# INTRODUCTORY CIRCUIT ANALYSIS

6th Edition

# INTRODUCTORY CIRCUIT ANALYSIS

Robert L. Boylestad

Merrill Publishing Company
A Bell & Howell Information Company
Columbus / Toronto / London / Melbourne

**To Else Marie,**
**Eric, Alison, and Stacey**

Cover Photo: Marcel Eskenazy/PhotoEdit

Published by Merrill Publishing Company
A Bell & Howell Information Company
Columbus, Ohio 43216

This book was set in Times Roman and Univers.

Administrative Editor: Dave Garza
Developmental Editor: Carol Thomas
Production Coordinator: Rex Davidson
Art Coordinator: Raydelle Clement
Cover Designers: Cathy Watterson and Brian Deep
Text Designer: Cindy Brunk
Photo Editor: Terry Tietz
Page Layout: Ellen Robison



Library of Congress Catalog Card Number: 89–61344
International Standard Book Number: 0–675–21181–6
Printed in the United States of America
1 2 3 4 5 6 7 8 9—95 94 93 92 91 90

# Preface

The longer my tenure in the educational community, the more obvious the fact that the literature supporting a particular subject area will always have room for improvement in content, clarity, and direction. How often, through recent editions, have I felt that the text was now in its best form—that areas of improvement would be difficult to define. However, the classroom experience with the changing needs of the student and the industrial community continue to define areas that require attention. There was a period when my role in the educational process was almost totally administrative and void of lecture responsibilities—a period of time when it became quite obvious that student interaction is an ingredient of paramount importance in maintaining a sense of need. The publisher's field staff, suggestions received from my peers, and reviews enlisted by Merrill Publishing are important and appreciated components of the revision process, but the methods must be class-tested and sensitive to the student's reaction—considered valid only when the result is improved clarity and successful absorption by the student.

In recent years there is the added pressure of computer literacy at an early stage in student development. I can no longer complete an edition and assume all the needs have now been met to the best of my ability. In fact, I'm sure as this new edition is released a file for the seventh edition will begin to develop. I have now fully accepted the time-honored paradoxical phrase that "one constant in life is that change will occur."

Virtually all the material covered in the 5th edition is included in this edition. There are no deletions, simply new material and improved descriptions of a number of critical areas. The most obvious addition is the PSPICE coverage. I believe the approach is unique in that there is sufficient content in the text to actually expect the student to write the input files for the problems that appear at the end of each chapter. Appendix A also deals primarily with the mechanics of using PSPICE. The computer coverage is unique in that it compares the use of a language (BASIC) to that of a software package, revealing the benefits of each and which one is more appropriate for particular areas of application. The chosen programs should clearly reveal the benefits associated with becoming familiar with computer techniques.

Another obvious change is the use of units throughout most of the calculations. Proper units of measurement with each quantity are maintained with an awareness of the impact of each term in a series of calculations.

The subject of superconductors is treated for the first time to establish an awareness of the research effort and current applications. Superconductivity is a phenomenon that will have significant impact on this field in the years to come. The coverage includes sufficient detail to understand a number of problems that have to be overcome and how each step forward can affect future areas of application.

An additional chapter (Chapter 21) on dB, filters, and Bode plots is introduced to permit an isolation of the important area of resonant circuits. Resonance is certainly one of the most important concepts presented in the ac sequence and deserves all the clarity possible. The new chapter also attempts to remove some of the initial fear students have for the application and use of logarithms, whether it be on a graph or in calculations. A number of practical examples are provided, followed by their use in the analysis of a variety of filters. The Bode plot sections are expanded in response to several requests from current users.

An increased number of exercises have been added to each chapter throughout the text. In general, they are more difficult and challenging, in response to the needs of some institutions. The most difficult problems of each chapter continue to be identified by an asterisk (*). In addition, each section in the list of problems now has the name of the section to clearly identify the subject matter to be tested.

The use of four colors is a milestone for literature in this field. The intent, however, is to use color functionally by improving the presentation and clarity of the text material and artwork.

Although less obvious on a brief scan, a number of important areas are rewritten and the number of examples increased. In particular, note the voltage and current divider rule, potentiometer control, superposition, nodal analysis, Thevenin's theorem, average value, frequency response, resonance, and so on. In general, therefore, changes and improved presentations appear throughout the text and may not be obvious until the subject is covered.

The package of ancillary materials has grown substantially. Particularly unique to the offering is a video covering two specific areas, Thevenin's theorem and series-parallel circuits. The video was developed in a classroom setting by the author to bring the presentation as close as possible to the actual experience. A computer disk is also available with programs of the text and solutions to particularly difficult problems. The computer test bank provides a self-testing mechanism for the student. The instructor's manual is expanded to include suggested course outlines with problem selections and additional comments on the other ancillary materials.

Selected experiments of the laboratory manual are replaced or improved to reflect suggestions from current users. In addition, four computer laboratory experiments are added to provide a format for those institutions that include "hands-on" experience in the early semesters. A solutions manual for the laboratory experiments will also be available to provide the results expected without having to run every experiment. Although the package continues to grow, each element meets an important need. I am interested in the general response to some of the innovative components of this edition.

## Acknowledgements

A number of individuals were particularly helpful in shaping the content and direction of this edition. The reviews were provided by R. C. Adler, DeVry Institute of Technology, Toronto, Canada; William Barnes, County College of Morris; John Barstow, Jr., Union County College; Robert Cox, Wentworth Institute of Technology; George Crossman, Broward Community College; Pete Culwell, DeVry Institute of Technology, Atlanta; Harry J. Franz, Pennsylvania State University, Beaver Campus; Frank Gergelyi, Metropolitan Technical Institute; L. W. Heselton, Mohawk Valley Community College; Robert Koval, Salem Community College; Clay Laster, San Antonio College; Fred Lewellen, University of Houston; Vince Loizzo, DeVry Institute of Technology, Chicago; Linda Miller, Florissant Valley Community College; Rich Pfile, Indiana University, Purdue University, Indianapolis, David Phillips, Purdue University; David Robinson, University of Akron; Earl Schoenwetter, California State Polytechnic Univ.-Pomona; Bill Stabosz, Delaware Technical and Community College; Don Trotter, Forsythe Technical Institute; Melvin Vye, The University of Akron; Dr. Robert Weiler, Capitol College; and Dewey Yaeger, Oklahoma State Technical Institute.

These reviews were well beyond that normally expected with all the demands on our daily lives—thank you. A special note of thanks to Professor Fred Lewellen, University of Houston, for developing and implementing the test bank. Without the assistance of Ken Brown, NAVA, Inc., and Jim Bryant, DeVry Institute of Technology, Columbus, Ohio, the video production would never have run so smoothly and been accomplished in such a short period of time.

There are two individuals whom I can always count on to respond to my ideas, make suggestions, provide whatever support I need, or just be there when frustrations develop. Through the years and all the editions, I cannot thank my good friends Professor Louis Nashelsky and Professor Leon Katz enough.

The staff at Merrill Publishing is a highly talented and supportive team. I have personally met most of the individuals who have worked on this edition and deeply appreciate the "family" atmosphere that pervades the home office. In particular I thank Steve Helba and Dave Garza for guiding the text through the many phases, Carol Thomas for managing the developmental details, Rex Davidson for his production control, Raydelle Clement for coordinating the art program, Cindy Brunk for creating the text design, and Aynn Titchenal for development of advertising materials.

And lastly, but certainly not least, I must thank Professor Joseph B. Aidala for giving me the opportunity to start my teaching career some 26 years ago and for continually pointing out areas that can stand improvement or modification.

# Contents

## CHAPTER 21 Decibels, Filters, and Bode Plots 829



## CHAPTER 22 Pulse Waveforms and the *R-C* Response 887



## CHAPTER 23 Polyphase Systems 917



## CHAPTER 24 Nonsinusoidal Circuits 965



## CHAPTER 25 Transformers 991

# 1
# Introduction

## 1.1 THE ELECTRICAL/ELECTRONICS INDUSTRY

The foundation of modern-day society is particularly sensitive to a few areas of development, research, and interest. In recent years, it has become obvious that the electrical/electronics industry is one area that will have a broad impact on future development in a host of activities that affect our life style, general health, and capabilities. Can you think of a field today, even those headstrong to minimize technical ties, that does not, at the very least, seek to broaden its horizons through the use of some technical innovation such as recording, duplication, computing, or data-handling instrumentation?

Every facet of our lives seems touched by developments that appear to surface at an ever increasing rate. For the layperson, the most obvious improvement of recent years has been the reduced size of electrical/electronics systems. TVs are now small enough to be hand held and have a battery capability which allows them to be more portable. Computers with significant memory capacity are now as small as portable typewriters. The size of radios is limited simply by the ability to read the numbers on the face of the dial. Hearing aids are no longer visible, and pacemakers are significantly smaller and more reliable. All the reduction in size is due primarily to a marvelous development of the last few decades—the integrated circuit (IC). First developed in the late 1950s, the IC industry has now reached a point where it can cut 1-micrometer lines. Consider that some 25,000 of these lines would fit within 1 in. Try to visualize breaking down an inch into 100 divisions and then consider 1000 or 25,000 divisions—an incredible achievement.

The integrated circuit of Fig. 1.1 has over 68,000 transistors in addition to thousands of other elements, yet is only about 1/4 in. on each side.

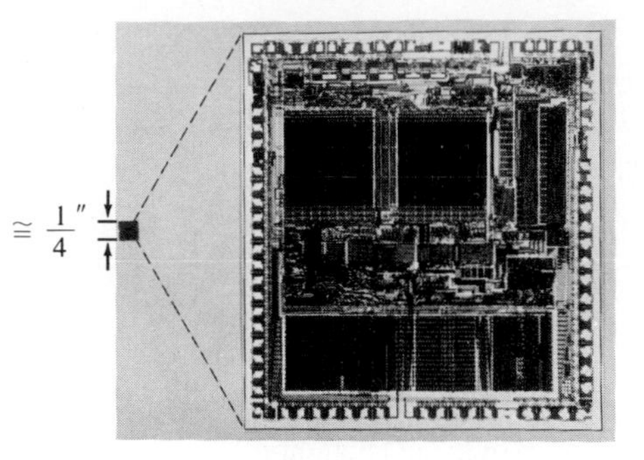


**FIG. 1.1**
*Integrated circuit. (Courtesy of Motorola Semiconductor Products)*

It is natural to wonder what the limits to growth may be when we consider the changes over the last few decades. Rather than following a steady growth curve that would be somewhat predictable, the industry is subject to surges that revolve around significant developments in the field. Present indications are that the level of miniaturization will continue but at a more moderate pace. Interest has turned toward increasing the quality and yield levels (percent of good integrated circuits in the production process).

History reveals that there have been peaks and valleys in industry growth but that revenues continue to rise at a steady rate and funds set aside for research and development continue to command an increasing share of the budget. The field changes at a rate that requires constant retraining of employees from the entry to the director level. Many companies have instituted their own training programs and have encouraged local universities to develop programs which will insure that the latest concepts and procedures are brought to the attention of their employees. A period of relaxation could be disastrous to a company dealing in competitive products.

No matter what the pressures on an individual in this field may be to keep up with the latest technology, there is one saving grace that becomes immediately obvious: Once a concept or procedure is clearly and correctly understood, it will bear fruit throughout the career of the individual at any level of the industry. For example, once a fundamental equation such as Ohm's law (Chapter 4) is understood, it will not be *replaced* by another equation as more advanced theory is considered. It is a relationship of fundamental quantities that can have application in the most advanced setting. In addition, once a procedure or method of analysis is understood, it usually can be applied to a wide (if not infinite) variety of problems, making it unnecessary to learn a different technique for each slight variation in the system. The content of this text is such that every morsel of information will have application in more advanced courses. It will not be replaced by a different set of equations and procedures unless required by the specific area of application. Even then, the new procedures will usually be an expanded application of concepts already presented in the text.

It is therefore paramount that the material presented in this introductory course be clearly and precisely understood. It is the foundation for the material to follow and will be applied throughout your working days in this growing and exciting field.

## 1.2 A BRIEF HISTORY

In the sciences, once a hypothesis is proven and accepted, it becomes one of the building blocks of that area of study, permitting further investigation and development. Naturally, the more pieces of a puzzle available, the more obvious the avenue toward a possible solution. In fact, history demonstrates that a single development may provide the key that will result in a mushroom effect that brings the science to a new plateau of understanding and impact.

If the opportunity presents itself, it would be time well spent to read one of the many publications reviewing the history of this field. Space requirements are such that only a brief review can be provided here. There are many more contributors than could be listed, and their efforts have often provided important keys to the solution of some very important concepts.

As noted earlier, there were periods characterized by what appeared to be an explosion of interest and development in particular areas. As you will see from the discussion of the late 1700s and the early 1800s, inventions, discoveries, and theories came fast and furiously. Each new concept broadened the possible areas of application until it becomes almost impossible to trace developments without picking a particular area of interest and following it through. In the review, as you read about the development of the radio, TV, and computer, keep in mind that similar progressive steps were occurring in the areas of the telegraph, the telephone, power generation, the phonograph, appliances, and so on.

There is a tendency when reading about the great scientists, inventors, and innovators to believe their contribution was a totally individual effort. In many instances, this was not the case. In fact, many of the great contributors were friends or associates and provided support and encouragement in their efforts to investigate various theories. At the very least, they were aware of one another's efforts to the degree possible in the days when a letter was often the best form of communication. In particular, note the closeness of the dates during periods of rapid development. One contributor seemed to spur on the efforts of the others or possibly provided the key needed to continue with the area of interest.

In the early stages, the contributors were not electrical, electronic, or computer engineers as we know them today. In most cases, they were physicists, chemists, mathematicians, or even philosophers. In addition, they were not from one or two communities of the Old World. The home country of many of the major contributors listed below is provided to show that almost every established community had some impact on the development of the fundamental laws of electrical circuits.

As you proceed through the remaining chapters of the text, you will find that a number of the units of measurement bear the name of major contributors in those areas—*volt* after Count Alessandro Volta, *ampere* after André Ampère, *ohm* after Georg Ohm, and so forth—fitting recognition for their important contribution to the birth of a major field of study.

Time charts indicating a limited number of major developments are provided in Fig. 1.2, primarily to identify specific periods of rapid development and to reveal how far we have come in the last few decades. In essence, the current state of the art is a result of efforts that began in earnest some 250 years ago, with progress in the last 100 years almost exponential.

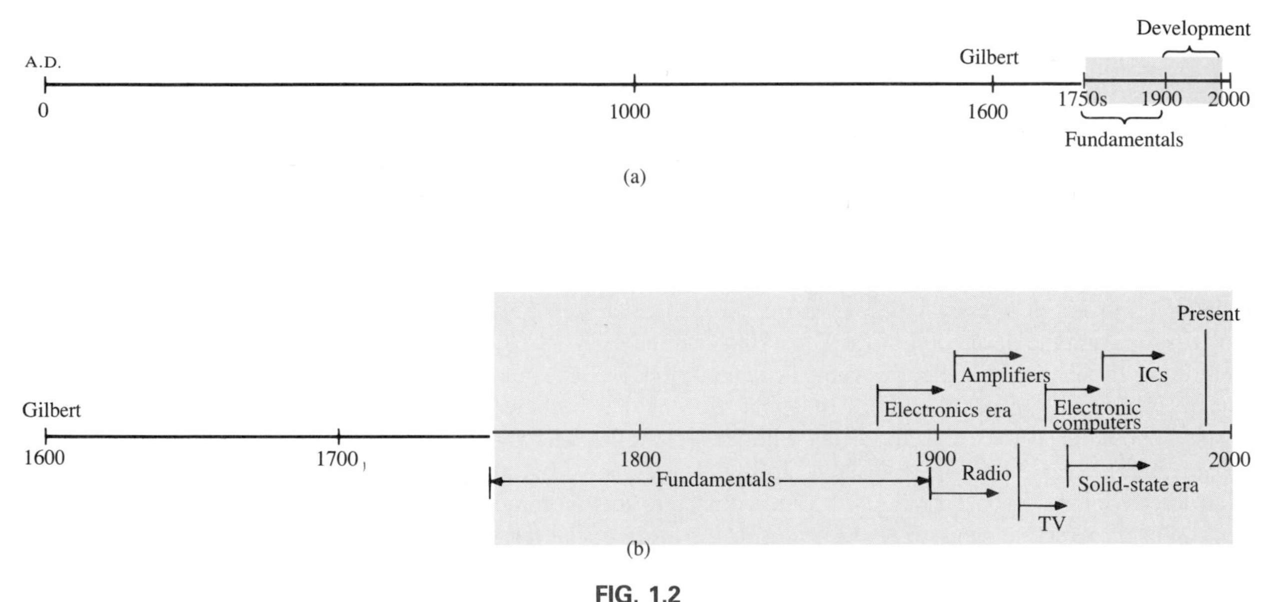


**FIG. 1.2**
*Time charts. (a) Long-range; (b) expanded.*

As you read through the following brief review, try to sense the growing interest in the field and the enthusiasm and excitement that must have accompanied each new revelation. Although you may find some of the terms used in the review new and essentially meaningless, the remaining chapters will explain them thoroughly.

## The Beginning

The phenomenon of static electricity has been toyed with since antiquity. The Greeks called the fossil resin substance so often used to demonstrate the effects of static electricity *elektron,* but no extensive study was made of the subject until William Gilbert researched the event in 1600. In the years to follow, there was a continuing investigation of electrostatic charge by a number of individuals such as Otto von Guericke, who developed the first machine to generate large amounts of charge, and Stephen Gray, who was able to transmit electrical charge over long distances on silk threads. Charles DuFay demonstrated that charges either attract or repel each other, leading him to believe that there were two types of charge—a theory we subscribe to today with our defined positive and negative charges.

There are many who believe the true beginnings of the electrical era lie with the efforts of Pieter van Musschenbroek and Benjamin Franklin. In 1745, van Musschenbroek introduced the *Leyden jar* for the storage of electrical charge (the first capacitor) and demonstrated electrical shock (and therefore the power of this new form of energy). Franklin used the Leyden jar some seven years later to establish that lightning is simply an electrical discharge, and expanded on a number of other important theories including the definition of the two types of charge as *positive* and *negative*. From this point on, new discoveries and theories seemed to occur at an increasing rate as the number of individuals performing research in the area grew.

In 1784, Charles Coulomb demonstrated in Paris that the force between charges is inversely related to the square of the distance between the charges. In 1791, Luigi Galvani, Professor of Anatomy at the University of Bologna, in Italy, performed experiments revealing that electricity is present in every animal. The first *voltaic cell* with its ability to produce electricity through the chemical action of a metal dissolving in an acid was developed by another Italian, Alessandro Volta, in 1799.

The fever pitch continued into the early 1800s with Hans Christian Oersted, a Swedish professor of physics, announcing in 1820 a relationship between magnetism and electricity that serves as the foundation for the theory of *electromagnetism* as we know it today. In the same year, a French physicist, André Ampère, demonstrated that there are magnetic effects around every current-carrying conductor and that current-carrying conductors can attract and repel each other just like magnets. In the period 1826 to 1827, a German physicist, Georg Ohm, introduced an important relationship between potential, current, and resistance which we now refer to as Ohm's law. In 1831, an English physicist, Michael Faraday, demonstrated his theory of *electromagnetic induction,* whereby a changing current in one coil can induce a changing current in another coil even though the two coils are not directly connected. Professor Faraday also did extensive work on a storage device he called the condenser, which we refer to as a capacitor today. He introduced the idea of adding a dielectric between the plates of a capacitor to increase the storage capacity (Chapter 10). James Clerk Maxwell, a Scottish professor of natural philosophy, performed extensive mathematical analyses to develop what are currently called *Maxwell's equations,* which support the efforts of Faraday linking electric and magnetic effects. Maxwell also developed the *electromagnetic theory of light* in 1862, which among other things revealed that electromagnetic waves travel through air at the velocity of light (186,000 miles per second or $3 \times 10^8$ meters per second). In 1888, a German physicist, Heinrich Rudolph Hertz, through experimentation with lower-frequency electromagnetic waves (microwaves), substantiated Maxwell's predictions and equations. In the mid 1800s, Professor Gustav Robert Kirchhoff introduced a series of laws of voltages and currents that find application at every level and area of this field (Chapters 5 and 6). In 1895, another German physicist, Wilhelm Röntgen, discovered electromagnetic waves of high frequency commonly called X rays today.

By the end of the 1800s, a significant number of the fundamental equations, laws, and relationships had been established and various fields of study including electronics, power generation, and calculating equipment started to develop in earnest.

## The Age of Electronics

**Radio** The true beginning of the electronics era is open to debate and is sometimes attributed to efforts by early scientists in applying potentials across evacuated glass envelopes. However, many trace the beginning to Thomas Edison, who added a metallic electrode to the vacuum of the tube and discovered that a current was established between the metal electrode and the filament when a positive voltage was applied to the metal electrode. The phenomenon, demonstrated in 1883, was referred to as the *Edison effect*. In the period to follow, the transmission of radio waves and the development of the radio received widespread attention. In 1887, Heinrich Hertz in his efforts to verify Maxwell's equations transmitted radio waves for the first time in his laboratory. In 1896, an Italian scientist, Guglielmo Marconi (often called the father of the radio), demonstrated that telegraph signals could be sent through the air over long distances (2.5 kilometers) using a grounded antenna. In the same year, Aleksandr Popov sent what might have been the first radio message some 300 yards. The message was the name ''*Heinrich Hertz*'' in respect for Hertz's earlier contributions. In 1901, Marconi established radio communication across the Atlantic.

In 1904, John Ambrose Fleming expanded on the efforts of Edison to develop the first diode, commonly called *Fleming's valve*—actually the first of the *electronic devices*. The device had a profound impact on the design of detectors in the receiving section of radios. In 1906, Lee De Forest added a third element to the vacuum structure and created the first amplifier, the triode. Shortly thereafter, in 1912, Edwin Armstrong built the first regenerative circuit to improve receiver capabilities and then used the same contribution to develop the first nonmechanical oscillator. By 1915 radio signals were being transmitted across the United States, and in 1918 Armstrong applied for a patent for the superheterodyne circuit employed in virtually every TV and radio to permit amplification at one frequency rather than at the full range of incoming signals. The major components of the modern-day radio were now in place and sales in radios grew from a few million dollars in the early 1920s to over one billion by the 1930s. The 1930s were truly the golden years of radio, with a wide range of productions for the listening audience.

**Television** The 1930s were also the true beginnings of the television era, although development on the picture tube began in earlier years with Paul Nipkow and his *electrical telescope* in 1884 and John Baird and his long list of successes including the transmission of TV pictures over telephone lines in 1927 and over radio waves in 1928, and simultaneous transmission of pictures and sound in 1930. In 1932, NBC installed the first commercial TV antenna on top of the Empire State Building, and RCA began regular broadcasting in 1939. The war slowed development and sales, but in the mid 1940s the number of sets grew from a few thousand to a few million. Color TV became popular in the early 1960s.

**Computers** The earliest computer system can be traced back to Blaise Pascal in 1642 with his mechanical machine for adding and subtracting numbers. In 1673 Gottfried Wilhelm von Leibniz used the *Leibniz wheel* to add multiplication and division to the range of operations, and in 1823 Charles Babbage developed the *Difference Engine* to add the mathematical operations of sine, cosine, logs, and a number of others. In the years to follow, improvements were made, but the system remained primarily mechanical until the 1930s when electromechanical systems using components such as relays were introduced. It was not until the 1940s that totally electronic systems became the new wave. It is interesting to note that even though IBM was formed in 1924, it did not enter the computer industry until 1937. An entirely electronic system known as *Eniac* was dedicated at the University of Pennsylvania in 1946. It contained 18,000 tubes and weighed 30 tons but was several times faster than most electromechanical systems. Although other vacuum tube systems were built, it was not until the birth of the solid-state era that computer systems experienced a major change in size, speed, and capability.

## The Solid-State Era

In 1947, physicists William Shockley, John Bardeen, and Walter H. Brattain of Bell Telephone Laboratories demonstrated the point-contact *transistor* (Fig. 1.3), an amplifier constructed entirely of solid-state materials with no requirement for a vacuum, glass envelope, or heater voltage for the filament. Although reluctant at first due to the vast amount of material available on the design, analysis, and synthesis of tube networks, the industry eventually accepted this new technology as the wave of the future. In 1958 the first *integrated circuit* (IC) was developed at Texas Instruments, and in 1961 the first commercial integrated circuit was manufactured by the Fairchild Corporation.

It is impossible to properly review the entire history of the electrical/electronics field in a few pages. The effort here, both through the discussion and the time graphs of Fig. 1.2, was to reveal the amazing progress of this field in the last 50 years. The growth appears to be truly exponential since the early 1900s, raising the interesting question as to where we go from here. The time chart suggests that the next few decades will probably contain a number of important innovative contributions that may cause an even faster growth curve than we are now experiencing.

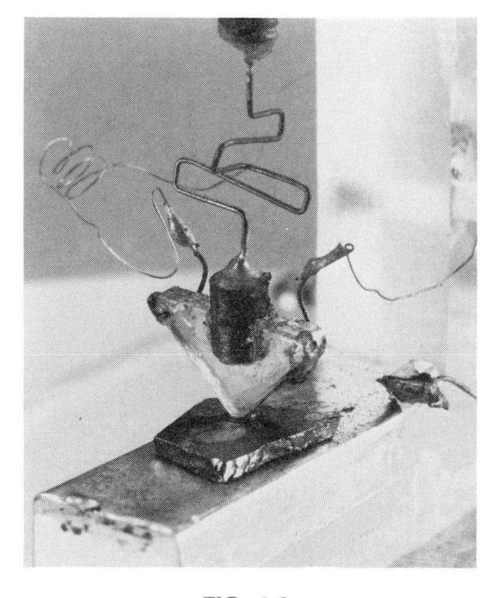

**FIG. 1.3**
*(Courtesy of AT&T, Bell Laboratories)*

$\sum_{I}^{S}$

## 1.3 UNITS OF MEASUREMENT

It is vital that the importance of units of measurement be understood and appreciated early in the development of a technically oriented background. Too frequently their effect on the most basic substitution is ignored. Consider, for example, the following very fundamental physics equation:

$$\boxed{v = \frac{d}{t}} \qquad \begin{aligned} v &= \text{velocity} \\ d &= \text{distance} \\ t &= \text{time} \end{aligned} \tag{1.1}$$

Assume, for the moment, that the following data are obtained:

$$d = 4000 \text{ ft}$$
$$t = 1 \text{ min}$$

and $v$ is desired in miles per hour. Often, without a second thought or consideration, the numerical values are simply substituted into the equation, with the result here that

$$v = \frac{d}{t} = \frac{4000 \text{ ft}}{1 \text{ min}} = \cancel{4000 \text{ mi/h}}$$

As indicated above, the solution is incorrect. If the result is desired in *miles per hour,* the unit of measurement for distance must be *miles,* and that for time, *hours.* In a moment, when the problem is analyzed properly, the extent of the error will demonstrate the importance of ensuring that

***the numerical value substituted into an equation must have the unit of measurement specified by the equation.***

The next question is normally how to convert the distance and time to the proper unit of measurement. A method will be presented in a later section of this chapter, but for now it is given that

$$1 \text{ mi} = 5280 \text{ ft}$$
$$4000 \text{ ft} = 0.7576 \text{ mi}$$
$$1 \text{ min} = \tfrac{1}{60} \text{ h} = 0.0167 \text{ h}$$

Substituting into Eq. (1.1), we have

$$v = \frac{d}{t} = \frac{0.7576 \text{ mi}}{0.0167 \text{ h}} = 45.37 \text{ mi/h}$$

which is significantly different from the result obtained before.

To complicate the matter further, suppose the distance is given in kilometers, as is now the case on many road signs. First, we must realize that the prefix *kilo* stands for a multiplier of 1000 (to be introduced in Section 1.5), and then we must find the conversion factor between kilometers and miles. If this conversion factor is not readily available, we must be able to make the conversion between units using the conversion factors between meters and feet or inches as described in Section 1.6.

Before substituting numerical values into an equation, try to mentally establish a reasonable range of solutions for comparison purposes. For instance, if a car travels 4000 ft in 1 min, does it seem reasonable that the speed would be 4000 mi/h? Obviously not! This self-checking procedure is particularly important in this day of the hand-held calculator when ridiculous results may be accepted simply because they appear on the digital display of the instrument.

Finally,

***if a unit of measurement is applicable to a result or piece of data, then it must be applied to the numerical value.***

To state that $v = 45.37$ without including the unit of measurement *mi/h* is meaningless.

Equation (1.1) is not a difficult one. A simple algebraic manipulation will result in the solution for any one of the three variables. However, in light of the number of questions arising from this equation, the reader may wonder if the difficulty associated with an equation will increase at the same rate as the number of terms in the equation. In the broad sense, this will not be the case. There is, of course, more room for a mathematical error with a more complex equation, but once the proper system of units is chosen and each term properly found in that system, there should be very little added difficulty associated with an equation requiring an increased number of mathematical calculations.

In review, before substituting numerical values into an equation, be absolutely sure of the following:

1. Each quantity has the proper unit of measurement as defined by the equation.
2. The proper magnitude of each quantity as determined by the defining equation is substituted.
3. Each quantity is in the same system of units (or as defined by the equation).
4. The magnitude of the result is of a reasonable nature when compared to the level of the substituted quantities.
5. The proper unit of measurement is applied to the result.

## 1.4 SYSTEMS OF UNITS

In the past, the *systems of units* most commonly used were the English and metric, as outlined by Table 1.1. Note that while the English system is based on a single standard, the metric is subdivided into two interrelated standards: the MKS and CGS. Fundamental quantities of these systems are compared in Table 1.1 along with their abbreviations. The MKS and CGS systems draw their names from the units of measurement used with each system; the MKS system uses *M*eters, *K*ilograms, and *S*econds, while the CGS system uses *C*entimeters, *G*rams, and *S*econds.

**TABLE 1.1**
*Comparison of the English and metric systems of units.*

| English | Metric | | SI |
|---|---|---|---|
| | MKS | CGS | |
| *Length:*<br>Yard (yd)<br>(0.914 m) | Meter (m)<br>(39.37 in.)<br>(100 cm) | Centimeter (cm)<br>(2.54 cm = 1 in.) | Meter (**m**) |
| *Mass:*<br>Slug<br>(14.6 kg) | Kilogram (kg)<br>(1000 g) | Gram (g) | Kilogram (**kg**) |
| *Force:*<br>Pound (lb)<br>(4.45 N) | Newton (N)<br>(100,000 dynes) | Dyne | Newton (**N**) |
| *Temperature:*<br>Fahrenheit (°F)<br>$\left(= \frac{9}{5}°C + 32\right)$ | Celsius or<br>Centigrade (°C)<br>$\left(= \frac{5}{9}(°F - 32)\right)$ | Centigrade (°C) | Kelvin (**K**)<br>K = 273.15 + °C |
| *Energy:*<br>Foot-pound (ft-lb)<br>(1.356 joules) | Newton-meter (N-m)<br>or Joule (J)<br>(0.7378 ft-lb) | Dyne-centimeter or Erg<br>(1 joule = $10^7$ ergs) | Joule (**J**) |
| *Time:*<br>Second (s) | Second (s) | Second (s) | Second (**s**) |

Understandably, the use of more than one system of units in a world that finds itself continually shrinking in size, due to advanced technical developments in communications and transportation, would introduce unnecessary complications to the basic understanding of any technical data. The need for a standard set of units to be adopted by all nations has become increasingly obvious. The International Bureau of Weights and Measures located at Sèvres, France, has been the host for the General Conference of Weights and Measures, attended by representatives from all nations of the world. In 1960, the General Conference adopted a system called Le Système International d'Unités (International System of Units), which has the international abbreviation SI. Since then, it has been adopted by the Institute of Electrical and Electronic Engineers, Inc. (IEEE) in 1965 and by the United States of America Standards Institute in 1967 as a standard for all scientific and engineering literature.

The inevitable changeover to the metric system has already resulted in the use of *both* miles per hour (mi/h) and kilometers per hour (km/h) on some road signs and the distribution of English-to-metric conversion charts as advertising literature by some firms. In fact, calculators are now available that are designed specifically to convert from one system to the other.

For comparison, the SI units of measurement and their abbreviations appear in Table 1.1. These abbreviations are those usually applied to each unit of measurement, and they were carefully chosen to be the most effective. Therefore, it is important that they be used whenever applicable to insure universal understanding. Note the similarities of the SI system to the MKS system. This text will employ, whenever possible and practical, all of the major units and abbreviations of the SI system in an effort to support the need for a universal system. For those readers requiring further information on the SI system, a complete kit has been assembled for general distribution by the American Society for Engineering Education (ASEE).*

Figure 1.4 should help the reader develop some feeling for the relative magnitudes of the units of measurement of each system of units.

*American Society for Engineering Education (ASEE), One Dupont Circle, Suite 400, Washington, D.C. 20036.

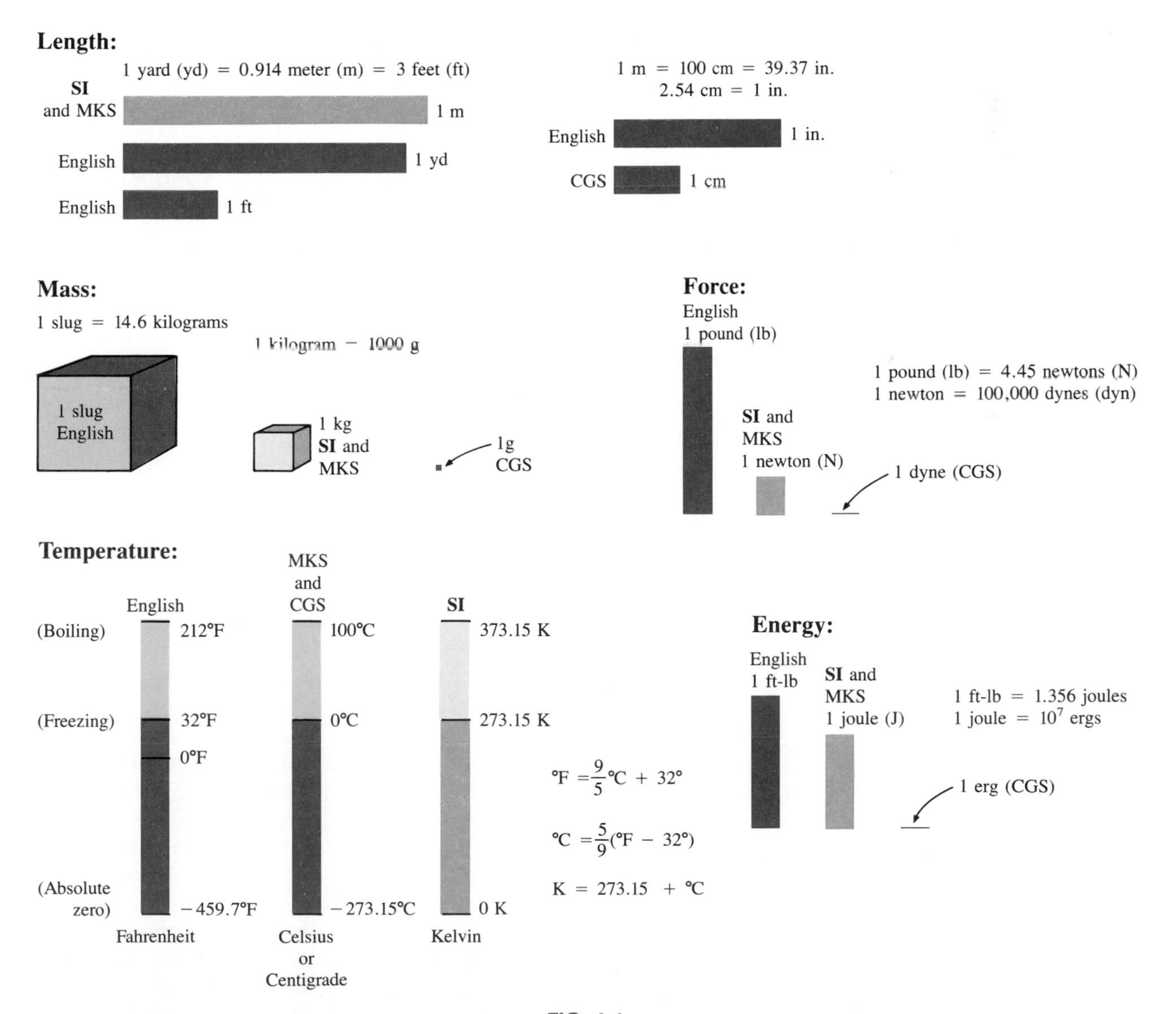


**FIG. 1.4**

*Comparison of units of the various systems of units.*

Note in the figure the relatively small magnitude of the units of measurement for the CGS system.

A standard exists for each unit of measurement of each system. The standards of some units are quite interesting.

The meter was originally defined in 1790 to be 1/10,000,000 the distance between the equator and either pole at sea level, a length preserved on a platinum-iridium bar at the International Bureau of Weights and Measures at Sèvres, France.

***The meter is now defined with reference to the speed of light in a vacuum, which is 299,792,458 m/s.***

***The kilogram is defined as a mass equal to 1000 times the mass of one cubic centimeter of pure water at 4°C.***

This standard is preserved in the form of a platinum-iridium cylinder in Sèvres.

The *second* was originally defined as 1/86,400 of the mean solar day. However, since Earth's rotation is slowing down by almost 1 second every 10 years,

***the second was redefined in 1967 as 9,192,631,770 periods of the electromagnetic radiation emitted by a particular transition of cesium atom.***

## 1.5 POWERS OF TEN

It should be apparent from the relative magnitude of the various units of measurement that very large and very small numbers will frequently be encountered in the study of the sciences. To ease the difficulty of mathematical operations with numbers of extreme size, *powers of ten* are usually employed. This notation takes full advantage of the mathematical properties of powers of 10. The notation used to represent numbers that are integer powers of 10 is as follows:

$$
\begin{aligned}
1 &= 10^0 & 1/10 &= 0.1 = 10^{-1} \\
10 &= 10^1 & 1/100 &= 0.01 = 10^{-2} \\
100 &= 10^2 & 1/1000 &= 0.001 = 10^{-3} \\
1000 &= 10^3 & 1/10{,}000 &= 0.0001 = 10^{-4}
\end{aligned}
$$

In particular, note that $10^0 = 1$, and, in fact, any quantity to the zero power is 1 ($x^0 = 1$, $1000^0 = 1$, and so on). Also, note that the numbers in the list that are greater than 1 are associated with positive powers of ten and numbers in the list that are less than 1 are associated with negative powers of ten.

A quick method of determining the proper power of 10 is to place a caret mark to the right of the numeral 1 wherever it may occur; then count from this point to the number of places to the right or left before arriving at the decimal point. Moving to the right indicates a positive power of 10, while moving to the left indicates a negative power. For example,

$$10{,}000.0 = 1\underbrace{0{,}0\,0\,0}_{1\ 2\ 3\ 4}. = 10^{+4}$$

$$0.00001 = 0.\underbrace{0\,0\,0\,0\,1}_{5\ 4\ 3\ 2\ 1} = 10^{-5}$$

Since some of these powers of 10 appear frequently, a written form of abbreviation has been adopted (as indicated in Table 1.2), which when written in conjunction with the unit of measurement eliminates the need to include the power of 10 in numerical form.

**TABLE 1.2**

| Power of 10 | Prefix | Abbreviation |
|---|---|---|
| $10^{12}$ | Tera | T |
| $10^{9}$ | Giga | G |
| $10^{6}$ | Mega | M |
| $10^{3}$ | Kilo | k |
| $10^{-3}$ | Milli | m |
| $10^{-6}$ | Micro | $\mu$ |
| $10^{-9}$ | Nano | n |
| $10^{-12}$ | Pico | p |

## EXAMPLES

$$\begin{aligned}
1{,}000{,}000 \text{ ohms} &= 1 \times 10^{6} \text{ ohms} \\
&= 1 \text{ megohm } (\text{M}\Omega) \\
100{,}000 \text{ meters} &= 100 \times 10^{3} \text{ meters} \\
&= 100 \text{ kilometers (km)} \\
0.0001 \text{ second} &= 0.1 \times 10^{-3} \text{ second} \\
&= 0.1 \text{ millisecond (ms)} \\
0.000001 \text{ farad} &= 1 \times 10^{-6} \text{ farad} \\
&= 1 \text{ microfarad } (\mu\text{F})
\end{aligned}$$

Some important mathematical equations and relationships applying to powers of 10 are listed below along with a few examples. In each case, $n$ and $m$ can be any positive or negative real number.

$$\frac{1}{10^{n}} = 10^{-n} \qquad \frac{1}{10^{-n}} = 10^{n} \tag{1.2}$$

Equation (1.2) clearly reveals that shifting a power of 10 from the denominator to the numerator, or the reverse, requires simply changing the sign of the power.

## EXAMPLES

$$\frac{1}{1000} = \frac{1}{10^{+3}} = 10^{-3}$$

$$\frac{1}{0.00001} = \frac{1}{10^{-5}} = 10^{+5}$$

The product of powers of 10:

$$(10^{n})(10^{m}) = 10^{(n+m)} \tag{1.3}$$

**EXAMPLES**

$$(1000)(10{,}000) = (10^3)(10^4) = 10^{(3+4)} = 10^7$$
$$(0.00001)(100) = (10^{-5})(10^2) = 10^{(-5+2)} = 10^{-3}$$

The division of powers of 10:

$$\frac{10^n}{10^m} = 10^{(n-m)} \tag{1.4}$$

**EXAMPLES**

$$\frac{100{,}000}{100} = \frac{10^5}{10^2} = 10^{(5-2)} = 10^3$$
$$\frac{1000}{0.0001} = \frac{10^3}{10^{-4}} = 10^{(3-(-4))} = 10^{(3+4)} = 10^7$$

Note in the last example the use of parentheses to insure the proper sign is established between operators.

The power of powers of 10:

$$(10^n)^m = 10^{(nm)} \tag{1.5}$$

**EXAMPLES**

$$(100)^4 = (10^2)^4 = 10^{(2)(4)} = 10^8$$
$$(1000)^{-2} = (10^3)^{-2} = 10^{(3)(-2)} = 10^{-6}$$
$$(0.01)^{-3} = (10^{-2})^{-3} = 10^{(-2)(-3)} = 10^6$$

## Basic Arithmetic Operations

Let us now examine the use of powers of ten to perform some basic arithmetic operations.

**Addition and subtraction** In order to perform addition or subtraction using powers of ten, the power of ten must be the same for each term.

That is,

$$A \times 10^n \pm B \times 10^n = (A \pm B) \times 10^n \tag{1.6}$$

### EXAMPLES

$$\begin{aligned}6300 + 75{,}000 &= (6.3)(1000) + (75)(1000)\\ &= 6.3 \times 10^3 + 75 \times 10^3\\ &= (6.3 + 75) \times 10^3\\ &= \mathbf{81.3 \times 10^3}\end{aligned}$$

$$\begin{aligned}0.00096 - 0.000086 &= (96)(0.00001) - (8.6)(0.00001)\\ &= 96 \times 10^{-5} - 8.6 \times 10^{-5}\\ &= (96 - 8.6) \times 10^{-5}\\ &= \mathbf{87.4 \times 10^{-5}}\end{aligned}$$

**Multiplication** In general,

$$(A \times 10^n)(B \times 10^m) = (A)(B) \times 10^{n+m} \quad \textbf{(1.7)}$$

revealing that the operations with the powers of ten can be separated from the operation with the multipliers.

### EXAMPLES

$$\begin{aligned}(0.0002)(0.000007) &= [(2)(0.0001)][(7)(0.000001)]\\ &= (2 \times 10^{-4})(7 \times 10^{-6})\\ &= (2)(7) \times (10^{-4})(10^{-6})\\ &= \mathbf{14 \times 10^{-10}}\end{aligned}$$

$$\begin{aligned}(340{,}000)(0.00061) &= (3.4 \times 10^5)(61 \times 10^{-5})\\ &= (3.4)(61) \times (10^5)(10^{-5})\\ &= 207.4 \times 10^0\\ &= \mathbf{207.4}\end{aligned}$$

**Division** In general,

$$\frac{A \times 10^n}{B \times 10^m} = \frac{A}{B} \times 10^{n-m} \quad \textbf{(1.8)}$$

revealing again that the operations with the powers of ten can be separated from the same operation with the multipliers.

### EXAMPLES

$$\begin{aligned}\frac{0.00047}{0.002} &= \frac{47 \times 10^{-5}}{2 \times 10^{-3}} = \left(\frac{47}{2}\right) \times \left(\frac{10^{-5}}{10^{-3}}\right)\\ &= \mathbf{23.5 \times 10^{-2}}\end{aligned}$$

$$\begin{aligned}\frac{690{,}000}{0.00000013} &= \frac{69 \times 10^4}{13 \times 10^{-8}} = \left(\frac{69}{13}\right) \times \left(\frac{10^4}{10^{-8}}\right)\\ &= \mathbf{5.31 \times 10^{12}}\end{aligned}$$

**Powers** In general,

$$(A \times 10^n)^m = A^m \times 10^{nm} \tag{1.9}$$

which again permits the separation of the operation with the powers of ten from the multipliers.

---

**EXAMPLES**

$$(0.00003)^3 = (3 \times 10^{-5})^3 = (3)^3 \times (10^{-5})^3$$
$$= \mathbf{27 \times 10^{-15}}$$
$$(90{,}800{,}000)^2 = (9.08 \times 10^7)^2 = (9.08)^2 \times (10^7)^2$$
$$= \mathbf{82.4464 \times 10^{14}}$$

---

The following examples include units of measurement.

---

**EXAMPLES**

a. 41,200 m is equivalent to $41.2 \times 10^3$ m = **41.2 km.**
b. 0.00956 J is equivalent to $9.56 \times 10^{-3}$ J = **9.56 mJ.**
c. 0.000768 s is equivalent to $768 \times 10^{-6}$ s = **768 μs.**
d. $\dfrac{8400 \text{ m}}{0.06} = \dfrac{8.4 \times 10^3 \text{ m}}{6 \times 10^{-2}} = \left(\dfrac{8.4}{6}\right) \times \left(\dfrac{10^3}{10^{-2}}\right) \text{m}$
$= 1.4 \times 10^5 \text{ m} = 140 \times 10^3 \text{ m} = \mathbf{140\ km}$
e. $(0.0003)^4 \text{ s} = (3 \times 10^{-4})^4 \text{ s} = 81 \times 10^{-16} \text{ s}$
$= 0.0081 \times 10^{-12} \text{ s} = \mathbf{0.0081\ ps}$

---

To demonstrate the amount of work saved and the reduced possibility of error that result by using powers of 10, consider finding the solution to the last example in the following manner:

$$\begin{array}{r} 0.0003 \\ \times\ 0.0003 \\ \hline 0.00000009 \\ \times\ 0.0003 \\ \hline 0.000000000027 \\ \times\ 0.0003 \\ \hline 0.0000000000000081 \end{array} = 81 \times 10^{-16} \text{ s} = \mathbf{0.0081\ ps}$$

**Scientific Notation** If the multiplier is limited to a number between 1 and 10, the number is said to be in *scientific notation*. It is not a required standard, however, since it would negate the use of some of the important prefixes introduced earlier.

### EXAMPLES

$$50{,}000 = \mathbf{5.0 \times 10^5}$$
$$0.00064 = \mathbf{6.4 \times 10^{-4}}$$

## 1.6 CONVERSION BETWEEN LEVELS

It is often necessary to convert from one power of ten to another. For instance, if a meter measures kilohertz (kHz), it may be necessary to find the corresponding level in megahertz (MHz), or if time is measured in milliseconds (ms), it may be necessary to find the corresponding time in microseconds ($\mu$s) for a graphical plot. The process is not a difficult one if we simply keep in mind that an increase or decrease in the power of ten must be associated with the opposite effect on the multiplying factor. The procedure is best described by a few examples.

### EXAMPLES

a. Convert 20 kHz to MHz.

***Solution:*** In the power-of-ten format:

$$20 \text{ kHz} = 20 \times 10^3 \text{ Hz}$$

The conversion requires that we find thc multiplying factor to appear in the space below:

$$20 \times 10^3 \text{ Hz} \Rightarrow \_\_ \times 10^6 \text{ Hz} \quad (\text{increase by } 3)$$

Since the power of ten will be *increased* by a factor of *three,* the multiplying factor must be *decreased* by moving the decimal point *three* places to the left as shown below:

$$020. = 0.02 \quad (3)$$

and $$20 \times 10^3 \text{ Hz} = 0.02 \times 10^6 \text{ Hz} = \mathbf{0.02\ MHz}$$

b. Convert 0.10 ms to microseconds.

***Solution:*** In the power-of-ten format:

$$0.01 \text{ ms} = 0.01 \times 10^{-3} \text{ s}$$

and $$0.01 \times 10^{-3} \text{ s} = \_\_ \times 10^{-6} \text{ s} \quad (\text{reduce by } 3)$$

Since the power of ten will be *reduced* by a factor of three, the multiplying factor must be *increased* by moving the decimal point three places to the right, as follows:

$$0.\underbrace{010}_{3} = 10$$

and $$0.01 \times 10^{-3}\ \text{s} = 10 \times 10^{-6}\ \text{s} = \mathbf{10\ \mu s}$$

There is a tendency when comparing $-3$ to $-6$ to think the power of ten has increased, but keep in mind when making your judgement about increasing or decreasing the magnitude of the multiplier that $10^{-6}$ is a great deal smaller than $10^{-3}$.

c. Convert 0.002 km to millimeters.

***Solution:***

$$0.002 \times 10^{3}\ \text{m} \Rightarrow \underline{\quad} \times 10^{-3}\ \text{m} \quad (\text{reduce by } 6)$$

In this example we have to be very careful because the difference between $+3$ and $-3$ is a factor of 6, requiring that the multiplying factor be modified as follows:

$$0.\underbrace{002000}_{6} = 2000$$

and $$0.002 \times 10^{3}\ \text{m} = 2000 \times 10^{-3}\ \text{m} = \mathbf{2000\ mm}$$

## 1.7 CONVERSION WITHIN AND BETWEEN SYSTEMS OF UNITS

The conversion within and between systems of units is a process that cannot be avoided in the study of any technical field. It is an operation, however, that is performed incorrectly so often that this section was included to provide one approach which, if applied properly, will lead to the correct result.

There is more than one method to perform the conversion process. In fact, some people prefer to determine mentally whether the conversion factor is multiplied or divided. This approach is acceptable for some elementary conversions but is risky with more complex operations.

The procedure to be described here is best introduced by examining a relatively simple problem such as converting inches to meters. Specifically, let us convert 48 in. (4 ft) to meters.

If we multiply the 48 in. by a factor of 1, the magnitude of the quantity remains the same:

$$48\ \text{in.} = 48\ \text{in.}(1) \tag{1.10}$$

Let us now look at the conversion factor, which is the following for this example:

$$1\ \text{m} = 39.37\ \text{in.}$$

Dividing both sides of the conversion factor by 39.37 in. will result in

$$\frac{1\ \text{m}}{39.37\ \text{in.}} = \frac{\cancel{39.37\ \text{in.}}}{\cancel{39.37\ \text{in.}}} = (1)$$

Note that the end result is that the ratio 1 m/39.37 in. equals 1, as it should since they are equal quantities. If we now substitute this factor (1) into Eq. (1.6), we obtain

$$48\ \text{in.}(1) = 48\ \cancel{\text{in.}}\left(\frac{1\ \text{m}}{39.37\ \cancel{\text{in.}}}\right)$$

which results in the cancellation of inches as a unit of measurement and leaves meters as the unit of measure. In addition, since the 39.37 is in the denominator, it must be divided into the 48 to complete the operation:

$$\frac{48}{39.37}\ \text{m} = \mathbf{1.219\ m}$$

Let us now review the method, which has the following sequence of steps:

1. Multiply the quantity to be converted by the factor (1).
2. Set up the conversion factor to form a numerical value of (1) with the unit of measurement to be removed in the denominator.
3. Perform the required mathematics to obtain the proper magnitude for the remaining unit of measurement.

---

**EXAMPLES** Convert 6.8 min to seconds.

*Step 1:* $6.8\ \text{min}(1)$

*Step 2:* $\left(\dfrac{60\ \text{s}}{1\ \text{min}}\right) = (1)$

*Step 3:* $6.8\ \cancel{\text{min}}\left(\dfrac{60\ \text{s}}{1\ \cancel{\text{min}}}\right) = (6.8)(60)\ \text{s}$

$= \mathbf{408\ s}$

Convert 0.24 m to centimeters.

*Step 1:* $0.24\ \text{m}(1)$

*Step 2:* $\left(\dfrac{100\ \text{cm}}{1\ \text{m}}\right) = (1)$

*Step 3:* $0.24\ \cancel{\text{m}}\left(\dfrac{100\ \text{cm}}{1\ \cancel{\text{m}}}\right) = (0.24)(100)\ \text{cm}$

$= \mathbf{24\ cm}$

---

The product (1)(1) or (1)(1)(1) is still 1. Using this fact, we can perform a series of conversions in the same operation.

**EXAMPLES** Determine the number of minutes in half a day.

$$0.5\ \cancel{\text{day}}\left(\frac{24\ \cancel{\text{h}}}{1\ \cancel{\text{day}}}\right)\left(\frac{60\ \text{min}}{1\ \cancel{\text{h}}}\right) = (0.5)(24)(60)\ \text{min}$$
$$= \mathbf{720\ min}$$

Convert 1/4 in. to millimeters.

$$\tfrac{1}{4}\ \cancel{\text{in.}}\left(\frac{1\ \cancel{\text{m}}}{39.37\ \cancel{\text{in.}}}\right)\left(\frac{10^3\ \text{mm}}{1\ \cancel{\text{m}}}\right) = \frac{0.25}{39.37}(10^3)\ \text{mm}$$
$$= \mathbf{6.35\ mm}$$

The following examples are variations of the above in practical situations.

**EXAMPLES** In Europe, the speed limit is posted in kilometers per hour. How fast in miles per hour is 100 km/h?

$$\left(\frac{100\ \text{km}}{\text{h}}\right)(1)(1)(1)(1)$$
$$= \left(\frac{100\ \cancel{\text{km}}}{\text{h}}\right)\left(\frac{1000\ \cancel{\text{m}}}{1\ \cancel{\text{km}}}\right)\left(\frac{39.37\ \cancel{\text{in.}}}{1\ \cancel{\text{m}}}\right)\left(\frac{1\ \cancel{\text{ft}}}{12\ \cancel{\text{in.}}}\right)\left(\frac{1\ \text{mi}}{5280\ \cancel{\text{ft}}}\right)$$
$$= \frac{(100)(1000)(39.37)}{(12)(5280)}\ \frac{\text{mi}}{\text{h}}$$
$$= \mathbf{62.14\ mi/h}$$

Determine the speed in miles per hour of a competitor who can run a 4-min mile. Inverting the factor 4 mi/1 mi to 1 mi/4 min, we can proceed as follows:

$$\left(\frac{1\ \text{mi}}{4\ \cancel{\text{min}}}\right)\left(\frac{60\ \cancel{\text{min}}}{\text{h}}\right) = \frac{60}{4}\ \text{mi/h} = \mathbf{15\ mi/h}$$

## 1.8 SYMBOLS

Throughout the text, various symbols will be employed that the reader may not have had occasion to use. Some are defined in Table 1.3, and others will be defined in the text as the need arises.

**TABLE 1.3**

| Symbol | Meaning |
|---|---|
| $\neq$ | Not equal to<br>$6.12 \neq 6.13$ |
| $>$ | Greater than<br>$4.78 > 4.20$ |
| $\gg$ | Much greater than<br>$840 \gg 16$ |
| $<$ | Less than<br>$430 < 540$ |
| $\ll$ | Much less than<br>$0.002 \ll 46$ |
| $\geq$ | Greater than or equal to<br>$x \geq y$ is satisfied for $y = 3$ and $x > 3$ or $x = 3$ |
| $\leq$ | Less than or equal to<br>$x \leq y$ is satisfied for $y = 3$ and $x < 3$ or $x = 3$ |
| $\cong$ | Approximately equal to<br>$3.14159 \cong 3.14$ |
| $\Sigma$ | Sum of<br>$\Sigma\,(4 + 6 + 8) = 18$ |
| $\lvert\ \rvert$ | Absolute magnitude of<br>$\lvert a \rvert = 4$, where $a = -4$ or $+4$ |
| $\therefore$ | Therefore<br>$x = \sqrt{4} \qquad \therefore x = \pm 2$ |

## 1.9 CONVERSION TABLES

Conversion tables such as those appearing in Appendix B can be very useful when time does not permit the application of methods described in this chapter. However, even though such tables appear easy to use, frequent errors occur because the operations appearing at the head of the table are not properly performed. In any case, when using such tables, try to establish mentally some order of magnitude for the quantity to be determined as compared to the magnitude of the quantity in its original set of units. This simple operation should prevent a number of the impossible results that may occur if the conversion operation is improperly applied.

For example, consider the following from such a conversion table:

| To convert from | To | Multiply by |
|---|---|---|
| Miles | Meters | $1.609 \times 10^3$ |

A conversion of 2.5 mi to meters would require that we multiply 2.5 by the conversion factor. That is,

$$2.5 \text{ mi}(1.609 \times 10^3) = 4.0225 \times 10^3 \text{ m}$$

A conversion from 4000 m to miles would require a division process:

$$\frac{4000 \text{ m}}{1.609 \times 10^3} = 2486.02 \times 10^{-3} = 2.48602 \text{ mi}$$

In each of the above, there should have been little difficulty realizing that 2.5 mi would convert to a few thousand meters, and 4000 m would be only a few miles. As indicated above, this kind of prior thinking will eliminate the possibility of ridiculous conversion results.

## 1.10 COMPUTER ANALYSIS AND DESIGN

The use of computers in the educational process has been growing at a very rapid rate in the past few years. There are very few texts at this introductory level that now fail to include some discussion of current popular computer techniques. In fact, the very accreditation of a technology program may be a function of the depth to which computer methods are incorporated in the program.

There is no question that a basic knowledge of computer methods is one that the graduating student should carry away from a 2-year or 4-year program. Industry is now expecting students to have a basic knowledge of computer jargon and some hands-on experience.

For many students the thought of having to learn how to use a computer will result in an insecure, uncomfortable feeling normally associated with outright fear. Be assured, however, that through the proper learning experience and exposure, the computer can become a very "friendly," useful, and supportive "tool" in the development and application of your technical skills in a professional environment.

For the new student of computers, there are two general directions that can be taken to develop the necessary computer skills: The study of languages or software packages.

### Languages

There are a number of languages that provide a direct line of communication with the computer and the operations it can perform. A language is a set of symbols, letters, words, or statements that the user can enter into the computer. The computer system will "understand" these entries and perform them in the order established by a series of commands called a *program*. The program tells the computer what to do on a sequential line-by-line basis in the same order a student would perform the calculations in longhand fashion. The computer can respond only to the commands entered by the user. This requires that the programmer fully understand the sequence of operations and calculations required to obtain a particular solution. In other words, the computer can only respond to the user's input—it does not have some mysterious way of providing solutions unless told how to obtain those solutions. A lengthy analysis can result in a program having hundreds or thousands of lines. Once written, the program has to be carefully checked to be sure the results have meaning and are valid for an expected range of input variables. Writing a program can, therefore, be a long, tedious process, but keep in mind that once the program is tested and true, it can be stored in memory for future use. The user can be assured that any future results obtained have a high degree of accuracy but require a minimum expenditure of energy and time. Some of the popular languages applied in the electrical/electronics field today include BASIC, PASCAL, FORTRAN, and C. Each has its own set of commands and statements to communicate with the computer, but each can be used to perform the same type of analysis. This text uses BASIC (Beginning All-purpose Symbolic Instruction Code) because the commands and statements compare directly with similar directives in the English language. The BASIC language permits an understanding of the format, subsets, and calculations of a program with a minimum of prior exposure and explanation. There is absolutely no suggestion that the coverage of BASIC in this text is sufficient to become adept at writing programs in BASIC. The purpose here is simply to expose the student to the general characteristics of the program and how the computer can be used effectively to analyze and design networks. A proper exposure to BASIC would require a course in itself or at least a very supportive structure provided by an educational institution. A sample program in BASIC is shown in Fig. 4.27. Note the use of words such as PRINT to tell the computer to print out the results or statement, the use of INPUT to request data from the user, the letters REM from the word REMark to specify comments to appear on the output, and the word END to terminate the program. More is said about particular statements and commands in the chapters to follow.

## Software Packages

The second approach to computer analysis avoids the need to know a particular language; in fact, the user may not be aware of which language was used to write the programs within the package. All that is required is a knowledge of how to provide the input about a system into the computer; the package will solve for specific unknowns of the system. The individual steps toward a solution are beyond the needs of the user—all the user needs is an idea of how to get the network parameters into the computer and how to extract the results. Herein lies one of the concerns of the author with packaged programs—whether a student has the ability to obtain a result for a particular analysis without knowing or understanding the steps leading to that solution. It is imperative that the student realize that the computer should be used as a tool to assist the user—it must not be allowed to control the scope and potential of the user! Therefore, as we progress through the chapters of the text, be sure concepts are clearly understood before turning to the computer for support and efficiency.

Each software package has a *menu,* which defines the range of application of the package. Once entered into the computer, the system is preprogrammed to perform all the functions appearing in the menu. The user simply has to provide the parameters of the network and signal the computer that all the data are in and a solution is desired. The package will then generate and print the results for the desired unknowns. Be aware, however, that if a particular type of analysis is requested that is not on the menu, the software package cannot provide the desired results. The package is limited solely to those maneuvers developed by the team of programmers that developed the software package. In such situations the user must turn to another software package or write a program using one of the languages listed above.

In broad terms, if a software package is available to perform a particular analysis, then it should be used rather than developing routines. Most popular software packages are the result of many hours of effort by teams of programmers with years of experience. However, if the results are not in the desired format or the software package does not provide all the desired results, then the user's innovative talents should be put to use to develop a software package. As noted above, any program the user writes that passes the tests of range and accuracy can be considered a software package of his or her authorship for future use.

The software package to be employed in this text is PSPICE, which is an educational version of a larger commercial version referred to simply as SPICE (Simulation Program with Integrated Circuit Emphasis). A photograph of the educational package as received from the MicroSim Corporation appears in Fig. 1.5. It includes a user's manual,

**FIG. 1.5**
*(Courtesy of MicroSim Corp.)*

floppy disks, and a demo package. Appendix A of this text was designed to provide a condensed listing of important information required to apply the SPICE package successfully to a variety of networks appearing in this text. The scope of the coverage was limited to the content of this text to permit as much detail as possible for the range of coverage. The PSPICE menu is an extensive one and can perform most of the procedures described in the text. In those instances where the capability was not built in, the author has turned to the BASIC language and a detailed program. In fact, in numerous instances the same analysis is performed using BASIC and PSPICE to demonstrate the salient differences between the application of a language and a software package. Again, it is important to realize that the coverage of PSPICE in this text is only a surface treatment to supplement a more detailed coverage in another course or a structured supportive ancillary of the present course. In other words, the content of this text will require additional support and guidance from your instructor or a concurrent computer course or laboratory session.

Other software packages available include BREADBOARD, which has a slightly different menu from PSPICE but has the added capability of being able to draw the network on the computer screen when the network parameters are entered.

Computer simulation and methods are an important, integral part of the text and should not be treated as superfluous material of the lowest priority. Once a basic concept is understood, take the time to investigate computer methods and start to develop a familiarity with the terminology and basic format of a program or utilization of a software package. It will be time well spent in preparation for the instruction you will eventually receive on computer systems. As noted earlier the material presented on computers in this text is not complete in itself but requires enhancement through either a concurrent instructional program or instructor-provided information. Sufficient content is included, however, to develop a first level of familiarity with the application of computer techniques to the analysis of electrical/electronic systems.

## PROBLEMS

Note: More difficult problems are denoted by an asterisk (*) throughout the text.

### SECTION 1.2 A Brief History

1. Visit your local library (at school or home) and describe the extent to which it provides literature and computer support for the technologies—in particular, electricity, electronics, electromagnetics, and computers.
2. Choose an area of particular interest in this field and write a very brief report on the history of the subject.
3. Choose an individual of particular importance in this field and write a very brief review of his or her life and important contributions.

## SECTION 1.3 Units of Measurement

**4.** Determine the distance in feet traveled by a car moving at 50 mi/h for 1 min.

**5.** How many hours would it take a person to walk 12 mi if the average pace is 15 min/mile?

## SECTION 1.4 Systems of Units

**6.** Are there any relative advantages associated with the metric system as compared to the English system with respect to length, mass, force, and temperature? If so, explain.

**7.** Which of the four systems of units appearing in Table 1.1 has the smallest units for length, mass, and force? When would this system be used most effectively?

***8.** Which system of Table 1.1 is closest in definition to the SI system? How are the two systems different? Why do you think the units of measurement for the SI system were chosen as listed in Table 1.1? Give the best reasons you can without referencing additional literature.

**9.** What is room temperature (68°F) in the MKS, CGS, and SI systems?

**10.** How many foot-pounds of energy are associated with 1000 J?

**11.** How many centimeters are there in 1/2 yd?

## SECTION 1.5 Powers of Ten

**12.** Express the following numbers as powers of 10:
**a.** 10,000 **b.** 0.0001
**c.** 1000 **d.** 1,000,000
**e.** 0.0000001 **f.** 0.00001

**13.** Using only those powers of 10 listed in Table 1.2, express the following numbers in what seems to you the most logical form for future calculations:
**a.** 15,000 **b.** 0.03000
**c.** 7,400,000 **d.** 0.0000068
**e.** 0.00040200 **f.** 0.0000000002

Perform each of the following operations and express the result as a power of 10:

**14.** **a.** (100)(100) **b.** (0.01)(1000)
**c.** $(10^3)(10^6)$ **d.** (1000)(0.00001)
**e.** $(10^{-6})(10,000,000)$ **f.** $(10,000)(10^{-8})(10^{35})$

**15.** **a.** $\frac{100}{1000}$ **b.** $\frac{0.01}{100}$
**c.** $\frac{10,000}{0.00001}$ **d.** $\frac{0.0000001}{100}$
**e.** $\frac{10^{38}}{0.000100}$ **f.** $\frac{(100)^{1/2}}{0.01}$

**16.** **a.** $(100)^3$ **b.** $(0.0001)^{1/2}$
**c.** $(10,000)^8$ **d.** $(0.00000010)^9$

$\sum_{I}^{S}$

Perform each of the following operations and express the result in scientific notation:

**17.** **a.** $(-0.001)^2$ **b.** $\frac{(100)(10^{-4})}{10}$

**c.** $\frac{(0.01)^2(100)}{10,000}$ **d.** $\frac{(10^2)(10,000)}{0.001}$

**e.** $\frac{(0.0001)^3(100)}{1,000,000}$ ***f.** $\frac{[(100)(0.01)]^{-3}}{[(100)^2][0.001]}$

***18.** **a.** $\frac{(300)^2(100)}{10^4}$ **b.** $[(40,000)^2][(20)^{-3}]$

**c.** $\frac{(60,000)^2}{(0.02)^2}$ **d.** $\frac{(0.000027)^{1/3}}{210,000}$

**e.** $\frac{[(4000)^2][300]}{0.02}$

**f.** $[(0.000016)^{1/2}][(100,000)^5][0.02]$

**g.** $\frac{[(0.003)^3][(0.00007)^2][(800)^2]}{[(100)(0.0009)]^{1/2}}$ (a challenge)

## SECTION 1.6 Conversion Between Levels

**19.** Convert 2000 μs to milliseconds.

**20.** Convert 0.04 ms to microseconds.

**21.** Convert 0.06 μF to nanofarads.

**22.** Convert 8400 ps to microseconds.

***23.** Convert 0.006 km to millimeters.

***24.** Convert $260 \times 10^3$ mm to kilometers.

## SECTION 1.7 Conversion Within and Between Systems of Units

Convert the following:

**25.** **a.** 1.5 min to seconds
**b.** 0.04 h to seconds
**c.** 0.05 s to microseconds
**d.** 0.16 m to millimeters
**e.** 0.00000012 s to nanoseconds
**f.** 3,620,000 s to days
**g.** 1020 mm to meters

**26.** **a.** 0.1 μF (microfarad) to picofarads
**b.** 0.467 km to meters
**c.** 63.9 mm to centimeters
**d.** 69 cm to kilometers
**e.** 3.2 h to milliseconds
**f.** 0.016 mm to μm
**g.** 60 sq cm ($cm^2$) to square meters ($m^2$)

***27.** **a.** 100 in. to meters
**b.** 4 ft to meters
**c.** 6 lb to newtons
**d.** 60,000 dyn to pounds
**e.** 150,000 cm to feet
**f.** 0.002 mi to meters (5280 ft = 1 mi)
**g.** 7800 m to yards

28. What is a mile in feet, yards, meters, and kilometers?
29. Calculate the speed of light in miles per hour using the defined speed of Section 1.4.
30. Find the velocity in miles per hour of a mass that travels 50 ft in 20 s.
31. How long in seconds will it take a car traveling at 100 mi/h to travel the length of a football field (100 yd)?
32. Convert 6 mi/h to meters per second.
33. If an athlete can row at a rate of 50 m/min, how many days would it take to cross the Atlantic (3000 mi)?
34. How long would it take a runner to complete a 10-km race if a pace of 6.5 min/mi were maintained?
35. Quarters are about 1 in. in diameter. How many would be required to stretch from one end of a football field to the other (100 yd)?
36. Compare the total time in hours to cross the United States (3000 mi) at an average speed of 55 mi/h versus an average speed of 65 mi/h. What is your reaction to the total time required versus the safety factor?
*37. Find the distance in meters that a mass traveling at 600 cm/s will cover in 0.016 h.
*38. If you were able to climb two steps per second, how long would it take to climb the Empire State Building if there are 102 floors and each floor is 12 ft? Each step is about 9 in.
*39. If the height of the Empire State Building were a horizontal distance, how long would it take a runner who can run 6-min miles to cover the distance? Use the data of Problem 38 and compare results. Gravity is certainly a factor with which to be reckoned.

## SECTION 1.9 Conversion Tables

40. Using Appendix A, determine the number of
    a. Btu in 5 joules of energy.
    b. cubic meters in 24 ounces of a liquid.
    c. seconds in 1.4 days.
    d. pints in 1 cubic meter of a liquid.

## SECTION 1.10 Computer Analysis and Design

41. Investigate the availability of computer courses and computer time in your curriculum. Which languages are commonly used and which software packages are popular?
42. Develop a list of five popular computer languages with a few characteristics of each. Why do you think some languages are better for the analysis of electric circuits than others?

## GLOSSARY

**BASIC** A language that employs familiar English phrases to direct the operation of a computer.

**CGS system** The system of units employing the *C*entimeter, *G*ram, and *S*econd as its fundamental units of measure.

**Difference Engine** One of the first mechanical calculators.

**Edison effect** Establishing a flow of charge between two elements in an evacuated tube.

**Electromagnetism** The relationship between magnetic and electrical effects.

**Eniac** The first totally electronic computer.

**Fleming's valve** The first of the electronic devices, the diode.

**Integrated circuit (IC)** A subminiature structure containing a vast number of electronic devices designed to perform a particular set of functions.

**Joule (J)** A unit of measurement for energy in the SI or MKS system. Equal to 0.7378 foot-pound in the English system and $10^7$ ergs in the CGS system.

**Kelvin (K)** A unit of measurement for temperature in the SI system. Equal to 273.15 + °C in the MKS and CGS systems.

**Kilogram (kg)** A unit of measure for mass in the SI and MKS systems. Equal to 1000 grams in the CGS system.

**Language** A communication link between user and computer to define the operations to be performed and the results to be displayed or printed.

**Leyden jar** One of the first charge storage devices.

**Menu** A computer generated list of choices for the user to determine the next operation to be performed.

**Meter (m)** A unit of measure for length in the SI and MKS systems. Equal to 1.094 yards in the English system and 100 centimeters in the CGS system.

**MKS system** The system of units employing the *M*eter, *K*ilogram, and *S*econd as its fundamental units of measure.

**Newton (N)** A unit of measurement for force in the SI and MKS systems. Equal to 100,000 dynes in the CGS system.

**Pound (lb)** A unit of measurement for force in the English system. Equal to 4.45 newtons in the SI or MKS system.

**Program** A sequential list of commands, instructions, etc. to perform a specified task using a computer.

**PSPICE** A software package designed to analyze a variety of dc, ac and transient electrical and electronic systems.

**Scientific notation** A method for describing very large and very small numbers through the use of powers of 10, which requires that the multiplier be a number between 1 and 10.

**Second (s)** A unit of measurement for time in the SI, MKS, English, and CGS systems.

**SI system** The system of units adopted by the IEEE in 1965 and the USASI in 1967 as the International System of Units (*S*ystème *I*nternational d'Unités).

**Slug** A unit of measure for mass in the English system. Equal to 14.6 kilograms in the SI or MKS system.

**Software package** A computer program designed to perform specific analysis and design operations or generate results in a particular format.

**Static electricity** Stationary charge in a state of equilibrium.

**Transistor** The first semiconductor amplifier.

**Voltaic cell** A storage device that converts chemical to electrical energy.

# 2
# Current and Voltage

## 2.1 ATOMS AND THEIR STRUCTURE

A basic understanding of the fundamental concepts of current and voltage requires a degree of familiarity with the atom and its structure. The simplest of all atoms is the hydrogen atom, made up of two basic particles, the *proton* and the *electron,* in the relative positions shown in Fig. 2.1(a). The *nucleus* of the hydrogen atom is the proton, a positively

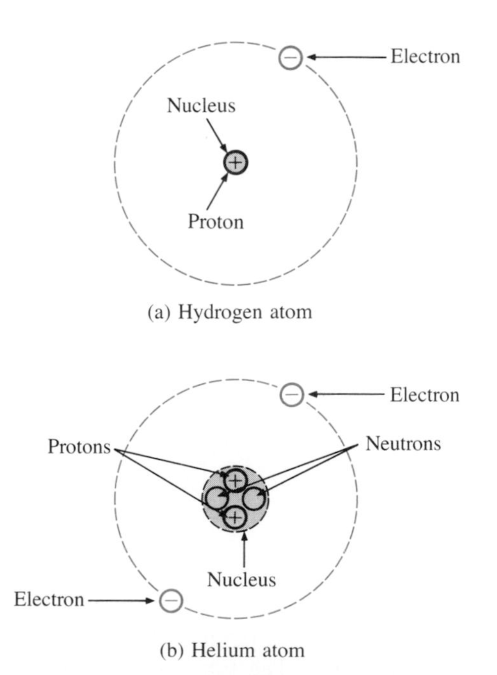


**FIG. 2.1**
*The hydrogen and helium atoms.*

charged particle. *The orbiting electron carries a negative charge that is equal in magnitude to the positive charge of the proton.* In all other elements, the nucleus also contains *neutrons,* which are slightly heavier than protons and have no electrical charge. The helium atom, for example, has two neutrons in addition to two electrons and two protons as shown in Fig. 2.1(b). *In all neutral atoms the number of electrons is equal to the number of protons.* The mass of the electron is $9.11 \times 10^{-28}$ g, and that of the proton and neutron is $1.672 \times 10^{-24}$ g. The mass of the proton (or neutron) is therefore approximately 1836 times that of the electron. The radii of the proton, neutron, and electron are all of the order of magnitude of $2 \times 10^{-15}$ m.

For the hydrogen atom, the radius of the smallest orbit followed by the electron is about $5 \times 10^{-11}$ m. The radius of this orbit is approximately 25,000 times that of the basic constituents of the atom. This is approximately equivalent to a sphere the size of a dime rotating about another sphere of the same size more than a quarter of a mile away.

Different atoms will have various numbers of electrons in the concentric shells about the nucleus. The first shell, which is closest to the nucleus, can contain only two electrons. If an atom should have three electrons, the third must go to the next shell. The second shell can contain a maximum of eight electrons, the third 18, and the fourth 32, as determined by the equation $2n^2$, where $n$ is the shell number. These shells are usually denoted by a number ($n = 1, 2, 3, \ldots$) or letter ($n = k, l, m, \ldots$).

Each shell is then broken down into subshells, where the first subshell can contain a maximum of two electrons, the second subshell six electrons, the third 10 electrons, and the fourth 14, as shown in Fig. 2.2. The subshells are usually denoted by the letters $s$, $p$, $d$, and $f$, in that order, outward from the nucleus.

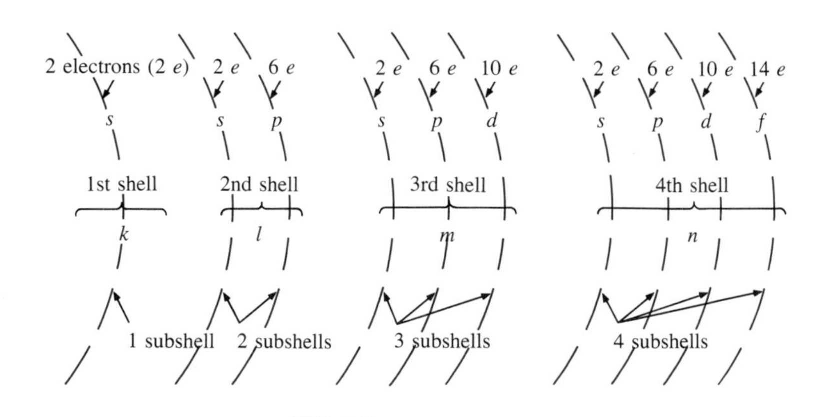


**FIG. 2.2**
*Shells and subshells of the atomic structure.*

It has been determined by experimentation that *unlike charges attract, and like charges repel.* The force of attraction or repulsion between two charged bodies $Q_1$ and $Q_2$ can be determined by Coulomb's law:

$$F \text{ (attraction or repulsion)} = \frac{kQ_1Q_2}{r^2} \qquad \textbf{(2.1)}$$

where $F$ is in newtons, $k$ = constant = $9.0 \times 10^9$, $Q_1$ and $Q_2$ are the charges in coulombs (to be introduced in Section 2.2), and $r$ is the distance in meters between the two charges. In particular, note the squared $r$ term in the denominator, resulting in rapidly decreasing levels of $F$ for increasing values of $r$.

In the atom, therefore, electrons will repel each other, and protons and electrons will attract each other. Since the nucleus consists of many positive charges (protons), a strong attractive force exists for the electrons in orbits close to the nucleus [note the effects of a large charge $Q$ and a small distance $r$ in Eq. (2.1)]. As the distance between the nucleus and the orbital electrons increases, the binding force diminishes until it reaches its lowest level at the outermost subshell (largest $r$). Due to the weaker binding forces, less energy must be expended to remove an electron from an outer subshell than from an inner subshell. Also, it is generally true that electrons are more readily removed from atoms having outer subshells that are incomplete *and,* in addition, possess few electrons. These properties of the atom that permit the removal of electrons under certain conditions are essential if motion of charge is to be created. Without this motion, this text could venture no further—our basic quantities rely on it.

*Copper* is the most commonly used metal in the electrical/electronics industry. An examination of its atomic structure will help identify why it has such widespread applications. The copper atom (Fig. 2.3) has one more electron than needed to complete the first three shells. This incomplete outermost subshell, possessing only one electron, and the distance

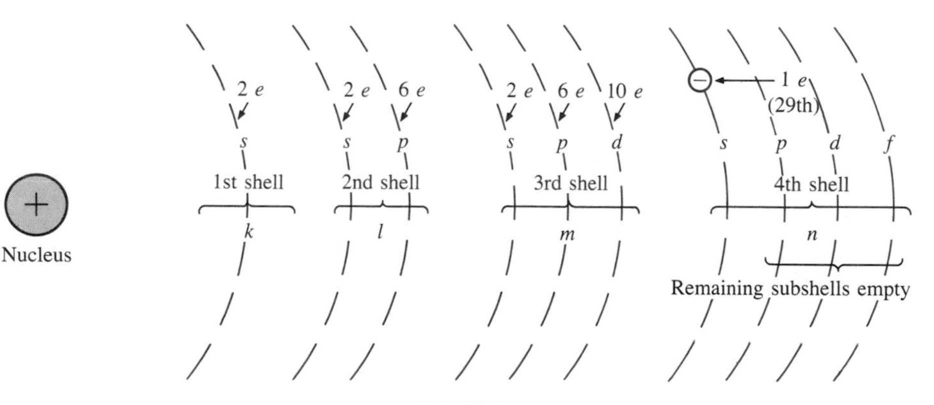


**FIG. 2.3**
*The copper atom.*

between this electron and the nucleus, reveal that the twenty-ninth electron is loosely bound to the copper atom. If this twenty-ninth electron gains sufficient energy from the surrounding medium to leave its parent atom, it is called a *free electron*. In one cubic inch of copper at room temperature there are approximately $1.4 \times 10^{+24}$ free electrons. Copper also has the advantage of being able to be drawn into long thin wires (ductility) or worked into many different shapes (malleability). Other metals that exhibit the same properties as copper, but to a different degree, are silver, gold, platinum, and aluminum. Gold is used extensively in integrated circuits where the performance level and amount of material required balance the cost factor. Aluminum has found some commercial use but suffers from being more temperature sensitive (expansion and contraction) than copper.

## 2.2 CURRENT

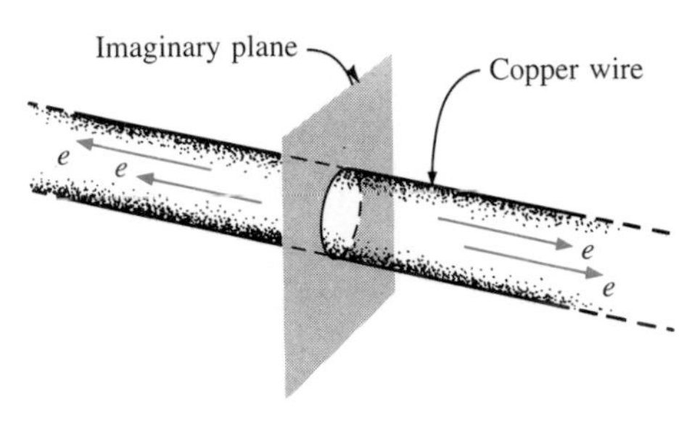


FIG. 2.4

Consider a short length of copper wire cut with an imaginary perpendicular plane, producing the circular cross section shown in Fig. 2.4. At room temperature with no external forces applied, there exists within the copper wire the random motion of free electrons created by the thermal energy that the electrons gain from the surrounding medium. When an atom loses its free electron, it acquires a net positive charge and is referred to as a *positive ion*. The free electron is able to move within these positive ions and leave the general area of the parent atom, while the positive ions only oscillate in a mean fixed position. For this reason,

***the free electron is the charge carrier in a copper wire or in any other solid conductor of electricity.***

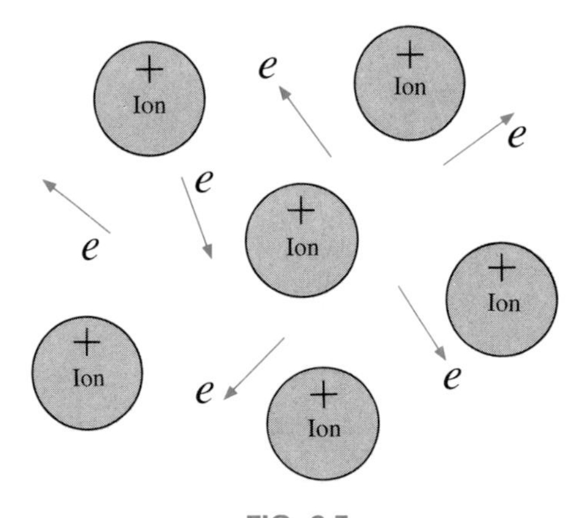


FIG. 2.5
*Random motion of free electrons in an atomic structure.*

An array of positive ions and free electrons is depicted in Fig. 2.5. Within this array, the free electrons find themselves continually gaining or losing energy by virtue of their changing direction and velocity. Some of the factors responsible for this random motion include (1) the collisions with positive ions and other electrons, (2) the attractive forces for the positive ions, and (3) the force of repulsion that exists between electrons. This random motion of free electrons is such that over a period of time, the number of electrons moving to the right across the circular cross section of Fig. 2.4 is exactly equal to the number passing over to the left.

***With no external forces applied, the net flow of charge in a conductor in any one direction is zero.***

Let us now connect this copper wire between two battery terminals as shown in Fig. 2.6. The battery, at the expense of chemical energy, places a net positive charge on one terminal and a net negative charge on the other. The instant the wire is connected between these two terminals, the free electrons of the copper wire will drift toward the positive

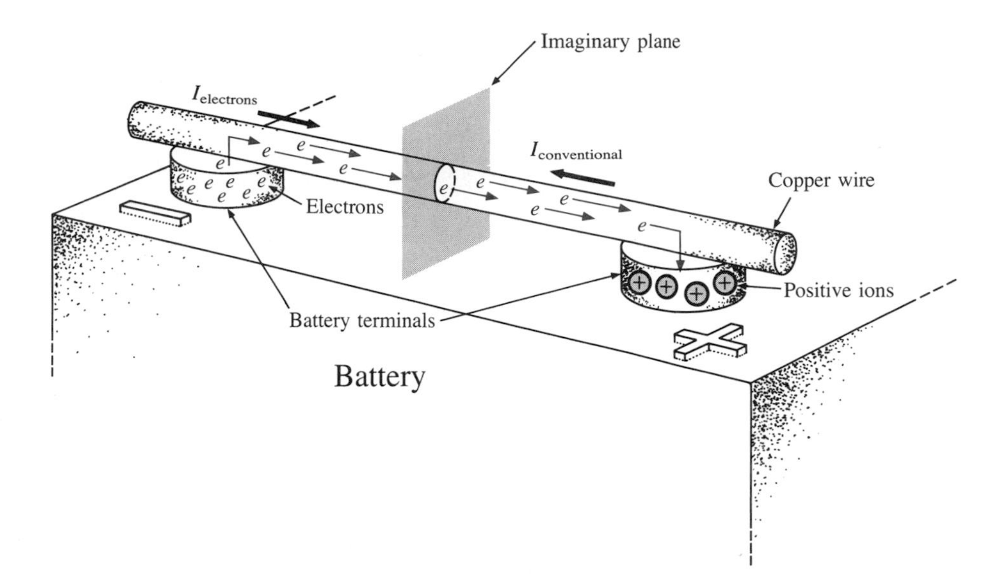

**FIG. 2.6**

terminal, while the positive ions will simply oscillate in a mean fixed position. The negative terminal is a supply of electrons to be drawn from when the electrons of the copper wire drift toward the positive terminal. The chemical activity of the battery will absorb the electrons at the positive terminal and maintain a steady supply of electrons at the negative terminal.

If $6.242 \times 10^{18}$ *electrons* drift at uniform velocity through the imaginary circular cross section of Fig. 2.6 in 1 *second,* the flow of charge, or *current,* is said to be 1 *ampere* (A). The discussion of Chapter 1 revealed that this is an enormous number of electrons passing through the surface in 1 second. The current associated with only a few electrons per second would be inconsequential and of little practical value. To establish numerical values that permit immediate comparisons between levels, a coulomb (C) of charge was defined as the total charge associated with $6.242 \times 10^{18}$ electrons. The charge associated with one electron can then be determined from

$$\text{Charge/electron} = Q_e = \frac{1\text{ C}}{6.242 \times 10^{18}} = 1.6 \times 10^{-19}\text{ C}$$

The current in amperes can now be calculated using the following equation:

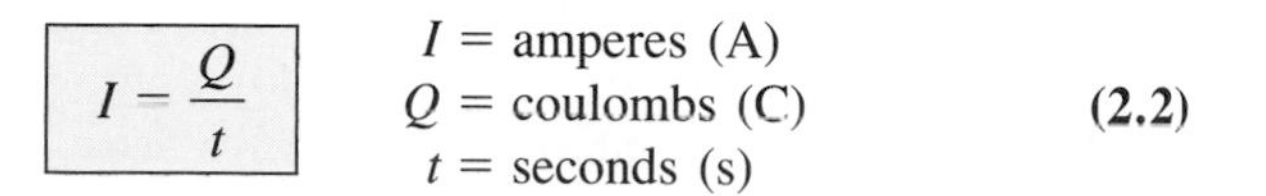

$$I = \frac{Q}{t} \qquad \begin{aligned} I &= \text{amperes (A)} \\ Q &= \text{coulombs (C)} \\ t &= \text{seconds (s)} \end{aligned} \qquad \textbf{(2.2)}$$

The capital letter $I$ was chosen from the French word for current: *intensité*. The SI abbreviation for each quantity in Eq. (2.2) is provided to the right of the equation. The equation clearly reveals that for equal time intervals, the more charge that flows through the wire, the heavier the current.

Through algebraic manipulations, the other two quantities can be determined as follows:

$$\boxed{Q = It} \qquad \text{(coulombs, C)} \qquad \textbf{(2.3)}$$

and

$$\boxed{t = \frac{Q}{I}} \qquad \text{(seconds, s)} \qquad \textbf{(2.4)}$$

---

**EXAMPLE 2.1.** The charge flowing through the imaginary surface of Fig. 2.6 is 0.16 C every 64 ms. Determine the current in amperes.

***Solution:*** Eq. (2.2):

$$I = \frac{Q}{t} = \frac{0.16\ \text{C}}{64 \times 10^{-3}\ \text{s}} = \frac{160 \times 10^{-3}\ \text{C}}{64 \times 10^{-3}\ \text{s}} = \mathbf{2.50\ A}$$

---

**EXAMPLE 2.2.** Determine the time required for $4 \times 10^{16}$ electrons to pass through the imaginary surface of Fig. 2.6 if the current is 5 mA.

***Solution:*** Determine $Q$:

$$4 \times 10^{16}\ \cancel{\text{electrons}} \left( \frac{1\ \text{coulomb}}{6.242 \times 10^{18}\ \cancel{\text{electrons}}} \right) = 0.641 \times 10^{-2}\ \text{C}$$

$$= 0.00641\ \text{C} = 6.41\ \text{mC}$$

Calculate $t$ [Eq. (2.4)]:

$$t = \frac{Q}{I} = \frac{6.41 \times 10^{-3}\ \text{C}}{5 \times 10^{-3}\ \text{A}} = \mathbf{1.282\ s}$$

---

A second glance at Fig. 2.6 will reveal that two directions of charge flow have been indicated. One is called *conventional flow* while the other is called *electron flow*. This text will deal only with conventional flow for a variety of reasons, including the fact that it is the most widely used at educational institutions and in industry, is employed in the design of all electronic device symbols, and is the popular choice for all major computer software packages. The flow controversy is a result of an assumption made at the time electricity was discovered that the positive charge was the moving particle in metallic conductors. Be assured that the choice of conventional flow will not create great difficulty and confusion in the chapters to follow. Once the direction of $I$ is established, the issue is dropped and the analysis can continue.

## 2.3 VOLTAGE

The flow of charge described in the previous section is established by an external ''pressure'' derived from the energy that a mass has by virtue of its position: *potential energy*.

*Energy*, by definition, is the *capacity to do work*. If a mass ($m$) is raised to some height ($h$) above a reference plane, it has a measure of potential energy expressed in *joules* (J) that is determined by

$$\boxed{\text{Potential energy (PE)} = mgh} \qquad \text{(joules, J)} \qquad \textbf{(2.5)}$$

where $g$ is the gravitational acceleration (9.754 m/s$^2$). This mass now has the ability to do work such as crush an object placed on the reference plane. If the weight is raised further, it has an increased measure of potential energy and can do additional work. There is an obvious *difference in potential* between the two heights above the reference plane.

In the battery of Fig. 2.6, the internal chemical action will establish (through an expenditure of energy) an accumulation of negative charges (electrons) on one terminal (the negative terminal) and positive charges (positive ions) on the other (the positive terminal). A ''positioning'' of the charges has been established that will result in a *potential difference* between the terminals. If a conductor is connected between the terminals of the battery, the electrons at the negative terminal have sufficient potential energy to overcome collisions with other particles in the conductor and the repulsion from similar charges to reach the positive terminal to which they are attracted.

Charge can be raised to a higher potential level through the expenditure of energy from an external source, or it can lose potential energy as it travels through an electrical system. In any case, by definition:

***A potential difference of 1 volt (V) exists between two points if 1 joule (J) of energy is exchanged in moving 1 coulomb (C) of charge between the two points.***

Pictorially, if one joule of energy (1 J) is required to move the one coulomb (1 C) of charge of Fig. 2.7 from position $x$ to position $y$, the

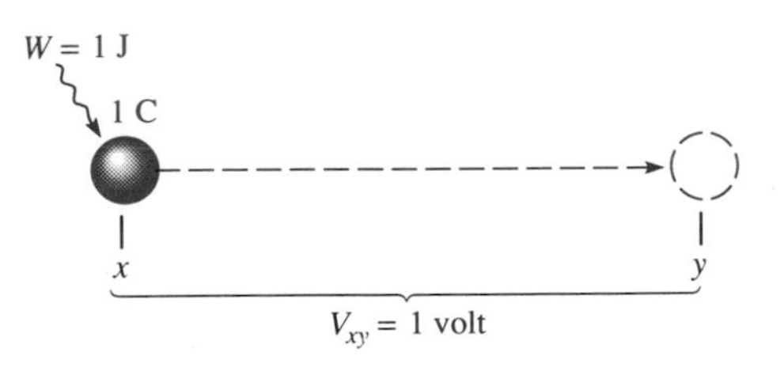


**FIG. 2.7**
*Defining the unit of measurement for voltage.*

potential difference or voltage between the two points is one volt (1 V). If the energy required to move the 1 C of charge increases to 12 J due to additional opposing forces then the potential difference will increase to

12 V. Voltage is therefore an indication of how much energy is involved in moving a charge between two points in an electrical system. Conversely, the higher the voltage rating of an energy source such as a battery, the more energy available to move charge through the system. Note in the above discussion that two points are always involved when talking about voltage or potential difference. In the future, therefore, it is very important to keep in mind that

***a potential difference or voltage is always measured between two points in the system. Changing either point may change the potential difference between the two points under investigation.***

In general, the potential difference between two points is determined by

$$V = \frac{W}{Q} \quad \text{(volts)} \tag{2.6}$$

Through algebraic manipulations, we have

$$W = QV \quad \text{(joules)} \tag{2.7}$$

and

$$Q = \frac{W}{V} \quad \text{(coulombs)} \tag{2.8}$$

---

**EXAMPLE 2.3.** Find the potential difference between two points in an electrical system if 60 J of energy are expended by a charge of 20 C between these two points.

***Solution:*** Eq. (2.6):

$$V = \frac{W}{Q} = \frac{60 \text{ J}}{20 \text{ C}} = \mathbf{3\ V}$$

---

---

**EXAMPLE 2.4.** Determine the energy expended moving a charge of 50 $\mu$C through a potential difference of 6 V.

***Solution:*** Eq. (2.7):

$$W = QV = (50 \times 10^{-6} \text{ C})(6 \text{ V}) = 300 \times 10^{-6} \text{ J} = \mathbf{300\ \mu J}$$

---

Notation plays a very important role in the analysis of electrical and electronic systems. To distinguish between sources of voltage (batteries and the like) and losses in potential across dissipative elements, the following notation will be used:

$E$ for voltage sources (volts)
$V$ for voltage drops (volts)

In summary, the applied potential difference (in volts) of a voltage source in an electric circuit is the "pressure" to set the system in motion and "cause" the flow of charge or current through the electrical system. A mechanical analogy of the applied voltage is the pressure applied to the water in a main. The resulting flow of water through the system is likened to the flow of charge through an electric circuit. Without the applied pressure from the spigot, the water will simply sit in the hose, just as the electrons of a copper wire do not have a general direction without an applied voltage.

## 2.4 FIXED (dc) SUPPLIES

The terminology *dc* employed in the heading of this section is an abbreviation for *direct current,* which encompasses the various electrical systems in which there is a *unidirectional* ("one direction") flow of charge. A great deal more will be said about this terminology in the chapters to follow. For now, we will consider only those supplies that provide a fixed voltage or current.

### dc Voltage Sources

Since the dc voltage source is the more familiar of the two types of supplies, it will be examined first. The symbol used for all dc voltage supplies in this text appears in Fig. 2.8. The relative lengths of the bars indicate the terminals they represent.

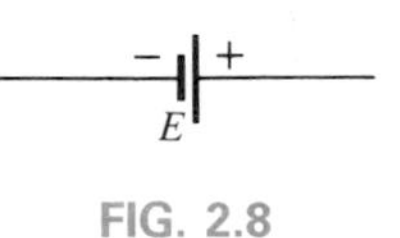

FIG. 2.8

Dc voltage sources can be divided into three broad categories: (1) batteries (chemical action), (2) generators (electromechanical), and (3) power supplies (rectification).

**Batteries** For the layperson, the battery is the most common of the dc sources. By definition, a battery (derived from the expression "battery of cells") consists of a combination of two or more similar *cells,* a cell being the fundamental source of electrical energy developed through the conversion of chemical or solar energy. All cells can be divided into the *primary* or *secondary* types. The secondary is rechargeable, whereas the primary is not. That is, the chemical reaction of the secondary cell can be reversed to restore its capacity. The two most common rechargeable batteries are the lead-acid unit (used primarily in automobiles) and the nickel-cadmium battery (used in calculators, tools, photoflash units, shavers, and so on). The obvious advantage of the rechargeable unit is the reduced costs associated with not having to continually replace discharged primary cells.

All of the cells appearing in this chapter except the *solar cell,* which absorbs energy from incident light in the form of photons, establish a

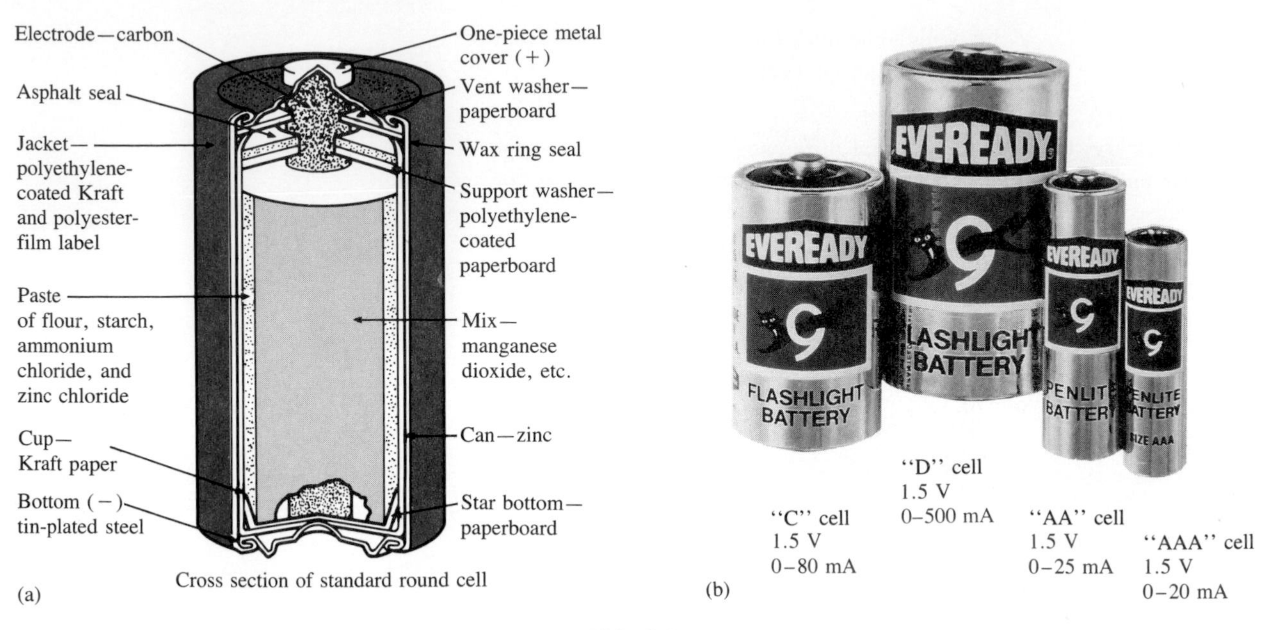


**FIG. 2.9**

*Carbon-zinc primary battery. (a) Construction; (b) appearance and ratings. (Courtesy of Eveready Batteries)*

potential difference at the expense of chemical energy. In addition, each has a positive and a negative *electrode* and an *electrolyte* to complete the circuit between electrodes within the battery. The electrolyte is the contact element and the source of ions for conduction between the terminals.

The popular carbon-zinc primary battery uses a zinc can as its negative electrode, a manganese dioxide mix and carbon rod as its positive electrode, and an electrolyte that is a mix of ammonium and zinc chlorides, flour, and starch, as shown in Fig. 2.9. Figure 2.10 shows a number of other types of primary units with an area of application and a rating to be considered later in this section.

(a) Lithiode™ lithium-iodine cell
2.8 V, 870 mAh
Long-life power sources with printed circuit board mounting capability

(b) Lithium-iodine pacemaker cell
2.8 V, 2.0 Ah

(c) Eveready transistor battery
9 V, 450 mAh

**FIG. 2.10**

*Primary cells. (Parts (a) and (b) courtesy of Catalyst Research Corp.; part (c) courtesy of Eveready Batteries)*

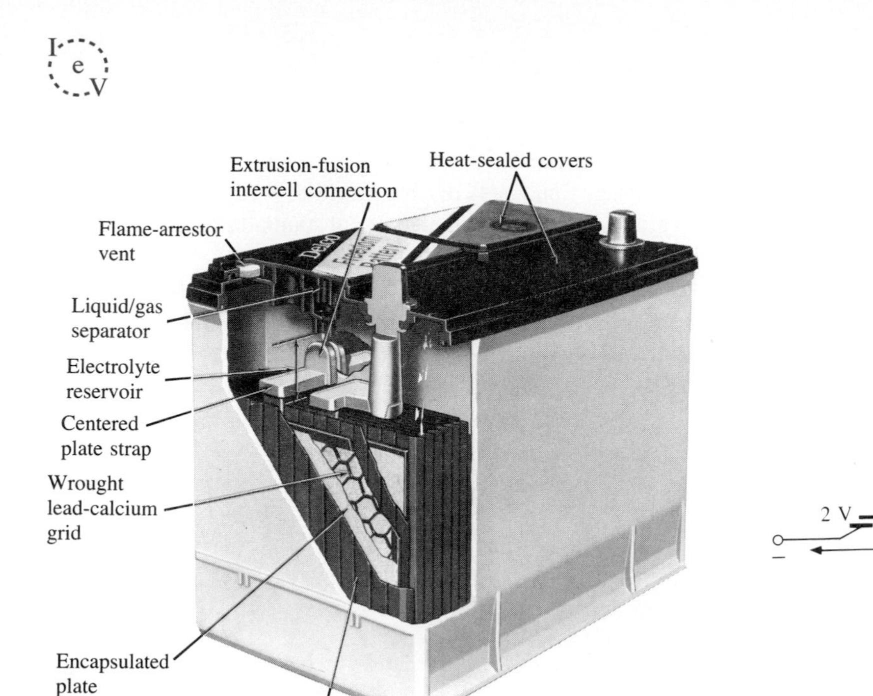


**FIG. 2.11**
*Maintenance-free 12-V lead-acid battery.* (*Courtesy of Delco-Remy, a division of General Motors Corp.*)

For the secondary lead-acid unit appearing in Fig. 2.11, the electrolyte is sulfuric acid and the electrodes are spongy lead (Pb) and lead peroxide ($PbO_2$). When a load is applied to the battery terminals, there is a transfer of electrons from the spongy lead electrode to the lead peroxide electrode through the load. This transfer of electrons will continue until the battery is completely discharged. The discharge time is determined by how diluted the acid has become and how heavy the coating of lead sulfate is on each plate. The state of discharge of a lead storage cell can be determined by measuring the specific gravity of the electrolyte with a hydrometer. The specific gravity of a substance is defined to be the ratio of the weight of a given volume of the substance to the weight of an equal volume of water at 4°C. For fully charged batteries, the specific gravity should be somewhere between 1.28 and 1.30. When the specific gravity drops to about 1.1, the battery should be recharged.

Since the lead storage cell is a secondary cell, it can be recharged at any point during the discharge phase simply by applying an external dc source across the cell that will pass current through the cell in a direction opposite to that in which the cell supplied current to the load. This will remove the lead sulfate from the plates and restore the concentration of sulfuric acid.

The output of a lead storage cell over most of the discharge phase is about 2 V. In the commercial lead storage batteries used in the automobile, the 12 V can be produced by six cells in series, as shown in Fig. 2.11. The use of a grid made from a wrought lead-calcium alloy strip rather than the lead-antimony cast grid commonly used has resulted in maintenance-free batteries such as that appearing in the same figure.

The lead-antimony structure was susceptible to corrosion, overcharge, gassing, water usage, and self-discharge. Improved design with the lead-calcium grid has either eliminated or substantially reduced most of these problems.

The nickel-cadmium battery is a rechargeable battery that has been receiving enormous interest and development in recent years. A number of such batteries manufactured by the Union Carbide Corporation and the General Electric Company appear in Fig. 2.12. The internal con-

(a)

Eveready® BH 500 cell
1.2 V, 500 mAh
*App:* Where vertical height is severe limitation

(b)

Printed circuit board mountable battery
2.4 V, 70 mAh

(c)

**FIG. 2.12**
*Rechargeable nickel-cadmium batteries. (Parts (a) and (b) courtesy of Eveready Batteries; part (c) courtesy of General Electric Co.)*

struction of the cylindrical-type cell appears in Fig. 2.13. In the fully charged condition the positive electrode is nickel hydroxide [$Ni(OH)_2$]; the negative electrode, metallic cadmium (Cd); and the electrolyte, potassium hydroxide (KOH). The oxidation (increased oxygen content) of the negative electrode occurring simultaneously with the reduction of the positive electrode provides the required electrical energy. The separator is required to isolate the two electrodes and maintain the location of the electrolyte. The advantage of such cells is that the active materials go through a change in oxidation state necessary to establish the required ion level without a change in the physical state. This establishes an excellent recovery mechanism for the recharging phase.

A high-density, 40-W solar cell appears in Fig. 2.14 with some of its associated data and areas of application. Since the maximum available wattage in an average bright sunlit day is 100 mW/cm$^2$ and conversion efficiencies are currently between 10% and 14%, the maximum available power per square centimeter from most commercial units is between 10 mW and 14 mW. For a square meter, however, the return

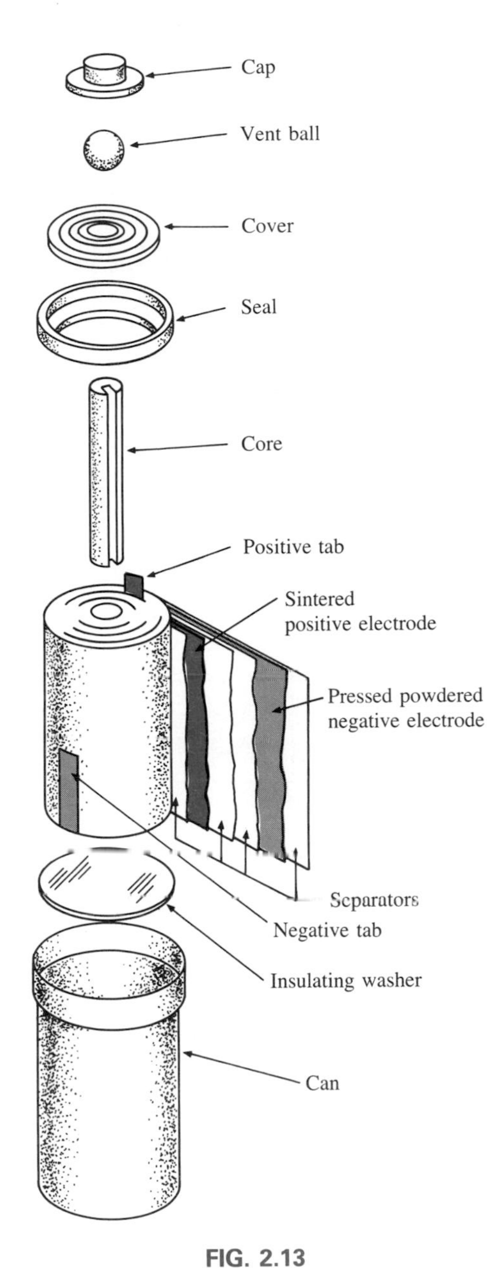


**FIG. 2.13**
*Internal structure of the cylindrical-type nickel-cadmium rechargeable cell. (Courtesy of Eveready Batteries)*

40-W, high-density solar module
100-mm × 100-mm (4″ × 4″) square cells are used to provide maximum power in a minimum of space. The 33 series cell module provides a strong 12-V battery charging current for a wide range of temperatures (−40°C to 60°C)

**FIG. 2.14**
*Solar module. (Courtesy of Motorola Semiconductor Products)*

would be 100 W to 140 W. A more detailed description of the solar cell will appear in your electronics courses. For now it is important to realize that a fixed illumination of the solar cell will provide a fairly steady dc voltage for driving various loads, from watches to automobiles.

Batteries have a capacity rating given in ampere-hours (Ah) or milliampere-hours (mAh). Some of these ratings are included in the above figures. A battery with an ampere-hour rating of 100 will theoretically provide a steady current of 1 A for 100 h, 2 A for 50 h, 10 A for 10 h, and so on, as determined by the following equation:

$$\text{Life (hours)} = \frac{\text{ampere-hour rating (Ah)}}{\text{amperes drawn (A)}} \tag{2.9}$$

Two factors that affect this rating, however, are the temperature and the rate of discharge. The disc-type EVEREADY® BH 500 cell appearing in Fig. 2.12 has the terminal characteristics appearing in Fig. 2.15. Figure 2.15 reveals that

***the capacity of a dc battery decreases with increase in the current demand***

and

***the capacity of a dc battery decreases at relatively (compared to room temperature) low and high temperatures.***

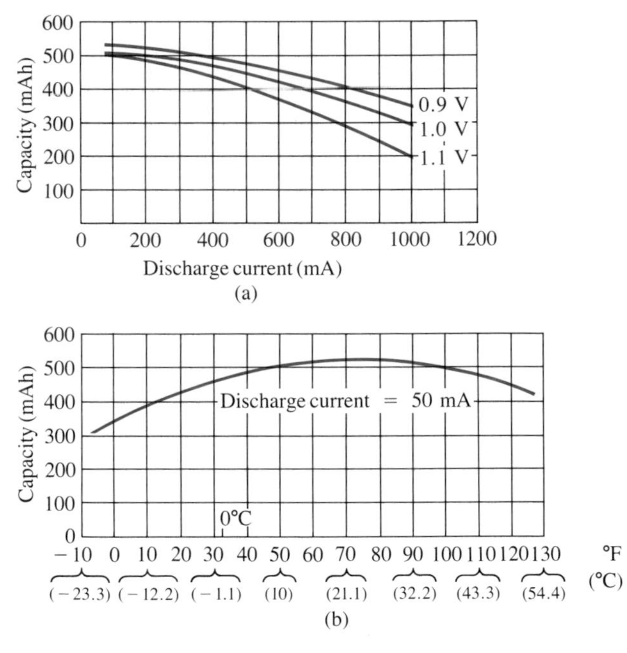

**FIG. 2.15**

*EVEREADY® BH 500 cell characteristics. (a) Capacity vs. discharge current; (b) capacity vs. temperature. (Courtesy of Eveready Batteries)*

For the 1-V unit of Fig. 2.15(a), the rating is above 500 mAh at a discharge current of 100 mA but drops to 300 mAh at about 1 A. For a unit that is less than $1\frac{1}{2}$ in. in diameter and less than 1/2 in. in thickness, however, these are excellent terminal characteristics. Figure 2.15(b) reveals that the maximum mAh rating (at a current drain of 50 mA) occurs at about 75°F (≅ 24°C), or just above average room temperature. Note how the curve drops to the right and left of this maximum value. We are all aware of the reduced "strength" of a battery at low temperatures. Note that it has dropped to almost 300 mAh at −20°C.

Another curve of interest appears in Fig. 2.16. It provides the expected cell voltage at a particular drain over a period of hours of use. It is noteworthy that the loss in hours between 50 mA and 100 mA is much greater than between 100 mA and 150 mA, even though the increase in current is the same between levels. In general,

***the terminal voltage of a dc battery decreases with the length of the discharge time at a particular drain current.***

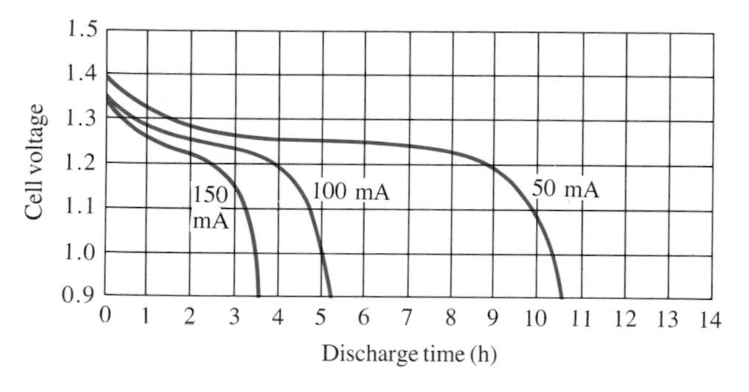


**FIG. 2.16**
*EVEREADY® BH 500 cell discharge curves.* (*Courtesy of Eveready Batteries*)

---

**EXAMPLE 2.5.**

a. Determine the capacity in milliampere-hours for the 0.9-V BH 500 cell of Fig. 2.15(a) if the discharge current is 600 mA.
b. At what temperature will the mAh rating of the cell of Fig. 2.15(b) be 90% of its maximum value if the discharge current is 50 mA?

***Solutions:***

a. From Fig. 2.15(a), the capacity at 600 mA is about 450 mAh. Thus, from Eq. (2.9),

$$\text{Life} = \frac{450\text{ mAh}}{600\text{ mA}} = 0.75\text{ h} = \mathbf{45\text{ min}}$$

b. From Fig. 2.15(b), the maximum is approximately 520 mAh. The 90% level is therefore 468 mAh, which occurs just above freezing, or **1°C,** and at the higher temperature of **45°C.**

---

**Generators** The dc generator is quite different, both in construction (Fig. 2.17) and in mode of operation, from the battery. When the shaft of the generator is rotating at the nameplate speed due to the applied torque of some external source of mechanical power, a voltage of rated value will appear across the external terminals. The terminal voltage and power-handling capabilities of the dc generator are typically higher than those of most batteries, and its lifetime is determined only by its construction. Commercially used dc generators are typically of the

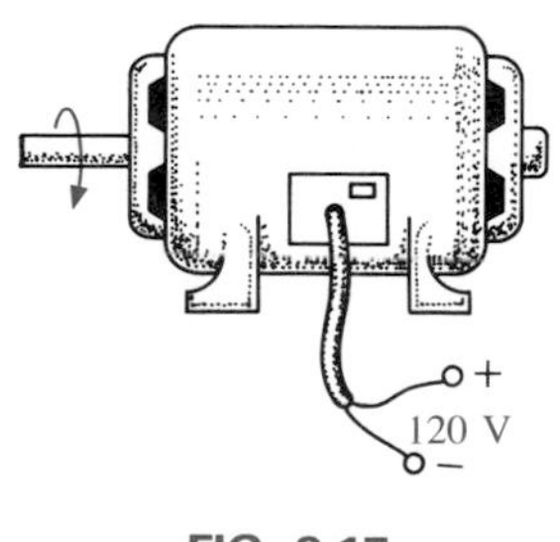


**FIG. 2.17**
*dc generator.*

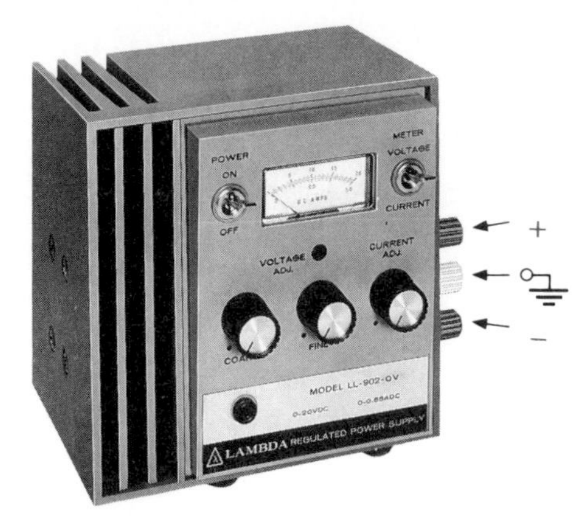


**FIG. 2.18**
*dc laboratory supply. (Courtesy of Lambda Electronics Corp.)*

120-V or 240-V variety. As pointed out earlier in this section, for the purposes of this text no distinction will be made between the symbol for a battery and a generator.

**Power supplies** The dc supply encountered most frequently in the laboratory employs the rectification and filtering processes as its means toward obtaining a steady dc voltage. By this process, a time-varying voltage (such as ac voltage available from a home outlet) is converted to one of a fixed magnitude. This process will be covered in detail in the basic electronics courses. A dc laboratory supply of this type appears in Fig. 2.18.

Most dc laboratory supplies have a regulated, adjustable voltage output with three available terminals, as indicated in Figs. 2.18 and 2.19(a). The symbol for ground or zero potential (the reference) is also shown in Fig. 2.19(a). If 10 volts above ground potential are required, then the connections are made as shown in Fig. 2.19(b). If 15 volts below ground potential are required, then the connections are made as shown in Fig. 2.19(c). If connections are as shown in Fig. 2.19(d), we say we have a ''floating'' voltage of 5 volts since the reference level is not included. Seldom is the configuration of Fig. 2.19(d) employed since it fails to protect the operator by providing a direct low resistance

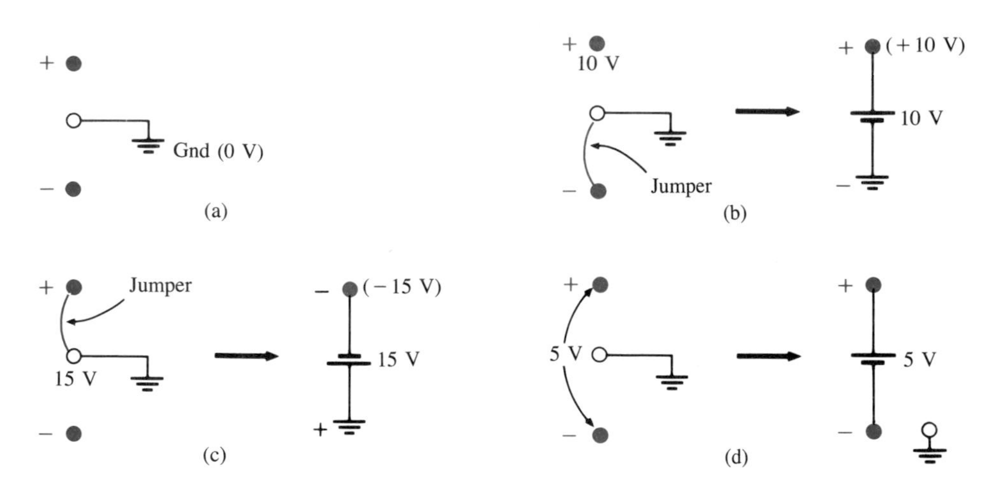


**FIG. 2.19**
*Possible output connections of a dc laboratory supply.*

path to ground and to establish a common ground for the system. In any case, the positive and negative terminals must be part of any circuit configuration.

## dc Current Sources

The wide variety of types of and applications for the dc voltage source has resulted in its becoming a rather familiar device, the characteristics of which are understood, at least basically, by the layperson. For example, it is common knowledge that a 12-V car battery has a terminal voltage (at least approximately) of 12 V even though the current drain by the automobile may vary under different operating conditions. In other words, *a dc voltage source ideally will provide a fixed terminal voltage even though the current drain may vary,* as depicted in Fig. 2.20(a). A dc current source is the dual of the voltage source. That is,

***the current source will, ideally, supply a fixed current to a load even though there will be variations in the terminal voltage as determined by the load,***

as depicted in Fig. 2.20(b). (Do not become alarmed if the concept of a current source is strange and somewhat confusing at this point. It will be covered in great detail in later chapters.)

The introduction of semiconductor devices such as the transistor has accounted in large measure for the increasing interest in current sources. A representative commercially available dc current source appears in Fig. 2.21.

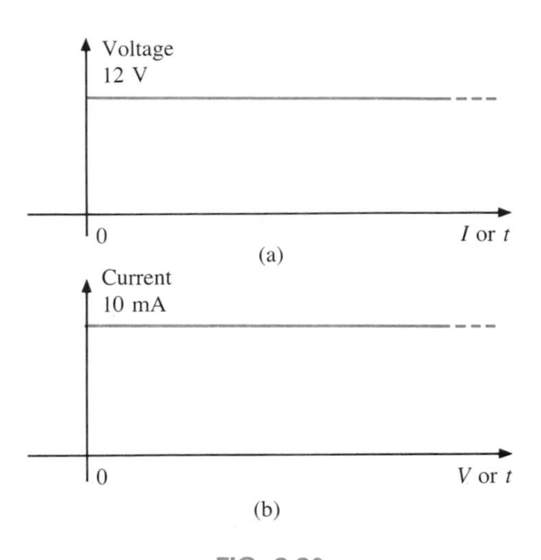

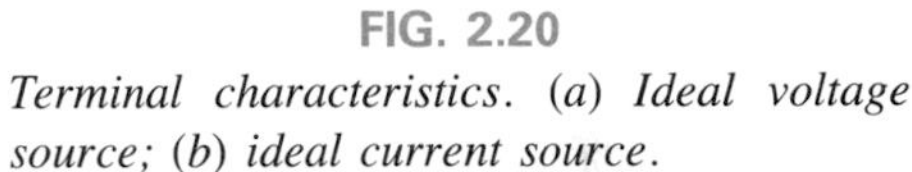


FIG. 2.20
*Terminal characteristics. (a) Ideal voltage source; (b) ideal current source.*

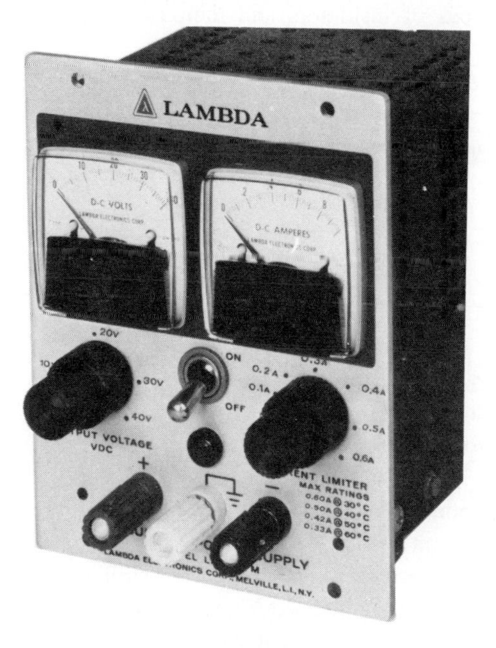


FIG. 2.21
*dc current source. (Courtesy of Lambda Electronics Corp.)*

## 2.5 CONDUCTORS AND INSULATORS

Different wires placed across the same two battery terminals will allow different amounts of charge to flow between the terminals. Many factors, such as density, mobility, and stability of the material, account for these variations in charge flow. In general, however,

***conductors are those materials that permit a generous flow of electrons with very little external force (voltage) applied.***

**TABLE 2.1**

*Relative conductivity of various materials.*

| Metal | Relative Conductivity (%) |
|---|---|
| Silver | 105 |
| Copper | **100** |
| Gold | 70.5 |
| Aluminum | 61 |
| Tungsten | 31.2 |
| Nickel | 22.1 |
| Iron | 14 |
| Constantan | 3.52 |
| Nichrome | 1.73 |
| Calorite | 1.44 |

In addition,

***good conductors typically have only one electron in the valence (most distant from the nucleus) ring.***

Since copper is used most frequently, it serves as the standard of comparison for the relative conductivity in Table 2.1. Note that aluminum, which has seen some commercial use, has only 61% of the conductivity level of copper, but keep in mind that this must be weighed against the cost and weight factors.

***Insulators are those materials that have very few free electrons and require a large applied potential (voltage) to establish a measureable current level.***

A common use of insulating material is for covering current-carrying wire, which, if uninsulated, could cause dangerous side effects. Power-line repair people wear rubber gloves and stand on rubber mats as safety measures when working on high-voltage transmission lines. A number of different types of insulators and their applications appear in Fig. 2.22.

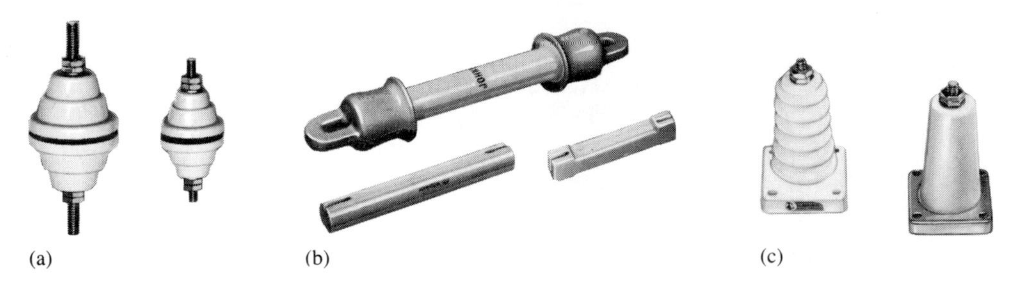

(a) (b) (c)

**FIG. 2.22**

*Insulators. (a) Insulated thru-panel bushings; (b) antenna strain insulators; (c) porcelain stand-off insulators. (Courtesy of Herman H. Smith, Inc.)*

It must be pointed out, however, that even the best insulator will break down (permit charge to flow through it) if a sufficiently large potential is applied across it. The breakdown strengths of some common insulators are listed in Table 2.2. According to this table, for insulators with the same geometric shape, it would require 270/30 = 9 times as much potential to pass current through rubber as compared to air and approximately 67 times as much voltage to pass current through mica as through air.

**TABLE 2.2**

*Breakdown strength of some common insulators.*

| Material | Average Breakdown Strength (kV/cm) |
|---|---|
| Air | 30 |
| Porcelain | 70 |
| Oils | 140 |
| Bakelite | 150 |
| Rubber | 270 |
| Paper (Paraffin-coated) | 500 |
| Teflon | 600 |
| Glass | 900 |
| Mica | 2000 |

## 2.6 SEMICONDUCTORS

***Semiconductors are a specific group of elements that exhibit characteristics that lie between those of insulators and conductors.***

The term *semi,* included in the terminology, has the dictionary definition of *half, partial,* or *between* as defined by its use. The entire electronics industry is dependent on this class of materials, since the electronic devices and integrated circuits (ICs) are constructed of semiconductor materials. Although *silicon* (Si) is the most extensively employed material, *germanium* (Ge) and *gallium arsenide* (GaAs) are also used in a number of important devices.

***Semiconductor materials typically have four electrons in the outermost valence ring.***

Semiconductors are further characterized as being photoconductive and having a negative temperature coefficient. Photoconductivity is a phenomenon where the photons (small packages of energy) from incident light can increase the carrier density in the material and thereby the charge flow level. A negative temperature coefficient reveals that the resistance (a characteristic to be described in detail in the next chapter) will decrease with increase in temperature (opposite to that of most conductors). A great deal more will be said about semiconductors in the chapters to follow and in your basic electronics courses.

## 2.7 AMMETERS AND VOLTMETERS

It is important to be able to measure the current and voltage levels in the network in order to check its operation, isolate malfunctions, and investigate effects impossible to predict on paper. As the names imply, *ammeters* are used to measure current levels, and *voltmeters,* the potential difference between two points. If the current levels are usually of the order of milliamperes, the instrument will be referred to as a milliammeter, and if in the microampere range, as a microammeter. Similar statements can be made for voltage levels. Throughout the industry, voltage levels are measured more frequently than current levels primarily because the former does not require that the network connections be disturbed.

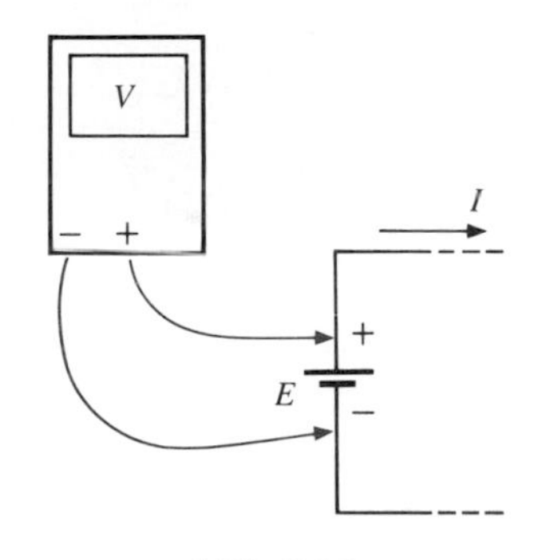

**FIG. 2.23**
*Voltmeter connection for an up-scale reading.*

The potential difference between two points can be measured by simply connecting the leads of the meter *across the two points* as indicated in Fig. 2.23. An up-scale reading is obtained by placing the positive lead of the meter to the point of higher potential of the network and the common or negative lead to the point of lower potential. The reverse connection will result in a negative reading or a below-zero indication.

Ammeters are connected in the *same branch* in which the current is to be measured, as shown in Fig. 2.24. Since ammeters measure the rate of flow of charge, the meter must be placed in the network such that the charge will flow through the meter. The only way this can be accomplished is to open the branch in which the current is to be measured and place the meter between the two resulting terminals. For the network of Fig. 2.24, the source lead must be disconnected from the network and the ammeter inserted as shown. An up-scale reading will be obtained if the polarities on the terminals of the ammeter are such that the current of the network enters the positive terminal or terminal at the higher potential.

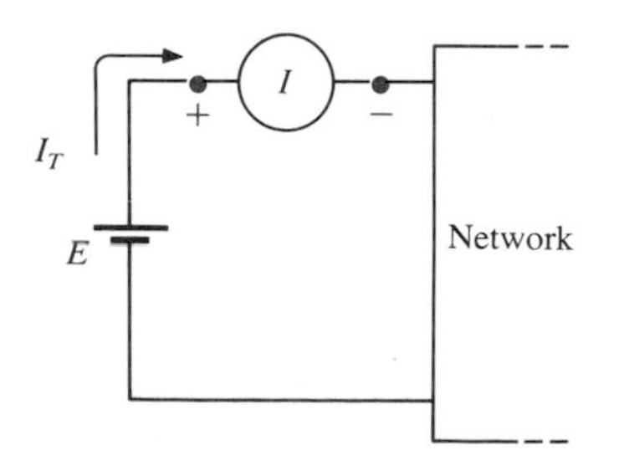

**FIG. 2.24**
*Ammeter connection for an up-scale reading.*

The introduction of any meter into the network raises a concern about whether the meter will affect the behavior of the network. This question and others will be examined in Chapters 5 and 6 after additional terms and concepts have been introduced.

Instruments exist that are designed to measure just current or just voltage levels. However, the most common laboratory meters include the *Volt-Ohm-Milliammeter* (VOM) and the *Digital Multimeter* (DMM) of Figs. 2.25 and 2.26, respectively. Both instruments will measure

FIG. 2.25
*Volt-Ohm-Milliammeter (VOM). (Courtesy of Simpson Electric Co.)*

FIG. 2.26
*Digital Multimeter (DMM). (Courtesy of John Fluke Mfg. Co. Inc.)*

voltage and current and a third quantity, resistance, to be introduced in the next chapter. The VOM uses an analog scale, which requires interpreting the position of a pointer on a continuous scale, while the DMM provides a display of numbers with decimal point accuracy determined by the chosen scale. Comments on the characteristics and use of a variety of meters will be made throughout the text. However, the major study of meters will be left for the laboratory sessions.

## 2.8 COMPUTER INPUT OF dc VOLTAGE SOURCES

In an effort to develop a familiarity with the PSPICE software package and the BASIC language, the early chapters of the text will review how some of the basic elements of a network are entered into the computer. In this chapter, since the independent voltage source was introduced, we review the procedure for entering the information.

The key word in entering any data into a computer is *format*. The data *must* be entered in a specified manner, or it will be rejected or, worse yet, misinterpreted. There is also no room for sloppiness when entering data. One misplaced comma or a wrong letter can invalidate entered data or completely change their magnitude.

Neither PSPICE nor BASIC distinguishes between upper- and lowercase letters, but, since computer programs are traditionally written in uppercase, we use uppercase throughout this text. Some of the following comments may be a repetition of material appearing in Appendix A, but they are repeated for completeness. In addition, keep in mind throughout the text that the computer content is not meant to make you an expert at using PSPICE or BASIC but simply to provide a surface treatment of each for familiarity and comparison purposes. For both approaches, literature on each should be consulted using those avenues available to you through your educational institution.

## PSPICE

In PSPICE all the network elements are entered in a specific format followed by a statement that indicates what output information you require. You will find that the first few applications of PSPICE will result in very short listings of input data, letting the software package do all the necessary maneuvers and calculations.

In PSPICE every element is defined between terminals called *nodes*. At this stage in your development suffice it to say that a node is simply a connection point between one element and another. Obviously, therefore, since every dc voltage source has two terminals to be connected in the network, it has two nodes to be specified. The basic format for entering a dc voltage source into PSPICE is the following:

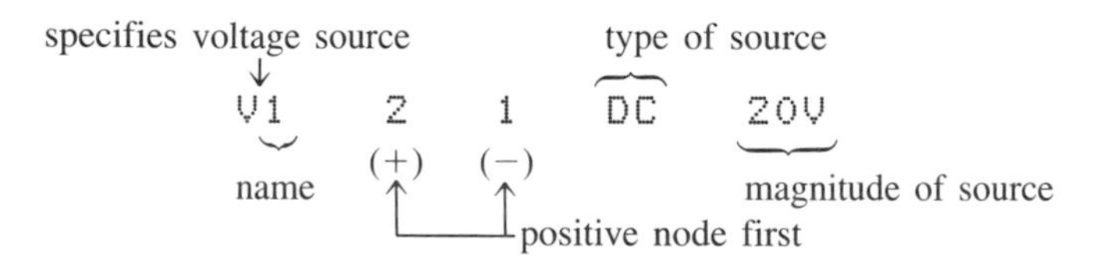


The input must begin with the uppercase letter V followed by whatever *name* you may want to give the source. Typically the name is limited to 8 characters, which can include numbers and letters. The next number is the node corresponding with the positive side of the battery, as shown in Fig. 2.27(a). Node numbers can have any value between (and including) 0 and 9999 but do not have to be sequential in their use as long as the network is properly defined. The DC reveals that it is a dc source, and the 20 V is its magnitude. The unit V is unnecessary but is included for clarity. In PSPICE any letters that follow a number are ignored, but they are often included for the user and future reference. The spacing between items is not important, but an even spacing provides a pleasing format. The basic format is certainly not that hard to follow and should present little difficulty in entering or reading. For instance, the following defines the voltage source of Fig. 2.27(b).

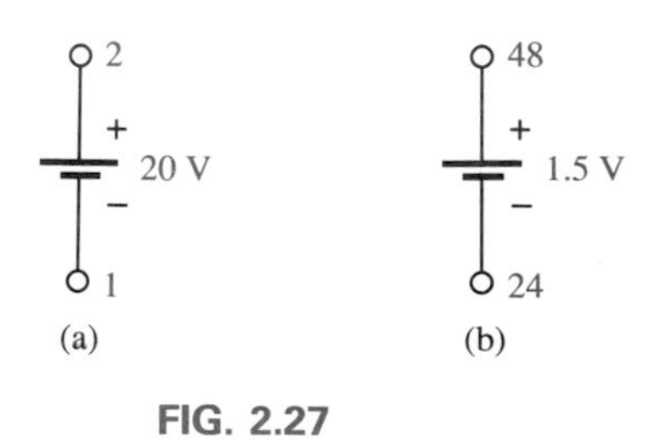


**FIG. 2.27**

```
VDCELL    48    24    DC    1.5V
```

## BASIC

In BASIC a line of the program must have a designated location (address) followed by the word INPUT to specify that a quantity is to be entered into the computer to be operated on. The word INPUT is followed immediately by an identification statement within quotes that specifies the quantity to be entered. The trailing quotation marks are followed by a semicolon and the quantity to be entered, as shown below:

program line or location
↓
```
120     INPUT "Voltage,E = ";E
```
operation — call for statement — quantity to be entered

When the program is *run*, the following statement will appear on the screen and printout:

```
Voltage,E = ?
```

The statement within the quotes will appear exactly as entered on line 120 (errors and all) with the addition of the question mark. There is no need to add the question mark; it is included automatically as part of the INPUT statement. It is now up to the user to enter the value of E so the computer can move on to the next line of the program. When entering line 120 into the computer, the simple error of typing a colon (:) rather than a semicolon (;) or misspelling INPUT will invalidate the command—everything must be exactly as specified in the defining format. You may want to turn to Section 4.8 to review other INPUT entries using BASIC.

It is virtually impossible to cover all the nuances with regard to the input of dc voltages sources in PSPICE and BASIC in this short section. However, the above are valid input entries and can be referenced in the programs and software runs to follow. This section is simply a foundation for further investigation and an opportunity to familiarize yourself with the basic format of each entry.

## PROBLEMS

### SECTION 2.1

**1.** The number of orbiting electrons in aluminum and silver is 13 and 47, respectively. Draw the electronic configuration, including all the shells and subshells, and discuss briefly why each is a good conductor.

**2.** Find the force of attraction between a proton and an electron separated by a distance equal to the radius of the smallest orbit followed by an electron ($5 \times 10^{-11}$ m) in a hydrogen atom.

3. Find the force of attraction in newtons between the charges $Q_1$ and $Q_2$ in Fig. 2.28 when
   a. $r = 1$ m  b. $r = 3$ m
   c. $r = 10$ m
   (Note how quickly the force drops with increase in $r$.)

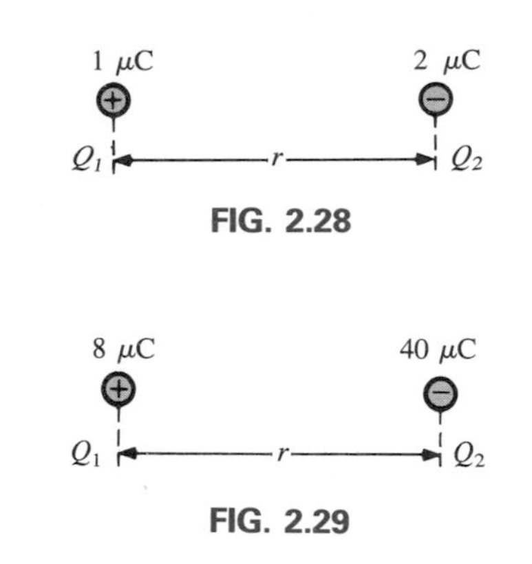


FIG. 2.28

FIG. 2.29

*4. Find the force of repulsion in newtons between $Q_1$ and $Q_2$ in Fig. 2.29 when
   a. $r = 1$ mi  b. $r = 0.01$ m
   c. $r = 1/16$ in.

5. Determine the distance between two charges of 20 μC if the force between the two charges is $3.6 \times 10^4$ N.

6. Two charged bodies, $Q_1$ and $Q_2$, when separated by a distance of 2 m, experience a force of repulsion equal to 1.8 N.
   a. What will the force of repulsion be when they are 10 m apart?
   b. If the ratio $Q_1/Q_2 = 1/2$, find $Q_1$ and $Q_2$ ($r = 10$ m).

7. Plot the force of repulsion between two charges of 1 μC as a function of the distance between the charges. Use a range of $r$ extending from 1 mm to 0.5 cm. Choose an appropriate scale for the vertical and horizontal axes.

## SECTION 2.2

8. Find the current in amperes if 650 C of charge pass through a wire in 50 s.

9. If 465 C of charge pass through a wire in 2.5 min, find the current in amperes.

10. If a current of 40 A exists for 1 min, how many coulombs of charge have passed through the wire?

11. How many coulombs of charge pass through a lamp in 2 min if the current is constant at 750 mA?

12. If the current in a conductor is constant at 2 mA, how much time is required for $4600 \times 10^{-6}$ C to pass through the conductor?

13. If $21.847 \times 10^{+18}$ electrons pass through a wire in 7 s, find the current.

14. How many electrons pass through a conductor in 1 min if the current is 1 A?

15. Will a fuse rated at 1 A ''blow'' if 86 C pass through it in 1.2 min?

*16. If $0.784 \times 10^{+18}$ electrons pass through a wire in 643 ms, find the current.

*17. Which would you prefer?
   a. A penny for every electron that passes through a wire in 0.01 μs at a current of 2 mA, or
   b. A dollar for every electron that passes through a wire in 1.5 ns if the current is 100 μA.

## SECTION 2.3 Voltage

**18.** What is the voltage between two points if 96 mJ of energy are required to move $50 \times 10^{18}$ electrons between the two points?

**19.** If the potential difference between two points is 42 V, how much work is required to bring 6 C from one point to the other?

**20.** Find the charge $Q$ that requires 96 J of energy to be moved through a potential difference of 16 V.

**21.** How much charge passes through a battery of 22.5 V if the energy expended is 90 J?

**22.** If a conductor with a current of 200 mA passing through it converts 40 J of electrical energy into heat in 30 s, what is the potential drop across the conductor?

***23.** Charge is flowing through a conductor at the rate of 420 C/min. If 742 J of electrical energy are converted to heat in 30 s, what is the potential drop across the conductor?

***24.** The potential difference between two points in an electric circuit is 24 V. If 0.4 J of energy were dissipated in a period of 5 ms, what would the current be between the two points?

## SECTION 2.4 Fixed (dc) Supplies

**25.** What current will a battery with an Ah rating of 200 theoretically provide for 40 h?

**26.** What is the Ah rating of a battery that can provide 0.8 A for 76 h?

**27.** For how many hours will a battery with an Ah rating of 32 theoretically provide a current of 1.28 A?

**28.** Find the mAh rating of the EVEREADY® BH 500 battery at 100°F and 0°C at a discharge current of 50 mA using Fig. 2.15(b).

**29.** Find the mAh rating of the 1.0 V EVEREADY® BH 500 battery if the current drain is 550 mA using Fig. 2.15(a). How long will it supply this current?

**30.** For how long can 50 mA be drawn from the battery of Fig. 2.16 before its terminal voltage drops below 1 V? Determine the number of hours at a drain current of 150 mA, and compare the ratio of drain current to the resulting ratio of hours of availability.

**31.** A standard 12-V car battery has an ampere-hour rating of 40 Ah, whereas a heavy-duty battery has a rating of 60 Ah. How would you compare the energy levels of each and the available current for starting purposes?

***32.** Using the relevant equations of the past few sections, determine the available energy from the Eveready transistor battery of Fig. 2.10.

*33. A portable TV using an 8-V, 3-Ah rechargeable battery can operate for a period of about 5.5 h. What is the average current drawn during this period? What is the energy expended by the battery in joules?

34. Discuss briefly the difference between the three types of dc voltage supplies (batteries, rectification, and generators).

35. Compare the characteristics of a dc current source with those of a dc voltage source. How are they similar and how are they different?

## SECTION 2.5 Conductors and Insulators

36. Discuss two properties of the atomic structure of copper that make it a good conductor.

37. Name two materials not listed in Table 2.1 that are good conductors of electricity.

38. Explain the terms *insulator* and *breakdown strength*.

39. List three uses of insulators not mentioned in Section 2.5.

## SECTION 2.6 Semiconductors

40. What is a semiconductor? How does it compare with a conductor and insulator?

41. Consult a semiconductor electronics text and note the extensive use of germanium and silicon semiconductor materials. Review the characteristics of each material.

## SECTION 2.7 Ammeters and Voltmeters

42. What are the significant differences in the way ammeters and voltmeters are connected?

43. If an ammeter reads 2.5 A for a period of 4 min, determine the charge that has passed through the meter.

44. Between two points in an electric circuit, a voltmeter reads 12.5 V for a period of 20 s. If the current measured by an ammeter is 10 mA, determine the energy expended and the charge that flowed between the two points.

## SECTION 2.8 Computer Input of dc Voltage Sources

45. Using PSPICE, what is the computer input for a 9-V dc battery connected between terminals 6 and 7 with the positive terminal on node 7? Choose any name for the supply you prefer.

46. For BASIC,
    a. How would you request the magnitude of the dc battery of the previous exercise using line 60 and a battery labeled VEE?
    b. How would the request for VEE appear?

## GLOSSARY

**Ammeter** An instrument designed to read the current through elements in series with the meter.

**Ampere (A)** The SI unit of measurement applied to the flow of charge through a conductor.

**Ampere-hour rating** The rating applied to a source of energy that will reveal how long a particular level of current can be drawn from that source.

**Cell** A fundamental source of electrical energy developed through the conversion of chemical or solar energy.

**Conductors** Materials that permit a generous flow of electrons with very little voltage applied.

**Copper** A material possessing physical properties that make it particularly useful as a conductor of electricity.

**Coulomb (C)** The fundamental SI unit of measure for charge. It is equal to the charge carried by $6.242 \times 10^{18}$ electrons.

**Coulomb's law** An equation defining the force of attraction or repulsion between two charges.

**dc current source** A source that will provide a fixed current level even though the load to which it is applied may cause its terminal voltage to change.

**dc generator** A source of dc voltage available through the turning of the shaft of the device by some external means.

**Direct current** Current in which the magnitude does not change over a period of time.

**Ductility** The property of a material that allows it to be drawn into long thin wires.

**Electrolytes** The contact element and the source of ions between the electrodes of the battery.

**Electron** The particle with negative polarity that orbits the nucleus of an atom.

**Free electron** An electron unassociated with any particular atom, relatively free to move through a crystal lattice structure under the influence of external forces.

**Insulators** Materials in which a very high voltage must be applied to produce any measurable current flow.

**Malleability** The property of a material that allows it to be worked into many different shapes.

**Neutron** The particle having no electrical charge, found in the nucleus of the atom.

**Node** A terminal point between elements of a network.

**Nucleus** The structural center of an atom which contains both protons and neutrons.

**Positive ion** An atom having a net positive charge due to the loss of one of its negatively charged electrons.

**Potential difference** The difference in potential between two points in an electrical system.

**Potential energy** The energy that a mass possesses by virtue of its position.

**Primary cell** Sources of voltage that cannot be recharged.

**Proton** The particle of positive polarity found in the nucleus of the atom.

**Rectification** The process by which an ac signal is converted to one which has an average dc level.

**Secondary cell** Sources of voltage that can be recharged.

**Semiconductor** A material having a conductance value between that of an insulator and that of a conductor. Of significant importance in the manufacture of semiconductor electronic devices.

**Solar cell** Sources of voltage available through the conversion of light energy (photons) into electrical energy.

**Specific gravity** The ratio of the weight of a given volume of a substance to the weight of an equal volume of water at 4°C.

**Volt (V)** The unit of measurement applied to the difference in potential between two points. If one joule of energy is required to move one coulomb of charge between two points, the difference in potential is said to be one volt.

**Voltmeter** An instrument designed to read the voltage across an element or between any two points in a network.

# 3 Resistance

## 3.1 INTRODUCTION

The flow of charge through any material encounters an opposing force similar in many respects to mechanical friction. This opposition, due to the collisions between electrons and between electrons and other atoms in the material, *which converts electrical energy into heat,* is called the *resistance* of the material. The unit of measurement of resistance is the *ohm,* for which the symbol is Ω, the capital Greek letter omega. The circuit symbol for resistance appears in Fig. 3.1 with the graphic abbreviation for resistance ($R$).

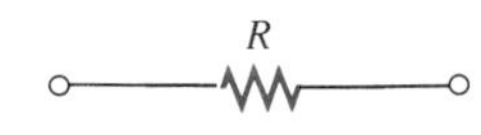


**FIG. 3.1**
*Resistance symbol and notation.*

The resistance of any material with a uniform cross-sectional area is determined by the following four factors:

1. Material
2. Length
3. Cross-sectional area
4. Temperature

The chosen material, with its unique molecular structure, will react differentially to pressures to establish current through its core. Conductors that permit a generous flow of charge with little external pressure will have low resistance levels, while insulators will have high resistance characteristics.

As one might expect, the longer the path the charge must pass through, the higher the resistance level, whereas the larger the area (and therefore available room), the lower the resistance. Resistance is thus directly proportional to length and inversely proportional to area.

As the temperature of most conductors increases, the increased motion of the particles within the molecular structure makes it increasingly difficult for the ''free'' carriers to pass through, and the resistance level increases.

At a fixed temperature of 20°C (room temperature), the resistance is related to the other three factors by

$$R = \rho \frac{l}{A} \quad \text{(ohms, } \Omega\text{)} \tag{3.1}$$

where $\rho$ (Greek letter rho) is a characteristic of the material called the *resistivity*, $l$ is the length of the sample, and $A$ is the cross-sectional area of the sample.

The units of measurement substituted into Eq. (3.1) are related to the application. For circular wires, units of measurement are usually defined as in Section 3.2. For most other applications involving important areas such as integrated circuits, the units are as defined in Section 3.4.

## 3.2 RESISTANCE: CIRCULAR WIRES

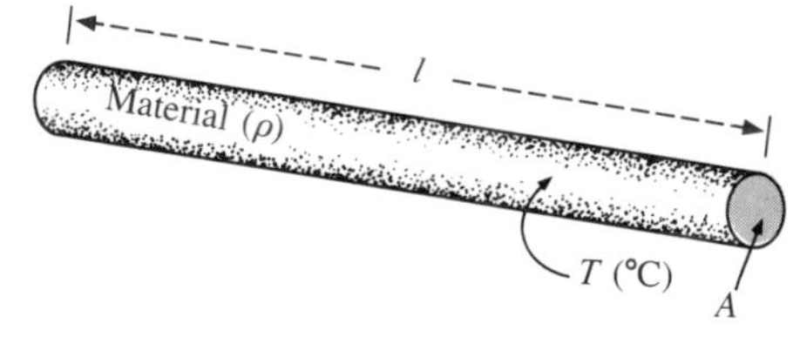


**FIG. 3.2**

For a circular wire, the quantities appearing in Eq. (3.1) are defined by Fig. 3.2. For two wires of the same physical size at the same temperature, as shown in Fig. 3.3(a), the relative resistances will be determined solely by the material. As indicated in Fig. 3.3(b), an increase in length will result in an increased resistance for similar areas, material, and temperature. Increased area [Fig. 3.3(c)] for remaining similar determining variables will result in a decrease in resistance. Finally, increased temperature [Fig. 3.3(d)] for metallic wires of identical construction and material will result in an increased resistance.

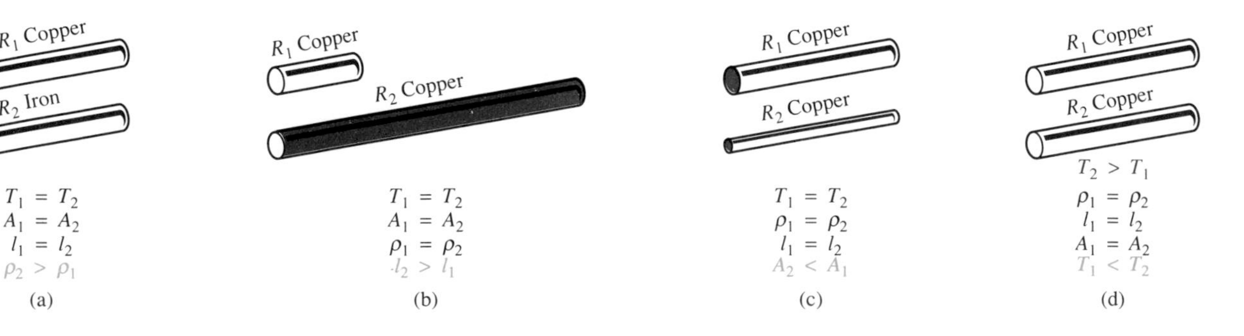


**FIG. 3.3**
*Cases in which $R_2 > R_1$.*

For circular wires, the quantities of Eq. (3.1) have the following units:

$\rho$—CM-ohms/ft at $T = 20°C$
$l$—feet
$A$—circular mils (CM)

Note that the area of the conductor is measured in *circular mils* and *not* in square meters, inches, and so on, as determined by the equation

$$\text{Area (circle)} = \pi r^2 = \frac{\pi d^2}{4} \qquad \begin{array}{l} r = \text{radius} \\ d = \text{diameter} \end{array} \qquad (3.2)$$

By definition:

$$\mathbf{1\ mil} = \frac{\mathbf{1}}{\mathbf{1000}}\ \mathbf{in.}$$

or

$$\mathbf{1000\ mils = 1\ in.}$$

A square mil will appear as shown in Fig. 3.4(a). By definition, *a wire that has a diameter of 1 mil, as shown in Fig. 3.4(b), has an area of 1 circular mil (CM).* One square mil was superimposed on the 1-CM area of Fig. 3.4(b) to show clearly that the square mil has a larger surface area than the circular mil.

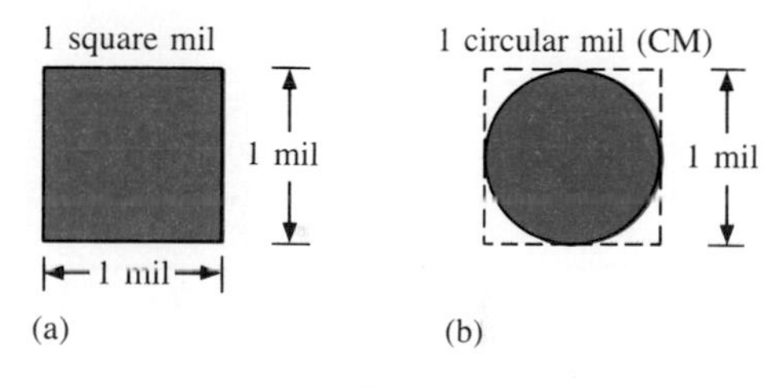

**FIG. 3.4**

Applying the above definition to a wire having a diameter of 1 mil, we have

$$A = \frac{\pi d^2}{4} = \frac{\pi}{4}(1)^2 = \frac{\pi}{4}\text{ sq mils} \overset{\text{by definition}}{\equiv} 1\text{ CM}$$

Therefore,

$$\mathbf{1\ CM} = \frac{\boldsymbol{\pi}}{\mathbf{4}}\ \mathbf{sq\ mils}$$

or

$$\mathbf{1\ sq\ mil} = \frac{\mathbf{4}}{\boldsymbol{\pi}}\ \mathbf{CM}$$

For conversion purposes,

$$\begin{aligned} \text{CM} &= \left(\frac{4}{\pi}\right) \times (\text{no. of sq mils}) \\ \text{sq miles} &= \left(\frac{\pi}{4}\right) \times (\text{no. of CM}) \end{aligned} \qquad (3.3)$$

For a wire with a diameter of $N$ mils (where $N$ can be any positive number),

$$A = \frac{\pi d^2}{4} = \frac{\pi N^2}{4} \text{ sq mils}$$

Substituting the fact that $4/\pi$ CM = 1 sq mil, we have

$$A = \frac{\pi N^2}{4} \text{ (sq mils)} = \left(\frac{\pi N^2}{4}\right)\left(\frac{4}{\pi} \text{ CM}\right) = N^2 \text{ CM}$$

Since $d = N$, the area in circular mils is simply equal to the diameter in mils square; that is,

$$A_{CM} = (d_{mils})^2 \quad \textbf{(3.4)}$$

Therefore, in order to find the area in circular mils, the diameter must first be converted to mils. Since 1 mil = 0.001 in., if the diameter is given in inches, simply move the decimal point three places to the right. For example,

$$0.123 \text{ in.} = 123.0 \text{ mils}$$

If in fractional form, first convert to decimal form and then proceed as above. For example,

$$\frac{1}{8} \text{ in.} = 0.125 \text{ in.} = 125 \text{ mils}$$

The constant $\rho$ (resistivity) is different for every material. Its value is the resistance of a length of wire 1 ft by 1 mil in diameter, measured at

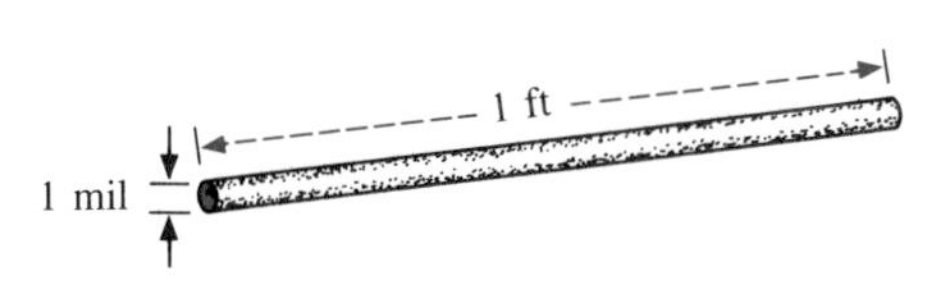


**FIG. 3.5**

20°C (Fig. 3.5). The unit of measurement for $\rho$ can be determined from Eq. (3.1) as follows:

$$R = \rho\frac{l}{A}$$

$$\text{Ohms} = \rho\frac{\text{ft}}{\text{CM}}$$

$$\text{Units of } \rho = \frac{\textbf{CM-}\boldsymbol{\Omega}}{\textbf{ft}}$$

The resistivity $\rho$ is also measured in ohms per mil-foot as determined by Fig. 3.5, or *ohm-meters* in the SI system of units.

Some typical values of $\rho$ are listed in Table 3.1.

**TABLE 3.1**
*The resistivity of various materials.*

| Material | $\rho\left(\frac{\text{CM-}\Omega}{\text{ft}}\right)$ @ 20°C |
|---|---|
| Silver | 9.9 |
| Copper | **10.37** |
| Gold | 14.7 |
| Aluminum | 17.0 |
| Tungsten | 33.0 |
| Nickel | 47.0 |
| Iron | 74.0 |
| Constantan | 295.0 |
| Nichrome | 600.0 |
| Calorite | 720.0 |
| Carbon | 21,000.0 |

**EXAMPLE 3.1.** What is the resistance of a 100-ft length of copper wire with a diameter of 0.020 in. at 20°C?

***Solution:***

$$\rho = 10.37 \frac{\text{CM-}\Omega}{\text{ft}} \qquad 0.020 \text{ in.} = 20 \text{ mils}$$

$$A_{\text{CM}} = (d_{\text{mils}})^2 = (20 \text{ mils})^2 = 400 \text{ CM}$$

$$R = \rho\frac{l}{A} = \frac{(10.37 \text{ CM-}\Omega/\text{ft})(100 \text{ ft})}{400 \text{ CM}}$$

$$R = \mathbf{2.59\ \Omega}$$

**EXAMPLE 3.2.** An undetermined number of feet of wire have been used from the carton of Fig. 3.6. Find the length of the remaining copper wire if it has a diameter of 1/16 in. and a resistance of 0.5 Ω.

***Solution:***

$$\rho = 10.37 \text{ CM-}\Omega/\text{ft} \qquad \frac{1}{16} \text{ in.} = 0.0625 \text{ in.} = 62.5 \text{ mils}$$

$$A_{\text{CM}} = (d_{\text{mils}})^2 = (62.5 \text{ mils})^2 = 3906.25 \text{ CM}$$

$$R = \rho\frac{l}{A} \Rightarrow l = \frac{RA}{\rho} = \frac{(0.5\ \Omega)(3906.25 \text{ CM})}{10.37 \dfrac{\text{CM-}\Omega}{\text{ft}}} = \frac{1953.125}{10.37}$$

$$l = \mathbf{188.34\ ft}$$

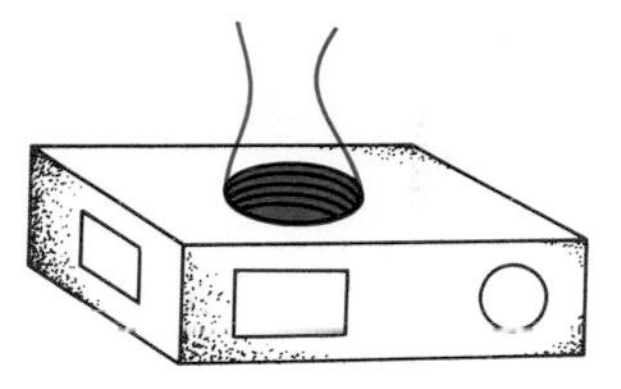

FIG. 3.6

**EXAMPLE 3.3.** What is the resistance of a copper bus-bar as used in the power distribution panel of a high-rise office building with the dimensions indicated in Fig. 3.7?

***Solution:***

$$A_{\text{CM}} \begin{cases} 5.0 \text{ in.} = 5000 \text{ mils} \\ \dfrac{1}{2} \text{ in.} = 500 \text{ mils} \\ A = (5000 \text{ mils})(500 \text{ mils}) = 2.5 \times 10^6 \text{ sq mils} \\ \quad = 2.5 \times 10^6 \cancel{\text{sq mils}}\left(\dfrac{4/\pi \text{ CM}}{1 \cancel{\text{ sq mil}}}\right) \\ A = 3.185 \times 10^6 \text{ CM} \end{cases}$$

$$R = \rho\frac{l}{A} = \frac{(10.37 \text{ CM-}\Omega/\text{ft})(3 \text{ ft})}{3.185 \times 10^6 \text{ CM}} = \frac{31.110}{3.185 \times 10^6}$$

$$R = \mathbf{9.768 \times 10^{-6}\ \Omega}$$

(quite small, 0.000009768 Ω)

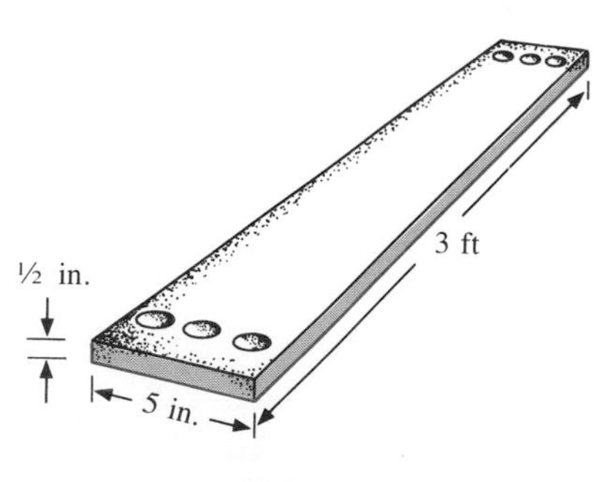


FIG. 3.7

## 3.3 WIRE TABLES

The wire table was designed primarily to standardize the size of wire produced by manufacturers throughout the United States. As a result, the manufacturer has a larger market and the consumer knows that standard wire sizes will always be available. The table was designed to assist the user in every way possible; it usually includes such data as the cross-sectional area in circular mils, diameter in mils, ohms per 1000 feet at 20°C, and weight per 1000 feet.

The American Wire Gage (AWG) sizes are given in Table 3.2 for solid round copper wire. A column indicating the maximum allowable current in amperes, as determined by the National Fire Protection Association, has also been included.

**TABLE 3.2**
*American Wire Gage (AWG) sizes.*

| | AWG # | Area (CM) | Ω/1000 ft at 20°C | Maximum Allowable Current for RHW Insulation (A)* |
|---|---|---|---|---|
| (4/0) | 0000 | 211,600 | 0.0490 | 230 |
| (3/0) | 000 | 167,810 | 0.0618 | 200 |
| (2/0) | 00 | 133,080 | 0.0780 | 175 |
| (1/0) | 0 | 105,530 | 0.0983 | 150 |
| | 1 | 83,694 | 0.1240 | 130 |
| | 2 | 66,373 | 0.1563 | 115 |
| | 3 | 52,634 | 0.1970 | 100 |
| | 4 | 41,742 | 0.2485 | 85 |
| | 5 | 33,102 | 0.3133 | — |
| | 6 | 26,250 | 0.3951 | 65 |
| | 7 | 20,816 | 0.4982 | — |
| | 8 | 16,509 | 0.6282 | 50 |
| | 9 | 13,094 | 0.7921 | — |
| | 10 | 10,381 | 0.9989 | 30 |
| | 11 | 8,234.0 | 1.260 | — |
| | 12 | 6,529.0 | 1.588 | 20 |
| | 13 | 5,178.4 | 2.003 | — |
| | 14 | 4,106.8 | 2.525 | 15 |
| | 15 | 3,256.7 | 3.184 | |
| | 16 | 2,582.9 | 4.016 | |
| | 17 | 2,048.2 | 5.064 | |
| | 18 | 1,624.3 | 6.385 | |
| | 19 | 1,288.1 | 8.051 | |
| | 20 | 1,021.5 | 10.15 | |
| | 21 | 810.10 | 12.80 | |
| | 22 | 642.40 | 16.14 | |
| | 23 | 509.45 | 20.36 | |
| | 24 | 404.01 | 25.67 | |
| | 25 | 320.40 | 32.37 | |

| AWG # | Area (CM) | Ω/1000 ft at 20°C | Maximum Allowable Current for RHW Insulation (A)* |
|---|---|---|---|
| 26 | 254.10 | 40.81 | |
| 27 | 201.50 | 51.47 | |
| 28 | 159.79 | 64.90 | |
| 29 | 126.72 | 81.83 | |
| 30 | 100.50 | 103.2 | |
| 31 | 79.70 | 130.1 | |
| 32 | 63.21 | 164.1 | |
| 33 | 50.13 | 206.9 | |
| 34 | 39.75 | 260.9 | |
| 35 | 31.52 | 329.0 | |
| 36 | 25.00 | 414.8 | |
| 37 | 19.83 | 523.1 | |
| 38 | 15.72 | 659.6 | |
| 39 | 12.47 | 831.8 | |
| 40 | 9.89 | 1049.0 | |


*Not more than three conductors in raceway, cable, or direct burial.

The chosen sizes have an interesting relationship: For every drop in three gage numbers the area is doubled, and for every drop in 10 gage numbers the area increases by a factor of 10.

Examining Eq. (3.1), we note also that *doubling the area cuts the resistance in half, and increasing the area by a factor of 10 decreases the resistance to 1/10 the original,* everything else kept constant.

The actual sizes of some of the gage wires listed in Table 3.2 are shown in Fig. 3.8 with a few of their areas of application. A few examples using Table 3.2 follow.

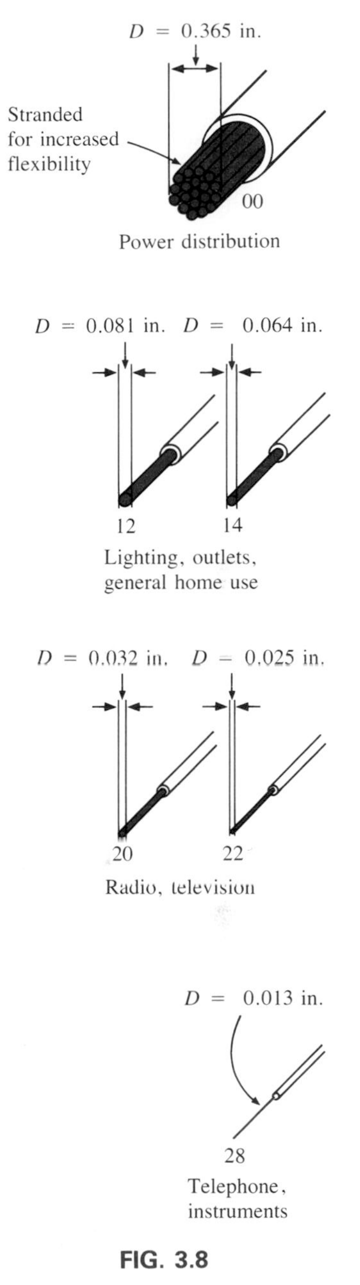


**FIG. 3.8**

**EXAMPLE 3.4.** Find the resistance of 650 ft of #8 copper wire ($T$ = 20°C).

***Solution:*** For #8 copper wire (solid), Ω/1000 ft at 20°C = 0.6282 Ω, and

$$650\,\cancel{\text{ft}}\left(\frac{0.6282\ \Omega}{1000\,\cancel{\text{ft}}}\right) = \mathbf{0.408\ \Omega}$$

**EXAMPLE 3.5.** What is the diameter, in inches, of a #12 copper wire?

***Solution:*** For #12 copper wire (solid), $A = 6529.9$ CM, and

$$d_{\text{mils}} = \sqrt{A_{\text{CM}}} = \sqrt{6529.9 \text{ CM}} \cong 80.81 \text{ mils}$$

$$d = \mathbf{0.0808\ in.} \text{ (or close to 1/12 in.)}$$

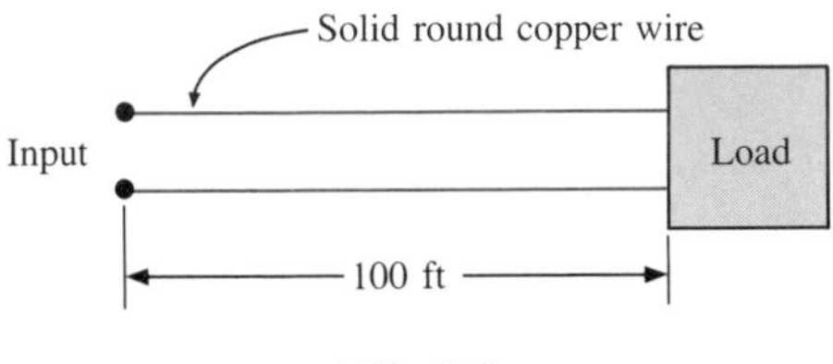


FIG. 3.9

**EXAMPLE 3.6.** For the system of Fig. 3.9, the total resistance of *each* power line cannot exceed 0.025 Ω, and the maximum current to be drawn by the load is 95 A. What gage wire should be used?

***Solution:***

$$R = \rho\frac{l}{A} \Rightarrow A = \rho\frac{l}{R} = \frac{(10.37 \text{ CM-}\Omega/\text{ft})(100 \text{ ft})}{0.025\ \Omega} = 41{,}480 \text{ CM}$$

Using the wire table, we choose the wire with the next largest area, which is #4, to satisfy the resistance requirement. We note, however, that 95 A must flow through the line. This specification requires that #3 wire be used, since the #4 wire can carry a maximum current of only 85 A.

## 3.4 RESISTANCE: METRIC UNITS

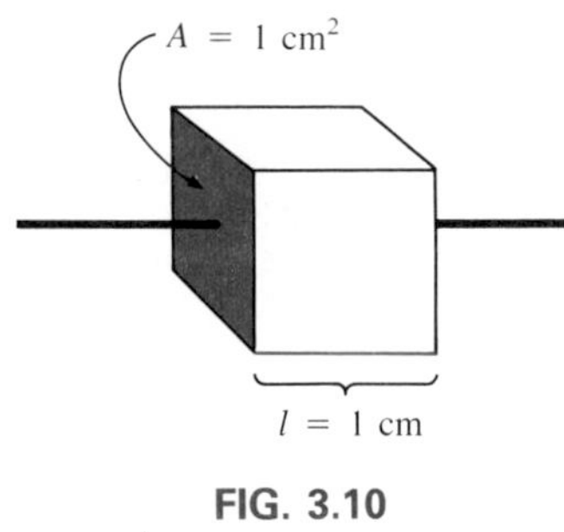


FIG. 3.10
*Defining ρ.*

The design of resistive elements for a variety of areas of application including thin-film resistors and integrated circuits uses metric units for the quantities of Eq. (3.1). In SI units, the resistivity would be measured in ohm-meters, the area in square meters, and the length in meters. However, the meter is generally too large a unit of measure for most applications, and so the centimeter is usually employed. The resulting dimensions for Eq. (3.1) are therefore

$\rho$—ohm-centimeters
$l$—centimeters
$A$—square centimeters

The units for $\rho$ can be derived from

$$R = \rho\frac{l}{A} \Rightarrow \rho = \frac{RA}{l} = \frac{\Omega\text{-cm}^2}{\text{cm}} = \Omega\text{-cm}$$

The resistivity of a material is actually the resistance of a sample such as appearing in Fig. 3.10. Table 3.3 provides a list of values of $\rho$ in ohm-centimeters.

Note that the area now is expressed in square centimeters, which can be determined using the basic equation $A = \pi d^2/4$, eliminating the need to work with circular mils, the special unit of measure associated with circular wires.

**TABLE 3.3**
*Resistivity (ρ) of various materials in ohm-centimeters.*

| Material | ρ |
|---|---|
| Silver | $1.645 \times 10^{-6}$ |
| Copper | $1.723 \times 10^{-6}$ |
| Gold | $2.443 \times 10^{-6}$ |
| Aluminum | $2.825 \times 10^{-6}$ |
| Tungsten | $5.485 \times 10^{-6}$ |
| Nickel | $7.811 \times 10^{-6}$ |
| Iron | $12.299 \times 10^{-6}$ |
| Tantalum | $15.54 \times 10^{-6}$ |
| Nichrome | $99.72 \times 10^{-6}$ |
| Tin oxide | $250 \times 10^{-6}$ |
| Carbon | $3500 \times 10^{-6}$ |

**EXAMPLE 3.7.** Determine the resistance of 100 ft of 28 copper telephone wire if the diameter is 0.0126 in.

***Solution:*** Unit conversions:

$$l = 100\,\cancel{\text{ft}}\left(\frac{12\,\cancel{\text{in.}}}{1\,\cancel{\text{ft}}}\right)\left(\frac{2.54\text{ cm}}{1\,\cancel{\text{in.}}}\right) = 3048\text{ cm}$$

$$d = 0.0126\text{ in.}\left(\frac{2.54\text{ cm}}{1\text{ in.}}\right) = 0.032\text{ cm}$$

Therefore,

$$A = \frac{\pi d^2}{4} = \frac{(3.1416)(0.032\text{ cm})^2}{4} = 8.04 \times 10^{-4}\text{ cm}^2$$

$$R = \rho\frac{l}{A} = \frac{(1.724 \times 10^{-6}\ \Omega\text{-cm})(3048\text{ cm})}{8.04 \times 10^{-4}\text{ cm}^2} \cong \mathbf{6.5\ \Omega}$$

Using the units for circular wires and Table 3.2 for the area of a #28 wire, we find

$$R = \rho\frac{l}{A} = \frac{(10.37\text{ CM-}\Omega/\text{ft})(100\text{ ft})}{159.79\text{ CM}} \cong \mathbf{6.5\ \Omega}$$

**EXAMPLE 3.8.** Determine the resistance of the thin-film resistor of Fig. 3.11 if the sheet resistance $R_S$ (defined by $R_S = \rho/d$) is 100 Ω.

***Solution:*** For deposited materials of the same thickness, the sheet resistance factor is usually employed in the design of thin-film resistors.

Equation (3.1) can be written

$$R = \rho\frac{l}{A} = \rho\frac{l}{dw} = \left(\frac{\rho}{d}\right)\left(\frac{l}{w}\right) = R_S\frac{l}{w}$$

where $l$ is the length of the sample and $w$ is the width. Substituting into the above equation yields

$$R = R_S\frac{l}{w} = \frac{(100\ \Omega)(0.6\text{ cm})}{0.3\text{ cm}} = \mathbf{200\ \Omega}$$

as one might expect since $l = 2w$.

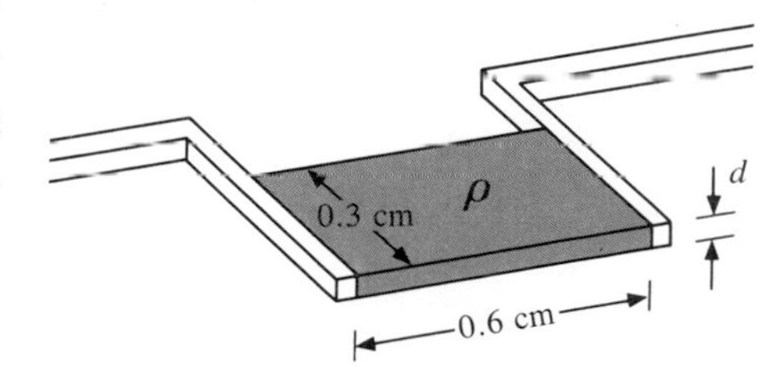


**FIG. 3.11**
*Thin-film resistor (note Fig. 3.22).*

The conversion factor between resistivity in circular mil-ohms per foot and ohm-centimeters is the following:

$$\boxed{\rho\ (\Omega\text{-cm}) = (1.662 \times 10^{-7}) \times (\text{value in CM-}\Omega/\text{ft})}$$

For example, for copper $\rho = 10.37$ CM-Ω/ft:

$$\begin{aligned}\rho\ (\Omega\text{-cm}) &= 1.662 \times 10^{-7}(10.37\text{ CM-}\Omega/\text{ft})\\ &= 1.723 \times 10^{-6}\ \Omega\text{-cm}\end{aligned}$$

as indicated in Table 3.3.

The resistivity in IC design is typically in ohm-centimeter units, although tables often provide $\rho$ in ohm-meters or microhm-centimeters. Using the conversion technique of Chapter 1, we find that the conversion factor between ohm-centimeters and ohm-meters is the following:

$$1.723 \times 10^{-6}\ \Omega\text{-}\cancel{\text{cm}} \left[\frac{1\text{ m}}{100\ \cancel{\text{cm}}}\right] = \frac{1}{100}[1.723 \times 10^{-6}]\ \Omega\text{-m}$$

or the value in ohm-meters is 1/100 the value in ohm-centimeters, and

$$\boxed{\rho\ (\Omega\text{-m}) = \left(\frac{1}{100}\right) \times (\text{value in } \Omega\text{-cm})}$$

Similarly:

$$\boxed{\rho\ (\mu\Omega\text{-cm}) = (10^{6}) \times (\text{value in } \Omega\text{-cm})}$$

For comparison purposes, typical values of $\rho$ in ohm-centimeters for conductors, semiconductors, and insulators are provided in Table 3.4.

**TABLE 3.4**
*Comparing levels of ρ in Ω-cm.*

| Conductor | Semiconductor | Insulator |
|---|---|---|
| Copper: $1.723 \times 10^{-6}$ | Ge 50 | In general: $10^{15}$ |
| | Si $200 \times 10^{3}$ | |
| | GaAs $70 \times 10^{6}$ | |

In particular, note the power of ten difference between conductors and insulators ($10^{21}$)—a difference of huge proportions. There is a significant difference is levels of $\rho$ for the list of semiconductors, but the power of ten difference between the conductor and insulator levels is at least $10^{6}$ for each of the semiconductors listed.

## 3.5 TEMPERATURE EFFECTS

Temperature has a significant effect on the resistance of conductors, semiconductors, and insulators.

### Conductors

In conductors there is a generous number of free electrons, and any introduction of thermal energy will have little impact on the total number of free carriers. In fact the thermal energy will only increase the intensity of the random motion of the particles within the material and make it increasingly difficult for a general drift of electrons in any one direction to be established. The result is that

***for good conductors, an increase in temperature will result in an increase in the resistance level. Consequently, conductors have a positive temperature coefficient.***

## Semiconductors

In semiconductors an increase in temperature will impart a measure of thermal energy to the system that will result in an increase in the number of free carriers in the material for conduction. The result is that

***for semiconductor materials, an increase in temperature will result in a decrease in the resistance level. Consequently, semiconductors have negative temperature coefficients.***

The thermistor and photoconductive cell of Sections 3.10 and 3.11 of this chapter are excellent examples of semiconductor devices with negative temperature coefficients.

## Insulators

***As with semiconductors, an increase in temperature will result in a decrease in the resistance of an insulator. The result is a negative temperature coefficient.***

## Inferred Absolute Temperature

Figure 3.12 reveals that for copper (and most other metallic conductors), the resistance increases almost linearly (in a straight-line relationship) with increase in temperature.

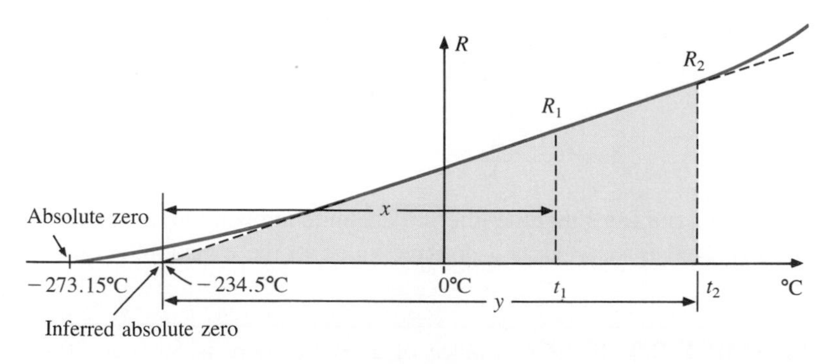


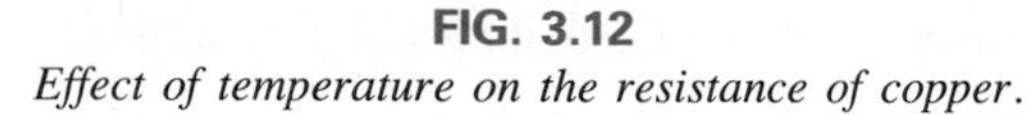

**FIG. 3.12**
*Effect of temperature on the resistance of copper.*

Since temperature can have such a pronounced effect on the resistance of a conductor, it is important that we have some method of determining the resistance at any temperature within operating limits. An equation for this purpose can be obtained by approximating the curve of

Fig. 3.12 by the straight dashed line that intersects the temperature scale at −234.5°C. Although the actual curve extends to *absolute zero* (−273.15°C, or 0 K), the straight-line approximation is quite accurate for the normal operating temperature range. At two different temperatures, $t_1$ and $t_2$, the resistance of copper is $R_1$ and $R_2$, as indicated on the curve. Using a property of similar triangles, we may develop a mathematical relationship between these values of resistances at different temperatures. Let $x$ equal the distance from −234.5°C to $t_1$ and $y$ the distance from −234.5°C to $t_2$, as shown in Fig. 3.12. From similar triangles,

$$\frac{x}{R_1} = \frac{y}{R_2}$$

or

$$\boxed{\frac{234.5 + t_1}{R_1} = \frac{234.5 + t_2}{R_2}} \qquad \textbf{(3.5)}$$

The temperature of −234.5°C is called the *inferred absolute temperature* of copper. For different conducting materials, the intersection of the straight-line approximation will occur at different temperatures. A few typical values are listed in Table 3.5.

**TABLE 3.5**
*Inferred absolute temperatures.*

| | |
|---|---|
| Silver | −243 |
| Copper | **−234.5** |
| Gold | −274 |
| Aluminum | −236 |
| Tungsten | −204 |
| Nickel | −147 |
| Iron | −162 |
| Nichrome | −2,250 |
| Constantan | −125,000 |

The minus sign does not appear with the inferred absolute temperature on either side of Eq. (3.5) because $x$ and $y$ are the *total distances* from −234.5°C to $t_1$ and $t_2$, respectively, and therefore are always positive quantities. For $t_1$ and $t_2$ less than zero, $x$ and $y$ are less than −234.5°C and the distances are the differences between the inferred absolute temperature and the temperature of interest.

Equation (3.5) can easily be adapted to any material by inserting the proper inferred absolute temperature. It may therefore be written as follows:

$$\boxed{\frac{|T| + t_1}{R_1} = \frac{|T| + t_2}{R_2}} \qquad \textbf{(3.6)}$$

where $|T|$ indicates that the inferred absolute temperature of the material involved is inserted as a positive value in the equation.

**EXAMPLE 3.9.** If the resistance of a copper wire is 50 Ω at 20°C, what is its resistance at 100°C (boiling point of water)?

***Solution:*** Eq. (3.5):

$$\frac{234.5°\text{C} + 20°\text{C}}{50\ \Omega} = \frac{234.5°\text{C} + 100°\text{C}}{R_2}$$

$$R_2 = \frac{(50\ \Omega)(334.5°\text{C})}{254.5°\text{C}} = \mathbf{65.72\ \Omega}$$

**EXAMPLE 3.10.** If the resistance of a copper wire at freezing (0°C) is 30 Ω, what is its resistance at −40°C?

***Solution:*** Eq. (3.5):

$$\frac{234.5°\text{C} + 0}{30\ \Omega} = \frac{234.5°\text{C} - 40°\text{C}}{R_2}$$

$$R_2 = \frac{(30\ \Omega)(194.5°\text{C})}{234.5°\text{C}} = \mathbf{24.88\ \Omega}$$

**EXAMPLE 3.11.** If the resistance of an aluminum wire at room temperature (20°C) is 100 mΩ (measured by a milliohmmeter), at what temperature will its resistance increase to 120 mΩ?

***Solution:*** Eq. (3.5):

$$\frac{236°\text{C} + 20°\text{C}}{100\ \text{m}\Omega} = \frac{236°\text{C} + t_2}{120\ \text{m}\Omega}$$

and

$$t_2 = 120\ \text{m}\Omega\left(\frac{256°\text{C}}{100\ \text{m}\Omega}\right) - 236°\text{C}$$

$$t_2 = \mathbf{71.2°C}$$

## Temperature Coefficient of Resistance

There is a second popular equation for calculating the resistance of a conductor at different temperatures. Defining

$$\alpha_1 = \frac{1}{|T| + t_1}$$

as the *temperature coefficient of resistance* at a temperature $t_1$, we have

$$\boxed{R_2 = R_1[1 + \alpha_1(t_2 - t_1)]} \qquad \textbf{(3.7)}$$

The values of $\alpha_1$ for different materials at a temperature of 20°C have been evaluated, and a few are listed in Table 3.6.

Equation (3.7) can be written in the following form:

$$\alpha_1 = \frac{1}{R_1}\overbrace{\left(\frac{R_2 - R_1}{t_2 - t_1}\right)}^{m\,=\,\text{slope of the curve}\,=\,\frac{\Delta y}{\Delta x}}$$

Referring to Fig. 3.12, we find that the temperature coefficient is directly proportional to the slope of the curve, so the greater the slope of the curve, the greater the value of $\alpha_1$. We can then conclude that *the higher the value of $\alpha_1$, the greater the rate of change of resistance with temperature*. Referring to Table 3.5, we find that copper is more sensi-

**TABLE 3.6**
*Temperature coefficient of resistance for various conductors at 20°C.*

| Material | Temperature Coefficient ($\alpha_1$) |
|---|---|
| Silver | 0.0038 |
| Copper | **0.00393** |
| Gold | 0.0034 |
| Aluminum | 0.00391 |
| Tungsten | 0.005 |
| Nickel | 0.006 |
| Iron | 0.0055 |
| Constantan | 0.000008 |
| Nichrome | 0.00044 |

tive to temperature variations than silver, gold, or aluminum, although the differences are quite small.

## PPM/°C

For resistors, as for conductors, resistance changes with change in temperature. The specification is normally provided in parts per million per degree Celsius (PPM/°C), providing an immediate indication of the sensitivity level of the resistor to temperature. For resistors, a 5000 PPM level is considered high, whereas 20 PPM is quite low. A 1000-PPM/°C characteristic reveals that a 1° change in temperature will result in a change in resistance equal to 1000 PPM, or 1000/1,000,000 = 1/1000 of its nameplate value—not a significant change for most applications. However, a 10° change would result in a change equal to 1/100 (1%) of its nameplate value, which is becoming significant. The concern, therefore, lies not only with the PPM level but with the range of expected temperature variation.

In equation form, the change in resistance is given by

$$\Delta R = \frac{R_{\text{nominal}}}{10^6}(\text{PPM})(\Delta T) \qquad (3.8)$$

where $R_{\text{nominal}}$ is the nameplate value of the resistor at room temperature and $\Delta T$ is the change in temperature from the reference level of 25°C.

---

**EXAMPLE 3.12.** For a 1-kΩ carbon composition resistor with a PPM of 2500, determine the resistance at 60°C.

***Solution:***

$$\Delta R = \frac{1000\ \Omega}{10^6}(2500)(60°\text{C} - 25°\text{C})$$
$$= 87.5\ \Omega$$

$$\text{and } R = R_{\text{nominal}} + \Delta R = 1000\ \Omega + 87.5\ \Omega$$
$$= \mathbf{1087.5\ \Omega}$$

---

# 3.6 SUPERCONDUCTORS

## Introduction

There is no question that the field of electricity/electronics has to be one of the most exciting of the 20th century. Even though new developments appear almost weekly from extensive research and development activities, every once in a while there is some very special step forward that has the whole field at the edge of its seat waiting to see what might develop in the near future. Such a level of excitement and interest surrounds the research drive to develop a room-temperature *superconductor*—an advance that will rival the introduction of semiconductor devices such as the transistor (to replace tubes), wireless communication, or the electric light. The implications of such a development are so far reaching that it is difficult to forecast the vast impact it will have on the entire field.

The intensity of the research effort throughout the world today to develop a room-temperature superconductor is described by some researchers as "unbelieveable, contagious, exciting, and demanding" but an adventure in which they treasure the opportunity to be involved. Progress in the field since 1986 hints that room-temperature superconductors may be a reality by the year 2000 or perhaps before this text passes through the production phase. It is indeed an exciting era full of growing anticipation! Why this interest in superconductors? What are they all about? In a nutshell,

***superconductors are conductors of electric charge that, for all practical purposes, have zero resistance.***

## Cooper Effect

In a conventional conductor, electrons travel at approximately 2% the speed of light (about 1000 mi/s). Einstein's theory of relativity suggests that the maximum speed of information transmission is the speed of light, or 186,000 mi/s. Beyond the speed of electronic flow, however, there is an obvious opportunity for improvement in the speed of transmission using techniques such as superconductivity. The relatively slow speed of conventional conduction is due to collisions with other atoms in the material, repulsive forces between electrons (like charges repel), thermal agitation that results in indirect paths due to the increased motion of the neighboring atoms, impurities in the conductor, and so on. In the superconductive state, there is a pairing of electrons denoted by the *Cooper effect,* in which electrons travel in pairs and help each other maintain a significantly higher velocity through the medium. In some ways this is like "drafting" by competitive cyclists or runners. There is an oscillation of energy between partners or even "new" partners (as the need arises) to ensure passage through the conductor at the highest possible velocity with the least total expenditure of energy.

## Ceramics

Even though the concept of superconductivity first surfaced in 1911, it was not until 1986 that the possibility of superconductivity at room temperature became a renewed goal of the research community. For some 74 years superconductivity could be established only at temperatures colder than 23 K (Kelvin temperature is universally accepted as the unit of measurement for temperature for superconductive effects. Recall that $K = 273.15° + °C$, so a temperature of 23 K is $-250°C$, or $-418°F$). In 1986, however, physicists Alex Muller and George Bednorz of the IBM Zurich Research Center found a ceramic material, lanthanum barium copper oxide, that exhibited superconductivity at

30 K. Although it would not appear to be a significant step forward, it introduced a new direction to the research effort and spurred others to improve on the new standard. In October 1987 both scientists received the Nobel prize for their contribution to an important area of development.

In just a few short months, Professors Paul Chu of the University of Houston and Man Kven Wu of the University of Alabama raised the temperature to 95 K using a superconductor of yttrium barium copper oxide. The result was a level of excitement in the scientific community that brought research in the area to a new level of effort and investment. The major impact of such a discovery was that liquid nitrogen (boiling point of 77 K) could now be used to bring the material down to the required temperature rather than liquid helium, which boils at 4 K. The result is a tremendous saving in the cooling expense, since liquid helium is at least ten times more expensive than liquid nitrogen. Pursuing the same direction, some success has been achieved at 125 K (February 1988) and 162 K (August 1988) using a thallium compound (unfortunately, however, thallium is a very poisonous substance). The time chart of Fig. 3.13 clearly reveals the tremendous change in the success

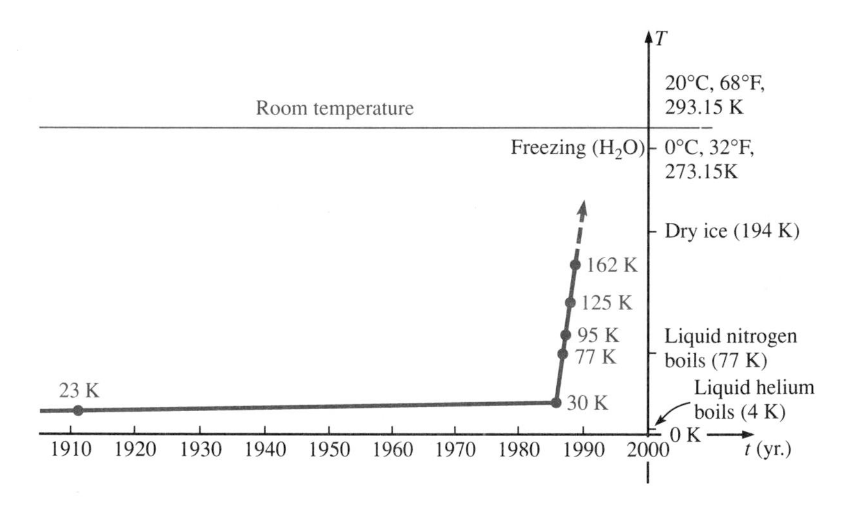


**FIG. 3.13**
*Rising temperatures of superconductors.*

curve since 1911 and also suggests that room-temperature success in the relatively near future is a strong possibility. However, the compound will probably not be one that has already surfaced and may in fact be of a totally different nature.

The fact that ceramics have provided the recent breakthrough in superconductivity is probably a surprise, when you consider that they are also an important class of insulators. However, the ceramics that exhibit the characteristics of superconductivity are compounds that include copper, oxygen, and rare earth elements such as yttrium, lanthanum, and thallium. There are also indicators that the current compounds may be limited to a maximum temperature of 200 K (about 100 K short of room temperature) leaving the door wide open to innovative approaches to compound selection. The temperature at which a superconductor reverts

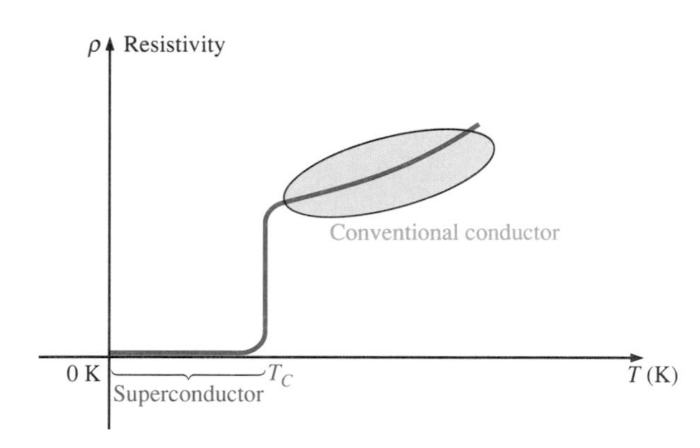


**FIG. 3.14**
*Defining the critical temperature $T_c$.*

back to the characteristics of a conventional conductor is called the *critical temperature,* denoted by $T_c$. Note in Fig. 3.14 how the resistivity level changes abruptly at $T_c$. The sharpness of the transition region is a function of the purity of the sample. Long listings of critical temperatures for a variety of tested compounds can be found in reference material devoted to providing tables of a wide variety to support research in physics, chemistry, geology, and related fields. Two such publications include the CRC (The Chemical Rubber Co.) *Handbook of Tables for Applied Engineering Science* and the CRC *Handbook of Chemistry and Physics*.

Even though ceramic compounds have established higher transition temperatures, there is concern about their brittleness and current density limitations. In the area of integrated circuit manufacturing, current density levels must equal or exceed 1 MA/cm$^2$, or 1 million amperes through a cross-sectional area about one-half the size of a dime. Recently IBM attained a level of 4 MA/cm$^2$ at 77 K, permitting the use of superconductors in the design of some new generation high speed computers.

## TIB Relationship

There are, in fact, three factors linked together in the development of a superconductor for extensive practical applications at room temperature—*temperature, current density,* and *magnetic field strength* (Chapter 11). As one element (such as temperature) is pushed to its limit, the other two (in this case, current density and magnetic flux density) will eventually drop off rather sharply, as shown in Fig. 3.15. In particular, note the temperature $T_1$ in Fig. 3.15, which defines current density and magnetic field strength less than maximum values. In addition, both continue to drop rather rapidly as the temperature is increased toward its critical value. Fortunately, the magnetic field strength currently available at workable superconductor temperatures is high enough (in excess of 250*T* at low temperatures and 100*T* at higher temperatures) for the majority of applications. The major concern for current materials lies in ensuring sufficient current density at the temperature of interest.

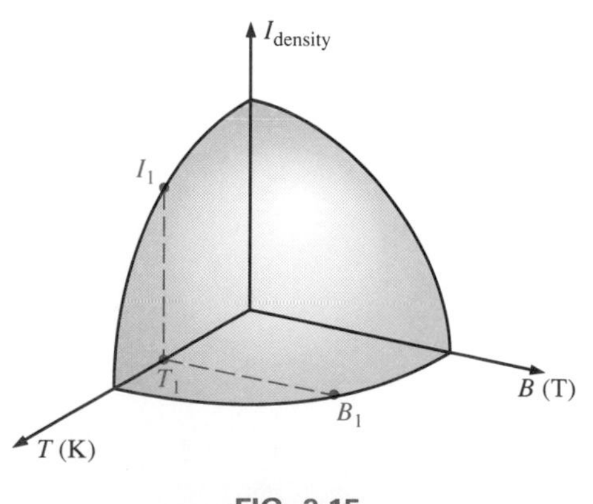


**FIG. 3.15**
*The TIB relationship.*

## Meissner Effect

A true physical indicator of whether superconduction has been established is beautifully demonstrated by the Meissner effect (Fig. 3.16(a)). At temperatures above the critical temperature, magnetic lines of force can pass through a conductor, as shown in Fig. 3.16(a). When the temperature of the conductor has been brought down to a level where superconductivity can be established, external magnetic lines of force cannot pass through the superconductor and the magnet will levitate (float above) the superconductor, as shown in Fig. 3.16(b), with a photograph of the actual phenomenon shown in Fig. 3.16(c). The inability of external magnetic flux lines to pass through a material in the superconducting state has extensive applications in sensors to be described in the applications section.

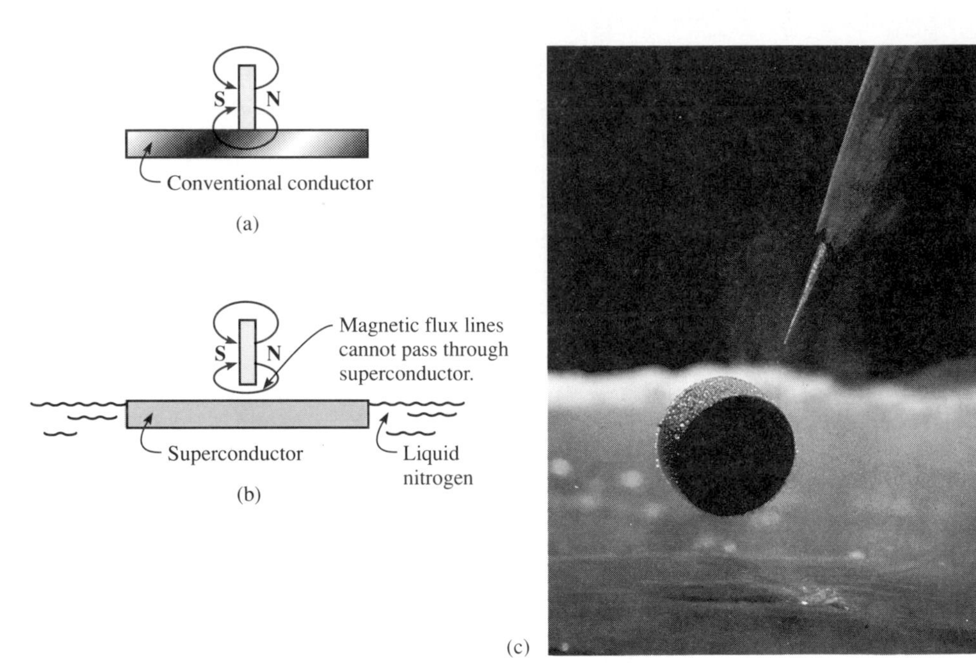


**FIG. 3.16**
*Demonstrating the Meissner effect. (Courtesy of IBM)*

## Applications

Although room-temperature success has not been attained (as of the publication of this text) there are applications for some of the superconductors developed thus far. It is simply a matter of balancing the additional cost against the results obtained or deciding whether any results at all can be obtained without the use of this zero-resistance state. Some research efforts require high-energy accelerators or strong magnets attainable only with superconductive materials. Superconductivity is currently applied in the design of 300-mi/h Meglev trains (trains that ride on a cushion of air established by opposite magnetic poles), in nuclear magnetic resonance imaging systems to obtain cross-sectional images of the brain (and other parts of the body), and in the design of computers with operating speeds four times that of conventional systems.

Through the use of the Josephson effect (left to the student as a research activity), there are magnetic field detectors known as SQUIDs (superconducting quantum interference devices) that can measure magnetic fields thousands of times smaller than conventional methods. Applications of such devices range from medicine to geology. Since the human skull distorts electric currents and not magnetic fields, SQUIDs can be used to detect extremely small magnetic fields that can provide important diagnostic information about the patient. In geology they can be used to detect magnetic fields that reveal the presence of specific minerals or even oil and water.

The range of future uses for superconductors is a function of the success physicists have in raising the operating temperature. However, it would appear that it is only a matter of time (the eternal optimist) before magnetically levitated trains increase in number, improved medical diagnostic equipment is available, computers operate at much higher speeds, high-efficiency power and storage systems are available, and transmission systems operate at very high efficiency levels due to this area of developing interest. Only time will reveal the impact that this new direction will have on the quality of life.

## 3.7 TYPES OF RESISTORS

### Fixed Resistors

Resistors are made in many forms, but all belong in either of two groups: fixed or variable. The most common of the low-wattage, fixed-type resistors is the molded carbon composition resistor. The basic construction is shown in Fig. 3.17.

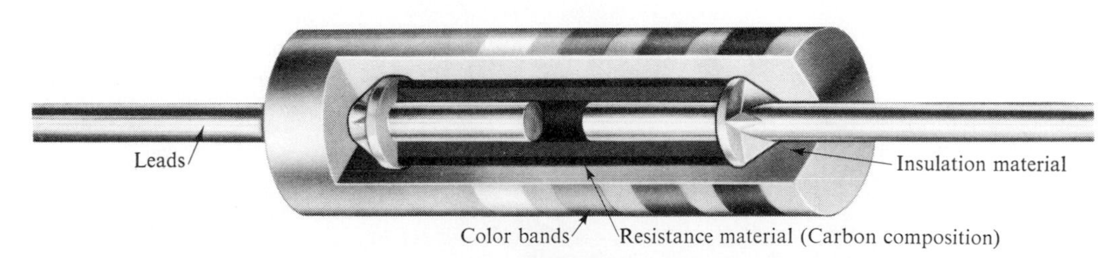


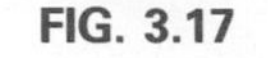

**FIG. 3.17**
*Fixed composition resistor. (Courtesy of Ohmite Manufacturing Co.)*

The relative sizes of all fixed and variable resistors change with the wattage (power) rating, increasing in size for increased wattage ratings in order to withstand the higher currents and dissipation losses. The relative sizes of the molded composition resistors for different wattage ratings are shown in Fig. 3.18. Resistors of this type are readily available in values ranging from 2.7 Ω to 22 MΩ.

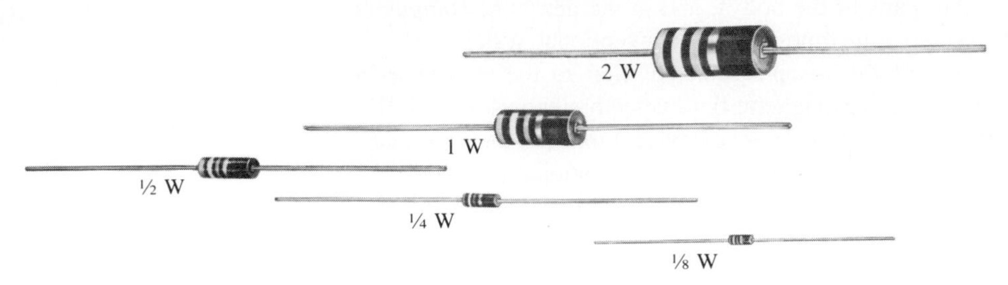


**FIG. 3.18**
*Fixed composition resistors of different wattage ratings. (Courtesy of Ohmite Manufacturing Co.)*

The temperature-versus-resistance curve for a 10,000-Ω and 0.5-MΩ composition-type resistor is shown in Fig. 3.19. Note the small

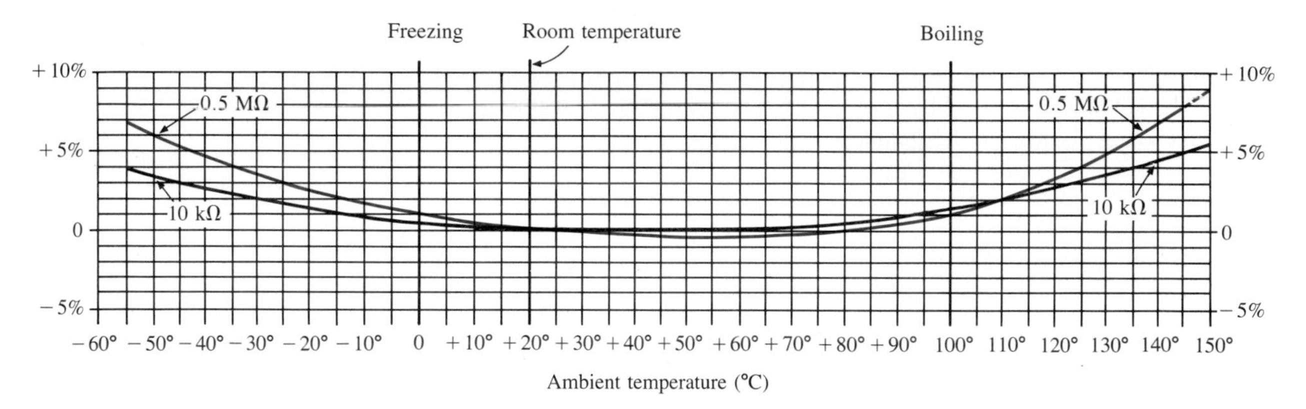


**FIG. 3.19**
*Curves showing percent temporary resistance changes from +25°C values. (Courtesy of Allen-Bradley Co.)*

percent resistance change in the normal temperature operating range. Several other types of fixed resistors are shown in Fig. 3.20.

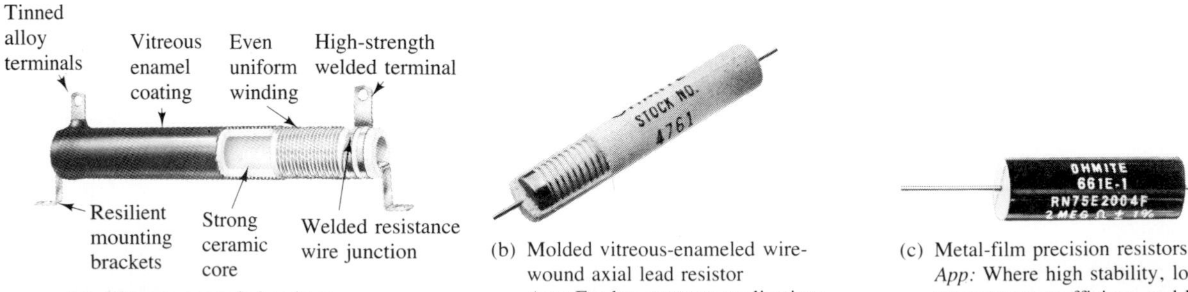


(a) Vitreous-enameled resistor
*App:* All types of equipment

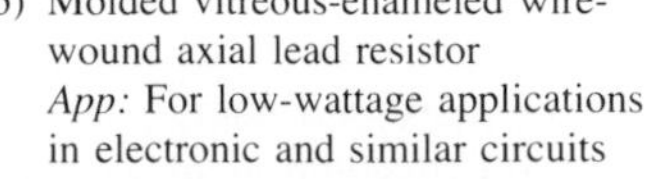

(b) Molded vitreous-enameled wire-wound axial lead resistor
*App:* For low-wattage applications in electronic and similar circuits

(c) Metal-film precision resistors
*App:* Where high stability, low temperature coefficient, and low noise level desired

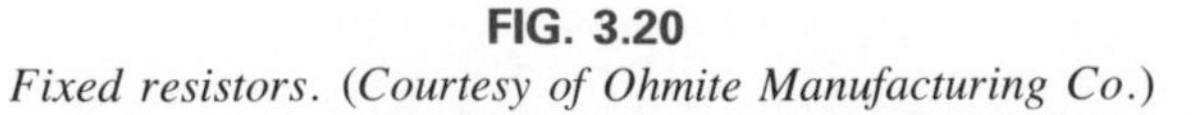

**FIG. 3.20**
*Fixed resistors. (Courtesy of Ohmite Manufacturing Co.)*

The miniaturization of parts—used quite extensively in computers—requires that resistances of different values be placed in very small packages. Two steps leading to the packaging of three resistors in a single module are shown in Fig. 3.21.

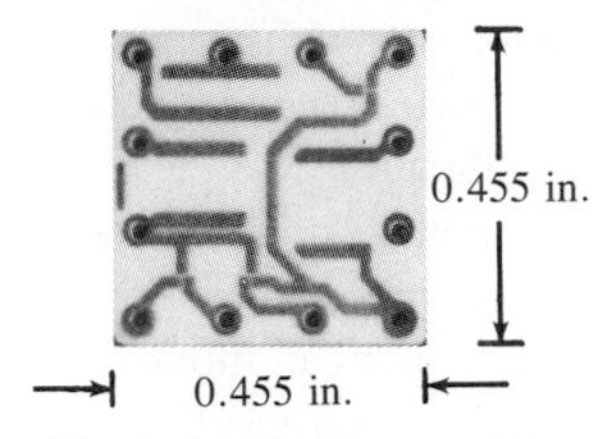

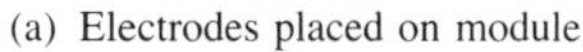


(a) Electrodes placed on module

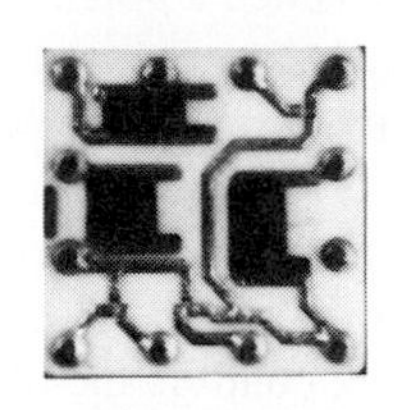

(b) Resistance applied and adjusted to desired value by air-abrasion techniques

(c) Module completely encased

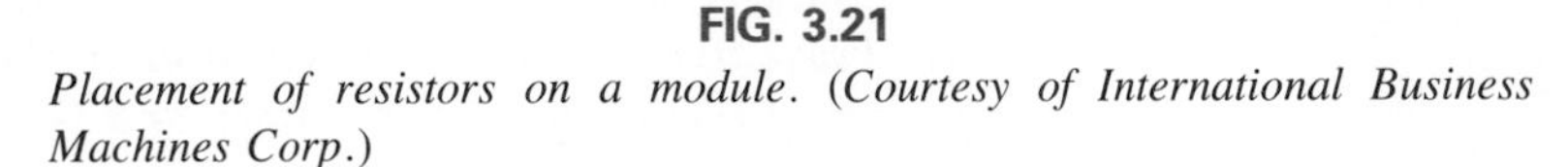

**FIG. 3.21**
*Placement of resistors on a module. (Courtesy of International Business Machines Corp.)*

For use with printed circuit boards, fixed resistor networks in a variety of configurations are available in miniature packages, such as shown in Fig. 3.22 with a photograph of the casing and pins. The LDP is a

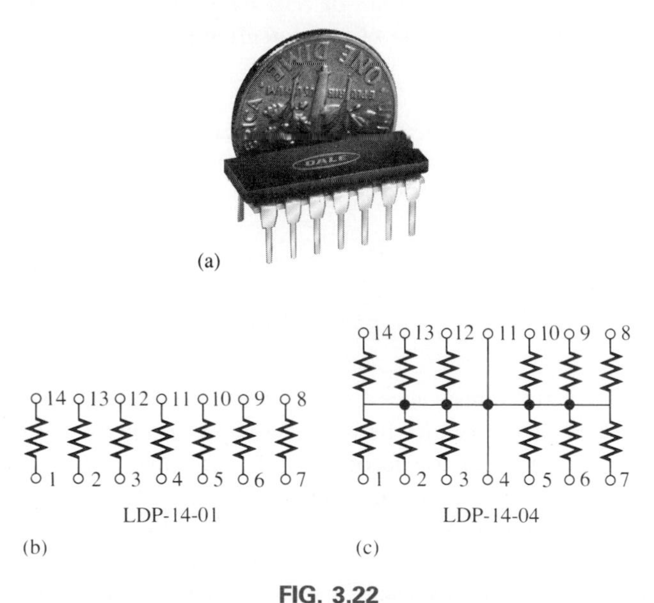


**FIG. 3.22**
*Resistor configuration microcircuit. (Courtesy of Dale Electronics, Inc.)*

coding for the production series, while the second number, 14, is the number of pins. The last two digits indicate the internal circuit configuration. The resistance range for the discrete elements in each chip is 10 Ω to 10 MΩ.

## Variable Resistors

Variable resistors, as the name implies, have a terminal resistance that can be varied by turning a dial, knob, screw, or whatever seems appro-

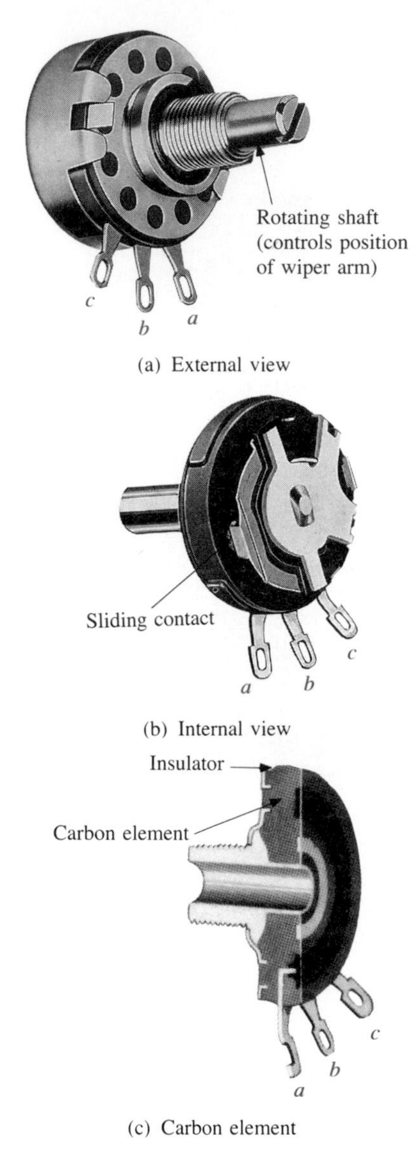


(a) External view

(b) Internal view

(c) Carbon element

**FIG. 3.23**
*Molded composition type potentiometer. (Courtesy of Allen-Bradley Co.)*

priate for the application. They can have two or three terminals, but most have three terminals. If the two- or three-terminal device is used as a variable resistor, it is usually referred to as a *rheostat*. If the three-terminal device is used for controlling potential levels, it is then commonly called a *potentiometer*. Even though a three-terminal device can be used as a rheostat or potentiometer (depending on how it is connected), it is typically called a potentiometer when listed in trade magazines or requested for a particular application.

Most potentiometers have three terminals in the relative positions shown in Fig. 3.23(a). The knob, dial, or screw in the center of the housing controls the motion of a contact that can move along the resistive element connected between the outer two terminals.

***The resistance between the outside terminals (a) and (c) of Fig. 3.23 is always fixed at the full rated value of the potentiometer, irrespective of the position of the wiper arm (b).***

In other words, the resistance between terminals $a$ and $c$ of Fig. 3.23 for a 1-MΩ potentiometer will always be 1 MΩ no matter how we turn the control element.

***The resistance between the wiper arm and either outside terminal can be varied from a minimum of zero ohms to a maximum value equal to the full rated value of the potentiometer.***

In addition,

***the sum of the resistances between the wiper arm and each outside terminal must equal the full rated resistance level of the potentiometer.***

Specifically:

$$R_{ac} = R_{ab} + R_{bc} \quad \textbf{(3.9)}$$

Therefore, as the resistance from the wiper arm to one outside contact increases, the resistance between the wiper arm and the other outside terminal must decrease accordingly. For example, if $R_{ab}$ of a 1-kΩ potentiometer is 200 Ω, then the resistance $R_{bc}$ must be 800 Ω. If $R_{ab}$ is further decreased to 50 Ω, then $R_{bc}$ must increase to 950 Ω, and so on.

The symbol for a three-terminal potentiometer appears in Fig. 3.24(a). When used as a variable resistor (or rheostat), the potentiometer is connected as shown in Fig. 3.24(b). The universally accepted symbol for a rheostat appears in Fig. 3.24(b).

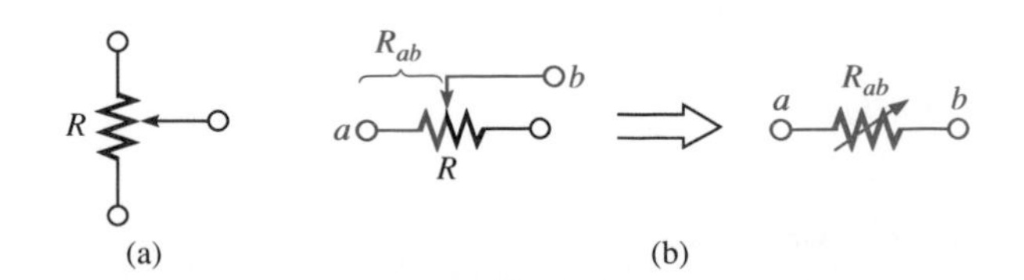


**FIG. 3.24**
*Potentiometer: (a) Symbol; (b) used on a rheostat.*

When the device is used as a potentiometer, the connections are as shown in Fig. 3.25. It can be used to control the level of $V_{ab}$, $V_{bc}$, or

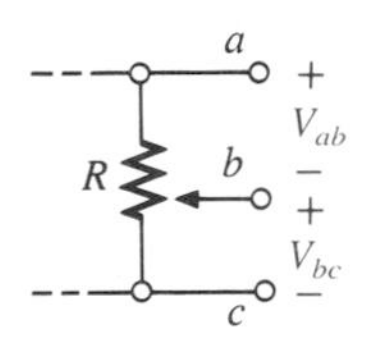


**FIG. 3.25**
*Potentiometer controlling potential levels.*

both, depending on the application. Further discussion of the potentiometer in a loaded situation can be found in the chapters that follow.

Figure 3.26 shows both a linear and a tapered type of potentiometer. In the linear type of Fig. 3.26(a), the number of turns of the high-resistance wire per unit length of the core is uniform; therefore, the resistance will vary linearly with the position of the rotating contact. One-half of a turn will result in half the total resistance between either stationary lug and the moving contact. Three-quarters of a turn will establish three-quarters of the total across two terminals and one-quarter between the other stationary lug and the moving contact. If the number of turns is not uniform as in the tapered unit of Fig. 3.26(b), the resistance will vary nonlinearly with the position of the rotating contact. That is, a quarter turn may result in less or more than one-quarter the total resistance between a stationary lug and the moving contact. Potentiometers of both types in Fig. 3.26 are made in all sizes, with a range of maximum values from 200 Ω to 50,000 Ω.

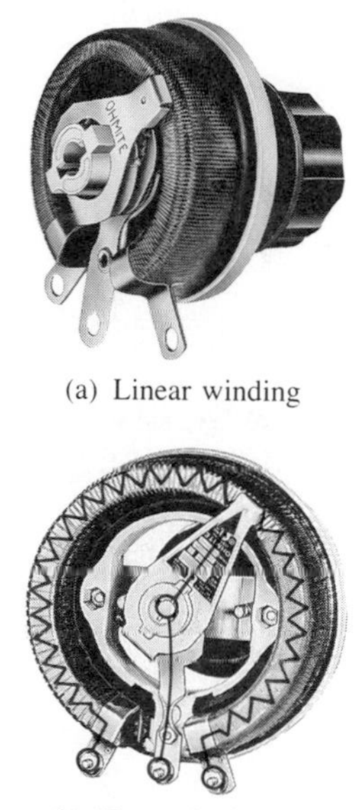

(a) Linear winding

(b) Tapered winding

**FIG. 3.26**
*Wirewound vitreous enameled potentiometers. (Courtesy of Ohmite Manufacturing Co.)*

The molded composition linear potentiometer of Fig. 3.23 is the type used in circuits with smaller power demands than the one previously described. It is smaller in size but has maximum values ranging from 20 Ω to 22 MΩ.

The resistance of the screw-drive linear variable resistor of Fig. 3.27 is determined by the position of the contact arm, which can be moved by using the handwheel. The stationary terminal used with the movable

**FIG. 3.27**
*Screw-drive rheostat. (Courtesy of James G. Biddle Co.)*

contact determines whether the resistance increases or decreases with movement of the contact arm.

## 3.8 COLOR CODING AND STANDARD RESISTOR VALUES

A wide variety of resistors, fixed or variable, are large enough to have their resistance in ohms printed on the casing. There are some, however, that are too small to have numbers printed on them, so a system of color coding is used. For the fixed molded composition resistor, four or five color bands are printed on one end of the outer casing as shown in Fig. 3.28. Each color has the numerical value indicated in Table 3.7. The color bands are always read from the end that has the band closest to it, as shown in Fig. 3.28. The first and second bands represent the first and second digits, respectively. The third band determines the power-of-10 multiplier for the first two digits (actually the number of zeros that follow the second digit), or a multiplying factor determined by the gold and silver bands. The fourth band is the manufacturer's tolerance, which is an indication of the precision by which the resistor was made. If the fourth band is omitted, the tolerance is assumed to be ±20%. The fifth band is a reliability factor which gives the percentage of failure per 1000 hours of use. For instance, a 1% failure rate would reveal that one out of every 100 (or 10 out of every 1000) will fail to fall within the tolerance range after 1000 hours of use.

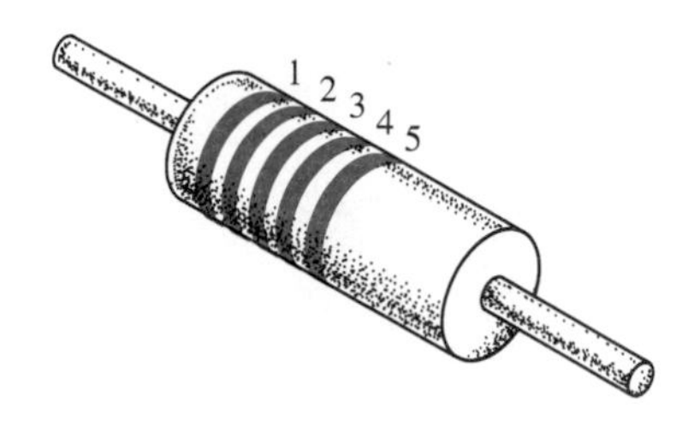


**FIG. 3.28**
*Color coding—fixed molded composition resistor.*

**TABLE 3.7**
*Resistor color coding.*

| Bands 1–3 | Band 3 | Band 4 | Band 5 |
|---|---|---|---|
| 0 Black | 0.1 Gold } multiplying | 5% Gold | 1% Brown |
| 1 Brown | 0.01 Silver } factors | 10% Silver | 0.1% Red |
| 2 Red | | 20% No band | 0.01% Orange |
| 3 Orange | | | 0.001% Yellow |
| 4 Yellow | | | |
| 5 Green | | | |
| 6 Blue | | | |
| 7 Violet | | | |
| 8 Gray | | | |
| 9 White | | | |

**EXAMPLE 3.13.** Find the range in which a resistor having the following color bands must exist to satisfy the manufacturer's tolerance:

a.

| 1st band | 2nd band | 3rd band | 4th band | 5th band |
|---|---|---|---|---|
| Gray | Red | Black | Gold | Brown |
| 8 | 2 | 0 | ±5% | 1% |

**82 Ω ± 5% (1% reliability)**

Since 5% of 82 = 4.10, the resistor should be within the range 82 Ω ± 4.10 Ω, or *between 77.90 and 86.10 Ω.*

b.

| 1st band | 2nd band | 3rd band | 4th band | 5th band |
|---|---|---|---|---|
| Orange | White | Gold | Silver | No color |
| 3 | 9 | 0.1 | ±10% | |

**3.9 Ω ± 10% = 3.9 ± 0.39 Ω**

The resistor should lie somewhere *between 3.51 and 4.29 Ω.*

Throughout the text material, resistor values in the network will be chosen to reduce the mathematical complexity of finding the solution. It was felt that the procedure or analysis technique was of primary importance and the mathematical exercise secondary. Many of the values appearing in the text are not *standard values*. That is, they are available only through special request. In the problem sections, however, standard values were frequently employed to make them more familiar and demonstrate their effect on the required calculations. A list of readily available standard values appears in Table 3.8. All the resistors appearing in Table 3.8 are available with 5% tolerance. Those in boldface are available with 5% and 10% tolerances while those in color are available with 5%, 10%, and 20% tolerances.

**TABLE 3.8**
*Standard values of commercially available resistors.*

| Ohms (Ω) | | | | | Kilohms (kΩ) | | Megohms (MΩ) | |
|---|---|---|---|---|---|---|---|---|
| 0.10 | 1.0 | 10 | 100 | 1000 | 10 | 100 | 1.0 | 10.0 |
| 0.11 | 1.1 | 11 | 110 | 1100 | 11 | 110 | 1.1 | 11.0 |
| **0.12** | **1.2** | **12** | **120** | **1200** | **12** | **120** | **1.2** | **12.0** |
| 0.13 | 1.3 | 13 | 130 | 1300 | 13 | 130 | 1.3 | 13.0 |
| 0.15 | 1.5 | 15 | 150 | 1500 | 15 | 150 | 1.5 | 15.0 |
| 0.16 | 1.6 | 16 | 160 | 1600 | 16 | 160 | 1.6 | 16.0 |
| **0.18** | **1.8** | **18** | **180** | **1800** | **18** | **180** | **1.8** | **18.0** |
| 0.20 | 2.0 | 20 | 200 | 2000 | 20 | 200 | 2.0 | 20.0 |
| 0.22 | 2.2 | 22 | 220 | 2200 | 22 | 220 | 2.2 | 22.0 |
| 0.24 | 2.4 | 24 | 240 | 2400 | 24 | 240 | 2.4 | |
| **0.27** | **2.7** | **27** | **270** | **2700** | **27** | **270** | **2.7** | |
| 0.30 | 3.0 | 30 | 300 | 3000 | 30 | 300 | 3.0 | |
| 0.33 | 3.3 | 33 | 330 | 3300 | 33 | 330 | 3.3 | |
| 0.36 | 3.6 | 36 | 360 | 3600 | 36 | 360 | 3.6 | |
| **0.39** | **3.9** | **39** | **390** | **3900** | **39** | **390** | **3.9** | |
| 0.43 | 4.3 | 43 | 430 | 4300 | 43 | 430 | 4.3 | |
| 0.47 | 4.7 | 47 | 470 | 4700 | 47 | 470 | 4.7 | |
| 0.51 | 5.1 | 51 | 510 | 5100 | 51 | 510 | 5.1 | |
| **0.56** | **5.6** | **56** | **560** | **5600** | **56** | **560** | **5.6** | |
| 0.62 | 6.2 | 62 | 620 | 6200 | 62 | 620 | 6.2 | |
| 0.68 | 6.8 | 68 | 680 | 6800 | 68 | 680 | 6.8 | |
| 0.75 | 7.5 | 75 | 750 | 7500 | 75 | 750 | 7.5 | |
| **0.82** | **8.2** | **82** | **820** | **8200** | **82** | **820** | **8.2** | |
| 0.91 | 9.1 | 91 | 910 | 9100 | 91 | 910 | 9.1 | |

## 3.9 CONDUCTANCE

By finding the reciprocal of the resistance of a material, we have a measure of how well the material will conduct electricity. The quantity is called *conductance,* has the symbol $G$, and is measured in *siemens* (S).

In equation form, conductance is

$$G = \frac{1}{R} \qquad \text{(siemens, S)} \qquad \mathbf{(3.10)}$$

A resistance of 1 M$\Omega$ is equivalent to a conductance of $10^{-6}$ S, and a resistance of 10 $\Omega$ is equivalent to a conductance of $10^{-1}$ S. The larger the conductance, therefore, the less the resistance and the greater the conductivity.

In equation form, the conductance is determined by

$$G = \frac{A}{\rho l} \qquad \text{(S)} \qquad \mathbf{(3.11)}$$

indicating that increasing the area or decreasing either the length or the resistivity will increase the conductance.

---

**EXAMPLE 3.14.** What is the relative increase or decrease in conductivity of a conductor if the area is reduced by 80% and the length is increased by 40%? The resistivity is fixed.

***Solution:*** Setting up a ratio

$$\frac{G_i = \dfrac{A_i}{\rho_i l_i}}{G_n = \dfrac{A_n}{\rho_n l_n}}$$

with the subscript $i$ for initial value and $n$ for new value. Since $\rho_i = \rho_n$ (same material):

$$\frac{G_i}{G_n} = \frac{A_i l_n}{A_n l_i}$$

with $A_n = (1/5)A_i$ and $l_n = 1.4l_i$, resulting in

$$\frac{G_i}{G_n} = \frac{(A_i)(1.4l_i)}{(0.2A_i)(l_i)} = \frac{1.4}{0.2} = 7$$

and

$$\mathbf{G_n = \frac{1}{7}G_i}$$

---

## 3.10 OHMMETERS

The *ohmmeter* is an instrument used to perform the following tasks and a number of other useful functions:

1. Measure the resistance of individual or combined elements
2. Detect open-circuit (high-resistance) and short-circuit (low-resistance) situations
3. Check continuity of network connections and identify wires of a multilead cable
4. Test some semiconductor (electronic) devices

For most applications, the ohmmeters used most frequently appear as part of a VOM or DMM such as appearing in Figs. 2.25 and 2.26. The details of the internal circuitry and the method of using the meter will be left primarily for the laboratory exercise. In general, however, the resistance of a resistor can be measured by simply connecting the two leads of the meter across the resistor as shown in Fig. 3.29. There is no need to be concerned about which lead goes on which end; the result will be the same in either case since resistors offer the same resistance to the flow of charge (current) in either direction. If the VOM is employed, a switch must be set to the proper resistance range and a nonlinear scale (usually the top scale of the meter) must be properly read to obtain the resistance value. The DMM also requires choosing the best scale setting for the resistance to be measured but the result appears as a numerical display with the proper placement of the decimal point as determined by the chosen scale. When measuring the resistance of a single resistor, it is usually best to remove the resistor from the network before making the measurement. If this is difficult or impossible, at least one end of the resistor must not be connected to the network or the reading may include the effects of the other elements of the system.

If the two leads of the meter are touching in the ohmmeter mode, the resulting resistance is obviously zero. A connection can be checked as shown in Fig. 3.30 by simply hooking up the meter to either side of the connection. If the resistance is zero, the connection is secure. If other than zero, it could be a weak connection and, if infinite, there is no connection at all.

If one wire of a harness is known, a second can be found as shown in Fig. 3.31. Simply connect the end of the known lead to the end of any other lead. When the ohmmeter indicates zero ohms (or very low resistance), the second lead has been identified. The above procedure can also be used to determine the first known lead by simply connecting the

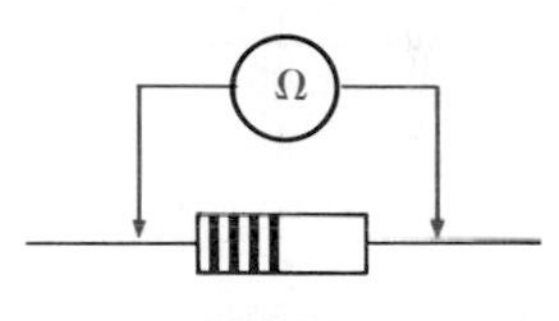


**FIG. 3.29**
*Measuring the resistance of a single element.*

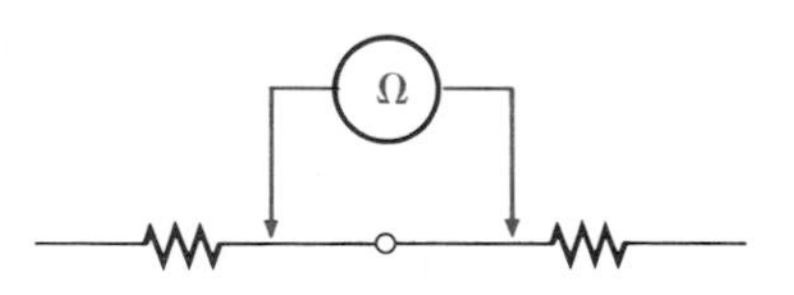


**FIG. 3.30**
*Checking the continuity of a connection.*

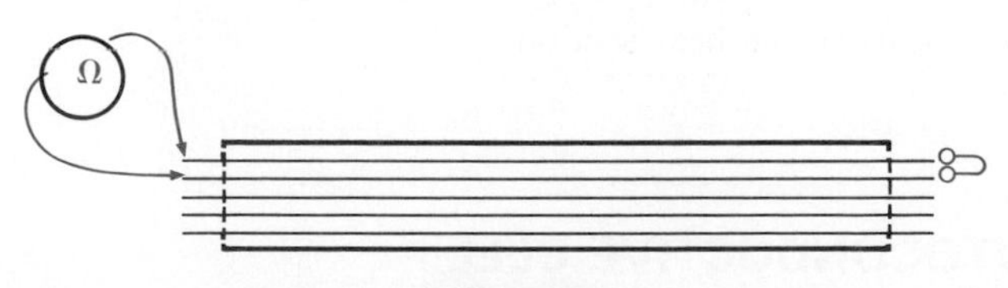


**FIG. 3.31**
*Identifying the leads of a multilead cable.*

meter to any wire at one end and then touching all the leads at the other end until a zero-ohm indication is obtained.

Preliminary measurements of the condition of some electronic devices such as the diode and transistor can be made using the ohmmeter. The meter can also be used to identify the terminals of such devices.

One important note about the use of any ohmmeter:

***Never hook up an ohmmeter to a live circuit!***

The reading will be meaningless and you may damage the instrument. The ohmmeter section of any meter is designed to pass a small sensing current through the resistance to be measured. A large external current could damage the movement and would certainly throw off the calibration of the instrument. In addition,

***never store an ohmmeter in the resistance mode.***

The two leads of the meter could touch and the small sensing current could drain the internal battery. VOMs should be stored with the selector switch on the highest voltage range, and DMMs in the off position.

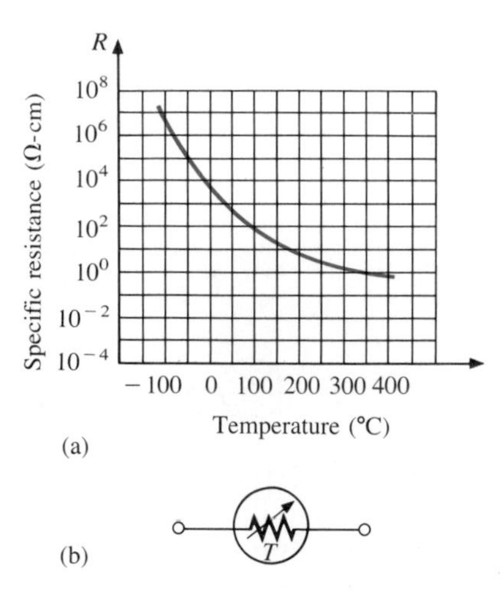


**FIG. 3.32**
*Thermistor. (a) Characteristics; (b) symbol.*

## 3.11 THERMISTORS

The *thermistor* is a two-terminal semiconductor device whose resistance, as the name suggests, is temperature sensitive. A representative characteristic appears in Fig. 3.32 with the graphic symbol for the device. Note the nonlinearity of the curve and the drop in resistance from about 5000 Ω to 100 Ω for an increase in temperature from 20°C to 100°C. The decrease in resistance with increase in temperature indicates a negative temperature coefficient.

The temperature of the device can be changed internally or externally. An increase in current through the device will raise its temperature, causing a drop in its terminal resistance. Any externally applied heat source will result in an increase in its body temperature and a drop in resistance. This type of action (internal or external) lends itself well to control mechanisms. A number of different types of thermistors are shown in Fig. 3.33. Materials employed in the manufacture of thermistors include oxides of cobalt, nickel, strontium, and manganese.

Note the use of a log scale in Fig. 3.32 for the vertical axis. The log scale permits the display of a wider range of specific resistance levels than a linear scale such as the horizontal axis. Note that it extends from 0.0001 Ω-cm to 100,000,000 Ω-cm over a very short interval. The log scale is used for both the vertical and the horizontal axis of Fig. 3.34, which appears in the next section.

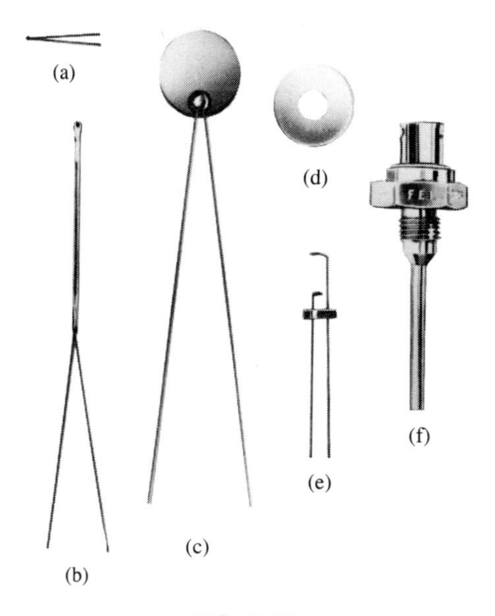


**FIG. 3.33**
*Thermistors. (a) Beads; (b) glass probe; (c) disc; (d) washer; (e) specially mounted bead; (f) special probe assembly. (Courtesy of Fenwal Electronics, Inc.)*

## 3.12 PHOTOCONDUCTIVE CELL

The *photoconductive cell* is a two-terminal semiconductor device whose terminal resistance is determined by the intensity of the incident light on its exposed surface. As the applied illumination increases in intensity, the energy state of the surface electrons and atoms increases, with a resultant increase in the number of "free carriers" and a corresponding drop in resistance. A typical set of characteristics and its graphic symbol appear in Fig. 3.34. Note the negative illumination coefficient. A number of cadmium sulfide photoconductive cells appear in Fig. 3.35.

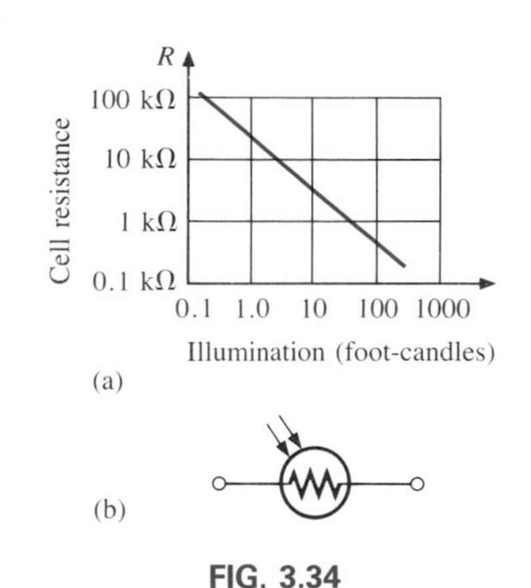


**FIG. 3.34**
*Photoconductive cell. (a) Characteristics; (b) symbol.*

## 3.13 VARISTORS

*Varistors* are voltage-dependent, nonlinear resistors used to suppress high-voltage transients. That is, their characteristics are such as to limit the voltage that can appear across the terminals of a sensitive device or system. A typical set of characteristics appears in Fig. 3.36(a) along

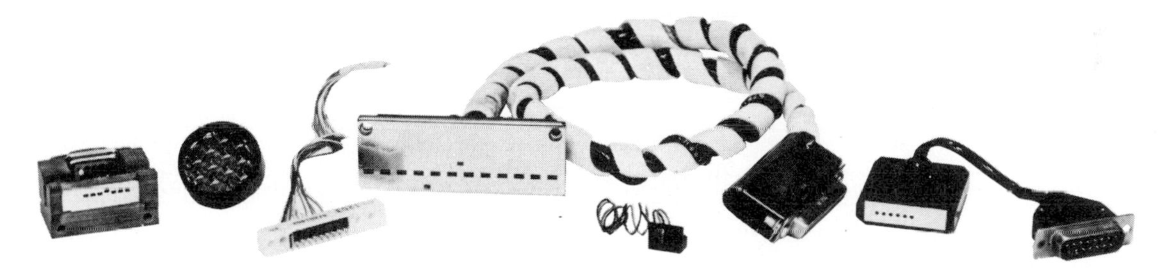

**FIG. 3.35**
*Photoconductive cells. (Courtesy of International Rectifier)*

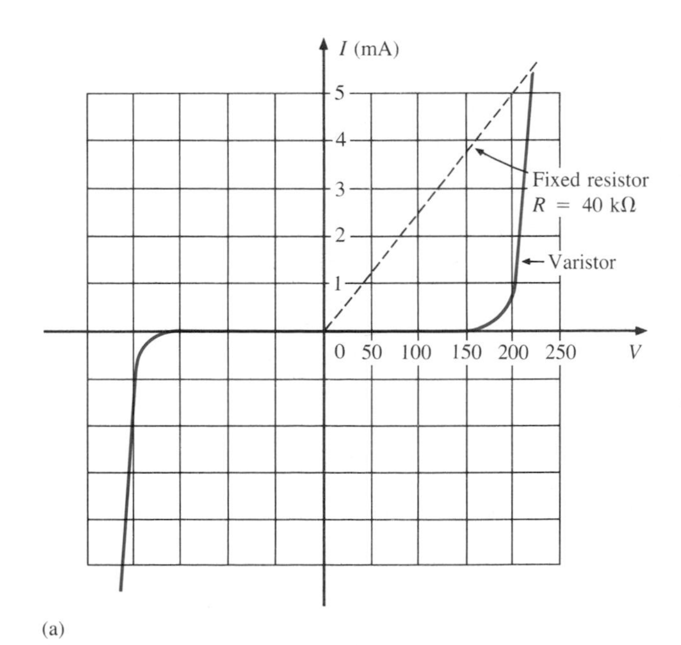

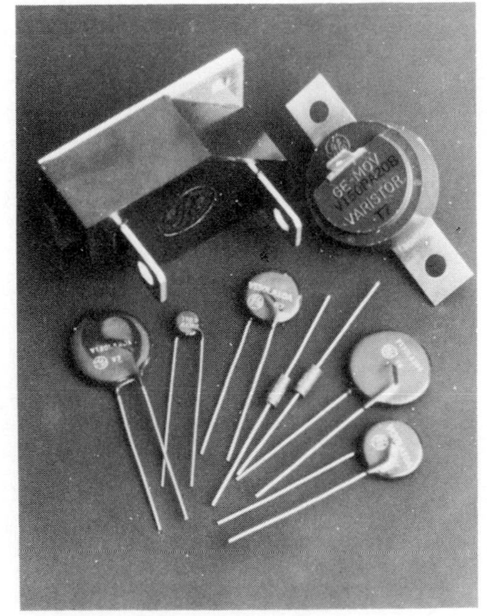


(b)

**FIG. 3.36**
*Varistors. (a) Characteristics; (b) photograph. (Courtesy of General Electric Co.)*

with a linear resistance characteristic for comparison purposes. Note that at a particular "firing voltage," the current rises rapidly but the voltage is limited to a level just above this firing potential. In other words, the magnitude of the voltage that can appear across this device cannot exceed that level defined by its characteristics. Through proper design techniques this device can therefore limit the voltage appearing across sensitive regions of a network. The current is simply limited by the network to which it is connected. A photograph of a number of commerical units appears in Fig. 3.36(b).

## 3.14 COMPUTER INPUT OF RESISTIVE ELEMENTS

The input of resistive elements will give us a chance to review the power-of-ten notation. Virtually all computer systems use power-of-ten notation, although some specify scientific notation (decimal point to the right of the first whole number ($5.67 \times 10^3$) and others ask for floating-point notation (decimal point at any location, such as $0.567 \times 10^4$ or $567. \times 10^1$). Both PSPICE and BASIC use the floating-point notation, permitting any one of the following forms for the number 1000 (using the uppercase letter *E* to designate power-of-ten format).

1000 1000.0 1E3 .001E6 0.1E4 10E2

Negative numbers simply have a negative sign in front of the number.

### PSPICE

In PSPICE specific suffixes (following the number) are given to frequently used powers of ten, as listed below. Since uppercase letters are specified for a particular purpose, you must be very careful in their use for other applications.

$$
\begin{aligned}
\text{F} &= 10^{-15} \\
\text{P} &= 10^{-12} \\
\text{N} &= 10^{-9} \\
\text{U} &= 10^{-6} \\
\text{M} &= 10^{-3} \\
\text{K} &= 10^{+3} \\
\text{MEG} &= 10^{+6} \\
\text{G} &= 10^{+9} \\
\text{T} &= 10^{+12}
\end{aligned}
$$

Using the above notation, the following are equivalent:

2E6 2MEG 2E3K .002G 2000K

Note, in particular, that milli and mega are differentiated by using an uppercase letter M for milli and uppercase letters MEG for mega.

The format for the input of a 2-kΩ resistor in PSPICE is the following:

```
              assumed polarities for V_R4
required      (+)   (-)
   ↓
  R4           4     3     2K
  name                     magnitude
```

The entry is very similar to that of a voltage source except the first letter must now be capital *R*. The name follows immediately using numbers or letters and then the node assignment. Even though resistors do not have polarity, as does a dc voltage source, the voltage across the resistor will have a particular polarity. In many cases the actual polarity will not be known when the circuit is entered, but an assumption is made that the output will confirm or reverse. For the above entry it is assumed that if the voltage across R4 is determined, node 4 will be at a higher potential than node 3. If correct, the output magnitude will have no sign at all (signifying +), but if the actual polarity is the reverse, the output will include a negative sign. In any case, do not be overly concerned about the polarity when the resistor values are entered. Make a reasonable assumption about the polarity and then let the computer package determine the actual result.

## BASIC

In BASIC, resistors are input using the same format as applied to dc voltage sources. The following is a listing for the same resistor entered above using PSPICE.

```
line number or location
 36     INPUT "R4=";R4
                item requested
```

The only difference is the comment within the quotation marks and the resistor name after the semicolon. When the program is run, the computer will request the magnitude of R4; after it is entered as 2E3, the computer will continue on to the next line of the program.

## PROBLEMS

### SECTION 3.2 Resistance: Circular Wires

**1.** Convert the following to mils:

**a.** 0.5 in. **b.** 0.01 in.
**c.** 0.004 in. **d.** 1 in.
**e.** 0.02 ft **f.** 0.01 cm

2. Calculate the area in circular mils (CM) of wires having the following diameters:
   **a.** 0.050 in. **b.** 0.016 in.
   **c.** 0.30 in. **d.** 0.1 cm
   **e.** 0.003 ft **f.** 0.0042 m

3. The area in circular mils is
   **a.** 1600 CM **b.** 900 CM
   **c.** 40,000 CM **d.** 625 CM
   **e.** 7.75 CM **f.** 81 CM
   What is the diameter of each wire in inches?

4. What is the resistance of a copper wire 200 ft long and 0.01 inch in diameter ($T = 20°C$)?

5. Find the resistance of a silver wire 50 yd long and 0.0045 inch in diameter ($T = 20°C$).

6. **a.** What is the area in circular mils of an aluminum conductor that is 80 ft long with a resistance of 2.5 Ω?
   **b.** What is its diameter in inches?

7. A 2.2-Ω resistor is to be made of nichrome wire. If the available wire is 1/32 in. in diameter, how much wire is required?

8. **a.** What is the area in circular mils of a copper wire that has a resistance of 2.5 Ω and is 300 ft long ($T = 20°C$)?
   **b.** Without working out the numerical solution, determine whether the area of an aluminum wire will be smaller or larger than that of the copper wire. Explain.
   **c.** Repeat (b) for a silver wire.

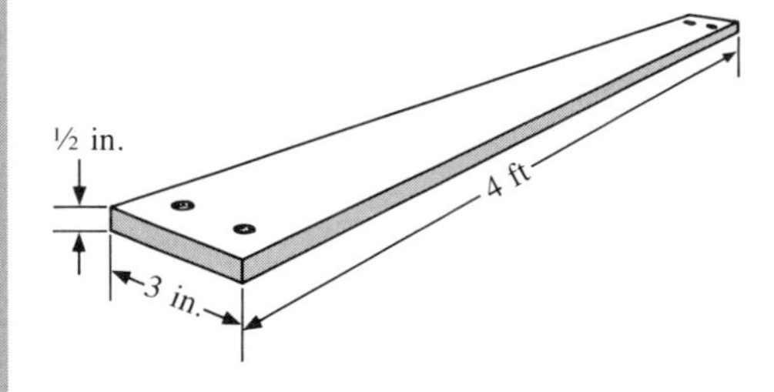

FIG. 3.37

9. In Fig. 3.37, three conductors of different materials are presented.
   **a.** Without working out the numerical solution, which section would appear to have the most resistance? Explain.
   **b.** Find the resistance of each section and compare with the result of (a) ($T = 20°C$).

10. A wire 1000 ft long has a resistance of 0.5 kΩ and an area of 94 CM. Of what material is the wire made ($T = 20°C$)?

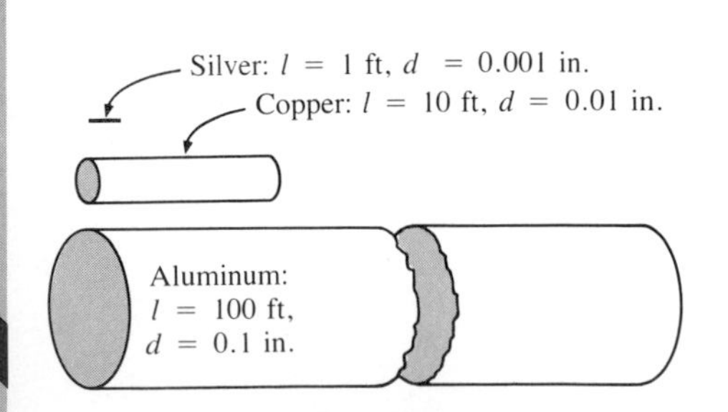


FIG. 3.38

*11. **a.** What is the resistance of a copper bus-bar with the dimensions shown ($T = 20°C$) in Fig. 3.38?
   **b.** Repeat (a) for aluminum and compare the results.
   **c.** Without working out the numerical solution, determine whether the resistance of the bar (aluminum or copper) will increase or decrease with increase in length. Explain your answer.
   **d.** Repeat (c) for increase in cross-sectional area.

12. Determine the increase in resistance of a copper conductor if the area is reduced by a factor of 4 and the length doubled. The original resistance was 0.2 Ω. The temperature remains fixed.

*13. What is the new resistance level of a copper wire if the length is changed from 200 ft to 100 yd, the area is changed from 40,000 CM to 0.04 in.$^2$ and the original resistance was 800 m$\Omega$?

## SECTION 3.3 Wire Tables

14. a. Using Table 3.2, find the resistance of 450 ft of #11 and #14 AWG wires.
    b. Compare the resistances of the two wires.
    c. Compare the areas of the two wires.

15. a. Using Table 3.2, find the resistance of 1800 ft of #8 and #18 AWG wires.
    b. Compare the resistances of the two wires.
    c. Compare the areas of the two wires.

16. a. For the system of Fig. 3.39, the resistance of each line cannot exceed 0.006 $\Omega$, and the maximum current drawn by the load is 110 A. What gage wire should be used?
    b. Repeat (a) for a maximum resistance of 0.003 $\Omega$, $d$ = 30 ft, and a maximum current of 110 A.

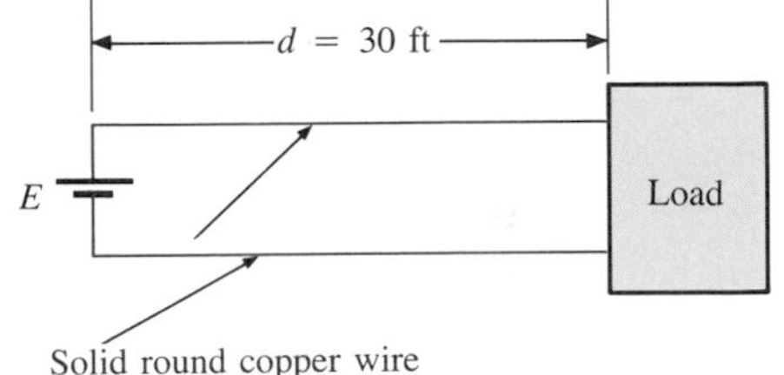


FIG. 3.39

*17. a. From Table 3.2, determine the maximum permissible current density (A/CM) for an AWG #0000 wire.
    b. Convert the result of (a) to A/in.$^2$.
    c. Using the result of (b), determine the cross-sectional area required to carry a current of 5000 A.

## SECTION 3.4 Resistance: Metric Units

18. Using metric units, determine the length of a copper wire that has a resistance of 0.2 $\Omega$ and a diameter of 1/10 in.

19. Repeat Problem 8 using metric units. That is, convert the given dimensions to metric units before determining the resistance.

20. If the sheet resistance of a tin oxide sample is 100 $\Omega$, what is the thickness of the oxide layer?

21. Determine the width of a carbon resistor having a sheet resistance of 150 $\Omega$ if the length is 1/2 in. and the resistance is 500 $\Omega$.

*22. Derive the conversion factor between $\rho$ (CM-$\Omega$/ft) and $\rho$ ($\Omega$-cm) by
    a. Solving for $\rho$ for the wire of Fig. 3.40 in CM-$\Omega$/ft.
    b. Solving for $\rho$ for the same wire of Fig. 3.40 in $\Omega$-cm by making the necessary conversions.
    c. Use the equation $\rho_2 = k\rho_1$ to determine the conversion factor $k$ if $\rho_1$ is the solution of part (a) and $\rho_2$ the solution of part (b).

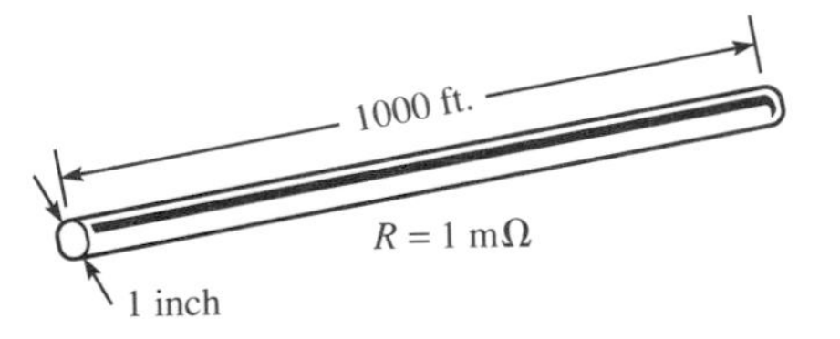


FIG. 3.40

## SECTION 3.5 Temperature Effects

23. The resistance of a copper wire is 2 $\Omega$ at 10°C. What is its resistance at 60°C?

24. The resistance of an aluminum bus-bar is 0.02 $\Omega$ at 0°C. What is its resistance at 100°C?

25. The resistance of a copper wire is 4 $\Omega$ at 70°F. What is its resistance at 32°F?

26. The resistance of a copper wire is 0.76 Ω at 30°C. What is its resistance at −40°C?

27. If the resistance of a silver wire is 0.04 Ω at −30°C, what is its resistance at 0°C?

*28. a. The resistance of a copper wire is 0.002 Ω at room temperature (68°F). What is its resistance at 32°F (freezing) and 212°F (boiling)?
  b. For (a), determine the change in resistance for each 10° change in temperature between room temperature and 212°F.

29. a. The resistance of a copper wire is 0.92 Ω at 4°C. At what temperature (°C) will it be 1.06 Ω?
  b. At what temperature will it be 0.15 Ω?

*30. a. If the resistance of a 1000-ft length of copper wire is 10 Ω at room temperature (20°C), what will its resistance be at 50 K (Kelvin units) using Eq. (3.6)?
  b. Repeat part (a) for a temperature of 38.65 K. Comment on the results obtained by reviewing the curve of Fig. 3.12.
  c. What is the temperature of absolute zero in Fahrenheit units?

31. Find the values of $\alpha_1$ for copper and aluminum at 20°C, and compare them with those given in Table 3.5.

32. Using Eq. (3.7), find the resistance of a copper wire at 16°C if its resistance at 20°C is 0.4 Ω.

33. a. Find the value of $\alpha_1$ at $t_1 = 40°C$ for copper.
  b. Using the result of (a), find the resistance of a copper wire at 75°C if its resistance is 0.3 Ω at 40°C.

34. A 22-Ω wire-wound resistor is rated at +20 PPM for a temperature range of −10°C to +75°C. Determine its resistance at 65°C.

35. Determine the PPM rating of the 10-kΩ carbon composition resistor of Fig. 3.19 using the resistance level determined at 90°C.

## SECTION 3.6 Superconductors

36. Visit your local library and find a table listing the critical temperatures for a variety of materials. List at least five materials with the critical temperatures that are not mentioned in this text. Choose a few materials that have relatively high critical temperatures.

*37. a. Using the curve of Fig. 3.13, what month of what year (make your best approximation) would you expect superconductors at room temperature to be a certainty.
  b. What is the approximate increase in temperature per year ($\Delta T/\Delta t$) using 30 K temperature as a starting point?
  c. Using the results of part (b) and a starting point of October 1986, determine when room-temperature superconductors may become a reality. How does your result compare with the solution of part (a)?

***38.** Using the required 1-MA/cm$^2$ density level for IC manufacturing, what would the resulting current be through a #12 house wire? Compare the result obtained with the allowable limit of Table 3.2.

***39.** Research the SQUID magnetic field detector and review its basic mode of operation and an application or two.

## SECTION 3.7 Types of Resistors

**40. a.** What is the approximate increase in size from a 1-W to a 2-W carbon resistor?

**b.** What is the approximate increase in size from a 1/2-W to a 2-W carbon resistor?

**41.** If the 10-kΩ resistor of Fig. 3.19 is exactly 10 kΩ at room temperature, what is its approximate resistance at −30°C and 100°C (boiling)?

**42.** Repeat Problem 41 at a temperature of 100°F.

**43.** If the resistance between the outside terminals of a linear potentiometer is 10 kΩ, what is its resistance between the wiper (movable) arm and an outside terminal if the resistance between the wiper arm and the other outside terminal is 3.5 kΩ?

**44.** If the wiper arm of a linear potentiometer is one-quarter the way around the contact surface, what is the resistance between the wiper arm and each terminal if the total resistance is 25 kΩ?

***45.** Show the connections required to establish 4 kΩ between the wiper arm and one outside terminal of a 10-kΩ potentiometer while having only zero ohms between the other outside terminal and the wiper arm.

## SECTION 3.8 Color Coding

**46.** Find the range in which a resistor having the following color bands must exist to satisfy the manufacturer's tolerance:

| | 1st band | 2nd band | 3rd band | 4th band |
|---|---|---|---|---|
| **a.** | green | blue | orange | gold |
| **b.** | red | red | brown | silver |
| **c.** | brown | black | black | — |

**47.** Is there an overlap in coverage between 20% resistors? That is, determine the tolerance range for a 10-Ω 20% resistor and a 15-Ω 20% resistor and note whether their tolerance ranges overlap.

**48.** Repeat Problem 37 for 10% resistors of the same value.

**49.** Repeat Problem 37 for a 47-Ω 20% resistor and a 68-Ω 20% resistor.

## SECTION 3.9 Conductance

**50.** Find the conductance of each of the following resistances:

**a.** 0.086 Ω **b.** 4000 Ω

**c.** 0.05 MΩ

Compare the three results.

**51.** Find the conductance of 1000 ft of #18 AWG wire made of
  **a.** copper
  **b.** aluminum
  **c.** iron

***52.** The conductance of a wire is 100 S. If the area of the wire is increased by a factor of 2/3 and the length is reduced by the same factor, find the new conductance of the wire if the temperature remains fixed.

## SECTION 3.10 Ohmmeters

**53.** How would you check the status of a fuse with an ohmmeter?

**54.** How would you determine the on and off states of a switch using an ohmmeter?

**55.** How would you use an ohmmeter to check the status of a light bulb?

## SECTION 3.11 Thermistors

***56.** **a.** Find the resistance of the thermistor having the characteristics of Fig. 3.32 at −50°C, 50°C, and 200°C. Note that it is a log scale. If necessary, consult a reference with an expanded log scale.
  **b.** Does the thermistor have a positive or negative temperature coefficient?
  **c.** Is the coefficient a fixed value for the range −100°C to 400°C? Why?
  **d.** What is the approximate rate of change of $\rho$ with temperature at 100°C?

## SECTION 3.12 Photoconductive Cell

***57.** **a.** Using the characteristics of Fig. 3.34, determine the resistance of the photoconductive cell at 10 and 100 foot-candle illumination. As in Problem 56, note that it is a log scale.
  **b.** Does the cell have a positive or negative illumination coefficient?
  **c.** Is the coefficient a fixed value for the range 0.1 to 1000 foot-candles? Why?
  **d.** What is the approximate rate of change of $R$ with illumination at 10 foot-candles?

## SECTION 3.13 Varistors

**58.** **a.** Referring to Fig. 3.36(a), find the terminal voltage of the device at 0.5, 1, 3, and 5 mA.
  **b.** What is the total change in voltage for the indicated range of current levels?
  **c.** Compare the ratio of maximum to minimum current levels above to the corresponding ratio of voltage levels.

### SECTION 3.14 Computer Input of Resistive Elements

**59.** What is the PSPICE input for a 22-Ω resistor labeled RBB between terminals 9 (+) and 10 (−)?

**60.** Repeat Exercise 59 for a 50,000-Ω and 1.2-MΩ resistor.

**61. a.** In BASIC how would one request the value of RBB from Exercise 59? Use line 1020.
**b.** How would the program request the value of RBB?

## GLOSSARY

**Absolute zero** The temperature at which all molecular motion ceases; −273.15°C.

**Circular mil (CM)** The cross-sectional area of a wire having a diameter of one mil.

**Color coding** A technique employing bands of color to indicate the resistance levels and tolerance of resistors.

**Conductance** ($G$) An indication of the relative ease with which current can be established in a material. It is measured in siemens (S).

**Cooper effect** The "pairing" of electrons as they travel through a medium.

**Inferred absolute temperature** The temperature through which a straight-line approximation for the actual resistance-versus-temperature curve will intersect the temperature axis.

**Meissner effect** The levitation of a magnet above a superconductor.

**Negative temperature coefficient of resistance** The value which reveals that the resistance of a material will decrease with increase in temperature.

**Ohm (Ω)** The unit of measurement applied to resistance.

**Ohmmeter** An instrument for measuring resistance levels.

**Photoconductive cell** A two-terminal semiconductor device whose terminal resistance is determined by intensity of the incident light on its exposed surface.

**Positive temperature coefficient of resistance** The value which reveals that the resistance of a material will increase with increase in temperature.

**Potentiometer** A three-terminal device through which potential levels can be varied in a linear or nonlinear manner.

**PPM/°C** Temperature sensitivity of a resistor in parts per million per degree Celsius.

**Resistance** A measure of the opposition to the flow of charge through a material.

**Resistivity** ($\rho$) A constant of proportionality between the resistance of a material and its physical dimensions.

**Rheostat** An element whose terminal resistance can be varied in a linear or nonlinear manner.

**Sheet resistance** Defined by $\rho/d$ for thin-film and integrated circuit design.

**SQUID** Superconducting quantum interference device.

**Superconductor** Conductors of electric charge that have for all practical purposes zero ohms.

**Thermistor** A two-terminal semiconductor device whose resistance is temperature sensitive.

**Varistor** A voltage-dependent, nonlinear resistor used to suppress high-voltage transients.

# 4 Ohm's Law, Power, and Energy

## 4.1 OHM'S LAW

Consider the following relationship:

$$\text{Effect} = \frac{\text{cause}}{\text{opposition}} \qquad \textbf{(4.1)}$$

Every conversion of energy from one form to another can be related to this equation. In electric circuits, the *effect* we are trying to establish is the flow of charge, or *current*. The *potential difference* between two points is the *cause* (''pressure''), and the opposition is the *resistance* encountered.

Substituting these terms into Eq. (4.1) results in

$$\text{Current} = \frac{\text{potential difference}}{\text{resistance}}$$

and

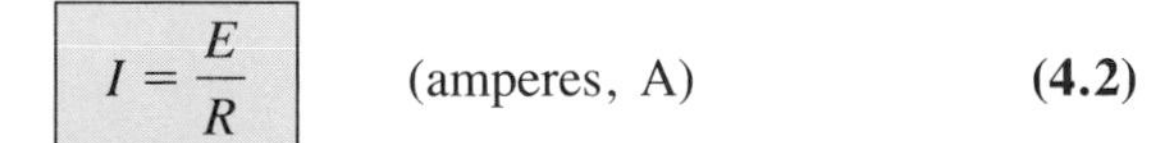

$$I = \frac{E}{R} \qquad \text{(amperes, A)} \qquad \textbf{(4.2)}$$

Equation (4.2), known as *Ohm's law,* clearly reveals that for a fixed resistance the greater the voltage (or pressure) across a resistor, the more the current, and the more the resistance for the same voltage, the less the current. In other words, the current is proportional to the applied voltage and inversely proportional to the resistance.

By simple mathematical manipulations, the voltage and resistance can be found in terms of the other two quantities:

$$\boxed{E = IR} \quad \text{(volts, V)} \tag{4.3}$$

$$\boxed{R = \frac{E}{I}} \quad \text{(ohms, }\Omega\text{)} \tag{4.4}$$

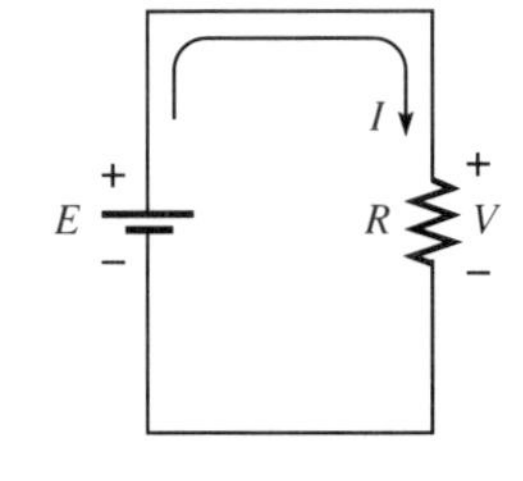


**FIG. 4.1**
*Basic circuit.*

The three quantities of Eqs. (4.2) through (4.4) are defined by the simple circuit of Fig. 4.1. The current $I$ of Eq. (4.2) results from applying a dc supply of $E$ volts across a network having a resistance $R$. Equation (4.3) determines the voltage $E$ required to establish a current $I$ through a network with a total resistance $R$, and Eq. (4.4) provides the resistance of a network that results in a current $I$ due to an impressed voltage $E$.

Note in Fig. 4.1 that the voltage source ''pressures'' current in a direction that passes from the negative to positive terminal of the battery. This will always be the case for single-source circuits. The effect of more than one source in the network will be examined in the chapters to follow. The symbol for the voltage of the battery (a source of electrical energy) is the uppercase letter $E$, whereas the loss in potential energy across the resistor is given the symbol $V$. The polarity of the voltage drop across the resistor is as defined by the applied source because the two terminals of the battery are connected directly across the resistive element.

---

**EXAMPLE 4.1.** Determine the current resulting from the application of a 9-V battery across a network with a resistance of 2.2 Ω.

***Solution:*** Eq. (4.2):

$$I = \frac{E}{R} = \frac{9\text{ V}}{2.2\ \Omega} = \mathbf{4.09\ A}$$

---

**EXAMPLE 4.2.** Calculate the resistance of a 60-W bulb if a current of 500 mA results from an applied voltage of 120 V.

***Solution:*** Eq. (4.4):

$$R = \frac{E}{I} = \frac{120\text{ V}}{500 \times 10^{-3}\text{ A}} = \mathbf{240\ \Omega}$$

---

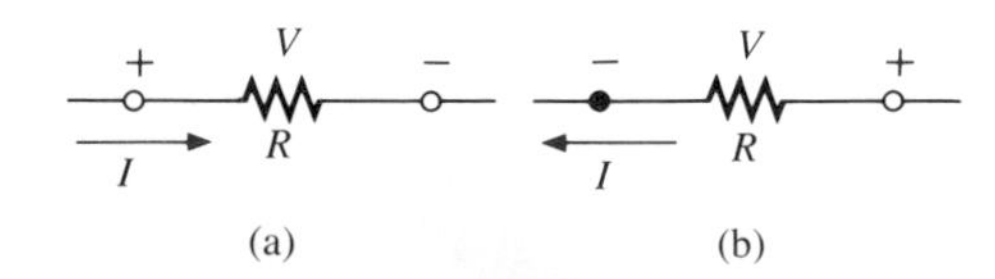


**FIG. 4.2**
*Defining polarities.*

For an isolated resistive element, the polarity of the voltage drop is as shown in Fig. 4.2(a) for the indicated current direction. A reversal in current will reverse the polarity, as shown in Fig. 4.2(b). In general, the flow of charge is from a high (+) to a low (−) potential. Polarities as established by current direction will become increasingly important in the analysis to follow.

**EXAMPLE 4.3.** Calculate the current through the 2-kΩ resistor of Fig. 4.3 if the voltage drop across it is 16 V.

***Solution:***

$$I = \frac{V}{R} = \frac{16\text{ V}}{2 \times 10^3\ \Omega} = \mathbf{8\ mA}$$

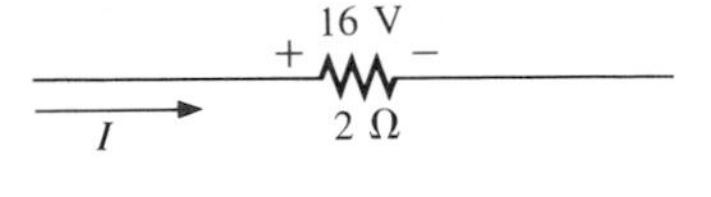

**FIG. 4.3**

**EXAMPLE 4.4.** Calculate the voltage that must be applied across the soldering iron of Fig. 4.4 to establish a current of 1.5 A through the iron if its internal resistance is 80 Ω.

***Solution:***

$$E = IR = (1.5\text{ A})(80\ \Omega) = \mathbf{120\ V}$$

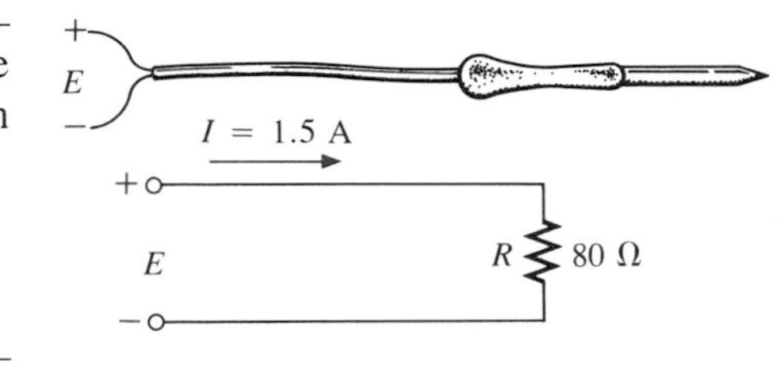

**FIG. 4.4**

## 4.2 PLOTTING OHM'S LAW

Graphs, characteristics, plots, and the like, play an important role in every technical field as a mode through which the broad picture of the behavior or response of a system can be conveniently displayed. It is therefore critical to develop the skills necessary both to read data and to plot them in such a manner that they can be interpreted easily.

For most sets of characteristics of electronic devices, the current is represented by the vertical axis (ordinate), and the voltage by the horizontal axis (abscissa), as shown in Fig. 4.5. First note that the vertical

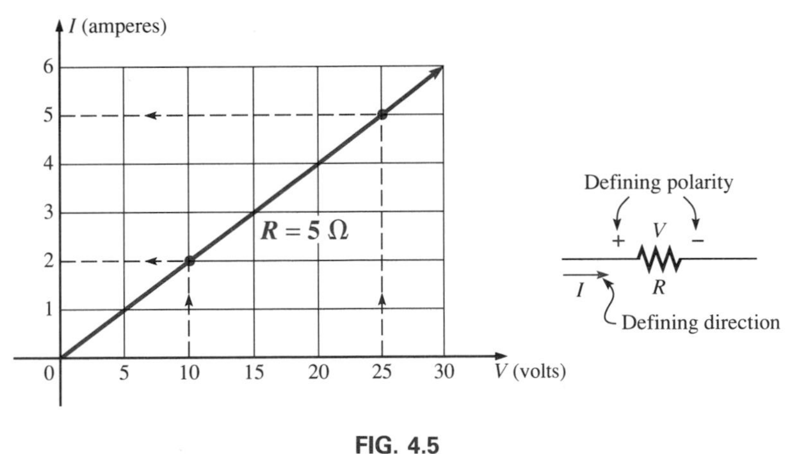


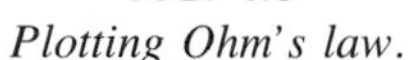
**FIG. 4.5**
*Plotting Ohm's law.*

axis is in amperes and the horizontal axis in volts. For some plots, *I* may be in milliamperes, microamperes, or whatever is appropriate for the range of interest. The same is true for the levels of voltage on the horizontal axis. Note also that the chosen parameters require that the spacing between numerical values of the vertical axis be different from that of the horizontal axis. The linear (straight-line) graph reveals that

the resistance is not changing with current or voltage level; rather, it is a fixed quantity throughout. The current direction and the voltage polarity appearing to the right of Fig. 4.5 are the defined direction and polarity for the provided plot. If the current direction is opposite to the defined direction, the region below the horizontal axis is the region of interest for the current $I$. If the voltage polarity is opposite to that defined, the region to the left of the current axis is the region of interest. For the standard fixed resistor, the first quadrant, or region, of Fig. 4.5 is the only region of interest. There are a number of devices, however, that you will encounter in your electronics courses that will use the other quadrants of a graph.

Once a graph such as Fig. 4.5 is developed, the current or voltage at any level can be found from the other quantity by simply using the resulting plot. For instance, at $V = 25$ V, if a vertical line is drawn on Fig. 4.5 to the curve as shown, the resulting current can be found by drawing a horizontal line over to the current axis, where a result of 5 A is obtained. Similarly, at $V = 10$ V, a vertical line to the plot and a horizontal line to the current axis will result in a current of 2 A, as determined by Ohm's law.

If the resistance of a plot is unknown, it can be determined at any point on the plot, since a straight line indicates a fixed resistance. At any point on the plot, find the resulting current and voltage and simply substitute into the following equation:

$$R_{dc} = \frac{V}{I} \tag{4.5}$$

To test Eq. 4.5 consider a point on the plot where $V = 20$ V and $I = 4$ A. The resulting resistance is $R_{dc} = V/I = 20\text{ V}/4\text{ A} = 5\ \Omega$. For comparison purposes, a 1-Ω and 10-Ω resistor were plotted on the graph of Fig. 4.6. Note that the less the resistance, the steeper the slope (closer to the vertical axis) of the curve.

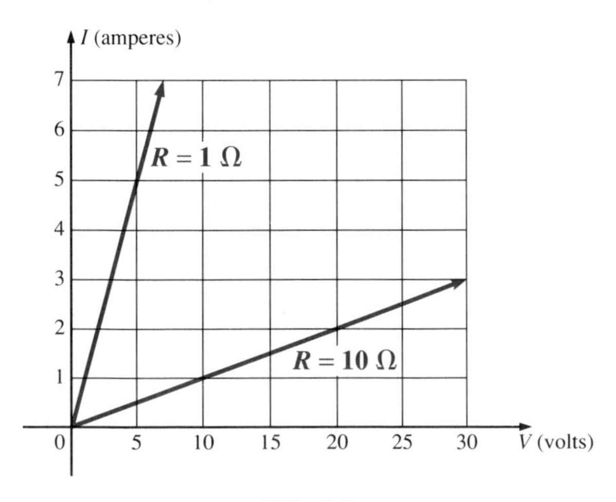

FIG. 4.6

If we write Ohm's law in the following manner and relate it to the basic straight-line equation

$$\begin{array}{ccccccc} I & = & \frac{1}{R} & \cdot & E & + & 0 \\ \downarrow & & \downarrow & & \downarrow & & \downarrow \\ y & = & m & \cdot & x & + & b \end{array}$$

we find that the slope is equal to 1 divided by the resistance value, as indicated by the following:

$$m = \text{slope} = \frac{\Delta y}{\Delta x} = \frac{\Delta I}{\Delta V} = \frac{1}{R} \qquad \textbf{(4.6)}$$

where $\Delta$ signifies a small finite change in the variable.

Equation (4.6) clearly reveals that the greater the resistance, the less the slope. If written in the following form, Eq. (4.6) can be used to determine the resistance from the linear curve:

$$R = \frac{\Delta V}{\Delta I} \quad \text{(ohms)} \qquad \textbf{(4.7)}$$

The equation states that by choosing a particular $\Delta V$ (or $\Delta I$), the corresponding $\Delta I$ (or $\Delta V$, respectively) can be obtained from the graph, as shown in Fig. 4.7, and the resistance can be determined. If the plot is

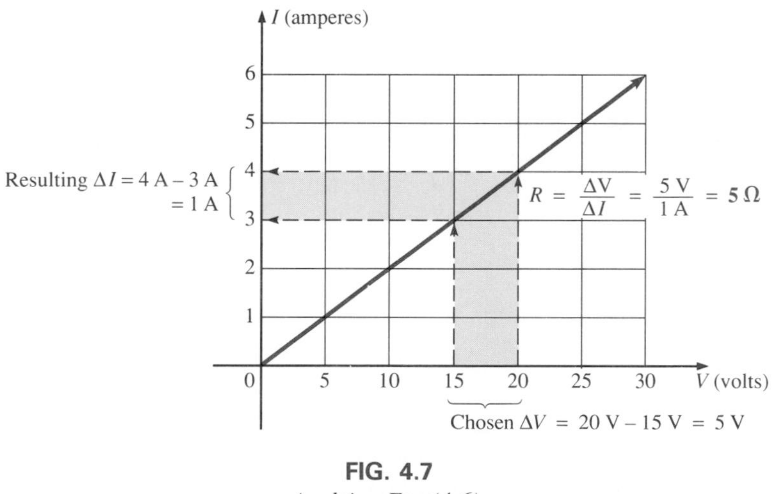


**FIG. 4.7**
*Applying Eq. (4.6).*

a straight line, Eq. (4.7) will provide the same result no matter where the equation is applied. However, if the plot curves at all the resistance will change.

**EXAMPLE 4.5.** Determine the resistance associated with the curve of Fig. 4.8 using Eqs. (4.5) and (4.7), and compare results.

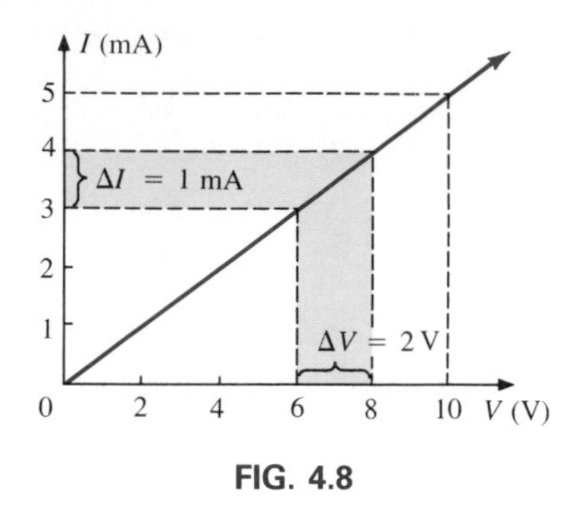


**FIG. 4.8**

***Solution:*** At $V = 6$ V, $I = 3$ mA and

$$R_{dc} = \frac{V}{I} = \frac{6\text{ V}}{3\text{ mA}} = \mathbf{2\ k\Omega}$$

For the interval between 6 V and 8 V,

$$R = \frac{\Delta V}{\Delta I} = \frac{2\text{ V}}{1\text{ mA}} = \mathbf{2\ k\Omega}$$

The results are equivalent.

Before leaving the subject, let us first investigate the characteristics of a very important semiconductor device called the *diode,* which will be examined in detail in the basic electronics courses. This device will ideally act like a low resistance path to current in one direction and a high resistance path to current in the reverse direction, much like a switch that will pass current in only one direction. A typical set of characteristics appears in Fig. 4.9. Without any mathematical calcula-

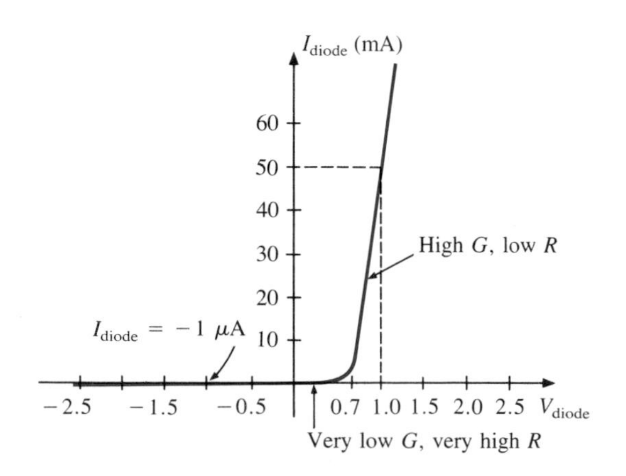


**FIG. 4.9**
*Semiconductor diode characteristic.*

tions, the closeness of the characteristic to the voltage axis for negative values of applied voltage indicates that this is the low conductance (high resistance, switch opened) region. Note that this region extends to approximately 0.7 V positive. However, for values of applied voltage greater than 0.7 V, the vertical rise in the characteristics indicates a high conductivity (low resistance, switch closed) region. Application of Ohm's law will now verify the above conclusions.

At $V = +1$ V,

$$R_{\text{diode}} = \frac{V}{I} = \frac{1\text{ V}}{50\text{ mA}} = \frac{1}{50 \times 10^{-3}}$$
$$= 20\ \Omega$$

(relatively low value for most applications)

At $V = -1$ V,

$$R_{\text{diode}} = \frac{V}{I} = \frac{1\text{ V}}{1\ \mu\text{A}}$$
$$= 1\text{ M}\Omega$$

## 4.3 POWER

*Power* is an indication of how much work (the conversion of energy from one form to another) can be accomplished in a specified amount of time, that is, a *rate* of doing work. For instance, a large motor has more power than a small motor because it can convert more electrical energy into mechanical energy in the same period of time. Since converted energy is measured in *joules* (J) and time in seconds (s), power is measured in joules/second (J/s). The electrical unit of measurement for power is the watt (W), defined by

$$1\text{ watt (W)} = 1\text{ joule/second (J/s)} \tag{4.8}$$

In equation form, power is determined by

$$P = \frac{W}{t} \qquad \text{(watts, W, or joules/second, J/s)} \tag{4.9}$$

with the energy $W$ measured in joules and the time $t$ in seconds.

Throughout the text, the abbreviation for energy ($W$) can be distinguished from that for the watt (W) by the fact that one is in italics while the other is in roman. In fact, all variables in the dc section appear in italics while the units appear in roman.

The unit of measurement, the watt, is derived from the surname of James Watt, who was instrumental in establishing the standards for

power measurements. He introduced the *horsepower* (hp) as a measure of the average power of a strong dray horse over a full working day. It is approximately 50% more than can be expected from the average horse. The horsepower and watt are related in the following manner:

$$\boxed{1 \text{ horsepower} \cong 746 \text{ watts}}$$

The power delivered to, or absorbed by, an electrical device or system can be found in terms of the current and voltage by first substituting Eq. (2.7) into Eq. (4.9):

$$P = \frac{W}{t} = \frac{QV}{t} = V\frac{Q}{t}$$

But

$$I = \frac{Q}{t}$$

so that

$$\boxed{P = VI} \qquad \text{(watts)} \qquad \textbf{(4.10)}$$

By direct substitution of Ohm's law, the equation for power can be obtained in two other forms:

$$P = VI = V\left(\frac{V}{R}\right)$$

and

$$\boxed{P = \frac{V^2}{R}} \qquad \text{(watts)} \qquad \textbf{(4.11)}$$

or

$$P = VI = (IR)I$$

and

$$\boxed{P = I^2R} \qquad \text{(watts)} \qquad \textbf{(4.12)}$$

The result is that the power absorbed by the resistor of Fig. 4.10 can be found directly depending on the information available. In other

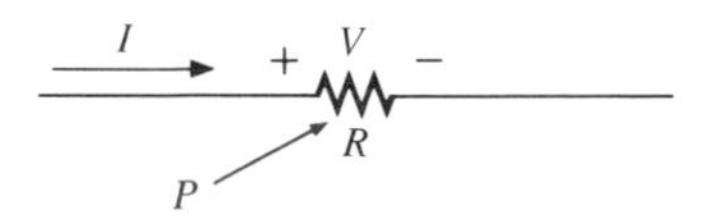


FIG. 4.10

words, if the current and resistance are known, it pays to use Eq. (4.12) directly, and if $V$ and $I$ are known, Eq. (4.10) is appropriate. It saves having to apply Ohm's law before determining the power.

Power can be delivered or absorbed as defined by the polarity of the voltage and direction of the current. For all dc voltage sources, power is being *delivered* by the source if the current has the direction appearing in Fig. 4.11(a). Note that the current has the same direction as established by the source in a single-source network. If the current direction and polarity are as shown in Fig. 4.11(b) due to a multisource network, the battery is absorbing power much as when a battery is being charged.

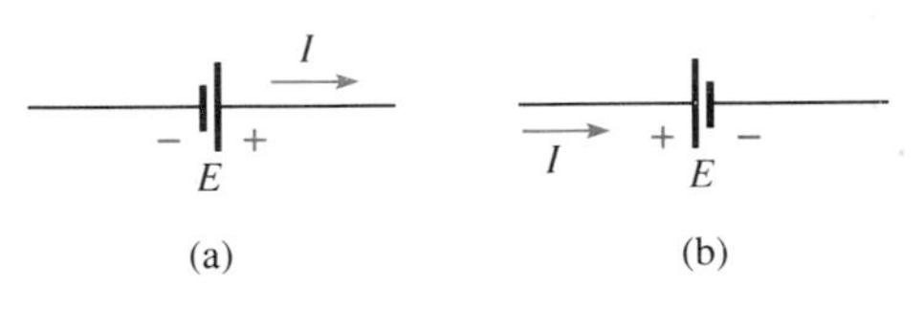


**FIG. 4.11**
*Battery power. (a) Supplied; (b) absorbed.*

For resistive elements, all the power delivered is dissipated in the form of heat because the voltage polarity is defined by the current direction (and vice versa) and current will always enter the terminal of higher potential corresponding with the absorbing state of Fig. 4.11(b). A reversal of the current direction in Fig. 4.10 will also reverse the polarity of the voltage across the resistor and match the conditions of Fig. 4.11(b).

The magnitude of the power delivered or absorbed by a battery is given by

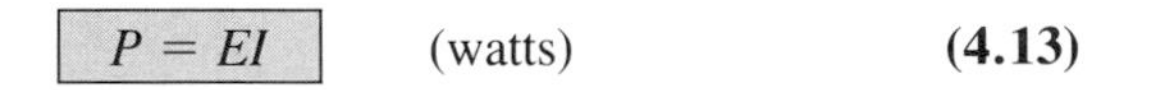

$$P = EI \qquad \text{(watts)} \qquad \textbf{(4.13)}$$

with $E$ the battery terminal voltage and $I$ the current through the source.

**EXAMPLE 4.6.** Find the power delivered to the dc motor of Fig. 4.12.

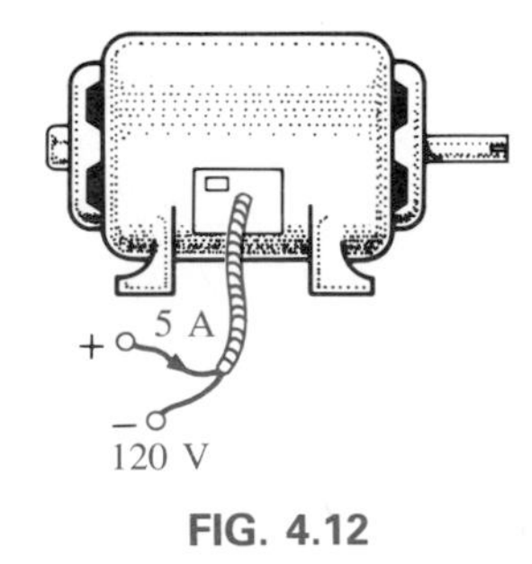


**FIG. 4.12**

***Solution:***

$$P = VI = (120\text{ V})(5\text{ A}) = 600\text{ W} = \mathbf{0.6\ kW}$$

**EXAMPLE 4.7.** What is the power dissipated by a 5-Ω resistor if the current is 4 A?

***Solution:***

$$P = I^2R = (4\text{ A})^2(5\ \Omega) = \mathbf{80\ W}$$

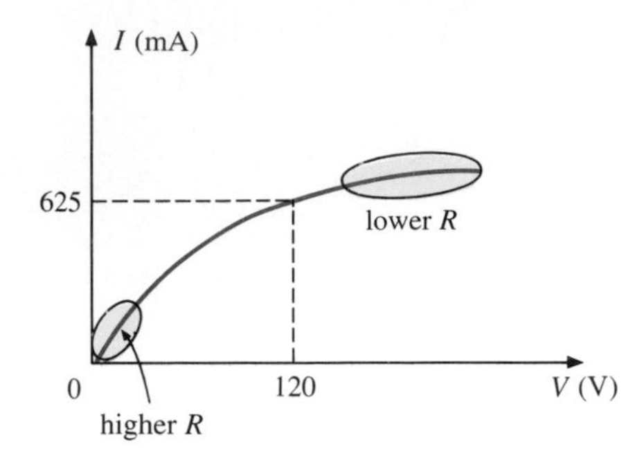


FIG. 4.13

**EXAMPLE 4.8.** The $I$-$V$ characteristics of a light bulb are provided in Fig. 4.13. Note the nonlinearity of the curve, indicating a wide range in resistance of the bulb with applied voltage as defined by the discussion of Section 4.2. If the rated voltage is 120 V, find the wattage rating of the bulb. Also calculate the resistance of the bulb under rated conditions.

***Solution:*** At 120 V,

$$I = 0.625 \text{ A}$$

and

$$P = VI = (120 \text{ V})(0.625 \text{ A}) = \mathbf{75\ W}$$

At 120 V,

$$R = \frac{V}{I} = \frac{120 \text{ V}}{0.625 \text{ A}} = \mathbf{192\ \Omega}$$

Sometimes the power is given and the current or voltage must be determined. Through algebraic manipulations, an equation for each variable is derived as follows:

$$P = I^2R \Rightarrow I^2 = \frac{P}{R}$$

and

$$\boxed{I = \sqrt{\frac{P}{R}}} \qquad \text{(amperes)} \qquad \textbf{(4.14)}$$

$$P = \frac{V^2}{R} \Rightarrow V^2 = PR$$

and

$$\boxed{V = \sqrt{PR}} \qquad \text{(volts)} \qquad \textbf{(4.15)}$$

**EXAMPLE 4.9.** Determine the current through a 5-k$\Omega$ resistor when the power dissipated by the element is 20 mW.

***Solution:*** Equation (4.14):

$$I = \sqrt{\frac{P}{R}} = \sqrt{\frac{20 \times 10^{-3} \text{ W}}{5 \times 10^3\ \Omega}} = \sqrt{4 \times 10^{-6}} = 2 \times 10^{-3}$$

$$= \mathbf{2\ mA}$$

## 4.4 WATTMETERS

As one might expect, instruments exist that can measure the power delivered by a source and to a dissipative element. One such instrument appears in Fig. 4.14. Since power is a function of both the current and

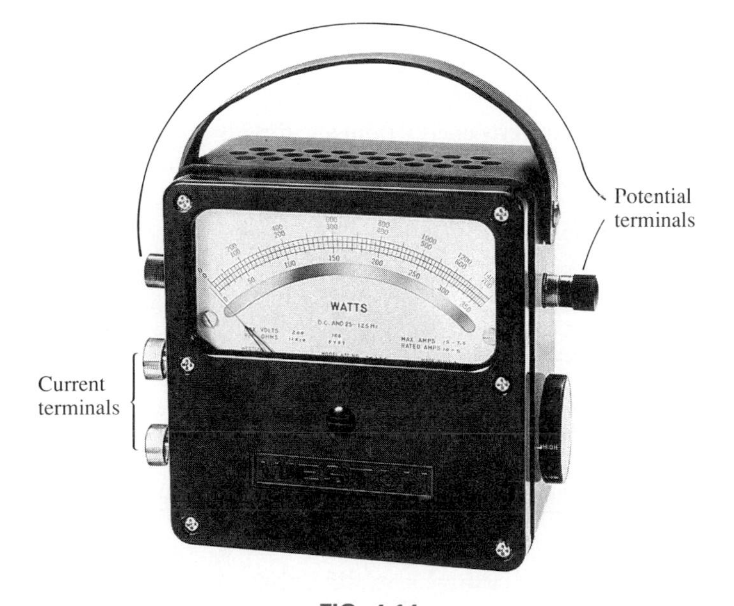


**FIG. 4.14**
*Wattmeter. (Courtesy of Electrical Instrument Service, Inc.)*

the voltage levels, four terminals must be connected as shown in Fig. 4.15 to measure the power to the resistor $R$.

If the current coils (CC) and potential coils (PC) of the wattmeter are connected as shown in Fig. 4.15, there will be an up-scale reading on

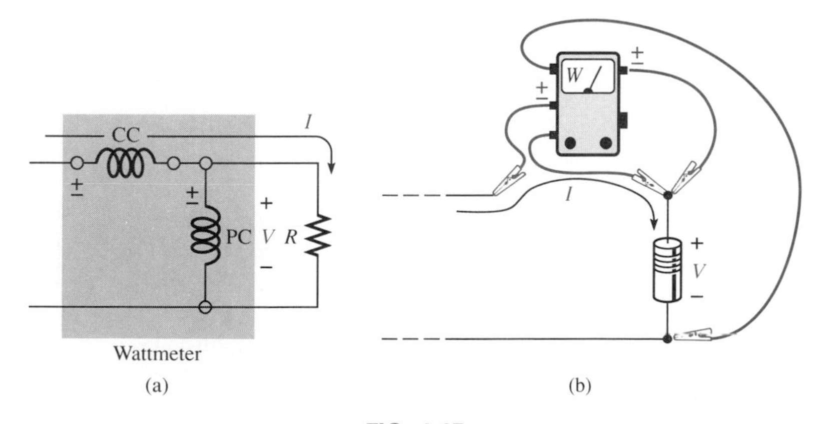


**FIG. 4.15**
*Wattmeter connections.*

the wattmeter. A reversal of either coil will result in a below-zero indication. Three voltage terminals may be available on the voltage side to permit a choice of voltage levels. On most wattmeters, the current terminals are physically larger than the voltage terminals for safety reasons and to insure a solid connection.

## 4.5 EFFICIENCY

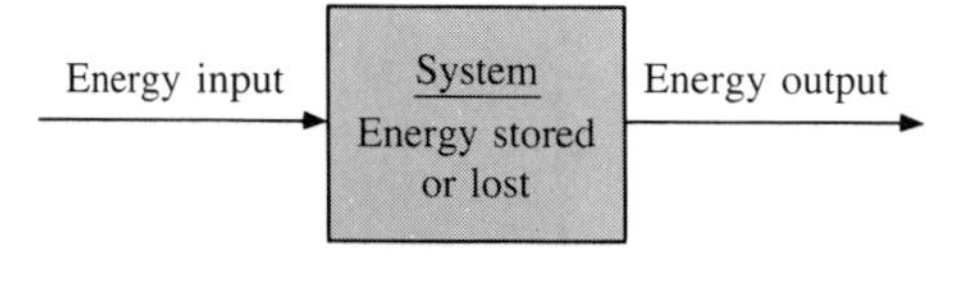


FIG. 4.16

Any system that converts energy from one form to another can be represented by the block diagram of Fig. 4.16 with an energy input and output terminal.

Conservation of energy requires that

Energy input = energy output + energy lost or stored in the system

Dividing both sides of the relationship by $t$ gives

$$\frac{W_{\text{in}}}{t} = \frac{W_{\text{out}}}{t} + \frac{W_{\text{lost or stored by the system}}}{t}$$

Since $P = W/t$, we have the following:

$$P_i = P_o + P_{\text{lost or stored}} \tag{4.16}$$

The efficiency ($\eta$) of the system is then determined by the following equation:

$$\text{Efficiency} = \frac{\text{power output}}{\text{power input}}$$

and

$$\eta = \frac{P_o}{P_i} \tag{4.17}$$

where $\eta$ (lowercase Greek letter eta) is a decimal number. Expressed as a percentage,

$$\eta = \frac{P_o}{P_i} \times 100\% \tag{4.18}$$

In terms of the input and output energy, the efficiency in percent is given by

$$\eta = \frac{W_o}{W_i} \times 100\% \tag{4.19}$$

The maximum possible efficiency is 100%, which occurs when $P_o = P_i$, or when the power lost or stored in the system is zero. Obviously, the greater the internal losses of the system in generating the necessary output power or energy, the lower the net efficiency.

---

**EXAMPLE 4.10.** A 2-hp motor operates at an efficiency of 75%. What is the power input in watts? If the input current is 9.05 A, what is the input voltage?

***Solution:***

$$\eta = \frac{P_o}{P_i} \times 100\%$$

$$0.75 = \frac{(2 \text{ hp})(746 \text{ W/hp})}{P_i}$$

and

$$P_i = \frac{1492 \text{ W}}{0.75} = \mathbf{1989.33 \text{ W}}$$

$$P = EI \quad \text{or} \quad E = \frac{P}{I} = \frac{1990 \text{ W}}{9.05 \text{ A}} = 219.82 \text{ V} \cong \mathbf{220 \text{ V}}$$

---

**EXAMPLE 4.11.** What is the output in horsepower of a motor with an efficiency of 80% and an input current of 8 A at 120 V?

***Solution:***

$$\eta = \frac{P_o}{P_i} \times 100\%$$

$$0.80 = \frac{P_o}{(120 \text{ V})(8 \text{ A})}$$

and

$$P_o = (0.80)(120 \text{ V})(8 \text{ A}) = 768 \text{ W}$$

with

$$768 \cancel{\text{W}}\left(\frac{1 \text{ hp}}{746 \cancel{\text{W}}}\right) = \mathbf{1.029 \text{ hp}}$$

---

**EXAMPLE 4.12.** What is the efficiency in percent of a system in which the input energy is 50 J and the output energy is 42.5 J?

***Solution:***

$$\eta = \frac{W_o}{W_i} \times 100\% = \frac{42.5 \text{ J}}{50 \text{ J}} \times 100\% = \mathbf{85\%}$$

---

The very basic components of a generating (voltage) system are depicted in Fig. 4.17. The source of mechanical power is a structure such

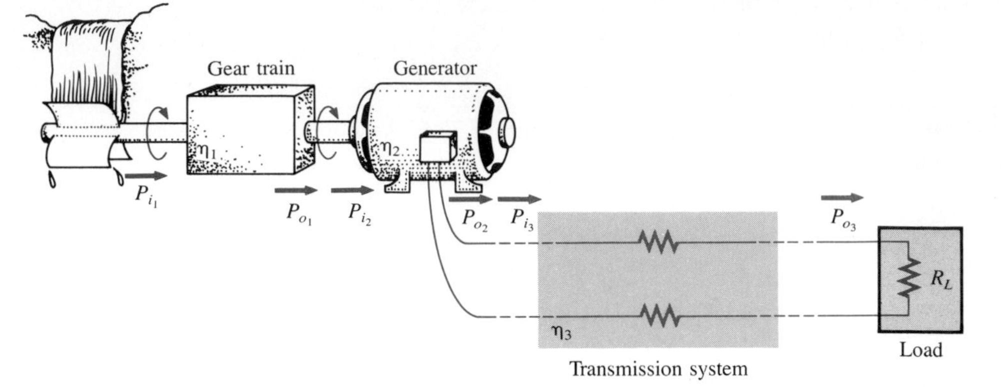


**FIG. 4.17**
*Basic components of a generating system.*

as a paddlewheel that is turned by the water rushing over the dam. The gear train will then insure that the rotating member of the generator is turning at rated speed. The output voltage must then be fed through a transmission system to the load. For each component of the system, an input and output power have been indicated. The efficiency of each system is given by

$$\eta_1 = \frac{P_{o_1}}{P_{i_1}} \qquad \eta_2 = \frac{P_{o_2}}{P_{i_2}} \qquad \eta_3 = \frac{P_{o_3}}{P_{i_3}}$$

If we form the product of these three efficiencies,

$$\eta_1 \cdot \eta_2 \cdot \eta_3 = \frac{P_{o_1}}{P_{i_1}} \cdot \frac{P_{o_2}}{P_{i_2}} \cdot \frac{P_{o_3}}{P_{i_3}}$$

and substitute the fact that $P_{i_2} = P_{o_1}$ and $P_{i_3} = P_{o_2}$, we find that the quantities indicated above will cancel, resulting in $P_{o_3}/P_{i_1}$, which is a measure of the efficiency of the entire system. In general, for the representative cascaded system of Fig. 4.18,

$$\boxed{\eta_{\text{total}} = \eta_1 \cdot \eta_2 \cdot \eta_3 \cdot \cdots \cdot \eta_n} \qquad \textbf{(4.20)}$$

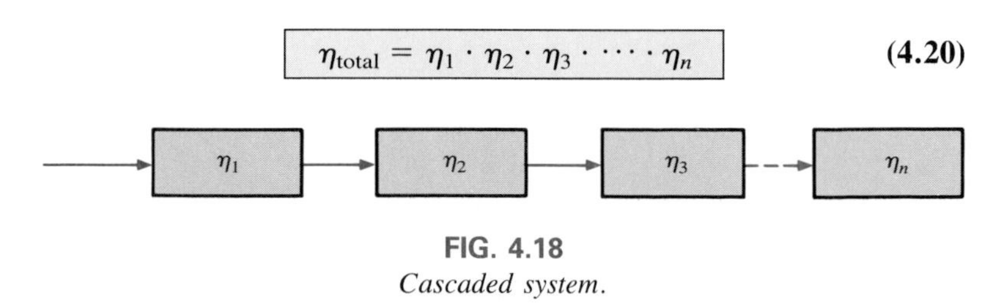


**FIG. 4.18**
*Cascaded system.*

**EXAMPLE 4.13.** Find the overall efficiency of the system of Fig. 4.17 if $\eta_1 = 90\%$, $\eta_2 = 85\%$, and $\eta_3 = 95\%$.

***Solution:***

$$\eta_T = \eta_1 \cdot \eta_2 \cdot \eta_3 = (0.90)(0.85)(0.95) = 0.727, \text{ or } \mathbf{72.7\%}$$

**EXAMPLE 4.14.** If the efficiency $\eta_1$ drops to 40%, find the new overall efficiency and compare the result with that obtained in Example 4.13.

***Solution:***

$$\eta_T = \eta_1 \cdot \eta_2 \cdot \eta_3 = (0.40)(0.85)(0.95) = 0.323, \text{ or } \mathbf{32.3\%}$$

Certainly 32.3% is noticeably less than 72.7%. The total efficiency of a cascaded system is therefore determined primarily by the lowest efficiency (weakest link) and is less than (or equal to if the remaining efficiences are 100%) the least efficient link of the system.

## 4.6 ENERGY

In order for power, which is the rate of doing work, to produce an energy conversion of any form, it must be *used over a period of time*. For example, a motor may have the horsepower to run a heavy load, but unless the motor is *used* over a period of time, there will be no energy conversion. In addition, the longer the motor is used to drive the load, the greater will be the energy expended.

The energy lost or gained by any system is therefore determined by

$$\boxed{W = Pt} \qquad \text{(wattseconds, Ws, or joules)} \qquad \mathbf{(4.21)}$$

Since power is measured in watts (or joules per second) and time in seconds, the unit of energy is the *wattsecond* or *joule* as indicated above. The wattsecond, however, is too small a quantity for most practical purposes, so the *watthour* (Wh) and *kilowatthour* (kWh) were defined, as follows:

$$\boxed{\text{Energy (Wh)} = \text{power (W)} \times \text{time (h)}} \qquad \mathbf{(4.22)}$$

$$\boxed{\text{Energy (kWh)} = \frac{\text{power (W)} \times \text{time (h)}}{1000}} \qquad \mathbf{(4.23)}$$

Note that the energy in kilowatthours is simply the energy in watthours divided by 1000.

To develop some sense for the kilowatthour energy level, consider that 1 kWh is the energy dissipated by a 100-W bulb in 10 h.

The *kilowatthour meter* is an instrument for measuring the energy supplied to the residential or commercial user of electricity. It is normally connected directly to the lines at a point just prior to entering the power distribution panel of the building. A typical set of dials is shown in Fig. 4.17 with a photograph of a kilowatthour meter. As indicated, each power of 10 below a dial is in kilowatthours. The more rapidly the aluminum disc rotates, the greater the energy demand. The dials are connected through a set of gears to the rotation of this disc.

**FIG. 4.19**
*Kilowatthour meter. (Courtesy of Westinghouse Electric Corp.)*

**EXAMPLE 4.15.** For the dial positions of Fig. 4.19, calculate the electricity bill if the previous reading was 4650 kWh and the average cost is 8¢ per kilowatthour.

***Solution:***

$$5360\text{ kWh} - 4650\text{ kWh} = 710\text{ kWh used}$$

$$710\,\cancel{\text{kWh}}\left(\frac{8¢}{\cancel{\text{kWh}}}\right) = \mathbf{\$56.80}$$

**EXAMPLE 4.16.** How much energy (in kilowatthours) is required to light a 60-W bulb continuously for 1 year (365 days)?

***Solution:***

$$W = \frac{Pt}{1000} = \frac{(60\text{ W})(24\text{ h/day})(365\text{ days})}{1000} = \frac{525{,}600\text{ Wh}}{1000} = \mathbf{525.60\text{ kWh}}$$

**EXAMPLE 4.17.** How long can a 205-W television set be on before using more than 4 kWh of energy?

***Solution:***

$$W = \frac{Pt}{1000} = \frac{(205\text{ W})(t)}{1000} \Rightarrow t\text{ (hours)} = \frac{(W)(1000)}{P} = \frac{(4\text{ kWh})(1000)}{205\text{ W}} = \mathbf{19.51\text{ h}}$$

**EXAMPLE 4.18.** What is the cost of using a 5-hp motor for 3 h if the cost is 8¢ per kilowatthour?

***Solution:***

$$W\text{ (kilowatthours)} = \frac{Pt}{1000} = \frac{(5\text{ hp} \times 746\text{ W/hp})(3\text{ h})}{1000} = 11.2\text{ kWh}$$

$$\text{Cost} = (11.2\text{ kWh})(8¢/\text{kWh}) = \mathbf{89.6¢}$$

**EXAMPLE 4.19.** What is the total cost of using the following at 8¢ per kilowatthour?

a. a 1200-W toaster for 30 min
b. six 50-W bulbs for 4 h
c. a 400-W washing machine for 45 min
d. a 4800-W electric clothes dryer for 20 min

***Solution:***

$$W = \frac{(1200\text{ W})(\frac{1}{2}\text{ h}) + (6)(50\text{ W})(4\text{ h}) + (400\text{ W})(\frac{3}{4}\text{ h}) + (4800\text{ W})(\frac{1}{3}\text{ h})}{1000}$$

$$= \frac{600\text{ Wh} + 1200\text{ Wh} + 300\text{ Wh} + 1600\text{ Wh}}{1000} = \frac{3700\text{ Wh}}{1000}$$

$$W = 3.7\text{ kWh}$$

$$\text{Cost} = (3.7\text{ kWh})(8¢/\text{kWh}) = \mathbf{29.6¢}$$

The chart in Fig. 4.20 shows the average cost per kilowatthour as compared to the kilowatthours used per customer. Note that the cost

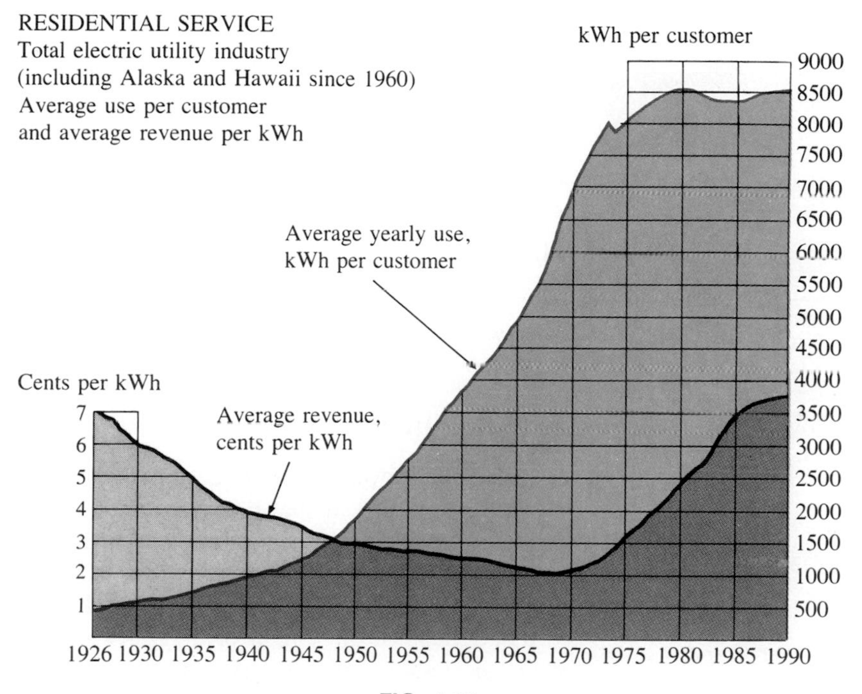


**FIG. 4.20**
(*Courtesy of Edison Electric Institute*)

today is just above the level of 1926 but that the average customer uses more than 20 times as much electrical energy in a year. Keep in mind that the chart of Fig. 4.20 is the average cost across the nation. Some states have average rates of 3¢ or 4¢ per kilowatthour, whereas others pay over 11¢ per kilowatthour.

Table 4.1 lists some common household items with their typical wattage ratings. It might prove interesting for the reader to calculate the cost of operating some of these appliances over a period of time using the preceding chart to find the cost per kilowatthour.

**TABLE 4.1**
*Typical wattage ratings of some common household appliances.*

| Appliance | Wattage Rating | Appliance | Wattage Rating |
|---|---|---|---|
| Air conditioner | 860 | Microwave oven | 800 |
| Blow dryer | 1,300 | Phonograph | 75 |
| Cassette player/ | | Projector | 1,200 |
| recorder | 5 | Radio | 70 |
| Clock | 2 | Range (self-cleaning) | 12,200 |
| Clothes dryer (electric) | 4,800 | Refrigerator (automatic | |
| Coffee maker | 900 | defrost) | 1,800 |
| Dishwasher | 1,200 | Shaver | 15 |
| Fan: | | Stereo equipment | 110 |
| Portable | 90 | Sun lamp | 280 |
| Window | 200 | Toaster | 1,200 |
| Heater | 1,322 | Trash compactor | 400 |
| Heating equipment: | | TV (color) | 150 |
| Furnace fan | 320 | Videocassette recorder | 110 |
| Oil-burner motor | 230 | Washing machine | 400 |
| Iron, dry or steam | 1,100 | Water heater | 2,500 |

Courtesy of General Electric Co.

(a)

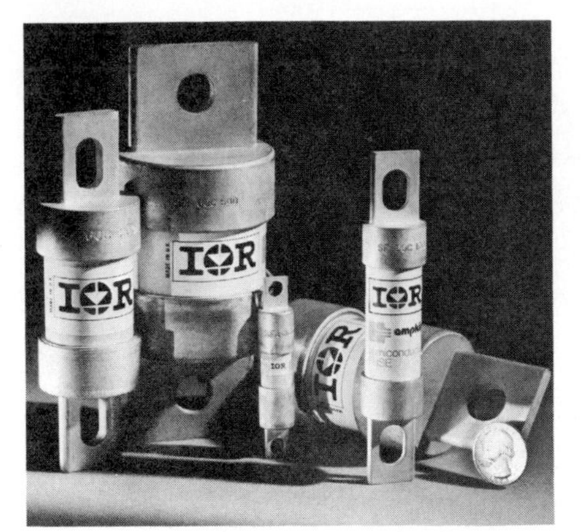

(b)

**FIG. 4.21**
*Bimetallic fuses. (Part (a) courtesy of Bussman Manufacturing Co.; part (b) courtesy of International Rectifier Corp.)*

**FIG. 4.22**
*Circuit breakers. (Courtesy of Potter and Brumfield Division, AMF, Inc.)*

## 4.7 CIRCUIT BREAKERS AND FUSES

The incoming power to any large industrial plant, heavy equipment, simple circuit in the home, or meters used in the laboratory must be limited to insure that the current through the lines is not above the rated value. Otherwise, the electric or electronic equipment may be damaged or dangerous side effects such as fire or smoke may result. To limit the current level, fuses or circuit breakers are installed where the power enters the installation, such as in the panel in the basement of most homes at the point where the outside feeder lines enter the dwelling. The fuse (depicted in Fig. 4.21) has an internal bimetallic conductor through which the current will pass; it will begin to melt if the current through the system exceeds the rated value printed on the casing. Of course, if it melts through, the current path is broken and the load in its path protected.

In homes and industrial plants built in recent years, the fuse has been replaced by circuit breakers such as those appearing in Fig. 4.22. When the current exceeds rated conditions, an electromagnet in the device will have sufficient strength to draw the connecting metallic link in the breaker out of the circuit and open the current path. When conditions have been corrected, the breaker can be reset and used again, unlike the fuse, which has to be replaced.

## 4.8 COMPUTER APPLICATION

This section will begin to reveal the differences between the use of a language and software package by examining the use of Ohm's law in its three forms. In BASIC it is possible to write a program that will ask the user which quantity he or she would like to calculate, ask for the other two known quantities, and solve for the desired unknown—in fact, the program to be described shortly will perform all the above to a high degree of accuracy and reliability. In PSPICE, however, the package is designed primarily to solve for nodal voltages, from which it can calculate current and power levels. There is no built-in facility to question the user about his or her interests or to input specific equations into the analysis routines—they are cast in stone within the package. In total, therefore, the software package is designed to perform specific tasks and provide output data in a variety of forms, whereas a language can be used to establish an interactive mode with the user and provide data in any format desired by the user.

The PSPICE analysis to follow will be limited to finding the current through a resistor using Ohm's law, whereas the BASIC program will determine *I*, *V*, or *R* depending on the input data.

### PSPICE

The earlier sections on computers were limited to the input format for dc voltage sources and resistors. We now look at the basic format of the *input file* (the name given to the lines of information fed into the computer) and how the data will appear in the *output file* (the manner in which the package will display the desired results).

The input file *must* contain three components as appearing in Fig. 4.23—the *title line, network description,* and the *.END statement*. As indicated in Fig. 4.23, the first line *must* be the title line. It is constructed of characters (letters, numbers, and so on) in any format that normally catalogues the file for future reference and reveals the analysis to be performed. The next necessary component of the input file is the network description. Elements can be entered in any order as long as the nodes are properly defined. It is always advisable to label all the nodes

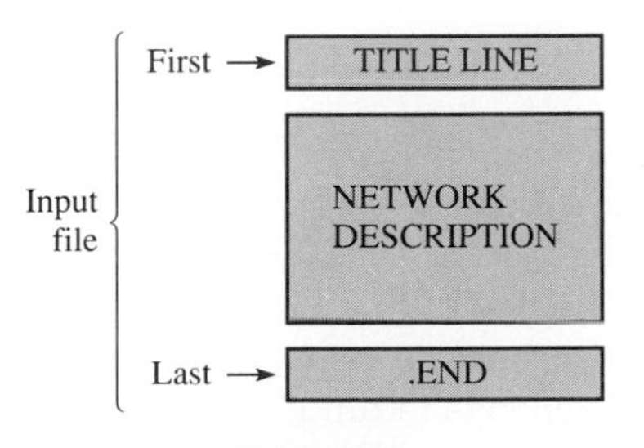


**FIG. 4.23**
*PSPICE input file.*

of the network before beginning the input process. It is vitally important, however, to remember that

***in every network one node must be designated as the reference node and assigned the number 0. It is usually the node connected to ground potential (zero volts) with all other nodal voltages taken with reference to this defined reference level.***

For example, the voltage output by the computer for node 1 is the voltage from node 1 to ground. A voltage V(8) is the voltage from node 8 to node 0 (or ground potential). Unless specified, PSPICE will automatically generate the nodal voltages as defined above.

The last line of any input file must be the .END command, to let the software package know that the network is fully described and all the information the package requires has been provided. *Don't forget* the period before the END statement. Its absence will invalidate the entry.

Let us now look at the analysis of the simple network of Fig. 4.24 with one voltage source and a resistor. There are two nodes (or connection points) between the two elements. As required, one is specified as the reference level 0 (in this case, the ground or zero-volt level) and the other, node 1. Note the relative simplicity of the input file, with a title line, two element entries, and an .END statement.

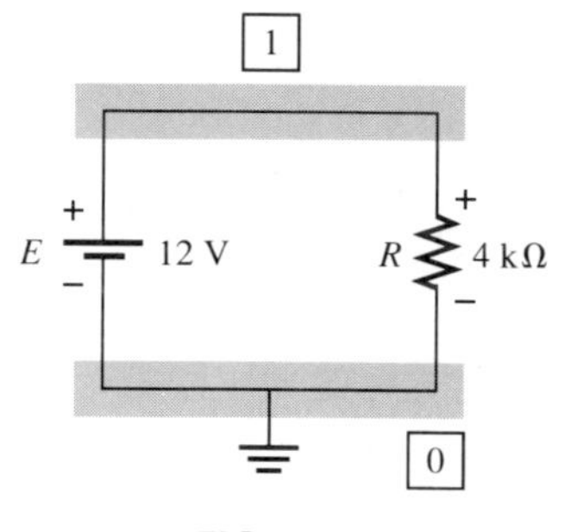

FIG. 4.24

```
Chapter 4-Ohm's Law
VE 1 0 12V
R 1 0 4K
.END
```

Once the input file is entered, PSPICE can be run, and the output file of Fig. 4.25 will result. Note how the title line and network description have been placed within the framework of a heading controlled by PSPICE. The title CIRCUIT DESCRIPTION simply heads the network description to follow and the SMALL-SIGNAL BIAS SOLUTION simply reveals that a solution is to follow. The heading SMALL SIGNAL will have more meaning when you begin your electronics courses, but for now simply relate the phrase BIAS SOLUTION to mean dc solution.

```
************************ Evaluation PSpice (January 1989) ******* 20:41:04 *******

 Chapter 4 - Ohm's Law

 ****     CIRCUIT DESCRIPTION

*****************************************************************************


VE 1 0 12V
R  1 0 4K
.END
```

```
************************ Evaluation PSpice (January 1989) ******* 20:41:04 *******


 Chapter 4 - Ohm's Law

 ****     SMALL SIGNAL BIAS SOLUTION       TEMPERATURE =   27.000 DEG C

*****************************************************************************


 NODE    VOLTAGE      NODE    VOLTAGE      NODE    VOLTAGE      NODE    VOLTAGE

(    1)   12.0000



    VOLTAGE SOURCE CURRENTS
    NAME            CURRENT

    VE            -3.000E-03

    TOTAL POWER DISSIPATION   3.60E-02  WATTS


            JOB CONCLUDED

            TOTAL JOB TIME            5.33
```

**FIG. 4.25**

In the solution, note that there is only one nodal voltage provided, since there is only one in the original network description. The current provided is the voltage source current, but since the current through the battery of Fig. 4.24 is the same as that through the resistor, we have the

desired current through the resistor. For all elements of a network, PSPICE defines the direction of the current as from the + node to the − node. Since the current through the source is opposite to this defined direction, a minus sign appears in the printout. Using powers of ten, the resulting current is 3 mA. In this network, since the power supplied by the battery is equal to the power absorbed by the resistor, the power level listed can be applied to either element. Its magnitude is 36 mW. The final line of the output file provides the time in seconds required to perform the necessary analysis.

A savings in paper and space can be accomplished by adding a control line .OPTIONS NOPAGE to the input file, as shown in Fig. 4.26. Note that the second heading in the output file was reduced to simply one line without a repeat of the title line PSPICE specifics.

```
************************ Evaluation PSpice (January 1989) ******* 15:23:07 *******

Chapter 4 - Ohm's Law

****     CIRCUIT DESCRIPTION

****************************************************************************

VE 1 0 12V
R  1 0 4K
.OPTIONS NOPAGE
.END

****     SMALL SIGNAL BIAS SOLUTION        TEMPERATURE =   27.000 DEG C

 NODE   VOLTAGE     NODE   VOLTAGE     NODE   VOLTAGE     NODE   VOLTAGE

(    1)   12.0000

    VOLTAGE SOURCE CURRENTS
    NAME          CURRENT

    VE          -3.000E-03

    TOTAL POWER DISSIPATION   3.60E-02  WATTS

          JOB CONCLUDED

          TOTAL JOB TIME           6.15
```

FIG. 4.26

## BASIC

The first complete program in BASIC is provided in Fig. 4.27. Everything you see is either entered into the computer or printed out by the computer except the brackets at the left of the line numbers with the printed explanation of the function performed by those sections. First note that the commands, statements, and so on, found on each line use

```
 10 REM ***** PROGRAM 4-1 *****
 20 REM ********************************************
 30 REM Program demonstrates selecting various forms
 40 REM of equations
 50 REM ********************************************
 60 REM
```

Equation Selection

```
100 PRINT:PRINT "Select which form of Ohm's law equation "
110 PRINT "you wish to use."
120 PRINT
130 PRINT TAB(10);"(1)  V=I*R"
140 PRINT TAB(10);"(2)  I=V/R"
150 PRINT TAB(10);"(3)  R=V/I"
160 PRINT TAB(20);
170 INPUT "choice=";C
180 IF C<1 OR C>3 THEN GOTO 100
190 ON C GOSUB 400,600,800
200 PRINT:PRINT
```

Continue?

```
210 INPUT "More (YES or NO)";A$
220 IF A$="YES" THEN 100
230 PRINT "Have a good day"
240 END
```

$V = IR$

```
400 REM Accept input of I,R and output V
410 PRINT:PRINT "Enter the following data:"
420 INPUT "I=";I
430 INPUT "R=";R
440 V=I*R
450 PRINT "Voltage is ";V;"volts"
460 RETURN
```

$I = \frac{V}{R}$

```
600 REM Accept input of V,R and output I
610 PRINT "Enter the following data:"
620 INPUT "V=";V
630 INPUT "R=";R
640 I=V/R
650 PRINT "Current is ";I;"amperes"
660 RETURN
```

$R = \frac{V}{I}$

```
800 REM Accept input of V,I and output R
810 PRINT "Enter the following data:"
820 INPUT "V=";V
830 INPUT "I=";I
840 R=V/I
850 PRINT "Resistance is ";R;"ohms"
860 RETURN

READY

RUN


Select which form of Ohm's law equation
you wish to use.

          (1)  V=I*R
          (2)  I=V/R
          (3)  R=V/I
                    choice=? 2
```

```
Enter the following data:
V=? 12

R=? 4E3

Current is  3E-03 amperes


More (YES or NO)? YES


Select which form of Ohm's law equation
you wish to use.

          (1) V=I*R
          (2) I=V/R
          (3) R=V/I
                    choice=? 1


Enter the following data:
I=? 2E-3

R=? 5.6E3

Voltage is  11.2 volts


More (YES or NO)? YES


Select which form of Ohm's law equation
you wish to use.

          (1) V=I*R
          (2) I=V/R
          (3) R=V/I
                    choice=? 3

Enter the following data:
V=? 48

I=? 0.025

Resistance is  1920 ohms


More (YES or NO)? NO

Have a good day


READY
```

**FIG. 4.27**
*Program 4.1.*

English words and phrases to indicate the operation to be performed. The REM statement (from the word *REMark*) simply indicates that a descriptive statement is being made about the program. The PRINT statement tells the system to print the characters between the quotes on the screen. The INPUT commands request the value of the variable appearing on the same line. The equations on lines 440, 640, and 840 carry out the operation to be performed in each module defined by the brackets at the left of the program. Lines 400 through 460 will request $I$ and $R$ and calculate the voltage. Line 450 will print the solution, while line 460 will ''return'' the program to line 210 to determine if a second

calculation is to be performed. The module from line 600 to line 660 will calculate the current *I*, and the module from line 800 to line 860 will determine the resistance from the input voltage and current. Lines 100 through 200 permit a selection of the form of Ohm's law to be applied. Three runs of the program are provided in the figure to reveal the format of the request for data and the output response.

If you start with line (location) 10 and perform the operations indicated for each succeeding line, you will find that the program proceeds in much the same manner as if done by longhand. For instance, line 170 asks for your choice of which form of Ohm's law you would like to apply from lines 130 through 160. If the choice is C=2 (I=V/R), line 190 will define line 600 as the next line to be performed. At 600 a sequence of lines follows that calculates I and prints out the result. The RETURN statement brings us back to line 200 for a two-line space (PRINT:PRINT), followed by the question of whether we want to continue. If so, we return to line 100 for a second choice of equations.

This longer section on computers begins to clarify the difference between the use of a language and a software package. Using a language requires that the entire program be written line by line, although a program does provide options on the type of analysis that can be performed and an interactive mode between user and machine. Using software provides the benefits of the many hours of writing and testing performed by a team of experienced programmers but limits the type of analysis we can perform and specifies the form of the input and output information.

## PROBLEMS

### SECTION 4.1 Ohm's Law

1. What is the potential drop across a 6-Ω resistor if the current through it is 2.5 A?
2. What is the current through a 72-Ω resistor if the voltage drop across it is 12 V?
3. How much resistance is required to limit the current to 1.5 mA if the potential drop across the resistor is 6 V?
4. At starting, what is the current drain on a 12-V car battery if the resistance of the starting motor is 0.056 Ω?
5. If the current through a 0.02-MΩ resistor is 3.6 μA, what is the voltage drop across the resistor?
6. If a voltmeter has an internal resistance of 15 kΩ, find the current through the meter when it reads 62 V.
7. If a refrigerator draws 2.2 A at 120 V, what is its resistance?
8. If a clock has an internal resistance of 7.5 kΩ, find the current through the clock if it is plugged into a 120-V outlet.
9. A washing machine is rated at 4.2 A at 120 V. What is its internal resistance?

10. If a soldering iron draws 0.76 A at 120 V, what is its resistance?

11. The input current to a transistor is 20 $\mu$A. If the applied (input) voltage is 24 mV, determine the input resistance of the transistor.

12. The internal resistance of a dc generator is 0.5 $\Omega$. Determine the loss in terminal voltage across this internal resistance if the current is 15 A.

*13. **a.** If an electric heater draws 9.5 A when connected to a 120-V supply, what is the internal resistance of the heater?
**b.** Using the basic relationships of Chapter 2, how much energy is converted in 1 h?

## SECTION 4.2 Plotting Ohm's Law

14. Plot the linear curves of a 100-$\Omega$ and 0.5-$\Omega$ resistor on the graph of Fig. 4.5.

15. Sketch the characteristics of a device that has an internal resistance of 20 $\Omega$ up to 10 V and an internal resistance of 2 $\Omega$ for higher voltages. Use the axes of Fig. 4.5.

16. Plot the linear curves of a 2-k$\Omega$ and 50-k$\Omega$ resistor on the graph of Fig. 4.5. Use a horizontal scale that extends from 0 V to 20 V and a vertical axis scaled off in milliamperes.

17. What is the change in voltage across a 2-k$\Omega$ resistor established by a change in current of 400 mA through the resistor?

*18. **a.** Using the axis of Fig. 4.9, sketch the characteristics of a device that has an internal resistance of 500 $\Omega$ up to 1 V and 50 $\Omega$ between 1 and 2 V. Its resistance then changes to $-20$ $\Omega$ for higher voltages. The result is a set of characteristics very similar to an electronic device called a tunnel diode.
**b.** Using the above characteristics, determine the resulting current at voltages of 0.7 V, 1.5 V and 2.5 V.

## SECTION 4.3 Power

19. If 420 J of energy are absorbed by a resistor in 7 min, what is the power to the resistor?

20. The power to a device is 40 joules per second (J/s). How long will it take to deliver 640 J?

21. **a.** How many joules of energy does a 2-W nightlight dissipate in 8 h?
**b.** How many kilowatthours does it dissipate?

22. A resistor of 10 $\Omega$ has charge flowing through it at the rate of 300 coulombs per minute (C/min). How much power is dissipated?

23. How long will a steady current of 2 A have to exist in a resistor that has 3 V across it to dissipate 12 J of energy?

24. What is the power delivered by a 6-V battery if the charge flows at the rate of 48 C/min?

25. The current through a 4-Ω resistor is 7 mA. What is the power delivered to the resistor?

26. The voltage drop across a 3-Ω resistor is 9 mV. What is the power input to the resistor?

27. If the power input to a 4-Ω resistor is 64 W, what is the current through the resistor?

28. A 1/2-W resistor has a resistance of 1000 Ω. What is the maximum current that it can safely handle?

29. A 2.2-kΩ resistor in a stereo system dissipates 42 mW of power. What is the voltage across the resistor?

30. A dc battery can deliver 45 mA at 9 V. What is the power rating?

31. What are the ''hot'' resistance level and current rating of a 120-V, 100-W bulb?

32. What are the internal resistance and voltage rating of a 450-W automatic washer that draws 3.75 A?

33. A 20-kΩ resistor has a rating of 100 W. What are the maximum current and the maximum voltage that can be applied to the resistor?

*34. **a.** Plot power versus current for a 100-Ω resistor. Use a power scale from 0 to 1 W and a current scale from 0 to 100 mA with divisions of 0.1 W and 10 mA, respectively.
**b.** Is the curve linear or nonlinear?
**c.** Using the resulting plot, determine the current at a power level of 500 mW.

*35. A small portable black-and-white TV draws 0.455 A at 9 V.
**a.** What is the power rating of the TV?
**b.** What is the internal resistance of the TV?
**c.** What is the energy converted in 6 h of typical battery life?

*36. **a.** If a home is supplied with a 120-V, 100-A service, find the maximum power capability.
**b.** Can the homeowner safely operate the following loads at the same time?
1. a 5-hp motor
2. a 3000-W clothes dryer
3. a 2400-W electric range
4. a 1000-W steam iron

## SECTION 4.5 Efficiency

37. What is the efficiency of a motor that has an output of 0.5 hp with an input of 450 W?

38. The motor of a power saw is rated 68.5% efficient. If 1.8 hp are required to cut a particular piece of lumber, what is the current drawn from a 120-V supply?

39. What is the efficiency of a dryer motor that delivers 1 hp when the input current and voltage are 4 A and 220 V, respectively?

40. If an electric motor having an efficiency of 87% and operating off a 220-V line delivers 3.6 hp, what input current does the motor draw?

41. A motor is rated to deliver 2 hp.
   a. If it runs on 110 V and is 90% efficient, how many watts does it draw from the power line?
   b. What is the input current?
   c. What is the input current if the motor is only 70% efficient?

42. An electric motor used in an elevator system has an efficiency of 90%. If the input voltage is 220 V, what is the input current when the motor is delivering 15 hp?

43. A 2-hp motor drives a sanding belt. If the efficiency of the motor is 87% and that of the sanding belt 75% due to slippage, what is the overall efficiency of the system?

44. If two systems in cascade each have an efficiency of 80% and the input energy is 60 J, what is the output energy?

45. The overall efficiency of two systems in cascade is 72%. If the efficiency of one is 0.9, what is the efficiency in percent of the other?

*46. If the total input and output power of two systems in cascade are 400 W and 128 W, respectively, what is the efficiency of each system if one has twice the efficiency of the other?

47. a. What is the total efficiency of three systems in cascade with efficiencies of 98%, 87%, and 21%?
   b. If the system with the least efficiency (21%) were removed and replaced by one with an efficiency of 90%, what would be the percent increase in total efficiency?

48. a. Perform the following conversions:
   1 watthour to joules
   1 kilowatthour to joules
   b. Based on the results of part a, discuss when it is more appropriate to use one unit versus the other.

## SECTION 4.6 Energy

49. A 10-Ω resistor is connected across a 15-V battery.
   a. How many joules of energy will it dissipate in 1 min?
   b. If the resistor is left connected for 2 min instead of 1 min, will the energy used increase? Will the power dissipation level increase?

50. How much energy in kilowatthours is required to keep a 230-W oil-burner motor running 12 h a week for 5 months?

51. How long can a 1500-W heater be on before using more than 10 kWh of energy?

52. How much does it cost to use a 30-W radio for 3 h at 8¢ per kilowatthour?

**53. a.** In 10 h an electrical system converts 500 kWh of electrical energy into heat. What is the power level of the system?

**b.** If the applied voltage is 208 V what is the current drawn from the supply?

**c.** If the efficiency of the system is 82%, how much energy is lost or stored in 10 h?

**54. a.** At 8¢ per kilowatthour, how long can one play a 250-W color TV for $1?

**b.** For $1, how long can one use a 4.8-kW dryer?

**c.** Compare the results of parts a and b and comment on the effect of the wattage level on the relative cost of using an appliance.

**55.** What is the total cost of using the following at 8¢ per kilowatthour?

**a.** an 860-W air conditioner for 24 h
**b.** a 4800-W clothes dryer for 30 min
**c.** a 400-W washing machine for 1 h
**d.** a 1200-W dishwasher for 45 min

***56.** What is the total cost of using the following at 8¢ per kilowatthour?

**a.** a 110-W stereo set for 4 h
**b.** a 1200-W projector for 20 min
**c.** a 60-W tape recorder for 1.5 h
**d.** a 150-W color television set for 3 h 45 min

## SECTION 4.8 Computer Application

**57.** Write an input file for a network like the one shown in Fig. 4.24 with $E = 400$ mV and $R = 0.04$ MΩ. Use the appropriate control line to limit the amount of paper required for the output file.

**58.** Write an input file for a network like the one shown in Fig. 4.24, except reverse the polarity of the battery. Use $E = 0.02$ V and $R = 240$ Ω. Keep the paper used for the output file to a minimum.

**59.** Write a program to calculate the cost of using five different appliances for varying lengths of time if the cost is 7¢ per kilowatthour.

**60.** Request $I$, $R$, and $t$ and determine $V$, $P$, and $W$. Print out the results with the proper units.

**61.** Given a resistance in kilohms, tabulate the voltage and power to the resistor for a range of current extending from 1 mA to 10 mA in increments of 1 mA.

**62.** Tabulate the time (in hours) and the cost (at 8¢ per kilowatthour) for the use of a particular system for $T = 1$ to $N$ hours in increments of 1 h. $N$ is an input quantity in the range 2–10.

## GLOSSARY

**Circuit breaker** A two-terminal device designed to insure that current levels do not exceed safe levels. If "tripped," it can be reset with a switch or a reset button.

**Diode** A semiconductor device whose behavior is much like that of a simple switch; that is, it will pass current ideally in only one direction when operating within specified limits.

**Efficiency** ($\eta$) A ratio of output to input power that provides immediate information about the energy-converting characteristics of a system.

**Energy** ($W$) A quantity whose change in state is determined by the product of the rate of conversion ($P$) and the period involved ($t$). It is measured in joules (J) or wattseconds (Ws).

**Fuse** A two-terminal device whose sole purpose is to insure that current levels in a circuit do not exceed safe levels.

**Horsepower (hp)** Equivalent to 746 watts in the electrical system.

**Input file** The lines of information fed into the computer to define the system to be analyzed and the operations to be performed.

**Kilowatthour meter** An instrument for measuring kilowatthours of energy supplied to a residential or commercial user of electricity.

**Ohm's law** An equation that establishes a relationship among the current, voltage, and resistance of an electrical system.

**Output file** The manner in which the results of a computer run are displayed.

**Power** An indication of how much work can be done in a specified amount of time; a *rate* of doing work. It is measured in joules/second (J/s) or watts (W).

**Wattmeter** An instrument capable of measuring the power delivered to an element by sensing both the voltage across the element and the current through the element.

# 5
# Series Circuits

## 5.1 INTRODUCTION

Two types of current are readily available to the consumer today. One is *direct current* (dc), in which ideally the flow of charge (current) does not change in magnitude or direction. The other is *sinusoidal alternating current* (ac), in which the flow of charge is continually changing in magnitude and direction. The next few chapters are an introduction to circuit analysis purely from a dc approach. The methods and concepts will be discussed in detail for direct current; and thus, when possible, a short discussion will suffice to cover any variations we might encounter when we consider ac in the later chapters.

The battery of Fig. 5.1, by virtue of the potential difference between its terminals, has the ability to cause (or "pressure") charge to flow

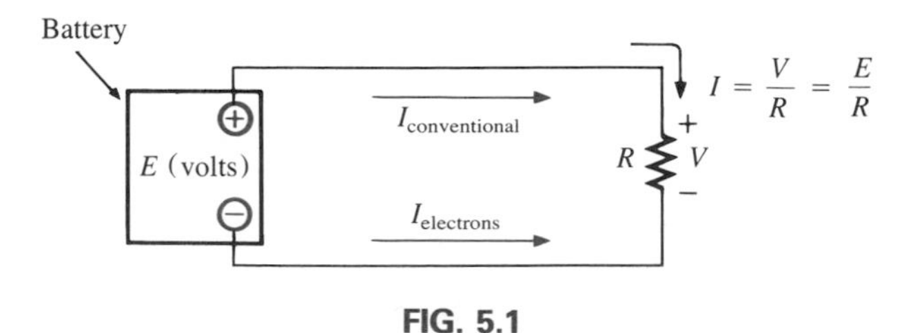


FIG. 5.1

through the simple circuit. The positive terminal attracts the electrons through the wire at the same rate at which electrons are supplied by the negative terminal. As long as the battery is connected in the circuit and maintains its terminal characteristics, the current (dc) through the circuit will not change in magnitude or direction.

If we consider the wire to be an ideal conductor (that is, having no opposition to flow), the potential difference $V$ across the resistor will equal the applied voltage of the battery: $V$ (volts) $= E$ (volts).

The current is limited only by the resistor $R$. The higher the resistance, the less the current, and conversely, as determined by Ohm's law.

By convention as discussed in Chapter 2, the direction of $I$ as shown in Fig. 5.1 is opposite to that of electron flow. Also, the uniform flow of charge dictates that the direct current $I$ be the same everywhere in the circuit. By following the direction of conventional flow, we notice that there is a rise in potential across the battery ($-$ to $+$), and a drop in potential across the resistor ($+$ to $-$). For single-voltage-source dc circuits, conventional flow always passes from a low potential to a high potential when passing through a voltage source, as shown in Fig. 5.2. However, conventional flow always passes from a high to a low potential when passing through a resistor for any number of voltage sources in the same circuit, as shown in Fig. 5.3.

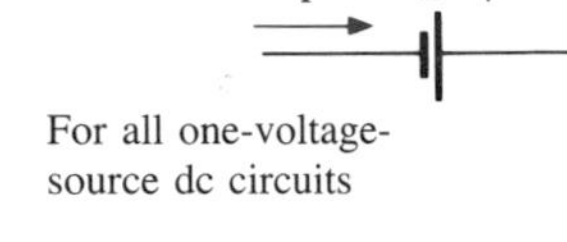


For all one-voltage-source dc circuits

FIG. 5.2

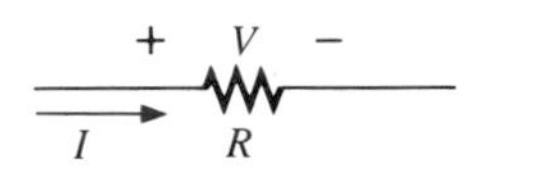


For any combination of voltage sources in the same dc circuit

FIG. 5.3

The circuit of Fig. 5.1 is the simplest possible configuration. This chapter and the chapters to follow will add elements to the system in a very specific manner to introduce a range of concepts that will form a major part of the foundation required to analyze the most complex system. Be aware that the laws, rules, and so on, introduced in Chapters 5 and 6 will be used throughout your studies of electrical, electronic, or computer systems. They will not be dropped for a more advanced set as you progress to more sophisticated material. It is therefore critical that the concepts be understood thoroughly and that the various procedures and methods be applied with confidence.

## 5.2 SERIES CIRCUITS

A *circuit* consists of any number of elements joined at terminal points, providing at least one closed path through which charge can flow. The circuit of Fig. 5.4(a) has three elements joined at three terminal points ($a$, $b$, and $c$) to provide a closed path for the current $I$.

***Two elements are in series if they have only one point in common that is not connected to other current-carrying elements of the network.***

In Fig. 5.4(a), the resistors $R_1$ and $R_2$ are in series because they have *only* point $b$ in common. The other ends of the resistors are connected elsewhere in the circuit. For the same reason, the battery $E$ and resistor $R_1$ are in series (terminal $a$ in common) and the resistor $R_2$ and the battery $E$ are in series (terminal $c$ in common). Since all the elements are in series, the network is called a *series circuit*. Some common examples of series connections include the tying of small pieces of rope together to form a longer rope, the connecting of pipes to get water from one point to another, and the joining of hands of a group in a circle.

***In a series circuit, the current is the same through each series element.***

For the circuit of Fig. 5.4(a), therefore, the current $I$ through each resistor is the same as that through the battery.

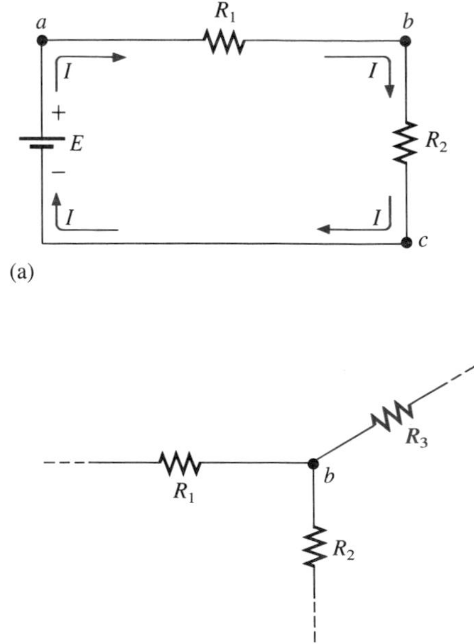


FIG. 5.4

If a third element, $R_3$, is added at terminal $b$ as shown in Fig. 5.4(b), the resistors $R_1$ and $R_2$ are no longer in series because an additional element has been connected between the common terminal and some other point in the network.

A *branch* of a circuit is any portion of the circuit that has one or more elements in series. In Fig. 5.4(a), the resistor $R_1$ forms one branch of the circuit, the resistor $R_2$ another, and the battery $E$ a third.

The total resistance of the circuit is determined by simply adding the values of the various resistors. In Fig. 5.4(a), for example, the total resistance ($R_T$) is equal to $R_1 + R_2$. Note that the total resistance is actually the resistance "seen" by the battery as it "looks" into the series combination of elements as shown in Fig. 5.5.

In general, to find the total resistance of $N$ resistors in series, the following equation is applied:

$$R_T = R_1 + R_2 + R_3 + \cdots + R_N \quad (\Omega) \qquad \textbf{(5.1)}$$

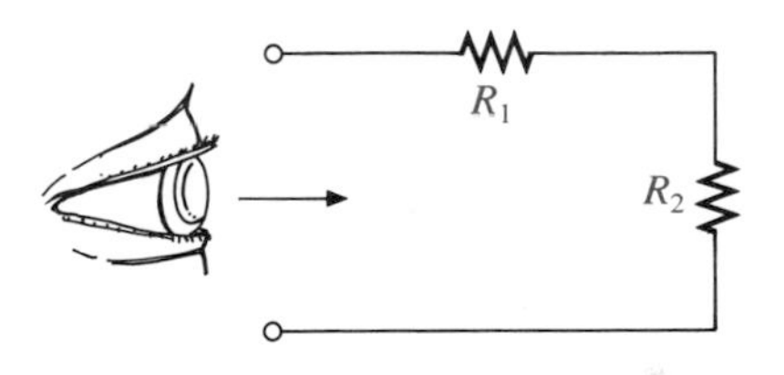


**FIG. 5.5**
*Resistance "seen" by source.*

Once the total resistance is known the circuit of Fig. 5.4(a) can be redrawn as shown in Fig. 5.6 and the current can be determined from Ohm's law:

$$I = \frac{E}{R_T} \quad (\text{A}) \qquad \textbf{(5.2)}$$

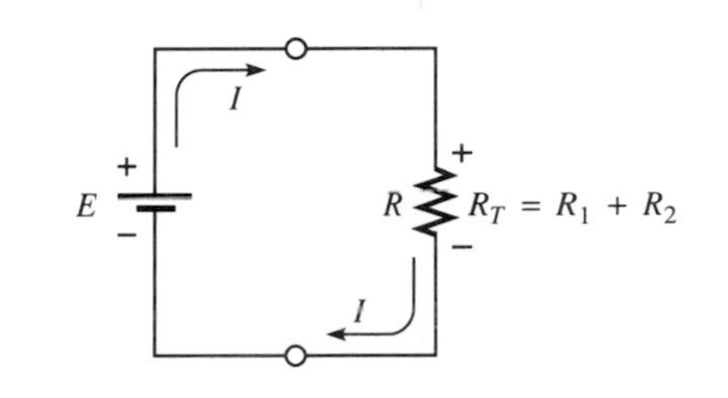


**FIG. 5.6**

The voltage across each element can also be determined using Ohm's law:

$$V_1 = IR_1,\ V_2 = IR_2,\ V_3 = IR_3,\ \cdots,\ V_N = IR_N \quad (\text{V}) \qquad \textbf{(5.3)}$$

The power delivered to each resistor can then be determined using any one of three equations as listed below for $R_1$.

$$P_1 = V_1 I_1 = I_1^2 R_1 = \frac{V_1^2}{R_1} \quad (\text{W}) \qquad \textbf{(5.4)}$$

The power delivered by the source is

$$P_{\text{del}} = EI \quad (\text{W}) \qquad \textbf{(5.5)}$$

For any combination of series resistors,

$$P_{\text{del}} = P_1 + P_2 + P_3 + \cdots + P_N \qquad \textbf{(5.6)}$$

revealing that the power delivered by the source is equal to the power dissipated by the resistors.

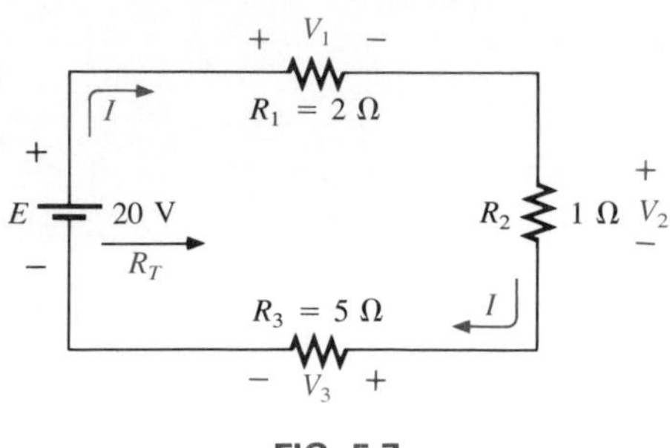


FIG. 5.7

**EXAMPLE 5.1.**

a. Find the total resistance for the series circuit of Fig. 5.7.
b. Calculate the current $I$.
c. Determine the voltages $V_1$, $V_2$, and $V_3$.
d. Calculate the power dissipated by $R_1$, $R_2$, and $R_3$.
e. Determine the power delivered by the source and compare to the sum of the power levels of part d.

***Solutions:***

a. $R_T = R_1 + R_2 + R_3 = 2\ \Omega + 1\ \Omega + 5\ \Omega = \mathbf{8\ \Omega}$

b. $I = \dfrac{E}{R_T} = \dfrac{20\text{ V}}{8\ \Omega} = \mathbf{2.5\ A}$

c. $V_1 = IR_1 = (2.5\text{ A})(2\ \Omega) = \mathbf{5\ V}$
$V_2 = IR_2 = (2.5\text{ A})(1\ \Omega) = \mathbf{2.5\ V}$
$V_3 = IR_3 = (2.5\text{ A})(5\ \Omega) = \mathbf{12.5\ V}$

d. $P_1 = V_1I_1 = (5\text{ V})(2.5\text{ A}) = \mathbf{12.5\ W}$
$P_2 = I_2^2R_2 = (2.5\text{ A})^2(1\ \Omega) = \mathbf{6.25\ W}$
$P_3 = V_3^2/R_3 = (12.5\text{ V})^2/5\ \Omega = \mathbf{31.25\ W}$

e. $P_{del} = EI = (20\text{ V})(2.5\text{ A}) = \mathbf{50\ W}$
$P_{del} = P_1 + P_2 + P_3$
$50\text{ W} = 12.5\text{ W} + 6.25\text{ W} + 31.25\text{ W}$
$50\text{ W} = 50\text{ W}$ (checks)

To find the total resistance of $N$ resistors of the same value in series, simply multiply the value of *one* of the resistors by the number in series. That is,

$$R_T = NR \tag{5.7}$$

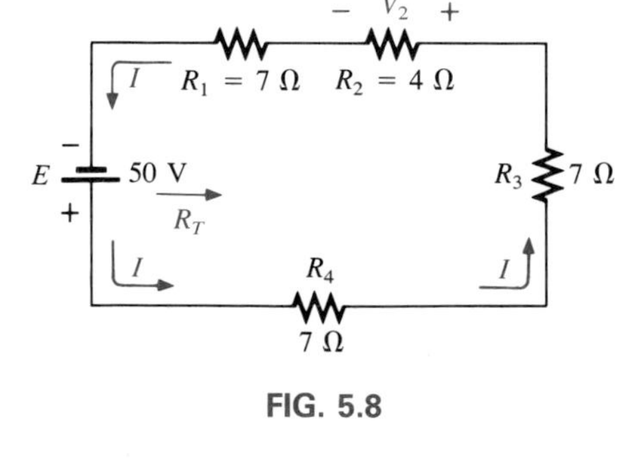


FIG. 5.8

**EXAMPLE 5.2.** Determine $R_T$, $I$, and $V_2$ for the circuit of Fig. 5.8.

***Solution:*** Note the current direction as established by the battery and the polarity of the voltage drops across $R_2$ as determined by the current direction.

Since $R_1 = R_3 = R_4$,

$$R_T = NR_1 + R_2 = (3)(7\ \Omega) + 4\ \Omega = 21\ \Omega + 4\ \Omega = \mathbf{25\ \Omega}$$

$$I = \frac{E}{R_T} = \frac{50\text{ V}}{25\ \Omega} = \mathbf{2\ A}$$

$$V_2 = IR_2 = (2\text{ A})(4\ \Omega) = \mathbf{8\ V}$$

Examples 5.1 and 5.2 are straightforward substitution-type problems that are relatively easy to solve with some practice. Example 5.3, however, is evidence of another type of problem which requires a firm grasp of the fundamental equations and an ability to identify which equation to use first. The best preparation for this type of exercise is simply to work through as many problems of this kind as possible.

---

**EXAMPLE 5.3.** Given $R_T$ and $I$, calculate $R_1$ and $E$ for the circuit of Fig. 5.9.

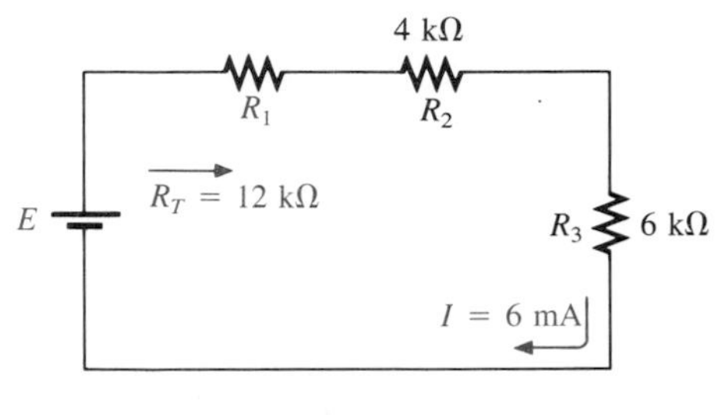

FIG. 5.9

***Solution:***

$$R_T = R_1 + R_2 + R_3$$
$$12\text{ k}\Omega = R_1 + 4\text{ k}\Omega + 6\text{ k}\Omega$$
$$R_1 = 12\text{ k}\Omega - 10\text{ k}\Omega = \mathbf{2\text{ k}\Omega}$$
$$E = IR_T = (6 \times 10^{-3}\text{ A})(12 \times 10^3\ \Omega) = \mathbf{72\text{ V}}$$

---

## 5.3 VOLTAGE SOURCES IN SERIES

Voltage sources can be connected in series as shown in Fig. 5.10 to increase or decrease the total voltage applied to a system. The net voltage is determined simply by summing the sources with the same polarity and subtracting the total of the sources with the opposite "pressure." The net polarity is the polarity of the larger sum.

In Fig. 5.10(a), for example, the sources are all "pressuring" current to the right, so the net voltage is

$$E_T = E_1 + E_2 + E_3 = 10\text{ V} + 6\text{ V} + 2\text{ V} = 18\text{ V}$$

as shown in the figure. In Fig. 5.8(b), however, the greater "pressure" is to the left, with a net voltage of

$$E_T = E_2 + E_3 - E_1 = 9\text{ V} + 3\text{ V} - 4\text{ V} = 8\text{ V}$$

and the polarity shown in the figure.

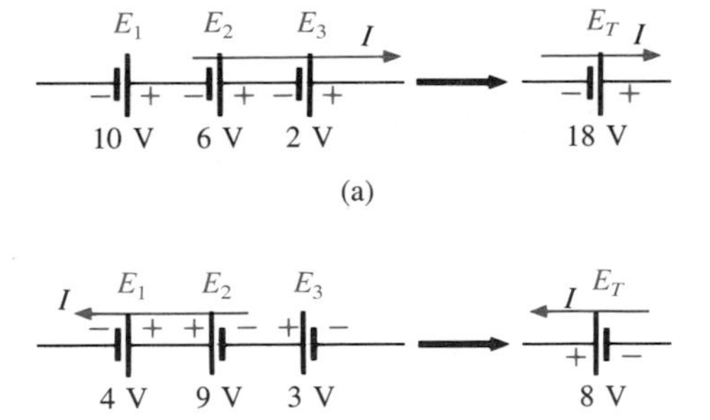


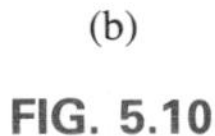
FIG. 5.10

## 5.4 KIRCHHOFF'S VOLTAGE LAW

***Kirchhoff's voltage law (KVL) states that the algebraic sum of the potential rises and drops around a closed loop (or path) is zero.***

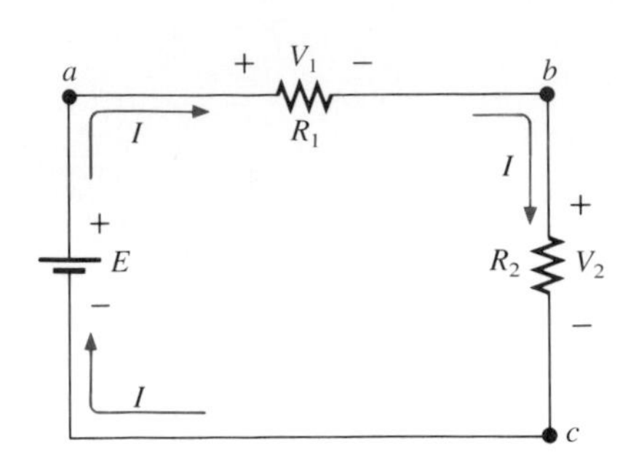


FIG. 5.11

A *closed loop* is any continuous connection of branches that allows us to trace a path that leaves a point in one direction and returns to that same point from another direction without leaving the circuit. In Fig. 5.11, by following the current, we can trace a continuous path that leaves point $a$ through $R_1$ and returns through $E$ without leaving the circuit. Therefore, *abca* is a closed loop. In order for us to be able to apply Kirchhoff's voltage law, the summation of potential rises and drops must be made in one direction around the closed loop.

For uniformity, the clockwise (CW) direction will be used throughout the text for all applications of Kirchhoff's voltage law. Be aware, however, that the same result will be obtained if the counterclockwise (CCW) direction is chosen and the law applied correctly.

A plus sign is assigned to a potential rise (− to +) and a minus sign to a potential drop (+ to −). If we follow the current in Fig. 5.11 from point $a$, we first encounter a potential drop $V_1$ (+ to −) across $R_1$ and then another potential drop $V_2$ across $R_2$. Continuing through the voltage source, we have a potential rise $E$ (− to +) before returning to point $a$. In symbolic form, where $\Sigma$ represents summation, $\circlearrowright$ the closed loop, and $V$ the potential drops and rises, we have

$$\Sigma_{\circlearrowright} V = 0 \qquad \text{(Kirchhoff's voltage law in symbolic form)} \tag{5.8}$$

which for the circuit of Fig. 5.11 yields (clockwise direction, following the current $I$)

$$+E - V_1 - V_2 = 0$$

or

$$E = V_1 + V_2$$

revealing that the potential impressed on the circuit by the battery is equal to the potential drops within the circuit. To go a step further, the sum of the potential rises will equal the sum of the potential drops around a closed path. Therefore, another way of stating Kirchhoff's voltage law is the following:

$$\Sigma_{\circlearrowright} V_{\text{rises}} = \Sigma_{\circlearrowright} V_{\text{drops}} \tag{5.9}$$

The text will emphasize the use of Eq. (5.8), however.

If the loop were taken in the counterclockwise direction, the following would result:

$$\Sigma_{\circlearrowright} V = 0$$
$$-E + V_2 + V_1 = 0$$

or, as before,

$$E = V_1 + V_2$$

**EXAMPLE 5.4.** Determine the unknown voltages for the networks of Figs. 5.12 and 5.13 using Kirchhoff's voltage law.

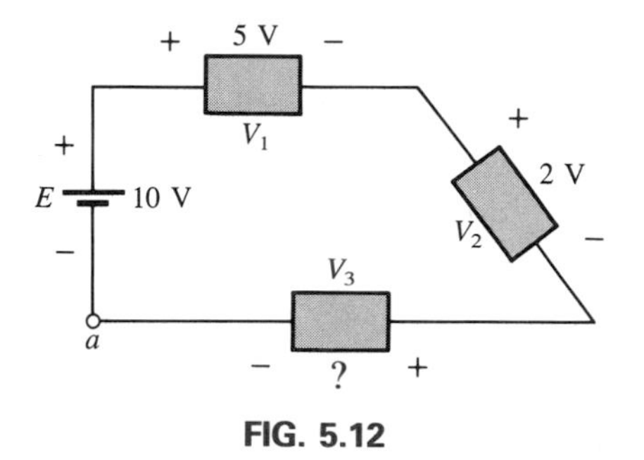


**FIG. 5.12**

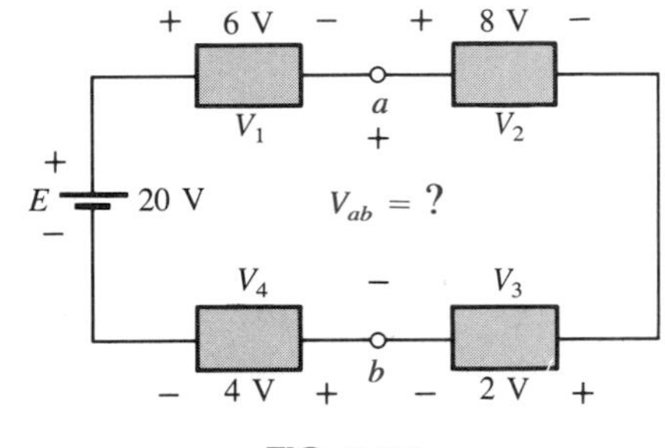


**FIG. 5.13**

***Solution:*** In each case, note that we do not need to know whether there is a voltage source or dissipative element in each container. Once the magnitude and polarity of the voltage for each element are known, Kirchhoff's voltage law can be applied.

For Fig. 5.12, applying Kirchhoff's voltage law in the clockwise direction starting at point $a$ will result in

$$E - V_1 - V_2 - V_3 = 0$$

and

$$V_3 = E - V_1 - V_2 = 10 - 5 \text{ V} - 2 \text{ V} = 10 \text{ V} - 7 \text{ V} = \mathbf{3\ V}$$

In the case of Fig. 5.13, the voltage to be determined is not across a single element but between two points in the network. There are two routes to follow toward a solution.

First, Kirchhoff's voltage law can be applied in the clockwise direction around a closed loop including the voltage source $E$. That is,

$$+E - V_1 - V_{ab} - V_4 = 0$$

and

$$V_{ab} = E - V_1 - V_4 = 20 \text{ V} - 6 \text{ V} - 4 \text{ V} = 20 \text{ V} - 10 \text{ V} = \mathbf{10\ V}$$

The other alternative is to apply the law in a clockwise direction around a closed path that includes only $V_2$ and $V_3$. That is,

$$+V_{ab} - V_2 - V_3 = 0$$

and

$$V_{ab} = V_2 + V_3 = 8 \text{ V} + 2 \text{ V} = \mathbf{10\ V}$$

Note that both routes will result in the same solution.

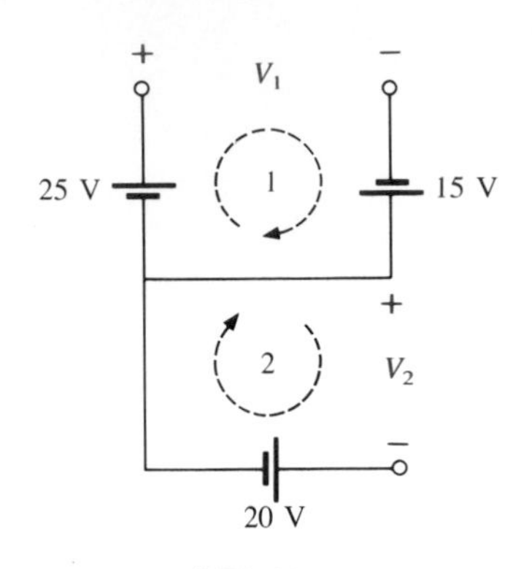


FIG. 5.14

**EXAMPLE 5.5.** Find $V_1$ and $V_2$ for the network of Fig. 5.14.

***Solution:*** For path 1,

$$+25 \text{ V} - V_1 + 15 \text{ V} = 0$$

and

$$V_1 = \mathbf{40 \text{ V}}$$

For path 2,

$$-20 \text{ V} - V_2 = 0$$

and

$$V_2 = \mathbf{-20 \text{ V}}$$

The minus sign simply indicates that the actual polarities of the potential difference are opposite the assumed polarity indicated in Fig. 5.14.

Note in Fig. 5.14 that a potential difference can exist between two points not connected by an element. In other words, a potential difference can exist between two points even though there may not be a current or connecting element between the two points.

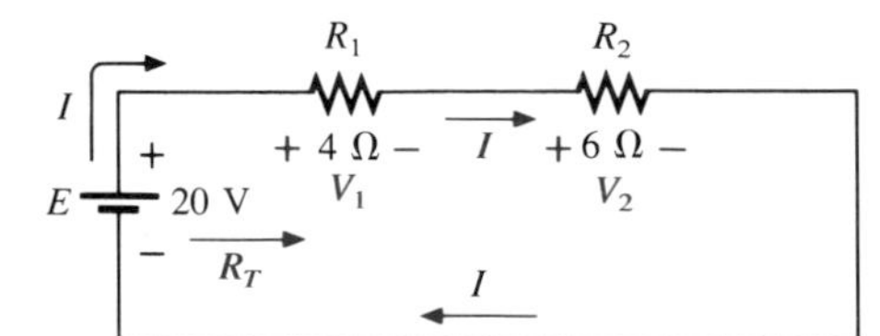

FIG. 5.15

**EXAMPLE 5.6.** For the circuit of Fig. 5.15.

a. Find $R_T$.
b. Find $I$.
c. Find $V_1$ and $V_2$.
d. Find the power to the 4-Ω and 6-Ω resistors.
e. Find the power delivered by the battery, and compare it to that dissipated by the 4-Ω and 6-Ω resistors combined.
f. Verify Kirchhoff's voltage law (clockwise direction).

***Solutions:***

a. $R_T = R_1 + R_2 = 4\ \Omega + 6\ \Omega = \mathbf{10\ \Omega}$

b. $I = \dfrac{E}{R_T} = \dfrac{20 \text{ V}}{10\ \Omega} = \mathbf{2 \text{ A}}$

c. $V_1 = IR_1 = (2 \text{ A})(4\ \Omega) = \mathbf{8 \text{ V}}$
$V_2 = IR_2 = (2 \text{ A})(6\ \Omega) = \mathbf{12 \text{ V}}$

d. $P_{4\Omega} = \dfrac{V_1^2}{R_1} = \dfrac{(8 \text{ V})^2}{4} = \dfrac{64}{4} = \mathbf{16 \text{ W}}$

$P_{6\Omega} = I^2R_2 = (2 \text{ A})^2(6\ \Omega) = (4)(6) = \mathbf{24 \text{ W}}$

e. $P_E = EI = (20 \text{ V})(2 \text{ A}) = \mathbf{40 \text{ W}}$
$P_E = P_{4\Omega} + P_{6\Omega}$
$40 \text{ W} = 16 \text{ W} + 24 \text{ W}$
$40 \text{ W} = 40 \text{ W}$ (checks)

f. $\Sigma_{\circlearrowright} V = +E - V_1 - V_2 = 0$
$E = V_1 + V_2$
$20 \text{ V} = 8 \text{ V} + 12 \text{ V}$
$20 \text{ V} = 20 \text{ V}$ (checks)

**EXAMPLE 5.7.** For the circuit of Fig. 5.16:

a. Determine $V_2$ using Kirchhoff's voltage law.
b. Determine $I$.
c. Find $R_1$ and $R_3$.

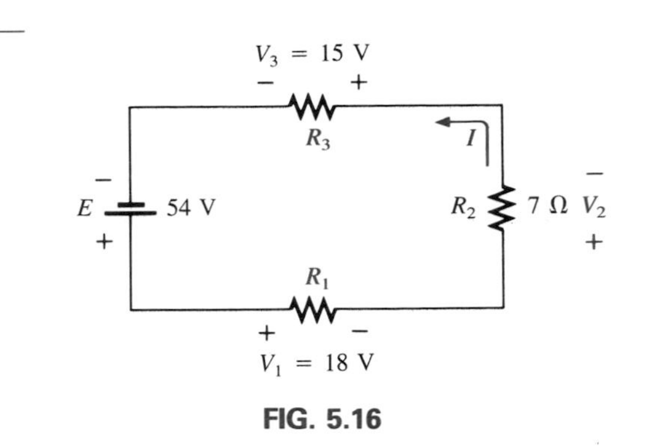


**FIG. 5.16**

***Solutions:***

a. Kirchhoff's voltage law (clockwise direction):

$$-E + V_3 + V_2 + V_1 = 0$$

or

$$E = V_1 + V_2 + V_3$$

and

$$V_2 = E - V_1 - V_3 = 54 \text{ V} - 18 \text{ V} - 15 \text{ V} = \mathbf{21\ V}$$

b. $I = \dfrac{V_2}{R_2} = \dfrac{21 \text{ V}}{7\ \Omega} = \mathbf{3\ A}$

c. $R_1 = \dfrac{V_1}{I} = \dfrac{18 \text{ V}}{3 \text{ A}} = \mathbf{6\ \Omega}$

$R_3 = \dfrac{V_3}{I} = \dfrac{15 \text{ V}}{3 \text{ A}} = \mathbf{5\ \Omega}$

## 5.5 INTERCHANGING SERIES ELEMENTS

The elements of a series circuit can be interchanged without affecting the total resistance, current, or power to each element. For instance, the network of Fig. 5.17 can be redrawn as shown in Fig. 5.18 without

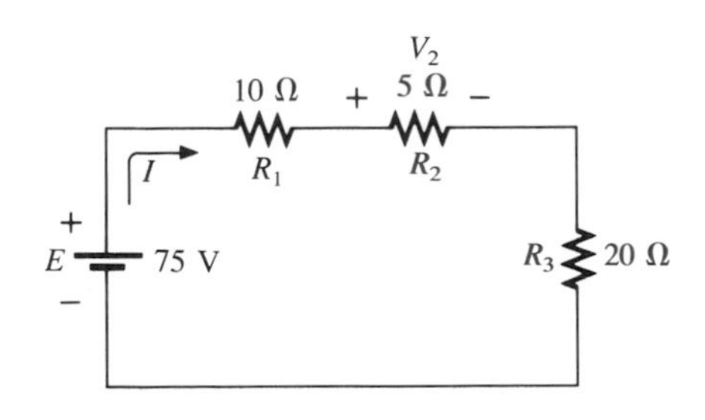

**FIG. 5.17**

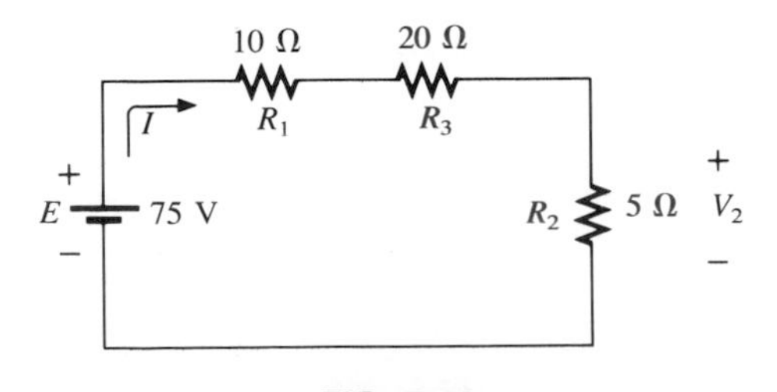

**FIG. 5.18**

affecting $I$ or $V_2$. Total resistance $R_T$ is 25 Ω in both cases, and $I = 75\text{ V}/25\ \Omega = 3\text{ A}$. The voltage $V_2 = IR_2 = (3\text{ A})(5\ \Omega) = 15\text{ V}$ for both configurations.

---

**EXAMPLE 5.8.** Determine $I$ and the voltage across the 7-Ω resistor for the network of Fig. 5.19.

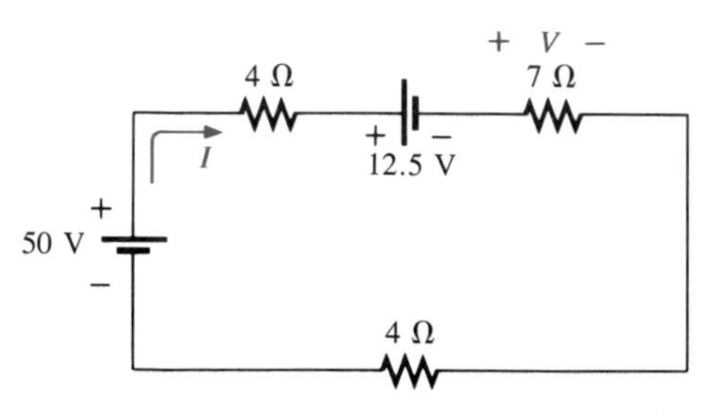

FIG. 5.19

***Solution:*** Network redrawn in Fig. 5.20.

$$R_T = (2)(4\ \Omega) + 7\ \Omega = 15\ \Omega$$

$$I = \frac{E}{R_T} = \frac{37.5\text{ V}}{15\ \Omega} = \mathbf{2.5\ A}$$

$$V_{7\Omega} = IR = (2.5\text{ A})(7\ \Omega) = \mathbf{17.5\ V}$$

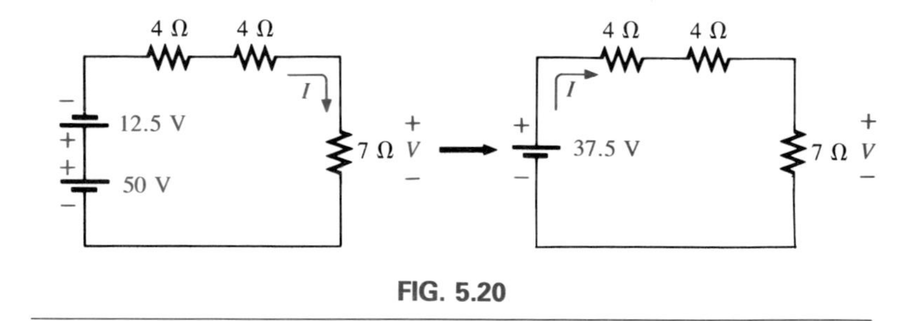


FIG. 5.20

---

## 5.6 VOLTAGE DIVIDER RULE

In a series circuit,

***the voltage across the resistive elements will divide as the magnitude of the resistance levels***

and

***the sum of the voltage drops across the series resistors will equal the applied voltage.***

For example, in Fig. 5.21 the applied voltage of 20 V is equal to the sum of the voltage drops across the three series resistors. That is,

$$E = 20\text{ V} = 12\text{ V} + 6\text{ V} + 2\text{ V}$$
$$20\text{ V} = 20\text{ V}$$

Note also that since $R_1$ is twice $R_2$, $V_{R_1} = 12$ V is twice $V_{R_2} = 6$ V. In addition, since $R_1$ is six times $R_3$, $V_{R_1} = 12$ V is six times $V_{R_3} = 2$ V. Further, since $R_2$ is three times $R_3$, the voltage $V_{R_2} = 6$ V is three times $V_{R_3} = 2$ V.

Based on this, a first glance at the series network of Fig. 5.22 should suggest that the major part of the applied voltage will appear across the 1-MΩ resistor and very little across the 100-Ω resistor. In fact 1 MΩ = (1000)1 kΩ = (10,000)100 Ω, revealing that $V_1 = 1000V_2 = 10{,}000V_3$.

Solving for the current and then the three voltage levels will result in

$$I = \frac{E}{R_T} = \frac{100\text{ V}}{1{,}001{,}100\ \Omega} \cong 99.89\ \mu\text{A}$$

and

$$V_1 = IR_1 = (99.89\ \mu\text{A})(1\text{ M}\Omega) = \mathbf{99.89\text{ V}}$$
$$V_2 = IR_2 = (99.89\ \mu\text{A})(1\text{ k}\Omega) = \mathbf{99.89\text{ mV}} = 0.09989\text{ V}$$
$$V_3 = IR_3 = (99.89\ \mu\text{A})(100\ \Omega) = \mathbf{9.989\text{ mV}} = 0.009989\text{ V}$$

clearly substantiating the above conclusions. For the future, therefore, use this discussion to estimate the share of the input voltage across series elements to act as a check against the actual calculations or to simply obtain an estimate with a minimum of effort.

In the above discussion the current was determined before the voltages of the network were determined. There is, however, a method referred to as the *voltage divider rule* that permits determining the voltage levels without first finding the current. The rule can be derived by analyzing the network of Fig. 5.23.

The total resistance is

$$R_T = R_1 + R_2$$

resulting in

$$I = \frac{E}{R_T}$$

and

$$V_1 = IR_1 = \left(\frac{E}{R_T}\right)R_1 = \frac{R_1E}{R_T}$$

with

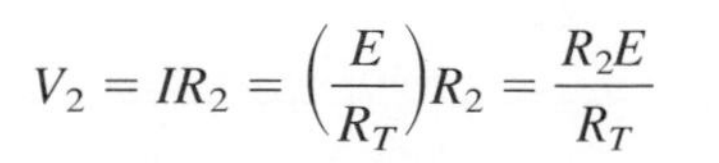

$$V_2 = IR_2 = \left(\frac{E}{R_T}\right)R_2 = \frac{R_2E}{R_T}$$

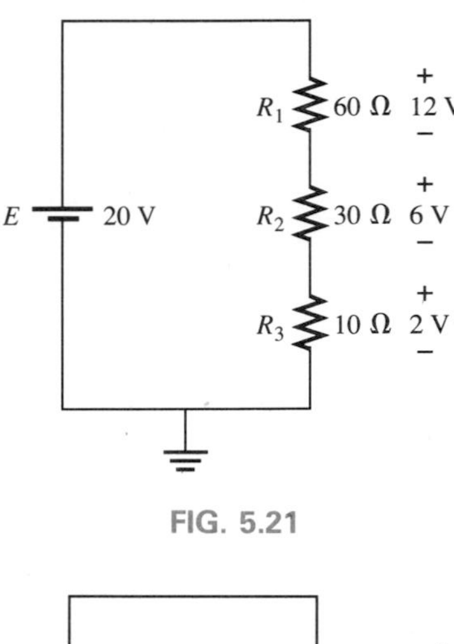


FIG. 5.21

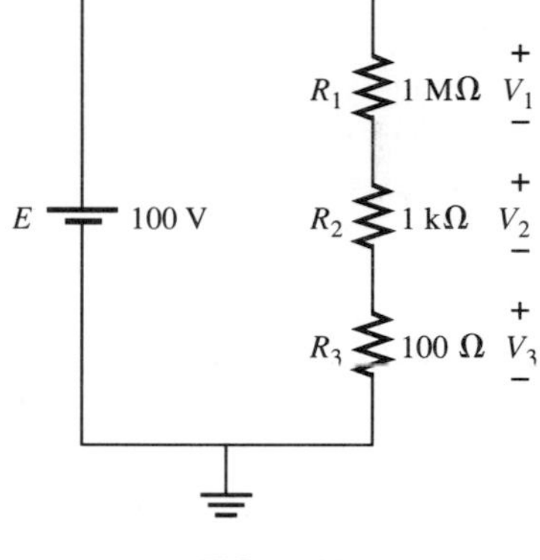

FIG. 5.22

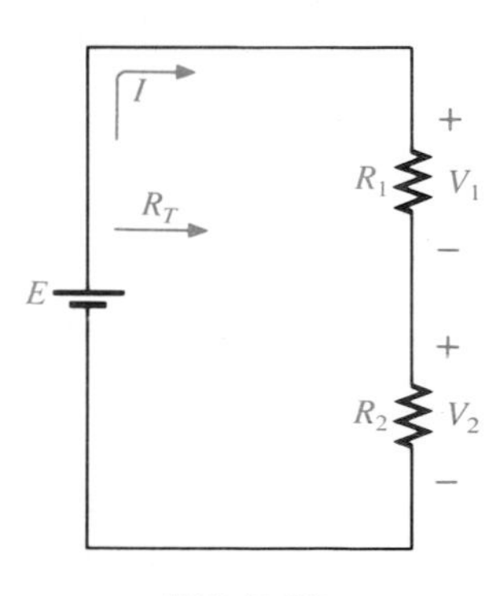


FIG. 5.23

Note that the format for $V_1$ and $V_2$ is

$$\boxed{V_x = \frac{R_x E}{R_T}} \qquad \text{(voltage divider rule)} \qquad \textbf{(5.10)}$$

where $V_x$ is the voltage across $R_x$, $E$ is the impressed voltage across the series elements, and $R_T$ is the total resistance of the series circuit.

In words, the *voltage divider rule* states that

***the voltage across a resistor in a series circuit is equal to the value of that resistor times the total impressed voltage across the series elements divided by the total resistance of the series elements.***

**EXAMPLE 5.9.** Determine the voltage $V_1$ for the network of Fig. 5.24.

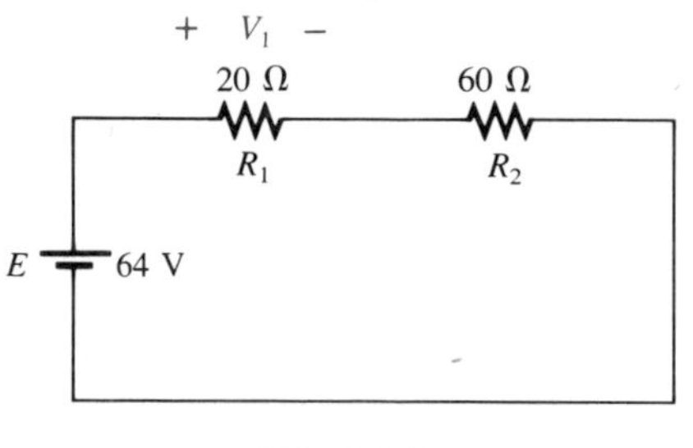


FIG. 5.24

***Solution:*** Eq. (5.10):

$$V_1 = \frac{R_1 E}{R_T} = \frac{R_1 E}{R_1 + R_2} = \frac{(20\ \Omega)(64\ \text{V})}{20\ \Omega + 60\ \Omega} = \frac{1280\ \text{V}}{80} = \mathbf{16\ V}$$

**EXAMPLE 5.10.** Using the voltage divider rule, determine the voltages $V_1$ and $V_3$ for the series circuit of Fig. 5.25.

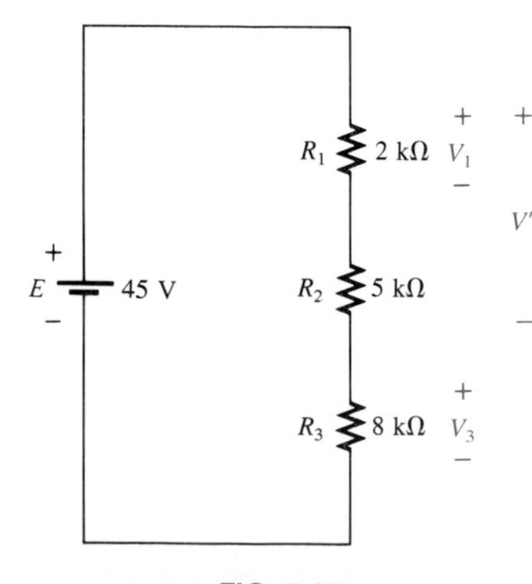

FIG. 5.25

***Solution:***

$$V_1 = \frac{R_1 E}{R_T} = \frac{(2\ \text{k}\Omega)(45\ \text{V})}{2\ \text{k}\Omega + 5\ \text{k}\Omega + 8\ \text{k}\Omega} = \frac{(2\ \text{k}\Omega)(45\ \text{V})}{15\ \text{k}\Omega}$$

$$= \frac{(2 \times 10^3\ \Omega)(45\ \text{V})}{15 \times 10^3\ \Omega} = \frac{90\ \text{V}}{15} = \mathbf{6\ V}$$

$$V_3 = \frac{R_2 E}{R_T} = \frac{(8\ \text{k}\Omega)(45\ \text{V})}{15\ \text{k}\Omega} = \frac{(8 \times 10^3\ \Omega)(45\ \text{V})}{15 \times 10^3\ \Omega}$$

$$= \frac{360\ \text{V}}{15} = \mathbf{24\ V}$$

The rule can be extended to the voltage across two or more series elements if the resistance in the numerator of Eq. (5.10) is expanded to include the total resistance of the series elements that the voltage is to be found across ($R'$). That is,

$$V' = \frac{R'E}{R_T} \quad \text{(volts)} \qquad \textbf{(5.11)}$$

**EXAMPLE 5.11.** Determine the voltage $V'$ in Fig. 5.25 across resistors $R_1$ and $R_2$.

***Solution:***

$$V' = \frac{R'E}{R_T} = \frac{(2\text{ k}\Omega + 5\text{ k}\Omega)(45\text{ V})}{15\text{ k}\Omega} = \frac{(7\text{ k}\Omega)(45\text{ V})}{15\text{ k}\Omega} = \mathbf{21\ V}$$

There is also no need for the voltage $E$ in the equation to be the source voltage of the network. For example, if $V$ is the total voltage across a number of series elements such as shown in Fig. 5.26, then

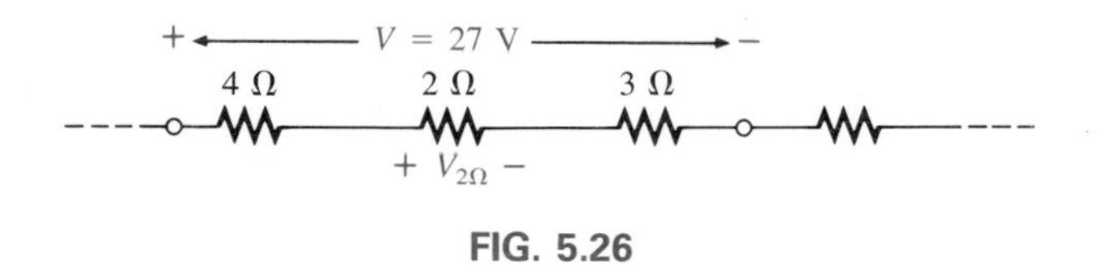

FIG. 5.26

$$V_{2\Omega} = \frac{(2\ \Omega)(27\text{ V})}{4\ \Omega + 2\ \Omega + 3\ \Omega} = \frac{54\text{ V}}{9} = \mathbf{6\ V}$$

**EXAMPLE 5.12.** Design the voltage divider of Fig. 5.27 such that $V_{R_1} = 4V_{R_2}$.

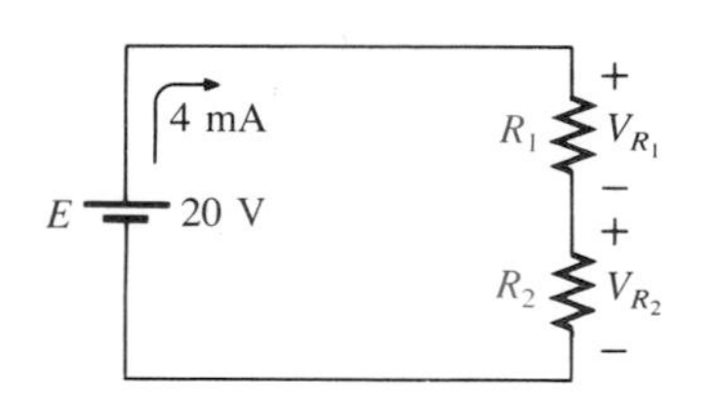

FIG. 5.27

***Solution:*** The total resistance is defined by

$$R_T = \frac{E}{I} = \frac{20\text{ V}}{4\text{ mA}} = 5\text{ k}\Omega$$

Since $V_{R_1} = 4V_{R_2}$,

$$R_1 = 4R_2$$

Thus,

$$R_T = R_1 + R_2 = 4R_2 + R_2 = 5R_2$$

and

$$5R_2 = 5\text{ k}\Omega$$
$$R_2 = \mathbf{1\ k\Omega}$$

and

$$R_1 = 4R_2 = \mathbf{4\ k\Omega}$$

## 5.7 NOTATION

Notation will play an increasingly important role in the analysis to follow. It is important, therefore, that we begin to consider some of the notation used throughout the industry.

### Voltage Sources and Ground

Except for a few special cases, electrical and electronic systems are grounded for reference and safety purposes. The symbol for the ground connection appears in Fig. 5.28 with its defined potential level—zero volts. None of the circuits discussed thus far have contained the ground connection. If Fig. 5.4(a) were redrawn with a grounded supply, it might appear as shown in Fig. 5.29(a) or (b) or (c). In either case, it is understood that the negative terminal of the battery and the bottom of

FIG. 5.28
*Ground potential.*

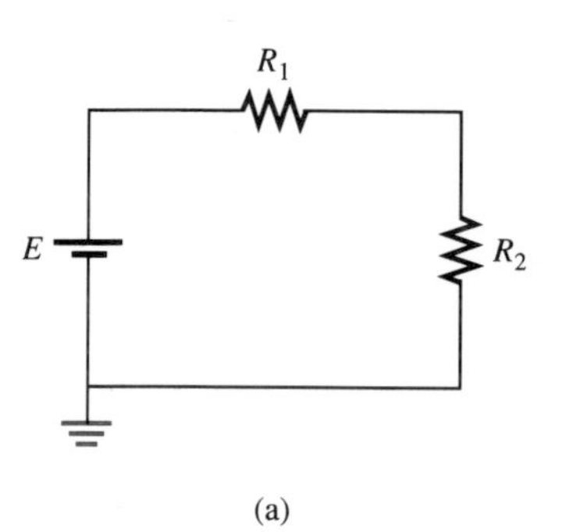


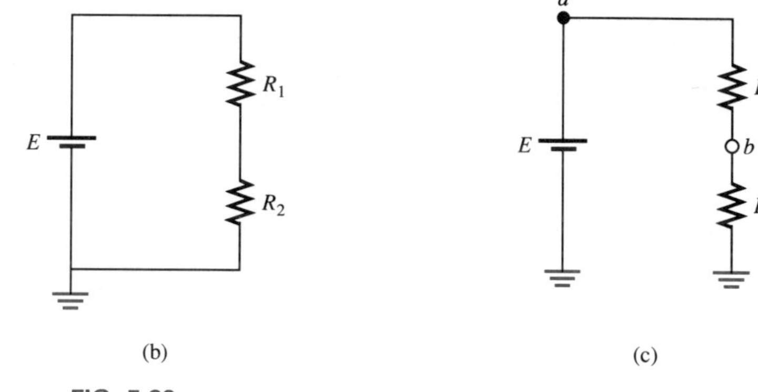


FIG. 5.29

the resistor $R_2$ are at ground potential. Although Fig. 5.29(c) shows no connection between the two grounds, it is recognized that such a connection exists for the continuous flow of charge. If $E = 12$ V, then point $a$ is 12 V positive with respect to ground potential and 12 V exist across the series combination of resistors $R_1$ and $R_2$. If a voltmeter placed from point $b$ to ground reads 4 V, then the voltage across $R_2$ is 4 V, with the higher potential at point $b$.

On large schematics where space is at a premium and clarity is important, voltage sources may be indicated as shown in Figs. 5.30(a) and

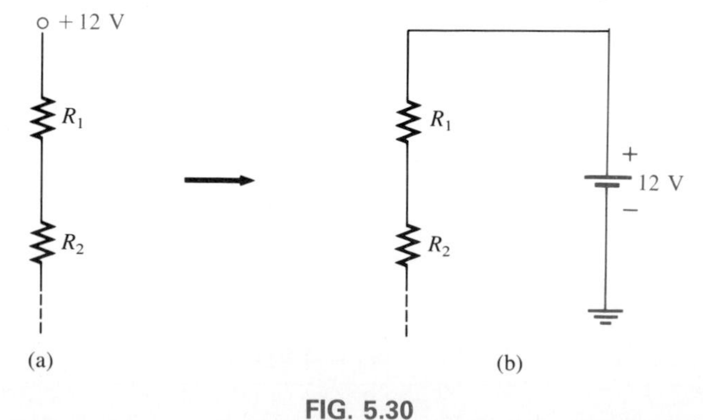

FIG. 5.30

5.31(a) rather than as illustrated in Figs. 5.30(b) and 5.31(b). In addition, potential levels may be indicated as in Fig. 5.32, to permit a rapid

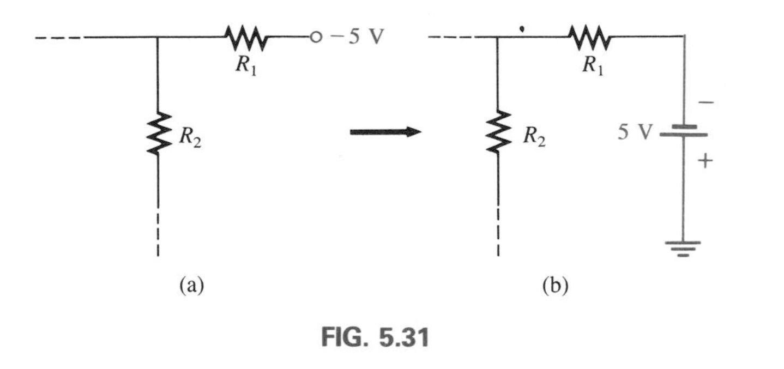


FIG. 5.31

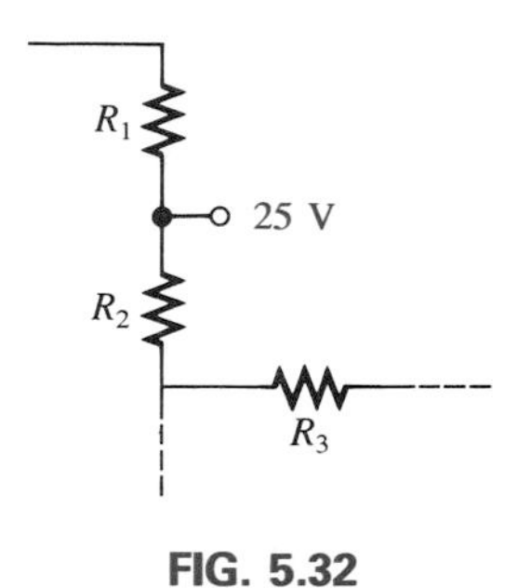


FIG. 5.32

check of the potential levels at various points in a network with respect to ground to insure the system is operating properly.

## Double-Subscript Notation

The fact that voltage is an *across* variable and exists between two points has resulted in a double-subscript notation that defines the first subscript as the higher potential. In Fig. 5.33(a), the two points that define the

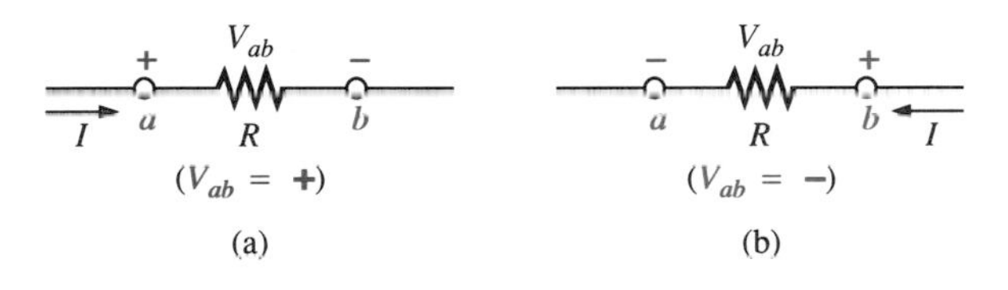


FIG. 5.33
*Defining the sign for double-subscript notation.*

voltage across the resistor $R$ are denoted by $a$ and $b$. Since $a$ is the first subscript for $V_{ab}$, point $a$ must have a higher potential than point $b$ if $V_{ab}$ is to have a positive value. If in fact, point $b$ is at a higher potential than point $a$, $V_{ab}$ will have a negative value, as indicated in Fig. 5.33(b).

In summary:

***The double-subscript notation $V_{ab}$ specifies point a as the higher potential. If this is not the case, a negative sign must be associated with the magnitude of $V_{ab}$.***

## Single-Subscript Notation

If point $b$ of the notation $V_{ab}$ is specified as ground potential (zero volts), then a single-subscript notation can be employed, which provides the voltage at a point with respect to ground.

In Fig. 5.34, $V_a$ is the voltage from point $a$ to ground. In this case it is obviously 10 V, since it is right across the source voltage $E$. The voltage $V_b$ is the voltage from point $b$ to ground. Because it is directly across the 4-Ω resistor, $V_b = 4$ V.

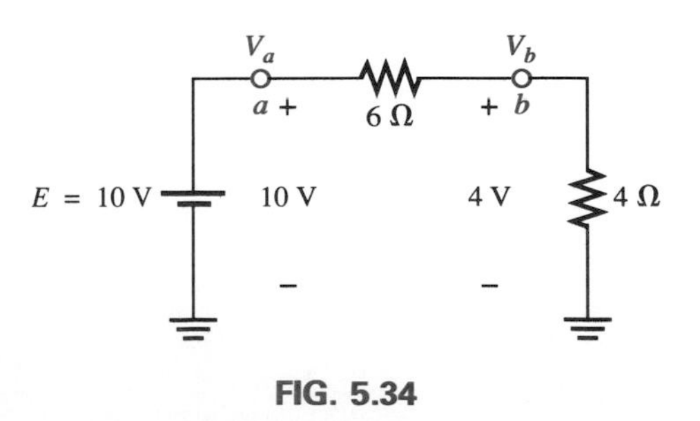


FIG. 5.34

In summary:

***The single-subscript notation $V_a$ specifies the voltage at point a with respect to ground (zero volts). If the voltage is less than zero volts, a negative sign must be associated with the magnitude of $V_a$.***

## General Comments

A particularly useful relationship can now be established that will have extensive applications in the analysis of electronic circuits. For the above notational standards, the following relationship exists:

$$\boxed{V_{ab} = V_a - V_b} \tag{5.12}$$

In other words, if the voltage at points $a$ and $b$ is known with respect to ground, then the voltage $V_{ab}$ can be determined using the above equation. In Fig. 5.34, for example,

$$V_{ab} = V_a - V_b = 10\text{ V} - 4\text{ V} = 6\text{ V}$$

**EXAMPLE 5.13.** Find the voltage $V_{ab}$ for the conditions of Fig. 5.35.

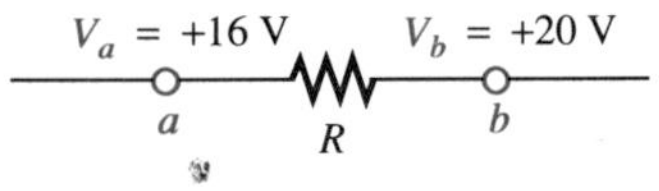

FIG. 5.35

***Solution:*** Applying Eq. (5.12):

$$V_{ab} = V_a - V_b = 16\text{ V} - 20\text{ V} = \mathbf{-4\ V}$$

Note the negative sign to reflect the fact that point $b$ is at a higher potential than point $a$.

**EXAMPLE 5.14.** Find the voltage $V_a$ for the configuration of Fig. 5.36.

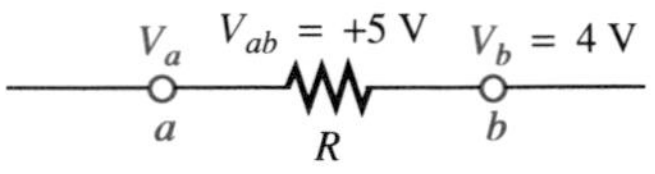

FIG. 5.36

***Solution:*** Applying Eq. (5.12):

$$V_{ab} = V_a - V_b$$

and

$$V_a = V_{ab} + V_b = 5\text{ V} + 4\text{ V} = \mathbf{9\ V}$$

**EXAMPLE 5.15.** Find the voltage $V_{ab}$ for the configuration of Fig. 5.37.

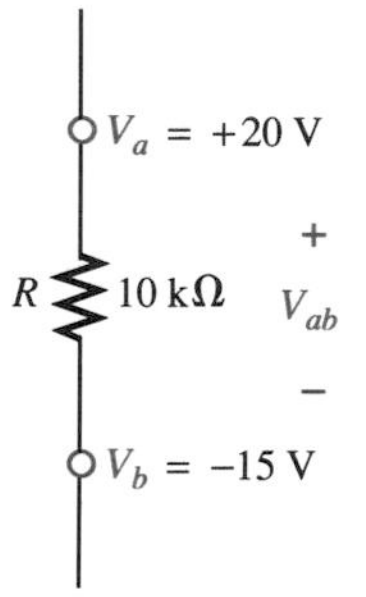

FIG. 5.37

***Solution:*** Applying Eq. (5.12):

$$V_{ab} = V_a - V_b = 20 \text{ V} - (-15 \text{ V}) = 20 \text{ V} + 15 \text{ V}$$
$$= \mathbf{35 \text{ V}}$$

Note in this case the care that must be taken with the signs when applying the equation. The voltage is dropping from a high level of +20 V to a negative voltage of −15 V. As shown in Fig. 5.38, this represents a drop in voltage of 35 V. In some ways it's like going from a positive checking balance of $20 to owing $15; the total expenditure is $35.

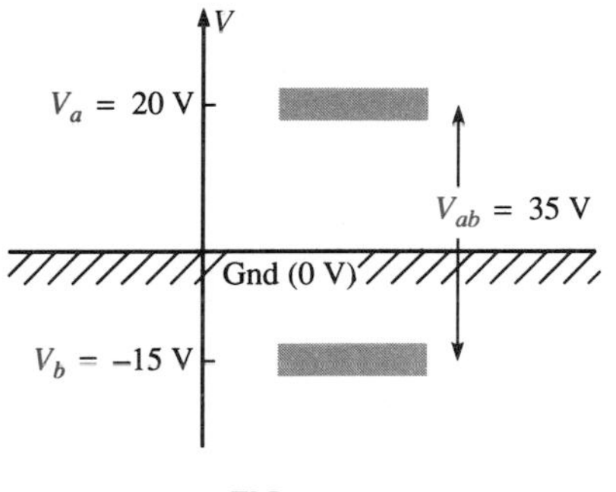

FIG. 5.38

**EXAMPLE 5.16.** Find the voltages $V_b$, $V_c$, and $V_{ac}$ for the network of Fig. 5.39.

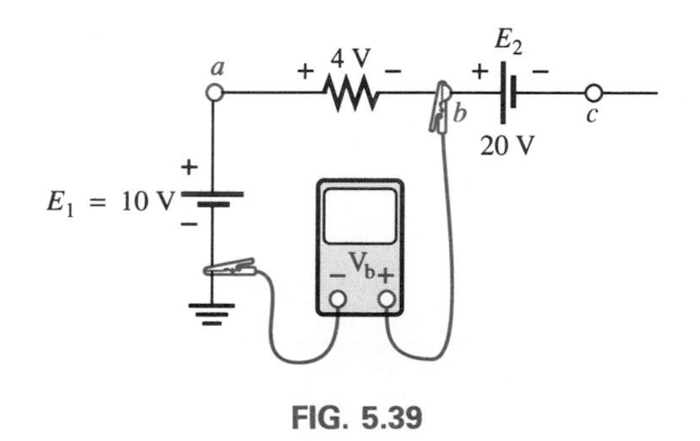


FIG. 5.39

***Solution:*** Starting at ground potential (zero volts), we proceed through a rise of 10 V to reach point $a$ and then pass through a drop in potential of 4 V to point $b$. The result is that the meter will read

$$V_b = +10 \text{ V} - 4 \text{ V} = \mathbf{6 \text{ V}}$$

as clearly demonstrated by Fig. 5.40.

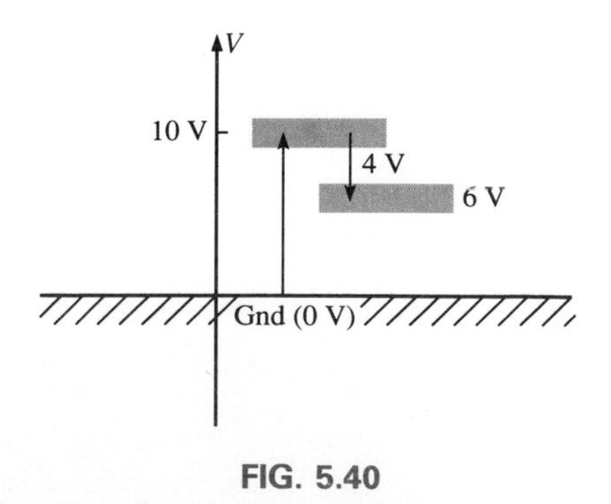


FIG. 5.40

If we then proceed to point $c$, there is an additional drop of 20 V, resulting in

$$V_c = V_b - 20\text{ V} = 6\text{ V} - 20\text{ V} = \mathbf{-14\ V}$$

as shown in Fig. 5.41.

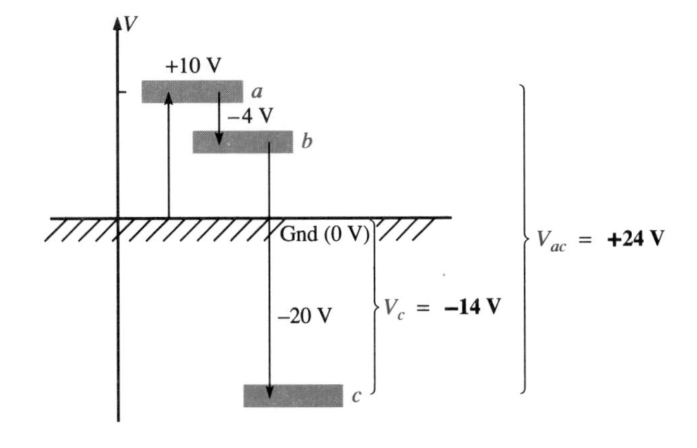

FIG. 5.41

The voltage $V_{ac}$ can be obtained using Eq. (5.12) or by simply referring to Fig. 5.41:

$$\begin{aligned} V_{ac} &= V_a - V_c = 10\text{ V} - (-14\text{ V}) \\ &= \mathbf{24\ V} \end{aligned}$$

---

**EXAMPLE 5.17.** Determine $V_{ab}$, $V_{cb}$, and $V_c$ for the network of Fig. 5.42.

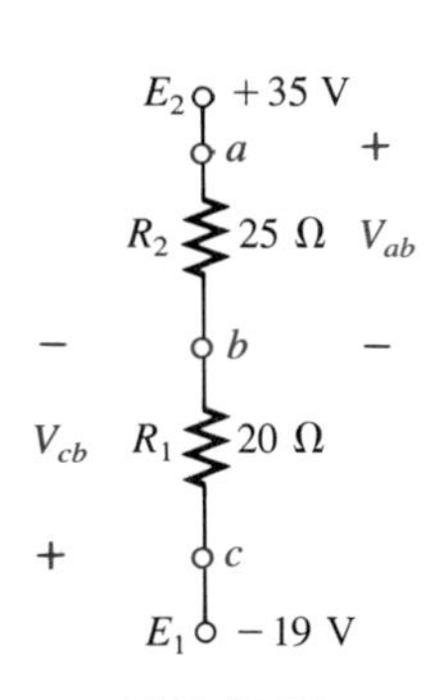


FIG. 5.42

***Solution:*** There are two ways to approach this problem. First is to sketch the diagram of Fig. 5.43 and note that there is a 54-V drop across

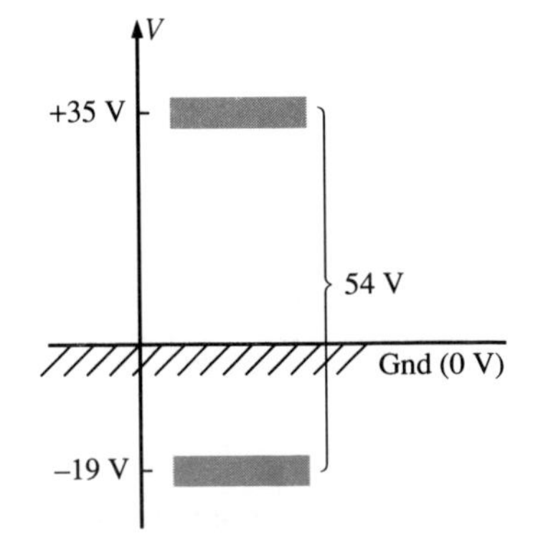

FIG. 5.43

the series resistors $R_1$ and $R_2$. The current can then be determined using Ohm's law and the voltage levels as follows:

$$I = \frac{54\ \text{V}}{45\ \Omega} = 1.2\ \text{A}$$

$$V_{ab} = IR_2 = (1.2\ \text{A})(25\ \Omega) = \mathbf{30\ V}$$

$$V_{cb} = -IR_1 = -(1.2\ \text{A})(20\ \Omega) = \mathbf{-24\ V}$$

$$V_c = E_1 = \mathbf{-19\ V}$$

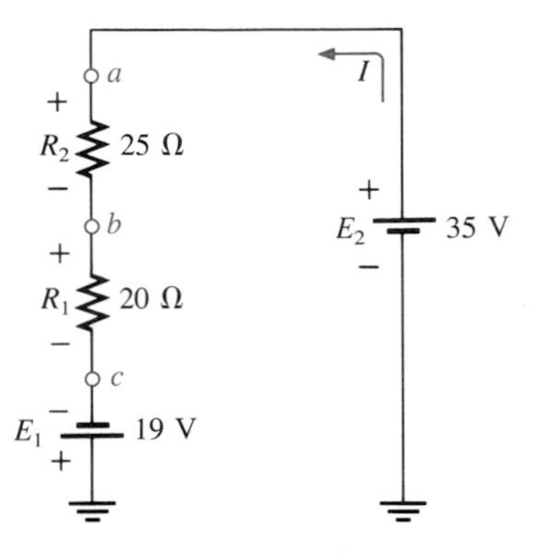

FIG. 5.44

The other approach is to redraw the network as shown in Fig. 5.44 to clearly establish the aiding effect of $E_1$ and $E_2$ and solve the resulting series circuit.

$$I = \frac{E_1 + E_2}{R_T} = \frac{19\ \text{V} + 35\ \text{V}}{45\ \Omega} = \frac{54\ \text{V}}{45\ \Omega} = 1.2\ \text{A}$$

and

$$V_{ab} = \mathbf{30\ V}, \qquad V_{cb} = \mathbf{-24\ V}, \qquad V_c = \mathbf{-19\ V}$$

---

**EXAMPLE 5.18.** Using the voltage divider rule, determine the voltages $V_1$ and $V_2$ of Fig. 5.45.

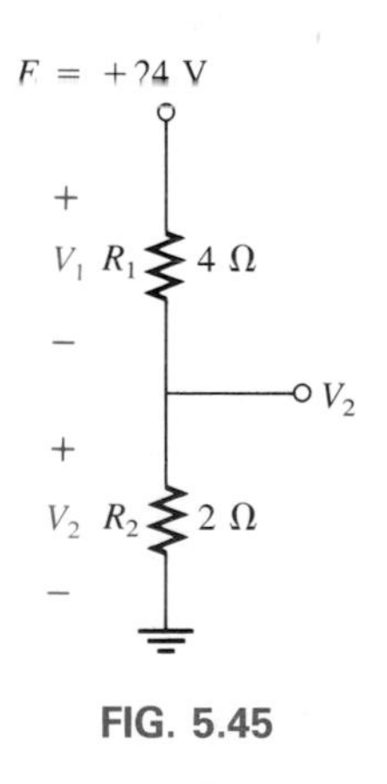


FIG. 5.45

***Solution:*** Redrawing the network with the standard battery symbol will result in the network of Fig. 5.46. Applying the voltage divider rule,

$$V_1 = \frac{R_1E}{R_1 + R_2} = \frac{(4\ \Omega)(24\ \text{V})}{4\ \Omega + 2\ \Omega} = \mathbf{16\ V}$$

$$V_2 = \frac{R_2E}{R_1 + R_2} = \frac{(2\ \Omega)(24\ \text{V})}{4\ \Omega + 2\ \Omega} = \mathbf{8\ V}$$

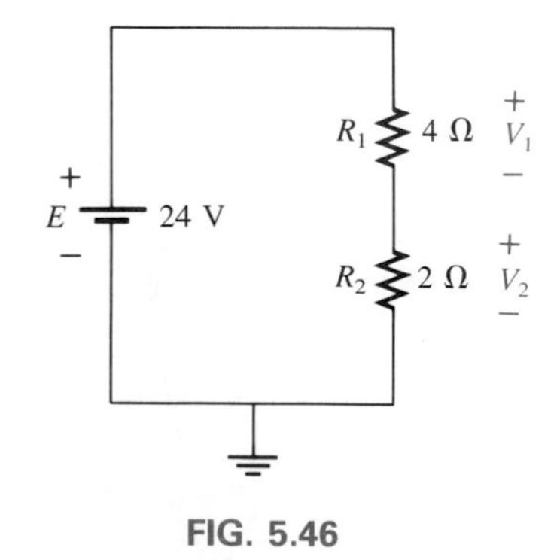


FIG. 5.46

**EXAMPLE 5.19.** For the network of Fig. 5.47:

a. Calculate $V_{ab}$.
b. Determine $V_b$.
c. Calculate $V_c$.

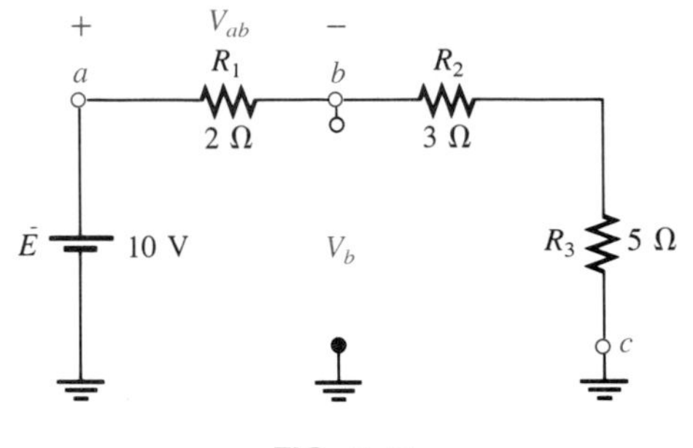

FIG. 5.47

***Solutions:***

a. Voltage divider rule:

$$V_{ab} = \frac{R_1 E}{R_T} = \frac{(2\ \Omega)(10\ \text{V})}{2\ \Omega + 3\ \Omega + 5\ \Omega} = \mathbf{+2\ V}$$

b. Voltage divider rule:

$$V_b = V_{R_2} + V_{R_3} = \frac{(R_2 + R_3)E}{R_T} = \frac{(3\ \Omega + 5\ \Omega)(10\ \text{V})}{10\ \Omega} = \mathbf{8\ V}$$

or

$$V_b = V_a - V_{ab} = E - V_{ab} = 10\ \text{V} - 2\ \text{V} = \mathbf{8\ V}$$

c. $V_c$ = ground potential = **0 V**

## 5.8 INTERNAL RESISTANCE OF VOLTAGE SOURCES

Every source of voltage, whether it be a generator, battery, or laboratory supply as shown in Fig. 5.48(a), will have some internal resistance.

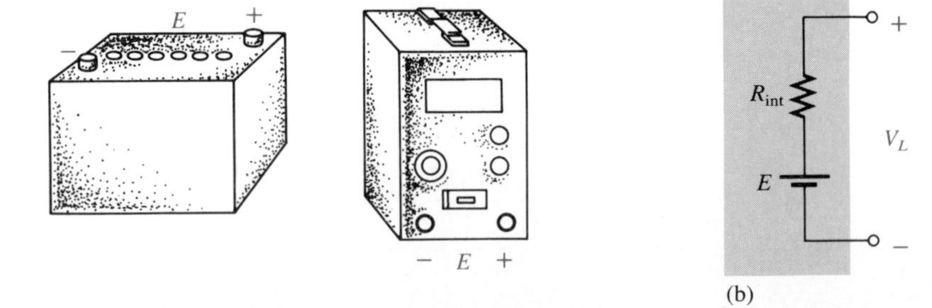


FIG. 5.48

The equivalent circuit of any source of voltage will therefore appear as shown in Fig. 5.48(b). In this section, we will examine the effect of the internal resistance on the output voltage so that any unexpected changes in terminal characteristics can be explained.

In all the circuit analyses to this point, the ideal voltage source (no internal resistance) was used [see Fig. 5.49(a)]. The ideal voltage

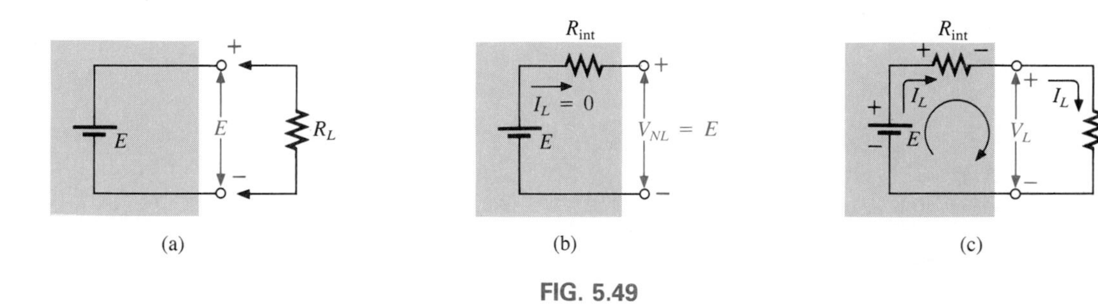


**FIG. 5.49**

source has no internal resistance and an output voltage of $E$ volts with no load or full load. In the practical case [Fig. 5.49(b)], where we consider the effects of the internal resistance, the output voltage will be $E$ volts only when no-load ($I_L = 0$) conditions exist. When a load is connected [Fig. 5.49(c)], the output voltage of the voltage source will decrease due to the voltage drop across the internal resistance.

By applying Kirchhoff's voltage law around the indicated loop of Fig. 5.49(c), we obtain

$$E - I_L R_{\text{int}} - V_L = 0$$

or, since

$$E = V_{NL}$$

we have

$$V_{NL} - I_L R_{\text{int}} - V_L = 0$$

and

$$\boxed{V_L = V_{NL} - I_L R_{\text{int}}} \qquad \textbf{(5.13)}$$

If the value of $R_{\text{int}}$ is not available, it can be found by first solving for $R_{\text{int}}$ in the equation just derived for $V_L$. That is,

$$R_{\text{int}} = \frac{V_{NL} - V_L}{I_L} = \frac{V_{NL}}{I_L} - \frac{I_L R_L}{I_L}$$

and

$$\boxed{R_{\text{int}} = \frac{V_{NL}}{I_L} - R_L} \qquad \textbf{(5.14)}$$

A plot of the output voltage versus current appears in Fig. 5.50(b) for the circuit in Fig. 5.50(a). Note that an increase in load demand ($I_L$)

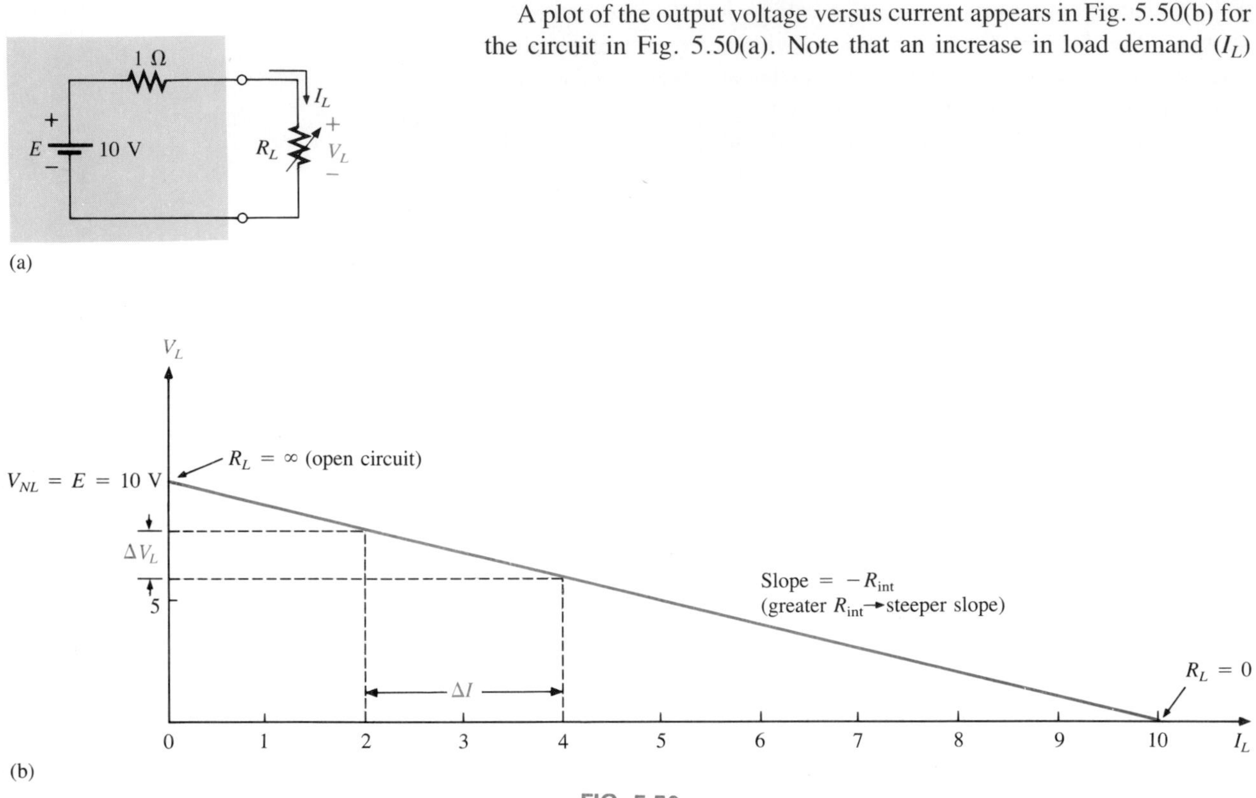

FIG. 5.50

increases the current through, and thereby the voltage drop across, the internal resistance of the source, resulting in a decrease in terminal voltage. Eventually, as the load resistance approaches zero ohms, all the generated voltage will appear across the internal resistance and none at the output terminals. The steeper the slope of the curve of Fig. 5.50(b), the greater the internal resistance. In fact, for any chosen interval of voltage or current, the magnitude of the internal resistance is given by

$$R_{int} = \frac{\Delta V_L}{\Delta I_L} \tag{5.15}$$

where Δ signifies finite change.

For the chosen interval of $2 \rightarrow 4$ A ($\Delta I_L = 2$ A) on Fig. 5.50(b), $\Delta V_L$ is 2 V, and so $R_{int} = 2/2 = 1\ \Omega$, as indicated in Fig. 5.50(a).

A direct consequence of the loss in output voltage is a loss in power delivered to the load. Multiplying both sides of Eq. (5.13) by the current $I_L$ in the circuit, we obtain

$$\underset{\text{Power to load}}{I_L V_L} = \underset{\text{Power output by battery}}{I_L V_{NL}} - \underset{\text{Power loss in the form of heat}}{I_L^2 R_{int}} \tag{5.16}$$

**EXAMPLE 5.20.** Before a load is applied, the terminal voltage of the power supply of Fig. 5.51 is set to 40 V. When a load of 500 Ω is attached, as shown in Fig. 5.51(b), the terminal voltage drops to 36 V. What happened to the remainder of the no-load voltage, and what is the internal resistance of the source?

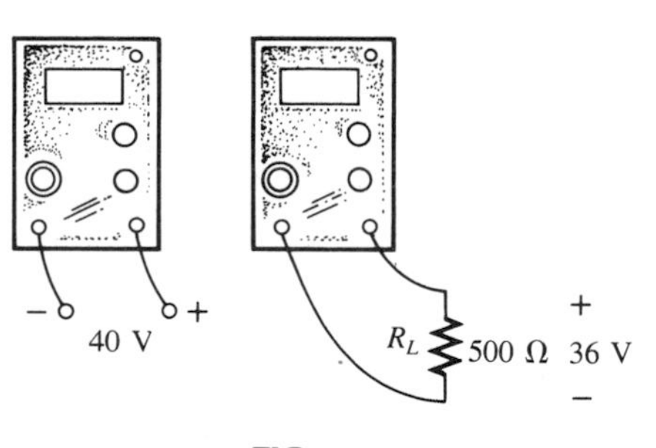


FIG. 5.51

***Solution:*** The difference of 40 V − 36 V = 4 V now appears across the internal resistance of the source. The load current is 36 V/0.5 kΩ = 72 mA. Applying Eq. (5.14),

$$R_{\text{int}} = \frac{V_{NL}}{I_L} - R_L = \frac{40\text{ V}}{72\text{ mA}} - 0.5\text{ k}\Omega$$
$$= 555.55\ \Omega - 500\ \Omega = \mathbf{55.55\ \Omega}$$

**EXAMPLE 5.21.** The battery of Fig. 5.52 has an internal resistance of 2 Ω. Find the voltage $V_L$ and the power lost to the internal resistance if the applied load is a 13-Ω resistor.

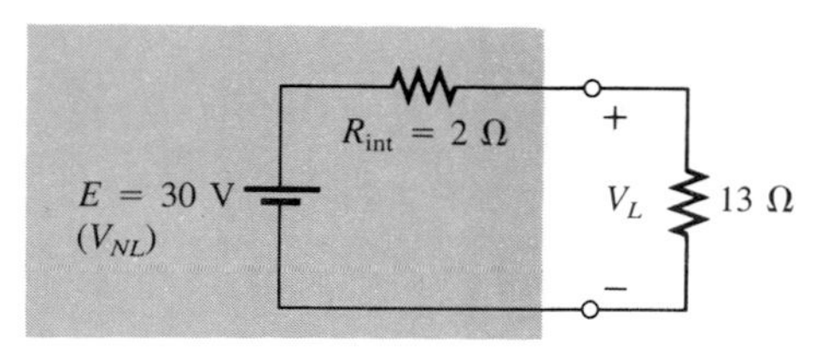


FIG. 5.52

***Solution:***

$$I_L = \frac{30\text{ V}}{2\ \Omega + 13\ \Omega} = \frac{30\text{ V}}{15\ \Omega} = 2\text{ A}$$
$$V_L = V_{NL} - I_L R_{\text{int}} = 30\text{ V} - (2\text{ A})(2\ \Omega) = \mathbf{26\ V}$$
$$P_{\text{lost}} = I_L^2 R_{\text{int}} = (2\text{ A})^2(2\ \Omega) = (4)(2) = \mathbf{8\ W}$$

**EXAMPLE 5.22.** The terminal characteristics of a dc generator appear in Fig. 5.53. Rated (full-load) conditions are indicated at 120 V, 8 A.

a. Calculate the average internal resistance of the supply.
b. At what load current will the terminal voltage drop to 100 V?

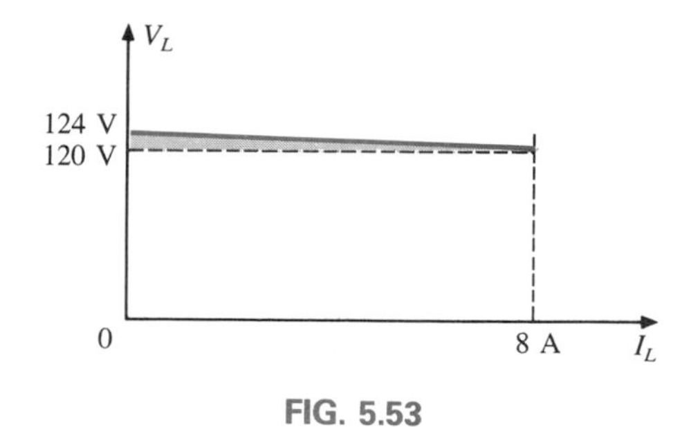


FIG. 5.53

***Solutions:***

a. $R_{\text{int}} = \dfrac{\Delta V_L}{\Delta I_L} = \dfrac{4\text{ V}}{8\text{ A}} = \mathbf{0.5\ \Omega}$

b. From Eq. (5.15),

$$\Delta I_L = \frac{\Delta V_L}{R_{\text{int}}}$$

Since

$$\Delta V_L = 124\text{ V} - 100\text{ V} = 24\text{ V}$$

then

$$\Delta I_L = \frac{\Delta V_L}{R_{\text{int}}} = \frac{24\text{ V}}{0.5\ \Omega} = 48\text{ A} \qquad \text{(from zero amps)}$$

so that

$$I_L = \mathbf{48\ A}$$

## 5.9 VOLTAGE REGULATION

For any supply, ideal conditions dictate that for the range of load demand ($I_L$), the terminal voltage remain fixed in magnitude. In other words, if a supply is set for 12 V, it is desirable that it maintain this terminal voltage even though the current demand on the supply may vary. A measure of how close a supply will come to ideal conditions is given by the voltage regulation characteristic. By definition, the voltage regulation of a supply between the limits of full-load and no-load conditions (Fig. 5.54) is given by the following:

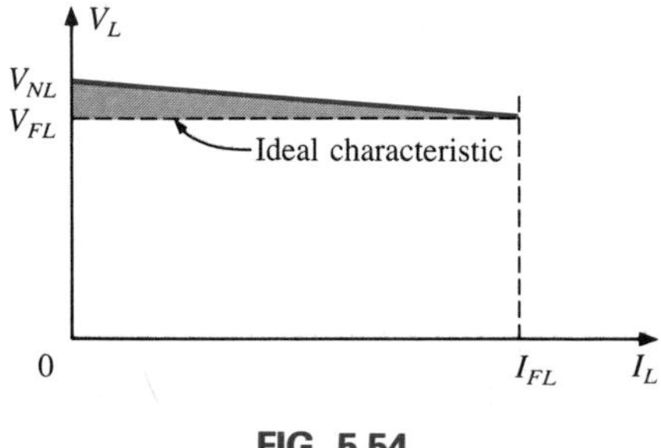


FIG. 5.54

$$\text{Voltage regulation } (VR)\% = \frac{V_{NL} - V_{FL}}{V_{FL}} \times 100\% \tag{5.17}$$

For ideal conditions, $V_{FL} = V_{NL}$ and $VR\% = 0$. Therefore, *the smaller the voltage regulation, the less the variation in terminal voltage with change in load.*

It can be shown with a short derivation that the voltage regulation is also given by

$$VR\% = \frac{R_{\text{int}}}{R_L} \times 100\% \tag{5.18}$$

In other words, the smaller the internal resistance for the same load, the smaller the regulation and more ideal the output.

---

**EXAMPLE 5.23.** Calculate the voltage regulation of a supply having the characteristics of Fig. 5.53.

***Solution:***

$$VR\% = \frac{V_{NL} - V_{FL}}{V_{FL}} \times 100\% = \frac{124 \text{ V} - 120 \text{ V}}{120 \text{ V}} \times 100\%$$

$$= \frac{4}{120} \times 100\% = \mathbf{3.33\%}$$

---

**EXAMPLE 5.24.** Determine the voltage regulation of the supply of Fig. 5.52.

***Solution:***

$$VR\% = \frac{R_{\text{int}}}{R_L} \times 100\% = \frac{2\ \Omega}{13\ \Omega} \times 100\% \cong \mathbf{15.38\%}$$

## 5.10 AMMETERS: LOADING EFFECTS

In Chapter 2, it was noted that ammeters are inserted in the branch in which the current is to be measured. We now realize that such a condition specifies that

***ammeters are placed in series with the branch in which the current is to be measured***

as shown in Fig. 5.55.

If the ammeter is to have minimal impact on the behavior of the network, its resistance should be very small (ideally zero ohms) compared to the other series elements of the branch such as the resistor $R$ of Fig. 5.55. If the meter resistance approaches or exceeds 10% of $R$, it would naturally have a significant impact on the current level it is measuring. It is also noteworthy that the resistances of the separate current scales of the same meter are usually not the same. In fact, the meter resistance normally increases with decreasing current levels. However, for the majority of situations one can simply assume that the internal ammeter resistance is small enough compared to the other circuit elements that it can be ignored.

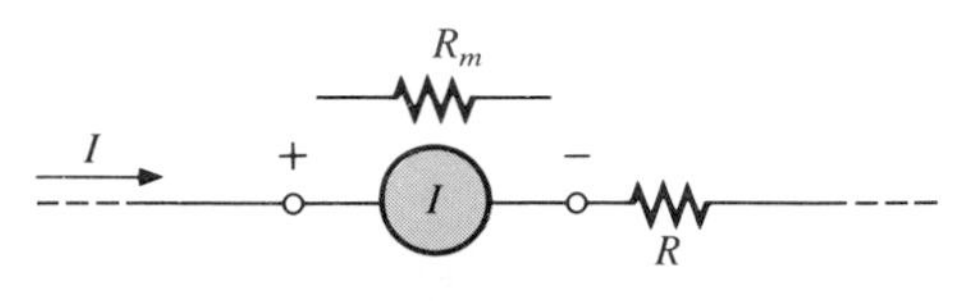


**FIG. 5.55**
*Series connection of an ammeter.*

## 5.11 COMPUTER ANALYSIS

Let us now build on the foundation established in the last chapter to analyze series dc circuits. A new command or two may be introduced, but in general, the input file for PSPICE will remain fairly simple, and in fact, the BASIC program will be shorter than encountered in Chapter 4.

### PSPICE

If you want to limit the output file of an analysis using PSPICE to specific values or control the format of the data, the following command statement must be included somewhere between the title line and the .END statement:

.DC    Source Name    Starting Value    Final Value    Increment

First note that after the required .DC notation (to specify dc analysis), the remainder of the line is concerned with the applied dc source. The .DC command permits the analysis of a dc network for a *range* of source voltages that extend from the starting value to the final value in increments that are under the user's control.

If the analysis is for a fixed source, the following format can be employed where the starting and final values are the same. The increment is required in the format but is ignored by the package. Any number can be used for the increment, but we will use the number 1 throughout this text.

.DC    Source Name    Starting Value    Starting Value    1

It is important to realize that once entered, the .DC command overrides the line of the network description for the dc voltage source. However, the voltage source must still appear in the network description.

Once the .DC command is employed, output data can be obtained only by specifying the quantities desired and the format. The following command will provide an output file for a dc analysis:

.PRINT DC Desired Quantities

Note the continued use of the period at the beginning of important control lines of the input file.

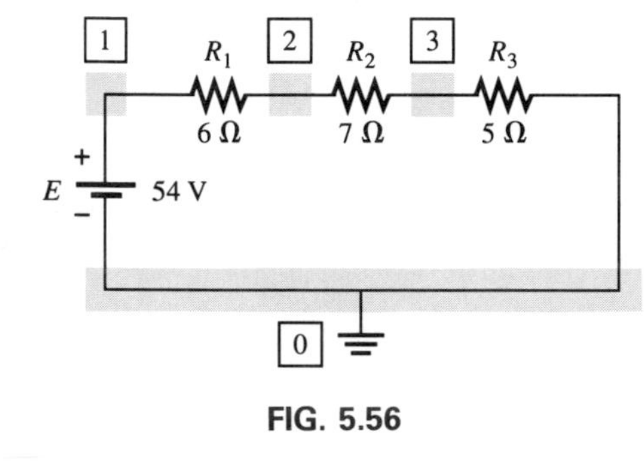


FIG. 5.56

Let us now use the above to analyze the series network of Fig. 5.56. First the nodes are labeled with the ground connection, defined in our reference with a 0 label. The title line and circuit description follow the formats defined in the previous chapters. Now, however, we want to limit our output to the voltages $V_2$ and $V_{R_2}$ and the series current $I$. This control of the output file requires the use of the .DC analysis line, which specifies the name of the source as VE, the beginning and ending values at 54 V (a fixed value), and the 1 for completeness. The .PRINT statement specifies the desired output values V(2), V(R2), and I(R1). Obviously, since it is a series circuit, I(R1) = I(R2) = I(R3). In addition V(R2) = V(2, 3). The .OPTIONS NOPAGE command continues to

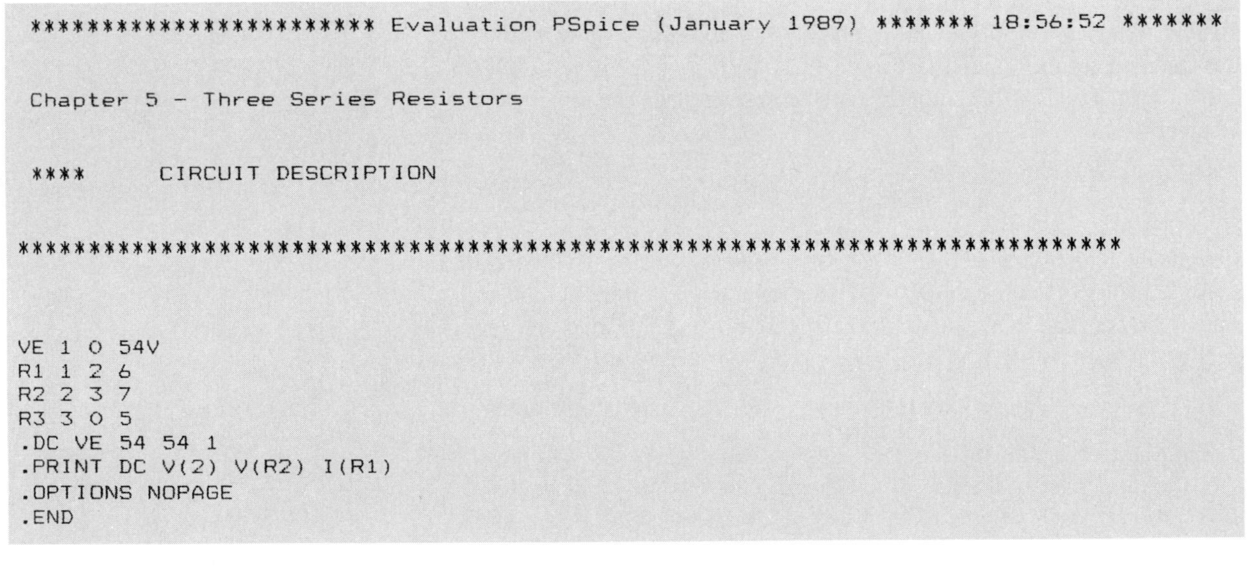

```
************************ Evaluation PSpice (January 1989) ******* 18:56:52 *******

Chapter 5 - Three Series Resistors

****      CIRCUIT DESCRIPTION

******************************************************************************

VE 1 0 54V
R1 1 2 6
R2 2 3 7
R3 3 0 5
.DC VE 54 54 1
.PRINT DC V(2) V(R2) I(R1)
.OPTIONS NOPAGE
.END

****      DC TRANSFER CURVES                    TEMPERATURE =   27.000 DEG C

  VE            V(2)          V(R2)         I(R1)

  5.400E+01     3.600E+01     2.100E+01     3.000E+00
```

FIG. 5.57

minimize paper usage. The output file appears in Fig. 5.57 with the requested values in the order appearing in the .PRINT command. The voltage $V_2$ is 36 V, $V_{R_2} = 21$ V, and $I = E/R_T = 54\ \text{V}/18\ \Omega = 3$ A, which will all be verified by the BASIC program to follow.

## BASIC

The BASIC program of Fig. 5.59 will provide a general solution for the series dc network of Fig. 5.58. When the program is run, it will request the network parameters and provide the output appearing in the run provided. Note that the results are the same as obtained for the PSPICE run. The total resistance $R_T$ is calculated on line 150; $I$, on line 180, the voltage across each resistor, on line 230; and the power to each resistor, on line 250. In addition, Kirchhoff's voltage law is applied on line 300

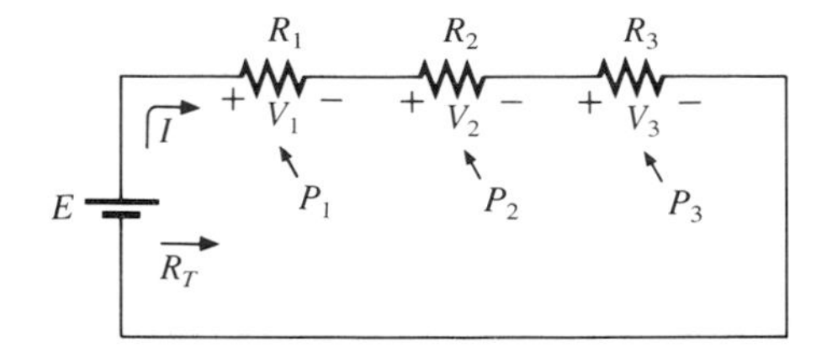


**FIG. 5.58**
*Network for Program 5.1.*

```
   10 REM ***** PROGRAM 5-1 *****
   20 REM ***********************************************************
   30 REM Analysis of a series resistor network
   40 REM ***********************************************************
   50 REM
  100 PRINT:PRINT "Enter resistor values for up to 3 resistors"
  110 PRINT "in series (enter 0 if no resistor):"
  120 INPUT "R1=";R1
  130 INPUT "R2=";R2
  140 INPUT "R3=";R3
RT 150 RT=R1+R2+R3
  160 PRINT:PRINT "The total resistance is RT=";RT;"ohms"
  170 PRINT:INPUT "Enter value of supply voltage, E=";E
I  180 I=E/RT
  190 PRINT
  200 PRINT "Supply current is, I=";I;"amperes"
  210 PRINT
  220 PRINT "The voltage drop across each resistor is:"
Vx 230 V1=I*R1:V2=I*R2 :V3=I*R3
  240 PRINT "V1=";V1;"volts   V2=";V2;"volts    V3=";V3;"volts"
Px 250 P1=I^2*R1 :P2=I^2*R2 :P3=I^2*R3
  260 PRINT
  270 PRINT "The power dissipated by each resistor is:"
  280 PRINT "P1=";P1;"watts","P2=";P2;"watts","P3=";P3;"watts"
  290 PRINT
  300 PRINT "Total voltage around loop is, V1+V2+V3=";V1+V2+V3;"volts"
  310 PRINT "and total power dissipated, P1+P2+P3=";P1+P2+P3;"watts"
  320 END

READY

RUN


Enter resistor values for up to 3 resistors
in series (enter 0 if no resistor):
R1=? 6

R2=? 7

R3=? 5
```

```
The total resistance is RT= 18 ohms

Enter value of supply voltage, E=? 54


Supply current is, I= 3 amperes

The voltage drop across each resistor is:
V1= 18 volts    V2= 21 volts    V3= 15 volts

The power dissipated by each resistor is:
P1= 54 watts    P2= 63 watts    P3= 45 watts

Total voltage around loop is, V1+V2+V3= 54 volts
and total power dissipated, P1+P2+P3= 162 watts
```

**FIG. 5.59**
*Program 5.1.*

to show that the applied voltage equals the sum of the voltage drops. The total power supplied or dissipated is then calculated on line 310.

At first exposure the PSPICE program is the most direct route for obtaining a solution of the series circuit of Fig. 5.56 because of the rather short input file. In BASIC, however, the entire program has to be written and tested before the network data can be entered. Of course, once the program is written, it can be placed in memory for future use; keep in mind that BASIC permits a greater range of choice regarding the output selection and the format.

## PROBLEMS

### SECTION 5.2 Series Circuits

**1.** Find the total resistance and current $I$ for each circuit of Fig. 5.60.

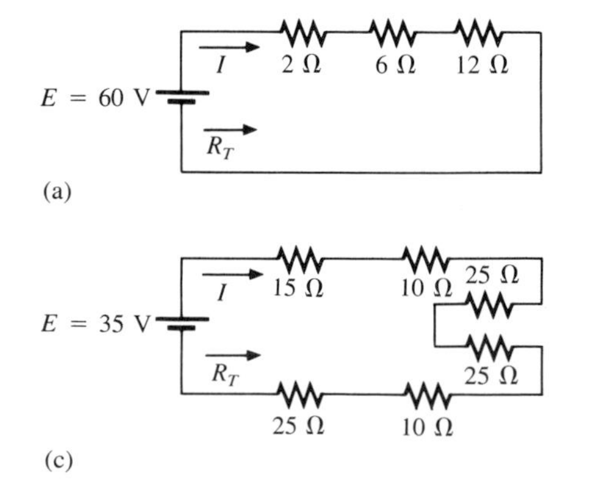


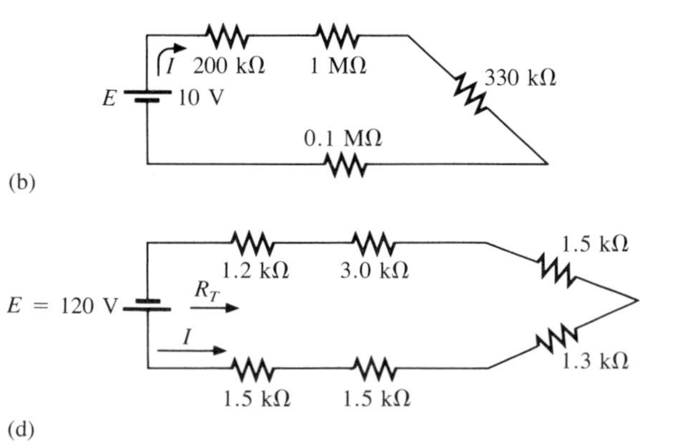


**FIG. 5.60**

2. For the circuits of Fig. 5.61, the total resistance is specified. Find the unknown resistances and the current $I$ for each circuit.

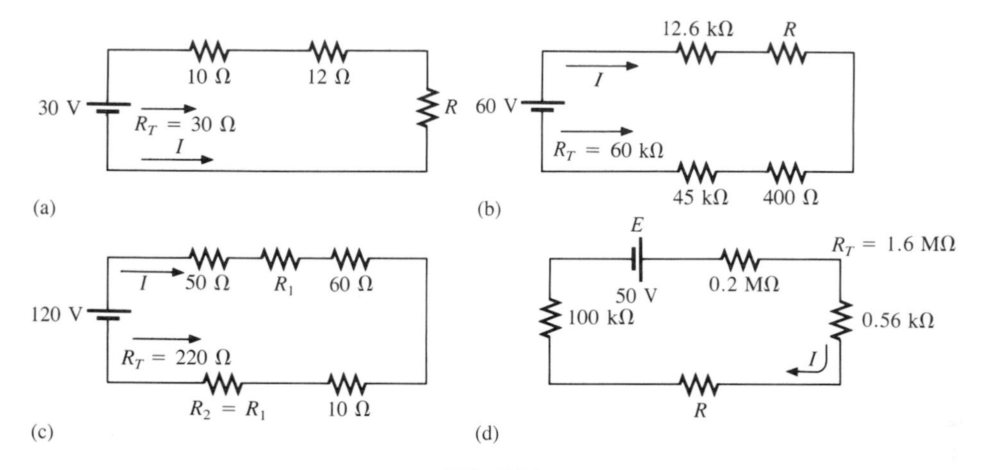

FIG. 5.61

3. Find the applied voltage $E$ necessary to develop the current specified in each network of Fig. 5.62.

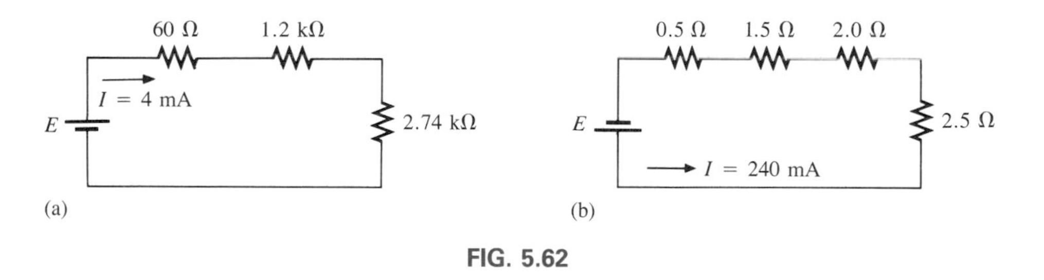


FIG. 5.62

*4. For each network of Fig. 5.63, determine the current $I$, the unknown resistance, and the voltage across each element.

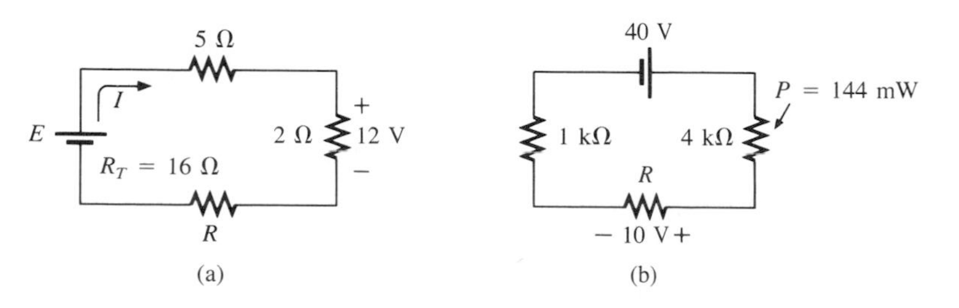


FIG. 5.63

## SECTION 5.3 Voltage Sources in Series

5. Determine the current $I$ for each network of Fig. 5.64. Before solving for $I$, redraw each network with a single voltage source.

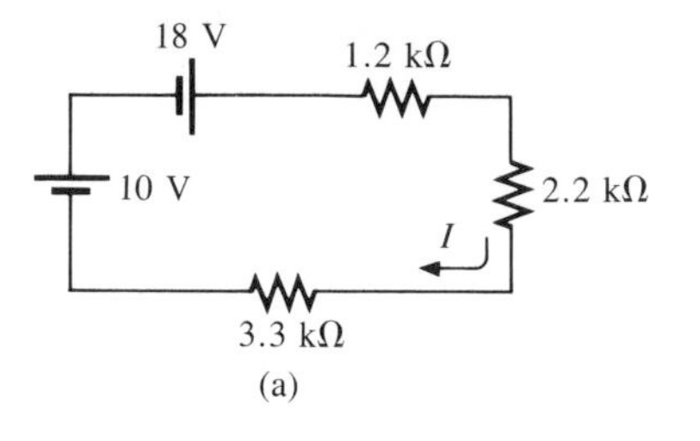


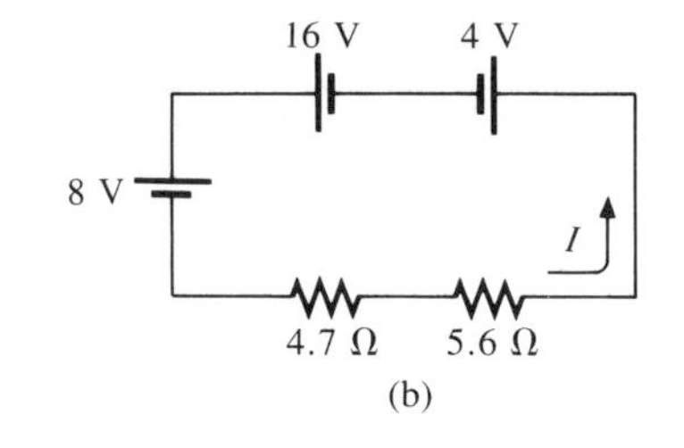


FIG. 5.64

*6. Find the unknown voltage source and resistor for the networks of Fig. 5.65. Also indicate the direction of the resulting current.

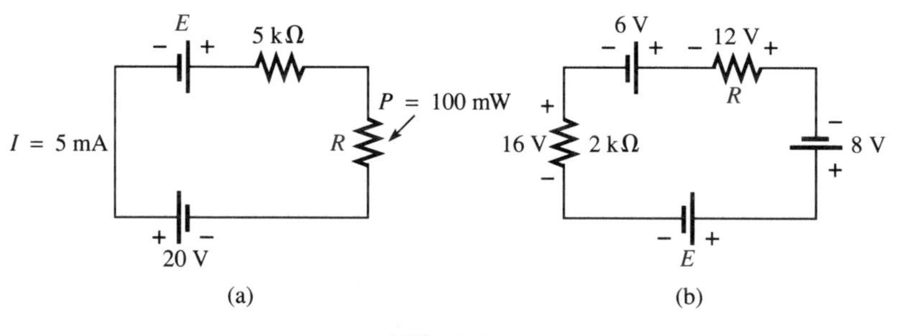

FIG. 5.65

## SECTIONS 5.4 AND 5.5 Kirchhoff's Voltage Law

7. Find $V_{ab}$ with polarity for the circuits of Fig. 5.66. Each box can contain a load or a power supply, or a combination of both.

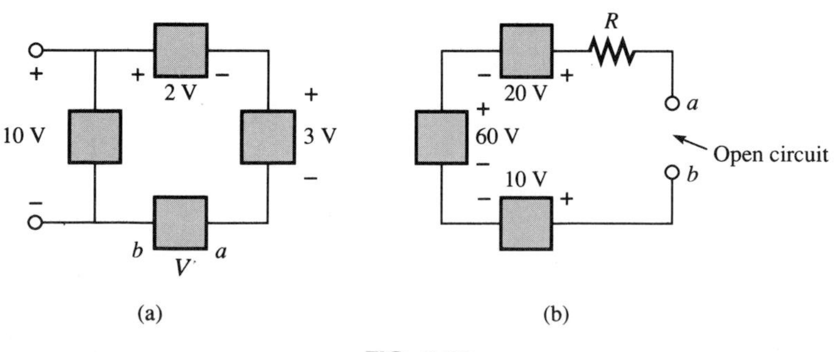


FIG. 5.66

8. Although the networks of Fig. 5.67 are not simply series circuits, determine the unknown voltages using Kirchhoff's voltage law.

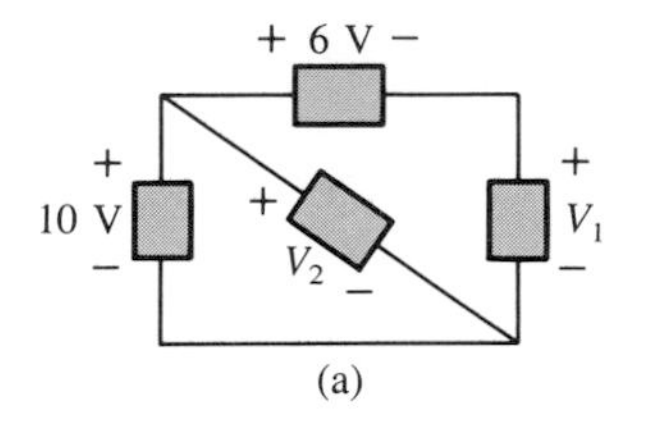


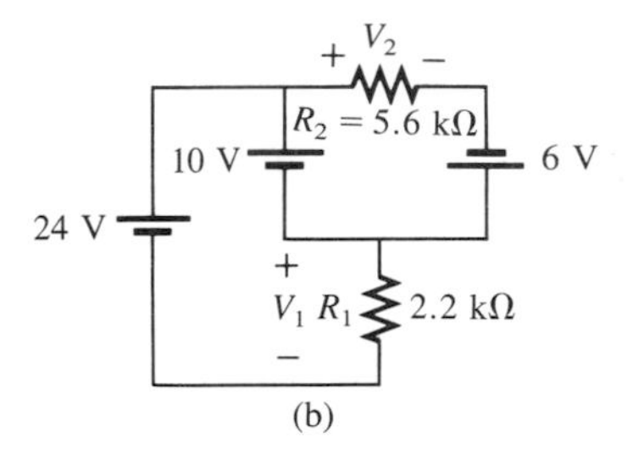


FIG. 5.67

9. Determine the current $I$ and the voltage $V_1$ for the network of Fig. 5.68.

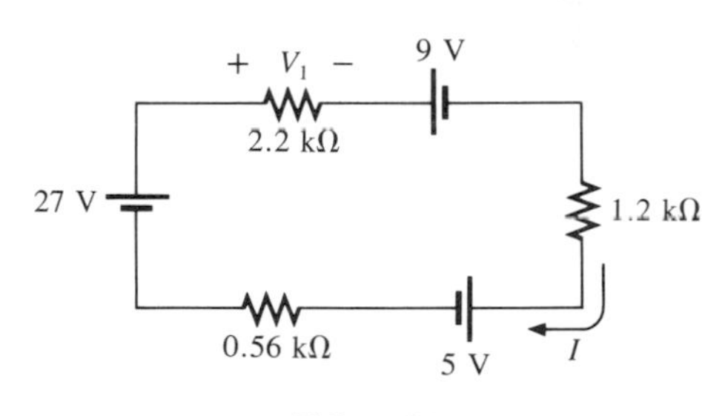


FIG. 5.68

10. For the circuit of Fig. 5.69.
   a. Find the total resistance, current, and unknown voltage drops.
   b. Verify Kirchhoff's voltage law around the closed loop.
   c. Find the power dissipated by each resistor, and note whether the power delivered is equal to the power dissipated.
   d. If the resistors are available with wattage ratings of 1/2, 1, and 2 W, what minimum wattage rating can be used for each resistor in this circuit?

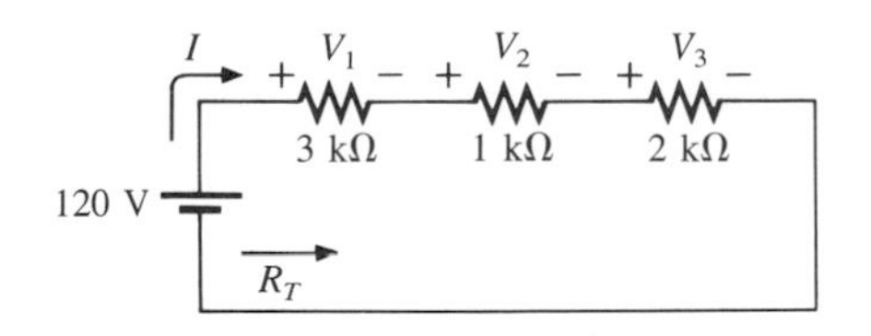


FIG. 5.69

11. Repeat Problem 9 for the circuit of Fig. 5.70.

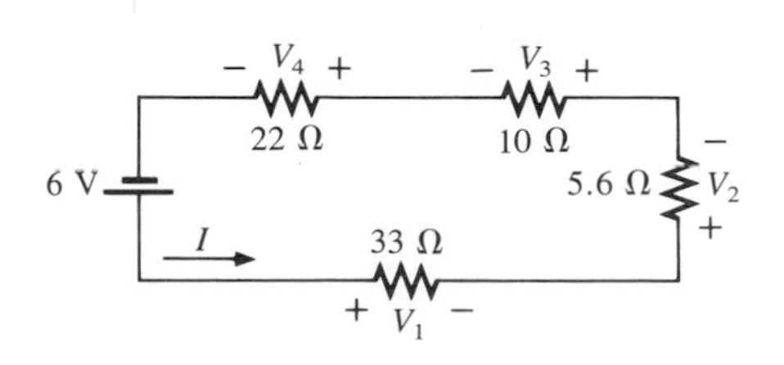

FIG. 5.70

*12. Find the unknown quantities in the circuits of Fig. 5.71 using the information provided.

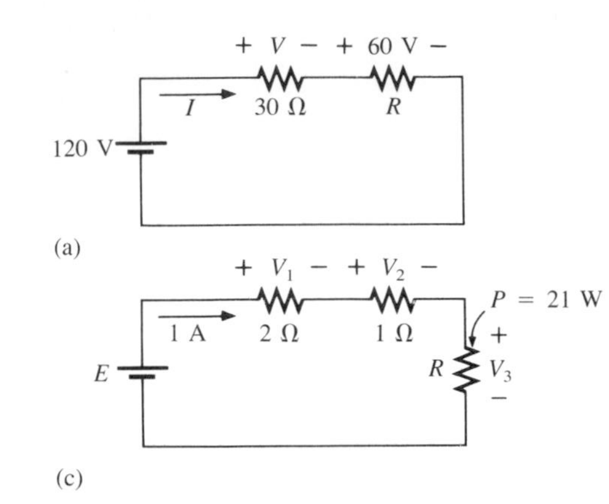


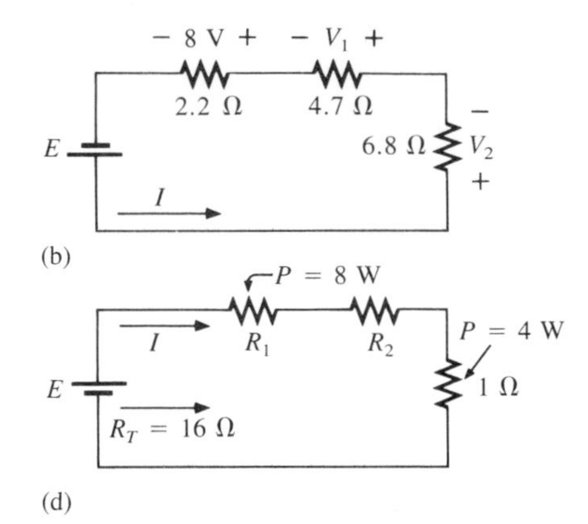

FIG. 5.71

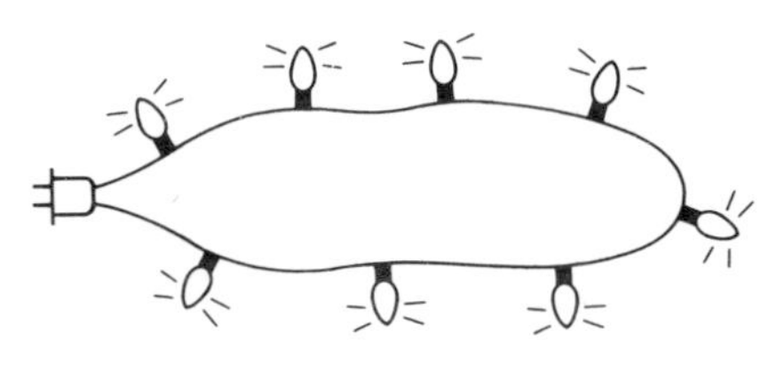

FIG. 5.72

13. There are eight Christmas tree lights connected in series as shown in Fig. 5.72.
   a. If the set is connected to a 120-V source, what is the current through the bulbs if each bulb has an internal resistance of $28\frac{1}{8}\ \Omega$?
   b. Determine the power delivered to each bulb.
   c. Calculate the voltage drop across each bulb.
   d. If one bulb burns out (that is, the filament opens), what is the effect on the remaining bulbs?

## SECTION 5.6 Voltage Divider Rule

14. Using the voltage divider rule, find $V_{ab}$ (with polarity) for the circuits of Fig. 5.73.

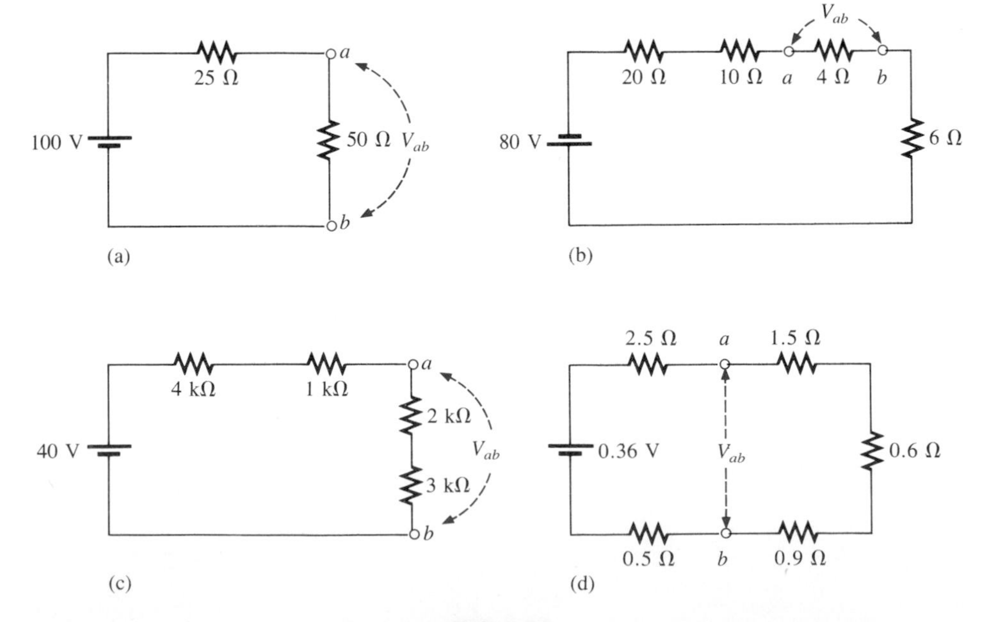

FIG. 5.73

15. Find the unknown quantities using the voltage divider rule and the information provided for the circuits of Fig. 5.74.

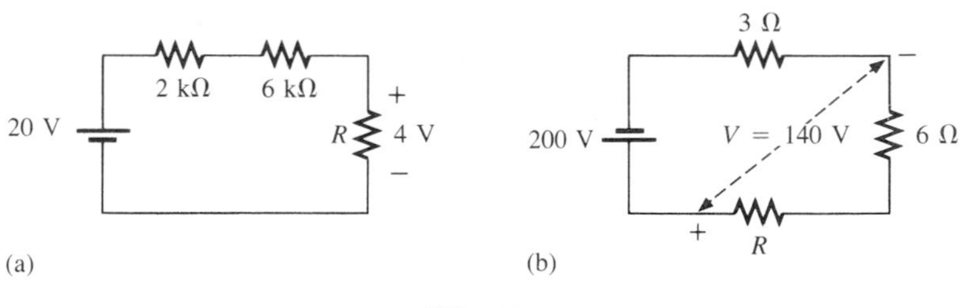


FIG. 5.74

16. Design the voltage divider of Fig. 5.75 such that $V_{R_1} = (1/5)V_{R_2}$ if $I = 4$ mA.

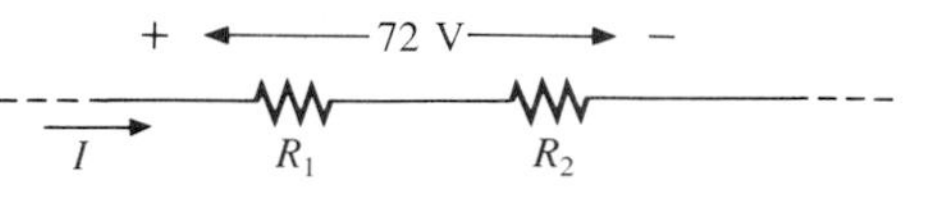


FIG. 5.75

17. Find the voltage across each resistor of Fig. 5.76 if $R_1 = 2R_3$ and $R_2 = 7R_3$.

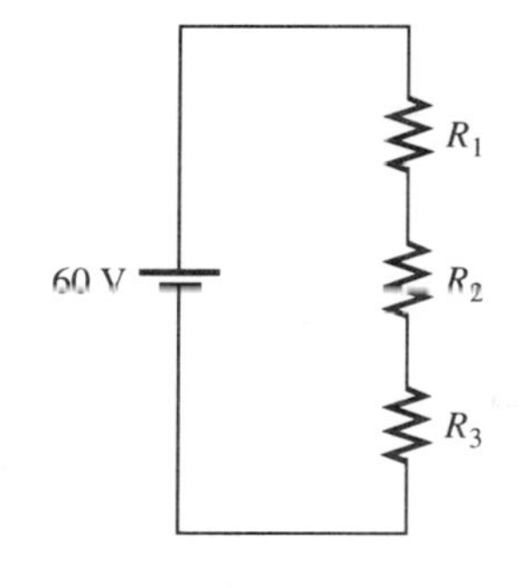

FIG. 5.76

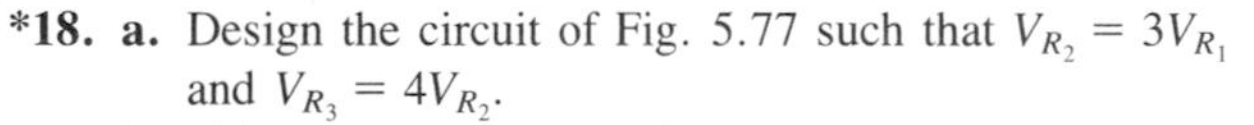

*18. a. Design the circuit of Fig. 5.77 such that $V_{R_2} = 3V_{R_1}$ and $V_{R_3} = 4V_{R_2}$.
  b. If the current $I$ is reduced to 10 $\mu$A, what are the new values of $R_1$, $R_2$, and $R_3$? How do they compare to the results of part a?

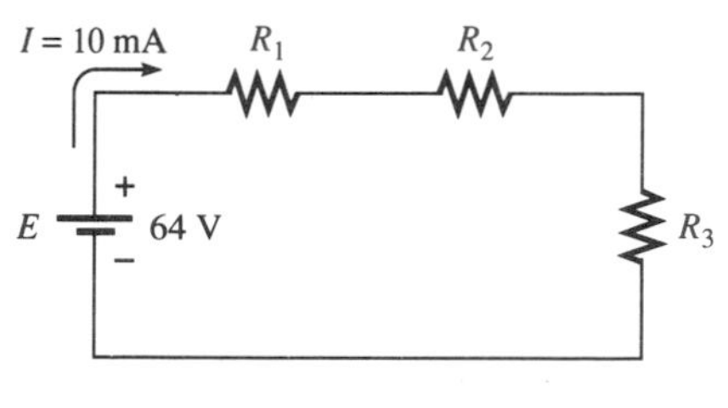

FIG. 5.77

## SECTION 5.7 Notation

19. a. Determine the voltages $V_a$ and $V_b$ for the networks of Fig. 5.78.
  b. Determine the voltage $V_{ab}$ for each network of Fig. 5.78.

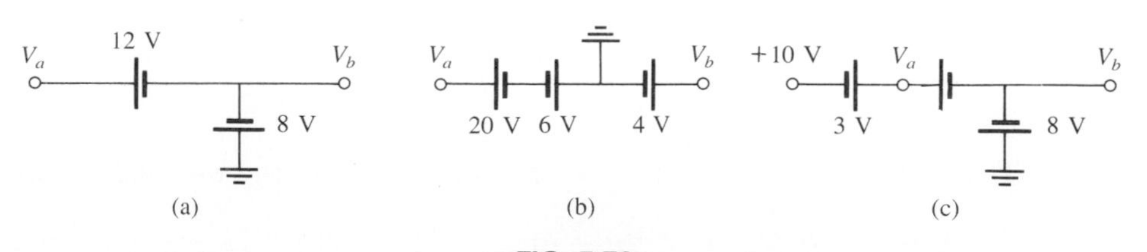


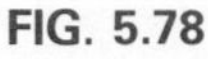
FIG. 5.78

**20.** Determine the voltages $V_a$ and $V_1$ for the networks of Fig. 5.79.

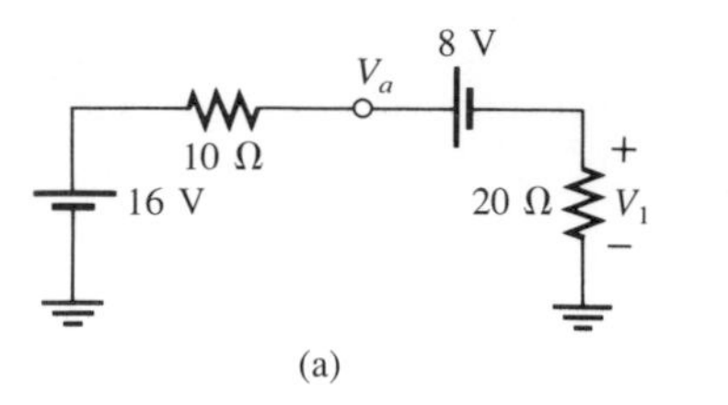


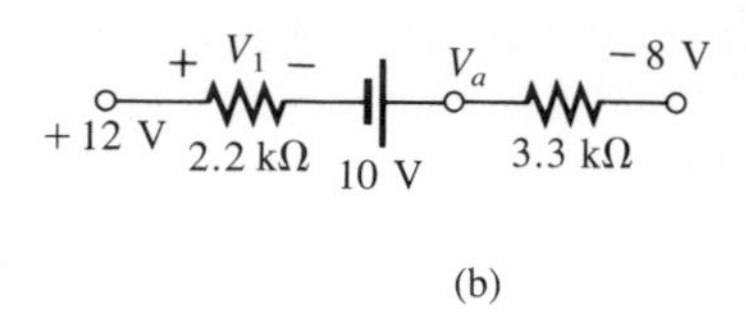

FIG. 5.79

***21.** For the network of Fig. 5.80, determine the voltages:
- **a.** $V_a$, $V_b$, $V_c$, $V_d$, $V_e$
- **b.** $V_{ab}$, $V_{dc}$, $V_{cb}$
- **c.** $V_{ac}$, $V_{db}$

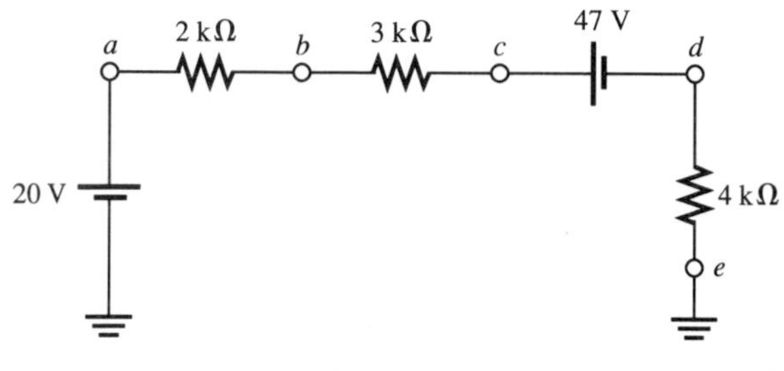


FIG. 5.80

***22.** For the network of Fig. 5.81, determine the voltages:
- **a.** $V_a$, $V_b$, $V_c$, $V_d$
- **b.** $V_{ab}$, $V_{cb}$, $V_{cd}$
- **c.** $V_{ad}$, $V_{ca}$

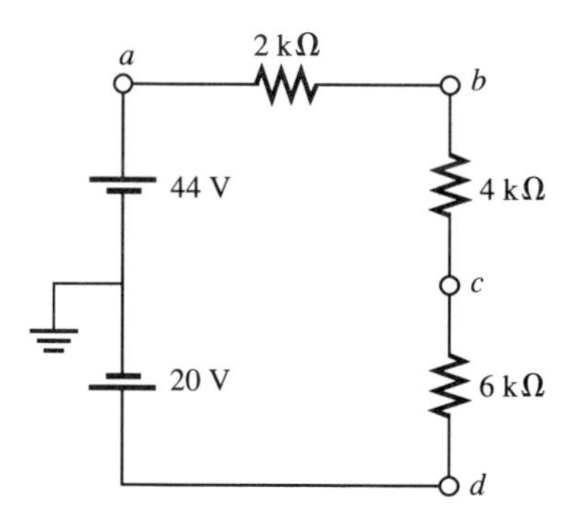


FIG. 5.81

## SECTION 5.8 Internal Resistance of Voltage Sources

**23.** Find the internal resistance of a battery that has a no-load output voltage of 60 V and supplies a current of 2 A to a load of 28 Ω.

**24.** Find the voltage $V_L$ and the power loss in the internal resistance for the configuration of Fig. 5.82.

**25.** Find the internal resistance of a battery that has a no-load output voltage of 6 V and supplies a current of 10 mA to a load of 1/2 kΩ.

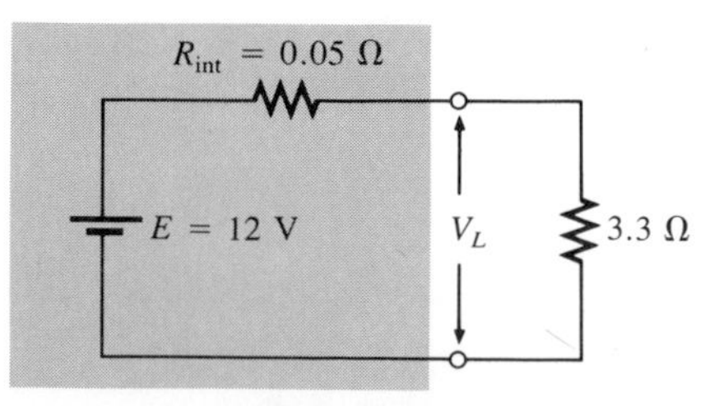


FIG. 5.82

## SECTION 5.9 Voltage Regulation

**26.** Determine the voltage regulation for the battery of Problem 23.

**27.** Calculate the voltage regulation for the supply of Fig. 5.82.

## SECTION 5.11 Computer Analysis

### PSPICE

**28.** Write an input file to provide the voltages and current for the network of Fig. 5.69.

**29.** Write an input file to solve for the voltages across the resistors of Fig. 5.81.

**30.** Write an input file that will only print out the current for the network of Fig. 5.60(c).

### BASIC

**31.** Write a program to determine the total resistance of any number of resistors in series.

**32.** Write a program that will apply the voltage divider rule to either resistor of a series circuit with a single source and two series resistors.

**33.** Write a program to tabulate the current and power to the resistor $R_L$ of the network of Fig. 5.83 for a range of values for $R_L$ from 1 Ω to 20 Ω. Print out the value of $R_L$ that results in maximum power to $R_L$.

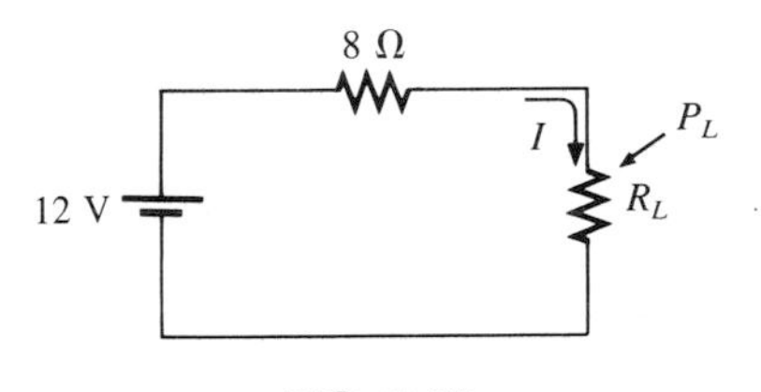


FIG. 5.83

## GLOSSARY

**Branch** The portion of a circuit consisting of one or more elements in series.

**Circuit** A combination of a number of elements joined at terminal points providing at least one closed path through which charge can flow.

**Closed loop** Any continuous connection of branches that allows tracing of a path that leaves a point in one direction and returns to that same point from another direction without leaving the circuit.

**Conventional current flow** A defined direction for the flow of charge in an electrical system that is opposite to that of the motion of electrons.

**Electron flow** The flow of charge in an electrical system having the same direction as the motion of electrons.

**Internal resistance** The inherent resistance found internal to any source of energy.

**Kirchhoff's voltage law** Law which states that the algebraic sum of the potential rises and drops around a closed loop (or path) is zero.

**Series circuit** A circuit configuration in which the elements have only one point in common and each terminal is not connected to a third, current-carrying element.

**Voltage divider rule** A method by which a voltage in a series circuit can be determined without first calculating the current in the circuit.

**Voltage regulation** (*VR*) A value, given as a percent, that provides an indication of the change in terminal voltage of a supply with change in load demand.

# 6

# Parallel Circuits

## 6.1 INTRODUCTION

There are two network configurations that form the framework for some of the most complex network structures. A clear understanding of each will pay enormous dividends as more complex methods and networks are examined. The series connection was discussed in detail in the last chapter. We will now examine the *parallel* connection and all the methods and laws associated with this important configuration.

## 6.2 PARALLEL ELEMENTS

***Two elements, branches, or networks are in parallel if they have two points in common.***

In Fig. 6.1, for example, elements 1 and 2 have terminals $a$ and $b$ in common; they are therefore in parallel.

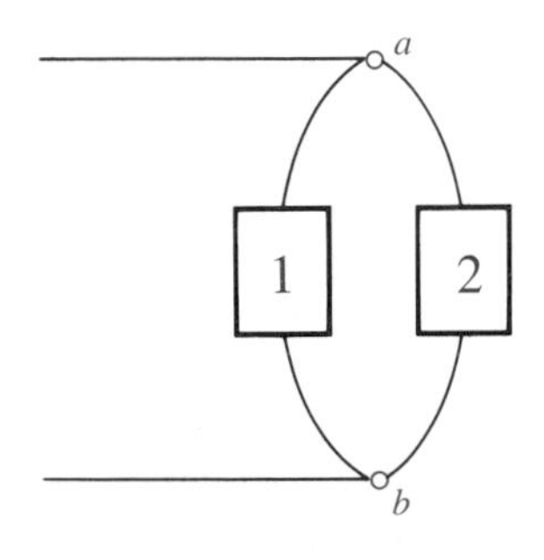


**FIG. 6.1**
*Parallel elements.*

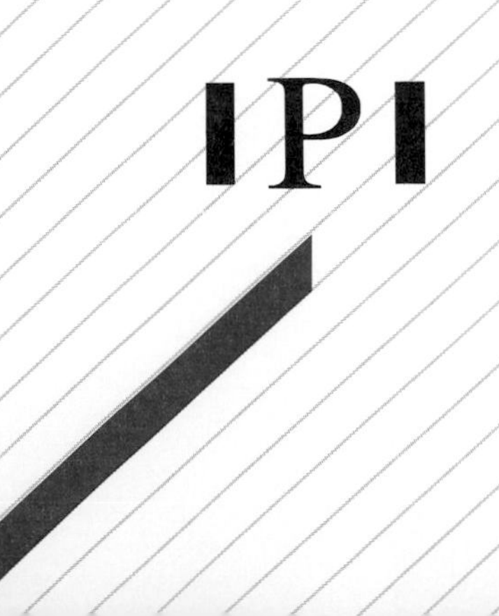

**IPI**

In Fig. 6.2, all the elements are in parallel because they satisfy the above criterion. Three configurations are provided to demonstrate how the parallel networks can be drawn. Do not let the squaring of the connection at the top and bottom of Figs. 6.2(a) and (b) cloud the fact that all the elements are connected to one terminal point at top and bottom as shown in Fig. 6.2(c).

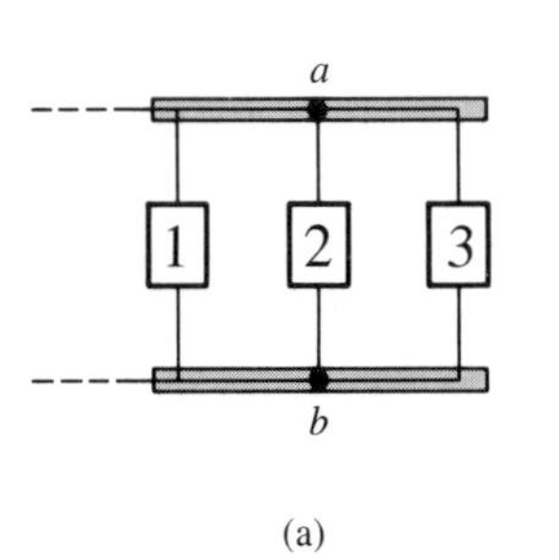


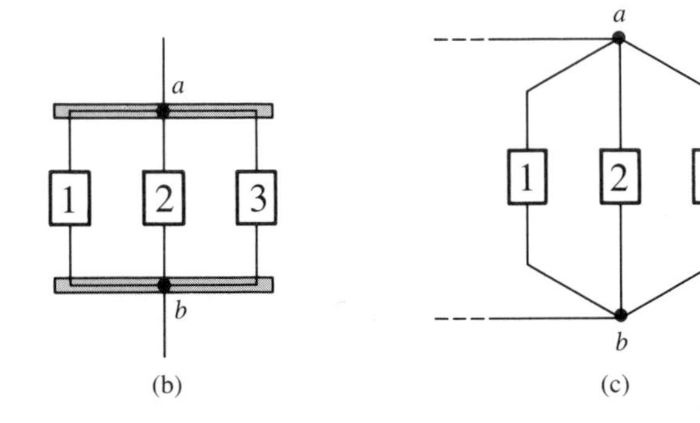


**FIG. 6.2**
*Parallel configurations.*

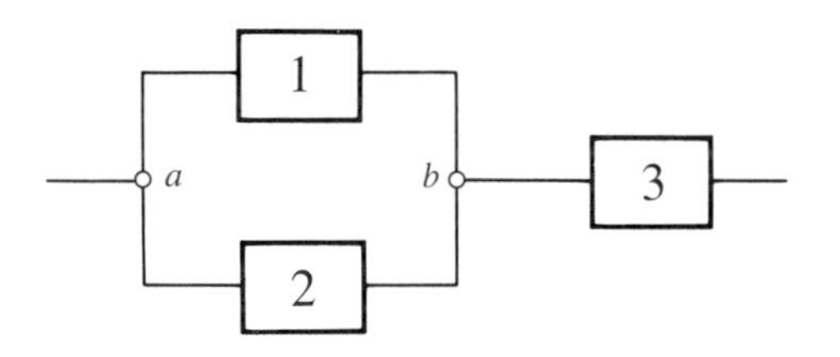


**FIG. 6.3**

In Fig. 6.3, elements 1 and 2 are in parallel because they have terminals $a$ and $b$ in common. The parallel combination of 1 and 2 is then in series with element 3 due to the common terminal point $b$.

In Fig. 6.4, elements 1 and 2 are in series due to the common point $a$, but the series combination of 1 and 2 is in parallel with element 3 as defined by the common terminal connections at $b$ and $c$.

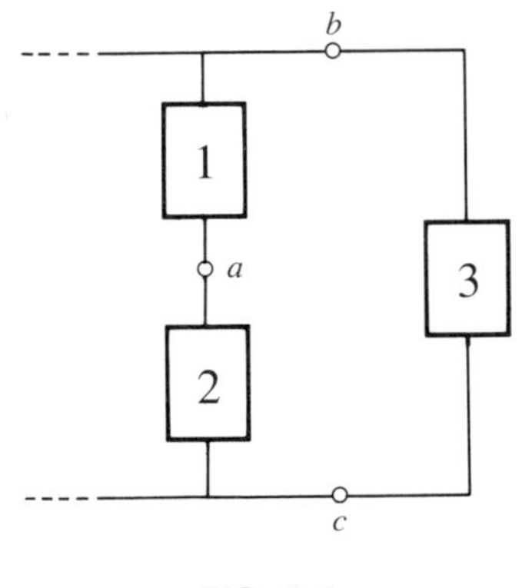

**FIG. 6.4**

In Figs. 6.1 through 6.4, the numbered boxes were used as a general symbol representing either single resistive elements, batteries, or complex network configurations.

Common examples of parallel elements include the rungs of a ladder, the tying of more than one rope between two points to increase the strength of the connection, and the use of pipes between two points to split the water between the two points at a ratio determined by the area of the pipes.

## 6.3 TOTAL CONDUCTANCE AND RESISTANCE

For series resistors, the total resistance is the sum of the resistor values.

***For parallel elements, the total conductance is the sum of the individual conductances.***

That is, for the parallel network of Fig. 6.5, we write

$$G_T = G_1 + G_2 + G_3 + \cdots + G_N \tag{6.1}$$

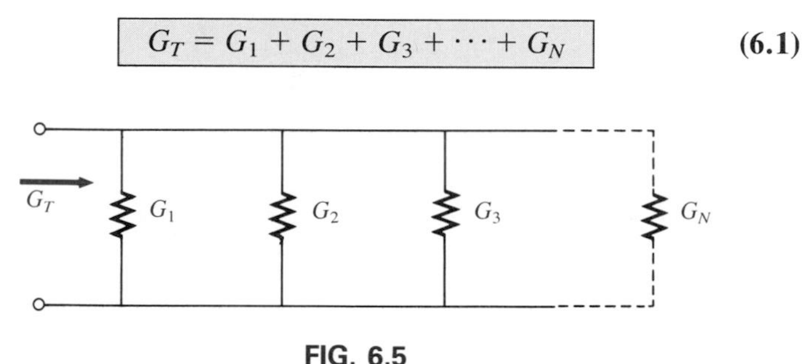


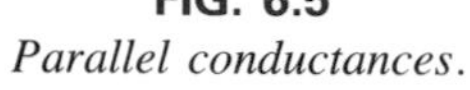

**FIG. 6.5**
*Parallel conductances.*

Since increasing levels of conductance will establish higher current levels, the more terms appearing in Eq. (6.1), the higher the input current level. In other words, as the number of resistors in parallel increases, the input current level will increase for the same applied voltage the opposite effect of increasing the number of resistors in series.

Substituting resistor values for the network of Fig. 6.5 will result in the network of Fig. 6.6. Since $G = 1/R$, the total resistance for the network can be determined by direct substitution into Eq. (6.1):

$$\frac{1}{R_T} = \frac{1}{R_1} + \frac{1}{R_2} + \frac{1}{R_3} + \cdots + \frac{1}{R_N} \tag{6.2}$$

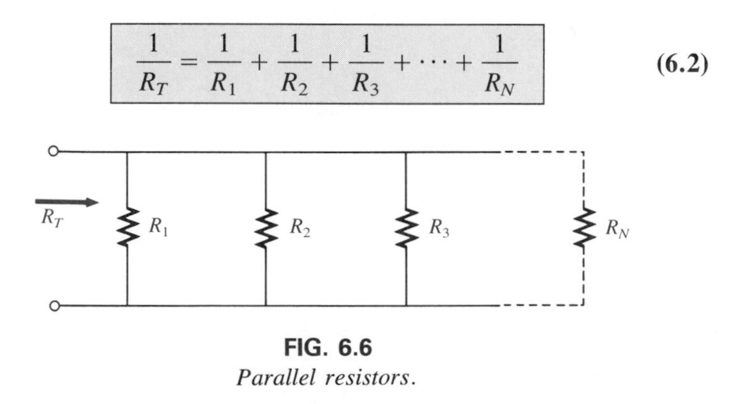


**FIG. 6.6**
*Parallel resistors.*

Note that the equation is for 1 divided by the total resistance rather than the total resistance. Once the sum of the terms to the right of the equal sign has been determined, it will then be necessary to divide the result into 1 to determine the total resistance. The following examples will demonstrate the additional calculations introduced by the inverse relationship.

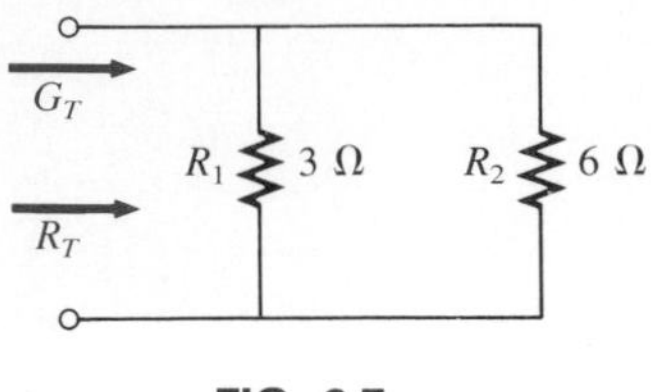


FIG. 6.7

**EXAMPLE 6.1.** Determine the total conductance and resistance for the parallel network of Fig. 6.7.

***Solution:***

$$G_T = G_1 + G_2 = \frac{1}{3\ \Omega} + \frac{1}{6\ \Omega} = 0.333\ \text{S} + 0.167\ \text{S} = \mathbf{0.5\ S}$$

and

$$R_T = \frac{1}{G_T} = \frac{1}{0.5\ \text{S}} = \mathbf{2\ \Omega}$$

**EXAMPLE 6.2.** Determine the effect on the total conductance and resistance of the network of Fig. 6.7 if another resistor of 10 Ω were added in parallel with the other elements.

***Solution:***

$$G_T = 0.5\ \text{S} + \frac{1}{10\ \Omega} = 0.5\ \text{S} + 0.1\ \text{S} = \mathbf{0.6\ S}$$

$$R_T = \frac{1}{G_T} = \frac{1}{0.6\ \text{S}} \cong \mathbf{1.67\ \Omega}$$

Note, as mentioned above, that adding additional terms increases the conductance level and decreases the resistance level.

**EXAMPLE 6.3.** Determine the total resistance for the network of Fig. 6.8.

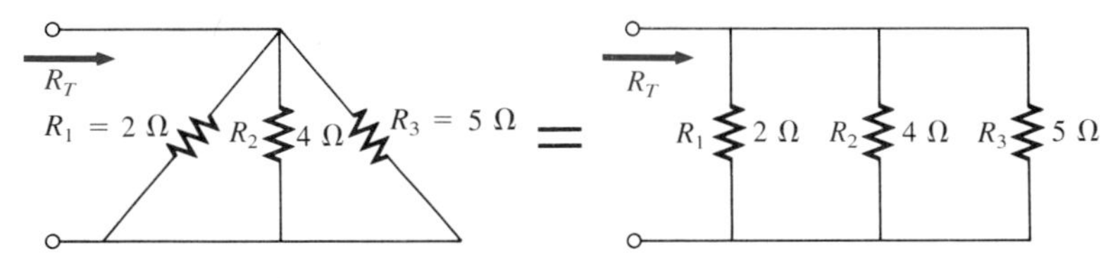


FIG. 6.8

***Solution:***

$$\frac{1}{R_T} = \frac{1}{R_1} + \frac{1}{R_2} + \frac{1}{R_3}$$

$$= \frac{1}{2\ \Omega} + \frac{1}{4\ \Omega} + \frac{1}{5\ \Omega} = 0.5\ \text{S} + 0.25\ \text{S} + 0.2\ \text{S}$$

$$= 0.95\ \text{S}$$

and

$$R_T = \frac{1}{0.95\ \text{S}} = \mathbf{1.053\ \Omega}$$

The above examples demonstrate an interesting and useful (for checking purposes) characteristic of parallel resistors:

***The total resistance of parallel resistors is always less than the value of the smallest resistor.***

In addition, the wider the spread in numerical value between two parallel resistors, the closer the total resistance will be to the smaller resistor. For instance, the total resistance of 3 Ω in parallel with 6 Ω is 2 Ω, as demonstrated in Example 6.1. However, the total resistance of 3 Ω in parallel with 60 Ω is 2.85 Ω, which is much closer to the value of the smaller resistor.

For *equal* resistors in parallel, the equation becomes significantly easier to apply. For $N$ equal resistors in parallel, Eq. (6.2) becomes

$$\frac{1}{R_T} = \underbrace{\frac{1}{R} + \frac{1}{R} + \frac{1}{R} + \cdots + \frac{1}{R}}_{N}$$

$$= N\left(\frac{1}{R}\right)$$

and

$$\boxed{R_T = \frac{R}{N}} \qquad \textbf{(6.3)}$$

In other words, the total resistance of $N$ equal parallel resistors is the resistance of *one* resistor divided by the number ($N$) of parallel elements.

For conductance levels, we have

$$\boxed{G_T = NG} \qquad \textbf{(6.4)}$$

**EXAMPLE 6.4.**

a. Find the total resistance of the network of Fig. 6.9.
b. Calculate the total resistance for the network of Fig. 6.10.

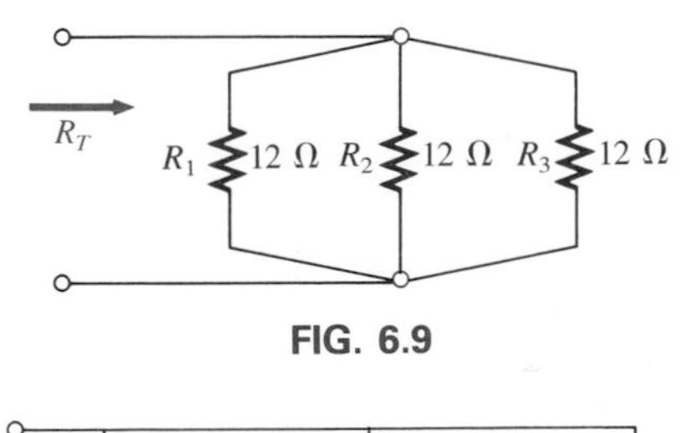


**FIG. 6.9**

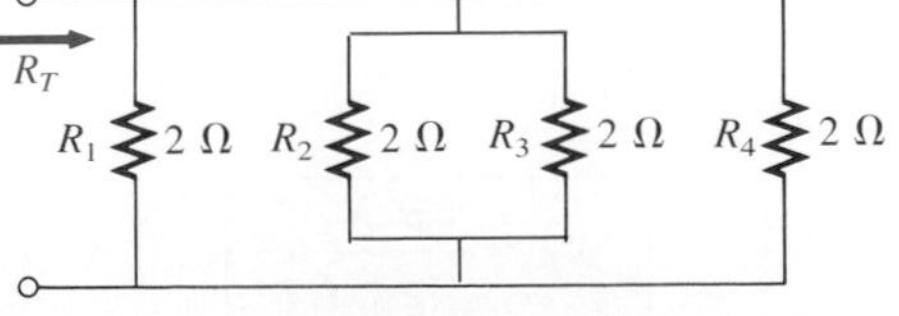


**FIG. 6.10**

***Solutions:***

a. Fig. 6.9 redrawn in Fig. 6.11:

$$R_T = \frac{R}{N} = \frac{12\ \Omega}{3} = \mathbf{4\ \Omega}$$

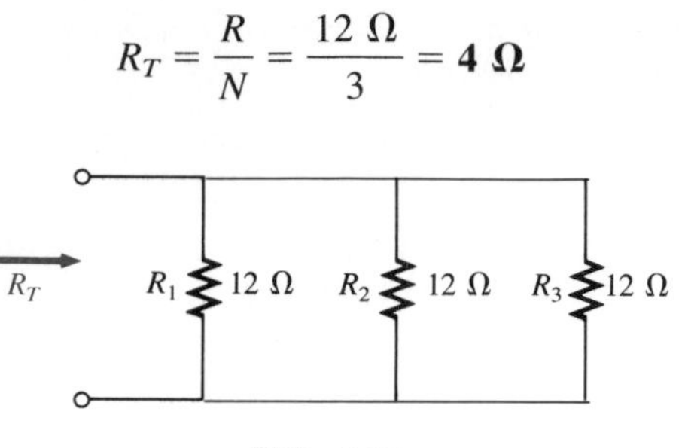

**FIG. 6.11**

b. Fig. 6.10 redrawn in Fig. 6.12:

$$R_T = \frac{R}{N} = \frac{2\ \Omega}{4} = \mathbf{0.5\ \Omega}$$

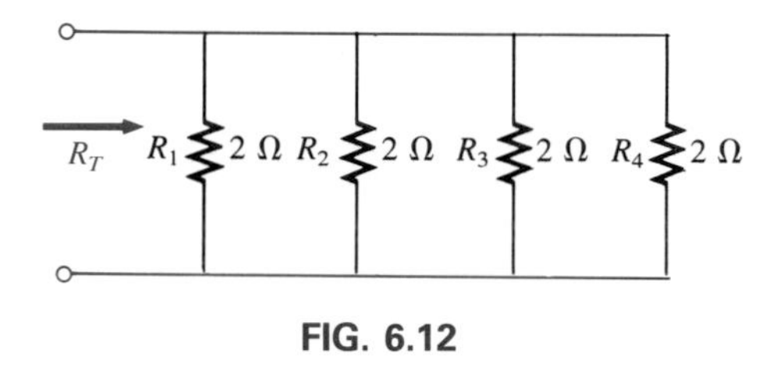


**FIG. 6.12**

---

In the vast majority of situations, only two or three parallel resistive elements need be combined. With this in mind, the following equations were developed to reduce the negative effects of the inverse relationship when determining $R_T$.

For two parallel resistors, we write

$$\frac{1}{R_T} = \frac{1}{R_1} + \frac{1}{R_2}$$

Multiplying the top and bottom of each term of the right side of the equation by the other resistor will result in

$$\frac{1}{R_T} = \left(\frac{R_2}{R_2}\right)\frac{1}{R_1} + \left(\frac{R_1}{R_1}\right)\frac{1}{R_2} = \frac{R_2}{R_1R_2} + \frac{R_1}{R_1R_2}$$

$$= \frac{R_2 + R_1}{R_1R_2}$$

and

$$\boxed{R_T = \frac{R_1R_2}{R_1 + R_2}} \qquad \textbf{(6.5)}$$

In words,

***the total resistance of two parallel resistors is the product of the two divided by their sum.***

For three parallel resistors, the equation becomes

$$R_T = \frac{R_1R_2R_3}{R_1R_2 + R_1R_3 + R_2R_3} \quad \textbf{(6.6)}$$

with the denominator showing all the possible product combinations of the resistors taken two at a time.

---

**EXAMPLE 6.5.** Repeat Example 6.1 using Eq. (6.5).

***Solution:***

$$R_T = \frac{R_1R_2}{R_1 + R_2} = \frac{(3\ \Omega)(6\ \Omega)}{3\ \Omega + 6\ \Omega} = \frac{18\ \Omega}{9} = \mathbf{2\ \Omega}$$

---

**EXAMPLE 6.6.** Repeat Example 6.3 using Eq. (6.6).

***Solution:***

$$\begin{aligned} R_T &= \frac{R_1R_2R_3}{R_1R_2 + R_1R_3 + R_2R_3} \\ &= \frac{(2\ \Omega)(4\ \Omega)(5\ \Omega)}{(2\ \Omega)(4\ \Omega) + (2\ \Omega)(5\ \Omega) + (4\ \Omega)(5\ \Omega)} \\ &= \frac{40\ \Omega}{8 + 10 + 20} = \frac{40\ \Omega}{38} = \mathbf{1.053\ \Omega} \end{aligned}$$

---

Recall that series elements can be interchanged without affecting the magnitude of the total resistance or current. In parallel networks,

***parallel elements can be interchanged without changing the total resistance or input current.***

Note in the next example how redrawing the network can often clarify which operations and equations should be applied.

---

**EXAMPLE 6.7.** Calculate the total resistance of the parallel network of Fig. 6.13.

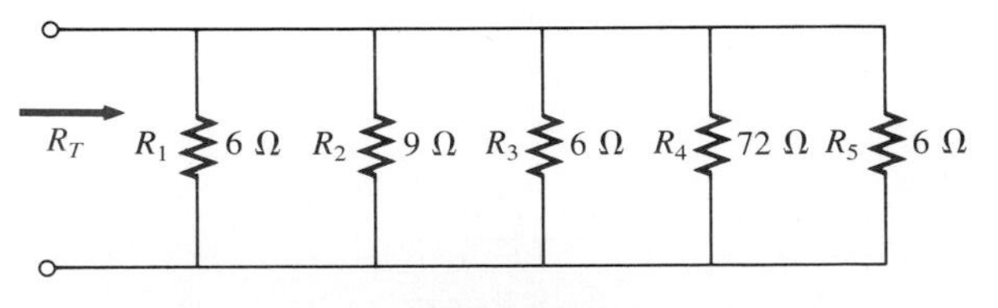


**FIG. 6.13**

**Solution:** Network redrawn in Fig. 6.14:

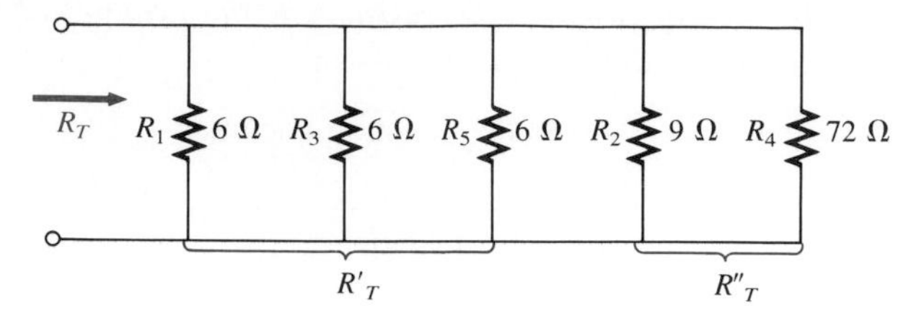

FIG. 6.14

$$R'_T = \frac{R}{N} = \frac{6\ \Omega}{3} = 2\ \Omega$$

$$R''_T = \frac{R_1R_2}{R_1 + R_2} = \frac{(9\ \Omega)(72\ \Omega)}{9\ \Omega + 72\ \Omega} = \frac{648\ \Omega}{81} = 8\ \Omega$$

and

$$R_T = R'_T \parallel R''_T$$

(in parallel with)

$$R_T = \frac{R'_TR''_T}{R'_T + R''_T} = \frac{(2\ \Omega)(8\ \Omega)}{2\ \Omega + 8\ \Omega} = \frac{16\ \Omega}{10} = \mathbf{1.6\ \Omega}$$

The preceding examples show direct substitution, where once the proper equation is defined it is only a matter of plugging in the numbers and performing the required algebraic maneuvers. The next two examples have a design orientation, where specific network parameters are defined and the circuit elements have to be determined.

**EXAMPLE 6.8.** Determine the value of $R_2$ in Fig. 6.15 to establish a total resistance of 9 kΩ.

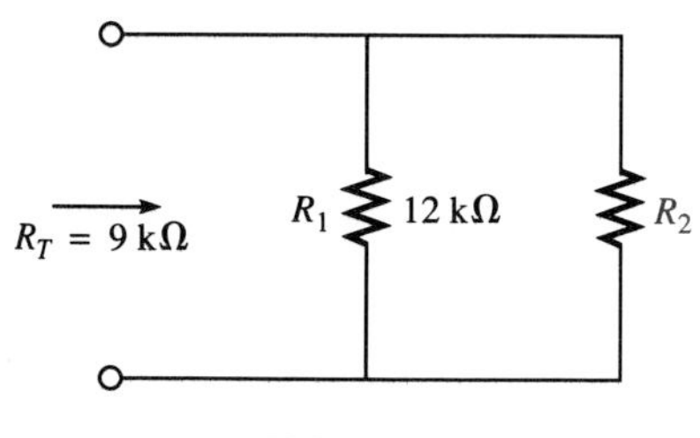


FIG. 6.15

**Solution:**

$$R_T = \frac{R_1R_2}{R_1 + R_2}$$

$$R_T(R_1 + R_2) = R_1R_2$$

$$R_TR_1 + R_TR_2 = R_1R_2$$

$$R_TR_1 = R_1R_2 - R_TR_2$$

$$R_TR_1 = (R_1 - R_T)R_2$$

and

$$\boxed{R_2 = \frac{R_TR_1}{R_1 - R_T}} \qquad \textbf{(6.7)}$$

Substituting values:

$$R_2 = \frac{(9\ \text{k}\Omega)(12\ \text{k}\Omega)}{12\ \text{k}\Omega - 9\ \text{k}\Omega}$$

$$= \frac{108\ \text{k}\Omega}{3} = \mathbf{36\ k\Omega}$$

---

**EXAMPLE 6.9.** Determine the values of $R_1$, $R_2$, and $R_3$ in Fig. 6.16 if $R_1 = 2R_2$ and $R_3 = 2R_2$ and the total resistance is 16 kΩ.

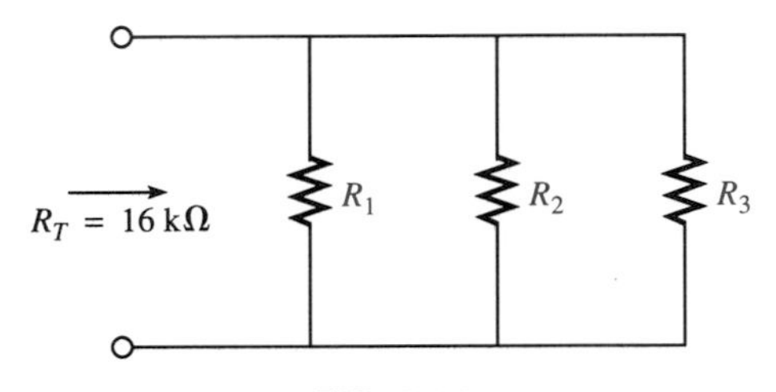

FIG. 6.16

***Solution:***

$$\frac{1}{R_T} = \frac{1}{R_1} + \frac{1}{R_2} + \frac{1}{R_3}$$

$$\frac{1}{16\ \text{k}\Omega} = \frac{1}{R_1} + \frac{1}{2R_1} + \frac{1}{4R_1}$$

since

$$R_3 = 2R_2 = 2(2R_1) = 4R_1$$

and

$$\frac{1}{16\ \text{k}\Omega} = \frac{1}{R_1} + \frac{1}{2}\left(\frac{1}{R_1}\right) + \frac{1}{4}\left(\frac{1}{R_1}\right)$$

$$\frac{1}{16\ \text{k}\Omega} = 1.75\left(\frac{1}{R_1}\right)$$

with

$$R_1 = 1.75(16\ \text{k}\Omega) = \mathbf{28\ k\Omega}$$

---

## 6.4 PARALLEL NETWORKS

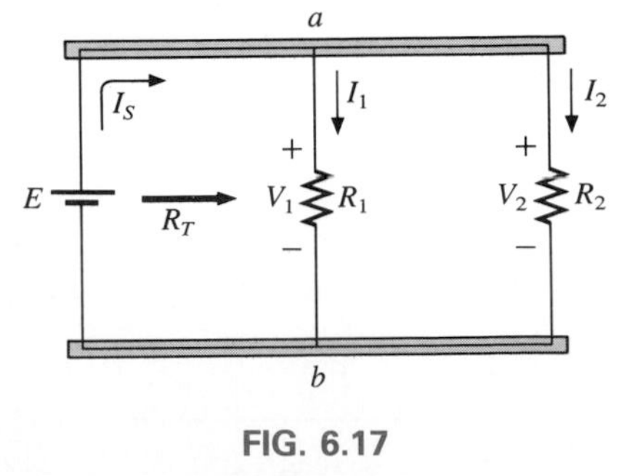


FIG. 6.17

The network of Fig. 6.17 is the simplest of parallel networks. All the elements have terminals $a$ and $b$ in common. The total resistance is determined by $R_T = R_1R_2/(R_1 + R_2)$, and the source current by

$I_s = E/R_T$. Since the terminals of the battery are connected directly across the resistors $R_1$ and $R_2$, the following should be fairly obvious:

***The voltage across parallel elements is the same.***

Using this fact will result in

$$V_1 = V_2 = E$$

and

$$I_1 = \frac{V_1}{R_1} = \frac{E}{R_1}$$

with

$$I_2 = \frac{V_2}{R_2} = \frac{E}{R_2}$$

If we take the equation for the total resistance and multiply both sides by the applied voltage, we obtain

$$E\left(\frac{1}{R_T}\right) = E\left(\frac{1}{R_1} + \frac{1}{R_2}\right)$$

and

$$\frac{E}{R_T} = \frac{E}{R_1} + \frac{E}{R_2}$$

Substituting the Ohm's law relationships appearing above, we find that the source current

$$I_s = I_1 + I_2$$

permitting the following conclusion:

***For single-source parallel networks, the source current is equal to the sum of the individual branch currents.***

The power dissipated by the resistors and delivered by the source can be determined from

$$P_1 = V_1 I_1 = I_1^2 R_1 = \frac{V_1^2}{R_1}$$

$$P_2 = V_2 I_2 = I_2^2 R_2 = \frac{V_2^2}{R_2}$$

$$P_s = E I_T = I_T^2 R_T = \frac{E^2}{R_T}$$

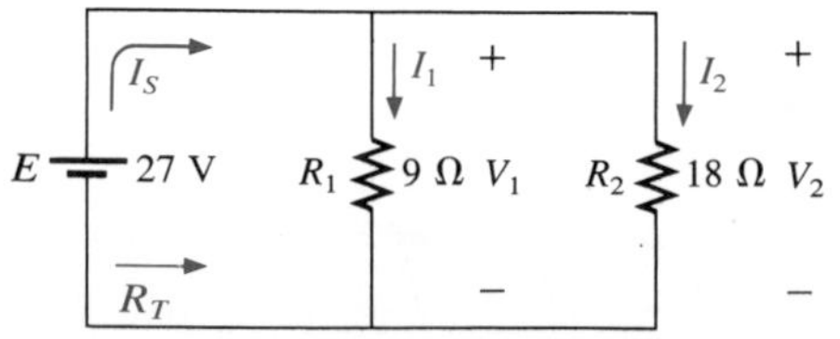

FIG. 6.18

**EXAMPLE 6.10.** For the parallel network of Fig. 6.18:

a. Calculate $R_T$.
b. Determine $I_s$.
c. Calculate $I_1$ and $I_2$ and demonstrate that $I_s = I_1 + I_2$.
d. Determine the power to each resistive load.
e. Determine the power delivered by the source and compare it to the total power dissipated by the resistive elements.

***Solutions:***

a. $R_T = \dfrac{R_1R_2}{R_1 + R_2} = \dfrac{(9\ \Omega)(18\ \Omega)}{9\ \Omega + 18\ \Omega} = \dfrac{162\ \Omega}{27} = \mathbf{6\ \Omega}$

b. $I_s = \dfrac{E}{R_T} = \dfrac{27\ \text{V}}{6\ \Omega} = \mathbf{4.5\ A}$

c.
$$I_1 = \frac{V_1}{R_1} = \frac{E}{R_1} = \frac{27\ \text{V}}{9\ \Omega} = 3\ \text{A}$$
$$I_2 = \frac{V_2}{R_2} = \frac{E}{R_2} = \frac{27\ \text{V}}{18\ \Omega} = 1.5\ \text{A}$$
$$I_s = I_1 + I_2$$
$$4.5\ \text{A} = 3\ \text{A} + 1.5\ \text{A}$$
$$4.5\ \text{A} = 4.5\ \text{A} \quad \text{(checks)}$$

d. $P_1 = V_1I_1 = EI_1 = (27\ \text{V})(3\ \text{A}) = \mathbf{81\ W}$
$P_2 = V_2I_2 = EI_2 = (27\ \text{V})(1.5\ \text{A}) = \mathbf{40.5\ W}$

e. $P_s = EI_s = (27\ \text{V})(4.5\ \text{A}) = \mathbf{121.5\ W}$
$P_s = P_1 + P_2 = 81\ \text{W} + 40.5\ \text{W} = \mathbf{121.5\ W}$

---

**EXAMPLE 6.11.** Given the information provided in Fig. 6.19:

a. Determine $R_3$.
b. Calculate $E$.
c. Find $I_s$.
d. Find $I_2$.
e. Determine $P_2$.

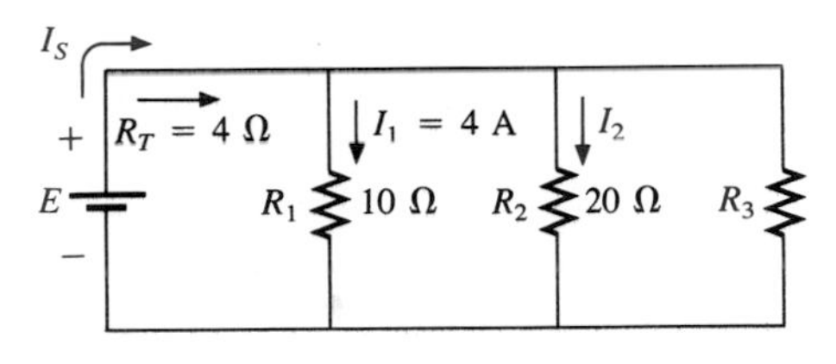


**FIG. 6.19**

***Solutions:***

a.
$$\frac{1}{R_T} = \frac{1}{R_1} + \frac{1}{R_2} + \frac{1}{R_3}$$
$$\frac{1}{4\ \Omega} = \frac{1}{10\ \Omega} + \frac{1}{20\ \Omega} + \frac{1}{R_3}$$
$$0.25\ \text{S} = 0.1\ \text{S} + 0.05\ \text{S} + \frac{1}{R_3}$$
$$0.25\ \text{S} = 0.15\ \text{S} + \frac{1}{R_3}$$
$$\frac{1}{R_3} = 0.1\ \text{S}$$
$$R_3 = \frac{1}{0.1\ \text{S}} = \mathbf{10\ \Omega}$$

b. $E = V_1 = I_1R_1 = (4\ \text{A})(10\ \Omega) = \mathbf{40\ V}$

c. $I_s = \dfrac{E}{R_T} = \dfrac{40\ \text{V}}{4\ \Omega} = \mathbf{10\ A}$

d. $I_2 = \dfrac{V_2}{R_2} = \dfrac{E}{R_2} = \dfrac{40\ \text{V}}{20\ \Omega} = \mathbf{2\ A}$

e. $P_2 = I_2^2R_2 = (2\ \text{A})^2(20\ \Omega) = \mathbf{80\ W}$

## 6.5 KIRCHHOFF'S CURRENT LAW

Kirchhoff's voltage law provides an important relationship between voltage levels around any closed loop of a network. We now consider Kirchhoff's current law, which provides an equally important relationship between current levels at any junction.

***Kirchhoff's current law (KCL) states that the algebraic sum of the currents entering and leaving a junction is zero.***

In other words,

***the sum of the currents entering a junction must equal the sum of the currents leaving the junction.***

In equation form:

$$\Sigma I_{\text{entering}} = \Sigma I_{\text{leaving}} \tag{6.8}$$

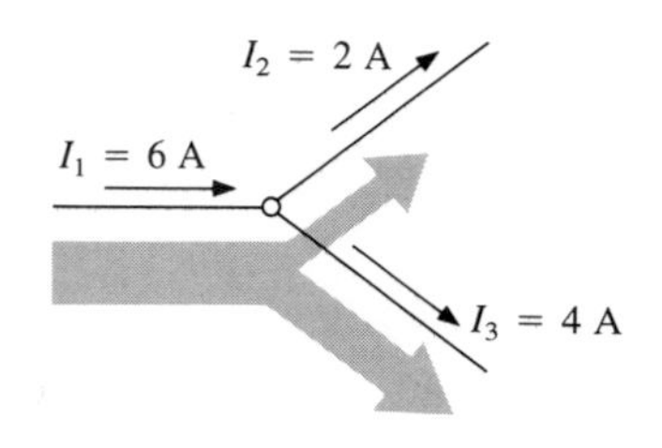


**FIG. 6.20**
*Demonstrating Kirchhoff's current law.*

In Fig. 6.20,

$$\Sigma I_{\text{entering}} = \Sigma I_{\text{leaving}}$$
$$6\text{ A} = 2\text{ A} + 4\text{ A}$$
$$6\text{ A} = 6\text{ A} \qquad \text{(checks)}$$

In the next two examples, unknown currents can be determined by applying Kirchhoff's current law. Simply remember to place all current levels entering a junction to the left of the equals sign and the sum of all currents leaving a junction to the right of the equals sign. The water-in-the-pipe analogy is an excellent one for supporting and clarifying the preceding law. Quite obviously, the sum total of the water entering a junction must equal the total of the water leaving the exit pipes.

**EXAMPLE 6.12.** Determine the current $I_4$ of Fig. 6.21 using Kirchhoff's current law.

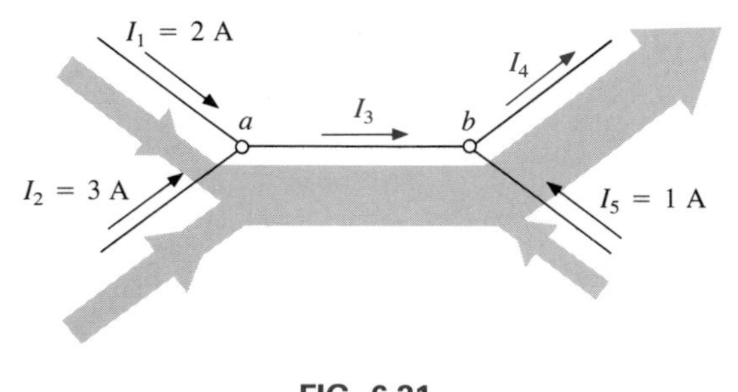


**FIG. 6.21**
*Figure for Example 6.12.*

***Solution:*** We must first work with junction *a*, since the only unknown is $I_3$. At junction *b* there are two unknowns and both cannot be determined from one application of the law.

At *a*:

$$\Sigma I_{\text{entering}} = \Sigma I_{\text{leaving}}$$
$$I_1 + I_2 = I_3$$
$$2\text{ A} + 3\text{ A} = I_3$$
$$I_3 = \mathbf{5\ A}$$

At *b*:

$$\Sigma I_{\text{entering}} = \Sigma I_{\text{leaving}}$$
$$I_3 + I_5 = I_4$$
$$5\text{ A} + 1\text{ A} = I_4$$
$$I_4 = \mathbf{6\ A}$$

---

**EXAMPLE 6.13.** Determine $I_1$, $I_3$, $I_4$, and $I_5$ for the network of Fig. 6.22.

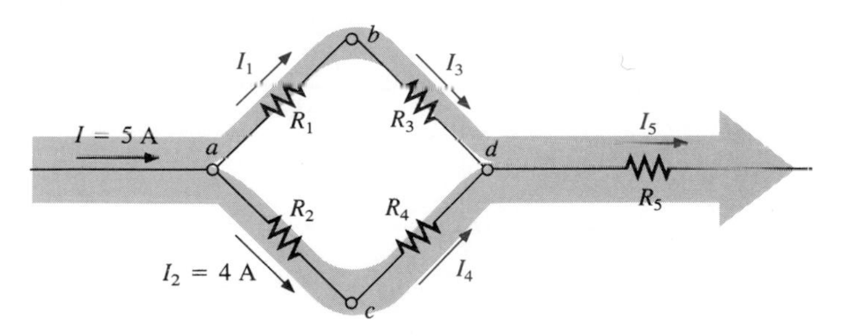


**FIG. 6.22**
*Network for Example 6.13.*

***Solution:*** At *a*:

$$\Sigma I_{\text{entering}} = \Sigma I_{\text{leaving}}$$
$$I = I_1 + I_2$$
$$5\text{ A} = I_1 + 4\text{ A}$$

Subtracting 4 A from both sides gives

$$5\text{ A} - 4\text{ A} = I_1 + 4\text{ A} - 4\text{ A}$$
$$I_1 = 5\text{ A} - 4\text{ A} = \mathbf{1\ A}$$

At *b*:

$$\Sigma I_{\text{entering}} = \Sigma I_{\text{leaving}}$$
$$I_1 = I_3 = \mathbf{1\ A}$$

as it should, since $R_1$ and $R_3$ are in series and the current is the same in series elements.

At *c:*

$$I_2 = I_4 = \mathbf{4\ A}$$

for the same reasons given for junction *b*.

At *d:*

$$\Sigma I_{\text{entering}} = \Sigma I_{\text{leaving}}$$
$$I_3 + I_4 = I_5$$
$$1\ \text{A} + 4\ \text{A} = I_5$$
$$I_5 = \mathbf{5\ A}$$

If we enclose the entire network, we find that the current entering is $I = 5$ A; the net current leaving from the far right is $I_5 = 5$ A. The two must be equal, since the net current entering any system must equal that leaving.

---

**EXAMPLE 6.14.** Determine the currents $I_3$ and $I_5$ of Fig. 6.23 through applications of Kirchhoff's current law.

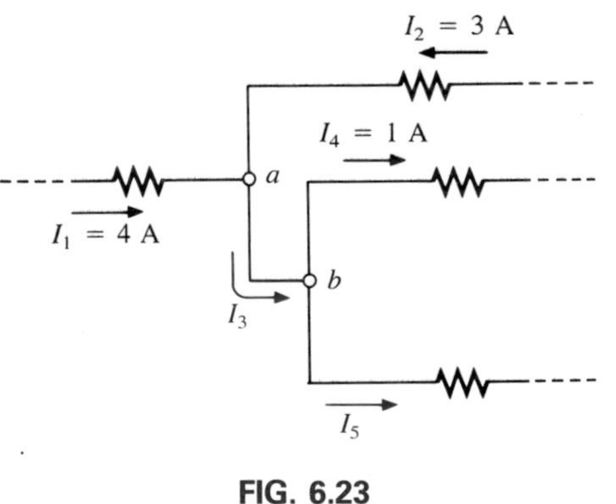


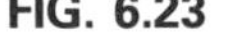

**FIG. 6.23**

***Solution:*** Note that since node *b* has two unknown quantities and node *a* only one, we must first apply Kirchhoff's current law to node *a*. The result can then be applied to node *b*. For node *a*,

$$I_1 + I_2 = I_3$$
$$4\ \text{A} + 3\ \text{A} = I_3$$

and

$$I_3 = \mathbf{7\ A}$$

For node *b*,

$$I_3 = I_4 + I_5$$
$$7\ \text{A} = 1\ \text{A} + I_5$$

and

$$I_5 = 7\ \text{A} - 1\ \text{A} = \mathbf{6\ A}$$

**EXAMPLE 6.15.** Find the magnitude and direction of the currents $I_3$, $I_4$, $I_6$, and $I_7$ for the network of Fig. 6.24. Even though the elements are

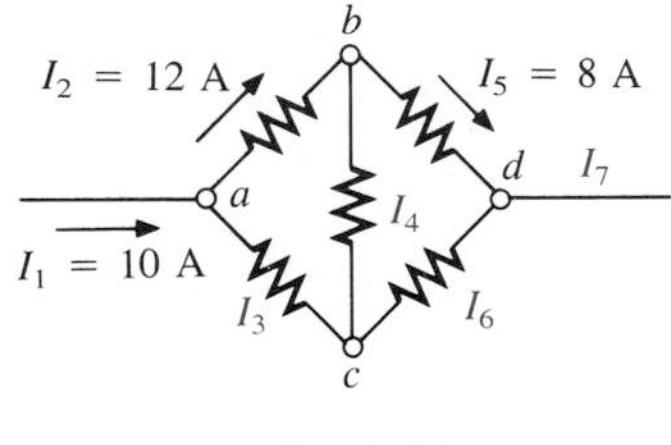

**FIG. 6.24**

not in series or parallel, Kirchhoff's current law can be applied to determine all the unknown currents.

***Solution:*** Considering the overall system, we know that the current entering must equal that leaving. Therefore,

$$I_7 = I_1 = \mathbf{10\ A}$$

Since 10 A are entering node *a* and 12 A are leaving, $I_3$ must be supplying current to the node. Applying Kirchhoff's current law at node *a*,

$$I_1 + I_3 = I_2$$
$$10\ \text{A} + I_3 = 12\ \text{A}$$

and

$$I_3 = 12\ \text{A} - 10\ \text{A} = \mathbf{2\ A}$$

At node *b*, since 12 A are entering and 8 A are leaving, $I_4$ must be leaving. Therefore,

$$I_2 = I_4 + I_5$$
$$12\ \text{A} = I_4 + 8\ \text{A}$$

and

$$I_4 = 12\ \text{A} - 8\ \text{A} = \mathbf{4\ A}$$

At node *c*, $I_3$ is leaving at 2 A and $I_4$ is entering at 4 A, requiring that $I_6$ be leaving. Applying Kirchhoff's current law at node *c*,

$$I_4 = I_3 + I_6$$
$$4\ \text{A} = 2\ \text{A} + I_6$$

and

$$I_6 = 4\ \text{A} - 2\ \text{A} = \mathbf{2\ A}$$

As a check at node *d*,

$$I_5 + I_6 = I_7$$
$$8\ \text{A} + 2\ \text{A} = 10\ \text{A}$$
$$10\ \text{A} = 10\ \text{A} \quad \text{(checks)}$$

Looking back at Example 6.10, we find that the current entering the top node is 4.5 A and the current leaving the node is $I_1 + I_2 = 3\text{ A} + 1.5\text{ A} = 4.5\text{ A}$. For Example 6.9, we have

$$I_T = I_1 + I_2 + I_3$$
$$10\text{ A} = 4\text{ A} + 2\text{ A} + I_3$$

and

$$I_3 = 10\text{ A} - 6\text{ A} = 4\text{ A}$$

## 6.6 CURRENT DIVIDER RULE

As the name suggests, the *current divider rule* (CDR) will determine how the current entering a set of parallel branches will split between the elements.

***For two parallel elements of equal value, the current will divide equally.***

***For parallel elements with different values, the smaller the resistance, the greater the share of input current.***

The current divider rule will be derived through the use of the representative network of Fig. 6.25. The input current $I$ equals $V/R_T$, where

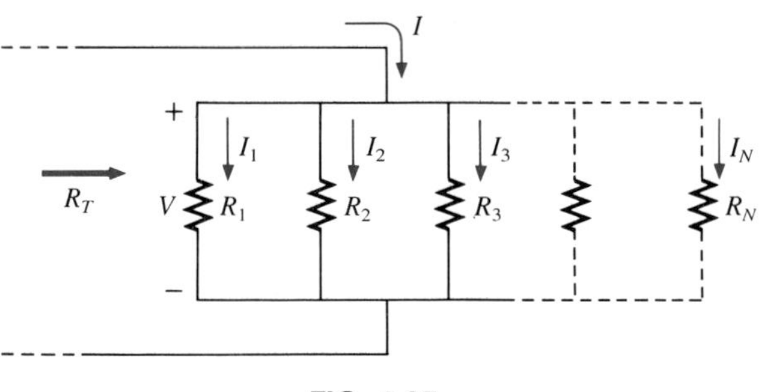


FIG. 6.25

$R_T$ is the total resistance of the parallel branches. Substituting $V = I_x R_x$ into the above equation, where $I_x$ refers to the current through a parallel branch of resistance $R_x$, we have

$$I = \frac{V}{R_T} = \frac{I_x R_x}{R_T}$$

and

$$\boxed{I_x = \frac{R_T}{R_x} I} \tag{6.9}$$

which is the general form for the current divider rule. In words, the current through any parallel branch is equal to the product of the *total* resistance of the parallel branches and the input current divided by the resistance of the branch through which the current is to be determined.

For the current $I_1$,

$$I_1 = \frac{R_T}{R_1} I$$

and for $I_2$,

$$I_2 = \frac{R_T}{R_2} I$$

and so on.

For the particular case of *two parallel resistors,* as shown in Fig. 6.26,

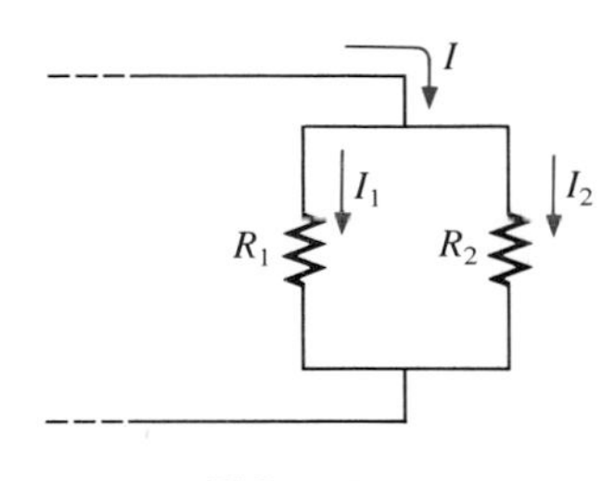


FIG. 6.26

$$R_T = \frac{R_1 R_2}{R_1 + R_2}$$

and

$$I_1 = \frac{R_T}{R_1} I = \frac{\dfrac{R_1 R_2}{R_1 + R_2}}{R_1} I$$

and

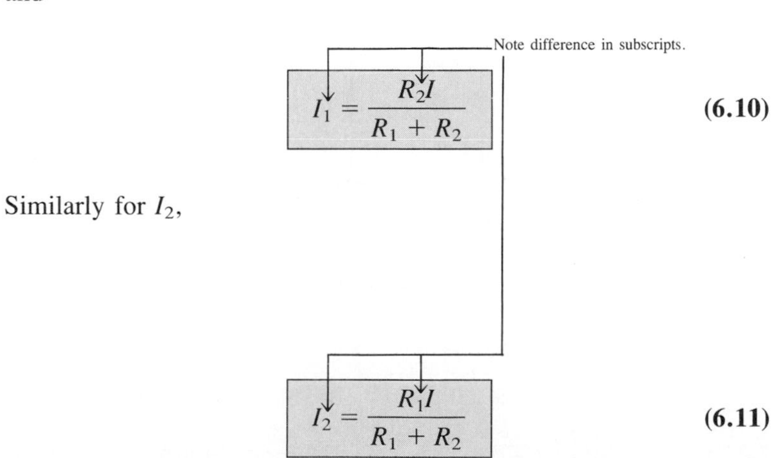


$$I_1 = \frac{R_2 I}{R_1 + R_2} \qquad (6.10)$$

Similarly for $I_2$,

$$I_2 = \frac{R_1 I}{R_1 + R_2} \qquad (6.11)$$

In words, for two parallel branches, the current through either branch is equal to the product of the *other* parallel resistor and the input current divided by the *sum* (not total parallel resistance) of the two parallel resistances.

---

**EXAMPLE 6.16.** Determine the current $I_2$ for the network of Fig. 6.27 using the current divider rule.

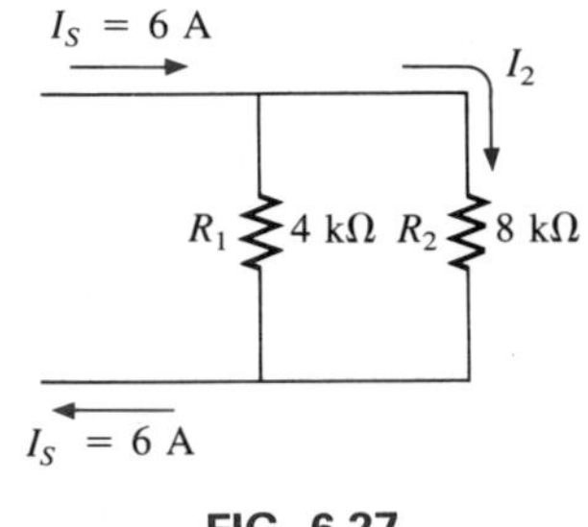


FIG. 6.27

***Solution:***

$$I_2 = \frac{R_1 I_s}{R_1 + R_2} = \frac{(4\ \text{k}\Omega)(6\ \text{A})}{4\ \text{k}\Omega + 8\ \text{k}\Omega} = \frac{4}{12}(6\ \text{A}) = \frac{1}{3}(6\ \text{A})$$
$$= \mathbf{2\ A}$$

---

**EXAMPLE 6.17.** Find the current $I_1$ for the network of Fig. 6.28.

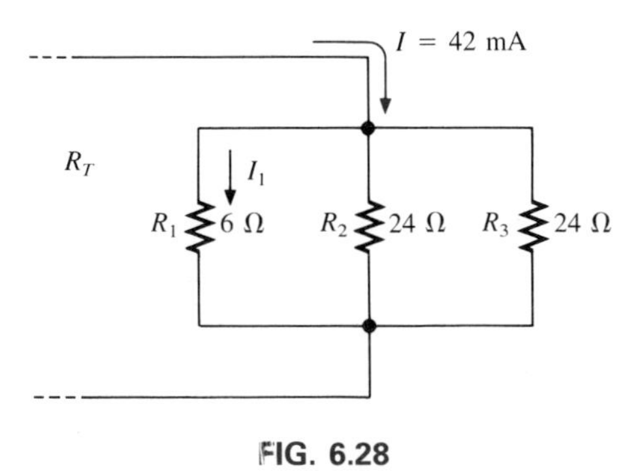


FIG. 6.28

***Solution:*** By Eq. (6.9),

$$R_T = 6\ \Omega \parallel 24\ \Omega \parallel 24\ \Omega = 6\ \Omega \parallel 12\ \Omega = 4\ \Omega$$

$$I_1 = \frac{R_T}{R_1} I = \frac{(4\ \Omega)(42 \times 10^{-3}\ \text{A})}{6\ \Omega} = \mathbf{28\ mA}$$

**EXAMPLE 6.18.** Determine the magnitude of the currents $I_1$, $I_2$, and $I_3$ for the network of Fig. 6.29.

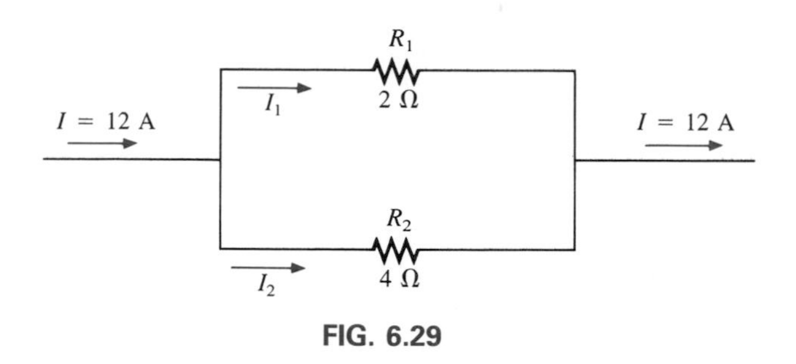


FIG. 6.29

***Solution:*** By Eq. (6.10), the current divider rule,

$$I_1 = \frac{R_2 I}{R_1 + R_2} = \frac{(4\ \Omega)(12\ \text{A})}{2\ \Omega + 4\ \Omega} = \mathbf{8\ A}$$

Applying Kirchhoff's current law,

$$I = I_1 + I_2$$

and

$$I_2 = I - I_1 = 12\ \text{A} - 8\ \text{A} = \mathbf{4\ A}$$

or, using the current divider rule again,

$$I_2 = \frac{R_1 I}{R_1 + R_2} = \frac{(2\ \Omega)(12\ \text{A})}{2\ \Omega + 4\ \Omega} = \mathbf{4\ A}$$

The total current entering the parallel branches must equal that leaving. Therefore,

$$I_3 = I = \mathbf{12\ A}$$

or

$$I_3 = I_1 + I_2 = 8\ \text{A} + 4\ \text{A} = \mathbf{12\ A}$$

**EXAMPLE 6.19.** Determine the resistance $R_1$ to effect the division of current in Fig. 6.30.

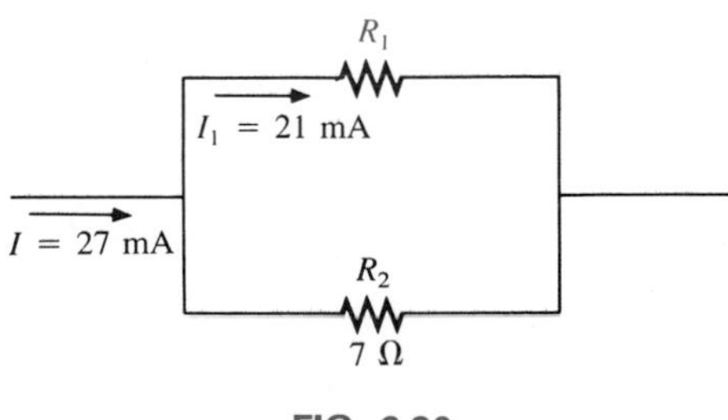


FIG. 6.30

***Solution:*** Applying the current divider rule,

$$I_1 = \frac{R_2 I}{R_1 + R_2}$$

and

$$(R_1 + R_2)I_1 = R_2 I$$
$$R_1 I_1 + R_2 I_1 = R_2 I$$
$$R_1 I_1 = R_2 I - R_2 I_1$$
$$R_1 = \frac{R_2(I - I_1)}{I_1}$$

Substituting values:

$$R_1 = \frac{7\ \Omega(27\text{ mA} - 21\text{ mA})}{21\text{ mA}}$$

$$= 7\ \Omega\left(\frac{6}{21}\right) = \frac{42\ \Omega}{21}$$

$$= \mathbf{2\ \Omega}$$

An alternative approach is

$$I_2 = I - I_1 \qquad \text{(Kirchhoff's current law)}$$
$$= 27\text{ mA} - 21\text{ mA} = 6\text{ mA}$$
$$V_2 = I_2R_2 = (6\text{ mA})(7\ \Omega) = 42\text{ mV}$$
$$V_1 = I_1R_1 = V_2 = 42\text{ mV}$$

and

$$R_1 = \frac{V_1}{I_1} = \frac{42\text{ mV}}{21\text{ mA}} = \mathbf{2\ \Omega}$$

---

From the examples just described, note the following:

***Current seeks the path of least resistance.***

That is,

1. More current passes through the smaller of two parallel resistors.
2. The current entering any number of parallel resistors divides into these resistors as the inverse ratio of their ohmic values. This relationship is depicted in Fig. 6.31.

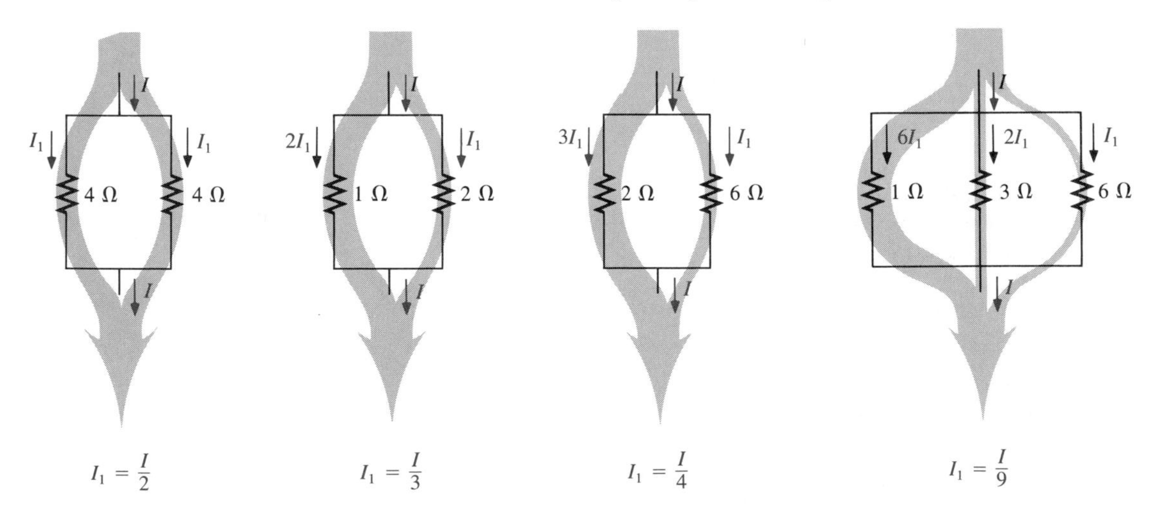


**FIG. 6.31**
*Current division through parallel branches.*

A mechanical analogy often used to describe this division of current is the flow of water through pipes. The water represents the flow of charge, and the tubes or pipes represent conductors. In this analogy, the greater the resistance of the corresponding electrical element, the smaller the area of the tubing.

The total current $I$ in Fig. 6.32(a) divides equally between the two equal resistors. The analogy just described is shown to the right. Obviously, for two pipes of equal diameter, the water will divide equally.

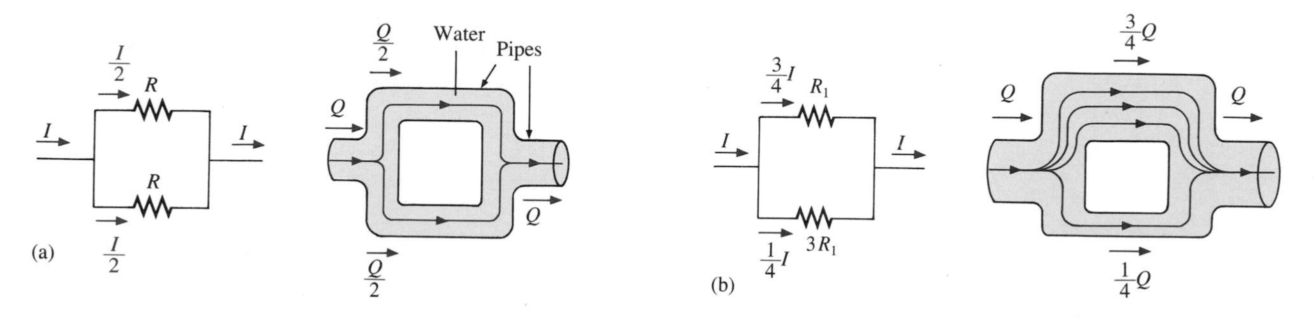

**FIG. 6.32**

In Fig. 6.32(b), one resistor is three times the other, resulting in the current dividing as shown. Its mechanical analogy is shown in the adjoining figure. Again, it should be obvious that three times as much water (current) will pass through one pipe as through the other. In both Figs. 6.32(a) and (b), the total water (current) entering the parallel systems from the left will equal that leaving to the right.

## 6.7 VOLTAGE SOURCES IN PARALLEL

Voltage sources are placed in parallel as shown in Fig. 6.33 only if they have the same voltage rating. Otherwise, Kirchhoff's voltage law

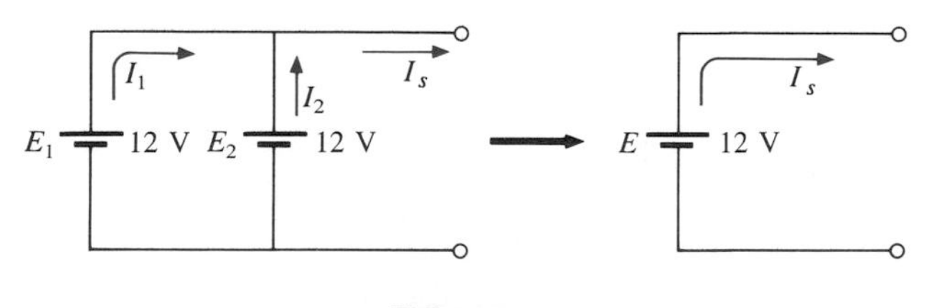


**FIG. 6.33**
*Parallel voltage sources.*

would be violated around the internal loop of the two batteries. The primary reason for placing two or more batteries in parallel of the same terminal voltage would be to increase the current rating (and, therefore, the power rating) of the source. As shown in Fig. 6.33, the current rating of the combination is determined by $I_s = I_1 + I_2$ at the same terminal voltage. The resulting power rating is twice that available with one supply.

## 6.8
## OPEN AND SHORT CIRCUITS

Open circuits and short circuits can often cause more confusion and difficulty in the analysis of a system than standard series or parallel configurations. This will become more obvious in the chapters to follow when we apply some of the methods and theorems.

An *open circuit* is simply two isolated terminals not connected by an element of any kind. Consider the battery of Fig. 6.34. An open circuit

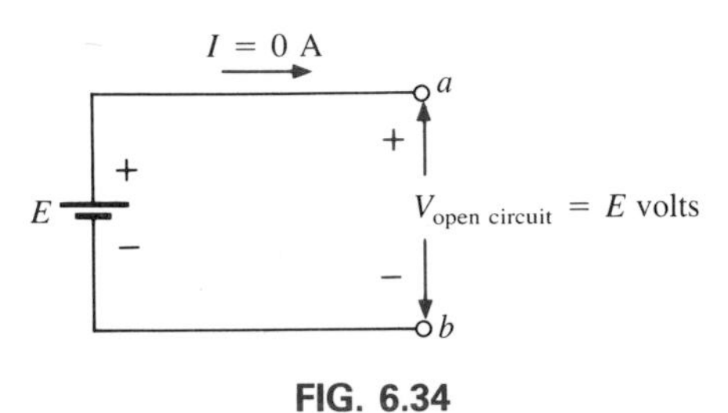

FIG. 6.34

exists between terminals $a$ and $b$. There is a voltage of $E$ volts between the two terminals, but the current between the two is zero due to the absence of a closed path for the flow of charge. In general:

***An open circuit can have a potential difference (voltage) across its terminals but the current is always zero amperes.***

A *short circuit* is a direct connection of zero ohms across an element or combination of elements. In Fig. 6.35(a), the current through the 2-Ω

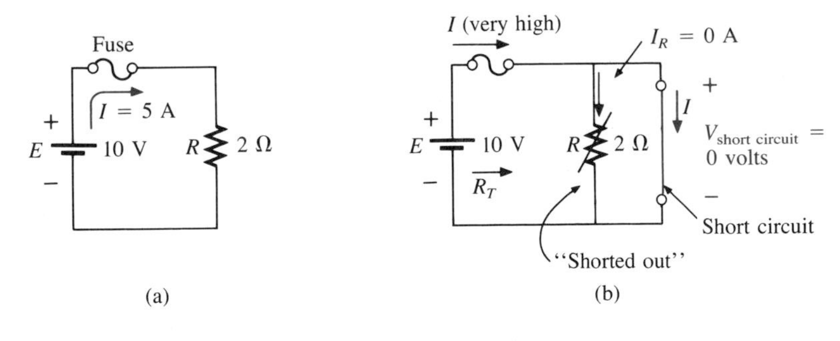

FIG. 6.35

resistor is 5 A. If a short circuit is established across the 2-Ω resistor as shown in Fig. 6.35(b) due to a faulty wire, connection, or other unexpected circumstance, the total resistance of the circuit $R_T$ is now $R_T = (2\ \Omega)(0)/(2\ \Omega + 0) = 0\ \Omega$ and the current will rise to very high levels. The 2-Ω resistor has effectively been "shorted out" by the low-resistance connection. The maximum current is limited only by the circuit breaker or fuse in series with the source. The resulting high current is often the cause of fire or smoke if the protective device fails to respond quickly enough. Since the resistance of a short circuit is zero ohms, there is no voltage drop across a short circuit, as determined by Ohm's law ($V = IR$). In general:

***A short circuit can carry a current of any level but the potential difference (voltage) across its terminals is always zero volts.***

**EXAMPLE 6.20.** Determine the voltage $V_{ab}$ for the network of Fig. 6.36.

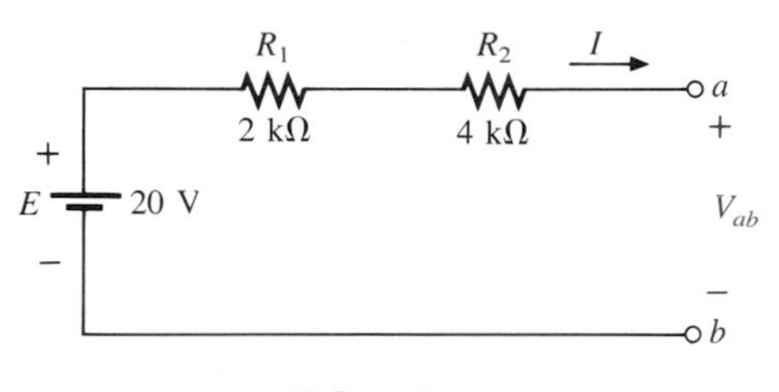


FIG. 6.36

***Solution:*** The open circuit requires that $I$ be zero amperes. The voltage drop across both resistors is therefore zero volts since $V = IR = (0)R = 0$ V. Applying Kirchhoff's voltage law around the closed loop,

$$V_{ab} = E = \mathbf{20\ V}$$

**EXAMPLE 6.21.** Determine the voltages $V_{ab}$ and $V_{cd}$ for the network of Fig. 6.37.

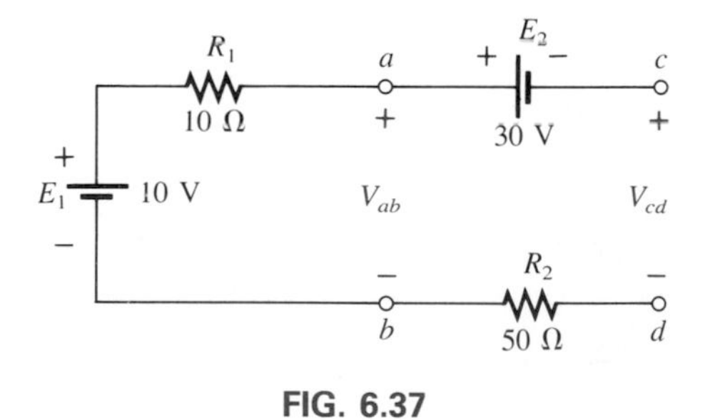


FIG. 6.37

***Solution:*** The current through the system is zero amperes due to the open circuit resulting in a 0-V drop across each resistor. Both resistors can therefore be replaced by short circuits as shown in Fig. 6.38. The voltage $V_{ab}$ is then directly across the 10-V battery, and

$$V_{ab} = E_1 = \mathbf{10\ V}$$

The voltage $V_{cd}$ requires an application of Kirchhoff's voltage law:

$$+E_1 - E_2 - V_{cd} = 0$$

or

$$V_{cd} = E_1 - E_2 = 10\ \text{V} - 30\ \text{V} = \mathbf{-20\ V}$$

The negative sign in the solution simply indicates that the actual voltage $V_{cd}$ has the opposite polarity of that appearing in Fig. 6.37.

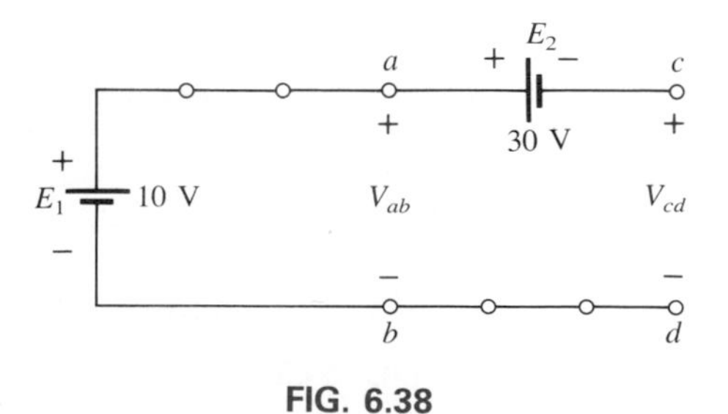


FIG. 6.38

**EXAMPLE 6.22.** Calculate the current $I$ and the voltage $V$ for the network of Fig. 6.39.

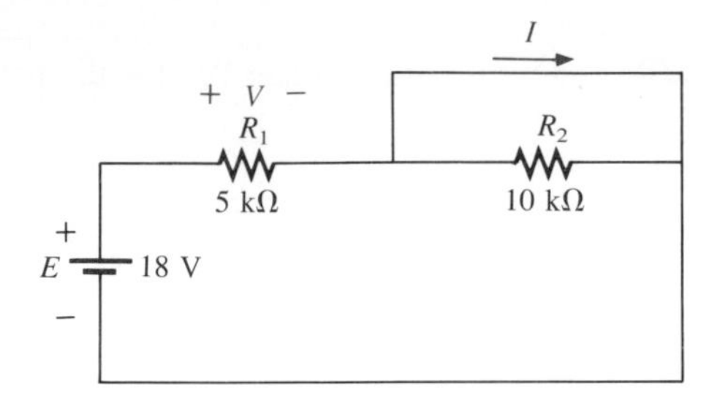

FIG. 6.39

***Solution:*** The 10-Ω resistor has been effectively shorted out, resulting in the equivalent network of Fig. 6.40. Using Ohm's law,

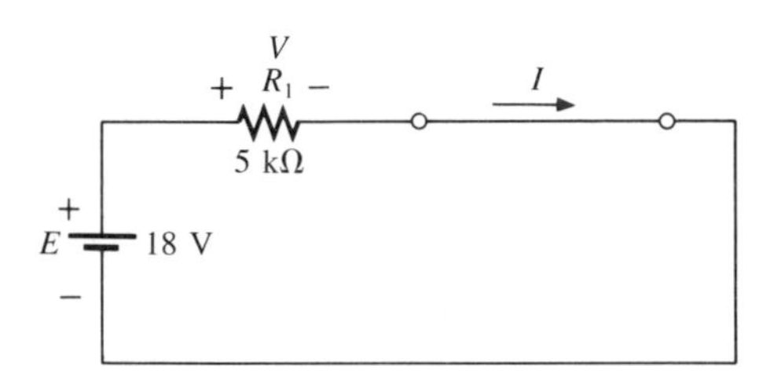


FIG. 6.40

$$I = \frac{E}{R_1} = \frac{18\text{ V}}{5\text{ k}\Omega} = \mathbf{3.6\text{ mA}}$$

and

$$V = E = \mathbf{18\text{ V}}$$

**EXAMPLE 6.23.** Determine $V$ and $I$ for the network of Fig. 6.41 if the resistor $R_2$ is shorted out.

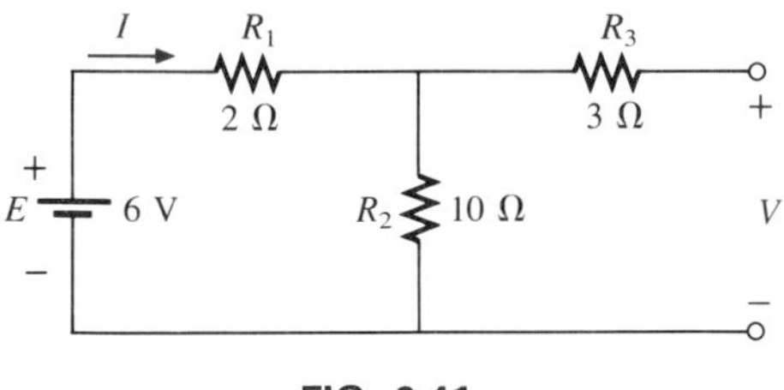


FIG. 6.41

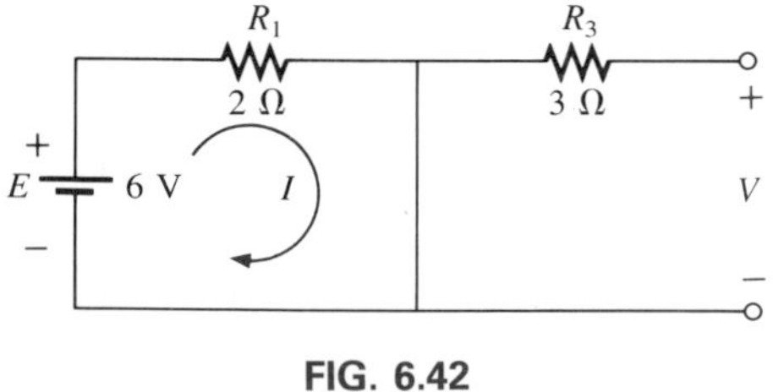


FIG. 6.42

***Solution:*** The redrawn network appears in Fig. 6.42. The current through the 3-Ω resistor is zero due to the open circuit causing all of the current $I$ to pass through the short circuit. Since $V_{3\Omega} = IR = (0)R =$ 0 V, the voltage $V$ is directly across the short, and

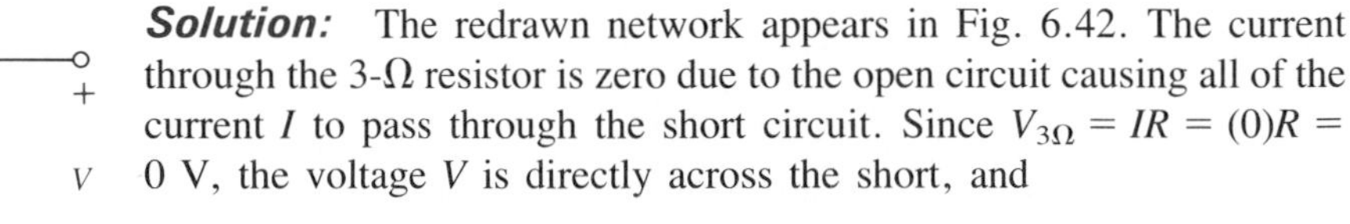

$$V = \mathbf{0\text{ V}}$$

with

$$I = \frac{E}{R_1} = \frac{6\text{ V}}{2\ \Omega} = \mathbf{3\text{ A}}$$

## 6.9 VOLTMETERS: LOADING EFFECT

In Chapter 2, it was noted that voltmeters are always placed across an element to measure the potential difference. We now realize that this connection is synonymous with placing the voltmeter in parallel with the element. The insertion of a meter in parallel with a resistor results in a combination of parallel resistors as shown in Fig. 6.43. Since the

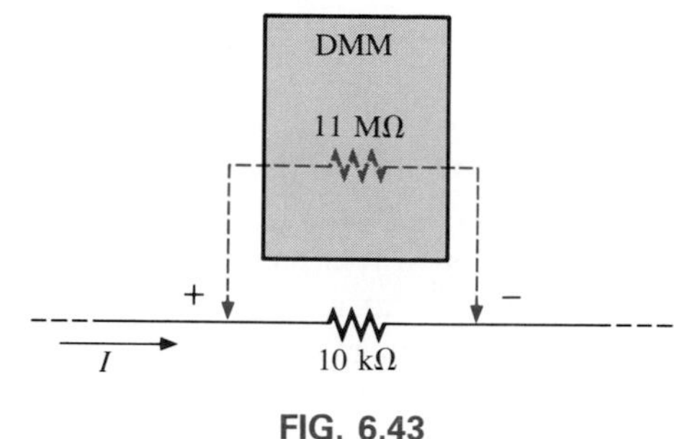


**FIG. 6.43**
*Voltmeter loading.*

resistance of two parallel branches is always less than the smaller parallel resistance, the resistance of the voltmeter should be as large as possible (ideally infinite). In Fig. 6.43, a DMM with an internal resistance of 11 MΩ is measuring the voltage across a 10-kΩ resistor. The total resistance of the combination is

$$R_T = 10\ \text{k}\Omega \parallel 11\ \text{M}\Omega = \frac{(10^4)(11 \times 10^6)}{10^4 + (11 \times 10^6)} = 9.99\ \text{k}\Omega$$

and we find that the network is essentially undisturbed. However, if we use a VOM with an internal resistance of 50 kΩ on the 2.5-V scale, the parallel resistance is

$$R_T = 10\ \text{k}\Omega \parallel 50\ \text{k}\Omega = \frac{(10^4)(50 \times 10^3)}{10^4 + (50 \times 10^3)} = 8.33\ \text{k}\Omega$$

and the behavior of the network will be altered somewhat since the 10-kΩ resistor will now appear to be 8.33 kΩ to the rest of the network.

The loading of a network by the insertion of meters is not to be taken lightly, especially in research efforts where accuracy is a primary consideration. It is good practice always to check the meter resistance level against the resistive elements of the network before making measurements. A factor of 10 between resistance levels will usually provide fairly accurate meter readings for a wide range of applications.

Most DMMs have internal resistance levels in excess of 10 MΩ on all voltage scales, while the internal resistance of VOMs is sensitive to the chosen scale. To determine the resistance of each scale setting of a VOM in the voltmeter mode, simply multiply the maximum voltage of the scale setting by the ohm/volt (Ω/V) rating of the meter, normally found at the bottom of the face of the meter.

For a typical ohm/volt rating of 20,000, the 2.5-V scale would have an internal resistance of

$$(2.5)(20{,}000) = 50\ \text{k}\Omega$$

while for the 100-V scale, it would be

$$(100)(20{,}000) = 2\ \text{M}\Omega$$

and for the 250-V scale,

$$(250)(20{,}000) = 5\ \text{M}\Omega$$

---

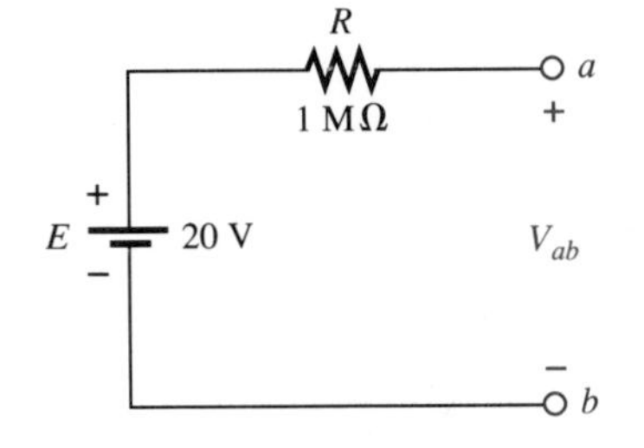

FIG. 6.44

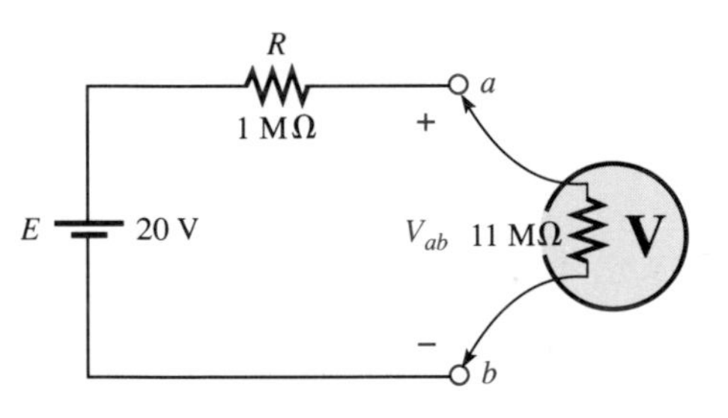

FIG. 6.45

**EXAMPLE 6.24.** For the relatively simple network of Fig. 6.44,

a. What is the open-circuit voltage $V_{ab}$.
b. What will a DMM indicate if it has an internal resistance of 11 MΩ. Compare to the results of part a.
c. Repeat part b for a VOM with an Ω/V rating of 20,000 on the 100-V scale.

***Solutions:***

a. $V_{ab} = \mathbf{20\ V}$.
b. The meter will complete the circuit as shown in Fig. 6.45 and using the voltage divider rule,

$$V_{ab} = \frac{11\ \text{M}\Omega(20\ \text{V})}{11\ \text{M}\Omega + 1\ \text{M}\Omega} = \mathbf{18.33\ V}$$

c. For the VOM the internal resistance of the meter is

$$R_m = 100\ \text{V}(20{,}000\ \Omega/\text{V}) = 2\ \text{M}\Omega$$

and

$$V_{ab} = \frac{2\ \text{M}\Omega(20\ \text{V})}{2\ \text{M}\Omega + 1\ \text{M}\Omega} = \mathbf{13.33\ V}$$

revealing the need to carefully consider the internal resistance of the meter in some instances.

---

## 6.10 COMPUTER ANALYSIS

The computer analysis of parallel dc networks is very similar to that applied to series circuits in the previous chapter. However, since the voltage across parallel elements is the same, the currents are the only quantities to be determined. The analysis in this chapter and some of those to follow will be limited solely to PSPICE due primarily to space considerations and the growing interest in PSPICE as an educational tool. However, if you prefer an extended coverage of BASIC as applied to parallel networks, there are numerous references available on the subject.*

## PSPICE

The input file for the parallel network of Fig. 6.46 is provided in Fig. 6.47(a). Note again the use of the .DC control line to specify the output file desired. The .PRINT line specifies only $I_1$ and $I_2$ for the output file and the .OPTIONS NOPAGE line continues to minimize wasted space. The results obtained for the parameters entered support the results obtained in Example 6.10.

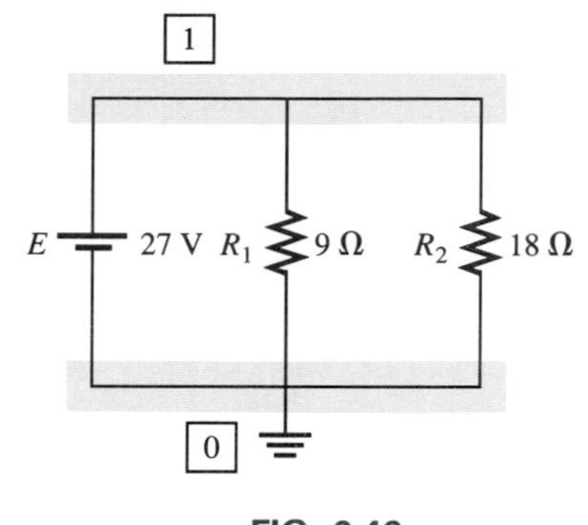


FIG. 6.46

```
************************ Evaluation PSpice (January 1989) ******* 19:02:07 *******

Chapter 6 - Two Parallel Resistors

****     CIRCUIT DESCRIPTION

**************************************************************************

VE 1 0 27V
R1 1 0 9
R2 1 0 18
.DC VE 27 27 1
.PRINT DC I(R1) I(R2)
.OPTIONS NOPAGE
.END

****     DC TRANSFER CURVES                    TEMPERATURE =   27.000 DEG C

  VE           I(R1)        I(R2)

  2.700E+01    3.000E+00    1.500E+00
```

FIG. 6.47(a)

The second input file (Fig. 6.47(b)) is different from the above only in that the dc battery voltage is incremented from 20 V to 30 V in increments of 1 V. Note how the currents increase in magnitude with increase in voltage and how the results obtained at $E = 27\text{ V} = 2.700\text{E}{+}01$ match those obtained above.

*L. Nashelsky and R. Boylestad, *BASIC Applied to Circuit Analysis,* Columbus, Ohio: Merrill, 1984.

```
************************ Evaluation PSpice (January 1989) ******* 09:40:56 *******

Chapter 6 - Two Parallel Resistors

****     CIRCUIT DESCRIPTION

****************************************************************************

VE 1 0 27V
R1 1 0 9
R2 1 0 18
.DC VE 20 30 1
.PRINT DC I(R1) I(R2)
.OPTIONS NOPAGE
.END
```

```
****     DC TRANSFER CURVES                    TEMPERATURE =   27.000 DEG C


 VE           I(R1)        I(R2)

  2.000E+01   2.222E+00   1.111E+00
  2.100E+01   2.333E+00   1.167E+00
  2.200E+01   2.444E+00   1.222E+00
  2.300E+01   2.556E+00   1.278E+00
  2.400E+01   2.667E+00   1.333E+00
  2.500E+01   2.778E+00   1.389E+00
  2.600E+01   2.889E+00   1.444E+00
  2.700E+01   3.000E+00   1.500E+00
  2.800E+01   3.111E+00   1.556E+00
  2.900E+01   3.222E+00   1.611E+00
  3.000E+01   3.333E+00   1.667E+00
```

**FIG. 6.47(b)**

## PROBLEMS

### SECTION 6.2 Parallel Elements

1. For each configuration of Fig. 6.48, determine which elements are in series or parallel.

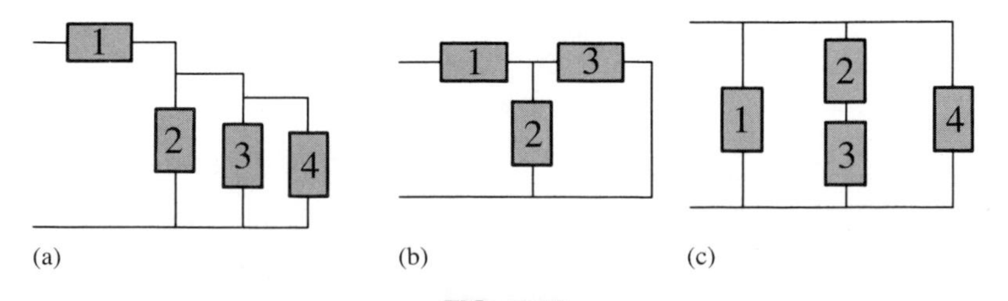


**FIG. 6.48**

2. For the network of Fig. 6.49,
   a. Which elements are in parallel?
   b. Which elements are in series?
   c. Which branches are in parallel?

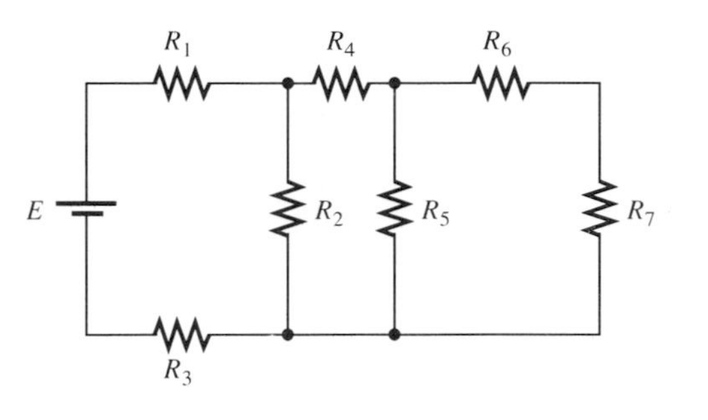

FIG. 6.49

## SECTION 6.3 Total Conductance and Resistance

3. Find the total conductance and resistance for the networks of Fig. 6.50.

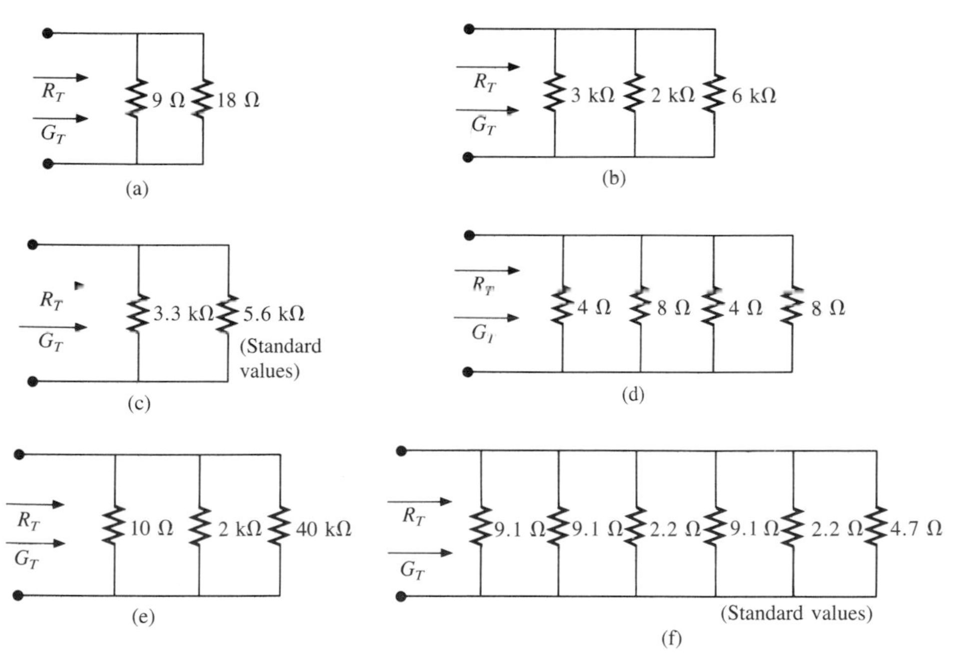


FIG. 6.50

4. The total conductance of the networks of Fig. 6.51 is specified. Find the value in ohms of the unknown resistances.

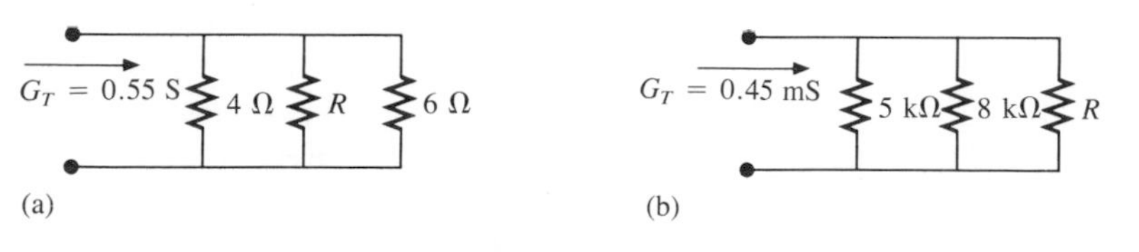


FIG. 6.51

5. The total resistance of the circuits of Fig. 6.52 is specified. Find the value in ohms of the unknown resistances.

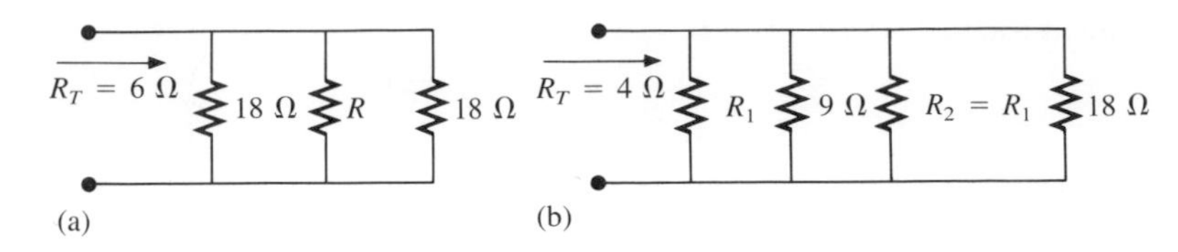

FIG. 6.52

*6. Determine the unknown resistors of Fig. 6.53 given the fact that $R_2 = 5R_1$ and $R_3 = \frac{1}{2}R_1$

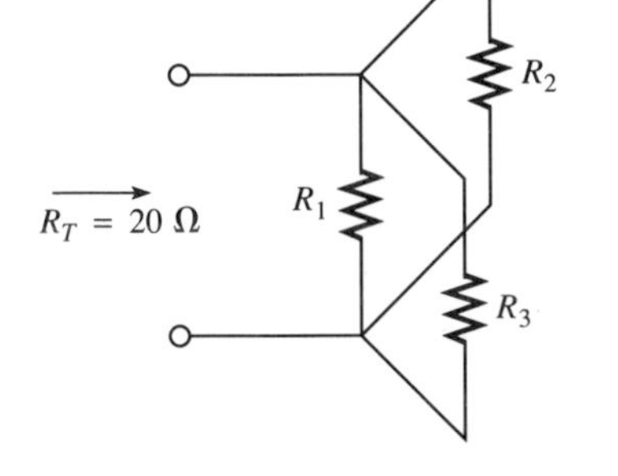


FIG. 6.53

*7. Determine $R_1$ for the network of Fig. 6.54.

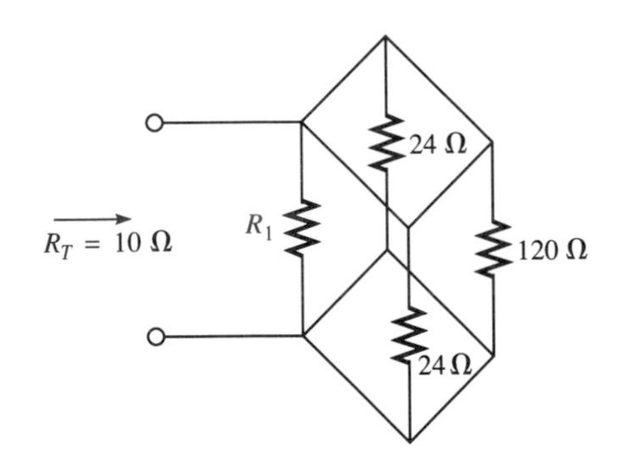


FIG. 6.54

## SECTION 6.4 Parallel Networks

8. For the network of Fig. 6.55:
   a. Find the total conductance and resistance.
   b. Determine $I_s$ and the current through each parallel branch.
   c. Verify that the source current equals the sum of the parallel branch currents.
   d. Find the power dissipated by each resistor, and note whether the power delivered is equal to the power dissipated.
   e. If the resistors are available with wattage ratings of 1/2, 1, 2, and 50 W, what is the minimum wattage rating for each resistor?

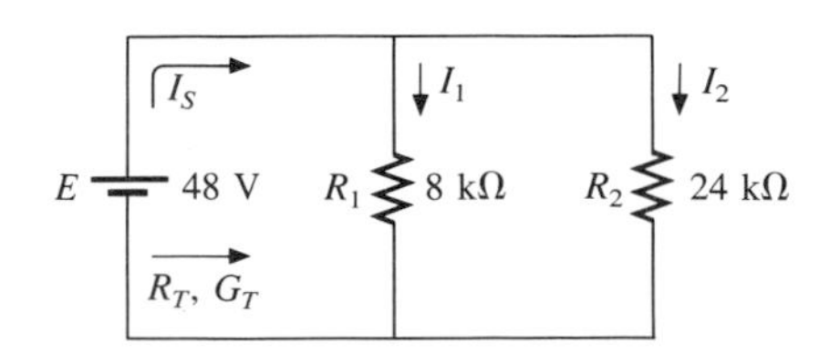


FIG. 6.55

**9.** Repeat Problem 8 for the network of Fig. 6.56.

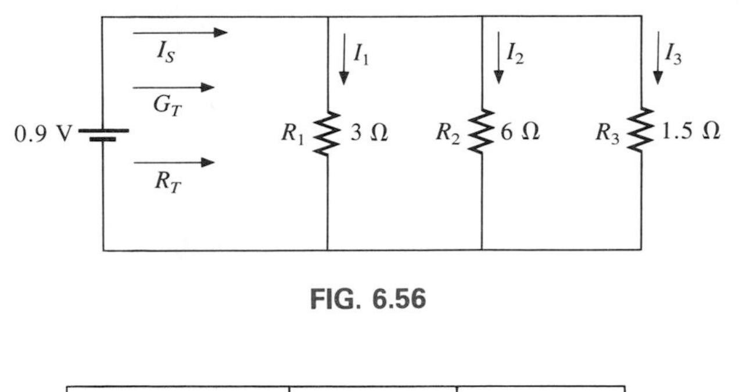


FIG. 6.56

**10.** Repeat Problem 8 for the network of Fig. 6.57 constructed of standard resistor values.

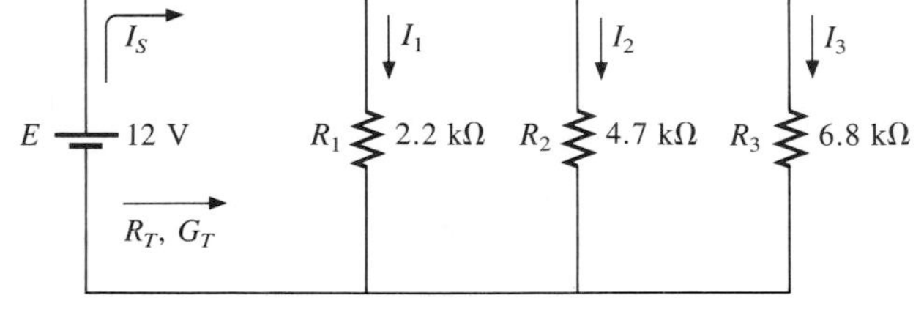


FIG. 6.57

**11.** There are eight Christmas tree lights connected in parallel as shown in Fig. 6.58.

**a.** If the set is connected to a 120-V source, what is the current through each bulb if each bulb has an internal resistance of 1.8 kΩ?

**b.** Determine the total resistance of the network.

**c.** Find the power delivered to each bulb.

**d.** If one bulb burns out (that is, the filament opens), what is the effect on the remaining bulbs?

**e.** Compare the parallel arrangement of Fig. 6.58 to the series arrangement of Fig. 5.72. What are the relative advantages and disadvantages of the parallel system as compared to the series arrangement?

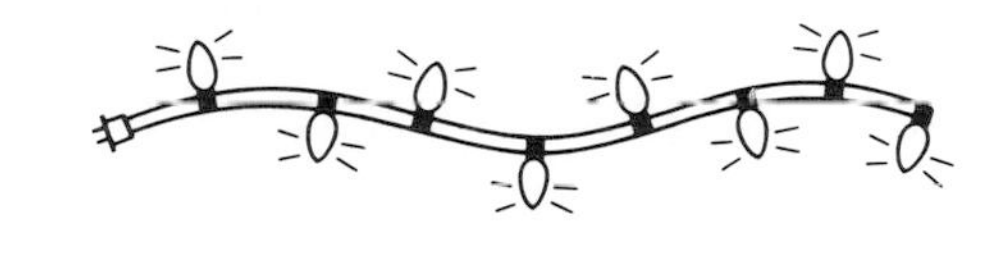

FIG. 6.58

**12.** A portion of a residential service to a home is depicted in Fig. 6.59.

**a.** Determine the current through each parallel branch of the network.

**b.** Calculate the current drawn from the 120-V source. Will the 20-A circuit breaker trip?

**c.** What is the total resistance of the network?

**d.** Determine the power supplied by the 120-V source. How does it compare to the total power of the load?

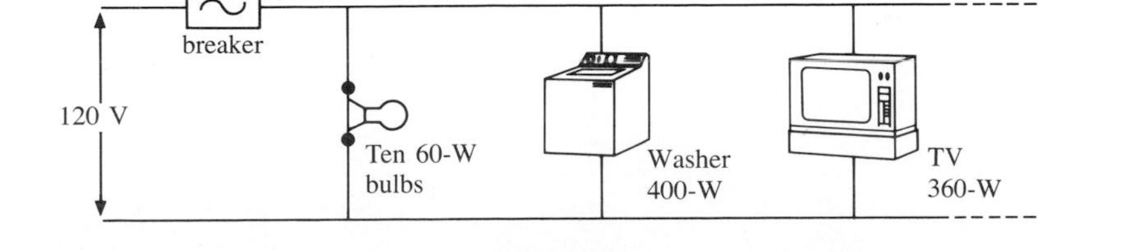


FIG. 6.59

13. Determine the currents $I_1$ and $I_s$ for the networks of Fig. 6.60.

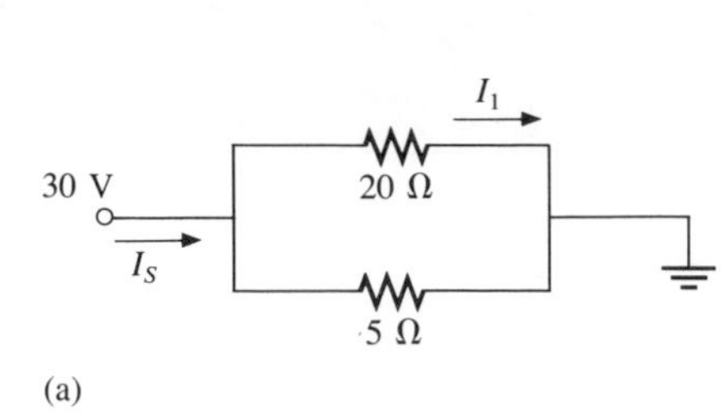


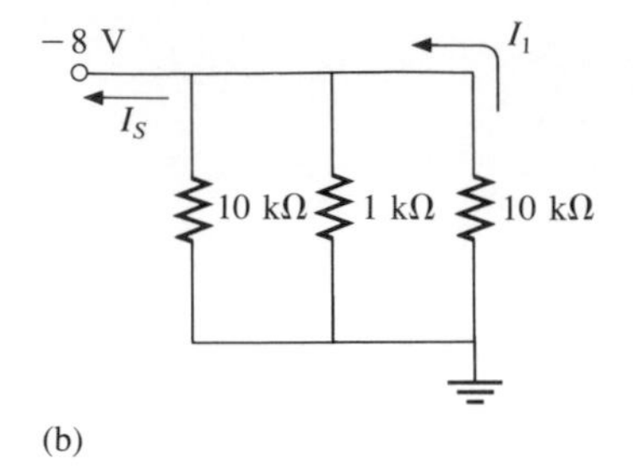

FIG. 6.60

*14. Determine the power delivered by the dc battery in Fig. 6.61.

FIG. 6.61

*15. For the network of Fig. 6.62,
   a. Find the source current $I_s$.
   b. Calculate the power dissipated by the 4-Ω resistor.
   c. Find the current $I_2$.

FIG. 6.62

## SECTION 6.5 Kirchhoff's Current Law

16. Find all unknown currents and their directions in the circuits of Fig. 6.63.

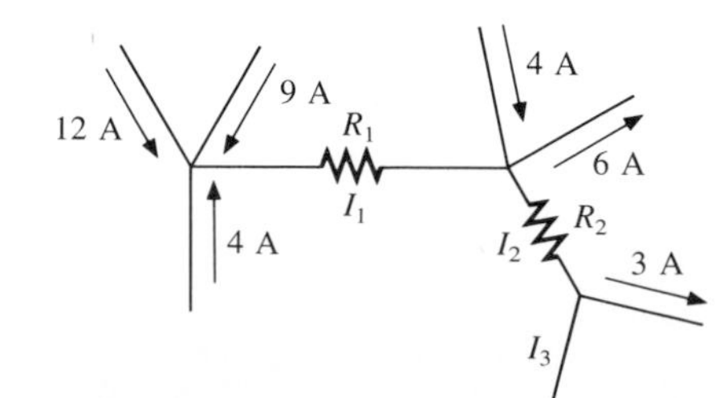


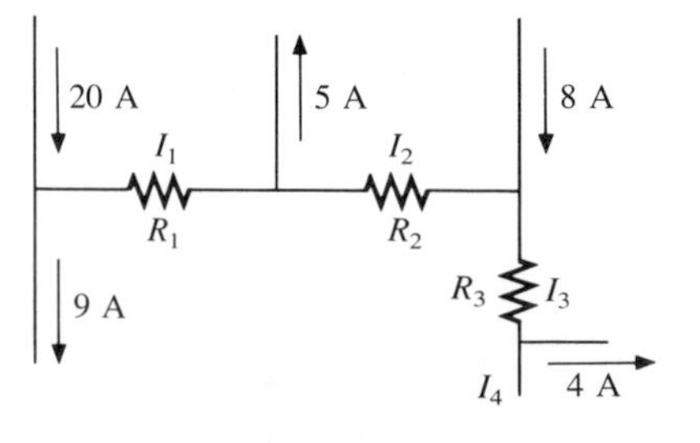


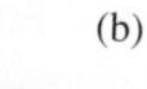

FIG. 6.63

**17.** Using Kirchhoff's current law, determine the unknown currents for the networks of Fig. 6.64.

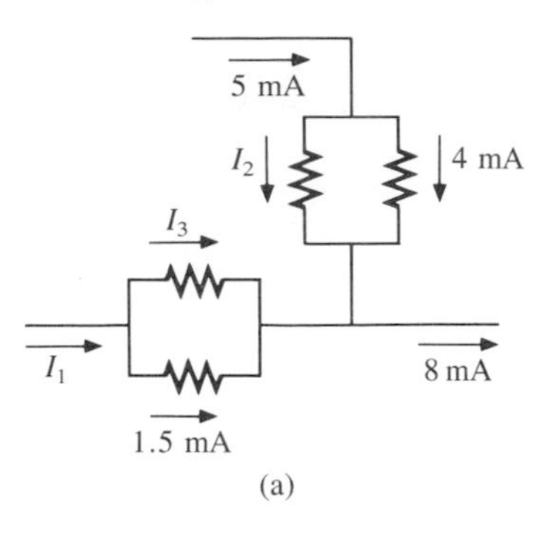


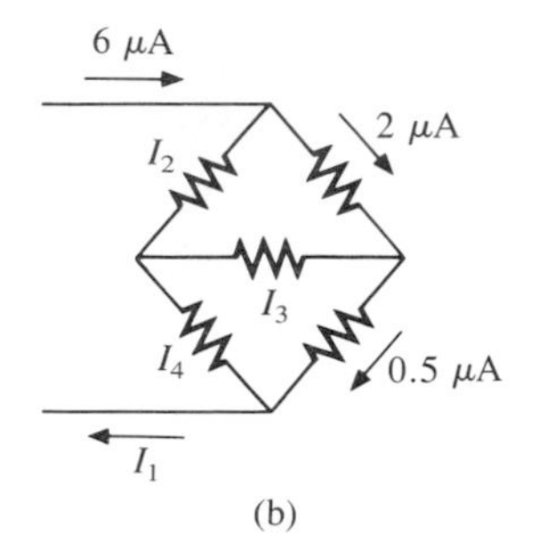

**FIG. 6.64**

***18.** Find the unknown quantities for the circuits of Fig. 6.65 using the information provided.

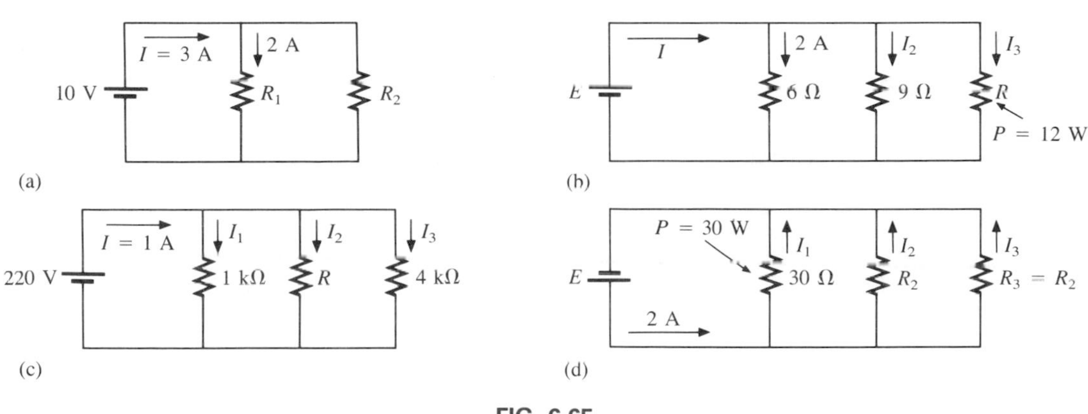

**FIG. 6.65**

## SECTION 6.6 Current Divider Rule

**19.** Using the current divider rule, find the unknown currents for the networks of Fig. 6.66.

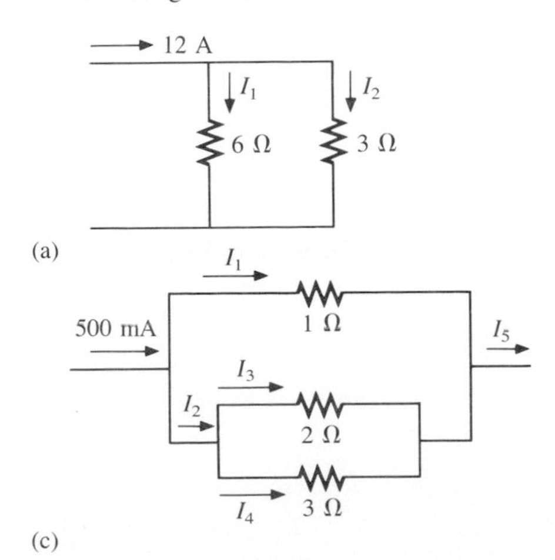


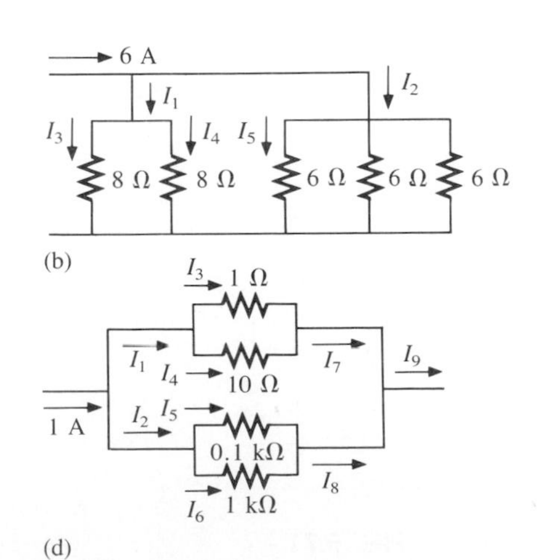


**FIG. 6.66**

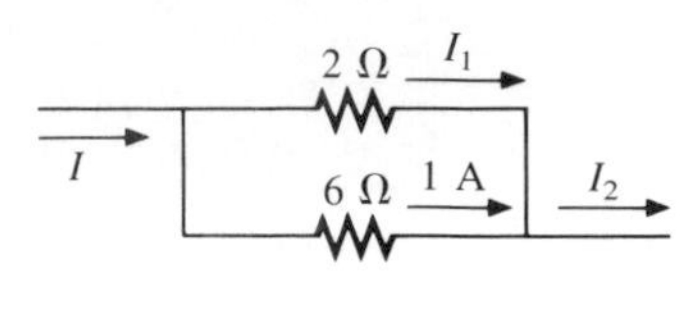

(a)

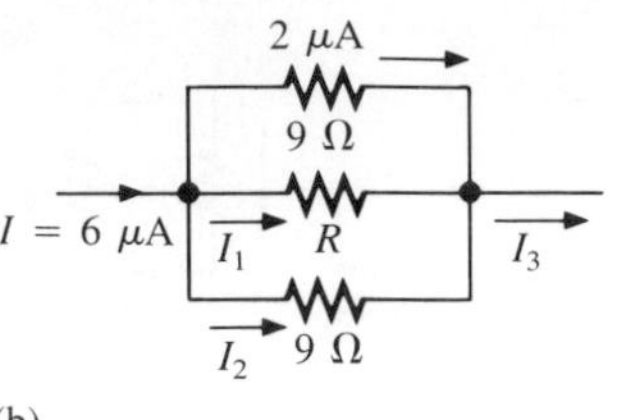


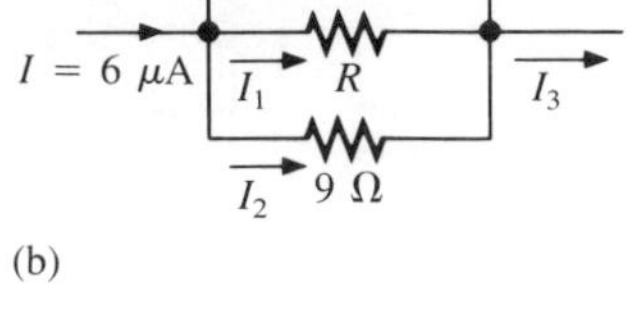

(b)

FIG. 6.67

**20.** Find the unknown quantities using the information provided for the networks of Fig. 6.67.

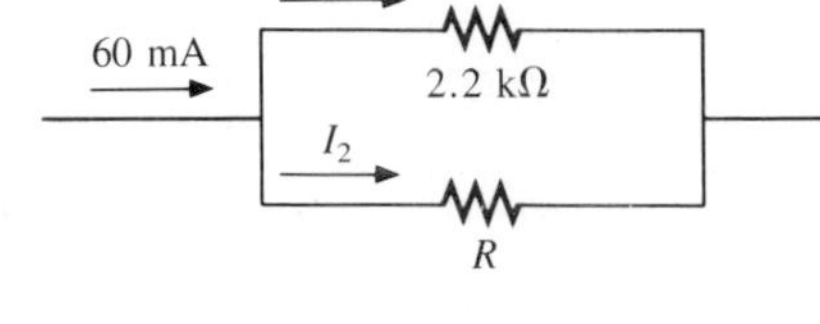


FIG. 6.68

***21.** Calculate the resistor $R$ for the network of Fig. 6.68 that will insure the current $I_1 = 3I_2$.

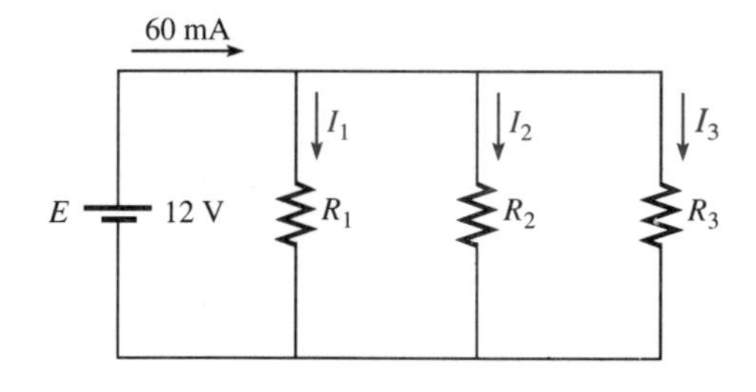

FIG. 6.69

***22.** Design the network of Fig. 6.69 such that $I_2 = 4I_1$ and $I_3 = 3I_2$.

## SECTION 6.7 Voltage Sources in Parallel

**23.** Determine the currents $I_1$ and $I_2$ for the network of Fig. 6.70.

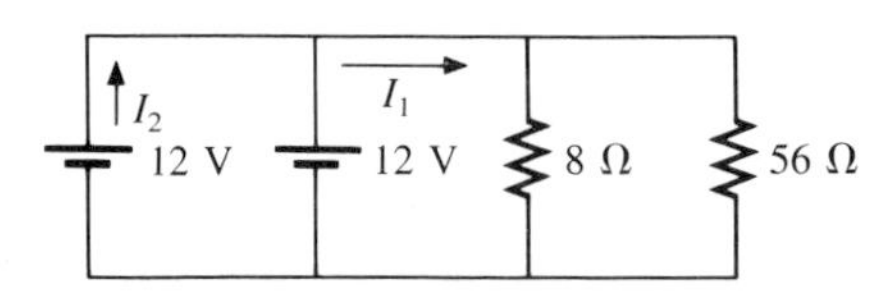

FIG. 6.70

## SECTION 6.8 Open and Short Circuits

**24.** For the network of Fig. 6.71:

**a.** Determine $I_s$ and $V_L$.

**b.** Determine $I_s$ if $R_L$ is shorted out.

**c.** Determine $V_L$ if $R_L$ is replaced by an open circuit.

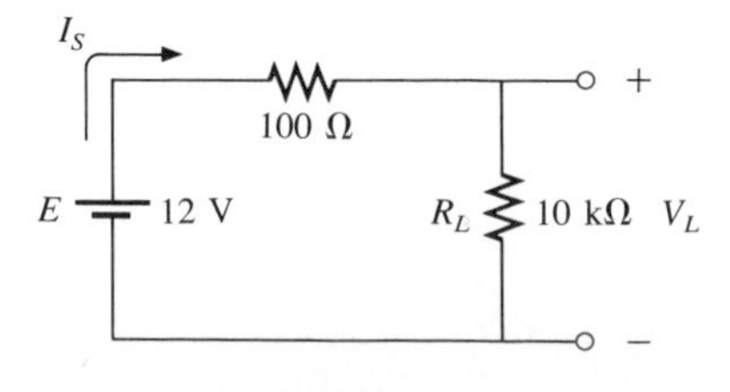

FIG. 6.71

**25.** For the network of Fig. 6.72:
  **a.** Determine the open-circuit voltage $V_L$.
  **b.** Place a short circuit across the output terminals and determine the current through the short circuit.
  **c.** If the 2.2-kΩ resistor is short circuited, what is the new value of $V_L$?
  **d.** Repeat part (b) with the 4.7-kΩ resistor replaced by an open circuit.

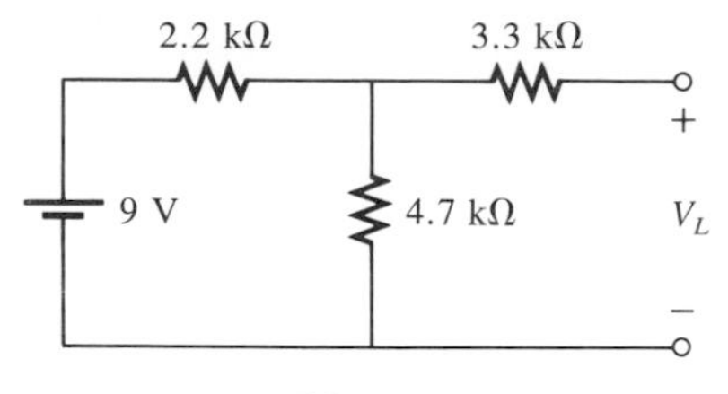

FIG. 6.72

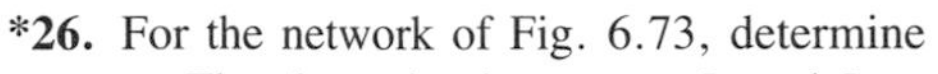

***26.** For the network of Fig. 6.73, determine
  **a.** The short-circuit currents $I_1$ and $I_2$.
  **b.** The voltages $V_1$ and $V_2$.
  **c.** The source current $I_s$.

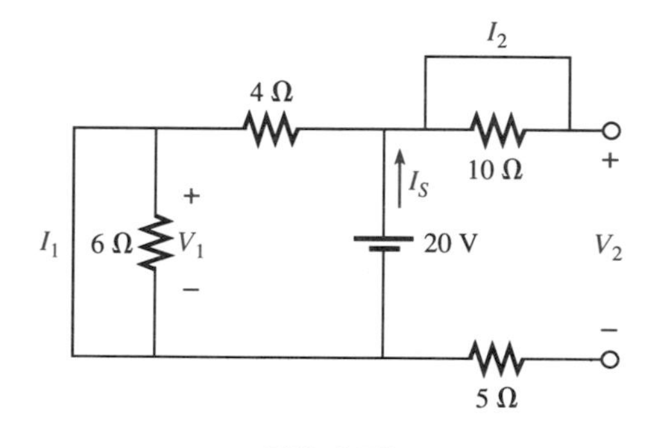


FIG. 6.73

## SECTION 6.9 Voltmeters: Loading Effect

**27.** For the network of Fig. 6.74:
  **a.** Determine the voltage $V_2$.
  **b.** Determine the reading of a DMM having an internal resistance of 11 MΩ when used to measure $V_2$.
  **c.** Repeat part (b) with a VOM having an ohm/volt rating of 20,000 using the 10-V scale. Compare the results of parts (b) and (c).
  **d.** Repeat part (c) with $R_1 = 100$ kΩ and $R_2 = 200$ kΩ.
  **e.** Based on the above can you make any general conclusions about the use of a voltmeter?

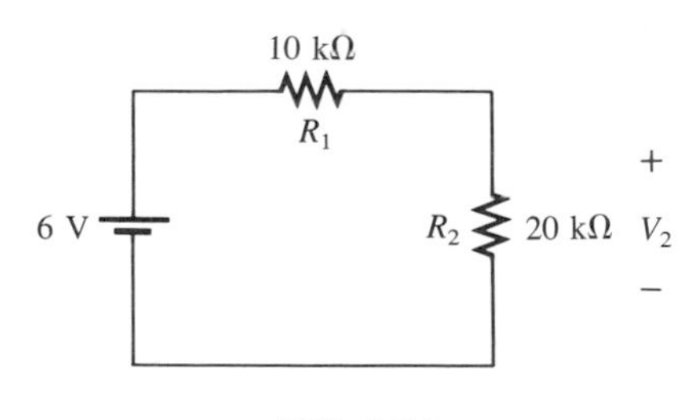


FIG. 6.74

## SECTION 6.10 Computer Analysis

### PSPICE

**28.** Write an input file to solve for the current through each resistor of Fig. 6.57.

**29.** Write an input file to find the source current and the current through each resistor of Fig. 6.61.

**30.** Write an input file to find the current through each resistor of Fig. 6.62.

### BASIC

**31.** Write a program to determine the total resistance and conductance of any number of elements in parallel.

**32.** Write a program to provide a complete solution of a parallel network with a single source and two parallel resistors. That is, print out the total resistance, input current, current through each branch, and power to each element.

**33.** Write a program that will tabulate the voltage $V_2$ of Fig. 6.74 measured by a VOM with an internal resistance of 200 kΩ as $R_2$ varies from 10 kΩ to 200 kΩ in increments of 10 kΩ.

## GLOSSARY

**Current divider rule** A method by which the current through parallel elements can be determined without first finding the voltage across those parallel elements.

**Kirchhoff's current law** The law which states that the algebraic sum of the currents entering and leaving a node is zero.

**Node** A junction of two or more branches.

**Ohm/volt rating** A rating used to determine both the current sensitivity of the movement and the internal resistance of the meter.

**Open circuit** The absence of a direct connection between two points in a network.

**Parallel circuit** A circuit configuration in which the elements have two points in common.

**Short circuit** A direct connection of low resistive value that can significantly alter the behavior of an element or system.

# 7
# Series-Parallel Networks

## 7.1 SERIES-PARALLEL NETWORKS

A firm understanding of the basic principles associated with series and parallel circuits is a sufficient background to approach most series-parallel networks (a network being a combination of any number of series and parallel elements) with *one* source of voltage. Multisource networks are considered in detail in Chapters 8 and 9. In general,

***series-parallel networks are networks that contain both series and parallel circuit configurations.***

One can become proficient in the analysis of series-parallel networks only through exposure, practice, and experience. In time the path to the desired unknown becomes more obvious as one recalls similar configurations and the frustration resulting from choosing the wrong approach. There are a few steps that can be helpful in getting started on the first few exercises, although the value of each will only become apparent with experience.

1. Take a moment to study the problem ''in total'' and make a brief mental sketch of the overall approach you plan to use. The result may be time- and energy-saving shortcuts.
2. Next examine each branch independently before tying them together in series-parallel combinations. This will usually simplify the network and possibly reveal a direct approach toward obtaining one or more desired unknowns. It also eliminates many of the errors that might result due to the lack of a systematic approach.
3. Redraw the network as often as possible with the reduced branches and undisturbed unknown quantities to maintain clarity and provide the reduced networks for the trip back from the source.

4. When you have a solution, check that it is reasonable by considering the magnitudes of the energy source and the elements in the network. If it does not seem reasonable, either solve the circuit using another approach or check over your work very carefully.

The block diagram approach will be employed to emphasize the fact that combinations of elements, not simply single resistive elements, can be in series or parallel. The approach will also reveal the number of seemingly different networks that have the same basic structure and therefore can involve similar analysis techniques.

The analysis of series-parallel dc networks with a single source usually requires that the resistance "seen" by the source ($R_T$) be determined by combining series and parallel elements until a single equivalent resistance remains. The source current ($I_S$) can then be calculated and the remaining currents of the network determined by working back through the network.

Initially, there will be some concern about identifying series and parallel elements and branches and choosing the best procedure to follow toward a solution. However, as you progress through the examples and try a few problems, a common path toward most solutions will surface that can actually make the analysis of such systems an interesting, enjoyable experience.

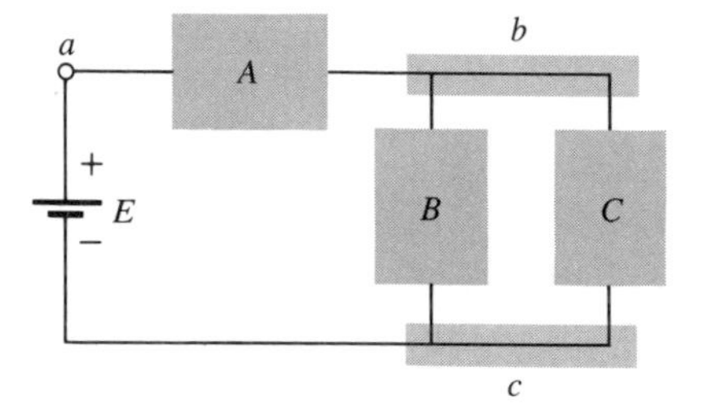


FIG. 7.1

In Fig. 7.1, blocks $B$ and $C$ are in parallel (points $b$ and $c$ in common), and the voltage source $E$ is in series with block $A$ (point $a$ in common). The parallel combination of $B$ and $C$ is also in series with $A$ and the voltage source $E$ due to the common points $b$ and $c$, respectively.

To insure that the analysis to follow is as clear and uncluttered as possible, the following notation will be used for series and parallel combinations of elements. For series resistors $R_1$ and $R_2$, a comma will be inserted between their subscript notation as shown here:

$$R_{1,2} = R_1 + R_2$$

For parallel resistors $R_1$ and $R_2$, the parallel symbol will be inserted between their subscript notation as follows:

$$R_{1\|2} = R_1 \parallel R_2 = \frac{R_1 R_2}{R_1 + R_2}$$

---

**EXAMPLE 7.1.** If each block of Fig. 7.1 were a single resistive element, the network of Fig. 7.2 might result.

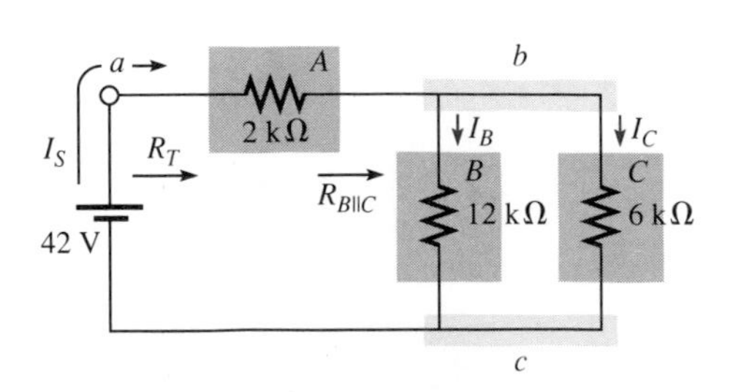


FIG. 7.2

The parallel combination of $R_B$ and $R_C$ results in

$$R_{B\|C} = R_B \parallel R_C = \frac{(12\text{ k}\Omega)(6\text{ k}\Omega)}{12\text{ k}\Omega + 6\text{ k}\Omega} = 4\text{ k}\Omega$$

The equivalent resistance $R_{B\|C}$ is then in series with $R_A$ and the total resistance "seen" by the source is

$$\begin{aligned} R_T &= R_A + R_{B\|C} \\ &= 2\text{ k}\Omega + 4\text{ k}\Omega = \mathbf{6\text{ k}\Omega} \end{aligned}$$

The result is an equivalent network, as shown in Fig. 7.3, permitting the determination of the source current $I_S$.

$$I_S = \frac{E}{R_T} = \frac{42\text{ V}}{6\text{ k}\Omega} = \mathbf{7\text{ mA}}$$

and, since the source and $R_A$ are in series,

$$I_A = I_S = 7\text{ mA}$$

We can then use the equivalent network of Fig. 7.4 to determine $I_B$ and $I_C$ using the current divider rule:

$$I_B = \frac{6\text{ k}\Omega(I_S)}{6\text{ k}\Omega + 12\text{ k}\Omega} = \frac{6}{18}I_S = \frac{1}{3}(7\text{ mA}) = \mathbf{2\tfrac{1}{3}\text{ mA}}$$

$$I_C = \frac{12\text{ k}\Omega(I_S)}{12\text{ k}\Omega + 6\text{ k}\Omega} = \frac{12}{18}I_S = \frac{2}{3}(7\text{ mA}) = \mathbf{4\tfrac{2}{3}\text{ mA}}$$

or, applying Kirchhoff's current law,

$$I_C = I_S - I_B = 7\text{ mA} - 2\frac{1}{3}\text{ mA} = \frac{21}{3}\text{ mA} - \frac{7}{3}\text{ mA}$$
$$= \frac{14}{3}\text{ mA} = \mathbf{4\tfrac{2}{3}\text{ mA}}$$

Note in the preceding solution that we worked back to the source to obtain the source current or total current supplied by the source. The remaining unknowns were then determined by working back through the network to find the other unknowns.

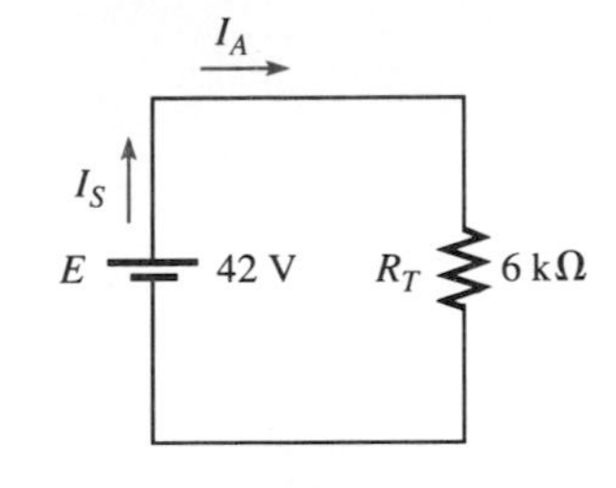

FIG. 7.3

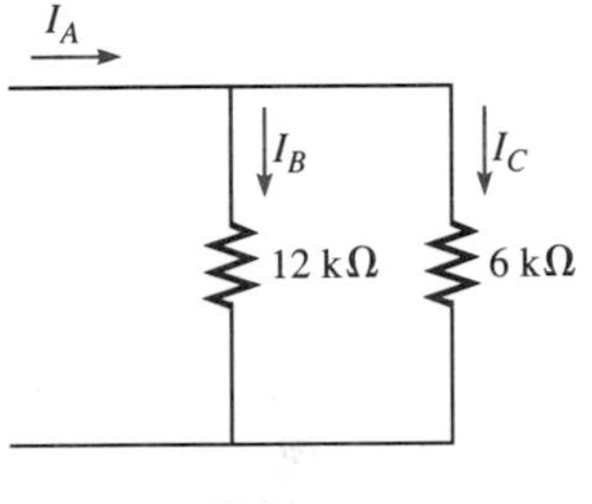

FIG. 7.4

**EXAMPLE 7.2.** It is also possible that the blocks $A$, $B$, and $C$ of Fig. 7.1 contain the elements and configurations of Fig. 7.5. Recalling Step 2, for branch

$$A\text{:}\quad R_A = 4\ \Omega$$

$$B\text{:}\quad R_B = R_2 \parallel R_3 = R_{2\parallel 3} = \frac{R}{N} = \frac{4\ \Omega}{2} = 2\ \Omega$$

$$C\text{:}\quad R_C = R_4 + R_5 = R_{4,5} = 0.5\ \Omega + 1.5\ \Omega = 2\ \Omega$$

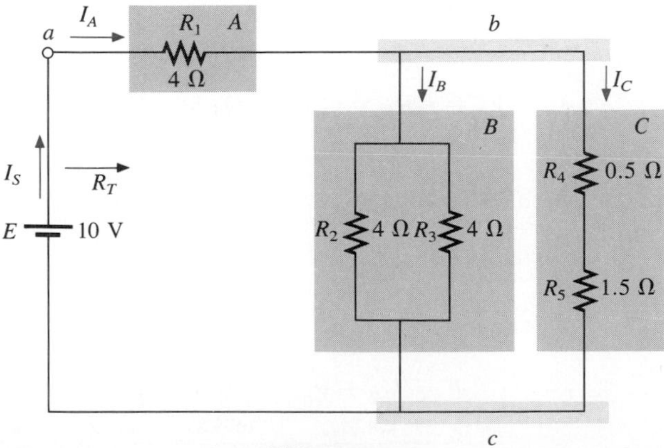

FIG. 7.5

Branches $B$ and $C$ are still in parallel and

$$R_{B\|C} = \frac{R}{N} = \frac{2\ \Omega}{2} = 1\ \Omega$$

with

$$R_T = R_A + R_{B\|C} = 4\ \Omega + 1\ \Omega = \mathbf{5\ \Omega}$$

(Note the similarity between this equation and that obtained for Example 7.1.)

and

$$I_S = \frac{E}{R_T} = \frac{10\ \text{V}}{5\ \Omega} = \mathbf{2\ A}$$

We can find the currents $I_A$, $I_B$, and $I_C$ using the reduction of the network of Fig. 7.5 (recall Step 3) as found in Fig. 7.6. Note that $I_A$, $I_B$,

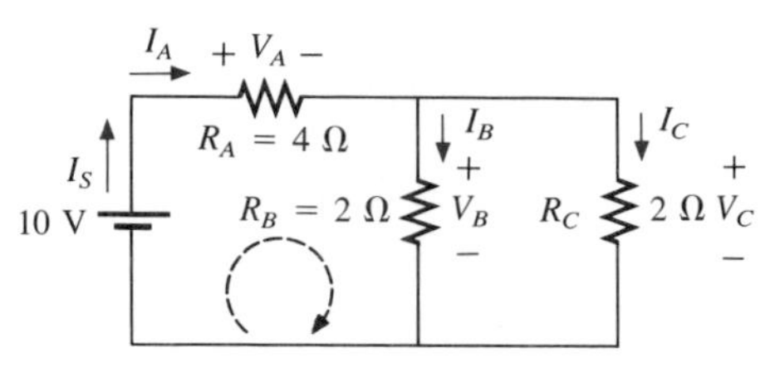


**FIG. 7.6**

and $I_C$ are the same in Figs. 7.5 and 7.6 and therefore also appear in Fig. 7.6. In other words, the currents $I_A$, $I_B$, and $I_C$ of Fig. 7.6 will have the same magnitude as the same currents of Fig. 7.5.

$$I_A = I_S = \mathbf{2\ A}$$

and

$$I_B = I_C = \frac{I_A}{2} = \frac{I_S}{2} = \frac{2\ \text{A}}{2} = \mathbf{1\ A}$$

Returning to the network of Fig. 7.5, we have

$$I_{R_2} = I_{R_3} = \frac{I_B}{2} = \mathbf{0.5\ A}$$

The voltages $V_A$, $V_B$, and $V_C$ from either figure are

$$V_A = I_A R_A = (2\ \text{A})(4\ \Omega) = \mathbf{8\ V}$$
$$V_B = I_B R_B = (1\ \text{A})(2\ \Omega) = \mathbf{2\ V}$$
$$V_C = V_B = \mathbf{2\ V}$$

Applying Kirchhoff's voltage law for the loop indicated in Fig. 7.6, we obtain

$$\Sigma_{\circlearrowright} V = E - V_A - V_B = 0$$

or

$$E = V_A + V_B = 8\ \text{V} + 2\ \text{V}$$
$$10\ \text{V} = 10\ \text{V} \qquad \text{(checks)}$$

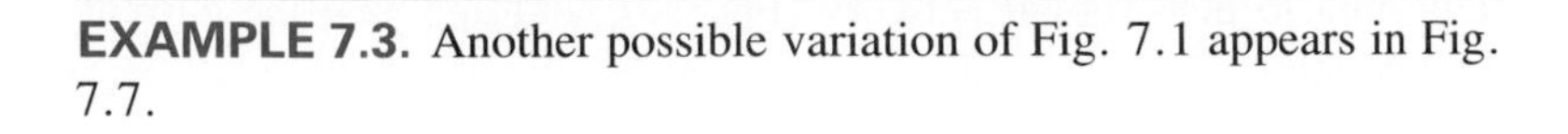

**EXAMPLE 7.3.** Another possible variation of Fig. 7.1 appears in Fig. 7.7.

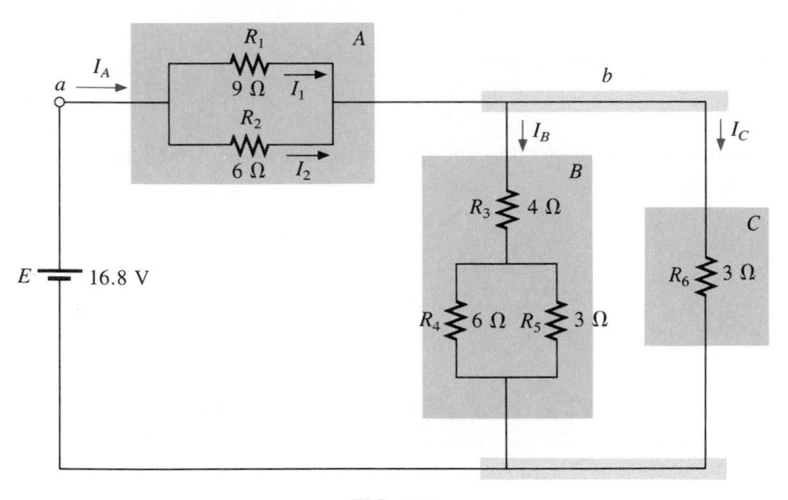


FIG. 7.7

$$R_A = R_{1\|2} = \frac{(9\ \Omega)(6\ \Omega)}{9\ \Omega + 6\ \Omega} = \frac{54\ \Omega}{15} = 3.6\ \Omega$$

$$R_B = R_3 + R_{4\|5} = 4\ \Omega + \frac{(6\ \Omega)(3\ \Omega)}{6\ \Omega + 3\ \Omega} = 4\ \Omega + 2\ \Omega = 6\ \Omega$$

$$R_C = 3\ \Omega$$

The network of Fig. 7.7 can then be redrawn in reduced form as shown in Fig. 7.8. Note the similarities between this circuit and those of Figs. 7.2 and 7.6.

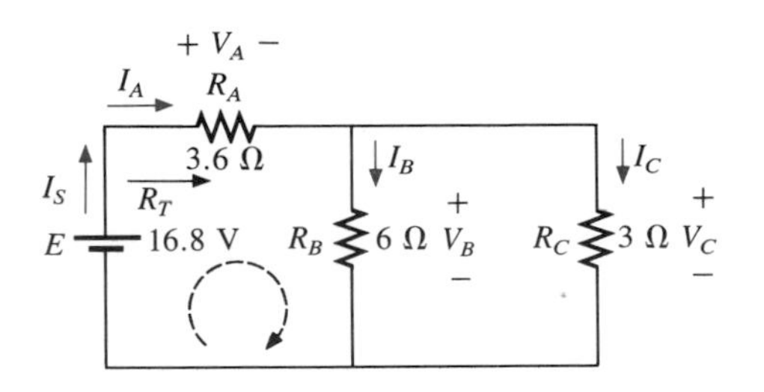


FIG. 7.8

$$R_T = R_A + R_{B\|C} = 3.6\ \Omega + \frac{(6\ \Omega)(3\ \Omega)}{6\ \Omega + 3\ \Omega}$$

$$= 3.6\ \Omega + 2\ \Omega = \mathbf{5.6\ \Omega}$$

$$I_S = \frac{E}{R_T} = \frac{16.8\ \text{V}}{5.6\ \Omega} = \mathbf{3\ A}$$

$$I_A = I_S = \mathbf{3\ A}$$

Applying the current divider rule yields

$$I_B = \frac{R_C I_A}{R_C + R_B} = \frac{(3\ \Omega)(3\ \text{A})}{3\ \Omega + 6\ \Omega} = \frac{9\ \text{A}}{9} = \mathbf{1\ A}$$

By Kirchhoff's current law,

$$I_C = I_A - I_B = 3\ \text{A} - 1\ \text{A} = \mathbf{2\ A}$$

By Ohm's law,

$$V_A = I_A R_A = (3\ \text{A})(3.6\ \Omega) = \mathbf{10.8\ V}$$

$$V_B = I_B R_B = V_C = I_C R_C = (2\ \text{A})(3\ \Omega) = \mathbf{6\ V}$$

Returning to the original network (Fig. 7.7) and applying the current divider rule,

$$I_1 = \frac{R_2 I_A}{R_2 + R_1} = \frac{(6\ \Omega)(3\ \text{A})}{6\ \Omega + 9\ \Omega} = \frac{18\ \text{A}}{15} = \mathbf{1.2\ A}$$

By Kirchhoff's current law,

$$I_2 = I_A - I_1 = 3\ \text{A} - 1.2\ \text{A} = \mathbf{1.8\ A}$$

Kirchhoff's voltage law for the loop indicated in the reduced network (Fig. 7.8) is

$$\Sigma_{\circlearrowright} V = E - V_A - V_B = 0$$
$$E = V_A + V_B$$
$$16.8\ \text{V} = 10.8\ \text{V} + 6\ \text{V}$$
$$16.8\ \text{V} = 16.8\ \text{V} \quad \text{(checks)}$$

Figures 7.2, 7.5, and 7.7 are only a few of the infinite variety of configurations that the network can assume starting with the basic arrangement of Fig. 7.1. They were included in our discussion to emphasize the importance of considering each branch independently before finding the solution for the network as a whole.

There are a variety of ways in which the blocks of Fig. 7.1 can be arranged. In fact, there is no limit on the number of series-parallel configurations that can appear within a given network. In reverse, the block diagram approach can be effectively used to reduce the apparent complexity of a system by identifying the major series and parallel components of the network. This approach will be demonstrated in the next few examples.

## 7.2 DESCRIPTIVE EXAMPLES

**EXAMPLE 7.4.** Find the current $I_4$ and the voltage $V_2$ for the network of Fig. 7.9.

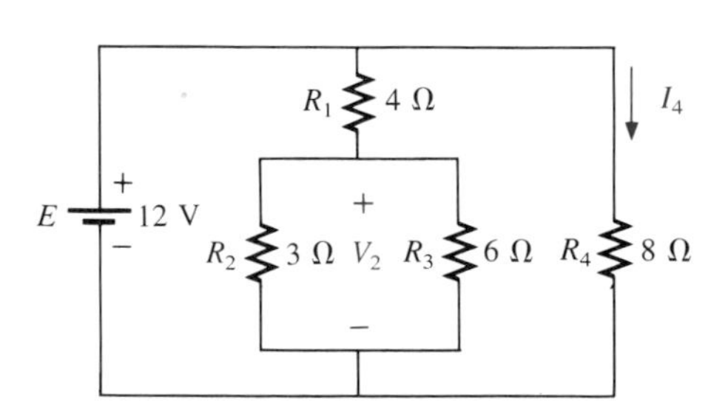


FIG. 7.9

***Solution:*** In this case, particular unknowns are requested instead of a complete solution. It would, therefore, be a waste of time to find all the currents and voltages of the network. The method employed should concentrate on obtaining only the unknowns requested. With the block diagram approach, the network has the basic structure of Fig. 7.10,

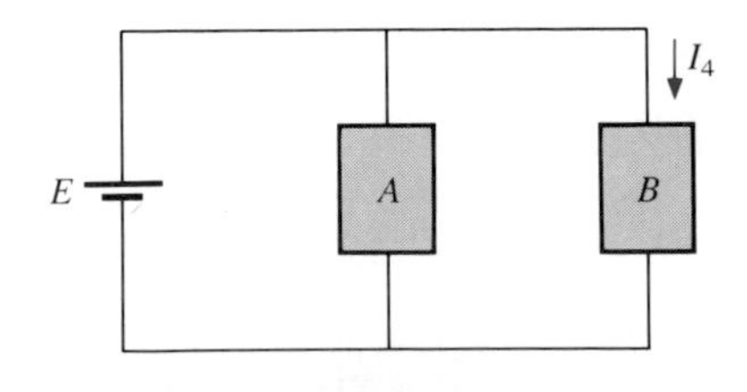


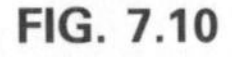
FIG. 7.10

clearly indicating that the three branches are in parallel and the voltage across $A$ and $B$ is the supply voltage. The current $I_4$ is now immediately obvious as simply the supply voltage divided by the resultant resistance for $B$. If desired, block $A$ could be broken down further as shown in Fig. 7.11 to identify $C$ and $D$ as series elements with the voltage $V_2$ capable of being determined using the voltage divider rule once the resistance of $C$ and $D$ is reduced to a single value. This is an example of how a mental sketch of the approach might be made before applying laws, rules, and so on, to avoid dead ends and growing frustration.

Applying Ohm's law,

$$I_4 = \frac{E}{R_B} = \frac{E}{R_4} = \frac{12\ \text{V}}{8\ \Omega} = \mathbf{1.5\ A}$$

Combining the resistors $R_2$ and $R_3$ of Fig. 7.9 will result in

$$R_D = R_2 \parallel R_3 = 3\ \Omega \parallel 6\ \Omega = \frac{(3\ \Omega)(6\ \Omega)}{3\ \Omega + 6\ \Omega} = \frac{18\ \Omega}{9} = 2\ \Omega$$

and, applying the voltage divider rule,

$$V_2 = \frac{R_D E}{R_D + R_C} = \frac{(2\ \Omega)(12\ \text{V})}{2\ \Omega + 4\ \Omega} = \frac{24\ \text{V}}{6} = \mathbf{4\ V}$$

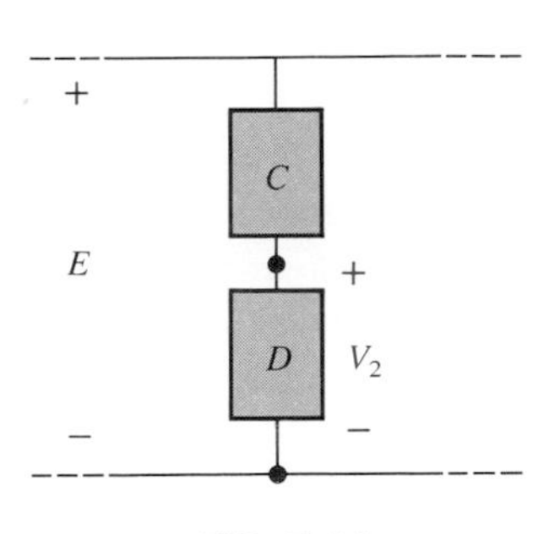


FIG. 7.11

**EXAMPLE 7.5.** Find the indicated currents and voltages for the network of Fig. 7.12.

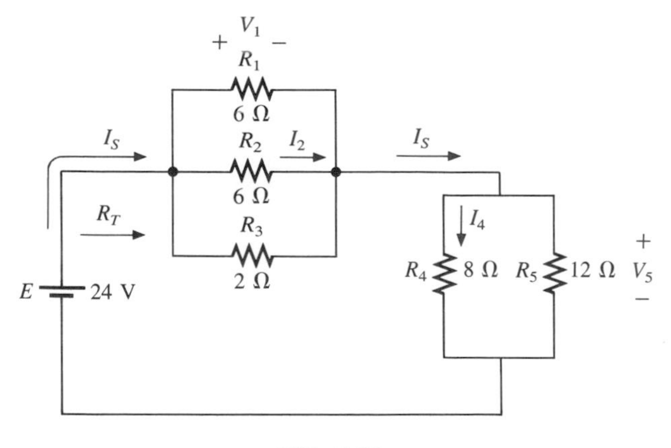

FIG. 7.12

***Solution:*** Again, only specific unknowns are requested. When the network is redrawn, it will be particularly important to note which unknowns are preserved and which will have to be determined using the original configuration. The block diagram of the network may appear as shown in Fig. 7.13, clearly revealing that $A$ and $B$ are in series. Note in this form the number of unknowns that have been preserved. The voltage $V_1$ will be the same across the three parallel branches of Fig. 7.12,

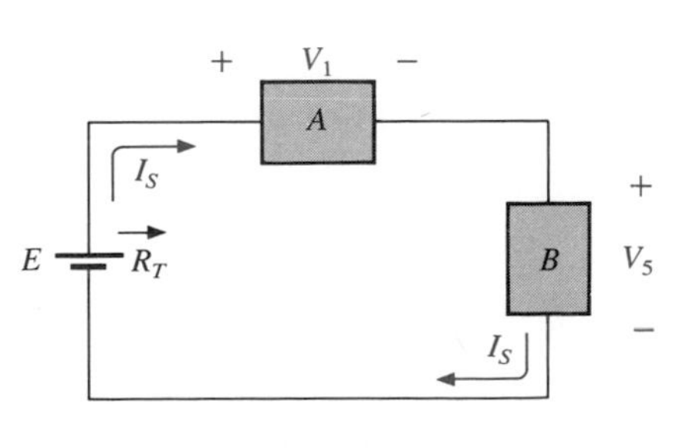

FIG. 7.13

and $V_5$ will be the same across $R_4$ and $R_5$. The unknown currents $I_2$ and $I_4$ are lost, since they represent the currents through only one of the parallel branches. However, once $V_1$ and $V_5$ are known, the required currents can be found using Ohm's law.

$$R_{1\|2} = \frac{R}{N} = \frac{6\ \Omega}{2} = 3\ \Omega$$

$$R_A = R_{1\|2\|3} = \frac{(3\ \Omega)(2\ \Omega)}{3\ \Omega + 2\ \Omega} = \frac{6\ \Omega}{5} = 1.2\ \Omega$$

$$R_B = R_{4\|5} = \frac{(8\ \Omega)(12\ \Omega)}{8\ \Omega + 12\ \Omega} = \frac{96\ \Omega}{20} = 4.8\ \Omega$$

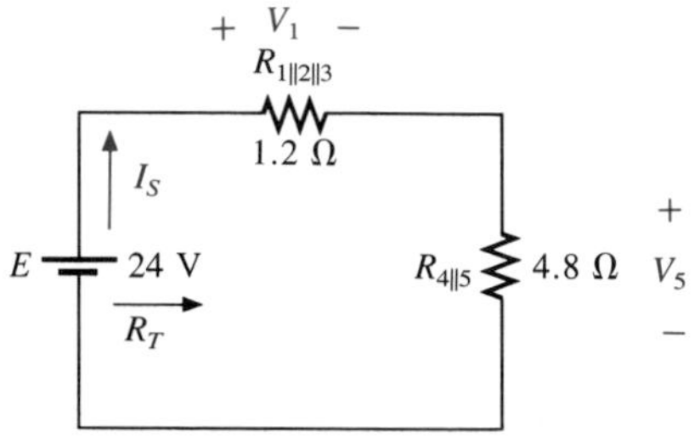

The reduced form of Fig. 7.12 will then appear as shown in Fig. 7.14, and

FIG. 7.14

$$R_T = R_{1\|2\|3} + R_{4\|5} = 1.2\ \Omega + 4.8\ \Omega = \mathbf{6\ \Omega}$$

$$I_S = \frac{E}{R_T} = \frac{24\text{ V}}{6\ \Omega} = \mathbf{4\ A}$$

with

$$V_1 = I_S R_{1\|2\|3} = (4\text{ A})(1.2\ \Omega) = \mathbf{4.8\ V}$$

$$V_5 = I_S R_{4\|5} = (4\text{ A})(4.8\ \Omega) = \mathbf{19.2\ V}$$

Applying Ohm's law,

$$I_4 = \frac{V_5}{R_4} = \frac{19.2\text{ V}}{8\ \Omega} = \mathbf{2.4\ A}$$

$$I_2 = \frac{V_2}{R_2} = \frac{V_1}{R_2} = \frac{4.8\text{ V}}{6\ \Omega} = \mathbf{0.8\ A}$$

---

The next example demonstrates that unknown voltages do not have to be across elements but can exist between any two points in a network. In addition, the importance of redrawing the network in a more familiar form is clearly revealed by the analysis to follow.

---

**EXAMPLE 7.6.**

a. Find the voltages $V_1$, $V_3$, and $V_{ab}$ for the network of Fig. 7.15.
b. Calculate the source current $I_S$.

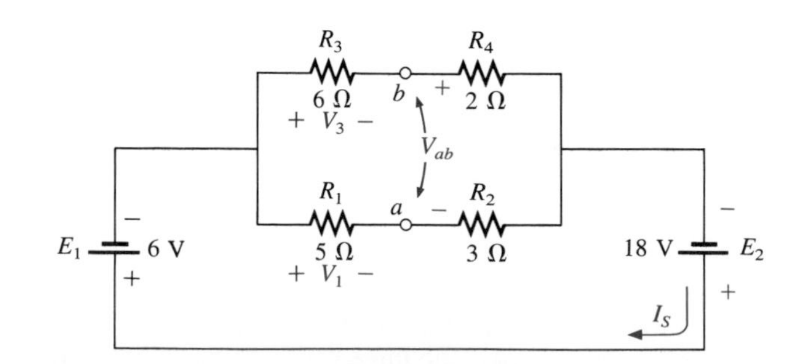


FIG. 7.15

***Solutions:*** This is one of those situations where it might be best to redraw the network before beginning the analysis. Since combining both sources will not affect the unknowns, the network is redrawn as shown in Fig. 7.16, establishing a parallel network with the total source voltage across each parallel branch. The net source voltage is the difference between the two with the polarity of the larger.

a. Note the similarities with Fig. 7.11, permitting the use of the voltage divider rule to determine $V_1$ and $V_3$:

$$V_1 = \frac{R_1 E}{R_1 + R_2} = \frac{(5\ \Omega)(12\ \text{V})}{5\ \Omega + 3\ \Omega} = \frac{60\ \text{V}}{8} = \mathbf{7.5\ V}$$

$$V_3 = \frac{R_3 E}{R_3 + R_4} = \frac{(6\ \Omega)(12\ \text{V})}{6\ \Omega + 2\ \Omega} = \frac{72\ \text{V}}{8} = \mathbf{9\ V}$$

The open-circuit voltage $V_{ab}$ is determined by applying Kirchhoff's voltage law around the indicated loop of Fig. 7.16 in the clockwise direction starting at terminal $a$.

$$+V_1 - V_3 + V_{ab} = 0$$

and

$$V_{ab} = V_3 - V_1 = 9\ \text{V} - 7.5\ \text{V} = \mathbf{1.5\ V}$$

b. By Ohm's law,

$$I_1 = \frac{V_1}{R_1} = \frac{7.5\ \text{V}}{5\ \Omega} = 1.5\ \text{A}$$

$$I_3 = \frac{V_3}{R_3} = \frac{9\ \text{V}}{6\ \Omega} = 1.5\ \text{A}$$

By Kirchhoff's current law,

$$I_S = I_1 + I_3 = 1.5\ \text{A} + 1.5\ \text{A} = \mathbf{3\ A}$$

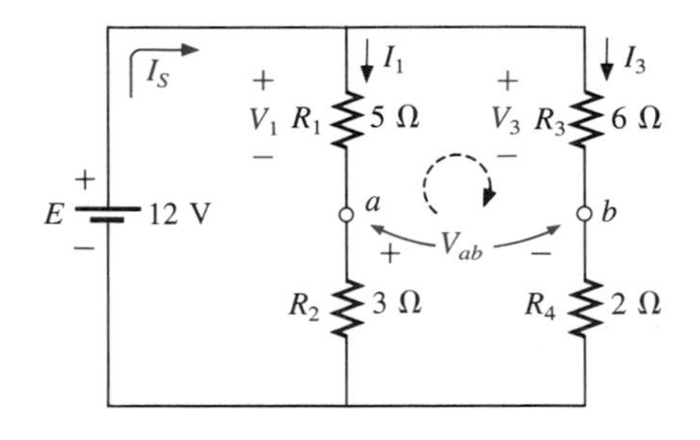


FIG. 7.16

**EXAMPLE 7.7.** For the network of Fig. 7.17, determine the voltages $V_1$ and $V_2$ and the current $I$.

***Solution:*** It would indeed be difficult to analyze the network in the form of Fig. 7.17 with the symbolic notation for the sources and the reference or ground connection in the upper left-hand corner of the diagram. However, when the network is redrawn as shown in Fig. 7.18,

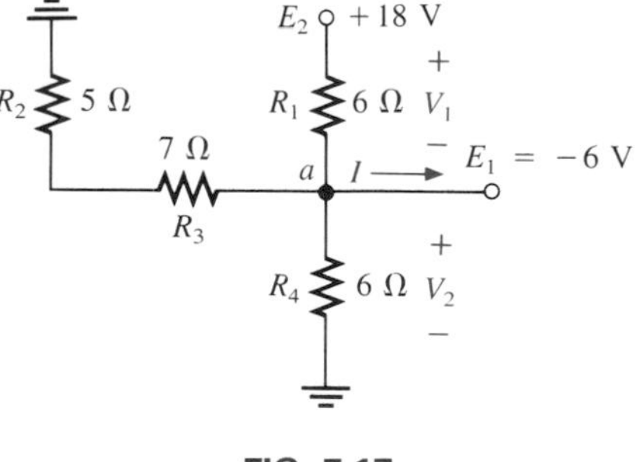


FIG. 7.17

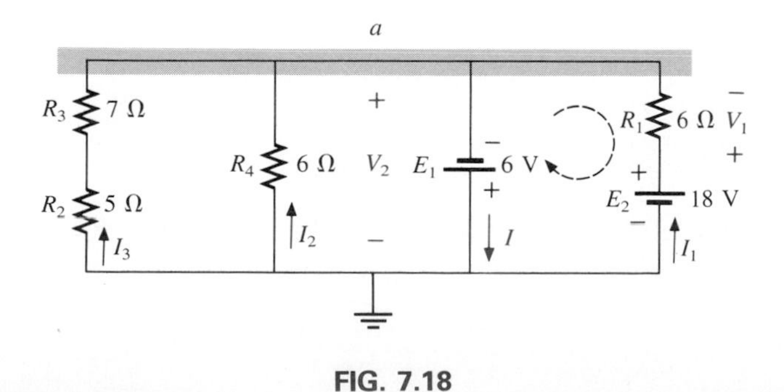


FIG. 7.18

the unknowns and the relationship between branches become significantly clearer. Note the common connection of the grounds and the replacing of the terminal notation by actual supplies.

It is now obvious that

$$V_2 = -E_1 = \mathbf{-6\ V}$$

The minus sign simply indicates that the chosen polarity for $V_2$ in Fig. 7.17 is opposite to the actual voltage. Applying Kirchhoff's voltage law to the loop indicated, we obtain

$$-E_1 + V_1 - E_2 = 0$$

and

$$V_1 = E_2 + E_1 = 18\ \text{V} + 6\ \text{V} = \mathbf{24\ V}$$

Applying Kirchhoff's current law to node $a$ yields

$$\begin{aligned} I &= I_1 + I_2 + I_3 \\ &= \frac{V_1}{R_1} + \frac{E_1}{R_4} + \frac{E_1}{R_2 + R_3} \\ &= \frac{24\ \text{V}}{6\ \Omega} + \frac{6\ \text{V}}{6\ \Omega} + \frac{6\ \text{V}}{12\ \Omega} \\ &= 4\ \text{A} + 1\ \text{A} + 0.5\ \text{A} \\ I &= \mathbf{5.5\ A} \end{aligned}$$

The next example is clear evidence of the fact that techniques learned in the current chapters will have far-reaching applications and will not be dropped for improved methods. Even though the transistor has not been introduced in this text, the dc levels of a transistor network can be examined using the basic rules and laws introduced in the early chapters of this text.

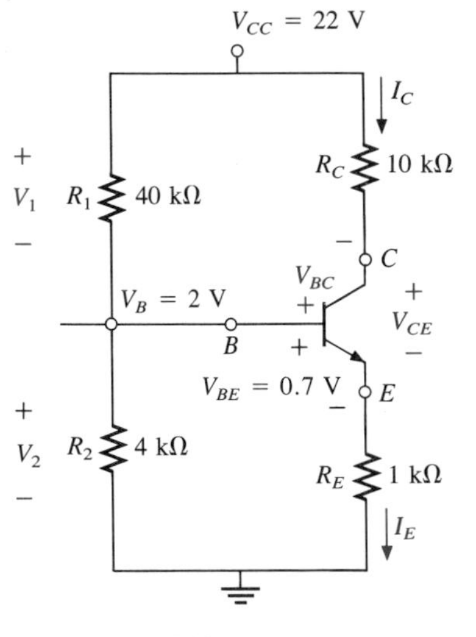

FIG. 7.19

**EXAMPLE 7.8.** For the transistor configuration of Fig. 7.19, in which $V_B$ and $V_{BE}$ have been provided:

a. Determine the voltage $V_E$ and the current $I_E$.
b. Calculate $V_1$.
c. Determine $V_{BC}$ using the fact that the approximation $I_C = I_E$ is often applied to transistor networks.
d. Calculate $V_{CE}$ using the information obtained in parts (a) through (c).

***Solutions:***

a. From Fig. 7.19, we find

$$V_2 = V_B = 2\ \text{V}$$

Writing Kirchhoff's voltage law around the lower loop yields

$$V_2 - V_{BE} - V_E = 0$$

or

$$V_E = V_2 - V_{BE} = 2\text{ V} - 0.7\text{ V} = \mathbf{1.3\text{ V}}$$

and

$$I_E = \frac{V_E}{R_E} = \frac{1.3\text{ V}}{1000\ \Omega} = \mathbf{1.3\text{ mA}}$$

b. The 22-V dc source can be split as shown in Fig. 7.20 as an aid in the required analysis. The section of interest can then be sketched as shown in Fig. 7.21, and Kirchhoff's law will result in

$$V_{CC} - V_1 - V_B = 0$$

or

$$V_1 = V_{CC} - V_B = 22\text{ V} - 2\text{ V} = \mathbf{20\text{ V}}$$

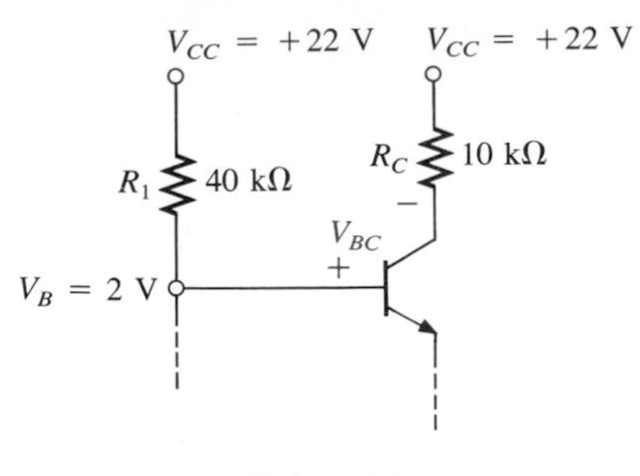

FIG. 7.20

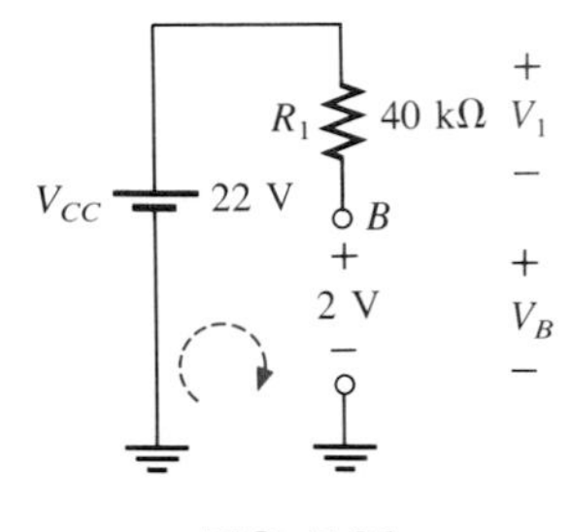


FIG. 7.21

c. Redrawing the section of the network of immediate interest will result in Fig. 7.22, where Kirchhoff's voltage law yields

$$V_B - V_{BC} + V_C - V_{CC} = 0$$

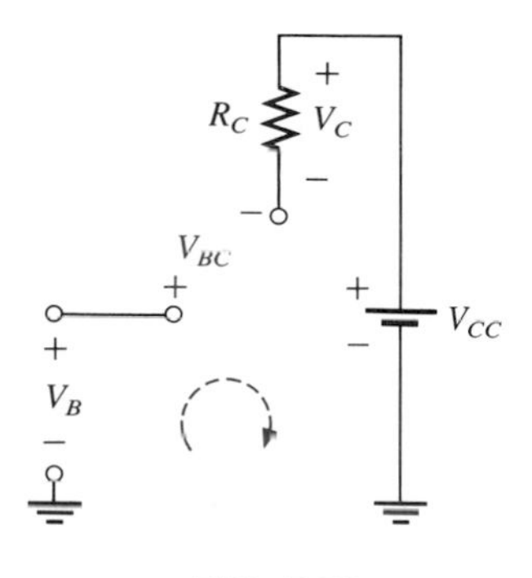

FIG. 7.22

or

$$V_{BC} = V_C + V_B - V_{CC}$$

with

$$V_C = I_C R_C = (1.3\text{ mA})(10\text{ k}\Omega) = 13\text{ V}$$

and

$$V_{BC} = 13\text{ V} + 2\text{ V} - 22\text{ V} = \mathbf{-7\text{ V}}$$

d. The appropriate section appears in Fig. 7.23. Application of Kirchhoff's voltage law will result in

$$V_E + V_{CE} + V_C = V_{CC}$$

or

$$V_{CE} = V_{CC} - V_C - V_E = 22 - 13 - 1.3 = \mathbf{7.7\text{ V}}$$

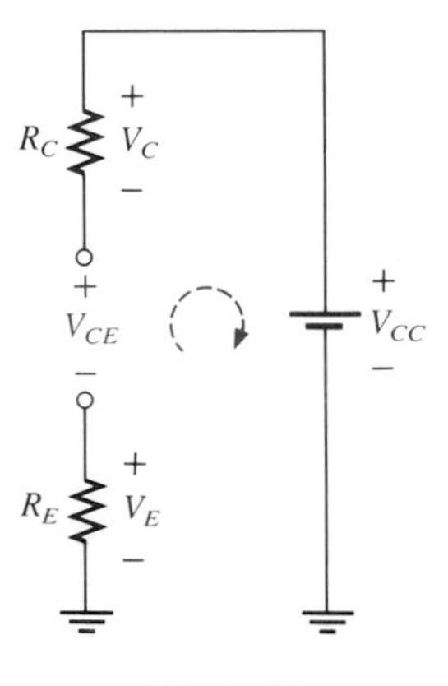


FIG. 7.23

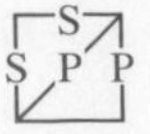

**EXAMPLE 7.9.** Calculate the indicated currents and voltage of Fig. 7.24.

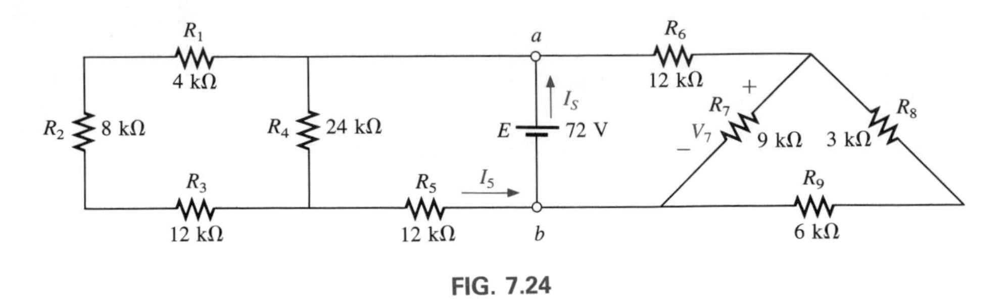


FIG. 7.24

***Solution:*** Redrawing the network after combining series elements yields Fig. 7.25, and

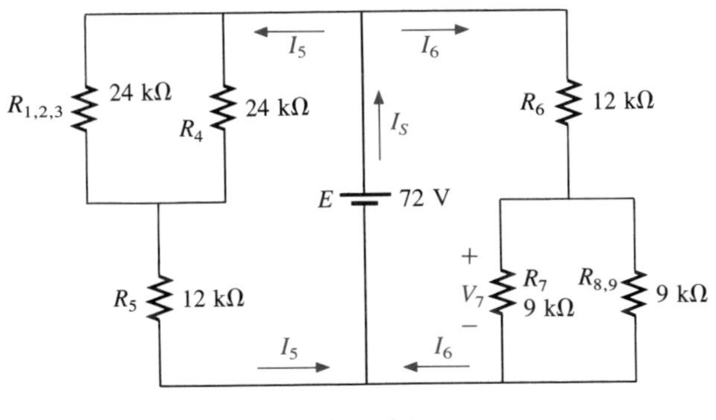


FIG. 7.25

$$I_5 = \frac{E}{R_{(1,2,3)\|4} + R_5} = \frac{72 \text{ V}}{12 \text{ k}\Omega + 12 \text{ k}\Omega} = \frac{72 \text{ V}}{24 \text{ k}\Omega} = \mathbf{3 \text{ mA}}$$

with

$$V_7 = \frac{R_{7\|(8,9)}E}{R_{7\|(8,9)} + R_6} = \frac{(4.5 \text{ k}\Omega)(72 \text{ V})}{4.5 \text{ k}\Omega + 12 \text{ k}\Omega} = \frac{324 \text{ V}}{16.5} = \mathbf{19.6 \text{ V}}$$

$$I_6 = \frac{V_7}{R_{7\|(8,9)}} = \frac{19.6 \text{ V}}{4.5 \text{ k}\Omega} = \mathbf{4.35 \text{ mA}}$$

and

$$I_S = I_5 + I_6 = 3 + 4.35 = \mathbf{7.35 \text{ mA}}$$

Since the potential difference between points *a* and *b* of Fig. 7.24 is fixed at *E* volts, the circuit to the right or left is unaffected if the network is reconstructed as shown in Fig. 7.26.

We can find each quantity required, except $I_S$, by analyzing each circuit independently. To find $I_S$, we must find the source current for each circuit and add it as in the above solution; that is, $I_S = I_5 + I_6$.

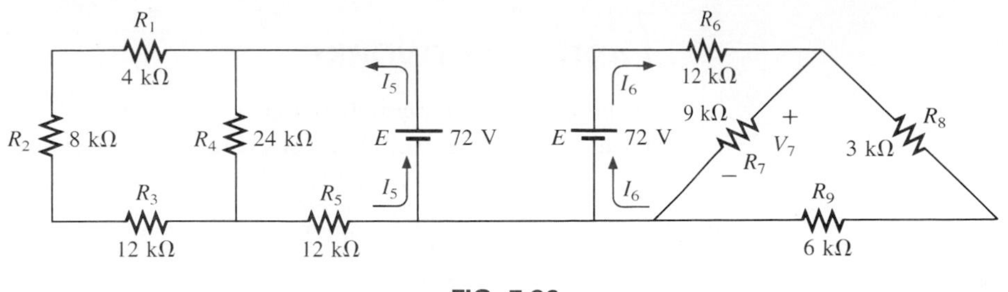

FIG. 7.26

**EXAMPLE 7.10.** This example will demonstrate the power of Kirchhoff's voltage law by determining the voltages $V_1$, $V_2$, and $V_3$ for the network of Fig. 7.27. For path 1 of Fig. 7.28,

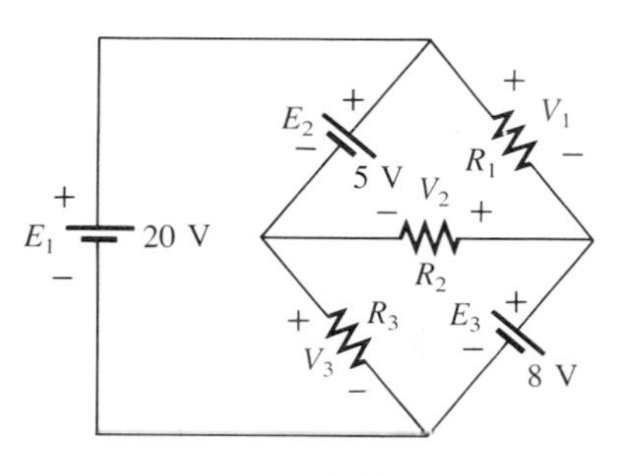


FIG. 7.27

$$E_1 - V_1 - E_3 = 0$$

and

$$V_1 = E_1 - E_3 = 20 \text{ V} - 8 \text{ V} = \mathbf{12\ V}$$

For path 2,

$$E_2 - V_1 - V_2 = 0$$

and

$$V_2 = E_2 - V_1 = 5 \text{ V} - 12 \text{ V} = \mathbf{-7\ V}$$

indicating that $V_2$ has a magnitude of 7 V but a polarity opposite to that appearing in Fig. 7.27. For path 3,

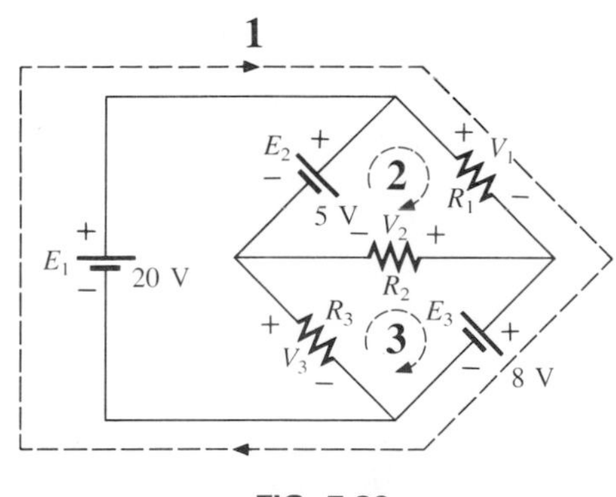


FIG. 7.28

$$V_3 + V_2 - E_3 = 0$$

and

$$V_3 = E_3 - V_2 = 8 \text{ V} - (-7 \text{ V}) = 8 \text{ V} + 7 \text{ V} = \mathbf{15\ V}$$

Note that the polarity of $V_2$ was maintained as originally assumed, requiring that $-7$ V be substituted for $V_2$.

## 7.3 LADDER NETWORKS

A three-section *ladder* network appears in Fig. 7.29. The reason for the terminology is quite obvious for the repetitive structure. There are basically two approaches used to solve networks of this type.

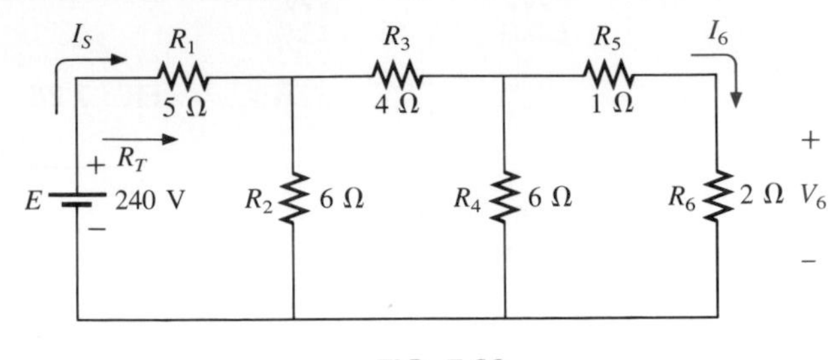

FIG. 7.29

### Method 1

Calculate the total resistance and resulting source current and then work back through the ladder until the desired current or voltage is obtained. This method is now employed to determine $V_6$ in Fig. 7.29.

Combining parallel and series elements as shown in Fig. 7.30 will result in the reduced network of Fig. 7.31, and

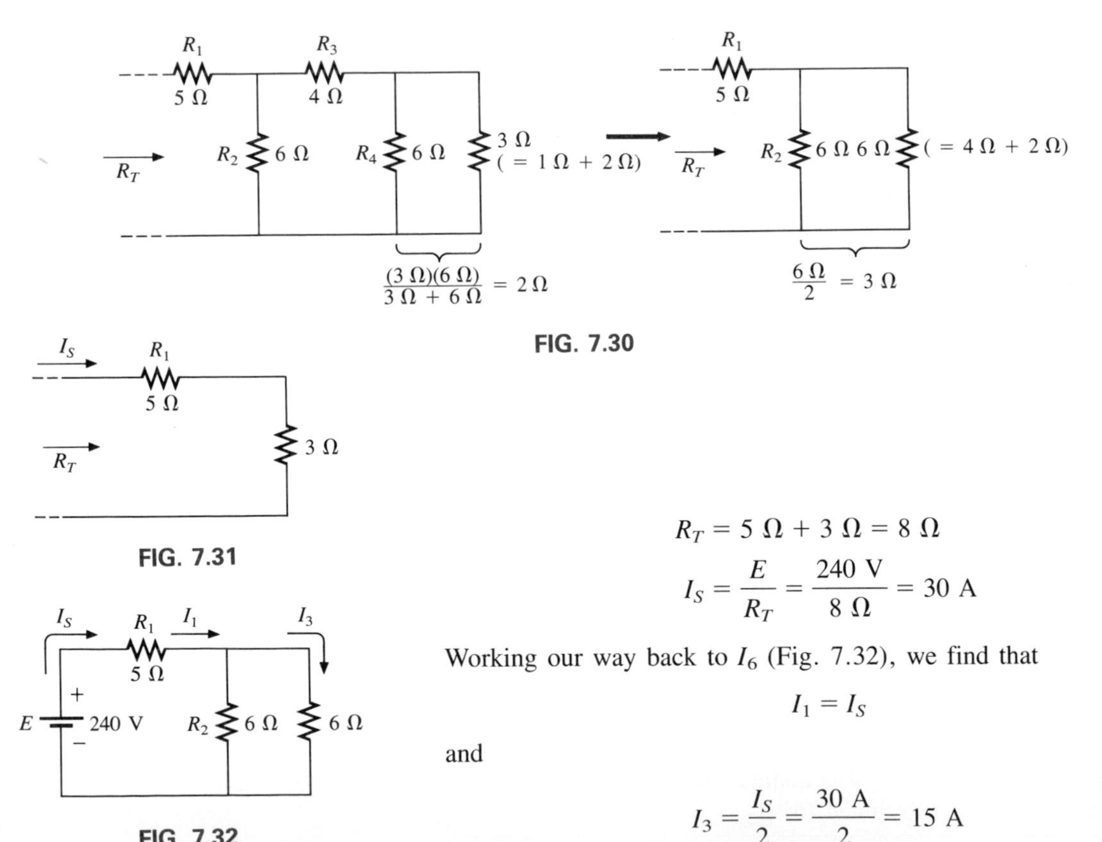


FIG. 7.30

FIG. 7.31

$$R_T = 5\ \Omega + 3\ \Omega = 8\ \Omega$$

$$I_S = \frac{E}{R_T} = \frac{240\ \text{V}}{8\ \Omega} = 30\ \text{A}$$

Working our way back to $I_6$ (Fig. 7.32), we find that

$$I_1 = I_S$$

and

$$I_3 = \frac{I_S}{2} = \frac{30\ \text{A}}{2} = 15\ \text{A}$$

FIG. 7.32

and, finally (Fig. 7.33),

$$I_6 = \frac{6I_3}{6+3} = \frac{6}{9}(15 \text{ A}) = 10 \text{ A}$$

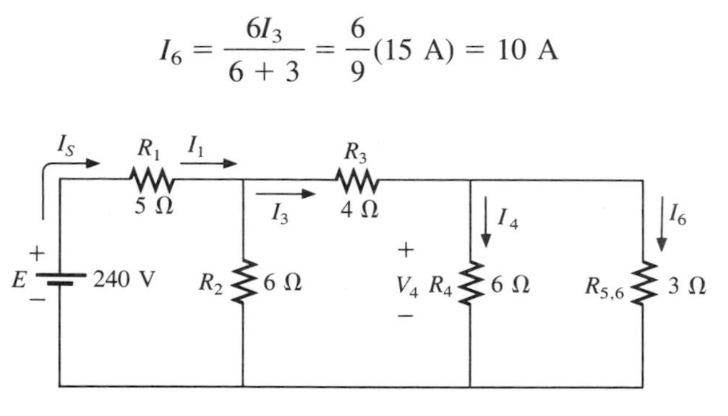

FIG. 7.33

and

$$V_6 = I_6R_6 = (10 \text{ A})(2\ \Omega) = \mathbf{20\ V}$$

## Method 2

Assign a letter symbol to the last branch current and work back through the network to the source, maintaining this assigned current or other current of interest. The desired current can then be found directly. This method can best be described through the analysis of the same network considered above, redrawn in Fig. 7.34.

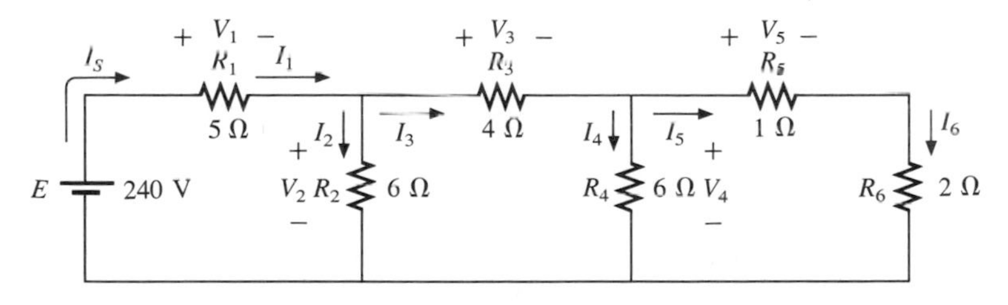


FIG. 7.34

The assigned notation for the current through the final branch is $I_6$:

$$I_6 = \frac{V_4}{R_5 + R_6} = \frac{V_4}{1+2} = \frac{V_4}{3}$$

or

$$V_4 = 3I_6$$

so that

$$I_4 = \frac{V_4}{R_4} = \frac{3I_6}{6} = 0.5I_6$$

and

$$I_3 = I_4 + I_6 = 0.5I_6 + I_6 = 1.5I_6$$
$$V_3 = I_3R_3 = (1.5I_6)(4) = 6I_6$$

Also,

$$V_2 = V_3 + V_4 = 6I_6 + 3I_6 = 9I_6$$

so that

$$I_2 = \frac{V_2}{R_2} = \frac{9I_6}{6} = 1.5I_6$$

and

$$I_S = I_2 + I_3 = 1.5I_6 + 1.5I_6 = 3I_6$$

with

$$V_1 = I_1R_1 = I_SR_1 = 5I_S$$

so that

$$E = V_1 + V_2 = 5I_S + 9I_6$$
$$= (5)(3I_6) + 9I_6 = 24I_6$$

and

$$I_6 = \frac{E}{24\ \Omega} = \frac{240\ \text{V}}{24\ \Omega} = 10\ \text{A}$$

with

$$V_6 = I_6R_6 = (10\ \text{A})(2\ \Omega) = \mathbf{20\ V}$$

as was obtained using method 1.

## 7.4 POTENTIOMETER LOADING

For the unloaded potentiometer of Fig. 7.35, the output voltage is determined by the voltage divider rule with $R_T$ in the figure representing the

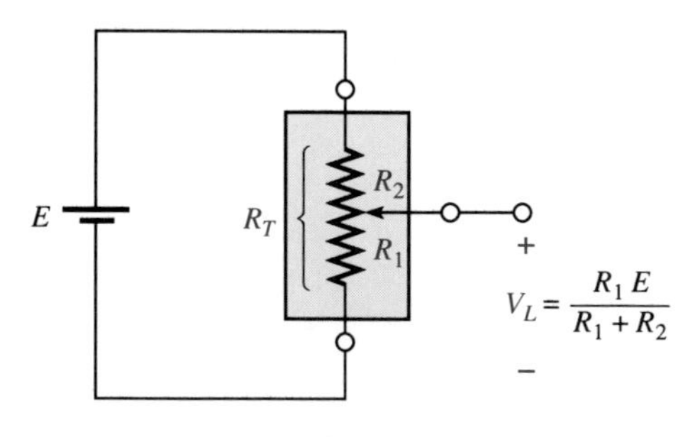


**FIG. 7.35**
*Unloaded potentiometer.*

total resistance of the potentiometer. Too often it is assumed that the voltage across a load connected to the wiper arm is determined solely by the potentiometer and the effect of the load can be ignored. This is definitely not the case, as is demonstrated in the next few paragraphs.

When a load is applied as shown in Fig. 7.36, the output voltage $V_L$ is now a function of the magnitude of the load applied, since $R_1$ is not as shown in Fig. 7.35 but is instead the parallel combination of $R_1$ and $R_L$.

The output voltage is now:

$$V_L = \frac{R'E}{R' + R_2} \qquad \text{with } R' = R_1 \parallel R_L \tag{7.1}$$

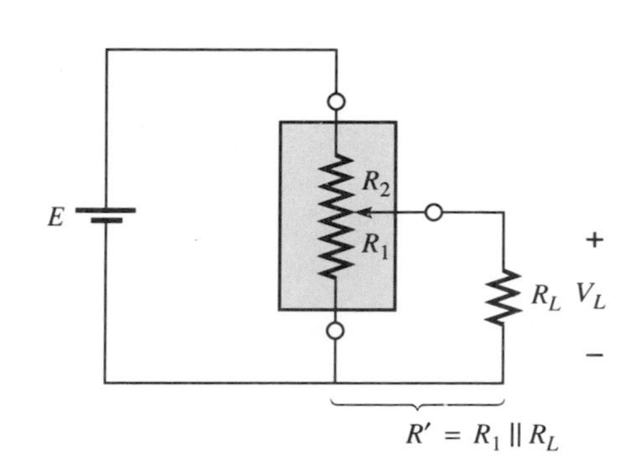


**FIG. 7.36**
*Loaded potentiometer.*

If it is desired to have good control of the output voltage $V_L$ through the controlling dial, knob, screw, or whatever, it is advisable to choose a load or potentiometer that satisfies the following relationship:

$$\boxed{R_L \geq R_T} \tag{7.2}$$

For example, if we disregard Eq. (7.2) and choose a 1-MΩ potentiometer with a 100-Ω load and set the wiper arm to $\frac{1}{10}$ the total resistance, as shown in Fig. 7.37, then

$$R' = 100 \text{ k}\Omega \parallel 100\ \Omega = 99.9\ \Omega$$

and

$$V_L = \frac{99.9\ \Omega(10 \text{ V})}{99.9\ \Omega + 900 \text{ k}\Omega} \cong 0.001 \text{ V} = 1 \text{ mV}$$

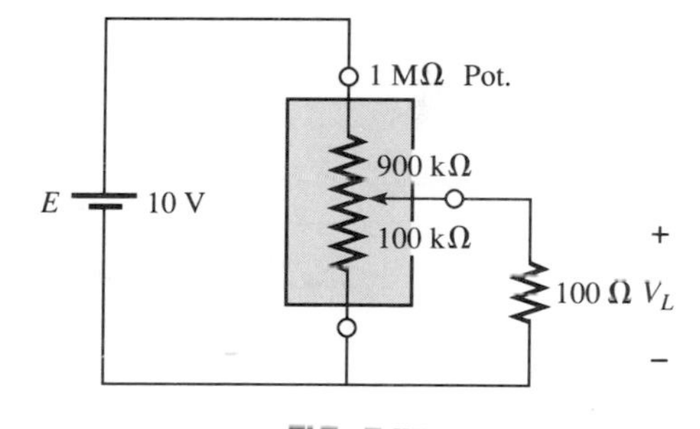


**FIG. 7.37**
$R_T > R_L$

which is extremely small compared to the total available voltage of 10 V.

In fact, if we move the wiper arm to the midpoint,

$$R' = 500 \text{ k}\Omega \parallel 100\ \Omega = 99.89\ \Omega$$

and

$$V_L = \frac{(99.89\ \Omega)(10 \text{ V})}{99.89\ \Omega + 500 \text{ k}\Omega} \cong 0.002 \text{ V} = 2 \text{ mV}$$

or only 0.002 V/10 V = 1/5000 of the available voltage.

In other words, the available voltage $V_L$ is only a very small part of the total applied voltage, even though the wiper arm has moved halfway through the available resistance.

Even at $R_1$ = 900 kΩ, $V_L$ is simply 0.01 V, or 1/1000 of the available voltage.

Using the reverse situation of $R_T$ = 100 Ω and $R_L$ = 1 MΩ and the wiper arm at the $\frac{1}{10}$ position, as in Fig. 7.38, we find

$$R' = 10\ \Omega \parallel 1 \text{ M}\Omega \cong 10\ \Omega$$

and

$$V_L = \frac{10\ \Omega(10 \text{ V})}{10\ \Omega + 90\ \Omega} = 1 \text{ V}$$

as desired.

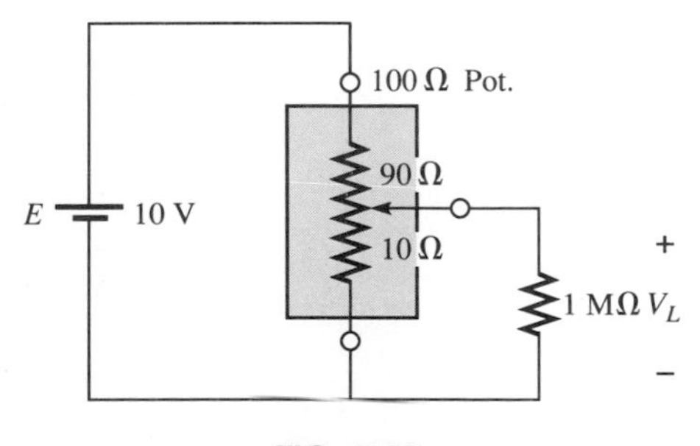


**FIG. 7.38**
$R_L > R_T$

For the lower limit of $R_L = R_T = 100\ \Omega$, as defined by Eq. (7.2) and the halfway position of Fig. 7.36,

$$R' = 50\ \Omega \parallel 100\ \Omega = 33.33\ \Omega$$

and

$$V_L = \frac{33.33\ \Omega(10\ \text{V})}{33.33\ \Omega + 50\ \Omega} \cong 40\ \text{V}$$

It may not be the ideal level of 50 V, but at least 40% of the voltage $E$ has been achieved at the halfway position rather than the 0.02% obtained with $R_L = 100\ \Omega$ and $R_T = 1\ \text{M}\Omega$.

In general, therefore, try to establish a situation for potentiometer control in which Eq. (7.2) is satisfied to the highest degree possible.

Someone might suggest that we make $R_T$ as small as possible to bring the percent result as close to the ideal as possible. Keep in mind, however, that the potentiometer has a power rating, and for networks such as Fig. 7.38, $P_{\text{max}} \cong E^2/R_T = (10\ \text{V})^2/100\ \Omega = 1\ \text{W}$. If $R_T$ is reduced to $10\ \Omega$, $P_{\text{max}} = (10\ \text{V})^2/10\ \Omega = 10\ \text{W}$, which would require a *much larger* unit.

---

**EXAMPLE 7.11.** Find the voltages $V_1$ and $V_2$ for the loaded potentiometer of Fig. 7.39.

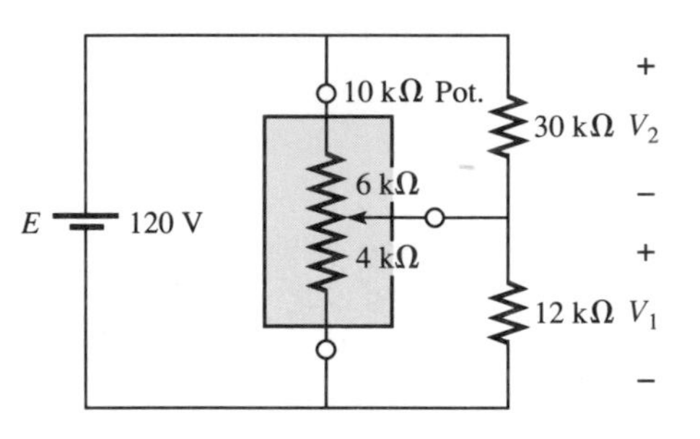


**FIG. 7.39**
*Example 7.11.*

***Solution:***

Ideal (no load):

$$V_1 = \frac{4\ \text{k}\Omega(120\ \text{V})}{10\ \text{k}\Omega} = 48\ \text{V}$$

$$V_2 = \frac{6\ \text{k}\Omega(120\ \text{V})}{10\ \text{k}\Omega} = 72\ \text{V}$$

Loaded:

$$R' = 4\ \text{k}\Omega \parallel 12\ \text{k}\Omega = 3\ \text{k}\Omega$$

$$R'' = 6\ \text{k}\Omega \parallel 30\ \text{k}\Omega = 5\ \text{k}\Omega$$

$$V_1 = \frac{3\ \text{k}\Omega(120\ \text{V})}{8\ \text{k}\Omega} = \mathbf{45\ V}$$

$$V_2 = \frac{5\ \text{k}\Omega(120\ \text{V})}{8\ \text{k}\Omega} = \mathbf{75\ V}$$

The ideal and loaded voltage levels are so close that the design can be considered a good one for the applied loads. A slight variation in the position of the wiper arm will establish the ideal voltage levels across the two loads.

---

## 7.5 AMMETER, VOLTMETER, AND OHMMETER DESIGN

Now that the fundamentals of series, parallel, and series-parallel networks have been introduced, we are prepared to investigate the fundamental design of an ammeter, voltmeter, and ohmmeter. Our design of each will employ the d'Arsonval movement of Fig. 7.40. The movement consists basically of an iron-core coil mounted on bearings between a permanent magnet. The helical springs limit the turning motion of the coil and provide a path for the current to reach the coil. When a current is passed through the movable coil, the fluxes of the coil and permanent magnet will interact to develop a torque on the coil which will cause it to rotate on its bearings. The movement is adjusted to indicate zero deflection on a meter scale when the current through the coil is zero. The direction of current through the coil will then determine whether the pointer will display an up-scale or below-zero indication. For this reason, ammeters and voltmeters have an assigned polarity on their terminals to insure an up-scale reading.

**FIG. 7.40**
*d'Arsonval movement. (Courtesy of Weston Instruments, Inc.)*

D'Arsonval movements are usually rated by current and resistance. The specifications of a typical movement might be 1 mA, 50 Ω. The 1 mA is the *current sensitivity* (*CS*) of the movement, which is the current required for a full-scale deflection. It will be denoted by the symbol $I_{CS}$. The 50 Ω represents the internal resistance ($R_m$) of the movement. A common notation for the movement and its specifications is provided in Fig. 7.41.

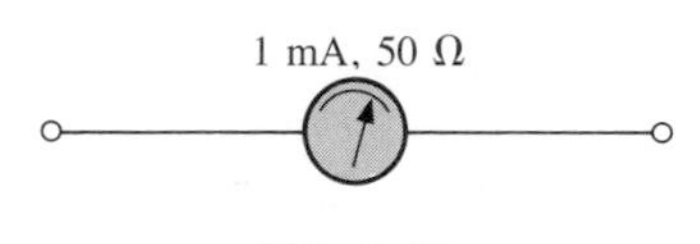


**FIG. 7.41**
*Movement notation.*

## The Ammeter

The maximum current that the d'Arsonval movement can read independently is equal to the current sensitivity of the movement. However, higher currents can be measured if additional circuitry is introduced. This additional circuitry, as shown in Fig. 7.42, results in the basic construction of an ammeter.

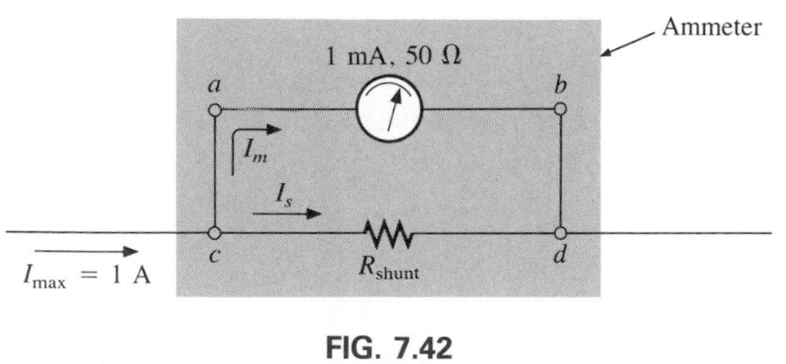


**FIG. 7.42**
*Basic ammeter.*

The resistance $R_{shunt}$ is chosen for the ammeter of Fig. 7.42 to allow 1 mA to flow through the movement when a maximum current of 1 A enters the ammeter. If less than 1 A should flow through the ammeter, the movement will have less than 1 mA flowing through it and will indicate less than full-scale deflection.

Since the voltage across parallel elements must be the same, the potential drop across *a-b* in Fig. 7.42 must equal that across *c-d;* that is,

$$(1 \text{ mA})(50 \ \Omega) = R_{shunt} I_S$$

Also, $I_S$ must equal 1 A − 1 mA = 999 mA if the current is to be limited to 1 mA through the movement (Kirchhoff's current law). Therefore,

$$(1 \text{ mA})(50 \ \Omega) = R_{\text{shunt}}(999 \text{ mA})$$

$$R_{\text{shunt}} = \frac{(1 \text{ mA})(50 \ \Omega)}{999 \text{ mA}}$$

$$\cong 0.05 \ \Omega$$

In general,

$$R_{\text{shunt}} = \frac{R_m I_{CS}}{I_{\max} - I_{CS}} \tag{7.3}$$

One method of constructing a multirange ammeter is shown in Fig. 7.43, where the rotary switch determines the $R_{\text{shunt}}$ to be used for the

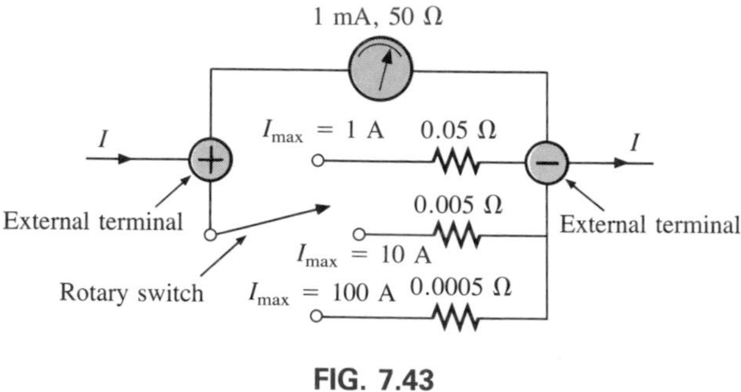


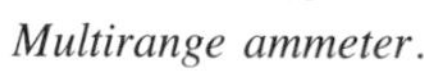
**FIG. 7.43**
*Multirange ammeter.*

maximum current indicated on the face of the meter. Most meters employ the same scale for various values of maximum current. If you read 375 on the 0–5 mA scale with the switch on the 5 setting, the current is 3.75 mA; on the 50 setting, the current is 37.5 mA; and so on.

## The Voltmeter

A variation in the additional circuitry will permit the use of the d'Arsonval movement in the design of a voltmeter. The 1-mA, 50-Ω movement can also be rated as a 50-mV (1 mA × 50 Ω), 50-Ω movement, indicating that the maximum voltage that the movement can measure independently is 50 mV. The millivolt rating is sometimes referred to as the *voltage sensitivity* (*VS*). The basic construction of the voltmeter is shown in Fig. 7.44.

The $R_{\text{series}}$ is adjusted to limit the current through the movement to 1 mA when the maximum voltage is applied across the voltmeter. A lesser voltage would simply reduce the current in the circuit, and thereby the deflection of the movement.

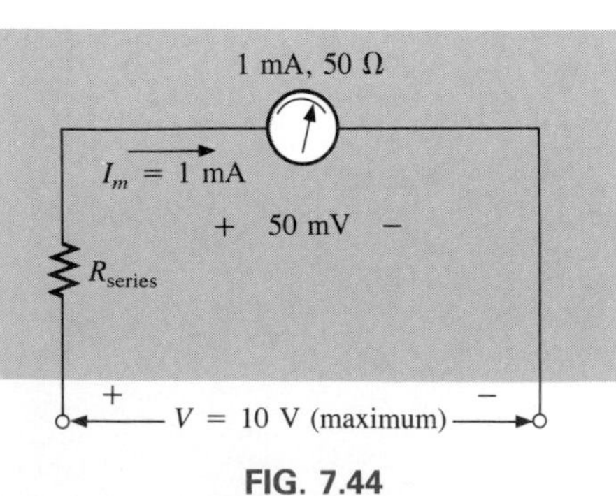


**FIG. 7.44**
*Basic voltmeter.*

Applying Kirchhoff's voltage law around the closed loop of Fig. 7.44, we obtain

$$[10\text{ V} - (1\text{ mA})(R_{\text{series}})] - 50\text{ mV} = 0$$

or

$$R_{\text{series}} = \frac{10\text{ V} - (50\text{ mV})}{1\text{ mA}} = 9950\ \Omega$$

In general,

$$\boxed{R_{\text{series}} = \frac{V_{\text{max}} - V_{VS}}{I_{CS}}} \qquad (7.4)$$

One method of constructing a multirange voltmeter is shown in Fig. 7.45. If the rotary switch is at 10 V, $R_{\text{series}} = 9.950\text{ k}\Omega$; at 50 V, $R_{\text{series}} = 40\text{ k}\Omega + 9.950\text{ k}\Omega = 49.950\text{ k}\Omega$; and at 100 V, $R_{\text{series}} = 50\text{ k}\Omega + 40\text{ k}\Omega + 9.950\text{ k}\Omega = 99.950\text{ k}\Omega$.

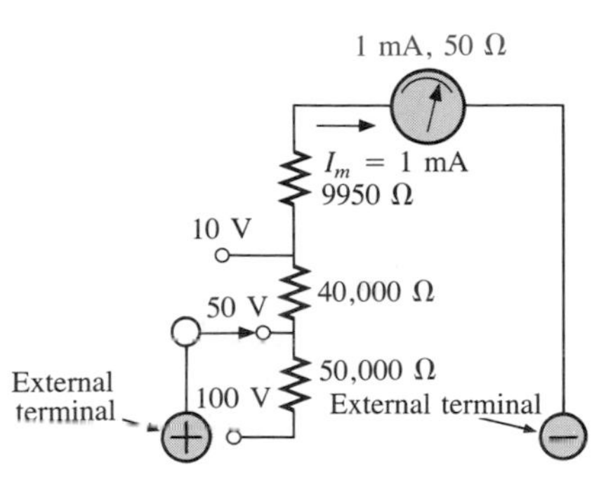


**FIG. 7.45**
*Multirange voltmeter.*

## The Ohmmeter

In general, ohmmeters are designed to measure resistance in the low, mid-, or high range. The most common is the *series ohmmeter*, designed to read resistance levels in the midrange. It employs the series configuration of Fig. 7.46. The design is quite different from that of the ammeter or voltmeter in that it will show a full-scale deflection for zero ohms and no deflection for infinite resistance.

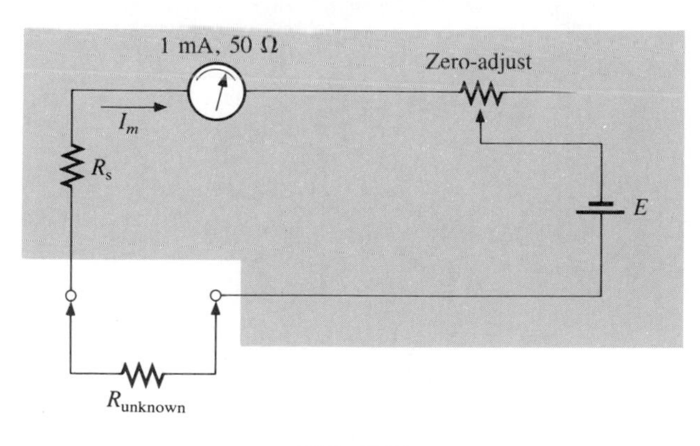


**FIG. 7.46**
*Series ohmmeter.*

To determine the series resistance $R_s$, the external terminals are shorted (a direct connection of zero ohms between the two) to simulate zero ohms, and the zero-adjust is set to half its maximum value. The resistance $R_s$ is then adjusted to allow a current equal to the current sensitivity of the movement (1 mA) to flow in the circuit. The zero-adjust is set to half its value so that any variation in the components of the meter that may produce a current more or less than the current sensitivity can be compensated for. The current $I_m$ is

$$I_m \text{ (full scale)} = I_{CS} = \frac{E}{R_s + R_m + \dfrac{\text{zero-adjust}}{2}} \qquad \textbf{(7.5)}$$

and

$$R_s = \frac{E}{I_{CS}} - R_m - \frac{\text{zero-adjust}}{2} \qquad \textbf{(7.6)}$$

If an unknown resistance is then placed between the external terminals, the current will be reduced, causing a deflection less than full scale. If the terminals are left open, simulating infinite resistance, the pointer will not deflect since the current through the circuit is zero.

An instrument designed to read very low values of resistance appears in Fig. 7.47. Because of its low range capability, the network design

**FIG. 7.47**
*Milliohmmeter. (Courtesy of Keithley Instruments, Inc.)*

must be a great deal more sophisticated than described above. It employs electronic components that eliminate the inaccuracies introduced by lead and contact resistances. It is similar to the above system in the sense that it is completely portable and does require a dc battery to establish measurement conditions. Note the special leads designed to limit any introduced resistance levels. The maximum scale setting can be set as low as 0.00352 (3.52 mΩ).

The Megger tester is an instrument for measuring very high resistance values. The term *Megger* is derived from the fact that the device measures resistance values in the megohm range. Its primary function is to test the insulation found in power transmission systems, electrical machinery, transformers, and so on. To measure the high-resistance values, a high dc voltage is established by a hand-driven generator. If the shaft is rotated above some set value, the output of the generator will be fixed at one selectable voltage, typically 250, 500, or 1000 V. A photograph of the commercially available Megger tester is shown in Fig. 7.48. The unknown resistance is connected between the terminals marked *Line* and *Earth*. For this instrument, the range is zero to 2000 MΩ.

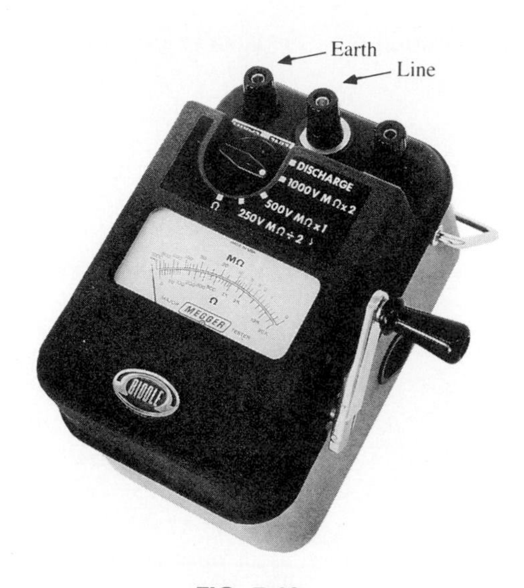


**FIG. 7.48**
*The Megger® tester. (Courtesy of James G. Biddle Co.)*

## 7.6 COMPUTER ANALYSIS

The series-parallel network to be analyzed using PSPICE is the same network examined in Example 7.2 and redrawn in Fig. 7.49. A program in BASIC would require a solution of the network that would sequentially calculate $R_T$, followed by the source current and then the best path back to the desired unknowns. The resulting program would require a

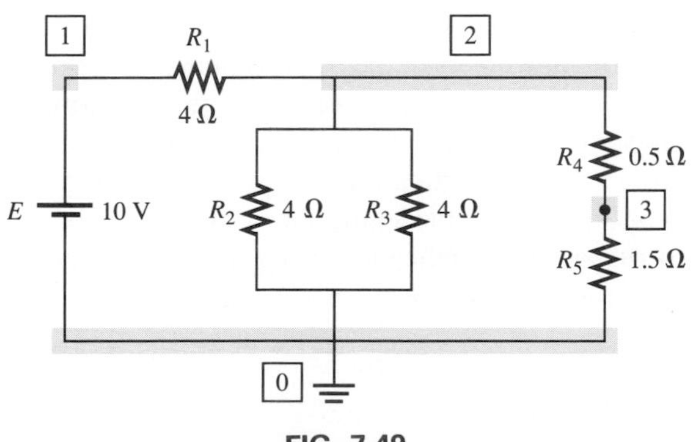


**FIG. 7.49**

number of lines and would obviously take some time to write and test. The beauty of the PSPICE package now becomes more obvious, since the input file (Fig. 7.50) is not that difficult or lengthy, and all the

```
************************* Evaluation PSpice (January 1989) ******* 19:04:07 *******

Chapter 7 - Multi R Network

****      CIRCUIT DESCRIPTION

****************************************************************************

VE 1 0 10V
R1 1 2 4
R2 2 0 4
R3 2 0 4
R4 2 3 0.5
R5 3 0 1.5
.DC VE 10 10 1
.PRINT DC I(R1) I(R2) V(3,0)
.OPTIONS NOPAGE
.END

****      DC TRANSFER CURVES                    TEMPERATURE =   27.000 DEG C

  VE            I(R1)         I(R2)         V(3,0)

   1.000E+01    2.000E+00    5.000E-01    1.500E+00
```

**FIG. 7.50**

desired results would be available in a few seconds of computer time. As an instructor, however, my concern remains that a student can use the package to obtain the solution yet have no idea (or a very weak understanding of) how to solve the network longhand. Do not let computer techniques place the "cart before the horse." Always try to have some understanding of the basic maneuvers a computer is going to go through before simply accepting the computer output as gospel. You do not have to be an expert in the subject area, but at least be aware of the general process and the implication of the results obtained.

The output file reveals that

$$I_{R_1} = I_E = 2\text{ A}, \qquad I_{R_2} = I_{R_3} = 0.5\text{ A}, \quad \text{and} \quad V_{R_5} = 1.5\text{ V}$$

as obtained in Example 7.2.

# PROBLEMS

## SECTION 7.2 Descriptive Examples

1. For the network of Fig. 7.51:
   a. Does $I = I_3 = I_6$? Explain.
   b. If $I = 5$ A and $I_1 = 2$ A, find $I_2$.
   c. Does $I_1 + I_2 = I_4 + I_5$? Explain.
   d. If $V_1 = 6$ V and $E = 10$ V, find $V_2$.
   e. If $R_1 = 3\ \Omega$, $R_2 = 2\ \Omega$, $R_3 = 4\ \Omega$, and $R_4 = 1\ \Omega$, what is $R_T$?
   f. If the resistors have the values given in part (e) and $E = 10$ V, what is the value of $I$ in amperes?
   g. Using values given in parts (e) and (f), find the power delivered by the battery $E$ and dissipated by the resistors $R_1$ and $R_2$.

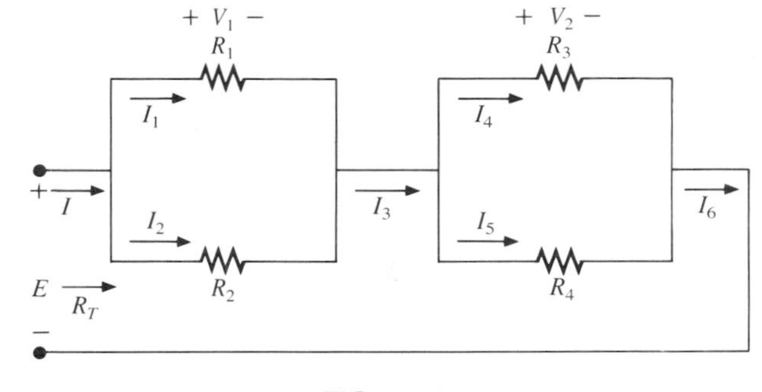


FIG. 7.51

2. For the network of Fig. 7.52:
   a. Calculate $R_T$.
   b. Determine $I$ and $I_1$.
   c. Find $V_3$.

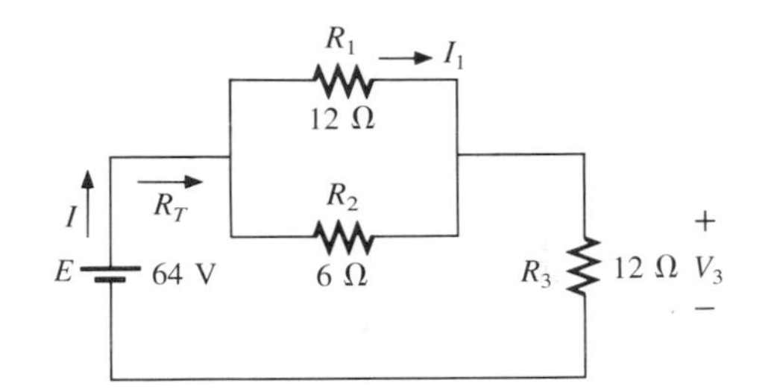

FIG. 7.52

3. For the network of Fig. 7.53:
   a. Determine $R_T$.
   b. Find $I_S$, $I_1$, and $I_2$.
   c. Calculate $V_a$.

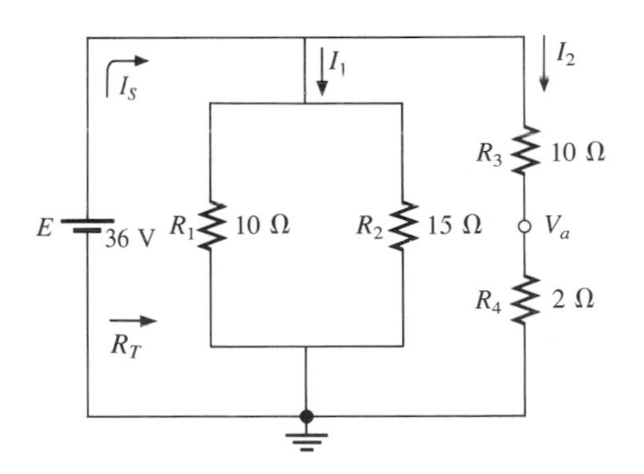

FIG. 7.53

4. Determine the currents $I_1$ and $I_2$ for the network of Fig. 7.54.

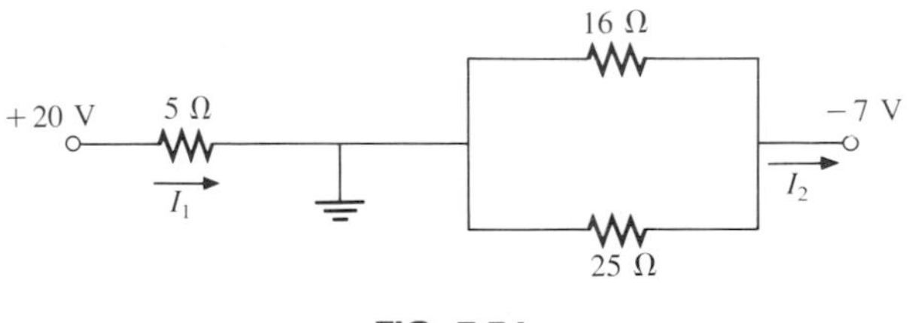

FIG. 7.54

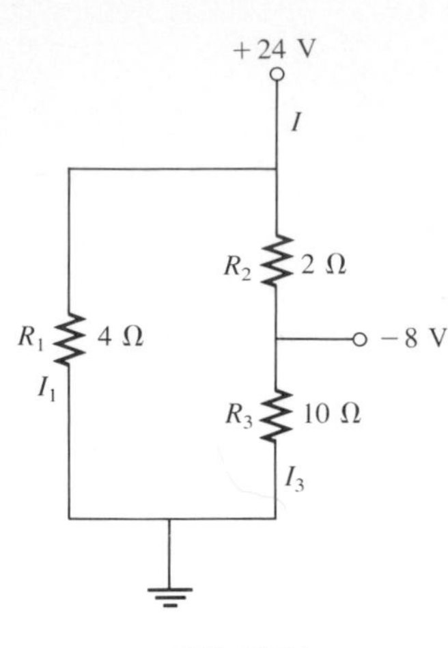

FIG. 7.55

5. **a.** Find the currents $I$, $I_1$, and $I_3$ for the network of Fig. 7.55.
   **b.** Indicate their direction on Fig. 7.55.

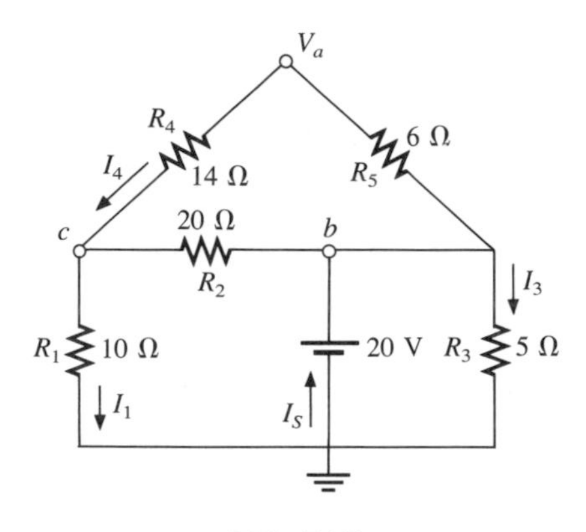

FIG. 7.56

*6. For the network of Fig. 7.56:
   **a.** Determine the currents $I_S$, $I_1$, $I_3$, and $I_4$.
   **b.** Calculate $V_a$ and $V_{bc}$.

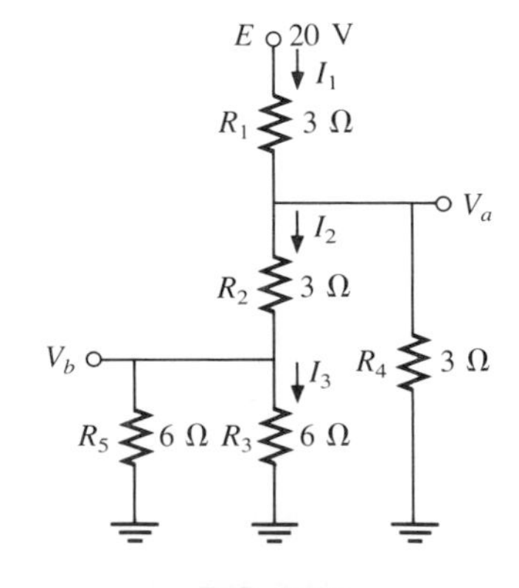


FIG. 7.57

7. For the network of Fig. 7.57:
   **a.** Determine the current $I_1$.
   **b.** Calculate the currents $I_2$ and $I_3$.
   **c.** Determine the voltage levels $V_a$ and $V_b$.

**8.** For the network of Fig. 7.58:

  **a.** Find the currents $I$ and $I_6$.
  **b.** Find the voltages $V_1$ and $V_5$.
  **c.** Find the power delivered to the 6-kΩ resistor.

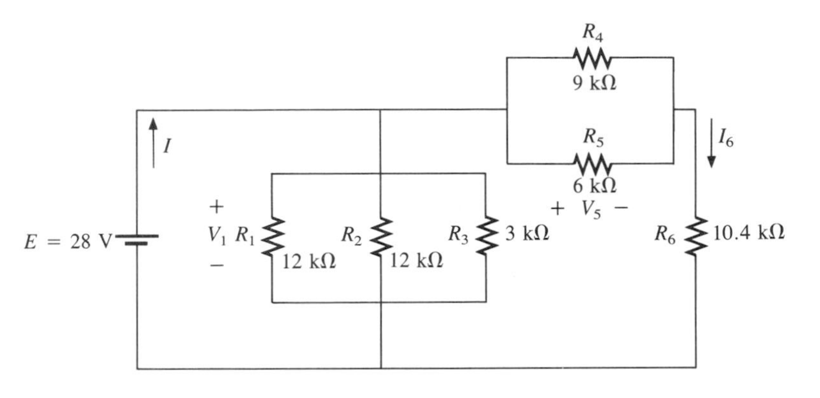


FIG. 7.58

***9.** For the series-parallel network of Fig. 7.59:

  **a.** Find the current $I$.
  **b.** Find the currents $I_3$ and $I_9$.
  **c.** Find the current $I_8$.
  **d.** Find the voltage $V_{ab}$.

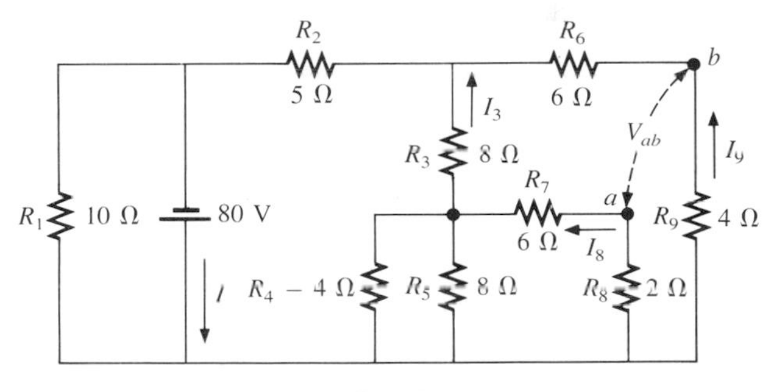

FIG. 7.59

***10.** Determine the dc levels for the transistor network of Fig. 7.60 using the fact that $V_{BE} = 0.7$ V, $V_E = 2$ V, and $I_C = I_E$. That is,

  **a.** Determine $I_E$ and $I_C$.
  **b.** Calculate $I_B$.
  **c.** Determine $V_B$ and $V_C$.
  **d.** Find $V_{CE}$ and $V_{BC}$.

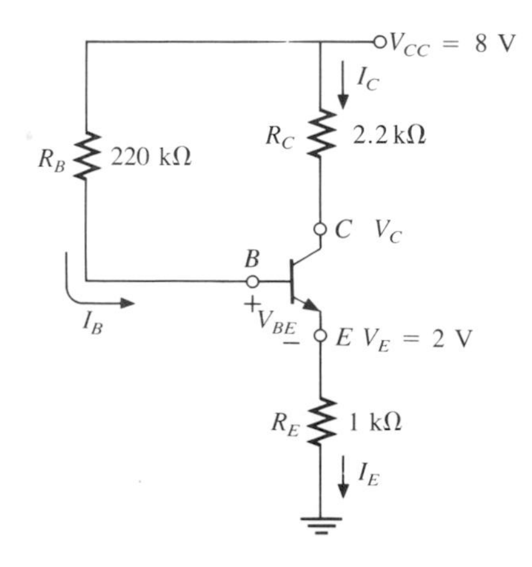

FIG. 7.60

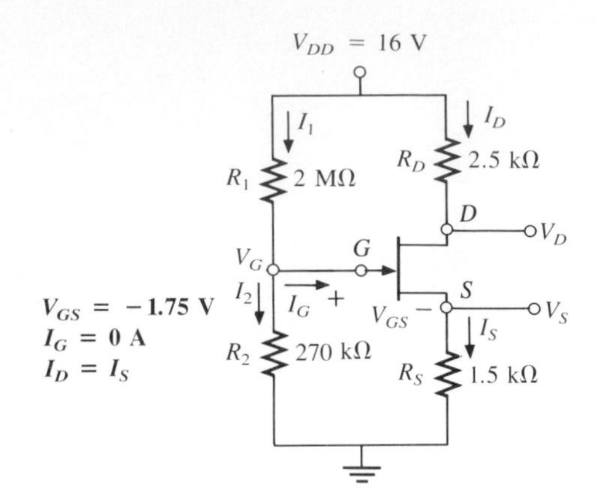


FIG. 7.61

***11.** The network of Fig. 7.61 is the basic biasing arrangement for the *field-effect transistor* (FET), a device of increasing importance in electronic design. (*Biasing* simply means the application of dc levels to establish a particular set of operating conditions.) Even though you may be unfamiliar with the FET, you can perform the following analysis using only the basic laws introduced in this chapter and the information provided on the diagram.

**a.** Determine the voltages $V_G$ and $V_S$.
**b.** Find the currents $I_1$, $I_2$, $I_D$, and $I_S$.
**c.** Determine $V_{DS}$.
**d.** Calculate $V_{DG}$.

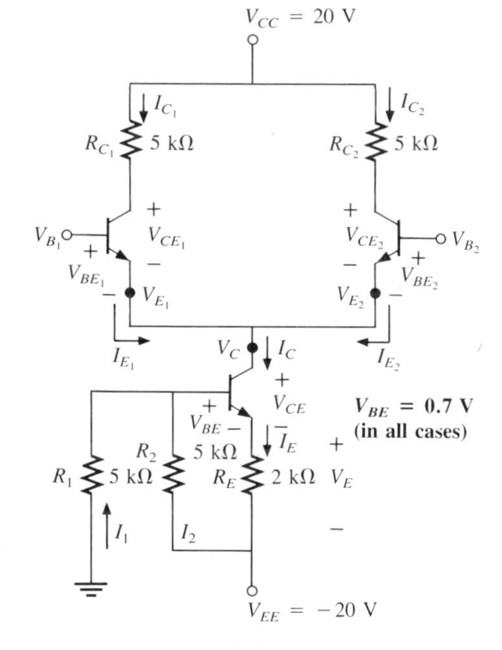

FIG. 7.62

***12.** The *difference amplifier* of Fig. 7.62 is a compound configuration that will establish an output (in the ac domain) that is the difference between the two input signals. Using the concepts introduced in this chapter and previous chapters, determine the following dc levels:

**a.** $V_{R_E}$ (given that $I_1$ = 2 mA).
**b.** $I_2$ (magnitude and direction).
**c.** $I_{E_1}$ and $I_{E_2}$ using the fact that $I_C = I_E$ and $I_{E_1} = I_{E_2}$ (balanced system).
**d.** $V_C$, $V_{E_1}$, and $V_{E_2}$ if $V_{CE}$ = 10.7 V.
**e.** $V_{B_1}$ and $V_{B_2}$ using the information obtained above.
**f.** $V_{CE_1}$ and $V_{CE_2}$ if $I_{C_1} = I_{E_1}$ and $I_{C_2} = I_{E_2}$.

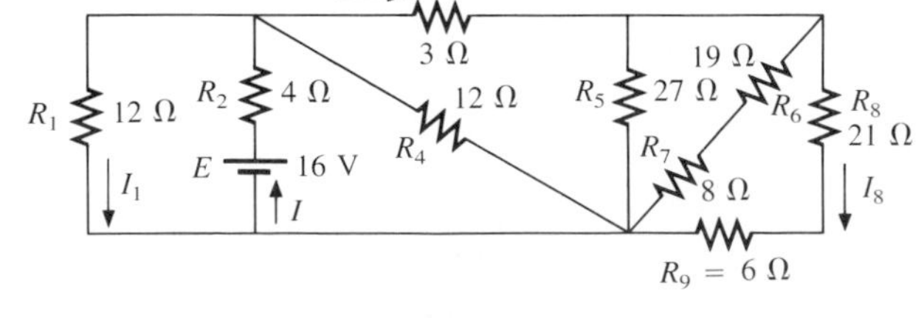

FIG. 7.63

***13.** For the series-parallel configuration of Fig. 7.63:

**a.** Find the current $I$.
**b.** Find the currents $I_1$, $I_3$, and $I_8$.
**c.** Find the power delivered to the 21-Ω resistor.

**14.** For the network of Fig. 7.64:

**a.** Determine the current $I$.

**b.** Find $V$.

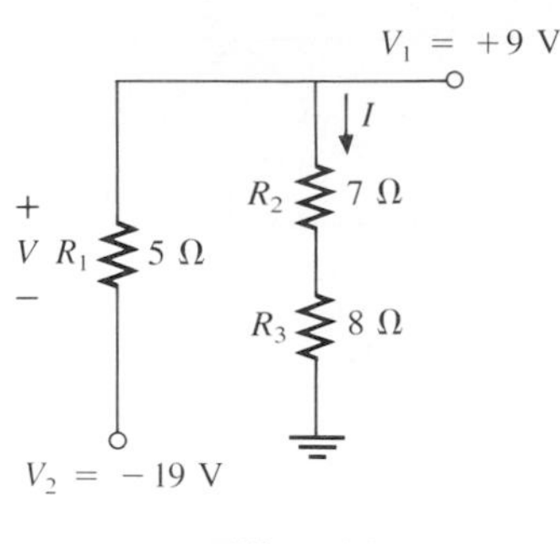

**FIG. 7.64**

**15.** For the configuration of Fig. 7.65:

**a.** Find the currents $I_2$, $I_6$, and $I_8$.

**b.** Find the voltages $V_4$ and $V_8$.

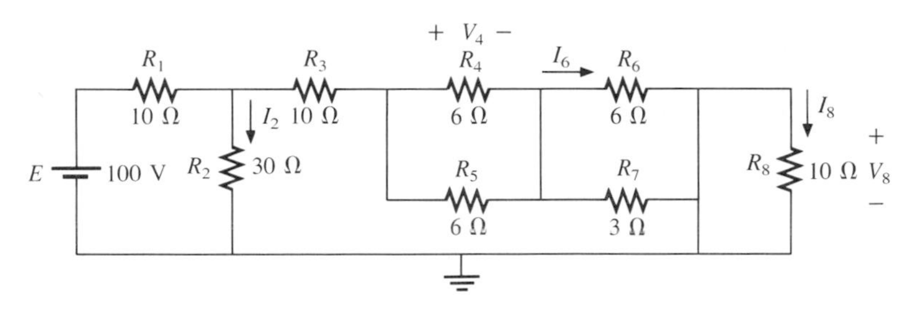


**FIG. 7.65**

***16.** For the network of Fig. 7.66, find the resistance $R_3$ if the current through it is 2 A.

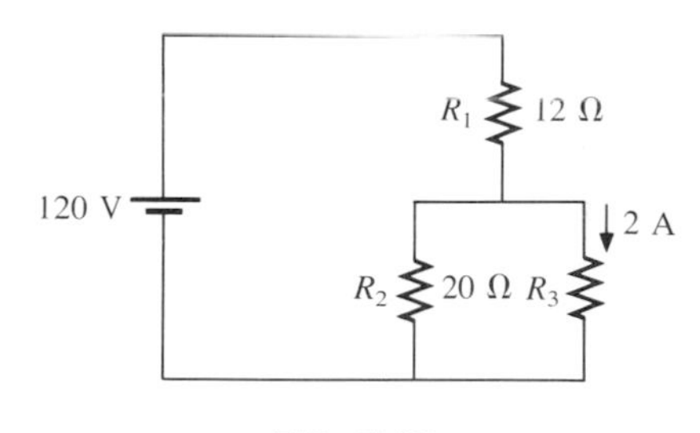


**FIG. 7.66**

**17.** Determine the voltage $V$ and the current $I$ for the network of Fig. 7.67.

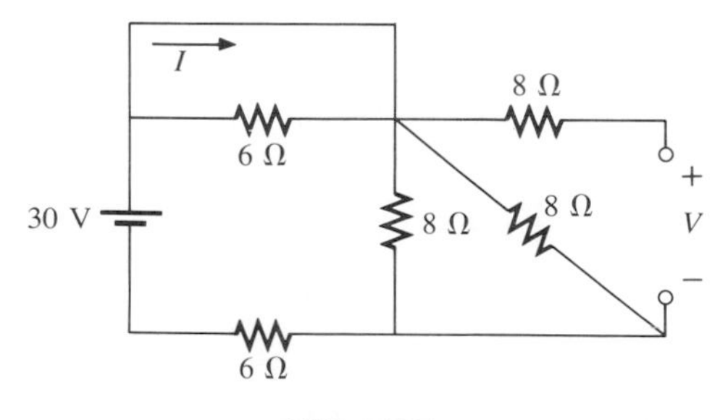

**FIG. 7.67**

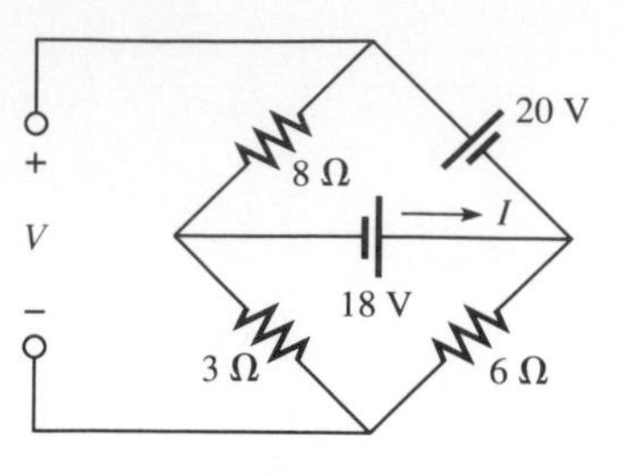


FIG. 7.68

***18.** For the network of Fig. 7.68:
**a.** Determine the current $I$.
**b.** Calculate the open-circuit voltage $V$.

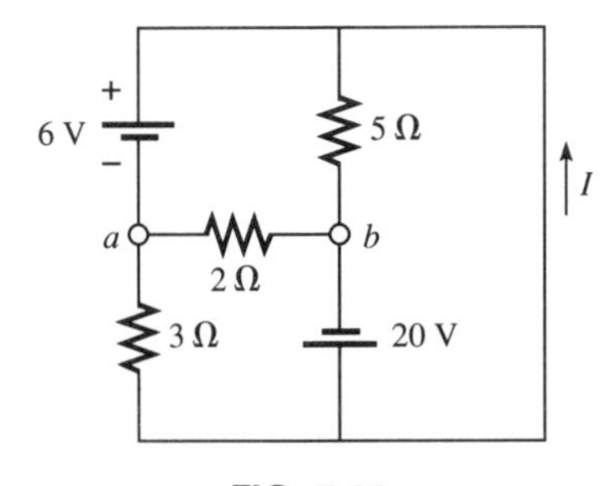


FIG. 7.69

***19.** For the network of Fig. 7.69:
**a.** Determine the voltage $V_{ab}$.
**b.** Calculate the short-circuit current $I$.

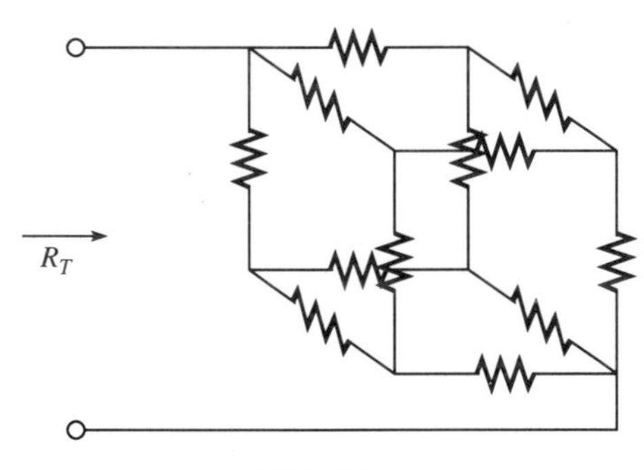


FIG. 7.70

***20.** If all the resistors of the following cube (Fig. 7.70) are 10 Ω, what is the total resistance? *Hint:* Make some basic assumptions about current division through the cube.

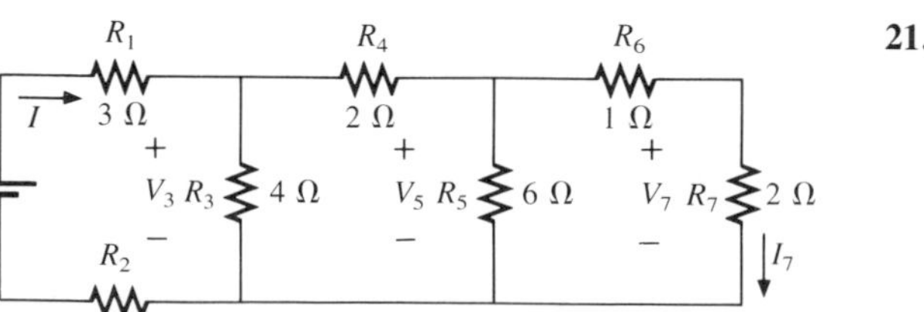


FIG. 7.71

## SECTION 7.3 Ladder Networks

**21.** For the ladder network of Fig. 7.71:
**a.** Find the current $I$.
**b.** Find the current $I_7$.
**c.** Determine the voltages $V_3$, $V_5$, and $V_7$.
**d.** Calculate the power delivered to $R_7$ and compare it to the power delivered by the 240-V supply.

**22.** For the ladder network of Fig. 7.72:

**a.** Determine $R_T$.

**b.** Calculate $I$.

**c.** Find $I_8$.

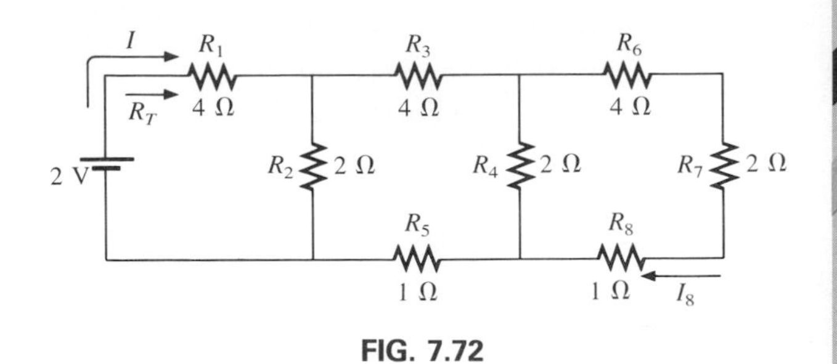


FIG. 7.72

***23.** For the multiple ladder configuration of Fig. 7.73:

**a.** Determine $I$.

**b.** Calculate $I_4$.

**c.** Find $I_6$.

**d.** Find $I_{10}$.

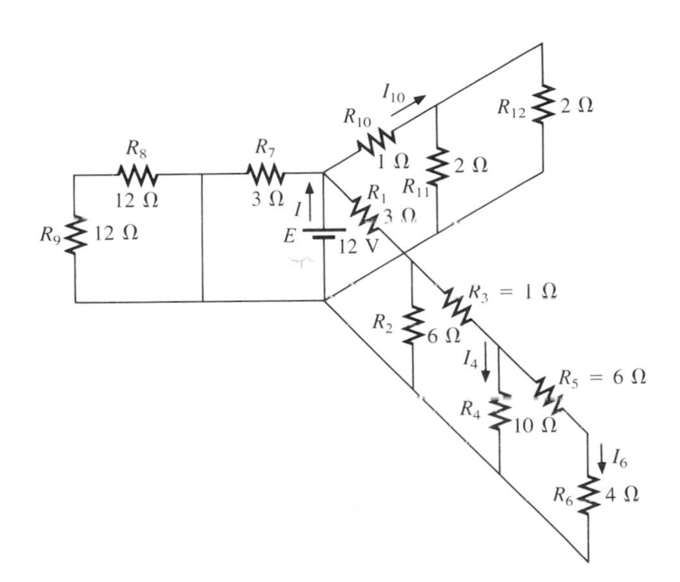


FIG. 7.73

## SECTION 7.4 Potentiometer Loading

***24.** For the system of Fig. 7.74:

**a.** At first exposure, does the design appear to be a good one?

**b.** In the absence of the 10-kΩ load, what are the values of $R_1$ and $R_2$ to establish 3 V across $R_2$?

**c.** Determine the values of $R_1$ and $R_2$ when the load is applied and compare to the results of part (b).

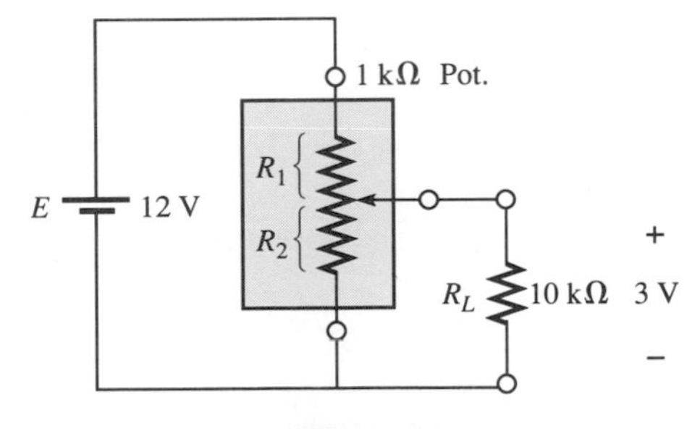

FIG. 7.74

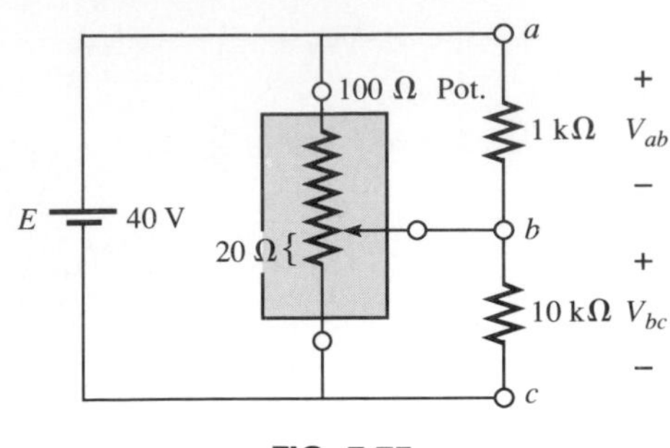


FIG. 7.75

***25.** For the potentiometer of Fig. 7.75:

**a.** What are the voltages $V_{ab}$ and $V_{bc}$ with no load applied?

**b.** What are the voltages $V_{ab}$ and $V_{bc}$ with the indicated loads applied?

**c.** What is the power dissipated by the potentiometer under the loaded conditions of Fig. 7.75?

**d.** What is the power dissipated by the potentiometer with no loads applied? Compare to the results of part (c).

## SECTION 7.5 Ammeter, Voltmeter, and Ohmmeter Design

**26.** A d'Arsonval movement is rated 1 mA, 100 Ω.

**a.** What is the current sensitivity?

**b.** Design a 20-A ammeter using the above movement. Show the circuit and component values.

**27.** Using a 50-μA, 1000-Ω d'Arsonval movement, design a multirange milliammeter having scales of 25, 50, and 100 mA. Show the circuit and component values.

**28.** A d'Arsonval movement is rated 50 μA, 1000 Ω.

**a.** Design a 15-V dc voltmeter. Show the circuit and component values.

**b.** What is the ohm/volt rating of the voltmeter?

**29.** Using a 1-mA, 100-Ω d'Arsonval movement, design a multirange voltmeter having scales of 5, 50, and 500 V. Show the circuit and component values.

**30.** A digital meter has an internal resistance of 10 MΩ on its 0.5-V range. If you had to build a voltmeter with a d'Arsonval movement, what current sensitivity would you need if the meter were to have the same internal resistance on the same voltage scale?

***31. a.** Design a series ohmmeter using a 100-μA, 1000-Ω movement, a zero-adjust with a maximum value of 2 kΩ, a battery of 3 V, and a series resistor whose value is to be determined.

**b.** Find the resistance required for full-scale, $\frac{3}{4}$-scale, $\frac{1}{2}$-scale, and $\frac{1}{4}$-scale deflection.

**c.** Using the results of part (b), draw the scale to be used with the ohmmeter.

**32.** Describe the basic construction and operation of the Megger.

## SECTION 7.6 Computer Analysis

### PSPICE

**33.** Write the input file for the network of Fig. 7.7 and request $I_1$, $I_3$, and $I_6$.

**34.** Write the input file for the network of Fig. 7.15 and print out $V_1$, $V_3$, and $V_{ab}$.

**35.** Write the input file for the network of Fig. 7.17 and print out $V_1$, $V_2$, and $I$.

**36.** Write the input file for the network of Fig. 7.29 and print out $V_6$ and $I_6$.

### BASIC

**37.** Write a program that will find the complete solution for the network of Fig. 7.5. That is, given all the parameters of the network, calculate the current, voltage, and power to each element.

**38.** Write a program to find all the quantities of Example 7.8 given the network parameters.

***39.** Write a program to find $R_T$, $I_T$, and $I_6$ for the network of Fig. 7.29 given the network parameters.

## GLOSSARY

**d'Arsonval movement** An iron-core coil mounted on bearings between a permanent magnet. A pointer connected to the movable core indicates the strength of the current passing through the coil.

**Field-effect transistor (FET)** A three-terminal electronic device capable of amplifying time-varying signals.

**Ladder network** A network that consists of a cascaded set of series-parallel combinations and has the appearance of a ladder.

**Megger® tester** An instrument for measuring very high resistance levels, such as in the megohm range.

**Series ohmmeter** A resistance-measuring instrument in which the movement is placed in series with the unknown resistance.

**Series-parallel network** A network consisting of a combination of both series and parallel branches.

**Shunt ohmmeter** A resistance measuring instrument in which the movement is placed in parallel with the unknown resistance.

**Transistor** A three-terminal semiconductor electronic device that can be used for amplification and switching purposes.

# 8

# Methods of Analysis and Selected Topics (dc)

## 8.1 INTRODUCTION

The circuits described in the previous chapters had only one source or two or more sources in series or parallel present. The step-by-step procedure outlined in those chapters cannot be applied if two or more sources in the same network are not in series or parallel. There will be an interaction of sources that will not permit the reduction technique used in Chapter 7 to find such quantities as the total resistance and source current.

Methods of analysis have been developed that allow us to approach, in a systematic manner, a network with any number of sources in any arrangement. Fortunately, these methods can also be applied to networks with only one source. The methods to be discussed in detail in this chapter include *branch-current analysis, mesh analysis,* and *nodal analysis*. Each can be applied to the same network. The "best" method cannot be defined by a set of rules but can be determined only by acquiring a firm understanding of the relative advantages of each. All of the methods will be described for *linear bilateral* networks. The term *linear* indicates that the characteristics of the network elements (such as the resistors) are independent of the voltage across or current through them. The second term, *bilateral,* refers to the fact that there is no change in the behavior or characteristics of an element if the current through or voltage across the element is reversed. Before discussing these methods, we shall consider the current source and the use of determinants. At the end of the chapter we shall consider bridge networks and Δ-Y and Y-Δ conversions.

Chapter 9 will present the important theorems of network analysis that can also be employed to solve networks with more than one source.

$N_A$

## 8.2 CURRENT SOURCES

The concept of the current source was introduced in Section 2.4 with the photograph of a commercially available unit. We must now investigate its characteristics in greater detail so that we can properly determine its effect on the networks to be examined in this chapter.

The current source is often referred to as the *dual* of the voltage source. That is, while a battery supplies a *fixed* voltage and the source current can vary, the current source supplies a *fixed* current to the branch in which it is located, while its terminal voltage may vary as determined by the network to which it is applied. Note from the above that *duality* simply implies an interchange of current and voltage to distinguish the characteristics of one source from the other.

The increasing interest in the current source is due fundamentally to semiconductor devices such as the transistor. In the basic electronics courses, you will find that the transistor is a current-controlled device. In the physical model (equivalent circuit) of a transistor used in the analysis of transistor networks, there appears a current source as indicated in Fig. 8.1. The symbol for a current source appears in Fig. 8.1.

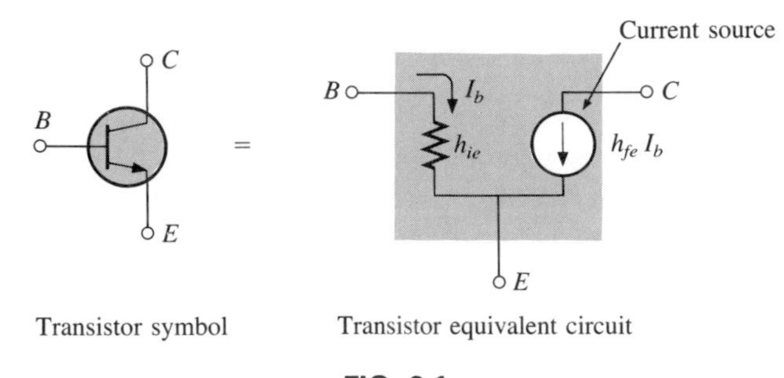

FIG. 8.1

The direction of the arrow within the circle indicates the direction in which current is being supplied.

For further comparison, the terminal characteristics of a *dc* voltage and current source are presented in Fig. 8.2. Note that for the voltage

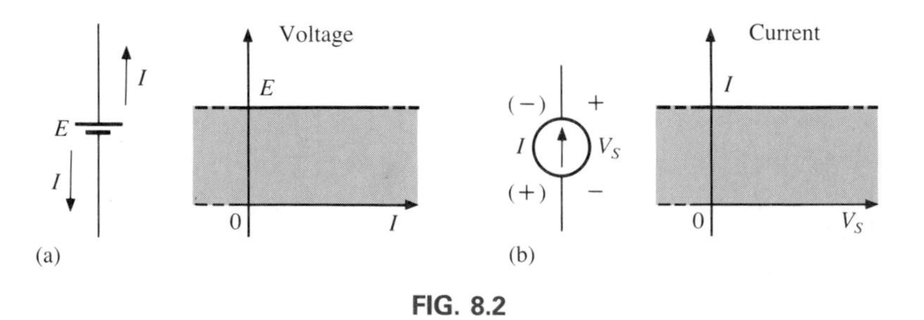


FIG. 8.2

source the terminal voltage is fixed at $E$ volts for the range of current values. For the region to the right of the voltage axis, the current will have one direction through the source, while to the left of the same axis it is reversed. In other words, as indicated in the associated figure of Fig. 8.2(a), the terminal voltage of a source is unaffected by the direction of current through the source. The characteristics of the current source, shown in Fig. 8.2(b), indicate that the current source will supply a fixed current even though the voltage across the source may vary in magnitude or reverse its polarity. This is indicated in the associated figure of Fig. 8.2(b). For the voltage source, the current direction will be determined by the remaining elements of the network. For all one-voltage-source networks it will have the direction indicated to the right of the battery in Fig. 8.2(a). For the current source, the network to which it is connected will also determine the magnitude and polarity of the voltage across the source. For all single-current-source networks, it will have the polarity indicated to the right of the current source in Fig. 8.2(b).

In review:

***A current source determines the current in the branch in which it is located***

and

***the voltage across a current source is a function of the network to which it is applied.***

---

**EXAMPLE 8.1.** Find the voltage $V_S$ and the current $I_1$ for the circuit of Fig. 8.3.

***Solution:***

$$I_1 = I = \mathbf{10\ mA}$$
$$V_S = V_1 = I_1R_1 = (10\ \text{mA})(20\ \text{k}\Omega) = \mathbf{200\ V}$$

---

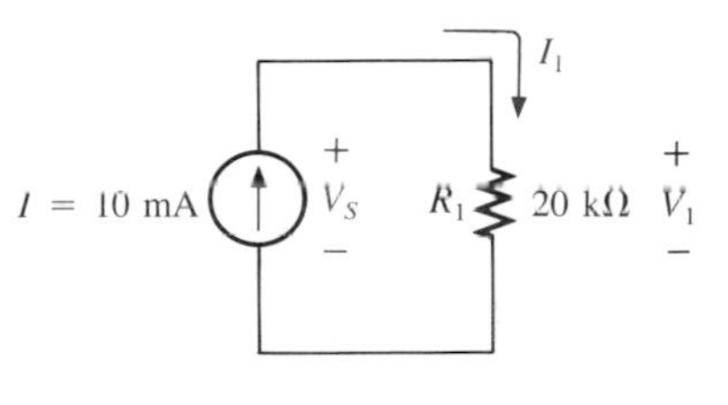

FIG. 8.3

---

**EXAMPLE 8.2.** Calculate the voltages $V_1$, $V_2$, and $V_S$ for the circuit of Fig. 8.4.

***Solution:***

$$V_1 = I_1R_1 = IR_1 = (5\ \text{A})(2\ \Omega) = \mathbf{10\ V}$$
$$V_2 = I_2R_2 = IR_2 = (5\ \text{A})(3\ \Omega) = \mathbf{15\ V}$$

Applying Kirchhoff's voltage law in the clockwise direction, we obtain

$$\Sigma_{\circlearrowright} V = +V_S - V_1 - V_2 = 0$$

or

$$V_S = V_1 + V_2 = 10\ \text{V} + 15\ \text{V}$$

and

$$V_S = \mathbf{25\ V}$$

Note the polarity of $V_S$ for the single-source circuit.

---

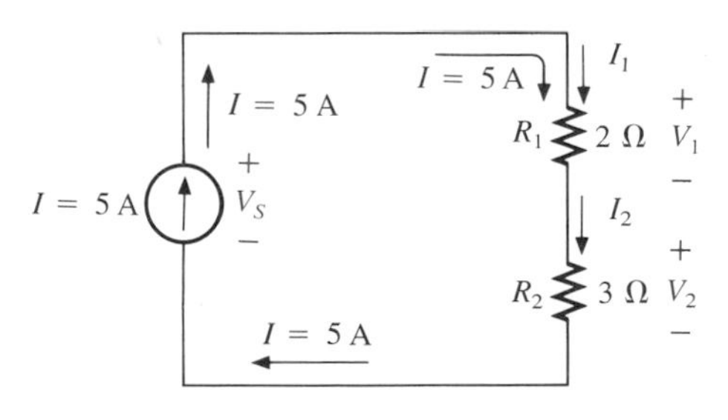


FIG. 8.4

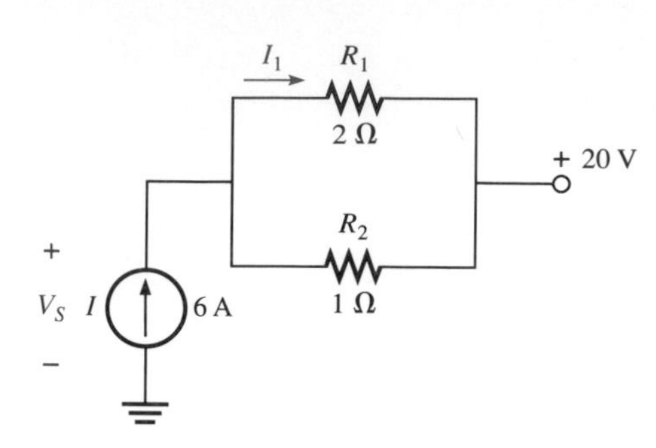


FIG. 8.5

**EXAMPLE 8.3.** Determine the current $I_1$ and the voltage $V_S$ for the network of Fig. 8.5.

***Solution:*** Using the current divider rule:

$$I_1 = \frac{R_2 I}{R_2 + R_1} = \frac{(1\ \Omega)(6\ \text{A})}{1\ \Omega + 2\ \Omega} = \mathbf{2\ A}$$

The voltage $V_1$ is

$$V_1 = I_1 R_1 = (2\ \text{A})(2\ \Omega) = 4\ \text{V}$$

and, applying Kirchhoff's voltage law,

$$+V_S + V_1 - 20\ \text{V} = 0$$

and

$$V_S = V_1 + 20\ \text{V} = 4\ \text{V} + 20\ \text{V} = \mathbf{24\ V}$$

Note the polarity of $V_S$ as determined by the multisource network.

## 8.3 SOURCE CONVERSIONS

The current source described in the previous section is called an *ideal source* due to the absence of any internal resistance. In reality, all sources—whether they be voltage or current—have some internal resistance in the relative positions shown in Figs. 8.6 and 8.7. For the

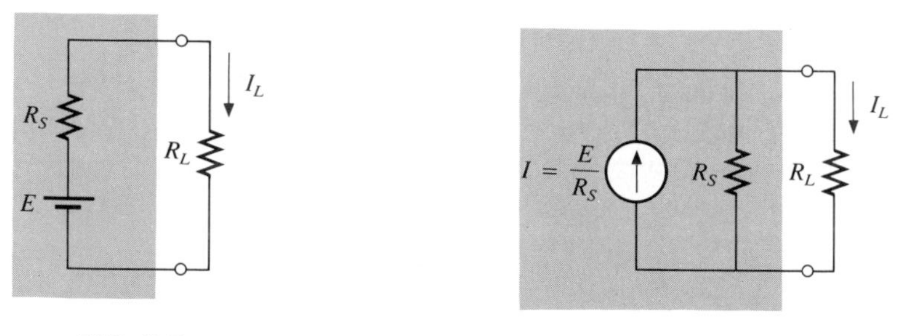

FIG. 8.6

FIG. 8.7

voltage source, if $R_S = 0\ \Omega$ or is so small compared to any series resistor that it can be ignored, then we have an ''ideal'' voltage source. For the current source, if $R_S = \infty\ \Omega$ or is large enough compared to other parallel elements that it can be ignored, then we have an ''ideal'' current source.

If the internal resistance is included with either source, then that source can be converted to the other type using the procedure to be described in this section. Since it is often advantageous to make such a maneuver, this entire section is devoted to being sure the steps are understood. It is important to realize, however, as we proceed through this section that

***source conversions are* equivalent *only at their external terminals.***

The internal characteristics of each are quite different.

We want the equivalence to ensure that the applied load of Figs. 8.6 and 8.7 will receive the same current, voltage, and power from each source and in effect not know, or care, which source is present.

In Fig. 8.6 if we solve for the load current $I_L$, we obtain

$$I_L = \frac{E}{R_S + R_L} \tag{8.1}$$

If we multiply this by a factor of 1, which we can choose to be $R_S/R_S$, we obtain

$$I_L = \frac{(1)E}{R_S + R_L} = \frac{(R_S/R_S)E}{R_S + R_L} = \frac{R_S(E/R_S)}{R_S + R_L} = \frac{R_S I}{R_S + R_L} \tag{8.2}$$

If we define $I = E/R_S$, Eq. (8.2) is the same as that obtained by applying the current divider rule to the network of Fig. 8.7. The result is an equivalence between the networks of Fig. 8.6 and Fig. 8.7 that simply requires that $I = E/R_S$ and the series resistor $R_S$ of Fig. 8.6 be placed in parallel, as in Fig. 8.7. The validity of this is demonstrated in the first example of this section.

For clarity, the equivalent sources *as far as terminals a and b are concerned* are repeated in Fig. 8.8 with the equations for converting in

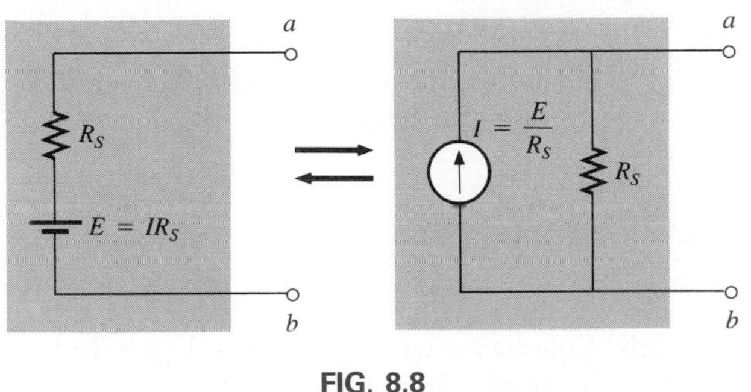


**FIG. 8.8**
*Source conversion.*

either direction. Note, as just indicated, that the resistor $R_S$ is the same in each source, only its position changes. The current of the current source or the voltage of the voltage source is determined using Ohm's law and the parameters of the other configuration. It was pointed out in some detail in Chapter 6 that every source of voltage has some internal series resistance. *For the current source, some internal parallel resistance will always exist in the practical world.* However, in many cases, it is an excellent approximation to drop the internal resistance of a source due to the magnitude of the elements of the network to which it is applied. For this reason, in the analyses to follow, voltage sources may appear without a series resistor, and current sources may appear without a parallel resistance. Realize, however, that in order to perform a conversion from one type of source to another, a voltage source must have a resistor in series with it, and a current source must have a resistor in parallel.

**EXAMPLE 8.4.** Convert the voltage source of Fig. 8.9 to a current source and calculate the current through the 4-Ω load for each source.

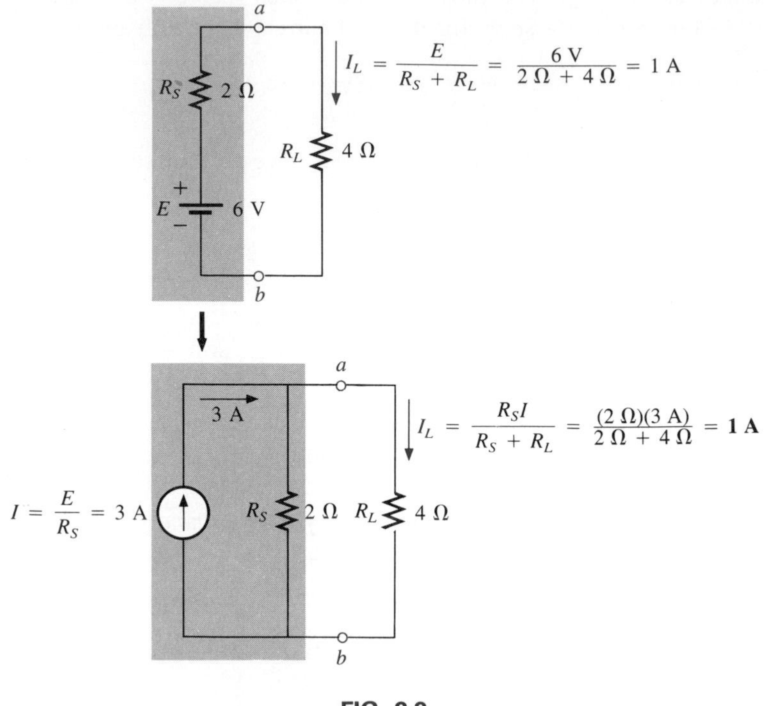


FIG. 8.9

***Solution:*** See the right side of Fig. 8.9.

**EXAMPLE 8.5.** Convert the current source of Fig. 8.10 to a voltage source and find the current through the load for each source.

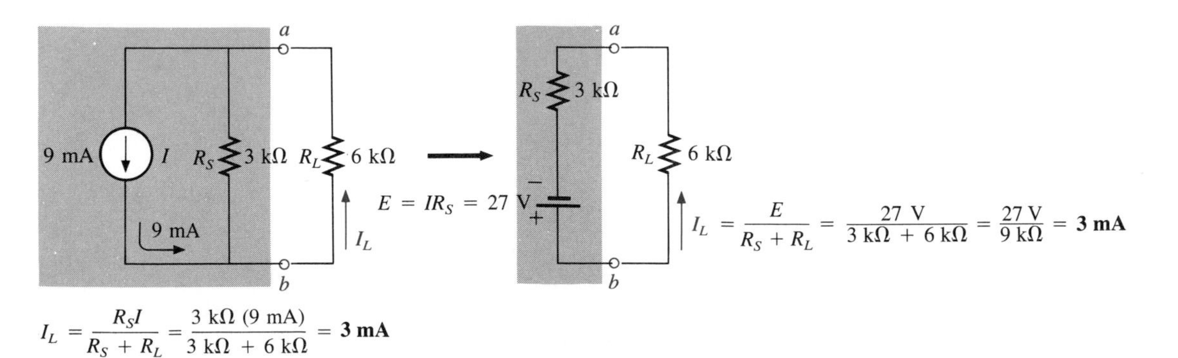

FIG. 8.10

***Solution:*** See the right side of Fig. 8.10.

## 8.4 CURRENT SOURCES IN PARALLEL

If two or more current sources are in parallel, they may all be replaced by one current source having the magnitude and direction of the resultant, which can be found by summing the currents in one direction and subtracting the sum of the currents in the opposite direction. The new parallel resistance is determined by methods described in the discussion of parallel resistors in Chapter 5. Consider the following examples.

---

**EXAMPLE 8.6.** Reduce the left sides of Figs. 8.11 and 8.12 to a minimum number of elements.

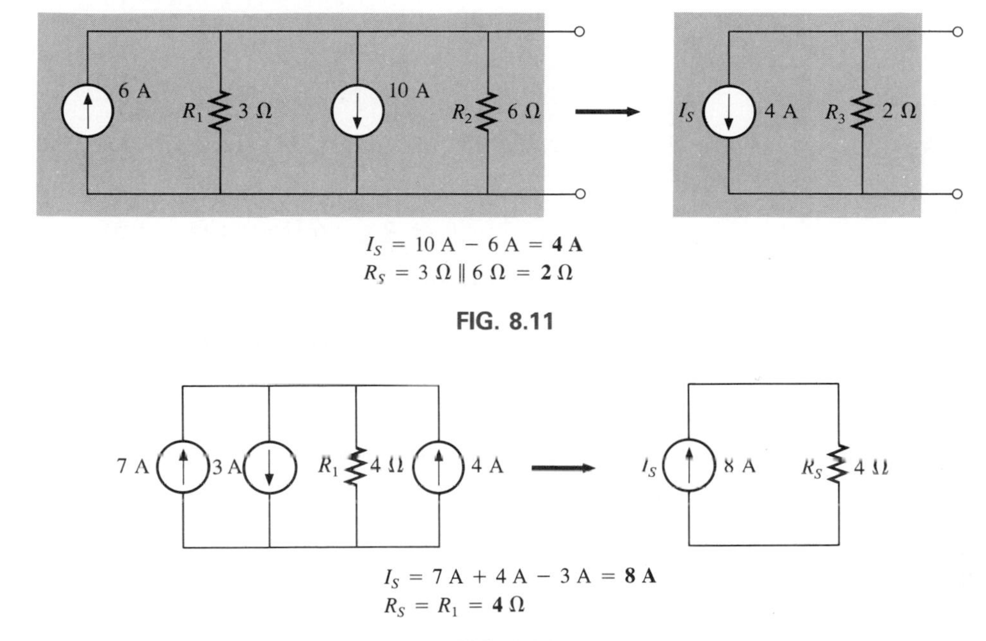


$I_S = 10\text{ A} - 6\text{ A} = \mathbf{4\ A}$
$R_S = 3\ \Omega \parallel 6\ \Omega = \mathbf{2\ \Omega}$

FIG. 8.11

$I_S = 7\text{ A} + 4\text{ A} - 3\text{ A} = \mathbf{8\ A}$
$R_S = R_1 = \mathbf{4\ \Omega}$

FIG. 8.12

***Solution:*** See the right side of the figures.

---

**EXAMPLE 8.7.** Reduce the network of Fig. 8.13 to a single current source and calculate the current through $R_L$.

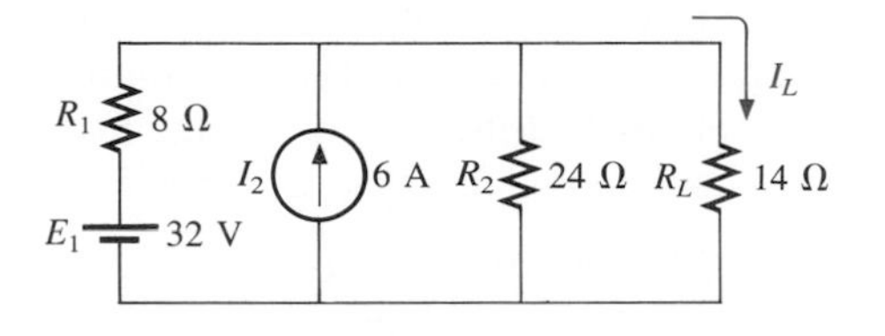


FIG. 8.13

***Solution:*** In this example, the voltage source will first be converted to a current source as shown in Fig. 8.14. Combining current sources,

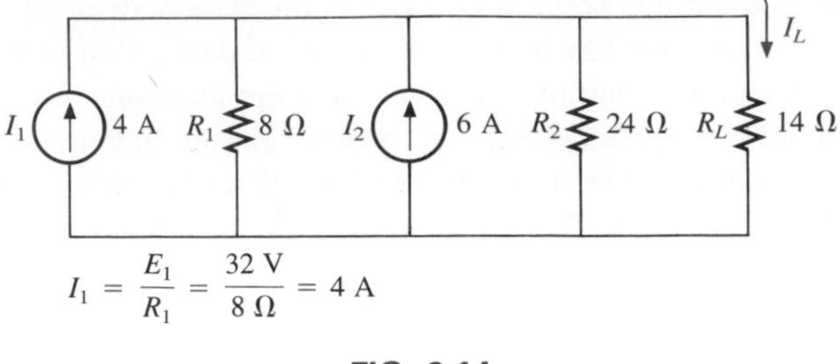


FIG. 8.14

$$I_S = I_1 + I_2 = 4\text{ A} + 6\text{ A} = \mathbf{10\ A}$$

and

$$R_S = R_1 \parallel R_2 = 8\ \Omega \parallel 24\ \Omega = \mathbf{6\ \Omega}$$

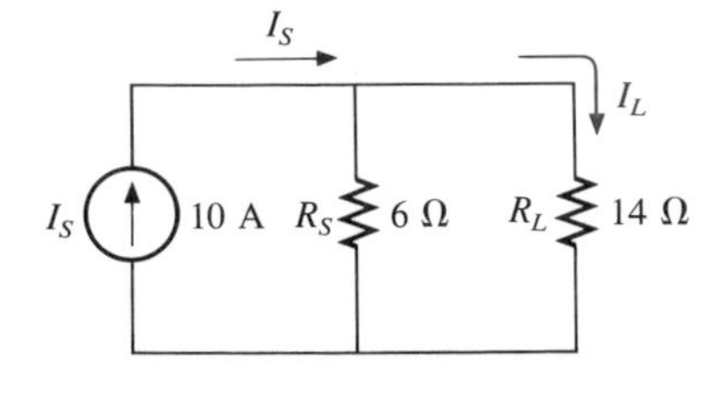


FIG. 8.15

Applying the current divider rule to the resulting network of Fig. 8.15,

$$I_L = \frac{R_S I_S}{R_S + R_L} = \frac{(6\ \Omega)(10\text{ A})}{6\ \Omega + 14\ \Omega} = \frac{60\text{ A}}{20} = \mathbf{3\ A}$$

---

**EXAMPLE 8.8.** Determine the current $I_2$ in the network of Fig. 8.16.

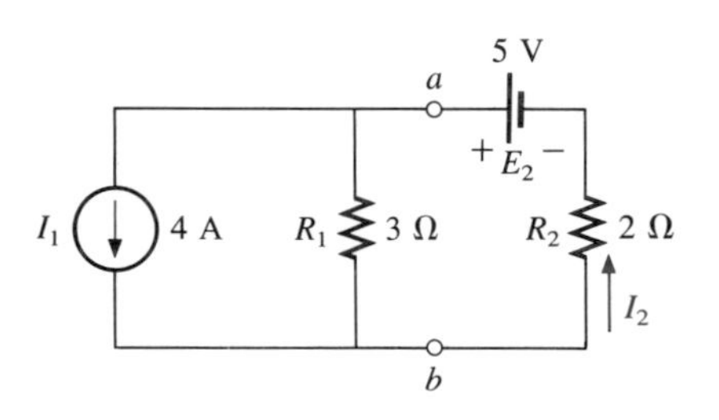

FIG. 8.16

***Solution:*** Although it might appear that the network cannot be solved using methods introduced thus far, one source conversion as shown in Fig. 8.17 will result in a simple series circuit:

$$E_S = I_1 R_1 = (4\text{ A})(3\ \Omega) = 12\text{ V}$$

and

$$R_S = R_1 = 3\ \Omega$$

and

$$I_2 = \frac{E_S + E_2}{R_S + R_2} = \frac{12\text{ V} + 5\text{ V}}{3\ \Omega + 2\ \Omega} = \frac{17\text{ A}}{5} = \mathbf{3.4\ A}$$

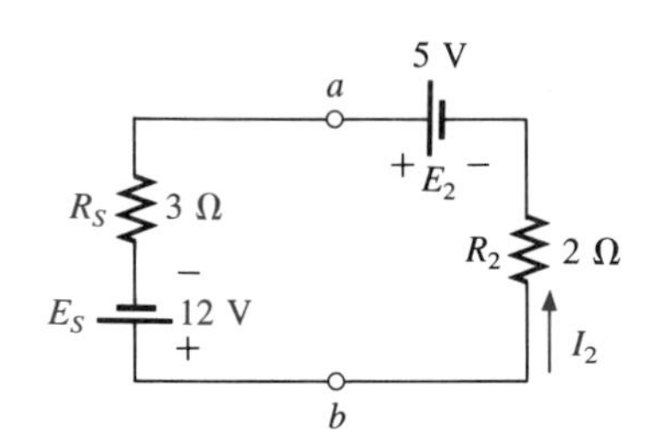


FIG. 8.17

## 8.5 CURRENT SOURCES IN SERIES

The current through any branch of a network can be only single-valued. For the situation indicated at point *a* in Fig. 8.18, we find by application of Kirchhoff's current law that the current leaving that point is greater than that entering—an impossible situation. Therefore,

***current sources of different current ratings are not connected in series,***

just as voltage sources of different voltage ratings are not connected in parallel.

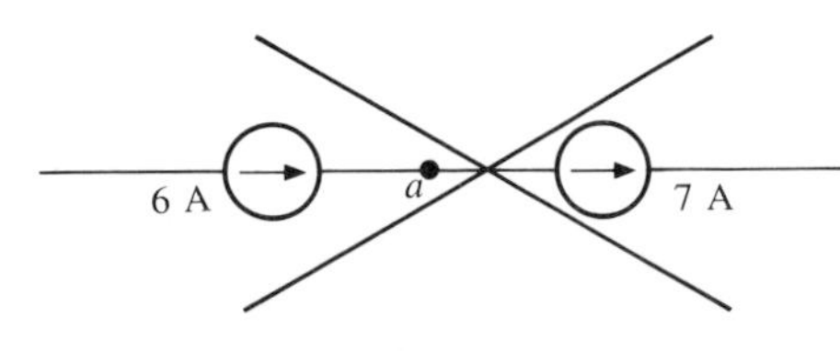


FIG. 8.18

## 8.6 BRANCH-CURRENT ANALYSIS

We will now consider the first in a series of methods for solving networks with two or more sources. Once this method is mastered, there is no linear bilateral dc network for which a solution cannot be found. Keep in mind that networks with two isolated voltage sources cannot be solved using the approach of Chapter 7. For further evidence of this fact, try solving for the unknown elements of Example 8.9 using the methods introduced in Chapter 7. The network of Fig. 8.21 can be solved using the source conversions described in the last section, but the method to be described in this section has applications far beyond the configuration of this network. The most direct introduction to a method of this type is to list the series of steps required for its application. There are four steps, as indicated below. Before continuing, understand that this method will produce the current through each branch of the network, the *branch current*. Once this is known, all other quantities, such as voltage or power, can be determined.

1. Assign a distinct current of *arbitrary* direction to each branch of the network.
2. Indicate the polarities for each resistor *as determined by the assumed current direction*.
3. Apply Kirchhoff's voltage law around each closed independent loop of the network.

The best way to determine how many times Kirchhoff's voltage law will have to be applied is to determine the number of "windows" in the network. The network of Example 8.9 has a definite similarity to the two-window configuration of Fig. 8.19(a). The result is a need to apply Kirchhoff's voltage law twice. For networks with three windows, as shown in Fig. 8.19(b), three applications of Kirchhoff's voltage law are required, and so on.

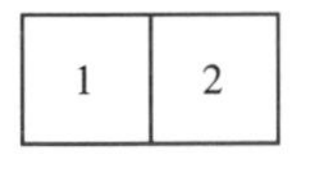


(a)

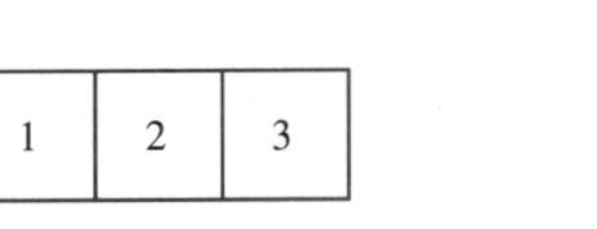


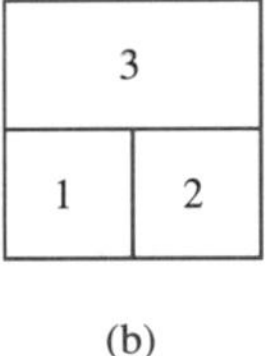


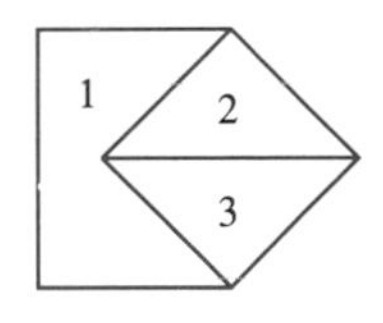


(b)

**FIG. 8.19**
*Determining the number of independent closed loops.*

4. Apply Kirchhoff's current law at the minimum number of nodes that will include all the branch currents of the network.

The minimum number is one less than the number of independent nodes of the network. For the purposes of this analysis a node is a junction of two or more branches, where a branch is any combination of series elements. Figure 8.20 defines the number of applications of Kirchhoff's current law for each configuration of Fig. 8.19.

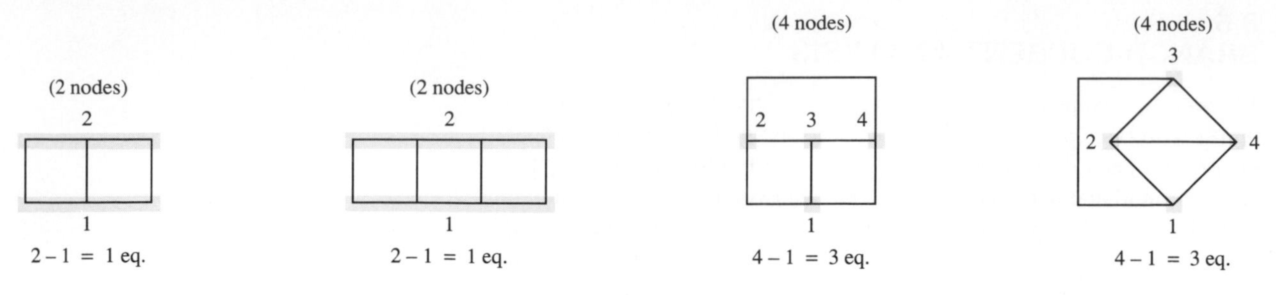


**FIG. 8.20**
*Determining the number of applications of Kirchhoff's current law required.*

5. Solve the resulting simultaneous linear equations for assumed branch currents.

It is assumed that the use of determinants to solve for the currents $I_1$, $I_2$, and $I_3$ is understood and is a part of the student's mathematical background. If not, a detailed explanation of the procedure is provided in Appendix C.

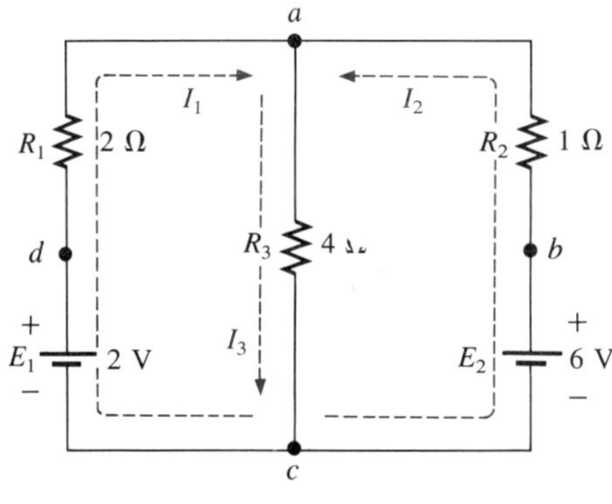

**FIG. 8.21**

**EXAMPLE 8.9.** Apply the branch-current method to the network of Fig. 8.21.

***Solution:***

*Step 1:* Since there are three distinct branches (*cda*, *cba*, *ca*), three currents of arbitrary directions ($I_1$, $I_2$, $I_3$) are chosen as indicated in Fig. 8.21. The current directions for $I_1$ and $I_2$ were chosen to match the "pressure" applied by sources $E_1$ and $E_2$, respectively. Since both $I_1$ and $I_2$ enter node *a*, $I_3$ is leaving.

*Step 2:* Polarities for each resistor are drawn to agree with assumed current directions as indicated in Fig. 8.22.

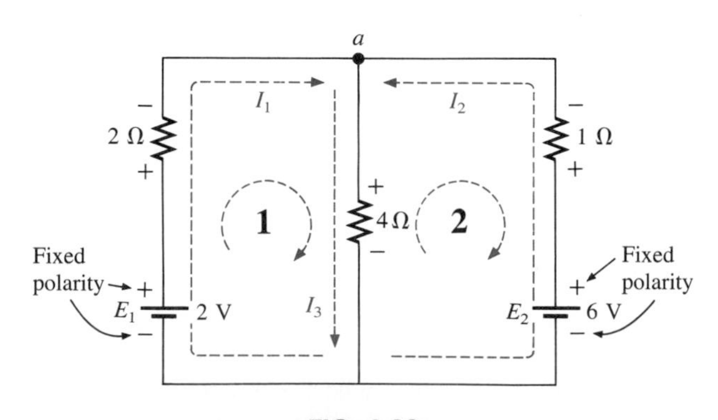


**FIG. 8.22**

*Step 3:* Kirchhoff's voltage law is applied around each closed loop (1 and 2) in the clockwise direction:

Rise in potential — Drop in potential

$$\text{loop 1:}\quad \Sigma_{\circlearrowright} V = \underbrace{+2\text{ V}}_{\substack{\text{Battery}\\ \text{potential}}} - \underbrace{(2\ \Omega)I_1}_{\substack{\text{Voltage drop}\\ \text{across 2-}\Omega\\ \text{resistor}}} - \underbrace{(4\ \Omega)I_3}_{\substack{\text{Voltage drop}\\ \text{across 4-}\Omega\\ \text{resistor}}} = 0$$

$$\text{loop 2:}\quad \Sigma_{\circlearrowright} V = (4\ \Omega)I_3 + (1\ \Omega)I_2 - 6\text{ V} = 0$$

*Step 4:* Applying Kirchhoff's current law at node *a* (in a two-node network, the law is applied or only one node),

$$I_1 + I_2 = I_3$$

*Step 5:* There are three equations and three unknowns (units removed for clarity):

$$\begin{aligned} 2 - 2I_1 - 4I_3 &= 0 \\ 4I_3 + 1I_2 - 6 &= 0 \\ I_1 + I_2 &= I_3 \end{aligned} \qquad \text{Rewritten:}\quad \begin{aligned} 2I_1 + 0 + 4I_3 &= +2 \\ 0 + I_2 + 4I_3 &= +6 \\ I_1 + I_2 - I_3 &= 0 \end{aligned}$$

Using third-order determinants, we have

$$I_1 = \frac{\begin{vmatrix} 2 & 0 & 4 \\ 6 & 1 & 4 \\ 0 & 1 & -1 \end{vmatrix}}{D = \begin{vmatrix} 2 & 0 & 4 \\ 0 & 1 & 4 \\ 1 & 1 & -1 \end{vmatrix}} = -1\text{ A}$$

A negative sign in front of a branch current indicates only that the actual current is in the direction opposite to that assumed.

$$I_2 = \frac{\begin{vmatrix} 2 & 2 & 4 \\ 0 & 6 & 4 \\ 1 & 0 & -1 \end{vmatrix}}{D} = 2\text{ A}$$

$$I_3 = \frac{\begin{vmatrix} 2 & 0 & 2 \\ 0 & 1 & 6 \\ 1 & 1 & 0 \end{vmatrix}}{D} = 1\text{ A}$$

Instead of using third-order determinants, we could reduce the three equations to two by substituting the third equation in the first and second equations:

$$\left.\begin{aligned} 2 - 2I_1 - 4(\overbrace{I_1 + I_2}^{I_3}) &= 0 \\ 4(\overbrace{I_1 + I_2}^{I_3}) + I_2 - 6 &= 0 \end{aligned}\right\} \quad \begin{aligned} 2 - 2I_1 - 4I_1 - 4I_2 &= 0 \\ 4I_1 + 4I_2 + I_2 - 6 &= 0 \end{aligned}$$

or

$$\begin{aligned} -6I_1 - 4I_2 &= -2 \\ +4I_1 + 5I_2 &= +6 \end{aligned}$$

Multiplying through by $-1$ in the top equation yields

$$\begin{aligned} 6I_1 + 4I_2 &= +2 \\ 4I_1 + 5I_2 &= +6 \end{aligned}$$

and using determinants,

$$I_1 = \frac{\begin{vmatrix} 2 & 4 \\ 6 & 5 \end{vmatrix}}{\begin{vmatrix} 6 & 4 \\ 4 & 5 \end{vmatrix}} = \frac{10 - 24}{30 - 16} = \frac{-14}{14} = \mathbf{-1\ A}$$

$$I_2 = \frac{\begin{vmatrix} 6 & 2 \\ 4 & 6 \end{vmatrix}}{14} = \frac{36 - 8}{14} = \frac{28}{14} = \mathbf{2\ A}$$

$$I_3 = I_1 + I_2 = -1 + 2 = \mathbf{1\ A}$$

It is now important that the impact of the results obtained is understood. The currents $I_1$, $I_2$, and $I_3$ are the actual currents in the branches in which they were defined. A negative sign in the solution simply reveals that the actual current has the opposite direction than initially defined—the magnitude is correct. Once the actual current directions and their magnitudes are inserted in the original network, the various voltages and power levels can be determined. For Example 8.9, the actual current directions and their magnitudes have been entered on the original network in Fig. 8.23. Note that the current through the series

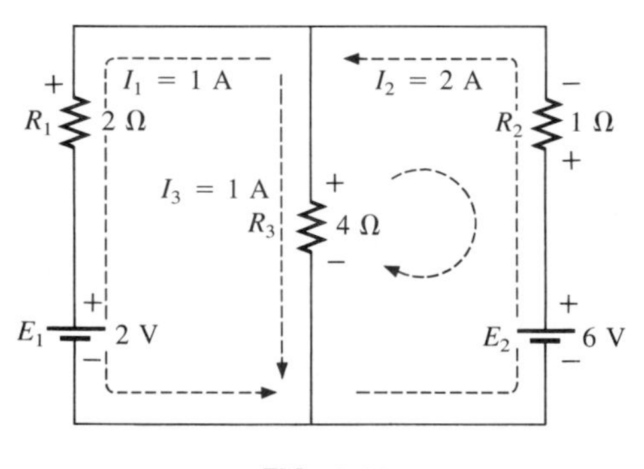


**FIG. 8.23**

elements $R_1$ and $E_1$ is 1 A, the current through $R_3$, 1 A, and the current through the series elements $R_2$ and $E_2$, 2 A. Due to the minus sign in the solution, the direction of $I_1$ is opposite to that shown in Fig. 8.21. The voltage across any resistor can now be found using Ohm's law, and the power delivered by either source or to any one of the three resistors can be found using the appropriate power equation.

Applying Kirchhoff's voltage law around the loop indicated in Fig. 8.23,

$$\Sigma_{\circlearrowleft} V = +(4\ \Omega)I_3 + (1\ \Omega)I_2 - 6\ \text{V} = 0$$

or

$$(4\ \Omega)I_3 + (1\ \Omega)I_2 = 6\ \text{V}$$

and

$$\begin{aligned}(4\ \Omega)(1\ \text{A}) + (1\ \Omega)(2\ \text{A}) &= 6\ \text{V}\\ 4\ \text{V} + 2\ \text{V} &= 6\ \text{V}\\ 6\ \text{V} &= 6\ \text{V} \quad \text{(checks)}\end{aligned}$$

---

**EXAMPLE 8.10.** Apply branch-current analysis to the network of Fig. 8.24.

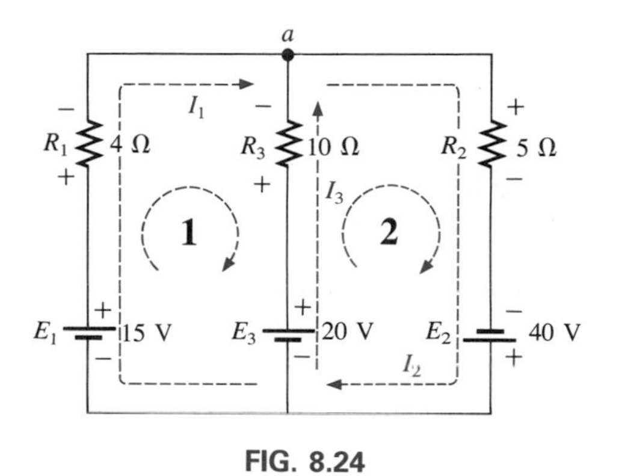


FIG. 8.24

***Solution:*** Again, the current directions were chosen to match the "pressure" of each battery. The polarities are then added and Kirchhoff's voltage law is applied around each closed loop in the clockwise direction. The result is as follows:

$$\begin{aligned}&\text{loop 1:}\quad +15\ \text{V} - (4\ \Omega)I_1 + (10\ \Omega)I_3 - 20\ \text{V} = 0\\ &\text{loop 2:}\quad +20\ \text{V} - (10\ \Omega)I_3 - (5\ \Omega)I_2 + 40\ \text{V} = 0\end{aligned}$$

Applying Kirchhoff's current law at node $a$,

$$I_1 + I_3 = I_2$$

Substituting the third equation into the other two yields (with units removal for clarity):

$$\left.\begin{aligned}15 - 4I_1 + 10I_3 - 20 &= 0\\ 20 - 10I_3 - 5(I_1 + I_3) + 40 &= 0\end{aligned}\right\}\ \text{Substituting for } I_2 \text{ (since it occurs only once in the two equations)}$$

or

$$\begin{aligned}-4I_1 + 10I_3 &= 5\\ -5I_1 - 15I_3 &= -60\end{aligned}$$

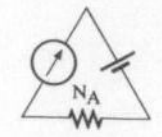

Multiplying the lower equation by −1, we have

$$\begin{aligned} -4I_1 + 10I_3 &= 5 \\ 5I_1 + 15I_3 &= 60 \end{aligned}$$

$$I_1 = \frac{\begin{vmatrix} 5 & 10 \\ 60 & 15 \end{vmatrix}}{\begin{vmatrix} -4 & 10 \\ 5 & 15 \end{vmatrix}} = \frac{75 - 600}{-60 - 50} = \frac{-525}{-110} = \mathbf{4.773\ A}$$

$$I_3 = \frac{\begin{vmatrix} -4 & 5 \\ 5 & 60 \end{vmatrix}}{-110} = \frac{-240 - 25}{-110} = \frac{-265}{-110} = \mathbf{2.409\ A}$$

$$I_2 = I_1 + I_3 = 4.773 + 2.409 = \mathbf{7.182\ A}$$

revealing that the assumed directions were the actual directions, with $I_2$ equal to the sum of $I_1$ and $I_3$.

## 8.7 MESH ANALYSIS (GENERAL APPROACH)

The second method of analysis to be described is called *mesh analysis*. The term *mesh* is derived from the similarities in appearance between the closed loops of a network and a wire mesh fence. Although this approach is on a more sophisticated plane than the branch-current method, it incorporates many of the ideas just developed. Of the two methods, mesh analysis is the one more frequently applied today. Branch-current analysis is introduced as a stepping stone to mesh analysis because branch currents are initially more ''real'' to the student than the loop currents employed in mesh analysis. Essentially, the mesh-analysis approach simply eliminates the need to substitute the results of Kirchhoff's current law into the equations derived from Kirchhoff's voltage law. It is now accomplished in the initial writing of the equations. The systematic approach outlined below should be followed when applying this method.

1. Assign a distinct current in the clockwise direction to each independent closed loop of the network. It is not absolutely necessary to choose the clockwise direction for each loop current. In fact, any direction can be chosen for each loop current with no loss in accuracy as long as the remaining steps are followed properly. However, by choosing the clockwise direction as a standard, we can develop a shorthand method (Section 8.8) for writing the required equations that will save time and possibly prevent some common errors.

This first step is most effectively accomplished by placing a loop current *within* each ''window'' of the network, as demonstrated in the previous section to insure they are all independent. There are a variety of other loop currents that can be assigned. In each case, however, be sure that the information carried by any one loop equation is not included in a combination of the other network equations. This is the crux of the terminology: *independent*. No matter how you choose your loop currents, the number of loop currents required is always equal to the number of windows of a planar (no-crossovers) network. On occasion a network may appear to be nonplanar. However, a redrawing of the network may reveal that it is, in fact, planar. Such may be the case in one or two problems at the end of the chapter.

Before continuing to the next step, let us insure that the concept of a loop current is clear. For the network of Fig. 8.25, the loop current $I_1$ is the branch current of the branch containing the 2-Ω resistor and 2-V battery. The current through the 4-Ω resistor is not $I_1$, however, since there is also a loop current $I_2$ through it. Since they have opposite directions, $I_{4\Omega}$ equals $I_1 - I_2$, as pointed out in the example to follow. In other words, *a loop current is a branch current only when it is the only loop current assigned to that branch.*

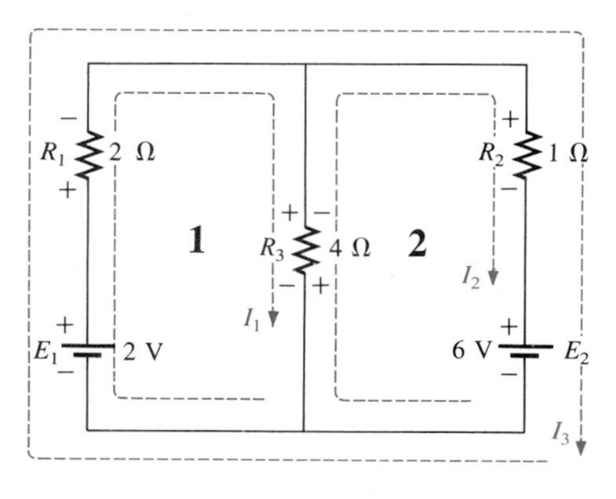


FIG. 8.25

2. Indicate the polarities *within* each loop for each resistor as determined by the assumed direction of loop current for that loop. Note the requirement that the polarities be placed within each loop. This requires, as shown in Fig. 8.25, that the 4-Ω resistor have two sets of polarities across it.
3. Apply Kirchhoff's voltage law around each closed loop in the clockwise direction. Again, the clockwise direction was chosen to establish uniformity and prepare us for the method to be introduced in the next section.
   a. If a resistor has two or more assumed currents through it, the total current through the resistor is the assumed current of the loop in which Kirchhoff's voltage law is being applied, plus the assumed currents of the other loops passing through in the *same* direction, minus the assumed currents through in the *opposite* direction.
   b. The polarity of a voltage source is unaffected by the direction of the assigned loop currents.
4. Solve the resulting simultaneous linear equations for the assumed loop currents.

**EXAMPLE 8.11.** Consider the same basic network as in Example 8.9 of the preceding section, now appearing in Fig. 8.25.

***Solution:***

*Step 1:* Two loop currents ($I_1$ and $I_2$) are assigned in the clockwise direction in the windows of the network. A third loop ($I_3$) could have been included around the entire network but the information carried by this loop is already included in the other two.

*Step 2:* Polarities are drawn within each window to agree with assumed current directions. Note that for this case the polarities across the 4-Ω resistor are the opposite for each loop current.

*Step 3:* Kirchhoff's voltage law is applied around each loop in the clockwise direction. Keep in mind as this step is performed that the law

is concerned only with the magnitude and polarity of the voltages around the closed loop and not with whether a voltage rise or drop is due to a battery or resistive element. The voltage across each resistor is determined by $V = IR$, and for a resistor with more than one current through it, the current is the loop current of the loop being examined plus or minus the other loop currents as determined by their directions. If clockwise applications of Kirchhoff's voltage law are always chosen, the other loop currents will always be subtracted from the loop current of the loop being analyzed.

loop 1: $+E_1 - V_1 - V_3 = 0$ (clockwise starting at point $a$)

$$+2\text{ V} - (2\ \Omega)I_1 - \overbrace{(4\ \Omega)(\underbrace{I_1 - I_2}_{\substack{\text{Total current}\\ \text{through}\\ 4\text{-}\Omega\text{ resistor}}})}^{\substack{\text{Voltage drop across}\\ 4\text{-}\Omega\text{ resistor}}} = 0$$

Subtracted since $I_2$ is opposite in direction to $I_1$.

loop 2: $-V_3 - V_2 - E_2 = 0$ (clockwise starting at point $b$)

$$-(4\ \Omega)(I_2 - I_1) - (1\ \Omega)I_2 - 6\text{ V} = 0$$

*Step 4:* The equations are then rewritten as follows (without units for clarity):

$$\begin{aligned} \text{loop 1:} &\quad +2 - 2I_1 - 4I_1 + 4I_2 = 0 \\ \text{loop 2:} &\quad -4I_2 + 4I_1 - 1I_2 - 6 = 0 \end{aligned}$$

and

$$\begin{aligned} \text{loop 1:} &\quad +2 - 6I_1 + 4I_2 = 0 \\ \text{loop 2:} &\quad -5I_2 + 4I_1 - 6 = 0 \end{aligned}$$

or

$$\begin{aligned} \text{loop 1:} &\quad -6I_1 + 4I_2 = -2 \\ \text{loop 2:} &\quad +4I_1 - 5I_2 = +6 \end{aligned}$$

Applying determinants will result in

$$I_1 = \mathbf{-1\ A} \quad \text{and} \quad I_2 = \mathbf{-2\ A}$$

The minus signs indicate that the currents have a direction opposite to that indicated by the assumed loop current.

The actual current through the 2-V source and 2-Ω resistor is therefore 1 A in the other direction, and the current through the 6-V source and 1-Ω resistor is 2 A in the opposite direction indicated on the circuit. The current through the 4-Ω resistor is determined by the following equation from the original network:

$$\begin{aligned} \text{loop 1:} \quad I_{4\Omega} &= I_1 - I_2 = -1\text{ A} - (-2\text{ A}) = -1\text{ A} + 2\text{ A} \\ &= \mathbf{1\ A} \qquad \text{(in the direction of } I_1\text{)} \end{aligned}$$

The outer loop ($I_3$) and *one* inner loop (either $I_1$ or $I_2$) would also have produced the correct results. This approach, however, will often lead to errors since the loop equations may be more difficult to write. The best method of picking these loop currents is usually to use the window approach.

**EXAMPLE 8.12.** Find the current through each branch of the network of Fig. 8.26.

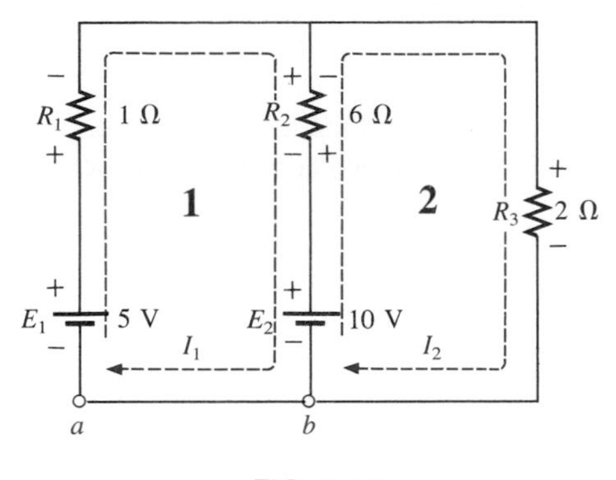

**FIG. 8.26**

***Solution:***

*Steps 1 and 2* are as indicated in the circuit. Note that the polarities of the 6-Ω resistor are different for each loop current.

*Step 3:* Kirchhoff's voltage law is applied around each closed loop in the clockwise direction:

loop 1: $+E_1 - V_1 - V_2 - E_2 = 0$ (clockwise starting at point $a$)

$$+5\text{ V} - (1\ \Omega)I_1 - (6\ \Omega)(I_1 - I_2) - 10\text{ V} = 0$$

$I_2$ flows through the 6-Ω resistor in the direction opposite to $I_1$.

loop 2: $E_2 - V_2 - V_3 = 0$ (clockwise starting at point $b$)

$$+10\text{ V} - (6\ \Omega)(I_2 - I_1) - (2\ \Omega)I_2 = 0$$

The equations are rewritten as

$$\left.\begin{aligned} 5 - I_1 - 6I_1 + 6I_2 - 10 = 0 \\ 10 - 6I_2 + 6I_1 - 2I_2 = 0 \end{aligned}\right\} \begin{aligned} -7I_1 + 6I_2 &= 5 \\ +6I_1 - 8I_2 &= -10 \end{aligned}$$

$$I_1 = \frac{\begin{vmatrix} 5 & 6 \\ -10 & -8 \end{vmatrix}}{\begin{vmatrix} -7 & 6 \\ 6 & -8 \end{vmatrix}} = \frac{-40 + 60}{56 - 36} = \frac{20}{20} = \mathbf{1\ A}$$

$$I_2 = \frac{\begin{vmatrix} -7 & 5 \\ 6 & -10 \end{vmatrix}}{20} = \frac{70 - 30}{20} = \frac{40}{20} = \mathbf{2\ A}$$

Since $I_1$ and $I_2$ are positive and flow in opposite directions through the 6-Ω resistor and 10-V source, the total current in this branch is equal to the difference of the two currents in the direction of the larger:

$$I_2 > I_1 \qquad (2\text{ A} > 1\text{ A})$$

Therefore, $I_{R_2} = I_2 - I_1 = 2\text{ A} - 1\text{ A} = \mathbf{1\ A}$ in the direction of $I_2$.

It is sometimes impractical to draw all the branches of a circuit at right angles to one another. The next example demonstrates how a portion of a network may appear due to various constraints. The method of analysis does not change with this change in configuration.

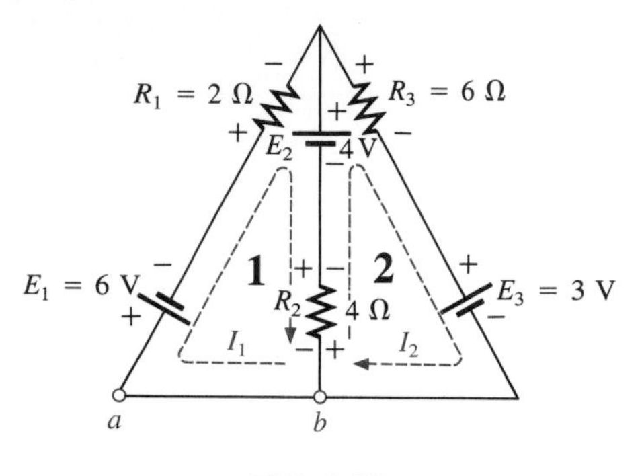


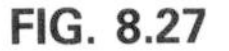

FIG. 8.27

**EXAMPLE 8.13.** Find the branch currents of the network of Fig. 8.27.

***Solution:***

*Steps 1 and 2* are as indicated in the circuit.

*Step 3:* Kirchhoff's voltage law is applied around each closed loop:

loop 1: $-E_1 - I_1R_1 - E_2 - V_2 = 0$ (clockwise from point $a$)

$$-6\text{ V} - (2\ \Omega)I_1 - 4\text{ V} - (4\ \Omega)(I_1 - I_2) = 0$$

loop 2: $-V_2 + E_2 - V_3 - E_3 = 0$ (clockwise from point $b$)

$$-(4\ \Omega)(I_2 - I_1) + 4\text{ V} - (6\ \Omega)(I_2) - 3\text{ V} = 0$$

which are rewritten as

$$\left.\begin{array}{r} -10 - 4I_1 - 2I_1 + 4I_2 = 0 \\ +1 + 4I_1 - 4I_2 - 6I_2 = 0 \end{array}\right\} \begin{array}{r} -6I_1 + 4I_2 = +10 \\ +4I_1 - 10I_2 = -1 \end{array}$$

or, by multiplying the top equation by $-1$, we obtain

$$\begin{array}{r} 6I_1 - 4I_2 = -10 \\ 4I_1 - 10I_2 = -1 \end{array}$$

and

$$I_1 = \frac{\begin{vmatrix} -10 & -4 \\ -1 & -10 \end{vmatrix}}{\begin{vmatrix} 6 & -4 \\ 4 & -10 \end{vmatrix}} = \frac{100 - 4}{-60 + 16} = \frac{96}{-44} = \mathbf{-2.182\ A}$$

$$I_2 = \frac{\begin{vmatrix} 6 & -10 \\ 4 & -1 \end{vmatrix}}{-44} = \frac{-6 + 40}{-44} = \frac{34}{-44} = \mathbf{-0.773\ A}$$

The current in the 4-Ω resistor and 4-V source for loop 1 is

$$\begin{aligned} I_1 - I_2 &= -2.182\text{ A} - (-0.773\text{ A}) \\ &= -2.182\text{ A} + 0.773\text{ A} \\ &= \mathbf{-1.409\ A} \end{aligned}$$

revealing that it is 1.409 A in a direction opposite (due to the minus sign) to $I_1$ in loop 1.

It may happen that current sources will be present in the network to which we wish to apply mesh analysis. To maintain the standard procedure, the first step will then be to convert all current sources to voltage sources as in the next example. The example will also use standard 20% resistor values to demonstrate how the analysis can become clouded by a more extensive mathematical exercise.

**EXAMPLE 8.14.** Using mesh analysis, determine the current through the 9-V battery for the network of Fig. 8.28.

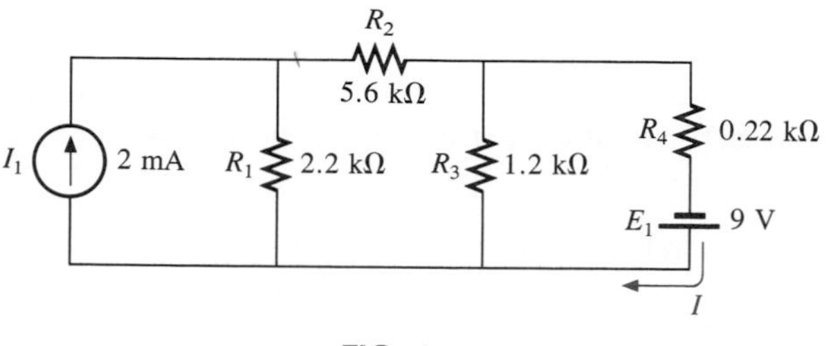


**FIG. 8.28**

***Solution:*** The current source is first converted to a voltage source as shown in Fig. 8.29 followed by a combination of series resistive elements as shown in Fig. 8.30.

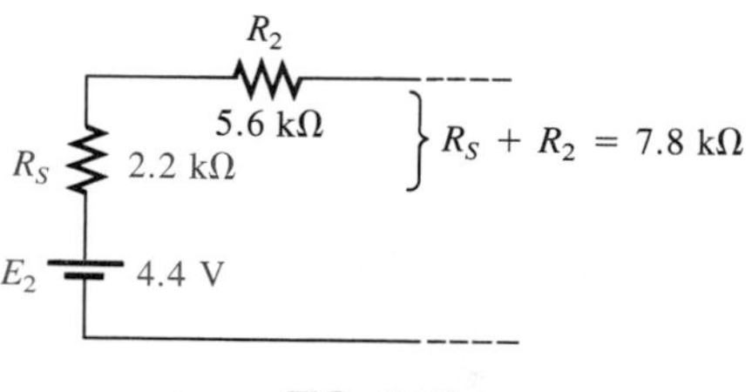


**FIG. 8.29**

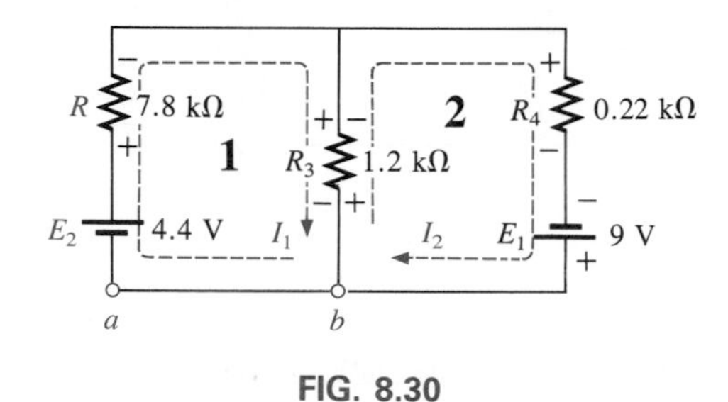


**FIG. 8.30**

The loop currents were chosen as shown in Fig. 8.30 and Kirchhoff's voltage law was applied around each closed loop in the clockwise direction:

loop 1: $+E_2 - I_1 R - V_3 = 0$ (clockwise from point $a$)

$$4.4\text{ V} - 7.8\text{ k}\Omega(I_1) - 1.2\text{ k}\Omega(I_1 - I_2) = 0$$

loop 2: $-V_3 - I_2 R_4 + E_1 = 0$ (clockwise from point $b$)

$$-1.2\text{ k}\Omega(I_2 - I_1) - 0.22\text{ k}\Omega(I_2) + 9\text{ V} = 0$$

which are rewritten as

$$\begin{aligned} 9\text{ k}\Omega(I_1) - 1.2\text{ k}\Omega(I_2) &= 4.4 \\ -1.2\text{ k}\Omega(I_1) + 1.42\text{ k}\Omega(I_2) &= 9 \end{aligned}$$

and

$$I_{9\text{V}} = I_2 = \frac{\begin{vmatrix} 9\text{ k}\Omega & 4.4 \\ -1.2\text{ k}\Omega & 9 \end{vmatrix}}{\begin{vmatrix} 9\text{ k}\Omega & -1.2\text{ k}\Omega \\ -1.2\text{ k}\Omega & +1.42\text{ k}\Omega \end{vmatrix}}$$

$$= \frac{(9)(9\text{ k}\Omega) - (4.4)(-1.2\text{ k}\Omega)}{(9\text{ k}\Omega)(1.42\text{ k}\Omega) - (-1.2\text{ k}\Omega)(-1.2\text{ k}\Omega)}$$

$$= \frac{86.28}{11.34\text{ k}\Omega} = \mathbf{7.608\text{ mA}}$$

In the previous example, if the currents through $R_1$ and $R_2$ were desired, the current source can remain if it is clearly understood that *the loop current through the current source must have the magnitude and direction of the current source.*

---

**EXAMPLE 8.15.** Find the currents through the resistors $R_1$ and $R_2$ in Fig. 8.28.

***Solution:*** As indicated in Fig. 8.31, three loop currents must now be defined and based on the preceding discussion, $I_1$ = **2 mA.**

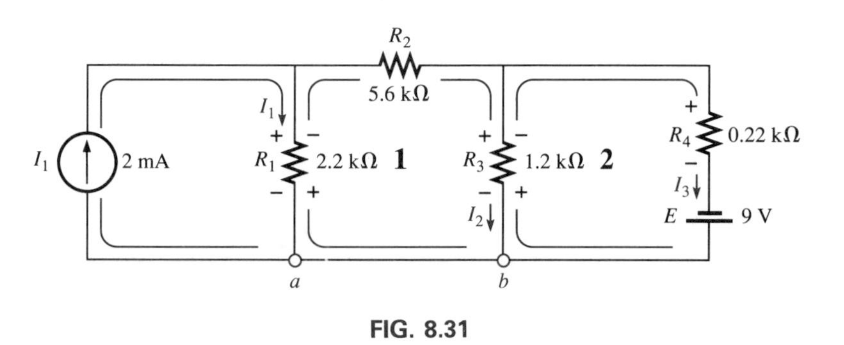


**FIG. 8.31**

Applying Kirchhoff's voltage law around loop 1 in the clockwise direction starting at point *a* will yield:

$$-V_1 - V_2 - V_3 = 0$$
$$-(2.2 \text{ k}\Omega)(I_2 - I_1) - (5.6 \text{ k}\Omega)I_2 - (1.2 \text{ k}\Omega)(I_2 - I_3) = 0$$

For loop 2, if we apply Kirchhoff's voltage law in the clockwise direction starting at point *b*, we obtain:

$$-V_3 - V_4 + 9 \text{ V} = 0$$
$$-1.2 \text{ k}\Omega(I_3 - I_2) - (0.22 \text{ k}\Omega)I_3 + 9 \text{ V} = 0$$

Rewriting both equations and reducing to the simplest form results in:

$$-9 \text{ k}\Omega\ I_2 + 1.2 \text{ k}\Omega\ I_3 = -4.4$$
$$1.2 \text{ k}\Omega\ I_2 - 1.42 \text{ k}\Omega\ I_3 = -9$$

Solving for $I_2$ will result in

$$I_2 = \mathbf{1.503\ mA}$$

and

$$I_{R_1} = I_1 - I_2 \qquad \text{(direction of } I_1\text{)}$$
$$= 2 \text{ mA} - 1.503 \text{ mA}$$
$$= \mathbf{0.497\ mA}$$

with

$$I_{R_2} = I_2 = \mathbf{1.503\ mA}$$

---

## 8.8 MESH ANALYSIS (FORMAT APPROACH)

Now that the basis for the mesh-analysis approach has been established, in this section we will consider a technique for writing the mesh equations more rapidly and usually with fewer errors. As an aid in introducing the procedure, the network of Example 8.12 (Fig. 8.26) has been redrawn in Fig. 8.32 with the assigned loop currents. (Note that each loop current has a clockwise direction.)

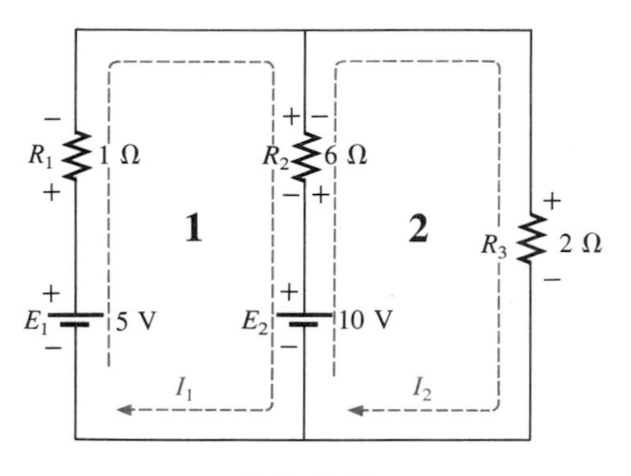


**FIG. 8.32**

The equations obtained are

$$-7I_1 + 6I_2 = 5$$
$$6I_1 - 8I_2 = -10$$

which can also be written as

$$7I_1 - 6I_2 = -5$$
$$8I_2 - 6I_1 = 10$$

and expanded as

| Col. 1 | | Col. 2 | Col. 3 |
|---|---|---|---|
| $(1 + 6)I_1$ | $-$ | $6I_2$ | $= (5 - 10)$ |
| $(2 + 6)I_2$ | $-$ | $6I_1$ | $= 10$ |

Note in the above equations that column 1 is composed of a loop current times the sum of the resistors through which that loop current passes. Column 2 is the product of the resistors common to another loop current times that other loop current. Note that in each equation this column is subtracted from column 1. Column 3 is the *algebraic* sum of the voltage sources through which the loop current of interest passes. A source is assigned a positive sign if the loop current passes from the negative to the positive terminal, and a negative value if the polarities are reversed. The comments above are correct only for a standard direction of loop current in each window, the chosen being the clockwise direction.

The above statements can be extended to develop the following *format approach* to mesh analysis:

1. Assign a loop current to each independent closed loop (as in the previous section) in a *clockwise* direction.
2. The number of required equations is equal to the number of chosen independent closed loops. Column 1 of each equation is formed by summing the resistance values of those resistors through which the loop current of interest passes and multiplying the result by that loop current.
3. We must now consider the mutual terms which, as noted in the examples above, are always subtracted from the first column. A mutual term is simply any resistive element having an additional loop current passing through it. It is possible to have more than one mutual term if the loop current of interest has an element in common with more than one other loop current. This will be demonstrated in an example to follow. Each term is the product

of the mutual resistor and the other loop current passing through the same element.

4. The column to the right of the equality sign is the algebraic sum of the voltage sources through which the loop current of interest passes. Positive signs are assigned to those sources of voltage having a polarity such that the loop current passes from the negative to the positive terminal. A negative sign is assigned to those potentials for which the reverse is true.
5. Solve the resulting simultaneous equations for the desired loop currents.

Let us now consider a few examples.

---

**EXAMPLE 8.16.** Write the mesh equations for the network of Fig. 8.33 and find the current through the 7-Ω resistor.

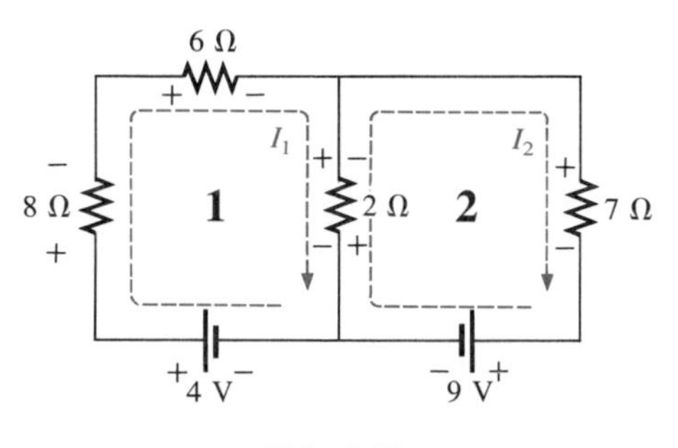

FIG. 8.33

***Solution:***
*Step 1:* As indicated in Fig. 8.33, each assigned loop current has a clockwise direction.
*Steps 2 to 4:*

$$\begin{aligned} I_1&: \quad (8\ \Omega + 6\ \Omega + 2\ \Omega)I_1 - (2\ \Omega)I_2 = 4\ \text{V} \\ I_2&: \quad (7\ \Omega + 2\ \Omega)I_2 - (2\ \Omega)I_1 = -9\ \text{V} \end{aligned}$$

and

$$\begin{aligned} 16I_1 - 2I_2 &= 4 \\ 9I_2 - 2I_1 &= -9 \end{aligned}$$

which for determinants are

$$\begin{aligned} 16I_1 - 2I_2 &= 4 \\ -2I_1 + 9I_2 &= -9 \end{aligned}$$

and

$$I_2 = I_{7\Omega} = \frac{\begin{vmatrix} 16 & 4 \\ -2 & -9 \end{vmatrix}}{\begin{vmatrix} 16 & -2 \\ -2 & 9 \end{vmatrix}} = \frac{-144 + 8}{144 - 4} = \frac{-136}{140}$$

$$= \mathbf{-0.971\ A}$$

---

**EXAMPLE 8.17.** Write the mesh equations for the network of Fig. 8.34.

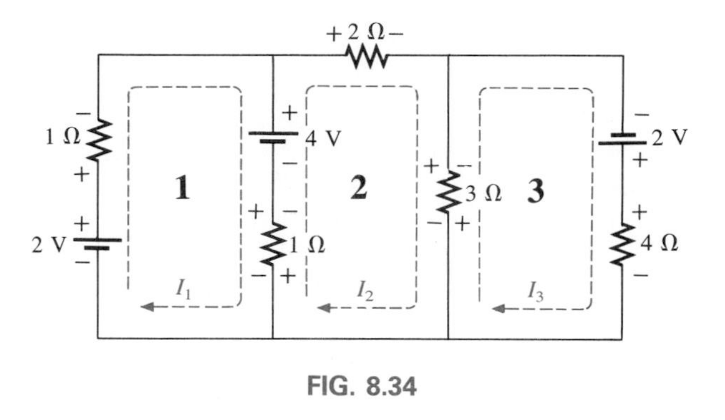


FIG. 8.34

***Solution:***

*Step 1:* Each window is assigned a loop current in the clockwise direction:

$I_1$ does not pass through an element mutual with $I_3$.

$$
\begin{aligned}
I_1:&\quad (1\ \Omega + 1\ \Omega)I_1 - (1\ \Omega)I_2 + 0 = 2\ \text{V} - 4\ \text{V}\\
I_2:&\quad (1\ \Omega + 2\ \Omega + 3\ \Omega)I_2 - (1\ \Omega)I_1 - (3\ \Omega)I_3 = 4\ \text{V}\\
I_3:&\quad (3\ \Omega + 4\ \Omega)I_3 - (3\ \Omega)I_2 + 0 = 2\ \text{V}
\end{aligned}
$$

$I_3$ does not pass through an element mutual with $I_1$.

Summing terms yields

$$
\begin{aligned}
2I_1 - I_2 + 0 &= -2\\
6I_2 - I_1 - 3I_3 &= 4\\
7I_3 - 3I_2 + 0 &= 2
\end{aligned}
$$

which are rewritten for determinants as

$$
\begin{aligned}
2I_1 - I_2 + 0 &= -2\\
-I_1 + 6I_2 - 3I_3 &= 4\\
0 - 3I_2 + 7I_3 &= 2
\end{aligned}
$$

which compares directly with the equations obtained in that example.

Note that the coefficients of the $a$ and $b$ diagonals are equal. This *symmetry* about the $c$ axis will always be true for equations written using the format approach. It is a check on whether the equations were obtained correctly. Note the symmetry of the equations of Example 8.17, as demonstrated by the dashed lines.

We will now consider a network with only one source of voltage to point out that mesh analysis can be used to advantage in other than multisource networks.

**EXAMPLE 8.18.** Find the current through the 10-Ω resistor of the network of Fig. 8.35.

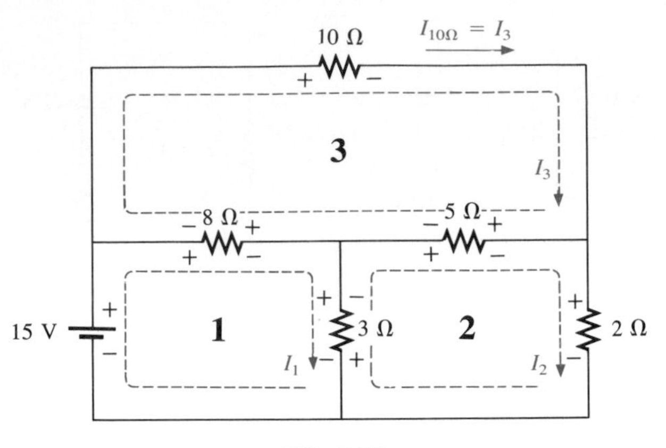

FIG. 8.35

***Solution:***

$$
\begin{aligned}
I_1\text{:} \quad & (8\ \Omega + 3\ \Omega)I_1 - (8\ \Omega)I_3 - (3\ \Omega)I_2 = 15\text{ V} \\
I_2\text{:} \quad & (3\ \Omega + 5\ \Omega + 2\ \Omega)I_2 - (3\ \Omega)I_1 - (5\ \Omega)I_3 = 0 \\
I_3\text{:} \quad & (8\ \Omega + 10\ \Omega + 5\ \Omega)I_3 - (8\ \Omega)I_1 - (5\ \Omega)I_2 = 0
\end{aligned}
$$

$$
\begin{aligned}
11I_1 - 8I_3 - 3I_2 &= 15 \\
10I_2 - 3I_1 - 5I_3 &= 0 \\
23I_3 - 8I_1 - 5I_2 &= 0
\end{aligned}
$$

or

$$
\begin{aligned}
11I_1 - 3I_2 - 8I_3 &= 15 \\
-3I_1 + 10I_2 - 5I_3 &= 0 \\
-8I_1 - 5I_2 + 23I_3 &= 0
\end{aligned}
$$

and

$$
I_3 = I_{10\Omega} = \frac{\begin{vmatrix} 11 & -3 & 15 \\ -3 & 10 & 0 \\ -8 & -5 & 0 \end{vmatrix}}{\begin{vmatrix} 11 & -3 & -8 \\ -3 & 10 & -5 \\ -8 & -5 & 23 \end{vmatrix}} = \mathbf{1.220\ A}
$$

## 8.9 NODAL ANALYSIS (GENERAL APPROACH)

Recall from the development of loop analysis that the general network equations were obtained by applying Kirchhoff's voltage law around each closed loop. We will now employ Kirchhoff's current law to develop a method referred to as *nodal analysis*.

A *node* is defined as a junction of two or more branches. If we now define one node of any network as a reference (that is, a point of zero potential or ground), the remaining nodes of the network will all have a fixed potential relative to this reference. For a network of $N$ nodes, therefore, there will exist $(N - 1)$ nodes with a fixed potential relative to the assigned reference node. Equations relating these nodal voltages can be written by applying Kirchhoff's current law at each of the $(N - 1)$ nodes. To obtain the complete solution of a network, these nodal voltages are then evaluated in the same manner in which loop currents were found in loop analysis.

The nodal analysis method is applied as follows:

1. Determine the number of nodes within the network.
2. Pick a reference node and label each remaining node with a subscripted value of voltage: $V_1$, $V_2$, and so on.
3. Assume a direction of current for each branch.
4. Apply Kirchhoff's current law at each node except the reference.
5. Solve the resulting equations for the nodal voltages.

**EXAMPLE 8.19.** Apply nodal analysis to the network of Fig. 8.36.

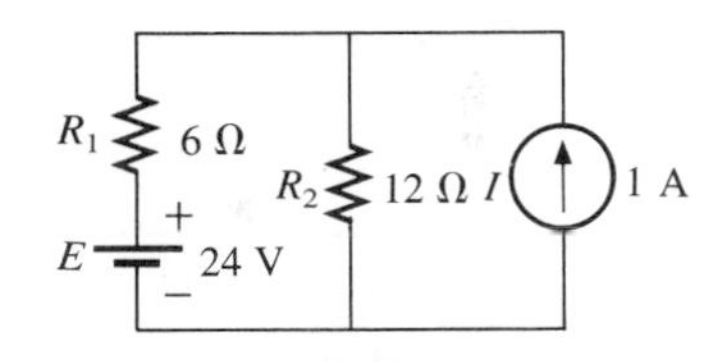

FIG. 8.36

***Solution:***

*Steps 1 and 2:* The network has two nodes, as shown in Fig. 8.37. The lower node is defined as the reference node at ground potential (zero volts) and the other $V_1$, the voltage from node 1 to ground.

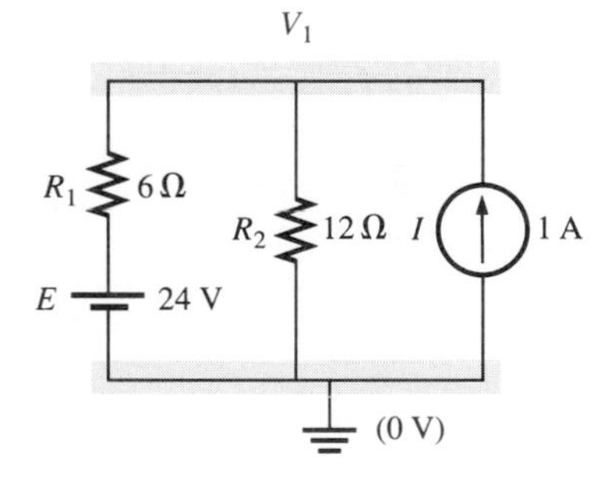

FIG. 8.37

*Step 3:* The assumed directions of $I_1$ and $I_2$ are shown on Fig. 8.38. Note the resulting polarities of the voltages across the resistors in the same branches. Quite obviously, the current in the branch with the current source has the magnitude of the current source with the direction indicated in the graphic symbol.

*Step 4:* Applying Kirchhoff's current law at node 1:

$$I_1 + I = I_2$$

The current $I_2$ is related to $V_1$ by Ohm's law:

$$I_2 = \frac{V_1}{R_2}$$

with the voltage across $R_1$ equal to $E - V_1$, resulting in a current through $R_1$ that can be determined by Ohm's law:

$$I_1 = \frac{V_{R_1}}{R_1} = \frac{(E - V_1)}{R_1}$$

Substituting into the preceding Kirchhoff's current law equation:

$$\frac{(E - V_1)}{R_1} + I = \frac{V_1}{R_2}$$

*Step 5:* The resulting equation is then solved to determine $V_1$.

$$\frac{E}{R_1} - \frac{V_1}{R_1} + I = \frac{V_1}{R_2}$$

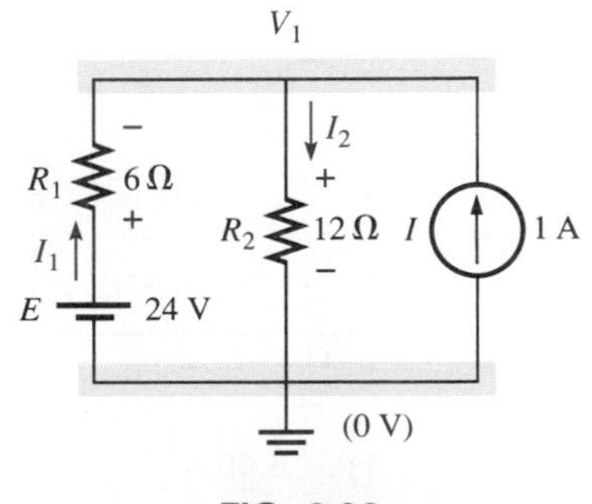

FIG. 8.38

and

$$V_1\left(\frac{1}{R_1} + \frac{1}{R_2}\right) = \frac{E}{R_1} + I$$

Substituting numerical values,

$$V_1\left(\frac{1}{6\ \Omega} + \frac{1}{12\ \Omega}\right) = \frac{24\ \text{V}}{6\ \Omega} + 1\ \text{A} = 4\ \text{A} + 1\ \text{A}$$

$$V_1\left(\frac{1}{4\ \Omega}\right) = 5\ \text{A}$$

$$V_1 = \mathbf{20\ V}$$

The currents $I_1$ and $I_2$ can then be determined using the preceding equations:

$$I_1 = \frac{E - V_1}{R_1} = \frac{24\ \text{V} - 20\ \text{V}}{6\ \Omega} = \frac{4\ \text{V}}{6\ \Omega}$$

$$= \mathbf{0.667\ A}$$

$$I_2 = \frac{V_1}{R_2} = \frac{20\ \text{V}}{12\ \Omega} = \mathbf{1.667\ A}$$

---

**EXAMPLE 8.20.** Apply nodal analysis to the network of Fig. 8.39.

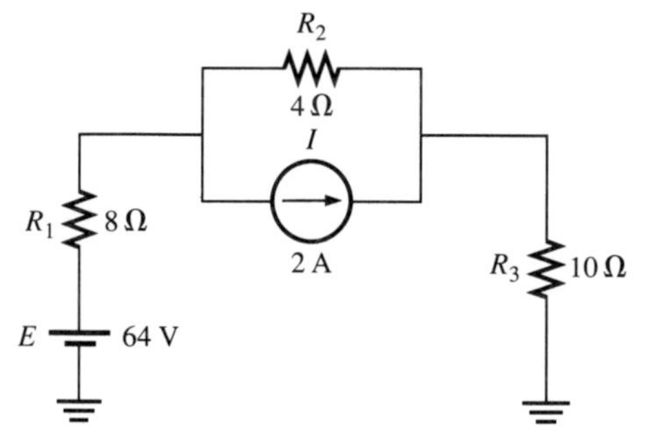

FIG. 8.39

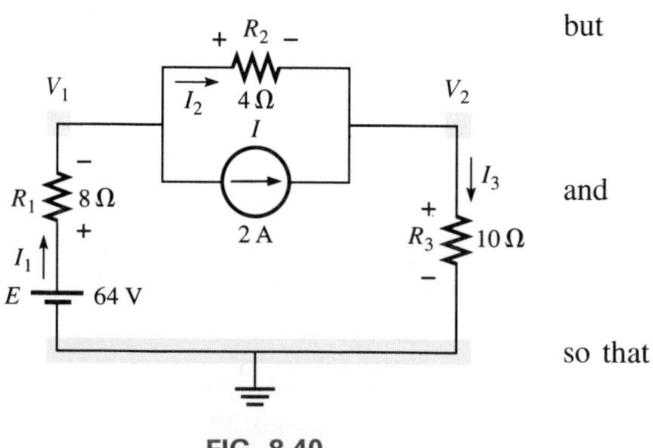


FIG. 8.40

***Solution:***

*Steps 1 and 2:* The network has three nodes, as defined in Fig. 8.40, with the bottom node again defined as the reference node (at ground potential, or zero volts) and the other nodes $V_1$ and $V_2$.

*Step 3:* The assumed current directions are also shown in Fig. 8.40 with the polarities they define across the resistors of the branch.

*Step 4:* Since there are two subscripted nodes, Kirchhoff's current law must be applied twice.

At node 1:

$$I_1 = I_2 + 2\ \text{A}$$

but

$$I_1 = \frac{V_{R_1}}{R_1} = \frac{E - V_1}{R_1} = \frac{64\ \text{V} - V_1}{8\ \Omega}$$

and

$$I_2 = \frac{V_{R_2}}{R_2} = \frac{V_1 - V_2}{4\ \Omega}$$

so that

$$\frac{(64\ \text{V} - V_1)}{8\ \Omega} = \frac{(V_1 - V_2)}{4\ \Omega} + 2\ \text{A}$$

Rearranging the terms results in

$$V_1\left(\frac{1}{8\ \Omega} + \frac{1}{4\ \Omega}\right) - \frac{1}{4\ \Omega}V_2 = +6\ \text{A}$$

At node 2:

$$I_2 + 2\ \text{A} = I_3$$

and

$$\frac{(V_1 - V_2)}{4\ \Omega} + 2\ \text{A} = \frac{V_2}{10\ \Omega}$$

Rearranging the terms results in

$$V_2\left(\frac{1}{4\ \Omega} + \frac{1}{10\ \Omega}\right) - \frac{1}{4\ \Omega}V_1 = +2\ \text{A}$$

*Step 5:* The two equations are then written as follows

$$\begin{aligned} \left(\frac{1}{8\ \Omega} + \frac{1}{4\ \Omega}\right)V_1 \qquad &- \frac{1}{4\ \Omega}V_2 = +6\ \text{A} \\ -\frac{1}{4\ \Omega}V_1 + \left(\frac{1}{4\ \Omega} + \frac{1}{10\ \Omega}\right)V_2 &= +2\ \text{A} \end{aligned}$$

or

$$\begin{aligned} 0.375V_1 - 0.25V_2 &= 6 \\ -0.25V_1 + 0.35V_2 &= 2 \end{aligned}$$

Using determinants,

$$V_1 = \mathbf{37.818\ V}$$

and

$$V_2 = \mathbf{32.727\ V}$$

The current is

$$I_1 = \frac{(64\ \text{V} - 37.818\ \text{V})}{8\ \Omega} = \mathbf{3.273\ A}$$

$$I_2 = \frac{(37.818\ \text{V} - 32.727\ \text{V})}{4\ \Omega} = \mathbf{1.273\ A}$$

$$I_3 = \frac{32.727\ \text{V}}{10\ \Omega} = \mathbf{3.2727\ A}$$

The signs of $V_1$ and $V_2$ reveal that both potentials are positive with respect to ground. In addition, $V_1$ is at a higher potential than $V_2$, so $I_2$ does in fact have the direction assumed. In fact, the assumed directions for $I_1$ and $I_3$ were correct also. In most cases the nodal voltages will clearly reveal the actual direction of current flow. Simply assume any direction for the currents and the resulting voltage levels will reveal the accuracy of the choice.

**EXAMPLE 8.21.** Determine the nodal voltages for the network of Fig. 8.41.

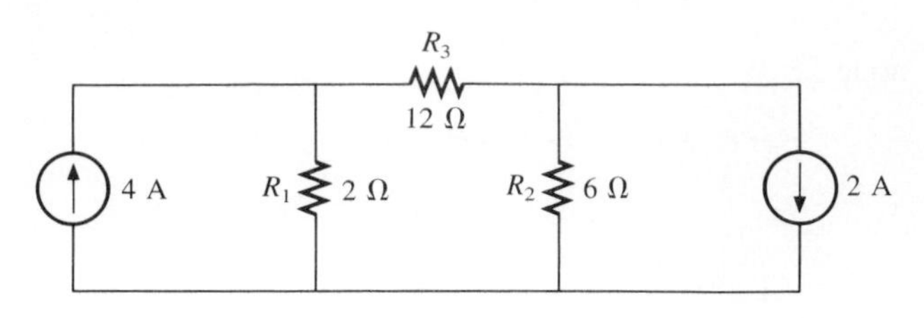

FIG. 8.41

***Solution:***

*Steps 1 to 3:* See Fig. 8.42.

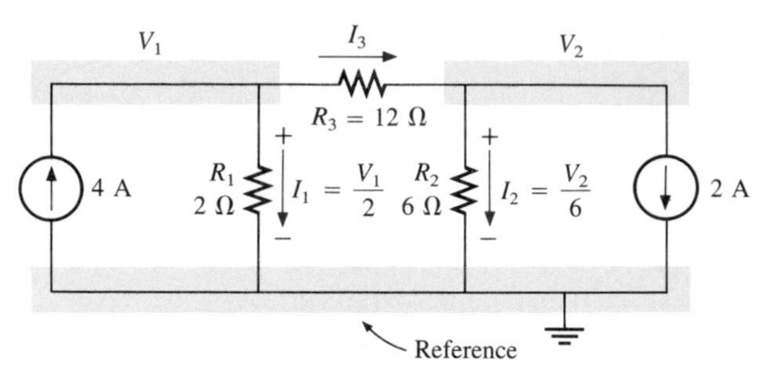

FIG. 8.42

*Step 4:* The resulting nodal equations are

$$\text{node 1:}\quad 4\text{ A} - I_1 - I_3 = 0$$
$$\text{node 2:}\quad I_3 - I_2 - 2\text{ A} = 0$$

Expanding in terms of $V_1$ and $V_2$, we have

$$\text{node 1:}\quad 4\text{ A} - \frac{V_1}{2\ \Omega} - \frac{(V_1 - V_2)}{12\ \Omega} = 0$$
$$\text{node 2:}\quad \frac{(V_1 - V_2)}{12\ \Omega} - \frac{V_2}{6\ \Omega} - 2\text{ A} = 0$$

or

$$V_1\left(\frac{1}{2\ \Omega} + \frac{1}{12\ \Omega}\right) - V_2\left(\frac{1}{12\ \Omega}\right) = +4\text{ A}$$
$$V_2\left(\frac{1}{12\ \Omega} + \frac{1}{6\ \Omega}\right) - V_1\left(\frac{1}{12\ \Omega}\right) = -2\text{ A} \qquad \textbf{(8.3)}$$

producing

$$\left.\begin{aligned} \frac{7}{12}V_1 - \frac{1}{12}V_2 &= +4 \\ -\frac{1}{12}V_1 + \frac{3}{12}V_2 &= -2 \end{aligned}\right\} \quad \begin{aligned} 7V_1 - \phantom{3}V_2 &= 48 \\ -1V_1 + 3V_2 &= -24 \end{aligned}$$

and

$$V_1 = \frac{\begin{vmatrix} 48 & -1 \\ -24 & 3 \end{vmatrix}}{\begin{vmatrix} 7 & -1 \\ -1 & 3 \end{vmatrix}} = \frac{120}{20} = \mathbf{+6\ V}$$

$$V_2 = \frac{\begin{vmatrix} 7 & 48 \\ -1 & -24 \end{vmatrix}}{20} = \frac{-120}{20} = \mathbf{-6\ V}$$

Since $V_1$ is greater than $V_2$, the assumed direction of $I_3$ was correct. Its value is

$$I_3 = \frac{V_1 - V_2}{12\ \Omega} = \frac{6\text{ V} - (-6\text{ V})}{12\ \Omega} = \frac{12\text{ V}}{12\ \Omega} = \mathbf{1\ A}$$

$$I_1 = \frac{V_1}{2\ \Omega} = \frac{6\text{ V}}{2\ \Omega} = \mathbf{3\ A}$$

$$I_2 = \frac{V_2}{6\ \Omega} = \frac{-6\text{ V}}{6\ \Omega} = \mathbf{-1\ A}$$

A negative sign indicates that the current in the original network has the opposite direction.

## 8.10 NODAL ANALYSIS (FORMAT APPROACH)

A close examination of Eq. (8.3) appearing in Example 8.21 reveals that the subscripted voltage at the node in which Kirchhoff's current law is applied is multiplied by the sum of the conductances attached to that node. Note also that the other nodal voltages within the same equation are multiplied by the negative of the conductance between the two nodes. The current sources are represented to the right of the equal sign with a positive sign if they supply current to the node and with a negative sign if they draw current from the node.

These conclusions can be expanded to include networks with any number of nodes. This will allow us to write nodal equations rapidly and in a form that is convenient for the use of determinants. A major requirement, however, is that all voltage sources first be converted to current sources before the procedure is applied. Note the parallelism between the following four steps of application and those required for mesh analysis in Section 8.8:

1. Choose a reference node and assign a subscripted voltage label to the $(N - 1)$ remaining nodes of the network.
2. The number of equations required for a complete solution is equal to the number of subscripted voltages $(N - 1)$. Column 1 of each equation is formed by summing the conductances tied to the node of interest and multiplying the result by that subscripted nodal voltage.
3. We must now consider the mutual terms which, as noted in the preceding example, are always subtracted from the first column. It is possible to have more than one mutual term if the nodal voltage of current interest has an element in common with more than one other nodal voltage. This will be demonstrated in an example to follow. Each mutual term is the product of the mutual conductance and the other nodal voltage tied to that conductance.
4. The column to the right of the equality sign is the algebraic sum of the current sources tied to the node of interest. A current source is assigned a positive sign if it supplies current to a node and a negative sign if it draws current from the node.
5. Solve the resulting simultaneous equations for the desired voltages.

Let us now consider a few examples.

---

**EXAMPLE 8.22.** Write the nodal equations for the network of Fig. 8.43.

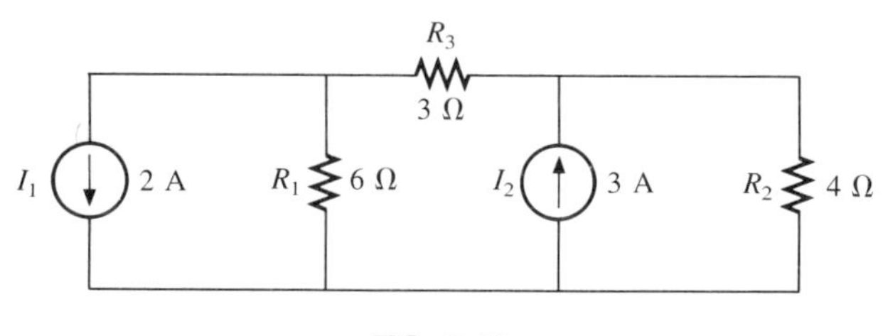

**FIG. 8.43**

***Solution:***
*Step 1:* The figure is redrawn with assigned subscripted voltages in Fig. 8.44.
*Steps 2 to 4:*

$$V_1: \quad \underbrace{\left(\frac{1}{6\,\Omega} + \frac{1}{3\,\Omega}\right)}_{\substack{\text{Sum of} \\ \text{conductances} \\ \text{connected} \\ \text{to node 1}}} V_1 - \underbrace{\left(\frac{1}{3\,\Omega}\right)}_{\substack{\text{Mutual} \\ \text{conductance}}} V_2 = \overset{\substack{\text{Drawing current} \\ \text{from node 1} \\ \downarrow}}{-2\text{ A}}$$

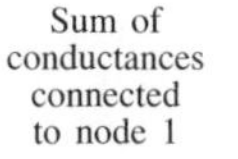

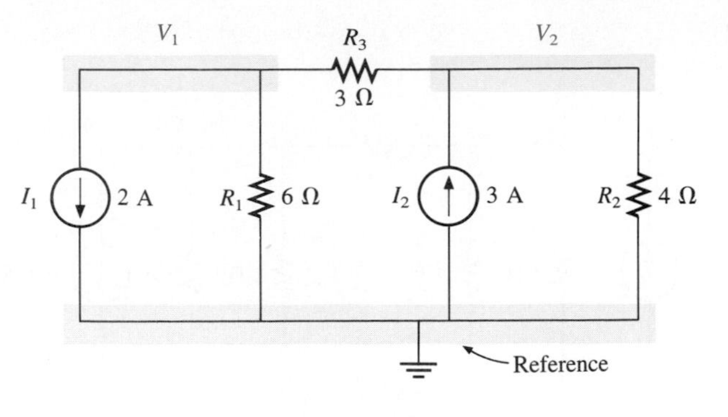


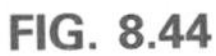
FIG. 8.44

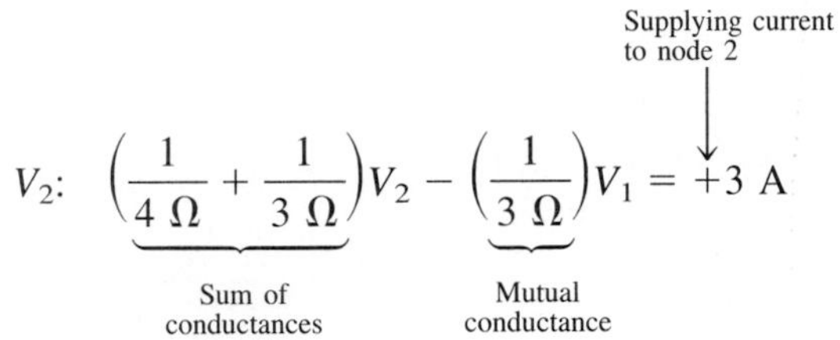


$$V_2\text{:}\quad \left(\frac{1}{4\ \Omega} + \frac{1}{3\ \Omega}\right)V_2 - \left(\frac{1}{3\ \Omega}\right)V_1 = +3\ \text{A}$$

and

$$\frac{1}{2}V_1 - \frac{1}{3}V_2 = -2$$

$$-\frac{1}{3}V_1 + \frac{7}{12}V_2 = 3$$

---

**EXAMPLE 8.23.** Find the voltage across the 3-Ω resistor of Fig. 8.45 by nodal analysis.

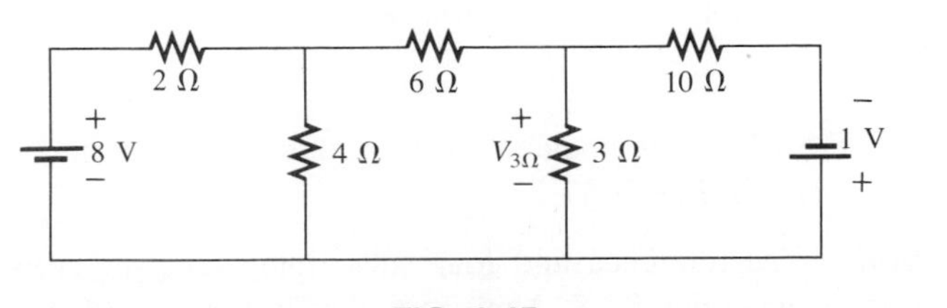


FIG. 8.45

***Solution:*** Converting sources and choosing nodes (Fig. 8.46), we have

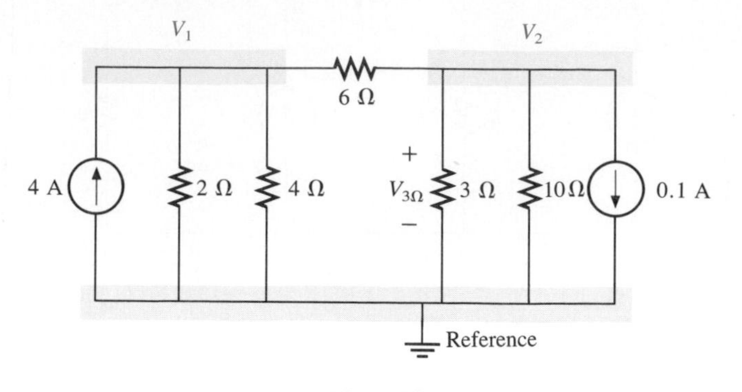


FIG. 8.46

$$\left(\frac{1}{2\ \Omega}+\frac{1}{4\ \Omega}+\frac{1}{6\ \Omega}\right)V_1-\left(\frac{1}{6\ \Omega}\right)V_2 = +4\text{ A}$$

$$\left(\frac{1}{10\ \Omega}+\frac{1}{3\ \Omega}+\frac{1}{6\ \Omega}\right)V_2-\left(\frac{1}{6\ \Omega}\right)V_1 = -0.1\text{ A}$$

$$\frac{11}{12}V_1-\frac{1}{6}V_2 = 4$$

$$-\frac{1}{6}V_1+\frac{3}{5}V_2 = -0.1$$

resulting in

$$11V_1 - 2V_2 = +48$$
$$-5V_1 + 18V_2 = -3$$

and

$$V_2 = V_{3\Omega} = \frac{\begin{vmatrix} 11 & 48 \\ -5 & -3 \end{vmatrix}}{\begin{vmatrix} 11 & -2 \\ -5 & 18 \end{vmatrix}} = \frac{-33+240}{198-10} = \frac{207}{188} = \mathbf{1.101\ V}$$

As demonstrated for mesh analysis, nodal analysis can also be a very useful technique for solving networks with only one source.

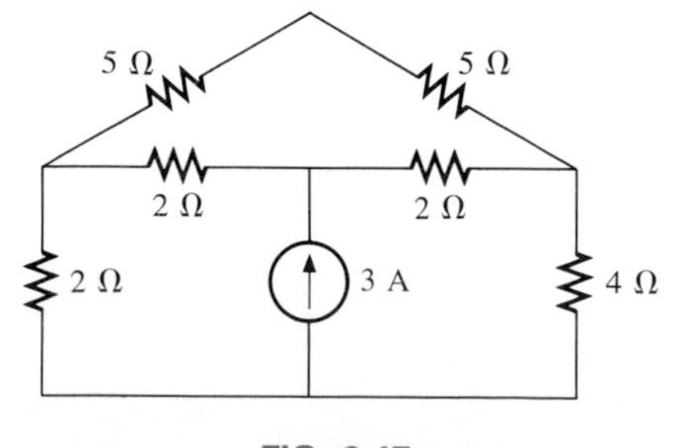


FIG. 8.47

**EXAMPLE 8.24.** Using nodal analysis, determine the potential across the 4-Ω resistor in Fig. 8.47.

***Solution:*** The reference and four subscripted voltage levels were chosen as shown in Fig. 8.48. A moment of reflection should reveal that for any difference in potential between $V_1$ and $V_3$, the current through and the potential drop across each 5-Ω resistor will be the same. Therefore, $V_4$ is simply a midvoltage level between $V_1$ and $V_3$ and is known if $V_1$ and $V_3$ are available. We will therefore not include it in a nodal voltage and will redraw the network as shown in Fig. 8.49. Understand, however, that $V_4$ can be included if desired, although four nodal voltages will result rather than the three to be obtained in the solution of this problem.

$$V_1: \quad \left(\frac{1}{2\ \Omega} + \frac{1}{2\ \Omega} + \frac{1}{10\ \Omega}\right)V_1 - \left(\frac{1}{2\ \Omega}\right)V_2 - \left(\frac{1}{10\ \Omega}\right)V_3 = 0$$

$$V_2: \quad \left(\frac{1}{2\ \Omega} + \frac{1}{2\ \Omega}\right)V_2 - \left(\frac{1}{2\ \Omega}\right)V_1 - \left(\frac{1}{2\ \Omega}\right)V_3 = 3\text{ A}$$

$$V_3: \quad \left(\frac{1}{10\ \Omega} + \frac{1}{2\ \Omega} + \frac{1}{4\ \Omega}\right)V_3 - \left(\frac{1}{2\ \Omega}\right)V_2 - \left(\frac{1}{10\ \Omega}\right)V_1 = 0$$

which are rewritten as

$$\begin{aligned} 1.1V_1 - 0.5V_2 - 0.1V_3 &= 0 \\ V_2 - 0.5V_1 - 0.5V_3 &= 3 \\ 0.85V_3 - 0.5V_2 - 0.1V_1 &= 0 \end{aligned}$$

For determinants,

$$\begin{aligned} 1.1V_1 - 0.5V_2 - 0.1V_3 &= 0 \\ -0.5V_1 + 1V_2 - 0.5V_3 &= 3 \\ -0.1V_1 - 0.5V_2 + 0.85V_3 &= 0 \end{aligned}$$

Before continuing, note the symmetry about the major diagonal in the equation above. Recall a similar result for mesh analysis. Examples 8.22 and 8.23 also exhibit this property in the resulting equations. Keep this thought in mind as a check on future applications of nodal analysis.

$$V_3 = V_{4\Omega} = \frac{\begin{vmatrix} 1.1 & -0.5 & 0 \\ -0.5 & +1 & 3 \\ -0.1 & -0.5 & 0 \end{vmatrix}}{\begin{vmatrix} 1.1 & -0.5 & -0.1 \\ -0.5 & +1 & -0.5 \\ -0.1 & -0.5 & +0.85 \end{vmatrix}} = \mathbf{4.645\ V}$$

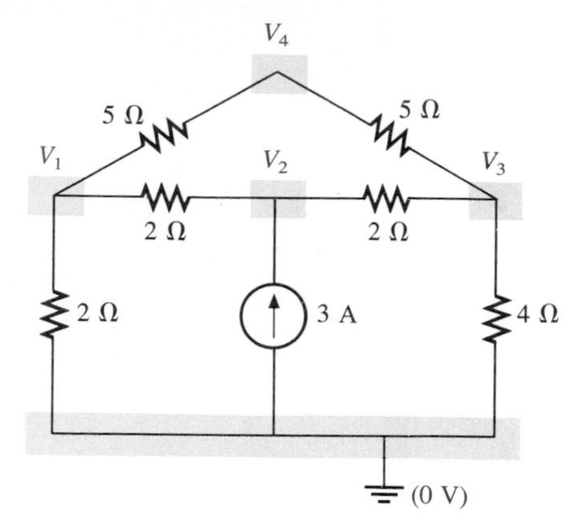

FIG. 8.48

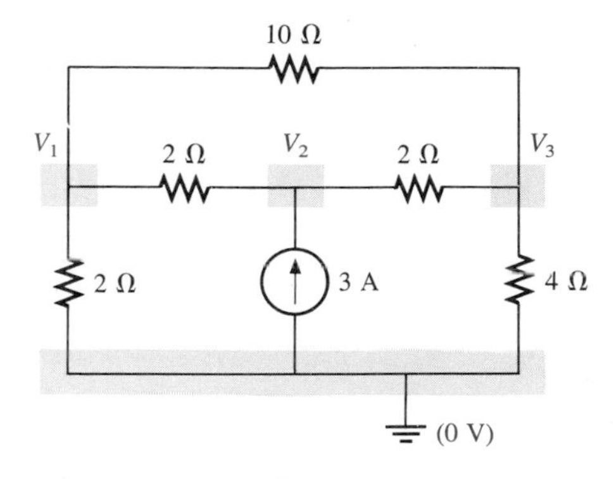

FIG. 8.49

Another example with only one source involves a ladder network.

**EXAMPLE 8.25.** Write the nodal equations and find the voltage across the 2-Ω resistor for the network of Fig. 8.50.

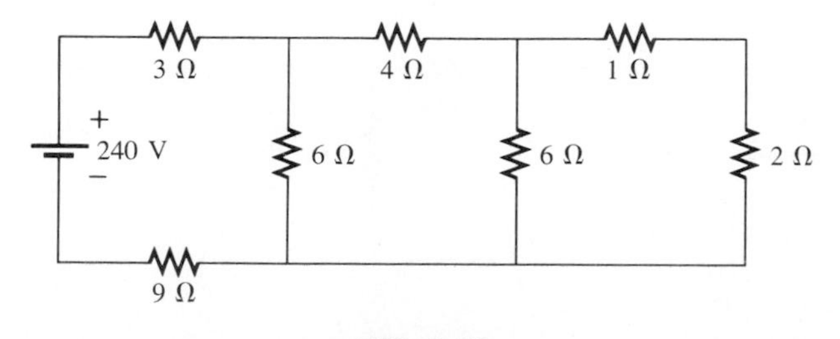

FIG. 8.50

***Solution:*** The nodal voltages are chosen as shown in Fig. 8.51.

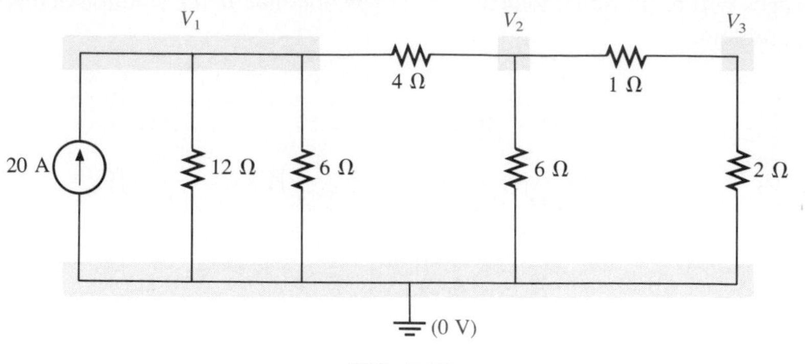

FIG. 8.51

$$V_1: \quad \left(\frac{1}{12\ \Omega} + \frac{1}{6\ \Omega} + \frac{1}{4\ \Omega}\right)V_1 - \left(\frac{1}{4\ \Omega}\right)V_2 + \ 0 = 20$$

$$V_2: \quad \left(\frac{1}{4\ \Omega} + \frac{1}{6\ \Omega} + \frac{1}{1\ \Omega}\right)V_2 - \left(\frac{1}{4\ \Omega}\right)V_1 - \left(\frac{1}{1\ \Omega}\right)V_3 = 0$$

$$V_3: \quad \left(\frac{1}{1} + \frac{1}{2\ \Omega}\right)V_3 - \left(\frac{1}{1\ \Omega}\right)V_2 + \ 0 = 0$$

and

$$\begin{aligned} 0.5V_1 - 0.25V_2 + \quad 0 &= 20 \\ -0.25V_1 + \frac{17}{12}V_2 - \quad 1V_3 &= 0 \\ 0 - \quad 1V_2 + 1.5V_3 &= 0 \end{aligned}$$

Note the symmetry present about the major axis. Application of determinants reveals that

$$V_3 = V_{2\Omega} = \mathbf{10.667\ V}$$

Another example of general interest is the bridge network. This will be set aside for the section on bridge networks to follow in this chapter.

There are various situations in which it may appear impossible to apply nodal analysis. However, by eliminating components or introducing new components, we can often resolve this problem. For example, in the network of Fig. 8.52(a), the voltage source cannot be converted to a current source, since it does not have a resistance in series with it.

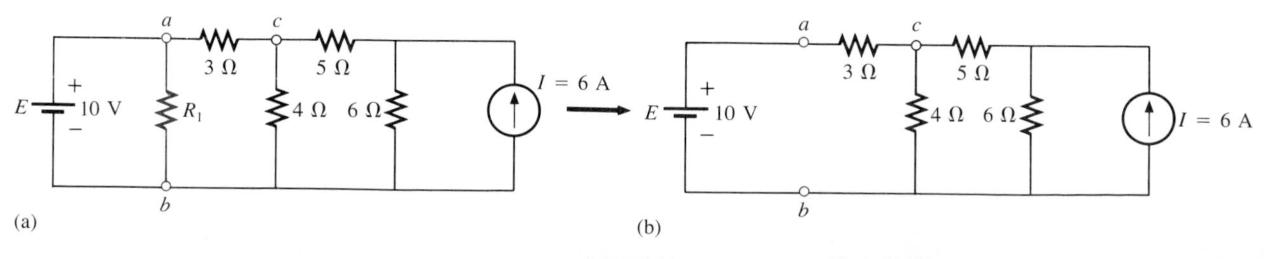


FIG. 8.52

Note that the voltage $V_{ab}$ across the resistor $R_1$ is always the source voltage $E = 10$ V, no matter what value the resistance $R_1$ may be. The resistor $R_1$, therefore, does not affect the voltage $V_{ab}$, $V_{ac}$, or any other voltage within the network. Since we are solving only for voltage levels when we apply nodal analysis, the resistor $R_1$ can be removed without affecting our solution. Once the nodal analysis is complete (all nodal voltages known), the resistor $R_1$ must be considered if the current through it or through the battery is desired. After the resistor $R_1$ is removed, the circuit is as shown in the diagram of Fig. 8.52(b), and we can readily arrive at a solution by nodal analysis after we convert the source.

## 8.11 BRIDGE NETWORKS

This section introduces the bridge network, a configuration that has a multitude of applications. In the chapters to follow, it will be employed in both dc and ac meters. In the electronics courses it will be encountered early in the discussion of rectifying circuits employed in converting a varying signal to one of a steady nature (such as dc). There are a

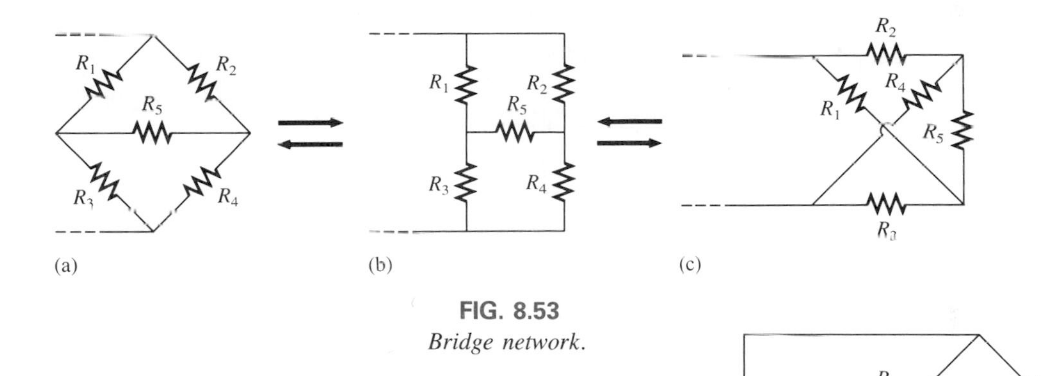


**FIG. 8.53**
*Bridge network.*

number of other areas of application that require some knowledge of ac networks which will be discussed later.

The bridge network may appear in one of three forms as indicated in Fig. 8.53. The network of Fig. 8.53(c) is also called a symmetrical lattice network if $R_2 = R_3$ and $R_1 = R_4$. Figure 8.53(c) is an excellent example of how a planar network can be made to appear nonplanar. For the purposes of investigation, let us examine the network of Fig. 8.54 using mesh and nodal analysis.

**FIG. 8.54**

Mesh analysis (Fig. 8.55) yields

$$
\begin{aligned}
(3\ \Omega + 4\ \Omega + 2\ \Omega)I_1 - (4\ \Omega)I_2 - (2\ \Omega)I_3 &= 20\ \text{V} \\
(4\ \Omega + 5\ \Omega + 2\ \Omega)I_2 - (4\ \Omega)I_1 - (5\ \Omega)I_3 &= 0 \\
(2\ \Omega + 5\ \Omega + 1\ \Omega)I_3 - (2\ \Omega)I_1 - (5\ \Omega)I_2 &= 0
\end{aligned}
$$

and

$$
\begin{aligned}
9I_1 - 4I_2 - 2I_3 &= 20 \\
-4I_1 + 11I_2 - 5I_3 &= 0 \\
-2I_1 - 5I_2 + 8I_3 &= 0
\end{aligned}
$$

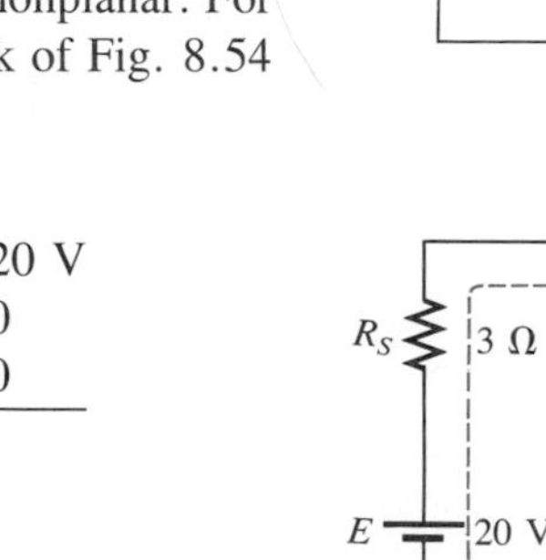

**FIG. 8.55**

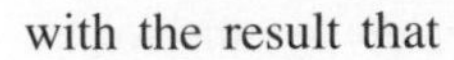

with the result that

$$I_1 = \mathbf{4\ A}$$
$$I_2 = \mathbf{2\tfrac{2}{3}\ A}$$
$$I_3 = \mathbf{2\tfrac{2}{3}\ A}$$

The net current through the 5-Ω resistor is

$$I_{5\Omega} = I_2 - I_3 = 2\tfrac{2}{3} - 2\tfrac{2}{3} = \mathbf{0\ A}$$

Nodal analysis (Fig. 8.56) yields

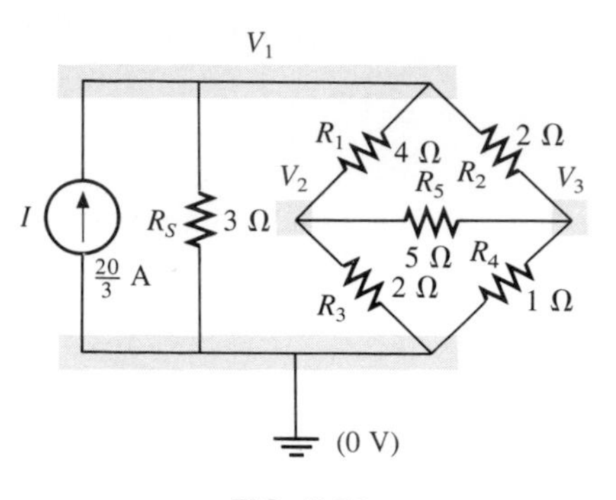


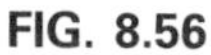

FIG. 8.56

$$\left(\frac{1}{3\ \Omega} + \frac{1}{4\ \Omega} + \frac{1}{2\ \Omega}\right)V_1 - \left(\frac{1}{4\ \Omega}\right)V_2 - \left(\frac{1}{2\ \Omega}\right)V_3 = \frac{20}{3}\ \text{A}$$
$$\left(\frac{1}{4\ \Omega} + \frac{1}{2\ \Omega} + \frac{1}{5\ \Omega}\right)V_2 - \left(\frac{1}{4\ \Omega}\right)V_1 - \left(\frac{1}{5\ \Omega}\right)V_3 = 0$$
$$\left(\frac{1}{5\ \Omega} + \frac{1}{2\ \Omega} + \frac{1}{1\ \Omega}\right)V_3 - \left(\frac{1}{2\ \Omega}\right)V_1 - \left(\frac{1}{5\ \Omega}\right)V_2 = 0$$

and

$$\left(\frac{1}{3\ \Omega} + \frac{1}{4\ \Omega} + \frac{1}{2\ \Omega}\right)V_1 - \left(\frac{1}{4\ \Omega}\right)V_2 - \left(\frac{1}{2\ \Omega}\right)V_3 = \frac{20}{3}\ \text{A}$$
$$-\left(\frac{1}{4\ \Omega}\right)V_1 + \left(\frac{1}{4\ \Omega} + \frac{1}{2\ \Omega} + \frac{1}{5\ \Omega}\right)V_2 - \left(\frac{1}{5\ \Omega}\right)V_3 = 0$$
$$-\left(\frac{1}{2\ \Omega}\right)V_1 - \left(\frac{1}{5\ \Omega}\right)V_2 + \left(\frac{1}{5\ \Omega} + \frac{1}{2\ \Omega} + \frac{1}{1\ \Omega}\right)V_3 = 0$$

Note the symmetry of the solution. The results are

$$V_1 = \mathbf{8\ V}$$
$$V_2 = \mathbf{2\tfrac{2}{3}\ V}$$
$$V_3 = \mathbf{2\tfrac{2}{3}\ V}$$

and the voltage across the 5-Ω resistor is

$$V_{5\Omega} = V_2 - V_3 = 2\tfrac{2}{3} - 2\tfrac{2}{3} = \mathbf{0\ V}$$

Since $V_{5\Omega} = 0$ V, we can insert a short in place of the bridge arm without affecting the network behavior. (Certainly $V = IR = I \cdot (0) = 0$ V.) In Fig. 8.57, a short circuit has replaced the resistor $R_5$, and the

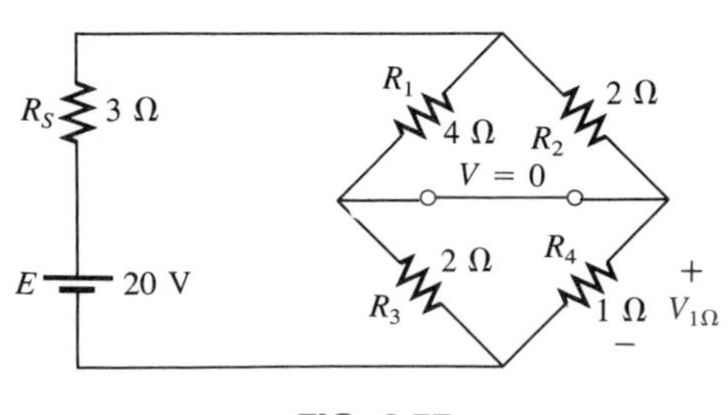

FIG. 8.57

voltage across $R_4$ is to be determined. The network is redrawn in Fig. 8.58, and

$$V_{1\Omega} = \frac{(2\ \Omega \parallel 1\ \Omega)20\ \text{V}}{(2\ \Omega \parallel 1\ \Omega) + (4\ \Omega \parallel 2\ \Omega) + 3\ \Omega} \qquad \text{(voltage divider rule)}$$

$$= \frac{\frac{2\ \Omega}{3}(20\ \text{V})}{\frac{2\ \Omega}{3} + \frac{8\ \Omega}{6} + 3\ \Omega} = \frac{\frac{2\ \Omega}{3}(20\ \text{V})}{\frac{2\ \Omega}{3} + \frac{4\ \Omega}{3} + \frac{9\ \Omega}{3}}$$

$$= \frac{2\ \Omega(20\ \text{V})}{2\ \Omega + 4\ \Omega + 9\ \Omega} = \frac{40\ \text{V}}{15} = \mathbf{2\tfrac{2}{3}\ V}$$

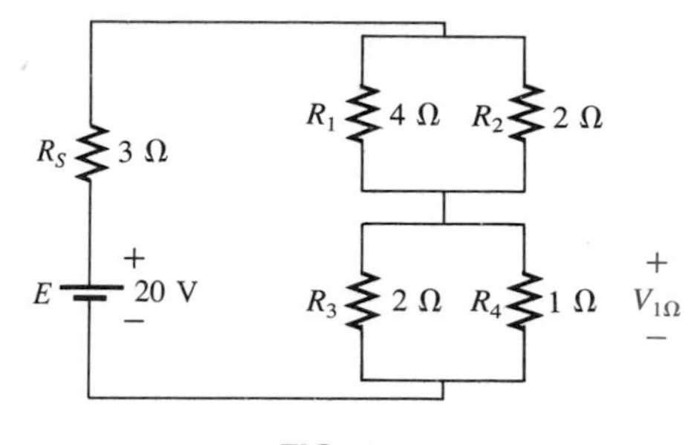


FIG. 8.58

as obtained earlier.

We found through mesh analysis that $I_{5\Omega} = 0$ A, which has as its equivalent an open circuit as shown in Fig. 8.59. (Certainly $I = V/R = 0/(\infty\ \Omega) = 0$ A.) The voltage across the resistor $R_4$ will again be determined and compared with the result above.

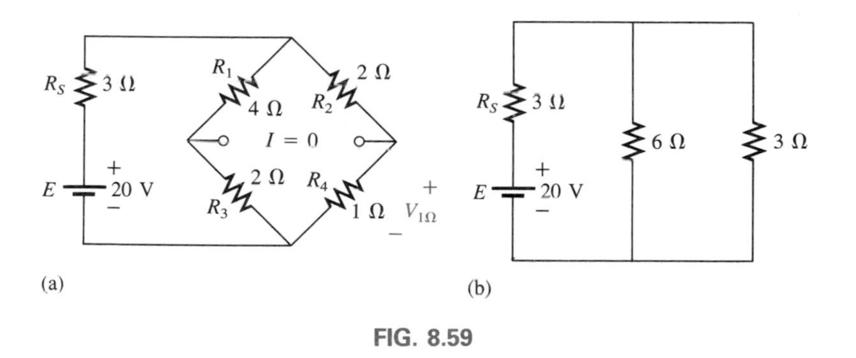


FIG. 8.59

The network is redrawn after combining series elements as shown in Fig. 8.59(b), and

$$V_{3\Omega} = \frac{(6\ \Omega \parallel 3\ \Omega)(20\ \text{V})}{6\ \Omega \parallel 3\ \Omega + 3\ \Omega} = \frac{2\ \Omega(20\ \text{V})}{2\ \Omega + 3\ \Omega} = 8\ \text{V}$$

and

$$V_{1\Omega} = \frac{1\ \Omega(8\ \text{V})}{1\ \Omega + 2\ \Omega} = \frac{8\ \text{V}}{3} = \mathbf{2\tfrac{2}{3}\ V}$$

as above.

The condition $V_{5\Omega} = 0$ V or $I_{5\Omega} = 0$ A exists only for a particular relationship between the resistors of the network. Let us now derive this relationship using the network of Fig. 8.60, in which it is indicated that $I = 0$ A and $V = 0$ V. Note that resistor $R_S$ of the network of Fig. 8.59 will not appear in the following analysis.

The bridge network is said to be *balanced* when the condition of $I = 0$ A or $V = 0$ V exists.

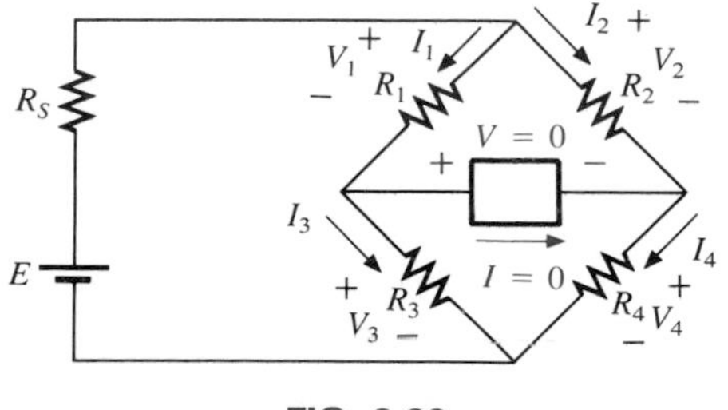


FIG. 8.60

If $V = 0$ V (short circuit between $a$ and $b$), then

$$V_1 = V_2$$

and

$$I_1R_1 = I_2R_2$$

or

$$I_1 = \frac{I_2R_2}{R_1}$$

In addition, when $V = 0$ V,

$$V_3 = V_4$$

and

$$I_3R_3 = I_4R_4$$

If we set $I = 0$ A, then $I_3 = I_1$ and $I_4 = I_2$ with the result that the above equation becomes

$$I_1R_3 = I_2R_4$$

Substituting for $I_1$ from above yields

$$\left(\frac{I_2R_2}{R_1}\right)R_3 = I_2R_4$$

or, rearranging, we have

$$\boxed{\frac{R_1}{R_3} = \frac{R_2}{R_4}} \quad \textbf{(8.4)}$$

This conclusion states that if the ratio of $R_1$ to $R_3$ is equal to that of $R_2$ to $R_4$, the bridge will be balanced, and $I = 0$ A or $V = 0$ V. A method of memorizing this form is indicated in Fig. 8.61.

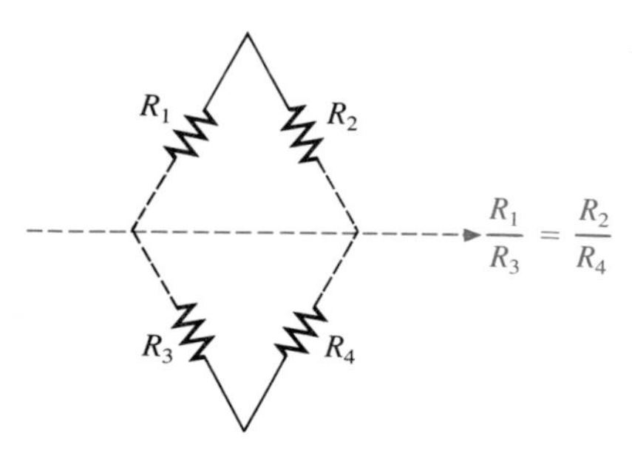


FIG. 8.61

For the example above, $R_1 = 4\ \Omega$, $R_2 = 2\ \Omega$, $R_3 = 2\ \Omega$, $R_4 = 1\ \Omega$, and

$$\frac{R_1}{R_3} = \frac{R_2}{R_4} \Rightarrow \frac{4\ \Omega}{2\ \Omega} = \frac{2\ \Omega}{1\ \Omega} = 2$$

The emphasis in this section has been on the balanced situation. Understand that if the ratio is not satisfied, there will be a potential drop across the balance arm and a current through it. The methods just described (mesh and nodal analysis) will yield any and all potentials or currents desired, just as they were applied for the balanced situation.

## 8.12 Y-Δ (T-$\pi$) AND Δ-Y ($\pi$-T) CONVERSIONS

Circuit configurations are often encountered in which the resistors do not appear to be in series or parallel. Under these conditions, it may be necessary to convert the circuit from one form to another in order to solve for any unknown quantities if mesh or nodal analysis is not applied. Two circuit configurations that often account for these difficulties are the wye (Y) and delta (Δ), depicted in Fig. 8.62(a). They are also referred to as the tee (T) and pi (π), respectively, as indicated in Fig. 8.62(b). Note that the pi is actually an inverted delta.

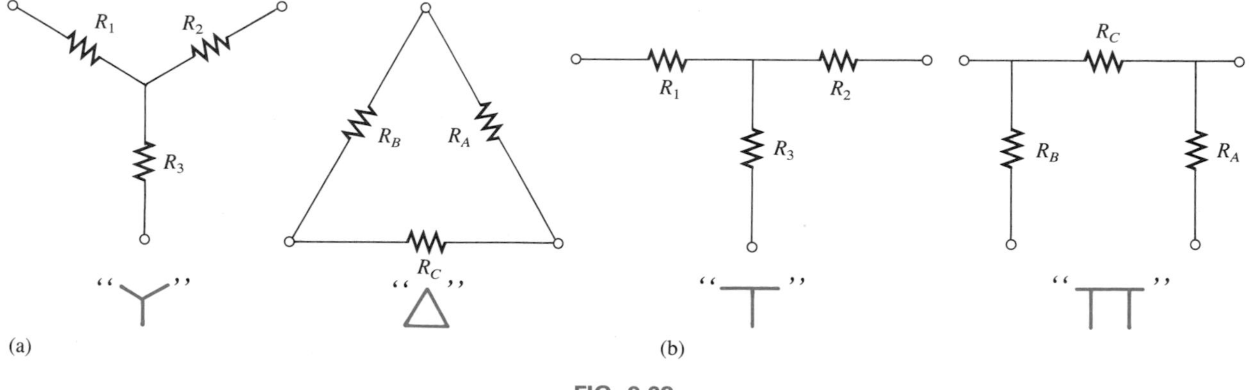

FIG. 8.62

The purpose of this section is to develop the equations for converting from Δ to Y or vice versa. This type of conversion will normally lead to a network that can be solved using techniques such as described in Chapter 7. In other words, in Fig. 8.63, with terminals $a$, $b$, and $c$ held fast, if the wye (Y) configuration were desired *instead of* the inverted delta (Δ) configuration, all that would be necessary is a direct application of the equations to be derived. The phrase *instead of* is emphasized to insure that it is understood that only one of these configurations is to appear at one time between the indicated terminals.

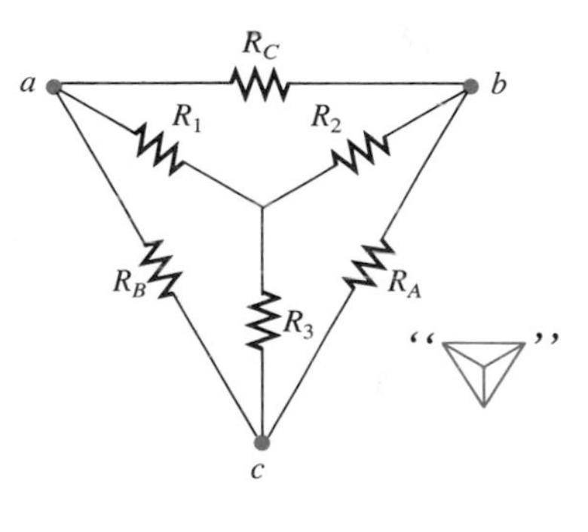

FIG. 8.63

It is our purpose (referring to Fig. 8.63) to find some expression for $R_1$, $R_2$, and $R_3$ in terms of $R_A$, $R_B$, and $R_C$ and vice versa that will ensure that the resistance between any two terminals of the Y configuration will be the same with the Δ configuration inserted in place of the Y configuration (and vice versa). If the two circuits are to be equivalent, the total resistance between any two terminals must be the same. Consider terminals $a$-$c$ in the Δ-Y configurations of Fig. 8.64.

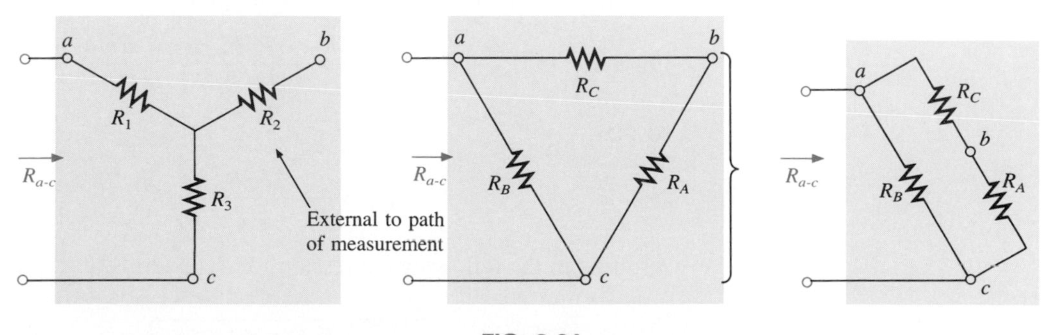


FIG. 8.64

Let us first assume that we want to convert the Δ ($R_A$, $R_B$, $R_C$) to the Y ($R_1$, $R_2$, $R_3$). This requires that we have a relationship for $R_1$, $R_2$, and $R_3$ in terms of $R_A$, $R_B$, and $R_C$. If the resistance is to be the same between terminals *a-c* for both the Δ and the Y, the following must be true:

$$R_{a\text{-}c}\ (\mathrm{Y}) = R_{a\text{-}c}\ (\Delta)$$

so that

$$R_{a\text{-}c} = R_1 + R_3 = \frac{R_B(R_A + R_C)}{R_B + (R_A + R_C)} \tag{8.5a}$$

Using the same approach for *a-b* and *b-c*, we obtain the following relationships:

$$R_{a\text{-}b} = R_1 + R_2 = \frac{R_C(R_A + R_B)}{R_C + (R_A + R_B)} \tag{8.5b}$$

and

$$R_{b\text{-}c} = R_2 + R_3 = \frac{R_A(R_B + R_C)}{R_A + (R_B + R_C)} \tag{8.5c}$$

Subtracting Eq. (8.5a) from Eq. (8.5b), we have

$$(R_1 + R_2) - (R_1 + R_3) = \left(\frac{R_C R_B + R_C R_A}{R_A + R_B + R_C}\right) - \left(\frac{R_B R_A + R_B R_C}{R_A + R_B + R_C}\right)$$

so that

$$R_2 - R_3 = \frac{R_A R_C - R_B R_A}{R_A + R_B + R_C} \tag{8.5d}$$

Subtracting Eq. (8.5d) from Eq. (8.5c) yields

$$(R_2 + R_3) - (R_2 - R_3) = \left(\frac{R_A R_B + R_A R_C}{R_A + R_B + R_C}\right) - \left(\frac{R_A R_C - R_B R_A}{R_A + R_B + R_C}\right)$$

so that

$$2R_3 = \frac{2R_B R_A}{R_A + R_B + R_C}$$

resulting in the following expression for $R_3$ in terms of $R_A$, $R_B$, and $R_C$:

$$R_3 = \frac{R_A R_B}{R_A + R_B + R_C} \tag{8.6a}$$

Following the same procedure for $R_1$ and $R_2$, we have

$$R_1 = \frac{R_B R_C}{R_A + R_B + R_C} \tag{8.6b}$$

and

$$R_2 = \frac{R_A R_C}{R_A + R_B + R_C} \tag{8.6c}$$

**Note that each resistor of the Y is equal to the product of the resistors in the two closest branches of the Δ divided by the sum of the resistors in the Δ.**

To obtain the relationships necessary to convert from a Y to a Δ, first divide Eq. (8.6a) by Eq. (8.6b):

$$\frac{R_3}{R_1} = \frac{(R_A R_B)/(R_A + R_B + R_C)}{(R_B R_C)/(R_A + R_B + R_C)} = \frac{R_A}{R_C}$$

or

$$R_A = \frac{R_C R_3}{R_1}$$

Then divide Eq. (8.6a) by Eq. (8.6c):

$$\frac{R_3}{R_2} = \frac{(R_A R_B)/(R_A + R_B + R_C)}{(R_A R_C)/(R_A + R_B + R_C)} = \frac{R_B}{R_C}$$

or

$$R_B = \frac{R_3 R_C}{R_2}$$

Substituting for $R_A$ and $R_B$ in Eq. (8.6c) yields

$$R_2 = \frac{(R_C R_3/R_1)R_C}{(R_3 R_C/R_2) + (R_C R_3/R_1) + R_C}$$

$$= \frac{(R_3/R_1)R_C}{(R_3/R_2) + (R_3/R_1) + 1}$$

Placing these over a common denominator, we obtain

$$R_2 = \frac{(R_3 R_C/R_1)}{(R_1 R_2 + R_1 R_3 + R_2 R_3)/(R_1 R_2)}$$

$$= \frac{R_2 R_3 R_C}{R_1 R_2 + R_1 R_3 + R_2 R_3}$$

and

$$R_C = \frac{R_1 R_2 + R_1 R_3 + R_2 R_3}{R_3} \tag{8.7a}$$

We follow the same procedure for $R_B$ and $R_A$:

$$R_A = \frac{R_1R_2 + R_1R_3 + R_2R_3}{R_1} \tag{8.7b}$$

and

$$R_B = \frac{R_1R_2 + R_1R_3 + R_2R_3}{R_2} \tag{8.7c}$$

**Note that the value of each resistor of the Δ is equal to the sum of the possible product combinations of the resistances of the Y divided by the resistance of the Y farthest from the resistor to be determined.**

Let us consider what would occur if all the values of a Δ or Y were the same. If $R_A = R_B = R_C$, Eq. (8.6a) would become (using $R_A$ only)

$$R_3 = \frac{R_AR_B}{R_A + R_B + R_C} = \frac{R_AR_A}{R_A + R_A + R_A} = \frac{R_A^2}{3R_A} = \frac{R_A}{3}$$

and, following the same procedure,

$$R_1 = \frac{R_A}{3} \qquad R_2 = \frac{R_A}{3}$$

In general, therefore,

$$R_Y = \frac{R_\Delta}{3} \tag{8.8a}$$

or

$$R_\Delta = 3R_Y \tag{8.8b}$$

which indicates that *for a Y of three equal resistors, the value of each resistor of the Δ is equal to three times the value of any resistor of the Y.* If only two elements of a Y or a Δ are the same, the corresponding Δ or Y of each will also have two equal elements. The converting of equations will be left as an exercise for the reader.

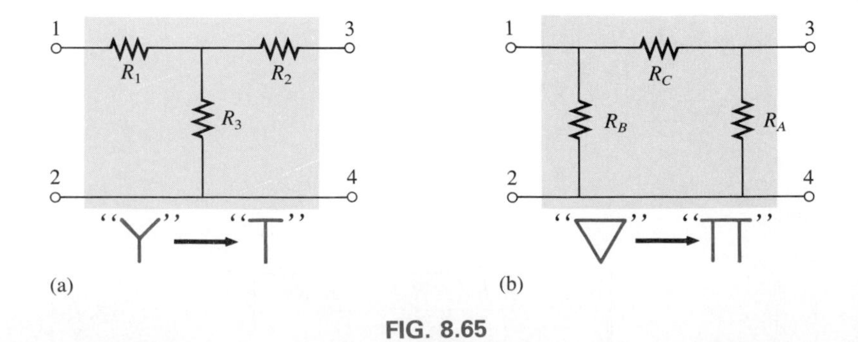


FIG. 8.65

The Y and the Δ will often appear as shown in Fig. 8.65. They are then referred to as a *tee* (T) and *pi* (π) network. The equations used to convert from one form to the other are exactly the same as those developed for the Y and Δ transformation.

**EXAMPLE 8.26.** Convert the Δ of Fig. 8.66 to a Y.

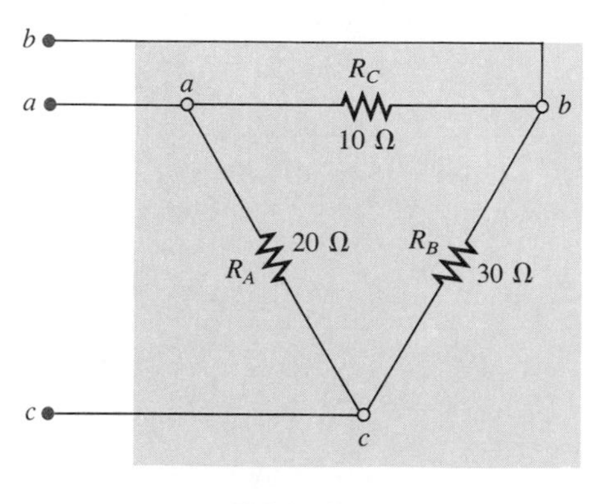


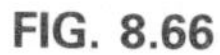
FIG. 8.66

***Solution:***

$$R_1 = \frac{R_B R_C}{R_A + R_B + R_C} = \frac{(20\ \Omega)(10\ \Omega)}{20\ \Omega + 30\ \Omega + 10\ \Omega} = \frac{200\ \Omega}{60} = \mathbf{3\tfrac{1}{3}\ \Omega}$$

$$R_2 = \frac{R_A R_C}{R_A + R_B + R_C} = \frac{(30\ \Omega)(10\ \Omega)}{60\ \Omega} = \frac{300\ \Omega}{60} = \mathbf{5\ \Omega}$$

$$R_3 = \frac{R_A R_B}{R_A + R_B + R_C} = \frac{(20\ \Omega)(30\ \Omega)}{60\ \Omega} = \frac{600\ \Omega}{60} = \mathbf{10\ \Omega}$$

The equivalent network is shown in Fig. 8.67.

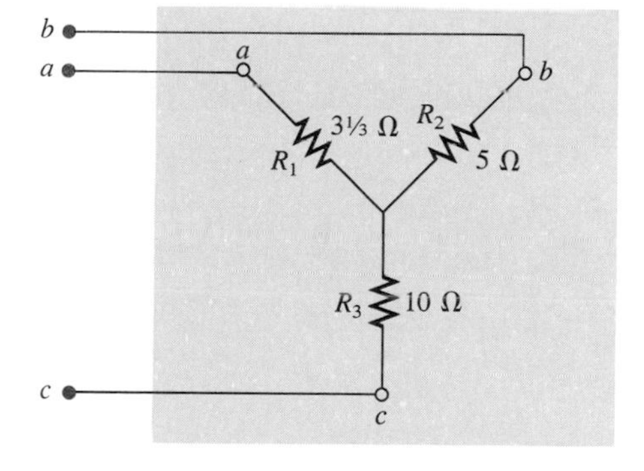


FIG. 8.67

**EXAMPLE 8.27.** Convert the Y of Fig. 8.68 to a Δ.

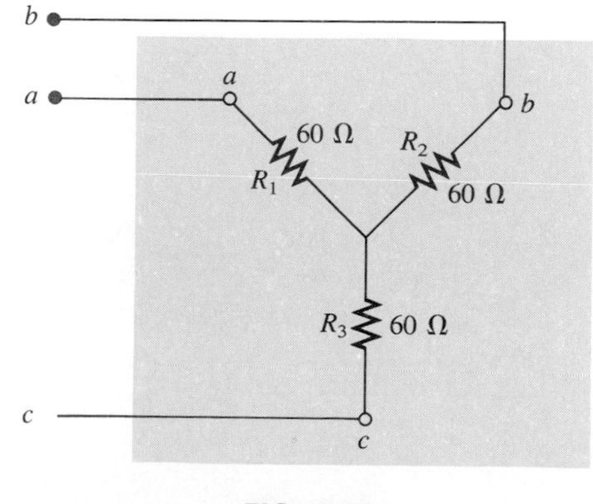


FIG. 8.68

***Solution:***

$$R_A = \frac{R_1R_2 + R_1R_3 + R_2R_3}{R_1}$$

$$= \frac{(60\ \Omega)(60\ \Omega) + (60\ \Omega)(60\ \Omega) + (60\ \Omega)(60\ \Omega)}{60\ \Omega}$$

$$= \frac{3600\ \Omega + 3600\ \Omega + 3600\ \Omega}{60} = \frac{10{,}800\ \Omega}{60}$$

$$R_A = \mathbf{180\ \Omega}$$

However, the three resistors for the Y are equal, permitting the use of Eq. (8.8) and yielding

$$R_\Delta = 3R_Y = 3(60\ \Omega) = 180\ \Omega$$

and

$$R_B = R_C = \mathbf{180\ \Omega}$$

The equivalent network is shown in Fig. 8.69.

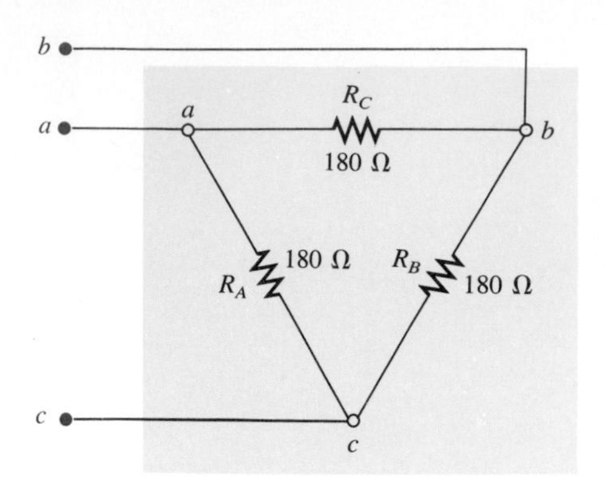

FIG. 8.69

**EXAMPLE 8.28.** Find the total resistance of the network of Fig. 8.70, where $R_A = 3\ \Omega$, $R_B = 3\ \Omega$, and $R_C = 6\ \Omega$.

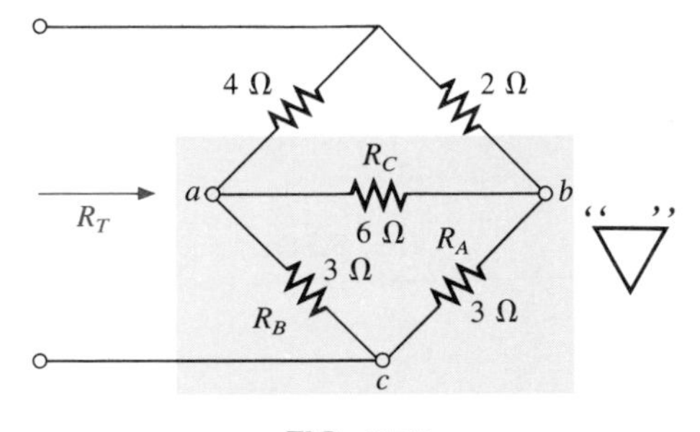


FIG. 8.70

***Solution:***

Two resistors of the Δ were equal; therefore, two resistors of the Y will be equal.

$$R_1 = \frac{R_B R_C}{R_A + R_B + R_C} = \frac{(3\ \Omega)(6\ \Omega)}{3\ \Omega + 3\ \Omega + 6\ \Omega} = \frac{18\ \Omega}{12} = \mathbf{1.5\ \Omega}$$

$$R_2 = \frac{R_A R_C}{R_A + R_B + R_C} = \frac{(3\ \Omega)(6\ \Omega)}{12\ \Omega} = \frac{18\ \Omega}{12} = \mathbf{1.5\ \Omega}$$

$$R_3 = \frac{R_A R_B}{R_A + R_B + R_C} = \frac{(3\ \Omega)(3\ \Omega)}{12\ \Omega} = \frac{9\ \Omega}{12} = \mathbf{0.75\ \Omega}$$

Replacing the Δ by the Y, as shown in Fig. 8.71, yields

$$R_T = 0.75\ \Omega + \frac{(4\ \Omega + 1.5\ \Omega)(2\ \Omega + 1.5\ \Omega)}{(4\ \Omega + 1.5\ \Omega) + (2\ \Omega + 1.5\ \Omega)}$$

$$= 0.75\ \Omega + \frac{(5.5\ \Omega)(3.5\ \Omega)}{5.5\ \Omega + 3.5\ \Omega}$$

$$= 0.75\ \Omega + 2.139\ \Omega$$

$$R_T = \mathbf{2.889\ \Omega}$$

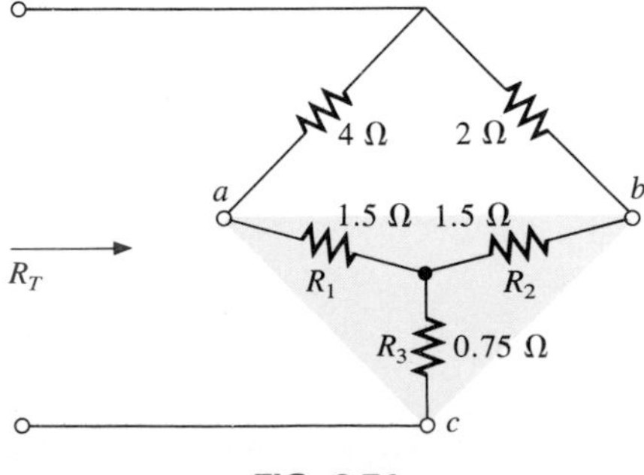


FIG. 8.71

**EXAMPLE 8.29.** Find the total resistance of the network of Fig. 8.72.

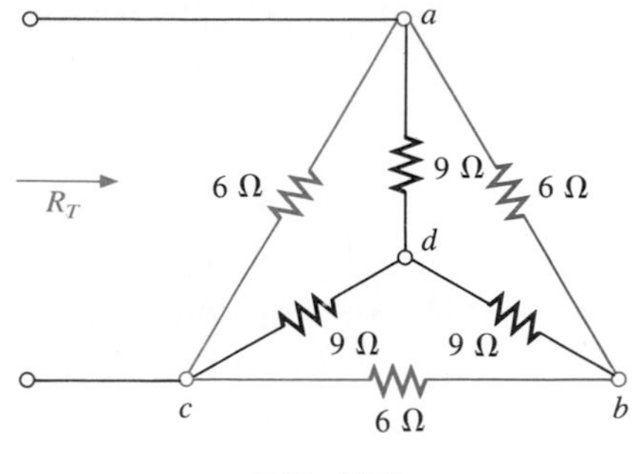

FIG. 8.72

***Solution:*** Since all the resistors of the Δ or Y are the same, Eqs. (8.8a) and (8.8b) can be used to convert either form to the other.

a. *Converting the Δ to a Y*. Note: When this is done, the resulting $d'$ of the new Y will be the same as the point $d$ shown in the original figure, only because both systems are "balanced." That is, the resistance in each branch of each system has the same value:

$$R_Y = \frac{R_\Delta}{3} = \frac{6\ \Omega}{3} = 2\ \Omega \qquad \text{(Fig. 8.73)}$$

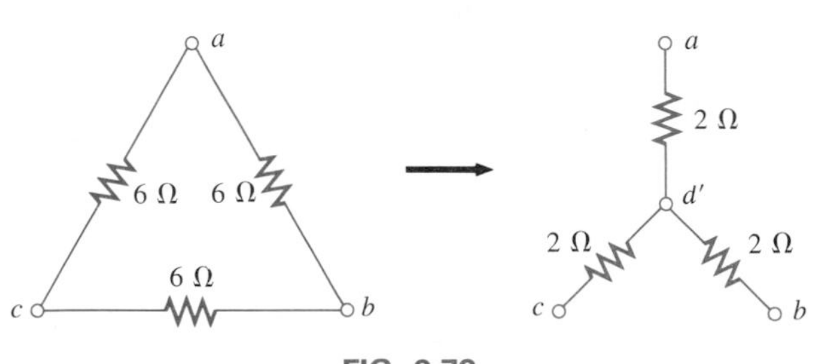


FIG. 8.73

The network then appears as shown in Fig. 8.74.

$$R_T = 2\left[\frac{(2\ \Omega)(9\ \Omega)}{2\ \Omega + 9\ \Omega}\right] = \mathbf{3.2727\ \Omega}$$

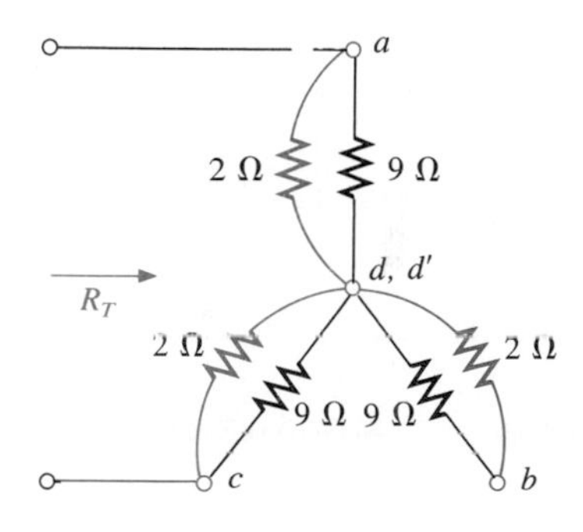

FIG. 8.74

b. *Converting the Y to a Δ*.

$$R_\Delta = 3R_Y = (3)(9\ \Omega) = 27\ \Omega \qquad \text{(Fig. 8.75)}$$

$$R'_T = \frac{(6\ \Omega)(27\ \Omega)}{6\ \Omega + 27\ \Omega} = \frac{162\ \Omega}{33} = 4.9091\ \Omega$$

$$R_T = \frac{R'_T(R'_T + R'_T)}{R'_T + (R'_T + R'_T)} = \frac{R'_T 2R'_T}{3R'_T} = \frac{2R'_T}{3}$$

$$= \frac{2(4.9091\ \Omega)}{3} = \mathbf{3.2727\ \Omega}$$

which checks with the previous solution.

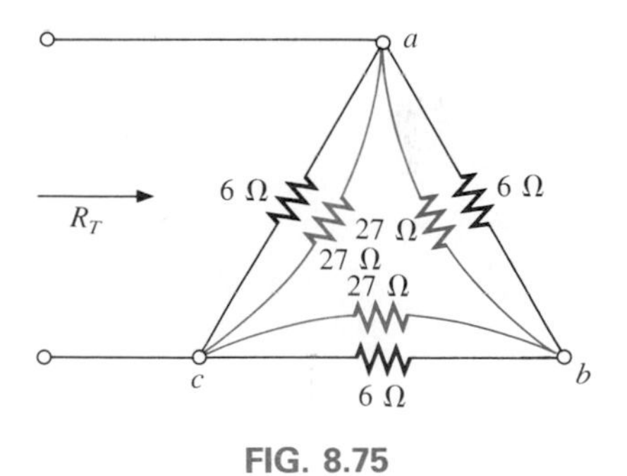


FIG. 8.75

## 8.13 COMPUTER METHODS

The multisource network will now be examined using PSPICE and BASIC. Although PSPICE is primarily interested in determining the nodal voltages, it can also print out the mesh or branch currents in the output file. In BASIC either mesh or nodal analysis can be applied to generate the general solution of a network and then through the proper command statements the other quantities of interest can be determined.

## PSPICE

The network appearing in Example 8.9 (Fig. 8.21) is redrawn in Fig. 8.76 with the nodes defined for a PSPICE input file. The input file of Fig. 8.77 is requesting the nodal voltage $V_2$ and the branch currents $I_1$, $I_2$, and $I_3$. Since the current through $R_1$ is the same as the mesh current of that window, $I_1$ is both the branch and mesh current. The same is true for $I_2$ in the other window. The results reveal that $V_2 = V_{R_3} = +4$ V, that the branch and mesh current through $R_1$ is 1 A from node 1 to node 2, and that the branch and mesh current through $R_2$ is 2 A from node 3 to 2. The branch current through $R_3$ is 1 A from node 2 to our reference node (0). Keep in mind that a positive result for a current in PSPICE indicates that the flow direction is from the first to second nodes of the input file for the resistor between the two nodes. A negative sign indicates the reverse. The results, in total, as obtained using PSPICE are the same as those obtained in Example 8.9.

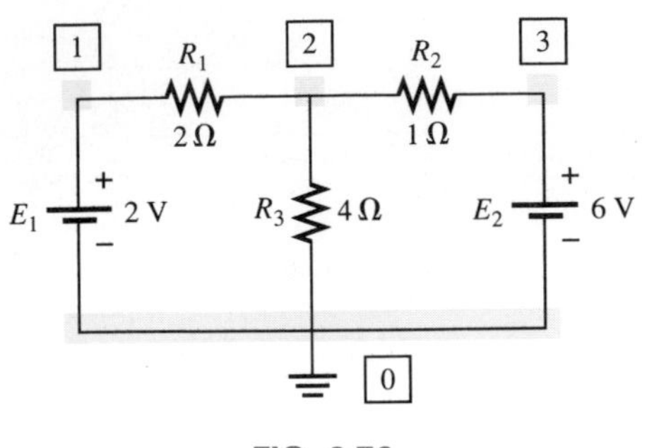


FIG. 8.76

```
************************ Evaluation PSpice (January 1989) ******* 17:48:40 *******

Chapter 8 - Two loop circuit

 ****     CIRCUIT DESCRIPTION

******************************************************************************

VE1  1 0 2V
VE2  3 0 6V
R1   2 1 2
R2   2 3 1
R3   2 0 4
.DC VE1 2 2 1
.PRINT DC V(2) I(R1) I(R2) I(R3)
.OPTIONS NOPAGE
.END

 ****     DC TRANSFER CURVES                    TEMPERATURE =   27.000 DEG C

  VE1          V(2)          I(R1)         I(R2)         I(R3)

   2.000E+00    4.000E+00    1.000E+00   -2.000E+00    1.000E+00
```

FIG. 8.77

The next PSPICE input file determines the nodal voltages for the same network of Example 8.23 (Fig. 8.45), repeated in Fig. 8.78. The

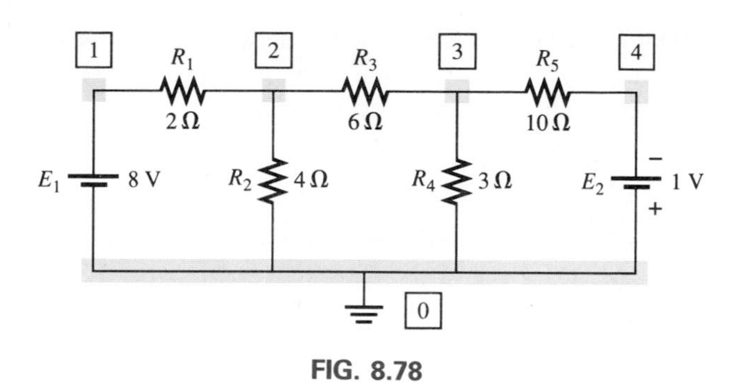


FIG. 8.78

input file appears in Fig. 8.79, with a request for the unknown nodal voltages $V_2$ and $V_3$. The output file reveals that $V_2$ is at a higher poten-

```
************************ Evaluation PSpice (January 1989) ******* 17:49:21 *******

Chapter 8 - Three loop circuit

****     CIRCUIT DESCRIPTION

******************************************************************************

VE1 1 0 8V
VE2 4 0 -1V
R1  1 2 2
R2  2 0 4
R3  2 3 6
R4  3 0 3
R5  3 4 10
.DC VE1 8 8 1
.PRINT DC V(2) V(3)
.OPTIONS NOPAGE
.END

****     DC TRANSFER CURVES                    TEMPERATURE =   27.000 DEG C

  VE1          V(2)         V(3)

  8.000E+00    4.564E+00    1.101E+00
```

FIG. 8.79

tial than $V_3$, establishing a current from node 2 to node 3 through the resistor $R_3$. If desired, the current through $R_3$ could be obtained by a simple application of Ohm's law and a voltage equal to the difference between the two levels.

The input file of Fig. 8.81 is for the network of Fig. 8.80, which appeared as Fig. 8.46. It is the same network as that just examined but with the voltage sources converted to current sources. There are only

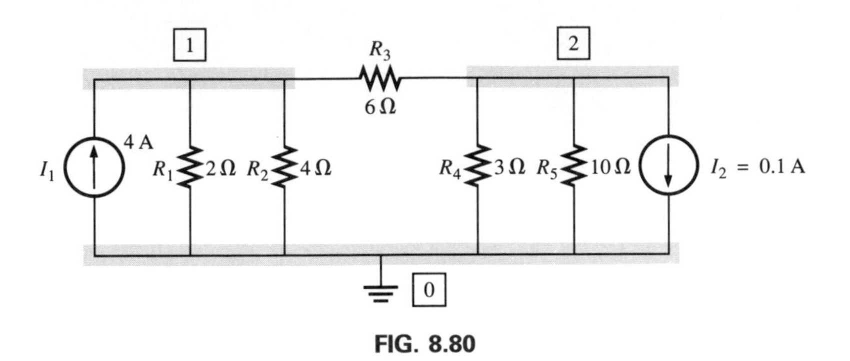


FIG. 8.80

```
************************ Evaluation PSpice (January 1989) ******* 17:49:55 *******

Chapter 8 - Three loop circuit

****      CIRCUIT DESCRIPTION

******************************************************************************

IS1 0 1 4
IS2 2 0 .1
R1  1 0 2
R2  1 0 4
R3  1 2 6
R4  2 0 3
R5  2 0 10
.DC IS2 .1 .1 1
.PRINT DC V(1) V(2)
.OPTIONS NOPAGE
.END

****      DC TRANSFER CURVES                TEMPERATURE =   27.000 DEG C

  IS2          V(1)         V(2)

  1.000E-01    4.564E+00    1.101E+00
```

FIG. 8.81

two nodal voltages to be determined for Fig. 8.80, but note that the result for the voltage across the 4-Ω resistor is the same for either approach and corresponds exactly with the result of Example 8.23.

## BASIC

The natural sequence of steps in either the mesh or nodal format approach simplifies the writing of the BASIC program required to solve single- or multisource networks.

A BASIC program to analyze the basic two-loop network of Fig. 8.82 is provided in Fig. 8.83. The resulting network equations are

$$I_1(R_1 + R_3) - I_2R_3 = E_1$$
$$-I_1R_3 + I_2(R_2 + R_3) = -E_2$$

with the mesh currents calculated using determinants on lines 200 through 220. Note that the program requests the network parameters on lines 130 through 170 and prints out the result, as directed by lines 230 through 250. Since the values of the network are not specified in Fig. 8.82 or in the program, once the program is run and the parameters are

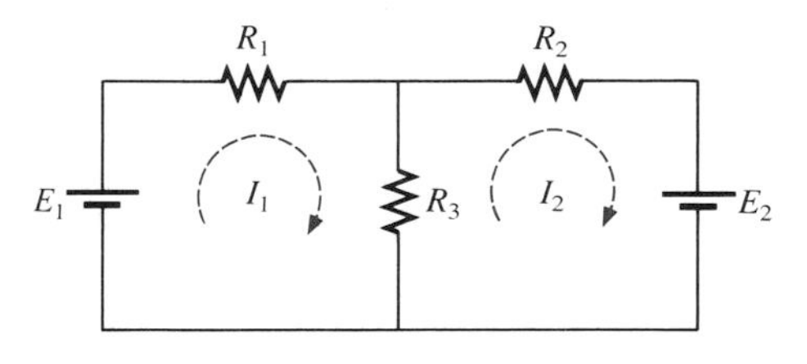


FIG. 8.82

```
        10 REM ***** PROGRAM 8-1 *****
        20 REM ********************************************
        30 REM Program to evaluate the loop currents for a
        40 REM 2-loop network.
        50 REM ********************************************
        60 REM
       100 PRINT "For a 2-loop network"
       110 PRINT "enter the following data:"
       120 PRINT
      [130 INPUT "R1=";R1
      |140 INPUT "R2=";R2
Input |150 INPUT "R3=";R3
      |160 INPUT "Voltage, E1=";E1
      [170 INPUT "Voltage, E2=";E2
       180 PRINT
      [190 REM Calculate I1 and I2
Calc. |200 D=R1*R2+R1*R3+R2*R3
      |210 I1=(E1*(R2+R3)-E2*R3)/D
      [220 I2=(-E2*(R1+R3)+E1*R3)/D
      [230 PRINT "The loop currents are:"
Output|240 PRINT "I1=";I1;"amps"
      [250 PRINT "I2=";I2;"amps"
       260 END

READY

For a 2-loop network
enter the following data:

R1=? 2

R2=? 1

R3=? 4

Voltage, E1=? 2

Voltage, E2=? 6
```

```
The loop currents are:
I1=-1 amps
I2=-2 amps


READY

RUN


For a 2-loop network
enter the following data:

R1=? 1E3

R2=? 2.2E3

R3=? 3.3E3

Voltage, E1=? -5.4

Voltage, E2=? 8.6


The loop currents are:
I1=-4.5517E-03 amps
I2=-4.2947E-03 amps


READY
```

FIG. 8.83
*Program 8.1.*

requested, any values can be substituted and a correct solution will be obtained. In PSPICE, however, a new input file must be made up for each change of input parameters. The question is then whether it takes longer to input the parameters using BASIC or input a whole new file in PSPICE. In many cases it will probably be a toss-up, but remember that it takes a great deal more time to write the BASIC program initially than to obtain a solution with PSPICE. Obviously, we can debate this issue from many viewpoints. The important thing, however, is that you realize the trade-offs with each and have a working knowledge of both software packages and an appropriate language.

The first run employed the same values as Example 8.11, whereas the second run includes resistors in the kilohm range and a reversed source.

## PROBLEMS

### SECTION 8.2 Current Sources

1. Find the voltage $V_{ab}$ (with polarity) for the circuit of Fig. 8.84.

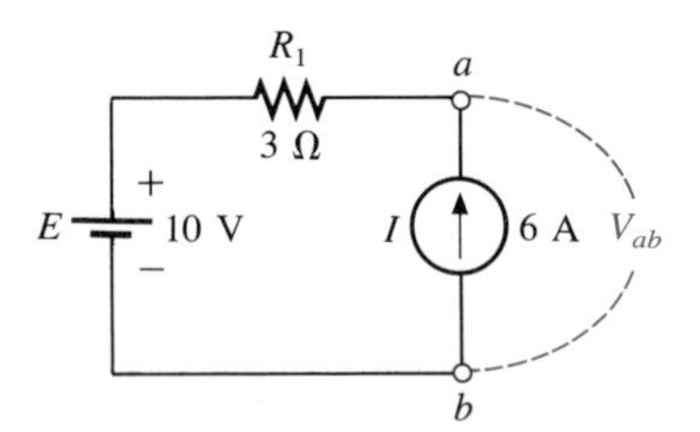

FIG. 8.84

**2.** For the network of Fig. 8.85:
  **a.** Find the voltages $V_S$ and $V_4$.
  **b.** Find the current $I_2$.

FIG. 8.85

**3.** For the network of Fig. 8.86:
  **a.** Find the currents $I_1$ and $I_S$.
  **b.** Find the voltages $V_S$ and $V_3$.

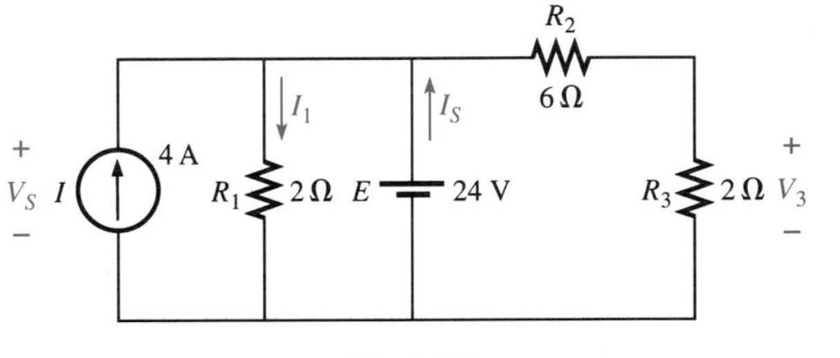


FIG. 8.86

**4.** Find the voltage $V_3$ and the current $I_2$ for the network of Fig. 8.87.

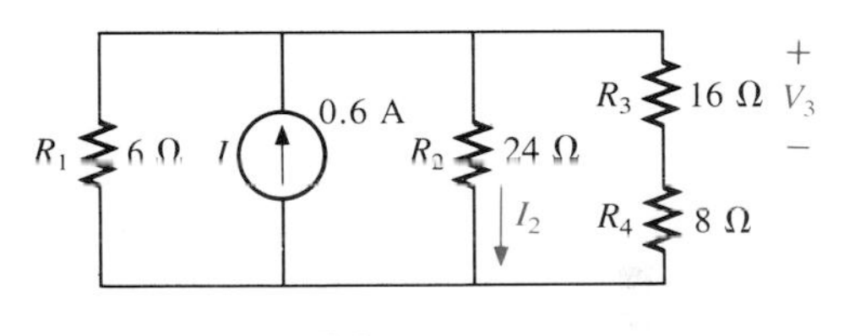


FIG. 8.87

## SECTION 8.3 Source Conversions

**5.** Convert the voltage sources of Fig. 8.88 to current sources.

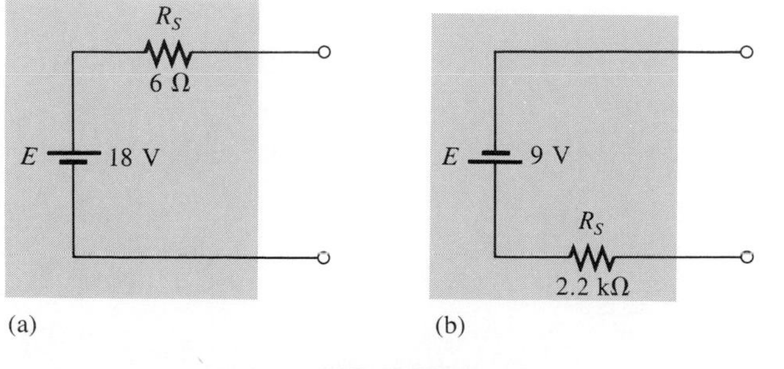


FIG. 8.88

6. Convert the current sources of Fig. 8.89 to voltage sources.

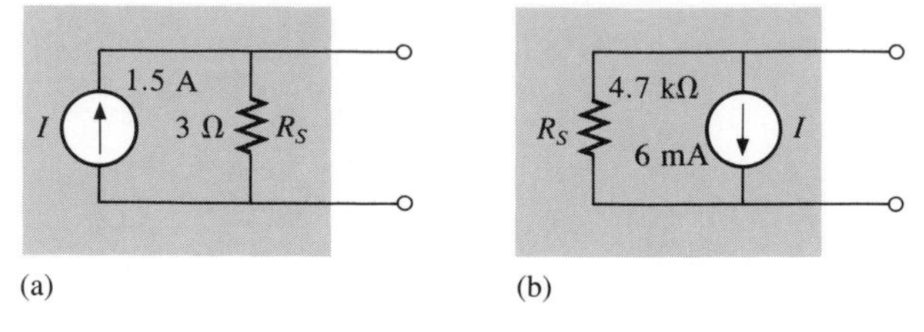

FIG. 8.89

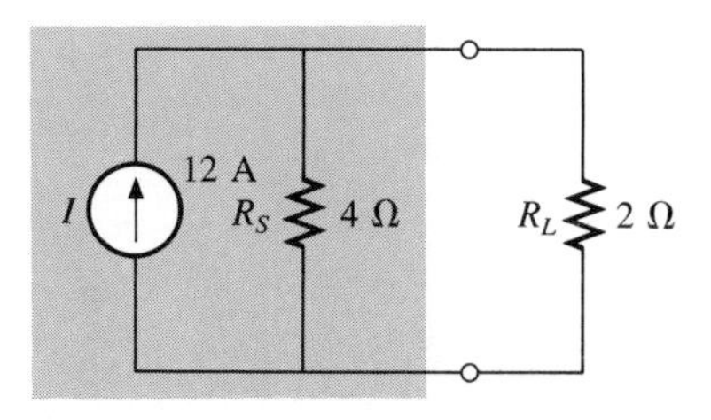


FIG. 8.90

7. For the network of Fig. 8.90:
   a. Find the current through the 2-Ω resistor.
   b. Convert the current source and 4-Ω resistor to a voltage source, and again solve for the current in the 2-Ω resistor. Compare the results.

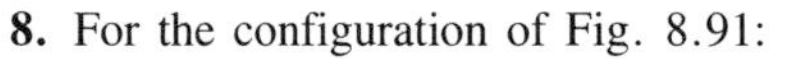

8. For the configuration of Fig. 8.91:
   a. Convert the current source and 6.8-Ω resistor to a voltage source.
   b. Find the magnitude and direction of the current $I_1$.
   c. Find the voltage $V_{ab}$ and the polarity of points $a$ and $b$.

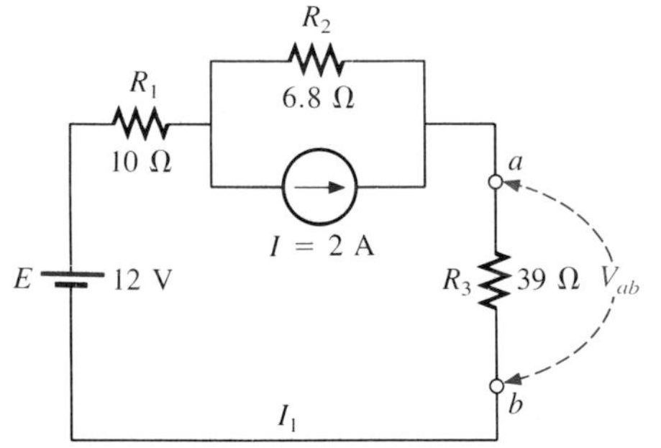

FIG. 8.91

## SECTION 8.4 Current Sources in Parallel

9. Find the voltage $V_2$ and the current $I_1$ for the network of Fig. 8.92.

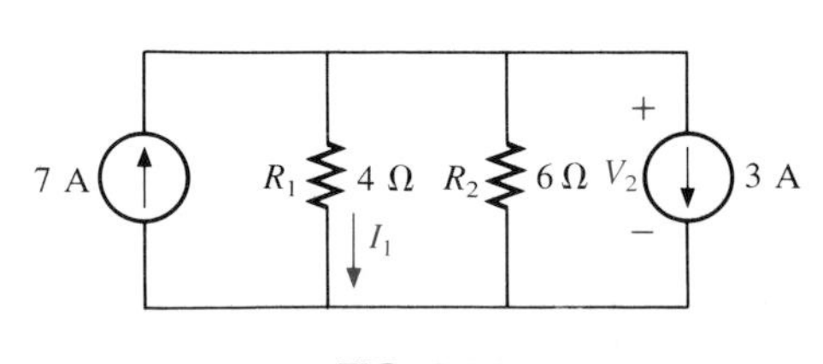


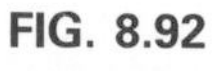
FIG. 8.92

**10. a.** Convert the voltage sources of Fig. 8.93 to current sources.

**b.** Find the voltage $V_{ab}$ and the polarity of points $a$ and $b$.

**c.** Find the magnitude and direction of the current $I$.

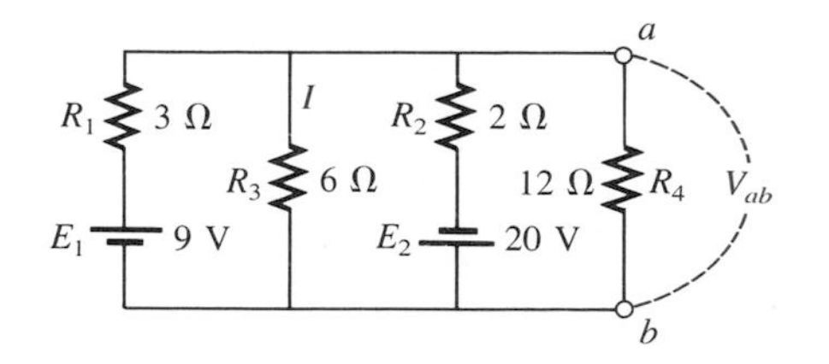


FIG. 8.93

**11.** For the network of Fig. 8.94:

**a.** Convert the voltage source to a current source.

**b.** Reduce the network to a single current source and determine the voltage $V_1$.

**c.** Using the results of part (b), determine $V_2$.

**d.** Calculate the current $I_2$.

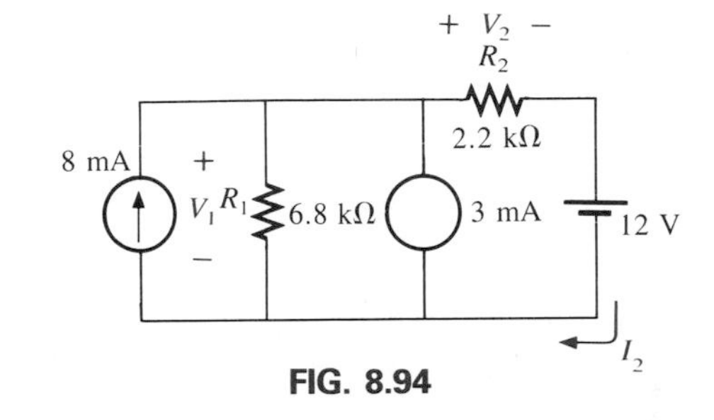


FIG. 8.94

## SECTION 8.6 Branch-Current Analysis

**12.** Using branch-current analysis, find the current through each resistor of each network of Fig. 8.95.

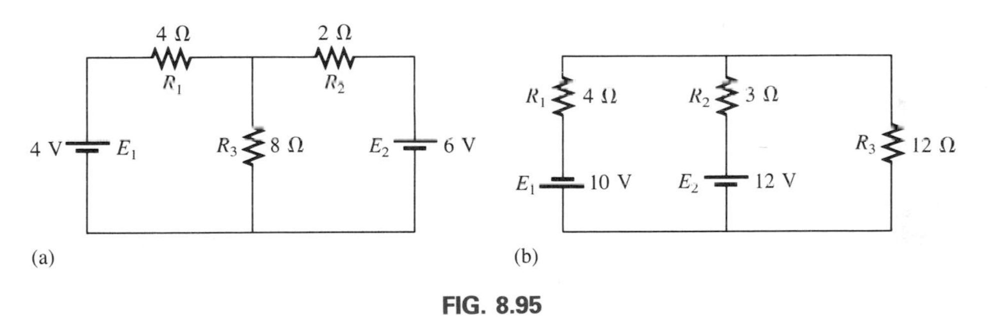


FIG. 8.95

***13.** Using branch-current analysis, find the current through each resistor of each network of Fig. 8.96.

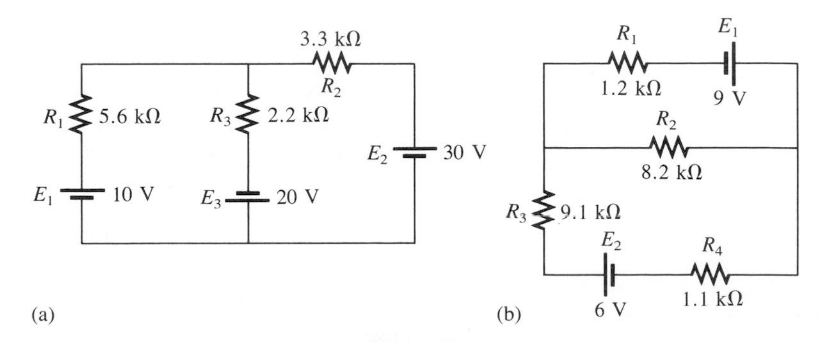


FIG. 8.96

***14.** For the networks of Fig. 8.97:

**a.** Determine the current $I_2$ using branch-current analysis.

**b.** Find the voltage $V_{ab}$.

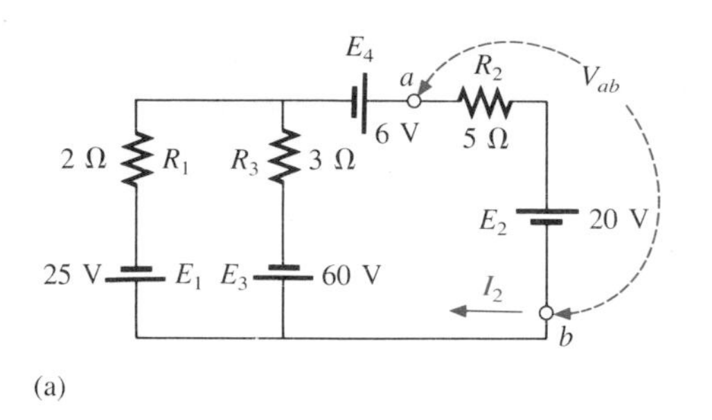


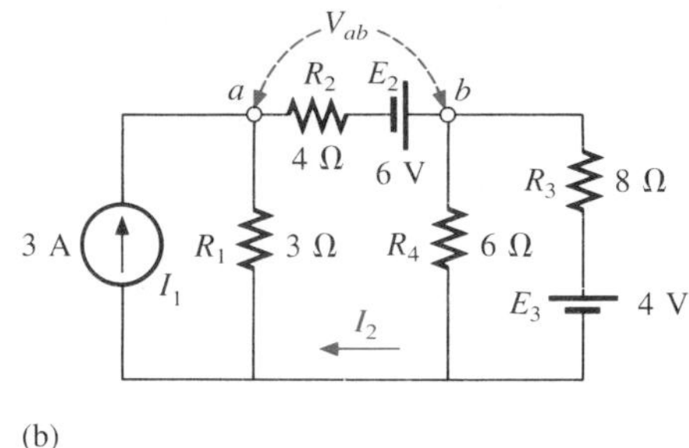

FIG. 8.97

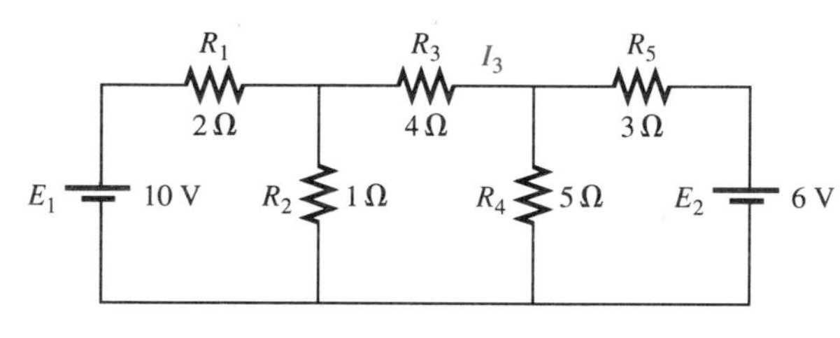


FIG. 8.98

***15.** For the network of Fig. 8.98:

**a.** Write the equations necessary to solve for the branch currents.

**b.** By substitution of Kirchhoff's current law, reduce the set to three equations.

**c.** Rewrite the equations in a format that could be solved using third-order determinants.

**d.** Solve for the branch current through the resistor $R_3$.

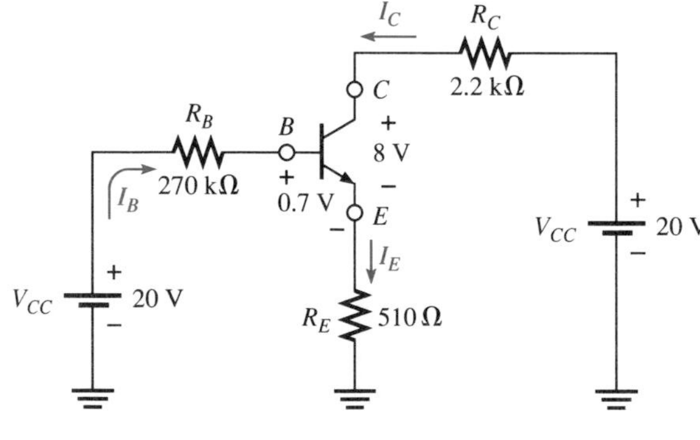

FIG. 8.99

***16.** For the transistor configuration of Fig. 8.99:

**a.** Solve for the currents $I_B$, $I_C$, and $I_E$ using the fact that $V_{BE} = 0.7$ V and $V_{CE} = 8$ V.

**b.** Find the voltages $V_B$, $V_C$, and $V_E$ with respect to ground.

**c.** What is the ratio of output current $I_C$ to input current $I_B$?

(*Note:* In transistor analysis this ratio is referred to as the dc beta of the transistor ($B_{dc}$).)

## SECTION 8.7 Mesh Analysis (General Approach)

**17.** Repeat Problem 12 using mesh analysis.

**18.** Repeat Problem 13 using mesh analysis.

***19.** Repeat Problem 14 using mesh analysis.

***20.** **a.** Repeat Problem 15 using mesh analysis.

**b.** Based on the results of part (a), how would you compare the application of mesh analysis to the branch-current method?

***21.** Repeat Problem 16 using mesh analysis.

*22. Using mesh analysis, determine the current through the 5-Ω resistor for each network of Fig. 8.100.

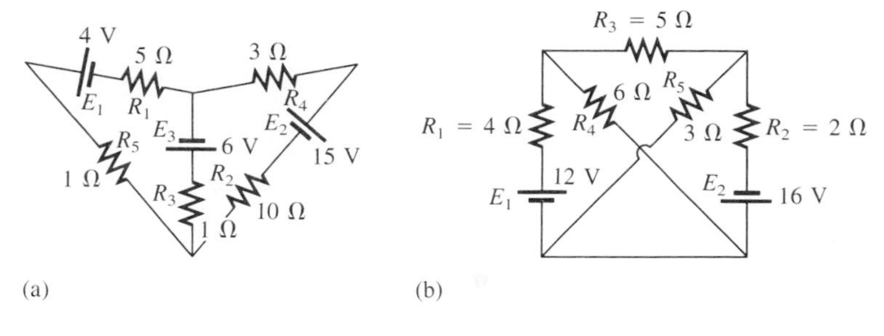


FIG. 8.100

*23. Write the mesh equations for each of the networks of Fig. 8.101, and, using determinants, solve for the loop currents in each circuit.

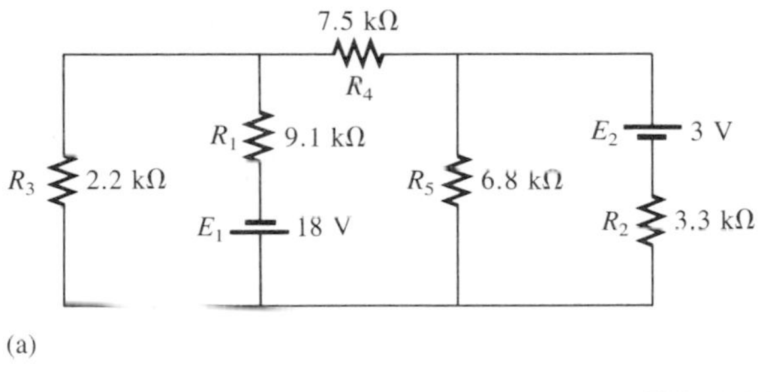


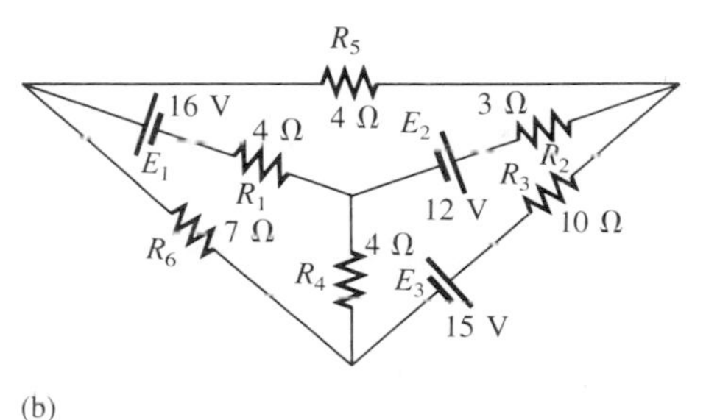

FIG. 8.101

*24. Repeat Problem 23 for the networks of Fig. 8.102.

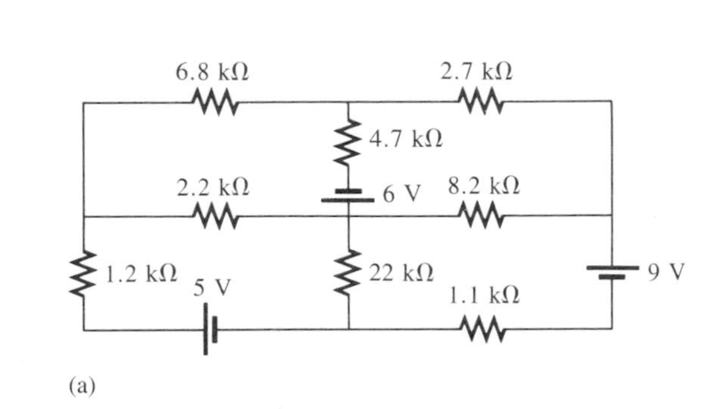


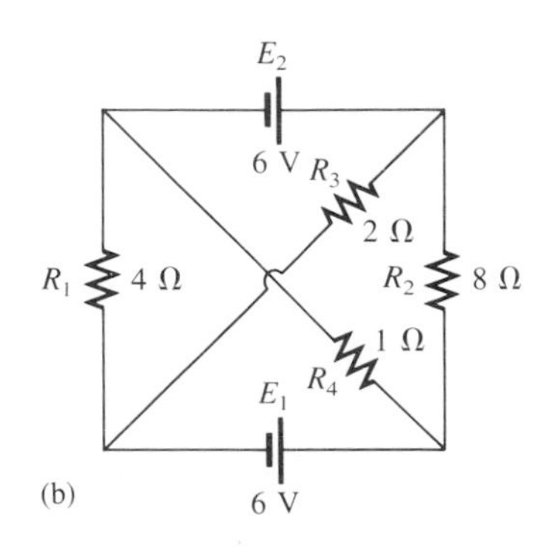

FIG. 8.102

## SECTION 8.8 Mesh Analysis (Format Approach)

25. Using the format approach, write the mesh equations for the networks of Fig. 8.95. Is symmetry present? Using determinants, solve for the mesh currents.

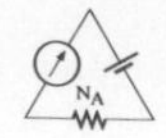

**26.** Repeat Problem 25 for the networks of Fig. 8.96.

***27.** Repeat Problem 25 for the networks of Fig. 8.97.

***28.** Repeat Problem 25 for the network of Fig. 8.98 and compare the effort required with the branch-current method.

**29.** Repeat Problem 25 for the transistor network of Fig. 8.99.

**30.** Repeat Problem 25 for the networks of Fig. 8.100.

***31.** Repeat Problem 25 for the networks of Fig. 8.101.

***32.** Repeat Problem 25 for the networks of Fig. 8.102 but do not solve for the mesh currents.

## SECTION 8.10 Nodal Analysis (Format Approach)

**33.** Using the format approach, write the nodal equations for the networks of Fig. 8.103, and, using determinants, solve for the nodal voltages. Is symmetry present?

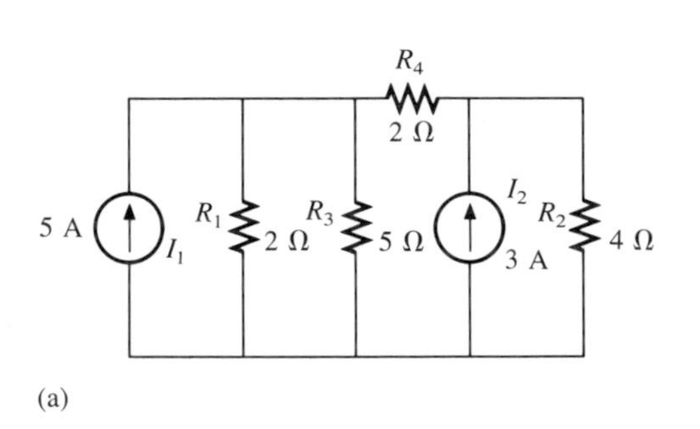


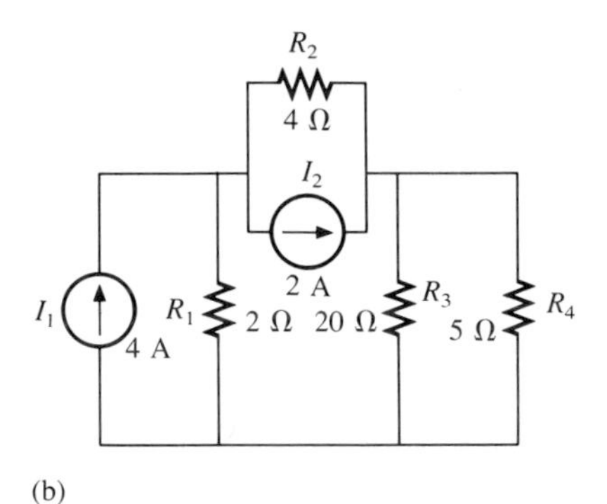


FIG. 8.103

***34.** Repeat Problem 33 for the networks of Fig. 8.104.

**35.** Repeat Problem 33 for the networks of Fig. 8.101.

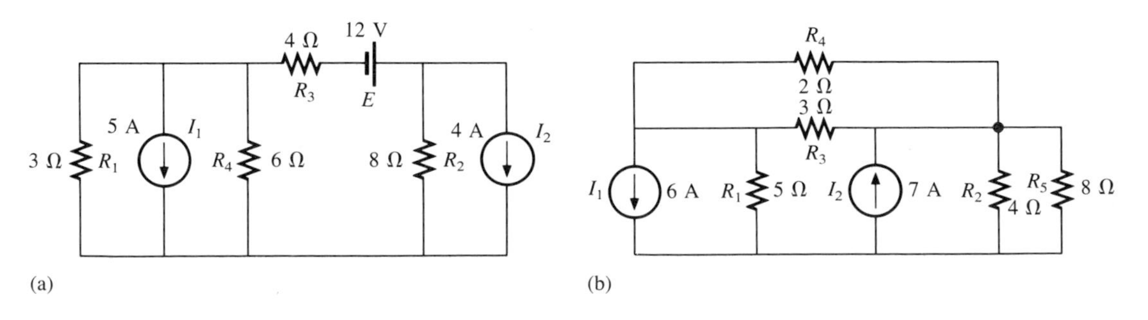


FIG. 8.104

***36.** For the networks of Fig. 8.105, using the format approach, write the nodal equations and solve for the nodal voltages.

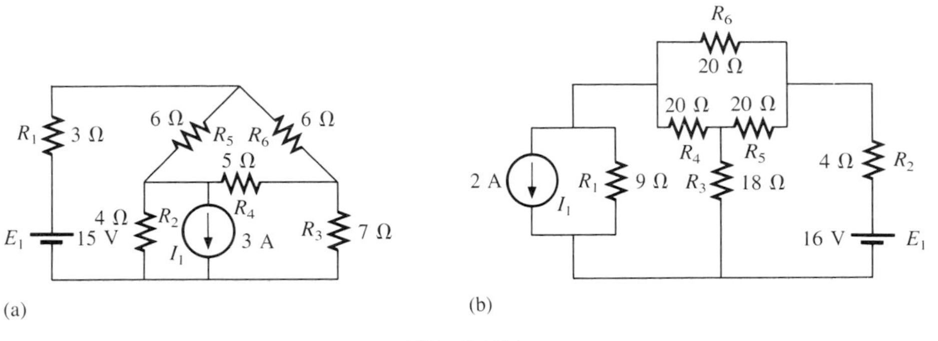


**FIG. 8.105**

***37.** Repeat Problem 36 for the networks of Fig. 8.106.

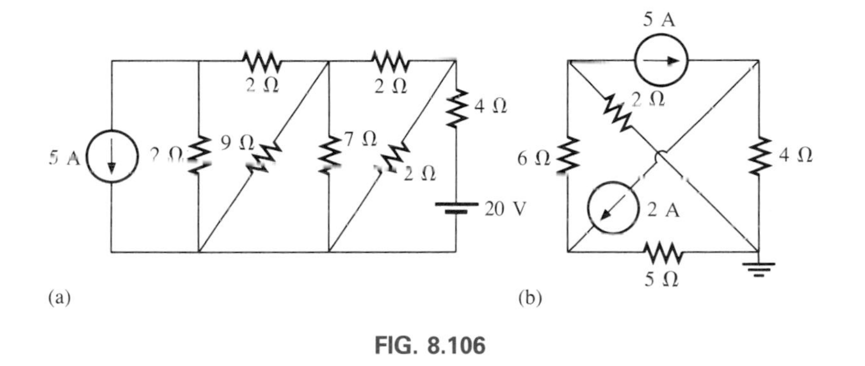


**FIG. 8.106**

## SECTION 8.11 Bridge Networks

***38.** For the bridge network of Fig. 8.107:

**a.** Using the format approach, write the mesh equations.
**b.** Determine the current through $R_5$.
**c.** Is the bridge balanced?
**d.** Is Eq. (8.4) satisfied?

***39.** For the network of Fig. 8.107:

**a.** Using the format approach, write the nodal equations.
**b.** Determine the voltage across $R_5$.
**c.** Is the bridge balanced?
**d.** Is Eq. (8.4) satisfied?

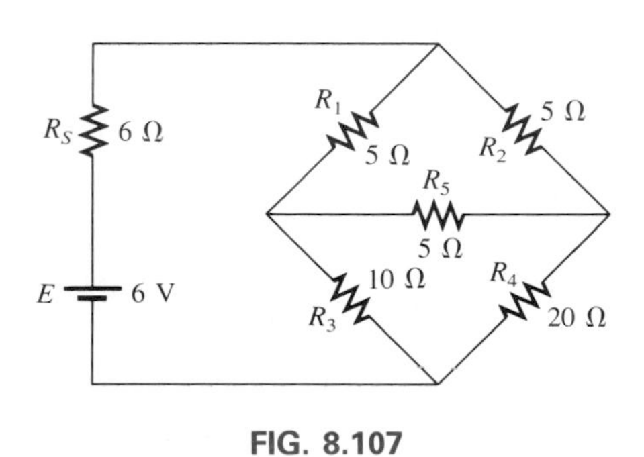


**FIG. 8.107**

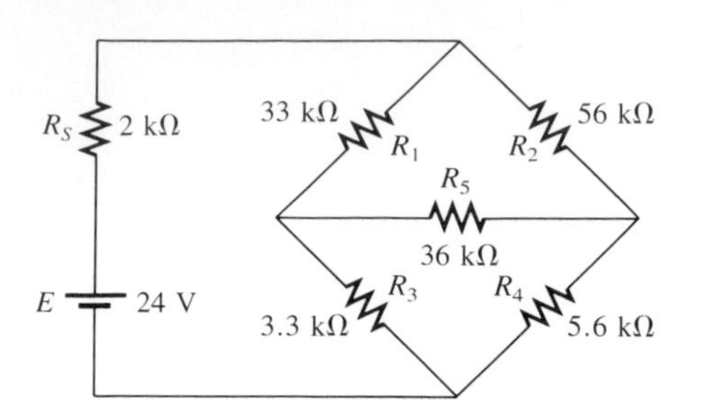

FIG. 8.108

40. Repeat Problem 38 for the network of Fig. 8.108.

41. Repeat Problem 39 for the network of Fig. 8.108.

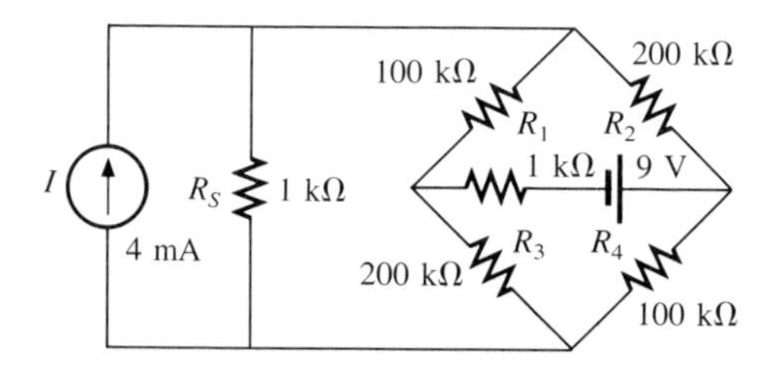


FIG. 8.109

42. Write the nodal equations for the bridge configuration of Fig. 8.109. Use the format approach.

*43. Determine the current through the resistor $R_S$ of each network of Fig. 8.110 using either mesh or nodal analysis. Discuss why you chose one method over the other.

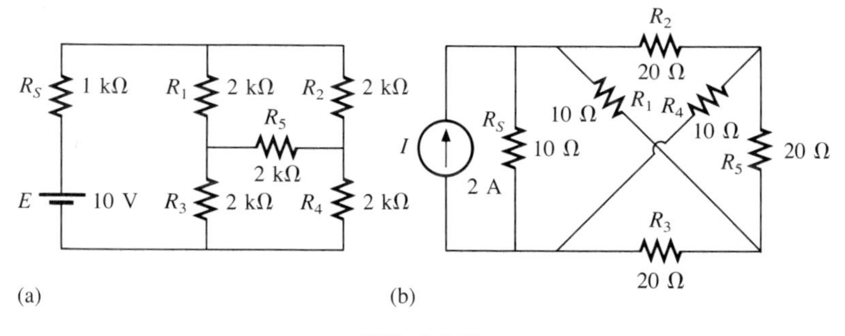


FIG. 8.110

## SECTION 8.12 Y-Δ (T-π) and Δ-Y (π-T) Conversions

44. Using a Δ-Y or Y-Δ conversion, find the current $I$ in each of the networks of Fig. 8.111.

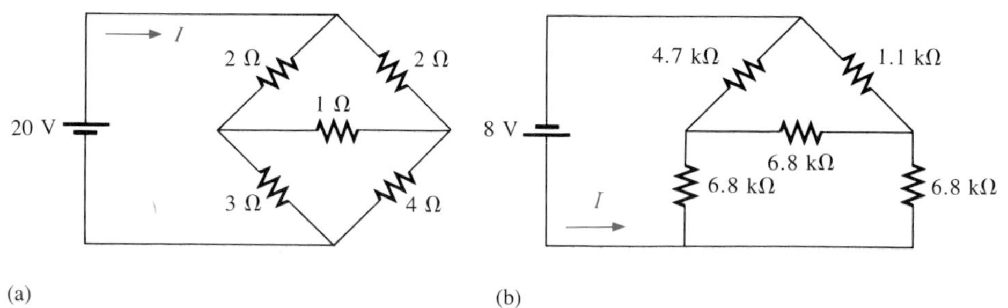


FIG. 8.111

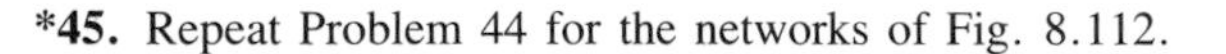

***45.** Repeat Problem 44 for the networks of Fig. 8.112.

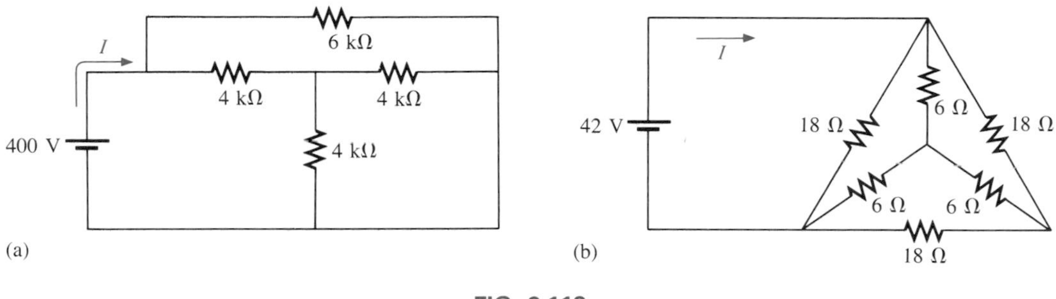

FIG. 8.112

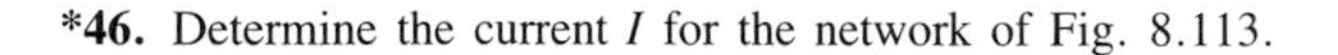

***46.** Determine the current $I$ for the network of Fig. 8.113.

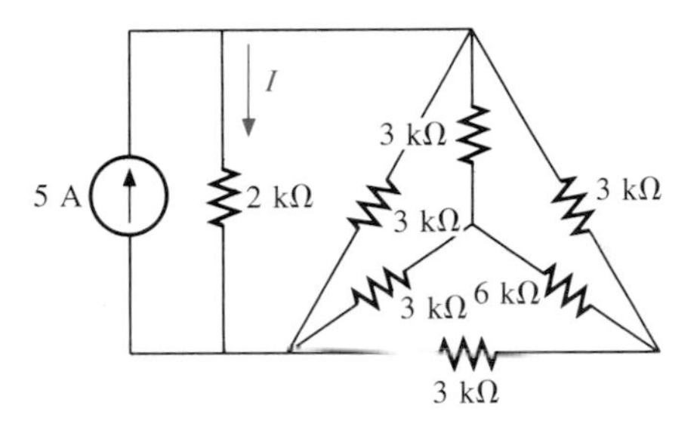

FIG. 8.113

***47. a.** Replace the T configuration of Fig. 8.114 (composed of 6 kΩ resistors) with a $\pi$ configuration.

**b.** Solve for the source current $I_{S_1}$.

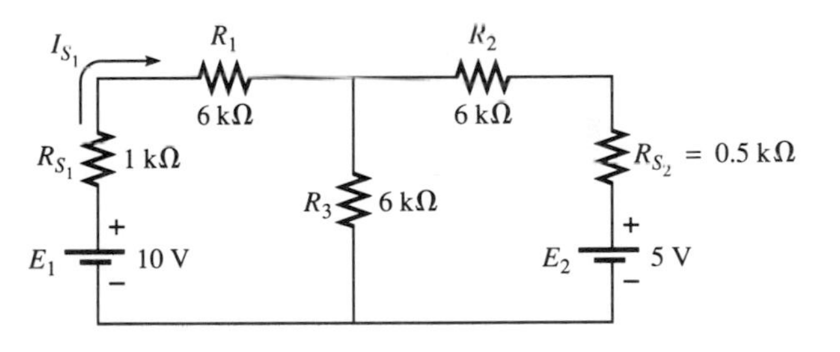

FIG. 8.114

***48. a.** Replace the $\pi$ configuration of Fig. 8.115 (composed of 3 kΩ resistors) with a T configuration.

**b.** Solve for the source current $I_{S_1}$.

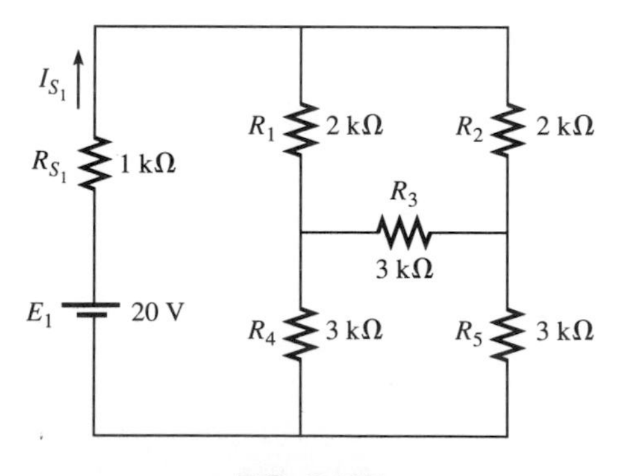


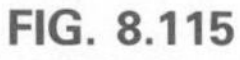

FIG. 8.115

## SECTION 8.13 Computer Methods

### PSPICE

**49.** Write the PSPICE input file for the network of Fig. 8.95(b) that will print out the three branch currents.

**50.** Write the PSPICE input file for the network of Fig. 8.101 that will print out the voltage $V_{R_4}$ and current $I_{R_2}$.

**51.** Write the PSPICE input file for the network of Fig. 8.105 that will print out $V_{R_3}$.

***52.** Write the PSPICE input file for the network of Fig. 8.107 that will print out the voltage $V_{R_5}$.

***53.** Write the PSPICE input file for the network of Fig. 8.110 that will print out $I_{R_1}$.

***54.** Write the PSPICE input file for the network of Fig. 8.115 that will print out $I_{S_1}$.

### BASIC

**55.** Write a program to perform a source conversion of either type. That is, given a voltage source, convert to a current source, and given a current source, convert to a voltage source.

**56.** Given two simultaneous equations, write a program to solve for the unknown variables.

***57.** Write a program to solve for both mesh currents of the network of Fig. 8.25 (for any component values) using mesh analysis and determinants.

***58.** Write a program to solve for the nodal voltages of the network of Fig. 8.40 (for any component values) using nodal analysis and determinants.

***59.** Write a program to determine the total resistance (resistance ''seen'' by the source) for a bridge configuration of any component values.

## GLOSSARY

**Branch-current method** A technique for determining the branch currents of a multiloop network.

**Bridge network** A network configuration typically having a diamond appearance in which no two elements are in series or parallel.

**Current sources** Sources that supply a fixed current to a network and have a terminal voltage dependent on the network to which they are applied.

**Delta (Δ), pi (π) configuration** A network structure that consists of three branches and has the appearance of the Greek letter delta (Δ) or pi (π).

**Determinants method** A mathematical technique for finding the unknown variables of two or more simultaneous linear equations.

**Mesh analysis** A technique for determining the mesh (loop) currents of a network that results in a reduced set of equations compared to the branch-current method.

**Mesh (loop) current** A labeled current assigned to each distinct closed loop of a network that can, individually or in combination with other mesh currents, define all of the branch currents of a network.

**Nodal analysis** A technique for determining the node voltages of a network.

**Node** A junction of two or more branches in a network.

**Wye (Y), tee (T) configuration** A network structure that consists of three branches and has the appearance of the capital letter Y or T.

# 9
# Network Theorems

## 9.1 INTRODUCTION

This chapter will introduce the important fundamental theorems of network analysis. Included are the *superposition, Thevenin's, Norton's, maximum power transfer, substitution, Millman's,* and *reciprocity* theorems. We will consider a number of areas of application for each. A thorough understanding of each theorem is important because a number will be applied repeatedly in the material to follow.

## 9.2 SUPERPOSITION THEOREM

The superposition theorem, like the methods of the last chapter, can be used to find the solution to networks with two or more sources that are not in series or parallel. The most obvious advantage of this method is that it does not require the use of a mathematical technique such as determinants to find the required voltages or currents. Instead, each source is treated independently, and the algebraic sum is found to determine a particular unknown quantity of the network.

The superposition theorem states the following:

***The current through, or voltage across, an element in a linear bilateral network is equal to the algebraic sum of the currents or voltages produced independently by each source.***

When applying the theorem it is possible to consider the effects of two sources at the same time and reduce the number of neworks that have to be analyzed, but, in general,

$$\text{Number of networks to be analyzed} = \text{Number of independent sources} \tag{9.1}$$

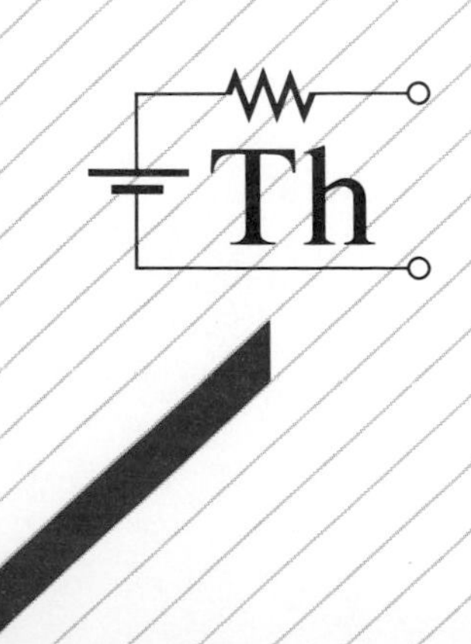

To consider the effects of each source independently requires that sources be removed and replaced without affecting the final result. To remove a voltage source when applying this theorem, the difference in potential between the terminals of the voltage source must be set to zero (short circuited); removing a current source requires that its terminals be opened (open circuit). Any internal resistance or conductance associated with the displaced sources is not eliminated but must still be considered.

Figure 9.1 reviews the various substitutions required when removing an ideal source, and Fig. 9.2 reviews the substitutions with practical sources that have an internal resistance.

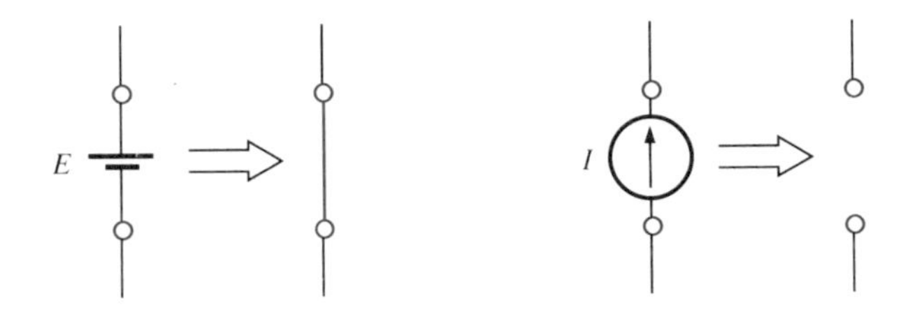


**FIG. 9.1**
*Removing the effects of ideal sources.*

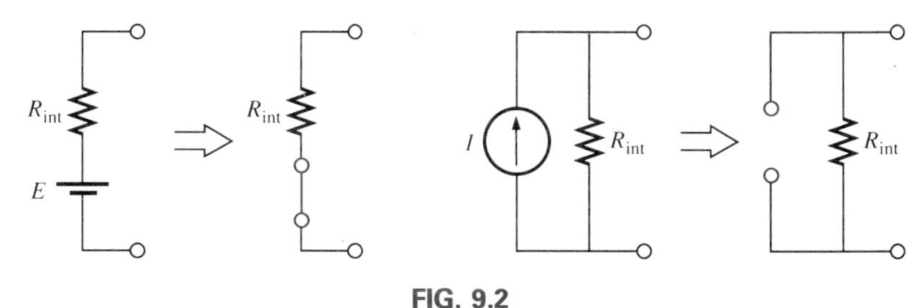


**FIG. 9.2**
*Removing the effects of practical sources.*

The total current through any portion of the network is equal to the algebraic sum of the currents produced independently by each source. That is, for a two-source network, if the current produced by one source is in one direction, while that produced by the other is in the opposite direction through the same resistor, *the resulting current is the difference of the two and has the direction of the larger*. If the individual currents are in the same direction, *the resulting current is the sum of two in the direction of either current*. This rule holds true for the voltage across a portion of a network as determined by polarities, and it can be extended to networks with any number of sources.

The superposition principle is not applicable to power effects since the power loss in a resistor varies as the square (nonlinear) of the current or voltage. For this reason, the power to an element cannot be calculated until the total current through (or voltage across) the element has been determined by superposition. This will be demonstrated in Example 9.3.

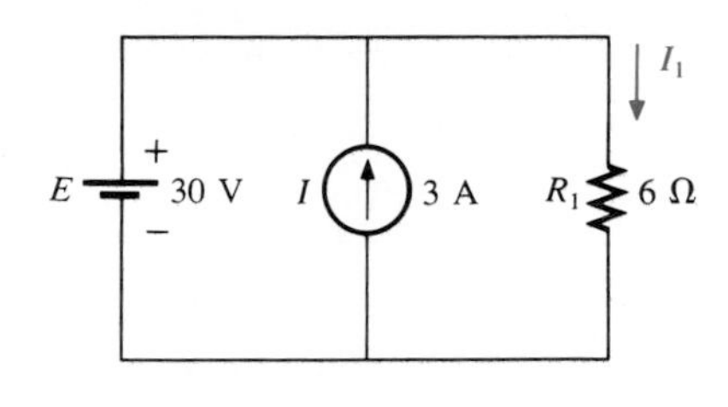


FIG. 9.3

**EXAMPLE 9.1.** Determine $I_1$ for the network of Fig. 9.3.

***Solution:*** Setting $E = 0$ V for the network of Fig. 9.3 results in the network of Fig. 9.4(a), where a short-circuit equivalent has replaced the 30-V source.

As shown in Fig. 9.4(a), the source current will choose the short-circuit path, and $I'_1 = 0$ A. If we applied the current divider rule,

$$I'_1 = \frac{R_{sc}I}{R_{sc} + R_1} = \frac{(0\ \Omega)I}{0\ \Omega + 6\ \Omega} = 0\ \text{A}$$

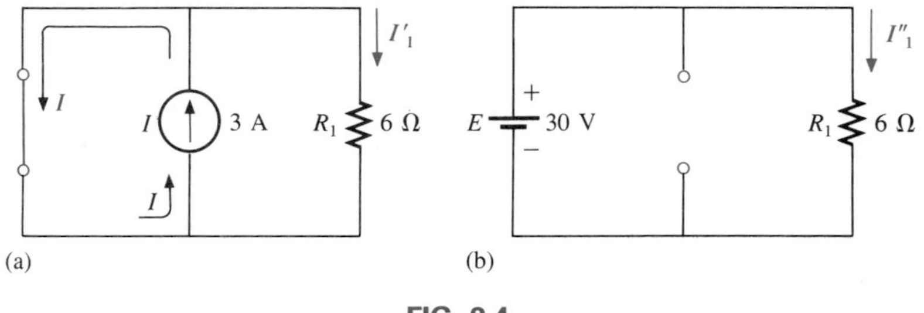


FIG. 9.4

Setting $I$ to zero amperes will result in the network of Fig. 9.4(b) with the current source replaced by an open circuit. Applying Ohm's law,

$$I''_1 = \frac{E}{R_1} = \frac{30\ \text{V}}{6\ \Omega} = 5\ \text{A}$$

Since $I'_1$ and $I''_1$ have the same defined direction in Figs. 9.4(a) and (b), the current $I_1$ is the sum of the two, and

$$I_1 = I'_1 + I''_1 = 0\ \text{A} + 5\ \text{A} = \mathbf{5\ A}$$

Note in this case that the current source has no effect on the current through the 6-Ω resistor since the voltage across the resistor must be fixed at 30 V because they are parallel elements.

**EXAMPLE 9.2.** Using superposition, determine the current through the 4-Ω resistor of Fig. 9.5. Note that this is a two-source network of the type considered in Chapter 8.

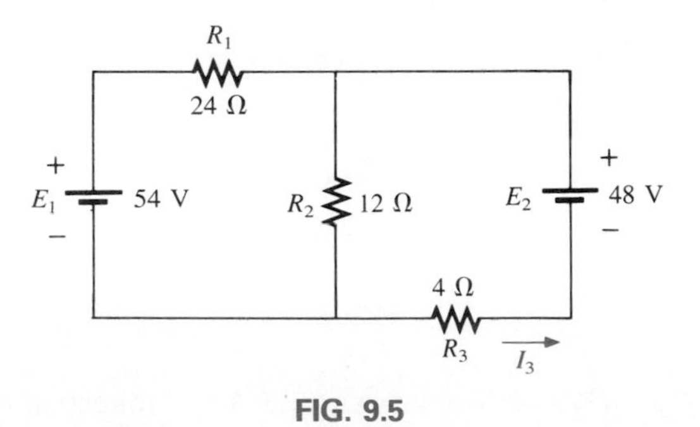


FIG. 9.5

***Solution:*** *Considering the effects of a* 54-V *source* (*Fig*. 9.6):

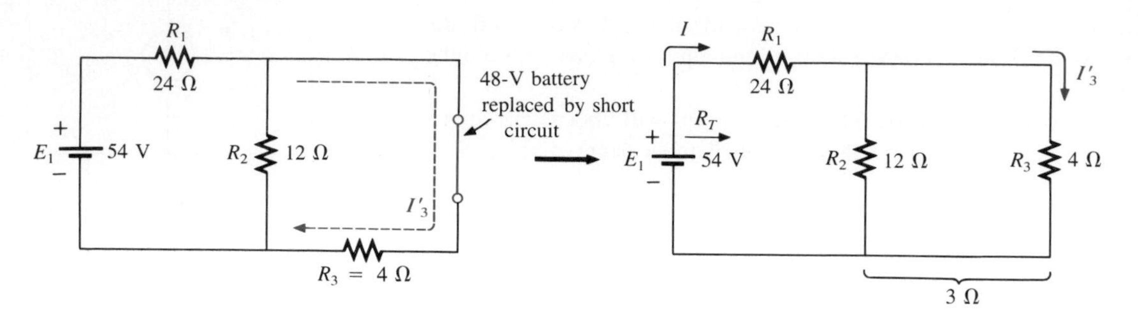

FIG. 9.6

$$R_T = R_1 + R_2 \parallel R_3 = 24\ \Omega + 12\ \Omega \parallel 4\ \Omega = 24\ \Omega + 3\ \Omega = 27\ \Omega$$

$$I = \frac{E_1}{R_T} = \frac{54\text{ V}}{27\ \Omega} = 2\text{ A}$$

By the current divider rule,

$$I'_3 = \frac{R_2 I}{R_2 + R_3} = \frac{(12\ \Omega)(2\text{ A})}{12\ \Omega + 4\ \Omega} = \frac{24\ \Omega}{16} = 1.5\text{ A}$$

*Considering the effects of the* 48-V *source* (*Fig*. 9.7):

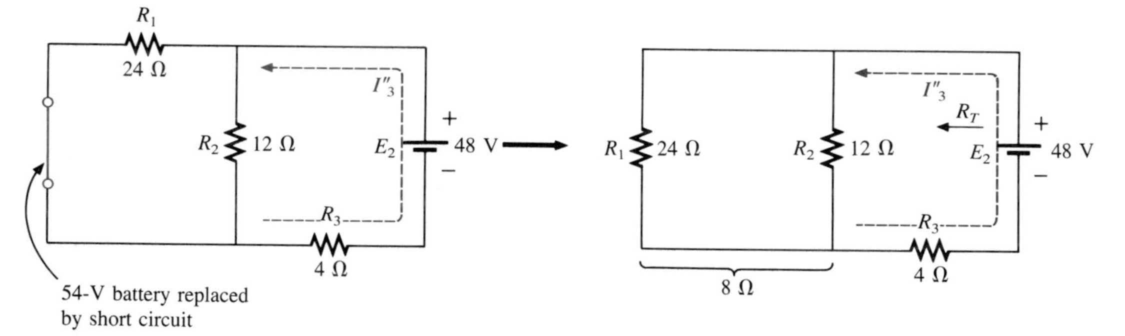


FIG. 9.7

$$R_T = R_3 + R_1 \parallel R_2 = 4\ \Omega + 24\ \Omega \parallel 12\ \Omega = 4\ \Omega + 8\ \Omega = 12\ \Omega$$

$$I''_3 = \frac{E_2}{R_T} = \frac{48\text{ V}}{12\ \Omega} = 4\text{ A}$$

The total current through the 4-Ω resistor (Fig. 9.8) is

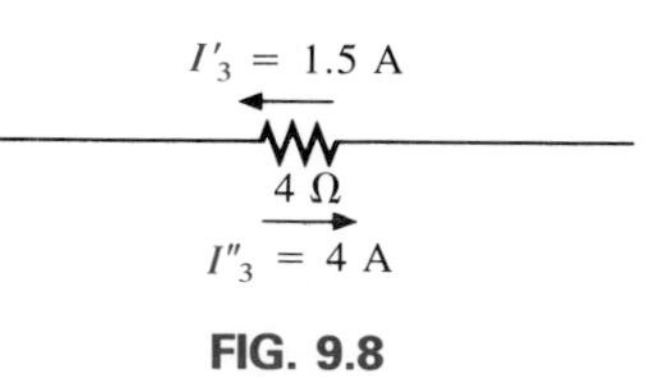


FIG. 9.8

$$I_3 = I''_3 - I'_3 = 4\text{ A} - 1.5\text{ A} = \mathbf{2.5\text{ A}} \qquad \text{(direction of } I''_3\text{)}$$

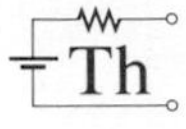




**EXAMPLE 9.3.** Using superposition, find the current through the 6-Ω resistor of the network of Fig. 9.9.

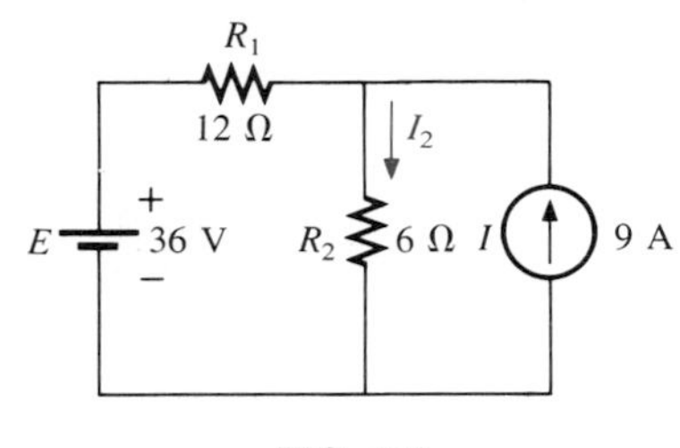


**FIG. 9.9**

***Solution:*** *Considering the effect of the* 36-V *source* (*Fig.* 9.10):

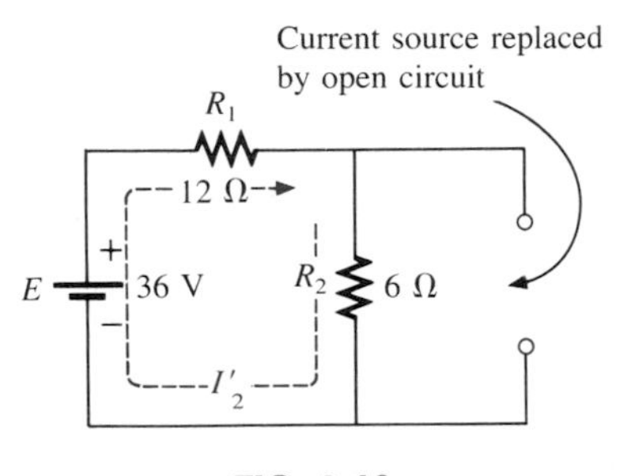

**FIG. 9.10**

$$I'_2 = \frac{E}{R_T} = \frac{E}{R_1 + R_2} = \frac{36\text{ V}}{12\ \Omega + 6\ \Omega} = 2\text{ A}$$

*Considering the effect of the* 9-A *source* (*Fig.* 9.11): Applying the current divider rule,

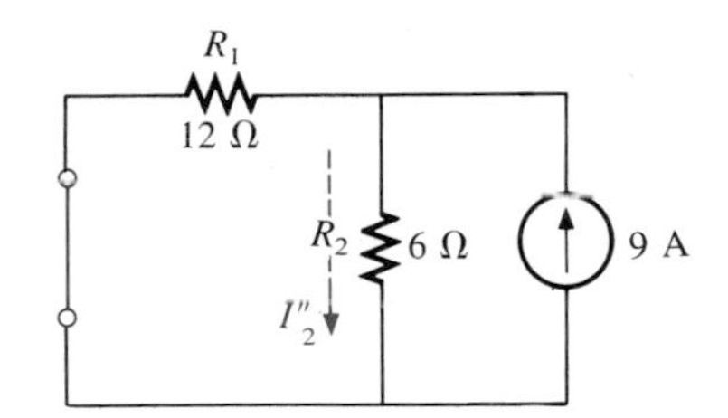


**FIG. 9.11**

$$I''_2 = \frac{R_1 I}{R_1 + R_2} = \frac{(12\ \Omega)(9\text{ A})}{12\ \Omega + 6\ \Omega} = \frac{108\text{ A}}{18} = 6\text{ A}$$

The total current through the 6-Ω resistor (Fig. 9.12) is

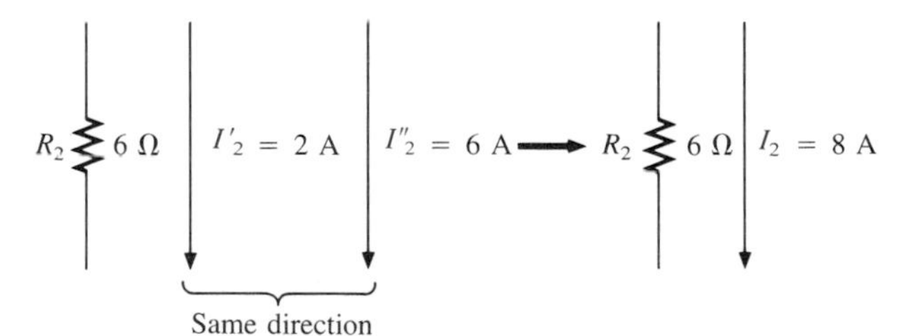

**FIG. 9.12**

$$I_2 = I'_2 + I''_2 = 2\text{ A} + 6\text{ A} = \mathbf{8\text{ A}}$$

The power to the 6-Ω resistor is

$$\text{Power} = I^2R = (8\text{ A})^2(6\ \Omega) = \mathbf{384\text{ W}}$$

The calculated power to the 6-Ω resistor due to each source, *misusing* the principle of superposition, is

$$P_1 = (I'_2)^2R = (2\text{ A})^2(6\ \Omega) = 24\text{ W}$$
$$P_2 = (I''_2)^2R = (6\text{ A})^2(6\ \Omega) = 216\text{ W}$$
$$P_1 + P_2 = 240\text{ W} \neq 384\text{ W}$$

This results because 2 + 6 = 8, but

$$(2)^2 + (6)^2 \neq (8)^2$$

As mentioned previously, the superposition principle is not applicable to power effects, since power is proportional to the square of the current or voltage ($I^2R$ or $V^2/R$).

Figure 9.13 is a plot of the power delivered to the 6-Ω resistor versus current.

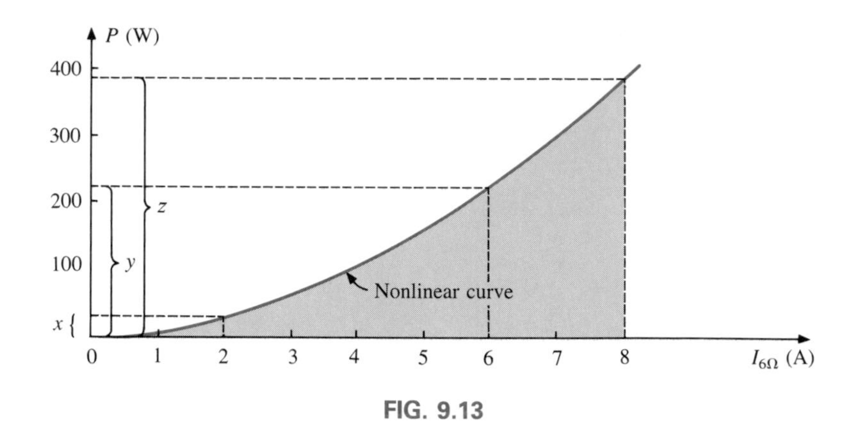


FIG. 9.13

Obviously, $x + y \neq z$, or 24 W + 216 W $\neq$ 384 W, and superposition does not hold. However, for a linear relationship, such as that between the voltage and current of the fixed-type 6-Ω resistor, superposition can be applied, as demonstrated by the graph of Fig. 9.14, where $a + b = c$, or 2 A + 6 A = 8 A.

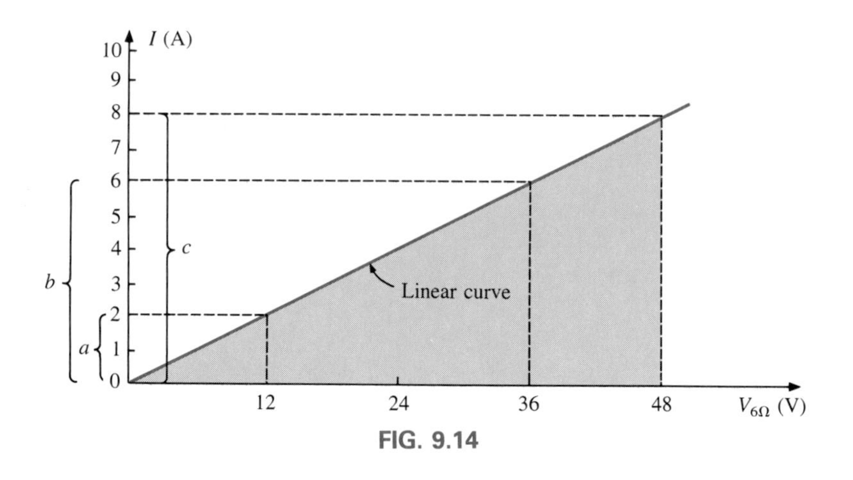


FIG. 9.14

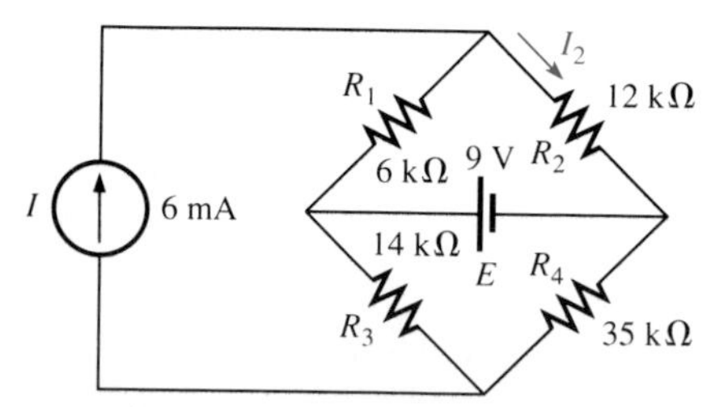


FIG. 9.15

**EXAMPLE 9.4.** Using the principle of superposition, find the current $I_2$ through the 12-kΩ resistor of Fig. 9.15.

***Solution:*** *Considering the effect of the* 6-mA *current source (Fig.* 9.16):

Current divider rule:

$$I'_2 = \frac{R_1 I}{R_1 + R_2} = \frac{(6\ \text{k}\Omega)(6\ \text{mA})}{6\ \text{k}\Omega + 12\ \text{k}\Omega} = 2\ \text{mA}$$

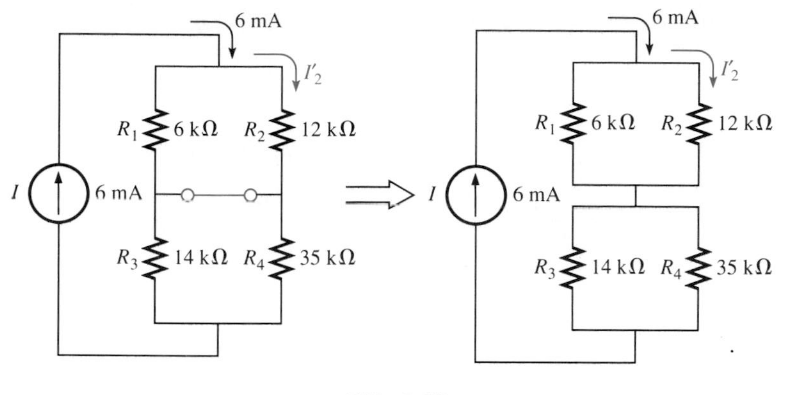


FIG. 9.16

*Considering the effect of the* 9-V *voltage source* (*Fig.* 9.17):

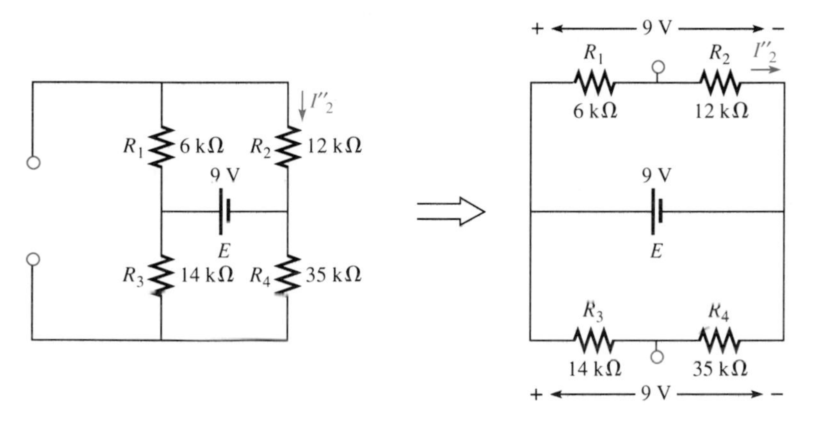

FIG. 9.17

$$I''_2 = \frac{E}{R_1 + R_2} = \frac{9\text{ V}}{6\text{ k}\Omega + 12\text{ k}\Omega} = 0.5\text{ mA}$$

Since $I'_2$ and $I''_2$ have the same direction through $R_2$, the desired current is the sum of the two:

$$\begin{aligned} I_2 &= I'_2 + I''_2 \\ &= 2\text{ mA} + 0.5\text{ mA} \\ &= \mathbf{2.5\text{ mA}} \end{aligned}$$

---

**EXAMPLE 9.5.** Find the current through the 2-Ω resistor of the network of Fig. 9.18. The presence of three sources will result in three different networks to be analyzed.

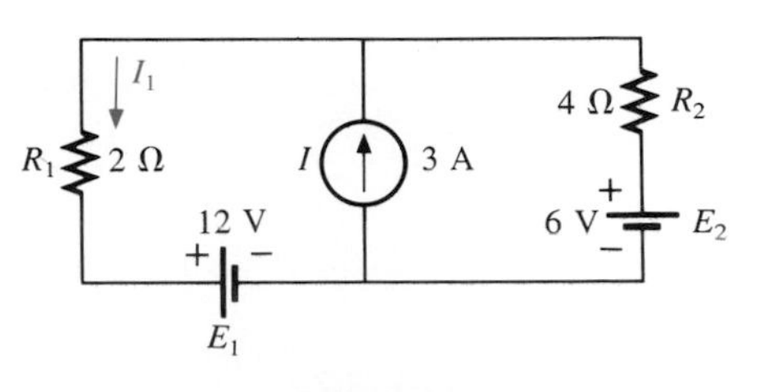


FIG. 9.18

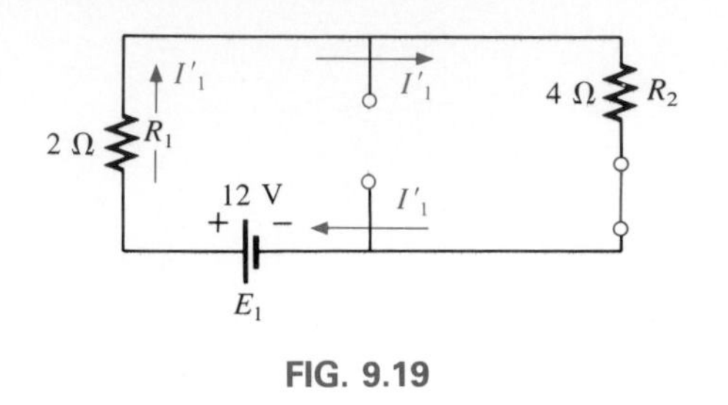


FIG. 9.19

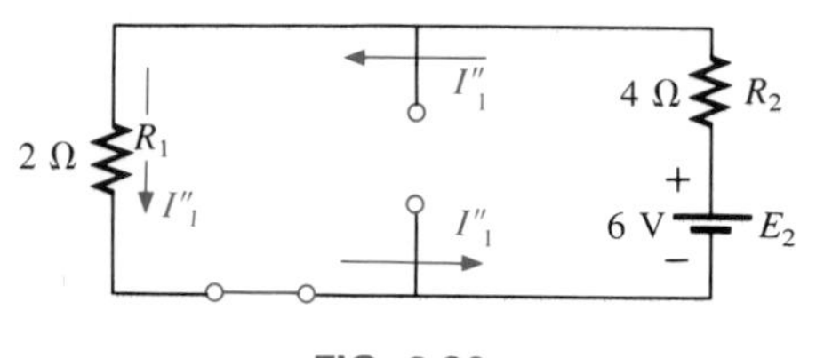


FIG. 9.20

***Solution:*** *Considering the effect of the* 12-V *source* (*Fig.* 9.19):

$$I'_1 = \frac{E_1}{R_1 + R_2} = \frac{12\text{ V}}{2\ \Omega + 4\ \Omega} = \frac{12\text{ V}}{6\ \Omega} = 2\text{ A}$$

*Considering the effect of the* 6-V *source* (*Fig.* 9.20):

$$I''_1 = \frac{E_2}{R_1 + R_2} = \frac{6\text{ V}}{2\ \Omega + 4\ \Omega} = \frac{6\text{ V}}{6\ \Omega} = 1\text{ A}$$

*Considering the effect of the* 3-A *source* (*Fig.* 9.21):

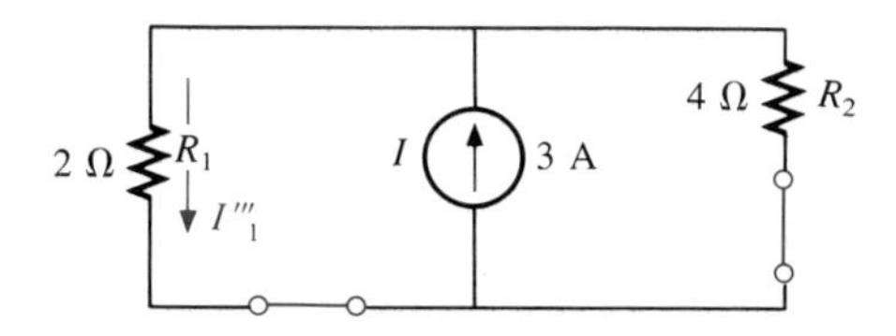

FIG. 9.21

Applying the current divider rule,

$$I'''_1 = \frac{R_2 I}{R_1 + R_2} = \frac{(4\ \Omega)(3\text{ A})}{2\ \Omega + 4\ \Omega} = \frac{12\text{ A}}{6} = 2\text{ A}$$

The total current through the 2-Ω resistor appears in Fig. 9.22, and

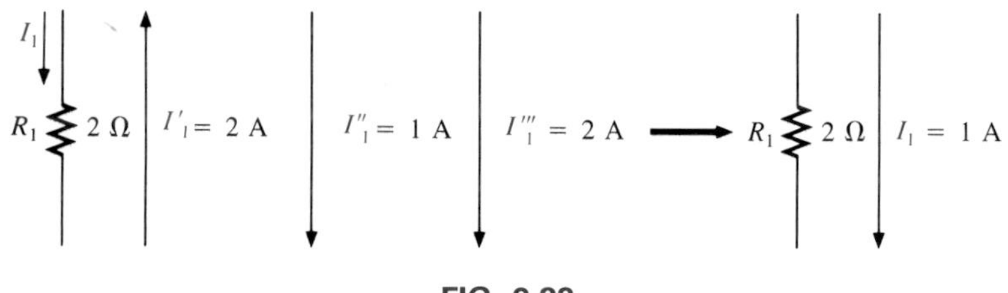

FIG. 9.22

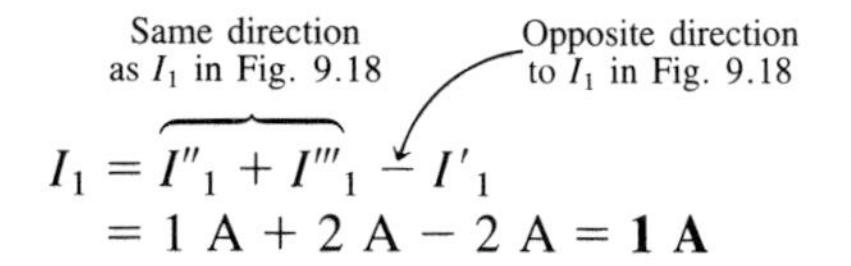


$$I_1 = I''_1 + I'''_1 - I'_1$$
$$= 1\text{ A} + 2\text{ A} - 2\text{ A} = \mathbf{1\text{ A}}$$

## 9.3 THEVENIN'S THEOREM

Thevenin's theorem states the following:

***Any two-terminal linear bilateral dc network can be replaced by an equivalent circuit consisting of a voltage source and a series resistor, as shown in Fig. 9.23.***

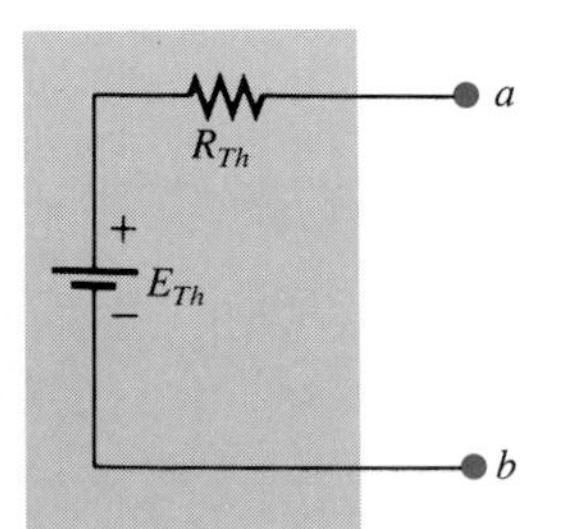


FIG. 9.23
*Thevenin equivalent circuit.*

In Fig. 9.24(a), for example, the network within the container has only two terminals available to the outside world, labeled *a* and *b*. It is

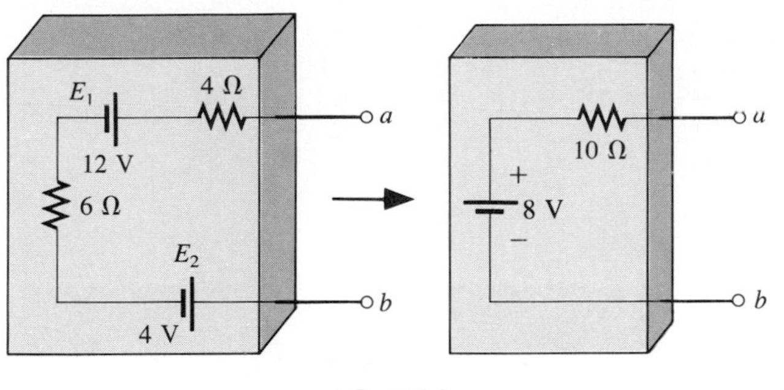


**FIG. 9.24**

possible using Thevenin's theorem to replace everything in the container by one source and one resistor, as shown in Fig. 9.24(b), and maintain the same terminal characteristics at terminals *a* and *b*. That is, any load connected to terminals *a* and *b* will not know whether it is hooked up to the network of Fig.9.24(a) or Fig. 9.24(b). The load will receive the same current, voltage, and power from either configuration of Fig. 9.24. Throughout the discussion to follow, however, always keep in mind that

***the Thevenin equivalent circuit provides an equivalence at the terminals only—the internal construction and characteristics of the original network and the Thevenin equivalent are usually quite different.***

For the network of Fig. 9.24(a), the Thevenin equivalent circuit can be found quite directly by simply combining the series batteries and resistors. Note the exact similarity of the network of Fig. 9.24(b) with the Thevenin configuration of Fig. 9.23. The method described below will allow us to extend the procedure just applied to more complex configurations and still end up with the relatively simple network of Fig. 9.23.

In most cases, there will be other elements connected to the right of terminals *a* and *b* in Fig. 9.24. To apply the theorem, however, the network to be reduced to the Thevenin equivalent form must be isolated as shown in Fig. 9.24 and the two ''holding'' terminals identified. Once the proper Thevenin equivalent circuit has been determined, the voltage, current, or resistance readings between the two ''holding'' terminals will be the same whether the original or Thevenin equivalent circuit is connected to the left of terminals *a* and *b* in Fig. 9.24. Any load connected to the right of terminals *a* and *b* of Fig. 9.24 will receive the same voltage or current with either network.

This theorem achieves two important objectives. First, as was true for all the methods previously described, it allows us to find any particular voltage or current in a linear network with one, two, or any other number of sources. Second, we can concentrate on a specific portion of a network by replacing the remaining network by an equivalent circuit.

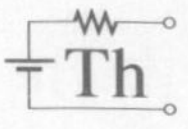


In Fig. 9.25, for example, by finding the Thevenin equivalent circuit for the network in the shaded area, we can quickly calculate the change in

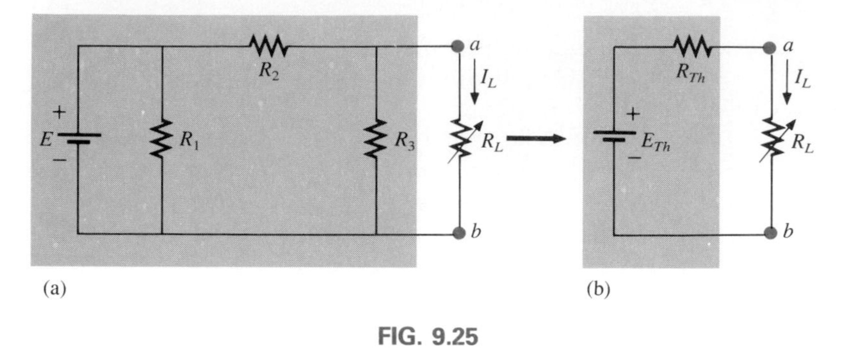


**FIG. 9.25**

current through or voltage across the variable resistor $R_L$ for the various values that it may assume. This will be demonstrated in Example 9.6.

Before we examine the steps involved in applying this theorem, it is important that an additional word be included here to insure that the implications of the Thevenin equivalent circuit are clear. In Fig. 9.25, the entire network, except $R_L$, is to be replaced by a single series resistor and battery as shown in Fig. 9.23. The values of these two elements of the Thevenin equivalent circuit must be chosen to insure that the resistor $R_L$ will react to the network of Fig. 9.25(a) in the same manner as to the network of Fig. 9.25(b). In other words, the current through or voltage across $R_L$ must be the same for either network for any value of $R_L$.

The following sequence of steps will lead to the proper value of $R_{Th}$ and $E_{Th}$.

**Preliminary:**

1. Remove that portion of the network across which the Thevenin equivalent circuit is to be found. In Fig. 9.25(a), this requires that the load resistor $R_L$ be temporarily removed from the network.
2. Mark the terminals of the remaining two-terminal network. (The importance of this step will become obvious as we progress through some complex networks.)

***$R_{Th}$***

3. Calculate $R_{Th}$ by first setting all sources to zero (voltage sources are replaced by short circuits and current sources by open circuits) and then finding the resultant resistance between the two marked terminals. (If the internal resistance of the voltage and/or current sources is included in the original network, it must remain when the sources are set to zero.)

***$E_{Th}$***

4. Calculate $E_{Th}$ by first returning all sources to their original position and finding the *open-circuit* voltage between the marked terminals. (This step is invariably the one that will lead to the most confusion and errors. In *all* cases, keep in mind that it is the *open-circuit* potential between the two terminals marked in step 2 above.)

**Conclusion:**

5. Draw the Thevenin equivalent circuit with the portion of the circuit previously removed replaced between the terminals of the equivalent circuit. This step is indicated by the placement of the resistor $R_L$ between the terminals of the Thevenin equivalent circuit as shown in Fig. 9.25(b).

---

**EXAMPLE 9.6.** Find the Thevenin equivalent circuit for the network in the shaded area of the network of Fig. 9.26. Then find the current through $R_L$ for values of 2 Ω, 10 Ω, and 100 Ω.

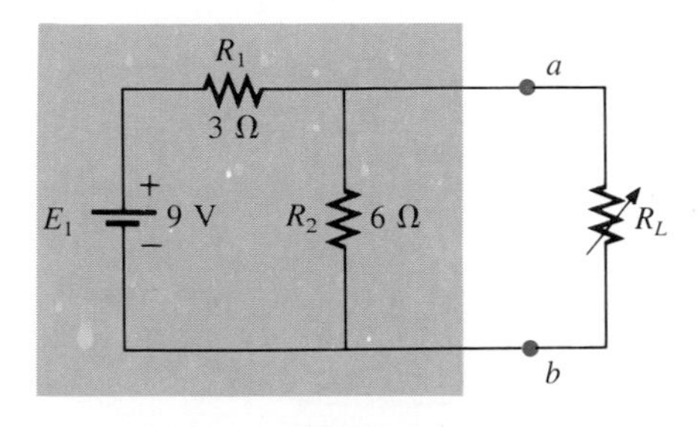

FIG. 9.26

***Solution:***

*Steps 1 and 2* produce the network of Fig. 9.27. Note that the load resistor $R_L$ has been removed and the two "holding" terminals defined as $a$ and $b$.

*Step 3:* Replacing the voltage source $E_1$ by a short-circuit equivalent yields the network of Fig. 9.28(a), where

$$R_{Th} = R_1 \parallel R_2 = \frac{(3\ \Omega)(6\ \Omega)}{3\ \Omega + 6\ \Omega} = \mathbf{2\ \Omega}$$

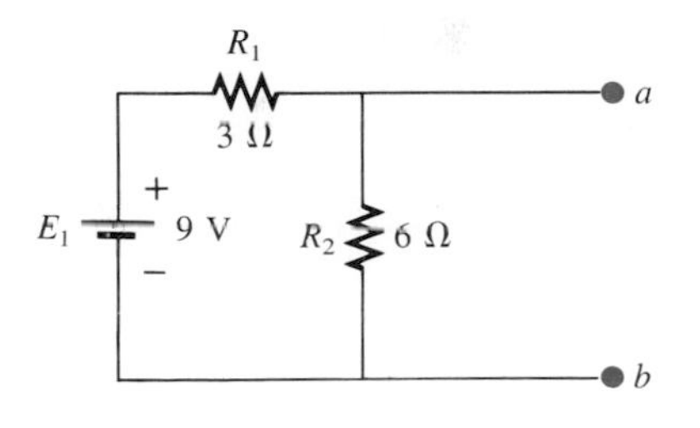


FIG. 9.27

The importance of the two marked terminals now begins to surface. They are the two terminals across which the Thevenin resistance is measured. It is no longer the total resistance as seen by the source as determined in the majority of problems of Chapter 7. If some difficulty develops when determining $R_{Th}$ with regard to whether the resistive elements are in series or parallel, consider recalling that the ohmmeter sends out a trickle current into a resistive combination and senses the level of the resulting voltage to establish the measured resistance level.

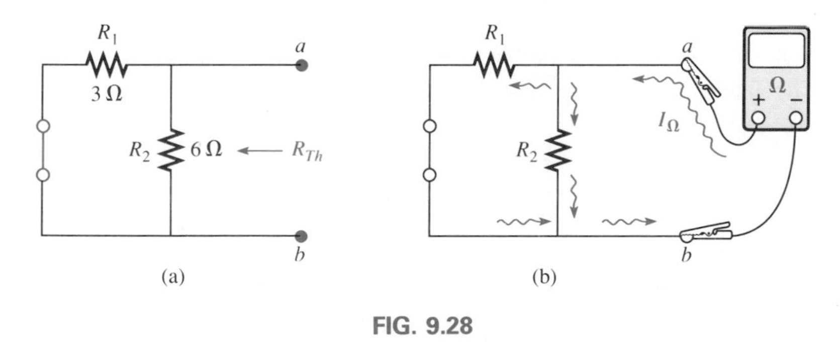


FIG. 9.28

In Fig. 9.28(b), the trickle current of the ohmmeter approaches the network through terminal $a$, and when it reaches the junction of $R_1$ and $R_2$, it splits as shown. The fact that the trickle current splits and then recombines at the lower node reveals that the resistors are in parallel as far as the ohmmeter reading is concerned. In essence, the path of the sensing current of the ohmmeter has revealed the manner in which the resistors are connected to the two terminals of interest and how the Thevenin resistance should be determined. Keep the above in mind as you work through the various examples of this section.

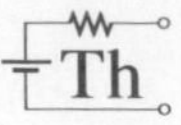


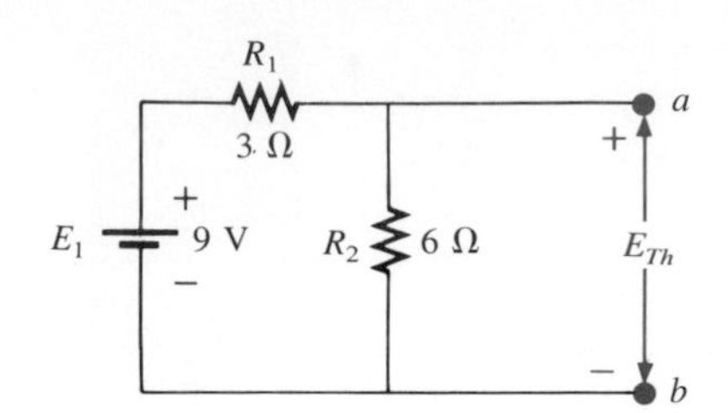


FIG. 9.29

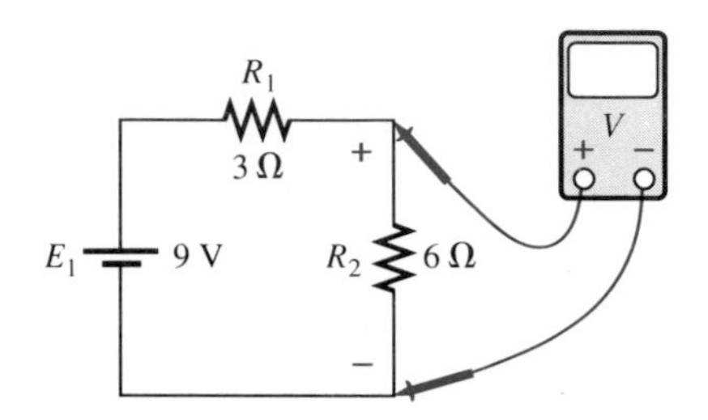

FIG. 9.30

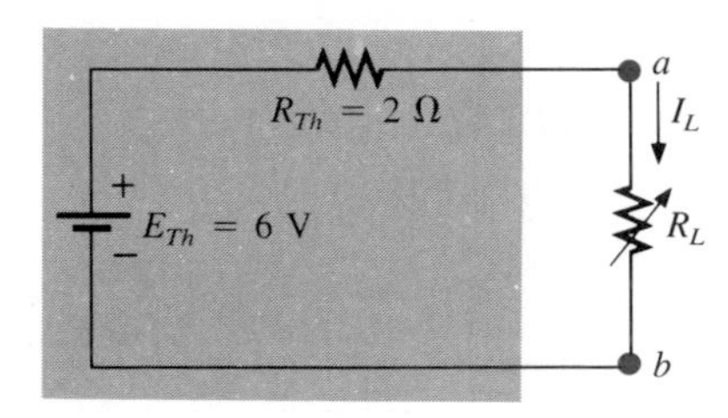


FIG. 9.31

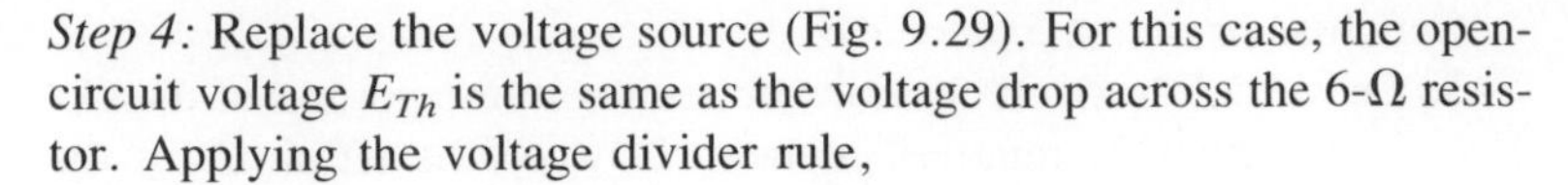

*Step 4:* Replace the voltage source (Fig. 9.29). For this case, the open-circuit voltage $E_{Th}$ is the same as the voltage drop across the 6-Ω resistor. Applying the voltage divider rule,

$$E_{Th} = \frac{R_2E_1}{R_2 + R_1} = \frac{(6\ \Omega)(9\ \text{V})}{6\ \Omega + 3\ \Omega} = \frac{54\ \text{V}}{9} = \mathbf{6\ V}$$

It is particularly important to recognize that $E_{Th}$ is the open-circuit potential between points $a$ and $b$. Remember that an open circuit can have any voltage across it but the current must be zero. In fact, the current through any element in series with the open circuit must be zero also. The use of a voltmeter to measure $E_{Th}$ appears in Fig. 9.30. Note that it is placed directly across the resistor $R_2$, since $E_{Th}$ and $V_{R_2}$ are in parallel.

*Step 5* (Fig. 9.31):

$$I_L = \frac{E_{Th}}{R_{Th} + R_L}$$

$$R_L = 2\ \Omega: \quad I_L = \frac{6\ \text{V}}{2\ \Omega + 2\ \Omega} = \mathbf{1.5\ A}$$

$$R_L = 10\ \Omega: \quad I_L = \frac{6\ \text{V}}{2\ \Omega + 10\ \Omega} = \mathbf{0.5\ A}$$

$$R_L = 100\ \Omega: \quad I_L = \frac{6\ \text{V}}{2\ \Omega + 100\ \Omega} = \mathbf{0.059\ A}$$

If Thevenin's theorem were unavailable, each change in $R_L$ would require that the entire network of Fig. 9.26 be reexamined to find the new value of $R_L$.

**EXAMPLE 9.7.** Find the Thevenin equivalent circuit for the network in the shaded area of the network of Fig. 9.32.

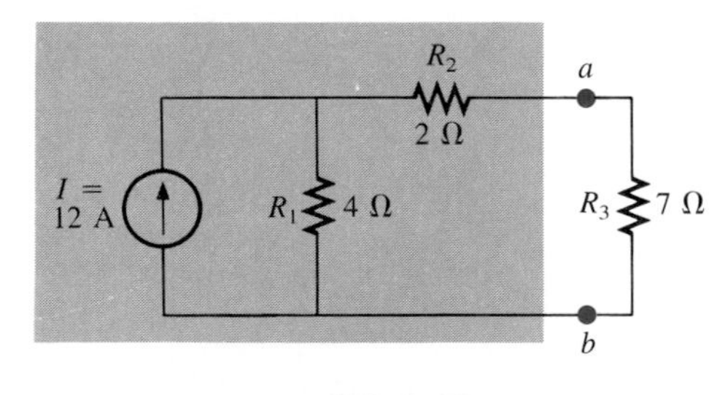


FIG. 9.32

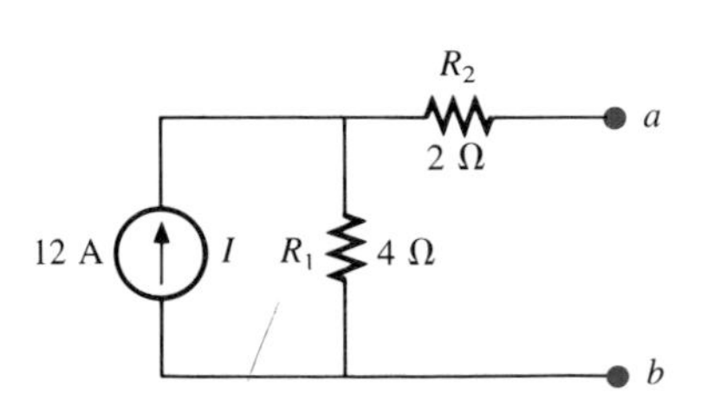


FIG. 9.33

***Solution:***

*Steps 1 and 2* are shown in Fig. 9.33.

*Step 3* is shown in Fig. 9.34. The current source has been replaced by an open-circuit equivalent and the resistance determined between terminals $a$ and $b$.

In this case an ohmmeter connected between terminals $a$ and $b$ would send out a sensing current that would flow directly through $R_1$ and $R_2$. The result is that $R_1$ and $R_2$ are in series and the Thevenin resistance is the sum of the two.

$$R_{Th} = R_1 + R_2 = 4\ \Omega + 2\ \Omega = \mathbf{6\ \Omega}$$

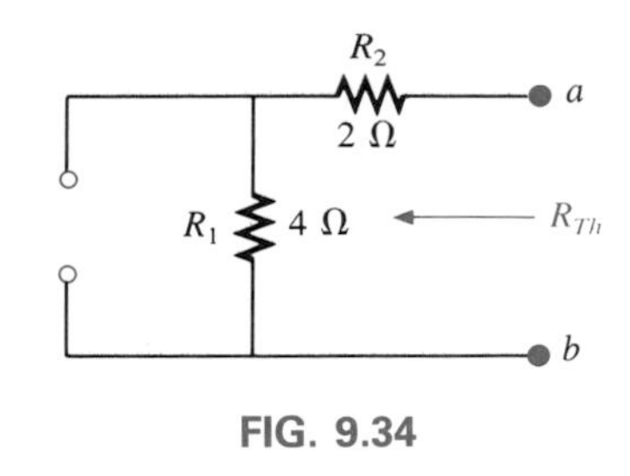


FIG. 9.34

*Step 4* (Fig. 9.35): In this case, since an open circuit exists between the two marked terminals, the current is zero between these terminals and through the 2-Ω resistor. The voltage drop across $R_2$ is therefore

$$V_2 = I_2R_2 = (0)R_2 = 0\ \text{V}$$

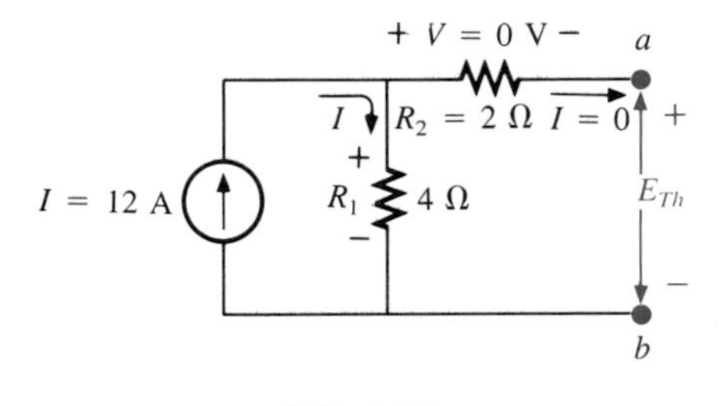


FIG. 9.35

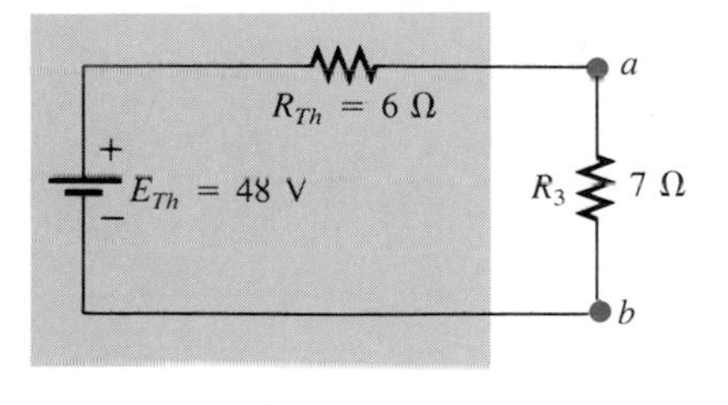

FIG. 9.36

and

$$E_{Th} = V_1 = I_1R_1 = IR_1 = (12\ \text{A})(4\ \Omega) = \mathbf{48\ V}$$

*Step 5* is shown in Fig. 9.36.

---

**EXAMPLE 9.8.** Find the Thevenin equivalent circuit for the network in the shaded area of the network of Fig. 9.37. Note in this example that

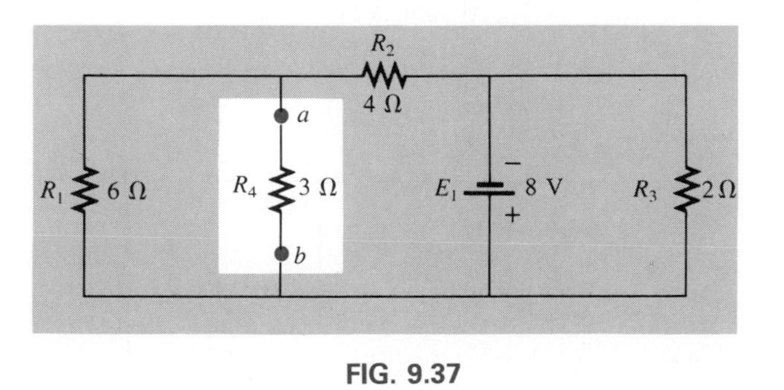


FIG. 9.37

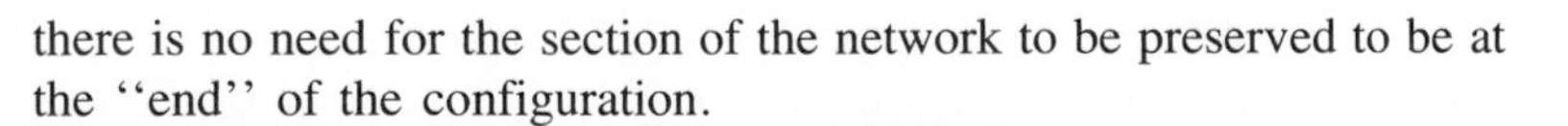

there is no need for the section of the network to be preserved to be at the ''end'' of the configuration.

***Solution:***

*Steps 1 and 2:* See Fig. 9.38.

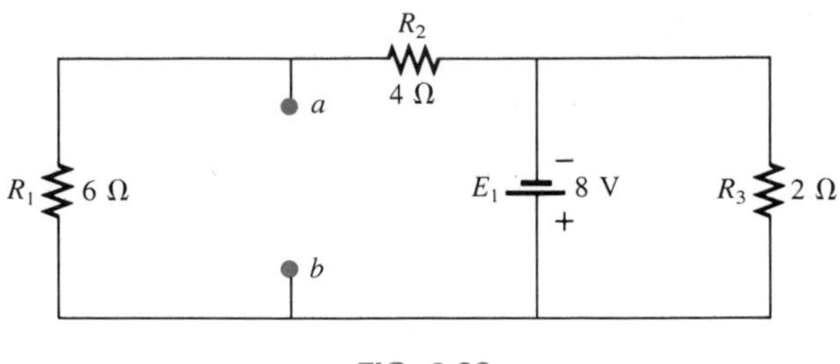

FIG. 9.38

*Step 3:* See Fig. 9.39. Steps 1 and 2 are relatively easy to apply, but now we must be careful to "hold" onto the terminals $a$ and $b$ as the Thevenin resistance and voltage are determined. In Fig. 9.39, all the

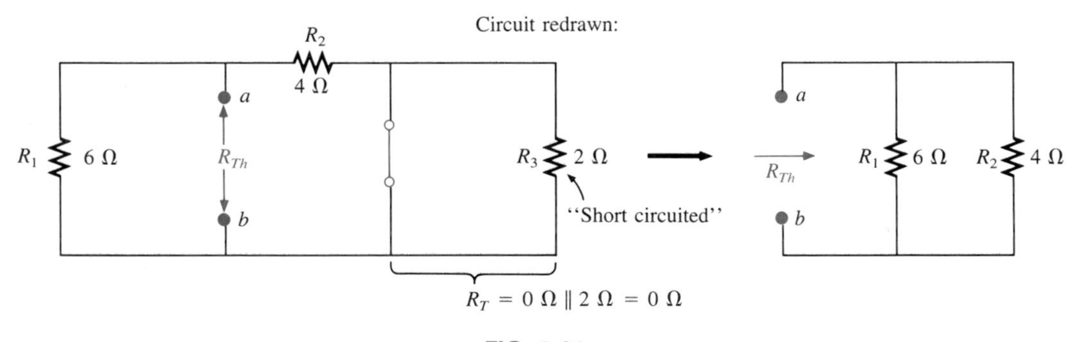

FIG. 9.39

remaining elements turn out to be in parallel, and the network can be redrawn as shown.

$$R_{Th} = R_1 \parallel R_2 = \frac{(6\ \Omega)(4\ \Omega)}{6\ \Omega + 4\ \Omega} = \frac{24\ \Omega}{10} = \mathbf{2.4\ \Omega}$$

*Step 4:* See Fig. 9.40. In this case, the network can be redrawn as

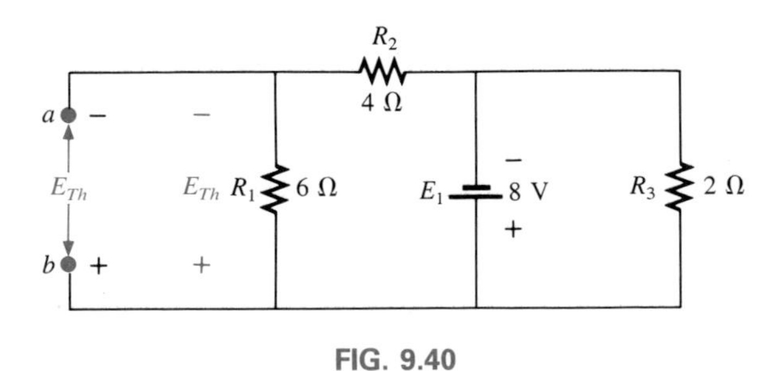


FIG. 9.40

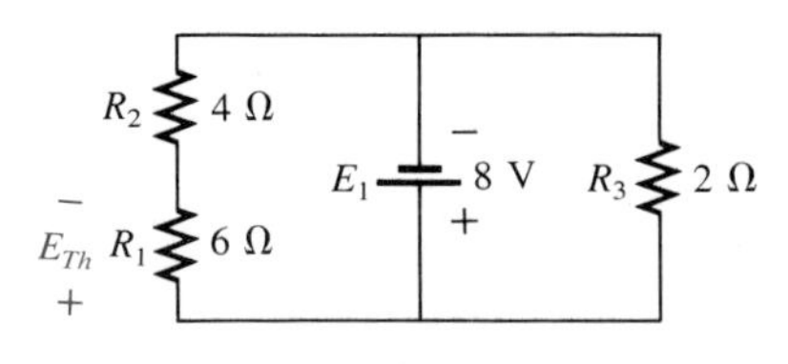


FIG. 9.41

shown in Fig. 9.41, and since the voltage is the same across parallel elements, the voltage across the series resistors $R_1$ and $R_2$ is $E_1$, or 8 V.

Applying the voltage divider rule,

$$E_{Th} = \frac{R_1E_1}{R_1 + R_2} = \frac{(6\ \Omega)(8\ \text{V})}{6\ \Omega + 4\ \Omega} = \frac{48\ \text{V}}{10} = \mathbf{4.8\ V}$$

*Step 5:* See Fig. 9.42.

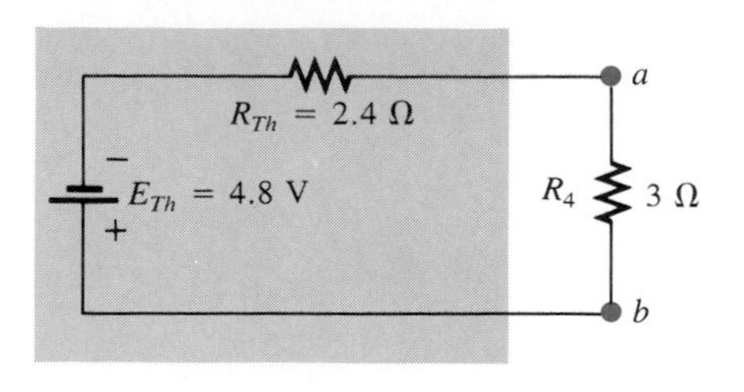

FIG. 9.42

The importance of marking the terminals should be obvious from Example 9.8. Note that there is no requirement that the Thevenin voltage have the same polarity as the equivalent circuit originally introduced.

**EXAMPLE 9.9.** Find the Thevenin equivalent circuit for the network in the shaded area of the bridge network of Fig. 9.43.

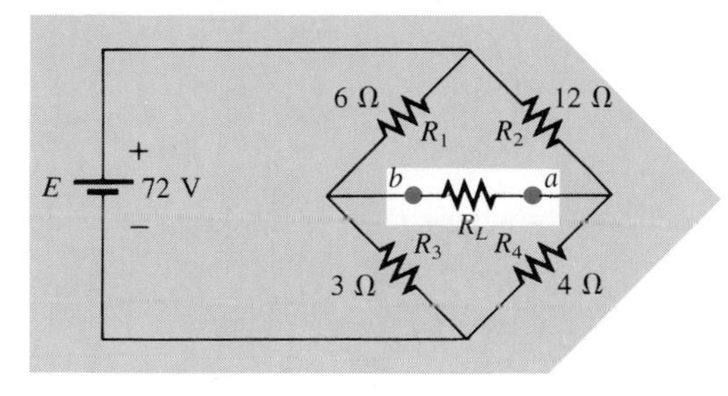

FIG. 9.43

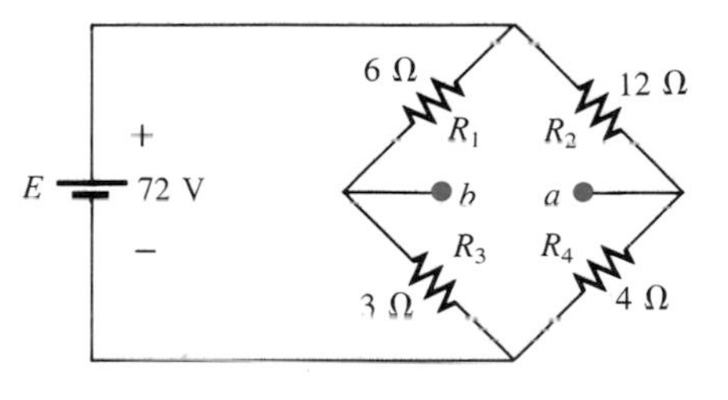

FIG. 9.44

***Solution:***

*Steps 1 and 2* are shown in Fig. 9.44.

*Step 3:* See Fig. 9.45. In this case, the short-circuit replacement of the voltage source $E$ provides a direct connection between $c$ and $c'$ of Fig. 9.45(a), permitting a ''folding'' of the network around the horizontal line of $a$-$b$ to produce the configuration of Fig. 9.45(b).

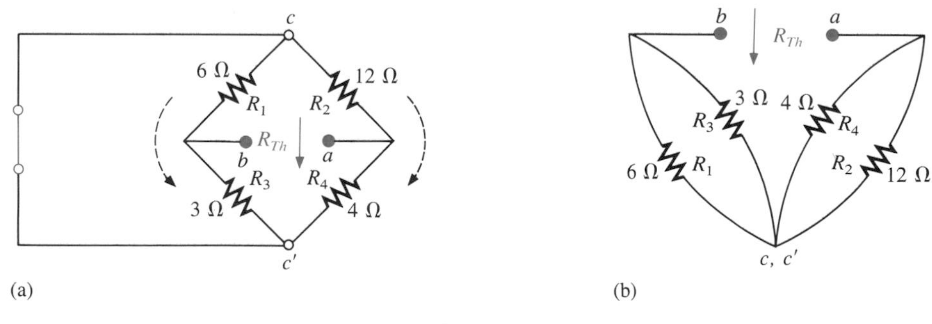

FIG. 9.45

$$\begin{aligned} R_{Th} = R_{a\text{-}b} &= R_1 \parallel R_3 + R_2 \parallel R_4 \\ &= 6\ \Omega \parallel 3\ \Omega + 4\ \Omega \parallel 12\ \Omega \\ &= 2\ \Omega + 3\ \Omega = \mathbf{5\ \Omega} \end{aligned}$$

*Step 4:* The circuit is redrawn in Fig. 9.46. The absence of a direct connection between $a$ and $b$ results in a network with three parallel

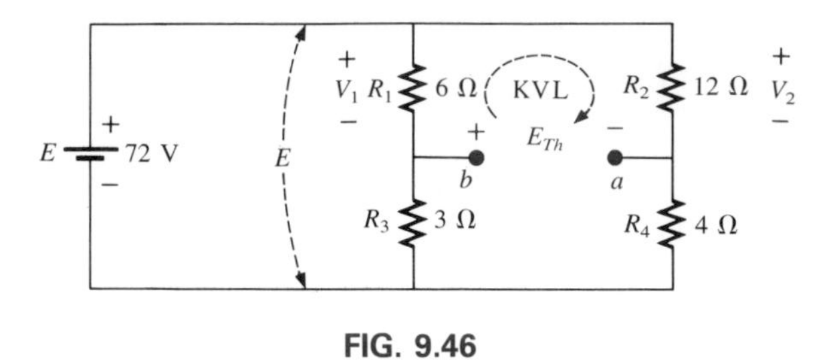


FIG. 9.46

branches. The voltages $V_1$ and $V_2$ can therefore be determined using the voltage divider rule:

$$V_1 = \frac{R_1 E}{R_1 + R_3} = \frac{(6\ \Omega)(72\ \text{V})}{6\ \Omega + 3\ \Omega} = \frac{432\ \text{V}}{9} = 48\ \text{V}$$

$$V_2 = \frac{R_2 E}{R_2 + R_4} = \frac{(12\ \Omega)(72\ \text{V})}{12\ \Omega + 4\ \Omega} = \frac{864\ \text{V}}{16} = 54\ \text{V}$$

Assuming the polarity shown for $E_{Th}$ and applying Kirchhoff's voltage law to the top loop in the clockwise direction will result in

$$\Sigma_{\circlearrowright} V = +E_{Th} + V_1 - V_2 = 0$$

and

$$E_{Th} = V_2 - V_1 = 54\ \text{V} - 48\ \text{V} = \mathbf{6\ V}$$

*Step 5* is shown in Fig. 9.47.

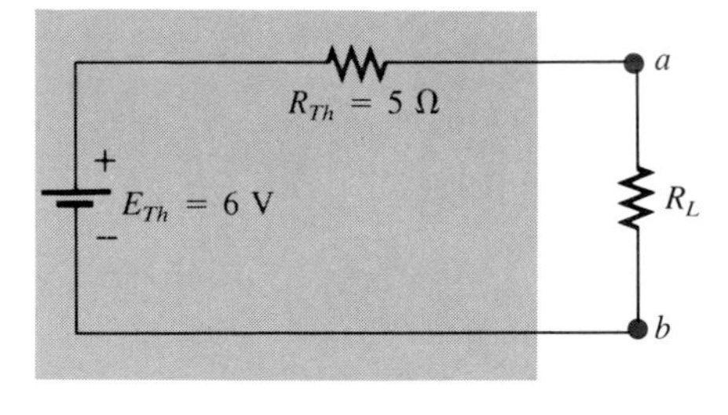


FIG. 9.47

---

Thevenin's theorem is not restricted to a single passive element, as shown in the preceding examples, but can be applied across sources, whole branches, portions of networks, or any circuit configuration, as shown in the following example. It is also possible that one of the methods previously described, such as mesh analysis or superposition, may have to be used to find the Thevenin equivalent circuit.

**EXAMPLE 9.10.** (Two sources) Find the Thevenin circuit for the network within the shaded area of Fig. 9.48.

***Solution:*** The network is redrawn and steps 1 and 2 are applied as shown in Fig. 9.49.

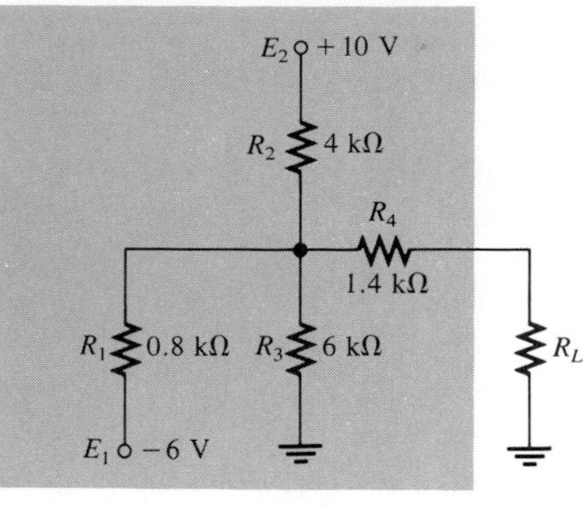


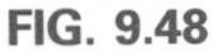
FIG. 9.48

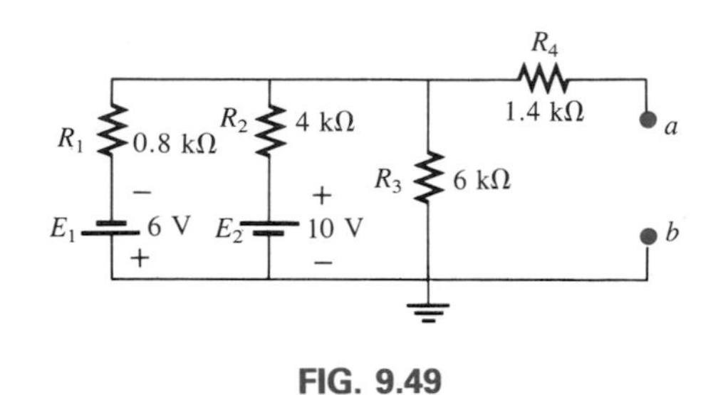


FIG. 9.49

*Step 3:* See Fig. 9.50.

$$\begin{aligned} R_{Th} &= R_4 + R_1 \parallel R_2 \parallel R_3 \\ &= 1.4\ \text{k}\Omega + 0.8\ \text{k}\Omega \parallel 4\ \text{k}\Omega \parallel 6\ \text{k}\Omega \\ &= 1.4\ \text{k}\Omega + 0.8\ \text{k}\Omega \parallel 2.4\ \text{k}\Omega \\ &= 1.4\ \text{k}\Omega + 0.6\ \text{k}\Omega \\ &= \mathbf{2\ k\Omega} \end{aligned}$$

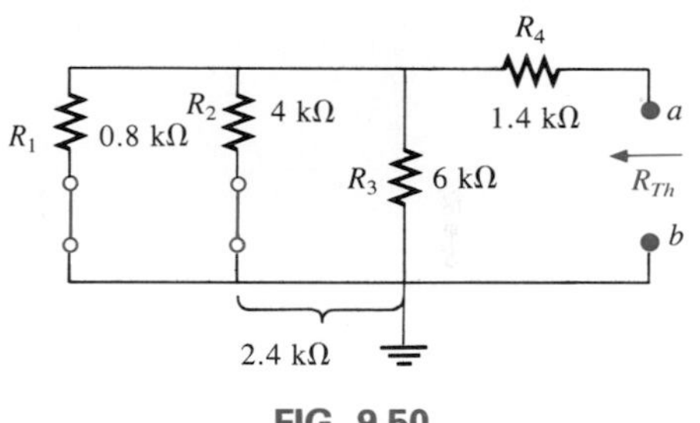

FIG. 9.50

*Step 4:* Applying superposition, we will consider the effects of the voltage source $E_1$ first. Note Fig. 9.51. The open circuit requires that $V_4 = I_4R_4 = (0)R_4 = 0$ V, and

$$E'_{Th} = V_3$$

$$R'_T = R_2 \parallel R_3 = 4\ \text{k}\Omega \parallel 6\ \text{k}\Omega = 2.4\ \text{k}\Omega$$

Applying the voltage divider rule,

$$V_3 = \frac{R'_T E_1}{R'_T + R_1} = \frac{(2.4\ \text{k}\Omega)(6\ \text{V})}{2.4\ \text{k}\Omega + 0.8\ \text{k}\Omega} = \frac{14.4\ \text{V}}{3.2} = 4.5\ \text{V}$$

and

$$E'_{Th} = V_3 = 4.5\ \text{V}$$

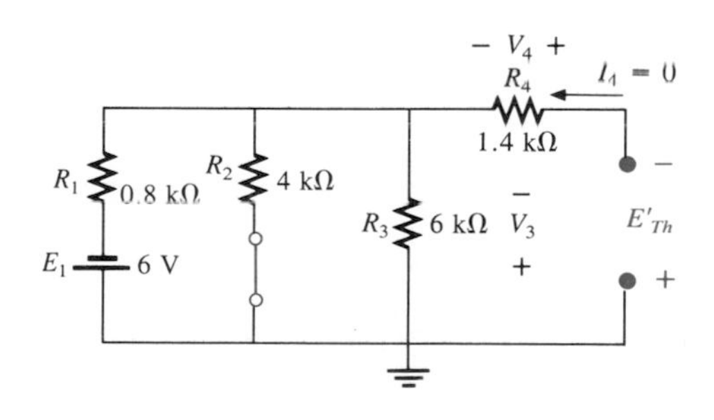


FIG. 9.51

For the source $E_2$, the network of Fig. 9.52 will result. Again, $V_4 = I_4R_4 = (0)R_4 = 0$ V, and

$$E''_{Th} = V_3$$

$$R'_T = R_1 \parallel R_3 = 0.8\ \text{k}\Omega \parallel 6\ \text{k}\Omega = 0.706\ \text{k}\Omega$$

and

$$V_3 = \frac{R'_T E_2}{R'_T + R_2} = \frac{(0.706\ \text{k}\Omega)(10\ \text{V})}{0.706\ \text{k}\Omega + 4\ \text{k}\Omega} = \frac{7.06\ \text{V}}{4.706} = 1.5\ \text{V}$$

and

$$E''_{Th} = V_3 = 1.5\ \text{V}$$

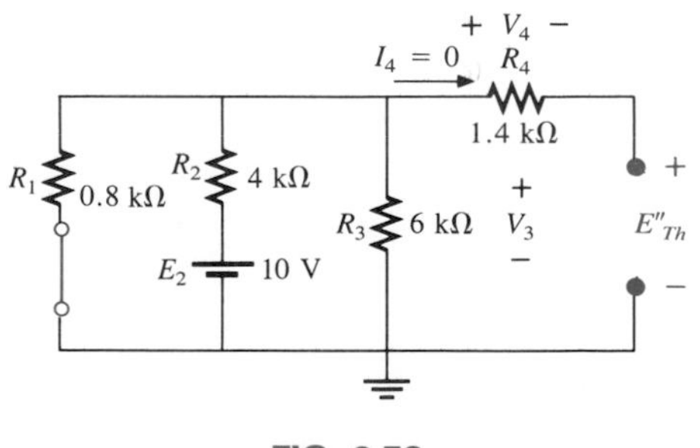


FIG. 9.52

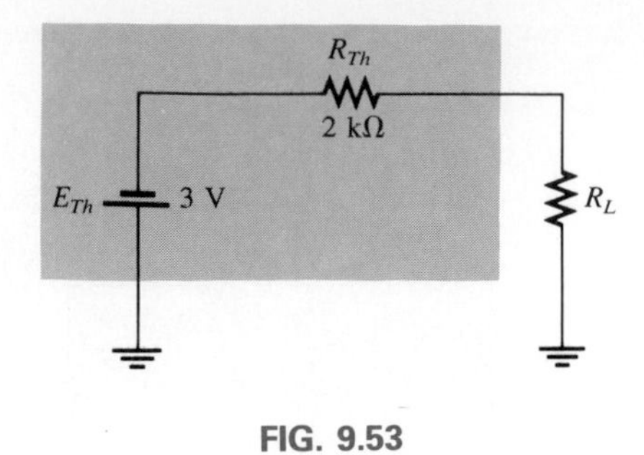


FIG. 9.53

Since $E'_{Th}$ and $E''_{Th}$ have opposite polarities,

$$
\begin{aligned}
E_{Th} &= E'_{Th} - E''_{Th} \\
&= 4.5 \text{ V} - 1.5 \text{ V} \\
&= \mathbf{3\ V} \qquad \text{(polarity of } E'_{Th}\text{)}
\end{aligned}
$$

*Step 5:* See Fig. 9.53.

---

For any physical network, the value of $E_{Th}$ can be determined experimentally by measuring the open-circuit voltage across the load terminals, as shown in Fig. 9.54; $E_{Th} = V_{ab}$. The value of $R_{Th}$ can then be

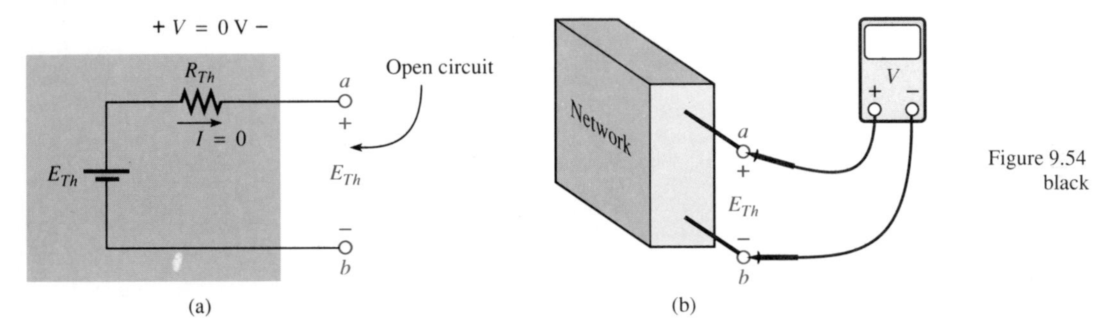


FIG. 9.54
*Determining $E_{Th}$ experimentally.*

determined by completing the network with an $R_L$ such as the potentiometer of Fig. 9.55(b). $R_L$ can then be varied until the voltage appearing

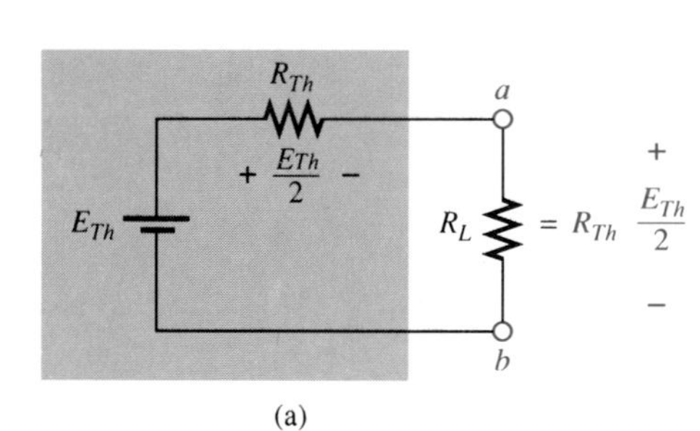


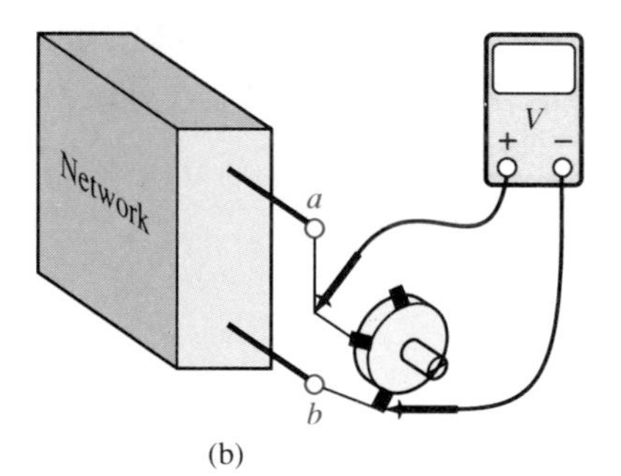


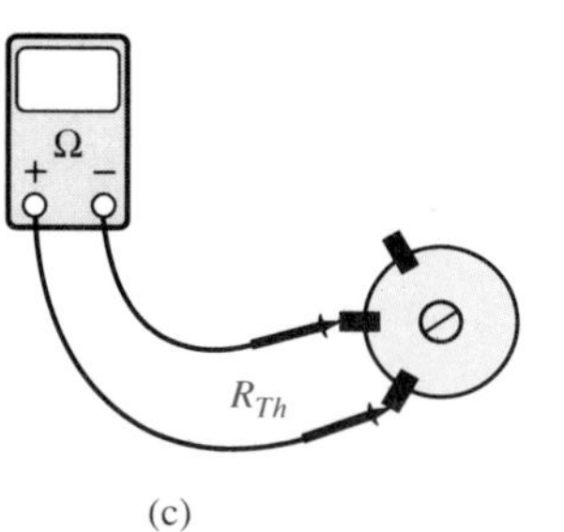


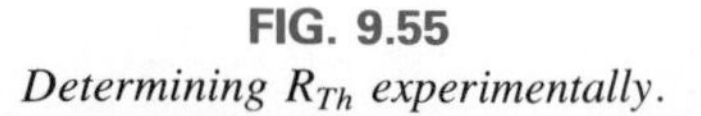

FIG. 9.55
*Determining $R_{Th}$ experimentally.*

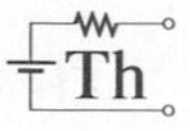



across the load is one-half the open-circuit value, or $V_L = E_{Th}/2$. For the series circuit of Fig. 9.55(a), when the load voltage is reduced to one-half the open-circuit level, the voltage across $R_{Th}$ and $R_L$ must be the same. If we read the value of $R_L$ [as shown in Fig. 9.55(c)] that resulted in the preceding calculations, we will also have the value of $R_{Th}$, since $R_L = R_{Th}$ if $V_L$ equals the voltage across $R_{Th}$.

## 9.4 NORTON'S THEOREM

It was demonstrated in Section 8.3 that every voltage source with a series internal resistance has a current source equivalent. The current source equivalent of the Thevenin network (which, you will note, satisfies the above conditions), as shown in Fig. 9.56, can be determined by Norton's theorem. It can also be found through the conversions of Section 8.3.

The theorem states the following:

***Any two-terminal linear bilateral dc network can be replaced by an equivalent circuit consisting of a current source and a parallel resistor as shown in Fig. 9.56.***

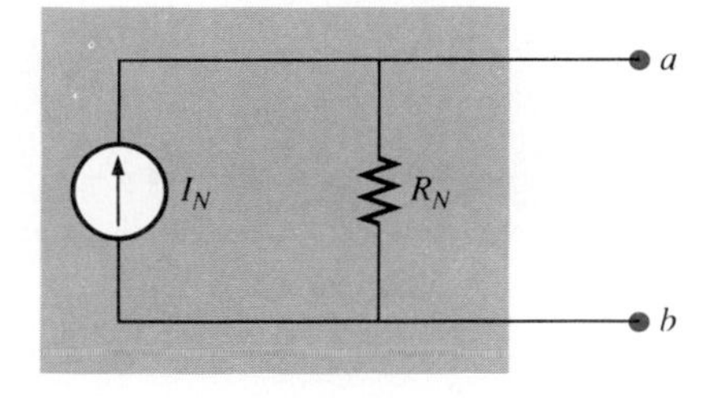


**FIG. 9.56**
*Norton equivalent circuit.*

The discussion of Thevenin's theorem with respect to the equivalent circuit can also be applied to the Norton equivalent circuit.

The steps leading to the proper values of $I_N$ and $R_N$ are now listed.

**Preliminary:**

1. Remove that portion of the network across which the Norton equivalent circuit is found.
2. Mark the terminals of the remaining two-terminal network.

***$R_N$:***

3. Calculate $R_N$ by first setting all sources to zero (voltage sources are replaced by short circuits and current sources by open circuits) and then finding the resultant resistance between the two marked terminals. (If the internal resistance of the voltage and/or current sources is included in the original network, it must remain when the sources are set to zero.) Since $R_N = R_{Th}$, the procedure and value obtained using the approach described for Thevenin's theorem will determine the proper value of $R_N$.

***$I_N$:***

4. Calculate $I_N$ by first returning all sources to their original position and then finding the *short-circuit* current between the marked terminals. It is the same current that would be measured by an ammeter placed between the marked terminals.

**Conclusion:**

5. Draw the Norton equivalent circuit with the portion of the circuit previously removed replaced between the terminals of the equivalent circuit.

The Norton and Thevenin equivalent circuits can also be found from each other by using the source transformation discussed earlier in this chapter and reproduced in Fig. 9.57.

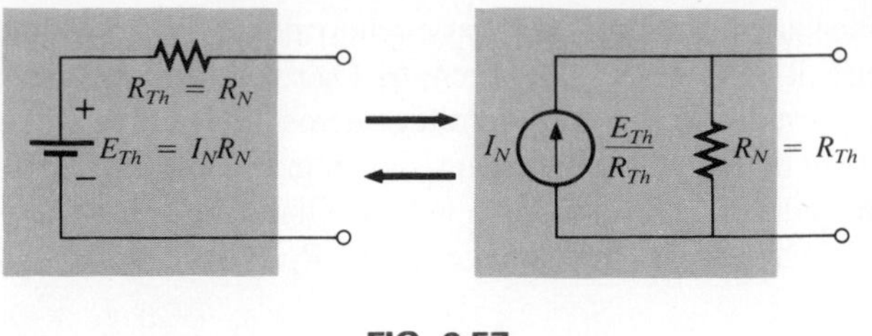

FIG. 9.57

**EXAMPLE 9.11.** Find the Norton equivalent circuit for the network in the shaded area of Fig. 9.58.

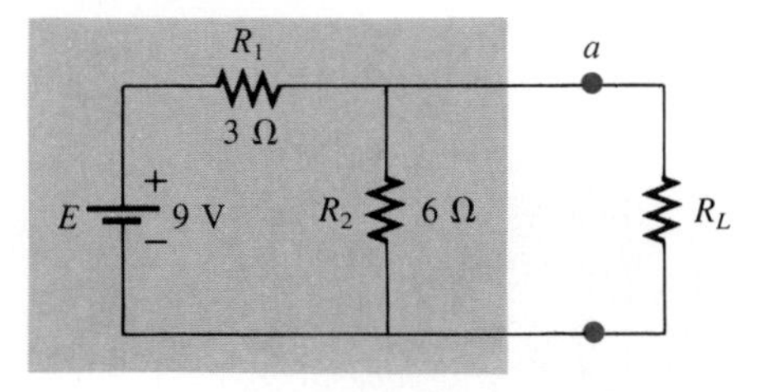


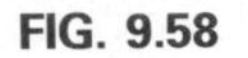
FIG. 9.58

***Solution:***

*Steps 1 and 2* are shown in Fig. 9.59.

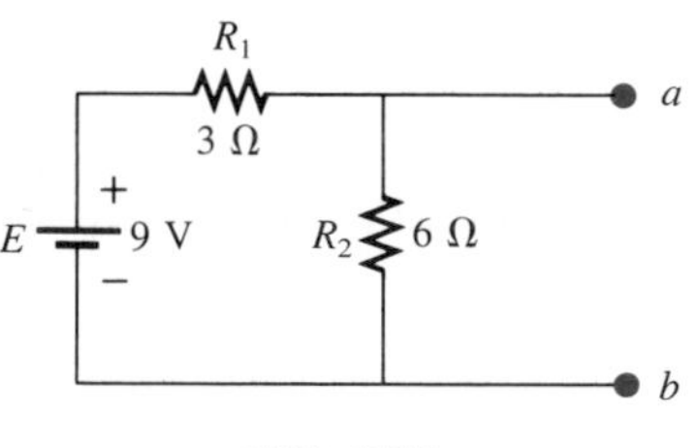

FIG. 9.59

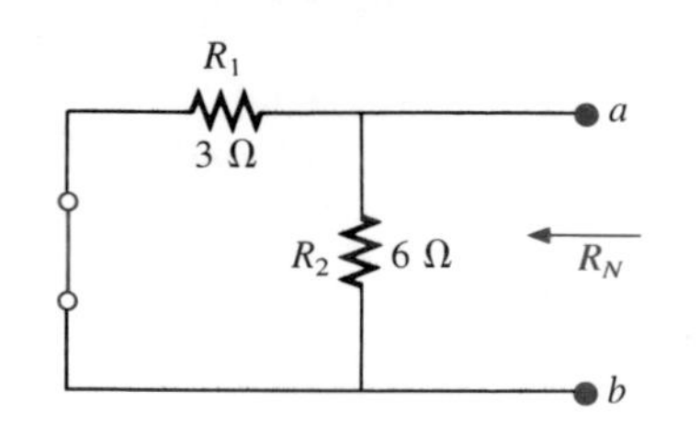

FIG. 9.60

*Step 3* is shown in Fig. 9.60, and

$$R_N = R_1 \parallel R_2 = 3\ \Omega \parallel 6\ \Omega = \frac{(3\ \Omega)(6\ \Omega)}{3\ \Omega + 6\ \Omega} = \frac{18\ \Omega}{9} = \mathbf{2\ \Omega}$$

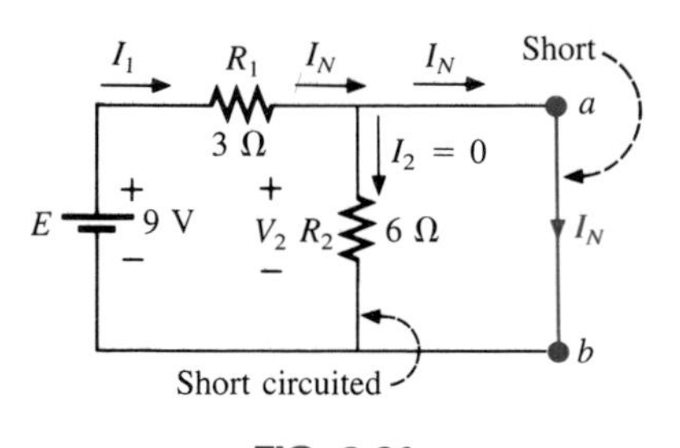


FIG. 9.61

*Step 4* is shown in Fig. 9.61, clearly indicating that the short-circuit connection between terminals $a$ and $b$ is in parallel with $R_2$ and eliminates its effect. $I_N$ is therefore the same as through $R_1$, and the full battery voltage appears across $R_1$ since

$$V_2 = I_2 R_2 = (0)6\ \Omega = 0\ \text{V}$$

Therefore,

$$I_N = \frac{E}{R_1} = \frac{9\ \text{V}}{3\ \Omega} = \mathbf{3\ A}$$

*Step 5:* See Fig. 9.62.

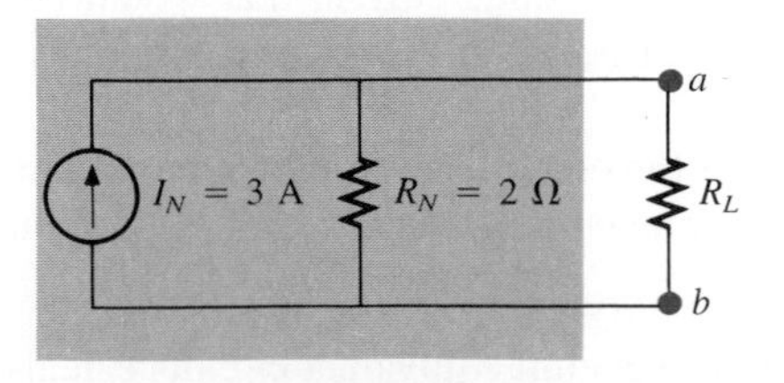

FIG. 9.62

This circuit is the same as the first one considered in the development of Thevenin's theorem. A simple conversion indicates that the Thevenin circuits are, in fact, the same (Fig. 9.63).

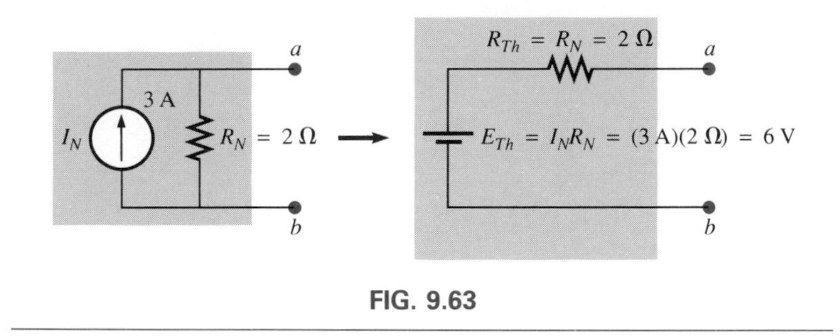


FIG. 9.63

**EXAMPLE 9.12.** Find the Norton equivalent circuit for the network external to the 9-Ω resistor in Fig. 9.64.

***Solution:***

*Steps 1 and 2:* See Fig. 9.65.

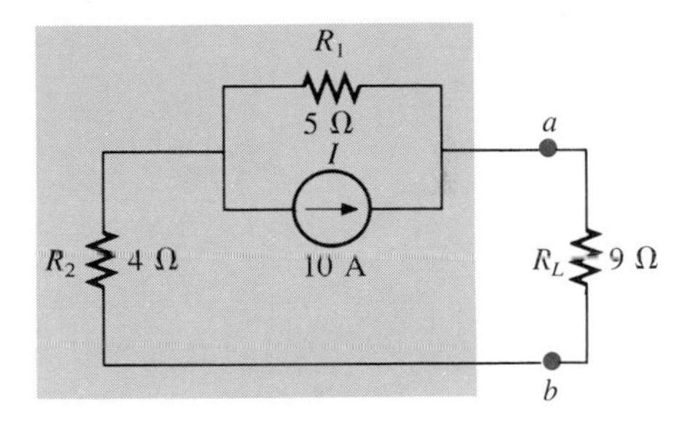

FIG. 9.64

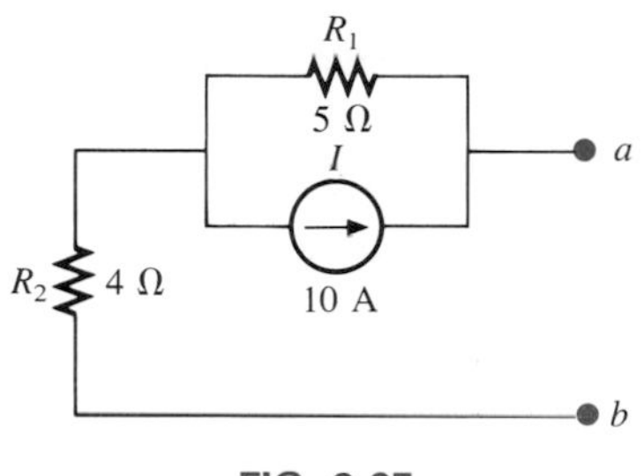

FIG. 9.65

*Step 3:* See Fig. 9.66, and

$$R_N = R_1 + R_2 = 5\ \Omega + 4\ \Omega = \mathbf{9\ \Omega}$$

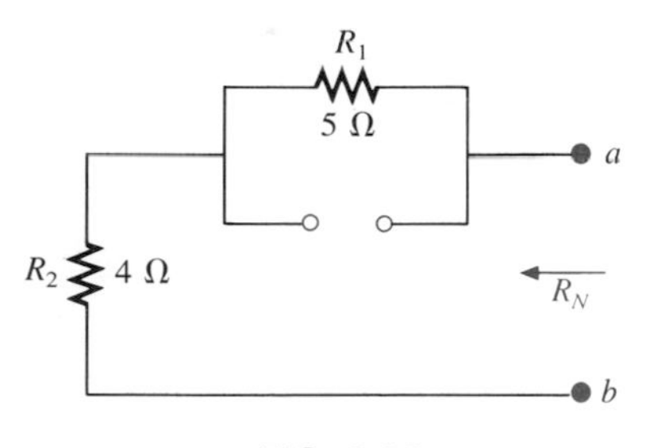


FIG. 9.66

*Step 4:* As shown in Fig. 9.67, the Norton current is the same as the current through the 4-Ω resistor. Applying the current divider rule,

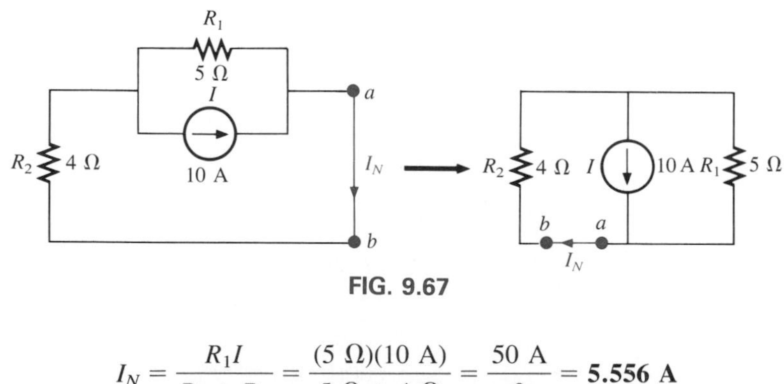


FIG. 9.67

$$I_N = \frac{R_1 I}{R_1 + R_2} = \frac{(5\ \Omega)(10\text{ A})}{5\ \Omega + 4\ \Omega} = \frac{50\text{ A}}{9} = \mathbf{5.556\ A}$$

*Step 5:* See Fig. 9.68.

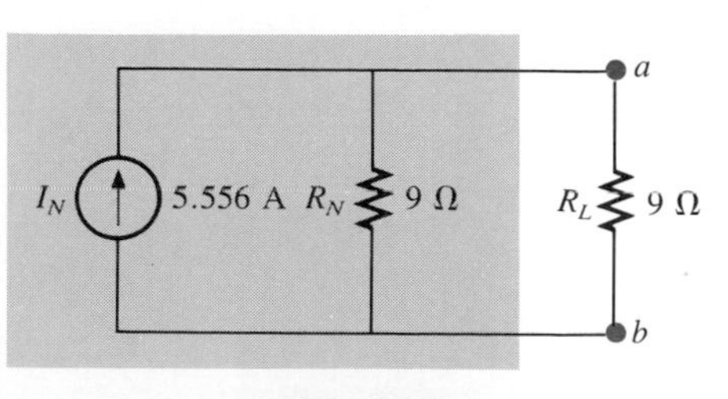


FIG. 9.68

**EXAMPLE 9.13.** (Two sources) Find the Norton equivalent circuit for the portion of the network to the left of *a-b* in Fig. 9.69.

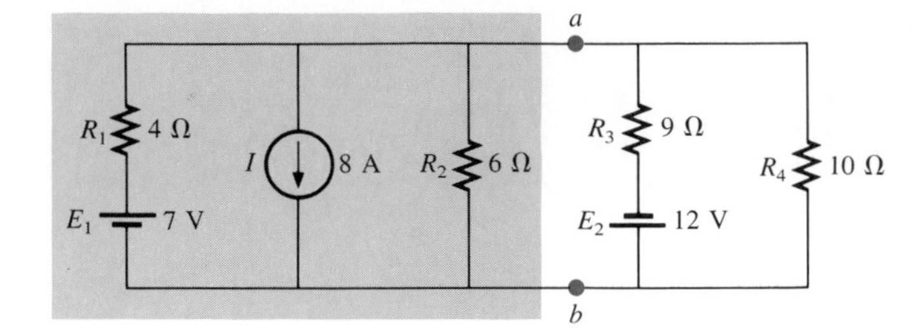

FIG. 9.69

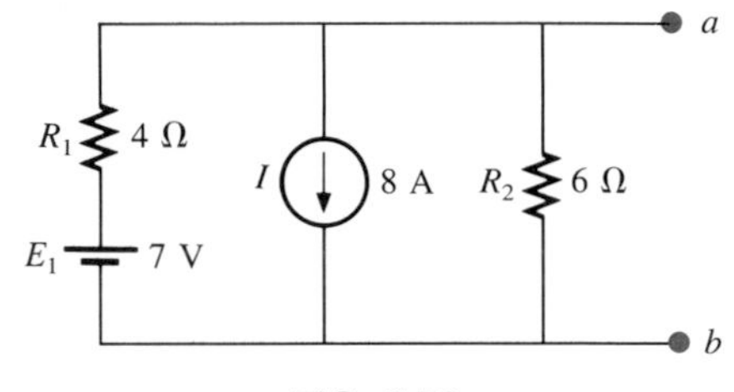

FIG. 9.70

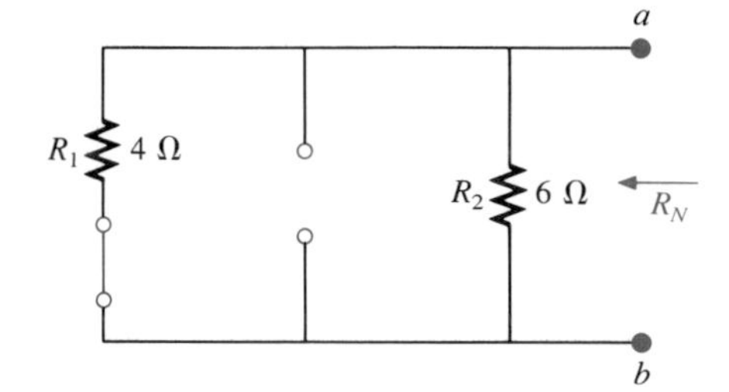

FIG. 9.71

***Solution:***

*Steps 1 and 2:* See Fig. 9.70.

*Step 3* is shown in Fig. 9.71, and

$$R_N = R_1 \parallel R_2 = 4\ \Omega \parallel 6\ \Omega = \frac{(4\ \Omega)(6\ \Omega)}{4\ \Omega + 6\ \Omega} = \frac{24\ \Omega}{10} = \mathbf{2.4\ \Omega}$$

*Step 4:* (Using superposition) For the 7-V battery (Fig. 9.72),

$$I'_N = \frac{E_1}{R_1} = \frac{7\ \text{V}}{4\ \Omega} = 1.75\ \text{A}$$

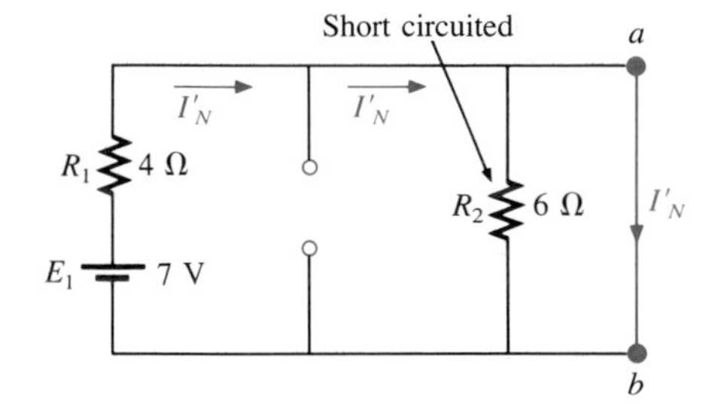


FIG. 9.72

For the 8-A source (Fig. 9.73), we find that both $R_1$ and $R_2$ have been "short circuited" by the direct connection between *a* and *b*, and

$$I''_N = I = 8\ \text{A}$$

The result is

$$I_N = I''_N - I'_N = 8\ \text{A} - 1.75\ \text{A} = \mathbf{6.25\ A}$$

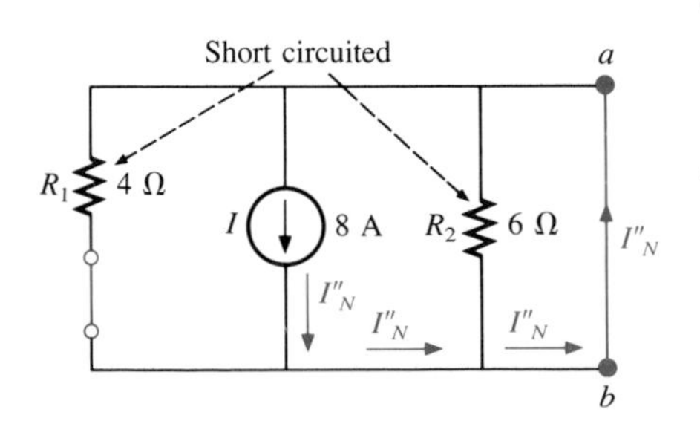


FIG. 9.73

*Step 5:* See Fig. 9.74.

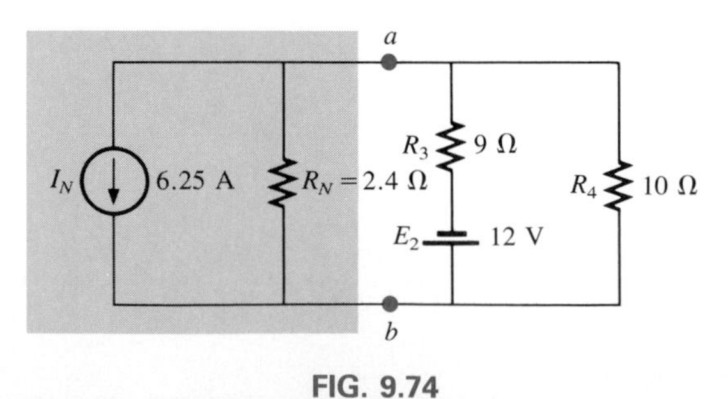


FIG. 9.74

## 9.5 MAXIMUM POWER TRANSFER THEOREM

The maximum power transfer theorem states the following:

***A load will receive maximum power from a linear bilateral dc network when its total resistive value is exactly equal to the Thevenin resistance of the network as "seen" by the load.***

For the network of Fig. 9.75, maximum power will be delivered to the load when

$$\boxed{R_L = R_{Th}} \qquad \textbf{(9.2)}$$

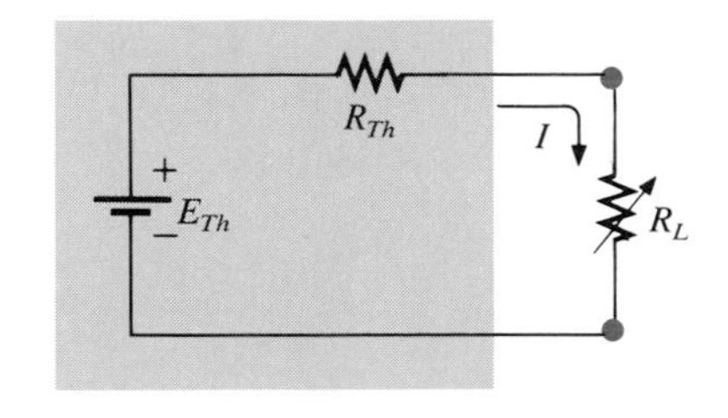


FIG. 9.75

From past discussions, we realize that a Thevenin equivalent circuit can be found across any element or group of elements in a linear bilateral dc network. Therefore, if we consider the case of the Thevenin equivalent circuit with respect to the maximum power transfer theorem, we are, in essence, considering the *total* effects of any network across a resistor $R_L$, such as in Fig. 9.75.

For the Norton equivalent circuit of Fig. 9.76, maximum power will be delivered to the load when

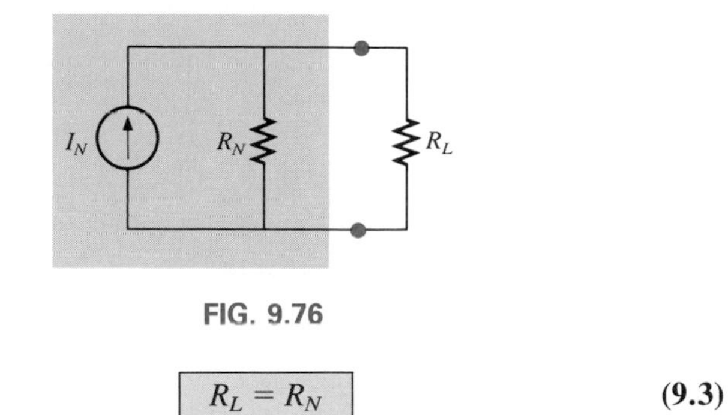


FIG. 9.76

$$\boxed{R_L = R_N} \qquad \textbf{(9.3)}$$

This result [Eq. (9.3)] will be used to its fullest advantage in the analysis of transistor networks where the most frequently applied transistor circuit model employs a current source rather than a voltage source.

For the network of Fig. 9.75,

$$I = \frac{E_{Th}}{R_{Th} + R_L}$$

and

$$P_L = I^2R_L = \left(\frac{E_{Th}}{R_{Th} + R_L}\right)^2 R_L$$

so that

$$P_L = \frac{E_{Th}^2 R_L}{(R_{Th} + R_L)^2}$$

For $E_{Th} = 4$ V and $R_{Th} = 5\ \Omega$, the powers to $R_L$ for different values of $R_L$ are tabulated in Table 9.1. A plot of these data (Fig. 9.77) clearly

TABLE 9.1

| $R_L$ (Ohms) | $P_L = \dfrac{16R_L}{(5 + R_L)^2}$ (Watts) | |
|---|---|---|
| 1 | 0.444 | |
| 2 | 0.653 | |
| 3 | 0.750 | Increase ↓ |
| 4 | 0.790 | |
| 5 | 0.800 ← | Maximum |
| 6 | 0.793 | |
| 7 | 0.778 | |
| 8 | 0.757 | Decrease ↓ |
| 9 | 0.735 | |
| 10 | 0.711 | |

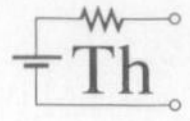


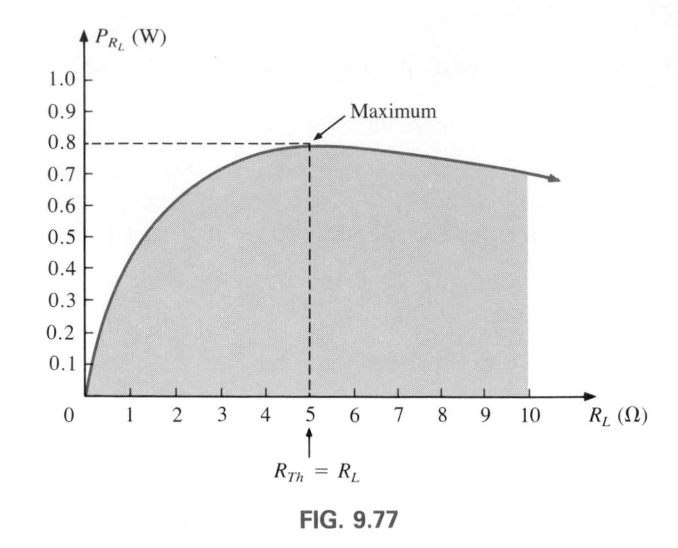


FIG. 9.77

reveals that maximum power is delivered to $R_L$ when it is exactly equal to $R_{Th}$.

Under maximum power transfer conditions, the operating efficiency is

$$\eta\% = \frac{P_o}{P_i} \times 100\% = \frac{V_L I_L}{E_{Th} I_L} \times 100\% = \frac{V_L}{E_{Th}} \times 100\%$$

with

$$R_L = R_{Th} \qquad V_L = \frac{E_{Th}}{2}$$

and

$$\eta\% = \frac{E_{Th}/2}{E_{Th}} \times 100\% = \frac{1}{2} \times 100\% = \mathbf{50\%}$$

revealing that only one-half of the power delivered by the source is getting to the load.

A relatively low efficiency of 50% can be tolerated in situations where power levels are relatively low, such as in a wide variety of electronic systems. However, when large power levels are involved, such as at generating stations, efficiencies of 50% would not be acceptable. In fact, a great deal of expense and research is dedicated to raising power-generating and transmission efficiencies a few percent. Raising an efficiency level of a 10-mega-kW power plant from 94% to 95% (a 1% increase) can save 0.1 mega-kW, or 100 million watts, of power—an enormous saving!

Consider a change in load levels from 5 Ω to 10 Ω. In Fig. 9.77, the power level has dropped only from 0.8 W to 0.711 W (an 11.125% drop), while the efficiency has increased to 66.7% (a 33.4% increase), as shown in Fig. 9.78. A balance point must be identified where the efficiency is sufficiently high without reducing the power to the load to too low a level.

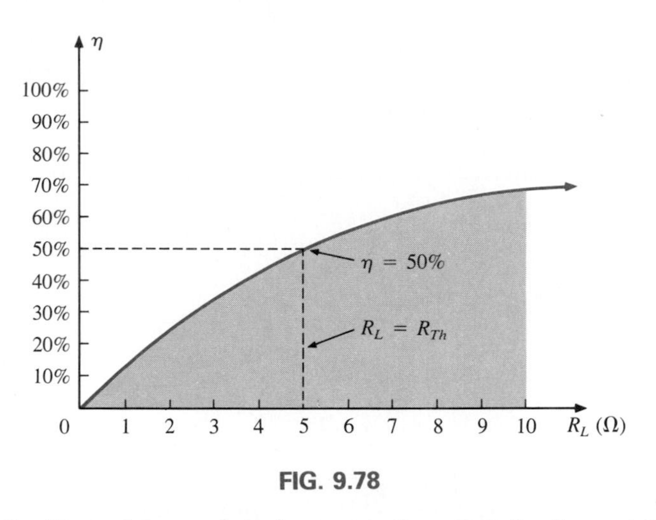


FIG. 9.78

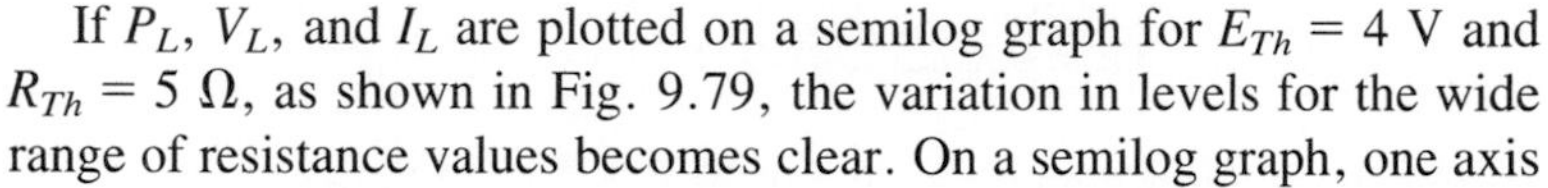

If $P_L$, $V_L$, and $I_L$ are plotted on a semilog graph for $E_{Th} = 4$ V and $R_{Th} = 5\ \Omega$, as shown in Fig. 9.79, the variation in levels for the wide range of resistance values becomes clear. On a semilog graph, one axis

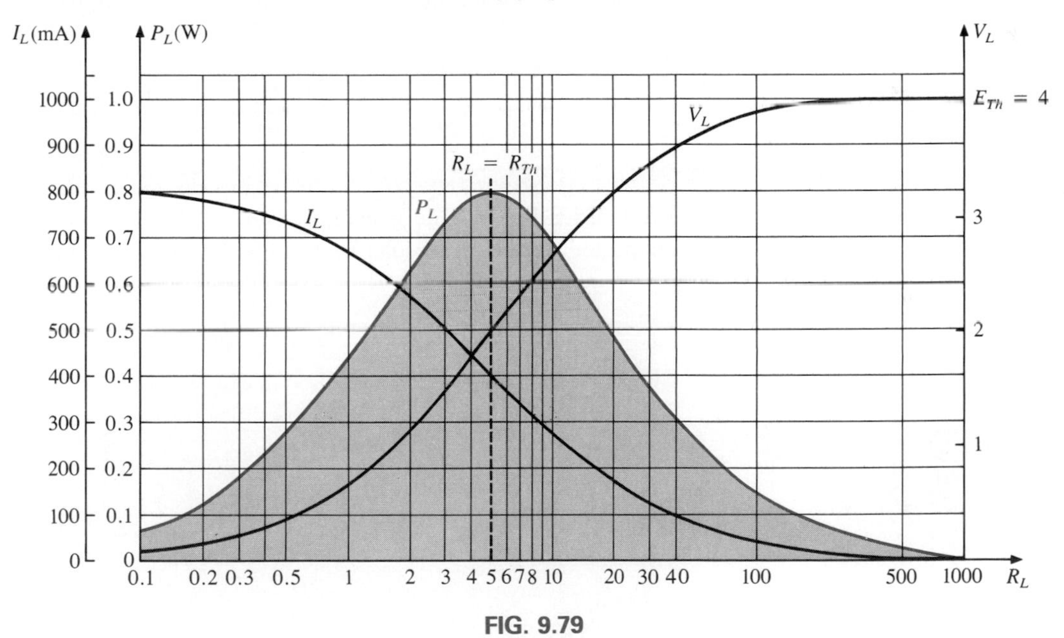


FIG. 9.79

is plotted on a log scale while the other employs a linear scale. Note that $P_L$ reaches only one maximum (at $R_L = R_{Th}$), that $V_L$ increases with increasing values of $R_L$ as determined by the voltage divider rule, and that $I_L$ drops with increasing levels of $R_L$ as controlled by Ohm's law. One obvious advantage of log scales is that they permit a wide variation in the magnitude of a parameter (such as $R_L$ in Figure 9.79). Log scales will be described in detail in Chapter 21.

Figure 9.79 reveals another important factor regarding power generation and transmission and the impact of efficiency levels. For values of $R_L$ greater than 100 Ω, the level of $V_L$ stays fairly constant—a desirable characteristic for power transmission. It would be quite disturbing to see our voltage level at the home or business vary with load ($R_L$), as shown

for values of $R_L$ from 0.1 Ω to 100 Ω. The combination of the fairly constant load voltage and higher efficiencies (as implied by Fig. 9.78) reveals that the far right region of Fig. 9.79 is the appropriate region for high-power-level situations.

The power delivered to $R_L$ under maximum power conditions ($R_L = R_{Th}$) is

$$I = \frac{E_{Th}}{R_{Th} + R_L} = \frac{E_{Th}}{2R_{Th}}$$

$$P_L = I^2 R_L = \left(\frac{E_{Th}}{2R_{Th}}\right)^2 R_{Th} = \frac{E_{Th}^2 R_{Th}}{4R_{Th}^2}$$

and

$$P_{L_{max}} = \frac{E_{Th}^2}{4R_{Th}} \qquad \text{(watts, W)} \qquad \textbf{(9.4)}$$

For the Norton circuit of Fig. 9.76,

$$P_{L_{max}} = \frac{I_N^2 R_N}{4} \qquad \text{(W)} \qquad \textbf{(9.5)}$$

---

**EXAMPLE 9.14.** A dc generator, battery, and laboratory supply are connected to a resistive load $R_L$ in Figs. 9.80(a), (b), and (c), respectively. For each, determine the value of $R_L$ for maximum power transfer to $R_L$.

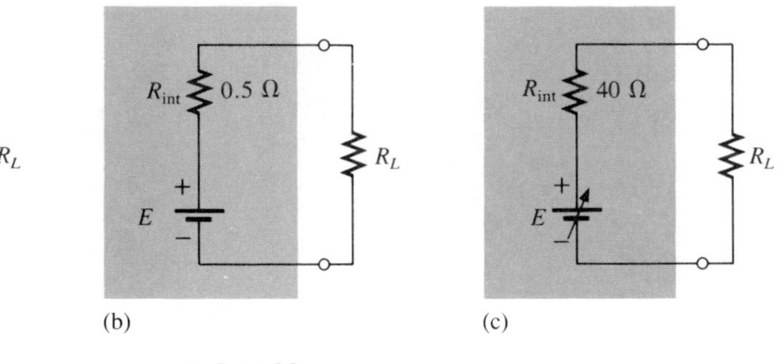

FIG. 9.80

***Solution:*** For the dc generator,

$$R_L = \mathbf{2.5\ \Omega}$$

For the battery,

$$R_L = \mathbf{0.5\ \Omega}$$

For the laboratory supply,

$$R_L = \mathbf{40\ \Omega}$$

---

The results of the preceding example reveal that the following modified form of the maximum power transfer theorem is valid:

***For loads connected directly to a dc voltage supply, maximum power will be delivered to the load when the load resistance is equal to the internal resistance of the source; that is, when***

$$\boxed{R_L = R_{\text{int}}} \tag{9.6}$$

**EXAMPLE 9.15.** Analysis of a transistor network resulted in the reduced configuration of Fig. 9.81. Determine the $R_L$ necessary to transfer maximum power to $R_L$, and calculate the power to $R_L$ under these conditions.

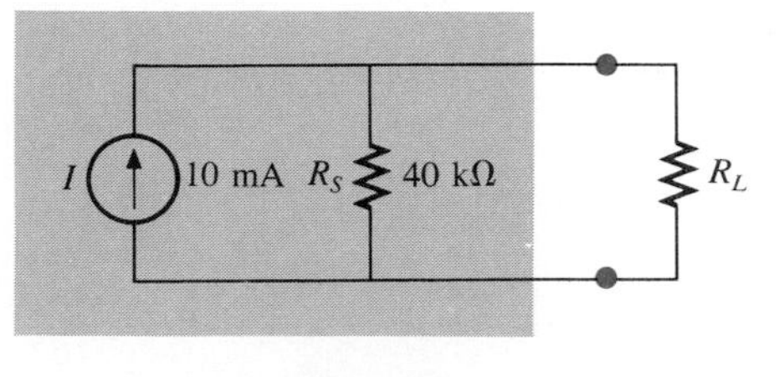


FIG. 9.81

***Solution:*** Equation (9.3):

$$R_L = R_S = \mathbf{40\ k\Omega}$$

Equation (9.5):

$$P_{L_{\text{max}}} = \frac{I_N^2 R_N}{4} = \frac{(10 \times 10^{-3}\ \text{A})^2(40\ \text{k}\Omega)}{4} = \mathbf{1\ W}$$

**EXAMPLE 9.16.** For the network of Fig. 9.82, determine the value of $R$ for maximum power to $R$, and calculate the power delivered under these conditions.

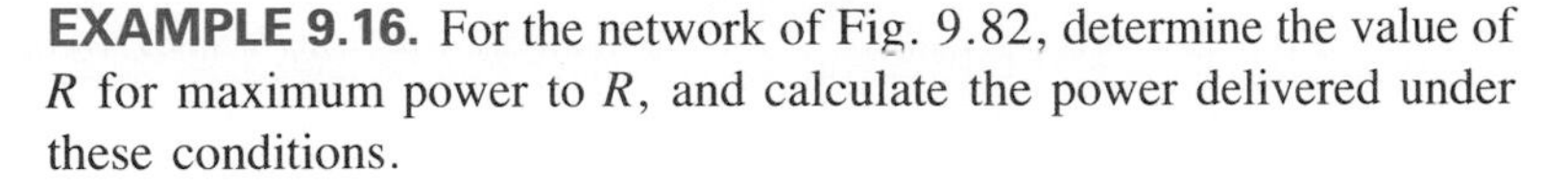

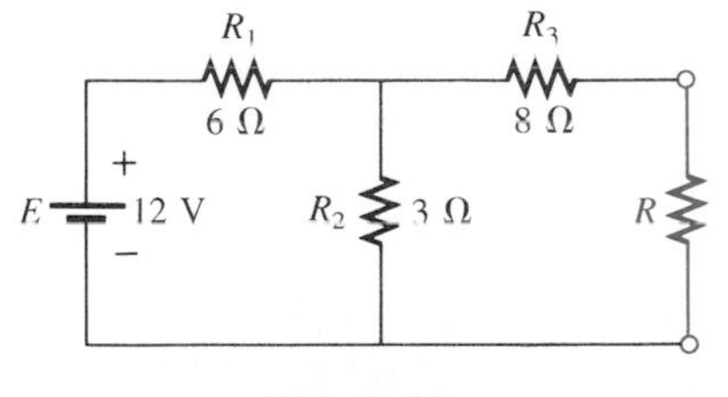


FIG. 9.82

***Solution:*** See Fig. 9.83;

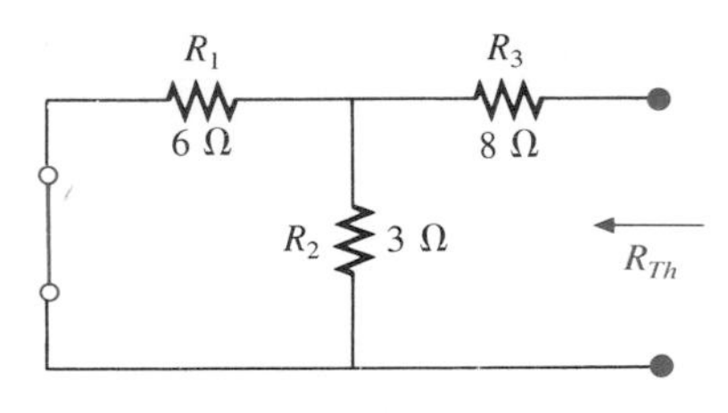

FIG. 9.83

$$R_{Th} = R_3 + R_1 \parallel R_2 = 8\ \Omega + \frac{(6\ \Omega)(3\ \Omega)}{6\ \Omega + 3\ \Omega} = 8\ \Omega + 2\ \Omega$$

and

$$R = R_{Th} = \mathbf{10\ \Omega}$$

See Fig. 9.84;

$$E_{Th} = \frac{R_2 E}{R_2 + R_1} = \frac{(3\ \Omega)(12\ \text{V})}{3\ \Omega + 6\ \Omega} = \frac{36\ \text{V}}{9} = \mathbf{4\ V}$$

and, by Eq. (9.4),

$$P_{L_{\text{max}}} = \frac{E_{Th}^2}{4R_{Th}} = \frac{(4\ \text{V})^2}{4(10\ \Omega)} = \mathbf{0.4\ W}$$

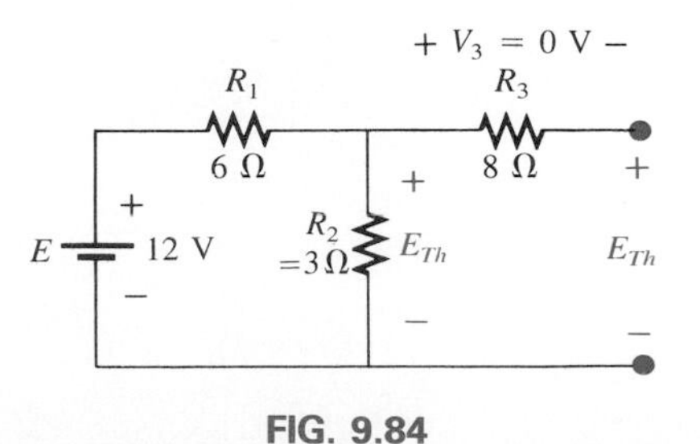


FIG. 9.84

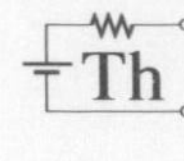


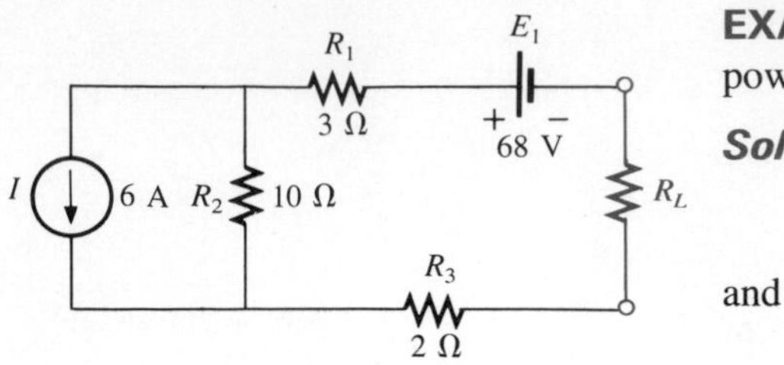

FIG. 9.85

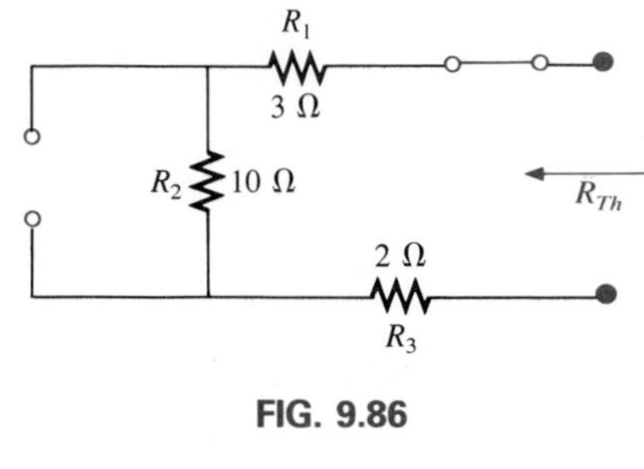


FIG. 9.86

**EXAMPLE 9.17.** Find the value of $R_L$ in Fig. 9.85 for maximum power to $R_L$, and determine the maximum power.

***Solution:*** See Fig. 9.86;

$$R_{Th} = R_1 + R_2 + R_3 = 3\ \Omega + 10\ \Omega + 2\ \Omega = 15\ \Omega$$

and

$$R_L = R_{Th} = \mathbf{15\ \Omega}$$

Note Fig. 9.87, where

$$V_1 = V_3 = 0\ \text{V}$$

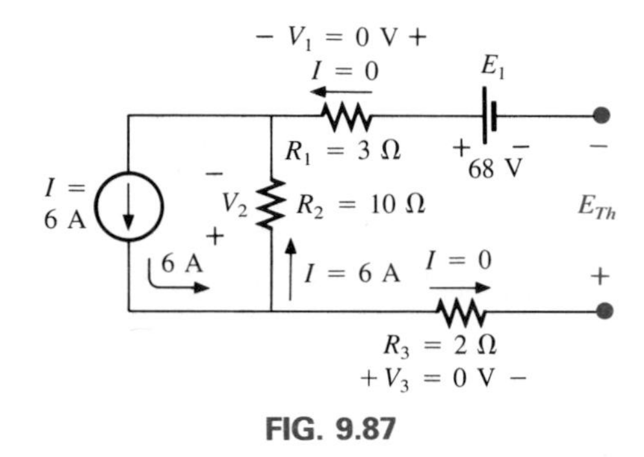


FIG. 9.87

and

$$V_2 = I_2R_2 = IR_2 = (6\ \text{A})(10\ \Omega) = 60\ \text{V}$$

Applying Kirchhoff's voltage law,

$$\Sigma_{\circlearrowright} V = -V_2 - E_1 + E_{Th} = 0$$

and

$$E_{Th} = V_2 + E_1 = 60\ \text{V} + 68\ \text{V} = 128\ \text{V}$$

Thus,

$$P_{L_{\text{max}}} = \frac{E_{Th}^2}{4R_{Th}} = \frac{(128\ \text{V})^2}{4(15\ \Omega)} = \mathbf{273.07\ W}$$

## 9.6 MILLMAN'S THEOREM

Through the application of Millman's theorem, any number of parallel voltage sources can be reduced to one. In Fig. 9.88, for example, the three voltage sources can be reduced to one. This would permit finding the current through or voltage across $R_L$ without having to apply a method such as mesh analysis, nodal analysis, superposition, and so on. The theorem can best be described by applying it to the network of Fig. 9.88. There are basically three steps included in its application.

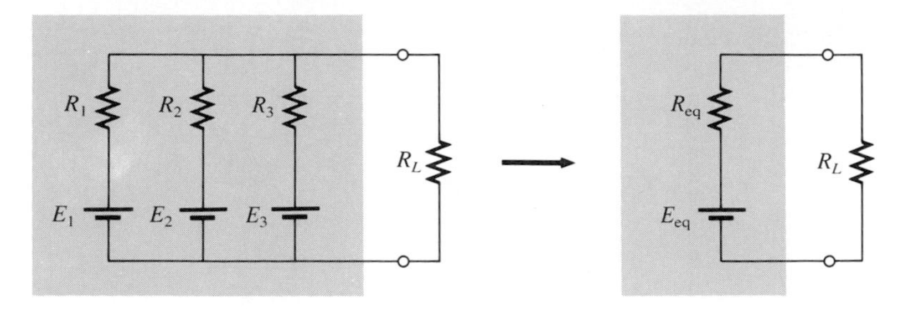


FIG. 9.88

*Step 1:* Convert all voltage sources to current sources as outlined in Section 8.3. This is performed in Fig. 9.89 for the network of Fig. 9.88.

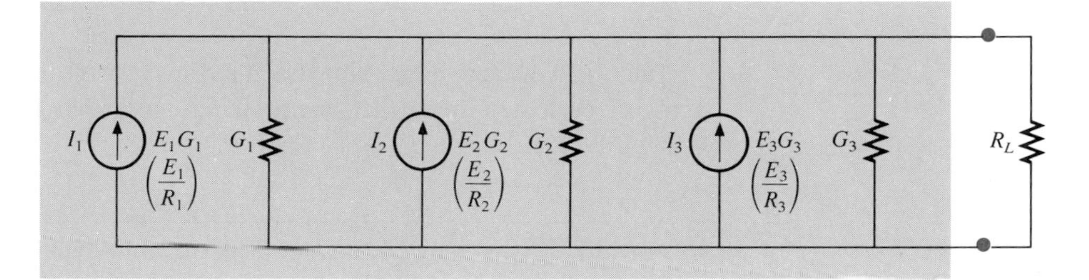


FIG. 9.89

*Step 2:* Combine parallel current sources as described in Section 8.4. The resulting network is shown in Fig. 9.90, where

$$I_T = I_1 + I_2 + I_3 \qquad G_T = G_1 + G_2 + G_3$$

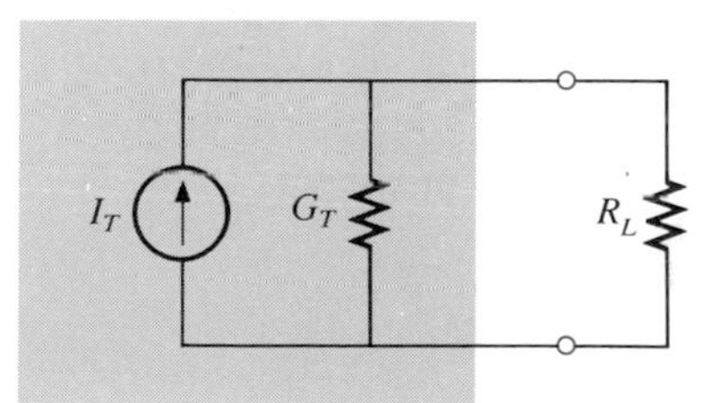


FIG. 9.90

*Step 3:* Convert the resulting current source to a voltage source, and the desired single-source network is obtained as shown in Fig. 9.91.

In general, Millman's theorem states that for any number of parallel voltage sources,

$$E_{eq} = \frac{I_T}{G_T} = \frac{\pm I_1 \pm I_2 \pm I_3 \pm \cdots \pm I_N}{G_1 + G_2 + G_3 + \cdots + G_N}$$

or

$$E_{eq} = \frac{\pm E_1G_1 \pm E_2G_2 \pm E_3G_3 \pm \cdots \pm E_NG_N}{G_1 + G_2 + G_3 + \cdots + G_N} \qquad \textbf{(9.7)}$$

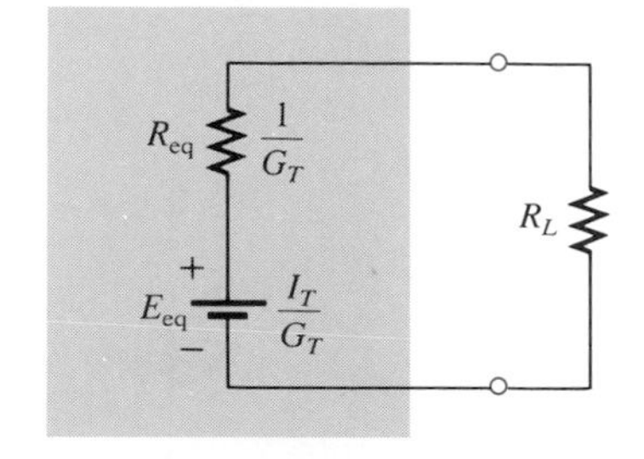


FIG. 9.91

The plus and minus signs appear in Eq. (9.7) to include those cases where the sources may not be supplying energy in the same direction. (Note Example 9.18.)

The equivalent resistance is

$$R_{eq} = \frac{1}{G_T} = \frac{1}{G_1 + G_2 + G_3 + \cdots + G_N} \qquad \textbf{(9.8)}$$

In terms of the resistance values,

$$E_{eq} = \frac{\pm \dfrac{E_1}{R_1} \pm \dfrac{E_2}{R_2} \pm \dfrac{E_3}{R_3} \pm \cdots \pm \dfrac{E_N}{R_N}}{\dfrac{1}{R_1} + \dfrac{1}{R_2} + \dfrac{1}{R_3} + \cdots + \dfrac{1}{R_N}} \tag{9.9}$$

and

$$R_{eq} = \frac{1}{\dfrac{1}{R_1} + \dfrac{1}{R_2} + \dfrac{1}{R_3} + \cdots + \dfrac{1}{R_N}} \tag{9.10}$$

The relatively few direct steps required may result in the student's applying each step rather than memorizing and employing Eqs. (9.7) through (9.10).

---

**EXAMPLE 9.18.** Using Millman's theorem, find the current through and voltage across the resistor $R_L$ of Fig. 9.92.

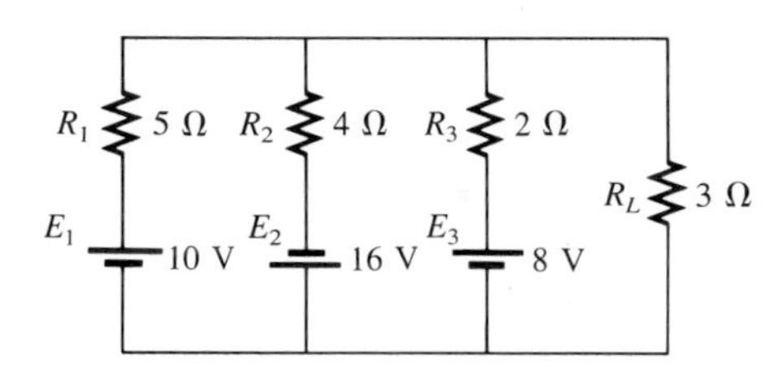

FIG. 9.92

***Solution:*** By Eq. (9.9),

$$E_{eq} = \frac{+\dfrac{E_1}{R_1} - \dfrac{E_2}{R_2} + \dfrac{E_3}{R_3}}{\dfrac{1}{R_1} + \dfrac{1}{R_2} + \dfrac{1}{R_3}}$$

The minus sign is used for $E_2/R_2$ because that supply has the opposite polarity of the other two. The chosen reference direction is therefore that of $E_1$ and $E_3$. The total conductance is unaffected by the direction, and

$$E_{eq} = \frac{+\dfrac{10\text{ V}}{5\ \Omega} - \dfrac{16\text{ V}}{4\ \Omega} + \dfrac{8\text{ V}}{2\ \Omega}}{\dfrac{1}{5\ \Omega} + \dfrac{1}{4\ \Omega} + \dfrac{1}{2\ \Omega}} = \frac{2\text{ A} - 4\text{ A} + 4\text{ A}}{0.2\text{ S} + 0.25\text{ S} + 0.5\text{ S}}$$

$$= \frac{2\text{ A}}{0.95\text{ S}} = \mathbf{2.105\ V}$$

with

$$R_{eq} = \frac{1}{\dfrac{1}{5\ \Omega} + \dfrac{1}{4\ \Omega} + \dfrac{1}{2\ \Omega}} = \frac{1}{0.95\text{ S}} = \mathbf{1.053\ \Omega}$$

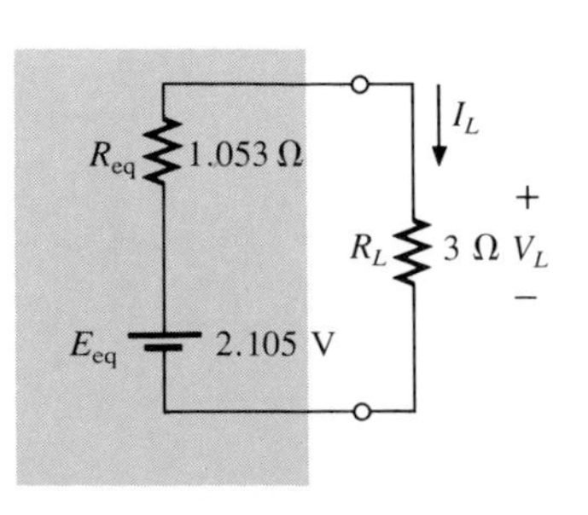


FIG. 9.93

The resultant source is shown in Fig. 9.93, and

$$I_L = \frac{2.105\text{ V}}{1.053\ \Omega + 3\ \Omega} = \frac{2.105\text{ V}}{4.053\ \Omega} = \mathbf{0.519\ A}$$

with

$$V_L = I_L R_L = (0.519\text{ A})(3\ \Omega) = \mathbf{1.557\ V}$$

**EXAMPLE 9.19.** Let us now consider the type of problem encountered in the introduction to mesh and nodal analysis in Chapter 8. Mesh analysis was applied to the network of Fig. 9.94 (Example 8.12). Let us now use Millman's theorem to find the current through the 2-Ω resistor and compare the results.

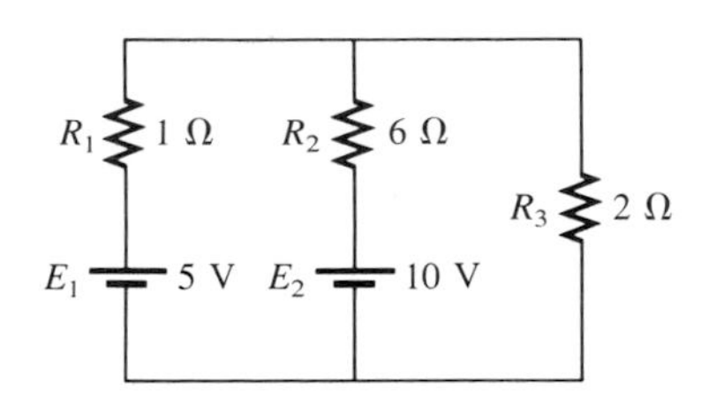


FIG. 9.94

***Solution:***

a. Let us first apply each step and, in the (b) solution, Eq. (9.9). Converting sources yields Fig. 9.95. Combining sources and parallel conductance branches (Fig. 9.96) yields

$$I_T = I_1 + I_2 = 5\text{ A} + \frac{5}{3}\text{ A} = \frac{15}{3}\text{ A} + \frac{5}{3}\text{ A} = \frac{20}{3}\text{ A}$$

$$G_T = G_1 + G_2 = 1\text{ S} + \frac{1}{6}\text{ S} = \frac{6}{6}\text{ S} + \frac{1}{6}\text{ S} = \frac{7}{6}\text{ S}$$

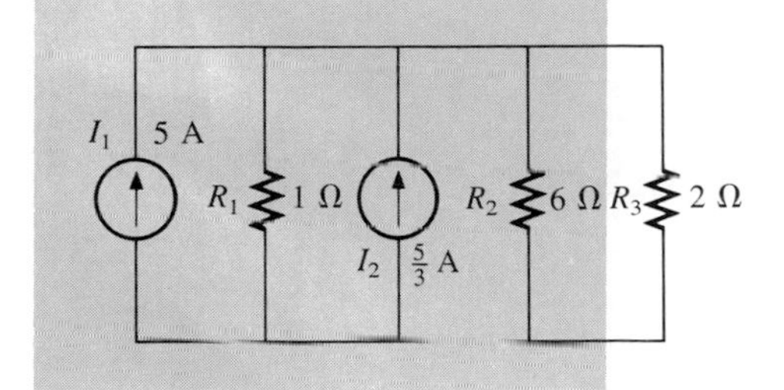


FIG. 9.95

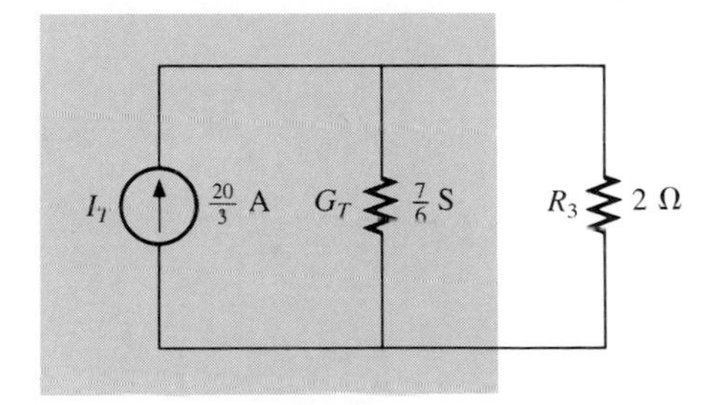


FIG. 9.96

Converting the current source to a voltage source (Fig. 9.97), we obtain

$$E_{eq} = \frac{I_T}{G_T} = \frac{\dfrac{20}{3}\text{ A}}{\dfrac{7}{6}\text{ S}} = \frac{(6)(20)}{(3)(7)}\text{ V} = \mathbf{\frac{40}{7}\ V}$$

and

$$R_{eq} = \frac{1}{G_T} = \frac{1}{\dfrac{7}{6}\text{ S}} = \mathbf{\frac{6}{7}\ \Omega}$$

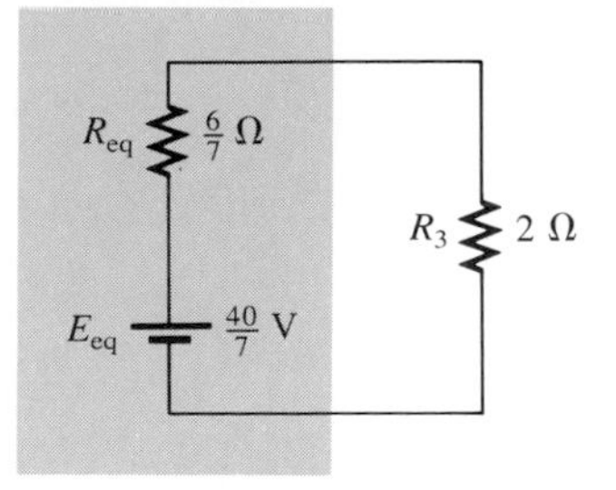


FIG. 9.97

so that

$$I_{2\Omega} = \frac{E_{eq}}{R_{eq} + R_3} = \frac{\dfrac{40}{7}\text{ V}}{\dfrac{6}{7}\ \Omega + 2\ \Omega} = \frac{\dfrac{40}{7}\text{ V}}{\dfrac{6}{7}\ \Omega + \dfrac{14}{7}\ \Omega} = \frac{40\text{ V}}{20\ \Omega} = \mathbf{2\ A}$$

which agrees with the result obtained in Example 8.18.

b. Let us now simply apply the proper equation, Eq. (9.9):

$$E_{eq} = \frac{+\dfrac{5\ \text{V}}{1\ \Omega} + \dfrac{10\ \text{V}}{6\ \Omega}}{\dfrac{1}{1\ \Omega} + \dfrac{1}{6\ \Omega}} = \frac{\dfrac{30\ \text{V}}{6\ \Omega} + \dfrac{10\ \text{V}}{6\ \Omega}}{\dfrac{6}{6\ \Omega} + \dfrac{1}{6\ \Omega}} = \mathbf{\frac{40}{7}\ V}$$

and

$$R_{eq} = \frac{1}{\dfrac{1}{1\ \Omega} + \dfrac{1}{6\ \Omega}} = \frac{1}{\dfrac{6}{6\ \Omega} + \dfrac{1}{6\ \Omega}} = \frac{1}{\dfrac{7}{6}\ \text{S}} = \mathbf{\frac{6}{7}\ \Omega}$$

which are the same values obtained above.

---

The dual of Millman's theorem is the combining of series current sources. The dual of Fig. 9.88 is Fig. 9.98.

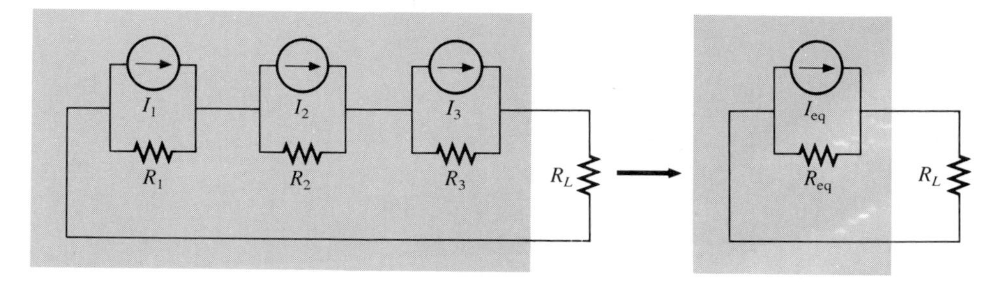


FIG. 9.98

It can be shown that $I_{eq}$ and $R_{eq}$, as shown in Fig. 9.98, are given by

$$I_{eq} = \frac{\pm I_1R_1 \pm I_2R_2 \pm I_3R_3}{R_1 + R_2 + R_3} \qquad \textbf{(9.11)}$$

and

$$R_{eq} = R_1 + R_2 + R_3 \qquad \textbf{(9.12)}$$

The derivation will appear as a problem at the end of the chapter.

## 9.7 SUBSTITUTION THEOREM

The substitution theorem states the following:

***If the voltage across and current through any branch of a dc bilateral network are known, this branch can be replaced by any***

***combination of elements that will maintain the same voltage across and current through the chosen branch.***

More simply, the theorem states that for branch equivalence, the terminal voltage and current must be the same. Consider the circuit of Fig. 9.99 in which the voltage across and current through the branch *a-b* are determined. Through the use of the substitution theorem, a number of equivalent *a-a′* branches are shown in Fig. 9.100.

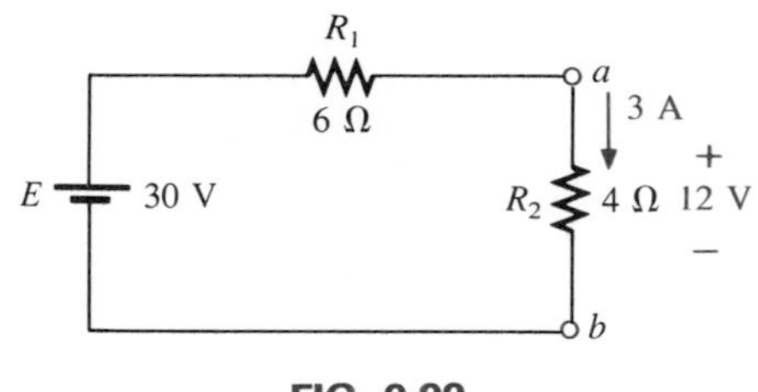


FIG. 9.99

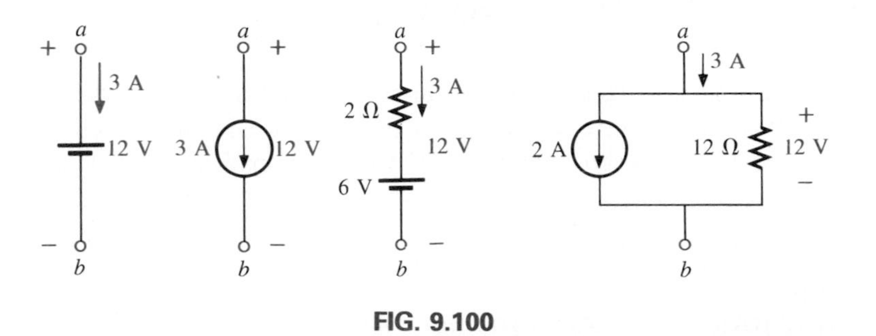


FIG. 9.100

Note that for each equivalent, the terminal voltage and current are the same. Also consider that the response of the remainder of the circuit of Fig. 9.99 is unchanged by substituting any one of the equivalent branches. As demonstrated by the single-source equivalents of Fig. 9.100, *a known potential difference and current in a network can be replaced by an ideal voltage source and current source, respectively.*

Understand that this theorem cannot be used to *solve* networks with two or more sources that are not in series or parallel. For it to be applied, a potential difference or current value must be known or found using one of the techniques discussed earlier. One application of the theorem is shown in Fig. 9.101. Note that in the figure the known potential difference $V$ was replaced by a voltage source, permitting the isolation of the portion of the network including $R_3$, $R_4$, and $R_5$. Recall that this was basically the approach employed in the analysis of the ladder network as we worked our way back toward the terminal resistance $R_5$.

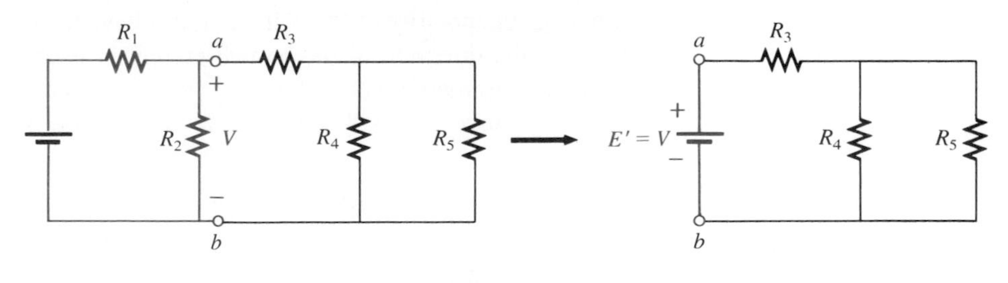


FIG. 9.101

The current source equivalence of the above is shown in Fig. 9.102, where a known current is replaced by an ideal current source, permitting the isolation of $R_4$ and $R_5$.

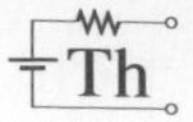


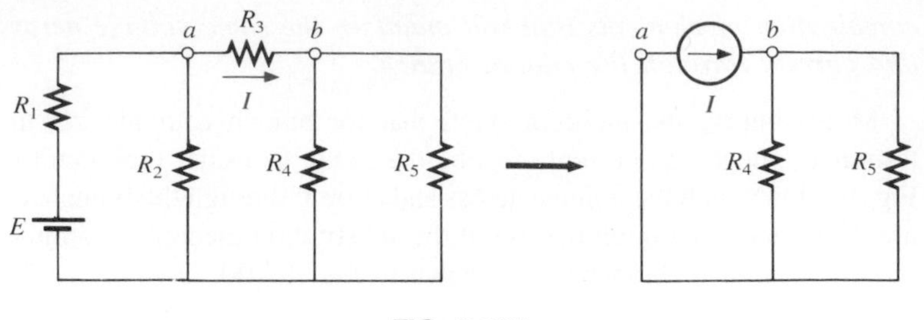


FIG. 9.102

You will also recall from the discussion of bridge networks that $V = 0$ and $I = 0$ were replaced by a short circuit and an open circuit, respectively. This substitution is a very specific application of the substitution theorem.

## 9.8 RECIPROCITY THEOREM

The reciprocity theorem is applicable only to single-source networks. It is, therefore, not a theorem employed in the analysis of multisource networks described thus far.

The theorem states the following:

***The current I in any branch of a network, due to a single voltage source E anywhere else in the network, will equal the current through the branch in which the source was originally located if the source is placed in the branch in which the current I was originally measured.***

In other words, the location of the voltage source and the resulting current may be interchanged without a change in current. The theorem requires that the polarity of the voltage source have the same correspondence with the direction of the branch current in each position.

In the representative network of Fig. 9.103(a), the current $I$ due to the voltage source $E$ was determined. If the position of each is inter-

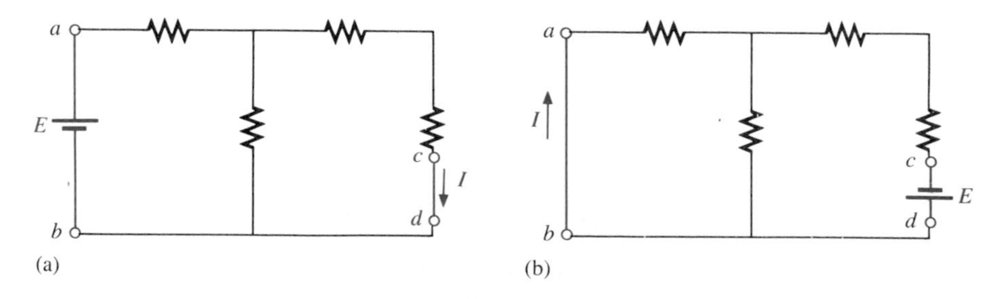


FIG. 9.103

changed as shown in Fig. 9.103(b), the current $I$ will be the same value as indicated. To demonstrate the validity of this statement and the theorem, consider the network of Fig. 9.104, in which values for the elements of Fig. 9.103(a) have been assigned.

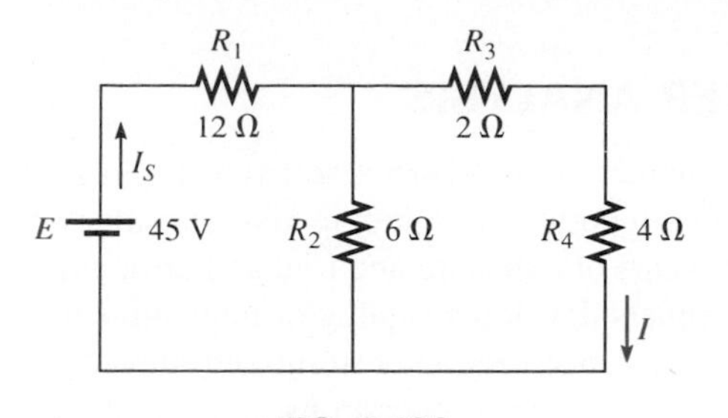


FIG. 9.104

The total resistance is

$$R_T = R_1 + R_2 \parallel (R_3 + R_4) = 12\ \Omega + 6\ \Omega \parallel (2\ \Omega + 4\ \Omega)$$
$$= 12\ \Omega + 6\ \Omega \parallel 6\ \Omega = 12\ \Omega + 3\ \Omega = 15\ \Omega$$

and

$$I_S = \frac{E}{R_T} = \frac{45\ \text{V}}{15\ \Omega} = 3\ \text{A}$$

with

$$I = \frac{3\ \text{A}}{2} = \mathbf{1.5\ A}$$

For the network of Fig. 9.105, which corresponds to that of Fig. 9.103(b), we find

$$R_T = R_4 + R_3 + R_1 \parallel R_2$$
$$= 4\ \Omega + 2\ \Omega + 12\ \Omega \parallel 6\ \Omega = 10\ \Omega$$

and

$$I_S = \frac{E}{R_T} = \frac{45\ \text{V}}{10\ \Omega} = 4.5\ \text{A}$$

so that

$$I = \frac{(6\ \Omega)(4.5\ \text{A})}{12\ \Omega + 6\ \Omega} = \frac{4.5\ \text{A}}{3} = \mathbf{1.5\ A}$$

which agrees with the above.

The uniqueness and power of such a theorem can best be demonstrated by considering a complex single-source network such as shown in Fig. 9.106.

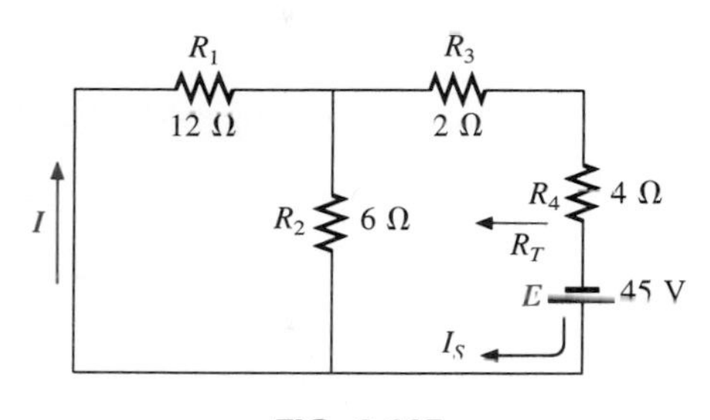


FIG. 9.105

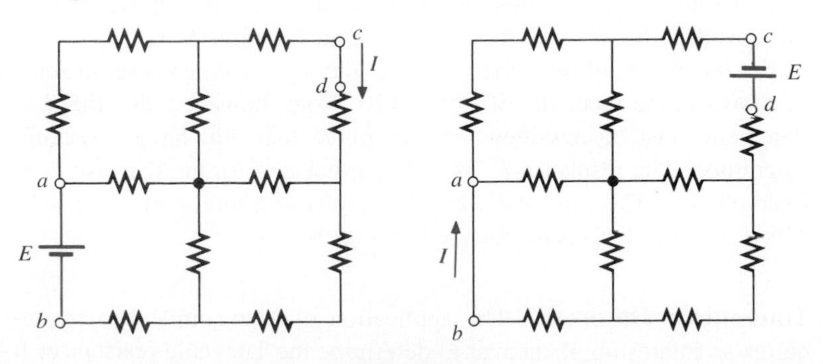


FIG. 9.106

## 9.9 COMPUTER ANALYSIS

Once the mechanics of applying a software package or language are understood, the opportunity to be creative and innovative presents itself. Through years of exposure and trial and error experiences, professional programmers develop a catalog of innovative techniques that are not only functional but very interesting and almost artistic in nature. Now that some of the basic operations associated with PSPICE have been introduced, a few innovative maneuvers will be made in the programs to follow. A program in BASIC will be included, but the analysis will follow a format similar to that encountered in earlier chapters, with the only major variation being the tabulation of a number of quantities for a changing variable.

### PSPICE

**Superposition** Let us now apply superposition to the network of Fig. 9.107, which appeared earlier as Example 9.3, to permit a comparison of the resulting solutions.

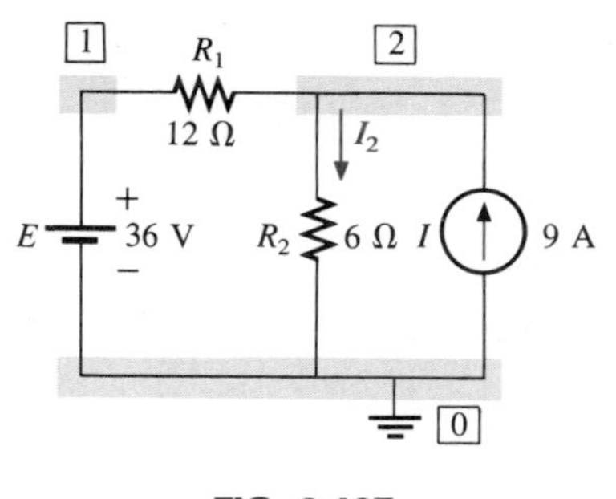


FIG. 9.107

The input file for the entire network is provided in Fig. 9.108 with the results for the current $I_{R_2}$.

In order to find the component of $I_{R_2}$ due solely to the voltage source, the only required modification of the input file of Fig. 9.108 is to set the current source to zero amps, as shown in the input file of Fig. 9.109. The result is that $I'_2 = 2$ A, as obtained in Example 9.3.

For the effects of the current source, the input voltage is set to zero, as shown in the input file of Fig. 9.110. Note, however, that the .DC statement must be rewritten for the source that will have a specific magnitude. The result for $I''_2$ is 6 A, further confirming the results of Example 9.3. The sum of $I'_2$ and $I''_2$ results in a total current of 8 A, which matches the result obtained in Fig. 9.108.

**Thevenin's Theorem** The application of Thevenin's theorem requires an interesting maneuver to determine the Thevenin resistance. It is a maneuver, however, that has application beyond Thevenin's theo-

```
************************ Evaluation PSpice (January 1989) ******* 10:27:15 *******

 CHAPTER 9, SUPERPOSITION

 ****     CIRCUIT DESCRIPTION

*******************************************************************************


VE 1 0 36V
R1 1 2 12
R2 2 0 6
IS 0 2 9A
.DC VE 36 36 1
.PRINT DC I(R2)
.OPTIONS NOPAGE
.END

 ****     DC TRANSFER CURVES                    TEMPERATURE =   27.000 DEG C


  VE            I(R2)

   3.600E+01    8.000E+00
```

FIG. 9.108

```
************************ Evaluation PSpice (January 1989) ******* 10:31:19 *******

 CHAPTER 9, SUPERPOSITION

 ****     CIRCUIT DESCRIPTION

*******************************************************************************


VE 1 0 36V
R1 1 2 12
R2 2 0 6
IS 0 2 0A
.DC VE 36 36 1
.PRINT DC I(R2)
.OPTIONS NOPAGE
.END

 ****     DC TRANSFER CURVES                    TEMPERATURE =   27.000 DEG C


  VE            I(R2)

   3.600E+01    2.000E+00
```

FIG. 9.109

```
************************ Evaluation PSpice (January 1989) ******* 10:33:25 *******

CHAPTER 9, SUPERPOSITION

****      CIRCUIT DESCRIPTION

******************************************************************************

VE 1 0 0V
R1 1 2 12
R2 2 0 6
IS 0 2 9A
.DC IS 9 9 1
.PRINT DC I(R2)
.OPTIONS NOPAGE
.END

****      DC TRANSFER CURVES                    TEMPERATURE =    27.000 DEG C

  IS              I(R2)

  9.000E+00     6.000E+00
```

FIG. 9.110

rem whenever a resistance level is required. The network to be analyzed appears in Fig. 9.111 and is the same analyzed in Example 9.10. Note

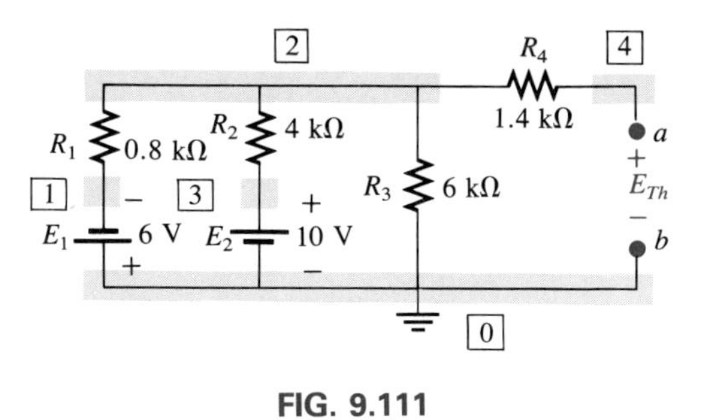


FIG. 9.111

the very large resistance in the input file of Fig. 9.112 between terminals 4 and 0 to simulate an open circuit. PSPICE does not recognize ''dangling nodes'' requiring at least one dc path to the reference level (in our case, ground). It also requires that each node appears twice in the input file (to accomplish the preceding for resistive circuits).

The output file of Fig. 9.112 reveals that the open-circuit voltage $E_{Th}$ is $-3$ V to match the solution of Example 9.10. The minus sign simply reveals that node 4 is at a lower potential than the reference node 0.

```
************************ Evaluation PSpice (January 1989) ******* 17:27:53 *******

Chapter 9 - Thevenin's Theorem E(th)

****       CIRCUIT DESCRIPTION

******************************************************************************


V1 0 1 DC 6V
V2 3 0 DC 10V
R1 1 2 .8K
R2 2 3 4K
R3 2 0 6K
R4 2 4 1.4K
RL 4 0 1E30
.DC V1 6 6 1
.PRINT DC V(4,0)
.OPTIONS NOPAGE
.END

****       DC TRANSFER CURVES                    TEMPERATURE =    27.000 DEG C


  V1            V(4,0)

   6.000E+00   -3.000E+00
```

FIG. 9.112

In order to determine $R_{Th}$, a current source of 1 A is applied to the network of Fig. 9.111, as shown in Fig. 9.113(a). In Fig. 9.113(b) the

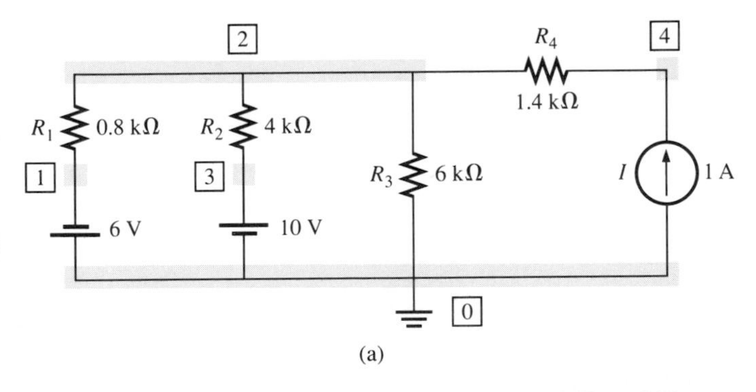


(a)

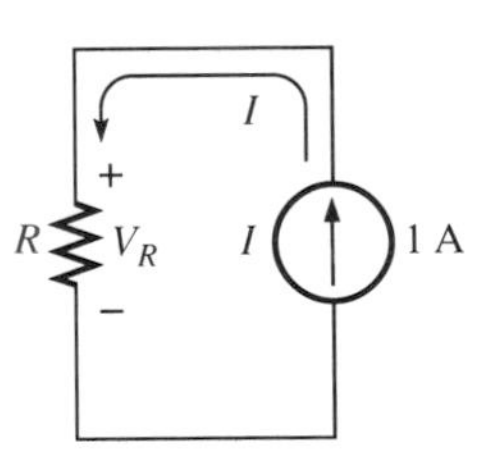


(b)

FIG. 9.113

resistance $R$ can be determined using Ohm's law and the resulting voltage $V_R$, as follows:

$$R = \frac{V_R}{I} = \frac{V_R}{1\text{ A}} \tag{9.13}$$

In Eq. (9.13), since $I = 1$ A the magnitude of $R$ in ohms is the same as the magnitude of the voltage $V_R$. Therefore, when V(4, 0) is requested in the input file of 9.114, you are in actuality also obtaining the magnitude of $R_{Th}$. The output file of Fig. 9.114 printed out a level of 2 kV for $V_R$, which we will read as $R_{Th} = 2$ kΩ, as obtained in Example 9.10.

```
************************ Evaluation PSpice (January 1989) ******* 17:38:34 *******

Chapter 9 - Thevenin's Theorem R(th)

****      CIRCUIT DESCRIPTION

****************************************************************************

V1 0 1 DC 0V
V2 3 0 DC 0V
I1 0 4 DC 1A
R1 1 2 .8K
R2 2 3 4K
R3 2 0 6K
R4 2 4 1.4K
.DC I1 1 1 1
.PRINT DC V(4,0)
.OPTIONS NOPAGE
.END

****      DC TRANSFER CURVES                    TEMPERATURE =   27.000 DEG C

  I1          V(4,0)

  1.000E+00   2.000E+03
```

FIG. 9.114

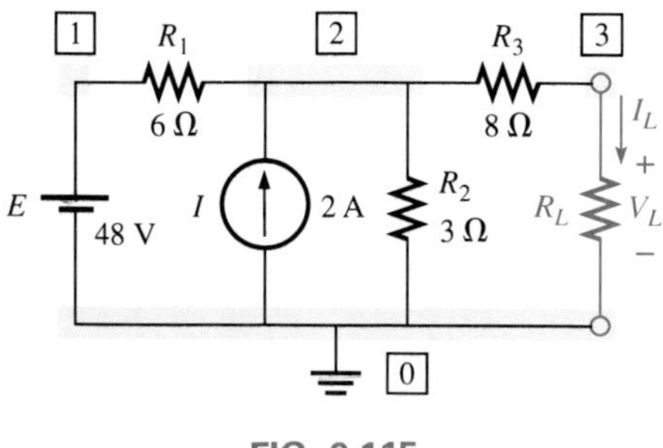


FIG. 9.115

## Load Variation

The next input file analyzes the fairly complicated network of Fig. 9.115 for a range of values for $R_L$. The first five lines of the input file of Fig. 9.116 are similar to the entries of previous examples, but the $R_L$ line specifies an initial value of the quantity to be varied. In this case the first value of $R_L$ for which $V_{R_L}$ and $I_{R_L}$ will be determined is 1 Ω. The .MODEL command is used to specify specific characteristics about the element (resistors, inductors, capacitors, semiconductor devices, and so on) for future reference in the analysis process. For the resistor it could be the tolerance or temperature coefficient. In this application specific characteristics are not provided but the command is required if $R_L$ is to

```
************************ Evaluation PSpice (January 1989) ******* 20:02:12 *******

 Chapter 9 - Load Variation

 ****     CIRCUIT DESCRIPTION

*****************************************************************************

VE 1 0 48V
IS 0 2 2A
R1 1 2 6
R2 2 0 3
R3 2 3 8
RL 3 0 RLOAD 1
.MODEL RLOAD RES
.DC RES RLOAD (R) 1 20 1
.PRINT DC V(RL) I(RL)
.PLOT DC V(RL) I(RL)
.OPTIONS NOPAGE
.END
```

```
 ****     DC TRANSFER CURVES                    TEMPERATURE =   27.000 DEG C

   R             V(RL)          I(RL)

   1.000E+00     1.818E+00      1.818E+00
   2.000E+00     3.333E+00      1.667E+00
   3.000E+00     4.615E+00      1.538E+00
   4.000E+00     5.714E+00      1.429E+00
   5.000E+00     6.667E+00      1.333E+00
   6.000E+00     7.500E+00      1.250E+00
   7.000E+00     8.235E+00      1.176E+00
   8.000E+00     8.889E+00      1.111E+00
   9.000E+00     9.474E+00      1.053E+00
   1.000E+01     1.000E+01      1.000E+00
   1.100E+01     1.048E+01      9.524E-01
   1.200E+01     1.091E+01      9.091E-01
   1.300E+01     1.130E+01      8.696E-01
   1.400E+01     1.167E+01      8.333E-01
   1.500E+01     1.200E+01      8.000E-01
   1.600E+01     1.231E+01      7.692E-01
   1.700E+01     1.259E+01      7.407E-01
   1.800E+01     1.286E+01      7.143E-01
   1.900E+01     1.310E+01      6.897E-01
   2.000E+01     1.333E+01      6.667E-01
```

```
****      DC TRANSFER CURVES                     TEMPERATURE =    27.000 DEG C


 LEGEND:

*: V(RL)
+: I(RL)

  R              V(RL)
(*)----------      0.0000E+00    5.0000E+00    1.0000E+01    1.5000E+01    2.0000E+01
(+)----------      5.0000E-01    1.0000E+00    1.5000E+00    2.0000E+00    2.5000E+00
                 - - - - - - - - - - - - - - - - - - - - - - - - - - - - - - - - -
 1.000E+00  1.818E+00 .    *          .            .     +      .            .
 2.000E+00  3.333E+00 .        *      .            .  +         .            .
 3.000E+00  4.615E+00 .           *.               .+           .            .
 4.000E+00  5.714E+00 .            .  *          + .            .            .
 5.000E+00  6.667E+00 .            .    *     +    .            .            .
 6.000E+00  7.500E+00 .            .       X       .            .            .
 7.000E+00  8.235E+00 .            .     +   *     .            .            .
 8.000E+00  8.889E+00 .            .   +       *   .            .            .
 9.000E+00  9.474E+00 .            .+            *.            .            .
 1.000E+01  1.000E+01 .            +               *            .            .
 1.100E+01  1.048E+01 .           +.               .*           .            .
 1.200E+01  1.091E+01 .          + .               . *          .            .
 1.300E+01  1.130E+01 .         +  .               .  *         .            .
 1.400E+01  1.167E+01 .        +   .               .   *        .            .
 1.500E+01  1.200E+01 .       +    .               .    *       .            .
 1.600E+01  1.231E+01 .      +     .               .     *      .            .
 1.700E+01  1.259E+01 .     +      .               .      *     .            .
 1.800E+01  1.286E+01 .     +      .               .      *     .            .
 1.900E+01  1.310E+01 .    +       .               .       *    .            .
 2.000E+01  1.333E+01 .   +        .               .        *   .            .
                 - - - - - - - - - - - - - - - - - - - - - - - - - - - - - - - - -
```

FIG. 9.116

be varied between specific limits. For input files of this nature it has the following format with the name being the only variable of choice by the user.

```
.MODEL    RLOAD    RES
          Name
```

The next command specifies the variable to be varied between the limits indicated on the same line. In this case $R_L$ will be varied from 1 Ω to 20 Ω in increments of 1 Ω. The basic format is the following with brackets beneath those quantities that can be controlled by the user.

```
.DC    RES    RLOAD    (R)      1          20      1
              Name           Beginning   Final  Increment
                               value     value
```

The .PRINT statement specifies $V_{R_L}$ and $I_{R_L}$ as output quantities and the .PLOT DC command requests that the data for the above be plotted for the range of $R_L$ values.

The output file of Fig. 9.116 lists the values of $R_L$, $V_{R_L}$, and $I_{R_L}$ for the range of values for $R_L$. Note that $V_{R_L}$ increased in magnitude but at a slower rate than $R_L$ resulting in a decreasing magnitude for $I_{R_L}$. The entire plot as shown was generated by PSPICE without any additional comment or calculations in the input file. The scale settings, choice of notation for $V_{R_L}$ (*) and $I_{R_L}$ (+), plotting routine, etc. are all part of the PSPICE software package. At the top of the plot, note the provided LEGEND for clarity and the range of $V_{R_L}$ from 0 to 20 V with $I_{R_L}$ from 0.5 A to 2.5 A. The values of $R_L$ are printed along the left hand edge to form the *x*-axis, while $V_{R_L}$ and $I_{R_L}$ form the *y*-axis (with the plot held on a 90° angle). The plot clearly reveals that $V_{R_L}$ increased with increasing values of $R_L$, while $I_{R_L}$ decreased in magnitude.

## .PROBE

The last command to be introduced in this chapter is the .PROBE command, which acts like a ''software oscilloscope.'' It provides a visual display of a particular variable with a number of controls on the output that are unavailable using the .PLOT command. The basic procedure for running .PROBE is covered in Appendix A, with the comments offered here limited to the output curves. The plot to be obtained is for the power to $R_L$ of the network of Fig. 9.75 for a range of $R_L$ from 1 Ω to 10 Ω in increments of 1 Ω. $E_{Th}$ is 4 V and $R_{Th}$ is 5 Ω, which should result in a plot similar to Fig. 9.77. If your system has the necessary Math Processor to run .PROBE all that is required is to call up .PROBE as indicated in the input file of Fig. 9.117. Note, however, that the input file does not specify the plot to be obtained. One of the important advantages of the .PROBE option is that a specific variable can be plotted or a quantity determined by a mathematical operation. In this case, since $P_L = V_L^2/R_L$, we will request a plot of $V(2) * V(2)/R$, result-

```
************************ Evaluation PSpice (January 1989) ******* 10:30:55 *******

Chapter 9 - POWER CIRCUIT

****      CIRCUIT DESCRIPTION

*****************************************************************************


VE 1 0 4V
R1 1 2 5
R2 2 0 RLOAD 1
.MODEL RLOAD RES
.DC RES RLOAD (R) 1 10 0.1
.PROBE
.OPTIONS NOPAGE
.END
```

FIG. 9.117

ing in the curve of Fig. 9.118. Note that the plot is an exact match of that obtained in Fig. 9.77 with a maximum power of 800 mW (0.8 W)

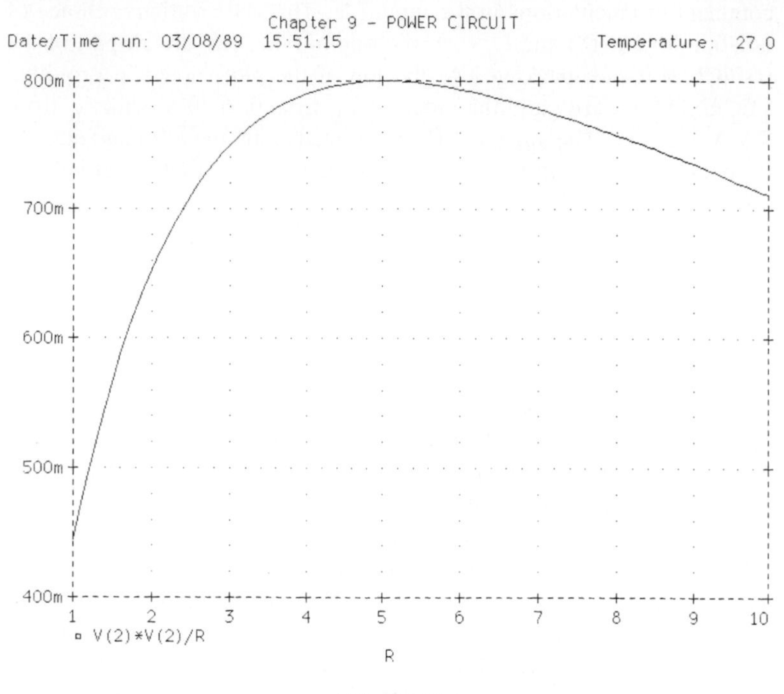


FIG. 9.118

at $R_L = 5\ \Omega$. Using the .PROBE option of changing the vertical scale setting, a magnified view of the peak region can be obtained as shown in Fig. 9.119.

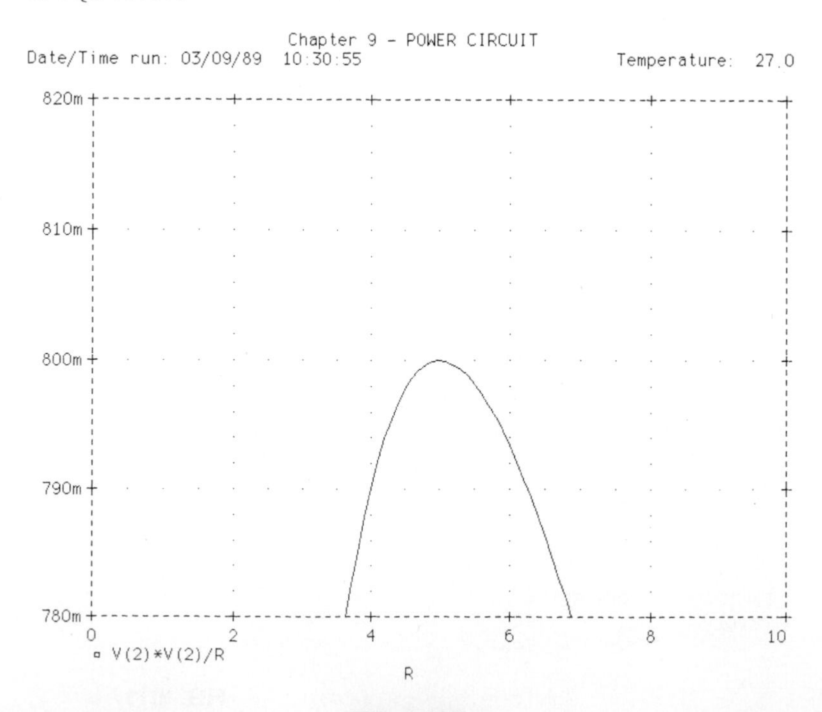


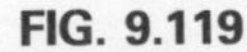
FIG. 9.119

## BASIC

The BASIC program of Fig. 9.120 will also analyze the network of Fig. 9.115 but will also provide the values of $E_{Th}$ and $R_{Th}$ and tabulate a list of output variables that extends from $R_{Th}/5$ to $2R_{Th}$ in increments of $R_{Th}/5$ (line 330). Since $R_{Th}$ is 10 Ω, the range of $R_L$ is from 2 Ω to 20 Ω in increments of 2 Ω. Lines 130 through 180 request the network parameters, whereas lines 200 and 210 calculate $R_{Th}$. $E_{Th}$ is determined by lines 220 through 260 using superposition, and both $E_{Th}$ and $R_{Th}$ are printed out by lines 270 through 290. Lines 310 and 320 provide a heading for the printout, with the TAB command simply specifying the spaces from the left edge of the paper and between columns. The range of $R_L$ is specified by line 330, and all the required calculations are

```
10 REM ***** PROGRAM 9-1 *****
20 REM ***********************************************
30 REM Program to tabulate changes in load levels for
40 REM a range of load values using Thevenin's theorem
50 REM ***********************************************
60 REM
100 PRINT "For the network of Fig. 9.115"
110 PRINT "enter the following data:"
120 PRINT
130 INPUT "R1=";R1 :REM Enter 0 if resistor non-existant
140 INPUT "R2=";R2 :REM Enter 1E30 if resistor non-existant
150 INPUT "R3=";R3 :REM Enter 0 if resistor non-existant
160 INPUT "RL=";RL
170 INPUT "Supply voltage, E=";E
180 INPUT "and supply current, I=";I
190 PRINT
200 REM Determine Rth
210 RT=R3+R1*R2/(R1+R2)
220 REM Use suerposition to determine Eth
230 E1=R2*E/(R1+R2)
240 I2=R2*I/(R1+R2)
250 E2=R1*R2*I/(R1+R2)
260 ET=E1+E2
270 PRINT "Using Thevenin's Theorem:"
280 PRINT "Rth=";RT;"ohms"
290 PRINT "and Eth=";ET;"volts"
300 PRINT
310 PRINT TAB(7);"RL";TAB(15);"IL";TAB(25);"VL";
320 PRINT TAB(35);"PL";TAB(45);"PD";TAB(55);"n%"
330 FOR RL=RT/5 TO 2*RT STEP RT/5
340 IL=ET/(RT+RL)
350 VL=IL*RL
360 PL=IL^2*RL
370 PD=ET*IL
380 N=100*PL/PD
390 IF RL=RT THEN PRINT "Rth=";
400 PRINT TAB(5);RL;TAB(13);IL;TAB(23);VL;
410 PRINT TAB(33);PL;TAB(43);PD;TAB(53);N
420 NEXT RL
430 END
RUN
For the network of Fig. 9.115
enter the following data:

R1=? 6
R2=? 3
R3=? 8
RL=? 10
Supply voltage, E=? 48
and supply current, I=? 2
```

Using Thevenin's Theorem:
Rth= 10 ohms
and Eth= 20 volts

| RL | IL | VL | PL | PD | n% |
|---|---|---|---|---|---|
| 2 | 1.666667 | 3.333333 | 5.555556 | 33.33333 | 16.66667 |
| 4 | 1.428572 | 5.714286 | 8.163265 | 28.57143 | 28.57143 |
| 6 | 1.25 | 7.5 | 9.375 | 25 | 37.5 |
| 8 | 1.111111 | 8.888889 | 9.876544 | 22.22222 | 44.44445 |
| Rth= 10 | 1 | 10 | 10 | 20 | 50 |
| 12 | .9090909 | 10.90909 | 9.917354 | 18.18182 | 54.54545 |
| 14 | .8333333 | 11.66667 | 9.722221 | 16.66667 | 58.33333 |
| 16 | .7692308 | 12.30769 | 9.467456 | 15.38462 | 61.53847 |
| 18 | .7142858 | 12.85714 | 9.183674 | 14.28572 | 64.28571 |
| 20 | .6666667 | 13.33333 | 8.888889 | 13.33333 | 66.66666 |

Ok

FIG. 9.120

performed by lines 340 through 360. Line 390 specifies that if the value of $R_L$ for a particular loop is equal to $R_{Th}$, then the comment "$R_{Th}$=" should be added to the printout as shown. The magnitude of all the quantities is printed out by lines 400 and 410, with line 420 sending the program back to line 330 to repeat the calculations for the next value of $R_L$.

Note that the maximum power is obtained at $R_L = R_{Th} = 10\ \Omega$ and the efficiency is 50% under maximum power transfer conditions. PD is the power delivered by the source, whereas PL is the power to the load.

## PROBLEMS

### SECTION 9.2 Superposition Theorem

1. **a.** Using superposition, find the current through each resistor of the network of Fig. 9.121.
   **b.** Find the power delivered to $R_1$ for each source.
   **c.** Find the power delivered to $R_1$ using the total current through $R_1$.
   **d.** Does superposition apply to power effects? Explain.

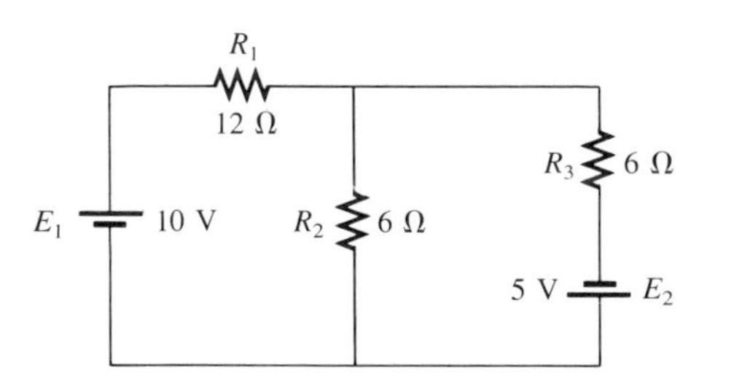


FIG. 9.121

2. Using superposition, find the current $I$ through the 10-Ω resistor for each of the networks of Fig. 9.122.

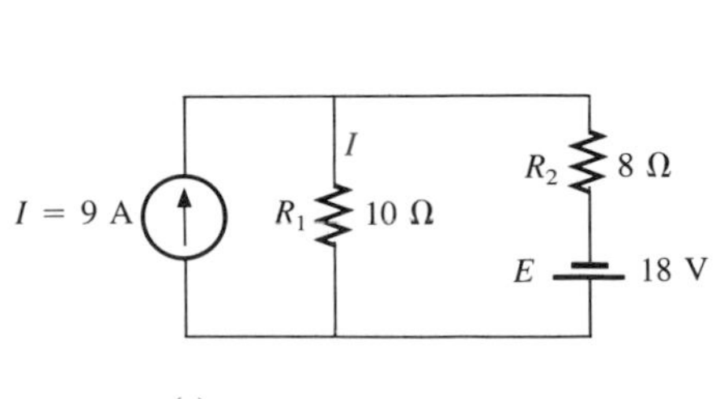

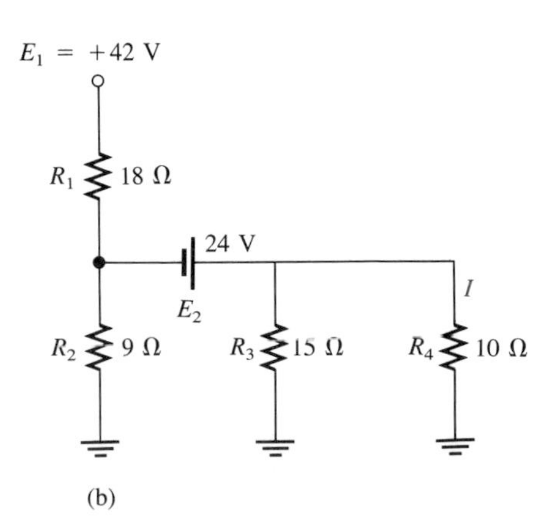


FIG. 9.122

***3.** Using superposition, find the current through $R_1$ for each network of Fig. 9.123.

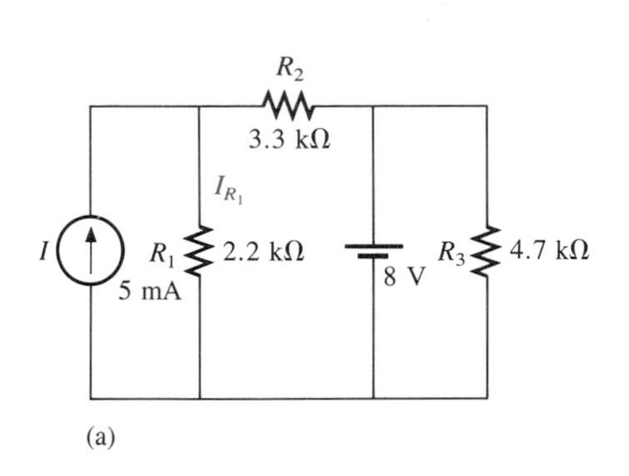


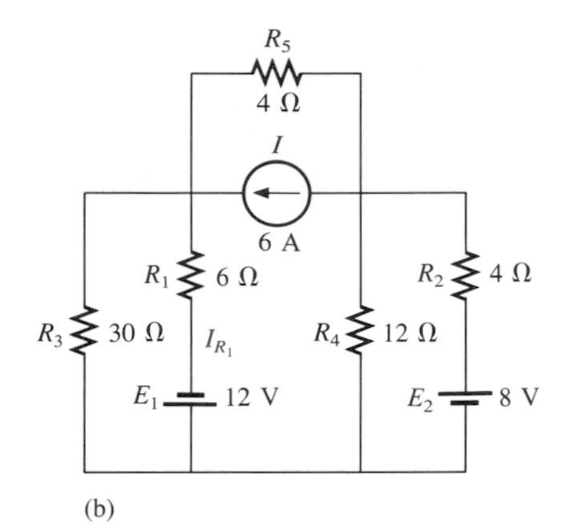

**FIG. 9.123**

**4.** Using superposition, find the voltage $V_2$ for the network of Fig. 9.124.

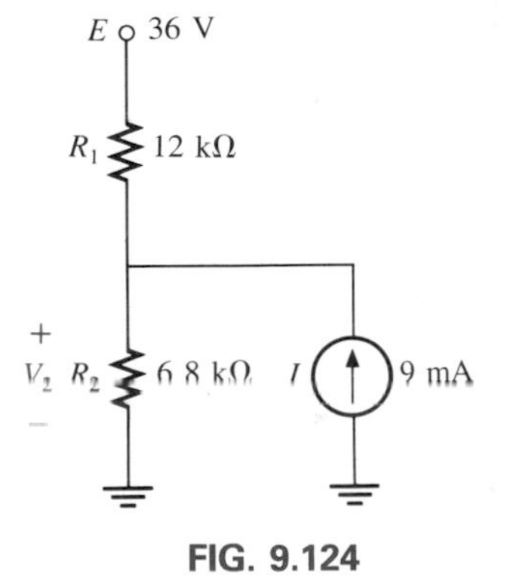


**FIG. 9.124**

## SECTION 9.3 Thevenin's Theorem

**5. a.** Find the Thevenin equivalent circuit for the network external to the resistor $R$ of Fig. 9.125.

**b.** Find the current through $R$ when $R$ is 2, 30, and 100 Ω.

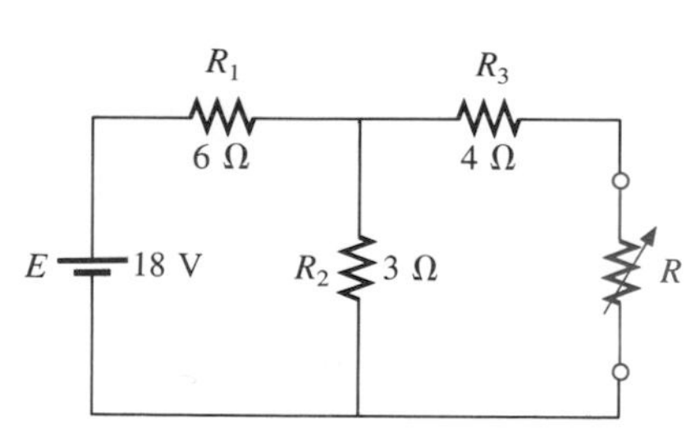

**FIG. 9.125**

**6. a.** Find the Thevenin equivalent circuit for the network external to the resistor $R$ in each of the networks of Fig. 9.126.

**b.** Find the power delivered to $R$ when $R$ is 2 Ω and 100 Ω.

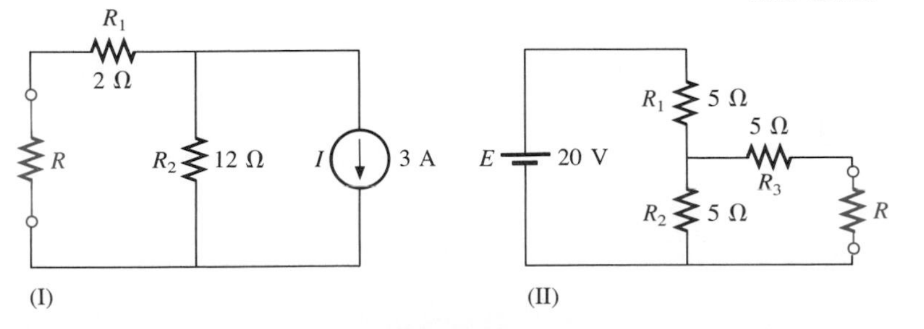


**FIG. 9.126**

**7.** Repeat Problem 6a for the networks of Fig. 9.127.

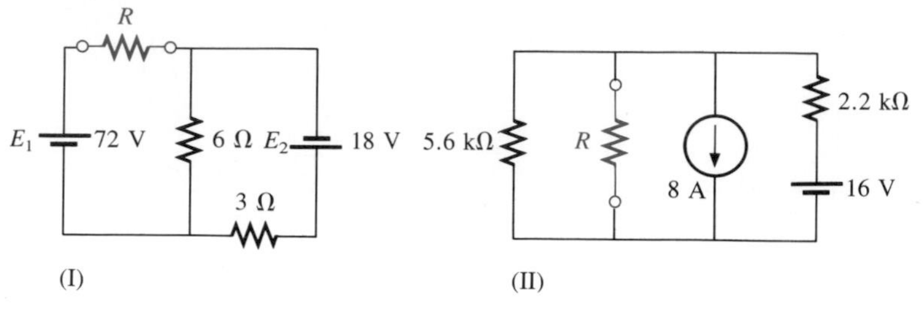

FIG. 9.127

***8.** Repeat Problem 6 for the networks of Fig. 9.128.

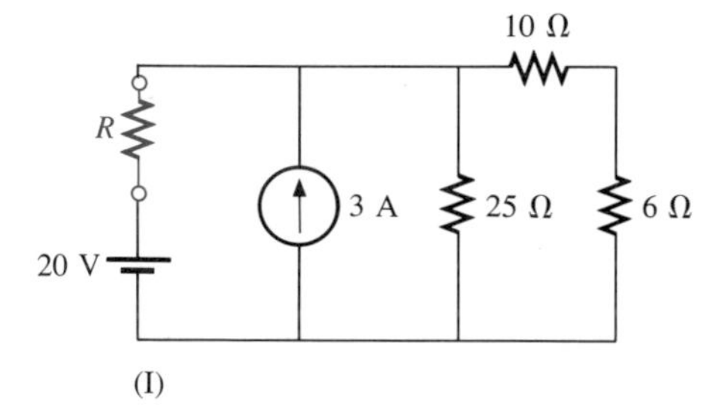


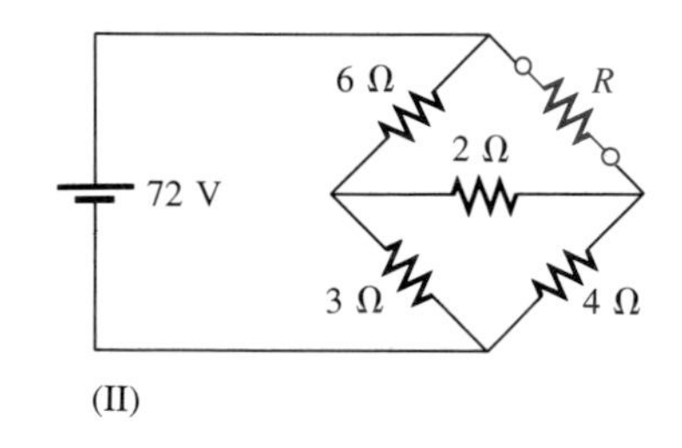


FIG. 9.128

***9.** Find the Thevenin equivalent circuit for the portions of the networks of Fig. 9.129 external to points $a$ and $b$.

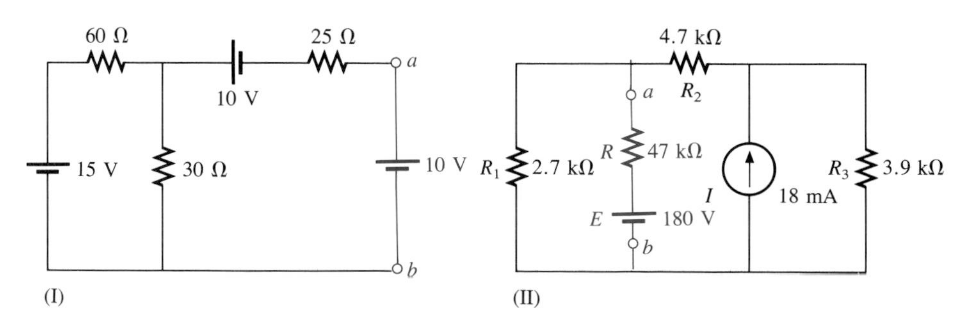


FIG. 9.129

***10.** Determine the Thevenin equivalent circuit for the network external to the resistor $R$ in both networks of Fig. 9.130.

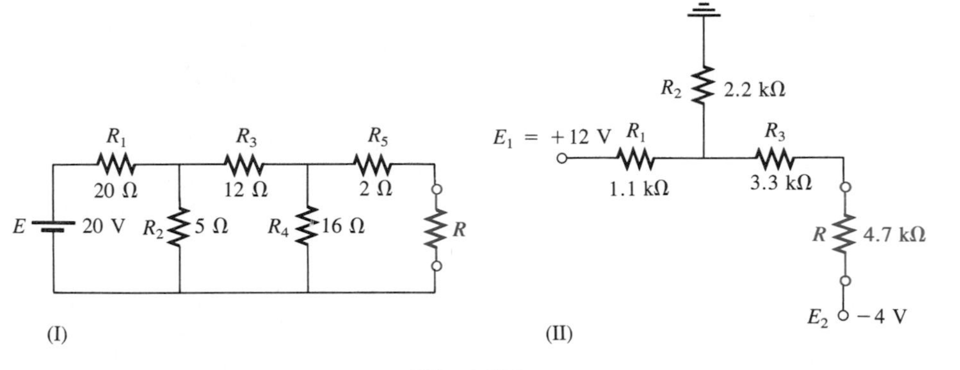

FIG. 9.130

***11.** For the network of Fig. 9.131, find the Thevenin equivalent circuit for the network external to the load resistor $R_L$.

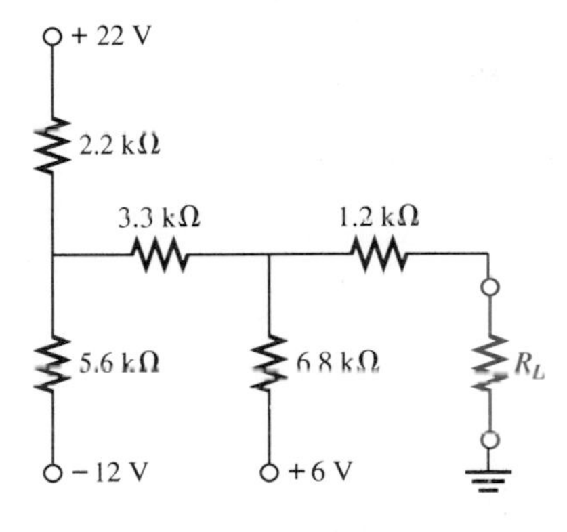

FIG. 9.131

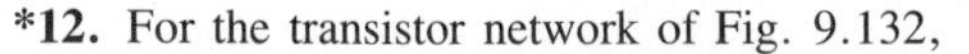

***12.** For the transistor network of Fig. 9.132,

**a.** Find the Thevenin equivalent circuit for that portion of the network to the left of the base ($B$) terminal.

**b.** Using the fact that $I_C = I_E$ and $V_{CE} = 8$ V, determine the magnitude of $I_E$.

**c.** Using the results of parts (a) and (b), calculate the base current $I_B$ if $V_{BE} = 0.7$ V.

**d.** What is the voltage $V_C$?

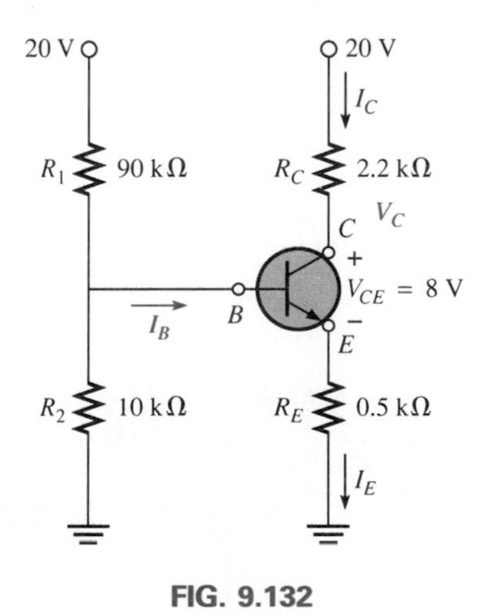


FIG. 9.132

## SECTION 9.4 Norton's Theorem

**13.** Find the Norton equivalent circuit for the network external to the resistor $R$ in each network of Fig. 9.126 by

**a.** following the procedure outlined in the text.

**b.** converting the Thevenin equivalent circuit to a Norton equivalent circuit if available from a previous assignment.

14. Repeat Problem 13 for the networks of Fig. 9.127.

*15. Repeat Problem 13 for the networks of Fig. 9.128.

16. Find the Norton equivalent circuit for the portions of the networks of Fig. 9.129 external to branch *a-b* by
   a. following the procedure outlined in the text.
   b. converting the Thevenin equivalent circuit to a Norton equivalent circuit if available from a previous assignment.

*17. Repeat Problem 16 for the resistor $R$ of Fig. 9.130.

18. Find the Norton equivalent circuit for the portions of the networks of Fig. 9.133 external to branch *a-b*.

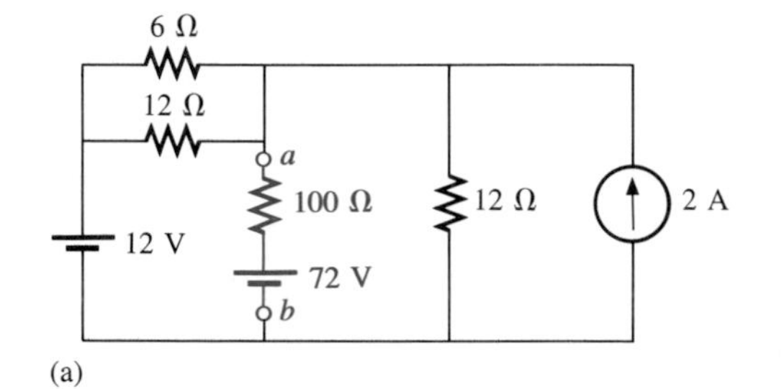


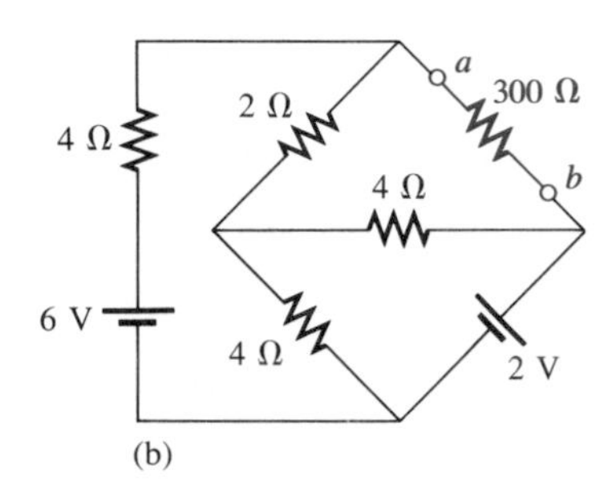

FIG. 9.133

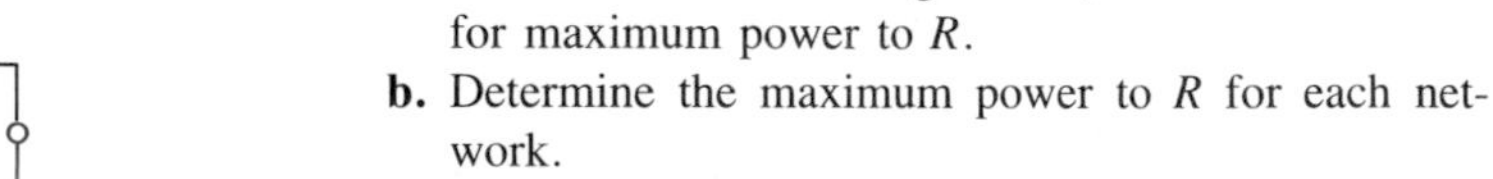
## SECTION 9.5 Maximum Power Transfer Theorem

19. a. For each network of Fig. 9.126, find the value of $R$ for maximum power to $R$.
   b. Determine the maximum power to $R$ for each network.

20. Repeat Problem 19 for the networks of Fig. 9.127.

*21. Repeat Problem 19 for the networks of Fig. 9.128.

22. a. For the network of Fig. 9.134, determine the value of $R$ for maximum power to $R$.
   b. Determine the maximum power to $R$.
   c. Plot a curve of power to $R$ versus $R$ for $R$ equal to $\frac{1}{4}$, $\frac{1}{2}$, $\frac{3}{4}$, 1, $1\frac{1}{4}$, $1\frac{1}{2}$, $1\frac{3}{4}$, and 2 times the value obtained in part (a).

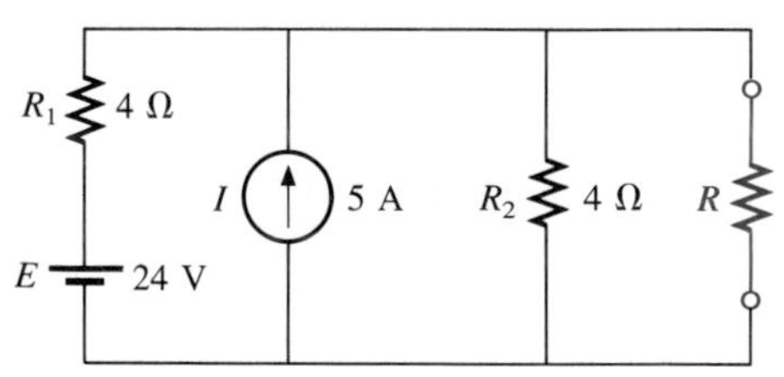


FIG. 9.134

*23. Find the resistance $R_1$ of Fig. 9.135 such that the resistor $R_4$ will receive maximum power. Think!

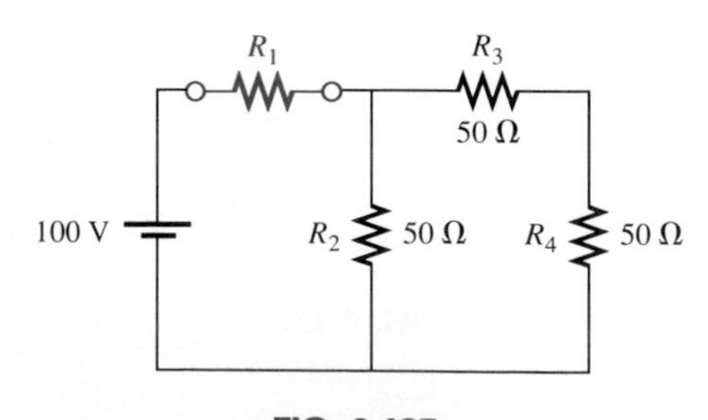


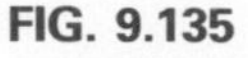
FIG. 9.135

*24. a. For the network of Fig. 9.136, determine the value $R_2$ for maximum power to $R_4$.

b. Is there a general statement that can be made about situations such as those presented here and in Problem 23?

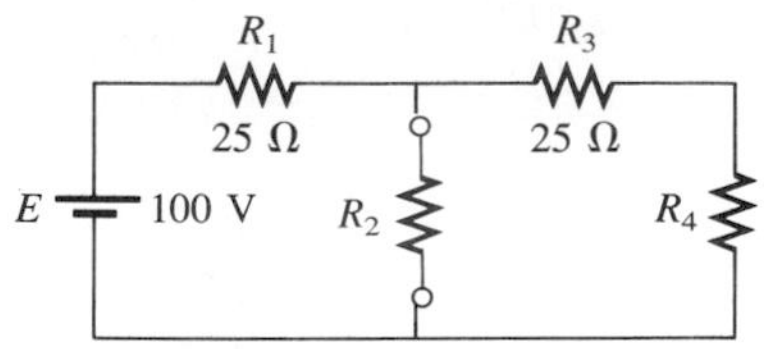

FIG. 9.136

*25. For the network of Fig. 9.137, determine the level of $R$ that will ensure maximum power to the 100-Ω resistor.

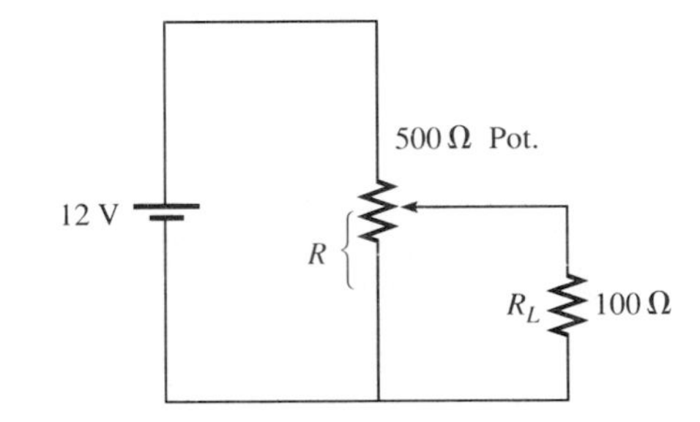

FIG. 9.137

## SECTION 9.6 Millman's Theorem

26. Using Millman's theorem, find the current through and voltage across the resistor $R_L$ of Fig. 9.138.

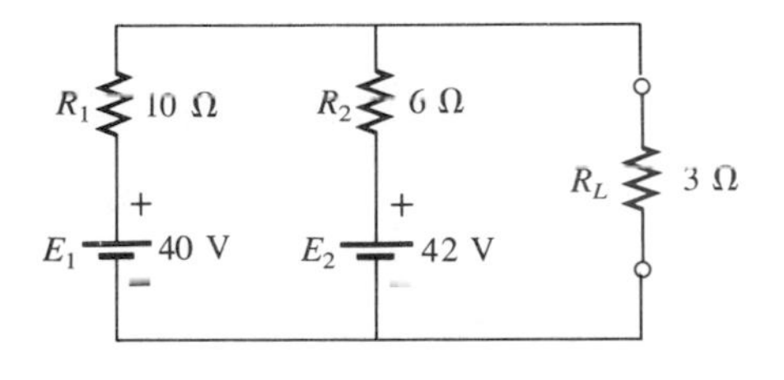

FIG. 9.138

27. Repeat Problem 26 for the network of Fig. 9.139.

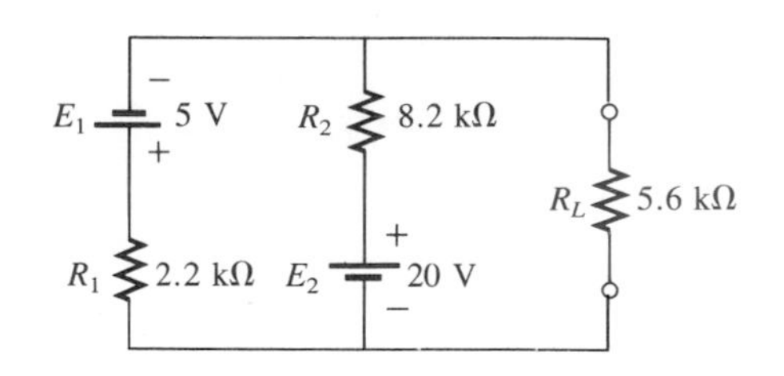

FIG. 9.139

28. Repeat Problem 26 for the network of Fig. 9.140.

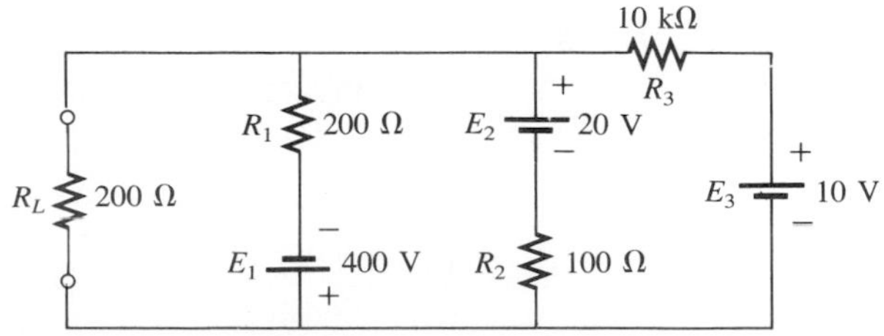

FIG. 9.140

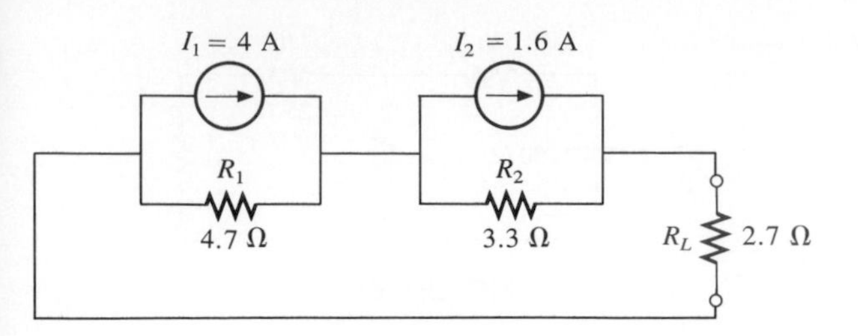


FIG. 9.141

**29.** Using the dual of Millman's theorem, find the current through and voltage across the resistor $R_L$ of Fig. 9.141.

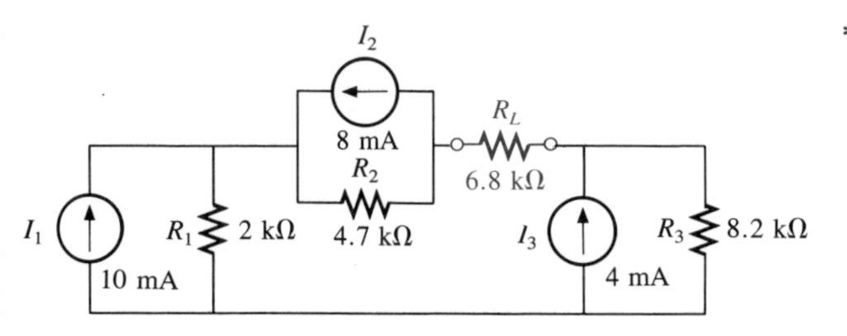


FIG. 9.142

***30.** Repeat Problem 29 for the network of Fig. 9.142.

## SECTION 9.7 Substitution Theorem

**31.** Using the substitution theorem, draw three equivalent branches for the branch $a$-$b$ of the network of Fig. 9.143.

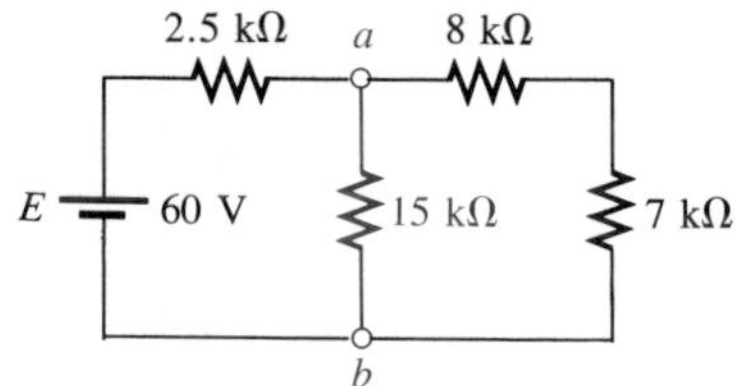
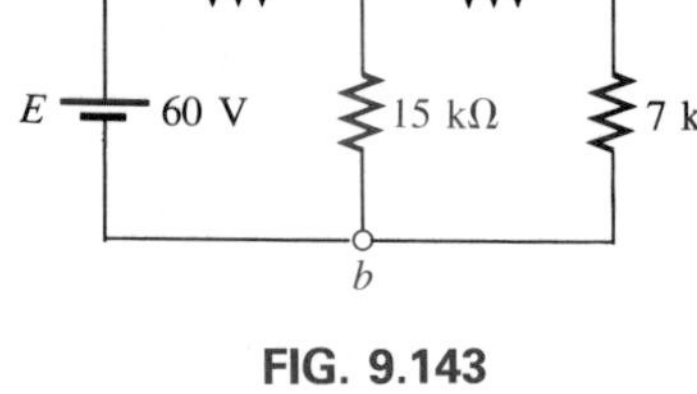


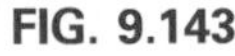
FIG. 9.143

**32.** Repeat Problem 31 for the network of Fig. 9.144.

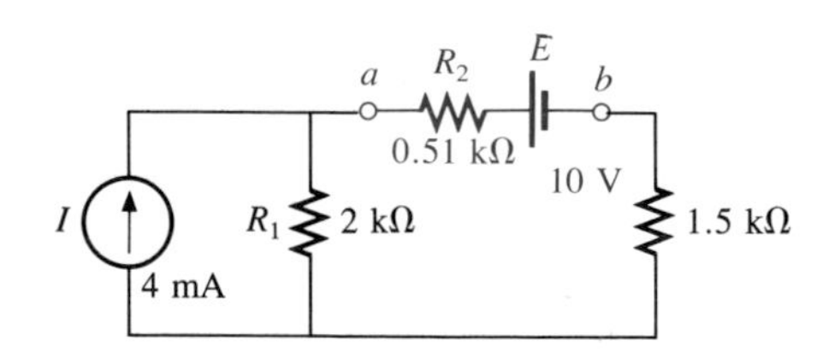

FIG. 9.144

***33.** Repeat Problem 31 for the network of Fig. 9.145. Be careful!

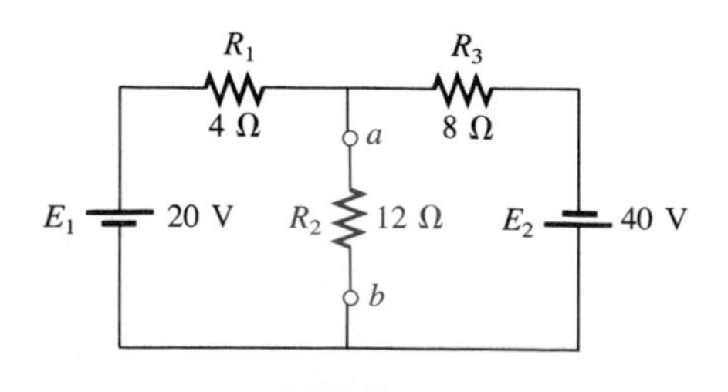


FIG. 9.145

## SECTION 9.8 Reciprocity Theorem

**34. a.** For the network of Fig. 9.146(a), determine the current $I$.

**b.** Repeat part (a) for the network of Fig. 9.146(b).

**c.** Is the reciprocity theorem satisfied?

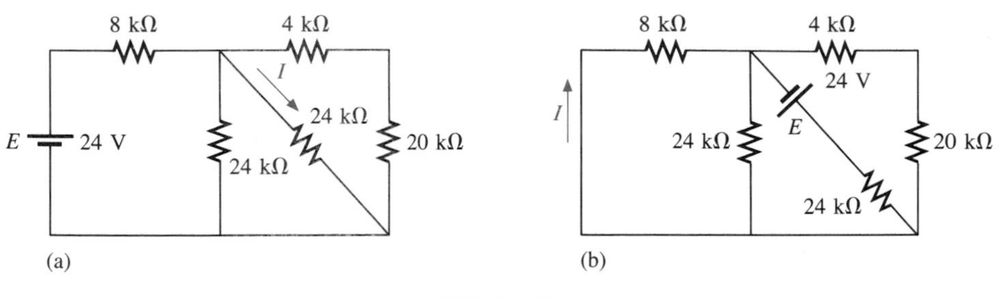

FIG. 9.146

**35.** Repeat Problem 34 for the networks of Fig. 9.147.

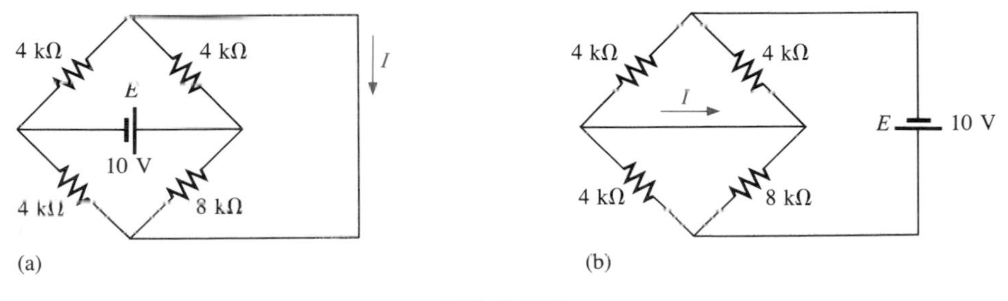

FIG. 9.147

**36. a.** Determine the voltage $V$ for the network of Fig. 9.148(a).

**b.** Repeat part (a) for the network of Fig. 9.148(b).

**c.** Is the dual of the reciprocity theorem satisfied?

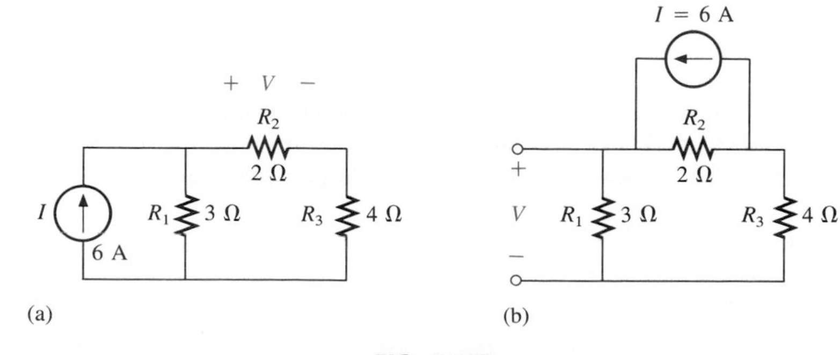


FIG. 9.148

### SECTION 9.9 Computer Analysis

#### PSPICE

**37.** Write the input files required to determine the current *I* and its components for the network of Fig. 9.122(a) using superposition.

**38.** Repeat Problem 37 for the network of Fig. 9.122(b).

***39.** Write the input files required to determine the voltage $V_2$ and its components for the network of Fig. 9.124 using superposition.

**40.** Write the input file to determine the Thevenin equivalent circuit for the network of Fig. 9.126(II) external to the resistor *R*.

***41.** Repeat Problem 40 for the network of Fig. 9.128(II).

***42.** Repeat Problem 40 for the network of Fig. 9.130(I).

**43.** Write the input files necessary to determine the Norton equivalent circuit for the network external to the resistor *R* in Fig. 9.128(I).

***44.** Repeat Problem 22 using PSPICE.

#### BASIC

**45.** Write a program to determine the current through the 10-Ω resistor of Fig. 9.122(a) (for any component values) using superposition.

**46.** Write a program to perform the analysis required for Problem 8, network (II) for any component values.

***47.** Write a program to perform the analysis of Problem 22 and tabulate the power to *R* for the values listed in part (c).

## GLOSSARY

**Maximum power transfer theorem** A theorem used to determine the load resistance necessary to insure maximum power transfer to the load.

**Millman's theorem** A method employing source conversions that will permit the determination of unknown variables in a multiloop network.

**Norton's theorem** A theorem that permits the reduction of any two-terminal linear dc network to one having a single current source and parallel resistor.

**.PROBE** A PSPICE command for obtaining output files that contain plots and data values not otherwise available.

**Reciprocity theorem** A theorem that states that for single-source networks, the current in any branch of a network, due to a single voltage source in the network, will equal the current through the branch in which the source was originally located if the source is placed in the branch in which the current was originally measured.

**Substitution theorem** A theorem that states that if the voltage across and current through any branch of a dc bilateral network are known, the branch can be replaced by any combination of elements that will maintain the same voltage across and current through the chosen branch.

**Superposition theorem** A network theorem that permits considering the effects of each source independently. The resulting current and/or voltage is the algebraic sum of the currents and/or voltages developed by each source independently.

**Thevenin's theorem** A theorem that permits the reduction of any two-terminal linear dc network to one having a single voltage source and series resistor.

# 10
# Capacitors

## 10.1 INTRODUCTION

Thus far, the only passive device appearing in the text has been the resistor. We will now consider two additional passive devices called the *capacitor* and the *inductor,* which are quite different from the resistor in purpose, operation, and construction.

Unlike the resistor, both elements display their total characteristics only when a change in voltage or current is made in the circuit in which they exist. In addition, if we consider the *ideal* situation, they do not dissipate energy like the resistor but store it in a form that can be returned to the circuit whenever required by the circuit design.

Proper treatment of each requires that we devote this entire chapter to the capacitor and Chapter 12 to the inductor. Since electromagnetic effects are a major consideration in the design of inductors, Chapter 11 on magnetic circuits will appear first.

## 10.2 THE ELECTRIC FIELD

Recall from Chapter 2 that a force of attraction or repulsion exists between two charged bodies. We shall now examine this phenomenon in greater detail by considering the electric field that exists in the region around any charged body. This electric field is represented by electric flux lines, which are drawn to indicate the strength of the electric field at any point around the charged body; that is, the denser the lines of flux, the stronger the electric field. In Fig. 10.1, the electric field strength is stronger at position *a* than at position *b* because the flux lines are denser at *a* than at *b*. The symbol for electric flux is the Greek letter

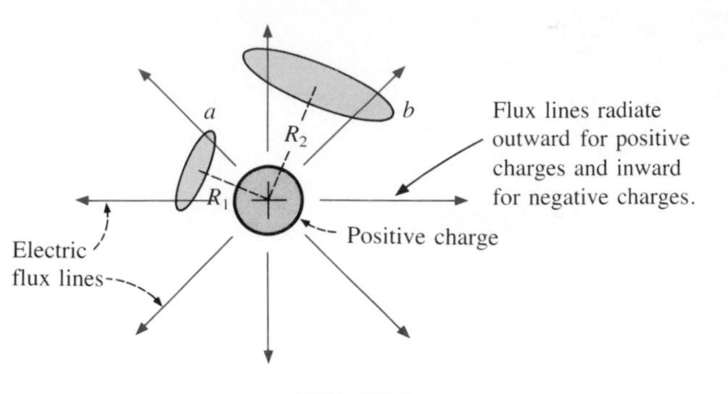


**FIG. 10.1**

$\psi$ (psi). The flux per unit area (flux density) is represented by the capital letter $D$ and is determined by

$$D = \frac{\psi}{A} \qquad \text{(flux/unit area)} \qquad \textbf{(10.1)}$$

The larger the charge $Q$ in coulombs, the greater the number of flux lines extending or terminating per unit area, independent of the surrounding medium. Twice the charge will produce twice the flux per unit area. The two can therefore be equated:

$$\psi \equiv Q \qquad \text{(coulombs, C)} \qquad \textbf{(10.2)}$$

By definition, the *electric field strength* at a point is the force acting on a unit positive charge at that point; that is,

$$\mathscr{E} = \frac{F}{Q} \qquad \text{(newtons/coulomb, N/C)} \qquad \textbf{(10.3)}$$

The force exerted on a unit positive charge ($Q_2 = 1$), by a charge $Q_1$, $r$ meters away, as determined by Coulomb's law is

$$F = \frac{kQ_1Q_2}{r^2} = \frac{kQ_1(1)}{r^2} = \frac{kQ_1}{r^2} \qquad (k = 9 \times 10^9)$$

Substituting this force $F$ into Eq. (10.3) yields

$$\mathscr{E} = \frac{F}{Q_2} = \frac{kQ_1/r^2}{1}$$

$$\mathscr{E} = \frac{kQ_1}{r^2} \qquad \text{(N/C)} \qquad \textbf{(10.4)}$$

We can therefore conclude that the electric field strength at any point distance $r$ from a point charge of $Q$ coulombs is directly proportional to the magnitude of the charge and inversely proportional to the distance squared from the charge. The squared term in the denominator will result in a rapid decrease in the strength of the electric field with distance from the point charge. In Fig. 10.1, substituting distances $R_1$ and $R_2$ into Eq. (10.4) will verify our previous conclusion that the electric field strength is greater at $a$ than at $b$.

***Electric flux lines always extend from a positively charged to a negatively charged body, always extend or terminate perpendicular to the charged surfaces, and never intersect.***

For two charges of similar and opposite polarities, the flux distribution would appear as shown in Fig. 10.2.

The attraction and repulsion between charges can now be explained in terms of the electric field and its flux lines. In Fig. 10.2(a), the flux lines are not interlocked but tend to act as a buffer, preventing attraction and causing repulsion. Since the electric field strength is stronger (flux lines denser) for each charge the closer we are to the charge, the more we try to bring the two charges together, the stronger will be the force of repulsion between them. In Fig. 10.2(b), the flux lines extending from the positive charge are terminated at the negative charge. A basic law of physics states that electric flux lines always tend to be as short as possible. The two charges will therefore be drawn to each other. Again, the closer the two charges, the stronger the attraction between the two charges due to the increased field strengths.

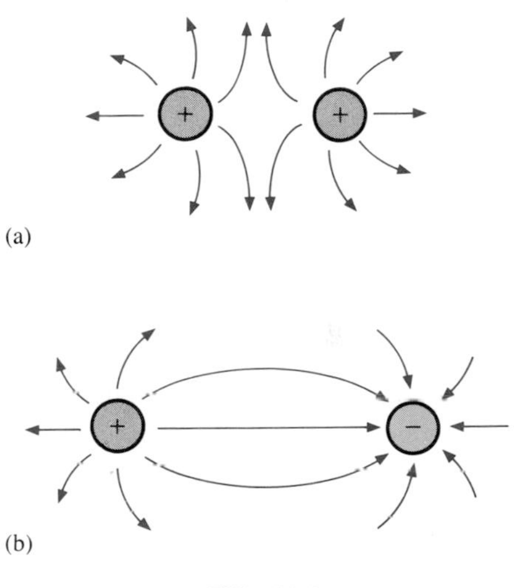


FIG. 10.2

## 10.3 CAPACITANCE

Up to this point we have considered only isolated positive and negative spherical charges, but the analysis can be extended to charged surfaces of any shape and size. In Fig. 10.3, for example, two parallel plates of

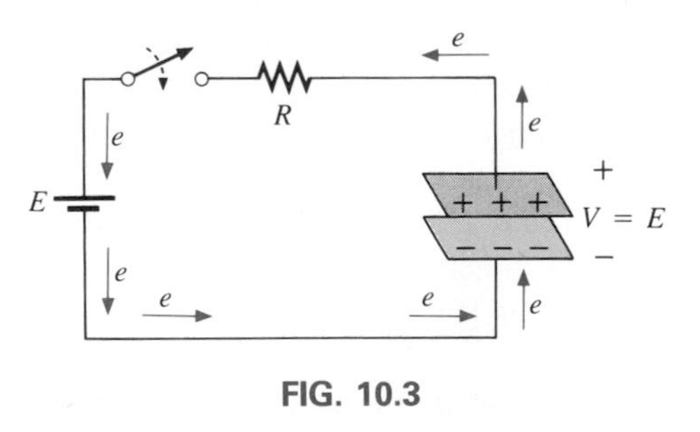


FIG. 10.3

a conducting material separated by an air gap have been connected through a switch and a resistor to a battery. If the parallel plates are initially uncharged and the switch is left open, no net positive or negative charge will exist on either plate. The instant the switch is closed, however, electrons are drawn from the upper plate through the resistor to the positive terminal of the battery. There will be a surge of current at first, limited in magnitude by the resistance present. The level of flow will then decline, as will be demonstrated in the sections to follow. This action creates a net positive charge on the top plate. Electrons are being repelled by the negative terminal through the lower conductor to the bottom plate at the same rate they are being drawn to the positive terminal. This transfer of electrons continues until the potential difference across the parallel plates is exactly equal to the battery voltage. The final result is a net positive charge on the top plate and a negative charge on the bottom plate, very similar in many respects to the two isolated charges of Fig. 10.2(b).

This element, constructed simply of two parallel conducting plates separated by an insulating material (in this case, air), is called a *capacitor*. *Capacitance* is a measure of a capacitor's ability to store charge on its plates—in other words, its storage capacity.

***A capacitor has a capacitance of 1 farad if 1 coulomb of charge is deposited on the plates by a potential difference of 1 volt across the plates.***

The farad is named after Michael Faraday, a nineteenth-century English chemist and physicist. The farad, however, is generally too large a measure of capacitance for most practical applications, so the microfarad ($10^{-6}$) or picofarad ($10^{-12}$) is more commonly used. Expressed as an equation, the capacitance is determined by

$$C = \frac{Q}{V} \qquad \begin{aligned} C &= \text{farads (F)} \\ Q &= \text{coulombs (C)} \\ V &= \text{volts (V)} \end{aligned} \tag{10.5}$$

Different capacitors for the same voltage across their plates will acquire greater or lesser amounts of charge on their plates. Hence the capacitors have a greater or lesser capacitance, respectively.

A cross-sectional view of the parallel plates is shown with the distribution of electric flux lines in Fig. 10.4(a). The number of flux lines per unit area ($D$) between the two plates is quite uniform. At the edges, the flux lines extend outside the common surface area of the plates, producing an effect known as *fringing*. This effect, which reduces the capacitance somewhat, can be neglected for most practical applications. For the analysis to follow, we will assume that all the flux lines leaving the positive plate will pass directly to the negative plate within the common surface area of the plates [Fig. 10.4(b)].

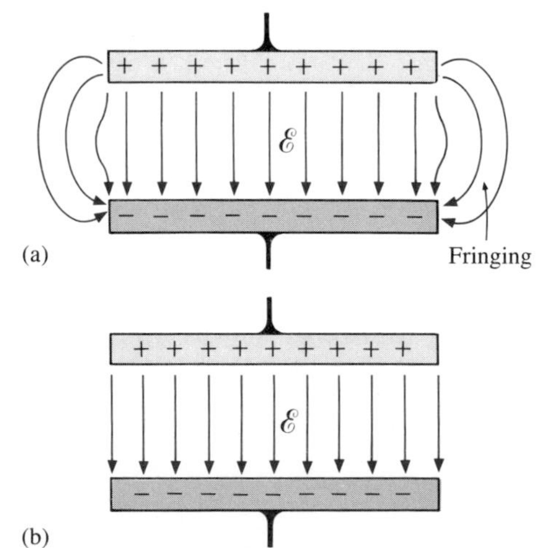

FIG. 10.4

If a potential difference of $V$ volts is applied across the two plates separated by a distance of $d$, the electric field strength between the plates is determined by

$$\mathscr{E} = \frac{V}{d} \qquad \text{(volts/meter, V/m)} \tag{10.6}$$

The uniformity of the flux distribution in Fig. 10.4(b) also indicates that the electric field strength is the same at any point between the two plates.

Many values of capacitance can be obtained for the same set of parallel plates by the addition of certain insulating materials between the plates. In Fig. 10.5(a), an insulating material has been placed between a set of parallel plates having a potential difference of $V$ volts across them.

Since the material is an insulator, the electrons within the insulator are unable to leave the parent atom and travel to the positive plate. The positive components (protons) and negative components (electrons) of each atom do shift, however [as shown in Fig. 10.5(a)], to form *dipoles*.

When the dipoles align themselves as shown in Fig. 10.5(a), the material is *polarized*. A close examination within this polarized material will indicate that the positive and negative components of adjoining dipoles are neutralizing the effects of each other [note the dashed area in Fig. 10.5(a)]. The layer of positive charge on one surface and the negative charge on the other are not neutralized, however, resulting in the establishment of an electric field within the insulator [$\mathscr{E}_{\text{dielectric}}$, Fig. 10.5(b)]. The net electric field between the plates ($\mathscr{E}_{\text{resultant}} = \mathscr{E}_{\text{air}} - \mathscr{E}_{\text{dielectric}}$) would therefore be reduced due to the insertion of the dielectric.

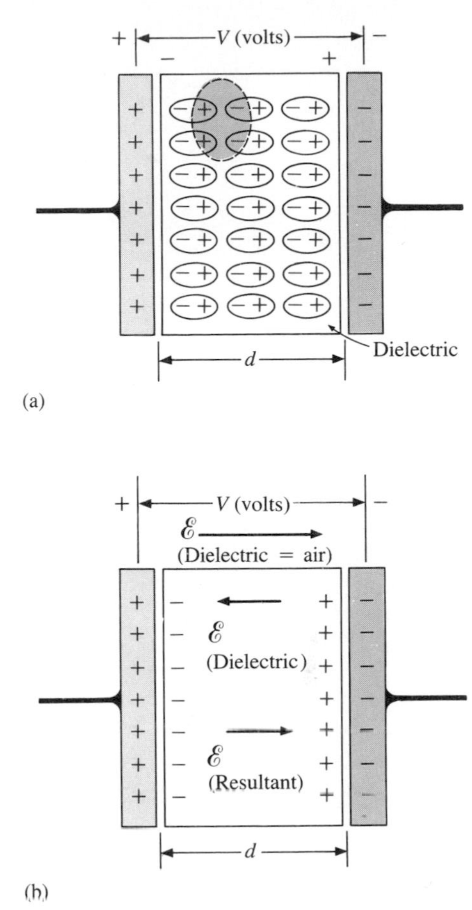

FIG. 10.5

The purpose of the dielectric, therefore, is to create an electric field to oppose the electric field set up by free charges on the parallel plates. For this reason, the insulating material is referred to as a *dielectric, di* for *opposing* and *electric* for *electric field*.

In either case—with or without the dielectric—if the potential across the plates is kept constant and the distance between the plates is fixed, the net electric field within the plates must remain the same, as determined by the equation $\mathscr{E} = V/d$. We just ascertained, however, that the net electric field between the plates would decrease with insertion of the dielectric for a fixed amount of free charge on the plates. To compensate and keep the net electric field equal to the value determined by $V$ and $d$, more charge must be deposited on the plates. [Look ahead to Eq. (10.11).] This additional charge for the same potential across the plates increases the capacitance, as determined by the following equation:

$$C\uparrow = \frac{Q\uparrow}{V}$$

For different dielectric materials between the same two parallel plates, different amounts of charge will be deposited on the plates. But $\psi \equiv Q$, so the dielectric is also determining the number of flux lines between the two plates and consequently the flux density ($D = \psi/A$) since $A$ is fixed.

The ratio of the flux density to the electric field intensity in the dielectric is called the *permittivity* of the dielectric:

$$\epsilon = \frac{D}{\mathscr{E}} \qquad \text{(farads/meter, F/m)} \qquad \textbf{(10.7)}$$

It is a measure of how easily the dielectric will "permit" the establishment of flux lines within the dielectric. The greater its value, the greater the amount of charge deposited on the plates, and, consequently, the greater the flux density for a fixed area.

For a vacuum, the value of $\epsilon$ (denoted by $\epsilon_o$) is $8.85 \times 10^{-12}$ F/m. The ratio of the permittivity of any dielectric to that of a vacuum is called the *relative permittivity,* $\epsilon_r$. It simply compares the permittivity of the dielectric to that of air. In equation form,

$$\boxed{\epsilon_r = \frac{\epsilon}{\epsilon_o}} \tag{10.8}$$

The value of $\epsilon$ for any material, therefore, is

$$\epsilon = \epsilon_r \epsilon_o$$

Note that $\epsilon_r$ is a dimensionless quantity. The relative permittivity, or *dielectric constant,* as it is often called, is provided in Table 10.1 for various dielectric materials.

**TABLE 10.1**
*Relative permittivity (dielectric constant) of various dielectrics.*

| Dielectric | $\epsilon_r$ (Average Values) |
|---|---|
| Vacuum | 1.0 |
| Air | 1.0006 |
| Teflon® | 2.0 |
| Paper, paraffined | 2.5 |
| Rubber | 3.0 |
| Transformer oil | 4.0 |
| Mica | 5.0 |
| Porcelain | 6.0 |
| Bakelite | 7.0 |
| Glass | 7.5 |
| Distilled water | 80.0 |
| Barium-strontium titanite (ceramic) | 7500.0 |

Substituting for $D$ and $\mathscr{E}$ in Eq. (10.7), we have

$$\epsilon = \frac{D}{\mathscr{E}} = \frac{\psi/A}{V/d} = \frac{Q/A}{V/d} = \frac{Qd}{VA}$$

But

$$C = \frac{Q}{V}$$

and therefore

$$\epsilon = \frac{Cd}{A}$$

and

$$C = \epsilon \frac{A}{d} \qquad \text{(F)} \qquad \textbf{(10.9)}$$

or

$$C = \epsilon_o \epsilon_r \frac{A}{d} = 8.85 \times 10^{-12} \epsilon_r \frac{A}{d} \qquad \text{(F)} \qquad \textbf{(10.10)}$$

where $A$ is the area in square meters of the plates, $d$ is the distance in meters between the plates, and $\epsilon_r$ is the relative permittivity. The capacitance, therefore, will be greater if the area of the plates is increased, or the distance between the plates is decreased, or the dielectric is changed so that $\epsilon_r$ is increased.

Solving for the distance $d$ in Eq. (10.9), we have

$$d = \frac{\epsilon A}{C}$$

and substituting into Eq. (10.6) yields

$$\mathscr{E} = \frac{V}{d} = \frac{V}{\epsilon A/C} = \frac{CV}{\epsilon A}$$

But $Q = CV$, and therefore

$$\mathscr{E} = \frac{Q}{\epsilon A} \qquad \text{(V/m)} \qquad \textbf{(10.11)}$$

which gives the electric field intensity between the plates in terms of the permittivity $\epsilon$, the charge $Q$, and the surface area $A$ of the plates. The ratio

$$\frac{C = \epsilon A/d}{C_o = \epsilon_o A/d} = \frac{\epsilon}{\epsilon_o} = \epsilon_r$$

or

$$C = \epsilon_r C_o \qquad \textbf{(10.12)}$$

which, in words, states that for the same set of parallel plates, the capacitance using a dielectric is $\epsilon_r$ times that obtained for a vacuum (or air, approximately) between the plates. This relationship between $\epsilon_r$ and the capacitances provides an excellent experimental method for finding the value of $\epsilon_r$ for various dielectrics.

**EXAMPLE 10.1.** Determine the capacitance of each capacitor on the right side of Fig. 10.6.

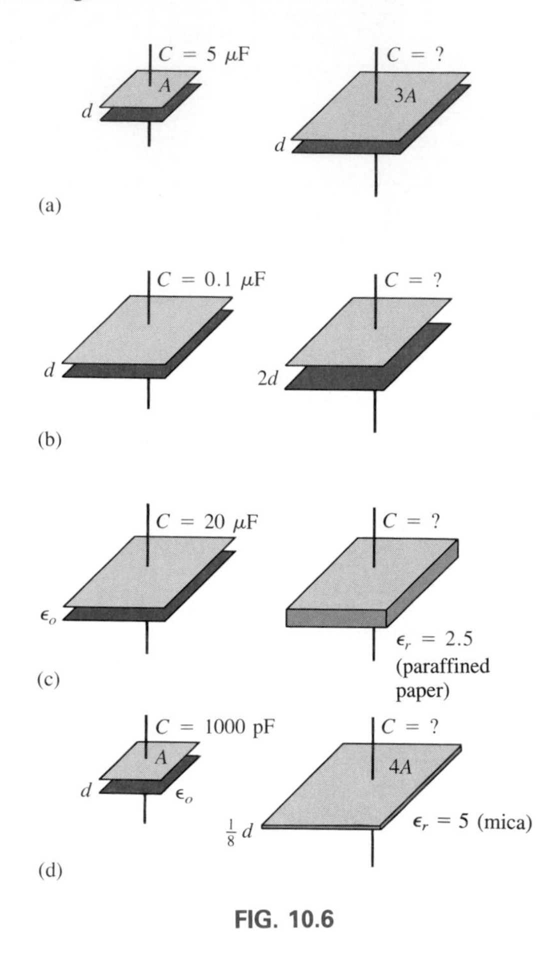


FIG. 10.6

***Solution:***

a. $C = 3(5\ \mu F) = \mathbf{15\ \mu F}$

b. $C = \frac{1}{2}(0.1\ \mu F) = \mathbf{0.05\ \mu F}$

c. $C = 2.5(20\ \mu F) = \mathbf{50\ \mu F}$

d. $C = (5)\frac{4}{(1/8)}(1000\ pF) = (160)(1000\ pF) = \mathbf{0.16\ \mu F}$

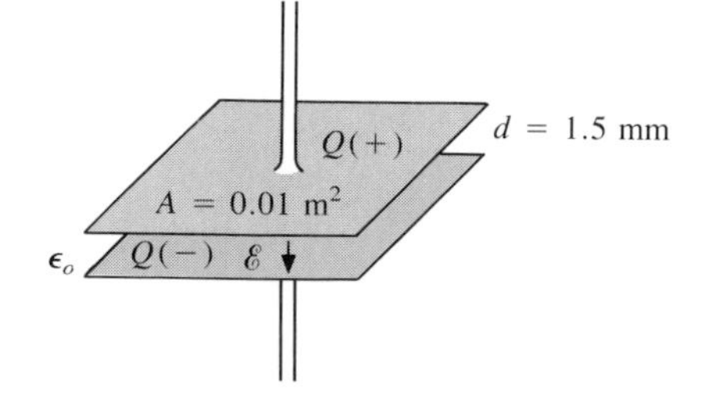


FIG. 10.7

**EXAMPLE 10.2.** For the capacitor of Fig. 10.7:

a. Determine the capacitance.
b. Determine the electric field strength between the plates if 450 V are applied across the plates.
c. Find the resulting charge on each plate.

***Solutions:***

a. $$C_o = \frac{\epsilon_o A}{d} = \frac{(8.85 \times 10^{-12}\text{ F/m})(0.01\text{ m}^2)}{1.5 \times 10^{-3}\text{ m}} = 59.0 \times 10^{-12}\text{ F}$$
$$= \mathbf{59\ pF}$$

b. $$\mathscr{E} = \frac{V}{d} = \frac{450\text{ V}}{1.5 \times 10^{-3}\text{ m}}$$
$$\cong \mathbf{300 \times 10^3\ V/m}$$

c. $$C = \frac{Q}{V}$$

or

$$Q = CV = (59.0 \times 10^{-12}\text{ F})(450\text{ V})$$
$$= 26.550 \times 10^{-9}\text{ C}$$
$$= \mathbf{26.55\ nC}$$

---

**EXAMPLE 10.3.** A sheet of mica 1.5 mm thick having the same area as the plates is inserted between the plates of Example 10.2.

a. Find the electric field strength between the plates.
b. Find the charge on each plate.
c. Find the capacitance.

***Solutions:***

a. $\mathscr{E}$ is fixed by

$$\mathscr{E} = \frac{V}{d} = \frac{450\text{ V}}{1.5 \times 10^{3}\text{ m}}$$
$$\cong \mathbf{300 \times 10^3\ V/m}$$

b. $$\mathscr{E} = \frac{Q}{\epsilon A}$$

or

$$Q = \epsilon \mathscr{E} A = \epsilon_r \epsilon_o \mathscr{E} A$$
$$= (5)(8.85 \times 10^{-12}\text{ F/m})(300 \times 10^3\text{ V/m})(0.01\text{ m}^2)$$
$$= 132.75 \times 10^{-9}\text{ C} = \mathbf{132.75\ nC}$$

(five times the amount for air between the plates)

c. $$C = \epsilon_r C_o$$
$$= (5)(59 \times 10^{-12}\text{ F}) = \mathbf{295\ pF}$$

---

## 10.4 DIELECTRIC STRENGTH

For every dielectric there is a potential that if applied across the dielectric will break the bonds within the dielectric and cause current to flow. The voltage required per unit length (electric field intensity) to establish conduction in a dielectric is an indication of its *dielectric strength* and is called the *breakdown voltage*. When breakdown occurs, the capacitor has characteristics very similar to those of a conductor. A typical example of breakdown is lightning, which occurs when the potential between the clouds and the earth is so high that charge can pass from one to the other through the atmosphere, which acts as the dielectric.

The average dielectric strengths for various dielectrics are tabulated in volts/mil in Table 10.2 (1 mil = 0.001 in.). The relative permittivity appears in parentheses to emphasize the importance of considering both factors in the design of capacitors. Take particular note of barium-strontium titanite and mica.

**TABLE 10.2**
*Dielectric strength of some dielectric materials.*

| Dielectric | Dielectric Strength (Average Value), in Volts/Mil | ($\epsilon_r$) |
|---|---|---|
| Air | 75 | (1.006) |
| Barium-strontium titanite (ceramic) | 75 | (7500) |
| Porcelain | 200 | (6.0) |
| Transformer oil | 400 | (4.0) |
| Bakelite | 400 | (7.0) |
| Rubber | 700 | (3.0) |
| Paper, paraffined | 1300 | (2.5) |
| Teflon | 1500 | (2.0) |
| Glass | 3000 | (7.5) |
| Mica | 5000 | (5.0) |

**EXAMPLE 10.4.** Find the maximum voltage that can be applied across a 0.2-$\mu$F capacitor having a plate area of 0.3 $m^2$. The dielectric is porcelain. Assume a linear relationship between the dielectric strength and the thickness of the dielectric.

***Solution:***

$$C = 8.85 \times 10^{-12}\epsilon_r \frac{A}{d}$$

or

$$d = \frac{8.85\epsilon_r A}{10^{12}C} = \frac{(8.85)(6)(0.3\text{ m}^2)}{(10^{12})(0.2 \times 10^{-6}\text{ F})} = 7.965 \times 10^{-5}\text{ m}$$

$$\cong \mathbf{79.65\ \mu m}$$

Converting millimeters to mils, we have

$$79.76\ \cancel{\mu m}\left(\frac{10^{-6}\ \cancel{m}}{\cancel{\mu m}}\right)\left(\frac{39.371\ \cancel{in.}}{\cancel{m}}\right)\left(\frac{1000\text{ mils}}{1\ \cancel{in.}}\right) = 3.136\text{ mils}$$

$$\text{Dielectric strength} = 200\text{ V/mil}$$

Therefore,

$$\left(\frac{200\ \text{V}}{\cancel{\text{mil}}}\right)(3.136\ \cancel{\text{mils}}) = \mathbf{627.20\ V}$$

## 10.5 LEAKAGE CURRENT

Up to this point, we have assumed that the flow of electrons will occur in a dielectric only when the breakdown voltage is reached. This is the ideal case. In actuality, there are free electrons in every dielectric due in part to impurities in the dielectric and forces within the material itself.

When a voltage is applied across the plates of a capacitor, a leakage current due to the free electrons flows from one plate to the other. The current is usually so small, however, that it can be neglected for most practical applications. This effect is represented by a resistor in parallel with the capacitor, as shown in Fig. 10.8(a), whose value is typically more than 100 megohms (MΩ). There are some capacitors, however, such as the electrolytic type, that have high leakage currents. When charged and then disconnected from the charging circuit, these capacitors lose their charge in a matter of seconds because of the flow of charge (leakage current) from one plate to the other [Fig. 10.8(b)].

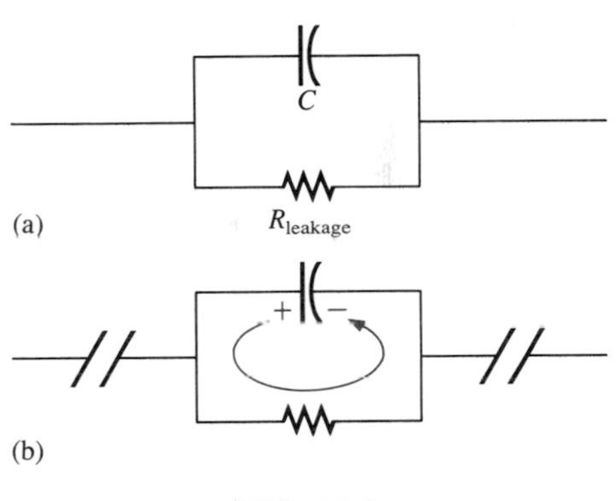


FIG. 10.8

## 10.6 TYPES OF CAPACITORS

Like resistors, all capacitors can be included under either of two general headings: fixed or variable. The symbol for a fixed capacitor is ⊥ and for a variable capacitor ⊥̸. The curved line represents the plate that is usually connected to the point of lower potential.

### Fixed Capacitors

Many types of fixed capacitors are available today. Some of the most common are the mica, ceramic, electrolytic, tantalum, and polyester-film capacitors. The typical *mica capacitor* consists basically of mica sheets separated by sheets of metal foil. The plates are connected to two electrodes, as shown in Fig. 10.9. The total area is the area of one sheet times the number of dielectric sheets. The entire system is encased in a plastic insulating material as shown in Fig. 10.10(a). The mica capacitor exhibits excellent characteristics under stress of temperature variations and high voltage applications (its dielectric strength is 5000 V/mil). Its leakage current is also very small ($R_{\text{leakage}}$ about 1000 MΩ).

Mica capacitors are typically between a few picofarads and 0.2 μF with voltages of 100 V or more. The color code for the mica capacitors of Fig. 10.10(a) can be found in Appendix D.

A second type of mica capacitor appears in Fig. 10.10(b). Note in particular the cylindrical unit in the bottom left-hand corner of the

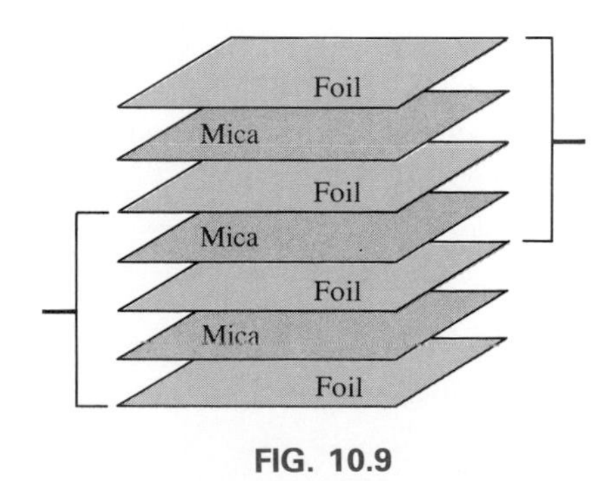


FIG. 10.9

(a) 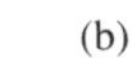 (b)

**FIG. 10.10**
*Mica capacitors. (Courtesy of Custom Electronics Inc.)*

figure. The ability to "roll" the mica to form the cylindrical shape is due to a process whereby the soluble contaminants in natural mica are removed, leaving a paperlike structure due to the cohesive forces in natural mica. It is commonly referred to as *reconstituted mica,* although the terminology does not mean "recycled" or "second-hand" mica. For some of the units in the photograph, different levels of capacitance are available between different sets of terminals.

The *ceramic capacitor* is made in many shapes and sizes, some of which are shown in Fig. 10.11. The basic construction, however, is about the same for each, as shown in Fig. 10.12. A ceramic base is

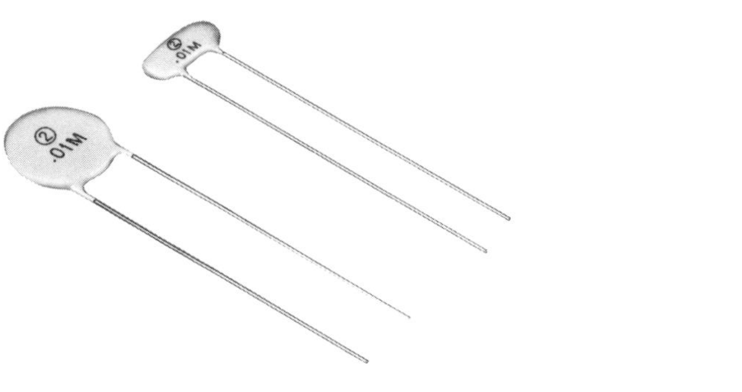


**FIG. 10.11**
*Ceramic disc capacitors. (Courtesy of Sprague Electric Co.)*

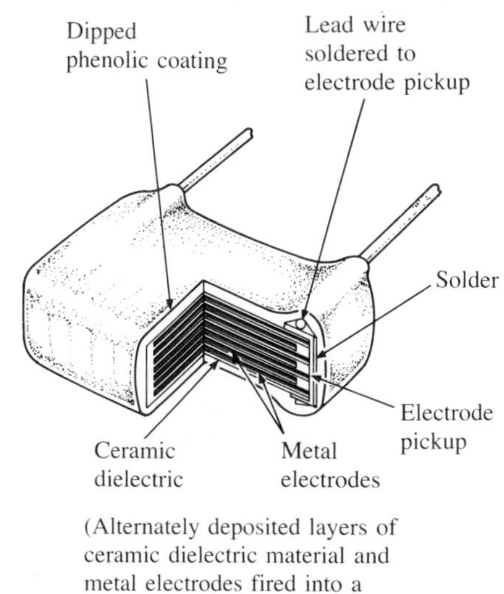


**FIG. 10.12**
*Multilayer, radial-lead ceramic capacitor.*

coated on two sides with a metal, such as copper or silver, to act as the two plates. The leads are then attached through electrodes to the plates. An insulating coating of ceramic or plastic is then applied over the plates and dielectric. Ceramic capacitors also have a very low leakage current ($R_{\text{leakage}}$ about 1000 MΩ) and can be used in both dc and ac networks. They can be found in values ranging from a few picofarads to perhaps 2 μF, with very high working voltages such as 5000 V or more.

In recent years there has been increasing interest in monolithic (single-structure) chip capacitors such as appearing in Fig. 10.13(a) due to their application on hybrid circuitry [networks using both discrete and integrated circuit (IC) components]. There has also been increasing use of microstrip (strip-line) circuitry such as appearing in Fig. 10.13(b). Note the small chips in this cutaway section. The *L* and *H* of Fig. 10.13(a) indicate the level of capacitance. For example, the letter *H* in black letters represents 16 units of capacitance (in picofarads), or 16 pF. If blue ink is used, a multiplier of 100 is applied, resulting in 1600 pF. Although the size is similar, the type of ceramic material controls the capacitance level.

(a)

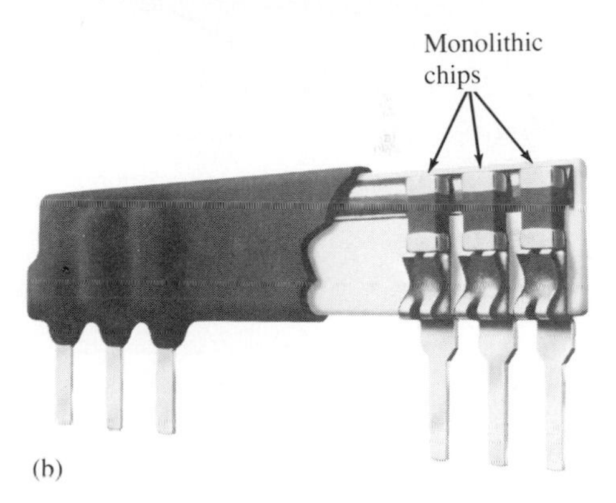


(b)

**FIG. 10.13**
*Monolithic chip capacitors. (Courtesy of Vitramon, Inc.)*

The *electrolytic capacitor* is used most commonly in situations where capacitances of the order of one to several thousand microfarads are required. They are designed primarily for use in networks where only dc voltages will be applied across the capacitor because they have good insulating characteristics (high leakage current) between the plates in one direction but take on the characteristics of a conductor in the other direction. There are electrolytic capacitors available that can be used in ac circuits (for starting motors) and in cases where the polarity of the dc voltage will reverse across the capacitor for short periods of time.

The basic construction of the electrolytic capacitor consists of a roll of aluminum foil coated on one side with an aluminum oxide, the aluminum being the positive plate and the oxide the dielectric. A layer of paper or gauze saturated with an electrolyte is placed over the aluminum oxide on the positive plate. Another layer of aluminum without the oxide coating is then placed over this layer to assume the role of the negative plate. In most cases the negative plate is connected directly to the aluminum container, which then serves as the negative terminal for external connections. Because of the size of the roll of aluminum foil, the overall area of this capacitor is large; and due to the use of an oxide as the dielectric, the distance between the plates is extremely small. The negative terminal of the electrolytic capacitor is usually the one with no visible identification on the casing. The positive is usually indicated by such designs as +, △, □, and so on. Due to the polarity requirement, the symbol for an electrolytic will normally appear as ⊥⊤⁺.

Associated with each electrolytic capacitor are the dc working voltage and the surge voltage. The *working voltage* is the voltage that can be applied across the capacitor for long periods of time without breakdown. The *surge voltage* is the maximum dc voltage that can be applied for a short period of time. Electrolytic capacitors are characterized as having low breakdown voltages and high leakage currents ($R_{\text{leakage}}$ about 1 M$\Omega$). Various types of electrolytic capacitors are shown in Fig. 10.14. They can be found in values extending from a few microfarads to several thousand microfarads and working voltages as high as 500 V. However, increased levels of voltage are normally associated with lower values of available capacitance.

There are fundamentally two types of *tantalum capacitors:* the *solid* and the *wet-slug*. In each case, tantalum powder of high purity is pressed into a rectangular or cylindrical shape as shown in Fig. 10.15.

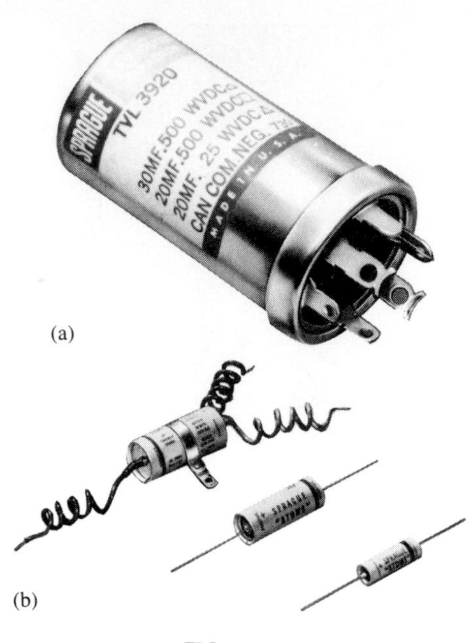


**FIG. 10.14**
*Electrolytic capacitors. (Courtesy of Sprague Electric Co.)*

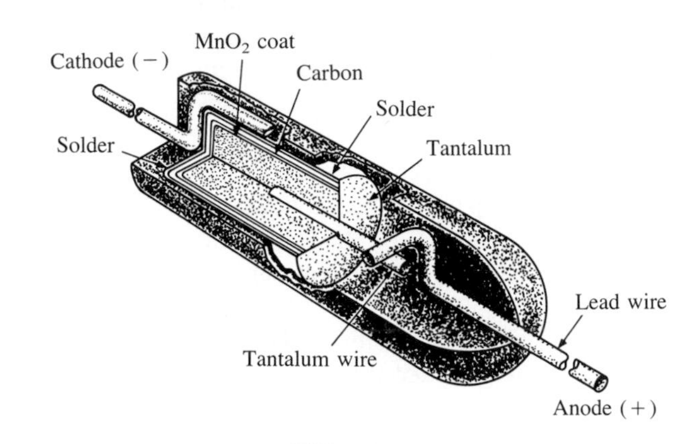


**FIG. 10.15**
*Tantalum capacitor. (Courtesy of Union Carbide Corp.)*

The anode (+) connection is then simply pressed into the resulting structures as shown in the figure. The resulting unit is then sintered (baked) in a vacuum at very high temperatures to establish a very porous material. The result is a structure with a very large surface area in a limited volume. Through immersion in an acid solution, a very thin manganese dioxide ($MnO_2$) coating is established on the large, porous surface area. An electrolyte is then added to establish contact between the surface area and the cathode, producing a solid tantalum capacitor. If an appropriate ''wet'' acid is introduced, it is called a *wet-slug* tantalum capacitor.

The last type of fixed capacitor to be introduced is the *polyester-film capacitor,* the basic construction of which is shown in Fig. 10.16. It consists simply of two metal foils separated by a strip of polyester material such as Mylar®. The outside layer of polyester is applied to act as an insulating jacket. Each metal foil is connected to a lead which extends either axially or radially from the capacitor. The rolled construction results in a large surface area, and the use of the plastic dielectric results in a very thin layer between the conducting surfaces.

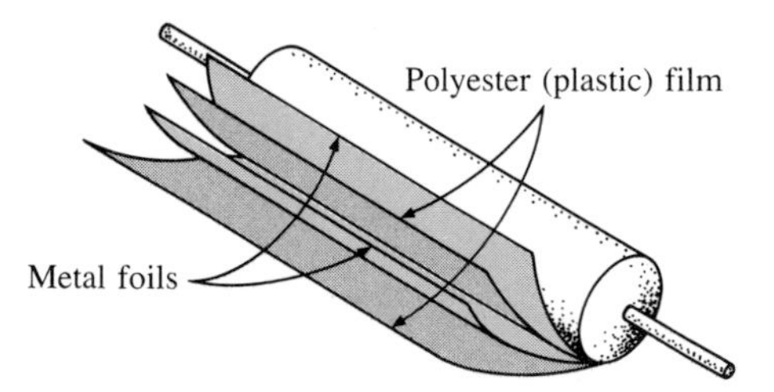


**FIG. 10.16**
*Polyester-film capacitor.*

Data such as capacitance and working voltage are printed on the outer wrapping if the polyester capacitor is large enough. Color coding is used on smaller devices (see Appendix E). A band (usually black) is sometimes printed near the lead that is connected to the outer metal foil. The lead nearest this band should always be connected to the point of lower potential. This capacitor can be used for both dc and ac networks.

Its leakage resistance is of the order of 100 MΩ. A typical polyester capacitor appears in Fig. 10.17. Polyester capacitors range in value from a few hundred picofarads to 10–20 μF, with working voltages as high as a few thousand volts.

**FIG. 10.17**
*Orange Drop® tubular capacitor. (Courtesy of Sprague Electric Co.)*

## Variable Capacitors

The most common of the variable-type capacitors is shown in Fig. 10.18. The dielectric for each capacitor is air. The capacitance in Fig. 10.18(a) is changed by turning the shaft at one end to vary the common area of the movable and fixed plates. The greater the common area, the larger the capacitance, as determined by Eq. (10.10). The capacitance of the trimmer capacitor in Fig. 10.18(b) is changed by turning the screw, which will vary the distance between the plates and thereby the capacitance.

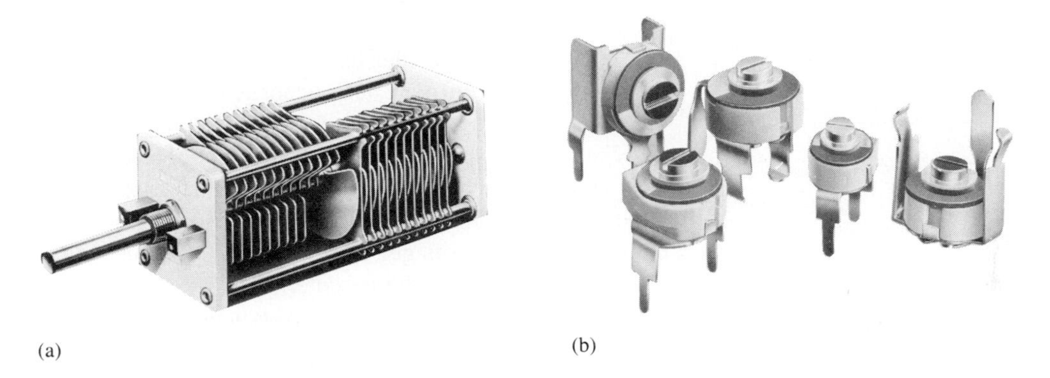

(a) (b)

**FIG. 10.18**
*Variable air capacitors. (Part (a) courtesy of James Millen Manufacturing Co.; part (b) courtesy of Johnson Manufacturing Co.)*

**FIG. 10.19**
*Digital reading capacitance meter. (Courtesy of Global Specialties Corp.)*

## Measurement and Testing

A digital reading capacitance meter appears in Fig. 10.19 that displays the level of capacitance by simply placing the capacitor between the provided clips with the proper polarity.

The best check of a capacitor is to use a meter designed to perform the necessary tests. However, an ohmmeter can identify those in which the dielectric has deteriorated (especially in paper and electrolytic capacitors). As the dielectric breaks down, the insulating qualities decrease to a point where the resistance between the plates drops to a relatively low level. After ensuring that the capacitor is fully discharged, place an ohmmeter across the capacitor, as shown in Fig. 10.20. In a polarized capacitor, the polarities of the meter should match those of the capacitor. A low resistance reading (zero ohms to a few hundred ohms) normally indicates a defective capacitor.

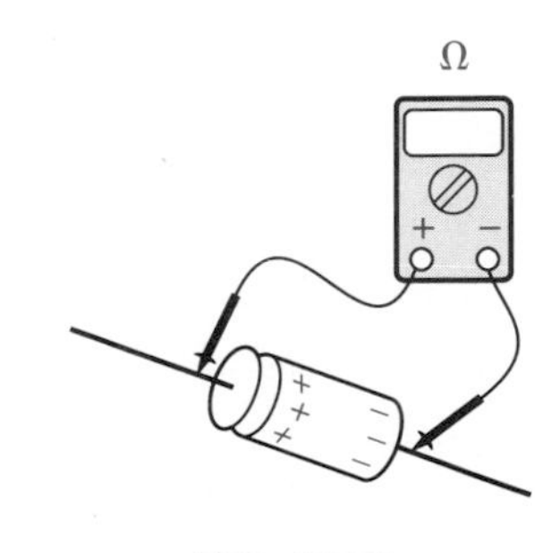


**FIG. 10.20**
*Checking the dielectric of an electrolytic capacitor.*

The above test of leakage is not all inclusive, since some capacitors will break down only when higher voltages are applied. The test, how-

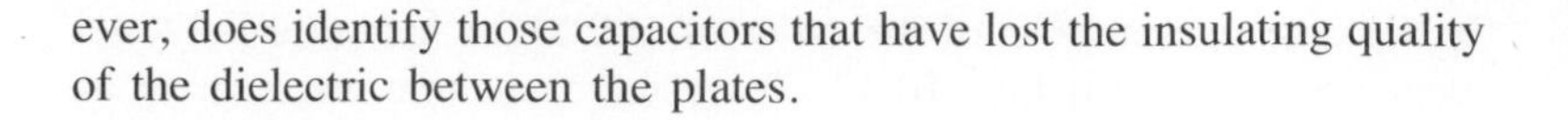
ever, does identify those capacitors that have lost the insulating quality of the dielectric between the plates.

## 10.7 TRANSIENTS IN CAPACITIVE NETWORKS: CHARGING PHASE

FIG. 10.21

Section 10.3 described how a capacitor acquires its charge. Let us now extend this discussion to include the potentials and current developed within the network of Fig. 10.21 following the closing of the switch (to position 1).

You will recall that the instant the switch is closed, electrons are drawn from the top plate and deposited on the bottom plate by the battery, resulting in a net positive charge on the top plate, and a negative charge on the bottom plate. The transfer of electrons is very rapid at first, slowing down as the potential across the capacitor approaches the applied voltage of the battery. When the voltage across the capacitor equals the battery voltage, the transfer of electrons will cease and the plates will have a net charge determined by $Q = CV_C = CE$.

Plots of the changing current and voltage appear in Figs. 10.22 and 10.23, respectively. When the switch is closed at $t = 0$ s, the current

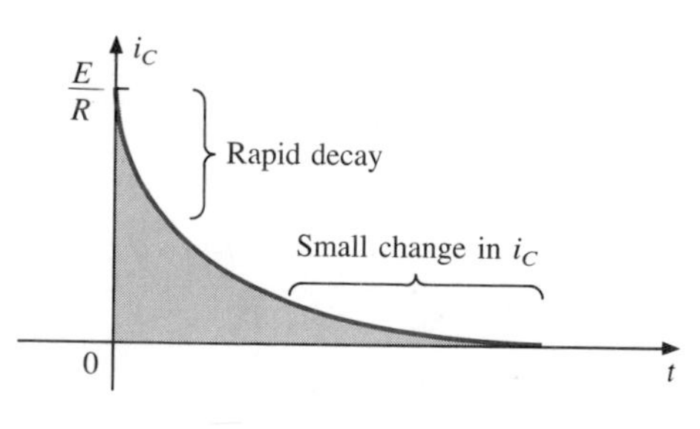


FIG. 10.22
*$i_C$ during the charging phase.*

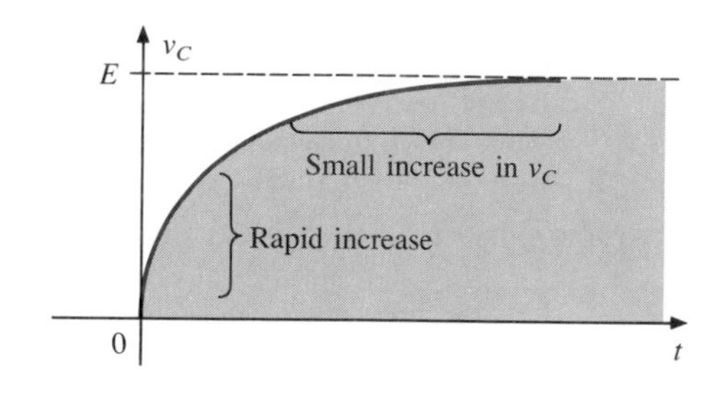


FIG. 10.23
*$v_C$ during the charging phase.*

jumps to a value limited only by the resistance of the network and then decays to zero as the plates are charged. Note the rapid decay in current level revealing that the amount of charge deposited on the plates per unit time is rapidly decaying also. Since the voltage across the plates is directly related to the charge on the plates by $v_C = q/C$, the rapid rate with which charge is initially deposited on the plates will result in a rapid increase in $v_C$. Obviously, as the rate of flow of charge ($I$) decreases, the rate of change in voltage will follow suit. Eventually, the flow of charge will stop, the current $I$ will be zero, and the voltage will cease to change in magnitude—the *charging phase* has passed. At this point the capacitor takes on the characteristics of an open circuit: a voltage drop across the plates without a flow of charge "between" the plates. As demonstrated in Fig. 10.24, the voltage across the capacitor

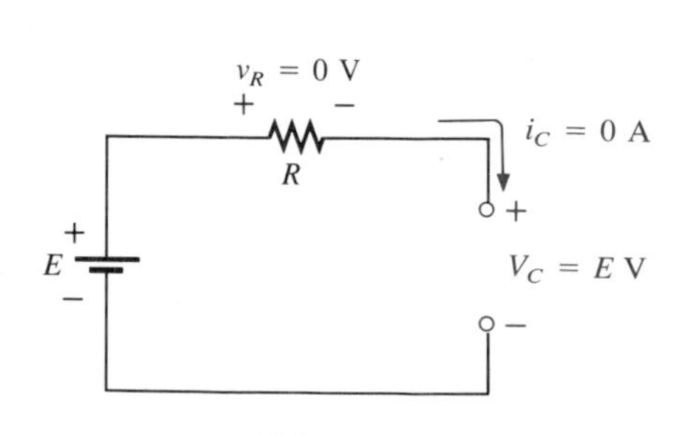


FIG. 10.24
*Open-circuit equivalent for a capacitor following the charging phase.*

is the source voltage since $i = i_C = i_R = 0$ A and $v_R = i_R R = (0)R = 0$ V. For all future analysis:

***A capacitor can be replaced by an open-circuit equivalent once the charging phase in a dc network has passed.***

Looking back at the instant the switch is closed, we can also surmise that a capacitor behaves like a short circuit the moment the switch is closed in a dc charging network, as shown in Fig. 10.25. The current $i = i_C = i_R = E/R$, and the voltage $v_C = E - v_R = E - i_R R = E - (E/R)R = E - E = 0$ V at $t = 0$ s.

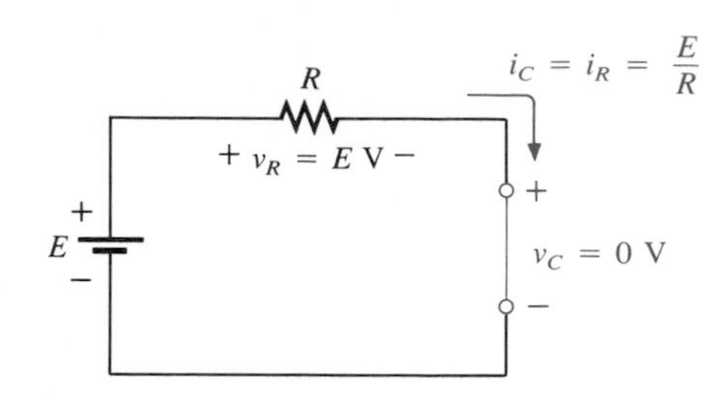


**FIG. 10.25**
*Short-circuit equivalent for a capacitor (switch closed, $t = 0$).*

Through the use of calculus, the following mathematical equation for the charging current $i_C$ can be obtained:

$$i_C = \frac{E}{R}e^{-t/RC} \tag{10.13}$$

The factor $e^{-t/RC}$ is an exponential function of the form $e^{-x}$, where $x = -t/RC$ and $e = 2.71828\ldots$ . A plot of $e^{-x}$ for $x \geq 0$ appears in Fig. 10.26. Exponentials are mathematical functions that all students of electrical, electronic, or computer systems must become very familiar with. They will appear throughout the analysis to follow in this course, and in succeeding courses.

Our current interest in the function $e^{-x}$ is limited to values of $x$ greater than zero, as noted by the curve of Fig. 10.22. All modern-day scientific calculators have the function $e^x$. To obtain $e^{-x}$, the sign of $x$ must be changed using the sign key before the exponential function is keyed in. The magnitude of $e^{-x}$ has been listed in Table 10.3 for a range of values of $x$. Note the rapidly decreasing magnitude of $e^{-x}$ with increasing value of $x$.

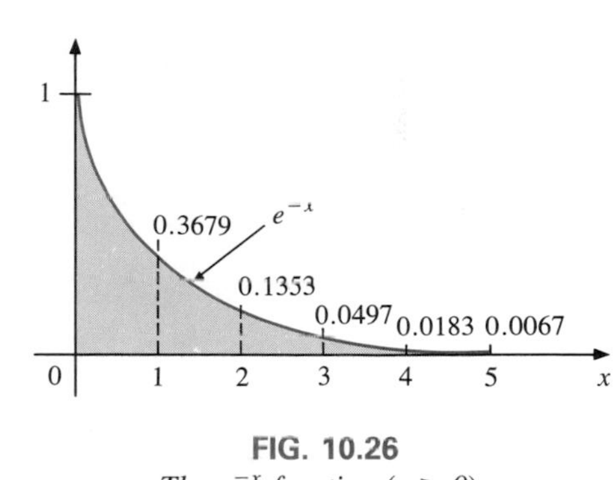


**FIG. 10.26**
*The $e^{-x}$ function ($x \geq 0$).*

**TABLE 10.3**
*Selected values of $e^{-x}$.*

| | |
|---|---|
| $x = 0$ | $e^{-x} = e^{-0} = \frac{1}{e^0} = \frac{1}{1} = 1$ |
| $x = 1$ | $e^{-1} = \frac{1}{e} = \frac{1}{2.71828\ldots} = 0.3679$ |
| $x = 2$ | $e^{-2} = \frac{1}{e^2} = 0.1353$ |
| $x = 5$ | $e^{-5} = \frac{1}{e^5} = 0.00674$ |
| $x = 10$ | $e^{-10} = \frac{1}{e^{10}} = 0.0000454$ |
| $x = 100$ | $e^{-100} = \frac{1}{e^{100}} = 3.72 \times 10^{-44}$ |

**TABLE 10.4**
*$i_C$ vs. $\tau$ (charging phase).*

| $t$ | Magnitude |
|---|---|
| 0 | 100% |
| $1\tau$ | 36.8% |
| $2\tau$ | 13.5% |
| $3\tau$ | 5.0% |
| $4\tau$ | 1.8% |
| $5\tau$ | 0.67% ← Less than 1% of maximum |
| $6\tau$ | 0.24% |

**TABLE 10.5**
*Change in $i_C$ between time constants.*

| | |
|---|---|
| $(0 \rightarrow 1)\tau$ | 63.2% |
| $(1 \rightarrow 2)\tau$ | 23.3% |
| $(2 \rightarrow 3)\tau$ | 8.6% |
| $(3 \rightarrow 4)\tau$ | 3.0% |
| $(4 \rightarrow 5)\tau$ | 1.2% |
| $(5 \rightarrow 6)\tau$ | 0.4% ← Less than 1% |

The factor $RC$ in Eq. (10.13) is called the *time constant* of the system and has the units of time as follows:

$$RC = \left(\frac{V}{I}\right)\left(\frac{Q}{V}\right) = \left(\frac{\not{V}}{\not{Q}/t}\right)\left(\frac{\not{Q}}{\not{V}}\right) = t$$

Its symbol is the Greek letter $\tau$ (tau), and its unit of measure is the second. Thus,

$$\boxed{\tau = RC} \qquad \text{(seconds, s)} \qquad \textbf{(10.14)}$$

If we substitute $\tau = RC$ into the exponential function $e^{-t/RC}$, we obtain $e^{-t/\tau}$. In one time constant, $e^{-t/\tau} = e^{-\tau/\tau} = e^{-1} = 0.3679$, or the function equals 36.79% of its maximum value of 1. At $t = 2\tau$, $e^{-t/\tau} = e^{-2\tau/\tau} = e^{-2} = 0.1353$, and the function has decayed to only 13.53% of its maximum value.

The magnitude of $e^{-t/\tau}$ and the percent change between time constants have been tabulated in Tables 10.4 and 10.5, respectively. Note that the current has dropped 63.2% (100% − 36.8%) in the first time constant but only 0.4% between the fifth and sixth time constants. The rate of change of $i_C$ is therefore quite sensitive to the time constant determined by the network parameters $R$ and $C$. For this reason, the universal time constant chart of Fig. 10.27 is provided to permit a more

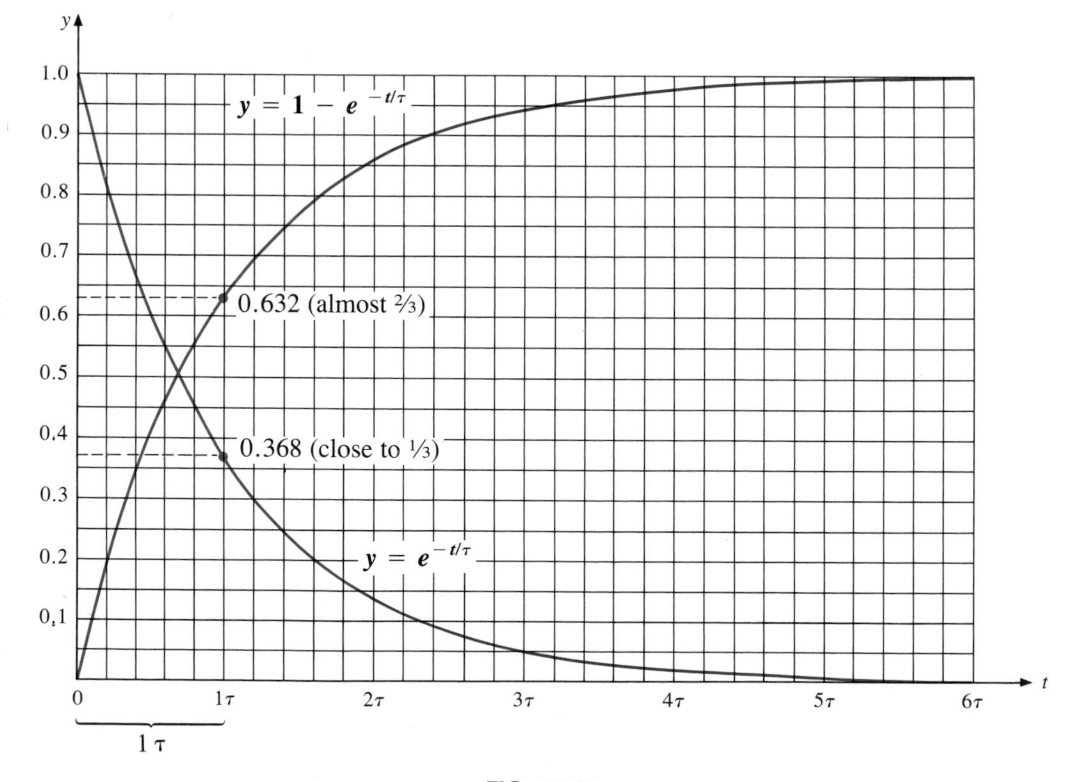


**FIG. 10.27**
*Universal time constant chart.*

accurate estimate of the value of the function $e^{-x}$ for specific time intervals related to the time constant. The term *universal* is used because the axes are not scaled to specific values.

Returning to Eq. (10.13), we find that the multiplying factor $E/R$ is the maximum value the current $i_C$ can attain, as shown in Fig. 10.22. Substituting $t = 0$ s into Eq. (10.13) yields

$$i_C = \frac{E}{R}e^{-t/RC} = \frac{E}{R}e^{-0} = \frac{E}{R}$$

verifying our earlier conclusion.

For increasing values of $t$, the magnitude of $e^{-t/\tau}$, and therefore the value of $i_C$, will decrease as shown in Fig. 10.28. Since the magnitude of $i_C$ is less than 1% of its maximum after five time constants, we will assume for future analysis that:

***The current $i_C$ of a capacitive network is essentially zero after five time constants of the charging phase have passed in a dc network.***

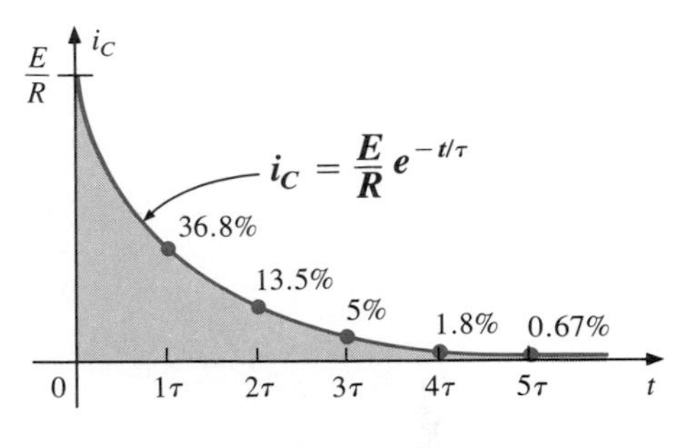


**FIG. 10.28**
*$i_C$ vs. t during the charging phase.*

Since $C$ is usually found in microfarads or picofarads, the time constant $\tau = RC$ will never be greater than a few seconds unless $R$ is very large.

Let us now turn our attention to the charging voltage across the capacitor. Through further mathematical analysis, the following equation for the voltage across the capacitor can be determined:

$$\boxed{v_C = E(1 - e^{-t/RC})} \qquad \textbf{(10.15)}$$

Note the presence of the same factor $e^{-t/RC}$ and the function $(1 - e^{-t/RC})$ appearing in Fig. 10.27. Since $e^{-t/\tau}$ is a decaying function, the factor $(1 - e^{-t/\tau})$ will grow toward a maximum value of 1 with time as shown in Fig. 10.27. In addition, since $E$ is the multiplying factor, we can conclude that for all practical purposes the voltage $v_C$ is $E$ volts after five time constants of the charging phase. A plot of $v_C$ versus $t$ is provided in Fig. 10.29.

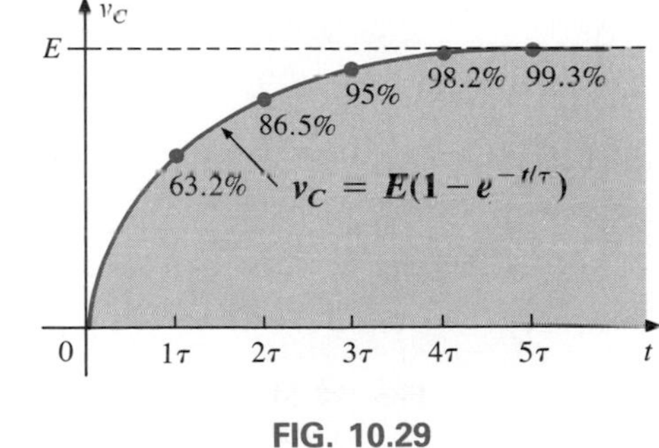


**FIG. 10.29**
*$v_C$ vs. t during the charging phase.*

If we keep $R$ constant and reduce $C$, the product $RC$ will decrease, and the rise time of five time constants will decrease. The change in transient behavior of the voltage $v_C$ is plotted in Fig. 10.30 for various

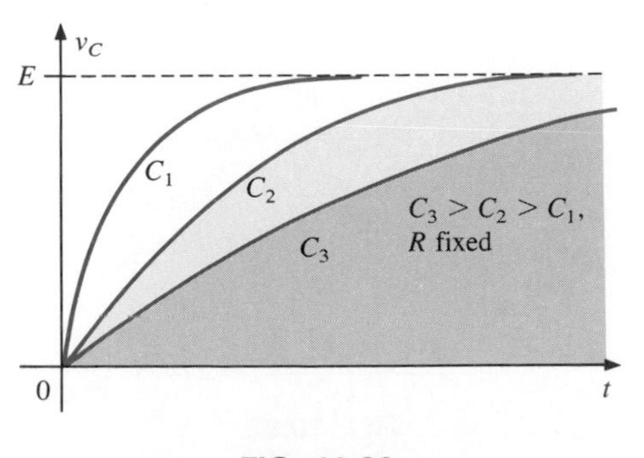

**FIG. 10.30**

values of $C$. The product $RC$ will always have some numerical value, even though it may be very small in some cases. For this reason,

***the voltage across a capacitor cannot change instantaneously.***

In fact, the capacitance of a network is also a measure of how much it will oppose a change in voltage across the network. The larger the capacitance, the larger the time constant and the longer it takes to charge up to its final value (curve of $C_3$ in Fig. 10.30). A lesser capacitance would permit the voltage to build up more quickly since the time constant is less (curve of $C_1$ in Fig. 10.30).

The rate at which charge is deposited on the plates during the charging phase can be found by substituting the following for $v_C$ in Eq. (10.15):

$$v_C = \frac{q}{C}$$

and

$$\boxed{q = Cv_C = CE(1 - e^{-t/\tau})} \quad \textit{charging} \qquad \textbf{(10.16)}$$

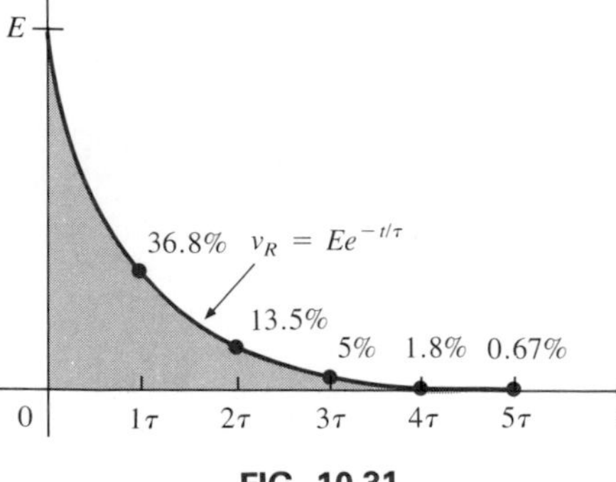

FIG. 10.31

indicating that the charging rate is very high during the first few time constants and less than 1% after five time constants.

The voltage across the resistor is determined by Ohm's law:

$$v_R = i_R R = Ri_C = R\frac{E}{R}e^{-t/\tau}$$

or

$$\boxed{v_R = Ee^{-t/\tau}} \qquad \textbf{(10.17)}$$

A plot of $v_R$ appears in Fig. 10.31.

---

**EXAMPLE 10.5.**

a. Find the mathematical expressions for the transient behavior of $v_C$, $i_C$, and $v_R$ for the circuit of Fig. 10.32 when the switch is moved to position 1. Plot the curves of $v_C$, $i_C$, and $v_R$.

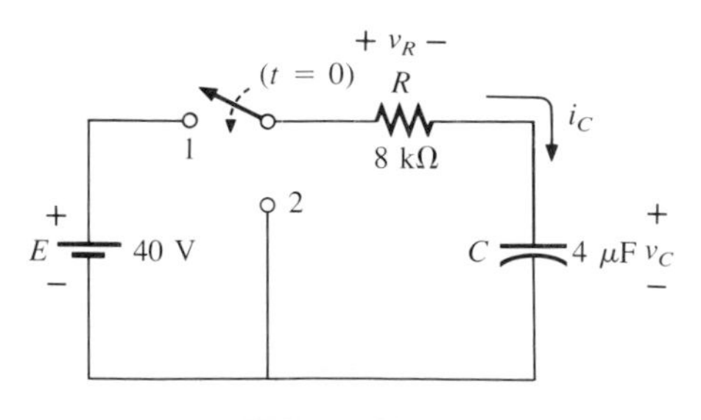

FIG. 10.32

b. How much time must pass before it can be assumed for all practical purposes that $i_C \cong 0$ A and $v_C \cong E$ volts?

***Solutions:***

a. $\tau = RC = (8 \times 10^3\ \Omega)(4 \times 10^{-6}\ \text{F}) = 32 \times 10^{-3}\ \text{s} = \mathbf{32\ ms}$

By Eq. (10.15),

$$v_C = E(1 - e^{-t/\tau}) = \mathbf{40(1 - e^{-t/(32\times10^{-3})})}$$

By Eq. (10.13),

$$i_C = \frac{E}{R}e^{-t/\tau} = \frac{40\ \text{V}}{8\ \text{k}\Omega}e^{-t/(32\times10^{-3})}$$
$$= \mathbf{(5 \times 10^{-3})e^{-t/(32\times10^{-3})}}$$

By Eq. (10.17),

$$v_R = Ee^{-t/\tau} = \mathbf{40e^{-t/(32\times10^{-3})}}$$

The curves appear in Fig. 10.33.

b. $5\tau = 5(32\ \text{ms}) = \mathbf{160\ ms}$

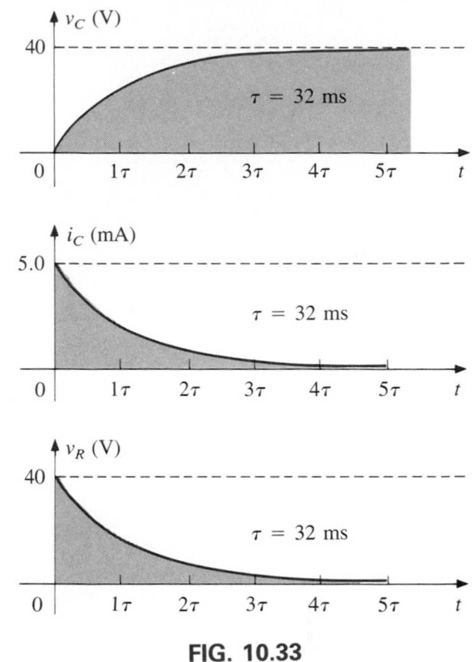


**FIG. 10.33**

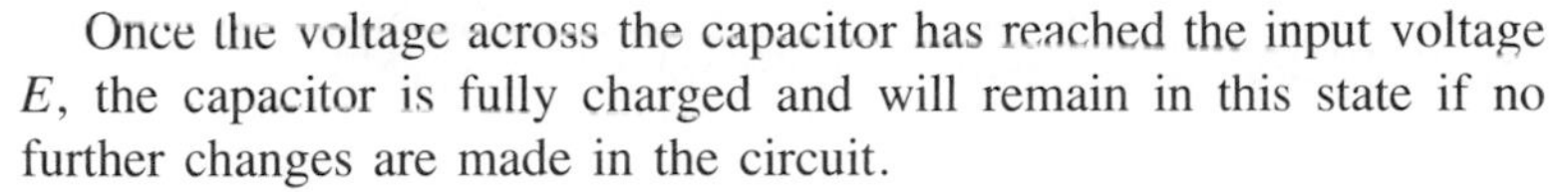

Once the voltage across the capacitor has reached the input voltage $E$, the capacitor is fully charged and will remain in this state if no further changes are made in the circuit.

If the switch of Fig. 10.21 is opened, as shown in Fig. 10.34(a), the capacitor will retain its charge for a period of time determined by its

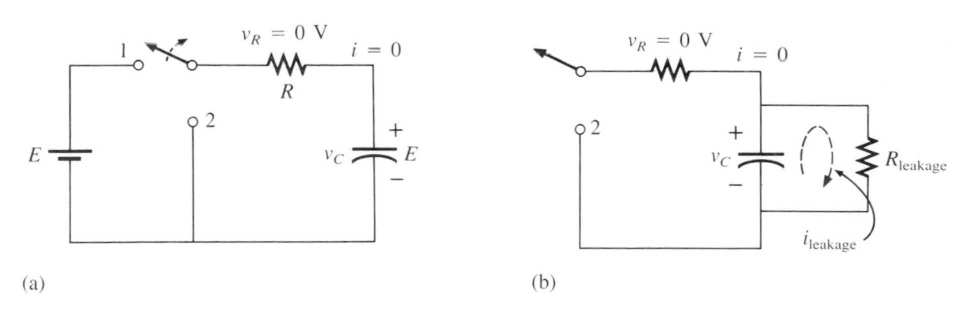


**FIG. 10.34**

leakage current. For capacitors such as the mica and ceramic, the leakage current ($i_{leakage} = v_C/R_{leakage}$) is very small, enabling the capacitor to retain its charge, and hence the potential difference across its plates, for a long period of time. For electrolytic capacitors, which have very high leakage currents, the capacitor will discharge more rapidly, as shown in Fig. 10.34(b). In any event, to insure that they are completely discharged, capacitors should be shorted by a lead or a screwdriver before they are handled.

## 10.8 DISCHARGE PHASE

The network of Fig. 10.21 is designed to both charge and discharge the capacitor. When the switch is placed in position 1, the capacitor will charge toward the supply voltage as described in the last section. At any point in the charging process, if the switch is moved to position 2, the capacitor will begin to discharge at a rate sensitive to the same time constant $\tau = RC$. The established voltage across the capacitor will create a flow of charge in the closed path that will eventually discharge the capacitor completely. In essence, the capacitor functions like a battery with a decreasing terminal voltage. Note in particular that the current $i_C$ has reversed direction, changing the polarity of the voltage across $R$.

If the capacitor had charged to the full battery voltage as indicated in Fig. 10.35, the equation for the decaying voltage across the capacitor would be the following:

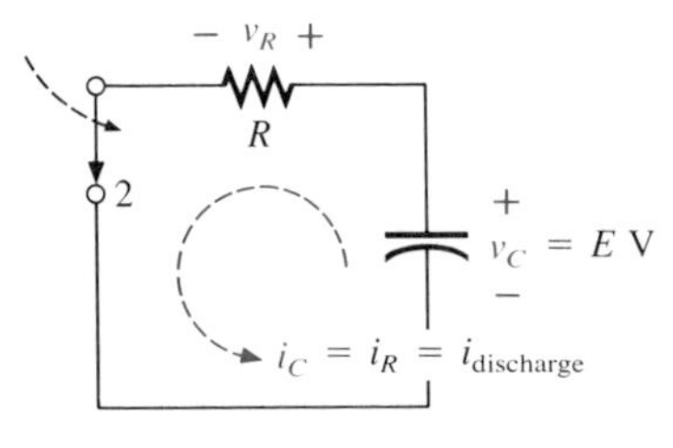


FIG. 10.35

$$\boxed{v_C = Ee^{-t/RC}} \quad \textit{discharging} \tag{10.18}$$

which employs the function $e^{-x}$ and the same time constant used above. The resulting curve will have the same shape as the curve for $i_C$ and $v_R$ in the last section. During the discharge phase, the current $i_C$ will also decrease with time as defined by the following equation:

$$\boxed{i_C = \frac{E}{R}e^{-t/RC}} \quad \textit{discharging} \tag{10.19}$$

The voltage $v_R = v_C$, and

$$\boxed{v_R = Ee^{-t/RC}} \quad \textit{discharging} \tag{10.20}$$

The complete discharge will occur, for all practical purposes, in five time constants. If the switch is moved between terminals 1 and 2 every five time constants, the waveshapes of Fig. 10.36 will result for $v_C$, $i_C$, and $v_R$. For each curve, the current direction and voltage polarities were defined by Fig. 10.21. Since the polarity of $v_C$ is the same for both the charging and discharging phases, the entire curve lies above the axis. The current $i_C$ reverses direction during the charging and discharging phases, producing a negative pulse for both the current and the voltage $v_R$. Note that the voltage $v_C$ never changes magnitude instantaneously but that the current $i_C$ has the ability to change instantaneously as demonstrated by its vertical rises and drops to maximum values.

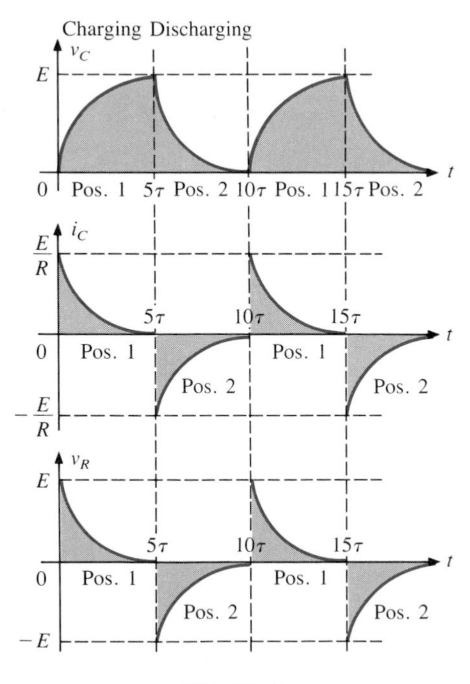


FIG. 10.36

**EXAMPLE 10.6.** After $v_C$ in Example 10.5 has reached its final value of 40 V, the switch is thrown into position 2 as shown in Fig. 10.37.

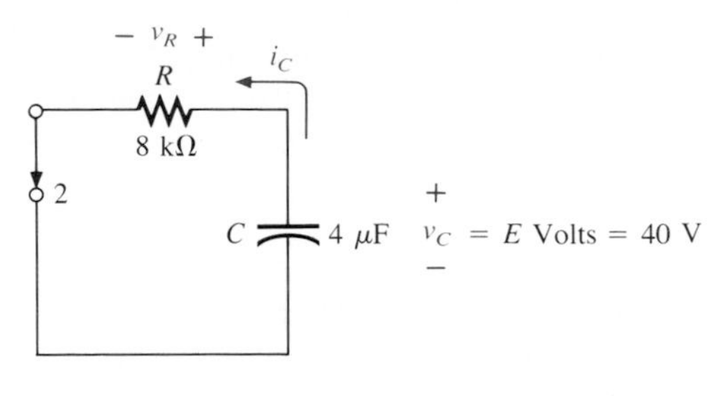

**FIG. 10.37**

Find the mathematical expressions for the transient behavior of $v_C$, $i_C$, and $v_R$ after the closing of the switch. Plot the curves for $v_C$, $i_C$, and $v_R$ using the defined directions and polarities of Fig. 10.32.

***Solution:***

$$\tau = 32 \text{ ms}$$

By Eq. (10.18),

$$v_C = Ee^{-t/\tau} = \mathbf{40}\boldsymbol{e}^{-t/(32\times 10^{-3})}$$

By Eq. (10.19),

$$i_C = \frac{E}{R}e^{-t/\tau} = \mathbf{(5 \times 10^{-3})}\boldsymbol{e}^{-t/(32\times 10^{-3})}$$

By Eq. (10.20),

$$v_R = Ee^{-t/\tau} = \mathbf{40}\boldsymbol{e}^{-t/(32\times 10^{-3})}$$

The curves appear in Fig. 10.38.

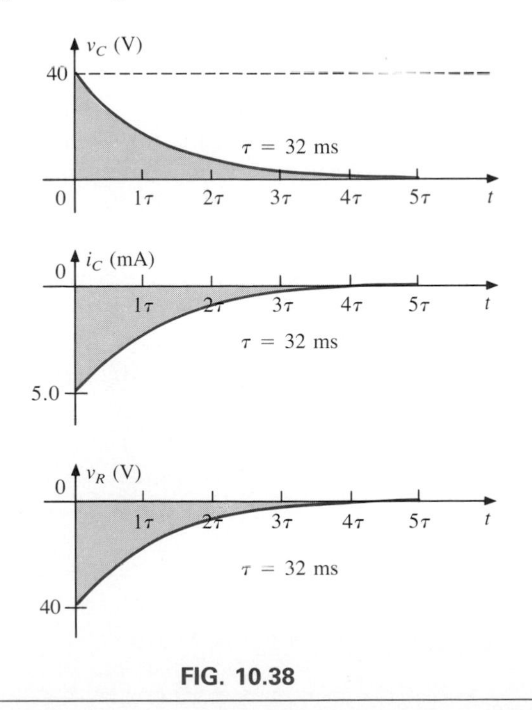


**FIG. 10.38**

The preceding discussion and examples apply to situations in which the capacitor charges to the battery voltage. If the charging phase is disrupted before reaching the supply voltage, the capacitive voltage will be less, and the equation for the discharging voltage $v_C$ will take on the form

$$v_C = V_i e^{-t/RC} \qquad (10.21)$$

where $V_i$ is the starting or *i*nitial voltage for the discharge phase. The equation for the decaying current is also modified by simply substituting $V_i$ for $E$. That is,

$$i_C = \frac{V_i}{R} e^{-t/\tau} = I_i e^{-t/\tau} \qquad (10.22)$$

Use of the above equations will be demonstrated in Examples 10.8 and 10.9.

**EXAMPLE 10.7.**

a. Find the mathematical expression for the transient behavior of the voltage across the capacitor of Fig. 10.39 if the switch is thrown into position 1 at $t = 0$ s.
b. Repeat part (a) for $i_C$.
c. Find the mathematical expressions for the response of $v_C$ and $i_C$ if the switch is thrown into position 2 at 30 ms (assuming the leakage resistance of the capacitor is ∞ ohms).
d. Find the mathematical expressions for the voltage $v_C$ and current $i_C$ if the switch is thrown into position 3 at $t = 48$ ms.
e. Plot the waveforms obtained in parts (a) through (d) on the same time axis for the voltage $v_C$ and the current $i_C$ using the defined polarity and current direction of Fig. 10.39.

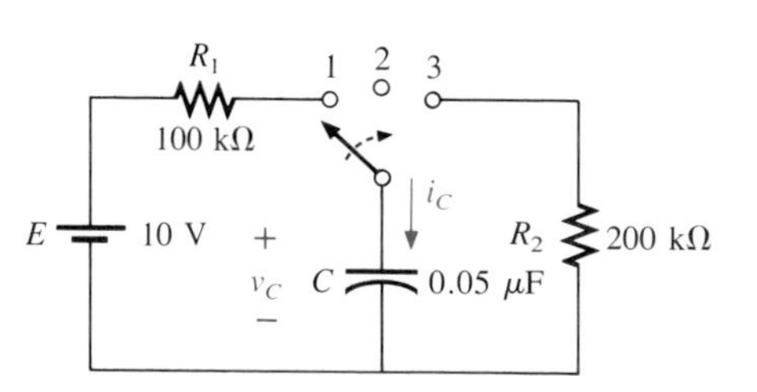


FIG. 10.39

***Solutions:***

a. Charging phase:

$$v_C = E(1 - e^{-t/\tau})$$

$$\tau = R_1 C = (100 \times 10^3\ \Omega)(0.05 \times 10^{-6}\ \text{F}) = 5 \times 10^{-3}\ \text{s} = 5\ \text{ms}$$

$$v_C = \mathbf{10(1 - e^{-t/(5\times 10^{-3})})}$$

b. $$i_C = \frac{E}{R_1} e^{-t/\tau} = \frac{10\ \text{V}}{100 \times 10^3\ \Omega} e^{-t/(5\times 10^{-3})}$$

$$i_C = \mathbf{(0.1 \times 10^{-3}) e^{-t/(5\times 10^{-3})}}$$

c. Storage phase:

$$v_C = E = \mathbf{10\ V}$$
$$i_C = \mathbf{0\ A}$$

d. Discharging phase:

$$v_C = Ee^{-t/\tau}$$
$$\tau = R_2C = (200 \times 10^3\ \Omega)(0.05 \times 10^{-6}\ \text{F}) = 10 \times 10^{-3}\ \text{s} = 10\ \text{ms}$$
$$v_C = \mathbf{10e^{-t/(10\times10^{-3})}}$$
$$i_C = \frac{E}{R_2}e^{-t/\tau} = \frac{10\ \text{V}}{200 \times 10^3\ \Omega}e^{-t/(10\times10^{-3})}$$
$$i_C = \mathbf{(0.05 \times 10^{-3})e^{-t/(10\times10^{-3})}}$$

e. See Fig. 10.40.

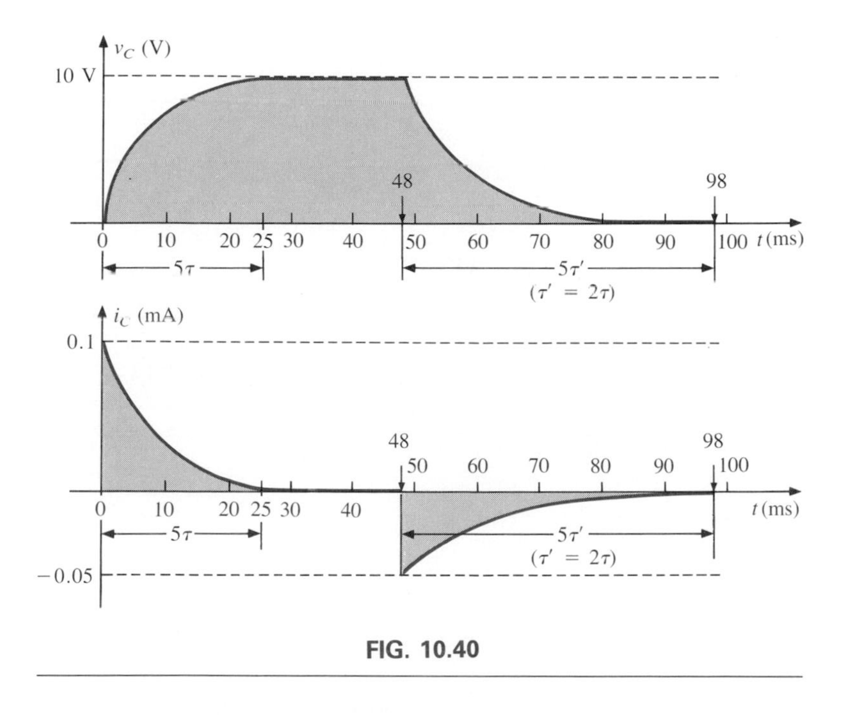


**FIG. 10.40**

**EXAMPLE 10.8.**

a. Find the mathematical expression for the transient behavior of the voltage across the capacitor of Fig. 10.41 if the switch is thrown into position 1 at $t = 0$ s.
b. Repeat part (a) for $i_C$.
c. Find the mathematical expression for the response of $v_C$ and $i_C$ if the switch is thrown into position 2 at $t = 1\tau$ of the charging phase.

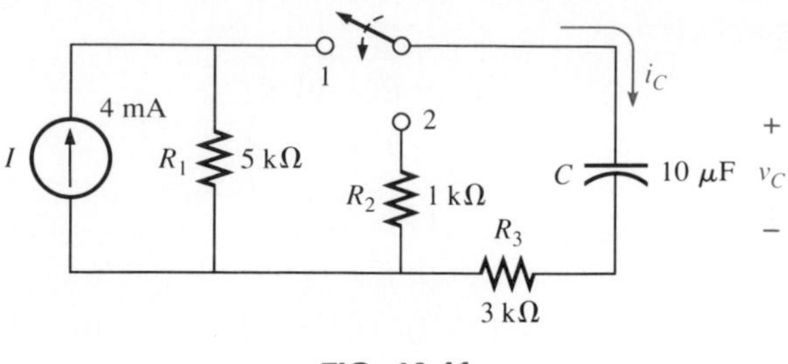

FIG. 10.41

d. Plot the waveforms obtained in parts (a) through (c) on the same time axis for the voltage $v_C$ and the current $i_C$ using the defined polarity and current direction of Fig. 10.41.

***Solutions:***

a. *Charging phase:* Converting the current source to a voltage source will result in the network of Fig. 10.42.

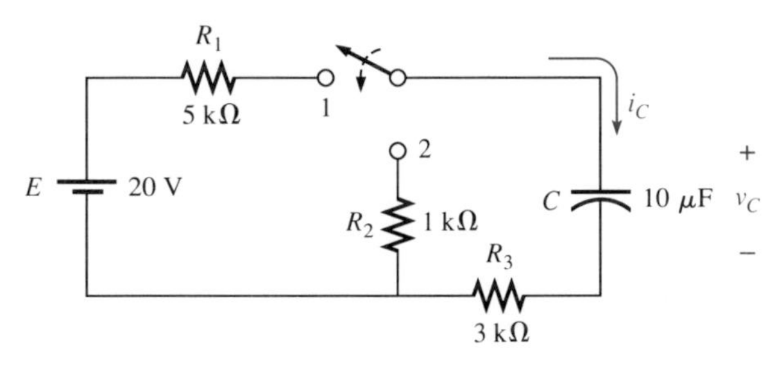

FIG. 10.42

$$v_C = E(1 - e^{-t/\tau_1})$$

$$\tau_1 = (R_1 + R_3)C = (5\text{ k}\Omega + 3\text{ k}\Omega)(10 \times 10^{-6}\text{ F})$$
$$= 80\text{ ms}$$

$$v_C = \mathbf{20(1 - e^{-t/(80\times10^{-3})})}$$

b. $$i_C = \frac{E}{R_1 + R_3}e^{-t/\tau_1}$$
$$= \frac{20\text{ V}}{8\text{ k}\Omega}e^{-t/(80\times10^{-3})}$$
$$i_C = \mathbf{(2.5 \times 10^{-3})e^{-t/(80\times10^{-3})}}$$

c. $v_C = V_i e^{-t/\tau_2}$

At $t = 1\tau$, $v_C = 0.632E = 0.632(20\text{ V}) = 12.64\text{ V}$.

$$\tau_2 = (R_2 + R_3)C = (1\text{ k}\Omega + 3\text{ k}\Omega)(10 \times 10^{-6}\text{ F})$$
$$= 40\text{ ms}$$

$$v_C = \mathbf{12.64\, e^{-t/(40\times10^{-3})}}$$

$$i_C = -I_i e^{-t/\tau_2}$$

At $t = 1\tau$, $i_C = (0.368)(2.5\text{ mA}) = 0.92\text{ mA}$.

$$i_C = \mathbf{-(0.92 \times 10^{-3})e^{-t/(40\times10^{-3})}}$$

d. See Fig. 10.03.

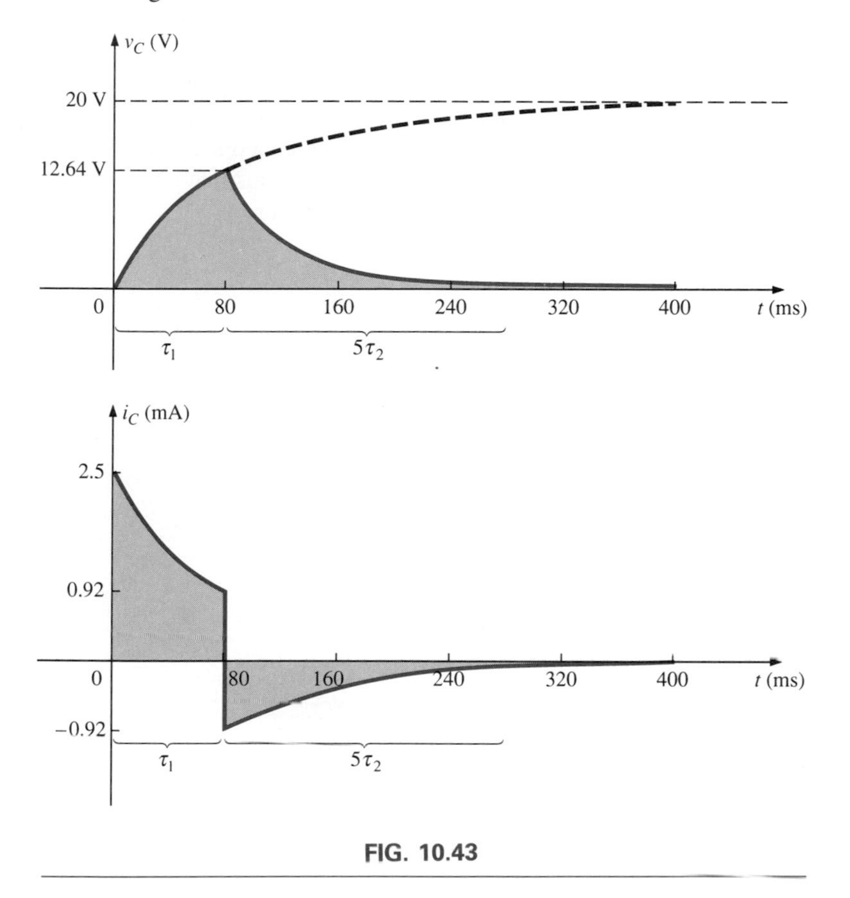


**FIG. 10.43**

## 10.9 INSTANTANEOUS VALUES

On occasion it will be necessary to determine the voltage or current at a particular instant of time. For example, if

$$v_C = 20(1 - e^{-t/(2\times10^{-3})})$$

the voltage $v_C$ may be required at $t = 5$ ms, which does not correspond to a particular value of $\tau$. Figure 10.27 reveals that $(1 - e^{-t/\tau})$ is approximately 0.93 at $t = 5\text{ ms} = 2.5\tau$, resulting in $v_C = 20(0.93) = 18.6$ V. Additional accuracy can be obtained simply by substituting $t = 5$ ms into the equation and solving for $v_C$ using a calculator or table to determine $e^{-2.5}$. Thus,

$$\begin{aligned} v_C &= 20(1 - e^{-5\text{ms}/2\text{ms}}) \\ &= 20(1 - e^{-2.5}) \\ &= 20(1 - 0.082) \\ &= 20(0.918) \\ &= \mathbf{18.36\ V} \end{aligned}$$

The results are close, but accuracy beyond the tenths' place is suspect using Fig. 10.27. The above procedure can also be applied to any other equation introduced in this chapter for currents or other voltages.

There are also occasions when the time to reach a particular voltage or current is required. The procedure is complicated somewhat by the use of natural logs ($\log_e$, or ln), but today's calculators are equipped to handle the operation with ease. There are two forms that require some development. First, consider the following sequence:

$$v_C = E(1 - e^{-t/\tau})$$

$$\frac{v_C}{E} = 1 - e^{-t/\tau}$$

$$1 - \frac{v_C}{E} = e^{-t/\tau}$$

$$\log_e\left(1 - \frac{v_C}{E}\right) = \log_e e^{-t/\tau}$$

$$\log_e\left(1 - \frac{v_C}{E}\right) = -\frac{t}{\tau}$$

and

$$t = -\tau \log_e\left(1 - \frac{v_C}{E}\right) \tag{10.23}$$

The second form is as follows:

$$v_C = Ee^{-t/\tau}$$

$$\frac{v_C}{E} = e^{-t/\tau}$$

$$\log_e\frac{v_C}{E} = \log_e e^{-t/\tau}$$

$$\log_e\frac{v_C}{E} = -\frac{t}{\tau}$$

and

$$t = -\tau \log_e\frac{v_C}{E} \tag{10.24}$$

For $i_C = (E/R)e^{-t/\tau}$,

$$t = -\tau \log_e\left(\frac{i_C R}{E}\right) \tag{10.25}$$

For example, suppose

$$v_C = 20(1 - e^{-t/(2\times 10^{-3})})$$

and the time to reach 10 V is required. Substituting into Eq. (10.23), we have

$$t = (-2 \times 10^{-3}) \log_e\left(1 - \frac{10}{20}\right)$$
$$= (-2 \times 10^{-3}) \log_e 0.5$$
$$= (-2 \times 10^{-3})(-0.693) \quad \leftarrow \boxed{\text{ln}} \text{ key on calculator}$$
$$= \mathbf{1.386\ ms}$$

Using Fig. 10.27 we find at $(1 - e^{-t/\tau}) = v_C/E = 0.5$ that $t \cong 0.7\tau = 0.7(2\text{ ms}) = 1.4\text{ ms}$, which is relatively close to the above.

## 10.10 $\tau = R_{Th}C$

Occasions will arise in which the network does not have the simple series form of Fig. 10.21. It will then be necessary to first find the Thevenin equivalent circuit for the network external to the capacitive element. $E_{Th}$ will then be the source voltage $E$ of Eqs. (10.15) through (10.20) and $R_{Th}$ the resistance $R$. The time constant is then $\tau = R_{Th}C$.

---

**EXAMPLE 10.9.** For the network of Fig. 10.44:

a. Find the mathematical expression for the transient behavior of the voltage $v_C$ and the current $i_C$ following the closing of the switch (position 1 at $t = 0$ s).
b. Find the mathematical expression for the voltage $v_C$ and current $i_C$ as a function of time if the switch is thrown into position 2 at $t = 9$ ms.
c. Draw the resultant waveforms of parts (a) and (b) on the same time axis.

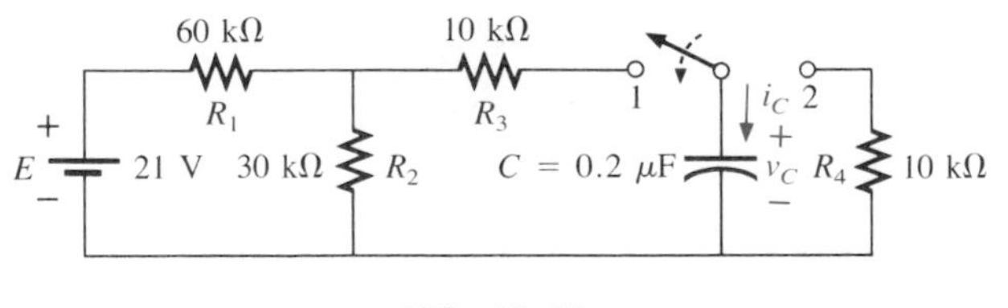


FIG. 10.44

***Solutions:***

a. Applying Thevenin's theorem to the 0.2-$\mu$F capacitor, we obtain Fig. 10.45:

$$R_{Th} = R_1 \parallel R_2 + R_3 = \frac{(60\text{ k}\Omega)(30\text{ k}\Omega)}{90\text{ k}\Omega} + 10\text{ k}\Omega$$
$$= 20\text{ k}\Omega + 10\text{ k}\Omega$$
$$R_{Th} = 30\text{ k}\Omega$$
$$E_{Th} = \frac{R_2E}{R_2 + R_1} = \frac{(30\text{ k}\Omega)(21\text{ V})}{30\text{ k}\Omega + 60\text{ k}\Omega} = \frac{1}{3}(21\text{ V}) = 7\text{ V}$$

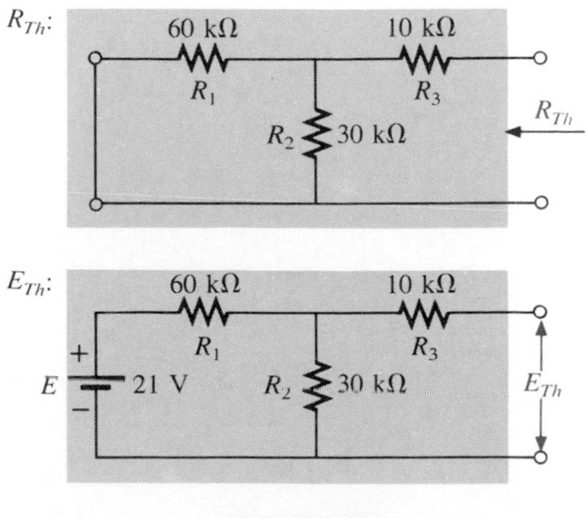


FIG. 10.45

The resultant Thevenin equivalent circuit with the capacitor replaced is shown in Fig. 10.46:

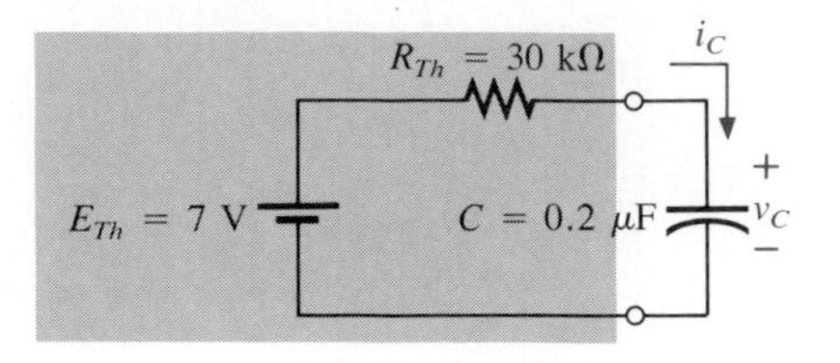

FIG. 10.46

$$v_C = E_{Th}(1 - e^{-t/\tau})$$
$$\tau = R_{Th}C = (30\ \text{k}\Omega)(0.2\ \mu\text{F})$$
$$= (30 \times 10^3\ \Omega)(0.2 \times 10^{-6}\ \text{F}) = 6 \times 10^{-3}\ \text{s}$$
$$\tau = 6\ \text{ms}$$
$$v_C = \mathbf{7(1 - e^{-t/(6\times 10^{-3})})}$$

and

$$i_C = \frac{E_{Th}}{R}e^{-t/RC}$$
$$= \frac{7\ \text{V}}{30\ \text{k}\Omega}e^{-t/(6\times 10^{-3})}$$
$$i_C = \mathbf{(0.233 \times 10^{-3})e^{-t/(6\times 10^{-3})}}$$

b. At $t = 9$ ms,

$$v_C = E_{Th}(1 - e^{-t/\tau}) = 7(1 - e^{-(9\times 10^{-3})/(6\times 10^{-3})})$$
$$= 7(1 - e^{-1.5}) = 7(1 - 0.223)$$
$$v_C = 7(0.777) = 5.44\ \text{V}$$

$$i_C = \frac{E_{Th}}{R}e^{-t/\tau} = (0.233 \times 10^{-3})e^{-1.5}$$
$$= (0.233 \times 10^{-3})(0.223)$$
$$i_C = 0.052 \times 10^{-3} = 0.052\ \text{mA}$$

By Eq. (10.21),

$$v_C = V_i e^{-t/\tau'}$$

with

$$\tau' = R_4C = (10 \times 10^3\ \Omega)(0.2 \times 10^{-6}\ \text{F}) = 2 \times 10^{-3}\ \text{s}$$
$$= 2\ \text{ms}$$

and

$$v_C = \mathbf{5.44e^{-t/(2\times 10^{-3})}}$$

By Eq. (10.22),

$$I_i = \frac{5.44\ \text{V}}{100\ \text{k}\Omega} = 0.054\ \text{mA}$$

and

$$i_C = -I_i e^{-t/\tau} = -(0.054 \times 10^{-3})e^{-t/(2\times 10^{-3})}$$

c. See Fig. 10.47.

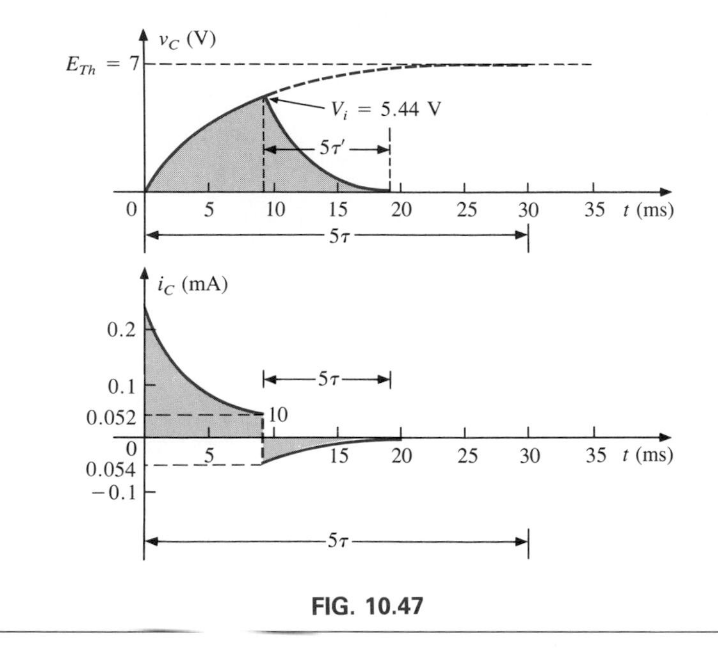


FIG. 10.47

**EXAMPLE 10.10.** The capacitor of Fig. 10.48 is initially charged to 80 V. Find the mathematical expression for $v_C$ after the closing of the switch.

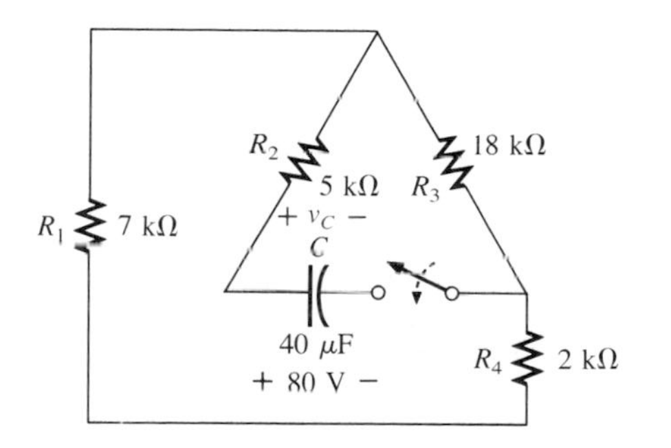


FIG. 10.48

***Solution:*** The network is redrawn in Fig. 10.49:

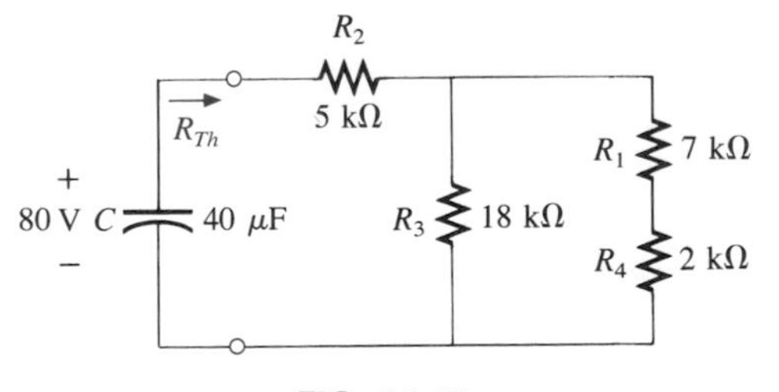


FIG. 10.49

$$R_{Th} = R_2 + R_3 \parallel (R_1 + R_4) = 5\text{ k}\Omega + 18\text{ k}\Omega \parallel (7\text{ k}\Omega + 2\text{ k}\Omega)$$
$$= 5\text{ k}\Omega + 18\text{ k}\Omega \parallel 9\text{ k}\Omega = 5\text{ k}\Omega + 6\text{ k}\Omega = 11\text{ k}\Omega$$

$$\tau = R_{Th}C = (11 \times 10^3\ \Omega)(40 \times 10^{-6}\text{ F})$$
$$= 440 \times 10^{-3}\text{ s} = 0.44\text{ s}$$

and

$$v_C = V_i e^{-t/\tau} = \mathbf{80}\boldsymbol{e}^{-t/0.44}$$

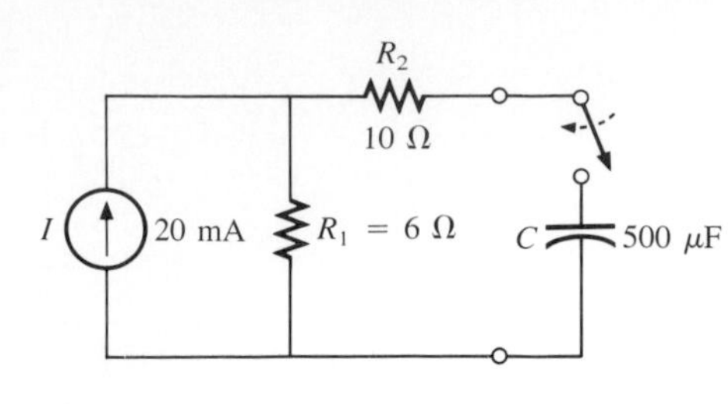


FIG. 10.50

**EXAMPLE 10.11.** For the network of Fig. 10.50, find the mathematical expression for the voltage $v_C$ after the closing of the switch (at $t = 0$).

***Solution:***

$$R_{Th} = R_1 + R_2 = 6\ \Omega + 10\ \Omega = 16\ \Omega$$
$$E_{Th} = V_1 + V_2 = IR_1 + 0$$
$$= (20 \times 10^{-3}\ \text{A})(6\ \Omega) = 120 \times 10^{-3}\ \text{V} = 0.12\ \text{V}$$

and

$$\tau = R_{Th}C = (16\ \Omega)(500 \times 10^{-6}\ \text{F}) = 8\ \text{ms}$$

so that

$$v_C = \mathbf{0.12(1 - e^{-t/(8 \times 10^{-3})})}$$

## 10.11 THE CURRENT $i_C$

The current $i_C$ associated with a capacitance $C$ is related to the voltage across the capacitor by

$$i_C = C\frac{dv_C}{dt} \qquad \textbf{(10.26)}$$

where $dv_C/dt$ is a measure of the change in $v_C$ in a vanishingly small period of time. The function $dv_C/dt$ is called the *derivative* of the voltage $v_C$ with respect to time $t$.

If the voltage fails to change at a particular instant, then

$$dv_C = 0$$

and

$$i_C = C\frac{dv_C}{dt} = 0$$

In other words, if the voltage across a capacitor fails to change with time, the current $i_C$ associated with the capacitor is zero. To take this a step further, the equation also states that the more rapid the change in voltage across the capacitor, the greater the resulting current.

In an effort to develop a clearer understanding of Eq. (10.26), let us calculate the average current associated with a capacitor for various voltages impressed across the capacitor. The average current is defined by the equation

$$i_{C\text{av}} = C\frac{\Delta v_C}{\Delta t} \qquad \textbf{(10.27)}$$

where Δ indicates a finite (measurable) change in charge, voltage, or time. The instantaneous current can be derived from Eq. (10.27) by letting $\Delta t$ become vanishingly small; that is,

$$i_{C\text{inst}} = \lim_{\Delta t \to 0} C\frac{\Delta v_C}{\Delta t} = C\frac{dv_C}{dt}$$

In the following example, the change in voltage $\Delta v_C$ will be considered for each slope of the voltage waveform. If the voltage increases with time, the average current is the change in voltage divided by the change in time, with a positive sign. If the voltage decreases with time, the average current is again the change in voltage divided by the change in time, but with a negative sign.

**EXAMPLE 10.12.** Find the waveform for the average current if the voltage across a 2-μF capacitor is as shown in Fig. 10.51.

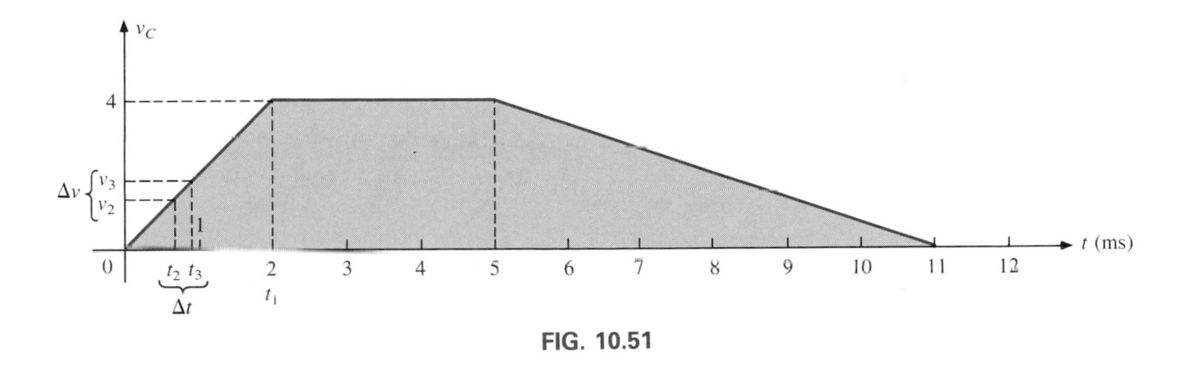


**FIG. 10.51**

***Solution:***

a. From 0 to 2 ms, the voltage increases linearly from 0 to 4 V, the change in voltage $\Delta v = 4\text{ V} - 0 = 4\text{ V}$ (with a positive sign, since the voltage increases with time). The change in time $\Delta t = 2\text{ ms} - 0 = 2\text{ ms}$, and

$$i_{C\text{av}} = C\frac{\Delta v_C}{\Delta t} = (2 \times 10^{-6}\text{ F})\left(\frac{4\text{ V}}{2 \times 10^{-3}\text{ s}}\right)$$
$$= 4 \times 10^{-3}\text{ A} = 4\text{ mA}$$

b. From 2 to 5 ms, the voltage remains constant at 4 V; the change in voltage $\Delta v = 0$. The change in time $\Delta t = 3$ ms, and

$$i_{C\text{av}} = C\frac{\Delta v_C}{\Delta t} = C\frac{0}{\Delta t} = 0$$

c. From 5 to 11 ms, the voltage decreases from 4 to 0 V. The change in voltage $\Delta v$ is therefore $4\text{ V} - 0 = 4\text{ V}$ (with a negative sign, since the voltage is decreasing with time). The change in time $\Delta t = 11 \text{ ms} - 5 \text{ ms} = 6 \text{ ms}$, and

$$i_{C\text{av}} = C\frac{\Delta v_C}{\Delta t} = -(2 \times 10^{-6} \text{ F})\left(\frac{4 \text{ V}}{6 \times 10^{-3} \text{ s}}\right)$$
$$= -1.33 \times 10^{-3} \text{ A} = -1.33 \text{ mA}$$

d. From 11 ms on, the voltage remains constant at 0 and $\Delta v = 0$, so $i_{C\text{av}} = 0$. The waveform for the average current for the impressed voltage is as shown in Fig. 10.52.

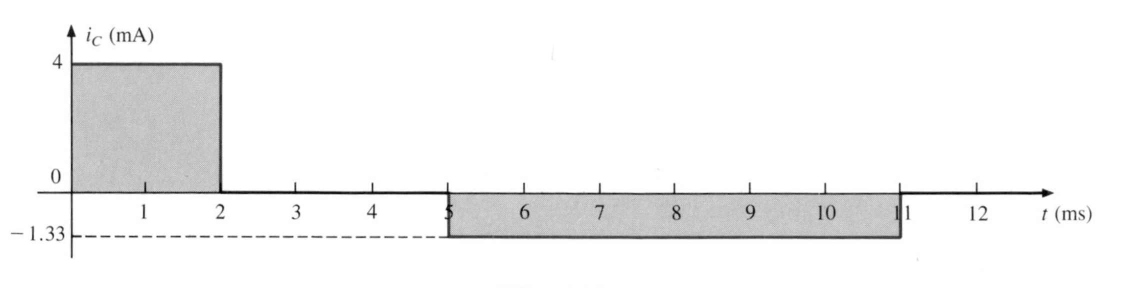

FIG. 10.52

In general, note in Example 10.12 that the steeper the slope, the greater the current, and when the voltage fails to change, the current is zero. In addition, the average value is the same as the instantaneous value at any point along the slope over which the average value was found. For example, if the interval $\Delta t$ is reduced from $0 \rightarrow t_1$ to $t_2 - t_3$ as noted in Fig. 10.51, $\Delta v/\Delta t$ is still the same. In fact, no matter how small the interval $\Delta t$, the slope will be the same, and therefore the current $i_{C\text{av}}$ will be the same. If we consider the limit as $\Delta t \rightarrow 0$, the slope will still remain the same, and therefore $i_{C\text{av}} = i_{C\text{inst}}$ at any instant of time between 0 and $t_1$. The same can be said about any portion of the voltage waveform that has a constant slope.

An important point to be gained from this discussion is that it is not the magnitude of the voltage across a capacitor that determines the current but rather how quickly the voltage *changes* across the capacitor. An applied steady dc voltage of 10,000 V would (ideally) not create any flow of charge (current), but a change in voltage of 1 V in a very brief period of time could create a significant current.

The method described above is only for waveforms with straight-line (linear) segments. For nonlinear (curved) waveforms, a method of calculus (differentiation) must be employed.

## 10.12 CAPACITORS IN SERIES AND PARALLEL

Capacitors, like resistors, can be placed in series and parallel. Increasing levels of capacitance can be obtained by placing capacitors in parallel, while decreasing levels can be obtained by placing capacitors in series.

For capacitors in series, the charge is the same on each capacitor (Fig. 10.53):

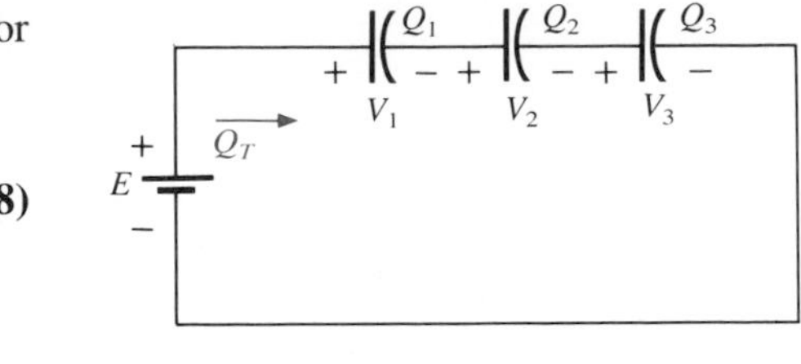


FIG. 10.53

$$Q_T = Q_1 = Q_2 = Q_3 \tag{10.28}$$

Applying Kirchhoff's voltage law around the closed loop gives

$$E = V_1 + V_2 + V_3$$

However,

$$V = \frac{Q}{C}$$

so that

$$\frac{Q_T}{C_T} = \frac{Q_1}{C_1} + \frac{Q_2}{C_2} + \frac{Q_3}{C_3}$$

Using Eq. (10.28) and dividing both sides by $Q$ yields

$$\frac{1}{C_T} = \frac{1}{C_1} + \frac{1}{C_2} + \frac{1}{C_3} \tag{10.29}$$

which is similar to the manner in which we found the total resistance of a parallel resistive circuit. The total capacitance of two capacitors in series is

$$C_T = \frac{C_1 C_2}{C_1 + C_2} \tag{10.30}$$

For capacitors in parallel as shown in Fig. 10.54, the voltage is the same across each capacitor, and the total charge is the sum of that on each capacitor:

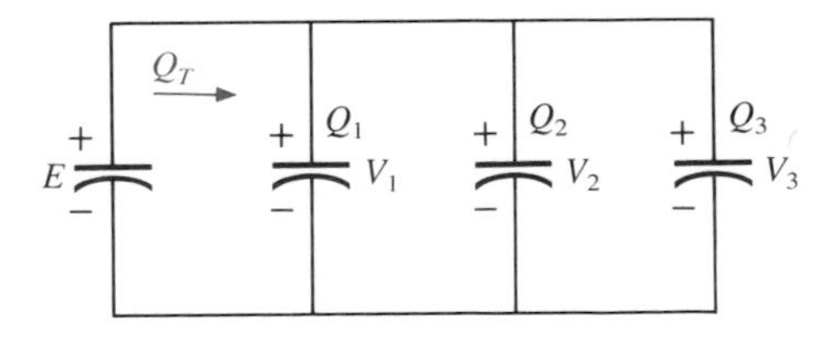


FIG. 10.54

$$Q_T = Q_1 + Q_2 + Q_3 \tag{10.31}$$

However,

$$Q = CV$$

Therefore,

$$C_T E = C_1 V_1 + C_2 V_2 + C_3 V_3$$

and

$$E = V_1 = V_2 = V_3$$

Thus,

$$C_T = C_1 + C_2 + C_3 \tag{10.32}$$

which is similar to the manner in which the total resistance of a series circuit is found.

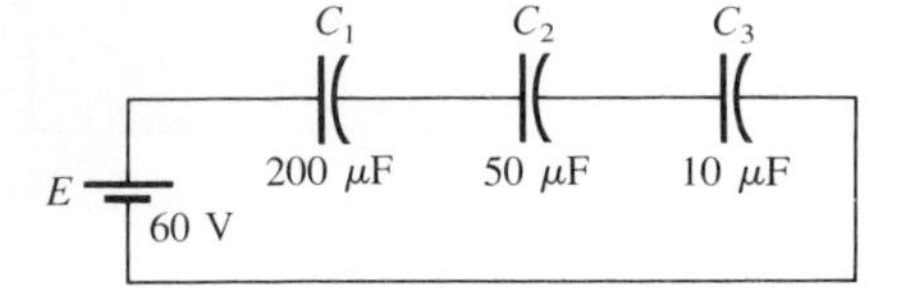


**FIG. 10.55**

**EXAMPLE 10.13.** For the circuit of Fig. 10.55:
a. Find the total capacitance.
b. Determine the charge on each plate.
c. Find the voltage across each capacitor.

***Solutions:***

a. $$\frac{1}{C_T} = \frac{1}{C_1} + \frac{1}{C_2} + \frac{1}{C_3}$$
$$= \frac{1}{200 \times 10^{-6}\ \text{F}} + \frac{1}{50 \times 10^{-6}\ \text{F}} + \frac{1}{10 \times 10^{-6}\ \text{F}}$$
$$= 0.005 \times 10^6 + 0.02 \times 10^6 + 0.1 \times 10^6$$
$$= 0.125 \times 10^6$$

and

$$C_T = \frac{1}{0.125 \times 10^6} = \mathbf{8\ \mu F}$$

b. $Q_T = Q_1 = Q_2 = Q_3$
$Q_T = C_T E = (8 \times 10^{-6}\ \text{F})(60\ \text{V}) = \mathbf{480\ \mu C}$

c. $$V_1 = \frac{Q_1}{C_1} = \frac{480 \times 10^{-6}\ \text{C}}{200 \times 10^{-6}\ \text{F}} = \mathbf{2.4\ V}$$
$$V_2 = \frac{Q_2}{C_2} = \frac{480 \times 10^{-6}\ \text{C}}{50 \times 10^{-6}\ \text{F}} = \mathbf{9.6\ V}$$
$$V_3 = \frac{Q_3}{C_3} = \frac{480 \times 10^{-6}\ \text{C}}{10 \times 10^{-6}\ \text{F}} = \mathbf{48.0\ V}$$

and

$$E = V_1 + V_2 + V_3 = 2.4\ \text{V} + 9.6\ \text{V} + 48\ \text{V}$$
$$= \mathbf{60\ V} \qquad \text{(checks)}$$

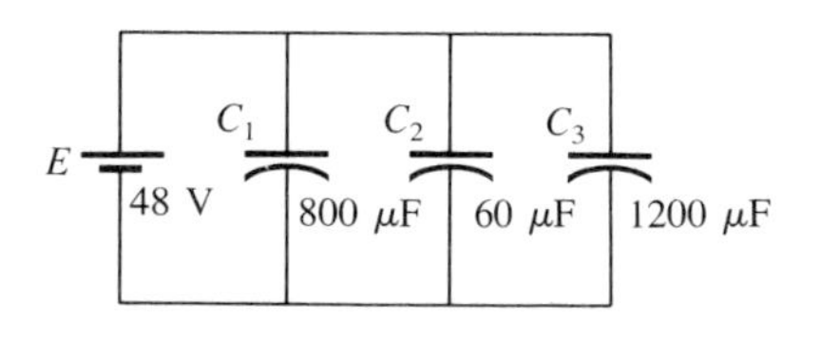


**FIG. 10.56**

**EXAMPLE 10.14.** For the network of Fig. 10.56:
a. Find the total capacitance.
b. Determine the charge on each plate.
c. Find the total charge.

***Solutions:***

a. $C_T = C_1 + C_2 + C_3 = 800\ \mu\text{F} + 60\ \mu\text{F} + 1200\ \mu\text{F}$
$= \mathbf{2060\ \mu F}$

b. $Q_1 = C_1 E = (800 \times 10^{-6}\ \text{F})(48\ \text{V}) = \mathbf{38.4\ mC}$
$Q_2 = C_2 E = (60 \times 10^{-6}\ \text{F})(48\ \text{V}) = \mathbf{2.88\ mC}$
$Q_3 = C_3 E = (1200 \times 10^{-6}\ \text{F})(48\ \text{V}) = \mathbf{57.6\ mC}$

c. $Q_T = Q_1 + Q_2 + Q_3 = 38.4\ \text{mC} + 2.88\ \text{mC} + 57.6\ \text{mC}$
$= \mathbf{98.88\ mC}$

**EXAMPLE 10.15.** Find the voltage across and charge on each capacitor for the network of Fig. 10.57.

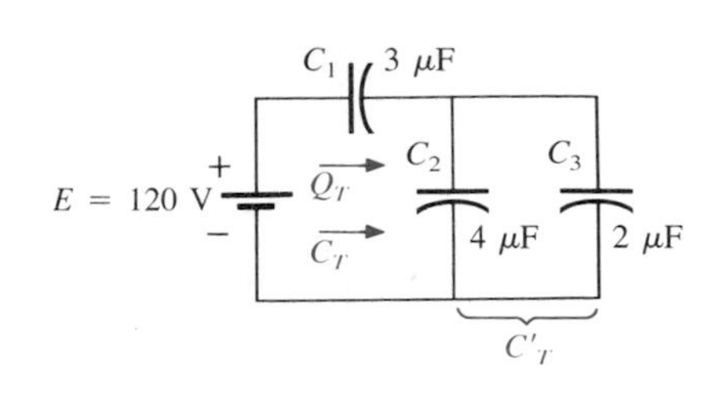


FIG. 10.57

***Solution:***

$$C'_T = C_2 + C_3 = 4\ \mu\text{F} + 2\ \mu\text{F} = 6\ \mu\text{F}$$

$$C_T = \frac{C_1 C'_T}{C_1 + C'_T} = \frac{(3\ \mu\text{F})(6\ \mu\text{F})}{3\ \mu\text{F} + 6\ \mu\text{F}} = 2\ \mu\text{F}$$

$$Q_T = C_T E = (2 \times 10^{-6}\ \text{F})(120\ \text{V}) = \mathbf{240\ \mu C}$$

An equivalent circuit (Fig. 10.58) has

$$Q_T = Q_1 = Q'_T$$

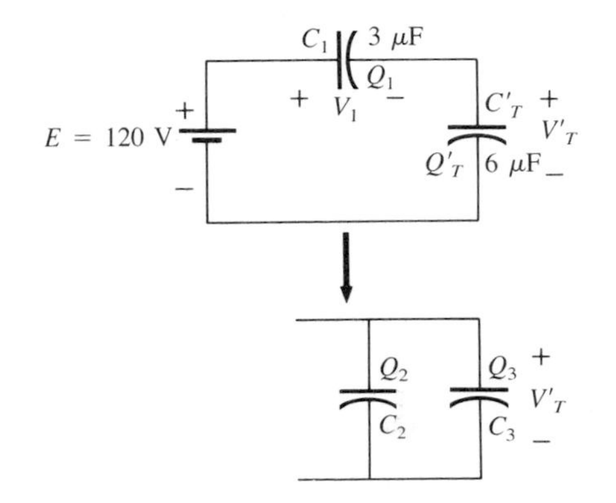


FIG. 10.58

and therefore

$$Q_1 = \mathbf{240\ \mu C}$$

and

$$V_1 = \frac{Q_1}{C_1} = \frac{240 \times 10^{-6}\ \text{C}}{3 \times 10^{-6}\ \text{F}} = \mathbf{80\ V}$$

$$Q'_T = 240\ \mu\text{C}$$

and therefore

$$V'_T = \frac{Q'_T}{C'_T} = \frac{240 \times 10^{-6}\ \text{C}}{6 \times 10^{-6}\ \text{F}} = \mathbf{40\ V}$$

and

$$Q_2 = C_2 V'_T = (4 \times 10^{-6}\ \text{F})(40\ \text{V}) = \mathbf{160\ \mu C}$$

$$Q_3 = C_3 V'_T = (2 \times 10^{-6}\ \text{F})(40\ \text{V}) = \mathbf{80\ \mu C}$$

**EXAMPLE 10.16.** Find the voltage across and charge on capacitor $C_1$ of Fig. 10.59 after it has charged up to its final value.

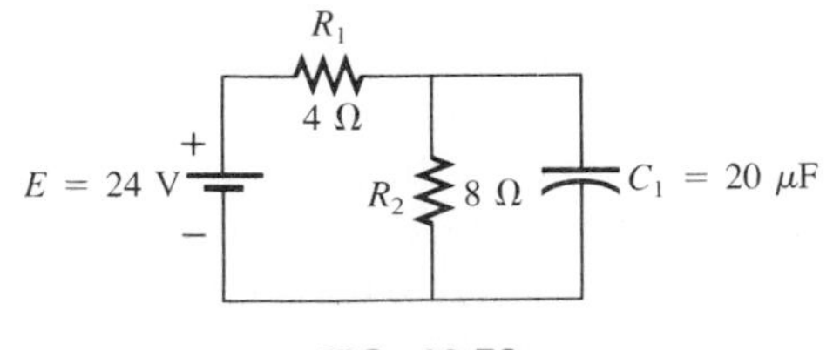


FIG. 10.59

***Solution:*** As previously discussed, the capacitor is effectively an open circuit for dc after charging up to its final value (Fig. 10.60). Therefore,

$$V_C = \frac{(8\ \Omega)(24\ \text{V})}{4\ \Omega + 8\ \Omega} = \mathbf{16\ V}$$

$$Q_1 = C_1 V_C = (20 \times 10^{-6}\ \text{F})(16\ \text{V}) = \mathbf{320\ \mu C}$$

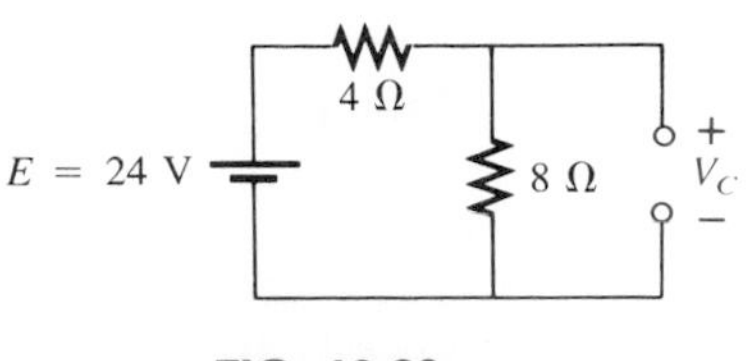

FIG. 10.60

**EXAMPLE 10.17.** Find the voltage across and charge on each capacitor of the network of Fig. 10.61 after each has charged up to its final value.

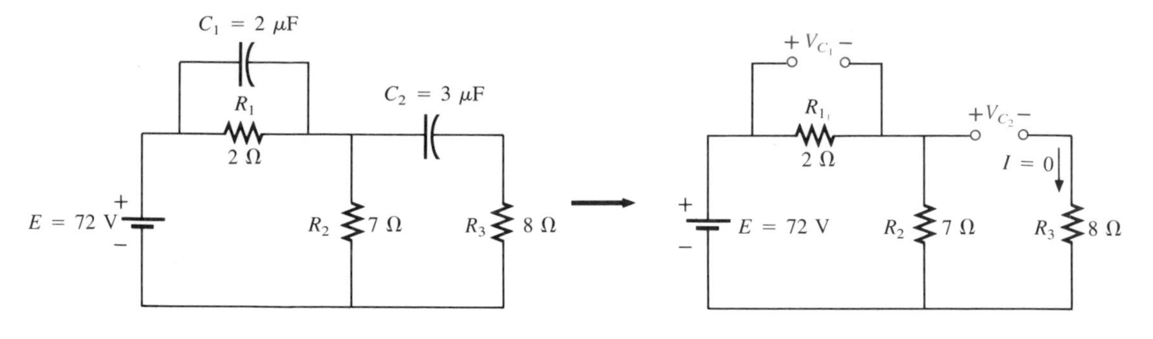

FIG. 10.61

***Solution:***

$$V_{C_2} = \frac{(7\ \Omega)(72\ \text{V})}{7\ \Omega + 2\ \Omega} = \mathbf{56\ V}$$

$$V_{C_1} = \frac{(2\ \Omega)(72\ \text{V})}{2\ \Omega + 7\ \Omega} = \mathbf{16\ V}$$

$$Q_1 = C_1V_{C_1} = (2 \times 10^{-6}\ \text{F})(16\ \text{V}) = \mathbf{32\ \mu C}$$

$$Q_2 = C_2V_{C_2} = (3 \times 10^{-6}\ \text{F})(56\ \text{V}) = \mathbf{168\ \mu C}$$

## 10.13 ENERGY STORED BY A CAPACITOR

The ideal capacitor does not dissipate any of the energy supplied to it. It stores the energy in the form of an electric field between the conducting surfaces. A plot of the voltage, current, and power to a capacitor during the charging phase is shown in Fig. 10.62. The power curve can be obtained by finding the product of the voltage and current at selected intervals of time and connecting the points obtained. (Note $p_1$ on the curve of Fig. 10.62.) The energy stored is represented by the shaded area under the power curve. Using calculus, we can determine the area under the curve:

$$W_C = \frac{1}{2}CE^2$$

FIG. 10.62

In general,

$$\boxed{W_C = \frac{1}{2}CV^2} \qquad \text{(J)} \qquad \mathbf{(10.33)}$$

where $V$ is the steady-state voltage across the capacitor. In terms of $Q$ and $C$,

$$W_C = \frac{1}{2}C\left(\frac{Q}{C}\right)^2$$

or

$$W_C = \frac{Q^2}{2C} \qquad \text{(J)} \qquad \textbf{(10.34)}$$

**EXAMPLE 10.18.** For the network of Fig. 10.61, determine the energy stored by each capacitor.

***Solution:*** For $C_1$,

$$W_C = \frac{1}{2}CV^2$$

$$= \frac{1}{2}(2 \times 10^{-6}\text{ F})(16\text{ V})^2 = (1 \times 10^{-6})(256)$$

$$= \mathbf{256\ \mu J}$$

For $C_2$,

$$W_C = \frac{1}{2}CV^2$$

$$= \frac{1}{2}(3 \times 10^{-6}\text{ F})(56\text{ V})^2 = (1.5 \times 10^{-6})(3136)$$

$$= \mathbf{4704\ \mu J}$$

Due to the squared term, note the difference in energy stored due to a higher voltage.

## 10.14 STRAY CAPACITANCES

In addition to the capacitors discussed so far in this chapter, there are stray capacitances that exist not through design but simply because two conducting surfaces are relatively close to each other. Two conducting wires in the same network will have a capacitive effect between them, as shown in Fig. 10.63(a). In electronic circuits, capacitance levels exist between conducting surfaces of the transistor as shown in Fig. 10.63(b). In Chapter 12 we will discuss another element called the *inductor* which will have capacitive effects between the windings [Fig. 10.63(c)]. Stray capacitances can often lead to serious errors in system design if not considered carefully.

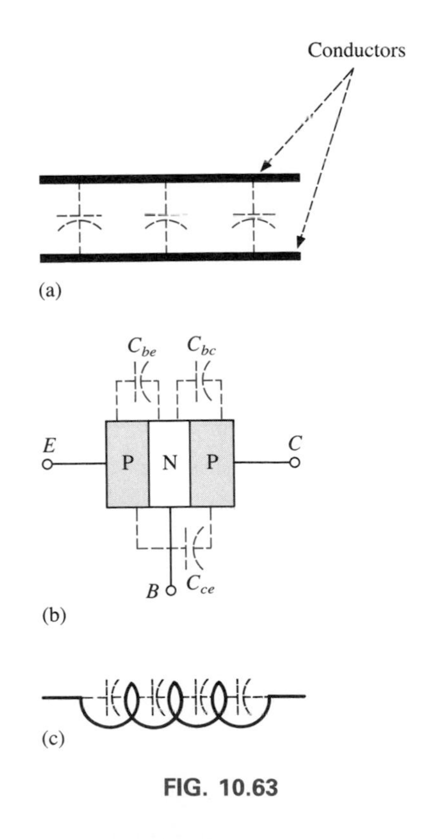


**FIG. 10.63**

## 10.15 COMPUTER ANALYSIS

Both PSPICE and BASIC can provide the transient response of an *RC* circuit. In BASIC the appropriate equations are employed to determine the voltage or current as it changes with time. A table of values can then be generated or a plot obtained using a plotting routine.

PSPICE has a specific command (.TRAN) that will determine the transient response over time. A plot of the response can then be obtained using the .PLOT or .PROBE commands.

### PSPICE

The .TRAN command determines the transient response of a network from $t = 0$ s to some final value. The format of the command is the following:

```
                          Final value of t
                          for calculations
.TRAN     TSTEP           TSTOP
          Time increment
          between determined values
Example:  .TRAN   5U   100U
```

TSTEP and TSTOP are specified by the user. TSTEP is the time interval between determined values and TSTOP is the final value of time ($t$) for which a magnitude is to be determined. The example specifies that the magnitude will be determined every 5 $\mu$s from 0 to 100 $\mu$s with the first level determined at $t = 5$ $\mu$s.

Capacitors are entered in much the same manner as resistors as demonstrated by the format below:

```
CBYPASS    6      7      100U     IC = 2V
Name       +      -
           Node   Node   Value    Initial value
```

The user fills in those quantities with brackets underneath although the initial condition entry (the voltage across the capacitor before the switching action occurs) can be omitted if the initial value is zero volts. The preceding entry is for a 100-$\mu$F bypass capacitor between nodes 6 and 7 with node 6 as the defined higher potential. In PSPICE capacitors have the additional limitation that they can not be entered as series elements. However, this problem can be circumvented by placing a very large resistance (equivalent to an open circuit for the network of interest) in parallel with one of the capacitors.

Let us now write the input file for the network of Fig. 10.64. Rather than use a switch to establish the 20 V across the network at $t = 0$ s, a pulse is applied that starts at $t = 0$ s, switching to 20 V for a period of time to exceed $5\tau$ of the network.

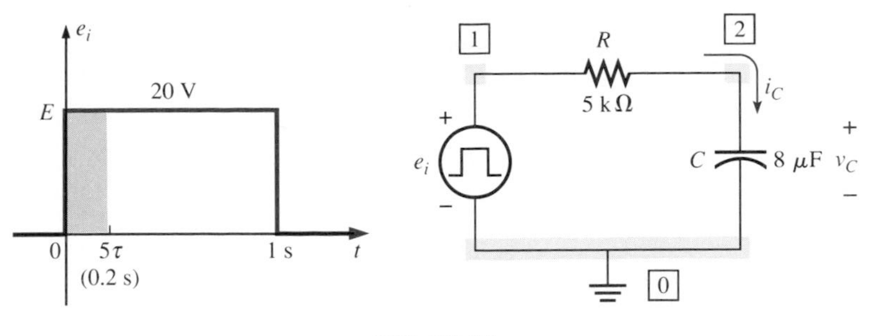


**FIG. 10.64**

In general, the format for an applied pulse has the following format:

| VPULSE | 3 | 2 | PULSE(2 | 8 | 1U | 10N | 20N | 50U | 100U) |
|---|---|---|---|---|---|---|---|---|---|
| Name | + Node | − Node | Initial level of pulse | Pulsed level | Delay time (DT) | Rise time (RT) | Fall time (FT) | Pulse width (PW) | Period of pulse (PER) |

All the quantities with a bracket below the entry are controlled by the user. Don't forget the parentheses at each end of the PULSE statement! The 2 V is the initial level of the pulse and the 8 V is the pulsed level. The delay time (1 $\mu$s) is the time interval following $t = 0$ s before the pulse changes levels. If desired it can be zero seconds. The rise and fall times are the time intervals required to change levels. The rise and fall times must be specified (other than zero seconds), or PSPICE will fall back to TSTEP of the .TRAN command for RT and FT. For the preceding example, the rise time is 10 ns and the fall time is 20 ns. The pulse width (50 $\mu$s) is the time interval at the pulsed level, and the period (100 $\mu$s) is the time between successive pulses. If the pulse width and period are unspecified, TSTOP of the .TRAN command will be employed for each.

For the input of Fig. 10.64, the statement for the pulse appears in the input file of Fig. 10.65. The pulse starts at $t = 0$ s (TD = 0) and has a rise and fall time of 1 ns, which are both significantly less than the time scale of the input or time constant of the network. The pulse width is 1 s, which is 5 times the period associated with $5\tau$ (200 ms) for this network and beyond the range of interest as defined by the .TRAN command. In other words, the 20-V level will be present for the period of interest defined by the 0–$5\tau$ interval.

The .PRINT statement of the input file specifies that the magnitude of the pulse waveform, $v_C$ and $i_C$, be tabulated for the .TRAN period of 10 ms to 200 ms at intervals of 10 ms. Note in the output file that the first value of each quantity is not determined until 10 ms have passed. Note also, however, that once past the 10-ms interval, 20 V are available at the input and that $v_C$ is very close to the 20-V level after $5\tau$ (or 200 ms). The current $i_C$ drops from 4 mA to 3.541 mA in the first 10 ms and then continues to drop to a very low level during the rest of the defined range of interest.

```
************************ Evaluation PSpice (January 1989) ******* 20:28:04 *******

 Chapter 10 - R-C Circuit Transient Analysis

 ****      CIRCUIT DESCRIPTION

*****************************************************************************

VE 1 0 PULSE(0 20V 0 1N 1N 1)
R  1 2 5K
C  2 0 8U
.TRAN 10M 200M
.PRINT TRAN V(1) V(2) I(C)
.OPTIONS NOPAGE
.END
```

```
 ****      TRANSIENT ANALYSIS                    TEMPERATURE =    27.000 DEG C

  TIME          V(1)          V(2)          I(C)

  0.000E+00     0.000E+00     0.000E+00     0.000E+00
  1.000E-02     2.000E+01     4.414E+00     3.117E-03
  2.000E-02     2.000E+01     7.862E+00     2.428E-03
  3.000E-02     2.000E+01     1.055E+01     1.890E-03
  4.000E-02     2.000E+01     1.264E+01     1.472E-03
  5.000E-02     2.000E+01     1.427E+01     1.145E-03
  6.000E-02     2.000E+01     1.554E+01     8.921E-04
  7.000E-02     2.000E+01     1.653E+01     6.943E-04
  8.000E-02     2.000E+01     1.730E+01     5.408E-04
  9.000E-02     2.000E+01     1.790E+01     4.209E-04
  1.000E-01     2.000E+01     1.836E+01     3.278E-04
  1.100E-01     2.000E+01     1.872E+01     2.551E-04
  1.200E-01     2.000E+01     1.901E+01     1.987E-04
  1.300E-01     2.000E+01     1.923E+01     1.547E-04
  1.400E-01     2.000E+01     1.940E+01     1.205E-04
  1.500E-01     2.000E+01     1.953E+01     9.376E-05
  1.600E-01     2.000E+01     1.963E+01     7.302E-05
  1.700E-01     2.000E+01     1.972E+01     5.683E-05
  1.800E-01     2.000E+01     1.978E+01     4.426E-05
  1.900E-01     2.000E+01     1.983E+01     3.445E-05
  2.000E-01     2.000E+01     1.987E+01     2.680E-05
```

**FIG. 10.65**

Replacing the .PRINT command by the .PROBE command will result in the beautiful plots of Fig. 10.66 for the input voltage, $v_C$ and $i_C$. Note how .PROBE assigns a notation for each of the plots and determines the appropriate scaling for the horizontal and vertical axes. As expected, $v_C$ rises toward 20 V in $5\tau$ and the input V(1) jumps to 20 V at zero milliseconds and continues at that level for the rest of the plot. The current $i_C$ starts at 4 mA and then decays exponentially to almost zero milliampere in the $5\tau$ period.

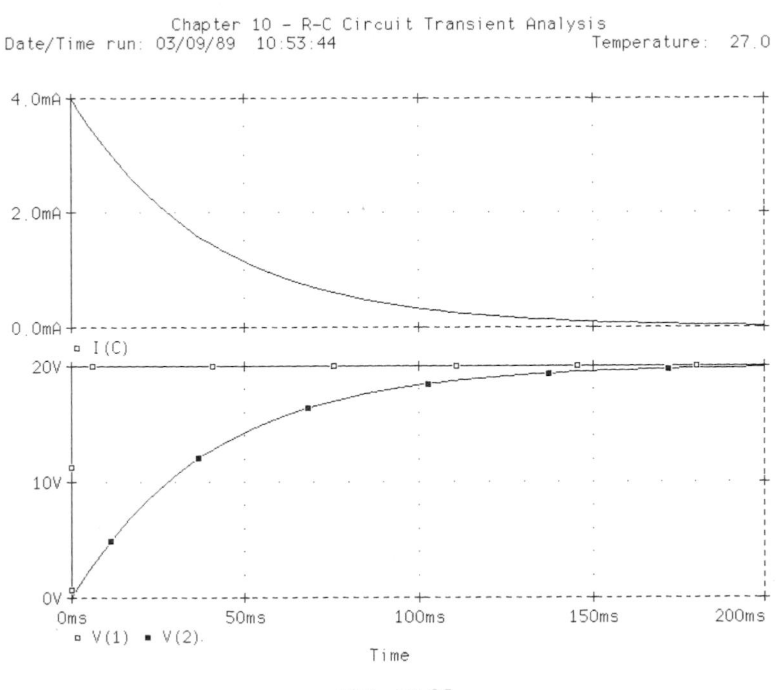


**FIG. 10.66**

# PROBLEMS

## SECTION 10.2 The Electric Field

1. Find the electric field strength at a point 2 m from a charge of 4 $\mu$C.
2. The electric field strength is 36 newtons/coulomb (N/C) at a point $r$ meters from a charge of 0.064 $\mu$C. Find the distance $r$.

## SECTION 10.3 Capacitance

3. Find the capacitance of a parallel plate capacitor if 1400 $\mu$C of charge are deposited on its plates when 20 V are applied across the plates.
4. How much charge is deposited on the plates of a 0.05-$\mu$F capacitor if 45 V are applied across the capacitor?
5. Find the electric field strength between the plates of a parallel plate capacitor if 100 mV are applied across the plates and the plates are 2 mm apart.
6. Repeat Problem 5 if the plates are separated by 4 mils.
7. A 4-$\mu$F parallel plate capacitor has 160 $\mu$C of charge on its plates. If the plates are 5 mm apart, find the electric field strength between the plates.
8. Find the capacitance of a parallel plate capacitor if the area of each plate is 0.075 $m^2$ and the distance between the plates is 1.77 mm. The dielectric is air.

9. Repeat Problem 8 if the dielectric is paraffin-coated paper.

10. Find the distance in mils between the plates of a 2-$\mu$F capacitor if the area of each plate is 0.09 $m^2$ and the dielectric is transformer oil.

11. The capacitance of a capacitor with a dielectric of air is 1200 pF. When a dielectric is inserted between the plates, the capacitance increases to 0.006 $\mu$F. Of what material is the dielectric made?

12. The plates of a parallel plate air capacitor are 0.2 mm apart, have an area of 0.08 $m^2$, and 200 V are applied across the plates.
 a. Determine the capacitance.
 b. Find the electric field intensity between the plates.
 c. Find the charge on each plate if the dielectric is air.

13. A sheet of Bakelite 0.2 mm thick having an area of 0.08 $m^2$ is inserted between the plates of Problem 12.
 a. Find the electric field strength between the plates.
 b. Determine the charge on each plate.
 c. Determine the capacitance.

## SECTION 10.4 Dielectric Strength

14. Find the maximum voltage ratings of the capacitors of Problems 12 and 13 assuming a linear relationship between the breakdown voltage and the thickness of the dielectric.

15. Find the maximum voltage that can be applied across a parallel plate capacitor of 0.006 $\mu$F. The area of one plate is 0.02 $m^2$ and the dielectric is mica. Assume a linear relationship between the dielectric strength and the thickness of the dielectric.

16. Find the distance in millimeters between the plates of a parallel plate capacitor if the maximum voltage that can be applied across the capacitor is 1250 V. The dielectric is mica. Assume a linear relationship between the breakdown strength and the thickness of the dielectric.

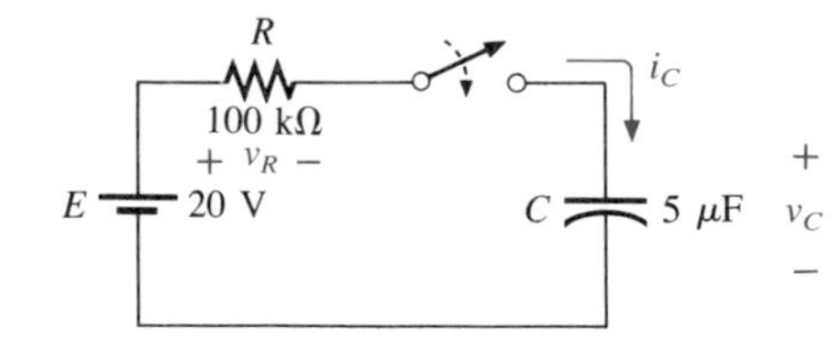

FIG. 10.67

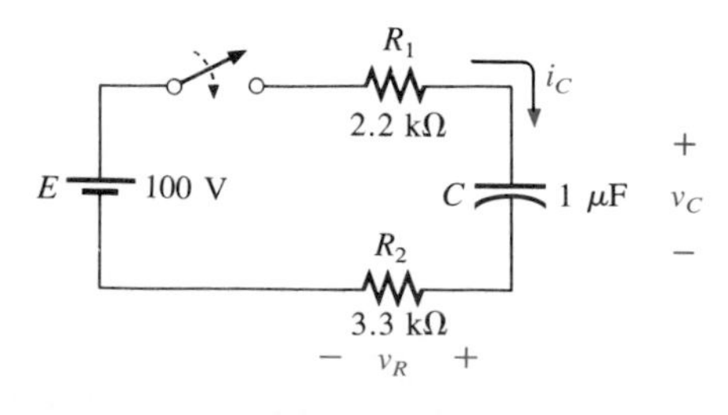


FIG. 10.68

## SECTION 10.7 Transients in Capacitive Networks: Charging Phase

17. For the circuit of Fig. 10.67:
 a. Determine the time constant of the circuit.
 b. Write the mathematical equation for the voltage $v_C$ following the closing of the switch.
 c. Determine the voltage $v_C$ after one, three, and five time constants.
 d. Write the equations for the current $i_C$ and the voltage $v_R$.
 e. Sketch the waveforms for $v_C$ and $i_C$.

18. Repeat Problem 17 for $R = 1\ M\Omega$ and compare results.

19. Repeat Problem 17 for the network of Fig. 10.68.

**20.** For the circuit of Fig. 10.69:

**a.** Determine the time constant of the circuit.

**b.** Write the mathematical expression for the voltage $v_C$ following the closing of the switch.

**c.** Write the mathematical expression for the current $i_C$ following the closing of the switch.

**d.** Sketch the waveforms of $v_C$ and $i_C$.

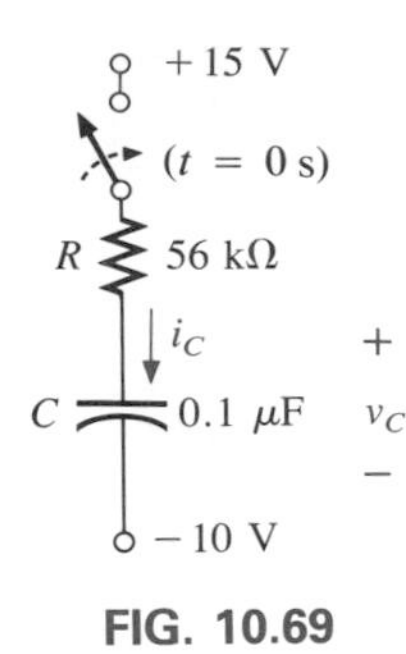


FIG. 10.69

## SECTION 10.8 Discharge Phase

**21.** For the circuit of Fig. 10.70:

**a.** Determine the time constant of the circuit when the switch is thrown into position 1.

**b.** Find the mathematical expression for the voltage across the capacitor after the switch is thrown into position 1.

**c.** Determine the mathematical expression for the current following the closing of the switch (position 1).

**d.** Determine the voltage $v_C$ and the current $i_C$ if the switch is thrown into position 2 at $t = 100$ ms.

**e.** Determine the mathematical expressions for the voltage $v_C$ and the current $i_C$ if the switch is thrown into position 3 at $t = 200$ ms.

**f.** Plot the waveforms of $v_C$ and $i_C$ for a period of time extending from $t = 0$ to $t = 300$ ms.

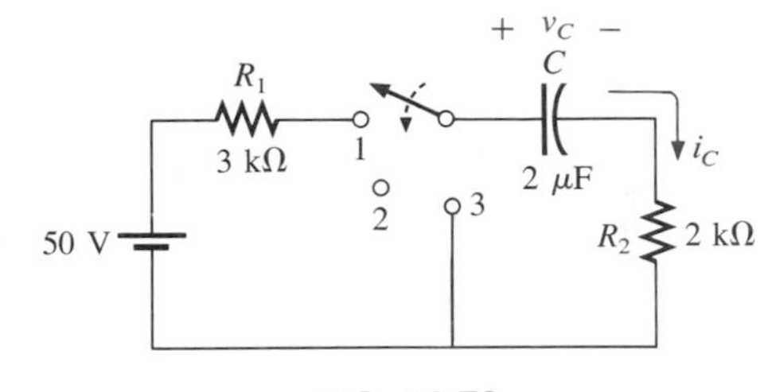


FIG. 10.70

**22.** Repeat Problem 21 for a capacitance of 20 $\mu$F.

***23.** For the network of Fig. 10.71:

**a.** Find the mathematical expression for the voltage across the capacitor after the switch is thrown into position 1.

**b.** Repeat part (a) for the current $i_C$.

**c.** Find the mathematical expressions for the voltage $v_C$ and current $i_C$ if the switch is thrown into position 2 at a time equal to five time constants of the charging circuit.

**d.** Plot the waveforms of $v_C$ and $i_C$ for a period of time extending from $t = 0$ to $t = 30$ $\mu$s.

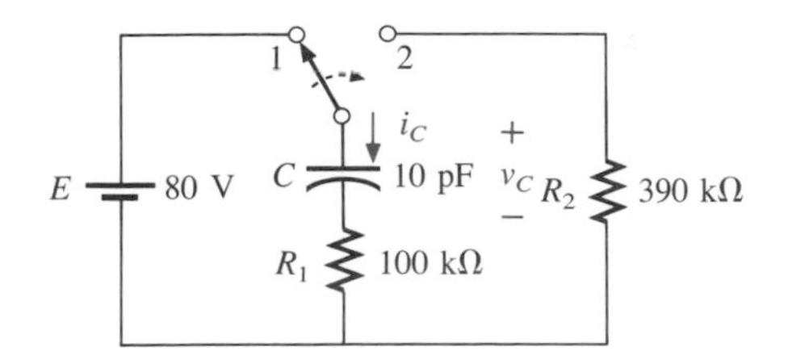


FIG. 10.71

**24.** The capacitor of Fig. 10.72 is initially charged to 40 V before the switch is closed. Write the expressions for the voltages $v_C$ and $v_R$ and the current $i_C$ for the decay phase.

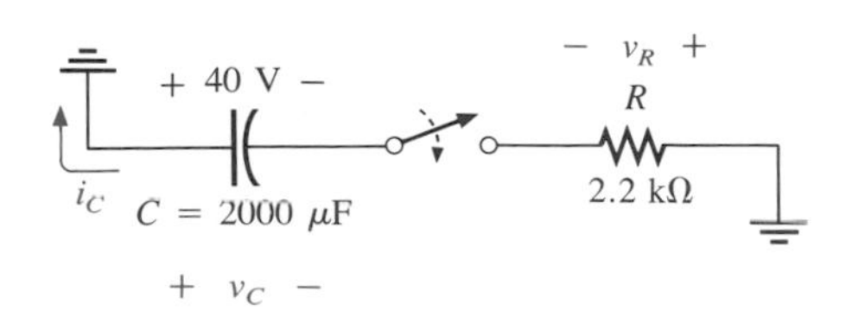

FIG. 10.72

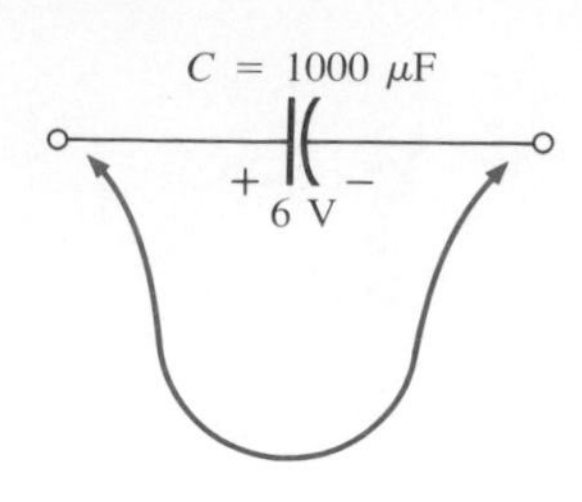

FIG. 10.73

25. The 1000-$\mu$F capacitor of Fig. 10.73 is charged to 6 V. To discharge the capacitor before further use, a wire with a resistance of 0.002 $\Omega$ is placed across the capacitor.
    **a.** How long will it take to discharge the capacitor?
    **b.** What is the peak value of the current?
    **c.** Based on the result of part (b), is a spark expected when contact is made with both ends of the capacitor?

## SECTION 10.9 Instantaneous Values

26. Given the expression $v_C = 8(1 - e^{-t/(20\times10^{-6})})$:
    **a.** Determine $v_C$ after five time constants.
    **b.** Determine $v_C$ after 10 time constants.
    **c.** Determine $v_C$ at $t = 5\ \mu$s.

27. For the situation of Problem 25, determine when the discharge current is one-half its maximum value if contact is made at $t = 0$ s.

28. For the network of Fig. 10.74, $V_L$ must be 8 V before the system is activated. If the switch is closed at $t = 0$ s, how long will it take for the system to be activated?

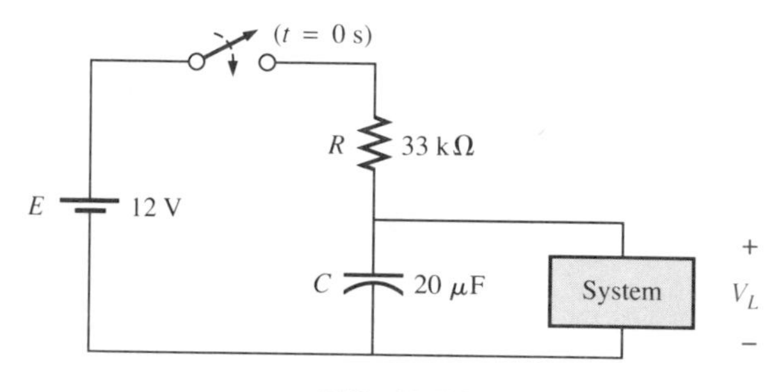

FIG. 10.74

*29. Design the network of Fig. 10.75 such that the system will turn on 10 s after the switch is closed.

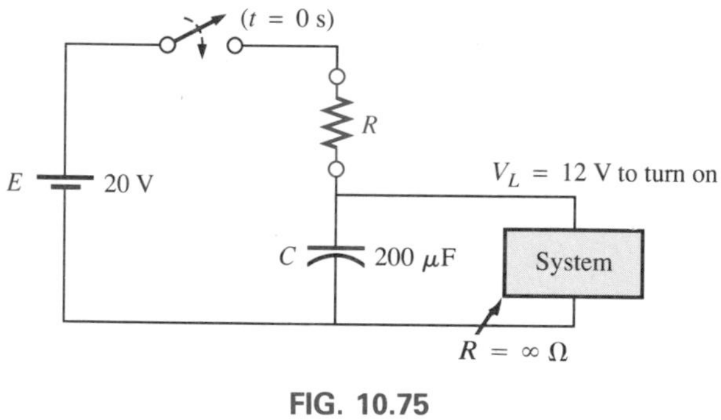

FIG. 10.75

*30. For the network of Fig. 10.76,
    **a.** Calculate $v_C$, $i_C$, and $v_{R_1}$ at 0.5 s and 1 s after the switch makes contact with position 1.
    **b.** The network sits in position 1 10 min before the switch is moved to position 2. How long after making contact with position 2 will it take for the current $i_C$ to drop to 8 $\mu$A and how much *longer* will it take for $v_C$ to drop to 10 V?

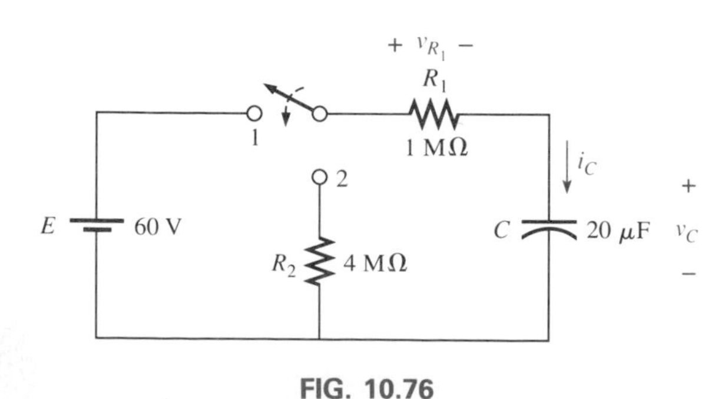


FIG. 10.76

## SECTION 10.10 $\tau = R_{Th}C$

**31.** For the circuit of Fig. 10.77:

**a.** Find the mathematical expressions for the transient behavior of the voltage $v_C$ and the current $i_C$ following the closing of the switch.

**b.** Sketch the waveforms of $v_C$ and $i_C$.

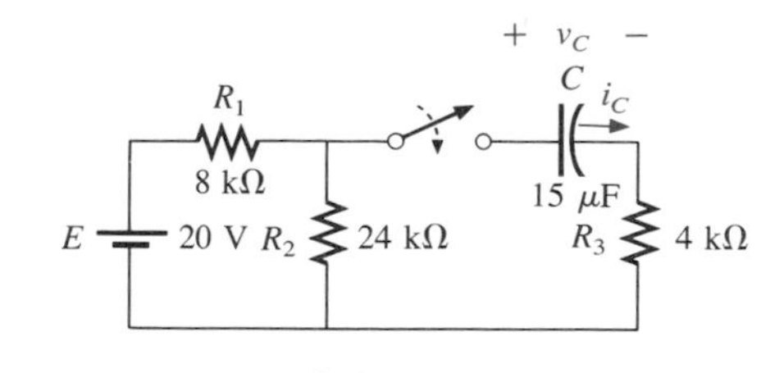

FIG. 10.77

***32.** Repeat Problem 31 for the circuit of Fig. 10.78.

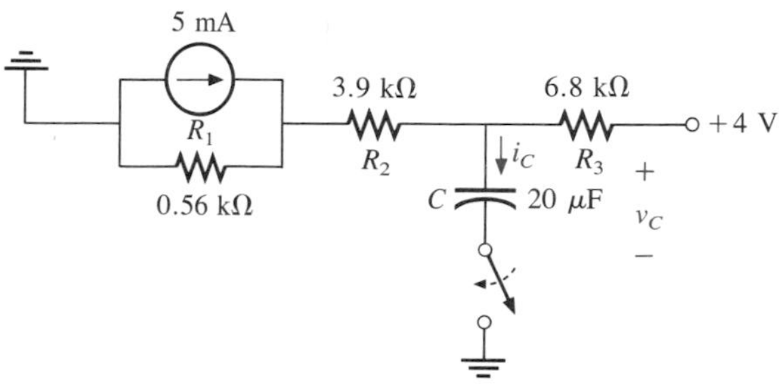

FIG. 10.78

## SECTION 10.11 The Current $i_C$

**33.** Find the waveform for the average current if the voltage across a 0.06-μF capacitor is as shown in Fig. 10.79.

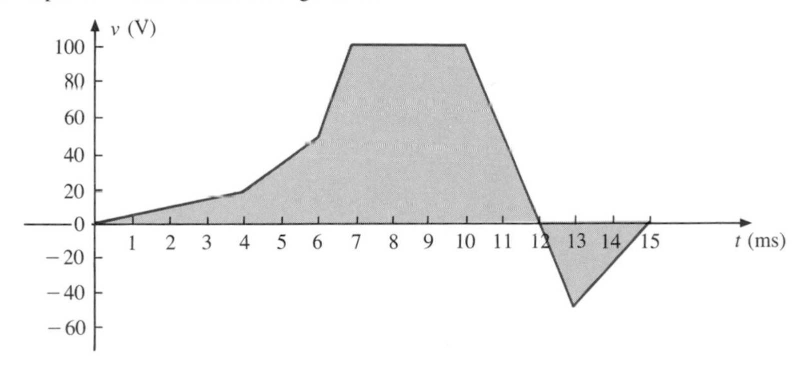


FIG. 10.79

**34.** Repeat Problem 33 for the waveform of Fig. 10.80.

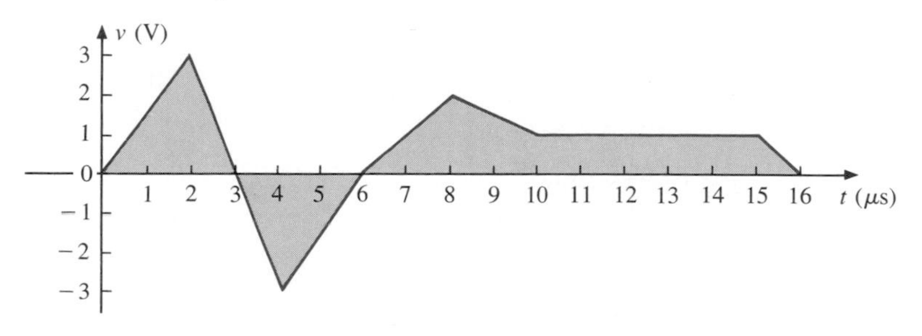


FIG. 10.80

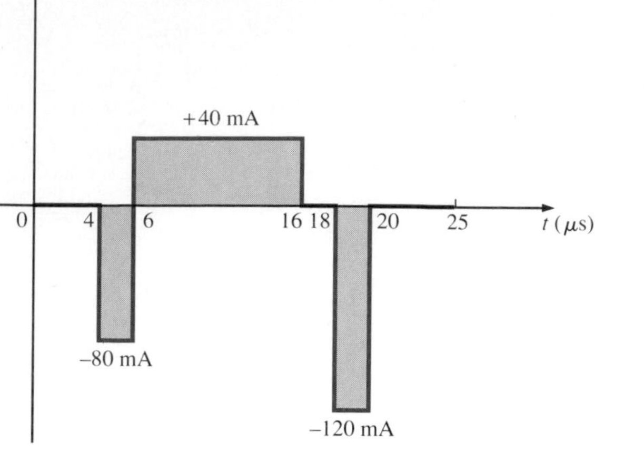


FIG. 10.81

***35.** Given the following waveform (Fig. 10.81) for the current of a 20-μF capacitor sketch the waveform of the voltage $v_C$ across the capacitor if $v_C = 0$ V at $t = 0$ s.

## SECTION 10.12 Capacitors in Series and Parallel

**36.** Find the total capacitance $C_T$ between points $a$ and $b$ of the circuits of Fig. 10.82.

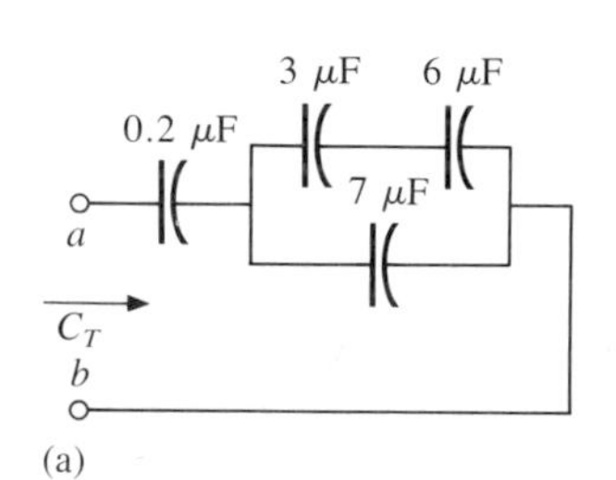


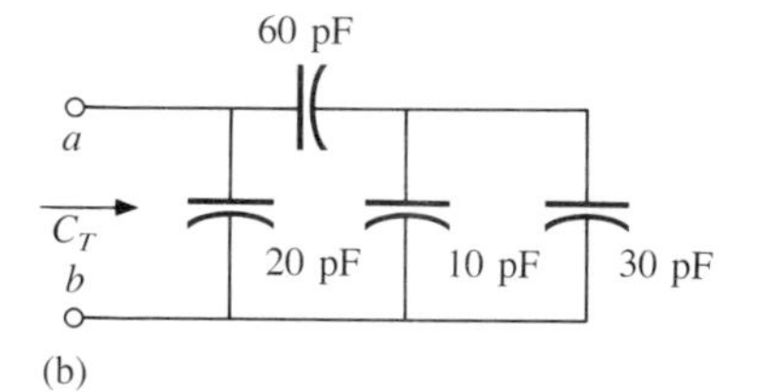

FIG. 10.82

**37.** Find the voltage across and charge on each capacitor for the circuits of Fig. 10.83.

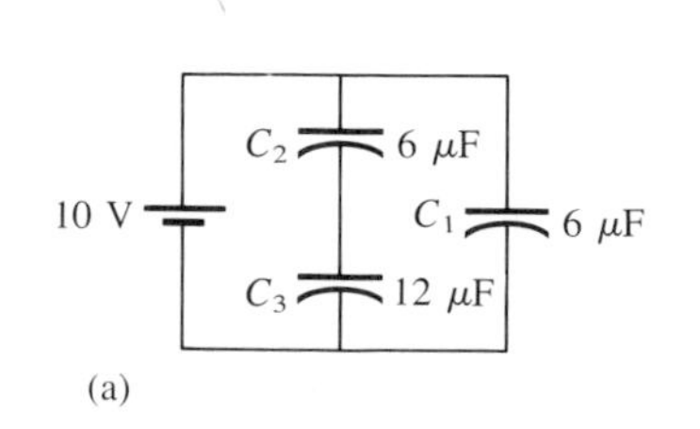


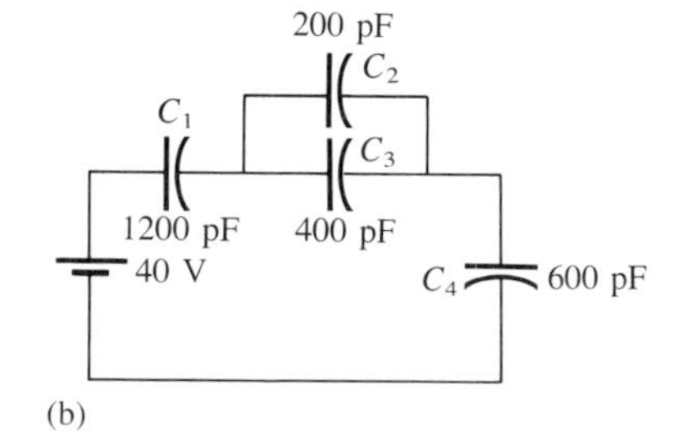


FIG. 10.83

***38.** For each configuration of Fig. 10.84, determine the voltage across each capacitor and the charge on each capacitor.

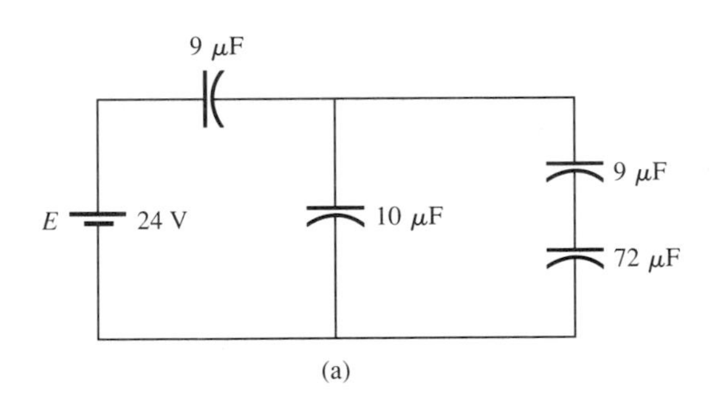


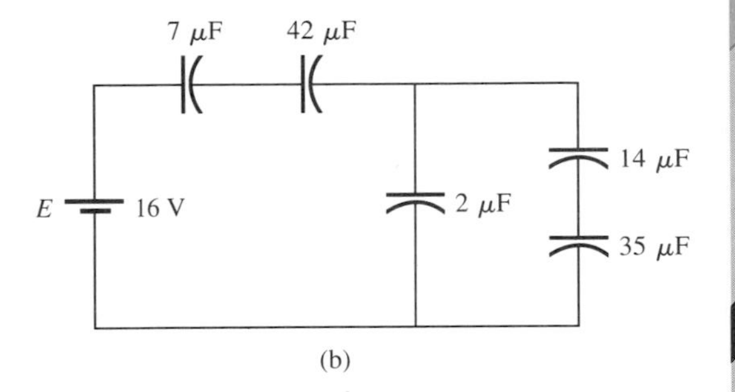

**FIG. 10.84**

***39.** For the network of Fig. 10.85, determine the following 100 ms after the switch is closed.

**a.** $V_{ab}$
**b.** $V_{ac}$
**c.** $V_{cb}$
**d.** $V_{da}$
**e.** If the switch is moved to position 2 one hour later, find the time required for $v_{R_2}$ to drop to 20 V.

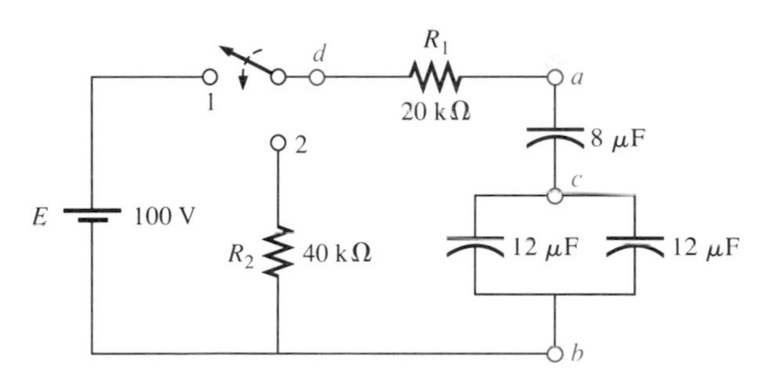

**FIG. 10.85**

**40.** For the circuits of Fig. 10.86, find the voltage across and charge on each capacitor after each capacitor has charged to its final value.

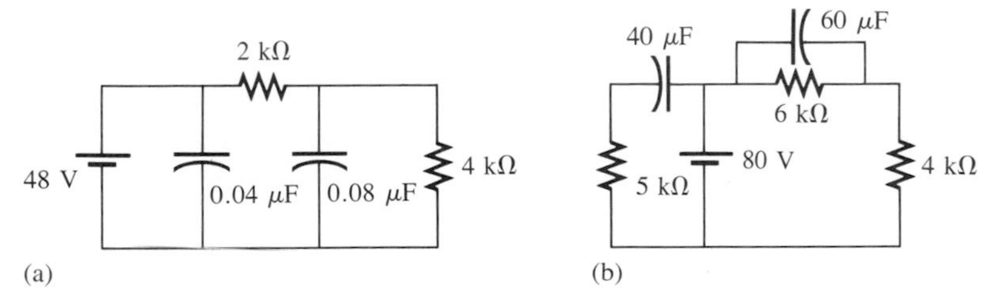


**FIG. 10.86**

## SECTION 10.13 Energy Stored by a Capacitor

**41.** Find the energy stored by a 120-pF capacitor with 12 V across its plates.

**42.** If the energy stored by a 6-$\mu$F capacitor is 1200 J, find the charge $Q$ on each plate of the capacitor.

***43.** An electronic flashgun has a 1000-$\mu$F capacitor which is charged to 100 V.

**a.** How much energy is stored by the capacitor?

**b.** What is the charge on the capacitor?

**c.** When the photographer takes a picture, the flash fires for 1/2000 s. What is the average current through the flashtube?

**d.** Find the power delivered to the flashtube.

**e.** After a picture is taken, the capacitor has to be recharged by a power supply which delivers a maximum current of 10 mA. How long will it take to charge the capacitor?

## SECTION 10.15 Computer Analysis

### PSPICE

**44.** Write the input file to obtain the waveforms of Fig. 10.33 for the network of Fig. 10.32.

**45.** Write the input file to obtain the transient waveforms of $v_C$ and $i_C$ for the network of Fig. 10.39 when the switch is moved to position 1.

**46.** Write the input file to obtain the waveforms of $v_C$ and $i_C$ for the network of Fig. 10.41 if the switch is moved to position 1 at $t = 0$ s.

***47.** Write the input file to obtain the waveforms of $v_C$ and $i_C$ for the network of Fig. 10.44 if the switch is moved to position 1 at $t = 0$ s.

### BASIC

**48.** Write a BASIC program to tabulate the voltage $v_C$ and current $i_C$ for the network of Fig. 10.41 for five time constants after the switch is moved to position 1 at $t = 0$ s. Use an increment of $\frac{1}{5}\tau$.

***49.** Write a program to write the mathematical expression for the voltage $v_C$ for the network of Fig. 10.44 for any element values when the switch is moved to position 1.

**50.** Write a program to tabulate the voltage of Problem 26 for each time constant from one to 20 time constants.

***51.** Given three capacitors in any series-parallel arrangement, write a program to determine the total capacitance. That is, determine the total number of possibilities and ask the user to identify the configuration and provide the capacitor values. Then calculate the total capacitance.

# GLOSSARY

**Breakdown voltage** Another term for *dielectric strength,* listed below.

**Capacitance** A measure of a capacitor's ability to store charge; measured in farads (F).

**Capacitive time constant** The product of resistance and capacitance that establishes the required time for the charging and discharging phases of a capacitive transient.

**Capacitive transient** The waveforms for the voltage and current of a capacitor that result during the charging and discharging phases.

**Capacitor** A fundamental electrical element having two conducting surfaces separated by an insulating material and having the capacity to store charge on its plates.

**Coulomb's law** An equation relating the force between two like or unlike charges.

**Dielectric** The insulating material between the plates of a capacitor that can have a pronounced effect on the charge stored on the plates of a capacitor.

**Dielectric constant** Another term for *relative permittivity,* listed below.

**Dielectric strength** An indication of the voltage required for unit length to establish conduction in a dielectric.

**Electric field strength** The force acting on a unit positive charge in the region of interest.

**Electric flux lines** Lines drawn to indicate the strength and direction of an electric field in a particular region.

**Fringing** An effect established by flux lines that do not pass directly from one conducting surface to another.

**Leakage current** The current that will result in the total discharge of a capacitor if the capacitor is disconnected from the charging network for a sufficient length of time.

**Permittivity** A measure of how well a dielectric will *permit* the establishment of flux lines within the dielectric.

**Relative permittivity** The permittivity of a material compared to that of air.

**Stray capacitance** Capacitances that exist not through design but simply because two conducting surfaces are relatively close to each other.

**Surge voltage** The maximum voltage that can be applied across the capacitor for very short periods of time.

**.TRAN** PSPICE command to specify a transient analysis.

**TSTEP** PSPICE entry to specify the interval between values to be determined.

**TSTOP** PSPICE entry to specify the final value for which function is to be determined.

**Working voltage** The voltage that can be applied across a capacitor for long periods of time without concern for dielectric breakdown.

# 11 Magnetic Circuits

## 11.1 INTRODUCTION

Magnetism plays an integral part in almost every electrical device used today in industry, research, or the home. Generators, motors, transformers, circuit breakers, televisions, computers, tape recorders, and telephones all employ magnetic effects to perform a variety of important tasks.

The compass, used by Chinese sailors as early as the second century A.D., relies on a *permanent magnet* for indicating direction. The permanent magnet is made of a material, such as steel or iron, that will remain magnetized for long periods of time without the need for an external source of energy.

In 1820, the Danish physicist Hans Christian Oersted discovered that the needle of a compass would deflect if brought near a current-carrying conductor. For the first time it was demonstrated that electricity and magnetism were related, and in the same year the French physicist André Marie Ampère performed experiments in this area and developed what is presently known as *Ampère's circuital law*. In subsequent years, men such as Michael Faraday, Karl Friedrich Gauss, and James Clerk Maxwell continued to experiment in this area and developed many of the basic concepts of *electromagnetism*—magnetic effects induced by the flow of charge, or current.

There is a great deal of similarity between the analyses of electric circuits and magnetic circuits. This will be demonstrated later in this chapter when we compare the basic equations and methods used to solve magnetic circuits with those used for electric circuits.

Difficulty in understanding methods used with magnetic circuits will often arise in simply learning to use the proper set of units, not because of the equations themselves. The problem exists because three different systems of units are still being used in the industry. To the extent practical, SI will be used throughout this chapter. For the CGS and English systems, a conversion table is provided in Appendix G.

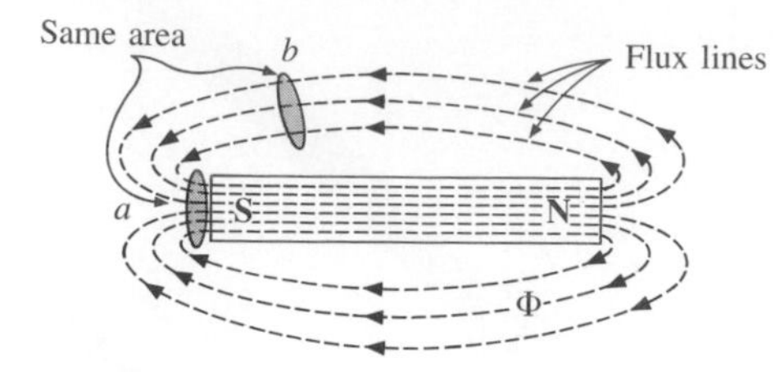


FIG. 11.1

## 11.2 MAGNETIC FIELDS

In the region surrounding a permanent magnet there exists a magnetic field, which can be represented by magnetic flux lines similar to electric flux lines. Magnetic flux lines, however, do not have origins or terminating points like electric flux lines but exist in continuous loops, as shown in Fig. 11.1. The symbol for magnetic flux is the Greek letter $\Phi$ (phi).

The magnetic flux lines radiate from the north pole to the south pole, returning to the north pole through the metallic bar. Note the equal spacing between the flux lines within the core and the symmetric distribution outside the magnetic material. These are additional properties of magnetic flux lines in homogeneous materials (that is, materials having uniform structure or composition throughout). It is also important to realize that the continuous magnetic flux line will strive to occupy as small an area as possible. This will result in magnetic flux lines of minimum length between the like poles, as shown in Fig. 11.2. The strength of a magnetic field in a particular region is directly related to the density of flux lines in that region. In Fig. 11.1, for example, the magnetic field strength at *a* is twice that at *b* since there are twice as many magnetic flux lines associated with the perpendicular plane at *a* than at *b*. Recall from childhood experiments how the strength of permanent magnets was always stronger near the poles.

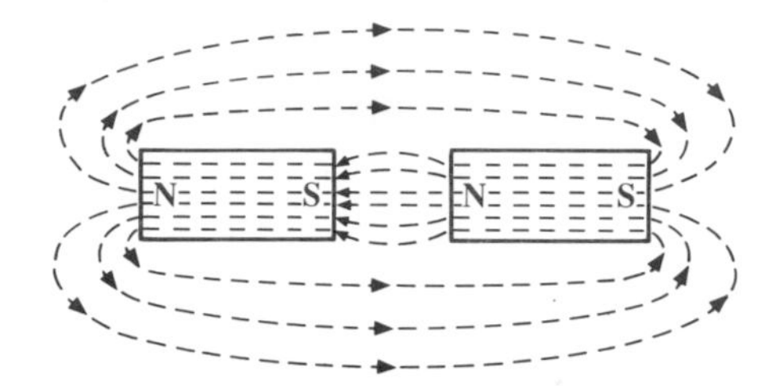


FIG. 11.2

If unlike poles of two permanent magnets are brought together, the magnets will attract, and the flux distribution will be as shown in Fig. 11.2. If like poles are brought together, the magnets will repel, and the flux distribution will be as shown in Fig. 11.3.

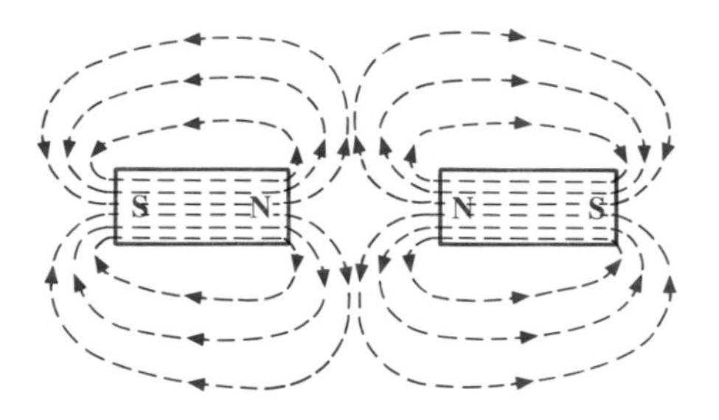

FIG. 11.3

If a nonmagnetic material, such as glass or copper, is placed in the flux paths surrounding a permanent magnet, there will be an almost unnoticeable change in the flux distribution (Fig. 11.4). However, if a magnetic material, such as soft iron, is placed in the flux path, the flux lines will pass through the soft iron rather than the surrounding air because flux lines pass with greater ease through magnetic materials than through air. This principle is put to use in the shielding of sensitive electrical elements and instruments that can be affected by stray magnetic fields (Fig. 11.5).

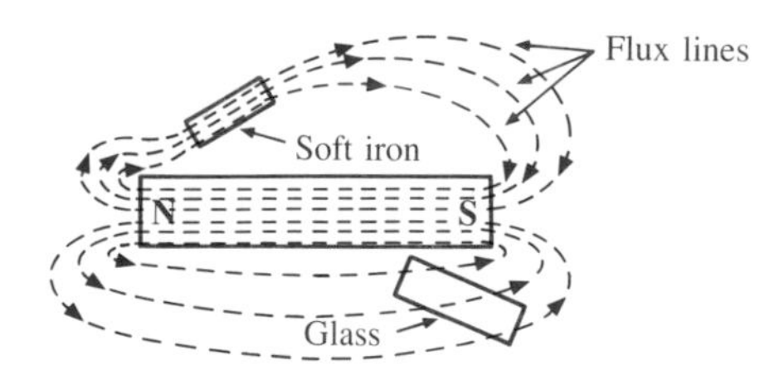


FIG. 11.4

As indicated in the introduction, a magnetic field (represented by concentric magnetic flux lines, as in Fig. 11.6) is present around every wire that carries an electric current. The direction of the magnetic flux

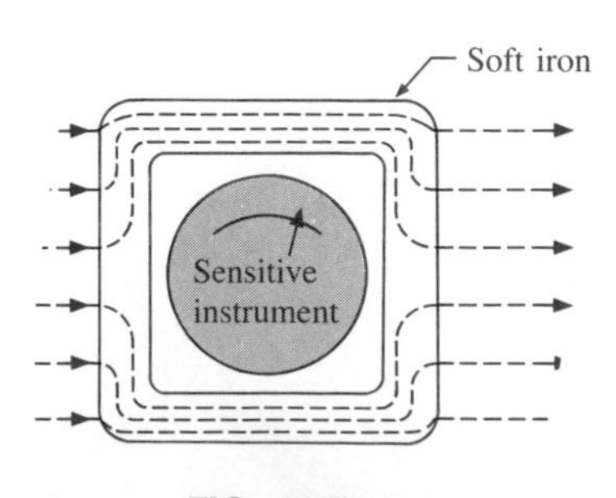


FIG. 11.5

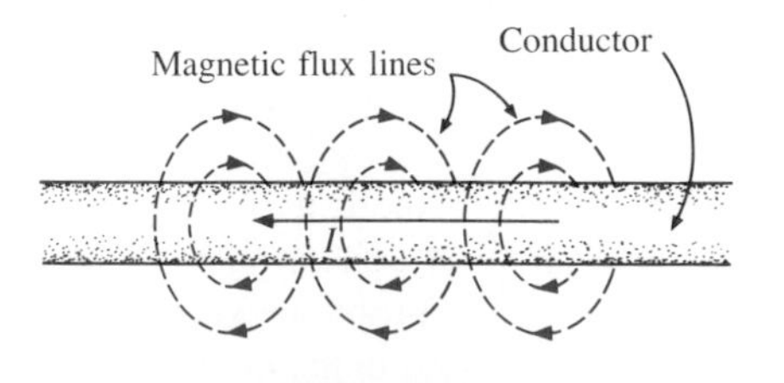


FIG. 11.6

lines can be found simply by placing the thumb of the *right* hand in the direction of *conventional* current flow and noting the direction of the fingers. (This method is commonly called the *right-hand rule*.) If the conductor is wound in a single-turn coil (Fig. 11.7), the resulting flux will flow in a common direction through the center of the coil.

A coil of more than one turn would produce a magnetic field that would exist in a continuous path through and around the coil (Fig. 11.8).

The flux distribution of the coil is quite similar to that of the permanent magnet. The flux lines leaving the coil from the left and entering to the right simulate a north and south pole, respectively. The principal difference between the two flux distributions is that the flux lines are more concentrated for the permanent magnet than for the coil. Also, since the strength of a magnetic field is determined by the density of the flux lines, the coil has a weaker field strength. The field strength of the coil can be effectively increased by placing certain materials, such as iron, steel, or cobalt, within the coil to increase the flux density within the coil. By increasing the field strength with the addition of the core, we have devised an *electromagnet* (Fig. 11.9) which, in addition to having all the properties of a permanent magnet, also has a field strength that can be varied by changing one of the component values (current, turns, and so on). Of course, current must pass through the coil of the electromagnet in order for magnetic flux to be developed, whereas there is no need for the coil or current in the permanent magnet. The direction of flux lines can be determined for the electromagnet (or in any core with a wrapping of turns) by placing the fingers of the right hand in the direction of current flow around the core. The thumb will then point in the direction of the north pole of the induced magnetic flux. This is demonstrated in Fig. 11.10. A cross section of the same electromagnet was included in the figure to introduce the convention for

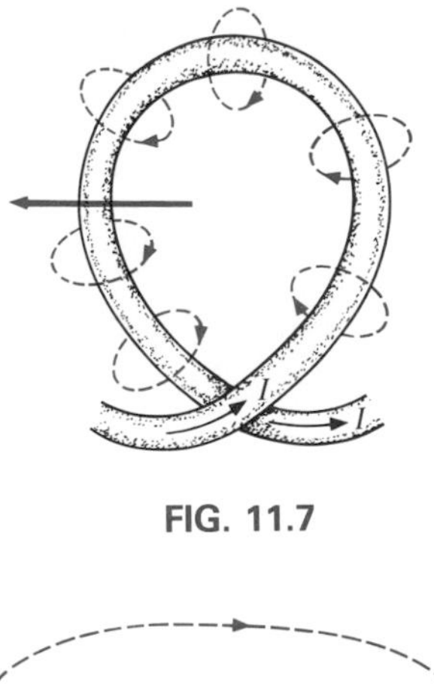


FIG. 11.7

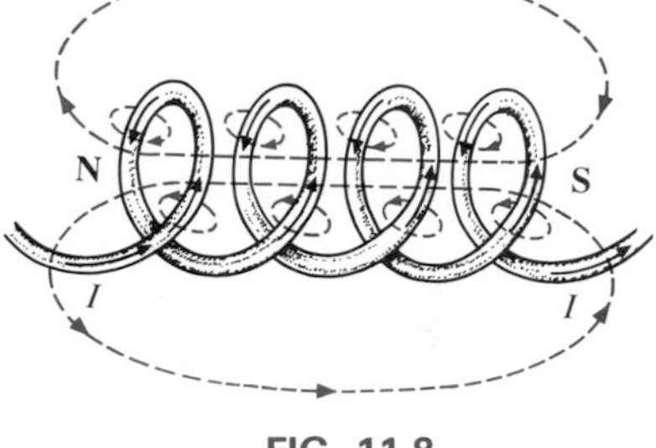


FIG. 11.8

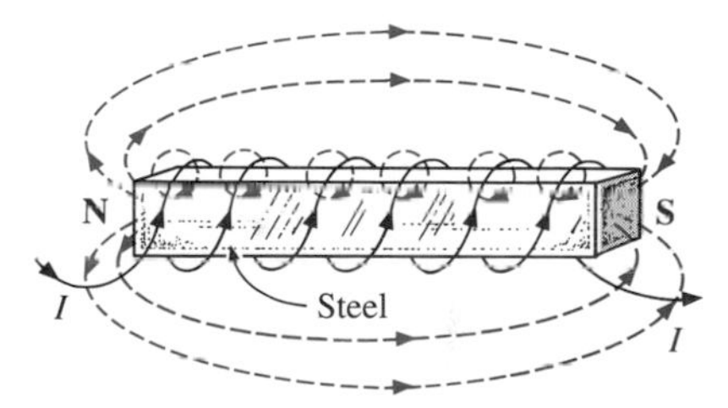


FIG. 11.9

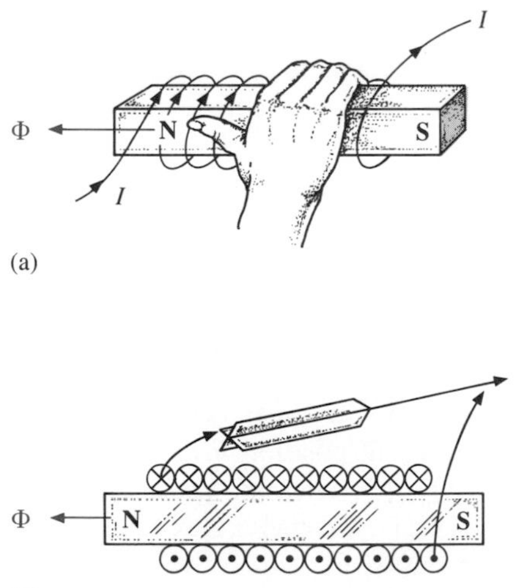


FIG. 11.10

directions perpendicular to the page. The cross and dot refer to the tail and head of the arrow, respectively.

Other areas of application for electromagnetic effects are shown in Fig. 11.11. The flux path for each is indicated in each figure.

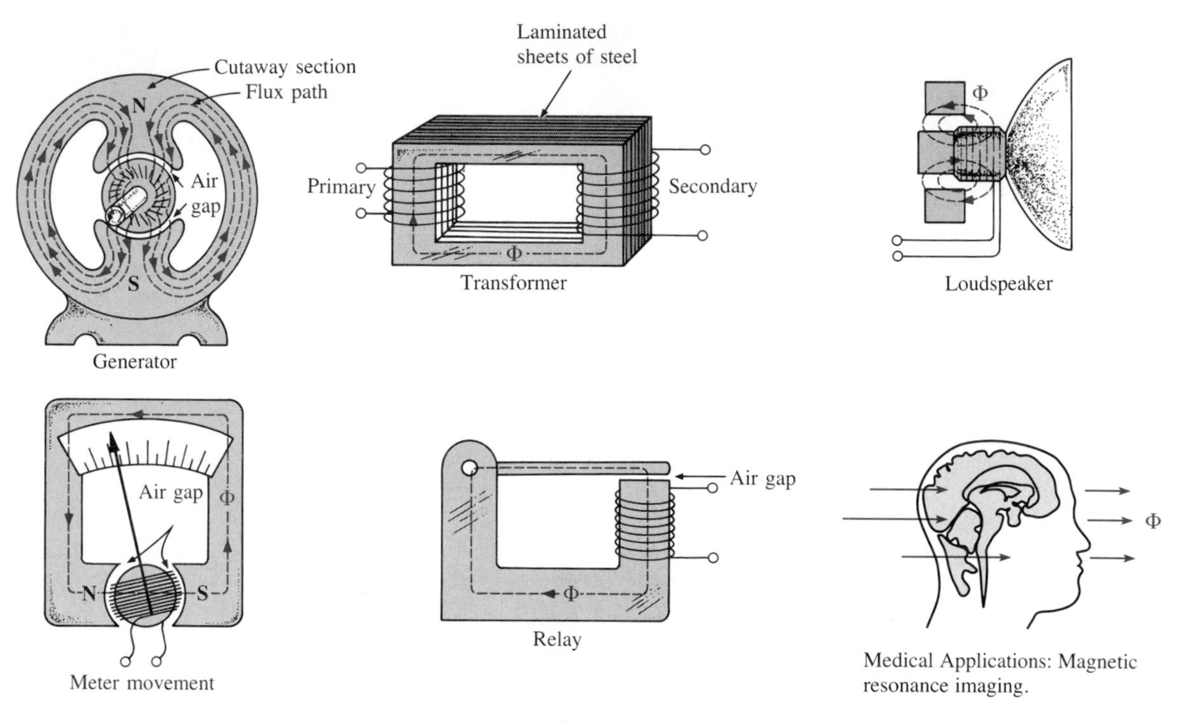


FIG. 11.11

## 11.3 FLUX DENSITY

In the SI system of units, magnetic flux is measured in webers (Wb) and has the symbol Φ. The number of flux lines per unit area is called the *flux density* and is denoted by the capital letter $B$. Its magnitude is determined by the following equation:

$$B = \frac{\Phi}{A} \qquad \begin{aligned} B &= \text{teslas (T)} \\ \Phi &= \text{webers (Wb)} \\ A &= \text{square meters (m}^2\text{)} \end{aligned} \tag{11.1}$$

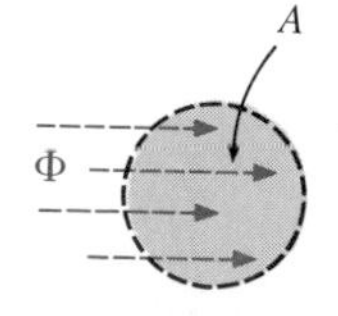


FIG. 11.12

where Φ is the number of flux lines passing through the area $A$ (Fig. 11.12). The flux density at position $a$ in Fig. 11.1 is twice that at $b$ because twice as many flux lines are passing through the same area.

As noted in Eq. (11.1), magnetic flux density in the SI system of units is measured in *teslas,* for which the symbol is T. By definition,

$$1 \text{ tesla (T)} = 1 \text{ Wb/m}^2$$

**EXAMPLE 11.1.** For the core of Fig. 11.13, determine the flux density $B$ in teslas.

***Solution:***

$$B = \frac{\Phi}{A} = \frac{6 \times 10^{-5}\ \text{Wb}}{1.2 \times 10^{-3}\ \text{m}^2} = \mathbf{5 \times 10^{-2}\ T}$$

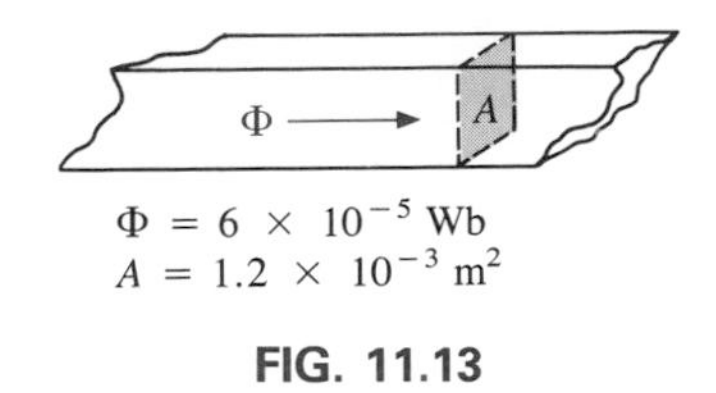


$\Phi = 6 \times 10^{-5}$ Wb
$A = 1.2 \times 10^{-3}\ \text{m}^2$

**FIG. 11.13**

**EXAMPLE 11.2.** In Fig. 11.13, if the flux density is 1.2 T and the area is 0.25 in.$^2$, determine the flux through the core.

***Solution:*** By Eq. (11.1),

$$\Phi = BA$$

However, converting 0.25 in.$^2$ to metric units,

$$A = 0.25\ \cancel{\text{in.}}^2\left(\frac{1\ \text{m}}{39.37\ \cancel{\text{in.}}}\right)\left(\frac{1\ \text{m}}{39.37\ \cancel{\text{in.}}}\right) = 1.613 \times 10^{-4}\ \text{m}^2$$

and

$$\begin{aligned}\Phi &= (1.2\ \text{T})(1.613 \times 10^{-4}\ \text{m}^2)\\ &= \mathbf{1.936 \times 10^{-4}\ Wb}\end{aligned}$$

An instrument designed to measure flux density in gauss (CGS system) appears in Fig. 11.14. Appendix G reveals that 1 T = $10^4$ gauss. The magnitude of the reading appearing on the face of the meter in Fig. 11.14 is therefore

$$1.964\ \cancel{\text{gauss}}\left(\frac{1\ \text{T}}{10^4\ \cancel{\text{gauss}}}\right) = 1.964 \times 10^{-4}\ \text{T}$$

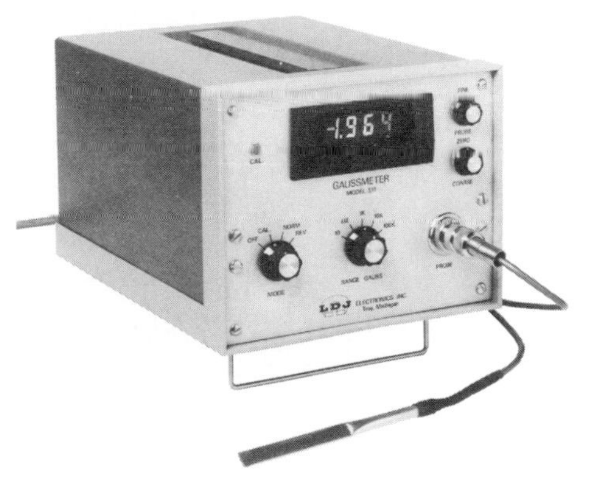


**FIG. 11.14**
*Digital display gaussmeter. (Courtesy of LDJ Electronics, Inc.)*

## 11.4 PERMEABILITY

If cores of different materials with the same physical dimensions are used in the electromagnet described in Section 11.2, the strength of the magnet will vary in accordance with the core used. This variation in strength is due to the greater or lesser number of flux lines passing through the core. Materials in which flux lines can readily be set up are said to be *magnetic* and to have *high permeability*. The permeability ($\mu$) of a material, therefore, is a measure of the ease with which magnetic flux lines can be established in the material. It is similar in many respects to conductivity in electric circuits. The permeability of free space $\mu_o$ (vacuum) is

$$\mu_o = 4\pi \times 10^{-7}\ \text{weber/ampere-meter}$$

As indicated above, $\mu$ has the units of webers/ampere. Practically speaking, the permeability of all nonmagnetic materials, such as copper, aluminum, wood, glass, and air, is the same as that for free space. Materials that have permeabilities slightly less than that of free space are said to be *diamagnetic,* and those with permeabilities slightly greater than that of free space are said to be *paramagnetic*. Magnetic materials, such as iron, nickel, steel, cobalt, and alloys of these metals, have permeabilities hundreds and even thousands of times that of free space. Materials with these very high permeabilities are referred to as *ferromagnetic*.

The ratio of the permeability of a material to that of free space is called its *relative permeability;* that is,

$$\mu_r = \frac{\mu}{\mu_o} \tag{11.2}$$

In general, for ferromagnetic materials, $\mu_r \geq 100$, and for nonmagnetic materials, $\mu_r = 1$.

Since $\mu_r$ is a variable, dependent on other quantities of the magnetic circuit, values of $\mu_r$ are not tabulated. Methods of calculating $\mu_r$ from the data supplied by manufacturers will be considered in a later section.

## 11.5 RELUCTANCE

The resistance of a material to the flow of charge (current) is determined for electric circuits by the equation

$$R = \rho\frac{l}{A} \qquad \text{(ohms, } \Omega\text{)}$$

The *reluctance* of a material to the setting up of magnetic flux lines in the material is determined by the following equation:

$$\mathcal{R} = \frac{l}{\mu A} \qquad \text{(rels, or At/Wb)} \tag{11.3}$$

where $\mathcal{R}$ is the reluctance, $l$ is the length of the magnetic path, and $A$ is its cross-sectional area. The $t$ in the units At/Wb is the number of turns of the applied winding. More will be said about ampere-turns (At) in the next section. Note that the resistance and reluctance are inversely proportional to the area, indicating that an increase in area will result in a reduction in each and an *increase* in the desired result: current and flux. For an increase in length the opposite is true, and the desired effect is reduced. The reluctance, however, is inversely proportional to the permeability, while the resistance is directly proportional to the resistivity. The larger the $\mu$ or smaller the $\rho$, the smaller the reluctance and resistance, respectively. Obviously, therefore, materials with high permeability, such as the ferromagnetics, have very small reluctances and will result in an increased measure of flux through the core. There is no widely accepted unit for reluctance, although the *rel* and the At/Wb are usually applied.

## 11.6 OHM'S LAW FOR MAGNETIC CIRCUITS

Recall the equation

$$\text{Effect} = \frac{\text{cause}}{\text{opposition}}$$

appearing in Chapter 4 to introduce Ohm's law for electric circuits. For magnetic circuits, the effect desired is the flux $\Phi$. The cause is the *magnetomotive force* (mmf) $\mathscr{F}$, which is the external force (or "pressure") required to set up the magnetic flux lines within the magnetic material. The opposition to the setting up of the flux $\Phi$ is the reluctance $\mathscr{R}$.

Substituting, we have

$$\boxed{\Phi = \frac{\mathscr{F}}{\mathscr{R}}} \qquad \textbf{(11.4)}$$

The magnetomotive force $\mathscr{F}$ is proportional to the product of the number of turns around the core (in which the flux is to be established) and the current through the turns of wire (Fig. 11.15). In equation form,

$$\boxed{\mathscr{F} = NI} \qquad \text{(ampere-turns, At)} \qquad \textbf{(11.5)}$$

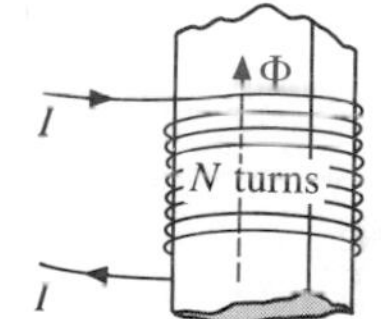

**FIG. 11.15**

The equation clearly indicates that an increase in the number of turns or the current through the wire will result in an increased "pressure" on the system to establish flux lines through the core.

Although there is a great deal of similarity between electric and magnetic circuits, one must continue to realize that the flux $\Phi$ is not a "flow" variable such as current in an electric circuit. Magnetic flux is established in the core through the alteration of the atomic structure of the core due to external pressure and is not a measure of the flow of some charged particles through the core.

## 11.7 MAGNETIZING FORCE

The magnetomotive force per unit length is called the *magnetizing force* ($H$). In equation form,

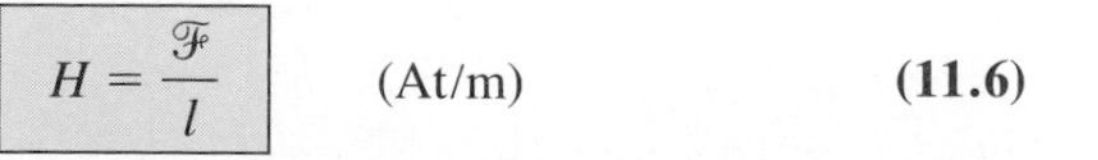

$$\boxed{H = \frac{\mathscr{F}}{l}} \qquad \text{(At/m)} \qquad \textbf{(11.6)}$$

Substituting for the magnetomotive force will result in

$$\boxed{H = \frac{NI}{l}} \qquad \text{(At/m)} \tag{11.7}$$

For the magnetic circuit of Fig. 11.16, if $NI = 40$ At and $l = 0.2$ m, then

$$H = \frac{NI}{l} = \frac{40}{0.2} = 200 \text{ At/m}$$

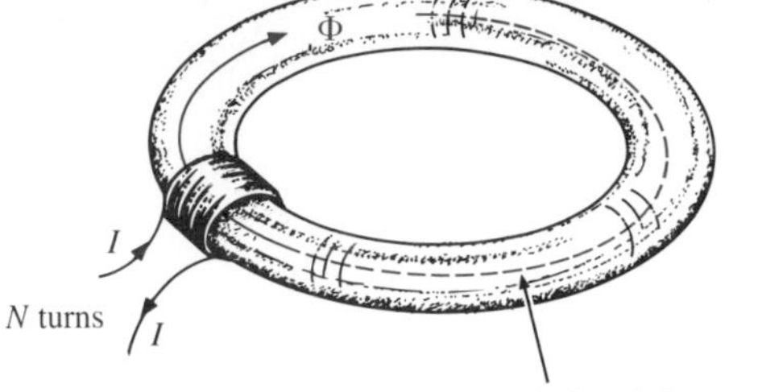


FIG. 11.16

In words, the result indicates that there are 200 At of "pressure" per meter to establish flux in the core.

Note in Fig. 11.16 that the direction of the flux Φ can be determined by placing the fingers of the right hand in the direction of current around the core and noting the direction of the thumb. It is interesting to realize that *the magnetizing force is independent of the type of core material*—it is determined solely by the number of turns, the current, and the length of the core.

The applied magnetizing force has a pronounced effect on the resulting permeability of a magnetic material. As the magnetizing force increases, the permeability rises to a maximum and then drops to a minimum, as shown in Fig. 11.17 for three commonly employed magnetic materials.

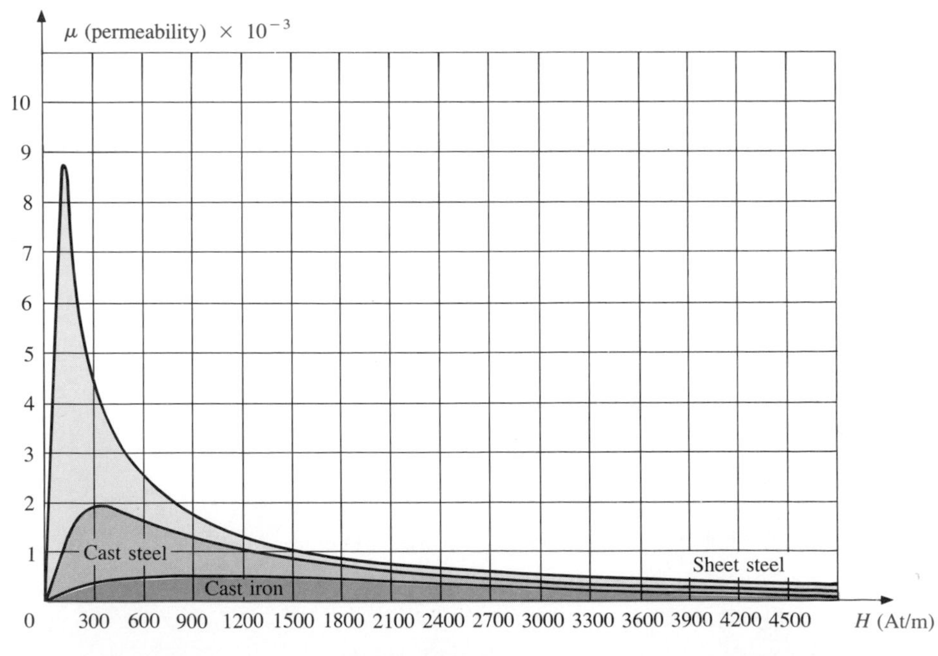

FIG. 11.17

The flux density and the magnetizing force are related by the following equation:

$$B = \mu H \tag{11.8}$$

This equation indicates that for a particular magnetizing force, the greater the permeability, the greater will be the induced flux density.

## 11.8 HYSTERESIS

A curve of the flux density $B$ versus the magnetizing force $H$ of a material is of particular importance to the engineer. Curves of this type can usually be found in manuals and descriptive pamphlets and brochures published by manufacturers of magnetic materials. A typical $B$-$H$ curve for a ferromagnetic material such as steel can be derived using the setup of Fig. 11.18.

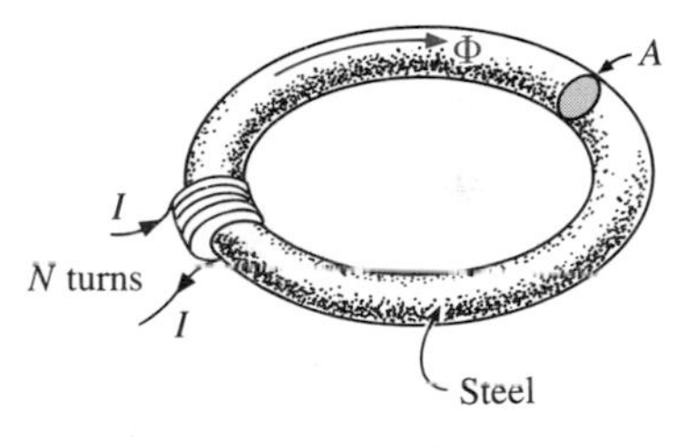


FIG. 11.18

The core is initially unmagnetized and the current $I = 0$. If the current $I$ is increased to some value above zero, the magnetizing force $H$ will increase to a value determined by

$$H\uparrow = \frac{NI\uparrow}{l}$$

The flux $\Phi$ and the flux density $B$ ($B = \Phi/A$) will also increase with the current $I$ (or $H$). If the material has no residual magnetism and the magnetizing force $H$ is increased from zero to some value $H_a$, the $B$-$H$ curve will follow the path shown in Fig. 11.19 between $o$ and $a$. If the

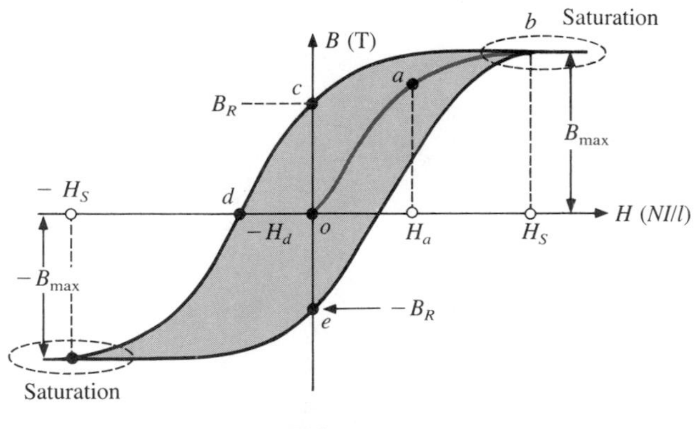


FIG. 11.19

magnetizing force $H$ is increased until saturation ($H_S$) occurs, the curve will continue as shown in the figure to point $b$. When saturation occurs, the flux density has, *for all practical purposes,* reached its maximum value. Any further increase in current through the coil increasing $H = NI/l$ will result in a very small increase in flux density $B$.

If the magnetizing force is reduced to zero by letting $I$ decrease to zero, the curve will follow the path of the curve between $b$ and $c$. The flux density $B_R$, which remains when the magnetizing force is zero, is called the *residual flux density*. It is this residual flux density that makes it possible to create permanent magnets. If the coil is now removed from the core of Fig. 11.18, the core will still have the magnetic properties determined by the residual flux density, a measure of its "retentivity." If the current $I$ is reversed, developing a magnetizing force, $-H$, the flux density $B$ will decrease with increase in $I$. Eventually, the flux density will be zero when $-H_d$ (the portion of curve from $c$ to $d$) is reached. The magnetizing force $-H_d$ required to "coerce" the flux density to reduce its level to zero is called the *coercive force,* a measure of the coercivity of the magnetic sample. As the force $-H$ is increased until saturation again occurs and is then reversed and brought back to zero, the path shown from $d$ to $e$ will result. If the magnetizing force is increased in the positive direction ($+H$), the curve will trace the path shown from $e$ to $b$. The entire curve represented by $bcded$ is called the *hysteresis curve* for the ferromagnetic material, from the Greek *hysterein,* meaning "to lag behind." The flux density $B$ *lagged* behind the magnetizing force $H$ during the entire plotting of the curve. When $H$ was zero at $c$, $B$ was not zero but had only begun to decline. Long after $H$ had passed through zero and had become equal to $-H_d$ did the flux density $B$ finally become equal to zero.

If the entire cycle is repeated, the curve obtained for the same core will be determined by the maximum $H$ applied. Three hysteresis loops for the same material for maximum values of $H$ less than the saturation value are shown in Fig. 11.20. In addition, the saturation curve is repeated for comparison purposes.

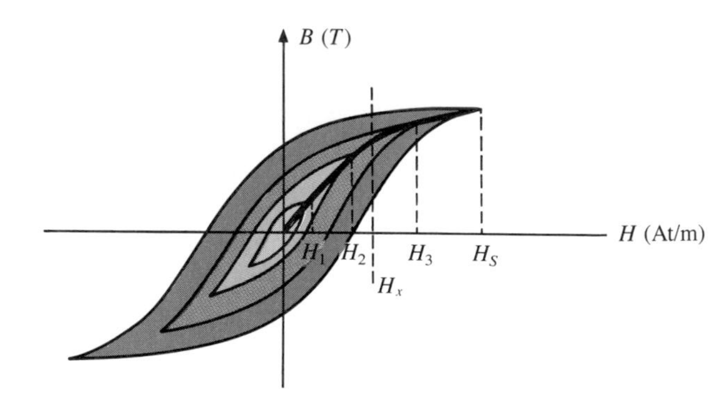

**FIG. 11.20**

Note from the various curves that for a particular value of $H$, say, $H_x$, the value of $B$ can vary widely, as determined by the history of the core. In an effort to assign a particular value of $B$ to each value of $H$, we compromise by connecting the tips of the hysteresis loops. The resulting curve, shown by the heavy, solid line in Fig. 11.20 and for various materials in Fig. 11.21, is called the *normal magnetization curve*. An expanded view of one region appears in Fig. 11.22.

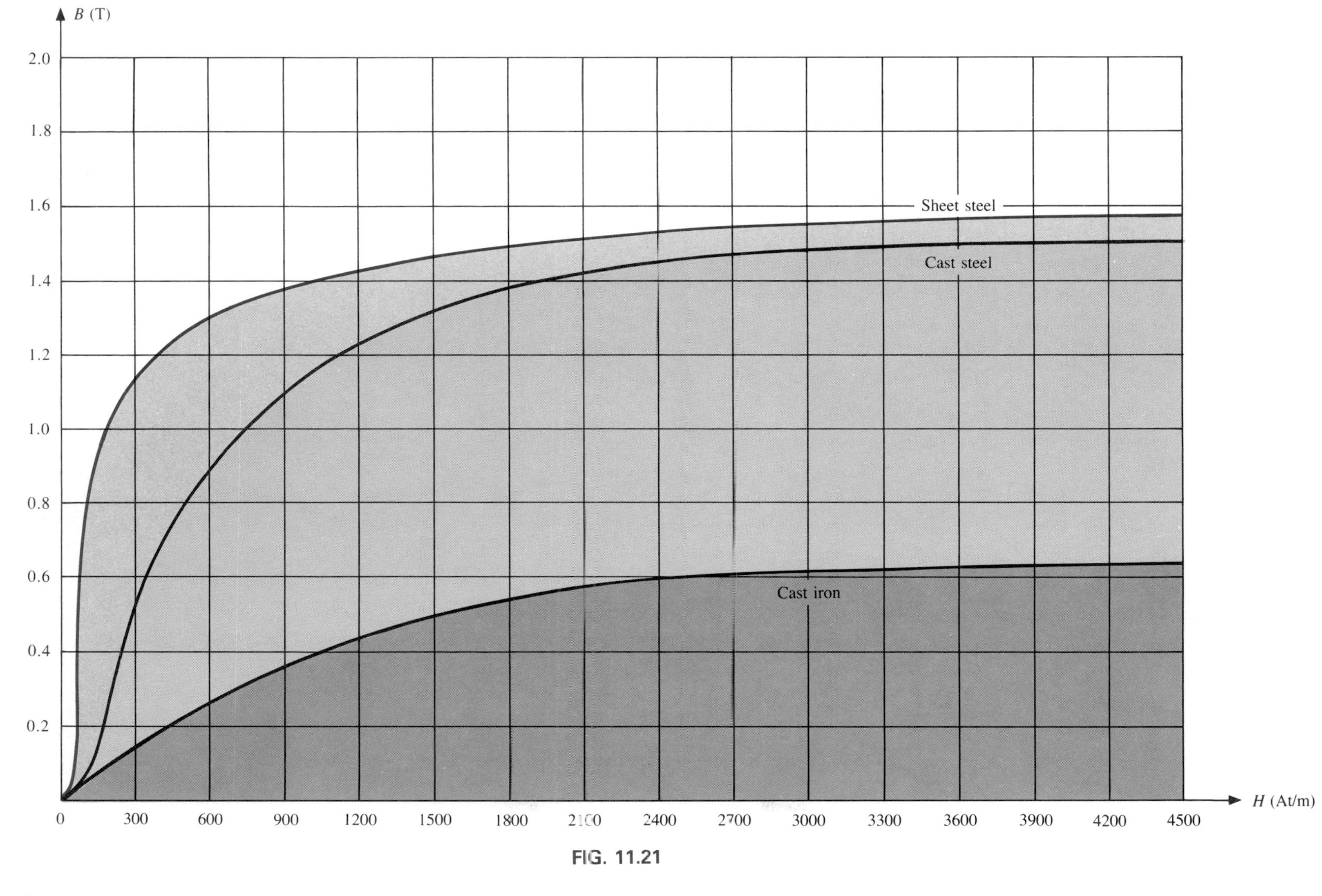
B (T)
2.0
1.8
1.6
1.4
1.2
1.0
0.8
0.6
0.4
0.2
0
300
600
900
1200
1500
1800
2100
2400
2700
3000
3300
3600
3900
4200
4500
H (At/m)
Sheet steel
Cast steel
Cast iron

FIG. 11.21

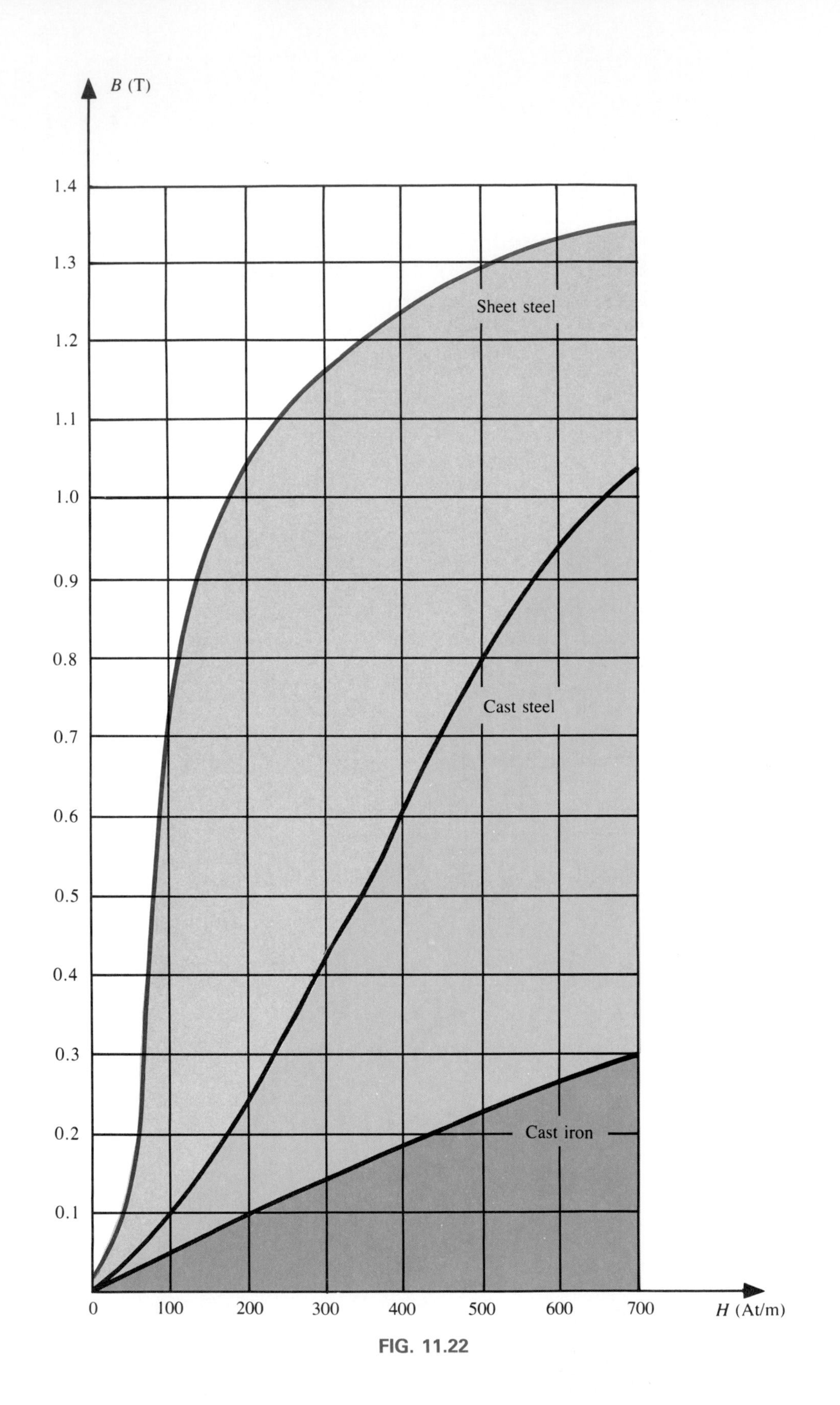
B (T)
1.4
1.3
1.2
1.1
1.0
0.9
0.8
0.7
0.6
0.5
0.4
0.3
0.2
0.1
0
100
200
300
400
500
600
700
H (At/m)
Sheet steel
Cast steel
Cast iron

FIG. 11.22

A comparison of Figs. 11.17 and 11.21 shows that for the same value of $H$, the value of $B$ is higher in Fig. 11.21 for the materials with the higher $\mu$ in Fig. 11.17. This is particularly obvious for low values of $H$. This correspondence between the two figures must exist, since $B = \mu H$. In fact, if in Fig. 11.21 we find $\mu$ for each value of $H$ using the equation $\mu = B/H$, we will obtain the curves of Fig. 11.17.

An instrument that will provide a plot of the $B$-$H$ curve for a magnetic sample appears in Fig. 11.23.

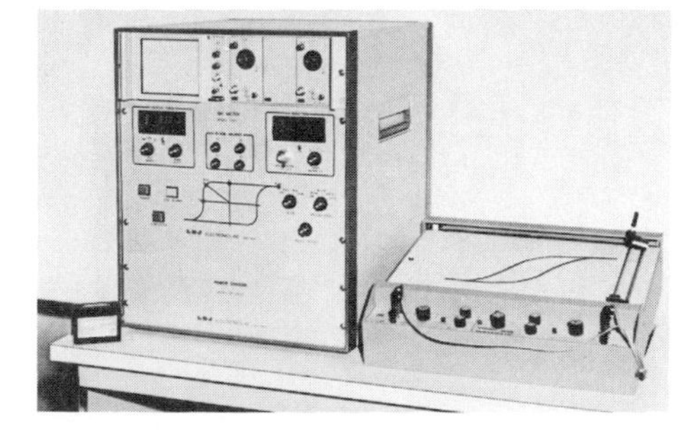

**FIG. 11.23**
*(Courtesy of LDJ Electronics, Inc.)*

It is interesting to note that the hysteresis curves of Fig. 11.20 have a *point symmetry* about the origin. That is, the inverted pattern to the left of the vertical axis is the same as that appearing to the right of the vertical axis. In addition, you will find that a further application of the same magnetizing forces to the sample will result in the same plot. For a current $I$ in $H = NI/l$ that will move between positive and negative maximums at a fixed rate, the same $B$-$H$ curve will result during each cycle. Such will be the case when we examine ac (sinusoidal) networks in the later chapters. The reversal of the field ($\Phi$) due to the changing current direction will result in a loss of energy that can best be described by first introducing the *domain theory of magnetism*.

Within each atom, the orbiting electrons (described in Chapter 2) are also spinning as they revolve around the nucleus. The atom, due to its spinning electrons, has a magnetic field associated with it. In nonmagnetic materials, the net magnetic field is effectively zero, since the magnetic fields due to the atoms of the material oppose each other. In magnetic materials such as iron and steel, however, the magnetic fields of groups of atoms numbering in the order of $10^{12}$ are aligned, forming very small bar magnets. This group of magnetically aligned atoms is called a *domain*. Each domain is a separate entity; that is, each domain is independent of the surrounding domains. For an unmagnetized sample of magnetic material, these domains appear in a random manner, such as shown in Fig. 11.24(a). The net magnetic field in any one direction is zero.

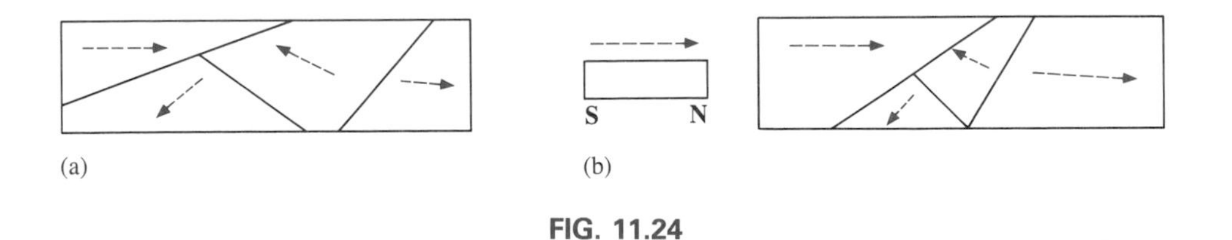


**FIG. 11.24**

When an external magnetizing force is applied, the domains that are nearly aligned with the applied field will grow at the expense of the less favorably oriented domains, such as shown in Fig. 11.24(b). Eventually, if a sufficiently strong field is applied, all of the domains will have the orientation of the applied magnetizing force, and any further increase in external field will not increase the strength of the magnetic flux through the core—a condition referred to as *saturation*. The elasticity of the above is evidenced by the fact that when the magnetizing force is removed, the alignment will be lost to some measure and the flux density will drop to $B_R$. In other words, the removal of the magnetizing force will result in the return of a number of misaligned domains within the core. The continued alignment of a number of the domains, however, accounts for our ability to create permanent magnets.

At a point just before saturation, the opposing unaligned domains are reduced to small cylinders of various shapes referred to as *bubbles*. These bubbles can be moved within the magnetic sample through the application of a *controlling* magnetic field. It is these magnetic bubbles that form the basis of the recently designed bubble memory system for computers.

## 11.9 AMPÈRE'S CIRCUITAL LAW

It was mentioned in the introduction to this chapter that there is a broad similarity between the analyses of electric and magnetic circuits. This has already been demonstrated to some extent for the quantities in Table 11.1.

**TABLE 11.1**

| | Electric Circuits | Magnetic Circuits |
|---|---|---|
| Cause | $E$ | $\mathscr{F}$ |
| Effect | $I$ | $\Phi$ |
| Opposition | $R$ | $\mathscr{R}$ |

If we apply the "cause" analogy to Kirchhoff's voltage law ($\Sigma_{\circlearrowright} V = 0$), we obtain the following:

$$\Sigma_{\circlearrowright} \mathscr{F} = 0 \quad \text{(for magnetic circuits)} \tag{11.9}$$

which, in words, states that the algebraic sum of the rises and drops of the mmf around a closed loop of a magnetic circuit is equal to zero; that is, the sum of the mmf rises equals the sum of the mmf drops around a closed loop.

Equation (11.9) is referred to as *Ampère's circuital law*. When it is applied to magnetic circuits, sources of mmf are expressed by the equation

$$\mathscr{F} = NI \quad \text{(At)} \tag{11.10}$$

The equation for the mmf drop across a portion of a magnetic circuit can be found by applying the relationships listed in Table 11.1. That is, for electric circuits,

$$V = IR$$

resulting in the following for magnetic circuits:

$$\mathscr{F} = \Phi\mathscr{R} \qquad \text{(At)} \qquad \mathbf{(11.11)}$$

where Φ is the flux passing through a section of the magnetic circuit and $\mathscr{R}$ is the reluctance of that section. The reluctance, however, is seldom calculated in the analysis of magnetic circuits. A more practical equation for the mmf drop is

$$\mathscr{F} = Hl \qquad \text{(At)} \qquad \mathbf{(11.12)}$$

as derived from Eq. (11.6), where $H$ is the magnetizing force on a section of a magnetic circuit and $l$ is the length of the section. As an example of Eq. (11.9), consider the magnetic circuit appearing in Fig. 11.25 constructed of three different ferromagnetic materials.

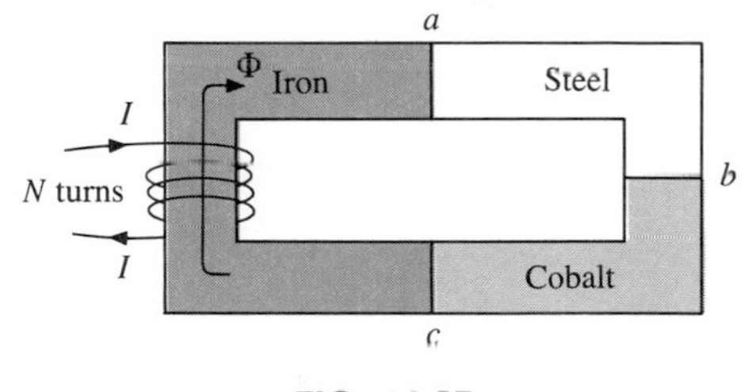


FIG. 11.25

Applying Ampère's circuital law, we have

$$\Sigma_{\circlearrowright} \mathscr{F} = 0$$

$$\underbrace{+NI}_{\text{rise}} - \underbrace{H_{ab}l_{ab}}_{\text{drop}} - \underbrace{H_{bc}l_{bc}}_{\text{drop}} - \underbrace{H_{ca}l_{ca}}_{\text{drop}} = 0$$

or

$$\underbrace{NI}_{\substack{\text{impressed}\\\text{mmf}}} = \underbrace{H_{ab}l_{ab} + H_{bc}l_{bc} + H_{ca}l_{ca}}_{\text{mmf drops}}$$

All of the terms of the equation are known except the magnetizing force for each portion of the magnetic circuit, which can be found by using the $B$-$H$ curve if the flux density $B$ is known.

## 11.10 THE FLUX Φ

If we continue to apply the relationships described in the previous section to Kirchhoff's current law, we will find that the sum of the fluxes entering a junction is equal to the sum of the fluxes leaving a junction; that is, for the circuit of Fig. 11.26,

$$\Phi_a = \Phi_b + \Phi_c \qquad \text{(at junction } a\text{)}$$

or

$$\Phi_b + \Phi_c = \Phi_a \qquad \text{(at junction } b\text{)}$$

both of which are equivalent.

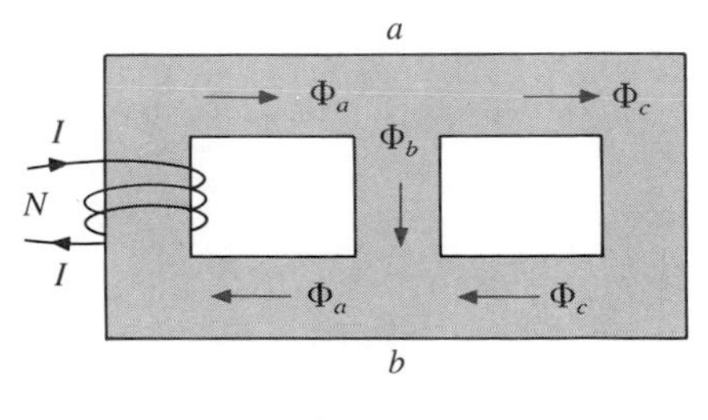


FIG. 11.26

## 11.11 SERIES MAGNETIC CIRCUITS: DETERMINING *NI*

We are now in a position to solve a few magnetic circuit problems, which are basically of two types. In one type, $\Phi$ is given, and the impressed mmf $NI$ must be computed. This is the type of problem encountered in the design of motors, generators, and transformers. In the other type, $NI$ is given, and the flux $\Phi$ of the magnetic circuit must be found. This type of problem is encountered primarily in the design of magnetic amplifiers and is more difficult since the approach is "hit or miss."

As indicated in earlier discussions, the value of $\mu$ will vary from point to point along the magnetization curve. This eliminates the possibility of finding the reluctance of each "branch" or the "total reluctance" of a network as was done for electric circuits where $\rho$ had a fixed value for any applied current or voltage. If the total reluctance could be determined, $\Phi$ could then be determined using the Ohm's law analogy for magnetic circuits.

For magnetic circuits, the level of $B$ or $H$ is determined from the other using the $B$-$H$ curve, and $\mu$ is seldom calculated unless asked for.

An approach frequently employed in the analysis of magnetic circuits is the *table* method. Before a problem is analyzed in detail, a table is prepared listing in the extreme left-hand column the various sections of the magnetic circuit. The columns on the right are reserved for the quantities to be found for each section. In this way, the individual doing the problem can keep track of what is required to complete the problem and also of what the next step should be. After a few examples, the usefulness of this method should become clear.

This section will consider only *series* magnetic circuits in which the flux $\Phi$ is the same throughout. In each example, the magnitude of the magnetomotive force is to be determined.

**EXAMPLE 11.3.** For the series magnetic circuit of Fig. 11.27:

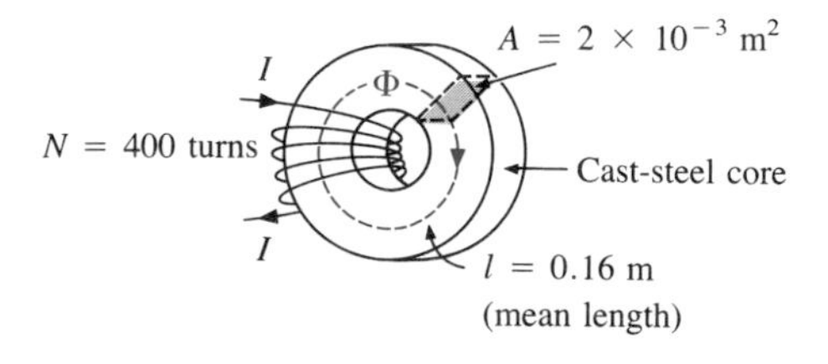

FIG. 11.27

a. Find the value of $I$ required to develop a magnetic flux of $\Phi = 4 \times 10^{-4}$ Wb.
b. Determine $\mu$ and $\mu_r$ for the material under these conditions.

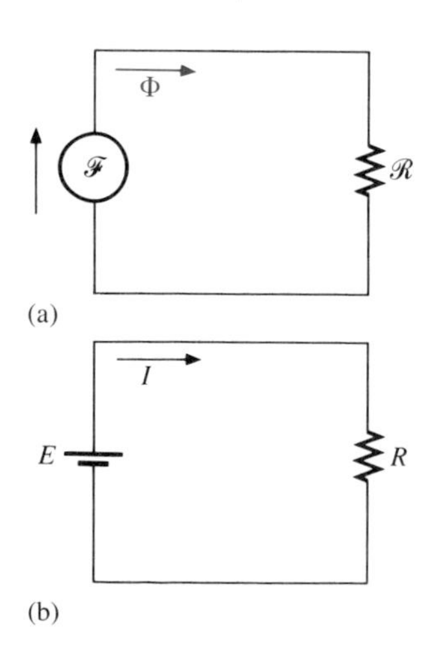


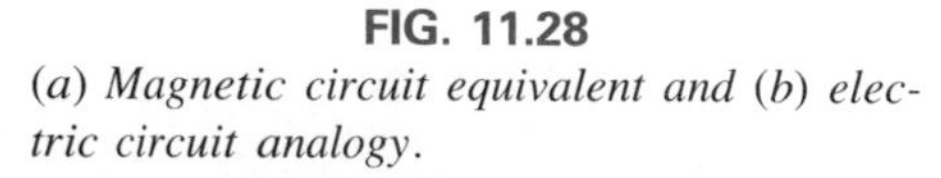
FIG. 11.28
*(a) Magnetic circuit equivalent and (b) electric circuit analogy.*

***Solutions:*** The magnetic circuit can be represented by the system shown in Fig. 11.28(a). The electric circuit analogy is shown in Fig. 11.28(b). Analogies of this type can be very helpful in the solution of

**TABLE 11.2**

| Section | $\Phi$ (Wb) | $A$ (m²) | $B$ (T) | $H$ (At/m) | $l$ (m) | $Hl$ (At) |
|---|---|---|---|---|---|---|
| One continuous section | $4 \times 10^{-4}$ | $2 \times 10^{-3}$ | | | 0.16 | |

magnetic circuits. Table 11.2 is for part (a) of this problem. The table is fairly trivial for this example but it does define the quantities to be found.

a. The flux density $B$ is

$$B = \frac{\Phi}{A} = \frac{4 \times 10^{-4}\ \text{Wb}}{2 \times 10^{-3}\ \text{m}^2} = 2 \times 10^{-1}\ \text{T} = 0.2\ \text{T}$$

Using the $B$-$H$ curves of Fig. 11.22, we can determine the magnetizing force $H$:

$$H\ (\text{cast steel}) = 110\ \text{At/m}$$

Applying Ampère's circuital law yields

$$NI = Hl$$

and

$$I = \frac{Hl}{N} = \frac{(110\ \text{At/m})(0.16\ \text{m})}{400\ \text{t}} = 0.044\ \text{A} = \mathbf{44\ mA}$$

(Recall that t represents turns.)

b. The permeability of the material can be found using Eq. (11.8):

$$\mu = \frac{B}{H} = \frac{0.2\ \text{T}}{110\ \text{At/m}} = \mathbf{1.818 \times 10^{-3}\ Wb/Am}$$

and the relative permeability is

$$\mu_r = \frac{\mu}{\mu_o} = \frac{1.818 \times 10^{-3}}{4\pi \times 10^{-7}} = \frac{1.818 \times 10^{-3}}{12.57 \times 10^{-7}} = \mathbf{1446.3}$$

**EXAMPLE 11.4.** The electromagnet of Fig. 11.29 has picked up a section of cast iron. Determine the current $I$ required to establish the indicated flux in the core.

***Solution:*** To be able to use Figs. 11.21 and 11.22, the dimensions must first be converted to the metric system. However, since the area is

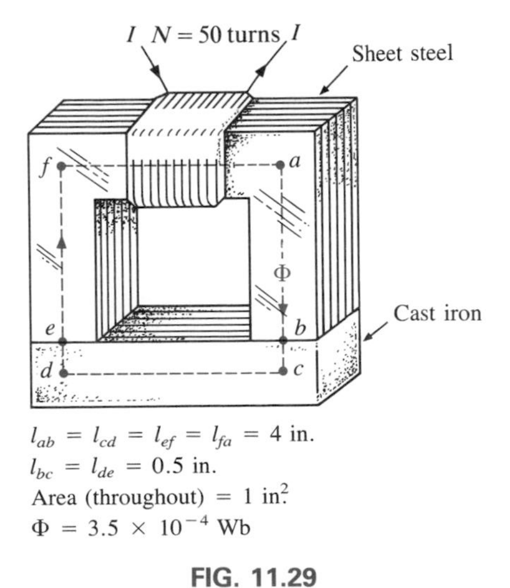


$l_{ab} = l_{cd} = l_{ef} = l_{fa} = 4$ in.
$l_{bc} = l_{de} = 0.5$ in.
Area (throughout) = 1 in.²
$\Phi = 3.5 \times 10^{-4}$ Wb

**FIG. 11.29**
*Electromagnet for Example 11.4.*

the same throughout, we can determine the length for each material rather than work with the individual sections:

$$l_{efab} = 4 \text{ in.} + 4 \text{ in.} + 4 \text{ in.} = 12 \text{ in.}$$

$$l_{bcde} = 0.5 \text{ in.} + 4 \text{ in.} + 0.5 \text{ in.} = 5 \text{ in.}$$

$$12 \text{ in.}\left(\frac{1 \text{ m}}{39.37 \text{ in.}}\right) = 304.8 \times 10^{-3} \text{ m}$$

$$5 \text{ in.}\left(\frac{1 \text{ m}}{39.37 \text{ in.}}\right) = 127 \times 10^{-3} \text{ m}$$

$$1 \text{ in.}^2\left(\frac{1 \text{ m}}{39.37 \text{ in.}}\right)\left(\frac{1 \text{ m}}{39.37 \text{ in.}}\right) = 6.452 \times 10^{-4} \text{ m}^2$$

The information available from the specifications of the problem has been inserted in Table 11.3. When the problem has been completed, each space will contain some information. Sufficient data to complete the problem can be found if we fill in each column from left to right. As the various quantities are calculated, they will be placed in a similar table found at the end of the example.

**TABLE 11.3**

| Section | $\Phi$ (Wb) | $A$ (m$^2$) | $B$ (T) | $H$ (At/m) | $l$ (m) | $Hl$ (At) |
|---|---|---|---|---|---|---|
| *efab* | $3.5 \times 10^{-4}$ | $6.452 \times 10^{-4}$ | | | $304.8 \times 10^{-3}$ | |
| *bcde* | $3.5 \times 10^{-4}$ | $6.452 \times 10^{-4}$ | | | $127 \times 10^{-3}$ | |

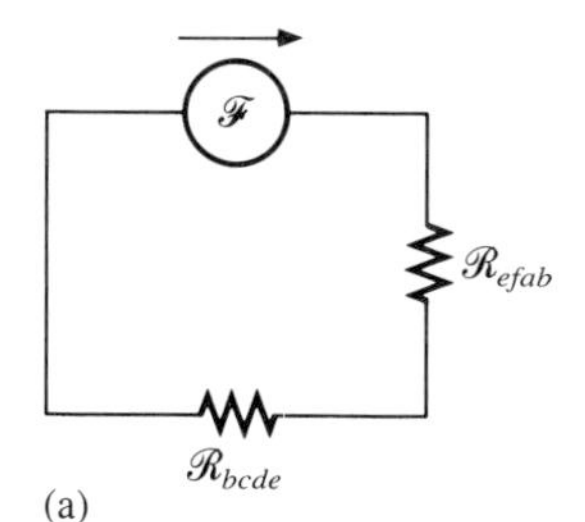


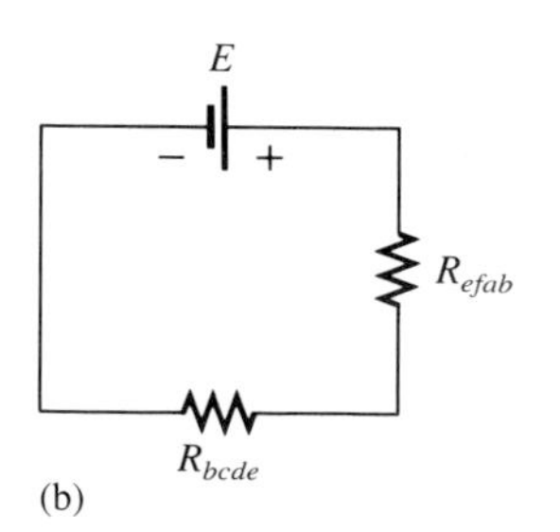


**FIG. 11.30**
*(a) Magnetic circuit equivalent and (b) electric circuit analogy for the electromagnetic of Fig. 11.29.*

The flux density for each section is

$$B = \frac{\Phi}{A} = \frac{3.5 \times 10^{-4} \text{ Wb}}{6.452 \times 10^{-4} \text{ m}^2} = 0.542 \text{ T}$$

and the magnetizing force is

$$H \text{ (sheet steel, Fig. 11.22)} \cong 60 \text{ At/m}$$

$$H \text{ (cast iron, Fig. 11.21)} \cong 1600 \text{ At/m}$$

Note the extreme difference in magnetizing force for each material for the required flux density. In fact, when we apply Ampère's circuital law, we will find that the sheet steel section could be ignored with a minimal error in the solution.

Determining $Hl$ for each section yields

$$H_{efab}l_{efab} = (60 \text{ At/m})(304.8 \times 10^{-3} \text{ m}) = 18.29 \text{ At}$$

$$H_{bcde}l_{bcde} = (1600 \text{ At/m})(127 \times 10^{-3} \text{ m}) = 203.2 \text{ At}$$

Inserting the above data in Table 11.3 will result in Table 11.4.

The magnetic circuit equivalent and the electric circuit analogy for the system of Fig. 11.29 appear in Fig. 11.30.

**TABLE 11.4**

| Section | $\Phi$ (Wb) | $A$ (m$^2$) | $B$ (T) | $H$ (At/m) | $l$ (m) | $Hl$ (At) |
|---|---|---|---|---|---|---|
| *efab* | $3.5 \times 10^{-4}$ | $6.452 \times 10^{-4}$ | 0.542 | 60 | $304.8 \times 10^{-3}$ | 18.29 |
| *bcde* | $3.5 \times 10^{-4}$ | $6.452 \times 10^{-4}$ | 0.542 | 1600 | $127 \times 10^{-3}$ | 203.2 |

Applying Ampère's circuital law,

$$NI = H_{efab}l_{efab} + H_{bcde}l_{bcde}$$
$$= 18.29 \text{ At} + 203.2 \text{ At} = 221.49 \text{ At}$$

and

$$(50 \text{ t})I = 221.49 \text{ At}$$

or

$$I = \frac{221.49 \text{ At}}{50 \text{ t}} = \mathbf{4.43\ A}$$

**EXAMPLE 11.5.** Determine the secondary current $I_2$ for the transformer of Fig. 11.31 if the flux in the core is $1.5 \times 10^{-5}$ Wb.

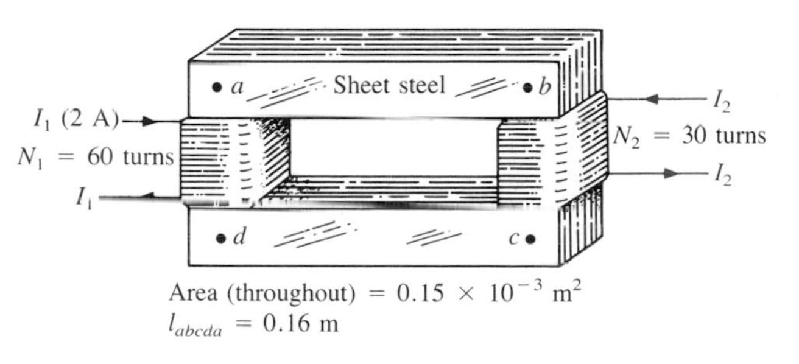


**FIG. 11.31**
*Transformer for Example 11.5.*

***Solution:*** This is the first example with two magnetizing forces to consider. In the analogies of Fig. 11.32 you will note that the resulting

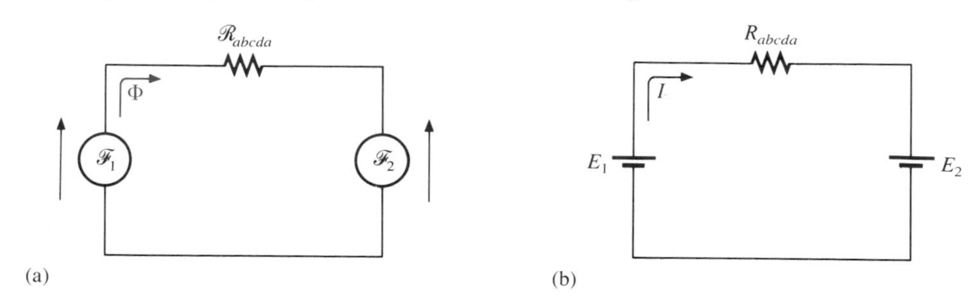


**FIG. 11.32**
*(a) Magnetic circuit equivalent and (b) electric circuit analogy for the transformer of Fig. 11.31.*

**TABLE 11.5**

| Section | Φ (Wb) | A (m²) | B (T) | H (At/m) | l (m) | Hl (At) |
|---|---|---|---|---|---|---|
| *abcda* | $1.5 \times 10^{-5}$ | $0.15 \times 10^{-3}$ | | | 0.16 | |

flux of each is opposing, just as the two sources of voltage are opposing in the electric circuit analogy.

The structural data appear in Table 11.5.

The flux density throughout is

$$B = \frac{\Phi}{A} = \frac{1.5 \times 10^{-5}\text{ Wb}}{0.15 \times 10^{-3}\text{ m}^2} = 10 \times 10^{-2}\text{ T} = 0.10\text{ T}$$

and

$$H\text{ (from Fig. 11.22)} \cong \frac{1}{7}(100\text{ At/m}) = 14.29\text{ At/m}$$

Applying Ampère's circuital law,

$$N_1I_1 - N_2I_2 = H_{abcda}l_{abcda}$$

$$(60\text{ t})(2\text{ A}) - (30\text{ t})(I_2) = (14.29\text{ At/m})(0.16\text{ m})$$

$$120\text{ At} - (30\text{ t})I_2 = 2.29\text{ At}$$

and

$$(30\text{ t})I_2 = 120\text{ At} - 2.29\text{ At}$$

or

$$I_2 = \frac{117.71\text{ At}}{30\text{ t}} = \mathbf{3.924\text{ A}}$$

---

For the analysis of most transformer systems, the equation $N_1I_1 = N_2I_2$ is employed. This would result in 4 A versus 3.924 A above. This difference is normally ignored, however, and the equation $N_1I_1 = N_2I_2$ considered exact.

Because of the nonlinearity of the *B-H* curve, *it is not possible to apply superposition to magnetic circuits;* that is, in the previous example, we cannot consider the effects of each source independently and then find the total effects by using superposition.

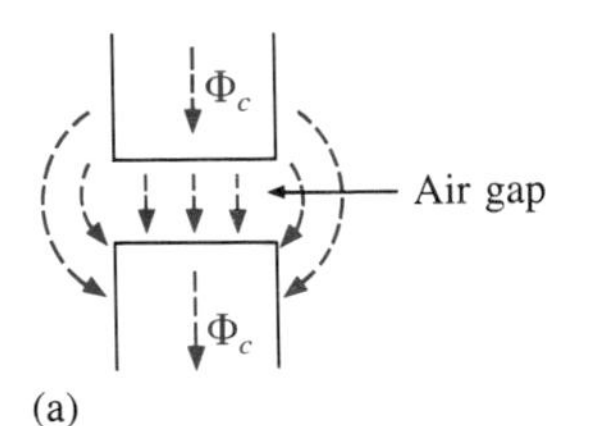


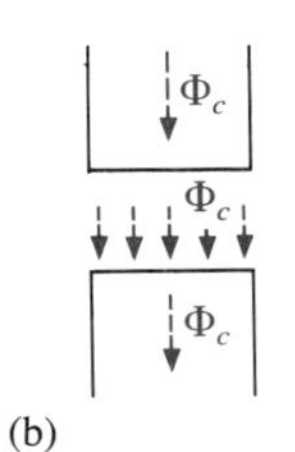


FIG. 11.33

## 11.12 AIR GAPS

Before continuing with the illustrative examples, let us consider the effects an air gap has on a magnetic circuit. Note the presence of air gaps in the magnetic circuits of the motor and meter of Fig. 11.11. The spreading of the flux lines outside the common area of the core for the air gap in Fig. 11.33(a) is known as *fringing*. For our purposes, we shall neglect this effect and assume the flux distribution to be as in Fig. 11.33(b).

The flux density of the air gap in Fig. 11.33(b) is given by

$$B_g = \frac{\Phi_g}{A_g} \qquad (11.13)$$

where, for our purposes,

$$\Phi_g = \Phi_{core}$$

and

$$A_g = A_{core}$$

For most practical applications, the permeability of air is taken to be equal to that of free space. The magnetizing force of the air gap is then determined by

$$H_g = \frac{B_g}{\mu_o} \qquad (11.14)$$

and the mmf drop across the air gap is equal to $H_g l_g$. An equation for $H_g$ is as follows:

$$H_g = \frac{B_g}{\mu_o} = \frac{B_g}{4\pi \times 10^{-7}}$$

and

$$H_g = (7.96 \times 10^5)B_g \quad \text{(At/m)} \qquad (11.15)$$

---

**EXAMPLE 11.6.** Find the value of $I$ required to establish a magnetic flux of $\Phi = 0.75 \times 10^{-4}$ Wb in the series magnetic circuit of Fig. 11.34.

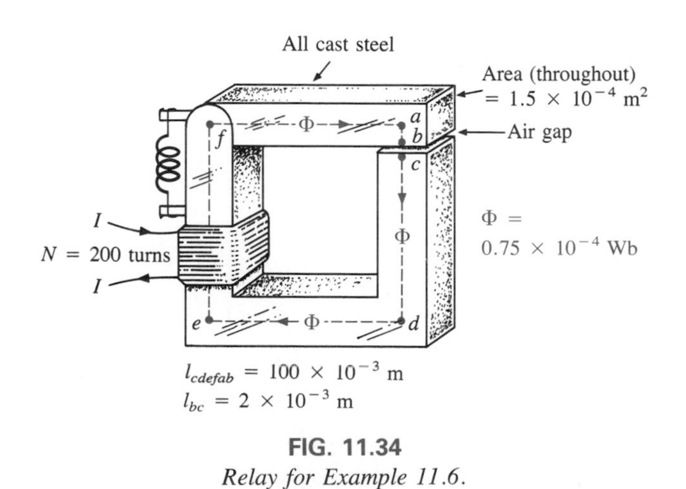


**FIG. 11.34**
*Relay for Example 11.6.*

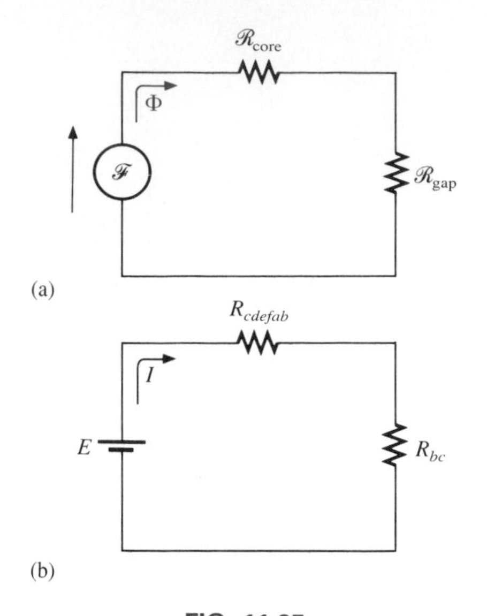


**FIG. 11.35**
*(a) Magnetic circuit equivalent and (b) electric circuit analogy for the relay of Fig. 11.34.*

***Solution:*** An equivalent magnetic circuit and its electric circuit analogy are shown in Fig. 11.35.

The flux density for each section is

$$B = \frac{\Phi}{A} = \frac{0.75 \times 10^{-4}\text{ Wb}}{1.5 \times 10^{-4}\text{ m}^2} = 0.5\text{ T}$$

From the *B-H* curves of Fig. 11.22,

$$H\text{ (cast steel)} \cong 280\text{ At/m}$$

Applying Eq. (11.15),

$$H_g = (7.96 \times 10^5)B_g = (7.96 \times 10^5)(0.5\text{ T}) = 3.98 \times 10^5\text{ At/m}$$

The mmf drops are

$$H_{core}l_{core} = (280\text{ At/m})(100 \times 10^{-3}\text{ m}) = 28\text{ At}$$
$$H_g l_g = (3.98 \times 10^5\text{ At/m})(2 \times 10^{-3}\text{ m}) = 796\text{ At}$$

Applying Ampère's circuital law,

$$NI = H_{core}l_{core} + H_g l_g$$
$$= 28\text{ At} + 796\text{ At}$$
$$(200\text{ t})I = 824\text{ At}$$
$$I = \mathbf{4.12\text{ A}}$$

Note from the above that the air gap requires the biggest share (by far) of the impressed *NI* due to the fact that air is nonmagnetic.

## 11.13 SERIES-PARALLEL MAGNETIC CIRCUITS

As one might expect, the close analogies between electric and magnetic circuits will eventually lead to series-parallel magnetic circuits similar in many respects to those encountered in Chapter 7. In fact, the electric circuit analogy will prove helpful in defining the procedure to follow toward a solution.

**EXAMPLE 11.7.** Determine the current *I* required to establish a flux of $1.5 \times 10^{-4}$ Wb in the section of the core indicated in Fig. 11.36.

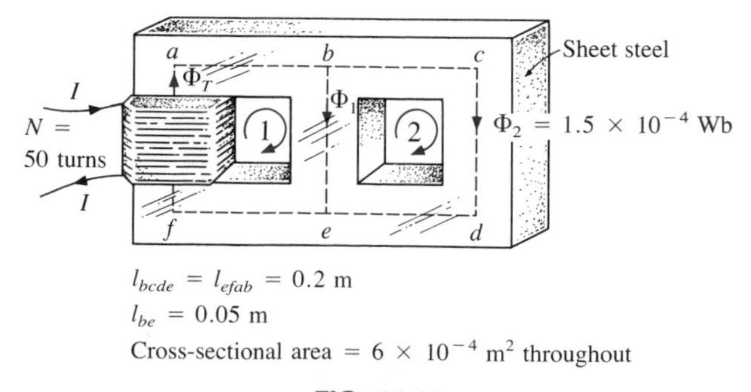


**FIG. 11.36**

***Solution:*** The equivalent magnetic circuit and the electric circuit analogy appear in Fig. 11.37. We have

$$B_2 = \frac{\Phi_2}{A} = \frac{1.5 \times 10^{-4}\ \text{Wb}}{6 \times 10^{-4}\ \text{m}^2} = 0.25\ \text{T}$$

From Fig. 11.22,

$$H_{bcde} \cong 30\ \text{At/m}$$

Applying Ampère's circuital law around loop 2 of Figs. 11.36 and 11.37,

$$\Sigma_{\circlearrowleft}\ \mathcal{F} = 0$$
$$H_{be}l_{be} - H_{bcde}l_{bcde} = 0$$
$$H_{be}(0.05\ \text{m}) - (30\ \text{At/m})(0.2\ \text{m}) = 0$$
$$H_{be} = \frac{6\ \text{At}}{0.05\ \text{m}} = 120\ \text{At/m}$$

From Fig. 11.22,

$$B_1 \cong 0.8\ \text{T}$$

and

$$\Phi_1 = B_1A = (0.8\ \text{T})(6 \times 10^{-4}\ \text{m}^2) = 4.8 \times 10^{-4}\ \text{Wb}$$

The results are then entered in Table 11.6.

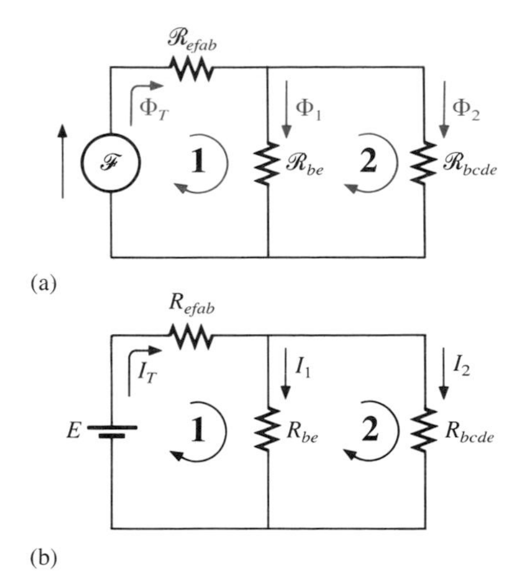


**FIG. 11.37**
*(a) Magnetic circuit equivalent and (b) electric circuit analogy for the series-parallel system of Fig. 11.36.*

**TABLE 11.6**

| Section | Φ (Wb) | A (m²) | B (T) | H (At/m) | l (m) | Hl (At) |
|---|---|---|---|---|---|---|
| *bcde* | $1.5 \times 10^{-4}$ | $6 \times 10^{-4}$ | 0.25 | 30 | 0.2 | 6 |
| *be* | $4.8 \times 10^{-4}$ | $6 \times 10^{-4}$ | 0.8 | 120 | 0.05 | 6 |
| *efab* | | $6 \times 10^{-4}$ | | | 0.2 | |

The table reveals that we must now turn our attention to section *efab:*

$$\Phi_T = \Phi_1 + \Phi_2 = 4.8 \times 10^{-4}\ \text{Wb} + 1.5 \times 10^{-4}\ \text{Wb}$$
$$= 6.3 \times 10^{-4}\ \text{Wb}$$
$$B = \frac{\Phi_T}{A} = \frac{6.3 \times 10^{-4}\ \text{Wb}}{6 \times 10^{-4}\ \text{m}^2}$$
$$= 1.05\ \text{T}$$

From Fig. 11.21,

$$H_{efab} \cong 200\ \text{At}$$

Applying Ampère's circuital law,

$$+NI - H_{efab}l_{efab} - H_{be}l_{be} = 0$$
$$NI = (200\ \text{At/m})(0.2\ \text{m}) + (120\ \text{At/m})(0.05\ \text{m})$$
$$(50\ \text{t})I = 40\ \text{At} + 6\ \text{At}$$
$$I = \frac{46\ \text{At}}{50\ \text{t}} = \mathbf{0.92\ A}$$

To demonstrate that $\mu$ is sensitive to the magnetizing force, the permeability of each section is determined as follows. For section *bcde,*

$$\mu = \frac{B}{H} = \frac{0.25 \text{ T}}{30 \text{ At/m}} = 0.0083$$

and

$$\mu_r = \frac{\mu}{\mu_o} = \frac{0.0083}{12.57 \times 10^{-7}} = \mathbf{6603}$$

For section *be,*

$$\mu = \frac{B}{H} = \frac{0.8 \text{ T}}{120 \text{ At/m}} = 0.0067$$

and

$$\mu_r = \frac{\mu}{\mu_o} = \frac{0.0067}{12.57 \times 10^{-7}} = \mathbf{5330}$$

For section *efab,*

$$\mu = \frac{B}{H} = \frac{1.05 \text{ T}}{200 \text{ At/m}} = 0.00525$$

and

$$\mu_r = \frac{\mu}{\mu_o} = \frac{0.00525}{12.57 \times 10^{-7}} = \mathbf{4176.6}$$

## 11.14 DETERMINING $\Phi$

The examples of this section are of the second type, where *NI* is given and the flux $\Phi$ must be found. This is a relatively straightforward problem if only one magnetic section is involved. Then

$$H = \frac{NI}{l} \qquad H \rightarrow B \text{ } (B\text{-}H \text{ curve})$$

and

$$\Phi = BA$$

For magnetic circuits with more than one section, there is no set order of steps that will lead to an exact solution for every problem on the first attempt. In general, however, we proceed as follows. We must find the impressed mmf for a *calculated guess* of the flux $\Phi$ and then compare this with the specified value of mmf. We can then make adjustments on our guess to bring it closer to the actual value. For most applications, a value within $\pm 5\%$ of the actual $\Phi$ or specified *NI* is acceptable.

We can make a reasonable guess at the value of $\Phi$ if we realize that the maximum mmf drop appears across the material with the smallest permeability if the length and area of each material are the same. As shown in Example 11.6, if there is an air gap in the magnetic circuit, there will be a considerable drop in mmf across the gap. As a starting point for problems of this type, therefore, we shall assume that the total mmf ($NI$) is across the section with the lowest $\mu$ or greatest $\mathcal{R}$ (if the other physical dimensions are relatively similar). This assumption gives a value of $\Phi$ that will produce a calculated $NI$ greater than the specified value. Then, after considering the results of our original assumption very carefully, we shall *cut* $\Phi$ and $NI$ by introducing the effects (reluctance) of the other portions of the magnetic circuit and *try* the new solution. For obvious reasons, this approach is frequently called the *cut and try* method.

**EXAMPLE 11.8.** Calculate the magnetic flux $\Phi$ for the magnetic circuit of Fig. 11.38.

***Solution:*** By Ampère's circuital law,

$$NI = H_{abcda} l_{abcda}$$

or

$$H_{abcda} = \frac{NI}{l_{abcda}} = \frac{(60\text{ t})(5\text{ A})}{0.3\text{ m}} = \frac{300\text{ At}}{0.3\text{ m}} = 1000\text{ At/m}$$

and

$$B_{abcda}\ (\text{from Fig. 11.22}) \cong 0.38\text{ T}$$

Since $B = \Phi/A$, we have

$$\Phi = BA = (0.38\text{ T})(2 \times 10^{-4}\text{ m}^2) = \mathbf{0.760 \times 10^{-4}\ Wb}$$

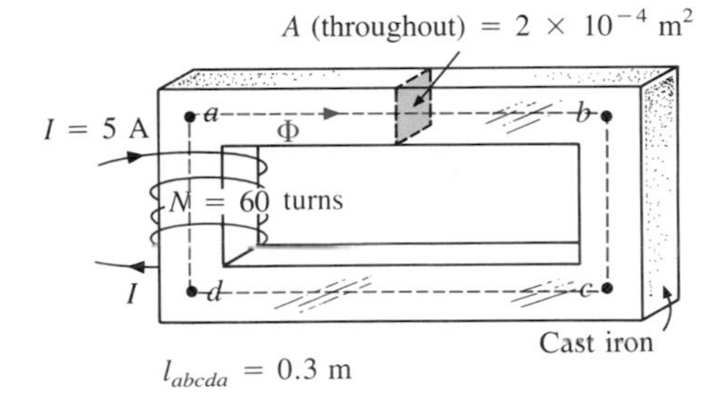

FIG. 11.38

**EXAMPLE 11.9.** Find the magnetic flux $\Phi$ for the series magnetic circuit of Fig. 11.39 for the specified impressed mmf.

***Solution:*** Assuming that the total impressed mmf $NI$ is across the air gap,

$$NI = H_g l_g$$

or

$$H_g = \frac{NI}{l_g} = \frac{400\text{ At}}{0.001\text{ m}} = 4 \times 10^5\text{ At/m}$$

and

$$B_g = \mu_o H_g = (4\pi \times 10^{-7})(4 \times 10^5\text{ At/m}) = 50.265 \times 10^{-2}\text{ T}$$

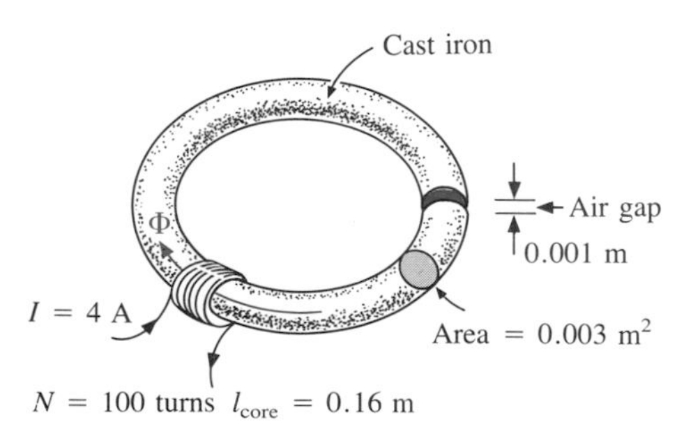


FIG. 11.39

The flux

$$\begin{aligned} \Phi_g &= \Phi_{core} = B_g A \\ &= (50.265 \times 10^{-2}\ \text{T})(0.003\ \text{m}^2) \\ \Phi_{core} &= 150.8 \times 10^{-5}\ \text{Wb} \end{aligned}$$

Using this value of $\Phi$, we can find $NI$. The data are inserted in Table 11.7.

**TABLE 11.7**

| Section | $\Phi$ (Wb) | $A$ (m²) | $B$ (T) | $H$ (At/m) | $l$ (m) | $Hl$ (At) |
|---|---|---|---|---|---|---|
| Core | $150.8 \times 10^{-5}$ | 0.003 | 0.50 | 1500 ($B$-$H$ curve) | 0.16 | |
| Gap | $150.8 \times 10^{-5}$ | 0.003 | 0.50 | $4 \times 10^5$ | 0.001 | 400 |

$$H_{core} l_{core} = (1500\ \text{At/m})(0.16\ \text{m}) = 240\ \text{At}$$

Applying Ampère's circuital law results in

$$\begin{aligned} NI &= H_{core} l_{core} + H_g l_g \\ &= 240\ \text{At} + 400\ \text{At} \\ NI &= 640\ \text{At} > 400\ \text{At} \end{aligned}$$

Since we neglected the reluctance of all the magnetic paths but the air gap, the calculated value is greater than the specified value. We must therefore reduce this value by including the effect of these reluctances. Since approximately $(640\ \text{At} - 400\ \text{At})/640\ \text{At} = 240\ \text{At}/640\ \text{At} \cong 37.5\%$ of our calculated value is above the desired value, let us reduce $\Phi$ by 37.5% and see how close we come to the impressed mmf of 400 At:

$$\begin{aligned} \Phi &= (1 - 0.375)(150.8 \times 10^{-5}\ \text{Wb}) \\ &= 94.25 \times 10^{-5}\ \text{Wb} \end{aligned}$$

See Table 11.8.

**TABLE 11.8**

| Section | $\Phi$ (Wb) | $A$ (m²) | $B$ (T) | $H$ (At/m) | $l$ (m) | $Hl$ (At) |
|---|---|---|---|---|---|---|
| Core | $94.25 \times 10^{-5}$ | 0.003 | | | 0.16 | |
| Gap | $94.25 \times 10^{-5}$ | 0.003 | | | 0.001 | |

$$B = \frac{\Phi}{A} = \frac{94.25 \times 10^{-5}\ \text{Wb}}{0.003\ \text{m}^2} = 31.42 \times 10^{-2}\ \text{T} \cong 0.31\ \text{T}$$

$$\begin{aligned} H_g l_g &= (7.96 \times 10^5) B_g l_g \\ &= (7.96 \times 10^5)(0.31\ \text{T})(0.001\ \text{m}) \\ &\cong 246.76\ \text{At} \end{aligned}$$

From $B$-$H$ curves,

$$\begin{aligned} H_{core} &\cong 730\ \text{At/m} \\ H_{core} l_{core} &= (730\ \text{At/m})(0.16\ \text{m}) = 116.8\ \text{At} \end{aligned}$$

Applying Ampère's circuital law yields

$$NI = H_{core}l_{core} + H_g l_g$$
$$= 116.8 \text{ At} + 297.45 \text{ At}$$
$$NI = \mathbf{414.25\ At} > 400 \text{ At} \quad \text{(but within } \pm 5\% \text{ and therefore acceptable)}$$

The solution is therefore

$$\Phi \cong \mathbf{94.25 \times 10^{-5}\ Wb}$$

## PROBLEMS

### SECTION 11.3 Flux Density

1. Using Appendix F, fill in the blanks in the following table. Indicate the units for each quantity.

| | $\Phi$ | $B$ |
|---|---|---|
| SI | $5 \times 10^{-4}$ Wb | $8 \times 10^{-4}$ T |
| CGS | ___ | ___ |
| English | ___ | ___ |

2. Repeat Problem 1 for the following table if area = 2 in.$^2$:

| | $\Phi$ | $B$ |
|---|---|---|
| SI | ___ | ___ |
| CGS | 60,000 maxwells | ___ |
| English | ___ | ___ |

3. For the electromagnet of Fig. 11.40:
   a. Find the flux density in the core.
   b. Sketch the magnetic flux lines and indicate their direction.
   c. Indicate the north and south poles of the magnet.

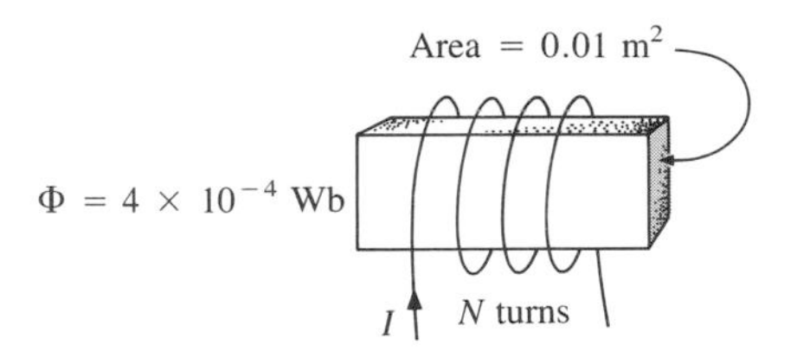


FIG. 11.40

### SECTION 11.5 Reluctance

4. Which section of Fig. 11.41 [(a), (b), or (c)] has the largest reluctance to the setting up of flux lines through its longest dimension?

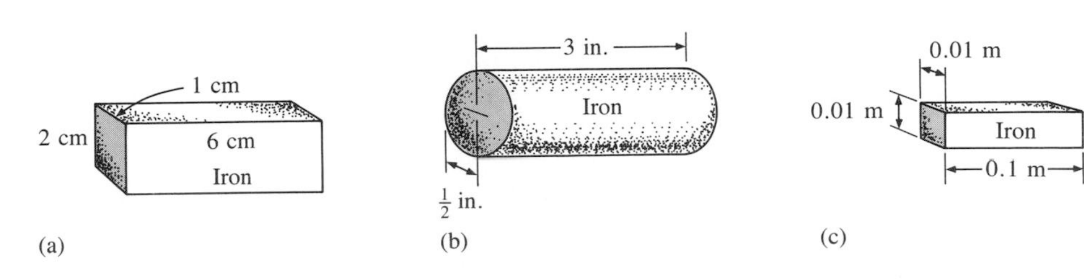


FIG. 11.41

## SECTION 11.6 Ohm's Law for Magnetic Circuits

5. Find the reluctance of a magnetic circuit if a magnetic flux $\Phi = 4.2 \times 10^{-4}$ Wb is established by an impressed mmf of 400 At.

6. Repeat Problem 5 for $\Phi = 72{,}000$ maxwells and an impressed mmf of 120 gilberts.

## SECTION 11.7 Magnetizing Force

7. Find the magnetizing force $H$ for Problem 5 in SI units if the magnetic circuit is 6 inches in length.

8. If a magnetizing force $H$ of 600 At/m is applied to a magnetic circuit, a flux density $B$ of $1200 \times 10^{-4}$ Wb/m$^2$ is established. Find the permeability $\mu$ of a material that will produce twice the original flux density for the same magnetizing force.

## SECTION 11.8 Hysteresis

9. For the series magnetic circuit of Fig. 11.42, determine the current $I$ necessary to establish the indicated flux.

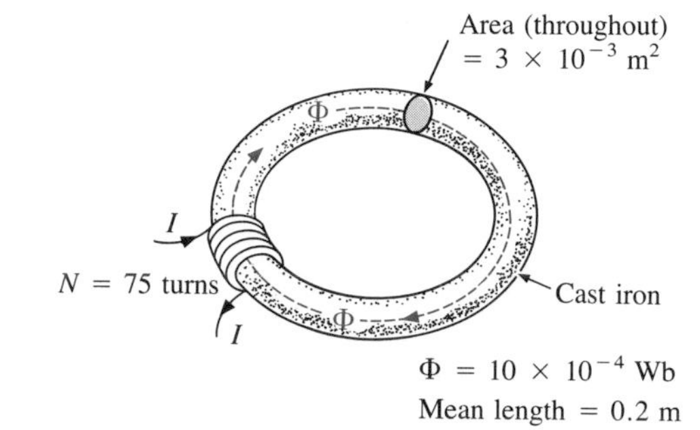

FIG. 11.42

10. Find the current necessary to establish a flux of $\Phi = 3 \times 10^{-4}$ Wb in the series magnetic circuit of Fig. 11.43.

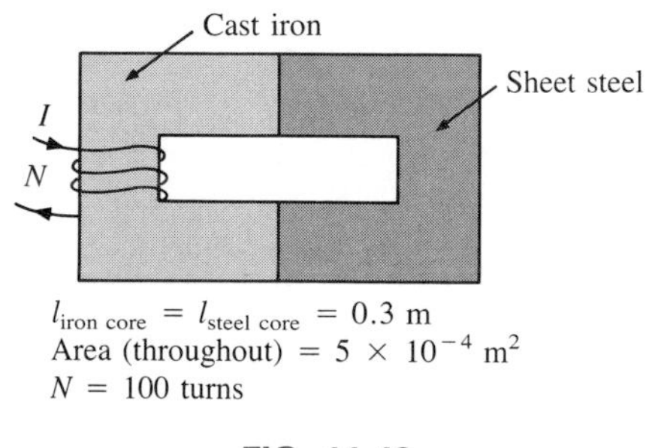

FIG. 11.43

11. a. Find the number of turns $N_1$ required to establish a flux $\Phi = 12 \times 10^{-4}$ Wb in the magnetic circuit of Fig. 11.44.
    b. Find the permeability $\mu$ of the material.

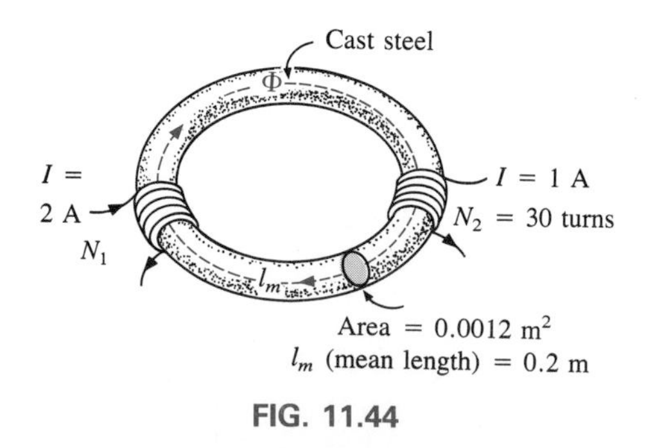


FIG. 11.44

**12. a.** Find the mmf ($NI$) required to establish a flux $\Phi = 80{,}000$ lines in the magnetic circuit of Fig. 11.45.

**b.** Find the permeability of each material.

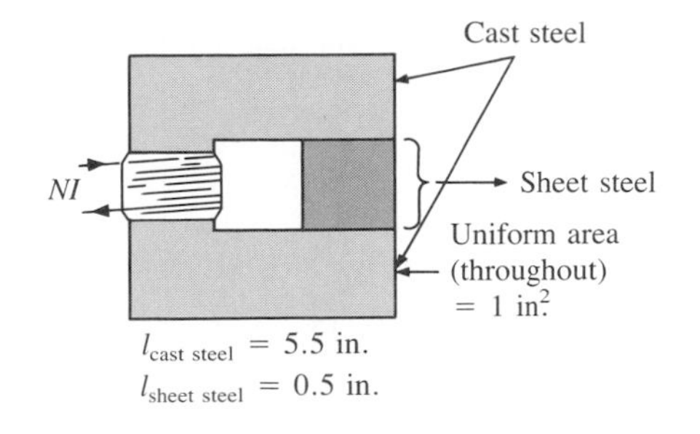


$l_{\text{cast steel}} = 5.5$ in.
$l_{\text{sheet steel}} = 0.5$ in.

**FIG. 11.45**

***13.** For the series magnetic circuit of Fig. 11.46 with two impressed sources of magnetic "pressure," determine the current $I$. Each applied mmf establishes a flux pattern in the clockwise direction.

$\Phi = 0.8 \times 10^{-4}$ Wb
$l_{\text{cast steel}} = 5.5$ in.
$l_{\text{cast iron}} = 2.5$ in.

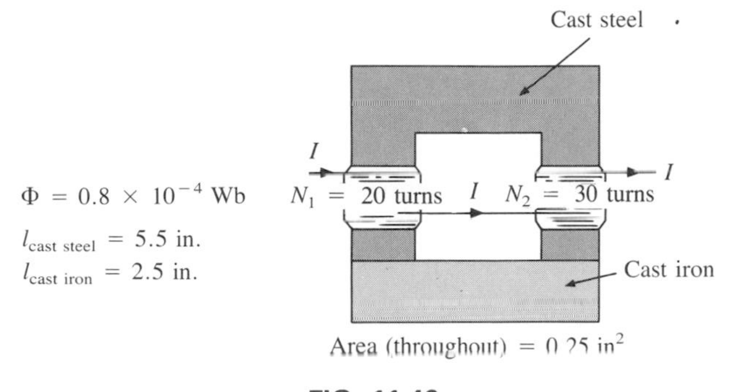

**FIG. 11.46**

## SECTION 11.12 Air Gaps

**14. a.** Find the current $I$ required to establish a flux $\Phi = 2.4 \times 10^{-4}$ Wb in the magnetic circuit of Fig. 11.47.

**b.** Compare the mmf drop across the air gap to that across the rest of the magnetic circuit. Discuss your results using the value of $\mu$ for each material.

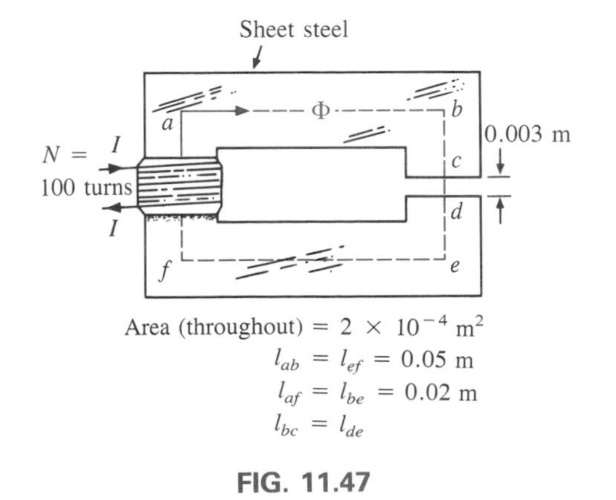


**FIG. 11.47**

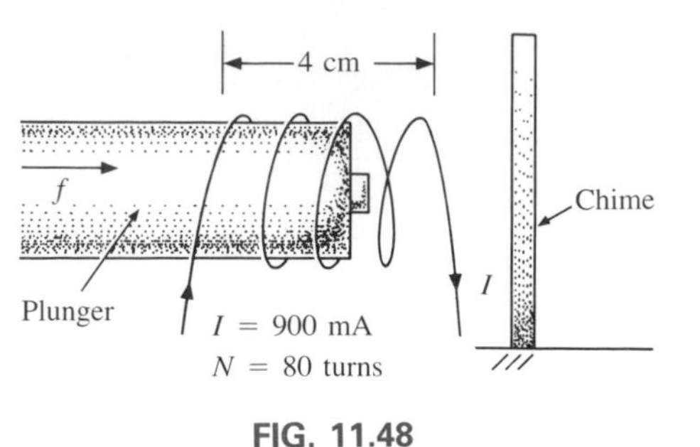


FIG. 11.48
*Door chime.*

***15.** The force carried by the plunger of the door chime of Fig. 11.48 is determined by

$$f = \frac{1}{2}NI\frac{d\phi}{dx} \qquad \text{(newtons)}$$

where $d\phi/dx$ is the rate of change of flux linking the coil as the core is drawn into the coil. The greatest rate of change of flux will occur when the core is 1/4 to 3/4 the way through. In this region, if $\Phi$ changes from $0.5 \times 10^{-4}$ Wb to $8 \times 10^{-4}$ Wb, what is the force carried by the plunger?

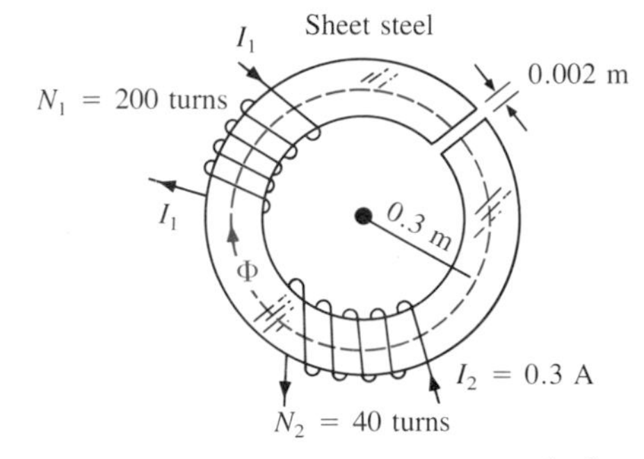

FIG. 11.49

**16.** Determine the current $I_1$ required to establish a flux of $\Phi = 2 \times 10^{-4}$ Wb in the magnetic circuit of Fig. 11.49.

***17. a.** A flux of $0.2 \times 10^{-4}$ Wb will establish sufficient attractive force for the armature of the relay of Fig. 11.50 to close the contacts. Determine the required current to establish this flux level if we assume the total mmf drop is across the air gap.

**b.** The force exerted on the armature is determined by the equation

$$F \text{ (newtons)} = \frac{1}{2} \cdot \frac{B_g^2 A}{\mu_o}$$

where $B_g$ is the flux density within the air gap and $A$ is the common area of the air gap. Find the force in newtons exerted when the flux $\Phi$ specified in part (a) is established.

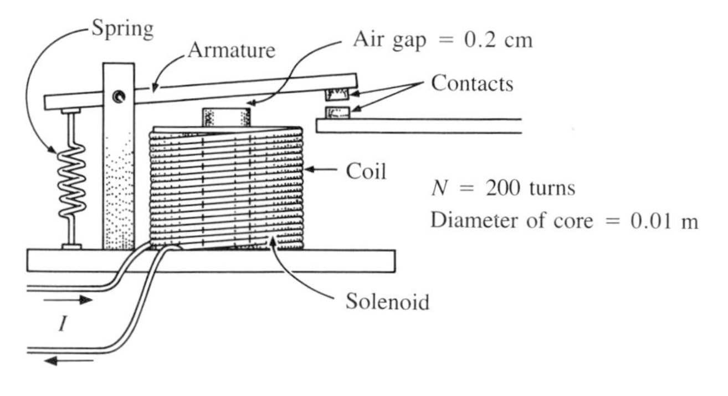

FIG. 11.50
*Relay.*

*18. For the series-parallel magnetic circuit of Fig. 11.51, find the value of $I$ required to establish a flux in the gap $\Phi_g = 2 \times 10^{-4}$ Wb.

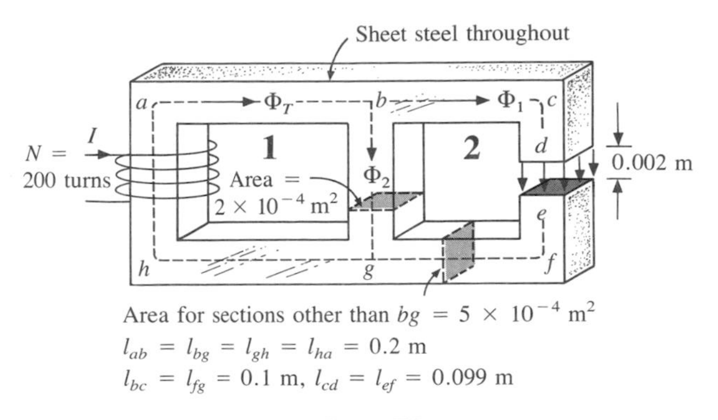

FIG. 11.51

## SECTION 11.14 Determining $\Phi$

19. Find the magnetic flux $\Phi$ established in the series magnetic circuit of Fig. 11.52.

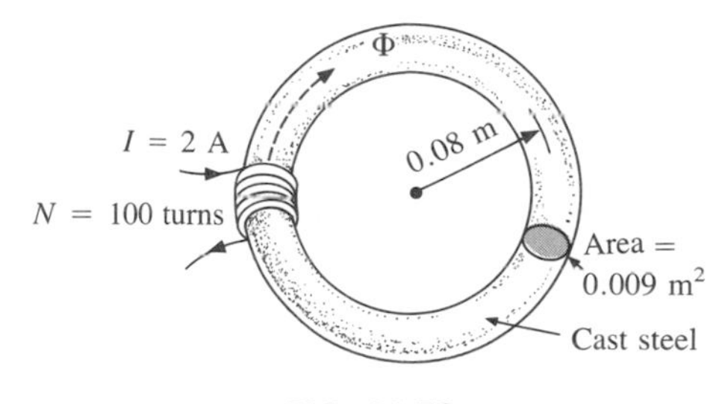

FIG. 11.52

*20. Determine the magnetic flux $\Phi$ established in the series magnetic circuit of Fig. 11.53.

*21. Note how closely the $B$-$H$ curve of cast steel in Fig. 11.21 matches the curve for the voltage across a capacitor as it charges from zero volts to its final value.

**a.** Using the equation for the charging voltage as a guide, write an equation for $T$ as a function of $H$ ($T = f(H)$) for cast steel.

**b.** Test the resulting equation at $H = 900$ At/m, 1800 At/m, and 2700 At/m.

**c.** Using the equation of part (a) derive an equation for $H$ in terms of $B$ ($H = f(T)$).

**d.** Test the resulting equation at $B = 1$ T and $B = 1.4$ T.

**e.** Using the result of part (c), perform the analysis of Example 11.3 and compare the results for the current $I$.

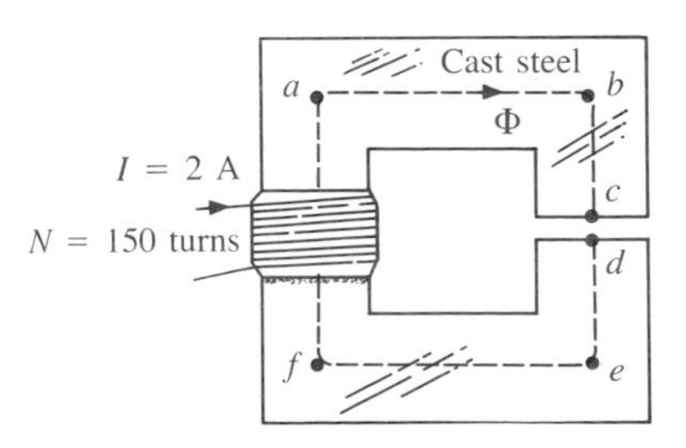


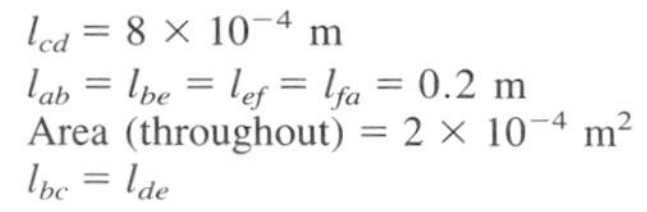

FIG. 11.53

## COMPUTER PROBLEMS

PSPICE is not designed to perform an analysis of magnetic circuits, but BASIC can be employed if we use mathematical representations of the $B$-$H$ curve of Fig. 11.21, as developed in Problem 21.

***22.** Using the results of Problem 21 write a BASIC program to perform the analysis of a core such as that shown in Example 11.3. That is, let the dimensions of the core and the applied turns be input variables requested by the program.

***23.** Using the results of Problem 21 develop a BASIC program to perform the analysis appearing in Example 11.9 for cast steel. A test routine will have to be developed to determine whether the results obtained are sufficiently close to the applied ampere-turns.

# GLOSSARY

**Ampère's circuital law** A law establishing the fact that the algebraic sum of the rises and drops of the mmf around a closed loop of a magnetic circuit is equal to zero.

**Diamagnetic materials** Materials that have permeabilities slightly less than that of free space.

**Domain** A group of magnetically aligned atoms.

**Electromagnetism** Magnetic effects introduced by the flow of charge or current.

**Ferromagnetic materials** Materials having permeabilities hundreds and thousands of times greater than that of free space.

**Flux density** ($B$) A measure of the flux per unit area perpendicular to a magnetic flux path. It is measured in teslas (T) or webers per square meter (Wb/m$^2$).

**Hysteresis** The lagging effect between flux density of a material and the magnetizing force applied.

**Magnetic flux lines** Lines of a continuous nature that reveal the strength and direction of the magnetic field.

**Magnetizing force** ($H$) A measure of the magnetomotive force per unit length of a magnetic circuit.

**Magnetomotive force** ($\mathscr{F}$) The ''pressure'' required to establish magnetic flux in a ferromagnetic material. It is measured in ampere-turns (At).

**Paramagnetic materials** Materials that have permeabilities slightly greater than that of free space.

**Permanent magnet** A material such as steel or iron that will remain magnetized for long periods of time without the aid of external means.

**Permeability** ($\mu$) A measure of the ease with which magnetic flux can be established in a material. It is measured in Wb/Am.

**Relative permeability** ($\mu_r$) The ratio of the permeability of a material to that of free space.

**Reluctance** ($\mathscr{R}$) A quantity determined by the physical characteristics of a material that will provide an indication of the ''reluctance'' of that material to the setting up of magnetic flux lines in the material. It is measured in rels or At/Wb.

# 12 Inductors

## 12.1 INTRODUCTION

We have examined the resistor and the capacitor in detail. In this chapter we shall consider a third element, the *inductor,* which has a number of response characteristics similar in many respects to those of the capacitor. In fact, some sections of this chapter will proceed parallel to those for the capacitor to emphasize the similarity that exists between the two elements.

## 12.2 FARADAY'S LAW OF ELECTROMAGNETIC INDUCTION

If a conductor is moved through a magnetic field so that it cuts magnetic lines of flux, a voltage will be induced across the conductor, as shown in Fig. 12.1. The greater the number of flux lines cut per unit time (by

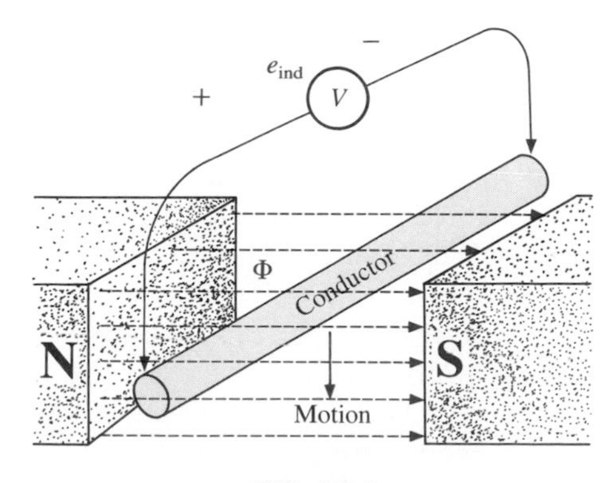


FIG. 12.1

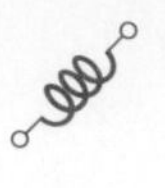

increasing the speed with which the conductor passes through the field), or the stronger the magnetic field strength (for the same traversing speed), the greater will be the induced voltage across the conductor. If the conductor is held fixed and the magnetic field is moved so that its flux lines cut the conductor, the same effect will be produced.

If a coil of $N$ turns is placed in the region of a changing flux, as in Fig. 12.2, a voltage will be induced across the coil as determined by *Faraday's law:*

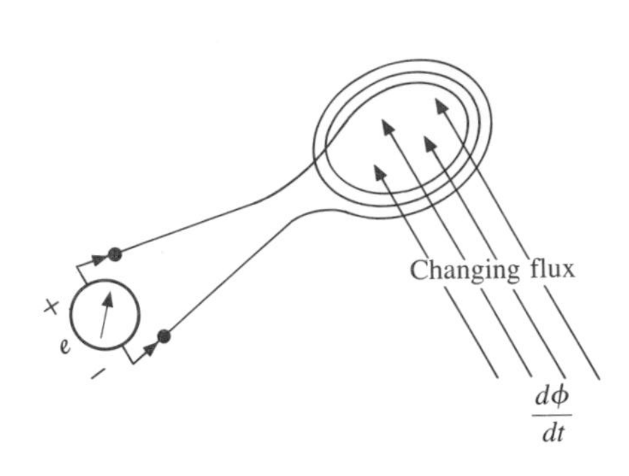


FIG. 12.2

$$e = N\frac{d\phi}{dt} \quad \text{(volts, V)} \qquad \textbf{(12.1)}$$

where $N$ represents the number of turns of the coil and $d\phi/dt$ is the instantaneous change in flux (in webers) linking the coil. The term *linking* refers to the flux within the turns of wire. The term *changing* simply indicates that either the strength of the field linking the coil changes in magnitude or the coil is moved through the field in such a way that the number of flux lines through the coil changes with time.

If the flux linking the coil ceases to change, such as when the coil simply sits still in a magnetic field of fixed strength, $d\phi/dt = 0$, and the induced voltage $e = N(d\phi/dt) = N(0) = 0$.

## 12.3 LENZ'S LAW

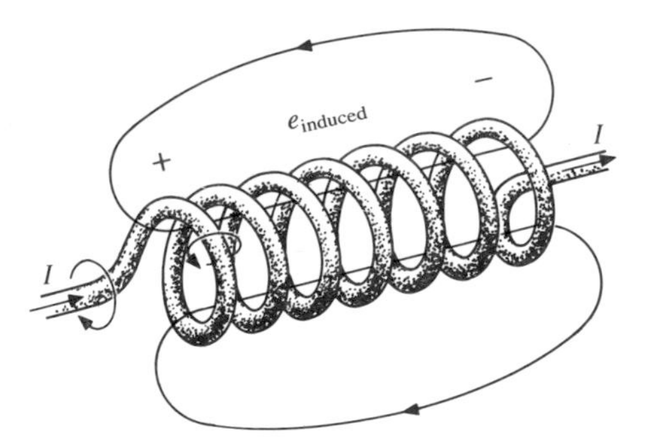


FIG. 12.3

In Section 11.2 it was shown that the magnetic flux linking a coil of $N$ turns with a current $I$ has the distribution of Fig. 12.3.

If the current increases in magnitude, the flux linking the coil also increases. It was shown in Section 12.2, however, that a changing flux linking a coil induces a voltage across the coil. For this coil, therefore, an induced voltage is developed *across* the coil due to the change in current *through* the coil. The polarity of this induced voltage tends to establish a current in the coil which produces a flux that will oppose any change in the original flux. In other words, the induced effect ($e_{ind}$) is a result of the increasing current through the coil. However, the resulting induced voltage will tend to establish a current that will oppose the increasing change in current through the coil. Keep in mind that this is all occurring simultaneously. The instant the current begins to increase in magnitude, there will be an opposing effect trying to limit the change. It is ''choking'' the change in current through the coil. Hence, the term *choke* is often applied to the inductor or coil. In fact, we will find shortly that the current through a coil cannot change instantaneously. A period of time determined by the coil and the resistance of the circuit is required before the inductor discontinues its opposition to a momentary change in current. Recall a similar situation for the voltage

across a capacitor in Chapter 10. The reaction above is true for increasing or decreasing levels of current through the coil. This effect is an example of a general principle known as *Lenz's law,* which states that

***an induced effect is always such as to oppose the cause that produced it.***

## 12.4 SELF-INDUCTANCE

The ability of a coil to oppose any change in current is a measure of the *self-inductance* $L$ of the coil. For brevity, the prefix *self* is usually dropped. Inductance is measured in henries (H), after the American physicist Joseph Henry.

*Inductors* are coils of various dimensions designed to introduce specified amounts of inductance into a circuit. The inductance of a coil varies directly with the magnetic properties of the coil. Ferromagnetic materials, therefore, are frequently employed to increase the inductance by increasing the flux linking the coil.

A close approximation, in terms of physical dimensions, for the inductance of the coils of Fig. 12.4 can be found using the following equation:

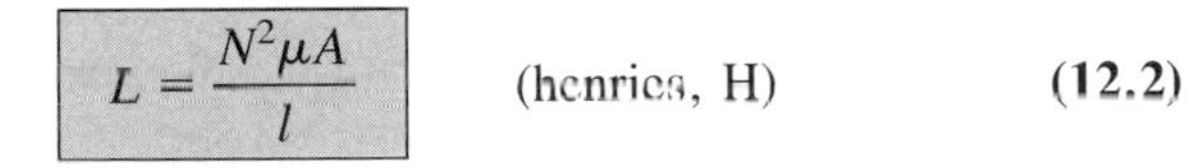

$$L = \frac{N^2\mu A}{l} \qquad \text{(henries, H)} \qquad \textbf{(12.2)}$$

where $N$ represents the number of turns, $\mu$ the permeability of the core (recall that $\mu$ is not a constant but depends on the level of $B$ and $H$ since $\mu = B/H$), $A$ the area of the core in square meters, and $l$ the mean length of the core in meters.

Substituting $\mu = \mu_r\mu_o$ into Eq. (12.2) yields

$$L = \frac{N^2\mu_r\mu_o A}{l} = \mu_r\frac{N^2\mu_o A}{l}$$

and

$$L = \mu_r L_o \qquad \textbf{(12.3)}$$

where $L_o$ is the inductance of the coil with an air core. In other words, the inductance of a coil with a ferromagnetic core is the relative permeability of the core times the inductance achieved with an air core.

Equations for the inductance of coils different from those shown above can be found in reference handbooks. Most of the equations are more complex than those just described.

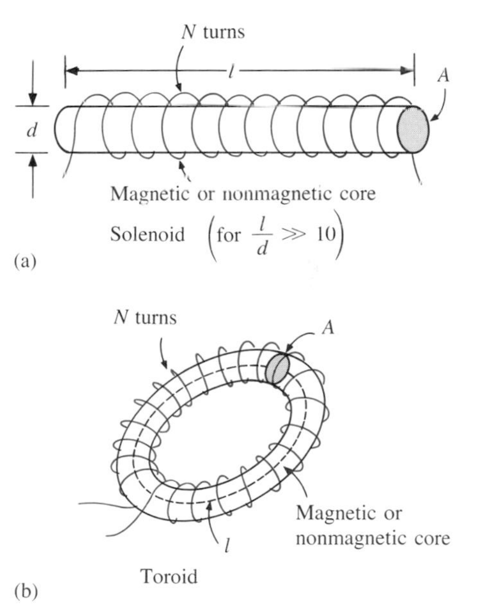


**FIG. 12.4**

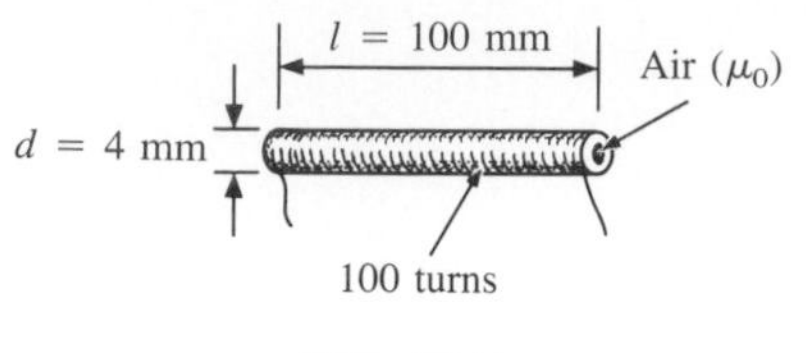

FIG. 12.5

**EXAMPLE 12.1.** Find the inductance of the air-core coil of Fig. 12.5.

***Solution:***

$$\mu = \mu_r\mu_o = (1)(\mu_o) = \mu_o$$

$$A = \frac{\pi d^2}{4} = \frac{(3.1416)(4 \times 10^{-3}\text{ m})^2}{4} = 12.57 \times 10^{-6}\text{ m}^2$$

$$L_o = \frac{N^2\mu_o A}{l} = \frac{(100\text{ t})^2(4\pi \times 10^{-7}\text{ Wb/A}\cdot\text{m})(12.57 \times 10^{-6}\text{ m}^2)}{0.1\text{ m}}$$

$$= \mathbf{1.58\ \mu H}$$

**EXAMPLE 12.2.** Repeat Example 12.1, but with an iron core and conditions such that $\mu_r = 2000$.

***Solution:*** By Eq. (12.3),

$$L = \mu_r L_o = (2000)(1.58 \times 10^{-6}\text{ H}) = \mathbf{3.16\ mH}$$

## 12.5 TYPES OF INDUCTORS

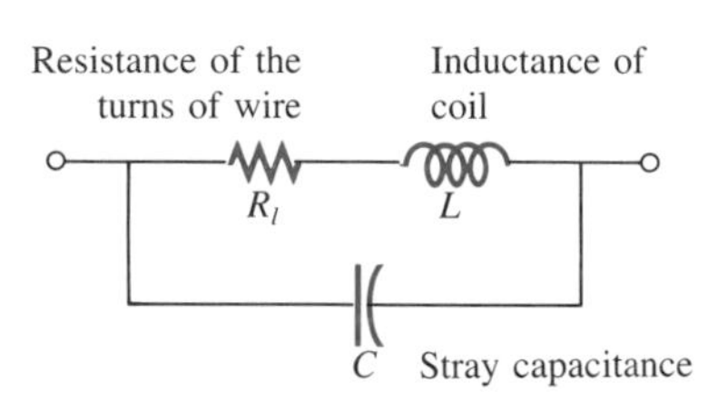

FIG. 12.6
*Complete equivalent model for an inductor.*

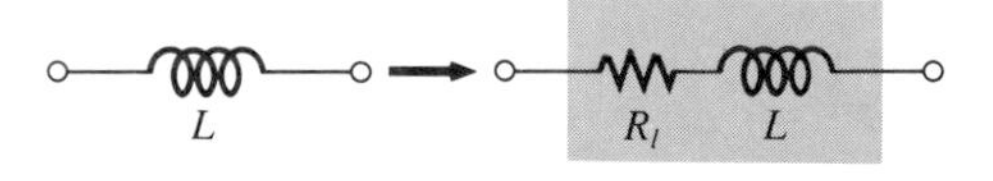

FIG. 12.7
*Practical equivalent model for an inductor.*

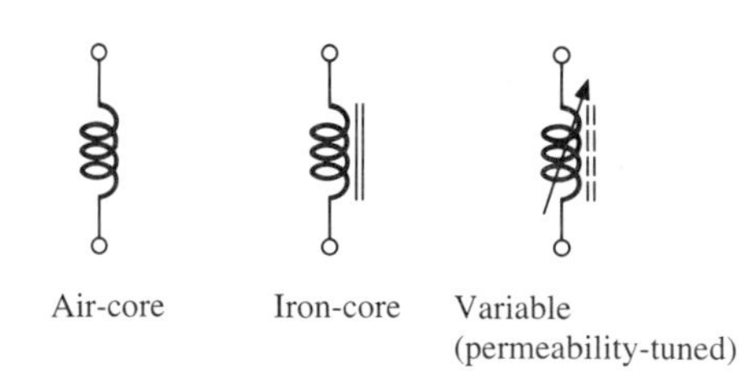

FIG. 12.8

Associated with every inductor are a resistance equal to the resistance of the turns and a stray capacitance due to the capacitance between the turns of the coil. To include these effects, the equivalent circuit for the inductor is as shown in Fig. 12.6. However, for most applications considered in this text, the stray capacitance appearing in Fig. 12.6 can be ignored, resulting in the equivalent model of Fig. 12.7.

The resistance $R_l$ can play an important role in the analysis of networks with inductive elements. For most applications, we have been able to treat the capacitor as an ideal element and maintain a high degree of accuracy. For the inductor, however, $R_l$ must often be included in the analysis and can have a pronounced effect on the response of a system (see Chapter 20, Resonance). The level of $R_l$ can extend from a few ohms to a few hundred ohms. Keep in mind that the longer or thinner the wire used in the construction of the inductor, the greater will be the dc resistance as determined by $R = \rho l/A$. Our initial analysis will treat the inductor as an ideal element. Once a general feeling for the response of the element is established, the effects of $R_l$ will be included.

The primary function of the inductor, however, is to introduce inductance—not resistance or capacitance—into the network. For this reason, the symbols employed for inductance are as shown in Fig. 12.8.

All inductors, like capacitors, can be listed under two general headings: *fixed* and *variable*. The fixed air-core and iron-core inductors were described in the last section. The permeability-tuned variable coil has a ferromagnetic shaft that can be moved within the coil to vary the flux linkages of the coil and thereby its inductance. Several fixed and variable inductors appear in Fig. 12.9.

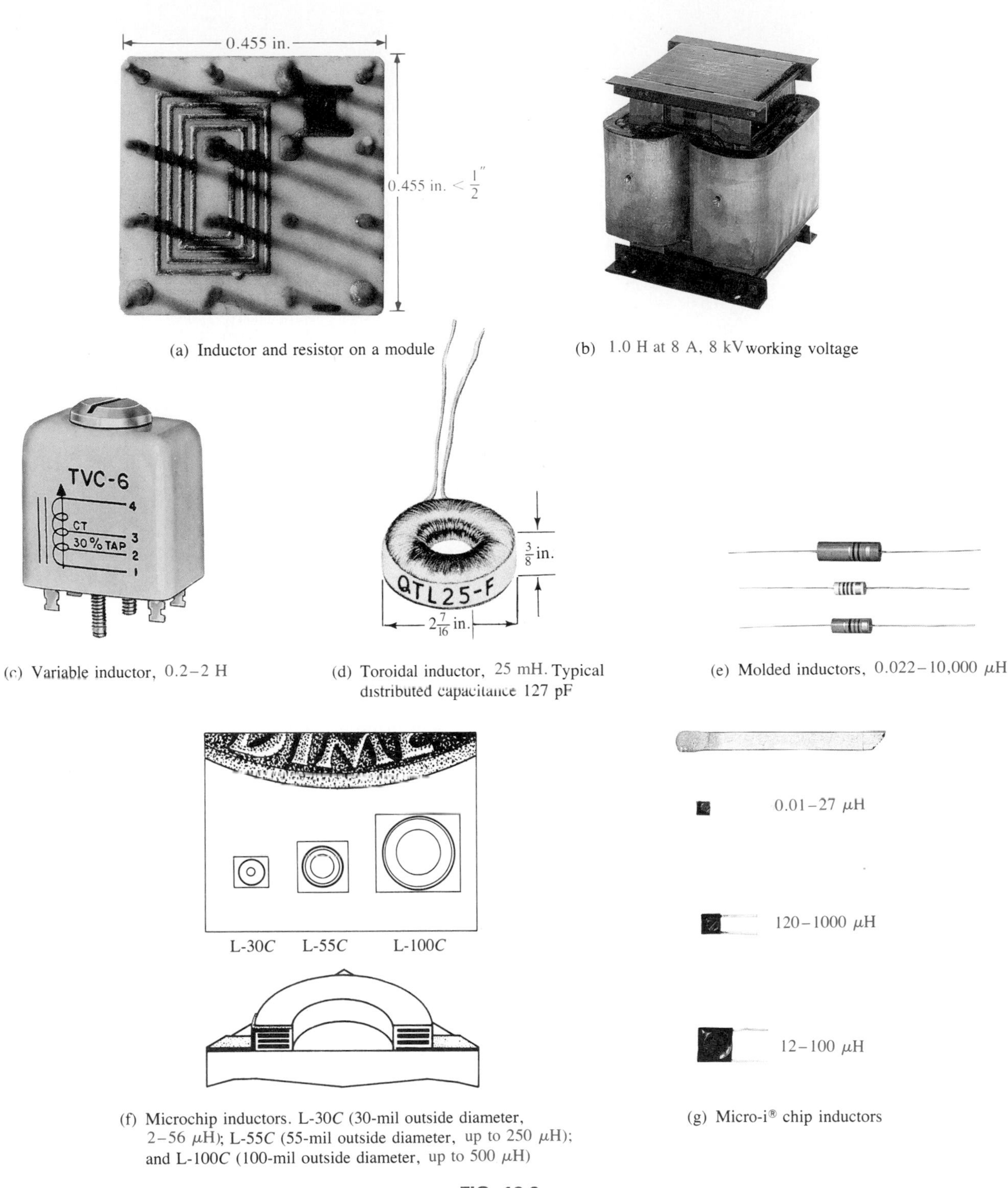


(a) Inductor and resistor on a module

(b) 1.0 H at 8 A, 8 kV working voltage

(c) Variable inductor, 0.2–2 H

(d) Toroidal inductor, 25 mH. Typical distributed capacitance 127 pF

(e) Molded inductors, 0.022–10,000 μH

(f) Microchip inductors. L-30*C* (30-mil outside diameter, 2–56 μH); L-55*C* (55-mil outside diameter, up to 250 μH); and L-100*C* (100-mil outside diameter, up to 500 μH)

(g) Micro-i® chip inductors

**FIG. 12.9**

*Various types of inductors. (Part (a) courtesy of International Business Machines Corp.; part (b) courtesy of Basler Electric Co.; part (c) courtesy of United Transformer Corp.; part (d) courtesy of Microtan Company, Inc.; part (e) courtesy of Delevan, Division of American Precision Industries, Inc.; part (f) courtesy of Thinco Division, Hull Corp.; part (g) courtesy of Delevan, Division of American Precision Industries, Inc.)*

The primary reasons for inductor failure are shorts that develop between the windings and open circuits in the windings due to factors such as excessive currents, overheating, and age. The open-circuit condition can easily be checked with an ohmmeter ($\infty$ ohms indication), but the short-circuit condition is harder because the resistance of many good inductors is relatively small and the shorting of a few windings will not adversely affect the total resistance. Of course, if one is aware of the typical resistance of the coil, it can be compared to the measured value. A short between the windings and the core can be checked by simply placing one lead of the meter on one wire (terminal) and the other on the core itself. An indication of zero ohms reflects a short between the two because the wire that makes up the winding has an insulation jacket throughout.

## 12.6 INDUCED VOLTAGE

The inductance of a coil is also a measure of the change in flux linking a coil due to a change in current through the coil; that is,

$$L = N\frac{d\phi}{di} \qquad \text{(H)} \qquad \textbf{(12.4)}$$

where $N$ is the number of turns, $\phi$ is the flux in webers, and $i$ is the current through the coil. The equation states that the larger the inductance of a coil (with $N$ fixed), the larger will be the instantaneous change in flux linking the coil due to an instantaneous change in current through the coil.

If we write Eq. (12.1) as

$$e_L = N\frac{d\phi}{dt} = \left(N\frac{d\phi}{di}\right)\left(\frac{di}{dt}\right)$$

and substitute Eq. (12.4), we then have

$$e_L = L\frac{di}{dt} \qquad \text{(V)} \qquad \textbf{(12.5)}$$

revealing that the magnitude of the voltage across an inductor is directly related to the inductance $L$ and the instantaneous rate of change of current through the coil. Obviously, therefore, the greater the *rate* of change of current through the coil, the greater will be the induced voltage. This certainly agrees with our earlier discussion of Lenz's law.

When induced effects are employed in the generation of voltages such as available from dc or ac generators, the symbol $e$ is appropriate for the induced voltage. However, in network analysis the voltage across an inductor will always have a polarity such as to oppose the source that produced it, and therefore the following notation will be used throughout the analysis to come:

$$v_L = L\frac{di}{dt} \tag{12.6}$$

If the current through the coil fails to change at a particular instant, the induced voltage across the coil will be zero. For dc applications, after the transient effect has passed, $di/dt = 0$, and the induced voltage is

$$v_L = L\frac{di}{dt} = L(0) = 0 \text{ V}$$

Recall that the equation for the current of a capacitor is the following:

$$i_C = C\frac{dv_C}{dt}$$

Note the similarity between this equation and Eq. (12.6). In fact, if we apply the duality $v \rightleftarrows i$ (that is, interchange the two) and $L \rightleftarrows C$ for capacitance and inductance, each equation can be derived from the other.

The average voltage across the coil is defined by the equation

$$v_{L_{av}} = L\frac{\Delta i}{\Delta t} \quad \text{(V)} \tag{12.7}$$

where $\Delta$ signifies finite change (a measurable change). Compare this to $i_C = C(\Delta v/\Delta t)$, and the meaning of $\Delta$ and application of this equation should be clarified from Chapter 10. An example follows.

---

**EXAMPLE 12.3.** Find the waveform for the average voltage across the coil if the current through a 4-mH coil is as shown in Fig. 12.10.

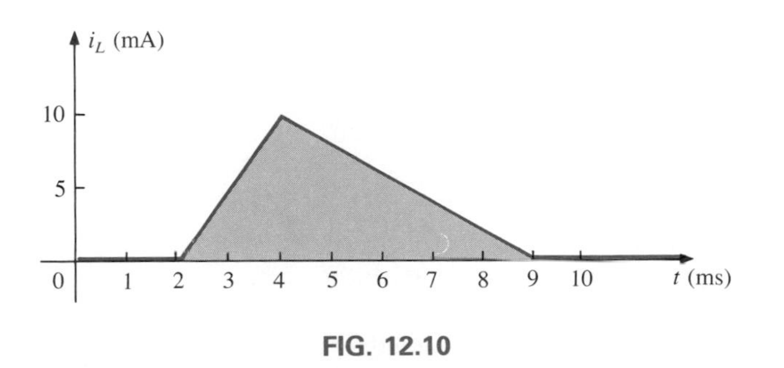


**FIG. 12.10**

***Solution:***

a. 0 to 2 ms: Since there is no change in current through the coil, there is no voltage induced across the coil; that is,

$$v_L = L\frac{\Delta i}{\Delta t} = L\frac{0}{\Delta t} = \mathbf{0}$$

b. 2 ms to 4 ms:

$$v_L = L\frac{\Delta i}{\Delta t} = (4 \times 10^{-3}\ \text{H})\left(\frac{10 \times 10^{-3}\ \text{A}}{2 \times 10^{-3}\ \text{s}}\right) = 20 \times 10^{-3}\ \text{V}$$
$$= \mathbf{20\ mV}$$

c. 4 ms to 9 ms:

$$v_L = L\frac{\Delta i}{\Delta t} = (-4 \times 10^{-3}\ \text{H})\left(\frac{10 \times 10^{-3}\ \text{A}}{5 \times 10^{-3}\ \text{s}}\right) = -8 \times 10^{-3}\ \text{V}$$
$$= \mathbf{-8\ mV}$$

d. 9 ms to ∞:

$$v_L = L\frac{\Delta i}{\Delta t} = L\frac{0}{\Delta t} = \mathbf{0}$$

The waveform for the average voltage across the coil is shown in Fig. 12.11. Note from the curve that

***the voltage across the coil is not determined solely by the magnitude of the change in current through the coil ($\Delta i$), but by the rate of change of current through the coil ($\Delta i/\Delta t$).***

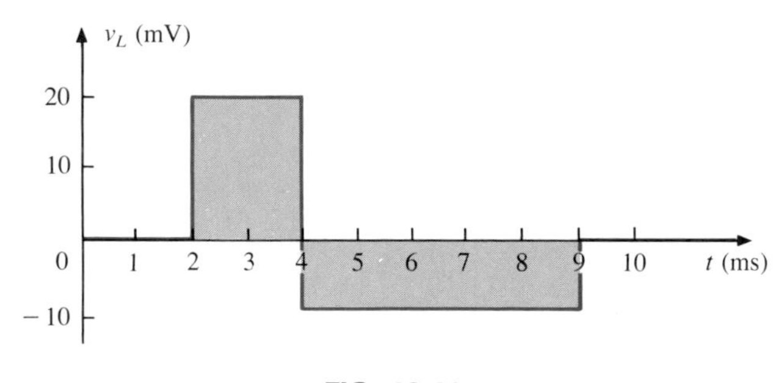


**FIG. 12.11**

A similar statement was made for the current of a capacitor due to a change in voltage across the capacitor.

---

A careful examination of Fig. 12.11 will also reveal that the area under the positive pulse from 2 ms to 4 ms equals the area under the negative pulse from 4 ms to 9 ms. In Section 12.13, we will find that the area under the curves represents the energy stored or released by the inductor. From 2 ms to 4 ms, the inductor is storing energy, while from

4 ms to 9 ms, the inductor is releasing the energy stored. For the full period zero to 10 ms, energy has simply been stored and released; there has been no dissipation as experienced for the resistive elements. Over a full cycle, both the ideal capacitor and inductor do not consume energy but simply store and release it in their respective forms.

## 12.7 R-L TRANSIENTS: STORAGE CYCLE

The changing voltages and current that result during the storing of energy in the form of a magnetic field by an inductor in a dc circuit can best be described using the circuit of Fig. 12.12. At the instant the switch is closed, the inductance of the coil will prevent an instantaneous change in current through the coil. The potential drop across the coil, $v_L$, will equal the impressed voltage $E$ as determined by Kirchhoff's voltage law since $v_R = iR = (0)R = 0$ V. The current $i_L$ will then build up from zero, establishing a voltage drop across the resistor and a corresponding drop in $v_L$. The current will continue to increase until the voltage across the inductor drops to zero volts and the full impressed voltage appears across the resistor. Initially, the current $i_L$ increases quite rapidly, followed by a continually decreasing rate until it reaches its maximum value of $E/R$.

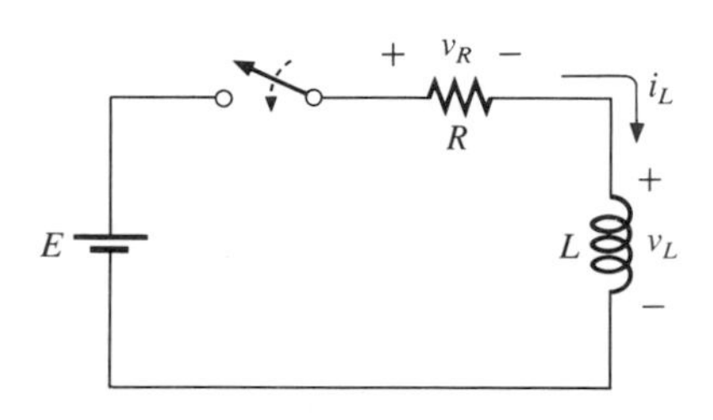


FIG. 12.12

You will recall from the discussion of capacitors that a capacitor has a short-circuit equivalent when the switch is first closed and an open-circuit equivalent when steady-state conditions are established. The inductor assumes the opposite equivalents for each stage. The instant the switch of Fig. 12.12 is closed, the equivalent network will appear as shown in Fig. 12.13. Note the correspondence with the earlier comments regarding the levels of voltage and current. The inductor obviously meets all the requirements for an open-circuit equivalent—$v_L = E$ volts, $i_L = 0$ A.

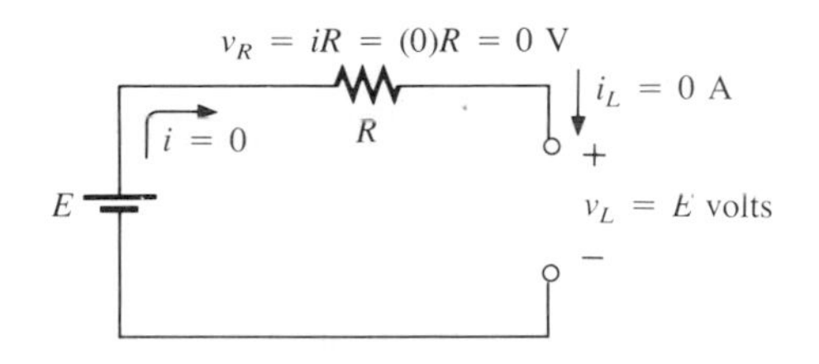


FIG. 12.13
*Circuit of Fig. 12.12 the instant the switch is closed.*

When steady-state conditions have been established and the storage phase is complete, the ''equivalent'' network will appear as shown in Fig. 12.14. The network clearly reveals that:

***An ideal inductor ($R_l = 0\ \Omega$) assumes a short-circuit equivalent in a dc network once steady-state conditions have been established.***

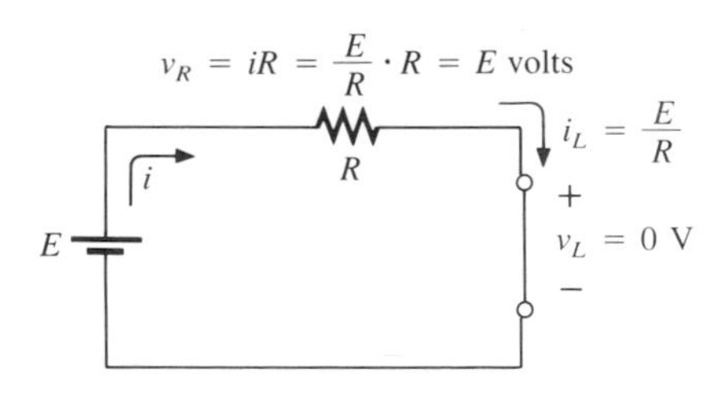


FIG. 12.14
*Circuit of Fig. 12.12 under steady-state conditions.*

Fortunately, the mathematical equations for the voltages and current for the storage phase are similar in many respects to those encountered for the *R-C* network. The experience gained with these equations in Chapter 10 will undoubtedly make the analysis of *R-L* networks somewhat easier to understand.

The equation for the current $i_L$ during the storage phase is the following:

$$i_L = I_m(1 - e^{-t/\tau}) = \frac{E}{R}(1 - e^{-t/(L/R)}) \tag{12.8}$$

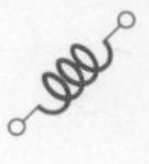

Note the factor $(1 - e^{-t/\tau})$, which also appeared for the voltage $v_C$ of a capacitor during the charging phase. A plot of the equation is given in Fig. 12.15, clearly indicating that the maximum steady-state value of $i_L$

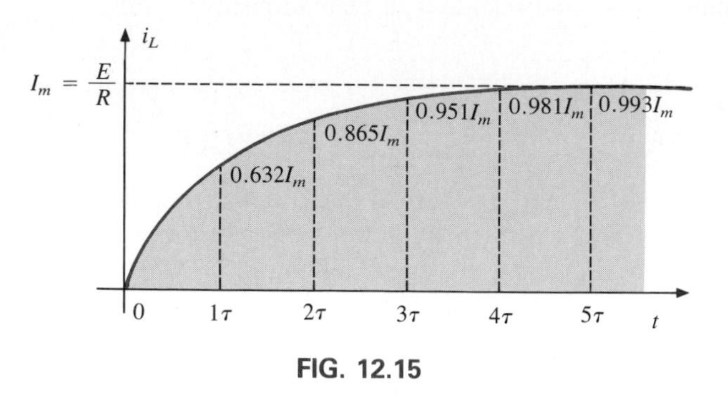


**FIG. 12.15**

is $E/R$, and that the rate of change in current decreases as time passes. The abscissa is scaled in time constants, with $\tau$ for inductive circuits defined by the following:

$$\tau = \frac{L}{R} \quad \text{(seconds, s)} \tag{12.9}$$

The fact that $\tau$ has the units of time can be verified by taking the equation for the induced voltage

$$v_L = L\frac{di}{dt}$$

and solving for $L$:

$$L = \frac{v_L}{di/dt}$$

which leads to the ratio

$$\tau = \frac{L}{R} = \frac{\dfrac{v_L}{di/dt}}{R} = \frac{v_L}{\dfrac{di}{dt}R} \Rightarrow \frac{V}{\dfrac{IR}{t}}$$

$$= \frac{\cancel{V}}{\dfrac{\cancel{V}}{t}} = t \text{ (seconds)}$$

Our experience with the factor $(1 - e^{-t/\tau})$ verifies the level of 63.2% after one time constant, 86.5% after two time constants, and so on. For convenience, Fig. 10.27 is repeated as Fig. 12.16 to evaluate the functions $(1 - e^{-t/\tau})$ and $e^{-t/\tau}$ at various values of $\tau$.

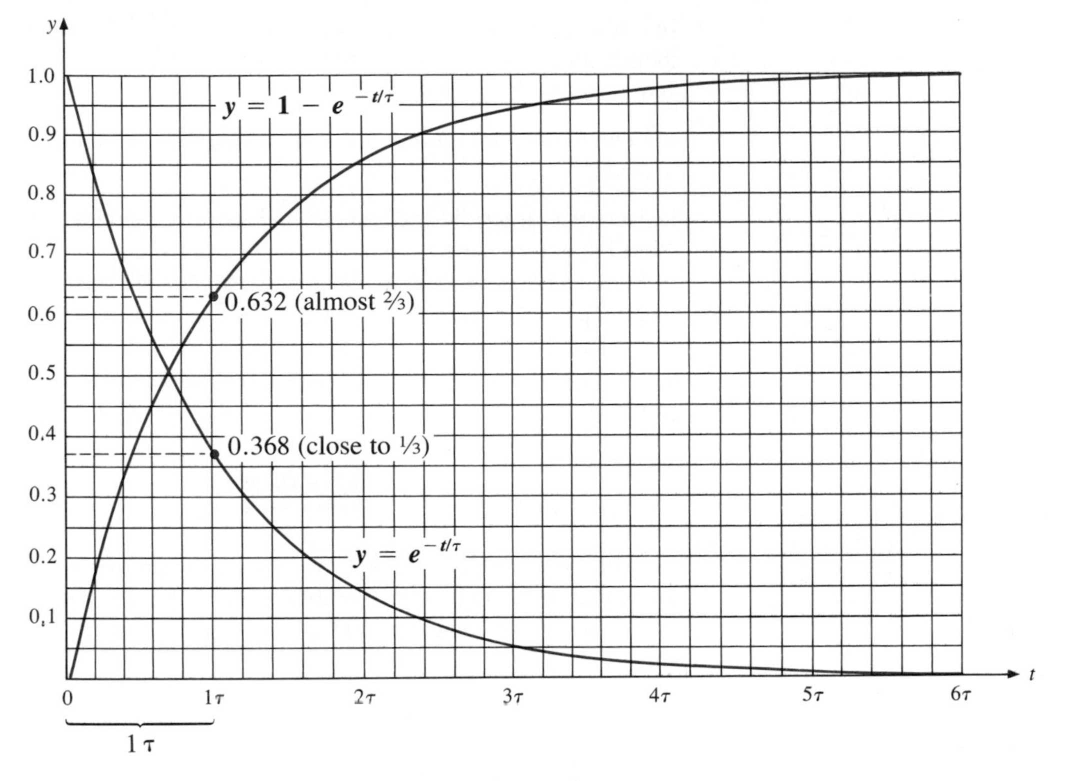


FIG. 12.16

If we keep $R$ constant and increase $L$, the ratio $L/R$ increases and the rise time increases. The change in transient behavior for the current $i_L$ is plotted in Fig. 12.17 for various values of $L$. Note again the duality

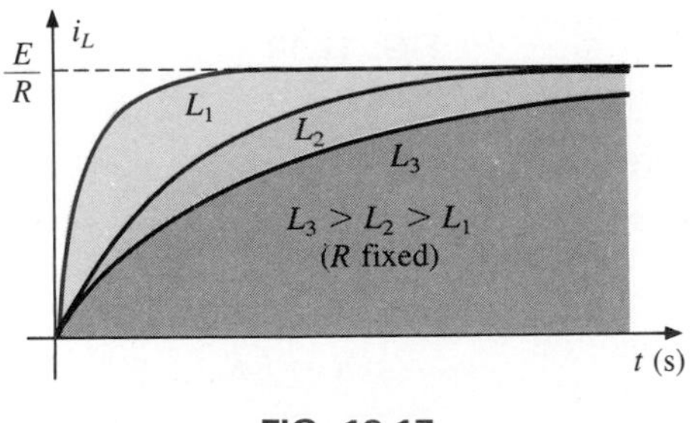


FIG. 12.17

between these curves and those obtained for the $R$-$C$ network in Fig. 10.30.

For most practical applications, we will assume that:

***The storage phase has passed and steady-state conditions have been established once a period of time equal to five time constants has occurred.***

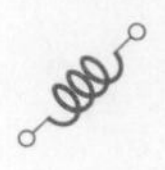

In addition, since $L/R$ will always have some numerical value even though it may be very small, the period $5\tau$ will always be greater than zero, confirming the fact that

***the current cannot change instantaneously in an inductive network.***

In fact, the larger the inductance, the more the circuit will oppose a rapid buildup in current level.

Figures 12.13 and 12.14 clearly reveal that the voltage across the coil jumps to $E$ volts when the switch is closed and decays to zero volts with time. The decay occurs in an exponential manner, and $v_L$ during the storage phase can be described mathematically by the following equation:

$$v_L = Ee^{-t/\tau} \quad \textbf{(12.10)}$$

A plot of $v_L$ appears in Fig. 12.18 with the time axis again divided into equal increments of $\tau$. Obviously, the voltage $v_L$ will decrease to zero volts at the same rate the current presses toward its maximum value.

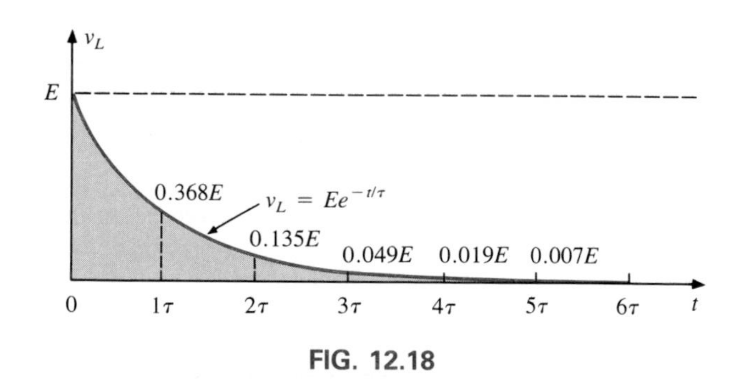


**FIG. 12.18**

***In five time constants, $i_L = E/R$, $v_L = 0$ V, and the inductor can be replaced by its short-circuit equivalent.***

Since

$$v_R = i_R R = i_L R$$

then

$$v_R = \left[\frac{E}{R}(1 - e^{-t/\tau})\right]R$$

and

$$v_R = E(1 - e^{-t/\tau}) \quad \textbf{(12.11)}$$

and the curve for $v_R$ will have the same shape as obtained for $i_L$.

**EXAMPLE 12.4.** Find the mathematical expressions for the transient behavior of $i_L$ and $v_L$ for the circuit of Fig. 12.19 after the closing of the switch. Sketch the resulting curves.

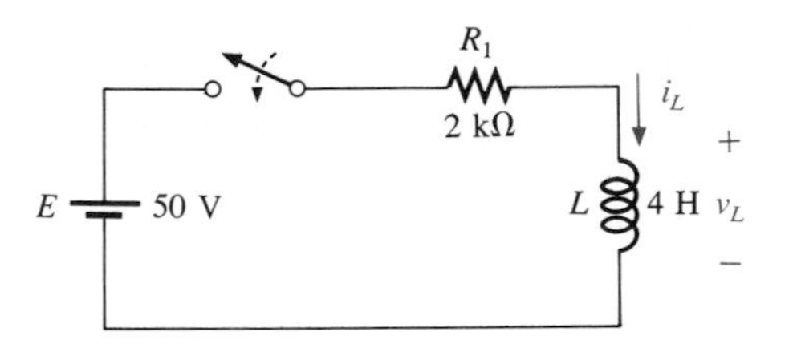


FIG. 12.19

***Solution:***

$$\tau = \frac{L}{R_1} = \frac{4\text{ H}}{2\text{ k}\Omega} = 2\text{ ms}$$

By Eq. (12.8),

$$I_m = \frac{E}{R_1} = \frac{50}{2\text{ k}\Omega} = 25 \times 10^{-3}\text{ A} = 25\text{ mA}$$

and

$$i_L = (\mathbf{25 \times 10^{-3}})(\mathbf{1} - e^{-t/(2\times 10^{-3})})$$

By Eq. (12.10),

$$v_L = \mathbf{50}e^{-t/(2\times 10^{-3})}$$

Both waveforms appear in Fig. 12.20.

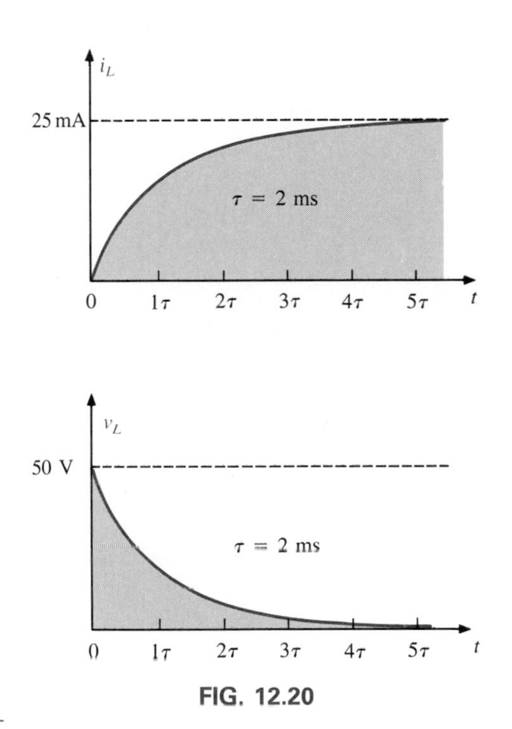


FIG. 12.20

## 12.8 *R-L* TRANSIENTS: DECAY PHASE

In the analysis of *R-C* circuits, we found that the capacitor could hold its charge and store energy in the form of an electric field for a period of time determined by the leakage factors. In *R-L* circuits, the energy is stored in the form of a magnetic field established by the current through the coil. Unlike the capacitor, however, an isolated inductor cannot continue to store energy since the absence of a closed path would cause the current to drop to zero, releasing the energy stored in the form of a magnetic field. If the switch of Fig. 12.12 were opened quickly, a spark would probably occur across the contacts due to the rapid change in current from a maximum of $E/R$ to zero amperes. The change in current $di/dt$ of the equation $v_L = L(di/dt)$ would establish a high voltage $v_L$ across the coil that would discharge across the points of the switch. This is the same mechanism as applied in the ignition system of a car to ignite the fuel in the cylinder. Some 25,000 volts are generated by the rapid decrease in ignition coil current that occurs when the switch in the system is opened. (In older systems, the "points" in the distributor served as the switch.) This inductive reaction is significant when you consider that the only independent source in a car is a 12-V battery.

If opening the switch to move it to another position will cause such a rapid discharge in stored energy, how can the decay phase of an *R-L* circuit be analyzed in much the same manner as for the *R-C* circuit? The solution is to use a network such as that appearing in Fig. 12.21. When the switch is closed, the voltage across the resistor $R_2$ is $E$ volts and the *R-L* branch will respond in the same manner as described above, with the same waveforms and levels. A Thevenin network of $E$ in parallel with $R_2$ would simply result in the source since $R_2$ would be shorted out by the short-circuit replacement of the voltage source $E$ when the Thevenin resistance is determined.

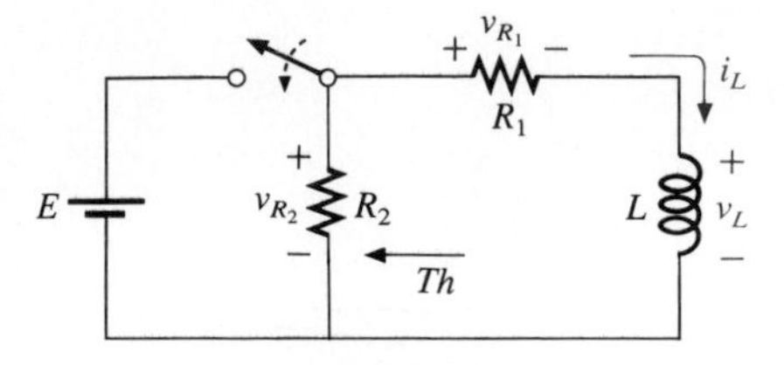


FIG. 12.21

After the storage phase has passed and steady-state conditions are established, the switch can be opened without the sparking effect or rapid discharge due to the resistor $R_2$, which provides a complete path for the current $i_L$. In fact, for clarity the discharge path is isolated in Fig. 12.22. The voltage $v_L$ across the inductor will reverse polarity and have a magnitude determined by

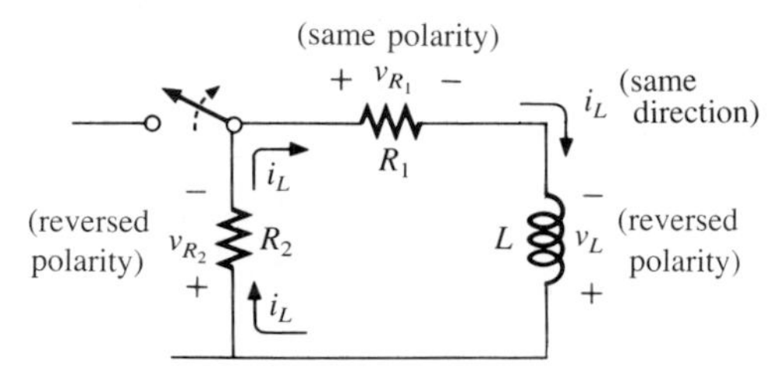


FIG. 12.22
*Network of Fig. 12.21 the instant the switch is opened.*

$$v_L = v_{R_1} + v_{R_2} \tag{12.12}$$

Recall that the voltage across an inductor can change instantaneously but the current cannot. The result is that the current $i_L$ must maintain the same direction and magnitude as shown in Fig. 12.22. Therefore, the instant after the switch is opened, $i_L$ is still $I_m = E/R_1$, and

$$\begin{aligned} v_L &= v_{R_1} + v_{R_2} = i_1R_1 + i_2R_2 \\ &= i_L(R_1 + R_2) = \frac{E}{R_1}(R_1 + R_2) = \left(\frac{R_1}{R_1} + \frac{R_2}{R_1}\right)E \end{aligned}$$

and

$$v_L = \left(1 + \frac{R_2}{R_1}\right)E \tag{12.13}$$

which is bigger than $E$ volts by the ratio $R_2/R_1$. In other words, when the switch is opened, the voltage across the inductor will jump instantaneously from $E$ to $[1 + (R_2/R_1)]E$ volts, with the reverse polarity.

As the inductor releases its stored energy, the voltage across the coil will decay to zero in the following manner:

$$v_L = V_i e^{-t/\tau'} \tag{12.14}$$

with

$$V_i = \left(1 + \frac{R_2}{R_1}\right)E$$

and

$$\tau' = \frac{L}{R_T} = \frac{L}{R_1 + R_2}$$

The current will decay from a maximum of $I_m = E/R_1$ to zero, in the following manner:

$$i_L = I_m e^{-t/\tau'} \tag{12.15}$$

with

$$I_m = \frac{E}{R_1} \quad \text{and} \quad \tau' = \frac{L}{R_1 + R_2}$$

The mathematical expression for the voltage across either resistor can then be determined using Ohm's law:

$$\begin{aligned} v_{R_1} &= i_{R_1} R_1 = i_L R_1 \\ &= I_m e^{-t/\tau'} R_1 \\ &= \frac{E}{R_1} R_1 e^{-t/\tau'} \end{aligned}$$

and

$$v_{R_1} = E e^{-t/\tau'} \tag{12.16}$$

The voltage $v_{R_1}$ has the same polarity as during the storage phase since the current $i_L$ has the same direction. The voltage $v_{R_2}$ is expressed as follows:

$$\begin{aligned} v_{R_2} &= i_{R_2} R_2 = i_L R_2 \\ &= I_m e^{-t/\tau'} R_2 \\ &= \frac{E}{R_1} R_2 e^{-t/\tau'} \end{aligned}$$

and

$$v_{R_2} = \frac{R_2}{R_1} E e^{-t/\tau'} \tag{12.17}$$

with the polarity indicated in Fig. 12.22.

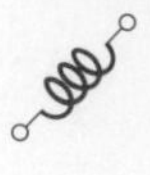

**EXAMPLE 12.5.** The resistor $R_2$ was added to the network of Fig. 12.19 as shown in Fig. 12.23.

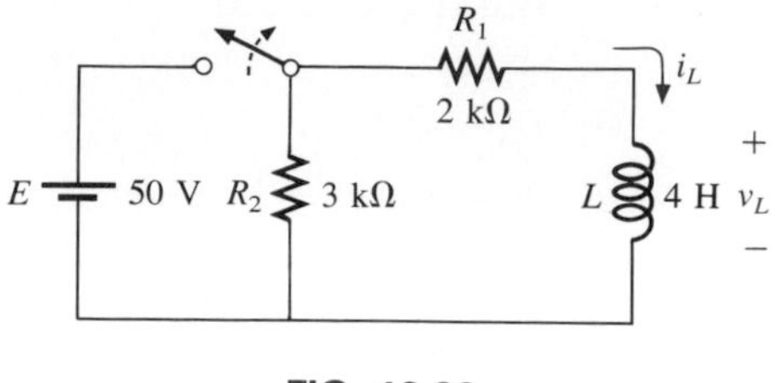


**FIG. 12.23**

a. Find the mathematical expressions for $i_L$, $v_L$, $v_{R_1}$, and $v_{R_2}$ after the storage phase has been completed and the switch is opened.
b. Sketch the waveforms for each voltage and current for both phases covered by this example and Example 12.4 if five time constants pass between phases. Use the defined polarities of Fig. 12.21.

***Solutions:***

a. $$\tau' = \frac{L}{R_1 + R_2} = \frac{4\text{ H}}{2\text{ k}\Omega + 3\text{ k}\Omega} = \frac{4\text{ H}}{5 \times 10^3\Omega} = 0.8 \times 10^{-3}\text{ s}$$
$$= 0.8\text{ ms}$$

By Eq. (12.14),

$$V_i = \left(1 + \frac{R_2}{R_1}\right)E = \left(1 + \frac{3\text{ k}\Omega}{2\text{ k}\Omega}\right)(50\text{ V}) = 125\text{ V}$$

and

$$v_L = V_i e^{-t/\tau'} = \mathbf{125}\boldsymbol{e}^{-t/(0.8\times10^{-3})}$$

By Eq. (12.15),

$$I_m = \frac{E}{R_1} = \frac{50\text{ V}}{2\text{ k}\Omega} = 25\text{ mA}$$

and

$$i_L = I_m e^{-t/\tau'} = \mathbf{(25 \times 10^{-3})}\boldsymbol{e}^{-t/(0.8\times10^{-3})}$$

By Eq. (12.16),

$$v_{R_1} = Ee^{-t/\tau'} = \mathbf{50}\boldsymbol{e}^{-t/(0.8\times10^{-3})}$$

By Eq. (12.17),

$$v_{R_2} = \frac{R_2}{R_1}Ee^{-t/\tau'} = \frac{3\text{ k}\Omega}{2\text{ k}\Omega}(50\text{ V})e^{-t/\tau'} = \mathbf{75}\boldsymbol{e}^{-t/(0.8\times10^{-3})}$$

b. See Fig. 12.24.

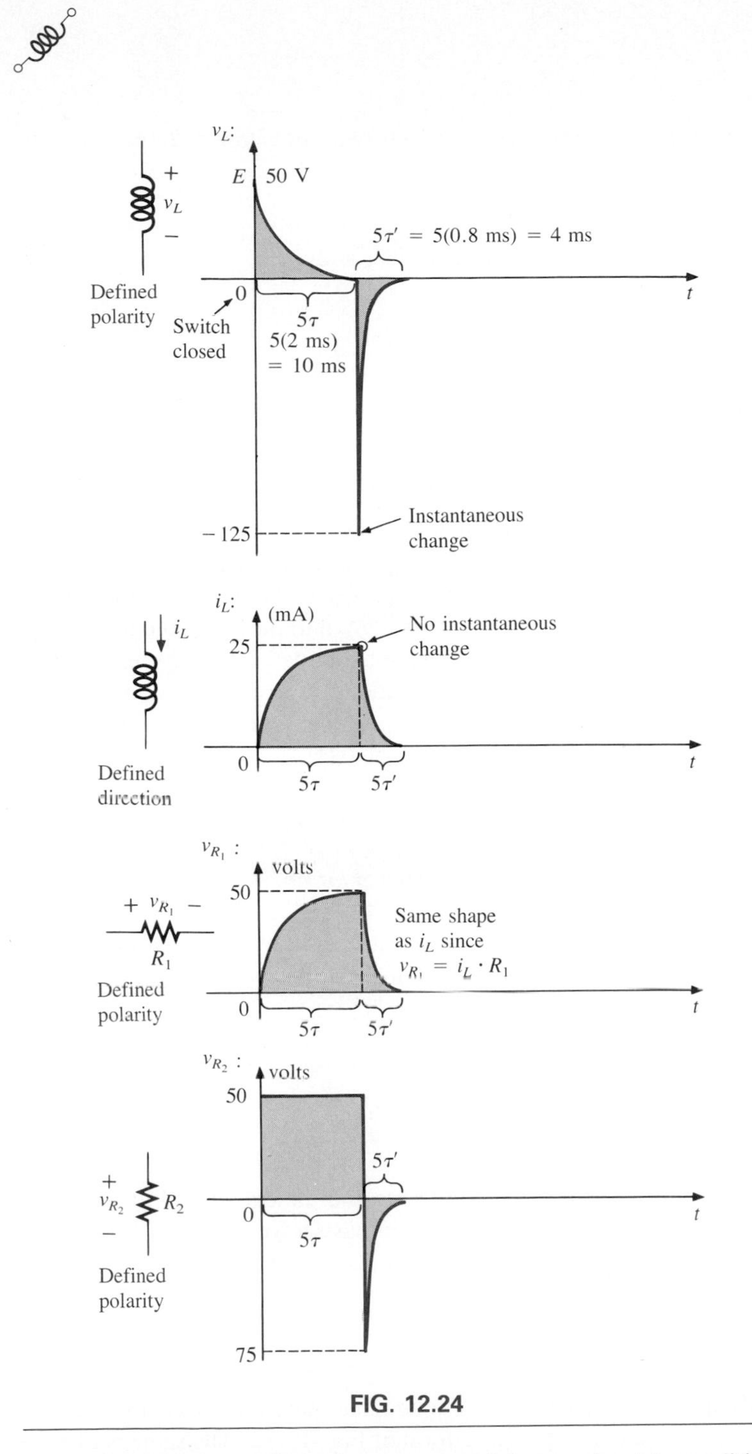


**FIG. 12.24**

In the preceding analysis, it was assumed that steady-state conditions were established during the charging phase and $I_m = E/R_1$, with $v_L = 0$ V. However, if the switch of Fig. 12.22 is opened before $i_L$ reaches its maximum value, the equation for the decaying current of Fig. 12.22 must change to

$$i_L = I_i e^{-t/\tau'} \tag{12.18}$$

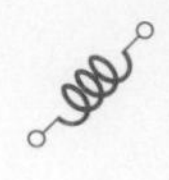

where $I_i$ is the starting or *i*nitial current. Equation (12.14) would be modified as follows:

$$v_L = V_i e^{-t/\tau'} \quad \textbf{(12.19)}$$

with

$$V_i = I_i(R_1 + R_2)$$

## 12.9 INSTANTANEOUS VALUES

The development presented in Section 10.9 for capacitive networks can also be applied to *R-L* networks to determine instantaneous voltages, currents, and time. The instantaneous values of any voltage or current can be determined by simply inserting $t$ into the equation and using a calculator or table to determine the magnitude of the exponential term.

The similarity between the equations $v_C = E(1 - e^{-t/\tau})$ and $i_L = I_m(1 - e^{-t/\tau})$ results in a derivation of the following for $t$ which is identical to that used to obtain Eq. (10.21):

$$t = -\tau \log_e\left(1 - \frac{i_L}{I_m}\right) \quad \textbf{(12.20)}$$

For the other form, the equation $v_C = Ee^{-t/\tau}$ is a close match with $v_L = Ee^{-t/\tau}$, permitting a derivation similar to that employed for Eq. (10.22):

$$t = -\tau \log_e \frac{v_L}{E} \quad \textbf{(12.21)}$$

The similarities between the above and the equations in Chapter 10 should make the equation for $t$ fairly easy to obtain.

## 12.10 $\tau = L/R_{Th}$

In Chapter 10 (Capacitors), we found that there are occasions when the circuit does not have the basic form of Fig. 12.12. The same is true for inductive networks. Again, it is necessary to find the Thevenin equivalent circuit before proceeding in the manner described in this chapter. Consider the following example.

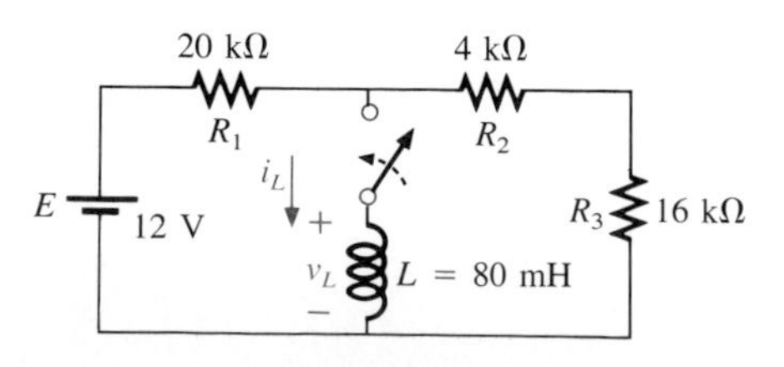


FIG. 12.25

**EXAMPLE 12.6.** For the network of Fig. 12.25:

a. Find the mathematical expression for the transient behavior of the current $i_L$ and the voltage $v_L$ after the closing of the switch.
b. Draw the resultant waveform for each.

***Solutions:***

a. Applying Thevenin's theorem to the 80-mH inductor (Fig. 12.26) yields

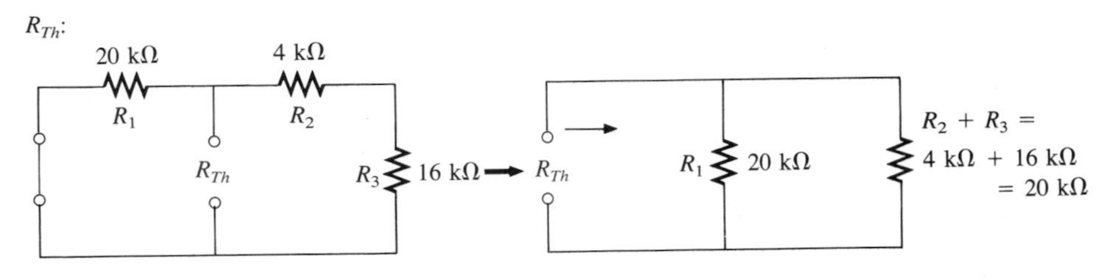

FIG. 12.26

$$R_{Th} = \frac{R}{N} = \frac{20\ \text{k}\Omega}{2} = 10\ \text{k}\Omega$$

Applying the voltage divider rule (Fig. 12.27),

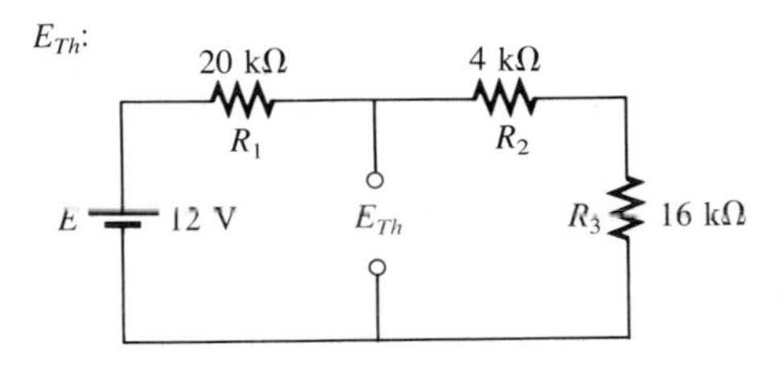

FIG. 12.27

$$E_{Th} = \frac{(R_2 + R_3)E}{R_1 + R_2 + R_3}$$

$$= \frac{(4\ \text{k}\Omega + 16\ \text{k}\Omega)(12\ \text{V})}{20\ \text{k}\Omega + 4\ \text{k}\Omega + 16\ \text{k}\Omega} = \frac{(20\ \text{k}\Omega)(12\ \text{V})}{40\ \text{k}\Omega} = 6\ \text{V}$$

The Thevenin equivalent circuit is shown in Fig. 12.28. Using Eq. (12.8),

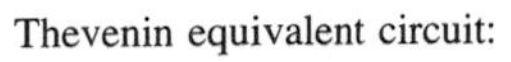
Thevenin equivalent circuit:

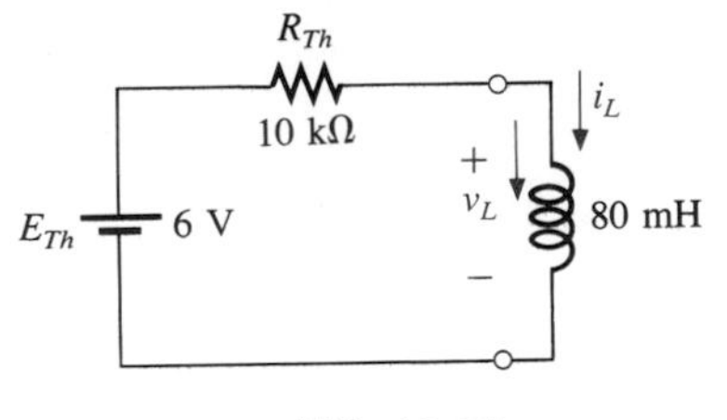


FIG. 12.28

$$i_L = \frac{E_{Th}}{R}(1 - e^{-t/\tau})$$

$$\tau = \frac{L}{R_{Th}} = \frac{80 \times 10^{-3}\ \text{H}}{10 \times 10^3\ \Omega} = 8 \times 10^{-6}\ \text{s}$$

$$I_m = \frac{E_{Th}}{R_{Th}} = \frac{6\ \text{V}}{10 \times 10^3\ \Omega} = 0.6 \times 10^{-3}\ \text{A}$$

and

$$\boldsymbol{i_L = (0.6 \times 10^{-3})(1 - e^{-t/(8\times10^{-6})})}$$

Using Eq. (12.10),

$$v_L = E_{Th}e^{-t/\tau}$$

so that

$$\boldsymbol{v_L = 6e^{-t/(8\times10^{-6})}}$$

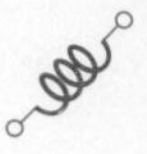

b. See Fig. 12.29.

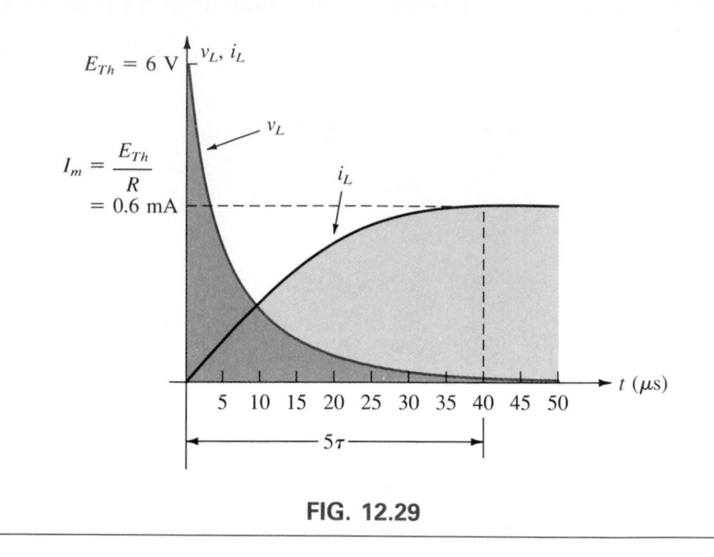


FIG. 12.29

## 12.11 INDUCTORS IN SERIES AND PARALLEL

Inductors, like resistors and capacitors, can be placed in series or parallel. Increasing levels of inductance can be obtained by placing inductors in series, while decreasing levels can be obtained by placing inductors in parallel.

For inductors in series, the total inductance is found in the same manner as the total resistance of resistors in series (Fig. 12.30):

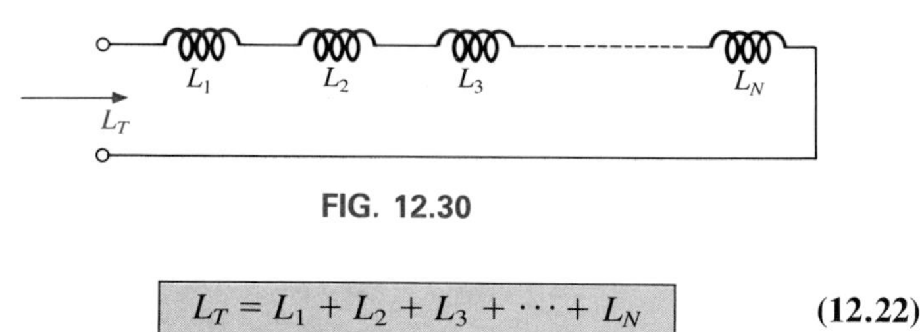


FIG. 12.30

$$L_T = L_1 + L_2 + L_3 + \cdots + L_N \quad (12.22)$$

For inductors in parallel, the total inductance is found in the same manner as the total resistance of resistors in parallel (Fig. 12.31):

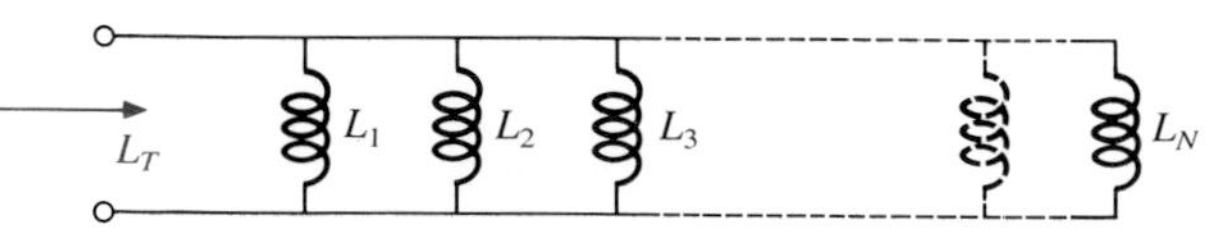


FIG. 12.31

$$\frac{1}{L_T} = \frac{1}{L_1} + \frac{1}{L_2} + \frac{1}{L_3} + \cdots + \frac{1}{L_N} \quad (12.23)$$

For two inductors in parallel,

$$L_T = \frac{L_1 L_2}{L_1 + L_2} \qquad (12.24)$$

## 12.12 R-L AND R-L-C CIRCUITS WITH dc INPUTS

We found in Section 12.7 that for all practical purposes, an inductor can be replaced by a short circuit in a dc circuit after a period of time greater than five time constants has passed. If in the following circuits we assume that all of the currents and voltages have reached their final values, the current through each inductor can be found by replacing each inductor by a short circuit. For the circuit of Fig. 12.32, for example,

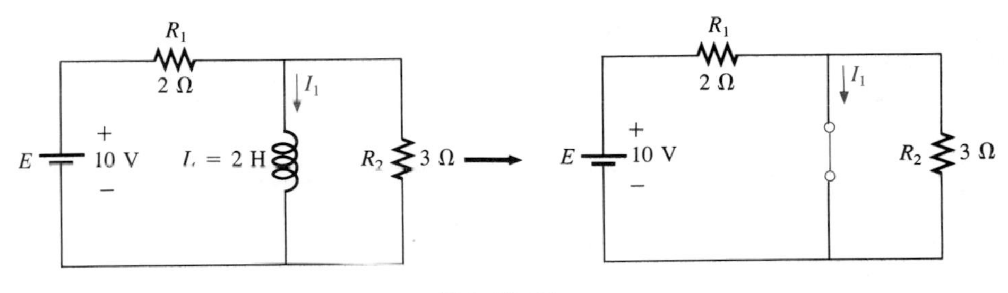


FIG. 12.32

$$I_1 = \frac{E}{R_1} = \frac{10}{2} = \mathbf{5\ A}$$

For the circuit of Fig. 12.33,

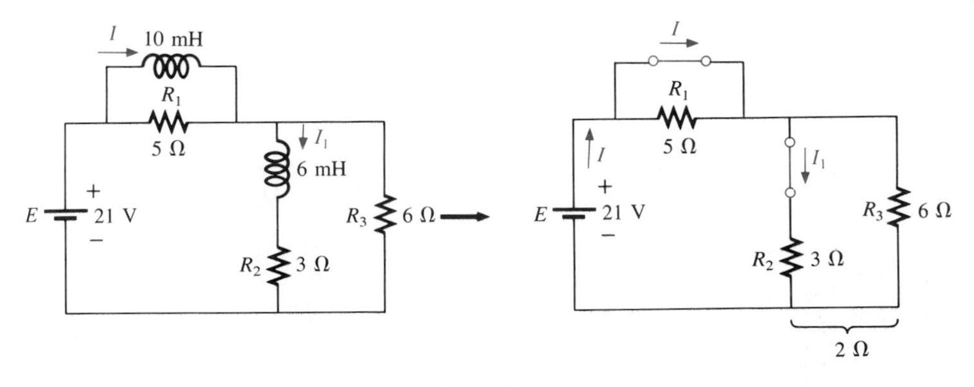


FIG. 12.33

$$I = \frac{E}{R_2 \parallel R_3} = \frac{21\ \text{V}}{2\ \Omega} = \mathbf{10.5\ A}$$

Applying the current divider rule,

$$I_1 = \frac{R_3 I}{R_3 + R_2} = \frac{(6\ \Omega)(10.5\ \text{A})}{6\ \Omega + 3\ \Omega} = \frac{63\ \text{A}}{9} = \mathbf{7\ A}$$

In the following examples we will assume that the voltage across the capacitors and the current through the inductors have reached their final values. Under these conditions, the inductors can be replaced by short circuits, and the capacitors by open circuits.

---

**EXAMPLE 12.7.** Find the current $I_L$ and the voltage $V_C$ for the network of Fig. 12.34.

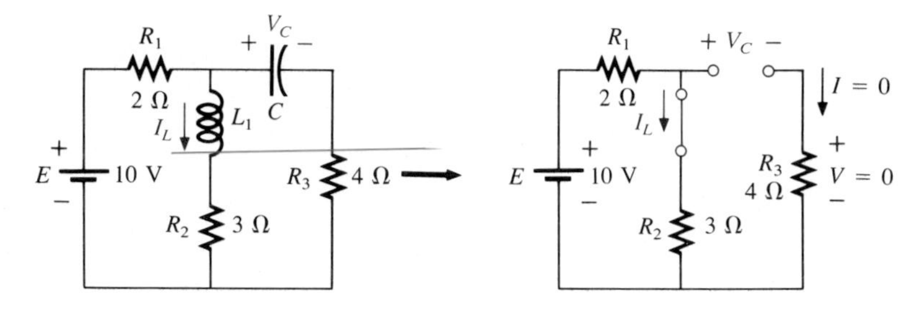


FIG. 12.34

***Solution:***

$$I_L = \frac{E}{R_1 + R_2} = \frac{10\ \text{V}}{5\ \Omega} = \mathbf{2\ A}$$

$$V_C = \frac{R_2 E}{R_2 + R_1} = \frac{(3\ \Omega)(10\ \text{V})}{3\ \Omega + 2\ \Omega} = \mathbf{6\ V}$$

---

**EXAMPLE 12.8.** Find the currents $I_1$ and $I_2$ and the voltages $V_1$ and $V_2$ for the network of Fig. 12.35.

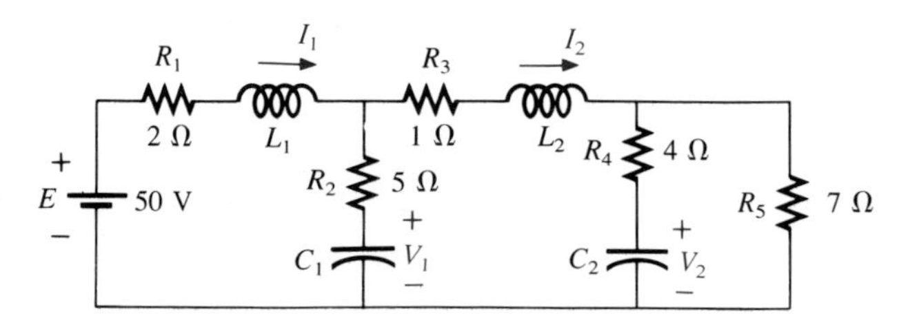


FIG. 12.35

***Solution:*** Note Fig. 12.36:

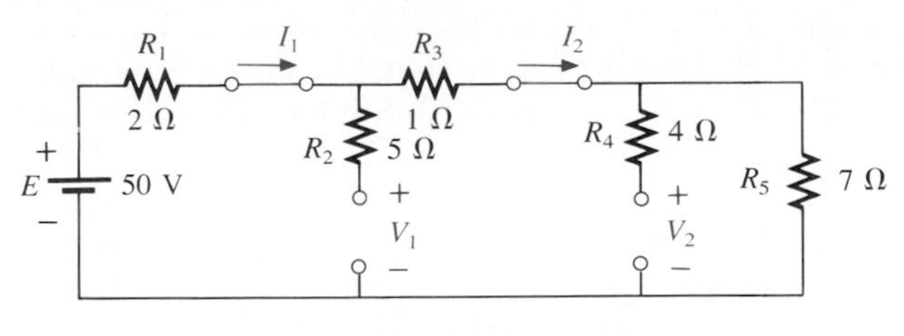


FIG. 12.36

$$I_1 = I_2$$

$$I_1 = \frac{E}{R_1 + R_3 + R_5} = \frac{50\text{ V}}{2\ \Omega + 1\ \Omega + 7\ \Omega} = \frac{50\text{ V}}{10\ \Omega} = \mathbf{5\text{ A}}$$

$$V_2 = I_2R_5 = (5\text{ A})(7\ \Omega) = \mathbf{35\text{ V}}$$

Applying the voltage divider rule,

$$V_1 = \frac{(R_3 + R_5)E}{R_1 + R_3 + R_5} = \frac{(1\ \Omega + 7\ \Omega)(50\text{ V})}{2\ \Omega + 1\ \Omega + 7\ \Omega} = \frac{(8\ \Omega)(50\text{ V})}{10\ \Omega} = \mathbf{40\text{ V}}$$

## 12.13 ENERGY STORED BY AN INDUCTOR

The ideal inductor, like the ideal capacitor, does not dissipate the electrical energy supplied to it. It stores the energy in the form of a magnetic field. A plot of the voltage, current, and power to an inductor is shown in Fig. 12.37 during the buildup of the magnetic field surrounding the

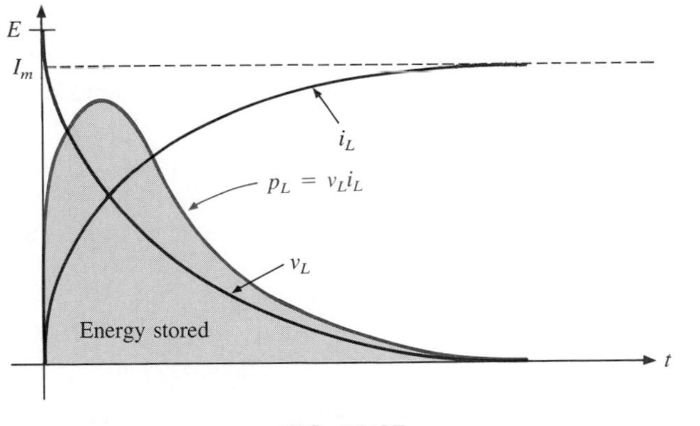


FIG. 12.37

inductor. The energy stored is represented by the shaded area under the power curve. Using calculus, we can show that the evaluation of the area under the curve yields

$$W_{\text{stored}} = \frac{1}{2}LI_m^2 \qquad \text{(joules, J)} \qquad \textbf{(12.25)}$$

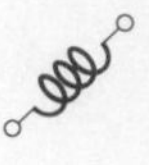

**EXAMPLE 12.9.** Find the energy stored by the inductor in the circuit of Fig. 12.38 when the current through it has reached its final value.

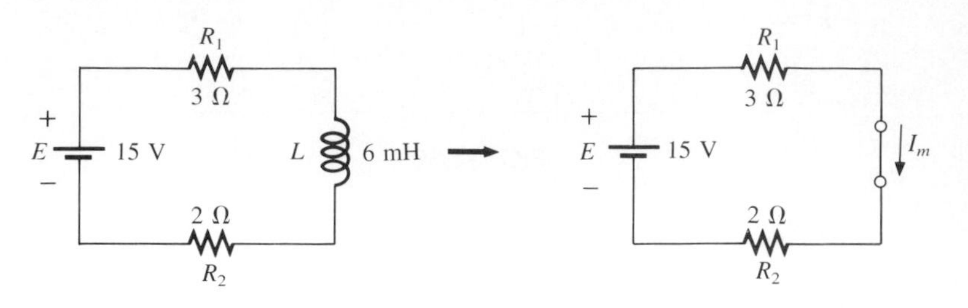


FIG. 12.38

***Solution:***

$$I_m = \frac{E}{R_1 + R_2} = \frac{15\text{ V}}{3\ \Omega + 2\ \Omega} = \frac{15\text{ V}}{5\ \Omega} = 3\text{ A}$$

$$W_{\text{stored}} = \frac{1}{2}LI_m^2 = \frac{1}{2}(6 \times 10^{-3}\text{ H})(3\text{ A})^2 = \frac{54}{2} \times 10^{-3}\text{ J}$$

$$= \mathbf{27\ mJ}$$

## 12.14 COMPUTER ANALYSIS

Both PSPICE and BASIC can provide the transient response for an *R-L* circuit. In BASIC, the appropriate equations are employed to determine the voltages and currents as they change with time. A table of values can then be generated or a plot obtained using a plotting routine.

For PSPICE, the .TRAN and PULSE commands introduced for capacitive networks will be employed to generate the desired waveforms.

### PSPICE

Inductors are entered in much the same manner as resistors and capacitors as demonstrated by the format below:

| LTOROID | 3 | 4 | 5M | IC = 2M |
|---|---|---|---|---|
| | + | − | | |
| Name | Node | Node | Value | Initial value |

The user fills in those quantities with brackets underneath, although the initial condition entry (the current through the inductor before the switching action occurs) can be omitted if the initial value is zero amperes. The above entry is for a 5-mH coil between nodes 3 and 4 with node 3 as the defined higher potential and an initial current of 2 mA. In PSPICE, inductors have the additional limitation that they cannot form a closed loop (like parallel inductors). However, this limitation can be circumvented by placing a small resistor (negligible compared to the parameters of the network) in series with one of the inductors (as in the example to follow).

The network to by analyzed appears in Fig. 12.39 with the input file in Fig. 12.40. Since two parallel inductors form a closed loop a 1-mΩ

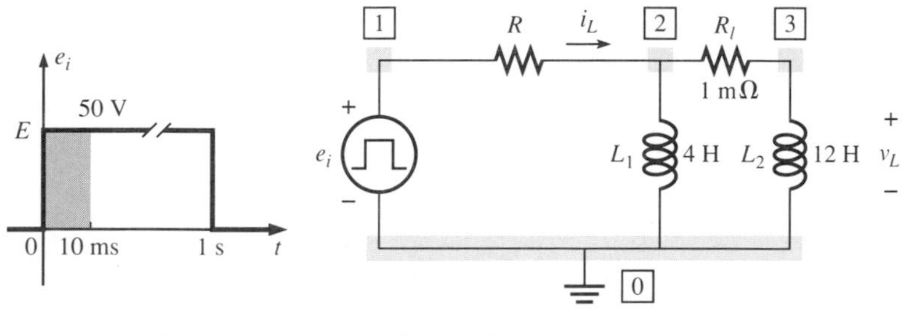


**FIG. 12.39**

```
************************ Evaluation PSpice (January 1989) ******* 20:31:54 *******

 Chapter 12 - R-L Circuit Transient Analysis

 ****      CIRCUIT DESCRIPTION

*******************************************************************************

VE 1 0 PULSE(0 50 0 1N 1N 1)
R  1 2 2K
RL 2 3 1M
L1 2 0 4H
L2 3 0 12H
.TRAN 0.5M 10M
.PROBE
.OPTIONS NOPAGE
.END
```

**FIG. 12.40**

resistor was placed in series with one of the coils before the nodes were defined. The input pulse is as defined for *R-C* circuits to simulate the closing of a switch at $t = 0$ s to establish 50 V across the network. The parallel combination of the inductors is $4\text{ H} \parallel 12\text{ H} = 3\text{ H}$ and $\tau = L/R = 3\text{ H}/2\text{ k}\Omega = 1.5\text{ ms}$. The .TRAN command is therefore established from 0.5 ms to 10 ms to provide at least three data points in each

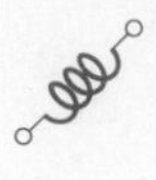

time constant interval. The .PROBE command then permits a request for V(3), which is $v_L$, and I(R), which is the total current $i_L$ through the parallel coils, as shown in the output file of Fig. 12.41. Note how $v_L$ approaches zero volts after $5\tau = 7.5$ ms and $i_L$ approaches its final value of $E/R = 50\ \text{V}/2\ \text{k}\Omega = 25$ mA in the same time interval. It is satisfying to watch the PSPICE and .PROBE combination present the excellent results of Fig. 12.41 with a minimum of effort on the part of the user.

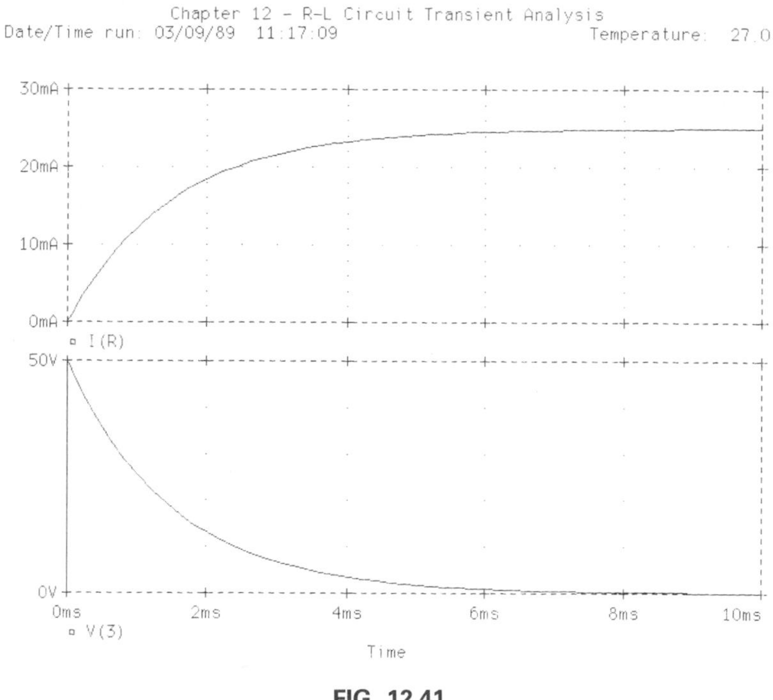


**FIG. 12.41**

## PROBLEMS

### SECTION 12.2 Faraday's Law of Electromagnetic Induction

1. If the flux linking a coil of 50 turns changes at a rate of 0.085 Wb/s, what is the induced voltage across the coil?
2. Determine the rate of change of flux linking a coil if 20 V are induced across a coil of 40 turns.
3. How many turns does a coil have if 42 mV are induced across the coil by a change of flux of 0.003 Wb/s?

## SECTION 12.4 Self-Inductance

4. Find the inductance $L$ in henries of the inductor of Fig. 12.42.
5. Repeat Problem 4 with $l = 4$ in. and $d = 0.25$ in.

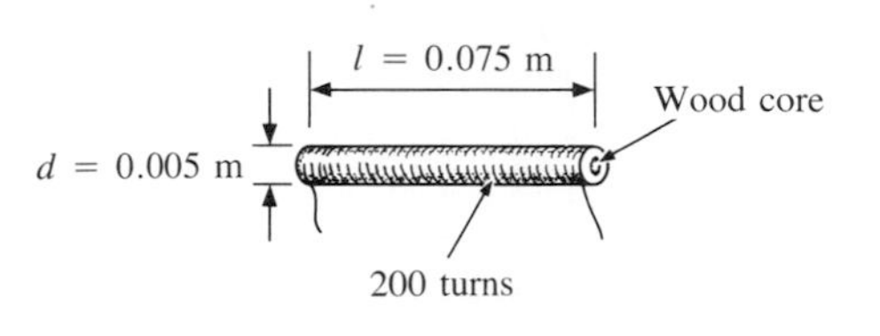


FIG. 12.42

6. a. Find the inductance $L$ in henries of the inductor of Fig. 12.43.
   b. Repeat part (a) if a ferromagnetic core is added having a $\mu_r$ of 2000.

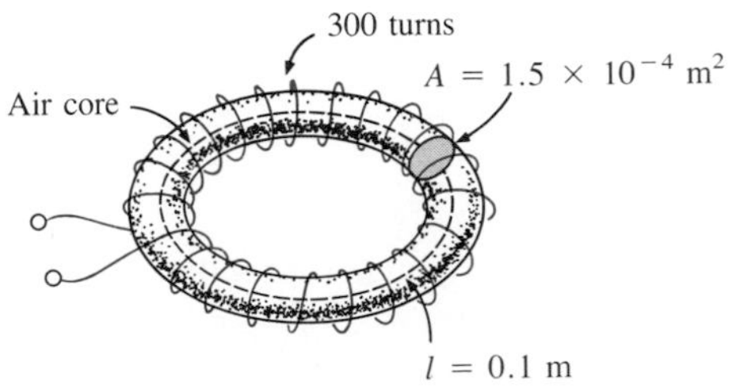

300 turns

$A = 1.5 \times 10^{-4}\ \text{m}^2$

Air core

$l = 0.1$ m

FIG. 12.43

## SECTION 12.6 Induced Voltage

7. Find the voltage induced across a coil of 5 H if the rate of change of current through the coil is
   a. 0.5 A/s
   b. 60 mA/s
   c. 0.04 A/ms
8. Find the induced voltage across a 50-mH inductor if the current through the coil changes at a rate of 0.1 mA/$\mu$s.
9. Find the waveform for the voltage induced across a 200-mH coil if the current through the coil is as shown in Fig. 12.44.

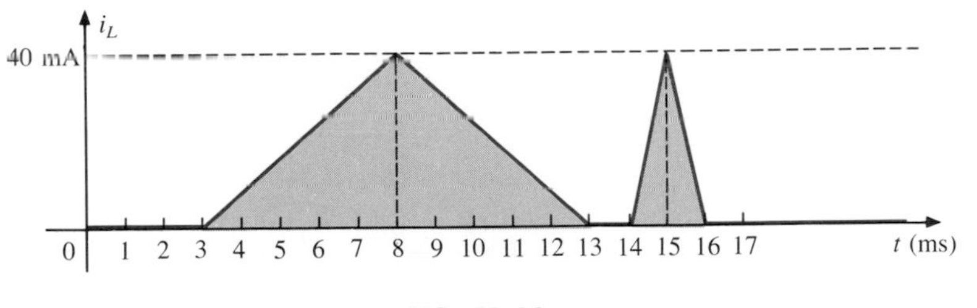


FIG. 12.44

10. Repeat Problem 9 for the waveform of Fig. 12.45.

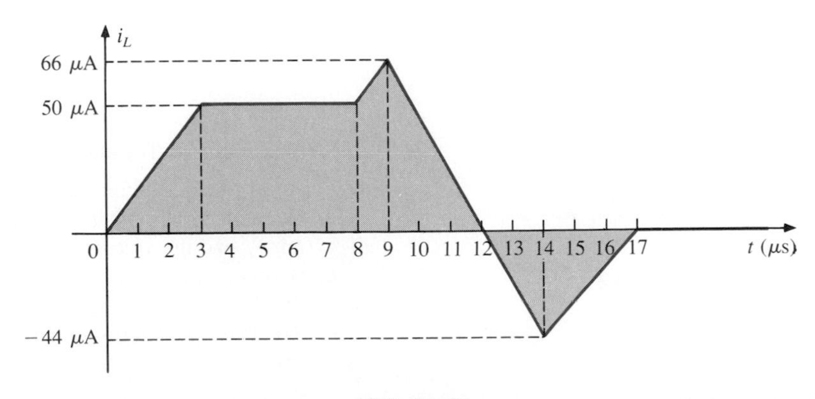


FIG. 12.45

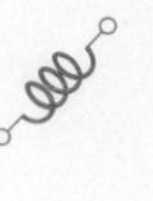

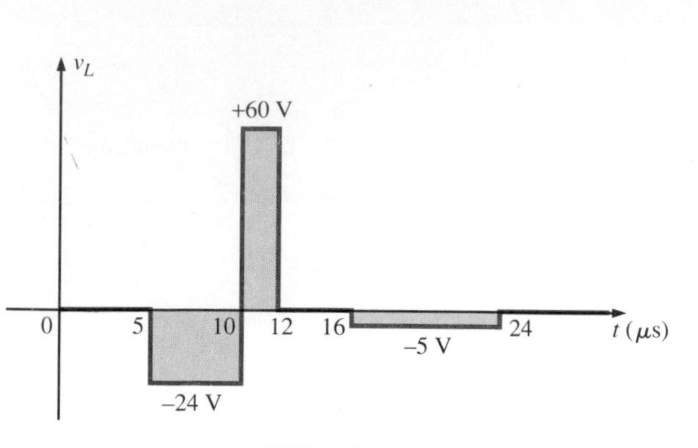

FIG. 12.46

*11. Find the waveform for the current of a 10-mH coil if the voltage across the coil follows the pattern of Fig. 12.46. The current $i_L$ is 4 mA at $t = 0$ s.

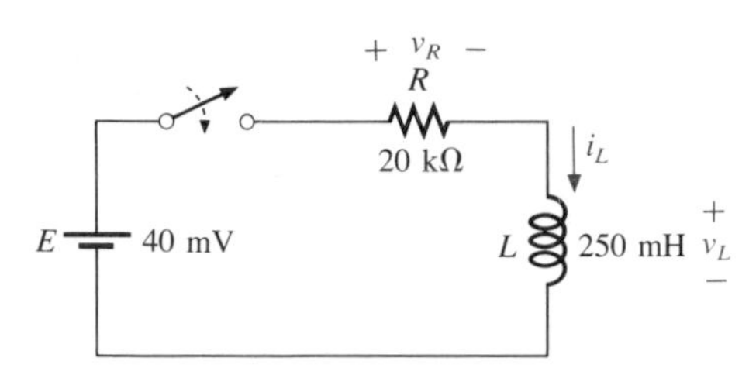


FIG. 12.47

## SECTION 12.7 *R-L* Transients: Storage Cycle

**12.** For the circuit of Fig. 12.47:
- **a.** Determine the time constant.
- **b.** Write the mathematical expression for the current $i_L$ after the switch is closed.
- **c.** Repeat part (b) for $v_L$ and $v_R$.
- **d.** Determine $i_L$ and $v_L$ at one, three, and five time constants.
- **e.** Sketch the waveforms of $i_L$, $v_L$, and $v_R$.

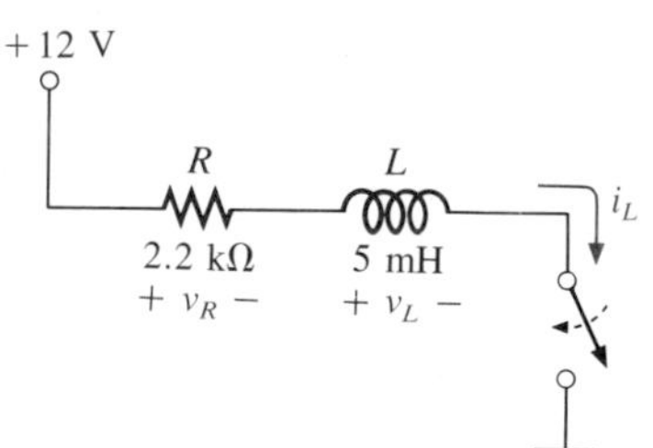


FIG. 12.48

**13.** Repeat Problem 12 for the network of Fig. 12.48.

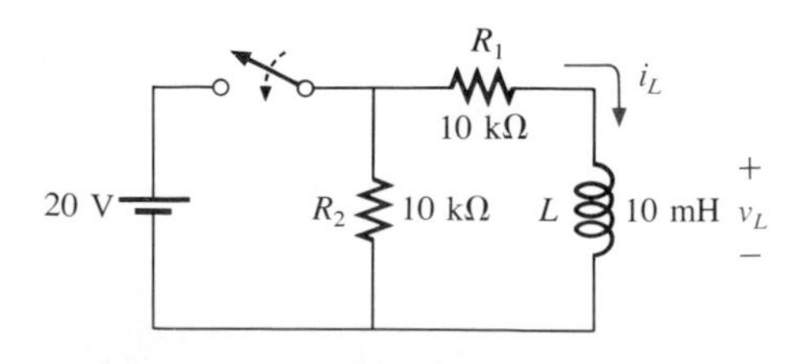


FIG. 12.49

## SECTION 12.8 *R-L* Transients: Decay Phase

**14.** For the network of Fig. 12.49:
- **a.** Determine the mathematical expressions for the current $i_L$ and the voltage $v_L$ when the switch is closed.
- **b.** Repeat part (a) if the switch is opened after a period of five time constants has passed.
- **c.** Sketch the waveforms of parts (a) and (b) on the same axis.

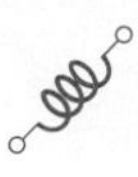



*15. **a.** Repeat Problem 14 for the network of Fig. 12.50.

**b.** Sketch the waveform for the voltage across $R_2$ for the same period of time encompassed by $i_L$ and $v_L$. Take careful note of the defined polarities and directions.

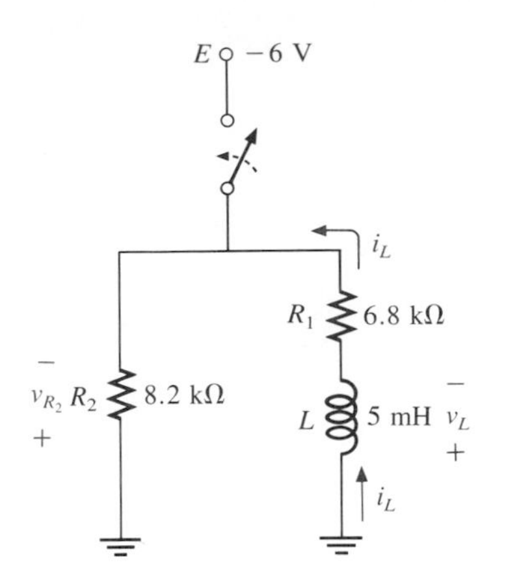

FIG. 12.50

*16. For the network of Fig. 12.51:

**a.** Determine the mathematical expressions for the current $i_L$ and the voltage $v_L$ following the closing of the switch.

**b.** Repeat part (a) if the switch is opened at $t = 1\ \mu s$.

**c.** Sketch the waveforms of parts (a) and (b) on the same axis.

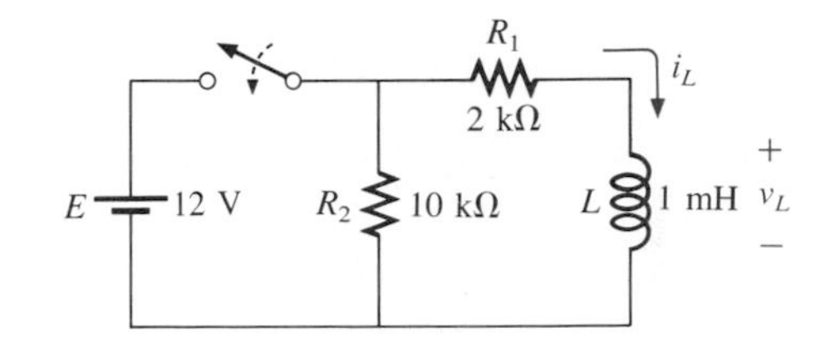


FIG. 12.51

## SECTIONS 12.9 AND 12.10 Instantaneous Values; $\tau = L/R_{Th}$

17. **a.** Determine the mathematical expressions for $i_L$ and $v_L$ following the closing of the switch in Fig. 12.52.

**b.** Determine $i_L$ and $v_L$ at $t = 100\ \mu s$.

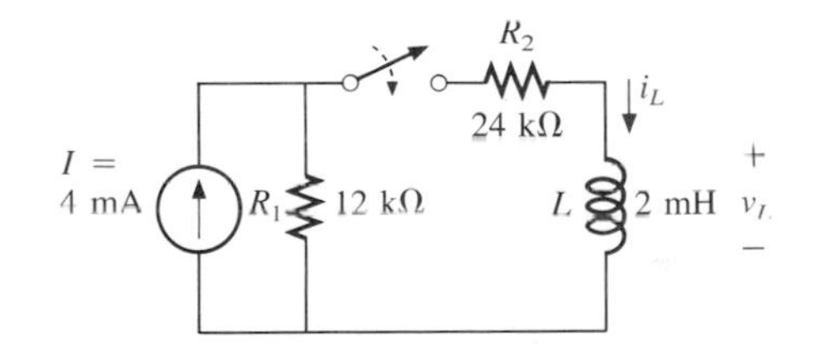

FIG. 12.52

*18. **a.** Determine the mathematical expressions for $i_L$ and $v_L$ following the closing of the switch in Fig. 12.53.

**b.** Calculate $i_L$ and $v_L$ at $t = 10\ \mu s$.

**c.** Write the mathematical expressions for the current $i_L$ and the voltage $v_L$ if the switch is opened at $t = 10\ \mu s$.

**d.** Sketch the waveforms of $i_L$ and $v_L$ for parts (a) and (c).

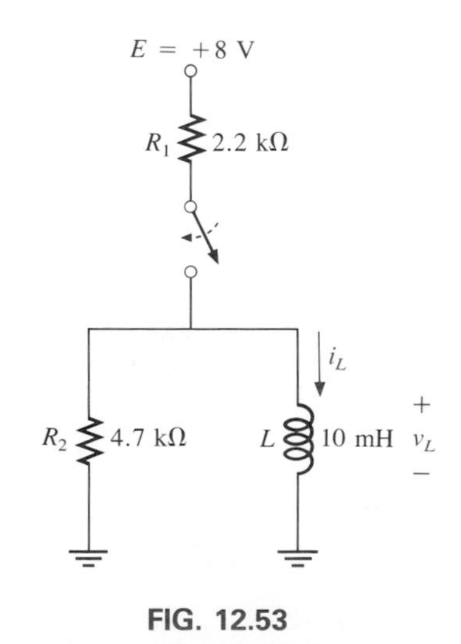


FIG. 12.53

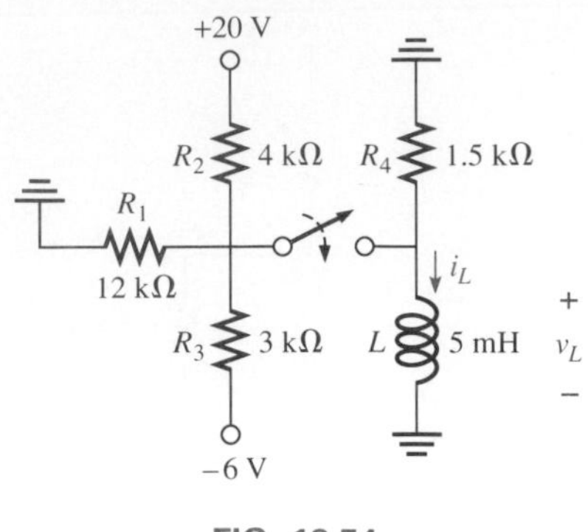


FIG. 12.54

*19. **a.** Determine the mathematical expressions for $i_L$ and $v_L$ following the closing of the switch in Fig. 12.54.
**b.** Determine $i_L$ and $v_L$ after two time constants of the storage phase.
**c.** Write the mathematical expressions for the current $i_L$ and the voltage $v_L$ if the switch is opened at the instant defined by part (b).
**d.** Sketch the waveforms of $i_L$ and $v_L$ for parts (a) and (c).

### SECTION 12.11 Inductors in Series and Parallel

**20.** Find the total inductance of the circuits of Fig. 12.55.

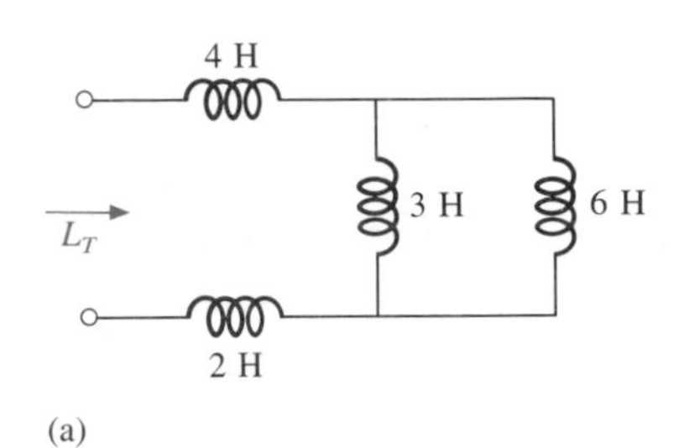


3.6 H
$L_T$
6 H
4 H
12 H
(b)

FIG. 12.55

**21.** Reduce the networks of Fig. 12.56 to the fewest elements.

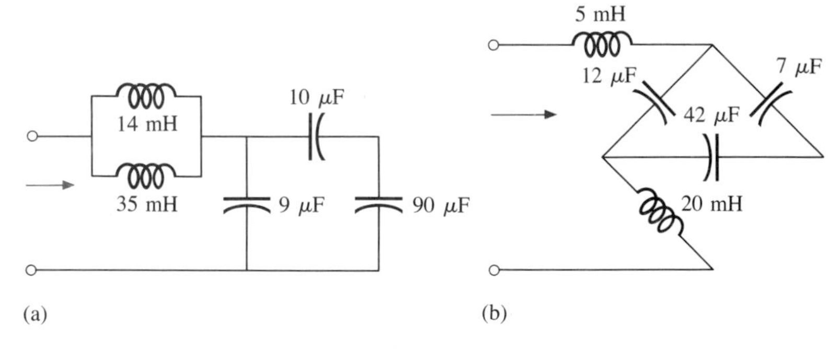

FIG. 12.56

20 V
$R_1$
5 kΩ
$R_2$ 20 kΩ
+
$v_L$
−
$L_1$
5 H
$i_L$
$L_2$ 6 H
$L_3$ 30 H
+
$v_{L_3}$
−

FIG. 12.57

*22. For the network of Fig. 12.57:
**a.** Find the mathematical expressions for the voltage $v_L$ and $i_L$ following the closing of the switch.
**b.** Sketch the waveforms of $v_L$ and $i_L$ obtained in part (a).
**c.** Determine the mathematical expression for the voltage $v_{L_3}$ following the closing of the switch and sketch the waveform.

## SECTION 12.12 *R-L* and *R-L-C* Circuits with dc Inputs

For Problems 23 through 25, assume that the voltage across each capacitor and the current through each inductor have reached their final values.

**23.** Find the voltages $V_1$ and $V_2$ and the current $I_1$ for the circuit of Fig. 12.58.

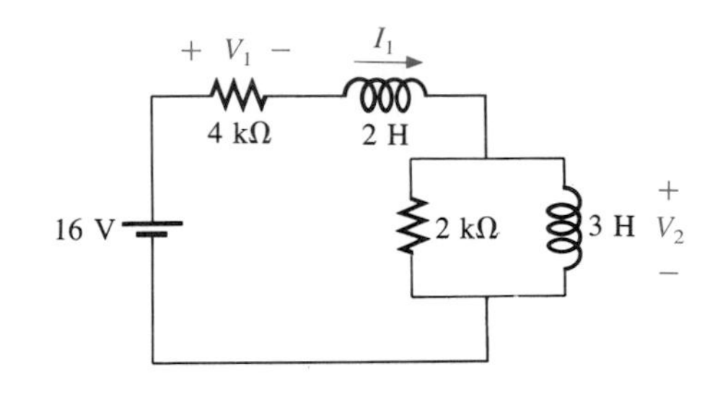


FIG. 12.58

**24.** Find the current $I_1$ and the voltage $V_1$ for the circuit of Fig. 12.59.

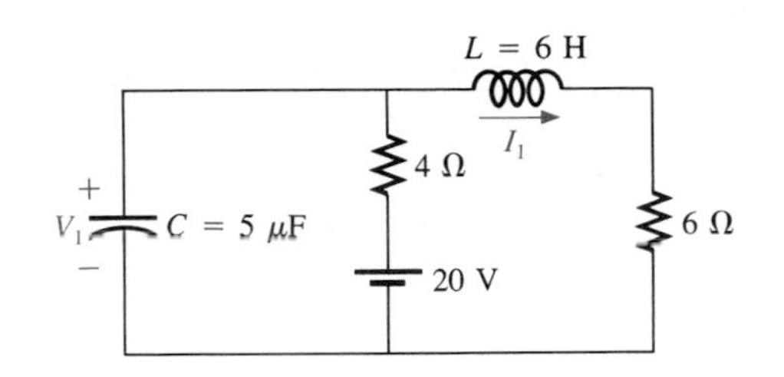

FIG. 12.59

**25.** Find the voltage $V_1$ and the current through each inductor in the circuit of Fig. 12.60.

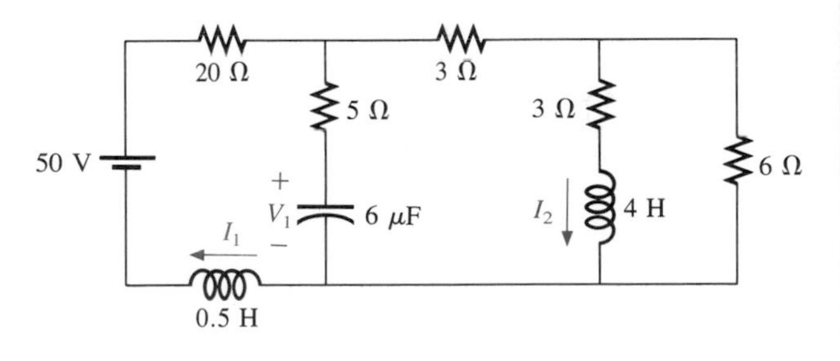


FIG. 12.60

## SECTION 12.13 Energy Stored by an Inductor

**26.** Find the energy stored in each inductor of Problem 23.

**27.** Find the energy stored in the capacitor and inductor of Problem 24.

**28.** Find the energy stored in each inductor of Problem 25.

## SECTION 12.14 Computer Analysis

### PSPICE

**29.** Write the input file to obtain a plot of $v_L$ and $i_L$ for the network of Fig. 12.52 following the closing of the switch.

***30.** Write the input file to obtain a plot of $v_L$ and $i_L$ for the network of Fig. 12.54 following the closing of the switch.

***31.** Write the input file to obtain a plot of $v_L$, $i_L$, and $v_{L_3}$ for the network of Fig. 12.57 following the closing of the switch.

### BASIC

**32.** Write a program to provide a general solution for the circuit of Fig. 12.12. That is, given the network parameters, generate the equations for $i_L$, $v_L$, and $v_R$.

33. Write a program that will provide a general solution for the storage and decay phase of the network of Fig. 12.49. That is, given the network values, generate the equations for $i_L$ and $v_L$ for each phase. In this case, assume the storage phase has passed through five time constants before the decay phase begins.
34. Repeat Problem 33 but assume the storage phase was not completed, requiring that the instantaneous values of $i_L$ and $v_L$ be determined when the switch is opened.

## GLOSSARY

**Choke** A term often applied to an inductor, due to the ability of an inductor to resist a change in current through it.

**Faraday's law** A law relating the voltage induced across a coil to the number of turns in the coil and the rate at which the flux linking the coil is changing.

**Inductor** A fundamental element of electrical systems constructed of numerous turns of wire around a ferromagnetic or air core.

**Lenz's law** A law stating that an induced effect is always such as to oppose the cause that produced it.

**Self-inductance** A measure of the ability of a coil to oppose any change in current through the coil and to store energy in the form of a magnetic field in the region surrounding the coil.

# 13

# Sinusoidal Alternating Waveforms

## 13.1 INTRODUCTION

The analysis thus far has been limited to dc networks, networks in which the currents or voltages are fixed in magnitude except for transient effects. We will now turn our attention to the analysis of networks in which the magnitude of the source varies in a set manner. Of particular interest is the time-varying voltage that is commercially available in large quantities and is commonly called the *ac voltage*. (The letters *ac* are an abbreviation for *alternating current*.) To be absolutely rigorous, the terminology *ac voltage* or *ac current* is not sufficient to describe the type of signal we will be analyzing. Each waveform of Fig. 13.1 is an

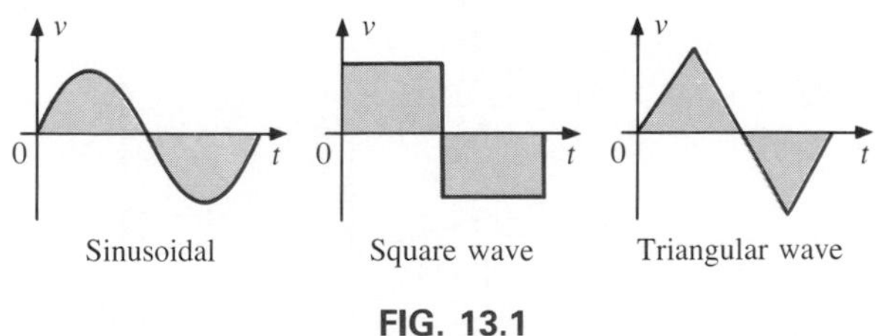


**FIG. 13.1**
*Alternating waveforms.*

alternating waveform available from commercial supplies. The term *alternating* indicates only that the waveform alternates between two prescribed levels in a set time sequence (Fig. 13.1). To be absolutely correct, the term *sinusoidal, square wave,* or *triangular* must also be applied. The pattern of particular interest is the *sinusoidal* ac voltage of Fig. 13.1. Since this type of signal is encountered in the vast majority of instances, the abbreviated phrases *ac voltage* and *ac current* are commonly applied without confusion. For the other patterns of Fig. 13.1, the descriptive term is always present, but frequently the *ac* abbreviation is dropped, resulting in the designation *square-wave* or *triangular* waveforms.

One of the important reasons for concentrating on the sinusoidal ac voltage is that it is the voltage generated by utilities throughout the world. Other reasons include its application throughout electrical, electronic, communication, and industrial systems. In addition, the chapters to follow will reveal that the waveform itself has a number of characteristics that will result in a unique response when it is applied to the basic electrical elements. The wide range of theorems and methods introduced for dc networks will also be applied to sinusoidal ac systems. Although the application of sinusoidal signals will raise the required math level, once the notation given in Chapter 14 is understood, most of the concepts introduced in the dc chapters can be applied to ac networks with a minimum of added difficulty.

The increasing number of computer systems used in the industrial community requires, at the very least, a brief introduction to the terminology employed with pulse waveforms and the response of some fundamental configurations to the application of such signals. Chapter 22 will serve such a purpose.

## 13.2 SINUSOIDAL ac VOLTAGE GENERATION

The characteristics of the sinusoidal voltage and current and their effect on the basic $R$, $L$, and $C$ elements will be described in some detail in this chapter and those to follow. Of immediate interest is the generation of sinusoidal voltages.

The terminology *ac generator* or *alternator* should not be new to most technically oriented students. It is an electromechanical device capable of converting mechanical power to electrical power. As shown in the very basic ac generator of Fig. 13.2, it is constructed of two main

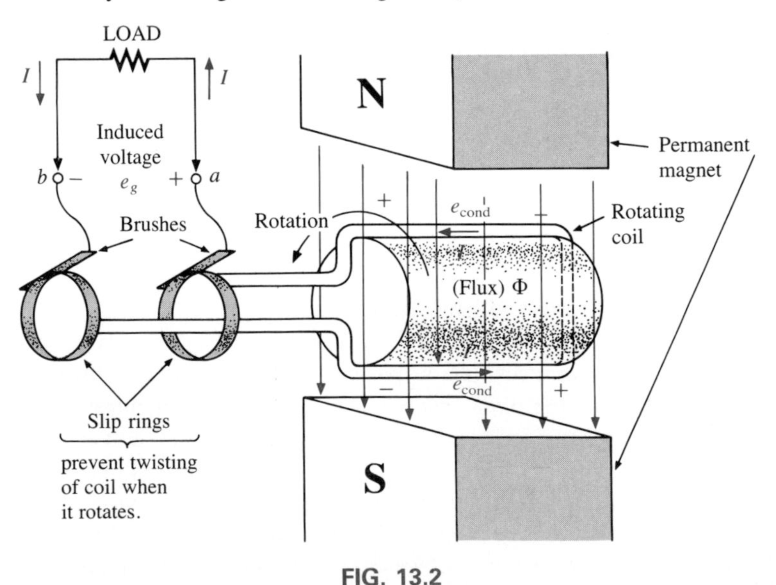


FIG. 13.2

components: the *rotor* (or armature, in this case) and the *stator*. As implied by the terminology, the rotor rotates within the framework of the stator, which is stationary. When the rotor is caused to rotate by some mechanical power such as is available from the forces of rushing water (dams) or steam-turbine engines, the conductors on the rotor will cut magnetic lines of force established by the poles of the stator, as shown in Fig. 13.2. The poles may be those of a permanent magnet or may result from the turns of wire around the ferromagnetic core of the pole through which a dc current is passed to establish the necessary magnetomotive force for the required flux density.

Dictated by Eq. (12.1), the length of conductor passing through the magnetic field will have a voltage induced across it as shown in Fig. 13.2. Note that the induced voltage across each conductor is additive, so the generated terminal voltage is the sum of the two induced voltages. Since the armature of Fig. 13.2 is rotating, and the output terminals $a$ and $b$ are connected to some fixed external load, there is the necessity for the *slip rings*. The slip rings are circular conducting surfaces that provide a path of conduction from the generated voltage to the load and prevent a twisting of the coil at $a$ and $b$ when the coil rotates. The induced voltage will have a polarity at terminals $a$ and $b$ and will develop a current $I$ having the direction indicated in Fig. 13.2. Note that the direction of $I$ is also the direction of increasing induced voltage within the generator.

A method will now be described for determining the direction of the resulting current or of the increasing induced voltage. For the generator, the thumb, forefinger, and middle finger are placed at right angles, as shown in Fig. 13.3. The thumb is placed in the direction of force or

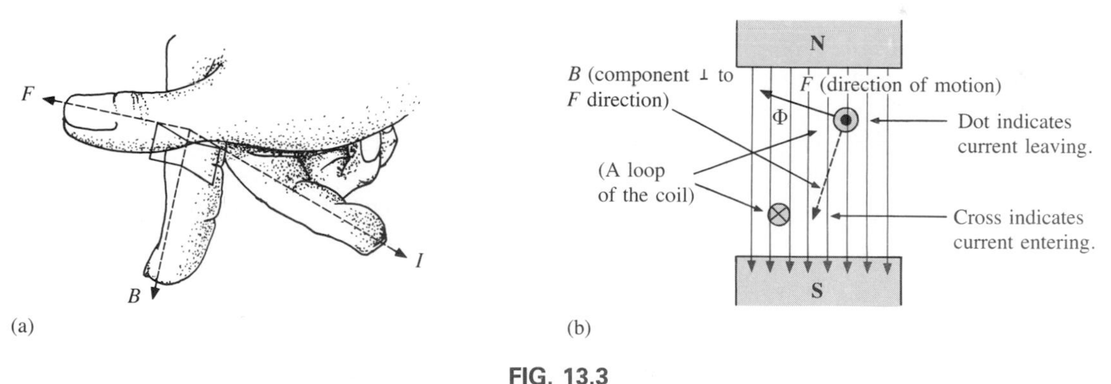


**FIG. 13.3**
*(a) Right-hand rule; (b) current directions as determined using the right-hand rule for the indicated position of the rotating coil.*

motion of a conductor, the index finger in the direction of the magnetic flux lines, and the middle finger in the direction of current flow resulting in the conductor if a load is attached. If a load is not attached, the middle finger indicates the direction of increasing induced voltage. The placement of the fingers is indicated in Fig. 13.3(a) for the top conductor of the rotor of Fig. 13.2 as it passes through the position indicated in Fig. 13.3(b). From this point on, we will assume that a load has been applied so that the current directions can be included using the dot (·)-cross (×) convention described in Chapter 11. Note that the resulting direction for the upper conductor is opposite that of the lower conductor. This is a necessary condition for the current $I$ in the series configuration. The reversal in the direction of motion (the thumb) in this region will result in the opposite direction for $I$.

Let us now consider a few representative positions of the rotating coil and determine the relative magnitude and polarity of the generated voltage at these positions. At the instant the coil passes through position 1 in Fig. 13.4(a), there are no flux lines being cut, and the induced

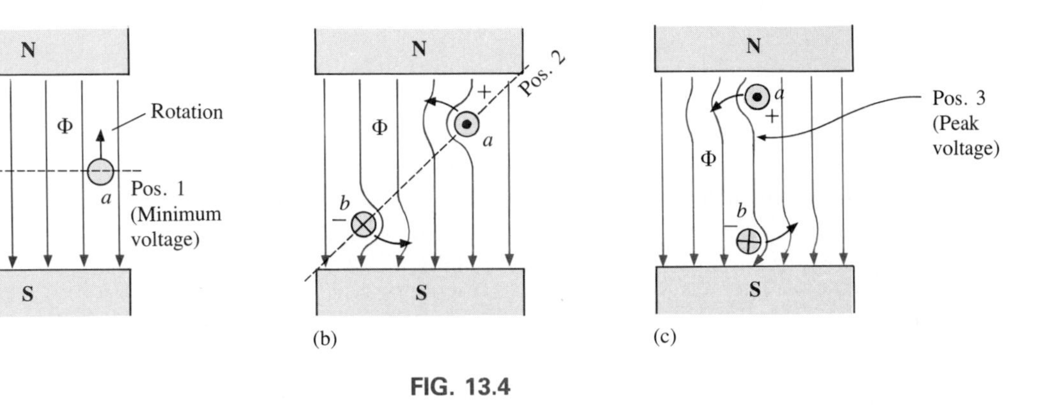


FIG. 13.4

voltage is zero. As the coil moves from position 1 to position 2, indicated in Fig. 13.4(b), the number of flux lines cut per unit time will increase, resulting in an increased induced voltage across the coil. For position 2, the resulting current direction and polarity of terminals $a$ and $b$ are indicated as determined by the right-hand rule. At position 3, the number of flux lines being cut per unit time is a maximum, resulting in a maximum induced voltage. The polarities and current direction are the same as at position 2.

As the coil continues to rotate toward position 4, indicated in Fig. 13.5(a), the polarity of the induced voltage and the current direction

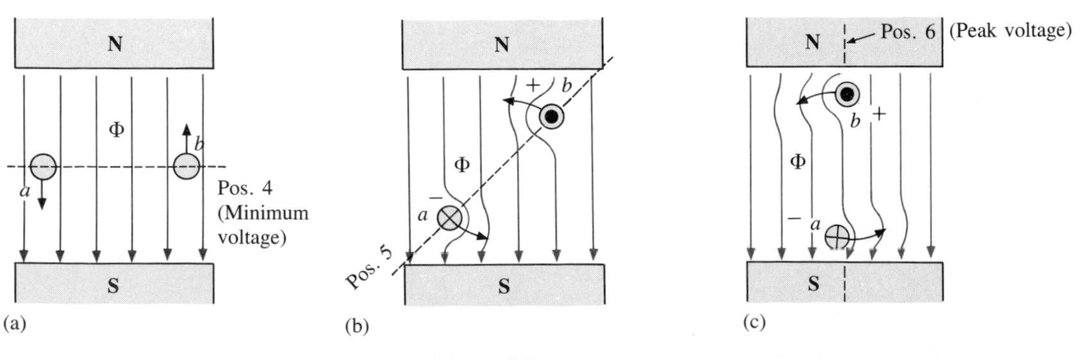


FIG. 13.5

remain the same, as shown in the figure, although the induced voltage will drop due to the reduced number of flux lines cut per unit time. At position 4, the induced voltage is again zero, since the number of flux lines cut per unit time has dropped to zero. As the coil now turns toward position 5, the magnitude of the induced voltage will again increase, but note the change in polarity for terminals *a* and *b* and the reversal of current direction in each conductor. The similarities between the coil positions of positions 2 and 5 [Fig. 13.5(b)], and of 3 and 6 [Fig. 13.5(c)], indicate that the magnitude of the induced voltage is the same although the polarity of *a-b* has reversed.

A continuous plot of the induced voltage $e_g$ appears in Fig. 13.6. The polarities of the induced voltage are shown for terminals *a* and *b* to the left of the vertical axis.

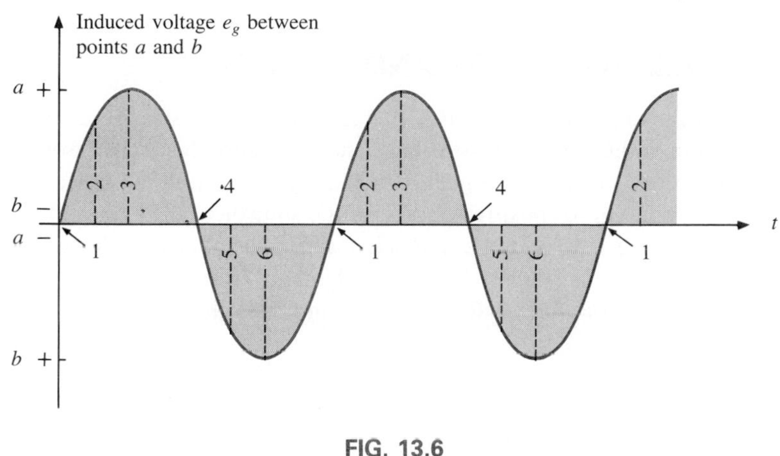


**FIG. 13.6**
*Sinusoidal waveform.*

Take a moment to relate the various positions to the resulting waveform of Fig. 13.6. This waveform will become very familiar in the discussions to follow. Note some of its obvious characteristics. As shown in the figure, if the coil is allowed to continue rotating, the generated voltage will repeat itself in equal intervals of time. Note also that the pattern is exactly the same below the axis as it is above, and that it changes continually with time (the horizontal axis). At the risk of being repetitious, let us again state that the waveform of Fig. 13.6 is the appearance of a *sinusoidal ac voltage*.

The function generator of Fig. 13.7, which employs semiconductor electronic components, will provide sinusoidal, square-wave, and triangular signals over a wide frequency range determined by the dial setting and the chosen frequency range.

If a sinusoidal signal with a frequency of 1000 Hz and a peak value of 10 V were required, the sinusoidal function switch would first be depressed as shown in the figure. Next, the dial would be set to 1 and the 1k range switch depressed as shown. The output frequency is the product of the dial position and the chosen range setting. The amplitude control would adjust the output until an ac voltmeter or oscilloscope

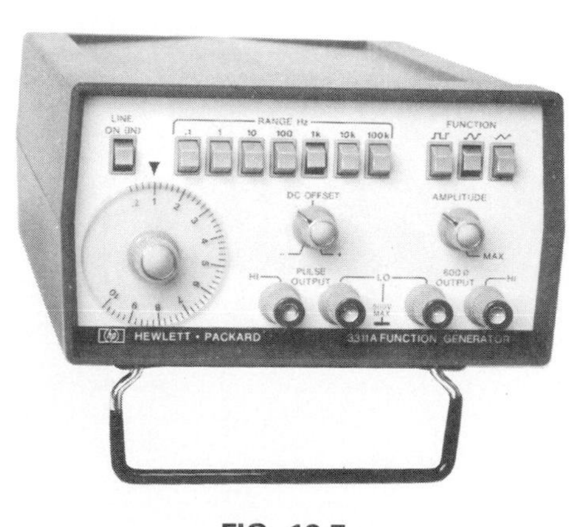


**FIG. 13.7**
*Function generator. (Courtesy of Hewlett Packard Co.)*

indicated an output with a peak value of 10 V. The same frequency could also have been set by choosing a dial position of 10 and pressing the 100 range switch.

The DC SET control at the center of the generator controls the dc level associated with the output ac signal. If OFF, it is zero volts. As indicated in Fig. 13.7, it is set on a slightly negative level. The 600-Ω label near the output terminals reveals that the generator has an internal resistance of 600-Ω, as discussed in Chapter 5. The HI and LO simply designate the relative magnitudes of the output voltage available at the indicated terminals. Since all the output ports are coax connectors, which have two internal conductors, only one terminal is associated with each output.

## 13.3 DEFINED POLARITIES AND DIRECTION

In the following analysis, we will find it necessary to establish a set of polarities for the sinusoidal ac voltage and a direction for the sinusoidal ac current. In each case, the polarity and current direction will be for an instant of time in the positive portion of the sinusoidal waveform. This is shown in Fig. 13.8 with the symbols for the sinusoidal ac voltage and current. A lowercase letter is employed for each to indicate that the quantity is time dependent; that is, its magnitude will change with time. The need for defining polarities and current direction will become quite obvious when we consider multisource ac networks. Note in the last sentence the absence of the term *sinusoidal* before the phrase *ac networks*. This will occur to an increasing degree as we progress; *sinusoidal* is to be understood unless otherwise indicated.

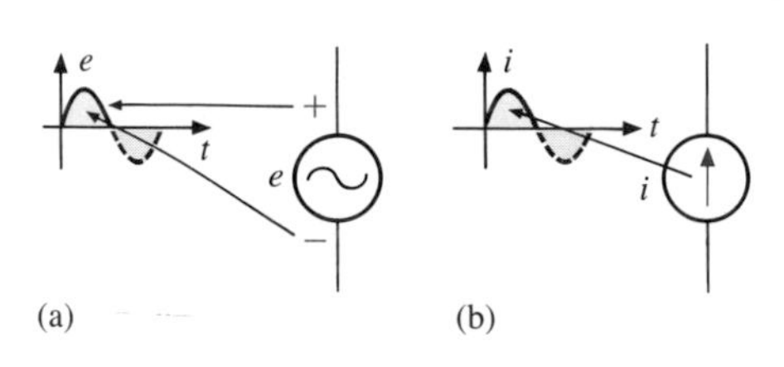


**FIG. 13.8**
*(a) Sinusoidal ac voltage sources; (b) sinusoidal ac current sources.*

## 13.4 DEFINITIONS

The sinusoidal waveform of Fig. 13.9 with its additional notation will now be used as a model in defining a few basic terms. These terms can, however, be applied to any alternating waveform. It is important to remember as you proceed through the various definitions that the vertical scaling is in volts or ampere and the horizontal scaling is *always* in units of time.

*Waveform:* The path traced by a quantity, such as the voltage in Fig. 13.9, plotted as a function of some variable such as time (as above), position, degrees, radians, temperature, and so on.

*Instantaneous value:* The magnitude of a waveform at any instant of time; denoted by lowercase letters ($e_1$, $e_2$).

*Peak amplitude:* The maximum value of a waveform as measured from its *average,* or *mean,* value, denoted by uppercase letters (such as $E_m$ for sources of voltage and $V_m$ for the voltage drop across a load). For the waveform of Fig. 13.9 the average value is zero volts and $E_m$ is as defined by the figure.

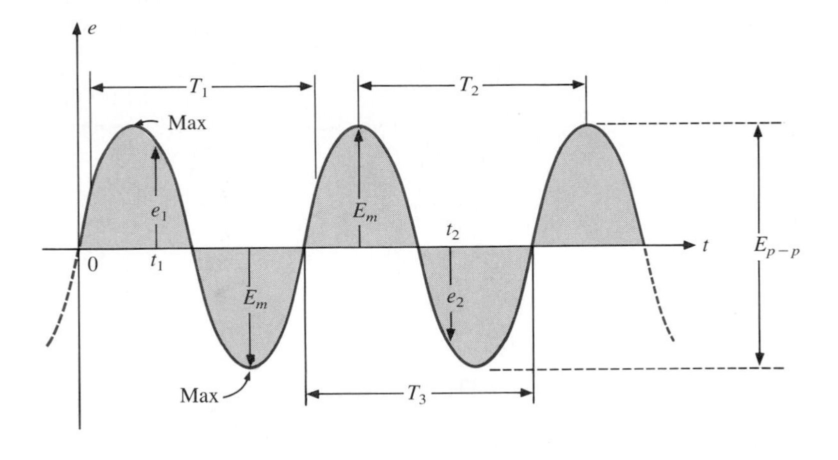


**FIG. 13.9**
*Sinusoidal voltage.*

*Peak value:* The maximum instantaneous value of a function as measured from the zero-volt level. For the waveform of Fig. 13.9 the peak amplitude and peak value are the same, since the average value of the function is zero volts.

*Peak-to-peak value:* Denoted by $E_{p\text{-}p}$ or $V_{p\text{-}p}$, the full voltage between positive and negative peaks of the waveform, that is, the sum of the magnitude of the positive and negative peaks.

*Periodic waveform:* A waveform that continually repeats itself after the same time interval. The waveform of Fig. 13.9 is a periodic waveform.

*Period (T):* The time interval between successive repetitions of a periodic waveform; the period $T_1 = T_2 = T_3$ in Fig. 13.9, so long as successive *similar points* of the periodic waveform are used in determining $T$.

*Cycle:* The portion of a waveform contained in *one period* of time. The cycles within $T_1$, $T_2$, and $T_3$ of Fig. 13.9 may appear different in Fig. 13.10, but they are all bounded by one period of time and therefore satisfy the definition of a cycle.

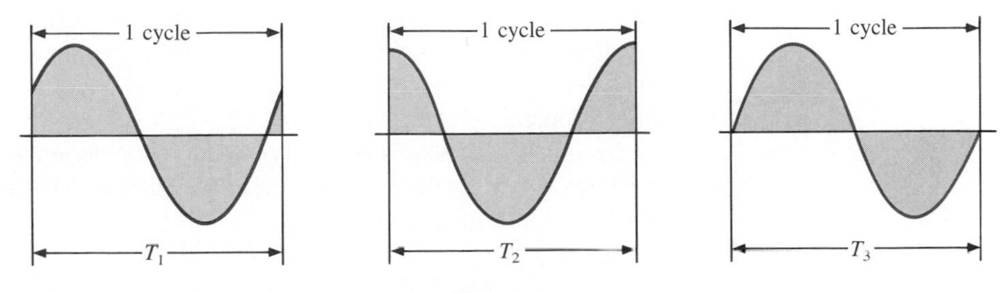


**FIG. 13.10**

*Frequency (f):* The number of cycles that occur in 1 second. The frequency of the waveform of Fig. 13.11(a) is 1 cycle per second, and for Fig. 13.11(b), $2\frac{1}{2}$ cycles per second. If a waveform of similar shape had a period of 0.5 second [Fig. 13.11(c)], the frequency would be 2 cycles per second.

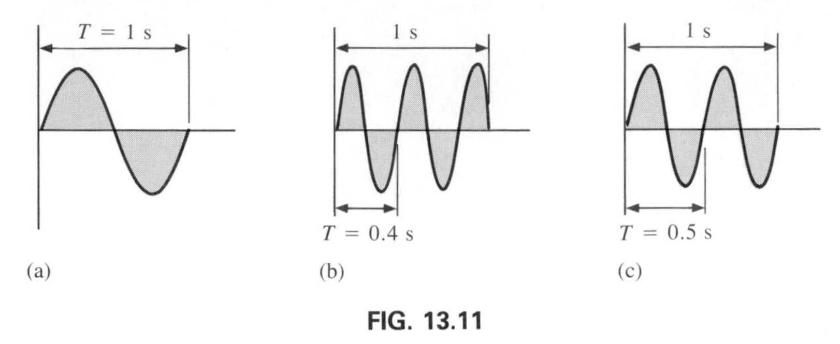


**FIG. 13.11**

The unit of measure for frequency is the *hertz* (Hz), where

$$1 \text{ hertz (Hz)} = 1 \text{ cycle per second (c/s)} \quad \textbf{(13.1)}$$

The unit hertz is derived from the surname of Heinrich Rudolph Hertz, who did original research in the area of alternating currents and voltages and their effect on the basic $R$, $L$, and $C$ elements. The frequency standard for North America is 60 Hz, while for Europe it is predominantly 50 Hz.

Using a log scale (described in detail in Chapter 21) a frequency spectrum from 1 to 1000 GHz can be scaled off on the same axis as shown in Fig. 13.12. A number of terms in the various spectrums are probably familiar to the reader from everyday experiences. Note that the audio range (human ear) extends from only 15 Hz to 20 kHz, but the transmission of radio signals can occur between 3 kHz and 300 GHz. The uniform process of defining the intervals of the radio frequency spectrum from VLF to EHF is quite evident from the length of the bars in the figure (although keep in mind that it is a log scale so the frequencies encompassed within each segment are quite different). Other frequencies of particular interest (TV, CB, microwave, etc.) are also included for reference purposes. Although it is numerically easy to talk about frequencies in the megahertz and gigahertz range, keep in mind that a frequency of 100 MHz, for instance, represents a sinusoidal waveform that passes through 100,000,000 cycles in only 1 s—an incredible number when we compare it to the 60 Hz of our conventional power sources.

Since the frequency is inversely related to the period—that is, as one increases the other decreases by an equal amount—the two can be related by the following equation:

$$f = \frac{1}{T} \qquad \begin{aligned} f &= \text{Hz} \\ T &= \text{seconds (s)} \end{aligned} \quad \textbf{(13.2)}$$

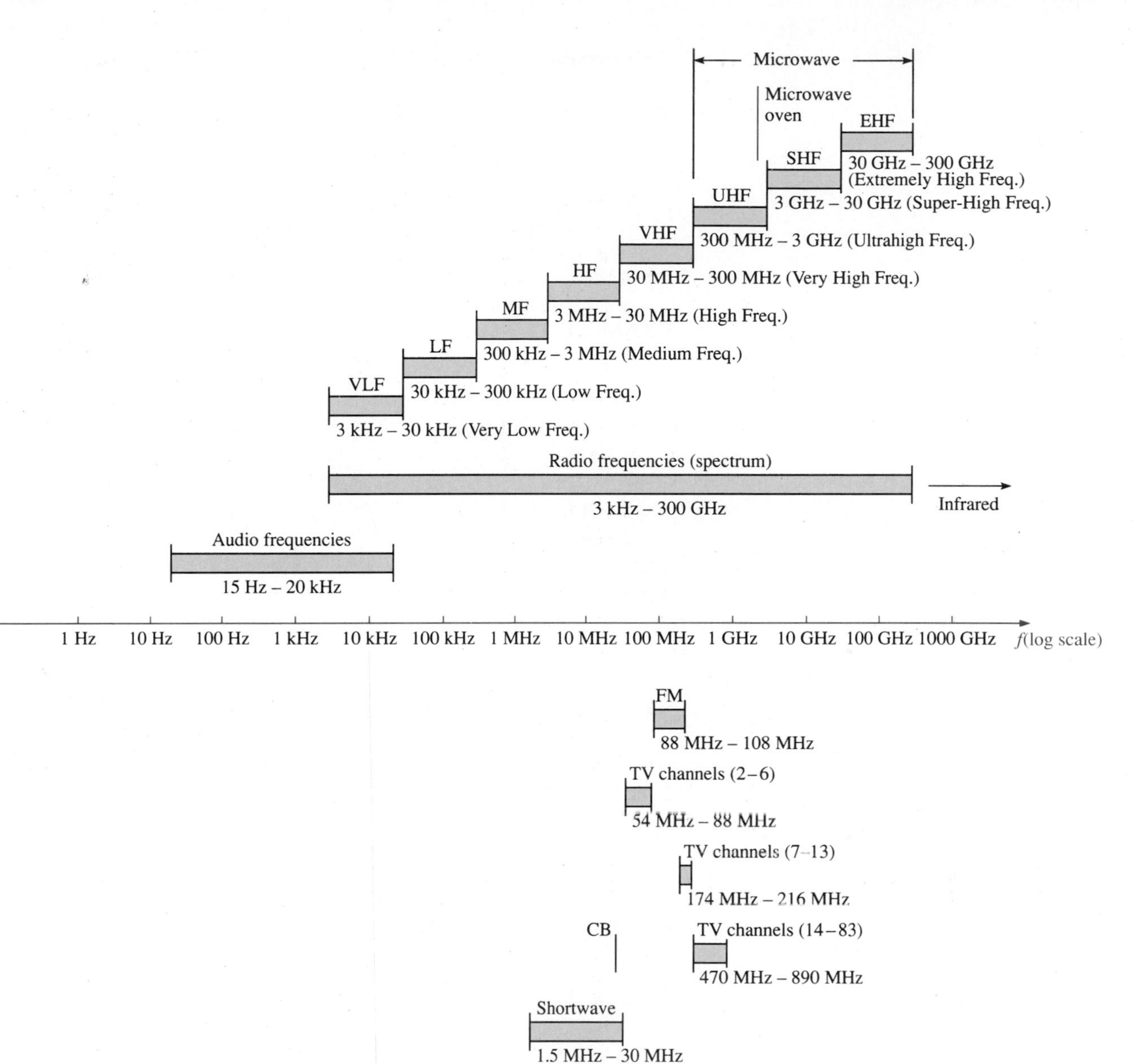


**FIG. 13.12**

or

$$T = \frac{1}{f} \qquad \textbf{(13.3)}$$

---

**EXAMPLE 13.1.** Find the period of a periodic waveform with a frequency of

a. 60 Hz
b. 1000 Hz

***Solutions:***

a. $T = \dfrac{1}{f} = \dfrac{1}{60\text{ Hz}} = 0.01667\text{ s or } \mathbf{16.67\text{ ms}}$

(a recurring value since 60 Hz is so prevalent)

b. $T = \dfrac{1}{f} = \dfrac{1}{1000\text{ Hz}} = 10^{-3}\text{ s} = \mathbf{1\text{ ms}}$

---

**EXAMPLE 13.2.** Determine the frequency of the waveform of Fig. 13.13.

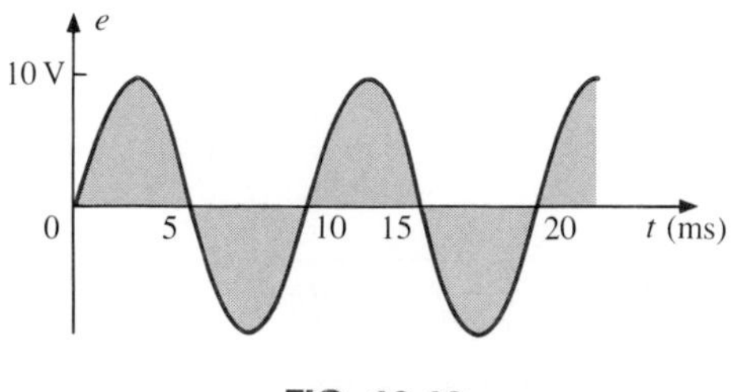

**FIG. 13.13**

***Solution:*** From the figure, $T = 10$ ms, and

$$f = \frac{1}{T} = \frac{1}{10 \times 10^{-3}\text{ s}} = \mathbf{100\text{ Hz}}$$

---

**EXAMPLE 13.3.** The oscilloscope is an instrument that will display alternating waveforms such as those described above. A sinusoidal pattern appears on the oscilloscope of Fig. 13.14 with the indicated vertical

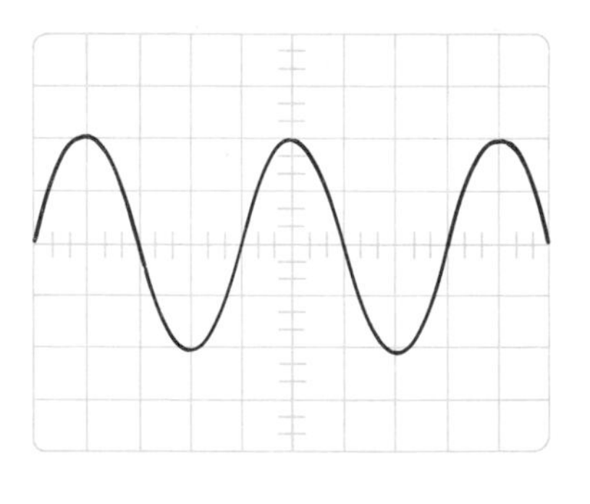

Vertical sensitivity = 0.1 V/cm

Horizontal sensitivity = 50 $\mu$s/cm

**FIG. 13.14**

and horizontal sensitivities. The vertical sensitivity defines the voltage associated with each vertical centimeter (cm) of the display. Virtually all oscilloscope screens are cut into a crosshatch pattern of lines separated by 1 cm in the vertical and horizontal directions. The horizontal sensitivity defines the time period associated with each horizontal centimeter of the display.

For the pattern of Fig. 13.14 and the indicated sensitivities, determine the period, frequency, and peak value of the waveform.

***Solution:*** One cycle spans 4 cm. The period is

$$T = 4(50\ \mu\text{s}) = \mathbf{200\ \mu s}$$

and the frequency is

$$f = \frac{1}{T} = \frac{1}{200 \times 10^{-6}\ \text{s}} = \mathbf{5\ kHz}$$

The vertical height above the horizontal axis encompasses 2 cm. Therefore,

$$V_m = (2\ \text{cm})(0.1\ \text{V/cm}) = \mathbf{0.2\ V}$$

## 13.5 THE SINE WAVE

The terms defined in the previous section can be applied to any type of periodic waveform, whether smooth or discontinuous. The sinusoidal waveform is of particular importance, however, since it lends itself readily to the mathematics and the physical phenomena associated with electric circuits. Consider the power of the following statement:

***The sine wave is the only alternating waveform whose shape is unaffected by the response characteristics of R, L, and C elements.***

In other words, if the voltage across (or current through) a resistor, coil, or capacitor is sinusoidal in nature, the resulting current (or voltage, respectively) for each will also have sinusoidal characteristics as shown in Fig. 13.15. If a square wave or a triangular wave were ap-

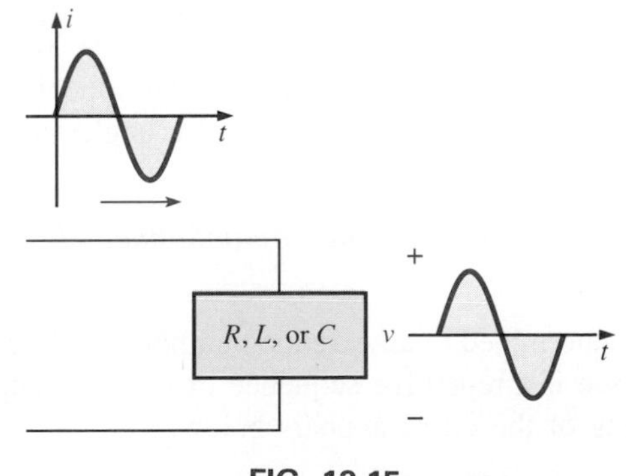


FIG. 13.15

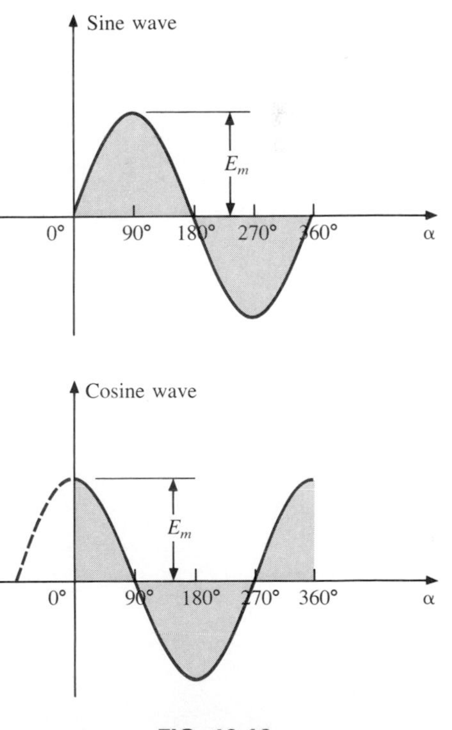


FIG. 13.16

plied, such would not be the case. It must be pointed out that the above statement is also applicable to the cosine wave since the waves differ only by a 90° shift on the horizontal axis, as shown in Fig. 13.16.

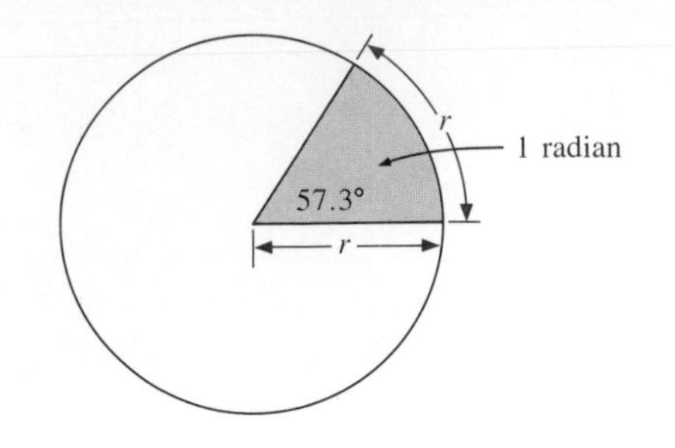


**FIG. 13.17**
*Defining the radian.*

The unit of measurement for the horizontal axis of Fig. 13.16 is the *degree*. A second unit of measurement frequently used is the *radian* (rad). It is defined by a quadrant of a circle such as in Fig. 13.17 where the distance subtended on the circumference equals the radius of the circle.

If we define $x$ as the number of intervals of $r$ (the radius) around the circumference of the circle, then

$$C = 2\pi r = x \cdot r$$

and we find

$$x = 2\pi$$

Therefore, there are $2\pi$ radians around a 360° circle, as shown in Fig. 13.18, and

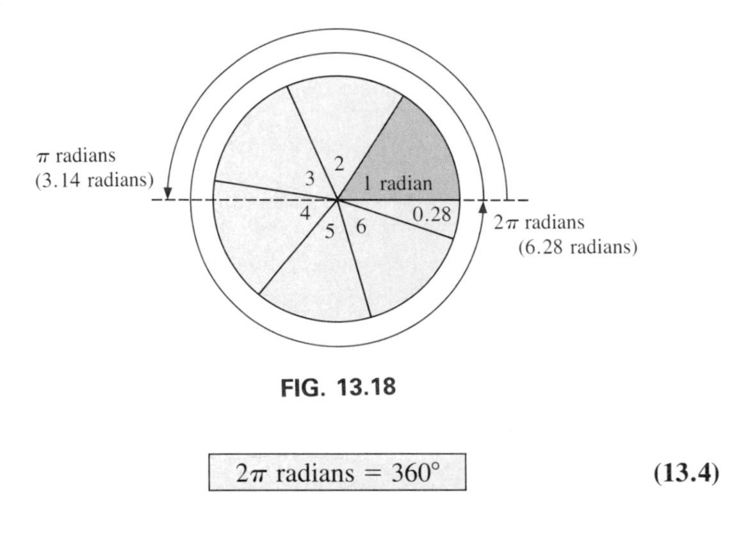


**FIG. 13.18**

$$\boxed{2\pi \text{ radians} = 360°} \tag{13.4}$$

with

$$\boxed{1 \text{ radian} \cong 57.3°} \tag{13.5}$$

A number of electrical formulas contain a multiplier of $\pi$. This is one reason it is sometimes preferable to measure angles in radians rather than in degrees.

***The quantity π is the ratio of the circumference of a circle to its diameter.***

$\pi$ has been determined to an extended number of places primarily in an attempt to see if a repetitive sequence of numbers appears. It does not. A sampling of the effort appears below:

$$\pi = 3.14159\ 26535\ 89793\ 23846\ 26433\ .\ .\ .$$

For our purposes, the following approximation will often be applied:

$$\boxed{\pi = 3.14} \tag{13.6}$$

For 180° and 360°, the two units of measurement are related as shown in Fig. 13.18. The conversion equations between the two are the following:

$$\text{Radians} = \left(\frac{\pi}{180^\circ}\right) \times (\text{degrees}) \tag{13.7}$$

$$\text{Degrees} = \left(\frac{180^\circ}{\pi}\right) \times (\text{radians}) \tag{13.8}$$

Applying these equations, we find

$$90^\circ\text{:}\quad \text{Radians} = \frac{\pi}{180^\circ}(90^\circ) = \boldsymbol{\frac{\pi}{2}}\ \textbf{rad}$$

$$30^\circ\text{:}\quad \text{Radians} = \frac{\pi}{180^\circ}(30^\circ) = \boldsymbol{\frac{\pi}{6}}\ \textbf{rad}$$

$$\frac{\pi}{3}\text{:}\quad \text{Degrees} = \frac{180^\circ}{\pi}\left(\frac{\pi}{3}\right) = \mathbf{60^\circ}$$

$$\frac{3\pi}{2}\text{:}\quad \text{Degrees} = \frac{180^\circ}{\pi}\left(\frac{3\pi}{2}\right) = \mathbf{270^\circ}$$

Using the radian as the unit of measurement for the abscissa, we would obtain a sine wave as shown in Fig. 13.19.

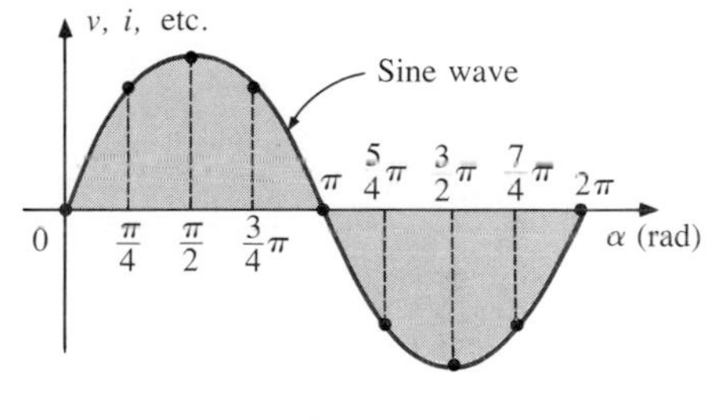


FIG. 13.19

It is of particular interest that the sinusoidal waveform can be derived from the length of the *vertical projection* of a radius vector rotating in a uniform circular motion about a fixed point. Starting as shown in Fig. 13.20(a) and plotting the amplitude (above and below zero) on the coordinates drawn to the right [Figs. 13.20(b) through (i)], we will trace a complete sinusoidal waveform after the radius vector has completed a 360° rotation about the center.

The velocity with which the radius vector rotates about the center, called the *angular velocity*, can be determined from the following equation:

$$\text{Angular velocity} = \frac{\text{distance (degrees or radians)}}{\text{time (seconds)}} \tag{13.9}$$

Substituting into Eq. (13.9) and assigning the Greek letter omega ($\omega$) to the angular velocity, we have

$$\omega = \frac{\alpha}{t} \tag{13.10}$$

and

$$\alpha = \omega t \tag{13.11}$$

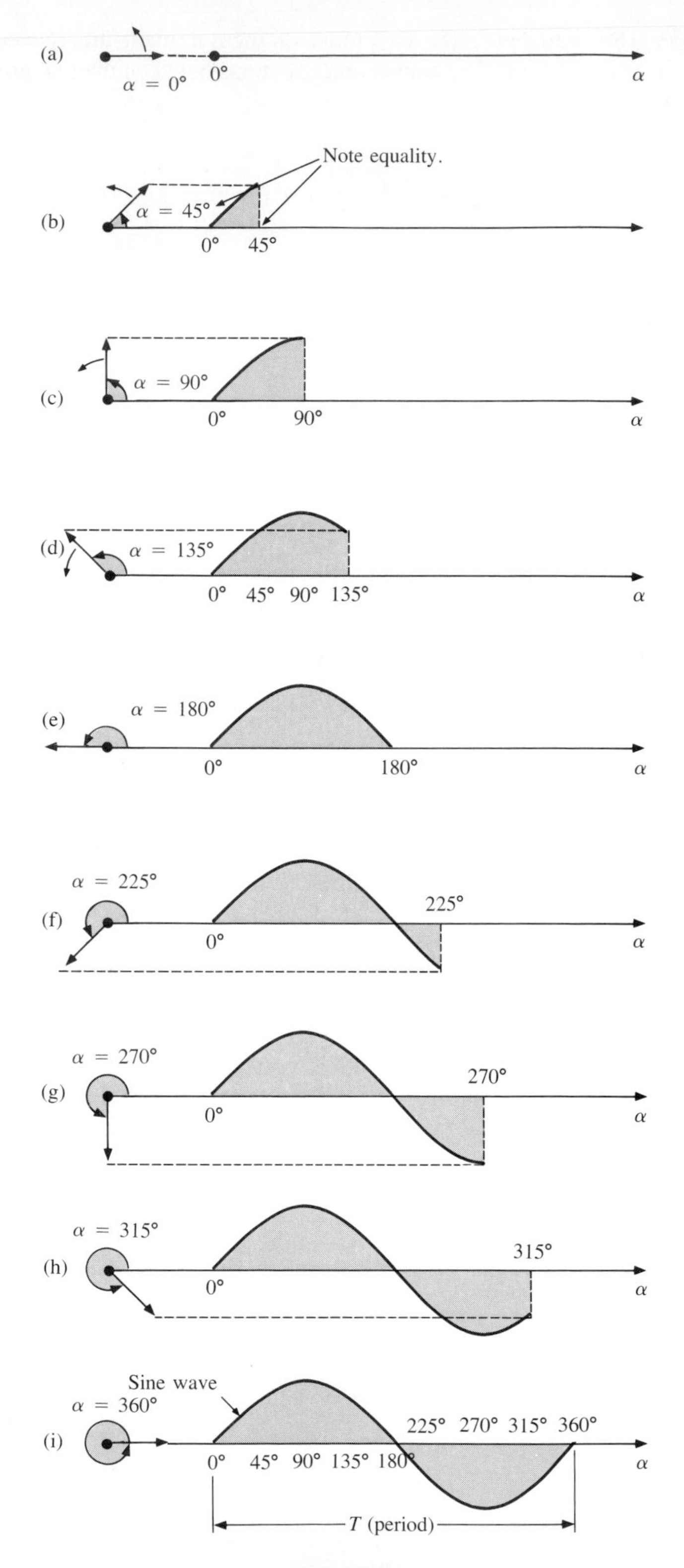
(a)
$\alpha = 0°$
0°
$\alpha$
Note equality.
(b)
$\alpha = 45°$
0° 45°
(c)
$\alpha = 90°$
0° 90°
$\alpha$
(d)
$\alpha = 135°$
0° 45° 90° 135°
$\alpha$
(e)
$\alpha = 180°$
0° 180°
$\alpha$
(f)
$\alpha = 225°$
225°
0°
$\alpha$
(g)
$\alpha = 270°$
270°
0°
$\alpha$
(h)
$\alpha = 315°$
315°
0°
$\alpha$
(i)
Sine wave
$\alpha = 360°$
225° 270° 315° 360°
0° 45° 90° 135° 180°
$\alpha$
T (period)

**FIG. 13.20**

Since $\omega$ is typically provided in radians per second, the angle $\alpha$ obtained using Eq. (13.11) is usually in radians. If $\alpha$ is required in degrees, Eq. (13.8) must be applied. The importance of remembering the above will become obvious in the examples to follow.

In Fig. 13.20, the time required to complete one revolution is equal to the period ($T$) of the sinusoidal waveform of Fig. 13.20(i). The radians subtended in this time interval are $2\pi$. Substituting, we have

$$\omega = \frac{2\pi}{T} \quad \text{(rad/s)} \tag{13.12}$$

In words, this equation states that the smaller the period of the sinusoidal waveform of Fig. 13.20(i), or the smaller the time interval before one complete cycle is generated, the greater must be the angular velocity of the rotating radius vector. Certainly this statement agrees with what we have learned thus far. We can now go one step further and apply the fact that the frequency of the generated waveform is inversely related to the period of the waveform; that is, $f = 1/T$. Thus,

$$\omega = 2\pi f \quad \text{(rad/s)} \tag{13.13}$$

This equation states that the higher the frequency of the generated sinusoidal waveform, the higher must be the angular velocity. Equations (13.12) and (13.13) are verified somewhat by Fig. 13.21, where for the same radius vector, $\omega$ = 100 rad/s and 500 rad/s.

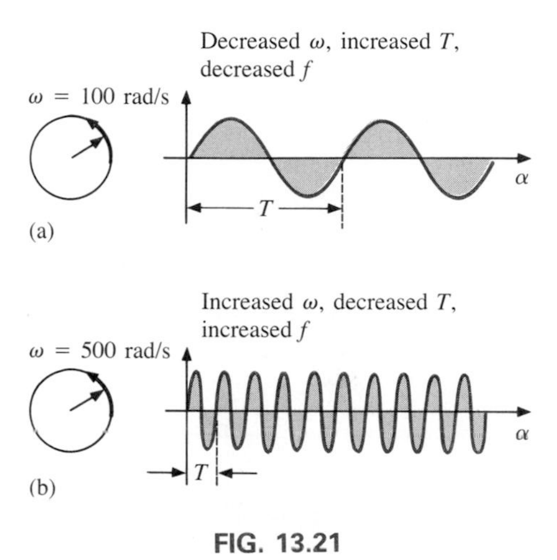


**FIG. 13.21**

**EXAMPLE 13.4.** Determine the angular velocity of a sine wave having a frequency of 60 Hz.

***Solution:***

$$\omega = 2\pi f = (6.28 \text{ rad})(60 \text{ Hz}) \cong \mathbf{377 \text{ rad/s}}$$

(a recurring value due to 60-Hz predominance)

**EXAMPLE 13.5.** Determine the frequency and period of the sine wave of Fig. 13.21(b).

***Solution:*** Since $\omega = 2\pi/T$,

$$T = \frac{2\pi}{\omega} = \frac{2\pi \text{ rad}}{500 \text{ rad/s}} = \frac{6.28 \text{ rad}}{500 \text{ rad/s}} = \mathbf{12.56 \text{ ms}}$$

and

$$f = \frac{1}{T} = \frac{1}{12.56 \times 10^{-3} \text{ s}} = \mathbf{79.62 \text{ Hz}}$$

**EXAMPLE 13.6.** Given $\omega = 200$ rad/s, determine how long it will take the sinusoidal waveform to pass through an angle of 90°.

***Solution:*** Eq. (13.11): $\alpha = \omega t$, and

$$t = \frac{\alpha}{\omega}$$

However, $\alpha$ must be substituted as $\pi/2$ (= 90°) since $\omega$ is in radians per second:

$$t = \frac{\alpha}{\omega} = \frac{\pi/2 \text{ rad}}{200 \text{ rad/s}} = \frac{\pi}{400} \text{ s} = \frac{3.14}{400} \text{ s} = \mathbf{7.85 \text{ ms}}$$

**EXAMPLE 13.7.** Find the angle a sinusoidal waveform of 60 Hz will pass through in a period of 5 ms.

***Solution:*** Eq. (13.11): $\alpha = \omega t$, or

$$\alpha = 2\pi f t = (6.28 \text{ rad})(60 \text{ Hz})(5 \times 10^{-3} \text{ s}) = \mathbf{1.884 \text{ rad}}$$

If not careful, one might be tempted to interpret the answer as 1.884°. However,

$$\alpha\ (°) = \frac{180°}{\pi \text{ rad}}(1.884 \text{ rad}) = \mathbf{108°}$$

## 13.6 GENERAL FORMAT FOR THE SINUSOIDAL VOLTAGE OR CURRENT

The basic mathematical format for the sinusoidal waveform is

$$\boxed{A_m \sin \alpha} \qquad \textbf{(13.14)}$$

where $A_m$ is the peak value of the waveform and $\alpha$ is the unit of measure for the horizontal axis as shown in Fig. 13.22.

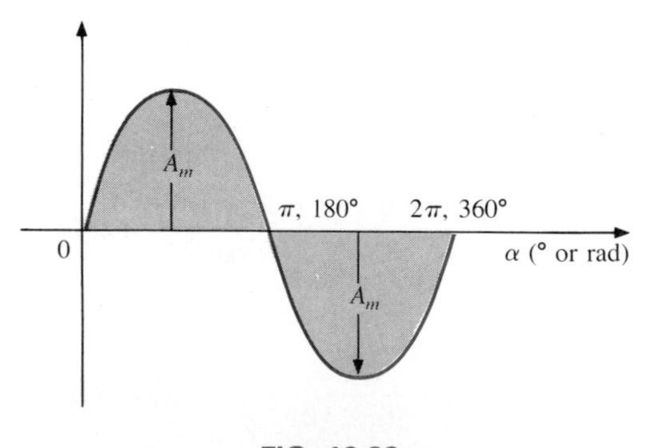


FIG. 13.22

The equation $\alpha = \omega t$ states that the angle $\alpha$ through which the rotating vector of Fig. 13.20 will pass is determined by the angular velocity of the rotating vector and the length of time the vector rotates. For example, for a particular angular velocity (fixed $\omega$), the longer the radius vector is permitted to rotate (that is, the greater the value of $t$), the greater will be the number of degrees or radians through which the vector will pass. Relating this statement to the sinusoidal waveform, for a particular angular velocity, the longer the time, the greater the number of cycles shown. For a fixed time interval, the greater the angular velocity, the greater the number of cycles generated.

Due to Eq. (13.11), the general format of a sine wave can also be written

$$A_m \sin \omega t \tag{13.15}$$

with $\omega t$ as the horizontal unit of measure.

For electrical quantities such as current and voltage, the general format is

$$i = I_m \sin \omega t = I_m \sin \alpha$$
$$e = E_m \sin \omega t = E_m \sin \alpha$$

where the capital letters with the subscript $m$ represent the amplitude and the lowercase letters $i$ and $e$ represent the instantaneous value of current or voltage, respectively, at any time $t$. This format is particularly important, since it presents the sinusoidal voltage or current as a function of time, which is the horizontal scale for the oscilloscope. Recall that the horizontal sensitivity of a scope is in time per centimeter and not degrees per centimeter.

**EXAMPLE 13.8.** Given $e = 5 \sin \alpha$, determine $e$ at $\alpha = 40°$ and $\alpha = 0.8\pi$.

***Solution:*** For $\alpha = 40°$,

$$e = 5 \sin 40° = 5(0.6428) = \mathbf{3.214\ V}$$

For $\alpha = 0.8\pi$,

$$\alpha\ (°) = \frac{180°}{\pi}(0.8\pi) = 144°$$

and

$$e = 5 \sin 144° = 5(0.5878) = \mathbf{2.939\ V}$$

The conversion to degrees will not be required for most modern-day scientific calculators, since they can perform the function directly. First be sure the calculator is in the RAD mode and then simply enter the radian measure and use the appropriate trigonometric key (sin, cos, tan, etc.).

The angle at which a particular voltage level is attained can be determined by rearranging the equation

$$e = E_m \sin \alpha$$

in the following manner:

$$\sin \alpha = \frac{e}{E_m}$$

which can be written

$$\alpha = \sin^{-1} \frac{e}{E_m} \tag{13.16}$$

Similarly, for a particular current level,

$$\alpha = \sin^{-1} \frac{i}{I_m} \tag{13.17}$$

The function $\sin^{-1}$ is available on all scientific calculators.

---

**EXAMPLE 13.9.**

a. Determine the angle at which the magnitude of the sinusoidal function $v = 10 \sin 377t$ is 4 V.
b. Determine the time at which the magnitude is attained.

***Solutions:***

a. Eq. (13.16):

$$\alpha_1 = \sin^{-1} \frac{e}{E_m} = \sin^{-1} \frac{4\text{ V}}{10\text{ V}} = \sin^{-1} 0.4 = \mathbf{23.578°}$$

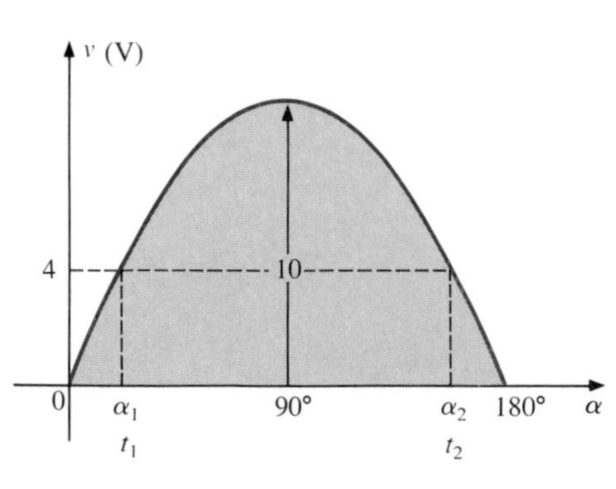

FIG. 13.23

However, Fig. 13.23 reveals that the magnitude of 4 V (positive) will be attained at two points between 0° and 180°. The second intersection is determined by

$$\alpha_2 = 180° - 23.578° = \mathbf{156.422°}$$

In general, therefore, keep in mind that Eqs. (13.16) and (13.17) will provide an angle with a magnitude between 0° and 90°.

b. Eq. (13.11): $\alpha = \omega t$, and so $t = \alpha/\omega$. However, $\alpha$ must be in radians. Thus,

$$\alpha\text{ (rad)} = \frac{\pi}{180°}(23.578°) = 0.411\text{ rad}$$

and

$$t_1 = \frac{\alpha}{\omega} = \frac{0.411\text{ rad}}{377\text{ rad/s}} = \mathbf{1.09\text{ ms}}$$

For the second intersection,

$$\alpha(\text{rad}) = \frac{\pi}{180^\circ}(156.422^\circ) = 2.73 \text{ rad}$$

$$t_2 = \frac{\alpha}{\omega} = \frac{2.73 \text{ rad}}{377 \text{ rad/s}} = 7.24 \text{ ms}$$

The sine wave can also be plotted against *time* on the horizontal axis. The time period for each interval can be determined from $t = \alpha/\omega$, but the most direct route is simply to find the period $T$ from $T = 1/f$ and break it up into the required intervals. This latter technique will be demonstrated in Example 13.10.

Before reviewing the example, take special note of the relative simplicity of the mathematical equation that can represent a sinusoidal waveform. Any alternating waveform whose characteristics differ from those of the sine wave cannot be represented by a single term, but may require two, four, six, or perhaps an infinite number of terms to be represented accurately. A further description of nonsinusoidal waveforms can be found in Chapter 24.

**EXAMPLE 13.10.** Sketch $e = 10 \sin 314t$ with the abscissa

a. angle ($\alpha$) in degrees.
b. angle ($\alpha$) in radians.
c. time ($t$) in seconds.

***Solutions:***

a. See Fig. 13.24. (Note that no calculations are required.)
b. See Fig. 13.25. (Once the relationship between degrees and radians is understood, there is again no need for calculations.)
c. 360°: $T = \dfrac{2\pi}{\omega} = \dfrac{6.28}{314} = 20 \text{ ms}$

180°: $\dfrac{T}{2} = \dfrac{20}{2} \times 10^{-3} = 10 \text{ ms}$

90°: $\dfrac{T}{4} = \dfrac{20}{4} \times 10^{-3} = 5 \text{ ms}$

30°: $\dfrac{T}{12} = \dfrac{20}{12} \times 10^{-3} = 1.67 \text{ ms}$

See Fig. 13.26.

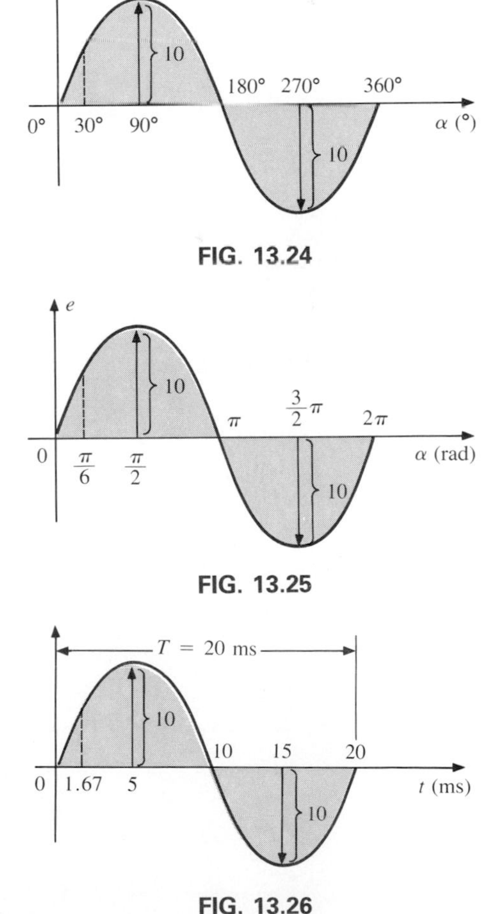


FIG. 13.24

FIG. 13.25

FIG. 13.26

**EXAMPLE 13.11.** Given $i = 6 \times 10^{-3} \sin 1000t$, determine $i$ at $t = 2$ ms.

***Solution:***

$$\alpha = \omega t = 1000t = (1000 \text{ rad/s})(2 \times 10^{-3} \text{ s}) = 2 \text{ rad}$$

$$\alpha\,(°) = \frac{180°}{\pi \text{ rad}}(2 \text{ rad}) = 114.59°$$

$$i = (6 \times 10^{-3})(\sin 114.59°$$
$$= (6 \text{ mA})(0.9093) = \mathbf{5.46 \text{ mA}}$$

## 13.7 PHASE RELATIONS

Thus far, we have considered only sine waves that have maxima at $\pi/2$ and $3\pi/2$, with a zero value at 0, $\pi$, and $2\pi$, as shown in Fig. 13.25. If the waveform is shifted to the right or left of 0°, the expression becomes

$$\boxed{A_m \sin(\omega t \pm \theta)} \qquad \textbf{(13.18)}$$

where $\theta$ is the angle in degrees or radians that the waveform has been shifted.

If the waveform passes through the horizontal axis with a *positive-going* (increasing with time) slope *before* 0°, as shown in Fig. 13.27, the expression is

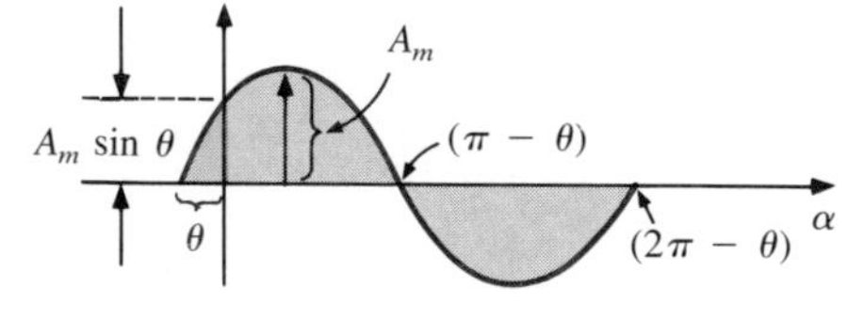


FIG. 13.27

$$\boxed{A_m \sin(\omega t + \theta)} \qquad \textbf{(13.19)}$$

At $\omega t = \alpha = 0°$, the magnitude is determined by $A_m \sin \theta$. If the waveform passes through the horizontal axis with a positive-going slope *after* 0°, as shown in Fig. 13.28, the expression is

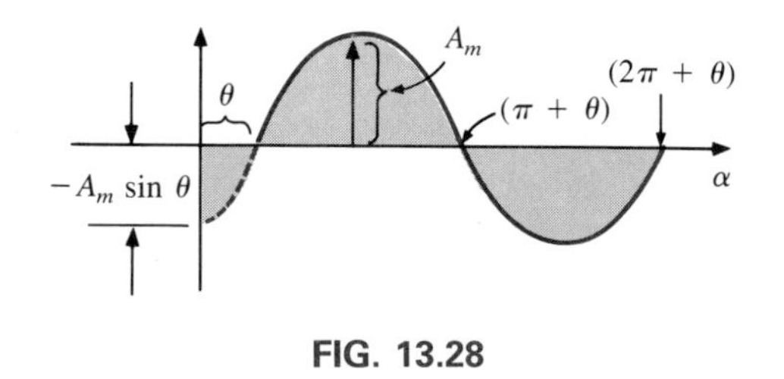


FIG. 13.28

$$\boxed{A_m \sin(\omega t - \theta)} \qquad \textbf{(13.20)}$$

And at $\omega t = \alpha = 0°$, the magnitude is $A_m \sin(-\theta)$, which by a trigonometric identity is $-A_m \sin \theta$.

If the waveform crosses the horizontal axis with a positive-going slope 90° ($\pi/2$) sooner, as shown in Fig. 13.29, it is called a *cosine wave*. That is,

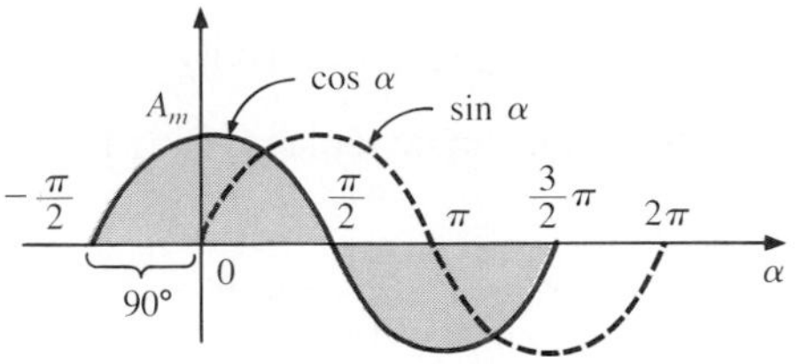

FIG. 13.29

$$\sin(\omega t + 90°) = \sin\left(\omega t + \frac{\pi}{2}\right) = \cos \omega t \qquad \textbf{(13.21)}$$

or

$$\sin \omega t = \cos(\omega t - 90°) = \cos\left(\omega t - \frac{\pi}{2}\right) \qquad \textbf{(13.22)}$$

The terms *lead* and *lag* are used to indicate the relationship between two sinusoidal waveforms of the *same frequency* plotted on the same set of axes. In Fig. 13.29, the cosine curve is said to *lead* the sine curve by 90°, and the sine curve is said to *lag* the cosine curve by 90°. The 90° is referred to as the phase angle between the two waveforms. In language commonly applied, the waveforms are *out of phase* by 90°. Note that the phase angle between the two waveforms is measured between those two points on the horizontal axis through which each passes with the *same slope*. If both waveforms cross the axis at the same point with the same slope, they are *in phase*.

A few additional geometric relations that may prove useful in applications involving sines or cosines in phase relationships are the following:

$$\begin{aligned}\sin(-\alpha) &= -\sin \alpha \\ \cos(-\alpha) &= \cos \alpha \\ -\sin(\alpha) &= \sin(\alpha \pm 180°) \\ -\cos(\alpha) &= \cos(\alpha \pm 180°)\end{aligned} \qquad \textbf{(13.23)}$$

If a sinusoidal expression should appear as

$$e = -E_m \sin \omega t$$

the negative sign is associated with the sine portion of the expression, not the peak value $E_m$. In other words, the expression, if not for convenience, would be written

$$e = E_m(-\sin \omega t)$$

Since

$$-\sin \omega t = \sin(\omega t \pm 180°)$$

the expression can also be written

$$e = E_m \sin(\omega t \pm 180°)$$

revealing that a negative sign can be replaced by a 180° change in phase angle (+ or −). That is,

$$e = -E_m \sin \omega t = E_m \sin(\omega t + 180°)$$
$$= E_m \sin(\omega t - 180°)$$

A plot of each will clearly show their equivalence. There are, therefore, two correct mathematical representations for the functions.

The *phase relationship* between two waveforms indicates which one leads or lags, and by how many degrees or radians.

---

**EXAMPLE 13.12.** What is the phase relationship between the sinusoidal waveforms of each of the following sets?

a. $v = 10 \sin(\omega t + 30°)$
   $i = 5 \sin(\omega t + 70°)$
b. $i = 15 \sin(\omega t + 60°)$
   $v = 10 \sin(\omega t - 20°)$
c. $i = 2 \cos(\omega t + 10°)$
   $v = 3 \sin(\omega t - 10°)$
d. $i = -\sin(\omega t + 30°)$
   $v = 2 \sin(\omega t + 10°)$
e. $i = -2 \cos(\omega t - 60°)$
   $v = 3 \sin(\omega t - 150°)$

***Solutions:***

a. See Fig. 13.30.
   ***i* leads *v* by 40°, or *v* lags *i* by 40°.**

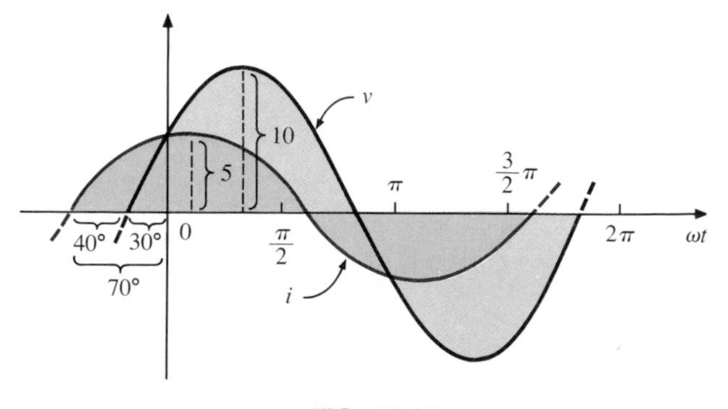


**FIG. 13.30**

b. See Fig. 13.31.
   ***i* leads *v* by 80°, or *v* lags *i* by 80°.**

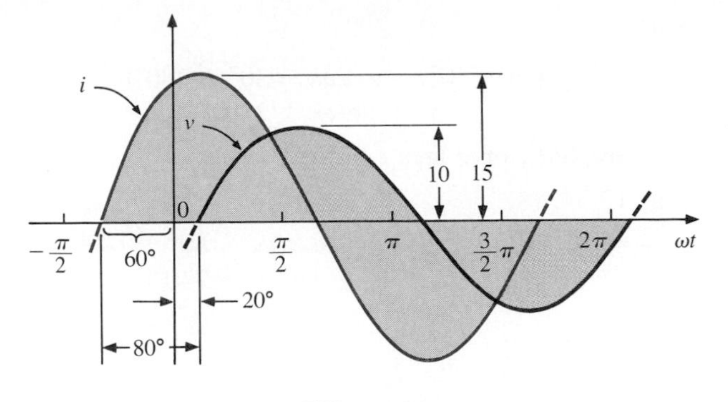

FIG. 13.31

c. See Fig. 13.32.

$$i = 2\cos(\omega t + 10°) = 2\sin(\omega t + 10° + 90°)$$
$$= 2\sin(\omega t + 100°)$$

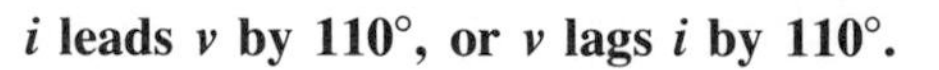
***i* leads *v* by 110°, or *v* lags *i* by 110°.**

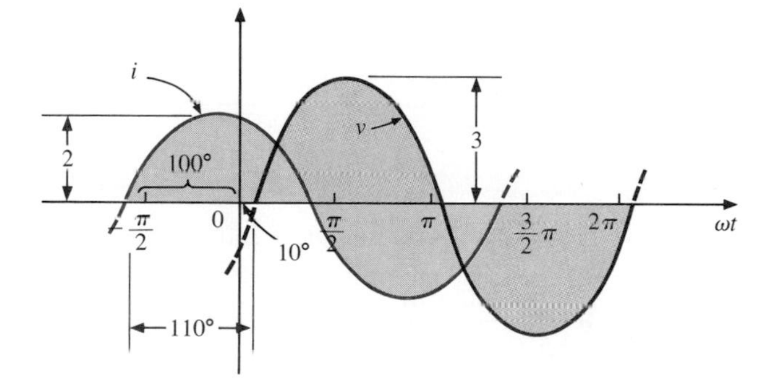

FIG. 13.32

d. See Fig. 13.33.

$$-\sin(\omega t + 30°) = \sin(\omega t + 30° \overset{\text{Note}}{-} 180°)$$
$$= \sin(\omega t - 150°)$$

***v* leads *i* by 160°, or *i* lags *v* by 160°.**

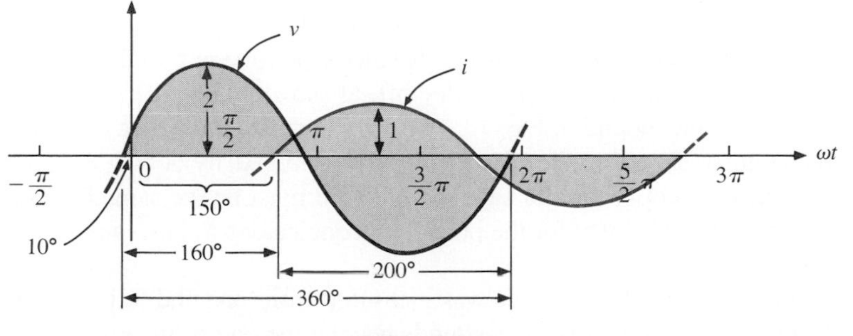


FIG. 13.33

Or using

$$-\sin(\omega t + 30°) = \sin(\omega t + 30° \overset{\text{Note}}{+} 180°)$$
$$= \sin(\omega t + 210°)$$

***i* leads *v* by 200°, or *v* lags *i* by 200°.**

e. See Fig. 13.34.

$$i = -2\cos(\omega t - 60°) = 2\cos(\omega t - 60° \overset{\text{By choice}}{-} 180°)$$
$$= 2\cos(\omega t - 240°)$$

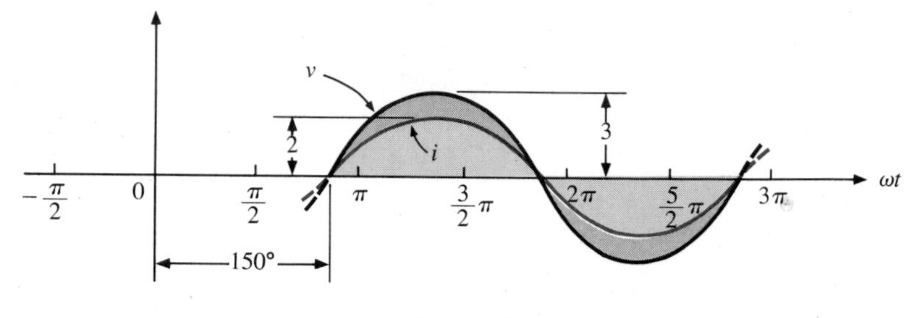

FIG. 13.34

However,

$$\cos\alpha = \sin(\alpha + 90°)$$

so that

$$2\cos(\omega t - 240°) = 2\sin(\omega t - 240° + 90°)$$
$$= 2\sin(\omega t - 150°)$$

***v* and *i* are in phase.**

## 13.8 AVERAGE VALUE

Even though the concept of the *average value* is an important one in most technical fields, its true meaning is often misinterpreted. In Fig. 13.35(a), for example, the average height of the sand may be required to determine the volume of sand available. The average height of the sand is that height obtained if the distance from one end to the other is maintained while the sand is leveled off, as shown in Fig. 13.35(b). The area under the mound of Fig. 13.35(a) will then equal the area under the rectangular shape of Fig. 13.35(b) as determined by $A = b \times h$. Of course, the depth (into the page) of the sand must be the same for Fig. 13.35(a) and 13.35(b) for the preceding conclusions to have any meaning.

In Fig. 13.35 the distance was measured from one end to the other. In Fig. 13.36(a) the distance extends beyond the end of the pile for the same original pile of Fig. 13.35. The situation could be one where a landscaper would like to know the average height of the sand if spread out over a distance such as defined in Fig. 13.36(a). The result of an increased distance is as shown in Fig. 13.36(b). The average height has

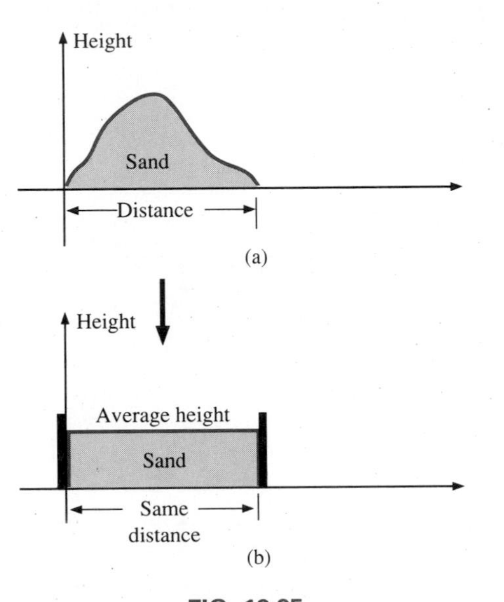


**FIG. 13.35**
*Defining average value.*

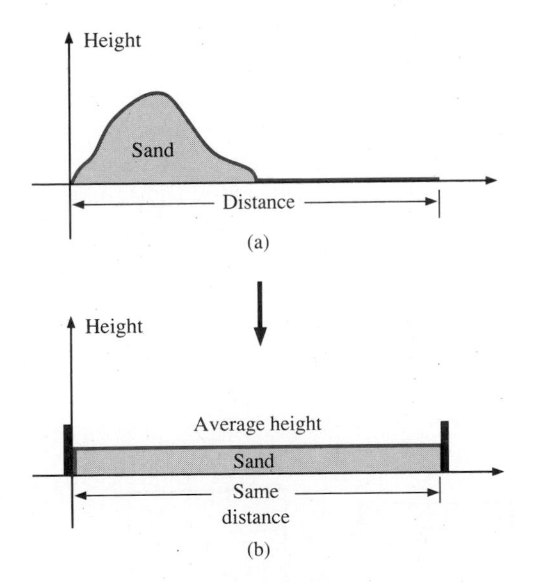


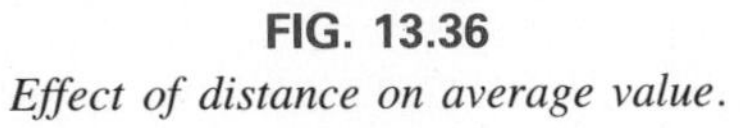

**FIG. 13.36**
*Effect of distance on average value.*

decreased compared to Fig. 13.35. Quite obviously, therefore, the longer the distance, the lower is the average value.

If the distance parameter includes a depression, as shown in Fig. 13.37(a), some of the sand will be used to fill the depression, resulting in an even lower average value for the landscaper, as shown in Fig. 13.37(b). For a sinusoidal waveform, the depression would have the same shape as the mound of sand (over one full cycle) resulting in an

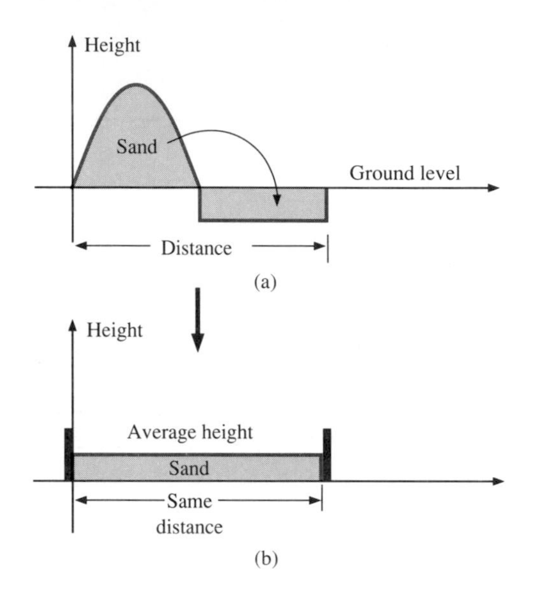


**FIG. 13.37**
*Effect of depressions (negative excursions) on average value.*

average value at ground level (or zero volts for a sinusoidal voltage over one full period).

After traveling a considerable distance by car, some drivers like to calculate their average speed for the entire trip. This is usually done by dividing the miles traveled by the hours required to drive that distance. For example, if a person traveled 180 mi in 5 h, his average speed was 180 mi/5 h or 36 mi/h. This same distance may have been traveled at various speeds for various intervals of time, as shown in Fig. 13.38.

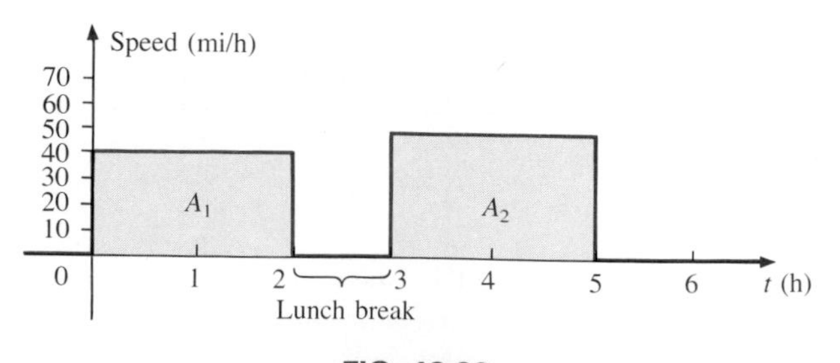

**FIG. 13.38**

By finding the total area under the curve for the 5 h and then dividing the area by 5 h (the total time for the trip), we obtain the same result of 36 mi/h; that is,

$$\text{Average speed} = \frac{\text{area under curve}}{\text{length of curve}} \tag{13.24}$$

$$= \frac{A_1 + A_2}{5\text{ h}}$$

$$= \frac{(40\text{ mi/h})(2\text{ h}) + (50\text{ mi/h})(2\text{ h})}{5\text{ h}}$$

$$= \frac{80}{5}\text{ mi/h}$$

$$= \mathbf{36\ mi/h}$$

Equation (13.24) can be extended to include any variable quantity, such as current or voltage, if we let $G$ denote the average value, as follows:

$$G\text{ (average value)} = \frac{\text{algebraic sum of areas}}{\text{length of curve}} \tag{13.25}$$

The algebraic sum of the areas must be determined, since some area contributions will be from below the horizontal axis. Areas above the axis will be assigned a positive sign, and those below a negative sign. A positive average value will then be above the axis, and a negative value below.

The average value of *any* current or voltage is the value indicated on a dc meter. In other words, over a complete cycle, the average value is the equivalent dc value. In the analysis of electronic circuits to be considered in a later course, both dc and ac sources of voltage will be applied to the same network. It will then be necessary to know or determine the dc (or average value) and ac components of the voltage or current in various parts of the system.

---

**EXAMPLE 13.13.** Determine the average value of the waveforms of Fig. 13.39.

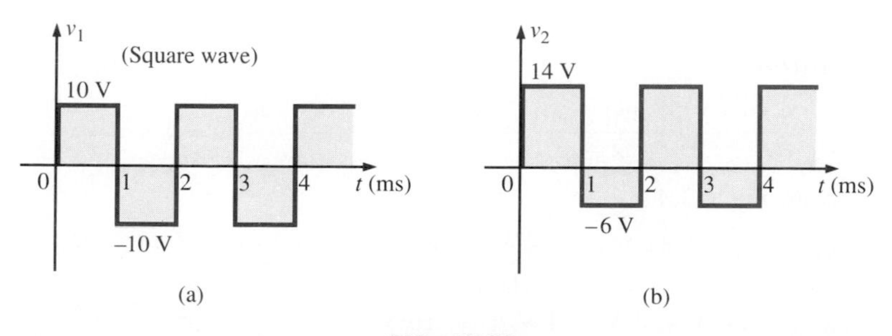


**FIG. 13.39**

***Solutions:***

a. By inspection, the area above the axis equals the area below over one cycle, resulting in an average value of zero volts.
Using Eq. (13.25):

$$G = \frac{(10\ \text{V})(1\ \text{ms}) - (10\ \text{V})(1\ \text{ms})}{2\ \text{ms}}$$

$$= \frac{0}{2\ \text{ms}} = \mathbf{0\ V}$$

b. Using Eq. (13.25):

$$G = \frac{(14\ \text{V})(1\ \text{ms}) - (6\ \text{V})(1\ \text{ms})}{2\ \text{ms}}$$

$$= \frac{14\ \text{V} - 6\ \text{V}}{2} = \frac{8\ \text{V}}{2}$$

$$= \mathbf{4\ V}$$

as shown in Fig. 13.40.

In reality, the waveform of Fig. 13.39(b) is simply the square wave of Fig. 13.39(a) with a dc shift of 4 V. That is,

$$v_2 = v_1 + 4\ \text{V}$$

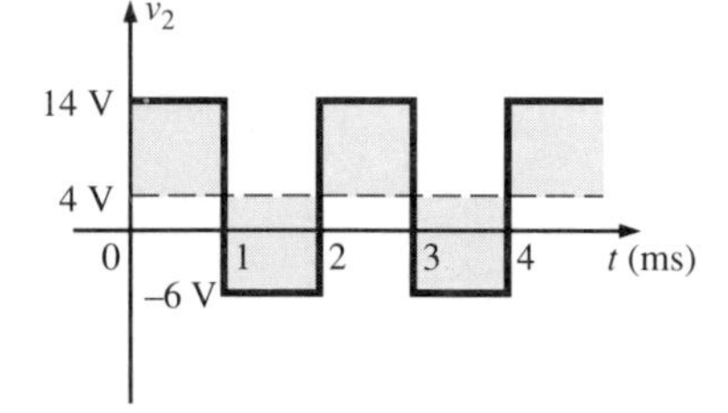


FIG. 13.40

**EXAMPLE 13.14.** Find the average values of the following waveforms over one full cycle:

a. Fig. 13.41.
b. Fig. 13.42.

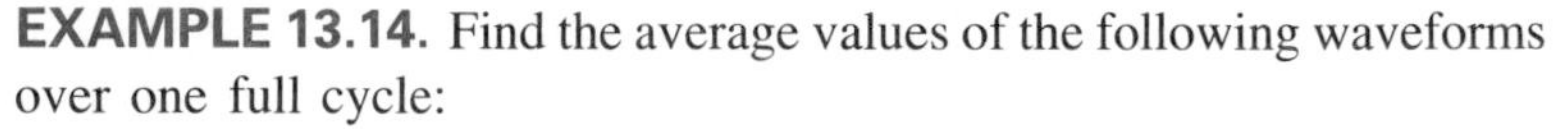

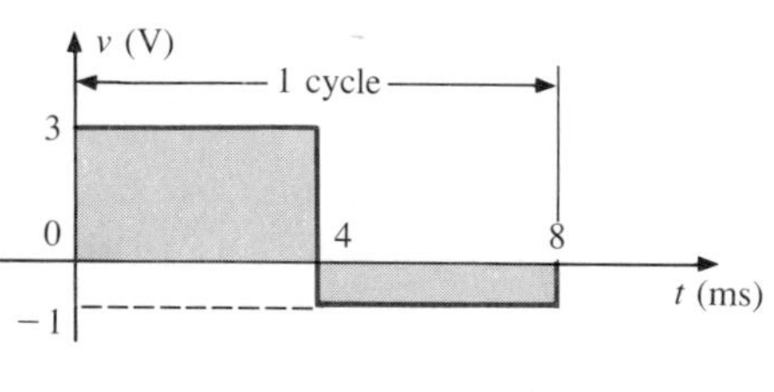


FIG. 13.41

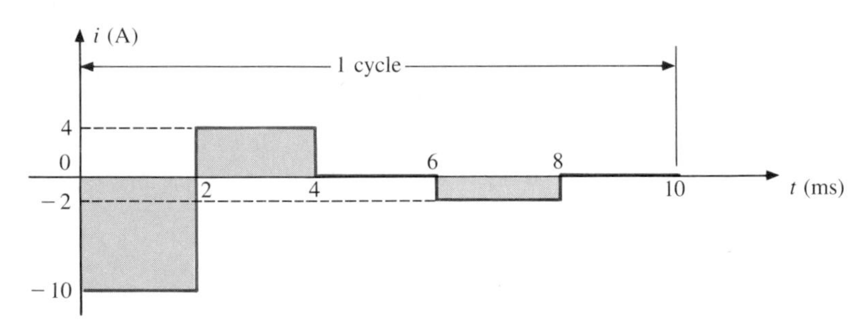


FIG. 13.42

***Solutions:***

a. $G = \dfrac{+(3\ \text{V})(4\ \text{ms}) - (1\ \text{V})(4\ \text{ms})}{8\ \text{ms}} = \dfrac{12\ \text{V} - 4\ \text{V}}{8} = \mathbf{1\ V}$

Note Fig. 13.43.

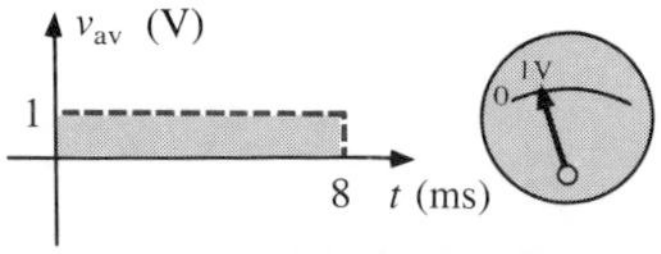

b. $G = \dfrac{-(10\ \text{V})(2\ \text{ms}) + (4\ \text{V})(2\ \text{ms}) - (2\ \text{V})(2\ \text{ms})}{10\ \text{ms}}$

$$= \frac{-20\ \text{V} + 8\ \text{V} - 4\ \text{V}}{10} = -\frac{16\ \text{V}}{10} = \mathbf{-\ 1.6\ V}$$

$v_{av}$ (V)
1
8 t (ms)
1 V
0

dc voltmeter (between 0 and 8 ms)

FIG. 13.43

Note Fig. 13.44.

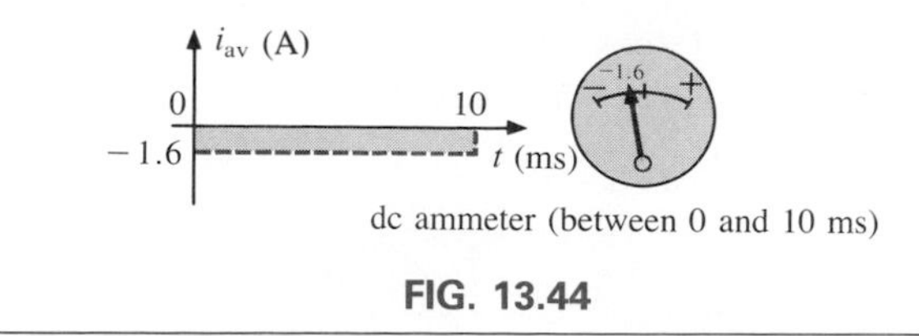


**FIG. 13.44**

We found the areas under the curves in the preceding example by using a simple geometric formula. If we should encounter a sine wave or any other unusual shape, however, we must find the area by some other means. We can obtain a good approximation of the area by attempting to reproduce the original wave shape using a number of small rectangles or other familiar shapes, the area of which we already know through simple geometric formulas. For example,

***the area of the positive (or negative) pulse of a sine wave is $2A_m$.***

Approximating this waveform by two triangles (Fig. 13.45), we obtain (using *area* = 1/2 *base* × *height* for the area of a triangle) a rough idea of the actual area:

$$\text{Area (shaded)} = 2\left(\frac{1}{2}bh\right) = 2\left[\left(\frac{1}{2}\right)\overbrace{\left(\frac{\pi}{2}\right)}^{b}\overbrace{(A_m)}^{h}\right] = \frac{\pi}{2}A_m$$
$$\cong 1.58A_m$$

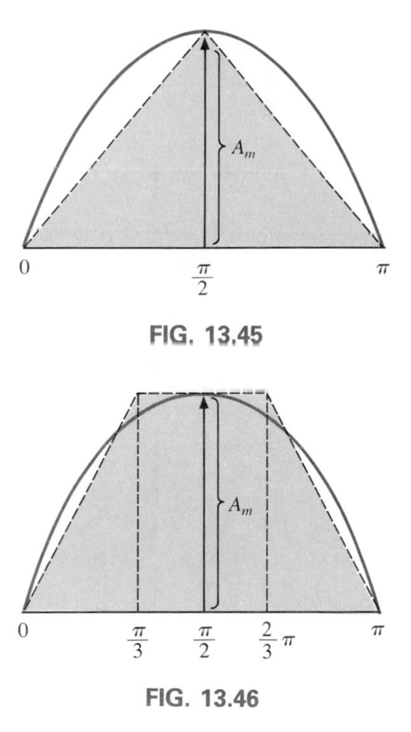


**FIG. 13.45**

**FIG. 13.46**

A closer approximation might be a rectangle with two similar triangles (Fig. 13.46):

$$\text{Area} = A_m\frac{\pi}{3} + 2\left(\frac{1}{2}bh\right) = A_m\frac{\pi}{3} + \frac{\pi}{3}A_m = \frac{2}{3}\pi A_m$$
$$= 2.094A_m$$

which is certainly close to the actual area. If an infinite number of forms were used, an exact answer of $2A_m$ could be obtained. For irregular waveforms, this method can be especially useful if data such as the average value are desired.

The procedure of calculus that gives the exact solution $2A_m$ is known as *integration*. Integration is presented here only to make the method recognizable to the reader; it is not necessary to be proficient in its use to continue with this text. It is a useful mathematical tool, however, and should be learned. Finding the area under the positive pulse of a sine wave using integration, we have

$$\text{Area} = \int_0^{\pi} A_m \sin \alpha \, d\alpha$$

where $\int$ is the sign of integration, $\pi$ and 0 are the limits of integration, $A_m \sin \alpha$ is the function to be integrated, and $d\alpha$ indicates that we are integrating with respect to $\alpha$.

Integrating, we obtain

$$\begin{aligned}\text{Area} &= A_m[-\cos\alpha]_0^\pi \\ &= -A_m(\cos\pi - \cos 0°) \\ &= -A_m[-1-(+1)] = -A_m(-2)\end{aligned}$$

$$\boxed{\text{Area} = 2A_m} \qquad (13.26)$$

Since we know the area under the positive (or negative) pulse, we can easily determine the average value of the positive (or negative) region of a sine wave pulse by applying Eq. (13.25):

$$G = \frac{2A_m}{\pi}$$

and

$$\boxed{G = 0.637A_m} \qquad (13.27)$$

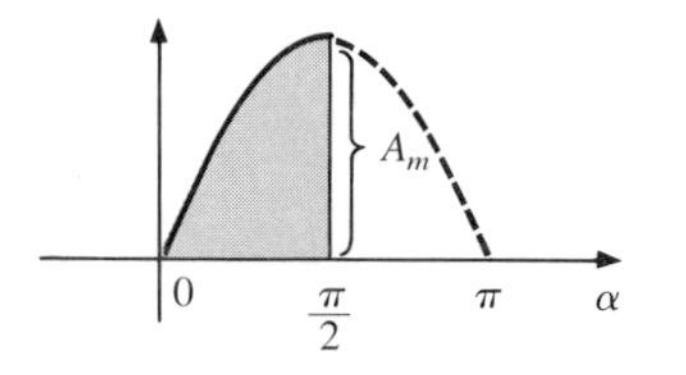


FIG. 13.47

For the waveform of Fig. 13.47,

$$G = \frac{(2A_m/2)}{\pi/2} = \frac{2A_m}{\pi} \qquad \text{(average the same as for a full pulse)}$$

**EXAMPLE 13.15.** Determine the average value of the sinusoidal waveform of Fig. 13.48.

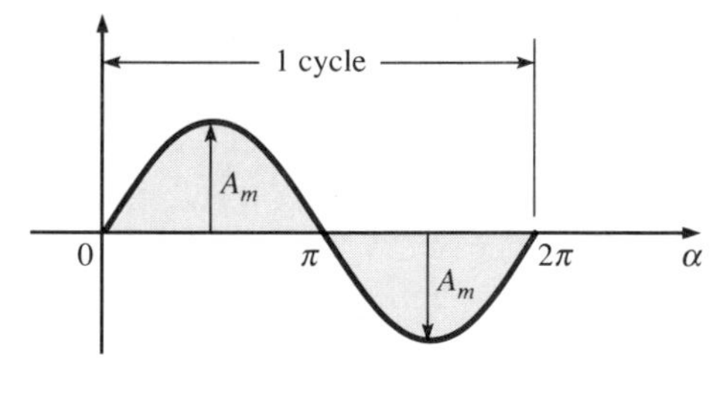


FIG. 13.48

***Solution:*** By inspection it is fairly obvious that

***the average value of a pure sinusoidal waveform over one full cycle is zero.***

Equation (13.25):

$$G = \frac{+2A_m - 2A_m}{2\pi} = \mathbf{0\ V}$$

**EXAMPLE 13.16.** Determine the average value of the waveform of Fig. 13.49.

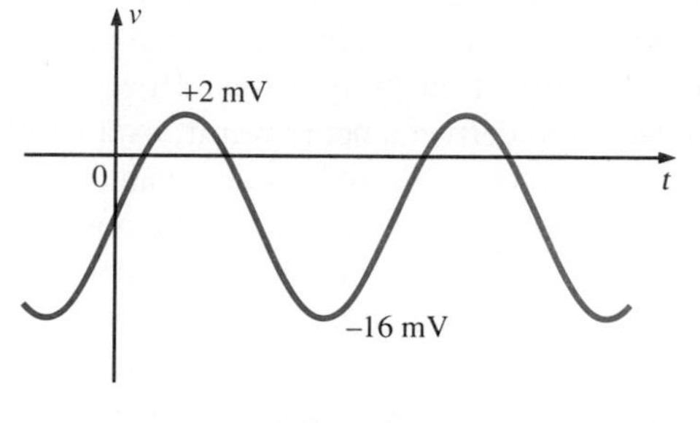


**FIG. 13.49**

***Solution:*** The peak-to-peak value of the sinusoidal function is 16 mV + 2 mV = 18 mV. The peak amplitude of the sinusoidal waveform is therefore 18 mV/2 = 9 mV. Counting down 9 mV from 2 mV (or 9 mV up from −16 mV) results in an average or dc level of −7 mV as noted by the dashed line of Fig. 13.49.

**EXAMPLE 13.17.** Determine the average value of the waveform of Fig. 13.50.

***Solution:***

$$G = \frac{2A_m + 0}{2\pi} = \frac{2(10\text{ V})}{2\pi} \cong \mathbf{3.18\ V}$$

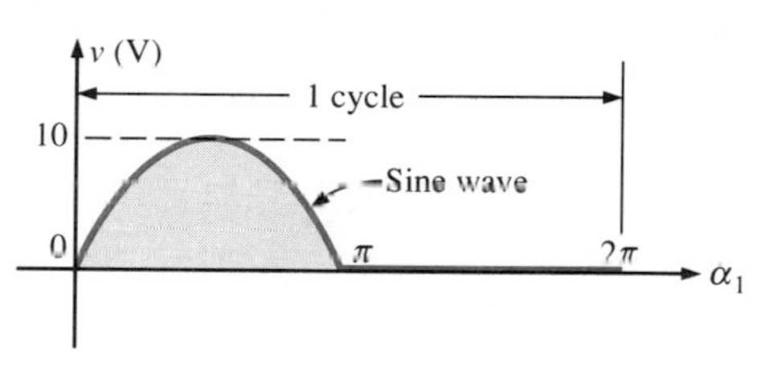


**FIG. 13.50**

**EXAMPLE 13.18.** For the waveform of Fig. 13.51 determine whether the average value is positive or negative and its approximate value.

***Solution:*** From the appearance of the waveform, the average value is positive and in the vicinity of 2 mV. Occasionally, judgments of this type will have to be made.

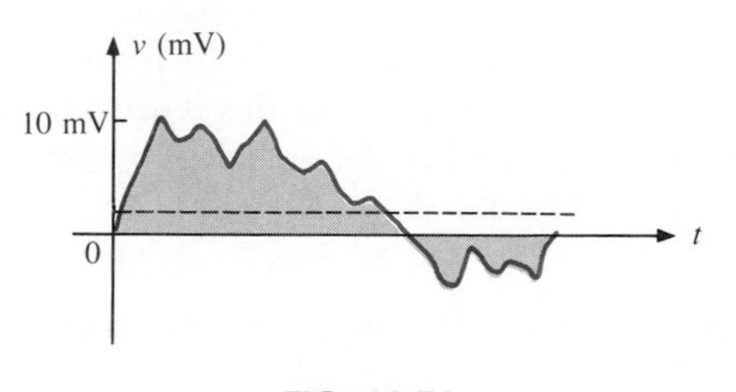


**FIG. 13.51**

## 13.9 EFFECTIVE VALUES

This section will begin to relate dc and ac quantities with respect to the power delivered to a load. It will help us determine the amplitude of a sinusoidal ac current required to deliver the same power as a particular dc current. The question frequently arises, How is it possible for a sinusoidal ac quantity to deliver a net power if, over a full cycle, the net current in any one direction is zero (average value = 0)? It would almost appear that the power delivered during the positive portion of the sinusoidal waveform is withdrawn during the negative portion, and since the two are equal in magnitude, the net power delivered is zero. However, understand that *irrespective of direction,* current of any magnitude through a resistor will deliver power *to that resistor*. In other words, during the positive or negative portions of a sinusoidal ac current, power is being delivered at *each instant of time* to the resistor. The power delivered at each instant will, of course, vary with the magnitude of the sinusoidal ac current, but there will be a net flow during either the positive or negative pulses with a net flow over the full cycle. The net power flow will equal twice that delivered by either the positive or negative regions of the sinusoidal quantity.

A fixed relationship between ac and dc voltages and currents can be derived from the experimental setup shown in Fig. 13.52. A resistor in

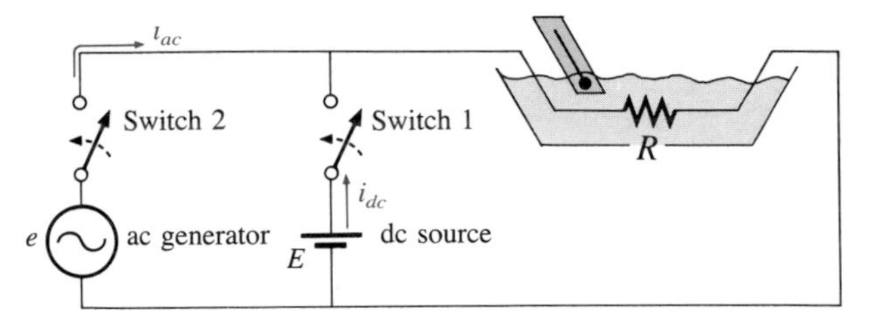


**FIG. 13.52**

a water bath is connected by switches to a dc and an ac supply. If switch 1 is closed, a dc current $I$, determined by the resistance $R$ and battery voltage $E$, will be established through the resistor $R$. The temperature reached by the water is determined by the dc power dissipated in the form of heat by the resistor.

If switch 2 is closed and switch 1 left open, the ac current through the resistor will have a peak value of $I_m$. The temperature reached by the water is now determined by the ac power dissipated in the form of heat by the resistor. The ac input is varied until the temperature is the same as that reached with the dc input. When this is accomplished, the average electrical power delivered to the resistor $R$ by the ac source is the same as that delivered by the dc source.

The power delivered by the ac supply at any instant of time is

$$P_{ac} = (i_{ac})^2R = (I_m \sin \omega t)^2R = (I_m^2 \sin^2\omega t)R$$

but

$$\sin^2\omega t = \frac{1}{2}(1 - \cos 2\omega t) \qquad \text{(trigonometric identity)}$$

Therefore,

$$P_{ac} = I_m^2\left[\frac{1}{2}(1 - \cos 2\omega t)\right]R$$

and

$$P_{ac} = \frac{I_m^2 R}{2} - \frac{I_m^2 R}{2}\cos 2\omega t \qquad \textbf{(13.28)}$$

The *average power* delivered by the ac source is just the first term, since the average value of a cosine wave is zero even though the wave may have twice the frequency of the original input current waveform. Equating the average power delivered by the ac generator to that delivered by the dc source,

$$P_{av(ac)} = P_{dc}$$

$$\frac{I_m^2 R}{2} = I_{dc}^2 R$$

and

$$I_m - \sqrt{2}I_{dc}$$

or

$$I_{dc} = \frac{I_m}{\sqrt{2}} = 0.707 I_m$$

which, in words, states that

***the equivalent dc value of a sinusoidal current or voltage is $1/\sqrt{2}$ or 0.707 of its maximum value.***

The equivalent dc value is called the effective value of the sinusoidal quantity.

In summary,

$$I_{eq\ dc} = I_{eff} = 0.707 I_m \qquad \textbf{(13.29)}$$

or

$$I_m = \sqrt{2}I_{eff} = 1.414 I_{eff} \qquad \textbf{(13.30)}$$

and

$$E_{eff} = 0.707 E_m \qquad \textbf{(13.31)}$$

or

$$E_m = \sqrt{2}E_{eff} = 1.414 E_{eff} \qquad \textbf{(13.32)}$$

As a simple numerical example, it would require an ac current with a peak value of $\sqrt{2}(10) = 14.14$ A to deliver the same power to the resistor in Fig. 13.52 as a dc current of 10 A. The effective value of any quantity plotted as a function of time can be found by using the following equation derived from the experiment just described:

$$I_{\text{eff}} = \sqrt{\frac{\int_0^T i^2\,dt}{T}} \tag{13.33}$$

or

$$I_{\text{eff}} = \sqrt{\frac{\text{area}(i^2)}{T}} \tag{13.34}$$

which, in words, states that to find the effective value, the function $i(t)$ must first be squared. After $i(t)$ is squared, the area under the curve is found by integration. It is then divided by $T$, the length of the cycle or period of the waveform, to obtain the average or *mean* value of the squared waveform. The final step is to take the *square root* of the mean value. This procedure gives us another designation for the effective value, the *root-mean-square* (rms) value.

**EXAMPLE 13.19.** Find the effective values of the sinusoidal waveforms in each part of Fig. 13.53.

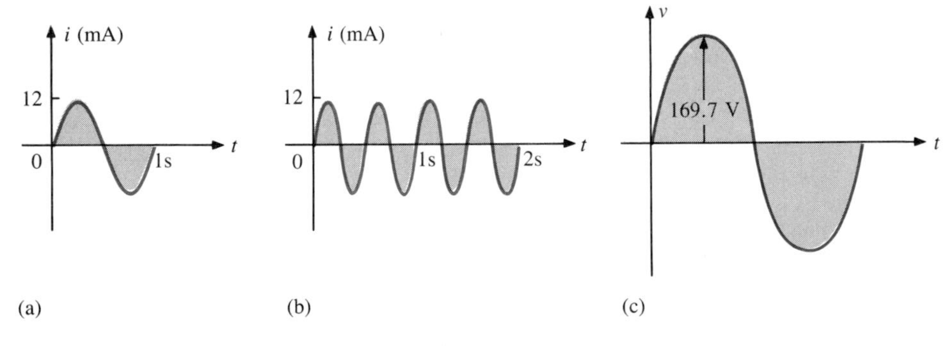

FIG. 13.53

***Solution:*** For part (a), $I_{\text{eff}} = 0.707(12 \times 10^{-3}\text{ A}) =$ **8.484 mA.** For part (b), again $I_{\text{eff}} =$ **8.484 mA.** Note that frequency did not change the effective value in (b) above as compared to (a). For part (c), $V_{\text{eff}} = 0.707(169.73\text{ V}) \cong$ **120 V,** as available from a home outlet.

**EXAMPLE 13.20.** The 120-V dc source of Fig. 13.54(a) delivers 3.6 W to the load. Determine the peak value of the applied voltage ($E_m$) and the current ($I_m$) if the ac source (Fig. 13.54(b)) is to deliver the same power to the load.

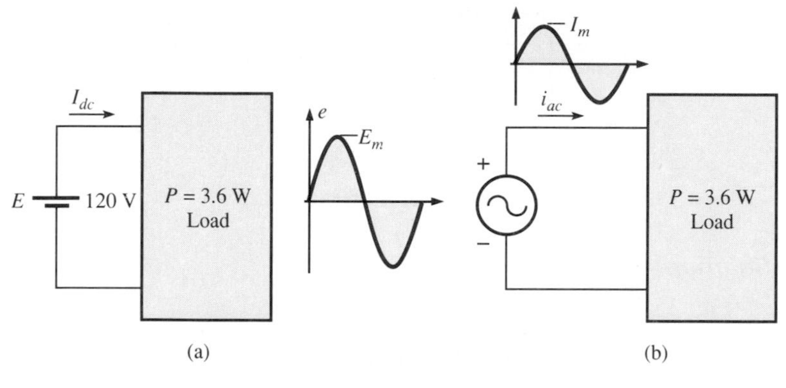

**FIG. 13.54**

***Solution:***

$$P_{dc} = V_{dc}I_{dc}$$

and

$$I_{dc} = \frac{P_{dc}}{V_{dc}} = \frac{3.6\text{ W}}{120\text{ V}} = 30\text{ mA}$$

$$I_m = \sqrt{2}I_{dc} = (1.414)(30\text{ mA}) = \mathbf{42.42\text{ mA}}$$

$$E_m = \sqrt{2}E_{dc} = (1.414)(120\text{ V}) = \mathbf{169.68\text{ V}}$$

**EXAMPLE 13.21.** Find the effective or rms value of the waveform of Fig. 13.55.

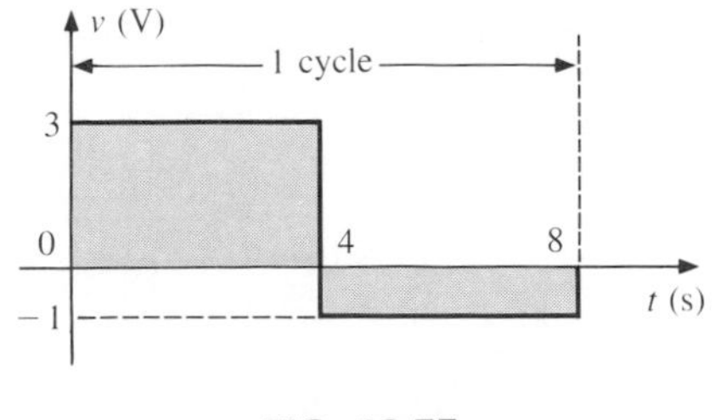


**FIG. 13.55**

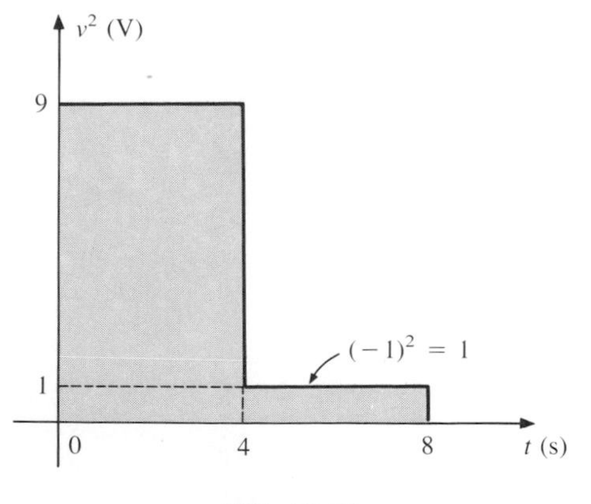


**FIG. 13.56**

***Solution:***

$v^2$ (Fig. 13.56):

$$V_{eff} = \sqrt{\frac{(9)(4) + (1)(4)}{8}} = \sqrt{\frac{40}{8}} = \mathbf{2.236\text{ V}}$$

**EXAMPLE 13.22.** Calculate the effective value of the voltage of Fig. 13.57.

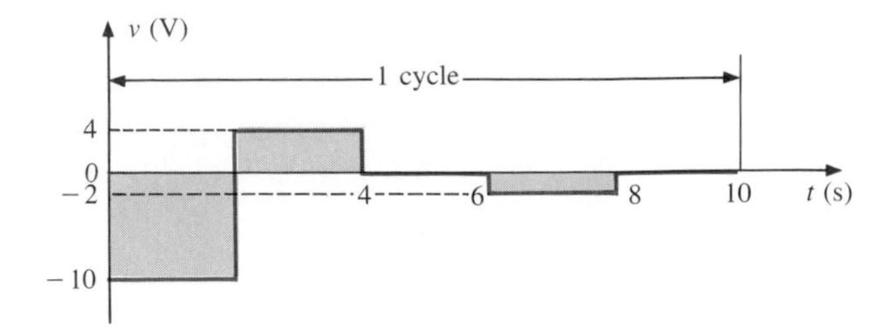

FIG. 13.57

***Solution:***

$v^2$ (Fig. 13.58):

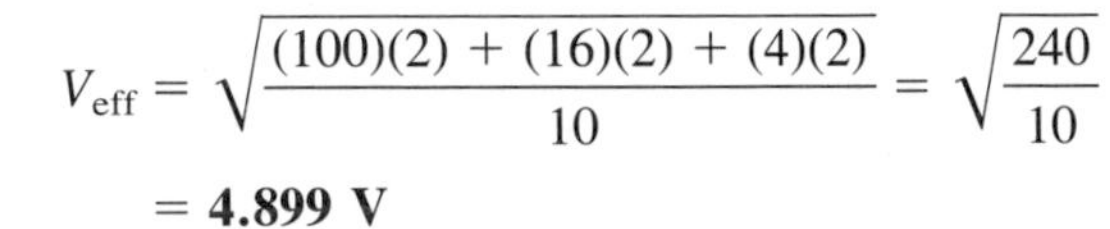

$$V_{\text{eff}} = \sqrt{\frac{(100)(2) + (16)(2) + (4)(2)}{10}} = \sqrt{\frac{240}{10}}$$

$$= \mathbf{4.899\ V}$$

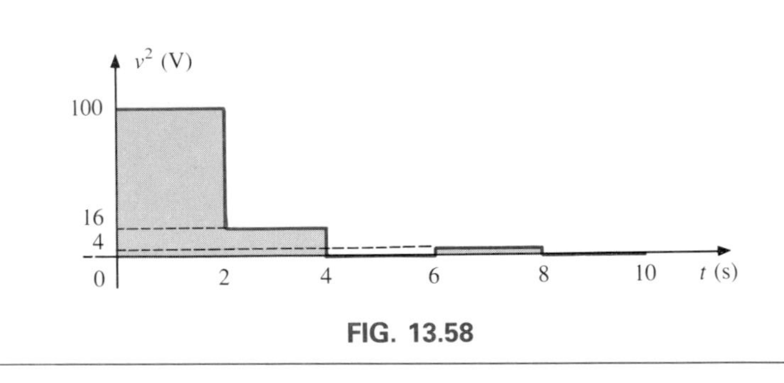


FIG. 13.58

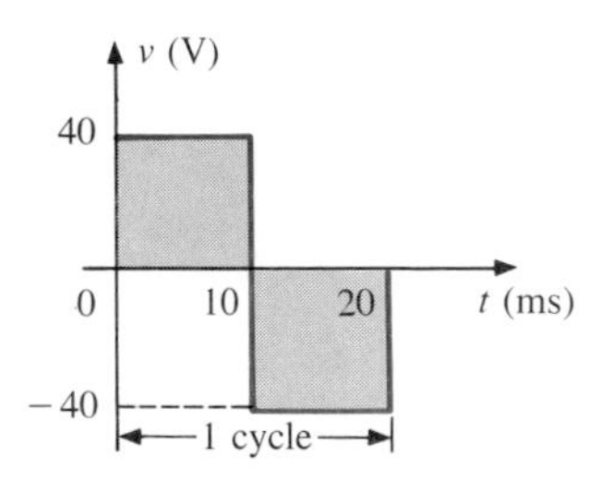

FIG. 13.59

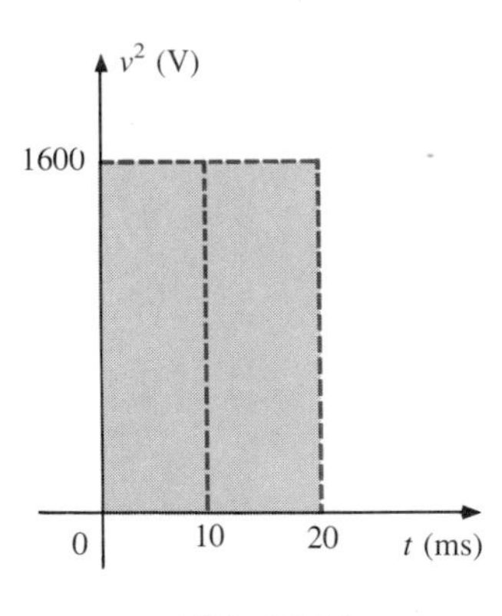


FIG. 13.60

**EXAMPLE 13.23.** Determine the average and effective values of the square wave of Fig. 13.59.

***Solution:*** By inspection, the average value is zero.

$v^2$ (Fig. 13.60):

$$V_{\text{eff}} = \sqrt{\frac{(1600)(10 \times 10^{-3}) + (1600)(10 \times 10^{-3})}{20 \times 10^{-3}}}$$

$$= \sqrt{\frac{32{,}000 \times 10^{-3}}{20 \times 10^{-3}}} = \sqrt{1600}$$

$$V_{\text{eff}} = \mathbf{40\ V}$$

(the maximum value of the waveform of Fig. 13.60)

The waveforms appearing in these examples are the same as those used in the examples on the average value. It might prove interesting to compare the effective and average values of these waveforms.

The effective values of sinusoidal quantities such as voltage or current will be represented by $E$ and $I$. These symbols are the same as those used for dc voltages and currents. To avoid confusion, the peak value of a waveform will always have a subscript $m$ associated with it: $I_m \sin \omega t$. *Caution:* When finding the effective value of the positive pulse of a sine wave, note that the squared area is *not* simply $(2A_m)^2 = 4A_m^2$; it must be found by a completely new integration. This will always be the case for any waveform that is not rectangular.

## 13.10 ac METERS AND INSTRUMENTS

The d'Arsonval movement employed in dc meters can also be used to measure sinusoidal voltages and currents if the *bridge rectifier* of Fig. 13.61 is placed between the signal to be measured and the average reading movement.

The bridge rectifier composed of four diodes (electronic switches) will convert the input signal of zero average value to one having an average value sensitive to the peak value of the input signal. The conversion process is well described in most basic electronics texts. Fundamentally, conduction is permitted through the diodes in such a manner as to convert the sinusoidal input of Fig. 13.62(a) to one having the appearance of Fig. 13.62(b). The negative portion of the input has been effectively "flipped over" by the bridge configuration. The resulting waveform of Fig. 13.62(b) is called a *full-wave rectified waveform.*

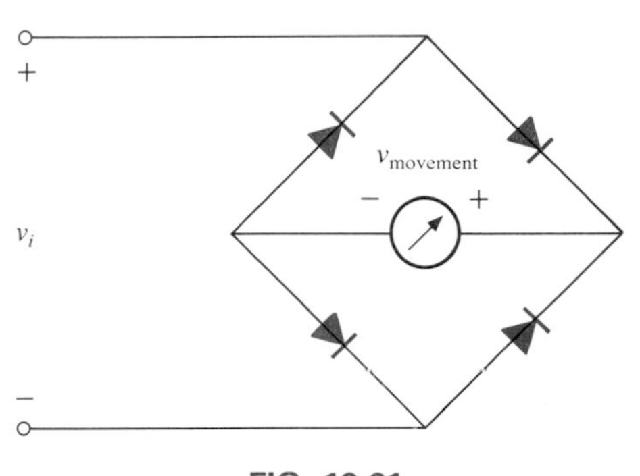

**FIG. 13.61**
*Full-wave bridge rectifier.*

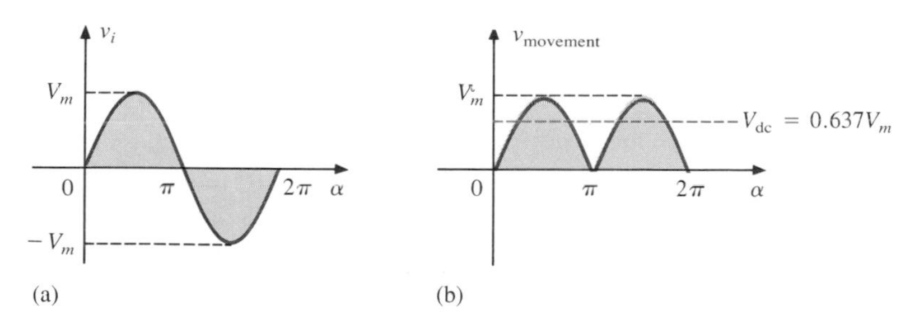

**FIG. 13.62**
*(a) Sinusoidal input; (b) full-wave rectified signal.*

The zero average value of Fig. 13.62(a) has been replaced by a pattern having an average value determined by

$$G = \frac{2V_m + 2V_m}{2\pi} = \frac{4V_m}{2\pi} = \frac{2V_m}{\pi} = 0.637V_m$$

The movement of the pointer will therefore be directly related to the peak value of the signal by the factor 0.637.

Forming the ratio between the rms and dc levels will result in

$$\frac{V_{\text{rms}}}{V_{\text{dc}}} = \frac{0.707V_m}{0.637V_m} \cong 1.11$$

revealing that the scale indication is 1.11 times the dc level measured by the movement. That is,

$$\boxed{\text{Meter indication} = 1.11(\text{dc or average value})} \quad \textit{full-wave} \qquad \textbf{(13.35)}$$

Some ac meters use a half-wave rectifier arrangement that results in the waveform of Fig. 13.63, which has half the average value of Fig. 13.62(b) over one full cycle. The result is

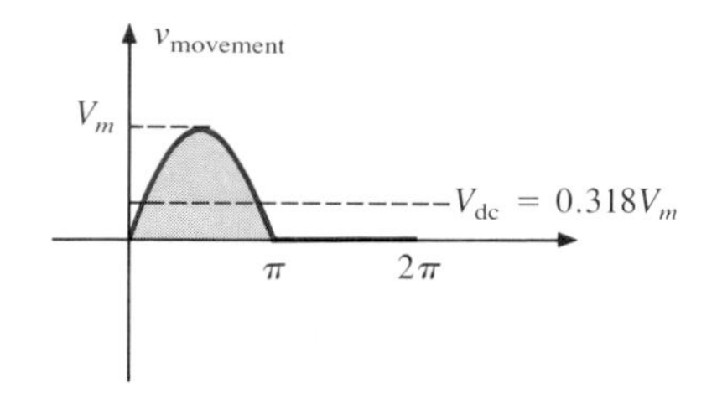


**FIG. 13.63**
*Half-wave rectified signal.*

$$\boxed{\text{Meter indication} = 2.22(\text{dc or average value})} \quad \textit{half-wave} \qquad \textbf{(13.36)}$$

A second movement, called the electrodynamometer movement (Fig. 13.64), can measure both ac and dc quantities without a change in internal circuitry. The movement can, in fact, read the effective value of any periodic or nonperiodic waveform, because a reversal in current direction reverses the fields of both the stationary and the movable coils, so the deflection of the pointer is always up-scale.

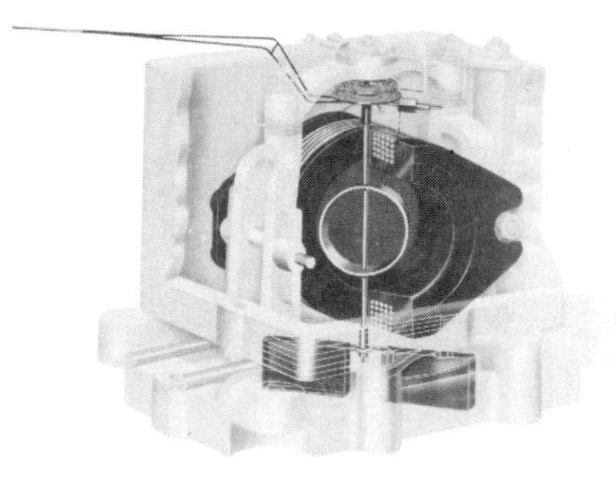

**FIG. 13.64**
*Electrodynamometer movement. (Courtesy of Weston Instruments, Inc.)*

The VOM, introduced in Chapter 2, can be used to measure both dc and ac voltages using a d'Arsonval movement and the proper switching networks. That is, when the meter is used for dc measurements, the dial setting will establish the proper series resistance for the chosen scale and permit the appropriate dc level to pass directly to the movement. For ac measurements, the dial setting will introduce a network that employs a full- or half-wave rectifier to establish a dc level. As discussed above, each setting is properly calibrated to indicate the desired quantity on the face of the instrument.

---

**EXAMPLE 13.24.** Determine the reading of each meter for each situation of Fig. 13.65.

***Solution:*** For part (a), situation (1): By Eq. (13.35),

$$\text{Meter indication} = 1.11(20 \text{ V}) = \mathbf{22.2 \text{ V}}$$

For part (a), situation (2):

$$V_{\text{rms}} = 0.707V_m = 0.707(20 \text{ V}) = \mathbf{14.14 \text{ V}}$$

For part (b), situation (1):

$$V_{\text{rms}} = V_{\text{dc}} = \mathbf{25 \text{ V}}$$

For part (b), situation (2):

$$V_{\text{rms}} = 0.707V_m = 0.707(15\text{ V}) \cong \mathbf{10.6\text{ V}}$$

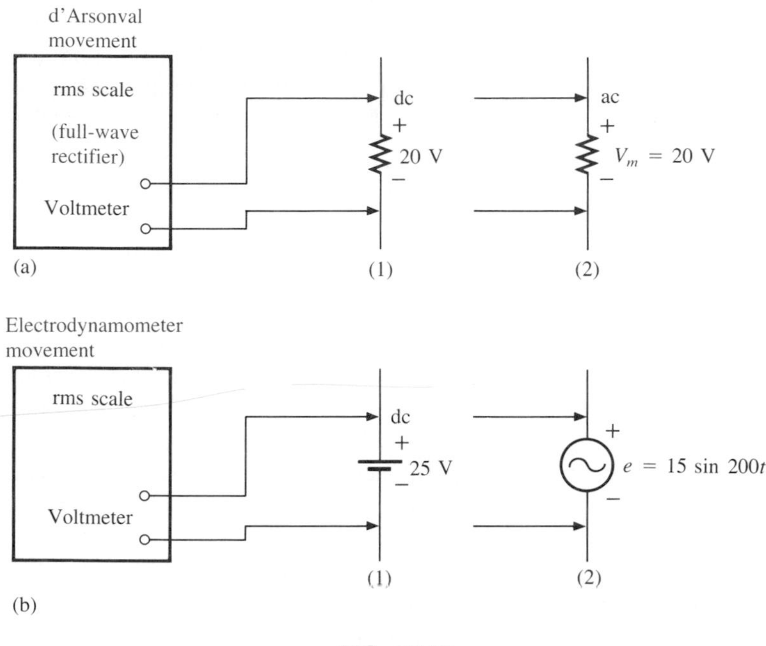


FIG. 13.65

Most DMMs employ a full-wave rectification system to convert the input ac signal to one with an average value. In fact, for the DMM of Fig. 2.26, the same scale factor of Eq. (13.35) is employed. That is, the average value is scaled up by a factor of 1.11 to obtain the rms value. In the digital meters, however, there are no moving parts such as in the d'Arsonval or electrodynamometer movements to display the signal level. Rather, the average value is sensed by a multiprocessor integrated circuit (IC), which in turn determines which digits should appear on the digital display.

Digital meters can also be used to measure nonsinusoidal signals, but the scale factor of each input waveform must first be known (normally provided by the manufacturer in the operator's manual). For instance, the scale factor for an average responding DMM on the ac rms scale will produce an indication for a square-wave input that is 1.11 times the peak value. For a triangular input, the response is 0.555 times the peak value. Obviously, for a sine wave input the response is 0.707 times the peak value.

For any instrument, it is always good practice to read (if only briefly) the operator's manual if it appears you will use the instrument on a regular basis.

For frequency measurements, the frequency counter of Fig. 13.66 provides a digital readout of the frequency or period of waveforms having a frequency range from 5 Hz to 80 MHz. In a period average mode, it can average the cycle time over 10, 100, or 1000 cycles. It has an input impedance of 1 MΩ and an internal rechargeable battery for

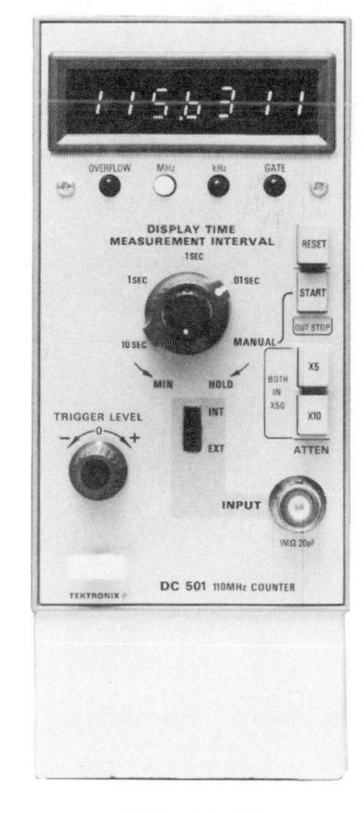


FIG. 13.66
*Frequency counter. (Courtesy of Tektronix, Inc.)*

portability. Note the high degree of accuracy available from the six-digit display.

The Amp-Clamp® of Fig. 13.67 is an instrument that can measure alternating current in the ampere range without having to open the circuit. The loop is opened by squeezing the "trigger"; then it is placed around the current-carrying conductor. Through transformer action, the level of current in rms units will appear on the appropriate scale. The accuracy of this instrument is ±3% of full scale at 60 Hz, and its scales have maximum values ranging from 6 A to 300 A. The addition of two leads as indicated in the figure permits its use as both a voltmeter and an ohmmeter.

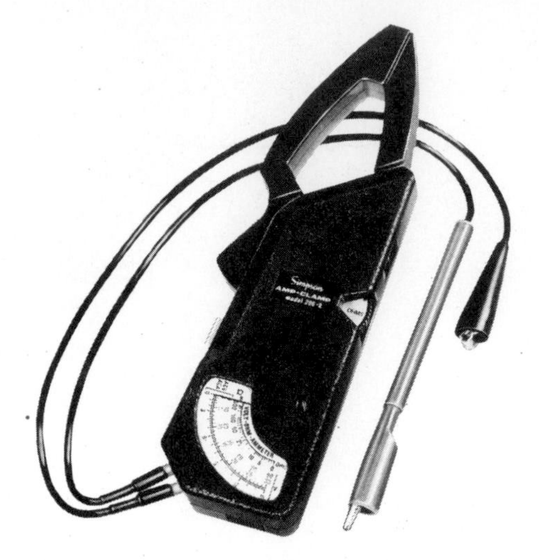

**FIG. 13.67**
Amp-Clamp®. *(Courtesy of Simpson Instruments, Inc.)*

One of the most versatile and important instruments in the electronics industry is the oscilloscope. It provides a display of the waveform on a cathode-ray tube to permit the detection of irregularities and the determination quantities such as magnitude, frequency, period, dc component, and so on. The unit of Fig. 13.68 is particularly interesting for two reasons: It is portable (working off internal batteries), and it is very small and lightweight. It weighs only 3.5 lb and is approximately 3″ × 5″ × 10″ in size. It has an input impedance of 1 MΩ and a time base that can be set for 5 μs to 500 ms per horizontal division. The vertical scale can be set to sensitivities extending from 1 mV to 50 V per division. This oscilloscope can also display two signals (dual trace) at the same time for magnitude and phase comparisons.

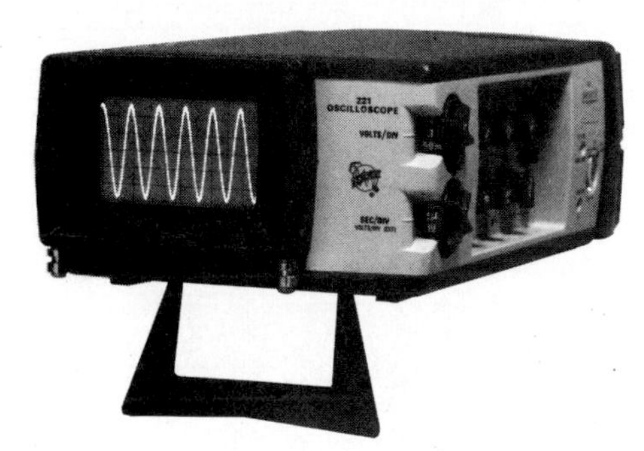

**FIG. 13.68**
*Miniscope. (Courtesy of Tektronix, Inc.)*

A student accustomed to watching TV might be confused when first introduced to an oscilloscope. There is, at least initially, an assumption that the oscilloscope is generating the waveform on the screen—much like a TV broadcast. However, it is important to clearly understand that

***an oscilloscope displays only those signals generated elsewhere and connected to the input terminals of the oscilloscope. The absence of an external signal will simply result in a horizontal line on the screen of the scope.***

On most modern-day oscilloscopes, there is a switch or knob with the choice DC/GND/AC as shown in Fig. 13.69(a) that is often ignored

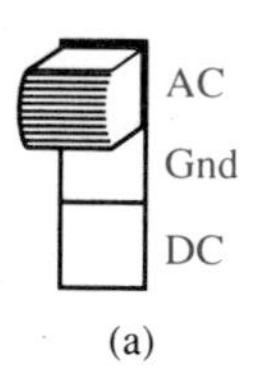


(a)

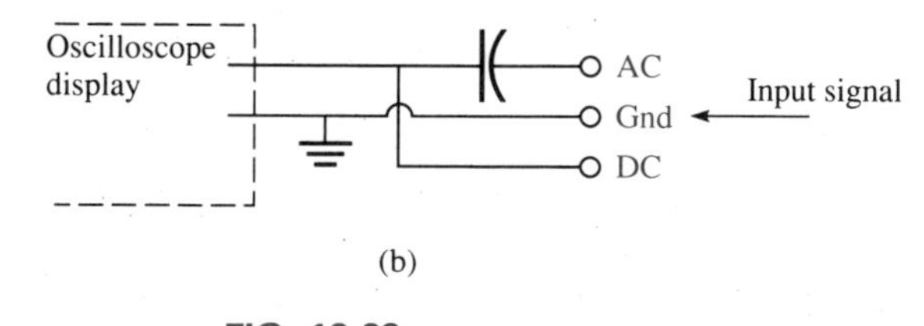


(b)

**FIG. 13.69**

or treated too lightly in the early stages of scope utilization. The effect of each position is fundamentally as shown in Fig. 13.69(b). In the DC mode the dc and ac components of the input signal can pass directly to the display. In the AC position the dc input is blocked by the capacitor, but the ac portion of the signal can pass on through to the screen. In the GND position the input signal is prevented from reaching the scope display by a direct ground connection, which reduces the scope display to a single horizontal line. The DC/AC option is an excellent way to determine the dc content of the input signal. All that is required is to note the vertical swing of the input signal when the switch is moved between the AC and DC positions. The lack of any change in the waveform indicates a lack of dc in the input signal, but an upward swing indicates a positive dc level and a downward shift a negative dc level. Using the vertical sensitivity (V/cm) permits a direct measurement of the dc content.

## PROBLEMS

### SECTION 13.4 Definitions

1. For the cycles of the periodic waveform shown in Fig. 13.70:
   a. Find the period $T$.
   b. How many cycles are shown?
   c. What is the frequency?
   *d. Determine the positive amplitude and peak-to-peak value (think!).

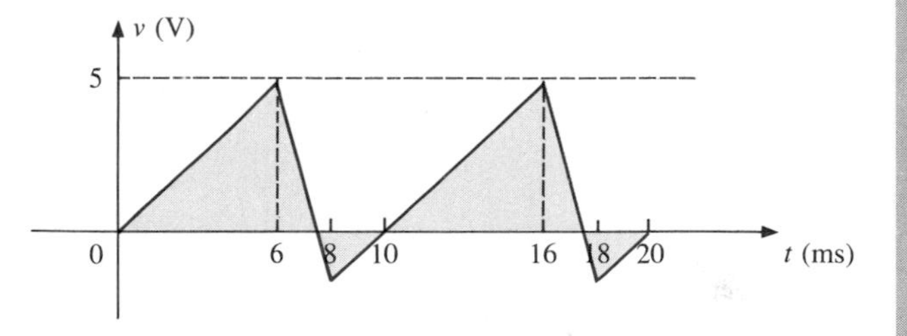

FIG. 13.70

2. Repeat Problem 1 for the periodic waveform of Fig. 13.71.

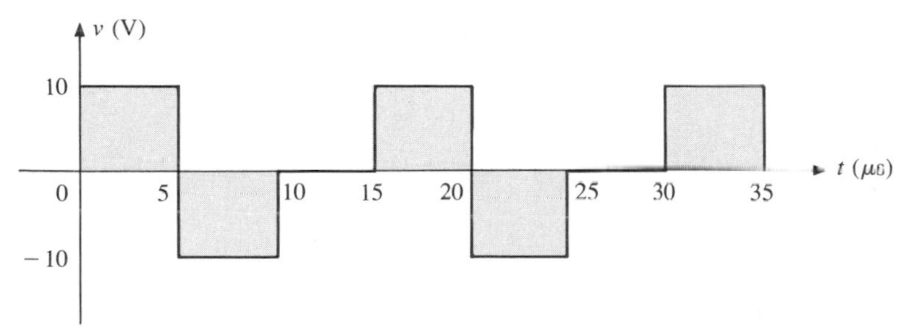

FIG. 13.71

3. Determine the period and frequency of the sawtooth waveform of Fig. 13.72.

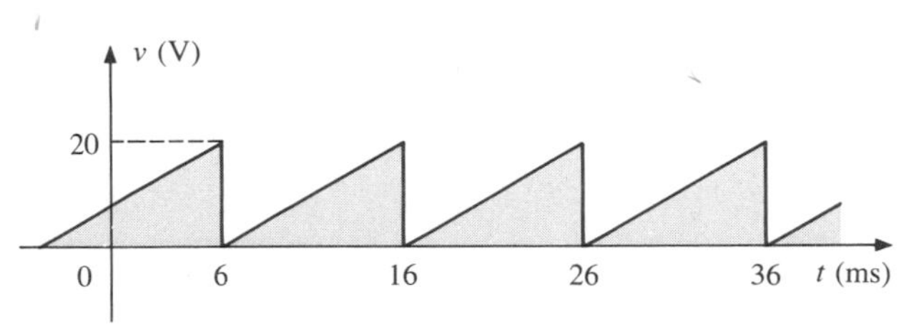


FIG. 13.72

4. Find the period of a periodic waveform whose frequency is
   a. 25 Hz
   b. 35 MHz
   c. 55 kHz
   d. 1 Hz
5. Find the frequency of a repeating waveform whose period is
   a. 1/60 s
   b. 0.01 s
   c. 34 ms
   d. 25 $\mu$s
6. Find the period of a sinusoidal waveform that completes 80 cycles in 24 ms.
7. If a periodic waveform has a frequency of 20 Hz, how long (in seconds) will it take to complete 5 cycles?

8. What is the frequency of a periodic waveform that completes 42 cycles in 6 s?

9. Sketch a periodic square wave like that appearing in Fig. 13.71 with a frequency of 20,000 Hz and a peak value of 10 mV.

10. For the oscilloscope pattern of Fig. 13.73,
    **a.** Determine the peak amplitude.
    **b.** Find the period.
    **c.** Calculate the frequency.
    Redraw the oscilloscope pattern if a +25-mV dc level were added to the input waveform.

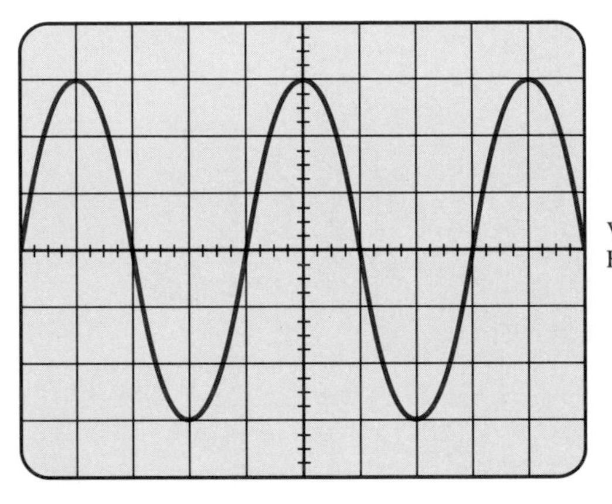


FIG. 13.73

## SECTION 13.5 The Sine Wave

11. Convert the following degrees to radians:

| | |
|---|---|
| **a.** 45° | **b.** 60° |
| **c.** 120° | **d.** 270° |
| **e.** 178° | **f.** 221° |

12. Convert the following radians to degrees:

| | |
|---|---|
| **a.** $\pi/4$ | **b.** $\pi/6$ |
| **c.** $\frac{1}{10}\pi$ | **d.** $\frac{7}{6}\pi$ |
| **e.** $3\pi$ | **f.** $0.55\pi$ |

13. Find the angular velocity of a waveform with a period of

| | |
|---|---|
| **a.** 2 s | **b.** 0.3 ms |
| **c.** 4 μs | **d.** 1/25 s |

14. Find the angular velocity of a waveform with a frequency of

| | |
|---|---|
| **a.** 50 Hz | **b.** 600 Hz |
| **c.** 2 kHz | **d.** 0.004 MHz |

15. Find the frequency and period of sine waves having an angular velocity of

| | |
|---|---|
| **a.** 754 rad/s | **b.** 8.4 rad/s |
| **c.** 6000 rad/s | **d.** 1/16 rad/s |

**16.** Given $f = 60$ Hz, determine how long it will take the sinusoidal waveform to pass through an angle of 45°.

**17.** If a sinusoidal waveform passes through an angle of 30° in 5 ms, determine the angular velocity of the waveform.

## SECTION 13.6 General Format for the Sinusoidal Voltage or Current

**18.** Find the amplitude and frequency of the following waves:
**a.** $20 \sin 377t$ **b.** $5 \sin 754t$
**c.** $10^6 \sin 10{,}000t$ **d.** $0.001 \sin 942t$
**e.** $-7.6 \sin 43.6t$ **f.** $1/42 \sin 6.28t$

**19.** Sketch $5 \sin 754t$ with the abscissa
**a.** angle in degrees.
**b.** angle in radians.
**c.** time in seconds.

**20.** Sketch $10^6 \sin 10{,}000t$ with the abscissa
**a.** angle in degrees.
**b.** angle in radians.
**c.** time in seconds.

**21.** Sketch $-7.6 \sin 43.6t$ with the abscissa
**a.** angle in degrees.
**b.** angle in radians.
**c.** time in seconds.

**22.** If $e = 300 \sin 157t$, how long (in seconds) does it take this waveform to complete 1/2 cycle?

**23.** Given $i = 0.5 \sin \alpha$, determine $i$ at $\alpha = 72°$.

**24.** Given $v = 20 \sin \alpha$, determine $v$ at $\alpha = 1.2\pi$.

***25.** Given $v = 30 \times 10^{-3} \sin \alpha$, determine the angles at which $v$ will be 6 mV.

***26.** If $v = 40$ V at $\alpha = 30°$ and $t = 1$ ms, determine the mathematical expression for the sinusoidal voltage.

## SECTION 13.7 Phase Relations

**27.** Sketch $\sin(377t + 60°)$ with the abscissa
**a.** angle in degrees.
**b.** angle in radians.
**c.** time in seconds.

**28.** Sketch the following waveforms:
**a.** $50 \sin(\omega t + 0°)$ **b.** $-20 \sin(\omega t + 2°)$
**c.** $5 \sin(\omega t + 60°)$ **d.** $4 \cos \omega t$
**e.** $2 \cos(\omega t + 10°)$ **f.** $-5 \cos(\omega t + 20°)$

**29.** Find the phase relationship between the waveforms of each set:
**a.** $v = 4 \sin(\omega t + 50°)$
$i = 6 \sin(\omega t + 40°)$
**b.** $v = 25 \sin(\omega t - 80°)$
$i = 5 \times 10^{-3} \sin(\omega t - 10°)$
**c.** $v = 0.2 \sin(\omega t - 60°)$
$i = 0.1 \sin(\omega t + 20°)$
**d.** $v = 200 \sin(\omega t - 210°)$
$i = 25 \sin(\omega t - 60°)$

*30. Repeat Problem 29 for the following sets:

a. $v = 2\cos(\omega t - 30^\circ)$
$i = 5\sin(\omega t + 60^\circ)$

b. $v = -1\sin(\omega t + 20^\circ)$
$i = 10\sin(\omega t - 70^\circ)$

c. $v = -4\cos(\omega t + 90^\circ)$
$i = -2\sin(\omega t + 10^\circ)$

31. Write the analytical expression for the waveforms of Fig. 13.74 with the phase angle in degrees.

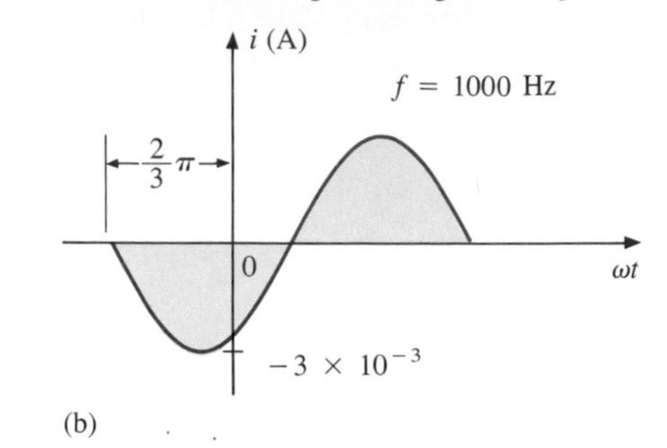


(b)

FIG. 13.74

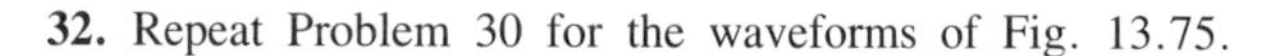

32. Repeat Problem 30 for the waveforms of Fig. 13.75.

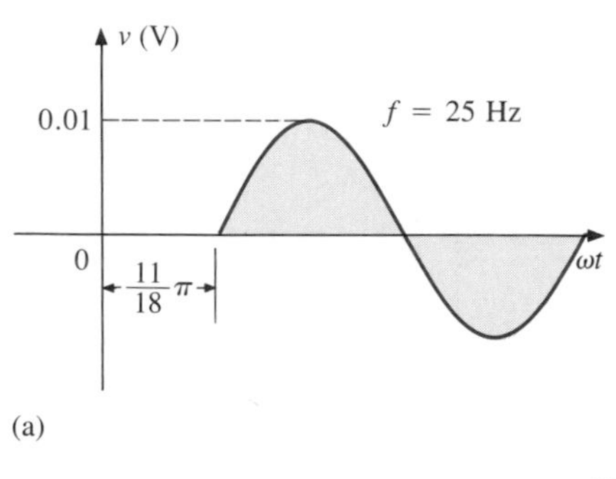


(a)

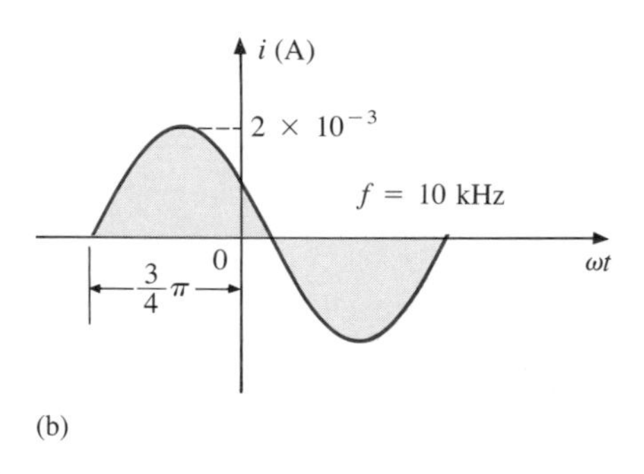


(b)

FIG. 13.75

## SECTION 13.8 Average Value

33. Find the average value of the periodic waveforms of Fig. 13.76 over one full cycle.

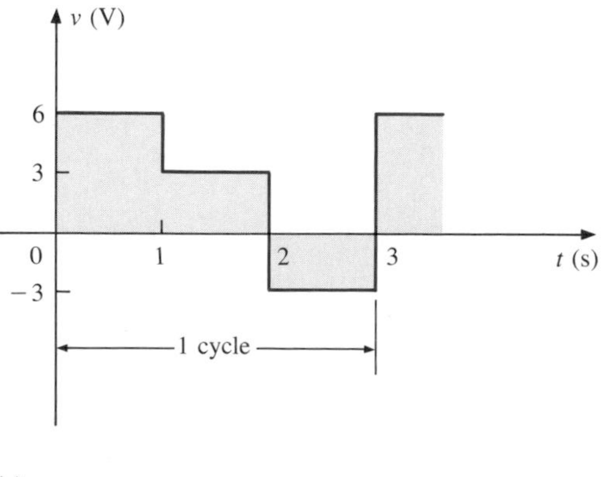


(a)

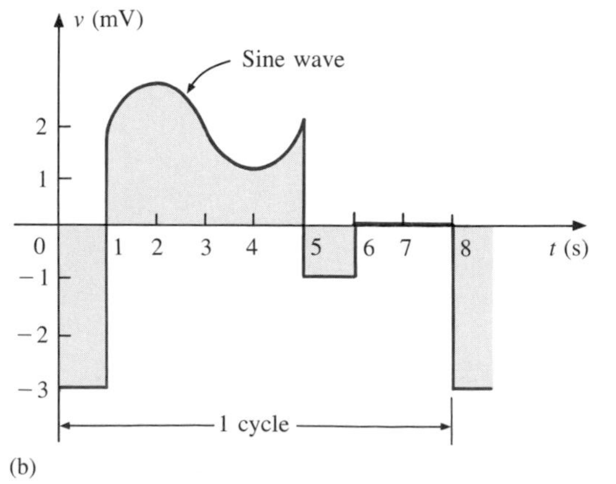


(b)

FIG. 13.76

**34.** Repeat Problem 33 for the waveforms of Fig. 13.77.

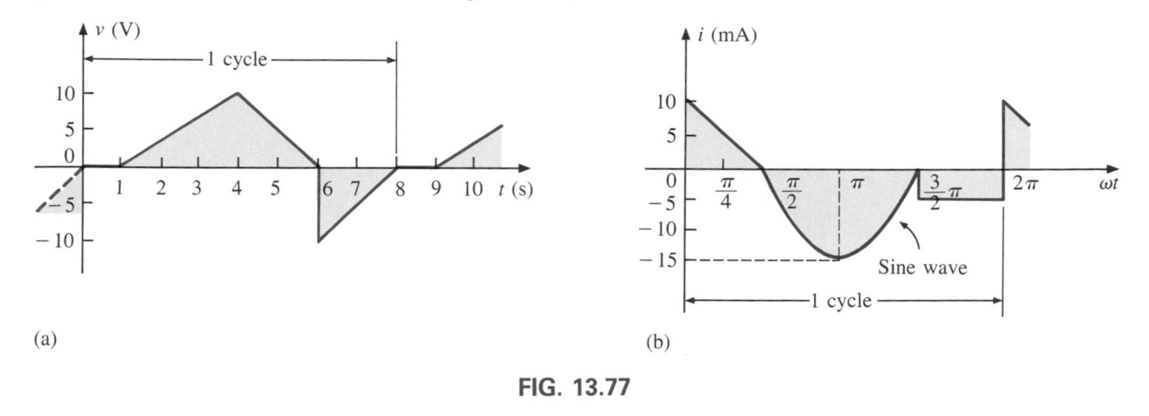


FIG. 13.77

***35.** **a.** By the method of approximation, using familiar geometric shapes, find the area under the curve of Fig. 13.78 from zero to 10 s. Compare your solution with the actual area of 5 volt-seconds (V-s).
**b.** Find the average value of the waveform from zero to 10 s.

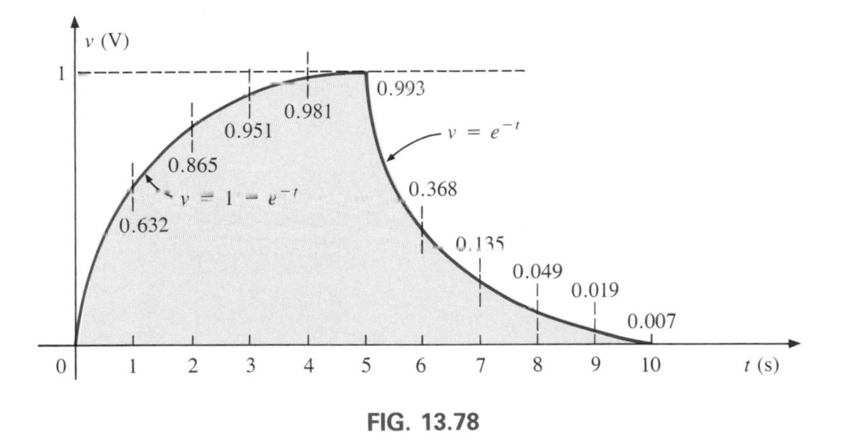


FIG. 13.78

## SECTION 13.9 Effective Values

**36.** Find the effective values of the following sinusoidal waveforms:
**a.** $v = 20 \sin 754t$
**b.** $v = 7.07 \sin 377t$
**c.** $i = 0.006 \sin(400t + 20°)$
**d.** $i = 16 \times 10^{-3} \sin(377t - 10°)$

**37.** Write the sinusoidal expressions for voltages and currents having the following effective values at a frequency of 60 Hz with zero phase shift:
**a.** 1.414 V **b.** 70.7 V
**c.** 0.06 A **d.** 24 $\mu$A

**38.** Find the effective value of the periodic waveform of Fig. 13.79 over one full cycle.

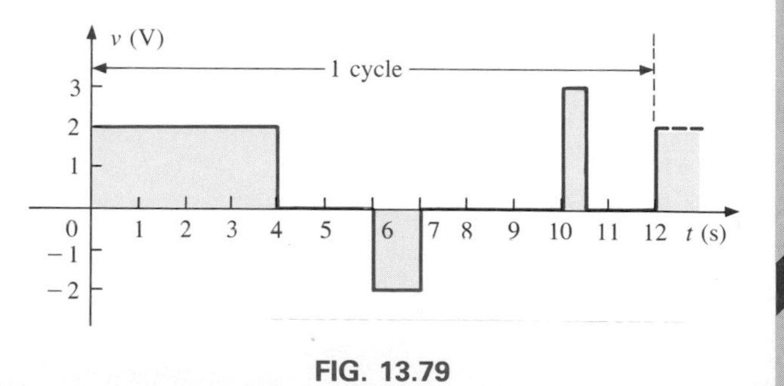


FIG. 13.79

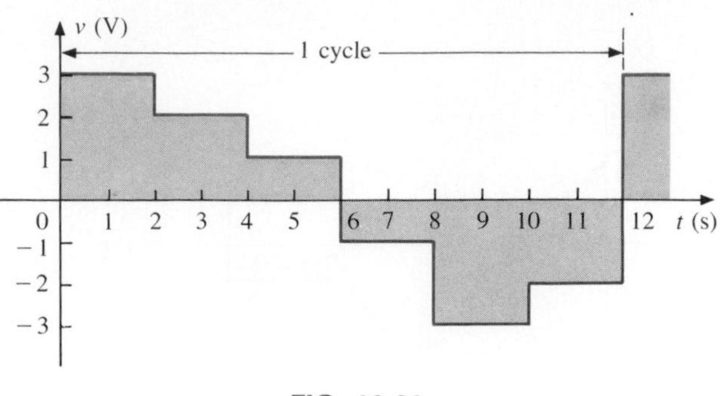

FIG. 13.80

39. Repeat Problem 38 for the waveform of Fig. 13.80.

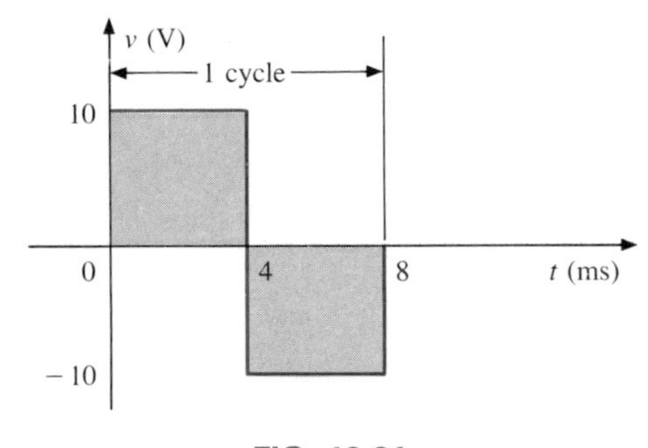


FIG. 13.81

40. What are the average and effective values of the square wave of Fig. 13.81?

41. What are the average and effective values of the waveform of Fig. 13.71?

42. What is the average value of the waveform of Fig. 13.72?

## SECTION 13.10 ac Meters and Instruments

43. Determine the reading of the meter for each situation of Fig. 13.82.

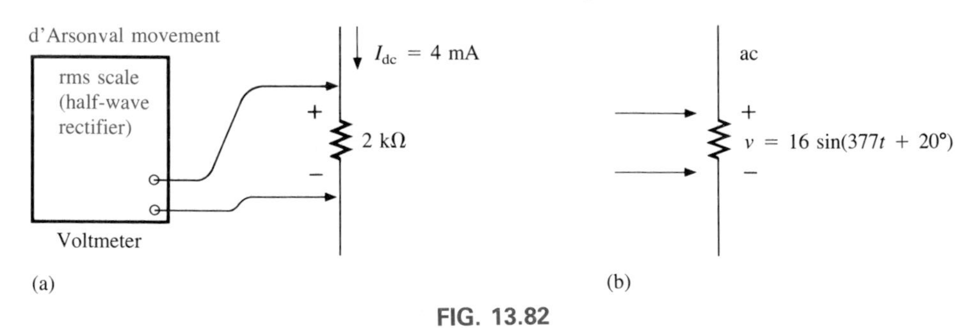


FIG. 13.82

### COMPUTER PROBLEMS

#### BASIC

**44.** Given a sinusoidal function, write a program to determine the effective value, frequency, and period.

**45.** Given two sinusoidal functions, write a program to determine the phase shift between the two waveforms, and indicate which is leading or lagging.

**46.** Given an alternating pulse waveform, write a program to determine the average and effective value of the waveform over one complete cycle.

## GLOSSARY

**Alternating waveform** A waveform that oscillates above and below a defined reference level.

**Amp-Clamp®** A clamp-type instrument that will permit noninvasive current measurements and that can be used as a conventional voltmeter or ohmmeter.

**Angular velocity** The velocity with which a radius vector projecting a sinusoidal function rotates about its center.

**Average value** The level of a waveform defined by the condition that the area enclosed by the curve above this level is exactly equal to the area enclosed by the curve below this level.

**Cycle** A portion of a waveform contained in one period of time.

**Effective value** The equivalent dc value of any alternating voltage or current.

**Electrodynamometer meters** Instruments that can measure both ac and dc quantities without a change in internal circuitry.

**Frequency** ($f$) The number of cycles of a periodic waveform that occur in one second.

**Frequency counter** An instrument that will provide a digital display of the frequency or period of a periodic time-varying signal.

**Instantaneous value** The magnitude of a waveform at any instant of time, denoted by lowercase letters.

**Oscilloscope** An instrument that will display, through the use of a cathode-ray tube, the characteristics of a time-varying signal.

**Peak-to-peak value** The magnitude of the total swing of a signal from positive to negative peaks. The sum of the absolute values of the positive and negative peak values.

**Peak value** The maximum value of a waveform, denoted by uppercase letters.

**Period** ($T$) The time interval between successive repetitions of a periodic waveform.

**Periodic waveform** A waveform that continually repeats itself after a defined time interval.

**Phase relationship** An indication of which of two waveforms leads or lags the other, and by how many degrees or radians.

**Radian** A unit of measure used to define a particular segment of a circle. One radian is approximately equal to 57.3°; $2\pi$ radians are equal to 360°.

**Rectifier-type ac meter** An instrument calibrated to indicate the effective value of a current or voltage through the use of a rectifier network and d'Arsonval-type movement.

**rms value** The root-mean-square or effective value of a waveform.

**Sinusoidal ac waveform** An alternating waveform of unique characteristics that oscillates with equal amplitude above and below a given axis.

**VOM** A multimeter with the capability to measure resistance and both ac and dc levels of current and voltage.

**Waveform** The path traced by a quantity, plotted as a function of some variable such as position, time, degrees, temperature, and so on.

# 14
# The Basic Elements and Phasors

θ

## 14.1 INTRODUCTION

The response of the basic $R$, $L$, and $C$ elements to a sinusoidal voltage and current will be examined in this chapter with special note of how frequency will affect the "opposing" characteristic of each element. Phasor notation will then be introduced to establish a method of analysis that permits a direct correspondence with a number of the methods, theorems, and concepts introduced in the dc chapters.

## 14.2 THE DERIVATIVE

It is fundamental to the understanding of the response of the basic $R$, $L$, and $C$ elements to a sinusoidal signal that the concept of the *derivative* be examined in some detail. It will not be necessary that you become proficient in the mathematical technique but simply that you understand the impact of a relationship defined by a derivative.

Recall from Section 10.11 that the derivative $dx/dt$ is defined as the rate of change of $x$ with respect to time. If $x$ fails to change at a particular instant, $dx = 0$, and the derivative is zero. For the sinusoidal waveform, $dx/dt$ is zero only at the positive and negative peaks ($\omega t = \pi/2$ and $\frac{3}{2}\pi$ in Fig. 14.1), since $x$ fails to change at these instants of time. The derivative $dx/dt$ is actually the slope of the graph at any instant of time.

A close examination of the sinusoidal waveform will also indicate that the greatest change in $x$ will occur at the instants $\omega t = 0$, $\pi$, and $2\pi$. The derivative is therefore a maximum at these points. At 0 and $2\pi$, $x$ increases at its greatest rate, and the derivative is given a positive sign since $x$ increases with time. At $\pi$, $dx/dt$ decreases at the same rate as it increases at 0 and $2\pi$, but the derivative is given a negative sign since $x$ decreases with time. Since the rate of change at 0, $\pi$, and $2\pi$ is the

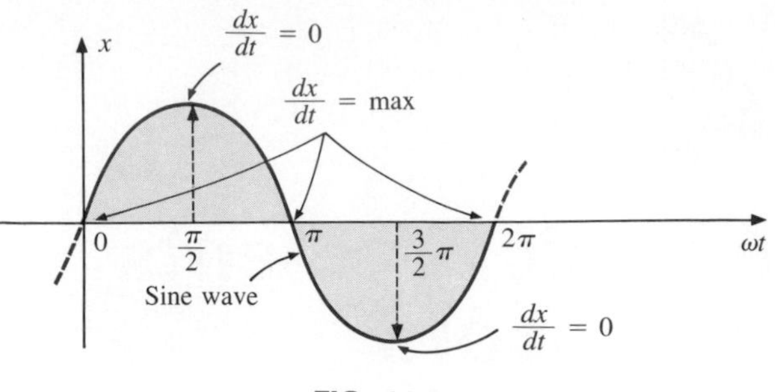

**FIG. 14.1**

same, the magnitude of the derivative at these points is the same also. For various values of $\omega t$ between these maxima and minima, the derivative will exist and have values from the minimum to the maximum inclusive. A plot of the derivative in Fig. 14.2 shows that the derivative of a sine wave is a cosine wave.

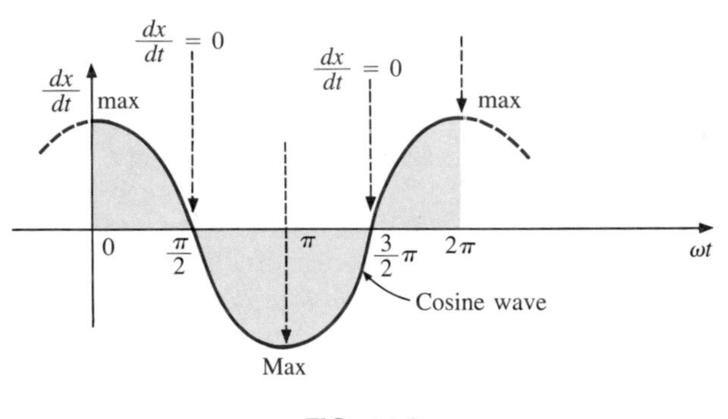

**FIG. 14.2**

The peak value of the cosine wave is directly related to the frequency of the original waveform. The higher the frequency, the steeper the slope at the abscissa, and the greater the value of $dx/dt$, as shown in Fig. 14.3.

Note in Fig. 14.3 that even though both waveforms have the same peak value, the sinusoidal function with the higher frequency produces the larger peak value for the derivative. In addition, note that the derivative has the same period and frequency as the original sinusoidal waveform.

For a sinusoidal voltage

$$e(t) = E_m \sin(\omega t \pm \theta)$$

the derivative can be found directly by differentiation (calculus) to produce the following:

$$\frac{d}{dt} e(t) = \omega E_m \cos(\omega t \pm \theta) = 2\pi f E_m \cos(\omega t \pm \theta) \tag{14.1}$$

clearly substantiating that the derivative of a sinusoidal function is a cosine wave and that the peak value is directly related to the frequency.

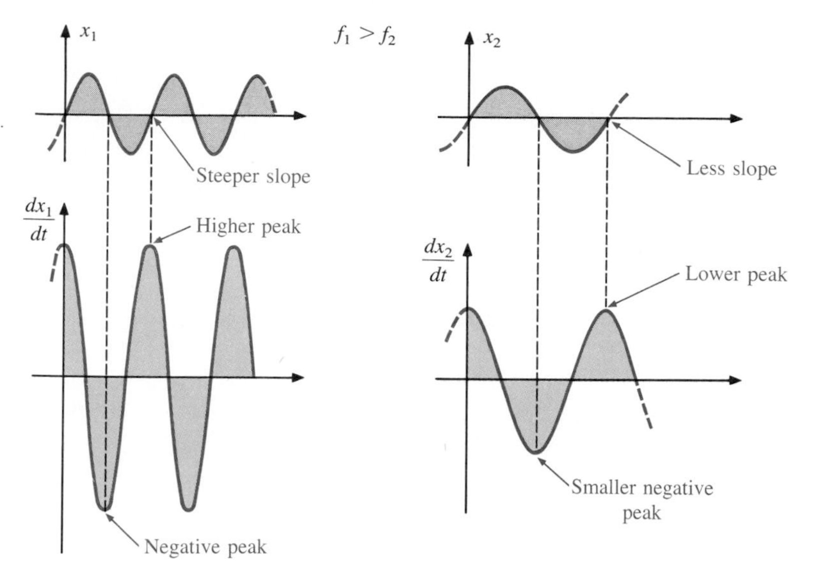


FIG. 14.3

By similar means, if

$$e(t) = E_m \cos(\omega t \pm \theta)$$

then

$$\begin{aligned} \frac{d}{dt} e(t) &= -\omega E_m \sin(\omega t \pm \theta) \\ &= -2\pi f E_m \sin(\omega t \pm \theta) \end{aligned} \quad \textbf{(14.2)}$$

## 14.3 RESPONSE OF BASIC *R, L,* AND *C* ELEMENTS TO A SINUSOIDAL VOLTAGE OR CURRENT

Using Ohm's law and the basic equations for the capacitor and inductor, we can now apply the sinusoidal voltage or current to the basic $R$, $L$, and $C$ elements.

### Resistor

For power-line frequencies and frequencies up to a few hundred kilohertz, resistance is, for all practical purposes, unaffected by the frequency of the applied sinusoidal voltage or current. For this frequency

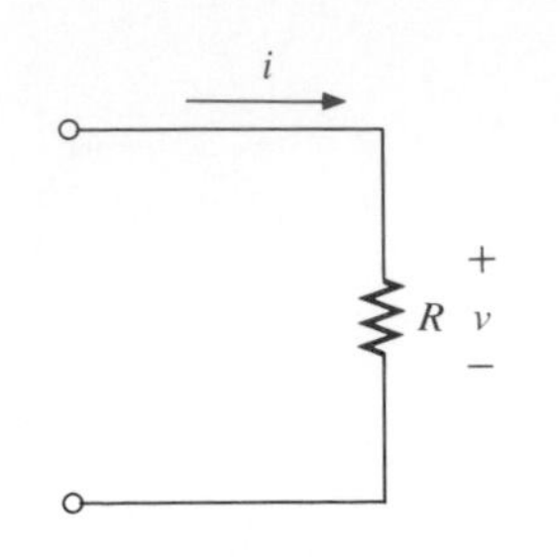

FIG. 14.4

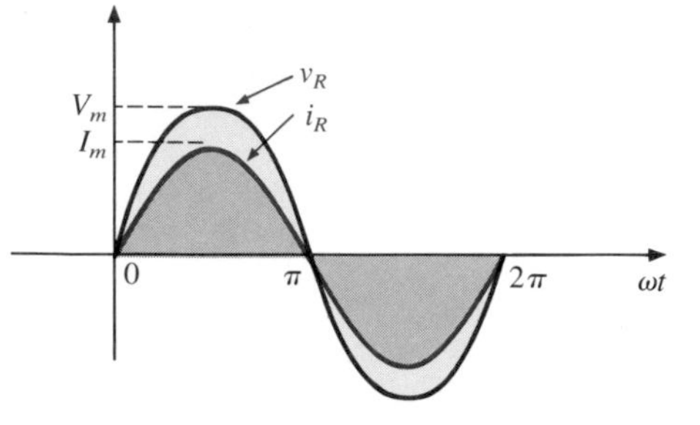


FIG. 14.5

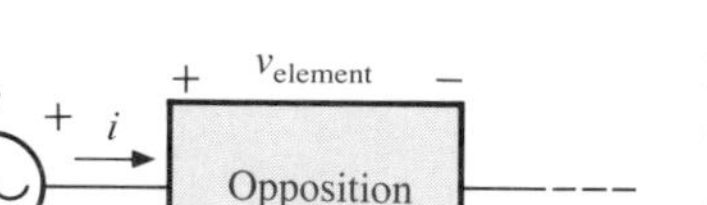


FIG. 14.6

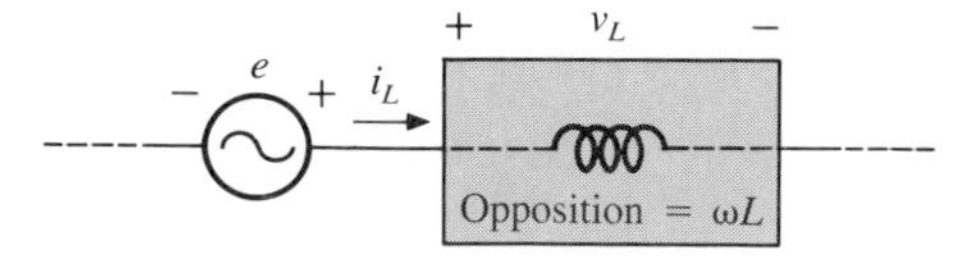


FIG. 14.7

region, the resistor $R$ of Fig. 14.4 can be treated as a constant, and Ohm's law applied as follows. For $v = V_m \sin \omega t$,

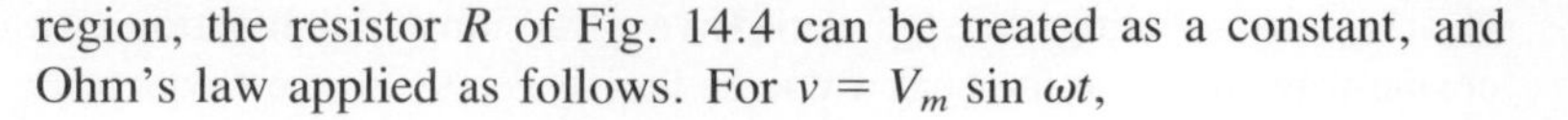

$$i = \frac{v}{R} = \frac{V_m \sin \omega t}{R} = \frac{V_m}{R} \sin \omega t = I_m \sin \omega t$$

where

$$I_m = \frac{V_m}{R} \tag{14.3}$$

In addition, for a given $i$,

$$v = iR = (I_m \sin \omega t)R = I_m R \sin \omega t = V_m \sin \omega t$$

where

$$V_m = I_m R \tag{14.4}$$

A plot of $v$ and $i$ in Fig. 14.5 reveals that

***for a purely resistive element, the voltage across and the current through the element are* in phase.**

## Inductor

For the series configuration of Fig. 14.6, the voltage appearing across the boxed-in element opposes the source $e$ and thereby reduces the magnitude of the current $i$. The magnitude of the voltage across the element is directly related to the opposition of the element to the flow of charge, or current $i$. For a resistive element, we have found that the opposition is its resistance and that $v_{element}$ and $i$ are determined by $v_{element} = iR$.

For the inductor, we found in Chapter 12 that the voltage across an inductor is directly related to the rate of change of current through the coil. Consequently, the higher the frequency, the greater will be the rate of change of current through the coil, and the greater the magnitude of the voltage. In addition, we found in the same chapter that the inductance of a coil will determine the rate of change of flux linking a coil for a particular change in current through the coil. The higher the inductance, the greater the rate of change of the flux linkages, and the greater the resulting voltage across the coil.

The inductive voltage, therefore, is directly related to the frequency (or, more specifically, the angular velocity of the sinusoidal ac current through the coil) and the inductance of the coil. For increasing values of $\omega$ and $L$ in Fig. 14.7, the magnitude of $v_L$ will increase as described above.

Utilizing the similarities between Figs. 14.6 and 14.7, we find that increasing levels of $v_L$ are directly related to increasing levels of opposition in Fig. 14.6. Since $v_L$ will increase with both $\omega$ and $L$, the opposition of an inductive element is defined as their product in Fig. 14.7.

We will now verify some of the preceding conclusions using a more mathematical approach and then define a few important quantities to be employed in the sections and chapters to follow.

For the inductor of Fig. 14.8, we know that

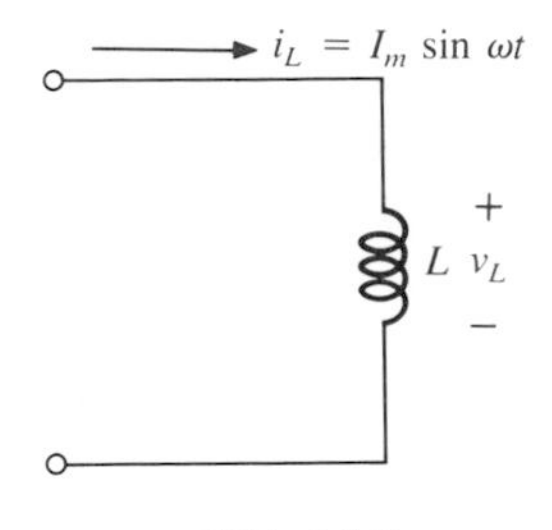

FIG. 14.8

$$v_L = L\frac{di_L}{dt}$$

and, applying differentiation,

$$\frac{di_L}{dt} = \frac{d}{dt}(I_m \sin \omega t) = \omega I_m \cos \omega t$$

Therefore,

$$v_L = L\frac{di_L}{dt} = L(\omega I_m \cos \omega t) = \omega L I_m \cos \omega t$$

or

$$v_L = V_m \sin(\omega t + 90°)$$

where

$$V_m = \omega L I_m$$

Note that the peak value of $v_L$ is directly related to $\omega$ and $L$ as predicted in the discussion above.

A plot of $v_L$ and $i_L$ in Fig. 14.9 reveals that

***for an inductor $v_L$ leads $i_L$ by 90°, or $i_L$ lags $v_L$ by 90°.***

If

$$i_L = I_m \sin(\omega t \pm \theta)$$

then

$$v_L = \omega L I_m \sin(\omega t \pm \theta + 90°)$$

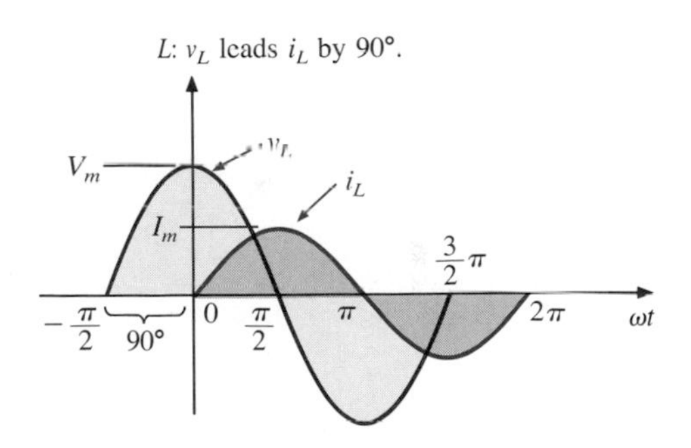


FIG. 14.9

The opposition to current developed by an inductor in a sinusoidal ac network can be found by applying Eq. (4.1):

$$\text{Effect} = \frac{\text{cause}}{\text{opposition}}$$

which for our purposes can be written

$$\text{Opposition} = \frac{\text{cause}}{\text{effect}}$$

Substituting values, we have

$$\text{Opposition} = \frac{V_m}{I_m} = \frac{\omega L I_m}{I_m} = \omega L$$

which agrees with the results obtained earlier.

The quantity $\omega L$, called the *reactance* (from the word *reaction*) of an inductor, is symbolically represented by $X_L$ and is measured in ohms; that is,

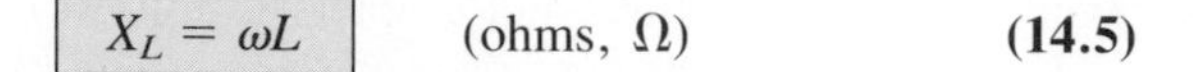

$$X_L = \omega L \qquad \text{(ohms, } \Omega\text{)} \qquad \textbf{(14.5)}$$

Inductive reactance is the opposition to the flow of current, which results in the continual interchange of energy between the source and the magnetic field of the inductor. In other words, reactance, unlike resistance (which dissipates energy in the form of heat), does not dissipate electrical energy.

## Capacitor

Let us now return to the series configuration of Fig. 14.6 and insert the capacitor as the element of interest. For the capacitor, however, we will determine $i$ for a particular voltage across the element. When this approach reaches its conclusion, the relationship between the voltage and current will be known, and the opposing voltage ($v_{\text{element}}$) can be determined for any sinusoidal current $i$.

Our investigation of the inductor revealed that the inductive voltage across a coil opposes the instantaneous change in current through the coil. For capacitive networks, the voltage across the capacitor is limited by the rate at which charge can be deposited on, or released by, the plates of the capacitor during the charging and discharging phases, respectively. In other words, an instantaneous change in voltage across a capacitor is opposed by the fact that there is an element of time required to deposit charge on (or release charge from) the plates of a capacitor, and $V = Q/C$.

Since capacitance is a measure of the rate at which a capacitor will store charge on its plates

***for a particular change in voltage across the capacitor, the greater the value of capacitance, the greater will be the resulting capacitive current.***

In addition, the fundamental equation relating the voltage across a capacitor to the current of a capacitor [$i = C\,(dv/dt)$] indicates that

***for a particular capacitance, the greater the rate of change of voltage across the capacitor, the greater the capacitive current.***

Certainly, an increase in frequency corresponds to an increase in the rate of change of voltage across the capacitor and to an increase in the current of the capacitor.

The current of a capacitor is therefore directly related to the frequency (or, again more specifically, the angular velocity) and the capacitance of the capacitor. An increase in either quantity will result in an increase in the current of the capacitor. For the basic configuration of Fig. 14.10, however, we are interested in determining the opposition of the capacitor as related to the resistance of a resistor and $\omega L$ for the

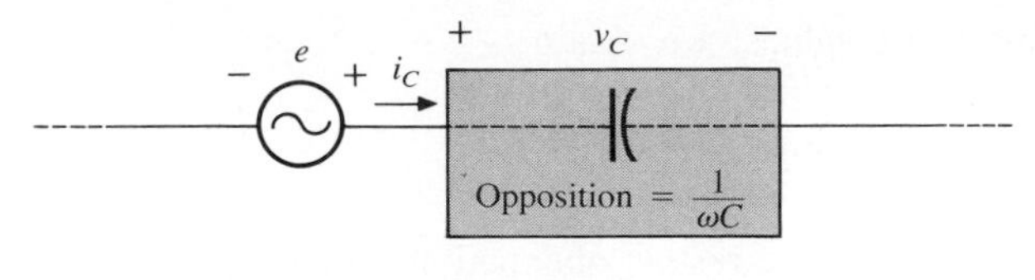


FIG. 14.10

inductor. Since an increase in current corresponds to a decrease in opposition, and $i_C$ is proportional to $\omega$ and $C$, the opposition of a capacitor is directly related to the reciprocal of $\omega C$, or $1/\omega C$, as shown in Fig. 14.10. In other words, the higher the angular velocity (or frequency) and capacitance, the less the opposition to the current $i_C$ or the lower the countervoltage of the capacitor ($v_C$), which limits the current $i_C$ as indicated by Fig. 14.6.

We will now verify, as we did for the inductor, some of these conclusions using a more mathematical approach. Certain important quantities used repeatedly in the following analysis will then be defined.

For the capacitor of Fig. 14.11, we know that

$$i_C = C\,\frac{dv_C}{dt}$$

and that

$$\frac{dv_C}{dt} = \frac{d}{dt}(V_m \sin \omega t) = \omega V_m \cos \omega t$$

Therefore,

$$i_C = C\,\frac{dv_C}{dt} = C(\omega V_m \cos \omega t) = \omega C V_m \cos \omega t$$

or

$$i_C = I_m \sin(\omega t + 90°)$$

where

$$I_m = \omega C V_m$$

Note that the peak value of $i_C$ is directly related to $\omega$ and $C$, as predicted in the discussion above.

A plot of $v_C$ and $i_C$ in Fig. 14.12 reveals that

***for a capacitor, $i_C$ leads $v_C$ by 90°, or $v_C$ lags $i_C$ by 90°.***

If

$$v_C = V_m \sin(\omega t \pm \theta)$$

then

$$i_C = \omega C V_m \sin(\omega t \pm \theta + 90°)$$

Applying

$$\text{Opposition} = \frac{\text{cause}}{\text{effect}}$$

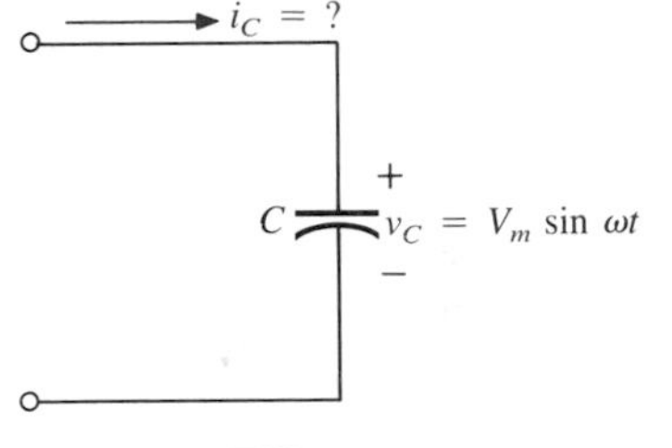


FIG. 14.11

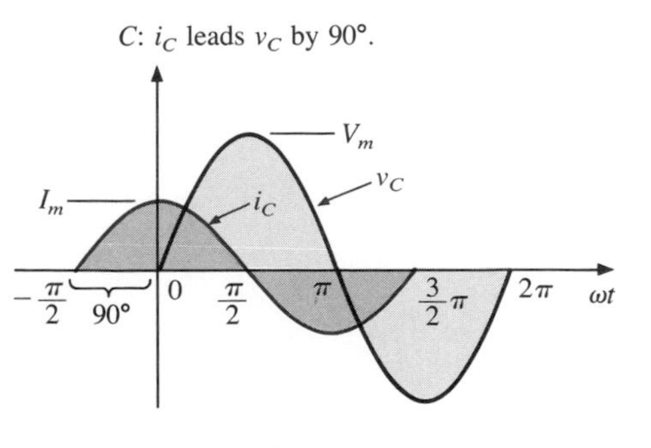


FIG. 14.12

and substituting values, we obtain

$$\text{Opposition} = \frac{V_m}{I_m} = \frac{V_m}{\omega C V_m} = \frac{1}{\omega C}$$

which agrees with the results obtained above.

The quantity $1/\omega C$, called the *reactance* of a capacitor, is symbolically represented by $X_C$ and is measured in ohms; that is,

$$X_C = \frac{1}{\omega C} \quad (\Omega) \tag{14.6}$$

Capacitive reactance is the opposition to the flow of charge, which results in the continual interchange of energy between the source and the electric field of the capacitor. Like the inductor, the capacitor does *not* dissipate energy in any form (ignoring the effects of the leakage resistance).

In the circuits just considered, the current was given in the inductive circuit, and the voltage in the capacitive circuit. This was done to avoid the use of integration in finding the unknown quantities. In the inductive circuit,

$$v_L = L\frac{di_L}{dt}$$

but

$$i_L = \frac{1}{L}\int v_L\,dt \tag{14.7}$$

In the capacitive circuit,

$$i_C = C\frac{dv_C}{dt}$$

but

$$v_C = \frac{1}{C}\int i_C\,dt \tag{14.8}$$

Shortly, we shall consider a method of analyzing ac circuits that will permit us to solve for an unknown quantity with sinusoidal input without having to use direct integration or differentiation.*

---

*A mnemonic phrase sometimes used to remember the phase relationship between the voltage and current of a coil and capacitor is ''*ELI* the *ICE* man.'' Note that the $L$ (inductor) has the *ER* before the $I$ ($e$ leads $i$ by 90°), and the $C$ (capacitor) has the $I$ before the $E$ ($i$ leads $e$ by 90°).

It is possible to determine whether a circuit with one or more elements is predominantly capacitive or inductive by noting the phase relationship between the input voltage and current.

***If the current leads the voltage, the circuit is predominantly capacitive, and if the voltage leads the current, it is predominantly inductive.***

Since we now have an equation for the reactance of an inductor or capacitor, we do not need to use derivatives or integration in the examples to be considered. Simply applying Ohm's law, $I_m = E_m/X_L$ (or $X_C$), and keeping in mind the phase relationship between the voltage and current for each element, will be sufficient to complete the examples.

---

**EXAMPLE 14.1.** The voltage across a resistor is indicated. Find the sinusoidal expression for the current if the resistor is 10 Ω. Sketch the $v$ and $i$ curves with the angle $\omega t$ as the abscissa.

a. $v = 100 \sin 377t$

b. $v = 25 \sin(377t + 60°)$

***Solutions:***

a. $i = \dfrac{v}{R} = \dfrac{100}{10} \sin 377t$

and

$$i = \mathbf{10 \sin 377}\boldsymbol{t}$$

The curves are sketched in Fig. 14.13.

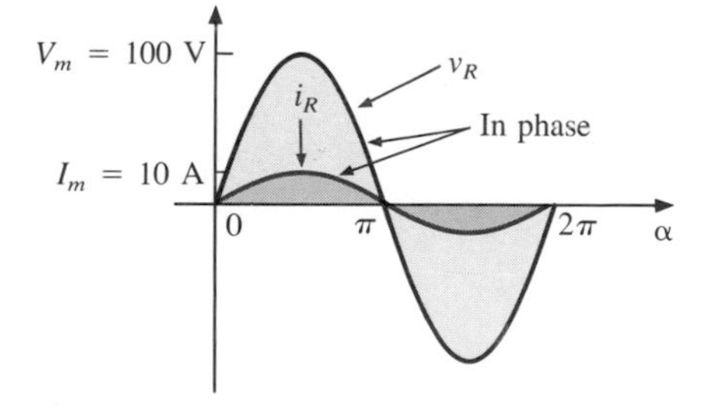


FIG. 14.13

b. $i = \dfrac{v}{R} = \dfrac{25}{10} \sin(377t + 60°)$

and

$$i = \mathbf{2.5 \sin(377}\boldsymbol{t}\mathbf{ + 60°)}$$

The curves are sketched in Fig. 14.14.

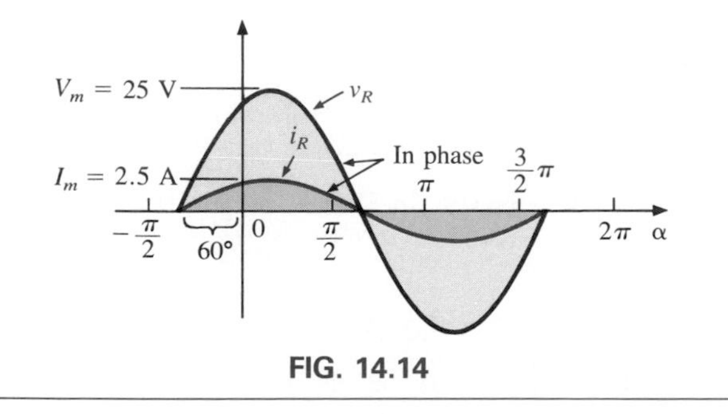


FIG. 14.14

---

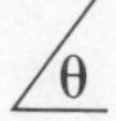


**EXAMPLE 14.2.** The current through a 5-Ω resistor is given. Find the sinusoidal expression for the voltage across the resistor for $i = 40 \sin(377t + 30°)$.

***Solution:***

$$v = iR = (40)(5) \sin(377t + 30°)$$

and

$$v = \mathbf{200 \sin(377t + 30°)}$$

**EXAMPLE 14.3.** The current through a 0.1-H coil is given. Find the sinusoidal expression for the voltage across the coil. Sketch the $v$ and $i$ curves.

a. $i = 10 \sin 377t$
b. $i = 7 \sin(377t - 70°)$

***Solutions:***

a. $X_L = \omega L = (377 \text{ rad/s})(0.1 \text{ H}) = 37.7 \ \Omega$
$V_m = I_m X_L = (10 \text{ A})(37.7 \ \Omega) = 377 \text{ V}$

and we know that for a coil, $v$ leads $i$ by 90°. Therefore,

$$v = \mathbf{377 \sin(377t + 90°)}$$

The curves are sketched in Fig. 14.15.

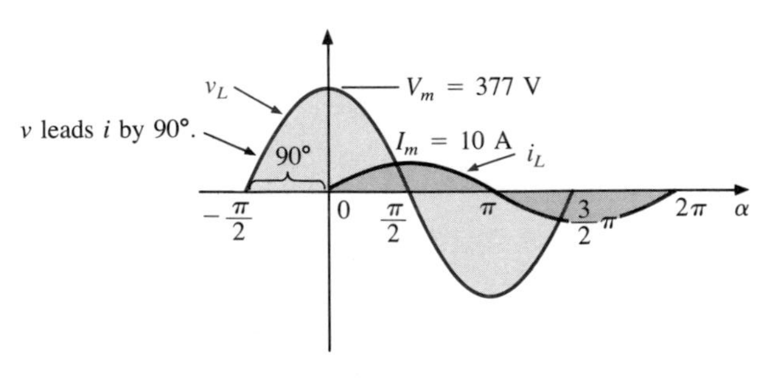


FIG. 14.15

b. $X_L = 37.7 \ \Omega$
$V_m = I_m X_L = (7 \text{ A})(37.7 \ \Omega) = 263.9 \text{ V}$
and we know that for a coil, $v$ leads $i$ by 90°. Therefore,

$$v = 263.9 \sin(377t - 70° + 90°)$$

and

$$v = \mathbf{263.9 \sin(377t + 20°)}$$

The curves are sketched in Fig. 14.16.

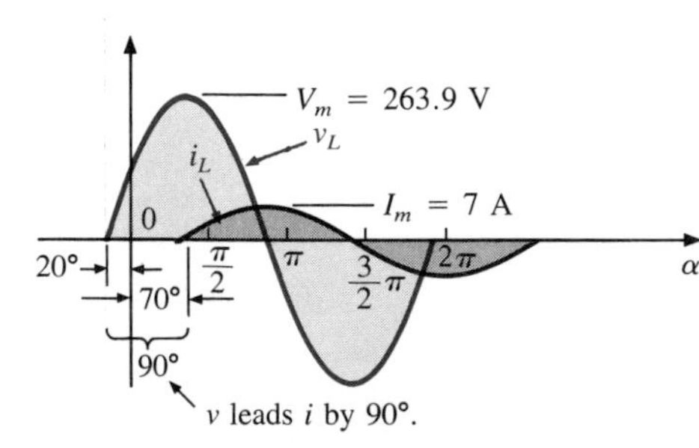


FIG. 14.16

**EXAMPLE 14.4.** The voltage across a 0.5-H coil is given. What is the sinusoidal expression for the current?

$$v = 100 \sin 20t$$

***Solution:***

$$X_L = \omega L = (20 \text{ rad/s})(0.5 \text{ H}) = 10 \ \Omega$$

$$I_m = \frac{V_m}{X_L} = \frac{100 \text{ V}}{10 \ \Omega} = 10 \text{ A}$$

and we know that $i$ lags $v$ by 90°. Therefore,

$$i = \mathbf{10 \sin(20}t\mathbf{ - 90°)}$$

**EXAMPLE 14.5.** The voltage across a 1-$\mu$F capacitor is given. What is the sinusoidal expression for the current? Sketch the $v$ and $i$ curves.

$$v = 30 \sin 400t$$

***Solution:***

$$X_C = \frac{1}{\omega C} = \frac{1}{(400 \text{ rad/s})(1 \times 10^{-6} \text{ F})} = \frac{10^6 \ \Omega}{400} = 2500 \ \Omega$$

$$I_m = \frac{V_m}{X_C} = \frac{30 \text{ V}}{2500 \ \Omega} = 0.0120 \text{ A} - 12 \text{ mA}$$

and we know that for a capacitor, $i$ leads $v$ by 90°. Therefore,

$$i = \mathbf{12 \times 10^{-3} \sin(400}t\mathbf{ + 90°)}$$

The curves are sketched in Fig. 14.17.

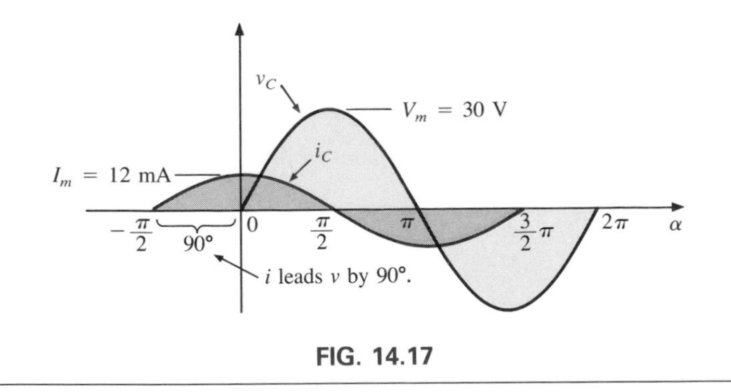


**FIG. 14.17**

**EXAMPLE 14.6.** The current through a 100-$\mu$F capacitor is given. Find the sinusoidal expression for the voltage across the capacitor.

$$i = 40 \sin(500t + 60°)$$

***Solution:***

$$X_C = \frac{1}{\omega C} = \frac{1}{(500 \text{ rad/s})(100 \times 10^{-6} \text{ F})} = \frac{10^6 \ \Omega}{5 \times 10^4} = \frac{10^2 \ \Omega}{5} = 20 \ \Omega$$

$$V_m = I_m X_C = (40 \text{ A})(20 \ \Omega) = 800 \text{ V}$$

and we know that for a capacitor, $v$ lags $i$ by 90°. Therefore,

$$v = 800 \sin(500t + 60° - 90°)$$

and

$$v = \mathbf{800 \sin(500t - 30°)}$$

---

**EXAMPLE 14.7.** For the following pairs of voltages and currents, indicate whether the element involved is a capacitor, inductor, or resistor, and determine the value of $C$, $L$, or $R$, if sufficient data are given (Fig. 14.18):

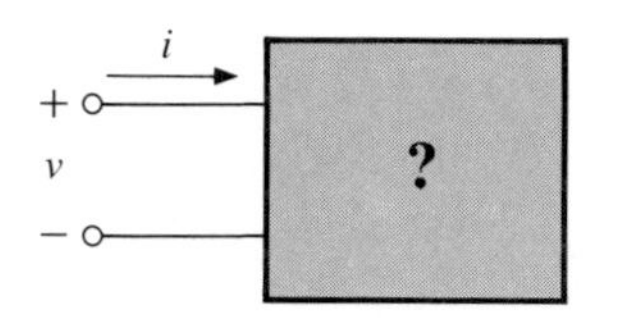


FIG. 14.18

a. $v = 100 \sin(\omega t + 40°)$
   $i = 20 \sin(\omega t + 40°)$
b. $v = 1000 \sin(377t + 10°)$
   $i = 5 \sin(377t - 80°)$
c. $v = 500 \sin(157t + 30°)$
   $i = 1 \sin(157t + 120°)$
d. $v = 50 \cos(\omega t + 20°)$
   $i = 5 \sin(\omega t + 110°)$

***Solutions:***

a. Since $v$ and $i$ are *in phase,* the element is a *resistor,* and

$$R = \frac{V_m}{I_m} = \frac{100 \text{ V}}{20 \text{ A}} = \mathbf{5 \ \Omega}$$

b. Since $v$ *leads* $i$ by 90°, the element is an *inductor,* and

$$X_L = \frac{V_m}{I_m} = \frac{1000 \text{ V}}{5 \text{ A}} = 200 \ \Omega$$

so that

$$X_L = \omega L = 200 \ \Omega$$

or

$$L = \frac{200 \ \Omega}{\omega} = \frac{200 \ \Omega}{377 \text{ rad/s}} = \mathbf{0.531 \text{ H}}$$

c. Since $i$ *leads* $v$ by 90°, the element is a *capacitor,* and

$$X_C = \frac{V_m}{I_m} = \frac{500 \text{ V}}{1 \text{ A}} = 500 \ \Omega$$

so that

$$X_C = \frac{1}{\omega C} = 500 \ \Omega$$

or

$$C = \frac{1}{\omega 500\ \Omega} = \frac{1}{(157\ \text{rad/s})(500\ \Omega)} = \mathbf{12.74\ \mu F}$$

d. $v = 50\cos(\omega t + 20°) = 50\sin(\omega t + 20° + 90°)$
$= 50\sin(\omega t + 110°)$

Since $v$ and $i$ are *in phase*, the element is a *resistor*, and

$$R = \frac{V_m}{I_m} = \frac{50\ \text{V}}{5\ \text{A}} = \mathbf{10\ \Omega}$$

For dc circuits, the frequency is zero, since the currents and voltages have constant magnitudes. The reactance of the coil for dc is, therefore,

$$X_L = 2\pi fL = 2\pi(0)L = 0\ \Omega$$

hence the short-circuit representation for the inductor in dc circuits (Chapter 12). At very high frequencies, $X_L\uparrow = 2\pi f\uparrow L$ is very large, and for some practical applications the inductor can be replaced by an open circuit.

The capacitor can be replaced by an open circuit in dc circuits since $f = 0$, and

$$X_C = \frac{1}{2\pi fC} = \frac{1}{2\pi(0)C} = \infty\ \Omega$$

once again substantiating our previous action (Chapter 10). At very high frequencies, for finite capacitances,

$$X_C\downarrow = \frac{1}{2\pi f\uparrow C}$$

is very small, and for some practical applications the capacitor can be replaced by a short circuit.

Table 14.1 reviews the preceding conclusions.

**TABLE 14.1**

*Effect of high and low frequencies on the circuit model of an inductor and capacitor.*

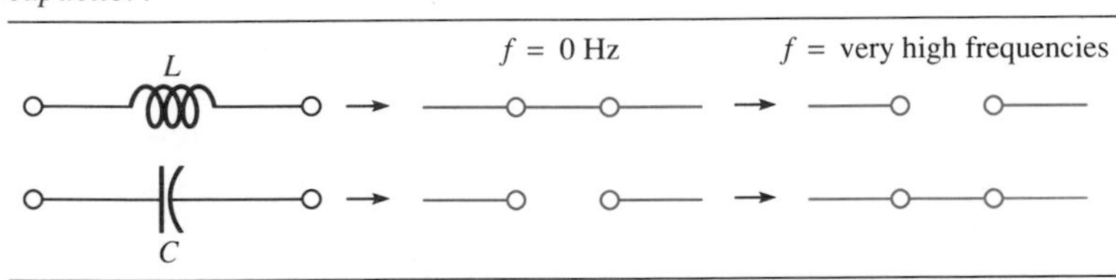


Now that we are familiar with phase relationships and understand how the elements affect the phase relationship between the applied voltage and resulting current, the use of the oscilloscope to measure the phase angle can be introduced. Recall from past discussions that the oscilloscope can be used only to display voltage levels versus time. However, now that we realize that the voltage across a resistor is in phase with the current through a resistor, we can consider the phase angle associated with the voltage across any resistor actually to be the phase angle of the current. For example, suppose we want to find the phase angle introduced by the unknown system of Fig. 14.19(a). In Fig. 14.19(b) a resistor was added to the input leads, and the two channels of a dual trace (most modern-day oscilloscopes can display two signals at the same time) were connected as shown. One channel will display the input voltage $v_i$, whereas the other will display $v_R$, as shown in Fig. 14.19(c). However, as noted before, since $v_R$ and $i_R$ are in phase, the phase angle appearing in Fig. 14.19(c) is also the phase angle between

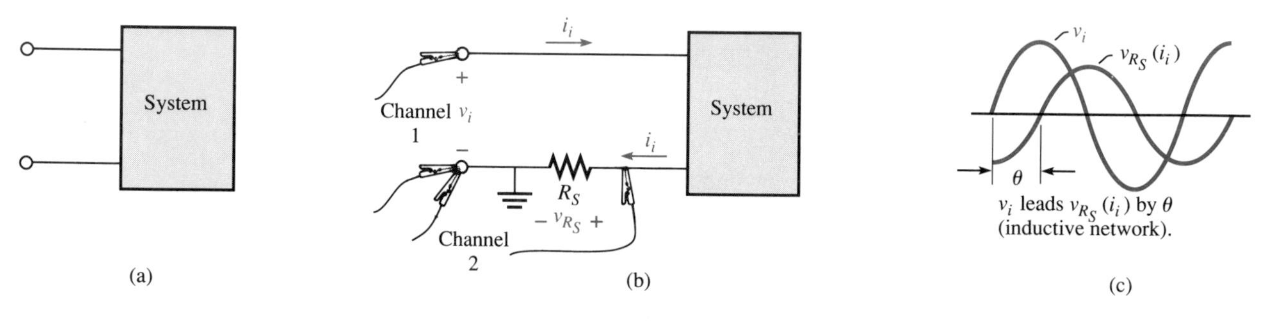


**FIG. 14.19**
*Using an oscilloscope to determine the phase angle of a system.*

$v_i$ and $i_i$. The addition of a ''sensing'' resistor, therefore, can be used to determine the phase angle introduced by the system and can be used to determine the magnitude of the resulting current. The details of the connections that must be made and how the actual phase angle is determined will be left for the laboratory experience.

## 14.4 FREQUENCY RESPONSE OF THE BASIC ELEMENTS

The analysis of Section 14.3 was limited to a particular applied frequency. What is the effect of varying the frequency on the level of opposition offered by a resistive, inductive, or capacitive element? We are aware from the last section that the inductive reactance increases with frequency while the capacitive reactance decreases. However, what is the pattern to this increase or decrease in opposition? Since applied signals may have frequencies extending from a few hertz to megahertz, it is important to be aware of the effect of frequency on the opposition level.

### *R*

Due to stray capacitance levels and lead inductance, the resistance of any resistor will change with frequency. However, the capacitive and inductive levels involved are so small that their real effect is not noticed until the megahertz range. The resistance-versus-frequency curves for a number of carbon composition resistors are provided in Fig. 14.20.

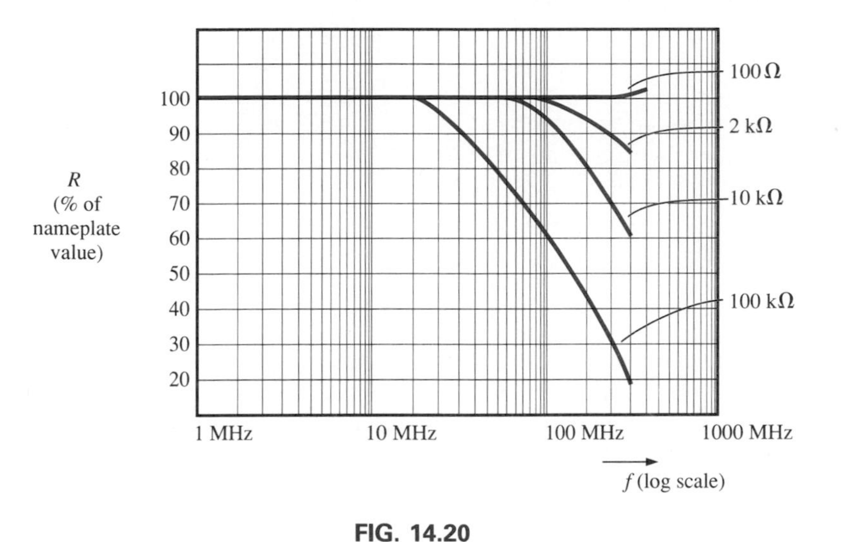


**FIG. 14.20**

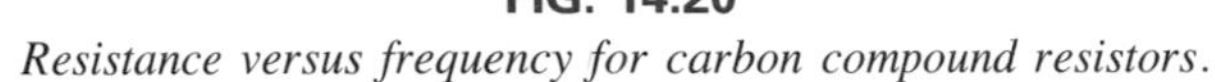

*Resistance versus frequency for carbon compound resistors.*

Note that the lower resistance levels seem to be less affected by the frequency level. The 100-Ω resistor is essentially stable up to about 300 MHz, whereas the 100-kΩ resistor starts its radical decline at about 15 MHz. Frequency, therefore, does have impact on the resistance of an element, but for our current frequency range of interest, we will assume the resistance-versus frequency plot of Fig. 14.21 (like Fig. 14.20 up to 15 MHz), which essentially specifies that the resistance level of a resistor is independent of frequency.

## *L*

The nameplate value of an inductor or capacitor will also change at very high frequencies for many of the same reasons already introduced for the resistor. However, for our current frequency range of interest it will be assumed that the nameplate values of inductors (5 mH, 10 $\mu$H, etc.) and capacitors (0.1 $\mu$F, 1000 pF, etc.) will remain fixed.

However, the equation

$$X_L = \omega L = 2\pi f L = 2\pi L f$$

is directly related to the straight-line equation

$$y = mx + b$$

with a slope ($m$) of $2\pi L$ and a $y$-intercept ($b$) of zero. $X_L$ is the $y$ variable and $f$ is the $x$ variable.

Quite obviously, the larger the inductance, the greater the slope for the same frequency range, as shown in Fig. 14.22. Keep in mind, as reemphasized by Fig. 14.22, that the opposition of an inductor at very

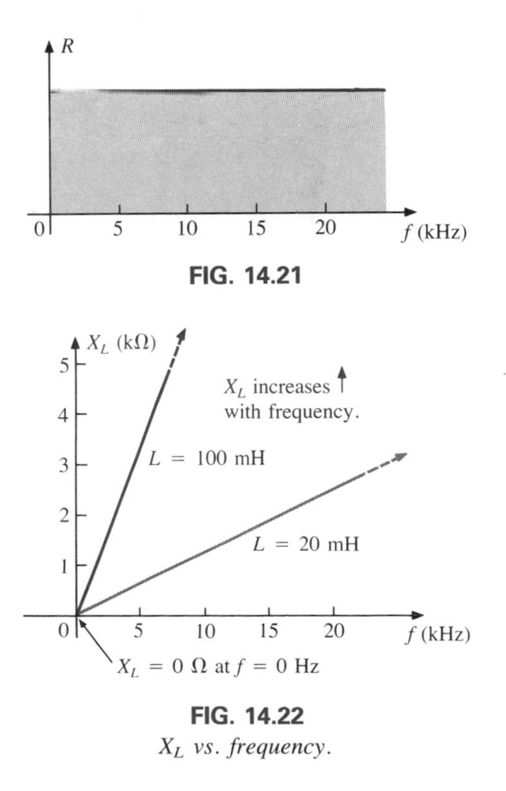


**FIG. 14.21**

**FIG. 14.22**

*$X_L$ vs. frequency.*

low frequencies approaches that of a short circuit, while at high frequencies the reactance approaches that of an open circuit.

For the capacitor, the reactance equation

$$X_C = \frac{1}{2\pi fC}$$

can be written

$$X_C f = \frac{1}{2\pi C}$$

which matches the basic format of a parabola,

$$yx = k$$

with the constant $k = 1/2\pi C$, $y = X_C$, and $x = f$.

At $f = 0$ Hz, the reactance of the capacitor is so large, as shown in Fig. 14.23, that it can be replaced by an open-circuit equivalent. As the

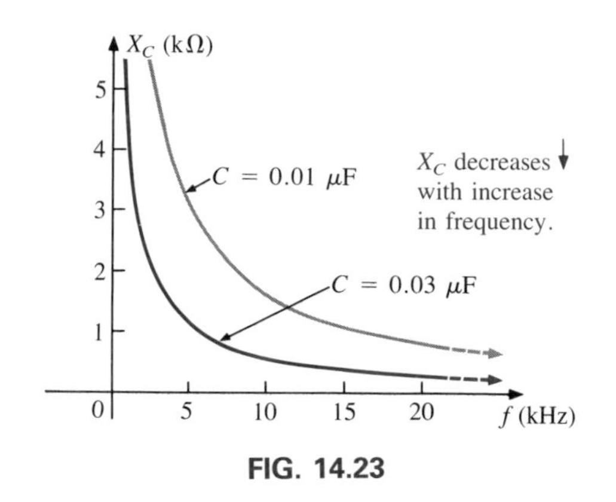


**FIG. 14.23**

frequency increases, the reactance decreases, until eventually a short-circuit equivalent would be appropriate. Note how an increase in capacitance causes the reactance to drop off more rapidly with frequency.

---

**EXAMPLE 14.8.** At what frequency will the reactance of a 200-mH inductor match the resistance level of a 5-kΩ resistor?

***Solution:*** The resistance remains constant at 5 kΩ for the frequency range of the inductor. Therefore,

$$R = 5000\ \Omega = X_L = 2\pi fL = 2\pi Lf$$
$$= (6.28\text{ rad})(200 \times 10^{-3}\text{ H})f = 1.256f$$

and

$$f = \frac{5000\text{ Hz}}{1.256} \cong \mathbf{3.98\ kHz}$$

---

**EXAMPLE 14.9.** At what frequency will an inductor of 5 mH have the same reactance as a capacitor of 0.1 $\mu$F?

***Solution:***

$$X_L = X_C$$

$$2\pi f L = \frac{1}{2\pi f C}$$

$$f^2 = \frac{1}{4\pi^2 LC}$$

and

$$f = \frac{1}{2\pi\sqrt{LC}} = \frac{1}{6.28 \text{ rad}\sqrt{(5 \times 10^{-3} \text{ H})(0.1 \times 10^{-6} \text{ F})}}$$

$$= \frac{1}{6.28\sqrt{5 \times 10^{-10}}} = \frac{1}{(6.28)(2.236 \times 10^{-5})}$$

$$f = \frac{10^5 \text{ Hz}}{14.04} \cong \mathbf{7.12\ kHz}$$

## 14.5 AVERAGE POWER AND POWER FACTOR

The instantaneous power to the load of Fig. 14.24 is

$$p = vi$$

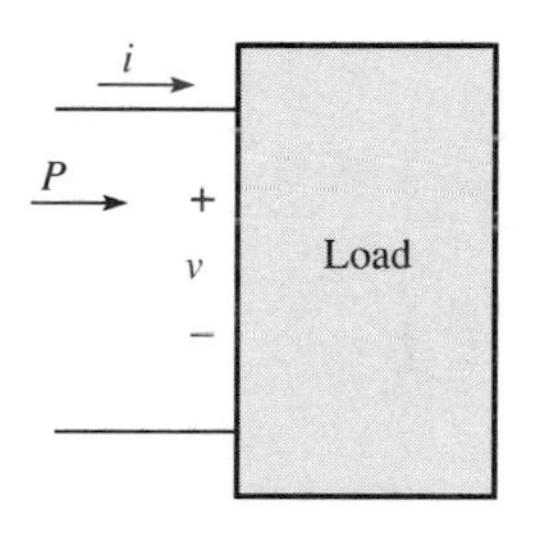


FIG. 14.24

If we consider the general case where

$$v = V_m \sin(\omega t + \beta) \quad \text{and} \quad i = I_m \sin(\omega t + \psi)$$

then

$$p = vi = V_m \sin(\omega t + \beta) I_m \sin(\omega t + \psi)$$
$$= V_m I_m \sin(\omega t + \beta) \sin(\omega t + \psi)$$

Using the trigonometric identity

$$\sin A \sin B = \frac{\cos(A - B) - \cos(A + B)}{2}$$

the function $\sin(\omega t + \beta) \sin(\omega t + \psi)$ becomes

$\sin(\omega t + \beta) \sin(\omega t + \psi)$

$$= \frac{\cos[(\omega t + \beta) - (\omega t + \psi)] - \cos[(\omega t + \beta) + (\omega t + \psi)]}{2}$$

$$= \frac{\cos(\beta - \psi) - \cos(2\omega t + \beta + \psi)}{2}$$

so that

$$p = \overbrace{\left[\frac{V_m I_m}{2}\cos(\beta - \psi)\right]}^{\text{fixed value}} - \overbrace{\left[\frac{V_m I_m}{2}\cos(2\omega t + \beta + \psi)\right]}^{\text{time-varying junction}}$$

A plot of $v$, $i$, and $p$ on the same set of axes is shown in Fig. 14.25.

Note that the second factor in the preceding equation is a cosine wave with an amplitude of $V_m I_m/2$, and a frequency twice that of the

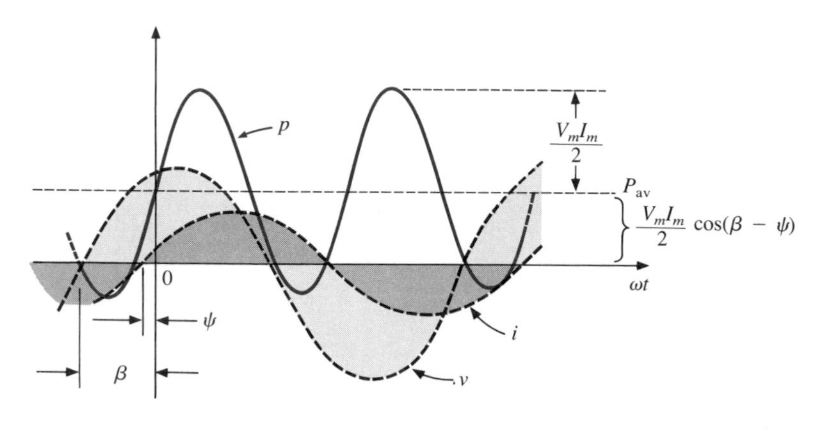


**FIG. 14.25**

voltage or current. The average value of this term is zero, producing no net transfer of energy in any one direction.

The first term in the preceding equation, however, has a constant magnitude (no time dependence) and therefore provides some net transfer of energy. This term is referred to as the *average power,* the reason for which is obvious from Fig. 14.25. The average power, or *real* power as it is sometimes called, is the power delivered to and dissipated by the load. It corresponds to the power calculations performed for dc networks. The angle $(\beta - \psi)$ is the phase angle between $v$ and $i$. Since $\cos(-\alpha) = \cos\alpha$

***the magnitude of average power delivered is independent of whether v leads i or i leads v.***

Defining $\theta$ as equal to $|\beta - \psi|$, where $|-|$ indicates that only the magnitude is important and the sign is immaterial, we have

$$P = \frac{V_m I_m}{2}\cos\theta \qquad \text{(watts, W)} \qquad \textbf{(14.9)}$$

where $P$ is the average power in watts. This equation can also be written

$$P = \left(\frac{V_m}{\sqrt{2}}\right)\left(\frac{I_m}{\sqrt{2}}\right)\cos\theta$$

Or, since

$$V_{\text{eff}} = \frac{V_m}{\sqrt{2}} \quad \text{and} \quad I_{\text{eff}} = \frac{I_m}{\sqrt{2}}$$

Eq. (14.9) becomes

$$\boxed{P = V_{\text{eff}} I_{\text{eff}} \cos \theta} \qquad \textbf{(14.10)}$$

Let us now apply Eqs. (14.9) and (14.10) to the basic $R$, $L$, and $C$ elements.

## Resistor

In a purely resistive circuit, since $v$ and $i$ are in phase, $|\beta - \psi| = \theta = 0°$, and $\cos \theta = \cos 0° = 1$, so that

$$\boxed{P = \frac{V_m I_m}{2} V_{\text{eff}} I_{\text{eff}}} \quad \text{(W)} \qquad \textbf{(14.11)}$$

Or, since

$$I_{\text{eff}} = \frac{V_{\text{eff}}}{R}$$

then

$$\boxed{P = \frac{V_{\text{eff}}^2}{R} = I_{\text{eff}}^2 R} \quad \text{(W)} \qquad \textbf{(14.12)}$$

## Inductor

In a purely inductive circuit, since $v$ leads $i$ by 90°, $|\beta - \psi| = \theta = 90°$. Therefore,

$$P = \frac{V_m I_m}{2} \cos 90° = \frac{V_m I_m}{2}(0) = \mathbf{0\ W}$$

***The average power or power dissipated by the ideal inductor (no associated resistance) is zero watts.***

## Capacitor

In a purely capacitive circuit, since $i$ leads $v$ by 90°, $|\beta - \psi| = \theta = |-90°| = 90°$. Therefore,

$$P = \frac{V_m I_m}{2} \cos(90°) = \frac{V_m I_m}{2}(0) = \mathbf{0}$$

***The average power or power dissipated by the ideal capacitor (no associated resistance) is zero watts.***

**EXAMPLE 14.10.** Find the average power dissipated in a circuit whose input current and voltage are the following:

$$i = 5 \sin(\omega t + 40°)$$
$$v = 10 \sin(\omega t + 40°)$$

***Solution:*** Since $v$ and $i$ are in phase, the circuit appears at the input terminals to be purely resistive. Therefore,

$$P = \frac{V_m I_m}{2} = \frac{(10 \text{ V})(5 \text{ A})}{2} = \mathbf{25 \text{ W}}$$

or

$$R = \frac{V_m}{I_m} = \frac{10 \text{ V}}{5 \text{ A}} = 2 \text{ }\Omega$$

and

$$P = \frac{V_{\text{eff}}^2}{R} = \frac{[(0.707)(10 \text{ V})]^2}{2} = \mathbf{25 \text{ W}}$$

or

$$P = I_{\text{eff}}^2 R = [(0.707)(5 \text{ A})]^2(2) = \mathbf{25 \text{ W}}$$

For the following examples, the circuit consists of a combination of resistances and reactances producing phase angles between the input current and voltage different from 0° or 90°.

**EXAMPLE 14.11.** Determine the average power delivered to networks having the following input voltage and current:

a. $v = 100 \sin(\omega t + 40°)$
   $i = 20 \sin(\omega t + 70°)$
b. $v = 150 \sin(\omega t - 70°)$
   $i = 3 \sin(\omega t - 50°)$

***Solutions:***

a. $V_m = 100, \ \beta = 40°$
   $I_m = 20, \ \psi = 70°$
   $\theta = |\beta - \psi| = |40° - 70°| = |-30°| = 30°$
   and

$$P = \frac{V_m I_m}{2} \cos \theta = \frac{(100 \text{ V})(20 \text{ A})}{2} \cos(30°) = (1000 \text{ W})(0.866)$$
$$= \mathbf{866 \text{ W}}$$

b. $V_m = 150 \text{ V}, \ \beta = -70°$
   $I_m = 3 \text{ A}, \ \psi = -50°$
   $\theta = |\beta - \psi| = |-70° - (-50°)|$
   $= |-70° + 50°| = |-20°| = 20°$

and

$$P = \frac{V_m I_m}{2} \cos\theta = \frac{(150\text{ V})(3\text{ A})}{2} \cos(20°) = (225\text{ W})(0.9397)$$
$$= \mathbf{211.43\ W}$$

## Power Factor

The frequency and the elements of the parallel network of Fig. 14.26(a) were chosen such that

$$i_1 = 50 \sin \omega t$$
$$i_2 = 50 \sin(\omega t - 90°)$$
$$i_3 = 50 \sin(\omega t + 90°)$$

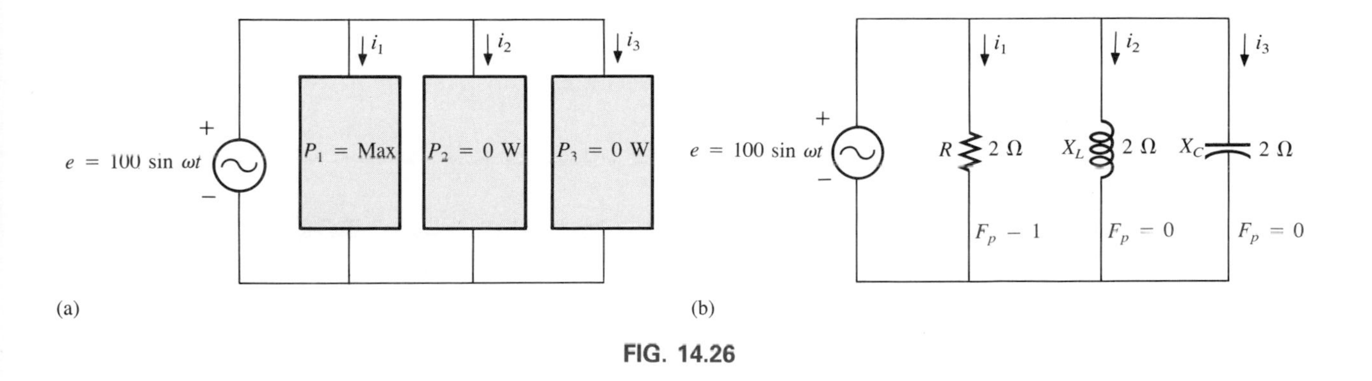


**FIG. 14.26**

Therefore, for each parallel branch, the peak values (or effective values) of the voltage and current are the same. Yet, as indicated in Fig. 14.26(a), the power to two of the branches is zero, and a maximum to the third. In our power equation, the only factor that accounts for this variation is cos $\theta$, related to the phase relationship between $v$ and $i$. This factor, cos $\theta$, is called the *power factor* and is symbolically represented by $F_p$; that is,

$$\text{Power factor} = F_p = \cos\theta \qquad \textbf{(14.13)}$$

The more reactive a load, the lower the power factor, and the smaller the average power delivered. The more resistive the load, the higher the power factor, and the greater the real power delivered. The elements within each load of Fig. 14.26(a) appear in Fig. 14.26(b). Note the power factor for each element in the figure. Low power factors are usually avoided, since a high current would be required to deliver any appreciable power. This higher current demand produces higher heating losses and, consequently, the system operates at a lower efficiency.

In terms of the average power and the terminal voltage and current,

$$F_p = \cos\theta = \frac{P}{V_{\text{eff}}I_{\text{eff}}} \tag{14.14}$$

The terms *leading* and *lagging* are often written in conjunction with the power factor. *They are defined by the current through the load.* If the current leads the voltage across a load, the load has a leading power factor. If the current lags the voltage across the load, the load has a lagging power factor. In other words,

***capacitive networks have leading power factors, and inductive networks have lagging power factors.***

---

**EXAMPLE 14.12.** Determine the power factors of the following loads, and indicate whether they are leading or lagging:

a. Fig. 14.27
b. Fig. 14.28

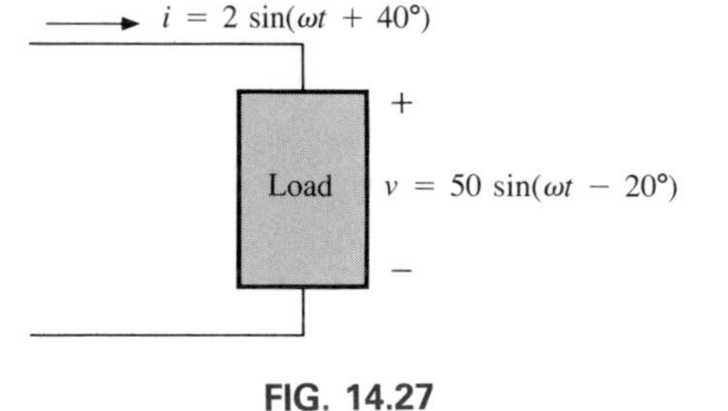


FIG. 14.27

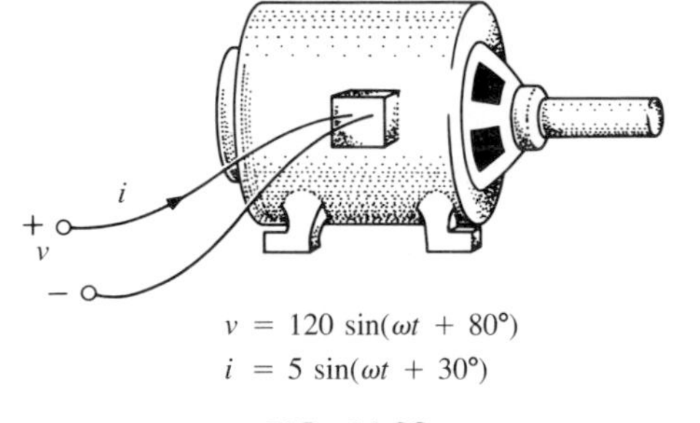

FIG. 14.28

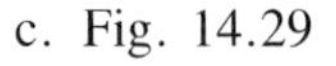

c. Fig. 14.29

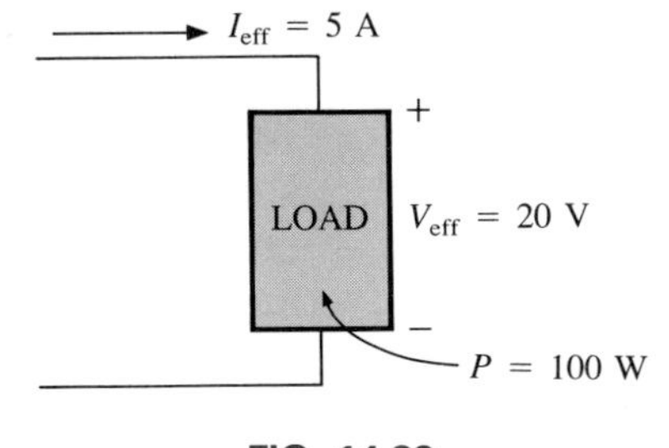


FIG. 14.29

***Solutions:***

a. $F_p = \cos\theta = \cos 60° = \mathbf{0.5\ leading}$

b. $F_p = \cos\theta = \cos 50° = \mathbf{0.6428\ lagging}$

c. $F_p = \cos\theta = \dfrac{P}{V_{\text{eff}}I_{\text{eff}}} = \dfrac{100\text{ W}}{(20\text{ V})(5\text{ A})} = \dfrac{100\text{ W}}{100\text{ W}} = \mathbf{1}$

The load is resistive, and $F_p$ is neither leading nor lagging.

---

## 14.6 COMPLEX NUMBERS

In our analysis of dc networks, we found it necessary to determine the algebraic sum of voltages and currents. Since the same will also be true for ac networks, the question arises, How do we determine the algebraic sum of two or more voltages (or currents) that are varying sinusoidally?

Although one solution would be to find the algebraic sum on a point-to-point basis (as shown in Section 14.12), this would be a long and tedious process in which accuracy would be directly related to the scale employed.

It is the purpose of this chapter to introduce a system of *complex numbers* which, when related to the sinusoidal ac waveform, will result in a technique for finding the algebraic sum of sinusoidal waveforms that is quick, direct, and accurate. In the following chapters, the technique will be extended to permit the analysis of sinusoidal ac networks in a manner very similar to that applied to dc networks. The methods and theorems as described for dc networks can then be applied to sinusoidal ac networks with little difficulty.

A *complex number* represents a point in a two-dimensional plane located with reference to two distinct axes. This point can also determine a radius vector drawn from the origin to the point. The horizontal axis is called the *real* axis, while the vertical axis is called the *imaginary* axis. Both are labeled in Fig. 14.30. For reasons that will be obvious later, the real axis is sometimes called the *resistance* axis, and the imaginary axis, the *reactance* axis. Every number from zero to $\pm\infty$ can be represented by some point along the real axis. Prior to the development of this system of complex numbers, it was believed that any number not on the real axis would not exist—hence the term *imaginary* for the vertical axis.

In the complex plane, the horizontal or real axis represents all positive numbers to the right of the imaginary axis and all negative numbers to the left of the imaginary axis. All positive imaginary numbers are represented above the real axis, and all negative imaginary numbers, below the real axis. The symbol $j$ (or sometimes $i$) is used to denote the imaginary component.

There are two forms used to represent a complex number: *rectangular* and *polar*. Each can represent a point in the plane or a radius vector drawn from the origin to that point.

## 14.7 RECTANGULAR FORM

The format for the rectangular form is

$$\boxed{\mathbf{C} = A + jB} \qquad \textbf{(14.15)}$$

as shown in Fig. 14.31.

**EXAMPLE 14.13.** Sketch the following complex numbers in the complex plane:

a. $\mathbf{C} = 3 + j4$
b. $\mathbf{C} = 0 - j6$
c. $\mathbf{C} = -10 - j20$

***Solutions:***

a. See Fig. 14.32.

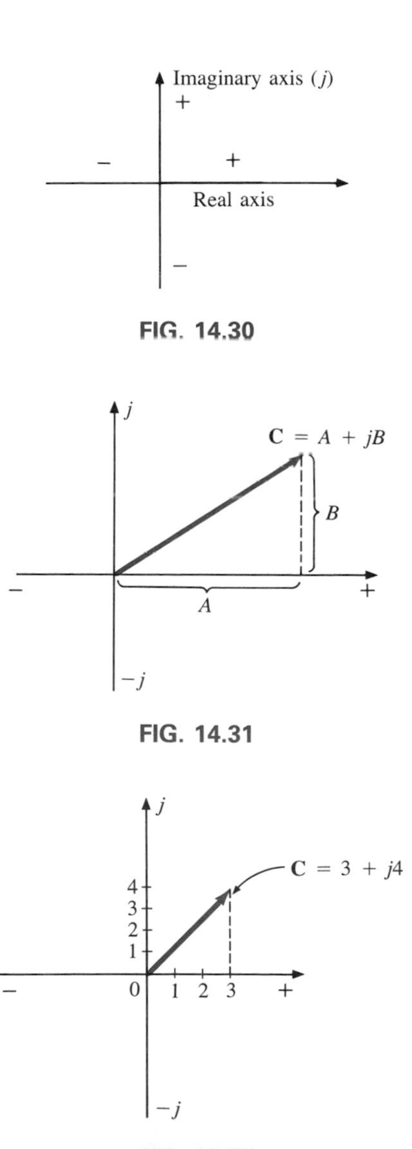


FIG. 14.30

FIG. 14.31

FIG. 14.32

b. See Fig. 14.33.
c. See Fig. 14.34.

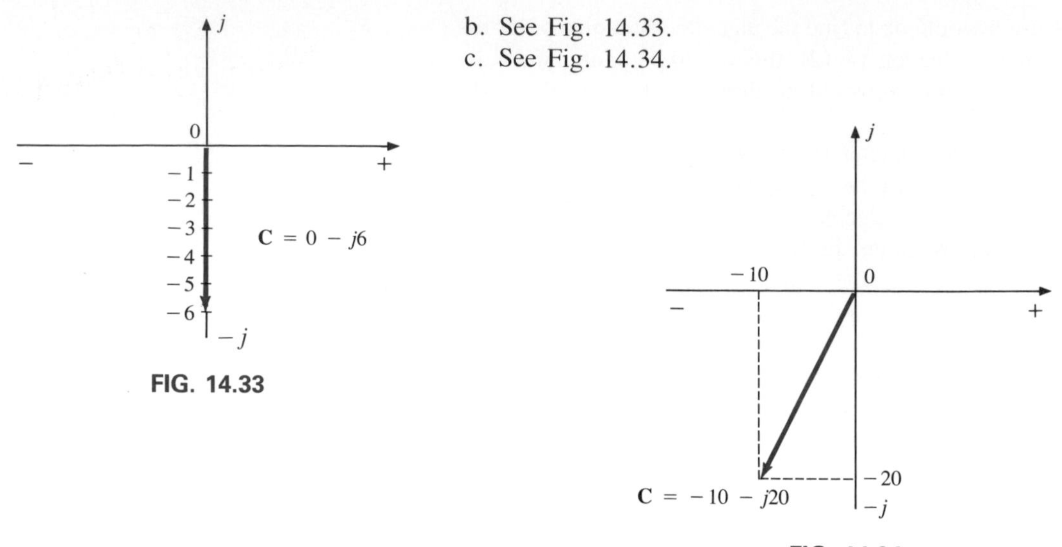


FIG. 14.33

FIG. 14.34

## 14.8 POLAR FORM

The format for the polar form is

$$\mathbf{C} = C \angle \theta \tag{14.16}$$

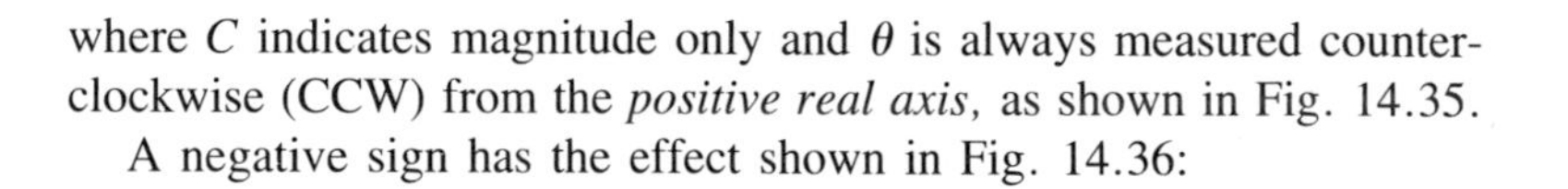

where $C$ indicates magnitude only and $\theta$ is always measured counterclockwise (CCW) from the *positive real axis,* as shown in Fig. 14.35.

A negative sign has the effect shown in Fig. 14.36:

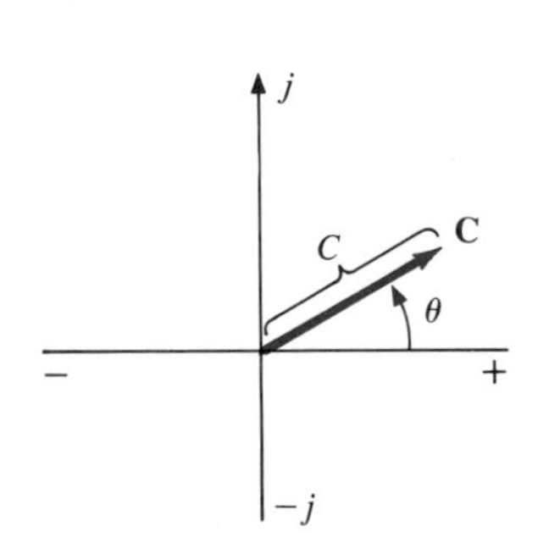

FIG. 14.35

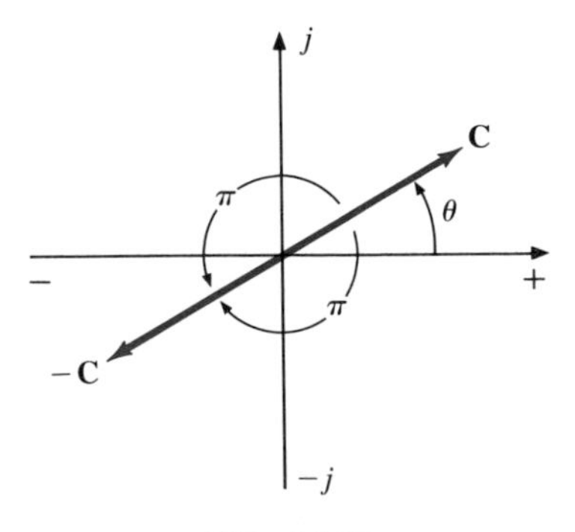

FIG. 14.36

$$-\mathbf{C} = -C \angle \theta = C \angle \theta \pm \pi \tag{14.17}$$

**EXAMPLE 14.14.** Sketch the following complex numbers in the complex plane:

a. $\mathbf{C} = 5 \angle 30^\circ$
b. $\mathbf{C} = 7 \angle 120^\circ$
c. $\mathbf{C} = -4.2 \angle 60^\circ$

***Solutions:***

a. See Fig. 14.37.

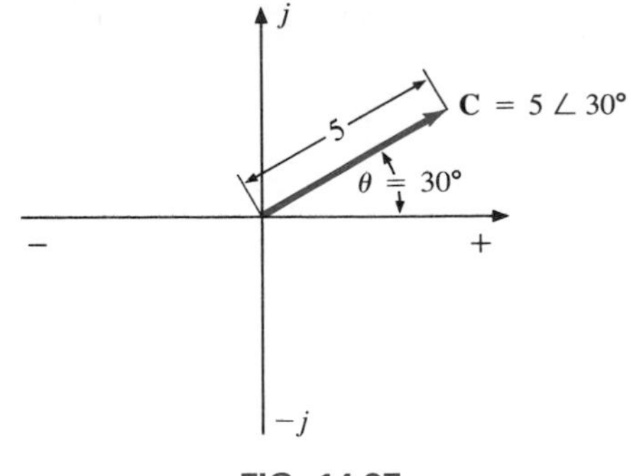

FIG. 14.37

b. See Fig. 14.38.

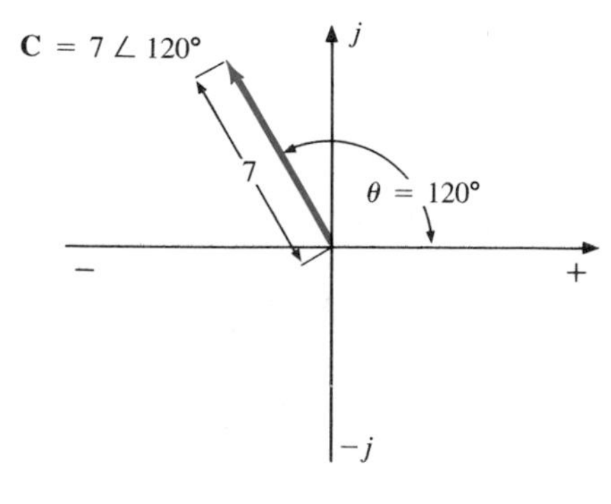

FIG. 14.38

c. See Fig. 14.39.

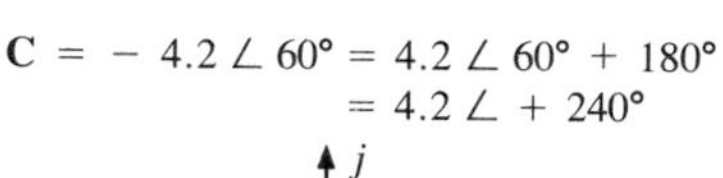


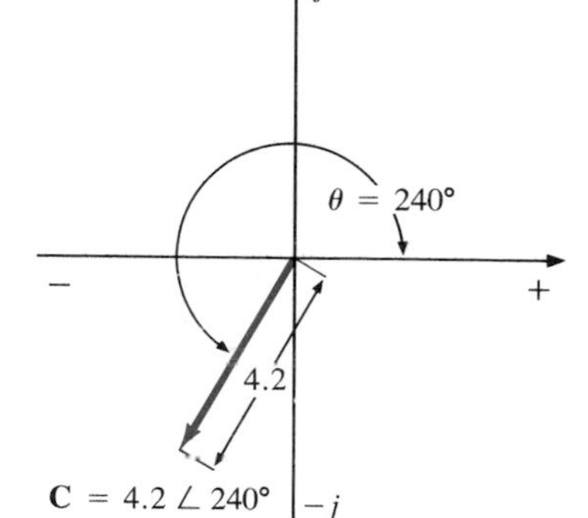


FIG. 14.39

## 14.9 CONVERSION BETWEEN FORMS

The two forms are related by the following equations.

### Rectangular to Polar

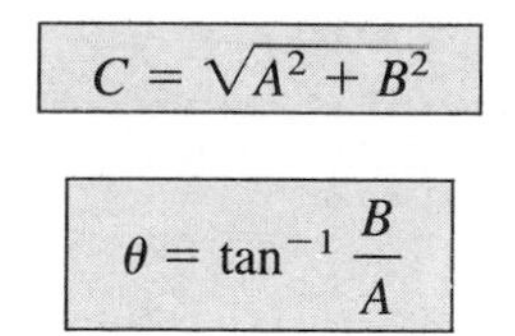

$$C = \sqrt{A^2 + B^2} \quad \textbf{(14.18)}$$

$$\theta = \tan^{-1} \frac{B}{A} \quad \textbf{(14.19)}$$

Note Fig. 14.40.

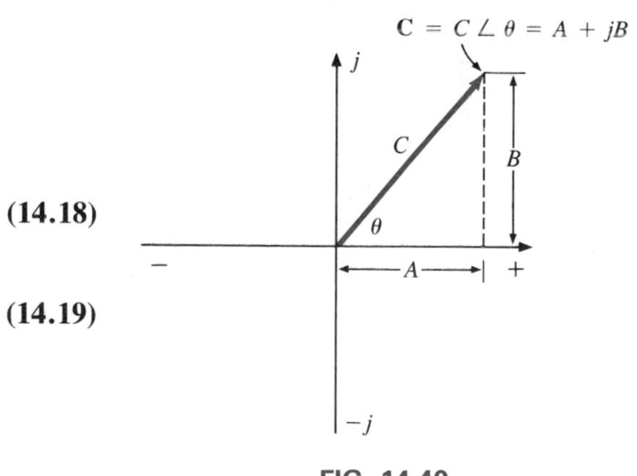


FIG. 14.40

### Polar to Rectangular

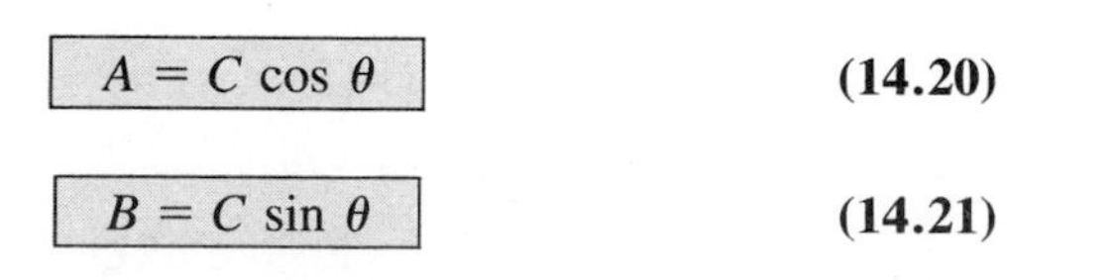

$$A = C \cos \theta \quad \textbf{(14.20)}$$

$$B = C \sin \theta \quad \textbf{(14.21)}$$

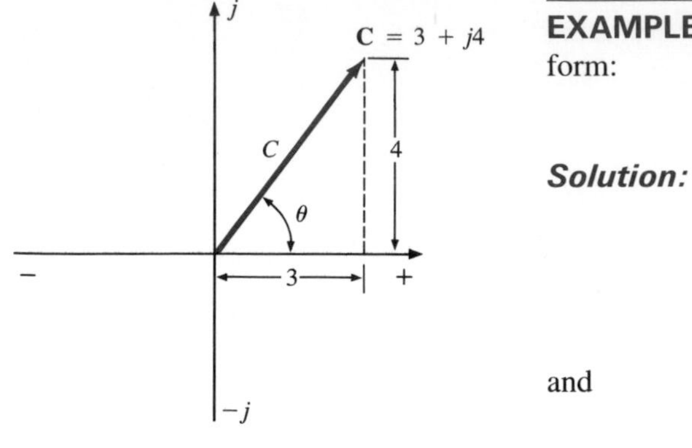


FIG. 14.41

**EXAMPLE 14.15.** Convert the following from rectangular to polar form:

$$\mathbf{C} = 3 + j4 \qquad \text{(Fig. 14.41)}$$

***Solution:***

$$C = \sqrt{(3)^2 + (4)^2} = \sqrt{25} = 5$$

$$\theta = \tan^{-1}\left(\frac{4}{3}\right) = 53.13°$$

and

$$\mathbf{C = 5 \angle 53.13°}$$

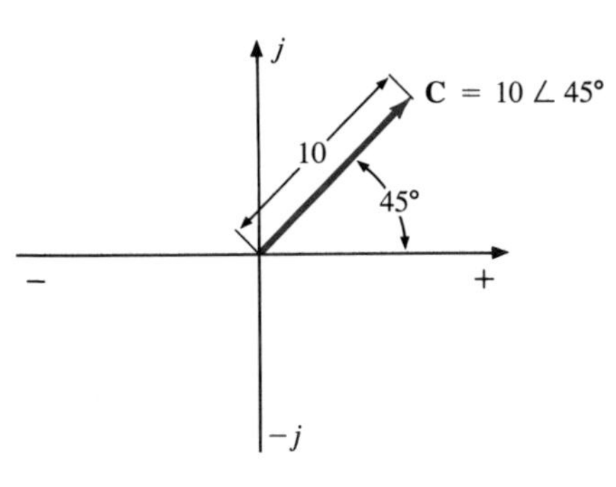


FIG. 14.42

**EXAMPLE 14.16.** Convert the following from polar to rectangular form:

$$\mathbf{C} = 10 \angle 45° \qquad \text{(Fig. 14.42)}$$

***Solution:***

$$A = 10 \cos 45° = (10)(0.707) = 7.07$$

$$B = 10 \sin 45° = (10)(0.707) = 7.07$$

and

$$\mathbf{C = 7.07 + \mathit{j}7.07}$$

If the complex number should appear in the second, third, or fourth quadrant, simply convert it in that quadrant, and carefully determine the proper angle to be associated with the magnitude of the vector.

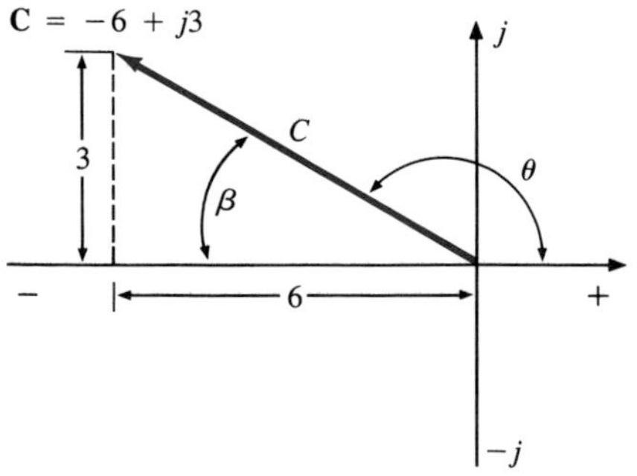

**EXAMPLE 14.17.** Convert the following from rectangular to polar form:

$$\mathbf{C} = -6 + j3 \qquad \text{(Fig. 14.43)}$$

***Solution:***

$$C = \sqrt{(6)^2 + (3)^2} = \sqrt{45} = 6.71$$

$$\beta = \tan^{-1}\left(\frac{3}{6}\right) = 26.57°$$

$$\theta = 180 - 26.57° = 153.43°$$

and

$$\mathbf{C = 6.71 \angle 153.43°}$$

FIG. 14.43

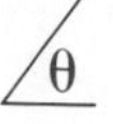



**EXAMPLE 14.18.** Convert the following from polar to rectangular form:

$$\mathbf{C} = 10 \angle 230^\circ \qquad \text{(Fig. 14.44)}$$

***Solution:***

$$A = C \cos \beta = 10 \cos(230^\circ - 180^\circ) = 10 \cos 50^\circ$$
$$= (10)(0.6428) = 6.428$$
$$B = C \sin \beta = 10 \sin 50^\circ = (10)(0.7660) = 7.660$$

and

$$\mathbf{C} = \mathbf{-6.428 - \mathit{j}7.660}$$

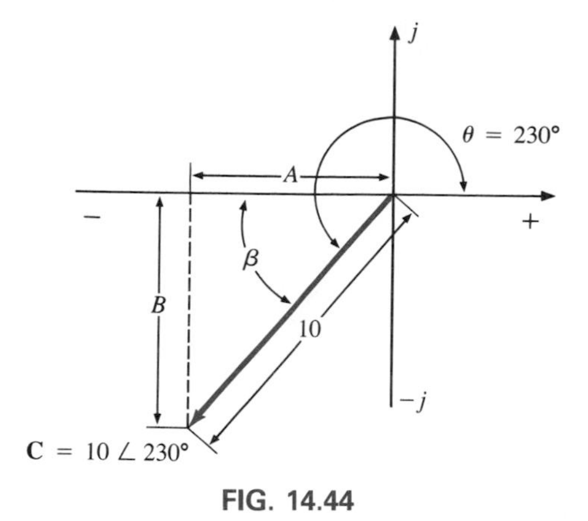


FIG. 14.44

## 14.10 MATHEMATICAL OPERATIONS WITH COMPLEX NUMBERS

Complex numbers lend themselves readily to the basic mathematical operations of addition, subtraction, multiplication, and division. A few basic rules and definitions must be understood before considering these operations.

Let us first examine the symbol $j$ associated with imaginary numbers. By definition,

$$j = \sqrt{-1} \qquad \textbf{(14.22)}$$

Thus,

$$j^2 = -1 \qquad \textbf{(14.23)}$$

and

$$j^3 = j^2 j = -1j = -j$$

with

$$j^4 = j^2 j^2 = (-1)(-1) = +1$$
$$j^5 = j$$

and so on. Further,

$$\frac{1}{j} = (1)\left(\frac{1}{j}\right) = \left(\frac{j}{j}\right)\left(\frac{1}{j}\right) = \frac{j}{j^2} = \frac{j}{-1}$$

and

$$\frac{1}{j} = -j \qquad \textbf{(14.24)}$$

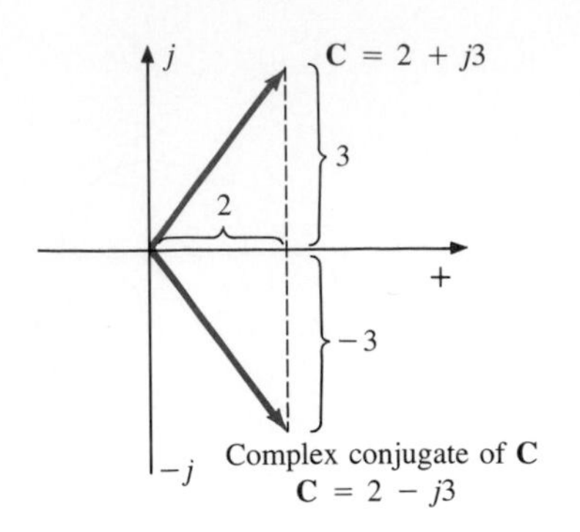


FIG. 14.45

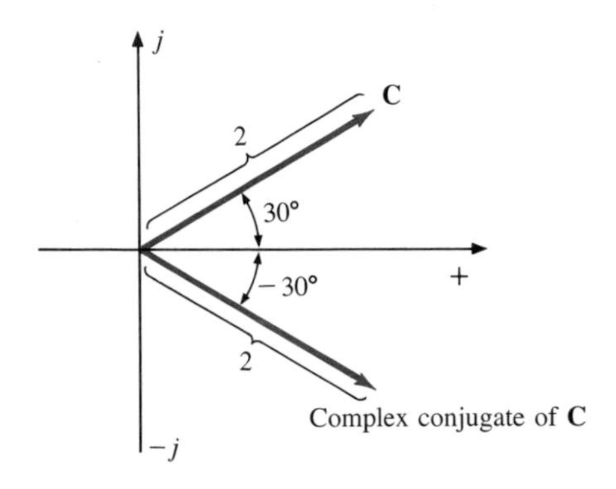


FIG. 14.46

## Complex Conjugate

The *conjugate* or *complex conjugate* of a complex number can be found by simply changing the sign of the imaginary part in the rectangular form or by using the negative of the angle of the polar form. For example, the conjugate of

$$\mathbf{C} = 2 + j3$$

is

$$2 - j3$$

as shown in Fig. 14.45. The conjugate of

$$\mathbf{C} = 2 \angle 30^\circ$$

is

$$2 \angle -30^\circ$$

as shown in Fig. 14.46.

## Reciprocal

The *reciprocal* of a complex number is 1 divided by the complex number. For example, the reciprocal of

$$\mathbf{C} = A + jB$$

is

$$\frac{1}{A + jB}$$

and of $C \angle \theta$,

$$\frac{1}{C \angle \theta}$$

We are now prepared to consider the four basic operations of *addition, subtraction, multiplication,* and *division* with complex numbers.

## Addition

To add two or more complex numbers, simply add the real and imaginary parts separately. For example, if

$$\mathbf{C}_1 = \pm A_1 \pm jB_1 \quad \text{and} \quad \mathbf{C}_2 = \pm A_2 \pm jB_2$$

then

$$\mathbf{C}_1 + \mathbf{C}_2 = (\pm A_1 \pm A_2) + j(\pm B_1 \pm B_2) \tag{14.25}$$

There is really no need to memorize the equation. Simply set one above the other and consider the real and imaginary parts separately, as shown in Example 14.19.

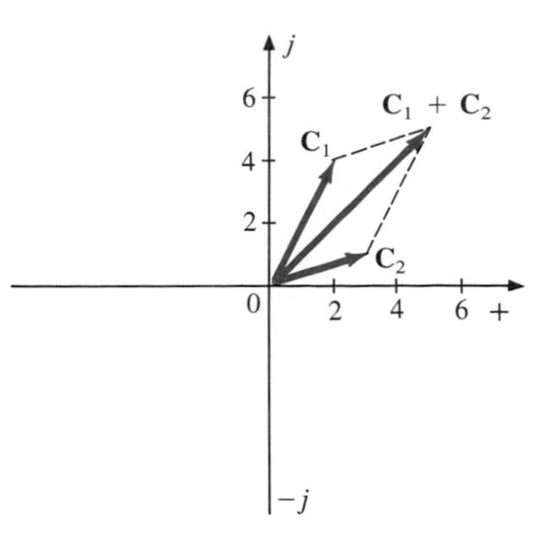

FIG. 14.47

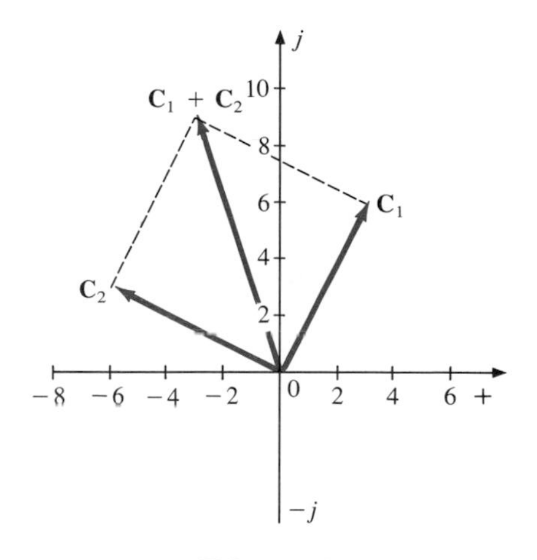


FIG. 14.48

---

**EXAMPLE 14.19.**

a. Add $\mathbf{C}_1 = 2 + j4$ and $\mathbf{C}_2 = 3 + j1$.
b. Add $\mathbf{C}_1 = 3 + j6$ and $\mathbf{C}_2 = -6 + j3$.

***Solutions:***

a. By Eq. (14.25),

$$\mathbf{C}_1 + \mathbf{C}_2 = (2 + 3) + j(4 + 1) = \mathbf{5 + j5}$$

Note Fig. 14.47. An alternative method is

$$\begin{array}{c} 2 + j4 \\ 3 + j1 \\ \hline \downarrow \quad \downarrow \\ \mathbf{5 + j5} \end{array}$$

b. By Eq. (14.25),

$$\mathbf{C}_1 + \mathbf{C}_2 = (3 - 6) + j(6 + 3) = \mathbf{-3 + j9}$$

Note Fig. 14.48. An alternative method is

$$\begin{array}{c} 3 + j6 \\ -6 + j3 \\ \hline \downarrow \quad \downarrow \\ \mathbf{-3 + j9} \end{array}$$

---

## Subtraction

In subtraction, the real and imaginary parts are again considered separately. For example, if

$$\mathbf{C}_1 = \pm A_1 \pm jB_1 \quad \text{and} \quad \mathbf{C}_2 = \pm A_2 \pm jB_2$$

then

$$\boxed{\mathbf{C}_1 - \mathbf{C}_2 = [\pm A_2 - (\pm A_2)] + j[\pm B_1 - (\pm B_2)]} \qquad \textbf{(14.26)}$$

Again, there is no need to memorize the equation if the alternative method of Example 14.20 is employed.

---

**EXAMPLE 14.20.**

a. Subtract $\mathbf{C}_2 = 1 + j4$ from $\mathbf{C}_1 = 4 + j6$.
b. Subtract $\mathbf{C}_2 = -2 + j5$ from $\mathbf{C}_1 = +3 + j3$.

***Solutions:***

a. By Eq. (14.26),

$$\mathbf{C}_1 - \mathbf{C}_2 = (4 - 1) + j(6 - 4) = \mathbf{3 + j2}$$

Note Fig. 14.49. An alternative method is

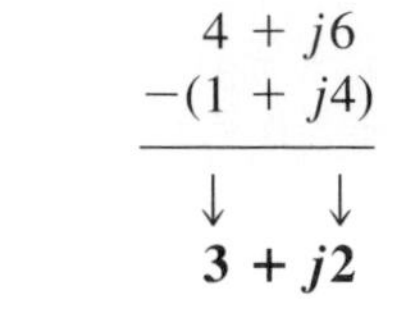


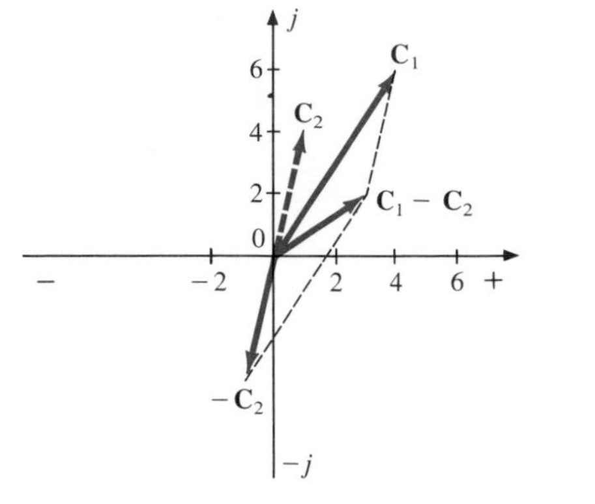

FIG. 14.49

b. By Eq. (14.26),

$$\mathbf{C}_1 - \mathbf{C}_2 = [3 - (-2)] + j(3 - 5) = \mathbf{5 - j2}$$

Note Fig. 14.50. An alternative method is

$$\begin{array}{r} 3 + j3 \\ -(-2 + j5) \\ \hline \downarrow \quad \downarrow \\ \mathbf{5 - j2} \end{array}$$

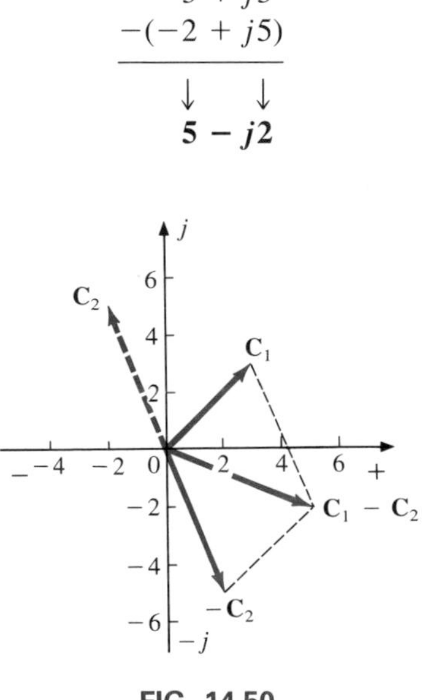

FIG. 14.50

***Addition or subtraction cannot be performed in polar form unless the complex numbers have the same angle θ or differ only by multiples of 180°.***

**EXAMPLE 14.21.**

$$2 \angle 45° + 3 \angle 45° = \mathbf{5 \angle 45°}$$

Note Fig. 14.51. Or

$$2 \angle 0° - 4 \angle 180° = \mathbf{6 \angle 0°}$$

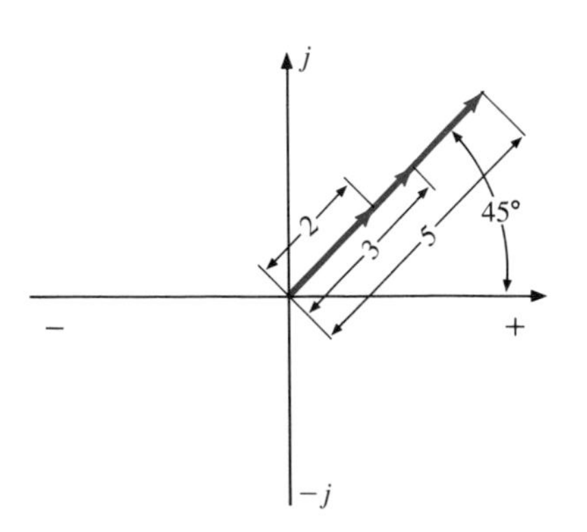


FIG. 14.51

Note Fig. 14.52.

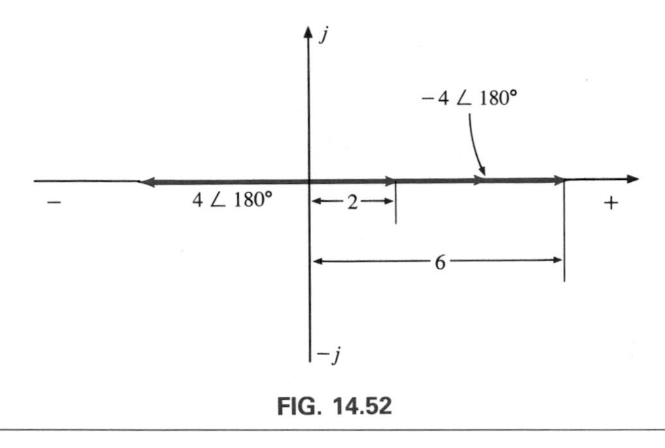


FIG. 14.52

## Multiplication

To multiply two complex numbers in *rectangular* form, multiply the real and imaginary parts of one in turn by the real and imaginary parts of the other. For example, if

$$\mathbf{C}_1 = A_1 + jB_1 \quad \text{and} \quad \mathbf{C}_2 = A_2 + jB_2$$

then

$$\begin{array}{rl} \mathbf{C}_1 \cdot \mathbf{C}_2 = & A_1 + jB_1 \\ & \underline{A_2 + jB_2} \\ & A_1A_2 + jB_1A_2 \\ & \underline{\quad + jA_1B_2 + j^2B_1B_2} \\ & A_1A_2 + j(B_1A_2 + A_1B_2) + B_1B_2(-1) \end{array}$$

and

$$\boxed{\mathbf{C}_1 \cdot \mathbf{C}_2 = (A_1A_2 - B_1B_2) + j(B_1A_2 + A_1B_2)} \qquad \textbf{(14.27)}$$

In Example 14.22(b), we obtain a solution without resorting to memorizing Eq. (14.27). Simply carry along the $j$ factor when multiplying each part of one vector with the real and imaginary parts of the other.

**EXAMPLE 14.22.**

a. Find $\mathbf{C}_1 \cdot \mathbf{C}_2$ if

$$\mathbf{C}_1 = 2 + j3 \quad \text{and} \quad \mathbf{C}_2 = 5 + j10$$

b. Find $\mathbf{C}_1 \cdot \mathbf{C}_2$ if

$$\mathbf{C}_1 = -2 - j3 \quad \text{and} \quad \mathbf{C}_2 = +4 - j6$$

***Solutions:***

a. Using the format above, we have

$$\mathbf{C}_1 \cdot \mathbf{C}_2 = [(2)(5) - (3)(10)] + j[(3)(5) + (2)(10)]$$
$$= \mathbf{-20 + j35}$$

b. Without using the format, we obtain

$$\begin{array}{l} -2 - j3 \\ +4 - j6 \\ \hline -8 - j12 \\ \quad + j12 + j^2 18 \\ \hline -8 + j(-12 + 12) - 18 \end{array}$$

and

$$\mathbf{C}_1 \cdot \mathbf{C}_2 = \mathbf{-26 = 26 \angle 180°}$$

In *polar* form, the magnitudes are multiplied and the angles added algebraically. For example, for

$$\mathbf{C}_1 = C_1 \angle \theta_1 \quad \text{and} \quad \mathbf{C}_2 = C_2 \angle \theta_2$$

we write

$$\boxed{\mathbf{C}_1 \cdot \mathbf{C}_2 = C_1 C_2 \angle \theta_1 + \theta_2} \qquad \textbf{(14.28)}$$

**EXAMPLE 14.23.**

a. Find $\mathbf{C}_1 \cdot \mathbf{C}_2$ if

$$\mathbf{C}_1 = 5 \angle 20° \quad \text{and} \quad \mathbf{C}_2 = 10 \angle 30°$$

b. Find $\mathbf{C}_1 \cdot \mathbf{C}_2$ if

$$\mathbf{C}_1 = 2 \angle -40° \quad \text{and} \quad \mathbf{C}_2 = 7 \angle +120°$$

***Solutions:***

a. $\mathbf{C}_1 \cdot \mathbf{C}_2 = (5)(10) \angle 20° + 30° = \mathbf{50 \angle 50°}$

b. $\mathbf{C}_1 \cdot \mathbf{C}_2 = (2)(7) \angle -40° + 120° = \mathbf{14 \angle +80°}$

To multiply a complex number in rectangular form by a real number requires that both the real part and the imaginary part be multiplied by the real number. For example,

$$(10)(2 + j3) = 20 + j30$$

and

$$50 \angle 0°(0 + j6) = j300 = 300 \angle 90°$$

## Division

To divide two complex numbers in *rectangular* form, multiply the numerator and denominator by the conjugate of the denominator and the resulting real and imaginary parts collected. That is, if

$$\mathbf{C}_1 = A_1 + jB_1 \quad \text{and} \quad \mathbf{C}_2 = A_2 + jB_2$$

then

$$\frac{\mathbf{C}_1}{\mathbf{C}_2} = \frac{(A_1 + jB_1)(A_2 - jB_2)}{(A_2 + jB_2)(A_2 - jB_2)}$$

$$= \frac{(A_1A_2 + B_1B_2) + j(A_2B_1 - A_1B_2)}{A_2^2 + B_2^2}$$

and

$$\boxed{\frac{\mathbf{C}_1}{\mathbf{C}_2} = \frac{A_1A_2 + B_1B_2}{A_2^2 + B_2^2} + j\frac{A_2B_1 - A_1B_2}{A_2^2 + B_2^2}} \qquad \textbf{(14.29)}$$

*The equation does not have to be memorized if the steps above used to obtain it are employed.* That is, first multiply the numerator by the complex conjugate of the denominator and separate the real and imaginary terms. Then divide each term by the real number obtained by multiplying the denominator by its conjugate.

---

**EXAMPLE 14.24.**

a. Find $\mathbf{C}_1/\mathbf{C}_2$ if

$$\mathbf{C}_1 - 1 + j4 \quad \text{and} \quad \mathbf{C}_2 = 4 + j5$$

b. Find $\mathbf{C}_1/\mathbf{C}_2$ if

$$\mathbf{C}_1 = -4 - j8 \quad \text{and} \quad \mathbf{C}_2 = +6 - j1$$

***Solutions:***

a. By Eq. (14.29),

$$\frac{\mathbf{C}_1}{\mathbf{C}_2} = \frac{(1)(4) + (4)(5)}{4^2 + 5^2} + j\frac{(4)(4) - (1)(5)}{4^2 + 5^2}$$

$$= \frac{24}{41} + \frac{j11}{41} \cong \mathbf{0.585 + \mathit{j}0.268}$$

b. Using an alternative method, we obtain

$$\begin{array}{l}
\quad -4 - j8 \\
\quad \underline{+6 + j1} \\
-24 - j48 \\
\quad\quad \underline{- j4 - j^2 8} \\
-24 - j52 + 8 = -16 - j52
\end{array}$$

$$\begin{array}{l}
+6 - j1 \\
\underline{+6 + j1} \\
36 + j6 \\
\quad \underline{- j6 - j^2 1} \\
36 + 0 + 1 = 37
\end{array}$$

and

$$\frac{\mathbf{C}_1}{\mathbf{C}_2} = \frac{-16}{37} - \frac{j52}{37} = \mathbf{-0.432 - \mathit{j}1.405}$$

---

To divide a complex number in rectangular form by a real number, both the real part and the imaginary part must be divided by the real number. For example,

$$\frac{8 + j10}{2} = 4 + j5$$

and

$$\frac{6.8 - j0}{2} = 3.4 - j0 = 3.4 \angle 0^\circ$$

In *polar* form, division is accomplished by simply dividing the magnitude of the numerator by the magnitude of the denominator and subtracting the angle of the denominator from that of the numerator. That is, for

$$\mathbf{C}_1 = C_1 \angle \theta_1 \quad \text{and} \quad \mathbf{C}_2 = C_2 \angle \theta_2$$

we write

$$\frac{\mathbf{C}_1}{\mathbf{C}_2} = \frac{C_1}{C_2} \angle \theta_1 - \theta_2 \tag{14.30}$$

---

**EXAMPLE 14.25.**

a. Find $\mathbf{C}_1/\mathbf{C}_2$ if

$$\mathbf{C}_1 = 15 \angle 10^\circ \quad \text{and} \quad \mathbf{C}_2 = 2 \angle 7^\circ$$

b. Find $\mathbf{C}_1/\mathbf{C}_2$ if

$$\mathbf{C}_1 = 8 \angle 120^\circ \quad \text{and} \quad \mathbf{C}_2 = 16 \angle -50^\circ$$

***Solutions:***

a. $\dfrac{\mathbf{C}_1}{\mathbf{C}_2} = \dfrac{15}{2} \angle 10^\circ - 7^\circ = \mathbf{7.5 \angle 3^\circ}$

b. $\dfrac{\mathbf{C}_1}{\mathbf{C}_2} = \dfrac{8}{16} \angle 120^\circ - (-50^\circ) = \mathbf{0.5 \angle 170^\circ}$

---

We obtain the *reciprocal* in the rectangular form by multiplying the numerator and denominator by the complex conjugate of the denominator:

$$\frac{1}{A + jB} = \left(\frac{1}{A + jB}\right)\left(\frac{A - jB}{A - jB}\right) = \frac{A - jB}{A^2 + B^2}$$

and

$$\frac{1}{A + jB} = \frac{A}{A^2 + B^2} - j\frac{B}{A^2 + B^2} \tag{14.31}$$

In the polar form, the reciprocal is

$$\frac{1}{C \angle \theta} = \frac{1}{C} \angle -\theta \tag{14.32}$$

Some concluding examples using the four basic operations follow.

---

**EXAMPLE 14.26.** Perform the following operations, leaving the answer in polar or rectangular form:

a. $$\frac{(2 + j3) + (4 + j6)}{(7 + j7) - (3 - j3)} = \frac{(2 + 4) + j(3 + 6)}{(7 - 3) + j(7 + 3)}$$
$$= \frac{(6 + j9)(4 - j10)}{(4 + j10)(4 - j10)}$$
$$= \frac{[(6)(4) + (9)(10)] + j[(4)(9) - (6)(10)]}{4^2 + 10^2}$$
$$= \frac{114 - j24}{116} = \mathbf{0.983 - j0.207}$$

b. $$\frac{(50 \angle 30^\circ)(5 + j5)}{10 \angle -20^\circ} = \frac{(50 \angle 30^\circ)(7.07 \angle 45^\circ)}{10 \angle -20^\circ} = \frac{353.5 \angle 75^\circ}{10 \angle -20^\circ}$$
$$= 35.35 \angle 75^\circ - (-20^\circ) = \mathbf{35.35 \angle 95^\circ}$$

c. $$\frac{(2 \angle 20^\circ)^2(3 + j4)}{8 - j6} = \frac{(2 \angle 20^\circ)(2 \angle 20^\circ)(5 \angle 53.13^\circ)}{10 \angle -36.87^\circ}$$
$$= \frac{(4 \angle 40^\circ)(5 \angle 53.13^\circ)}{10 \angle -36.87^\circ} = \frac{20 \angle 93.13^\circ}{10 \angle -36.87^\circ}$$
$$= 2 \angle 93.13^\circ - (-36.87^\circ) = \mathbf{2.0 \angle 130^\circ}$$

d. $$3 \angle 27^\circ - 6 \angle -40^\circ = (2.673 + j1.362) - (4.596 - j3.857)$$
$$= (2.673 - 4.596) + j(1.362 + 3.857)$$
$$= \mathbf{-1.923 + j5.219}$$

---

## 14.11 TECHNIQUES OF CONVERSION

For many years the technologist was dependent on tables, charts, and the slide rule to perform conversions between complex numbers. Today, calculators such as that shown in Fig. 14.53 are available that can perform the conversion to eight-place accuracy. In fact, this particular calculator is programmed to perform this particular operation by your pressing just a few buttons. The →R and →P refer to the rectangular and polar forms, respectively.

You will appreciate the speed and accuracy of calculator conversion after using other techniques of conversion. There are inexpensive calculators that have the necessary functions to perform the conversions

**FIG. 14.53**
*Scientific calculator. (Courtesy of Hewlett Packard Co.)*

using Eqs. (14.18) through (14.21). For those who seriously expect to stay in this field, a calculator with the proper functions would be a wise investment.

## 14.12 PHASORS

As noted earlier in this chapter, the addition of sinusoidal voltages and currents will frequently be required in the analysis of ac circuits. One indicated method of performing this operation is to place both sinusoidal waveforms on the same set of axes and add algebraically the magnitudes of each at every point along the abscissa, as shown for $c = a + b$ in Fig. 14.54. This, however, can be a long and tedious process with

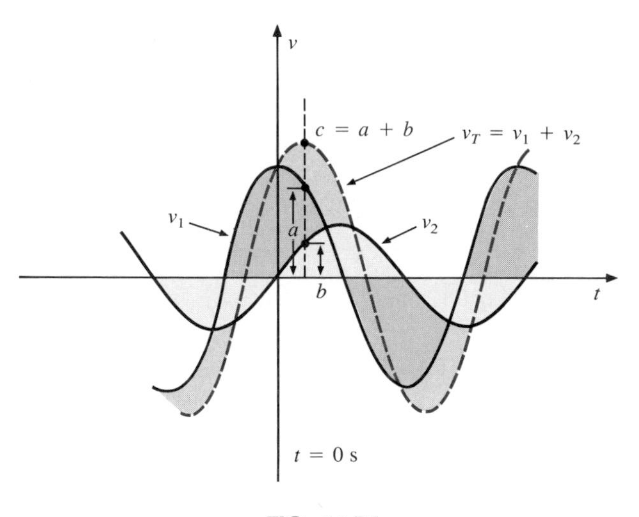

**FIG. 14.54**

limited accuracy. A shorter method uses the rotating radius vector shown in Fig. 13.20. This *radius vector,* having a *constant magnitude* (length) with *one end fixed at the origin,* is called a *phasor* when applied to electric circuits. During its rotational development of the sine wave, the phasor will, at the instant $t = 0$, have the positions shown in Fig. 14.55(a) for each waveform in Fig. 14.55(b).

Note in Fig. 14.55(b) that $v_2$ passes through the horizontal axis at $t = 0$, requiring that the radius vector in Fig. 14.55(a) be on the horizontal axis. Its length in Fig. 14.55(a) is equal to the peak value of the sinusoid as required by the radius vector of Fig. 13.20. The other sinusoid (actually a cosine wave) has passed through 90° of its rotation by the time $t = 0$ is reached and therefore has its maximum vertical projection as shown in Fig. 14.55(a). (Recall that 90° separates the minimum and maximum values of any sinusoidal function.) Since the vertical projection is a maximum, the peak value of the sinusoid that it will generate is also attained at $t = 0$, as shown in Fig. 14.55(b). Note also that $v_T = v_1$ at $t = 0$ since $v_2 = 0$ at this instant.

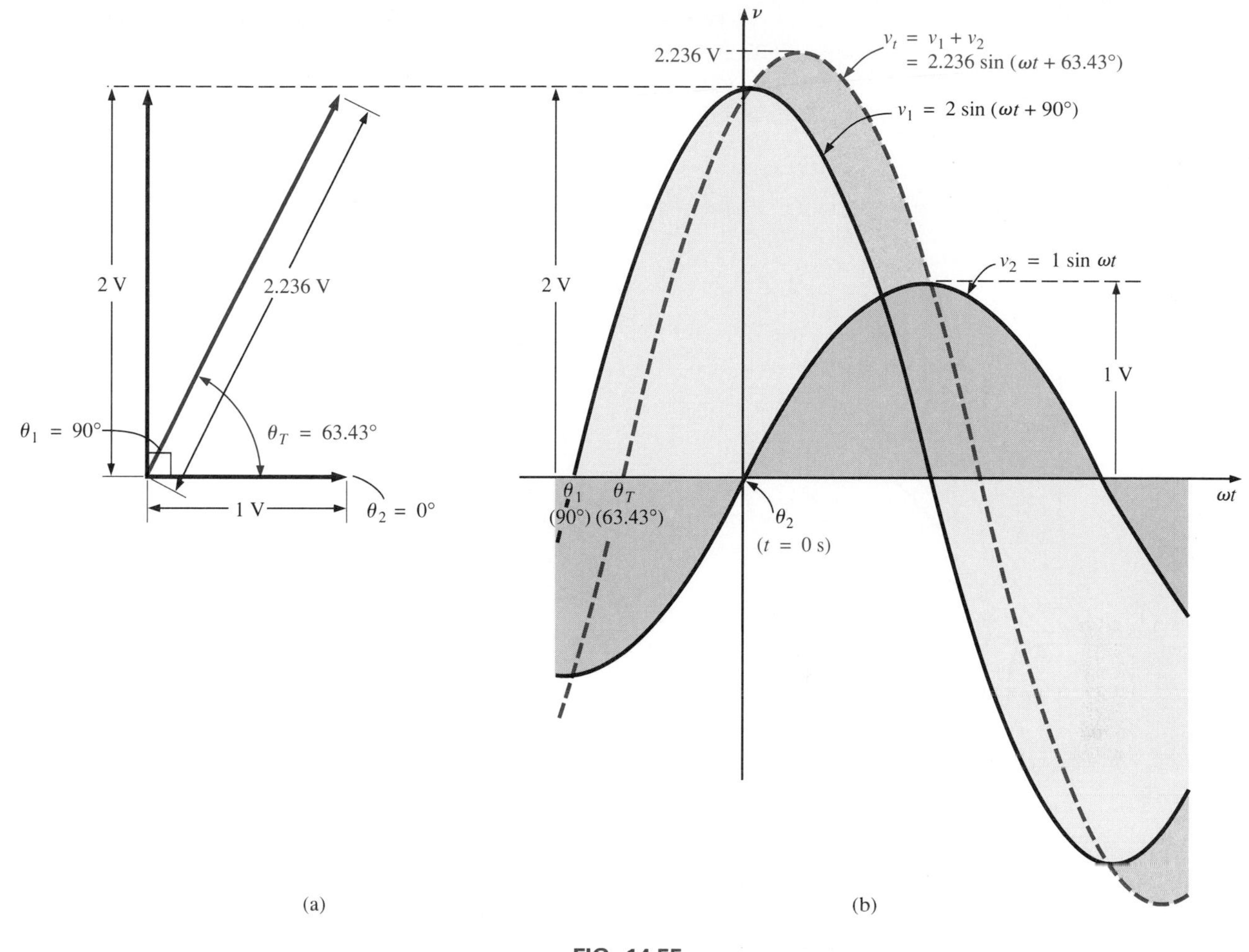


**FIG. 14.55**

It can be shown [see Fig. 14.55(a)] using the vector algebra described in Section 14.10 that

$$1 \text{ V} \angle 0° + 2 \text{ V} \angle 90° = 2.236 \text{ V} \angle 63.43°$$

In other words, if we convert $v_1$ and $v_2$ to the phasor form using

$$v = V_m \sin(\omega t \pm \theta) \Rightarrow V_m \angle \pm\theta$$

and add them using vector algebra, we can find the phasor form for $v_T$ with very little difficulty. It can then be converted to the time domain and plotted on the same set of axes as shown in Fig. 14.55(b). Figure 14.55(a), showing the magnitudes and relative positions of the various phasors, is called a *phasor diagram*. It is actually a ''snapshot'' of the phasors representing the sinusoidal waveforms at $t = 0$.

In the future, therefore, if the addition of two sinusoids is required, they should first be converted to the phasor domain and the sum found using complex algebra. The result can then be converted to the time domain if required.

The case of two sinusoidal functions having phase angles different from 0° and 90° appears in Fig. 14.56. Note again that the vertical height of the functions in Fig. 14.56(b) is determined by the rotational positions of the radius vectors in Fig. 14.56(a).

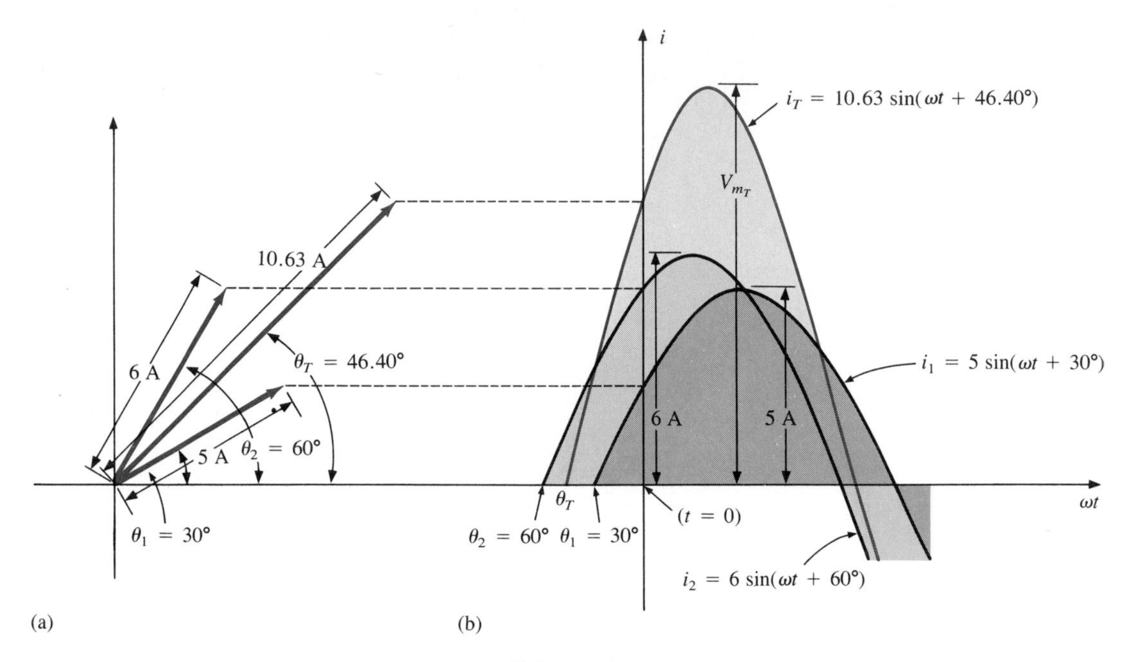

**FIG. 14.56**

Since the effective, rather than the peak, values are used almost exclusively in the analysis of ac circuits, the phasor will now be redefined for the purposes of practicality and uniformity as having a magnitude equal to the *effective value* of the sine wave it represents. The angle associated with the phasor will remain as previously described—the phase angle. The phasor diagram will therefore be as shown in Fig. 14.57, replacing that of Fig. 14.56(a).

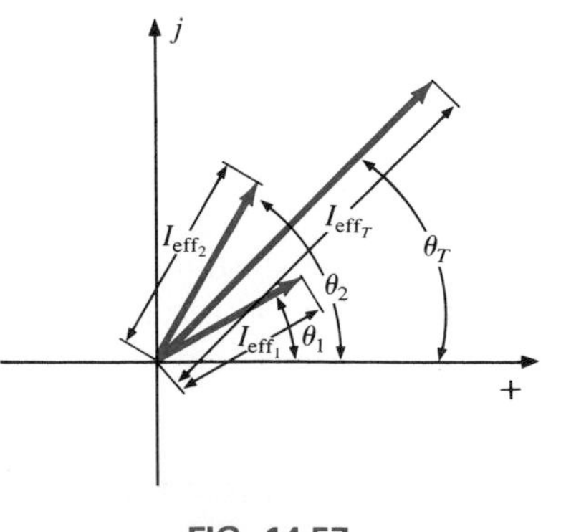

**FIG. 14.57**

In general, for all of the analyses to follow, the phasor form of a sinusoidal voltage or current will be

$$\mathbf{V} = V \angle\theta \quad \text{and} \quad \mathbf{I} = I \angle\theta$$

where $V$ and $I$ are effective values and $\theta$ is the phase angle. It should be pointed out that in phasor notation, the sine wave is always the reference, and the frequency is not represented. Phasor algebra for sinusoidal quantities is applicable only for waveforms having the *same frequency,* so that it can be carried along without any special notation.

**EXAMPLE 14.27.** Convert the following from the time to the phasor domain:

| Time Domain | Phasor Domain |
|---|---|
| a. $\sqrt{2}(50) \sin \omega t$ | $\mathbf{50 \angle 0^\circ}$ |
| b. $69.6 \sin(\omega t + 72^\circ)$ | $(0.707)(69.6) \angle 72^\circ = \mathbf{49.21 \angle 72^\circ}$ |
| c. $45 \cos \omega t$ | $(0.707)(45) \angle 90^\circ = \mathbf{31.82 \angle 90^\circ}$ |

**EXAMPLE 14.28.** Write the sinusoidal expression for the following phasors if the frequency is 60 Hz:

| Phasor Domain | Time Domain |
|---|---|
| a. $\mathbf{I} = 10 \angle 30^\circ$ | $i = \sqrt{2}(10) \sin(2\pi 60t + 30^\circ)$<br>and $i = \mathbf{14.14 \sin(377t + 30^\circ)}$ |
| b. $\mathbf{V} = 115 \angle -70^\circ$ | $v = \sqrt{2}(115) \sin(377t - 70^\circ)$<br>**and** $\mathbf{v = 162.6 \sin(377t - 70^\circ)}$ |

**EXAMPLE 14.29.** Find the input voltage of the circuit of Fig. 14.58 if

$$\left.\begin{aligned} v_a &= 50 \sin(377t + 30^\circ) \\ v_b &= 30 \sin(377t + 60^\circ) \end{aligned}\right\} \quad f = 60 \text{ Hz}$$

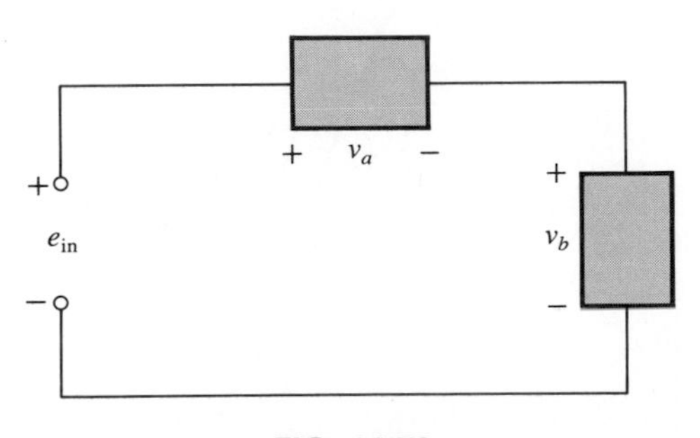


FIG. 14.58

***Solution:*** Applying Kirchhoff's voltage law, we have

$$e_{in} = v_a + v_b$$

Converting from the time to the phasor domain yields

$$v_a = 50 \sin(377t + 30°) \Rightarrow \mathbf{V}_a = 35.35 \text{ V} \angle 30°$$
$$v_b = 30 \sin(377t + 60°) \Rightarrow \mathbf{V}_b = 21.21 \text{ V} \angle 60°$$

Converting from polar to rectangular form for addition yields

$$\mathbf{V}_a = 35.35 \text{ V} \angle 30° = 30.61 + j17.68$$
$$\mathbf{V}_b = 21.21 \text{ V} \angle 60° = 10.61 + j18.37$$

Then

$$\mathbf{E}_{in} = \mathbf{V}_a + \mathbf{V}_b = (30.61 + j17.68) + (10.61 + j18.37)$$
$$= 41.22 + j36.05$$

Converting from rectangular to polar form, we have

$$\mathbf{E}_{in} = 41.22 + j36.05 = 54.76 \text{ V} \angle 41.17$$

Converting from the phasor to the time domain, we obtain

$$\mathbf{E}_{in} = 54.76 \text{ V} \angle 41.17° \Rightarrow e_{in} = \sqrt{2}(54.76) \sin(377t + 41.17°)$$

and

$$e_{in} = \mathbf{77.43 \sin(377t + 41.17°)}$$

A plot of the three waveforms is shown in Fig. 14.59. Note that at each instant of time, the sum of the two waveforms does in fact add up

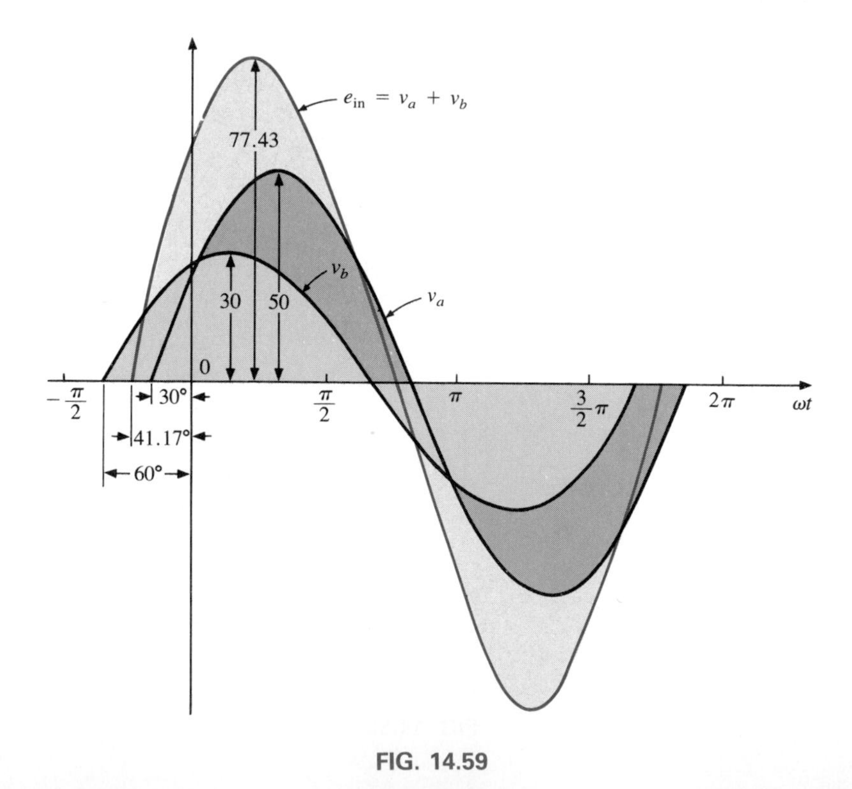


FIG. 14.59

to $e_{in}$. At $t = 0$ ($\omega t = 0$), $e_{in}$ is the sum of the two positive values, while at a value of $\omega t$ almost midway between $\pi/2$ and $\pi$, the sum of the positive value of $v_a$ and the negative value of $v_b$ results in $e_{in} = 0$.

---

**EXAMPLE 14.30.** Determine the current $i_2$ for the network of Fig. 14.60.

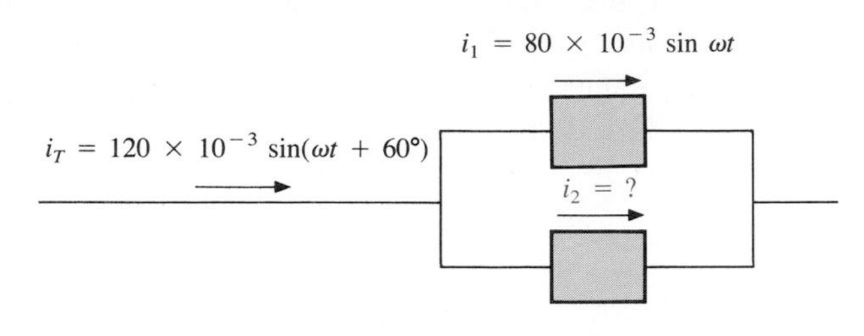


FIG. 14.60

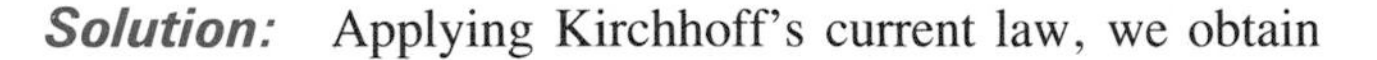

***Solution:*** Applying Kirchhoff's current law, we obtain

$$i_T = i_1 + i_2$$

or

$$i_2 = i_T - i_1$$

Converting from the time to the phasor domain yields

$$i_T = 120 \times 10^{-3} \sin(\omega t + 60°) \Rightarrow 84.84 \text{ mA} \angle 60°$$
$$i_1 = 80 \times 10^{-3} \sin \omega t \Rightarrow 56.56 \text{ mA} \angle 0°$$

Converting from polar to rectangular form for subtraction yields

$$\mathbf{I}_T = 84.84 \text{ mA} \angle 60° = 42.42 \times 10^{-3} + j73.47 \times 10^{-3}$$
$$\mathbf{I}_1 = 56.56 \text{ mA} \angle 0° = 56.56 \times 10^{-3} + j0$$

Then

$$\begin{aligned}\mathbf{I}_2 &= \mathbf{I}_T - \mathbf{I}_1 \\ &= (42.42 \times 10^{-3} + j73.47 \times 10^{-3}) - (56.56 \times 10^{-3} + j0)\end{aligned}$$

and

$$\mathbf{I}_2 = -14.14 \times 10^{-3} + j73.47 \times 10^{-3}$$

Converting from rectangular to polar form, we have

$$\mathbf{I}_2 = 74.82 \text{ mA} \angle 100.89°$$

Converting from the phasor to the time domain, we have

$$\mathbf{I}_2 = 74.82 \text{ mA} \angle 100.89° \Rightarrow$$
$$i_2 = \sqrt{2}(74.82 \times 10^{-3}) \sin(\omega t + 100.89°)$$

and

$$i_2 = \mathbf{105.8 \times 10^{-3} \sin(\omega t + 100.89°)}$$

A plot of the three waveforms appears in Fig. 14.61. The waveforms clearly indicate that $i_T = i_1 + i_2$.

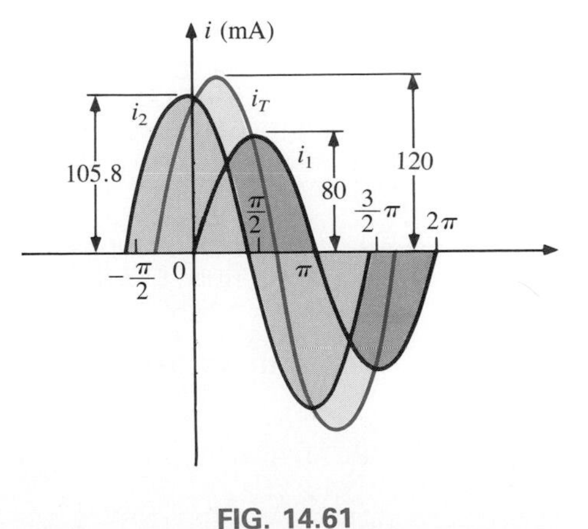


FIG. 14.61

## 14.13 COMPUTER METHODS: CONVERSION ROUTINE

Although PSPICE could be applied to perform some of the analysis described in this chapter, the true benefits of the package will not be evident until we investigate series and parallel ac networks in the next chapter.

BASIC, however, is not limited to the content of a specific software package and can be used to perform a variety of tasks other than solving for the important parameters of a network. For instance, the BASIC program of Fig. 14.62 is designed to perform a conversion from polar to

```
            10 REM ***** PROGRAM 14-1 *****
            20 REM ****************************************
            30 REM Program to perform selected conversions
            40 REM ****************************************
            50 REM
           100 PRINT
           110 PRINT "Enter (1) for rectangular to polar conversion"
           120 PRINT "      (2) for polar to rectangular conversion"
           130 PRINT TAB(20);
           140 INPUT "Choice=";C :REM C is choice 1 or 2
           150 IF C<0 OR C>2 THEN GOTO 110
           160 ON C GOSUB 200,300
           170 PRINT:INPUT "More(YES or NO)";A$
           180 IF A$="YES" THEN GOTO 100
           190 END
           200 REM Use rectangular to polar conversion module
           210 PRINT:PRINT:PRINT "Enter rectangular data:"
Input-     220 INPUT "X=";X :INPUT "Y=";Y
(Rect.)    230 GOSUB 2000
Output-    240 PRINT:PRINT "Polar form is";Z;"at an angle of";TH;"degrees"
(Polar)    250 RETURN
           300 REM Use polar to rectangular conversion
Input    [ 310 PRINT:PRINT "Enter polar data:":PRINT:INPUT "Z=";Z
(Polar)  [ 320 INPUT "Angle(degrees), TH=";TH
           330 GOSUB 2100
Output   [ 340 PRINT:PRINT "Rectangular form is";X;
(Rect.)  [ 350 IF Y>=0 THEN PRINT "+j";Y
         [ 360 IF Y<0 THEN PRINT "-j";ABS(Y)
           370 RETURN
         [2000 REM Module to convert from rectangular to polar form.
         [2010 REM Enter with X, Y - Return with Z, TH(eta)
Rect.    [2020 Z=SQR(X^2+Y^2)
 ↓       [2030 IF X<0 THEN TH=(180/3.14159)*ATN(Y/X)
Polar    [2040 IF X<0 THEN TH=180*SGN(Y)+(180/3.14159)*ATN(Y/X)
         [2050 IF X=0 THEN TH=90*SGN(Y)
         [2060 IF Y=0 THEN IF X<0 THEN TH=180
         [2070 RETURN
         [2100 REM Module to convert from polar to rectangular form.
Polar    [2110 REM Enter with Z, TH(eta) - return with X, Y
 ↓       [2120 X=Z*COS(TH*3.14159/180)
Rect.    [2130 Y=Z*SIN(TH*3.14159/180)
         [2140 RETURN

READY
```

```
Enter  (1)  for  rectangular  to  polar  conversion
       (2)  for  polar  to  rectangular  conversion
                     Choice=? 2


Enter  polar  data:

Z=? 5

Angle(degrees),  TH=? -53.13


Rectangular  form  is  3  -j  4

More(YES  or  NO)? YES


Enter  (1)  for  rectangular  to  polar  conversion
       (2)  for  polar  to  rectangular  conversion
                     Choice=? 1


Enter  rectangular  data:
X=? -10

Y=? 20


Polar  form  is  22.3607  at  an  angle  of  116.565  degrees

More(YES  or  NO)? YES


Enter  (1)  for  rectangular  to  polar  conversion
       (2)  for  polar  to  rectangular  conversion
                     Choice=? 2


Enter  polar  data:

Z=? 12

Angle(degrees),  TH=? 35


Rectangular  form  is  9.8298  +j  6.8829

More(YES  or  NO)? NO


READY
```

**FIG. 14.62**
*Program 14.1.*

rectangular form or rectangular to polar form. The input parameters of the rectangular form are entered on line 220, and the polar form on lines 310 and 320. Line 240 outputs the polar form, and lines 340 through 360 the rectangular form.

The rectangular-to-polar conversion routine appears on lines 2000 through 2070, while the polar-to-rectangular conversion appears on lines 2100 through 2140. Note on line 2020 the equation for the magnitude of the polar form $Z = \sqrt{X^2 + Y^2}$ and the testing of $X$ and $Y$ to determine the correct value of $\theta$ on lines 2030 through 2060. Lines 2120 and 2130 determine $X$ and $Y$ using the equations $X = Z \cos \theta$ and $Y = Z \sin \theta$, respectively. Note the need to convert the input angle in degrees (TH) to radians before the BASIC language can act on the SIN and COS functions.

The first run performs the following conversion:

$$5 \angle -53.13^\circ \Rightarrow 3 - j4$$

The second run performs a rectangular-to-polar conversion:

$$-10 + j20 = 22.3607 \angle 116.565^\circ$$

The last run performs a second polar-to-rectangular conversion:

$$12 \angle 35^\circ = 9.8298 + j6.8829$$

Now that Program 14.1 is available and in memory, it can be called up in the midst of any other ac analysis package to support the calculations being performed. In fact, chances are that the PSPICE software package uses a routine very similar to Program 14.1 to perform the conversions when required.

## PROBLEMS

### SECTION 14.2 The Derivative

1. Plot the following waveform versus time showing one clear complete cycle. Then determine the derivative of the waveform using Eq. (14.1), and sketch one complete cycle of the derivative directly under the original waveform. Compare the magnitude of the derivative at various points versus the slope of the original sinusoidal function.

$$v = 1 \sin 3.14t$$

2. Repeat Problem 1 for the following sinusoidal function and compare results. In particular, determine the frequency of the waveforms of Problems 1 and 2 and compare the magnitude of the derivative.

$$v = 1 \sin 15.7t$$

3. What is the derivative of each of the following sinusoidal expressions?
   **a.** $10 \sin 377t$
   **b.** $0.6 \cos 754t$
   **c.** $0.05 \cos(157t - 10^\circ)$
   **d.** $25 \cos(20t - 150^\circ)$

## SECTION 14.3 Response of Basic *R*, *L*, and *C* Elements to a Sinusoidal Voltage or Current

**4.** The voltage across a 5-Ω resistor is as indicated. Find the sinusoidal expression for the current. In addition, sketch the $v$ and $i$ curves with the abscissa in radians.
**a.** $150 \sin 377t$
**b.** $30 \sin(377t + 20°)$
**c.** $40 \cos(\omega t + 10°)$
**d.** $-80 \sin(\omega t + 40°)$

**5.** The current through a 7-kΩ resistor is as indicated. Find the sinusoidal expression for the voltage. In addition, sketch the $v$ and $i$ curves with the abscissa in radians.
**a.** $0.03 \sin 754t$
**b.** $2 \times 10^{-3} \sin(400t - 120°)$
**c.** $6 \times 10^{-6} \cos(\omega t - 2°)$
**d.** $-0.004 \cos(\omega t - 90°)$

**6.** Determine the inductive reactance (in ohms) of a 2-H coil for
**a.** dc
and for the following frequencies:
**b.** 25 Hz
**c.** 60 Hz
**d.** 2000 Hz
**e.** 100,000 Hz

**7.** Determine the inductance of a coil that has a reactance of
**a.** 20 Ω at $f = 2$ Hz.
**b.** 1000 Ω at $f = 60$ Hz.
**c.** 5280 Ω at $f = 1000$ Hz.

**8.** Determine the frequency at which a 10-H inductance has the following inductive reactances:
**a.** 50 Ω
**b.** 3770 Ω
**c.** 15.7 kΩ
**d.** 243 Ω

**9.** The current through a 20-Ω inductive reactance is given. What is the sinusoidal expression for the voltage? Sketch the $v$ and $i$ curves with the abscissa in radians.
**a.** $i = 5 \sin \omega t$
**b.** $i = 0.4 \sin(\omega t + 60°)$
**c.** $i = -6 \sin(\omega t - 30°)$
**d.** $i = 3 \cos(\omega t + 10°)$

**10.** The current through a 0.1-H coil is given. What is the sinusoidal expression for the voltage?
**a.** $30 \sin 30t$
**b.** $0.006 \sin 377t$
**c.** $5 \times 10^{-6} \sin(400t + 20°)$
**d.** $-4 \cos(20t - 70°)$

**11.** The voltage across a 50-Ω inductive reactance is given. What is the sinusoidal expression for the current? Sketch the $v$ and $i$ curves with the abscissa in radians.
**a.** $50 \sin \omega t$
**b.** $30 \sin(\omega t + 20°)$
**c.** $40 \cos(\omega t + 10°)$
**d.** $-80 \sin(377t + 40°)$

**12.** The voltage across a 0.2-H coil is given. What is the sinusoidal expression for the current?
**a.** $1.5 \sin 60t$
**b.** $0.016 \sin(t + 4°)$
**c.** $-4.8 \sin(0.05t + 50°)$
**d.** $9 \times 10^{-3} \cos(377t + 360°)$

**13.** Determine the capacitive reactance (in ohms) of a 5-$\mu$F capacitor for
**a.** dc
and for the following frequencies:
**b.** 60 Hz **c.** 120 Hz
**d.** 1800 Hz **e.** 24,000 Hz

**14.** Determine the capacitance in microfarads if a capacitor has a reactance of
**a.** 250 $\Omega$ at $f = 60$ Hz.
**b.** 55 $\Omega$ at $f = 312$ Hz.
**c.** 10 $\Omega$ at $f = 25$ Hz.

**15.** Determine the frequency at which a 50-$\mu$F capacitor has the following capacitive reactances:
**a.** 342 $\Omega$ **b.** 684 $\Omega$
**c.** 171 $\Omega$ **d.** 2000 $\Omega$

**16.** The voltage across a 2.5-$\Omega$ capacitive reactance is given. What is the sinusoidal expression for the current? Sketch the $v$ and $i$ curves with the abscissa in radians.
**a.** $100 \sin \omega t$ **b.** $0.4 \sin(\omega t + 20°)$
**c.** $8 \cos(\omega t + 10°)$ **d.** $-70 \sin(\omega t + 40°)$

**17.** The voltage across a 1-$\mu$F capacitor is given. What is the sinusoidal expression for the current?
**a.** $30 \sin 200t$ **b.** $90 \sin 377t$
**c.** $-120 \sin(374t + 30°)$ **d.** $70 \cos(800t - 20°)$

**18.** The current through a 10-$\Omega$ capacitive reactance is given. Write the sinusoidal expression for the voltage. Sketch the $v$ and $i$ curves with the abscissa in radians.
**a.** $i = 50 \sin \omega t$ **b.** $i = 40 \sin(\omega t + 60°)$
**c.** $i = -6 \sin(\omega t - 30°)$ **d.** $i = 3 \cos(\omega t + 10°)$

**19.** The current through a 0.5-$\mu$F capacitor is given. What is the sinusoidal expression for the voltage?
**a.** $0.20 \sin 300t$ **b.** $0.007 \sin 377t$
**c.** $0.048 \cos 754t$ **d.** $0.08 \sin(1600t - 80°)$

***20.** For the following pairs of voltages and currents, indicate whether the element involved is a capacitor, inductor, or resistor, and the value of $C$, $L$, or $R$ if sufficient data are given:
**a.** $v = 550 \sin(377t + 40°)$
$i = 11 \sin(377t - 50°)$
**b.** $v = 36 \sin(754t + 80°)$
$i = 4 \sin(754t + 170°)$
**c.** $v = 10.5 \sin(\omega t + 13°)$
$i = 1.5 \sin(\omega t + 13°)$

***21.** Repeat Problem 20 for the following pairs of voltages and currents:
**a.** $v = 2000 \sin \omega t$
$i = 5 \cos \omega t$
**b.** $v = 80 \sin(157t + 150°)$
$i = 2 \sin(157t + 60°)$
**c.** $v = 35 \sin(\omega t - 20°)$
$i = 7 \cos(\omega t - 110°)$

## SECTION 14.4 Frequency Response of the Basic Elements

**22.** Plot $X_L$ versus frequency for a 5-mH coil using a frequency range of zero to 100 kHz on a linear scale.

**23.** Plot $X_C$ versus frequency for a 1-$\mu$F capacitor using a frequency range of zero to 10 kHz on a linear scale.

**24.** At what frequency will the reactance of a 1-$\mu$F capacitor equal the resistance of a 2-k$\Omega$ resistor?

**25.** The reactance of a coil equals the resistance of a 10-k$\Omega$ resistor at a frequency of 5 kHz. Determine the inductance of the coil.

**26.** Determine the frequency at which a 1-$\mu$F capacitor and a 10-mH inductor will have the same reactance.

**27.** Determine the capacitance required to establish a capacitive reactance that will match that of a 2-mH coil at a frequency of 50 kHz.

## SECTION 14.5 Average Power and Power Factor

**28.** Find the average power loss in watts for each set of Problem 20.

**29.** Find the average power loss in watts for each set of Problem 21.

***30.** Find the average power loss and power factor for each of the circuits whose input current and voltage are as follows:

**a.** $v = 60 \sin(\omega t + 30°)$
$i = 15 \sin(\omega t + 60°)$

**b.** $v = -50 \sin(\omega t - 20°)$
$i = -2 \sin(\omega t + 40°)$

**c.** $v = 50 \sin(\omega t + 80°)$
$i = 3 \cos(\omega t + 20°)$

**d.** $v = 75 \sin(\omega t - 5°)$
$i = 0.08 \sin(\omega t - 35°)$

**31.** If the current through and voltage across an element are $i = 8 \sin(\omega t + 40°)$ and $v = 48 \sin(\omega t + 40°)$, respectively, compute the power by $I^2R$, $(V_m I_m/2) \cos \theta$, and $VI \cos \theta$, and compare answers.

**32.** A circuit dissipates 100 W (average power) at 150 V (effective input voltage) and 2 A (effective input current). What is the power factor? Repeat if the power is 0 W; 300 W.

***33.** The power factor of a circuit is 0.5 lagging. The power delivered in watts is 500. If the input voltage is $50 \sin(\omega t + 10°)$, find the sinusoidal expression for the input current.

**34.** In Fig. 14.63, $e = 30 \sin(377t + 20°)$.

**a.** What is the sinusoidal expression for the current?

**b.** Find the power loss in the circuit.

**c.** How long (in seconds) does it take the current to complete 6 cycles?

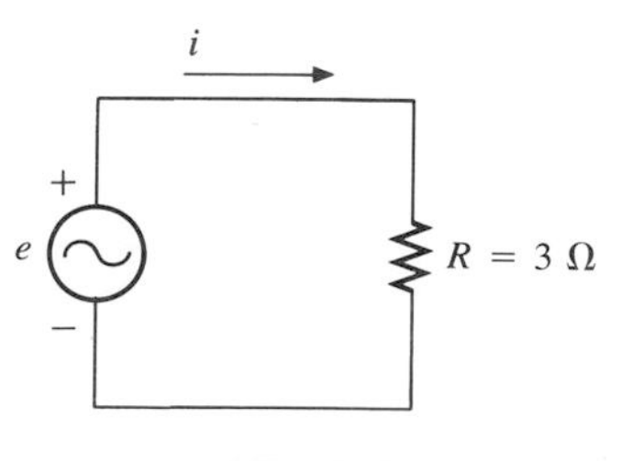


FIG. 14.63

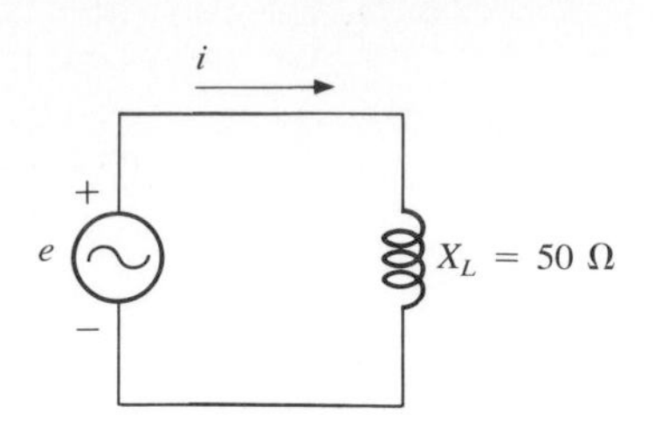


FIG. 14.64

35. In Fig. 14.64, $e = 100 \sin(157t + 30°)$.
    **a.** Find the sinusoidal expression for $i$.
    **b.** Find the value of the inductance $L$.
    **c.** Find the average power loss by the inductor.

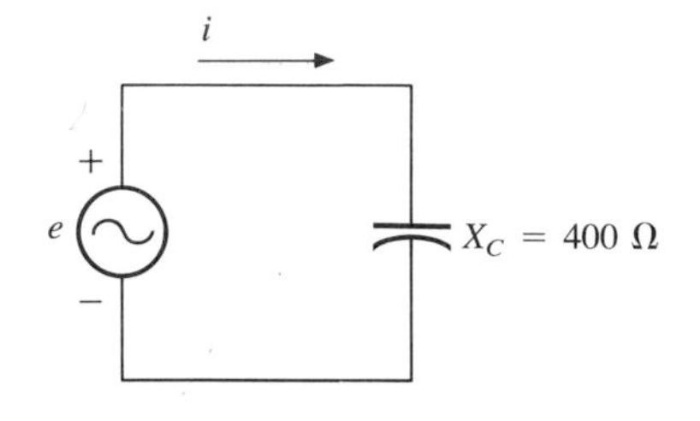


FIG. 14.65

36. In Fig. 14.65, $i = 3 \sin(377t - 20°)$.
    **a.** Find the sinusoidal expression for $e$.
    **b.** Find the value of the capacitance $C$ in microfarads.
    **c.** Find the average power loss in the capacitor.

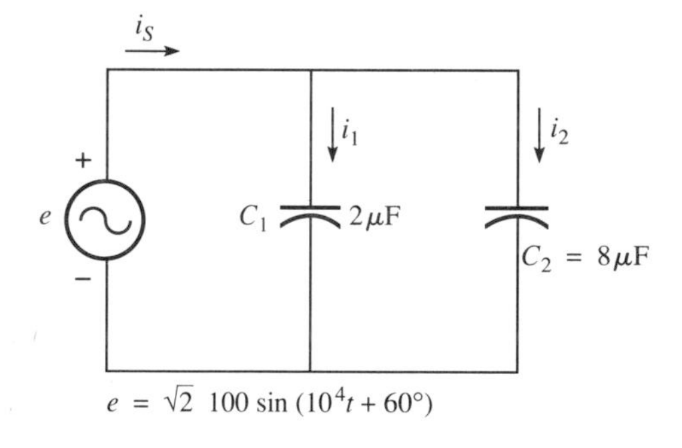


FIG. 14.66

*37. For the network of Fig. 14.66 and the applied signal:
    **a.** Determine $i_1$ and $i_2$.
    **b.** Find $i_S$.

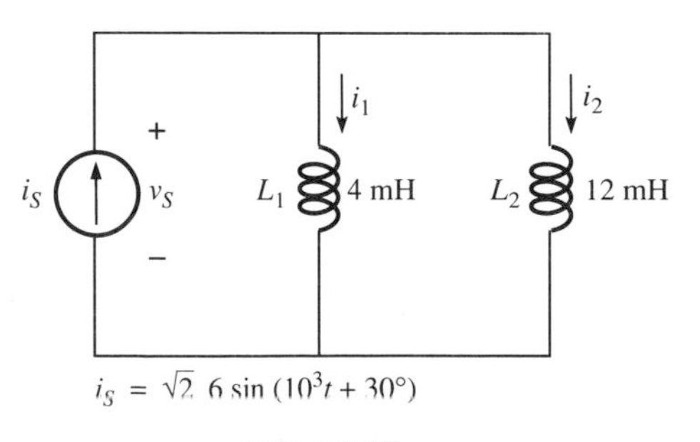


FIG. 14.67

*38. For the network of Fig. 14.67 and the applied source:
    **a.** Determine the source voltage $v_S$.
    **b.** Find the currents $i_1$ and $i_2$.

## SECTION 14.9 Conversion Between Forms

39. Convert the following from rectangular to polar form:
    **a.** $4 + j3$
    **b.** $2 + j2$
    **c.** $3.5 + j16$
    **d.** $100 + j800$
    **e.** $1000 + j400$
    **f.** $0.001 + j0.0065$
    **g.** $7.6 - j9$
    **h.** $-8 + j4$
    **i.** $-15 - j60$
    **j.** $+78 - j65$
    **k.** $-2400 + j3600$
    **l.** $5 \times 10^{-3} - j25 \times 10^{-3}$

**40.** Convert the following from polar to rectangular form:

**a.** $6 \angle 30°$ **b.** $40 \angle 80°$
**c.** $7400 \angle 70°$ **d.** $4 \times 10^{-4} \angle 8°$
**e.** $0.04 \angle 80°$ **f.** $0.0093 \angle 23°$
**g.** $65 \angle 150°$ **h.** $1.2 \angle 135°$
**i.** $500 \angle 200°$ **j.** $6320 \angle -35°$
**k.** $7.52 \angle -125°$ **l.** $0.008 \angle 310°$

**41.** Convert the following from rectangular to polar form:

**a.** $1 + j15$ **b.** $60 + j5$
**c.** $0.01 + j0.3$ **d.** $100 - j2000$
**e.** $-5.6 + j86$ **f.** $-2.7 - j38.6$

**42.** Convert the following from polar to rectangular form:

**a.** $13 \angle 5°$ **b.** $160 \angle 87°$
**c.** $7 \times 10^{-6} \angle 2°$ **d.** $8.7 \angle 177°$
**e.** $76 \angle -4°$ **f.** $396 \angle +265°$

## SECTION 14.10 Mathematical Operations with Complex Numbers

Perform the following operations.

**43.** Addition and subtraction (express your answers in rectangular form):

**a.** $(4.2 + j6.8) + (7.6 + j0.2)$
**b.** $(142 + j7) + (9.8 + j42) + (0.1 + j0.9)$
**c.** $(4 \times 10^{-6} + j76) + (7.2 \times 10^{-7} - j5)$
**d.** $(9.8 + j6.2) - (4.6 + j4.6)$
**e.** $(167 + j243) - (-42.3 - j68)$
**f.** $(-36.0 + j78) - (-4 - j6) + (10.8 - j72)$
**g.** $6 \angle 20° + 8 \angle 80°$
**h.** $42 \angle 45° + 62 \angle 60° - 70 \angle 120°$

**44.** Multiplication [express your answers in rectangular form for parts (i) through (1), and in polar form for parts (m) through (p)]:

**i.** $(2 + j3)(6 + j8)$
**j.** $(7.8 + j1)(4 + j2)(7 + j6)$
**k.** $(0.002 + j0.006)(-2 + j2)$
**l.** $(400 - j200)(-0.01 - j0.5)(-1 + j3)$
**m.** $(2 \angle 60°)(4 \angle 22°)$
**n.** $(6.9 \angle 8°)(7.2 \angle -72°)$
**o.** $0.002 \angle 120°)(0.5 \angle 200°)(40 \angle -60°)$
**p.** $(540 \angle -20°)(-5 \angle 180°)(6.2 \angle 0°)$

**45.** Division (express your answers in polar form):

**q.** $(42 \angle 10°)/(7 \angle 60°)$
**r.** $(0.006 \angle 120°)/(30 \angle -20°)$
**s.** $(4360 \angle -20°)/(40 \angle 210°)$
**t.** $(650 \angle -80°)/(8.5 \angle 360°)$
**u.** $(8 + j8)/(2 + j2)$
**v.** $(8 + j42)/(-6 + j60)$
**w.** $(0.05 + j0.25)/(8 - j60)$
**x.** $(-4.5 - j6)/(0.1 - j0.4)$

*46. Perform the following operations (express your answers in rectangular form):

a. $\dfrac{(4 + j3) + (6 - j8)}{(3 + j3) - (2 + j3)}$

b. $\dfrac{8 \angle 60^\circ}{(2 \angle 0^\circ) + (100 + j100)}$

c. $\dfrac{(6 \angle 20^\circ)(120 \angle -40^\circ)(3 + j4)}{2 \angle -30^\circ}$

d. $\dfrac{(0.4 \angle 60^\circ)^2(300 \angle 40^\circ)}{3 + j9}$

e. $\dfrac{(150 \angle 2^\circ)(4 \times 10^{-6} \angle 88^\circ)}{(0.002 \angle 10^\circ)^2(4 \angle 30^\circ)}$

## SECTION 14.12 Phasors

47. Express the following in phasor form:
   - a. $\sqrt{2}(100) \sin(\omega t + 30^\circ)$
   - b. $\sqrt{2}(0.25) \sin(157t - 40^\circ)$
   - c. $100 \sin(\omega t - 90^\circ)$
   - d. $42 \sin(377t + 0^\circ)$
   - e. $6 \times 10^{-6} \cos \omega t$
   - f. $3.6 \times 10^{-6} \cos(754t - 20^\circ)$

48. Express the following phasor currents and voltages as sine waves if the frequency is 60 Hz:
   - a. $\mathbf{I} = 40 \angle 20^\circ$ b. $\mathbf{V} = 120 \angle 0^\circ$
   - c. $\mathbf{I} = 8 \times 10^{-3} \angle 120^\circ$ d. $\mathbf{V} = 5 \angle 90^\circ$
   - e. $\mathbf{I} = 1200 \angle -120^\circ$ f. $\mathbf{V} = \dfrac{6000}{\sqrt{2}} \angle -180^\circ$

49. For the system of Fig. 14.68, find the sinusoidal expression for the unknown voltage $v_a$ if

$$e_{in} = 60 \sin(377t + 20^\circ)$$
$$v_b = 20 \sin 377t$$

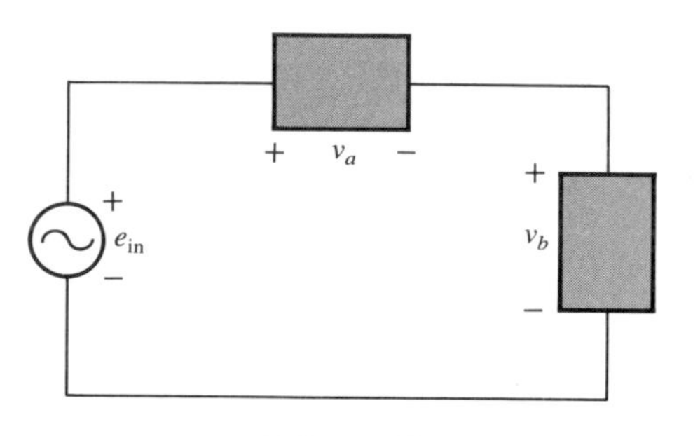

FIG. 14.68

50. For the system of Fig. 14.69, find the sinusoidal expression for the unknown current $i_1$ if

$$i_s = 20 \times 10^{-6} \sin(\omega t + 90^\circ)$$
$$i_2 = 6 \times 10^{-6} \sin(\omega t - 60^\circ)$$

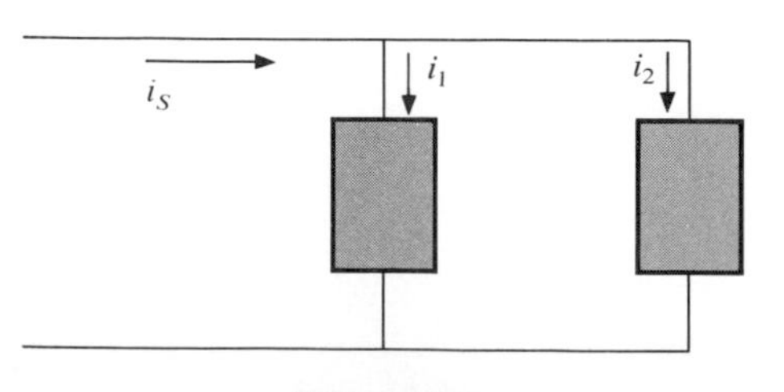


FIG. 14.69

**51.** Find the sinusoidal expression for the applied voltage $e$ for the system of Fig. 14.70 if

$$v_a = 60 \sin(\omega t + 30°)$$
$$v_b = 30 \sin(\omega t - 30°)$$
$$v_c = 40 \sin(\omega t + 120°)$$

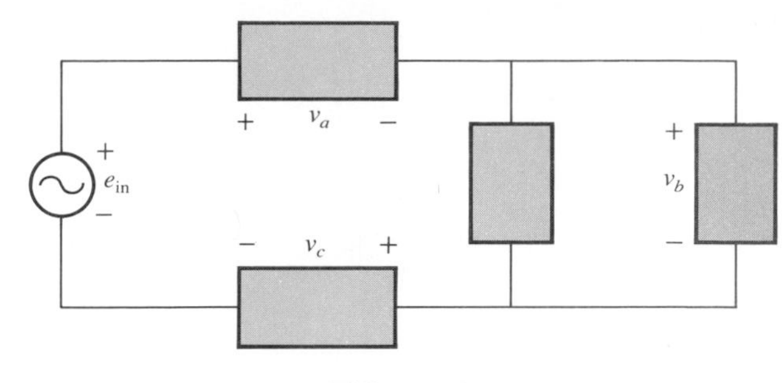

FIG. 14.70

**52.** Find the sinusoidal expression for the current $i_T$ for the system of Fig. 14.71 if

$$i_1 = 6 \times 10^{-3} \sin(377t + 180°)$$
$$i_2 = 8 \times 10^{-3} \sin 377t$$
$$i_3 = 2i_2$$

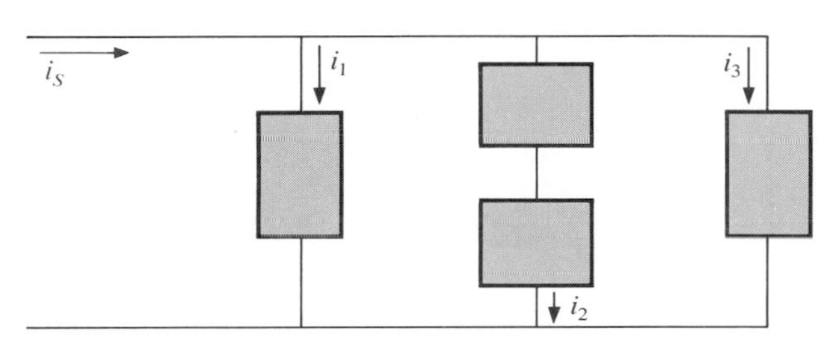

FIG. 14.71

## SECTION 14.13 Computer Methods: Conversion Routine

**53.** Given a sinusoidal function, write a program to print out the derivative.

**54.** Given the sinusoidal expression for the current, determine the expression for the voltage across a resistor, capacitor, or inductor, depending on the element involved. In other words, the program will ask which element is to be investigated and will then request the pertinent data to obtain the mathematical expression for the sinusoidal voltage.

**55.** Write a program to tabulate the reactance versus frequency for an inductor or capacitor for a specified frequency range.

**56.** Given the sinusoidal expression for the voltage and current of a load, write a program to determine the average power and power factor.

**57.** Given two sinusoidal functions, write a program to convert each to the phasor domain, add the two, and print out the sum in the phasor and time domains.

## GLOSSARY

**Average or real power** The power delivered to and dissipated by the load over a full cycle.

**Complex conjugate** A complex number defined by simply changing the sign of an imaginary component of a complex number in the rectangular form.

**Complex number** A number that represents a point in a two-dimensional plane located with reference to two distinct axes. It defines a vector drawn from the origin to that point.

**Derivative** The instantaneous rate of change of a function with respect to time or another variable.

**Leading and lagging power factors** An indication of whether a network is primarily capacitive or inductive in nature. Leading power factors are associated with capacitive networks, and lagging power factors with inductive networks.

**Phasor** A radius vector that has a constant magnitude at a fixed angle from the positive real axis and that represents a sinusoidal voltage or current in the vector domain.

**Phasor diagram** A "snapshot" of the phasors that represent a number of sinusoidal waveforms at $t = 0$.

**Polar form** A method of defining a point in a complex plane that includes a single magnitude to represent the distance from the origin, and an angle to reflect the counterclockwise distance from the positive real axis.

**Power factor** ($F_p$) An indication of how reactive or resistive an electrical system is. The higher the power factor, the greater the resistive component.

**Reactance** The opposition of an inductor or capacitor to the flow of charge that results in the continual exchange of energy between the circuit and magnetic field of an inductor or the electric field of a capacitor.

**Reciprocal** A format defined by 1 divided by the complex number.

**Rectangular form** A method of defining a point in a complex plane that includes the magnitude of the real component and the magnitude of the imaginary component, the latter component being defined by an associated letter $j$.

# 15
# Series and Parallel ac Circuits

## 15.1 INTRODUCTION

In this chapter, phasor algebra will be used to develop a quick, direct method for solving both the series and the parallel ac circuits. The close relationship that exists between this method for solving for unknown quantities and the approach used for dc circuits will become apparent after a few simple examples are considered. Once this association is established, many of the rules (current divider rule, voltage divider rule, and so on) for dc circuits can be readily applied to ac circuits.

### *SERIES ac CIRCUITS*

## 15.2 IMPEDANCE AND THE PHASOR DIAGRAM

In Chapter 14, we found, for the purely resistive circuit of Fig. 15.1, that $v$ and $i$ were in phase, and the magnitude

$$I_m = \frac{V_m}{R} \quad \text{or} \quad V_m = I_m R$$

In phasor form,

$$v = V_m \sin \omega t \Rightarrow \mathbf{V} = V \angle 0°$$

where $V = 0.707V_m$.

Applying Ohm's law and using phasor algebra, we have

$$\mathbf{I} = \frac{V \angle 0°}{R \angle \theta_R} = \frac{V}{R} \underline{/0° - \theta_R}$$

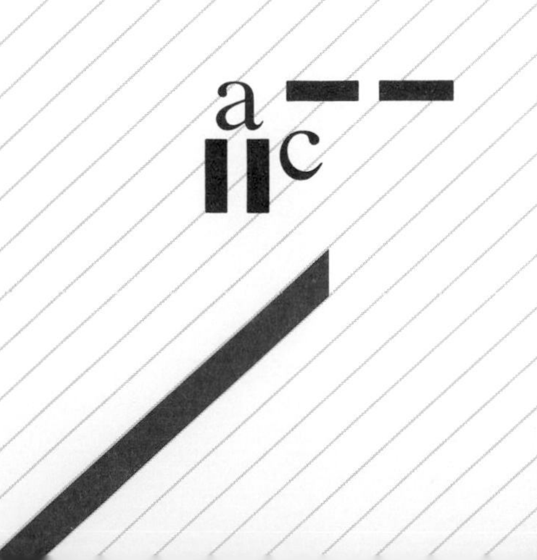

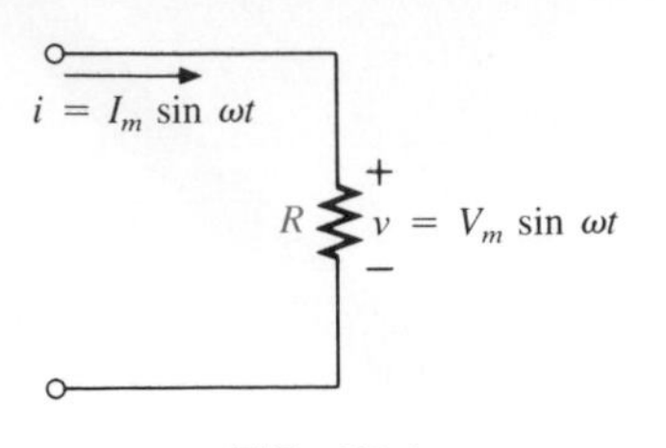


FIG. 15.1

Since $i$ and $v$ are in phase, the angle associated with $i$ also must be 0°. To satisfy this condition, $\theta_R$ must equal 0°. Substituting $\theta_R = 0°$, we find

$$\mathbf{I} = \frac{V \angle 0°}{R \angle 0°} = \frac{V}{R} \underline{/0° - 0°} = \frac{V}{R} \angle 0°$$

so that in the time domain,

$$i = \sqrt{2}\left(\frac{V}{R}\right) \sin \omega t$$

The complex number in the denominator of the above equation,

$$\boxed{\mathbf{R} = R \angle 0°} \tag{15.1}$$

does not represent a sinusoidal function in the phasor domain even though it has the same format. It is a radius vector in the complex plane that has a fixed magnitude $R$ at an angle of 0° (the positive real axis). The relative advantages of associating $\angle 0°$ with purely resistive elements is demonstrated in the following examples. You will find that it is no longer necessary to keep in mind that $v$ and $i$ are in phase. This fact was included when we associated an angle of 0° with $R$.

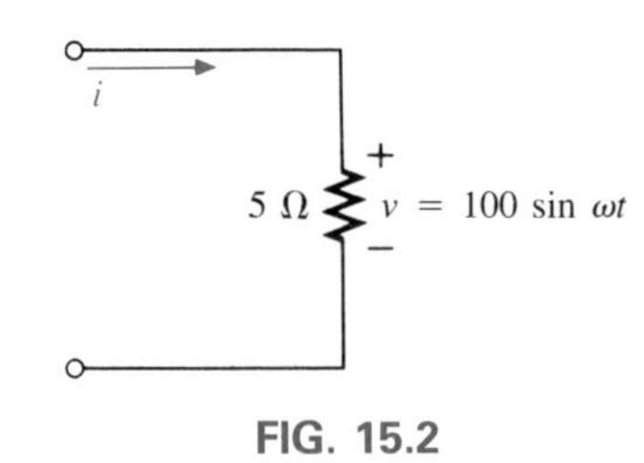


FIG. 15.2

**EXAMPLE 15.1.** Using phasor algebra, find the current $i$ for the circuit of Fig. 15.2. Sketch the waveforms of $v$ and $i$.

***Solution:*** Note Fig. 15.3:

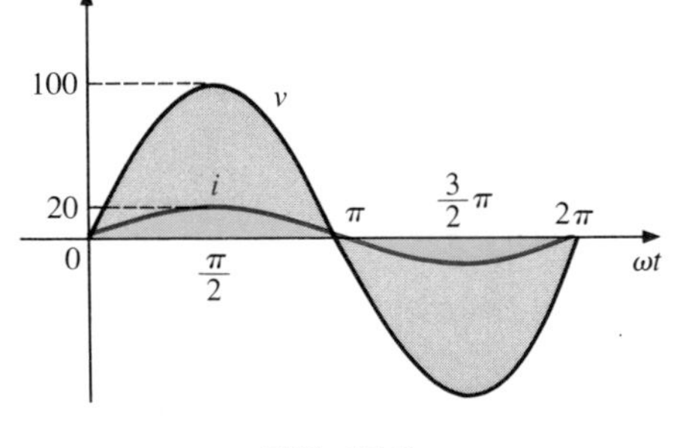


FIG. 15.3

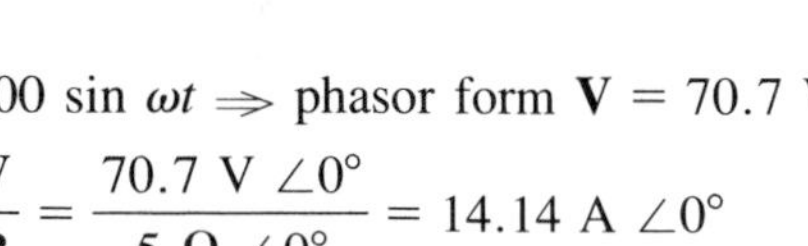

$$v = 100 \sin \omega t \Rightarrow \text{phasor form } \mathbf{V} = 70.7 \text{ V} \angle 0°$$

$$\mathbf{I} = \frac{\mathbf{V}}{\mathbf{R}} = \frac{70.7 \text{ V} \angle 0°}{5\ \Omega \angle 0°} = 14.14 \text{ A} \angle 0°$$

and

$$i = \sqrt{2}(14.14) \sin \omega t = \mathbf{20 \sin \omega t}$$

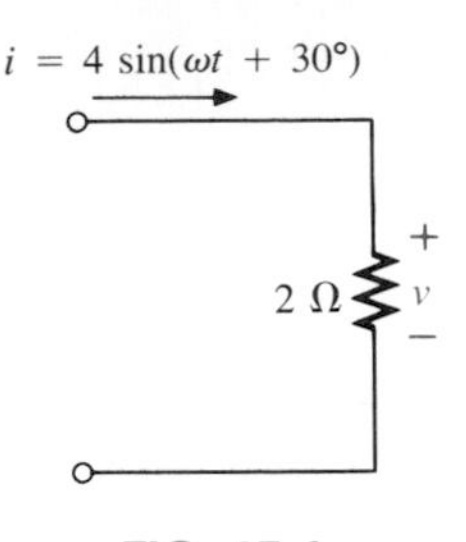


FIG. 15.4

**EXAMPLE 15.2.** Using phasor algebra, find the voltage $v$ for the circuit of Fig. 15.4. Sketch the waveforms of $v$ and $i$.

***Solution:*** Note Fig. 15.5:

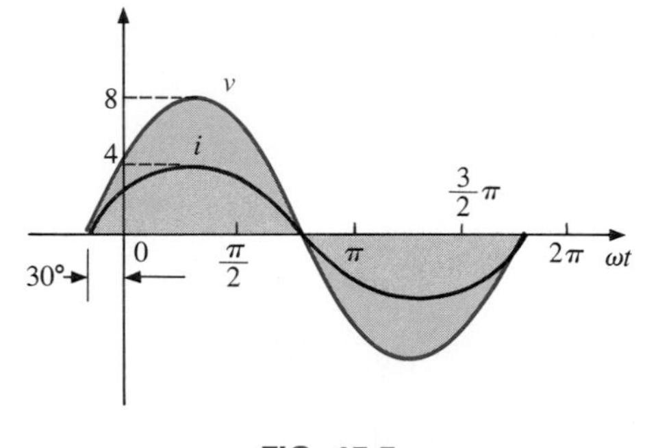

FIG. 15.5

$$i = 4 \sin(\omega t + 30°) \Rightarrow \text{phasor form } \mathbf{I} = 2.828 \text{ A } \angle 30°$$
$$\mathbf{V} = \mathbf{IR} = (2.828 \text{ A } \angle 30°)(2 \ \Omega \ \angle 0°) = 5.656 \text{ V } \angle 30°$$

and

$$v = \sqrt{2}(5.656) \sin(\omega t + 30°) = \mathbf{8.0 \sin(\boldsymbol{\omega} t + 30°)}$$

It is often helpful in the analysis of networks to have a *phasor diagram,* which shows at a glance the *magnitudes* and *phase relations* between the various quantities within the network. For example, the phasor diagrams of the circuits considered in the preceding examples would be as shown in Fig. 15.6. In both cases, it is immediately obvious that $v$ and $i$ are in phase, since they both have the same phase angle.

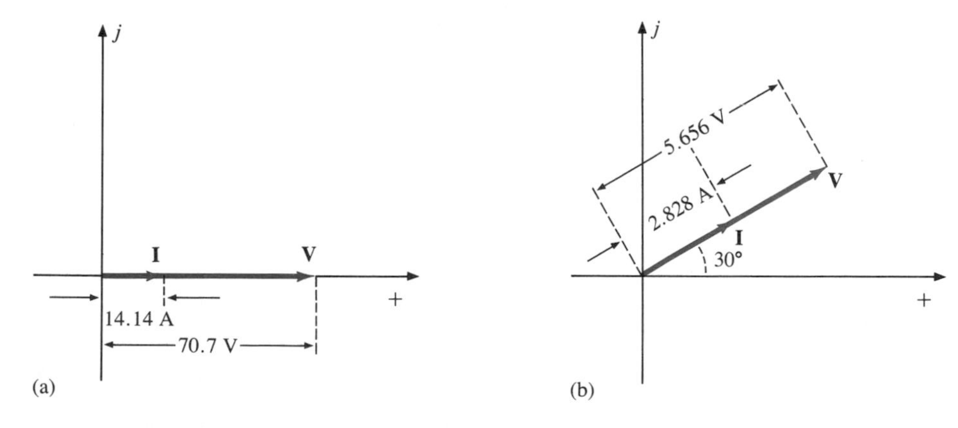

FIG. 15.6

For the pure inductor of Fig. 15.7, it was learned in Chapter 13 that the voltage leads the current by 90°, and that the reactance of the coil $X_L$ is determined by $\omega L$.

$$v = V_m \sin \omega t \Rightarrow \text{phasor form } \mathbf{V} = V \angle 0°$$

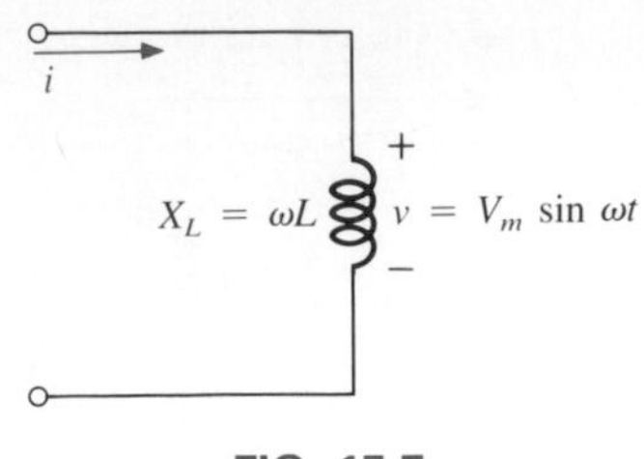


FIG. 15.7

By Ohm's law,

$$\mathbf{I} = \frac{V \angle 0^\circ}{X_L \angle \theta_L} = \frac{V}{X_L} \underline{/0^\circ - \theta_L}$$

Since $v$ leads $i$ by 90°, $i$ must have an angle of $-90^\circ$ associated with it. To satisfy this condition, $\theta_L$ must equal $+90^\circ$. Substituting $\theta_L = 90^\circ$, we obtain

$$\mathbf{I} = \frac{V \angle 0^\circ}{X_L \angle 90^\circ} = \frac{V}{X_L} \underline{/0^\circ - 90^\circ} = \frac{V}{X_L} \angle -90^\circ$$

so that in the time domain,

$$i = \sqrt{2}\left(\frac{V}{X_L}\right) \sin(\omega t - 90^\circ)$$

The complex number in the denominator of the preceding equation,

$$\boxed{\mathbf{X}_L = X_L \angle 90^\circ} \tag{15.2}$$

does not represent a sinusoidal function in the phasor domain even though it has the same format. It is a radius vector in the complex plane that has a fixed magnitude $X_L$ at an angle of 90°.

**EXAMPLE 15.3.** Using phasor algebra, find the current $i$ for the circuit of Fig. 15.8. Sketch the $v$ and $i$ curves.

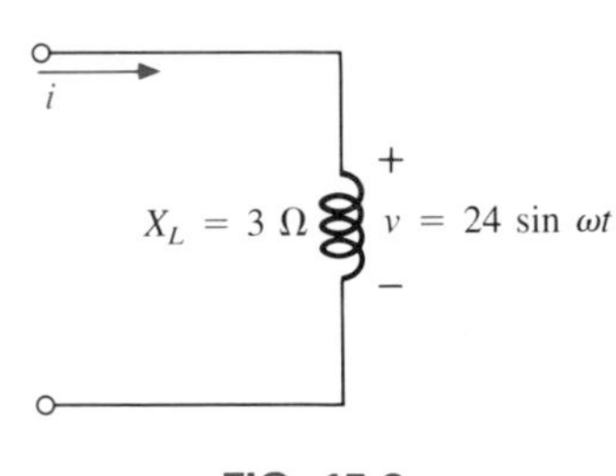


FIG. 15.8

***Solution:*** Note Fig. 15.9:

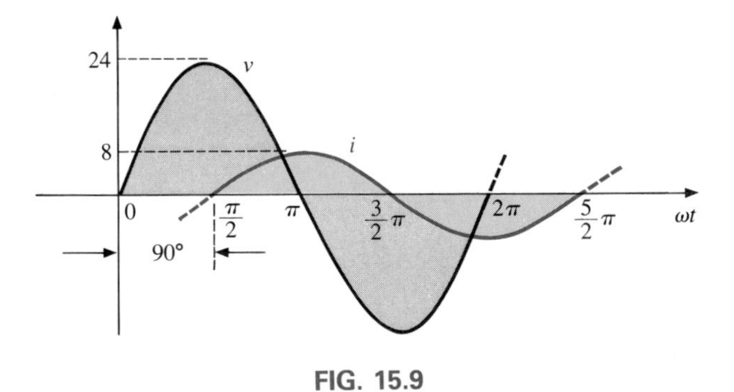


FIG. 15.9

$$v = 24 \sin \omega t \Rightarrow \text{phasor form } \mathbf{V} = 16.968 \text{ V} \angle 0^\circ$$

$$\mathbf{I} = \frac{\mathbf{V}}{\mathbf{X}_L} = \frac{16.968 \text{ V} \angle 0^\circ}{3\ \Omega \angle 90^\circ} = 5.656 \text{ A} \angle -90^\circ$$

and

$$i = \sqrt{2}(5.656) \sin(\omega t - 90^\circ) = \mathbf{8.0 \sin(\omega t - 90^\circ)}$$

**EXAMPLE 15.4.** Using phasor algebra, find the voltage $v$ for the circuit of Fig. 15.10. Sketch the $v$ and $i$ curves.

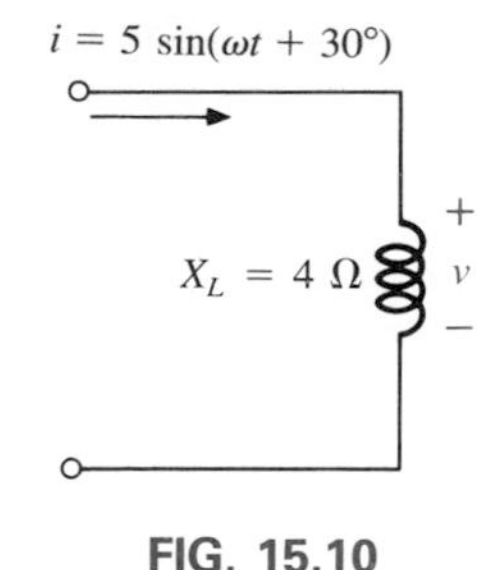


FIG. 15.10

***Solution:*** Note Fig. 15.11:

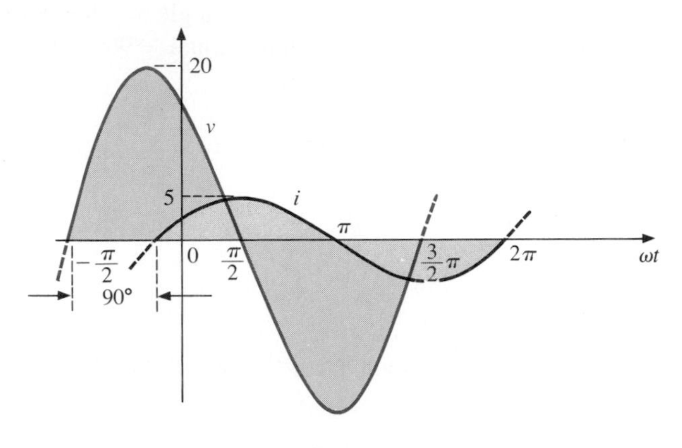

FIG. 15.11

$$i = 5 \sin(\omega t + 30°) \Rightarrow \text{phasor form } \mathbf{I} = 3.535 \text{ A } \angle 30°$$

$$\mathbf{V} = \mathbf{IX}_L = (3.535 \text{ A } \angle 30°)(4 \ \Omega \ \angle +90°) = 14.140 \text{ V } \angle 120°$$

and

$$v = \sqrt{2}(14.140) \sin(\omega t + 120°) = \mathbf{20 \sin(\omega t + 120°)}$$

The phasor diagrams for the two circuits of the preceding examples are shown in Fig. 15.12. Both indicate quite clearly that the voltage leads the current by 90°.

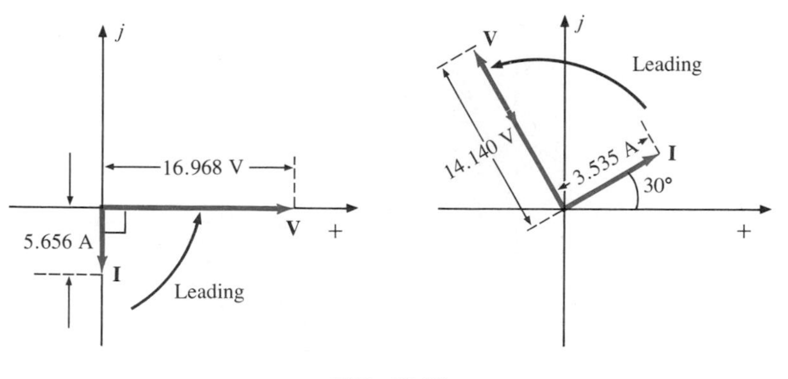


FIG. 15.12

For the pure capacitor of Fig. 15.13, it was learned in Chapter 13 that the current leads the voltage by 90°, and that the reactance of the capacitor $X_C$ is determined by $1/\omega C$.

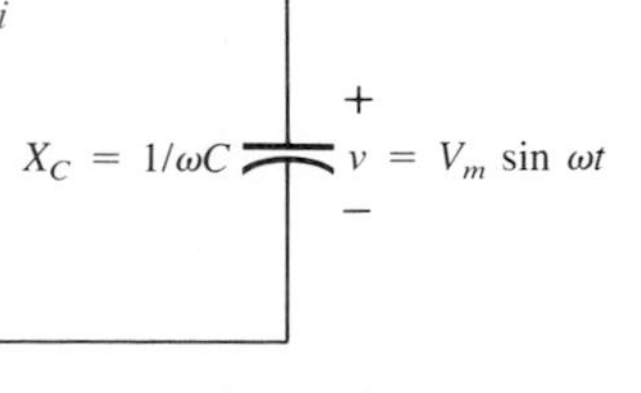


FIG. 15.13

$$v = V_m \sin \omega t \Rightarrow \text{phasor form } \mathbf{V} = V \angle 0°$$

Applying Ohm's law and using phasor algebra, we find

$$\mathbf{I} = \frac{V \angle 0^\circ}{X_C \angle \theta_C} = \frac{V}{X_C} \angle 0^\circ - \theta_C$$

Since we know $i$ leads $v$ by 90°, $i$ must have an angle of +90° associated with it. To satisfy this condition, $\theta_C$ must equal −90°. Substituting $\theta_C = -90^\circ$ yields

$$\mathbf{I} = \frac{V \angle 0^\circ}{X_C \angle -90^\circ} = \frac{V}{X_C} \angle 0^\circ - (-90^\circ) = \frac{V}{X_C} \angle 90^\circ$$

so, in the time domain,

$$i = \sqrt{2}\left(\frac{V}{X_C}\right) \sin(\omega t + 90^\circ)$$

Once more, the complex number in the denominator of the above equation,

$$\boxed{\mathbf{X}_C = X_C \angle -90^\circ} \qquad \textbf{(15.3)}$$

does not represent a sinusoidal function in the phasor domain even though it has the same format. It is a radius vector in the complex plane that has a fixed magnitude $X_C$ at an angle of −90°.

---

**EXAMPLE 15.5.** Using phasor algebra, find the current $i$ in the circuit of Fig. 15.14. Sketch the $v$ and $i$ curves.

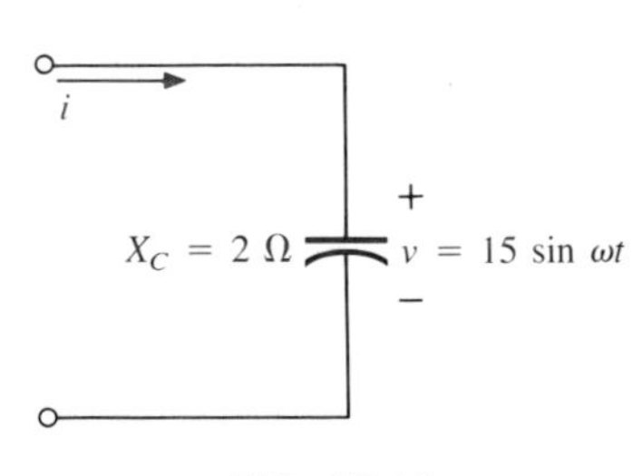

FIG. 15.14

***Solution:*** Note Fig. 15.15:

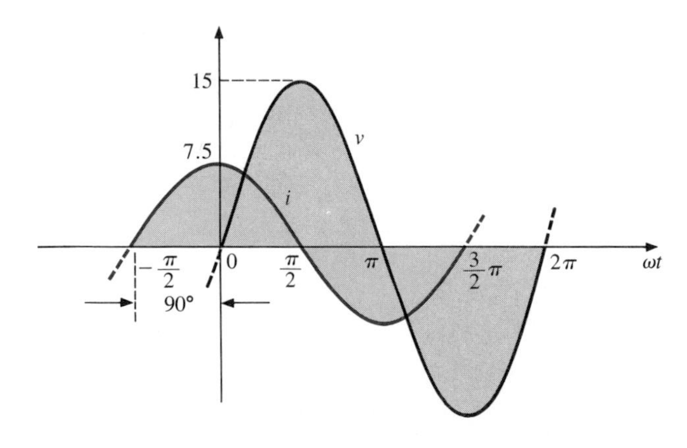


FIG. 15.15

$$v = 15 \sin \omega t \Rightarrow \text{phasor notation } \mathbf{V} = 10.605 \text{ V} \angle 0^\circ$$

$$\mathbf{I} = \frac{\mathbf{V}}{\mathbf{X}_C} = \frac{10.605 \text{ V} \angle 0^\circ}{2 \,\Omega \angle -90^\circ} = 5.303 \text{ A} \angle 90^\circ$$

and

$$i = \sqrt{2}(5.303)\ \sin(\omega t + 90°) = \mathbf{7.5\ sin(\omega t + 90°)}$$

**EXAMPLE 15.6.** Using phasor algebra, find the voltage $v$ in the circuit of Fig. 15.16. Sketch the $v$ and $i$ curves.

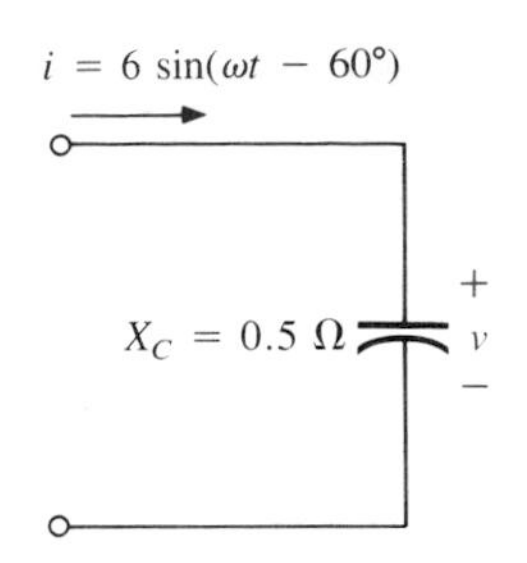

FIG. 15.16

***Solution:*** Note Fig. 15.17:

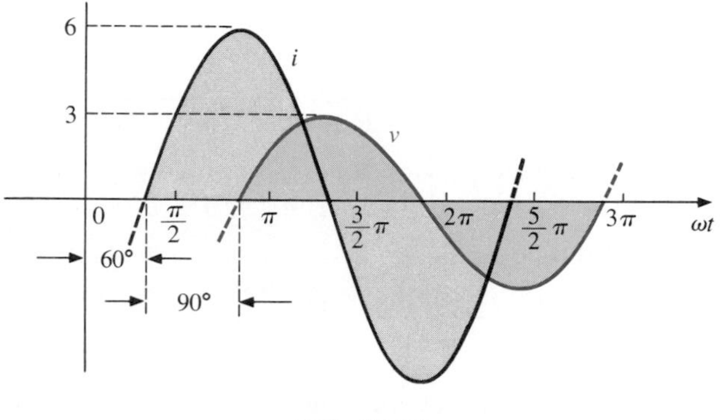

FIG. 15.17

$$i = 6\ \sin(\omega t - 60°) \Rightarrow \text{phasor notation } \mathbf{I} = 4.242\ \text{A} \angle -60°$$

$$\mathbf{V} = \mathbf{I}\mathbf{X}_C = (4.242\ \text{A} \angle -60°)(0.5\ \Omega \angle -90°) = 2.121\ \text{V} \angle -150°$$

and

$$v = \sqrt{2}(2.121)\ \sin(\omega t - 150°) = \mathbf{3.0\ sin(\omega t - 150°)}$$

The phasor diagrams for the two circuits of the preceding examples are shown in Fig. 15.18. Both indicate quite clearly that the current $i$ leads the voltage $v$ by 90°.

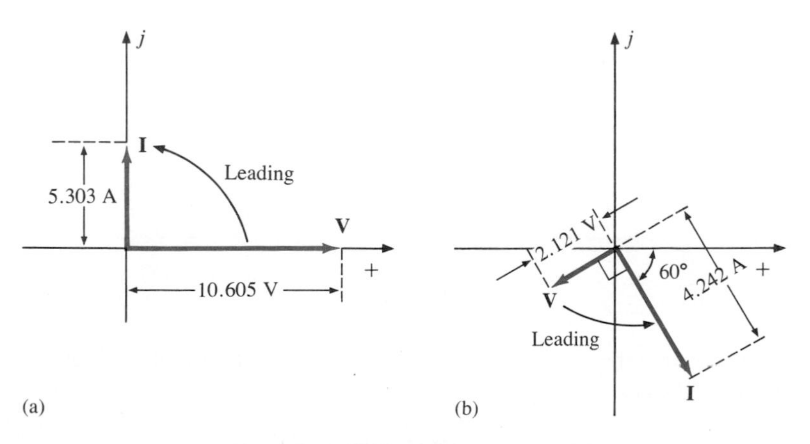


FIG. 15.18

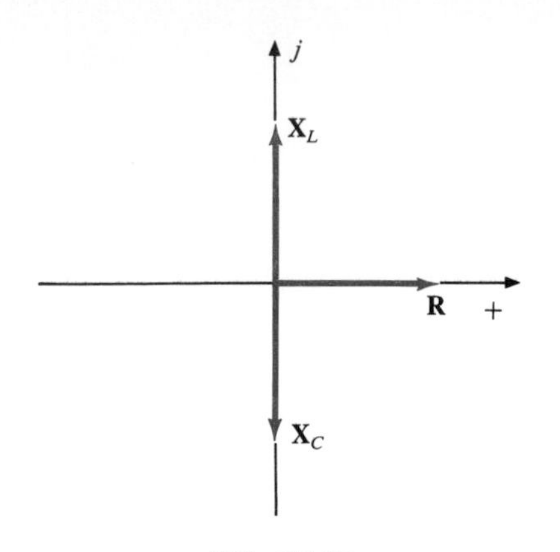


FIG. 15.19

A plot of resistance, inductive reactance, and capacitive reactance appears in Fig. 15.19. For any network, the resistance will *always* appear on the positive real axis, the inductive reactance on the positive imaginary axis, and the capacitive reactance on the negative imaginary axis.

Any *one or combination* of these elements in an ac circuit is called the *impedance* of the circuit. It is a measure of how much the circuit will *impede,* or hinder, the flow of current through it. The diagram of Fig. 15.19 is referred to as an *impedance diagram*. The symbol for impedance is $Z$.

For the individual elements,

$$\text{Resistance:} \quad \mathbf{Z} = \mathbf{R} = R \angle 0^\circ = R + j0 \qquad (15.4)$$

$$\text{Inductive reactance:} \quad \mathbf{Z} = \mathbf{X}_L = X_L \angle 90^\circ = 0 + jX_L \qquad (15.5)$$

$$\text{Capacitive reactance:} \quad \mathbf{Z} = \mathbf{X}_C = X_C \angle -90^\circ = 0 - jX_C \qquad (15.6)$$

***For any configuration (series, parallel, series-parallel, etc.), the angle associated with the total impedance is the angle by which the applied voltage leads the source current. For inductive networks, $\theta_T$ will be positive, whereas for capacitive networks, $\theta_T$ will be negative.***

## 15.3 SERIES CONFIGURATION

The overall properties of series ac circuits (Fig. 15.20) are the same as those for dc circuits; that is, *the current is the same through series*

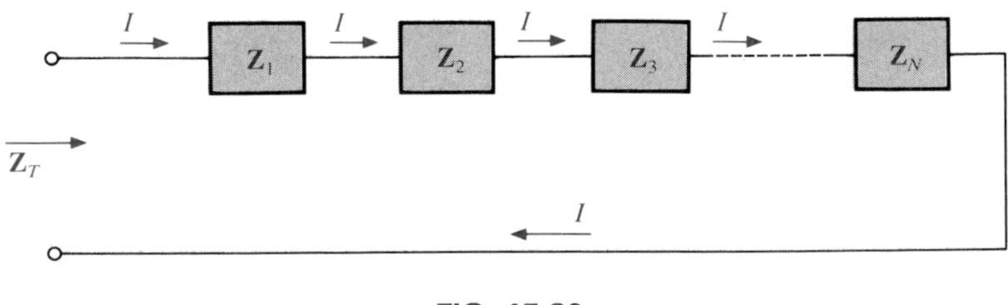


FIG. 15.20

*elements,* and the total impedance of a system is the sum of the individual impedances:

$$\mathbf{Z}_T = \mathbf{Z}_1 + \mathbf{Z}_2 + \mathbf{Z}_3 + \cdots + \mathbf{Z}_N \qquad (15.7)$$

**EXAMPLE 15.7.** Draw the impedance diagram for the circuit of Fig. 15.21 and find the total impedance.

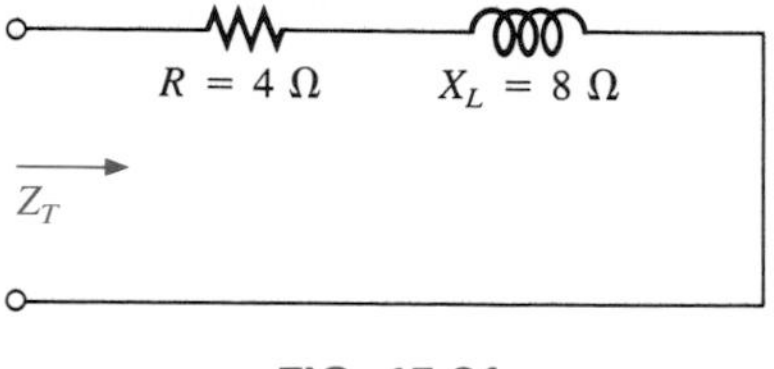


FIG. 15.21

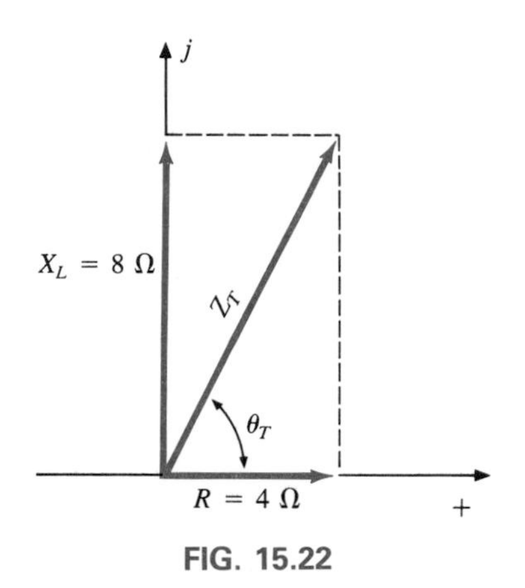


FIG. 15.22

***Solution:*** As indicated by Fig. 15.22, the input impedance can be found graphically from the impedance diagram by properly scaling the real and imaginary axes and finding the length of the resultant vector $Z_T$ and angle $\theta_T$. Or, by using vector algebra, we obtain

$$\begin{aligned}\mathbf{Z}_T &= \mathbf{Z}_1 + \mathbf{Z}_2\\ &= R\ \angle 0^\circ + X_L\ \angle 90^\circ\\ &= R + jX_L = 4\ \Omega + j8\ \Omega\\ \mathbf{Z}_T &= \mathbf{8.944\ \Omega\ \angle 63.43^\circ}\end{aligned}$$

**EXAMPLE 15.8.** Determine the input impedance to the series network of Fig. 15.23. Draw the impedance diagram.

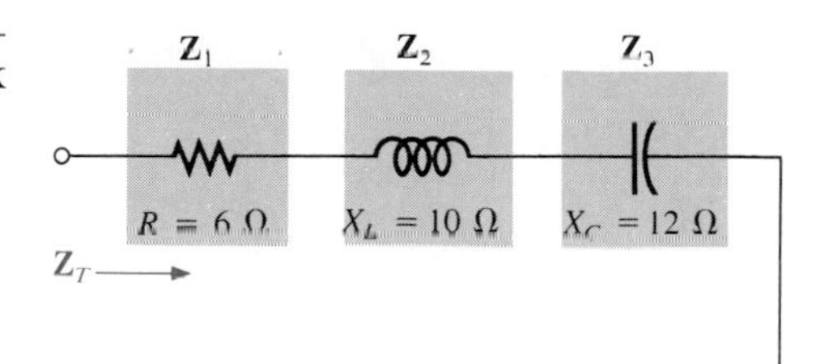


FIG. 15.23

***Solution:***

$$\begin{aligned}\mathbf{Z}_T &= \mathbf{Z}_1 + \mathbf{Z}_2 + \mathbf{Z}_3\\ &= R\ \angle 0^\circ + X_L\ \angle 90^\circ + X_C\ \angle -90^\circ\\ &= R + jX_L - jX_C\\ &= R + j(X_L - X_C) = 6\ \Omega + j(10\ \Omega - 12\ \Omega) = 6\ \Omega - j2\ \Omega\\ \mathbf{Z}_T &= \mathbf{6.325\ \Omega\ \angle -18.43^\circ}\end{aligned}$$

The impedance diagram appears in Fig. 15.24. Note that in this example, series inductive and capacitive reactances are in direct opposition. For the circuit of Fig. 15.23, if the inductive reactance were equal to the capacitive reactance, the input impedance would be purely resistive. We will have more to say about this particular condition in a later chapter.

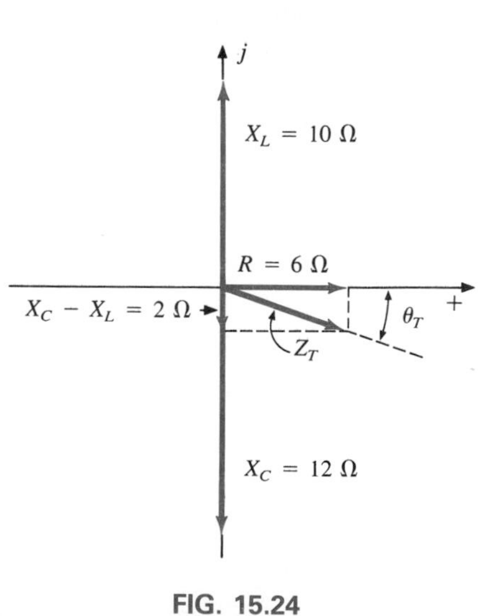


FIG. 15.24

In many of the circuits to be considered, $3 + j4 = 5\ \angle 53.13^\circ$ and $4 + j3 = 5\ \angle 36.87^\circ$ will be used quite frequently to insure that the approach is as clear as possible and not lost in mathematical complexity.

Let us now examine the *R-L*, *R-C*, and *R-L-C* series networks. Their basic nature dictates that they be examined in some detail. Numerical values were assigned to make the description as informative as possible.

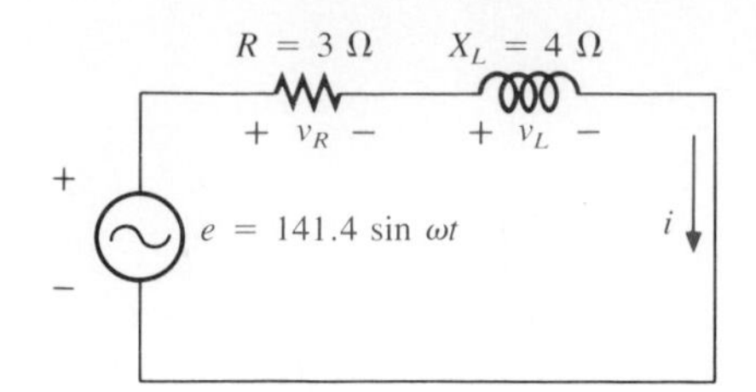


FIG. 15.25

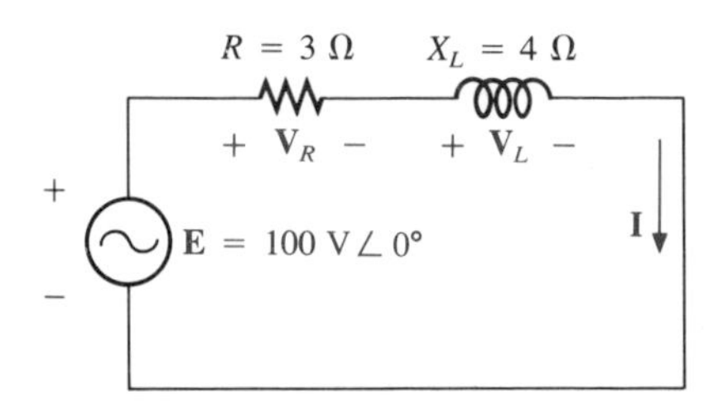

FIG. 15.26

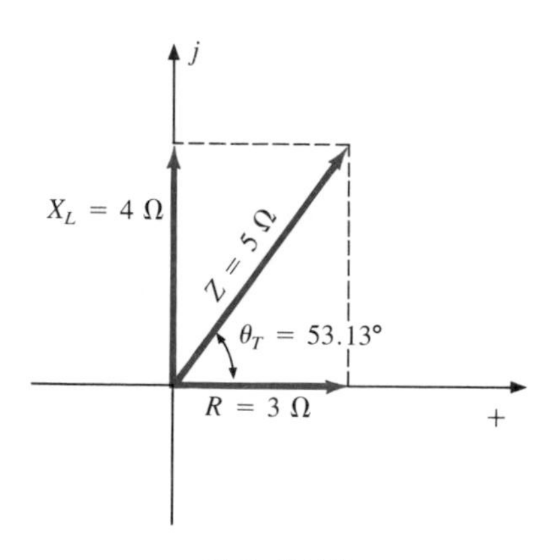

FIG. 15.27

***R-L*** (Fig. 15.25)

*Phasor notation:*

$$e = 141.4 \sin \omega t \Rightarrow \mathbf{E} = 100\ \text{V}\ \angle 0^\circ$$

Note Fig. 15.26.

$\mathbf{Z}_T$**:**

$$\mathbf{Z}_T = \mathbf{Z}_1 + \mathbf{Z}_2 = 3\ \Omega\ \angle 0^\circ + 4\ \Omega\ \angle 90^\circ = 3\ \Omega + j4\ \Omega$$

and

$$\mathbf{Z}_T = \mathbf{5\ \Omega\ \angle 53.13^\circ}$$

*Impedance diagram:* As shown in Fig. 15.27.

**I:**

$$\mathbf{I} = \frac{\mathbf{E}}{\mathbf{Z}_T} = \frac{100\ \text{V}\ \angle 0^\circ}{5\ \Omega\ \angle 53.13^\circ} = \mathbf{20\ A\ \angle -53.13^\circ}$$

$\mathbf{V}_R$, $\mathbf{V}_L$**:**

*Kirchhoff's voltage law:*

$$\Sigma_{\circlearrowleft} \mathbf{V} = \mathbf{E} - \mathbf{V}_R - \mathbf{V}_L = 0$$

or

$$\mathbf{E} = \mathbf{V}_R + \mathbf{V}_L$$

In rectangular form,

$$\mathbf{V}_R = 60\ \text{V}\ \angle -53.13^\circ = 36\ \text{V} - j48\ \text{V}$$
$$\mathbf{V}_L = 80\ \text{V}\ \angle +36.87^\circ = 64\ \text{V} + j48\ \text{V}$$

and

$$\mathbf{E} = \mathbf{V}_R + \mathbf{V}_L = (36\ \text{V} - j48\ \text{V}) + (64\ \text{V} + j48\ \text{V}) = 100\ \text{V} + j0 = 100\ \text{V}\ \angle 0^\circ$$

as applied.

*Phasor diagram:* Note that for the phasor diagram of Fig. 15.28, **I** is in phase with the voltage across the resistor and lags the voltage across the inductor by 90°.

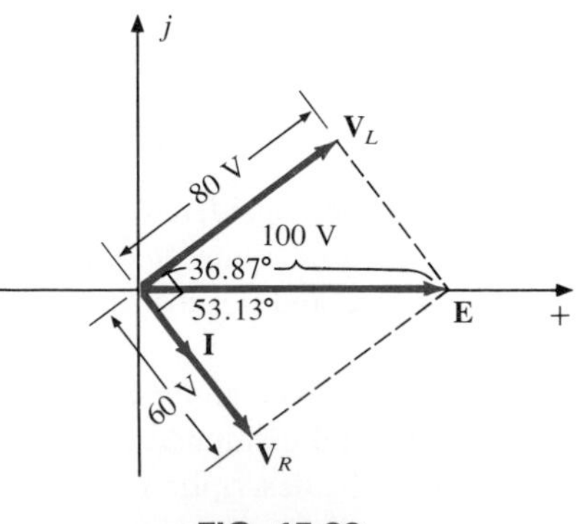


FIG. 15.28

*Power:* The total power in watts delivered to the circuit is

$$\begin{aligned} P_T &= EI \cos \theta_T \\ &= (100 \text{ V})(20 \text{ A}) \cos 53.13° = (2000 \text{ W})(0.6) \\ &= \mathbf{1200\ W} \end{aligned}$$

where $E$ and $I$ are effective values and $\theta_T$ is the phase angle between $E$ and $I$, or

$$\begin{aligned} P_T &= I^2R \\ &= (20 \text{ A})^2(3\ \Omega) = (400)(3) \\ &= \mathbf{1200\ W} \end{aligned}$$

where $I$ is the effective value, or, finally,

$$\begin{aligned} P_T = P_R + P_L &= V_R I \cos \theta_R + V_L I \cos \theta_L \\ &= (60 \text{ V})(20 \text{ A}) \cos 0° + (80 \text{ V})(20 \text{ A}) \cos 90° \\ &= 1200 \text{ W} + 0 \\ &= \mathbf{1200\ W} \end{aligned}$$

where $\theta_R$ is the phase angle between $V_R$ and $I$, and $\theta_L$ is the phase angle between $V_L$ and $I$.

*Power factor:* The power factor $F_p$ of the circuit is cos 53.13° = **0.6 lagging,** where 53.13° is the phase angle between **E** and **I.**

If we write the basic power equation $P = EI \cos \theta$ in the following form:

$$\cos \theta = \frac{P}{EI}$$

where $E$ and $I$ are the input quantities and $P$ is the power delivered to the network, and then perform the following substitutions from the basic series ac circuit:

$$\cos \theta = \frac{P}{EI} = \frac{I^2R}{EI} = \frac{IR}{E} = \frac{R}{E/I} = \frac{R}{Z_T}$$

we find

$$\boxed{F_p = \cos \theta_T = \frac{R}{Z_T}} \qquad \textbf{(15.8)}$$

Reference to Fig. 15.27 also indicates that $\theta$ is the impedance angle $\theta_T$ as written in Eq. (15.8), further supporting the fact that the impedance angle $\theta_T$ is also the phase angle between the input voltage and current for a series ac circuit. To determine the power factor, it is necessary only to form the ratio of the total resistance to the magnitude of the input impedance.

For the case at hand,

$$F_p = \cos \theta = \frac{R}{Z_T} = \frac{3\ \Omega}{5\ \Omega} = \mathbf{0.6\ lagging}$$

as found above.

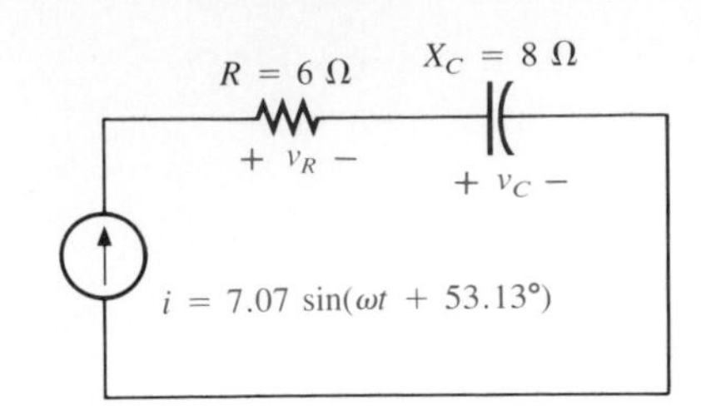


FIG. 15.29

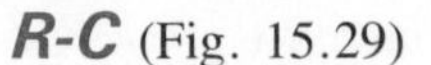

## *R-C* (Fig. 15.29)

*Phasor notation:*

$$i = 7.07 \sin(\omega t + 53.13°) \Rightarrow \mathbf{I} = 5 \text{ A} \angle 53.13°$$

Note Fig. 15.30.

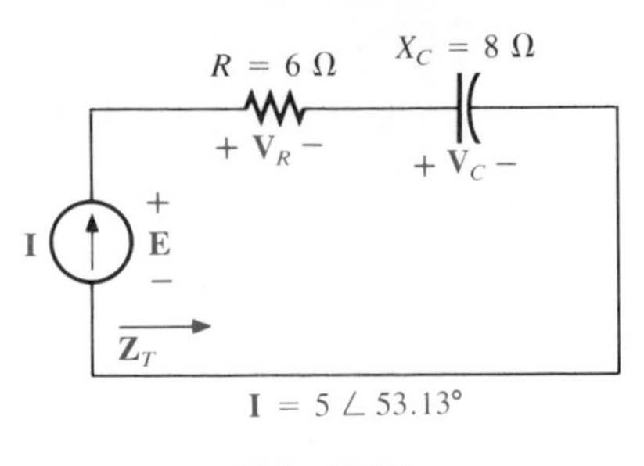

FIG. 15.30

FIG. 15.31

$\mathbf{Z}_T$:

$$\mathbf{Z}_T = \mathbf{Z}_1 + \mathbf{Z}_2 = 6 \ \Omega \angle 0° + 8 \ \Omega \angle -90° = 6 \ \Omega - j8 \ \Omega$$

and

$$\mathbf{Z}_T = \mathbf{10 \ \Omega \angle -53.13°}$$

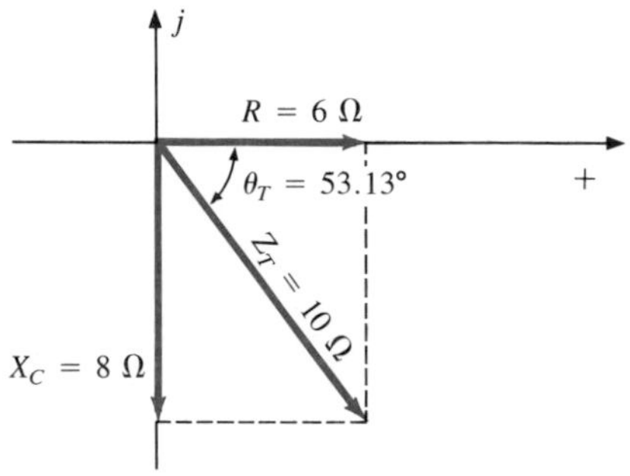

*Impedance diagram:* As shown in Fig. 15.31.

**E:**

$$\mathbf{E} = \mathbf{I}\mathbf{Z}_T = (5 \text{ A} \angle 53.13°)(10 \ \Omega \angle -53.13°) = \mathbf{50 \ V \angle 0°}$$

$\mathbf{V}_R$, $\mathbf{V}_C$:

$$\mathbf{V}_R = \mathbf{I}\mathbf{R} = (5 \text{ A} \angle 53.13°)(6 \ \Omega \angle 0°) = \mathbf{30 \ V \angle 53.13°}$$
$$\mathbf{V}_C = \mathbf{I}\mathbf{X}_C = (5 \text{ A} \angle 53.13°)(8 \ \Omega \angle -90°) = \mathbf{40 \ V \angle -36.87°}$$

*Kirchhoff's voltage law:*

$$\Sigma_{\circlearrowleft} \mathbf{V} = \mathbf{E} - \mathbf{V}_R - \mathbf{V}_C = 0$$

or

$$\mathbf{E} = \mathbf{V}_R + \mathbf{V}_C$$

which can be verified by vector algebra as demonstrated for the *R-L* circuit.

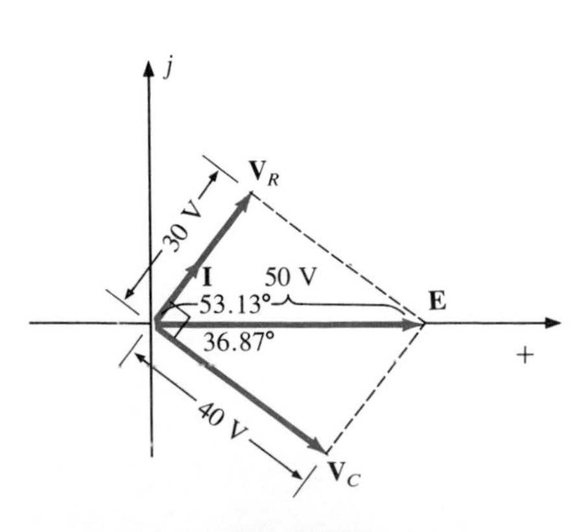


FIG. 15.32

*Phasor diagram:* Note on the phasor diagram of Fig. 15.32 that the current **I** is in phase with the voltage across the resistor and leads the voltage across the capacitor by 90°.

*Time domain:* In the time domain,

$$e = \sqrt{2}(50) \sin \omega t = \mathbf{70.70 \sin \omega t}$$
$$v_R = \sqrt{2}(30) \sin(\omega t + 53.13°) = \mathbf{42.42 \sin(\omega t + 53.13°)}$$
$$v_C = \sqrt{2}(40) \sin(\omega t - 36.87°) = \mathbf{56.56 \sin(\omega t - 36.87°)}$$

A plot of all of the voltages and the current of the circuit appears in Fig. 15.33. Note again that $i$ and $v_R$ are in phase and that $v_C$ lags $i$ by 90°.

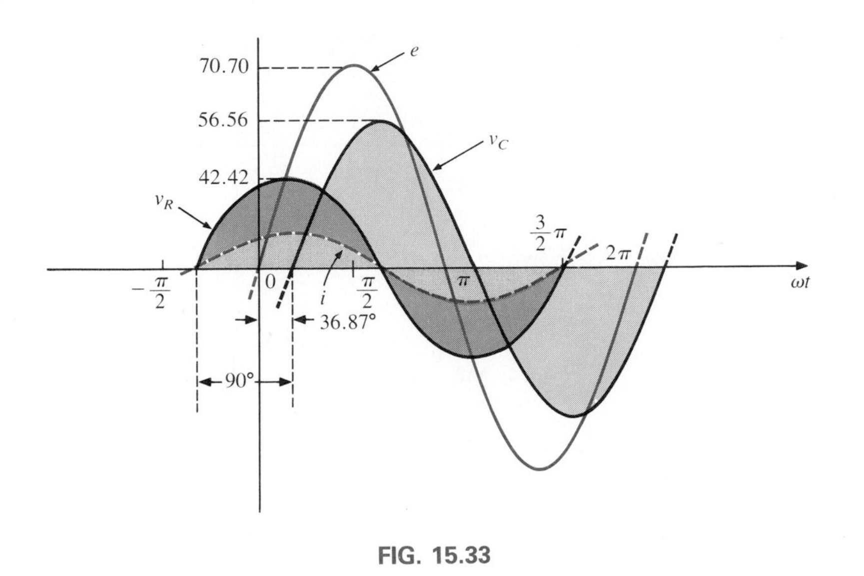


FIG. 15.33

*Power:* The total power in watts delivered to the circuit is

$$P_T = EI \cos \theta_T = (50 \text{ V})(5 \text{ A}) \cos 53.13^\circ$$
$$= (250)(0.6) = \mathbf{150 \text{ W}}$$

or

$$P_T = I^2R = (5 \text{ A})^2(6 \ \Omega) = (25)(6)$$
$$= \mathbf{150 \text{ W}}$$

or, finally,

$$P_T = P_R + P_C = V_RI \cos \theta_R + V_CI \cos \theta_C$$
$$= (30 \text{ V})(5 \text{ A}) \cos 0^\circ + (40 \text{ V})(5 \text{ A}) \cos 90^\circ$$
$$= 150 \text{ W} + 0$$
$$= \mathbf{150 \text{ W}}$$

*Power factor:* The power factor of the circuit is

$$F_p = \cos \theta = \cos 53.13^\circ = \mathbf{0.6 \ leading}$$

Using Eq. (15.8), we obtain

$$F_p = \cos \theta = \frac{R}{Z_T} = \frac{6 \ \Omega}{10 \ \Omega}$$
$$= \mathbf{0.6 \ leading}$$

as determined above.

***R-L-C*** (Fig. 15.34)

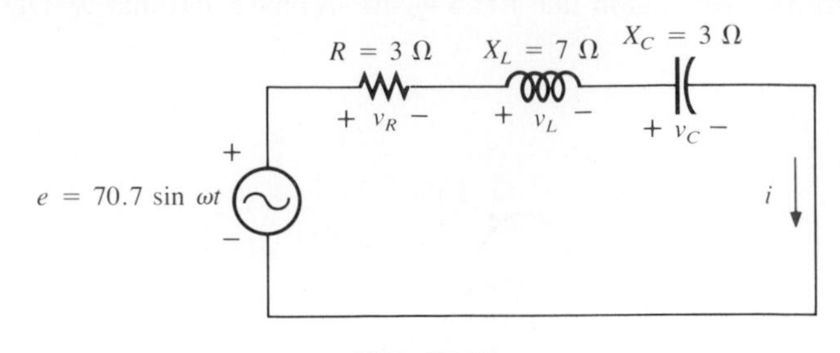

FIG. 15.34

*Phasor notation:* As shown in Fig. 15.35.

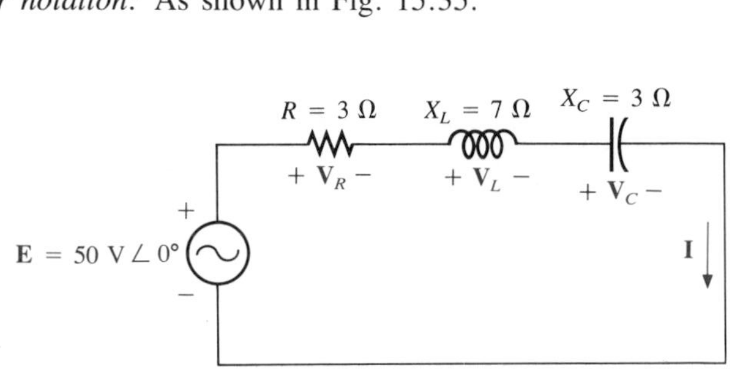


FIG. 15.35

$\mathbf{Z}_T$:

$$\mathbf{Z}_T = \mathbf{Z}_1 + \mathbf{Z}_2 + \mathbf{Z}_3 = R \angle 0° + X_L \angle 90° + X_C \angle -90°$$
$$= 3\ \Omega + j7\ \Omega - j3\ \Omega = 3\ \Omega + j4\ \Omega$$

and

$$\mathbf{Z}_T = \mathbf{5\ \Omega \angle 53.13°}$$

*Impedance diagram:* As shown in Fig. 15.36.

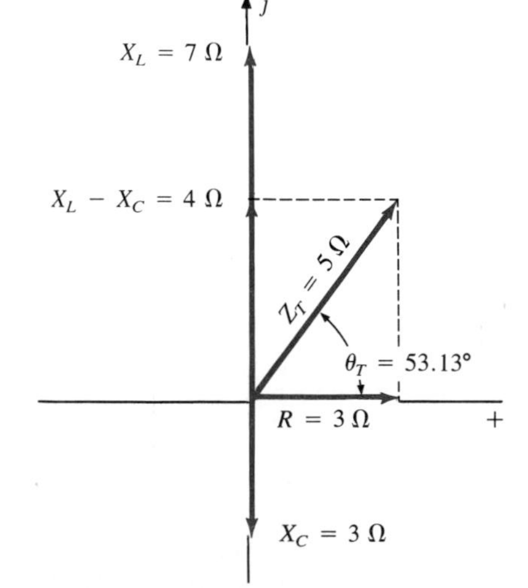


FIG. 15.36

**I:**

$$\mathbf{I} = \frac{\mathbf{E}}{\mathbf{Z}_T} = \frac{50\ \text{V} \angle 0°}{5\ \Omega \angle 53.13°} = \mathbf{10\ A \angle -53.13°}$$

$\mathbf{V}_R$, $\mathbf{V}_L$, $\mathbf{V}_C$:

$$\mathbf{V}_R = \mathbf{IR} = (10\ \text{A} \angle -53.13°)(3\ \Omega \angle 0°) = \mathbf{30\ V \angle -53.13°}$$
$$\mathbf{V}_L = \mathbf{IX}_L = (10\ \text{A} \angle -53.13°)(7\ \Omega \angle 90°) = \mathbf{70\ V \angle 36.87°}$$
$$\mathbf{V}_C = \mathbf{IX}_C = (10\ \text{A} \angle -53.13°)(3\ \Omega \angle -90°) = \mathbf{30\ V \angle -143.13°}$$

*Kirchhoff's voltage law:*

$$\Sigma_{\circlearrowleft} \mathbf{V} = \mathbf{E} - \mathbf{V}_R - \mathbf{V}_L - \mathbf{V}_C = 0$$

or

$$\mathbf{E} = \mathbf{V}_R + \mathbf{V}_L + \mathbf{V}_C$$

which can also be verified through vector algebra.

*Phasor diagram:* The phasor diagram of Fig. 15.37 indicates that the current **I** is in phase with the voltage across the resistor, lags the voltage across the inductor by 90°, and leads the voltage across the capacitor by 90°.

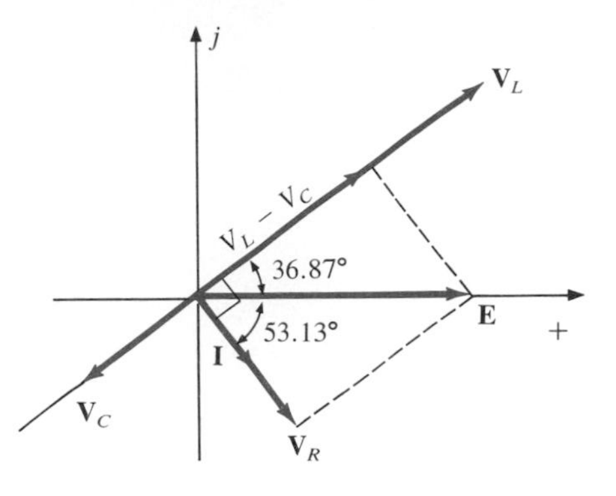


FIG. 15.37

*Time domain:*

$$i = \sqrt{2}(10)\sin(\omega t - 53.13^\circ) = \mathbf{14.14\,sin(\omega t - 53.13^\circ)}$$
$$v_R = \sqrt{2}(30)\sin(\omega t - 53.13^\circ) = \mathbf{42.42\,sin(\omega t - 53.13^\circ)}$$
$$v_L = \sqrt{2}(70)\sin(\omega t + 36.87^\circ) = \mathbf{98.98\,sin(\omega t + 36.87^\circ)}$$
$$v_C = \sqrt{2}(30)\sin(\omega t - 143.13^\circ) = \mathbf{42.42\,sin(\omega t - 143.13^\circ)}$$

A plot of all the voltages and the current of the circuit appears in Fig. 15.38.

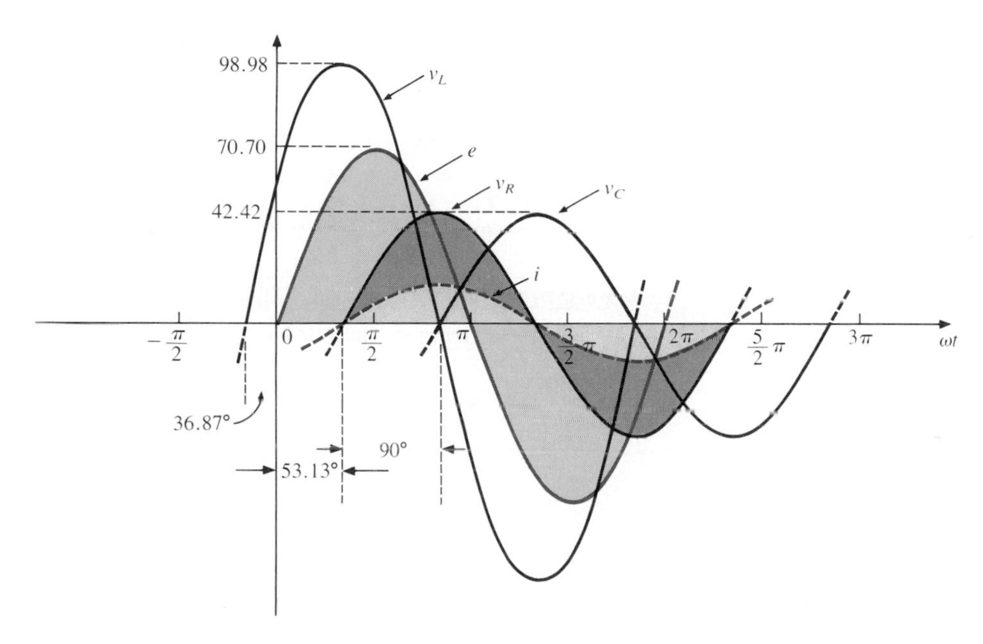

FIG. 15.38

*Power:* The total power in watts delivered to the circuit is

$$P_T = EI\cos\theta_T = (50\text{ V})(10\text{ I})\cos 53.13^\circ = (500)(0.6) = \mathbf{300\ W}$$

or

$$P_T = I^2R = (10\text{ A})^2(3\ \Omega) = (100)(3) = \mathbf{300\ W}$$

or

$$\begin{aligned}P_T &= P_R + P_L + P_C\\ &= V_RI\cos\theta_R + V_LI\cos\theta_L + V_CI\cos\theta_C\\ &= (30\text{ V})(10\text{ A})\cos 0^\circ + (70\text{ V})(10\text{ A})\cos 90^\circ\\ &\quad + (30\text{ V})(10\text{ A})\cos 90^\circ\\ &= (30\text{ V})(10\text{ A}) + 0 + 0 = \mathbf{300\ W}\end{aligned}$$

*Power factor:* The power factor of the circuit is

$$F_p = \cos \theta_T = \cos 53.13° = \mathbf{0.6\ lagging}$$

Using Eq. (15.8), we obtain

$$F_p = \cos \theta = \frac{R}{Z_T} = \frac{3\ \Omega}{5\ \Omega} = \mathbf{0.6\ lagging}$$

## 15.4 VOLTAGE DIVIDER RULE

The basic format for the voltage divider rule in ac circuits is exactly the same as that for dc circuits:

$$\mathbf{V}_x = \frac{\mathbf{Z}_x\mathbf{E}}{\mathbf{Z}_T} \tag{15.9}$$

where $\mathbf{V}_x$ is the voltage across one or more elements in series that have total impedance $\mathbf{Z}_x$, $\mathbf{E}$ is the total voltage appearing across the series circuit, and $\mathbf{Z}_T$ is the total impedance of the series circuit.

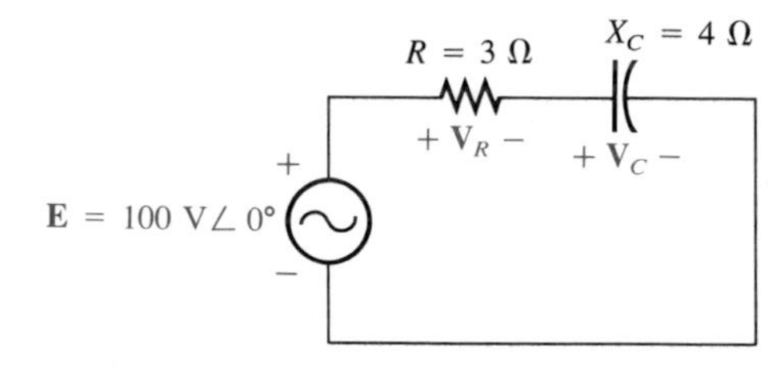


FIG. 15.39

**EXAMPLE 15.9.** Using the voltage divider rule, find the voltage across each element of the circuit of Fig. 15.39.

***Solution:***

$$\mathbf{V}_C = \frac{\mathbf{X}_C\mathbf{E}}{\mathbf{X}_C + \mathbf{R}} = \frac{(4\ \Omega\ \angle -90°)(100\ \text{V}\ \angle 0°)}{4\ \Omega\ \angle -90° + 3\ \Omega\ \angle 0°} = \frac{400\ \angle -90°}{3 - j4}$$

$$= \frac{400\ \angle -90°}{5\ \angle -53.13°} = \mathbf{80\ V\ \angle -36.87°}$$

$$\mathbf{V}_R = \frac{\mathbf{R}\mathbf{E}}{\mathbf{X}_C + \mathbf{R}} = \frac{(3\ \Omega\ \angle 0°)(100\ \text{V}\ \angle 0°)}{5\ \Omega\ \angle -53.13°} = \frac{300\ \angle 0°}{5\ \angle -53.13°}$$

$$= \mathbf{60\ V\ \angle +53.13°}$$

**EXAMPLE 15.10.** Using the voltage divider rule, find the unknown voltages $\mathbf{V}_R$, $\mathbf{V}_L$, $\mathbf{V}_C$, and $\mathbf{V}_1$ for the circuit of Fig. 15.40.

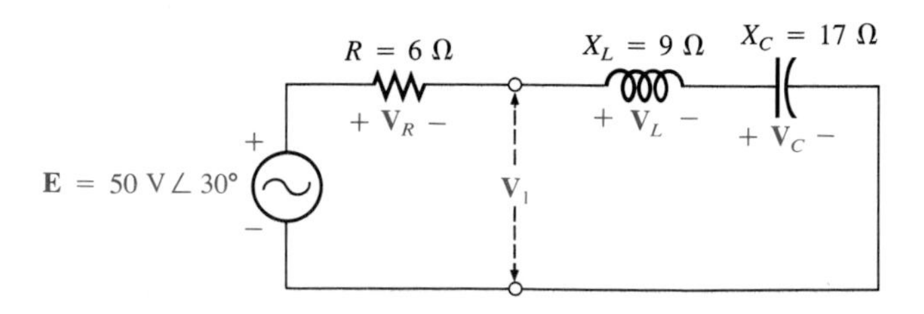


FIG. 15.40

***Solution:***

$$\mathbf{V}_R = \frac{\mathbf{RE}}{\mathbf{R} + \mathbf{X}_L + \mathbf{X}_C} = \frac{(6\ \Omega \angle 0°)(50\ \text{V} \angle 30°)}{6\ \Omega \angle 0° + 9\ \Omega \angle 90° + 17\ \Omega \angle -90°}$$

$$= \frac{300 \angle 30°}{6 + j9 - j17} = \frac{300 \angle 30°}{6 - j8}$$

$$= \frac{300 \angle 30°}{10 \angle -53.13°} = \mathbf{30\ V \angle 83.13°}$$

$$\mathbf{V}_L = \frac{\mathbf{X}_L\mathbf{E}}{\mathbf{Z}_T} = \frac{(9\ \Omega \angle 90°)(50\ \text{V} \angle 30°)}{10\ \Omega \angle -53.13°} = \frac{450 \angle 120°}{10 \angle -53.13°}$$

$$= \mathbf{45\ V \angle 173.13°}$$

$$\mathbf{V}_C = \frac{\mathbf{X}_C\mathbf{E}}{\mathbf{Z}_T} = \frac{(17\ \Omega \angle -90°)(50\ \text{V} \angle 30°)}{10\ \Omega \angle -53°} = \frac{850 \angle -60°}{10 \angle -53°}$$

$$= \mathbf{85\ V \angle -6.87°}$$

$$\mathbf{V}_1 = \frac{(\mathbf{X}_L + \mathbf{X}_C)\mathbf{E}}{\mathbf{Z}_T} = \frac{(9\ \Omega \angle 90° + 17\ \Omega \angle -90°)(50\ \text{V} \angle 30°)}{10\ \Omega \angle -53.13°}$$

$$= \frac{(8 \angle -90°)(50 \angle 30°)}{10 \angle -53.13°}$$

$$= \frac{400 \angle -60°}{10 \angle -53.13°} = \mathbf{40\ V \angle -6.87°}$$

---

**EXAMPLE 15.11.** For the circuit of Fig. 15.41:

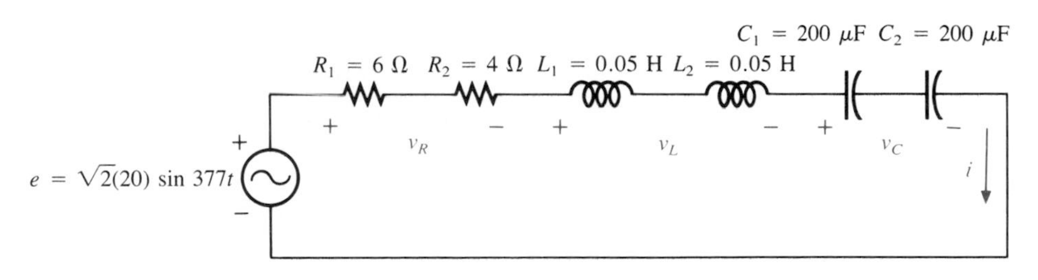


**FIG. 15.41**

a. Calculate $i$, $v_R$, $v_L$, and $v_C$ in phasor form.
b. Calculate the total power factor.
c. Calculate the average power delivered to the circuit.
d. Draw the phasor diagram.
e. Obtain the phasor sum of $\mathbf{V}_R$, $\mathbf{V}_L$, and $\mathbf{V}_C$, and show that it equals the input voltage **E.**
f. Find $\mathbf{V}_R$ and $\mathbf{V}_C$ using the voltage divider rule.

***Solutions:***

a. Combining common elements and finding the reactance of the inductor and capacitor, we obtain

$$R_T = 6\ \Omega + 4\ \Omega = 10\ \Omega$$

$$L_T = 0.05\text{ H} + 0.05\text{ H} = 0.1\text{ H}$$

$$C_T = \frac{200\ \mu\text{F}}{2} = 100\ \mu\text{F}$$

$$X_L = \omega L = (377\text{ rad/s})(0.1\text{ H}) = 37.70\ \Omega$$

$$X_C = \frac{1}{\omega C} = \frac{1}{(377\text{ rad/s})(100 \times 10^{-6}\text{ F})} = \frac{10^6\ \Omega}{37{,}700} = 26.53\ \Omega$$

Redrawing the circuit using phasor notation results in Fig. 15.42.

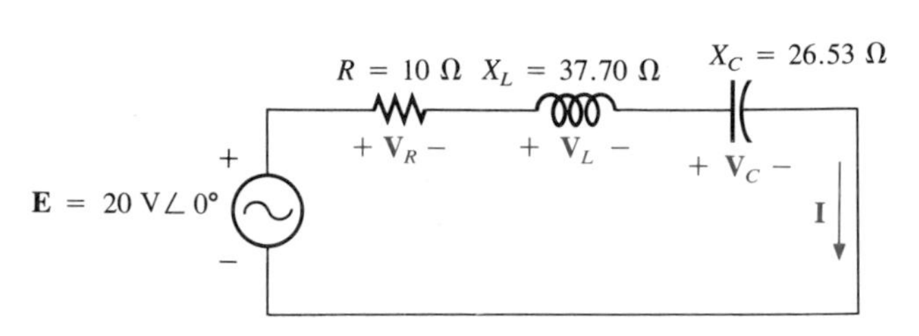

FIG. 15.42

For the circuit of Fig. 15.42,

$$\begin{aligned}\mathbf{Z}_T &= R\ \angle 0^\circ + X_L\ \angle 90^\circ + X_C\ \angle -90^\circ \\ &= 10\ \Omega + j37.70\ \Omega - j26.53\ \Omega \\ &= 10\ \Omega + j11.17\ \Omega = \mathbf{15\ \Omega\ \angle 48.16^\circ}\end{aligned}$$

The current **I** is

$$\mathbf{I} = \frac{\mathbf{E}}{\mathbf{Z}_T} = \frac{20\text{ V}\ \angle 0^\circ}{15\ \Omega\ \angle 48.16^\circ} = \mathbf{1.33\ A\ \angle -48.16^\circ}$$

The voltage across the resistor, inductor, and capacitor can be found using Ohm's law:

$$\begin{aligned}\mathbf{V}_R = \mathbf{IR} &= (1.33\text{ A}\ \angle -48.16^\circ)(10\ \Omega\ \angle 0^\circ) \\ &= \mathbf{13.30\ V\ \angle -48.16^\circ} \\ \mathbf{V}_L = \mathbf{IX}_L &= (1.33\text{ A}\ \angle -48.16^\circ)(37.70\ \Omega\ \angle 90^\circ) \\ &= \mathbf{50.14\ V\ \angle 41.84^\circ} \\ \mathbf{V}_C = \mathbf{IX}_C &= (1.33\text{ A}\ \angle -48.16^\circ)(26.53\ \Omega\ \angle -90^\circ) \\ &= \mathbf{35.28\ V\ \angle -138.16^\circ}\end{aligned}$$

b. The total power factor is determined by the angle between the applied voltage **E,** and the resulting current **I** is 48.16°:

$$F_p = \cos\theta = \cos 48.16^\circ = \mathbf{0.667\ lagging}$$

or

$$F_p = \cos\theta = \frac{R}{Z_T} = \frac{10\ \Omega}{15\ \Omega} = \mathbf{0.667\ lagging}$$

c. The total power in watts delivered to the circuit is

$$P_T = EI \cos \theta = (20 \text{ V})(1.33 \text{ A})(0.667) = \mathbf{17.74 \text{ W}}$$

d. The phasor diagram appears in Fig. 15.43.

e. The phasor sum of $\mathbf{V}_R$, $\mathbf{V}_L$, and $\mathbf{V}_C$ is

$$\mathbf{E} = \mathbf{V}_R + \mathbf{V}_L + \mathbf{V}_C$$
$$= 13.30 \text{ V} \angle -48.16° + 50.14 \text{ V} \angle 41.84° + 35.28 \text{ V} \angle -138.16°$$
$$\mathbf{E} = 13.30 \text{ V} \angle -48.16° + 14.86 \text{ V} \angle 41.84°$$

Therefore,

$$E = \sqrt{(13.30 \text{ V})^2 + (14.86 \text{ V})^2} = \mathbf{20 \text{ V}}$$

and

$$\theta_E = \mathbf{0°} \qquad \text{(from phasor diagram)}$$

and

$$\mathbf{E} = 20 \angle 0°$$

f. $$\mathbf{V}_R = \frac{\mathbf{RE}}{\mathbf{Z}_T} = \frac{(10 \ \Omega \angle 0°)(20 \text{ V} \angle 0°)}{15 \ \Omega \angle 48.16°} = \frac{200 \angle 0°}{15 \angle 48.16°}$$
$$= \mathbf{13.3 \text{ V} \angle -48.16°}$$

$$\mathbf{V}_C = \frac{\mathbf{X}_C\mathbf{E}}{\mathbf{Z}_T} = \frac{(26.5 \ \Omega \angle -90°)(20 \text{ V} \angle 0°)}{15 \ \Omega \angle 48.16°} = \frac{530.6 \angle -90°}{15 \angle 48.16°}$$
$$= \mathbf{35.37 \text{ V} \angle -138.16°}$$

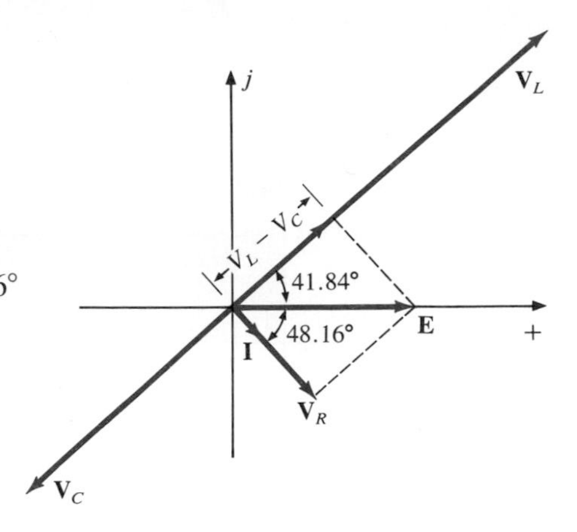


FIG. 15.43

## 15.5 FREQUENCY RESPONSE OF THE *R-C* CIRCUIT

Thus far, the analysis of series circuits has been limited to a particular frequency. We will now examine the effect of frequency on the response of an *R-C* series configuration such as that in Fig. 15.44.

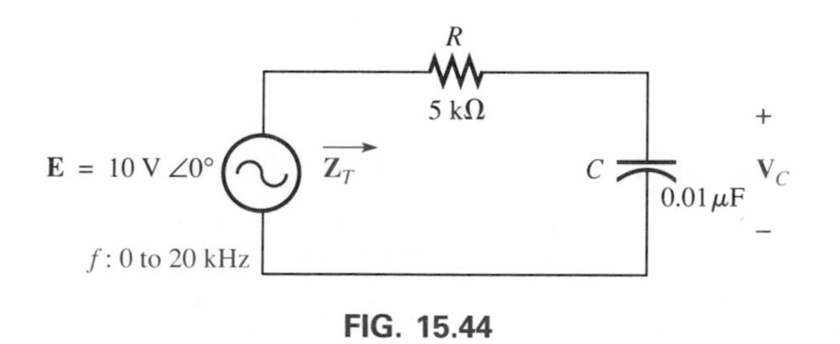


FIG. 15.44

The magnitude of the source is fixed at 10 V but the frequency range of analysis will extend from zero to 20 kHz.

**$\mathbf{Z}_T$:**

Let us first determine how the impedance of the circuit $\mathbf{Z}_T$ will vary with frequency for the specified frequency range of interest. Before getting into specifics, however, let us first develop a sense for what we should expect by noting the impedance versus frequency curve of each element as drawn in Fig. 15.45.

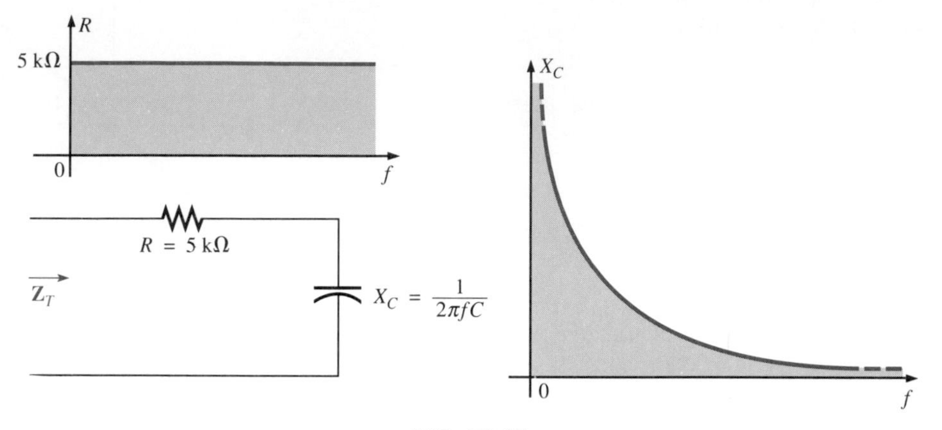

**FIG. 15.45**

At low frequencies the reactance of the capacitor will be quite high and considerably more than the level of the resistance $R$, suggesting that the total impedance will be primarily capacitive in nature. At high frequencies the reactance $X_C$ will drop below the $R = 5\text{-k}\Omega$ level and the network will start to shift toward one of a purely resistive nature (at 5 kΩ). The frequency at which $X_C = R$ can be determined in the following manner:

$$X_C = \frac{1}{2\pi f_1 C} = R$$

and

$$\boxed{f_1 = \frac{1}{2\pi RC}} \quad {}_{X_C = R} \qquad \textbf{(15.10)}$$

which for the network of interest is

$$f_1 = \frac{1}{2\pi(5\text{ k}\Omega)(0.01\ \mu\text{F})} \cong \mathbf{3183.1\ Hz}$$

For frequencies less than $f_1$, $X_C > R$ and for frequencies greater than $f_1$, $R > X_C$.

Now for the details. The total impedance is determined by the following equation:

$$\begin{aligned}\mathbf{Z}_T &= \mathbf{R} + \mathbf{X}_C \\ &= R - jX_C\end{aligned}$$

and

$$\boxed{\mathbf{Z}_T = Z_T \angle \theta_T = \sqrt{R^2 + X_C^2} \angle -\tan^{-1}\frac{X_C}{R}} \qquad \textbf{(15.11)}$$

The magnitude and angle of the total impedance can now be found at any frequency of interest by simply substituting into Eq. (15.11). The presence of the capacitor suggests that we start from a low frequency (100 Hz) and then open the spacing until we reach the upper limit of interest (20 kHz).

**$f$ = 100 Hz:**

$$X_C = \frac{1}{2\pi fC} = \frac{1}{2\pi(100\text{ Hz})(0.01\ \mu\text{F})} = 159.16\text{ k}\Omega$$

and

$$Z_T = \sqrt{R^2 + X_C^2} = \sqrt{(5\text{ k}\Omega)^2 + (159.16\text{ k}\Omega)^2} = 159.24\text{ k}\Omega$$

with

$$\theta_T = -\tan^{-1}\frac{X_C}{R} = -\tan^{-1}\frac{159.24\text{ k}\Omega}{5\text{ k}\Omega} = -\tan^{-1} 31.83 = -88.2^\circ$$

and

$$\mathbf{Z_T = 159.24\ k\Omega\ \angle -88.2^\circ}$$

which compares very closely with $\mathbf{X}_C = 159.16\text{ k}\Omega \angle -90^\circ$, which it would be if the circuit were purely capacitive ($R = 0\ \Omega$). Our assumption that the circuit is primarily capacitive at low frequencies is therefore confirmed.

**$f$ = 1 kHz:**

$$X_C = \frac{1}{2\pi fC} = \frac{1}{2\pi(1\text{ kHz})(0.01\ \mu\text{F})} = 15.92\text{ k}\Omega$$

and

$$Z_T = \sqrt{R^2 + X_C^2} = \sqrt{(5\text{ k}\Omega)^2 + (15.92\text{ k}\Omega)^2} = 16.69\text{ k}\Omega$$

with

$$\theta_T = -\tan^{-1}\frac{X_C}{R} = -\tan^{-1}\frac{16.69\text{ k}\Omega}{5\text{ k}\Omega} = -\tan^{-1} 3.34 = -73.32^\circ$$

and

$$\mathbf{Z_T = 16.69\ k\Omega\ \angle -73.32^\circ}$$

A noticeable drop in the magnitude has occurred and the impedance angle has dropped almost 17° from the purely capacitive level.

Continuing:

$$f = 5\text{ kHz:}\quad \mathbf{Z_T = 5.94\ k\Omega\ \angle -32.62^\circ}$$
$$f = 10\text{ kHz:}\quad \mathbf{Z_T = 5.25\ k\Omega\ \angle -17.75^\circ}$$
$$f = 15\text{ kHz:}\quad \mathbf{Z_T = 5.11\ k\Omega\ \angle -11.97^\circ}$$
$$f = 20\text{ kHz:}\quad \mathbf{Z_T = 5.06\ k\Omega\ \angle -9.09^\circ}$$

Note how close the magnitude of $Z_T$ at $f = 20$ kHz is to the resistance level of 5 kΩ. In addition, note how the phase angle is approaching that associated with a pure resistive network (0°).

A plot of $Z_T$ versus frequency in Fig. 15.46 completely supports our assumption based on the curves of Fig. 15.45. The plot of $\theta_T$ versus frequency in Fig. 15.47 further suggests the fact that the total impedance made a transition from one of a capacitive nature ($\theta_T = -90°$) to one with resistive characteristics ($\theta_T = 0°$).

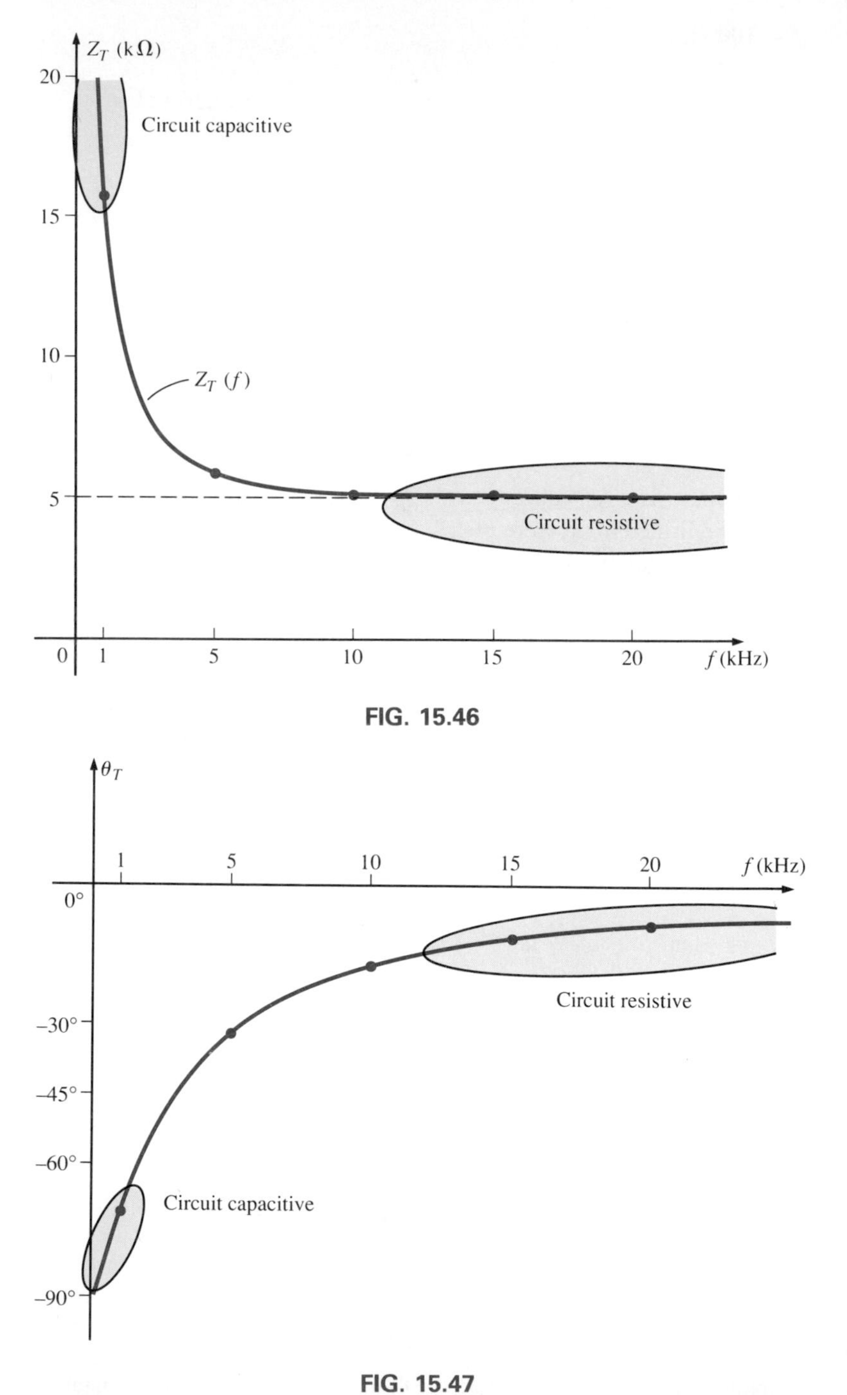


FIG. 15.46

FIG. 15.47

Applying the voltage divider rule to determine the voltage across the capacitor in phasor form yields

$$\mathbf{V}_C = \frac{\mathbf{X}_C\mathbf{E}}{\mathbf{R} + \mathbf{X}_C}$$

$$= \frac{(X_C \angle -90°)(E \angle 0°)}{R - jX_C} = \frac{X_CE \angle -90°}{R - jX_C}$$

$$= \frac{X_CE \angle -90°}{\sqrt{R^2 + X_C^2} \angle -\tan^{-1} X_C/R}$$

or

$$\mathbf{V}_C = V_C \angle \theta_C = \frac{X_CE}{\sqrt{R^2 + X_C^2}} \angle -90° + \tan^{-1}(X_C/R)$$

The magnitude of $\mathbf{V}_C$ is therefore determined by

$$V_C = \frac{X_CE}{\sqrt{R^2 + X_C^2}} \quad \textbf{(15.12)}$$

and the phase angle $\theta_C$ by which $\mathbf{V}_C$ leads $\mathbf{E}$ is given by

$$\theta = -90° + \tan^{-1}\frac{X_C}{R} \quad \textbf{(15.13)}$$

To determine the frequency response, $X_C$ must be calculated for each frequency of interest and inserted into Eqs. (15.12) and (15.13).

To begin our analysis, it makes good sense to consider the case of $f = 0$ Hz (dc conditions).

***f* = 0 Hz:**

$$X_C = \frac{1}{2\pi(0)C} = \frac{1}{0} \Rightarrow \text{very large value}$$

Applying the open-circuit equivalent for the capacitor based on the above calculation will result in the following:

$$\mathbf{V}_C = \mathbf{E} = 10 \text{ V} \angle 0°$$

If we apply Eq. (15.10), we find

$$X_C^2 \gg R^2$$

and

$$\sqrt{R^2 + X_C^2} \cong \sqrt{X_C^2} = X_C$$

and

$$V_C = \frac{X_CE}{\sqrt{R^2 + X_C^2}} = \frac{X_CE}{X_C} = E$$

with

$$\begin{aligned}\theta_C &= -90° + \tan^{-1}(\text{very large value})\\ &= -90° + 90°\\ &= 0°\end{aligned}$$

verifying the above conclusions.

***f* = 1 kHz:**

Applying Eq. (15.12):

$$X_C = \frac{1}{2\pi fC} = \frac{1}{(6.28)(1 \times 10^3\ \text{Hz})(0.01 \times 10^{-6}\ \text{F})} \cong \mathbf{15.92\ k\Omega}$$

$$\sqrt{R^2 + X_C^2} = \sqrt{(5\ \text{k}\Omega)^2 + (15.92\ \text{k}\Omega)^2} \cong 16.69\ \text{k}\Omega$$

and

$$V_C = \frac{X_C E}{\sqrt{R^2 + X_C^2}} = \frac{(15.92\ \text{k}\Omega)(10)}{16.69\ \text{k}\Omega} = \mathbf{9.54\ V}$$

Applying Eq. (15.13):

$$\begin{aligned}\theta_C &= -90° + \tan^{-1}\frac{X_C}{R}\\ &= -90° + \tan^{-1}\frac{15.9\ \text{k}\Omega}{5\ \text{k}\Omega} = -90° + \tan^{-1} 31.84\\ &= \mathbf{-17.46°}\end{aligned}$$

and

$$\mathbf{V}_C = \mathbf{9.53\ V} \angle \mathbf{-17.46°}$$

As expected, the high reactance of the capacitor at low frequencies has resulted in the major part of the applied voltage appearing across the capacitor.

If we plot the phasor diagrams for $f = 0$ Hz and $f = 1$ kHz, as shown in Fig. 15.48, we find that $\mathbf{V}_C$ is beginning a clockwise rotation with increase in frequency that will increase the angle $\theta$ and decrease the phase angle between **I** and **E.** Recall that for a purely capacitive network, **I** leads **E** by 90°. As the frequency increases, therefore, the capacitive reactance is decreasing, and eventually $R \gg X_C$ with $\theta =$ 90°, and the angle between **I** and **E** will approach 0°. Keep in mind as we proceed through the other frequencies that $\theta$ is the phase angle between $\mathbf{V}_C$ and **E** and that the magnitude of the angle by which **I** leads **E** is determined by

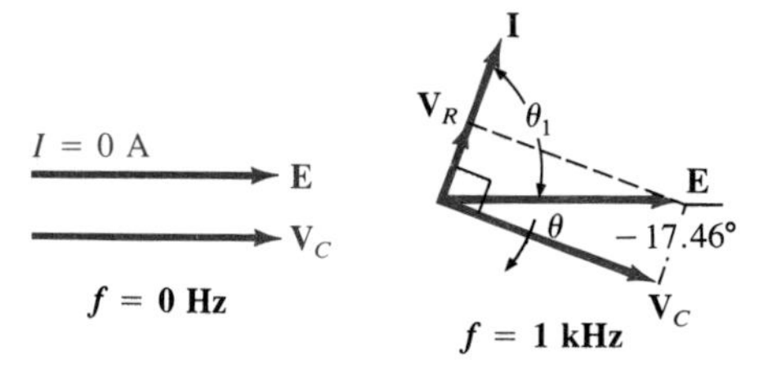

FIG. 15.48

$$|\theta_I| = 90° - |\theta_C| \tag{15.14}$$

***f* = 5 kHz:**

Applying Eq. (15.12):

$$X_C = \frac{1}{2\pi fC} = \frac{1}{(6.28)(5 \times 10^3\ \text{Hz})(0.01 \times 10^{-6}\ \text{F})} \cong \mathbf{3.18\ k\Omega}$$

Note the dramatic drop in $X_C$ from 1 kHz to 5 kHz. In fact, $X_C$ is now less than the resistance $R$ of the network, and the phase angle determined by $\tan^{-1}(X_C/R)$ must be less than 45°. Here,

$$V_C = \frac{X_C E}{\sqrt{R^2 + X_C^2}} = \frac{(3.18\ \text{k}\Omega)(10\ \text{V})}{\sqrt{(5\ \text{k}\Omega)^2 + (3.18\ \text{k}\Omega)^2}} = \mathbf{5.37\ V}$$

with

$$\theta_C = -90° + \tan^{-1}\frac{X_C}{R}$$

$$= -90° + \tan^{-1}\frac{3.2\ \text{k}\Omega}{5\ \text{k}\Omega} = -90° + \tan^{-1} 0.636$$

$$= \mathbf{-57.38°}$$

***f* = 10 kHz:**

$$X_C \cong 1.59\ \text{k}\Omega,\ V_C = \mathbf{3.03\ V},\ \theta_C = \mathbf{-72.34°}$$

***f* = 15 kHz:**

$$X_C \cong 1.06\ \text{k}\Omega,\ V_C = \mathbf{2.07\ V},\ \theta = \mathbf{78.02°}$$

***f* = 20 kHz:**

$$X_C \cong 795.78\ \Omega,\ V_C = \mathbf{1.57\ V},\ \theta = \mathbf{80.96°}$$

The phasor diagrams for $f = 5$ kHz and $f = 20$ kHz appear in Fig. 15.49 to show the continuing rotation of the $\mathbf{V}_C$ vector.

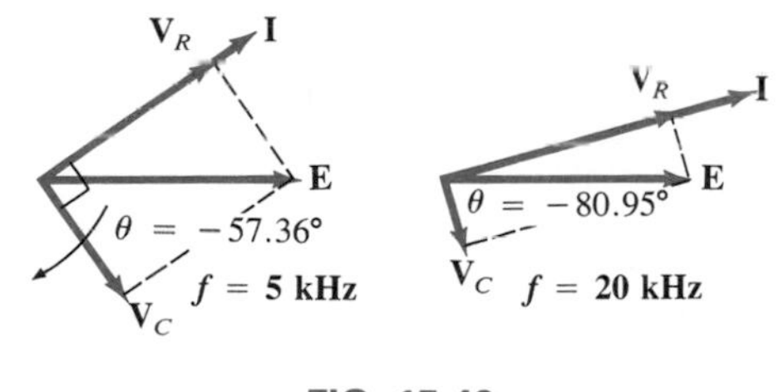


**FIG. 15.49**

Note also from Figs. 15.48 and 15.49 that the vector $\mathbf{V}_R$ and the current **I** have grown in magnitude with reduction in the capacitive reactance. Eventually, at very high frequencies $X_C$ will approach zero ohms and the short-circuit equivalent can be applied, resulting in $V_C = 0$ V and $\theta = 90°$, producing the phasor diagram of Fig. 15.50. The network is then purely resistive and the phase angle between **I** and **E** is zero degrees and $V_R$ and $I$ are their maximum values.

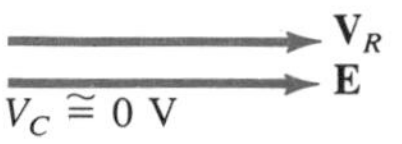

A plot of $V_C$ versus frequency appears in Fig. 15.51, and $\theta$ versus frequency in Fig. 15.52.

A plot of $V_R$ versus frequency would approach $E$ volts from 0 volts with increase in frequency, but remember $V_R \neq E - V_C$ due to vector relationship. The phase angle between **I** and **E** could be plotted directly from Fig. 15.52 using Eq. (15.14).

In Chapter 21, the analysis of this section will be extended to a much wider frequency range using a log axis for frequency. It will be demon-

$\mathbf{V}_R$
**E**
$V_C \cong 0$ V

***f* = very high frequencies**

**FIG. 15.50**

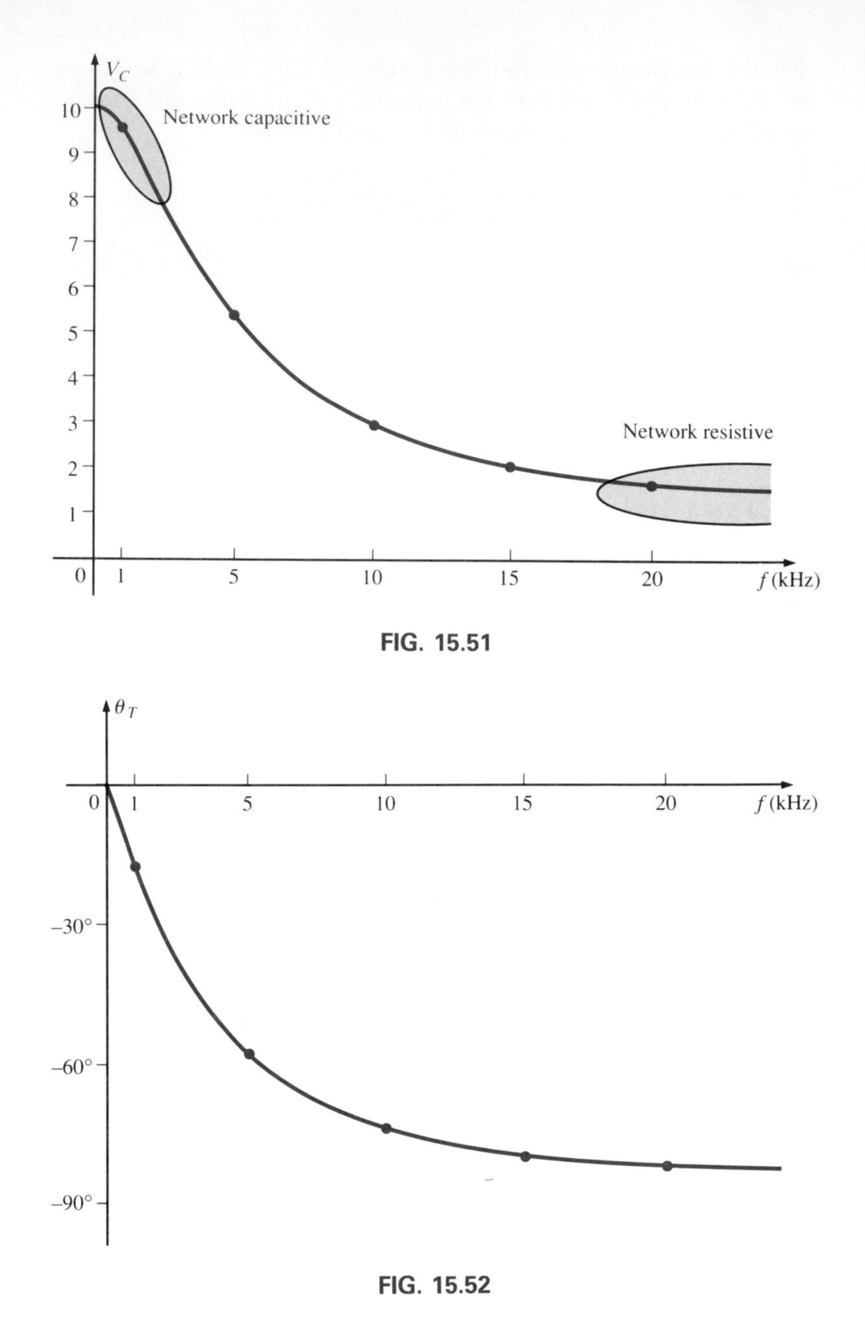


**FIG. 15.51**

**FIG. 15.52**

strated that an $R$-$C$ circuit such as that in Fig. 15.44 can be used as a filter to determine which frequencies will have the greatest impact on the stage to follow. From our current analysis, it is obvious that any network connected across the capacitor will receive the greatest potential level at low frequencies and be effectively ''shorted out'' at very high frequencies.

The analysis of a series $R$-$L$ circuit would proceed in much the same manner, except that $X_L$ and $V_L$ would increase with frequency and the angle between $\mathbf{I}$ and $\mathbf{E}$ would approach 90° (voltage leading the current) rather than 0°. If $\mathbf{V}_L$ were plotted versus frequency, $\mathbf{V}_L$ would approach

**E**, and $X_L$ would eventually attain a level at which the open-circuit equivalent would be appropriate.

## *PARALLEL ac CIRCUITS*

## 15.6 ADMITTANCE AND SUSCEPTANCE

The discussion for parallel ac circuits will be very similar to that for dc circuits. In dc circuits, *conductance* ($G$) was defined as equal to $1/R$. The total conductance of a parallel circuit was then found by adding the conductance of each branch. The total resistance $R_T$ is simply $1/G_T$.

In ac circuits, we define *admittance* ($Y$) as equal to $1/Z$. The unit of measure for admittance as defined by the SI system is *siemens,* which has the symbol S. Admittance is a measure of how well an ac circuit will *admit,* or allow, current to flow in the circuit. The larger its value, therefore, the heavier the current flow for the same applied potential. The total admittance of a circuit can also be found by finding the sum of the parallel admittances. The total impedance $Z_T$ of the circuit is then $1/Y_T$; that is, for the network of Fig. 15.53.

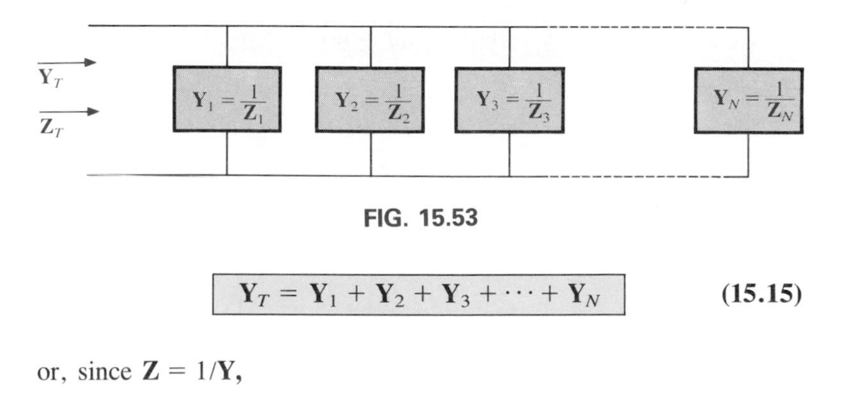


**FIG. 15.53**

$$\mathbf{Y}_T = \mathbf{Y}_1 + \mathbf{Y}_2 + \mathbf{Y}_3 + \cdots + \mathbf{Y}_N \tag{15.15}$$

or, since $\mathbf{Z} = 1/\mathbf{Y}$,

$$\frac{1}{\mathbf{Z}_T} = \frac{1}{\mathbf{Z}_1} + \frac{1}{\mathbf{Z}_2} + \frac{1}{\mathbf{Z}_3} + \cdots + \frac{1}{\mathbf{Z}_N} \tag{15.16}$$

For two impedances in parallel,

$$\frac{1}{\mathbf{Z}_T} = \frac{1}{\mathbf{Z}_1} + \frac{1}{\mathbf{Z}_2}$$

If the manipulations used in Chapter 6 to find the total resistance of two parallel resistors are now applied, the following similar equation will result:

$$\mathbf{Z}_T = \frac{\mathbf{Z}_1\mathbf{Z}_2}{\mathbf{Z}_1 + \mathbf{Z}_2} \tag{15.17}$$

For three parallel impedances,

$$\mathbf{Z}_T = \frac{\mathbf{Z}_1\mathbf{Z}_2\mathbf{Z}_3}{\mathbf{Z}_1\mathbf{Z}_2 + \mathbf{Z}_2\mathbf{Z}_3 + \mathbf{Z}_1\mathbf{Z}_3} \tag{15.18}$$

As pointed out in the introduction to this section, conductance is the reciprocal of resistance, and

$$\mathbf{Y} = \frac{1}{\mathbf{R}} = \frac{1}{R \angle 0^\circ} = G \angle 0^\circ$$

so that

$$\mathbf{G} = G \angle 0^\circ \tag{15.19}$$

The reciprocal of reactance ($1/X$) is called *susceptance* and is a measure of how *susceptible* an element is to the passage of current through it. Susceptance is also measured in *siemens,* and is represented by the capital letter $B$.

For the inductor,

$$\mathbf{Y} = \frac{1}{\mathbf{X}_L} = \frac{1}{X_L \angle 90^\circ} = \frac{1}{\omega L \angle 90^\circ} = \frac{1}{\omega L} \angle -90^\circ$$

Defining

$$B_L = \frac{1}{X_L} \qquad \text{(siemens, S)} \tag{15.20}$$

we have

$$\mathbf{B}_L = B_L \angle -90^\circ \tag{15.21}$$

Note that for inductance, an increase in frequency or inductance will result in a decrease in susceptance or, correspondingly, in admittance.

For the capacitor,

$$\mathbf{Y} = \frac{1}{\mathbf{X}_C} = \frac{1}{X_C \angle -90^\circ} = \frac{1}{1/\omega C \angle -90^\circ} = \omega C \angle 90^\circ$$

Defining

$$B_C = \frac{1}{X_C} \qquad \text{(S)} \tag{15.22}$$

we have

$$\mathbf{B}_C = B_C \angle 90^\circ \tag{15.23}$$

For the capacitor, therefore, an increase in frequency or capacitance will result in an increase in its susceptibility.

In summary, for parallel circuits,

Resistance:

$$\mathbf{Y} = \frac{1}{\mathbf{R}} = \mathbf{G} = G \angle 0° = G + j0 \tag{15.24}$$

Inductance:

$$\mathbf{Y} = \frac{1}{\mathbf{X}_L} = \mathbf{B}_L = B_L \angle -90° = 0 - jB_L \tag{15.25}$$

Capacitance:

$$\mathbf{Y} = \frac{1}{\mathbf{X}_C} = \mathbf{B}_C = B_C \angle 90° = 0 + jB_C \tag{15.26}$$

For parallel ac circuits, the *admittance diagram* is used with the three admittances represented as shown in Fig. 15.54.

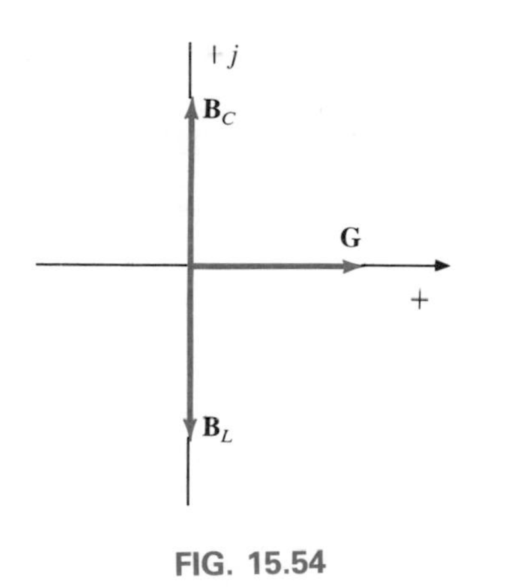


FIG. 15.54

Note in Fig. 15.54 that the conductance (like resistance) is on the positive real axis, while inductive and capacitive susceptances are still in direct opposition on the imaginary axis.

***For any configuration (series, parallel, series-parallel, etc.), the angle associated with the total admittance is the angle by which the source current leads the applied voltage. For inductive networks, $\theta_T$ is negative, whereas for capacitive networks, $\theta_T$ is positive.***

**EXAMPLE 15.12.** For the network of Fig. 15.55:

a. Find the admittance of each parallel branch.
b. Determine the input admittance.
c. Calculate the input impedance.
d. Draw the admittance diagram.

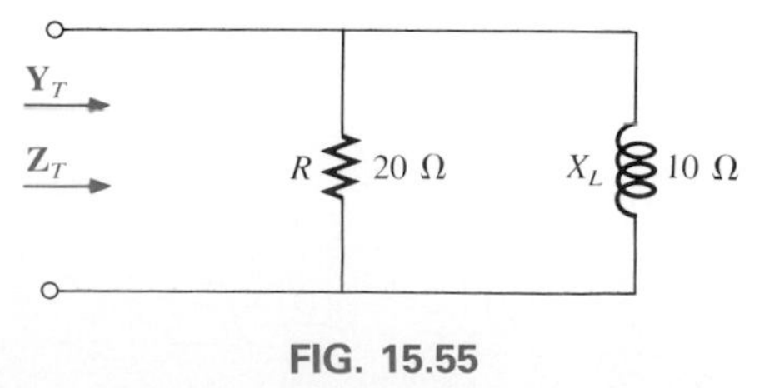


FIG. 15.55

***Solutions:***

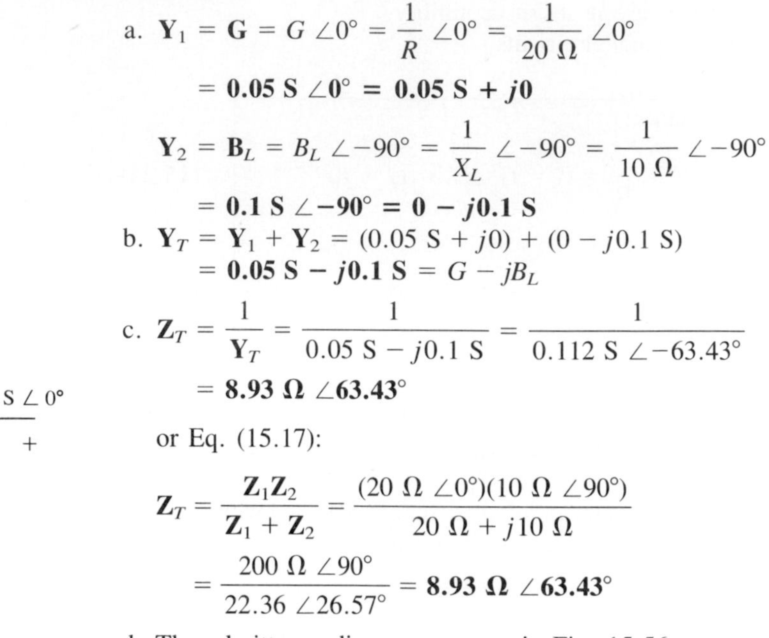

a. $\mathbf{Y}_1 = \mathbf{G} = G \angle 0° = \dfrac{1}{R} \angle 0° = \dfrac{1}{20\ \Omega} \angle 0°$

$= \mathbf{0.05\ S \angle 0° = 0.05\ S} + j\mathbf{0}$

$\mathbf{Y}_2 = \mathbf{B}_L = B_L \angle -90° = \dfrac{1}{X_L} \angle -90° = \dfrac{1}{10\ \Omega} \angle -90°$

$= \mathbf{0.1\ S \angle -90° = 0} - j\mathbf{0.1\ S}$

b. $\mathbf{Y}_T = \mathbf{Y}_1 + \mathbf{Y}_2 = (0.05\ \text{S} + j0) + (0 - j0.1\ \text{S})$

$= \mathbf{0.05\ S} - j\mathbf{0.1\ S} = G - jB_L$

c. $\mathbf{Z}_T = \dfrac{1}{\mathbf{Y}_T} = \dfrac{1}{0.05\ \text{S} - j0.1\ \text{S}} = \dfrac{1}{0.112\ \text{S} \angle -63.43°}$

$= \mathbf{8.93\ \Omega \angle 63.43°}$

or Eq. (15.17):

$$\mathbf{Z}_T = \frac{\mathbf{Z}_1\mathbf{Z}_2}{\mathbf{Z}_1 + \mathbf{Z}_2} = \frac{(20\ \Omega \angle 0°)(10\ \Omega \angle 90°)}{20\ \Omega + j10\ \Omega}$$

$$= \frac{200\ \Omega \angle 90°}{22.36 \angle 26.57°} = \mathbf{8.93\ \Omega \angle 63.43°}$$

d. The admittance diagram appears in Fig. 15.56.

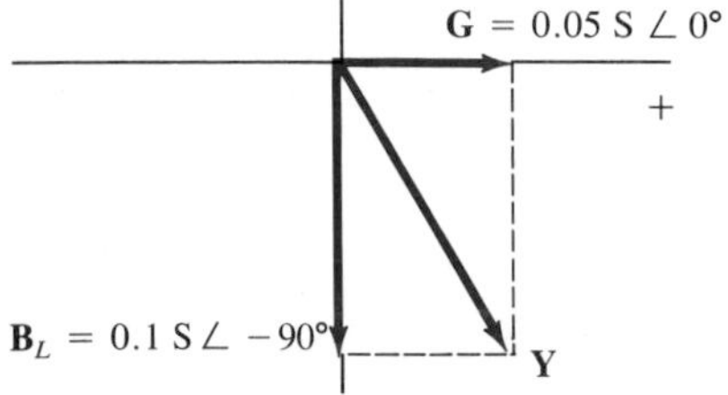


**FIG. 15.56**

**EXAMPLE 15.13.** Repeat Example 15.12 for the parallel network of Fig. 15.57.

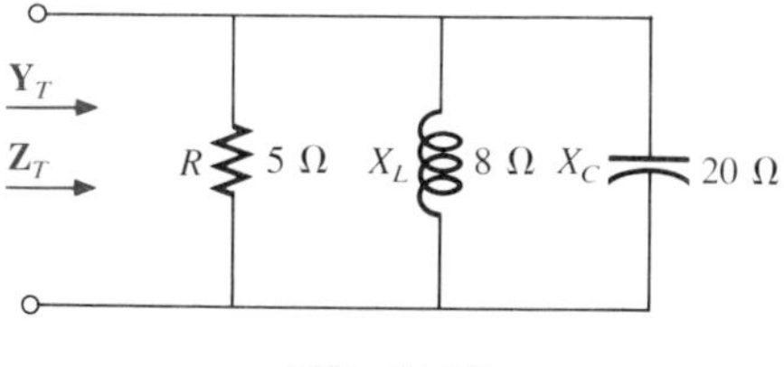


**FIG. 15.57**

***Solutions:***

a. $\mathbf{Y}_1 = \mathbf{G} = G \angle 0° = \dfrac{1}{R} \angle 0° = \dfrac{1}{5\ \Omega} \angle 0°$

$= \mathbf{0.2\ S \angle 0° = 0.2\ S} + j\mathbf{0}$

$\mathbf{Y}_2 = \mathbf{B}_L = B_L \angle -90° = \dfrac{1}{X_L} \angle -90° = \dfrac{1}{8\ \Omega} \angle -90°$

$= \mathbf{0.125\ S \angle -90° = 0} - j\mathbf{0.125\ S}$

$\mathbf{Y}_3 = \mathbf{B}_C = B_C \angle 90° = \dfrac{1}{X_C} \angle 90° = \dfrac{1}{20\ \Omega} \angle 90°$

$= \mathbf{0.050\ S \angle +90° = 0} + j\mathbf{0.050\ S}$

b. $\mathbf{Y}_T = \mathbf{Y}_1 + \mathbf{Y}_2 + \mathbf{Y}_3$
$= (0.2\text{ S} + j0) + (0 - j0.125\text{ S}) + (0 + j0.050\text{ S})$
$= 0.2\text{ S} - j0.075\text{ S} = \mathbf{0.2136\ S\ \angle{-20.56°}}$

c. $$\mathbf{Z}_T = \frac{1}{0.2136\text{ S}\ \angle{-20.56°}} = \mathbf{4.68\ \Omega\ \angle 20.56°}$$

or

$$\mathbf{Z}_T = \frac{\mathbf{Z}_1\mathbf{Z}_2\mathbf{Z}_3}{\mathbf{Z}_1\mathbf{Z}_2 + \mathbf{Z}_2\mathbf{Z}_3 + \mathbf{Z}_1\mathbf{Z}_3}$$

$$= \frac{(5\ \Omega\ \angle 0°)(8\ \Omega\ \angle 90°)(20\ \Omega\ \angle{-90°})}{(5\ \Omega\ \angle 0°)(8\ \Omega\ \angle 90°) + (8\ \Omega\ \angle 90°)(20\ \Omega\ \angle{-90°}) + (5\ \Omega\ \angle 0°)(20\ \Omega\ \angle{-90°})}$$

$$= \frac{800\ \angle 0°}{40\ \angle 90° + 160\ \angle 0° + 100\ \angle{-90°}}$$

$$= \frac{800}{160 + j40 - j100} = \frac{800}{160 - j60}$$

$$= \frac{800}{170.88\ \angle{-20.56°}}$$

$$= \mathbf{4.68\ \Omega\ \angle 20.56°}$$

d. The admittance diagram appears in Fig. 15.58.

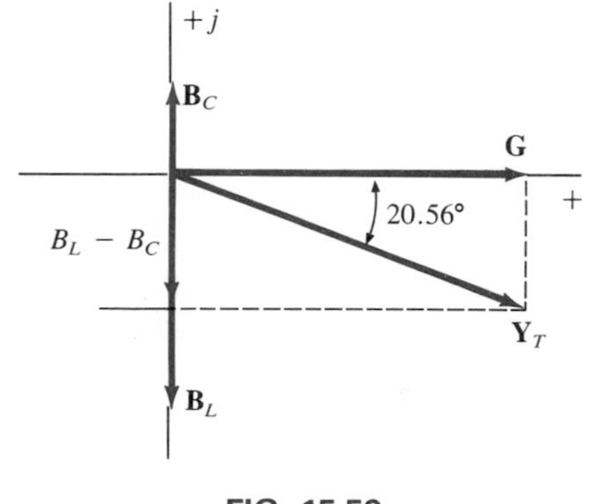


FIG. 15.58

On many occasions, the inverse relationship $\mathbf{Y}_T = 1/\mathbf{Z}_T$ or $\mathbf{Z}_T = 1/\mathbf{Y}_T$ will require that we divide the number 1 by a complex number having a real and an imaginary part. This division, if not performed in the polar form, requires that we multiply the numerator and denominator by the conjugate of the denominator, as follows:

$$\mathbf{Y}_T = \frac{1}{\mathbf{Z}_T} = \frac{1}{4\ \Omega + j6\ \Omega} = \left(\frac{1}{4\ \Omega + j6\ \Omega}\right)\frac{(4\ \Omega - j6\ \Omega)}{(4\ \Omega - j6\ \Omega)} = \frac{4 - j6}{4^2 + 6^2}$$

and

$$\mathbf{Y}_T = \frac{4}{52}\text{ S} - j\frac{6}{52}\text{ S}$$

To avoid this laborious task each time we want to find the reciprocal of a complex number in rectangular form, a format can be developed using the following complex number, which is symbolic of any impedance or admittance in the first or fourth quadrant:

$$\frac{1}{a_1 \pm jb_1} = \left(\frac{1}{a_1 \pm jb_1}\right)\left(\frac{a_1 \mp jb_1}{a_1 \mp jb_1}\right) = \frac{a_1 \mp jb_1}{a_1^2 + b_1^2}$$

or

$$\boxed{\frac{1}{a_1 \pm jb_1} = \frac{a_1}{a_1^2 + b_1^2} \mp \frac{b_1}{a_1^2 + b_1^2}} \quad \textbf{(15.27)}$$

Note that the denominator is simply the sum of the squares of each term. The sign is inverted between the real and imaginary parts. A few examples will develop some familiarity with the use of this equation.

**EXAMPLE 15.14.** Find the admittance of each set of series elements in Fig. 15.59.

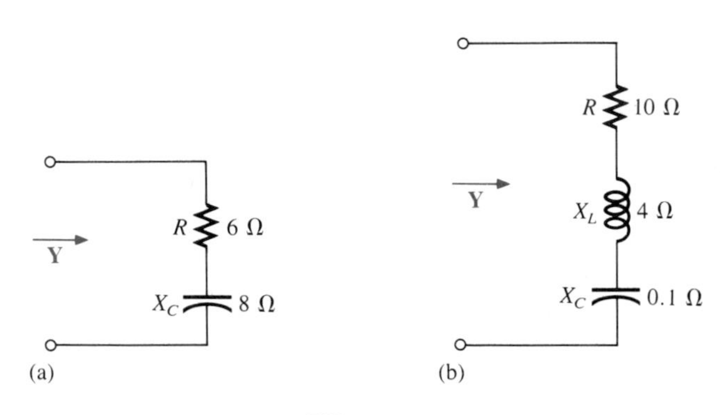

FIG. 15.59

***Solutions:***

a. $\mathbf{Z} = 6\ \Omega - j8\ \Omega$

$$\mathbf{Y} = \frac{1}{6\ \Omega - j8\ \Omega} = \frac{6}{(6)^2 + (8)^2} + j\frac{8}{(6)^2 + (8)^2}$$

$$= \frac{\mathbf{6}}{\mathbf{100}}\ \mathbf{S} + \boldsymbol{j}\frac{\mathbf{8}}{\mathbf{100}}\ \mathbf{S}$$

b. $\mathbf{Z} = 10\ \Omega + j4\ \Omega + (-j0.1\ \Omega) = 10\ \Omega + j3.9\ \Omega$

$$\mathbf{Y} = \frac{1}{\mathbf{Z}} = \frac{1}{10\ \Omega + j3.9\ \Omega} = \frac{10}{(10)^2 + (3.9)^2} - j\frac{3.9}{(10)^2 + (3.9)^2}$$

$$= \frac{10}{115.21} - j\frac{3.9}{115.21} = \mathbf{0.087\ S - \boldsymbol{j}0.034\ S}$$

## 15.7 *R-L, R-C,* AND *R-L-C* PARALLEL ac NETWORKS

### *R-L* (Fig. 15.60)

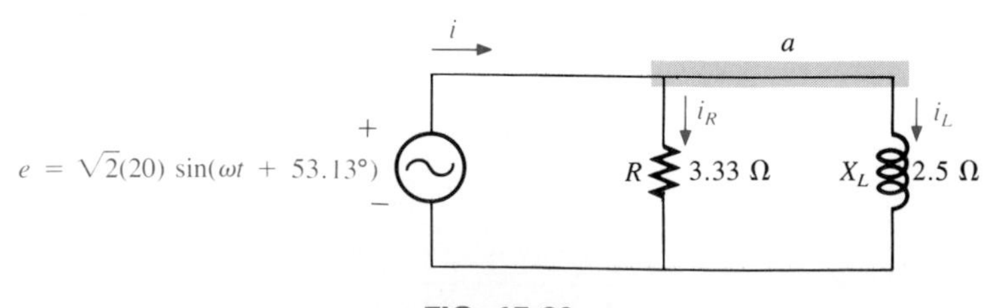


FIG. 15.60

*Phasor notation:* As shown in Fig. 15.61.

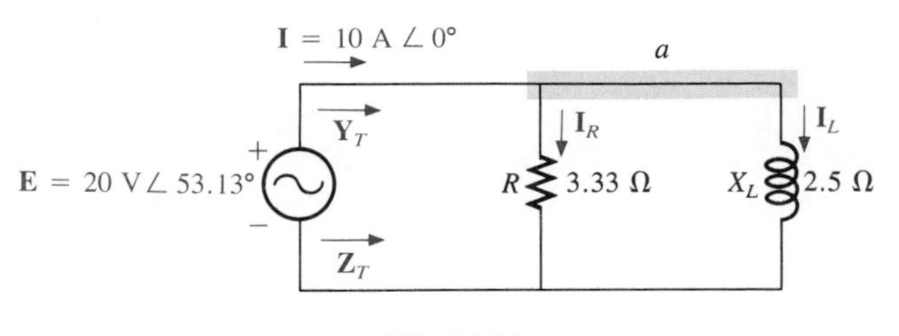


FIG. 15.61

**$\mathbf{Y}_T$, $\mathbf{Z}_T$:**

$$\mathbf{Y}_T = \mathbf{Y}_1 + \mathbf{Y}_2 = \mathbf{G} + \mathbf{B}_L = \frac{1}{3.33\ \Omega}\angle 0^\circ + \frac{1}{2.5\ \Omega}\angle -90^\circ$$

$$= 0.3\text{ S }\angle 0^\circ + 0.4\text{ S }\angle -90^\circ = 0.3\text{ S} - j0.4\text{ S}$$

$$= \mathbf{0.5\ S\ \angle -53.13^\circ}$$

$$\mathbf{Z}_T = \frac{1}{\mathbf{Y}_T} = \frac{1}{0.5\text{ S }\angle -53.13^\circ} = \mathbf{2\ \Omega\ \angle 53.13^\circ}$$

*Admittance diagram:* As shown in Fig. 15.62.

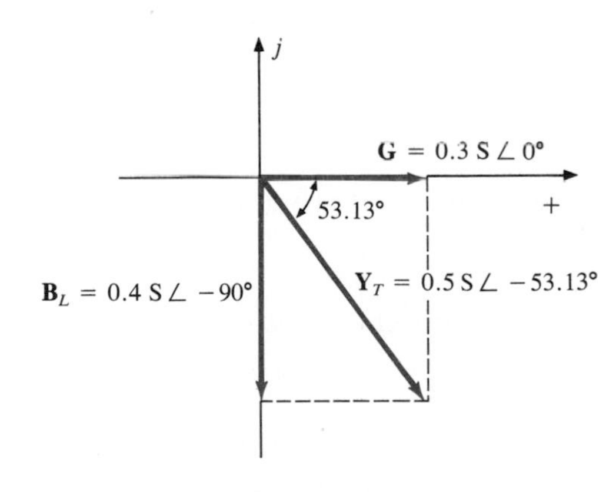


FIG. 15.62

**I:**

$$\mathbf{I} = \frac{\mathbf{E}}{\mathbf{Z}_T} = \mathbf{E}\mathbf{Y}_T = (20\text{ V }\angle 53.13^\circ)(0.5\text{ S }\angle -53.13^\circ) = \mathbf{10\ A\ \angle 0^\circ}$$

**$\mathbf{I}_R$, $\mathbf{I}_L$:**

$$\mathbf{I}_R = \frac{\mathbf{E}}{\mathbf{R}} = \mathbf{E}\mathbf{G} = (20\text{ V }\angle 53.13^\circ)(0.3\text{ S }\angle 0^\circ) = \mathbf{6\ A\ \angle 53.13^\circ}$$

$$\mathbf{I}_L = \frac{\mathbf{E}}{\mathbf{X}_L} = \mathbf{E}\mathbf{B}_L = (20\text{ V }\angle 53.13^\circ)(0.4\text{ S }\angle -90^\circ)$$

$$= \mathbf{8\ A\ \angle -36.87^\circ}$$

*Kirchhoff's current law:* At node *a*,

$$\mathbf{I} - \mathbf{I}_R - \mathbf{I}_L = 0$$

or

$$\mathbf{I} = \mathbf{I}_R + \mathbf{I}_L$$
$$10 \text{ A} \angle 0° = 6 \text{ A} \angle 53.13° + 8 \text{ A} \angle -36.87°$$
$$10 \text{ A} \angle 0° = (3.60 \text{ A} + j4.80 \text{ A}) + (6.40 \text{ A} - j4.80 \text{ A}) = 10 \text{ A} + j0$$

and

$$\mathbf{10 \text{ A} \angle 0° = 10 \text{ A} \angle 0°} \qquad \text{(checks)}$$

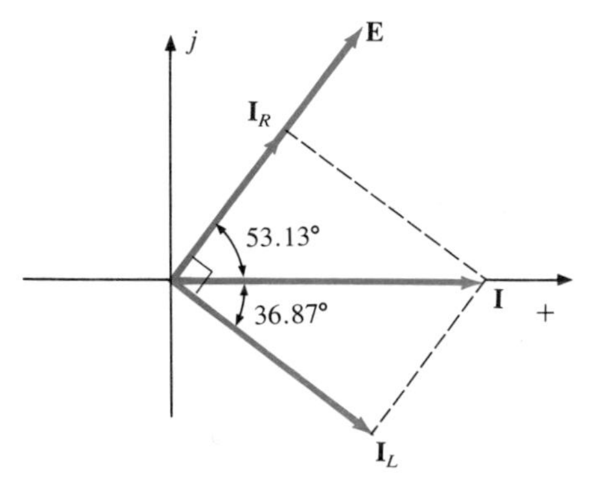

FIG. 15.63

*Phasor diagram:* The phasor diagram of Fig. 15.63 indicates that the applied voltage **E** is in phase with the current $\mathbf{I}_R$ and leads the current $\mathbf{I}_L$ by 90°.

*Power:* The total power in watts delivered to the circuit is

$$\begin{aligned} P_T &= EI \cos \theta_T \\ &= (20 \text{ V})(10 \text{ A}) \cos 53.13° = (200 \text{ W})(0.6) \\ &= \mathbf{120 \text{ W}} \end{aligned}$$

or

$$P_T = I^2R = \frac{V_R^2}{R} = V_R^2 G = (20 \text{ V})^2(0.3 \text{ S}) = \mathbf{120 \text{ W}}$$

or, finally,

$$\begin{aligned} P_T &= P_R + P_L = EI_R \cos \theta_R + EI_L \cos \theta_L \\ &= (20 \text{ V})(6 \text{ A}) \cos 0° + (20 \text{ V})(8 \text{ A}) \cos 90° = 120 \text{ W} + 0 \\ &= \mathbf{120 \text{ W}} \end{aligned}$$

*Power factor:* The power factor of the circuit is

$$F_p = \cos \theta_T = \cos 53.13° = \mathbf{0.6 \text{ lagging}}$$

or, through an analysis similar to that employed for a series ac circuit,

$$\cos \theta_T = \frac{P}{EI} = \frac{E^2/R}{EI} = \frac{EG}{I} = \frac{G}{I/V} = \frac{G}{Y_T}$$

and

$$\boxed{F_p = \cos \theta_T = \frac{G}{Y_T}} \qquad \mathbf{(15.28)}$$

where $G$ and $Y_T$ are the magnitudes of the total conductance and admittance of the parallel network. For this case,

$$F_p = \cos \theta_T = \frac{0.3 \text{ S}}{0.5 \text{ S}} = \mathbf{0.6 \text{ lagging}}$$

*Impedance approach:* The current **I** can also be found by first finding the total impedance of the network:

$$\mathbf{Z}_T = \frac{\mathbf{Z}_1\mathbf{Z}_2}{\mathbf{Z}_1 + \mathbf{Z}_2} = \frac{(3.33\ \Omega\ \angle 0^\circ)(2.5\ \Omega\ \angle 90^\circ)}{3.33\ \Omega\ \angle 0^\circ + 2.5\ \Omega\ \angle 90^\circ}$$

$$= \frac{8.325\ \angle 90^\circ}{4.164\ \angle 36.87^\circ} = \mathbf{2\ \Omega\ \angle 53.13^\circ}$$

And then, using Ohm's law, we obtain

$$\mathbf{I} = \frac{\mathbf{E}}{\mathbf{Z}_T} = \frac{20\ \text{V}\ \angle 53.13^\circ}{2\ \Omega\ \angle 53.13^\circ} = \mathbf{10\ A\ \angle 0^\circ}$$

### *R-C* (Fig. 15.64)

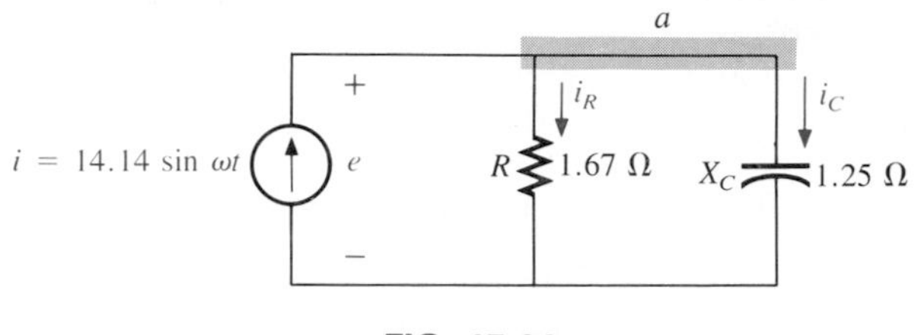


FIG. 15.64

*Phasor notation:* As shown in Fig. 15.65.

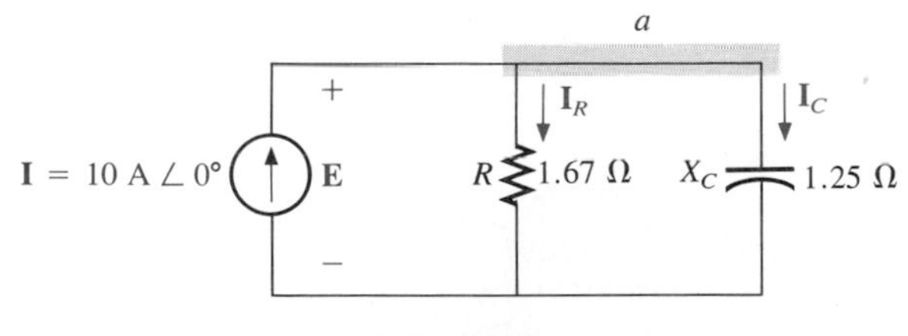


FIG. 15.65

**$\mathbf{Y}_T$, $\mathbf{Z}_T$:**

$$\mathbf{Y}_T = \mathbf{Y}_1 + \mathbf{Y}_2 = \mathbf{G} + \mathbf{B}_C = \frac{1}{1.67\ \Omega}\ \angle 0^\circ + \frac{1}{1.25\ \Omega}\ \angle 90^\circ$$

$$= 0.6\ \text{S}\ \angle 0^\circ + 0.8\ \text{S}\ \angle 90^\circ = 0.6\ \text{S} + j0.8\ \text{S} = \mathbf{1.0\ S\ \angle 53.13^\circ}$$

$$\mathbf{Z}_T = \frac{1}{\mathbf{Y}_T} = \frac{1}{1.0\ \text{S}\ \angle 53.13^\circ} = \mathbf{1\ \Omega\ \angle -53.13^\circ}$$

*Admittance diagram:* As shown in Fig. 15.66.

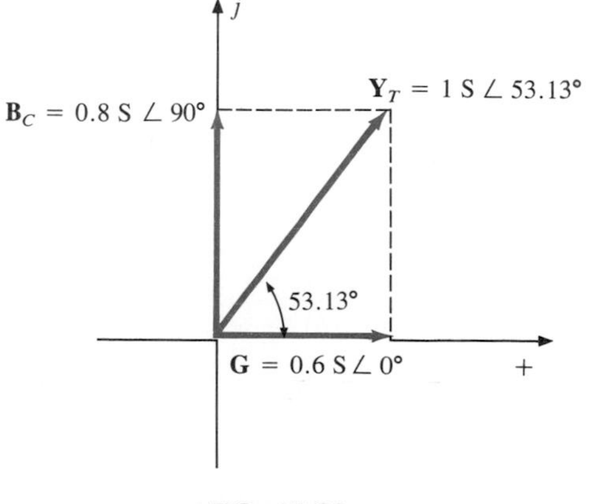


FIG. 15.66

**E:**

$$\mathbf{E} = \mathbf{I}\mathbf{Z}_T = \frac{\mathbf{I}}{\mathbf{Y}_T} = \frac{10\ \text{A}\ \angle 0^\circ}{1\ \text{S}\ \angle 53.13^\circ} = \mathbf{10\ V\ \angle -53.13^\circ}$$

$\mathbf{I}_R$, $\mathbf{I}_C$:

$$\mathbf{I}_R = \mathbf{E}\mathbf{G} = (10\ \text{V}\ \angle -53.13^\circ)(0.6\ \text{S}\ \angle 0^\circ) = \mathbf{6\ A\ \angle -53.13^\circ}$$
$$\mathbf{I}_C = \mathbf{E}\mathbf{B}_C = (10\ \text{V}\ \angle -53.13^\circ)(0.8\ \text{S}\ \angle 90^\circ) = \mathbf{8\ A\ \angle 36.87^\circ}$$

*Kirchhoff's current law:* At node $a$,

$$\mathbf{I} - \mathbf{I}_R - \mathbf{I}_C = 0$$

or

$$\mathbf{I} = \mathbf{I}_R + \mathbf{I}_C$$

which can also be verified (as for the *R-L* network) through vector algebra.

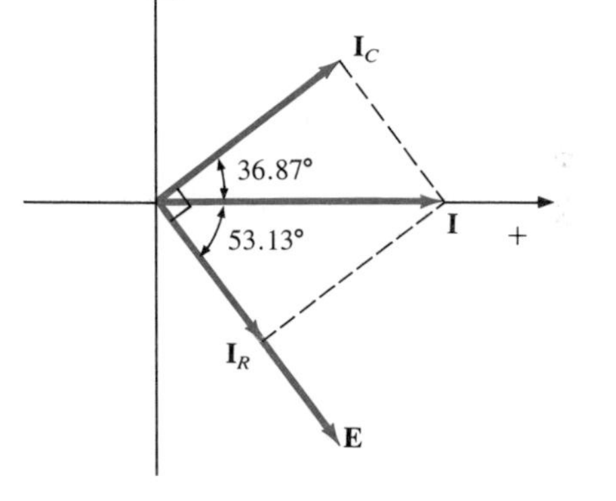


FIG. 15.67

*Phasor diagram:* The phasor diagram of Fig. 15.67 indicates that **E** is in phase with the current through the resistor $\mathbf{I}_R$ and lags the capacitive current $\mathbf{I}_C$ by 90°.

*Time domain:*

$$e = \sqrt{2}(10)\sin(\omega t - 53.13^\circ) = \mathbf{14.14\sin(\omega t - 53.13^\circ)}$$
$$i_R = \sqrt{2}(6)\sin(\omega t - 53.13^\circ) = \mathbf{8.48\sin(\omega t - 53.13^\circ)}$$
$$i_C = \sqrt{2}(8)\sin(\omega t + 36.87^\circ) = \mathbf{11.31\sin(\omega t + 36.87^\circ)}$$

A plot of all of the currents and the voltage appears in Fig. 15.68. Note that $e$ and $i_R$ are in phase and $e$ lags $i_C$ by 90°.

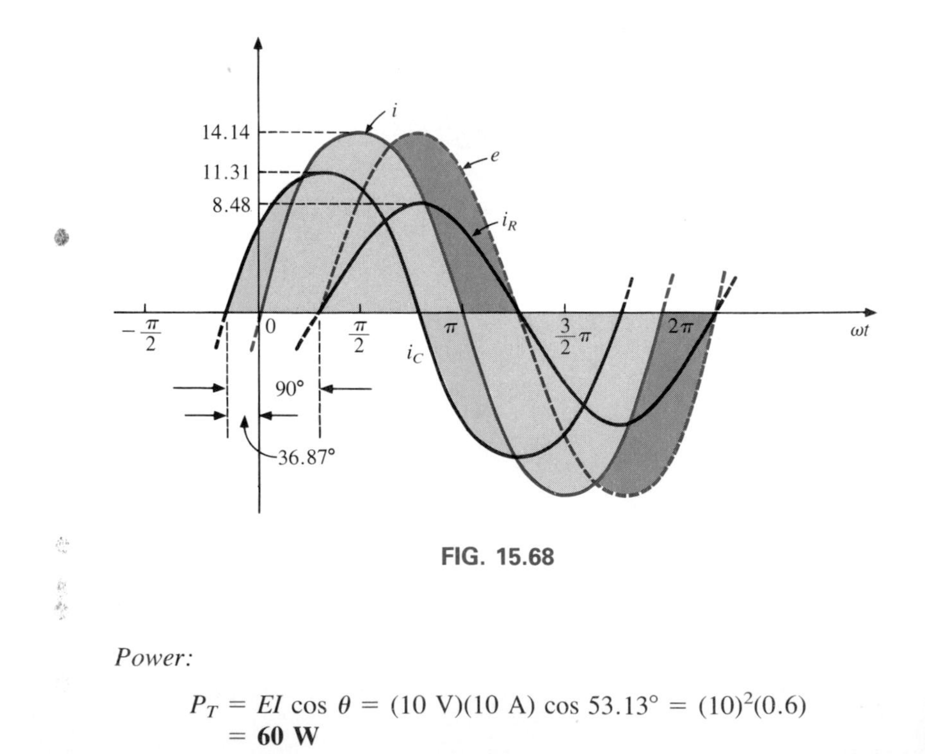


FIG. 15.68

*Power:*

$$P_T = EI\cos\theta = (10\ \text{V})(10\ \text{A})\cos 53.13^\circ = (10)^2(0.6)$$
$$= \mathbf{60\ W}$$

or

$$P_T = E^2G = (10\text{ V})^2(0.6\text{ S}) = \mathbf{60\text{ W}}$$

or, finally,

$$\begin{aligned}P_T = P_R + P_C &= EI_R\cos\theta_R + EI_C\cos\theta_C\\ &= (10\text{ V})(6\text{ A})\cos 0^\circ + (10\text{ V})(8\text{ A})\cos 90^\circ\\ &= \mathbf{60\text{ W}}\end{aligned}$$

*Power factor:* The power factor of the circuit is

$$F_p = \cos 53.13^\circ = \mathbf{0.6\text{ leading}}$$

Using Eq. (15.26), we have

$$F_p = \cos\theta_T = \frac{G}{Y_T} = \frac{0.6\text{ S}}{1.0\text{ S}} = \mathbf{0.6\text{ leading}}$$

*Impedance approach:* The voltage **E** can also be found by first finding the total impedance of the circuit:

$$\begin{aligned}\mathbf{Z}_T = \frac{\mathbf{Z}_1\mathbf{Z}_2}{\mathbf{Z}_1+\mathbf{Z}_2} &= \frac{(1.67\ \Omega\ \angle 0^\circ)(1.25\ \Omega\ \angle -90^\circ)}{1.67\ \Omega\ \angle 0^\circ + 1.25\ \Omega\ \angle -90^\circ}\\ &= \frac{2.09\ \angle -90^\circ}{2.09\ \angle -36.81^\circ} = \mathbf{1\ \Omega\ \angle -53.19^\circ}\end{aligned}$$

and then, using Ohm's law, we find

$$\mathbf{E} = \mathbf{I}\mathbf{Z}_T = (10\text{ A}\ \angle 0^\circ)(1\ \Omega\ \angle -53.19^\circ) = \mathbf{10\text{ V}\ \angle -53.19^\circ}$$

## *R-L-C* (Fig. 15.69)

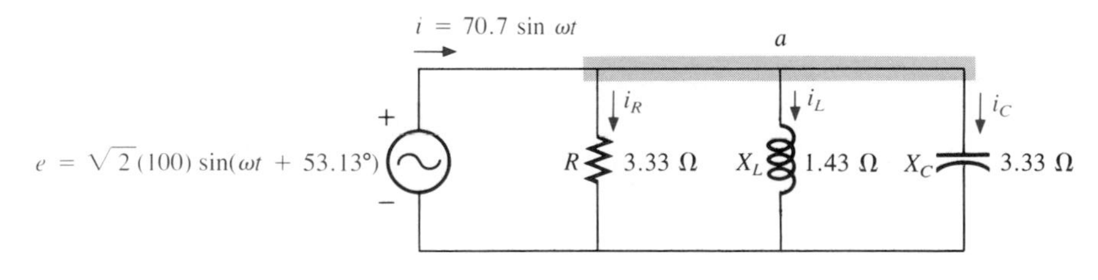


FIG. 15.69

*Phasor notation:* As shown in Fig. 15.70.

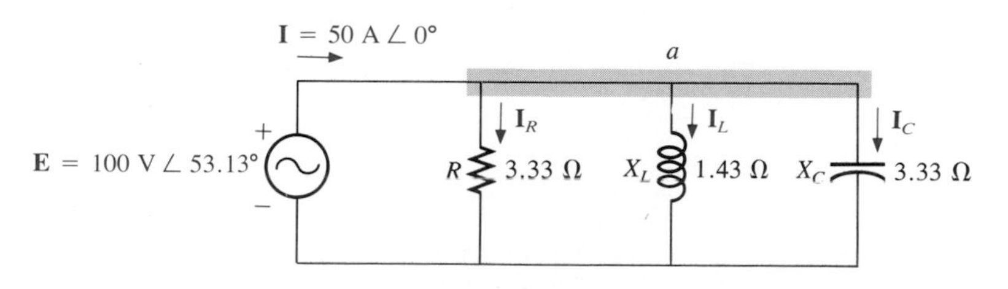


FIG. 15.70

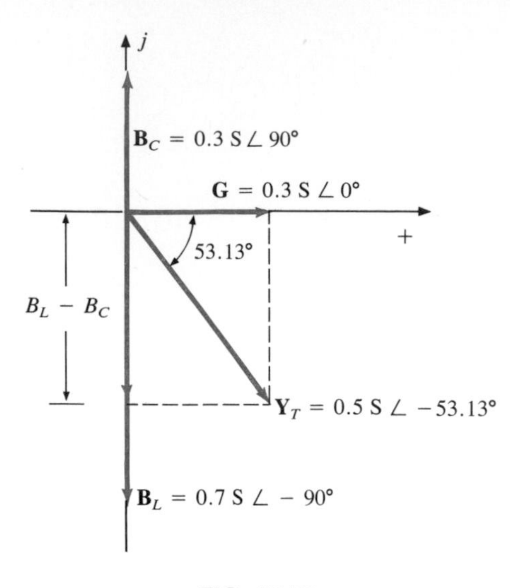

FIG. 15.71

$\mathbf{Y}_T$, $\mathbf{Z}_T$:

$$\mathbf{Y}_T = \mathbf{Y}_1 + \mathbf{Y}_2 + \mathbf{Y}_3 = \mathbf{G} + \mathbf{B}_L + \mathbf{B}_C$$
$$= \frac{1}{3.33\ \Omega}\angle 0^\circ + \frac{1}{1.43\ \Omega}\angle -90^\circ + \frac{1}{3.33\ \Omega}\angle 90^\circ$$
$$= 0.3\text{ S }\angle 0^\circ + 0.7\text{ S }\angle -90^\circ + 0.3\text{ S }\angle 90^\circ$$
$$= 0.3\text{ S} - j0.7\text{ S} + j0.3\text{ S}$$
$$= 0.3\text{ S} - j0.4\text{ S} = \mathbf{0.5\ S\ \angle -53.13^\circ}$$

$$\mathbf{Z}_T = \frac{1}{\mathbf{Y}_T} = \frac{1}{0.5\text{ S }\angle -53.13^\circ} = \mathbf{2\ \Omega\ \angle 53.13^\circ}$$

*Admittance diagram:* As shown in Fig. 15.71.

**I:**

$$\mathbf{I} = \frac{\mathbf{E}}{\mathbf{Z}_T} = \mathbf{E}\mathbf{Y}_T = (100\text{ V }\angle 53.13^\circ)(0.5\text{ S }\angle -53.13^\circ) = \mathbf{50\ A\ \angle 0^\circ}$$

$\mathbf{I}_R$, $\mathbf{I}_L$, $\mathbf{I}_C$:

$$\mathbf{I}_R = \mathbf{E}\mathbf{G} = (100\text{ V }\angle 53.13^\circ)(0.3\text{ S }\angle 0^\circ) = \mathbf{30\ A\ \angle 53.13^\circ}$$
$$\mathbf{I}_L = \mathbf{E}\mathbf{B}_L = (100\text{ V }\angle 53.13^\circ)(0.7\text{ S }\angle -90^\circ) = \mathbf{70\ A\ \angle -36.87^\circ}$$
$$\mathbf{I}_C = \mathbf{E}\mathbf{B}_C = (100\text{ V }\angle 53.13^\circ)(0.3\text{ S }\angle +90^\circ) = \mathbf{30\ A\ \angle 143.13^\circ}$$

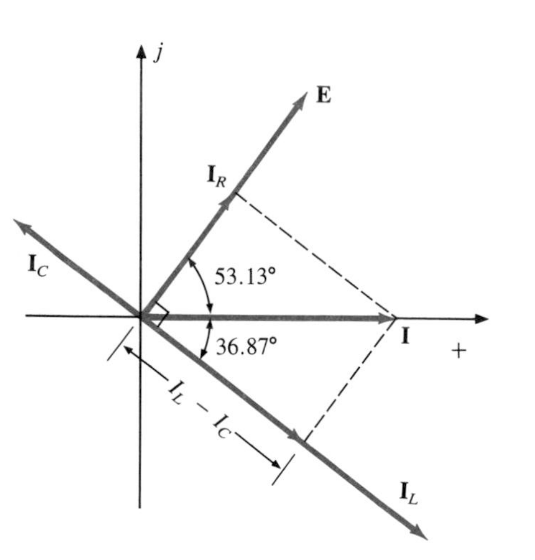


FIG. 15.72

*Kirchhoff's current law:* At node *a*,

$$\mathbf{I} - \mathbf{I}_R - \mathbf{I}_L - \mathbf{I}_C = 0$$

or

$$\mathbf{I} = \mathbf{I}_R + \mathbf{I}_L + \mathbf{I}_C$$

*Phasor diagram:* The phasor diagram of Fig. 15.72 indicates that the impressed voltage **E** is in phase with the current $\mathbf{I}_R$ through the resistor, leads the current $\mathbf{I}_L$ through the inductor by 90°, and lags the current $\mathbf{I}_C$ of the capacitor by 90°.

*Time domain:*

$$i = \sqrt{2}(50)\sin\omega t = \mathbf{70.70\ sin\ \omega t}$$
$$i_R = \sqrt{2}(30)\sin(\omega t + 53.13^\circ) = \mathbf{42.42\ sin(\omega t + 53.13^\circ)}$$
$$i_L = \sqrt{2}(70)\sin(\omega t - 36.87^\circ) = \mathbf{98.98\ sin(\omega t - 36.87^\circ)}$$
$$i_C = \sqrt{2}(30)\sin(\omega t + 143.13^\circ) = \mathbf{42.42\ sin(\omega t + 143.13^\circ)}$$

A plot of all of the currents and the impressed voltage appears in Fig. 15.73.

*Power:* The total power in watts delivered to the circuit is

$$P_T = EI\cos\theta = (100\text{ V})(50\text{ A})\cos 53.13^\circ = (5000)(0.6)$$
$$= \mathbf{3000\ W}$$

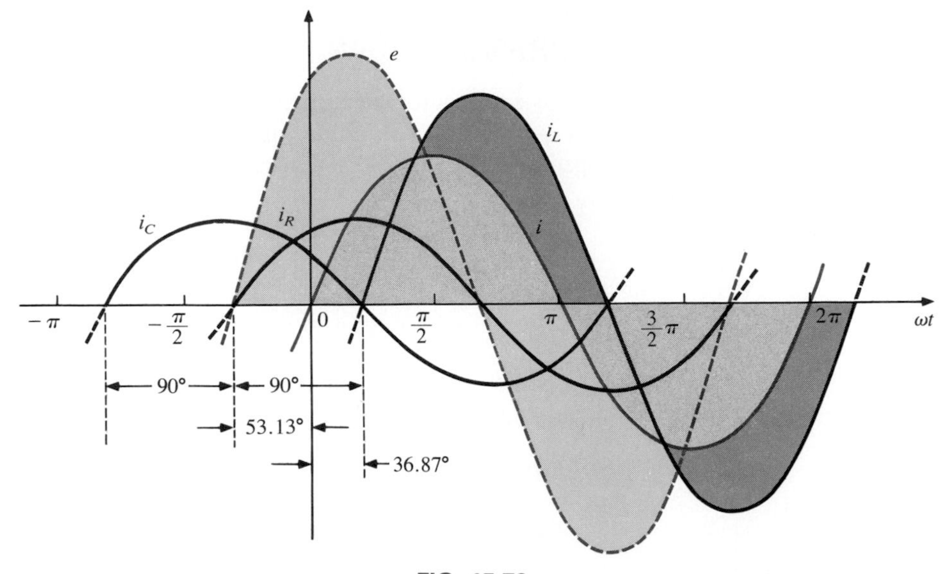

FIG. 15.73

or

$$P_T = E^2G = (100\text{ V})^2(0.3\text{ S}) = \mathbf{3000\ W}$$

or, finally,

$$\begin{aligned} P_T &= P_R + P_L + P_C \\ &= EI_R \cos \theta_R + EI_L \cos \theta_L + EI_C \cos \theta_C \\ &= (100\text{ V})(30\text{ A}) \cos 0^\circ + (100\text{ V})(70\text{ A}) \cos 90^\circ \\ &\quad + (100\text{ V})(30\text{ A}) \cos 90^\circ \\ &= 3000\text{ W} + 0 + 0 \\ &= \mathbf{3000\ W} \end{aligned}$$

*Power factor:* The power factor of the circuit is

$$F_p = \cos \theta_T = \cos 53.13^\circ = \mathbf{0.6\ lagging}$$

Using Eq. (15.26), we obtain

$$F_p = \cos \theta_T = \frac{G}{Y_T} = \frac{0.3\text{ S}}{0.5\text{ S}} = \mathbf{0.6\ lagging}$$

*Impedance approach:* The input current **I** can also be determined by first finding the total impedance in the following manner:

$$\mathbf{Z}_T = \frac{\mathbf{Z}_1\mathbf{Z}_2\mathbf{Z}_3}{\mathbf{Z}_1\mathbf{Z}_2 + \mathbf{Z}_2\mathbf{Z}_3 + \mathbf{Z}_1\mathbf{Z}_3} = \mathbf{2\ \Omega \angle 53.13^\circ}$$

and, applying Ohm's law, we obtain

$$\mathbf{I} = \frac{\mathbf{E}}{\mathbf{Z}_T} = \frac{100\text{ V} \angle 53.13^\circ}{2\ \Omega \angle 53.13^\circ} = \mathbf{50\ A \angle 0^\circ}$$

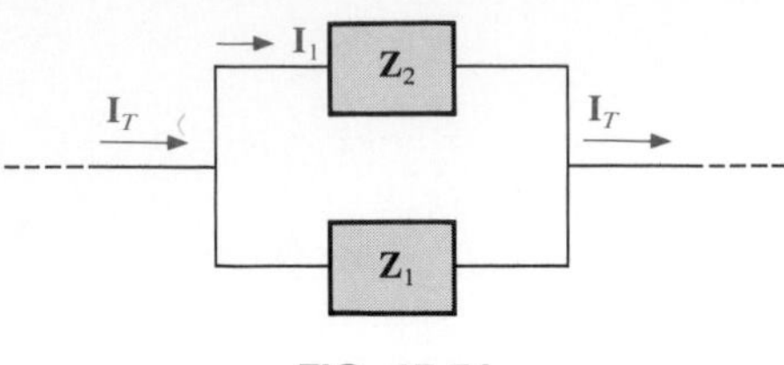


FIG. 15.74

## 15.8 CURRENT DIVIDER RULE

The basic format for the current divider rule in ac circuits is exactly the same as that for dc circuits; that is, for two parallel branches with impedances $\mathbf{Z}_1$ and $\mathbf{Z}_2$ as shown in Fig. 15.74,

$$\mathbf{I}_1 = \frac{\mathbf{Z}_2\mathbf{I}_T}{\mathbf{Z}_1 + \mathbf{Z}_2} \quad \text{or} \quad \mathbf{I}_2 = \frac{\mathbf{Z}_1\mathbf{I}_T}{\mathbf{Z}_1 + \mathbf{Z}_2} \tag{15.29}$$

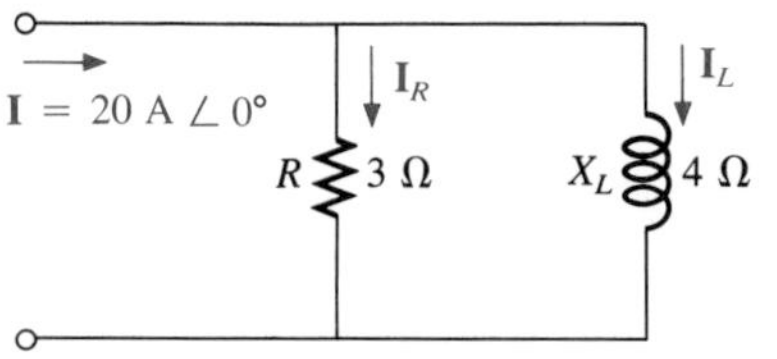

**EXAMPLE 15.15.** Using the current divider rule, find the current through each impedance of Fig. 15.75.

FIG. 15.75

***Solution:***

$$\mathbf{I}_R = \frac{\mathbf{X}_L\mathbf{I}_T}{\mathbf{R} + \mathbf{X}_L} = \frac{(4\ \Omega\ \angle 90^\circ)(20\ \text{A}\ \angle 0^\circ)}{3\ \Omega\ \angle 0^\circ + 4\ \Omega\ \angle 90^\circ} = \frac{80\ \angle 90^\circ}{5\ \angle 53.13^\circ}$$
$$= \mathbf{16\ A\ \angle 36.87^\circ}$$
$$\mathbf{I}_L = \frac{\mathbf{R}\mathbf{I}_T}{\mathbf{R} + \mathbf{X}_L} = \frac{(3\ \Omega\ \angle 0^\circ)(20\ \text{A}\ \angle 0^\circ)}{5\ \Omega\ \angle 53.13^\circ} = \frac{60\ \angle 0^\circ}{5\ \angle 53.13^\circ}$$
$$= \mathbf{12\ A\ \angle -53.13^\circ}$$

**EXAMPLE 15.16.** Using the current divider rule, find the current through each parallel branch of Fig. 15.76.

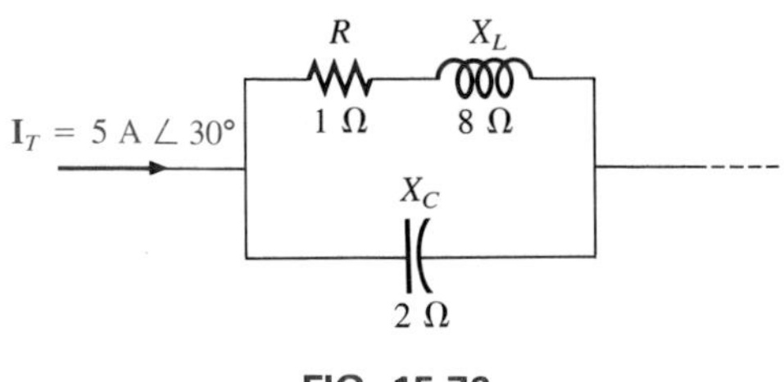

FIG. 15.76

***Solution:***

$$\mathbf{I}_{R\text{-}L} = \frac{\mathbf{X}_C\mathbf{I}_T}{\mathbf{X}_C + \mathbf{Z}_{R\text{-}L}} = \frac{(2\ \Omega\ \angle -90^\circ)(5\ \text{A}\ \angle 30^\circ)}{-j2\ \Omega + 1\ \Omega + j8\ \Omega} = \frac{10\ \angle -60^\circ}{1 + j6}$$
$$= \frac{10\ \angle -60^\circ}{6.083\ \angle 80.54^\circ} \cong \mathbf{1.644\ A\ \angle -140.54^\circ}$$
$$\mathbf{I}_C = \frac{\mathbf{Z}_{R\text{-}L}\mathbf{I}_T}{\mathbf{Z}_{R\text{-}L} + \mathbf{X}_C} = \frac{(1\ \Omega + j8\ \Omega)(5\ \text{A}\ \angle 30^\circ)}{6.08\ \Omega\ \angle 80.54^\circ}$$
$$= \frac{(8.06\ \angle 82.87^\circ)(5\ \angle 30^\circ)}{6.08\ \angle 80.54^\circ} = \frac{40.30\ \angle 112.87^\circ}{6.083\ \angle 80.54^\circ}$$
$$= \mathbf{6.625\ A\ \angle 32.33^\circ}$$

## 15.9 FREQUENCY RESPONSE OF THE PARALLEL *R-L* NETWORK

In Section 15.5 the frequency response of a series *R-C* circuit was analyzed. Let us now note the impact of frequency on the total impedance and inductive current for the parallel *R-L* network of Fig. 15.77 for a frequency range of 0 through 40 kHz.

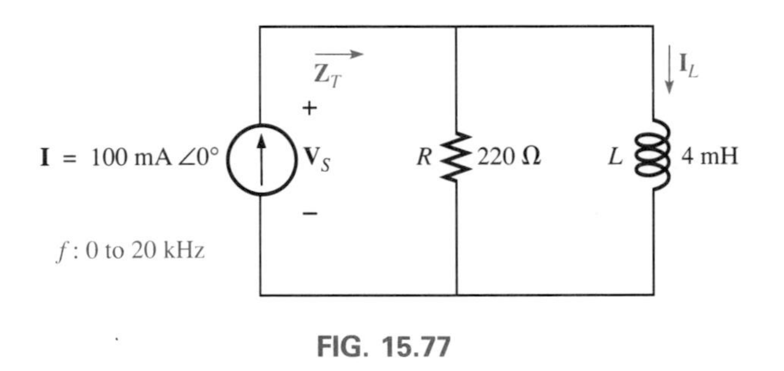


FIG. 15.77

### $\mathbf{Z}_T$:

Before getting into specifics, let us first develop a "sense" for the impact of frequency on the network of Fig. 15.77 by noting the impedance versus frequency curves of the individual elements, as shown in Fig. 15.78. The fact that the elements are now in parallel requires that

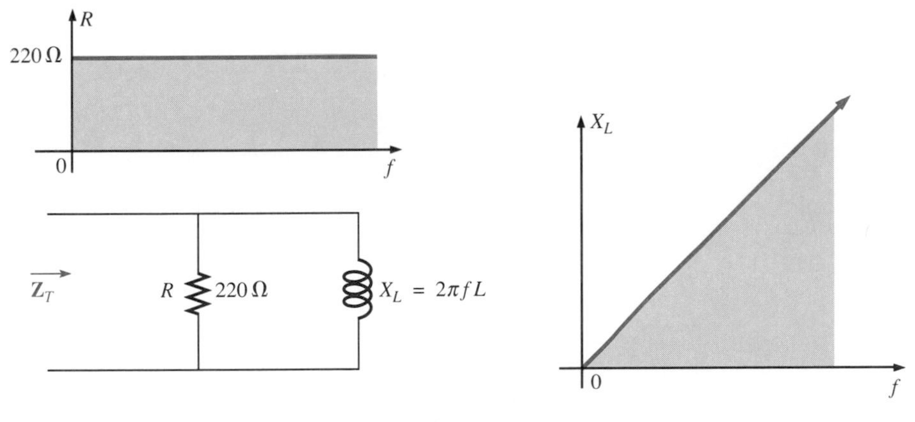

FIG. 15.78

we consider their characteristics in a different manner than occurred for the series *R-C* circuit of Section 15.5. Recall that for parallel elements, the element with the smallest impedance will have the greatest impact on the total impedance at that frequency. In Fig. 15.78, for example, $X_L$ is very small at low frequencies compared to *R*, establishing $X_L$ as the predominant factor in this frequency range. In other words, at low frequencies the network will be primarily inductive and the angle associated with the total impedance will be close to 90°, as with a pure inductor. As the frequency increases, $X_L$ will increase until it equals the impedance of the resistor (220 Ω). The frequency at which this situation occurs can be determined in the following manner:

$$X_L = 2\pi f_2 L = R$$

and

$$\boxed{f_2 = \frac{R}{2\pi L}} \tag{15.30}$$

which for the network of Fig. 15.77 is

$$f_2 = \frac{R}{2\pi L} = \frac{220\ \Omega}{2\pi(4 \times 10^{-3}\ \text{H})}$$
$$\cong \mathbf{8.75\ kHz}$$

falling within the frequency range of interest.

For frequencies less than $f_2$, $X_L < R$, and for frequencies greater than $f_2$, $X_L > R$. A general equation for the total impedance in vector form can be developed in the following manner:

$$\mathbf{Z}_T = \frac{\mathbf{R}\mathbf{X}_L}{\mathbf{R} + \mathbf{X}_L}$$
$$= \frac{(R \angle 0°)(X_L \angle 90°)}{R + jX_L} = \frac{RX_L \angle 90°}{\sqrt{R^2 + X_L^2} \angle \tan^{-1} X_L/R}$$

and

$$\boxed{\mathbf{Z}_T = Z_T \angle \theta = \frac{RX_L}{\sqrt{R^2 + X_L^2}} \angle 90° - \tan^{-1} X_L/R} \tag{15.31}$$

The magnitude and angle of the total impedance can now be found at any frequency of interest simply by substituting into Eq. (15.31).

***f* = 1 kHz:**

$$X_L = 2\pi f L = 2\pi(1\ \text{kHz})(4 \times 10^{-3}\ \text{H}) = 25.12\ \Omega$$

and

$$Z_T = \frac{RX_L}{\sqrt{R^2 + X_L^2}} = \frac{(220\ \Omega)(25.12\ \Omega)}{\sqrt{(220\ \Omega)^2 + (25.12\ \Omega)^2}} = 24.96\ \Omega$$

with

$$\theta_T = 90° - \tan^{-1}\frac{X_L}{R} = 90° - \tan^{-1}\frac{25.12\ \Omega}{220\ \Omega}$$
$$= 90° - \tan^{-1} 0.114 = 83.49°$$

and

$$\mathbf{Z}_T = \mathbf{24.96\ \Omega\ \angle 83.49°}$$

This value compares very closely with $X_L = 25.12\ \Omega\ \angle 90°$, which it would be if the network were purely inductive ($R = \infty\ \Omega$). Our assumption that the network is primarily inductive at low frequencies is therefore confirmed.

Continuing:

$$f = 5\text{ kHz:}\quad \mathbf{Z}_T = \mathbf{109.1\ \Omega\ \angle 60.23°}$$
$$f = 10\text{ kHz:}\quad \mathbf{Z}_T = \mathbf{165.5\ \Omega\ \angle 41.21°}$$
$$f = 15\text{ kHz:}\quad \mathbf{Z}_T = \mathbf{189.99\ \Omega\ \angle 30.28°}$$
$$f = 20\text{ kHz:}\quad \mathbf{Z}_T = \mathbf{201.53\ \Omega\ \angle 23.65°}$$
$$f = 30\text{ kHz:}\quad \mathbf{Z}_T = \mathbf{211.19\ \Omega\ \angle 16.27°}$$
$$f = 40\text{ kHz:}\quad \mathbf{Z}_T = \mathbf{214.91\ \Omega\ \angle 12.35°}$$

At $f = 40$ kHz, note how closely the magnitude of $Z_T$ has approached the resistance level of 220 Ω and how the associated angle with the total impedance is approaching zero degrees. The result is a network with terminal characteristics that are becoming more and more resistive as the frequency increases, which further confirms the earlier conclusions developed by the curves of Fig. 15.78.

Plots of $Z_T$ versus frequency in Fig. 15.79 and $\theta_T$ in Fig. 15.80 clearly reveal the transition from an inductive network to one that has

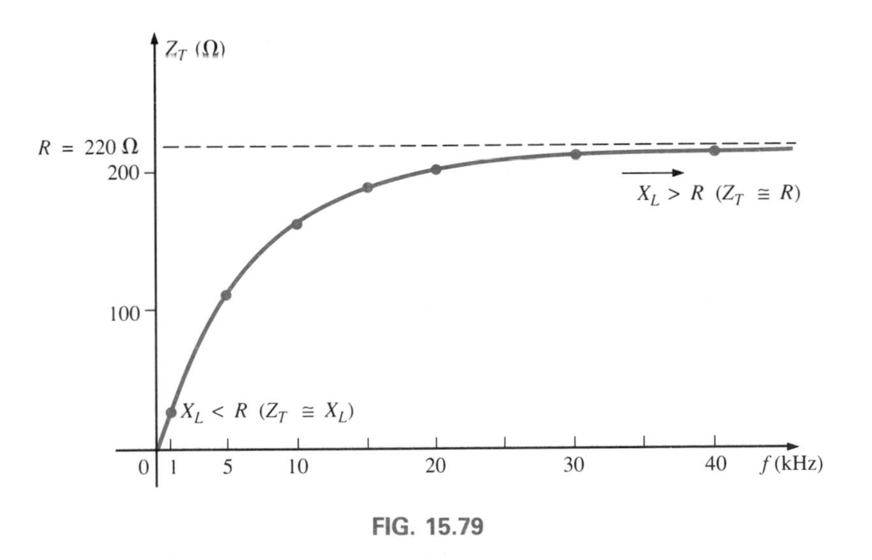


FIG. 15.79

resistive characteristics. Note that the transition frequency of 8.75 kHz occurs right in the middle of the knee of the curves for both $Z_T$ and $\theta_T$.

A review of Figs. 15.46 and 15.79 will reveal that a series $R$-$C$ and a parallel $R$-$L$ network will have an impedance level that approaches the resistance of the network at high frequencies. The capacitive circuit

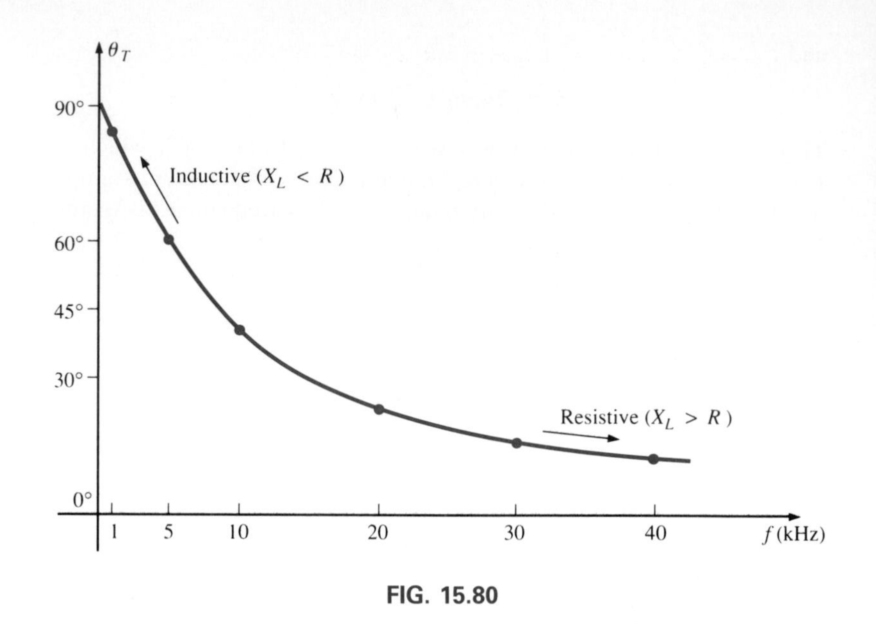


**FIG. 15.80**

approaches the level from above, whereas the inductive network does the same from below. For the series *R-L* circuit and the parallel *R-C* network, the total impedance will begin at the resistance level and then display the characteristics of the reactive elements at high frequencies.

**$\mathbf{I}_L$:**

Applying the current divider rule to the network of Fig. 15.77 will result in the following:

$$\mathbf{I}_L = \frac{\mathbf{R}\mathbf{I}}{\mathbf{R} + \mathbf{X}_L}$$

$$= \frac{(R \angle 0^\circ)(I \angle 0^\circ)}{R + jX_L} = \frac{RI \angle 0^\circ}{\sqrt{R^2 + X_L^2}} \angle \tan^{-1} X_L/R$$

and

$$\mathbf{I}_L = I_L \angle \theta_L = \frac{RI}{\sqrt{R^2 + X_L^2}} \angle -\tan^{-1} X_L/R$$

The magnitude of $I_L$ is therefore determined by

$$I_L = \frac{RI}{\sqrt{R^2 + X_L^2}} \tag{15.32}$$

and the phase angle $\theta_L$ by which $\mathbf{I}_L$ leads $\mathbf{I}$ is given by

$$\theta_L = -\tan^{-1} \frac{X_L}{R} \tag{15.33}$$

Because $\theta_L$ is always negative, the magnitude of $\theta_L$ is, in actuality, the angle by which $\mathbf{I}_L$ lags $\mathbf{I}$.

To begin our analysis, let us first consider the case of $f = 0$ Hz (dc conditions).

**$f = 0$ Hz:**

$$X_L = 2\pi fL = 2\pi(0 \text{ Hz})L = 0\ \Omega$$

Applying the short-circuit equivalent for the inductor in Fig. 15.77 would result in

$$\mathbf{I}_L = \mathbf{I} = 100 \text{ mA} \angle 0°$$

as appearing in Figs. 15.81 and 15.82.

**$f = 1$ kHz:**

Applying Eq. (15.32):

$$X_L = 2\pi fL = 2\pi(1 \text{ kHz})(4 \text{ mH}) = 25.12\ \Omega$$

and

$$\sqrt{R^2 + X_L^2} = \sqrt{(220\ \Omega)^2 + (25.12\ \Omega)^2} = 221.43\ \Omega$$

and

$$I_L = \frac{RI}{\sqrt{R^2 + X_L^2}} = \frac{(220\ \Omega)(100 \text{ mA})}{221.43\ \Omega} = 99.35 \text{ mA}$$

with

$$\theta_L = \tan^{-1}\frac{X_L}{R} = -\tan^{-1}\frac{25.12\ \Omega}{220\ \Omega} = -\tan^{-1} 0.114 = -6.51°$$

and

$$\mathbf{I}_L = \mathbf{99.35 \text{ mA} \angle -6.51°}$$

The result is a current $\mathbf{I}_L$ that is still very close to the source current $\mathbf{I}$ in both magnitude and phase.

Continuing:

$$f = 5 \text{ kHz:} \quad \mathbf{I}_L = \mathbf{86.84 \text{ mA} \angle -29.72°}$$
$$f = 10 \text{ kHz:} \quad \mathbf{I}_L = \mathbf{65.88 \text{ mA} \angle -48.79°}$$
$$f = 15 \text{ kHz:} \quad \mathbf{I}_L = \mathbf{50.43 \text{ mA} \angle -59.72°}$$
$$f = 20 \text{ kHz:} \quad \mathbf{I}_L = \mathbf{40.11 \text{ mA} \angle -66.35°}$$
$$f = 30 \text{ kHz:} \quad \mathbf{I}_L = \mathbf{28.02 \text{ mA} \angle -73.73°}$$
$$f = 40 \text{ kHz:} \quad \mathbf{I}_L = \mathbf{21.38 \text{ mA} \angle -77.65°}$$

The plot of the magnitude of $I_L$ versus frequency is provided in Fig. 15.81 and reveals that the current through the coil dropped from its maximum of 100 mA to almost 20 mA at 40 kHz. As the reactance of the coil increased with frequency, more of the source current chose the lower-resistance path of the resistor. The magnitude of the phase angle between $\mathbf{I}_L$ and $\mathbf{I}$ is approaching 90° with increase in frequency, as

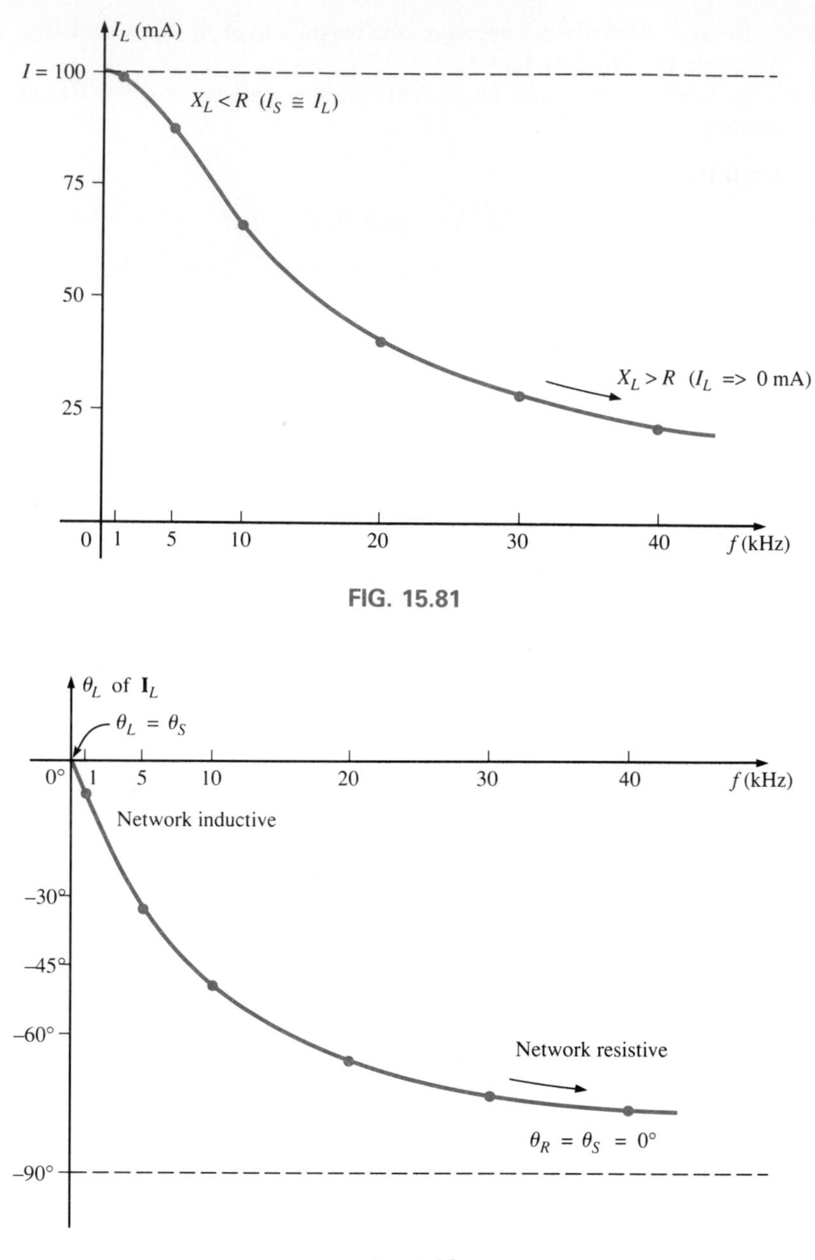


FIG. 15.81



FIG. 15.82

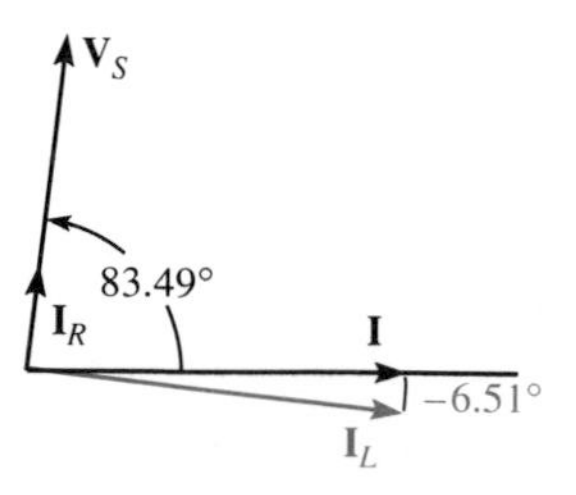


FIG. 15.83
*f = 1 kHz.*

shown in Fig. 15.82, leaving its initial value of zero degrees at $f = 0$ Hz far behind.

At $f = 1$ kHz the phasor diagram of the network appears as shown in Fig. 15.83. First note that the magnitude and phase angle of $\mathbf{I}_L$ is very close to that of $\mathbf{I}$. Since the voltage across a coil must lead the current through a coil by 90°, the voltage $\mathbf{V}_S$ appears as shown. The voltage across a resistor is in phase with the current through the resistor, resulting in the direction of $\mathbf{I}_R$ shown in Fig. 15.83. Of course, at this frequency $R > X_L$, and the current $I_R$ is relatively small in magnitude.

At $f = 40$ kHz the phasor diagram changes to that appearing in Fig. 15.84. Note that now $\mathbf{I}_R$ and $\mathbf{I}$ are close in magnitude and phase because $X_L > R$. The magnitude of $\mathbf{I}_L$ has dropped to very low levels and the phase angle associated with $\mathbf{I}_L$ is approaching $-90°$. The network is now more ''resistive'' as compared to its ''inductive'' characteristics at low frequencies.

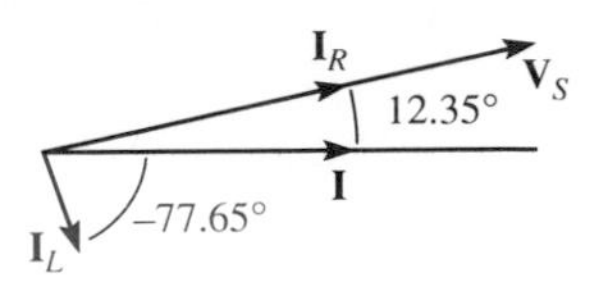


**FIG. 15.84**
*f = 40 kHz.*

The analysis of a parallel *R-C* or *R-L-C* network would proceed in much the same manner, with the inductive impedance predominating at low frequencies and the capacitive reactance at high frequencies. Keep in mind through the analysis that the impedance with the least impedance is the predominant factor in a parallel network, whereas the reverse is true in a series circuit.

## 15.10 EQUIVALENT CIRCUITS

In a series ac circuit, the total impedance of two or more elements in series is often equivalent to an impedance that can be achieved with fewer elements of different values, the elements and their values being determined by the frequency applied. This is also true for parallel circuits. For the circuit of Fig. 15.85(a),

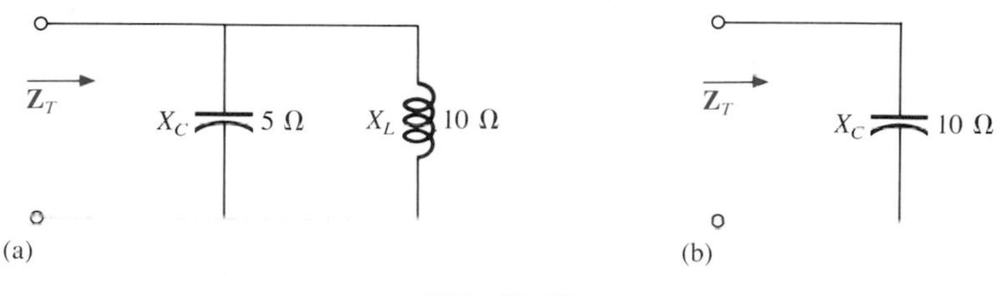


**FIG. 15.85**

$$\mathbf{Z}_T = \frac{\mathbf{X}_C\mathbf{X}_L}{\mathbf{X}_C + \mathbf{X}_L} = \frac{(5\ \Omega\ \angle -90°)(10\ \Omega\ \angle 90°)}{5\ \Omega\ \angle -90° + 10\ \Omega\ \angle 90°} = \frac{50\ \angle 0°}{5\ \angle 90°}$$
$$= 10\ \Omega\ \angle -90°$$

The total impedance at the frequency applied is equivalent to a capacitor with a reactance of 10 Ω, as shown in Fig. 15.85(b). Always keep in mind that this equivalence is true only at the applied frequency. If the frequency changes, the reactance of each element changes, and the equivalent circuit will change—perhaps from capacitive to inductive in the above example.

Another interesting development appears if the impedance of a parallel circuit, such as the one of Fig. 15.86(a), is found in rectangular coordinates. In this case,

$$\mathbf{Z}_T = \frac{\mathbf{X}_L\mathbf{R}}{\mathbf{X}_L + \mathbf{R}} = \frac{(4\ \Omega\ \angle 90°)(3\ \Omega\ \angle 0°)}{4\ \Omega\ \angle 90° + 3\ \Omega\ \angle 0°}$$
$$= \frac{12\ \angle 90°}{5\ \angle 53.13°} = 2.40\ \Omega\ \angle 36.87°$$
$$= 1.920\ \Omega + j1.440\ \Omega$$

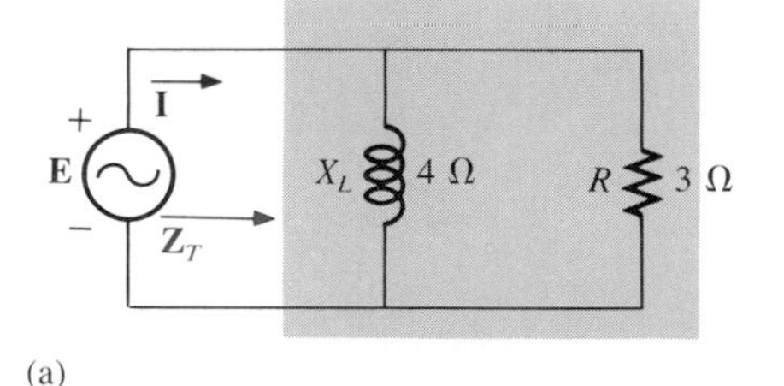


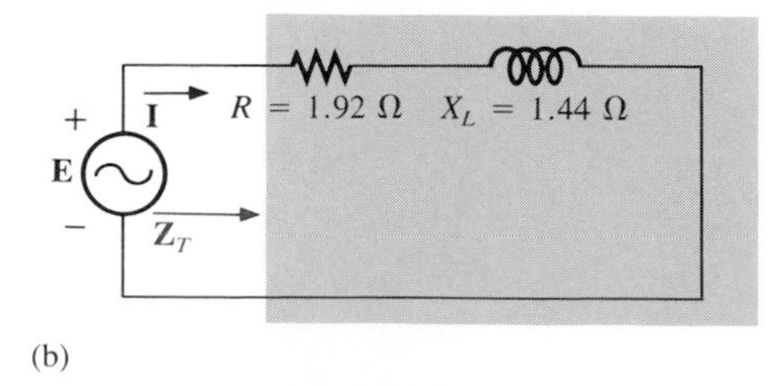

**FIG. 15.86**

which is the impedance of a series circuit with a resistor of 1.92 Ω and an inductive reactance of 1.44 Ω, as shown in Fig. 15.86(b).

The current **I** will be the same in each circuit of Fig. 15.85 or Fig. 15.86 if the same input voltage **E** is applied. For a parallel circuit of one resistive element and one reactive element, the series circuit with the same input impedance will always be composed of one resistive and one reactive element. The impedance of each element of the series circuit will be different from that of the parallel circuit, but the reactive elements will always be of the same type; that is, an *R-L* circuit and an *R-C* parallel circuit will have an equivalent *R-L* and *R-C* series circuit, respectively. The same is true when converting from a series to a parallel circuit. In the discussion to follow, keep in mind that the term *equivalent*

***refers only to the fact that for the same applied potential, the same impedance and input current will result.***

To formulate the equivalence between the series and parallel circuits, the equivalent series circuit for a resistor and reactance in parallel can be found by determining the total impedance of the circuit in rectangular form; that is, for the circuit of Fig. 15.87(a),

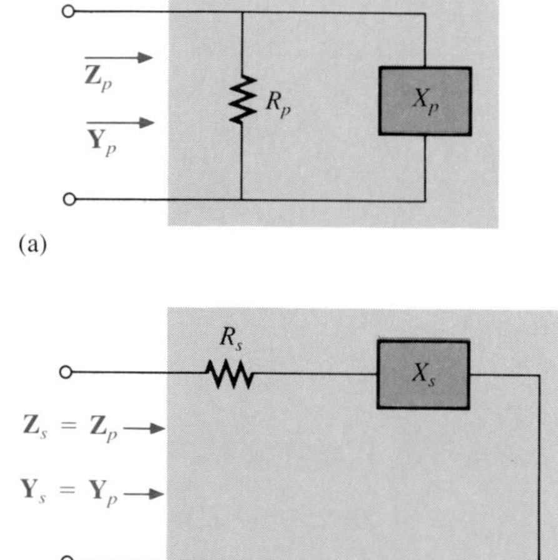


FIG. 15.87

$$\mathbf{Y}_p = \frac{1}{R_p} + \frac{1}{\pm jX_p}$$

and

$$\mathbf{Z}_p = \frac{1}{\mathbf{Y}_p} = \frac{1}{(1/R_p) \mp j(1/X_p)}$$
$$= \frac{1/R_p}{(1/R_p)^2 + (1/X_p)^2} \pm j\frac{1/X_p}{(1/R_p)^2 + (1/X_p)^2}$$

Multiplying the numerator and denominator of each term by $R_p^2X_p^2$ results in

$$\mathbf{Z}_p = \frac{R_pX_p^2}{X_p^2 + R_p^2} \pm j\frac{R_p^2X_p}{X_p^2 + R_p^2}$$
$$= R_s \pm jX_s \qquad \text{[Fig. 15.87(b)]}$$

and

$$\boxed{R_s = \frac{R_pX_p^2}{X_p^2 + R_p^2}} \qquad \textbf{(15.34)}$$

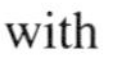

with

$$\boxed{X_s = \frac{R_p^2X_p}{X_p^2 + R_p^2}} \qquad \textbf{(15.35)}$$

For the network of Fig. 15.86,

$$R_s = \frac{R_p X_p^2}{X_p^2 + R_p^2} = \frac{(3\ \Omega)(4\ \Omega)^2}{(4\ \Omega)^2 + (3\ \Omega)^2} = \frac{48\ \Omega}{25} = \mathbf{1.920\ \Omega}$$

and

$$X_s = \frac{R_p^2 X_p}{X_p^2 + R_p^2} = \frac{(3\ \Omega)^2(4\ \Omega)}{(4\ \Omega)^2 + (3\ \Omega)^2} = \frac{36\ \Omega}{25} = \mathbf{1.440\ \Omega}$$

which agrees with the previous result.

The equivalent parallel circuit for a circuit with a resistor and reactance in series can be found by simply finding the total admittance of the system in rectangular form; that is, for the circuit of Fig. 15.87(b),

$$\mathbf{Z}_s = R_s \pm jX_s$$

$$\mathbf{Y}_s = \frac{1}{\mathbf{Z}_s} = \frac{1}{R_s \pm jX_s} = \frac{R_s}{R_s^2 + X_s^2} \mp j\frac{X_s}{R_s^2 + X_s^2}$$

$$= G_p \mp jB_p = \frac{1}{R_p} \mp j\frac{1}{X_p} \qquad \text{[Fig. 15.87(a)]}$$

or

$$\boxed{R_p = \frac{R_s^2 + X_s^2}{R_s}} \qquad (15.36)$$

with

$$\boxed{X_p = \frac{R_s^2 + X_s^2}{X_s}} \qquad (15.37)$$

For the above example,

$$R_p = \frac{R_s^2 + X_s^2}{R_s} = \frac{(1.92\ \Omega)^2 + (1.44\ \Omega)^2}{1.92\ \Omega} = \frac{5.76\ \Omega}{1.92} = \mathbf{3.0\ \Omega}$$

and

$$X_p = \frac{R_s^2 + X_s^2}{X_s} = \frac{5.76\ \Omega}{1.44} = \mathbf{4.0\ \Omega}$$

as shown in Fig. 15.86(a).

**EXAMPLE 15.17.** Determine the series equivalent circuit for the network of Fig. 15.88.

***Solution:***

$$R_p = 8\ \text{k}\Omega$$

$$X_p\ \text{(resultant)} = |X_L - X_C| = |9\ \text{k}\Omega - 4\ \text{k}\Omega|$$
$$= 5\ \text{k}\Omega$$

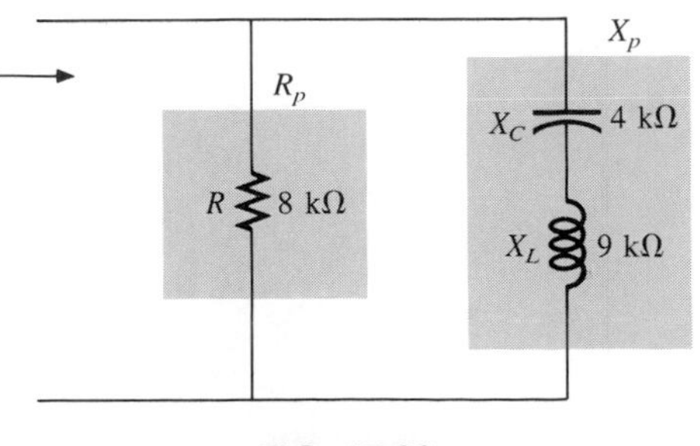


**FIG. 15.88**

and

$$R_s = \frac{R_p X_p^2}{X_p^2 + R_p^2} = \frac{(8\ \text{k}\Omega)(5\ \text{k}\Omega)^2}{(5\ \text{k}\Omega)^2 + (8\ \text{k}\Omega)^2} = \frac{200\ \text{k}\Omega}{89} = \mathbf{2.247\ k\Omega}$$

with

$$X_s = \frac{R_p^2 X_p}{X_p^2 + R_p^2} = \frac{(8\ \text{k}\Omega)^2(5\ \text{k}\Omega)}{(5\ \text{k}\Omega)^2 + (8\ \text{k}\Omega)^2} = \frac{320\ \text{k}\Omega}{89}$$

$$= \mathbf{3.596\ k\Omega} \qquad \textbf{(inductive)}$$

The equivalent series circuit appears in Fig. 15.89.

FIG. 15.89

**EXAMPLE 15.18.** For the network of Fig. 15.90:

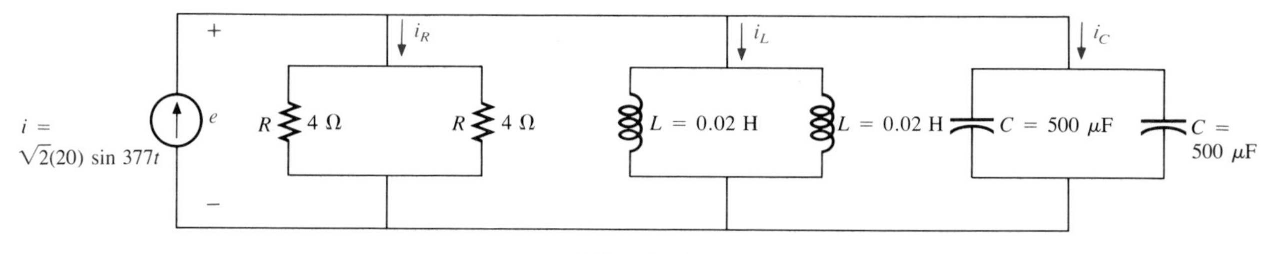

FIG. 15.90

a. Compute $e$, $i_R$, $i_L$, and $i_C$ in phasor form.
b. Compute the total power factor.
c. Compute the total power delivered to the network.
d. Draw the phasor diagram.
e. Obtain the phasor sum of $\mathbf{I}_R$, $\mathbf{I}_L$, and $\mathbf{I}_C$, and show that it equals $\mathbf{I}$.
f. Compute the impedance of the parallel combination of $\mathbf{X}_L$ and $\mathbf{X}_C$, and then find $\mathbf{I}_R$ by the current divider rule.
g. Determine the equivalent series circuit as far as the total impedance and current $\mathbf{I}$ are concerned.

***Solutions:***

a. Combining common elements and finding the reactance of the inductor and capacitor, we obtain

$$R_T = \frac{R}{2} = \frac{4\ \Omega}{2} = 2\ \Omega$$

$$L_T = \frac{0.02\ \text{H}}{2} = 0.01\ \text{H}$$

$$C_T = 500\ \mu\text{F} + 500\ \mu\text{F} = 1000\ \mu\text{F}$$

$$X_L = \omega L = (377\ \text{rad/s})(0.01\ \text{H}) = 3.77\ \Omega$$

$$X_C = \frac{1}{\omega C} = \frac{1}{(377\ \text{rad/s})(10^3 \times 10^{-6}\ \text{F})} = 2.65\ \Omega$$

The network is redrawn in Fig. 15.91 with phasor notation. The total admittance is

$$\mathbf{Y}_T = \mathbf{Y}_1 + \mathbf{Y}_2 + \mathbf{Y}_3 = \mathbf{G} + \mathbf{B}_L + \mathbf{B}_C$$

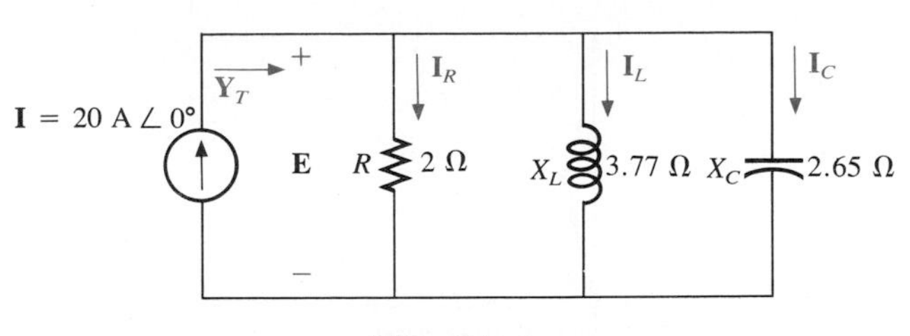


FIG. 15.91

$$= \frac{1}{2\ \Omega}\angle 0° + \frac{1}{3.77\ \Omega}\angle -90° + \frac{1}{2.65\ \Omega}\angle +90°$$
$$= 0.5\ \text{S}\ \angle 0° + 0.265\ \text{S}\ \angle -90° + 0.377\ \text{S}\ \angle +90°$$
$$= 0.5\ \text{S} - j0.265\ \text{S} + j0.377\ \text{S}$$
$$= 0.5\ \text{S} + j0.112\ \text{S} = \mathbf{0.512\ S\ \angle 12.63°}$$

The input voltage is

$$\mathbf{E} = \frac{\mathbf{I}}{\mathbf{Y}_T} = \frac{20\ \text{A}\ \angle 0°}{0.512\ \text{S}\ \angle 12.63°}$$
$$= \mathbf{39.06\ V\ \angle -12.63°}$$

The current through the resistor, inductor, and capacitor can be found using Ohm's law as follows:

$$\mathbf{I}_R - \mathbf{EG} - (39.06\ \text{V}\ \angle -12.63°)(0.5\ \text{S}\ \angle 0°)$$
$$= \mathbf{19.53\ \angle -12.63°}$$
$$\mathbf{I}_L = \mathbf{EB}_L = (39.06\ \text{V}\ \angle -12.63°)(0.265\ \text{S}\ \angle -90°)$$
$$= \mathbf{10.35\ \angle -102.63°}$$
$$\mathbf{I}_C = \mathbf{EB}_C = (39.06\ \text{V}\ \angle -12.63°)(0.377\ \text{S}\ \angle +90°)$$
$$= \mathbf{14.73\ \angle +77.37°}$$

b. The total power factor is

$$F_p = \cos\theta = \frac{G}{Y_T} = \frac{0.5\ \text{S}}{0.512\ \text{S}} = \mathbf{0.977}$$

c. The total power in watts delivered to the circuit is

$$P_T = I_R^2 R = E^2 G = (39.06\ \text{V})^2(0.5\ \text{S})$$
$$= \mathbf{762.84\ W}$$

d. The phasor diagram is shown in Fig. 15.92.
e. The phasor sum of $\mathbf{I}_R$, $\mathbf{I}_L$, and $\mathbf{I}_C$ is

$$I = I_R + \mathbf{I}_L + \mathbf{I}_C$$
$$= 19.53\ \text{A}\ \angle -12.63° + 10.35\ \text{A}\ \angle -102.63°$$
$$+ 14.73\ \text{A}\ \angle +77.37°$$
$$= 19.53\ \text{A}\ \angle -12.63° + 4.38\ \text{A}\ \angle +77.37°$$
$$I_T = \sqrt{(19.53\ \text{A})^2 + (4.38\ \text{A})^2} = 20\ \text{A}$$

and $\theta_T$ (from phasor diagram) = 0°. Therefore,

$$\mathbf{I}_T = \mathbf{20\ A\ \angle 0°}$$

which agrees with the original input.

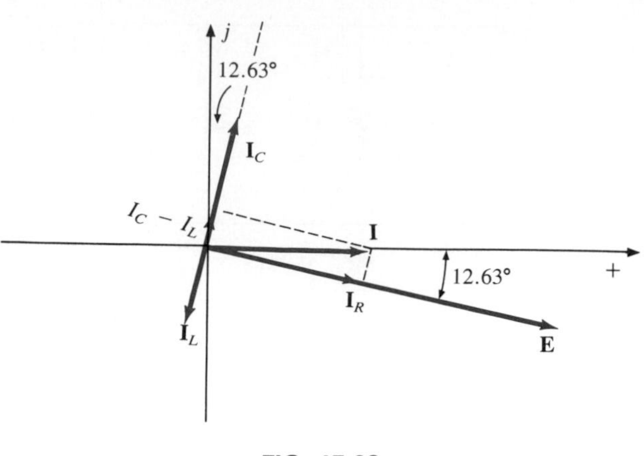


FIG. 15.92

f. The impedance of the parallel combination of $\mathbf{X}_L$ and $\mathbf{X}_C$ is

$$\mathbf{Z}_{T_1} = \frac{\mathbf{X}_L\mathbf{X}_C}{\mathbf{X}_L + \mathbf{X}_C} = \frac{(3.77\ \Omega\ \angle 90^\circ)(2.65\ \Omega\ \angle -90^\circ)}{3.77\ \Omega\ \angle 90^\circ + 2.65\ \Omega\ \angle -90^\circ}$$

$$= \frac{10\ \angle 0^\circ}{1.12\ \angle 90^\circ}$$

$$= 8.93\ \Omega\ \angle -90^\circ$$

The current $\mathbf{I}_R$, using the current divider rule, is

$$\mathbf{I}_R = \frac{\mathbf{Z}_{T_1}\mathbf{I}}{\mathbf{Z}_{T_1} + \mathbf{R}} = \frac{(8.93\ \Omega\ \angle -90^\circ)(20\ \text{A}\ \angle 0^\circ)}{8.93\ \Omega\ \angle -90^\circ + 2\ \Omega\ \angle 0^\circ}$$

$$= \frac{178.60\ \angle -90^\circ}{9.15\ \angle -77.37^\circ} = \mathbf{19.52\ A\ \angle -12.63^\circ}$$

g. $$\mathbf{Z}_T = \frac{1}{\mathbf{Y}_T} = \frac{1}{0.512\ \text{S}\ \angle 12.63^\circ} = \mathbf{1.95\ \Omega\ \angle -12.63^\circ}$$

which, in rectangular form, is $1.90\ \Omega - j0.427\ \Omega$, and

$$C = \frac{1}{\omega X_C} = \frac{1}{(377\ \text{rad/s})(0.427\ \Omega)} = 0.006212\ \text{F} = \mathbf{6212\ \mu F}$$

The series circuit appears in Fig. 15.93. Since $Z_{T_1}$ is available from part (f) above, we can apply Eqs. (15.34) and (15.35). That is, $Z_{T_1} = X_p$, and

$$R_s = \frac{R_p X_p^2}{X_p^2 + R_p^2} = \frac{(2\ \Omega)(8.93\ \Omega)^2}{(8.93\ \Omega)^2 + (2\ \Omega)^2} = \frac{159.49\ \Omega}{83.74} = \mathbf{1.90\ \Omega}$$

with

$$X_s = \frac{R_p^2 X_p}{X_p^2 + R_p^2} = \frac{(2\ \Omega)^2(8.93\ \Omega)}{(8.93\ \Omega)^2 + (2\ \Omega)^2} = \frac{35.72\ \Omega}{83.74} = \mathbf{0.427\ \Omega}$$

as obtained above.

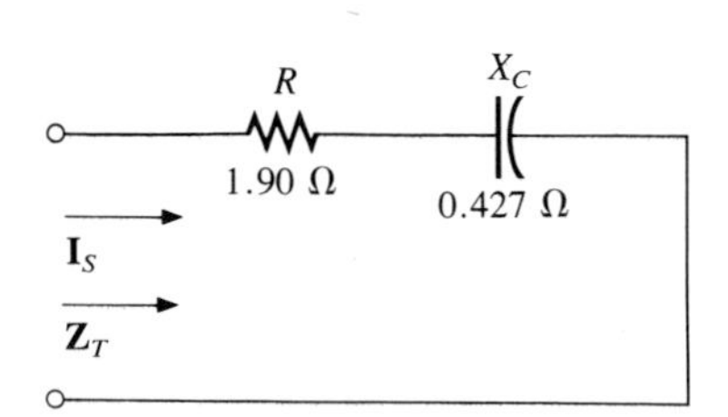


FIG. 15.93

## 15.11 COMPUTER ANALYSIS

We have now covered sufficient material to make effective use of the PSPICE software package. Although all the examples to appear in this section can be solved using BASIC, this section will be limited to PSPICE and the commands required to apply the package to ac systems.

## PSPICE

As with the analysis just described, there is a great deal of similarity between the application of PSPICE to dc and ac systems. The basic format of the input file is similar with passive elements ($R$, $L$, and $C$) entered in exactly the same manner. The two elements that have the most impact on the command statements are the frequency or frequencies of interest and the phase angles applied and obtained, as demonstrated by the following commands.

### Independent ac Sources

The term *independent* (as described more than once in the text) simply means that the magnitude of the source is independent of any other parameters of the network. Since effective values are the most frequently applied and measured, the magnitude associated with all sinusoidal voltages and currents will be the effective value. The format for an independent ac voltage source is the following:

| VSOURCE | 1 | 0 | AC | 20V | 0 |
|---|---|---|---|---|---|
| | + | − | | Magnitude | Phase angle |
| Name | Node | Node | | ACMAG | ACPHASE |

As with the commands for dc networks, the quantities with brackets are those controlled by the user. The others must appear as shown above in the same relative positions.

For a current source $\mathbf{I}_L = 10\text{ mA} \angle 30°$ (whose arrow within the graphic symbol points from node 3 to 4), the following command would be included in the input file:

```
I2    3    4    AC    10MA    30
```

If the ACMAG entry is left blank, a default value of 1 V is applied, whereas the absence of an ACPHASE entry will result in a default value of zero degrees.

Note that the command statement for an independent ac source does not include the applied frequency. The effect of frequency is relegated to the next command.

### .AC Command

Recall the importance of the .DC statement for specifying sweep values and permitting control of the output format. The .AC command provides the same control for ac networks, specifying the frequency of frequencies of interest. The basic format of the command is the following:

```
.AC    LIN    11        1KH        2KH
```

$\underbrace{11}_{\text{Number of data points}}$ $\underbrace{1KH}_{\text{Starting frequency}}$ $\underbrace{2KH}_{\text{Final frequency}}$

As before, the quantities with brackets are controlled by the user. The LIN notation is an abbreviation for the word *linear,* which specifies that the increment between frequencies to be applied be a fixed value. In other words, the spacing between frequencies must be the same from the initial to final values. The choice of the number of data points is an important one if the output file is to contain the desired results. In this light, let us take a moment to be absolutely sure we understand how to determine the number of data points. In Fig. 15.94, the initial ($f_I$) and

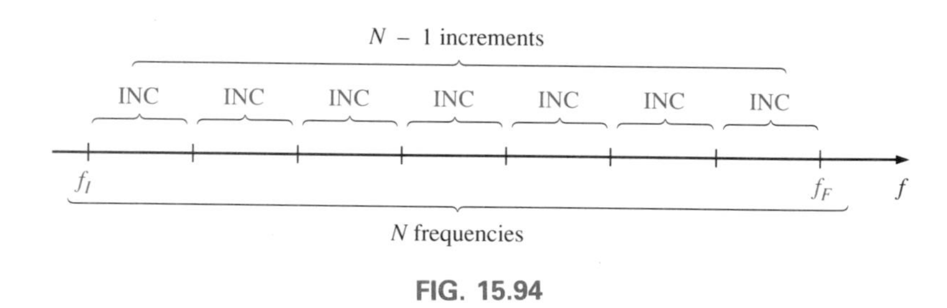


**FIG. 15.94**

final ($f_F$) frequencies have been specified. Let us define $N$ as the total number of data points to be determined between $f_I$ and $f_F$, *including* $f_I$ and $f_F$. Defining INC as the increment between frequencies will result in $N - 1$ increments to progress from $f_I$ to $f_F$. That is,

$$f_I + (N - 1)(\text{INC}) = f_F$$

Solving for the resulting increment:

$$\boxed{\text{INC (in Hz)} = \frac{f_F - f_I}{N - 1}} \qquad \textbf{(15.38)}$$

For the preceding command statement,

$$\text{INC} = \frac{2000\text{ Hz} - 1000\text{ Hz}}{11 - 1} = \frac{1000\text{ Hz}}{10} = 100\text{ Hz}$$

Fig. 15.95 provides a list of the frequencies to be applied for the preceding command. Note that there are 11 distinct frequencies, and the increment between each is 100 Hz.

If the analysis is to be performed at a fixed frequency such as 1 kHz, the command statement will be entered as

```
.AC    LIN    1    1KH    1KH
```

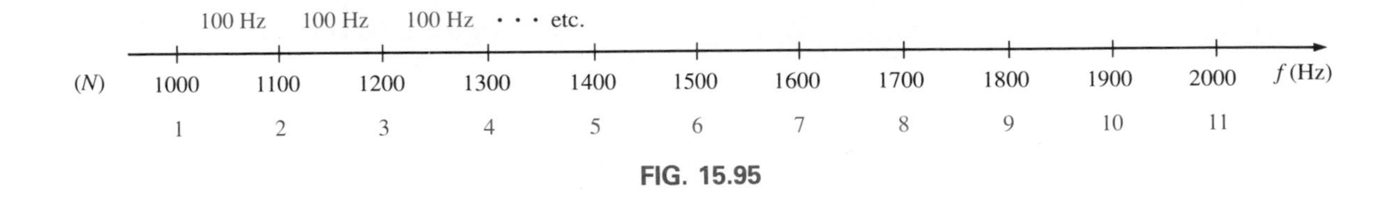


FIG. 15.95

## .PRINT Command

The last command required to perform a basic ac analysis using PSPICE is the .PRINT statement. It has the following general format:

```
.PRINT    AC    VM(2)        VP(2)
```

VM(2): Magnitude of the voltage V(2); VP(2): Phase angle of the voltage V(2)

The quantities within the brackets are controlled by the user and can define a voltage or current. The uppercase letter M specifies that the magnitude of the voltage V(2) be printed and the uppercase letter P specifies that the phase angle of the same voltage appear in the output file.

## The Inductor

The first ac system to be examined using the PSPICE package is the isolated inductor of Fig. 15.96(a). However, due to a restriction of

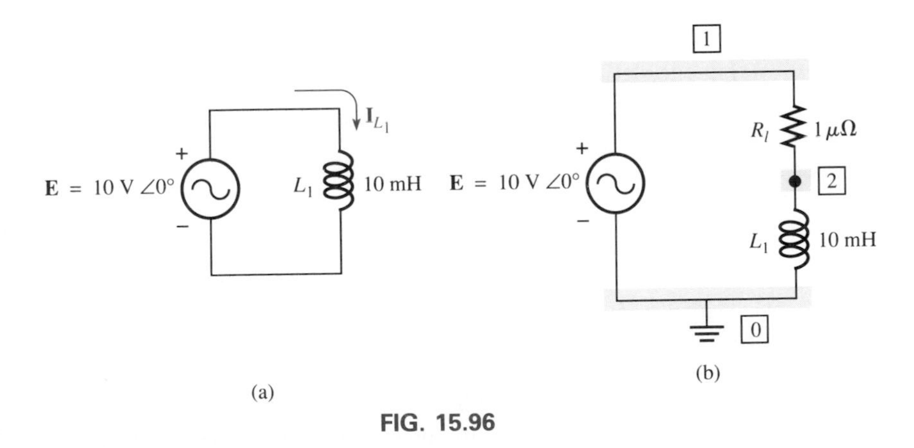


FIG. 15.96

PSPICE that does not permit a loop of inductors and/or voltage sources (parallel combination) a resistor must be added in series with the inductor, as shown in Fig. 15.96(b). The magnitude of the resistor ($1 \times 10^{-6}\ \Omega$) is sufficiently small to be ignored for the frequency range of interest, and the voltages $V_1$ and $V_2$ are essentially the same. The first three lines of the input file of Fig. 15.97 should now be clear based on the preceding discussion. The .AC line specifies an analysis at a fixed

```
************************ Evaluation PSpice (January 1989) ******* 16:30:02 *******

Chapter 15 - AC Inductor Circuit (Fixed Frequency)

****     CIRCUIT DESCRIPTION

**********************************************************************************

VE 1 0 AC 20V 0
RL 1 2 1M
L1 2 0 10MH
.AC LIN 1 1KH 1KH
.PRINT AC IM(L1) IP(L1)
.OPTIONS NOPAGE
.END
```

```
****     AC ANALYSIS                              TEMPERATURE =   27.000 DEG C

  FREQ          IM(L1)       IP(L1)

   1.000E+03    3.183E-01   -9.000E+01
```

**FIG. 15.97**

frequency of 1 kHz. The .PRINT command requests both the magnitude and phase angle of the current through the coil. Before reviewing the output file, let us first perform the analysis in longhand:

$$X_L = 2\pi fL = 2\pi(1\text{ kHz})(10\text{ mH}) = 62.83\ \Omega$$

and

$$\mathbf{I}_L = \frac{\mathbf{E}}{\mathbf{X}_L} = \frac{20\text{ V}\ \angle 0^\circ}{62.83\ \Omega\ \angle 90^\circ} = \mathbf{318.32\ mA\ \angle -90^\circ}$$

As expected the results of the PSPICE analysis are the same as obtained earlier. The FREQ is 1 kHz, the magnitude of $\mathbf{I}_L$ is 0.3183 A = 318.3 mA, and the phase angle is $-90^\circ$.

The next input file (Fig. 15.98) will give us a chance to use the .AC command to specify a range of frequencies for the analysis of the same circuit of Fig. 15.96. The analysis will be performed for a frequency range of 50 Hz to 2 kHz in increments of

$$\frac{2000\text{ Hz} - 50\text{ Hz}}{40 - 1} = \frac{1950\text{ Hz}}{39} = 50\text{ Hz}$$

The .PRINT and .OPTIONS NOPAGE commands are as in the first example. The output file of Fig. 15.98 provides a table of frequencies ranging from 50 Hz to 2 kHz with increments of 50 Hz, as specified.

```
************************ Evaluation PSpice (January 1989) ******* 09:03:07 *******

Chapter 15 - AC Inductor Circuit (Table)

****     CIRCUIT DESCRIPTION

****************************************************************************


VE 1 0 AC 20V 0
RL 1 2 1U
L1 2 0 10MH
.AC LIN 40 50 2KH
.PRINT AC IM(L1) IP(L1)
.OPTIONS NOPAGE
.END
```

```
****     AC ANALYSIS                        TEMPERATURE =   27.000 DEG C


  FREQ         IM(L1)       IP(L1)


   5.000E+01    6.366E+00   -9.000E+01
   1.000E+02    3.183E+00   -9.000E+01
   1.500E+02    2.122E+00   -9.000E+01
   2.000E+02    1.592E+00   -9.000E+01
   2.500E+02    1.273E+00   -9.000E+01
   3.000E+02    1.061E+00   -9.000E+01
   3.500E+02    9.095E-01   -9.000E+01
   4.000E+02    7.958E-01   -9.000E+01
   4.500E+02    7.074E-01   -9.000E+01
   5.000E+02    6.366E-01   -9.000E+01
   5.500E+02    5.787E-01   -9.000E+01
   6.000E+02    5.305E-01   -9.000E+01
   6.500E+02    4.897E-01   -9.000E+01
   7.000E+02    4.547E-01   -9.000E+01
   7.500E+02    4.244E-01   -9.000E+01
   8.000E+02    3.979E-01   -9.000E+01
   8.500E+02    3.745E-01   -9.000E+01
   9.000E+02    3.537E-01   -9.000E+01
   9.500E+02    3.351E-01   -9.000E+01
   1.000E+03    3.183E-01   -9.000E+01
   1.050E+03    3.032E-01   -9.000E+01
   1.100E+03    2.894E-01   -9.000E+01
   1.150E+03    2.768E-01   -9.000E+01
   1.200E+03    2.653E-01   -9.000E+01
   1.250E+03    2.546E-01   -9.000E+01
   1.300E+03    2.449E-01   -9.000E+01
   1.350E+03    2.358E-01   -9.000E+01
   1.400E+03    2.274E-01   -9.000E+01
   1.450E+03    2.195E-01   -9.000E+01
   1.500E+03    2.122E-01   -9.000E+01
   1.550E+03    2.054E-01   -9.000E+01
   1.600E+03    1.989E-01   -9.000E+01
   1.650E+03    1.929E-01   -9.000E+01
   1.700E+03    1.872E-01   -9.000E+01
   1.750E+03    1.819E-01   -9.000E+01
   1.800E+03    1.768E-01   -9.000E+01
   1.850E+03    1.721E-01   -9.000E+01
   1.900E+03    1.675E-01   -9.000E+01
   1.950E+03    1.632E-01   -9.000E+01
   2.000E+03    1.592E-01   -9.000E+01
```

FIG. 15.98

Since the inductor is the only element present, the phase angle remains at −90° throughout the frequency range. The reactance of the coil increases with frequency, so the magnitude of the current $\mathbf{I}_L$ will obviously decrease with increasing frequency. Its initial value is 6.366 A, and its final value for the given frequency range is .1592 A—a significant drop.

The last run for the inductor of Fig. 15.96 is a .PROBE plot of the magnitude of the current $\mathbf{I}_L$ with the same frequency range explored before. Note that the only change in the input file of Fig. 15.99 is to

```
************************ Evaluation PSpice (January 1989) ******* 09:06:34 ******

Chapter 15 - AC Inductor Circuit (Plot)

****     CIRCUIT DESCRIPTION

******************************************************************************

VE 1 0 AC 20V 0
RL 1 2 1U
L1 2 0 10MH
.AC LIN 40 50 2KH
.PROBE
.OPTIONS NOPAGE
.END
```

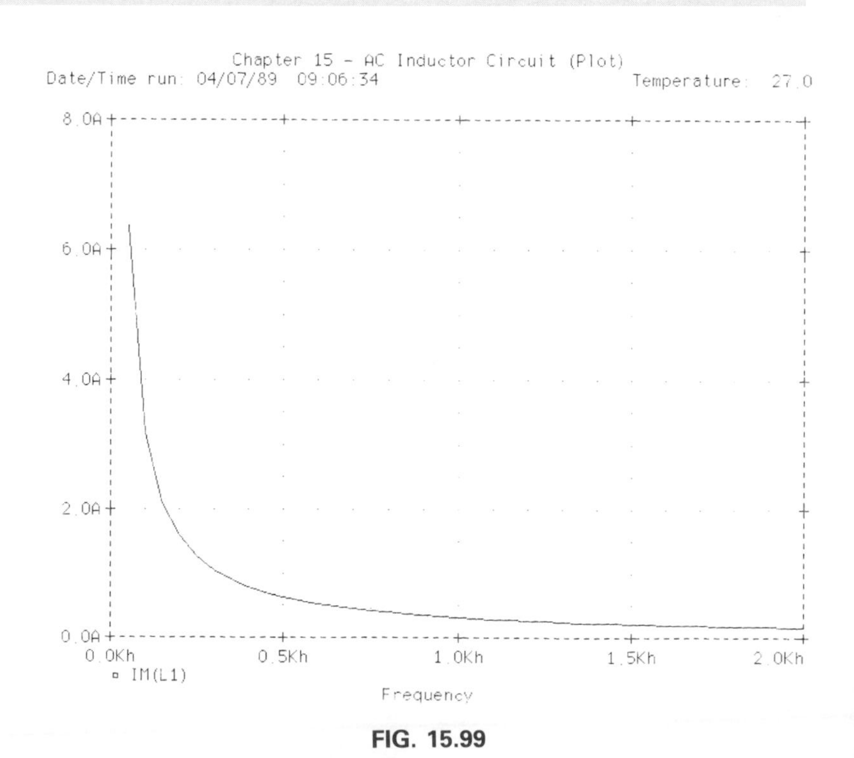


FIG. 15.99

replace the .PRINT statement by the .PROBE command. A LIN run was requested for the range of interest (linear scale for frequency) for IM(L1), producing the plot of Fig. 15.99. Note that its values at 50 Hz and 2 kHz match the values appearing in the output file of Fig. 15.98.

## Series *R-C* Circuit

The series *R-C* circuit of Fig. 15.44 analyzed in detail in Section 15.5 will now be examined using PSPICE. Although the network (Fig. 15.100) is more complex than the pure inductive circuit of Fig. 15.96,

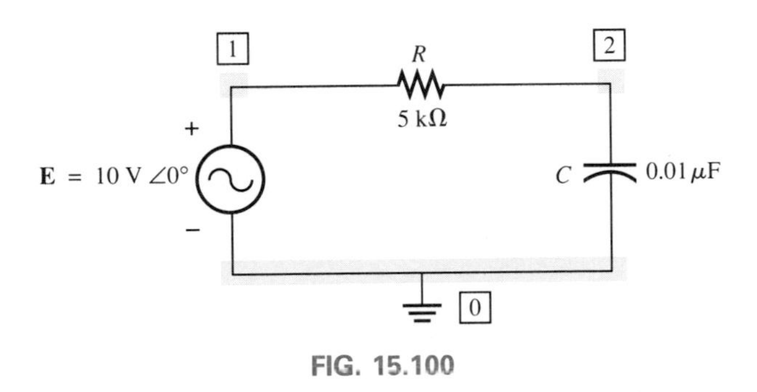


FIG. 15.100

the commands just introduced are sufficient to provide a detailed analysis of the frequency response of the circuit. The input file of Fig. 15.101 includes a source **E** = 10 V ∠0°, a resistor of 5 kΩ, and a capacitor of 0.01 μF. The analysis will take place from $f = 1$ kHz to 20 kHz in increments of (20 kHz − 1 kHz)/(20 − 1) = 1 kHz. The .PRINT statement requests the magnitude and phase angle of $\mathbf{V}_C$ as the applied frequency moves through the specified range.

Note the similarities between the tabulated values for the magnitude and phase angle and the plots of Figs. 15.51 and 15.52. In addition,

```
************************ Evaluation PSpice (January 1989) ******* 16:43:05 *******

Chapter 15 - Series RC Circuit - AC Analysis (Table)

****     CIRCUIT DESCRIPTION

*****************************************************************************

VE 1 0 AC 10V 0
R 1 2 5K
C 2 0 0.01UF
.AC LIN 20 1KH 20KH
.PRINT AC VM(C) VP(C)
.OPTIONS NOPAGE
.END
```

```
****     AC ANALYSIS                             TEMPERATURE =   27.000 DEG C


 FREQ         VM(C)        VP(C)


  1.000E+03    9.540E+00   -1.744E+01
  2.000E+03    8.467E+00   -3.214E+01
  3.000E+03    7.277E+00   -4.330E+01
  4.000E+03    6.227E+00   -5.149E+01
  5.000E+03    5.370E+00   -5.752E+01
  6.000E+03    4.687E+00   -6.205E+01
  7.000E+03    4.139E+00   -6.555E+01
  8.000E+03    3.697E+00   -6.830E+01
  9.000E+03    3.334E+00   -7.052E+01
  1.000E+04    3.033E+00   -7.234E+01
  1.100E+04    2.780E+00   -7.386E+01
  1.200E+04    2.564E+00   -7.514E+01
  1.300E+04    2.378E+00   -7.624E+01
  1.400E+04    2.217E+00   -7.719E+01
  1.500E+04    2.076E+00   -7.802E+01
  1.600E+04    1.951E+00   -7.875E+01
  1.700E+04    1.840E+00   -7.939E+01
  1.800E+04    1.741E+00   -7.997E+01
  1.900E+04    1.652E+00   -8.049E+01
  2.000E+04    1.572E+00   -8.096E+01
```

**FIG. 15.101**

note the correspondence between the magnitudes and angles in the output file compared to the calculated values of Section 15.5. Imagine the time saved and the accuracy maintained using the PSPICE software package versus the longhand approach—a very definite plus for the package approach, providing, of course, the theory is already clearly and correctly understood!

The next input file (Fig. 15.102) provides a plot of $V_C$ and $\theta_C$ for the circuit of Fig. 15.100 using the .PROBE command in place of the

```
************************ Evaluation PSpice (January 1989) ******* 09:33:20 *******

 Chapter 15 - Series RC Circuit - AC Analysis (Plot)

 ****     CIRCUIT DESCRIPTION

******************************************************************************


VE 1 0 AC 10V 0
R 1 2 5K
C 2 0 0.01UF
.AC LIN 100 100H 20KH
.PROBE
.OPTIONS NOPAGE
.END
```

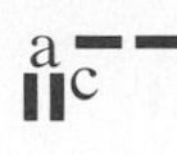



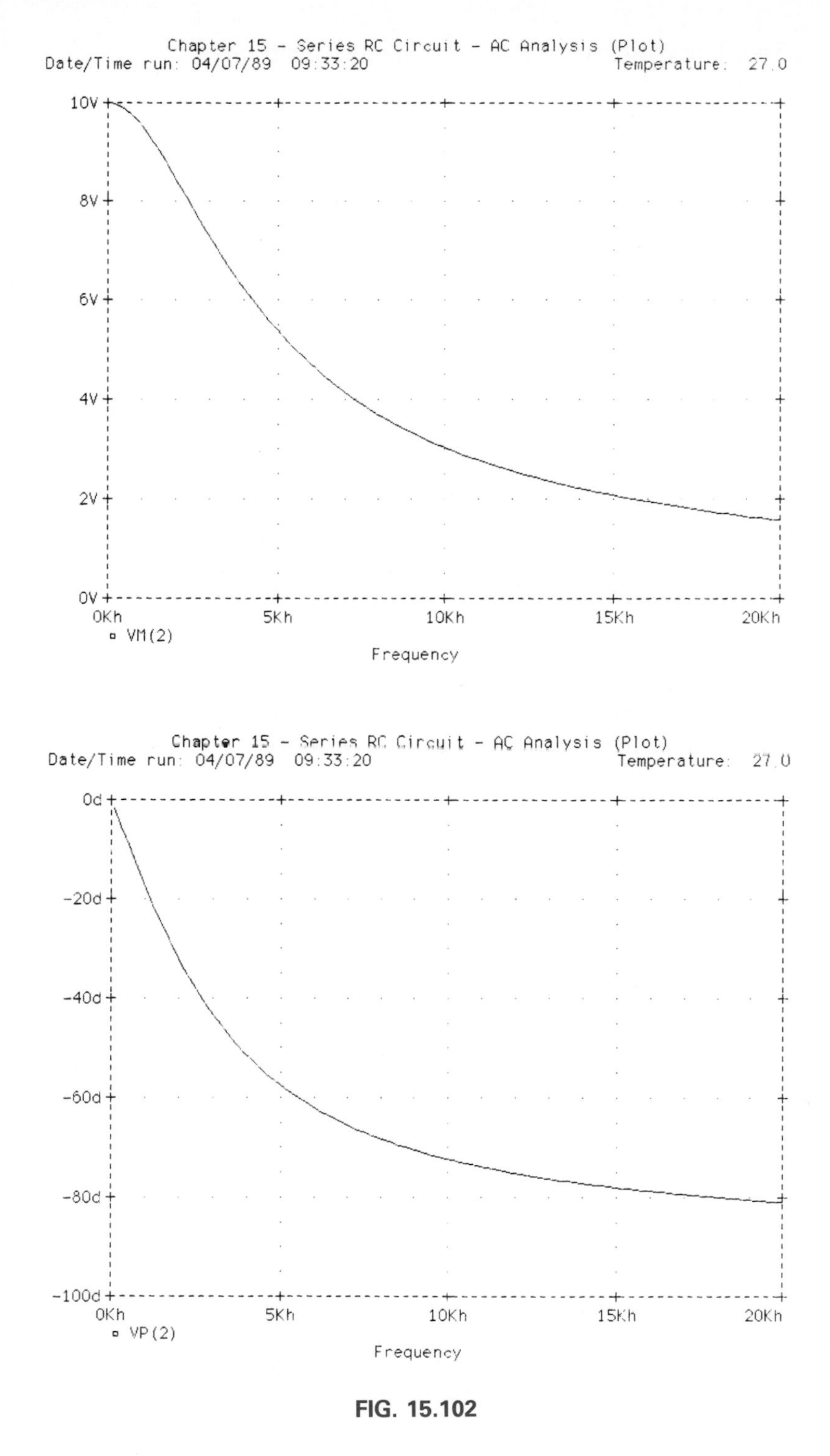


**FIG. 15.102**

.PRINT command. Compare the resulting curves with those obtained in Figs. 15.51 and 15.52.

## Total Impedance-Parallel *R-L-C* Network

The total impedance of the network of Fig. 15.103 will be tabulated and plotted using the .PLOT command, which has the following basic format:

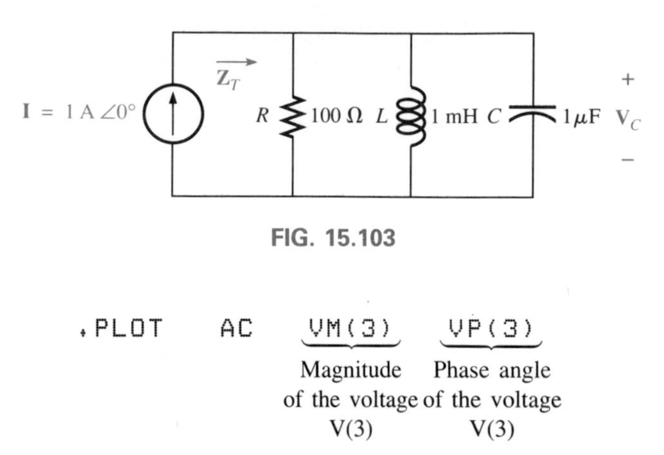


FIG. 15.103

```
.PLOT     AC     VM(3)      VP(3)
```

| .PLOT | AC | VM(3) | VP(3) |
|---|---|---|---|
| | | Magnitude of the voltage V(3) | Phase angle of the voltage V(3) |

The input file of Fig. 15.104 requests a plot of the magnitude of the voltage across the capacitor with frequency. However, since the voltage across parallel elements is the same, it is also the voltage across the input terminals of the network. Choosing a current source of 1 A will result in a magnitude of $\mathbf{V}_C$ that will equal the magnitude of the total impedance of the network, since

$$|Z_T| = \left|\frac{V_{\text{network}}}{I}\right| = \left|\frac{V_C}{1\text{ A}}\right| \equiv |V_C|$$

```
************************ Evaluation PSpice (January 1989) ******* 17:04:29 *******

 Chapter 15 - Impedance - RLC Parallel Network

 ****      CIRCUIT DESCRIPTION

*************************************************************************

IS 1 0 AC 1 0
R 1 0 100
L 1 0 1MH
C 1 0 1UF
.AC LIN 40 .5KH 20KH
.PLOT AC VM(C)
.OPTIONS NOPAGE
.END
```

```
****     AC ANALYSIS                           TEMPERATURE =   27.000 DEG C


   FREQ        VM(C)


(*)----------    1.0000E+00   1.0000E+01   1.0000E+02   1.0000E+03   1.0000E+04
                      - - - - - - - - - - - - - - - - - - - - - - - - - - - - - -
5.000E+02  3.171E+00 .      *     .            .            .            .
1.000E+03  6.527E+00 .          * .            .            .            .
1.500E+03  1.029E+01 .            .*           .            .            .
2.000E+03  1.476E+01 .            .  *         .            .            .
2.500E+03  2.041E+01 .            .    *       .            .            .
3.000E+03  2.806E+01 .            .     *      .            .            .
3.500E+03  3.918E+01 .            .       *    .            .            .
4.000E+03  5.636E+01 .            .         *  .            .            .
4.500E+03  8.156E+01 .            .           *.            .            .
5.000E+03  9.991E+01 .            .            *            .            .
5.500E+03  8.718E+01 .            .            *            .            .
6.000E+03  6.669E+01 .            .          * .            .            .
6.500E+03  5.216E+01 .            .         *  .            .            .
7.000E+03  4.259E+01 .            .        *   .            .            .
7.500E+03  3.601E+01 .            .       *    .            .            .
8.000E+03  3.127E+01 .            .      *     .            .            .
8.500E+03  2.770E+01 .            .     *      .            .            .
9.000E+03  2.492E+01 .            .     *      .            .            .
9.500E+03  2.268E+01 .            .    *       .            .            .
1.000E+04  2.085E+01 .            .    *       .            .            .
1.050E+04  1.931E+01 .            .   *        .            .            .
1.100E+04  1.800E+01 .            .   *        .            .            .
1.150E+04  1.687E+01 .            .  *         .            .            .
1.200E+04  1.589E+01 .            .  *         .            .            .
1.250E+04  1.502E+01 .            .  *         .            .            .
1.300E+04  1.425E+01 .            .  *         .            .            .
1.350E+04  1.357E+01 .            . *          .            .            .
1.400E+04  1.295E+01 .            . *          .            .            .
1.450E+04  1.238E+01 .            . *          .            .            .
1.500E+04  1.187E+01 .            .*           .            .            .
1.550E+04  1.140E+01 .            .*           .            .            .
1.600E+04  1.097E+01 .            .*           .            .            .
1.650E+04  1.058E+01 .            .*           .            .            .
1.700E+04  1.021E+01 .            .*           .            .            .
1.750E+04  9.866E+00 .            *            .            .            .
1.800E+04  9.548E+00 .            *            .            .            .
1.850E+04  9.251E+00 .            *            .            .            .
1.900E+04  8.972E+00 .            *            .            .            .
1.950E+04  8.711E+00 .            *            .            .            .
2.000E+04  8.465E+00 .            *            .            .            .
                      - - - - - - - - - - - - - - - - - - - - - - - - - - - - - -
```

**FIG. 15.104**

The .AC LIN line extends from $f = 0.5$ kHz to 20 kHz in increments of (20 kHz − 0.5 kHz)/(40 − 1) = 0.5 kHz.

Note in the output file that the increment is indeed 500 Hz and that the peak value of $Z_T$ is 100 Ω, which matches the value of the parallel resistor. The maximum value of $\mathbf{Z}_T$ will occur at a frequency determined by the following relationship:

$$X_L = X_C$$

$$2\pi fL = \frac{1}{2\pi fC}$$

and

$$f = \frac{1}{2\pi\sqrt{LC}} = \frac{1}{2\pi\sqrt{(1\text{ mH})(1\ \mu\text{F})}} = \mathbf{5.03\ kHz}$$

as confirmed by the plot. At low frequencies the inductive reactance establishes a low total impedance, whereas at high frequencies the capacitive reactance causes a reduction in the magnitude of $\mathbf{Z}_T$. At 0.5 kHz, $Z_T = 3.171\ \Omega$, peaking at 100 Ω at about 5 kHz, and dropping to 8.465 Ω at 20 kHz.

The obvious advantage of the .PLOT command over the .PROBE command is the tabulated values for specific frequencies next to the resulting plot.

## PROBLEMS

### SECTION 15.2 Impedance and the Phasor Diagram

1. Express the impedances of Fig. 15.105 in both polar and rectangular form.

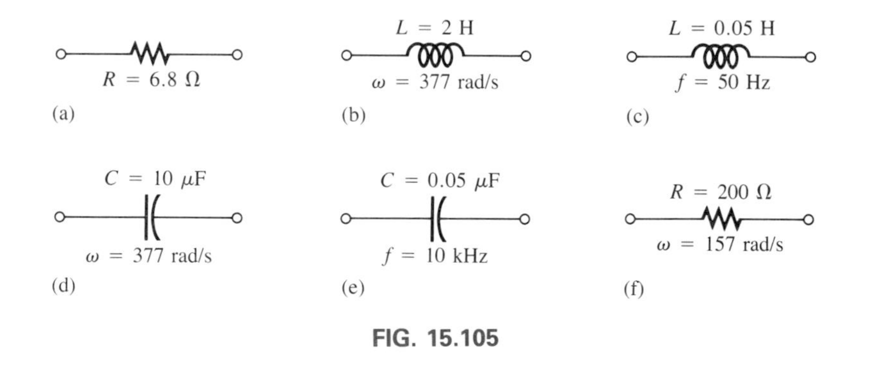


FIG. 15.105

2. Find the current $i$ for the elements of Fig. 15.106 using phasor algebra. Sketch the waveforms for $v$ and $i$ on the same set of axes.

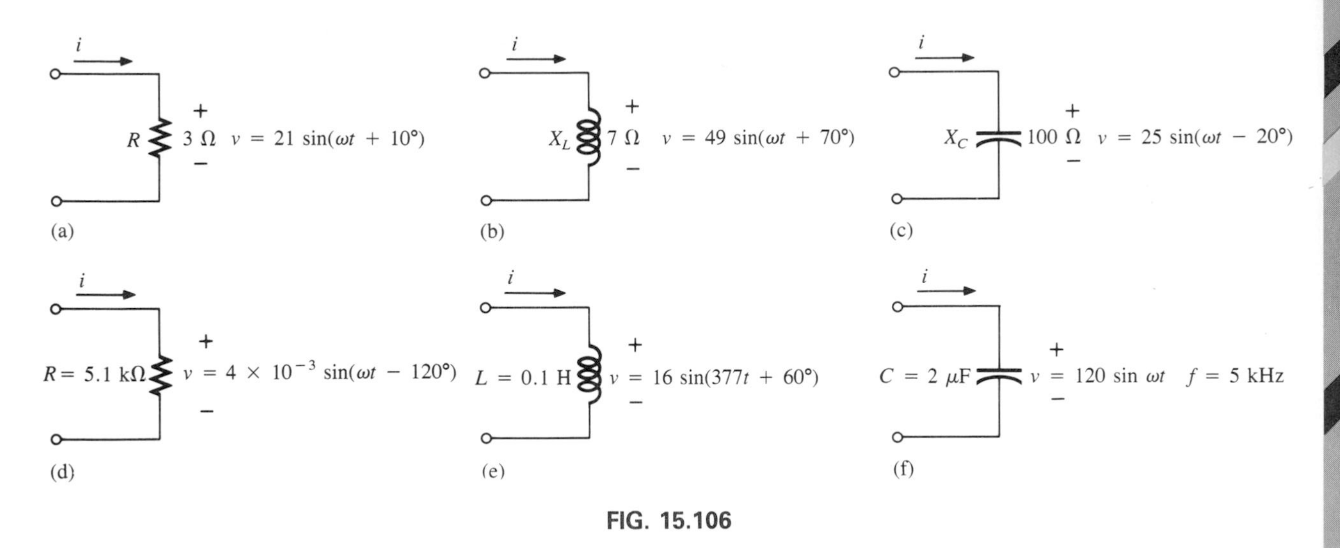


FIG. 15.106

3. Find the voltage $v$ for the elements of Fig. 15.107 using phasor algebra. Sketch the waveforms of $v$ and $i$ on the same set of axes.

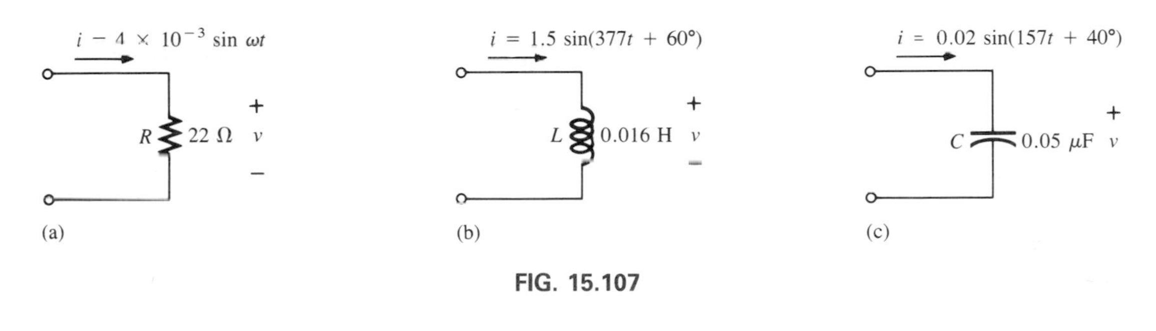


FIG. 15.107

## SECTION 15.3 Series Configuration

4. Calculate the total impedance of the circuits of Fig. 15.108. Express your answer in rectangular and polar form, and draw the impedance diagram.

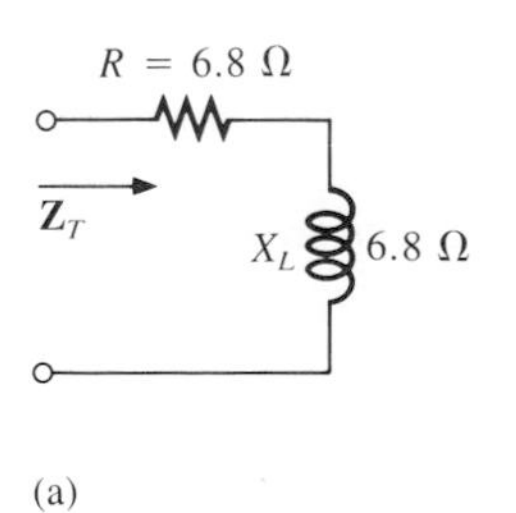


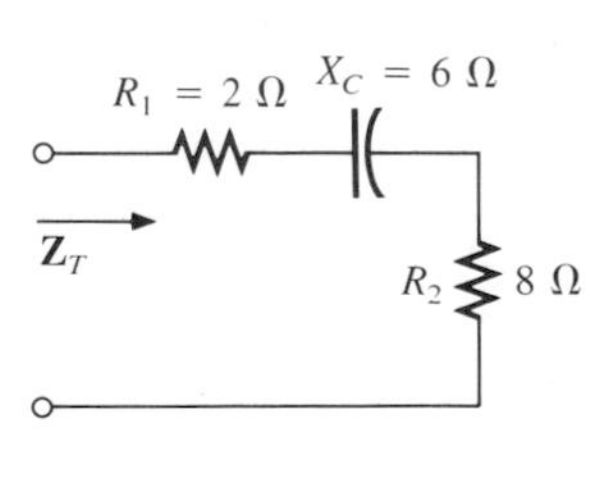


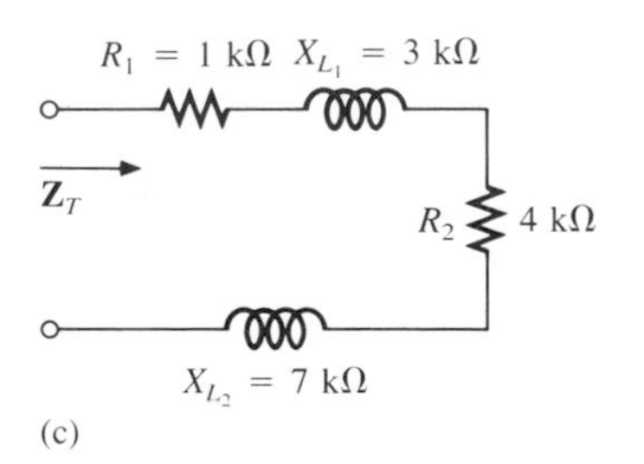


FIG. 15.108

5. Repeat Problem 4 for the circuits of Fig. 15.109.

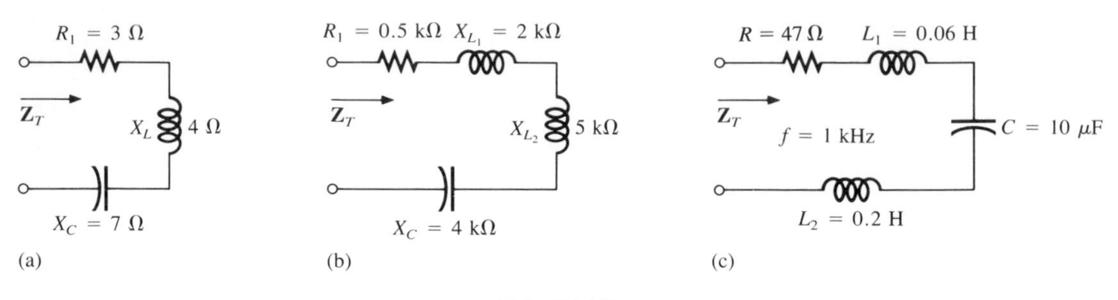

FIG. 15.109

6. Find the type and impedance in ohms of the series circuit elements that must be in the closed container of Fig. 15.110 in order for the indicated voltages and currents to exist at the input terminals. (Find the simplest series circuit that will satisfy the indicated conditions.)

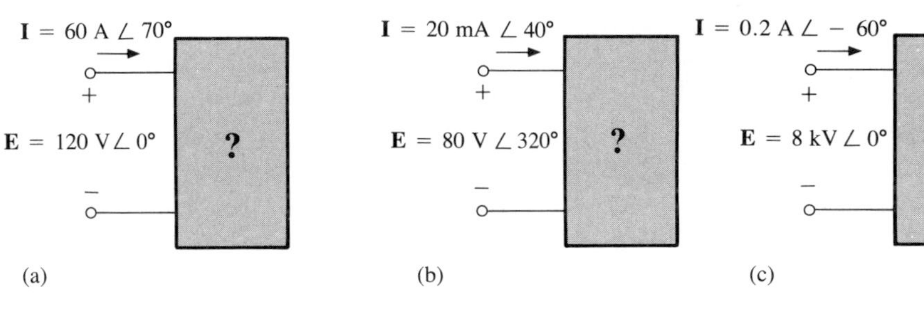


FIG. 15.110

7. For the circuit of Fig. 15.111:
   a. Find the total impedance $\mathbf{Z}_T$ in polar form.
   b. Draw the impedance diagram.
   c. Find the current **I** and the voltages $\mathbf{V}_R$ and $\mathbf{V}_L$ in phasor form.
   d. Draw the phasor diagram of the voltages **E**, $\mathbf{V}_R$, and $\mathbf{V}_L$, and the current **I**.
   e. Verify Kirchhoff's voltage law around the closed loop.
   f. Find the average power delivered to the circuit.
   g. Find the power factor of the circuit and indicate whether it is leading or lagging.
   h. Find the sinusoidal expressions for the voltages and current if the frequency is 60 Hz.
   i. Plot the waveforms for the voltages and current on the same set of axes.
8. Repeat Problem 7 for the circuit of Fig. 15.112, replacing $\mathbf{V}_L$ by $\mathbf{V}_C$ in parts (c) and (d).

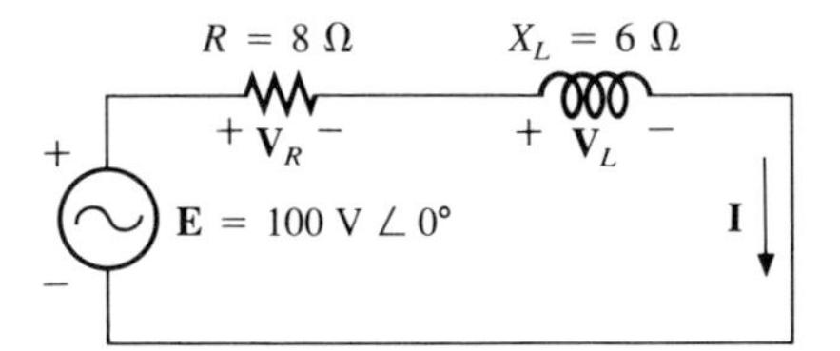


FIG. 15.111

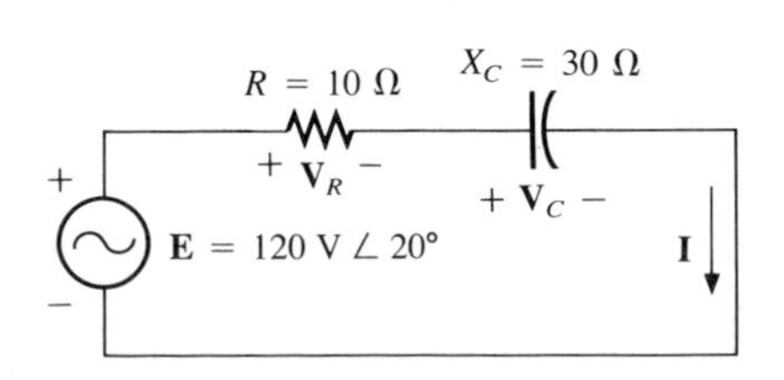


FIG. 15.112

9. Given the network of Fig. 15.113:
   a. Determine $\mathbf{Z}_T$.
   b. Find $\mathbf{I}$.
   c. Calculate $\mathbf{V}_R$ and $\mathbf{V}_L$.
   d. Find $P$ and $F_p$.

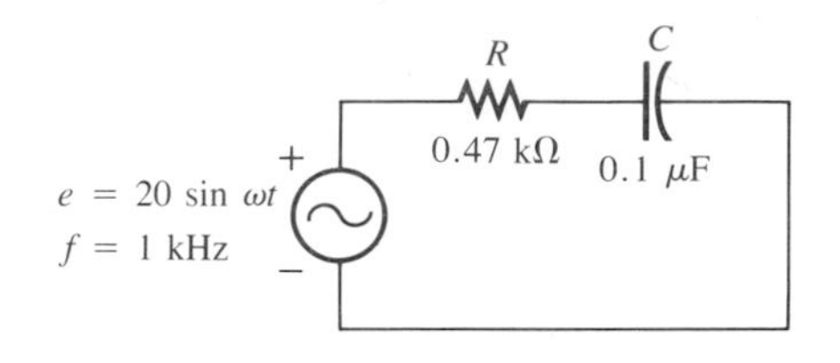

FIG. 15.113

10. For the circuit of Fig. 15.114:
    a. Find the total impedance $\mathbf{Z}_T$ in polar form.
    b. Draw the impedance diagram.
    c. Find the value of $C$ in microfarads and $L$ in henries.
    d. Find the current $i$ and the voltages $v_R$, $v_L$, and $v_C$ in phasor form.
    e. Draw the phasor diagram of the voltages $\mathbf{E}$, $\mathbf{V}_R$, $\mathbf{V}_L$, and $\mathbf{V}_C$, and the current $\mathbf{I}$.
    f. Verify Kirchhoff's voltage law around the closed loop.
    g. Find the average power delivered to the circuit.
    h. Find the power factor of the circuit and indicate whether it is leading or lagging.
    i. Find the sinusoidal expressions for the voltages and current.
    j. Plot the waveforms for the voltages and current on the same set of axes.

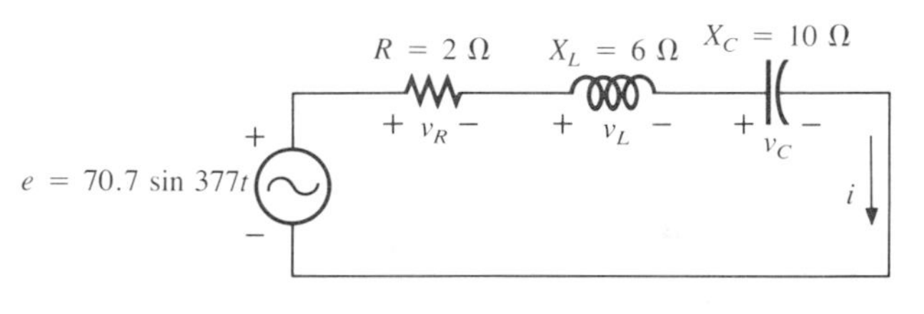


FIG. 15.114

11. Repeat Problem 10 for the circuit of Fig. 15.115.

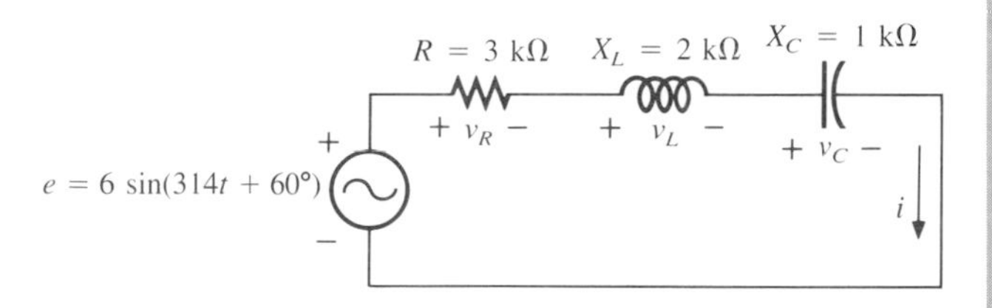


FIG. 15.115

## SECTION 15.4 Voltage Divider Rule

12. Calculate the voltages $\mathbf{V}_1$ and $\mathbf{V}_2$ for the circuit of Fig. 15.116 in phasor form using the voltage divider rule.

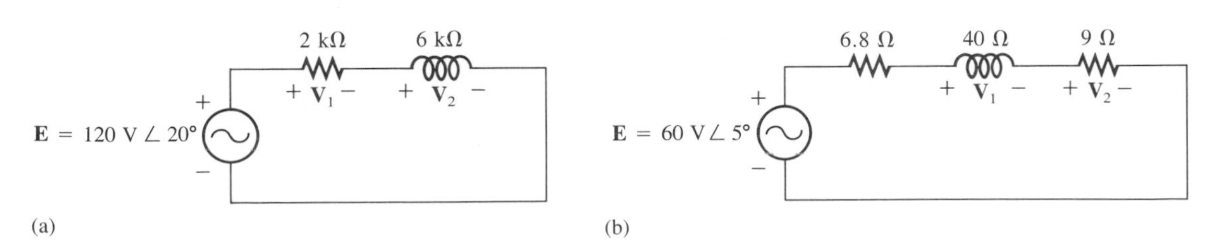


FIG. 15.116

**13.** Repeat Problem 12 for the circuits of Fig. 15.117.

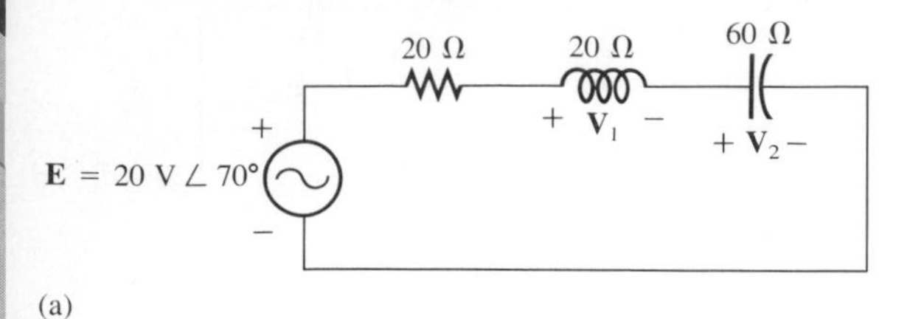


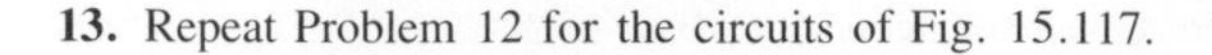


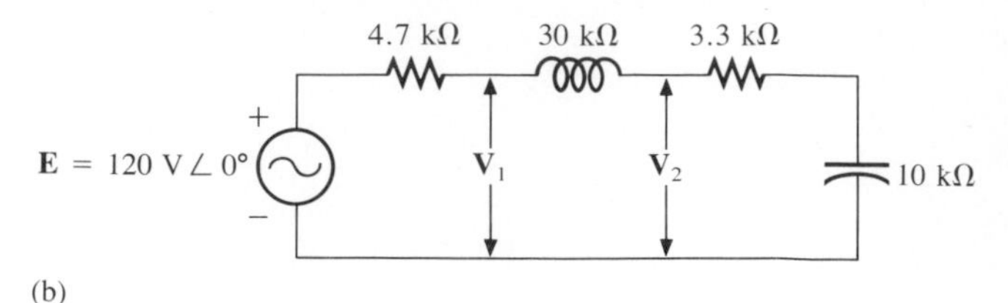

FIG. 15.117

***14.** For the circuit of Fig. 15.118:

**a.** Determine $i$, $v_R$, and $v_C$ in phasor form.

**b.** Calculate the total power factor and indicate whether it is leading or lagging.

**c.** Calculate the average power delivered to the circuit.

**d.** Draw the impedance diagram.

**e.** Draw the phasor diagram of the voltages **E**, $\mathbf{V}_R$, and $\mathbf{V}_C$, and the current **I**.

**f.** Find the voltages $\mathbf{V}_R$ and $\mathbf{V}_C$ using the voltage divider rule, and compare with part (a) above.

**g.** Draw the equivalent series circuit of the above as far as the total impedance and the current $i$ are concerned.

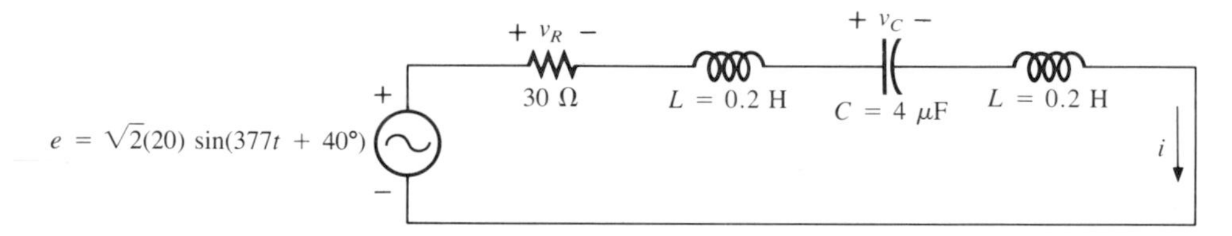

FIG. 15.118

***15.** Repeat Problem 14 if the capacitance is changed to 1000 μF.

**16.** An electrical load has a power factor of 0.8 lagging. It dissipates 8 kW at a voltage of 200 V. Calculate the impedance of this load in rectangular coordinates.

***17.** Find the series element or elements that must be in the enclosed container of Fig. 15.119 to satisfy the following conditions:

**a.** Average power to circuit = 300 W.

**b.** Circuit has a lagging power factor.

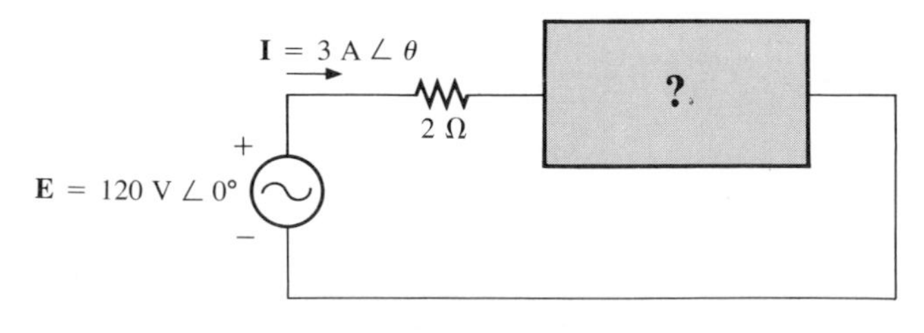


FIG. 15.119

## SECTION 15.5 Frequency Response of the *R-C* Circuit

***18.** For the circuit of Fig. 15.120:
  **a.** Plot $Z_T$ and $\theta_T$ versus frequency for a frequency range of zero to 20 kHz.
  **b.** Plot $V_L$ versus frequency for the frequency range of part (a).
  **c.** Plot $\theta_L$ versus frequency for the frequency range of part (a).
  **d.** Plot $V_R$ versus frequency for the frequency range of part (a).

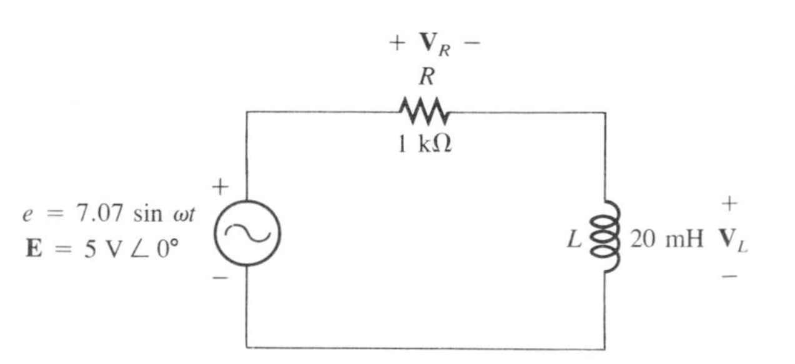

FIG. 15.120

***19.** For the circuit of Fig. 15.121:
  **a.** Plot $Z_T$ and $\theta_T$ versus frequency for a frequency range of zero to 10 kHz.
  **b.** Plot $V_C$ versus frequency for the frequency range of part (a).
  **c.** Plot $\theta_C$ versus frequency for the frequency range of part (a).
  **d.** Plot $V_R$ versus frequency for the frequency range of part (a)

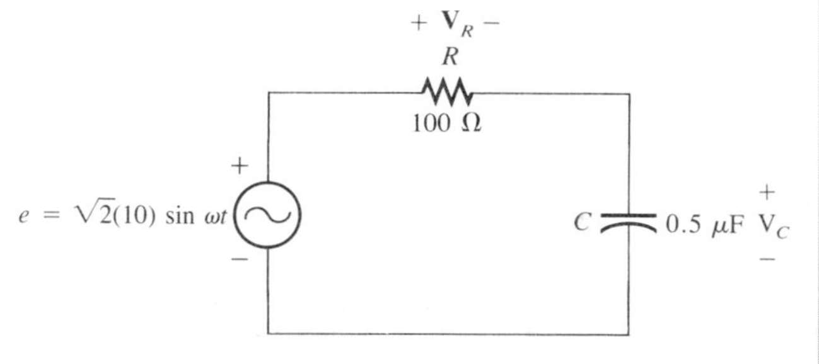

FIG. 15.121

***20.** For the series *R-L-C* circuit of Fig. 15.122:
  **a.** Plot $Z_T$ and $\theta_T$ versus frequency for a frequency range of zero to 20 kHz in increments of 1 kHz.
  **b.** Plot $V_C$ (magnitude only) versus frequency for the same frequency range of part (a).
  **c.** Plot $I$ (magnitude only) versus frequency for the same frequency range of part (a).

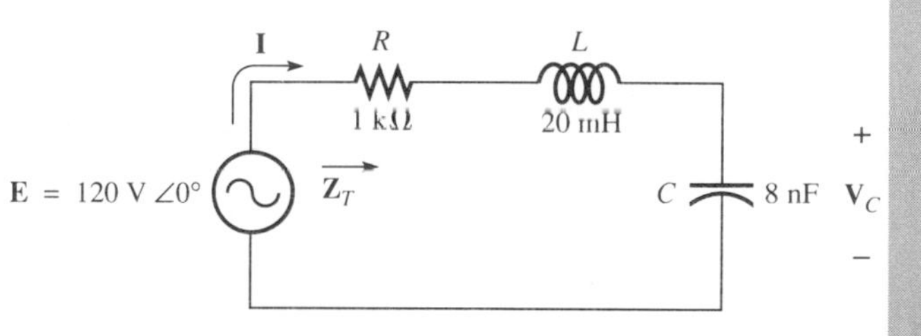

FIG. 15.122

## SECTION 15.6 Admittance and Susceptance

**21.** Find the total admittance and impedance of the circuits of Fig. 15.123. Identify the values of conductance and susceptance, and draw the admittance diagram.

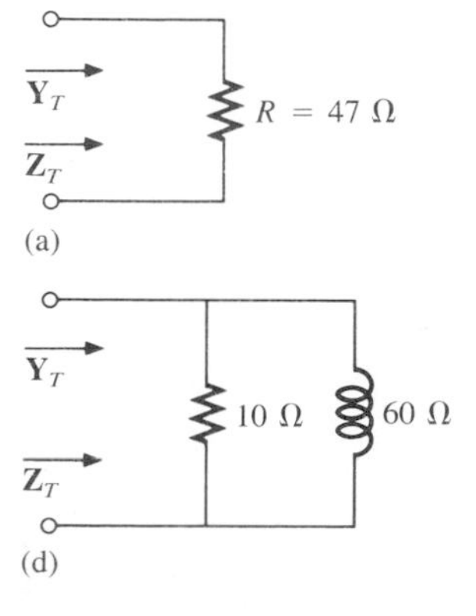


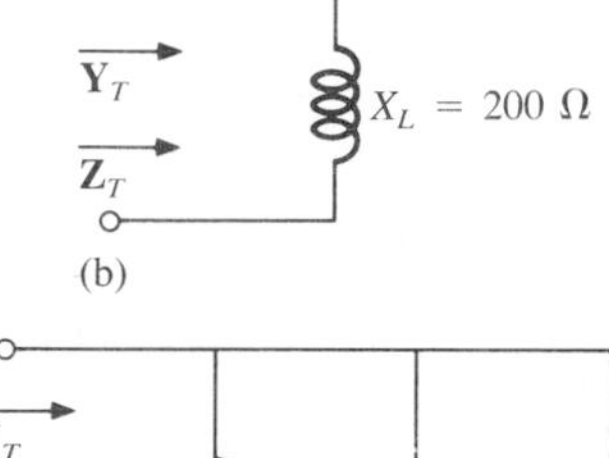


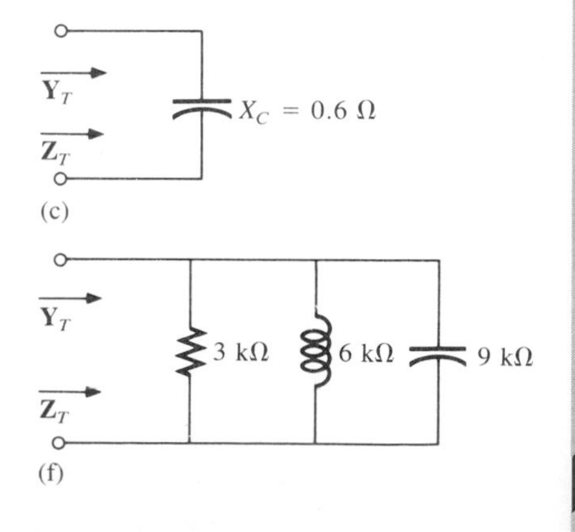


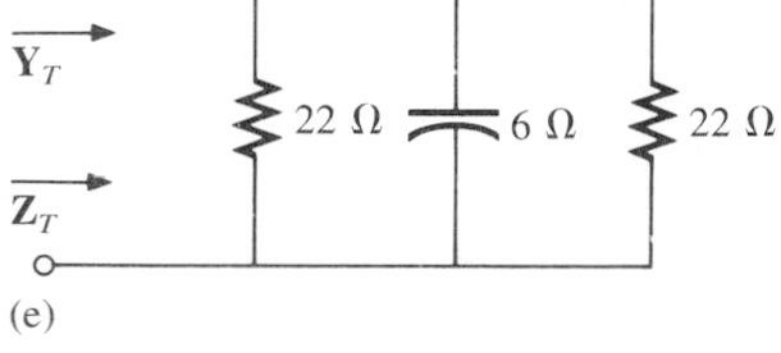


FIG. 15.123

22. Repeat Problem 21 for the networks of Fig. 15.124.

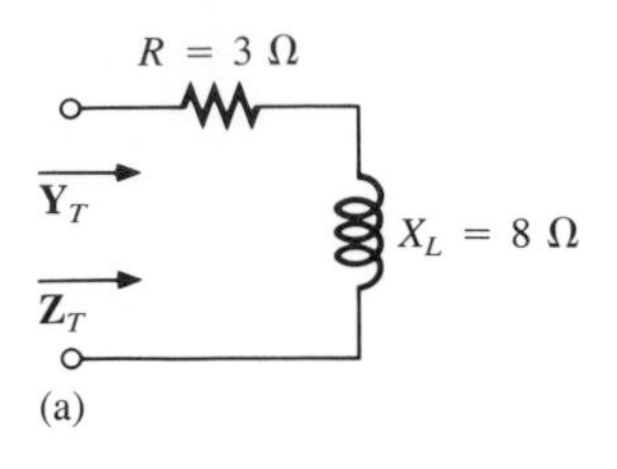


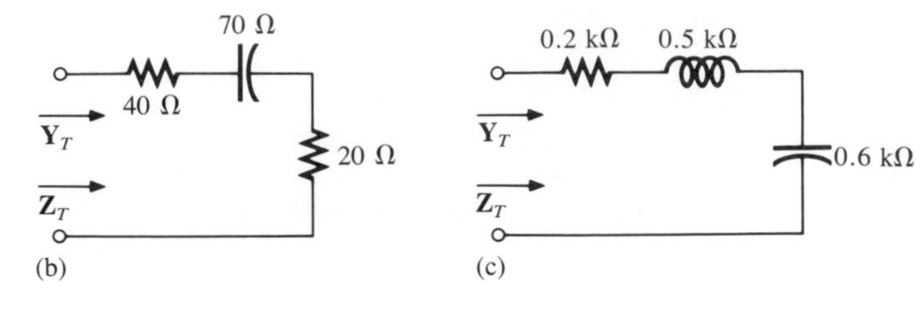


FIG. 15.124

23. Repeat Problem 6 for the parallel circuit elements that must be in the closed container for the same voltage and current to exist at the input terminals. (Find the simplest parallel circuit that will satisfy the conditions indicated.)

## SECTION 15.7 *R-L, R-C,* and *R-L-C* Parallel ac Networks

24. For the circuit of Fig. 15.125:
    a. Find the total admittance $\mathbf{Y}_T$ in polar form.
    b. Draw the admittance diagram.
    c. Find the voltage **E** and the currents $\mathbf{I}_R$ and $\mathbf{I}_L$ in phasor form.
    d. Draw the phasor diagram of the currents $\mathbf{I}_S$, $\mathbf{I}_R$, and $\mathbf{I}_L$, and the voltage **E.**
    e. Verify Kirchhoff's current law at one node.
    f. Find the average power delivered to the circuit.
    g. Find the power factor of the circuit and indicate whether it is leading or lagging.
    h. Find the sinusoidal expressions for the currents and voltage if the frequency is 60 Hz.
    i. Plot the waveforms for the currents and voltage on the same set of axes.

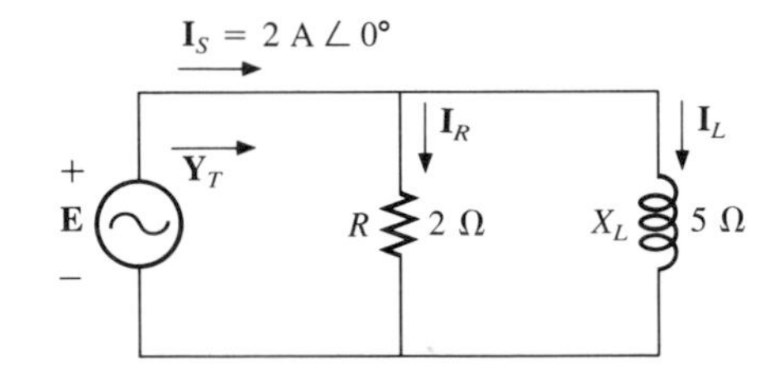


FIG. 15.125

25. Repeat Problem 24 for the circuit of Fig. 15.126, replacing $\mathbf{I}_L$ by $\mathbf{I}_C$ in parts (c) and (d).

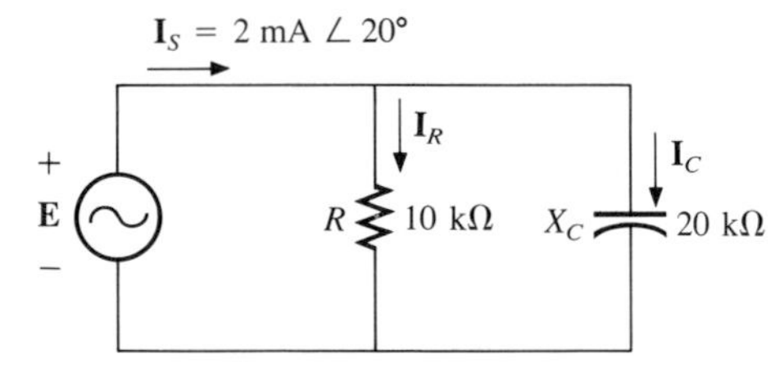


FIG. 15.126

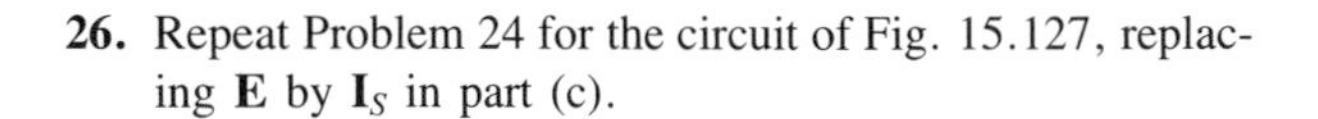
26. Repeat Problem 24 for the circuit of Fig. 15.127, replacing **E** by $\mathbf{I}_S$ in part (c).

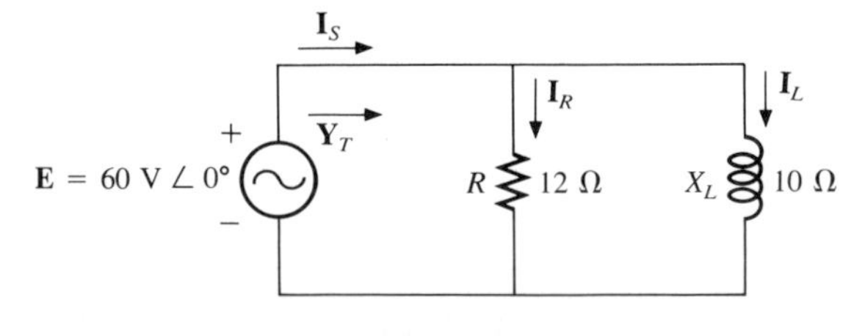


FIG. 15.127

**27.** For the circuit of Fig. 15.128:

**a.** Find the total admittance $\mathbf{Y}_T$ in polar form.
**b.** Draw the admittance diagram.
**c.** Find the value of $C$ in microfarads and $L$ in henries.
**d.** Find the voltage $e$ and currents $i_R$, $i_L$, and $i_C$ in phasor form.
**e.** Draw the phasor diagram of the currents $\mathbf{I}_S$, $\mathbf{I}_R$, $\mathbf{I}_L$, and $\mathbf{I}_C$, and the voltage $\mathbf{E}$.
**f.** Verify Kirchhoff's current law at one node.
**g.** Find the average power delivered to the circuit.
**h.** Find the power factor of the circuit and indicate whether it is leading or lagging.
**i.** Find the sinusoidal expressions for the currents and voltage.
**j.** Plot the waveforms for the currents and voltage on the same set of axes.

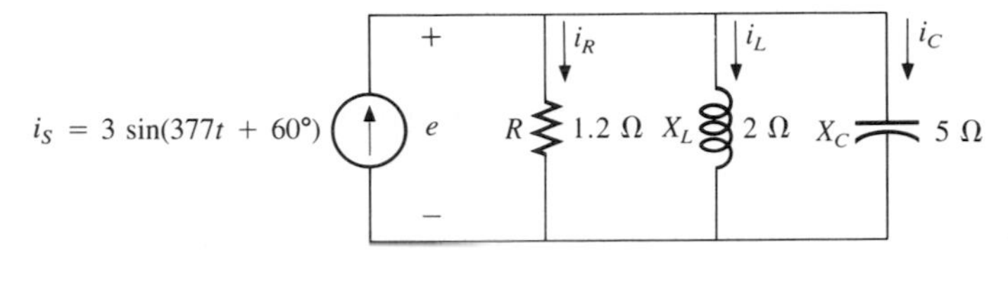


FIG. 15.128

**28.** Repeat Problem 27 for the circuit of Fig. 15.129.

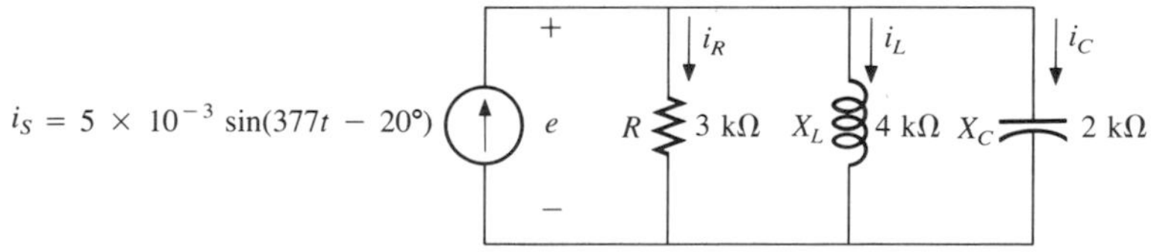

FIG. 15.129

**29.** Repeat Problem 27 for the circuit of Fig. 15.130, replacing $e$ by $i_S$ in part (d).

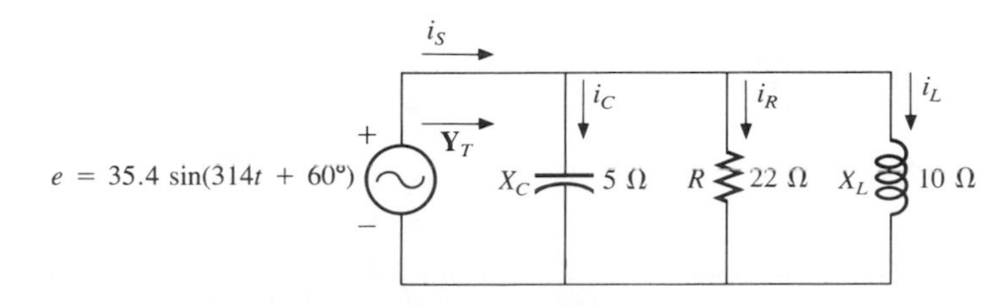


FIG. 15.130

## SECTION 15.8 Current Divider Rule

**30.** Calculate the currents $\mathbf{I}_1$ and $\mathbf{I}_2$ of Fig. 15.131 in phasor form using the current divider rule.

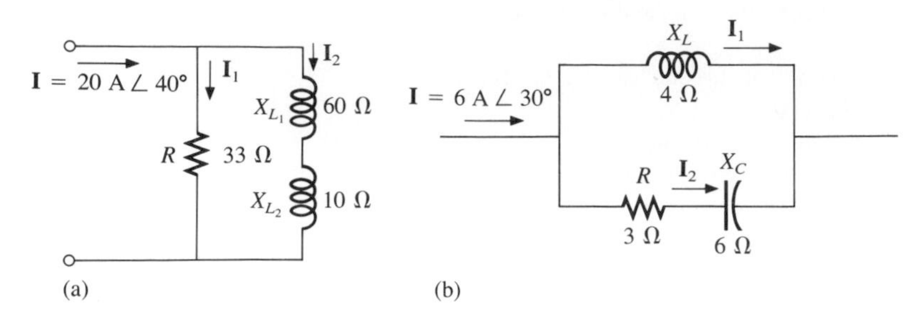

FIG. 15.131

## SECTION 15.9 Frequency Response of the Parallel *R-L* Network

***31.** For the parallel *R-C* network of Fig. 15.132:

**a.** Plot $Z_T$ and $\theta_T$ versus frequency for a frequency range of zero to 4 kHz.

**b.** Plot $V_C$ versus frequency for the frequency range of part (a).

**c.** Plot $I_R$ versus frequency for the frequency range of part (a).

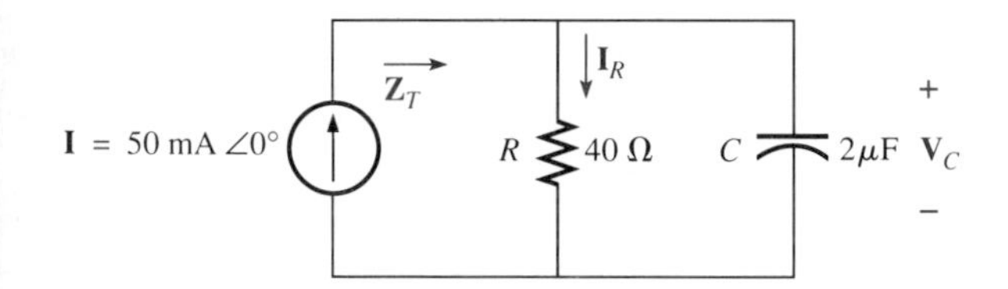


FIG. 15.132

***32.** For the parallel *R-L* network of Fig. 15.133:

**a.** Plot $Z_T$ and $\theta_T$ versus frequency for a frequency range of zero to 10 kHz.

**b.** Plot $I_L$ versus frequency for the frequency range of part (a).

**c.** Plot $I_R$ versus frequency for the frequency range of part (a).

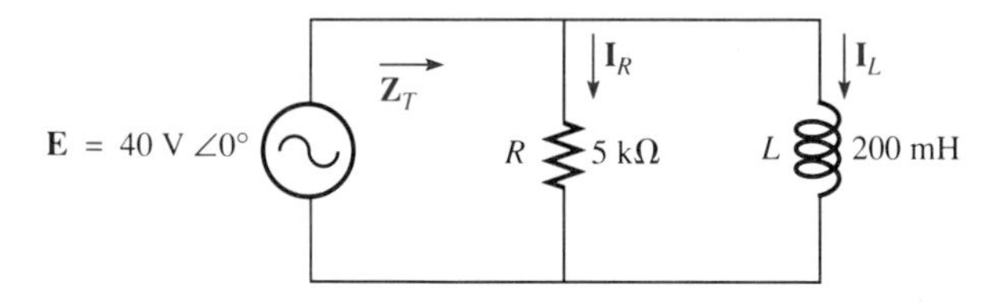


FIG. 15.133

**33.** Plot $Y_T$ and $\theta_T$ (of $\mathbf{Y}_T = Y_T \angle \theta_T$) for a frequency range of zero to 4 kHz for the network of Fig. 15.132.

**34.** Plot $Y_T$ and $\theta_T$ (of $\mathbf{Y}_T = Y_T \angle \theta_T$) for a frequency range of zero to 10 kHz for the network of Fig. 15.133.

***35.** For the parallel *R-L-C* network of Fig. 15.134:

**a.** Plot $Y_T$ and $\theta_T$ (of $\mathbf{Y}_T = Y_T \angle \theta_T$) for a frequency range of zero to 20 kHz.

**b.** Repeat part (a) for $Z_T$ and $\theta_T$ (of $\mathbf{Z}_T = Z_T \angle \theta_T$).

**c.** Plot $V_C$ versus frequency for the frequency range of part (a).

**d.** Plot $I_L$ versus frequency for the frequency range of part (a).

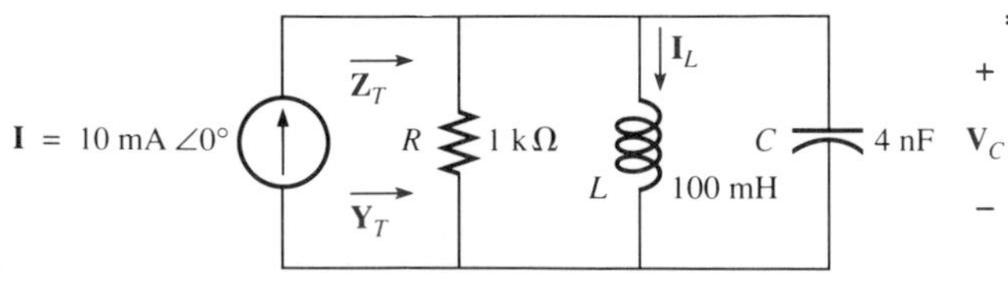


FIG. 15.134

## SECTION 15.10 Equivalent Circuits

**36.** For the series circuits of Fig. 15.135, find a parallel circuit that will have the same total impedance ($\mathbf{Z}_T$).

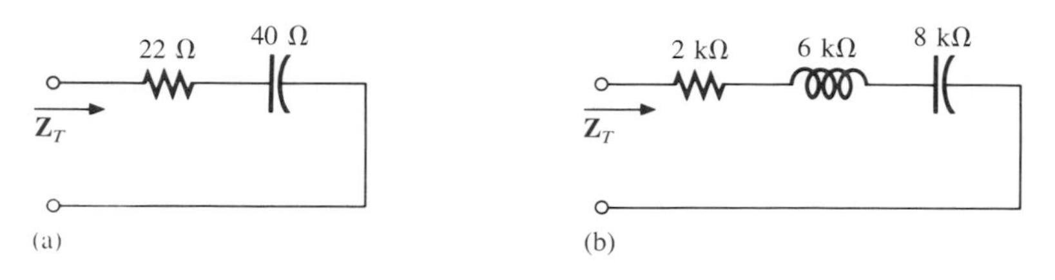

**FIG. 15.135**

**37.** For the parallel circuits of Fig. 15.136, find a series circuit that will have the same total impedance.

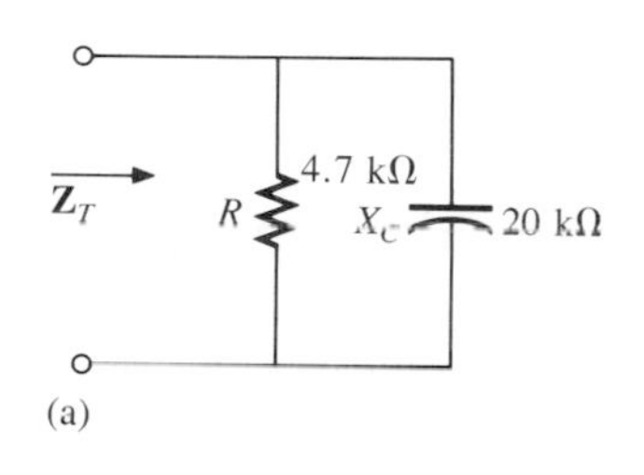


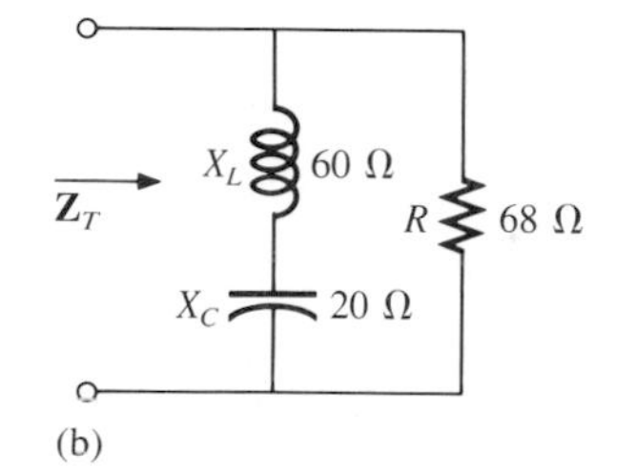

**FIG. 15.136**

**38.** For the network of Fig. 15.137:

**a.** Calculate $e$, $i_R$, and $i_L$ in phasor form.

**b.** Calculate the total power factor and indicate whether it is leading or lagging.

**c.** Calculate the average power delivered to the circuit.

**d.** Draw the admittance diagram.

**e.** Draw the phasor diagram of the currents $\mathbf{I}_S$, $\mathbf{I}_R$, and $\mathbf{I}_L$, and the voltage $\mathbf{E}$.

**f.** Find the current $\mathbf{I}_C$ for each capacitor using only Kirchhoff's current law.

**g.** Find the series circuit of one resistive and reactive element that will have the same impedance as the original circuit.

***39.** Repeat Problem 28 if the inductance is changed to 1 H.

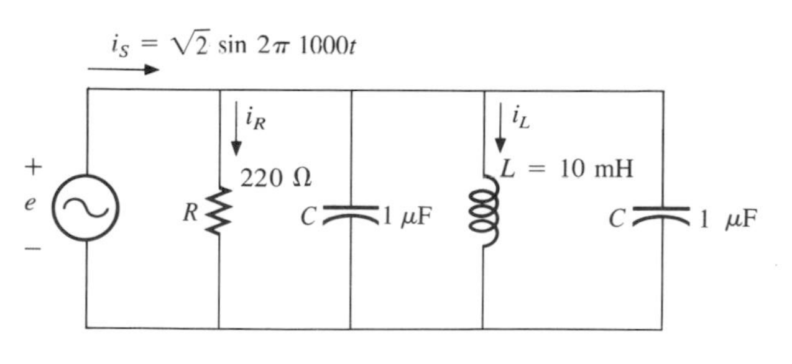

**FIG. 15.137**

**40.** Find the element or elements that must be in the closed container of Fig. 15.138 to satisfy the following conditions. (Find the simplest parallel circuit that will satisfy the indicated conditions.)

**a.** Average power to the circuit = 3000 W.

**b.** Circuit has a lagging power factor.

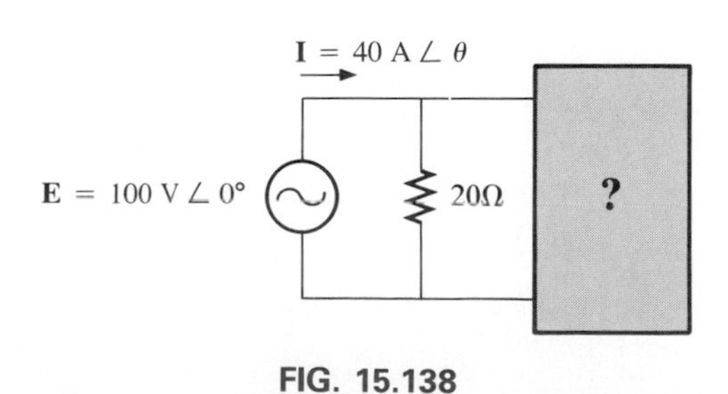


**FIG. 15.138**

## SECTION 15.11 Computer Analysis

### PSPICE

**41.** Write the input file to calculate the magnitude and phase angle of $\mathbf{V}_R$, $\mathbf{V}_L$, and $\mathbf{I}$ for the network of Fig. 15.111. Use a frequency of 1 kHz to determine the appropriate value of $L$ for the analysis.

**42.** Write the input file to calculate the magnitude and phase angle of $\mathbf{V}_1$ and $\mathbf{V}_2$ for the network of Fig. 15.117 using a frequency of 1 kHz to determine the appropriate value of $L$ and $C$ for the analysis.

**43.** For the network of Fig. 15.120:
   **a.** Write the input file to print the magnitude and angle of $\mathbf{Z}_T$ for a frequency range of 0 to 20 kHz in increments of 500 Hz.
   **b.** Repeat part (a) for $\mathbf{V}_L$.

***44.** For the network of Fig. 15.122:
   **a.** Write the input file to print the magnitude and angle of $\mathbf{Z}_T$ for a frequency range of zero to 20 kHz in increments of 1 kHz.
   **b.** Repeat part (a) for $\mathbf{I}_C$.

**45.** Write the input file to calculate the magnitude and phase angle of $\mathbf{I}_R$, $\mathbf{I}_L$, $\mathbf{I}_C$, and $\mathbf{E}$ for the network of Fig. 15.128 using any appropriate frequency to determine the value of $L$ and $C$ for the calculations.

**46.** For the network of Fig. 15.132:
   **a.** Write the input file to print out the magnitude and angle of $\mathbf{Z}_T$ for a frequency range of zero to 4 kHz with increments of 200 Hz.
   **b.** Repeat part (a) for $\mathbf{V}_C$.

***47.** For the network of Fig. 15.133:
   **a.** Write the input file to print out the magnitude and angle of $\mathbf{Y}_T$ for a frequency range of zero to 20 kHz in increments of 500 Hz.
   **b.** Repeat part (a) for $\mathbf{I}_R$.

### BASIC

**48.** Write a program to generate the sinusoidal expression for the current of a resistor, inductor, or capacitor given the value of $R$, $L$, or $C$ and the applied voltage in sinusoidal form.

**49.** Given the impedance of each element in rectangular form, write a program to determine the total impedance in rectangular form of any number of series elements.

**50.** Given two phasors in polar form in the first quadrant, write a program to generate the sum of the two phasors in polar form.

**51.** Given a series or parallel network as shown in Fig. 15.87, write a program to generate the parallel or series equivalent.

# GLOSSARY

**Admittance** A measure of how easily a network will ''admit'' the passage of current through that system. It is measured in siemens, abbreviated S, and is represented by the capital letter *Y*.

**Admittance diagram** A vector display that clearly depicts the magnitude of the admittance of the conductance, capacitive susceptance, and inductive susceptance, and the magnitude and angle of the total admittance of the system.

**Current divider rule** A method by which the current through either of two parallel branches can be determined in an ac network without first finding the voltage across the parallel branches.

**Equivalent circuits** For every series ac network there is a parallel ac network (and vice versa) that will be ''equivalent'' in the sense that the input current and impedance are the same.

**Impedance diagram** A vector display that clearly depicts the magnitude of the impedance of the resistive, reactive, and capacitive components of a network, and the magnitude and angle of the total impedance of the system.

**Parallel ac circuits** A connection of elements in an ac network in which all of the elements have two points in common. The voltage is the same across each element.

**Phasor diagram** A vector display that provides at a glance the magnitude and phase relationships among the various voltages and currents of a network.

**Series ac configuration** A connection of elements in an ac network in which no two impedances have more than one terminal in common and the current is the same through each element.

**Susceptance** A measure of how ''susceptible'' an element is to the passage of current through it. It is measured in siemens, abbreviated S, and is represented by the capital letter *B*.

**Voltage divider rule** A method through which the voltage across one element of a series of elements in an ac network can be determined without first having to find the current through the elements.

# 16

# Series-Parallel ac Networks

## 16.1 INTRODUCTION

In this chapter, we shall utilize the fundamental concepts of the previous chapter to develop a technique for solving series-parallel ac networks. A brief review of Chapter 7 may be helpful before considering these networks, since the approach here will be quite similar to that undertaken earlier. The circuits to be discussed will have only one source of energy, either potential or current. Networks with two or more sources will be considered in Chapters 17 and 18, using methods previously described for dc circuits.

In general, when working with series-parallel ac networks, consider the following approach:

1. Redraw the network employing block impedances to combine obvious series and parallel elements to reduce the network to one that clearly reveals the fundamental structure of the system.
2. Study the problem and make a brief mental sketch of the overall approach you plan to use. Doing this may result in time- and energy-saving shortcuts.
3. After the overall approach has been determined, consider each branch involved in your method independently before tying them together in series-parallel combinations.
4. When you have arrived at a solution, check to see that it is reasonable by considering the magnitudes of the energy source and the elements in the circuit. If not, either solve the network using another approach, or check over your work very carefully.

## 16.2 ILLUSTRATIVE EXAMPLES

**EXAMPLE 16.1.** For the network of Fig. 16.1:

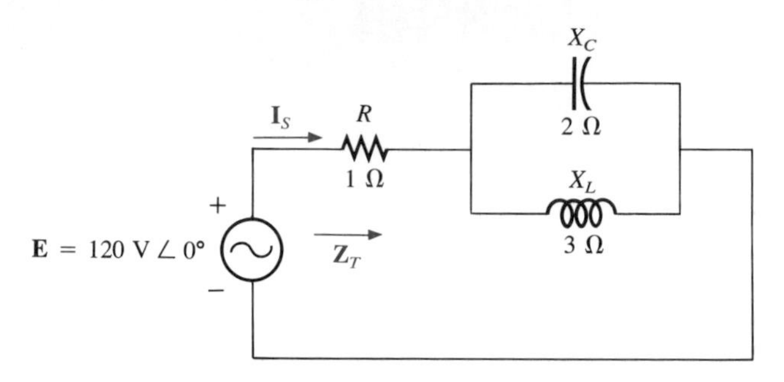

FIG. 16.1

a. Calculate $\mathbf{Z}_T$.
b. Determine $\mathbf{I}_S$.
c. Find $\mathbf{I}_C$.
d. Calculate $\mathbf{V}_R$ and $\mathbf{V}_C$.
e. Compute the power delivered.
f. Find $F_p$ of the network.

For convenience, the network is redrawn in Fig. 16.2. It is good practice when working with complex networks to represent impedances

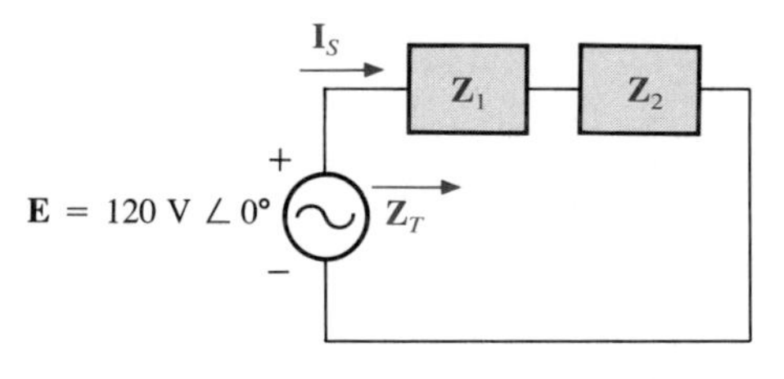

FIG. 16.2

in the manner indicated. When the unknown quantity is found in terms of these subscripted impedances, the numerical values can then be substituted to find the magnitude of the unknown. This will usually save time and prevent calculation errors.

***Solutions:***

a. $\mathbf{Z}_1 = R\ \angle 0^\circ = 1\ \Omega\ \angle 0^\circ$

$$\mathbf{Z}_2 = \mathbf{X}_C \parallel \mathbf{X}_L = \frac{\mathbf{X}_C\mathbf{X}_L}{\mathbf{X}_C + \mathbf{X}_L} = \frac{(2\ \Omega\ \angle -90^\circ)(3\ \Omega\ \angle 90^\circ)}{-j2\ \Omega + j3\ \Omega}$$

$$= \frac{6\ \angle 0^\circ}{j1} = \frac{6\ \angle 0^\circ}{1\ \angle 90^\circ}$$

$$= 6\ \Omega\ \angle -90^\circ$$

and

$$\mathbf{Z}_T = \mathbf{Z}_1 + \mathbf{Z}_2 = 1\ \Omega - j6\ \Omega = \mathbf{6.08\ \Omega\ \angle -80.54°}$$

b. $\mathbf{I}_S = \dfrac{\mathbf{E}}{\mathbf{Z}_T} = \dfrac{120\ \text{V}\ \angle 0°}{6.08\ \Omega\ \angle -80.5°} = \mathbf{19.74\ A\ \angle 80.54°}$

c. By the current divider rule,

$$\mathbf{I}_C = \frac{\mathbf{X}_L\mathbf{I}_S}{\mathbf{X}_L + \mathbf{X}_C} = \frac{(3\ \Omega\ \angle 90°)(19.74\ \text{A}\ \angle 80.54°)}{1\ \Omega\ \angle 90°}$$

$$= \frac{59.22\ \angle 170.54°}{1\ \angle 90°} = \mathbf{59.22\ A\ \angle 80.54°}$$

d. $\mathbf{V}_R = \mathbf{I}_S\mathbf{Z}_1 = (19.74\ \text{A}\ \angle 80.54°)(1\ \Omega\ \angle 0°) = \mathbf{19.74\ V\ \angle 80.54°}$
$\mathbf{V}_C = \mathbf{I}_S\mathbf{Z}_2 = (19.74\ \text{A}\ \angle 80.54°)(6\ \Omega\ \angle -90°)$
$= \mathbf{118.44\ V\ \angle -9.46°}$

e. $P_{\text{del}} = I_T^2R = (19.74\ \text{A})^2(1\ \Omega) = \mathbf{389.67\ W}$

f. $F_p = \cos\theta = \cos 80.54° =$ **0.164 leading,** indicating a very reactive network. That is, the closer $F_p$ is to zero, the more reactive and less resistive is the network.

---

**EXAMPLE 16.2.** For the network of Fig. 16.3:

a. If **I** is 50 A $\angle 30°$, calculate $\mathbf{I}_1$ using the current divider rule.
b. Repeat part (a) for $\mathbf{I}_2$.
c. Verify Kirchhoff's current law at one node.

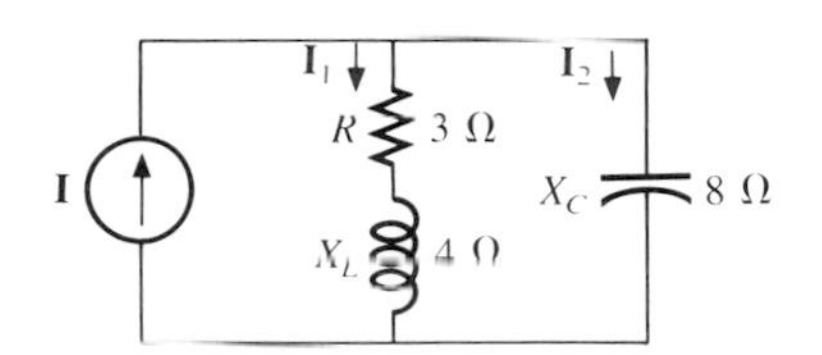


FIG. 16.3

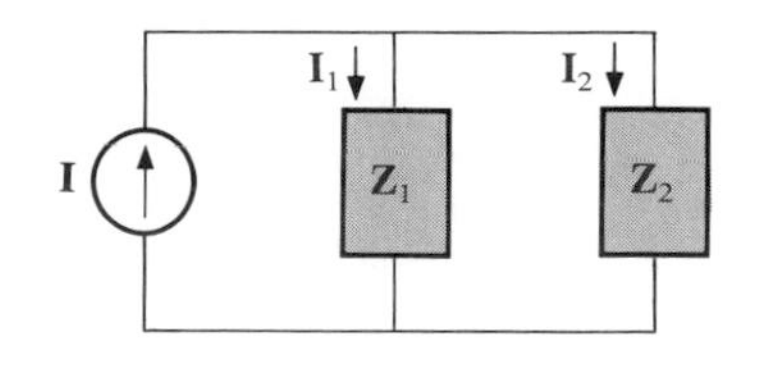


FIG. 16.4

***Solutions:***

a. Redrawing the circuit as in Fig. 16.4, we have

$$\mathbf{Z}_1 = R + jX_L = 3\ \Omega + j4\ \Omega = 5\ \Omega\ \angle 53.13°$$
$$\mathbf{Z}_2 = -jX_C = -j8\ \Omega = 8\ \Omega\ \angle -90°$$

Using the current divider rule yields

$$\mathbf{I}_1 = \frac{\mathbf{Z}_2\mathbf{I}}{\mathbf{Z}_2 + \mathbf{Z}_1} = \frac{(8\ \Omega\ \angle -90°)(50\ \text{A}\ \angle 30°)}{(-j8\ \Omega) + (3\ \Omega + j4\ \Omega)} = \frac{400\ \angle -60°}{3 - j4}$$

$$= \frac{400\ \angle -60°}{5\ \angle -53.13°} = \mathbf{80\ A\ \angle -6.87°}$$

b. $\mathbf{I}_2 = \dfrac{\mathbf{Z}_1\mathbf{I}}{\mathbf{Z}_2 + \mathbf{Z}_1} = \dfrac{(5\ \Omega\ \angle 53.13°)(50\ \text{A}\ \angle 30°)}{5\ \Omega\ \angle -53.13°} = \dfrac{250\ \angle 83.13°}{5\ \angle -53.13°}$
$= \mathbf{50\ A\ \angle 136.26°}$

c.
$$\mathbf{I} = \mathbf{I}_1 + \mathbf{I}_2$$
$$50\ \text{A}\ \angle 30° = 80\ \text{A}\ \angle -6.87° + 50\ \text{A}\ \angle 136.26°$$
$$= (79.43 - j9.57) + (-36.12 + j34.57)$$
$$= 43.31 + j25.0$$
$$50\ \text{A}\ \angle 30° = 50\ \text{A}\ \angle 30° \qquad \text{(checks)}$$

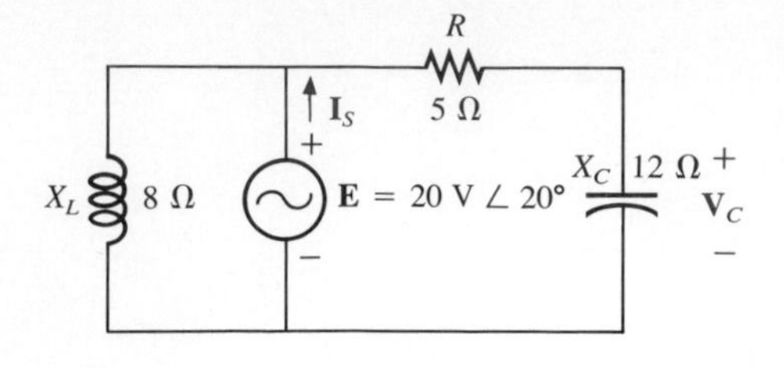


FIG. 16.5

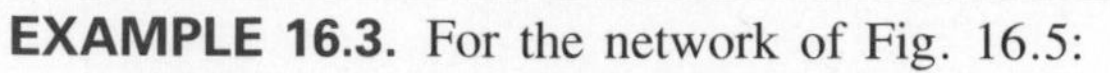

**EXAMPLE 16.3.** For the network of Fig. 16.5:

a. Calculate the voltage $\mathbf{V}_C$ using the voltage divider rule.
b. Calculate the current $\mathbf{I}_S$.

***Solutions:***

a. The network is redrawn as shown in Fig. 16.6, with

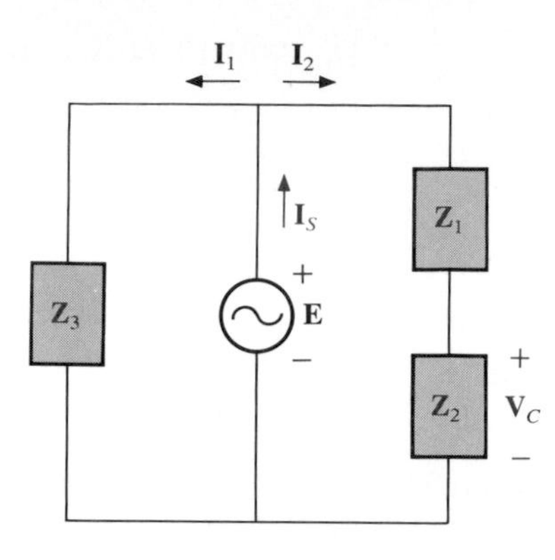


FIG. 16.6

$$\mathbf{Z}_1 = 5\ \Omega = 5\ \Omega\ \angle 0^\circ$$
$$\mathbf{Z}_2 = -j12\ \Omega = 12\ \Omega\ \angle -90^\circ$$
$$\mathbf{Z}_3 = +j8\ \Omega = 8\ \Omega\ \angle 90^\circ$$

Since $\mathbf{V}_C$ is desired, we will not combine $R$ and $X_C$ into a single block impedance. Note also how Fig. 16.6 clearly reveals that $\mathbf{E}$ is the total voltage across the series combination of $\mathbf{Z}_1$ and $\mathbf{Z}_2$, permitting the use of the voltage divider rule to calculate $\mathbf{V}_C$:

$$\mathbf{V}_C = \frac{\mathbf{Z}_2\mathbf{E}}{\mathbf{Z}_1 + \mathbf{Z}_2} = \frac{(12\ \Omega\ \angle -90^\circ)(20\ \text{V}\ \angle 20^\circ)}{5\ \Omega - j12\ \Omega} = \frac{240\ \angle -70^\circ}{13\ \angle -67.38^\circ}$$
$$= \mathbf{18.46\ V\ \angle -2.62^\circ}$$

b. $$\mathbf{I}_1 = \frac{\mathbf{E}}{\mathbf{Z}_3} = \frac{20\ \text{V}\ \angle 20^\circ}{8\ \Omega\ \angle 90^\circ} = 2.5\ \text{A}\ \angle -70^\circ$$

$$\mathbf{I}_2 = \frac{\mathbf{E}}{\mathbf{Z}_1 + \mathbf{Z}_2} = \frac{20\ \text{V}\ \angle 20^\circ}{13\ \Omega\ \angle -67.38^\circ} = 1.54\ \text{A}\ \angle 87.38^\circ$$

and

$$\mathbf{I}_S = \mathbf{I}_1 + \mathbf{I}_2$$
$$= 2.5\ \text{A}\ \angle -70^\circ + 1.54\ \text{A}\ \angle 87.38^\circ$$
$$= (0.86 - j2.35) + (0.07 + j1.54)$$
$$\mathbf{I}_S = 0.93 - j0.81 = \mathbf{1.23\ A\ \angle -41.05^\circ}$$

**EXAMPLE 16.4.** For Fig. 16.7:

a. Calculate the current $\mathbf{I}$.
b. Find the voltage $\mathbf{V}_{ab}$.

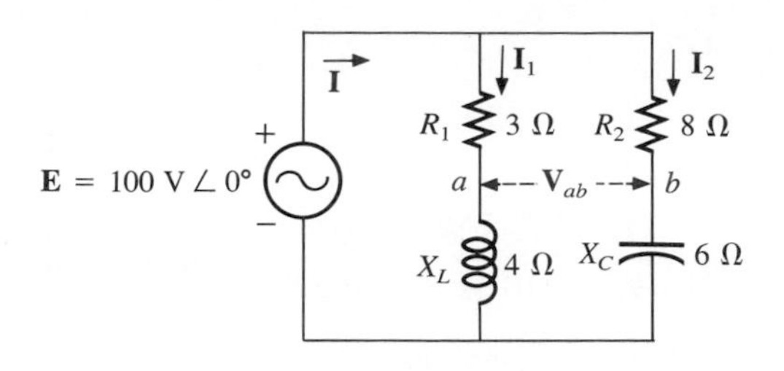


FIG. 16.7

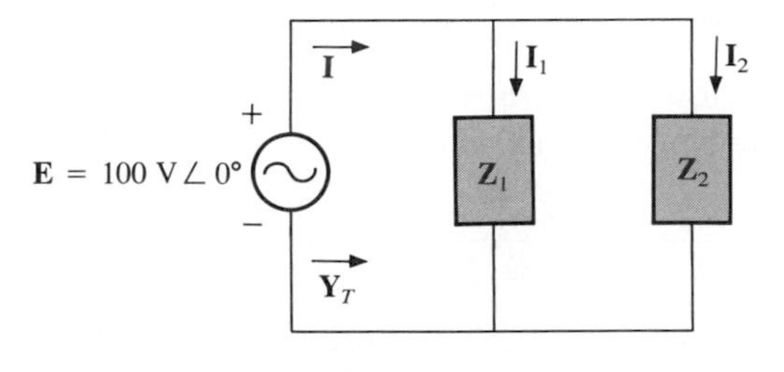


FIG. 16.8

***Solutions:***

a. Redrawing the circuit as in Fig. 16.8, we obtain

$$\mathbf{Z}_1 = R_1 + jX_L = 3\ \Omega + j4\ \Omega = 5\ \Omega\ \angle 53.13^\circ$$
$$\mathbf{Z}_2 = R_2 - jX_C = 8\ \Omega - j6\ \Omega = 10\ \Omega\ \angle -36.87^\circ$$
$$\mathbf{Y}_T = \mathbf{Y}_1 + \mathbf{Y}_2$$
$$= \frac{1}{\mathbf{Z}_1} + \frac{1}{\mathbf{Z}_2}$$
$$= \frac{1}{5\ \Omega\ \angle 53.13^\circ} + \frac{1}{10\ \Omega\ \angle -36.87^\circ}$$
$$= 0.2\ \text{S}\ \angle -53.13^\circ + 0.1\ \text{S}\ \angle 36.87^\circ$$
$$= (0.12\ \text{S} - j0.16\ \text{S}) + (0.08\ \text{S} + j0.06\ \text{S})$$
$$\mathbf{Y}_T = 0.2\ \text{S} - j0.1\ \text{S} = \mathbf{0.224\ S\ \angle -26.57^\circ}$$

By Ohm's law,

$$\mathbf{I} = \frac{\mathbf{E}}{\mathbf{Z}_T} = \mathbf{E}\mathbf{Y}_T = (100\ \text{V}\ \angle 0^\circ)(0.224\ \text{S}\ \angle -26.57^\circ)$$
$$= \mathbf{22.4\ A\ \angle -26.57^\circ}$$

Using another approach, we find the total impedance:

$$\mathbf{Z}_T = \frac{\mathbf{Z}_1\mathbf{Z}_2}{\mathbf{Z}_1 + \mathbf{Z}_2} = \frac{(5\ \Omega\ \angle 53.13^\circ)(10\ \Omega\ \angle -36.87^\circ)}{(3\ \Omega + j4\ \Omega) + (8\ \Omega - j6\ \Omega)}$$
$$= \frac{50\ \angle 16.26^\circ}{11 - j2} = \frac{50\ \angle 16.26^\circ}{11.18\ \angle -10.30^\circ}$$
$$= \mathbf{4.472\ \Omega\ \angle 26.56^\circ}$$

and

$$\mathbf{I} = \frac{\mathbf{E}}{\mathbf{Z}_T} = \frac{100\ \text{V}\ \angle 0^\circ}{4.472\ \Omega\ \angle 26.56^\circ} = \mathbf{22.36\ \angle -26.56^\circ}$$

b. By Ohm's law,

$$\mathbf{I}_1 = \frac{\mathbf{E}}{\mathbf{Z}_1} = \frac{100\text{ V}\angle 0^\circ}{5\ \Omega\angle 53.13^\circ} = \mathbf{20\ A\ \angle -53.13^\circ}$$

$$\mathbf{I}_2 = \frac{\mathbf{E}}{\mathbf{Z}_2} = \frac{100\text{ V}\angle 0^\circ}{10\ \Omega\angle -36.87^\circ} = \mathbf{10\ A\ \angle 36.87^\circ}$$

Returning to Fig. 16.7, we have

$$\mathbf{V}_{R_1} = \mathbf{I}_1\mathbf{R}_1 = (20\text{ A}\angle -53.13^\circ)(3\ \Omega\angle 0^\circ) = \mathbf{60\ V\ \angle -53.13^\circ}$$
$$\mathbf{V}_{R_2} = \mathbf{I}_2\mathbf{R}_2 = (10\text{ A}\angle +36.87^\circ)(8\ \Omega\angle 0^\circ) = \mathbf{80\ V\ \angle +36.87^\circ}$$

Instead of using the two steps just shown, $\mathbf{V}_{R_1}$ or $\mathbf{V}_{R_2}$ could have been determined in one step using the voltage divider rule:

$$\mathbf{V}_{R_1} = \frac{(3\ \Omega\angle 0^\circ)(100\text{ V}\angle 0^\circ)}{3\ \Omega\angle 0^\circ + 4\ \Omega\angle 90^\circ} = \frac{300\angle 0^\circ}{5\angle 53.13^\circ} = \mathbf{60\ V\ \angle -53.13^\circ}$$

In order to find $\mathbf{V}_{ab}$, Kirchhoff's voltage law must be applied around the loop (Fig. 16.9) consisting of the 3-Ω and 8-Ω resistors. By Kirchhoff's voltage law,

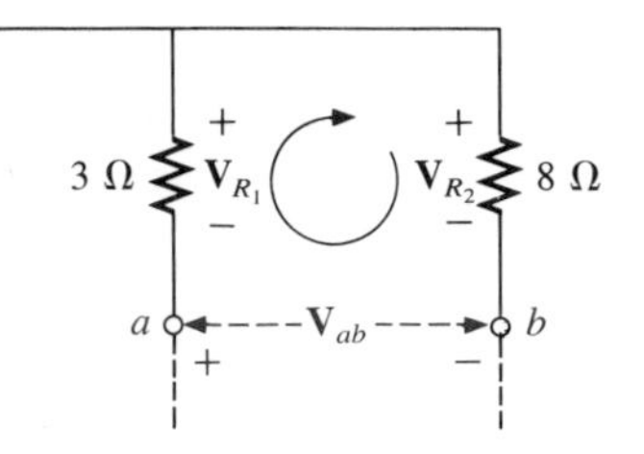


FIG. 16.9

$$\mathbf{V}_{ab} + \mathbf{V}_{R_1} - \mathbf{V}_{R_2} = 0$$

or

$$\begin{aligned}\mathbf{V}_{ab} &= \mathbf{V}_{R_2} - \mathbf{V}_{R_1}\\ &= 80\text{ V}\angle 36.87^\circ - 60\text{ V}\angle -53.13^\circ\\ &= (64 + j48) - (36 - j48)\\ &= 28 + j96\\ \mathbf{V}_{ab} &= \mathbf{100\ V\ \angle 73.74^\circ}\end{aligned}$$

**EXAMPLE 16.5.** For the network of Fig. 16.10:

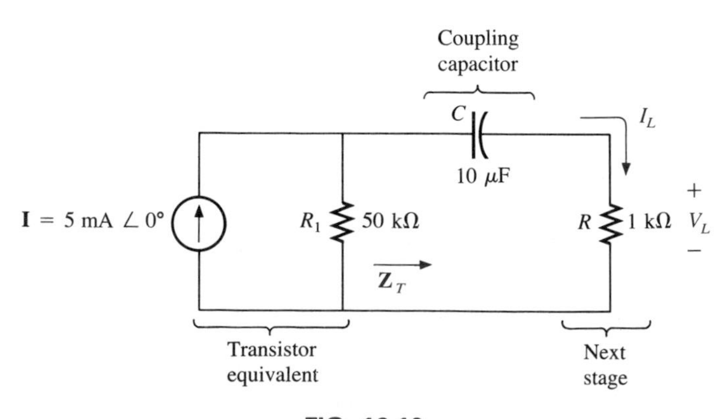

FIG. 16.10

a. Find $\mathbf{Z}_T$ and compare it to $R_1 = 50\text{ k}\Omega$ at $f = 0.1$ kHz and $f = 20$ kHz.

b. Determine $\mathbf{I}_L$ at both frequencies and compare it to $\mathbf{I}$, keeping in mind the results of part (a).
c. Calculate $\mathbf{V}_L$ at each frequency.

(The type of network found in Fig. 16.10 will appear frequently in the analysis of transistor networks. The coupling capacitor is designed to be an open circuit for dc and as low an impedance as possible for the frequencies of interest to insure that $\mathbf{V}_L$ is a maximum value. The frequencies chosen in the example represent the low and high ends of the audio range.)

***Solutions:***

a. $f = 0.1\text{ kHz} = 100\text{ Hz}$:

$$X_C = \frac{1}{2\pi fC} = \frac{1}{(6.28)(100\text{ Hz})(10 \times 10^{-6}\text{ F})} = 159.24\ \Omega$$

$$\mathbf{Z}_T = R - jX_C = 1000\ \Omega - j159.24\ \Omega = \mathbf{1012.6\ \Omega\ \angle -9.05^\circ}$$

$f = 20\text{ kHz}$:

$$X_C = \frac{1}{2\pi fC} = \frac{1}{(6.28)(20 \times 10^3\text{ Hz})(10 \times 10^{-6}\text{ F})} = 0.796\ \Omega$$

$$\mathbf{Z}_T = R - jX_C = 1000\ \Omega - j0.796\ \Omega = 1000\ \Omega\ \angle -0.046^\circ$$
$$\cong \mathbf{1000\ \Omega\ \angle 0^\circ} = R$$

Note the dramatic change in $X_C$ with frequency. Obviously, the higher the frequency, the better the short-circuit approximation for $X_C$ for ac conditions.

b. $f = 100\text{ Hz}$:

$$\mathbf{I}_L = \frac{R_1\mathbf{I}}{R_1 + R - jX_C} = \frac{(50 \times 10^3\ \Omega\ \angle 0^\circ)(5 \times 10^{-3}\text{ A}\ \angle 0^\circ)}{50{,}000\ \Omega + 1000\ \Omega - j159.24\ \Omega}$$
$$= \frac{250\ \angle 0^\circ}{51{,}000 - j159.24} \cong \frac{250\ \angle 0^\circ}{51{,}000\ \angle 0^\circ} = \mathbf{4.9\text{ mA}\ \angle 0^\circ}$$

$f = 20\text{ kHz}$:

$$\mathbf{I}_L = \frac{R_1\mathbf{I}}{R_1 + R - jX_C} = \frac{(50 \times 10^3\ \Omega\ \angle 0^\circ)(5 \times 10^{-3}\text{ A}\ \angle 0^\circ)}{50{,}000\ \Omega + 1000\ \Omega - j0.796\ \Omega}$$
$$= \frac{250\ \angle 0^\circ}{51{,}000\ \angle 0^\circ} = \mathbf{4.9\text{ mA}\ \angle 0^\circ}$$

Note that at both frequencies, the effect of $X_C$ on $\mathbf{I}_L$ is negligible.

c. Since $\mathbf{I}_L = 4.9\text{ mA}\ \angle 0^\circ$ at both frequencies,

$$\mathbf{V}_L = \mathbf{I}_L\mathbf{R} = (4.9\text{ mA}\ \angle 0^\circ)(1\text{ k}\Omega\ \angle 0^\circ) = \mathbf{4.9\text{ V}}$$

*Ideal conditions* require that $R_1 = \infty\ \Omega$ (open circuit) and $X_C = 0\ \Omega$, resulting in

$$\mathbf{V}_L = \mathbf{I}_L\mathbf{R} = (5\text{ mA}\ \angle 0^\circ)(1\text{ k}\Omega\ \angle 0^\circ) = \mathbf{5\text{ V}}\ \angle 0^\circ$$

The percent difference in magnitude is

$$\frac{5\text{ V} - 4.9\text{ V}}{5\text{ V}} \times 100\% = \mathbf{2\%}$$

**EXAMPLE 16.6.** For the network of Fig. 16.11:

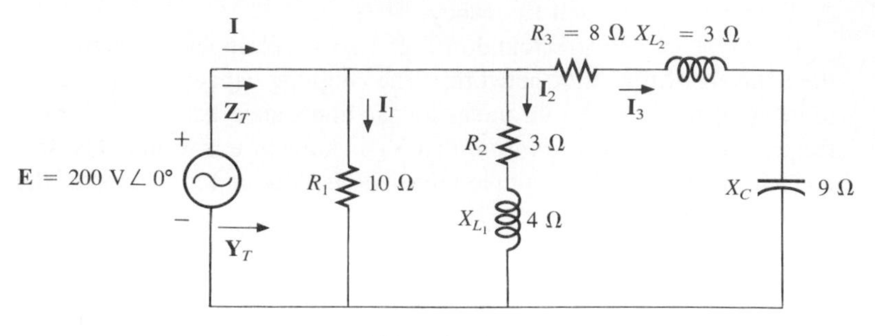

FIG. 16.11

a. Compute **I.**
b. Find $\mathbf{I}_1$, $\mathbf{I}_2$, and $\mathbf{I}_3$.
c. Verify Kirchhoff's current law by showing that

$$\mathbf{I} = \mathbf{I}_1 + \mathbf{I}_2 + \mathbf{I}_3$$

d. Find the total impedance of the circuit.

***Solutions:***

a. Redrawing the circuit as in Fig. 16.12, we obtain

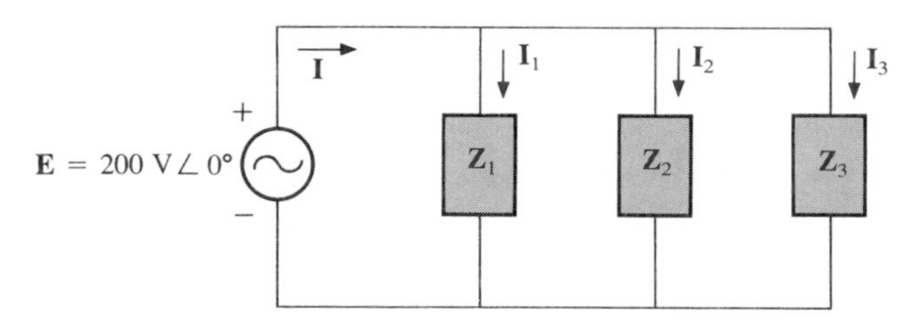


FIG. 16.12

$$\mathbf{Z}_1 = R_1 = 10\ \Omega\ \angle 0°$$
$$\mathbf{Z}_2 = R_2 + jX_{L_1} = 3\ \Omega + j4\ \Omega$$
$$\mathbf{Z}_3 = R_3 + jX_{L_2} - jX_C = 8\ \Omega + j3\ \Omega - j9\ \Omega = 8\ \Omega - j6\ \Omega$$

The total admittance is

$$\mathbf{Y}_T = \mathbf{Y}_1 + \mathbf{Y}_2 + \mathbf{Y}_3$$
$$= \frac{1}{\mathbf{Z}_1} + \frac{1}{\mathbf{Z}_2} + \frac{1}{\mathbf{Z}_3} = \frac{1}{10\ \Omega} + \frac{1}{3\ \Omega + j4\ \Omega} + \frac{1}{8\ \Omega - j6\ \Omega}$$

However,

$$\frac{1}{\mathbf{Z}_2} = \frac{1}{3\ \Omega + j4\ \Omega} = \frac{3}{(3)^2 + (4)^2} - j\frac{4}{(3)^2 + (4)^2} = \frac{3}{25}\ \text{S} - j\frac{4}{25}\ \text{S}$$

and

$$\frac{1}{\mathbf{Z}_3} = \frac{1}{8\ \Omega - j6\ \Omega} = \frac{8}{(8)^2 + (6)^2} + j\frac{6}{(8)^2 + (6)^2} = \frac{8}{100}\ \text{S} + j\frac{6}{100}\ \text{S}$$

Therefore,

$$\mathbf{Y}_1 = \frac{1}{10}\text{ S} + j0 = \frac{10}{100}\text{ S} + j0$$

$$\mathbf{Y}_2 = \frac{3}{25}\text{ S} - j\frac{4}{25}\text{ S} = \frac{12}{100}\text{ S} - j\frac{16}{100}\text{ S}$$

$$\mathbf{Y}_3 = \frac{8}{100}\text{ S} + j\frac{6}{100}\text{ S} = \frac{8}{100}\text{ S} + j\frac{6}{100}\text{ S}$$

$$\mathbf{Y}_T = \mathbf{Y}_1 + \mathbf{Y}_2 + \mathbf{Y}_3 = \frac{30}{100}\text{ S} - j\frac{10}{100}\text{ S}$$

$$\mathbf{I} = \mathbf{E}\mathbf{Y}_T = 200\text{ V }\angle 0°\left(\frac{30}{100}\text{ S} - j\frac{10}{100}\text{ S}\right) = \mathbf{60 - \mathit{j}20}$$
$$= \mathbf{63.25\text{ A }\angle -18.43°}$$

b. Since the voltage is the same across parallel branches,

$$\mathbf{I}_1 = \frac{\mathbf{E}}{\mathbf{Z}_1} = \frac{200\text{ V }\angle 0°}{10\ \Omega\ \angle 0°} = \mathbf{20\text{ A }\angle 0°}$$

$$\mathbf{I}_2 = \frac{\mathbf{E}}{\mathbf{Z}_2} = \frac{200\text{ V }\angle 0°}{5\ \Omega\ \angle 53.13°} = \mathbf{40\text{ A }\angle -53.13°}$$

$$\mathbf{I}_3 = \frac{\mathbf{E}}{\mathbf{Z}_3} = \frac{200\text{ V }\angle 0°}{10\ \Omega\ \angle -36.87°} = \mathbf{20\text{ A }\angle +36.87°}$$

c.
$$\mathbf{I} = \mathbf{I}_1 + \mathbf{I}_2 + \mathbf{I}_3$$
$$60 - j20 = 20\ \angle 0° + 40\ \angle -53.13° + 20\ \angle +36.87°$$
$$= (20 + j0) + (24 - j32) + (16 + j12)$$
$$60 - j20 = 60 - j20 \qquad \text{(checks)}$$

d.
$$\mathbf{Z}_T = \frac{1}{\mathbf{Y}_T} = \frac{1}{0.3\text{ S} - j0.1\text{ S}}$$
$$= \frac{0.3}{(0.3)^2 + (0.1)^2} + j\frac{0.1}{(0.3)^2 + (0.1)^2}$$
$$= \frac{0.3}{0.1} + j\frac{0.1}{0.1} = 3\ \Omega + j\ \Omega = \mathbf{3.16\ \Omega\ \angle 18.43°}$$

---

**EXAMPLE 16.7.** For the network of Fig. 16.13:

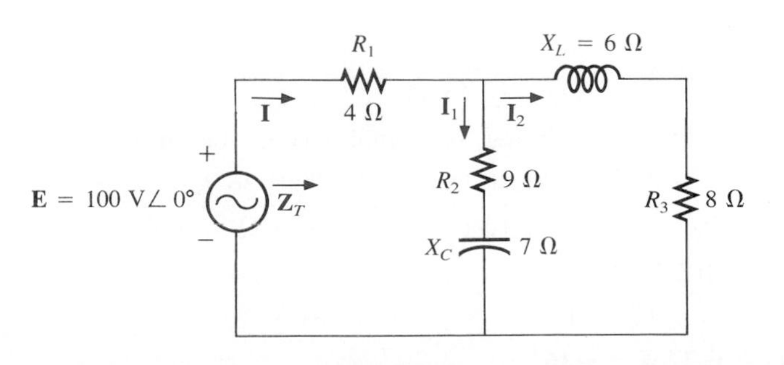


**FIG. 16.13**

a. Calculate the total impedance $\mathbf{Z}_T$.
b. Compute $\mathbf{I}$.
c. Find the total power factor.
d. Calculate $\mathbf{I}_1$ and $\mathbf{I}_2$.
e. Find the average power delivered to the circuit.

***Solutions:***

a. Redrawing the circuit as in Fig. 16.14, we have

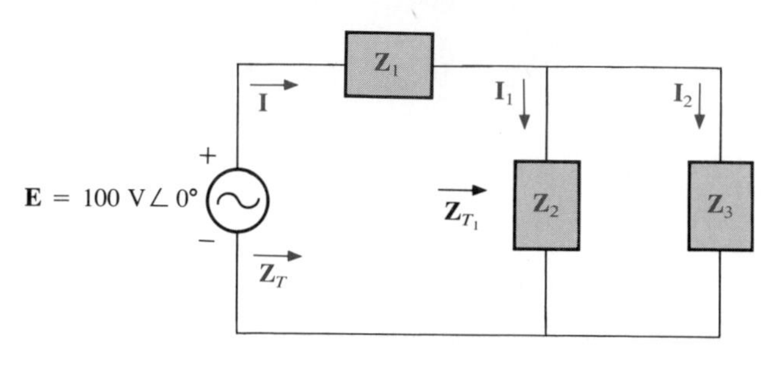

FIG. 16.14

$$\mathbf{Z}_1 = R_1 = 4\ \Omega\ \angle 0°$$
$$\mathbf{Z}_2 = R_2 - jX_C = 9\ \Omega - j7\ \Omega = 11.40\ \Omega\ \angle -37.87°$$
$$\mathbf{Z}_3 = R_3 + jX_L = 8\ \Omega + j6\ \Omega = 10\ \Omega\ \angle +36.87°$$

The total impedance is

$$\mathbf{Z}_T = \mathbf{Z}_1 + \mathbf{Z}_{T_1}$$
$$= \mathbf{Z}_1 + \frac{\mathbf{Z}_2\mathbf{Z}_3}{\mathbf{Z}_2 + \mathbf{Z}_3}$$
$$= 4\ \Omega + \frac{(11.4\ \Omega\ \angle -37.87°)(10\ \Omega\ \angle 36.87°)}{(9\ \Omega - j7\ \Omega) + (8\ \Omega + j6\ \Omega)}$$
$$= 4 + \frac{114\ \angle -1.00°}{17.03\ \angle -3.37°}$$
$$= 4 + 6.69\ \angle 2.37°$$
$$= 4 + 6.68 + j0.28$$
$$\mathbf{Z}_T = 10.68\ \Omega + j0.28\ \Omega = \mathbf{10.68\ \Omega\ \angle 1.5°}$$

b. $$\mathbf{I} = \frac{\mathbf{E}}{\mathbf{Z}_T} = \frac{100\ \text{V}\ \angle 0°}{10.68\ \Omega\ \angle 1.5°} = \mathbf{9.36\ A\ \angle -1.5°}$$

c. $$F_p = \cos\theta_T = \frac{R}{Z_T} = \frac{10.68\ \Omega}{10.68\ \Omega} = \mathbf{1}$$

(interesting, considering the complexity of the network)

d. $$\mathbf{I}_2 = \frac{\mathbf{Z}_2\mathbf{I}}{\mathbf{Z}_2 + \mathbf{Z}_3} = \frac{(11.40\ \Omega\ \angle -37.87°)(9.36\ \text{A}\ \angle -1.5°)}{(9\ \Omega - j7\ \Omega) + (8\ \Omega + j6\ \Omega)}$$
$$= \frac{106.7\ \angle -39.37°}{17 - j1} = \frac{106.7\ \angle -39.37°}{17.03\ \angle -3.37°}$$
$$\mathbf{I}_2 = \mathbf{6.27\ A\ \angle -36°}$$

Applying Kirchhoff's current law yields

$$\mathbf{I} = \mathbf{I}_1 + \mathbf{I}_2$$

or

$$\begin{aligned}\mathbf{I}_1 &= \mathbf{I} - \mathbf{I}_2 \\ &= (9.36 \angle -1.5^\circ) - (6.27 \angle -36^\circ) \\ &= (9.36 - j0.25) - (5.07 - j3.69) \\ \mathbf{I}_1 &= 4.29 + j3.44 = \mathbf{5.5\ A\ \angle 38.72^\circ}\end{aligned}$$

e. $$\begin{aligned}P_T &= EI \cos \theta_T \\ &= (100\ \text{V})(9.36\ \text{A}) \cos 1.5^\circ \\ &= (936)(1) \\ P_T &= \mathbf{936\ W}\end{aligned}$$

## 16.3 LADDER NETWORKS

Ladder networks were discussed in some detail in Chapter 7. This section will simply apply the first method described in Section 7.3 to the general sinusoidal ac ladder network of Fig. 16.15. The current $\mathbf{I}_6$ is desired.

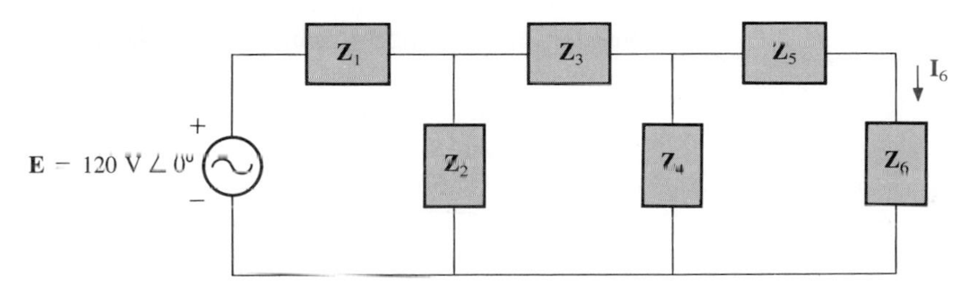


FIG. 16.15

Impedances $\mathbf{Z}_T$, $\mathbf{Z'}_T$, and $\mathbf{Z''}_T$ and currents $\mathbf{I}_1$ and $\mathbf{I}_3$ are defined in Fig. 16.16:

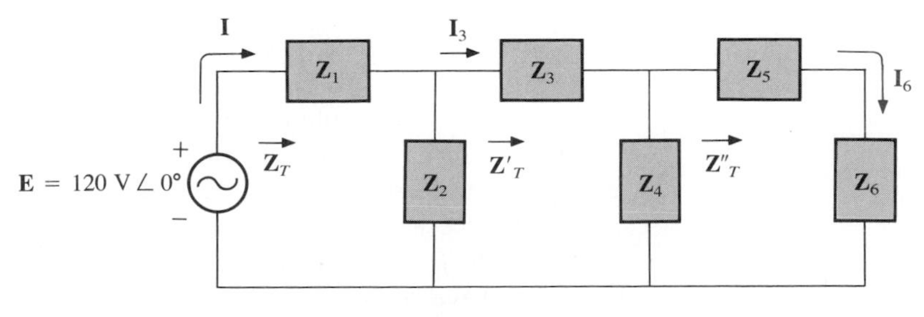


FIG. 16.16

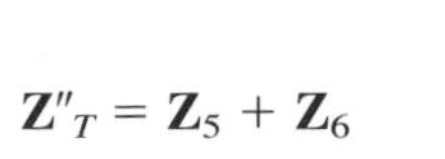

$$\mathbf{Z''}_T = \mathbf{Z}_5 + \mathbf{Z}_6$$

and

$$\mathbf{Z'}_T = \mathbf{Z}_3 + \mathbf{Z}_4 \parallel \mathbf{Z''}_T$$

with

$$\mathbf{Z}_T = \mathbf{Z}_1 + \mathbf{Z}_2 \parallel \mathbf{Z}'_T$$

Then

$$\mathbf{I} = \frac{\mathbf{E}}{\mathbf{Z}_T}$$

and

$$\mathbf{I}_3 = \frac{\mathbf{Z}_2\mathbf{I}}{\mathbf{Z}_2 + \mathbf{Z}'_T}$$

with

$$\mathbf{I}_6 = \frac{\mathbf{Z}_4\mathbf{I}_3}{\mathbf{Z}_4 + \mathbf{Z}''_T}$$

## 16.4 COMPUTER ANALYSIS

The content of this chapter reveals the extensive number of calculations required to solve for some variables. A BASIC program would require frequent use of the conversion subroutine introduced earlier to convert from one form to the other before the fundamental laws of electric circuits can be applied. In this light, PSPICE is a very definite asset, since it appears that all that is required is the correct entry of the network configuration and a few command lines to produce the desired results in a matter of seconds. The first run of this section is for the network of Fig. 16.17, which appeared as Example 16.4. Review the number of steps required to determine the requested unknown quantities in the example versus the time required to obtain a solution with PSPICE—a total of 9.88 seconds once the input file was entered!

There are a few factors associated with the use of packaged programs that have not been touched on in the earlier chapters and that deserve discussion at this point. First, we have to assume the package was properly written and can handle any nuances that a circuit configuration may introduce. In other words, no matter what elements or frequency are entered and in whatever configuration, the package will provide a *correct* solution. Any package can provide a solution—but more importantly, is it the correct solution? For the PSPICE software package, the background of its development and the extensive use it is enjoying in recent years suggest that it is of the highest caliber and for the analysis of the type we are currently interested in, fully trustworthy. Let us assume PSPICE is absolutely bug-free. What are the other concerns? How often have you used a calculator, obtained a result, and assumed the answer was correct for future calculations—only to find way down the line that the result was incorrect due to a wrong entry in the calculator sequence. The same thing can happen with PSPICE. An incorrect entry in the input file can result in a solution that is totally erroneous, but unless we have a "sense" for what to expect and some idea of the maneuvers the program is going through, we will have no way to "check" the results. For the future, therefore, be skeptical of all results—check and recheck the input file, compare the solutions to the magnitudes involved, and evaluate the results based on your own experience with similar systems. This is not to downgrade PSPICE but to simply ensure that the types of errors often encountered using a calculator are not repeated using the package approach.

The input file for Fig. 16.17 is provided in Fig. 16.18. If you compare the network of Fig. 16.7 with that of Fig. 16.17, you will find that

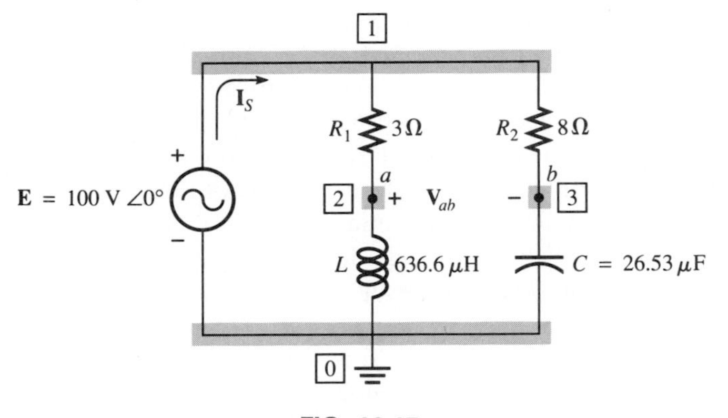


FIG. 16.17

```
************************* Evaluation PSpice (January 1989) ******* 17:08:53 *******

Chapter 16 - Series-Parallel Network (I) - AC Analysis

****     CIRCUIT DESCRIPTION

*****************************************************************************

VE 1 0 AC 100V 0
R1 1 2 3
L 2 0 636.6UH
R2 1 3 8
C 3 0 26.53UF
.AC LIN 1 1KH 1KH
.PRINT AC VM(2,3) VP(2,3) IM(VE) IP(VE)
.OPTIONS NOPAGE
.END

****     AC ANALYSIS                      TEMPERATURE =   27.000 DEG C

  FREQ        VM(2,3)     VP(2,3)     IM(VE)      IP(VE)

  1.000E+03   1.000E+02   7.374E+01   2.236E+01   1.534E+02
```

FIG. 16.18

the reactances of the elements were provided in ohms in Fig. 16.7 and in their individual units of measurement in Fig. 16.17. This maneuver was required because PSPICE calculates the reactance level at the applied frequency and does not permit the entry of reactance levels in ohms. In order to duplicate the analysis of Example 16.4, a frequency of 1 kHz was chosen, and the required inductor ($L = X_L/2\pi f$) and capacitor ($C = 1/2\pi f X_C$) were determined using the appropriate equation.

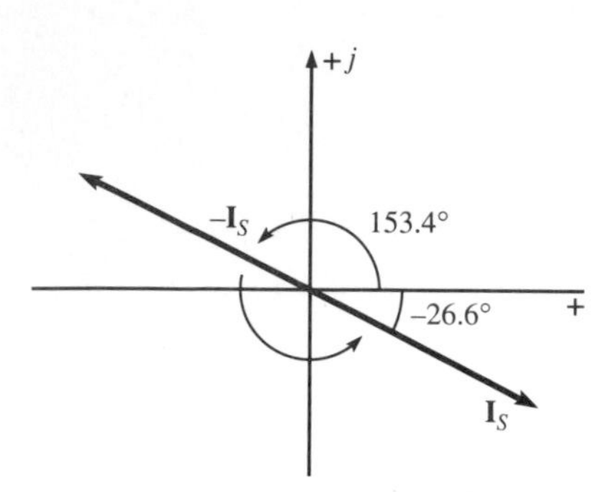


FIG. 16.19

In Example 16.4 $\mathbf{V}_{ab}$ was found to be 100 V $\angle 73.74°$, matching the result for V(2, 3) in the output file of Fig. 16.18. For the source current, Example 16.4 provided a result of $\mathbf{I}_S = 22.36\text{ A} \angle -26.56°$, whereas PSPICE produced a result of 22.36 A $\angle 153.4°$. The magnitude is the same, but the phase angle appears to be totally different. This difference in phase angle is due to the fact that current through an independent source (dc or ac) is defined by PSPICE to flow from the positive to negative node. In the example, the currents $\mathbf{I}_S$ of Figs. 16.7 and 16.17 have the opposite direction, requiring that a 180° phase shift be introduced to the PSPICE solution, as shown in Fig. 16.19. Now note the close correlation with the result of Example 16.4 where $\mathbf{I}_S = 22.36\text{ A} \angle -26.56°$.

The next application of PSPICE is to the series-parallel network of Fig. 16.20, which is Problem 10 in the text. Although the solutions are provided here, they should provide added incentive to obtain the same results longhand.

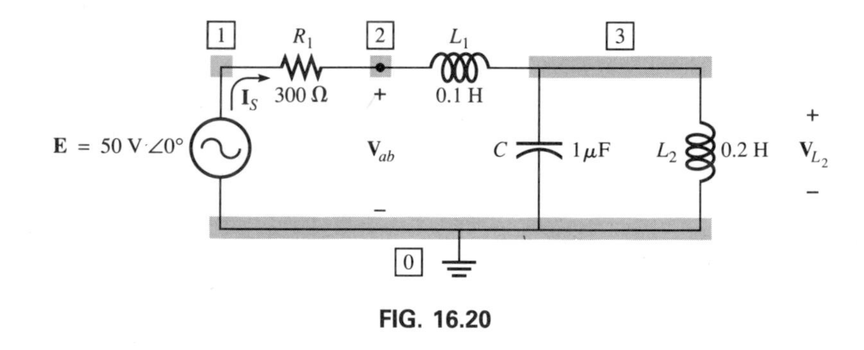


FIG. 16.20

The input file is provided in Fig. 16.21 with a request for $\mathbf{I}_S$, $\mathbf{V}_{L_2}$ and $\mathbf{V}_2$. The current $\mathbf{I}_S$ is requested as I(Rl), since I(R1) will be the same as $\mathbf{I}_T$ (series configuration) and I(Rl) will have the same direction as $\mathbf{I}_S$, avoiding the 180° phase shift introduced by asking for the current through the source. The solution for $\mathbf{I}_S$ obtained in the longhand manner is 93 mA $\angle -56.07°$, comparing very closely to the 93.01 mA $\angle -56.08°$ obtained with PSPICE. The longhand solution for $\mathbf{V}_{L_2}$ is 16.931 V $\angle 213.93°$ versus 16.95 V $\angle -143.1°$ for PSPICE. The results are very close when one changes the phase angle for the PSPICE solution to a positive number. That is, $\mathbf{V}_{L_2} = 16.95\text{ V} \angle +360° - 143.1° = 16.95\text{ V} \angle 213.9°$. The result for $\mathbf{V}_2$ using PSPICE is exactly the same as obtained in the longhand manner: $\mathbf{V}_2 = \mathbf{V}_{ab} = 41.49\text{ V} \angle 33.92°$.

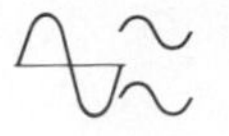



```
************************ Evaluation PSpice (January 1989) ******* 17:11:17 *******

Chapter 16 - Series-Parallel Network (II) - AC Analysis

****     CIRCUIT DESCRIPTION

******************************************************************************


VE 1 0 AC 50V 0
R1 1 2 300
L1 2 3 0.1H
C  3 0 1UF
L2 3 0 0.2H
.AC LIN 1 1KH 1KH
.PRINT AC IM(R1) IP(R1) VM(L2) VP(L2) VM(2) VP(2)
.OPTIONS NOPAGE
.END

****     AC ANALYSIS                        TEMPERATURE =   27.000 DEG C


 FREQ        IM(R1)       IP(R1)       VM(L2)       VP(L2)       VM(2)

  1.000E+03   9.301E-02  -5.608E+01   1.695E+01  -1.461E+02   4.149E+01

****     AC ANALYSIS                        TEMPERATURE =   27.000 DEG C


 FREQ        VP(2)

  1.000E+03   3.392E+01
```

FIG. 16.21

## PROBLEMS

### SECTION 16.2 Illustrative Examples

1. For the series-parallel network of Fig. 16.22:
   a. Calculate $\mathbf{Z}_T$.
   b. Determine **I**.
   c. Determine $\mathbf{I}_1$.
   d. Find $\mathbf{I}_2$ and $\mathbf{I}_3$.
   e. Find $\mathbf{V}_L$.

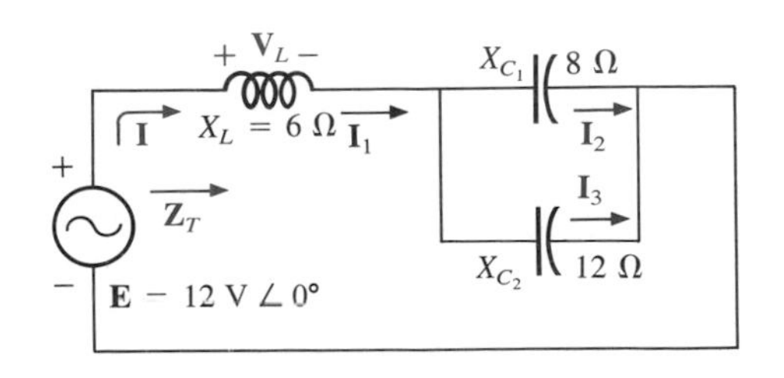


FIG. 16.22

2. For the network of Fig. 16.23:
   **a.** Find the total impedance $\mathbf{Z}_T$.
   **b.** Determine the current $\mathbf{I}_S$.
   **c.** Calculate $\mathbf{I}_C$ using the current divider rule.
   **d.** Calculate $\mathbf{V}_L$ using the voltage divider rule.

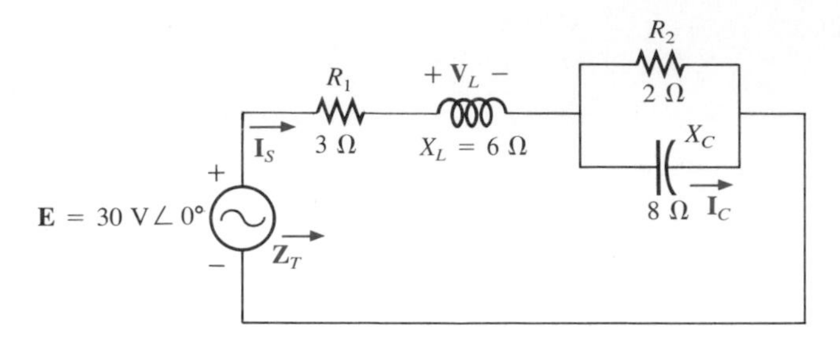

FIG. 16.23

3. For the network of Fig. 16.24:
   **a.** Find the total impedance $\mathbf{Z}_T$ and the total admittance $\mathbf{Y}_T$.
   **b.** Find the current $\mathbf{I}_S$.
   **c.** Calculate $\mathbf{I}_2$ using the current divider rule.
   **d.** Calculate $\mathbf{V}_C$.
   **e.** Calculate the average power delivered to the network.

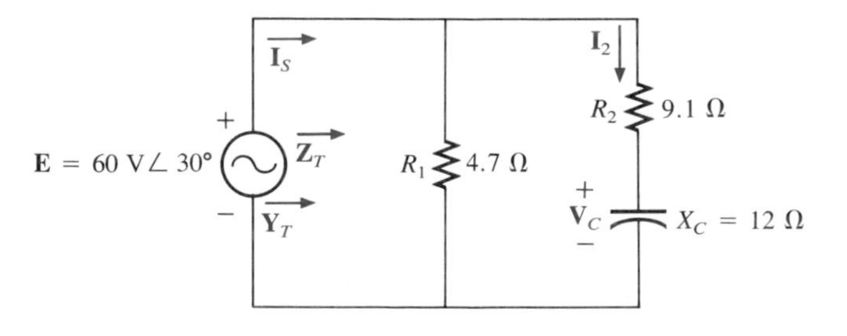


FIG. 16.24

4. For the network of Fig. 16.25:
   **a.** Find the total impedance $\mathbf{Z}_T$.
   **b.** Calculate the voltage $\mathbf{V}_2$ and the current $\mathbf{I}_L$.
   **c.** Find the power factor of the network.

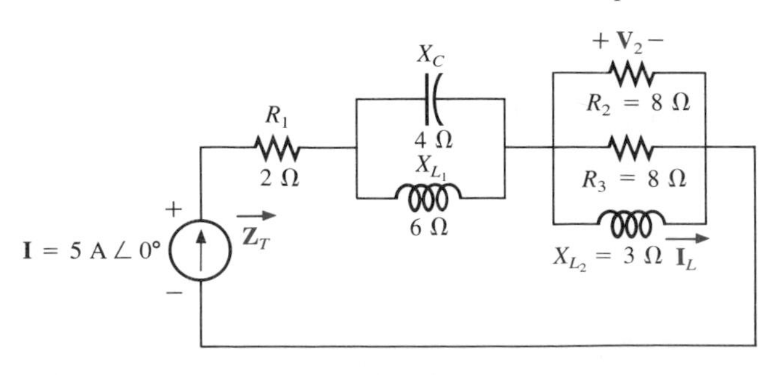


FIG. 16.25

5. For the network of Fig. 16.26:
   a. Find the current **I**.
   b. Find the voltage $\mathbf{V}_C$.
   c. Find the average power delivered to the network.

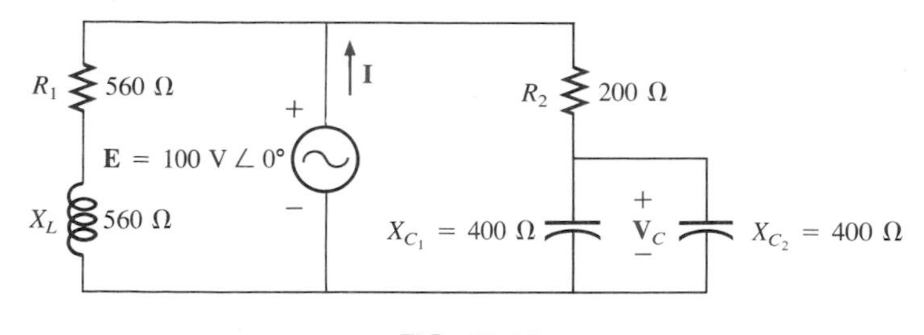


FIG. 16.26

*6. For the network of Fig. 16.27:
   a. Find the current $\mathbf{I}_1$.
   b. Calculate the voltage $\mathbf{V}_C$ using the voltage divider rule.
   c. Find the voltage $\mathbf{V}_{ab}$.

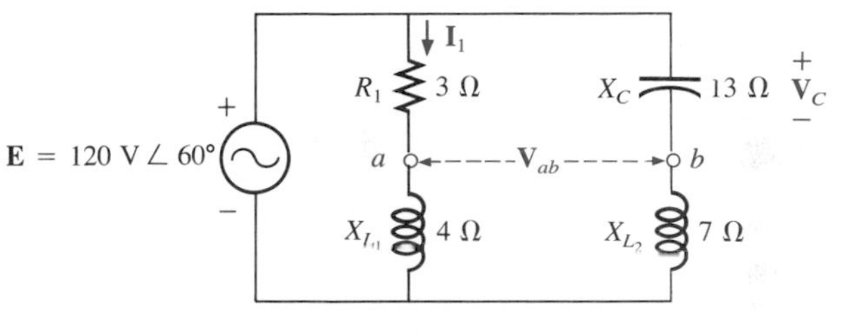

FIG. 16.27

*7. For the network of Fig. 16.28:
   a. Find the current $\mathbf{I}_1$.
   b. Find the voltage $\mathbf{V}_1$.
   c. Calculate the average power delivered to the network.

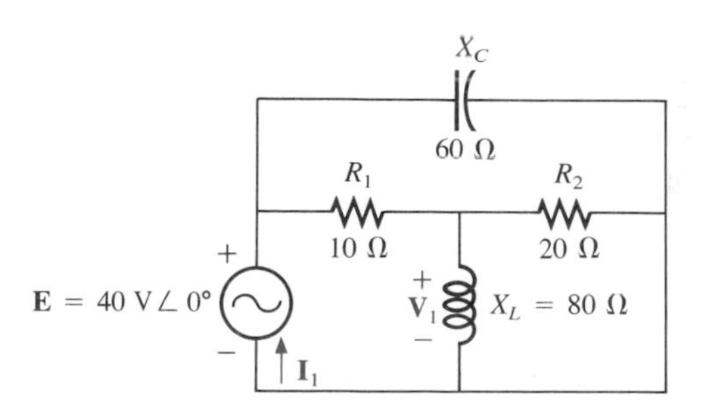

FIG. 16.28

8. For the network of Fig. 16.29:
   a. Find the total impedance $\mathbf{Z}_T$ and the admittance $\mathbf{Y}_T$.
   b. Find the currents $\mathbf{I}_1$, $\mathbf{I}_2$, and $\mathbf{I}_3$.
   c. Verify Kirchhoff's current law by showing that $\mathbf{I}_S = \mathbf{I}_1 + \mathbf{I}_2 + \mathbf{I}_3$.
   d. Find the power factor of the network and indicate whether it is leading or lagging.

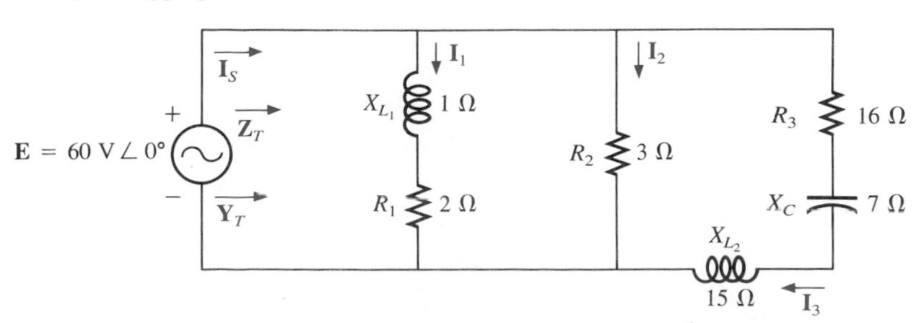


FIG. 16.29

***9.** For the network of Fig. 16.30:

**a.** Find the total admittance $\mathbf{Y}_T$.

**b.** Find the voltages $\mathbf{V}_1$ and $\mathbf{V}_2$.

**c.** Find the current $\mathbf{I}_3$.

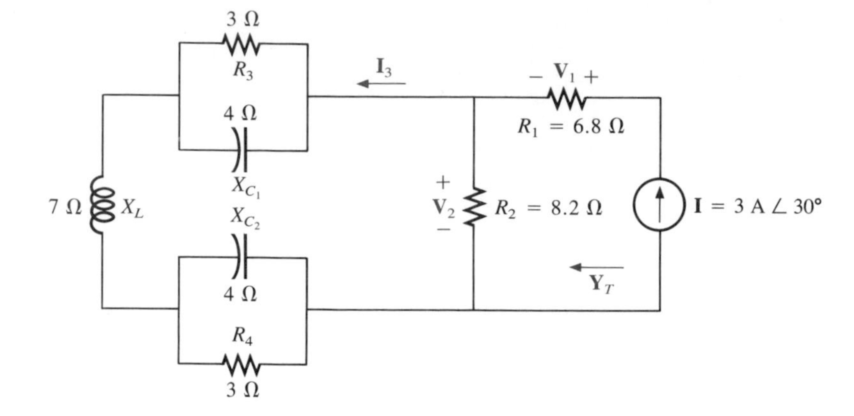

FIG. 16.30

***10.** For the network of Fig. 16.31:

**a.** Find the total impedance $\mathbf{Z}_T$ and the admittance $\mathbf{Y}_T$.

**b.** Find the current $i_S$ in phasor form.

**c.** Find the currents $i_1$ and $i_2$ in phasor form.

**d.** Find the voltages $v_1$ and $v_{ab}$ in phasor form.

**e.** Find the average power delivered to the network.

**f.** Find the power factor of the network and indicate whether it is leading or lagging.

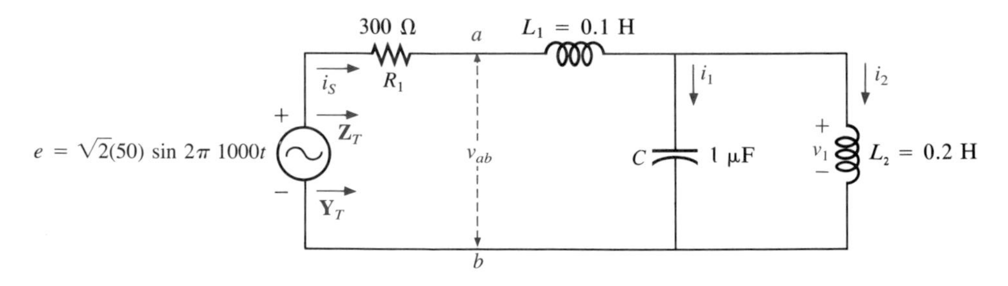


FIG. 16.31

***11.** Find the current **I** for the network of Fig. 16.32.

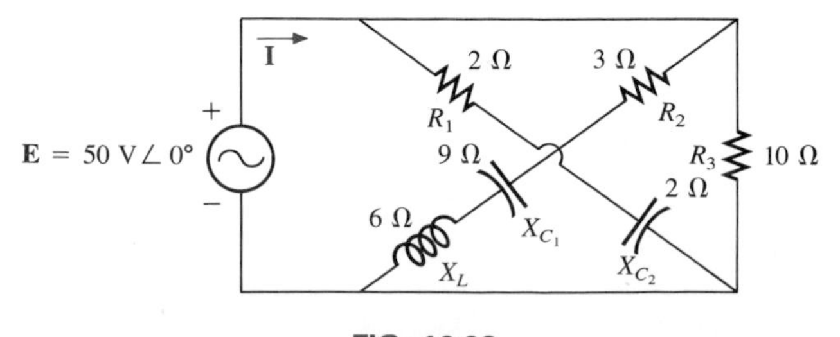


FIG. 16.32

## SECTION 16.3 Ladder Networks

**12.** Find the current $\mathbf{I}_5$ for the network of Fig. 16.33. Note the effect of one reactive element on the resulting calculations.

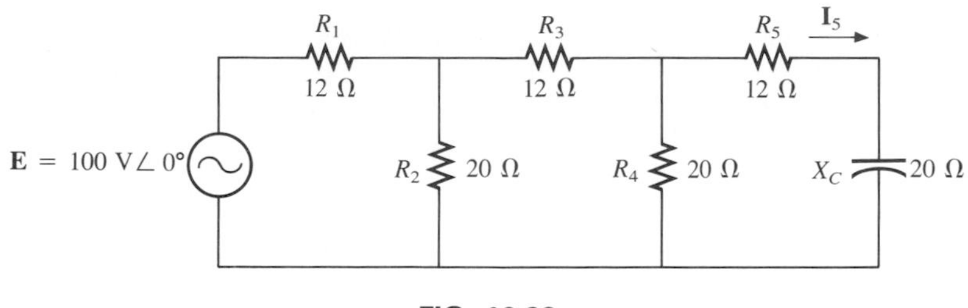


**FIG. 16.33**

**13.** Find the average power delivered to $R_4$ in Fig. 16.34.

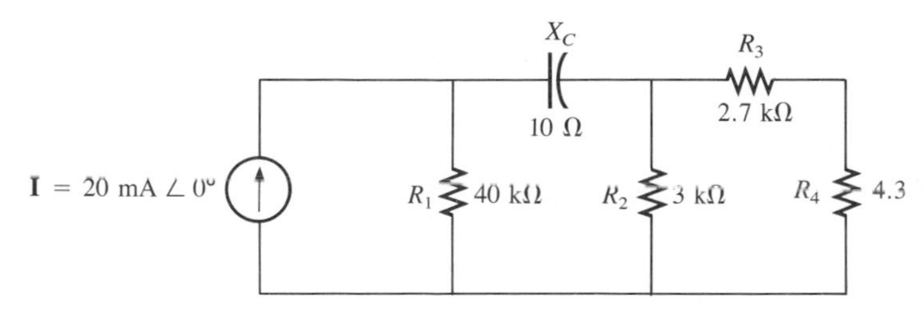


**FIG. 16.34**

**14.** Find the current $\mathbf{I}_1$ for the network of Fig. 16.35.

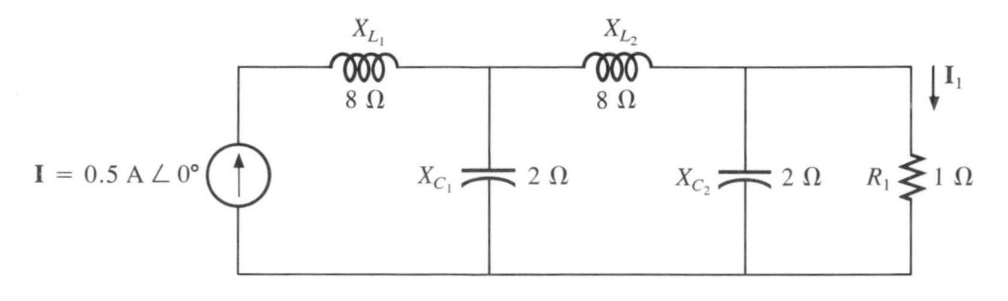


**FIG. 16.35**

## SECTION 16.4 Computer Analysis

### PSPICE

For Problems 15 through 19, use a frequency of 1 kHz to determine the inductive and capacitive levels required for the input files. In each case write the required input file.

***15.** Repeat Problem 1 using PSPICE. A separate input file must be written for part (a).

***16.** Repeat Problem 2 using PSPICE. A separate input file must be written for part (a).

*17. Repeat Problem 7, parts (a) and (b), using PSPICE.

*18. Repeat Problem 11 using PSPICE.

*19. Repeat Problem 14 using PSPICE.

**BASIC**

**20.** Write a program to provide a general solution to Problem 1. That is, given the reactance of each element, generate a solution for parts (a) through (e).

**21.** Given the network of Fig. 16.24, write a program to generate a solution for parts (a) and (b) of Problem 2. Use the values given.

**22.** Generate a program to obtain a general solution for the network of Fig. 16.26 for the questions asked in parts (a) through (c) of Problem 2. That is, given the resistance and reactance of the elements, determine the requested current, voltage, and power.

## GLOSSARY

**Ladder network** A repetitive combination of series and parallel branches that has the appearance of a ladder.

**Series-parallel ac network** A combination of series and parallel branches in the same network configuration. Each branch may contain any number of elements whose impedance is dependent on the applied frequency.

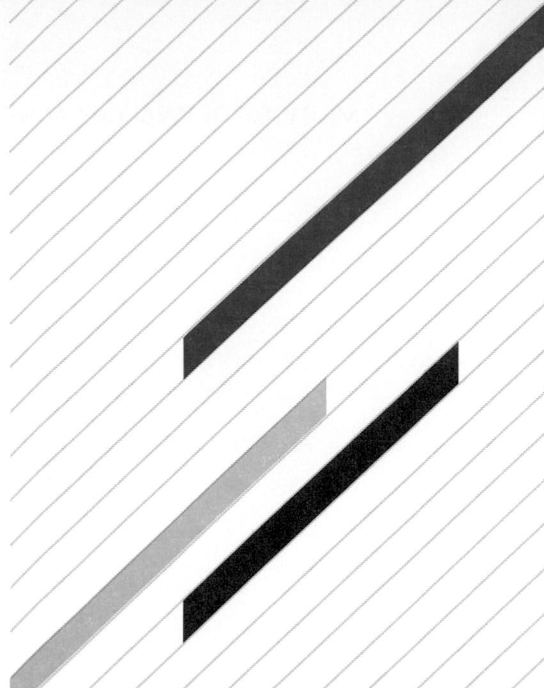

# 17 Methods of Analysis and Selected Topics (ac)

## 17.1 INTRODUCTION

For networks with two or more sources that are not in series or parallel, the methods described in the last two chapters cannot be applied. Rather, such methods as mesh analysis or nodal analysis must be employed. Since these methods were discussed in detail for dc circuits in Chapter 8, this chapter will consider the variations required to apply these methods to ac circuits.

The branch-current method will not be discussed again, since it falls within the framework of mesh analysis. In addition to the methods mentioned above, the bridge network and Δ-Y, Y-Δ conversions will also be discussed for ac circuits.

Before we examine these topics, however, we must consider the subject of independent and controlled sources.

## 17.2 INDEPENDENT VERSUS DEPENDENT (CONTROLLED) SOURCES

In the previous chapters, each source appearing in the analysis of dc or ac networks was an *independent source* such as $E$ and $I$ (or $\mathbf{E}$ and $\mathbf{I}$) in Fig. 17.1.

**FIG. 17.1**
*Independent sources.*

*The term* **independent** *specifies that the magnitude of the source is* **independent of the network** *to which it is applied and that it displays its terminal characteristics even if completely isolated.*

*A* **dependent** *or* **controlled source** *is one whose magnitude is determined (or controlled) by a current or voltage of the system in which it appears.*

There are currently two symbols used for controlled sources. One simply uses the independent symbol with an indication of the controlling element, as shown in Fig. 17.2. In Fig. 17.2(a), the magnitude and phase of the voltage are controlled by a voltage **V** elsewhere in the system with the magnitude further controlled by the constant $k_1$. In Fig. 17.2(b), the magnitude and phase of the current source are controlled by a current **I** elsewhere in the system with the magnitude further con-

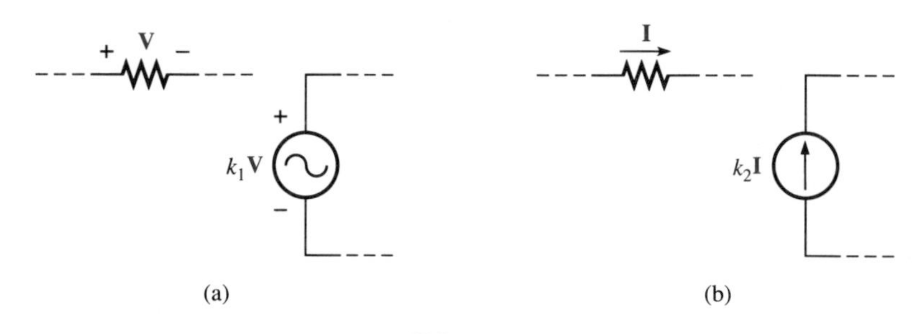


**FIG. 17.2**
*Controlled or dependent sources.*

trolled by the constant $k_2$. To distinguish between the dependent and independent sources, the notation of Fig. 17.3 was introduced. In recent years a number of respected publications on circuit analysis have accepted the notation of Fig. 17.3, although a number of excellent publi-

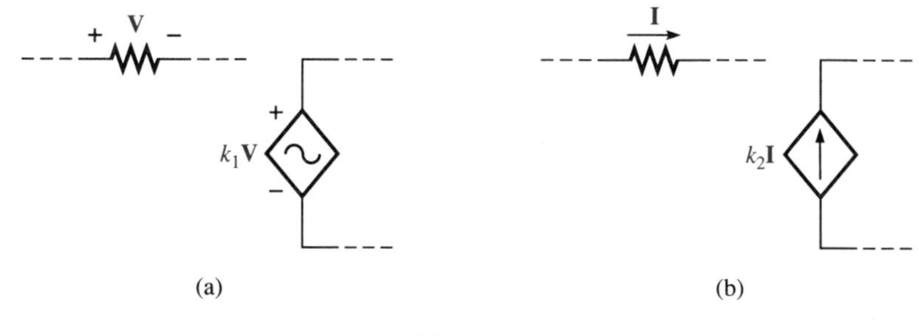

**FIG. 17.3**

cations in the area of electronics continue to use the symbol of Fig. 17.2, especially in the circuit modeling for a variety of electronic devices such as the transistor and FET. This text will use both with an effort toward using the symbol must commonly applied to the area of investigation and to ensure that when the student encounters either symbol, he or she will be aware of its characteristics.

Possible combinations for controlled sources are indicated in Fig. 17.4. Note that the magnitude of current sources or voltage sources can be controlled by a voltage and a current, respectively. Unlike with the independent source, isolation such that **V** or **I** = 0 in Fig. 17.4(a) will result in the short-circuit or open-circuit equivalent as indicated in Fig. 17.4(b). Note that the type of representation under these conditions is controlled by whether it is a current source or a voltage source, not by the controlling agent (**V** or **I**).

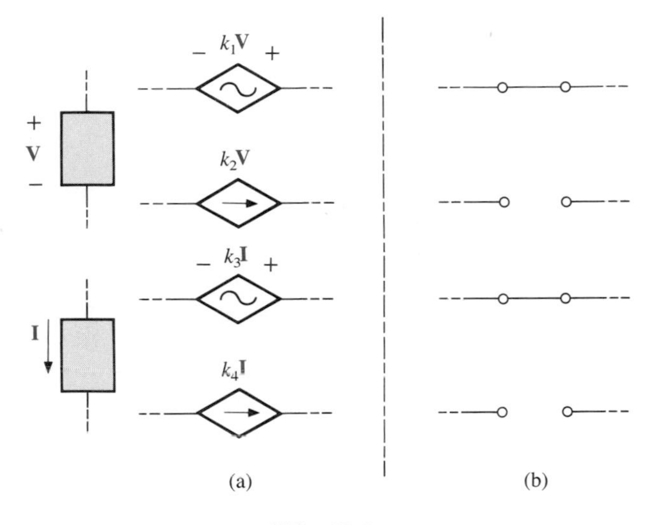


**FIG. 17.4**
*Conditions of V = 0 V and I = 0 A for a controlled source.*

## 17.3 SOURCE CONVERSIONS

When applying the methods to be discussed, it may be necessary to convert a current source to a voltage source, or a voltage source to a current source. This can be accomplished in much the same manner as for dc circuits, except now we shall be dealing with phasors and impedances instead of just real numbers and resistors.

In general, the format for converting from one to the other is as shown in Fig. 17.5.

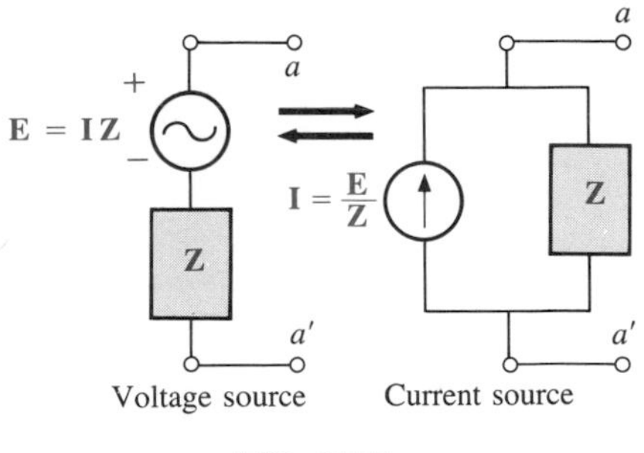


**FIG. 17.5**

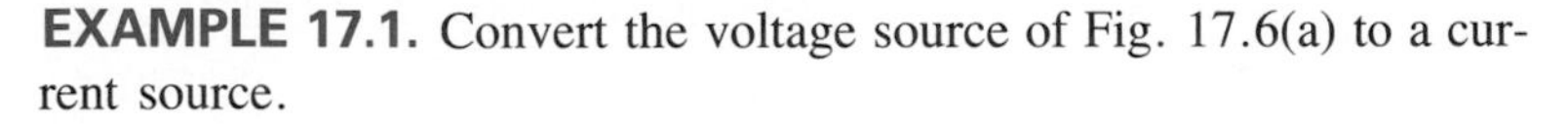

**EXAMPLE 17.1.** Convert the voltage source of Fig. 17.6(a) to a current source.

***Solution:***

$$\mathbf{I} = \frac{\mathbf{E}}{\mathbf{Z}} = \frac{100\text{ V }\angle 0^\circ}{5\ \Omega\ \angle 53.13^\circ}$$
$$= \mathbf{20\text{ A }\angle -53.13^\circ} \qquad \text{[Fig. 17.6(b)]}$$

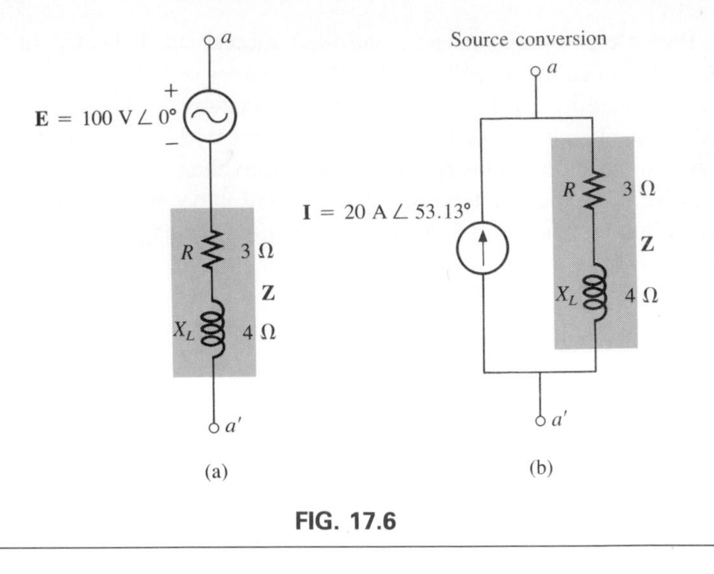


FIG. 17.6

**EXAMPLE 17.2.** Convert the current source of Fig. 17.7(a) to a voltage source.

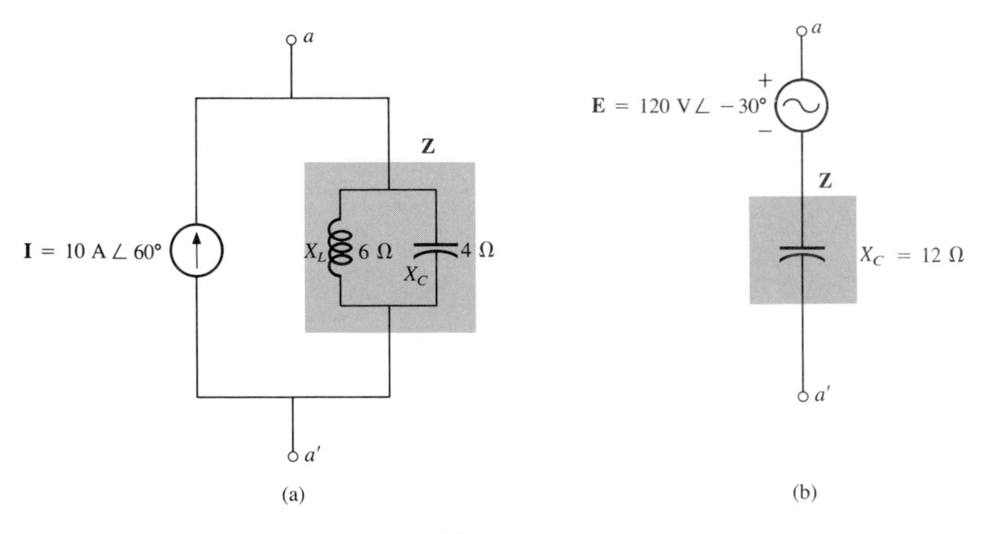


FIG. 17.7

***Solution:***

$$\mathbf{Z} = \frac{\mathbf{X}_C \mathbf{X}_L}{\mathbf{X}_C + \mathbf{X}_L} = \frac{(4\ \Omega\ \angle -90^\circ)(6\ \Omega\ \angle 90^\circ)}{-j4\ \Omega + j6\ \Omega} = \frac{24\ \angle 0^\circ}{2\ \angle 90^\circ}$$

$$= \mathbf{12\ \Omega\ \angle -90^\circ} \qquad \text{[Fig. 17.7(b)]}$$

$$\mathbf{E} = \mathbf{IZ} = (10\ \text{A}\ \angle 60^\circ)(12\ \Omega\ \angle -90^\circ)$$

$$= \mathbf{120\ V\ \angle -30^\circ} \qquad \text{[Fig. 17.7(b)]}$$

For dependent sources, the direct conversion of Fig. 17.5 can be applied if the controlling variable (**V** or **I** in Fig. 17.4) is not determined by a portion of the network to which the conversion is to be applied. For example, in Figs. 17.8 and 17.9, **V** and **I**, respectively, are controlled

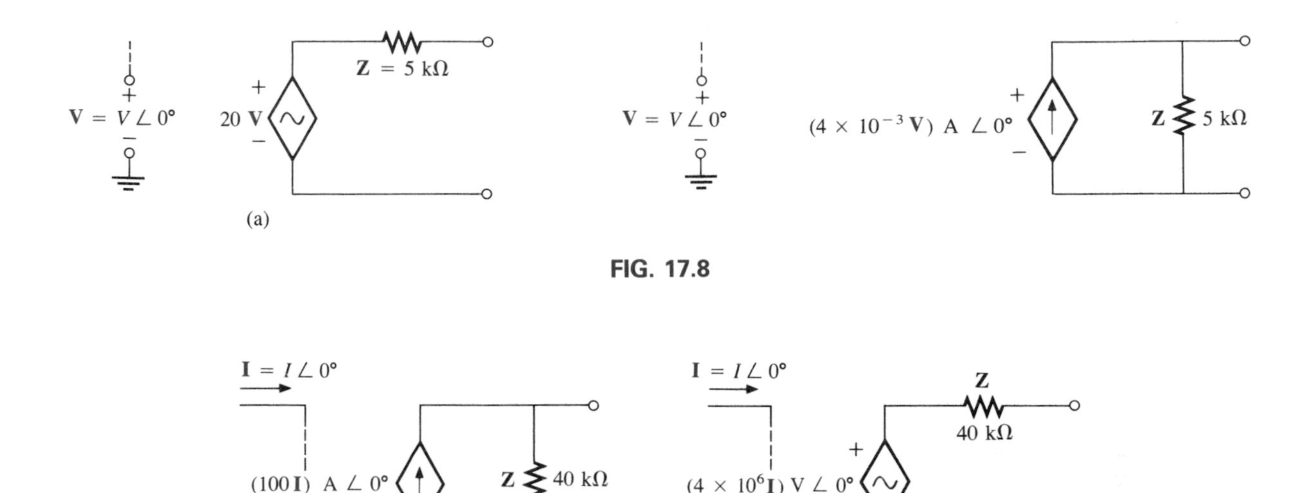


**FIG. 17.8**

(a) (b)

**FIG. 17.9**

by an external portion of the network. Conversions of the other kind, where **V** and **I** are controlled by a portion of the network to be converted, will be considered in Sections 18.3 and 18.4.

---

**EXAMPLE 17.3.** Convert the voltage source of Fig. 17.8(a) to a current source.

***Solution:***

$$\mathbf{I} = \frac{\mathbf{E}}{\mathbf{Z}} = \frac{(20\mathbf{V})\ \text{V}\ \angle 0^\circ}{5\ \text{k}\Omega\ \angle 0^\circ}$$
$$= \mathbf{(4 \times 10^{-3}\ V)\ A\ \angle 0^\circ} \qquad \text{[Fig. 17.8(b)]}$$

---

**EXAMPLE 17.4.** Convert the current source of Fig. 17.9(a) to a voltage source.

***Solution:***

$$\mathbf{E} = \mathbf{IZ} = [(100\mathbf{I})\ \text{A}\ \angle 0^\circ][40\ \text{k}\Omega\ \angle 0^\circ]$$
$$= \mathbf{(4 \times 10^{6} I)\ V\ \angle 0^\circ} \qquad \text{[Fig. 17.9(b)]}$$

---

## 17.4 MESH ANALYSIS (FORMAT APPROACH)

The first method to be considered is mesh analysis using the format approach, introduced in Section 8.9. The steps for applying this method are repeated here with changes for its use in ac circuits:

1. Assign a loop current to each independent closed loop (as in the previous section) in a *clockwise* direction.
2. The number of required equations is equal to the number of chosen independent closed loops. Column 1 of each equation is formed by simply summing the impedance values of those impedances through which the loop current of interest passes and multiplying the result by that loop current.
3. We must now consider the mutual terms that are always subtracted from the terms in the first column. It is possible to have more than one mutual term if the loop current of interest has an element in common with more than one other loop current. Each mutual term is the product of the mutual impedance and the other loop current passing through the same element.
4. The column to the right of the equality sign is the algebraic sum of the voltage sources through which the loop current of interest passes. Positive signs are assigned to those sources of voltage having a polarity such that the loop current passes from the negative to the positive terminal. A negative sign is assigned to those potentials for which the reverse is true.
5. Solve resulting simultaneous equations for the desired loop currents.

Note that the only change in the above as compared to its appearance in Chapter 8 (dc circuits) was to replace the word *resistor* by the word *impedance*. This and the use of phasors instead of real numbers will be the only major changes in applying these methods to ac circuits.

The technique is applied as above for all networks with independent sources or networks with dependent sources where the controlling variable is not a part of the network under investigation. If the controlling variable is part of the network being examined, additional care must be taken when applying the above steps.

---

**EXAMPLE 17.5.** Using mesh analysis, find the current $\mathbf{I}_1$ in Fig. 17.10.

***Solution:*** When applying these methods to ac circuits, it is good practice to represent the resistors and reactances (or combinations thereof) by subscripted impedances. When the total solution is found in terms of these subscripted impedances, the numerical values can be substituted to find the unknown quantities.

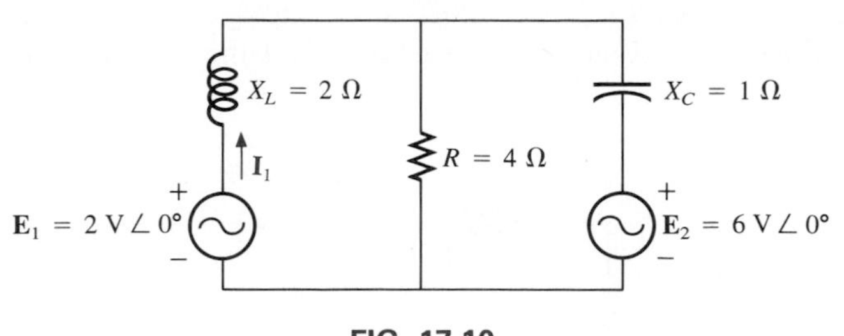

FIG. 17.10

The network is redrawn in Fig. 17.11 with subscripted impedances:

$$\mathbf{Z}_1 = +jX_L = +j2\ \Omega \qquad \mathbf{E}_1 = 2\ \text{V}\ \angle 0°$$
$$\mathbf{Z}_2 = R = 4\ \Omega \qquad \mathbf{E}_2 = 6\ \text{V}\ \angle 0°$$
$$\mathbf{Z}_3 = -jX_C = -j1\ \Omega$$

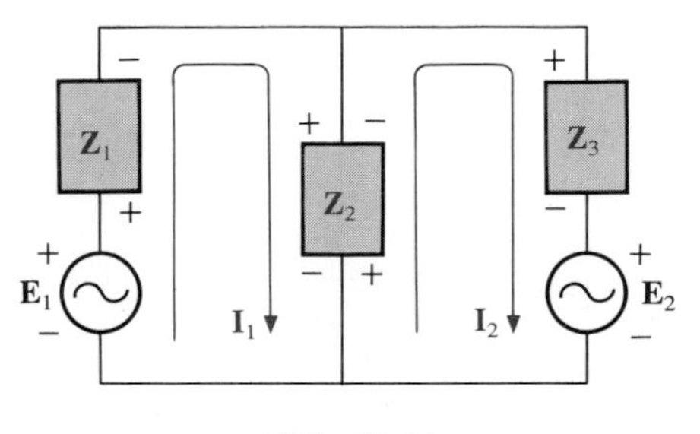


FIG. 17.11

*Step 1* is as indicated in Fig. 17.11.
*Steps 2 to 4:*

$$\mathbf{I}_1(\mathbf{Z}_1 + \mathbf{Z}_2) - \mathbf{I}_2\mathbf{Z}_2 = \mathbf{E}_1$$
$$\mathbf{I}_2(\mathbf{Z}_2 + \mathbf{Z}_3) - \mathbf{I}_1\mathbf{Z}_2 = -\mathbf{E}_2$$

which are rewritten as

$$\begin{aligned} \mathbf{I}_1(\mathbf{Z}_1 + \mathbf{Z}_2) - \mathbf{I}_2\mathbf{Z}_2 \qquad\qquad &= \mathbf{E}_1 \\ -\mathbf{I}_1\mathbf{Z}_2 \qquad + \mathbf{I}_2(\mathbf{Z}_2 + \mathbf{Z}_3) &= -\mathbf{E}_2 \end{aligned}$$

*Step 5:* Using determinants, we obtain

$$\mathbf{I}_1 = \frac{\begin{vmatrix} \mathbf{E}_1 & -\mathbf{Z}_2 \\ -\mathbf{E}_2 & \mathbf{Z}_2 + \mathbf{Z}_3 \end{vmatrix}}{\begin{vmatrix} \mathbf{Z}_1 + \mathbf{Z}_2 & -\mathbf{Z}_2 \\ -\mathbf{Z}_2 & \mathbf{Z}_2 + \mathbf{Z}_3 \end{vmatrix}} = \frac{\mathbf{E}_1(\mathbf{Z}_2 + \mathbf{Z}_3) - \mathbf{E}_2(\mathbf{Z}_2)}{(\mathbf{Z}_1 + \mathbf{Z}_2)(\mathbf{Z}_2 + \mathbf{Z}_3) - (\mathbf{Z}_2)^2} = \frac{(\mathbf{E}_1 - \mathbf{E}_2)\mathbf{Z}_2 + \mathbf{E}_1\mathbf{Z}_3}{\mathbf{Z}_1\mathbf{Z}_2 + \mathbf{Z}_1\mathbf{Z}_3 + \mathbf{Z}_2\mathbf{Z}_3}$$

Substituting numerical values yields

$$\mathbf{I}_1 = \frac{(2\ \text{V} - 6\ \text{V})(4\ \Omega) + (2\ \text{V})(-j\ \Omega)}{(+j2\ \Omega)(4\ \Omega) + (+j2\ \Omega)(-j2\ \Omega) + (4\ \Omega)(-j2\ \Omega)}$$
$$= \frac{-16 - j2}{j8 - j^2 2 - j4}$$
$$= \frac{-16 - j2}{2 + j4} = \frac{16.12\ \angle -172.87°}{4.47\ \angle 63.43°}$$
$$= 3.61\ \text{A}\ \angle -236.30° \quad \text{or} \quad 3.61\ \text{A}\ \angle 123.70°$$

Therefore,

$$\mathbf{I}_1 = \mathbf{3.61\ A}\ \angle \mathbf{123.70°}$$

**EXAMPLE 17.6.** Using mesh analysis, find the current $\mathbf{I}_2$ in Fig. 17.12.

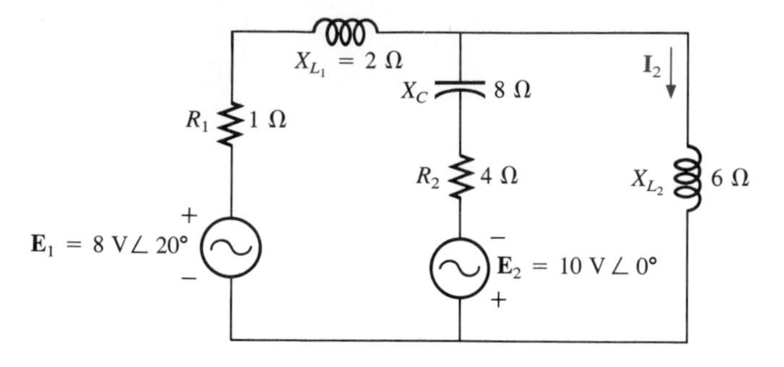

FIG. 17.12

***Solution:*** The network is redrawn in Fig. 17.13:

$$\mathbf{Z}_1 = R_1 + jX_{L_1} = 1\ \Omega + j2\ \Omega \qquad \mathbf{E}_1 = 8\ \text{V}\ \angle 20°$$
$$\mathbf{Z}_2 = R_2 - jX_C = 4\ \Omega - j8\ \Omega \qquad \mathbf{E}_2 = 10\ \text{V}\ \angle 0°$$
$$\mathbf{Z}_3 = +jX_{L_2} = +j6\ \Omega$$

Note the reduction in complexity of the problem with the substitution of the subscripted impedances.

*Step 1* is as indicated in Fig. 17.13.

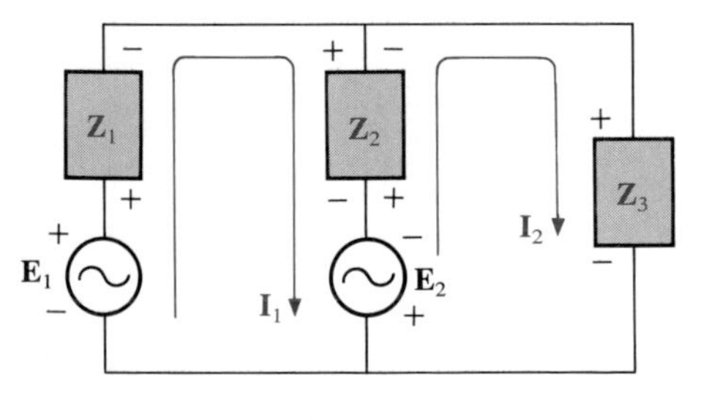


FIG. 17.13

*Steps 2 to 4:*

$$\begin{aligned} \mathbf{I}_1(\mathbf{Z}_1 + \mathbf{Z}_2) - \mathbf{I}_2\mathbf{Z}_2 &= \mathbf{E}_1 + \mathbf{E}_2 \\ \mathbf{I}_2(\mathbf{Z}_2 + \mathbf{Z}_3) - \mathbf{I}_1\mathbf{Z}_2 &= -\mathbf{E}_2 \end{aligned}$$

which are rewritten as

$$\begin{aligned} \mathbf{I}_1(\mathbf{Z}_1 + \mathbf{Z}_2) - \mathbf{I}_2\mathbf{Z}_2 \qquad\qquad &= \mathbf{E}_1 + \mathbf{E}_2 \\ -\mathbf{I}_1\mathbf{Z}_2 \qquad + \mathbf{I}_2(\mathbf{Z}_2 + \mathbf{Z}_3) &= -\mathbf{E}_2 \end{aligned}$$

*Step 5:* Using determinants, we have

$$\mathbf{I}_2 = \frac{\begin{vmatrix} \mathbf{Z}_1 + \mathbf{Z}_2 & \mathbf{E}_1 + \mathbf{E}_2 \\ -\mathbf{Z}_2 & -\mathbf{E}_2 \end{vmatrix}}{\begin{vmatrix} \mathbf{Z}_1 + \mathbf{Z}_2 & -\mathbf{Z}_2 \\ -\mathbf{Z}_2 & \mathbf{Z}_2 + \mathbf{Z}_3 \end{vmatrix}}$$

$$= \frac{-(\mathbf{Z}_1 + \mathbf{Z}_2)\mathbf{E}_2 + \mathbf{Z}_2(\mathbf{E}_1 + \mathbf{E}_2)}{(\mathbf{Z}_1 + \mathbf{Z}_2)(\mathbf{Z}_2 + \mathbf{Z}_3) - \mathbf{Z}_2^2}$$

$$= \frac{-\mathbf{Z}_1\mathbf{E}_2 + \mathbf{Z}_2\mathbf{E}_1}{\mathbf{Z}_1\mathbf{Z}_2 + \mathbf{Z}_1\mathbf{Z}_3 + \mathbf{Z}_2\mathbf{Z}_3}$$

Substituting numerical values yields

$$\mathbf{I}_2 = \frac{-(1\ \Omega + j2\ \Omega)(10\ \text{V}\ \angle 0^\circ) + (4\ \Omega - j8\ \Omega)(8\ \text{V}\ \angle 20^\circ)}{(1\ \Omega + j2\ \Omega)(4\ \Omega - j8\ \Omega) + (1\ \Omega + j2\ \Omega)(+j6\ \Omega) + (4\ \Omega - j8\ \Omega)(+j6\ \Omega)}$$

$$= \frac{-(10 + j20) + (4 - j8)(7.52 + j2.74)}{20 + (j6 - 12) + (j24 + 48)}$$

$$= \frac{-(10 + j20) + (52.0 - j49.20)}{56 + j30} = \frac{+42.0 - j69.20}{56 + j30}$$

$$= \frac{80.95\ \angle -58.74^\circ}{63.53\ \angle 28.18^\circ} = 1.27\ \text{A}\ \angle -86.92^\circ$$

Therefore,

$$\mathbf{I}_2 = \mathbf{1.27\ A\ \angle -86.92^\circ}$$

---

**EXAMPLE 17.7.** Write the mesh equations for the network of Fig. 17.14. Do not solve.

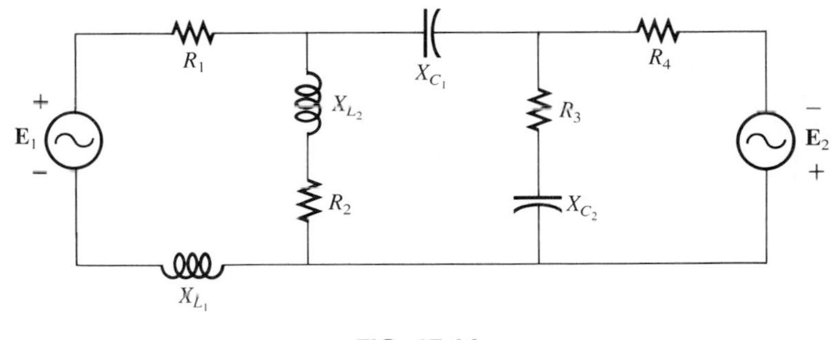


FIG. 17.14

***Solution:*** The network is redrawn in Fig. 17.15. Again note the reduced complexity and increased clarity by use of subscripted impedances:

$$\mathbf{Z}_1 = R_1 + jX_{L_1}$$
$$\mathbf{Z}_2 = R_2 + jX_{L_2}$$
$$\mathbf{Z}_3 = jX_{C_1}$$
$$\mathbf{Z}_4 = R_3 - jX_{C_2}$$
$$\mathbf{Z}_5 = R_4$$

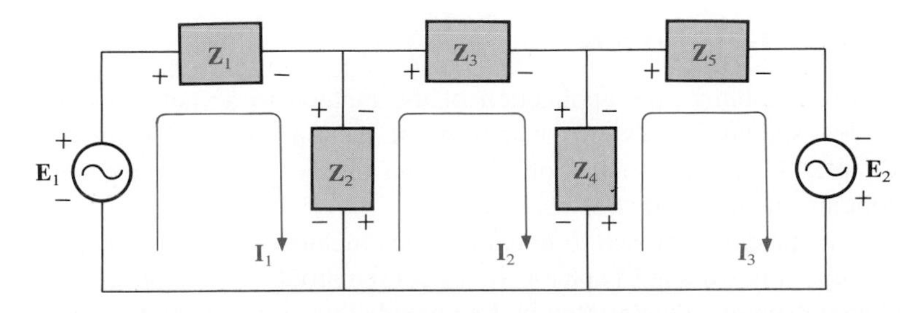


FIG. 17.15

and

$$\begin{aligned} \mathbf{I}_1(\mathbf{Z}_1 + \mathbf{Z}_2) - \mathbf{I}_2\mathbf{Z}_2 &= \mathbf{E}_1 \\ \mathbf{I}_2(\mathbf{Z}_2 + \mathbf{Z}_3 + \mathbf{Z}_4) - \mathbf{I}_1\mathbf{Z}_2 - \mathbf{I}_3\mathbf{Z}_4 &= 0 \\ \mathbf{I}_3(\mathbf{Z}_4 + \mathbf{Z}_5) - \mathbf{I}_2\mathbf{Z}_4 &= \mathbf{E}_2 \end{aligned}$$

or

$$\begin{array}{llll} \mathbf{I}_1(\mathbf{Z}_1 + \mathbf{Z}_2) & - \mathbf{I}_2(\mathbf{Z}_2) & + 0 & = \mathbf{E}_1 \\ \mathbf{I}_1(\mathbf{Z}_2) & - \mathbf{I}_2(\mathbf{Z}_2 + \mathbf{Z}_3 + \mathbf{Z}_4) & + \mathbf{I}_3(\mathbf{Z}_4) & = 0 \\ 0 & - \mathbf{I}_2(\mathbf{Z}_4) & + \mathbf{I}_3(\mathbf{Z}_4 + \mathbf{Z}_5) & = \mathbf{E}_2 \end{array}$$

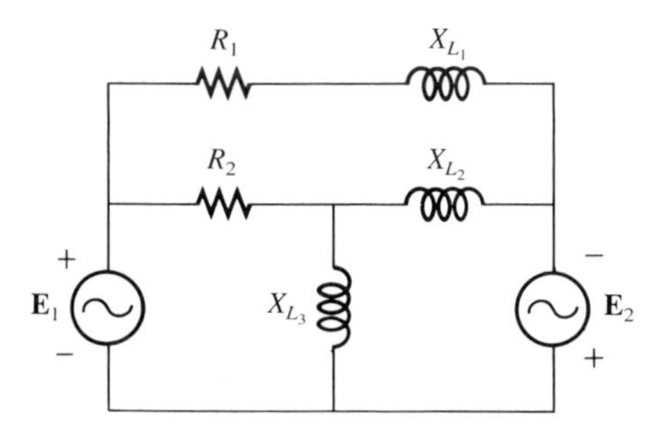

FIG. 17.16

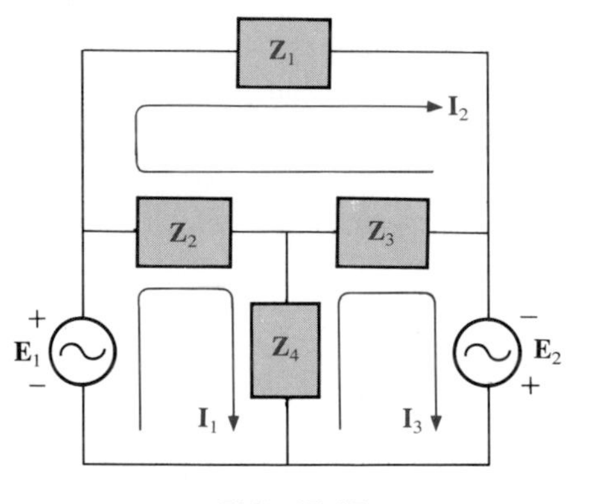


FIG. 17.17

**EXAMPLE 17.8.** Using the format approach, write the mesh equations for the network of Fig. 17.16.

***Solution:*** The network is redrawn as shown in Fig. 17.17, where

$$\begin{aligned} \mathbf{Z}_1 &= R_1 + jX_{L_1} \\ \mathbf{Z}_2 &= R_2 \\ \mathbf{Z}_3 &= jX_{L_2} \\ \mathbf{Z}_4 &= jX_{L_3} \end{aligned}$$

and

$$\begin{aligned} \mathbf{I}_1(\mathbf{Z}_2 + \mathbf{Z}_4) - \mathbf{I}_2\mathbf{Z}_2 - \mathbf{I}_3\mathbf{Z}_4 &= \mathbf{E}_1 \\ \mathbf{I}_2(\mathbf{Z}_1 + \mathbf{Z}_2 + \mathbf{Z}_3) - \mathbf{I}_1\mathbf{Z}_2 - \mathbf{I}_3\mathbf{Z}_3 &= 0 \\ \mathbf{I}_3(\mathbf{Z}_3 + \mathbf{Z}_4) - \mathbf{I}_2\mathbf{Z}_3 - \mathbf{I}_1\mathbf{Z}_4 &= \mathbf{E}_2 \end{aligned}$$

or

$$\begin{array}{llll} \mathbf{I}_1(\mathbf{Z}_2 + \mathbf{Z}_4) & - \mathbf{I}_2\mathbf{Z}_2 & - \mathbf{I}_3\mathbf{Z}_4 & = \mathbf{E}_1 \\ -\mathbf{I}_1\mathbf{Z}_2 & + \mathbf{I}_2(\mathbf{Z}_1 + \mathbf{Z}_2 + \mathbf{Z}_3) & - \mathbf{I}_3\mathbf{Z}_3 & = 0 \\ -\mathbf{I}_1\mathbf{Z}_4 & - \mathbf{I}_2\mathbf{Z}_3 & + \mathbf{I}_3(\mathbf{Z}_3 + \mathbf{Z}_4) & = \mathbf{E}_2 \end{array}$$

Note the symmetry *about* the diagonal axis. That is, note the location of $-\mathbf{Z}_2$, $-\mathbf{Z}_4$, and $-\mathbf{Z}_3$ off the diagonal.

## 17.5 NODAL ANALYSIS

### General Approach

Before examining the application of the method to ac networks, the student should first review the appropriate sections on nodal analysis in Chapter 8, since the content of this section will be limited to the general conclusions of Chapter 8.

The general approach to nodal analysis includes the same sequence of steps appearing in Chapter 8. In fact, throughout this section the only change from the dc coverage will be to substitute impedance for resistance and admittance for conductance in the general procedure.

The fundamental steps are the following:

1. Determine the number of nodes within the network.
2. Pick a reference node and label each remaining node with a subscripted value of voltage: $\mathbf{V}_1$, $\mathbf{V}_2$, and so on.
3. Assume a direction of current for each branch.
4. Apply Kirchhoff's current law at each node except the reference.
5. Solve the resulting equations for the nodal voltages.

A few examples will refresh your memory about the content of Chapter 8 and the general approach to a nodal analysis solution.

**EXAMPLE 17.9.** Determine the nodal voltages for the network of Fig. 17.18.

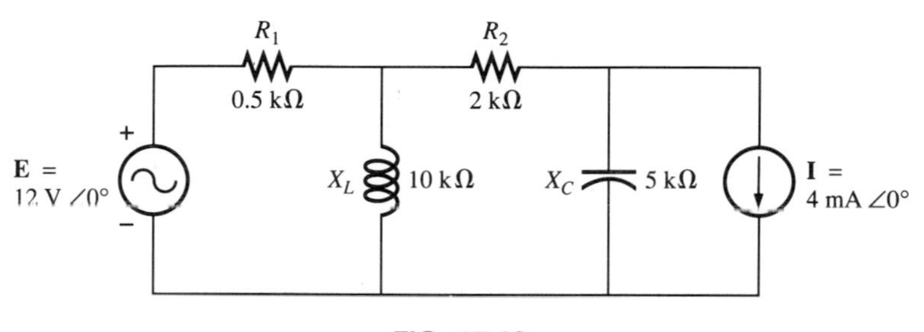


**FIG. 17.18**

***Solution:*** Choosing nodes, defining the impedances, and choosing (arbitrary) the current directions will result in the network of Fig. 17.19.

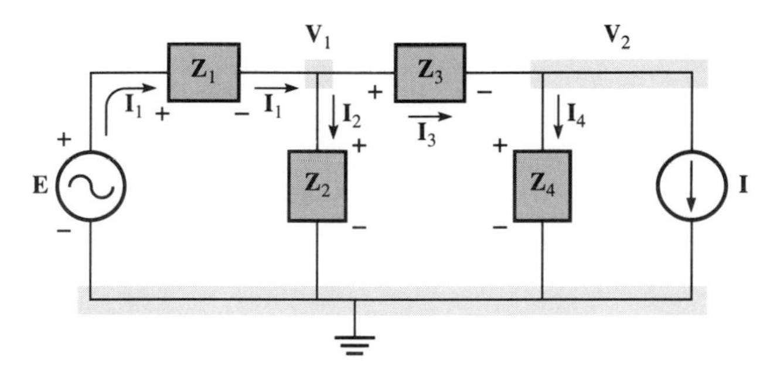


**FIG. 17.19**

At node 1:

$$\mathbf{I}_1 = \mathbf{I}_2 + \mathbf{I}_3$$

Substituting:

$$\frac{\mathbf{E} - \mathbf{V}_1}{\mathbf{Z}_1} = \frac{\mathbf{V}_1}{\mathbf{Z}_2} + \frac{(\mathbf{V}_1 - \mathbf{V}_2)}{\mathbf{Z}_3}$$

Rearranging:

$$\mathbf{V}_1\left[\frac{1}{\mathbf{Z}_1} + \frac{1}{\mathbf{Z}_2} + \frac{1}{\mathbf{Z}_3}\right] - \mathbf{V}_2\left[\frac{1}{\mathbf{Z}_3}\right] = \frac{\mathbf{E}}{\mathbf{Z}_1} \qquad \textbf{(17.1)}$$

At node 2:

$$\mathbf{I}_3 = \mathbf{I}_4 + \mathbf{I}$$

Substituting:

$$\frac{\mathbf{V}_1 - \mathbf{V}_2}{\mathbf{Z}_3} = \frac{\mathbf{V}_2}{\mathbf{Z}_4} + \mathbf{I}$$

Rearranging:

$$\mathbf{V}_2\left[\frac{1}{\mathbf{Z}_3} + \frac{1}{\mathbf{Z}_4}\right] - \mathbf{V}_1\left[\frac{1}{\mathbf{Z}_3}\right] = -\mathbf{I} \qquad \textbf{(17.2)}$$

Grouping equations:

$$\begin{aligned}
\mathbf{V}_1\left[\frac{1}{\mathbf{Z}_1} + \frac{1}{\mathbf{Z}_2} + \frac{1}{\mathbf{Z}_3}\right] - \mathbf{V}_2\left[\frac{1}{\mathbf{Z}_3}\right] &= \frac{\mathbf{E}}{\mathbf{Z}_1} \\
\mathbf{V}_1\left[\frac{1}{\mathbf{Z}_3}\right] - \mathbf{V}_2\left[\frac{1}{\mathbf{Z}_3} + \frac{1}{\mathbf{Z}_4}\right] &= \mathbf{I}
\end{aligned}$$

$$\frac{1}{\mathbf{Z}_1} + \frac{1}{\mathbf{Z}_2} + \frac{1}{\mathbf{Z}_3} = \frac{1}{0.5\text{ k}\Omega} + \frac{1}{j10\text{ k}\Omega} + \frac{1}{2\text{ k}\Omega} = 2.5\text{ mS }\angle -2.29^\circ$$

$$\frac{1}{\mathbf{Z}_3} + \frac{1}{\mathbf{Z}_4} = \frac{1}{2\text{ k}\Omega} + \frac{1}{-j5\text{ k}\Omega} = 0.539\text{ mS }\angle 21.80^\circ$$

and

$$\begin{aligned}
\mathbf{V}_1[2.5\text{ mS }\angle -2.229^\circ] - \mathbf{V}_2[0.5\text{ mS }\angle 0^\circ] &= 24\text{ mA }\angle 0^\circ \\
\mathbf{V}_1[0.5\text{ mS }\angle 0^\circ] - \mathbf{V}_2[0.539\text{ mS }\angle 21.80^\circ] &= 4\text{ mA }\angle 0^\circ
\end{aligned}$$

with

$$\mathbf{V}_1 = \frac{\begin{vmatrix} 24\text{ mA }\angle 0^\circ & -0.5\text{ mS }\angle 0^\circ \\ 4\text{ mA }\angle 0^\circ & -0.539\text{ mS }\angle 21.80 \end{vmatrix}}{\begin{vmatrix} 2.5\text{ mS }\angle -2.29^\circ & -0.5\text{ mS }\angle 0^\circ \\ 0.5\text{ mS }\angle 0^\circ & -0.539\text{ mS }\angle 21.80^\circ \end{vmatrix}}$$

$$= \frac{(24\text{ mA }\angle 0^\circ)(-0.539\text{ mS }\angle 21.80^\circ) + (0.5\text{ mS }\angle 0^\circ)(4\text{ mA }\angle 0^\circ)}{(2.5\text{ mS }\angle -2.29^\circ)(-0.539\text{ mS }\angle 21.80^\circ) + (0.5\text{ mS }\angle 0^\circ)(0.5\text{ mS }\angle 0^\circ)}$$

$$= \frac{-12.94 \times 10^{-6}\text{ V }\angle 21.80^\circ + 2 \times 10^{-6}\text{ V }\angle 0^\circ}{-1.348 \times 10^{-6}\ \angle 19.51^\circ + 0.25 \times 10^{-6}\ \angle 0^\circ}$$

$$= \frac{-(12.01 + j4.81) \times 10^{-6}\text{ V} + 2 \times 10^{-6}\text{ V}}{-(1.271 + j0.45) \times 10^{-6} + 0.25 \times 10^{-6}}$$

$$= \frac{-10.01\text{ V} - j4.81\text{ V}}{-1.021 - j0.45} = \frac{11.106\text{ V }\angle -154.33^\circ}{1.116\ \angle -156.21^\circ}$$

$\mathbf{V}_1 = \mathbf{9.95\ V\ \angle 1.88^\circ}$

Similarly

$$\mathbf{V}_2 = \mathbf{1.828\ V\ \angle -12.49^\circ}$$

## Format Approach

A close examination of Eqs. (17.1) and (17.2) in Example 17.9 will reveal that they are the same equations that would have been obtained using the format approach introduced in Chapter 8. Recall that the approach required that the voltage source first be converted to a current source, but the writing of the equations was quite direct and minimized and chances of an error due to a lost sign or missing term.

The sequence of steps required to apply the format approach is the following:

1. Choose a reference node and assign a subscripted voltage label to the $(N - 1)$ remaining independent nodes of the network.
2. The number of equations required for a complete solution is equal to the number of subscripted voltages $(N - 1)$. Column 1 of each equation is formed by summing the admittances tied to the node of interest and multiplying the result by that subscripted nodal voltage.
3. The mutual terms are always subtracted from the terms of the first column. It is possible to have more than one mutual term if the nodal voltage of interest has an element in common with more than one other nodal voltage. Each mutual term is the product of the mutual admittance and the other nodal voltage tied to that admittance.
4. The column to the right of the equality sign is the algebraic sum of the current sources tied to the node of interest. A current source is assigned a positive sign if it supplies current to a node, and a negative sign if it draws current from the node.
5. Solve resulting simultaneous equations for the desired nodal voltages. The comments offered for mesh analysis regarding independent and dependent sources apply here also.

**EXAMPLE 17.10.** Using nodal analysis, find the voltage across the 4-Ω resistor in Fig. 17.20.

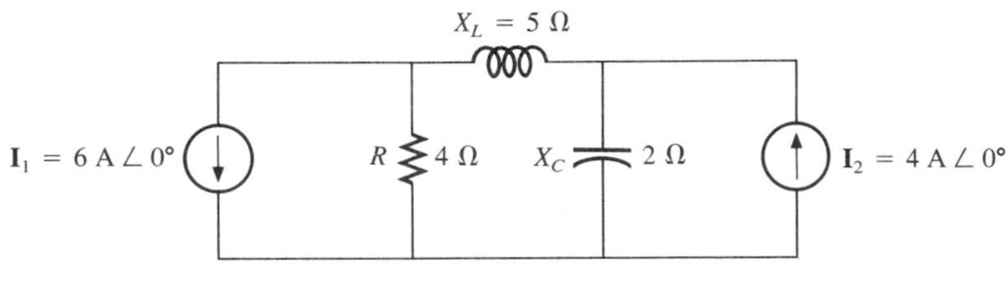


FIG. 17.20

***Solution:*** Choosing nodes (Fig. 17.21) and writing the nodal equations, we have

$$\mathbf{Z}_1 = R = 4\ \Omega \qquad \mathbf{Z}_2 = jX_L = j5\ \Omega \qquad \mathbf{Z}_3 = -jX_C = -j2\ \Omega$$

$$\mathbf{V}_1(\mathbf{Y}_1 + \mathbf{Y}_2) - \mathbf{V}_2(\mathbf{Y}_2) = -\mathbf{I}_1$$
$$\mathbf{V}_2(\mathbf{Y}_3 + \mathbf{Y}_2) - \mathbf{V}_1(\mathbf{Y}_2) = +\mathbf{I}_2$$

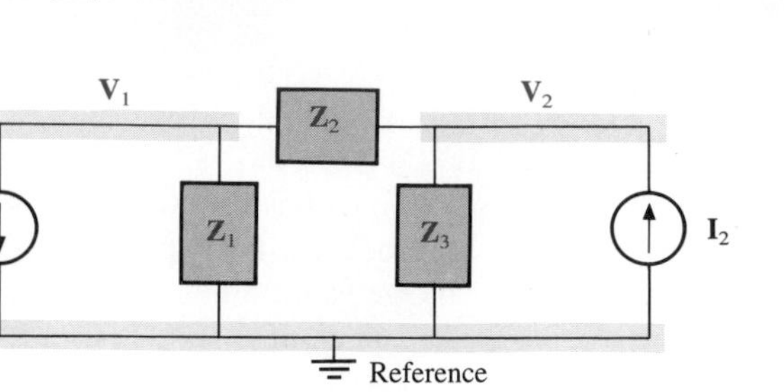

FIG. 17.21

or

$$\begin{aligned} \mathbf{V}_1(\mathbf{Y}_1 - \mathbf{Y}_2) - \mathbf{V}_2(\mathbf{Y}_2) \qquad\qquad &= -\mathbf{I}_1 \\ -\mathbf{V}_1(\mathbf{Y}_2) \qquad + \mathbf{V}_2(\mathbf{Y}_3 + \mathbf{Y}_2) &= +\mathbf{I}_2 \end{aligned}$$

$$\mathbf{Y}_1 = \frac{1}{\mathbf{Z}_1} \qquad \mathbf{Y}_2 = \frac{1}{\mathbf{Z}_2} \qquad \mathbf{Y}_3 = \frac{1}{\mathbf{Z}_3}$$

Using determinants yields

$$\begin{aligned} \mathbf{V}_1 &= \frac{\begin{vmatrix} -\mathbf{I}_1 & -\mathbf{Y}_2 \\ +\mathbf{I}_2 & \mathbf{Y}_3 + \mathbf{Y}_2 \end{vmatrix}}{\begin{vmatrix} \mathbf{Y}_1 + \mathbf{Y}_2 & -\mathbf{Y}_2 \\ -\mathbf{Y}_2 & \mathbf{Y}_3 + \mathbf{Y}_2 \end{vmatrix}} \\ &= \frac{-(\mathbf{Y}_3 + \mathbf{Y}_2)\mathbf{I}_1 + \mathbf{I}_2\mathbf{Y}_2}{(\mathbf{Y}_1 + \mathbf{Y}_2)(\mathbf{Y}_3 + \mathbf{Y}_2) - \mathbf{Y}_2^2} \\ &= \frac{-(\mathbf{Y}_3 + \mathbf{Y}_2)\mathbf{I}_1 + \mathbf{I}_2\mathbf{Y}_2}{\mathbf{Y}_1\mathbf{Y}_3 + \mathbf{Y}_2\mathbf{Y}_3 + \mathbf{Y}_1\mathbf{Y}_2} \end{aligned}$$

Substituting numerical values, we have

$$\begin{aligned} \mathbf{V}_1 &= \frac{-[(1/-j2\ \Omega) + (1/j5\ \Omega)]6\ \text{A}\ \angle 0^\circ + 4\ \text{A}\ \angle 0^\circ(1/j5\ \Omega)}{(1/4\ \Omega)(1/-j2\ \Omega) + (1/j5\ \Omega)(1/-j2\ \Omega) + (1/4\ \Omega)(1/j5\ \Omega)} \\ &= \frac{-(+j0.5 - j0.2)6\ \angle 0^\circ + 4\ \angle 0^\circ(-j0.2)}{(1/-j8) + (1/10) + (1/j20)} \\ &= \frac{(-0.3\ \angle 90^\circ)(6\ \angle 0^\circ) + (4\ \angle 0^\circ)(0.2\ \angle -90^\circ)}{j0.125 + 0.1 - j0.05} \\ &= \frac{-1.8\ \angle 90^\circ + 0.8\ \angle -90^\circ}{0.1 + j0.075} \\ &= \frac{2.6\ \angle -90^\circ}{0.125\ \angle 36.87^\circ} \\ &= \mathbf{20.80\ V}\ \angle \mathbf{-126.87^\circ} \end{aligned}$$

**EXAMPLE 17.11.** Write the nodal equations for the network of Fig. 17.22. In this case, a voltage source appears in the network.

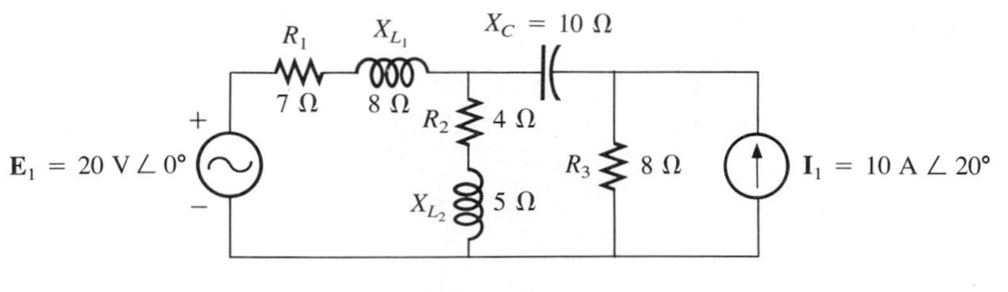

FIG. 17.22

***Solution:*** The circuit is redrawn in Fig. 17.23, where

$$\mathbf{Z}_1 = R_1 + jX_{L_1} = 7\ \Omega + j8\ \Omega \qquad \mathbf{E}_1 = 20\ \text{V}\ \angle 0°$$
$$\mathbf{Z}_2 = R_2 + jX_{L_2} = 4\ \Omega + j5\ \Omega \qquad \mathbf{I}_1 = 10\ \text{A}\ \angle 20°$$
$$\mathbf{Z}_3 = -jX_C = -j10\ \Omega$$
$$\mathbf{Z}_4 = R_3 = 8\ \Omega$$

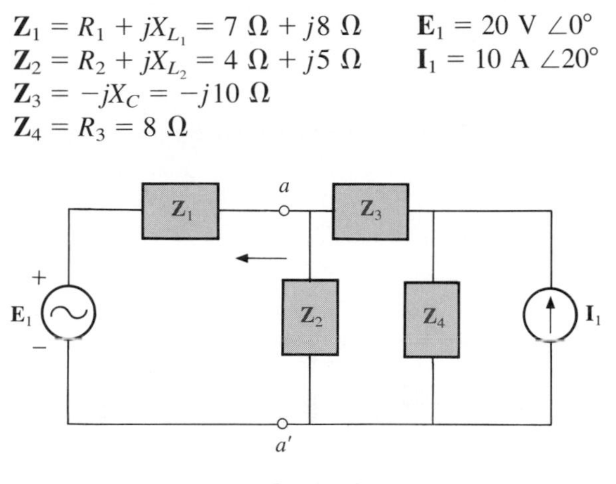


FIG. 17.23

Converting the voltage source to a current source and choosing nodes, we obtain Fig. 17.24. Note the ''neat'' appearance of the network using

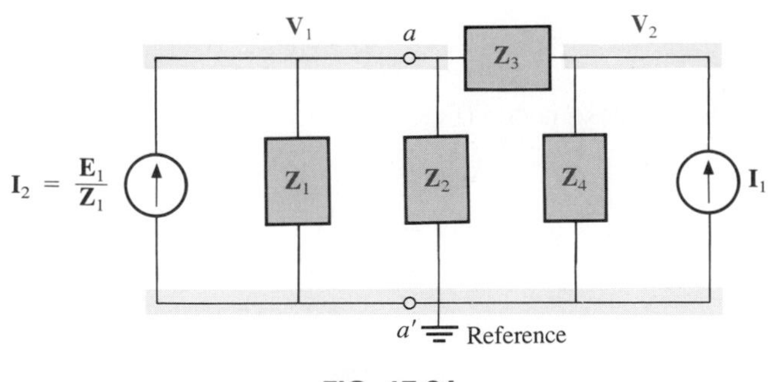


FIG. 17.24

the subscripted impedances. Working directly with Fig. 17.22 would be difficult and may produce errors.

Write the nodal equations:

$$\mathbf{V}_1(\mathbf{Y}_1 + \mathbf{Y}_2 + \mathbf{Y}_3) - \mathbf{V}_2(\mathbf{Y}_3) = +\mathbf{I}_2$$
$$\mathbf{V}_2(\mathbf{Y}_3 + \mathbf{Y}_4) - \mathbf{V}_1(\mathbf{Y}_3) = +\mathbf{I}_1$$

$$\mathbf{Y}_1 = \frac{1}{\mathbf{Z}_1} \qquad \mathbf{Y}_2 = \frac{1}{\mathbf{Z}_2} \qquad \mathbf{Y}_3 = \frac{1}{\mathbf{Z}_3} \qquad \mathbf{Y}_4 = \frac{1}{\mathbf{Z}_4}$$

which are rewritten as

$$\begin{array}{ll} \mathbf{V}_1(\mathbf{Y}_1 + \mathbf{Y}_2 + \mathbf{Y}_3) - \mathbf{V}_2(\mathbf{Y}_3) & = +\mathbf{I}_2 \\ -\mathbf{V}_1(\mathbf{Y}_3) \qquad\qquad + \mathbf{V}_2(\mathbf{Y}_3 + \mathbf{Y}_4) & = +\mathbf{I}_1 \end{array}$$

$$\mathbf{Y}_1 = \frac{1}{7\ \Omega + j8\ \Omega} \qquad \mathbf{Y}_2 = \frac{1}{4\ \Omega + j5\ \Omega}$$

$$\mathbf{Y}_3 = \frac{1}{-j10\ \Omega} \qquad \mathbf{Y}_4 = \frac{1}{8\ \Omega}$$

$$\mathbf{I}_2 = \frac{20\ \text{V}\ \angle 0^\circ}{7\ \Omega + j8\ \Omega} \qquad \mathbf{I}_1 = 10\ \text{A}\ \angle 20^\circ$$

**EXAMPLE 17.12.** Write the nodal equations for the network of Fig. 17.25.

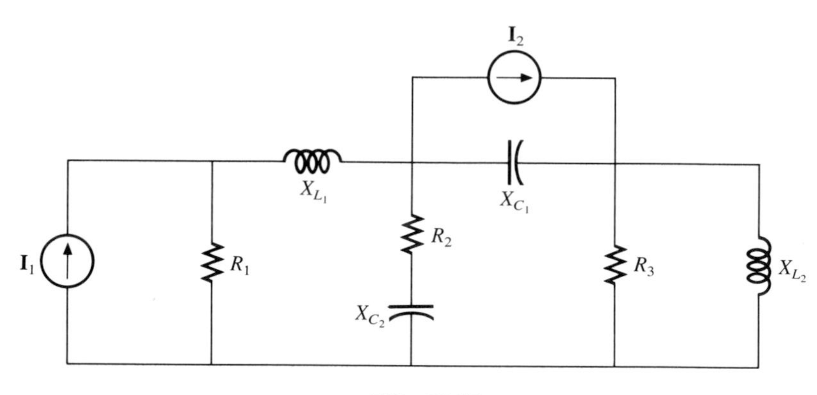

FIG. 17.25

***Solution:*** Choose nodes (Fig. 17.26):

$$\mathbf{Z}_1 = R_1 \qquad \mathbf{Z}_2 = jX_{L_1} \qquad \mathbf{Z}_3 = R_2 - jX_{C_2}$$
$$\mathbf{Z}_4 = -jX_{C_1} \qquad \mathbf{Z}_5 = R_3 \qquad \mathbf{Z}_6 = jX_{L_2}$$

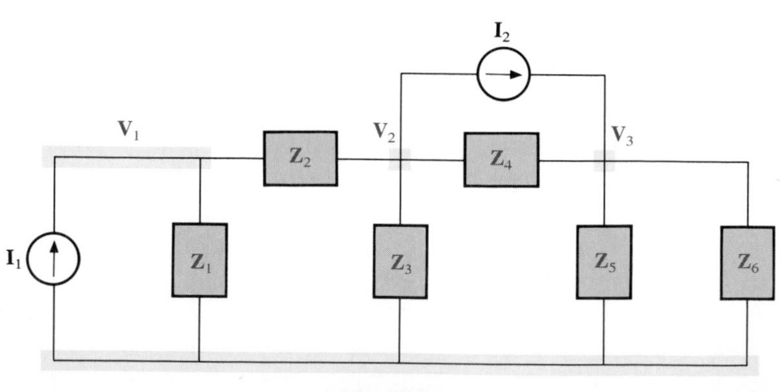


FIG. 17.26

and write nodal equations:

$$\begin{aligned}
\mathbf{V}_1(\mathbf{Y}_1 + \mathbf{Y}_2) - \mathbf{V}_2(\mathbf{Y}_2) &= +\mathbf{I}_1 \\
\mathbf{V}_2(\mathbf{Y}_2 + \mathbf{Y}_3 + \mathbf{Y}_4) - \mathbf{V}_1(\mathbf{Y}_2) - \mathbf{V}_3(\mathbf{Y}_4) &= -\mathbf{I}_2 \\
\mathbf{V}_3(\mathbf{Y}_4 + \mathbf{Y}_5 + \mathbf{Y}_6) - \mathbf{V}_2(\mathbf{Y}_4) &= +\mathbf{I}_2
\end{aligned}$$

which are rewritten as

$$\begin{array}{llll}
\mathbf{V}_1(\mathbf{Y}_1 + \mathbf{Y}_2) & - \mathbf{V}_2(\mathbf{Y}_2) & + 0 & = +\mathbf{I}_1 \\
-\mathbf{V}_1(\mathbf{Y}_2) & + \mathbf{V}_2(\mathbf{Y}_2 + \mathbf{Y}_3 + \mathbf{Y}_4) & - \mathbf{V}_3(\mathbf{Y}_4) & = -\mathbf{I}_2 \\
0 & - \mathbf{V}_2(\mathbf{Y}_4) & + \mathbf{V}_3(\mathbf{Y}_4 + \mathbf{Y}_5 + \mathbf{Y}_6) & = +\mathbf{I}_2
\end{array}$$

$$\mathbf{Y}_1 = \frac{1}{R_1} \qquad \mathbf{Y}_2 = \frac{1}{jX_{L_1}} \qquad \mathbf{Y}_3 = \frac{1}{R_2 - jX_{C_2}}$$

$$\mathbf{Y}_4 = \frac{1}{-jX_{C_1}} \qquad \mathbf{Y}_5 = \frac{1}{R_3} \qquad \mathbf{Y}_6 = \frac{1}{jX_{L_2}}$$

Note the symmetry about the diagonal for this example and those preceding it in this section.

**EXAMPLE 17.13.** Apply nodal analysis to the network of Fig. 17.27. Determine the voltage $\mathbf{V}_L$.

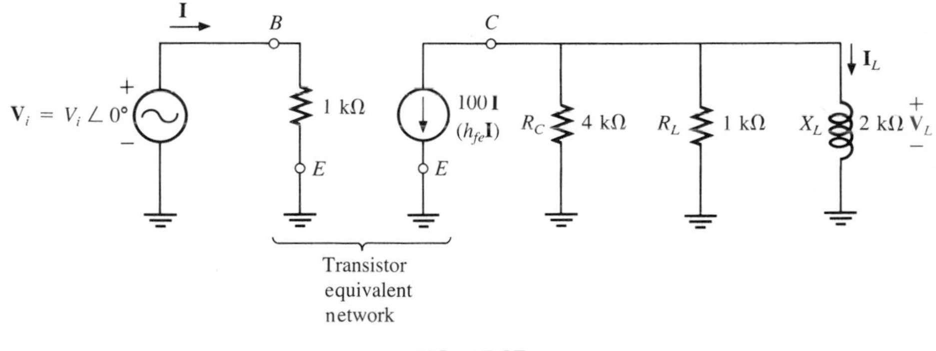


FIG. 17.27

***Solution:*** In this case there is no need for a source conversion. The network is redrawn in Fig. 17.28 with the chosen node voltage and subscripted impedances.

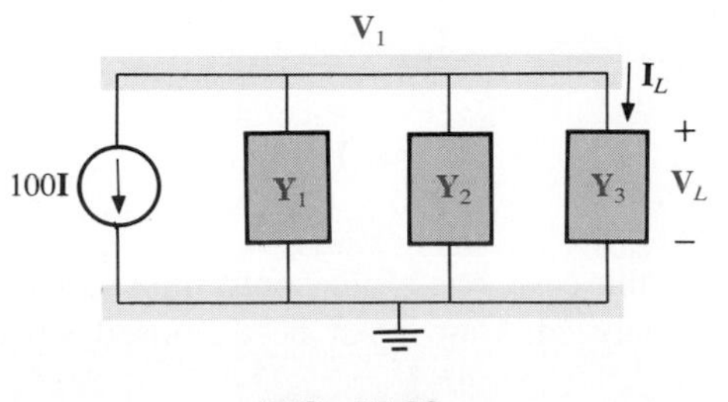


FIG. 17.28

Apply the format approach:

$$\mathbf{Y}_1 = \frac{1}{\mathbf{Z}_1} = \frac{1}{4\text{ k}\Omega} = 0.25\text{ mS }\angle 0° = \mathbf{G}_1$$

$$\mathbf{Y}_2 = \frac{1}{\mathbf{Z}_2} = \frac{1}{1\text{ k}\Omega} = 1\text{ mS }\angle 0° = \mathbf{G}_2$$

$$\mathbf{Y}_3 = \frac{1}{\mathbf{Z}_3} = \frac{1}{2\text{ k}\Omega \angle 90°} = 0.5\text{ mS }\angle -90°$$

$$= -j0.5\text{ mS} = \mathbf{B}_L$$

$$\mathbf{V}_1\text{:} (\mathbf{Y}_1 + \mathbf{Y}_2 + \mathbf{Y}_3)\mathbf{V}_1 = -100\mathbf{I}$$

and

$$\mathbf{V}_1 = \frac{-100\mathbf{I}}{\mathbf{Y}_1 + \mathbf{Y}_2 + \mathbf{Y}_3}$$

$$= \frac{-100\mathbf{I}}{0.25\text{ mS} + 1\text{ mS} - j0.5\text{ mS}}$$

$$= \frac{-100 \times 10^3\mathbf{I}}{1.25 - j0.5} = \frac{-100 \times 10^3\mathbf{I}}{1.3463 \angle -21.80°}$$

$$= -74.28 \times 10^3\mathbf{I} \angle 21.80°$$

$$= -74.28 \times 10^3\left(\frac{\mathbf{V}_i}{1\text{ k}\Omega}\right) \angle 21.80°$$

$$\mathbf{V}_1 = \mathbf{V}_L = \mathbf{-(74.28V}_i\mathbf{)\ V \angle 21.80°}$$

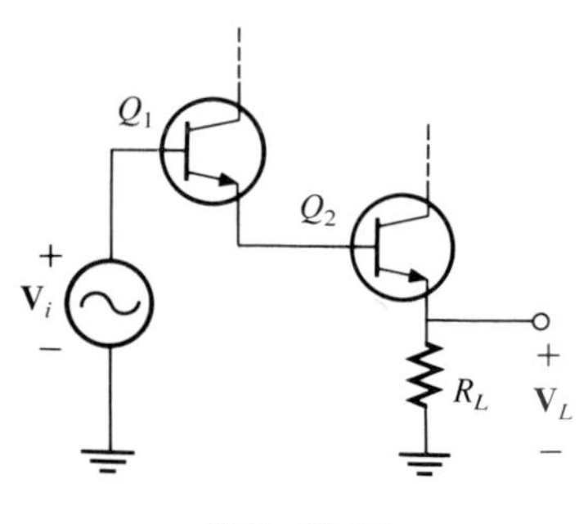

FIG. 17.29

**EXAMPLE 17.14.** The transistor configuration of Fig. 17.29 will result in a network very similar in appearance to Fig. 17.30 when the equivalent circuits are substituted. The quantities $h_1$ and $h_2$ are characteristic constants of the transistors. Determine $\mathbf{V}_L$ for the network of Fig. 17.30. The resistance values were chosen for clarity and are not

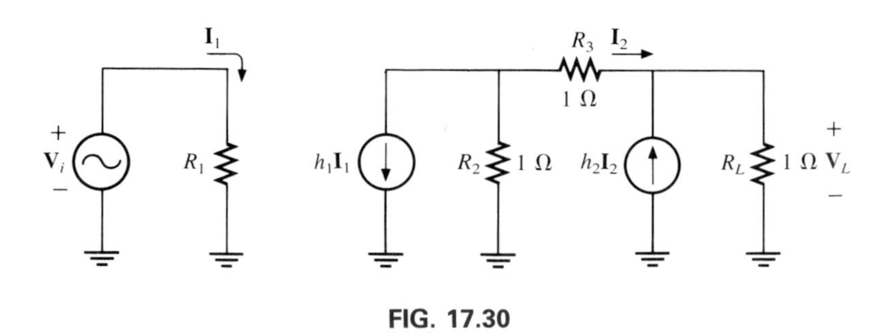


FIG. 17.30

typical values. In this case, one of the controlling variables is part of the network to be analyzed. Care must be exercised when applying the method.

***Solution:*** The network is redrawn in Fig. 17.31. Note that the controlling variable is defined by a separate network:

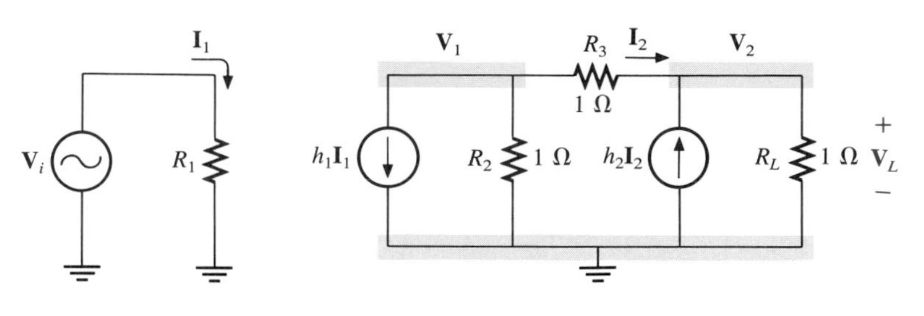

**FIG. 17.31**

$$\mathbf{I}_2 = \frac{\mathbf{V}_1 - \mathbf{V}_2}{R_3} = \frac{\mathbf{V}_1 - \mathbf{V}_2}{1\ \Omega} = \mathbf{V}_1 - \mathbf{V}_2$$

and

$$\mathbf{I}_1 = \frac{\mathbf{V}_i}{R_1} = \frac{\mathbf{V}_i}{1\ \Omega} = \mathbf{V}_i$$

Applying the format approach, we have

$$\begin{aligned} \mathbf{V}_1&: \mathbf{V}_1(1\text{ S} + 1\text{ S}) - (1\text{ S})\mathbf{V}_2 = -h_1\mathbf{I}_1 \\ \mathbf{V}_2&: \mathbf{V}_2(1\text{ S} + 1\text{ S}) - (1\text{ S})\mathbf{V}_1 = h_2\mathbf{I}_2 \end{aligned}$$

and

$$\begin{aligned} 2\mathbf{V}_1 - \mathbf{V}_2 &= -h_1\mathbf{V}_i \\ 2\mathbf{V}_2 - \mathbf{V}_1 &= h_2(\mathbf{V}_1 - \mathbf{V}_2) = h_2\mathbf{V}_1 - h_2\mathbf{V}_2 \end{aligned}$$

so that

$$2\mathbf{V}_1 - \mathbf{V}_2 = -h_1\mathbf{V}_i$$

and

$$-\mathbf{V}_1 - h_2\mathbf{V}_1 + 2\mathbf{V}_2 + h_2\mathbf{V}_2 = 0$$

Or

$$2\mathbf{V}_1 - \mathbf{V}_2 = -h_1\mathbf{V}_i$$

with

$$-(1 + h_2)\mathbf{V}_1 + (2 + h_2)\mathbf{V}_2 = 0$$

and

$$\mathbf{V}_2 = \mathbf{V}_L = \frac{\begin{vmatrix} 2 & -h_1\mathbf{V}_i \\ -(1 + h_2) & 0 \end{vmatrix}}{\begin{vmatrix} 2 & -1 \\ -(1 + h_2) & (2 + h_2) \end{vmatrix}} = \frac{-h_1(1 + h_2)\mathbf{V}_i}{2(2 + h_2) - (1 + h_2)} = \frac{-h_1(1 + h_2)\mathbf{V}_i}{4 + 2h_2 - 1 - h_2}$$

so that

$$\mathbf{V}_L = \left[-\frac{\mathbf{h}_1(\mathbf{1} + \mathbf{h}_2)}{\mathbf{3} + \mathbf{h}_2}\right]\mathbf{V}_i$$

For $h_1 = h_2 = 100$ (typical),

$$\mathbf{V}_L = \frac{(-100)(101)}{3 + 100}\mathbf{V}_i \cong -98\mathbf{V}_i$$

## 17.6 BRIDGE NETWORKS (ac)

The basic bridge configuration was discussed in some detail in Section 8.11 for dc networks. We now continue to examine bridge networks by considering those that have reactive components and a sinusoidal ac voltage or current applied.

We will first analyze various familiar forms of the bridge network using mesh analysis and nodal analysis (the format approach). The balance conditions will be investigated throughout the section.

Apply *mesh analysis* to the network of Fig. 17.32. The network is redrawn in Fig. 17.33, where

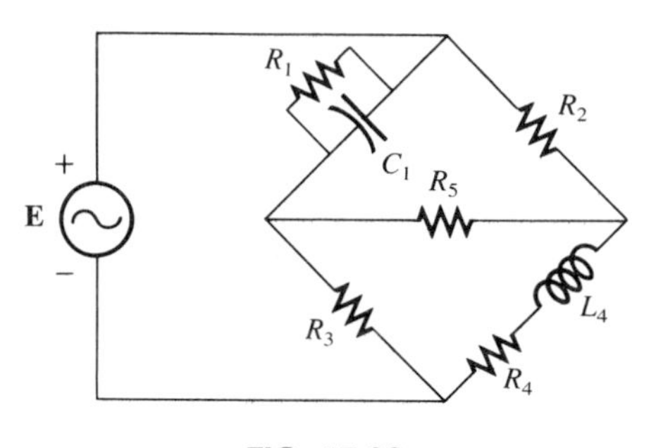


**FIG. 17.32**
*Maxwell bridge.*

$$\mathbf{Z}_1 = \frac{1}{\mathbf{Y}_1} = \frac{1}{G_1 + jB_C} = \frac{G_1}{G_1^2 + B_C^2} - j\frac{B_C}{G_1^2 + B_C^2}$$

$$\mathbf{Z}_2 = R_2 \qquad \mathbf{Z}_3 = R_3 \qquad \mathbf{Z}_4 = R_4 + jX_L \qquad \mathbf{Z}_5 = R_5$$

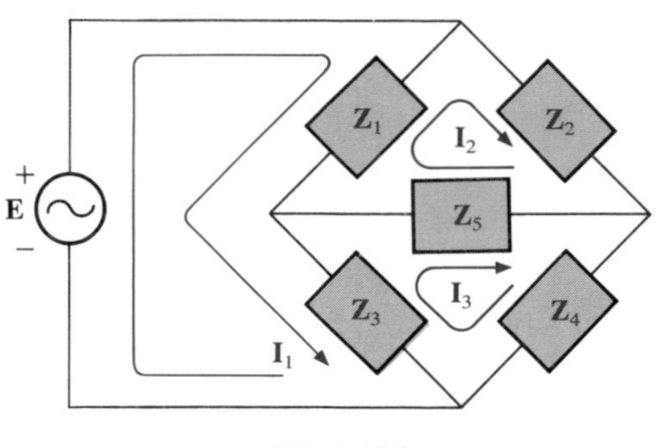


**FIG. 17.33**

Apply the format approach:

$$\begin{aligned}
(\mathbf{Z}_1 + \mathbf{Z}_3)\mathbf{I}_1 \qquad\qquad\qquad\qquad &- (\mathbf{Z}_1)\mathbf{I}_2 - (\mathbf{Z}_3)\mathbf{I}_3 = \mathbf{E}\\
(\mathbf{Z}_1 + \mathbf{Z}_2 + \mathbf{Z}_5)\mathbf{I}_2 &- (\mathbf{Z}_1)\mathbf{I}_1 - (\mathbf{Z}_5)\mathbf{I}_3 = 0\\
(\mathbf{Z}_3 + \mathbf{Z}_4 + \mathbf{Z}_5)\mathbf{I}_3 &- (\mathbf{Z}_3)\mathbf{I}_1 - (\mathbf{Z}_5)\mathbf{I}_2 = 0
\end{aligned}$$

which are rewritten as

$$\begin{aligned}
\mathbf{I}_1(\mathbf{Z}_1 + \mathbf{Z}_3) - \mathbf{I}_2\mathbf{Z}_1 - \mathbf{I}_3\mathbf{Z}_3 &= \mathbf{E}\\
-\mathbf{I}_1\mathbf{Z}_1 + \mathbf{I}_2(\mathbf{Z}_1 + \mathbf{Z}_2 + \mathbf{Z}_5) - \mathbf{I}_3\mathbf{Z}_5 &= 0\\
-\mathbf{I}_1\mathbf{Z}_3 - \mathbf{I}_2\mathbf{Z}_5 + \mathbf{I}_3(\mathbf{Z}_3 + \mathbf{Z}_4 + \mathbf{Z}_5) &= 0
\end{aligned}$$

Note the symmetry about the diagonal of the above equations. For balance, $\mathbf{I}_{\mathbf{Z}_5} = 0$ A, and

$$\mathbf{I}_{\mathbf{Z}_5} = \mathbf{I}_2 - \mathbf{I}_3 = 0$$

From the above equations,

$$\mathbf{I}_2 = \frac{\begin{vmatrix} \mathbf{Z}_1 + \mathbf{Z}_3 & \mathbf{E} & -\mathbf{Z}_3 \\ -\mathbf{Z}_1 & 0 & -\mathbf{Z}_5 \\ -\mathbf{Z}_3 & 0 & (\mathbf{Z}_3 + \mathbf{Z}_4 + \mathbf{Z}_5) \end{vmatrix}}{\begin{vmatrix} \mathbf{Z}_1 + \mathbf{Z}_3 & -\mathbf{Z}_1 & -\mathbf{Z}_3 \\ -\mathbf{Z}_1 & (\mathbf{Z}_1 + \mathbf{Z}_2 + \mathbf{Z}_3) & -\mathbf{Z}_5 \\ -\mathbf{Z}_3 & -\mathbf{Z}_5 & (\mathbf{Z}_3 + \mathbf{Z}_4 + \mathbf{Z}_5) \end{vmatrix}}$$
$$= \frac{\mathbf{E}(\mathbf{Z}_1\mathbf{Z}_3 + \mathbf{Z}_1\mathbf{Z}_4 + \mathbf{Z}_1\mathbf{Z}_5 + \mathbf{Z}_3\mathbf{Z}_5)}{\Delta}$$

where $\Delta$ signifies the determinant of the denominator (or coefficients). Similarly,

$$\mathbf{I}_3 = \frac{\mathbf{E}(\mathbf{Z}_1\mathbf{Z}_3 + \mathbf{Z}_3\mathbf{Z}_2 + \mathbf{Z}_1\mathbf{Z}_5 + \mathbf{Z}_3\mathbf{Z}_5)}{\Delta}$$

and

$$\mathbf{I}_{\mathbf{Z}_5} = \mathbf{I}_2 - \mathbf{I}_3 = \frac{\mathbf{E}(\mathbf{Z}_1\mathbf{Z}_4 - \mathbf{Z}_3\mathbf{Z}_2)}{\Delta}$$

For $\mathbf{I}_{\mathbf{Z}_5} = 0$, the following must be satisfied (for a finite $\Delta$ not equal to zero):

$$\boxed{\mathbf{Z}_1\mathbf{Z}_4 = \mathbf{Z}_3\mathbf{Z}_2} \qquad \mathbf{I}_{\mathbf{Z}_5} = 0 \qquad \mathbf{(17.3)}$$

This condition will be analyzed in greater depth later in this section.

Applying *nodal analysis* to the network of Fig. 17.34 will result in the configuration of Fig. 17.35, where

$$\mathbf{Y}_1 = \frac{1}{\mathbf{Z}_1} = \frac{1}{R_1 - jX_C} \qquad \mathbf{Y}_2 = \frac{1}{\mathbf{Z}_2} = \frac{1}{R_2}$$
$$\mathbf{Y}_3 = \frac{1}{\mathbf{Z}_3} = \frac{1}{R_3} \qquad \mathbf{Y}_4 = \frac{1}{\mathbf{Z}_4} = \frac{1}{R_4 + jX_L} \qquad \mathbf{Y}_5 = \frac{1}{R_5}$$

and

$$\begin{aligned} (\mathbf{Y}_1 + \mathbf{Y}_2)\mathbf{V}_1 - (\mathbf{Y}_1)\mathbf{V}_2 - (\mathbf{Y}_2)\mathbf{V}_3 &= \mathbf{I} \\ (\mathbf{Y}_1 + \mathbf{Y}_3 + \mathbf{Y}_5)\mathbf{V}_2 - (\mathbf{Y}_1)\mathbf{V}_1 - (\mathbf{Y}_5)\mathbf{V}_3 &= 0 \\ (\mathbf{Y}_2 + \mathbf{Y}_4 + \mathbf{Y}_5)\mathbf{V}_3 - (\mathbf{Y}_2)\mathbf{V}_1 - (\mathbf{Y}_5)\mathbf{V}_2 &= 0 \end{aligned}$$

which are rewritten as

$$\begin{aligned} \mathbf{V}_1(\mathbf{Y}_1 + \mathbf{Y}_2) \quad - \mathbf{V}_2\mathbf{Y}_1 \qquad\qquad & - \mathbf{V}_3\mathbf{Y}_2 &= \mathbf{I} \\ -\mathbf{V}_1\mathbf{Y}_1 \quad + \mathbf{V}_2(\mathbf{Y}_1 + \mathbf{Y}_3 + \mathbf{Y}_5) & - \mathbf{V}_3\mathbf{Y}_5 &= 0 \\ -\mathbf{V}_1\mathbf{Y}_2 \quad - \mathbf{V}_2\mathbf{Y}_5 \qquad\qquad & + \mathbf{V}_3(\mathbf{Y}_2 + \mathbf{Y}_4 + \mathbf{Y}_5) &= 0 \end{aligned}$$

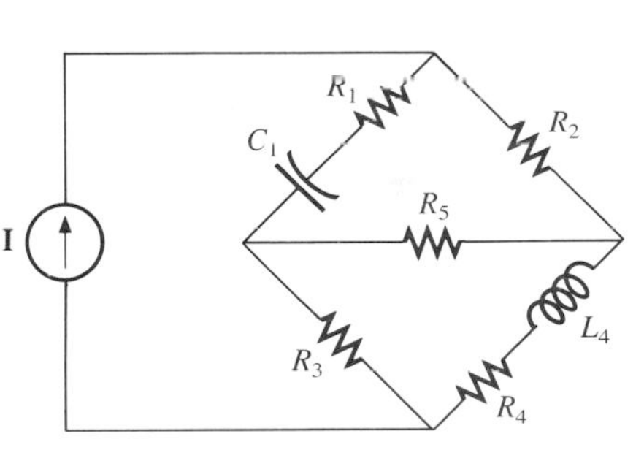


FIG. 17.34
*Hay bridge.*

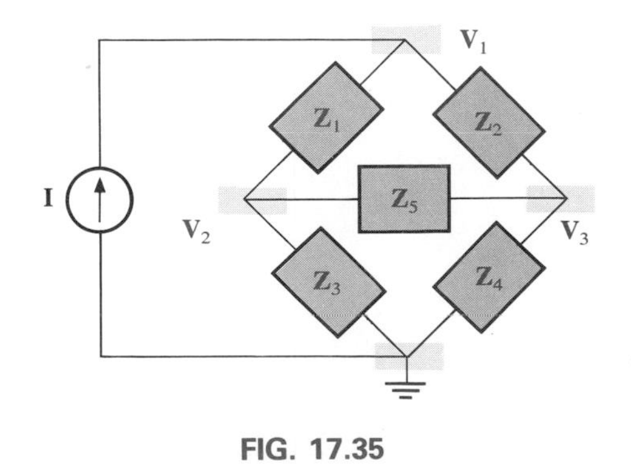


FIG. 17.35

Again, note the symmetry about the diagonal axis. For balance, $\mathbf{V}_{\mathbf{Z}_5} = 0$ V, and

$$\mathbf{V}_{\mathbf{Z}_5} = \mathbf{V}_2 - \mathbf{V}_3 = 0$$

From the above equations,

$$\mathbf{V}_2 = \frac{\begin{vmatrix} \mathbf{Y}_1 + \mathbf{Y}_2 & \mathbf{I} & -\mathbf{Y}_2 \\ -\mathbf{Y}_1 & 0 & -\mathbf{Y}_5 \\ -\mathbf{Y}_2 & 0 & (\mathbf{Y}_2 + \mathbf{Y}_4 + \mathbf{Y}_5) \end{vmatrix}}{\begin{vmatrix} \mathbf{Y}_1 + \mathbf{Y}_2 & -\mathbf{Y}_1 & -\mathbf{Y}_2 \\ -\mathbf{Y}_1 & (\mathbf{Y}_1 + \mathbf{Y}_3 + \mathbf{Y}_5) & -\mathbf{Y}_5 \\ -\mathbf{Y}_2 & -\mathbf{Y}_5 & (\mathbf{Y}_2 + \mathbf{Y}_4 + \mathbf{Y}_5) \end{vmatrix}}$$

$$= \frac{\mathbf{I}(\mathbf{Y}_1\mathbf{Y}_3 + \mathbf{Y}_1\mathbf{Y}_4 + \mathbf{Y}_1\mathbf{Y}_5 + \mathbf{Y}_3\mathbf{Y}_5)}{\Delta}$$

Similarly,

$$\mathbf{V}_3 = \frac{\mathbf{I}(\mathbf{Y}_1\mathbf{Y}_3 + \mathbf{Y}_3\mathbf{Y}_2 + \mathbf{Y}_1\mathbf{Y}_5 + \mathbf{Y}_3\mathbf{Y}_5)}{\Delta}$$

Note the similarities between the above equations and those obtained for mesh analysis. Then

$$\mathbf{V}_{\mathbf{Z}_5} = \mathbf{V}_2 - \mathbf{V}_3 = \frac{\mathbf{I}(\mathbf{Y}_1\mathbf{Y}_4 - \mathbf{Y}_3\mathbf{Y}_2)}{\Delta}$$

For $\mathbf{V}_{\mathbf{Z}_5} = 0$, the following must be satisfied for a finite $\Delta$ not equal to zero:

$$\boxed{\mathbf{Y}_1\mathbf{Y}_4 = \mathbf{Y}_3\mathbf{Y}_2} \qquad \mathbf{V}_{\mathbf{Z}_5} = 0 \qquad \textbf{(17.4)}$$

However, substituting $\mathbf{Y}_1 = 1/\mathbf{Z}_1$, $\mathbf{Y}_2 = 1/\mathbf{Z}_2$, $\mathbf{Y}_3 = 1/\mathbf{Z}_3$, and $\mathbf{Y}_4 = 1/\mathbf{Z}_4$, we have

$$\frac{1}{\mathbf{Z}_1\mathbf{Z}_4} = \frac{1}{\mathbf{Z}_3\mathbf{Z}_2}$$

or

$$\boxed{\mathbf{Z}_1\mathbf{Z}_4 = \mathbf{Z}_3\mathbf{Z}_2} \qquad \mathbf{V}_{\mathbf{Z}_5} = 0$$

corresponding with Eq. (17.3) obtained earlier.

Let us now investigate the balance criteria in more detail by considering the network of Fig. 17.36, where it is specified that **I**, **V** = 0.

Since **I** = 0,

$$\boxed{\mathbf{I}_1 = \mathbf{I}_3} \qquad \textbf{(17.5a)}$$

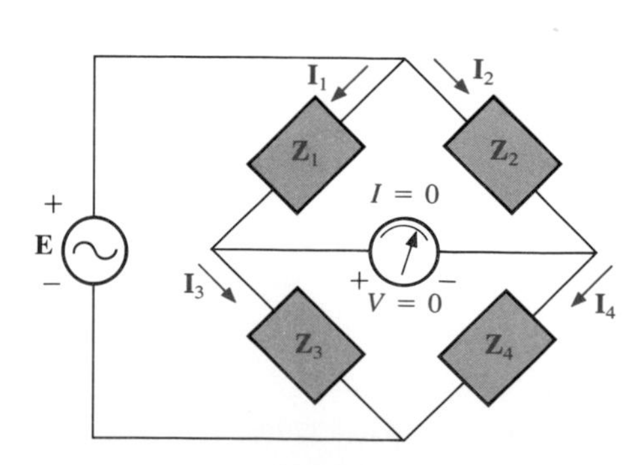

FIG. 17.36

and

$$\mathbf{I}_2 = \mathbf{I}_4 \qquad (17.5b)$$

In addition, for $\mathbf{V} = 0$,

$$\mathbf{I}_1\mathbf{Z}_1 = \mathbf{I}_2\mathbf{Z}_2 \qquad (17.5c)$$

and

$$\mathbf{I}_3\mathbf{Z}_3 = \mathbf{I}_4\mathbf{Z}_4 \qquad (17.5d)$$

Substituting the current relations above into Eq. (17.5d), we have

$$\mathbf{I}_1\mathbf{Z}_3 = \mathbf{I}_2\mathbf{Z}_4$$

and

$$\mathbf{I}_2 = \frac{\mathbf{Z}_3}{\mathbf{Z}_4}\mathbf{I}_1$$

Substituting this relationship for $\mathbf{I}_2$ into Eq. (17.5c) yields

$$\mathbf{I}_1\mathbf{Z}_1 = \left(\frac{\mathbf{Z}_3}{\mathbf{Z}_4}\mathbf{I}_1\right)\mathbf{Z}_2$$

and

$$\mathbf{Z}_1\mathbf{Z}_4 = \mathbf{Z}_2\mathbf{Z}_3$$

as obtained above. Rearranging, we have

$$\frac{\mathbf{Z}_1}{\mathbf{Z}_3} = \frac{\mathbf{Z}_2}{\mathbf{Z}_4} \qquad (17.6)$$

corresponding with Eq. (8.4) for dc resistive networks.

For the network of Fig. 17.34, which is referred to as a *Hay bridge* when $\mathbf{Z}_5$ is replaced by a sensitive galvanometer,

$$\mathbf{Z}_1 = R_1 - jX_C$$
$$\mathbf{Z}_2 = R_2$$
$$\mathbf{Z}_3 = R_3$$
$$\mathbf{Z}_4 = R_4 + jX_L$$

This particular network is used for measuring the resistance and inductance of coils in which the resistance is a small fraction of the reactance $X_L$.

Substitute into Eq. (17.6) in the following form:

$$\mathbf{Z}_2\mathbf{Z}_3 = \mathbf{Z}_4\mathbf{Z}_1$$
$$R_2R_3 = (R_4 + jX_L)(R_1 - jX_C)$$

or

$$R_2R_3 = R_1R_4 + j(R_1X_L - R_4X_C) + X_CX_L$$

so that

$$R_2R_3 + j0 = (R_1R_4 + X_CX_L) + j(R_1X_L - R_4X_C)$$

In order for the equations to be equal, *the real and imaginary parts must be equal*. Therefore, for a balanced Hay bridge,

$$\boxed{R_2R_3 = R_1R_4 + X_CX_L} \tag{17.7a}$$

and

$$\boxed{0 = R_1X_L - R_4X_C} \tag{17.7b}$$

or substituting

$$X_L = \omega L$$

$$X_C = \frac{1}{\omega C}$$

we have

$$X_CX_L = \left(\frac{1}{\omega C}\right)(\omega L) = \frac{L}{C}$$

and

$$R_2R_3 = R_1R_4 + \frac{L}{C}$$

with

$$R_1\omega L = \frac{R_4}{\omega C}$$

Solving for $R_4$ in the last equation yields

$$R_4 = \omega^2 LCR_1$$

and substituting into the previous equation, we have

$$R_2R_3 = R_1(\omega^2 LCR_1) + \frac{L}{C}$$

Multiply through by $C$ and factor:

$$CR_2R_3 = L(\omega^2C^2R_1^2 + 1)$$

and

$$\boxed{L = \frac{CR_2R_3}{1 + \omega^2C^2R_1^2}} \tag{17.8a}$$

with further algebra yielding

$$R_4 = \frac{\omega^2 C^2 R_1 R_2 R_3}{1 + \omega^2 C^2 R_1^2} \qquad \textbf{(17.8b)}$$

Equations (17.7) and (17.8) are the balance conditions for the Hay bridge. Note that each is frequency dependent. For different frequencies, the resistive and capacitive elements must vary for a particular coil to achieve balance. For a coil placed in the Hay bridge as shown in Fig. 17.35, the resistance and inductance of the coil can be determined by Eqs. (17.8a) and (17.8b) when balance is achieved.

The bridge of Fig. 17.32 is referred to as a *Maxwell bridge* when $\mathbf{Z}_5$ is replaced by a sensitive galvanometer. This setup is used for inductance measurements when the resistance of the coil is large enough not to require a Hay bridge.

Application of Eq. (17.6) will yield the following results for the inductance and resistance of the inserted coil:

$$L = CR_2R_3 \qquad \textbf{(17.9)}$$

$$R_4 = \frac{R_2R_3}{R_1} \qquad \textbf{(17.10)}$$

The derivation of these equations is quite similar to that employed for the Hay bridge. Keep in mind that the real and imaginary parts must be equal.

One remaining popular bridge is the *capacitance comparison bridge* of Fig. 17.37. An unknown capacitance and its associated resistance can be determined using this bridge. Application of Eq. (17.6) will yield the following results:

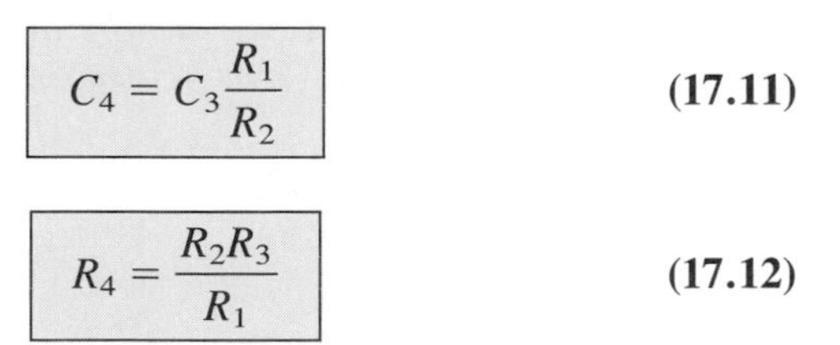

$$C_4 = C_3\frac{R_1}{R_2} \qquad \textbf{(17.11)}$$

$$R_4 = \frac{R_2R_3}{R_1} \qquad \textbf{(17.12)}$$

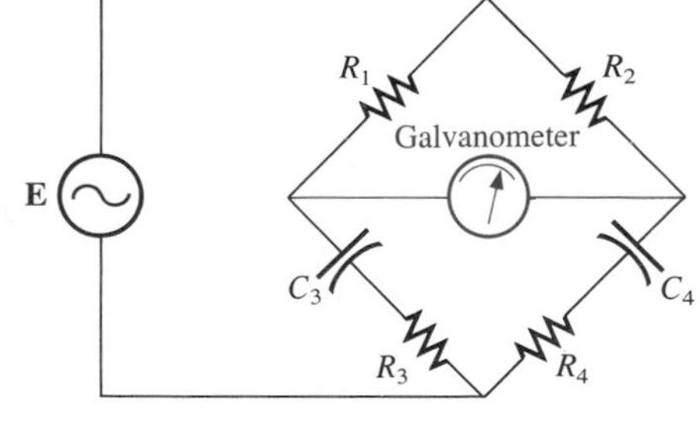


**FIG. 17.37**
*Capacitance comparison bridge.*

The derivation of these equations will appear as a problem at the end of the chapter.

## 17.7 Δ-Y, Y-Δ CONVERSIONS

The Δ-Y, Y-Δ (or π-T, T-π as defined in Section 8.12) conversions for ac circuits will not be derived here since the development corresponds

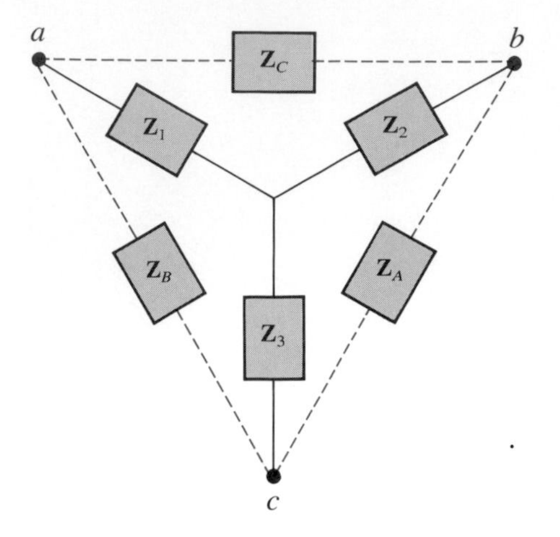


FIG. 17.38
*Δ-Y configuration.*

exactly with that for dc circuits. Taking the Δ-Y configuration shown in Fig. 17.38, we find the general equations for the impedances of the Y in terms of those for the Δ:

$$\mathbf{Z}_1 = \frac{\mathbf{Z}_B\mathbf{Z}_C}{\mathbf{Z}_A + \mathbf{Z}_B + \mathbf{Z}_C} \tag{17.13}$$

$$\mathbf{Z}_2 = \frac{\mathbf{Z}_A\mathbf{Z}_C}{\mathbf{Z}_A + \mathbf{Z}_B + \mathbf{Z}_C} \tag{17.14}$$

$$\mathbf{Z}_3 = \frac{\mathbf{Z}_A\mathbf{Z}_B}{\mathbf{Z}_A + \mathbf{Z}_B + \mathbf{Z}_C} \tag{17.15}$$

For the impedances of the Δ in terms of those for the Y, the equations are

$$\mathbf{Z}_B = \frac{\mathbf{Z}_1\mathbf{Z}_2 + \mathbf{Z}_1\mathbf{Z}_3 + \mathbf{Z}_2\mathbf{Z}_3}{\mathbf{Z}_2} \tag{17.16}$$

$$\mathbf{Z}_A = \frac{\mathbf{Z}_1\mathbf{Z}_2 + \mathbf{Z}_1\mathbf{Z}_3 + \mathbf{Z}_2\mathbf{Z}_3}{\mathbf{Z}_1} \tag{17.17}$$

$$\mathbf{Z}_C = \frac{\mathbf{Z}_1\mathbf{Z}_2 + \mathbf{Z}_1\mathbf{Z}_3 + \mathbf{Z}_2\mathbf{Z}_3}{\mathbf{Z}_3} \tag{17.18}$$

***Note that each impedance of the Y is equal to the product of the impedances in the two closest branches of the Δ, divided by the sum of the impedances in the Δ; and the value of each impedance of the Δ is equal to the sum of the possible product combinations of the impedances of the Y, divided by the impedances of the Y farthest from the impedance to be determined.***

Drawn in different forms (Fig. 17.39), they are also referred to as the T and $\pi$ configurations.

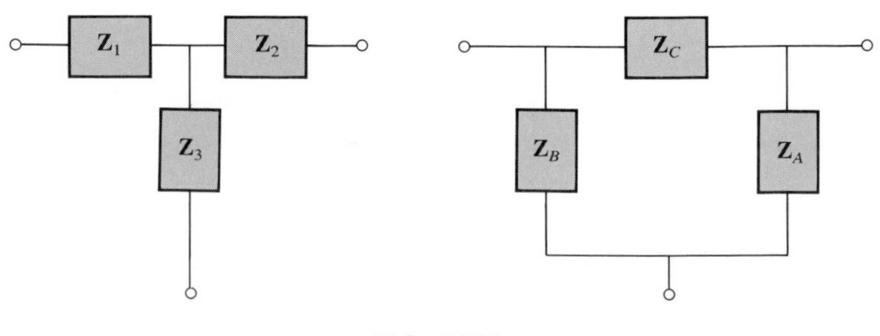


FIG. 17.39

In the study of dc networks, we found that if all of the resistors of the Δ or Y were the same, the conversion from one to the other could be accomplished using the equation

$$R_\Delta = 3R_Y \quad \text{or} \quad R_Y = \frac{R_\Delta}{3}$$

For ac networks,

$$\mathbf{Z}_\Delta = 3\mathbf{Z}_Y \quad \text{or} \quad \mathbf{Z}_Y = \frac{\mathbf{Z}_\Delta}{3} \tag{17.19}$$

Be careful when using this simplified form. It is not sufficient for all the impedances of the Δ or Y to be of the same magnitude: *The angle associated with each must also be the same.*

---

**EXAMPLE 17.15.** Find the total impedance $\mathbf{Z}_T$ of the network of Fig. 17.40.

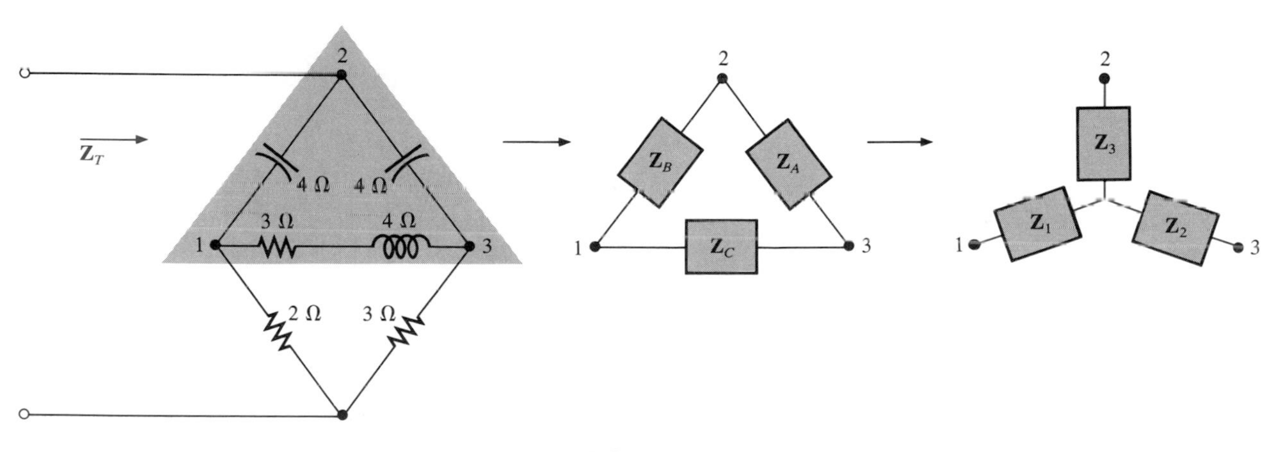

**FIG. 17.40**

***Solution:***

$$\mathbf{Z}_B = -j4 \qquad \mathbf{Z}_A = -j4 \qquad \mathbf{Z}_C = 3 + j4$$

$$\mathbf{Z}_1 = \frac{\mathbf{Z}_B\mathbf{Z}_C}{\mathbf{Z}_A + \mathbf{Z}_B + \mathbf{Z}_C} = \frac{(-j4\ \Omega)(3\ \Omega + j4\ \Omega)}{(-j4\ \Omega) + (-j4\ \Omega) + (3\ \Omega + j4\ \Omega)}$$

$$= \frac{(4\ \angle -90^\circ)(5\ \angle 53.13^\circ)}{3 - j4} = \frac{20\ \angle -36.87^\circ}{5\ \angle -53.13^\circ}$$

$$= 4\ \Omega\ \angle 16.13^\circ = 3.84\ \Omega + j1.11\ \Omega$$

$$\mathbf{Z}_2 = \frac{\mathbf{Z}_A\mathbf{Z}_C}{\mathbf{Z}_A + \mathbf{Z}_B + \mathbf{Z}_C} = \frac{(-j4\ \Omega)(3\ \Omega + j4\ \Omega)}{5\ \Omega\ \angle -53.13^\circ}$$

$$= 4\ \Omega\ \angle 16.13^\circ = 3.84\ \Omega + j1.11\ \Omega$$

Recall from the study of dc circuits that if two branches of the Y or Δ were the same, the corresponding Δ or Y, respectively, would also have two similar branches. In this example, $\mathbf{Z}_A = \mathbf{Z}_B$. Therefore, $\mathbf{Z}_1 = \mathbf{Z}_2$, and

$$\mathbf{Z}_3 = \frac{\mathbf{Z}_A\mathbf{Z}_B}{\mathbf{Z}_A + \mathbf{Z}_B + \mathbf{Z}_C} = \frac{(-j4\ \Omega)(-j4\ \Omega)}{5\ \Omega\ \angle -53.13^\circ}$$
$$= \frac{16\ \angle -180^\circ}{5\ \angle -53.13^\circ} = 3.2\ \Omega\ \angle -126.87^\circ = -1.92\ \Omega - j2.56\ \Omega$$

Replace the Δ by the Y (Fig. 17.41):

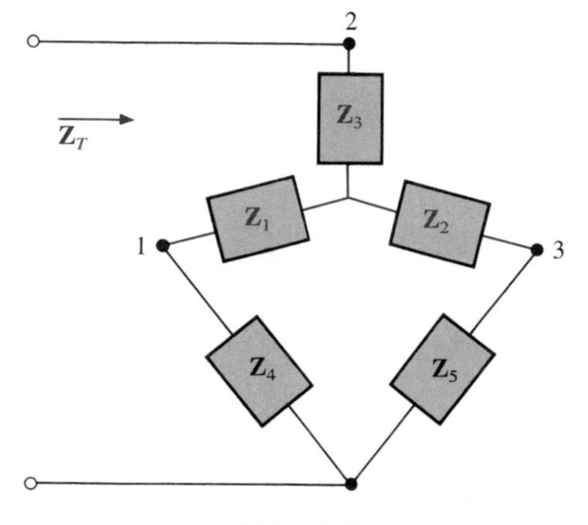

FIG. 17.41

$$\mathbf{Z}_1 = 3.84\ \Omega + j1.11\ \Omega \qquad \mathbf{Z}_2 = 3.84\ \Omega + j1.11\ \Omega$$
$$\mathbf{Z}_3 = -1.92\ \Omega - j2.56\ \Omega \qquad \mathbf{Z}_4 = 2\ \Omega \qquad \mathbf{Z}_5 = 3\ \Omega$$

Impedances $\mathbf{Z}_1$ and $\mathbf{Z}_4$ are in series:

$$\mathbf{Z}_{T_1} = \mathbf{Z}_1 + \mathbf{Z}_4 = 3.84\ \Omega + j1.11\ \Omega + 2\ \Omega = 5.84\ \Omega + j1.11\ \Omega$$
$$= 5.94\ \Omega\ \angle 10.76^\circ$$

Impedances $\mathbf{Z}_2$ and $\mathbf{Z}_5$ are in series:

$$\mathbf{Z}_{T_2} = \mathbf{Z}_2 + \mathbf{Z}_5 = 3.84\ \Omega + j1.11\ \Omega + 3\ \Omega = 6.84\ \Omega + j1.11\ \Omega$$
$$= 6.93\ \Omega\ \angle 9.22^\circ$$

Impedances $\mathbf{Z}_{T_1}$ and $\mathbf{Z}_{T_2}$ are in parallel:

$$\mathbf{Z}_{T_3} = \frac{\mathbf{Z}_{T_1}\mathbf{Z}_{T_2}}{\mathbf{Z}_{T_1} + \mathbf{Z}_{T_2}} = \frac{(5.94\ \Omega\ \angle 10.76^\circ)(6.93\ \Omega\ \angle 9.22^\circ)}{5.84\ \Omega + j1.11\ \Omega + 6.84\ \Omega + j1.11\ \Omega}$$
$$= \frac{41.16\ \angle 19.98^\circ}{12.68 + j2.22} = \frac{41.16\ \angle 19.98^\circ}{12.87\ \angle 9.93^\circ} = 3.198\ \Omega\ \angle 10.05^\circ$$
$$= 3.15\ \Omega + j0.56\ \Omega$$

Impedances $\mathbf{Z}_3$ and $\mathbf{Z}_{T_3}$ are in series. Therefore,

$$\mathbf{Z}_T = \mathbf{Z}_3 + \mathbf{Z}_{T_3} = -1.92\ \Omega - j2.56\ \Omega + 3.15\ \Omega + j0.56\ \Omega$$
$$= 1.23\ \Omega - j2.0\ \Omega = \mathbf{2.35\ \Omega\ \angle -58.41^\circ}$$

**EXAMPLE 17.16.** Using both the Δ-Y and Y-Δ transformations, find the total impedance $\mathbf{Z}_T$ for the network of Fig. 17.42.

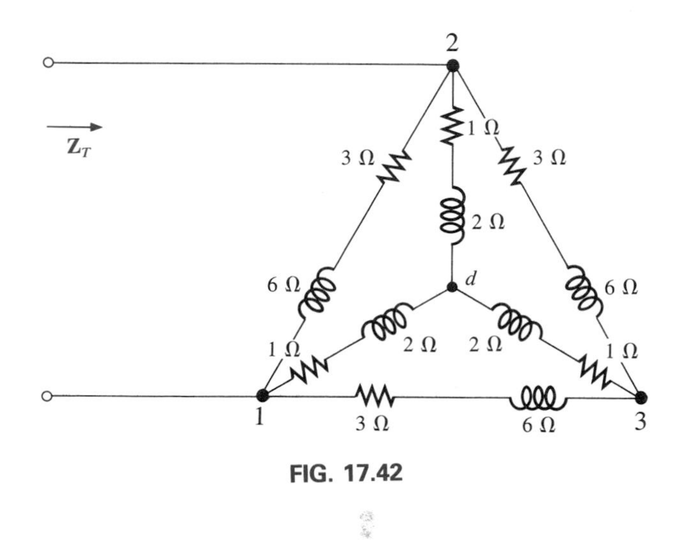


**FIG. 17.42**

***Solution:*** *Using the Δ-Y transformation,* we obtain Fig. 17.43. In this case, since both systems are balanced (same impedance in each

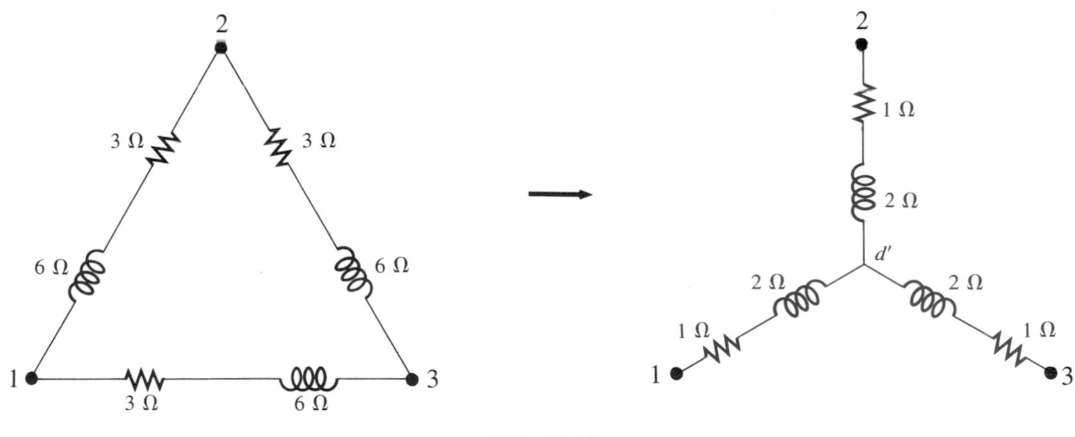


**FIG. 17.43**

branch), the center point $d'$ of the transformed Δ will be the same as the point $d$ of the original Y:

$$\mathbf{Z}_Y = \frac{\mathbf{Z}_\Delta}{3} = \frac{3\ \Omega + j6\ \Omega}{3} = 1\ \Omega + j2\ \Omega$$

and (Fig. 17.44)

$$\mathbf{Z}_T = 2\left(\frac{1\ \Omega + j2\ \Omega}{2}\right) = \mathbf{1\ \Omega + \mathit{j}2\ \Omega}$$

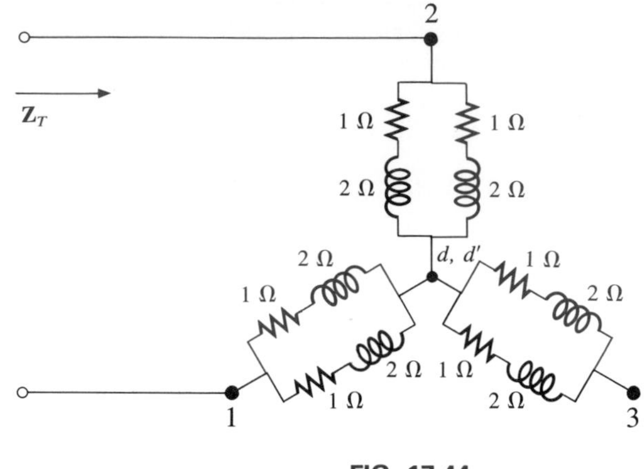


FIG. 17.44

*Using the Y-Δ transformation* (Fig. 17.45), we obtain

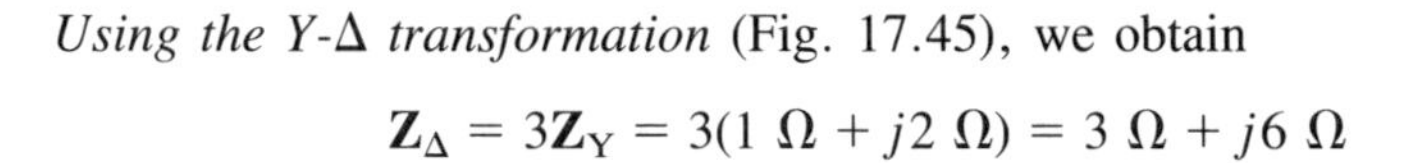

$$\mathbf{Z}_\Delta = 3\mathbf{Z}_Y = 3(1\ \Omega + j2\ \Omega) = 3\ \Omega + j6\ \Omega$$

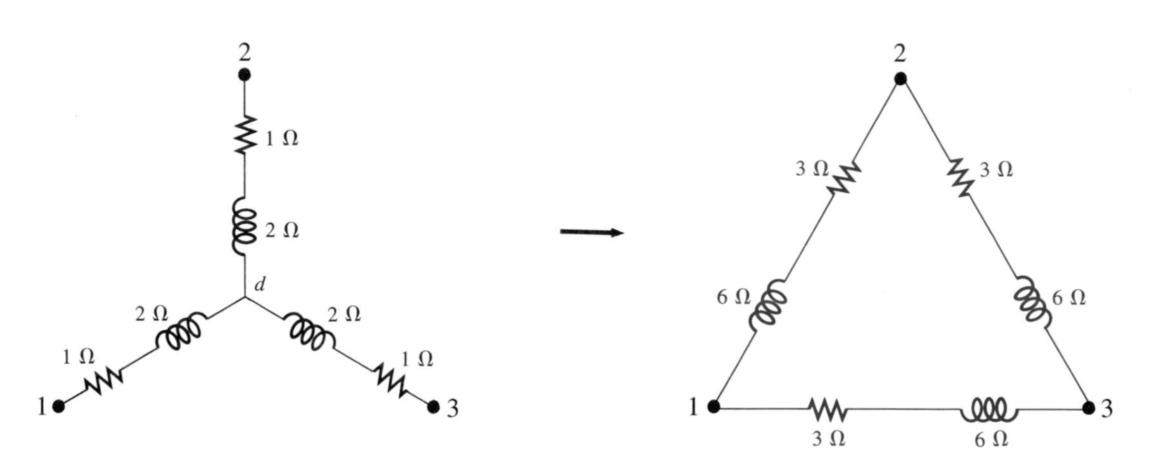

FIG. 17.45

Each resulting parallel combination in Fig. 17.46 will have the following impedance:

$$\mathbf{Z}' = \frac{3\ \Omega + j6\ \Omega}{2} = 1.5\ \Omega + j3\ \Omega$$

and

$$\mathbf{Z}_T = \frac{\mathbf{Z}'(2\mathbf{Z}')}{\mathbf{Z}' + 2\mathbf{Z}'} = \frac{2(\mathbf{Z}')^2}{3\mathbf{Z}'} = \frac{2\mathbf{Z}'}{3}$$

$$= \frac{2(1.5\ \Omega + j3\ \Omega)}{3} = \mathbf{1\ \Omega + \mathit{j}2\ \Omega}$$

which compares with the above result.

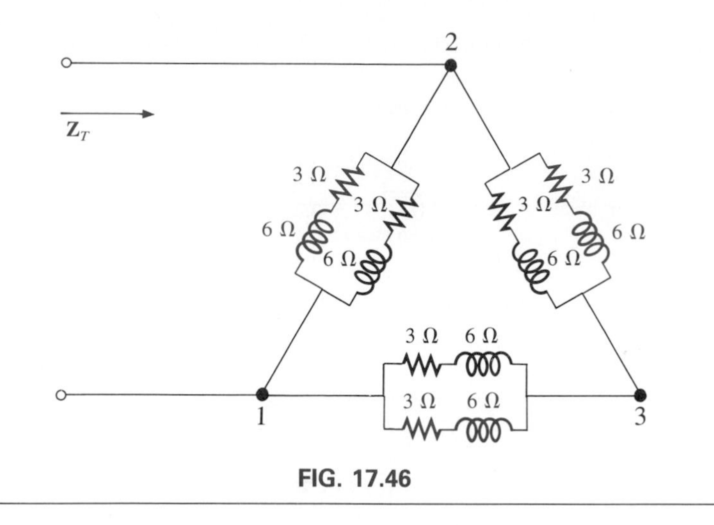


FIG. 17.46

## 17.8 COMPUTER ANALYSIS

### BASIC

The nodal voltages and mesh currents for any network appearing in this chapter can be obtained using BASIC to determine the solution to the equations that result from each configuration. Once established, the parameters of a network can be changed and the solution quickly determined by the BASIC program. The sequence, therefore, is to find a general solution for the network in terms of the network impedances and applied sources and then apply BASIC to determine the solution using vector algebra. Of course, any change in the original configuration will result in a different set of equations and require the writing of a new BASIC program.

### PSPICE

In this case, the obvious advantage of PSPICE is that a general solution does not have to be obtained before the software package can be applied. Through proper input of the network parameters, the software package can determine the relationships among parameters and print out the desired quantities. There is no need specifically to convey to the package which elements are in series or parallel and assist in the generation of a general solution. The input file is sufficient to provide this information to the package, from which it generates and prints out a solution for the user.

The first application of PSPICE in this section is to determine the nodal voltages for the network of Example 17.10 and compare solutions. The network is redrawn in Fig. 17.47 with the defined nodes and

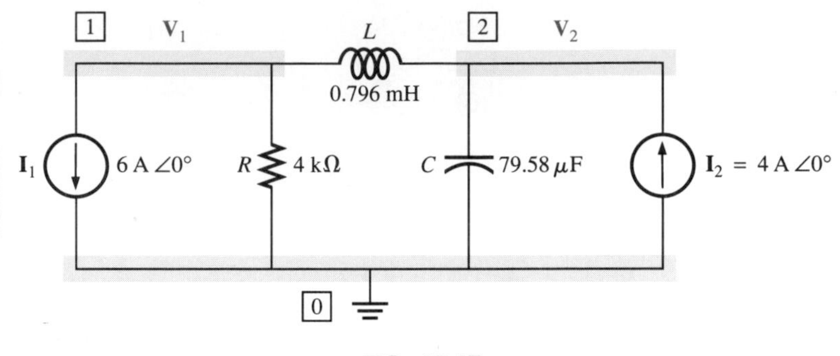

FIG. 17.47

parameter values as determined at a frequency of 1 kHz from the reactance levels of Fig. 17.20. The input file of Fig. 17.48 requires five lines for the network parameters, an .AC line to define the analysis and frequency, and a .PRINT statement to output the desired results. Note that $\mathbf{V}_1 = 20.80\text{ V} \angle -126.9°$ is exactly the same as obtained in the longhand fashion of Example 17.10. In addition, $\mathbf{V}_2 = 8.617\text{ V} \angle -150.9°$.

```
************************ Evaluation PSpice (January 1989) ******* 17:32:04 *******

 Chapter 17 - Nodal Analysis - AC

 ****     CIRCUIT DESCRIPTION

**************************************************************************


I1 1 0 AC 6 0
I2 0 2 AC 4 0
R  1 0 4
L  1 2 0.796MH
C  2 0 79.58UF
.AC LIN 1 1KH 1KH
.PRINT AC VM(1) VP(1) VM(2) VP(2)
.OPTIONS NOPAGE
.END


 ****     AC ANALYSIS                          TEMPERATURE =   27.000 DEG C


  FREQ         VM(1)        VP(1)        VM(2)        VP(2)


   1.000E+03   2.080E+01   -1.269E+02   8.617E+00   -1.509E+01
```

FIG. 17.48

The second application is to the bridge network of Fig. 17.49, which is Problem 24. The voltage across the bridge arm was requested to

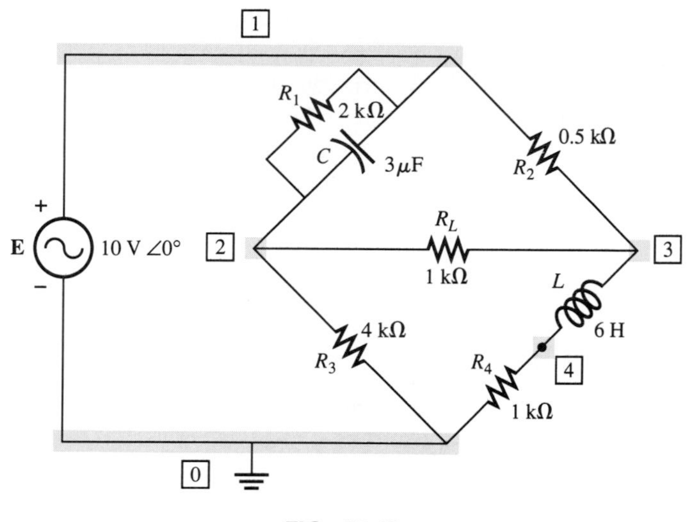


**FIG. 17.49**

determine if a balance condition exists, as defined by $|V| = 0$ V. The input file of Fig. 17.50 is quite lengthy but follows the same format as

```
************************ Evaluation PSpice (January 1989) ******* 17:40:06 *******

Chapter 17 - Bridge Network - AC

****     CIRCUIT DESCRIPTION

**************************************************************************


VE 0 1 AC 10V 0
R1 1 2 2K
R2 1 3 0.5K
R3 2 0 4K
R4 4 0 1K
RL 2 3 1K
L  3 4 6H
C  1 2 3UF
.AC LIN 1 159.15H 159.15H
.PRINT AC VM(2,3) VP(2,3)
.OPTIONS NOPAGE
.END

****     AC ANALYSIS                        TEMPERATURE =   27.000 DEG C

 FREQ         VM(2,3)      VP(2,3)

 1.592E+02    1.831E-15   -1.660E+02
```

**FIG. 17.50**

previously defined. The frequency applied was determined from the equation

$$f = \frac{\omega}{2\pi} = \frac{1000 \text{ rad/s}}{2\pi \text{ rad}} \cong 159.15 \text{ Hz}$$

The result of $18.31 \times 10^{-15}$ V certainly satisfies the balance condition, and the system is in a balanced state. There is a phase angle present but it is immaterial when associated with such a small current level. The rms-reading galvanometer will obviously reflect a current of zero amperes.

# PROBLEMS

## SECTION 17.2 Independent Versus Dependent (Controlled) Sources

1. Discuss, in your own words, the difference between a controlled and an independent source.

## SECTION 17.3 Source Conversions

2. Convert the voltage sources of Fig. 17.51 to current sources.

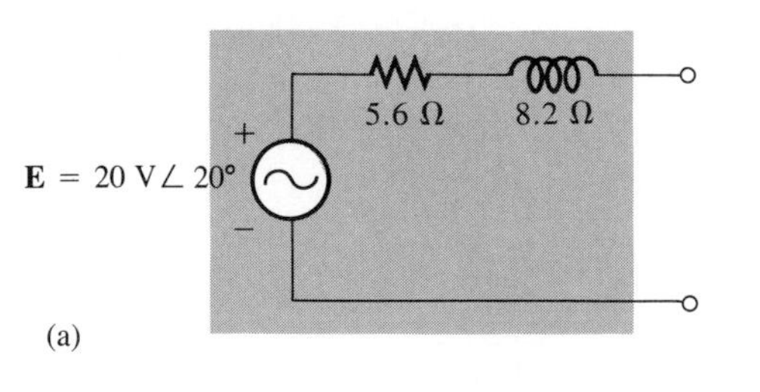


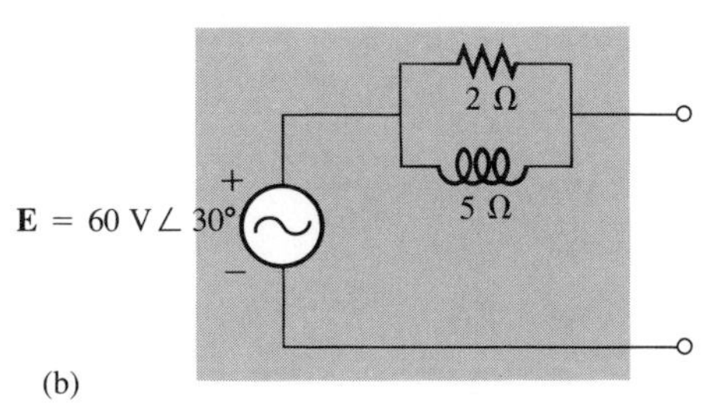


FIG. 17.51

3. Convert the current sources of Fig. 17.52 to voltage sources.

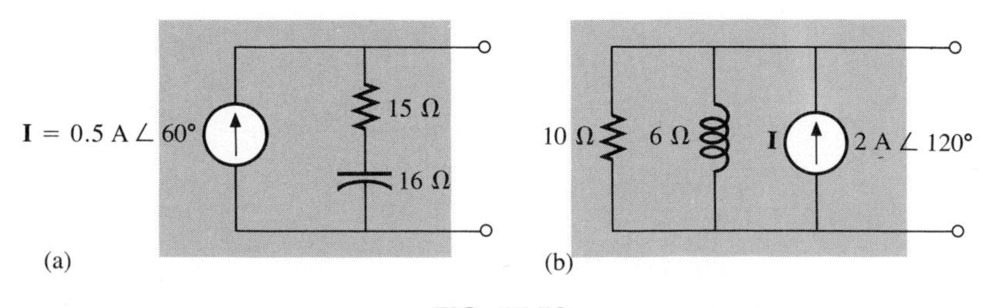


FIG. 17.52

4. Convert the voltage source of Fig. 17.53(a) to a current source and the current source of Fig. 17.53(b) to a voltage source.

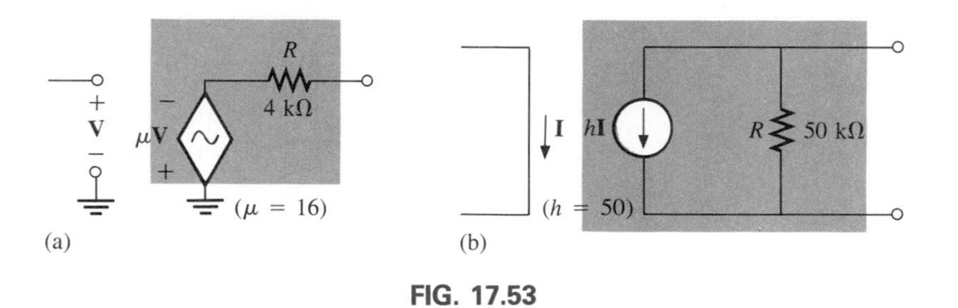


FIG. 17.53

## SECTION 17.4 Mesh Analysis (Format Approach)

5. Write the mesh equations for the networks of Fig. 17.54. Determine the current through the resistor $R_1$.

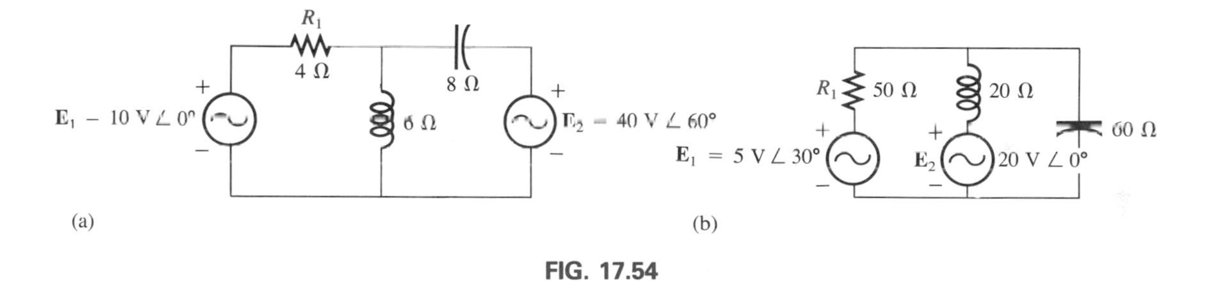


FIG. 17.54

6. Repeat Problem 5 for the networks of Fig. 17.55.

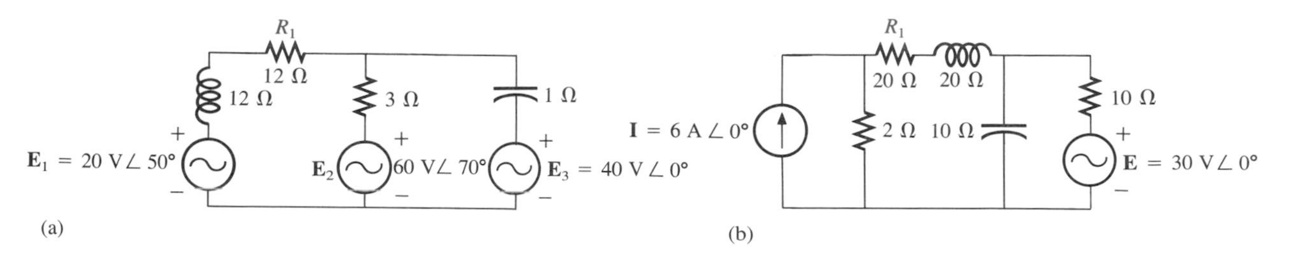


FIG. 17.55

*7. Repeat Problem 5 for the networks of Fig. 17.56.

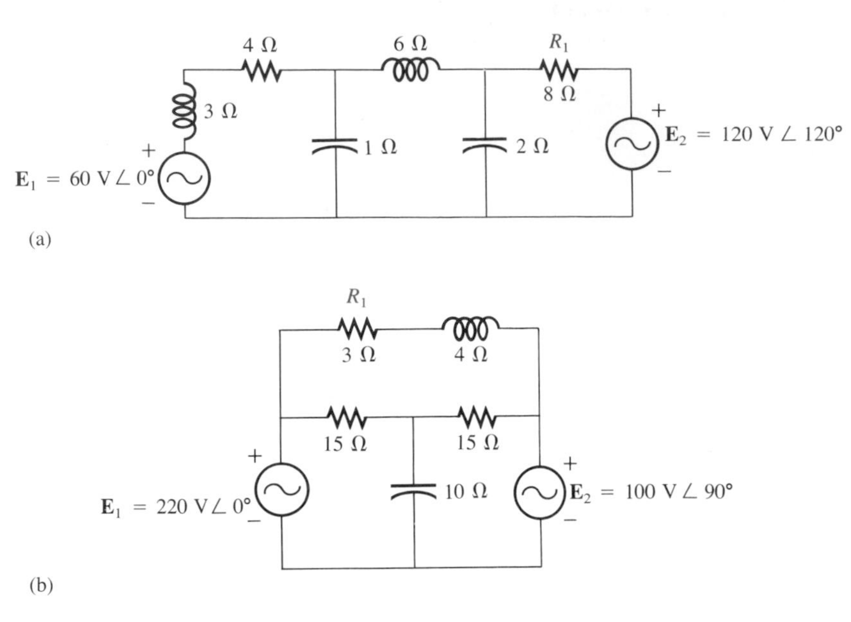


FIG. 17.56

*8. Repeat Problem 5 for the networks of Fig. 17.57.

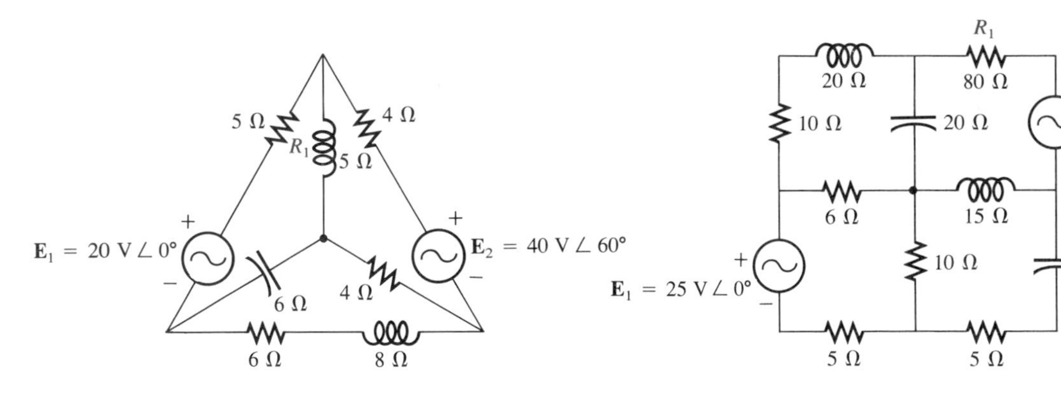

FIG. 17.57

9. Using the mesh-analysis format approach, determine the current $\mathbf{I}_L$ (in terms of $\mathbf{V}$) for the network of Fig. 17.58.

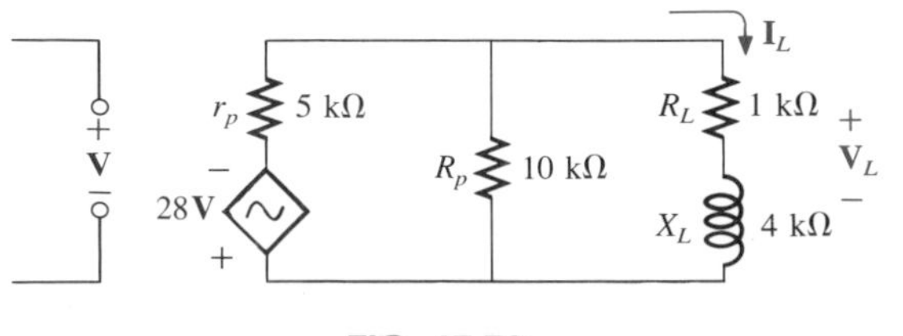


FIG. 17.58

*10. Using the mesh-analysis format approach, determine the current $\mathbf{I}_L$ (in terms of $\mathbf{I}$) for the network of Fig. 17.59.

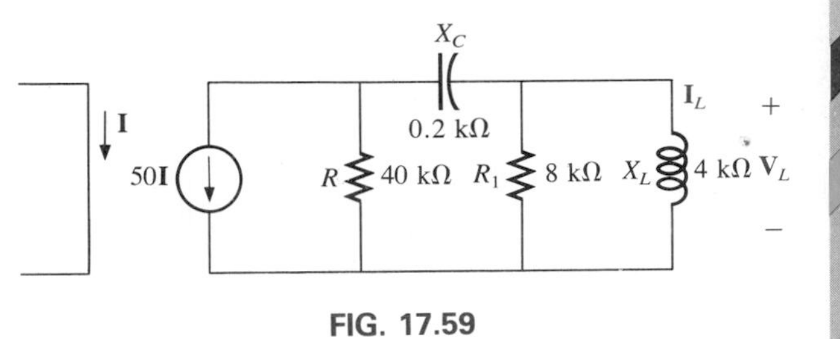


FIG. 17.59

## SECTION 17.5 Nodal Analysis

*11. Determine the nodal voltages for the networks of Fig. 17.60 using the general approach.

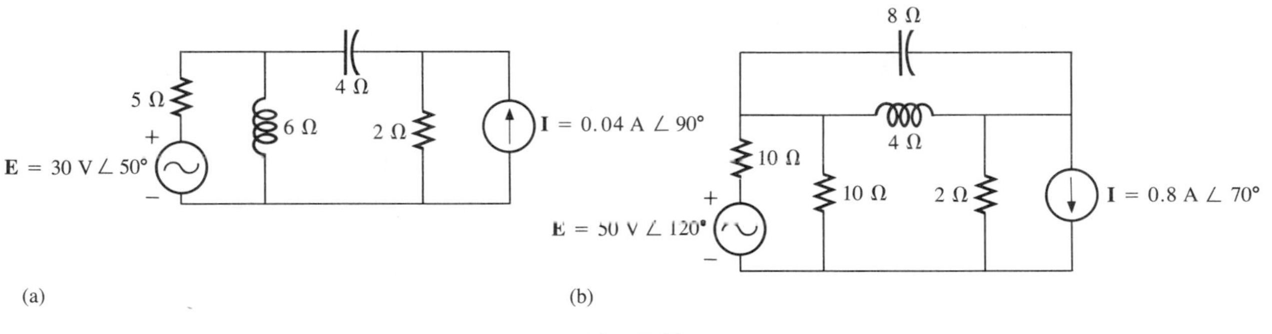

FIG. 17.60

12. Repeat Problem 11 for the networks of Fig. 17.61.

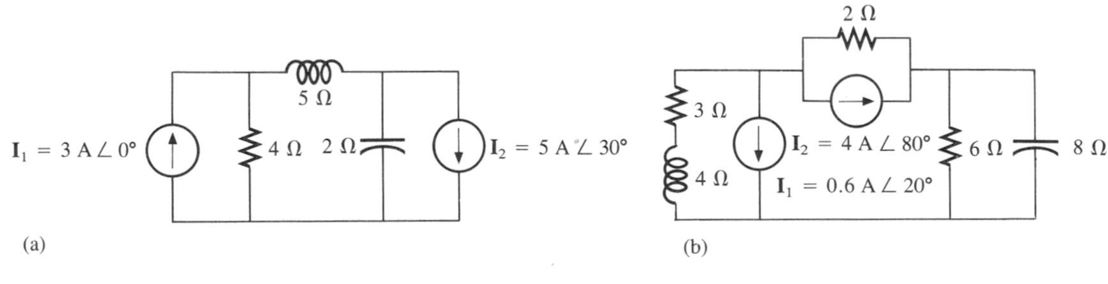

FIG. 17.61

*13. Determine the nodal voltages for the network of Fig. 17.60(a) using the format approach.

*14. Repeat Problem 13 for the network of Fig. 17.60(b).

15. Repeat Problem 13 for the network of Fig. 17.61(a).

16. Repeat Problem 13 for the network of Fig. 17.61(b).

***17.** Determine the nodal voltages for the networks of Fig. 17.62.

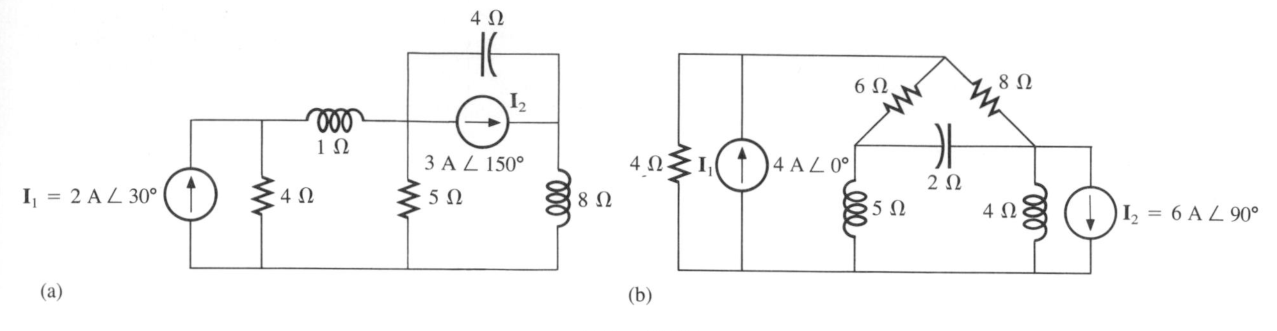

FIG. 17.62

**18.** Using the nodal-analysis approach, repeat Problem 9 (determine $\mathbf{V}_L$ in terms of $\mathbf{V}$).

***19.** Using the nodal-analysis approach, repeat Problem 10 (determine $\mathbf{V}_L$ in terms of $\mathbf{I}$).

***20.** For the network of Fig. 17.63:
- **a.** Write the nodal equations.
- **b.** Is symmetry present? If not, why?
- **c.** Determine $\mathbf{V}_L$ in terms of $\mathbf{E}_i$.

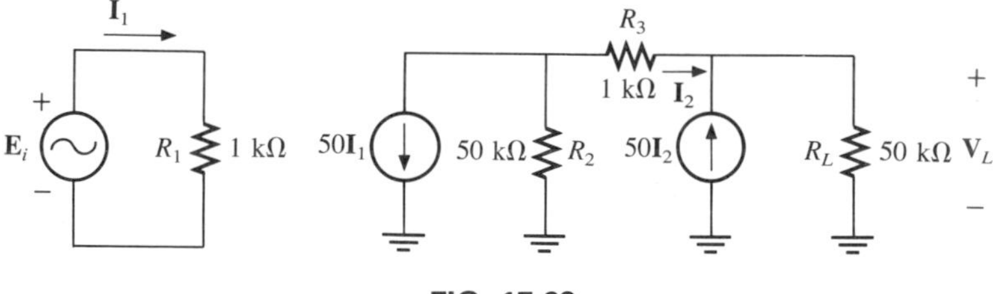


FIG. 17.63

## SECTION 17.6 Bridge Networks (ac)

**21.** For the bridge network of Fig. 17.64:
- **a.** Is the bridge balanced?
- **b.** Using mesh analysis, determine the current through the capacitive reactance.
- **c.** Using nodal analysis, determine the voltage across the capacitive reactance.

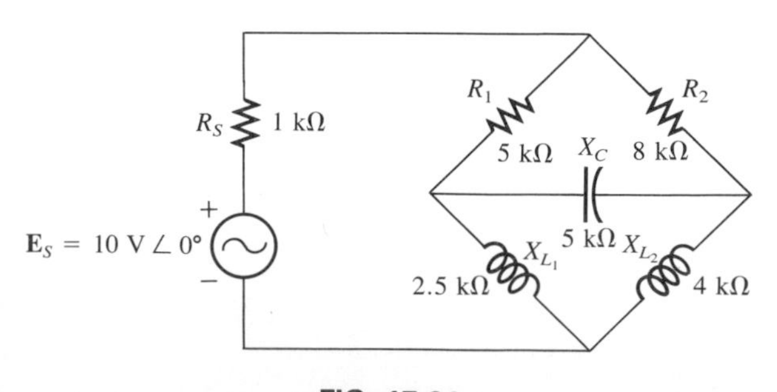


FIG. 17.64

**22.** Repeat Problem 21 for the bridge of Fig. 17.65.

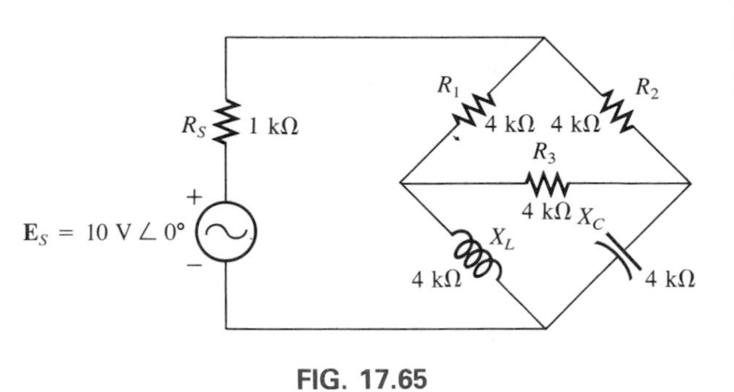


FIG. 17.65

**23.** The Hay bridge of Fig. 17.66 is balanced. Using Eq. (17.3), determine the unknown inductance $L_x$ and resistance $R_x$.

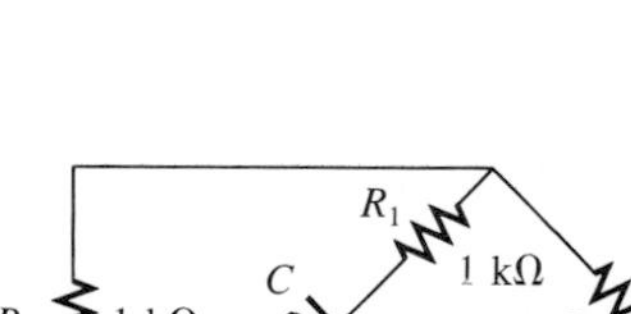
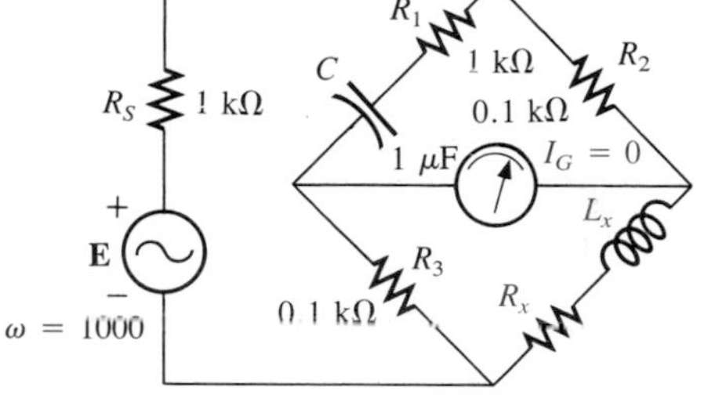

FIG. 17.66

**24.** Determine whether the Maxwell bridge of Fig. 17.67 is balanced (ω = 1000).

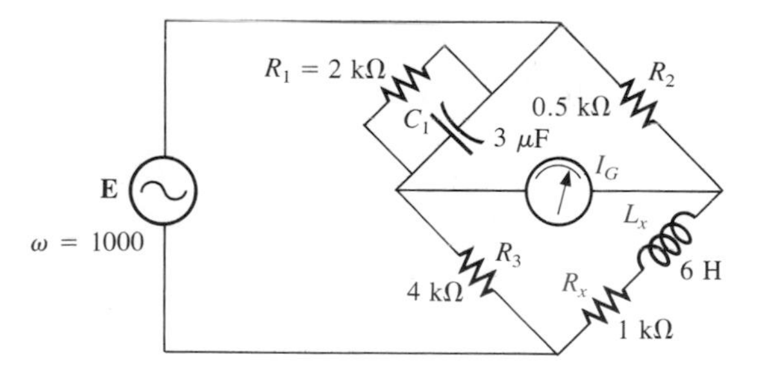

FIG. 17.67

**25.** Derive the balance equations (17.11) and (17.12) for the capacitance comparison bridge.

**26.** Determine the balance equations for the inductance bridge of Fig. 17.68.

FIG. 17.68

## SECTION 17.7 Δ-Y, Y-Δ Conversions

**27.** Using the Δ-Y or Y-Δ conversion, determine the current **I** for the networks of Fig. 17.69.

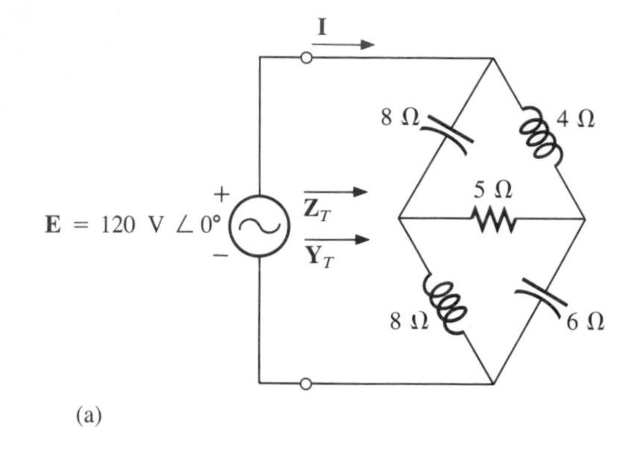


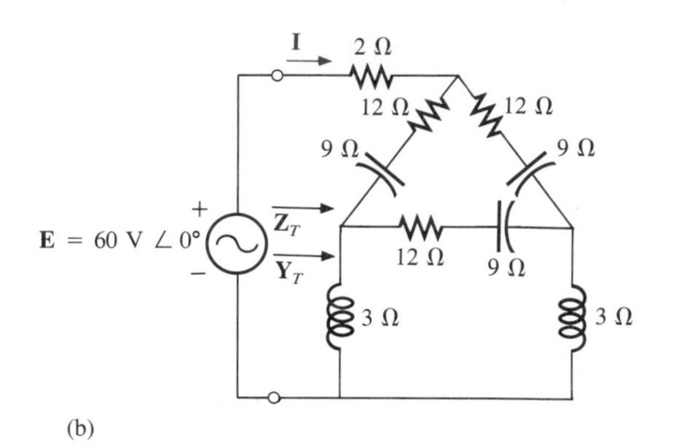

FIG. 17.69

**28.** Repeat Problem 27 for the networks of Fig. 17.70. ($\mathbf{E}$ = 100 V ∠0° in each case.)

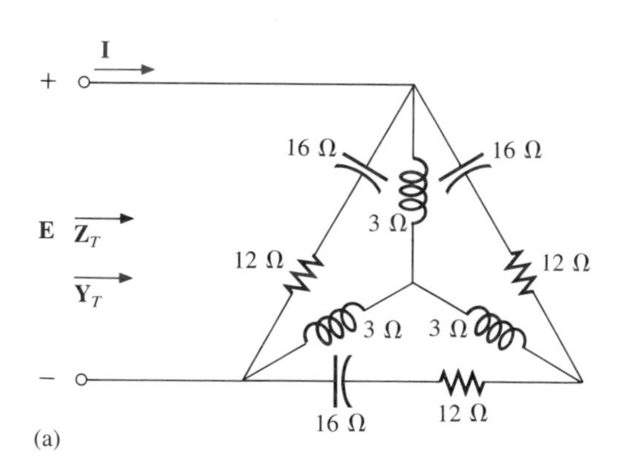


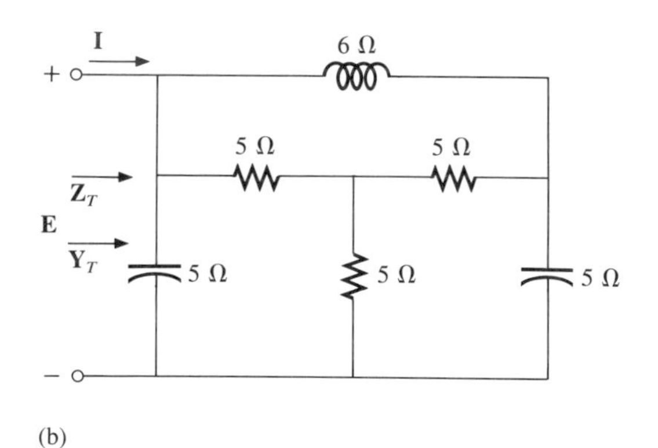


FIG. 17.70

## SECTION 17.8 Computer Analysis

### PSPICE

**29.** Write the input file to determine the mesh currents for the network of Fig. 17.55 using PSPICE.

***30.** Repeat Problem 29 for the network of Fig. 17.56(a).

***31.** Repeat Problem 29 for the network of Fig .17.56(b).

**32.** Write the input file to determine the nodal voltages for the network of Fig. 17.60(a).

***33.** Repeat Problem 32 for the network of Fig. 17.61(b).

***34.** Repeat Problem 32 for the network of Fig. 17.65 using a frequency of 1 kHz to determine the element values.

**BASIC**

**35.** Write a computer program that will provide a general solution for the network of Fig. 17.10. That is, given the reactance of each element and the parameters of the source voltages, generate a solution in phasor form for both mesh currents.

**36.** Repeat Problem 35 for the nodal voltages of Fig. 17.20.

**37.** Given a bridge composed of series impedances in each branch, write a program to test the balance condition as defined by Eq. (17.6).

## GLOSSARY

**Bridge network** A network configuration having the appearance of a diamond in which no two branches are in series or parallel.

**Capacitance comparison bridge** A bridge configuration having a galvanometer in the bridge arm that is used to determine an unknown capacitance and associated resistance.

**Delta (Δ) configuration** A network configuration having the appearance of the capital Greek letter delta.

**Dependent (controlled) source** A source whose magnitude and/or phase angle is determined (controlled) by a current or voltage of the system in which it appears.

**Hay bridge** A bridge configuration used for measuring the resistance and inductance of coils in those cases where the resistance is a small fraction of the reactance of the coil.

**Independent source** A source whose magnitude is independent of the network to which it is applied. It displays its terminal characteristics even if completely isolated.

**Maxwell bridge** A bridge configuration used for inductance measurements when the resistance of the coil is large enough not to require a Hay bridge.

**Mesh analysis** A method through which the loop (or mesh) currents of a network can be determined. The branch currents of the network can then be determined directly from the loop currents.

**Nodal analysis** A method through which the node voltages of a network can be determined. The voltage across each element can then be determined through application of Kirchhoff's voltage law.

**Source conversion** The changing of a voltage source to a current source, or vice versa, which will result in the same terminal behavior of the source. In other words, the external network is unaware of the change in sources.

**Wye (Y) configuration** A network configuration having the appearance of the capital letter Y.

# 18 Network Theorems (ac)

## 18.1 INTRODUCTION

This chapter will parallel Chapter 9, which dealt with network theorems as applied to dc networks. It would be time well spent to review each theorem in Chapter 9 before beginning this chapter, as many of the comments offered there will not be repeated.

Due to the need for developing confidence in the application of the various theorems to networks with controlled (dependent) sources, some sections have been divided into two parts: independent sources and dependent sources.

Theorems to be considered in detail include the superposition theorem, Thevenin and Norton theorems, and the maximum power theorem. The substitution and reciprocity theorems and Millman's theorem are not discussed in detail here, since a review of Chapter 9 will enable you to apply them to sinusoidal ac networks with little difficulty.

## 18.2 SUPERPOSITION THEOREM

You will recall from Chapter 9 that the superposition theorem eliminated the need for solving simultaneous linear equations by considering the effects of each source independently. To consider the effects of each source, we had to remove the remaining sources. This was accomplished by setting voltage sources to zero (short-circuit representation) and current sources to zero (open-circuit representation). The current through, or voltage across, a portion of the network produced by each source was then added algebraically to find the total solution for the current or voltage.

The only variation in applying this method to ac networks with independent sources is that we will now be working with impedances and phasors instead of just resistors and real numbers.

The superposition theorem is not applicable to power effects in ac networks, since we are still dealing with a nonlinear relationship. It can be applied to networks with sources of different frequencies only if the total response for *each* frequency is found independently and the results are expanded in a nonsinusoidal expression as appearing in Chapter 24.

One of the most frequent applications of the superposition theorem is to electronic systems in which the dc and ac analysis are treated separately and the total solution is the sum of the two. It is an important application of the theorem because the impact of the reactive elements changes dramatically in response to the two types of independent sources. In addition, the dc analysis of an electronic system can often define important parameters for the ac analysis. The fourth example will demonstrate the impact of the applied source on the general configuration of the network.

We will first consider networks with only independent sources to provide a close association with the analysis of Chapter 9.

## Independent Sources

**EXAMPLE 18.1.** Using the superposition theorem, find the current **I** through the 4-Ω reactance ($X_{L_2}$) of Fig. 18.1.

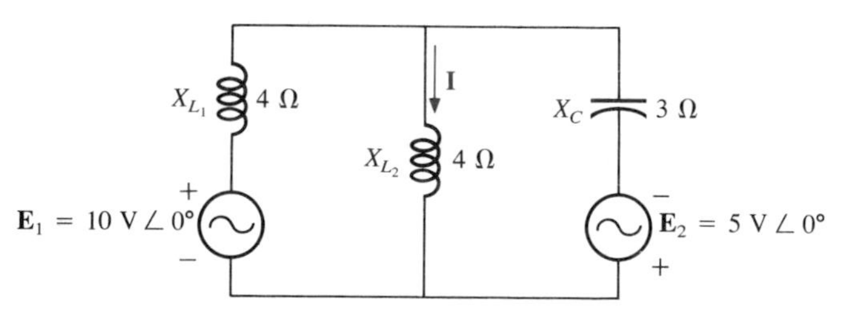

FIG. 18.1

***Solution:*** For the redrawn circuit (Fig. 18.2),

$$\mathbf{Z}_1 = +jX_{L_1} = j4\ \Omega$$
$$\mathbf{Z}_2 = +jX_{L_2} = j4\ \Omega$$
$$\mathbf{Z}_3 = -jX_C = -j3\ \Omega$$

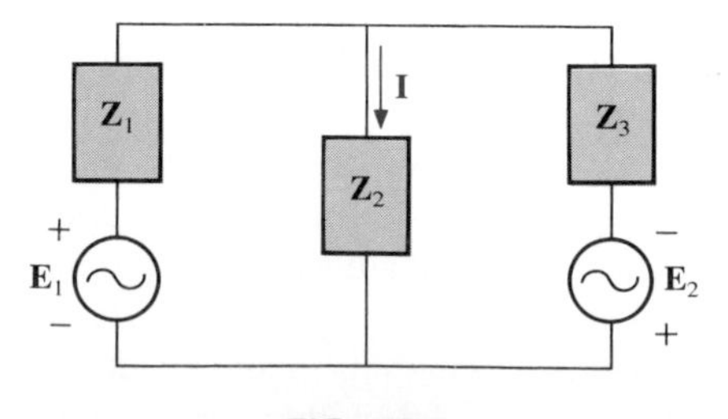


FIG. 18.2

Considering the effects of the voltage source $\mathbf{E}_1$ (Fig. 18.3), we have

$$\mathbf{Z}_{2\|3} = \frac{\mathbf{Z}_2\mathbf{Z}_3}{\mathbf{Z}_2 + \mathbf{Z}_3} = \frac{(j4\ \Omega)(-j3\ \Omega)}{j4\ \Omega - j3\ \Omega} = \frac{12\ \Omega}{j} = -j12\ \Omega$$
$$= 12\ \Omega\ \angle -90^\circ$$

$$\mathbf{I}_{T_1} = \frac{\mathbf{E}_1}{\mathbf{Z}_{2\|3} + \mathbf{Z}_1} = \frac{10\ \text{V}\ \angle 0^\circ}{-j12\ \Omega + j4\ \Omega} = \frac{10\ \text{V}\ \angle 0^\circ}{8\ \Omega\ \angle -90^\circ}$$
$$= 1.25\ \text{A}\ \angle 90^\circ$$

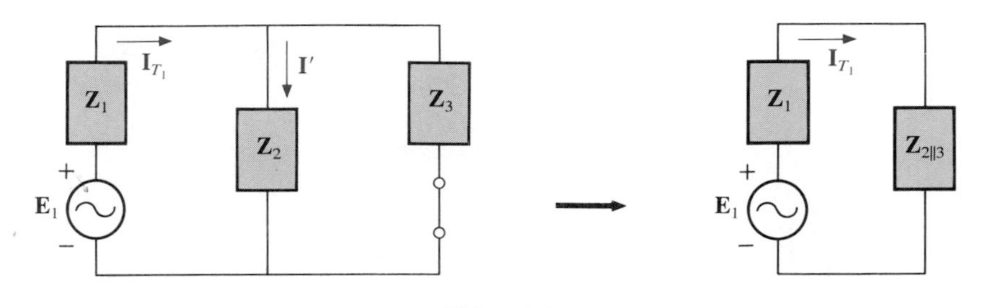

FIG. 18.3

and

$$\mathbf{I}' = \frac{\mathbf{Z}_3\mathbf{I}_{T_1}}{\mathbf{Z}_2 + \mathbf{Z}_3} \quad \text{(current divider rule)}$$
$$= \frac{(-j3\ \Omega)(j1.25\ \text{A})}{j4\ \Omega - j3\ \Omega} = \frac{3.75\ \text{A}}{j} = 3.75\ \text{A}\ \angle -90^\circ$$

Considering the effects of the voltage source $\mathbf{E}_2$ (Fig. 18.4), we have

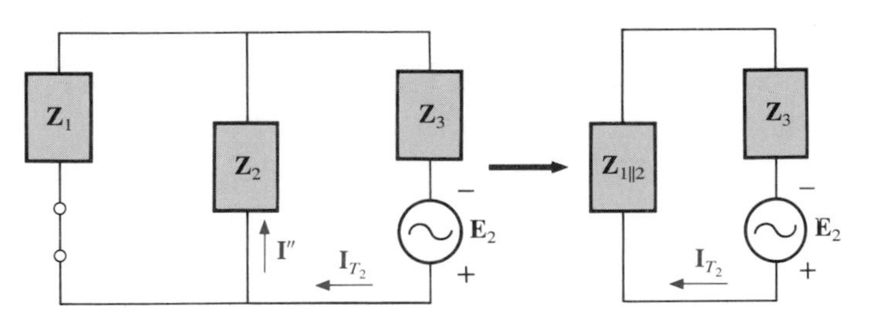

FIG. 18.4

$$\mathbf{Z}_{1\|2} = \frac{\mathbf{Z}_1}{N} = \frac{j4\ \Omega}{2} = j2\ \Omega$$

$$\mathbf{I}_{T_2} = \frac{\mathbf{E}_2}{\mathbf{Z}_{1\|2} + \mathbf{Z}_3} = \frac{5\ \text{V}\ \angle 0^\circ}{j2\ \Omega - j3\ \Omega} = \frac{5\ \text{V}\ \angle 0^\circ}{1\ \Omega\ \angle -90^\circ} = 5\ \text{A}\ \angle 90^\circ$$

and

$$\mathbf{I}'' = \frac{\mathbf{I}_{T_2}}{2} = 2.5\ \text{A}\ \angle 90^\circ$$

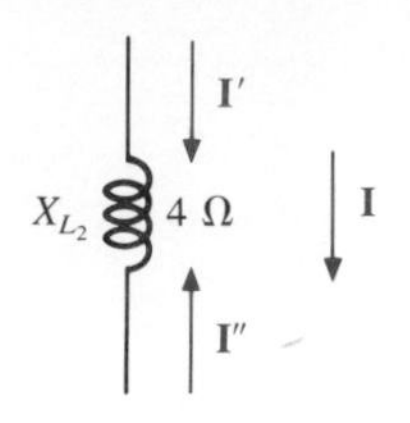


FIG. 18.5

The total current through the 4-Ω reactance $X_{L_2}$ (Fig. 18.5) is

$$\mathbf{I} = \mathbf{I}' - \mathbf{I}''$$
$$= 3.75 \text{ A } \angle -90° - 2.50 \text{ A } \angle 90° = -j3.75 \text{ A} - j2.50 \text{ A}$$
$$= -j6.25 \text{ A}$$
$$\mathbf{I} = \mathbf{6.25 \text{ A } \angle -90°}$$

**EXAMPLE 18.2.** Using superposition, find the current **I** through the 6-Ω resistor of Fig. 18.6.

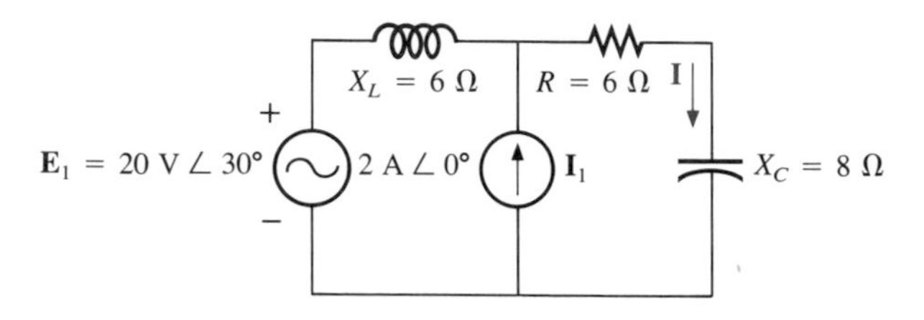

FIG. 18.6

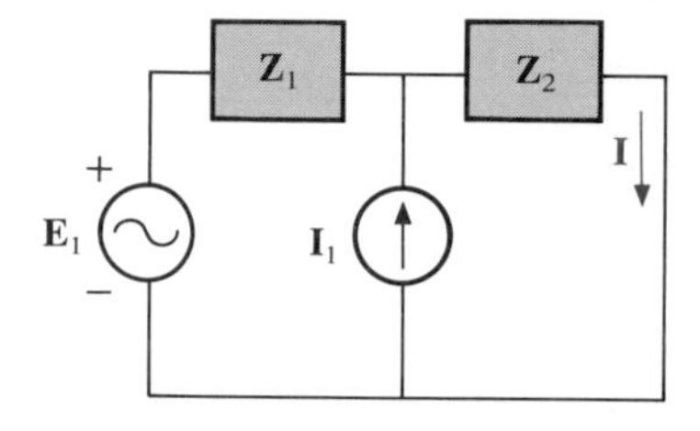


FIG. 18.7

***Solution:*** For the redrawn circuit (Fig. 18.7),

$$\mathbf{Z}_1 = j6\ \Omega \qquad \mathbf{Z}_2 = 6 - j8\ \Omega$$

Consider the effects of the current source (Fig. 18.8). Applying the current divider rule, we have

$$\mathbf{I}' = \frac{\mathbf{Z}_1\mathbf{I}_1}{\mathbf{Z}_1 + \mathbf{Z}_2} = \frac{(j6\ \Omega)(2\text{ A})}{j6\ \Omega + 6\ \Omega - j8\ \Omega} = \frac{j12}{6 - j2}$$
$$= \frac{12\ \angle 90°}{6.32\ \angle -18.43°}$$
$$\mathbf{I}' = 1.9 \text{ A } \angle 108.43°$$

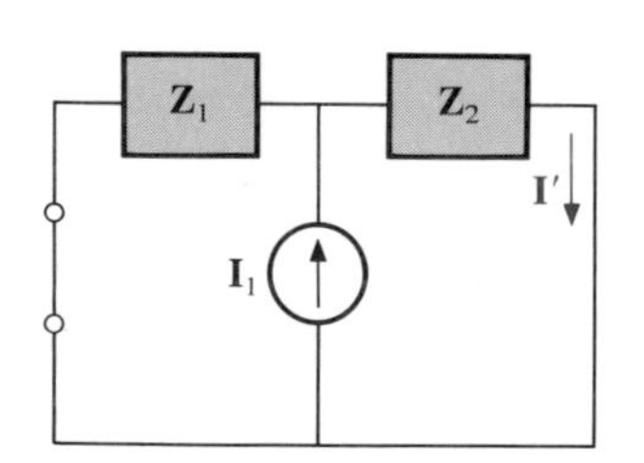


FIG. 18.8

Consider the effects of the voltage source (Fig. 18.9). Applying Ohm's law gives us

$$\mathbf{I}'' = \frac{\mathbf{E}_1}{\mathbf{Z}_T} = \frac{\mathbf{E}_1}{\mathbf{Z}_1 + \mathbf{Z}_2} = \frac{20 \text{ V } \angle 30°}{6.32\ \Omega\ \angle -18.43°}$$
$$= 3.16 \text{ A } \angle 48.43°$$

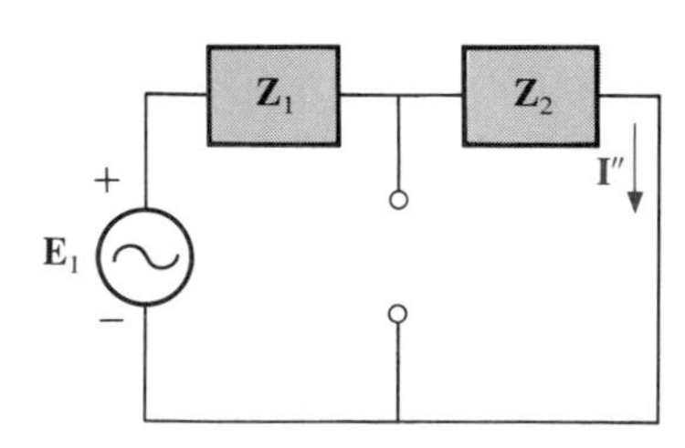

FIG. 18.9

The total current through the 6-Ω resistor (Fig. 18.10) is

$$\mathbf{I} = \mathbf{I}' + \mathbf{I}''$$
$$= 1.9 \text{ A } \angle 108.43° + 3.16 \text{ A } \angle 48.43°$$
$$= (-0.60 \text{ A} + j1.80 \text{ A}) + (2.10 \text{ A} + j2.36 \text{ A})$$
$$= 1.50 \text{ A} + j4.16 \text{ A}$$
$$\mathbf{I} = \mathbf{4.42 \text{ A } \angle 70.2°}$$

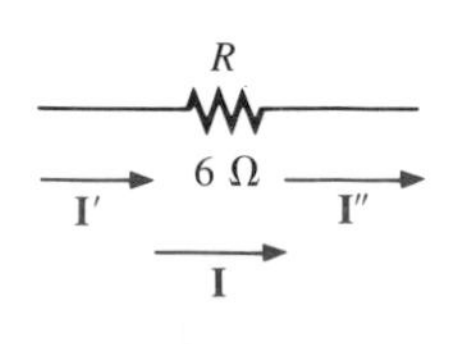


FIG. 18.10

**EXAMPLE 18.3.** Using superposition, find the voltage across the 6-Ω resistor in Fig. 18.6. Check the results against $\mathbf{V}_{6\Omega} = \mathbf{I}(6)$, where **I** is the current found through the 6-Ω resistor in the previous example.

***Solution:*** For the current source,

$$\mathbf{V'}_{6\Omega} = \mathbf{I'}(6\ \Omega) = (1.9\ \text{A}\ \angle 108.43^\circ)(6\ \Omega) = 11.4\ \text{V}\ \angle 108.43^\circ$$

For the voltage source,

$$\mathbf{V''}_{6\Omega} = \mathbf{I''}(6) = (3.16\ \text{A}\ \angle 48.43^\circ)(6\ \Omega) = 18.96\ \text{V}\ \angle 48.43^\circ$$

The total voltage across the 6-Ω resistor (Fig. 18.11) is

$$\begin{aligned}\mathbf{V}_{6\Omega} &= \mathbf{V'}_{6\Omega} + \mathbf{V''}_{6\Omega} \\ &= 11.4\ \text{V}\ \angle 108.43^\circ + 18.96\ \text{V}\ \angle 48.43^\circ \\ &= (-3.60\ \text{V} + j10.82\ \text{V}) + (12.58\ \text{V} + j14.18\ \text{V}) \\ &= 8.98\ \text{V} + j25.0\ \text{V} \\ \mathbf{V}_{6\Omega} &= \mathbf{26.5\ V\ \angle 70.2^\circ}\end{aligned}$$

Checking the result, we have

$$\begin{aligned}\mathbf{V}_{6\Omega} &= \mathbf{I}(6\ \Omega) = (4.42\ \text{A}\ \angle 70.2^\circ)(6\ \Omega) \\ &= \mathbf{26.5\ V\ \angle 70.2^\circ} \qquad \text{(checks)}\end{aligned}$$

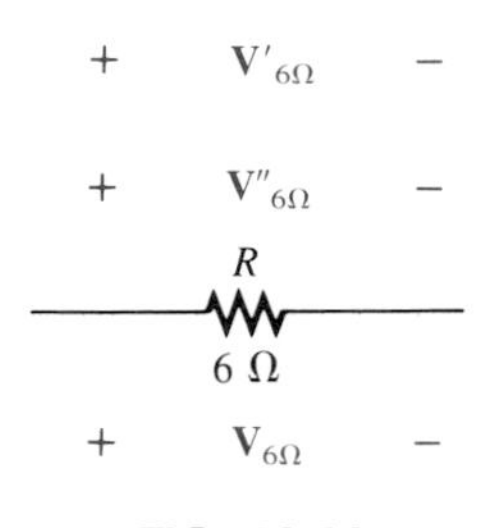


FIG. 18.11

**EXAMPLE 18.4.** For the network of Fig. 18.12, determine the voltage $v_3$ using superposition.

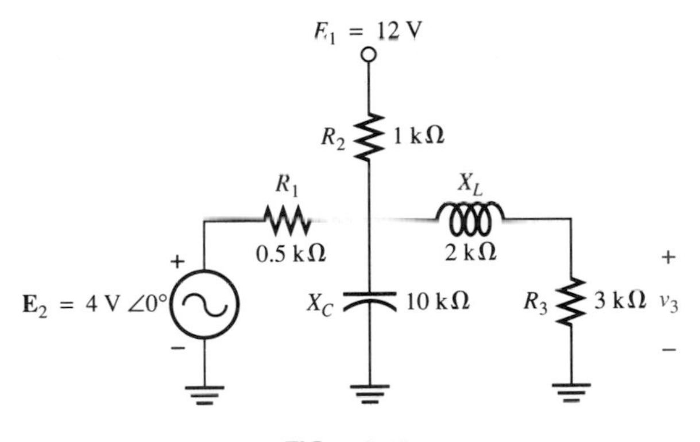


FIG. 18.12

***Solution:*** For the dc source, recall that for dc analysis, in the steady state the capacitor can be replaced by an open-circuit equivalent and the inductor by a short-circuit equivalent. The result is the network of Fig. 18.13.

The resistors $R_1$ and $R_3$ are then in parallel and the voltage $V_3$ can be determined using the voltage divider rule:

$$R' = R_1 \parallel R_3 = 0.5\ \text{k}\Omega \parallel 3\ \text{k}\Omega = 0.429\ \text{k}\Omega$$

and

$$\begin{aligned}V_3 &= \frac{R'E_1}{R' + R_2} \\ &= \frac{(0.429\ \text{k}\Omega)(12\ \text{V})}{0.429\ \text{k}\Omega + 1\ \text{k}\Omega} = \frac{5.148\ \text{V}}{1.429} \\ V_3 &\cong \mathbf{3.6\ V}\end{aligned}$$

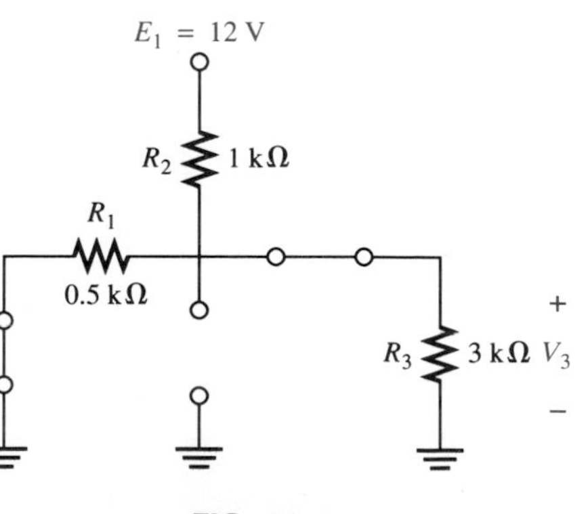


FIG. 18.13

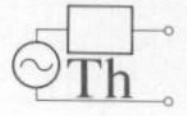

For ac analysis, the dc source is set to zero and the network is redrawn, as shown in Fig. 18.14.

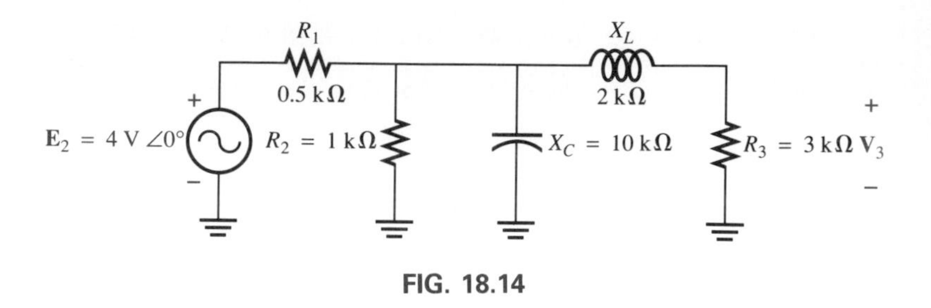


FIG. 18.14

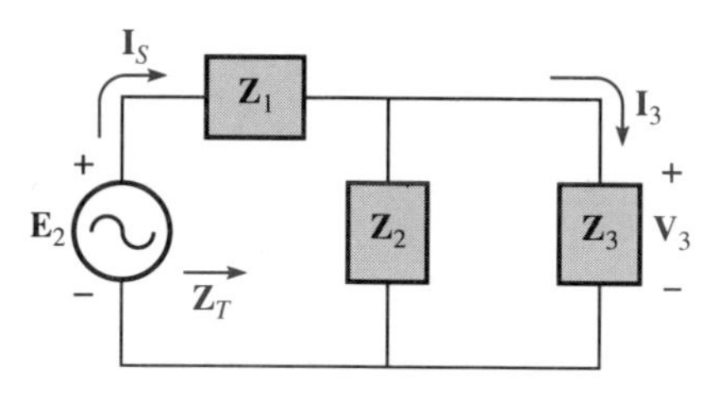


FIG. 18.15

The block impedances are then defined as in Fig. 18.15 and series-parallel techniques applied as follows:

$$\mathbf{Z}_1 = 0.5 \text{ k}\Omega \angle 0°$$

$$\mathbf{Z}_2 = \mathbf{R}_2 \parallel \mathbf{X}_C = \frac{(1 \text{ k}\Omega \angle 0°)(10 \text{ k}\Omega \angle -90°)}{1 \text{ k}\Omega - j10 \text{ k}\Omega} = \frac{10 \text{ k}\Omega \angle -90°}{10.05 \angle -84.29°}$$
$$= 0.995 \text{ k}\Omega \angle -5.71°$$

$$\mathbf{Z}_3 = R_3 + jX_L = 3 \text{ k}\Omega + j2 \text{ k}\Omega = 3.61 \text{ k}\Omega \angle 33.69°$$

and

$$\mathbf{Z}_T = \mathbf{Z}_1 + \mathbf{Z}_2 \parallel \mathbf{Z}_3 = 0.5 \text{ k}\Omega + (0.995 \text{ k}\Omega \angle -5.71°) \parallel (3.61 \text{ k}\Omega \angle 33.69°)$$
$$= 1.312 \text{ k}\Omega \angle 1.57°$$

$$\mathbf{I}_S = \frac{\mathbf{E}_2}{\mathbf{Z}_T} = \frac{4 \text{ V} \angle 0°}{1.312 \text{ k}\Omega \angle 1.57°} = 3.05 \text{ mA} \angle -1.57°$$

Current divider rule:

$$\mathbf{I}_3 = \frac{\mathbf{Z}_2\mathbf{I}_S}{\mathbf{Z}_2 + \mathbf{Z}_3} = \frac{(0.995 \text{ k}\Omega \angle -5.71°)(3.05 \text{ mA} \angle -1.57°)}{0.995 \text{ k}\Omega \angle -5.71° + 3.61 \text{ k}\Omega \angle 33.69°}$$
$$= 0.686 \text{ mA} \angle -32.74°$$

with

$$\mathbf{V}_3 = \mathbf{I}_3\mathbf{R}_3$$
$$= (0.686 \text{ mA} \angle -32.74°)(3 \text{ k}\Omega \angle 0°)$$
$$= \mathbf{2.06 \text{ V} \angle -32.74°}$$

The total solution:

$$v_3 = v_3 \text{ (dc)} + v_3 \text{ (ac)}$$
$$= 3.6 \text{ V} + 2.06 \text{ V} \angle -32.74°$$
$$v_3 = \mathbf{3.6 + 2.91 \sin(\omega t - 32.74°)}$$

The result is a sinusoidal voltage having a peak value of 2.91 V riding on an average value of 3.6 V, as shown in Fig. 18.16.

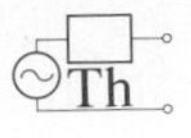



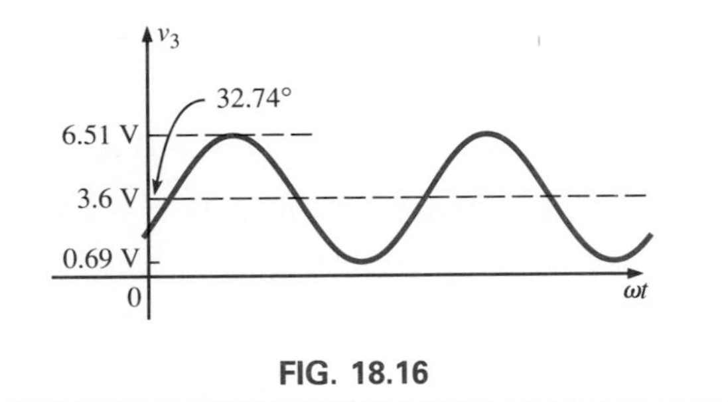


FIG. 18.16

## Dependent Sources

For dependent sources in which the controlling variable is not determined by the network to which the superposition theorem is to be applied, the application of the theorem is basically the same as for independent sources. The solution obtained will simply be in terms of the controlling variables.

**EXAMPLE 18.5.** Using the superposition theorem, determine the current $\mathbf{I}_2$ for the network of Fig. 18.17. The quantities $\mu$ and $h$ are constants.

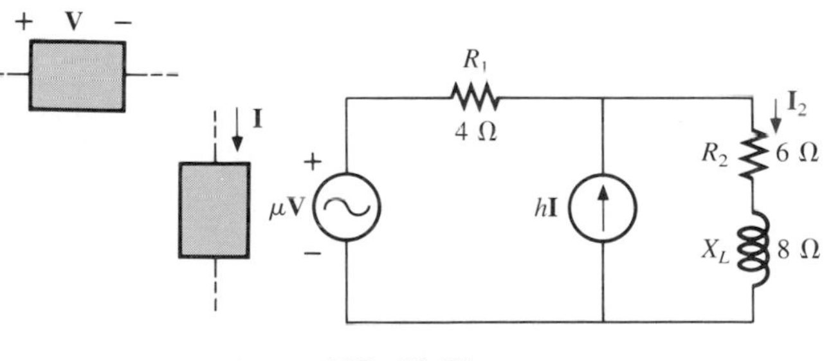


FIG. 18.17

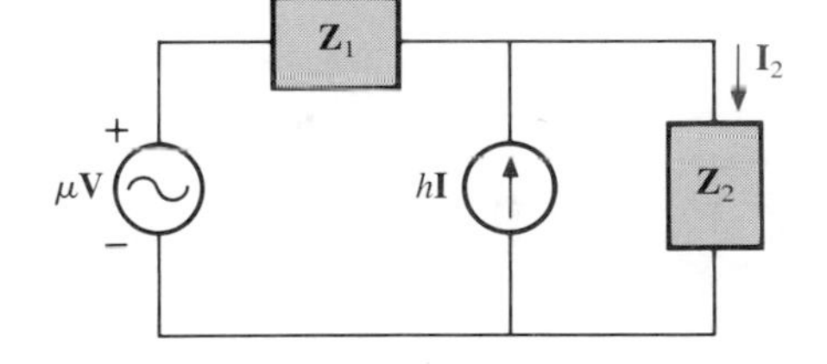


FIG. 18.18

***Solution:*** With a portion of the system redrawn (Fig. 18.18),

$$\mathbf{Z}_1 = R_1 = 4\ \Omega \qquad \mathbf{Z}_2 = R_2 + jX_L = 6\ \Omega + j8\ \Omega$$

For the voltage source (Fig. 18.19),

$$\mathbf{I}' = \frac{\mu\mathbf{V}}{\mathbf{Z}_1 + \mathbf{Z}_2} = \frac{\mu\mathbf{V}}{4\ \Omega + 6\ \Omega + j8\ \Omega} = \frac{\mu\mathbf{V}}{10\ \Omega + j8\ \Omega}$$

$$= \frac{\mu\mathbf{V}}{12.8\ \Omega\ \angle 38.66^\circ} = (0.078\ \mu\mathbf{V})\ \text{A}\ \angle -38.66^\circ$$

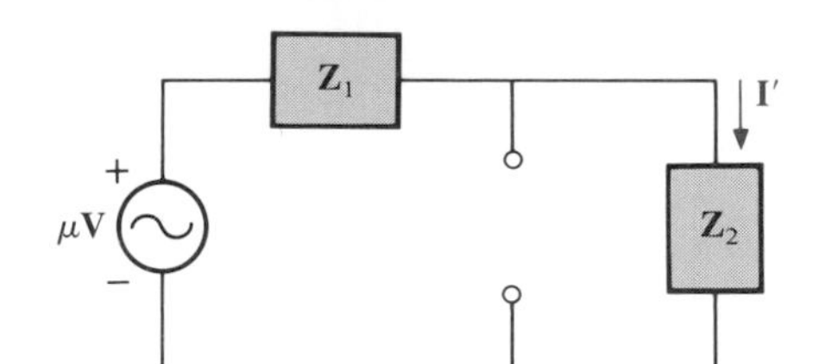


FIG. 18.19

For the current source (Fig. 18.20),

$$\mathbf{I}'' = \frac{\mathbf{Z}_1(h\mathbf{I})}{\mathbf{Z}_1 + \mathbf{Z}_2} = \frac{(4\ \Omega)(h\mathbf{I})}{12.8\ \Omega\ \angle 38.66^\circ} = 4(0.078)h\mathbf{I}\ \angle -38.66^\circ$$

$$= (0.312h\mathbf{I})\ \text{A}\ \angle -38.66^\circ$$

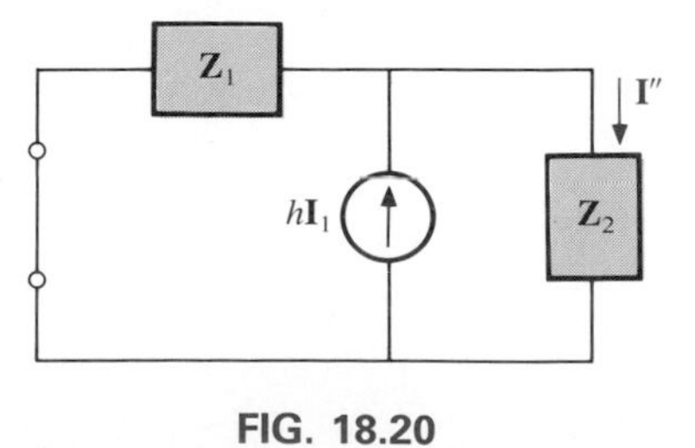


FIG. 18.20

The current $\mathbf{I}_2$ is

$$\mathbf{I}_2 = \mathbf{I}' + \mathbf{I}''$$
$$= \mathbf{(0.078\ \mu V)\ A\ \angle -38.66^\circ + (0.312\mathit{h}I)\ A\ \angle -38.66^\circ}$$

For $\mathbf{V} = 10\text{ V}\ \angle 0^\circ$, $\mathbf{I} = 20\text{ mA}\ \angle 0^\circ$, $\mu = 20$, $h = 100$,

$$\mathbf{I}_2 = 0.078(20)(10)\text{ A}\ \angle -38.66^\circ$$
$$+\ 0.312(100)(20 \times 10^{-3})\text{ A}\ \angle -38.66^\circ$$
$$= 15.60\text{ A}\ \angle -38.66^\circ + 0.62\text{ A}\ \angle -38.66^\circ$$
$$\mathbf{I}_2 = \mathbf{16.22\ A\ \angle -38.66^\circ}$$

For dependent sources in which the controlling variable is determined by the network to which the theorem is to be applied, the dependent source cannot be set to zero unless the controlling variable is also zero. For networks containing dependent sources such as indicated in Example 18.5 and dependent sources of the type just introduced above, the superposition theorem is applied for each independent source and each dependent source not having a controlling variable in the portions of the network under investigation. It must be reemphasized that dependent sources are not sources of energy in the sense that if all independent sources are removed from a system, all currents and voltages must be zero.

**EXAMPLE 18.6.** Determine the current $\mathbf{I}_L$ through the resistor $R_L$ of Fig. 18.21.

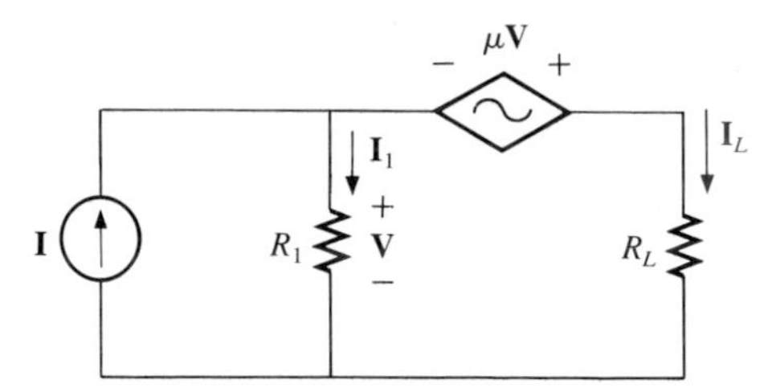


FIG. 18.21

***Solution:*** Note that the controlling variable $\mathbf{V}$ is determined by the network to be analyzed. From the above discussions, it is understood that the dependent source cannot be set to zero unless $\mathbf{V}$ is zero. If we set $\mathbf{I}$ to zero, the network lacks a source of voltage, and $\mathbf{V} = 0$ with $\mu\mathbf{V} = 0$. The resulting $\mathbf{I}_L$ under this condition is zero. Obviously, therefore, the network must be analyzed as it appears in Fig. 18.21, with the result that neither source can be eliminated, as is normally done using the superposition theorem.

Applying Kirchhoff's voltage law, we have

$$\mathbf{V}_L = \mathbf{V} + \mu\mathbf{V} = (1 + \mu)\mathbf{V}$$

and

$$\mathbf{I}_L = \frac{(1 + \mu)\mathbf{V}}{R_L}$$

The result, however, must be found in terms of $\mathbf{I}$ since $\mathbf{V}$ and $\mu\mathbf{V}$ are only dependent variables.

Applying Kirchhoff's current law gives us

$$\mathbf{I} = \mathbf{I}_1 + \mathbf{I}_L = \frac{\mathbf{V}}{R_1} + \frac{(1 + \mu)\mathbf{V}}{R_L}$$

and

$$\mathbf{I} = \mathbf{V}\left(\frac{1}{R_1} + \frac{1 + \mu}{R_L}\right)$$

or

$$\mathbf{V} = \frac{\mathbf{I}}{(1/R_1) + [(1 + \mu)/R_L]}$$

Substituting into the above yields

$$\mathbf{I}_L = \frac{(1 + \mu)\mathbf{V}}{R_L} = \frac{(1 + \mu)}{R_L}\left(\frac{\mathbf{I}}{(1/R_1) + [(1 + \mu)/R_L]}\right)$$

Therefore,

$$\mathbf{I}_L = \frac{(1 + \mu)R_1\mathbf{I}}{R_L + (1 + \mu)R_1}$$

## 18.3 THEVENIN'S THEOREM

Thevenin's theorem, as stated for sinusoidal ac circuits, is changed only to include the term *impedance* instead of *resistance;* that is,

***any two-terminal linear ac network can be replaced by an equivalent circuit consisting of a voltage source and an impedance in series* as shown in Fig. 18.22.**

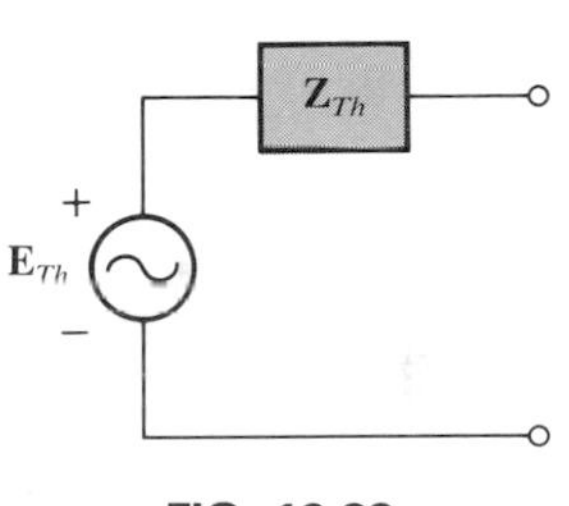


FIG. 18.22

Since the reactances of a circuit are frequency dependent, the Thevenin circuit found for a particular network is applicable only at *one* frequency.

The steps required to apply this method to dc circuits are repeated here with changes for sinusoidal ac circuits. As before, the only change is the replacement of the term *resistance* by *impedance*. Again, dependent and independent sources will be treated separately.

The last example of the independent source section will include a network with a dc and ac source to establish the groundwork for possible use in the electronics area.

### Independent Sources

1. Remove that portion of the network across which the Thevenin equivalent circuit is to be found.
2. Mark (○, •, and so on) the terminals of the remaining two-terminal network.
3. Calculate $\mathbf{Z}_{Th}$ by first setting all voltage and current sources to zero (short circuit and open circuit, respectively) and then finding the resulting impedance between the two marked terminals.
4. Calculate $\mathbf{E}_{Th}$ by first replacing the voltage and current sources and then finding the open-circuit voltage between the marked terminals.
5. Draw the Thevenin equivalent circuit with the portion of the circuit previously removed replaced between the terminals of the Thevenin equivalent circuit.

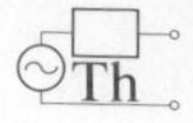


**EXAMPLE 18.7.** Find the Thevenin equivalent circuit for the network external to resistor $R$ in Fig. 18.23.

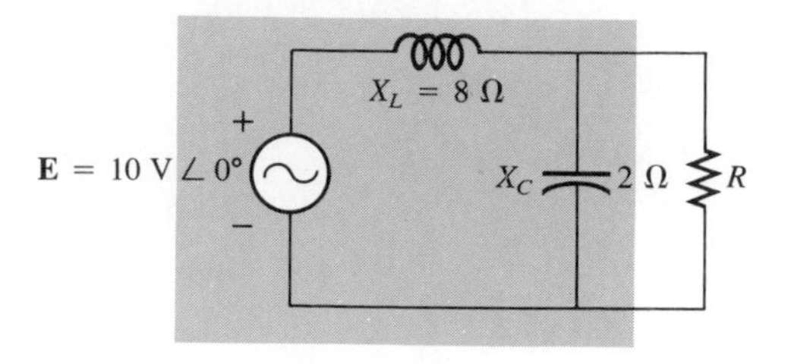

FIG. 18.23

***Solution:***
*Steps 1 and 2* (Fig. 18.24):

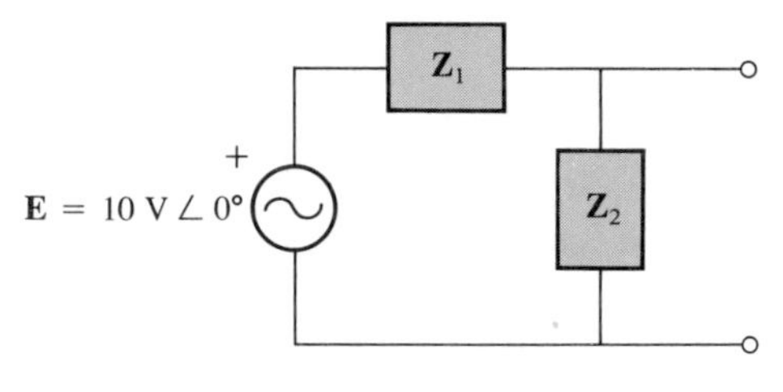

FIG. 18.24

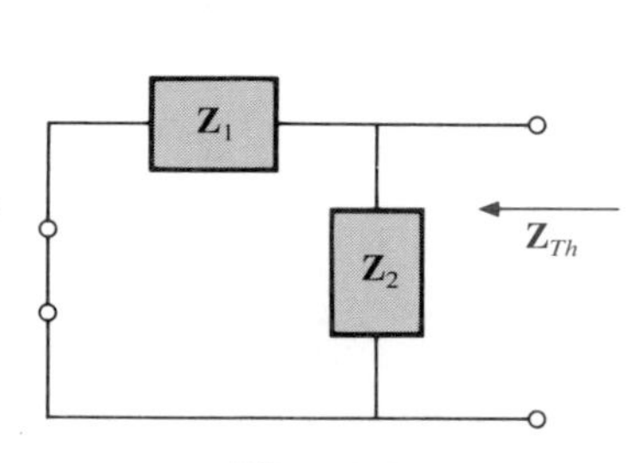

FIG. 18.25

$$\mathbf{Z}_1 = jX_L = j8\ \Omega \qquad \mathbf{Z}_2 = -jX_C = -j2\ \Omega$$

*Step 3* (Fig. 18.25):

$$\mathbf{Z}_{Th} = \frac{\mathbf{Z}_1\mathbf{Z}_2}{\mathbf{Z}_1 + \mathbf{Z}_2} = \frac{(j8\ \Omega)(-j2\ \Omega)}{j8\ \Omega - j2\ \Omega} = \frac{-j^2 16}{j6} = \frac{16}{6\ \angle 90°}$$
$$= \mathbf{2.67\ \Omega\ \angle{-90°}}$$

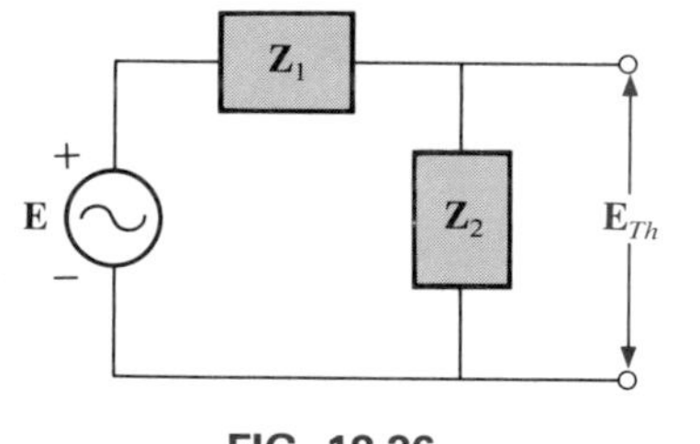


FIG. 18.26

*Step 4* (Fig. 18.26):

$$\mathbf{E}_{Th} = \frac{\mathbf{Z}_2\mathbf{E}}{\mathbf{Z}_1 + \mathbf{Z}_2} \qquad \text{(voltage divider rule)}$$
$$= \frac{(-j2\ \Omega)(10\ \text{V})}{j8\ \Omega - j2\ \Omega} = \frac{-j20}{j6} = \mathbf{3.33\ V\ \angle{-180°}}$$

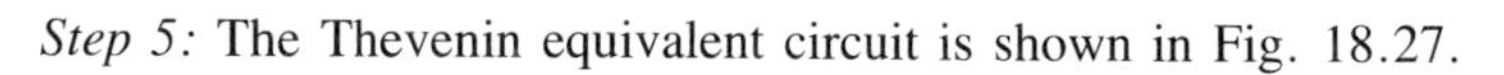
*Step 5:* The Thevenin equivalent circuit is shown in Fig. 18.27.

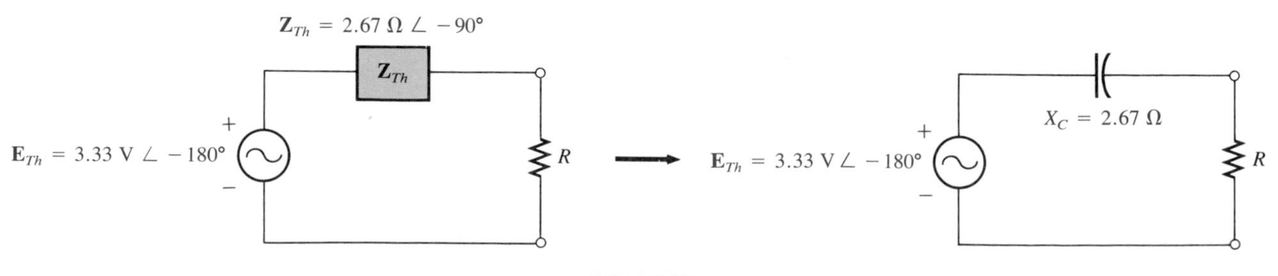


FIG. 18.27

**EXAMPLE 18.8.** Find the Thevenin equivalent circuit for the network external to branch *a-a'* in Fig. 18.28.

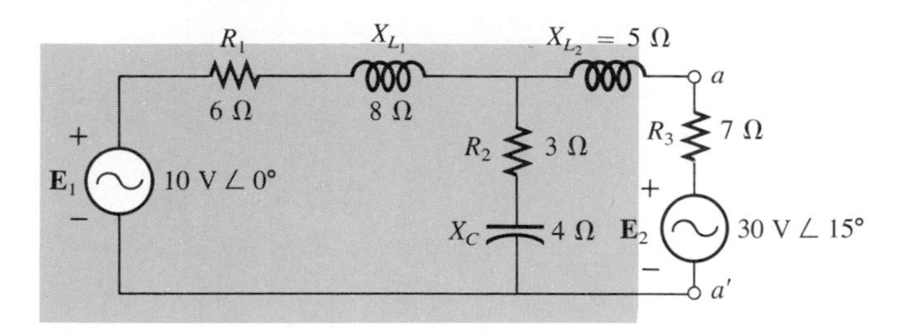


FIG. 18.28

***Solution:***

*Steps 1 and 2* (Fig. 18.29): Note the reduced complexity with subscripted impedances:

$$\mathbf{Z}_1 = R_1 + jX_{L_1} = 6\ \Omega + j8\ \Omega$$
$$\mathbf{Z}_2 = R_2 - jX_C = 3\ \Omega - j4\ \Omega$$
$$\mathbf{Z}_3 = +jX_{L_2} = j5\ \Omega$$

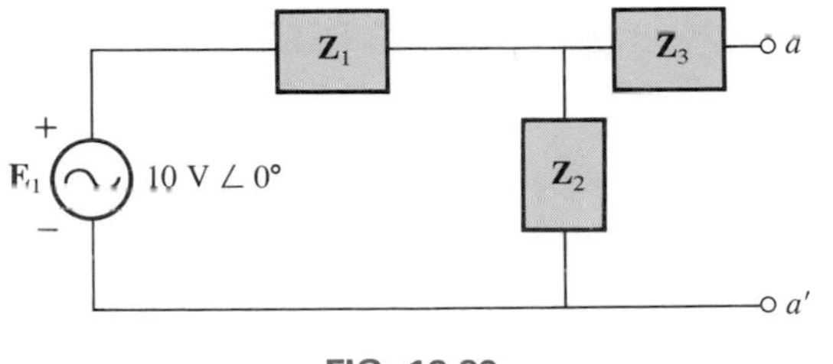

FIG. 18.29

*Step 3* (Fig. 18.30):

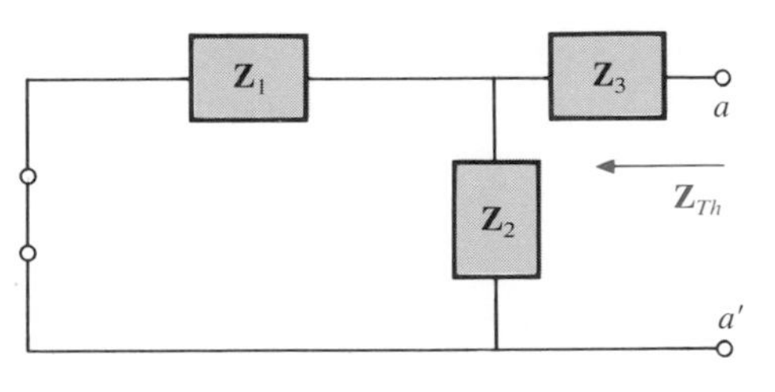

FIG. 18.30

$$\mathbf{Z}_{Th} = \mathbf{Z}_3 + \frac{\mathbf{Z}_1\mathbf{Z}_2}{\mathbf{Z}_1 + \mathbf{Z}_2} = j5\ \Omega + \frac{(10\ \Omega\ \angle 53.13^\circ)(5\ \Omega\ \angle -53.13^\circ)}{(6\ \Omega + j8\ \Omega) + (3\ \Omega - j4\ \Omega)}$$
$$= j5 + \frac{50\ \angle 0^\circ}{9 + j4} = j5 + \frac{50\ \angle 0^\circ}{9.85\ \angle 23.96^\circ}$$
$$= j5 + 5.08\ \angle -23.96^\circ = j5 + 4.64 - j2.06$$
$$\mathbf{Z}_{Th} = \mathbf{4.64\ \Omega + j2.94\ \Omega = 5.49\ \Omega\ \angle 32.36^\circ}$$

*Step 4* (Fig. 18.31): Since *a-a'* is an open circuit, $\mathbf{I}_{\mathbf{Z}_3} = 0$. Then $\mathbf{E}_{Th}$ is the voltage drop across $\mathbf{Z}_2$:

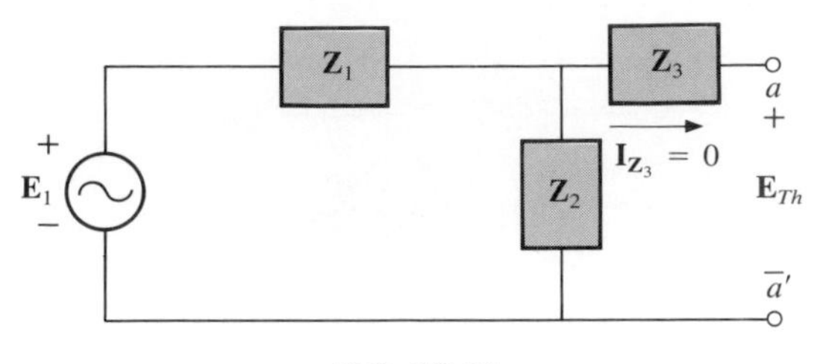

FIG. 18.31

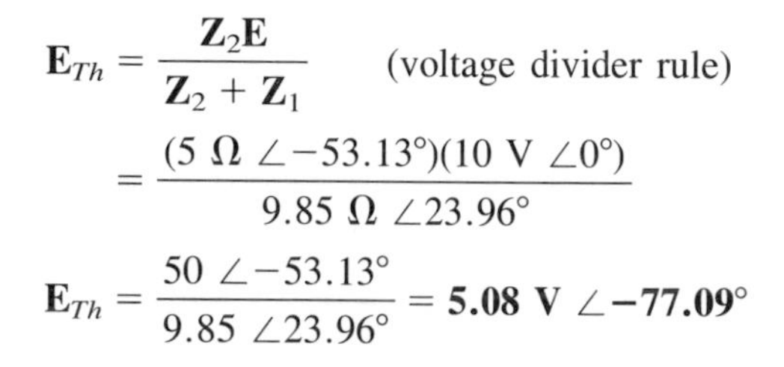

$$\mathbf{E}_{Th} = \frac{\mathbf{Z}_2\mathbf{E}}{\mathbf{Z}_2 + \mathbf{Z}_1} \qquad \text{(voltage divider rule)}$$

$$= \frac{(5\ \Omega\ \angle -53.13^\circ)(10\ \text{V}\ \angle 0^\circ)}{9.85\ \Omega\ \angle 23.96^\circ}$$

$$\mathbf{E}_{Th} = \frac{50\ \angle -53.13^\circ}{9.85\ \angle 23.96^\circ} = \mathbf{5.08\ V\ \angle -77.09^\circ}$$

*Step 5:* The Thevenin equivalent circuit is shown in Fig. 18.32.

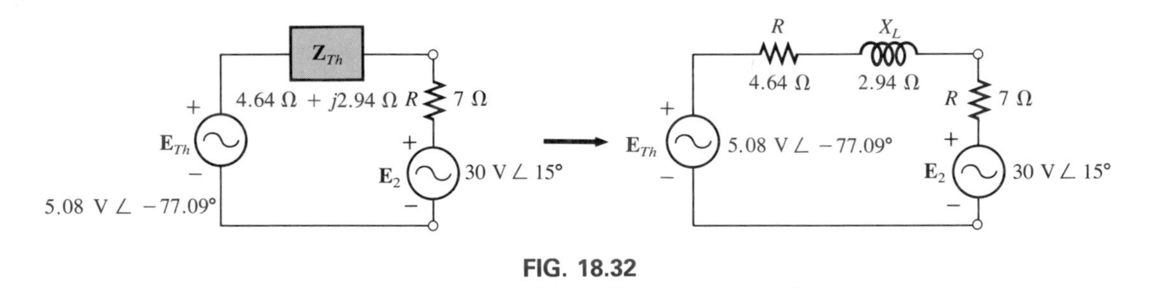


FIG. 18.32

The next example demonstrates how superposition is applied to electronic circuits to permit a separation of the dc and ac analysis. The fact that the controlling variable in this analysis is not in the portion of the network connected directly to the terminals of interest permits an analysis of the network in the same manner as applied above for independent sources.

**EXAMPLE 18.9.** Determine the Thevenin equivalent circuit for the transistor network external to the resistor $R_L$ in the following network (Fig. 18.33) and then determine $\mathbf{V}_L$.

***Solution:*** Applying superposition.

*DC conditions:*

Substituting the open-circuit equivalent for the coupling capacitor $C_2$ will isolate the dc source and the resulting currents from the load resistor. The result is for dc conditions that $V_L = 0$ V. Although the output

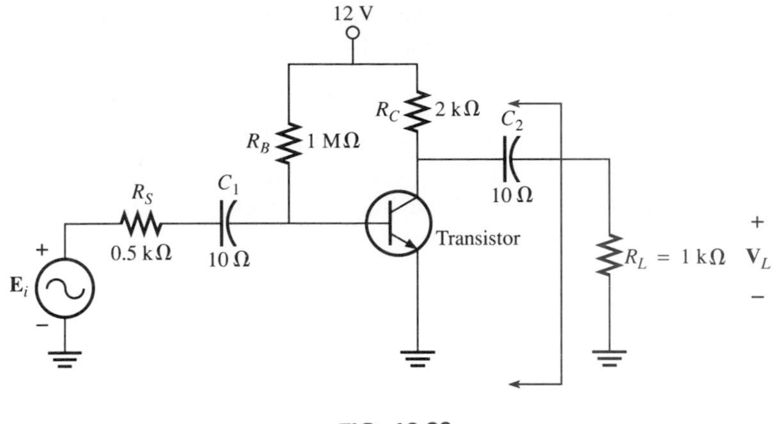


FIG. 18.33

dc voltage is zero, the application of the dc voltage is important to the basic operation of the transistor in a number of important ways, one of which is to determine the parameters of the "equivalent circuit" to appear in the ac analysis to follow.

*AC conditions:*

For the ac analysis an equivalent circuit is substituted for the transistor as established by the dc conditions above that will behave like the actual transistor. A great deal more will be said about equivalent circuits and the operations performed to obtain the network of Fig. 18.34, but for now let us limit our attention to the manner in which the Thevenin equivalent circuit is obtained. Note in Fig. 18.34 that the equivalent

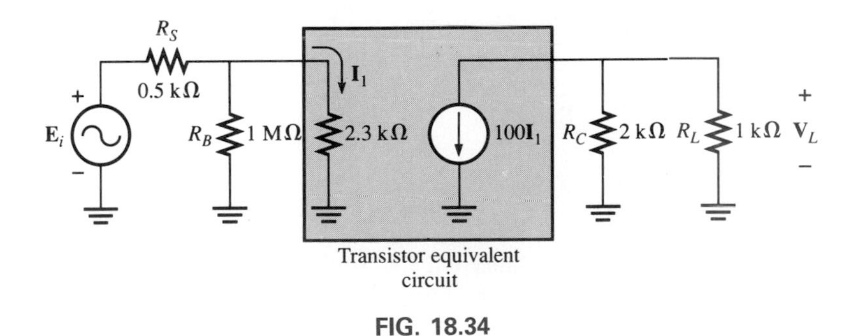


FIG. 18.34

circuit includes a resistor of 2.3 kΩ and a controlled current source whose magnitude is determined by the product of a factor of 100 and the current $I_1$ in another part of the network.

Note in Fig. 18.34 the absence of the coupling capacitors for the ac analysis. In general, coupling capacitors are designed to be open circuits for dc and short circuits for ac analysis. The short-circuit equivalent is valid because the other impedances in series with the coupling capacitors are so much larger in magnitude that the effect of the coupling capacitors can be ignored. Both $R_B$ and $R_C$ are now tied to ground because the dc source was set to zero volts (superposition) and replaced by a short-circuit equivalent to ground.

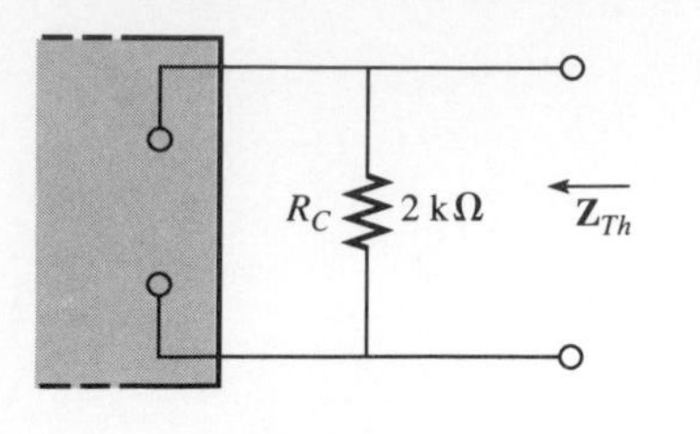

FIG. 18.35

For the analysis to follow, the effect of the resistor $R_B$ will be ignored, since it is so much larger than the parallel 2.3-kΩ resistor.

$\mathbf{Z}_{Th}$:

When $\mathbf{E}_i$ is set to zero volts, the current $\mathbf{I}_1$ will be zero amperes, and the controlled source $100\mathbf{I}_1$ will be zero amperes also. The result is an open-circuit equivalent for the source as appearing in Fig. 18.35.

It is fairly obvious from Fig. 18.35 that

$$\mathbf{Z}_{Th} = \mathbf{2\ k\Omega}$$

$\mathbf{E}_{Th}$:

For $\mathbf{E}_{Th}$ the current $\mathbf{I}_1$ of Fig. 18.34 will be

$$\mathbf{I}_1 = \frac{\mathbf{E}_i}{R_S + 2.3\ \text{k}\Omega} = \frac{\mathbf{E}_i}{0.5\ \text{k}\Omega + 2.3\ \text{k}\Omega} = \frac{\mathbf{E}_i}{2.8\ \text{k}\Omega}$$

and

$$100\mathbf{I}_1 = (100)\left(\frac{\mathbf{E}_i}{2.8\ \text{k}\Omega}\right) = 35.71 \times 10^{-3}\mathbf{E}_i$$

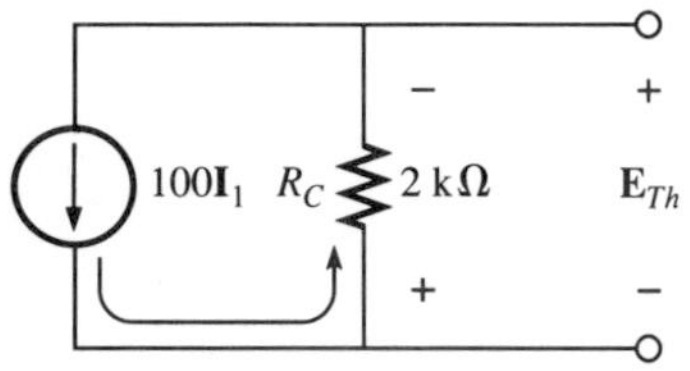

FIG. 18.36

Referring to Fig. 18.36, we find that

$$\begin{aligned}\mathbf{E}_{Th} &= -(100\mathbf{I}_1)\mathbf{R}_C \\ &= -(35.71 \times 10^{-3}\mathbf{E}_i)(2 \times 10^3\ \Omega) \\ \mathbf{E}_{Th} &= -71.42\mathbf{E}_i\end{aligned}$$

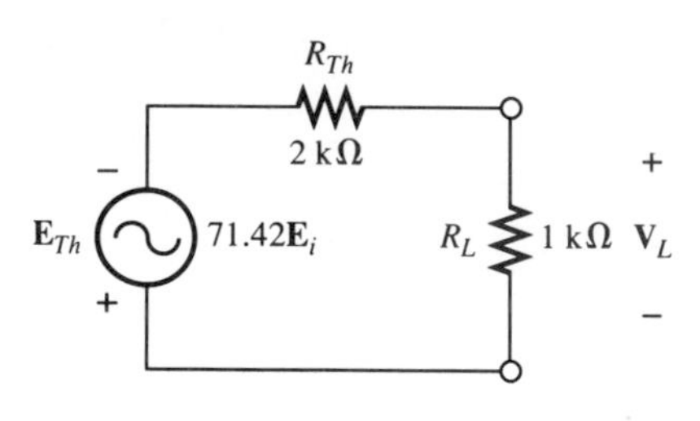

FIG. 18.37

The Thevenin equivalent circuit appears in Fig. 18.37 with the original load $R_L$.

The output voltage $V_L$:

$$\mathbf{V}_L = \frac{-\mathbf{R}_L\mathbf{E}_{Th}}{\mathbf{R}_L + \mathbf{Z}_{Th}} = \frac{-(1\ \text{k}\Omega)(71.42\ \mathbf{E}_i)}{1\ \text{k}\Omega + 2\ \text{k}\Omega}$$

and

$$\mathbf{V}_L = \mathbf{-23.81E}_i$$

revealing that the output voltage is 23.81 times the applied voltage with a phase shift of 180° due to the minus sign.

## Dependent Sources

For dependent sources with a controlling variable not in the network under investigation, the procedure indicated above can be applied. However, for dependent sources of the other type, where the controlling variable is part of the network to which the theorem is to be applied, another approach must be employed. The necessity for a different approach will be demonstrated in an example to follow. The method is not limited to dependent sources of the latter type. It can also be applied to any dc or sinusoidal ac network. However, for networks of independent sources, the method of application employed in Chapter 9 and the first portion of this section is generally more direct, with usual savings in time and errors.

The new approach to Thevenin's theorem can best be introduced at this stage in the development by considering the Thevenin equivalent circuit of Fig. 18.38(a). As indicated in Fig. 18.38(b), the open-circuit terminal voltage ($\mathbf{E}_{oc}$) of the Thevenin equivalent circuit is the Thevenin equivalent voltage. That is,

$$\mathbf{E}_{oc} = \mathbf{E}_{Th} \tag{18.1}$$

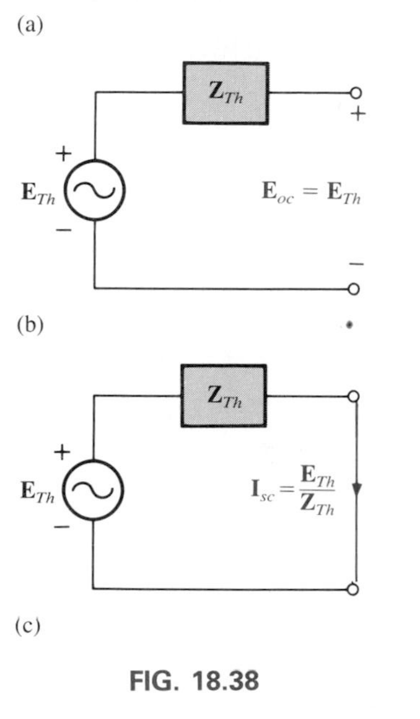

FIG. 18.38

If the external terminals are short circuited as in Fig. 18.38(c), the resulting short-circuit current is determined by

$$\mathbf{I}_{sc} = \frac{\mathbf{E}_{Th}}{\mathbf{Z}_{Th}} \tag{18.2}$$

or, rearranged,

$$\mathbf{Z}_{Th} = \frac{\mathbf{E}_{Th}}{\mathbf{I}_{sc}}$$

and

$$\mathbf{Z}_{Th} = \frac{\mathbf{E}_{oc}}{\mathbf{I}_{sc}} \tag{18.3}$$

Equations (18.1) and (18.3) indicate that for any linear bilateral dc or ac network with or without dependent sources of any type, if the open-circuit terminal voltage of a portion of a network can be determined along with the short-circuit current between the same two terminals, the Thevenin equivalent circuit is effectively known. A few examples will make the method quite clear. The advantage of the method, which was stressed earlier in this section for independent sources, should now be more obvious. The current $\mathbf{I}_{sc}$, which is necessary to find $\mathbf{Z}_{Th}$, is in general more difficult to obtain since all of the sources are present.

There is a third approach to the Thevenin equivalent circuit that is also useful from a practical viewpoint. The Thevenin voltage is found as in the two previous methods. However, the Thevenin impedance is obtained by applying a source of voltage to the terminals of interest and determining the source current as indicated in Fig. 18.39. For this method, the source voltage of the original network is set to zero. The Thevenin impedance is then determined by the following equation:

$$\mathbf{Z}_{Th} = \frac{\mathbf{E}_{g}}{\mathbf{I}_{g}} \tag{18.4}$$

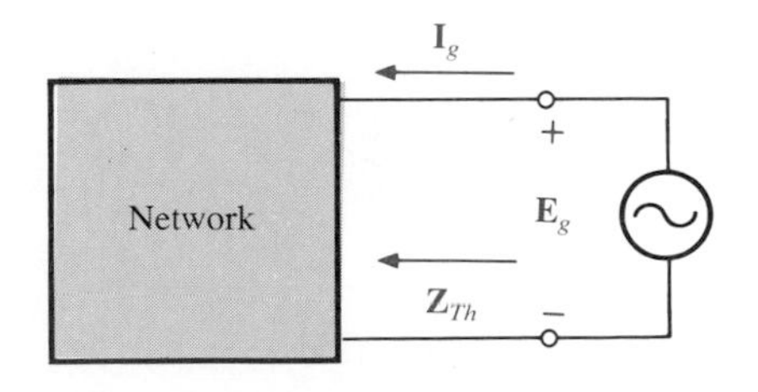

FIG. 18.39
*Determining $\mathbf{Z}_{Th}$.*

Note that for each technique, $\mathbf{E}_{Th} = \mathbf{E}_{oc}$, but the Thevenin impedance is found in different ways.

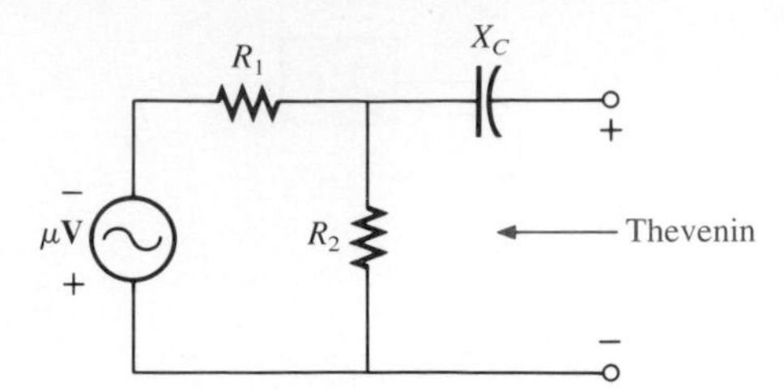

FIG. 18.40

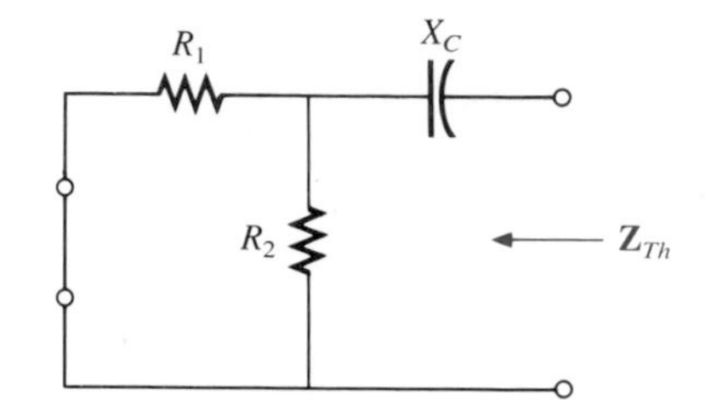

FIG. 18.41

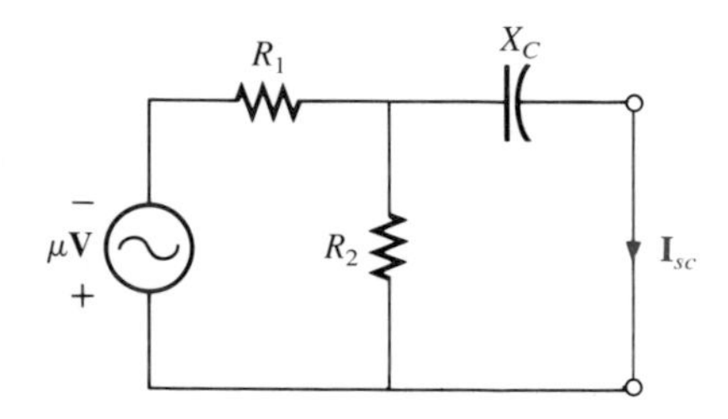


FIG. 18.42

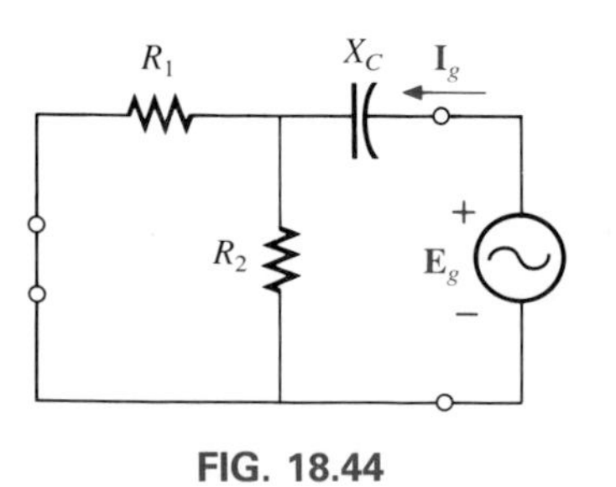


FIG. 18.44

**EXAMPLE 18.10.** Using each of the three techniques described in this section, determine the Thevenin equivalent circuit for the network of Fig. 18.40.

***Solution:*** Since for each approach the Thevenin voltage is found in exactly the same manner, it will be determined first. From Fig. 18.40, where $\mathbf{I}_{X_C} = 0$,

Due to the polarity for **V** and defined terminal polarities

$$\mathbf{V}_{R_1} = \mathbf{E}_{Th} = \mathbf{E}_{oc} = -\frac{R_2(\mu\mathbf{V})}{R_1 + R_2} = -\frac{\boldsymbol{\mu R_2 \mathbf{V}}}{\boldsymbol{R_1 + R_2}}$$

The following three methods for determining the Thevenin impedance appear in the order in which they were introduced in this section.

*Method 1* (Fig. 18.41):

$$\mathbf{Z}_{Th} = \boldsymbol{R_1 \parallel R_2 - jX_C}$$

*Method 2* (Fig. 18.42): Converting the voltage source to a current source (Fig. 18.43), we have (current divider rule)

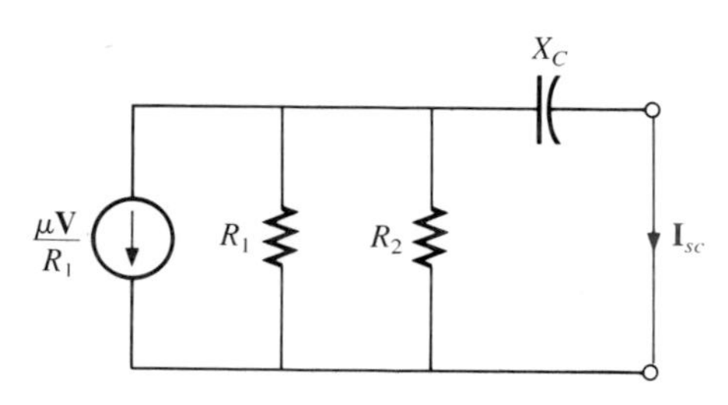


FIG. 18.43

$$\mathbf{I}_{sc} = \frac{-(R_1 \parallel R_2)\dfrac{\mu\mathbf{V}}{R_1}}{(R_1 \parallel R_2) - jX_C} = \frac{-\dfrac{R_1R_2}{R_1 + R_2}\left(\dfrac{\mu\mathbf{V}}{R_1}\right)}{(R_1 \parallel R_2) - jX_C}$$

$$= \frac{\dfrac{-\mu R_2\mathbf{V}}{R_1 + R_2}}{(R_1 \parallel R_2) - jX_C}$$

and

$$\mathbf{Z}_{Th} = \frac{\mathbf{E}_{oc}}{\mathbf{I}_{sc}} = \frac{\dfrac{-\mu R_2\mathbf{V}}{R_1 + R_2}}{\dfrac{\dfrac{-\mu R_2\mathbf{V}}{R_1 + R_2}}{(R_1 \parallel R_2) - jX_C}} = \frac{1}{\dfrac{1}{(R_1 \parallel R_2) - jX_C}}$$

$$= \boldsymbol{R_1 \parallel R_2 - jX_C}$$

*Method* 3 (Fig. 18.44):

$$\mathbf{I}_g = \frac{\mathbf{E}_g}{(R_1 \parallel R_2) - jX_C}$$

and

$$\mathbf{Z}_{Th} = \frac{\mathbf{E}_g}{\mathbf{I}_g} = R_1 \parallel R_2 - jX_C$$

In each case, the Thevenin impedance is the same. The resulting Thevenin equivalent circuit is shown in Fig. 18.45.

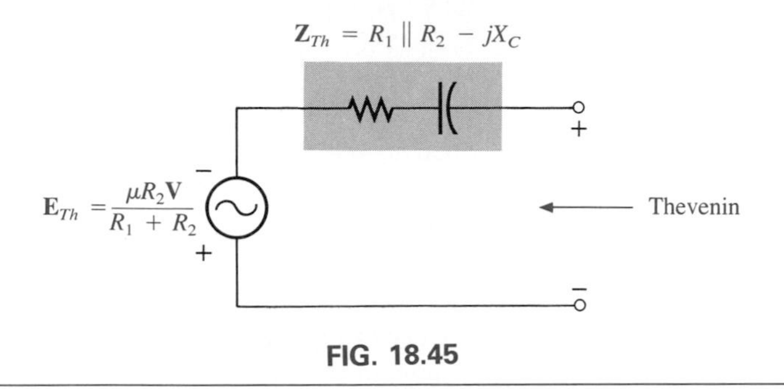


FIG. 18.45

**EXAMPLE 18.11.** Repeat Example 18.10 for the network of Fig. 18.46.

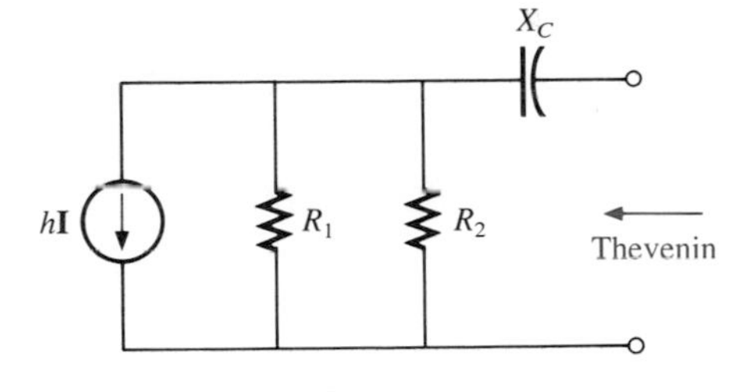


FIG. 18.46

***Solution:*** From Fig. 18.46, $\mathbf{E}_{Th}$ is

$$\mathbf{E}_{Th} = \mathbf{E}_{oc} = -h\mathbf{I}(R_1 \parallel R_2) = \frac{hR_1R_2\mathbf{I}}{R_1 + R_2}$$

*Method 1* (Fig. 18.47):

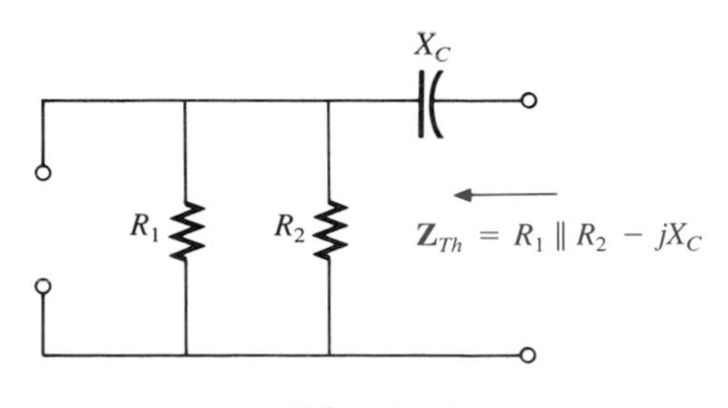

FIG. 18.47

$$\mathbf{Z}_{Th} = R_1 \parallel R_2 - jX_C$$

Note the similarity between this solution and that obtained for the previous example.

*Method 2* (Fig. 18.48):

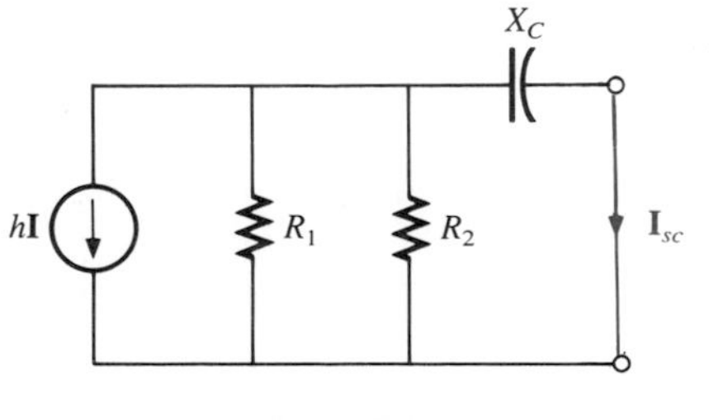


FIG. 18.48

$$\mathbf{I}_{sc} = \frac{-(R_1 \parallel R_2)h\mathbf{I}}{(R_1 \parallel R_2) - jX_C}$$

and

$$\mathbf{Z}_{Th} = \frac{\mathbf{E}_{oc}}{\mathbf{I}_{sc}} = \frac{-h\mathbf{I}(R_1 \parallel R_2)}{\dfrac{-(R_1 \parallel R_2)h\mathbf{I}}{(R_1 \parallel R_2) - jX_C}} = R_1 \parallel R_2 - jX_C$$

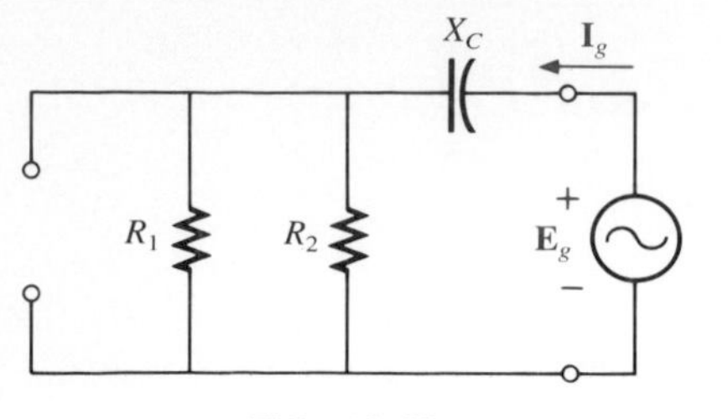

FIG. 18.49

*Method 3* (Fig. 18.49):

$$\mathbf{I}_g = \frac{\mathbf{E}_g}{(R_1 \parallel R_2) - jX_C}$$

and

$$\mathbf{Z}_{Th} = \frac{\mathbf{E}_g}{\mathbf{I}_g} = \boldsymbol{R_1 \parallel R_2 - jX_C}$$

---

The following example has a dependent source that will not permit the use of the method described in the beginning of this section for independent sources. All three methods will be applied, however, so that the results can be compared.

---

**EXAMPLE 18.12.** For the network of Fig. 18.50 (introduced in Example 18.5), determine the Thevenin equivalent circuit between the indicated terminals using each method described in this section. Compare your results.

FIG. 18.50

***Solution:*** First, using Kirchhoff's voltage law, $\mathbf{E}_{Th}$ (which is the same for each method) is written

$$\mathbf{E}_{Th} = \mathbf{V} + \mu\mathbf{V} = (1 + \mu)\mathbf{V}$$

However,

$$\mathbf{V} = \mathbf{I}R_1$$

so

$$\mathbf{E}_{Th} = \boldsymbol{(1 + \mu)\mathbf{I}R_1}$$

$\mathbf{Z}_{Th}$**:**

*Method 1* (Fig. 18.51): Since $\mathbf{I} = 0$, $\mathbf{V}$ and $\mu\mathbf{V} = 0$, and

~~$\mathbf{Z}_{Th} = R_1$~~ (incorrect)

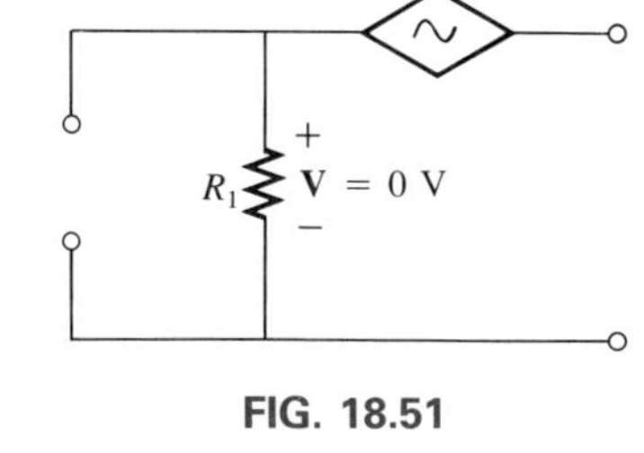

FIG. 18.51

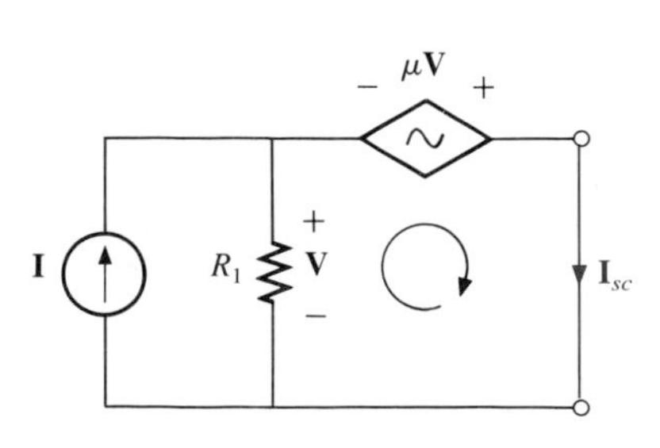

FIG. 18.52

*Method 2* (Fig. 18.52): Kirchhoff's voltage law around the indicated loop gives us

$$\mathbf{V} + \mu\mathbf{V} = 0$$

and

$$\mathbf{V}(1 + \mu) = 0$$

Since $\mu$ is a positive constant, the above equation can be satisfied only when $\mathbf{V} = 0$. Substitution of this result into Fig. 18.52 will yield the configuration of Fig. 18.53, and

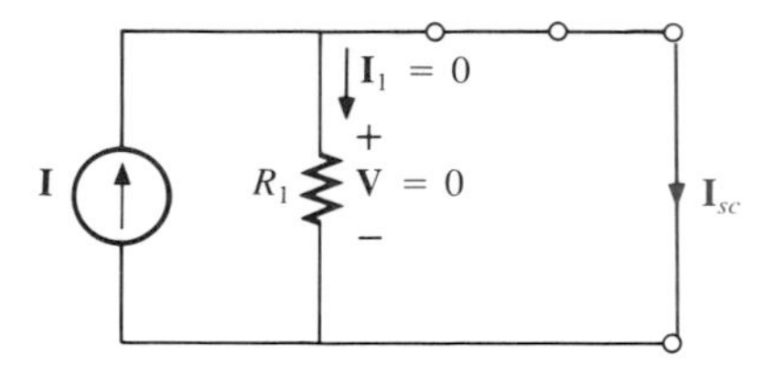


FIG. 18.53

$$\mathbf{I}_{sc} = \mathbf{I}$$

with

$$\mathbf{Z}_{Th} = \frac{\mathbf{E}_{oc}}{\mathbf{I}_{sc}} = \frac{(1 + \mu)\mathbf{I}R_1}{\mathbf{I}} = \mathbf{(1 + \mu)}\boldsymbol{R}_1 \qquad \text{(correct)}$$

*Method 3* (Fig. 18.54):

$$\mathbf{E}_g = \mathbf{V} + \mu\mathbf{V} = (1 + \mu)\mathbf{V}$$

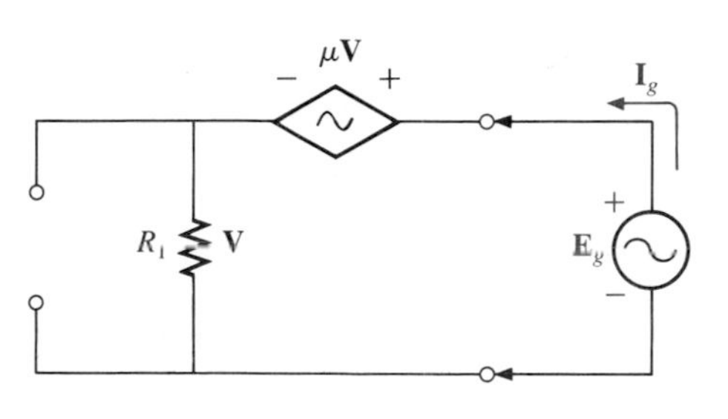


FIG. 18.54

or

$$\mathbf{V} = \frac{\mathbf{E}_g}{1 + \mu}$$

and

$$\mathbf{I}_g = \frac{\mathbf{V}}{R_1} = \frac{\mathbf{E}_g}{(1 + \mu)R_1}$$

and

$$\mathbf{Z}_{Th} = \frac{\mathbf{E}_g}{\mathbf{I}_g} = \mathbf{(1 + \mu)}\boldsymbol{R}_1 \qquad \text{(correct)}$$

The Thevenin equivalent circuit appears in Fig. 18.55, and

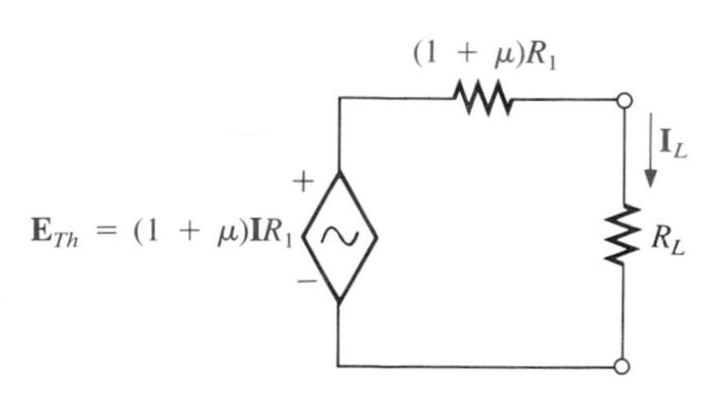

FIG. 18.55

$$\mathbf{I}_L = \frac{\mathbf{(1 + \mu)}\boldsymbol{R}_1\mathbf{I}}{\boldsymbol{R}_L + \mathbf{(1 + \mu)}\boldsymbol{R}_1}$$

which compares with the result of Example 18.5.

---

The network of Fig. 18.56 is the basic configuration of the transistor equivalent circuit applied most frequently today. Needless to say, it is necessary to know its characteristics and be adept in its use. Note that there is a controlled voltage and current source, each controlled by variables in the configuration.

---

**EXAMPLE 18.13.** Determine the Thevenin equivalent circuit for the indicated terminals of the network of Fig. 18.56.

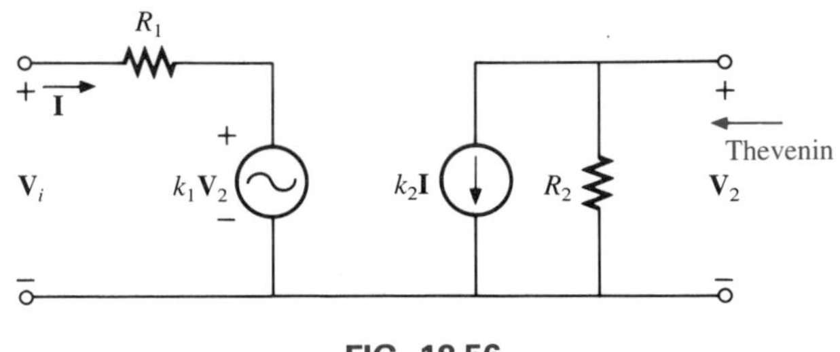

FIG. 18.56

***Solution:*** Apply the second method introduced in this section.

$\mathbf{E}_{Th}$:

$$\mathbf{E}_{oc} = \mathbf{V}_2$$

$$\mathbf{I} = \frac{\mathbf{V}_i - k_1\mathbf{V}_2}{R_1} = \frac{\mathbf{V}_i - k_1\mathbf{E}_{oc}}{R_1}$$

and

$$\mathbf{E}_{oc} = -k_2\mathbf{I}R_2 = -k_2R_2\left(\frac{\mathbf{V}_i - k_1\mathbf{E}_{oc}}{R_1}\right)$$

$$= \frac{-k_2R_2\mathbf{V}_i}{R_1} + \frac{k_1k_2R_2\mathbf{E}_{oc}}{R_1}$$

or

$$\mathbf{E}_{oc}\left(1 - \frac{k_1k_2R_2}{R_1}\right) = \frac{-k_2R_2\mathbf{V}_i}{R_1}$$

and

$$\mathbf{E}_{oc}\left(\frac{R_1 - k_1k_2R_2}{\cancel{R_1}}\right) = \frac{-k_2R_2\mathbf{V}_i}{\cancel{R_1}}$$

so

$$\boxed{\mathbf{E}_{oc} = \frac{-k_2R_2\mathbf{V}_i}{R_1 - k_1k_2R_2} = \mathbf{E}_{Th}} \qquad \textbf{(18.5)}$$

$\mathbf{I}_{sc}$:

For the network of Fig. 18.57, where

$$\mathbf{V}_2 = 0 \qquad k_1\mathbf{V}_2 = 0 \qquad \mathbf{I} = \frac{\mathbf{V}_i}{R_1}$$

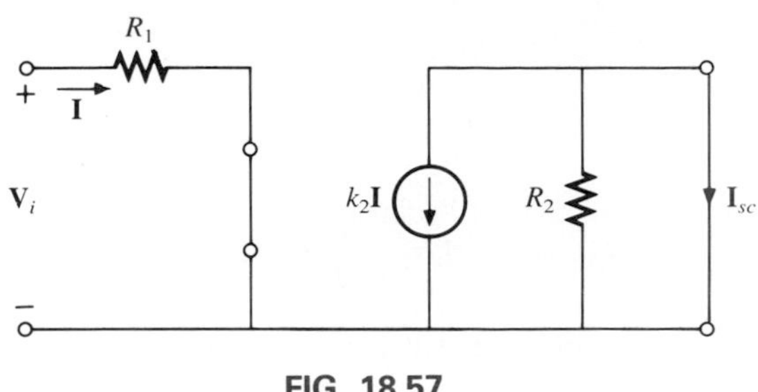


FIG. 18.57

and

$$\mathbf{I}_{sc} = -k_2\mathbf{I} = \frac{-k_2\mathbf{V}_i}{R_1}$$

so

$$\mathbf{Z}_{Th} = \frac{\mathbf{E}_{oc}}{\mathbf{I}_{sc}} = \frac{\dfrac{-k_2R_2\mathbf{V}_i}{R_1 - k_1k_2R_2}}{\dfrac{-k_2\mathbf{V}_i}{R_1}} = \frac{R_1R_2}{R_1 - k_1k_2R_2}$$

and

$$\mathbf{Z}_{Th} = \frac{R_2}{1 - \dfrac{k_1k_2R_2}{R_1}} \qquad \textbf{(18.6)}$$

Frequently, the approximation $k_1 \cong 0$ is applied. Then, the Thevenin voltage and impedance are

$$\mathbf{E}_{Th} = \frac{-k_2R_2\mathbf{V}_i}{R_1} \qquad k_1 = 0 \qquad \textbf{(18.7)}$$

$$\mathbf{Z}_{Th} = R_2 \qquad k_1 = 0 \qquad \textbf{(18.8)}$$

Apply $\mathbf{Z}_{Th} = \mathbf{E}_g/\mathbf{I}_g$ to the network of Fig. 18.58, where

$$\mathbf{I} = \frac{-k_1\mathbf{V}_2}{R_1}$$

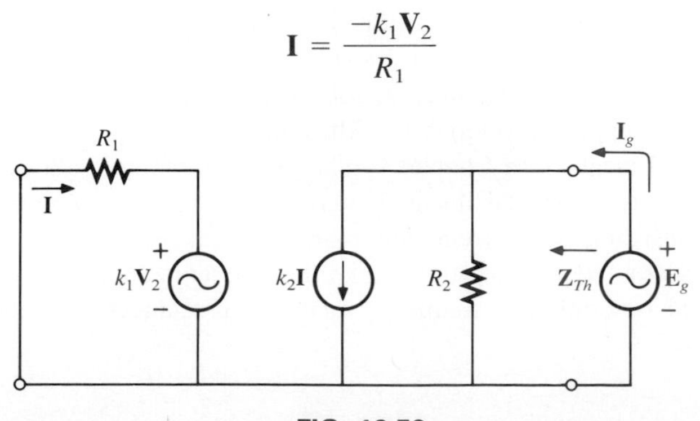


FIG. 18.58

But

$$\mathbf{V}_2 = \mathbf{E}_g$$

so

$$\mathbf{I} = \frac{-k_1\mathbf{E}_g}{R_1}$$

Applying Kirchhoff's current law, we have

$$\mathbf{I}_g = k_2\mathbf{I} + \frac{\mathbf{E}_g}{R_2} = k_2\left(-\frac{k_1\mathbf{E}_g}{R_1}\right) + \frac{\mathbf{E}_g}{R_2}$$

$$= \mathbf{E}_g\left(\frac{1}{R_2} - \frac{k_1k_2}{R_1}\right)$$

and

$$\frac{\mathbf{I}_g}{\mathbf{E}_g} = \frac{R_1 - k_1k_2R_2}{R_1R_2}$$

or

$$\mathbf{Z}_{Th} = \frac{\mathbf{E}_g}{\mathbf{I}_g} = \frac{\boldsymbol{R_1R_2}}{\boldsymbol{R_1 - k_1k_2R_2}}$$

as obtained above.

---

The last two methods presented in this section were applied only to networks in which the magnitudes of the controlled sources were dependent on a variable within the network for which the Thevenin equivalent circuit was to be obtained. Understand that both of those methods can also be applied to any dc or sinusoidal ac network containing only independent sources or dependent sources of the other kind.

## 18.4 NORTON'S THEOREM

The three methods described for Thevenin's theorem will each be altered to permit their use with Norton's theorem. Since the Thevenin and Norton impedances are the same for a particular network, certain portions of the discussion will be quite similar to those encountered in the previous section. We will first consider independent sources and the approach developed in Chapter 9, followed by dependent sources and the new techniques developed for Thevenin's theorem.

You will recall from Chapter 9 that Norton's theorem allows us to replace any two-terminal linear bilateral ac network by an equivalent circuit consisting of a current source and impedance, as in Fig. 18.59.

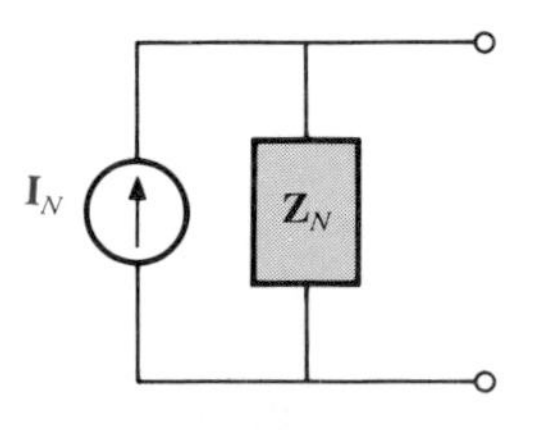

FIG. 18.59

The Norton equivalent circuit, like the Thevenin equivalent circuit, is applicable at only one frequency since the reactances are frequency dependent.

## Independent Sources

The procedure outlined below to find the Norton equivalent of a sinusoidal ac network is changed (from that in Chapter 9) in only one respect: to replace the term *impedance* with the term *resistance*.

1. Remove that portion of the network across which the Norton equivalent circuit is to be found.
2. Mark (○, •, and so on) the terminals of the remaining two-terminal network.
3. Calculate $\mathbf{Z}_N$ by first setting all voltage and current sources to zero (short circuit and open circuit, respectively) and then finding the resulting impedance between the two marked terminals.
4. Calculate $\mathbf{I}_N$ by first replacing the voltage and current sources and then finding the short-circuit current between the marked terminals.
5. Draw the Norton equivalent circuit with the portion of the circuit previously removed replaced between the terminals of the Norton equivalent circuit.

The Norton and Thevenin equivalent circuits can be found from each other by using the source transformation shown in Fig. 18.60. The source transformation is applicable for any Thevenin or Norton equivalent circuit determined from a network with any combination of independent or dependent sources.

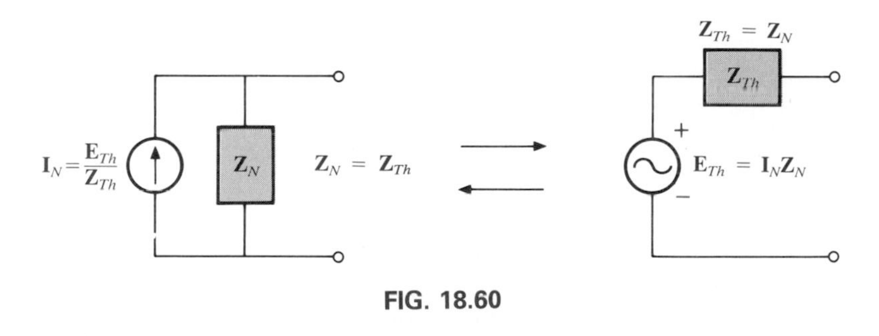


FIG. 18.60

**EXAMPLE 18.14.** Determine the Norton equivalent circuit for the network external to the 6-Ω resistor of Fig. 18.61.

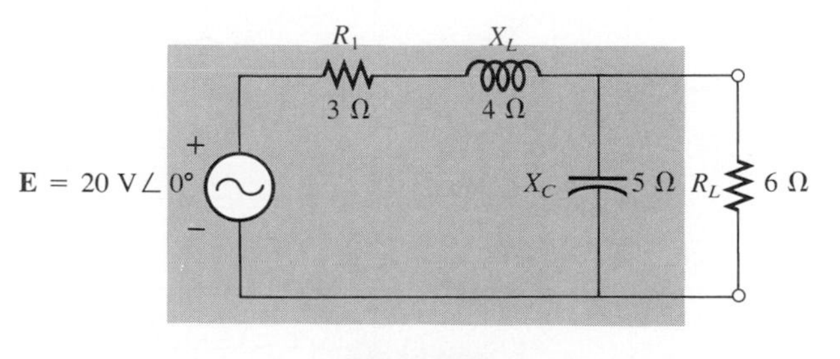


FIG. 18.61

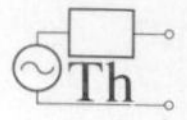


***Solution:***

*Steps 1 and 2* (Fig. 18.62):

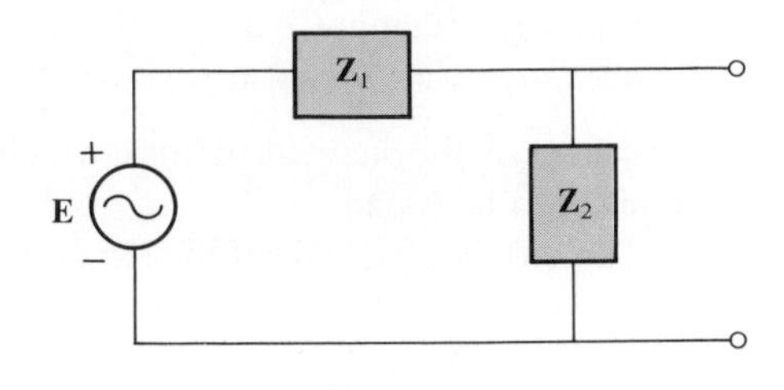

$$\mathbf{Z}_1 = R_1 + jX_L = 3\ \Omega + j4\ \Omega = 5\ \Omega\ \angle 53.13^\circ$$
$$\mathbf{Z}_2 = -jX_C = -j5\ \Omega$$

*Step 3* (Fig. 18.63):

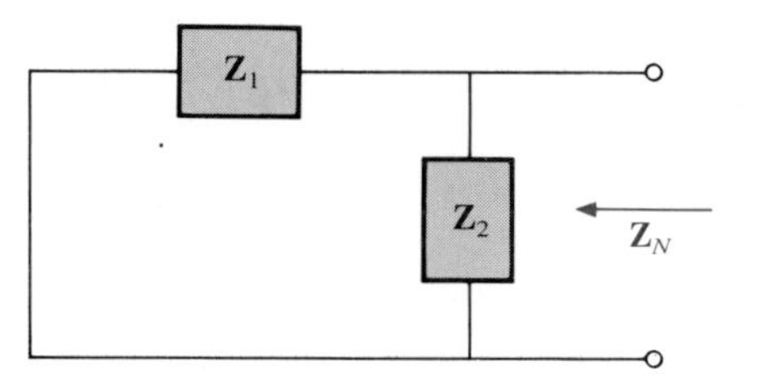


FIG. 18.63

$$\mathbf{Z}_N = \frac{\mathbf{Z}_1\mathbf{Z}_2}{\mathbf{Z}_1 + \mathbf{Z}_2} = \frac{(5\ \Omega\ \angle 53.13^\circ)(5\ \Omega\ \angle -90^\circ)}{3\ \Omega + j4\ \Omega - j5\ \Omega} = \frac{25\ \angle -36.87^\circ}{3 - j1}$$
$$= \frac{25\ \angle -36.87^\circ}{3.16\ \angle -18.43^\circ} = 7.91\ \Omega\ \angle -18.44^\circ = \mathbf{7.50\ \Omega - j2.50\ \Omega}$$

*Step 4* (Fig. 18.64):

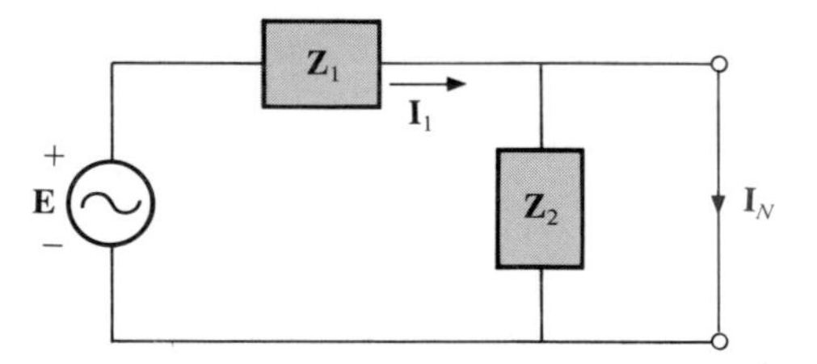

$$\mathbf{I}_N = \mathbf{I}_1 = \frac{\mathbf{E}}{\mathbf{Z}_1} = \frac{20\ \text{V}\ \angle 0^\circ}{5\ \Omega\ \angle 53.13^\circ} = \mathbf{4\ A\ \angle -53.13^\circ}$$

*Step 5:* The Norton equivalent circuit is shown in Fig. 18.65.

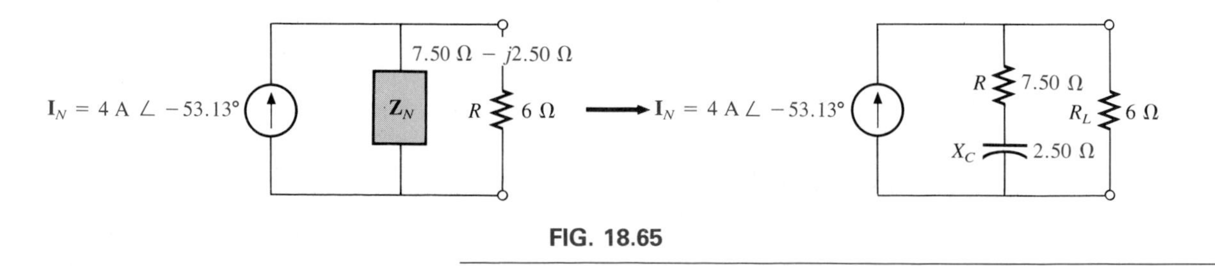


FIG. 18.65

---

**EXAMPLE 18.15.** Find the Norton equivalent circuit for the network external to the 7-Ω capacitive reactance in Fig. 18.66.

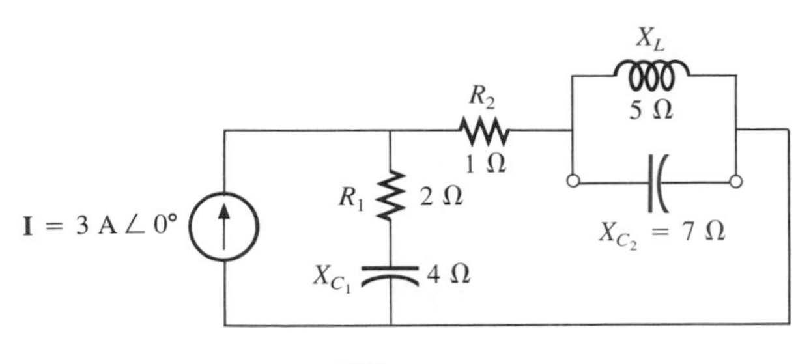

FIG. 18.66

***Solution:***

*Steps 1 and 2* (Fig. 18.67):

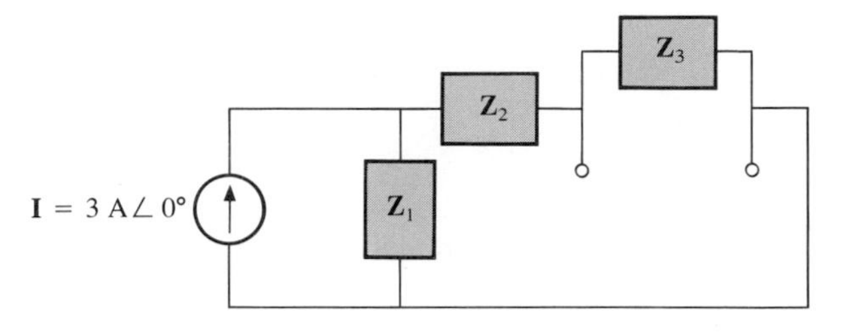


FIG. 18.67

$$\mathbf{Z}_1 = R_1 - jX_{C_1} = 2\ \Omega - j4\ \Omega$$
$$\mathbf{Z}_2 = R_2 = 1\ \Omega$$
$$\mathbf{Z}_3 = +jX_L = j5\ \Omega$$

*Step 3* (Fig. 18.68):

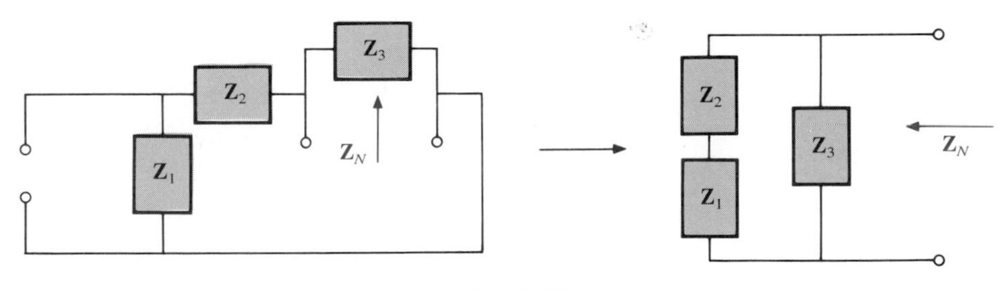


FIG. 18.68

$$\mathbf{Z}_N = \frac{\mathbf{Z}_3(\mathbf{Z}_1 + \mathbf{Z}_2)}{\mathbf{Z}_3 + (\mathbf{Z}_1 + \mathbf{Z}_2)}$$

$$\mathbf{Z}_1 + \mathbf{Z}_2 = 2\ \Omega - j4\ \Omega + 1\ \Omega = 3\ \Omega - j4\ \Omega = 5\ \Omega \angle -53.13°$$

$$\mathbf{Z}_N = \frac{(5\ \Omega \angle 90°)(5\ \Omega \angle -53.13°)}{j5\ \Omega + 3\ \Omega - j4\ \Omega} = \frac{25 \angle 36.87°}{3 + j1}$$

$$= \frac{25 \angle 36.87°}{3.16 \angle +18.43°}$$

$$\mathbf{Z}_N = 7.91\ \Omega \angle 18.44° = \mathbf{7.50\ \Omega + \mathit{j}2.50\ \Omega}$$

*Step 4* (Fig. 18.69):

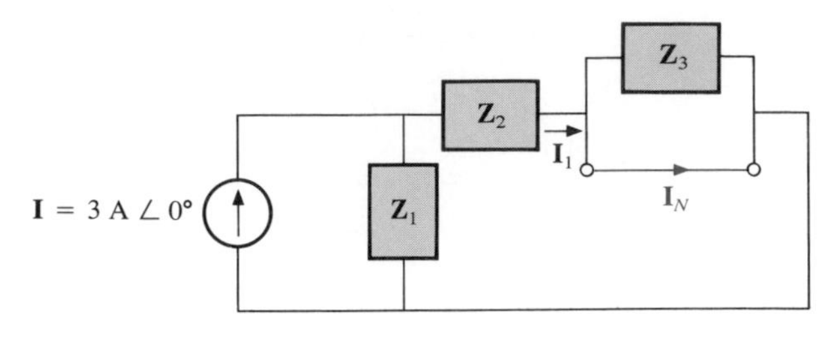

FIG. 18.69

$$\mathbf{I}_N = \mathbf{I}_1 = \frac{\mathbf{Z}_1\mathbf{I}}{\mathbf{Z}_1 + \mathbf{Z}_2} \qquad \text{(current divider rule)}$$

$$= \frac{(2\ \Omega - j4\ \Omega)(3\ \text{A})}{3\ \Omega - j4\ \Omega} = \frac{6 - j12}{5\ \angle -53.13^\circ} = \frac{13.4\ \angle -63.43^\circ}{5\ \angle -53.13^\circ}$$

$$\mathbf{I}_N = \mathbf{2.68\ A\ \angle -10.3^\circ}$$

*Step 5:* The Norton equivalent circuit is shown in Fig. 18.70.

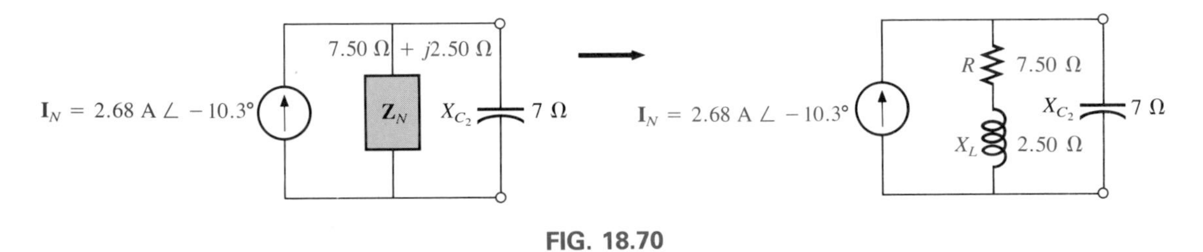


FIG. 18.70

---

**EXAMPLE 18.16.** Find the Thevenin equivalent circuit for the network external to the 7-Ω capacitive reactance in Fig. 18.66.

***Solution:*** Using the conversion between sources (Fig. 18.71), we obtain

$$\mathbf{Z}_{Th} = \mathbf{Z}_N = \mathbf{7.50\ \Omega + \mathit{j}2.50\ \Omega}$$

$$\mathbf{E}_{Th} = \mathbf{I}_N\mathbf{Z}_N = (2.68\ \text{A}\ \angle -10.3^\circ)(7.91\ \Omega\ \angle 18.44^\circ)$$
$$= \mathbf{21.2\ V\ \angle 8.14^\circ}$$

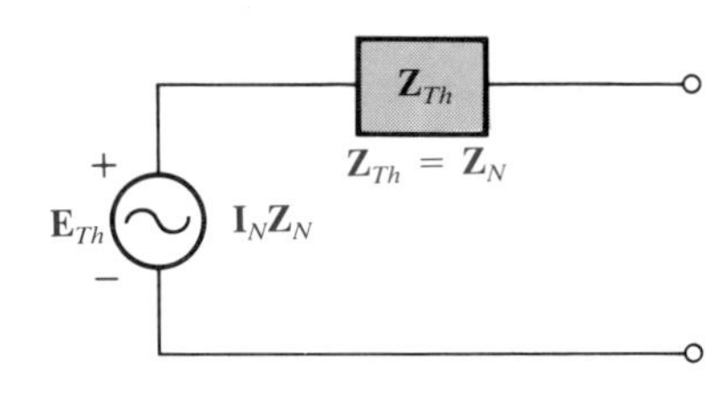

FIG. 18.71

The Thevenin equivalent circuit is shown in Fig. 18.72.

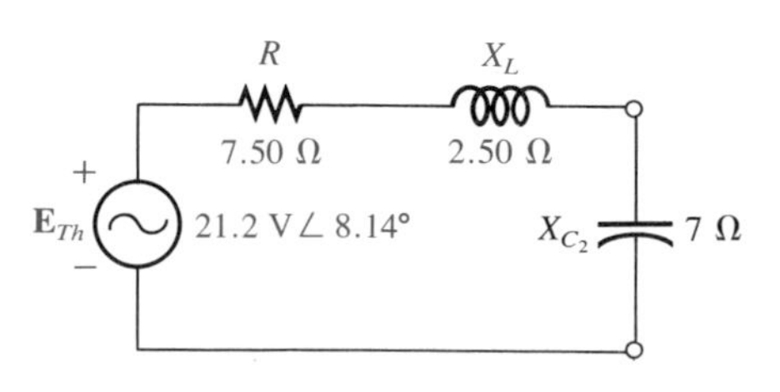

FIG. 18.72

## Dependent Sources

As stated for Thevenin's theorem, dependent sources in which the controlling variable is not determined by the network for which the Norton equivalent circuit is to be found do not alter the procedure outlined above.

For dependent sources of the other kind, one of the following procedures must be applied. Both of these procedures can also be applied to networks with any combination of independent sources and dependent sources not controlled by the network under investigation.

The Norton equivalent circuit appears in Fig. 18.73(a). In Fig. 18.73(b), we find that

$$\mathbf{I}_{sc} = \mathbf{I}_N \tag{18.9}$$

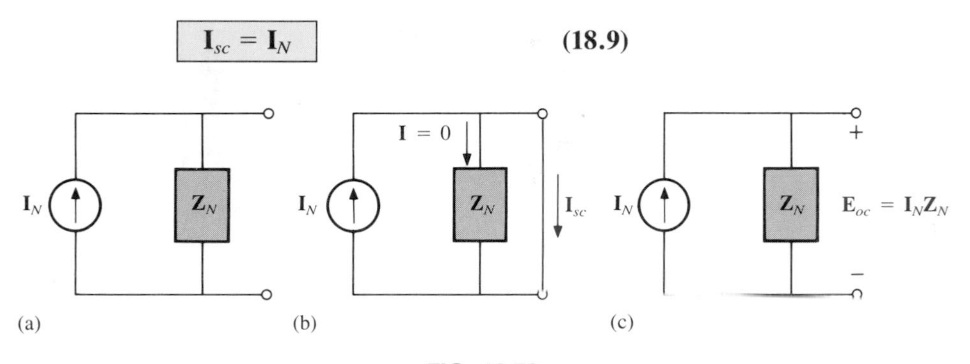


FIG. 18.73

and in Fig. 18.73(c) that

$$\mathbf{E}_{oc} = \mathbf{I}_N\mathbf{Z}_N$$

Or, rearranging, we have

$$\mathbf{Z}_N = \frac{\mathbf{E}_{oc}}{\mathbf{I}_N}$$

and

$$\mathbf{Z}_N = \frac{\mathbf{E}_{oc}}{\mathbf{I}_{sc}} \tag{18.10}$$

The Norton impedance can also be determined by applying a source of voltage $\mathbf{E}_g$ to the terminals of interest and finding the resulting $\mathbf{I}_g$, as shown in Fig. 18.74. All independent sources and dependent sources not controlled by a variable in the network of interest are set to zero, and

$$\mathbf{Z}_N = \frac{\mathbf{E}_g}{\mathbf{I}_g} \tag{18.11}$$

For this latter approach, the Norton current is still determined by the short-circuit current.

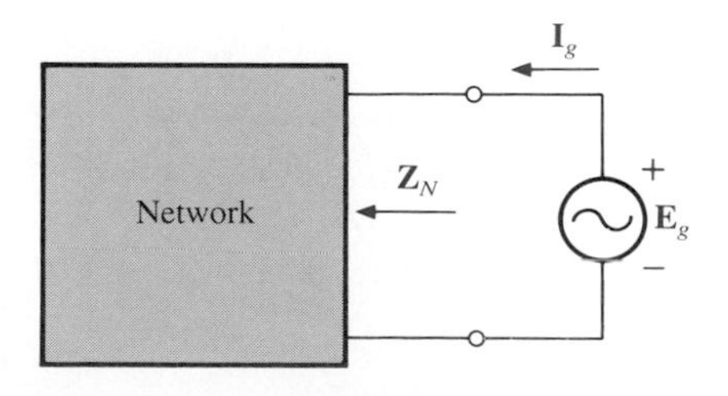


FIG. 18.74

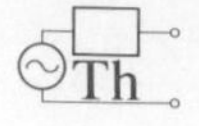


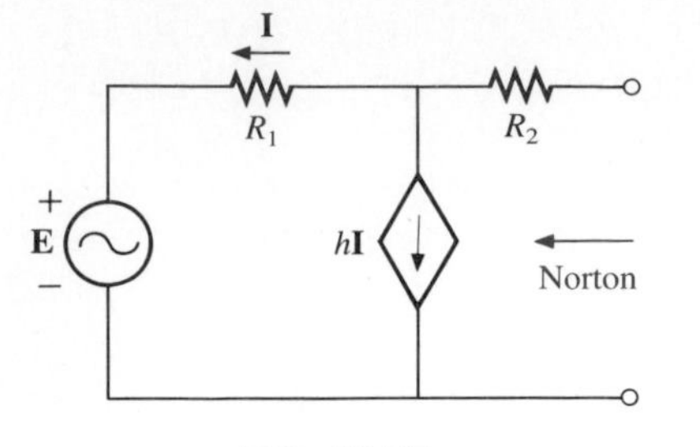


FIG. 18.75

**EXAMPLE 18.17.** Using each method described for dependent sources, find the Norton equivalent circuit for the network of Fig. 18.75.

***Solution:***

$\mathbf{I}_N$:

For each method, $\mathbf{I}_N$ is determined in the same manner. From Fig. 18.76, using Kirchhoff's current law, we have

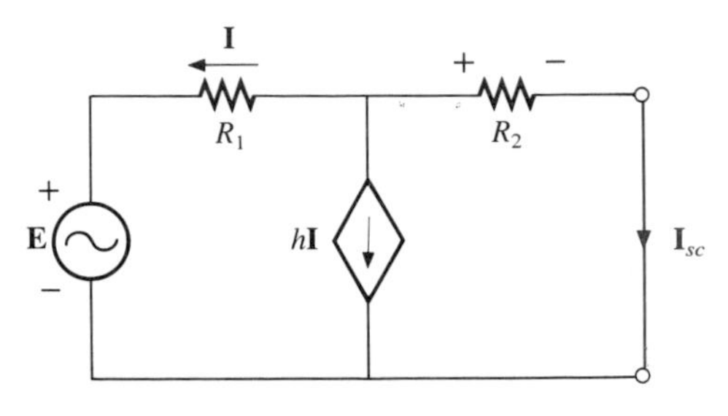


FIG. 18.76

$$0 = \mathbf{I} + h\mathbf{I} + \mathbf{I}_{sc}$$

or

$$\mathbf{I}_{sc} = -(1 + h)\mathbf{I}$$

Applying Kirchhoff's voltage law gives us

$$\mathbf{E} + \mathbf{I}R_1 - \mathbf{I}_{sc}R_2 = 0$$

and

$$\mathbf{I}R_1 = \mathbf{I}_{sc}R_2 - \mathbf{E}$$

or

$$\mathbf{I} = \frac{\mathbf{I}_{sc}R_2 - \mathbf{E}}{R_1}$$

so

$$\mathbf{I}_{sc} = -(1 + h)\mathbf{I} = -(1 + h)\left(\frac{\mathbf{I}_{sc}R_2 - \mathbf{E}}{R_1}\right)$$

or

$$R_1\mathbf{I}_{sc} = -(1 + h)\mathbf{I}_{sc}R_2 + (1 + h)\mathbf{E}$$

$$\mathbf{I}_{sc}[R_1 + (1 + h)R_2] = (1 + h)\mathbf{E}$$

$$\mathbf{I}_{sc} = \frac{(1 + h)\mathbf{E}}{R_1 + (1 + h)R_2} = \mathbf{I}_N$$

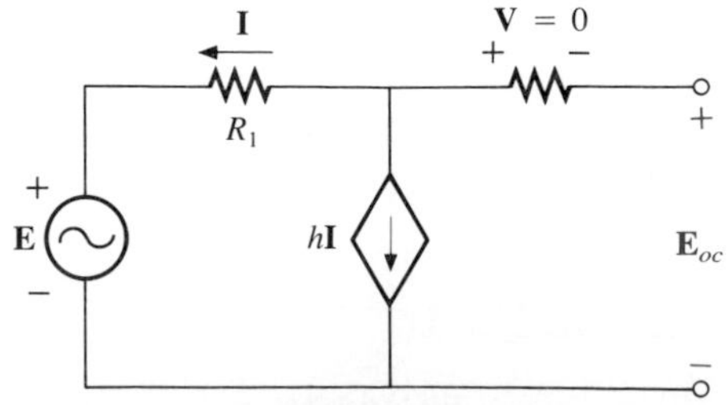

FIG. 18.77

$\mathbf{Z}_N$:

*Method 1:* $\mathbf{E}_{oc}$ is determined from the network of Fig. 18.77. By Kirchhoff's current law,

$$0 = \mathbf{I} + h\mathbf{I} \quad \text{or} \quad \mathbf{I}(h + 1) = 0$$

For $h$, a positive constant $\mathbf{I}$ must equal zero to satisfy the above. Therefore,

$$\mathbf{I} = 0 \quad \text{and} \quad h\mathbf{I} = 0$$

and

$$\mathbf{E}_{oc} = \mathbf{E}$$

with

$$\mathbf{Z}_N = \frac{\mathbf{E}_{oc}}{\mathbf{I}_{sc}} = \frac{\mathbf{E}}{\dfrac{(1 + h)\mathbf{E}}{R_1 + (1 + h)R_2}} = \frac{\boldsymbol{R_1 + (1 + h)R_2}}{\boldsymbol{(1 + h)}}$$

*Method 2:* Note Fig. 18.78. By Kirchhoff's current law,

$$\mathbf{I}_g = \mathbf{I} + h\mathbf{I} = (1 + h)\mathbf{I}$$

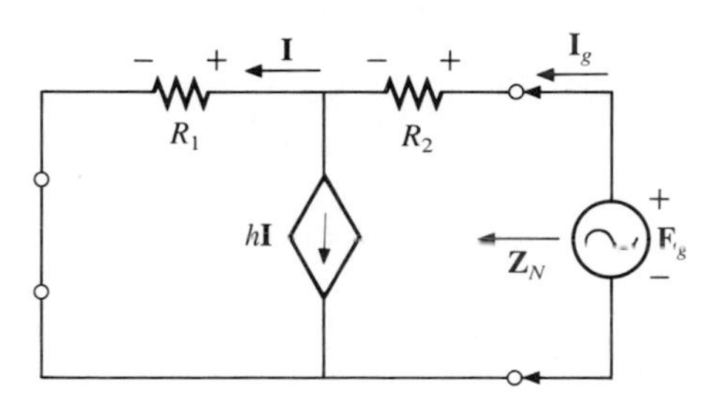


**FIG. 18.78**

By Kirchhoff's voltage law,

$$\mathbf{E}_g - \mathbf{I}_g R_2 - \mathbf{I}R_1 = 0$$

or

$$\mathbf{I} = \frac{\mathbf{E}_g - \mathbf{I}_g R_2}{R_1}$$

Substituting, we have

$$\mathbf{I}_g = (1 + h)\mathbf{I} = (1 + h)\left(\frac{\mathbf{E}_g - \mathbf{I}_g R_2}{R_1}\right)$$

and

$$\mathbf{I}_g R_1 = (1 + h)\mathbf{E}_g - (1 + h)\mathbf{I}_g R_2$$

so

$$\mathbf{E}_g(1 + h) = \mathbf{I}_g[R_1 + (1 + h)R_2]$$

or

$$\mathbf{Z}_N = \frac{\mathbf{E}_g}{\mathbf{I}_g} = \frac{\boldsymbol{R_1 + (1 + h)R_2}}{\boldsymbol{1 + h}}$$

which agrees with the above.

**EXAMPLE 18.18.** Find the Norton equivalent circuit for the network configuration of Fig. 18.56.

***Solution:*** By source conversion,

$$\mathbf{I}_N = \frac{\mathbf{E}_{Th}}{\mathbf{Z}_{Th}} = \frac{\dfrac{-k_2R_2\mathbf{V}_i}{R_1 - k_1k_2R_2}}{\dfrac{R_1R_2}{R_1 - k_1k_2R_2}}$$

and

$$\mathbf{I}_N = \frac{-k_2\mathbf{V}_i}{R_1} \tag{18.12}$$

which is $\mathbf{I}_{sc}$ as determined in that example, and

$$\mathbf{Z}_N = \mathbf{Z}_{Th} = \frac{R_2}{1 - \dfrac{k_1k_2R_2}{R_1}} \tag{18.13}$$

For $k_1 \cong 0$, we have

$$\mathbf{I}_N = \frac{-k_2\mathbf{V}_i}{R_1} \qquad k_1 = 0 \tag{18.14}$$

$$\mathbf{Z}_N = R_2 \qquad k_1 = 0 \tag{18.15}$$

## 18.5 MAXIMUM POWER TRANSFER THEOREM

When applied to ac circuits, the maximum power transfer theorem states that

***maximum power will be delivered to a load when the load impedance is the conjugate of the Thevenin impedance across its terminals.***

That is, for Fig. 18.79, for maximum power transfer to the load,

$$Z_L = Z_{Th} \quad \text{and} \quad \theta_L = -\theta_{Th_Z} \tag{18.16}$$

or, in rectangular form,

$$R_L = R_{Th} \quad \text{and} \quad \pm jX_L = \mp jX_{Th} \tag{18.17}$$

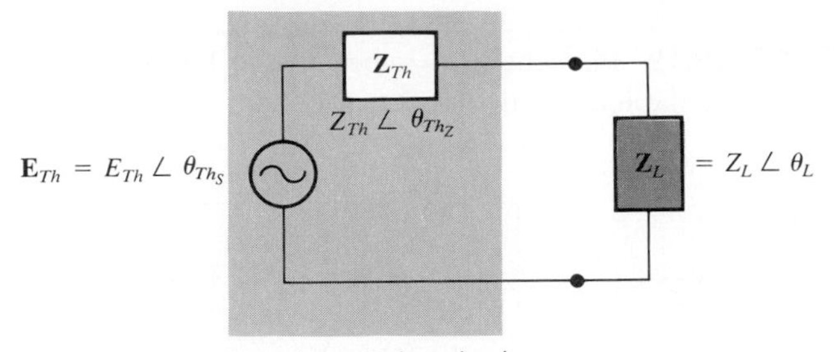


**FIG. 18.79**

The conditions just mentioned will make the total impedance of the circuit appear purely resistive, as indicated in Fig. 18.80:

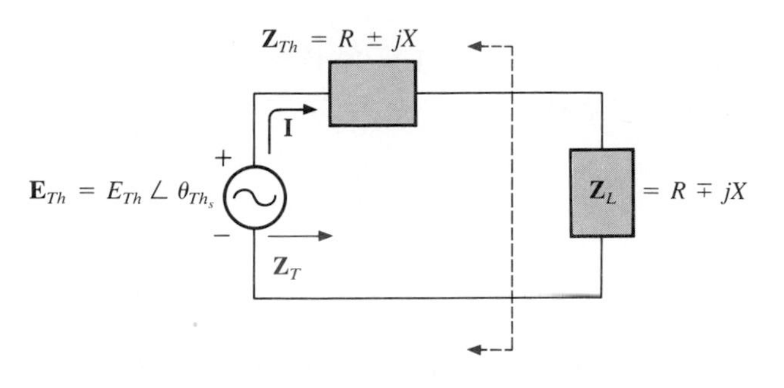


**FIG. 18.80**

$$\mathbf{Z}_T = (R \pm jX) + (R \mp jX)$$

and

$$\boxed{\mathbf{Z}_T = 2R} \qquad \textbf{(18.18)}$$

Since the circuit is purely resistive, the power factor of the circuit under maximum power conditions is 1. That is,

$$\boxed{F_p = 1} \qquad \text{(maximum power transfer)} \qquad \textbf{(18.19)}$$

The magnitude of the current **I** of Fig. 18.80 is

$$I = \frac{E_{Th}}{Z_T} = \frac{E_{Th}}{2R}$$

The maximum power to the load is

$$P_{\max} = I^2 R = \left(\frac{E_{Th}}{2R}\right)^2 R$$

and

$$\boxed{P_{\max} = \frac{E_{Th}^2}{4R}} \qquad \textbf{(18.20)}$$

**EXAMPLE 18.19.** Find the load impedance in Fig. 18.81 for maximum power to the load, and find the maximum power.

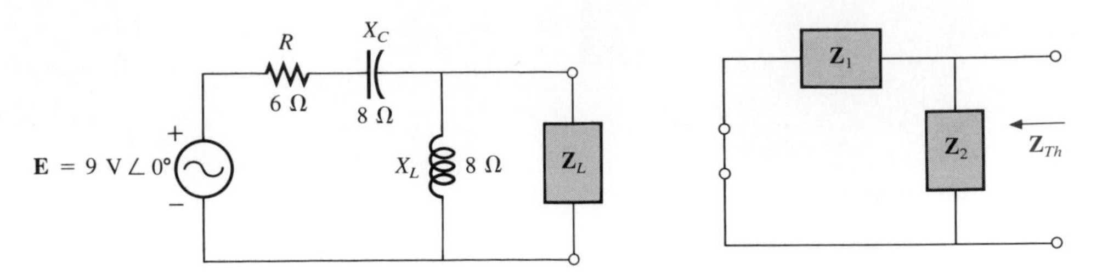

FIG. 18.81

***Solution:***

$$\mathbf{Z}_1 = R - jX_C = 6\ \Omega - j8\ \Omega = 10\ \Omega\ \angle -53.13^\circ$$
$$\mathbf{Z}_2 = +jX_L = j8\ \Omega$$
$$\mathbf{Z}_{Th} = \frac{\mathbf{Z}_1\mathbf{Z}_2}{\mathbf{Z}_1 + \mathbf{Z}_2} = \frac{(10\ \Omega\ \angle -53.13^\circ)(8\ \Omega\ \angle 90^\circ)}{6\ \Omega - j8\ \Omega + j8\ \Omega} = \frac{80\ \angle 36.87^\circ}{6\ \angle 0^\circ}$$
$$= 13.33\ \Omega\ \angle 36.87^\circ = 10.66\ \Omega + j8\ \Omega$$

and

$$\mathbf{Z}_L = 13.3\ \Omega\ \angle -36.87^\circ = \mathbf{10.66\ \Omega - j8\ \Omega}$$

In order to find the maximum power, we must first find $\mathbf{E}_{Th}$ (Fig. 18.82), as follows:

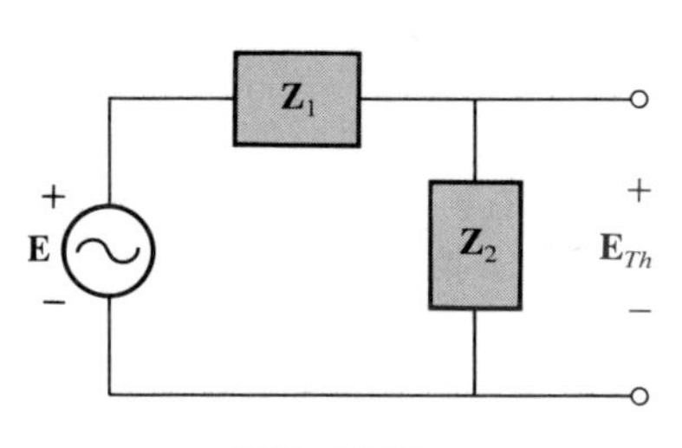


FIG. 18.82

$$\mathbf{E}_{Th} = \frac{\mathbf{Z}_2\mathbf{E}}{\mathbf{Z}_2 + \mathbf{Z}_1} \qquad \text{(voltage divider rule)}$$
$$= \frac{(8\ \Omega\ \angle 90^\circ)(9\ \text{V}\ \angle 0^\circ)}{j8\ \Omega + 6\ \Omega - j8\ \Omega} = \frac{72\ \angle 90^\circ}{6\ \angle 0^\circ} = 12\ \text{V}\ \angle 90^\circ$$

Then

$$P_{\max} = \frac{E_{Th}^2}{4R} = \frac{(12\ \text{V})^2}{4(10.66\ \Omega)} = \frac{144}{42.64} = \mathbf{3.38\ W}$$

**EXAMPLE 18.20.** Find the load impedance in Fig. 18.83 for maximum power to the load, and find the maximum power.

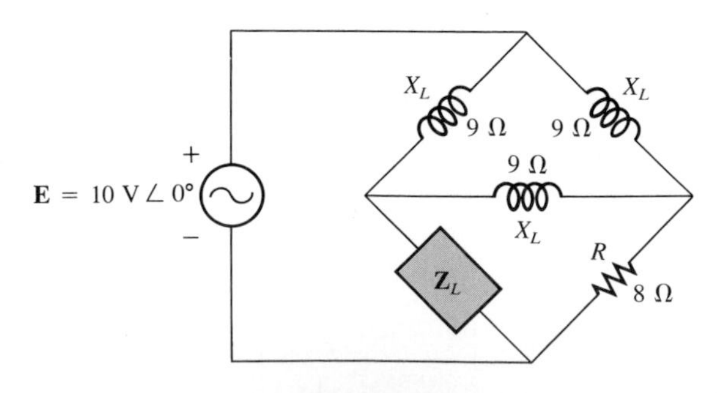

FIG. 18.83

***Solution:*** First we must find $\mathbf{Z}_{Th}$ (Fig. 18.84).

$$\mathbf{Z}_1 = +jX_L = j9\ \Omega \qquad \mathbf{Z}_2 = R = 8\ \Omega$$

Converting from a $\Delta$ to a Y (Fig. 18.85), we have

$$\mathbf{Z}'_1 = \frac{\mathbf{Z}_1}{3} = j3\ \Omega \qquad \mathbf{Z}_2 = 8\ \Omega$$

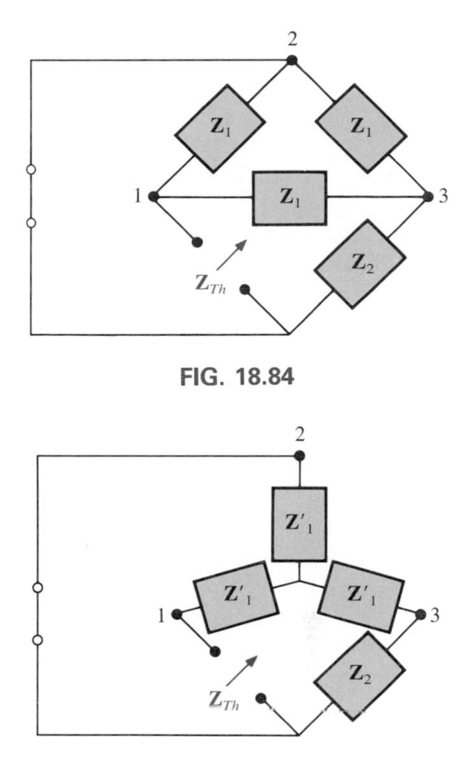

FIG. 18.85

The redrawn circuit (Fig. 18.86) shows

$$\begin{aligned}\mathbf{Z}_{Th} &= \mathbf{Z}'_1 + \frac{\mathbf{Z}'_1(\mathbf{Z}'_1 + \mathbf{Z}_2)}{\mathbf{Z}'_1 + (\mathbf{Z}'_1 + \mathbf{Z}_2)}\\ &= j3\ \Omega + \frac{3\ \Omega\ \angle 90°(j3\ \Omega + 8\ \Omega)}{j6\ \Omega + 8\ \Omega}\\ &= j3 + \frac{(3\ \angle 90°)(8.54\ \angle 20.56°)}{10\ \angle 36.87°}\\ &= j3 + \frac{25.62\ \angle 110.56°}{10\ \angle 36.87°} = j3 + 2.56\ \angle 73.69°\\ &= j3 + 0.72 + j2.46\\ \mathbf{Z}_{Th} &= 0.72\ \Omega + j5.46\ \Omega\end{aligned}$$

and

$$\mathbf{Z}_L = \mathbf{0.72\ \Omega - j5.46\ \Omega}$$

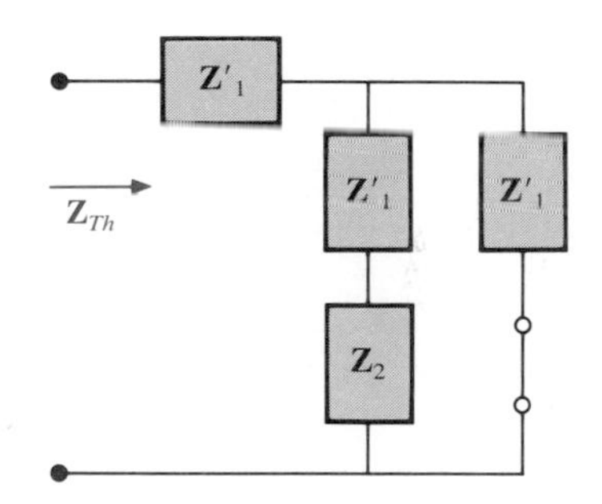

FIG. 18.86

For $\mathbf{E}_{Th}$, use the modified circuit of Fig. 18.87 with the voltage source replaced in its original position. Since $I_1 = 0$, $\mathbf{E}_{Th}$ is the voltage across the series impedance of $\mathbf{Z}'_1$ and $\mathbf{Z}_2$. Using the voltage divider rule gives us

$$\begin{aligned}\mathbf{E}_{Th} &= \frac{(\mathbf{Z}'_1 + \mathbf{Z}_2)\mathbf{E}}{\mathbf{Z}'_1 + \mathbf{Z}_2 + \mathbf{Z}'_1} = \frac{(j3\ \Omega + 8\ \Omega)(10\ \text{V}\ \angle 0°)}{8\ \Omega + j6\ \Omega}\\ &= \frac{(8.54\ \angle 20.56°)(10\ \angle 0°)}{10\ \angle 36.87°}\\ \mathbf{E}_{Th} &= 8.54\ \text{V}\ \angle -16.31°\end{aligned}$$

and

$$\begin{aligned}P_{\max} &= \frac{E_{Th}^2}{4R} = \frac{(8.54\ \text{V})^2}{4(0.72\ \Omega)} = \frac{72.93}{2.88}\\ &= \mathbf{25.32\ W}\end{aligned}$$

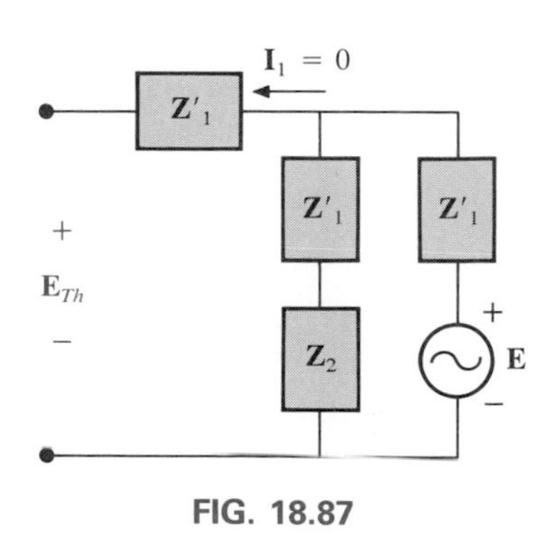

FIG. 18.87

If the load resistance is adjustable but the magnitude of the load reactance cannot be set equal to the magnitude of the Thevenin reactance then the maximum power *that can be delivered* to the load will occur when the load reactance is made as close to the Thevenin reactance as possible and the load resistance is set to the following value:

$$R_L = \sqrt{R_{Th}^2 + (X_{Th} + X_{\text{load}})^2} \qquad \textbf{(18.21)}$$

where each reactance carries a positive sign if inductive and a negative sign if capacitive.

The power delivered will be determined by

$$P = E_{Th}^2/4R_{av} \tag{18.22}$$

where

$$R_{av} = \frac{R_{Th} + R_L}{2} \tag{18.23}$$

The derivation of the above equations is given in Appendix H of the text. The following example demonstrates the use of the above.

---

**EXAMPLE 18.21.** For the network of Fig. 18.88:

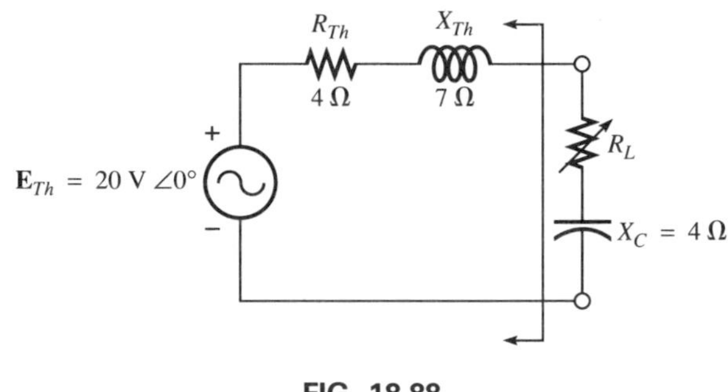

**FIG. 18.88**

a. Determine the value of $R_L$ for maximum power to the load if the load reactance is fixed at 4 Ω.
b. Find the power delivered to the load under the conditions of part (a).
c. Find the maximum power to the load if the load reactance is made adjustable to any value and compare the result to part (b) above.

***Solutions:***

a. Eq. (18.21):

$$\begin{aligned} R_L &= \sqrt{R_{Th}^2 + (X_{Th} + X_{load})^2} \\ &= \sqrt{(4\ \Omega)^2 + (7\ \Omega - 4\ \Omega)^2} \\ &= \sqrt{16 + 9} = \sqrt{25} \end{aligned}$$

$$R_L = \mathbf{5\ \Omega}$$

b. Eq. (18.23):

$$R_{av} = \frac{R_{Th} + R_L}{2} = \frac{4\ \Omega + 5\ \Omega}{2}$$

$$= \mathbf{4.5\ \Omega}$$

Eq. (18.22):

$$P = \frac{E_{Th}^2}{4R_{av}}$$
$$= \frac{(20 \text{ V})^2}{4(4.5 \ \Omega)} = \frac{400}{18} \text{ W}$$
$$\cong \mathbf{22.22 \ W}$$

c. For

$$\mathbf{Z}_L = 4 \ \Omega - j7 \ \Omega$$
$$P_{max} = \frac{E_{Th}^2}{4R_{Th}} = \frac{(20 \text{ V})^2}{4(4 \ \Omega)}$$
$$= \mathbf{25 \ W}$$

exceeding the result of part (b) by 2.78 W.

## 18.6 SUBSTITUTION, RECIPROCITY, AND MILLMAN'S THEOREMS

As indicated in the introduction to this chapter, the substitution and reciprocity theorems and Millman's theorem will not be considered here in detail. A careful review of Chapter 9 will enable you to apply these theorems to sinusoidal ac networks with little difficulty. A number of problems in the use of these theorems appear in the problem section.

## 18.7 COMPUTER ANALYSIS

The computer analysis of this chapter will be limited to an application of Thevenin's theorem to a network with independent sources and the writing of an input file for a network with controlled sources. The extended use of controlled sources in the analysis of electronic systems places a high priority on the introduction of the subject. There are essentially four types of controlled sources in electronic systems with which a student should become familiar: current-controlled current sources (CCCS), voltage-controlled current sources (VCCS), current-controlled voltage sources (CCVS), and voltage-controlled voltage sources (VCVS). Each has a controlling variable that must be properly listed in the input file to insure the proper magnitude and phase for the controlled source. In BASIC an analysis of the network as appearing in this chapter would have to be applied—simply an extension of the longhand approach to a specific configuration. PSPICE permits a fairly standard entry of network parameters with the software package able to perform the proper analysis and provide the desired solution. This section will be limited to the PSPICE approach to permit increased coverage of the controlled source input format.

## PSPICE

**Thevenin's Theorem** The network to be analyzed using Thevenin's theorem is the same as shown in Fig. 18.28 of the text. In order to enter the proper inductive and capacitive levels for the network of Fig. 18.89(a), a frequency of 1 kHz was chosen and the nameplate values calculated for the reactance levels of Fig. 18.28.

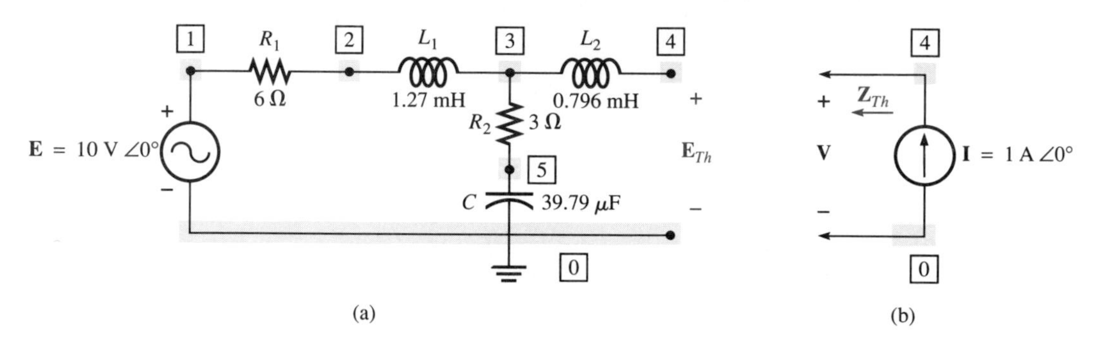


FIG. 18.89

```
************************ Evaluation PSpice (January 1989) ******* 10:45:10 *******

Chapter 18 - Thevenin's Theorem - AC - E(Th)

****     CIRCUIT DESCRIPTION

*****************************************************************************

VE 1 0 AC 10V 0
R1 1 2 6
L1 2 3 1.27MH
L2 3 4 0.796MH
R2 3 5 3
C 5 0 39.79UF
RL 4 0 1E30
.AC LIN 1 1KH 1KH
.PRINT AC VM(4) VP(4)
.OPTIONS NOPAGE
.END

****     AC ANALYSIS                         TEMPERATURE =   27.000 DEG C

 FREQ          VM(4)        VP(4)

  1.000E+03    5.081E+00   -7.698E+01
```

FIG. 18.90

**$E_{Th}$:**

For $\mathbf{E}_{Th}$ the input file appears in Fig. 18.90 with the values defined in Fig. 18.90(a) and an applied frequency of 1 kHz. Both the magnitude and phase angle of the Thevenin voltage between nodes 4 and 0 were requested.

The result of $\mathbf{E}_{Th} = 5.081 \text{ V} \angle -76.98°$ is a close match with the result of Example 18.8 ($\mathbf{E}_{Th} = 5.08 \text{ V} \angle -77.09°$).

**$Z_{Th}$:**

For $\mathbf{Z}_{Th}$ the technique applied to dc systems was applied with the source set to zero volts and current source of 1 A applied between nodes 4 and 0, as shown in Fig. 18.90(b). Since $\mathbf{Z}_T = \mathbf{V}/\mathbf{I}$ and $\mathbf{I} = 1 \text{ A} \angle 0°$, the magnitude and angle of $\mathbf{V}$ will be the same as that for $\mathbf{Z}_T$. The input file appears in Fig. 18.91, with the result of $\mathbf{V} = 5.493 \text{ V} \angle 32.42°$ revealing that $\mathbf{Z}_{Th} = 5.493 \ \Omega \angle 32.42°$, which is a very close match with the impedance determined in Example 18.8 ($\mathbf{Z}_{Th} = 5.49 \ \Omega \angle 32.36°$).

```
************************* Evaluation PSpice (January 1989) ******* 10:46:55 *******

 Chapter 18 - Thevenin's Theorem - AC - Z(Th)

 ****     CIRCUIT DESCRIPTION

*******************************************************************************

VE 1 0 AC 0V 0
R1 1 2 6
L1 2 3 1.27MH
L2 3 4 0.796MH
R2 3 5 3
C 5 0 39.79UF
RL 4 0 1E30
II 0 4 AC 1 0
.AC LIN 1 1KH 1KH
.PRINT AC VM(4) VP(4)
.OPTIONS NOPAGE
.END

 ****     AC ANALYSIS                     TEMPERATURE =   27.000 DEG C

  FREQ         VM(4)        VP(4)

   1.000E+03    5.493E+00    3.242E+01
```

FIG. 18.91

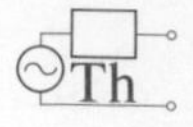

**Controlled sources** The hybrid equivalent circuit for a transistor has both a current-controlled current source (CCCS) and a voltage-controlled voltage source (VCVS), as shown in Fig. 18.92. As noted

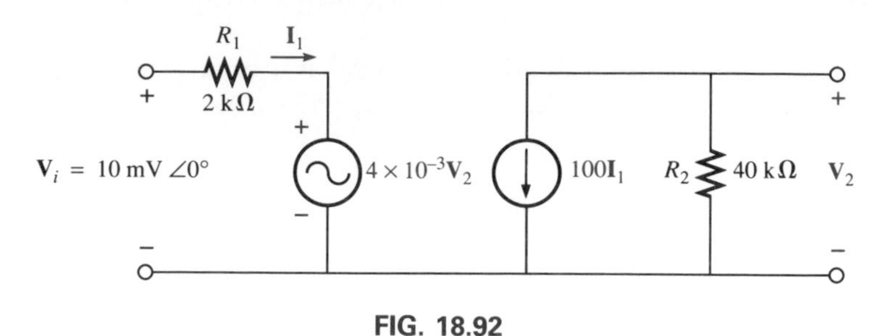


FIG. 18.92

earlier in this chapter, the concept of equivalent circuits and, in particular, the hybrid equivalent circuit, will all be discussed in detail in your electronics courses. The intent here is to learn how to input controlled sources using PSPICE to ensure the correct results are obtained.

The magnitude and angle of the controlled current source of Fig. 18.92 are determined by a multiplying factor (in this case 100) and a current $\mathbf{I}_1$ that flows through the resistor $R_1$ in the direction indicated. The magnitude and angle of the controlled voltage source are determined by a multiplying factor of $4 \times 10^{-3}$ and the output voltage $\mathbf{V}_2$ defined by nodes 4 and 0. Note in passing that the controlled current source is part of the output circuit, whereas the controlling variable is part of the input circuit. The reverse is true for the controlled voltage source, which is actually a ''feedback'' of the output voltage to the input circuit.

The format for CCCS and VCVS controlled sources are as follows.

**Current-Controlled Current Sources (CCCS)** For CCCS sources it is particularly important that the direction of the controlling current be properly entered. In Fig. 18.92, if the controlling current $\mathbf{I}_1$ were entered with a direction opposite to that of Fig. 18.92, the result would be meaningless.

In PSPICE a controlling current has a direction defined by an independent source in series with the branch in which the controlling current is defined. Keep in mind, however, that in PSPICE the current through an independent source has the direction shown in Fig. 18.93—from the + to − potential.

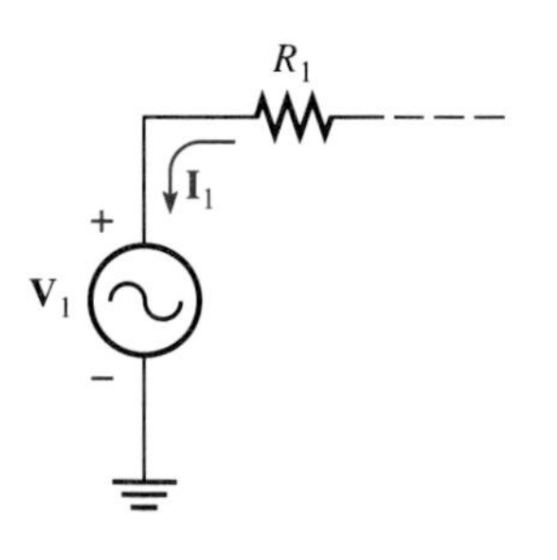


FIG. 18.93

In Fig. 18.92 this would result in a wrong direction for the controlling current $\mathbf{I}_1$. A correct direction can be obtained in PSPICE by introducing an independent sensing source of zero volts in the branch in which $\mathbf{I}_1$ is defined, as shown in Fig. 18.94. Note that the direction of $\mathbf{I}_1$ corresponds with the current direction defined for the ''dummy'' source VSENSE. The fact that the magnitude of the dummy source is zero volts removes any concern about it affecting the overall behavior of the network. Of course, if $\mathbf{I}_1$ had the opposite direction, the voltage source $\mathbf{V}_i$ could have been used to define the direction of $\mathbf{I}_1$ for the input file.

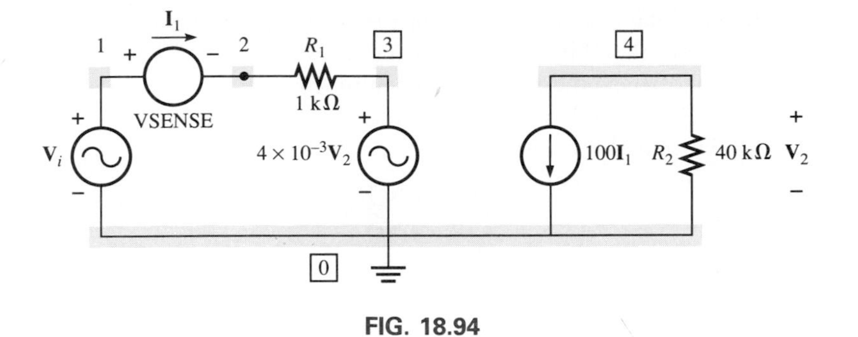


FIG. 18.94

In review:

***The direction of a controlling current is defined by an independent source is series with the branch in which the controlling current is defined. The absence of an independent source of the presence of a source with the wrong polarity requires the insertion of an independent source with the correct polarity and a magnitude of zero volts.***

The general format for a CCCS is as follows.

**Current-Controlled Current Source (CCCS)**

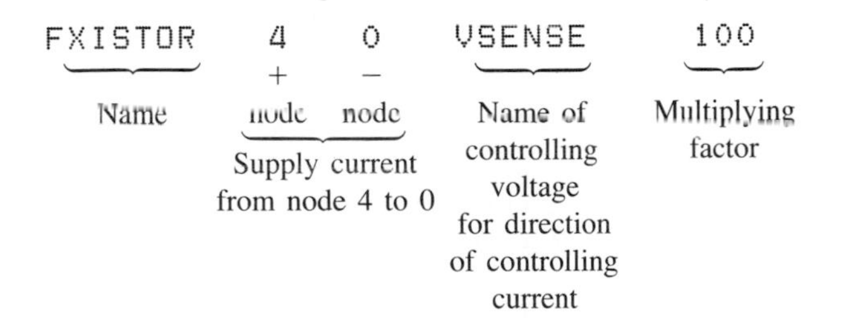


Note in the preceding format that a single letter F defines a CCCS and that a voltage is defining the direction of $\mathbf{I}_1$, as discussed earlier. Of course, the use of VSENSE in the CCCS command will require that VSENSE appear as an independent source in the input file.

**Voltage-Controlled Voltage Source (VCVS)** The input format for a VCVS is easier to grasp because the polarity of the controlling voltage source is entered as defined in Fig. 18.92. The general format is the following:

| EFBACK | 3 | 0 | 4 | 0 | 4M |
|---|---|---|---|---|---|
| | + | − | + | − | |
| Name | node | node | node | node | Multiplying factor |
| | For controlled source | | Of controlling voltage | | |

Again note the use of a single letter E to define a VCVS and the relative simplicity of writing the input line compared to the CCCS.

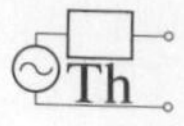


Before looking at the entire input file, let us take a moment to analyze the network of Fig. 18.92 longhand (as we would in BASIC) to be sure we understand the impact of a controlled source and have an answer to compare with the output of the PSPICE run.

For the controlling variables:

$$\mathbf{I}_1 = \frac{\mathbf{V}_i - 4 \times 10^{-3}\mathbf{V}_2}{1\ \text{k}\Omega}$$

and

$$\mathbf{V}_2 = -(100\mathbf{I}_1)(40\ \text{k}\Omega)$$

Substituting:

$$\mathbf{V}_2 = -(100)\left(\frac{\mathbf{V}_i - 4 \times 10^{-3}\mathbf{V}_2}{1\ \text{k}\Omega}\right)(40\ \text{k}\Omega)$$

$$\mathbf{V}_2 = -(4000)[\mathbf{V}_i - 4 \times 10^{-3}\mathbf{V}_2] = -4000\mathbf{V}_i + 16\mathbf{V}_2$$

and

$$-15\mathbf{V}_2 = -4000\mathbf{V}_i$$

or

$$\mathbf{V}_2 = \frac{-4000\mathbf{V}_i}{15} = -266.67\mathbf{V}_i$$

but

$$\mathbf{V}_i = 10\ \text{mV}\ \angle 0^\circ$$

and

$$\mathbf{V}_2 = -(266.67)(10 \times 10^{-3})\ \text{V}\ \angle 0^\circ$$
$$\mathbf{V}_2 = \mathbf{-2.67\ V\ \angle 0^\circ = 2.67\ V\ \angle 180^\circ}$$

The input file for Fig. 18.94 appears in Fig. 18.95, with the two controlled sources ER and FF and the independent source VSENSE to define the correct direction for $\mathbf{I}_1$. Even though there are no reactive elements in the network, a frequency is defined to establish the .AC command. Note how closely the result for $\mathbf{V}_2$ in the output file corresponds with the preceding solution.

Although not appearing in Fig. 18.92, the format of the remaining two types of controlled sources are now defined.

### Voltage-Controlled Current Source (VCCS)

| GFET | 2 | 3 | 5 | 8 | 4M |
|---|---|---|---|---|---|
| | + | − | + | − | |
| Name | node | node | node | node | Multiplying factor (units: siemens) |
| | Source current from node 2 to node 3 | | Defined by polarity of controlling voltage | | |

```
************************ Evaluation PSpice (January 1989) ******* 10:38:24 *******

Chapter 18 - Transistor Amplifier - Controlled Source

****     CIRCUIT DESCRIPTION

******************************************************************************

VI 1 0 AC 10MV 0
VSENSE 1 2 0V
R1 2 3 1K
ER 3 0 4 0 4M
FF 4 0 VSENSE 100
R2 4 0 40K
.AC LIN 1 1KH 1KH
.PRINT AC VM(4,0)
.OPTIONS NOPAGE
.END
```

```
****     AC ANALYSIS                          TEMPERATURE =   27.000 DEG C

 FREQ         VM(4,0)

  1.000E+03   2.667E+00
```

FIG. 18.95

### Current-Controlled Voltage Source (CCVS)

| HCCVS | 6 | 5 | VSENSE | 4 |
|---|---|---|---|---|
| Name | + node | − node | Defines direction of controlling variable | Multiplying factor (units: ohms) |
| | Polarity of controlled source | | | |

## PROBLEMS

### SECTION 18.2 Superposition Theorem

1. Using superposition, for each network of Fig. 18.96, determine the current through the inductance $X_L$.

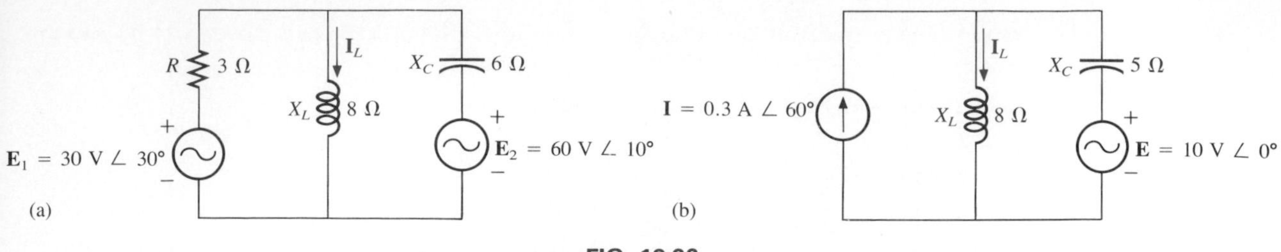

FIG. 18.96

*2. Repeat Problem 1 for the networks of Fig. 18.97.

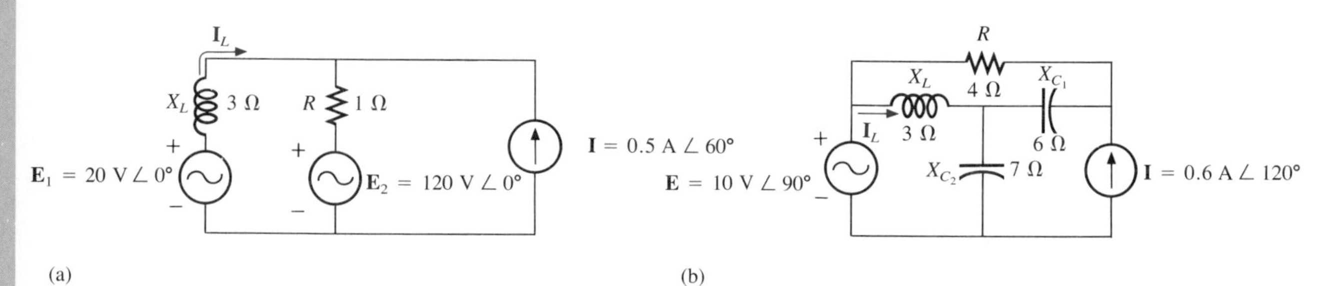

FIG. 18.97

*3. Using superposition, find the current **I** for the network of Fig. 18.98.

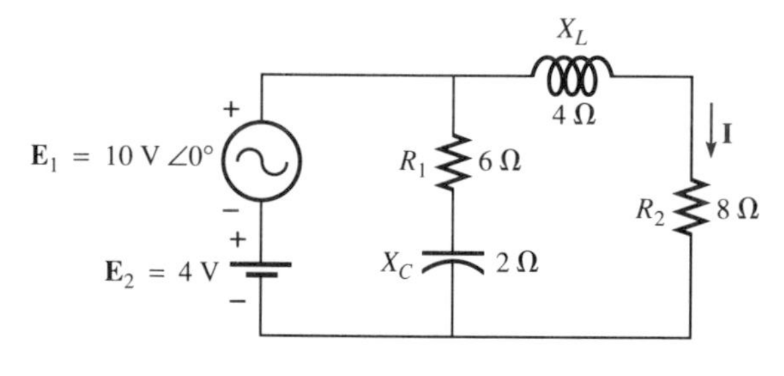


FIG. 18.98

4. Using superposition, find the voltage $\mathbf{V}_C$ for the network of Fig. 18.99.

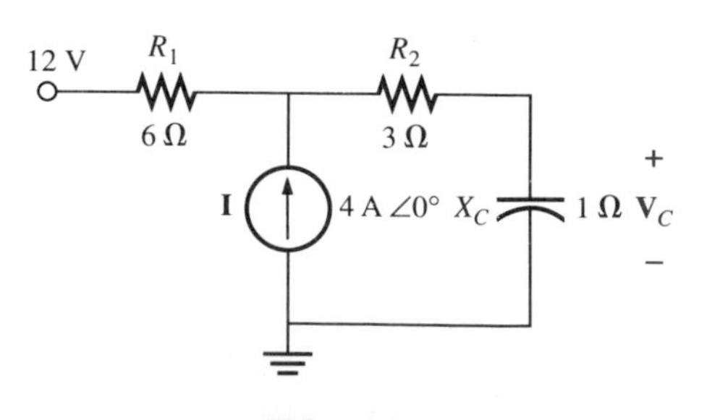


FIG. 18.99

*5. Using superposition, find the current **I** for the network of Fig. 18.100.

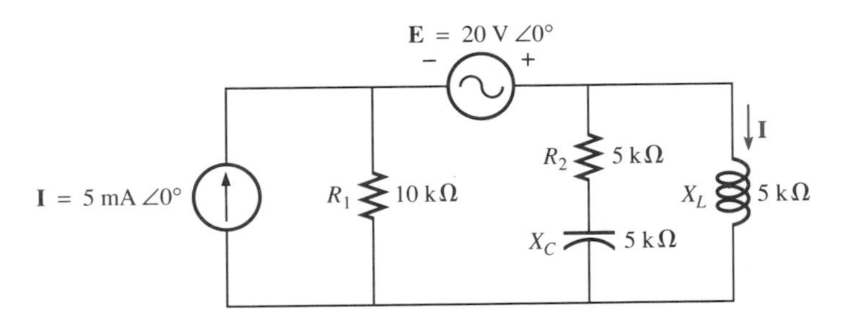


FIG. 18.100

6. Using superposition, for the network of Fig. 18.101, determine the current $\mathbf{I}_L$ ($h = 100$).

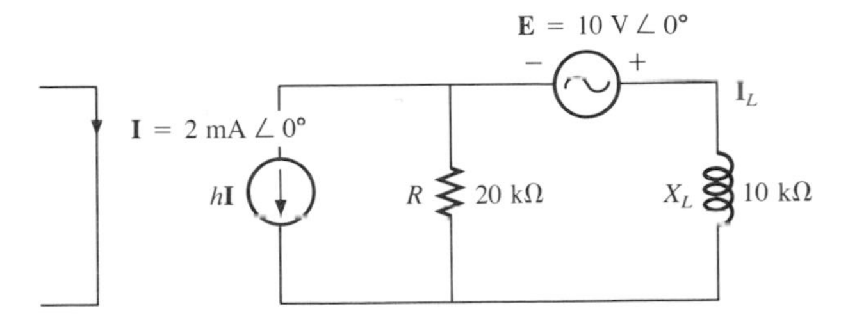


FIG. 18.101

7. Using superposition, for the network of Fig. 18.102, determine the voltage $\mathbf{V}_L$ ($\mu = 20$).

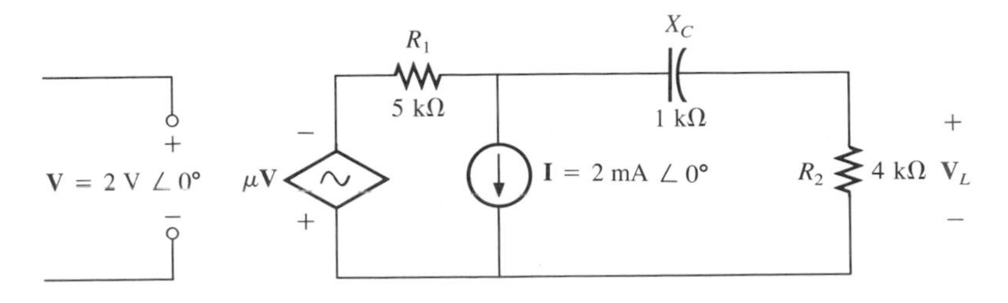


FIG. 18.102

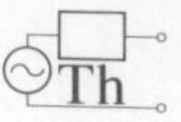


***8.** Using superposition, determine the current $\mathbf{I}_L$ for the network of Fig. 18.103 ($\mu = 20$; $h = 100$).

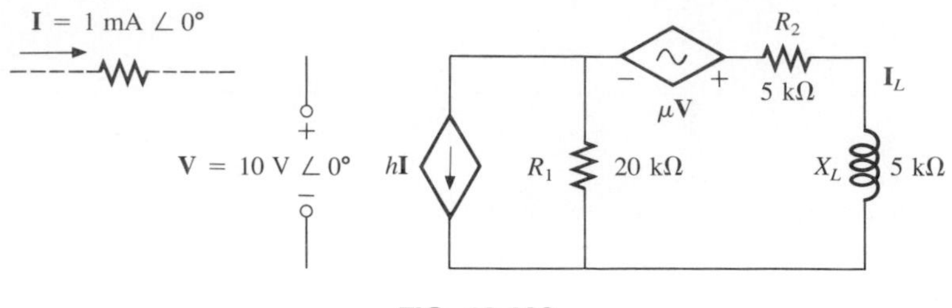

FIG. 18.103

***9.** Determine $\mathbf{V}_L$ for the network of Fig. 18.104 ($h = 50$).

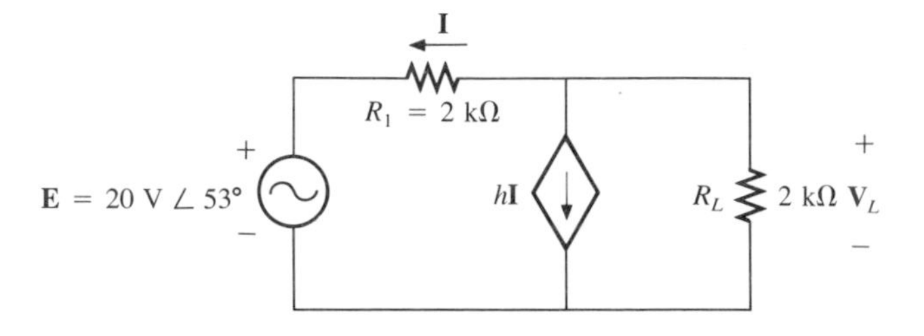


FIG. 18.104

***10.** Calculate the current **I** for the network of Fig. 18.105.

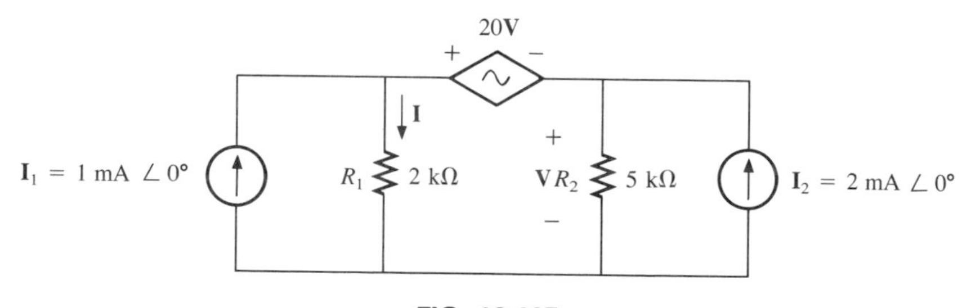


FIG. 18.105

## SECTION 18.3 Thevenin's Theorem

**11.** Find the Thevenin equivalent circuit for the portions of the networks of Fig. 18.106 external to the elements between points $a$ and $b$.

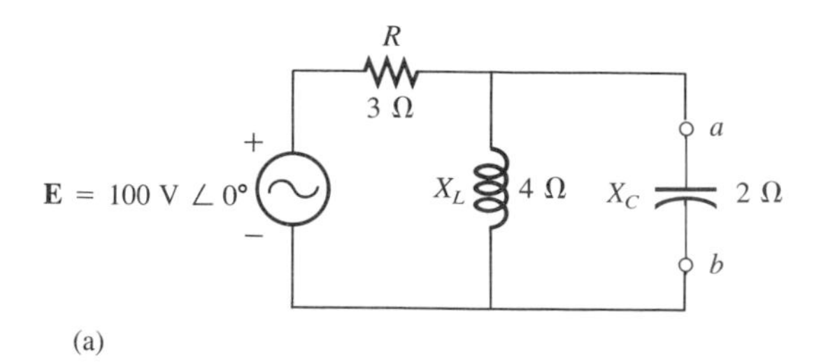


(a)

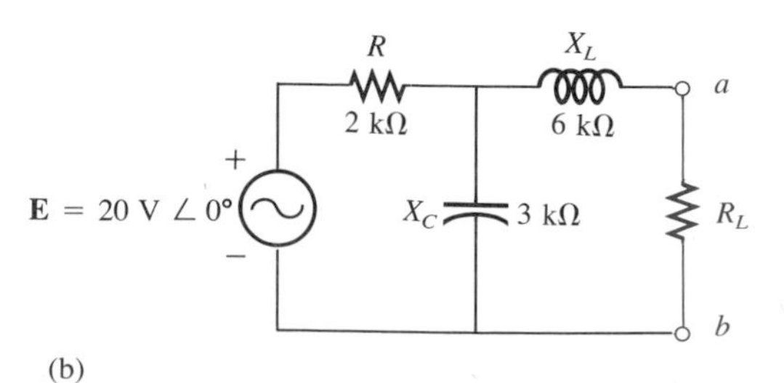


(b)

FIG. 18.106

***12.** Find the Thevenin equivalent circuit for the portions of the networks of Fig. 18.107 external to the elements between points $a$ and $b$.

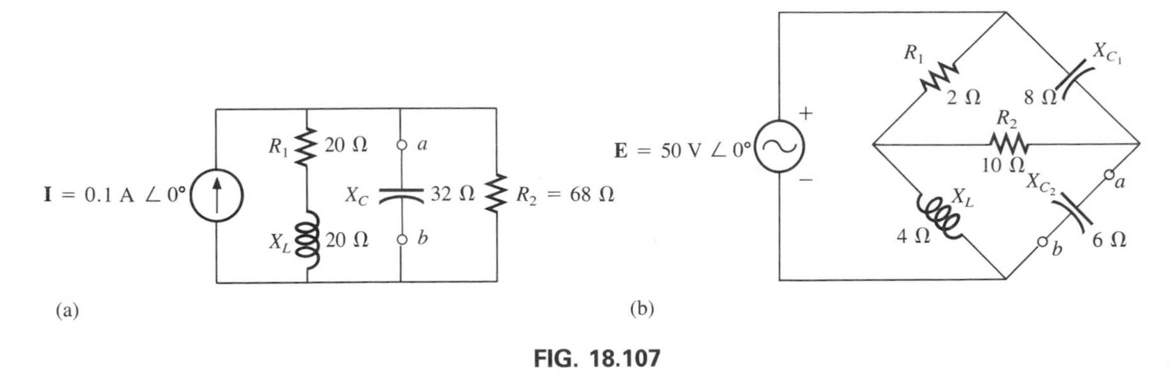


**FIG. 18.107**

***13.** Find the Thevenin equivalent circuit for the portions of the networks of Fig. 18.108 external to the elements between points $a$ and $b$.

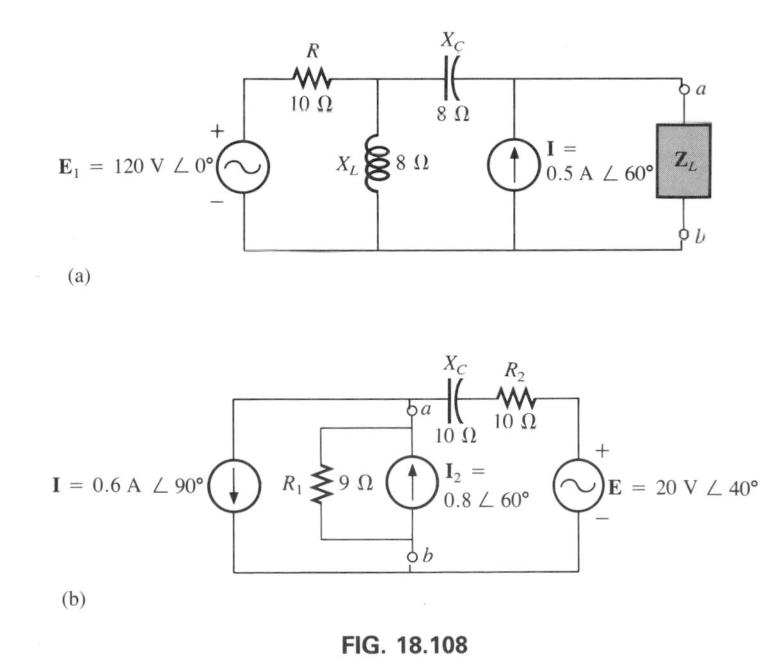


**FIG. 18.108**

***14. a.** Find the Thevenin equivalent circuit for the network external to the resistor $R_2$ in Fig. 18.98.

**b.** Using the results of part (a) determine the current **I** of the same figure.

**15. a.** Find the Thevenin equivalent circuit for the network external to the capacitor of Fig. 18.99.

**b.** Using the results of part (a) determine the voltage $\mathbf{V}_C$ for the same figure.

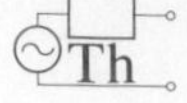


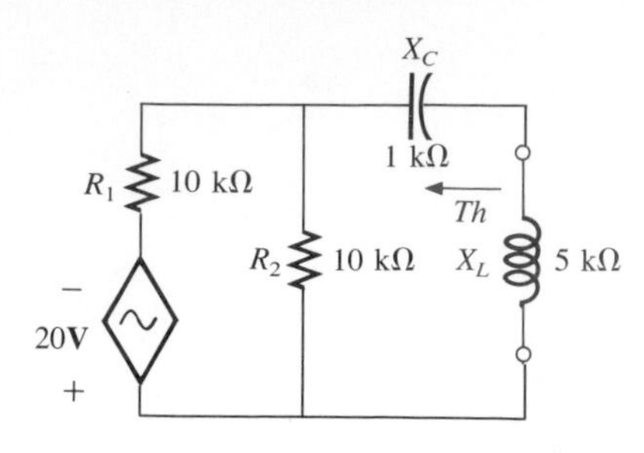

FIG. 18.109

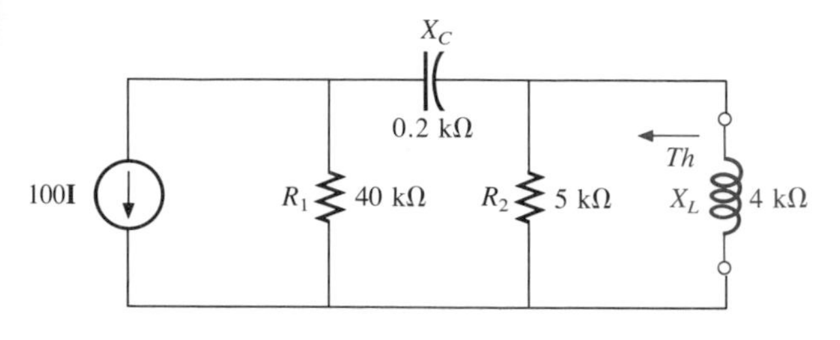


FIG. 18.110

***16. a.** Find the Thevenin equivalent circuit for the network external to the inductor of Fig. 18.100.

**b.** Using the results of part (a) determine the current **I** of the same figure.

**17.** Determine the Thevenin equivalent circuit for the network external to the 5-kΩ inductive reactance of Fig. 18.109 (in terms of **V**).

**18.** Determine the Thevenin equivalent circuit for the network external to the 4-kΩ inductive reactance of Fig. 18.110 (in terms of **I**).

**19.** Find the Thevenin equivalent circuit for the network external to the 10-kΩ inductive reactance of Fig. 18.101.

**20.** Determine the Thevenin equivalent circuit for the network external to the 4-kΩ resistor of Fig. 18.102.

***21.** Find the Thevenin equivalent circuit for the network external to the 5-kΩ inductive reactance of Fig. 18.103.

***22.** Determine the Thevenin equivalent circuit for the network external to the 2-kΩ resistor of Fig. 18.104.

***23.** Find the Thevenin equivalent circuit for the network external to the resistor $R_1$ of Fig. 18.105.

## SECTION 18.4 Norton's Theorem

**24.** Find the Norton equivalent circuit for the portions of the networks of Fig. 18.106 external to the elements between points $a$ and $b$.

**25.** Repeat Problem 24 for the networks of Fig. 18.107.

***26.** Repeat Problem 24 for the networks of Fig. 18.108.

***27.** Find the Norton equivalent circuit for the portions of the networks of Fig. 18.111 external to the elements between points $a$ and $b$.

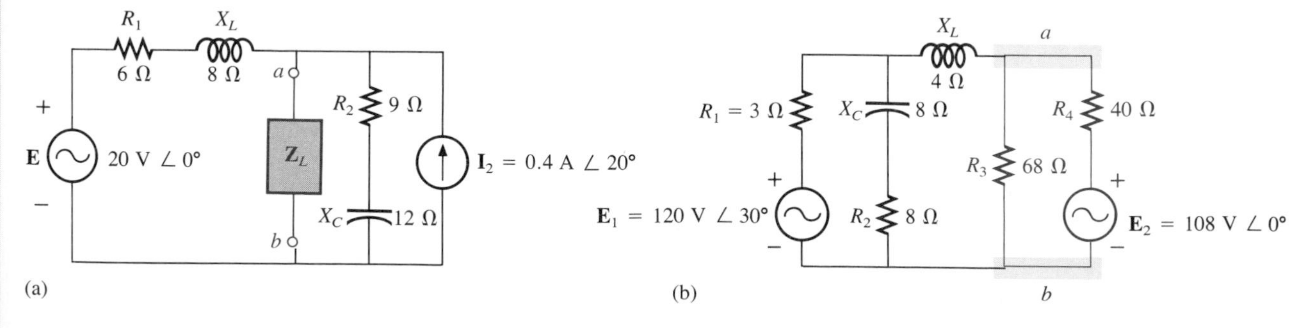

FIG. 18.111

***28. a.** Find the Norton equivalent circuit for the network external to the resistor $R_2$ in Fig. 18.98.

**b.** Using the results of part (a) determine the current **I** of the same figure.

*29. **a.** Find the Norton equivalent circuit for the network external to the capacitor of Fig. 18.99.
**b.** Using the results of part (a) determine the voltage $\mathbf{V}_C$ for the same figure.

*30. **a.** Find the Norton equivalent circuit for the network external to the inductor of Fig. 18.100.
**b.** Using the results of part (a) determine the current **I** of the same figure.

**31.** Determine the Norton equivalent circuit for the network external to the 5-kΩ inductive reactance of Fig. 18.109.

**32.** Determine the Norton equivalent circuit for the network external to the 4-kΩ inductive reactance of Fig. 18.110.

**33.** Find the Norton equivalent circuit for the network external to the 4-kΩ resistor of Fig. 18.102.

*34. Find the Norton equivalent circuit for the network external to the 5-kΩ inductive reactance of Fig. 18.103.

*35. For the network of Fig. 18.112, find the Norton equivalent circuit for the network external to the 2-kΩ resistor.

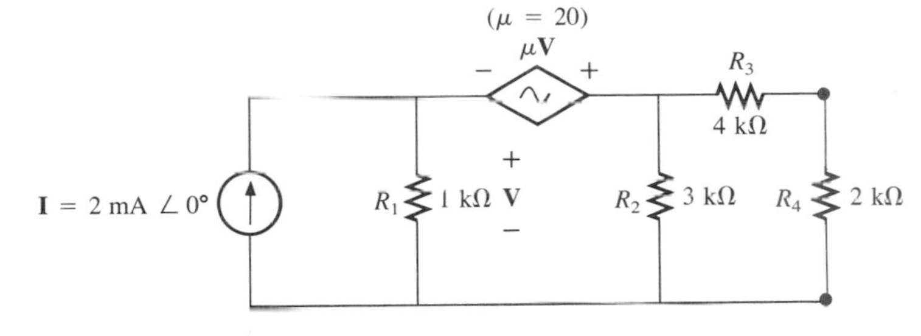

FIG. 18.112

*36. Find the Norton equivalent circuit for the network external to the $\mathbf{I}_1$ current source of Fig. 18.105.

## SECTION 18.5 Maximum Power Transfer Theorem

**37.** Find the load impedance $\mathbf{Z}_L$ for the networks of Fig. 18.113 for maximum power to the load, and find the maximum power to the load.

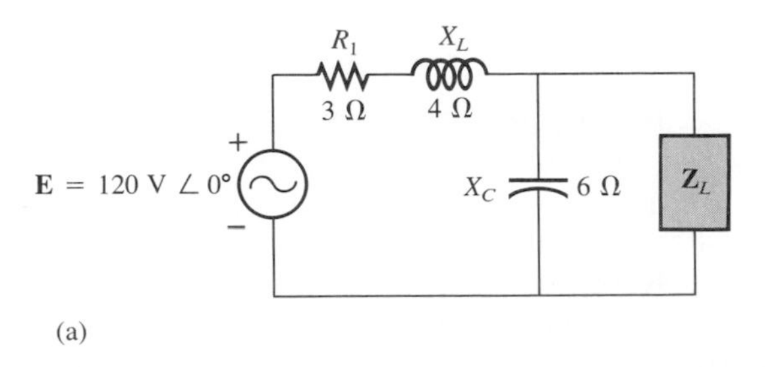

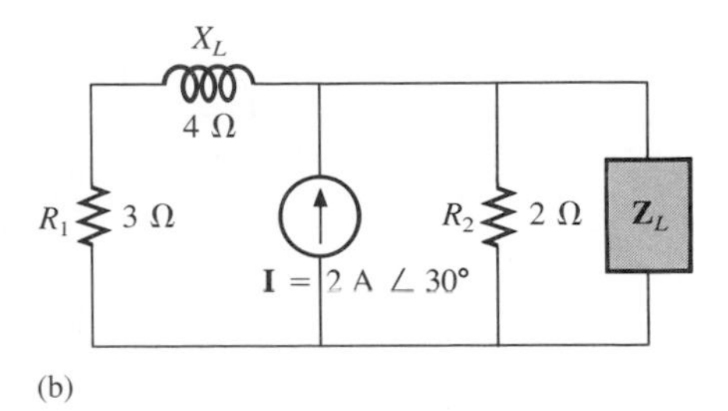


FIG. 18.113

*38. Find the load impedance $\mathbf{Z}_L$ for the networks of Fig. 18.114 for maximum power to the load, and find the maximum power to the load.

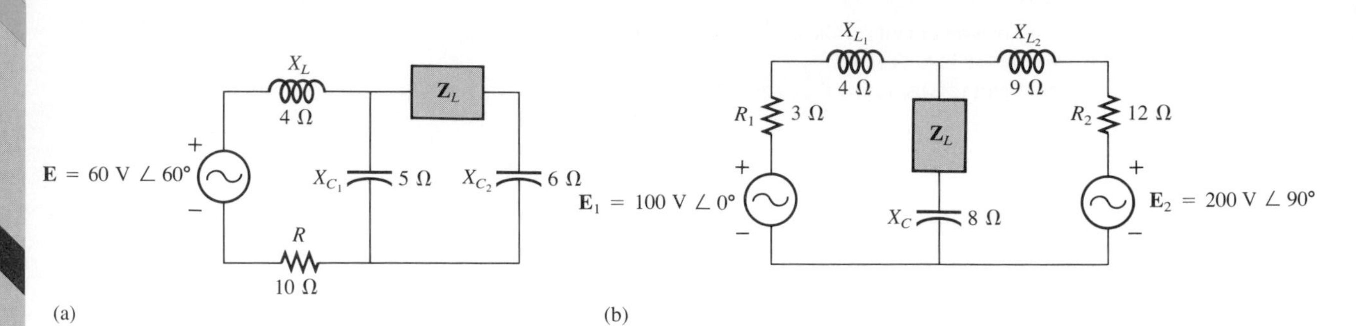


FIG. 18.114

39. Find the load impedance $R_L$ for the network of Fig. 18.115 for maximum power to the load, and find the maximum power to the load.

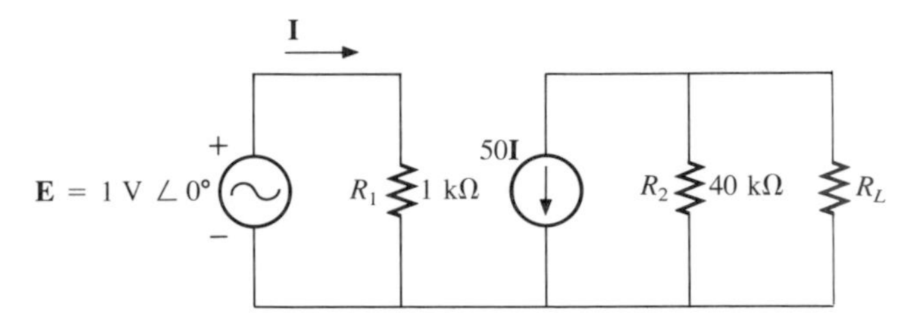

FIG. 18.115

*40. a. Determine the load impedance to replace the resistor $R_2$ of Fig. 18.98 to ensure maximum power to the load.
b. Using the results of part (a), determine the maximum power to the load.

*41. a. Determine the load impedance to replace the capacitor $X_C$ of Fig. 18.99 to ensure maximum power to the load.
b. Using the results of part (a), determine the maximum power to the load.

*42. a. Determine the load impedance to replace the inductor $X_L$ of Fig. 18.100 to ensure maximum power to the load.
b. Using the results of part (a), determine the maximum power to the load.

43. a. For the network of Fig. 18.116, determine the value of $R_L$ that will result in maximum power to the load.
b. Using the results of part (a), determine the maximum power delivered.

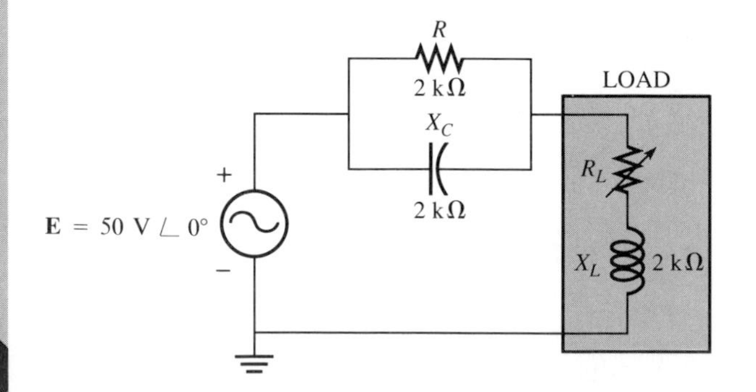


FIG. 18.116

*44. **a.** For the network of Fig. 18.117 determine the level of capacitance that will ensure maximum power to the load if the range of capacitance is limited to 1 to 5 nF.
**b.** Using the results of part (a), determine the value of $R_L$ that will ensure maximum power to the load.
**c.** Using the results of parts (a) and (b) determine the maximum power to the load.

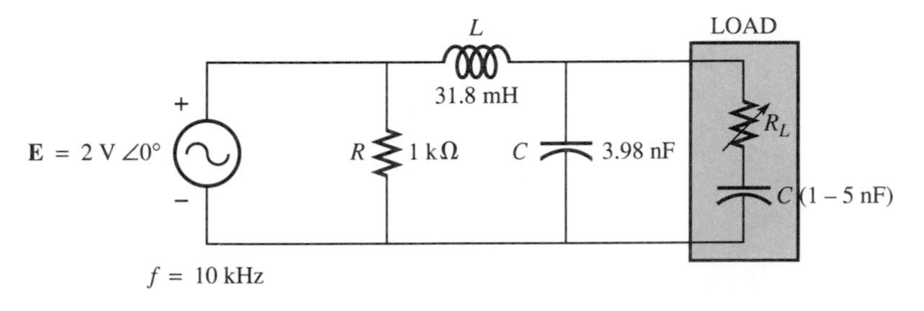

FIG. 18.117

## SECTION 18.6 Substitution, Reciprocity, and Millman's Theorems

**45.** For the network of Fig. 18.118, determine two equivalent branches through the substitution theorem for the branch *a-b*.

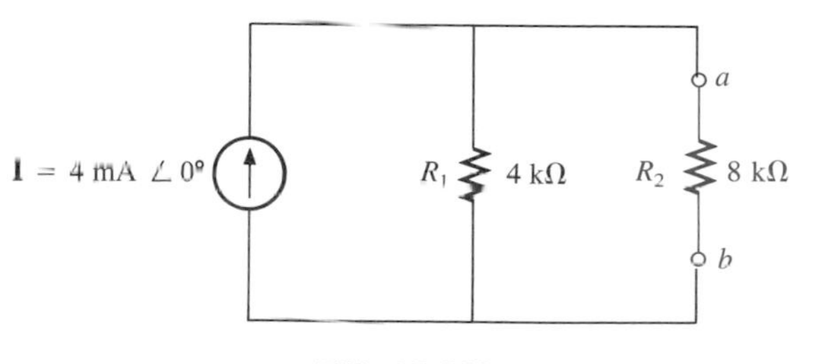

FIG. 18.118

**46. a.** For the network of Fig. 18.119(a), find the current **I**.
**b.** Repeat part (a) for the network of Fig. 18.119(b).
**c.** Do the results of parts (a) and (b) compare?

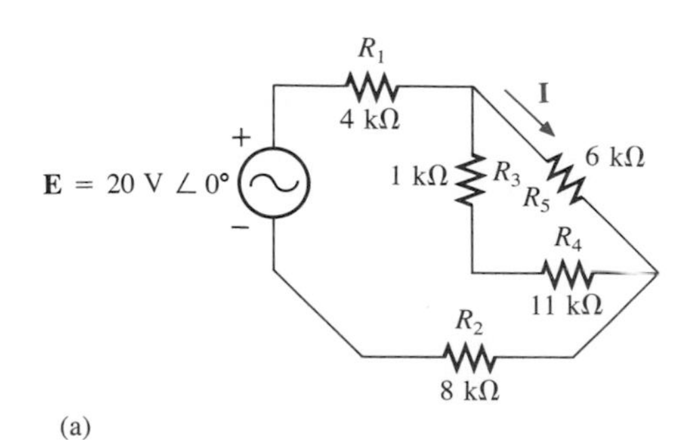


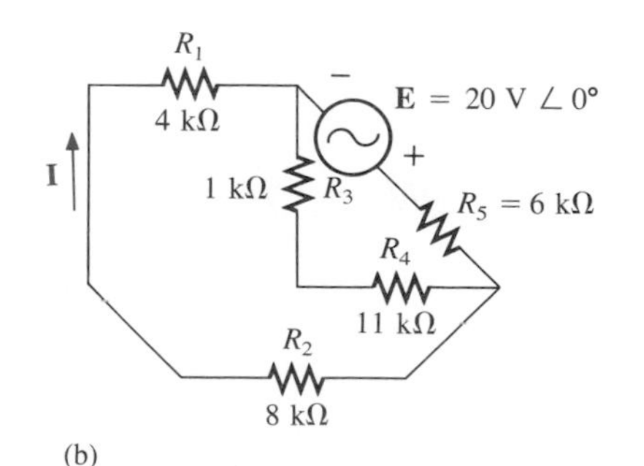


FIG. 18.119

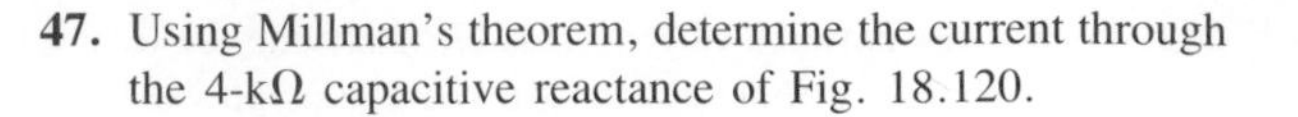

**47.** Using Millman's theorem, determine the current through the 4-kΩ capacitive reactance of Fig. 18.120.

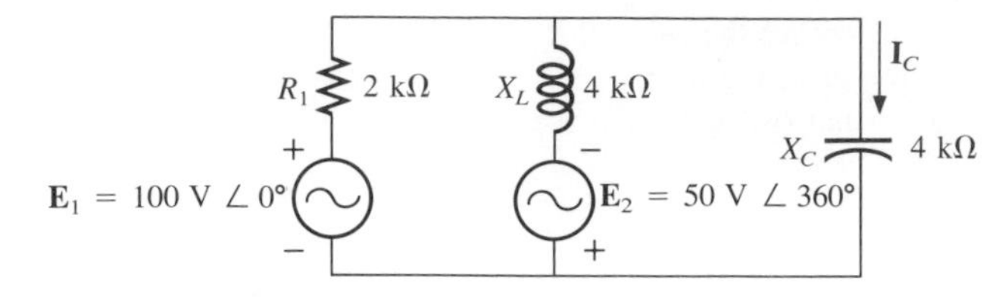

FIG. 18.120

## SECTION 18.7 Computer Analysis

### PSPICE

**48.** Write the input file for the network of Fig. 18.1 that will provide a format for a solution using superposition.

**49.** Write the input file for the network of Fig. 18.12 that will provide a format for a solution using superposition.

***50.** Write the input file to determine the Thevenin voltage and impedance for the network of Fig. 18.107(b).

***51.** Write the input file to determine the Thevenin voltage and impedance for the network of Fig. 18.50 if $R_1 =$ 2 kΩ, $\mu = 50$ and $\mathbf{I} = 10$ mA $\angle 0°$. Compare with the solution obtained using the results of Example 18.12.

***52.** Write the input file to determine the Norton equivalent circuit for the network of Example 18.15.

**53.** Substituting the values of $\mathbf{Z}_L$ for maximum power in the network of Fig. 18.83, write an input file to determine the current through $R_L$.

### BASIC

**54.** Given the network of Fig. 18.1, write a program to determine a general solution for the current **I** using superposition. That is, given the reactance of the same network elements, determine **I** for voltage sources of any magnitude but the same angle.

**55.** Given the network of Fig. 18.23, write a program to determine the Thevenin voltage and impedance for any level of reactance for each element and any magnitude of voltage for the voltage source. The angle of the voltage source should remain at zero degrees.

**56.** Given the configuration of Fig. 18.121, demonstrate that maximum power is delivered to the load when $X_C = X_L$ by tabulating the power to the load for $X_C$ varying from 0.1 kΩ to 2 kΩ in increments of 0.1 kΩ.

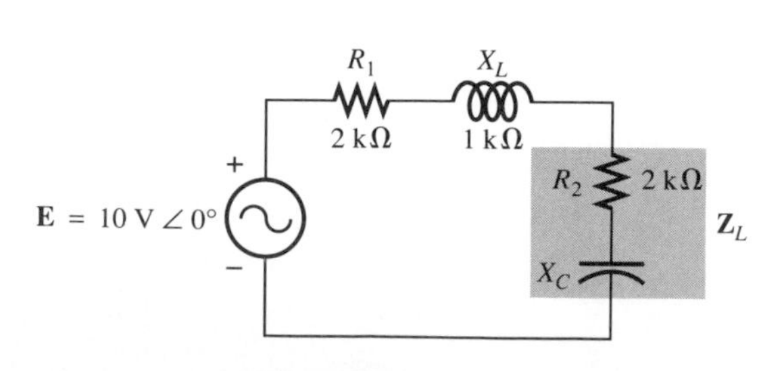


FIG. 18.121

## GLOSSARY

**Current-controlled current source (CCCS)** A current source whose parameters are controlled by a current elsewhere in the system.

**Current-controlled voltage source (CCVS)** A voltage source whose parameters are controlled by a voltage elsewhere in the system.

**Maximum power transfer theorem** A theorem used to determine the load impedance necessary to ensure maximum power to the load.

**Millman's theorem** A method employing voltage-to-current source conversions that will permit the determination of unknown variables in a multiloop network.

**Norton's theorem** A theorem that permits the reduction of any two-terminal linear ac network to one having a single current source and parallel impedance. The resulting configuration can then be employed to determine a particular current or voltage in the original network or to examine the effects of a specific portion of the network on a particular variable.

**Reciprocity theorem** A theorem stating that for single-source networks, the magnitude of the current in any branch of a network, due to a single voltage source anywhere else in the network, will equal the magnitude of the current through the branch in which the source was originally located if the source is placed in the branch in which the current was originally measured.

**Substitution theorem** A theorem stating that if the voltage across and current through any branch of an ac bilateral network are known, the branch can be replaced by any combination of elements that will maintain the same voltage across and current through the chosen branch.

**Superposition theorem** A method of network analysis that permits considering the effects of each source independently. The resulting current and/or voltage is the phasor sum of the currents and/or voltages developed by each source independently.

**Thevenin's theorem** A theorem that permits the reduction of any two-terminal linear ac network to one having a single voltage source and series impedance. The resulting configuration can then be employed to determine a particular current or voltage in the original network or to examine the effects of a specific portion of the network on a particular variable.

**Voltage-controlled current source (VCCS)** A current source whose parameters are controlled by a voltage elsewhere in the system.

**Voltage-controlled voltage source (VCVS)** A voltage source whose parameters are controlled by a voltage elsewhere in the system.

# 19
# Power (ac)

## 19.1 INTRODUCTION

The discussion of power in Chapter 14 included only the average power delivered to an ac network. We will now examine the total power equation in a slightly different form and introduce two additional types of power: *apparent* and *reactive*.

Let us define, for the configuration of Fig. 19.1,

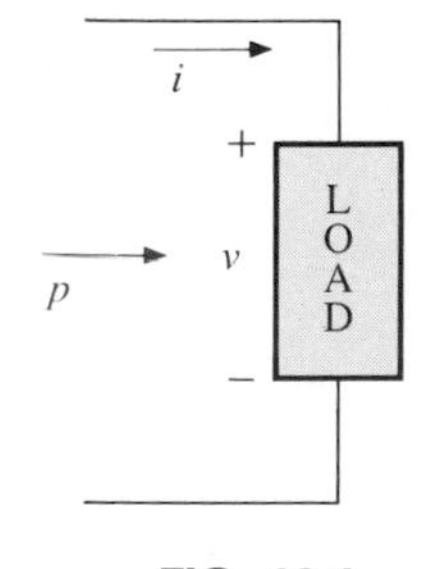


FIG. 19.1

$$v = V_m \sin(\omega t + \theta)$$

and

$$i = I_m \sin \omega t$$

where $\theta$ is the phase angle by which $v$ leads $i$, and since

$$\mathbf{Z} = \frac{V \angle\theta}{I \angle 0^\circ} = \frac{V}{I} \angle\theta = Z \angle\theta$$

it is also the angle associated with the total impedance of the load of Fig. 19.1.

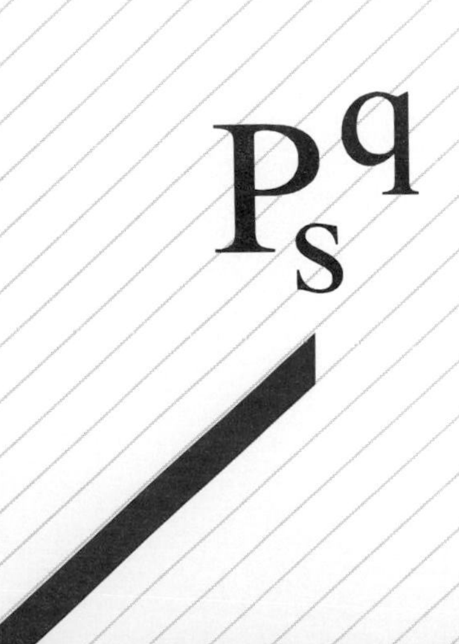

The power delivered to the load of Fig. 19.1 at any instant of time is determined by

$$p = vi$$

Substituting into the above, we have

$$p = V_m I_m \sin \omega t \sin(\omega t + \theta)$$

If we now apply a number of trigonometric identities, we will find that the power equation can also be written

$$\boxed{p = VI \cos \theta (1 - \cos 2\omega t) + VI \sin \theta (\sin 2\omega t)} \qquad \textbf{(19.1)}$$

where $V$ and $I$ are effective values.

It would appear initially that nothing has been gained by putting the equation in this form. However, the usefulness of the form of Eq. (19.1) will be demonstrated in the following sections. The derivation of Eq. (19.1) from the initial form will appear as an assignment at the end of the chapter.

If Eq. (19.1) is expanded to the form

$$p = \underbrace{VI \cos \theta}_{\text{Average}} - \underbrace{VI \cos \theta}_{\text{Peak}} \cos \underbrace{2\omega t}_{2x} + \underbrace{VI \sin \theta}_{\text{Peak}} \sin \underbrace{2\omega t}_{2x}$$

there are two obvious points that can be made. First, the average power still appears as an isolated term that is time independent. Second, both terms that follow vary at a frequency twice that of the applied voltage or current with peak values having a very similar format.

In an effort to insure completeness and order in presentation, each basic element ($R$, $L$, and $C$) will first be treated separately.

## 19.2 RESISTIVE CIRCUIT

For a purely resistive circuit (such as that in Fig. 19.2), $v$ and $i$ are in phase, and $\theta = 0°$. Substituting $\theta = 0°$ into Eq. (19.1), we obtain

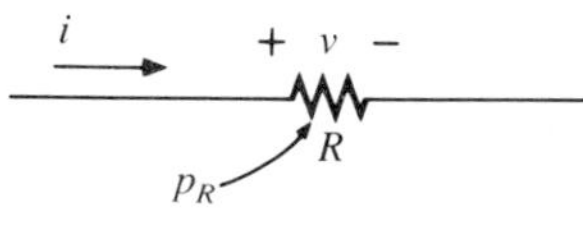


FIG. 19.2

$$\begin{aligned} p_R &= VI \cos(0°)(1 - \cos 2\omega t) + VI \sin(0°) \sin 2\omega t \\ &= VI(1 - \cos 2\omega t) + 0 \end{aligned}$$

or

$$\boxed{p_R = VI - VI \cos 2\omega t} \qquad \textbf{(19.2)}$$

where $VI$ is the average or dc term and $-VI \cos 2\omega t$ is a negative cosine wave with twice the frequency of either input quantity ($v$ or $i$) and a peak value of $VI$.

Plotting the waveform for $p_R$ (Fig. 19.3), we see that

$$T_1 = \text{period of input quantities}$$

$$T_2 = \text{period of power curve } p_R$$

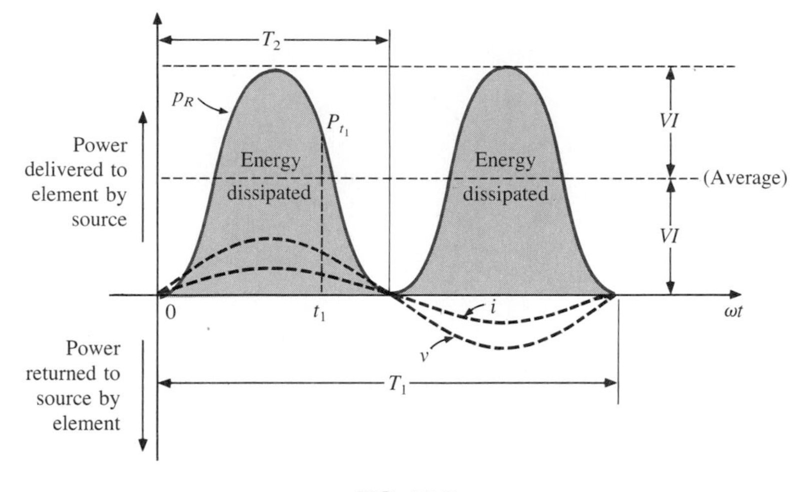


**FIG. 19.3**

Note that in Fig. 19.3 the power curve passes through two cycles about its average value of $VI$ for each cycle of either $v$ or $i$ ($T_1 = 2T_2$ or $f_2 = 2f_1$). Consider also that since the peak and average values of the power curve are the same, the curve is always above the horizontal axis. This indicates that

***the total power delivered to a resistor will be dissipated in the form of heat.***

The power returned to the source is represented by the portion of the curve below the axis, which in this case is zero. The power dissipated at any instant of time $t_1$ by the resistor can be found by substituting $t_1$ into Eq. (19.2). The average power from Eq. (19.2), or Fig. 19.3, is $VI$; or, as a summary,

$$P = VI = \frac{V_m I_m}{2} = I^2R = \frac{V^2}{R} \qquad \text{(watts, W)} \qquad \textbf{(19.3)}$$

as derived in Chapter 14.

The energy dissipated by the resistor ($W_R$) *over one full cycle of the power curve* (Fig. 19.3) can be found using the following equation:

$$W_R = \int_0^{T_2} p_R \, dt \qquad \text{(joules, J)}$$
$$= \text{area under the power curve from } 0 \text{ to } T_2 \text{ (period of } p_R) \qquad \textbf{(19.4)}$$

The area under the curve = (average value) × (length of the curve), and

$$W_R = (VI) \times (T_2)$$

or

$$\boxed{W_R = VIT_2} \qquad \text{(J)} \qquad \textbf{(19.5)}$$

or, since $T_2 = 1/f_2$, where $f_2$ is the frequency of the $p_R$ curve,

$$\boxed{W_R = \frac{VI}{f_2}} \qquad \text{(J)} \qquad \textbf{(19.6)}$$

## 19.3 APPARENT POWER

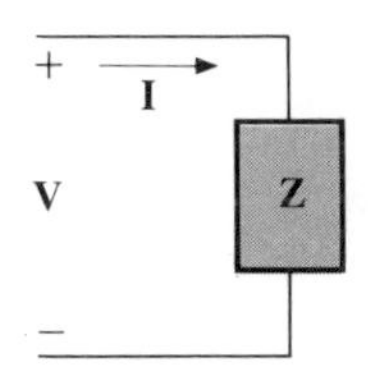


FIG. 19.4

From our analysis of dc networks (and resistive elements above), it would seem *apparent* that the power delivered to the load of Fig. 19.4 is simply determined by the product of the applied voltage and current, with no concern for the components of the load. That is, $P = VI$. However, we found in Chapter 14 that the power factor ($\cos\theta$) of the load will have a pronounced effect on the power dissipated, less pronounced for more reactive loads. Although the product of the voltage and current is not always the power delivered, it is a power rating of significant usefulness in the description and analysis of sinusoidal ac networks and in the maximum rating of a number of electrical components and systems. It is called the *apparent power* and is represented symbolically by $S$.* Since it is simply the product of voltage and current, its units are *volt-amperes,* for which the abbreviation is VA. Its magnitude is determined by

$$\boxed{S = VI} \qquad \text{(VA)} \qquad \textbf{(19.7)}$$

or, since

$$V = IZ \quad \text{and} \quad I = \frac{V}{Z}$$

then

$$\boxed{S = I^2Z} \qquad \text{(VA)} \qquad \textbf{(19.8)}$$

and

$$\boxed{S = \frac{V^2}{Z}} \qquad \text{(VA)} \qquad \textbf{(19.9)}$$

The average power to a system is defined by the equation

$$P = VI\cos\theta$$

---

*Prior to 1968, the symbol for apparent power was the more descriptive $P_a$.

However,

$$S = VI$$

Therefore,

$$\boxed{P = S\cos\theta} \qquad \text{(W)} \qquad \textbf{(19.10)}$$

or the power factor of a system $F_p$ is

$$\boxed{F_p = \cos\theta = \frac{P}{S}} \qquad \textbf{(19.11)}$$

The power factor of a circuit, therefore, is the ratio of the average power to the apparent power. For a purely resistive circuit, we have

$$P = VI = S$$

and

$$F_p = \cos\theta = \frac{P}{S} = 1$$

In general, electric equipment is rated in volt-amperes (VA) or in kilovolt-amperes (kVA), not in watts. By knowing the volt-ampere rating and the rated voltage of a device, we can readily determine the *maximum* current rating. For example, a device rated at 10 kVA at 200 V has a maximum current rating of $I = 10{,}000\text{ VA}/200\text{ V} = 50\text{ A}$ when operated under rated conditions. The volt-ampere rating of a piece of equipment is equal to the wattage rating only when the $F_p$ is 1. It is therefore a maximum power dissipation rating. This condition exists only when the total impedance of a system $Z\angle\theta$ is such that $\theta = 0°$.

The exact current demand of a device, when used under normal operating conditions, could be determined if the wattage rating and power factor were given instead of the volt-ampere rating. However, the power factor is sometimes not available, or it may vary with the load.

The reason for rating electrical equipment in kilovolt-amperes rather than in kilowatts is obvious from the configuration of Fig. 19.5. The

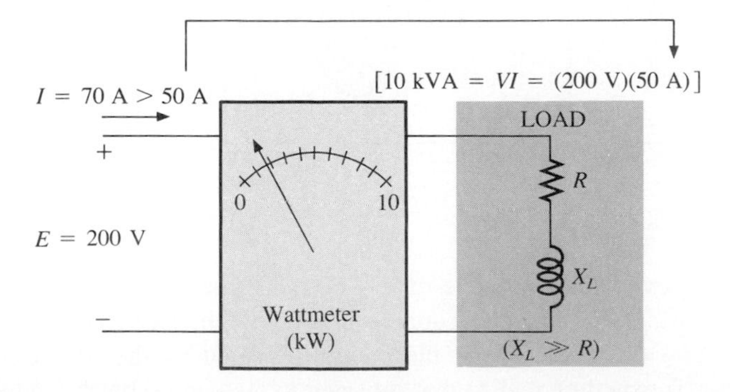


**FIG. 19.5**

load has an apparent power rating of 10 kVA and a current rating of 50 A at the applied voltage, 200 V. As indicated, the current demand is above the rated value and could damage the load element, yet the reading on the wattmeter is very low since the load is highly reactive. In other words, the wattmeter reading is an indication not of the current drawn but simply of the watts dissipated. Theoretically, if the load were purely reactive, the wattmeter reading could be zero, and the device burning up due to the high current demand.

## 19.4 INDUCTIVE CIRCUIT AND REACTIVE POWER

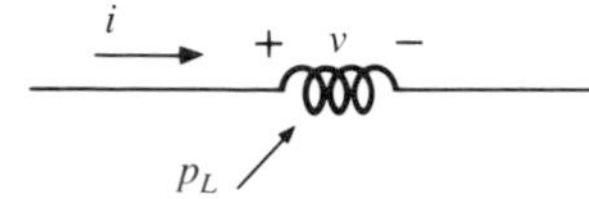


FIG. 19.6

For a purely inductive circuit (such as that in Fig. 19.6), $v$ leads $i$ by 90°. Therefore, in Eq. (19.1), $\theta = 90°$. Substituting $\theta = 90°$ into Eq. (19.1) yields

$$\begin{aligned} p_L &= VI\cos(90°)(1 - \cos 2\omega t) + VI\sin(90°)(\sin 2\omega t) \\ &= 0 + VI\sin 2\omega t \end{aligned}$$

or

$$\boxed{p_L = VI\sin 2\omega t} \qquad \textbf{(19.12)}$$

where $VI \sin 2\omega t$ is a sine wave with twice the frequency of either input quantity ($v$ or $i$) and a peak value of $VI$. Note the absence of an average or constant term in the equation.

Plotting the waveform for $p_L$ (Fig. 19.7), we obtain

$$T_1 = \text{period of either input quantity}$$
$$T_2 = \text{period of } p_L \text{ curve}$$

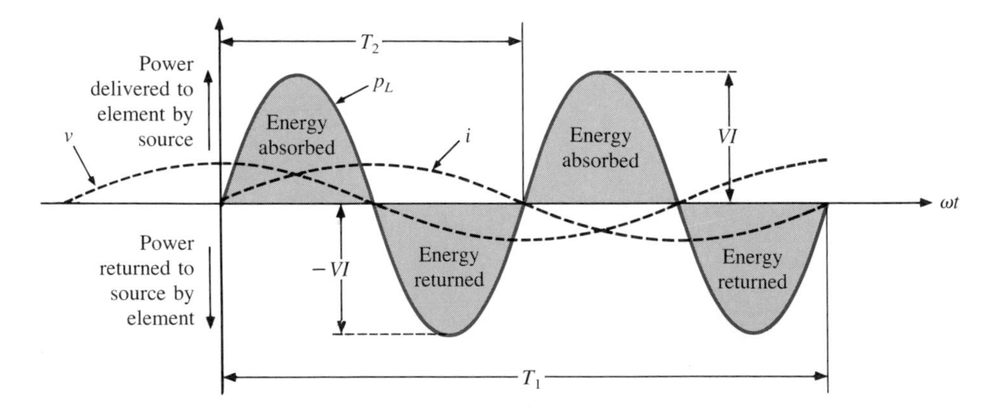

FIG. 19.7

Note that over one full cycle of $p_L$ ($T_2$), the area above the horizontal axis in Fig. 19.7 is exactly equal to that below the axis. This indicates that over a full cycle of $p_L$, the power delivered by the source to the inductor is exactly equal to that returned to the source by the inductor.

***The net flow of power to the pure inductor is zero over a full cycle, and no energy is lost in the transaction.***

The power absorbed or returned by the inductor at any instant of time $t_1$ can be found simply by substituting $t_1$ into Eq. (19.12). The peak value of the curve $VI$ is defined as the *reactive power* associated with a pure inductor.

In general, the reactive power associated with any circuit is defined to be $VI \sin \theta$, a factor appearing in the second term of Eq. (19.1). Note that it is the peak value of that term of the total power equation that produces no net transfer of energy. The symbol for reactive power is $Q$, and its unit of measure is the *volt-ampere reactive* (VAR).* The $Q$ is derived from the quadrature (90°) relationship between the various powers to be discussed in detail in a later section. Therefore,

$$\boxed{Q = VI \sin \theta} \qquad \text{(VAR)} \qquad \textbf{(19.13)}$$

where $\theta$ is the phase angle between $V$ and $I$.

For the inductor,

$$\boxed{Q_L = VI} \qquad \text{(VAR)} \qquad \textbf{(19.14)}$$

or, since $V = IX_L$ or $I = V/X_L$,

$$\boxed{Q_L = I^2 X_L} \qquad \text{(VAR)} \qquad \textbf{(19.15)}$$

or

$$\boxed{Q_L = \frac{V^2}{X_L}} \qquad \text{(VAR)} \qquad \textbf{(19.16)}$$

The apparent power associated with an inductor is $S = VI$, and the average power is $P = 0$, as noted in Fig. 19.7. The power factor is therefore

$$F_p = \cos \theta = \frac{P}{S} = \frac{0}{VI} = 0$$

If the average power is zero, and the energy supplied is returned within one cycle, why is reactive power of any significance? The reason is not obvious but can be explained using the curve of Fig. 19.7. At every instant of time along the power curve that the curve is above the axis (positive), energy must be supplied to the inductor even though it will be returned during the negative portion of the cycle. This power requirement during the positive portion of the cycle requires that the generating plant provide this energy during that interval. Therefore, the

---

*Prior to 1968, the symbol for reactive power was the more descriptive $P_q$.

effect of reactive elements such as the inductor can be to raise the power requirement of the generating plant even though the reactive power is not dissipated but simply ''borrowed.'' The increased power demand during these intervals is a cost factor that must be passed on to the industrial consumer. In fact, most larger users of electrical energy pay for the apparent power demand rather than the watts dissipated since the volt-amperes used are sensitive to the reactive power requirement (see Section 19.6). In other words, the closer the power factor of an industrial outfit is to 1, the more efficient is the plant's operation, since it is limiting its use of ''borrowed'' power.

The energy stored by the inductor during the positive portion of the cycle (Fig. 19.7) is equal to that returned during the negative portion and can be determined using the integral

$$W_L = \int_{T_2/2}^{T_2} p_L \, dt \qquad \text{(J)}$$

$$= \text{area under the power curve from } T_2/2 \text{ to } T_2$$

Recall from Chapter 14 that the average value of the positive portion of a sinusoid equals 2(peak value/$\pi$), and that the area under any curve equals (average value) × (length of the curve):

$$W_L = \left(\frac{2VI}{\pi}\right) \times \left(\frac{T_2}{2}\right)$$

or

$$\boxed{W_L = \frac{VIT_2}{\pi}} \qquad \text{(J)} \qquad \textbf{(19.17)}$$

or, since $T_2 = 1/f_2$, where $f_2$ is the frequency of the $p_L$ curve, we have

$$\boxed{W_L = \frac{VI}{\pi f_2}} \qquad \text{(J)} \qquad \textbf{(19.18)}$$

Since the frequency $f_2$ of the power curve is twice that of the input quantity, if we substitute the frequency $f_1$ of the input voltage or current, Eq. (19.18) becomes

$$W_L = \frac{VI}{\pi(2f_1)}$$

or

$$\boxed{W_L = \frac{VI}{\omega_1}} \qquad \text{(J)} \qquad \textbf{(19.19)}$$

However,

$$V = IX_L = I\omega_1 L$$

so

$$W_L = \frac{(I\omega_1 L)I}{\omega_1}$$

or

$$\boxed{W_L = LI^2} \quad \text{(J)} \qquad \textbf{(19.20)}$$

## 19.5 CAPACITIVE CIRCUIT

For a purely capacitive circuit (such as that in Fig. 19.8), $i$ leads $v$ by 90°. Therefore, in Eq. (19.1), $\theta = -90°$. Substituting $\theta = -90°$ into Eq. (19.1), we obtain

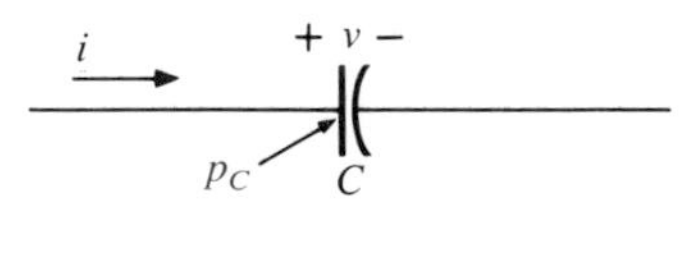

FIG. 19.8

$$\begin{aligned} p_C &= VI\cos(-90°)(1 - \cos 2\omega t) + VI\sin(-90°)(\sin 2\omega t) \\ &= 0 - VI\sin 2\omega t \end{aligned}$$

or

$$\boxed{p_C = -VI\sin 2\omega t} \qquad \textbf{(19.21)}$$

where $-VI\sin 2\omega t$ is a negative sine wave with twice the frequency of either input ($v$ or $i$) and a peak value of $VI$. Again, note the absence of an average or constant term.

Plotting the waveform for $p_C$ (Fig. 19.9) gives us

$$T_1 = \text{period of either input quantity}$$
$$T_2 = \text{period of } p_C \text{ curve}$$

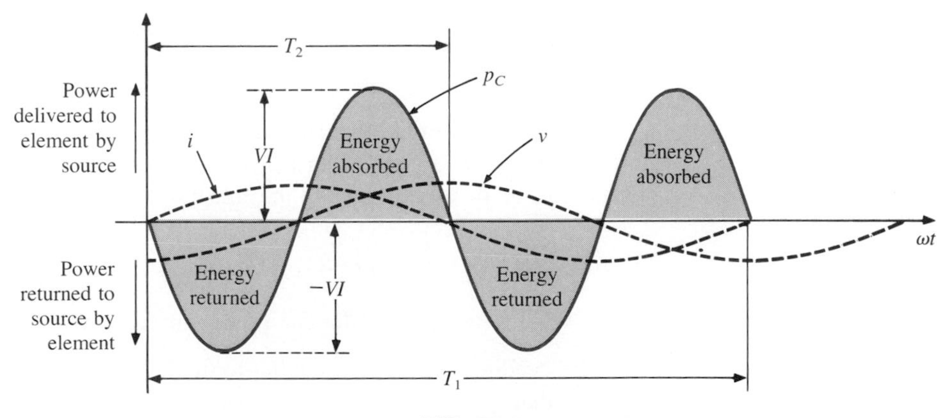


FIG. 19.9

Note that the same situation exists here for the $p_C$ curve as existed for the $p_L$ curve. The power delivered by the source to the capacitor is exactly equal to that returned to the source by the capacitor over one full cycle.

***The net flow of power to the pure capacitor is zero over a full cycle,***

and no energy is lost in the transaction. The power absorbed or returned by the capacitor at any instant of time $t_1$ can be found by substituting $t_1$ into Eq. (19.21).

The reactive power associated with the capacitor is equal to the peak value of the $p_C$ curve, as follows:

$$\boxed{Q_C = VI} \qquad \text{(VAR)} \qquad \textbf{(19.22)}$$

but, since $V = IX_C$ and $I = V/X_C$, the reactive power to the capacitor can also be written

$$\boxed{Q_C = I^2X_C} \qquad \text{(VAR)} \qquad \textbf{(19.23)}$$

and

$$\boxed{Q_C = \frac{V^2}{X_C}} \qquad \text{(VAR)} \qquad \textbf{(19.24)}$$

The apparent power associated with the capacitor is

$$\boxed{S = VI} \qquad \text{(VA)} \qquad \textbf{(19.25)}$$

and the average power is $P = 0$, as noted from Eq. (19.21) or Fig. 19.9. The power factor is therefore

$$F_p = \cos\theta = \frac{P}{S} = \frac{0}{VI} = 0$$

The energy stored by the capacitor during the positive portion of the cycle (Fig. 19.9) is equal to that returned during the negative portion and can be determined using the integral

$$W_C = \int_0^{T_2/2} p_C\,dt \qquad \text{(J)}$$

$$= \text{area under the power curve from 0 to } T_2/2$$

Proceeding in a manner similar to that used for the inductor, we can show that

$$\boxed{W_C = \frac{VIT_2}{\pi}} \qquad \text{(J)} \qquad \textbf{(19.26)}$$

or, since $T_2 = 1/f_2$, where $f_2$ is the frequency of the $p_C$ curve,

$$W_C = \frac{VI}{\pi f_2} \quad \text{(J)} \qquad \textbf{(19.27)}$$

In terms of the frequency $f_1$ of the input quantities $v$ and $i$,

$$W_C = \frac{VI}{\omega_1} \quad \text{(J)} \qquad \textbf{(19.28)}$$

and

$$W_C = CV^2 \quad \text{(J)} \qquad \textbf{(19.29)}$$

## 19.6 THE POWER TRIANGLE

The three quantities—average power, apparent power, and reactive power—can be related in the vector domain by

$$\mathbf{S} = \mathbf{P} + \mathbf{Q} \qquad \textbf{(19.30)}$$

with

$$\mathbf{P} = P \angle 0° \qquad \mathbf{Q}_L = Q_L \angle 90° \qquad \mathbf{Q}_C = Q_C \angle -90°$$

For an inductive load, the *phasor power* **S**, as it is often called, is defined by

$$\mathbf{S} = P + jQ_L$$

as shown in Fig. 19.10.

The 90° shift in $Q_L$ from $P$ is the source of another term for reactive power: *quadrature power*.

For a capacitive load, the phasor power **S** is defined by

$$\mathbf{S} = P - jQ_C$$

as shown in Fig. 19.11.

If a network has both capacitive and inductive elements, the reactive component of the power triangle will be determined by the *difference* between the reactive power delivered to each. If $Q_L > Q_C$, the resultant power triangle will be similar to Fig. 19.10. If $Q_C > Q_L$, the resultant power triangle will be similar to Fig. 19.11.

That the total reactive power is the difference between the reactive powers of the inductive and capacitive elements can be demonstrated by considering Eqs. (19.12) and (19.21). Using these equations, the reactive power delivered to each reactive element has been plotted for a

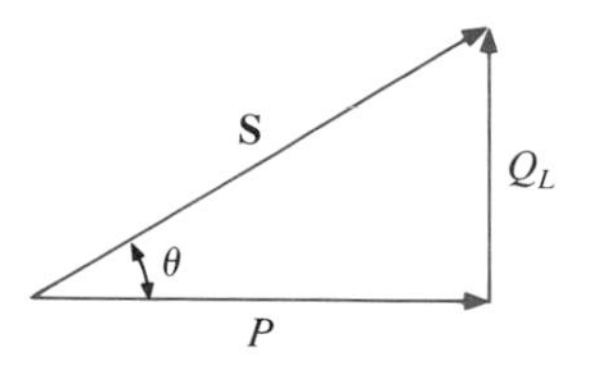


**FIG. 19.10**
*Power diagram for inductive loads.*

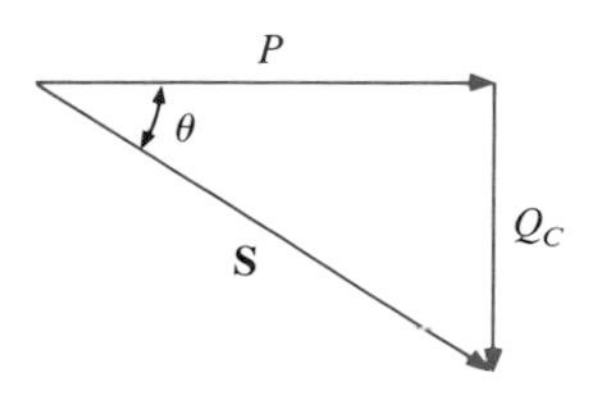


**FIG. 19.11**
*Power diagram for capacitive loads.*

series $L$-$C$ circuit on the same set of axes in Fig. 19.12. The reactive elements were chosen such that $X_L > X_C$. Note that the power curve for

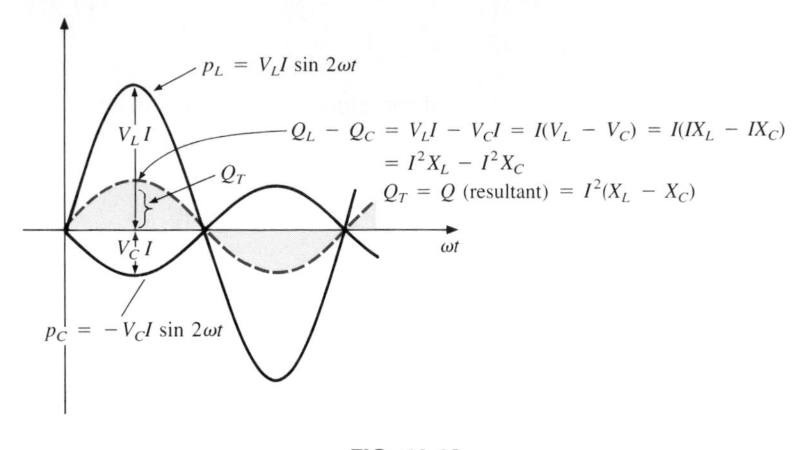

FIG. 19.12

each is exactly 180° out of phase. The curve for the resultant reactive power is therefore determined by the algebraic resultant of the two at each instant of time. Since the reactive power is defined as the peak value, the reactive component of the power triangle is as indicated in the figure: $I^2(X_L - X_C)$.

An additional verification can be derived by first considering the impedance diagram of a series $R$-$L$-$C$ circuit (Fig. 19.13). If we multiply each radius vector by the current squared ($I^2$), we obtain the results shown in Fig. 19.14, which is the power triangle for a predominantly inductive circuit.

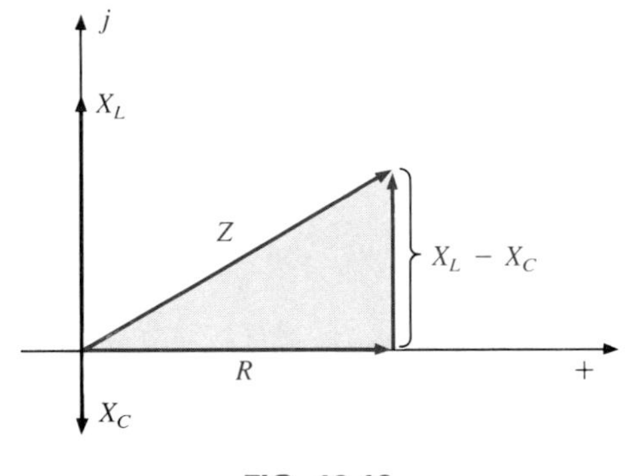


FIG. 19.13

Since the reactive power and average power are always angled 90° to each other, the three powers are related by the Pythagorean theorem; that is,

$$S^2 = P^2 + Q^2 \tag{19.31}$$

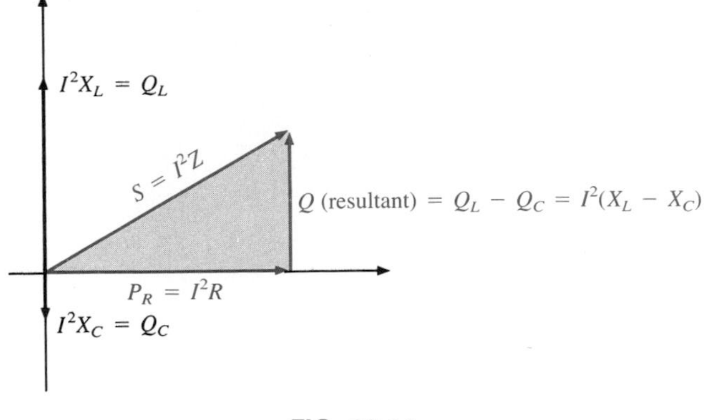


FIG. 19.14

Therefore, the third power can always be found if the other two are known.

It is particularly interesting that the equation

$$\mathbf{S} = \mathbf{VI}^* \tag{19.32}$$

will provide the vector form of the apparent power of a system. Here, **V** is the voltage across the system and **I*** is the complex conjugate of the current.

Consider, for example, the simple *R-L* circuit of Fig. 19.15, where

$$\mathbf{I} = \frac{\mathbf{V}}{\mathbf{Z}_T} = \frac{10 \text{ V} \angle 0^\circ}{3\ \Omega + j4\ \Omega} = \frac{10 \text{ V} \angle 0^\circ}{5\ \Omega \angle 53.13^\circ} = 2 \text{ A} \angle -53.13^\circ$$

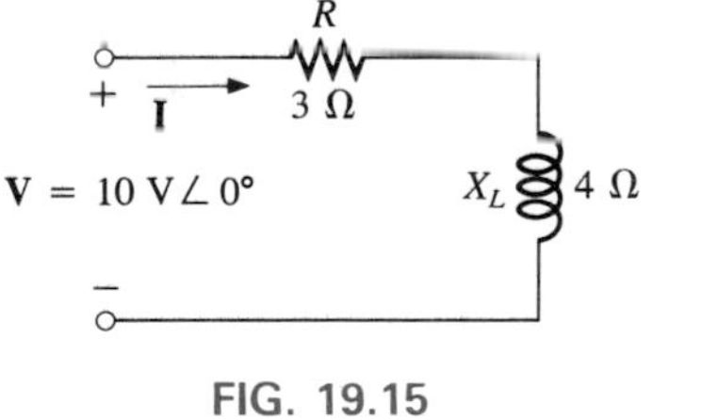


FIG. 19.15

The real power (the term *real* being derived from the positive real axis of the complex plane) is

$$P = I^2R = (2 \text{ A})^2(3\ \Omega) = 12 \text{ W}$$

and the reactive power is

$$Q_L = I^2X_L = (2 \text{ A})^2(4\ \Omega) = 16 \text{ VAR}$$

with

$$\mathbf{S} = P + jQ_L = 12 \text{ W} + j16 \text{ VAR} = 20 \text{ VA} \angle 53.13^\circ$$

as shown in Fig. 19.16. Applying Eq. (19.32) yields

$$\mathbf{S} = \mathbf{VI}^* = (10 \text{ V} \angle 0^\circ)(2 \text{ A} \angle +53.13^\circ) = 20 \text{ VA} \angle 53.13^\circ$$

as obtained above.

The angle $\theta$ associated with **S** and appearing in Figs. 19.10, 19.11, and 19.16 is the power-factor angle of the network. Since

$$P = VI \cos \theta$$

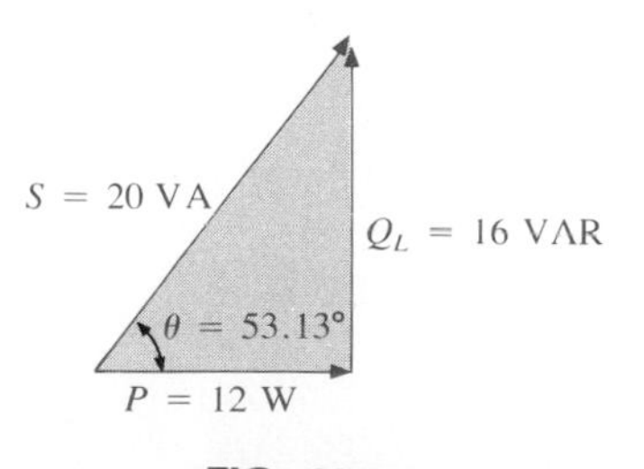

FIG. 19.16

or

$$P = S \cos \theta$$

then

$$F_p = \cos\theta = \frac{P}{S} \tag{19.33}$$

## 19.7 THE TOTAL *P*, *Q*, AND *S*

The total number of watts, volt-amperes reactive, volt-amperes, and the power factor of any system can be found using the following procedure:

1. Find the real power and reactive power for each branch of the circuit.
2. The total real power of the system ($P_T$) is then the sum of the average power delivered to each branch.
3. The total reactive power ($Q_T$) is the difference between the reactive power of the inductive loads and that of the capacitive loads.
4. The total apparent power is $S_T = \sqrt{P_T^2 + Q_T^2}$.
5. The total power factor is $P_T/S_T$.

There are two important points in the above tabulation. First, the total apparent power must be determined from the total average and reactive powers and *cannot* be determined from the apparent powers of each branch. Second, and more important, it is *not necessary* to consider the series-parallel arrangement of branches. In other words, the total real, reactive, or apparent power is independent of whether the loads are in series, parallel, or series-parallel. The following examples will demonstrate the relative ease with which all the quantities of interest can be found.

---

**EXAMPLE 19.1.** Find the total number of watts, volt-amperes reactive, volt-amperes, and the power factor $F_p$ of the network in Fig. 19.17. Draw the power triangle and find the current in phasor form.

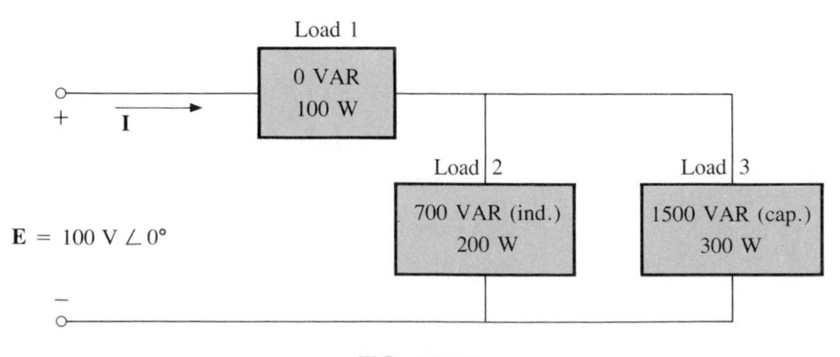

FIG. 19.17

***Solution:*** Use a table:

| Load | W | VAR | VA |
|---|---|---|---|
| 1 | 100 | 0 | 100 |
| 2 | 200 | 700 (ind.) | $\sqrt{(200)^2 + (700)^2} = 728.0$ |
| 3 | 300 | 1500 (cap.) | $\sqrt{(300)^2 + (1500)^2} = 1529.71$ |
| | $P_T =$ **600** Total power dissipated | $Q_T =$ **800 (cap.)** Resultant reactive power of network | $S_T = \sqrt{(600)^2 + (800)^2} =$ **1000** (Note that $S_T \neq$ sum of each branch: $1000 \neq 100 + 728 + 1529.71$) |

$$F_p = \frac{P_T}{S_T} = \frac{600 \text{ W}}{1000 \text{ VA}} = \mathbf{0.6 \text{ leading (cap.)}}$$

The power triangle is shown in Fig. 19.18.

Since $S_T = VI = 1000$ VA, $I = 1000$ VA/100 V $= 10$ A; and since $\theta$ of $\cos\theta = F_p$ is the angle between the input voltage and current,

$$\mathbf{I} = 10 \text{ A } \angle +53.13^\circ$$

The plus sign is associated with the phase angle since the circuit is predominantly capacitive.

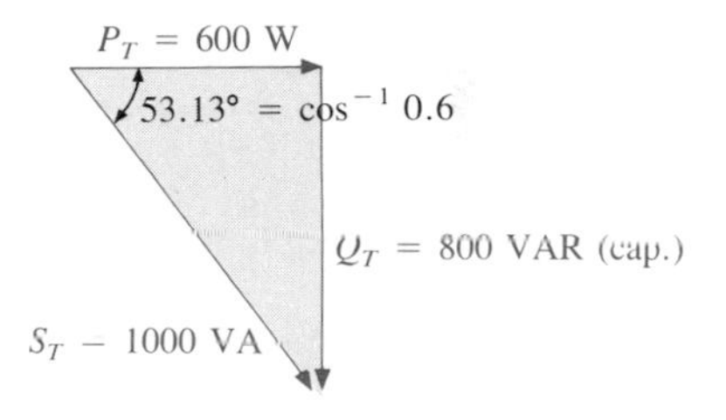

**FIG. 19.18**

**EXAMPLE 19.2.**

a. Find the total number of watts, volt-amperes reactive, volt-amperes, and the power factor $F_p$ for the network of Fig. 19.19.

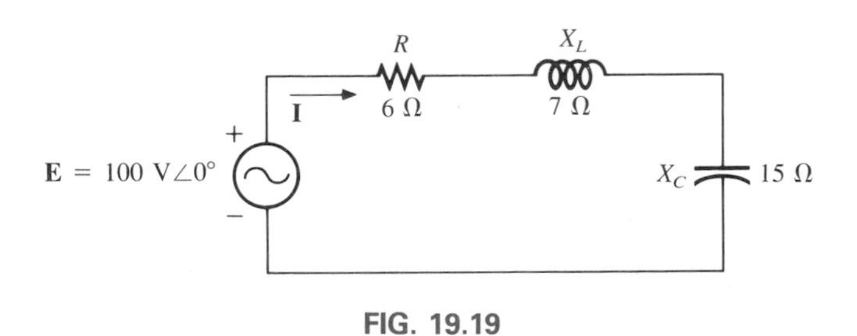


**FIG. 19.19**

b. Sketch the power triangle.
c. Find the energy dissipated by the resistor over one full cycle of the input voltage if the frequency of the input quantities is 60 Hz.
d. Find the energy stored in, or returned by, the capacitor or inductor over 1/2 cycle of the power curve for each if the frequency of the input quantities is 60 Hz.

***Solutions:***

a. $\mathbf{I} = \dfrac{\mathbf{E}}{\mathbf{Z}_T} = \dfrac{100 \text{ V} \angle 0°}{6\ \Omega + j7\ \Omega - j15\ \Omega} = \dfrac{100 \text{ V} \angle 0°}{10\ \Omega \angle -53.13°} = 10 \text{ A} \angle 53.13°$

$\mathbf{V}_R = \mathbf{IR} = (10 \text{ A} \angle 53.13°)(6\ \Omega \angle 0°) = 60 \text{ V} \angle 53.13°$
$\mathbf{V}_L = \mathbf{IX}_L = (10 \text{ A} \angle 53.13°)(7\ \Omega \angle 90°) = 70 \text{ V} \angle 143.13°$
$\mathbf{V}_C = \mathbf{IX}_C = (10 \text{ A} \angle 53.13°)(15\ \Omega \angle -90°) = 150 \text{ V} \angle -36.87°$

$P_T = EI \cos\theta = (100 \text{ V})(10 \text{ A}) \cos 53.13° = \mathbf{600\ W}$
$= I^2R = (10 \text{ A})^2(6\ \Omega) = \mathbf{600\ W}$
$= \dfrac{V_R^2}{R} = \dfrac{(60 \text{ V})^2}{6} = \mathbf{600\ W}$

$S_T = EI = (100 \text{ V})(10 \text{ A}) = \mathbf{1000\ VA}$
$= I^2Z_T = (10 \text{ A})^2(10\ \Omega) = \mathbf{1000\ VA}$
$= \dfrac{E^2}{Z_T} = \dfrac{(100 \text{ V})^2}{10\ \Omega} = \mathbf{1000\ VA}$

$Q_T = EI \sin\theta = (100 \text{ V})(10 \text{ A}) \sin 53.13° = \mathbf{800\ VAR}$
$= Q_C - Q_L$
$= I^2(X_C - X_L) = (10 \text{ A})^2(15\ \Omega - 7\ \Omega) = \mathbf{800\ VAR}$

$Q_T = \dfrac{V_C^2}{X_C} - \dfrac{V_L^2}{X_L} = \dfrac{(150 \text{ V})^2}{15\ \Omega} - \dfrac{(70 \text{ V})^2}{7\ \Omega}$
$= 1500 \text{ VAR} - 700 \text{ VAR} = \mathbf{800\ VAR}$

$F_p = \dfrac{P_T}{S_T} = \dfrac{600 \text{ W}}{1000 \text{ VA}} = \mathbf{0.6\ leading\ (cap.)}$

b. The power triangle is as shown in Fig. 19.20.

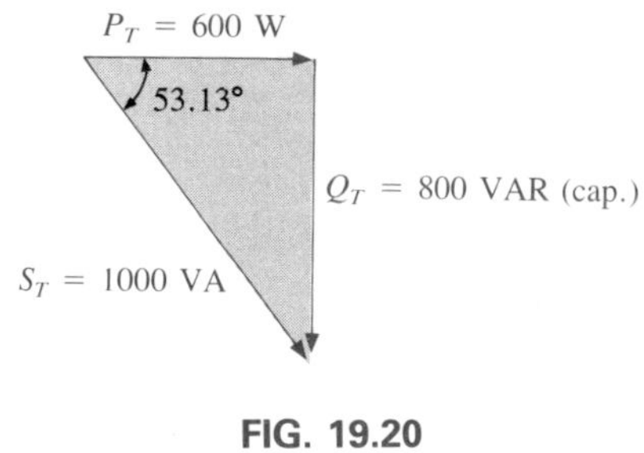

FIG. 19.20

c. $W_R = 2\left(\dfrac{V_RI}{f_2}\right) = 2\left(\dfrac{V_RI}{2f_1}\right) = \dfrac{V_RI}{f_1} = \dfrac{(60 \text{ V})(10 \text{ A})}{60 \text{ Hz}} = \mathbf{10\ J}$

d. $W_L = \dfrac{V_LI}{2\pi f_1} = \dfrac{(70 \text{ V})(10 \text{ A})}{(6.28)(60 \text{ Hz})} = \dfrac{700 \text{ J}}{377} = \mathbf{1.86\ J}$

$W_C = \dfrac{V_CI}{2\pi f_1} = \dfrac{(150 \text{ V})(10 \text{ A})}{377 \text{ rad/s}} = \dfrac{1500 \text{ J}}{377} = \mathbf{3.98\ J}$

**EXAMPLE 19.3.** For the system of Fig. 19.21,

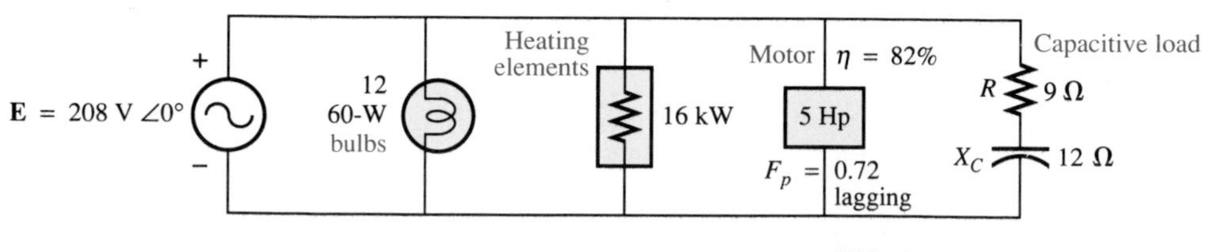

FIG. 19.21

a. Find the average power, apparent power, reactive power, and $F_p$ for each branch.
b. Find the total number of watts, volt-amperes reactive, volt-amperes, and the power factor of the system. Sketch the power triangle.
c. Find the source current $I$.

### *Solutions:*

a. *Bulbs:*

Total dissipation of applied power

$$P_1 = 12(60\text{W}) = \mathbf{720\ W}$$
$$Q_1 = \mathbf{0\ VAR}$$
$$S_1 = P_1 = \mathbf{720\ VA}$$
$$F_{p_1} = \mathbf{1}$$

*Heating elements:*

Total dissipation of applied power

$$P_2 = \mathbf{6.4\ kW}$$
$$Q_2 = \mathbf{0\ VAR}$$
$$S_2 = P_2 = \mathbf{6.4\ kVA}$$
$$F_{p_2} = \mathbf{1}$$

*Motor:*

$$\eta = \frac{P_o}{P_i} \Rightarrow P_i = \frac{P_o}{\eta} = \frac{5(746\ \text{W})}{0.82} = \mathbf{4548.78\ W} = P_3$$

$$F_p = \mathbf{0.72\ lagging}$$

$$P_3 = S_3 \cos\theta \Rightarrow S_3 = \frac{P_3}{\cos\theta} = \frac{4548.78\ \text{W}}{0.72} = \mathbf{6317.75\ VA}$$

Also, $\theta = \cos^{-1} 0.72 = 43.95°$, so that

$$Q_3 = S_3 \sin\theta = (6317.75\ \text{VA})(\sin 43.95°)$$
$$= (6317.75\ \text{VA})(0.694) = \mathbf{4384.71\ VAR}$$

*Capacitive load:*

$$\mathbf{I} = \frac{\mathbf{E}}{\mathbf{Z}} = \frac{208\ \text{V}\ \angle 0°}{9\ \Omega - j12\ \Omega} = \frac{208\ \text{V}\ \angle 0°}{15\ \Omega\ \angle -53.13°} = 13.87\ \text{A}\ \angle 53.13°$$

$$P_4 = I^2R = (13.87\ \text{A})^2 \cdot 9\ \Omega = \mathbf{1731.39\ W}$$

$$Q_4 = I^2X_C = (13.87\ \text{A})^2 \cdot 12\ \Omega = \mathbf{2308.52\ VAR}$$

$$S_4 = \sqrt{P_4^2 + Q_4^2} = \sqrt{(1731.39\ \text{W})^2 + (2308.52\ \text{VAR})^2}$$
$$= \mathbf{2885.65\ VA}$$

$$F_p = \frac{P_4}{S_4} = \frac{1731.39\ \text{W}}{2885.65\ \text{VA}} = \mathbf{0.6\ leading}$$

b. $P_T = P_1 + P_2 + P_3 + P_4$
$= 720 \text{ W} + 6400 \text{ W} + 4548.78 \text{ W} + 1731.39 \text{ W}$
$\cong \mathbf{13.4 \text{ kW}}$

$Q_T = \pm Q_1 \pm Q_2 \pm Q_3 \pm Q_4$
$= 0 + 0 + 4384.71 \text{ VAR} - 2308.52 \text{ VAR}$
$= \mathbf{2076.19 \text{ VAR (Ind.)}}$

$S_T = \sqrt{P_T^2 + Q_T^2} = \sqrt{(13.4 \text{ kW})^2 + (2076.19 \text{ VAR})^2}$
$= \mathbf{13{,}559.89 \text{ VA}}$

$$F_p = \frac{P_T}{S_T} = \frac{13.4 \text{ kW}}{13{,}559.89 \text{ VA}} = \mathbf{0.988 \text{ lagging}}$$

$\theta = \cos^{-1} 0.988 = 8.89°$
Note Fig. 19.22.

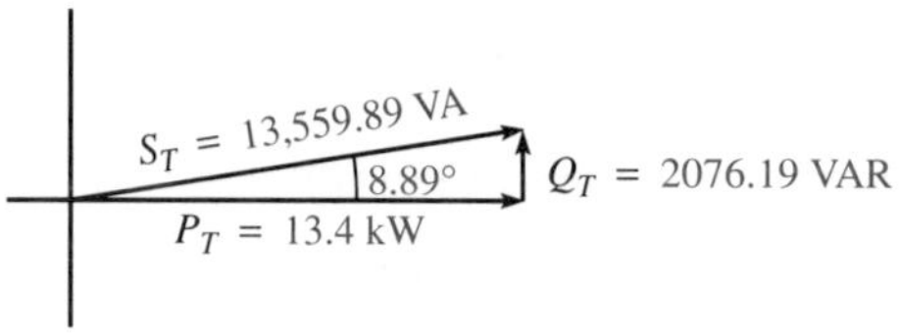

**FIG. 19.22**

c. $S_T = EI \Rightarrow I = \dfrac{S_T}{E} = \dfrac{13{,}559.89 \text{ VA}}{208 \text{ V}} = 65.19 \text{ A}$

Lagging power factor: **E** leads **I** by 8.89° and

$$\mathbf{I} = \mathbf{65.19 \text{ A} \angle -8.89°}$$

---

**EXAMPLE 19.4.** An electrical device is rated 5 kVA, 100 V at a 0.6 power-factor lag. What is the impedance of the device in rectangular coordinates?

***Solution:***

$$S = EI = 5000 \text{ VA}$$

Therefore,

$$I = \frac{5000 \text{ VA}}{100 \text{ V}} = 50 \text{ A}$$

For $F_p = 0.6$, we have

$$\theta = \cos^{-1} 0.6 = 53.13°$$

Since the power factor is lagging, the circuit is predominantly inductive, and **I** lags **E.** Or, for $\mathbf{E} = 100 \text{ V} \angle 0°$,

$$\mathbf{I} = 50 \text{ A} \angle -53.13°$$

However,

$$\mathbf{Z}_T = \frac{\mathbf{E}}{\mathbf{I}} = \frac{100 \text{ V} \angle 0°}{50 \text{ A} \angle -53.13°} = 2\ \Omega \angle 53.13° = \mathbf{1.2\ \Omega + \mathit{j}1.6\ \Omega}$$

which is the impedance of the circuit of Fig. 19.23.

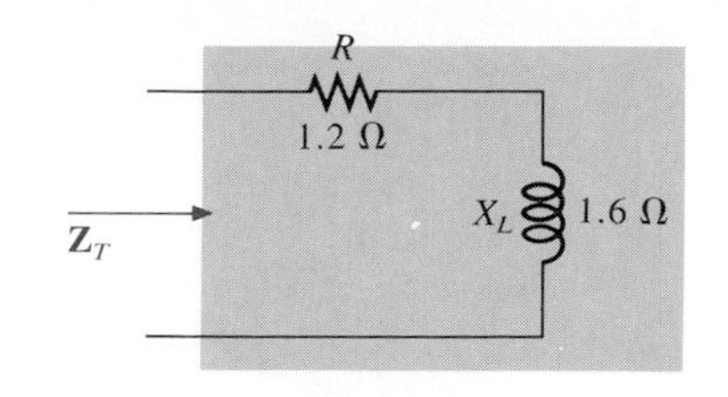

FIG. 19.23

## 19.8 POWER-FACTOR CORRECTION

The design of any power transmission system is very sensitive to the magnitude of the current in the lines as determined by the applied loads. Increased currents result in increased power losses (by a squared factor since $P = I^2R$) in the transmission lines due to the resistance of the lines. Heavier currents also require larger conductors, increasing the amount of copper needed for the system, and, quite obviously, they require increased generating capacities by the utility company.

Every effort must therefore be made to keep current levels at a minimum. Since the line voltage of a transmission system is fixed, the apparent power is directly related to the current level. In turn, the smaller the net apparent power, the smaller the current drawn from the supply. Minimum current is therefore drawn from a supply when $S = P$ and $Q_T = 0$. Note the effect of decreasing levels of $Q_T$ on the length (and magnitude) of $S$ in Fig. 19.24 for the same real power. Note also that the power-factor angle approaches zero degrees and $F_p$ approaches 1, revealing that the network is appearing more and more resistive at the input terminals.

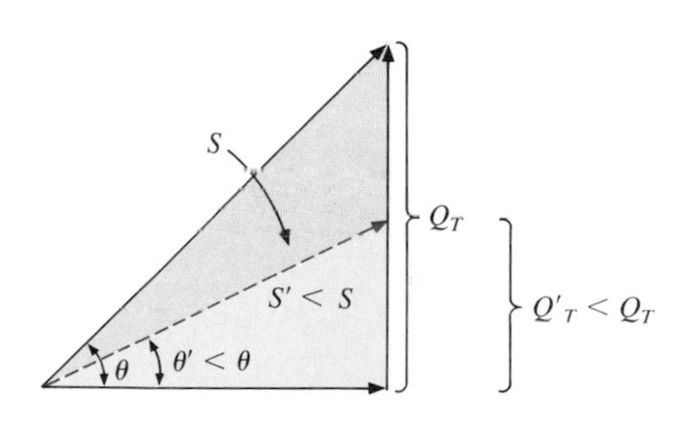


FIG. 19.24

The process of introducing reactive elements to bring the power factor closer to unity is called *power-factor correction*. Since most loads are inductive, the process normally involves introducing elements with capacitive terminal characteristics having the sole purpose of improving the power factor.

In Fig. 19.25(a), for instance, an inductive load is drawing a current $I_L$ that has a real and imaginary component. In Fig. 19.25(b) a capacitive load was added in parallel with the original load to raise the power factor of the total system to the unity power-factor level. Note that by

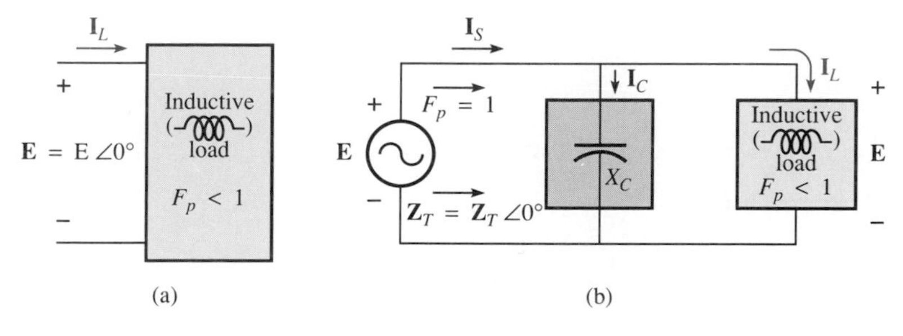


FIG. 19.25

placing all the elements in parallel the load still receives the same terminal voltage and draws the same current $I_L$. In other words, the load is unaware and unconcerned about whether it is hooked up as shown in Fig. 19.25(a) or Fig. 19.25(b).

Solving for the source current in Fig. 19.25(b):

$$\begin{aligned}\mathbf{I}_S &= \mathbf{I}_C + \mathbf{I}_L \\ &= -jI_C + I_L \text{ (real)} + jI_L \text{ (imag.)} \\ &= I_L \text{ (real)} + j(I_L \text{ (imag.)} - I_C)\end{aligned}$$

If $X_C$ is chosen such that $I_C = I_L$ (imag.)

$$\mathbf{I}_S = I_L \text{ (real)} + j(0)$$

and

$$\mathbf{I}_S = I_L \text{ (real)} \angle 0°$$

The result is a source current whose magnitude is simply equal to the real part of the load current, which can be considerably less than the magnitude of the load current of Fig. 19.25(a). In addition, since the phase angle associated with both the applied voltage and the source current is the same, the system appears ''resistive'' at the input terminals and all the power supplied is absorbed, creating maximum efficiency for a generating utility.

---

**EXAMPLE 19.5.** A 5-hp motor with a 0.6 lagging power factor and an efficiency of 92% is connected to a 208-V, 60-Hz supply. What level of capacitance in parallel with the motor will raise the power factor of the combined system to unity?

***Solution:*** First we establish the power triangle for the 5-hp motor. Since 1 hp = 746 W,

$$P_o = 5 \text{ hp} = 5(746 \text{ W}) = 3730 \text{ W}$$

and

$$P_i \text{ (drawn from the line)} = \frac{P_o}{\eta} = \frac{3730 \text{ W}}{0.92} = 4054.35 \text{ W}$$

Also,

$$F_p = \cos\theta = 0.6$$

and

$$\theta = \cos^{-1} 0.6 = 53.13°$$

Then we use

$$\tan\theta = \frac{Q_L}{P_i}$$

to obtain

$$\begin{aligned}Q_L &= P_i \tan\theta = (4054.35 \text{ W}) \tan 53.13° \\ &= (4054.35 \text{ W})(1.333) = 5404.45 \text{ VAR}\end{aligned}$$

and

$$S = \sqrt{P_i^2 + Q_L^2} = \sqrt{(4054.35 \text{ W})^2 + (5404.45 \text{ VAR})^2}$$
$$= 6756.17 \text{ VA}$$

The power triangle appears in Fig. 19.26.

A net unity power-factor level is established by introducing a capacitive reactive power level of 5404.45 VAR to balance $Q_L$. Since

$$Q_C = \frac{V^2}{X_C}$$

then

$$X_C = \frac{V^2}{Q_C}$$

and

$$X_C = \frac{(208 \text{ V})^2}{5404.45 \text{ VAR}} = 8 \ \Omega$$

and

$$C = \frac{1}{2\pi f X_C} \qquad (\text{from } X_C = 1/2\pi fC)$$
$$= \frac{1}{(6.28)(60 \text{ Hz})(8 \ \Omega)} = \mathbf{332 \ \mu F}$$

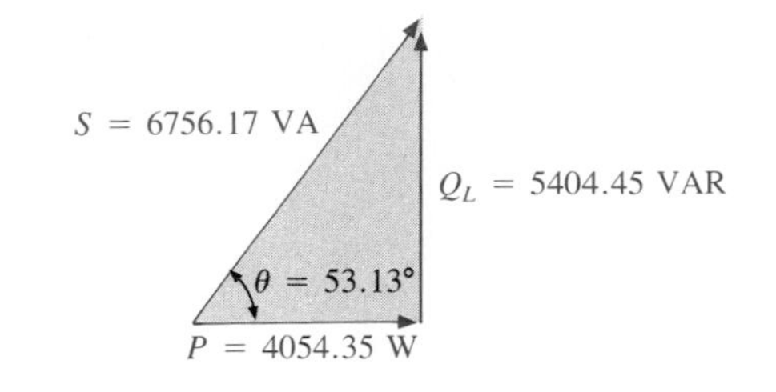


FIG. 19.26

**At 0.6$F_p$,**

$$S = VI = 6756.17 \text{ VA}$$

and

$$I = \frac{S}{V} = \frac{6756.17 \text{ VA}}{208 \text{ V}} = \mathbf{32.48 \text{ A}}$$

**At unity $F_p$,**

$$S = VI = 4054.35 \text{ VA}$$

and

$$I = \frac{S}{V} = \frac{4054.35 \text{ VA}}{208 \text{ V}} = \mathbf{19.49 \text{ A}}$$

producing a 40% reduction in supply current.

**EXAMPLE 19.6.** A small industrial plant has a 10-kW heating load and a 20-kVA inductive load due to a bank of induction motors. The heating elements are considered purely resistive ($F_p = 1$) and the induction motors have a lagging power factor of 0.7. If the supply is 1000 V at 60 Hz, determine the capacitive element required to raise the power factor to 0.95.

***Solution:*** For the induction motors,

$$S = VI = 20 \text{ kVA}$$
$$P = VI \cos\theta = (20 \times 10^3 \text{ VA})(0.7) = 14 \times 10^3 \text{ W}$$
$$\theta = \cos^{-1} 0.7 \cong 45.6^\circ$$

and

$$Q_L = VI \sin\theta = (20 \times 10^3 \text{ VA})(0.714) = 14.28 \times 10^3 \text{ VAR}$$

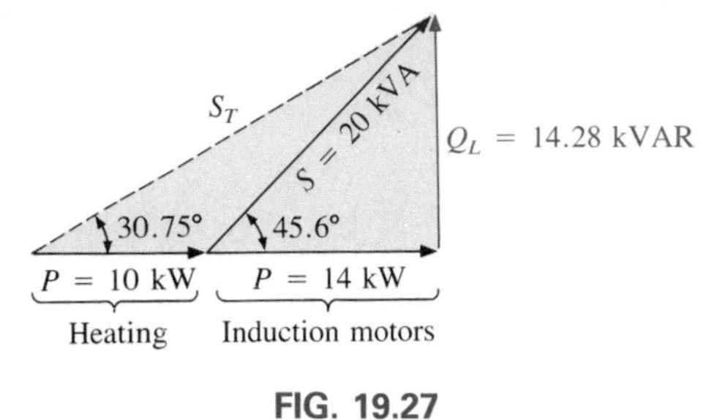


FIG. 19.27

The power triangle for the total system appears in Fig. 19.27.

Note the addition of real powers and the increased value of $S$:

$$S_T = \sqrt{(24 \text{ kW})^2 + (14.28 \text{ kVAR})^2} = 27.93 \text{ kVA}$$

with

$$I = \frac{S_T}{V} = \frac{27.93 \text{ kVA}}{1000 \text{ V}} = 27.93 \text{ A}$$

A power factor of 0.95 results in an angle between $S$ and $P$ of

$$\theta = \cos^{-1} 0.95 = 18.19^\circ$$

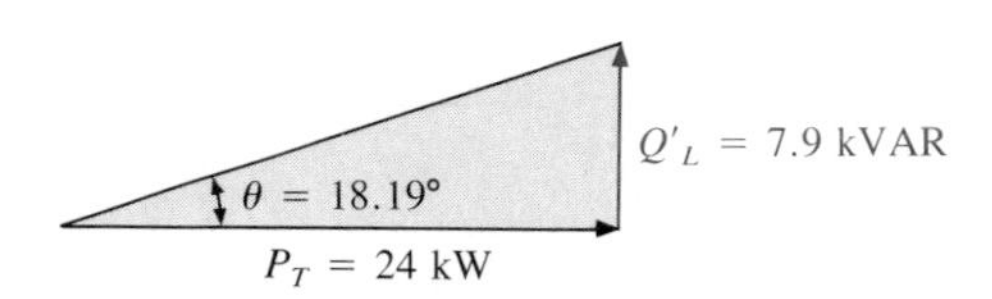

FIG. 19.28

changing the power triangle to the following (Fig. 19.28):

$$\tan\theta = \frac{Q'_L}{P_T} \Rightarrow Q'_L = P_T \tan\theta = (24 \times 10^3 \text{ W})(\tan 18.19^\circ)$$
$$= (24 \times 10^3 \text{ W})(0.329) = 7.9 \text{ kVAR}$$

The inductive reactive power must therefore be reduced by

$$Q_L - Q'_L = 14.28 \text{ kVAR} - 7.9 \text{ kVAR} = 6.38 \text{ kVAR}$$

Therefore, $Q_C = 6.38$ kVAR, and using

$$Q_C = \frac{V^2}{X_C}$$

we obtain

$$X_C = \frac{V^2}{Q_C} = \frac{(10^3 \text{ V})^2}{6.38 \times 10^3 \text{ VAR}} = 156.74\ \Omega$$

and

$$C = \frac{1}{2\pi f X_C} = \frac{1}{(6.28)(60 \text{ Hz})(156.74\ \Omega)} = \mathbf{16.93\ \mu F}$$

## 19.9 THE WATTMETER

The wattmeter, as the name suggests, is an instrument designed to read the power to an element or network. It employs an electrodynamometer-type movement or solid-state electronic system to measure the power in a dc or ac network. It can, in fact, be used to measure the wattage of any circuit with a periodic or nonperiodic input.

In an electrodynamometer movement, a moving coil rotates in a magnetic field produced by the current of a stationary coil. The fluxes of the stationary and movable coils interact to develop a torque on the pointer connected to the movable coil. In the wattmeter configuration (Fig. 19.29), the current in the stationary coils is the line current, while

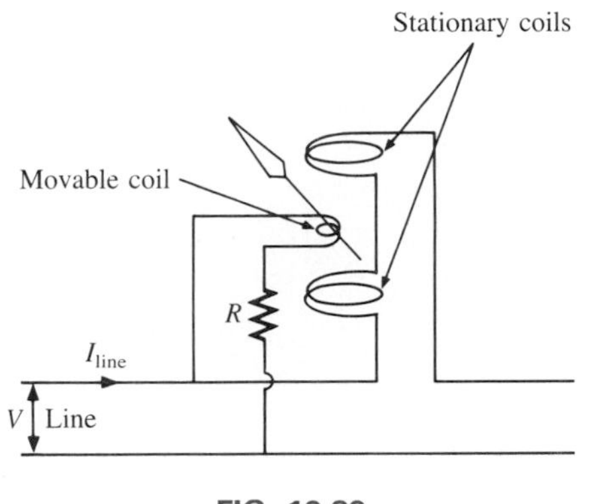


FIG. 19.29

the current in the moving coil is derived from the line voltage. The instrument then indicates power in watts on a linear scale. A typical wattmeter using an electrodynamometer movement appears in Fig. 19.30.

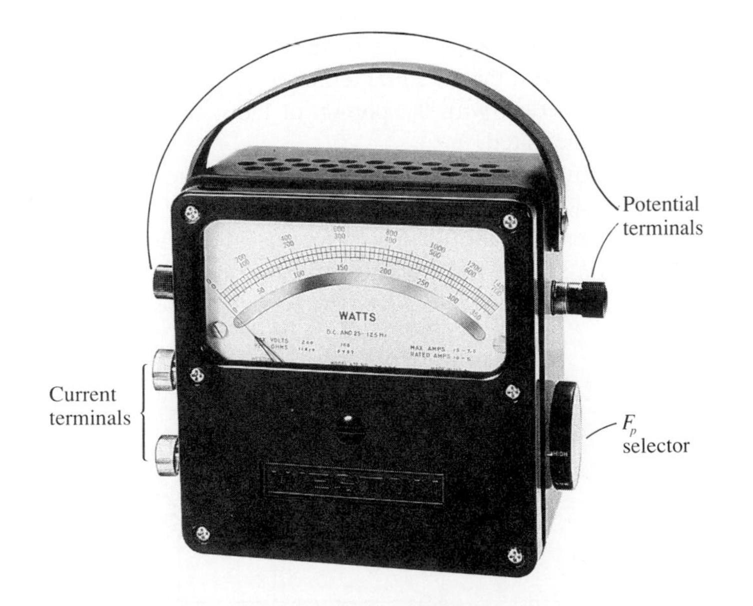


FIG. 19.30
*Wattmeter. (Courtesy of Electrical Instrument Service, Inc.)*

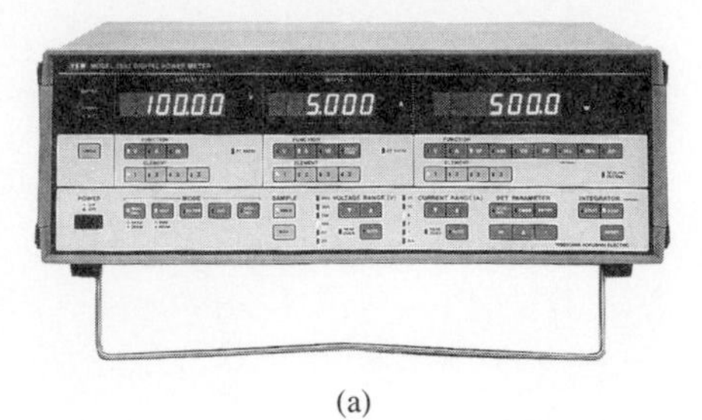


(a)

(b)

**FIG. 19.31**
(a) *Digital wattmeter*. (*Courtesy of Yokogawa Corporation of America*) (b) *Nuwatt Digital Powermeter*. (*Courtesy of AEMC Corporation*)

The digital display wattmeter of Fig. 19.31(a) employs a sophisticated electronic package to sense the voltage and current levels and through the use of an analog-to-digital conversion unit display the proper digits on the display. The Nuwatt Digital Powermeter of Fig. 19.31(b) can measure power levels extending from 0.1 W to 2 MW. The system can handle input currents up to 3000 A and voltages up to 700 V and works on single- or three-phase (Chapter 23) systems.

For an up-scale deflection, the electrodynamometer wattmeter is connected as shown in Fig. 19.32. Some electrodynamometer wattmeters will always give a wattage reading that is higher than that actually delivered to the load. They are high by the amount of power consumed by the potential coil ($V_{pc}^2/R_{pc}$). This correction is important and should be considered with every set of data. Many wattmeters are designed to compensate for this correction, and therefore they eliminate the need for any other adjustment in the reading. The wattmeter is always connected with the potential terminals in parallel, and the current terminals in series, with the portion of the network to which the power is being measured.

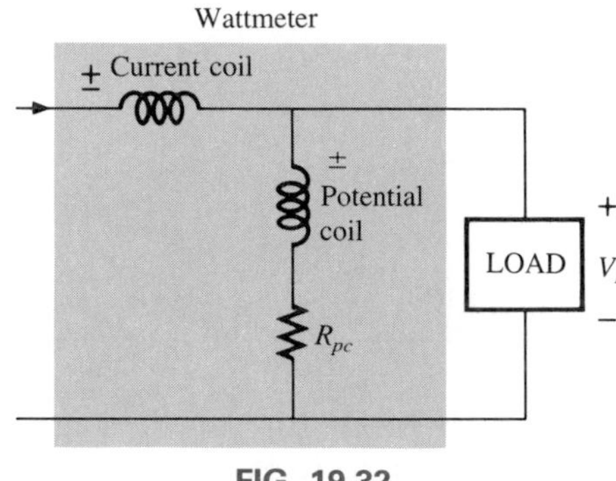


**FIG. 19.32**

The power delivered to $R_1$ in Fig. 19.33 can be found by connecting the electrodynamometer wattmeter as shown in Fig. 19.33(a). To find the power delivered to the total network, it should be connected as shown in Fig. 19.33(b). The connections for the digital meter are fundamentally the same as for the electrodynamometer meter.

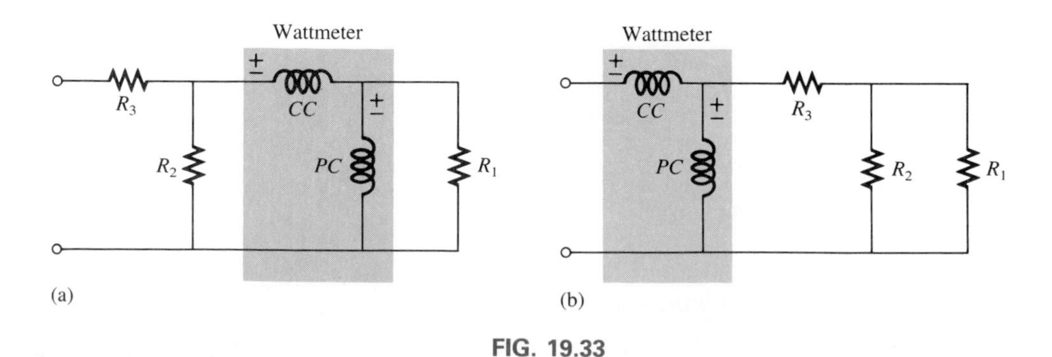


**FIG. 19.33**

When using the electrodynamometer wattmeter, the operator must take care not to exceed the current, voltage, or wattage rating. The product of the voltage and current ratings may or may not equal the wattage rating. In the high-power-factor wattmeter, the product of the voltage and current ratings is usually equal to the wattage rating, or at least 80% of it. For a low-power-factor wattmeter, the product of the current and voltage ratings is much greater than the wattage rating. For obvious reasons, the low-power-factor meter is used only in circuits with low power factors (total impedance highly reactive). Typical ratings for high-power-factor (HPF) and low-power-factor (LPF) meters are shown in Table 19.1. Meters of both high and low power factors have an accuracy of 0.5% to 1% of full scale.

**TABLE 19.1**

| Meter | Current Ratings | Voltage Ratings | Wattage Ratings |
|---|---|---|---|
| HPF | 2.5 A<br>5.0 A | 150 V<br>300 V | 1500/750/375 |
| LPF | 2.5 A<br>5.0 A | 150 V<br>300 V | 300/150/75 |

## 19.10 EFFECTIVE RESISTANCE

The resistance of a conductor as determined by the equation $R = \rho(l/A)$ is often called the *dc*, *ohmic*, or *geometric* resistance. It is a constant quantity determined only by the material used and its physical dimensions. In ac circuits, the actual resistance of a conductor (called the *effective* resistance) differs from the dc resistance because of the varying currents and voltages which introduce effects not present in dc circuits.

These effects include radiation losses, skin effect, eddy currents, and hysteresis losses. The first two effects apply to any network, while the latter two are concerned with the additional losses introduced by the presence of ferromagnetic materials in a changing magnetic field.

### Experimental Procedure

The effective resistance of an ac circuit cannot be measured by the ratio $V/I$, since this ratio is now the impedance of a circuit that may have both resistance and reactance. The effective resistance can be found, however, by using the power equation $P = I^2R$, where

$$R_{\text{eff}} = \frac{P}{I^2} \qquad \textbf{(19.34)}$$

A wattmeter and an ammeter are therefore necessary for measuring the effective resistance of an ac circuit.

## Radiation Losses

Let us now examine the various losses in greater detail. The radiation loss is the loss of energy in the form of electromagnetic waves during the transfer of energy from one element to another. This loss in energy requires that the input power be larger to establish the same current $I$, causing $R$ to increase as determined by Eq. (19.34). At a frequency of 60 Hz, the effects of radiation losses can be completely ignored. However, at radio frequencies, this is an important effect and may in fact become the main effect in an electromagnetic device such as an antenna.

## Skin Effect

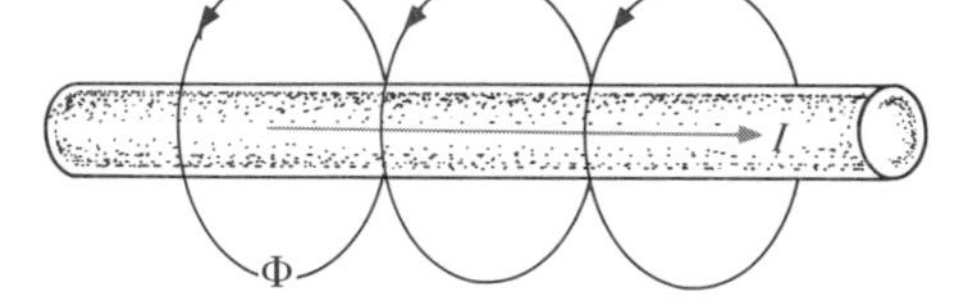


FIG. 19.34

The explanation of skin effect requires the use of some basic concepts previously described. It will be remembered from Chapter 11 that a magnetic field exists around every current-carrying conductor (Fig. 19.34). Since the amount of charge flowing in ac circuits changes with time, the magnetic field surrounding the moving charge (current) also changes. Recall also that a wire placed in a changing magnetic field will have an induced voltage across its terminals as determined by Faraday's law $e = N\ (d\phi/dt)$. The higher the frequency of the changing flux as determined by an alternating current, the greater the induced voltage will be.

For a conductor carrying alternating current, the changing magnetic field surrounding the wire links the wire itself, thus developing within the wire an induced voltage that opposes the original flow of charge or current. These effects are more pronounced at the center of the conductor than at the surface since the center is linked by the changing flux inside the wire as well as that outside the wire. As the frequency of the applied signal increases, the flux linking the wire will change at a greater rate. An increase in frequency will therefore increase the counter induced voltage at the center of the wire to the point where the current will, for all practical purposes, flow on the surface of the conductor. At 60 Hz, the effects of skin effect are almost noticeable. However, at radio frequencies, the skin effect is so pronounced that large conductors are frequently made hollow since the center part is relatively ineffective. The skin effect, therefore, reduces the effective area through which the current can flow, and causes the resistance of the conductor, given by the equation $R\uparrow = \rho(l/A\downarrow)$, to increase.

## Hysteresis and Eddy Current Losses

As mentioned earlier, hysteresis and eddy current losses will appear when a ferromagnetic material is placed in the region of a changing magnetic field. To describe eddy current losses in greater detail, we will consider the effects of an alternating current passing through a coil wrapped around a ferromagnetic core. As the alternating current passes through the coil, it will develop a changing magnetic flux $\Phi$ linking both the coil and the core which will develop an induced voltage within the core as determined by Faraday's law. This induced voltage and the geometric resistance of the core $R_C = \rho(l/A)$ cause currents to be developed within the core, $i_{core} = (e_{ind}/R_C)$, called *eddy currents*. The currents flow in circular paths, as shown in Fig. 19.35, changing direction with the applied ac potential.

The eddy current losses are determined by

$$P_{eddy} = i^2_{eddy}R_{core}$$

The magnitude of these losses is determined primarily by the type of core used. If the core is nonferromagnetic—and has a high resistivity like wood or air—the eddy current losses can be neglected. In terms of the frequency of the applied signal and the magnetic field strength produced, the eddy current loss is proportional to the square of the frequency times the square of the magnetic field strength:

$$P_{eddy} \propto f^2B^2$$

Eddy current losses can be reduced if the core is constructed of thin, laminated sheets of ferromagnetic material insulated from one another and aligned parallel to the magnetic flux. Such construction reduces the magnitude of the eddy currents by placing more resistance in their path.

Hysteresis losses were described in Section 11.8. You will recall that in terms of the frequency of the applied signal and the magnetic field strength produced, the hysteresis loss is proportional to the frequency to the 1st power times the magnetic field strength to the $n$th power:

$$P_{hys} \propto f^1B^n$$

where $n$ can vary from 1.4 to 2.6, depending on the material under consideration.

Hysteresis losses can be effectively reduced by the injection of small amounts of silicon into the magnetic core, constituting some 2% or 3% of the total composition of the core. This must be done carefully, however, as too much silicon makes the core brittle and difficult to machine into the shape desired.

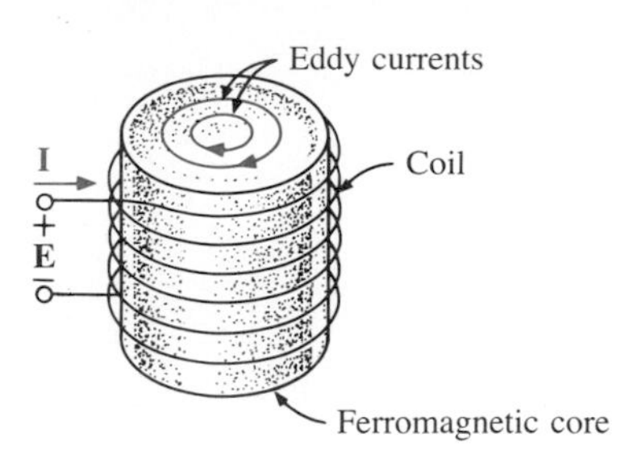


FIG. 19.35

**EXAMPLE 19.7.**

a. An air-core coil is connected to a 120-V, 60-Hz source as shown in Fig. 19.36. The current is found to be 5 A, and a wattmeter reading

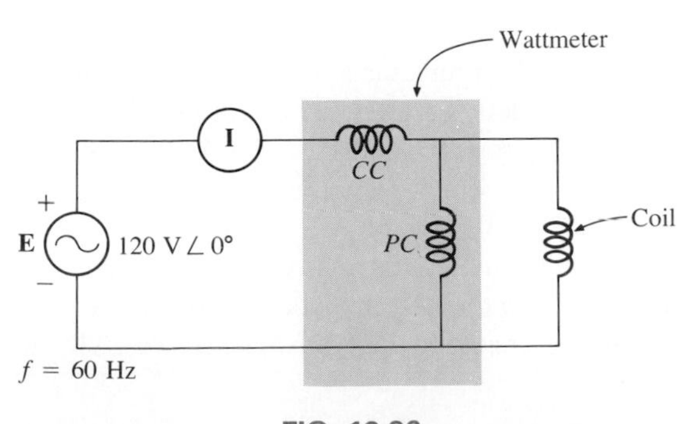

FIG. 19.36

$P_S^q$

of 75 W is observed. Find the effective resistance and the inductance of the coil.

b. A brass core is then inserted in the coil, and the ammeter reads 4 A and the wattmeter reads 80 W. Calculate the effective resistance of the core. To what do you attribute the increase in value over that of part (a)?

c. If a solid iron core is inserted in the coil, the current is found to be 2 A, and the wattmeter reads 52 W. Calculate the resistance and the inductance of the coil. Compare these values to those of part (a), and account for the changes.

***Solutions:***

a. $R = \dfrac{P}{I^2} = \dfrac{75\text{ W}}{(5\text{ A})^2} = \mathbf{3\ \Omega}$

$$Z_T = \frac{E}{I} = \frac{120\text{ V}}{5\text{ A}} = 24\ \Omega$$

$$X_L = \sqrt{Z_T^2 - R^2} = \sqrt{(24\ \Omega)^2 - (3\ \Omega)^2} = 23.81\ \Omega$$

and

$$X_L = 2\pi f L$$

or

$$L = \frac{X_L}{2\pi f} = \frac{23.81\ \Omega}{377\text{ rad/s}} = \mathbf{63.16\ mH}$$

b. $R = \dfrac{P}{I^2} = \dfrac{80\text{ W}}{(4\text{ A})^2} = \dfrac{80}{16} = \mathbf{5\ \Omega}$

The brass core has less reluctance than the air core. Therefore a greater magnetic flux density $B$ will be created in it. Since $P_{eddy} \propto f^2B^2$, and $P_{hys} \propto f^1B^n$, as the flux density increases, the core losses and the effective resistance increase.

c. $R = \dfrac{P}{I^2} = \dfrac{52\text{ W}}{(2\text{ A})^2} = \dfrac{52}{4} = \mathbf{13\ \Omega}$

$$Z_T = \frac{E}{I} = \frac{120\text{ V}}{2\text{ A}} = 60\ \Omega$$

$$X_L = \sqrt{Z_T^2 - R^2} = \sqrt{(60\ \Omega)^2 - (13\ \Omega)^2} = 58.57\ \Omega$$

$$L = \frac{X_L}{2\pi f} = \frac{58.57\ \Omega}{377\text{ rad/s}} = \mathbf{155.36\ mH}$$

The iron core has less reluctance than the air or brass cores. Therefore a greater magnetic flux density $B$ will be developed in the core. Again, since $P_{eddy} \propto f^2B^2$, and $P_{hys} \propto f^1B^n$, the increased flux density will cause the core losses and the effective resistance to increase.

Since the inductance $L$ is related to the change in flux by the equation $L = N\ (d\phi/di)$, the inductance will be greater for the iron core because the changing flux linking the core will increase.

## 19.11 COMPUTER ANALYSIS

PSPICE does not lend itself well to some of the important calculations to be performed in this chapter and therefore is not included in this computer analysis section. PSPICE can calculate power levels using the .PROBE routine but cannot tabulate real and reactive power levels using simply the .PRINT or .PLOT commands.

```
                 10 REM *****  PROGRAM 19-1  *****
                 20 REM ******************************************
                 30 REM This program calculates the total real,
                 40 REM reactive and apparent power of a network
                 50 REM with five individual loads.
                 60 REM ******************************************
                 70 REM
                100 DIM P(5),Q(5),S(5)
                110 PRINT "This program calculates the total real,"
                120 PRINT "reactive and apparent power of a network"
                130 PRINT "with five individual loads."
                140 PRINT
                150 PRINT "Input the following data:"
                160 PRINT "(use negative sign for capacitive vars)"
                170 PRINT
        Input [ 180 INPUT "E=";E
              [ 190 INPUT "at an angle=";EA
                200 PRINT
              [ 210 FOR I=1 TO 5
              [ 220 PRINT "For";I;"  ";
        Input [ 230 INPUT "P(watts)=";P(I)
              [ 240 PRINT TAB(8);
              [ 250 INPUT "Q(vars)=";Q(I)
       P,Pq   [ 260 PT=PT+P(I)
              [ 270 QT=QT+Q(I)
                280 NEXT I
                290 PRINT
              [ 300 PRINT "The apparent power associated with each load"
              [ 310 PRINT "is the following:"
              [ 320 PRINT
        Pa    [ 330 FOR I=1 TO 5
              [ 340 S(I)=SQR(P(I)^2+Q(I)^2)
              [ 350 PRINT "S";I;"=";S(I)
              [ 360 NEXT I
              [ 370 ST=SQR(PT^2+QT^2)
                380 PRINT:PRINT
              [ 390 PRINT "Total real power, PT=";PT;"watts"
        Power [ 400 PRINT
       Output [ 410 PRINT "Total reactive power, QT=";QT;"vars"
              [ 420 PRINT
              [ 430 PRINT "Total apparent power, ST=";ST;"VA"
              [ 440 FP=PT/ST
              [ 450 TH=-57.296*ATN(QT/PT)
              [ 460 IF QT>0 THEN IA=EA-TH
              [ 470 IF QT<0 THEN IA=EA+TH
        Fp    [ 480 PRINT
              [ 490 PRINT "Power factor angle=";IA;"degrees"
              [ 500 PRINT
              [ 510 PRINT "Power factor=";FP;
              [ 520 IF QT>0 THEN PRINT "(lagging)"
              [ 530 IF QT<0 THEN PRINT "(leading)"
              [ 540 I=ST/E
        IT    [ 550 PRINT
              [ 560 PRINT "Input current:";I;"at an angle of";IA;"degrees"
                570 END

            READY
```

```
RUN

This program calculates the total real,
reactive and apparent power of a network
with five individual loads.

Input the following data:
(use negative sign for capacitive vars)

E=? 50

at an angle=? 60


For 1     P(watts)=? 200

          Q(vars)=? 100

For 2     P(watts)=? 200

          Q(vars)=? 100

For 3     P(watts)=? 100

          Q(vars)=? -200

For 4     P(watts)=? 100

          Q(vars)=? -200

For 5     P(watts)=? 0

          Q(vars)=? 0


The apparent power associated with each load
is the following:

S 1 = 224
S 2 = 224
S 3 = 224
S 4 = 224
S 5 = 0


Total real power, PT= 600 watts

Total reactive power, QT=-200 vars

Total apparent power, ST= 632 VA

Power factor angle= 78 degrees

Power factor= 1 (leading)

Input current: 13 at an angle of 78 degrees
```

**FIG. 19.37**
*Program 19.1.*

BASIC, however, can perform a number of interesting calculations, such as determining the total real, reactive, and apparent power levels in addition to the power-factor and input current. Program 19.1 of Fig. 19.37 can handle up to five different loads of varying power-factor angles.

The program is limited to five individual loads as defined by lines 110 through 130. Lines 260 and 270 determine the total real and reactive power using a loop routine that begins on line 210 and ends on line

280. The apparent power for each load is determined on line 340, and the total apparent power on line 370. The results are printed out by lines 390 through 430, and the power factor is determined by lines 440 through 530. The input current is then determined by lines 540 through 560.

A run of the program for four loads is provided with parameter values that permit a relatively easy check of the program through review of the results.

## PROBLEMS

### SECTION 19.7 The Total *P, Q,* and *S*

1. For the network of Fig. 19.38:
   a. Find the average power delivered to each element.
   b. Find the reactive power for each element.
   c. Find the apparent power for each element.
   d. Find the total number of watts, volt-amperes reactive, volt-amperes, and the power factor $F_p$ of the circuit.
   e. Sketch the power triangle.
   f. Find the energy dissipated by the resistor over one full cycle of the input voltage.
   g. Find the energy stored or returned by the capacitor and the inductor over 1/2 cycle of the power curve for each.

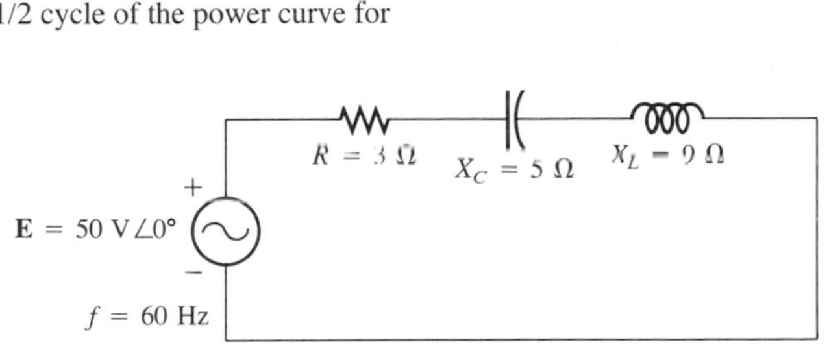


FIG. 19.38

2. For the system of Fig. 19.39:
   a. Find the total number of watts, volt-amperes reactive, volt-amperes, and the power factor $F_p$.
   b. Draw the power triangle.
   c. Find the current $\mathbf{I}_S$.

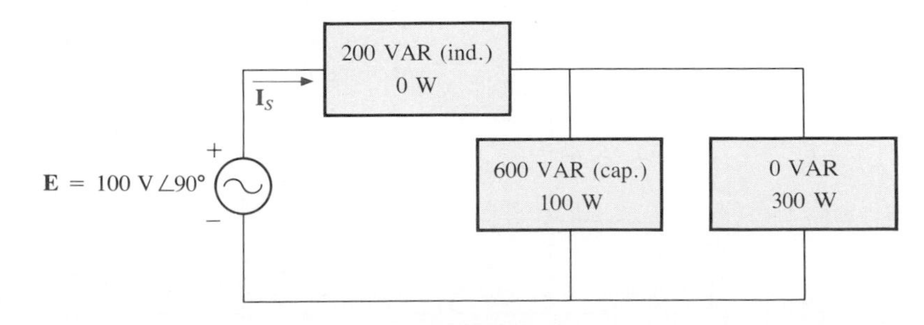


FIG. 19.39

**3.** Repeat Problem 2 for the system of Fig. 19.40.

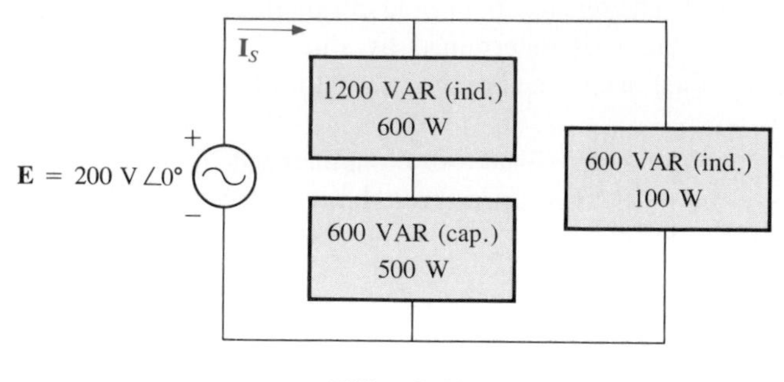

FIG. 19.40

**4.** Repeat Problem 2 for the system of Fig. 19.41.

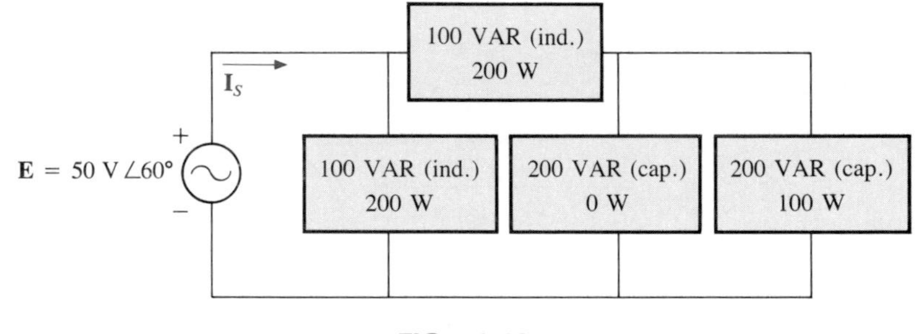

FIG. 19.41

**5.** For the circuit of Fig. 19.42:
  **a.** Find the average, reactive, and apparent power for the 20-Ω resistor.
  **b.** Repeat part (a) for the 10-Ω inductive reactance.
  **c.** Find the total number of watts, volt-amperes reactive, volt-amperes, and the power factor $F_p$.
  **d.** Find the current $\mathbf{I}_S$.

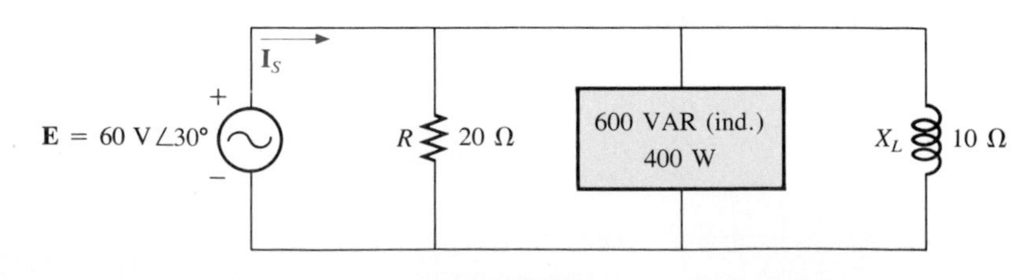


FIG. 19.42

**6.** Repeat Problem 1 for the circuit of Fig. 19.43.

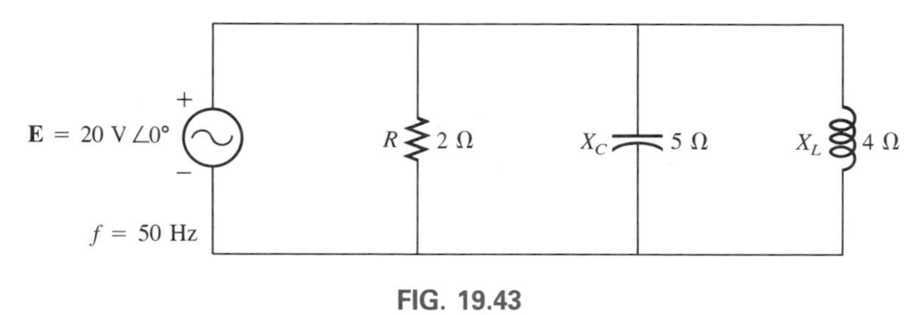


FIG. 19.43

**7.** Repeat Problem 1 for the circuit of Fig. 19.44.

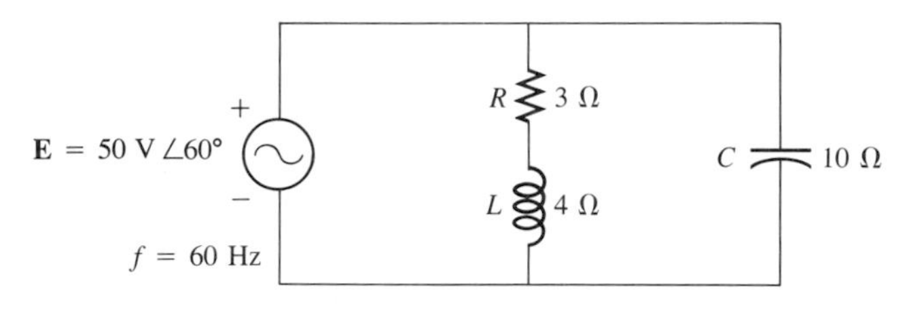

FIG. 19.44

***8.** Repeat Problem 1 for the circuit of Fig. 19.45.

**9.** An electrical system is rated 10 kVA, 200 V at a 0.5 leading power factor.

**a.** Determine the impedance of the system in rectangular coordinates.

**b.** Find the average power delivered to the system.

**10.** Repeat Problem 9 for an electrical system rated 5 kVA, 120 V at a 0.8 lagging power factor.

***11.** For the system of Fig. 19.46:

**a.** Find the total number of watts, volt-amperes reactive, volt-amperes, and $F_p$.

**b.** Find the current $\mathbf{I}_S$.

**c.** Draw the power triangle.

**d.** Find the type elements and their impedance in ohms within each electrical box. (Assume that all elements within the boxes are in series.)

**e.** Verify that the result of part (b) is correct by finding the current $\mathbf{I}_S$ using only the input voltage **E** and the results of part (d). Compare the value of $\mathbf{I}_S$ with that obtained for part (b).

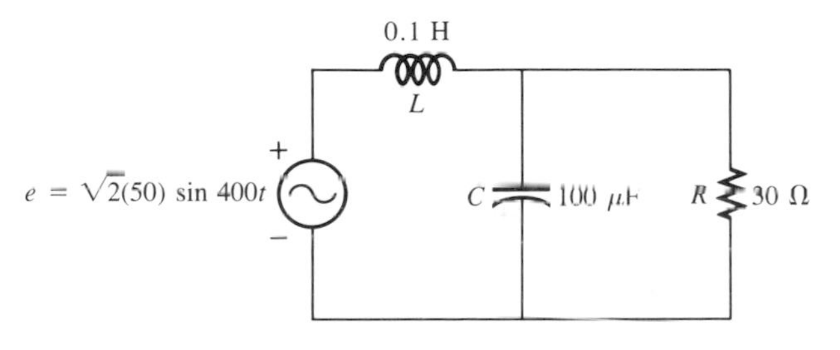


FIG. 19.45

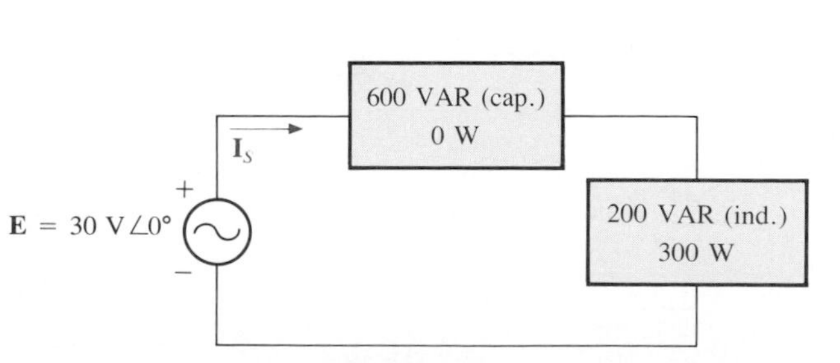


FIG. 19.46

*12. For the system of Fig. 19.47:

**a.** Find the total number of watts, volt-amperes reactive, volt-amperes, and $F_p$.

**b.** Find the current **I**.

**c.** Find the type elements and their impedance in each box. (Assume that the elements within each box are in series.)

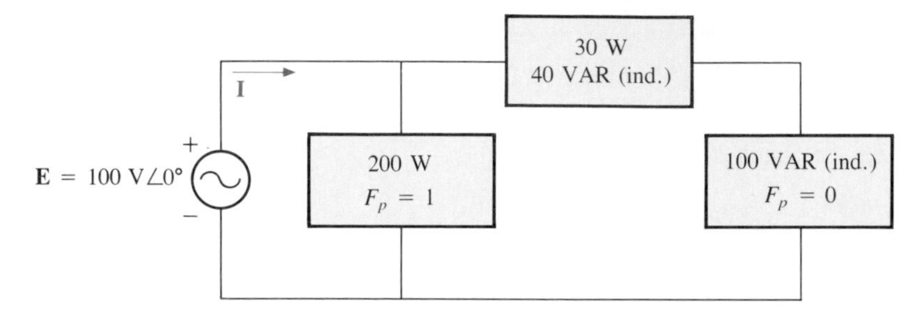

FIG. 19.47

13. For the circuit of Fig. 19.48:

**a.** Find the total number of watts, volt-amperes reactive, volt-amperes, and $F_p$.

**b.** Find the voltage **E**.

**c.** Find the type elements and their impedance in each box. (Assume that the elements within each box are in series.)

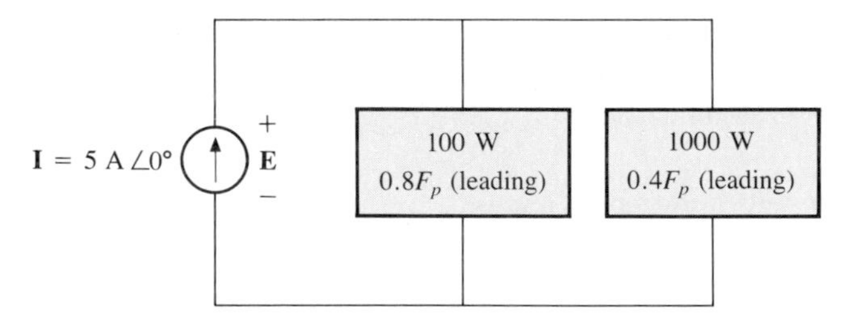


FIG. 19.48

*14. Repeat Problem 11 for the system of Fig. 19.49.

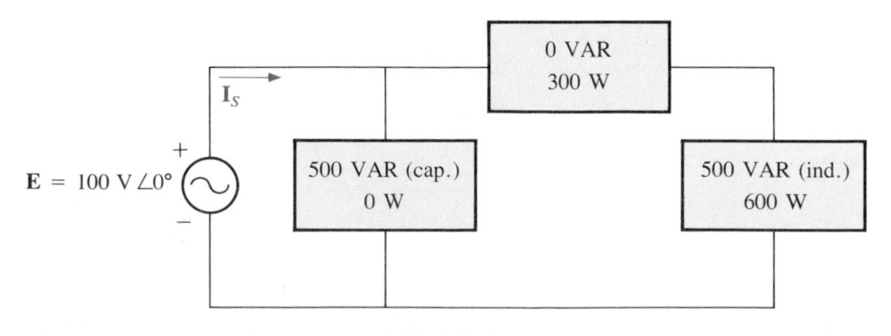


FIG. 19.49

## SECTION 19.8 Power-Factor Correction

**15.** The lighting and motor loads of a small factory establish a 10-kVA power demand at a 0.7 lagging power factor on a 208-V, 60-Hz supply. Determine what level of capacitance in parallel with the load will raise the power factor level to
  **a.** unity
  **b.** 0.9

**16.** The load on a 1200-V, 60-Hz supply is 5 kW (resistive), 8 kVAR (inductive), and 2 kVAR (capacitive).
  **a.** Determine the $F_p$ of the combined loads.
  **b.** Find the total kilovolt-amperes.
  **c.** Find the current drawn from the supply.
  **d.** Calculate the capacitance necessary to establish a unity power factor.
  **e.** Find the current drawn from the supply at unity power factor.

**17.** The loading of a factory on a 1000-V, 60-Hz system includes:

20-kW heating (unity power factor)

10-kW ($P_i$) induction motors (0.7 lagging power factor)

5-kW lighting (0.85 lagging power factor)

  **a.** Determine the total kilovolt-amperes.
  **b.** Find the total $F_p$.
  **c.** Derive the net reactive power.
  **d.** Find the current drawn from the supply.
  **e.** Calculate the capacitive contribution necessary to establish unity power factor for the total load.
  **f.** Calculate the current drawn from the supply under unity power-factor conditions.

## SECTION 19.9 The Wattmeter

**18.** **a.** A wattmeter is connected with its current coil as shown in Fig. 19.50 and the potential coil across points *f-g*. What does the wattmeter read?
  **b.** Repeat part (a) with the potential coil (*PC*) across *a-b*, *b-c*, *a-c*, *a-d*, *c-d*, *d-e*, and *f-e*.

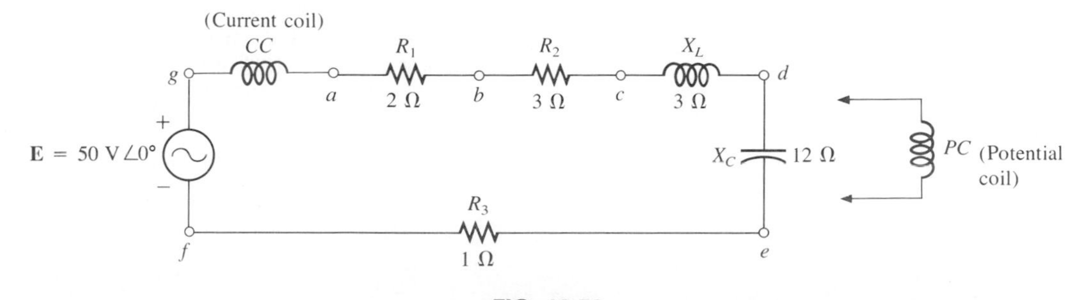


FIG. 19.50

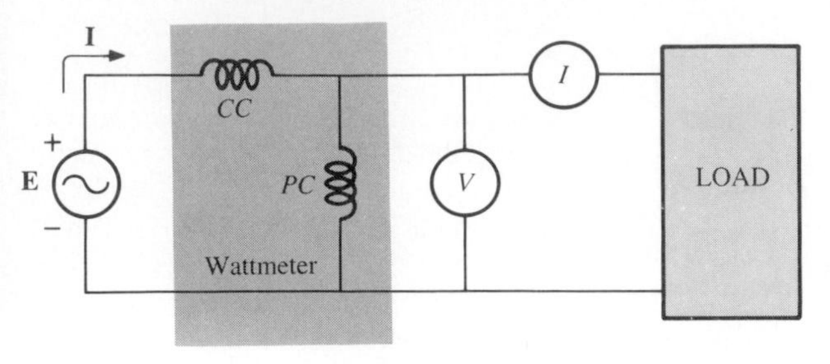

FIG. 19.51

19. The voltage source of Fig. 19.51 delivers 660 VA at 120 V with a supply current that lags the voltage by a power factor of 0.6.
    a. Determine the voltmeter, ammeter, and wattmeter readings.
    b. Find the load impedance in rectangular form.

## SECTION 19.10 Effective Resistance

20. a. An air-core coil is connected to a 200-V, 60-Hz source. The current is found to be 4 A, and a wattmeter reading of 80 W is observed. Find the effective resistance and the inductance of the coil.
    b. A brass core is inserted in the coil. The ammeter reads 3 A and the wattmeter reads 90 W. Calculate the effective resistance of the core. Explain the increase over the value of part (a).
    c. If a solid iron core is inserted in the coil, the current is found to be 2 A, and the wattmeter reads 60 W. Calculate the resistance and inductance of the coil. Compare these values to the values of part (a), and account for the changes.
21. a. The inductance of an air-core coil is 0.08 H, and the effective resistance is 4 Ω when a 60-V, 50-Hz source is connected across the coil. Find the current passing through the coil and the reading of a wattmeter across the coil.
    b. If a brass core is inserted in the coil, the effective resistance increases to 7 Ω, and the wattmeter reads 30 W. Find the current passing through the coil and the inductance of the coil.
    c. If a solid iron core is inserted in the coil, the effective resistance of the coil increases to 10 Ω, and the current decreases to 1.7 A. Find the wattmeter reading and the inductance of the coil.

## SECTION 19.11 Computer Analysis

22. Write a program that provides a general solution for the network of Fig. 19.19. That is, given the resistance or reactance of each element and the source voltage at zero degrees, calculate the real, reactive, and apparent power of the system.

23. Write a program that will demonstrate the effect of increasing reactive power on the power factor of a system. Tabulate the real power, reactive power, and power factor of the system for a fixed real power and a reactive power that starts at 10% of the real power and continues through to five times the real power in increments of 10% of the real power.

24. Write a program that will provide a general solution for exercises like that defined by Example 19.5.

# GLOSSARY

**Apparent power** The power delivered to a load without consideration of the effects of a power-factor angle of the load. It is determined solely by the product of the terminal voltage and current of the load.

**Average (real) power** The delivered power that is dissipated in the form of heat by a network or system.

**Eddy currents** Small, circular currents in a paramagnetic core, causing an increase in the power losses and the effective resistance of the material.

**Effective resistance** The resistance value that includes the effects of radiation losses, skin effect, eddy currents, and hysteresis losses.

**Hysteresis losses** Losses in a magnetic material introduced by changes in the direction of the magnetic flux within the material.

**Power-factor correction** The addition of reactive components (typically capacitive) to establish a system power factor closer to unity.

**Radiation losses** The loss of energy in the form of electromagnetic waves during the transfer of energy from one element to another.

**Reactive power** The power associated with reactive elements that provides a measure of the energy associated with setting up the magnetic and electric fields of inductive and capacitive elements, respectively.

**Skin effect** At high frequencies, a counter induced voltage builds up at the center of a conductor resulting in an increased flow near the surface (skin) of the conductor and a great reduction near the center. As a result, resistance increases as determined by the basic equation for resistance in terms of the geometric shape of the conductor.

# 20
# Resonance

## 20.1 INTRODUCTION

This chapter will introduce the very important resonant (or tuned) circuit, which is fundamental to the operation of a wide variety of electrical and electronic systems in use today. The resonant circuit is a combination of $R$, $L$, and $C$ elements having a frequency response characteristic as shown in Fig. 20.1. Note in the figure that the response

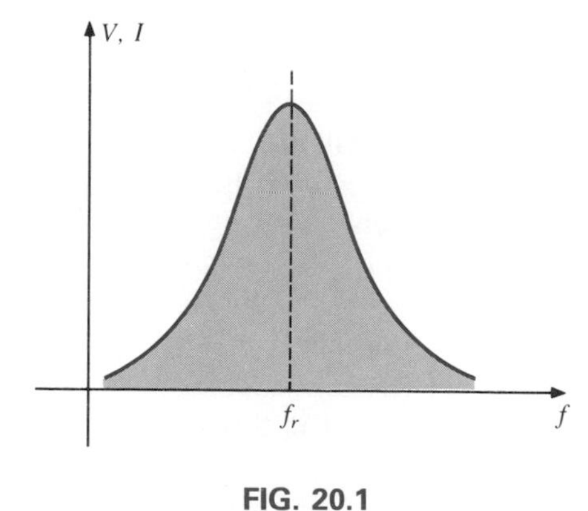


**FIG. 20.1**
*Resonance curve.*

is a maximum for the frequency $f_r$, decreasing to the right and left of this frequency. In other words, the resonant circuit selects a range of frequencies for which the response will be near or equal to the maximum. The frequencies to the far left or right are, for all practical purposes, nullified with respect to their effect on the system's response. The radio or television receiver has a response curve for each broadcast station of the type indicated in Fig. 20.1. When the receiver is set (or tuned) to a particular station, it is set on or near the frequency $f_r$ of Fig. 20.1.

Stations transmitting at frequencies to the far right or left of this resonant frequency are not carried through with significant power to affect the program of interest. The tuning process (setting the dial to $f_r$) as described above is the reason for the terminology *tuned circuit*. When the response is a maximum, the circuit is said to be in a state of *resonance,* with $f_r$ as the *resonant frequency*.

The concept of resonance is not limited to electrical or electronic systems. If mechanical impulses are applied to a mechanical system at the proper frequency, the system will enter a state of resonance in which sustained vibrations of very large amplitude will develop. The frequency at which this occurs is called the *natural frequency* of the system. The classic example of this effect was the Tacoma Narrows Bridge built in 1940 over Puget Sound in Washington State. It had a suspended span of 2800 feet. Four months after the bridge was completed, a 42-mi/h pulsating gale set the bridge into oscillations at its natural frequency. The amplitude of the oscillations increased to the point where the main span broke up and fell into the water below. It has since been replaced by the new Tacoma Narrows Bridge, completed in 1950.

The resonant electrical circuit *must* have both inductance and capacitance. In addition, resistance will always be present due either to the lack of ideal elements or to the control offered on the shape of the resonance curve. When resonance occurs due to the application of the proper frequency ($f_r$), the energy absorbed by one reactive element is the same as that released by another reactive element within the system. In other words, energy pulsates from one reactive element to the other. Therefore, once an ideal (pure $C$, $L$) system has reached a state of resonance, it requires no further reactive power since it is self-sustaining. In a practical circuit, there is some resistance associated with the reactive elements that will result in the eventual "damping" of the oscillations between reactive elements. In other words, the energy released by one reactive element will not be exactly equal to that absorbed by the other. Under resonant conditions, the total apparent power is equal to the average power dissipated by the resistive elements. *The average power absorbed by the system will also be a maximum at resonance,* just as the transfer of energy to the mechanical system above was a maximum at the natural frequency.

There are two types of resonant circuits: *series* and *parallel*. Each will be considered in some detail in this chapter.

## *SERIES RESONANCE*

## 20.2 SERIES RESONANT CIRCUIT

A resonant circuit (series or parallel) must have an inductive and capacitive element. A resistive element will always be present due to the resistance of the source ($R_s$), the internal resistance of the inductor ($R_l$), or an added resistance to control the shape of the response curve ($R_{\text{design}}$). The basic configuration for the series resonant circuit appears in Fig. 20.2(a) with the resistive elements listed above. The "cleaner" appearance of Fig. 20.2(b) is a result of combining the series resistive elements into one total value.

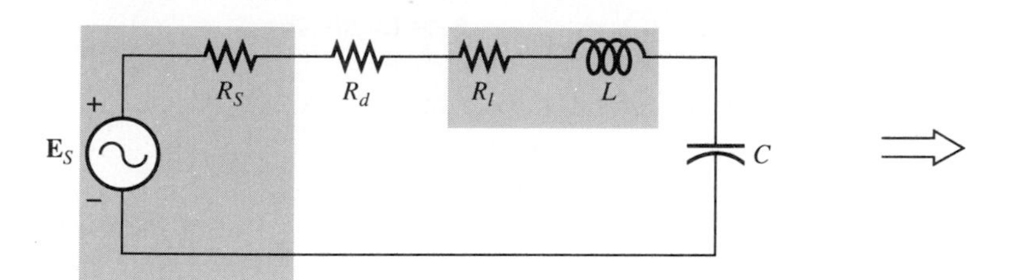


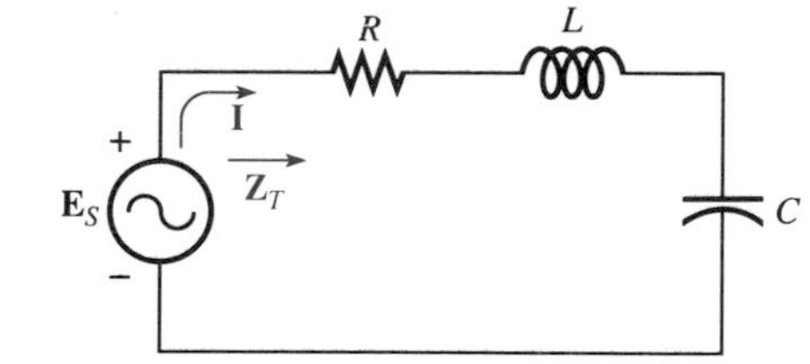


**FIG. 20.2**
*Series resonant circuit.*

That is,

$$R = R_s + R_l + R_d \tag{20.1}$$

The total impedance of this network at any frequency is determined by

$$\mathbf{Z}_T = R + jX_L - jX_C = R + j(X_L - X_C)$$

The resonant conditions described in the introduction will occur when

$$X_L = X_C \tag{20.2}$$

removing the reactive component from the total impedance equation. The total impedance at resonance is then simply

$$\mathbf{Z}_{T_s} = R \tag{20.3}$$

representing the minimum value of $\mathbf{Z}_T$ at any frequency. The subscript $s$ will be employed to indicate series resonant conditions.

The resonant frequency can be determined in terms of the inductance and capacitance by examining the defining equation for resonance [Eq. (20.2)]:

$$X_L = X_C$$

Substituting yields

$$\omega L = \frac{1}{\omega C} \quad \text{and} \quad \omega^2 = \frac{1}{LC}$$

and

$$\omega_s = \frac{1}{\sqrt{LC}} \tag{20.4}$$

or

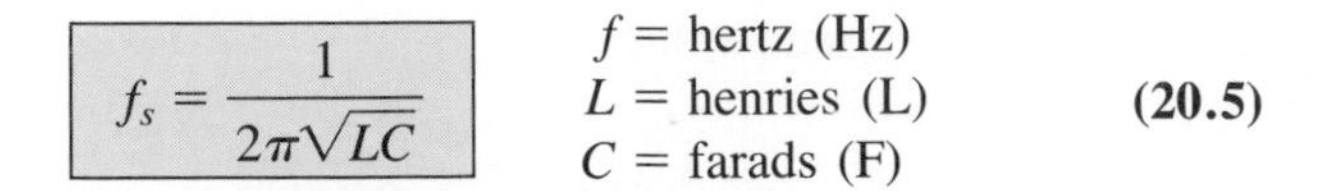

$$f_s = \frac{1}{2\pi\sqrt{LC}} \qquad \begin{aligned} f &= \text{hertz (Hz)} \\ L &= \text{henries (L)} \\ C &= \text{farads (F)} \end{aligned} \tag{20.5}$$

The current through the circuit at resonance is

$$\mathbf{I} = \frac{E \angle 0°}{R \angle 0°} = \frac{E}{R} \angle 0°$$

which you will note is the maximum current for the circuit of Fig. 20.2 for an applied voltage **E**, since $\mathbf{Z}_T$ is a minimum value. Consider also that *the input voltage and current are in phase at resonance.*

Since the current is the same through the capacitor and inductor, the voltage across each is equal in magnitude but 180° out of phase at resonance:

$$\left.\begin{aligned} \mathbf{V}_L &= \mathbf{IX}_L = (I \angle 0°)(X_L \angle 90°) = IX_L \angle 90° \\ \mathbf{V}_C &= \mathbf{IX}_C = (I \angle 0°)(X_C \angle -90°) = IX_C \angle -90° \end{aligned}\right\} \begin{aligned}&180° \\ &\text{out of} \\ &\text{phase}\end{aligned}$$

and, since $X_L = X_C$,

$$V_{L_s} = V_{C_s} \tag{20.6}$$

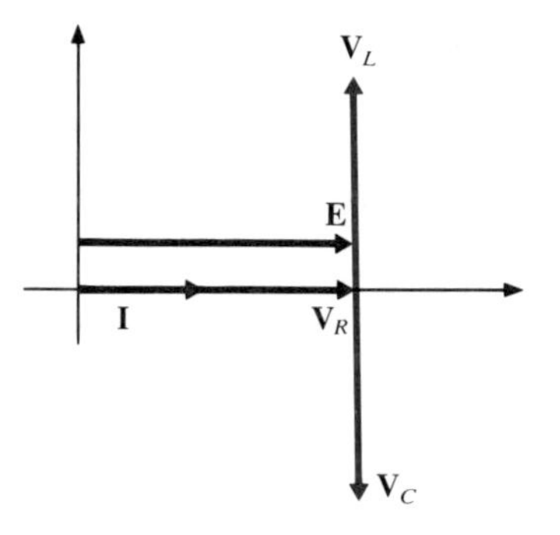


**FIG. 20.3**
*Phasor diagram for the series resonant circuit at resonance.*

Figure 20.3, a phasor diagram of the voltages and current, clearly indicates that the voltage across the resistor at resonance is the input voltage.

The average power to the resistor at resonance is equal to $I^2R$, and the reactive power to the capacitor and inductor are $I^2X_C$ and $I^2X_L$, respectively.

The power triangle at resonance (Fig. 20.4) shows that the total apparent power is equal to the average power dissipated by the resistor, since $Q_L = Q_C$. The power factor of the circuit at resonance is

$$F_p = \cos\theta = \frac{P}{S} = 1$$

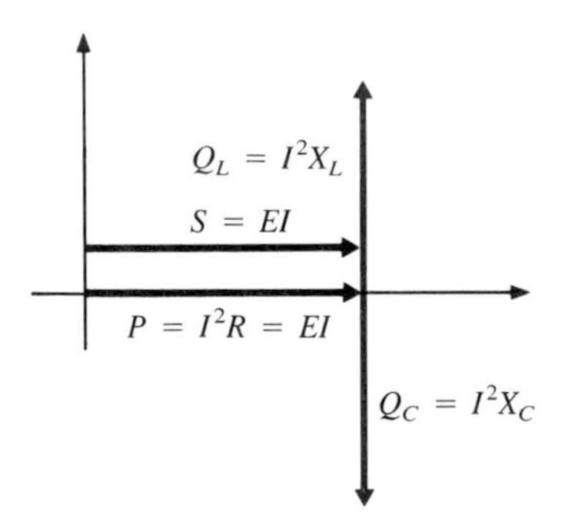


**FIG. 20.4**
*Power triangle for the series resonant circuit at resonance.*

Plotting the power curves of each element on the same set of axes (Fig. 20.5), we note that even though the total reactive power at any instant is equal to zero ($t = t'$), energy is still being absorbed and released by the inductor and capacitor at resonance.

A closer examination reveals that the energy absorbed by the inductor from time 0 to $t_1$ is the same as the energy being released by the capacitor from 0 to $t_1$. The reverse occurs from $t_1$ to $t_2$, and so on. Therefore, the total apparent power continues to be equal to the average power, even though the inductor and capacitor are absorbing and releasing energy. This condition occurs only at resonance. The slightest change in frequency introduces a reactive component into the power triangle which will increase the apparent power of the system above the average power dissipation, and resonance will no longer exist.

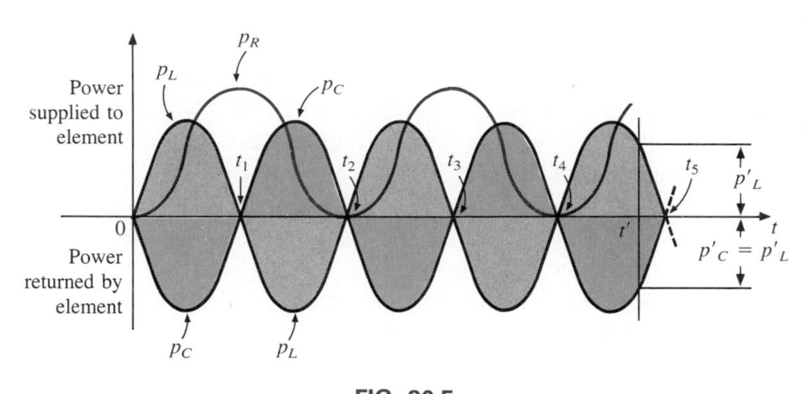


**FIG. 20.5**
*Power curves at resonance for the series resonant circuit.*

## 20.3 THE QUALITY FACTOR (Q)

The *quality factor* $Q$ of a series resonant circuit is defined as the ratio of the reactive power of either the inductor or the capacitor to the average power of the resistor at resonance; that is,

$$Q_s = \frac{\text{reactive power}}{\text{average power}} \quad \textbf{(20.7)}$$

The quality factor is also an indication of how much energy is placed in storage (continual transfer from one reactive element to the other) as compared to that dissipated. The lower the level of dissipation for the same reactive power, the larger the $Q_s$ factor and the more concentrated and intense the region of resonance.

Substituting for an inductive reactance in Eq. (20.7) at resonance gives us

$$Q_s = \frac{I^2 X_L}{I^2 R}$$

and

$$Q_s = \frac{X_L}{R} = \frac{\omega_s L}{R} \quad \textbf{(20.8)}$$

If the resistance $R$ is just the resistance of the coil ($R_l$), we can speak of the $Q$ of the coil, where

$$Q_{\text{coil}} = Q_l = \frac{X_L}{R_l} \qquad R = R_l \quad \textbf{(20.9)}$$

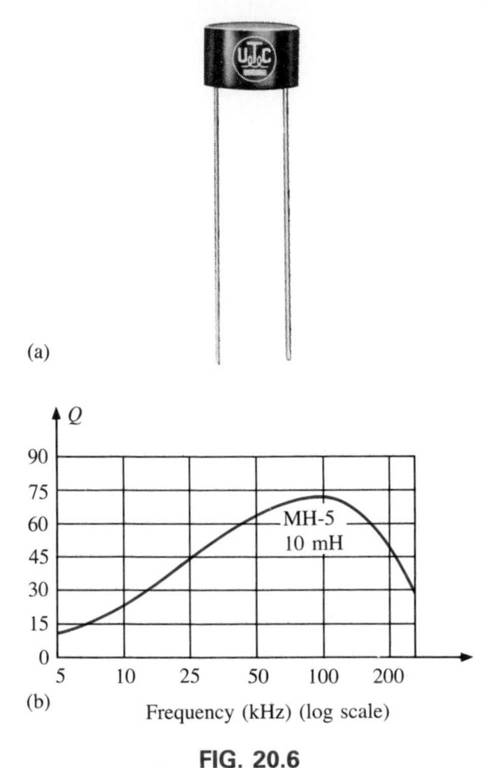


**FIG. 20.6**
*Q vs. frequency for a TRW/UTC 10-mH coil. (Courtesy of United Transformer Corp.)*

Since the quality factor of a coil is typically the information provided by manufacturers of inductors, it is often given the symbol $Q$ without an associated subscript. It would appear from Eq. (20.9) that $Q$ will increase linearly with frequency. That is, if the frequency doubles, then $Q$ will also increase by a factor of 2. This is approximately true for the low range to the midrange of frequencies such as shown for the coil of Fig. 20.6. Unfortunately, however, as the frequency increases, the effective resistance of the coil will also increase, due primarily to skin effect phenomena, and the resulting $Q$ will decrease. In addition, the capacitive effects between the windings will increase, further reducing the $Q$ of the coil. For this reason, the $Q$ of a coil must be specified at a particular frequency (usually at the maximum). For wide frequency applications, a plot of $Q$ versus frequency is often provided. The maximum $Q$ for most commercially available coils approaches 100.

If we substitute into Eq. (20.8) the fact that

$$\omega_s = 2\pi f_s$$

and

$$f_s = \frac{1}{2\pi\sqrt{LC}}$$

then

$$Q_s = \frac{\omega_s L}{R} = \frac{2\pi f_s L}{R} = \frac{2\pi}{R}\left(\frac{1}{2\pi\sqrt{LC}}\right)L$$

$$= \frac{L}{R}\left(\frac{1}{\sqrt{LC}}\right) = \left(\frac{\sqrt{L}}{\sqrt{L}}\right)\frac{L}{R\sqrt{LC}}$$

and

$$\boxed{Q_s = \frac{1}{R}\sqrt{\frac{L}{C}}} \qquad \textbf{(20.10)}$$

For series resonant circuits used in communication systems, $Q_s$ is usually greater than 1. By applying the voltage divider rule to the circuit of Fig. 20.2, we obtain

$$V_L = \frac{X_L E}{Z_T} = \frac{X_L E}{R} \qquad \text{(at resonance)}$$

and

$$\boxed{V_{L_s} = Q_s E} \qquad \textbf{(20.11)}$$

or

$$V_C = \frac{X_C E}{Z_T} = \frac{X_C E}{R}$$

and

$$V_{C_s} = Q_s E \quad (20.12)$$

Since $Q_s$ is usually greater than 1, the voltage across the capacitor or inductor of a series resonant circuit is usually greater than the input voltage. In fact, in many cases the $Q_s$ is so high that careful design and handling (including adequate insulation) are mandatory with respect to the voltage across the capacitor and inductor.

In the circuit of Fig. 20.7, for example, which is in the state of resonance,

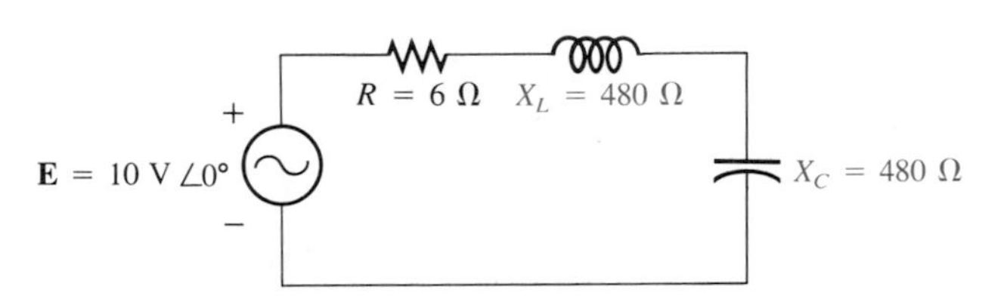


**FIG. 20.7**
*High-Q series resonant circuit.*

$$Q_s = \frac{X_L}{R} = \frac{480\ \Omega}{6\ \Omega} = 80$$

and

$$V_L = V_C = Q_s E = (80)(10\ \text{V}) = 800\ \text{V}$$

which is certainly a potential of significant magnitude.

## 20.4 $Z_T$ VERSUS FREQUENCY

The total impedance of the series *R-L-C* circuit of Fig. 20.2 at any frequency is determined by

$$\mathbf{Z}_T = R + jX_L - jX_C \quad \text{or} \quad \mathbf{Z}_T = R + j(X_L - X_C)$$

The magnitude of the impedance $\mathbf{Z}_T$ is

$$Z_T = \sqrt{R^2 + (X_L - X_C)^2}$$

The total-impedance-versus-frequency curve for the series resonant circuit of Fig. 20.2 can be found by applying the impedance-versus-frequency curve for each element of the equation just derived, written in the following form:

$$Z_T(f) = \sqrt{[R(f)]^2 + [X_L(f) - X_C(f)]^2} \quad (20.13)$$

where $Z_T(f)$ "means" the total impedance as a *function* of frequency. For the frequency range of interest, we will assume the resistance $R$

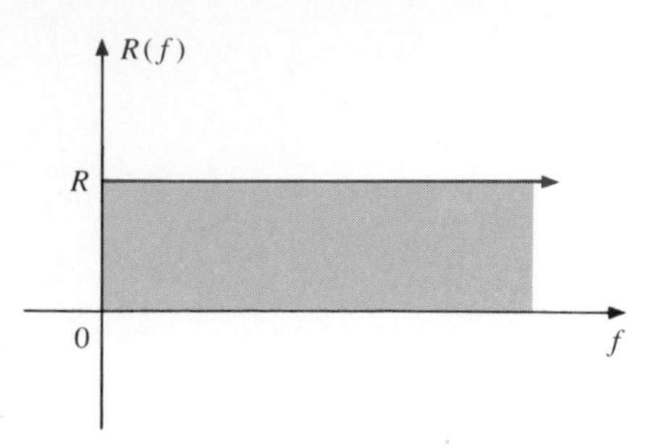


FIG. 20.8
*Resistance vs. frequency.*

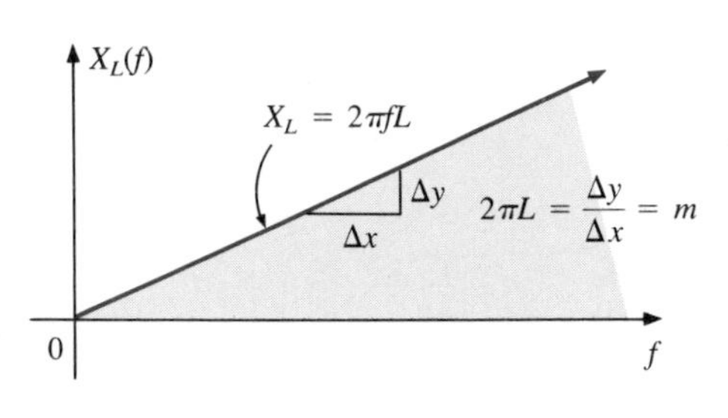


FIG. 20.9
*Inductive reactance vs. frequency.*

does not change with frequency, so its curve is a straight horizontal line with a magnitude $R$ above the frequency axis (Fig. 20.8). The curve for the inductance, as determined by the reactance equation, is a straight line intersecting the origin with a slope equal to the inductance of the coil. The mathematical expression for any straight line in a two-dimensional plane is given by

$$y = mx + b$$

Thus, for the coil,

$$\begin{array}{ccccc} X_L & = 2\pi fL + 0 = & \underbrace{(2\pi L)}(f) & + & 0 \\ \downarrow & & \downarrow \quad \downarrow & & \downarrow \\ y & = & m \quad \cdot \quad x & + & b \end{array}$$

(where $2\pi L$ is the slope), producing the results shown in Fig. 20.9.

For the capacitor,

$$X_C = \frac{1}{2\pi fC} \quad \text{or} \quad X_C f = \frac{1}{2\pi C}$$

which becomes $yx = k$, the equation for a hyperbola, where

$$y \text{ (variable)} = X_C$$
$$x \text{ (variable)} = f$$
$$k \text{ (constant)} = \frac{1}{2\pi C}$$

The hyperbolic curve for $X_C(f)$ is plotted in Fig. 20.10. In particular note its very large magnitude at low frequencies and its rapid drop-off as the frequency increases.

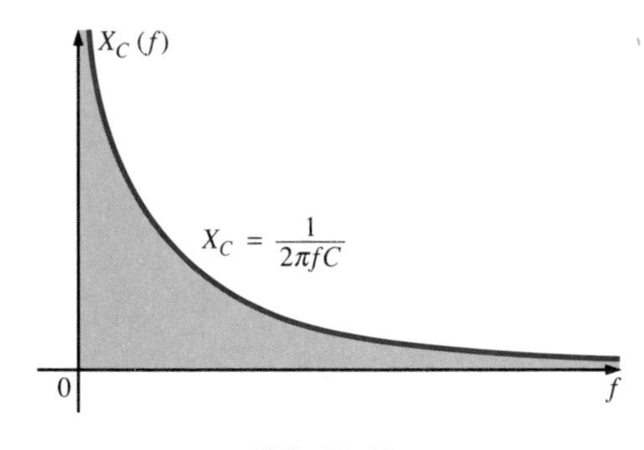


FIG. 20.10

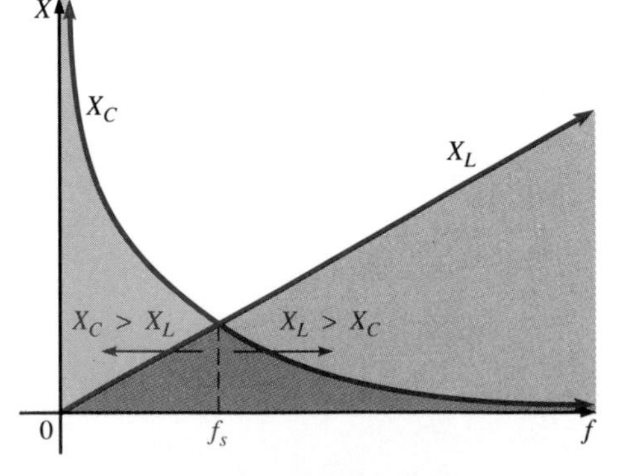


FIG. 20.11

If we place Figs. 20.9 and 20.10 on the same axis, we obtain the waveforms of Fig. 20.11. The condition of resonance is now clearly defined by the point of intersection, where $X_L = X_C$. For frequencies less than $f_s$, it is also quite clear that the network is primarily capacitive ($X_C > X_L$). For frequencies above the resonant condition, $X_L > X_C$ and the network is inductive.

Applying

$$Z_T(f) = \sqrt{[R(f)]^2 + [X_L(f) - X_C(f)]^2}$$
$$= \sqrt{[R(f)]^2 + [X(f)]^2}$$

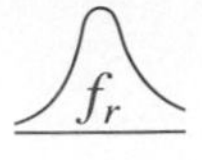




to the curves of Fig. 20.11 where $X(f) = X_L(f) - X_C(f)$, we obtain the curve for $Z_T(f)$ as shown in Fig. 20.12. The minimum impedance occurs at the resonant frequency and is equal to the resistance $R$. Note that the curve is not symmetrical about the resonant frequency (especially at higher values of $Z_T$).

The phase angle associated with the total impedance is

$$\theta = \tan^{-1} \frac{(X_L - X_C)}{R} \quad \textbf{(20.14)}$$

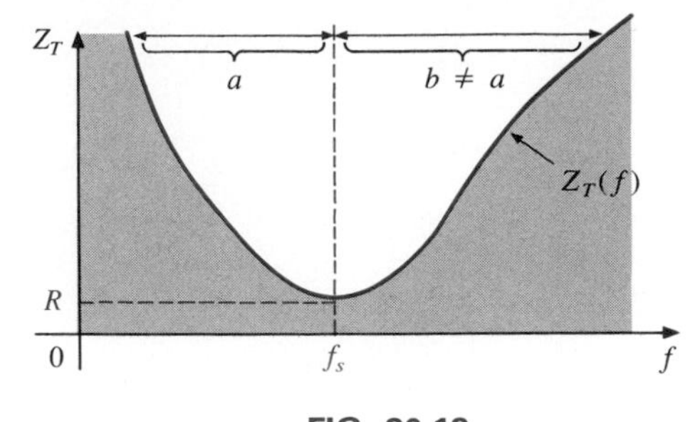


**FIG. 20.12**
$\mathbf{Z}_T$ *vs. frequency for the series resonant circuit.*

For the $\tan^{-1} x$ function, the larger $x$ is, the larger the angle $\theta$ (closer to 90°). However, one must also be aware that

$$\tan^{-1}(-x) = -\tan^{-1} x \quad \textbf{(20.15)}$$

At low frequencies, $X_C > X_L$ and $\theta$ will approach $-90°$ (capacitive) as shown in Fig. 20.13, whereas at high frequencies, $X_L > X_C$ and $\theta$ will approach 90°. In general, therefore, for a series resonant circuit:

| | |
|---|---|
| $f < f_s$: | network capacitive, **I** leads **E** |
| $f > f_s$: | network inductive, **E** leads **I** |
| $f = f_s$: | network resistive, **E** and **I** in phase |

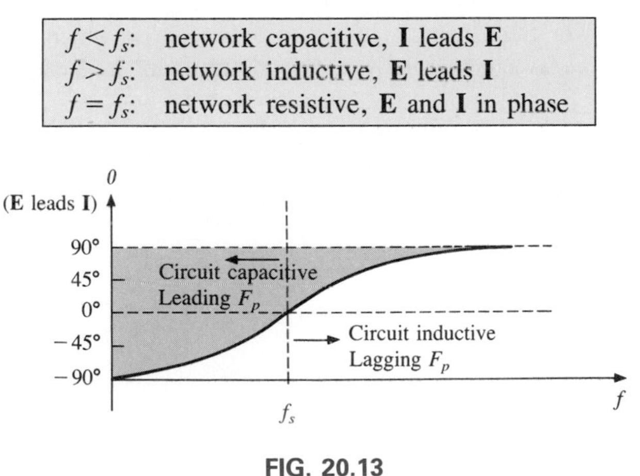


**FIG. 20.13**
*Phase plot for the series resonant circuit.*

## 20.5 SELECTIVITY

If we now plot the magnitude of the current $I = E/Z_T$ versus frequency for a *fixed* applied voltage $E$, we obtain the curve shown in Fig. 20.14, which rises from zero to a maximum value of $E/R$ (where $Z_T$ is a minimum) and then drops toward zero (as $Z_T$ increases) at a slower rate than it rose to its peak value. The curve is actually the inverse of the impedance-versus-frequency curve. Since the $Z_T$ curve is not absolutely symmetrical about the resonant frequency, the curve of the current versus frequency has the same property.

There is a definite range of frequencies at which the current is near its maximum value and the impedance is at a minimum. Those frequencies corresponding to 0.707 of the maximum current are called the *band frequencies, cutoff frequencies,* or *half-power frequencies.* They are indicated by $f_1$ and $f_2$ in Fig. 20.14. The range of frequencies between the two is referred to as the *bandwidth* (abbreviated BW) of the resonant circuit.

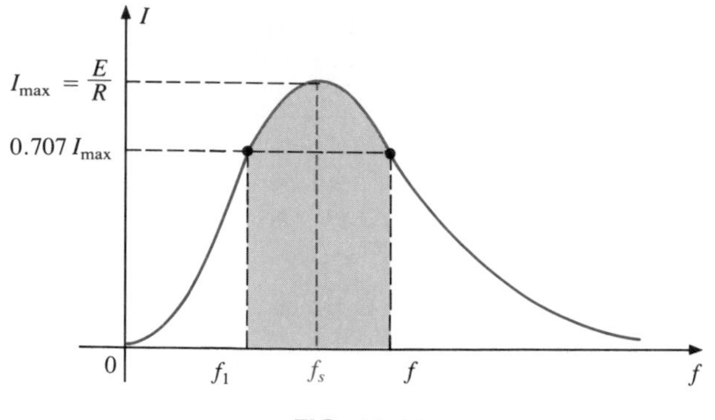


**FIG. 20.14**
*I vs. frequency for the series resonant circuit.*

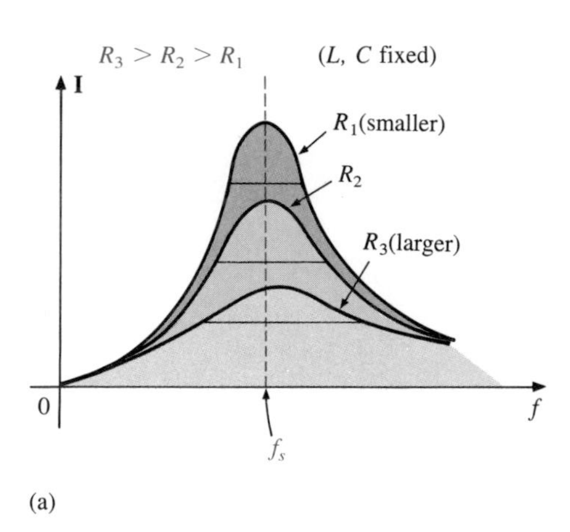


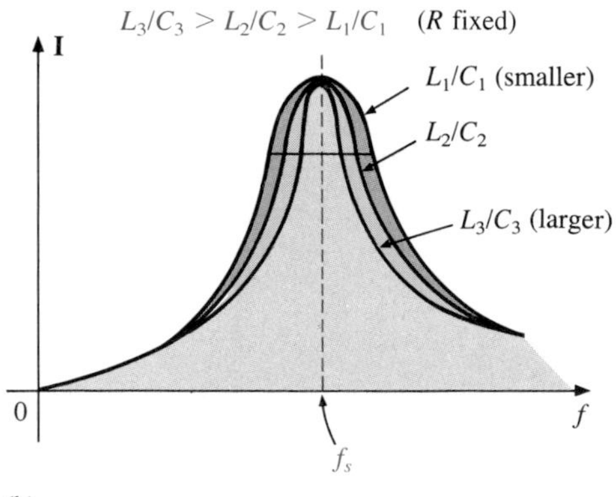


**FIG. 20.15**
*Effect of R, L, and C on the selectivity curve for the series resonant circuit.*

Half-power frequencies are those frequencies at which the power delivered is one-half that delivered at the resonant frequency; that is,

$$P_{HPF} = \frac{1}{2}P_{max} \tag{20.16}$$

The above condition is derived using the fact that

$$P_{max} = I_{max}^2 R$$

and

$$P_{HPF} = I^2R = (0.707I_{max})^2R = 0.5I_{max}^2R = \frac{1}{2}P_{max}$$

Since the resonant circuit is adjusted to select a band of frequencies, the curve of Fig. 20.14 is called the *selectivity curve*. The term is derived from the fact that one must be selective in choosing the frequency to insure that it is in the bandwidth. The smaller the bandwidth, the higher the selectivity. The shape of the curve, as shown in Fig. 20.15, depends on each element of the series *R-L-C* circuit. If the resistance is made smaller with a fixed inductance and capacitance, the bandwidth decreases and the selectivity increases. Similarly, if the ratio *L/C* increases with fixed resistance, the bandwidth again decreases with an increase in selectivity.

In terms of $Q_s$, if $R$ is larger for the same $X_L$, then $Q_s$ is less, as determined by the equation $Q_s = \omega_s L/R$.

***A small $Q_s$, therefore, is associated with a resonant curve with a large bandwidth and a small selectivity, while a large $Q_s$ indicates the opposite.***

***For circuits where $Q_s \geq 10$, a widely accepted approximation is that the resonant frequency bisects the bandwidth***

and that the resonant curve is symmetrical about the resonant frequency. These conditions are shown in Fig. 20.16, indicating that the cutoff frequencies are then equidistant from the resonant frequency.

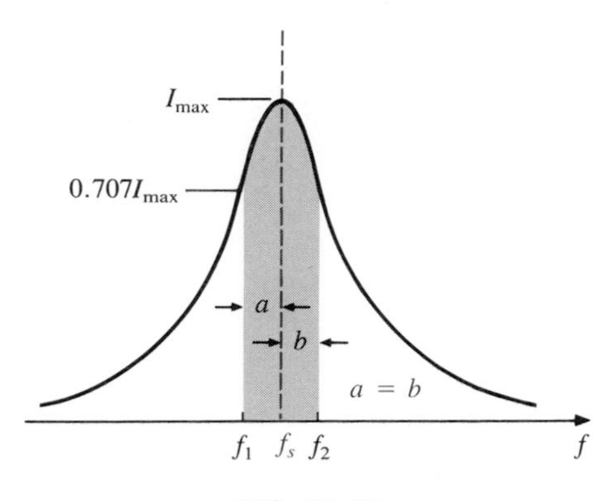


**FIG. 20.16**
*Approximate series resonance curve for $Q_s \geq 10$.*

For any $Q_s$, the preceding is not true. The cutoff frequencies $f_1$ and $f_2$ can be found for the general case (any $Q_s$) by first employing the fact that a drop in current to 0.707 of its resonant value corresponds to an increase in impedance equal to $1/0.707 = \sqrt{2}$ times the resonant value, which is $R$.

Substituting $\sqrt{2}R$ into the equation for the magnitude of $\mathbf{Z}_T$, we find that

$$\mathbf{Z}_T = \sqrt{R^2 + (X_L - X_C)^2}$$

becomes

$$\sqrt{2}R = \sqrt{R^2 + (X_L - X_C)^2}$$

or, squaring both sides, that

$$2R^2 = R^2 + (X_L - X_C)^2$$

and

$$R^2 = (X_L - X_C)^2$$

Taking the square root of both sides gives us

$$R = X_L - X_C$$

Let us first consider the case where $X_L > X_C$, which relates to $f_2$ or $\omega_2$. Substituting $\omega_2 L$ for $X_L$ and $1/\omega_2 C$ for $X_C$ and bringing both quantities to the left of the equal sign, we have

$$R - \omega_2 L + \frac{1}{\omega_2 C} = 0 \quad \text{or} \quad R\omega_2 - \omega_2^2 L + \frac{1}{C} = 0$$

which can be written

$$\omega_2^2 - \frac{R}{L}\omega_2 - \frac{1}{LC} = 0$$

Solving the quadratic, we have

$$\omega_2 = \frac{-(-R/L) \pm \sqrt{[-(R/L)]^2 - [-(4/LC)]}}{2}$$

and

$$\omega_2 = +\frac{R}{2L} \pm \frac{1}{2}\sqrt{\frac{R^2}{L^2} + \frac{4}{LC}}$$

with

$$f_2 = \frac{1}{2\pi}\left[\frac{R}{2L} + \frac{1}{2}\sqrt{\left(\frac{R}{L}\right)^2 + \frac{4}{LC}}\right] \quad \text{(Hz)} \qquad \textbf{(20.17)}$$

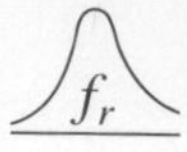


The negative sign in front of the second factor was dropped because $(1/2)\sqrt{R^2/L^2 + 4/LC}$ is always greater than $R/(2L)$. If it were not dropped, there would be a negative solution for the radian frequency $\omega$.

If we repeat the same procedure for $X_C > X_L$, which relates to $\omega_1$ or $f_1$ such that $Z_T = \sqrt{R^2 + (X_C - X_L)^2}$, the solution $f_1$ becomes

$$f_1 = \frac{1}{2\pi}\left[-\frac{R}{2L} + \frac{1}{2}\sqrt{\left(\frac{R}{L}\right)^2 + \frac{4}{LC}}\right] \quad \text{(Hz)} \qquad \textbf{(20.18)}$$

The bandwidth (BW) is

$$\text{BW} = f_2 - f_1 = \text{Eq. (20.17)} - \text{Eq. (20.18)}$$

and

$$\text{BW} = f_2 - f_1 = \frac{R}{2\pi L} \qquad \textbf{(20.19)}$$

Substituting $R/L = \omega_s/Q_s$ from $Q_s = \omega_s L/R$ and $1/2\pi = f_s/\omega_s$ from $\omega_s = 2\pi f_s$ gives us

$$\text{BW} = \frac{R}{2\pi L} = \left(\frac{f_s}{\omega_s}\right)\left(\frac{\omega_s}{Q_s}\right)$$

or

$$\text{BW} = \frac{f_s}{Q_s} \qquad \textbf{(20.20)}$$

which is a very convenient form, since it relates the bandwidth to the $Q_s$ of the circuit. As mentioned earlier, Eq. (20.20) verifies that the larger the $Q_s$, the smaller the bandwidth, and vice versa.

Written in a slightly different form, Eq. (20.20) becomes

$$\frac{f_2 - f_1}{f_s} = \frac{1}{Q_s} \qquad \textbf{(20.21)}$$

The ratio $(f_2 - f_1)/f_s$ is sometimes called the *fractional bandwidth,* providing an indication of the width of the bandwidth as compared to the resonant frequency.

It can also be shown through mathematical manipulations of the pertinent equations that the resonant frequency is related to the geometric mean of the band frequencies. That is,

$$f_s = \sqrt{f_1 f_2} \qquad \textbf{(20.22)}$$

## 20.6 $V_R$, $V_L$, AND $V_C$

Plotting the magnitude (effective value) of the voltages $\mathbf{V}_R$, $\mathbf{V}_L$, and $\mathbf{V}_C$ and the current $\mathbf{I}$ versus frequency for the series resonant circuit on the same set of axes, we obtain the curves shown in Fig. 20.17. Note that

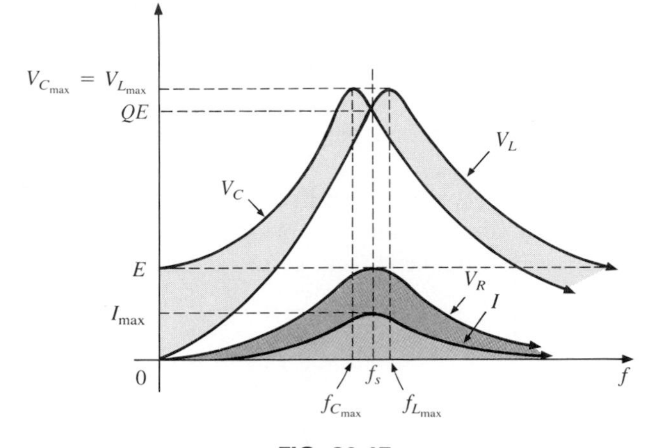


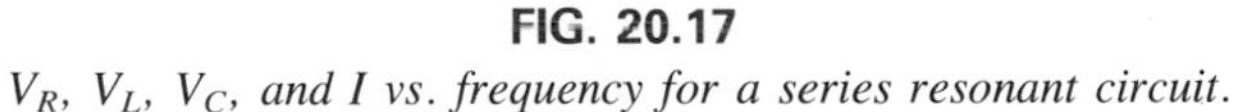

**FIG. 20.17**

*$V_R$, $V_L$, $V_C$, and $I$ vs. frequency for a series resonant circuit.*

the $V_R$ curve has the same shape as the $I$ curve and a peak value equal to the magnitude of the input voltage $E$. The $V_C$ curve builds up slowly at first from a value equal to the input voltage, since the reactance of the capacitor is infinite (open circuit) at zero frequency and the reactance of the inductor is zero (short circuit) at this frequency. As the frequency increases, $1/\omega C$ of the equation

$$V_C = IX_C = I\frac{1}{\omega C}$$

becomes smaller, but $I$ increases at a rate faster than that at which $1/\omega C$ drops. Therefore, $V_C$ rises and will continue to rise, due to the quickly rising current, until the frequency nears resonance. As it approaches the resonant condition, the rate of change of $I$ decreases. When this occurs, the factor $1/\omega C$, which decreased as the frequency rose, will overcome the rate of change of $I$, and $V_C$ will start to drop. The peak value will occur at a frequency just before resonance. After resonance, both $V_C$ and $I$ drop in magnitude, and $V_C$ approaches zero.

The higher the $Q_s$ of the circuit, the closer $f_{C_{max}}$ will be to $f_s$, and the closer $V_{C_{max}}$ will be to $Q_sE$. For circuits with $Q_s \geq 10$, $f_{C_{max}} \cong f_s$, and $V_{C_{max}} \cong Q_sE$.

The curve for $V_L$ increases steadily from zero to the resonant frequency, since both quantities $\omega L$ and $I$ of the equation $V_L = IX_L = \omega LI$ increase over this frequency range. At resonance, $I$ has reached its maximum value, but $\omega L$ is still rising. Therefore, $V_L$ will reach its maximum value after resonance. After reaching its peak value, the

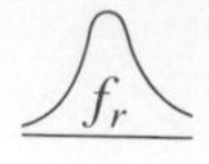


voltage $V_L$ will drop toward $E$, since the drop in $I$ will overcome the rise in $\omega L$. It approaches $E$ because $X_L$ will eventually be infinite, and $X_C$ will be zero.

As $Q_s$ of the circuit increases, the frequency $f_{L_{max}}$ drops toward $f_s$, and $V_{L_{max}}$ approaches $Q_sE$. For circuits with $Q_s \geq 10$, $f_{L_{max}} \cong f_s$, and $V_{L_{max}} \cong Q_sE$.

The $V_L$ curve has a greater magnitude than the $V_C$ curve for any frequency above resonance, and the $V_C$ curve has a greater magnitude than the $V_L$ curve for any frequency below resonance. This again verifies that the series *R-L-C* circuit is predominantly capacitive from zero to the resonant frequency and predominantly inductive for any frequency above resonance.

For the condition $Q_s \geq 10$, the curves of Fig. 20.17 will appear as shown in Fig. 20.18. Note that they each peak (on an approximate basis) at the resonant frequency and have a similar shape.

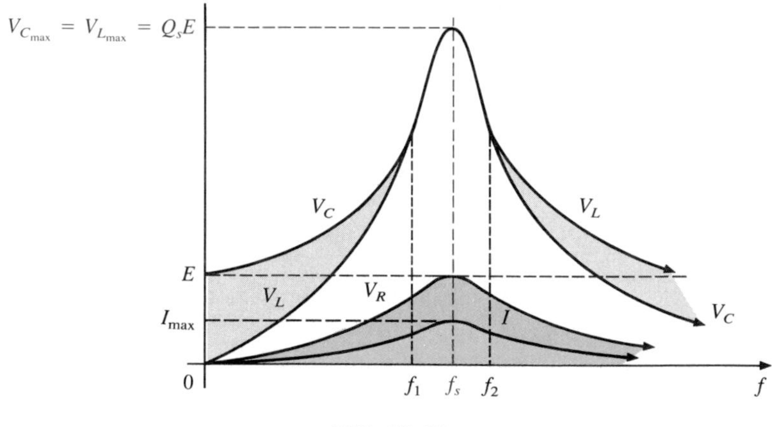


**FIG. 20.18**

*Approximate curves for $V_R$, $V_L$, $V_C$, and $I$ for a series resonant circuit where $Q_s \geq 10$.*

In review, for the series resonant circuit:

1. $V_C$ and $V_L$ are their maximum values at or near resonance (depending on $Q$).
2. At very low frequencies, $V_C$ is very close to the source voltage and $V_L$ is very close to zero volts, whereas at very high frequencies, $V_L$ approaches the source voltage and $V_C$ approaches zero volts.
3. Both $V_R$ and $I$ peak at the resonant frequency and have the same shape.

## 20.7 EXAMPLES (SERIES RESONANCE)

**EXAMPLE 20.1.** For the resonant circuit shown in Fig. 20.19, find $i$, $v_R$, $v_L$, and $v_C$ in phasor form. What is the $Q_s$ of the circuit? If the

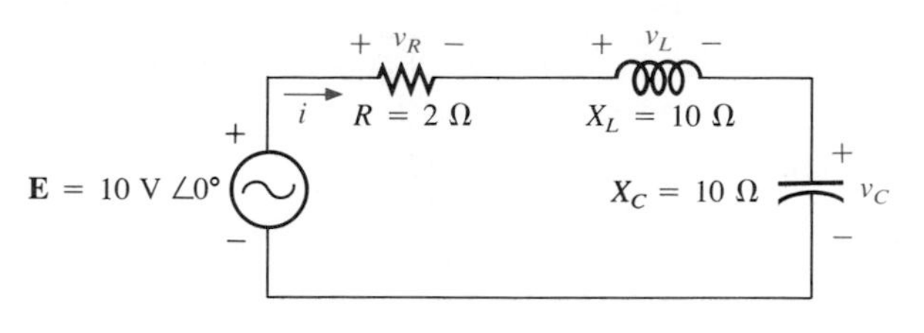


**FIG. 20.19**

resonant frequency is 5000 Hz, find the bandwidth. What is the power dissipated in the circuit at the half-power frequencies?

***Solution:***

$$\mathbf{Z}_{T_s} = R = 2\,\Omega$$

$$\mathbf{I} = \frac{\mathbf{E}}{\mathbf{Z}_{T_s}} = \frac{10\text{ V }\angle 0^\circ}{2\,\Omega\angle 0^\circ} = \mathbf{5\text{ A }\angle 0^\circ}$$

$$\mathbf{V}_R = \mathbf{E} = 10\text{ V }\angle 0^\circ$$

$$\mathbf{V}_L = \mathbf{I}\mathbf{X}_L = (5\text{ A }\angle 0^\circ)(10\,\Omega\angle 90^\circ) = \mathbf{50\text{ V }\angle 90^\circ}$$

$$\mathbf{V}_C = \mathbf{I}\mathbf{X}_C = (5\text{ A }\angle 0^\circ)(10\,\Omega\angle -90^\circ) = \mathbf{50\text{ V }\angle -90^\circ}$$

$$Q_s = \frac{X_L}{R} = \frac{10\,\Omega}{2\,\Omega} = \mathbf{5}$$

$$\text{BW} = f_2 - f_1 = \frac{f_s}{Q_s} = \frac{5000\text{ Hz}}{5} = \mathbf{1000\text{ Hz}}$$

$$P_{\text{HPF}} = \frac{1}{2}P_{\max} = \frac{1}{2}I_{\max}^2R = \left(\frac{1}{2}\right)(5\text{ A})^2(2\,\Omega) = \mathbf{25\text{ W}}$$

---

**EXAMPLE 20.2.** The bandwidth of a series resonant circuit is 400 Hz. If the resonant frequency is 4000 Hz, what is the value of $Q_s$? If $R = 10\,\Omega$, what is the value of $X_L$ at resonance? Find the inductance $L$ and capacitance $C$ of the circuit.

***Solution:***

$$\text{BW} = \frac{f_s}{Q_s} \quad \text{or} \quad Q_s = \frac{f_s}{\text{BW}} = \frac{4000\text{ Hz}}{400\text{ Hz}} = \mathbf{10}$$

$$Q_s = \frac{X_L}{R} \quad \text{or} \quad X_L = Q_sR = (10)(10\,\Omega) = \mathbf{100\,\Omega}$$

$$X_L = 2\pi f_sL \quad \text{or} \quad L = \frac{X_L}{2\pi f_s} = \frac{100\,\Omega}{(6.28)(4000\text{ Hz})} = \mathbf{3.98\text{ mH}}$$

$$X_C = \frac{1}{2\pi f_sC} \quad \text{or} \quad C = \frac{1}{2\pi f_sX_C} = \frac{1}{(6.28)(4000\text{ Hz})(100\,\Omega)} = \mathbf{0.398\ \mu F}$$

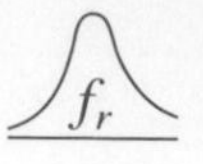


**EXAMPLE 20.3.** A series *R-L-C* circuit has a series resonant frequency of 12,000 Hz. If $R = 5\ \Omega$ and $X_L$ at resonance is $300\ \Omega$, find the bandwidth. Find the cutoff frequencies.

***Solution:***

$$Q_s = \frac{X_L}{R} = \frac{300\ \Omega}{5\ \Omega} = 60$$

$$\text{BW} = \frac{f_s}{Q_s} = \frac{12{,}000\ \text{Hz}}{60} = \mathbf{200\ Hz}$$

Since $Q_s \geq 10$, the bandwidth is bisected by $f_s$. Therefore,

$$f_2 = f_s + \frac{\text{BW}}{2} = 12{,}000\ \text{Hz} + 100\ \text{Hz} = \mathbf{12{,}100\ Hz}$$

$$f_1 = 12{,}000\ \text{Hz} - 100\ \text{Hz} = \mathbf{11{,}900\ Hz}$$

**EXAMPLE 20.4.**

a. Determine the $Q_s$ and bandwidth for the response curve of Fig. 20.20.
b. For $C = 0.1\ \mu\text{F}$, determine $L$ and $R$ for the series resonant circuit.

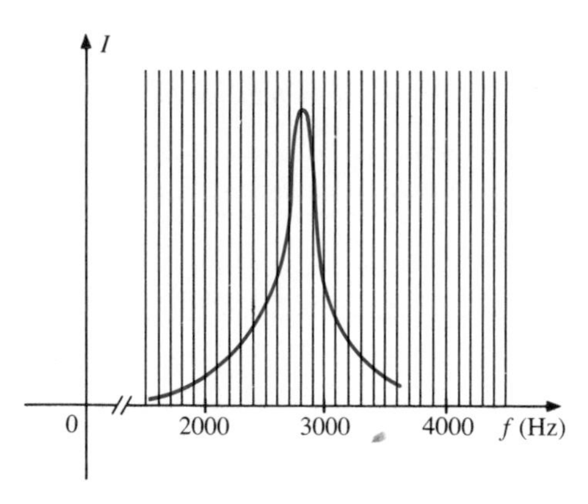


FIG. 20.20

***Solutions:***

a. The resonant frequency is 2800 Hz. At 0.707 times the peak value, BW = **200 Hz,** and

$$Q_s = \frac{f_s}{\text{BW}} = \frac{2800\ \text{Hz}}{200\ \text{Hz}} = \mathbf{14}$$

b. $f_s = \dfrac{1}{2\pi\sqrt{LC}}$ or $L = \dfrac{1}{4\pi^2 f_s^2 C}$

$$= \frac{1}{4\pi^2(2.8 \times 10^3\ \text{Hz})^2(0.1 \times 10^{-6}\ \text{F})}$$

$$= \frac{1}{30.951} = \mathbf{32.31\ mH}$$

$$Q_s = \frac{X_L}{R} \quad \text{or} \quad R = \frac{X_L}{Q_s} = \frac{2\pi(2800\ \text{Hz})(32.31 \times 10^{-3}\ \text{H})}{14}$$

$$= \mathbf{40.572\ \Omega}$$

**EXAMPLE 20.5.** A series *R-L-C* circuit is designed to resonate at $\omega_s = 10^5$ rad/s, have a bandwidth of $0.15\omega_s$, and draw 16 W from a 120-V source.

a. Determine the value of $R$.
b. Find the bandwidth in hertz.
c. Find the nameplate values of $L$ and $C$.
d. What is the $Q_s$ of the circuit?
e. Determine the fractional bandwidth.

***Solutions:***

a. $P = \dfrac{E^2}{R}$ and $R = \dfrac{E^2}{P} = \dfrac{(120\text{ V})^2}{16\text{ W}} = \mathbf{900\ \Omega}$

b. $f_s = \dfrac{\omega_s}{2\pi} = \dfrac{10^5\text{ rad/s}}{2\pi} = 15{,}915.49\text{ Hz}$

$\text{BW} = 0.15f_s = 0.15(15{,}915.49\text{ Hz}) = \mathbf{2387.32\ Hz}$

c. Eq. (20.19):

$$\text{BW} = \frac{R}{2\pi L} \quad \text{and} \quad L = \frac{R}{2\pi \text{BW}} = \frac{900\ \Omega}{2\pi(2387.32\text{ Hz})} = \mathbf{60\ mH}$$

$$f_s = \frac{1}{2\pi\sqrt{LC}} \quad \text{and} \quad C = \frac{1}{4\pi^2 f_s^2 L} = \frac{1}{4\pi^2(15{,}915.49\text{ Hz})^2(60 \times 10^{-3}\text{ H})}$$
$$= \mathbf{1.67\ nF}$$

d. $Q_s = \dfrac{X_L}{R} = \dfrac{2\pi f_s L}{R} = \dfrac{2\pi(15{,}915.49\text{ Hz})(60\text{ mH})}{R} = \mathbf{6.67}$

e. $\dfrac{f_2 - f_1}{f_s} = \dfrac{\text{BW}}{f_s} = \dfrac{1}{Q_s} = \dfrac{1}{6.67} = \mathbf{0.15}$

## *PARALLEL RESONANCE*

## 20.8 PARALLEL RESONANT CIRCUIT

The basic format of the series resonant circuit is a series *R-L-C* combination in series with an applied voltage source. The parallel resonant circuit has the basic configuration of Fig. 20.21, a parallel *R-L-C* combination in parallel with an applied current source.

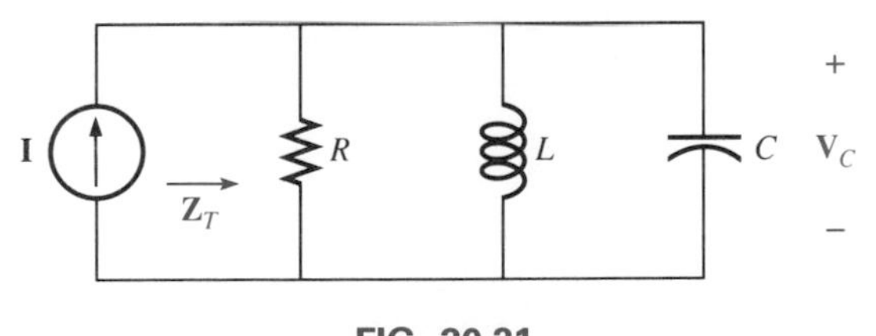


**FIG. 20.21**
*Ideal parallel resonant network.*

For the series circuit, the impedance was a minimum at resonance producing a significant current that resulted in a high output voltage for $\mathbf{V}_C$ and $\mathbf{V}_L$. For the parallel resonant circuit, the impedance is a maximum at resonance producing a significant voltage for $\mathbf{V}_C$ and $\mathbf{V}_L$ through the Ohm's law relationship ($\mathbf{V}_C = \mathbf{I}\mathbf{Z}_T$). For the network of Fig. 20.21, resonance will occur when $X_L = X_C$, and the resonant frequency will have the same format obtained for series resonance.

If the practical equivalent of Fig. 20.21 had the format of Fig. 20.21, the analysis would be as direct and lucid as that experienced for series

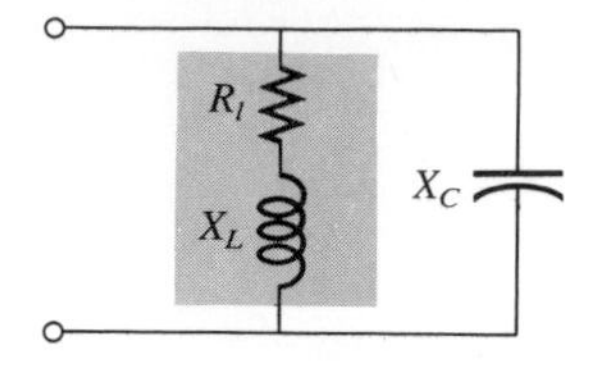

FIG. 20.22

resonance. However, in the practical world, the internal resistance of the coil must be placed in series with the inductor, as shown in Fig. 20.22. The resistance $R_l$ can no longer be included in a simple series or parallel combination with the source resistance and any other resistance added for design purposes. Even though $R_l$ is usually relatively small in magnitude compared with other resistance and reactance levels of the network, it does have an important impact on the parallel resonant condition, as will be demonstrated in the sections to follow. In other words, the network of Fig. 20.21 is an ideal situation that can only be assumed for specific network conditions.

Our first effort will be to find a parallel network equivalent (at the terminals) for the series $R$-$L$ branch of Fig. 20.22 using the technique introduced in Section 15.10. That is,

$$\mathbf{Z}_{R\text{-}L} = R_l + jX_L$$

and

$$\mathbf{Y}_{R\text{-}L} = \frac{1}{\mathbf{Z}_{R\text{-}L}} = \frac{1}{R_l + jX_L} = \frac{R_l}{R_l^2 + X_L^2} - j\frac{X_L}{R_l^2 + X_L^2}$$
$$= \frac{1}{\dfrac{R_l^2 + X_L^2}{R_l}} + \frac{1}{j\left(\dfrac{R_l^2 + X_L^2}{X_L}\right)} = \frac{1}{R_p} + \frac{1}{jX_{L_p}}$$

with

$$R_p = \frac{R_l^2 + X_L^2}{R_l} \qquad (20.23)$$

and

$$X_{L_p} = \frac{R_l^2 + X_L^2}{X_L} \qquad (20.24)$$

as shown in Fig. 20.23.

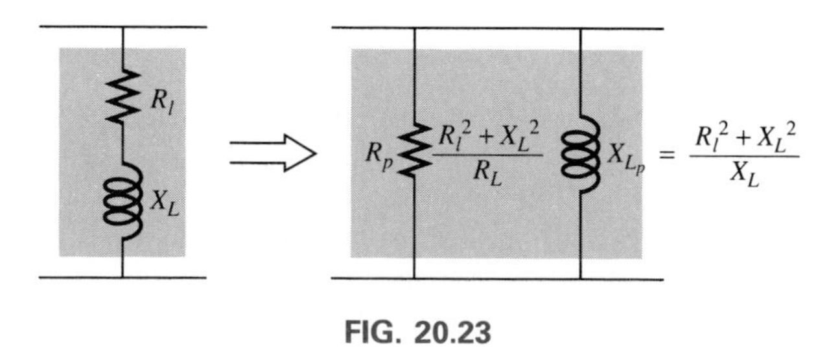

FIG. 20.23

Redrawing the network of Fig. 20.22 with the equivalent of Fig. 20.23 and a practical current source will result in the network of Fig. 20.24.

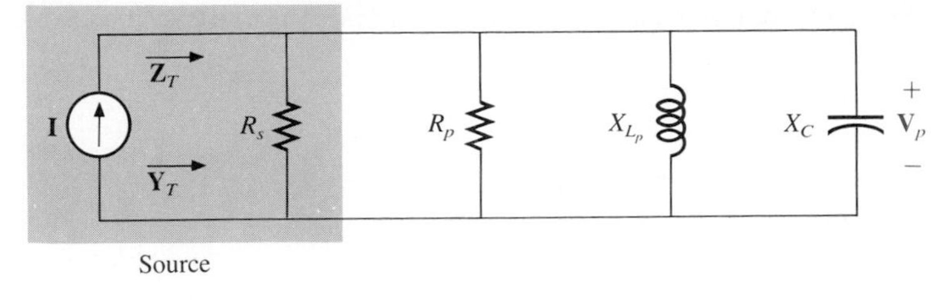


FIG. 20.24

If we define the parallel combination of $R_s$ and $R_p$ by the notation

$$R = R_s \parallel R_p \tag{20.25}$$

the network of Fig. 20.25 will result, which has the same format as the ideal configuration of Fig. 20.21.

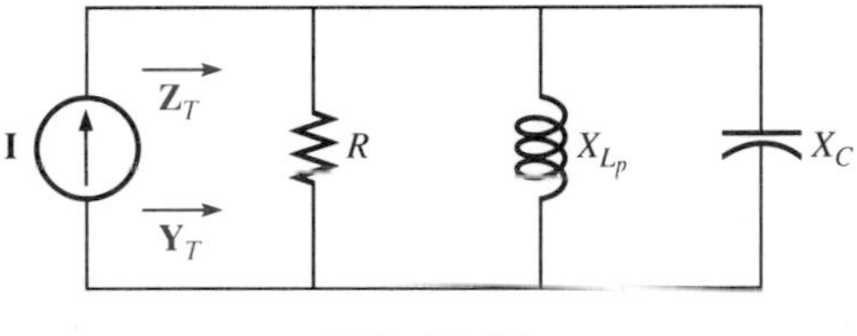


FIG. 20.25

We are now at a point where we can define the resonance conditions for the practical parallel resonant configuration. Recall for series resonance that the resonant frequency was the frequency at which the impedance was a minimum, the current a maximum, the input impedance purely resistive, and the network had a unity power factor. For parallel networks, since the resistance $R_p$ in our equivalent model is frequency dependent, the frequency at which maximum conditions are obtained is not the same as required for the unit power factor characteristic. Since both conditions are often used to define the resonant state, the frequency at which each occurs will be designated by different subscripts.

## Unity Power Factor, $f_p$

For the network of Fig. 20.24,

$$\mathbf{Y}_T = \frac{1}{\mathbf{R}} + \frac{1}{\mathbf{X}_{L_p}} + \frac{1}{\mathbf{X}_C}$$

$$= \frac{1}{R} - j\left(\frac{1}{X_{L_p}}\right) + j\left(\frac{1}{X_C}\right)$$

and

$$\mathbf{Y}_T = \frac{1}{R} + j\left(\frac{1}{X_C} - \frac{1}{X_{L_p}}\right) \tag{20.26}$$

For unity power factor, the reactive component must be zero as defined by

$$\frac{1}{X_C} - \frac{1}{X_{L_p}} = 0$$

Therefore,

$$\frac{1}{X_C} = \frac{1}{X_{L_p}}$$

and

$$\boxed{X_{L_p} = X_C} \tag{20.27}$$

Substituting for $X_{L_p}$ yields

$$\boxed{\frac{R_l^2 + X_L^2}{X_L} = X_C} \tag{20.28}$$

The resonant frequency, $f_p$, can now be determined from Eq. (20.28) as follows:

$$R_l^2 + X_L^2 = X_C X_L = \left(\frac{1}{\omega C}\right)\omega L = \frac{L}{C}$$

or

$$X_L^2 = \frac{L}{C} - R_l^2$$

with

$$2\pi f_p L = \sqrt{\frac{L}{C} - R_l^2}$$

and

$$f_p = \frac{1}{2\pi L}\sqrt{\frac{L}{C} - R_l^2}$$

Multiplying the top and bottom of the factor within the square-root sign by $C/L$ produces

$$f_p = \frac{1}{2\pi L}\sqrt{\frac{1 - R_l^2(C/L)}{C/L}} = \frac{1}{2\pi L\sqrt{C/L}}\sqrt{1 - \frac{R_l^2 C}{L}}$$

and

$$\boxed{f_p = \frac{1}{2\pi\sqrt{LC}}\sqrt{1 - \frac{R_l^2 C}{L}}} \tag{20.29}$$

or

$$f_p = f_s\sqrt{1 - \frac{R_l^2 C}{L}} \qquad \textbf{(20.30)}$$

where $f_p$ is the resonant frequency of a parallel resonant circuit (for $F_p = 1$) and $f_s$ is the resonant frequency as determined by $X_L = X_C$ for series resonance. Note that unlike a series resonant circuit, the resonant frequency $f_p$ is a function of resistance (in this case $R_l$). Note also, however, the absence of the source resistance $R_s$ in Eqs. (20.29) and (20.30). Since the factor $\sqrt{1 - (R_l^2C/L)}$ is less than one, $f_p$ is less than $f_s$. Recognize also that as the magnitude of $R_l$ approaches zero, $f_p$ rapidly approaches $f_s$.

## Maximum Impedance, $f_m$

At $f = f_p$ the input impedance of a parallel resonant circuit will be near its maximum value but not quite its maximum value due to the frequency dependence of $R_p$. The frequency at which maximum impedance will occur is defined by $f_m$ and is slightly more than $f_p$, as demonstrated in Fig. 20.26. The frequency $f_m$ is determined by differentiating (calculus) the general equation for $\mathbf{Z}_T$ with respect to frequency and then determining the frequency at which the resulting equation is equal to zero. The algebra is quite extensive and cumbersome and will not be included here. The resulting equation, however, is the following:

$$f_m = f_s\sqrt{1 - \frac{1}{4}\left(\frac{R_l^2 C}{L}\right)} \qquad \textbf{(20.31)}$$

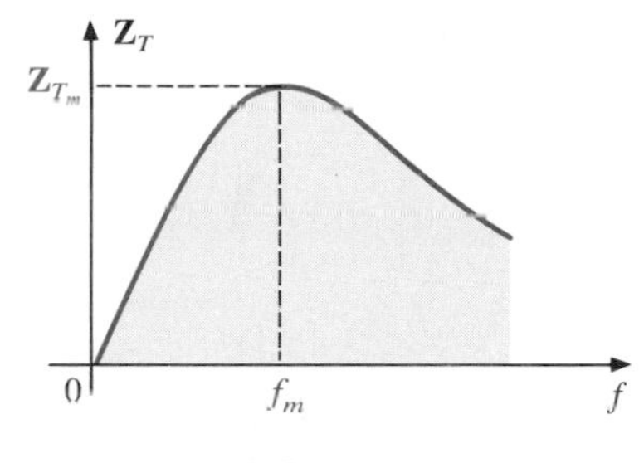

FIG. 20.26

*$\mathbf{Z}_T$ vs. frequency for the parallel resonant circuit.*

Note the similarities with Eq. (20.30). Since the square root factor of Eq. (20.31) is always more than the similar factor of Eq. (20.30), $f_m$ is always more than $f_p$. In general, therefore,

$$f_s > f_m > f_p \qquad \textbf{(20.32)}$$

Once $f_m$ is determined the network of Fig. 20.22 can be used to determine the magnitude and phase angle of the total impedance at the resonance condition defined by $f_m$.

That is,

$$\mathbf{Z}_{T_m} = \mathbf{Z}_{R-L} \parallel \mathbf{Z}_C \Big|_{f = f_m} \qquad \textbf{(20.33)}$$

## 20.9 SELECTIVITY CURVE FOR PARALLEL RESONANT CIRCUITS

The $\mathbf{Z}_T$-versus-frequency curve of Fig. 20.26 clearly reveals that a parallel resonant circuit exhibits maximum impedance at resonance ($f_m$), unlike the series resonant circuit which experiences minimum resistance levels at resonance.

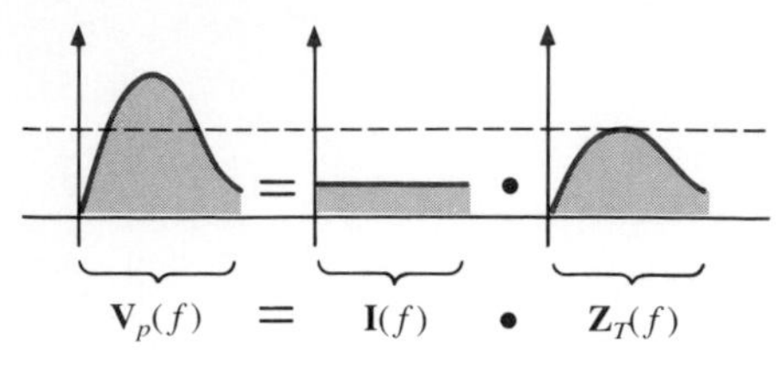


FIG. 20.27

Since the current $I$ of the current source is constant for any value of $Z_T$ or frequency the voltage across the parallel circuit will have the same shape as the total impedance $Z_T$, as shown in Fig. 20.27.

For the parallel circuit, the resonance curve of interest is that of the voltage $V_C$ across the capacitor. The reason for this interest in $V_C$ derives from electronic considerations that often place the capacitor at the input to another stage of a network.

Since the voltage across parallel elements is the same,

$$\boxed{\mathbf{V}_C = \mathbf{V}_p = \mathbf{I}\mathbf{Z}_T} \tag{20.34}$$

The resonant value of $\mathbf{V}_C$ is therefore determined by the value of $Z_{T_m}$ and the magnitude of the current source **I.**

The quality factor of the parallel resonant circuit continues to be determined by the ratio of the reactive power to the real power. That is,

$$Q = \frac{V_p^2/X_{L_p}}{V_p^2/R}$$

where $R = R_s \parallel R_p$, and $V_p$ is the voltage across the parallel branches. The result is

$$\boxed{Q = \frac{R}{X_{L_p}} = \frac{R_s \parallel R_p}{X_{L_p}}} \tag{20.35}$$

or since $X_{L_p} = X_C$ at resonance,

$$\boxed{Q = \frac{R}{X_C} = \frac{R_s \parallel R_p}{X_C}} \tag{20.36}$$

For the ideal current source ($R_s = \infty\ \Omega$) or when $R_s$ is sufficiently large compared to $R_p$, we can make the following approximation:

$$R = R_s \parallel R_p \cong R_p$$

and

$$Q = \frac{R_s \parallel R_p}{X_{L_p}} = \frac{R_p}{X_{L_p}} = \frac{(R_l^2 + X_L^2)/R_l}{(R_l^2 + X_L^2)/X_L}$$

so that

$$\boxed{Q = \frac{X_L}{R_l} = Q_l} \quad R_s \gg R_p \tag{20.37}$$

which is simply the quality factor $Q_l$ of the coil.

Depending on the frequency of interest, the quality factor can be defined as $Q_p$ or $Q_m$. Since $f_m > f_p$, $Q_m$ will always be greater than $Q_p$.

In general, the bandwidth is still related to the resonant frequency and the quality factor by

$$\text{BW} = f_2 - f_1 = \frac{f_r}{Q} \tag{20.38}$$

where $f_r$ and $Q$ can be determined by either $f_p$ or $f_m$.

The effect of $R_l$, $L$, and $C$ on the shape of the parallel resonance curve, as shown in Fig. 20.28 for the input impedance, is quite similar to their effect on the series resonance curve. Whether or not $R_l$ is zero, the parallel resonant circuit will frequently appear in a network schematic as shown in Fig. 20.28.

At resonance, an increase in $R_l$ or a decrease in the ratio $L/C$ will result in a decrease in the resonant impedance, with a corresponding increase in the current. The bandwidth of the resonance curves is given by Eq. (20.38). For increasing $R_l$ or decreasing $L$ (or $L/C$ for constant $C$), the bandwidth will increase as shown in Fig. 20.28.

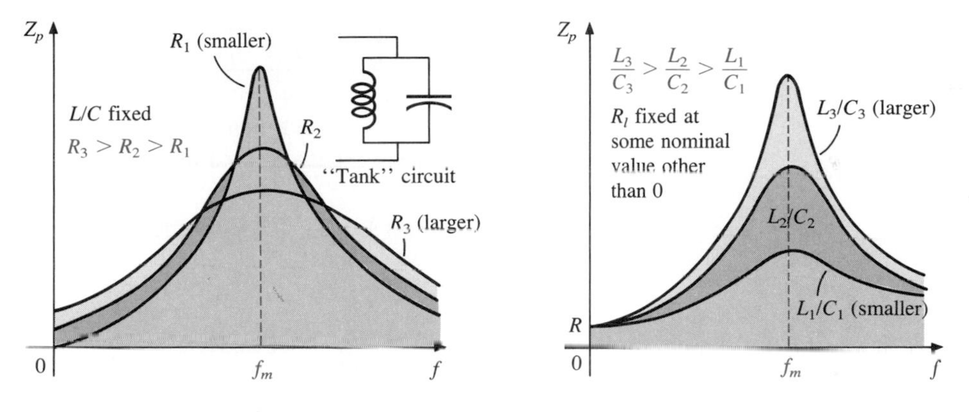


**FIG. 20.28**
*Effect of $R_l$, $L$, and $C$ on the parallel resonance curve.*

At low frequencies, the capacitive reactance is quite high, and the inductive reactance is low. Since the elements are in parallel, the total impedance at low frequencies will therefore be inductive. At high frequencies, the reverse is true, and the network is capacitive. At resonance ($f_p$), the network appears resistive. These facts lead to the phase plot of Fig. 20.29. Note that it is the inverse of that appearing for the

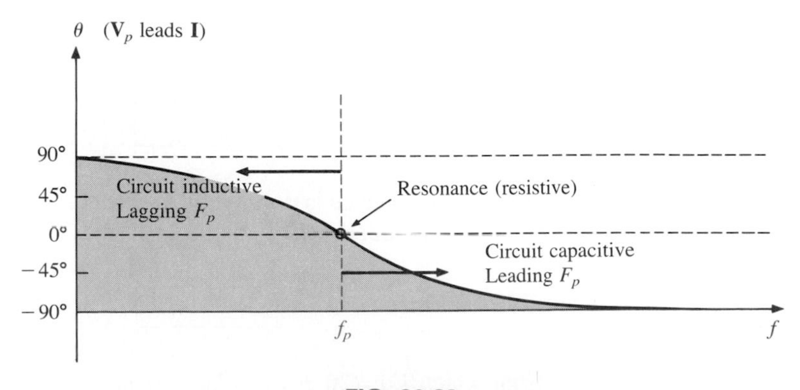


**FIG. 20.29**
*Phase plot for the parallel resonant circuit.*

series resonant circuit in that at low frequencies the series resonant circuit was capacitive and at high frequencies inductive.

## 20.10 EFFECT OF $Q_l \geq 10$

The content of the previous section may suggest that the analysis of parallel resonant circuits is significantly more complex than encountered for series resonant circuits. Fortunately, however, this is not the case since for the majority of parallel resonant circuits the quality factor of the coil $Q_l$ is sufficiently large to permit a number of approximations that simplify the required analysis.

First, and most important, if $Q_l \geq 10$ the equations for $f_s$, $f_m$, and $f_p$ are such that the following approximation can be employed:

$$\boxed{f_p = f_m = f_s} \quad Q_l \geq 10 \tag{20.39}$$

A direct result of the above equation is that for $Q_l \geq 10$ we can assume that maximum impedance and unity power factor occur at the same frequency.

### Resonant Frequency $f_p$

Using Eq. (20.39) and the equation for $f_s$, the equation for the resonant frequency of a parallel resonant frequency is defined by:

$$\boxed{f_p = \frac{1}{2\pi\sqrt{LC}}} \quad Q_l \geq 10 \tag{20.40}$$

### $X_{L_p}$

If we write

$$X_{L_p} = \frac{R_l^2 + X_L^2}{X_L} = \frac{R_l^2(X_L)}{X_L(X_L)} + X_L = \frac{X_L}{Q_l^2} + X_L$$

for $Q_l \geq 10$

$$\boxed{X_{L_p} = X_L} \quad Q_l \geq 10 \tag{20.41}$$

and since resonance is defined by $X_{L_p} = X_C$, the resulting condition for resonance is reduced to:

$$\boxed{X_L = X_C} \quad Q_l \geq 10 \tag{20.42}$$

from which Eq. (20.30) can be derived.

## $R_p$

$$R_p = \frac{R_l^2 + X_L^2}{R_l} = R_l + \frac{X_L^2}{R_l}\left(\frac{R_l}{R_l}\right) = R_l + \frac{X_L^2}{R_l^2}R_l$$
$$= R_l + Q_l^2 R_l = (1 + Q_l^2)R_l$$

For $Q_l \geq 10$, $1 + Q_l^2 \cong Q_l^2$ and

$$R_p \cong Q_l^2 R_l \qquad Q_l \geq 10 \tag{20.43}$$

Applying the approximations just derived to the network of Fig. 20.24 will result in the approximate equivalent circuit of Fig. 20.30, which is certainly a lot "cleaner" in general appearance.

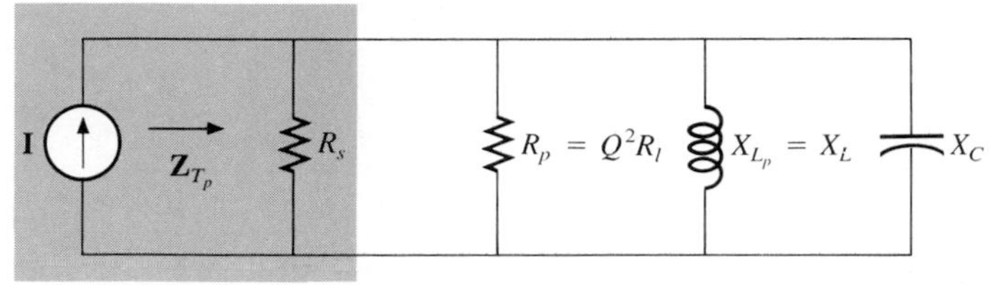


**FIG. 20.30**
*Approximate equivalent circuit for $Q_l \geq 10$.*

## $Z_{T_p}$

The total impedance at resonance is now defined by:

$$Z_{T_p} = R_s \,\|\, R_p \cong R_s \,\|\, Q_l^2 R_l \qquad Q_l \geq 10 \tag{20.44}$$

For an ideal current source ($R_s = \infty\ \Omega$) or if $R_s \gg R_p$, the equation reduces to

$$Z_{T_p} = Q_l^2 R_l \qquad Q_l \geq 10,\ R_s \gg R_p \tag{20.45}$$

## $Q_p$

The quality factor is now defined by

$$Q_p = \frac{R}{X_{L_p}} = \frac{R_s \,\|\, Q_l^2 R_l}{X_L} \tag{20.46}$$

or

$$Q_p = \frac{R}{X_C} = \frac{R_s \,\|\, Q_l^2 R_l}{X_C} \tag{20.47}$$

Quite obviously, therefore, $R_s$ does have an impact on the quality factor of the network and the shape of the resonant curves.

If an ideal current source ($R_s = \infty\ \Omega$) is employed or if $R_s \gg R_p$,

$$Q_p = \frac{R_s \parallel Q_l^2 R_l}{X_L} \cong \frac{Q_l^2 R_l}{X_L} = \frac{Q_l^2}{X_L/R_l} = \frac{Q_l^2}{Q_l}$$

and

$$\boxed{Q_p = Q_l} \qquad Q_l \geq 10,\ R_s \gg R_p \tag{20.48}$$

## BW

The bandwidth is defined by

$$\boxed{\text{BW} = f_2 - f_1 = \frac{f_p}{Q_p}} \tag{20.49}$$

By substituting $Q_p = (R_s \parallel Q^2 R_l)/X_C$ from above and performing a few algebraic manipulations it can be shown that

$$\boxed{\text{BW} = f_2 - f_1 \cong \frac{1}{2\pi}\left[\frac{R_l}{L} + \frac{1}{R_s C}\right]} \tag{20.50}$$

clearly revealing the impact of $R_s$ on the resulting bandwidth. Of course, if $R_s = \infty\ \Omega$ (ideal current source):

$$\boxed{\text{BW} = f_2 - f_1 \cong \frac{R_l}{2\pi L}} \qquad R_s = \infty\ \Omega \tag{20.51}$$

The equations resulting from the application of the condition $Q_l \geq 10$ are obviously a great deal easier to apply then those obtained earlier. For the future, therefore, the analysis of parallel resonant circuits should proceed as follows:

1. Determine $f_s$ to obtain some idea of the resonant frequency. Recall that for most situations, $f_s$, $f_m$, and $f_p$ are quite close together.
2. Calculate $Q_l$ using $f_s$ from above.
3. If $Q_l \geq 10$ use the approximate approach described above.
4. If $Q_l$ is in the range 4 to 10 the approximate approach can be employed if it is understood and accepted that the results will be a few percent off. Considering the typical variations from nameplate values for many of our components, the use of the approximations is quite valid for this range of $Q_l$.
5. For $Q_l$ in the neighborhood of 4 or less the analysis of Section 20.8 should be applied.

## 20.11 THE BRANCH CURRENTS $I_L$ AND $I_C$ AND SUMMARY TABLE

You will recall that for the series resonant circuit, $V_L = V_C = Q_l E$ at the resonant condition. A similar result can be obtained for the parallel resonant circuit if we carefully examine the network of Fig. 20.31. The current $\mathbf{I}_T$ is not the source current (due to $R_s$) but the current entering the tank circuit. Of course, for the condition $R_s \cong \infty\ \Omega$ (open circuit), which often occurs, $\mathbf{I}_T$ is equal to the source current $\mathbf{I}$.

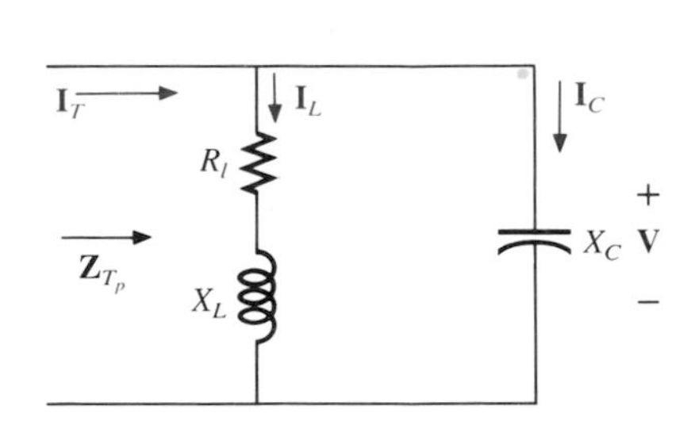


FIG. 20.31

At resonance, $\mathbf{Z}_{T_p} = Q_l^2 R_l$ (for $Q_l \geq 10$), as determined earlier, and through Ohm's law, $\mathbf{V} = \mathbf{I}_T \cdot Q_l^2 R_l$.

The current $\mathbf{I}_L$ is

$$\mathbf{I}_L = \frac{\mathbf{V}}{\mathbf{Z}_L} = \frac{\mathbf{I}_T Q_l^2 R_l}{R_l + jX_L}$$

Dividing by $R_l$ in the numerator and denominator gives us

$$\mathbf{I}_L = \frac{\mathbf{I}_T Q_l^2}{1 + j\dfrac{X_L}{R_l}} = \frac{\mathbf{I}_T Q_L^2}{1 + jQ_l}$$

The magnitude of $\mathbf{I}_L$ is given by

$$I_L = \frac{I_T Q_l^2}{\sqrt{1 + Q_l^2}}$$

which for $Q_l \geq 10$ becomes

$$I_L = \frac{I_T Q_l^2}{Q_l}$$

and

$$\boxed{I_L \cong Q_l I_T} \qquad (20.52)$$

In a parallel resonant circuit, therefore, the magnitude of the current through the inductive branch is $Q_l$ times the current entering the tank circuit (at resonance only). Furthermore, since

$$\mathbf{I}_C = \frac{\mathbf{V}}{-jX_C} = \frac{\mathbf{I}_T Q_l^2 R_l}{-jX_C}$$

if we divide by $R_l$, we get

$$\mathbf{I}_C = \frac{\mathbf{I}_T Q_l^2}{-j\dfrac{X_C}{R_l}}$$

However, at resonance, for $Q_l \geq 10$,

$$X_L = X_C \quad \text{and} \quad \frac{X_C}{R_l} = \frac{X_L}{R_l} = Q_l$$

and therefore

$$\mathbf{I}_C = \frac{\mathbf{I}_T Q_l^2}{-jQ_l}$$

The magnitude is

$$I_C = \frac{I_T Q_l^2}{Q_l}$$

and

$$\boxed{I_C \cong Q_l I_T} \qquad \textbf{(20.53)}$$

Table 20.1 is included as a review of the effect of $Q_l \geq 10$.

**TABLE 20.1**
*Parallel resonant circuit.*

| | Low $Q_l$ | High $Q_l$ | High $Q_l$ ($R_s \gg R_p$) |
|---|---|---|---|
| $f_s$ | $\dfrac{1}{2\pi\sqrt{LC}}$ | $\dfrac{1}{2\pi\sqrt{LC}}$ | $\dfrac{1}{2\pi\sqrt{LC}}$ |
| $f_p$ | $f_s\sqrt{1 - \dfrac{R_l^2C}{L}}$ | $\dfrac{1}{2\pi\sqrt{LC}}$ | $\dfrac{1}{2\pi\sqrt{LC}}$ |
| $f_m$ | $f_s\sqrt{1 - \dfrac{1}{4}\left[\dfrac{R_l^2C}{L}\right]}$ | $\dfrac{1}{2\pi\sqrt{LC}}$ | $\dfrac{1}{2\pi\sqrt{LC}}$ |
| $Z_{T_p}$ | $\mathbf{Z}_{R-L} \parallel \mathbf{Z}_C$ | $R_s \parallel Q_l^2R_l$ | $Q_l^2R_l$ |
| $Q_p$ | $\dfrac{R_s \parallel R_p}{X_{L_p}} = \dfrac{R_s \parallel R_p}{X_C}$ | $\dfrac{R_s \parallel Q_l^2R_l}{X_L} = \dfrac{R_s \parallel Q_l^2R_l}{X_C}$ | $Q_l$ |
| BW | $\dfrac{f_r}{Q}$ | $\dfrac{f_p}{Q_p}$ | $\dfrac{f_p}{Q_l}$ |
| $I_L, I_C$ | network analysis | $I_L = I_C = Q_pI_T$ | $I_L = I_C = Q_lI_T$ |

## 20.12 EXAMPLES (PARALLEL RESONANCE)

**EXAMPLE 20.6.** Given the parallel network of Fig. 20.32 with "ideal" elements:

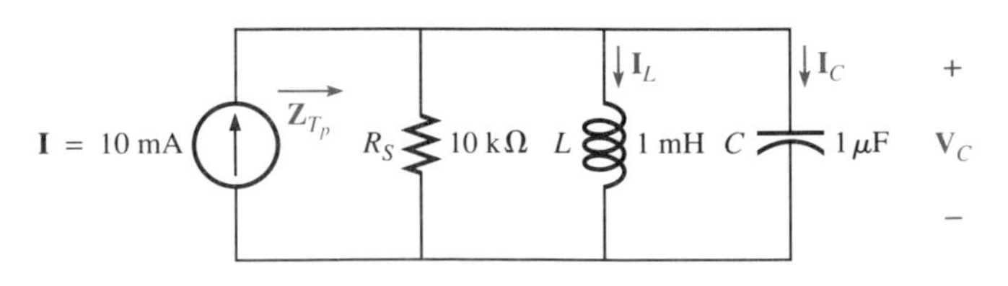


FIG. 20.32

a. Determine the resonant frequency.
b. Find the total impedance at resonance.

c. Calculate the quality factor and bandwidth of the system.
d. Find the voltage $V_C$ at resonance.
e. Determine the currents $I_L$ and $I_C$ at resonance.

***Solutions:***

a. The fact that $R_l$ is zero ohms results in a very high $Q_l$ $(= X_L/R_l)$, permitting the use of the following equation for $f_p$.

$$f_p = \frac{1}{2\pi\sqrt{LC}} = \frac{1}{2\pi\sqrt{(1\text{ mH})(1\ \mu\text{F})}}$$
$$= \mathbf{5.03\ kHz}$$

b. For the parallel reactive elements:

$$\mathbf{X}_T = \mathbf{X}_L \parallel \mathbf{X}_C = \frac{(X_L \angle 90^\circ)(X_C \angle -90^\circ)}{+j(X_L - X_C)}$$

but $X_L = X_C$ at resonance, resulting in a zero in the denominator of the equation and a very high impedance for $\mathbf{X}_T$ that can be approximated as an open circuit.

Therefore,

$$Z_{T_p} = \mathbf{R}_s \parallel \mathbf{X}_T \cong \mathbf{R}_s = \mathbf{10\ k\Omega}$$

c. $$Q_p = \frac{R_s}{X_{L_p}} = \frac{R_s}{2\pi f_p L} = \frac{10\text{ k}\Omega}{2\pi(5.03\text{ kHz})(1\text{ mH})} = \mathbf{316.41}$$

$$\text{BW} = \frac{f_p}{Q_p} = \frac{5.03\text{ kHz}}{316.41} = \mathbf{15.90\ Hz}$$

d. $V_C = IZ_{T_p} = (10\text{ mA})(10\text{ k}\Omega) = \mathbf{100\ V}$

e. $$I_L = \frac{V_L}{X_L} = \frac{V_C}{2\pi f_p L} = \frac{100\text{ V}}{2\pi(5.03\text{ kHz})(1\text{ mH})} = \frac{100\text{ V}}{31.6\ \Omega} = \mathbf{3.16\ A}$$

$$I_C = \frac{V_C}{X_C} = \frac{100\text{ V}}{31.6\ \Omega} = \mathbf{3.16\ A}$$

---

The preceding example demonstrates the impact of $R_l$ on the calculations associated with parallel resonance. The source impedance is the only factor to limit the input impedance and the level of $V_C$. In general, therefore, even though $R_l$ may be relatively small in magnitude compared to the impedance levels of the other elements of the network, it does have a significant impact on the response characteristics.

---

**EXAMPLE 20.7.** For the parallel resonant circuit of Fig. 20.33,

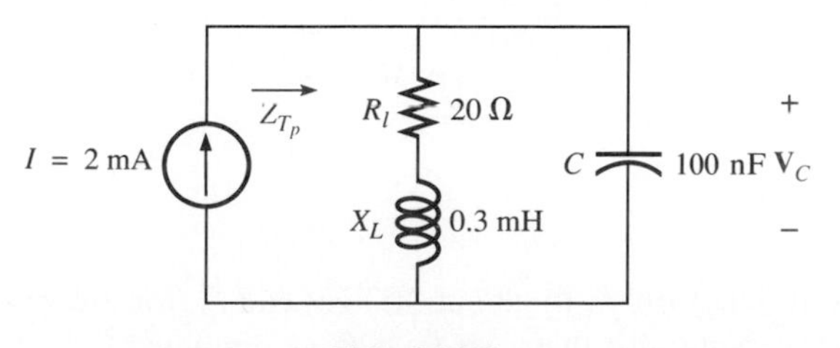


**FIG. 20.33**

a. Determine $f_s$, $f_m$, and $f_p$.
b. Calculate the maximum impedance and the magnitude of the voltage $V_C$ at $f_m$.
c. Determine the quality factors $Q_m$ and $Q_p$.
d. Calculate the bandwidth.
e. Compare the above results with those obtained using the equations associated with $Q_l \geq 10$.

***Solutions:***

a. $$f_s = \frac{1}{2\pi\sqrt{LC}} = \frac{1}{2\pi\sqrt{(0.3\text{ mH})(100\text{ nF})}} = \mathbf{29{,}057.58\text{ Hz}}$$

$$f_m = f_s\sqrt{1 - \frac{1}{4}\left[\frac{R_l^2C}{L}\right]}$$

$$= (29{,}057.58\text{ Hz})\sqrt{1 - \frac{1}{4}\left[\frac{(20\ \Omega)^2(100\text{ nF})}{0.3\text{ mH}}\right]}$$

$$= \mathbf{28{,}569.19\text{ Hz}}$$

$$f_p = f_s\sqrt{1 - \frac{R_l^2C}{L}} = (29{,}057.58\text{ Hz})\sqrt{1 - \left[\frac{(20\ \Omega)^2(100\text{ nF})}{0.3\text{ mH}}\right]}$$

$$= \mathbf{27{,}051.14\text{ Hz}}$$

b. $\mathbf{Z}_{T_m} = (R_l + jX_L) \parallel X_C$ at $f = f_m$

$X_L = 2\pi f_m L = 2\pi(28{,}569.19\text{ Hz})(0.3\text{ mH}) = 53{,}852\ \Omega$

$$X_C = \frac{1}{2\pi f_m C} = \frac{1}{2\pi(28{,}569.19\text{ Hz})(100\text{ nF})} = 55{,}709\ \Omega$$

$$R_l + jX_L = 20\ \Omega + j53.852\ \Omega = 57.446\ \Omega\ \angle 69.626^\circ$$

$$\mathbf{Z}_{T_m} = \frac{(57.446\ \Omega\ \angle 69.626^\circ)(55.709\ \Omega\ \angle -90^\circ)}{20\ \Omega + j53.852\ \Omega - j55.709\ \Omega}$$

$$= \mathbf{159.344\ \Omega\ \angle -15.069^\circ}$$

$$V_{C_{\max}} = IZ_{T_m} = (2\text{ mA})(159.344\ \Omega) = \mathbf{318.69\text{ mV}}$$

c. $R_s = \infty\ \Omega$, therefore, $Q = \dfrac{X_L}{R_l} = Q_l$

$f_m$:

$$Q_m = \frac{X_L}{R_l} = \frac{53.852\ \Omega}{20\ \Omega} = \mathbf{2.693}$$

$f_p$:

$$Q_p = \frac{X_L}{R_l} = \frac{2\pi(27{,}051.14\text{ Hz})(0.3\text{ mH})}{20\ \Omega} = \frac{50.990\ \Omega}{20\ \Omega} = \mathbf{2.55}$$

d. $$\text{BW} = \frac{f_m}{Q_m} = \frac{28{,}569.19\text{ Hz}}{2.693} = \mathbf{10{,}608.685\text{ Hz}}$$

e. For $Q_l \geq 10$

$$f_p = \mathbf{29{,}057.58\text{ Hz}}$$

which differs from $f_m$ by about 500 Hz and $f_p$ (for low $Q_l$) by about 2000 Hz. Note also that $f_s > f_m > f_p$ as predicted.

$$Z_{T_p} = Q_l^2 R_l$$

$$Q_l \text{ (at } f_s) = \frac{2\pi f_s L}{R_l} = \frac{2\pi(29{,}057.58 \text{ Hz})(0.3 \text{ mH})}{20\ \Omega} = 2.739$$

$$Z_{T_p} = (2.739)^2 \cdot 20\ \Omega = \mathbf{150.04\ \Omega \angle 0°}$$

(versus 159.344 Ω ∠−15.069° above)

$$V_{C_{max}} = IZ_{T_p} = (2 \text{ mA})(150.04\ \Omega) = \mathbf{300.08 \text{ mV}}$$

(versus 318.69 mV above)

$$\text{BW} = \frac{f_p}{Q_p} = \frac{29{,}057.58 \text{ Hz}}{2.739} = \mathbf{10{,}608.83 \text{ Hz}}$$

(versus 10,608.685 Hz above)

The results reveal that even for a low $Q$ system the approximate solutions are still in the same ball park as those obtained using the full equations. The primary difference is between $f_s$ *and* $f_p$ (about 7%), with the difference between $f_s$ and $f_m$ at less than 2%. For the future, using $f_s$ to determine $Q_l$ will certainly provide a measure of $Q_l$ that can be used to determine whether the approximate approach is appropriate.

---

**EXAMPLE 20.8.** For the network of Fig. 20.34:

a. Determine $Q_l$.
b. Determine $R_p$.
c. Calculate $Z_{T_p}$.
d. Find $C$ at resonance.
e. Find $Q_p$.
f. Calculate BW.

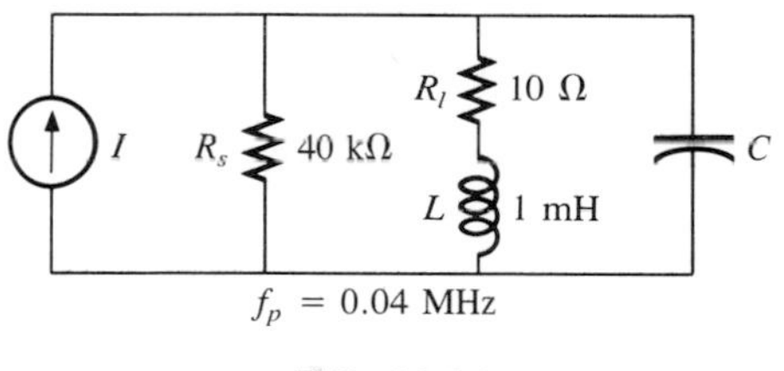


FIG. 20.34

***Solutions:***

a. $$Q_l = \frac{X_L}{R_l} = \frac{2\pi f_p L}{R_l} = \frac{(6.28)(0.04 \text{ MHz})(1 \text{ mH})}{10\ \Omega} = \mathbf{25.12}$$

b. $Q_l \geq 10$. Therefore,

$$R_p = Q_l^2 R_l = (25.12)^2(10\ \Omega) = \mathbf{6.31\ k\Omega}$$

c. $Z_{T_p} = R_s \parallel R_p = 40\ \text{k}\Omega \parallel 6.31\ \text{k}\Omega = \mathbf{5.45\ k\Omega}$

d. $Q_l \geq 10$. Therefore,

$$f_p = \frac{1}{2\pi\sqrt{LC}}$$

and

$$C = \frac{1}{L(f2\pi)^2} = \frac{1}{(1 \text{ mH})(0.04 \text{ MHz})^2 4\pi^2} = \mathbf{0.0159\ \mu F}$$

e. $Q_l \geq 10$. Therefore,

$$Q_p = \frac{R}{\omega_p L} = \frac{5.45\ \text{k}\Omega}{(6.28)(0.04 \text{ MHz})(1 \text{ mH})} = \mathbf{21.71}$$

f. $$\text{BW} = \frac{f_p}{Q_p} = \frac{0.04 \text{ MHz}}{21.71} = \mathbf{1.84\ kHz}$$

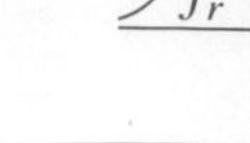


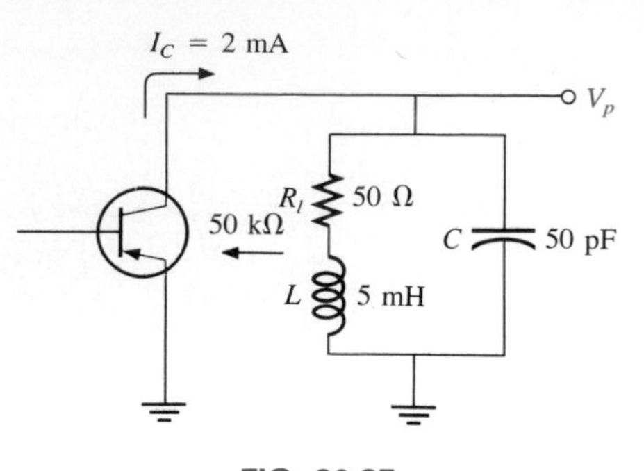


FIG. 20.35

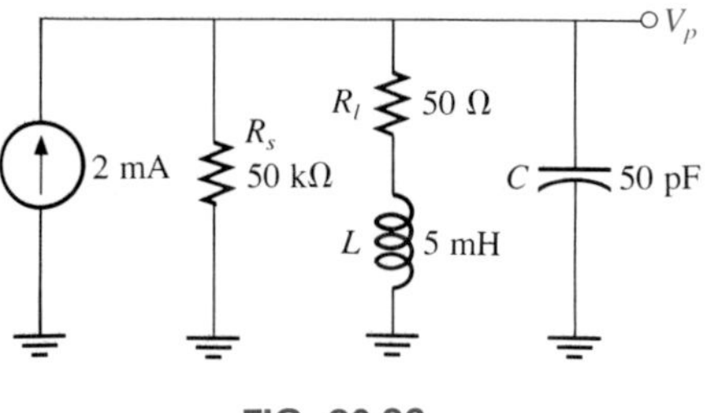


FIG. 20.36

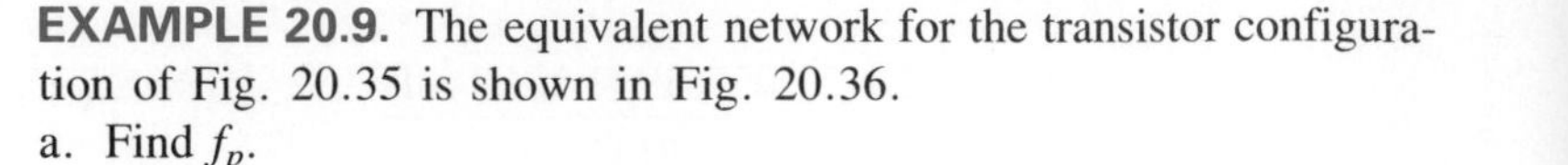

**EXAMPLE 20.9.** The equivalent network for the transistor configuration of Fig. 20.35 is shown in Fig. 20.36.

a. Find $f_p$.
b. Determine $Q_p$.
c. Calculate BW.
d. Determine $V_p$ at resonance.
e. Sketch the curve of $V_C$ versus frequency.

***Solutions:***

a. $f_s = \dfrac{1}{2\pi\sqrt{LC}} = \dfrac{1}{2\pi\sqrt{(5\text{ mH})(50\text{ pF})}} = 318.31\text{ kHz}$

$X_L = 2\pi f_s L = 2\pi(318.31\text{ kHz})(5\text{ mH}) = 10\text{ k}\Omega$

$Q_l = \dfrac{X_L}{R_l} = \dfrac{10\text{ k}\Omega}{50\ \Omega} = 200 > 10$

$f_p = f_s = \mathbf{318.31\ kHz}$

Using Eq. (20.30) would result in $\cong$ 318.5 kHz.

b. $Q_p = \dfrac{R_s \parallel R_p}{X_L}$

$R_p = Q_l^2 R_l = (200)^2 50\ \Omega = 2\text{ M}\Omega$

$Q_p = \dfrac{50\text{ k}\Omega \parallel 2\text{ M}\Omega}{10\text{ k}\Omega} = \dfrac{48.78\text{ k}\Omega}{10\text{ k}\Omega} = \mathbf{4.88}$

c. $\text{BW} = \dfrac{f_p}{Q_p} = \dfrac{318.31\text{ kHz}}{4.88} = \mathbf{65.23\ kHz}$

Using

$$\text{BW} = \frac{1}{2\pi}\left(\frac{R_l}{L} + \frac{1}{R_s C}\right) = \frac{1}{2\pi}\left[\frac{50\ \Omega}{5\text{ mH}} + \frac{1}{(50\text{ k}\Omega)(50\text{ pF})}\right]$$

$= \mathbf{65.25\ kHz}$

compares very favorably with the above solution.

d. $V_p = IZ_{T_p} = (2\text{ mA})(R_s \parallel R_p) = (2\text{ mA})(48.78\text{ k}\Omega) = \mathbf{97.56\ V}$

e. See Fig. 20.37.

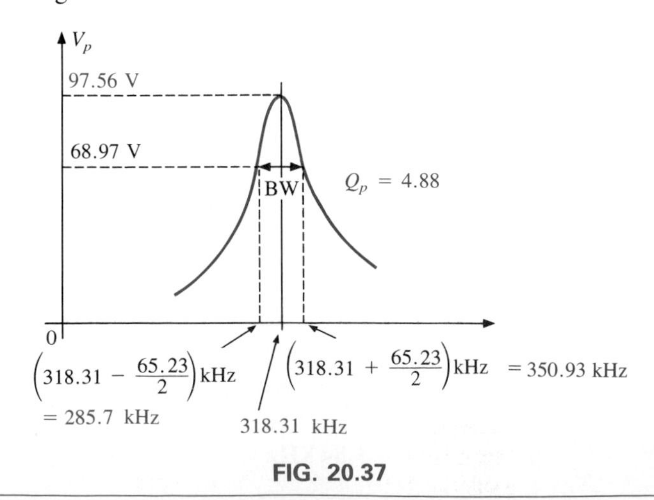


FIG. 20.37

**EXAMPLE 20.10.** Repeat Example 20.9 ignoring the effects of $R_s$, and compare results.

***Solution:***

a. $f_p$ is the same, **318.31 kHz.**

b. *For* $R_s = \infty\ \Omega$

$$Q_p = Q_l = \mathbf{200} \qquad \text{(versus 4.88)}$$

d. $\text{BW} = \dfrac{f_p}{Q_p} = \dfrac{318.31\text{ kHz}}{200} = \mathbf{1.592\ kHz} \qquad \text{(versus 65.23 kHz)}$

e. $Z_{T_p} = R_p = \mathbf{2\ M\Omega} \qquad \text{(versus 48.78 k}\Omega)$

f. $V_p = IZ_{T_p} = (2\text{ mA})(2\text{ M}\Omega) = \mathbf{4000\ V} \qquad \text{(versus 97.56 V)}$

The results obtained clearly reveal that the source resistance can have a significant impact on the response characteristics of a parallel resonant circuit.

**EXAMPLE 20.11.** Design a parallel resonant circuit to have the response curve of Fig. 20.38 using a 1-mH, 10-Ω inductor, and a current source with an internal resistance of 40 kΩ.

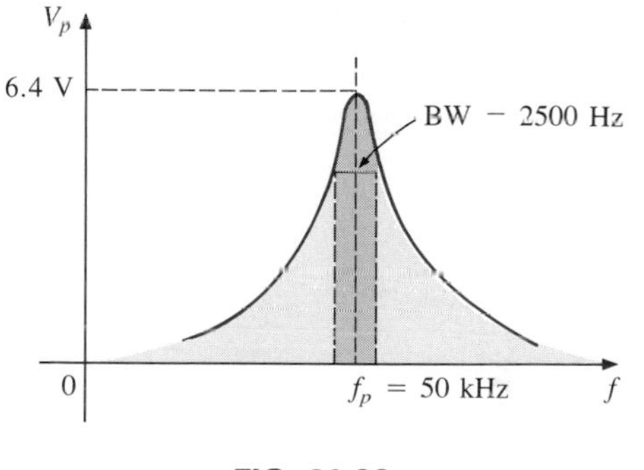


FIG. 20.38

***Solution:***

$$\text{BW} = \frac{f_p}{Q_p}$$

Therefore,

$$Q_p = \frac{f_p}{\text{BW}} = \frac{50{,}000\text{ Hz}}{2500\text{ Hz}} = 20$$

$$X_L = 2\pi f_p L = (6.28)(50\text{ kHz})(1\text{ mH}) = 314\ \Omega$$

and

$$Q_l = \frac{X_L}{R_l} = \frac{314\ \Omega}{10\ \Omega} = 31.4$$

$$R_p = Q_l^2 R_l = (31.4)^2(10\ \Omega) = 9859.6\ \Omega$$

$$Q_p = \frac{R}{X_L} = \frac{R_s \parallel 9859.6\ \Omega}{314\ \Omega} = 20$$

$$\frac{(R_s)(9859.6)}{R_s + 9859.6} = 6280$$

resulting in

$$R_s = 17.298\text{ k}\Omega$$

However, the source resistance was given as 40 kΩ. We must therefore add a parallel resistor ($R'$) that will reduce the 40 kΩ to approximately 17.298 kΩ. That is,

$$\frac{(40\text{ k}\Omega)(R')}{40\text{ k}\Omega + R'} = 17.298\text{ k}\Omega$$

Solving for $R'$:

$$R' = \mathbf{30.481\ k\Omega}$$

At resonance, $X_L = X_C$, and

$$X_C = \frac{1}{2\pi f_p C}$$

$$C = \frac{1}{2\pi f_p X_C}$$

$$= \frac{1}{2\pi(50\ \text{kHz})(314\ \Omega)}$$

and

$$C \cong \mathbf{0.01\ \mu F}$$

$$Z_{T_p} = R_s \parallel Q_l^2 R_l$$
$$= 17.298\ \text{k}\Omega \parallel 9859.6\ \Omega$$
$$= 6.28\ \text{k}\Omega$$

with

$$V_p = IZ_{T_p}$$

or

$$I = \frac{V_p}{Z_{T_p}} = \frac{10\ \text{V}}{6.28\ \text{k}\Omega} \cong \mathbf{1.6\ mA}$$

The network appears in Fig. 20.39.

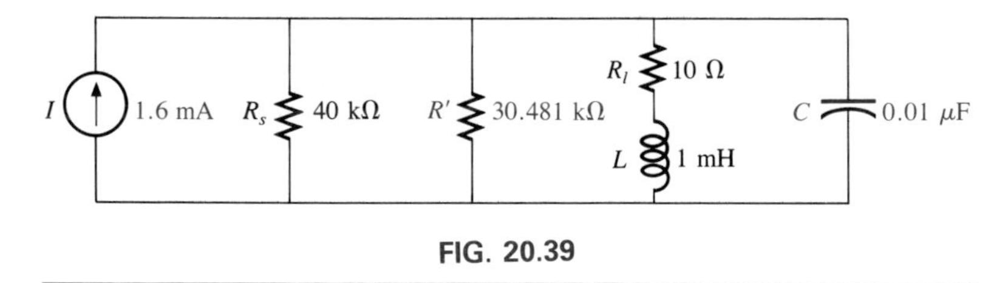


FIG. 20.39

## 20.13 COMPUTER ANALYSIS

### PSPICE

The beauty of the .PROBE response can be fully appreciated in the plots that result for series and parallel resonant circuits. The insertion of a few network parameters into the PSPICE file will result in a curve that immediately reveals the resonant frequency, the bandwidth of the response, and the relative quality factor. Also immediately available is the range of frequencies that will not receive sufficient power to significantly affect the response of the succeeding stage of the system.

The series resonant circuit of Fig. 20.40 is very similar to the circuit employed in Example 20.4 of the text with the addition of a voltage

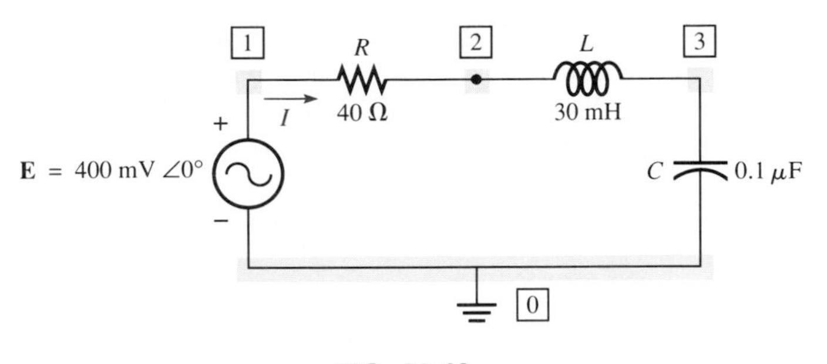


FIG. 20.40

source. The magnitude of the source was chosen to produce a maximum current of $I = 400\text{ mV}/40\ \Omega = 10\text{ mA}$ at resonance, and the reactive elements were chosen to have a resonant frequency of

$$f_s = \frac{1}{2\pi\sqrt{LC}} = \frac{1}{2\pi\sqrt{(30\text{ mH})(0.1\ \mu\text{F})}} \cong \mathbf{2.9\ kHz}$$

The quality factor is

$$Q_l = \frac{X_L}{R_l} = \frac{546.64\ \Omega}{40\ \Omega} = \mathbf{13.7}$$

and the bandwidth is

$$\text{BW} = \frac{f_s}{Q_l} = \frac{2.9\text{ kHz}}{13.7} \cong \mathbf{212\ Hz}$$

The input file of Fig. 20.41 requests 91 data points between 1 KHz and 10 KHz for an increment of

$$\text{Increment} = \frac{10\text{ kHz} - 1\text{ kHz}}{91 - 1} = \frac{9\text{ kHz}}{90} = \mathbf{100\ Hz}$$

```
************************ Evaluation PSpice (January 1989) ******* 10:04:05 *******

 Chapter 20 -Series Resonance

 ****      CIRCUIT DESCRIPTION

*****************************************************************************

VE 1 0 AC 400MV 0
R  1 2 40
L  2 3 30MH
C  3 0 0.1UF
.AC LIN 91 1KH 10KH
.PROBE
.OPTIONS NOPAGE
.END
```

FIG. 20.41

The output file of Fig. 20.42 clearly substantiates the above results with a resonant frequency of 2.9 kHz, a maximum current of 10 mA, and a bandwidth at 7.07 mA of about 200 Hz. The shape of the curve also suggests a quality factor greater than 10. In addition, note that the current level at 1 kHz and 8 kHz is about the same, and any frequency less than 1 kHz or greater than 8 kHz has a lower value.

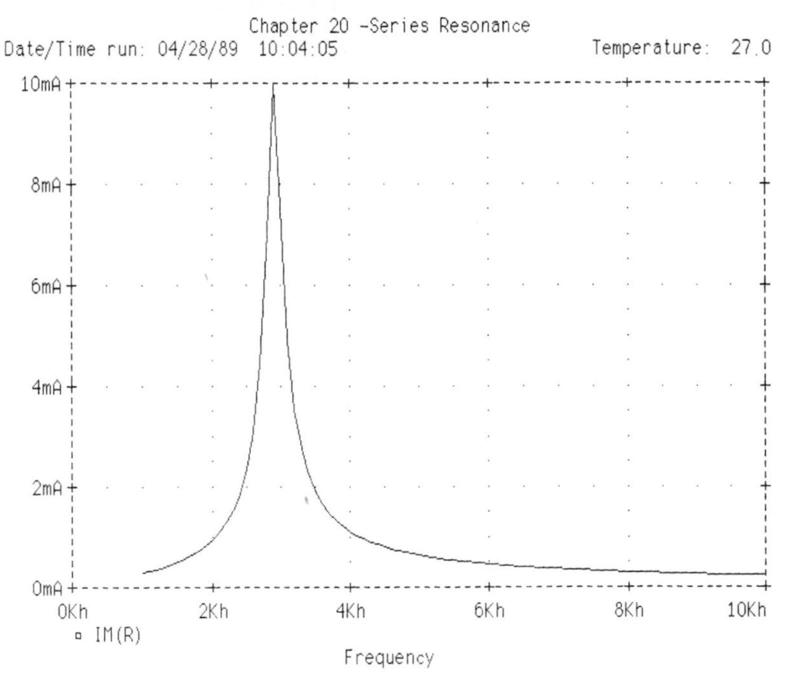

**FIG. 20.42**

The second run is for the parallel resonant circuit of Fig. 20.43, which appears as Example 20.10 in the text. The results of the example

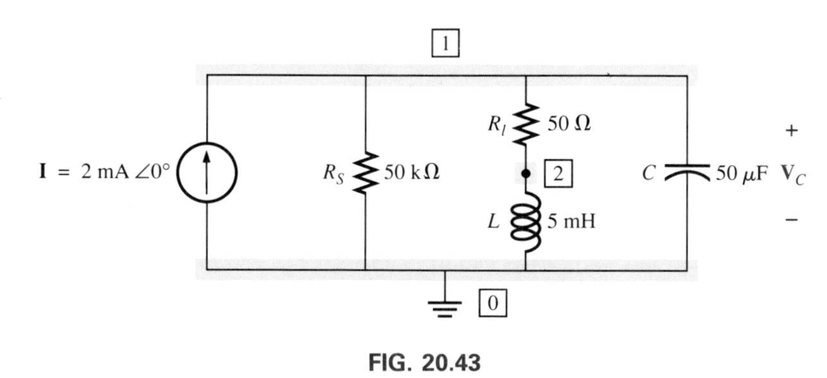


**FIG. 20.43**

were $V_p = 97.56$ V, $Q_p = 4.88$, BW $= 65.23$ kHz, and a resonant frequency of 318.31 kHz. The input file of Fig. 20.44 requests 81 data points for an increment of

$$\text{Increment} = \frac{500\text{ kHz} - 100\text{ kHz}}{81 - 1} = \frac{400\text{ kHz}}{80} = \mathbf{5\text{ kHz}}$$

```
************************ Evaluation PSpice (January 1989) ******* 10:11:33 *******

Chapter 20 - Parallel Resonance

****     CIRCUIT DESCRIPTION

******************************************************************************

IS 0 1 AC 2MA 0
RS 1 0 50K
RL 1 2 50
L  2 0 5MH
C  1 0 50PF
.AC LIN 81 100KH 500KH
.PROBE
.OPTIONS NOPAGE
.END
```

FIG. 20.44

The resulting .PROBE response of Fig. 20.45 reveals a resonant frequency near 320 kHz, a maximum voltage of about 97 V, and a bandwidth of about 65 kHz at 0.707(97 V) = 68.6 V. The shape of the curve for the frequency range also suggests a low quality factor ($\cong$ 5) and reveals that the voltage drops below 20 V for frequencies below 200 kHz and above 500 kHz.

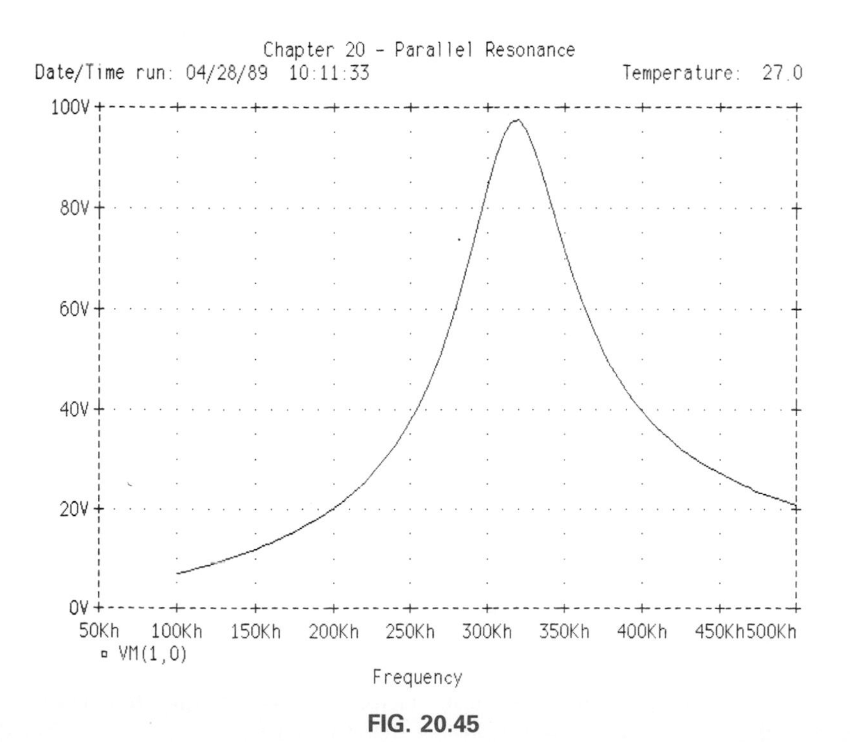


FIG. 20.45

## PROBLEMS

### SECTIONS 20.2 THROUGH 20.7 Series Resonance

**1.** Find the resonant $\omega_s$ and $f_s$ for the series circuit with the following parameters:
- **a.** $R = 10\,\Omega$, $L = 1\,\text{H}$, $C = 16\,\mu\text{F}$
- **b.** $R = 300\,\Omega$, $L = 0.5\,\text{H}$, $C = 0.16\,\mu\text{F}$
- **c.** $R = 20\,\Omega$, $L = 0.28\,\text{mH}$, $C = 7.46\,\mu\text{F}$

**2.** For the series circuit of Fig. 20.46:
- **a.** Find the value of $X_C$ for resonance.
- **b.** Find the current **I** and the voltages $\mathbf{V}_R$, $\mathbf{V}_L$, and $\mathbf{V}_C$ in phasor form at resonance.
- **c.** Draw the phasor diagram of the voltages and current.
- **d.** Sketch the power triangle for the circuit at resonance.
- **e.** Find the $Q_s$ of the circuit.

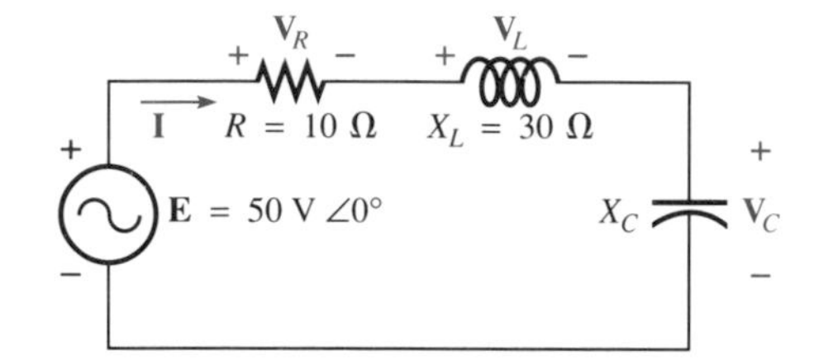


FIG. 20.46

**3.** Repeat Problem 2 for the circuit of Fig. 20.47.

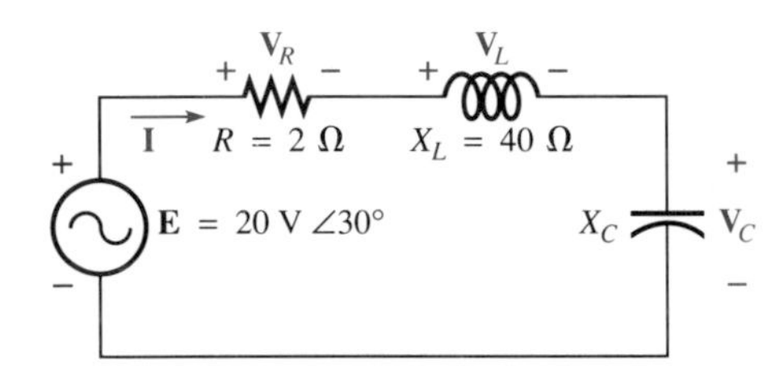


FIG. 20.47

**4.** For the circuit of Fig. 20.48:
- **a.** Find the value of $L$ in millihenries if the resonant frequency is 1800 Hz.
- **b.** Repeat Problem 2, parts (b) through (e).
- **c.** Calculate the cutoff frequencies.
- **d.** Find the bandwidth of the series resonant circuit.

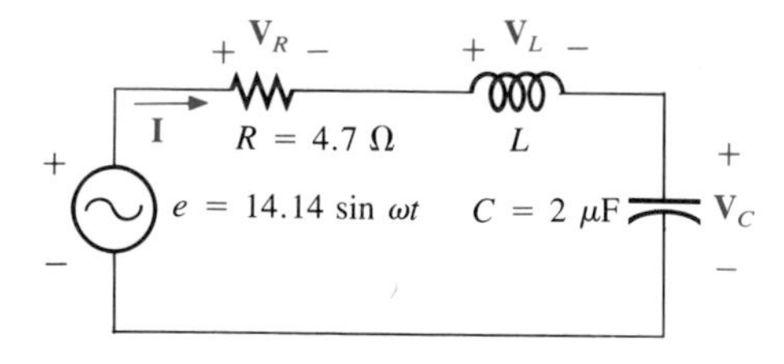


FIG. 20.48

**5.**
- **a.** Find the bandwidth of a series resonant circuit having a resonant frequency of 6000 Hz and a $Q_s$ of 15.
- **b.** Find the cutoff frequencies.
- **c.** If the resistance of the circuit at resonance is $3\,\Omega$, what are the values of $X_L$ and $X_C$ in ohms?
- **d.** What is the power dissipated at the half-power frequencies if the maximum current flowing through the circuit is 0.5 A?

**6.** A series circuit has a resonant frequency of 10 kHz. The resistance of the circuit is $5\,\Omega$, and $X_C$ at resonance is $200\,\Omega$.
- **a.** Find the bandwidth.
- **b.** Find the cutoff frequencies.
- **c.** Find $Q_s$.
- **d.** If the input voltage is $30\text{ V}\angle 0^\circ$, find the voltage across the coil and capacitor.
- **e.** Find the power dissipated at resonance.

7. a. The bandwidth of a series resonant circuit is 200 Hz. If the resonant frequency is 2000 Hz, what is the value of $Q_s$ for the circuit?
   b. If $R = 2\,\Omega$, what is the value of $X_L$ at resonance?
   c. Find the value of $L$ and $C$ at resonance.
   d. Find the cutoff frequencies.

8. The cutoff frequencies of a series resonant circuit are 5400 Hz and 6000 Hz.
   a. Find the bandwidth of the circuit.
   b. If $Q_s$ is 9.5, find the resonant frequency of the circuit.
   c. If the resistance of the circuit is $2\,\Omega$, find the value of $X_L$ and $X_C$ at resonance.
   d. Find the value of $L$ and $C$ at resonance.

*9. Design a series resonant circuit with an input voltage of $5\text{ V}\angle 0°$ to have the following specifications:
   a. a peak current of 500 mA at resonance
   b. a bandwidth of 120 Hz
   c. a resonant frequency of 8400 Hz
   Find the value of $L$ and $C$ and the cutoff frequencies.

*10. Design a series resonant circuit to have a bandwidth of 400 Hz using a coil with a $Q_l$ of 20 and a resistance of $2\,\Omega$. Find the value of $L$ and $C$ and the cutoff frequencies.

*11. A series resonant circuit is to resonate at $\omega_s = 2\pi \times 10^6$ rad/s and draw 20 W from a 120-V source at resonance. If the fractional bandwidth is 0.16,
   a. Determine the resonant frequency in hertz.
   b. Calculate the bandwidth in hertz.
   c. Determine the values of $R$, $L$, and $C$.
   d. Find the resistance of the coil if $Q_l = 80$.

*12. A series resonant circuit will resonate at a frequency of 1 MHz with a fractional bandwidth of 0.2. If the quality factor of the coil at resonance is 12.5 and its inductance is 100 $\mu$H, determine
   a. The resistance of the coil.
   b. The additional resistance required to establish the indicated fractional bandwidth.
   c. The required value of capacitance.

## SECTIONS 20.8 THROUGH 20.12 Parallel Resonance

13. For the "ideal" parallel resonant circuit of Fig. 20.49,
   a. Determine the resonant frequency.
   b. Find the voltage $V_C$ at resonance.
   c. Determine the currents $I_L$ and $I_C$ at resonance.
   d. Find $Q_p$.

14. For the parallel resonant circuit of Fig. 20.50,
   a. Find the resonant frequency.
   b. Find the voltage $V_C$ at resonance.
   c. Determine the power delivered by the source at resonance.
   d. Calculate the quality factor of the network, $Q_p$.

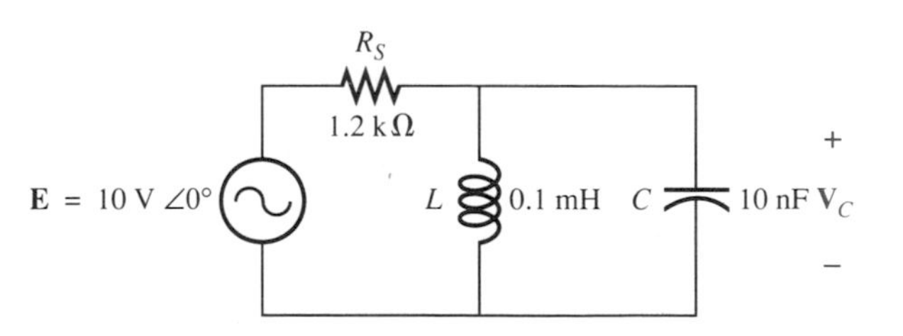

FIG. 20.49

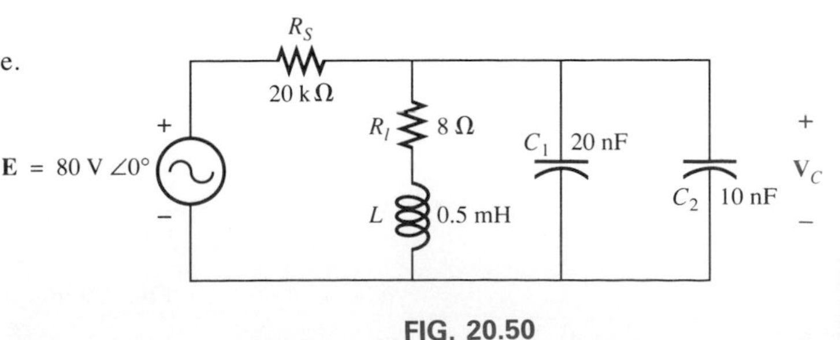

FIG. 20.50

15. For the circuit of Fig. 20.51:
   a. Find the value of $X_C$ at resonance ($f_p$).
   b. Find the total impedance $Z_T$ at resonance.
   c. Find the currents $I_L$ and $I_C$ at resonance.
   d. If the resonant frequency is 20,000 Hz, find the value of $L$ and $C$ at resonance.
   e. Find $Q_p$ and BW.

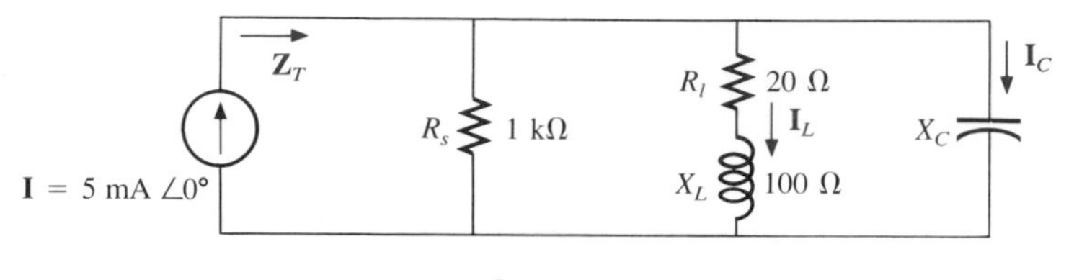


FIG. 20.51

16. Repeat Problem 15 for the circuit of Fig. 20.52.

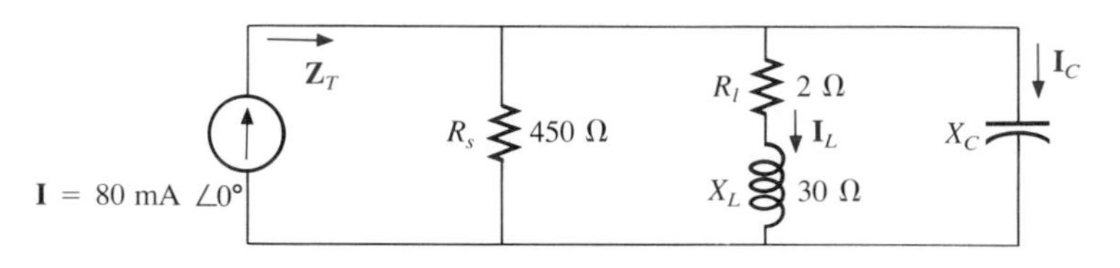

FIG. 20.52

17. For the circuit of Fig. 20.53:
   a. Find the resonant frequency.
   b. Find the value of $X_L$ and $X_C$ at resonance.
   c. Is the coil a high-$Q$ or low-$Q$ coil at resonance?
   d. Find the impedance $Z_{T_p}$ at resonance.
   e. Find the currents $I_L$ and $I_C$ at resonance.
   f. Calculate $Q_p$ and BW.

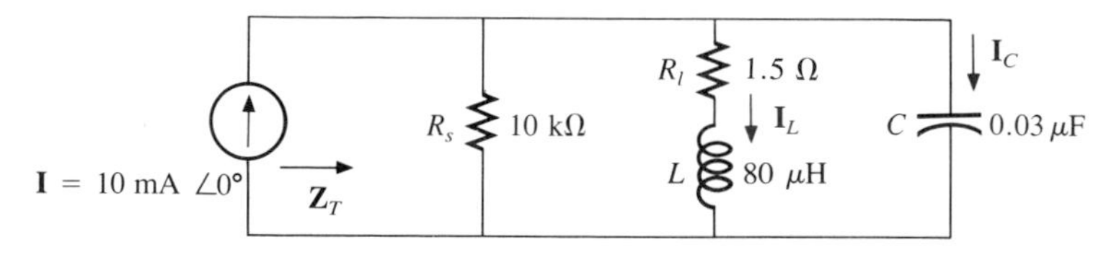


FIG. 20.53

18. Repeat Problem 17 for the circuit of Fig. 20.54.

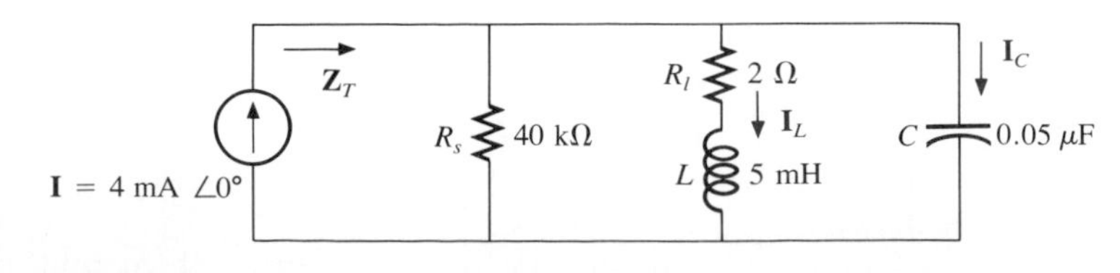


FIG. 20.54

**19.** It is desired that the impedance $Z_T$ of the circuit of Fig. 20.55 be a resistor of 50 k$\Omega$ at resonance.

**a.** Find the value of $X_L$.

**b.** Compute $X_C$.

**c.** Find the resonant frequency if $L = 16$ mH.

**d.** Find the value of $C$.

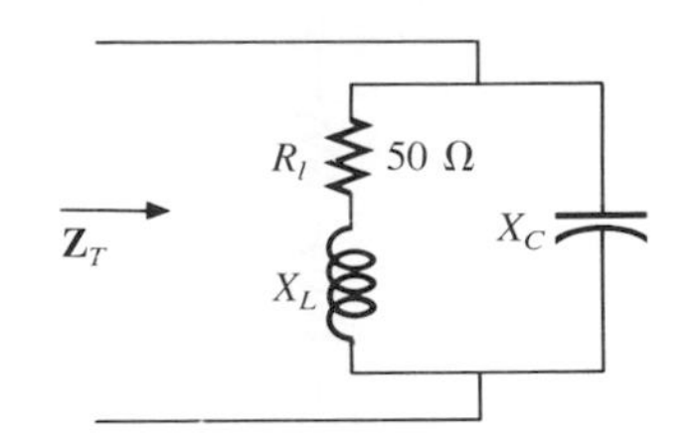

FIG. 20.55

**20.** For the network of Fig. 20.56:

**a.** Find $f_p$.

**b.** Calculate $V_C$ at resonance.

**c.** Determine the power absorbed at resonance.

**d.** Find BW.

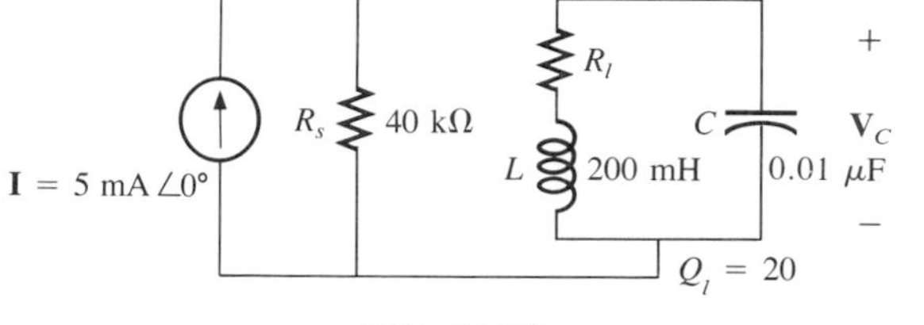

FIG. 20.56

***21.** For the network of Fig. 20.57, the following are specified:

$$f_p = 100\text{ kHz}$$
$$\text{BW} = 2500\text{ Hz}$$
$$L = 2\text{ mH}$$
$$Q_l = 80$$

Find $R_s$ and $C$.

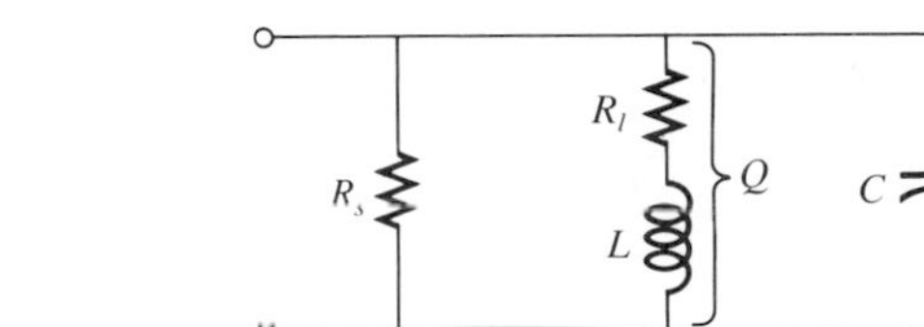


FIG. 20.57

***22.** For the network of Fig. 20.58:

**a.** Find the value of $X_L$ for resonance.

**b.** Find $Q_l$.

**c.** Find the resonant frequency if the bandwidth is 1000 Hz.

**d.** Find the maximum value of the voltage $V_C$.

**e.** Sketch the curve of $V_C$ versus frequency. Indicate its peak value, resonant frequency, and band frequencies.

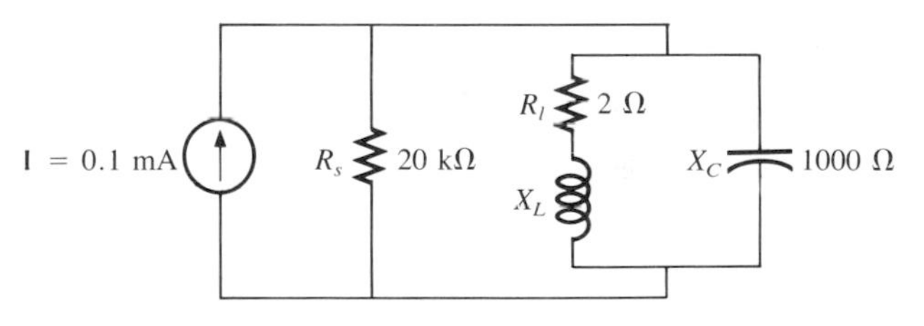

FIG. 20.58

***23.** Repeat Problem 22 for the network of Fig. 20.59.

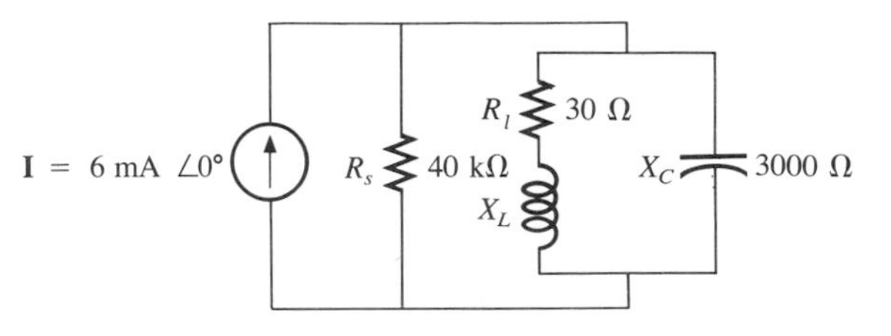

FIG. 20.59

***24.** Design the network of Fig. 20.60 to have the following characteristics:

**a.** a bandwidth of 500 Hz

**b.** $Q_p = 30$

**c.** $V_{C_{max}} = 1.8$ V

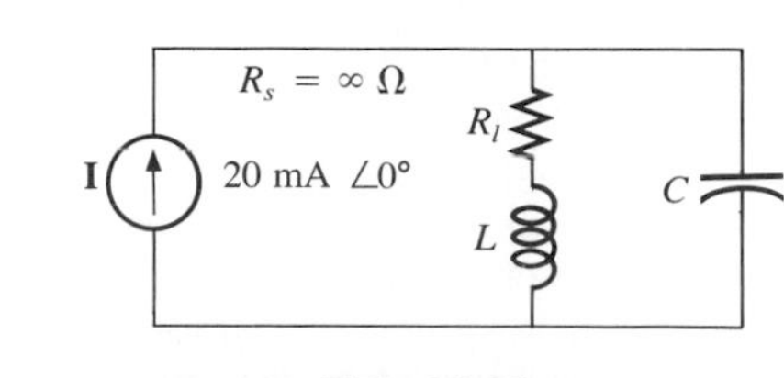


FIG. 20.60

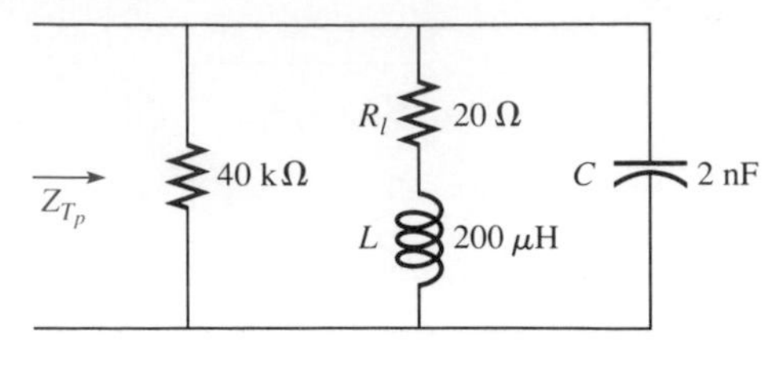


FIG. 20.61

*25. For the parallel resonant circuit of Fig. 20.61:
   **a.** Determine the resonant frequency.
   **b.** Find the total impedance at resonance.
   **c.** Find $Q_p$.
   **d.** Calculate the bandwidth.
   **e.** Repeat parts (a) through (d) for $L = 20\ \mu\text{H}$ and $C = 20$ nF.
   **f.** Repeat parts (a) through (d) for $L = 2\ \mu\text{H}$ and $C = 200$ nF.
   **g.** For the network of Fig. 20.61 and the parameters of parts (e) and (f), determine the ratio $L/C$.
   **h.** Do your results confirm the conclusions of Fig. 20.28 for changes in the $L/C$ ratio?

## SECTION 20.13 Computer Analysis

### BASIC

**26.** Write a program to tabulate the impedance and current of the network of Fig. 20.2 versus frequency for a frequency range extending from $0.1f_s$ to $2f_s$ in increments of $0.1f_s$. For the first run, use the parameters defined by Example 20.1.

**27.** Write a program to provide a general solution for the network of Fig. 20.34. That is, given the parameters appearing in Fig. 20.34, determine the quantities appearing in parts (a) through (f) of Example 20.8. For the first run, use the parameters appearing in Example 20.8, and compare results.

**28.** Write a program to provide a general solution for the network of Fig. 20.36. That is, determine the parameters requested in parts (a) through (e) of Example 20.9.

### PSPICE

**29.** Given a series *R-L-C* circuit with $R = 10\ \Omega$, $L = 3.98$ mH, and $C = 0.398\ \mu\text{F}$, write an input file to determine the voltages $V_R$, $V_L$, and $V_C$ and the current $I$ at resonance ($f_s = 2800$ Hz).

**30.** Write the input file to obtain a .PROBE response for the voltage $V_C$ for the network of Fig. 20.54.

# GLOSSARY

**Band (cutoff, half-power, corner) frequencies** Frequencies that define the points on the resonance curve that are 0.707 of the peak current or voltage value. In addition, they define the frequencies at which the power transfer to the resonant circuit will be half the maximum power level.

**Bandwidth** The range of frequencies between the band, cutoff, or half-power frequencies.

**Quality factor ($Q$)** A ratio that provides an immediate indication of the sharpness of the peak of a resonance curve. The higher the $Q$, the sharper the peak and the more quickly it drops off to the right and left of the resonant frequency.

**Resonance** A condition established by the application of a particular frequency (the resonant frequency) to a series or parallel *R-L-C* network. The transfer of power to the system is a maximum and, for frequencies above and below, the power transfer drops off to significantly lower levels.

**Selectivity** A characteristic of resonant networks directly related to the bandwidth of the resonant system. High selectivity is associated with small bandwidth (high $Q$'s), and low selectivity with larger bandwidths (low $Q$'s).

# 21 Decibels, Filters, and Bode Plots

dB

## 21.1 LOGARITHMS

The use of logarithms in industry is so extensive that a clear understanding of their purpose and use is an absolute necessity. At first exposure, logarithms often appear vague and mysterious due to the mathematical operations required to find the logarithm and antilogarithm using the longhand table approach that is typically taught in mathematics courses. However, virtually all of today's scientific calculators have the common and natural log functions, eliminating the complexity of applying logarithms and allowing us to concentrate on the positive characteristics of the function.

### Basic Relationships

Let us first examine the relationship between the variables of the logarithmic function. The mathematical expression

$$N = (b)^x$$

states that the number $N$ is equal to the base $b$ taken to the power $x$. A few examples:

$$100 = (10)^2$$
$$27 = (3)^3$$
$$54.6 = (e)^4 \qquad \text{where } e = 2.7183$$

If the question were to find the power $x$ to satisfy the equation

$$1200 = (10)^x$$

the value of $x$ could be determined using logarithms in the following manner:

$$x = \log_{10} 1200 = \mathbf{3.079}$$

Note that the logarithm was taken to the base 10—the number to be taken to the power of $x$. There is no limitation to the numerical value of the base except that tables and calculators are designed to handle either a base of 10 (common logarithm, log) or base $e = 2.7183$ (natural logarithm, ln). In review, therefore,

$$\text{If } N = (b)^x, \text{ then } x = \log_b N. \tag{21.1}$$

The base to be employed is a function of the area of application. If a conversion from one base to the other is required, the following equation can be applied:

$$\log_e x = 2.3 \log_{10} x \tag{21.2}$$

The content of this chapter is such that we will concentrate solely on the common logarithm. However, a number of the conclusions are also applicable to natural logarithms.

## Some Areas of Application

The following is a short list of the most common applications of the logarithmic function.

1. This chapter will demonstrate that the use of logarithms permits plotting the response of a system for a range of values that may otherwise be impossible or unwieldy with a linear scale.
2. Levels of power, voltage, and the like can be compared without dealing with very large or small numbers that often cloud the true impact of the difference in magnitudes.
3. There are a number of systems that respond to outside stimuli in a nonlinear logarithmic manner. The result is a mathematical model that permits a direct calculation of the response of the system to a particular input signal.
4. The response of a cascaded or compound system can be rapidly determined using logarithms if the gain of each stage is known on a logarithmic basis. This characteristic will be demonstrated in an example to follow.

## Graphs

Graph paper is available in the *semilog* or *log-log* variety. Semilog paper has only one log scale, with the other a linear scale. Both scales of log-log paper are log scales. A section of semilog paper appears in Fig. 21.1. Note the linear (even-spaced-interval) vertical scaling and the repeating intervals of the log scale.

The spacing of the log scale is determined by taking the common log (base 10) of the number. The scaling starts with 1, since $\log_{10} 1 = 0$. The distance between 1 and 2 is determined by $\log_{10} 2 = 0.3010$, or approximately 30% of the full distance of a log interval, as shown on

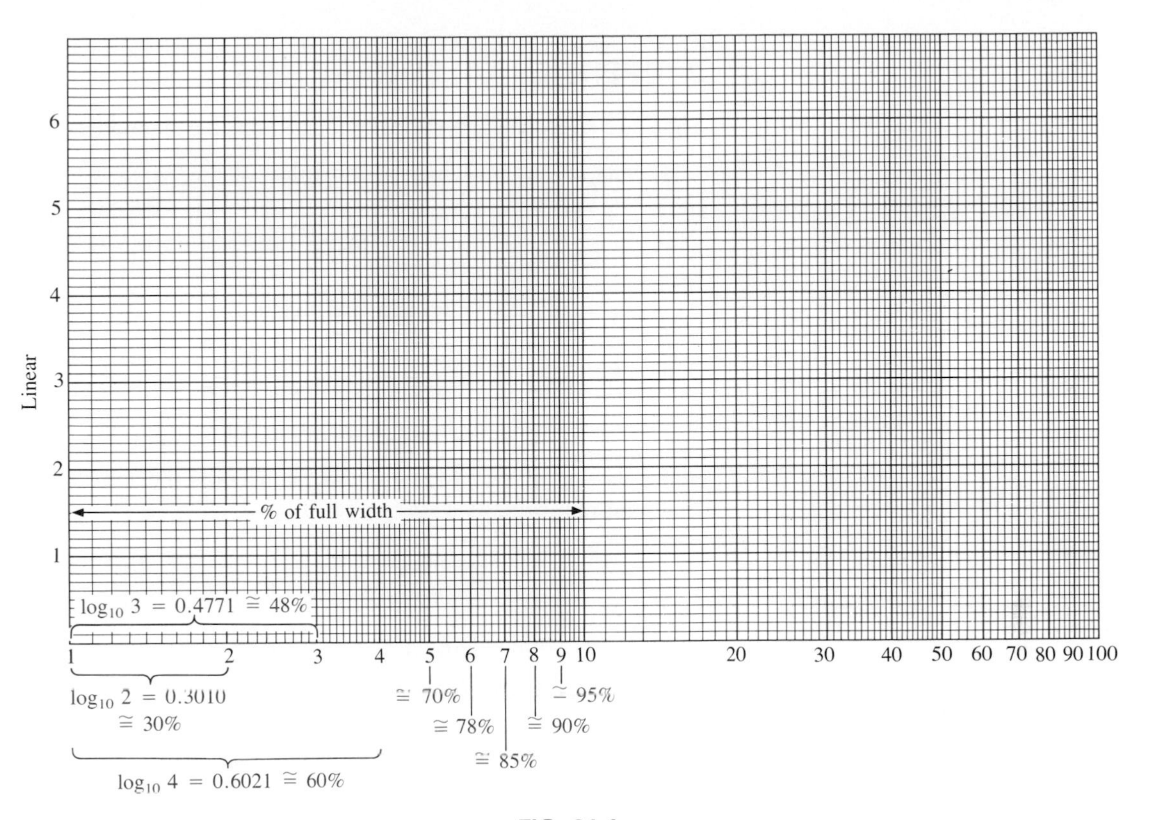


**FIG. 21.1**
*Semilog graph paper.*

the graph. The distance between 1 and 3 is determined by $\log_{10} 3 = 0.4771$, or about 48% of the full width. For future reference, keep in mind that almost 50% of the width of one log interval is represented by a 3 rather than by the 5 of a linear scale. This is particularly useful when the various lines of the graph are left unnumbered.

Note how the log scale becomes compressed at the high end of each interval. With increasing frequency levels assigned to each interval, a single graph can provide a frequency plot extending from 1 Hz to 1 MHz, as shown in Fig. 21.2 with particular reference to the 30%, 50%, and 70% levels of each interval.

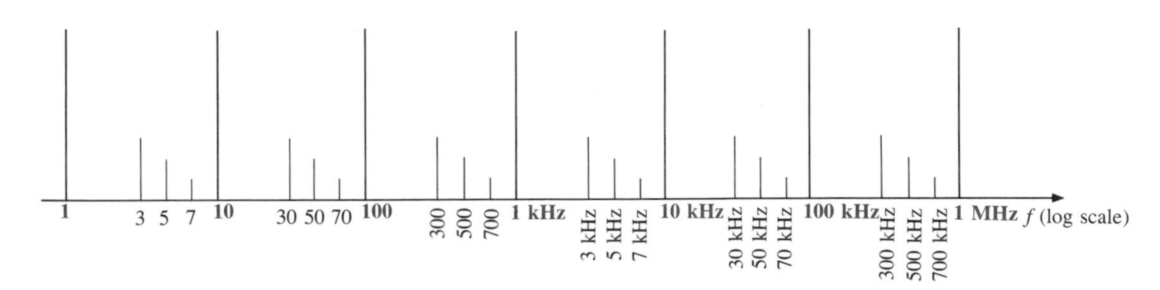


**FIG. 21.2**

## 21.2 PROPERTIES OF LOGARITHMS

There are a few characteristics of logarithms that should be emphasized.

**1.** The common or natural logarithm of the number 1 is 0.

$$\boxed{\log_{10} 1 = 0} \qquad \textbf{(21.3)}$$

just as $10^x = 1$ requires that $x = 0$.

**2.** The log of any number *less than* 1 is a *negative* number.

$$\log_{10} \tfrac{1}{2} = \log_{10} 0.5 = -3$$
$$\log_{10} \tfrac{1}{10} = \log_{10} 0.1 = -1$$

**3.** The log of the product of two numbers is the sum of the logs of the numbers.

$$\boxed{\log_{10} ab = \log_{10} a + \log_{10} b} \qquad \textbf{(21.4)}$$

**4.** The log of the quotient of two numbers is the log of the numerator minus the log of the denominator.

$$\boxed{\log_{10}\frac{a}{b} = \log_{10} a - \log_{10} b} \qquad \textbf{(21.5)}$$

**5.** The log of a number taken to a power is equal to the product of the power and the log of the number.

$$\boxed{\log_{10} a^n = n \log^{10} a} \qquad \textbf{(21.6)}$$

### Calculator Functions

On most calculators the log of a number is found by simply entering the number and pressing the $\boxed{\text{log}}$ or $\boxed{\text{ln}}$ key.

For example,

$$\log_{10} 80 = \boxed{8}\ \boxed{0}\ \boxed{\text{log}}$$

with a display of **1.903.**

For the reverse process, where $N$ is desired, the function $10^x$ is employed. On most calculators $10^x$ appears as a second function above the $\boxed{\text{log}}$ key. For the case of

$$0.6 = \log_{10} N$$

the following keys are employed:

$$\boxed{\cdot}\ \boxed{6}\ \boxed{\text{2ndF}}\ \boxed{10^x}$$

with a display of **3.981.** Checking: $\log^{10} 3.981 = 0.6$.

**EXAMPLE 21.1.** Evaluate each of the following logarithmic expressions.

a. $\log_{10} 0.004$
b. $\log_{10} 250{,}000$
c. $\log_{10}(0.08)(240)$
d. $\log_{10}\dfrac{1 \times 10^4}{1 \times 10^{-4}}$
e. $\log_{10}(10)^4$

***Solutions:***

a. $\mathbf{-2.398}$
b. $\mathbf{+5.398}$
c. $\log_{10}(0.08)(240) = \log_{10} 0.08 + \log_{10} 240 = -1.097 + 2.380 = \mathbf{1.283}$
d. $\log_{10}\dfrac{1 \times 10^4}{1 \times 10^{-4}} = \log_{10} 1 \times 10^4 - \log_{10} 1 \times 10^{-4} = 4 - (-4) = \mathbf{8}$
e. $\log_{10} 10^4 = 4 \log_{10} 10 = 4(1) = \mathbf{4}$

## 21.3 DECIBELS

### Power Gain

Two levels of power can be compared using a unit of measure called the *bel,* which is defined by the following equation:

$$\text{bels} = \log_{10}\frac{P_2}{P_1} \qquad \textbf{(21.7)}$$

However, to provide a unit of measure of *less* magnitude, a *decibel* is defined, where

$$1 \text{ bel} = 10 \text{ decibels (dB)} \qquad \textbf{(21.8)}$$

The result is the following important equation, which compares power levels $P_2$ and $P_1$ in decibels.

$$\text{dB} = 10 \log_{10}\frac{P_2}{P_1} \quad \text{(decibels, dB)} \qquad \textbf{(21.9)}$$

If the power levels are equal ($P_2 = P_1$), there is no change in power level, and dB $= 0$. If there is an increase in power level ($P_2 > P_1$), the resulting decibel level is positive. If there is a decrease in power level ($P_2 < P_1$), the resulting decibel level will be negative.

For the special case of $P_2 = 2P_1$, the gain in decibels is

$$\text{dB} = 10 \log_{10} \frac{P_2}{P_1} = 10 \log_{10} 2 = \mathbf{3\ dB}$$

Therefore, for a speaker system, a 3-dB increase in output would require that the power level be doubled. In the audio industry, it is a generally accepted rule that an increase in sound level is accomplished with 3-dB increments in the output level. In other words, a 1-dB increase is barely detectable, and a 2-dB increase just discernible. A 3-dB increase normally results in a readily detectable increase in sound level. A further increase in the sound level is normally accomplished by simply increasing the output level another 3 dB. If an 8-W system were in use, a 3-dB increase would require a 16-W output, whereas a further increase of 3 dB (a total of 6 dB) would require a 32-W system, as demonstrated by the calculations below:

$$\text{dB} = 10 \log_{10} \frac{P_2}{P_1} = 10 \log_{10} \frac{16}{8} = 10 \log_{10} 2 = \mathbf{3\ dB}$$

$$\text{dB} = 10 \log_{10} \frac{P_2}{P_1} = 10 \log_{10} \frac{32}{8} = 10 \log_{10} 4 = \mathbf{6\ dB}$$

For $P_2 = 10P_1$,

$$\text{dB} = 10 \log_{10} \frac{P_2}{P_1} = 10 \log_{10} 10 = 10(1) = \mathbf{10\ dB}$$

resulting in the unique situation where the power gain has the same magnitude as the decibel level.

For some applications, a reference level is established to permit a comparison of decibel levels from one situation to another. For communication systems a commonly applied reference level is

$$P_{\text{ref}} = 1\ \text{mW} \quad (\text{across a } 600\text{-}\Omega \text{ load})$$

Equation (21.9) is then typically written as

$$\text{dB}_m = 10 \log_{10} \frac{P}{1\ \text{mW}}\bigg|_{600\ \Omega} \tag{21.10}$$

Note the subscript $m$ to denote that the decibel level is determined with a reference level of 1 mW.

In particular, for $P = 40$ mW,

$$\text{dB}_m = 10 \log_{10} \frac{40\ \text{mW}}{1\ \text{mW}} = 10 \log_{10} 40 = 10(1.6) = \mathbf{16\ dBm}$$

whereas for $P = 4$ W,

$$\text{dB}_m = 10 \log_{10} \frac{4000\ \text{mW}}{1\ \text{mW}} = 10 \log_{10} 4000 = 10(3.6) = \mathbf{36\ dBm}$$

Even though the power level has increased by a factor of 4000 mW/40 mW = 100, the $dB_m$ increase is limited to 20 dBm. In time, the significance of $dB_m$ levels of 16 dB and 36 dBm will generate an immediate appreciation regarding the power levels involved. An increase of 20 dBm will also be associated with a significant gain in power levels.

## Voltage Gain

Decibels are also used to compare voltage levels for a system such as that shown in Fig. 21.3.

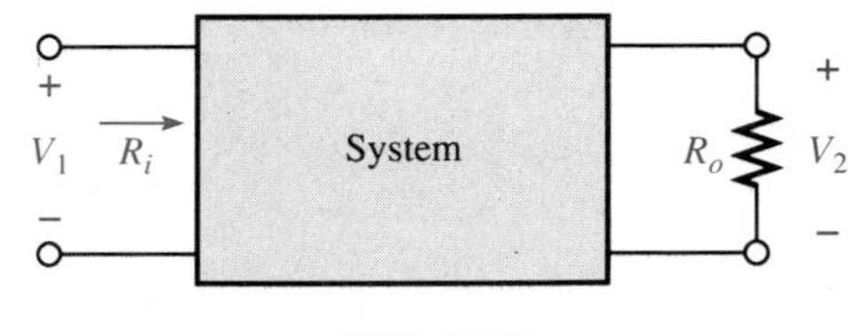


FIG. 21.3

For the input side,

$$P_i = \frac{V_1^2}{R_i}$$

and for the output,

$$P_o = \frac{V_2^2}{R_o}$$

Substituting into (21.9) will result in

$$\text{dB} = 10 \log_{10}\frac{P_2}{P_1} = 10 \log_{10}\frac{V_2^2/R_o}{V_1^2/R_i}$$

$$= 10 \log_{10}\frac{V_2^2/V_1^2}{R_o/R_i} = 10 \log_{10}\left(\frac{V_2}{V_1}\right)^2 - 10 \log_{10}\left(\frac{R_o}{R_i}\right)$$

and

$$\text{dB} = 20 \log_{10}\frac{V_2}{V_1} - 10 \log_{10}\frac{R_o}{R_i}$$

For the situation where $R_o = R_i$, a condition normally assumed when comparing voltage levels on a decibel basis, the second term of the preceding equation will drop out and

$$\text{dB}_v = 20 \log_{10}\frac{V_2}{V_1} \qquad \text{(dB)} \qquad \textbf{(21.11)}$$

Note the subscript $v$ to define the decibel level obtained.

---

**EXAMPLE 21.2.** Find the voltage gain in dB of a system where the applied signal is 2 mV and the output voltage is 1.2 V.

***Solution:***

$$\text{dB}_v = 20 \log_{10}\frac{V_2}{V_1} = 20 \log_{10}\frac{1.2\text{ V}}{2\text{ mV}} = 20 \log_{10} 600 = \mathbf{55.56\ dB}$$

for a voltage gain $A_v = V_o/V_i$ of 600.

---

**EXAMPLE** If a system has a voltage gain of 36 dB, find the applied voltage if the output voltage is 6.8 V.

***Solution:***

$$dB_v = 10\ \log_{10}\frac{V_2}{V_1}$$

$$36 = 20\ \log_{10}\frac{V_2}{V_1}$$

$$1.8 = \log_{10}\frac{V_2}{V_1}$$

$$\frac{V_2}{V_1} = 63.96$$

or

$$V_1 = \frac{V_2}{63.96} = \frac{6.8\ \text{V}}{63.96} = \mathbf{106.32\ mV}$$

TABLE 21.1

| $V_o/V_i$ | $dB = 20\ \log_{10}(V_o/V_i)$ |
|---|---|
| 1 | 0 dB |
| 2 | 6 dB |
| 10 | 20 dB |
| 20 | 26 dB |
| 100 | 40 dB |
| 1,000 | 60 dB |
| 100,000 | 100 dB |

Table 21.1 compares the magnitude of specific gains to the resulting decibel level. In particular, note that when voltage levels are compared, a doubling of the level results in a change of 6 dB rather than 3 dB as obtained for power levels.

In addition, note that an increase in gain from 1 to 100,000 results in a change in decibels that can easily be plotted on a single graph. Also note that doubling the gain (from 1 to 2 and 10 to 20) results in a 6-dB increase in the decibel level, while a change of 10 to 1 (from 1 to 10, 10 to 100, and so on) always results in a 20-dB increase in the decibel level.

## The Human Auditory Response

One of the most frequent applications of the decibel scale is in the communication and entertainment industries. The human ear does not respond in a linear fashion to changes in source power level. That is, a doubling of the audio power level from 1/2 W to 1 W does not result in a doubling of the loudness level for the human ear. In addition, a change from 5 W to 10 W will be received by the ear as the same change in sound intensity as experienced from 1/2 W to 1 W. In other words, the ratio between levels is the same in each case (1 W/0.5 W = 10 W/5 W = 2), resulting in the same decibel or logarithmic change defined by Eq. (21.6). The ear, therefore, responds in a logarithmic fashion to changes in audio power levels.

To establish a basis for comparison between audio levels, a reference level of 0.0002 microbars ($\mu$bar) was chosen, where 1 $\mu$bar is equal to the sound pressure of 1 dyne per square centimeter, or about 1 millionth of the normal atmospheric pressure at sea level. The 0.0002-$\mu$bar level is the threshold level of hearing. Using this reference level, the sound pressure level in decibels is defined by the following equation:

$$\mathrm{dB}_s = 20 \log_{10} \frac{P}{0.0002\ \mu\mathrm{bar}} \tag{21.12}$$

where $P$ is the sound pressure in microbars.

The decibel levels of Fig. 21.4 are defined by Eq. (21.12). Meters designed to measure audio levels are calibrated to the levels defined by Eq. (21.12) and shown in Fig. 21.4.

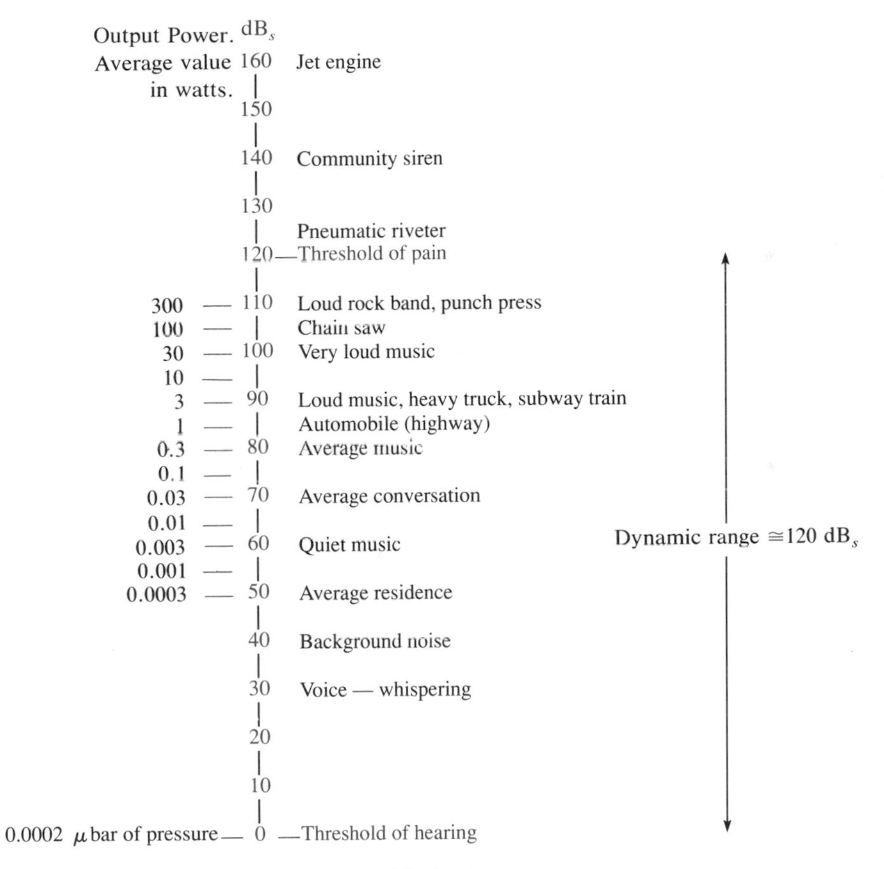


**FIG. 21.4**
*Typical sound levels and their decibel levels.*

A common question regarding audio levels is how much the power level of an acoustical source must be increased to double the sound level received by the human ear. The question is not as simple as it first seems, due to considerations such as the frequency content of the sound, the acoustical conditions of the surrounding area, the physical characteristics of the surrounding medium, and—of course—the unique characteristics of the human ear. However, a general conclusion can be formulated that has practicalvalue if we note the power levels of an acoustical source appearing to the left of Fig. 21.4. Each power level is associated with a particular decibel level, and a change of 10 dB in the scale corresponds with an increase or decrease in power by a factor of 10. For instance, a change from 90 db to 100 db is associated with a change in wattage from 3 W to 30 W. Through experimentation it has been found on an averge basis that the loudness level will double for every 10-db change in audio level—a conclusion somewhat verified by the examples to the right of Fig. 21.4. Using the fact that a 10-db change corresponds with a tenfold increase in power level supports the following conclusion (on an approximate basis): Through experimentation it has been found on an average basis that the loudness level will double for every 10-dB change in audio level.

***in order to double the sound level received by the human ear, the power rating of the acoustical source (in watts) must be increased by a factor of 10.***

In other words, doubling the sound level available from a 1-W acoustical source would require moving up to a 10-W source.

## 21.4 FILTERS

Any combination of passive (*R*, *L*, and *C*) and/or active (transistors or operational amplifiers) elements designed to select or reject a band of frequencies is called a *filter*. In communication systems, filters are employed to pass those frequencies containing the desired information and reject the remaining frequencies. In stereo systems, filters can be used to isolate particular bands of frequencies for increased or decreased emphasis by the output acoustical system (amplifier, speaker, etc.). Filters are employed to filter out any unwanted frequencies, commonly called *noise*, due to the nonlinear characteristics of some electronic devices or signals picked up from the surrounding medium.

In general, there are two classifications of filters:

1. *Passive filters* are those filters composed of series or parallel combinations of *R*, *L*, and *C* elements.
2. *Active filters* are filters that employ active devices such as transistors and operational amplifiers in combination with *R*, *L*, and *C* elements.

Since this text is limited to passive devices, the analysis of this chapter will be limited to passive filters. In addition, only the most fundamental forms will be examined in the next few sections. The subject of filters is a very broad one that continues to receive extensive research support from industry and the government as new communication systems are developed to meet the demands of increased volume and speed. There are courses and texts devoted solely to the analysis and design of filter systems that can become quite complex and sophisticated. In general, however, all filters belong to the four broad categories of *low-pass*, *high-pass*, *band-pass*, and *stop-band*, as depicted in

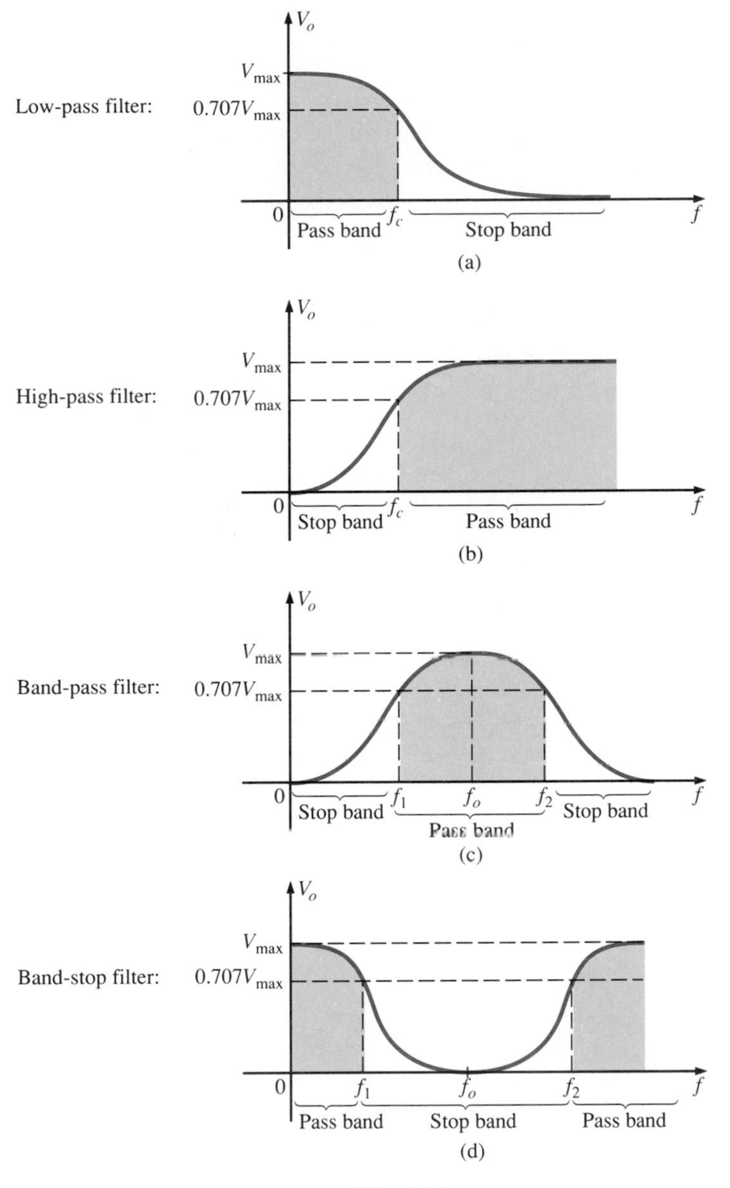


**FIG. 21.5**

Fig. 21.5. For each form there are critical frequencies that define the regions of pass- and stop-bands (often called *reject* bands). Any frequency in the pass-band will pass through to the next stage with at least 70.7% of the maximum output voltage. Recall the use of the 0.707 level to define the bandwidth of a series or parallel resonant circuit (both with the general shape of the pass-band filter). At least one example of each filter of Fig. 21.5 will be discussed in some detail in the sections to follow. Take particular note of the relative simplicity of some of the designs.

## 21.5 R-C LOW-PASS FILTER

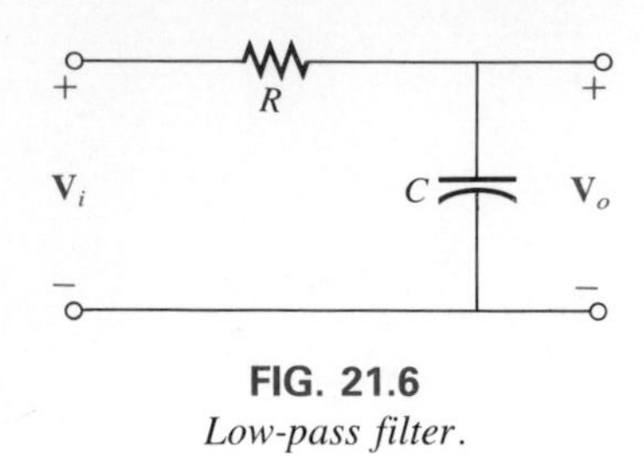


FIG. 21.6
*Low-pass filter.*

The *R-C* filter, incredibly simple in design, can be used as a low-pass or high-pass filter. If the output is taken off the capacitor, as shown in Fig. 21.6, it will respond as a low-pass filter. If the positions of the resistor and capacitor are interchanged and the output is taken off the resistor, the response will be that of a high-pass filter.

A glance at Fig. 21.5(a) reveals that the circuit should behave in a manner that will result in a high-level output for low frequencies and a declining level for frequencies above the critical value. Let us first examine the network at the frequency extremes of $f = 0$ Hz and very high frequencies to test the response of the circuit.

At $f = 0$ Hz,

$$X_C = \frac{1}{2\pi fC} = \infty\ \Omega$$

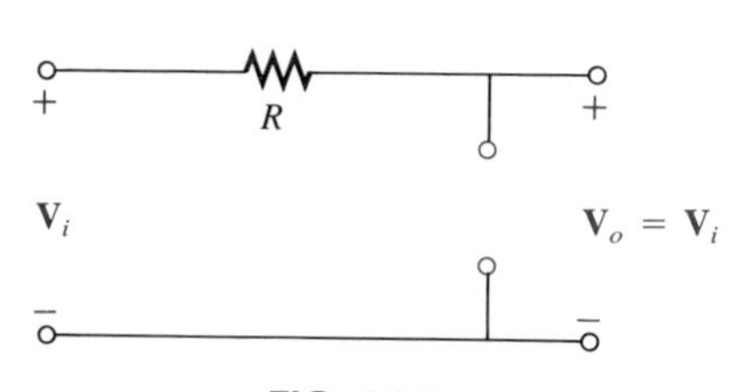


FIG. 21.7
*R-C low-pass filter at low frequencies.*

and the open-circuit equivalent can be substituted for the capacitor, as shown in Fig. 21.7, resulting in $V_o = V_i$.

At very high frequencies, the reactance is

$$X_C = \frac{1}{2\pi fC} \cong 0\ \Omega$$

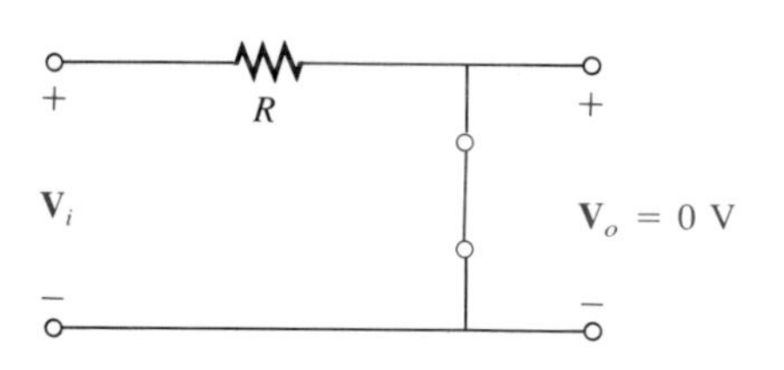


FIG. 21.8
*R-C low-pass filter at high frequencies.*

and the short-circuit equivalent can be substituted for the capacitor, as shown in Fig. 21.8, resulting in $V_o = 0$ V.

Using the extreme levels of $V_o$ and inserting a transition level will result in the characteristics of Fig. 21.9. Our next goal is now clearly defined: Find the frequency at which the transition takes place from a pass-band to stop-band.

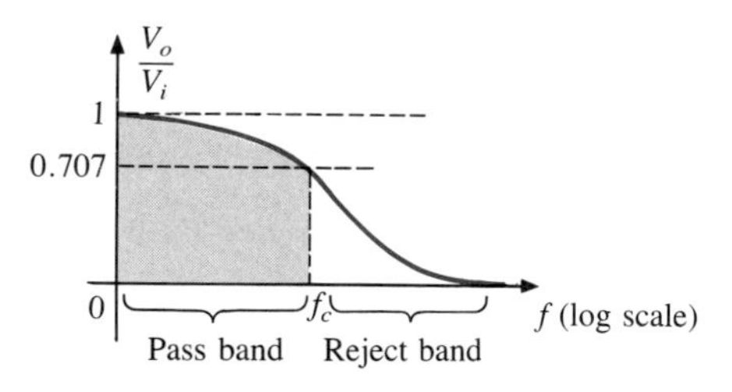


FIG. 21.9
*Low-pass filter characteristics.*

At any intermediate frequency, the output voltage can be determined using the voltage divider rule:

$$\mathbf{V}_o = \frac{\mathbf{X}_C \mathbf{V}_i}{\mathbf{R} + \mathbf{X}_C}$$

or

$$\frac{\mathbf{V}_o}{\mathbf{V}_i} = \frac{\mathbf{X}_C}{\mathbf{R} + \mathbf{X}_C} = \frac{X_C \angle -90^\circ}{R - jX_C} = \frac{X_C \angle -90^\circ}{\sqrt{R^2 + X_C^2}\ \angle -\tan^{-1}(X_C/R)}$$

and

$$\frac{\mathbf{V}_o}{\mathbf{V}_i} = \frac{X_C}{\sqrt{R^2 + X_C^2}} \angle -90^\circ + \tan^{-1}(X_C/R)$$

The magnitude of the ratio $V_o/V_i$ is therefore determined by

$$\boxed{\frac{V_o}{V_i} = \frac{X_C}{\sqrt{R^2 + X_C^2}} = \frac{1}{\sqrt{\left(\frac{R}{X_C}\right)^2 + 1}}} \quad \textbf{(21.13)}$$

and the phase angle, by

$$\boxed{\theta = -90^\circ + \tan^{-1}\frac{X_C}{R}} \quad \textbf{(21.14)}$$

For the special frequency at which $X_C = R$, the magnitude becomes

$$\frac{V_o}{V_i} = \frac{1}{\sqrt{\left(\frac{R}{X_C}\right)^2 + 1}} = \frac{1}{\sqrt{1 + 1}}$$

and

$$\boxed{\frac{V_o}{V_i} = \frac{1}{\sqrt{2}} = 0.707} \quad X_C = R, f = f_c$$

as shown in Fig. 21.9.

The frequency at which $X_C = R$ is determined by

$$\frac{1}{2\pi f_c C} = R$$

and

$$\boxed{f_c = \frac{1}{2\pi RC}} \quad \textbf{(21.15)}$$

The impact of Eq. (21.15) extends beyond its relative simplicity. For any low-pass filter, the application of any frequency less than $f_c$ will result in an output voltage $V_o$ that is at least 70.7% of the magnitude of the input signal. For any frequency above $f_c$, the output is less than 70.7% of the applied signal.

For the phase angle at high frequencies, $\tan^{-1}(X_C/R)$ approaches 0°, and

$$\theta = -90^\circ + \tan^{-1}\frac{X_C}{R} = -90^\circ + 0^\circ$$

$$= -90^\circ$$

At low frequencies, $\tan^{-1}(X_C/R)$ approaches 90°, and

$$\theta = -90^\circ + \tan^{-1}\frac{X_C}{R} = -90^\circ + 90^\circ$$
$$= 0^\circ$$

At $X_C = R$ or $f = f_c$, $\tan^{-1}(X_C/R) = \tan^{-1} 1 = 45^\circ$, and

$$\theta = -90^\circ + \tan^{-1}\frac{X_C}{R} = -90^\circ + 45^\circ$$
$$= -45^\circ$$

A plot of $\theta$ versus frequency results in the phase plot of Fig. 21.10.

The plot is of $\mathbf{V}_o$ leading $\mathbf{V}_i$, but since the phase angle is always negative, the phase plot of Fig. 21.11 is more appropriate. Note that a change in sign requires that the vertical axis be changed to the angle by which $\mathbf{V}_o$ lags $\mathbf{V}_i$. The low-pass *R-C* filter is therefore a *lagging network*.

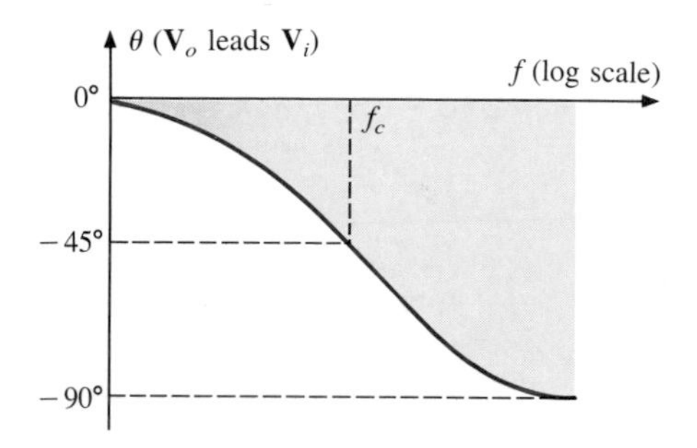


**FIG. 21.10**
*Phase response for the low-pass R-C filter.*

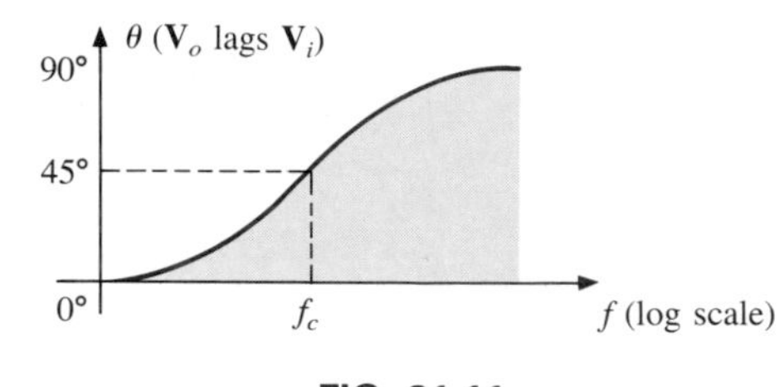


**FIG. 21.11**
*Phase response for the low-pass R-C filter.*

In summary, for the low-pass *R-C* filter:

$$f_c = \frac{1}{2\pi fC}$$

| | |
|---|---|
| For | $f < f_c$, $V_o > 0.707V_i$ |
| whereas for | $f > f_c$, $V_o < 0.707V_i$ |
| At $f_c$, | $\mathbf{V}_o$ lags $\mathbf{V}_i$ by 45°. |

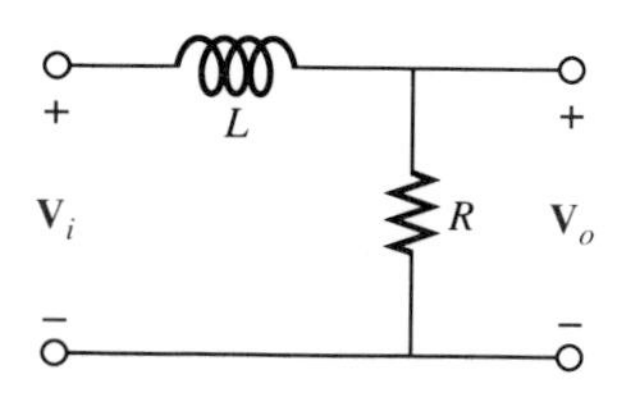


**FIG. 21.12**
*Low-pass R-L filter.*

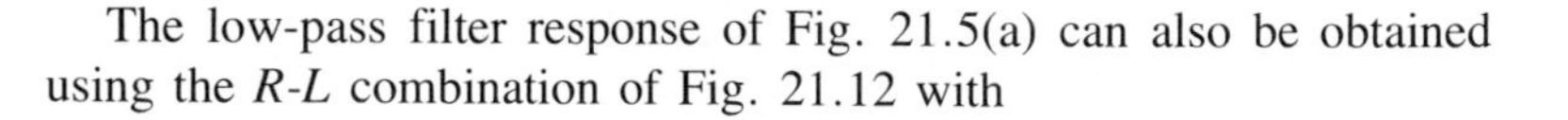

The low-pass filter response of Fig. 21.5(a) can also be obtained using the *R-L* combination of Fig. 21.12 with

$$f_c = \frac{R}{2\pi L} \tag{21.16}$$

In general, however, the $R$-$C$ combination is more popular due to the smaller size of capacitive elements and the nonlinearities associated with inductive elements. The details of the analysis of the low-pass $R$-$L$ will be left as an exercise for the reader.

---

**EXAMPLE 21.4.**

a. Sketch the output voltage $V_o$ versus frequency for the low-pass $R$-$C$ circuit of Fig. 21.13.
b. Determine the voltage $V_o$ at $f = 100$ kHz and 1 MHz and compare the results to the results obtained from the curve of part (a).

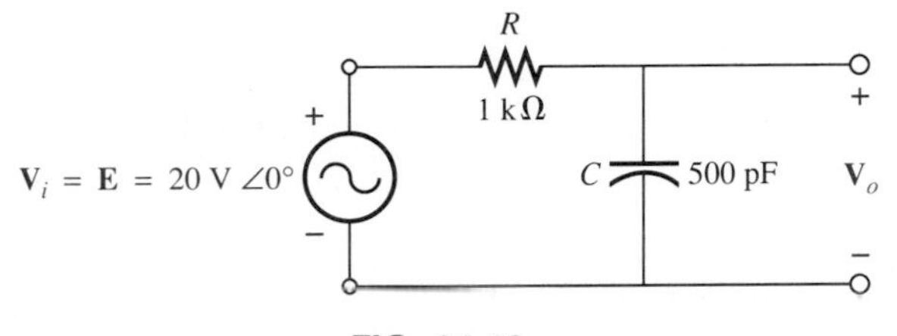

**FIG. 21.13**

***Solutions:***

a. Equation (21.15):

$$f_c = \frac{1}{2\pi RC} = \frac{1}{2\pi(1\ \text{k}\Omega)(500\ \text{pF})} = \mathbf{318.31\ kHz}$$

At $f_c$, $V_o = 0.707(20\ \text{V}) = 14.14$ V. See Fig. 21.14.

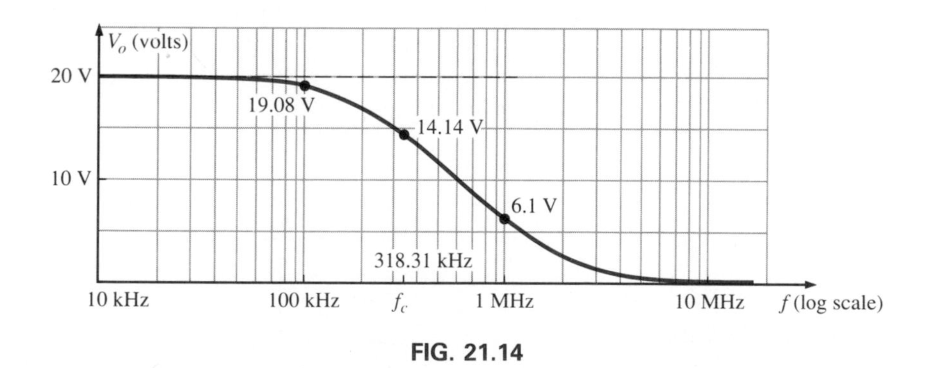


**FIG. 21.14**

b. Equation (21.13):

$$V_o = \frac{V_i}{\sqrt{\left(\dfrac{R}{X_C}\right)^2 + 1}}$$

At $f = 100$ kHz:

$$X_C = \frac{1}{2\pi fC} = \frac{1}{2\pi(100 \text{ kHz})(500 \text{ pF})} = 3.18 \text{ k}\Omega$$

and

$$V_o = \frac{20 \text{ V}}{\sqrt{\left(\dfrac{1 \text{ k}\Omega}{3.18 \text{ k}\Omega}\right)^2 + 1}} = \mathbf{19.08\ V}$$

At $f = 1$ mHz:

$$X_C = \frac{1}{2\pi fC} = \frac{1}{2\pi(1 \text{ MHz})(500 \text{ pF})} = 0.32 \text{ k}\Omega$$

and

$$V_o = \frac{20 \text{ V}}{\sqrt{\left(\dfrac{1 \text{ k}\Omega}{0.32 \text{ k}\Omega}\right)^2 + 1}} = \mathbf{6.1\ V}$$

Both levels are verified by Fig. 21.14.

## 21.6 *R-C* HIGH-PASS FILTER

As noted early in Section 21.5, a high-pass *R-C* filter can be constructed by simply reversing the positions of the capacitor and inductor, as shown in Fig. 21.15.

At very high frequencies the reactance of the capacitor is very small and the short-circuit equivalent can be substituted, as shown in Fig. 21.16. The result is $V_o = V_i$, or $V_o/V_i = 1$, as shown in the response characteristics of Fig. 21.17.

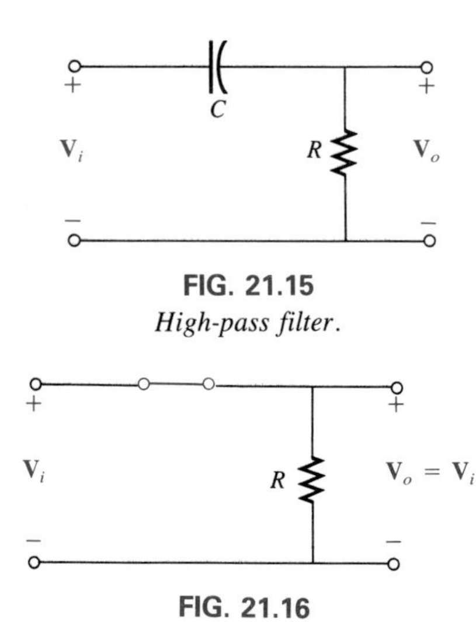


FIG. 21.15
*High-pass filter.*

FIG. 21.16
*R-C high-pass filter at very high frequencies.*

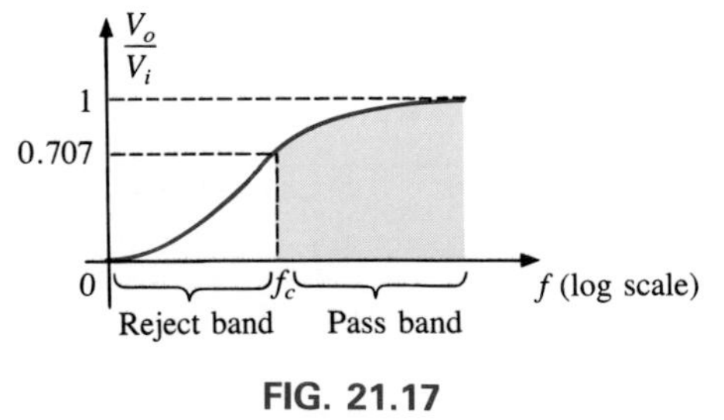


FIG. 21.17
*High-pass filter characteristics.*

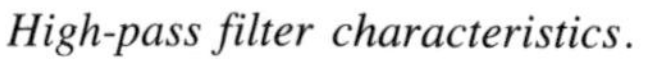

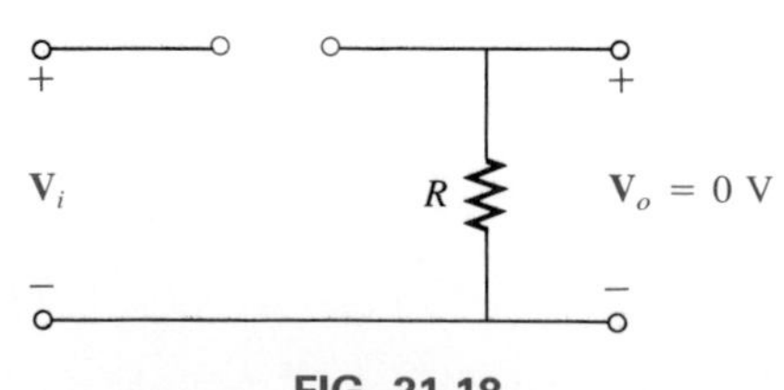


FIG. 21.18
*R-C high-pass filter at $f = 0$ Hz.*

At $f = 0$ Hz, the reactance of the capacitor is quite high, and the open-circuit equivalent can be substituted, as shown in Fig. 21.18. Also, $V_o = 0$ V, or $V_o/V_i = 0$, as shown in Fig. 21.17.

At any intermediate frequency, the output voltage can be determined using the voltage divider rule:

$$\mathbf{V}_o = \frac{\mathbf{R}\mathbf{V}_i}{\mathbf{R} + \mathbf{X}_C}$$

or

$$\frac{\mathbf{V}_o}{\mathbf{V}_i} = \frac{\mathbf{R}}{\mathbf{R} + \mathbf{X}_C} = \frac{R \angle 0°}{R - jX_C} = \frac{R \angle 0°}{\sqrt{R^2 + X_C^2} \angle -\tan^{-1}(X_C/R)}$$

and

$$\frac{\mathbf{V}_o}{\mathbf{V}_i} = \frac{R}{\sqrt{R^2 + X_C^2}} \angle \tan^{-1}(X_C/R)$$

The magnitude of the ratio $\mathbf{V}_o/\mathbf{V}_i$ is therefore determined by

$$\frac{V_o}{V_i} = \frac{R}{\sqrt{R^2 + X_C^2}} = \frac{1}{\sqrt{1 + \left(\frac{X_C}{R}\right)^2}} \tag{21.17}$$

and the phase angle $\theta$, by

$$\theta = \tan^{-1}\frac{X_C}{R} \tag{21.18}$$

For the frequency at which $X_C = R$, the magnitude becomes

$$\frac{V_o}{V_i} = \frac{1}{\sqrt{1 + \left(\frac{X_C}{R}\right)^2}} = \frac{1}{\sqrt{1 + 1}}$$

and

$$\frac{V_o}{V_i} = \frac{1}{\sqrt{2}} = 0.707$$

as shown in Fig. 21.17.

The frequency at which $X_C = R$ is determined by

$$X_C = \frac{1}{2\pi f_c C} = R$$

and

$$f_c = \frac{1}{2\pi RC} \tag{21.19}$$

For the high-pass *R-C* filter, the application of any frequency greater than $f_c$ will result in an output voltage $V_o$ that is at least 70.7% of the magnitude of the input signal. For any frequency below $f_c$, the output is less than 70.7% of the applied signal.

For the phase angle, high frequencies result in small values of $X_C$, and the ratio $X_C/R$ will approach zero with $\tan^{-1}(X_C/R)$ approaching 0°, as shown in Fig. 21.19. At low frequencies, the ratio $X_C/R$ becomes

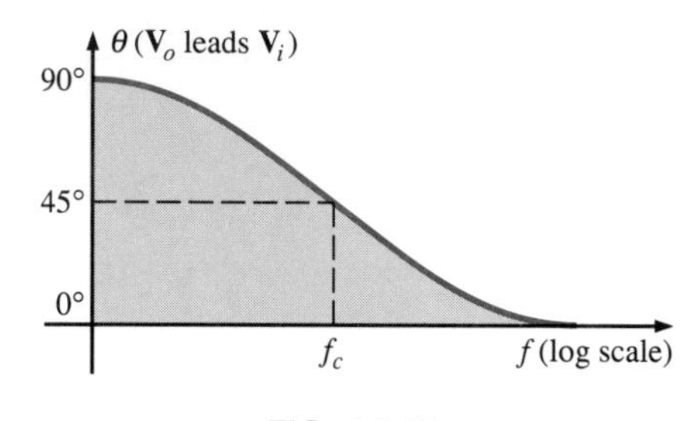


**FIG. 21.19**
*Phase-angle response for the high-pass R-C filter.*

quite large and $\tan^{-1}(X_C/R)$ approaches 90°. For the case $X_C = R$, $\tan^{-1}(X_C/R) = \tan^{-1} 1 = 45°$. Assigning a phase angle of 0° to $\mathbf{V}_i$ such that $\mathbf{V}_i = V_i \angle 0°$, the phase angle associated with $\mathbf{V}_o$ is $\theta$, resulting in $\mathbf{V}_o = V_o \angle \theta$ and revealing that $\theta$ is the angle by which $\mathbf{V}_o$ leads $\mathbf{V}_i$. Since the angle $\theta$ is the angle by which $\mathbf{V}_o$ leads $\mathbf{V}_i$ throughout the frequency range of Fig. 21.19, the high-pass *R-C* filter is referred to as a *leading network*.

In summary, for the high-pass *R-C* filter:

$$f_c = \frac{1}{2\pi RC}$$

| | |
|---|---|
| For | $f < f_c$, $V_o < 0.707V_i$ |
| whereas for | $f > f_c$, $V_o > 0.707V_i$ |
| At $f_c$, | $\mathbf{V}_o$ leads $\mathbf{V}_i$ by 45°. |

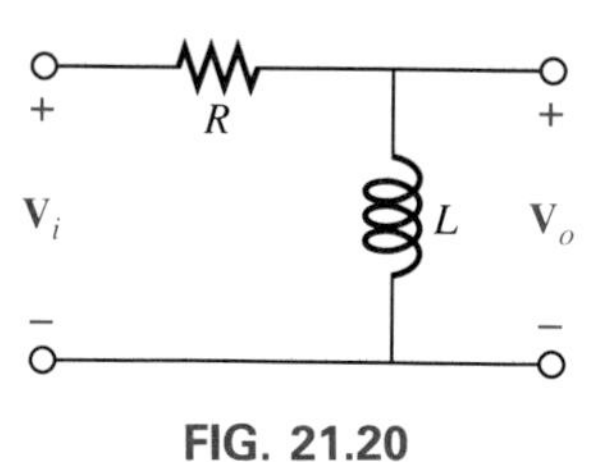


**FIG. 21.20**
*High-pass R-L filter.*

The high-pass filter response of Fig. 21.17 can also be obtained using the same elements of Fig. 21.12 but interchanging their positions, as shown in Fig. 21.20.

**EXAMPLE 21.5.** Given $R = 20\ \text{k}\Omega$ and $C = 1200$ pF.

a. Sketch the magnitude plot if the filter is used as both a high- and a low-pass filter.

b. Sketch the phase plot for both filters of part (a).
c. Determine the magnitude and phase of $\mathbf{V}_o/\mathbf{V}_i$ at $f = \frac{1}{2}f_1$ for the high-pass filter.

***Solutions:***

a. $f_c = \dfrac{1}{2\pi RC} = \dfrac{1}{(6.28)(20\text{ k}\Omega)(1200\text{ pF})}$

$= \mathbf{6634.82\ Hz}$

The magnitude plots appear in Fig. 21.21.

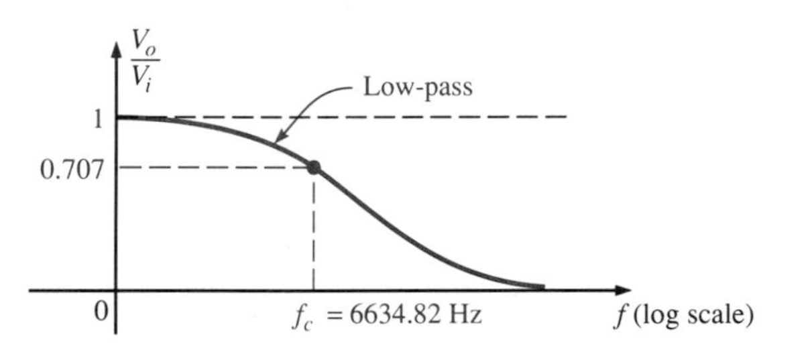


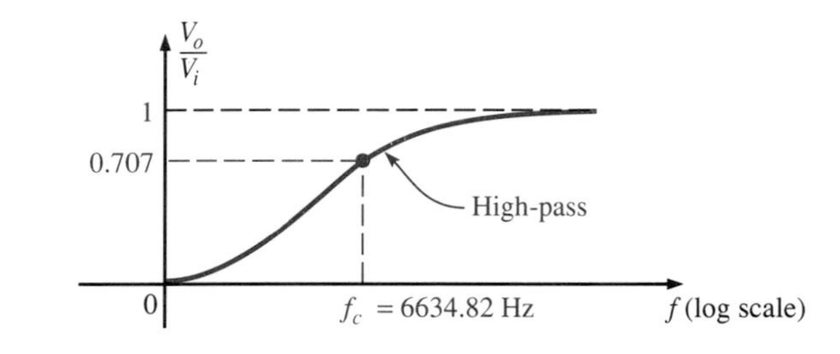

**FIG. 21.21**

b. The phase plots appear in Fig. 21.22.

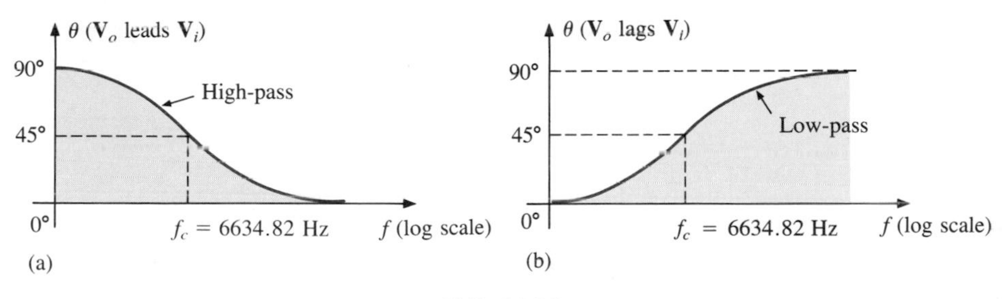

**FIG. 21.22**

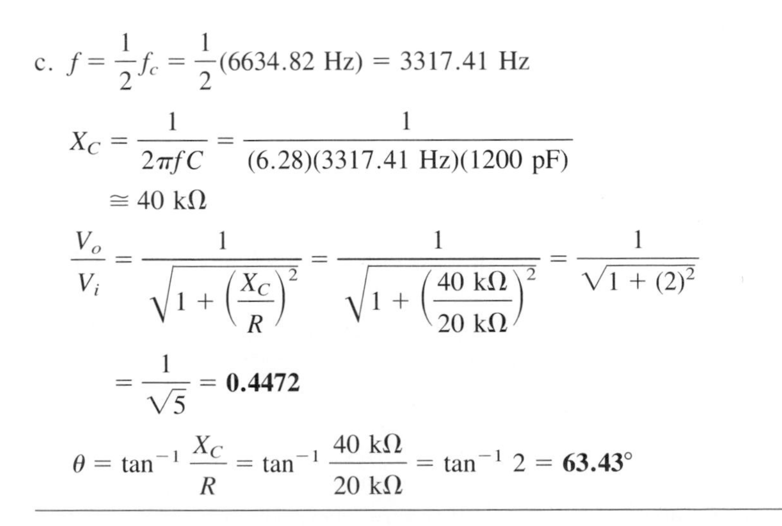

c. $f = \dfrac{1}{2}f_c = \dfrac{1}{2}(6634.82\text{ Hz}) = 3317.41\text{ Hz}$

$$X_C = \frac{1}{2\pi fC} = \frac{1}{(6.28)(3317.41\text{ Hz})(1200\text{ pF})}$$
$$\cong 40\text{ k}\Omega$$

$$\frac{V_o}{V_i} = \frac{1}{\sqrt{1 + \left(\dfrac{X_C}{R}\right)^2}} = \frac{1}{\sqrt{1 + \left(\dfrac{40\text{ k}\Omega}{20\text{ k}\Omega}\right)^2}} = \frac{1}{\sqrt{1 + (2)^2}}$$
$$= \frac{1}{\sqrt{5}} = \mathbf{0.4472}$$

$$\theta = \tan^{-1}\frac{X_C}{R} = \tan^{-1}\frac{40\text{ k}\Omega}{20\text{ k}\Omega} = \tan^{-1} 2 = \mathbf{63.43^\circ}$$

## 21.7 BAND-PASS FILTERS

There are a number of methods to establish the band-pass characteristic of Fig. 21.5(c). One method employs both a low-pass and high-pass filter in cascade, as shown in Fig. 21.23.

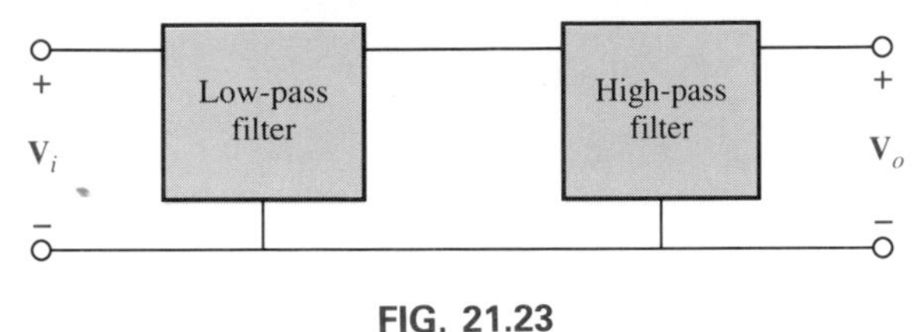


**FIG. 21.23**
*Band-pass filter.*

The components are chosen to establish a cutoff frequency for the low-pass filter that is higher than the critical frequency of the high-pass filter, as shown in Fig. 21.24. A frequency $f_1$ may pass through the low-pass filter at a relatively high level without affecting $V_o$ due to the

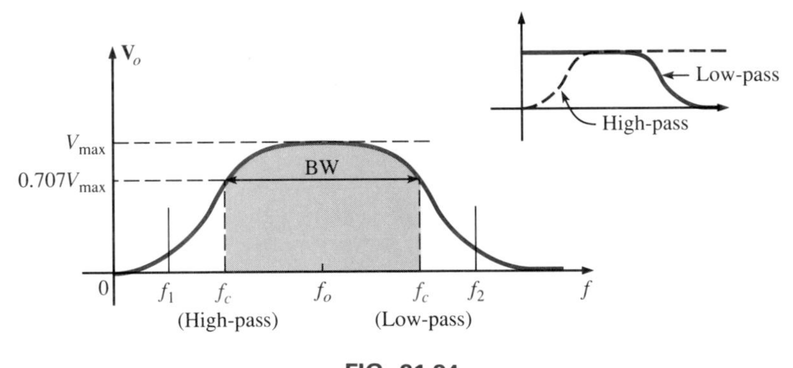


**FIG. 21.24**
*Band-pass characteristics.*

reject characteristics of the high-pass filter. A frequency $f_2$ would pass through the high-pass filter unmolested but is prohibited from reaching the high-pass filter by the low-pass characteristics. A frequency $f_o$ near the center of the pass-band will pass through both filters with very little degeneration.

The network of Example 21.6 will generate the characteristics of Fig. 21.24. However, for a circuit such as the one shown in Fig. 21.25, there is a loading between stages at each frequency that will affect the level of $\mathbf{V}_o$. Through proper design, the level of $\mathbf{V}_o$ may be very near $\mathbf{V}_i$ in the pass-band, but it will never be exactly equal. In addition, as the critical frequencies of each filter get closer and closer together to increase the quality factor of the response curve, the peak values within the pass-band will continue to drop.

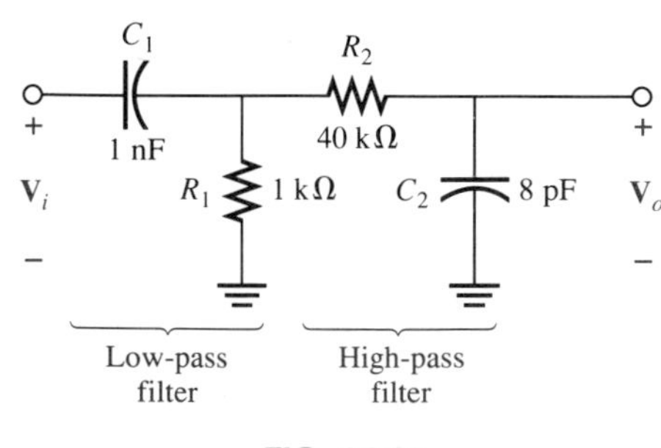


**FIG. 21.25**
*Band-pass filter.*

**EXAMPLE 21.6.** For the band-pass filter of Fig. 21.25,

a. Determine the critical frequencies for the low- and high-pass filters.
b. Using only the critical frequencies and assuming $V_o = V_i$ in the pass-band, sketch the ideal response characteristics and determine the bandwidth of the pass-band.
c. Determine the actual value of $V_o$ at the high-pass critical frequency and compare to the ideal value of $0.707V_i$.

***Solutions:***

a. Low-pass filter:

$$f_c = \frac{1}{2\pi R_1 C_1} = \frac{1}{2\pi(1\ \text{k}\Omega)(1\ \text{nF})} = \mathbf{159.15\ kHz}$$

High-pass filter:

$$f_c = \frac{1}{2\pi R_2 C_2} = \frac{1}{2\pi(40\ \text{k}\Omega)(8\ \text{pF})} = \mathbf{497.36\ kHz}$$

b. See Fig. 21.26.

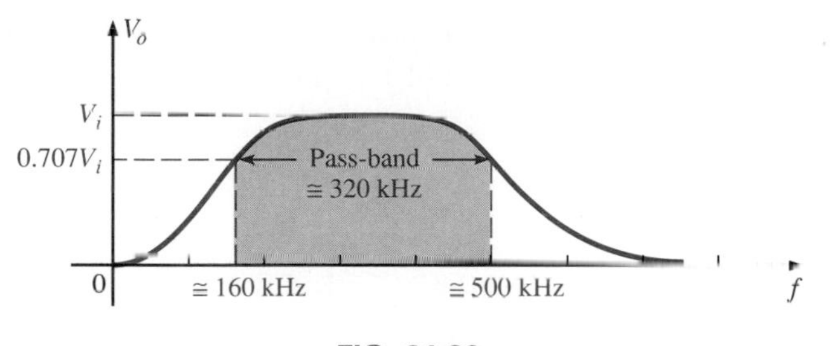

**FIG. 21.26**

c. At $f = 497.36$ kHz,

$$X_{C_1} = \frac{1}{2\pi f C_1} \cong 320\ \Omega$$

and

$$X_{C_2} = \frac{1}{2\pi f C_2} = R_2 = 40\ \text{k}\Omega$$

resulting in the network of Fig. 21.27.

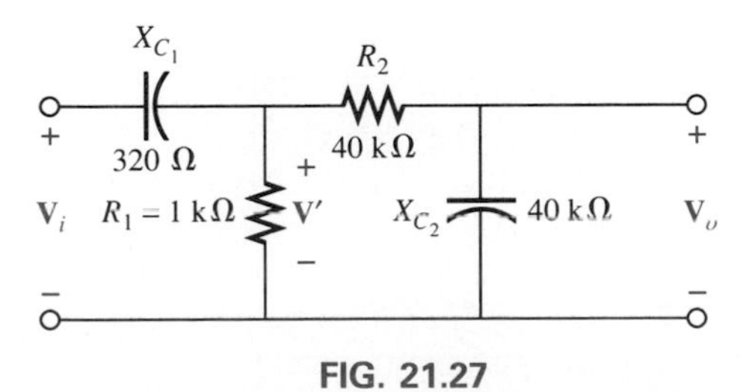


**FIG. 21.27**
*Network of Fig. 21.25 at $f = 497.36$ kHz.*

The magnitude of the series $R_2$-$X_{C_2}$ combination is so large compared to the parallel resistor $R_1$ that the loading effect on $R_1$ can be ignored as a good approximation.

The result is

$$\mathbf{V}' = \frac{\mathbf{R}_1\mathbf{V}_i}{\mathbf{R}_1 + \mathbf{X}_{C_1}} = \frac{(1\text{ k}\Omega \angle 0°)\mathbf{V}_i}{1\text{ k}\Omega - j0.32\text{ k}\Omega} = 0.952V_i \angle 17.74°$$

At $f = f_c$,

$$V_o = 0.707V' = 0.707(0.952V_i)$$

and

$$\frac{V_o}{V_i} = \mathbf{0.673}$$

which is very close to the ideal 0.707 level.

---

The band-pass response can also be obtained using the series and parallel resonant circuits discussed in Chapter 20. In each case, however, $V_o$ will not be equal to $V_i$ in the pass-band, but a frequency range in which $V_o$ will be equal to or greater than $0.707V_{max}$ can be defined.

For the series resonant circuit of Fig. 21.28, $X_L = X_C$ at resonance and

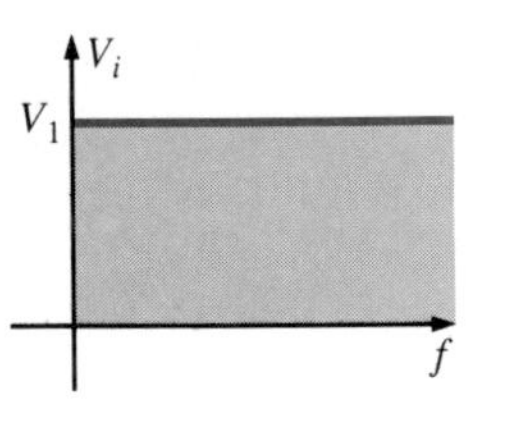


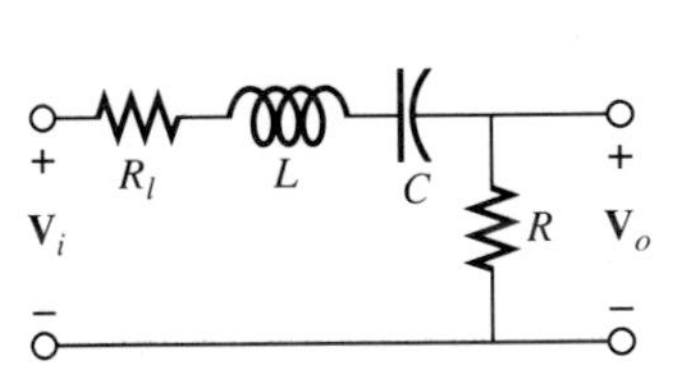


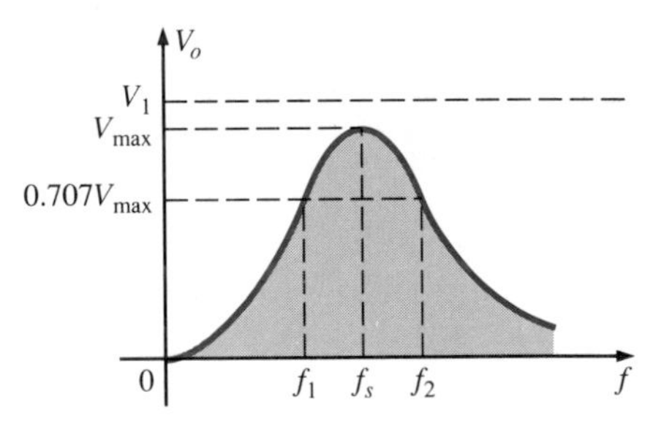


**FIG. 21.28**
*Series resonant band-pass filter.*

$$V_o = \frac{R}{R + R_l}V_i \qquad f = f_s \tag{21.20}$$

and

$$f_s = \frac{1}{2\pi\sqrt{LC}} \tag{21.21}$$

with

$$Q_s = \frac{X_L}{R + R_l} \tag{21.22}$$

and

$$\text{BW} = \frac{f_s}{Q_s} \tag{21.23}$$

For the parallel resonant circuit of Fig. 21.29, $Z_{T_p}$ is a maximum value at resonance and

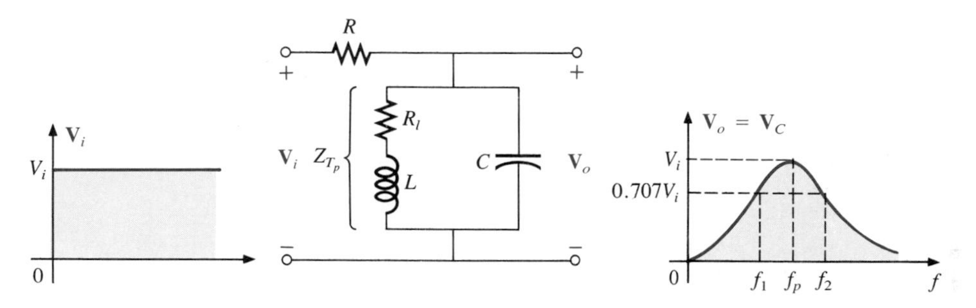


**FIG. 21.29**
*Parallel resonant band-pass filter.*

$$V_o = \frac{Z_{T_p} V_i}{Z_{T_p} + R} \quad f = f_p \tag{21.24}$$

with

$$Z_{T_p} = Q_l^2 R_l \quad Q_l \geq 10 \tag{21.25}$$

and

$$f_p = \frac{1}{2\pi\sqrt{LC}} \quad Q_l \geq 10 \tag{21.26}$$

For the parallel resonant circuit

$$Q_p = \frac{X_L}{R_l} \tag{21.27}$$

and

$$\text{BW} = \frac{f_p}{Q_p} \tag{21.28}$$

As a first approximation that is acceptable for most practical applications, it can be assumed that the resonant frequency bisects the bandwidth.

**EXAMPLE 21.7.** Determine the frequency response for the voltage $V_o$ for the series circuit of Fig. 21.30.

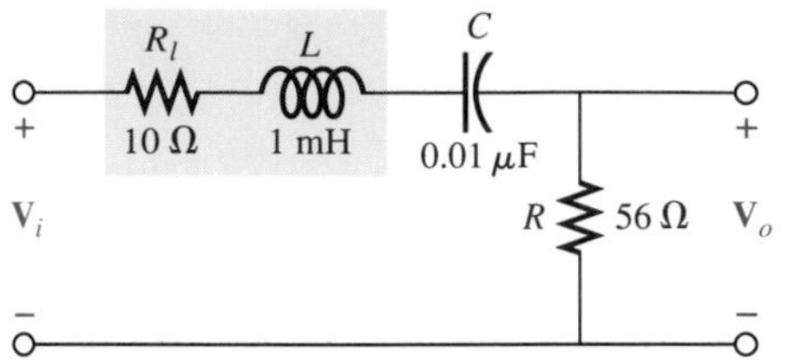

FIG. 21.30
*Series resonant band-pass filter.*

***Solution:***

$$f_s = \frac{1}{2\pi\sqrt{LC}} = \frac{1}{2\pi\sqrt{(1\text{ mH})(0.01\ \mu\text{F})}} = \mathbf{50{,}329.21\ Hz}$$

$$Q_s = \frac{X_L}{R + R_l} = \frac{2\pi(50{,}329.21\text{ Hz})(1\text{ mH})}{56\ \Omega + 10\ \Omega} = \mathbf{4.79}$$

At resonance:

$$V_o = \frac{RV_i}{R + R_l} = \frac{56\ \Omega(V_i)}{56\ \Omega + 10\ \Omega} = 0.848V_i$$

and

$$\frac{V_o}{V_i} = \mathbf{0.848}$$

Note Fig. 21.31.

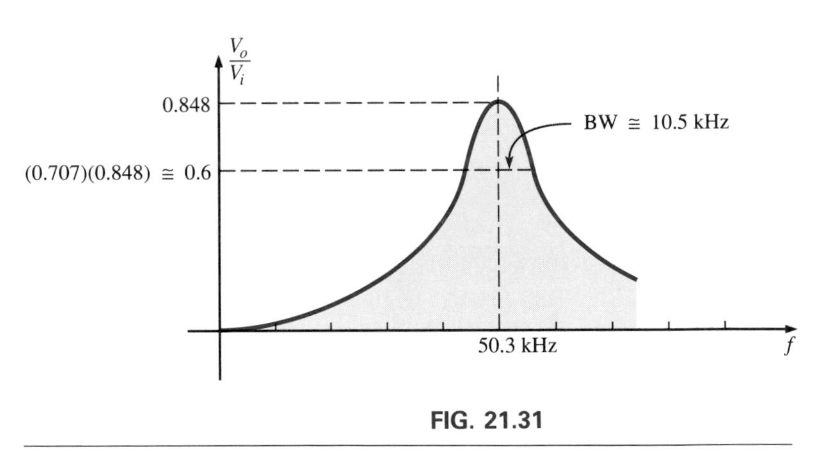


FIG. 21.31

## 21.8 BAND-STOP FILTERS

Band-stop filters can also be constructed using a low-pass and high-pass filter. However, rather than the cascaded configuration used for the pass-band filter, a parallel arrangement is required, as shown in Fig. 21.32. A low frequency $f_1$ can pass through the low-pass filter and a higher frequency $f_2$ can use the parallel path, as shown in Figs. 21.32 and 21.33. However, a frequency such as $f_o$ in the reject-band is higher than the low-pass critical frequency and lower than the high-pass critical

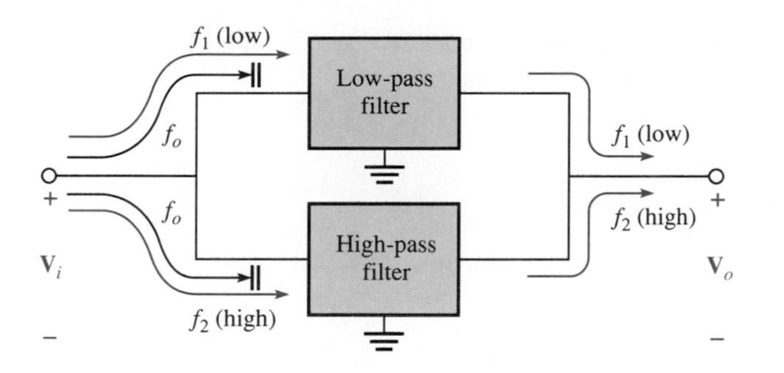


**FIG. 21.32**
*Band-stop filter.*

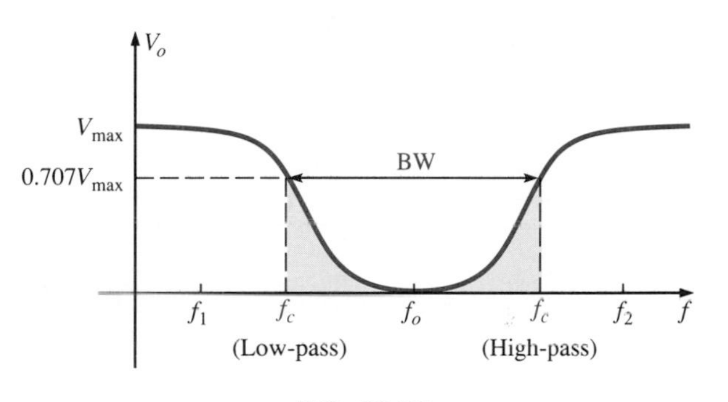


**FIG. 21.33**
*Band-stop characteristics.*

frequency and is therefore prevented from contributing to the levels of $V_o$ above $0.707V_{max}$.

Since the characteristics of a band-stop filter are the inverse of the pattern obtained for the band-pass filters, we can employ the fact that at any frequency the sum of the magnitudes of the two waveforms to the right of the equal sign in Fig. 21.34 will equal the applied voltage $V_i$.

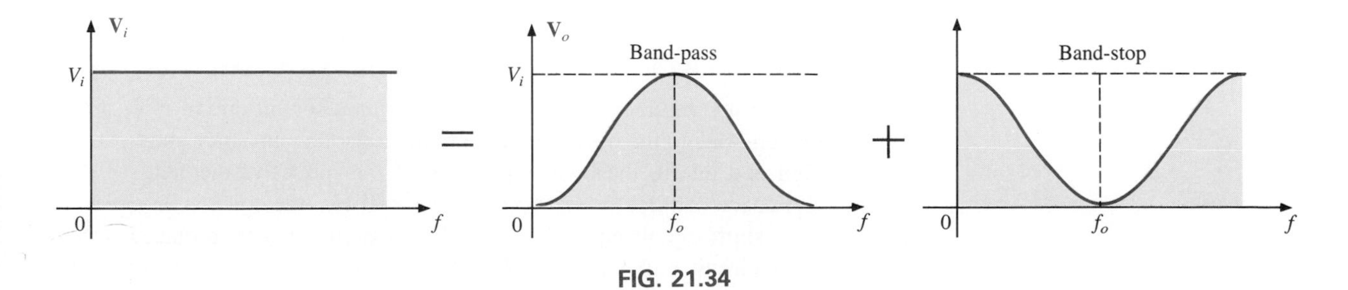


**FIG. 21.34**

For the band-pass filters of Figs. 21.28 and 21.29, therefore, if we take the output off the other series elements as shown in Figs. 21.35 and 21.36, a band-stop characteristic will be obtained, as required by Kirchhoff's voltage law.

For the series resonant circuit of Fig. 21.35, Eqs. (21.21)–(21.23) still apply, but now at resonance,

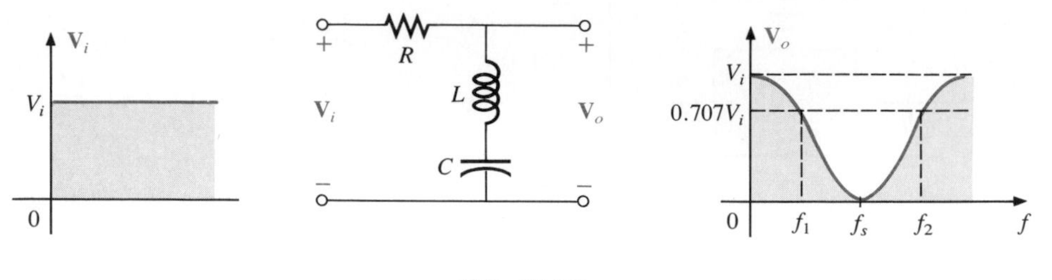

FIG. 21.35

$$V_o = \frac{R_l V_i}{R_l + R} \tag{21.29}$$

For the parallel resonant circuit of Fig. 21.36, Eqs. (21.25)–(21.28) are still applicable, but now at resonance,

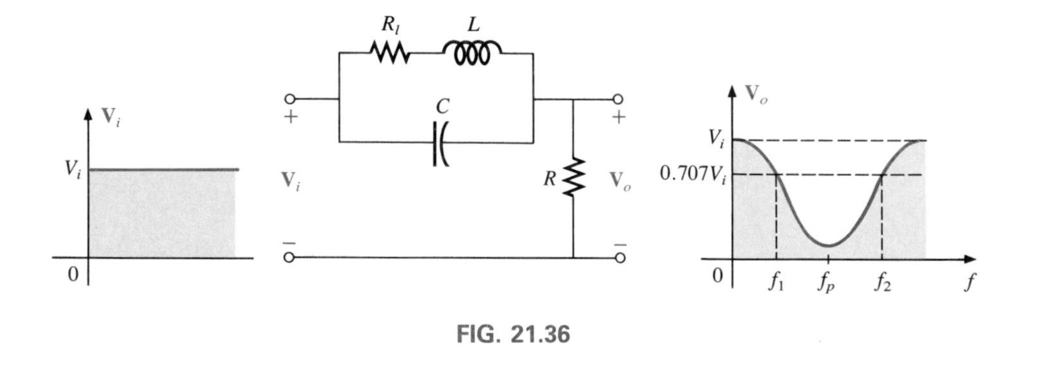


FIG. 21.36

$$V_o = \frac{RV_i}{R + Z_{T_p}} \tag{21.30}$$

The maximum value of $V_o$ for the series resonant circuit is $V_i$ at the low end due to the open-circuit equivalent for the capacitor and $V_i$ at the high end due to the high impedance of the inductive element.

For the parallel resonant circuit, at $f = 0$ Hz, the coil can be replaced by a short-circuit equivalent and the capacitor can be replaced by its open circuit and $V_o = RV_i/(R + R_l)$. At the high-frequency end, the capacitor approaches a short-circuit equivalent, and $V_o$ increases toward $V_i$.

The band-reject and band-pass characteristics of miniature filters manufactured by TRW/UTC inductive products appear in Fig. 21.37 along with a photograph of a typical unit. Note that the band-pass

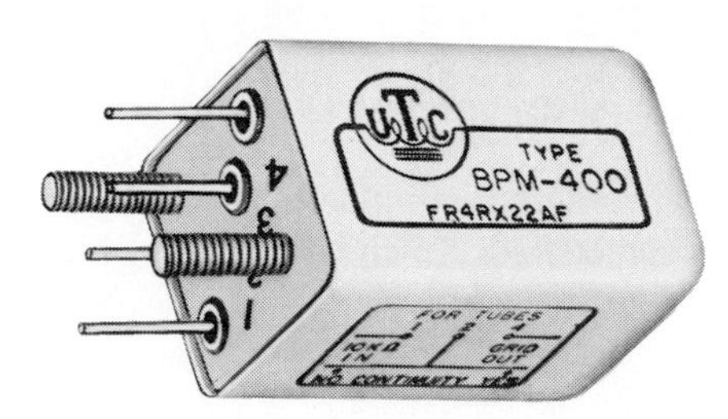


| Type No. | Center Frequency (Hz) | Pass Band (Less than 2 dB) (Hz) | Stop Band (More than 35 dB) | |
|---|---|---|---|---|
| | | | Below (Hz) | Above (Hz) |
| BPM 400 | 400 | 388–412 | 200 | 800 |
| BPM 440 | 440 | 427–453 | 220 | 880 |
| BPM 500 | 500 | 485–515 | 250 | 1,000 |
| BPM 600 | 600 | 582–618 | 300 | 1,200 |
| BPM 800 | 800 | 776–824 | 400 | 1,600 |
| BPM 1000 | 1,000 | 970–1,030 | 500 | 2,000 |
| BPM 1200 | 1,200 | 1,164–1,236 | 600 | 2,400 |
| BPM 1500 | 1,500 | 1,455–1,545 | 750 | 3,000 |
| BPM 1600 | 1,600 | 1,552–1,648 | 800 | 3,200 |
| BPM 2000 | 2,000 | 1,940–2,060 | 1,000 | 4,000 |
| BPM 2500 | 2,500 | 2,425–2,575 | 1,250 | 5,000 |
| BPM 3000 | 3,000 | 2,910–3,090 | 1,500 | 6,000 |
| BPM 3200 | 3,200 | 3,104–3,296 | 1,600 | 6,400 |
| BPM 4000 | 4,000 | 3,880–4,120 | 2,000 | 8,000 |
| BPM 4800 | 4,800 | 4,656–4,944 | 2,400 | 9,600 |
| BPM 5000 | 5,000 | 4,850–5,150 | 2,500 | 10,000 |
| BPM 6000 | 6,000 | 5,820–6,180 | 3,000 | 12,000 |
| BPM 8000 | 8,000 | 7,760–8,240 | 4,000 | 16,000 |
| BPM 10000 | 10,000 | 9,700–10,300 | 5,000 | 20,000 |
| BPM 20000 | 20,000 | 19,400–20,600 | 10,000 | 40,000 |

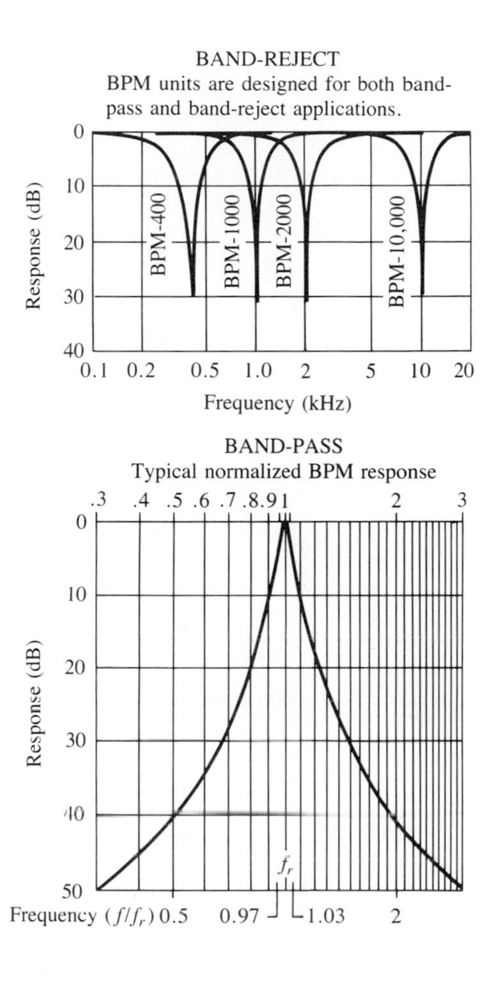


**FIG. 21.37**
*Band-pass and band-reject characteristics for BPM TRW/UTC filters. (Courtesy of United Transformer Corp.)*

characteristics are a universal set since the resonant frequency is undefined and the horizontal axis is the ratio $f/f_r$. When examining the curves, remember that the vertical scale is a measure of the attenuation of the input signal. That is, at 0 dB the output is equal to the input, while at $-30$ dB the output is significantly less than the input. Since each of these units can be used for the pass-band or stop-band function, data about each application are provided. Note that the pass band is inserted by the center frequency for each unit and the type number of the unit reflects the center frequency. The stop band is implying essentially zero response with the $-35$-dB criterion. Note that the center frequency does not bisect the stop band since the resonance curve is not symmetrical about $f_o$.

## 21.9 DOUBLE-TUNED FILTER

There are some network configurations that display both a pass-band and a band-stop characteristic, such as shown in Fig. 21.38. For the network of Fig. 21.38(a), the parallel resonant circuit will establish the band stop by resonating at the frequency not permitted to establish a $\mathbf{V}_L$. The greater part of the applied voltage **E** will appear across this parallel resonant circuit at this frequency due to its very high impedance compared with $R_L$.

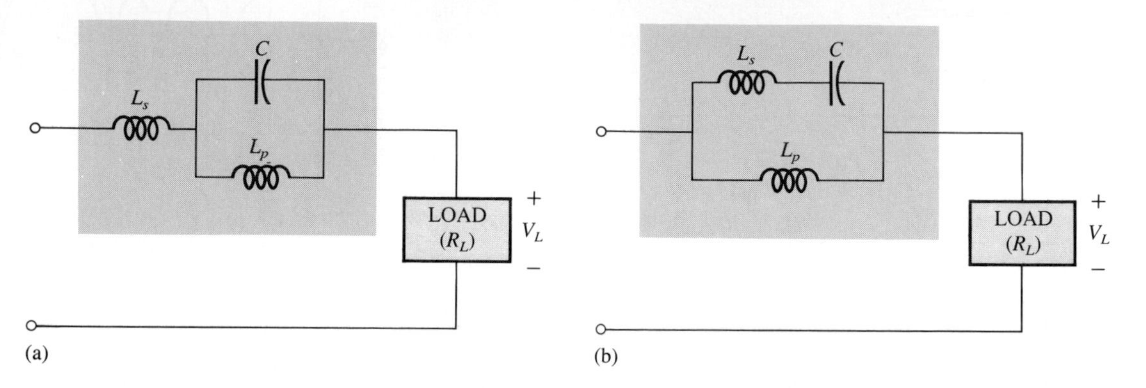


**FIG. 21.38**
*Double-tuned networks.*

For the pass band, the parallel resonant circuit is designed to be capacitive (inductive if $L_s$ is replaced by $C_s$). The inductance $L_s$ is chosen to cancel the effects of the resulting capacitive reactance at the resonant pass-band frequency of the tank circuit, thereby acting as a series resonant circuit. The applied voltage **E** will then appear across $R_L$ at this frequency.

For the network of Fig. 21.38(b), the series resonant circuit will still determine the pass band, acting as a very low impedance across the parallel inductor at resonance, establishing $\mathbf{V}_L = \mathbf{E}$. At the desired band-stop resonant frequency, the series resonant circuit is capacitive. The inductance $L_p$ is chosen to establish parallel resonance at the resonant band-stop frequency. The high impedance of the parallel resonant circuit will result in a very low load voltage $V_L$.

For rejected frequencies below the pass band, the networks should appear as shown in Fig. 21.38. For the reverse situation, $L_s$ in Fig. 21.38(a) and $L_p$ in Fig. 21.38(b) are replaced by capacitors.

---

**EXAMPLE 21.8.** For the network of Fig. 21.38(b), determine $L_s$ and $L_p$ for a capacitance $C$ of 500 pF if a frequency of 200 kHz is to be rejected and a frequency of 600 kHz accepted.

***Solution:*** For series resonance, we have

$$f_s = \frac{1}{2\pi\sqrt{LC}}$$

and

$$L_s = \frac{1}{4\pi^2 f_s^2 C} = \frac{1}{4(3.14)^2(600\text{ kHz})^2(500\text{ pF})}$$
$$= \mathbf{141\ \mu H}$$

At 200 kHz,

$$X_L = \omega L = 2\pi f_s L = (6.28)(200\text{ kHz})(141\ \mu\text{H})$$
$$= 177.1\ \Omega$$

and

$$X_C = \frac{1}{\omega C} = \frac{1}{(6.28)(200\text{ kHz})(500\text{ pF})}$$
$$= 1.59\text{ k}\Omega$$

For the series elements,

$$j(X_L - X_C) = j(177.1\ \Omega - 1592\ \Omega) = -j1414.8\ \Omega \quad \text{(cap.)}$$

At parallel resonance ($Q \geq 10$ assumed),

$$X_L = X_C$$

and

$$L_p = \frac{X_C}{\omega} = \frac{1412.9\ \Omega}{(6.28)(200\text{ kHz})} = \mathbf{1.125\ mH}$$

The frequency response for the preceding network appears as one of the examples of PSPICE in the last section of the chapter.

## 21.10 BODE PLOTS

There is a technique for sketching the frequency response of such factors as filters, amplifiers, and systems on a decibel scale that can save a great deal of time and effort and provide an excellent way to compare decibel levels at different frequency levels. The method employs a number of straight-line segments to form an envelope that can be used along with a few known points to sketch the actual frequency response.

To insure the derivation of the method is correctly and clearly understood, the first network to be analyzed will be examined in some detail. The second network will be treated in a more shorthand manner, and finally a method for quickly determining the response will be introduced.

### High-Pass *R-C* Filter

Let us start by examining the high-pass filter of Fig. 21.39. The high-pass filter was chosen as our starting point, since the frequencies of primary interest are at the low end of the frequency spectrum.

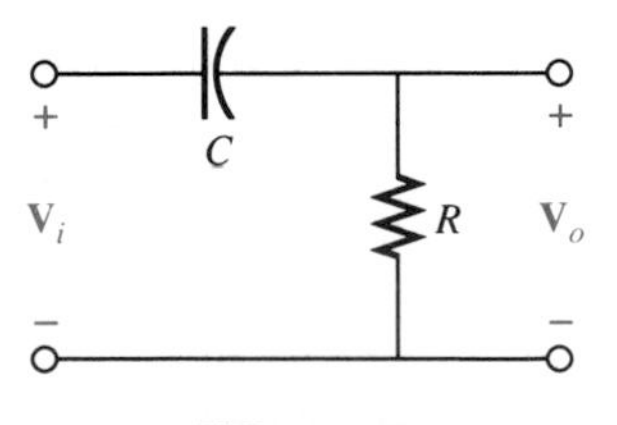


**FIG. 21.39**
*High-pass filter.*

The voltage gain of the system is given by:

$$\mathbf{A}_v = \frac{\mathbf{V}_o}{\mathbf{V}_i} = \frac{R}{R - jX_C} = \frac{1}{1 - j\dfrac{X_C}{R}} = \frac{1}{1 - j\dfrac{1}{2\pi fCR}}$$

$$= \frac{1}{1 - j\left(\dfrac{1}{2\pi RC}\right)\dfrac{1}{f}}$$

If we substitute

$$f_1 = \frac{1}{2\pi RC} \tag{21.31}$$

we find

$$\mathbf{A}_v = \frac{1}{1 - j(f_1/f)}$$

and

$$\mathbf{A}_v = \frac{\mathbf{V}_o}{\mathbf{V}_i} = A_v \angle\theta = \frac{1}{\sqrt{1 + (f_1/f)^2}} \angle \tan^{-1}(f_1/f) \tag{21.32}$$

providing an equation for the magnitude and phase of the high-pass filter in terms of the frequency levels.

Using Eq. (21.11),

$$A_{v_{\text{dB}}} = 20 \log_{10} A_v$$

and, substituting the magnitude component of Eq. (21.32),

$$A_{v_{\text{dB}}} = 20 \log_{10} \frac{1}{\sqrt{1 + (f_1/f)^2}} \tag{21.33}$$

In logs,

$$\log_{10} \frac{1}{x} = -\log_{10} x$$

resulting in

$$A_{v_{\text{dB}}} = -20 \log_{10} \sqrt{1 + \left(\frac{f_1}{f}\right)^2}$$

Further,

$$\log_{10} \sqrt{x} = \frac{1}{2} \log_{10} x$$

and

$$A_{v_{dB}} = -\frac{1}{2}(20)\log_{10}\left[1+\left(\frac{f_1}{f}\right)^2\right]$$
$$= -10\log_{10}\left[1+\left(\frac{f_1}{f}\right)^2\right]$$

For frequencies where $f \ll f_1$ or $(f_1/f)^2 \gg 1$,

$$1+\left(\frac{f_1}{f}\right)^2 \cong \left(\frac{f_1}{f}\right)^2$$

and

$$A_{v_{dB}} = -10\log_{10}\left(\frac{f_1}{f}\right)^2$$

but

$$\log_{10} x^2 = 2\log_{10} x$$

resulting in

$$\boxed{A_{v_{dB}} = -20\log_{10}\frac{f_1}{f}} \qquad f \ll f_1 \qquad \textbf{(21.34)}$$

First note the similarities between Eq. (21.34) and the basic equation for gain in decibels: $G_{dB} = 20\log_{10} V_o/V_i$. The comments regarding changes in decibel levels due to changes in $V_o/V_i$ can therefore be applied here also, except now a change in frequency by a 2:1 ratio will result in a −6-dB change in gain due to the negative sign in Eq. (21.34). A change in frequency by a 10:1 ratio will result in a −20-dB change in gain.

***Two frequencies separated by a 2:1 ratio are said to be an* octave *apart.***

***For Bode plots, a change in frequency by one octave will result in a 6-dB change in gain.***

***Two frequencies separated by a 10:1 ratio are said to be a* decade *apart.***

***For Bode plots, a change in frequency by one decade will result in a 20-dB change in gain.***

One may wonder about all the mathematical development to obtain an equation that initially appears confusing and of limited value. As specified, Eq. (21.34) is accurate only for frequency levels much less than $f_1$.

First, realize that the mathematical development of Eq. (21.34) will not have to be repeated for each configuration encountered. Second, the equation itself is seldom applied but simply used to define a straight line on a log plot that permits a sketch of the frequency response of a system with a minimum of effort and a high degree of accuracy.

To plot Eq. (21.34), consider the following levels:

For $f = f_1$, $f_1/f = 1$ and $-20 \log_{10} 1 = 0$ dB
For $f = f_1/2$, $f_1/f = 2$ and $-20 \log_{10} 2 = -6$ dB
For $f = f_1/4$, $f_1/f = 4$ and $-20 \log_{10} 4 = -12$ dB
For $f = f_1/10$, $f_1/f = 10$ and $-20 \log_{10} 10 = -20$ dB

A plot of these points on a log scale will result in the decibel plot of Fig. 21.40.

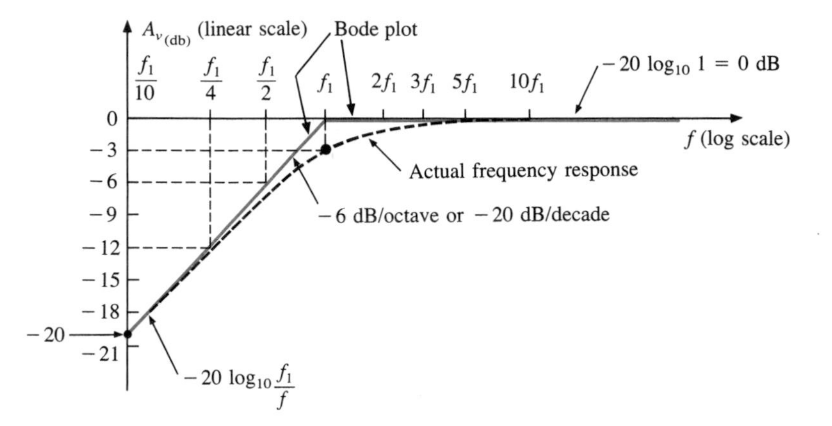


FIG. 21.40
*Bode plot for the low-frequency region.*

For the future, note that the resulting plot is a straight line intersecting the 0-dB line at $f_1$. It drops off to the left at a rate of $-6$ dB per octave or $-10$ dB per decade. In other words, once $f_1$ is determined, find $f_1/2$ and a plot point exists at $-6$ dB (or find $f_1/10$ and a plot point exists at $-20$ dB). The *actual* response curve will then pass through the $-3$-dB point at $f_1$ and approach the asymptote established by the straight line defined by Eq. (21.34). The curve will also approach an asymptote defined by $A_{v_{\text{dB}}} = 0$ dB, since at high frequencies

$$A_{v_{\text{dB}}} = 20 \log_{10} \frac{1}{\sqrt{1 + (f_1/f)^2}} \cong 20 \log_{10} \frac{1}{\sqrt{1 + 0}}$$
$$= 20 \log_{10} 1 = 0 \text{ dB}$$

The phase response can be determined from Eq. (21.32) where

$$\boxed{\theta = \tan^{-1} \frac{f_1}{f}} \qquad \textbf{(21.35)}$$

For frequencies where $f \ll f_1$, $\theta = \tan^{-1}(f_1/f)$ approaches 90°, and for frequencies where $f \gg f_1$, $\theta = \tan^{-1}(f_1/f)$ approaches 0°. At $f = f_1$,

$\theta = \tan^{-1}(f_1/f) = \tan^{-1} 1 = 45°$. A plot of $\theta$ versus frequency appears in Fig. 21.41.

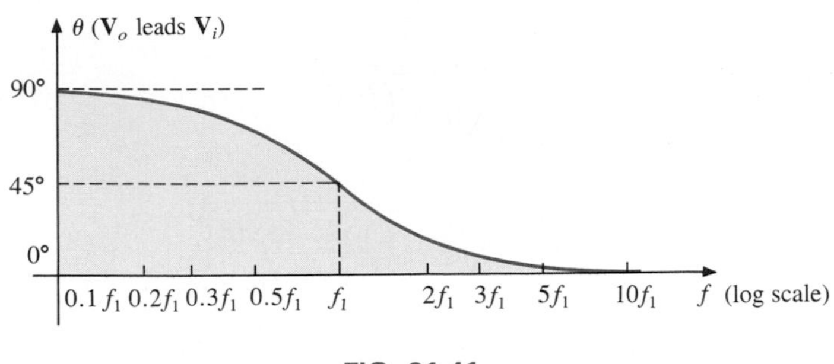


FIG. 21.41

---

### EXAMPLE 21.9.

a. Sketch the frequency response for the high-pass *R-C* filter of Fig. 21.42.
b. Determine the decibel level at $f = 1$ kHz.

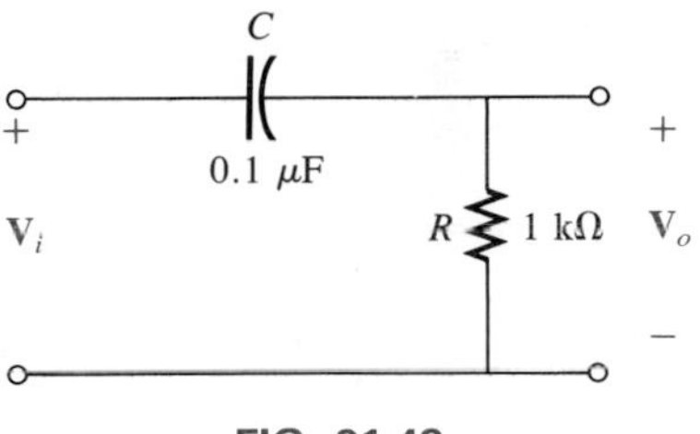


FIG. 21.42

***Solutions:***

a. $f_1 = \dfrac{1}{2\pi RC} = \dfrac{1}{(6.28)(1\text{ k}\Omega)(0.1\ \mu\text{F})} = \mathbf{1592.36\ Hz}$

The frequency $f_1$ is identified on the log scale as shown in Fig. 21.43. A straight line is then drawn from $f_1$ with a slope that will intersect $-20$ dB at $f_1/10 = 159.24$ Hz or $-6$ dB at $f_1/2 = 796.18$ Hz. The actual response curve can then be drawn through the $-3$-dB level at $f_1$ approaching the two asymptotes of Fig. 21.43.

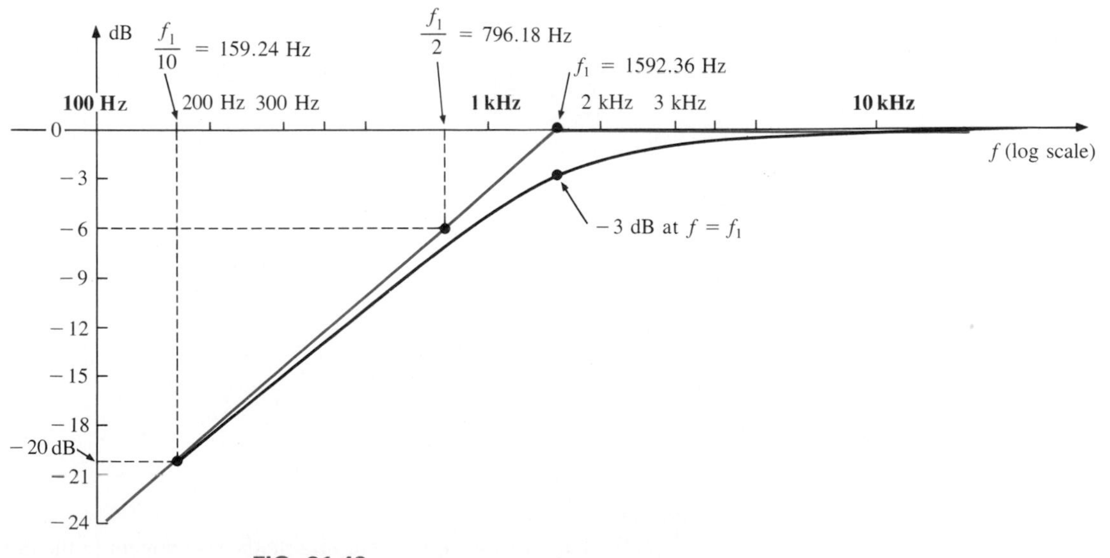


FIG. 21.43

Note in the preceding solution that there was no need to employ Eq. (21.34) or perform any extensive mathematical manipulations.

b. Eq. (21.33):

$$|A_{v_{\text{dB}}}| = 20 \log_{10} \frac{1}{\sqrt{1 + \left(\frac{f_1}{f}\right)^2}}$$

$$= 20 \log_{10} \frac{1}{\sqrt{1 + \left(\frac{1592.36 \text{ Hz}}{1000 \text{ Hz}}\right)^2}}$$

$$= 20 \log_{10} \frac{1}{\sqrt{1 + (1.592)^2}} = 20 \log_{10} 0.5318$$

$$= \mathbf{-5.485\ dB}$$

as verified by Fig. 21.43.

## Low-Pass *R-C* Filter

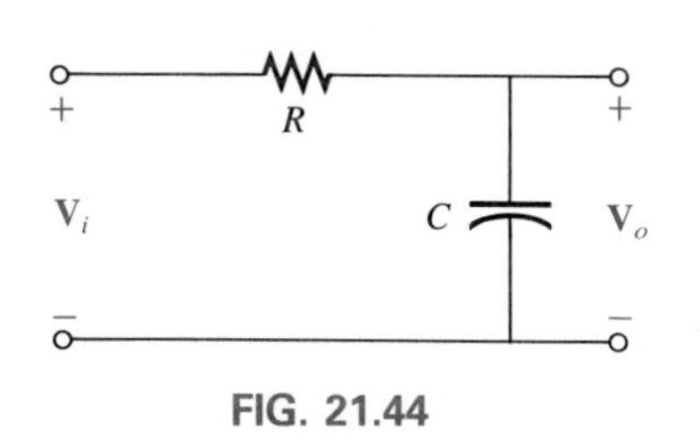

FIG. 21.44

For the low-pass filter of Fig. 21.44,

$$\mathbf{A}_v = \frac{\mathbf{V}_o}{\mathbf{V}_i} = \frac{\mathbf{X}_C}{\mathbf{R} + \mathbf{X}_C} = \frac{-jX_C}{R - jX_C} = \frac{1}{\frac{R}{-jX_C} + 1}$$

$$= \frac{1}{1 + j\frac{R}{X_C}} = \frac{1}{1 + j\frac{R}{\frac{1}{2\pi fC}}} = \frac{1}{1 + j\frac{f}{\frac{1}{2\pi RC}}}$$

and

$$\mathbf{A}_v = \frac{1}{1 + j(f/f_2)} \tag{21.36}$$

with

$$f_2 = \frac{1}{2\pi RC} \tag{21.37}$$

as defined earlier.

Note that now the sign of the imaginary component in the denominator is positive and $f_2$ appears in the denominator of the frequency ratio rather than in the numerator as in the case of $f_1$ for the high-pass filter.

In terms of magnitude and phase,

$$\mathbf{A}_v = \frac{\mathbf{V}_o}{\mathbf{V}_i} = A_v \angle\theta = \frac{1}{\sqrt{1 + (f/f_2)^2}} \angle -\tan^{-1}(f/f_2) \qquad \textbf{(21.38)}$$

An analysis similar to that performed for the high-pass filter will result in

$$A_{v_{dB}} = -20 \log_{10}\frac{f}{f_2} \qquad f \gg f_2 \qquad \textbf{(21.39)}$$

Note in particular that the equation is exact only for frequencies much greater than $f_2$, but a plot of Eq. (21.39) does provide an asymptote that performs the same function as the asymptote derived for the high-pass filter.

A plot of Eq. (21.39) appears in Fig. 21.45 for $f_2$ = 1 kHz. Note the 6-dB drop at $f = 2f_2$ and the 20-dB drop at $f = 10f_2$.

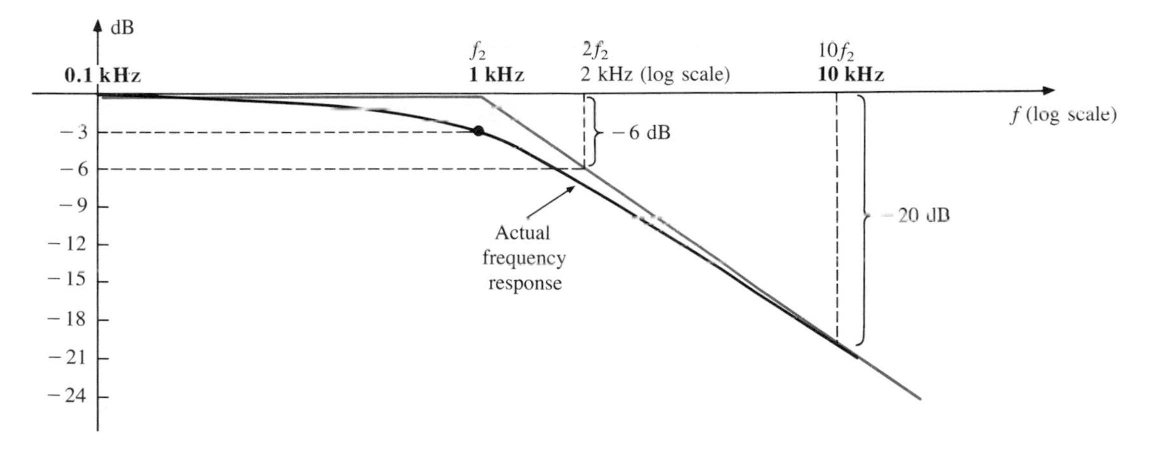


**FIG. 21.45**
*Bode plot for the high-frequency region.*

At $f \gg f_2$, the phase angle $\theta = -\tan^{-1}(f/f_2)$ approaches $-90°$, while for frequencies $f \ll f_2$, $\theta = -\tan^{-1}(f/f_2)$ approaches 0°. At $f = f_2$, $\theta = -\tan^{-1} 1 = -45°$, confirming the plot of Fig. 21.10.

Even though the preceding analysis has been limited solely to the *R-C* combination, the results obtained will have an impact on networks a great deal more complicated. One good example is the high- and low-frequency response of a standard transistor configuration. There are some capacitive elements in a practical transistor network that will affect the low-frequency response and others that will affect the high-frequency response. In the absence of the capacitive elements, the frequency response of a transistor would ideally stay level at the midband value. However, the coupling capacitors at low frequencies and the bypass and parasitic capacitors at high frequencies will define a bandwidth for numerous transistor configurations. In the low-frequency region, specific capacitors and resistors will form an $R$-$C$ combination that will define a low cutoff frequency. There are then other elements and capacitors forming a second $R$-$C$ combination that will define a high cutoff frequency. Once the cutoff frequencies are known, the $-3$-dB points are set and the bandwidth of the system can be determined.

## 21.11 SKETCHING THE BODE RESPONSE

We must now develop a procedure for finding the frequency response that reduces the mathematical complexity associated with defining the Bode plot.

### Low-Pass Filter

Recall that for the low-pass filter,

$$\mathbf{A}_v = \frac{-jX_C}{R - jX_C}$$

Multiplying numerator and denominator by $j$:

$$\mathbf{A}_v = \frac{(j)(-jX_C)}{(j)(R - jX_C)} = \frac{X_C}{jR + X_C}$$

Dividing numerator and denominator by $X_C$:

$$\frac{1}{1 + j\dfrac{R}{X_C}}$$

Substituting $X_C = 1/\omega C$:

$$\mathbf{A}_v = \frac{1}{1 + j\omega RC}$$

and

$$\mathbf{A}_v = A_v \angle\theta = \frac{1}{\sqrt{1 + (\omega RC)^2}} \angle -\tan^{-1} \omega RC \qquad \textbf{(21.40)}$$

When $\omega RC = 1$ or $f = 1/2\pi RC$,

$$A_{v_{\text{dB}}} = 20 \log_{10} \frac{1}{\sqrt{1 + (\omega RC)^2}} = 20 \log_{10} \frac{1}{\sqrt{1 + 1}}$$

$$= 20 \log_{10} \frac{1}{\sqrt{2}} = 20 \log_{10} 0.707 = -3\text{dB}$$

defining the cutoff frequency for the high end of the frequency spectrum and the corner frequency for the Bode plot.

In general, therefore, if the magnitude of the gain $A_v$ can be written as defined by Eq. (21.41):

$$A_v = \frac{1}{\sqrt{1 + x^2}} \qquad \textbf{(21.41)}$$

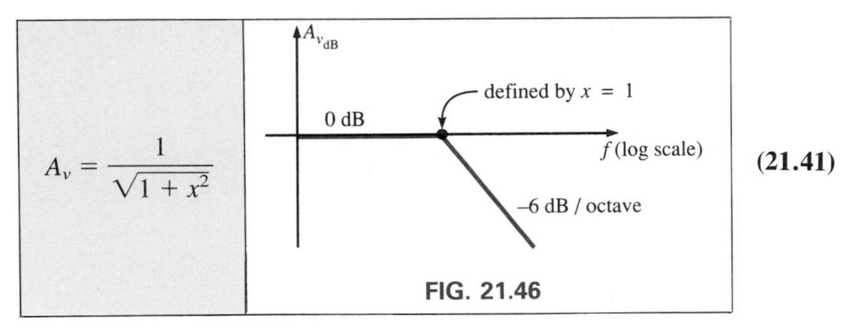


**FIG. 21.46**

A Bode plot for the gain in decibels composed of two asymptotes (one at zero decibels and one at $-6$ dB/octave) can be drawn as shown in Fig. 21.46. Note that the corner frequency is defined by the condition $x = 1 = \omega RC$ for this filter.

For future applications, if the magnitude of the gain of any system can be maneuvered into the basic format of Eq. (21.41), the asymptotes of Fig. 21.46 will form a straight-line approximation for the actual frequency response that will be $-3$ dB at a frequency defined by $x = 1$. Keep in mind that for some networks, $x$ may not have the simple form of $\omega RC$, but the condition $x = 1$ can still be applied.

## High-Pass Filter

For the high-pass filter,

$$\mathbf{A}_v = \frac{R}{R - jX_C}$$

Multiplying numerator and denominator by $j$:

$$\mathbf{A}_v = \frac{(j)R}{(j)(R - jX_C)} = \frac{jR}{jR + X_C}$$

Dividing numerator and denominator by $X_C$:

$$\mathbf{A}_v = \frac{jR/X_C}{jR/X_C + 1}$$

Substituting $X_C = 1/\omega C$ and rearranging:

$$\mathbf{A}_v = \frac{j\omega RC}{1 + j\omega RC}$$

with

$$\mathbf{A}_v = A_v \angle\theta = \frac{\omega RC}{\sqrt{1 + (\omega RC)^2}} \angle 90^\circ - \tan^{-1} \omega RC \qquad \textbf{(21.42)}$$

Since $\log_{10} M \cdot N - \log_{10} M + \log_{10} N$,

$$A_{v_{dB}} = 20 \log_{10} \frac{\omega RC}{\sqrt{1 + (\omega RC)^2}}$$

$$= 20 \log_{10} \omega RC + 20 \log_{10} \frac{1}{\sqrt{1 + (\omega RC)^2}}$$

Note the similarity of the second term with Eq. (21.40) for the low-pass filter.

Defining $x = \omega RC$:

$$A_{v_{\text{dB}}} = 20 \log_{10} x + 20 \log_{10} \frac{1}{\sqrt{1 + x^2}} \tag{21.43}$$

Each term of Eq. (21.43) will define asymptotes for the Bode plot. The first term will generate a bode plot with the characteristics appearing in Fig. 21.47.

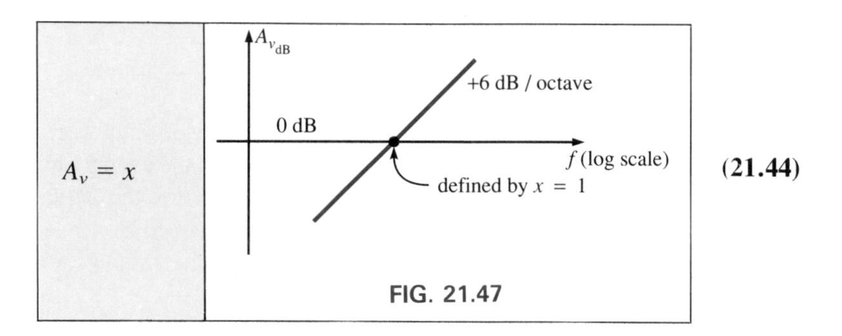


$$A_v = x \tag{21.44}$$

FIG. 21.47

For the high-pass filter, Eq. (21.43) specifies that the asymptotes for each term be added for specific frequency intervals to obtain the asymptotes that will define the high-frequency response. Fig. 21.48 reveals that the resulting plot is the same as obtained earlier.

For the interval 0 to the frequency defined by $x = 1$, the second term of Eq. (21.43) is zero decibels, and the sum is simply the first term of the equation. For higher frequencies, the +6-dB/octave slope of the first term and the −6-dB/octave slope of the second term will cancel the effects of each other, resulting in a zero-decibel asymptote, as shown in the adjoining figure.

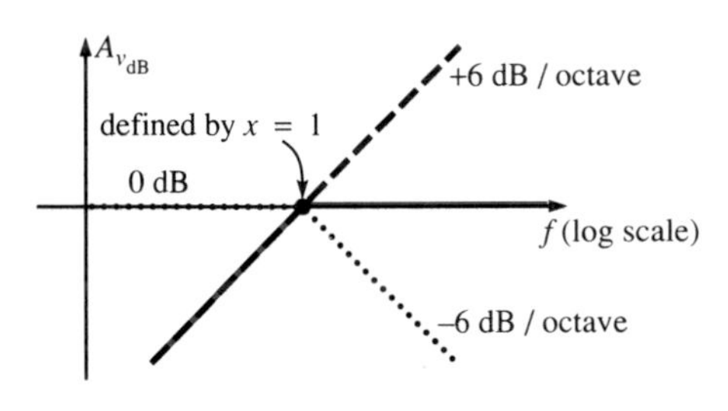


FIG. 21.48
*Asymptotic envelope for a high-pass filter.*

The net result of our analysis of the high-pass filter is another Bode plot for the expression of Eq. (21.44) that can be put to good use in other investigations.

The phase relationship as determined by

$$\boxed{\theta = 90° - \tan^{-1} \omega RC} \tag{21.45}$$

from Eq. (21.42) will result in the same plot as obtained for Eq. (21.32):

$$\theta = \tan^{-1}\frac{f_1}{f} = \tan^{-1}\frac{1}{2\pi fRC} = \tan^{-1}\frac{1}{\omega RC} \tag{21.46}$$

Through trigonometric identities it can be shown that

$$\theta = 90° - \tan^{-1} \omega RC = \tan^{-1}\frac{1}{\omega RC} \tag{21.47}$$

for the range of $\theta$ of interest.

## Low-Pass Filter with Limited Attenuation

The filter of Fig. 21.49 will limit the attenuation of the high frequencies to a level determined by the resistors $R_1$ and $R_2$.

Applying the voltage divider rule:

$$\mathbf{V}_o = \frac{(R_2 - jX_C)\mathbf{V}_i}{R_1 + R_2 - jX_C}$$

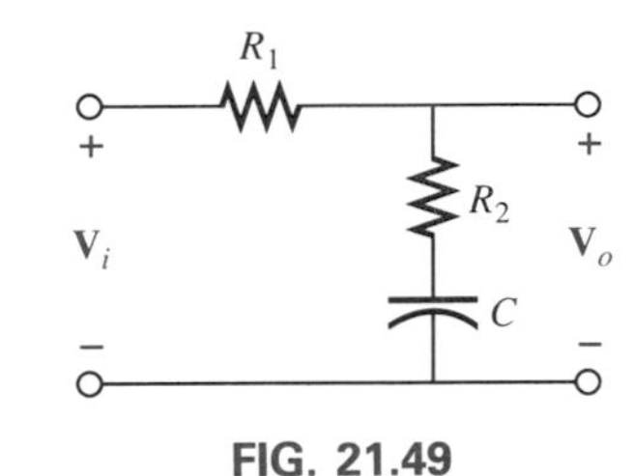


**FIG. 21.49**
*Low-pass filter with limited attenuation.*

and

$$\mathbf{A}_v = \frac{\mathbf{V}_o}{\mathbf{V}_i} = \frac{R_2 - jX_C}{R_1 + R_2 - jX_C} = \frac{R_2 - j\dfrac{1}{\omega C}}{R_1 + R_2 - j\dfrac{1}{\omega C}} = \frac{\omega R_2 C - j}{\omega(R_1 + R_2)C - j}$$

Multiplying numerator and denominator by $+j$:

$$\frac{(+j)(\omega R_2 C - j)}{(+j)(\omega(R_1 + R_2)C - j)} = \frac{j\omega R_2 C + 1}{j\omega(R_1 + R_2)C + 1}$$

and

$$\mathbf{A}_v = \frac{1 + j\omega R_2 C}{1 + j\omega(R_1 + R_2)C}$$

with

$$\mathbf{A}_v = A_v \angle\theta = \frac{\sqrt{1 + (\omega R_2 C)^2}}{\sqrt{1 + (\omega(R_1 + R_2)C)^2}} \angle \tan^{-1}\omega R_2 C - \tan^{-1}\omega(R_1 + R_2)C \qquad \textbf{(21.48)}$$

and

$$A_{v_{dB}} = 20\log_{10}\frac{\sqrt{1 + (\omega R_2 C)^2}}{\sqrt{1 + (\omega(R_1 + R_2)C)^2}}$$

$$= 20\log_{10}\sqrt{1 + (\omega R_2 C)^2} + 20\log_{10}\frac{1}{1 + (\omega(R_1 + R_2)C)^2}$$

Defining $x_1 = \omega R_2 C$ and $x_2 = \omega(R_1 + R_2)C$,

$$A_{v_{dB}} = 20\log_{10}\sqrt{1 + x_1^2} + 20\log_{10}\frac{1}{\sqrt{1 + x_2^2}} \qquad \textbf{(21.49)}$$

The second term has the same format as Eq. (21.41). The first term is new but has the same format as the denominator of the second term. The fact that the square root is in the numerator of the first term results in the asymptotes of Fig. 21.50.

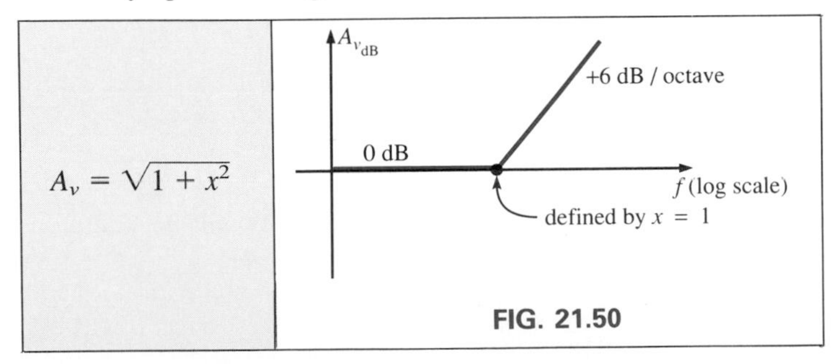


**FIG. 21.50**

For this filter the total asymptotic decibel level is determined by the sum of the two terms of Eq. (21.50) at each frequency. Plotting the asymptotes for each term of the preceding value of $A_{v_{\text{dB}}}$ on a frequency scale will result in the plot of Fig. 21.51.

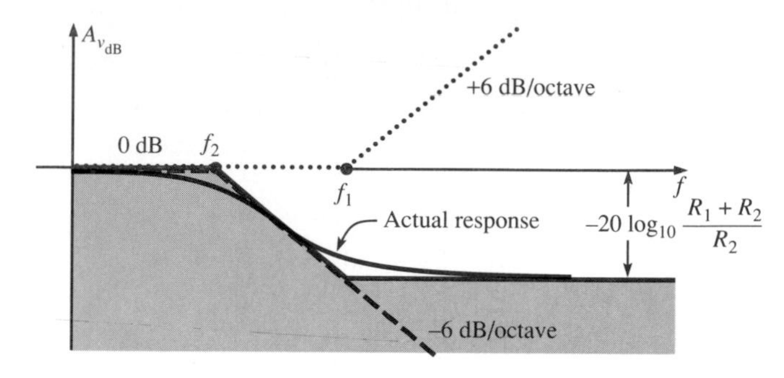


**FIG. 21.51**

The frequencies $f_1$ and $f_2$ are defined by

$$x_1 = 2\pi f_1 R_2 C = 1 \qquad \text{with } f_1 = \frac{1}{2\pi R_2 C}$$

and

$$x_2 = 2\pi f_2 (R_1 + R_2) C = 1 \qquad \text{with } f_2 = \frac{1}{2\pi (R_1 + R_2) C}$$

For frequencies from 0 to $f_2$, both asymptotes are zero decibels, resulting in a sum of zero decibels and the asymptote shown in Fig. 21.51. For frequencies from $f_2$ to $f_1$, one asymptote is at zero decibels and the other is at $-6$ dB/octave. The sum of the two is an asymptote dropping off at $-6$ dB/octave from the frequency $f_2$. For frequencies greater than $f_1$, the $+6$-dB and $-6$-dB asymptotes will cancel the effect of each other; the resulting decibel level is that defined by $f = f_1$.

The level at $f_1$ can be determined by noting that it is the same at $f = f_1$ as at very high frequencies. For the network of Fig. 21.49, if we apply a very high frequency the capacitor can be replaced by its short-circuit equivalent; applying the voltage divider rule:

$$\mathbf{V}_o = \frac{R_2 \mathbf{V}_i}{R_1 + R_2}$$

and

$$\mathbf{A}_v = \frac{\mathbf{V}_o}{\mathbf{V}_i} = \frac{R_2}{R_1 + R_2} \tag{21.51}$$

with

$$A_{v_{\text{dB}}} = 20 \log_{10} \frac{R_2}{R_1 + R_2}$$

but, since $R_2/(R_1 + R_2)$ is less than 1, Eq. (21.51) will be written as

$$A_{v_{\text{dB}}} = -20 \log_{10} \frac{R_1 + R_2}{R_2} \qquad f \geq f_1 \tag{21.52}$$

The resulting plot of Fig. 21.51 reveals that the actual curve follows the guide provided by the asymptotes but separates itself from each break frequency by 3 dB.

At each break frequency the output $\mathbf{V}_o$ will lag the input voltage by an angle less than (but relatively close to) 45°, as shown in Fig. 21.52 and defined by the following equation:

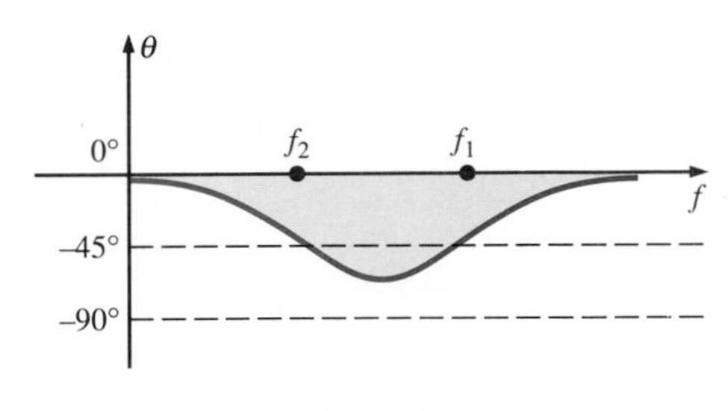


FIG. 21.52

$$\theta = \tan^{-1} \omega R_2 C - \tan^{-1} \omega (R_1 + R_2)C \tag{21.53}$$

For low frequencies the open-circuit equivalent for the capacitor results in $\mathbf{V}_o \cong \mathbf{V}_i$ and a 0° phase shift. At high frequencies, the resistive nature of the resulting network (with the capacitor replaced by its short-circuit equivalent) will again result in a 0° phase shift. The remaining points on the curve of Fig. 21.52 can be determined by simply plugging in specific frequencies into Eq. (21.53). However, it is also useful to know that the most dramatic changes in the phase angle occur when the dB plot of the magnitude also goes through its greatest change (such as at $f_1$ and $f_2$).

## High-Pass Filter with Limited Attenuation

The last filter to be analyzed in detail is the filter of Fig. 21.53, which will limit the low-frequency attentuation in much the same manner as just described for the low-pass filter.

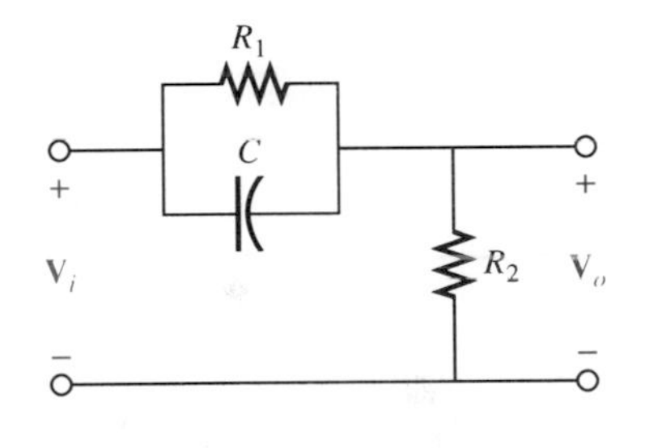


FIG. 21.53
*High-pass filter with limited attenuation.*

For the parallel elements:

$$\mathbf{R}_1 \parallel \mathbf{X}_C = \frac{(R_1)(-jX_C)}{R_1 - jX_C} = \frac{-jR_1X_C}{R_1 - jX_C}$$

Applying the voltage divider rule:

$$\mathbf{V}_o = \frac{R_2\mathbf{V}_i}{R_2 - j\dfrac{R_1X_C}{R_1 - jX_C}} = \frac{R_2(R_1 - jX_C)\mathbf{V}_i}{R_2(R_1 - jX_C) - jR_1X_C}$$

and

$$\mathbf{A}_v = \frac{\mathbf{V}_o}{\mathbf{V}_i} = \frac{R_1R_2 - jR_2X_C}{R_1R_2 - j(R_1 + R_2)X_C}$$

Multiplying numerator and denominator by $+j/R_2X_C$:

$$\frac{\left(\dfrac{+j}{R_2X_C}\right)(R_1R_2 - jR_2X_C)}{\left(\dfrac{+j}{R_2X_C}\right)(R_1R_2 - j(R_1 + R_2)X_C)} = \frac{1 + j\dfrac{R_1}{X_C}}{\dfrac{R_1 + R_2}{R_2} + j\dfrac{R_1}{X_C}}$$

Multiplying numerator and denominator by $R_2/(R_1 + R_2)$:

$$\frac{\left(\dfrac{R_2}{R_1 + R_2}\right)\left(1 + j\dfrac{R_1}{X_C}\right)}{\left(\dfrac{R_2}{R_1 + R_2}\right)\left(\dfrac{R_1 + R_2}{R_2} + j\dfrac{R_1}{X_C}\right)} = \frac{\dfrac{R_2}{R_1 + R_2}\left(1 + j\dfrac{R_1}{X_C}\right)}{1 + j\dfrac{R_1R_2}{(R_1 + R_2)X_C}}$$

Substituting $X_C = 1/\omega C$ results in

$$\mathbf{A}_v = \frac{\mathbf{V}_o}{\mathbf{V}_i} = \frac{R_2}{R_1 + R_2} \frac{1 + j\omega R_1 C}{\left[1 + j\omega\left(\dfrac{R_1 R_2}{R_1 + R_2}\right)C\right]} \tag{21.54}$$

with

$$\mathbf{A}_v = A_v \angle\theta = \frac{R_2}{R_1 + R_2} \cdot \frac{\sqrt{1 + (\omega R_1 C)^2}}{\sqrt{1 + \left[\omega\left(\dfrac{R_1 R_2}{R_1 + R_2}\right)C\right]^2}} \angle \tan^{-1} \omega R_1 C - \tan^{-1} \omega\left(\frac{R_1 R_2}{R_1 + R_2}\right)C \tag{21.55}$$

Before determining $A_{v_{\text{dB}}}$, the ratio $R_2/(R_1 + R_2)$ is less than 1, which suggests that we write the equation as $-20 \log_{10}(R_1 + R_2)/R_2$ rather than as $20 \log_{10} R_2/(R_1 + R_2)$, since the result will be a negative number. Therefore,

$$A_{v_{\text{dB}}} = -20 \log_{10}\frac{R_1 + R_2}{R_2} + 20 \log_{10} \sqrt{1 + (\omega R_1 C)^2} + 20 \log_{10}\frac{1}{\sqrt{1 + \left[\omega\left(\dfrac{R_1 R_2}{R_1 + R_2}\right)C\right]^2}} \tag{21.56}$$

Defining $x_1 = \omega R_1 C = 1$,

$$f_1 = \frac{1}{2\pi R_1 C}$$

and

$$x_2 = \omega\left(\frac{R_1 R_2}{R_1 + R_2}\right)C = 1$$

$$f_2 = \frac{1}{2\pi\left(\dfrac{R_1 R_2}{R_1 + R_2}\right)C}$$

The plot of Fig. 21.54 can then be established.

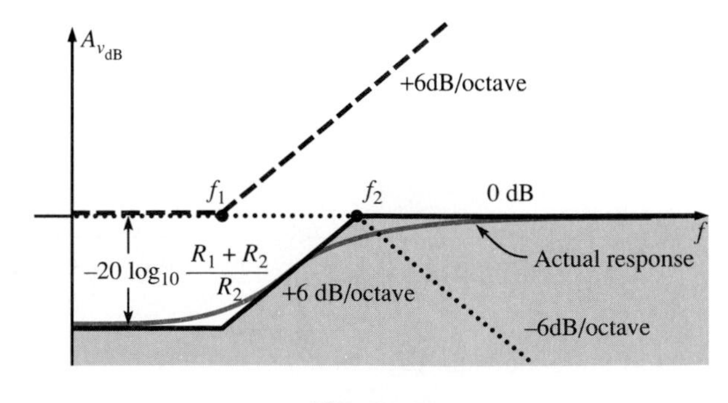

FIG. 21.54

For frequencies 0 to $f_1$, the asymptotes of the last two terms of Eq. (21.56) are 0 decibels and $A_{v_{\text{dB}}} = -20 \log_{10}(R_1 + R_2)/R_2$, as shown. At $f_1$ the second term of Eq. (21.56) starts to rise at +6 dB, whereas the last term remains at 0 dB. The result is a +6-dB rise from the low-frequency level. For this configuration it is then interesting to note that the +6-dB/octave curve passes through zero decibels at $f = f_2$. This result is verified by the following mathematical operations:

At $f = f_2$,

$$\omega R_1 C = 2\pi f_2 R_1 C = 2\pi \frac{1}{\left[2\pi\left(\frac{R_1 R_2}{R_1 + R_2}\right)C\right]} R_1 C = \frac{R_1 + R_2}{R_2}$$

Substituting into the second term of Eq. (21.56):

$$20 \log_{10} \sqrt{1 + (\omega R_1 C)^2} = 20 \log_{10} \sqrt{1 + \left(\frac{R_1 + R_2}{R_2}\right)^2}$$

$$= 10 \log_{10}\left[1 + \left(\frac{R_1 + R_2}{R_2}\right)^2\right]$$

Dropping the 1 to obtain the asymptotic plot,

$$10 \log_{10}\left[1 + \left(\frac{R_1 + R_2}{R_2}\right)^2\right] = 10 \log_{10}\left(\frac{R_1 + R_2}{R_2}\right)^2$$

$$= +20 \log_{10} \frac{R_1 + R_2}{R_2}$$

At $f = f_2$, therefore, the first and second terms of Eq. (21.56) will cancel, and the last term is 0 decibels. The result is $A_{v_{dB}} = 0$ dB at $f = f_2$, as shown in Fig. 21.54.

The low-frequency level can also be determined by replacing the capacitor by its open-circuit equivalent and applying the voltage divider rule:

$$\mathbf{V}_o = \frac{R_2 \mathbf{V}_i}{R_1 + R_2}$$

with

$$\mathbf{A}_v = \frac{\mathbf{V}_o}{\mathbf{V}_i} = \frac{R_2}{R_1 + R_2}$$

and

$$A_{v_{dB}} = 20 \log_{10} \frac{R_2}{R_1 + R_2} = -20 \log_{10} \frac{R_1 + R_2}{R_2}$$

For the high-frequency region, the capacitor can be replaced by its short-circuit equivalent, and

$$\mathbf{V}_o = \mathbf{V}_i$$

with

$$\mathbf{A}_v = \frac{\mathbf{V}_o}{\mathbf{V}_i} = 1$$

and

$$A_{v_{dB}} = 20 \log_{10} 1 = 0 \text{ dB}$$

The phase angle by which $\mathbf{V}_o$ leads $\mathbf{V}_i$ is given by

$$\theta = \tan^{-1} \omega R_1 C - \tan^{-1} \omega\left(\frac{R_1 R_2}{R_1 + R_2}\right)C \tag{21.57}$$

The phase plot for the leading network appears in Fig. 21.55, revealing—as in the previous network—that both the phase and magni-

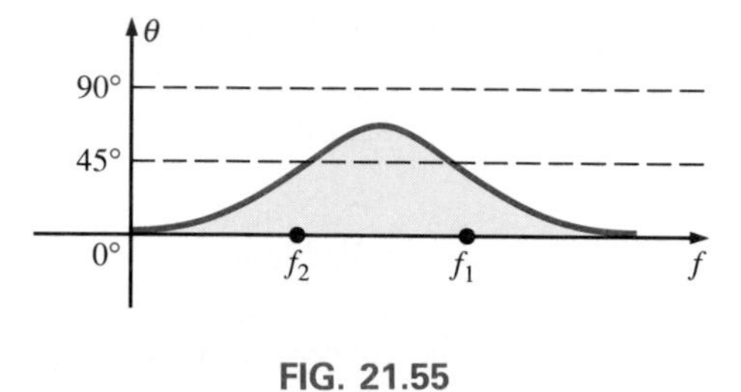


FIG. 21.55

tude plots experience their greatest changes in the same frequency intervals. At low frequencies the resistive nature of the network (capacitor = open-circuit) results in a zero-degree phase shift between $\mathbf{V}_o$ and $\mathbf{V}_i$. At the high end, the relationship $\mathbf{V}_o = \mathbf{V}_i$ dictates a zero-degree phase shift between the input and output quantities. At a frequency between $f_1$ and $f_2$, the angle will be a maximum, as noted in Fig. 21.55.

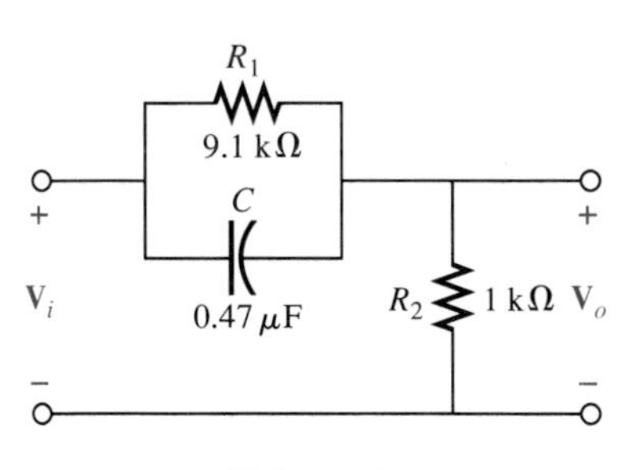

FIG. 21.56

**EXAMPLE 21.10.** For the filter of Fig. 21.56,

a. Sketch the curve of $A_{v_{\text{dB}}}$ versus frequency using a log scale.
b. Sketch the curve of $\theta$ versus frequency using a log scale.

***Solution:***

a. For the break frequencies:

$$f_1 = \frac{1}{2\pi R_1 C} = \frac{1}{2\pi(9.1\text{ k}\Omega)(0.47\ \mu\text{F})} = \mathbf{37.2\ Hz}$$

$$f_2 = \frac{1}{2\pi\left(\frac{R_1 R_2}{R_1 + R_2}\right)C} = \frac{1}{2\pi(0.9\text{ k}\Omega)(0.47\ \mu\text{F})} = \mathbf{376.25\ Hz}$$

The maximum low-level attenuation is

$$-20\log_{10}\frac{R_1 + R_2}{R_2} = -20\log_{10}\frac{9.1\text{ k}\Omega + 1\text{ k}\Omega}{1\text{ k}\Omega}$$

$$= -20\log_{10} 10.1 = \mathbf{-20.09\ dB}$$

The resulting plot appears in Fig. 21.57.

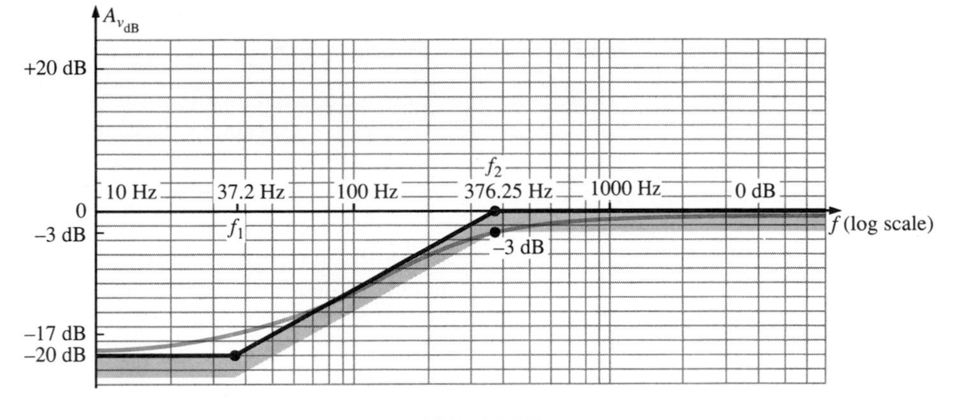


FIG. 21.57

b. For the break frequencies:

$f = f_1 = 37.2$ Hz:

$$\theta = \tan^{-1} 2\pi f R_1 C - \tan^{-1} 2\pi f\left(\frac{R_1 R_2}{R_1 + R_2}\right) C$$

$$= 45° - \tan^{-1} 2\pi(37.2 \text{ Hz})(0.9 \text{ k}\Omega)(0.47 \ \mu\text{F})$$
$$= 45° - 5.71°$$
$$= \mathbf{39.29°}$$

$f = f_2 = 376.26$ Hz:

$$\theta = \tan^{-1} 2\pi(376.25 \text{ Hz})(9.1 \text{ k}\Omega)(0.47 \ \mu\text{F}) - 45°$$
$$= 84.35° - 45°$$
$$= \mathbf{39.35°}$$

At a frequency midway between $f_1$ and $f_2$ on a log scale, for example, 120 Hz,

$$\theta = \tan^{-1} 2\pi(120 \text{ Hz})(9.1 \text{ k}\Omega)(0.47 \ \mu\text{F}) - \tan^{-1} 2\pi(120 \text{ Hz})(0.9 \text{ k}\Omega)(0.47 \ \mu\text{F})$$
$$= 72.77° - 17.69°$$
$$= \mathbf{55.08°}$$

The resulting phase plot appears in Fig. 21.58.

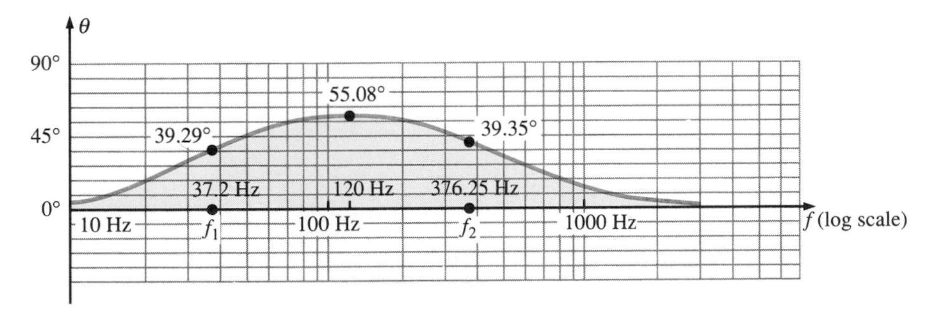


FIG. 21.58

## 21.12 COMPUTER ANALYSIS

This chapter provides an excellent opportunity to demonstrate the versatility of the PROBE option. By simply indicating the preference for each axis, the vertical and horizontal scales can be linear or log scales; in fact, a decibel scale can be chosen if desired.

The first input file of this section will investigate the impact of using a log-log scale to plot the current versus frequency for the inductor of Fig. 21.59 initially investigated in Chapter 15 with linear scales. By

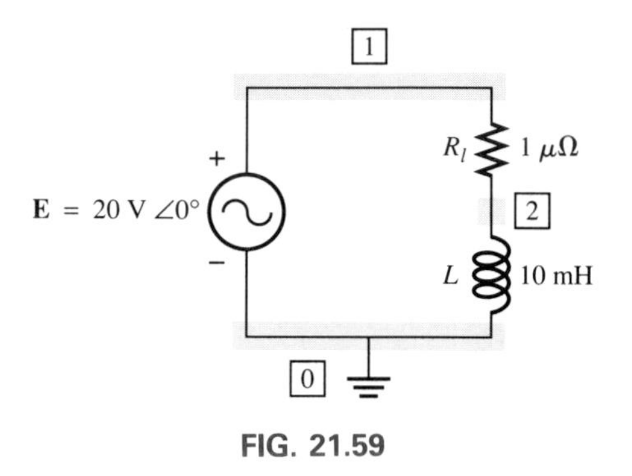


FIG. 21.59

choosing a log-log scale, the range of current and frequency values can go far beyond the plot of Chapter 15. The input file appears in Fig. 21.60 with the PROBE response in Fig. 21.61.

```
************************ Evaluation PSpice (Jan. 1988) ******* 16:12:11 *******

 Chapter 21 - Log-Log Plot (R-L network)

 ****     CIRCUIT DESCRIPTION

****************************************************************************

VE 1 0 AC 20V 0
RL 1 2 1U
L1 2 0 10MH
.AC LIN 100 1 100KH
.PROBE
.OPTIONS NOPAGE
.END
```

FIG. 21.60

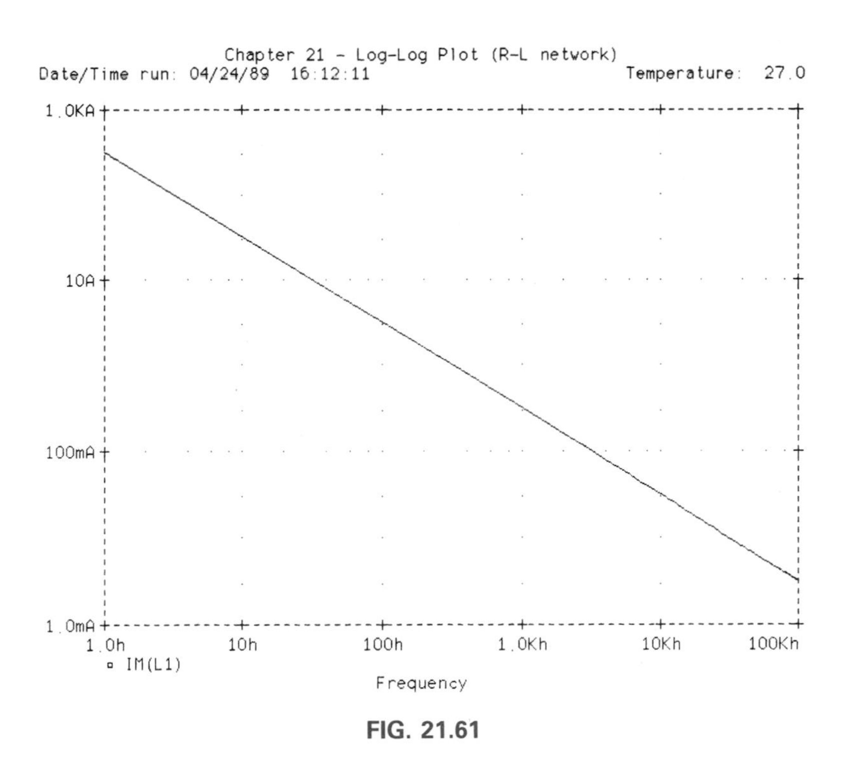


**FIG. 21.61**

The vertical scale of Fig. 21.61 must be read with particular care, since intermediate log values are not included. For instance, at $f =$ 1 Hz, the interval between 10 A and 1000 A must first be broken down, as shown in Fig. 21.62, to include the 100-A level. The intersection of the plot is then about 300 A on a log scale rather than 500 A if it were on a linear scale. Checking our conclusion, at $f = 1$ Hz,

$$X_L = 2\pi fL = 2\pi(1 \text{ Hz})(10 \text{ mH}) = 62.8 \text{ m}\Omega$$

and

$$I = \frac{E}{X_L} = \frac{20 \text{ V}}{62.8 \text{ m}\Omega} = 318.47 \text{ A}$$

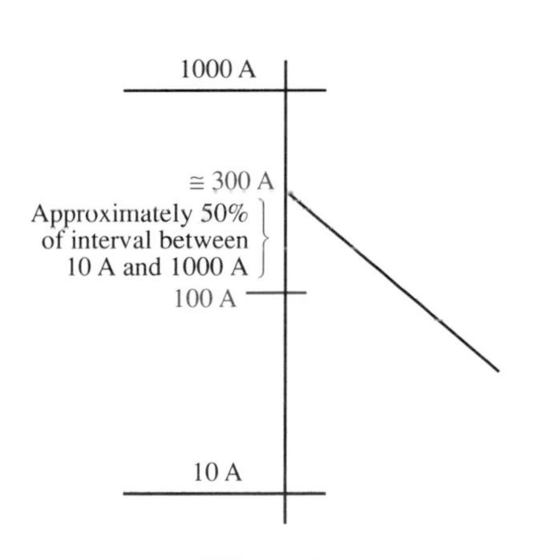


**FIG. 21.62**

At $f = 100$ kHz, the 10-mA level must be added between 1 mA and 100 mA before a level of about 3 mA can be estimated.

In particular, note the enormous range of current and frequency that can appear on the same plot. The resulting curve does not reflect the exponential relationship between the current and applied frequency, but it does reveal the level of current for a wide range of frequencies. Just imagine trying to plot the preceding graph using linear scales—especially if we use the interval designated for 1 mA to 100 mA for the full range of current values.

The frequency response of the *R-C* network of Fig. 21.63 will now be determined using the input file of Fig. 21.64. The PROBE response of Fig. 21.65 requested the magnitude of the output voltage versus

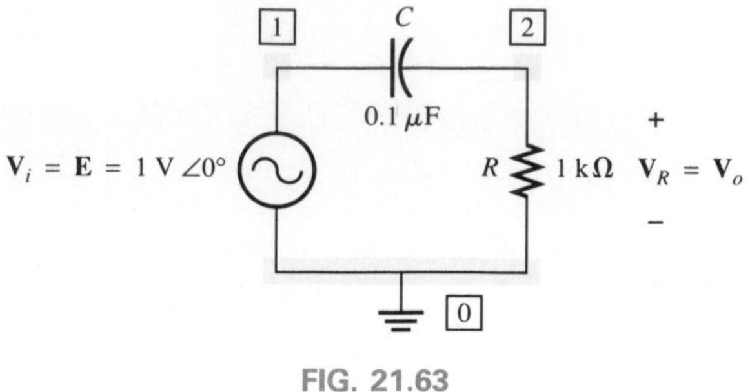


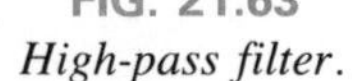

FIG. 21.63
*High-pass filter.*

```
************************ Evaluation PSpice (Jan. 1988) ******* 11:25:23 *******

 Chapter 21 - High Pass R-C Filter

 ****      CIRCUIT DESCRIPTION

*****************************************************************************


V 1 0 AC 1 0
C 1 2 0.1UF
R 2 0 1K
.AC LIN 101 100H 10KH
.PROBE
.OPTIONS NOPAGE
.END
```

FIG. 21.64

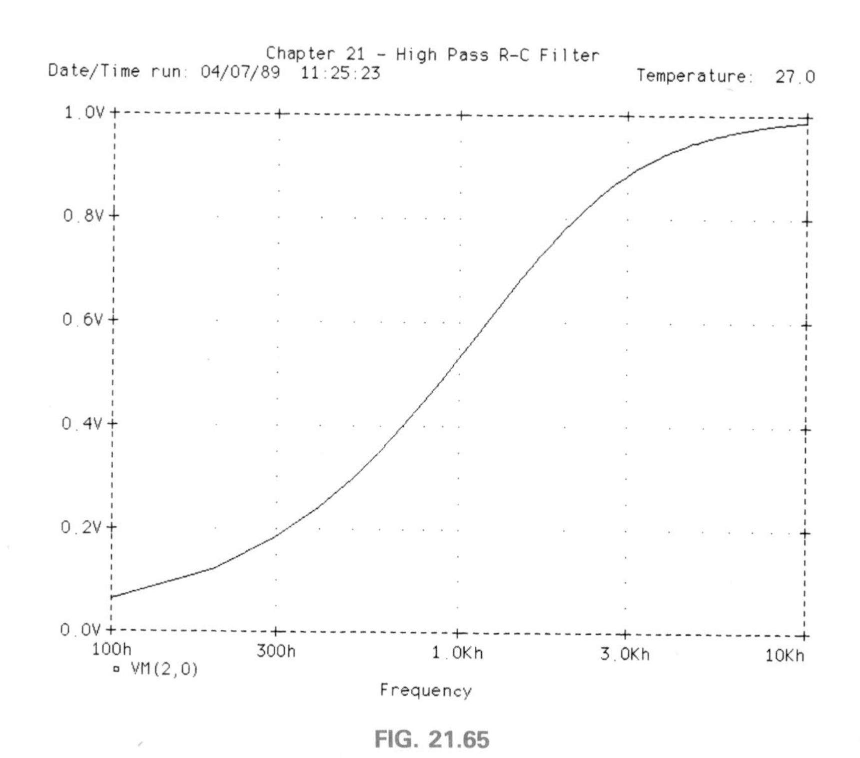


FIG. 21.65

frequency on a log scale. Quite obviously, as the frequency increased the magnitude of $V_o$ increased also, as required for a high-pass filter. The critical frequency is defined by 0.707 V, but reading the resulting frequency on the horizontal axis requires a bit of care due to the log scale. A careful breakdown of the region between 1 kHz and 3 kHz will result in a frequency of about 1.6 kHz versus the actual value of 1591.55 Hz. The phase plot of Fig. 21.66 reveals a phase shift of about 83° at $f$ = 100 Hz, 45° at $f_c$ = 1.6 kHz, and about 10° at 10 kHz, supporting the plot of Fig. 21.19 of the text.

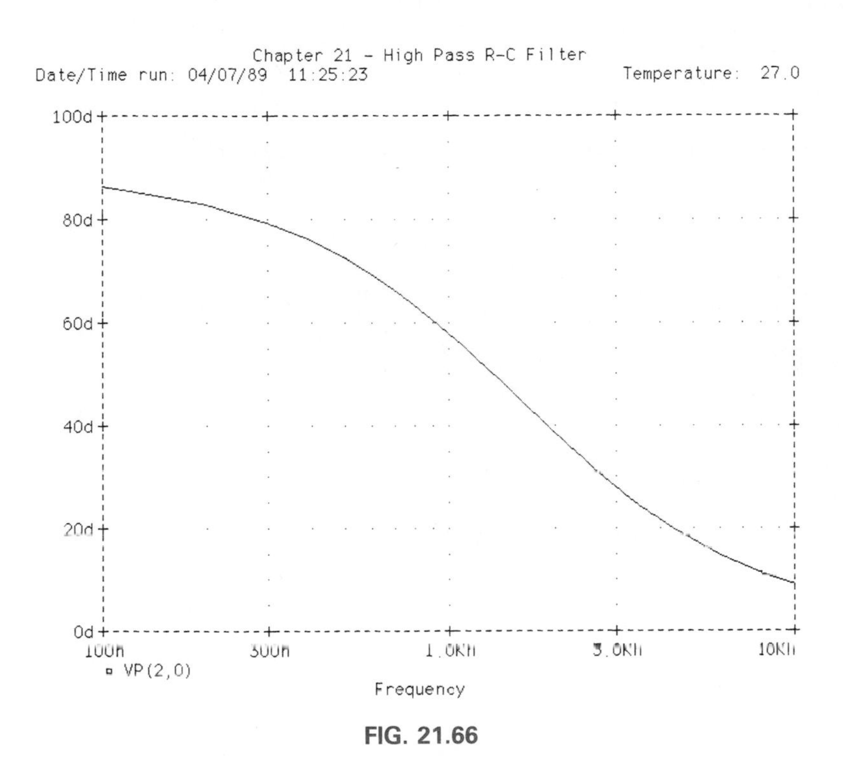


FIG. 21.66

The next input file will check our solution for the double-tuned filter of Example 21.8, given here as Fig. 21.67. The input file appears in

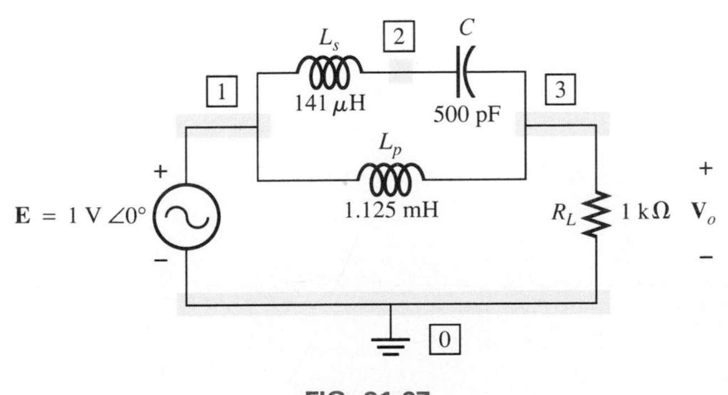


FIG. 21.67
*Double-tuned filter.*

Fig. 21.68, with the PROBE response in Fig. 21.69. Note how the filter response rejects the frequency at 0.2 MHz = 200 kHz and accepts the frequency at 0.6 MHz = 600 kHz, as required in the example. The quality factor of the series resonance reject response is obviously much higher than that obtained for the parallel resonant configuration.

```
************************ Evaluation PSpice (Jan. 1988) ******* 11:16:24 *******

 Chapter 21 - Double Tuned Filter

 ****     CIRCUIT DESCRIPTION

******************************************************************************

V  1 0 AC 1 0
L  1 2 141UH
C  2 3 500PF
LP 1 3 1.125MH
RL 3 0 1K
.AC LIN 91 100KH 1MEGH
.PROBE
.OPTIONS NOPAGE
.END
```

**FIG. 21.68**

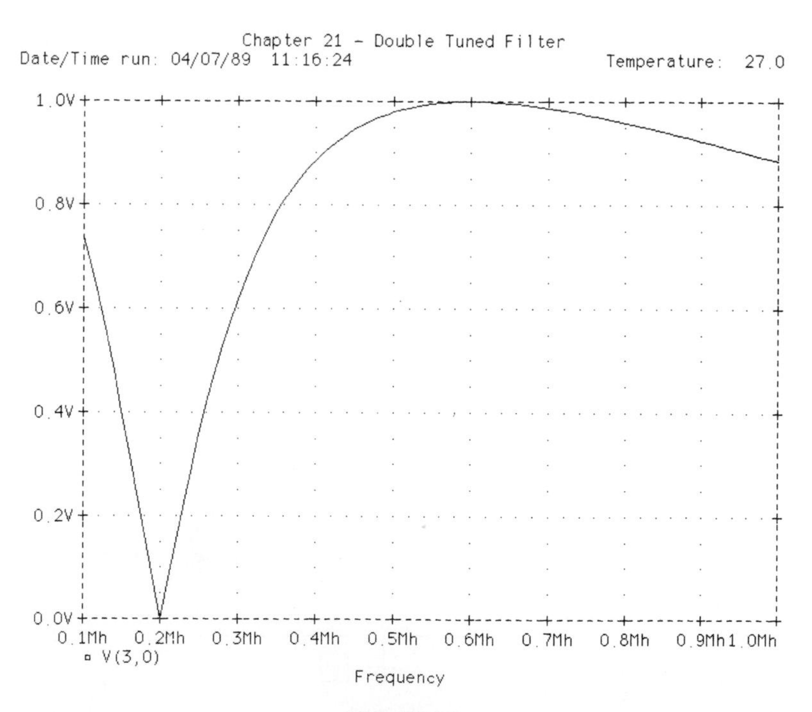

**FIG. 21.69**

The last input file of Fig. 21.70 will provide a Bode plot for the *R-C* network of Fig. 21.71. One of the options of the PROBE analysis is the DB ($x$) function appearing at the bottom of the response of Fig. 21.71. Note that the actual curve is 3-dB below the break frequency of 1.6 kHz. In addition, if we draw a straight line along the linear region

```
************************ Evaluation PSpice (January 1989) ******* 09:55:07 ****
***

 Bode Plot

 ****     CIRCUIT DESCRIPTION

*******************************************************************************


VI 1 0 AC 1V 0
C  1 2 0.1UF
R  2 0 1K
.AC LIN 100 100H 10KH
.PROBE
.OPTIONS NOPAGE
.END
```

FIG. 21.70

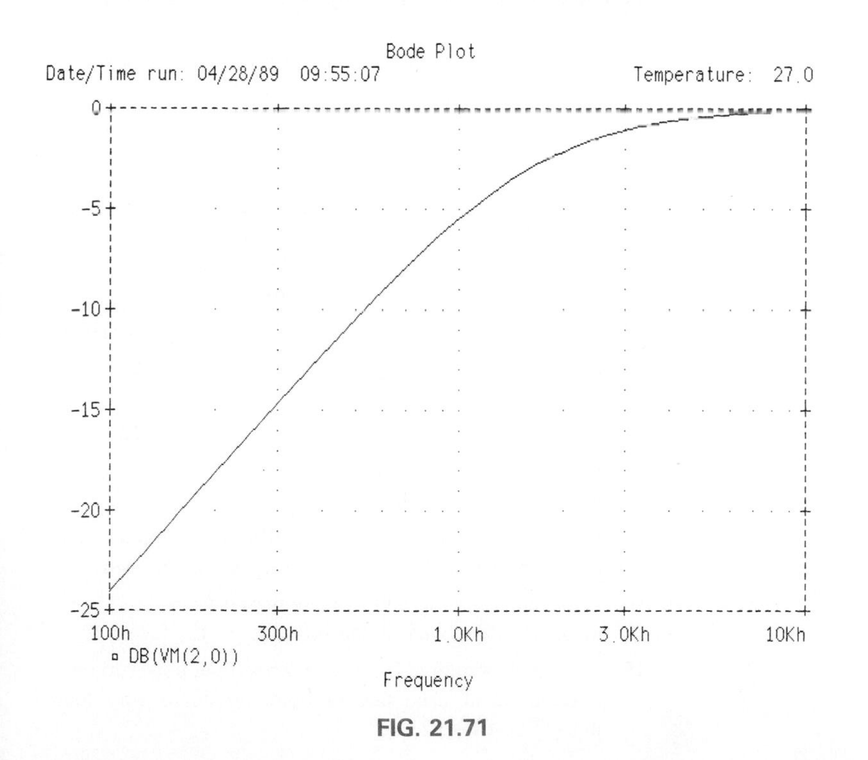


FIG. 21.71

of the low-frequency response between 100 Hz and 1 kHz (actually the low frequency asymptote), we will find that it will intersect −4 dB at 1 kHz, resulting in a 20-dB change (−24 dB to −4 dB) in 1 decade (100 Hz to 1 kHz).

## PROBLEMS

### SECTION 21.2 Properties of Logarithms

**1.** Determine $\log_{10} x$ for each value of $x$.
**a.** 100,000 **b.** 0.0001
**c.** $10^8$ **d.** $10^{-6}$
**e.** 20 **f.** 8643.4
**g.** 56,000 **h.** .318

**2.** Given $N = \log_{10} x$, determine $x$ for each value of $N$.
**a.** 3 **b.** 12
**c.** 0.2 **d.** 0.04
**e.** 10 **f.** 3.18
**g.** 1.001 **h.** 6.1

**3.** Determine $\log_e x$ for each value of $x$.
**a.** 100,000 **b.** 0.0001
**c.** 20 **d.** 8643.4
Compare with the solutions to Exercise 1.

**4.** Determine $\log_{10} 48 = \log_{10}(8)(6)$ and compare to $\log_{10} 8 + \log_{10} 6$.

**5.** Determine $\log_{10} 0.2 = \log_{10} 18/90$ and compare to $\log_{10} 18 - \log_{10} 90$.

**6.** Verify that $\log_{10} 0.5$ is equal to $-\log_{10} 1/0.5 = -\log_{10} 2$.

**7.** Find $\log_{10}(3)^3$ and compare with $3 \log_{10} 3$.

**8.** **a.** Determine the number of bels that relate power levels of $P_2 = 280$ mW and $P_1 = 4$ mW.
**b.** Determine the number of decibels for the power levels of part (a) and compare results.

**9.** A power level of 100 W is 6 dB above what power level?

**10.** If a 2-W speaker is replaced by one with a 40-W output, what is the increase in decibel level?

**11.** Determine the $\text{dB}_m$ level for an output power of 120 mW.

**12.** Find the $\text{dB}_v$ gain of an amplifier that raises the voltage level from 0.1 mW to 8.4 W.

**13.** Find the output voltage of an amplifier if the applied voltage is 20 mV and a $\text{dB}_v$ gain of 22 dB is attained.

**14.** If the sound pressure level is increased from 0.001 $\mu$bar to 0.016 $\mu$bar, what is the increase in $\text{dB}_s$ level?

**15.** What is the required increase in acoustical power to raise a sound level from that of quiet music to very loud music? Use Fig. 21.4.

16. Using semilog paper:
    a. Plot $X_L$ versus frequency for a 10-mH coil for a frequency range of 1 Hz to 1 MHz. Set the vertical scale by the maximum value of $X_L$.
    b. Plot $X_L$ versus frequency for a 1-$\mu$F capacitor for a frequency range of 10 Hz to 100 kHz. Choose an appropriate vertical scale for the frequency range of interest.

## SECTION 21.5 *R-C* Low-Pass Filter

17. For the *R-C* low-pass filter of Fig. 21.72,
    a. Sketch $V_o/V_i$ versus frequency using a log scale for the frequency axis. Determine $V_o/V_i$ at $0.1f_c$, $0.5f_c$, $f_c$, $2f_c$, and $10f_c$.
    b. Sketch $\theta$ of $\mathbf{V}_o = V_o \angle \theta$ versus frequency using a log scale for the frequency axis. Determine $\theta$ at the same frequencies listed in part (a).
18. Design an *R-C* low-pass filter to have a cutoff frequency of 500 Hz using a resistor of 1.2 kΩ. Then sketch the resulting magnitude and phase plots.

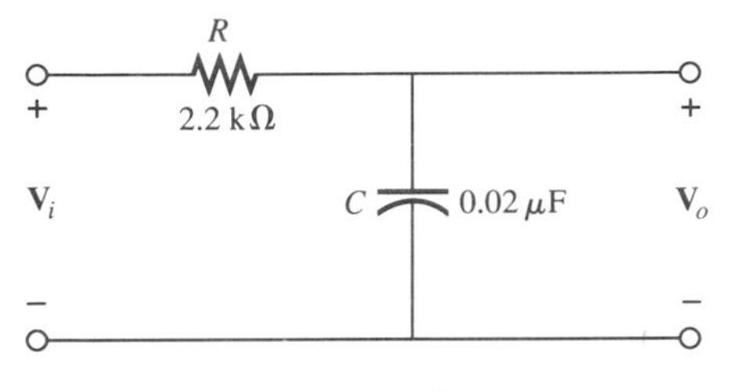

FIG. 21.72

## SECTION 21.6 *R-C* High-Pass Filter

19. For the high-pass *R-C* filter of Fig. 21.73,
    a. Determine the cutoff frequency.
    b. Plot $V_o/V_i$ on semilog paper for a frequency range of 1 Hz to 1 MHz.
    c. Plot the phase response for the frequency range of part (b).
20. Design a high-pass *R-C* filter having a cutoff frequency of 2 kHz, given a capacitor of 0.1 $\mu$F. Sketch the response $V_o/V_i$ on semilog paper for a frequency range of 10 Hz to 100 kHz. Use commercial resistor values (or at least the closest value available).

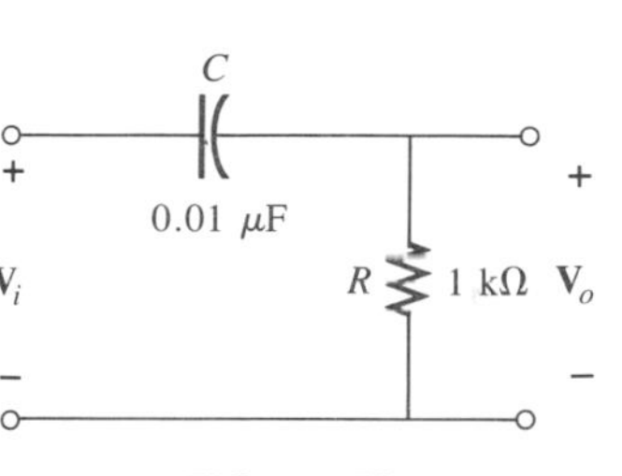


FIG. 21.73

## SECTION 21.7 Band-Pass Filters

21. For the band-pass filter of Fig. 21.74,
    a. Sketch the frequency response of $\mathbf{V}_o$ against a log scale extending from 10 Hz to 10 kHz.
    b. What is the bandwidth and center frequency?

*22. Design a band-pass filter such as appearing in Fig. 21.74 to have a low cutoff frequency of 40 kHz and a high cutoff frequency of 80 kHz.

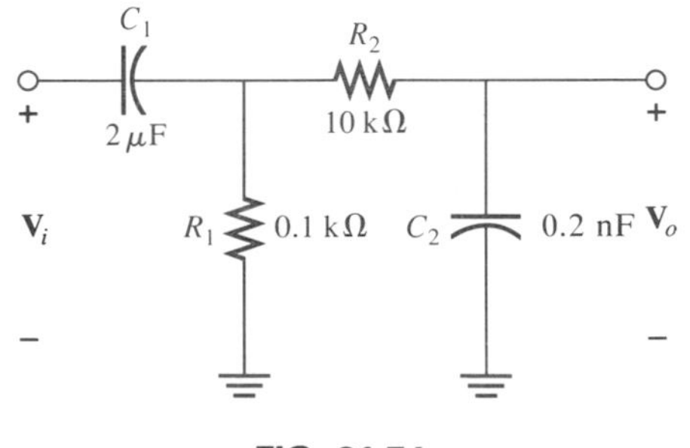

FIG. 21.74

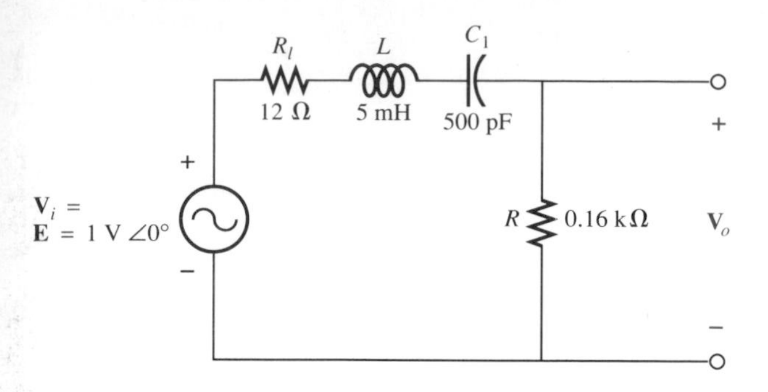

FIG. 21.75

23. For the band-pass filter of Fig. 21.75:
    a. Determine the frequency response of $V_o/V_i$ for a frequency range of 1 kHz to 1 MHz.
    b. Find the quality factor $Q_s$ and the bandwidth of the response.

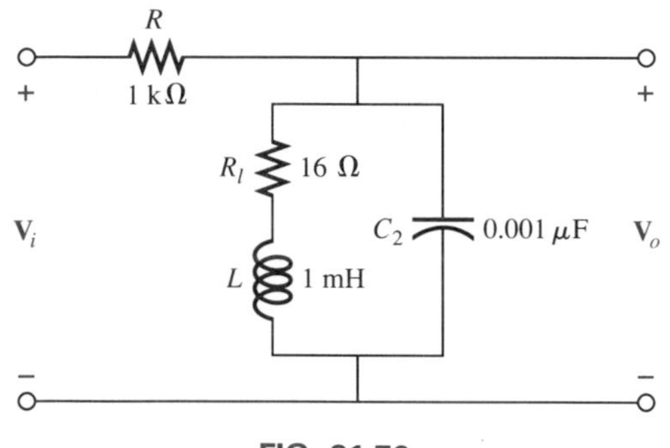


FIG. 21.76

24. For the band-pass filter of Fig. 21.76,
    a. Determine the frequency response of $V_o/V_i$ for a frequency range of 100 Hz to 1 MHz.
    b. Find the quality factor $Q_p$ and the bandwidth of the response.

## SECTION 21.8 Band-Stop Filters

*25. For the band-stop filter of Fig. 21.77:
    a. Determine $Q_s$.
    b. Find the bandwidth and half-power frequencies.
    c. Sketch the frequency characteristics of $V_o/V_i$.
    d. What is the effect on the curve of part (c) if a load of 2 kΩ is applied?

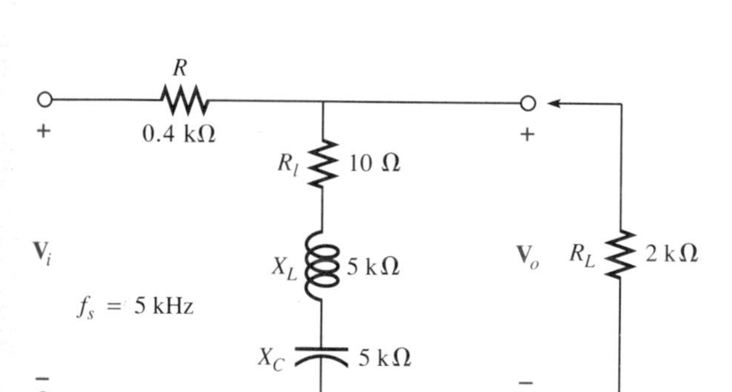

FIG. 21.77

*26. For the band-stop filter of Fig. 21.78:
    a. Determine $Q_p$ ($R_L = \infty$ Ω, an open circuit).
    b. Sketch the frequency characteristics of $V_o/V_i$.
    c. Find $Q_p$ (loaded) for $R_L$ = 100 kΩ, and indicate the effect of $R_L$ on the characteristics of part (b).
    d. Repeat part (c) for $R_L$ = 20 kΩ.

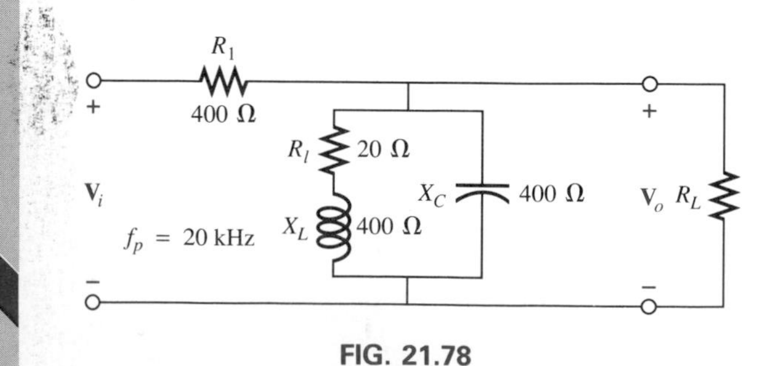


FIG. 21.78

## SECTION 21.9 Double-Tuned Filter

**27. a.** For the network of Fig. 21.38(a), if $L_p = 400\ \mu H$ ($Q > 10$), $L_s = 60\ \mu H$, and $C = 120$ pF, determine the rejected and accepted frequencies.

**b.** Sketch the response curve for part (a).

**28. a.** For the network of Fig. 21.38(b), if the rejected frequency is 30 kHz and the accepted is 100 kHz, determine the values of $L_s$ and $L_p$ ($Q > 10$) for a capacitance of 200 pF.

**b.** Sketch the response curve for part (a).

## SECTION 21.10 Bode Plots

**29. a.** Sketch the Bode plot for the frequency response of a high-pass *R-C* filter if $R = 0.47$ kΩ and $C = 0.05\ \mu F$.

**b.** Using the results of part (a), sketch the actual frequency response for the same frequency range.

**c.** Determine the decibel level at frequencies equal to one-half and twice the cutoff frequency.

**d.** Determine the gain $V_o/V_i$ at both frequencies of part (c).

**e.** Sketch the phase response for the same frequency range.

**30.** Repeat Problem 29 for an *R-C* low-pass filter if $R -$ 12 kΩ and C = 0.001 $\mu$F.

## SECTION 21.11 Sketching the Bode Response

**31.** For the filter of Fig. 21.79,

**a.** Sketch the curve of $A_{v_{dB}}$ versus frequency using a log scale.

**b.** Sketch the curve of $\theta$ versus frequency for the same frequency range as part (a).

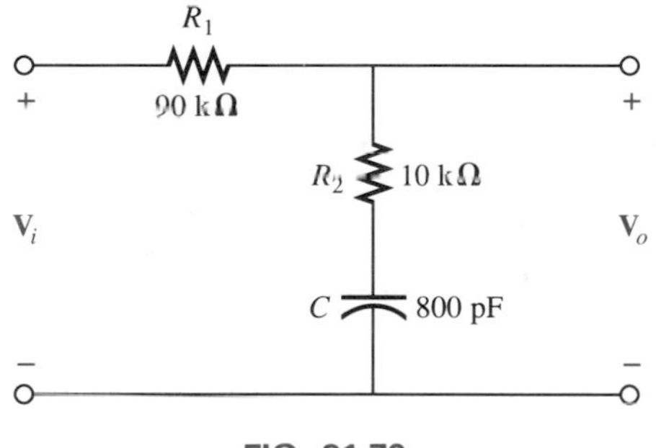

FIG. 21.79

**32.** For the filter of Fig. 21.80,

**a.** Sketch the curve of $A_{v_{dB}}$ versus frequency using a log scale.

**b.** Sketch the curve of $\theta$ versus frequency for the same frequency range as part (a).

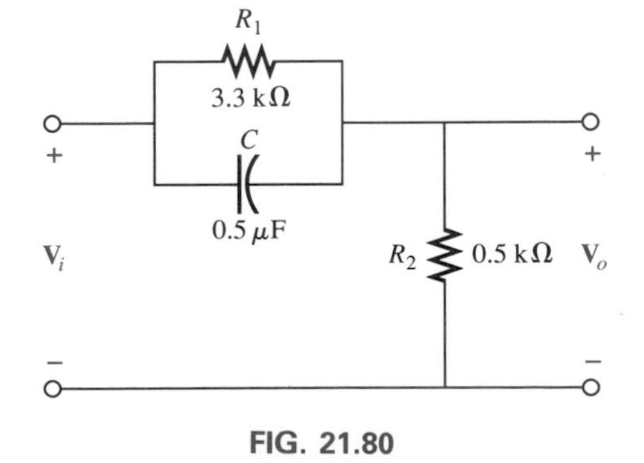


FIG. 21.80

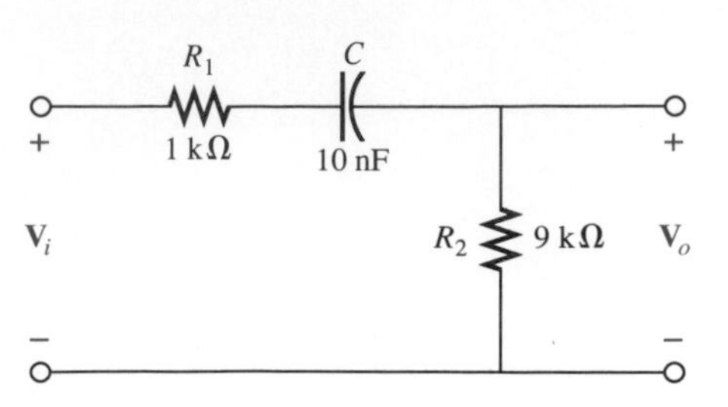

FIG. 21.81

*33. For the filter of Fig. 21.81,
   a. Sketch the curve of $A_{v_{dB}}$ versus frequency using a log scale.
   b. Sketch the curve of $\theta$ versus frequency for the same frequency range as part (a).

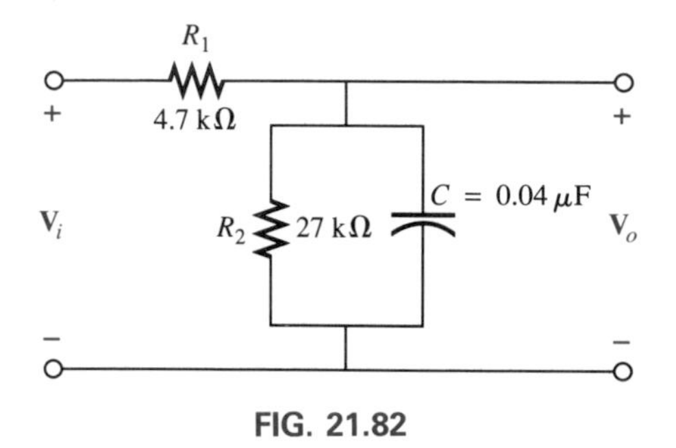


FIG. 21.82

*34. For the filter of Fig. 21.82,
   a. Sketch the curve of $A_{v_{dB}}$ versus frequency using a log scale.
   b. Sketch the curve of $\theta$ versus frequency for the same frequency range as part (a).

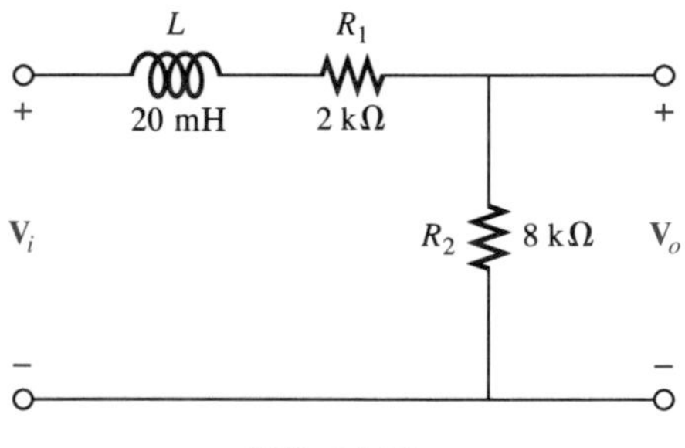

FIG. 21.83

*35. For the filter of Fig. 21.83,
   a. Sketch the curve of $A_{v_{dB}}$ versus frequency using a log scale.
   b. Sketch the curve of $\theta$ versus frequency for the same frequency range as part (a).

36. Sketch the Bode plot of the following function:

$$\mathbf{A}_v = \frac{0.05}{0.05 - j100/f}$$

37. Sketch the Bode plot of the following function:

$$\mathbf{A}_v = \frac{200}{200 + j0.1f}$$

38. Sketch the Bode plot of the following function:

$$\mathbf{A}_v = \frac{j\omega/1000}{(1 + j\omega/1000)(1 + j\omega/10{,}000)}$$

*39. Sketch the Bode plot of the following function:

$$\mathbf{A}_v = \frac{(1 + j\omega/1000)(1 + j\omega/2000)}{(1 + j\omega/3000)^2}$$

*40. Sketch the Bode plot of the following function:

$$\mathbf{A}_v = \frac{40(1 + j\omega/1000)}{(j\omega/1000)(1 + j\omega/5000)}$$

### SECTION 21.12

#### PSPICE

**41.** Setting $\mathbf{V}_i = 1\text{ V}\angle 0°$, write the input file to determine the frequency response of the magnitude of $\mathbf{V}_o$ for the network of Fig. 21.72.

**42.** Setting $\mathbf{V}_i = 1\text{ V}\angle 0°$, write the input file to obtain the frequency response of $\mathbf{V}_o$ for the magnitude of the network of Fig. 21.79.

**43.** Repeat Problem 42 for the frequency response of the phase angle associated with the output voltage.

**44.** Setting $\mathbf{V}_i = 1\text{ V}\angle 0°$, write the input file to obtain the frequency response of the magnitude of the output voltage $\mathbf{V}_o$ for the network of Fig. 21.74.

**45.** Setting $\mathbf{V}_i = 1\text{ V}\angle 0°$, write the input file to obtain the frequency response of the magnitude of the output voltage $\mathbf{V}_o$ for the network of Fig. 21.75.

#### BASIC

**46.** Write a program that will tabulate the gain of Eq. (21.13) versus frequency for a frequency range extending from $0.1f_1$ to $2f_1$ in increments of $0.1f_1$. Note whether $f = f_1$ when $V_o/V_i = 0.707$. Use $R = 1\text{ k}\Omega$ and $C = 500\text{ pF}$.

**47.** Repeat Problem 46 for the phase angle of Eq. (21.14). Note whether $\theta = 45°$ when $f = f_1$.

**48.** Write a program to tabulate $A_{v_{dB}}$ as determined by Eq. (21.33) and $A_{v_{dB}}$ as calculated by Eq. (21.34). For a frequency range extending from $0.01f_1$ to $f_1$ in increments of $0.01f_1$, compare the magnitudes, and note whether the values are closer when $f \ll f_1$ and whether $A_{v_{dB}} = -3\text{ dB}$ at $f = f_1$ for Eq. (21.33) and zero for Eq. (21.34).

## GLOSSARY

**Active filter** A filter that employs active devices such as transistors or operational amplifiers in combination with $R$, $L$, and $C$ elements.

**Band-stop filter** A network designed to reject (block) signals within a particular frequency range.

**Bode plot** A plot (envelope) of the frequency response of a system using straight-line segments called asymptotes.

**Decibel** A unit of measurement used to compare power levels.

**Double-tuned filter** A network having both a pass-band and a band-stop region.

**Filter** Networks designed to either pass or reject the transfer of signals at certain frequencies to a load.

**High-pass filter** A filter designed to pass high frequencies and reject low frequencies.

**Log-log paper** Graph paper with vertical and horizontal log scales.

**Low-pass filter** A filter designed to pass low frequencies and reject high frequencies.

**Micro-bar** A unit of measurement for sound pressure levels that permits comparing audio levels on a dB scale.

**Pass-band (band-pass) filter** A network designed to pass signals within a particular frequency range.

**Passive filter** A filter constructed of series, parallel, or series-parallel $R$, $L$, and $C$ elements.

**Semilog paper** Graph paper with one log scale and one linear scale.

# 22

# Pulse Waveforms and the *R-C* Response

## 22.1 INTRODUCTION

Our analysis thus far has been limited to alternating waveforms that vary in a sinusoidal manner. This chapter will introduce the basic terminology associated with the pulse waveform and examine the response of an *R-C* circuit to a square-wave input. The importance of the pulse waveform to the electrical/electronics industry cannot be overstated. A vast array of instrumentation, communication systems, computers, radar systems, and so on, all employ pulse signals to control operation, transmit data, and display information in a variety of formats.

The response of the networks described thus far to a pulse signal is quite different from that obtained for sinusoidal signals. In fact, we will be returning to the dc chapter on capacitors to retrieve a few fundamental concepts and equations that will help us in the analysis to follow. The content of this chapter is quite introductory in nature, designed simply to provide the fundamentals that will be helpful when the pulse waveform is encountered in specific areas of application.

## 22.2 IDEAL VERSUS ACTUAL

The *ideal* pulse of Fig. 22.1 has vertical sides, sharp corners, a flat peak characteristic, and starts instantaneously at $t_1$ and ends just as abruptly at $t_2$.

The waveform of Fig. 22.1 will be applied in the analysis to follow in this chapter and will probably appear in the initial investigation of areas of application beyond the scope of this text. Once the fundamental operation of a device, package, or system is clearly understood using ideal characteristics, the effect of a *true, actual,* or *practical* pulse must be considered. If an attempt were made to introduce all the differences

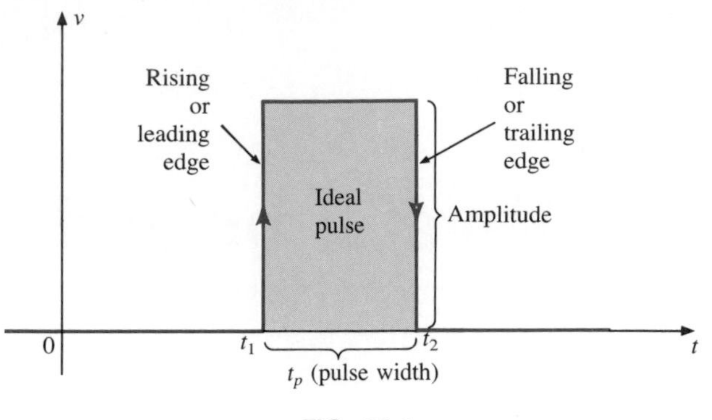


**FIG. 22.1**
*Ideal pulse waveform.*

between an ideal and actual pulse in a single figure, the result would probably be complex and confusing. A number of waveforms will therefore be used to define the critical parameters.

The reactive elements of a network, in their effort to prevent instantaneous changes in voltage (capacitor) and current (inductor), establish a slope to both edges of the pulse waveform as shown in Fig. 22.2. The *rising* edge of the waveform of Fig. 22.2 is defined as the edge that increases from a lower to a higher level. The *falling* edge is defined by the region or edge where the waveform decreases from a higher to a lower level. Since the rising edge is the first to be encountered, it is also called the *leading* edge. The falling edge always follows the leading edge and is therefore often called the *trailing* edge. Both regions are defined in Figs. 22.1 and 22.2.

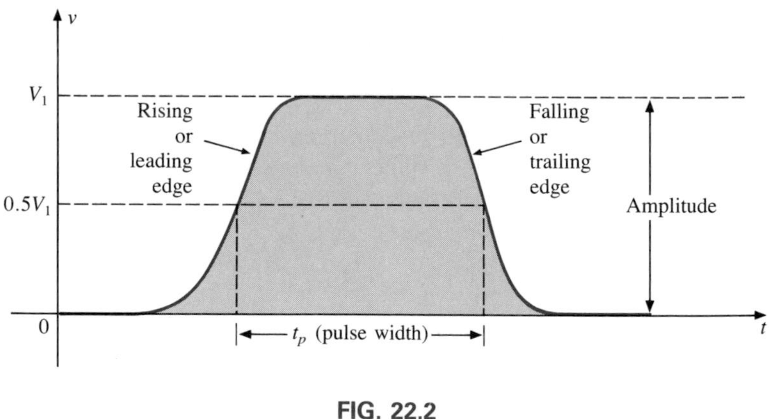


**FIG. 22.2**
*Actual pulse waveform.*

## Amplitude

For most applications, the amplitude of a pulse waveform is defined as the peak-to-peak value. Of course, if the waveforms all start and return to the zero-volt level, then the peak and peak-to-peak values are synonymous. For the purposes of this text, the amplitude of a pulse waveform is the peak-to-peak value, as illustrated in Figs. 22.1 and 22.2.

## Pulse Width

The *pulse width* ($t_p$), or *pulse duration,* is defined by a pulse level equal to 50% of the peak value. For the ideal pulse of Fig. 22.1, the pulse width is the same at any level, whereas $t_p$ for the waveform of Fig. 22.2 is a very specific value.

## Base-Line Voltage

The *base-line voltage* ($V_b$) is the voltage level from which the pulse is initiated. The waveforms of Figs. 22.1 and 22.2 both have a 0-V base-line voltage. In Fig. 22.3(a) the base-line voltage is 1 V, while in Fig. 22.3(b) the base-line voltage is −4 V.

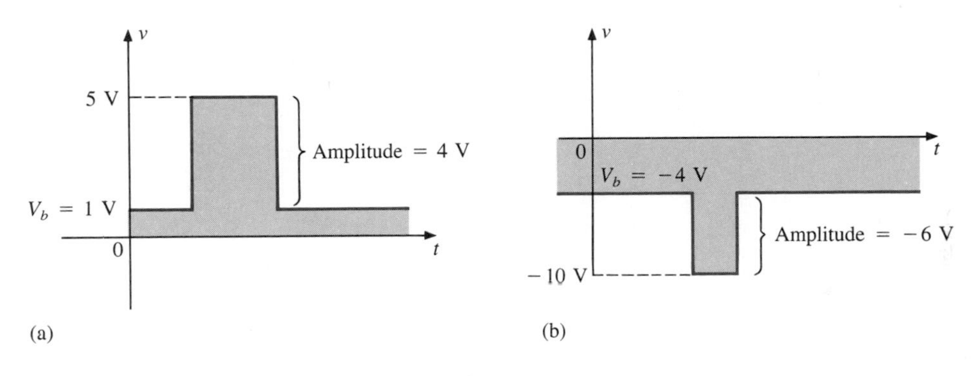


**FIG. 22.3**
*Defining the base-line voltage.*

## Positive-Going and Negative-Going Pulses

A *positive-going* pulse increases positively from the base-line voltage. Quite obviously, therefore, a *negative-going* pulse increases in the negative direction from the base-line voltage. The waveform of Fig. 22.3(a) is a positive-going pulse, whereas the waveform of Fig. 22.3(b) is a negative-going pulse.

Even though the base-line voltage of Fig. 22.4 is negative, the waveform is positive-going (with an amplitude of 10 V) since the voltage increased in the positive direction from the base-line voltage.

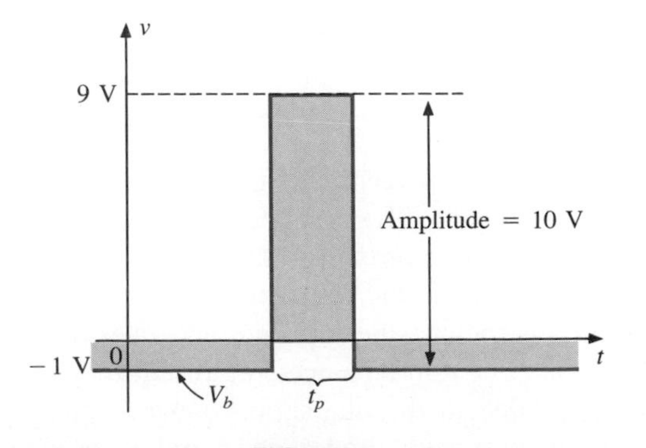


**FIG. 22.4**
*Positive-going pulse.*

## Rise Time ($t_r$) and Fall Time ($t_f$)

Of particular importance is the time required for the pulse to shift from one level to another. The *rounding* (defined in Fig. 22.5) that occurs at

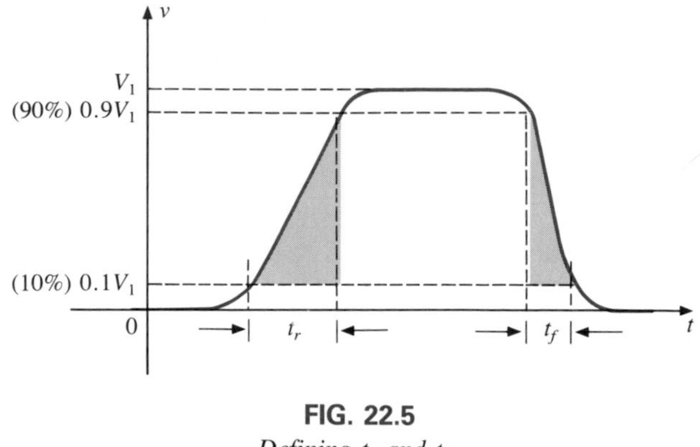


**FIG. 22.5**
*Defining $t_r$ and $t_f$.*

the beginning and end of each transition makes it difficult to define the exact point at which the rise time should be initiated and terminated. For this reason, the *rise* and *fall* times are defined by the 10% and 90% levels as indicated in Fig. 22.5. Note that there is no requirement that $t_r$ equal $t_f$.

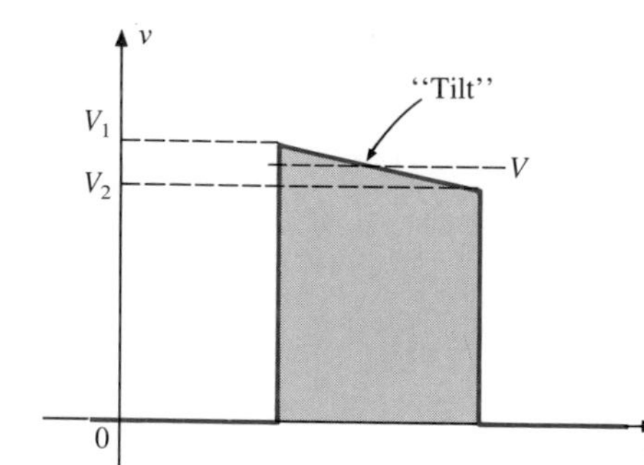


**FIG. 22.6**
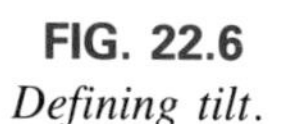
*Defining tilt.*

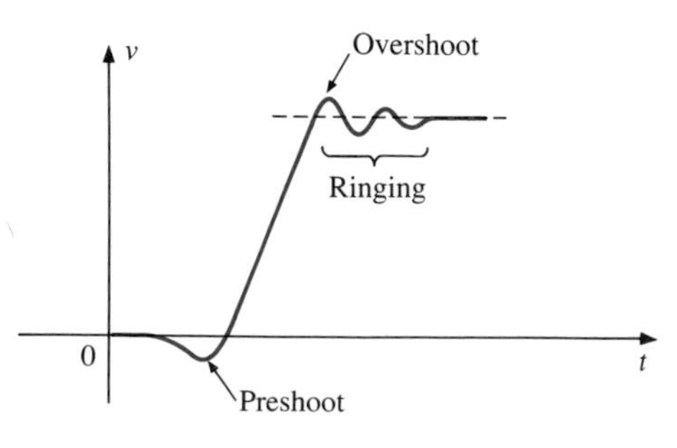


**FIG. 22.7**
*Defining preshoot, overshoot, and ringing.*

## Tilt

An undesirable but common distortion normally occurring due to a poor low-frequency response characteristic of the system through which a pulse has passed appears in Fig. 22.6. The drop in peak value is called *tilt, droop,* or *sag*. The percent tilt is defined by

$$\% \text{ tilt} = \frac{V_1 - V_2}{V} \times 100\% \tag{22.1}$$

where $V$ is the average value of the peak amplitude as determined by

$$V = \frac{V_1 + V_2}{2} \tag{22.2}$$

Naturally, the less the percent tilt or sag, the more ideal the pulse. Due to rounding, it may be difficult to define the values of $V_1$ and $V_2$. It is then necessary only to approximate the sloping region by a straight-line approximation and use the resulting values of $V_1$ and $V_2$.

Other distortions include the *preshoot* and *overshoot* appearing in Fig. 22.7, normally due to pronounced high-frequency effects of a system, and *ringing,* due to the interaction between the capacitive and inductive elements of a network at their natural or resonant frequency.

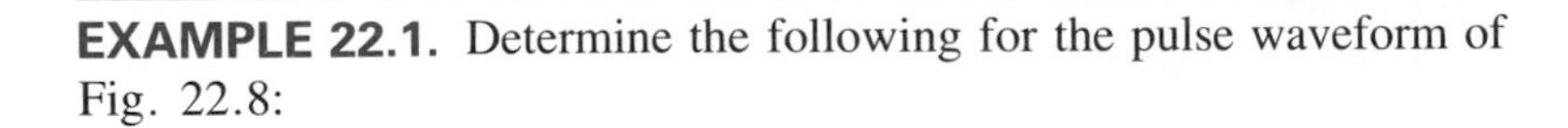

**EXAMPLE 22.1.** Determine the following for the pulse waveform of Fig. 22.8:

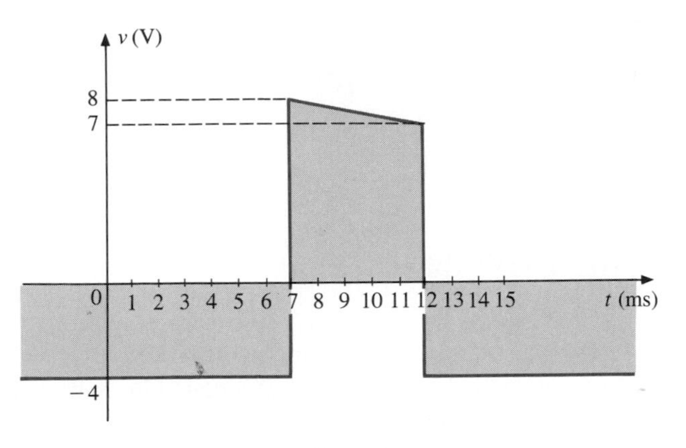


FIG. 22.8

a. positive- or negative-going?
b. base-line voltage
c. pulse width
d. maximum amplitude
e. tilt

***Solutions:***

a. **positive-going**
b. $V_b = \mathbf{-4\ V}$
c. $t_p = (12 - 7)\ \text{ms} = \mathbf{5\ ms}$
d. $V_{\text{max}} = 8\ \text{V} + 4\ \text{V} = \mathbf{12\ V}$
e. $V = \dfrac{V_1 + V_2}{2} = \dfrac{12\ \text{V} + 11\ \text{V}}{2} = \dfrac{23\ \text{V}}{2} = 11.5\ \text{V}$

$$\%\ \text{tilt} = \frac{V_1 - V_2}{V} \times 100\% = \frac{12\ \text{V} - 11\ \text{V}}{11.5\ \text{V}} \times 100\% = \mathbf{8.696\%}$$

(Remember, $V$ is defined by the average value of the peak amplitude.)

**EXAMPLE 22.2.** Determine the following for the pulse waveform of Fig. 22.9:

a. positive- or negative-going?
b. base-line voltage
c. tilt
d. amplitude
e. $t_p$
f. $t_r$ and $t_f$

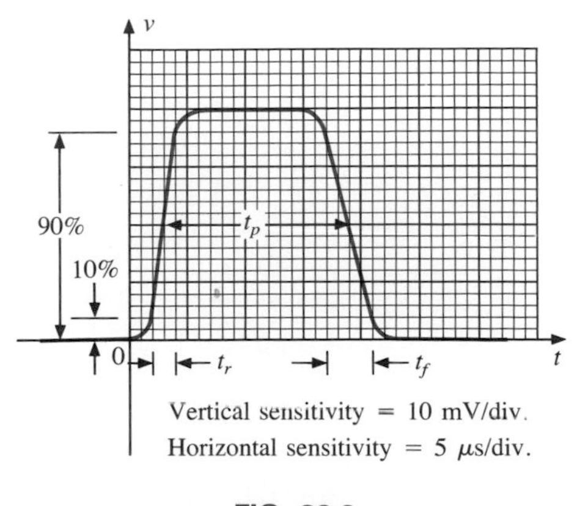


FIG. 22.9

***Solutions:***

a. **positive-going**
b. $V_b = \mathbf{0\ V}$
c. % tilt = **0%**
d. amplitude = (4 div.)(10 mV/div.) = **40 mV**
e. $t_p = (3.2\text{ div.})(5\ \mu\text{s/div.}) = \mathbf{16\ \mu s}$
f. $t_r = (0.4\text{ div.})(5\ \mu\text{s/div.}) = \mathbf{2\ \mu s}$
   $t_f = (0.8\text{ div.})(5\ \mu\text{s/div.}) = \mathbf{4\ \mu s}$

## 22.3 PULSE REPETITION RATE AND DUTY CYCLE

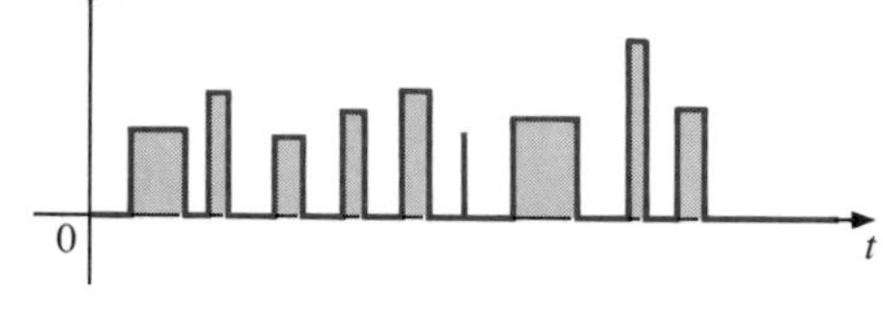

**FIG. 22.10**
*Pulse train.*

A series of pulses such as those appearing in Fig. 22.10 is called a *pulse train*. The varying widths and heights may contain information that can be decoded at the receiving end.

If the pattern repeats itself in a periodic manner as shown in Figs. 22.11(a) and (b), the result is called a *periodic pulse train*.

The *period* ($T$) of the pulse train is defined as the time differential between any two similar points on the pulse train as shown in Figs. 22.11(a) and (b).

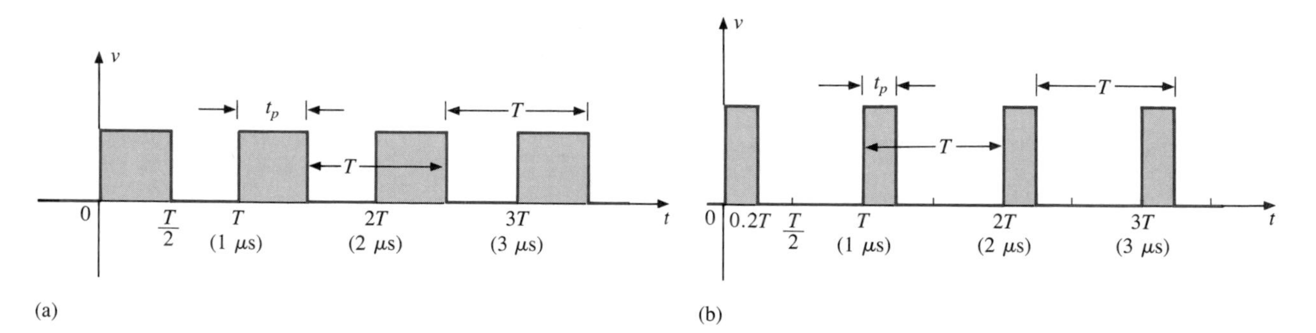

**FIG. 22.11**
*Periodic pulse trains.*

The *pulse repetition frequency* (prf), or *pulse repetition rate* (prr), is defined by

$$\text{prf (or prr)} = \frac{1}{T} \quad \text{(Hz or pulses/s)} \qquad \textbf{(22.3)}$$

Applying Eq. (22.3) to each waveform of Fig. 22.11 will result in the same pulse repetition frequency since the periods are the same. The result clearly reveals that the shape of the periodic pulse has nothing to do with the pulse repetition frequency. The pulse repetition frequency is determined solely by the period of the repeating pulse. The factor that will reveal how much of the period is encompassed by the pulse is called the *duty cycle,* defined as follows:

$$\text{Duty cycle} = \frac{\text{pulse width}}{\text{period}} \times 100\%$$

or

$$\text{Duty cycle} = \frac{t_p}{T} \times 100\% \qquad \textbf{(22.4)}$$

For Fig. 22.11(a) (a square-wave pattern),

$$\text{Duty cycle} = \frac{0.5T}{T} \times 100\% = \mathbf{50\%}$$

and for Fig. 22.11(b),

$$\text{Duty cycle} = \frac{0.2T}{T} \times 100\% = \mathbf{20\%}$$

The above results clearly reveal that the duty cycle provides a percent indication of the portion of the total period encompassed by the pulse waveform.

---

**EXAMPLE 22.3.** Determine the pulse repetition frequency and the duty cycle for the periodic pulse waveform of Fig. 22.12.

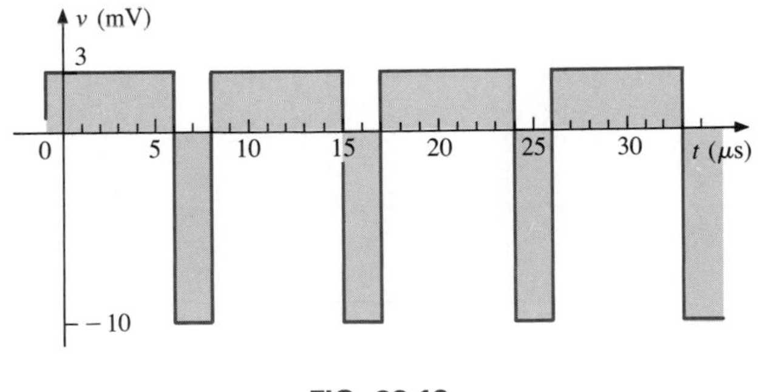

FIG. 22.12

***Solution:***

$$T = (15 - 6)\ \mu\text{s} = 9\ \mu\text{s}$$

$$\text{prf} = \frac{1}{T} = \frac{1}{9\ \mu\text{s}} \cong \mathbf{0.111\ MHz}$$

$$\text{Duty cycle} = \frac{t_p}{T} \times 100\% = \frac{(8 - 6)\ \mu\text{s}}{9\ \mu\text{s}} \times 100\%$$

$$= \frac{2}{9} \times 100\% \cong \mathbf{22.22\%}$$

---

**EXAMPLE 22.4.** Determine the pulse repetition frequency and the duty cycle for the periodic waveform of Fig. 22.13.

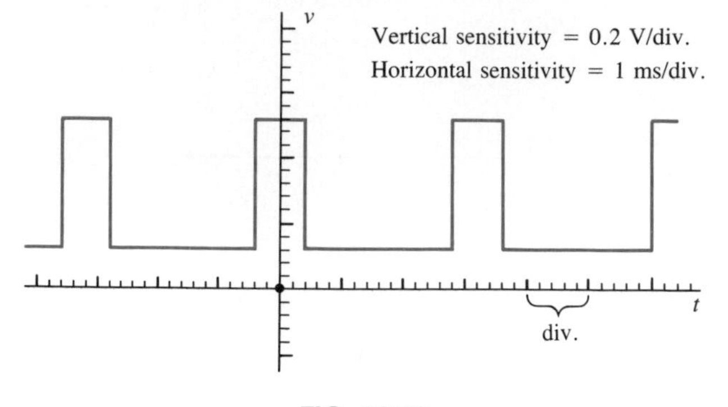

FIG. 22.13

***Solution:***

$$T = (3.2 \text{ div.})(1 \text{ ms/div.}) = 3.2 \text{ ms}$$

$$t_p = (0.8 \text{ div.})(1 \text{ ms/div.}) = 0.8 \text{ ms}$$

$$\text{prf} = \frac{1}{T} = \frac{1}{3.2 \text{ ms}} = \mathbf{312.5\ Hz}$$

$$\text{Duty cycle} = \frac{t_p}{T} \times 100\% = \frac{0.8 \text{ ms}}{3.2 \text{ ms}} \times 100\% = \mathbf{25\%}$$

**EXAMPLE 22.5.** Determine the pulse repetition rate and duty cycle for the trigger waveform of Fig. 22.14.

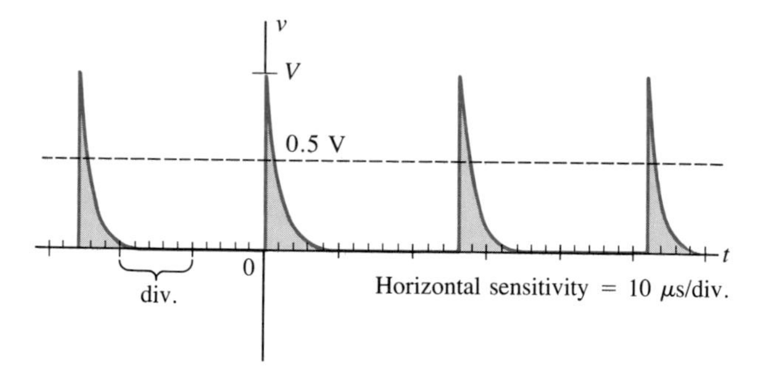

FIG. 22.14

***Solution:***

$$T = (2.6 \text{ div.})(10\ \mu\text{s/div.}) = 26\ \mu\text{s}$$

$$\text{prf} = \frac{1}{T} = \frac{1}{26\ \mu\text{s}} = \mathbf{38{,}462\ kHz}$$

$$t_p \cong (0.2 \text{ div.})(10\ \mu\text{s/div.}) = 2\ \mu\text{s}$$

$$\text{Duty cycle} = \frac{t_p}{T} \times 100\% = \frac{2\ \mu\text{s}}{26\ \mu\text{s}} \times 100\% = \mathbf{7.69\%}$$

## 22.4 AVERAGE VALUE

The average value of a pulse waveform can be determined using one of two methods. The first is the procedure outlined in Section 13.8 which can be applied to any alternating waveform. The second can be applied only to pulse waveforms since it utilizes terms specifically related to pulse waveforms. That is,

$$V_{av} = (\text{duty cycle})(\text{peak value}) + (1 - \text{duty cycle})(V_b) \tag{22.5}$$

In Eq. (22.5), the peak value is the maximum deviation from the reference or zero-volt level and the duty cycle is in decimal form. Eq. (22.5) does not include the effect of any tilt pulse waveforms with sloping sides.

**EXAMPLE 22.6.** Determine the average value for the periodic pulse waveform of Fig. 22.15.

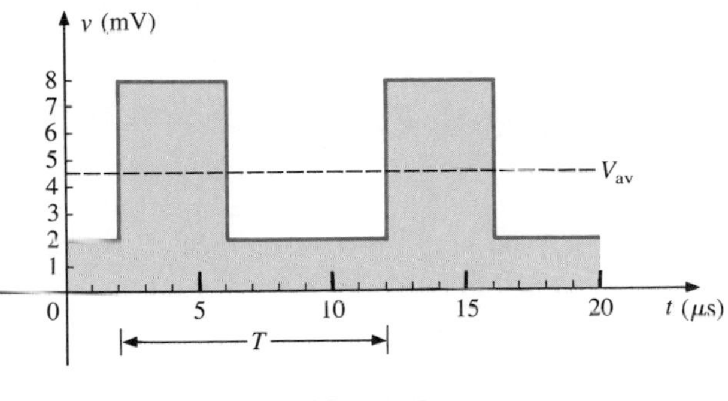

FIG. 22.15

***Solution:*** By the method of Section 13.8,

$$G = \frac{\text{area under curve}}{T}$$

$$T = (12 - 2)\ \mu\text{s} = 10\ \mu\text{s}$$

$$G = \frac{(8\text{ mV})(4\ \mu\text{s}) + (2\text{ mV})(6\ \mu\text{s})}{10\ \mu\text{s}} = \frac{32 \times 10^{-9} + 16 \times 10^{-9}}{10 \times 10^{-6}}$$

$$= \frac{44 \times 10^{-9}}{10 \times 10^{-6}} = \mathbf{4.4\ mV}$$

By Eq. (22.5),

$$V_b = +2\text{ mV}$$

$$\text{Duty cycle} = \frac{t_p}{T} = \frac{(6 - 2)\ \mu\text{s}}{10\ \mu\text{s}} = \frac{4}{10} = 0.4 \text{ (decimal form)}$$

Peak value (from 0-V reference) = 8 mV

$$
\begin{aligned}
V_{av} &= (\text{duty cycle})(\text{peak value}) + (1 - \text{duty cycle})(V_b) \\
&= (0.4)(8\ \text{mV}) + (1 - 0.4)(2\ \text{mV}) \\
&= 3.2\ \text{mV} + 1.2\ \text{mV} = \mathbf{4.4\ mV}
\end{aligned}
$$

as obtained above.

---

**EXAMPLE 22.7.** Given a periodic pulse waveform with a duty cycle of 28%, a peak value of 7 V, and a base-line voltage of −3 V:

a. Determine the average value.
b. Sketch the waveform.
c. Verify the result of part (a) using the method of Section 13.8.

***Solutions:***

a. By Eq. (22.5),

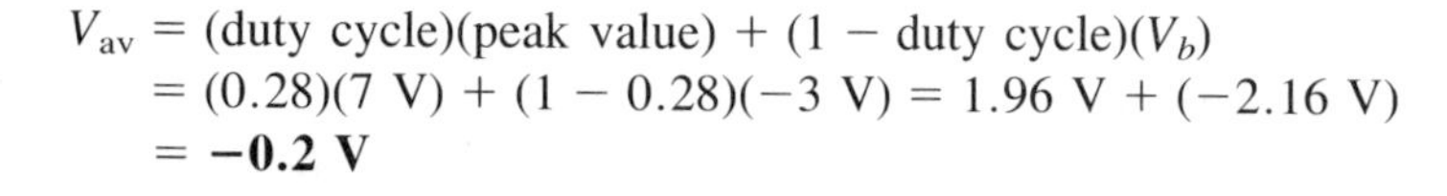

$$
\begin{aligned}
V_{av} &= (\text{duty cycle})(\text{peak value}) + (1 - \text{duty cycle})(V_b) \\
&= (0.28)(7\ \text{V}) + (1 - 0.28)(-3\ \text{V}) = 1.96\ \text{V} + (-2.16\ \text{V}) \\
&= \mathbf{-0.2\ V}
\end{aligned}
$$

b. See Fig. 22.16.

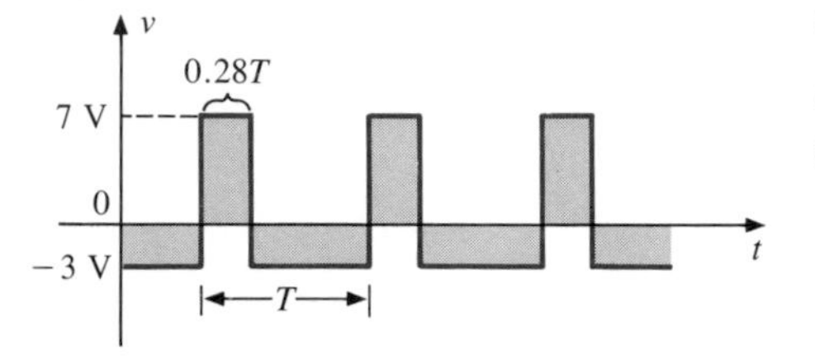

FIG. 22.16

c. $G = \dfrac{(7\ \text{V})(0.28T) - (3\ \text{V})(0.72T)}{T} = 1.96\ \text{V} - 2.16\ \text{V}$

$= \mathbf{-0.2\ V}$

as obtained above.

---

The average value (dc value) of any waveform can be easily determined using the oscilloscope. If the mode switch of the scope is set in the ac position, the average or dc component of the applied waveform will be blocked by an internal capacitor from reaching the screen. The pattern can be adjusted to establish the display of Fig. 22.17(a). If the mode switch is then placed in the dc position, the vertical shift (positive or negative) will reveal the average or dc level of the input signal, as shown in Fig. 22.17(b).

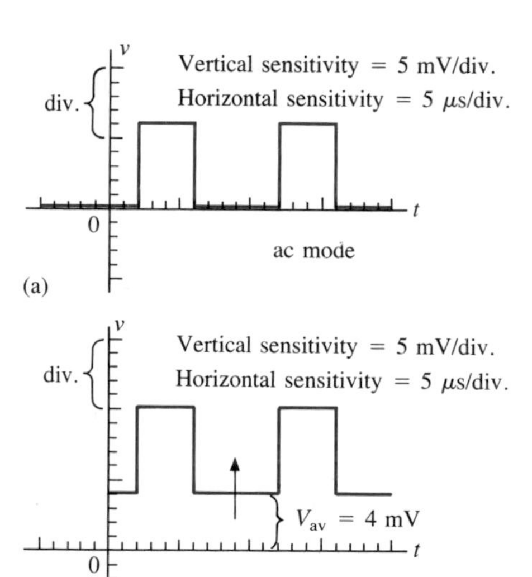


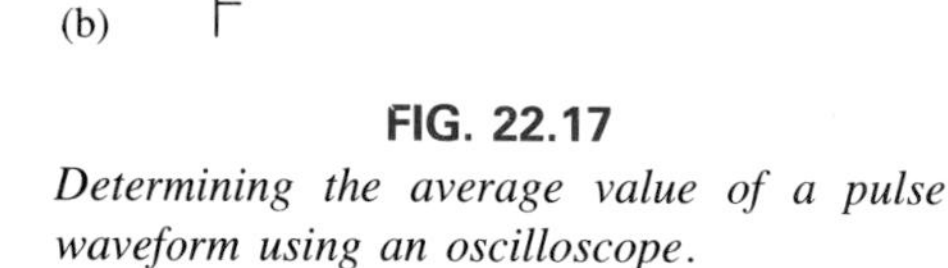

FIG. 22.17
*Determining the average value of a pulse waveform using an oscilloscope.*

## 22.5 *R-C* CIRCUITS WITH INITIAL VALUES

In the description of transients in an *R-C* circuit (Chapter 10), the capacitors were assumed to be initially uncharged and the final voltage was the supply voltage. The initial voltage $V_i$ was defined for those cases where the transient voltage did not reach the final value and the decay phase began with an initial voltage less than the supply voltage. In this section, a more general solution for the transient behavior of the *R-C* circuit will be provided that includes the effect of an initial voltage on the capacitor and a final voltage that may not be the supply voltage.

In Fig. 22.18, a capacitor has an initial voltage $V_i$ and will charge to a final voltage $V_f$. Using the conclusions of Chapter 10, we know that the equation for the transient waveform beginning at $t = 0$ s is $(V_f - V_i)(1 - e^{-t/RC})$. However, since $V_i$ is the same throughout the transient phase, it can simply be added to the above mathematical expression to generate an equation for the voltage $v_C$ of Fig. 22.18. That is,

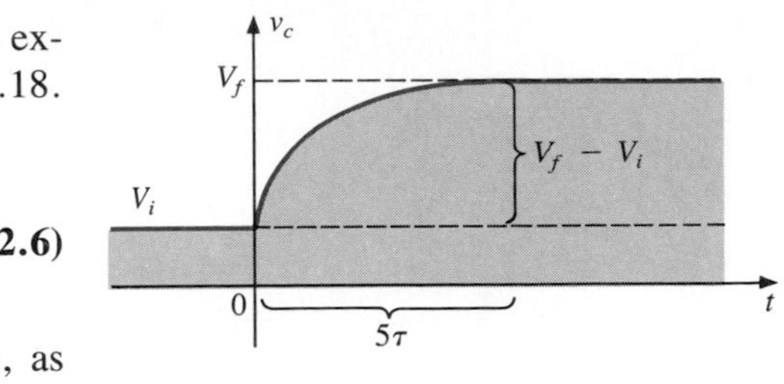


FIG. 22.18

$$\boxed{v_C = V_i + (V_f - V_i)(1 - e^{-t/RC})} \qquad \textbf{(22.6)}$$

For the case where $V_i$ equals zero volts and $V_f$ equals $E$ volts, as described in detail in Chapter 10,

$$v_C = 0 + (E - 0)(1 - e^{-t/RC}) = E(1 - e^{-t/RC})$$

For the case of Fig. 22.19, $V_i = -2$ V, $V_f = +5$ V, and

$$\begin{aligned} v_C &= V_i + (V_f - V_i)(1 - e^{-t/RC}) \\ &= -2\text{ V} + [5\text{ V} - (-2\text{ V})](1 - e^{-t/RC}) \\ v_C &= -2\text{ V} + 7\text{ V}(1 - e^{-t/RC}) \end{aligned}$$

For the case where $t = \tau = RC$,

$$\begin{aligned} v_C &= -2\text{ V} + 7\text{ V}(1 - e^{-t/\tau}) = -2\text{ V} + 7\text{ V}(1 - e^{-1}) \\ &= -2\text{ V} + 7\text{ V}(1 - 0.368) = -2\text{ V} + 7\text{ V}(0.632) \\ v_C &= 2.424\text{ V} \end{aligned}$$

as verified by Fig. 22.19.

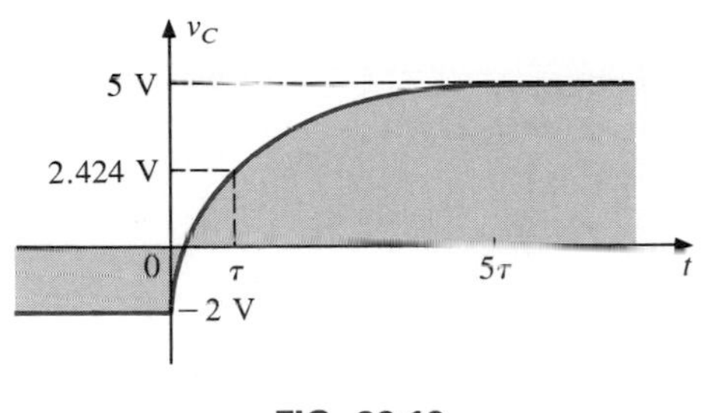

FIG. 22.19

---

**EXAMPLE 22.8.** The capacitor of Fig. 22.20 is initially charged to 2 V before the switch is closed. The switch is then closed.

a. Determine the mathematical expression for $v_C$.
b. Determine the mathematical expression for $i_C$.
c. Sketch the waveforms of $v_C$ and $i_C$.

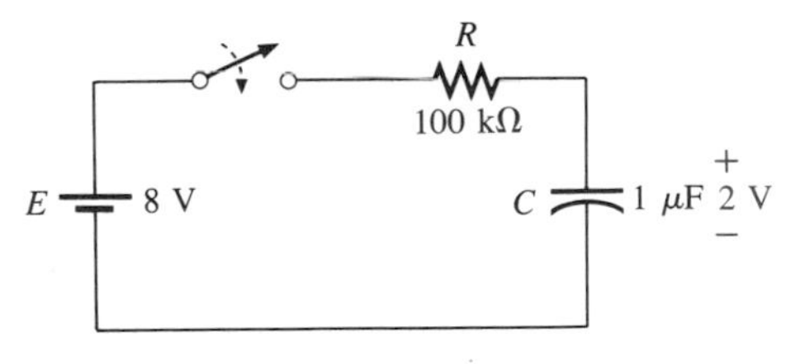


FIG. 22.20

***Solutions:***

a. $V_i = 2$ V
$V_f$ (after $5\tau$) $= E = 8$ V
$\tau = RC = (100\text{ k}\Omega)(1\ \mu\text{F}) = 100$ ms

By Eq. (22.6),

$$\begin{aligned} v_C &= V_i + (V_f - V_i)(1 - e^{-t/\tau}) \\ &= 2\text{ V} + (8\text{ V} - 2\text{ V})(1 - e^{-t/\tau}) \end{aligned}$$

and

$$\mathbf{v_C = 2\ V + 6\ V(1 - e^{-t/\tau})}$$

b. When the switch is first closed, the voltage across the capacitor cannot change instantaneously, and $V_R = E - V_i = 8\text{ V} - 2\text{ V} = 6\text{ V}$. The current therefore jumps to a level determined by Ohm's law:

$$I_{R_{\max}} = \frac{V_R}{R} = \frac{6\text{ V}}{100\text{ k}\Omega} = 0.06\text{ mA}$$

The current will then decay to zero amperes with the same time constant calculated in part (a), and

$$i_C = \mathbf{0.06\ mA}e^{-t/\tau}$$

c. See Fig. 22.21.

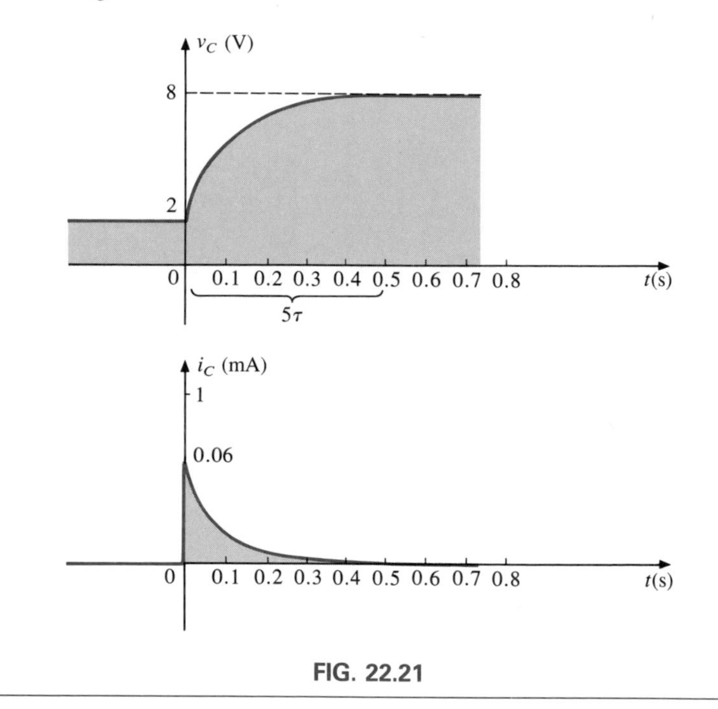


**FIG. 22.21**

**EXAMPLE 22.9.** Sketch $v_C$ for the step input shown in Fig. 22.22. Assume the $-4$ mV has been present for a period of time in excess of

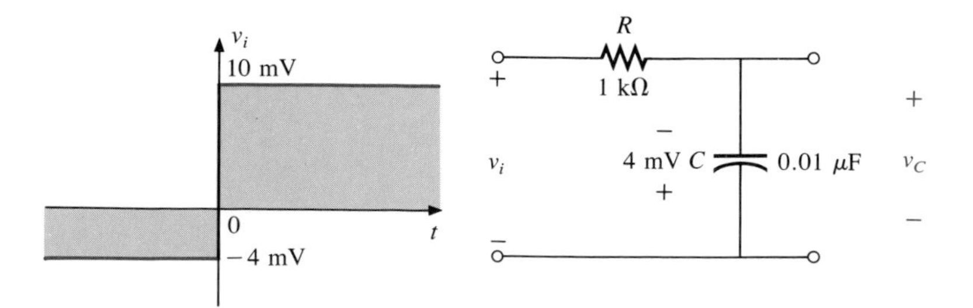

**FIG. 22.22**

five time constants of the network. Then determine when $v_C = 0$ V if the step changes levels at $t = 0$ s.

***Solution:***

$$V_i = -4 \text{ mV} \qquad V_f = 10 \text{ mV}$$
$$\tau = RC = (1 \text{ k}\Omega)(0.01 \ \mu\text{F}) = 10 \ \mu\text{s}$$

By Eq. (22.6),

$$\begin{aligned} v_C &= V_i + (V_f - V_i)(1 - e^{-t/\tau}) \\ &= -4 \text{ mV} + [10 \text{ mV} - (-4 \text{ mV})](1 - e^{-t/\tau}) \end{aligned}$$

and

$$v_C = \mathbf{-4\ mV + 14\ mV(1 - e^{-t/(10\ \mu s)})}$$

The waveform appears in Fig. 22.23.

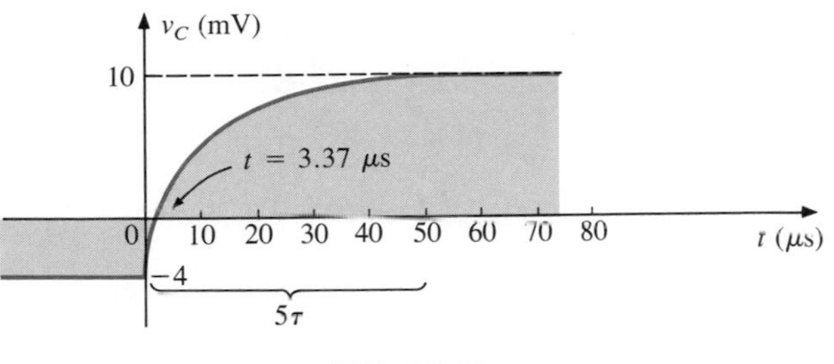

**FIG. 22.23**

Substituting $v_C = 0$ V into the above equation yields

$$v_C = 0 = -4 \text{ mV} + 14 \text{ mV}(1 - e^{-t/\tau})$$

and

$$\frac{4 \text{ mV}}{14 \text{ mV}} = 1 - e^{-t/\tau}$$

or

$$0.286 - 1 = -e^{-t/\tau}$$

and

$$0.714 = e^{-t/\tau}$$

but

$$\log_e 0.714 = \log_e(e^{-t/\tau}) = \frac{-t}{\tau}$$

and

$$t = -\tau \log_e 0.714 = -(10 \ \mu\text{s})(-0.337) = \mathbf{3.37\ \mu s}$$

as indicated in Fig. 22.23.

## 22.6 *R-C* RESPONSE TO SQUARE-WAVE INPUTS

The square wave of Fig. 22.24 is a particular form of pulse waveform. It has a duty cycle of 50% and an average value of zero volts, as calculated below:

$$\text{Duty cycle} = \frac{t_p}{T} \times 100\% = \frac{T/2}{T} \times 100\% = \mathbf{50\%}$$

$$V_{\text{av}} = \frac{(V_1)(T/2) + (-V_1)(T/2)}{T} = \frac{0}{T} = \mathbf{0\ V}$$

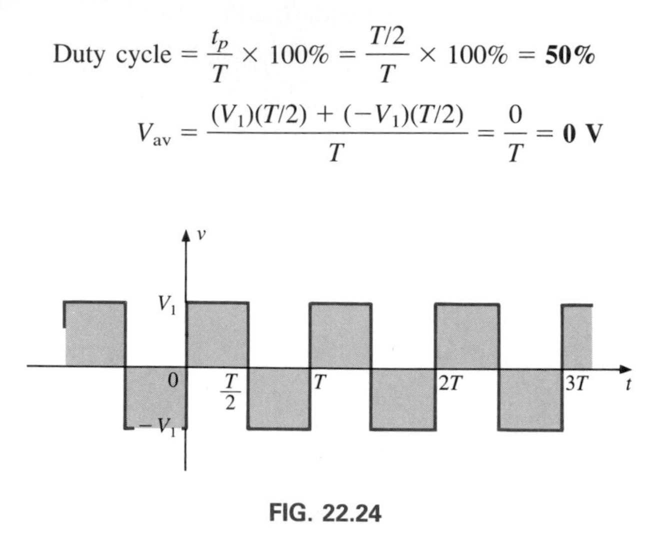


FIG. 22.24

The application of a dc voltage $V_1$ in series with the square wave of Fig. 22.24 can raise the base-line voltage from $-V_1$ to zero volts, and the average value to $V_1$ volts.

If a square wave such as developed in Fig. 22.25 is applied to an *R-C* circuit as shown in Fig. 22.26, the period of the square wave can have a pronounced effect on the resulting waveform for $v_C$.

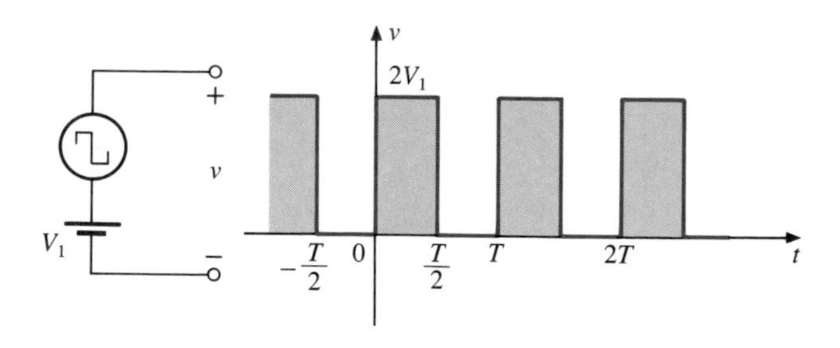

FIG. 22.25

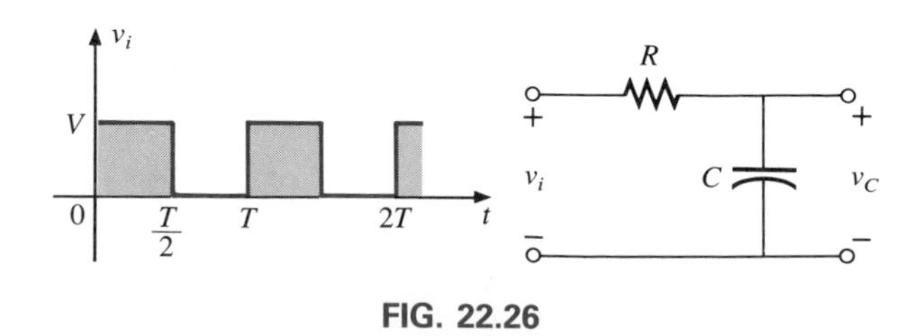


FIG. 22.26

*For the analysis to follow, we will assume that steady-state conditions will be established after a period of five time constants has passed.* The types of waveforms developed across the capacitor can then be separated into three fundamental types.

## $T/2 > 5\tau$

The condition $T/2 > 5\tau$, or $T > 10\tau$, establishes a situation where the capacitor can charge to its steady-state value in advance of $t = T/2$. The resulting waveforms for $v_C$ and $i_C$ will appear as shown in Fig. 22.27.

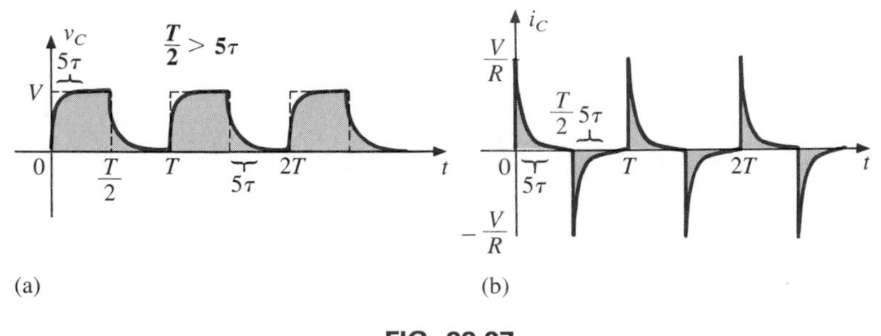


**FIG. 22.27**
*$v_C$ and $i_C$ for $T/2 > 5\tau$.*

Note how closely the voltage $v_C$ shadows the applied waveform and how $i_C$ is nothing more than a series of very sharp spikes. Note also that the change of $V_i$ from $V$ to zero volts during the trailing edge simply results in a rapid discharge of $v_C$ to zero volts. In essence, when $V_i = 0$ the capacitor and resistor are in parallel and the capacitor simply discharges through $R$ with a time constant equal to that encountered during the charging phase but with a direction of charge flow (current) opposite to that established during the charging phase.

## $T/2 = 5\tau$

If the frequency of the square wave is chosen such that $T/2 = 5\tau$ or $T = 10\tau$, the voltage $v_C$ will reach its final value just before beginning its discharge phase, as shown in Fig. 22.28. The voltage $v_C$ no longer resembles the square-wave input and, in fact, has some of the character-

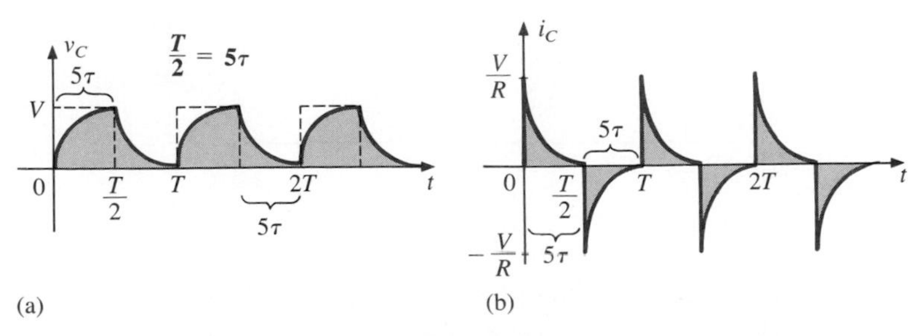


**FIG. 22.28**
*$v_C$ and $i_C$ for $T/2 = 5\tau$.*

istics of a triangular waveform. The increased time constant has resulted in a more rounded $v_C$, and $i_C$ has increased substantially in width to reveal the longer charging period.

### $T/2 < 5\tau$

If $T/2 < 5\tau$ or $T < 10\tau$, the voltage $v_C$ will not reach its final value during the first pulse (Fig. 22.29), and the discharge cycle will not

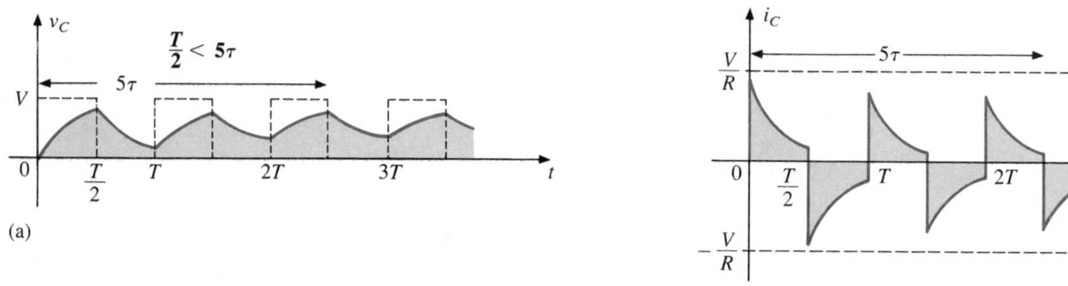


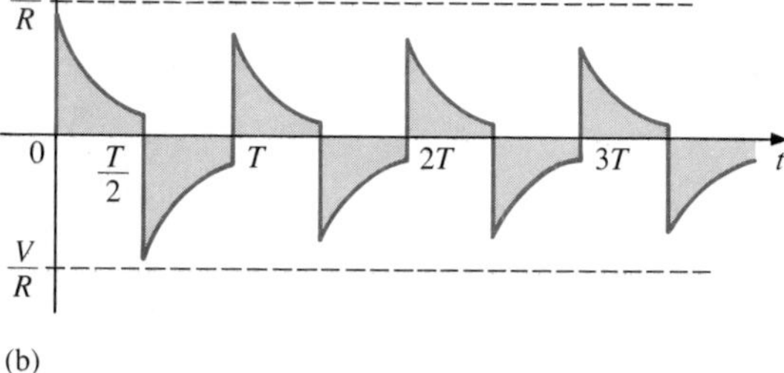


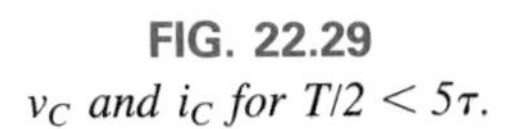

**FIG. 22.29**
*$v_C$ and $i_C$ for $T/2 < 5\tau$.*

return to zero volts. In fact, the initial value for each succeeding pulse will change until steady-state conditions are reached. In most instances, it is a good approximation to assume that steady-state conditions have been established in 5 cycles of the applied waveform.

As the frequency increases and the period decreases, there will be a flattening of the response for $v_C$ until a pattern like that in Fig. 22.30 results. Figure 22.30 begins to reveal an important conclusion regarding the response curve for $v_C$: Under steady-state conditions, the average value of $v_C$ will equal the average value of the applied square wave. Note in Figs. 22.29 and 22.30 how the waveform for $v_C$ approaches an average value of $V/2$.

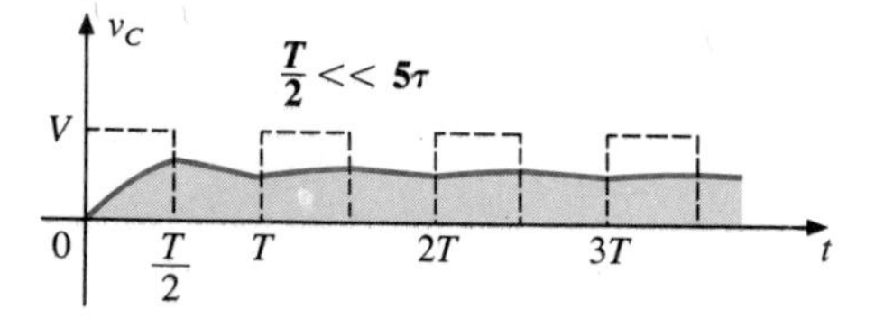


**FIG. 22.30**
*$v_C$ for $T/2 \ll 5\tau$ or $T \ll 10\tau$.*

---

**EXAMPLE 22.10.** The 1000-Hz square wave of Fig. 22.31 is applied to the *R-C* circuit of the same figure.

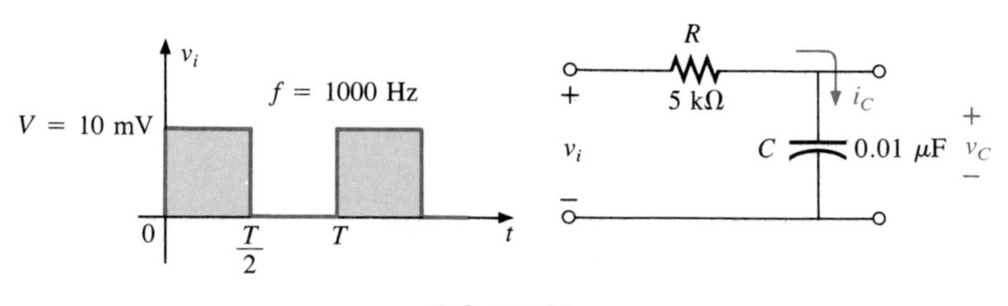

**FIG. 22.31**

a. Compare the pulse width of the square wave to the time constant of the circuit.
b. Sketch $v_C$.
c. Sketch $i_C$.

***Solutions:***

a. $T = \frac{1}{f} = \frac{1}{1000} = 1 \text{ ms}$

$t_p = \frac{T}{2} = 0.5 \text{ ms}$

$\tau = RC = (5 \times 10^3\ \Omega)(0.01 \times 10^{-6}\ \text{F}) = 0.05 \text{ ms}$

$\frac{t_p}{\tau} = \frac{0.5 \text{ ms}}{0.05 \text{ ms}} = 10$

and

$$\boldsymbol{t_p = 10\tau = \frac{T}{2}}$$

The result reveals that $v_C$ will charge to its final value in half the pulse width.

b. For the charging phase, $V_i = 0$ V and $V_f = 10$ mV, and

$$v_C = V_i + (V_f - V_i)(1 - e^{-t/\tau})$$
$$= 0 + (10 \text{ mV} - 0)(1 - e^{-t/\tau})$$

and

$$v_C = \mathbf{10\ mV(1 - e^{-t/\tau})}$$

For the discharge phase, $V_i = 10$ mV and $V_f = 0$ V, and

$$v_C = V_i + (V_f - V_i)(1 - e^{-t/\tau})$$
$$= 10 \text{ mV} + (0 - 10 \text{ mV})(1 - e^{-t/\tau})$$
$$v_C = 10 \text{ mV} - 10 \text{ mV} + 10 \text{ mV}(e^{-t/\tau})$$

and

$$v_C = \mathbf{10\ mV}\boldsymbol{e^{-t/\tau}}$$

The waveform for $v_C$ appears in Fig. 22.32.

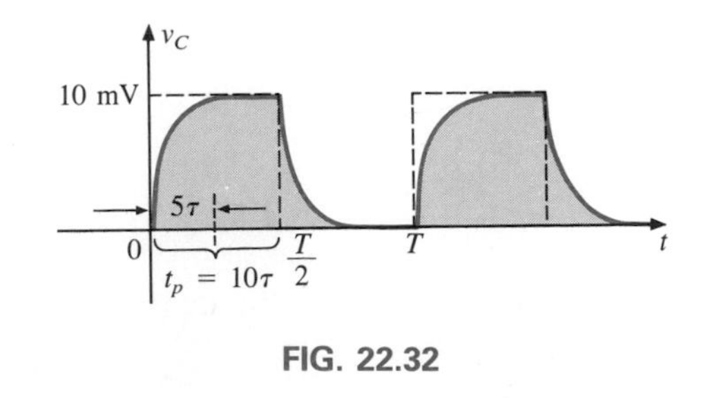


FIG. 22.32

c. For the charging phase at $t = 0$ s, $V_R = V$ and $I_{R_{max}} = V/R = 10 \text{ mV}/5 \text{ k}\Omega = 2\ \mu\text{A}$, and

$$i_C = I_{max}e^{-t/\tau} = \mathbf{2\ \mu A}\boldsymbol{e^{-t/\tau}}$$

For the discharge phase, the current will have the same mathematical formulation but the opposite direction, as shown in Fig. 22.33.

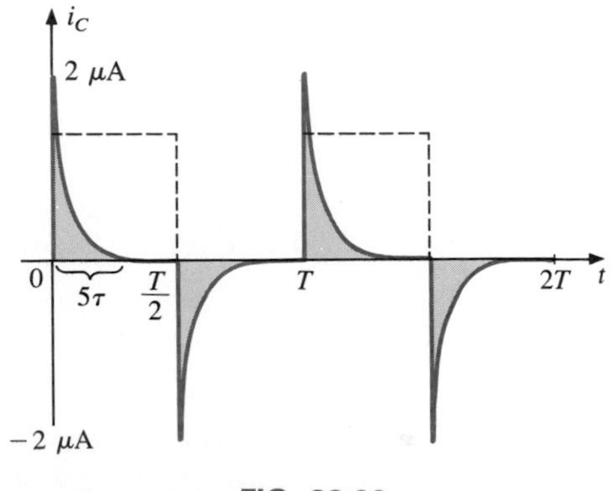


FIG. 22.33

**EXAMPLE 22.11.** Repeat Example 22.10 for $f = 10$ kHz.

***Solution:***

$$T = \frac{1}{f} = \frac{1}{10 \text{ kHz}} = 0.1 \text{ ms}$$

and

$$\frac{T}{2} = 0.05 \text{ ms}$$

with

$$\tau = t_p = \frac{T}{2} = 0.05 \text{ ms}$$

In other words, the pulse width is exactly equal to the time constant of the network. The voltage $v_C$ will not reach the final value before the first pulse of the square-wave input returns to zero volts.

For $t$ in the range $t = 0$ to $t = T/2$, $V_i = 0$ V and $V_f = 10$ mV, and

$$v_C = 10 \text{ mV}(1 - e^{-t/\tau})$$

At $t = \tau$, we recall from Chapter 10 that $v_C = 63.2\%$ of the final value. Substituting $t = \tau$ into the equation above yields

$$v_C = (10 \text{ mV})(1 - e^{-1}) = (10 \text{ mV})(1 - 0.368)$$
$$= (10 \text{ mV})(0.632) = 6.32 \text{ mV}$$

as shown in Fig. 22.34.

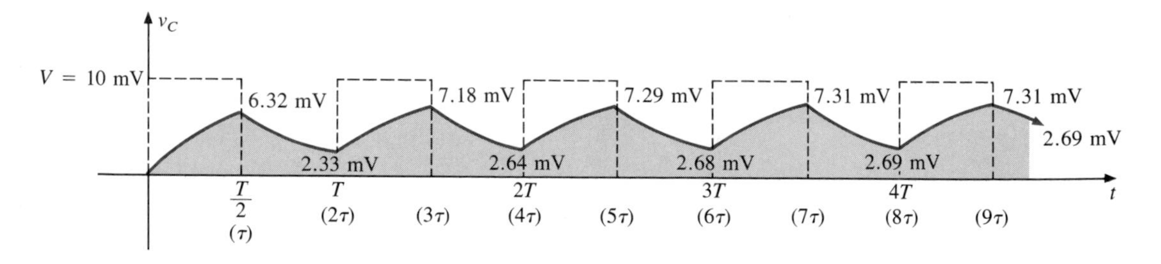


**FIG. 22.34**
*R-C response for $t_p = \tau = T/2$.*

For the discharge phase between $t = T/2$ and $T$, $V_i = 6.32$ mV and $V_f = 0$ V, and

$$v_C = V_i + (V_f - V_i)(1 - e^{-t/\tau})$$
$$= 6.32 \text{ mV} + (0 - 6.32 \text{ mV})(1 - e^{-t/\tau})$$
$$v_C = 6.32 \text{ mV}e^{-t/\tau}$$

with $t$ now being measured from $t = T/2$ in Fig. 22.34. In other words, for each interval of Fig. 22.34, the beginning of the transient waveform is defined as $t = 0$ s. The value of $v_C$ at $t = T$ is therefore determined by substituting $t = \tau$ into the above equation, and not $2\tau$ as defined by Fig. 22.34.

Substituting $t = \tau$,

$$\begin{aligned} v_C &= (6.32 \text{ mV})(e^{-1}) = (6.32 \text{ mV})(0.368) \\ &= 2.33 \text{ mV} \end{aligned}$$

as shown in Fig. 22.34.

For the next interval, $V_i = 2.33$ mV and $V_f = 10$ mV, and

$$\begin{aligned} v_C &= V_i + (V_f - V_i)(1 - e^{-t/\tau}) \\ &= 2.33 \text{ mV} + (10 \text{ mV} - 2.33 \text{ mV})(1 - e^{-t/\tau}) \\ v_C &= 2.33 \text{ mV} + 7.67 \text{ mV}(1 - e^{-t/\tau}) \end{aligned}$$

At $t = \tau$ (since $t = T = 2\tau$ is now $t = 0$ s for this interval),

$$\begin{aligned} v_C &= 2.33 \text{ mV} + 7.67 \text{ mV}(1 - e^{-1}) \\ &= 2.33 \text{ mV} + 4.85 \text{ mV} \\ v_C &= 7.18 \text{ mV} \end{aligned}$$

as shown in Fig. 22.34.

For the discharge interval, $V_i = 7.18$ mV and $V_f = 0$ V, and

$$\begin{aligned} v_C &= V_i + (V_f - V_i)(1 - e^{-t/\tau}) \\ &= 7.18 \text{ mV} + (0 - 7.18 \text{ mV})(1 - e^{-t/\tau}) \\ v_C &= 7.18 \text{ mV}e^{-t/\tau} \end{aligned}$$

At $t = \tau$ (measured from $3\tau$ of Fig. 22.34),

$$\begin{aligned} v_C &= (7.18 \text{ mV})(e^{-1}) = (7.18 \text{ mV})(0.368) \\ &= 2.64 \text{ mV} \end{aligned}$$

as shown in Fig. 22.34.

Continuing in the same manner, the remaining waveform for $v_C$ will be generated as depicted in Fig. 22.34. Note that repetition occurs after $t = 8\tau$ and the waveform has essentially reached steady-state conditions in a period of time less than $10\tau$, or 5 cycles of the applied square wave.

A closer look will reveal that both the peak and lower levels continued to increase until steady-state conditions were established. Since the exponential waveforms between $t = 4T$ and $t = 5T$ have the same time constant, the average value of $v_C$ can be determined from the steady-state 7.31-mV and 2.69-mV levels as follows:

$$V_{\text{av}} = \frac{7.31 \text{ mV} + 2.69 \text{ mV}}{2} = \frac{10 \text{ mV}}{2} = 5 \text{ mV}$$

which equals the average value of the applied signal as stated earlier in this section.

We can use the results of Fig. 22.34 to plot $i_C$.

At any instant of time,

$$v_i = v_R + v_C$$

or

$$v_R = v_i - v_C$$

and

$$i_R = i_C = \frac{v_i - v_C}{R}$$

At $t = 0^+$, $v_C = 0$ V, and

$$i_R = \frac{v_i - v_C}{R} = \frac{10 \text{ mV} - 0}{5 \text{ k}\Omega} = 2 \ \mu\text{A}$$

as shown in Fig. 22.35.

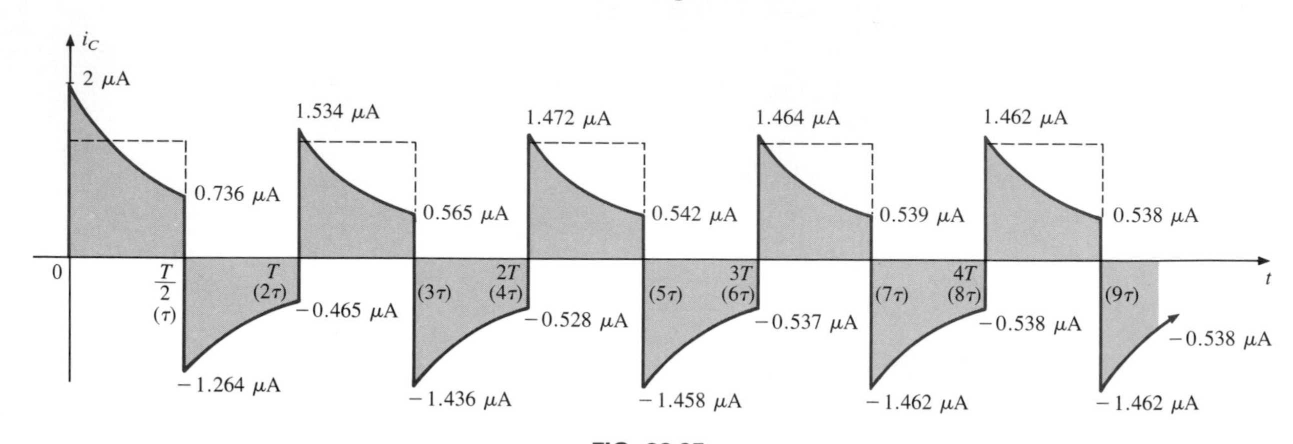

**FIG. 22.35**

As the charging process proceeds, the current $i_C$ will decay at a rate determined by

$$i_C = 2 \ \mu\text{A}e^{-t/\tau}$$

At $t = \tau$,

$$\begin{aligned} i_C &= (2 \ \mu\text{A})(e^{-\tau/\tau}) = (2 \ \mu\text{A})(e^{-1}) = (2 \ \mu\text{A})(0.368) \\ &= 0.736 \ \mu\text{A} \end{aligned}$$

as shown in Fig. 22.35.

For the trailing edge of the first pulse, the voltage across the capacitor cannot change instantaneously, resulting in the following when $v_i$ drops to zero volts:

$$i_C = i_R = \frac{v_i - v_C}{R} = \frac{0 - 6.32 \text{ mV}}{5 \text{ k}\Omega} = -1.264 \ \mu\text{A}$$

as illustrated in Fig. 22.35. The current will then decay as determined by

$$i_C = -1.264 \ \mu\text{A}e^{-t/\tau}$$

and at $t = \tau$ (actually $t = 2\tau$ in Fig. 22.35),

$$\begin{aligned} i_C &= (-1.264 \ \mu\text{A})(e^{-\tau/\tau}) = (-1.264 \ \mu\text{A})(e^{-1}) \\ &= (-1.264 \ \mu\text{A})(0.368) = -0.465 \ \mu\text{A} \end{aligned}$$

as shown in Fig. 22.35.

At $t = T$ ($t = 2\tau$), $v_C = 2.33$ mV and $v_i$ returns to 10 mV, resulting in

$$i_C = i_R = \frac{v_i - v_C}{R} = \frac{10 \text{ mV} - 2.33 \text{ mV}}{5 \text{ k}\Omega} = 1.534 \ \mu\text{A}$$

The equation for the decaying current is now

$$i_C = 1.534\ \mu\text{A}e^{-t/\tau}$$

and at $t = \tau$ (actually $t = 3\tau$ in Fig. 22.35),

$$i_C = (1.534\ \mu\text{A})(0.368) = 0.565\ \mu\text{A}$$

The process will continue until steady-state conditions are reached at the same time they were attained for $v_C$. Note in Fig. 22.35 that the positive peak current decreased toward steady-state conditions while the negative peak became more negative. It is also interesting and important to realize that the current waveform becomes symmetrical about the axis when steady-state conditions are established. The result is that the net average current over one cycle is zero, as it should be in a series *R-C* circuit. Recall from Chapter 10 that the capacitor under dc steady-state conditions can be replaced by an open-circuit equivalent, resulting in $I_C = 0$ A.

---

Although both examples provided above started with an uncharged capacitor, there is no reason that the same approach cannot be used effectively for initial conditions. Simply substitute the initial voltage on the capacitor as $V_i$ in Eq. (22.6) and proceed as above.

## 22.7 OSCILLOSCOPE ATTENUATOR AND COMPENSATING PROBE

The ×10 attenuator probe employed with oscilloscopes is designed to reduce the magnitude of the input voltage by a factor of 10. If the input impedance to a scope is 1 MΩ, the ×10 attenuator probe would have an internal resistance of 9 MΩ, as shown in Fig. 22.36(a).

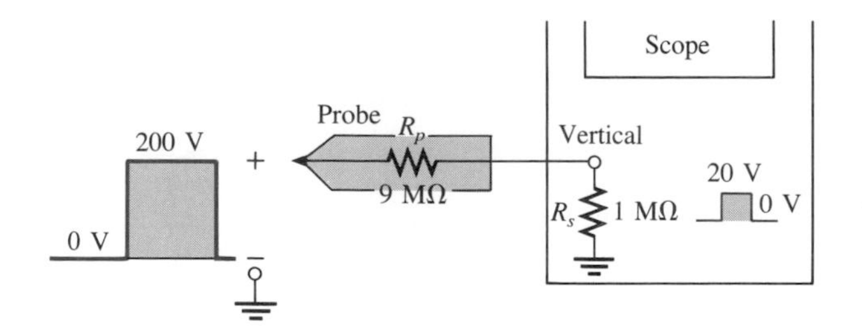


**FIG. 22.36**
*×10 attenuator probe.*

Applying the voltage divider rule,

$$V_{\text{scope}} = \frac{(1\ \text{M}\Omega)(V_i)}{1\ \text{M}\Omega + 9\ \text{M}\Omega} = \frac{1}{10}V_i$$

In addition to the internal resistance, oscilloscopes also have some internal input capacitance, represented by $C_s$ in Fig. 22.37.

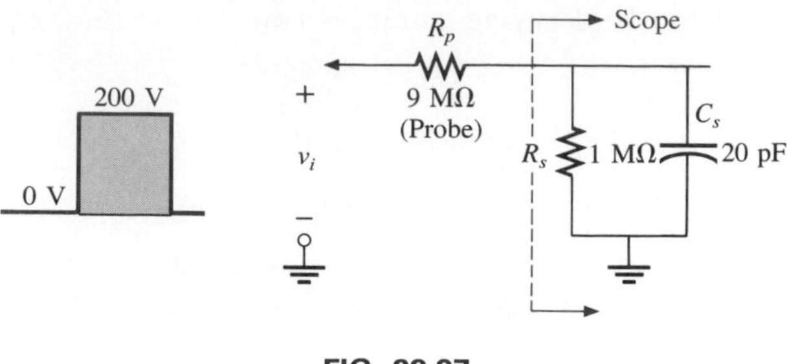

FIG. 22.37

For the analysis to follow, let us determine the Thevenin equivalent circuit for the capacitor $C_s$:

$$E_{Th} = \frac{(1\ \text{M}\Omega)(V_i)}{1\ \text{M}\Omega + 9\ \text{M}\Omega} = \frac{1}{10}V_i$$

and

$$R_{Th} = 9\ \text{M}\Omega \parallel 1\ \text{M}\Omega = 0.9\ \text{M}\Omega$$

The Thevenin network is shown in Fig. 22.38.

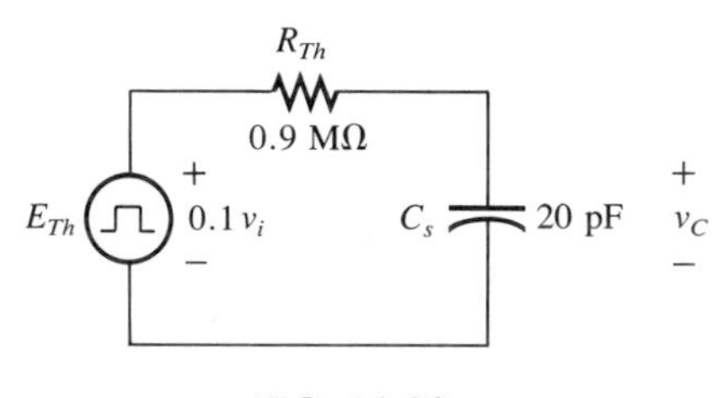


FIG. 22.38

For $v_i = 200$ V (peak),

$$E_{Th} = 0.1v_i = 20\ \text{V (peak)}$$

and for $v_C$, $V_f = 20$ V and $V_i = 0$ V, with

$$\tau = RC = (0.9 \times 10^6\ \Omega)(20 \times 10^{-12}\ \text{F}) = 18\ \mu\text{s}$$

For an applied frequency of 5 kHz,

$$T = \frac{1}{f} = 0.2\ \text{ms}$$

and

$$\frac{T}{2} = 0.1\ \text{ms} = 100\ \mu\text{s}$$

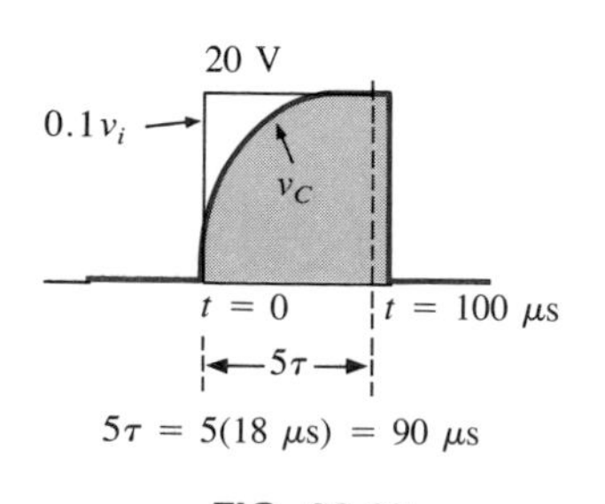


FIG. 22.39

The resulting waveform for $v_C$ appears in Fig. 22.39, clearly revealing a severe rounding distortion of the square wave and a poor representation of the applied signal.

To improve matters, a variable capacitor is often added in parallel with the resistance of the attenuator to establish a probe referred to as a *compensated attenuator*. In Chapter 25, it will be demonstrated that a square wave can be generated by a summation of sinusoidal signals of particular frequency and amplitude. If we therefore design a network such as shown in Fig. 22.40 that will insure that $V_{scope}$ is $\frac{1}{10}v_i$ for any

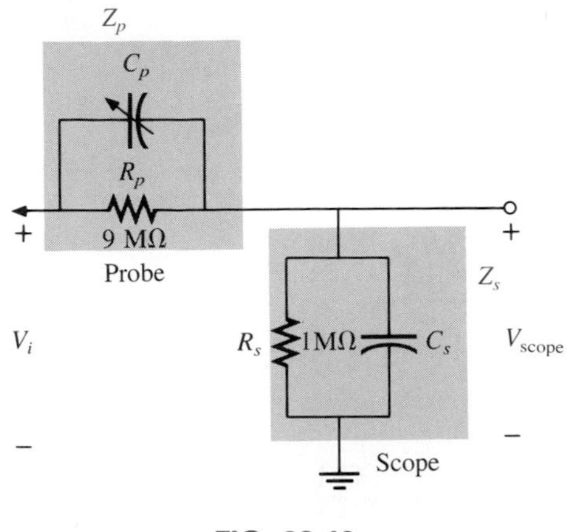


FIG. 22.40

frequency, then the rounding distortion will be removed and $V_{scope}$ will have the same appearance as $v_i$.

Applying the voltage divider rule to the network of Fig. 22.40,

$$\mathbf{V}_{scope} = \frac{\mathbf{Z}_s\mathbf{V}_i}{\mathbf{Z}_s + \mathbf{Z}_p} \tag{22.7}$$

If the parameters are chosen or adjusted such that

$$R_pC_p = R_sC_s \tag{22.8}$$

the phase angle of $\mathbf{Z}_s$ and $\mathbf{Z}_p$ will be the same and Eq. (22.7) will reduce to

$$\mathbf{V}_{scope} = \frac{R_s\mathbf{V}_i}{R_s + R_p} \tag{22.9}$$

which is insensitive to frequency.

In the laboratory, simply adjust the probe capacitance using a standard or known square-wave signal until the desired sharp corners of the square wave are obtained. If you avoid the calibration step, you may make a rounded signal look square since you assumed a square wave at the point of measurement.

Too much capacitance will result in an overshoot effect, while too little will continue to show the rounding effect.

## 22.8 COMPUTER ANALYSIS

In this section we will make full use of the .PROBE command in PSPICE to generate some of the waveforms appearing in this chapter. In particular, let us obtain the pulse response to the network of Fig. 22.41, which is also given as Fig. 22.31.

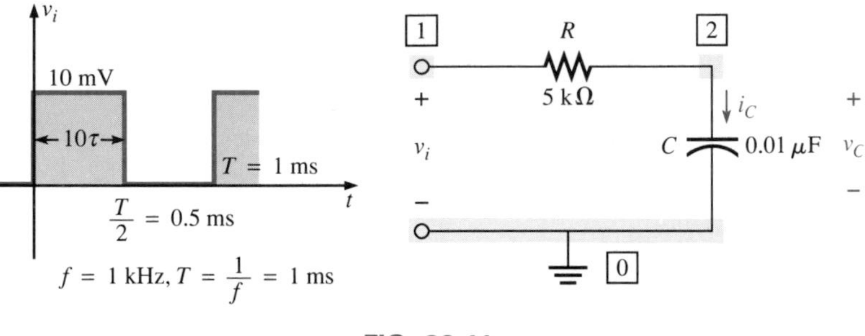


FIG. 22.41

For the applied frequency, the pulse is 10 mV for 0.5 ms and 0 V for the succeeding 0.5 ms. The first line of the input file of Fig. 22.42

```
************************ Evaluation PSpice (January 1989) ******* 15:35:18 *******

Chapter 22 - R-C Pulse Response

****     CIRCUIT DESCRIPTION

******************************************************************************

VI 1 0 PULSE(0V 10MV 0 1NS 1NS 0.5MS 1MS)
R  1 2 5K
C  2 0 0.01UF
.TRAN 0.05M 1M
.PROBE
.OPTIONS NOPAGE
.END
```

FIG. 22.42

specifies a 0-V to 10-mV transition with 0 s delay time. The next two 1-ns entries represent the rise and fall times, which—you recall—must be specified or they default back to TSTEP of the .TRAN command. The 0.5-ms entry is the length of the pulse at the 10-mV level and 1 ms specifies the period of the pulse. The .TRAN command will generate a data point every 0.05 ms, or 50 μs, up to 1 ms (1 M).

The .PROBE command was then used to obtain a plot of $v_i$ (nodal voltage $V_1$ of Fig. 22.41(b)) and $v_C$ (nodal voltage $V_2$ of Fig. 22.41(b)), as shown in Fig. 22.43. Note how the results for each plot match the

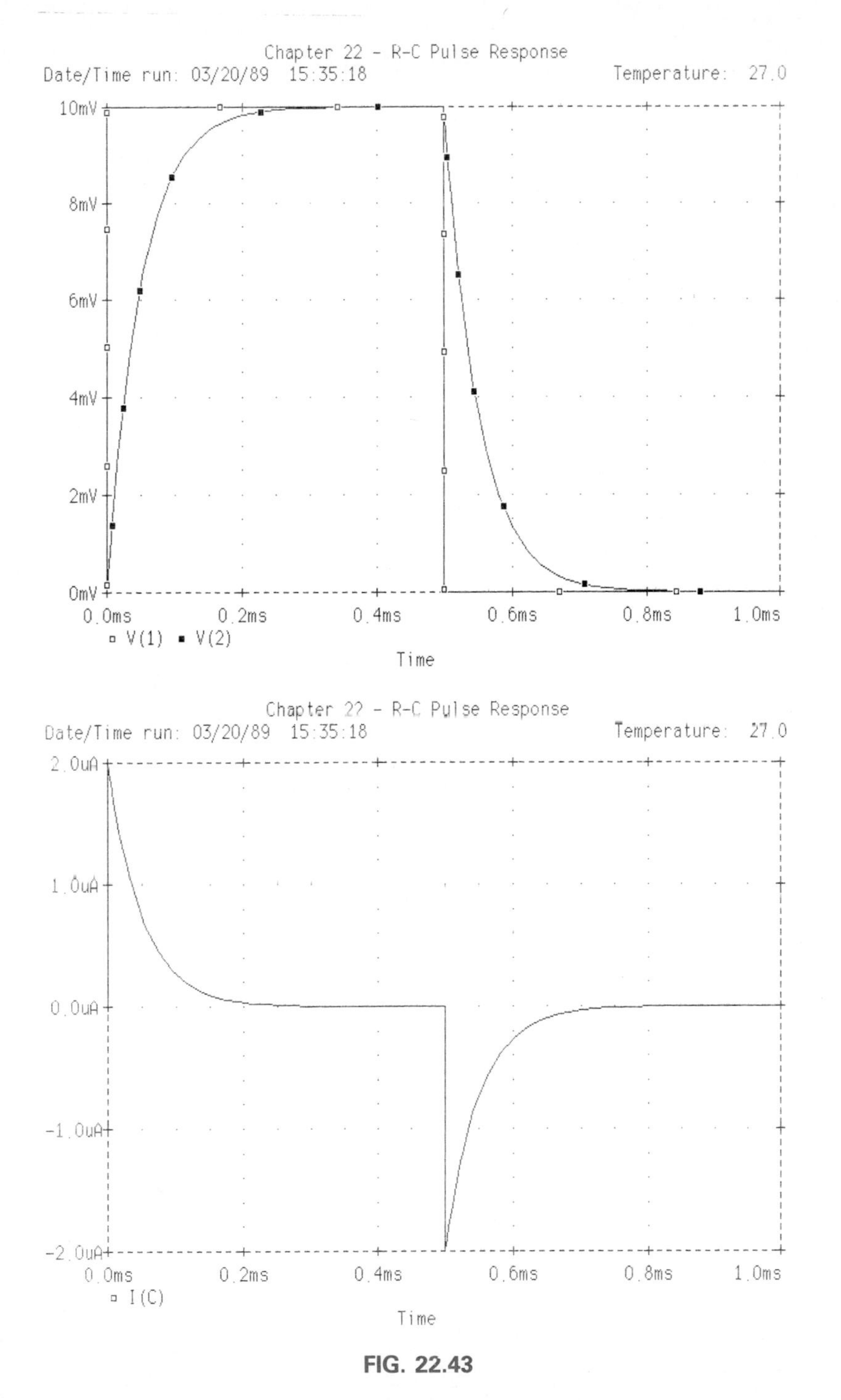


**FIG. 22.43**

curves appearing in Figs. 22.32 and 22.33. Note also how .PROBE specifies at the bottom of the plot the notation for each voltage, □ for $v_i$ and ■ for $v_C$. The plot of $i_C$ reveals a sensitivity of .PROBE to negative and positive values as it sets the horizontal axis in the middle of the page.

The second input file of Fig. 22.44 is for the applied pulse of Fig. 22.45, which has a frequency of 10 kHz (as applied in Example 22.11).

```
************************ Evaluation PSpice (January 1989) ******* 15:46:29 *******

 Chapter 22 - R-C Pulse Response

 ****     CIRCUIT DESCRIPTION

****************************************************************************


VI 1 0 PULSE(0V 10MV 0 1NS 1NS 0.05MS .1MS)
R  1 2 5K
C  2 0 0.01UF
.TRAN 0.005M .1M
.PROBE
.OPTIONS NOPAGE
.END
```

**FIG. 22.44**

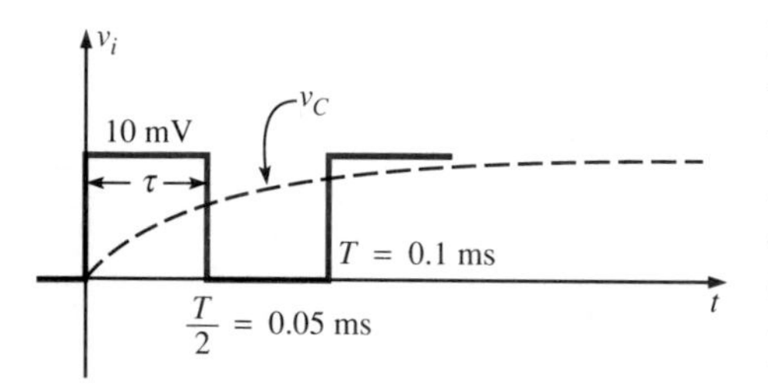

**FIG. 22.45**

In this case, the 10-mV pulse is present for a period of time matching the time constant of the network—a period insufficient (less than $5\tau$) for the voltage $v_C$ to climb to its final value. The result will be a level of $v_C$ at $t = 0.05$ ms less than 10 mV, as shown in the figure. The only changes to the input file of Fig. 22.42 are to change the pulse period to 0.1 ms ($2 \times 0.05$ ms) and to modify the .TRAN command with a data point every 0.005 ms = 5 $\mu$s for a period extending to 0.1 ms = 100 $\mu$s. A request for a plot of $v_i$ and $v_C$ resulted in the curves of Fig. 22.46, which compare very favorably with the curves of Fig. 22.34 for

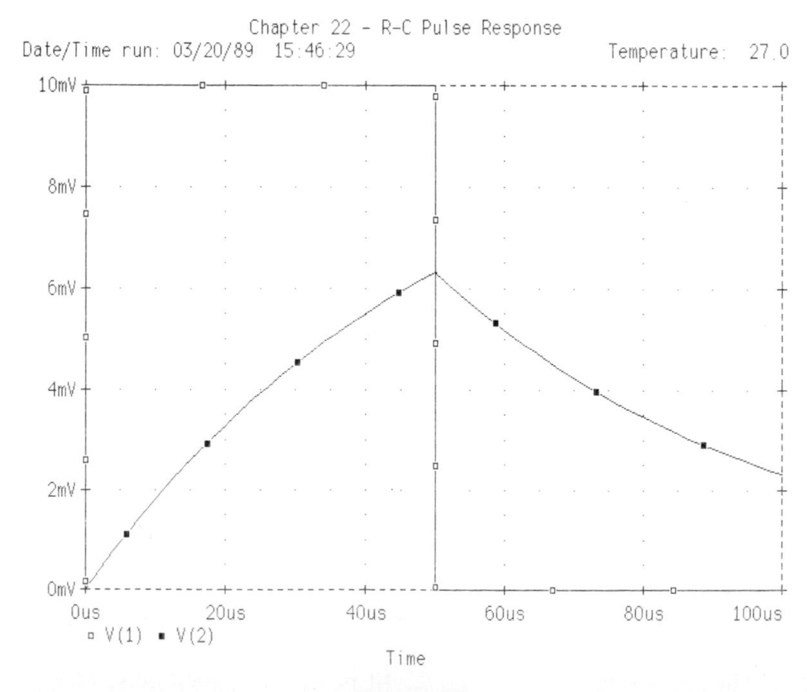

**FIG. 22.46**

the same conditions. The value of $v_C$ has risen to about 6.3 V in 50 $\mu$s and drops to about 2.3 V at $t = 100$ $\mu$s. Accuracy beyond the tenths place is difficult with the curves of Fig. 22.46, but keep in mind that the scaling can be changed if a higher level of accuracy is required for a particular region of the output waveforms.

## PROBLEMS

### SECTION 22.2 Ideal versus Actual

1. Determine the following for the pulse waveform of Fig. 22.47:
   - **a.** positive- or negative-going?
   - **b.** base-line voltage
   - **c.** pulse width
   - **d.** amplitude
   - **e.** % tilt

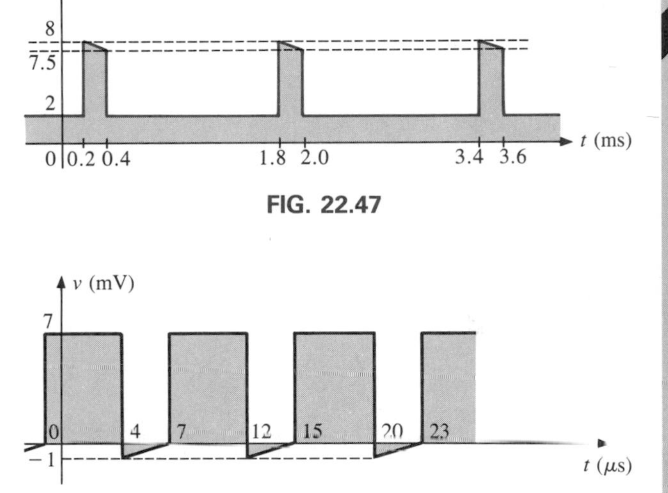


FIG. 22.47

FIG. 22.48

2. Repeat Problem 1 for the pulse waveform of Fig. 22.48.
3. Repeat Problem 1 for the pulse waveform of Fig. 22.49.
4. Determine the rise and fall time for the waveform of Fig. 22.49.
5. Sketch a pulse waveform that has a base-line voltage of −5 mV, a pulse width of 2 $\mu$s, an amplitude of 15 mV, a 10% tilt, a period of 10 $\mu$s, vertical sides, and is positive-going.
6. For the waveform of Fig. 22.50, established by straight-line approximations of the original waveform:
   - **a.** Determine the rise time.
   - **b.** Find the fall time.
   - **c.** Find the pulse width.
   - **d.** Calculate the frequency.

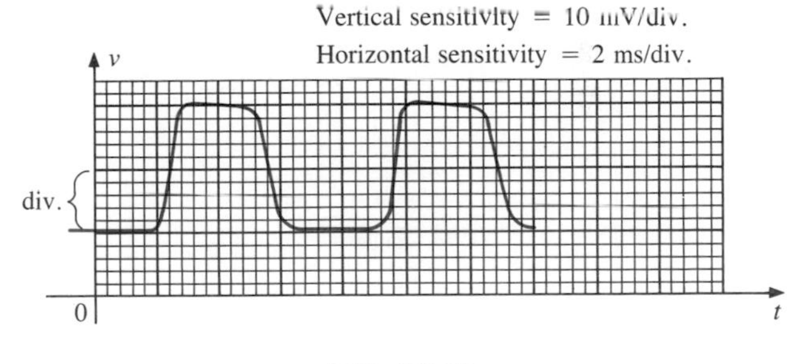

FIG. 22.49

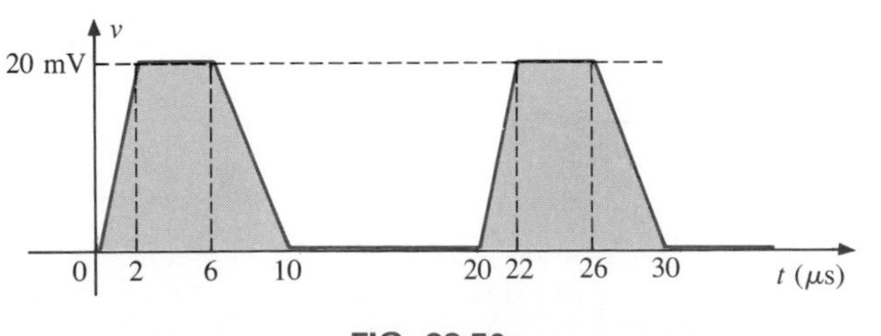


FIG. 22.50

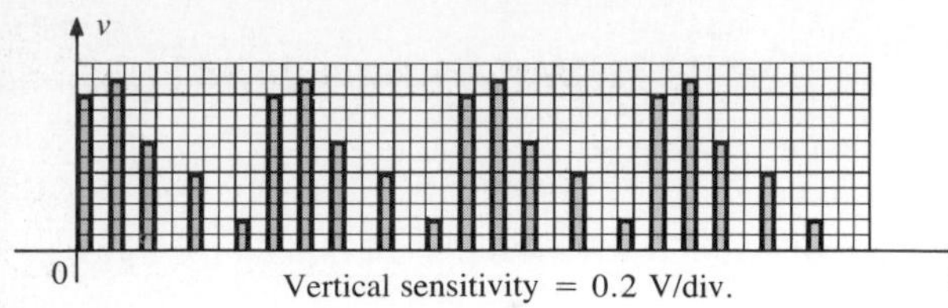


FIG. 22.51

7. For the waveform of Fig. 22.51:
   a. Determine the period.
   b. Find the frequency.
   c. Find the maximum and minimum amplitude.

## SECTION 22.3 Pulse Repetition Rate and Duty Cycle

8. Determine the pulse repetition frequency and duty cycle for the waveform of Fig. 22.47.
9. Repeat Problem 8 for the waveform of Fig. 22.48.
10. Repeat Problem 8 for the waveform of Fig. 22.49.

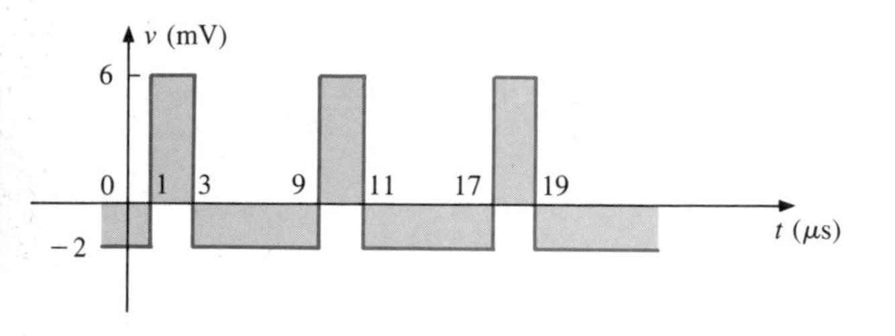


FIG. 22.52

## SECTION 22.4 Average Value

11. Determine the average value of the periodic pulse waveform of Fig. 22.52.
12. Determine the average value of the periodic pulse waveform of Fig. 22.47.
13. To the best accuracy possible, determine the average value of the waveform of Fig. 22.49.
14. Determine the average value of the waveform of Fig. 22.50.
15. Determine the average value of the periodic pulse train of Fig. 22.51.

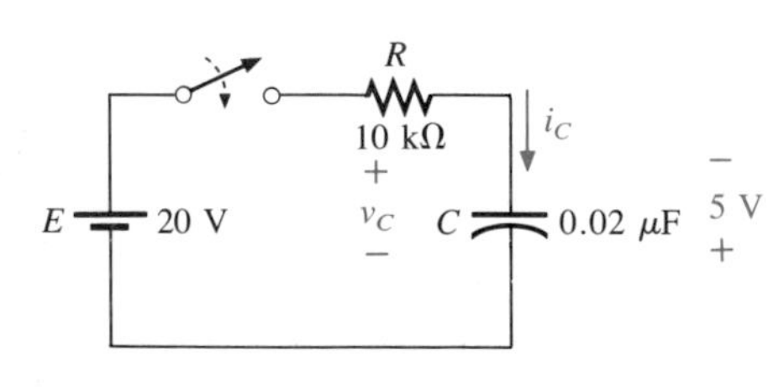


FIG. 22.53

## SECTION 22.5 *R-C* Circuits with Initial Values

16. The capacitor of Fig. 22.53 is initially charged to 5 V, with the polarity indicated in the figure. The switch is then closed at $t = 0$ s.
   a. What is the mathematical expression for the voltage $v_C$?
   b. Sketch $v_C$ versus $t$.
   c. What is the mathematical expression for the current $i_C$?
   d. Sketch $i_C$ versus $t$.
17. For the input voltage $v_i$ appearing in Fig. 22.54 sketch the waveform for $v_o$. Assume that steady-state conditions were established with $v_i = 8$ V.

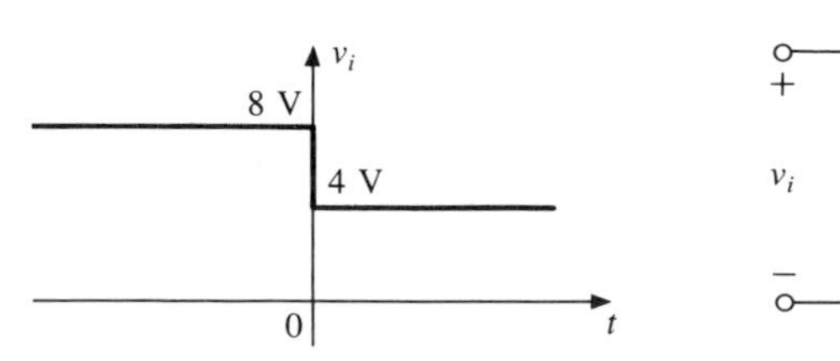


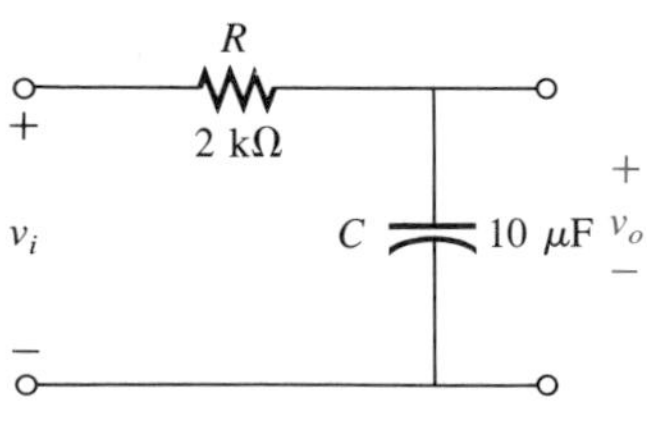


FIG. 22.54

18. The switch of Fig. 22.55 is in position 1 until steady-state conditions are established. Then the switch is moved (at $t = 0$ s) to position 2. Sketch the waveform for the voltage $v_C$.

19. Sketch the waveform for $i_C$ for Problem 18.

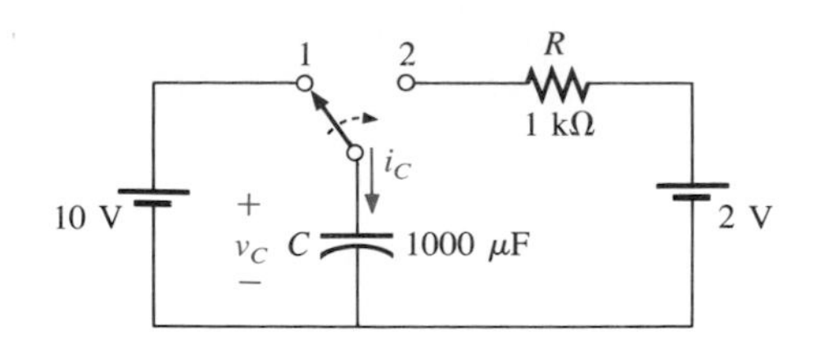

FIG. 22.55

## SECTION 22.6 *R-C* Response to Square-Wave Inputs

20. Sketch the voltage $v_C$ for the network of Fig. 22.56 due to the square-wave input of the same figure with a frequency of
   a. 500 Hz    b. 100 Hz
   c. 5000 Hz

21. Sketch the current $i_C$ for each frequency of Problem 20.

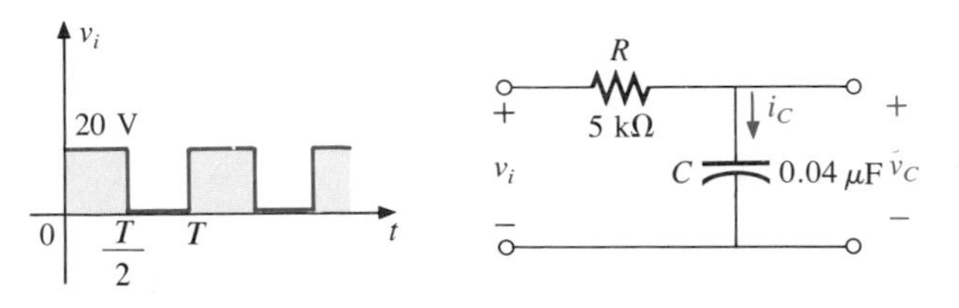

FIG. 22.56

22. Sketch the response ($v_C$) of the network of Fig. 22.56 to the square-wave input of Fig. 22.57.

23. If the capacitor of Fig. 22.56 is initially charged to 20 V, sketch the response ($v_C$) to the same input signal (of Fig. 22.56) at a frequency of 500 Hz.

24. Repeat Problem 23 if the capacitor is initially charged to $-10$ V.

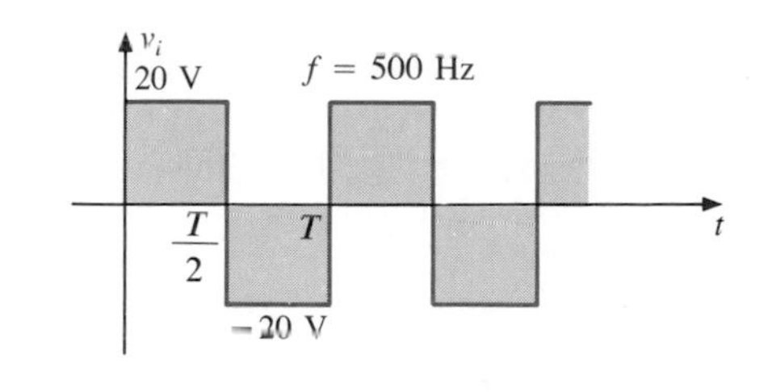

FIG. 22.57

## SECTION 22.7 Oscilloscope Attenuator and Compensating Probe

25. Given the network of Fig. 22.40 with $R_p = 9$ MΩ and $R_s = 1$ MΩ, find $\mathbf{V}_{scope}$ in polar form if $C_p = 2$ pF, $C_s = 18$ pF, and $v_i = \sqrt{2}(100) \sin 10{,}000t$. That is, determine $\mathbf{Z}_s$ and $\mathbf{Z}_p$ and substitute into Eq. (22.7) and compare the results obtained with Eq. (22.9). Is it verified that the phase angle of $\mathbf{Z}_s$ and $\mathbf{Z}_p$ is the same under the condition $R_pC_p = R_sC_s$?

26. Repeat Problem 25 at $\omega = 10^5$ rad/s.

## SECTION 22.8 Computer Analysis

### PSPICE

27. Write the input file to obtain the waveforms of $v_C$ and $v_i$ for the network of Fig. 22.50 for a frequency of 100 Hz.

28. Repeat Problem 27 for a frequency of 5000 Hz.

29. Write the input file to print out the plot $v_C$ versus time as requested in Problem 22. Use an interval of $\frac{1}{5}\tau$ for data points from 0 to $5\tau$.

30. Write the input file to print out and plot $i_C$ versus time for the conditions of Problem 23. Consult your PSPICE manual to determine how to handle initial conditions.

**BASIC**

**31.** Given a periodic pulse train such as that in Fig. 22.11, write a program to determine the average value, given the base-line voltage, peak value, and duty cycle.

**32.** Given the initial and final values and the network parameters ($R$ and $C$), write a program to tabulate the values of $v_C$ at each time constant (of the first five) of the transient phase.

**33.** For the case of $T/2 < 5\tau$ as defined by Fig. 22.29, write a program to determine the values of $v_C$ at each half-period of the applied square wave. Test the solution by entering the conditions of Example 22.10.

## GLOSSARY

**Actual (true, practical) pulse** A pulse waveform having a leading and trailing edge that are not vertical, along with other distortion effects such as tilt, ringing, or overshoot.

**Attenuator probe** A scope probe that will reduce the strength of the signal applied to the vertical channel of a scope.

**Base-line voltage** The voltage level from which a pulse is initiated.

**Compensated attenuator probe** A scope probe that can reduce the applied signal and balance the effects of the input capacitance of a scope on the signal to be displayed.

**Duty cycle** Reveals how much of a period is encompassed by the pulse waveform.

**Fall time** ($t_f$) The time required for the trailing edge of a pulse waveform to drop from the 90% to the 10% level.

**Ideal pulse** A pulse waveform characterized as having vertical sides, sharp corners, and a flat peak response.

**Negative-going pulse** A pulse that increases in the negative direction from the base-line voltage.

**Periodic pulse train** A sequence of pulses that repeats itself after a specific period of time.

**Positive-going pulse** A pulse that increases in the positive direction from the base-line voltage.

**Pulse amplitude** The peak-to-peak value of a pulse waveform.

**Pulse repetition frequency** The frequency of a periodic pulse train.

**Pulse train** A series of pulses that may have varying heights and widths.

**Pulse width** ($t_p$) The pulse width defined by the 50% voltage level.

**Rise time** ($t_r$) The time required for the leading edge of a pulse waveform to travel from the 10% to the 90% level.

**Square wave** A periodic pulse waveform with a 50% duty cycle.

**Step function** A waveform that abruptly changes from one level to another.

**Tilt** The drop in peak value across the pulse width of a pulse waveform.

# 23
# Polyphase Systems

## 23.1 INTRODUCTION

An ac generator designed to develop a single sinusoidal voltage for each rotation of the shaft (rotor) is referred to as a *single-phase generator*. If the number of coils on the rotor is increased in a specified manner, the result is a *polyphase generator,* which develops more than one ac phase voltage per rotation of the rotor. In this chapter, the three-phase system will be discussed in detail since it is the most frequently used for power transmission.

In general, the three-phase system is more economical for transmitting power at a fixed power loss than the single-phase system. This economy is due primarily to the reduction of the $I^2R$ losses of the transmission lines. A reduction in these losses permits the use of smaller conductors, which in turn reduces the weight of copper required.

The three-phase system is used in almost all commercial electric generators. This does not mean that single-phase and two-phase generating systems are obsolete. Most small emergency generators, such as the gasoline type, are one-phase generating systems.

One of the more common applications of the two-phase system is in servomechanisms, which are self-correcting control systems capable of detecting and adjusting their own operation. Servomechanisms are used in ships and aircraft to keep them on course automatically, or, in simpler devices such as a thermostatic circuit, to regulate heat output. In most cases, however, where single-phase and two-phase inputs are required, they are supplied by one and two phases of a three-phase generating system rather than generated independently.

The number of phase voltages that can be produced by a polyphase generator is not limited to three. Any number of phases can be obtained by spacing the windings for each phase at the proper angular position around the rotor. Some electrical systems operate more efficiently if more than three phases are used. One such system involves the process of rectification, which is used to convert alternating current to direct current. The greater the number of phases, the smoother the dc output of the system.

## 23.2 THE THREE-PHASE GENERATOR

The three-phase generator of Fig. 23.1(a) has three induction coils placed 120° apart on the rotor (armature), as shown symbolically by Fig. 23.1(b). Since the three coils have an equal number of turns, and each coil rotates with the same angular velocity, the voltage induced

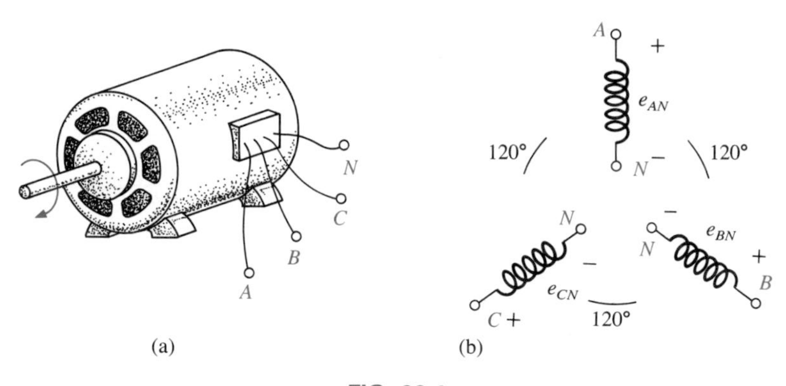


**FIG. 23.1**
*(a) Three-phase generator; (b) induced voltages of a three-phase generator.*

across each coil will have the same peak value, shape, and frequency. As the shaft of the generator is turned by some external means, the induced voltages $e_{AN}$, $e_{BN}$, and $e_{CN}$ will be generated simultaneously, as shown in Fig. 23.2. Note the 120° phase shift between waveforms and the similarities in appearance of the three sinusoidal functions.

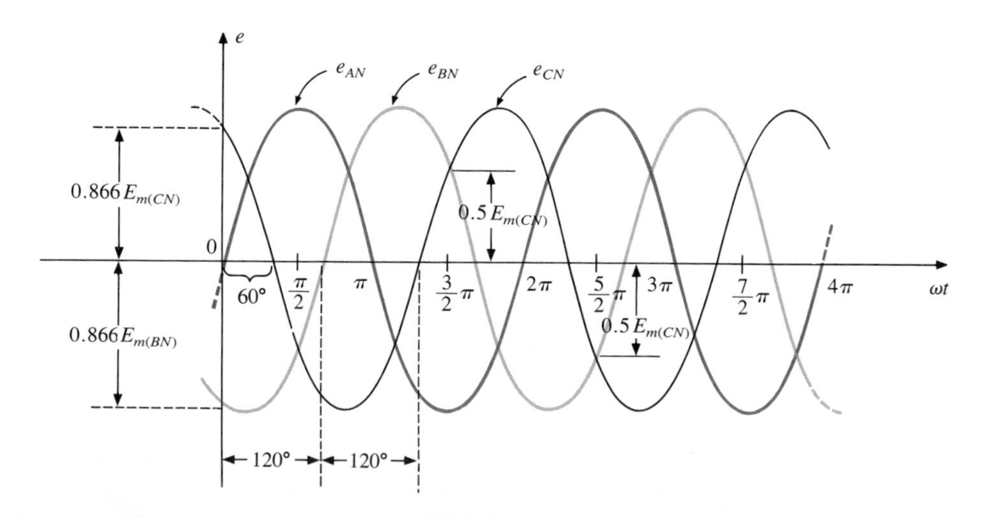


**FIG. 23.2**

In particular, note that *at any instant of time, the algebraic sum of the three phase voltages is zero*. This is shown at $\omega t = 0$ in Fig. 23.2, where it is also evident that *when one induced voltage is zero, the other two are 86.6% of their positive or negative maximums. In addition, when any two are equal in magnitude and sign (at $0.5E_m$), the remaining induced voltage has the opposite polarity and a peak value.*

The sinusoidal expression for each of the induced voltages of Fig. 23.2 is

$$\begin{aligned} e_{AN} &= E_{m(AN)} \sin \omega t \\ e_{BN} &= E_{m(BN)} \sin(\omega t - 120°) \\ e_{CN} &= E_{m(CN)} \sin(\omega t - 240°) = E_{m(CN)} \sin(\omega t + 120°) \end{aligned} \tag{23.1}$$

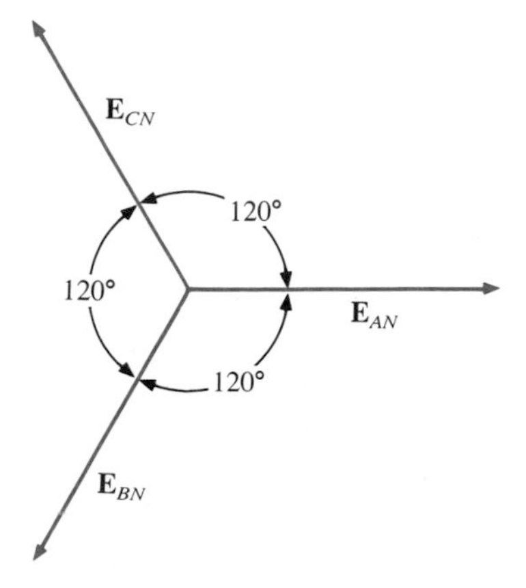

FIG. 23.3

The phasor diagram of the induced voltages is shown in Fig. 23.3, where

$$\begin{aligned} E_{AN} &= 0.707E_{m(AN)} \\ E_{BN} &= 0.707E_{m(BN)} \\ E_{CN} &= 0.707E_{m(CN)} \end{aligned}$$

and

$$\begin{aligned} \mathbf{E}_{AN} &= E_{AN} \angle 0° \\ \mathbf{E}_{BN} &= E_{BN} \angle -120° \\ \mathbf{E}_{CN} &= E_{CN} \angle +120° \end{aligned}$$

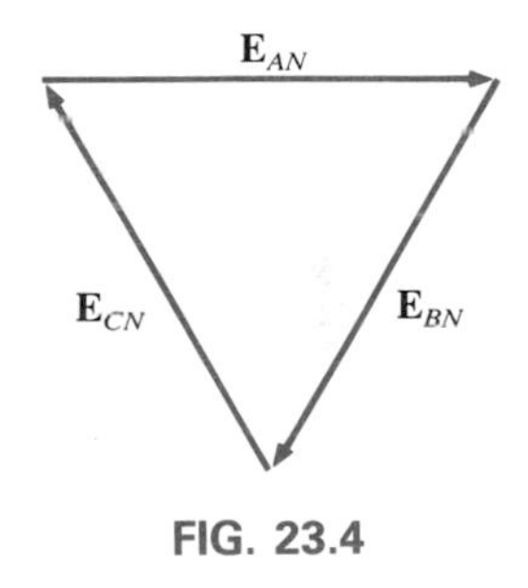


FIG. 23.4

By rearranging the phasors as shown in Fig. 23.4 and applying a law of vectors which states that *the vector sum of any number of vectors drawn such that the "head" of one is connected to the "tail" of the next, and that the head of the last vector is connected to the tail of the first is zero,* we can conclude that the phasor sum of the phase voltages in a three-phase system is zero. That is,

$$\Sigma\,(\mathbf{E}_{AN} + \mathbf{E}_{BN} + \mathbf{E}_{CN}) = 0 \tag{23.2}$$

## 23.3 THE Y-CONNECTED GENERATOR

If the three terminals denoted $N$ of Fig. 23.1(b) are connected together, the generator is referred to as a *Y-connected three-phase generator* (Fig. 23.5). As indicated in Fig. 23.5, the Y is inverted for ease of notation and for clarity. The point at which all the terminals are connected is called the *neutral point*. If a conductor is not attached from this point to the load, the system is called a *Y-connected, three-phase, three-wire generator*. If the neutral is connected, the system is a *Y-connected, three-phase, four-wire generator*. The function of the neutral will be discussed in detail when we consider the load circuit.

The three conductors connected from $A$, $B$, and $C$ to the load are called *lines*. For the Y-connected system, it should be obvious from

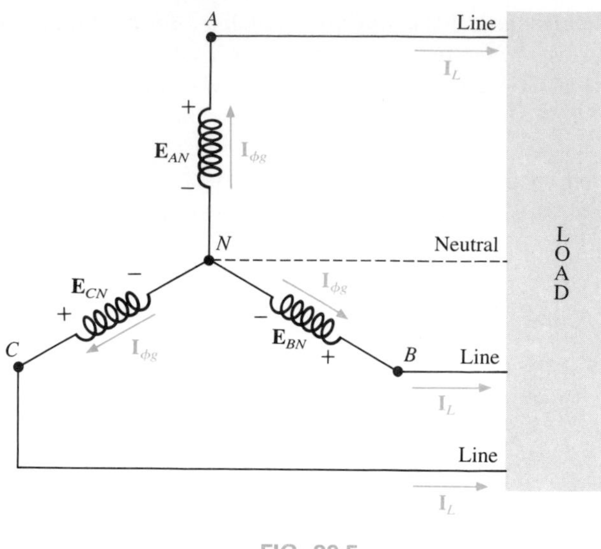


FIG. 23.5
*Y-connected generator.*

Fig. 23.5 that the line current equals the phase current for each phase; that is,

$$\mathbf{I}_L = \mathbf{I}_{\phi g} \tag{23.3}$$

The voltage from one line to another is called a *line voltage*. On the phasor diagram (Fig. 23.6) it is the phasor drawn from the end of one phase to another in the counterclockwise direction.

Applying Kirchhoff's voltage law around the indicated loop of Fig. 23.6, we obtain

$$\mathbf{E}_{AB} - \mathbf{E}_{AN} + \mathbf{E}_{BN} = 0$$

or

$$\mathbf{E}_{AB} = \mathbf{E}_{AN} - \mathbf{E}_{BN} = \mathbf{E}_{AN} + \mathbf{E}_{NB}$$

The phasor diagram is redrawn to find $\mathbf{E}_{AB}$ as shown in Fig. 23.7. Since each phase voltage when reversed ($\mathbf{E}_{NB}$) will bisect the other two,

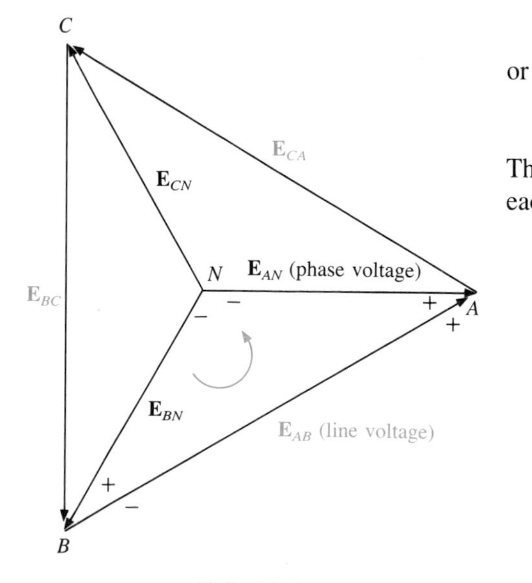


FIG. 23.6
*Line and phase voltages of the Y-connected three-phase generator.*

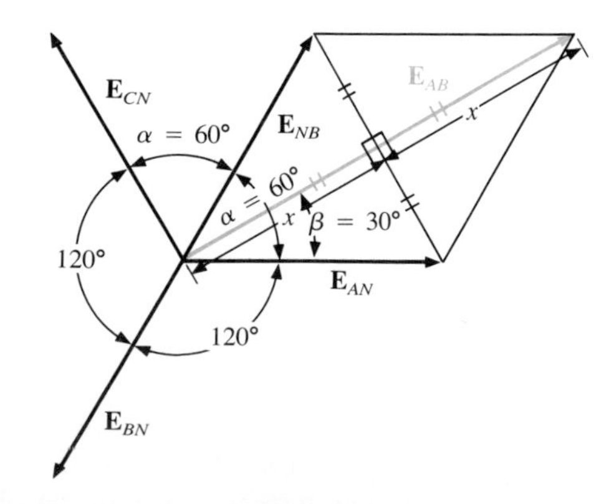


FIG. 23.7

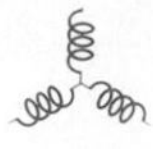



$\alpha = 60°$. The angle $\beta$ is 30°, since a line drawn from opposite ends of a rhombus will divide in half both the angle of origin and the opposite angle. Lines drawn between opposite corners of a rhombus will also bisect each other at right angles.

The length $x$ is

$$x = E_{AN} \cos 30° = \frac{\sqrt{3}}{2} E_{AN}$$

and

$$E_{AB} = 2x = (2)\frac{\sqrt{3}}{2} E_{AN} = \sqrt{3} E_{AN}$$

Noting from the phasor diagram that $\theta$ of $\mathbf{E}_{AB} = \beta = 30°$, the result is

$$\mathbf{E}_{AB} = E_{AB} \angle 30° = \sqrt{3} E_{AN} \angle 30°$$

In words, the magnitude of the line voltage of a Y-connected generator is $\sqrt{3}$ times the phase voltage:

$$\boxed{E_L = \sqrt{3} E_\phi} \tag{23.4}$$

In addition, the phase angle between any line voltage and the nearest phase voltage is 30°, as shown for $\mathbf{E}_{AB}$ and $\mathbf{E}_{AN}$ in Fig. 23.7 ($\beta = 30°$).

In sinusoidal notation,

$$e_{AB} = \sqrt{2} E_{AB} \sin(\omega t + 30°)$$

Repeating the same procedure for the other line voltages results in

$$e_{CA} = \sqrt{2} E_{CA} \sin(\omega t + 150°)$$

and

$$e_{BC} = \sqrt{2} E_{BC} \sin(\omega t + 270°)$$

The phasor diagram of the line and phase voltages is shown in Fig. 23.8. If the phasors representing the line voltages in Fig. 23.8(a) are rearranged slightly, they will form a closed loop [Fig. 23.8(b)]. There-

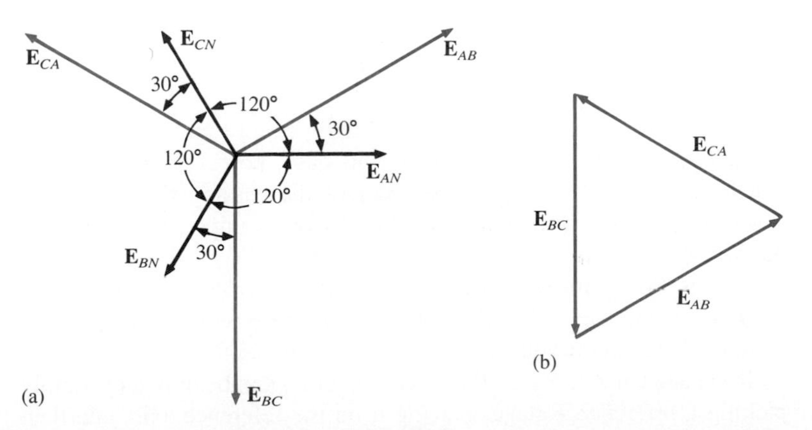


FIG. 23.8

fore, we can conclude that the sum of the line voltages is also zero; that is,

$$\Sigma\,(\mathbf{E}_{AB} + \mathbf{E}_{CA} + \mathbf{E}_{BC}) = 0 \tag{23.5}$$

## 23.4 PHASE SEQUENCE (Y-CONNECTED GENERATOR)

The phase sequence can be determined by the order in which the phasors representing the phase voltages pass through a fixed point on the phasor diagram if the phasors are rotated in a counterclockwise direction. For example, in Fig. 23.9 the phase sequence is *ABC*. However, since the fixed point can be chosen anywhere on the phasor diagram, the sequence can also be written as *BCA* or *CAB*. The phase sequence is quite important in the three-phase distribution of power. In a three-phase motor, for example, if two phase voltages are interchanged, the sequence will change and the direction of rotation of the motor will be reversed. Other effects will be described when we consider the loaded three-phase system.

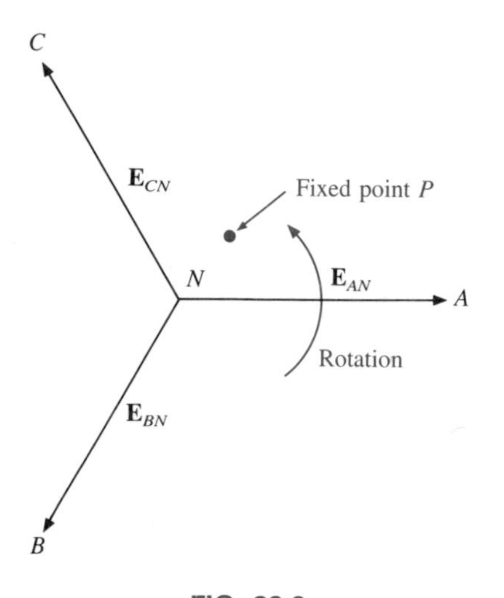


FIG. 23.9

The phase sequence can also be described in terms of the line voltages. Drawing the line voltages on a phasor diagram in Fig. 23.10, we

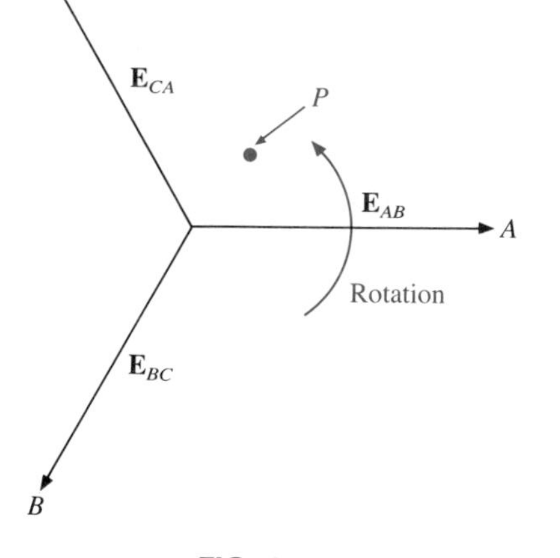


FIG. 23.10

are able to determine the phase sequence by again rotating the phasors in the counterclockwise direction. In this case, however, the sequence can be determined by noting the order of the passing first or second subscripts. In the system of Fig. 23.10, for example, the phase sequence of the first subscripts passing point *P* is *ABC*, and of the second subscripts *BCA*. But we know that *BCA* is equivalent to *ABC*, so the sequence is the same for each. Note that the phase sequence is the same as that of the phase voltages described in Fig. 23.9.

If the sequence is given, the phasor diagram can be drawn by simply picking a reference voltage, placing it on the reference axis, and then drawing the other voltages at the proper angular position. For a sequence of $ACB$, for example, we might choose $E_{AB}$ to be the reference [Fig. 23.11(a)] if we wanted the phasor diagram of the line voltages, or $E_{NA}$ for the phase voltages [Fig. 23.11(b)]. For the sequence indicated, the phasor diagrams would be as in Fig. 23.11. In phasor notation,

$$\text{Line voltages}\begin{cases}\mathbf{E}_{AB} = E_{AB}\ \angle 0^\circ & \text{(reference)}\\ \mathbf{E}_{CA} = E_{CA}\ \angle -120^\circ \\ \mathbf{E}_{BC} = E_{BC}\ \angle +120^\circ\end{cases}$$

$$\text{Phase voltages}\begin{cases}\mathbf{E}_{AN} = E_{AN}\ \angle 0^\circ & \text{(reference)}\\ \mathbf{E}_{CN} = E_{CN}\ \angle -120^\circ \\ \mathbf{E}_{BN} = E_{BN}\ \angle +120^\circ\end{cases}$$

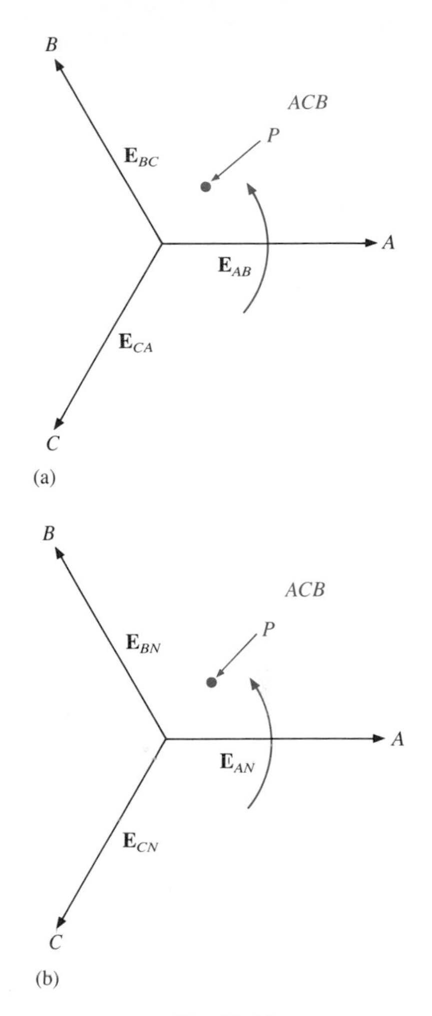

FIG. 23.11

## 23.5 THE Y-CONNECTED GENERATOR WITH A Y-CONNECTED LOAD

Loads connected to three-phase supplies are of two types: the Y and the Δ.

If a Y-connected load is connected to a Y-connected generator, the system is symbolically represented by Y-Y. The physical setup of such a system is shown in Fig. 23.12.

If the load is balanced, the neutral can be removed without affecting the circuit in any manner. That is, if

$$\mathbf{Z}_1 = \mathbf{Z}_2 = \mathbf{Z}_3$$

then $I_N$ will be zero. (This will be demonstrated in Example 23.1.) Note that in order to have a balanced load, the phase angle must also be the same for each impedance—a condition that was not necessary in dc circuits when we considered balanced systems.

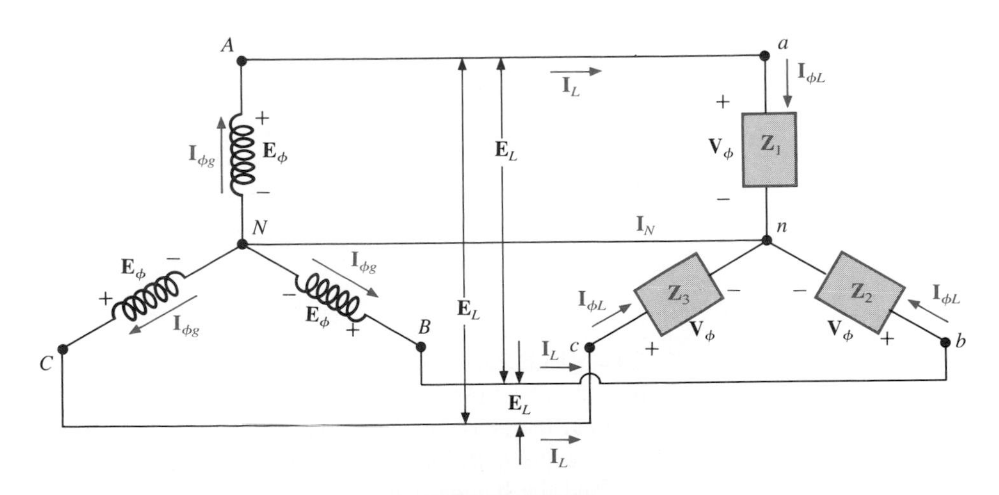


FIG. 23.12
*Y-connected generator with a Y-connected load.*

In practice, if a factory, for example, had only balanced three-phase loads, the absence of the neutral would have no effect, since ideally the system would always be balanced. The cost would therefore be less since the number of required conductors would be reduced. However, lighting and most other electrical equipment will use only one of the phase voltages, and even if the loading is designed to be balanced (as it should be), there will never be perfect continuous balancing, since lights and other electrical equipment will be turned on and off, upsetting the balanced condition. The neutral is therefore necessary to carry the resulting current away from the load and back to the Y-connected generator. This will be demonstrated when we consider unbalanced Y-connected systems.

We shall now examine the *four-wire Y-Y-connected system*. The current passing through each phase of the generator is the same as its corresponding line current, which in turn for a Y-connected load is equal to the current in the phase of the load to which it is attached:

$$\mathbf{I}_{\phi g} = \mathbf{I}_L = \mathbf{I}_{\phi L} \tag{23.6}$$

For a balanced or unbalanced load, since the generator and load have a common neutral point, then

$$\mathbf{V}_\phi = \mathbf{E}_\phi \tag{23.7}$$

In addition, since $\mathbf{I}_{\phi L} = \mathbf{V}_\phi/\mathbf{Z}_\phi$, the magnitude of the current in each phase will be equal for a balanced load and unequal for an unbalanced load. You will recall that for the Y-connected generator, the magnitude of the line voltage is equal to $\sqrt{3}$ times the phase voltage. This same relationship can be applied to a balanced or unbalanced four-wire Y-connected load:

$$E_L = \sqrt{3}V_\phi \tag{23.8}$$

For a voltage drop across a load element, the first subscript refers to that terminal through which the current enters the load element, and the second subscript to the terminal from which the current leaves. In other words, the first subscript is, by definition, positive with respect to the second for a voltage drop. Note Fig. 23.13, in which the standard double subscripts for a source of voltage and a voltage drop are indicated.

**EXAMPLE 23.1.** The phase sequence of the Y-connected generator in Fig. 23.13 is *ABC*.

a. Find the phase angles $\theta_2$ and $\theta_3$.
b. Find the magnitude of the line voltages.
c. Find the line currents.
d. Verify that since the load is balanced, $\mathbf{I}_N = 0$.

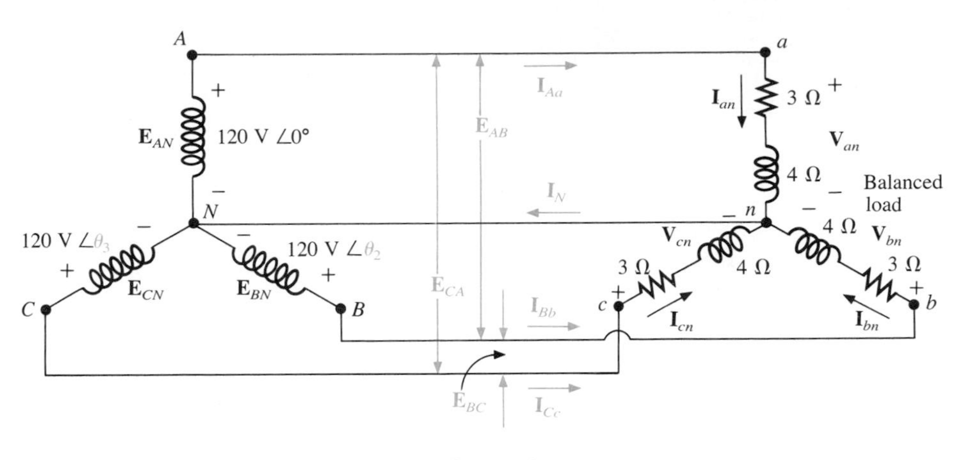


FIG. 23.13

*Solutions:*

a. For an $ABC$ phase sequence,

$$\theta_2 = \mathbf{-120^\circ} \quad \text{and} \quad \theta_3 = \mathbf{+120^\circ}$$

b. $E_L = \sqrt{3}E_\phi = (1.73)(120 \text{ V}) = 208$ V. Therefore,

$$E_{AB} = E_{BC} = E_{CA} = \mathbf{208\ V}$$

c. $\mathbf{V}_\phi = \mathbf{E}_\phi$. Therefore,

$$\mathbf{V}_{an} = \mathbf{E}_{AN} \qquad \mathbf{V}_{bn} = \mathbf{E}_{BN} \qquad \mathbf{V}_{cn} = \mathbf{E}_{CN}$$

$$\mathbf{I}_{\phi L} = \mathbf{I}_{an} = \frac{\mathbf{V}_{an}}{\mathbf{Z}_{an}} = \frac{120 \text{ V} \angle 0^\circ}{3\ \Omega + j4\ \Omega} = \frac{120 \text{ V} \angle 0^\circ}{5\ \Omega \angle 53.13^\circ}$$
$$= 24 \text{ A} \angle -53.13^\circ$$

$$\mathbf{I}_{bn} = \frac{\mathbf{V}_{bn}}{\mathbf{Z}_{bn}} = \frac{120 \text{ V} \angle -120^\circ}{5\ \Omega \angle 53.13^\circ} = 24 \text{ A} \angle -173.13^\circ$$

$$\mathbf{I}_{cn} = \frac{\mathbf{V}_{cn}}{\mathbf{Z}_{cn}} = \frac{120 \text{ V} \angle +120^\circ}{5\ \Omega \angle 53.13^\circ} = 24 \text{ A} \angle 66.87^\circ$$

and, since $\mathbf{I}_L = \mathbf{I}_{\phi L}$,

$$\mathbf{I}_{Aa} = \mathbf{I}_{an} = \mathbf{24\ A \angle -53.13^\circ}$$
$$\mathbf{I}_{Bb} = \mathbf{I}_{bn} = \mathbf{24\ A \angle -173.13^\circ}$$
$$\mathbf{I}_{Cc} = \mathbf{I}_{cn} = \mathbf{24\ A \angle 66.87^\circ}$$

d. Applying Kirchhoff's current law, we have

$$\mathbf{I}_N = \mathbf{I}_{Aa} + \mathbf{I}_{Bb} + \mathbf{I}_{Cc}$$

In rectangular form,

$$\begin{aligned}
\mathbf{I}_{Aa} &= 24 \text{ A} \angle -53.13^\circ &&= 14.40 \text{ A} - j19.20 \text{ A} \\
\mathbf{I}_{Bb} &= 24 \text{ A} \angle -173.13^\circ &&= -23.83 \text{ A} - j2.87 \text{ A} \\
\mathbf{I}_{Cc} &= 24 \text{ A} \angle 66.87^\circ &&= \underline{9.43 \text{ A} + j22.07 \text{ A}} \\
\Sigma\ (\mathbf{I}_{Aa} &+ \mathbf{I}_{Bb} + \mathbf{I}_{Cc}) &&= 0 + j0
\end{aligned}$$

and $\mathbf{I}_N$ is in fact equal to **zero,** as required for a balanced load.

## 23.6 THE Y-Δ SYSTEM

There is no neutral connection for the Y-Δ system of Fig. 23.14. Any variation in the impedance of a phase which produces an unbalanced system will simply vary the line and phase currents of the system.

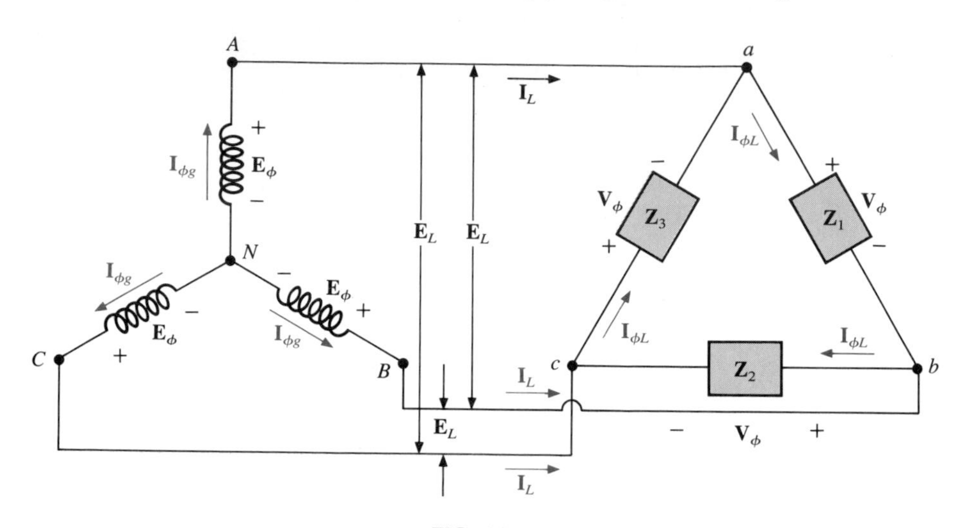


**FIG. 23.14**
*Y-connected generator with a Δ-connected load.*

For a balanced load,

$$\mathbf{Z}_1 = \mathbf{Z}_2 = \mathbf{Z}_3 \tag{23.9}$$

The voltage across each phase of the load is equal to the line voltage of the generator for a balanced or unbalanced load:

$$\mathbf{V}_\phi = \mathbf{E}_L \tag{23.10}$$

The relationship between the line currents and phase currents of a balanced Δ load can be found using an approach very similar to that used in Section 23.3 to find the relationship between the line voltages and phase voltages of a Y-connected generator. For this case, however, Kirchhoff's current law is employed instead of Kirchhoff's voltage law.

The results obtained are

$$I_L = \sqrt{3} I_\phi \tag{23.11}$$

and the phase angle between a line current and the nearest phase current is 30°. A more detailed discussion of this relationship between the line and phase currents of a Δ-connected system can be found in Section 23.7.

For a balanced load, the line currents will be equal in magnitude, as will the phase currents.

**EXAMPLE 23.2.** For the three-phase system of Fig. 23.15:

a. Find the phase angles $\theta_2$ and $\theta_3$.
b. Find the current in each phase of the load.
c. Find the magnitude of the line currents.

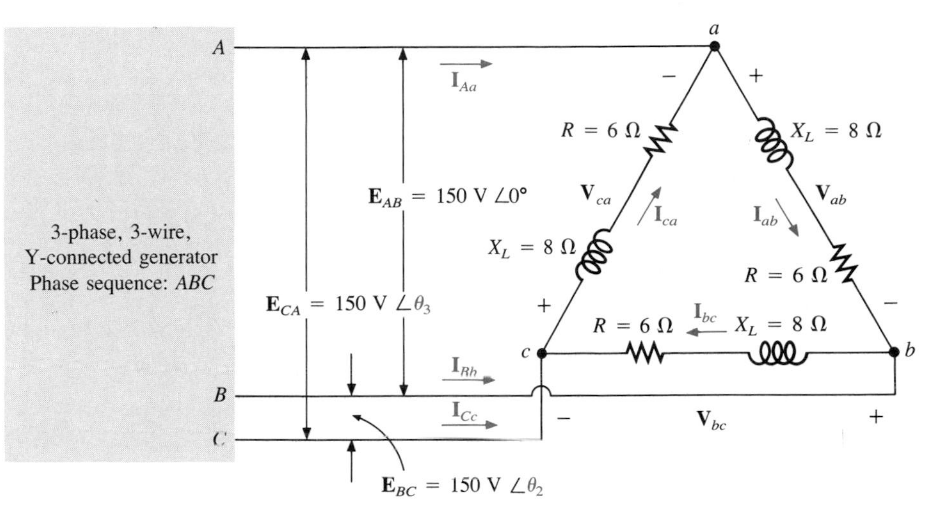


FIG. 23.15

***Solutions:***

a. For an *ABC* sequence,

$$\theta_2 = \mathbf{-120^\circ} \quad \text{and} \quad \theta_3 = \mathbf{+120^\circ}$$

b. $\mathbf{V}_\phi = \mathbf{E}_L$. Therefore,

$$\mathbf{V}_{ab} = \mathbf{E}_{AB} \qquad \mathbf{V}_{ca} = \mathbf{E}_{CA} \qquad \mathbf{V}_{bc} = \mathbf{E}_{BC}$$

The phase currents are

$$\mathbf{I}_{ab} = \frac{\mathbf{V}_{ab}}{\mathbf{Z}_{ab}} = \frac{150 \text{ V} \angle 0^\circ}{6\ \Omega + j8\ \Omega} = \frac{150 \text{ V} \angle 0^\circ}{10\ \Omega \angle 53.13^\circ} = \mathbf{15\ A \angle -53.13^\circ}$$

$$\mathbf{I}_{bc} = \frac{\mathbf{V}_{bc}}{\mathbf{Z}_{bc}} = \frac{150 \text{ V} \angle -120^\circ}{10\ \Omega \angle 53.13^\circ} = \mathbf{15\ A \angle -173.13^\circ}$$

$$\mathbf{I}_{ca} = \frac{\mathbf{V}_{ca}}{\mathbf{Z}_{ca}} = \frac{150 \text{ V} \angle +120^\circ}{10\ \Omega \angle 53.13^\circ} = \mathbf{15\ A \angle 66.87^\circ}$$

c. $I_L = \sqrt{3} I_\phi = (1.73)(15 \text{ A}) = 25.95 \text{ A}$. Therefore,

$$I_{Aa} = I_{Bb} = I_{Cc} = \mathbf{25.95\ A}$$

## 23.7 THE Δ-CONNECTED GENERATOR

If we rearrange the coils of the generator in Fig. 23.16(a) as shown in Fig. 23.16(b), the system is referred to as a *three-phase, three-wire, Δ-connected ac generator*. In this system, the phase and line voltages

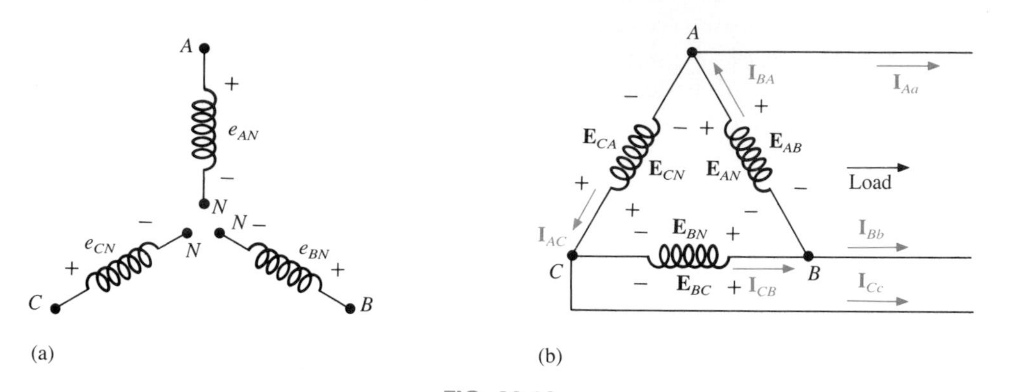


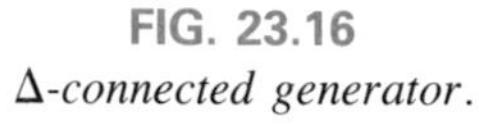

FIG. 23.16
*Δ-connected generator.*

are equivalent and equal to the voltage induced across each coil of the generator. That is,

$$\left.\begin{aligned} \mathbf{E}_{AB} &= \mathbf{E}_{AN} \quad \text{and} \quad e_{AN} = \sqrt{2}E_{AN}\sin\omega t \\ \mathbf{E}_{BC} &= \mathbf{E}_{BN} \quad \text{and} \quad e_{BN} = \sqrt{2}E_{BN}\sin(\omega t - 120^\circ) \\ \mathbf{E}_{CA} &= \mathbf{E}_{CN} \quad \text{and} \quad e_{CN} = \sqrt{2}E_{CN}\sin(\omega t + 120^\circ) \end{aligned}\right\} \begin{matrix}\text{Phase}\\ \text{sequence}\\ ABC\end{matrix}$$

or

$$\boxed{\mathbf{E}_L = \mathbf{E}_{\phi g}} \tag{23.12}$$

Note that only one voltage (magnitude) is available instead of the two available in the Y-connected system.

Unlike the line current for the Y-connected generator, the line current for the Δ-connected system is not equal to the phase current. The relationship between the two can be found by applying Kirchhoff's current law at one of the nodes and solving for the line current in terms of the phase currents. That is, at node $A$,

$$\mathbf{I}_{BA} = \mathbf{I}_{Aa} + \mathbf{I}_{AC}$$

or

$$\mathbf{I}_{Aa} = \mathbf{I}_{BA} - \mathbf{I}_{AC} = \mathbf{I}_{BA} + \mathbf{I}_{CA}$$

The phasor diagram is shown in Fig. 23.17 for a balanced load.

Using the same procedure to find the line current as was used to find the line voltage of a Y-connected generator produces the following general result:

$$\boxed{I_L = \sqrt{3}I_{\phi g}} \tag{23.13}$$

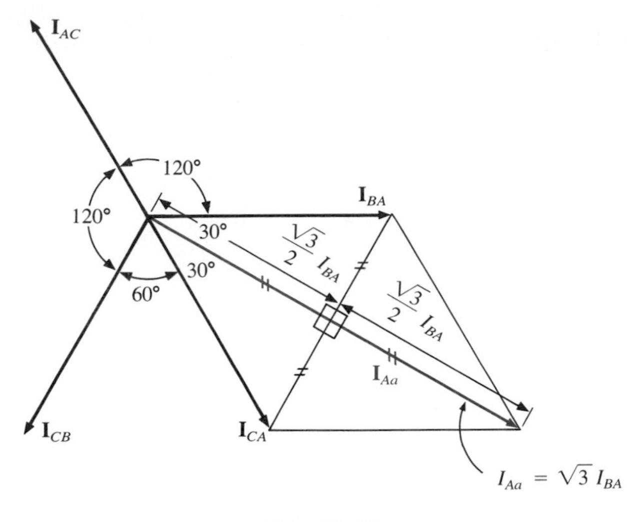


FIG. 23.17

The phase angle between a line current and the nearest phase current is 30°. The phasor diagram of the currents is shown in Fig. 23.18.

It can be shown in the same manner employed for the voltages of a Y-connected generator that the phasor sum of the line currents or phase currents for Δ-connected systems with balanced loads is zero.

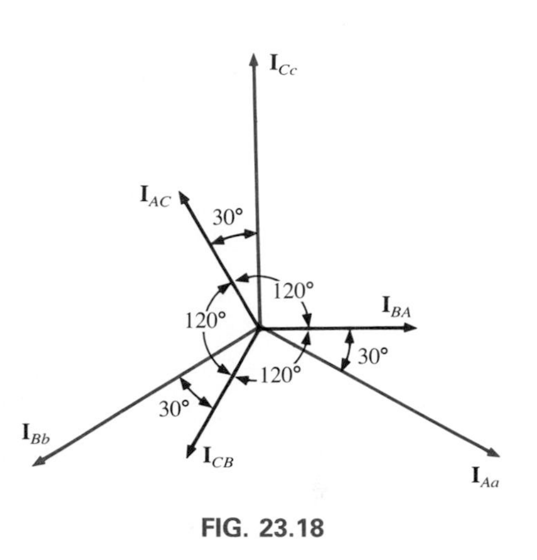


FIG. 23.18

## 23.8 PHASE SEQUENCE (Δ-CONNECTED GENERATOR)

Even though the line and phase voltages of a Δ-connected system are the same, it is standard practice to describe the phase sequence in terms of the line voltages. The method used is the same as that described for the line voltages of the Y-connected generator. For example, the phasor diagram of the line voltages for a phase sequence *ABC* is shown in Fig. 23.19. In drawing such a diagram, one must take care to have the sequence of the first and second subscripts the same.

In phasor notation,

$$\mathbf{E}_{AB} = E_{AB} \angle 0^\circ$$
$$\mathbf{E}_{BC} = E_{BC} \angle -120^\circ$$
$$\mathbf{E}_{CA} = E_{CA} \angle 120^\circ$$

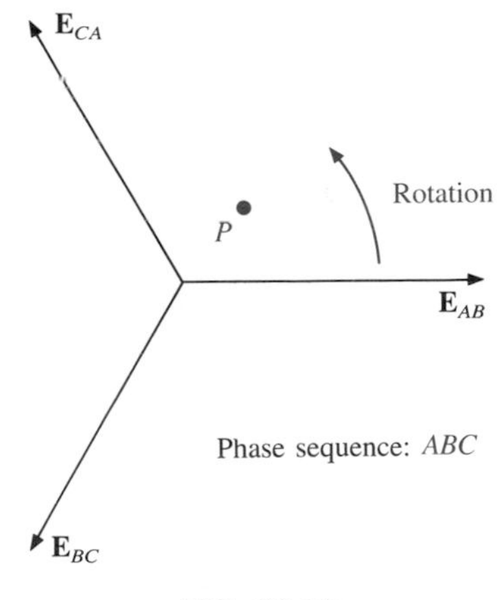

FIG. 23.19

## 23.9 THE Δ-Δ, Δ-Y THREE-PHASE SYSTEMS

The basic equations necessary to analyze either of the two systems (Δ-Δ, Δ-Y) have been presented at least once in this chapter. We will therefore proceed directly to two descriptive examples, one with a Δ-connected load and one with a Y-connected load.

**EXAMPLE 23.3.** For the Δ-Δ system shown in Fig. 23.20:

a. Find the phase angles $\theta_2$ and $\theta_3$ for the specified phase sequence.
b. Find the current in each phase of the load.
c. Find the magnitude of the line currents.

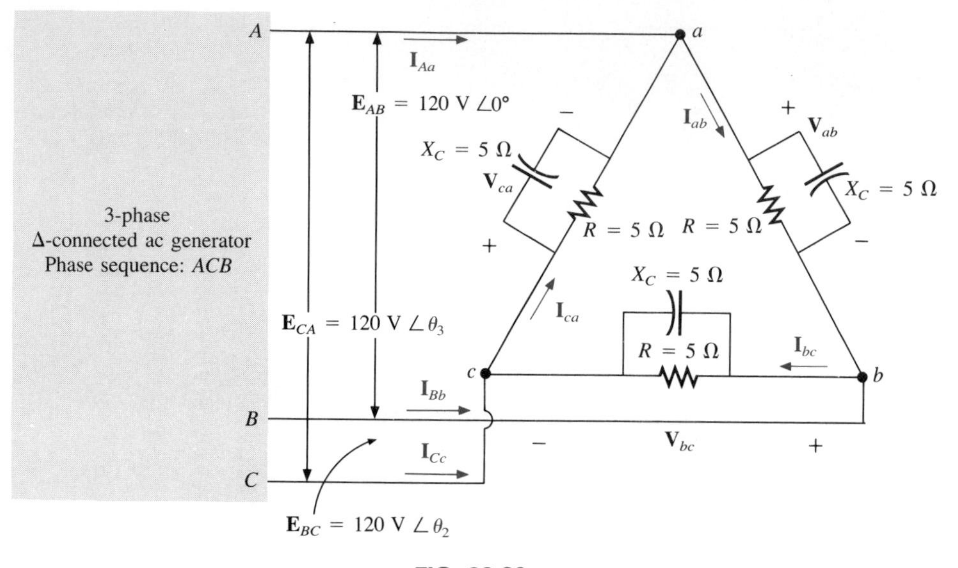


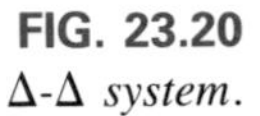

**FIG. 23.20**
*Δ-Δ system.*

***Solutions:***

a. For an *ACB* phase sequence,

$$\theta_2 = \mathbf{120^\circ} \quad \text{and} \quad \theta_3 = \mathbf{-120^\circ}$$

b. $\mathbf{V}_\phi = \mathbf{E}_L$. Therefore,

$$\mathbf{V}_{ab} = \mathbf{E}_{AB} \qquad \mathbf{V}_{ca} = \mathbf{E}_{CA} \qquad \mathbf{V}_{bc} = \mathbf{E}_{BC}$$

The phase currents are

$$\mathbf{I}_{ab} = \frac{\mathbf{V}_{ab}}{\mathbf{Z}_{ab}} = \frac{120\text{ V}\angle 0^\circ}{\dfrac{(5\ \Omega\angle 0^\circ)(5\ \Omega\angle -90^\circ)}{5\ \Omega - j5\ \Omega}} = \frac{120\text{ V}\angle 0^\circ}{\dfrac{25\ \Omega\angle -90^\circ}{7.07\angle -45^\circ}}$$

$$= \frac{120\text{ V}\angle 0^\circ}{3.54\ \Omega\angle -45^\circ} = \mathbf{33.9\ A\angle 45^\circ}$$

$$\mathbf{I}_{bc} = \frac{\mathbf{V}_{bc}}{\mathbf{Z}_{bc}} = \frac{120\text{ V}\angle 120^\circ}{3.54\ \Omega\angle -45^\circ} = \mathbf{33.9\ A\angle 165^\circ}$$

$$\mathbf{I}_{ca} = \frac{\mathbf{V}_{ca}}{\mathbf{Z}_{ca}} = \frac{120\text{ V}\angle -120^\circ}{3.54\ \Omega\angle -45^\circ} = \mathbf{33.9\ A\angle -75^\circ}$$

c. $I_L = \sqrt{3}I_\phi = (1.73)(34\text{ A}) = 58.82\text{ A}$. Therefore,

$$I_{Aa} = I_{Bb} = I_{Cc} = \mathbf{58.82\ A}$$

**EXAMPLE 23.4.** For the Δ-Y system shown in Fig. 23.21:

a. Find the voltage across each phase of the load.
b. Find the magnitude of the line voltages.

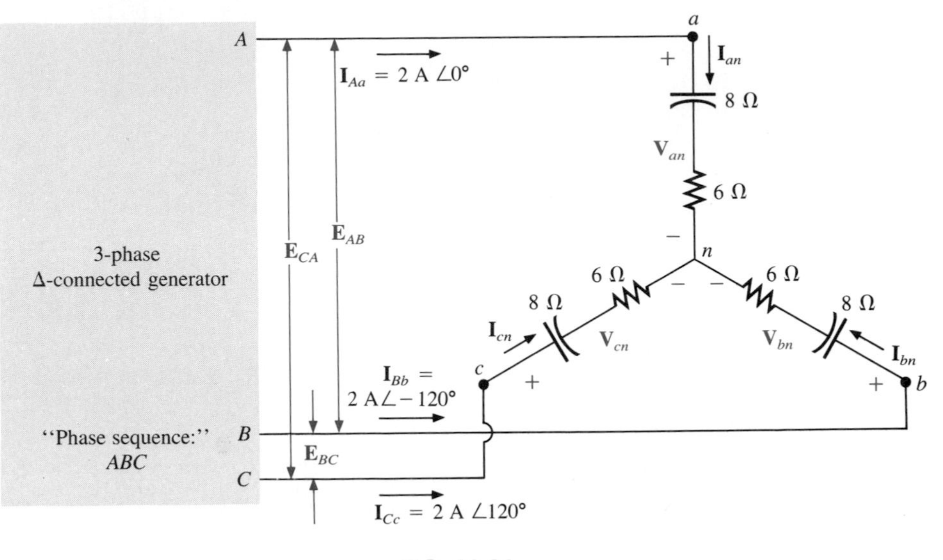


FIG. 23.21
*Δ-Y system.*

***Solutions:***

a. $\mathbf{I}_{\phi L} = \mathbf{I}_L$. Therefore,

$$\mathbf{I}_{an} = \mathbf{I}_{Aa} - 2\ \text{A} \angle 0°$$
$$\mathbf{I}_{bn} = \mathbf{I}_{Bb} = 2\ \text{A} \angle -120°$$
$$\mathbf{I}_{cn} = \mathbf{I}_{Cc} = 2\ \text{A} \angle 120°$$

The phase voltages are

$$\mathbf{V}_{an} = \mathbf{I}_{an}\mathbf{Z}_{an} = (2\ \text{A} \angle 0°)(10\ \Omega \angle -53.13°) = \mathbf{20\ V \angle -53.13°}$$
$$\mathbf{V}_{bn} = \mathbf{I}_{bn}\mathbf{Z}_{bn} = (2\ \text{A} \angle -120°)(10\ \Omega \angle -53.13°) = \mathbf{20\ V \angle -173.13°}$$
$$\mathbf{V}_{cn} = \mathbf{I}_{cn}\mathbf{Z}_{cn} = (2\ \text{A} \angle 120°)(10\ \Omega \angle -53.13°) = \mathbf{20\ V \angle 66.87°}$$

b. $E_L = \sqrt{3}V_\phi = (1.73)(20\ \text{V}) = 34.6\ \text{V}$. Therefore,

$$E_{BA} = E_{CB} = E_{AC} = \mathbf{34.6\ V}$$

## 23.10 POWER

### Y-Connected Balanced Load (Fig. 23.22)

**Average Power** The average power delivered to each phase can be determined by any one of Eqs. (23.14) through (23.16).

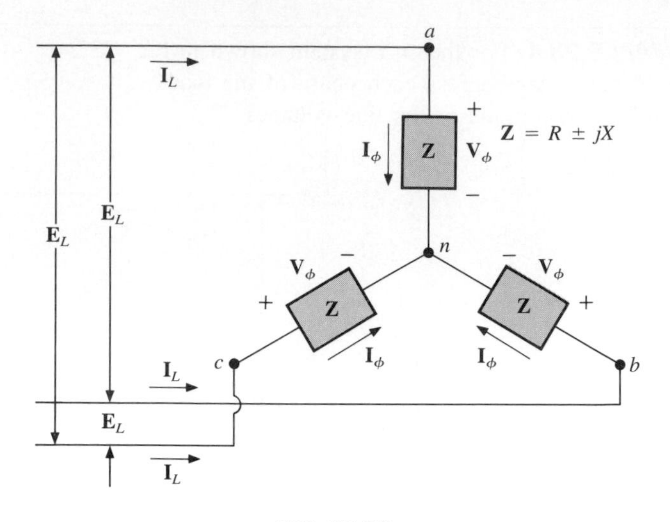

FIG. 23.22

$$P_\phi = V_\phi I_\phi \cos \theta_{I_\phi}^{V_\phi} = I_\phi^2 R_\phi = \frac{V_R^2}{R_\phi} \qquad \text{(watts, W)} \quad \mathbf{(23.14)}$$

where $\theta_{I_\phi}^{V_\phi}$ indicates that $\theta$ is the phase angle between $V_\phi$ and $I_\phi$. The total power to the balanced load is

$$P_T = 3P_\phi \qquad \text{(W)} \quad \mathbf{(23.15)}$$

or, since

$$V_\phi = \frac{E_L}{\sqrt{3}} \quad \text{and} \quad I_\phi = I_L$$

then

$$P_T = 3\frac{E_L}{\sqrt{3}} I_L \cos \theta_{I_\phi}^{V_\phi}$$

But

$$\left(\frac{3}{\sqrt{3}}\right)(1) = \left(\frac{3}{\sqrt{3}}\right)\left(\frac{\sqrt{3}}{\sqrt{3}}\right) = \frac{3\sqrt{3}}{3} = \sqrt{3}$$

Therefore,

$$P_T = \sqrt{3} E_L I_L \cos \theta_{I_\phi}^{V_\phi} = 3 I_L^2 R_\phi \qquad \text{(W)} \quad \mathbf{(23.16)}$$

**Reactive Power** The reactive power of each phase (in volt-amperes reactive) is

$$Q_\phi = V_\phi I_\phi \sin \theta_{I_\phi}^{V_\phi} = I_\phi^2 X_\phi = \frac{V_X^2}{X_\phi} \qquad \text{(VAR)} \qquad \textbf{(23.17)}$$

The total reactive power of the load is

$$Q_T = 3Q_\phi \qquad \text{(VAR)} \qquad \textbf{(23.18)}$$

or, proceeding in the same manner as above, we have

$$Q_T = \sqrt{3}E_L I_L \sin \theta_{I_\phi}^{V_\phi} = 3I_L^2 X_\phi \qquad \text{(VAR)} \qquad \textbf{(23.19)}$$

**Apparent Power** The apparent power of each phase is

$$S_\phi = V_\phi I_\phi \qquad \text{(VA)} \qquad \textbf{(23.20)}$$

The total apparent power of the load is

$$S_T = 3S_\phi \qquad \text{(VA)} \qquad \textbf{(23.21)}$$

or, as before,

$$S_T = \sqrt{3}E_L I_L \qquad \text{(VA)} \qquad \textbf{(23.22)}$$

**Power Factor** The power factor of the system is given by

$$F_p = \frac{P_T}{S_T} = \cos \theta \text{ (leading or lagging)} \qquad \textbf{(23.23)}$$

---

**EXAMPLE 23.5.** See Fig. 23.23. Here,

$$\mathbf{Z}_\phi = 3\ \Omega + j4\ \Omega = 5\ \Omega \angle 53.13^\circ$$

$$V_\phi = \frac{V_L}{\sqrt{3}} = \frac{173.2\ \text{V}}{1.732} = 100\ \text{V}$$

$$I_\phi = \frac{V_\phi}{Z_\phi} = \frac{100\ \text{V}}{5\ \Omega} = 20\ \text{A}$$

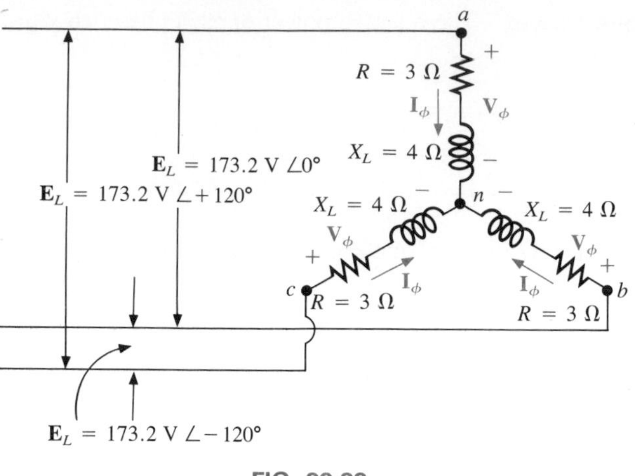

FIG. 23.23

The *average power* is

$$P_\phi = V_\phi I_\phi \cos \theta_{I_\phi}^{V_\phi} = (100 \text{ V})(20 \text{ A}) \cos 53.13° = (2000)(0.6) = \mathbf{1200\ W}$$

$$P_\phi = I_\phi^2 R_\phi = (20 \text{ A})^2(3\ \Omega) = (400)(3) = \mathbf{1200\ W}$$

$$P_\phi = \frac{V_R^2}{R_\phi} = \frac{(60 \text{ V})^2}{3\ \Omega} = \frac{3600}{3} = \mathbf{1200\ W}$$

$$P_T = 3P_\phi = (3)(1200 \text{ W}) = \mathbf{3600\ W}$$

or

$$P_T = \sqrt{3}E_L I_L \cos \theta_{I_\phi}^{V_\phi} = (1.732)(173.2 \text{ V})(20 \text{ A})(0.6) = \mathbf{3600\ W}$$

The *reactive power* is

$$Q_\phi = V_\phi I_\phi \sin \theta_{I_\phi}^{V_\phi} = (100 \text{ V})(20 \text{ A}) \sin 53.13° = (2000)(0.8) = \mathbf{1600\ VAR}$$

or

$$Q_\phi = I_\phi^2 X_\phi = (20 \text{ A})^2(4\ \Omega) = (400)(4) = \mathbf{1600\ VAR}$$

$$Q_T = 3Q_\phi = (3)(1600 \text{ VAR}) = \mathbf{4800\ VAR}$$

or

$$Q_T = \sqrt{3}E_L I_L \sin \theta_{I_\phi}^{V_\phi} = (1.732)(173.2 \text{ V})(20 \text{ A})(0.8) = \mathbf{4800\ VAR}$$

The *apparent power* is

$$S_\phi = V_\phi I_\phi = (100 \text{ V})(20 \text{ A}) = \mathbf{2000\ VA}$$

$$S_T = 3S_\phi = (3)(2000 \text{ VA}) = \mathbf{6000\ VA}$$

or

$$S_T = \sqrt{3}E_L I_L = (1.732)(173.2 \text{ V})(20 \text{ A}) = \mathbf{6000\ VA}$$

The *power factor* is

$$F_p = \frac{P_T}{S_T} = \frac{3600 \text{ W}}{6000 \text{ VA}} = \mathbf{0.6\ lagging}$$

## Δ-Connected Balanced Load (Fig. 23.24)

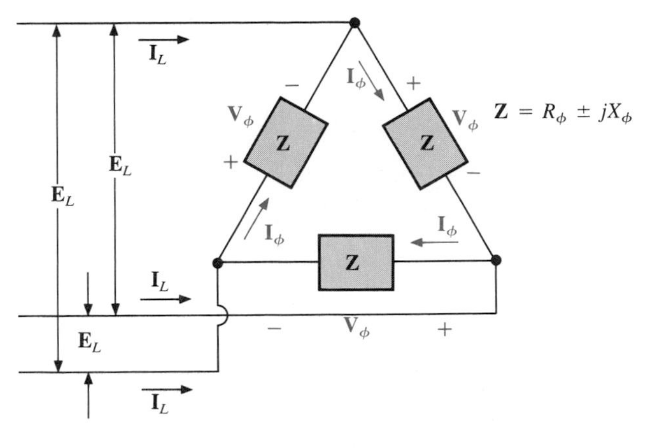

FIG. 23.24

### Average Power

$$P_\phi = V_\phi I_\phi \cos \theta_{I_\phi}^{V_\phi} = I_\phi^2 R_\phi = \frac{V_R^2}{R_\phi} \qquad \text{(W)} \qquad (23.24)$$

$$P_T = 3P_\phi \qquad \text{(W)} \qquad (23.25)$$

### Reactive Power

$$Q_\phi = V_\phi I_\phi \sin \theta_{I_\phi}^{V_\phi} = I_\phi^2 X_\phi = \frac{V_X^2}{X_\phi} \qquad \text{(VAR)} \qquad (23.26)$$

$$Q_T = 3Q_\phi \qquad \text{(VAR)} \qquad (23.27)$$

### Apparent Power

$$S_\phi = V_\phi I_\phi \qquad \text{(VA)} \qquad (23.28)$$

$$S_T = 3S_\phi = \sqrt{3}E_L I_L \qquad \text{(VA)} \qquad (23.29)$$

### Power Factor

$$F_p = \frac{P_T}{S_T} \qquad (23.30)$$

**EXAMPLE 23.6.** Determine the total watts, volt-amperes reactive, and volt-amperes for the network of Fig. 23.25. In addition, calculate the total power factor of the load.

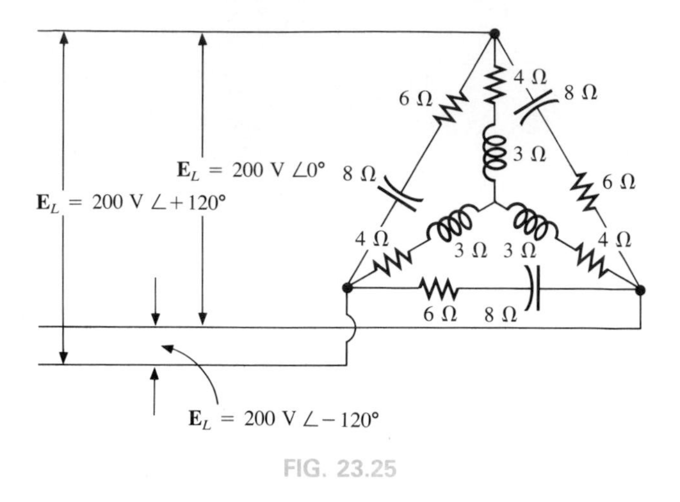


FIG. 23.25

*Solution:* Consider the Δ and Y separately.

**For the Δ:**

$$\mathbf{Z}_\Delta = 6\ \Omega - j8\ \Omega = 10\ \Omega\ \angle -53.13^\circ$$

$$I_\phi = \frac{E_L}{Z_\Delta} = \frac{200\ \text{V}}{10\ \Omega} = 20\ \text{A}$$

$$P_{T_\Delta} = 3I_\phi^2 R_\phi = (3)(20\ \text{A})^2(6\ \Omega) = \mathbf{7200\ W}$$

$$Q_{T_\Delta} = 3I_\phi^2 X_\phi = (3)(20\ \text{A})^2(8\ \Omega) = \mathbf{9600\ VAR\ (cap.)}$$

$$S_{T_\Delta} = 3V_\phi I_\phi = (3)(200\ \text{A})(20\ \Omega) = \mathbf{12{,}000\ VA}$$

**For the Y:**

$$\mathbf{Z}_Y = 4\ \Omega + j3\ \Omega = 5\ \Omega\ \angle 36.87^\circ$$

$$I_\phi = \frac{E_L/\sqrt{3}}{Z_Y} = \frac{200\ \text{V}/\sqrt{3}}{5\ \Omega} = \frac{116\ \text{V}}{5\ \Omega} = 23.12\ \text{A}$$

$$P_{T_Y} = 3I_\phi^2 R_\phi = (3)(23.12\ \text{A})^2(4\ \Omega) = \mathbf{6414.41\ W}$$

$$Q_{T_Y} = 3I_\phi^2 X_\phi = (3)(23.12\ \text{A})^2(3\ \Omega) = \mathbf{4810.81\ VAR\ (ind.)}$$

$$S_{T_Y} = 3V_\phi I_\phi = (3)(116\ \text{V})(23.12\ \text{A}) = \mathbf{8045.76\ VA}$$

**For the total load:**

$$P_T = P_{T_\Delta} + P_{T_Y} = 7200\ \text{W} + 6414.41\ \text{W} = \mathbf{13{,}614.41\ W}$$

$$Q_T = Q_{T_\Delta} - Q_{T_Y} = 9600\ \text{VAR (cap.)} - 4810.81\ \text{VAR (ind.)} = \mathbf{4789.19\ VAR\ (cap.)}$$

$$S_T = \sqrt{P_T^2 + Q_T^2} = \sqrt{(13{,}614.41\ \text{W})^2 + (4789.19\ \text{VAR})^2} = \mathbf{14{,}432.2\ VA}$$

$$F_p = \frac{P_T}{S_T} = \frac{13{,}614.41\ \text{W}}{14{,}432.20\ \text{VA}} = \mathbf{0.943\ leading}$$

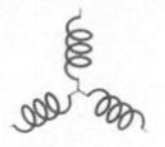

**EXAMPLE 23.7.** Each transmission line of the three-wire, three-phase system of Fig. 23.26 has an impedance of 15 Ω + *j*20 Ω. The system

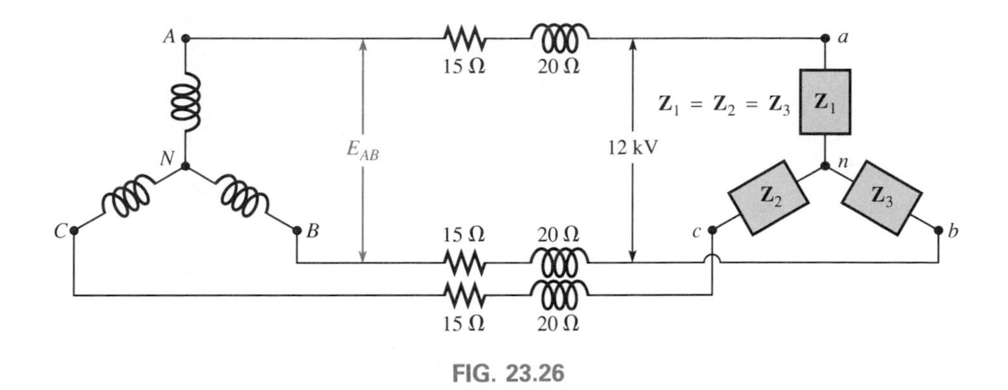


FIG. 23.26

delivers a total power of 160 kW at 12,000 V to a balanced three-phase load with a lagging power factor of 0.86.

a. Determine the magnitude of the line voltage $E_{AB}$ of the generator.
b. Find the power factor of the total load applied to the generator.
c. What is the efficiency of the system?

***Solutions:***

a. $V_\phi \text{ (load)} = \dfrac{V_L}{\sqrt{3}} = \dfrac{12{,}000 \text{ V}}{1.73} = 6936.42 \text{ V}$

$P_T \text{ (load)} = 3V_\phi I_\phi \cos\theta$

and

$$I_\phi = \frac{P_T}{3V_\phi \cos\theta} = \frac{160{,}000 \text{ W}}{3(6936.42 \text{ V})(0.86)} = \mathbf{8.94\ A}$$

Since $\theta = \cos^{-1} 0.86 = 30.68°$, assigning $\mathbf{V}_\phi$ an angle of 0° or $\mathbf{V}_\phi = V_\phi \angle 0°$, a lagging power factor results in

$$\mathbf{I}_\phi = 8.94 \text{ A} \angle -30.68°$$

For each phase, the system will appear as shown in Fig. 23.27, where

$$\mathbf{E}_{AN} - \mathbf{I}_\phi[\mathbf{Z}_{\text{line}}] - \mathbf{V}_\phi = 0$$

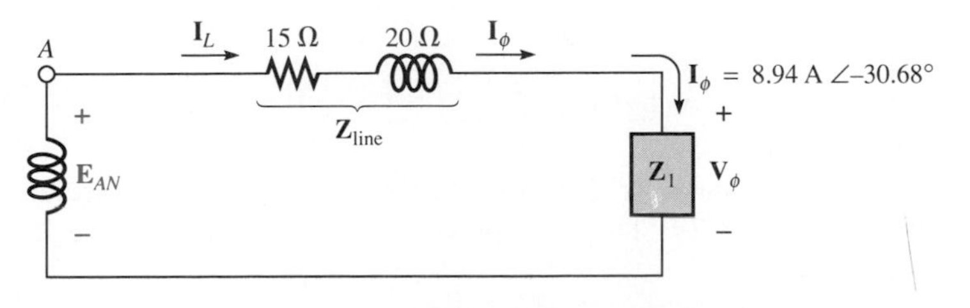


FIG. 23.27

or

$$\begin{aligned}\mathbf{E}_{AN} &= \mathbf{I}_\phi[\mathbf{Z}_{\text{line}}] + \mathbf{V}_\phi\\ &= (8.94\text{ A }\angle -30.68^\circ)(25\ \Omega\ \angle 53.13^\circ) + 6936.42\text{ V }\angle 0^\circ\\ &= 223.5\text{ V }\angle 22.45^\circ + 6936.42\text{ V }\angle 0^\circ\\ &= 206.56\text{ V} + j85.35\text{ V} + 6936.42\text{ V}\\ &= 7142.98\text{ V} + j85.35\text{ V}\\ &= 7143.5\text{ V }\angle 0.68^\circ\end{aligned}$$

then

$$\begin{aligned}E_{AB} &= \sqrt{3}E_{\phi g} = (1.73)(7143.5\text{ V})\\ &= \mathbf{12{,}358.26\ V}\end{aligned}$$

b. $$\begin{aligned}P_T &= P_{\text{load}} + P_{\text{lines}}\\ &= 160\text{ kW} + 3(I_L)^2R_{\text{line}}\\ &= 160\text{ kW} + 3(8.94\text{ A})^2 15\ \Omega\\ &= 160{,}000\text{ W} + 3596.55\text{ W}\\ &= 163{,}596.55\text{ W}\end{aligned}$$

and

$$P_T = \sqrt{3}V_L I_L \cos\theta_T$$

or

$$\cos\theta_T = \frac{P_T}{\sqrt{3}V_L I_L} = \frac{163{,}596.55\text{ W}}{(1.73)(12{,}358.26\text{ V})(8.94\text{ A})}$$

and

$$F_p = \mathbf{0.856} < 0.86 \text{ of load}$$

c. $$\begin{aligned}\eta &= \frac{P_o}{P_i} = \frac{P_o}{P_o + P_{\text{losses}}} = \frac{160\text{ kW}}{160\text{ kW} + 3596.55\text{ W}} = 0.978\\ &= \mathbf{97.8\%}\end{aligned}$$

## 23.11 THE THREE-WATTMETER METHOD

The power delivered to a balanced or an unbalanced four-wire Y-connected load can be found using three wattmeters in the manner shown in Fig. 23.28. Each wattmeter measures the power delivered to each phase. The potential coil of each wattmeter is connected parallel with the load, while the current coil is in series with the load. The total average power of the system can be found by summing the three wattmeter readings; that is,

$$P_{T_Y} = P_1 + P_2 + P_3 \tag{23.31}$$

For the load (balanced or unbalanced), the wattmeters are connected as shown in Fig. 23.29. The total power is again the sum of the three wattmeter readings:

$$P_{T_\Delta} = P_1 + P_2 + P_3 \tag{23.32}$$

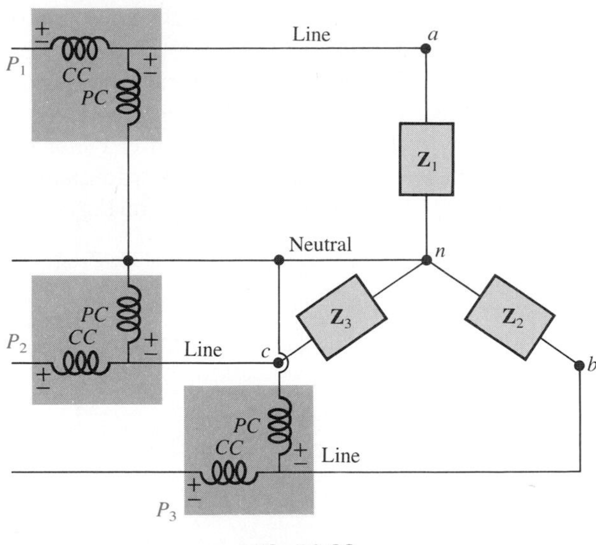


FIG. 23.28

FIG. 23.29

If in either of the cases just described the load is balanced, the power delivered to each phase will be the same. The total power is then just three times any one wattmeter reading.

## 23.12 THE TWO-WATTMETER METHOD

The power delivered to a three-phase, three-wire, Δ- or Y-connected balanced or unbalanced load can be found using only two wattmeters if the proper connection is employed and if the wattmeter readings are properly interpreted. The basic connections are shown in Fig. 23.30. One end of each potential coil is connected to the same line. The current coils are then placed in the remaining lines.

The connection shown in Fig. 23.31 will also satisfy the requirements. A third hookup is also possible, but this is left to the reader as an exercise.

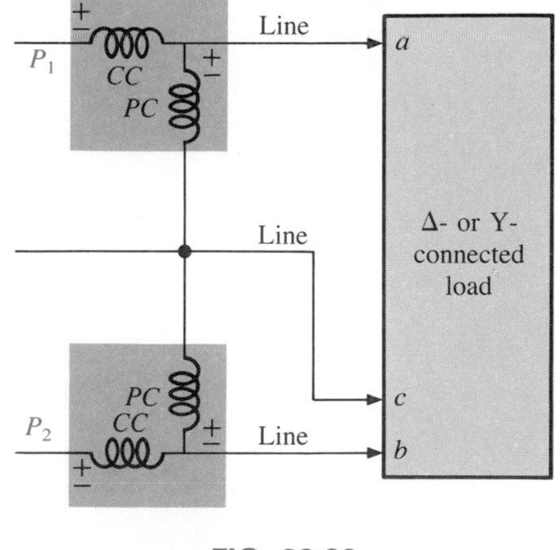


FIG. 23.30

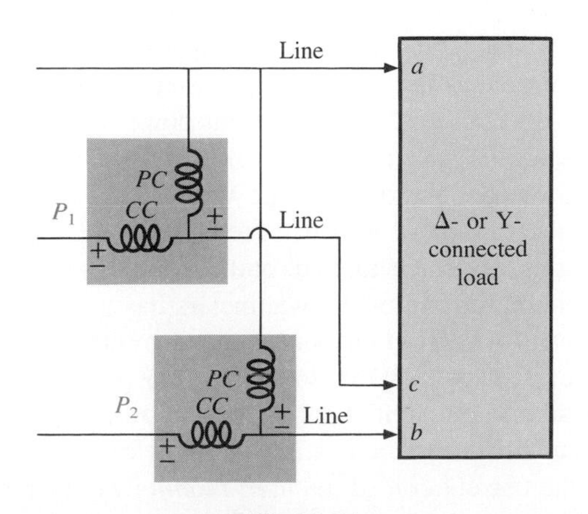


FIG. 23.31

The total power delivered to the load is the algebraic sum of the two wattmeter readings. For a *balanced* load, we will now consider two methods of determining whether the total power is the sum or the difference of the two wattmeter readings. The first method to be described requires that we know or be able to find the power factor (leading or lagging) of any one phase of the load. When this information has been obtained, it can be applied directly to the curve of Fig. 23.32.

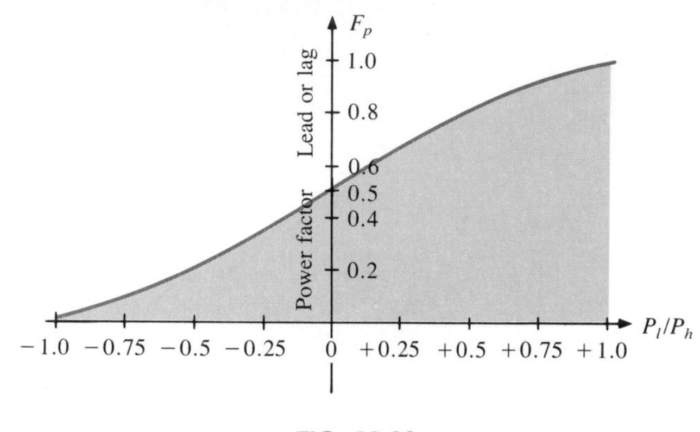


FIG. 23.32

The curve in Fig. 23.32 is a plot of the power factor of the load (phase) versus the ratio $P_l/P_h$, where $P_l$ and $P_h$ are the magnitudes of the lower- and higher-reading wattmeters, respectively. Note that for a power factor (leading or lagging) greater than 0.5, the ratio has a positive value. This indicates that both wattmeters are reading positive, and the total power is the sum of the two wattmeter readings; that is, $P_T = P_l + P_h$. For a power factor less than 0.5 (leading or lagging), the ratio has a negative value. This indicates that the smaller-reading wattmeter is reading negative, and the total power is the difference of the two wattmeter readings; that is, $P_T = P_h - P_l$.

A closer examination will reveal that when the power factor is 1 ($\cos 0° = 1$), corresponding to a purely resistive load, $P_l/P_h = 1$ or $P_l = P_h$, and both wattmeters will have the same wattage indication. At a power factor equal to 0 ($\cos 90° = 0$), corresponding to a purely reactive load, $P_l/P_h = -1$ or $P_l = -P_h$, and both wattmeters will again have the same wattage indication but with opposite signs. The transition from a negative to a positive ratio occurs when the power factor of the load is 0.5 or $\theta = \cos^{-1} 0.5 = 60°$. At this power factor, $P_l/P_h = 0$, so that $P_l = 0$, while $P_h$ will read the total power delivered to the load.

The second method for determining whether the total power is the sum or difference of the two wattmeter readings involves a simple laboratory test. For the test to be applied, both wattmeters must first have an up-scale deflection. If one of the wattmeters has a below-zero indication, an up-scale deflection can be obtained by simply reversing the leads of the current coil of the wattmeter. To perform the test, first remove the lead of the potential coil of the *low-reading* wattmeter from the line that has no current coil in it. Take this lead and touch it to the line that has the current coil of the *high-reading* wattmeter in it. If the pointer of the low-reading wattmeter deflects upward, the two wattmeter readings should be added. If the pointer deflects downward (below zero watts), the wattage reading of the low-reading wattmeter should be subtracted from that of the high-reading wattmeter.

For a *balanced system,* since

$$P_T = P_h \pm P_l = \sqrt{3}E_L I_L \cos \theta_{I_\phi}^{V_\phi}$$

the power factor of the load (phase) can be found from the wattmeter readings and the magnitude of the line voltage and current:

$$F_p = \cos \theta_{I_\phi}^{V_\phi} = \frac{P_h \pm P_l}{\sqrt{3}E_L I_L} \quad \textbf{(23.33)}$$

EXAMPLE 23.8. For the unbalanced delta-connected load of Fig. 23.33 with two properly connected wattmeters,

a. Determine the magnitude and angle of the phase currents.
b. Calculate the magnitude and angle of the line currents.
c. Determine the power reading of each wattmeter.
d. Calculate the total power absorbed by the load.
e. Compare the result of part (d) with the total power calculated using the phase currents and the resistive elements.

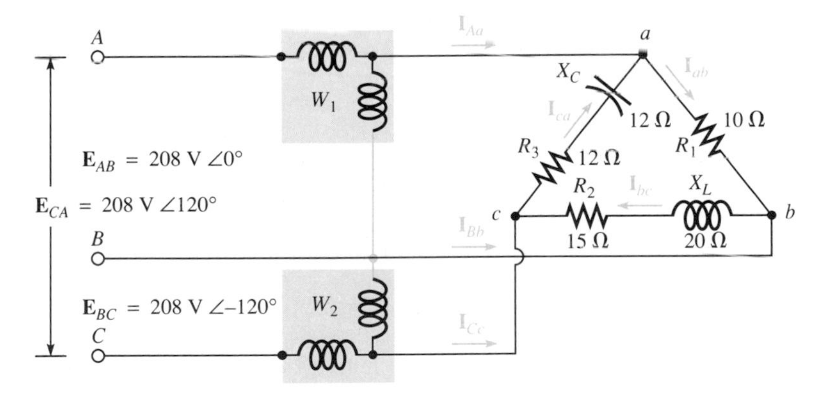


FIG. 23.33

*Solutions:*

a. $$\mathbf{I}_{ab} = \frac{\mathbf{V}_{ab}}{\mathbf{Z}_{ab}} = \frac{\mathbf{E}_{AB}}{\mathbf{Z}_{ab}} = \frac{208 \text{ V } \angle 0^\circ}{10\ \Omega \angle 0^\circ} = \mathbf{20.8\ A \angle 0^\circ}$$

$$\mathbf{I}_{bc} = \frac{\mathbf{V}_{bc}}{\mathbf{Z}_{bc}} = \frac{\mathbf{E}_{BC}}{\mathbf{Z}_{bc}} = \frac{208 \text{ V } \angle -120^\circ}{15\ \Omega + j20\ \Omega} = \frac{208 \text{ V } \angle -120^\circ}{25\ \Omega \angle 53.13^\circ}$$
$$= \mathbf{8.32\ A \angle -173.13^\circ}$$

$$\mathbf{I}_{ca} = \frac{\mathbf{V}_{ca}}{\mathbf{Z}_{ca}} = \frac{\mathbf{E}_{CA}}{\mathbf{Z}_{ca}} = \frac{208 \text{ V } \angle +120^\circ}{12\ \Omega + j12\ \Omega} = \frac{208 \text{ V } \angle +120^\circ}{16.97\ \Omega \angle -45^\circ}$$
$$= \mathbf{12.26\ A \angle 165^\circ}$$

b. $\mathbf{I}_{Aa} = \mathbf{I}_{ab} - \mathbf{I}_{ca}$
$= 20.8 \text{ A} \angle 0° - 12.26 \text{ A} \angle 165°$
$= 20.8 \text{ A} - (-11.84 \text{ A} + j3.17 \text{ A})$
$= 20.8 \text{ A} + 11.84 \text{ A} - j3.17 \text{ A} = 32.64 \text{ A} - j3.17 \text{ A}$
$= \mathbf{32.79 \text{ A} \angle -5.55°}$

$\mathbf{I}_{Bb} = \mathbf{I}_{bc} - \mathbf{I}_{ab}$
$= 8.32 \text{ A} \angle -173.13° - 20.8 \text{ A} \angle 0°$
$= (-8.26 \text{ A} - j1 \text{ A}) - 20.8 \text{ A}$
$= -8.26 \text{ A} - 20.8 \text{ A} - j1 \text{ A} = -29.06 \text{ A} - j1 \text{ A}$
$= \mathbf{29.08 \text{ A} \angle -178.03°}$

$\mathbf{I}_{Cc} = \mathbf{I}_{ca} - \mathbf{I}_{bc}$
$= 12.26 \text{ A} \angle 165° - 8.32 \text{ A} \angle -173.13°$
$= (-11.84 \text{ A} + j3.17 \text{ A}) - (-8.26 \text{ A} - j1 \text{ A})$
$= -11.84 \text{ A} + 8.26 \text{ A} + j(3.17 \text{ A} + 1 \text{ A}) = -3.58 \text{ A} + j4.17 \text{ A}$
$= \mathbf{5.5 \text{ A} \angle 130.65°}$

c. $P_1 = V_{ab} I_{Aa} \cos \theta_{\mathbf{I}_{Aa}}^{\mathbf{V}_{ab}}, \mathbf{V}_{ab} = 208 \text{ V} \angle 0°, \mathbf{I}_{Aa} = 32.79 \text{ A} \angle -5.55°$
$= (208 \text{ V})(32.79 \text{ A}) \cos 5.55°$
$= \mathbf{6788.35 \text{ W}}$

$\mathbf{V}_{bc} = \mathbf{E}_{BC} = 208 \text{ V} \angle -120°$
but $\mathbf{V}_{cb} = \mathbf{E}_{CB} = 208 \text{ V} \angle -120° + 180°$
$= 208 \text{ V} \angle 60°$
with $\mathbf{I}_{Cc} = 5.5 \text{ A} \angle 130.65°$

$P_2 = V_{cb} I_{Cc} \cos \theta_{\mathbf{I}_{Cc}}^{\mathbf{V}_{cb}}$
$= (208 \text{ V})(5.5 \text{ A}) \cos 70.65°$
$= \mathbf{379.1 \text{ W}}$

d. $P_T = P_1 + P_2 = 6788.35 \text{ W} + 379.1 \text{ W}$
$= \mathbf{7167.45 \text{ W}}$

e. $P_T = (I_{ab})^2 R_1 + (I_{bc})^2 R_2 + (I_{ca})^2 R_3$
$= (20.8 \text{ A})^2 10 \ \Omega + (8.32 \text{ A})^2 15 \ \Omega + (12.26 \text{ A})^2 12 \ \Omega$
$= 4326.4 \text{ W} + 1038.34 \text{ W} + 1803.69 \text{ W}$
$= \mathbf{7168.43 \text{ W}}$ (The slight difference is due to the level of accuracy carried through the calculations.)

## 23.13 UNBALANCED THREE-PHASE, FOUR-WIRE, Y-CONNECTED LOAD

For the three-phase, four-wire, Y-connected load of Fig. 23.34, conditions are such that *none* of the load impedances are equal. Since the neutral is a common point between the load and source, no matter what the impedance of each phase of the load, the voltage across each phase is the phase voltage of the generator:

$$\mathbf{V}_\phi = \mathbf{E}_\phi \qquad (23.34)$$

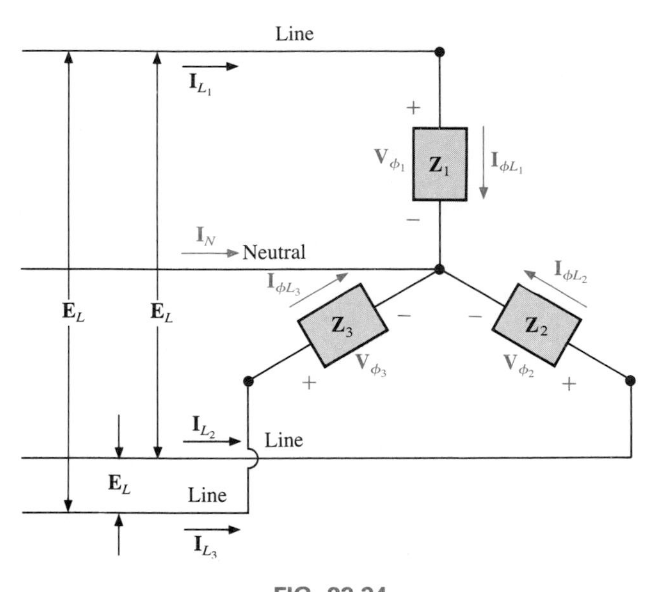


FIG. 23.34
*Unbalanced Y-connected load.*

The phase currents, therefore, can be determined by Ohm's law:

$$\mathbf{I}_{\phi_1} = \frac{\mathbf{V}_{\phi_1}}{\mathbf{Z}_1} = \frac{\mathbf{E}_{\phi_1}}{\mathbf{Z}_1}, \text{ and so on} \qquad \textbf{(23.35)}$$

The current in the neutral for any unbalanced system can then be found by applying Kirchhoff's current law at the common point *n*:

$$\mathbf{I}_N = \mathbf{I}_{\phi_1} + \mathbf{I}_{\phi_2} + \mathbf{I}_{\phi_3} = \mathbf{I}_{L_1} + \mathbf{I}_{L_2} + \mathbf{I}_{L_3} \qquad \textbf{(23.36)}$$

## 23.14 UNBALANCED THREE-PHASE, THREE-WIRE, Y-CONNECTED LOAD

For the system shown in Fig. 23.35, the required equations can be derived by first applying Kirchhoff's voltage law around each closed loop to produce

$$\mathbf{E}_{AB} - \mathbf{V}_{an} + \mathbf{V}_{bn} = 0$$
$$\mathbf{E}_{BC} - \mathbf{V}_{bn} + \mathbf{V}_{cn} = 0$$
$$\mathbf{E}_{CA} - \mathbf{V}_{cn} + \mathbf{V}_{an} = 0$$

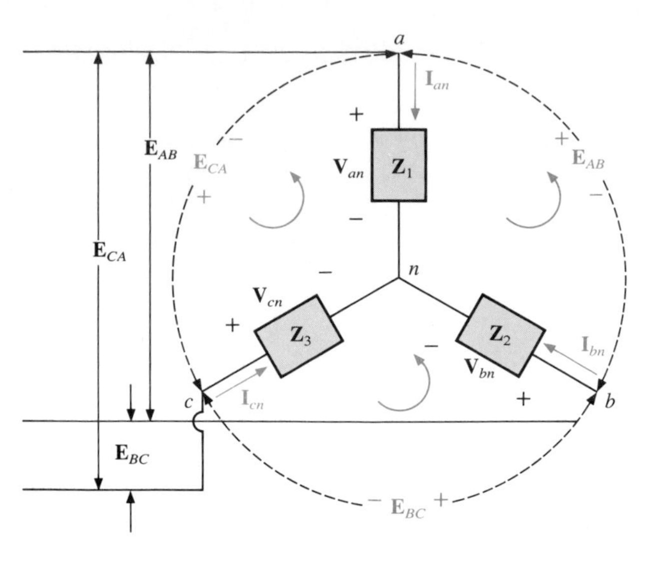

FIG. 23.35

Substituting, we have

$$\mathbf{V}_{an} = \mathbf{I}_{an}\mathbf{Z}_1 \qquad \mathbf{V}_{bn} = \mathbf{I}_{bn}\mathbf{Z}_2 \qquad \mathbf{V}_{cn} = \mathbf{I}_{cn}\mathbf{Z}_3$$

$$\mathbf{E}_{AB} = \mathbf{I}_{an}\mathbf{Z}_1 - \mathbf{I}_{bn}\mathbf{Z}_2 \tag{23.37a}$$

$$\mathbf{E}_{BC} = \mathbf{I}_{bn}\mathbf{Z}_2 - \mathbf{I}_{cn}\mathbf{Z}_3 \tag{23.37b}$$

$$\mathbf{E}_{CA} = \mathbf{I}_{cn}\mathbf{Z}_3 - \mathbf{I}_{an}\mathbf{Z}_1 \tag{23.37c}$$

Applying Kirchhoff's current law at node $n$ results in

$$\mathbf{I}_{an} + \mathbf{I}_{bn} + \mathbf{I}_{cn} = 0 \quad \text{and} \quad \mathbf{I}_{bn} = -\mathbf{I}_{an} - \mathbf{I}_{cn}$$

Substituting for $\mathbf{I}_{bn}$ in Eqs. (23.37a) and (23.37b) yields

$$\mathbf{E}_{AB} = \mathbf{I}_{an}\mathbf{Z}_1 - [-(\mathbf{I}_{an} + \mathbf{I}_{cn})]\mathbf{Z}_2$$

$$\mathbf{E}_{BC} = -(\mathbf{I}_{an} + \mathbf{I}_{cn})\mathbf{Z}_2 - \mathbf{I}_{cn}\mathbf{Z}_3$$

which are rewritten as

$$\mathbf{E}_{AB} = \mathbf{I}_{an}(\mathbf{Z}_1 + \mathbf{Z}_2) + \mathbf{I}_{cn}\mathbf{Z}_2$$

$$\mathbf{E}_{BC} = \mathbf{I}_{an}(-\mathbf{Z}_2) + \mathbf{I}_{cn}[-(\mathbf{Z}_2 + \mathbf{Z}_3)]$$

Using determinants, we have

$$\mathbf{I}_{an} = \frac{\begin{vmatrix} \mathbf{E}_{AB} & \mathbf{Z}_2 \\ \mathbf{E}_{BC} & -(\mathbf{Z}_2 + \mathbf{Z}_3) \end{vmatrix}}{\begin{vmatrix} \mathbf{Z}_1 + \mathbf{Z}_2 & \mathbf{Z}_2 \\ -\mathbf{Z}_2 & -(\mathbf{Z}_2 + \mathbf{Z}_3) \end{vmatrix}}$$

$$= \frac{-(\mathbf{Z}_2 + \mathbf{Z}_3)\mathbf{E}_{AB} - \mathbf{E}_{BC}\mathbf{Z}_2}{-\mathbf{Z}_1\mathbf{Z}_2 - \mathbf{Z}_1\mathbf{Z}_3 - \mathbf{Z}_2\mathbf{Z}_3 - \mathbf{Z}_2^2 + \mathbf{Z}_2^2}$$

$$\mathbf{I}_{an} = \frac{-\mathbf{Z}_2(\mathbf{E}_{AB} + \mathbf{E}_{BC}) - \mathbf{Z}_3\mathbf{E}_{AB}}{-\mathbf{Z}_1\mathbf{Z}_2 - \mathbf{Z}_1\mathbf{Z}_3 - \mathbf{Z}_2\mathbf{Z}_3}$$

Apply Kirchhoff's voltage law to the line voltages:

$$\mathbf{E}_{AB} + \mathbf{E}_{CA} + \mathbf{E}_{BC} = 0 \quad \text{or} \quad \mathbf{E}_{AB} + \mathbf{E}_{BC} = -\mathbf{E}_{CA}$$

Substituting for $(\mathbf{E}_{AB} + \mathbf{E}_{CB})$ in the above equation for $\mathbf{I}_{an}$:

$$\mathbf{I}_{an} = \frac{-\mathbf{Z}_2(-\mathbf{E}_{CA}) - \mathbf{Z}_3\mathbf{E}_{AB}}{-\mathbf{Z}_1\mathbf{Z}_2 - \mathbf{Z}_1\mathbf{Z}_3 - \mathbf{Z}_2\mathbf{Z}_3}$$

and

$$\mathbf{I}_{an} = \frac{\mathbf{E}_{AB}\mathbf{Z}_3 - \mathbf{E}_{CA}\mathbf{Z}_2}{\mathbf{Z}_1\mathbf{Z}_2 + \mathbf{Z}_1\mathbf{Z}_3 + \mathbf{Z}_2\mathbf{Z}_3} \qquad (23.38)$$

In the same manner, it can be shown that

$$\mathbf{I}_{cn} = \frac{\mathbf{E}_{CA}\mathbf{Z}_2 - \mathbf{E}_{BC}\mathbf{Z}_1}{\mathbf{Z}_1\mathbf{Z}_2 + \mathbf{Z}_1\mathbf{Z}_3 + \mathbf{Z}_2\mathbf{Z}_3} \qquad (23.39)$$

Substituting Eq. (23.39) for $\mathbf{I}_{cn}$ in the right-hand side of Eq. (23.37b), we obtain

$$\mathbf{I}_{bn} = \frac{\mathbf{E}_{BC}\mathbf{Z}_1 - \mathbf{E}_{AB}\mathbf{Z}_3}{\mathbf{Z}_1\mathbf{Z}_2 + \mathbf{Z}_1\mathbf{Z}_3 + \mathbf{Z}_2\mathbf{Z}_3} \qquad (23.40)$$

**EXAMPLE 23.9.** A *phase-sequence indicator* (Fig. 23.36) is an instrument that can determine the phase sequence of a polyphase circuit. The numbers 1-2-3 correspond to the terminals *A-B-C* described in this chapter.

A network that will perform the same function as the indicator of Fig. 23.36 appears in Fig. 23.37. As noted, the applied phase sequence

**FIG. 23.36**
*Phase-sequence indicator. (Courtesy of General Electric Co.)*

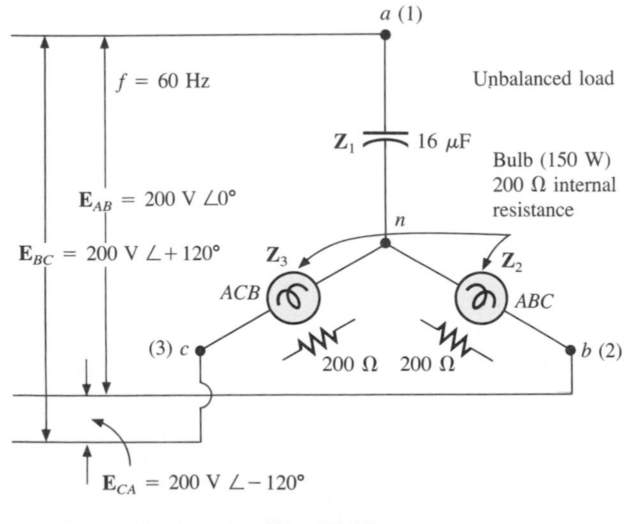


**FIG. 23.37**

is $ABC$. The bulb corresponding to this phase sequence will burn more brightly than the bulb indicating the $ACB$ sequence because a greater current is passing through the $ABC$ bulb. Calculating the phase currents will demonstrate that this situation does in fact exist:

$$Z_1 = X_C = \frac{1}{\omega C} = \frac{1}{(377 \text{ rad/s})(16 \times 10^{-6} \text{ F})} = 166 \ \Omega$$

By Eq. (23.39),

$$\mathbf{I}_{cn} = \frac{\mathbf{E}_{CA}\mathbf{Z}_2 - \mathbf{E}_{BC}\mathbf{Z}_1}{\mathbf{Z}_1\mathbf{Z}_2 + \mathbf{Z}_1\mathbf{Z}_3 + \mathbf{Z}_2\mathbf{Z}_3}$$

$$= \frac{(200 \text{ V} \angle 120^\circ)(200 \ \Omega \angle 0^\circ) - (200 \text{ V} \angle -120^\circ)(166 \ \Omega \angle -90^\circ)}{(166 \ \Omega \angle -90^\circ)(200 \ \Omega \angle 0^\circ) + (166 \ \Omega \angle -90^\circ)(200 \ \Omega \angle 0^\circ) + (200 \ \Omega \angle 0^\circ)(200 \ \Omega \angle 0^\circ)}$$

$$\mathbf{I}_{cn} = \frac{40{,}000 \text{ V} \angle 120^\circ + 33{,}200 \text{ V} \angle -30^\circ}{33{,}200 \ \Omega \angle -90^\circ + 33{,}200 \ \Omega \angle -90^\circ + 40{,}000 \ \Omega \angle 0^\circ}$$

Dividing the numerator and denominator by 1000 and converting both to the rectangular domain yields

$$\mathbf{I}_{cn} = \frac{(-20 + j34.64) + (28.75 - j16.60)}{40 - j66.4}$$

$$= \frac{8.75 + j18.04}{77.52 \angle -58.93^\circ} = \frac{20.05 \angle 64.13^\circ}{77.52 \angle -58.93^\circ}$$

$$\mathbf{I}_{cn} = \mathbf{0.259 \ A \angle 123.06^\circ}$$

By Eq. (23.40),

$$\mathbf{I}_{bn} = \frac{\mathbf{E}_{BC}\mathbf{Z}_1 - \mathbf{E}_{AB}\mathbf{Z}_3}{\mathbf{Z}_1\mathbf{Z}_2 + \mathbf{Z}_1\mathbf{Z}_3 + \mathbf{Z}_2\mathbf{Z}_3}$$

$$= \frac{(200 \text{ V} \angle -120^\circ)(166 \angle -90^\circ) - (200 \text{ V} \angle 0^\circ)(200 \angle 0^\circ)}{77.52 \times 10^3 \ \Omega \angle -58.93^\circ}$$

$$\mathbf{I}_{bn} = \frac{33{,}200 \text{ V} \angle -210^\circ - 40{,}000 \text{ V} \angle 0^\circ}{77.52 \times 10^3 \ \Omega \angle -58.93^\circ}$$

Dividing by 1000 and converting to the rectangular domain yields

$$\mathbf{I}_{bn} = \frac{-28.75 + j16.60 - 40.0}{77.52 \angle -58.93^\circ}$$

$$= \frac{-68.75 + j16.60}{77.52 \angle -58.93^\circ}$$

$$= \frac{70.73 \angle 166.43^\circ}{77.52 \angle -58.93^\circ}$$

$$\mathbf{I}_{bn} = \mathbf{0.91 \ A \angle 225.36^\circ}$$

and $I_{bn} > I_{cn}$ by a factor of more than 3:1. Therefore, the bulb indicating an $ABC$ sequence will burn more brightly due to the greater current. If the phase sequence were $ACB$, the reverse would be true.

# 23.15 COMPUTER ANALYSIS

## PSPICE

The application of PSPICE to three-phase systems requires that the user be particularly careful to insure the following:

1. A continuous loop of independent voltage sources is avoided.
2. All nodes have a dc path to ground.
3. A reference node is properly defined.

Writing an input file for the 3-$\phi$ balanced network of Fig. 23.38 requires that a resistor of relatively negligible value be placed between

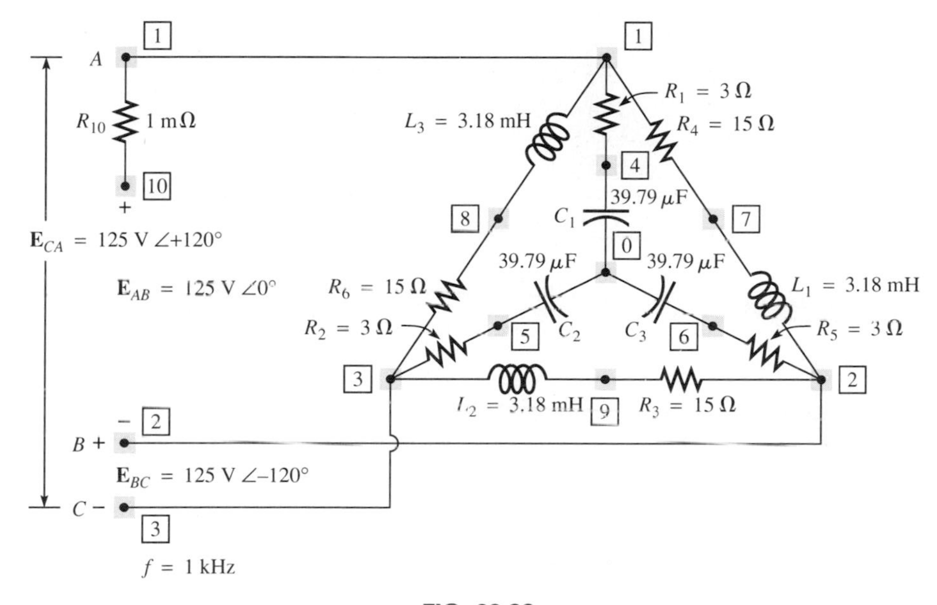

FIG. 23.38

nodes 1 and 2 to void the continuous loop of independent voltage sources $\mathbf{V}_{AB}$, $\mathbf{V}_{BC}$, and $\mathbf{V}_{CA}$. Note the choice of reference node at the center of the balanced load to establish a reference node for each branch voltage of the Y and each node of the Δ-connected load. To insure a dc path to ground for each node, a resistor of $10^{30}$ Ω is placed in parallel with each capacitor, as shown for R7, R8, and R9 in the input file of Fig. 23.39.

The network of Fig. 23.38 is the same as that appearing in Problem 42 of this chapter, with the inductor and capacitor values determined at a frequency of 1 kHz. Note in the printout that the magnitude of the branch current of each leg of the balanced Y is 1.443 A, with phase angles that differ by 120°. The magnitude of each phase current of the

```
************************ Evaluation PSpice (January 1989) ******* 14:08:11 *******

Chapter 23 - Three Phase System (balanced)

****     CIRCUIT DESCRIPTION

*****************************************************************************


V1 10 2 AC 125V 0DEG
R10 10 2 1M
V2 2 3 AC 125V -120DEG
V3 3 1 AC 125V +120DEG
R1 1 4 3
R2 3 5 3
R3 2 6 3
C1 4 0 39.79UF
C2 5 0 39.79UF
C3 6 0 39.79UF
R7 4 0 1E30
R8 6 0 1E30
R9 5 0 1E30
R4 1 7 15
R5 2 9 15
R6 3 8 15
L1 7 2 3.18MH
L2 9 3 3.18MH
L3 8 1 3.18MH
.AC LIN 1 1KH 1KH
.PRINT AC IM(R1) IM(R2) IM(R3)
.PRINT AC IP(R1) IP(R2) IP(R3)
.PRINT AC IM(R4) IM(R5) IM(R6)
.PRINT AC IP(R4) IP(R5) IP(R6)
.OPTIONS NOPAGE
.END
```

```
****     AC ANALYSIS                    TEMPERATURE =   27.000 DEG C


 FREQ        IM(R1)       IM(R2)       IM(R3)

  1.000E+03   1.443E+01    1.443E+01    1.443E+01


****     AC ANALYSIS                    TEMPERATURE =   27.000 DEG C


 FREQ        IP(R1)       IP(R2)       IP(R3)

  1.000E+03   2.313E+01    1.431E+02   -9.687E+01
```

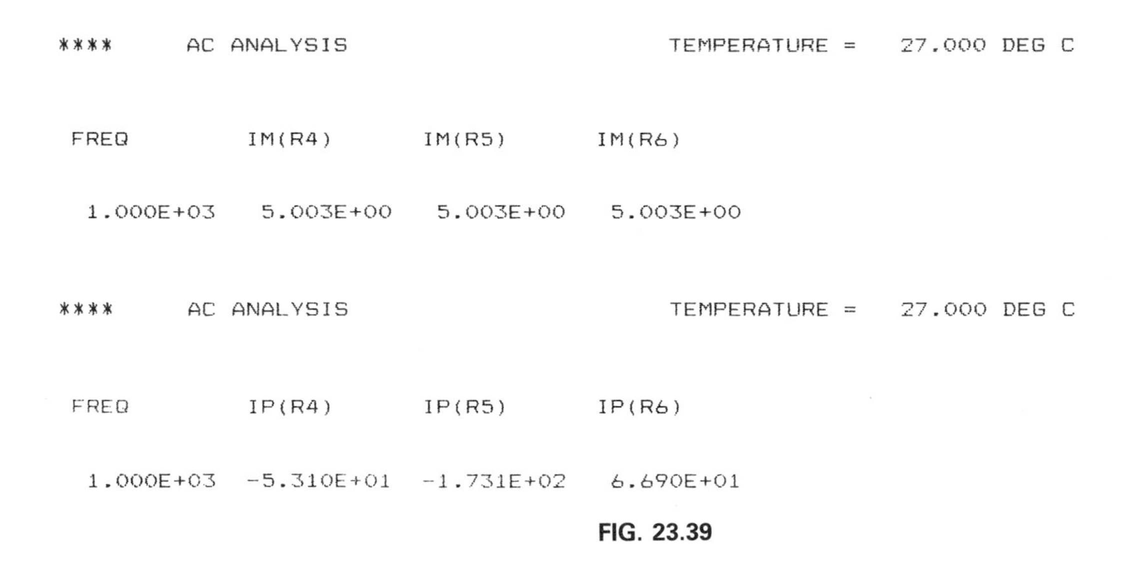

```
****     AC ANALYSIS                              TEMPERATURE =   27.000 DEG C


 FREQ         IM(R4)       IM(R5)       IM(R6)

  1.000E+03   5.003E+00    5.003E+00    5.003E+00


****     AC ANALYSIS                              TEMPERATURE =   27.000 DEG C


 FREQ         IP(R4)       IP(R5)       IP(R6)

  1.000E+03  -5.310E+01   -1.731E+02    6.690E+01
```

**FIG. 23.39**

delta is 5 A, with phase angles that also differ by 120°. The capacitive Y load establishes phase currents that lead the line voltages and the inductive delta establishes phase currents that lag the applied line voltages.

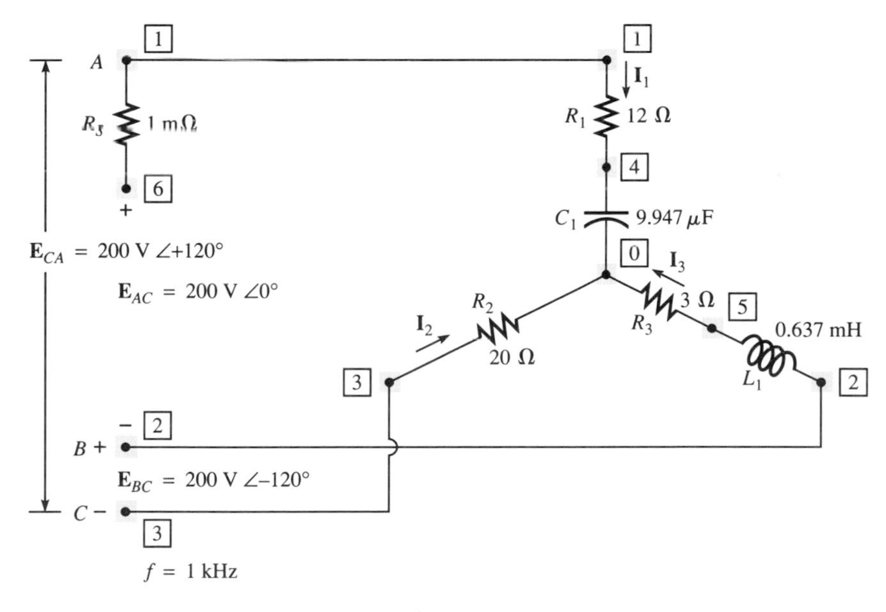


**FIG. 23.40**

The second 3-$\phi$ system to be analyzed by PSPICE is the unbalanced load of Fig. 23.40. Again, the input file of Fig. 23.41 requires a relatively small resistance between nodes 6 and 1 to eliminate the

```
************************ Evaluation PSpice (January 1989) ******* 17:35:35 *******

Chapter 23 - Three Phase Systems (unbalanced)

****     CIRCUIT DESCRIPTION

*****************************************************************************

V1 6 2 AC 200V 0
R21 6 1 1M
V2 2 3 AC 200V -120
V3 3 1 AC 200V -240
R1 1 4 12
C1 4 0 9.947UF
R2 3 0 20
R3 5 0 3
L3 2 5 0.637MH
.AC LIN 1 1KH 1KH
.PRINT AC IM(R1) IP(R1)
.PRINT AC IM(R2) IP(R2)
.PRINT AC IM(R3) IP(R3)
.OPTIONS NOPAGE
.END

****     AC ANALYSIS                        TEMPERATURE =   27.000 DEG C

 FREQ        IM(R1)      IP(R1)

  1.000E+03   1.071E+01   2.958E+01

****     AC ANALYSIS                        TEMPERATURE =   27.000 DEG C

 FREQ        IM(R2)      IP(R2)

  1.000E+03   6.512E+00   4.230E+01

****     AC ANALYSIS                        TEMPERATURE =   27.000 DEG C

 FREQ        IM(R3)      IP(R3)

  1.000E+03   1.712E+01  -1.456E+02
```

**FIG. 23.41**
*Input file.*

continuous loop of independent sources and a resistor of $10^{30}$ Ω must be placed across the capacitor to provide a dc path to ground for node 1.

The results are:

$$\mathbf{I}_1 = 10.71 \text{ A} \angle 29.58°$$
$$\mathbf{I}_2 = 6.512 \text{ A} \angle 42.3°$$
$$\mathbf{I}_3 = 17.12 \text{ A} \angle -145.6°$$

Note the different magnitudes for each phase current due to the unbalanced conditions and the fact that the phase angles are not out of phase by 120°, as in the balanced situation.

## BASIC

The BASIC program to follow analyzes the network of Fig. 23.42, which has the format of Fig. 23.38 but has the additional ability of

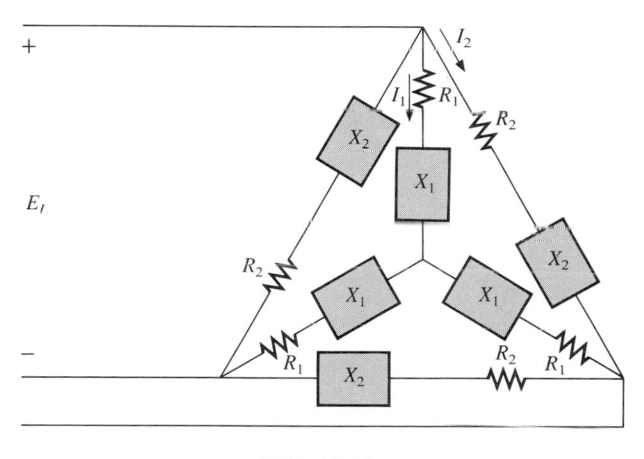


FIG. 23.42

being able to define the type of reactance for $X_1$ and $X_2$. The program of Fig. 23.43 will determine all the quantities of interest for the Δ-Y-connected load. As indicated by the INPUT bracket, lines 130 through 220 request the line voltage and the network parameters of the series-connected phase impedances. Inductive and capacitive reactances are distinguished by the sign entered on lines 180 and 220.

The calculations for the Y-connected loads are made by lines 250 through 360 using equations introduced in Section 23.10. Lines 380 through 490 perform a detailed analysis of the Δ section. The total real and reactive power are calculated by lines 510 through 570, and the total apparent power and the power factor by lines 580 through 630.

For comparison purposes, the run employed the network parameters used in Example 23.6.

```
 10 REM ***** PROGRAM  23-1 *****
 20 REM ******************************
 30 REM Program to analyze a 3-phase
 40 REM delta-wye load.
 50 REM ******************************
 60 REM
100 PRINT "This program analyzes the 3-phase"
110 PRINT "delta-wye load."
120 PRINT
130 PRINT "Enter the following network information:"
140 PRINT
150 INPUT "Line voltage, E=";EL
160 PRINT "For the series-connected wye load:"
170 INPUT "R=";R1
180 INPUT "X=";X1 :REM Enter negative sign if capacitive
190 PRINT
200 PRINT "And for the series-connected delta load:"
210 INPUT "R=";R2
220 INPUT "X=";X2 :REM Enter negative sign if capacitive
230 REM Now do calculations and print results
240 PRINT:PRINT
250 Z1=SQR(R1^2+X1^2)
260 I1=EL/(SQR(3)*Z1)
270 P1=3*I1^2*R1
280 Q1=3*I1^2*X1
290 S1=3*EL*I1/SQR(3)
300 PRINT "For the wye connection:"
310 PRINT "The phase current I=";I1;"amps"
320 PRINT "The total power dissipated is";P1;"watts"
330 PRINT "With a net reactive power of Q=";ABS(Q1);"vars";
340 IF SGN(X1)=-1 THEN PRINT "(cap.)"
350 IF SGN(X1)=1 THEN PRINT "(ind.)"
360 PRINT "and apparent power of:";S1;"VA"
370 PRINT:PRINT
380 Z2=SQR(R2^2+X2^2)
390 I2=EL/Z2
400 P2=3*I2^2*R2
410 Q2=3*I2^2*X2
420 S2=3*EL*I2
430 PRINT "For the delta connected load:"
440 PRINT "The phase current is, I=";I2;"amps"
450 PRINT "The total power dissipated is";P2;"watts"
460 PRINT "with a net reactive power of Q=";ABS(Q2);"vars";
470 IF SGN(X2)=-1 THEN PRINT "(cap.)"
480 IF SGN(X2)=1 THEN PRINT "(ind.)"
490 PRINT "and apparent power of:";S2;"VA"
500 PRINT:PRINT
510 PT=P1+P2
520 QT=Q1+Q2
530 PRINT "For the combined system:"
540 PRINT "The total power dissipated is";PT;"watts"
550 PRINT "and the net reactive power QT=";ABS(QT);"vars";
560 IF SGN(QT)=-1 THEN PRINT "(cap.)"
570 IF SGN(QT)=1 THEN PRINT "(ind.)"
580 ST=SQR(PT^2+QT^2)
590 FP=PT/ST
600 PRINT:PRINT "The total apparent power ST=";ST;"VA"
610 PRINT "with a network power factor Fp=";FP;
620 IF SGN(QT)=-1 THEN PRINT "(leading)"
630 IF SGN(QT)=1 THEN PRINT "(lagging)"
640 END

READY
```

Margin labels: Input (lines 130–220); $Y$ (lines 250–360); $\Delta$ (lines 380–490); $P_T$, $Q_T$ (lines 510–570); $S_T$, $F_p$ (lines 580–630)

```
RUN

This program analyzes the 3-phase
delta-wye load.

Enter the following network information:

Line voltage, E=? 200

For the series-connected wye load:
R=? 4

X=? 3


And for the series-connected delta load:
R=? 6

X=? -8



For the wye connection:
The phase current I= 23.094 amps
The total power dissipated is 6400 watts
With a net reactive power of Q= 4800 vars(ind.)
and apparent power of: 8000 VA


For the delta connected load:
The phase current is, I= 20 amps
The total power dissipated is 7200 watts
with a net reactive power of Q= 9600 vars(cap.)
and apparent power of: 1.2E+04 VA


For the combined system:
The total power dissipated is 1.36E+04 watts
and the net reactive power QT= 4800 vars(cap.)

The total apparent power ST= 1.4422E+04 VA
with a network power factor Fp= .943 (leading)


READY
```

**FIG. 23.43**
*Program 23.1.*

# PROBLEMS

## SECTION 23.5 The Y-Connected Generator with a Y-Connected Load

**1.** A balanced Y load having a 10-Ω resistance in each leg is connected to a three-phase, four-wire, Y-connected generator having a line voltage of 208 V. Calculate the magnitude of
   **a.** the phase voltage of the generator.
   **b.** the phase voltage of the load.
   **c.** the phase current of the load.
   **d.** the line current.

2. Repeat Problem 1 if each phase impedance is changed to a 12-Ω resistor in series with a 16-Ω capacitive reactance.

3. Repeat Problem 1 if each phase impedance is changed to a 10-Ω resistor in parallel with a 10-Ω capacitive reactance.

4. The phase sequence for the Y-Y system of Fig. 23.44 is *ABC*.
   **a.** Find the angles $\theta_2$ and $\theta_3$ for the specified phase sequence.
   **b.** Find the voltage across each phase impedance in phasor form.
   **c.** Find the current through each phase impedance in phasor form.
   **d.** Draw the phasor diagram of the currents found in part (c) and show that their phasor sum is zero.
   **e.** Find the magnitude of the line currents.
   **f.** Find the magnitude of the line voltages.

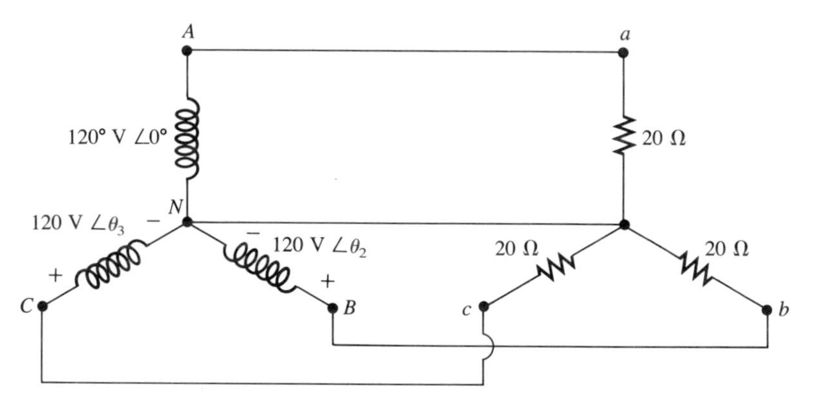

FIG. 23.44

5. Repeat Problem 4 if the phase impedances are changed to a 9-Ω resistor in series with a 12-Ω inductive reactance.

6. Repeat Problem 4 if the phase impedances are changed to a 6-Ω resistance in parallel with an 8-Ω capacitive reactance.

7. For the system of Fig. 23.45, find the magnitude of the unknown voltages and currents.

*8. Compute the magnitude of the voltage $E_{AB}$ for the balanced three-phase system of Fig. 23.46.

*9. For the Y-Y system of Fig. 23.47,
   **a.** Find the magnitude and angle associated with the voltages $\mathbf{E}_{AN}$, $\mathbf{E}_{BN}$, and $\mathbf{E}_{CN}$.
   **b.** Determine the magnitude and angle associated with each phase current of the load: $\mathbf{I}_{an}$, $\mathbf{I}_{bn}$, and $\mathbf{I}_{cn}$.
   **c.** Find the magnitude and phase angle of each line current: $\mathbf{I}_{Aa}$, $\mathbf{I}_{Bb}$, and $\mathbf{I}_{Cc}$.
   **d.** Determine the magnitude and phase angle of the voltage across each phase of the load: $\mathbf{V}_{an}$, $\mathbf{V}_{bn}$, and $\mathbf{V}_{cn}$.

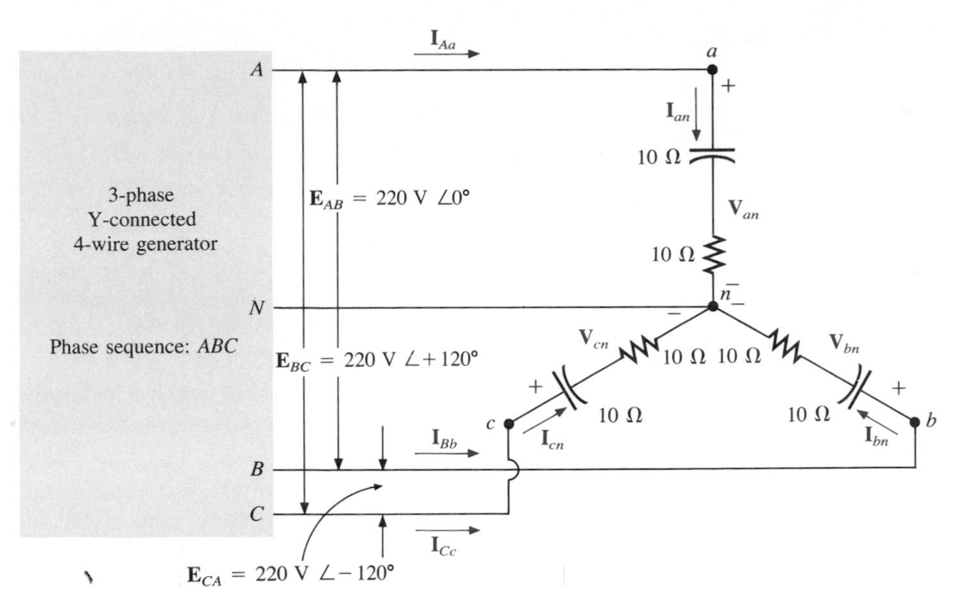


FIG. 23.45

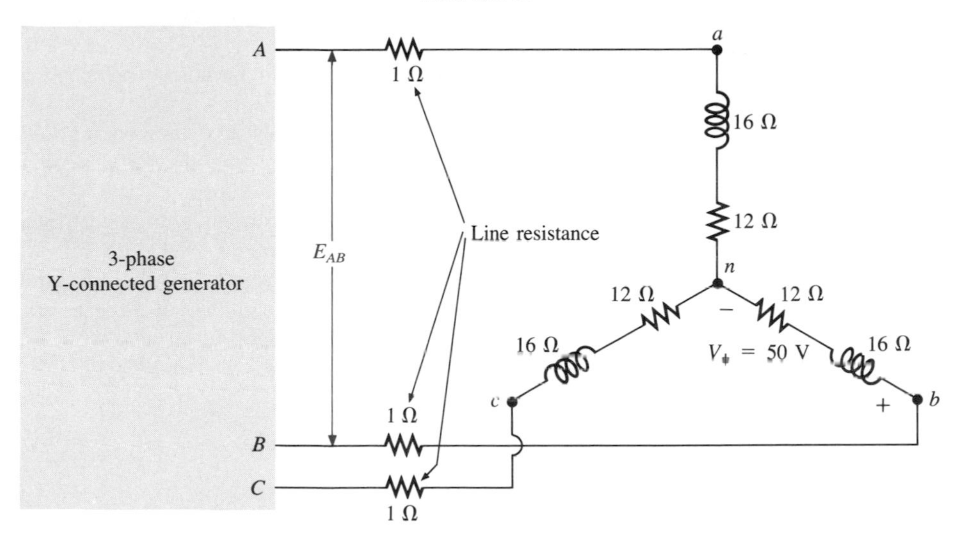

FIG. 23.46

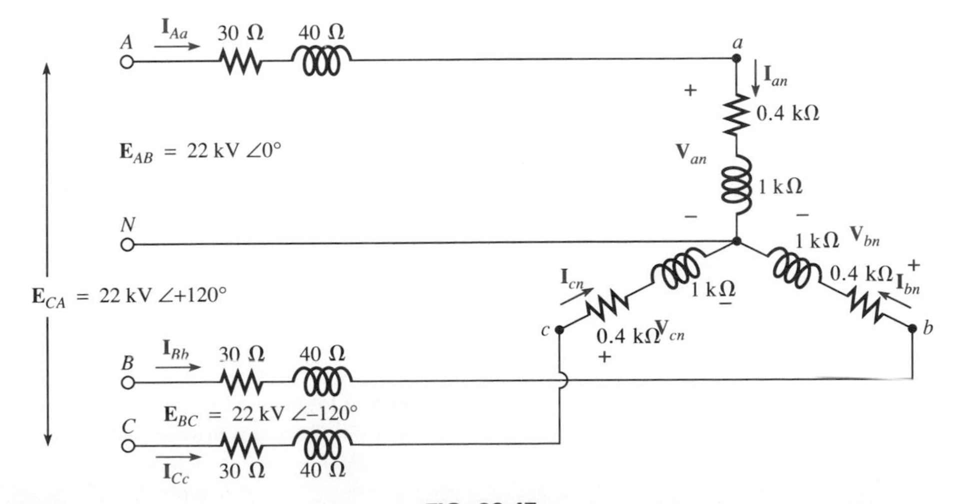


FIG. 23.47

## SECTION 23.6 The Y-Δ System

**10.** A balanced Δ load having a 20-Ω resistance in each leg is connected to a three-phase, three-wire, Y-connected generator having a line voltage of 208 V. Calculate the magnitude of

**a.** the phase voltage of the generator.
**b.** the phase voltage of the load.
**c.** the phase current of the load.
**d.** the line current.

**11.** Repeat Problem 10 if each phase impedance is changed to a 6.8-Ω resistor in series with a 14-Ω inductive reactance.

**12.** Repeat Problem 10 if each phase impedance is changed to an 18-Ω resistance in parallel with an 18-Ω capacitive reactance.

**13.** The phase sequence for the Y-Δ system of Fig. 23.48 is *ABC*.

**a.** Find the angles $\theta_2$ and $\theta_3$ for the specified phase sequence.
**b.** Find the voltage across each phase impedance in phasor form.
**c.** Draw the phasor diagram of the voltages found in part (b) and show that their sum is zero around the closed loop of the Δ load.
**d.** Find the current through each phase impedance in phasor form.
**e.** Find the magnitude of the line currents.
**f.** Find the magnitude of the generator phase voltages.

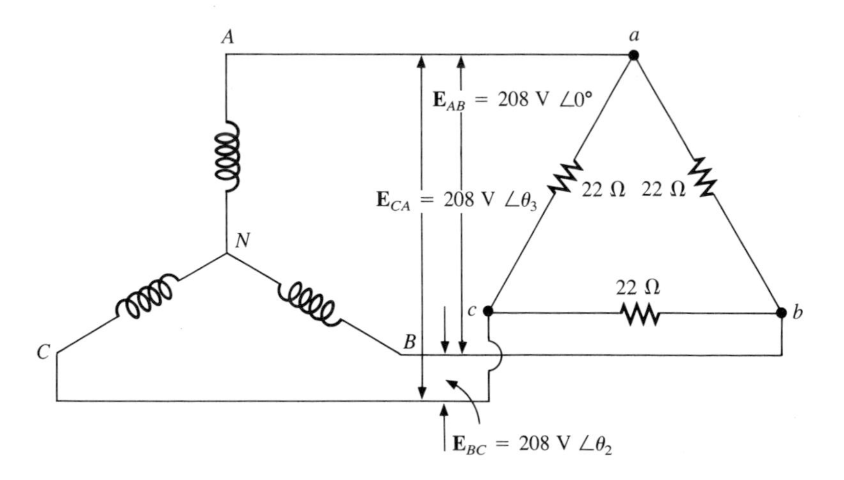


FIG. 23.48

**14.** Repeat Problem 13 if the phase impedances are changed to a 100-Ω resistor in series with a capacitive reactance of 100 Ω.

**15.** Repeat Problem 14 if the phase impedances are changed to a 3-Ω resistor in parallel with an inductive reactance of 4 Ω.

**16.** For the system of Fig. 23.49, find the magnitude of the unknown voltages and currents.

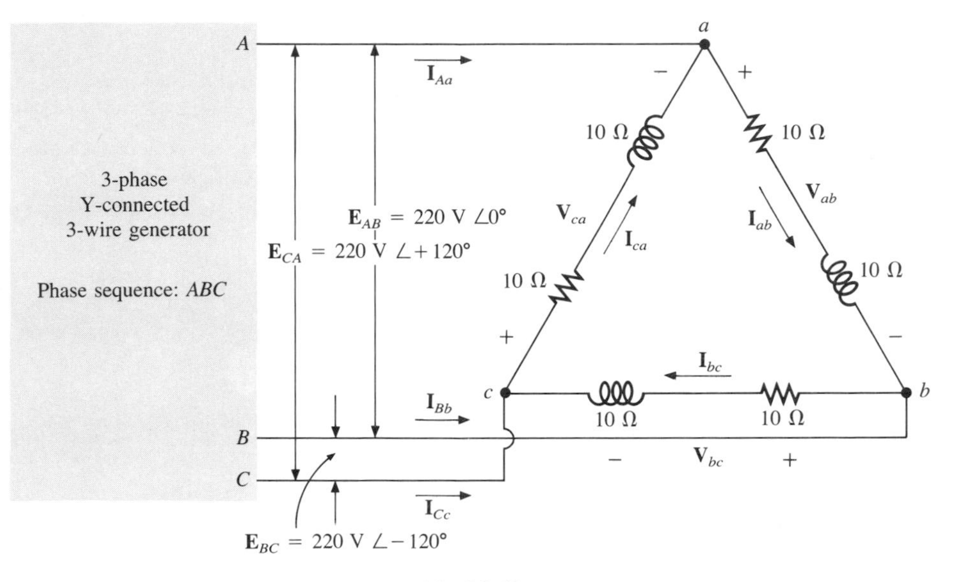

FIG. 23.49

***17.** For the Δ-connected load of Fig. 23.50,

**a.** Find the magnitude and angle of each phase current: $\mathbf{I}_{ab}$, $\mathbf{I}_{bc}$, $\mathbf{I}_{ca}$.

**b.** Calculate the magnitude and angle of each line current: $\mathbf{I}_{Aa}$, $\mathbf{I}_{Bb}$, $\mathbf{I}_{Cc}$.

**c.** Determine the magnitude and angle of the voltages $\mathbf{E}_{AB}$, $\mathbf{E}_{BC}$, and $\mathbf{E}_{CA}$.

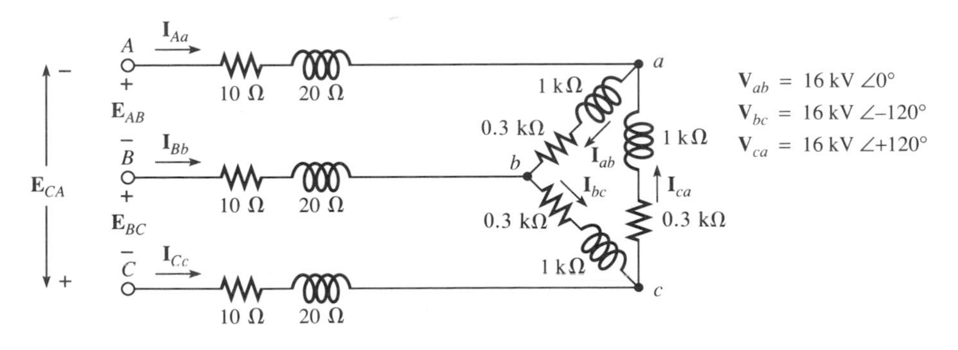


FIG. 23.50

## SECTION 23.9 The Δ-Δ, Δ-Y Three-Phase Systems

**18.** A balanced Y load having a 30-Ω resistance in each leg is connected to a three-phase Δ-connected generator having a line voltage of 208 V. Calculate the magnitude of
**a.** the phase voltage of the generator.
**b.** the phase voltage of the load.
**c.** the phase current of the load.
**d.** the line current.

**19.** Repeat Problem 18 if each phase impedance is changed to a 12-Ω resistor in series with a 12-Ω inductive reactance.

**20.** Repeat Problem 18 if each phase impedance is changed to a 15-Ω resistor in parallel with a 20-Ω capacitive reactance.

***21.** For the system of Fig. 23.51, find the magnitude of the unknown voltages and currents.

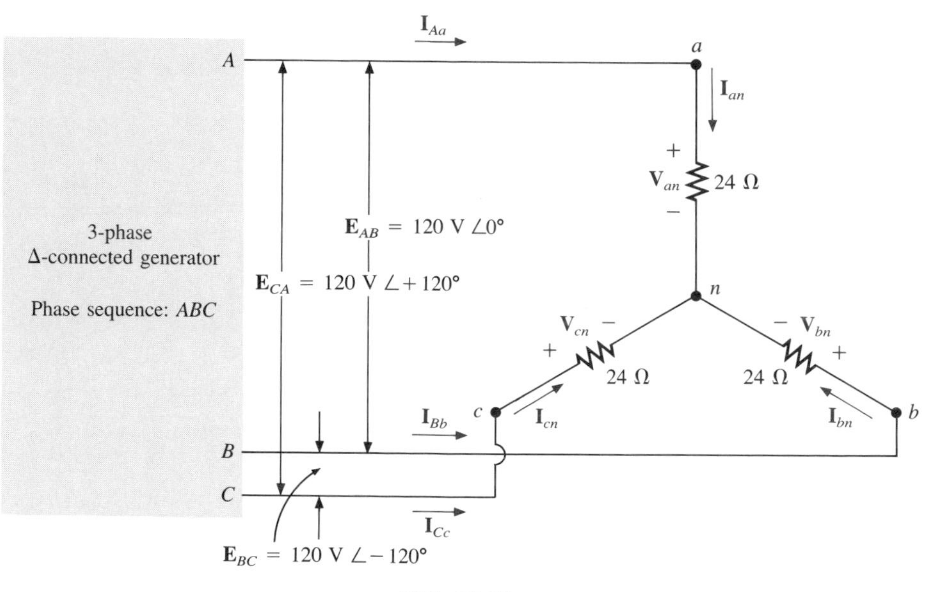

**FIG. 23.51**

**22.** Repeat Problem 21 if each phase impedance is changed to a 10-Ω resistor in series with a 20-Ω inductive reactance.

**23.** Repeat Problem 21 if each phase impedance is changed to a 20-Ω resistor in parallel with a 15-Ω capacitive reactance.

**24.** A balanced Δ load having a 220-Ω resistance in each leg is connected to a three-phase Δ-connected generator having a line voltage of 440 V. Calculate the magnitude of
**a.** the phase voltage of the generator.
**b.** the phase voltage of the load.
**c.** the phase current of the load.
**d.** the line current.

**25.** Repeat Problem 24 if each phase impedance is changed to a 12-Ω resistor in series with a 9-Ω capacitive reactance.

26. Repeat Problem 24 if each phase impedance is changed to a 22-Ω resistor in parallel with a 22-Ω inductive reactance.

27. The phase sequence for the Δ-Δ system of Fig. 23.52 is *ABC*.
   - **a.** Find the angles $\theta_2$ and $\theta_3$ for the specified phase sequence.
   - **b.** Find the voltage across each phase impedance in phasor form.
   - **c.** Draw the phasor diagram of the voltages found in part (b) and show that their phasor sum is zero around the closed loop of the Δ load.
   - **d.** Find the current through each phase impedance in phasor form.
   - **e.** Find the magnitude of the line currents.

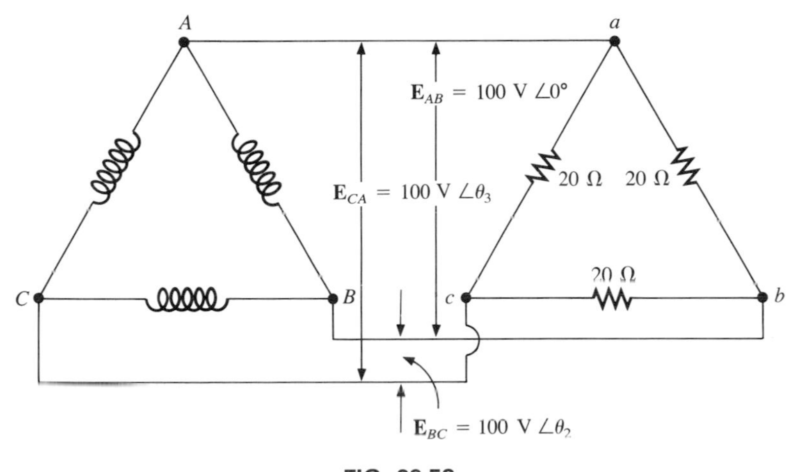

**FIG. 23.52**

28. Repeat Problem 25 if each phase impedance is changed to a 12-Ω resistor in series with a 16-Ω inductive reactance.

29. Repeat Problem 25 if each phase impedance is changed to a 20-Ω resistor in parallel with a 20-Ω capacitive reactance.

## SECTION 23.10 Power

30. Find the total watts, volt-amperes reactive, volt-amperes, and $F_p$ of the three-phase system of Problem 2.

31. Find the total watts, volt-amperes reactive, volt-amperes, and $F_p$ of the three-phase system of Problem 4.

32. Find the total watts, volt-amperes reactive, volt-amperes, and $F_p$ of the three-phase system of Problem 7.

33. Find the total watts, volt-amperes reactive, volt-amperes, and $F_p$ of the three-phase system of Problem 12.

34. Find the total watts, volt-amperes reactive, volt-amperes, and $F_p$ of the three-phase system of Problem 14.

35. Find the total watts, volt-amperes reactive, volt-amperes, and $F_p$ of the three-phase system of Problem 16.

36. Find the total watts, volt-amperes reactive, volt-amperes, and $F_p$ of the three-phase system of Problem 20.

37. Find the total watts, volt-amperes reactive, volt-amperes, and $F_p$ of the three-phase system of Problem 22.

38. Find the total watts, volt-amperes reactive, volt-amperes, and $F_p$ of the three-phase system of Problem 26.

39. Find the total watts, volt-amperes reactive, volt-amperes, and $F_p$ of the three-phase system of Problem 28.

40. A balanced three-phase Δ-connected load has a line voltage of 200 and a total power consumption of 4800 W at a lagging power factor of 0.8. Find the impedance of each phase in rectangular coordinates.

41. A balanced three-phase Y-connected load has a line voltage of 208 and a total power consumption of 1200 W at a leading power factor of 0.6. Find the impedance of each phase in rectangular coordinates.

*42. Find the total watts, volt-amperes reactive, volt-amperes, and $F_p$ of the system of Fig. 23.53.

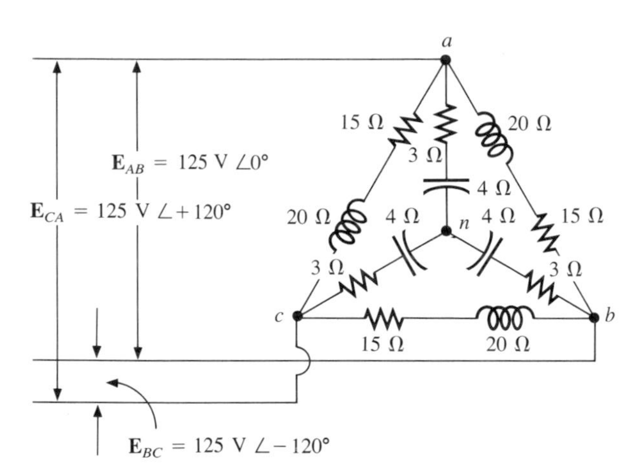

FIG. 23.53

*43. The Y-Y system of Fig. 23.54 has a balanced load and a line impedance $\mathbf{Z}_{\text{line}} = 4\ \Omega + j20\ \Omega$. If the line voltage at the generator is 16,000 V and the total power delivered to the load is 1200 kW at 80 A, determine each of the following;

a. The magnitude of each phase voltage of the generator.
b. The magnitude of the line currents.
c. The total power delivered by the source.
d. The power factor angle of the entire load "seen" by the source.
e. The magnitude and angle of the line current $\mathbf{I}_{Aa}$ if $\mathbf{E}_{AN} = E_{AN} \angle 0°$.
f. The magnitude and angle of the phase voltage $\mathbf{V}_{an}$.
g. The impedance of the load of each phase in rectangular coordinates.
h. The difference between the power factor of the load and the power factor of the entire system (including $\mathbf{Z}_{\text{line}}$).
i. The efficiency of the system.

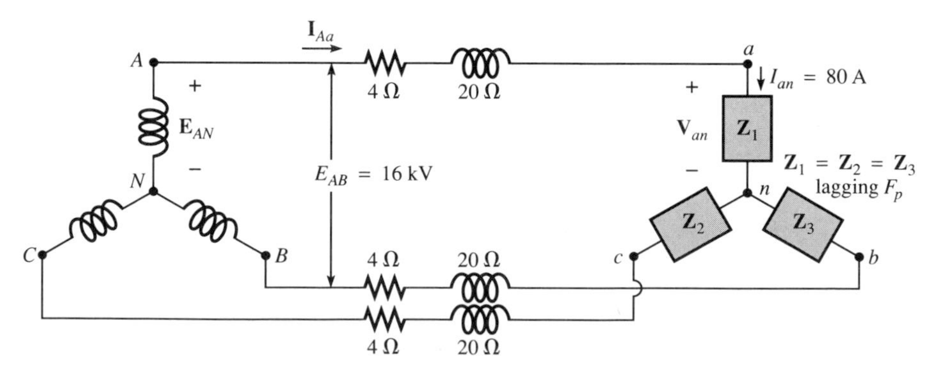

FIG. 23.54

## SECTION 23.11 The Three-Wattmeter Method

**44. a.** Sketch the connections required to measure the total watts delivered to the load of Fig. 23.45 using three wattmeters.

**b.** Determine the total wattage dissipation and the reading of each wattmeter.

**45.** Repeat Problem 44 for the network of Fig. 23.48.

## SECTION 23.12 The Two-Wattmeter Method

**46. a.** For the three-wire system of Fig. 23.55, properly connect a second wattmeter so that the two will measure the total power delivered to the load.

**b.** If one wattmeter has a reading of 200 W and the other a reading of 85 W, what is the total dissipation in watts if the total power factor is 0.8 leading?

**c.** Repeat part (b) if the total power factor is 0.2 lagging and $P_l = 100$ W.

**47.** Sketch three different ways that two wattmeters can be connected to measure the total power delivered to the load of Problem 16.

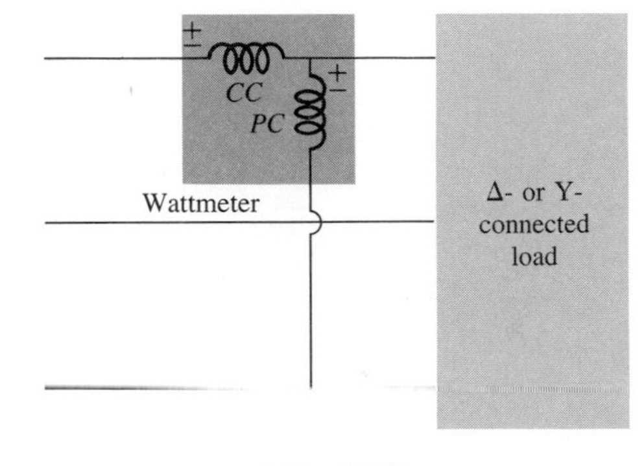


FIG. 23.55

***48.** For the Y-Δ system of Fig. 23.56,

**a.** Determine the magnitude and angle of the phase currents.

**b.** Find the magnitude and angle of the line currents.

**c.** Determine the reading of each wattmeter.

**d.** Find the total power delivered to the load.

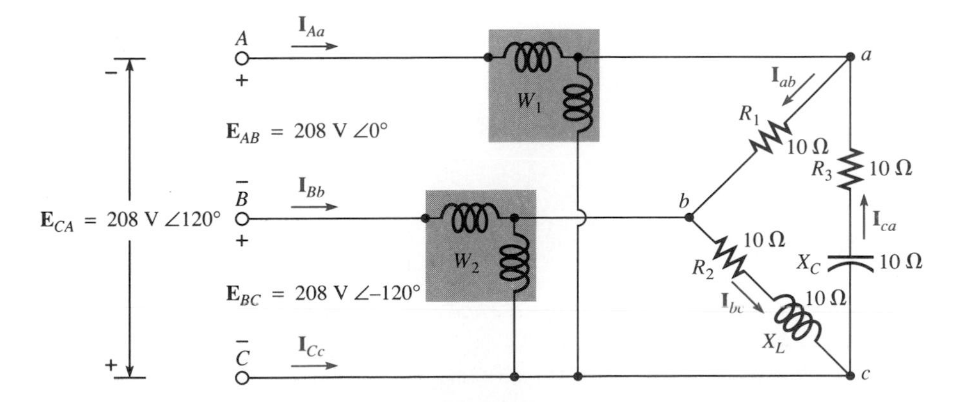


FIG. 23.56

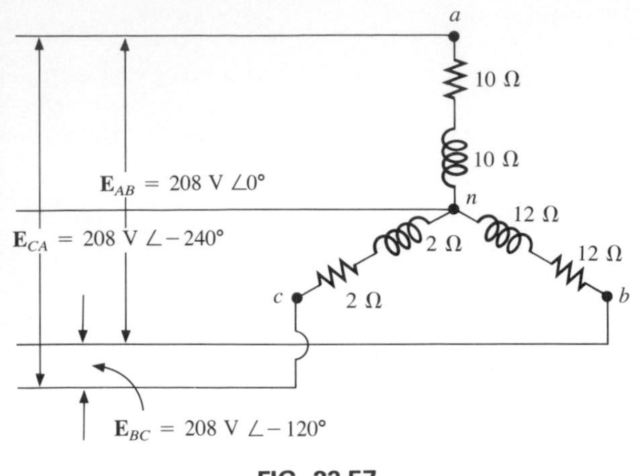


FIG. 23.57

## SECTION 23.13 Unbalanced Three-Phase, Four-Wire, Y-Connected Load

***49.** For the system of Fig. 23.57:

**a.** Calculate the magnitude of the voltage across each phase of the load.

**b.** Find the magnitude of the current through each phase of the load.

**c.** Find the total watts, volt-amperes reactive, volt-amperes, and $F_p$ of the system.

**d.** Find the phase currents in phasor form.

**e.** Using the results of part (c) determine the current $\mathbf{I}_N$.

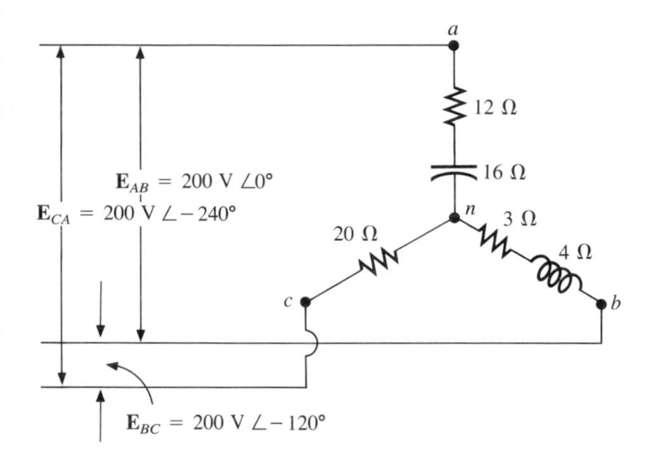


FIG. 23.58

## SECTION 23.14 Unbalanced Three-Phase, Three-Wire, Y-Connected Load

***50.** For the three-phase, three-wire system of Fig. 23.58, find the magnitude of the current through each phase of the load and the total watts, volt-amperes reactive, volt-amperes, and $F_p$ of the load.

## SECTION 23.15 Computer Analysis

### PSPICE

**51.** Write the input file to determine the phase voltages and currents for the network of Fig. 23.45.

**52.** Write the input file to determine the phase currents for the network of Fig. 23.49.

**53.** Write the input file to determine the phase voltages and currents for the load of Fig. 23.47.

**54.** Write the input file to determine the phase currents for the network of Fig. 23.56.

### BASIC

**55.** Given the magnitude of the line voltages and the impedance of each phase (in series or parallel), write a program to determine the magnitude of all the voltages and currents of a balanced Y-connected load.

**56.** Repeat Problem 55 for a Δ-connected load.

**57.** For a balanced Y-connected load, write a program to determine

**a.** the magnitude of the load currents and voltages.

**b.** the real, reactive, and apparent power to each phase.

**c.** the total real, reactive, and apparent power to the load.

**d.** the load power factor.

**58.** Repeat Problem 57 for a balanced Δ-connected load.

## GLOSSARY

**Delta (Δ)-connected generator** A three-phase generator having the three phases connected in the shape of the capital Greek letter delta (Δ).

**Line current** The current that flows from the generator to the load of a single-phase or polyphase system.

**Line voltage** The potential difference that exists between the lines of a single-phase or polyphase system.

**Neutral connection** The connection between the generator and the load that, under balanced conditions, will have zero current associated with it.

**Phase current** The current that flows through each phase of a single-phase or polyphase generator load.

**Phase sequence** The order in which the generated sinusoidal voltages of a polyphase generator will affect the load to which they are applied.

**Phase voltage** The voltage that appears between the line and neutral of a Y-connected generator and from line to line in a Δ-connected generator.

**Polyphase ac generator** An electromechanical source of ac power that generates more than one sinusoidal voltage per rotation of the rotor. The frequency generated is determined by the speed of rotation and the number of poles of the rotor.

**Single-phase ac generator** An electromechanical source of ac power that generates a single sinusoidal voltage having a frequency determined by the speed of rotation and the number of poles of the rotor.

**Three-wattmeter method** A method for determining the total power delivered to a three-phase load using three wattmeters.

**Two-wattmeter method** A method for determining the total power delivered to a Δ- or Y-connected three-phase load using only two wattmeters and considering the power factor of the load.

**Unbalanced polyphase load** A load not having the same impedance in each phase.

**WYE (Y)-connected generator** A three-phase source of ac power in which the three phases are connected in the shape of the letter Y.

# 24

# Nonsinusoidal Circuits

NON

## 24.1 INTRODUCTION

Any waveform that differs from the basic description of the sinusoidal waveform is referred to as *nonsinusoidal*. The most obvious are the dc, square-wave, triangular, or sawtooth voltage or current; others are voice patterns, such as those in Fig. 24.1, which were recorded as a person

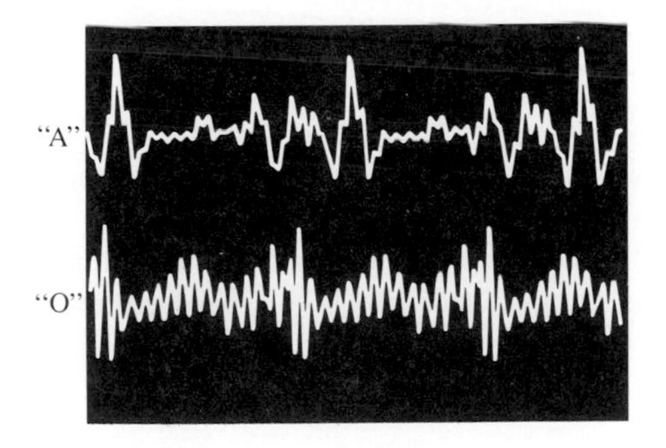


**FIG. 24.1**
*Voice patterns.*

held continuous ''A'' and ''O'' sounds. Note that the waveforms are almost periodic, but that a variation in tone causes a slight change in the waveform. Voice patterns are unique to people in much the same manner as fingerprints.

The output of many electrical and electronic devices will be a nonsinusoidal quantity if a sinusoidal input is applied. For example, the nonlinear characteristics of a diode are shown in Fig. 24.2(a). If a sinusoidal voltage is applied across the diode, the current passing through the diode will have the nonsinusoidal wave shape shown in Fig. 24.2(b). Note, however, that the nonsinusoidal output is also periodic: It has the

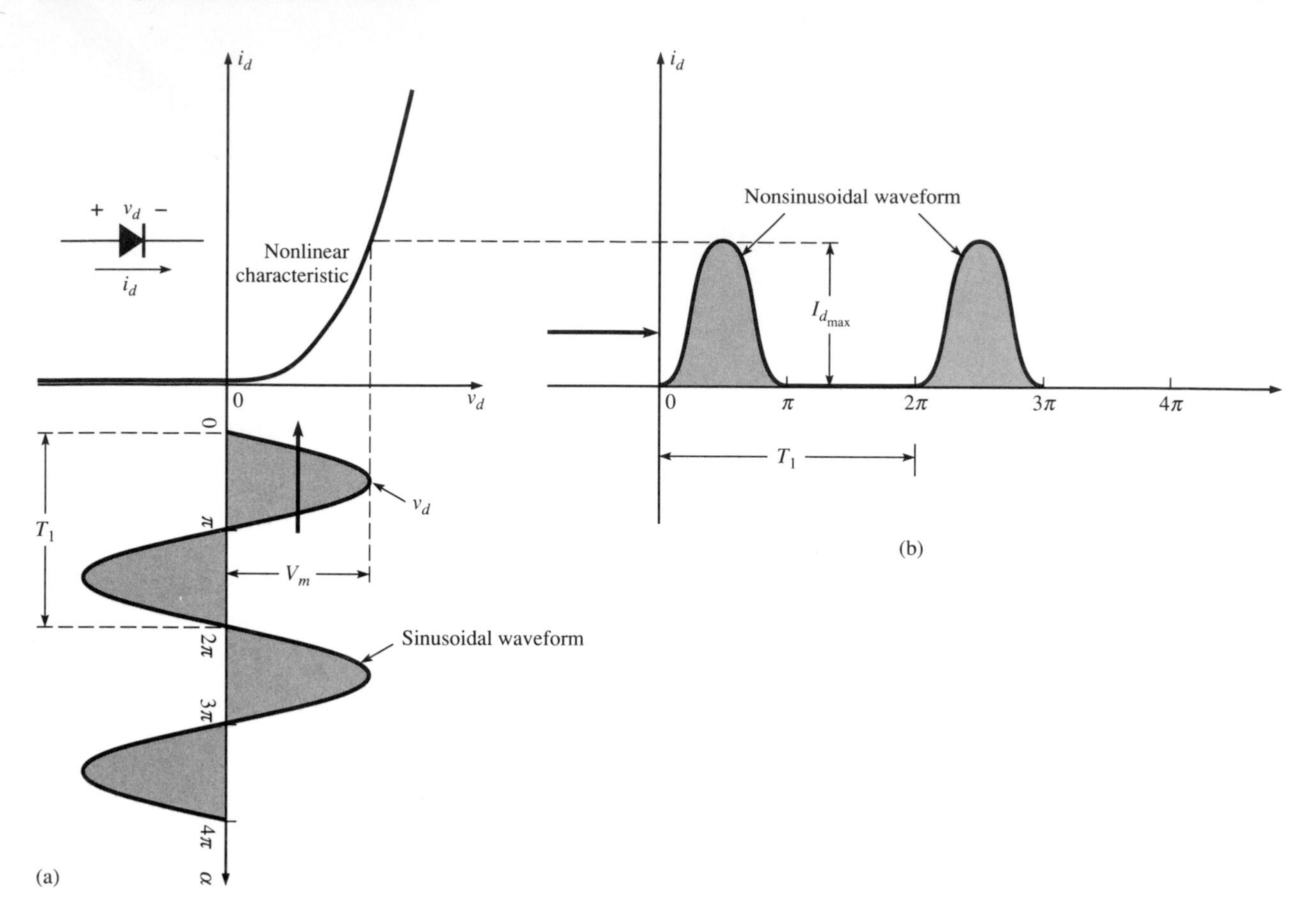


**FIG. 24.2**

same period as the input voltage. A periodic nonsinusoidal output will result for any nonlinear system that has no residual effects associated with it if a sinusoidal quantity is applied at the input.

In this chapter, we shall discuss one method of obtaining the response of a system to a periodic nonsinusoidal input.

## 24.2 FOURIER SERIES

*Fourier series* refers to a series of terms, developed in 1826 by Baron Jean Fourier, that can be used to represent a nonsinusoidal waveform. In the analysis of these waveforms, we solve for each term in the Fourier series:

$$f(\alpha) = \underbrace{A_0}_{\substack{\text{dc or} \\ \text{average value}}} + \underbrace{A_1 \cos \alpha + A_2 \cos 2\alpha + A_3 \cos 3\alpha + \cdots + A_n \cos n\alpha}_{\text{cosine terms}} + \underbrace{B_1 \sin \alpha + B_2 \sin 2\alpha + B_3 \sin 3\alpha + \cdots + B_n \sin n\alpha}_{\text{sine terms}} \tag{24.1}$$

Depending on the waveform, a large number of these terms may be required to approximate the waveform closely for the purpose of circuit analysis.

The Fourier series has three basic parts. The first is the dc term $A_0$, which is the average value of the waveform over one full cycle. The second is a series of cosine terms. There are no restrictions on the values or relative values of the amplitudes of these cosine terms, but each will have a frequency that is an integer multiple of the frequency of the first cosine term of the series. The third part is a series of sine terms. There are again *no* restrictions on the values or relative values of the amplitudes of these sine terms, but each will have a frequency that is an integer multiple of the frequency of the first sine term of the series. For a particular waveform, it is quite possible that all of the sine *or* cosine terms are zero. Characteristics of this type can be determined by simply examining the nonsinusoidal waveform and its position on the horizontal axis.

***If a waveform is symmetric about the vertical axis, it is called an* even *function or is said to have* axis symmetry.**

Fig. 24.3(a) is an example of such a waveform. For all even functions, the $B_{1\to n}$ constants will all be zero, and the function can be completely described by just the *dc and cosine terms*. Note that the cosine wave itself is also symmetrical about the vertical axis (ordinate) [Fig. 24.3(b)].

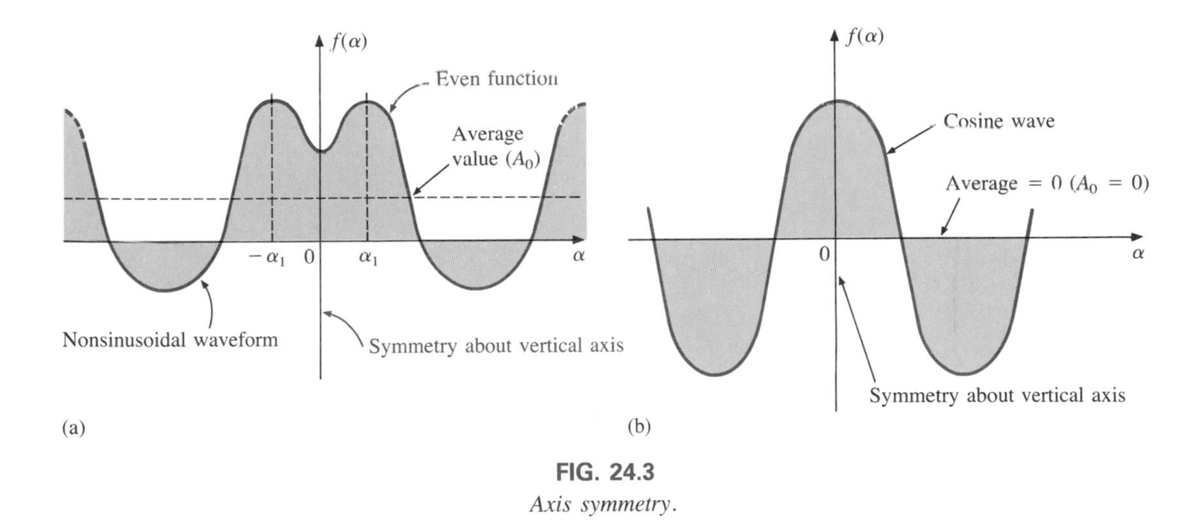


**FIG. 24.3**
*Axis symmetry.*

For both waveforms of Fig. 24.3, the following mathematical relationship is true:

$$f(\alpha) = f(-\alpha) \quad \text{(even function)} \qquad \textbf{(24.2)}$$

In words, it states that the magnitude of the function is the same at $+\alpha$ as at $-\alpha$ [$\alpha_1$ in Fig. 24.3(a)].

***If the waveform is such that its value for $+\alpha$ is the negative of that for $-\alpha$, it is called an* odd *function or is said to have* point symmetry *(about any point of intersection on the horizontal axis).***

Fig. 24.4(a) is an example of a waveform with point symmetry. For waveforms of this type all of the constants $A_{1\rightarrow n}$ will be zero. The function can then be represented by the *dc and sine terms*. Note that the sine wave itself is an odd function [Fig. 24.4(b)].

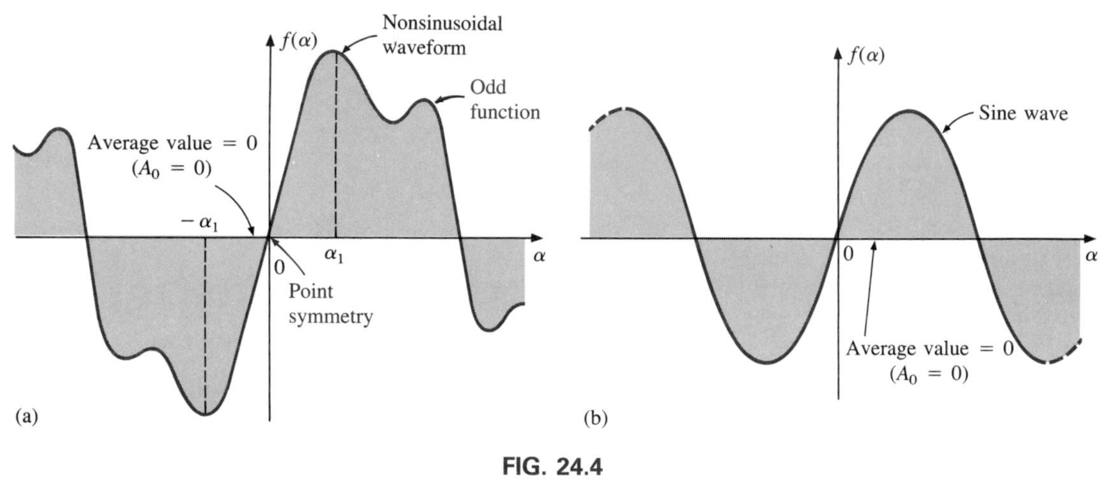


**FIG. 24.4**
*Point symmetry.*

For both waveforms of Fig. 24.4, the following mathematical relationship is true:

$$f(\alpha) = -f(-\alpha) \quad \text{(odd function)} \tag{24.3}$$

In words, it states that the magnitude of the function at $+\alpha$ is equal to the negative of the magnitude at $-\alpha$ [$\alpha_1$ in Fig. 24.4(a)].

The first term of the sine and cosine series is called the *fundamental component*. It represents the minimum frequency term required to represent a particular waveform, and it also has the same frequency as the waveform being represented. A fundamental term, therefore, must be present in any Fourier series representation. The other terms with higher-order frequencies (integer multiples of the fundamental) are called the *harmonic terms*. A term that has a frequency equal to twice the fundamental is the second harmonic; three times, the third harmonic; and so on.

If the waveform is such that

$$f(t) = f\left(\frac{T}{2} + t\right) \tag{24.4}$$

*the odd harmonics of the series of cosine and sine terms are zero.* Figure 24.5 is an example of this type of function.

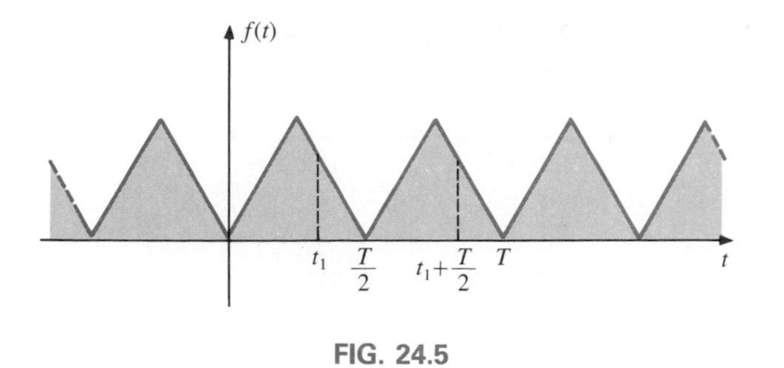

FIG. 24.5

Equation (24.4) states that the function repeats itself after each $T/2$ time interval ($t_1$ in Fig. 24.5). The waveform, however, will also repeat itself after each period $T$. In general, therefore, for a function of this type, if the period $T$ of the waveform is chosen to be twice that of the minimum period ($T/2$), the odd harmonics will all be zero.

If the waveform is such that

$$f(t) = -f\left(\frac{T}{2} + t\right) \tag{24.5}$$

the waveform is said to have *half wave* or *mirror symmetry* and *the even harmonics of the series of cosine and sine terms will be zero*. Figure 24.6 is an example of this type of function.

Equation (24.5) states that the waveform encompassed in one time interval $T/2$ will repeat itself in the next $T/2$ time interval, but in the negative sense ($t_1$ in Fig. 24.6). For example, the waveform of Fig. 24.6 from zero to $T/2$ will repeat itself in the time interval $T/2$ to $T$, but below the horizontal axis.

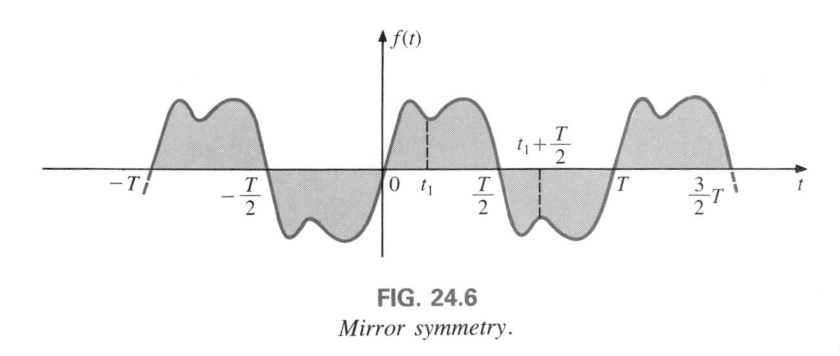

FIG. 24.6
*Mirror symmetry.*

An instrument for measuring harmonic frequencies and corresponding amplitudes is the wave analyzer shown in Fig. 24.7. It will generate the magnitude of each term of the Fourier series expansion of Eq. (24.1). For the sinusoidal and cosine terms, however, it will provide the rms values.

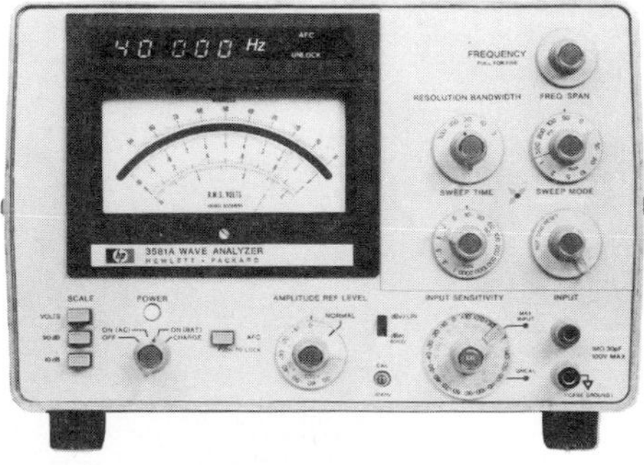

FIG. 24.7
*Wave analyzer. (Courtesy of Hewlett Packard Co.)*

The constants $A_0$, $A_{1\rightarrow n}$, $B_{1\rightarrow n}$ can be determined by using the following integral formulas:

$$A_0 = \frac{1}{T}\int_0^T f(t)\,dt = \frac{1}{2\pi}\int_0^{2\pi} f(\alpha)\,d\alpha \tag{24.6}$$

$$A_n = \frac{2}{T}\int_0^T f(t)\cos n\omega t\,dt = \frac{1}{\pi}\int_0^{2\pi} f(\alpha)\cos n\alpha\,d\alpha \tag{24.7}$$

$$B_n = \frac{2}{T}\int_0^T f(t)\sin n\omega t\,dt = \frac{1}{\pi}\int_0^{2\pi} f(\alpha)\sin n\alpha\,d\alpha \tag{24.8}$$

These equations have been presented for recognition purposes only; they will not be used in the following analysis.

The following examples will demonstrate the use of the equations and concepts introduced thus far in this chapter.

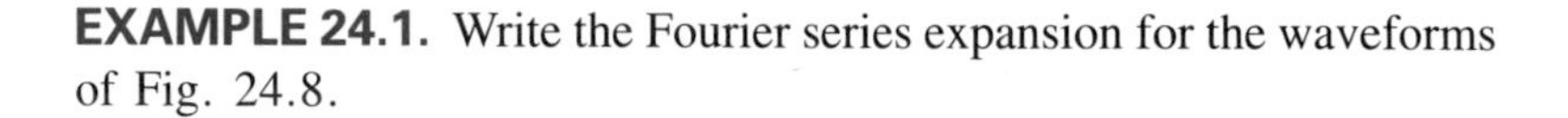

**EXAMPLE 24.1.** Write the Fourier series expansion for the waveforms of Fig. 24.8.

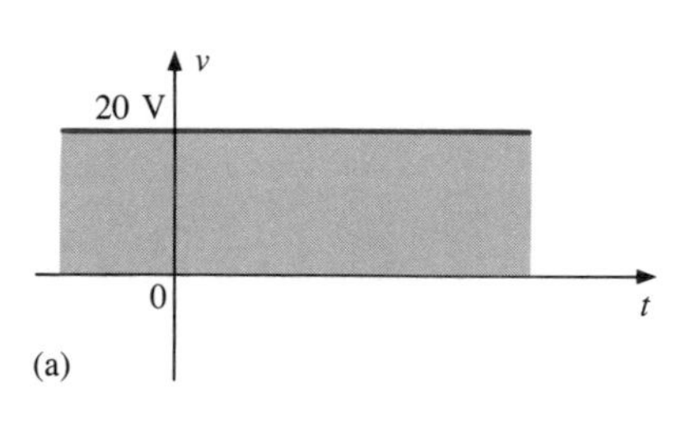


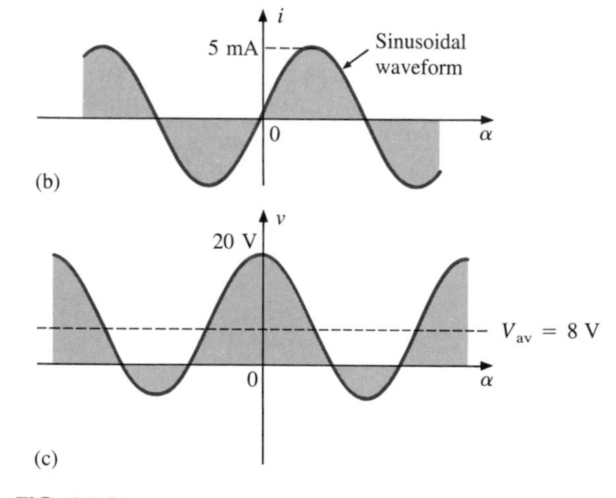


FIG. 24.8

***Solution:***

a. $A_0 = 20$, $A_{1\rightarrow n} = 0$, $B_{1\rightarrow n} = 0$
   $\boldsymbol{v = 20}$

b. $A_0 = 0$, $A_{1\rightarrow n} = 0$, $B_1 = 5 \times 10^{-3}$, $B_{2\rightarrow n} = 0$
   $\boldsymbol{i = 5 \times 10^{-3} \sin \alpha}$

c. $A_0 = 8$, $A_1 = 12$, $A_{2\rightarrow n} = 0$, $B_{1\rightarrow n} = 0$
   $\boldsymbol{v = 8 + 12 \cos \alpha}$

**EXAMPLE 24.2.** Sketch the following Fourier series expansion:

$$v = 2 + 1 \cos \alpha + 2 \sin \alpha$$

***Solution:*** Note Fig. 24.9.

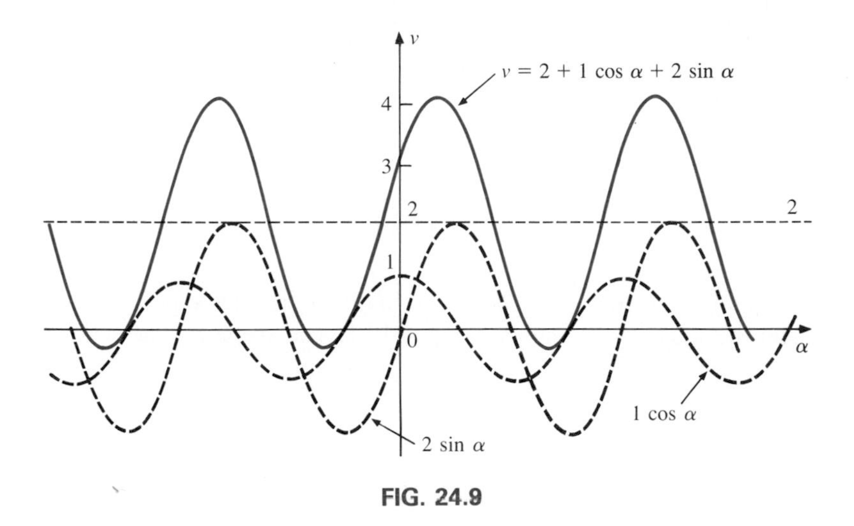


**FIG. 24.9**

The solution could be obtained graphically by first plotting all of the functions and considering a sufficient number of points on the horizontal axis; or phasor algebra could be employed as follows:

$$\begin{aligned} 1 \cos \alpha + 2 \sin \alpha &= 1 \text{ V} \angle 90^\circ + 2 \text{ V} \angle 0^\circ = j1 \text{ V} + 2 \text{ V} \\ &= 2 \text{ V} + j1 \text{ V} = 2.236 \text{ V} \angle 26.57^\circ \\ &= 2.236 \sin(\alpha + 26.57^\circ) \end{aligned}$$

and

$$v = 2 + 2.236 \sin(\alpha + 26.57^\circ)$$

which is simply the sine wave portion riding on a dc level of 2 V. That is, its positive maximum is 2 V + 2.236 V = 4.236 V, and its minimum is 2 V − 2.236 V = −0.236 V.

**EXAMPLE 24.3.** Sketch the following Fourier series expansion:

$$i = \sin \alpha + \sin 2\alpha$$

***Solution:*** See Fig. 24.10. Note that in this case the sum of the two sinusoidal waveforms of different frequencies is *not* a sine wave. Recall that complex algebra can be applied only to waveforms having the *same* frequency. In this case the solution is obtained graphically.

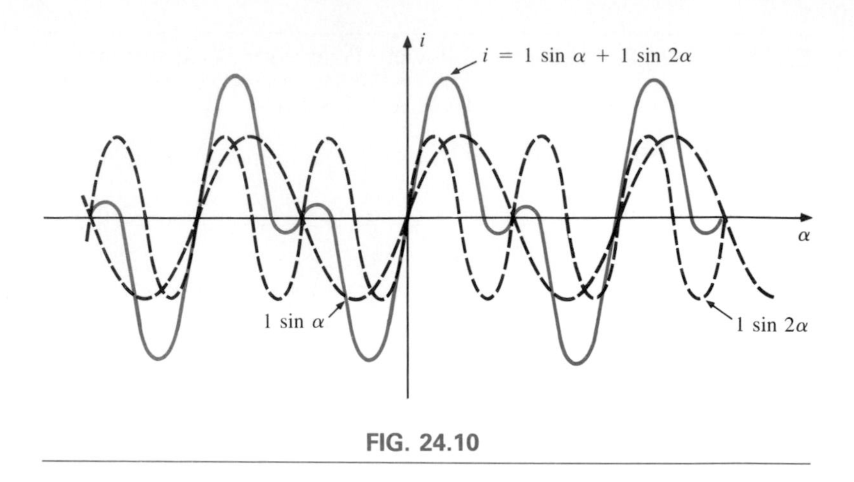


FIG. 24.10

As a further example in the use of the Fourier series approach, consider the square wave shown in Fig. 24.11. The average value is zero,

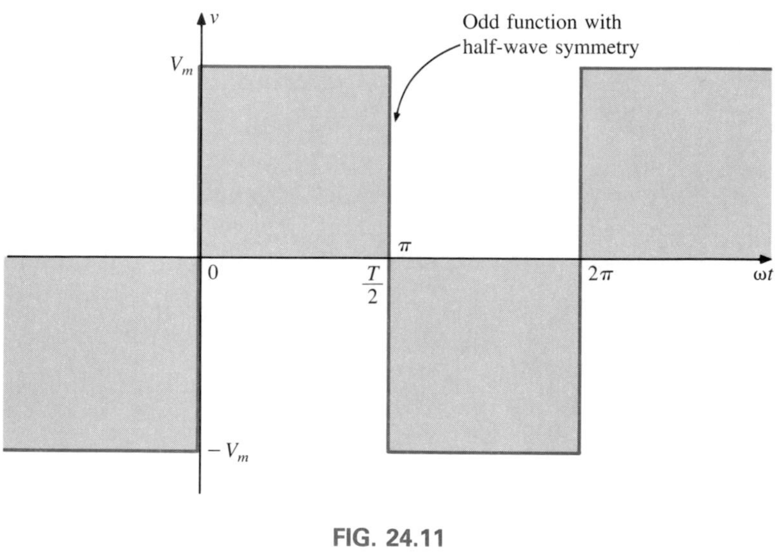


FIG. 24.11
*Square wave.*

so $A_0 = 0$. It is an odd function, so all the constants $A_{1\rightarrow n}$ equal zero; only sine terms will be present in the series expansion. Since the waveform satisfies the criteria for $f(t) = -f(T/2 + t)$, the even harmonics will also be zero.

The expression obtained after evaluating the various coefficients from Eq. (24.8) is

$$v = \frac{4}{\pi}V_m\left(\sin \omega t + \frac{1}{3}\sin 3\omega t + \frac{1}{5}\sin 5\omega t + \frac{1}{7}\sin 7\omega t + \cdots + \frac{1}{n}\sin n\omega t\right) \tag{24.9}$$

Note that the fundamental does indeed have the frequency of the square wave. If we add the fundamental and third harmonics, we obtain the results shown in Fig. 24.12.

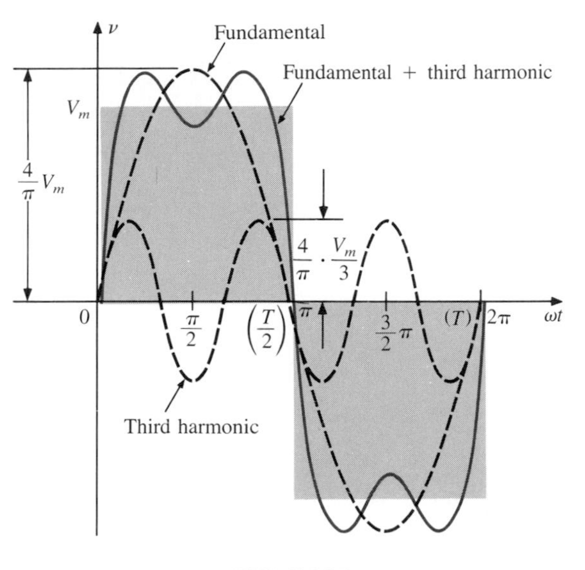


FIG. 24.12

Even with only the first two terms, the wave shape is beginning to look like a square wave. If we add the next two terms (Fig. 24.13), the width of the pulse increases, and the number of peaks increases.

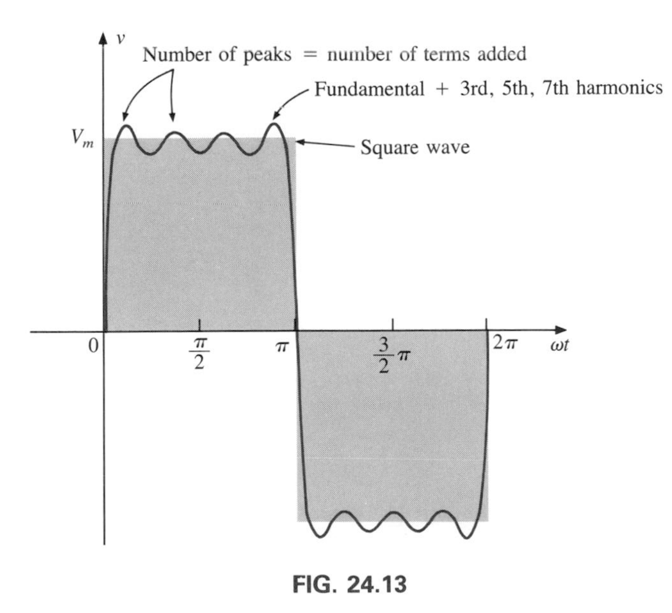


FIG. 24.13

As we continue to add terms, the series will better approximate the square wave. Note, however, that the amplitude of each succeeding term diminishes to the point at which it will be negligible compared with those of the first few terms. A good approximation would be to assume that the waveform is composed of the harmonics up to and including the ninth. Any higher harmonics would be less than one-tenth the fundamental. If the waveform just described were shifted above or below the horizontal axis, the Fourier series would be altered only by a change in the dc term. Figure 24.14(a), for example, is the sum of Figs.

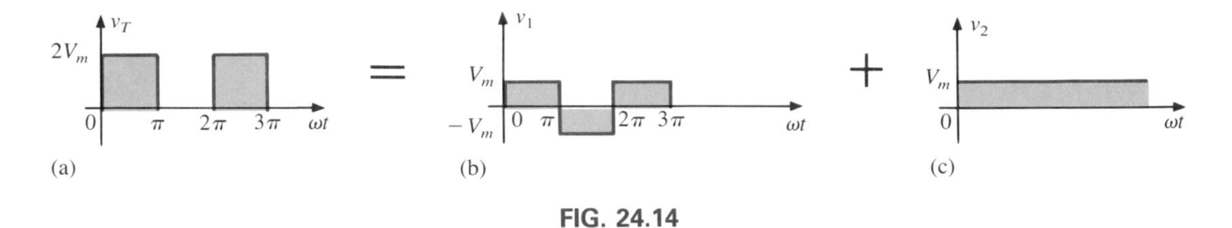


FIG. 24.14

24.14(b) and (c). The Fourier series for the complete waveform is therefore

$$v_T = V_m + \text{Eq. } (24.9)$$
$$= V_m + \frac{4}{\pi}V_m\left(\sin \omega t + \frac{1}{3}\sin 3\omega t + \frac{1}{5}\sin 5\omega t + \frac{1}{7}\sin 7\omega t + \cdots\right)$$

The equation for the pulsating waveform of Fig. 24.15(b) is

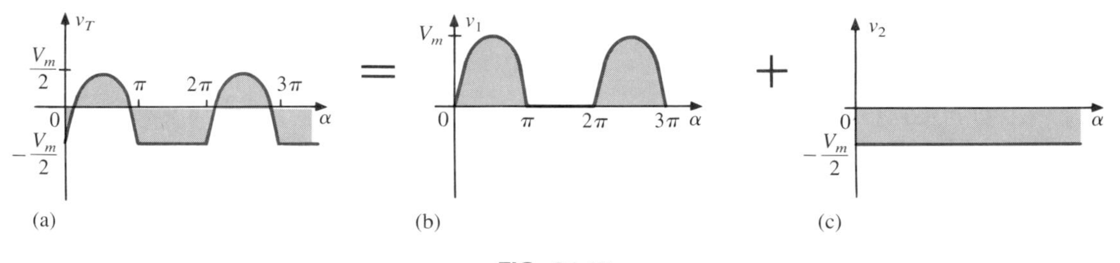


FIG. 24.15

$$v = 0.318V_m + 0.500V_m \sin \alpha - 0.212V_m \cos 2\alpha - 0.0424V_m \cos 4\alpha - \cdots \tag{24.10}$$

The waveform in Fig. 24.15(a) is the sum of the two in Figs. 24.15(b) and (c). The Fourier series for the waveform of Fig. 24.15(a) is therefore

$$v_T = -\frac{V_m}{2} + \text{Eq. } (24.10)$$
$$= \underbrace{(-0.500 + 0.318)}_{-0.182}V_m + 0.500V_m \sin \alpha - 0.212V_m \cos 2\alpha - 0.0424V_m \cos 4\alpha + \cdots$$

If either waveform were shifted to the right or left, the phase shift would be subtracted or added, respectively, from the sine and cosine terms. The dc term would not change with a shift to the right or left.

If the half-wave rectified signal is shifted 90° to the left, as in Fig. 24.16, the Fourier series becomes

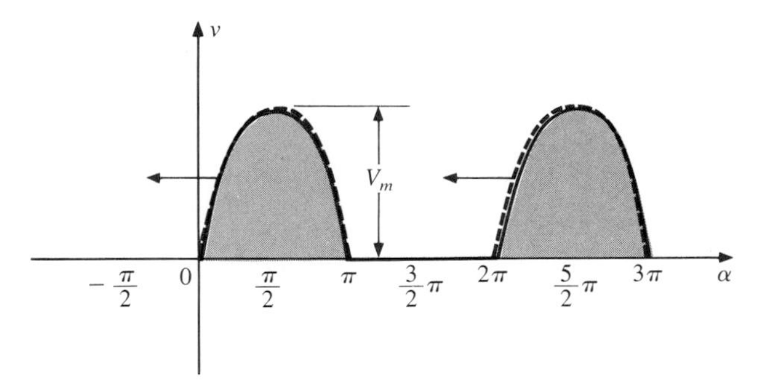


FIG. 24.16

$$v_o = 0.318V_m + 0.500V_m \underbrace{\sin(\alpha + 90°)}_{\cos\alpha} - 0.212V_m \cos 2(\alpha + 90°) - 0.0424V_m \cos 4(\alpha + 90°) + \cdots$$

$$= 0.318V_m + 0.500V_m \cos\alpha - 0.212V_m \cos(2\alpha + 180°) - 0.0424V_m \cos(4\alpha + 360°) + \cdots$$

$$v_o = 0.318V_m + 0.500V_m \cos\alpha + 0.212V_m \cos 2\alpha - 0.0424\ \mathrm{V_m} \cos 4\alpha + \cdots$$

## 24.3 CIRCUIT RESPONSE TO A NONSINUSOIDAL INPUT

The Fourier series representation of a nonsinusoidal input can be applied to a linear network using the principle of superposition. Recall that this theorem allowed us to consider the effects of each source of a circuit independently. If we replace the nonsinusoidal input by the terms of the Fourier series deemed necessary for practical considerations, we can use superposition to find the response of the network to each term (Fig. 24.17).

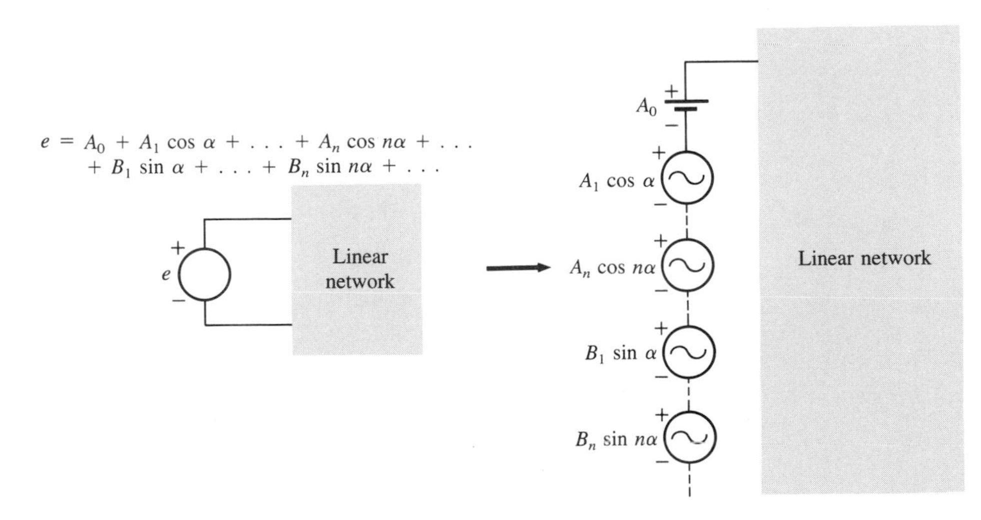


FIG. 24.17

The total response of the system is then the algebraic sum of the values obtained for each term. The major change between using this theorem for nonsinusoidal circuits and using it for the circuits previously described is that the frequency will be different for each term in the nonsinusoidal application. Therefore, the reactances

$$X_L = 2\pi fL \quad \text{and} \quad X_C = \frac{1}{2\pi fC}$$

will change for each term of the input voltage or current.

In Chapter 13, we found that the effective value of any waveform was given by

$$\sqrt{\frac{1}{T}\int_0^T [f(t)]^2\,dt}$$

If we apply this equation to the following Fourier series:

$$v(\alpha) = V_0 + V_{m_1}\cos\alpha + \cdots + V_{m_n}\cos n\alpha + V'_{m_1}\sin\alpha + \cdots + V'_{m_n}\sin n\alpha$$

then

$$V_{\text{eff}} = \sqrt{V_0^2 + \frac{V_{m_1}^2 + \cdots + V_{m_n}^2 + V'^2_{m_1} + \cdots + V'^2_{m_n}}{2}} \tag{24.11}$$

However, since

$$\frac{V_{m_1}^2}{2} = \left(\frac{V_{m_1}}{\sqrt{2}}\right)\left(\frac{V_{m_1}}{\sqrt{2}}\right) = (V_{1_{\text{eff}}})(V_{1_{\text{eff}}}) = V_{1_{\text{eff}}}^2$$

then

$$V_{\text{eff}} = \sqrt{V_0^2 + V_{1_{\text{eff}}}^2 + \cdots + V_{n_{\text{eff}}}^2 + V'^2_{1_{\text{eff}}} + \cdots + V'^2_{n_{\text{eff}}}} \tag{24.12}$$

Similarly, for

$$i(\alpha) = I_0 + I_{m_1}\cos\alpha + \cdots + I_{m_n}\cos n\alpha + I'_{m_1}\sin\alpha + \cdots + I'_{m_n}\sin n\alpha$$

we have

$$I_{\text{eff}} = \sqrt{I_0^2 + \frac{I_{m_1}^2 + \cdots + I_{m_n}^2 + I'^2_{m_1} + \cdots + I'^2_{m_n}}{2}} \tag{24.13}$$

and

$$I_{\text{eff}} = \sqrt{I_0^2 + I_{1_{\text{eff}}}^2 + \cdots + I_{n_{\text{eff}}}^2 + I'^2_{1_{\text{eff}}} + \cdots + I'^2_{n_{\text{eff}}}} \tag{24.14}$$

The total power delivered is the sum of that delivered by the corresponding terms of the voltage and current. In the following equations, all voltages and currents are effective values:

$$P_T = V_0 I_0 + V_1 I_1 \cos \theta_1 + \cdots + V_n I_n \cos \theta_n + \cdots \tag{24.15}$$

$$P_T = I_0^2 R + I_1^2 R + \cdots + I_n^2 R + \cdots \tag{24.16}$$

or

$$P_T = I_{\text{eff}}^2 R \tag{24.17}$$

with $I_{\text{eff}}$ as defined by Eq. (24.13), and, similarly,

$$P_T = \frac{V_{\text{eff}}^2}{R} \tag{24.18}$$

with $V_{\text{eff}}$ as defined by Eq. (24.11).

**EXAMPLE 24.4.** Determine the effective value of the waveform of Fig. 24.18.

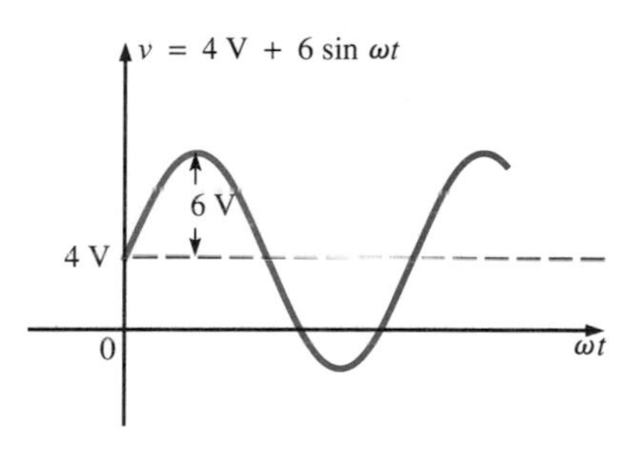


FIG. 24.18

***Solution:*** Eq. (24.12):

$$V_{\text{eff}} = \sqrt{V_0^2 + \frac{V_m^2}{2}}$$

$$= \sqrt{(4\text{ V})^2 + \frac{(6\text{ V})^2}{2}} = \sqrt{16 + \frac{36}{2}}\text{ V} = \sqrt{34}\text{ V}$$

and

$$V_{\text{eff}} = \mathbf{5.831\ V}$$

It is particularly interesting to note from the preceding example that the rms value of a waveform having both dc and ac components is not simply the sum of the effective values of each. In other words, there is a temptation in the absence of Eq. (24.12) to state that $V_{\text{eff}} = 4\text{ V} + 0.707(6\text{ V}) = 8.242\text{ V}$, which is incorrect and, in fact, exceeds the correct level by some 41%.

**EXAMPLE 24.5.** We learned in Chapter 13 that the effective value of a square wave is the peak value of the waveform. Let us test this result using the Fourier expansion and Eq. (24.11).

Determine the effective value of the square wave of Fig. 24.11 with $V_m = 20$ V using only the first four terms of the Fourier expansion and compare to the actual effective value of 20 V.

***Solution:***

$$v = \frac{4}{\pi}(20\text{ V})\sin\omega t + \frac{4}{\pi}\left(\frac{1}{3}\right)(20\text{ V})\sin 3\omega t + \frac{4}{\pi}\left(\frac{1}{5}\right)(20\text{ V})\sin 5\omega t + \frac{4}{\pi}\left(\frac{1}{7}\right)(20\text{ V})\sin 7\omega t$$

$$v = 25.465\sin\omega t + 8.488\sin 3\omega t + 5.093\sin 5\omega t + 3.638\sin 7\omega t$$

Eq. (24.11):

$$V_{\text{eff}} = \sqrt{V_0^2 + \frac{V_{m_1}^2 + V_{m_2}^2 + V_{m_3}^2 + V_{m_4}^2}{2}}$$

$$= \sqrt{(0\text{ V})^2 + \frac{(25.465\text{ V})^2 + (8.488\text{ V})^2 + (5.093\text{ V})^2 + (3.638\text{ V})^2}{2}}$$

$$= \mathbf{19.49\text{ V}}$$

The solution is about 0.5 V from the correct answer of 20 V. However, each additional term in the Fourier series will bring the result closer to the 20-V level. An infinite number would result in an exact solution of 20 V.

---

**EXAMPLE 24.6.** The input to the circuit of Fig. 24.19 is the following:

$$e = 12 + 10\sin 2t$$

a. Find the current $i$ and the voltages $v_R$ and $v_C$.
b. Find the effective values of $i$, $v_R$, and $v_C$.
c. Find the power delivered to the circuit.

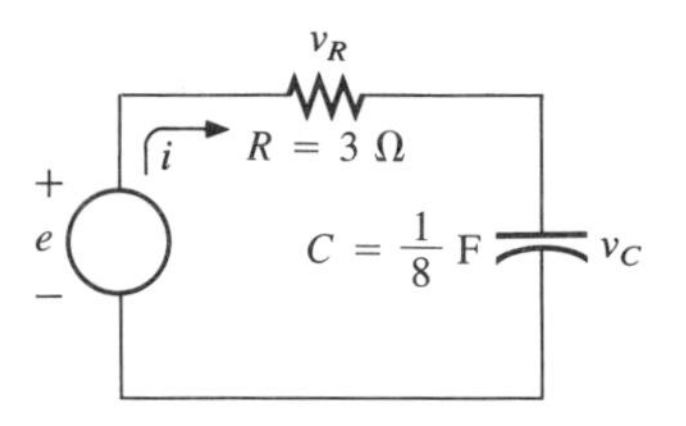


FIG. 24.19

***Solutions:***

a. Redraw the original circuit as shown in Fig. 24.20. Then apply superposition:

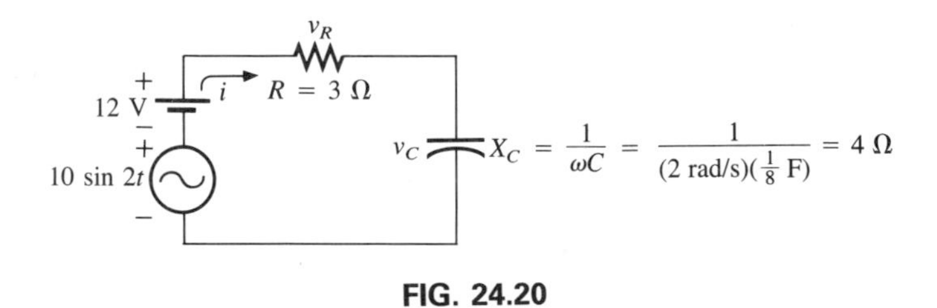


FIG. 24.20

1. *For the 12-V dc supply portion of the input,* $I = 0$ since the capacitor is an open circuit to dc when $v_C$ has reached its final (steady-state) value. Therefore,

$$V_R = IR = 0\text{ V}$$

and

$$V_C = 12\text{ V}$$

2. *For the ac supply,*

$$\mathbf{Z} = 3\ \Omega - j4\ \Omega = 5\ \Omega\ \angle -53.13^\circ$$

and

$$\mathbf{I} = \frac{\mathbf{E}}{\mathbf{Z}} = \frac{\frac{10}{\sqrt{2}}\ \text{V}\ \angle 0^\circ}{5\ \Omega\ \angle -53.13^\circ} = \frac{2}{\sqrt{2}}\ \text{A}\ \angle +53.13^\circ$$

$$\mathbf{V}_R = \mathbf{IR} = \left(\frac{2}{\sqrt{2}}\ \text{A}\ \angle +53.13^\circ\right)(3\ \Omega\ \angle 0^\circ) = \frac{6}{\sqrt{2}}\ \text{V}\ \angle +53.13^\circ$$

and

$$\mathbf{V}_C = \mathbf{IX}_C = \left(\frac{2}{\sqrt{2}}\ \text{A}\ \angle +53.13^\circ\right)(4\ \Omega\ \angle -90^\circ)$$

$$= \frac{8}{\sqrt{2}}\ \text{V}\ \angle -36.87^\circ$$

In the time domain,

$$i = \mathbf{0 + 2\ sin(2}t\ \mathbf{+ 53.13^\circ)}$$

Note that even though the dc term was present in the expression for the input voltage, the dc term for the current in this circuit is zero:

$$v_R = \mathbf{0 + 6\ sin(2}t\ \mathbf{+ 53.13^\circ)}$$

and

$$v_C = \mathbf{12 + 8\ sin(2}t\ \mathbf{- 36.87^\circ)}$$

b. Eq. (24.14): $I_{\text{eff}} = \sqrt{(0)^2 + \frac{(2\ \text{A})^2}{2}} = \sqrt{2} = \mathbf{1.414\ A}$

Eq. (24.12): $V_{R_{\text{eff}}} = \sqrt{(0)^2 + \frac{(6\ \text{V})^2}{2}} = \sqrt{18} = \mathbf{4.243\ V}$

Eq. (24.12): $V_{C_{\text{eff}}} = \sqrt{(12)^2 + \frac{(8\ \text{V})^2}{2}} = \sqrt{176} = \mathbf{13.267\ V}$

c. $P = I_{\text{eff}}^2 R = \left(\frac{2}{\sqrt{2}}\ \text{A}\right)^2 (3\ \Omega) = \mathbf{6\ W}$

---

**EXAMPLE 24.7.** Find the response of the circuit of Fig. 24.21 to the input shown.

$$e = 0.318E_m + 0.500E_m \sin \omega t - 0.212E_m \cos 2\omega t - 0.0424E_m \cos 4\omega t + \cdots$$

***Solution:*** For discussion purposes, only the first three terms will be used to represent $e$. Converting the cosine terms to sine terms and substituting for $E_m$ gives us

$$e = 63.60 + 100.0 \sin \omega t - 42.40 \sin(2\omega t + 90^\circ)$$

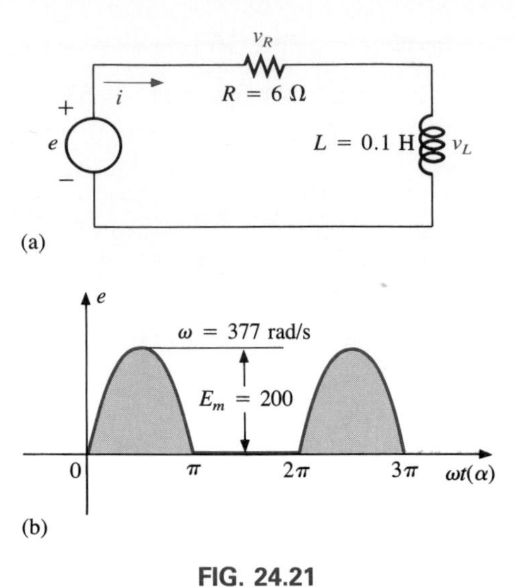

FIG. 24.21

Using phasor notation, the original circuit becomes as shown in Fig. 24.22. Apply superposition:

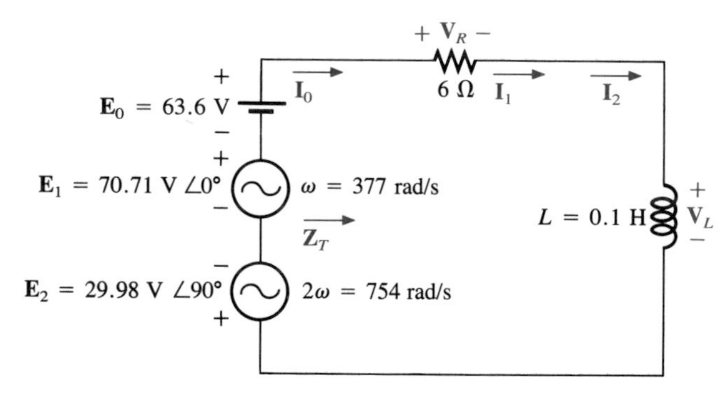

FIG. 24.22

*For the dc term* ($E_0 = 63.6$ V),

$$X_L = 0 \qquad \text{(short for dc)}$$

$$\mathbf{Z}_T = 6\ \Omega \angle 0° = \mathbf{R}$$

$$I_0 = \frac{E_0}{R} = \frac{63.6\text{ V}}{6\ \Omega} = 10.60\text{ A}$$

$$V_{R_0} = I_0 R = E_0 = 63.60\text{ V}$$

$$V_{L_0} = 0$$

The average power is

$$P_0 = I_0^2 R = (10.60\text{ A})^2(6\ \Omega) = 674.2\text{ W}$$

*For the fundamental term* ($\mathbf{E}_1 = 70.71\text{ V}\angle 0°$, $\omega = 377$),

$$X_{L_1} = \omega L = (377\text{ rad/s})(0.1\text{ H}) = 37.7\ \Omega$$
$$\mathbf{Z}_{T_1} = 6\ \Omega + j37.7\ \Omega = 38.17\ \Omega\angle 80.96°$$
$$\mathbf{I}_1 = \frac{\mathbf{E}_1}{\mathbf{Z}_{T_1}} = \frac{70.71\text{ V}\angle 0°}{38.17\ \Omega\angle 80.96°} = 1.85\text{ A}\angle -80.96°$$
$$\mathbf{V}_{R_1} = \mathbf{I}_1\mathbf{R} = (1.85\text{ A}\angle -80.96°)(6\ \Omega\angle 0°) = 11.10\text{ V}\angle -80.96°$$
$$\mathbf{V}_{L_1} = \mathbf{I}_1\mathbf{X}_L = (1.85\text{ A}\angle -80.96°)(37.7\ \Omega\angle 90°) = 69.75\text{ V}\angle 9.04°$$

The average power is

$$P_1 = I_1^2R = (1.85\text{ A})^2(6\ \Omega) = 20.54\text{ W}$$

*For the second harmonic* ($\mathbf{E}_2 = 29.98\text{ V}\angle -90°$, $\omega = 754$). The phase angle of $\mathbf{E}_2$ was changed to $-90°$ to give it the same polarity as the input voltages $\mathbf{E}_0$ and $\mathbf{E}_1$.

$$X_{L_2} = \omega L = (754\text{ rad/s})(0.1\text{ H}) = 75.4\ \Omega$$
$$\mathbf{Z}_{T_2} = 6\ \Omega + j75.4\ \Omega = 75.64\ \Omega\angle 85.45°$$
$$\mathbf{I}_2 = \frac{\mathbf{E}_2}{\mathbf{Z}_{T_2}} = \frac{29.98\text{ V}\angle -90°}{75.64\ \Omega\angle 85.45} = 0.396\text{ A}\angle -174.45°$$
$$\mathbf{V}_{R_2} = \mathbf{I}_2\mathbf{R} = (0.396\text{ A}\angle -174.45°)(6\ \Omega\angle 0°) = 2.38\text{ V}\angle -174.45°$$
$$\mathbf{V}_{L_2} = \mathbf{I}_2\mathbf{X}_{L_2} = (0.396\text{ A}\angle -174.45°)(75.4\ \Omega\angle 90°) = 29.9\text{ V}\angle -84.45°$$

The average power is

$$P_2 = I_2^2R = (0.396\text{ A})^2(6\ \Omega) = 0.941\text{ W}$$

The Fourier series expansion for $i$ is

$$i = \mathbf{10.6 + \sqrt{2}(1.85)\sin(377}t\ \mathbf{- 80.96°) + \sqrt{2}(0.396)\sin(754}t\ \mathbf{- 174.45°)}$$

and

$$I_{\text{eff}} = \sqrt{(10.6\text{ A})^2 + (1.85\text{ A})^2 + (0.396\text{ A})^2} = \mathbf{10.77\text{ A}}$$

The Fourier series expansion for $v_R$ is

$$v_R = \mathbf{63.6 + \sqrt{2}(11.10)\sin(377}t\ \mathbf{- 80.96°) + \sqrt{2}(2.38)\sin(754}t\ \mathbf{- 174.45°)}$$

and

$$V_{R_{\text{eff}}} = \sqrt{(63.6\text{ V})^2 + (11.10\text{ V})^2 + (2.38\text{ V})^2} = \mathbf{64.61\text{ V}}$$

The Fourier series expansion for $v_L$ is

$$v_L = \mathbf{\sqrt{2}(69.75)\sin(377}t\ \mathbf{+ 9.04°) + \sqrt{2}(29.93)\sin(754}t\ \mathbf{- 84.45°)}$$

and

$$V_{L_{\text{eff}}} = \sqrt{(69.75\text{ V})^2 + (29.93\text{ V})^2} = \mathbf{75.90\text{ V}}$$

The total average power is

$$P_T = I_{\text{eff}}^2R = (10.77\text{ A})^2(6\ \Omega) = \mathbf{695.96\text{ W}} = P_0 + P_1 + P_2$$

## 24.4 ADDITION AND SUBTRACTION OF NONSINUSOIDAL WAVEFORMS

The Fourier series expression for the waveform resulting from the addition or subtraction of two nonsinusoidal waveforms can be found using phasor algebra if the terms having the same frequency are considered separately.

For example, the sum of the following two nonsinusoidal waveforms is found using this method:

$$v_1 = 30 + 20 \sin 20t + \cdots + 5 \sin(60t + 30°)$$
$$v_2 = 60 + 30 \sin 20t + 20 \sin 40t + 10 \cos 60t$$

**1.** dc terms:

$$V_{T_o} = 30 \text{ V} + 60 \text{ V} = 90 \text{ V}$$

**2.** $\omega = 20$:

$$V_{T_{1(\max)}} = 30 \text{ V} + 20 \text{ V} = 50 \text{ V}$$

and

$$V_{T_1} = 50 \sin 20t$$

**3.** $\omega = 40$:

$$V_{T_2} = 20 \sin 40t$$

**4.** $\omega = 60$:

$$5 \sin(60t + 30°) = (0.707)(5) \text{ V} \angle 30° = 3.54 \text{ V} \angle 30°$$
$$10 \cos 60t = 10 \sin(60t + 90°) \Rightarrow (0.707)(10) \text{ V} \angle 90° = 7.07 \text{ V} \angle 90°$$
$$\mathbf{V}_{T_3} = 3.54 \text{ V} \angle 30° + 7.07 \text{ V} \angle 90° = 3.07 \text{ V} + j1.77 \text{ V} + j7.07 \text{ V} = 3.07 \text{ V} + j8.84 \text{ V}$$
$$\mathbf{V}_{T_3} = 9.36 \text{ V} \angle 70.85°$$

and

$$\mathbf{V}_{T_3} = 13.24 \sin(60t + 70.85°)$$

and

$$v_1 + v_2 = v_T = \mathbf{90 + 50 \sin 20t + 20 \sin 40t + 13.24 \sin(60t + 70.85°)}$$

## 24.5 COMPUTER ANALYSIS

All the mathematical operations appearing in this chapter can be applied using BASIC. PSPICE would be limited primarily to an analysis of the effect of each component of a Fourier expansion of the applied signal using superposition. In other words, a separate solution would be obtained at each applied frequency and then the total solution would be obtained in the longhand fashion.

## BASIC

To demonstrate the versatility of the language approach, BASIC was applied in Program 24.1 (Fig. 24.23) to obtain the Fourier series expansion resulting from the addition of two nonsinusoidal waveforms. The general format of the nonsinusoidal waveforms to be added appears on

```
 10 REM ***** PROGRAM 24-1 *****
 20 REM ********************************************
 30 REM Program to add two non-sinusoidal voltages.
 40 REM ********************************************
 50 REM
100 DIM V1(15),A1(15),V2(15),A2(15),VS(15),VA(15)
110 PRINT "This program add two non-sinusoidal voltages"
120 PRINT "of the form:"
130 PRINT "v(t)=VO + V1sin(wt+TH1) + V2sin(2wt+TH2) + V3sin(3wt+TH3)
140 PRINT
150 PRINT "Enter the following signal information:"
160 PRINT
170 INPUT "Radian frequency, w=";W
180 PRINT :PRINT "Enter the number of terms for the voltage"
190 PRINT "where N=dc + harmonics, i.e. N=1 + 4 = 5 for up "
200 PRINT "to and including the 4th harmonic."
210 PRINT:INPUT "Number of terms, N=";N
220 PRINT:PRINT "For the following, the number in the parentheses"
230 PRINT "refers to the term in the Fourier series expansion."
240 PRINT:INPUT "V1(0)=";V1(0)
250 INPUT "V2(0)=";V2(0)
260 FOR I=1 TO N-1
270 PRINT
280 PRINT "V1(";I;")=";
290 INPUT V1(I)
300 INPUT "at angle";A1(I)
310 PRINT "V2(";I;")=";
320 INPUT V2(I)
330 INPUT "at angle";A2(I)
340 NEXT I
350 PRINT
360 REM Calculate the sum terms
370 VS(0)=V1(0)+V2(0)
380 FOR I=1 TO N-1
390 A1=A1(I) :A2=A2(I)
400 XS=V1(I)*COS(A1/57.296)+V2(I)*COS(A2/57.296)
410 YS=V1(I)*SIN(A1/57.296)+V2(I)*SIN(A2/57.296)
420 VS(I)=SQR(XS^2+YS^2)
425 IF XS=0 THEN VA(I)=90 :GOTO 440
430 VA(I)=57.296*ATN(YS/XS)
440 NEXT I
450 PRINT
460 PRINT "Equation for the sum of the voltages is:"
470 PRINT "v(t)=v1(t) + v2(t) =";VS(0);
480 FOR I=1 TO N-1
490 PRINT TAB(21);
500 PRINT "+";VS(I);"sin(";W*I;"t";
510 IF VA(I)>=0 THEN PRINT "+";
520 PRINT VA(I);")"
530 NEXT I
540 END

READY
```

```
RUN


This program add two non-sinusoidal voltages
of the form:
v(t)=V0 + V1sin(wt+TH1) + V2sin(2wt+TH2) + V3sin(3wt+TH3) + ...

Enter the following signal information:

Radian frequency, w=? 60


Enter the number of terms for the voltage
where N=dc + harmonics, i.e. N=1 + 4 = 5 for up
to and including the 4th harmonic.

Number of terms, N=? 4


For the following, the number in the parentheses
refers to the term in the Fourier series expansion.

V1(0)=? 60

V2(0)=? 20


V1( 1 )=? 70

at angle? 0

V2( 1 )=? 30

at angle? 0


V1( 2 )=? 20

at angle? 90

V2( 2 )=? -20

at angle? 90


V1( 3 )=? 10

at angle? 60

V2( 3 )=? 5

at angle? 90



Equation for the sum of the voltages is:
v(t)=v1(t) + v2(t) = 80 + 100 sin( 60 t+ 0 )
                     + 0 sin( 120 t+ 90 )
                     + 14.5466 sin( 180 t+ 69.8961 )


READY
```

**FIG. 24.23**
*Program 24.1.*

line 130 of the program. The dc terms for each are requested on lines 240 and 250, and the harmonics on lines 260 through 340. Note on line 190 that the number of terms $N$ includes the dc term and the number of harmonics.

The algebraic addition of the dc terms occurs on line 370, while the sum of the sinusoidal terms is obtained by lines 380 through 440, using the indicated $I$ loop. As an example, consider the case of finding the sum of two fundamental frequency components for the provided printout. Line 390 will define A1 = A1(1) = $\theta_1$ = 30° and A2 = A2(1) = $\theta_2$ = 40°. Line 400 will then determine the sum of the real and imaginary components using the right-triangle relationships of Fig. 24.24. That is,

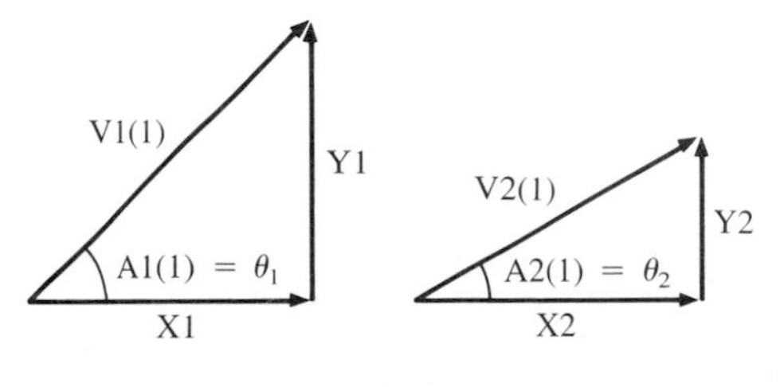


FIG. 24.24

$$\begin{aligned}\text{XS} = \text{X1} + \text{X2} &= \text{V1(1)} \cos \theta_1 + \text{V2(1)} \cos \theta_2 \\ &= \text{V1(1)} \cos \frac{\text{A1}}{57.296} + \text{V2(1)} \cos \frac{\text{A2}}{57.296} \\ &= 2 \cos \frac{30}{57.296} + 5 \cos \frac{40}{57.296}\end{aligned}$$

and, similarly,

$$\begin{aligned}\text{YS} = \text{Y1} + \text{Y2} &= \text{V1(1)} \sin \theta_1 + \text{V2(1)} \sin \theta_2 \\ &= 2 \sin \frac{30}{57.296} + 5 \sin \frac{40}{57.296}\end{aligned}$$

The magnitude and angle of the polar form of the sum are then determined by lines 420 and 430, respectively, and stored as VS(1) and VA(1). The same routine will then be repeated for each set of terms with the same frequency.

When the I loop is complete, the results will be printed out by lines 460 through 530, using an additional I loop from I = 1 to I = (N − 1).

Note in the program that the effective values are not employed in the calculations, because they were not to be printed out, and that the results are in the sinusoidal domain. The printout will also indicate the proper value of $\omega$ for each term and include each harmonic that appears in either or both of the nonsinusoidal signals.

The run shown here provides a solution of part (a) of Problem 16 in this chapter.

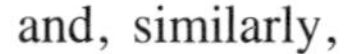

## PROBLEMS

### SECTION 24.2 Fourier Series

**1.** For the waveforms of Fig. 24.25, determine whether the following will be present in the Fourier series representation:

**a.** dc term
**b.** cosine terms
**c.** sine terms
**d.** even-ordered harmonics
**e.** odd-ordered harmonics

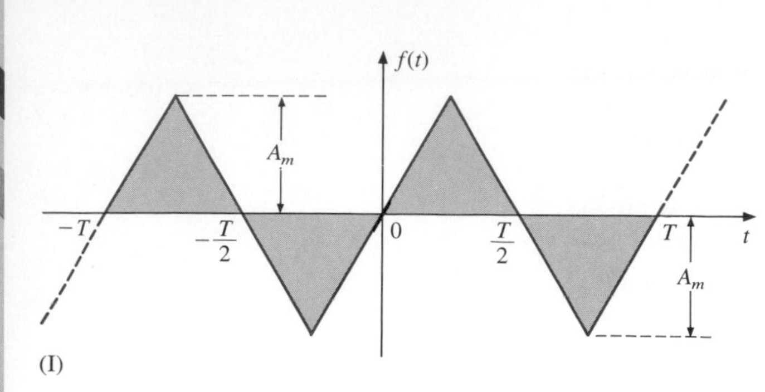


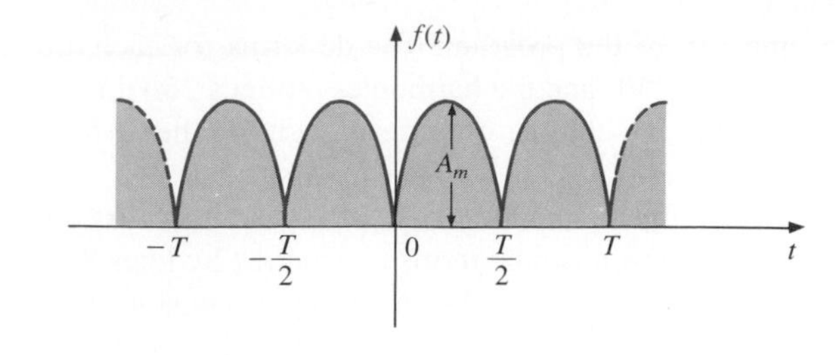


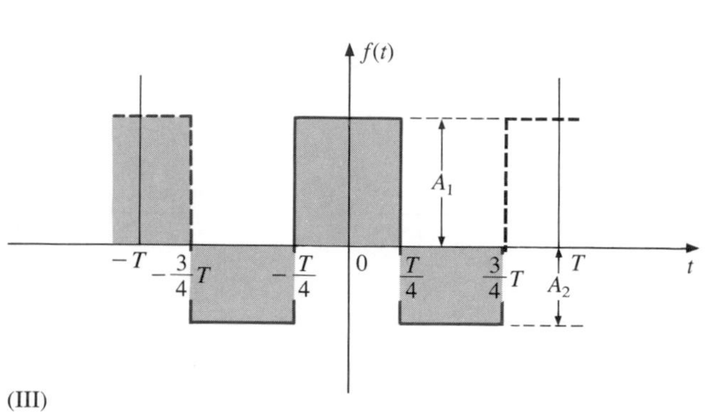


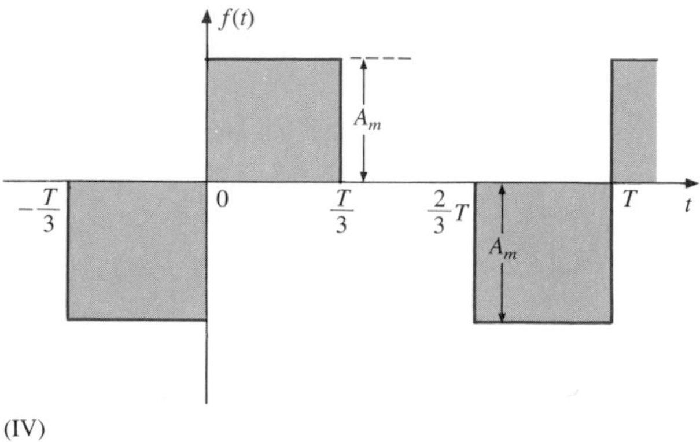

**FIG. 24.25**

**2.** If the Fourier series for the waveform of Fig. 24.26(a) is

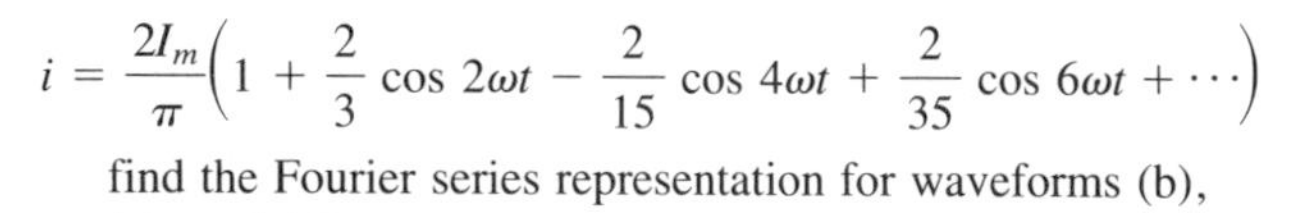

$$i = \frac{2I_m}{\pi}\left(1 + \frac{2}{3}\cos 2\omega t - \frac{2}{15}\cos 4\omega t + \frac{2}{35}\cos 6\omega t + \cdots\right)$$

find the Fourier series representation for waveforms (b), (c), and (d).

(a)

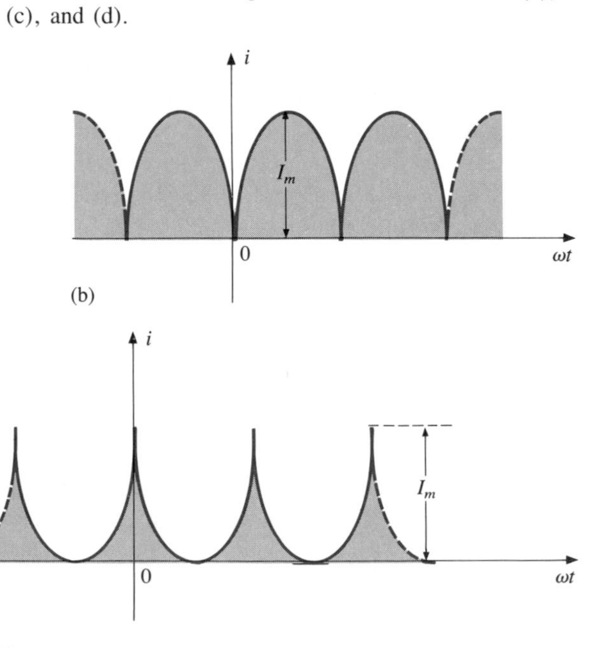


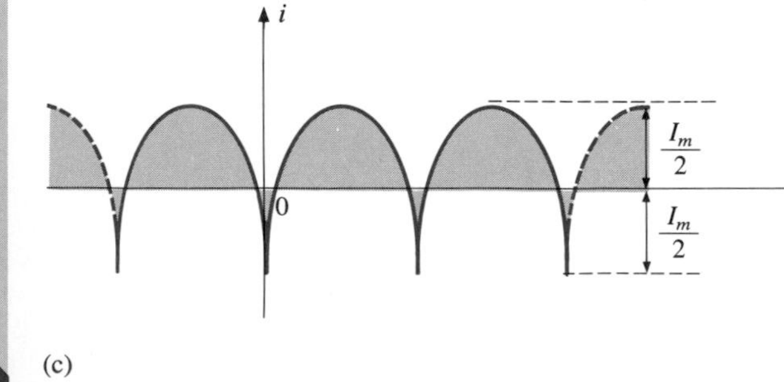


**FIG. 24.26**

3. Sketch the following nonsinusoidal waveforms with $\alpha = \omega t$ as the abscissa:
   a. $v = -4 + 2 \sin \alpha$ b. $v = (\sin \alpha)^2$
   c. $i = 2 - 2 \cos \alpha$
4. Sketch the following nonsinusoidal waveforms with $\alpha$ as the abscissa:
   a. $i = 3 \sin \alpha - 6 \sin 2\alpha$
   b. $v = 2 \cos 2\alpha + \sin \alpha$
5. Sketch the following nonsinusoidal waveforms with $\omega t$ as the abscissa:
   a. $i = 50 \sin \omega t + 25 \sin 3\omega t$
   b. $i = 50 \sin \alpha - 25 \sin 3\alpha$
   c. $i = 4 + 3 \sin \omega t + 2 \sin 2\omega t - 1 \sin 3\omega t$

## SECTION 24.3 Circuit Response to a Nonsinusoidal Input

6. Find the average and effective values of the following nonsinusoidal waves:
   a. $v = 100 + 50 \sin \omega t + 25 \sin 2\omega t$
   b. $i = 3 + 2 \sin(\omega t - 53°) + 0.8 \sin(2\omega t - 70°)$
7. Find the effective value of the following nonsinusoidal waves:
   a. $v = 20 \sin \omega t + 15 \sin 2\omega t - 10 \sin 3\omega t$
   b. $i = 6 \sin(\omega t + 20°) + 2 \sin(2\omega t + 30°) - 1 \sin(3\omega t + 60°)$
8. Find the total average power to a circuit whose voltage and current are as indicated in Problem 6.
9. Find the total average power to a circuit whose voltage and current are as indicated in Problem 7.
10. The Fourier series representation for the input voltage to the circuit of Fig. 24.27 is

$$e = 18 + 30 \sin 400t$$

   a. Find the nonsinusoidal expression for the current $i$.
   b. Calculate the effective value of the current.
   c. Find the expression for the voltage across the resistor.
   d. Calculate the effective value of the voltage across the resistor.
   e. Find the expression for the voltage across the reactive element.
   f. Calculate the effective value of the voltage across the reactive element.
   g. Find the average power delivered to the resistor.
11. Repeat Problem 10 for

$$e = 24 + 30 \sin 400t + 10 \sin 800t$$

12. Repeat Problem 10 for the following input voltage:

$$e = -60 + 20 \sin 300t - 10 \sin 600t$$

13. Repeat Problem 10 for the circuit of Fig. 24.28.

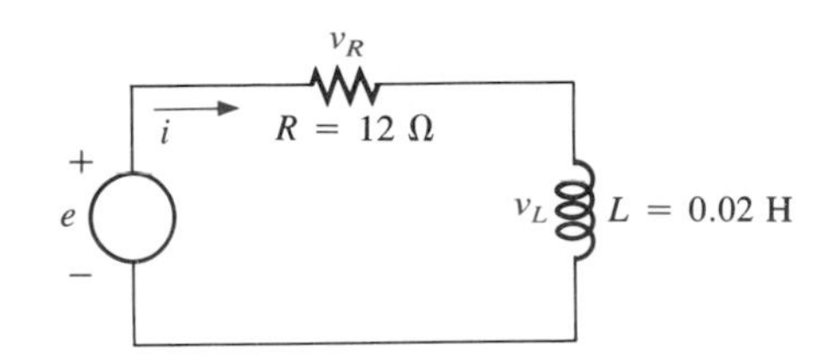


FIG. 24.27

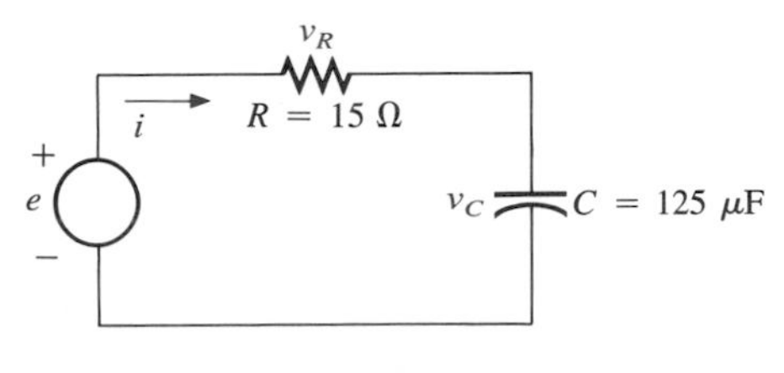

FIG. 24.28

***14.** The input voltage [Fig. 24.29(a)] to the circuit of Fig. 24.29(b) is a full-wave rectified signal having the following Fourier series expansion:

$$e = \frac{(2)(100)}{\pi}\left(1 + \frac{2}{3}\cos 2\omega t - \frac{2}{15}\cos 4\omega t + \frac{2}{53}\cos 6\omega t + \cdots\right)$$

where $\omega = 377$.

**a.** Find the Fourier series expression for the voltage $v_o$ using only the first three terms of the expression.

**b.** Find the effective value of $v_o$.

**c.** Find the average power delivered to the 1-kΩ resistor.

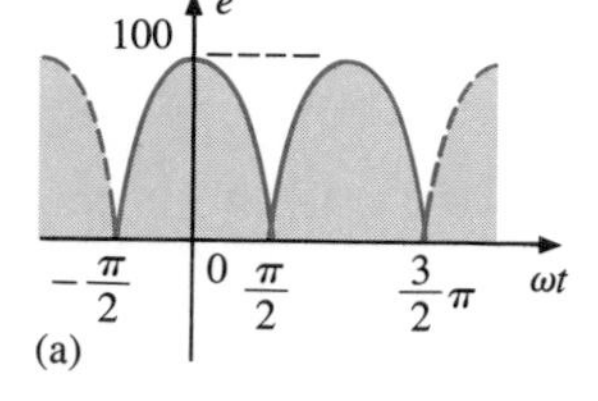


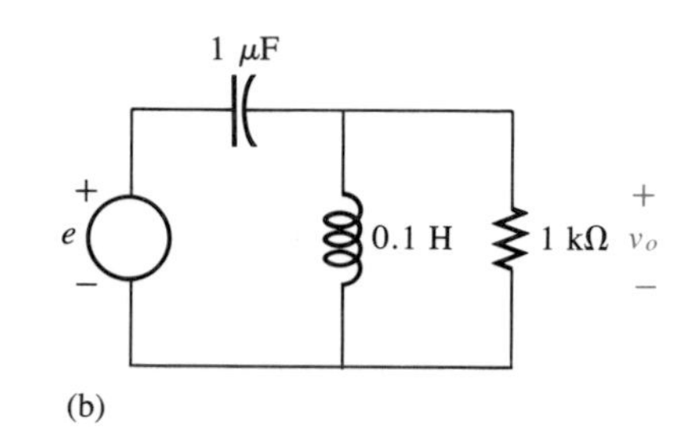


FIG. 24.29

***15.** Find the Fourier series expression for the voltage $v_o$ of Fig. 24.30.

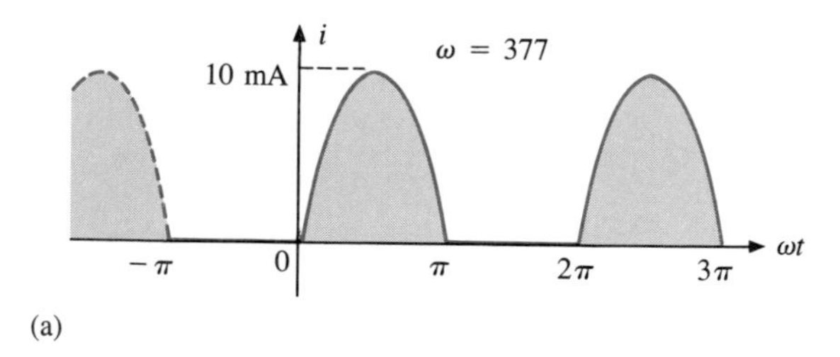


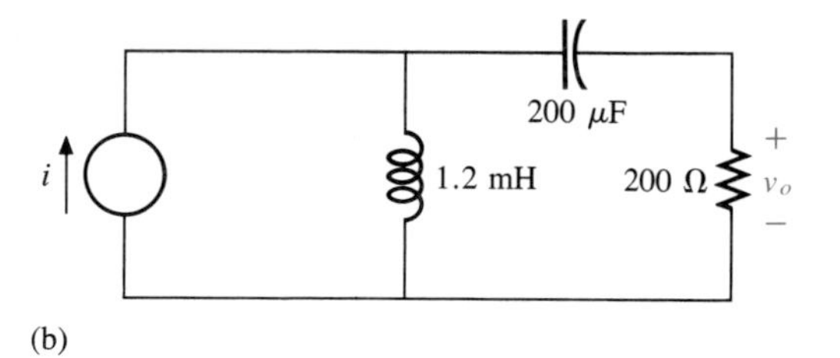


FIG. 24.30

## SECTION 24.4 Addition and Subtraction of Nonsinusoidal Waveforms

**16.** Perform the indicated operations on the following nonsinusoidal waveforms:

**a.** $[60 + 70 \sin \omega t + 20 \sin(2\omega t + 90°) + 10 \sin(3\omega t + 60°)] + [20 + 30 \sin \omega t - 20 \cos 2\omega t + 5 \cos 3\omega t]$

**b.** $[20 + 60 \sin \alpha + 10 \sin(2\alpha - 180°) + 5 \cos(3\alpha + 90°)] - [5 - 10 \sin \alpha + 4 \sin(3\alpha - 30°)]$

17. Find the nonsinusoidal expression for the current $i_S$ of the diagram of Fig. 24.31.

$$i_2 = 10 + 30 \sin 20t - 0.5 \sin(40t + 90°)$$
$$i_1 = 20 + 4 \sin(20t + 90°) + 0.5 \sin(40t + 30°)$$

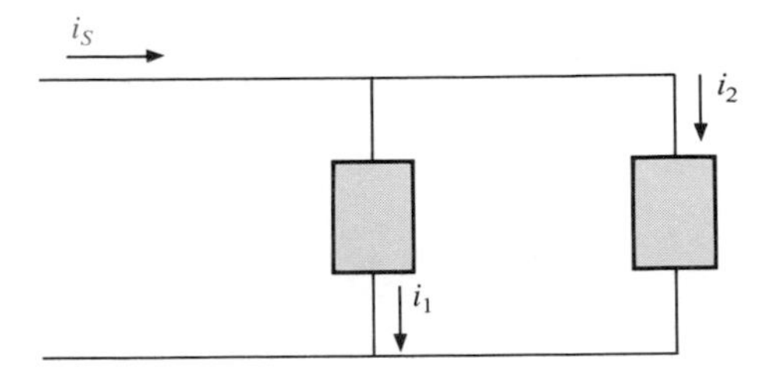

FIG. 24.31

18. Find the nonsinusoidal expression for the voltage $e$ of the diagram of Fig. 24.32.

$$v_1 = 20 - 200 \sin 600t + 100 \cos 1200t + 75 \sin 1800t$$
$$v_2 = -10 + 150 \sin(600t + 30°) + 50 \sin(1800t + 60°)$$

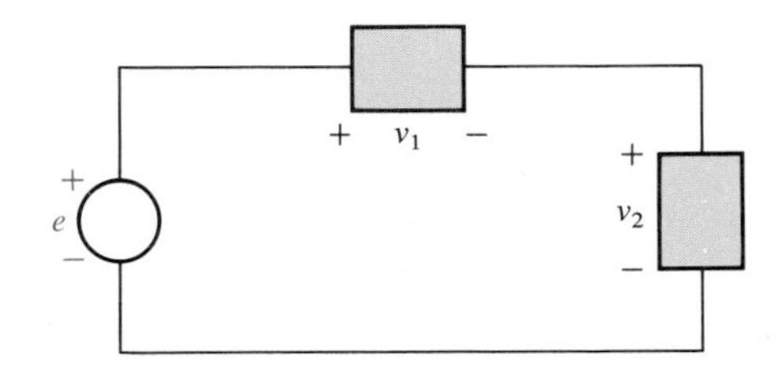

FIG. 24.32

## SECTION 24.5 Computer Analysis

### BASIC

19. Write a program to determine the sum of the first 10 terms of Eq. (24.9) at $\omega t = \pi/2$, $\pi$, and $(3/2)\pi$, and compare to the values determined by Fig. 24.11. That is, enter Eq. (24.9) into memory and calculate the sum of the terms at the points listed above.

20. Given any nonsinusoidal function, write a program that will determine the average and effective values of the waveform. The program should request the data required from the nonsinusoidal function.

21. Write a program that will provide a general solution for the network of Fig. 24.19 for a single dc and ac term in the applied voltage. In other words, the parameter values are given along with the particulars regarding the applied signal, and the nonsinusoidal expression for the current and each voltage is generated by the program.

## GLOSSARY

**Axis symmetry** A sinusoidal or nonsinusoidal function that has symmetry about the vertical axis.

**Even harmonics** The terms of the Fourier series expansion that have frequencies that are even multiples of the fundamental component.

**Fourier series** A series of terms, developed in 1826 by Baron Jean Fourier, that can be used to represent a nonsinusoidal function.

**Fundamental component** The minimum frequency term required to represent a particular waveform in the Fourier series expansion.

**Half-wave (mirror) symmetry** A sinusoidal or nonsinusoidal function that satisfies the relationship $f(t) = -f\left(\frac{T}{2} + t\right)$.

**Harmonic** The terms of the Fourier series expansion that have frequencies that are integer multiples of the fundamental component.

**Nonsinusoidal waveform** Any waveform that differs from the fundamental sinusoidal function.

**Odd harmonics** The terms of the Fourier series expansion that have frequencies that are odd multiples of the fundamental component.

**Point symmetry** A sinusoidal or nonsinusoidal function that satisfies the relationship $f(\alpha) = -f(-\alpha)$.

# 25
# Transformers

## 25.1 INTRODUCTION

Chapter 12 discussed the *self-inductance* of a coil. We shall now examine the *mutual inductance* that exists between coils of the same or different dimensions. Mutual inductance is a phenomenon basic to the operation of the transformer, an electrical device used today in almost every field of electrical engineering. This device plays an integral part in power distribution systems and can be found in many electronic circuits and measuring instruments. In this chapter, we will discuss three of the basic applications of a transformer. These are its ability to build up or step down the voltage or current, to act as an impedance matching device, and to isolate (no physical connection) one portion of a circuit from another. In addition, we will introduce the dot convention and consider the transformer equivalent circuit. The chapter will conclude with a word about the effect of mutual inductance on the mesh equations of a network.

## 25.2 MUTUAL INDUCTANCE

The transformer is constructed of two coils placed so that the changing flux developed by one will link the other as shown in Fig. 25.1. This will result in an induced voltage across each coil. To distinguish between the coils, we will apply the transformer convention that the coil to which the source is applied is called the *primary,* and the coil to which the load is applied is called the *secondary*. For the primary of the transformer of Fig. 25.1, an application of Faraday's law [Eq. (12.1)] will result in

$$e_p = N_p \frac{d\phi_p}{dt}$$

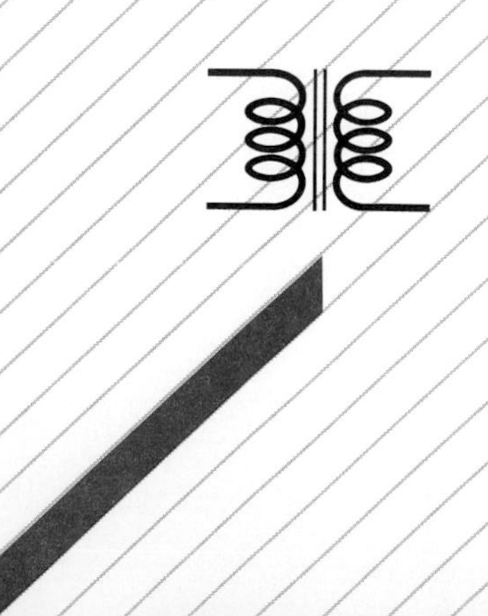

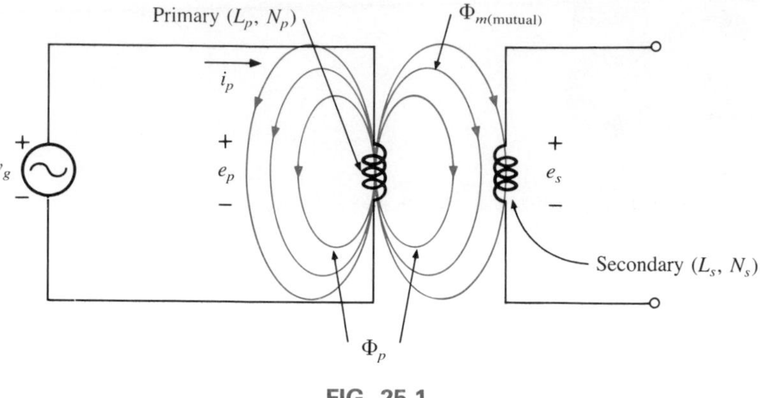


FIG. 25.1

or, from Eq. (12.5),

$$e_p = L_p \frac{di_p}{dt}$$

The magnitude of $e_s$, the voltage induced across the secondary, is determined by

$$e_s = N_s \frac{d\phi_m}{dt} \tag{25.1a}$$

where $\phi_m$ is the portion of the primary flux $\phi_p$ that links the secondary.

If all of the flux linking the primary links the secondary, then

$$\phi_m = \phi_p$$

and

$$e_s = N_s \frac{d\phi_p}{dt} \tag{25.1b}$$

The *coefficient of coupling* between two coils is determined by

$$k \text{ (coefficient of coupling)} = \frac{\phi_m}{\phi_p} \tag{25.2}$$

Since the maximum changing flux that can link the secondary is $\phi_p$, the coefficient of coupling between two coils can never be greater than 1. The coefficient of coupling between various coils is indicated in Fig. 25.2. Note that for the iron core, $k \cong 1$, while for the air core, $k$ is considerably less. Those coils with low coefficients of coupling are said to be *loosely coupled*.

For the secondary, we have

$$e_s = N_s \frac{d\phi_m}{dt} = N_s \frac{dk\phi_p}{dt}$$

and

$$e_s = kN_s \frac{d\phi_p}{dt} \qquad \textbf{(25.3)}$$

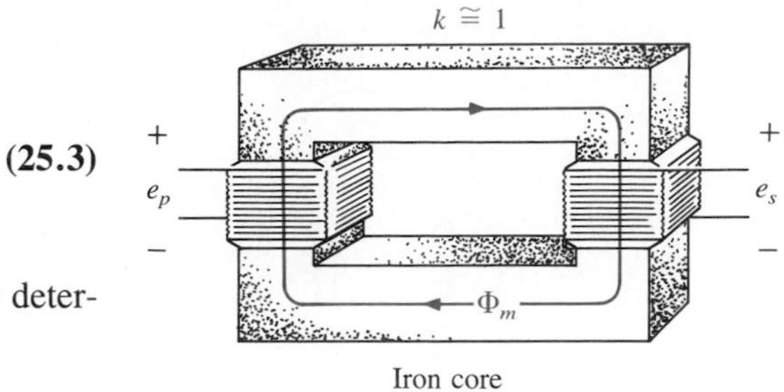


The mutual inductance between the two coils of Fig. 25.1 is determined by

$$M = N_s \frac{d\phi_m}{di_p} \qquad \text{(henries, H)} \qquad \textbf{(25.4a)}$$

or

$$M = N_p \frac{d\phi_p}{di_s} \qquad \text{(H)} \qquad \textbf{(25.4b)}$$

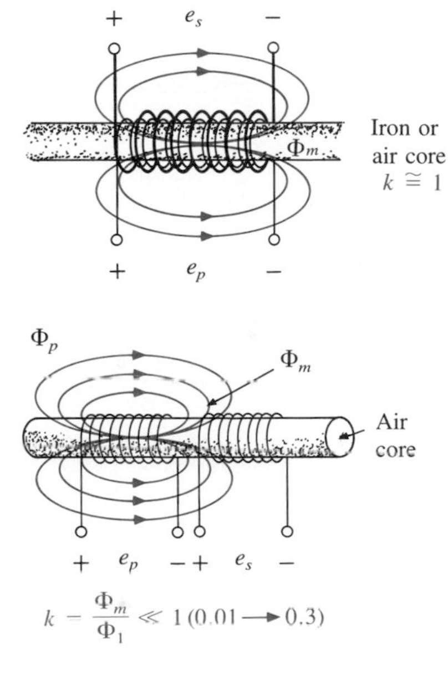


FIG. 25.2

Note in the above equations that the symbol for mutual inductance is the capital letter $M$, and that its unit of measurement, like that of self-inductance, is the *henry*. In words, Eqs. (25.4a) and (25.4b) state that the mutual inductance between two coils is proportional to the instantaneous change in flux linking one coil due to an instantaneous change in current through the other coil.

In terms of the inductance of each coil and the coefficient of coupling, the mutual inductance is determined by

$$M = k\sqrt{L_pL_s} \qquad \text{(H)} \qquad \textbf{(25.5)}$$

The greater the coefficient of coupling (greater flux linkages), or the greater the inductance of either coil, the higher the mutual inductance between the coils. Relate this fact to the configurations of Fig. 25.2.

The secondary voltage $e_s$ can also be found in terms of the mutual inductance if we rewrite Eq. (25.1a) as

$$e_s = N_s\left(\frac{d\phi_m}{di_p}\right)\left(\frac{di_p}{dt}\right)$$

and, since $M = N_s(d\phi_m/di_p)$, it can also be written

$$e_s = M\frac{di_p}{dt} \qquad \textbf{(25.6a)}$$

Similarly,

$$e_p = M\frac{di_s}{dt} \qquad \textbf{(25.6b)}$$

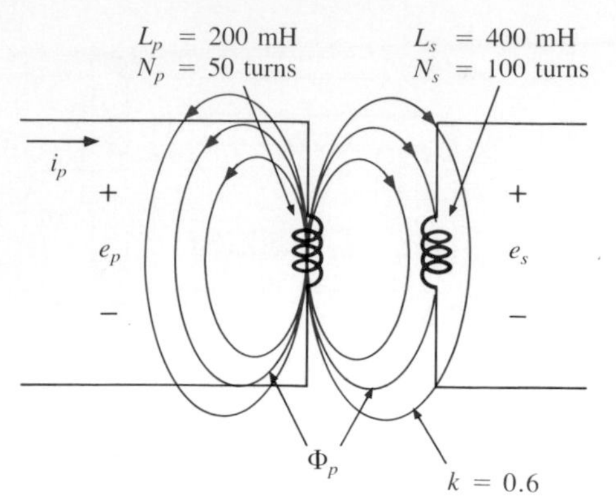


FIG. 25.3

**EXAMPLE 25.1.** For the transformer in Fig. 25.3:

a. Find the mutual inductance $M$.
b. Find the induced voltage $e_p$ if the flux changes at the rate of 450 mWb/s.
c. Find the induced voltage $e_s$ for the same rate of change indicated in part (b).
d. Find the induced voltages $e_p$ and $e_s$ if the current $i_p$ changes at the rate of 2 A/s.

***Solutions:***

a. $M = k\sqrt{L_pL_s} = 0.6\sqrt{(200 \text{ mH})(400 \text{ mH})}$
$= 0.6\sqrt{8 \times 10^{-2}} = (0.6)(2.828 \times 10^{-1}) = \mathbf{169.7\ mH}$

b. $e_p = N_p\dfrac{d\phi_p}{dt} = (50)(450 \text{ mWb/s}) = \mathbf{22.5\ V}$

c. $e_s = kN_s\dfrac{d\phi_p}{dt} = (0.6)(100)(450 \text{ mWb/s}) = \mathbf{27\ V}$

d. $e_p = L_p\dfrac{di_p}{dt} = (200 \text{ mH})(2 \text{ A/s}) = \mathbf{400\ mV}$

$e_s = M\dfrac{di_p}{dt} = (170 \text{ mH})(2 \text{ A/s}) = \mathbf{340\ mV}$

## 25.3 SERIES CONNECTION OF MUTUALLY COUPLED COILS

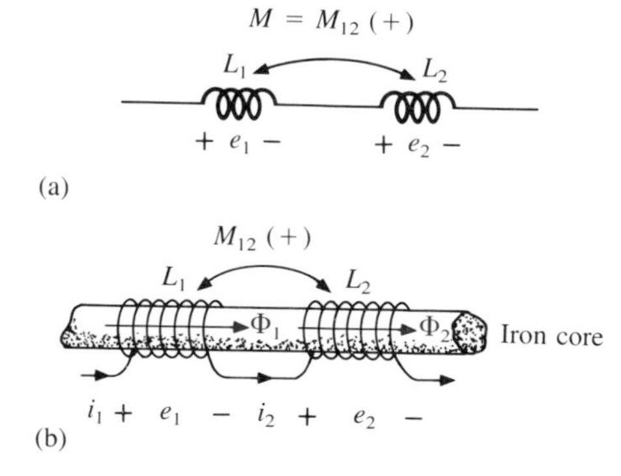


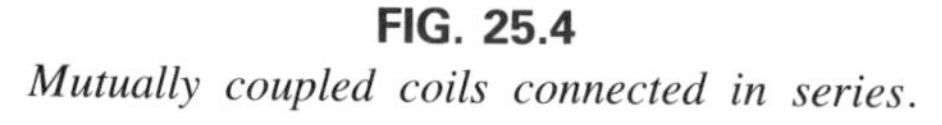

FIG. 25.4
*Mutually coupled coils connected in series.*

In Chapter 12, we found that the total inductance of series isolated coils was determined simply by the sum of the inductances. For two coils that are connected in series but also share the same flux linkages, such as those in Fig. 25.4(a), a mutual term is introduced that will alter the total inductance of the series combination. The physical picture of how the coils are connected is indicated in Fig. 25.4(b). An iron core is included, although the equations to be developed are for any two mutually coupled coils with any value of coefficient of coupling $k$. When referring to the voltage induced across the inductance $L_1$ (or $L_2$) due to the change in flux linkages of the inductance $L_2$ (or $L_1$, respectively), the mutual inductance is represented by $M_{12}$. This type of subscript notation is particularly important when there are two or more mutual terms.

Due to the presence of the mutual term, the induced voltage $e_1$ is composed of that due to the self-inductance $L_1$ and that due to the mutual inductance $M_{12}$. That is,

$$e_1 = L_1\frac{di_1}{dt} + M_{12}\frac{di_2}{dt}$$

However, since $i_1 = i_2 = i$,

$$e_1 = L_1\frac{di}{dt} + M_{12}\frac{di}{dt}$$

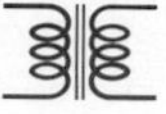



or

$$e_1 = (L_1 + M_{12})\frac{di}{dt} \qquad \textbf{(25.7a)}$$

and, similarly,

$$e_2 = (L_2 + M_{12})\frac{di}{dt} \qquad \textbf{(25.7b)}$$

For the series connection, the total induced voltage across the series coils, represented by $e_T$, is

$$e_T = e_1 + e_2 = (L_1 + M_{12})\frac{di}{dt} + (L_2 + M_{12})\frac{di}{dt}$$

or

$$e_T = (L_1 + L_2 + M_{12} + M_{12})\frac{di}{dt}$$

and the total inductance is

$$L_{T(+)} = L_1 + L_2 + 2M_{12} \qquad \text{(H)} \qquad \textbf{(25.8a)}$$

The subscript (+) was included to indicate that the mutual terms have a positive sign. If the coils were wound such as shown in Fig. 25.5, where $\phi_1$ and $\phi_2$ are in opposition, the induced voltage due to the

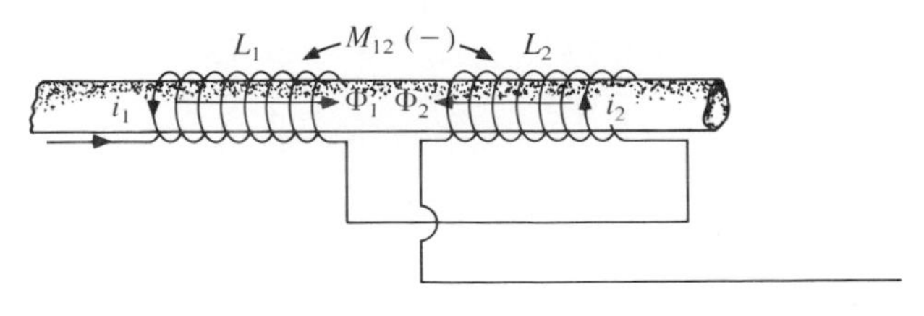


**FIG. 25.5**
*Mutually coupled coils connected in series with negative mutual inductance.*

mutual term would oppose that due to the self-inductance, and the total inductance would be determined by

$$L_{T(-)} = L_1 + L_2 - 2M_{12} \qquad \text{(H)} \qquad \textbf{(25.8b)}$$

Through Eqs. (25.8a) and (25.8b), the mutual inductance can be determined by

$$M = \frac{1}{4}(L_{T(+)} - L_{T(-)}) \qquad \textbf{(25.9)}$$

Equation (25.9) is very effective in determining the mutual inductance between two coils. It states that the mutual inductance is equal to one-quarter the difference between the total inductance with positive mutual inductance and with negative mutual inductance.

From the preceding, it should be clear that the mutual inductance will directly affect the magnitude of the voltage induced across a coil, since it will determine the net inductance of the coil. Further examination reveals that the sign of the mutual term for each coil of a coupled pair is the same. For $L_{T(+)}$ they were both positive, and for $L_{T(-)}$ they were both negative. On a network schematic where it is inconvenient to indicate the windings and the flux path, a system of dots is employed which will determine whether the mutual terms are to be positive or negative. The dot convention is shown in Fig. 25.6 for the series coils of Figs. 25.4 and 25.5.

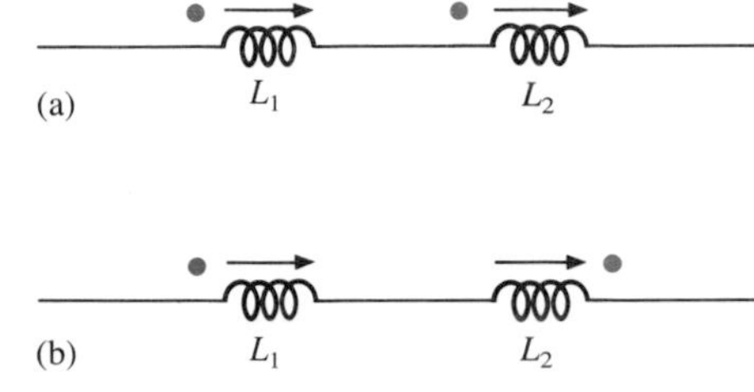


**FIG. 25.6**
*Dot convention for the series coils of (a) Fig. 25.4 and (b) Fig. 25.5.*

If the current through *each* of the mutually coupled coils is going away from (or toward) the dot as it *passes through the coil,* the mutual term will be positive, as shown for the case in Fig. 25.6(a). If the arrow indicating current direction through the coil is leaving the dot for one coil and entering the dot for the other, the mutual term is negative.

A few possibilities for mutually coupled transformer coils are indicated in Fig. 25.7(a). The sign of $M$ is indicated for each. When determining the sign, be sure to examine the current direction within the coil itself. In Fig. 25.7(b), one direction was indicated outside for one coil and through for the other. It initially might appear that the sign should be positive since both currents enter the dot, but the current *through* coil 1 is leaving the dot; hence a negative sign is in order.

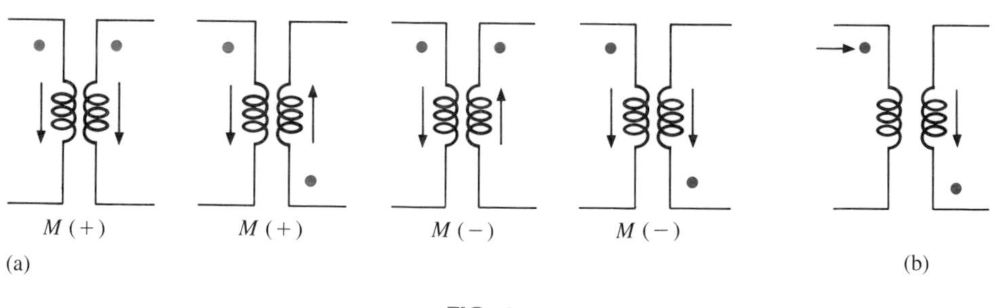


**FIG. 25.7**

The dot convention also reveals the polarity of the *induced* voltage across the mutually coupled coil. If the reference direction for the current *in* a coil leaves the dot, the polarity at the dot for the induced voltage of the mutually coupled coil is positive. In the first two figures of Fig. 25.7(a), the polarity at the dots of the induced voltages is positive. In the third figure of Fig. 25.7(a), the polarity at the dot of the right-hand coil is positive while the polarity at the dot of the left-hand coil is negative, since the current enters the dot (within the coil) of the right-hand coil. The comments for the third figure of Fig. 25.7(a) can also be applied to the last figure of Fig. 25.7(a).

**EXAMPLE 25.2.** Find the total inductance of the series coils of Fig. 25.8.

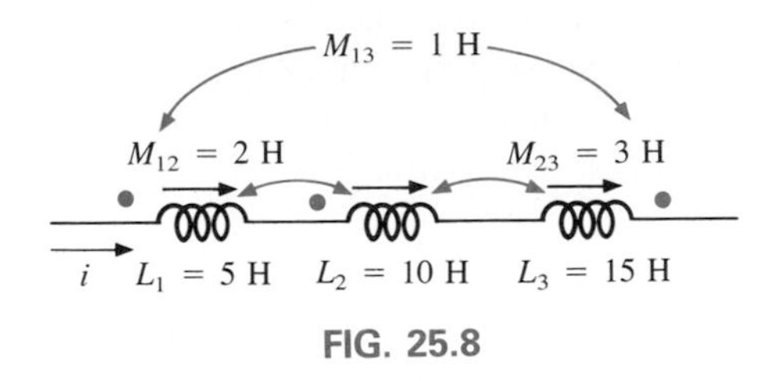


FIG. 25.8

***Solution:***

Current vectors leave dot.

Coil 1: $L_1 + M_{12} - M_{13}$

One current vector enters dot while one leaves.

Coil 2: $L_2 + M_{12} - M_{23}$

Coil 3: $L_3 - M_{23} - M_{13}$

Note that each has the same sign above.

and

$$\begin{aligned} L_T &= (L_1 + M_{12} - M_{13}) + (L_2 + M_{12} - M_{23}) + (L_3 - M_{23} - M_{13}) \\ &= L_1 + L_2 + L_3 + 2M_{12} - 2M_{23} - 2M_{13} \end{aligned}$$

Substituting values, we find

$$\begin{aligned} L_T &= 5\text{ H} + 10\text{ H} + 15\text{ H} + 2(2\text{ H}) - 2(3\text{ H}) - 2(1\text{ H}) \\ &= 34\text{ H} - 8\text{ H} = \mathbf{26\text{ H}} \end{aligned}$$

**EXAMPLE 25.3.** Write the mesh equations for the transformer network in Fig. 25.9.

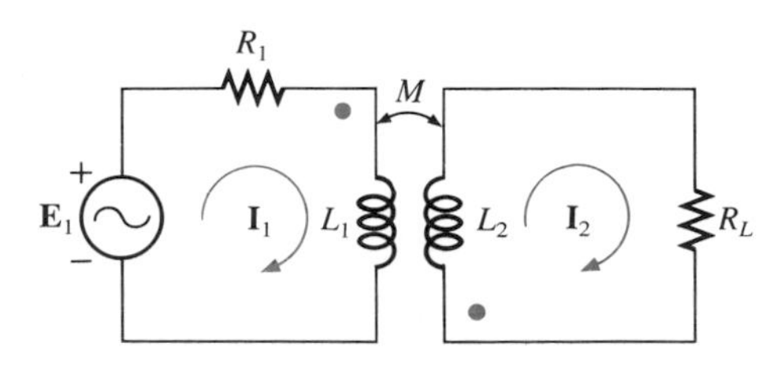


FIG. 25.9

***Solution:*** For each coil, the mutual term is positive, and the sign of $M$ in $\mathbf{X}_m = \omega M \angle 90^\circ$ is positive, as determined by the direction of $\mathbf{I}_1$ and $\mathbf{I}_2$. Thus,

$$\mathbf{E}_1 - \mathbf{I}_1\mathbf{R}_1 - \mathbf{I}_1\mathbf{X}_{L_1} - \mathbf{I}_2\mathbf{X}_m = 0$$

or

$$\mathbf{E}_1 - \mathbf{I}_1(\mathbf{R}_1 + \mathbf{X}_{L_1}) - \mathbf{I}_2\mathbf{X}_m = 0$$

For the other loop,

$$-\mathbf{I}_2\mathbf{X}_{L_2} - \mathbf{I}_1\mathbf{X}_m - \mathbf{I}_2\mathbf{R}_L = 0$$

or

$$\mathbf{I}_2(\mathbf{X}_{L_2} + \mathbf{R}_L) + \mathbf{I}_1\mathbf{X}_m = 0$$

## 25.4 THE IRON-CORE TRANSFORMER

An iron-core transformer under loaded conditions is shown in Fig. 25.10. The iron core will serve to increase the coefficient of coupling between the coils by increasing the mutual flux $\phi_m$. Recall from Chapter 11 that magnetic flux lines will always take the path of least reluctance, which in this case is the iron core.

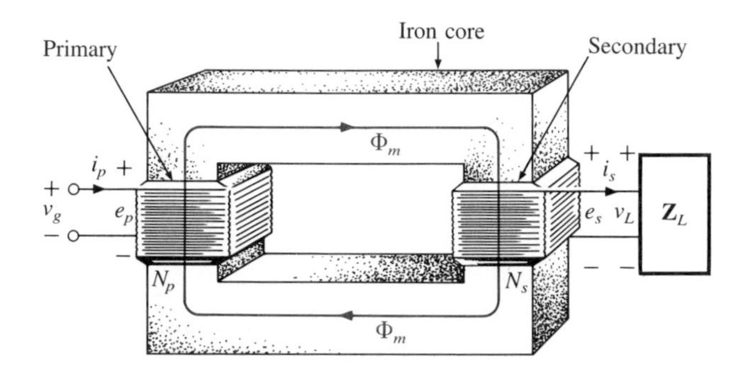


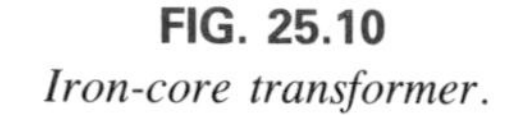
**FIG. 25.10**
*Iron-core transformer.*

We will assume in the analyses to follow in this chapter that all of the flux linking coil 1 will link coil 2. In other words, the coefficient of coupling is its maximum value, 1, and $\phi_m = \phi_1$. In addition, we will first analyze the transformer from an ideal viewpoint. That is, we will neglect such losses as the geometric or dc resistance of the coils, the leakage reactance due to the flux linking either coil that forms no part of $\phi_m$, and the hysteresis and eddy current losses. This is not to convey the impression, however, that we will be far from the actual operation of a transformer. Most transformers manufactured today can be considered almost ideal. The equations we will develop under ideal conditions will be, in general, a first approximation to the actual response which will never be off by more than a few percent. The losses will be considered in greater detail in Section 25.6.

When the current $i_1$ through the primary circuit of the iron-core transformer is a maximum, the flux $\phi_m$ linking both coils is also a maximum. In fact, the magnitude of the flux is directly proportional to the current through the primary windings. Therefore, the two are in phase, and for sinusoidal inputs, the magnitude of the flux will vary as a sinusoid also. That is, if

$$i_p = \sqrt{2}I_p \sin \omega t$$

then

$$\phi_m = \Phi_m \sin \omega t$$

The induced voltage across the primary due to a sinusoidal input can be determined by Faraday's law:

$$e_p = N_p \frac{d\phi_p}{dt} = N_p \frac{d\phi_m}{dt}$$

Substituting for $\phi_m$ gives us

$$e_p = N_p \frac{d}{dt}(\Phi_m \sin \omega t)$$

and differentiating, we obtain

$$e_p = \omega N_p \Phi_m \cos \omega t$$

or

$$e_p = \omega N_p \Phi_m \sin(\omega t + 90°)$$

indicating that the induced voltage $e_p$ leads the current through the primary coil by 90°.

The effective value of $e_p$ is

$$E_p = \frac{\omega N_p \Phi_m}{\sqrt{2}} = \frac{2\pi f N_p \Phi_m}{\sqrt{2}}$$

and

$$\boxed{E_p = 4.44 f N_p \Phi_m} \qquad \textbf{(25.10a)}$$

which is an equation for the effective value of the voltage across the primary coil in terms of the frequency of the input current or voltage, the number of turns of the primary, and the maximum value of the magnetic flux linking the primary.

For the case under discussion, where the flux linking the secondary equals that of the primary, if we repeat the procedure just described for the induced voltage across the secondary, we get

$$\boxed{E_s = 4.44 f N_s \Phi_m} \qquad \textbf{(25.10b)}$$

Dividing Eq. (25.10a) by Eq. (25.10b), as follows:

$$\frac{E_p}{E_s} = \frac{4.44 f N_p \Phi_m}{4.44 f N_s \Phi_m}$$

we obtain

$$\boxed{\frac{E_p}{E_s} = \frac{N_p}{N_s}} \qquad \textbf{(25.11)}$$

Note that the ratio of the magnitudes of the induced voltages is the same as the ratio of the corresponding turns.

If we consider that

$$e_p = N_p \frac{d\phi_m}{dt}$$

and

$$e_s = N_s \frac{d\phi_m}{dt}$$

and divide one by the other; that is,

$$\frac{e_p}{e_s} = \frac{N_p(d\phi_m/dt)}{N_s(d\phi_m/dt)}$$

then

$$\frac{e_p}{e_s} = \frac{N_p}{N_s}$$

The *instantaneous* values of $e_1$ and $e_2$ are therefore related by a constant determined by the turns ratio. Since their instantaneous magnitudes are related by a constant, the induced voltages are in phase, and Eq. (25.11) can be changed to include phasor notation; that is,

$$\boxed{\frac{\mathbf{E}_p}{\mathbf{E}_s} = \frac{N_p}{N_s}} \tag{25.12}$$

or, since $\mathbf{V}_g = \mathbf{E}_1$ and $\mathbf{V}_L = \mathbf{E}_2$ for the ideal situation,

$$\boxed{\frac{\mathbf{V}_g}{\mathbf{V}_L} = \frac{N_p}{N_s}} \tag{25.13}$$

The ratio $N_p/N_s$, usually represented by the lowercase letter $a$, is referred to as the *transformation ratio:*

$$\boxed{a = \frac{N_p}{N_s}} \tag{25.14}$$

If $a < 1$, the transformer is called a *step-up transformer,* since the voltage $E_s > E_p$:

$$\frac{E_p}{E_s} = \frac{N_p}{N_s} = a$$

so

$$E_s = \frac{E_p}{a}$$

and, if $a < 1$,

$$E_s > E_p$$

If $a > 1$, the transformer is called a *step-down transformer,* since $E_s < E_p$; that is,

$$E_p = aE_s$$

and, if $a > 1$, then

$$E_p > E_s$$

**EXAMPLE 25.4.** For the iron-core transformer of Fig. 25.11:

a. Find the maximum flux $\Phi_m$.

b. Find the number of turns $N_s$.

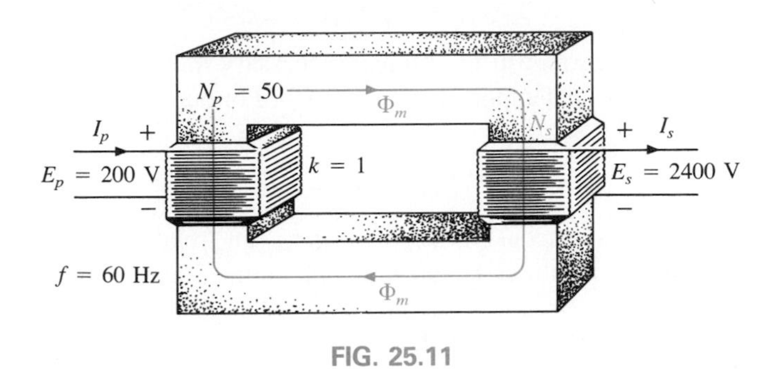


FIG. 25.11

***Solutions:***

a. $E_p = 4.44N_p f\Phi_m$

Therefore,

$$\Phi_m = \frac{E_p}{4.44N_p f} = \frac{200\text{ V}}{(4.44)(50\text{ t})(60\text{ Hz})}$$

and

$$\Phi_m = \mathbf{15.02\ mWb}$$

b. $\dfrac{E_p}{E_s} = \dfrac{N_p}{N_s}$

Therefore,

$$N_s = \frac{N_p E_s}{E_p} = \frac{(50\text{ t})(2400\text{ V})}{200\text{ V}}$$
$$= \mathbf{600\ turns}$$

The induced voltage across the secondary of the transformer of Fig. 25.10 will establish a current $i_s$ through the load $Z_L$ and the secondary windings. This current and the turns $N_s$ will develop an mmf $N_s i_s$ that would not be present under no-load conditions, since $i_s = 0$ and $N_s i_s = 0$. Under loaded or unloaded conditions, however, the net ampere-turns on the core produced by both the primary and the secondary must remain unchanged for the same flux $\phi_m$ to be established in the core. The flux $\phi_m$ must remain the same to have the same induced voltage across the primary to balance the voltage impressed across the primary. In order to counteract the mmf of the secondary, which is tending to change $\phi_m$, an additional current must flow in the primary. This current is called the *load component of the primary current* and is represented by the notation $i'_p$.

For the balanced or equilibrium condition,

$$N_p i'_p = N_s i_s$$

The total current in the primary under loaded conditions is

$$i_p = i'_p + i_{\phi_m}$$

where $i_{\phi_m}$ is the current in the primary necessary to establish the flux $\phi_m$. For most practical applications, $i'_p > i_{\phi_m}$. For our analysis, we will assume $i_p \cong i'_p$, so

$$N_p i_p = N_s i_s$$

Since the instantaneous values of $i_p$ and $i_s$ are related by the turns ratio, the phasor quantities $\mathbf{I}_p$ and $\mathbf{I}_s$ are also related by the same ratio:

$$N_p \mathbf{I}_p = N_s \mathbf{I}_s$$

or

$$\frac{\mathbf{I}_p}{\mathbf{I}_s} = \frac{N_s}{N_p} \tag{25.15}$$

Keep in mind that Eq. (25.15) holds true only if we neglect the effects of $i_{\phi_m}$. Otherwise, the magnitudes of $\mathbf{I}_p$ and $\mathbf{I}_s$ are not related by the turns ratio, and $\mathbf{I}_p$ and $\mathbf{I}_s$ are not in phase.

For the step-up transformer, $a < 1$, and the current in the secondary, $I_s = aI_p$, is less in magnitude than that in the primary. For a step-down transformer, the reverse is true.

## 25.5 REFLECTED IMPEDANCE AND POWER

In the previous sections, we found that

$$\frac{\mathbf{V}_g}{\mathbf{V}_L} = \frac{N_p}{N_s} \quad \text{and} \quad \frac{\mathbf{I}_p}{\mathbf{I}_s} = \frac{N_s}{N_p}$$

Dividing one by the other, we have

$$\frac{\mathbf{V}_g/\mathbf{V}_L \;=\; N_p/N_s \;=\; a}{\mathbf{I}_p/\mathbf{I}_s \;=\; N_s/N_p \;=\; 1/a}$$

or

$$\frac{\mathbf{V}_g/\mathbf{I}_p}{\mathbf{V}_L/\mathbf{I}_s} = a^2$$

and

$$\frac{\mathbf{V}_g}{\mathbf{I}_p} = a^2\frac{\mathbf{V}_L}{\mathbf{I}_s}$$

However, since

$$\mathbf{Z}_p = \frac{\mathbf{V}_g}{\mathbf{I}_p} \quad \text{and} \quad \mathbf{Z}_L = \frac{\mathbf{V}_L}{\mathbf{I}_s}$$

then

$$\boxed{\mathbf{Z}_p = a^2\mathbf{Z}_L} \qquad \textbf{(25.16)}$$

which in words states that the impedance of the primary circuit of an ideal transformer is the transformation ratio squared times the impedance of the load. If a transformer is used, therefore, an impedance can be made to appear larger or smaller at the primary by placing it in the secondary of a step-down ($a > 1$) or step-up ($a < 1$) transformer, respectively. Note that if the load is capacitive or inductive, the reflected impedance will also be capacitive or inductive.

For the ideal iron-core transformer,

$$\frac{E_p}{E_s} = a = \frac{I_s}{I_p}$$

or

$$\boxed{E_pI_p = E_sI_s} \qquad \textbf{(25.17)}$$

and

$$\boxed{P_{\text{in}} = P_{\text{out}}} \qquad \text{(ideal case)} \qquad \textbf{(25.18)}$$

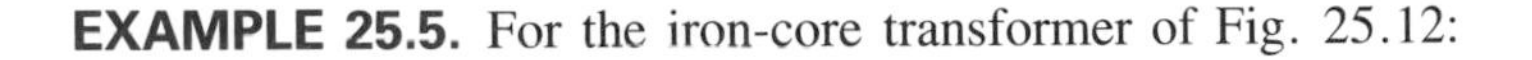

**EXAMPLE 25.5.** For the iron-core transformer of Fig. 25.12:

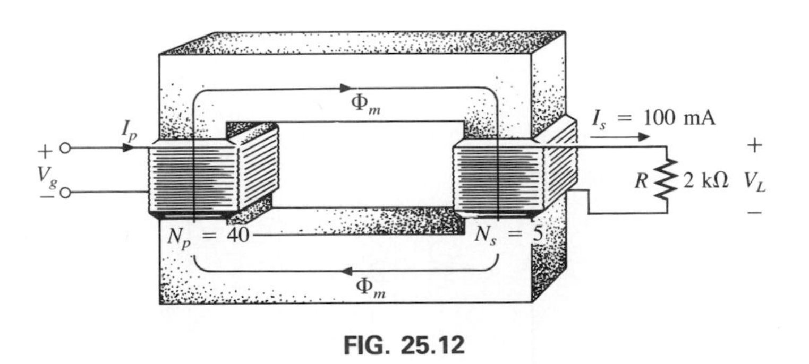


**FIG. 25.12**

a. Find the magnitude of the current in the primary and the impressed voltage across the primary.
b. Find the input resistance of the transformer.

**Solutions:**

a. $\dfrac{I_p}{I_s} = \dfrac{N_s}{N_p}$

or

$$I_p = \frac{N_s}{N_p} I_s = \left(\frac{5\text{ t}}{40\text{ t}}\right)(0.1\text{ A})$$

and

$$I_p = \mathbf{12.5\ mA}$$

$$V_L = I_s Z_L = (0.1\text{ A})(2\text{ k}\Omega) = 200\text{ V}$$

and

$$\frac{V_g}{V_L} = \frac{N_p}{N_s}$$

or

$$V_g = \frac{N_p}{N_s} V_L = \left(\frac{40\text{ t}}{5\text{ t}}\right)(200\text{ V})$$

and

$$V_g = \mathbf{1600\ V}$$

b. $Z_p = a^2 Z_L$

$$a = \frac{N_p}{N_s} = 8$$

$$Z_p = (8)^2(2\text{ k}\Omega) = R_p = \mathbf{128\ k\Omega}$$

---

**EXAMPLE 25.6.** For the speaker in Fig. 25.13 to receive maximum power from the circuit, the internal resistance of the speaker should be

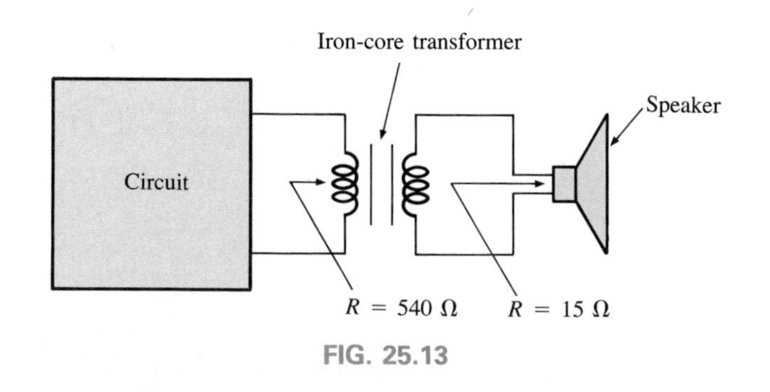


FIG. 25.13

540 Ω. If a transformer is used, the speaker resistance of 15 Ω can be made to appear 540 Ω at the primary. Find the transformation ratio

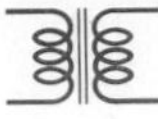



required and the number of turns in the primary if the secondary winding has 40 turns.

***Solution:***

$$Z_p = a^2 Z_L$$

or

$$a = \sqrt{\frac{Z_p}{Z_L}} = \sqrt{\frac{540\ \Omega}{15\ \Omega}} = \sqrt{36} = 6$$

Therefore,

$$a = 6 = \frac{N_p}{N_s}$$

or

$$N_p = 6N_s = 6(40\ \text{t})$$

and

$$N_p = \mathbf{240\ turns}$$

**EXAMPLE 25.7.** For the residential supply appearing in Fig. 25.14, determine (assuming a totally resistive load) the following:

a. the value of $R$ to insure a balanced load
b. the magnitude of $I_1$ and $I_2$
c. the line voltage $V_L$
d. the total power delivered
e. the turns ratio $N_1/N_2$

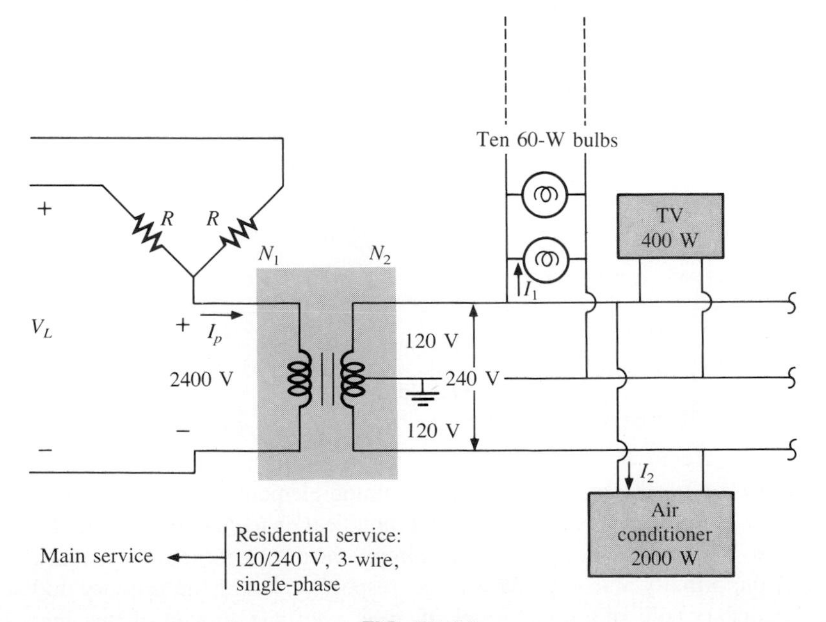


**FIG. 25.14**

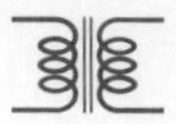

***Solutions:***

a. $P_T = 600\text{ W} + 400\text{ W} + 2000\text{ W} = 3000\text{ W}$

$P_{\text{in}} = P_{\text{out}}$

$V_pI_p = V_sI_s = 3000\text{ W}$ (purely resistive load)

$(2400\text{ V})I_p = 3000\text{ W}$

$I_p = 1.25\text{ A}$

$$R = \frac{V}{I} = \frac{2400\text{ V}}{1.25\text{ A}} = \mathbf{1920\ \Omega}$$

b. $P_1 = (10\text{ bulbs})(60\text{ W/bulb}) = 600\text{ W} = VI_1 = (120\text{ V})I_1$

and

$I_1 = \mathbf{5\ A}$

$P_2 = 2000\text{ W} = VI_2 = (240\text{ V})I_2$

and

$I_2 = \mathbf{8\tfrac{1}{3}\ A}$

c. $V_L = \sqrt{3}V_\phi = 1.73(2400\text{ V}) = \mathbf{4152\ V}$

d. $P_T = 3P_\phi = 3(3000\text{ W}) = \mathbf{9\ kW}$

e. $\dfrac{N_1}{N_2} = \dfrac{V_p}{V_s} = \dfrac{2400\text{ V}}{240\text{ V}} = \mathbf{10}$

## 25.6 EQUIVALENT CIRCUIT (IRON-CORE TRANSFORMER)

For the nonideal or practical iron-core transformer, the equivalent circuit appears as in Fig. 25.15. As indicated, part of this equivalent cir-

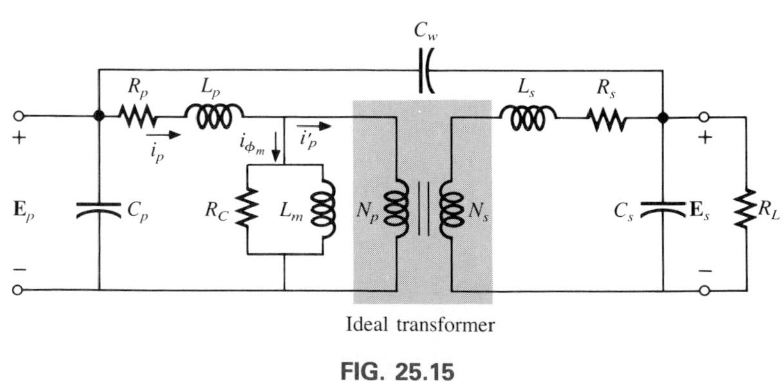


**FIG. 25.15**
*Equivalent circuit for the practical iron-core transformer.*

cuit is an ideal transformer. The remaining elements of Fig. 25.15 are those elements that contribute to the nonideal characteristics of the device. The resistances $R_p$ and $R_s$ are simply the dc or geometric resistance of the primary and secondary coils, respectively. For the primary and secondary coils of a transformer, there is a definite amount of flux that links each coil that does not pass through the core. This situation is shown in Fig. 25.16. This *leakage* flux, serving as a definite loss to the system since it employs an amount of input energy to be established but serves no useful purpose, is represented by an inductance $L_p$ in the primary circuit and an inductance $L_s$ in the secondary.

The resistance $R_C$ represents the hysteresis and eddy current losses (core losses) within the core due to an ac flux through the core. The inductance $L_m$ (magnetizing inductance) is the inductance associated with the magnetization of the core, that is, the establishing of the flux $\Phi_m$ in the core. The capacitances $C_p$ and $C_s$ are the lumped capacitances of the primary and secondary circuits, respectively, and $C_w$ represents the equivalent lumped capacitances between the windings of the transformer.

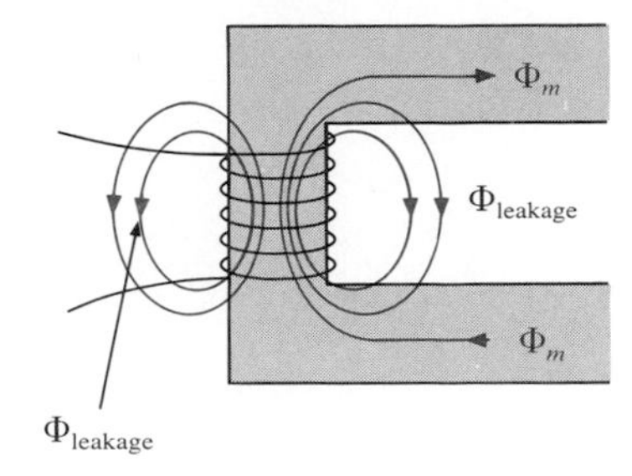


FIG. 25.16

Since $i'_p$ is normally considerably larger than $i_{\phi_m}$, we will ignore $i_{\phi_m}$ for the moment (set it equal to zero), resulting in the absence of $R_C$ and $L_m$ in the reduced equivalent circuit of Fig. 25.17. In addition, the

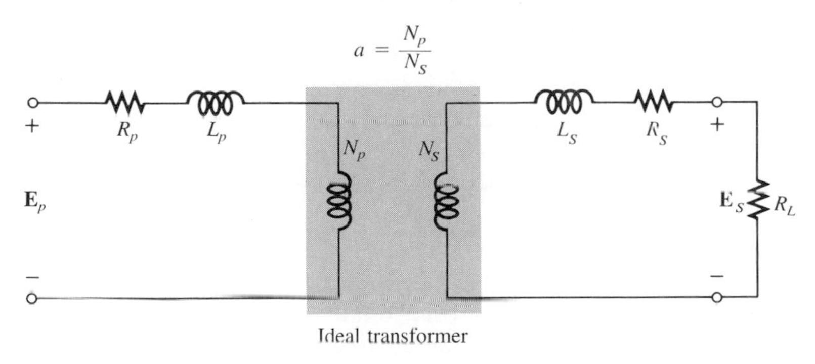


FIG. 25.17
*Reduced equivalent circuit for the nonideal iron-core transformer.*

capacitances $C_p$, $C_w$, and $C_s$ do not appear in the equivalent circuit of Fig. 25.17, since their reactance in the present frequency range of interest will not appreciably affect the transfer characteristics of the transformer.

If we now reflect the secondary circuit through the ideal transformer, as shown in Fig. 25.18(a), we will have the load and generator voltage

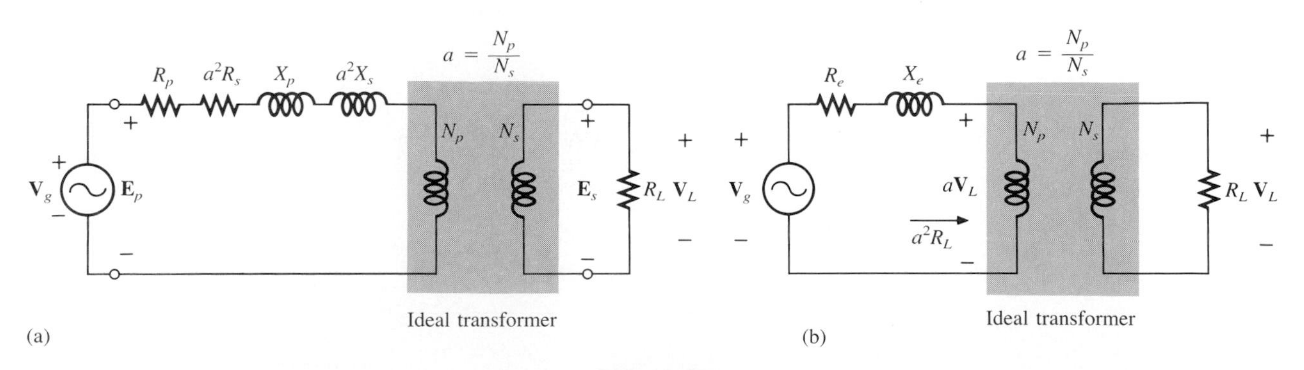


FIG. 25.18

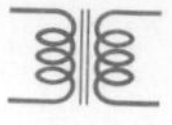

in the same physical circuit. The total resistance and inductive reactance are determined by

$$\boxed{R_{\text{equivalent}} = R_e = R_p + a^2R_s} \tag{25.19a}$$

and

$$\boxed{X_{\text{equivalent}} = X_e = X_p + a^2X_s} \tag{25.19b}$$

which result in the useful equivalent circuit of Fig. 25.18(b). The load voltage can be obtained directly from the circuit of Fig. 25.18(b) through the voltage divider rule:

$$\boxed{a\mathbf{V}_L = \frac{a^2R_L\mathbf{V}_g}{(R_e + a^2R_L) + jX_e}} \tag{25.20}$$

In a different light, the generator voltage necessary to establish a particular load voltage can also be determined through Eq. (25.20). The voltages across the elements of Fig. 25.18(b) have the phasor relationship indicated in Fig. 25.19(a). Note that the current is the reference phasor for drawing the phasor diagram. That is, the voltages across the resistive elements are *in phase* with the current phasor, while the voltage across the equivalent inductance leads the current by 90°. The primary voltage, by Kirchhoff's voltage law, is then the phasor sum of these voltages, as indicated in Fig. 25.19(a). For an inductive load, the phasor diagram appears in Fig. 25.19(b). Note that $a\mathbf{V}_L$ leads $\mathbf{I}$ by the power-factor angle of the load. The remainder of the diagram is then similar to that for a resistive load. (The phasor diagram for a capacitive load will be left to the reader as an exercise.)

The effect of $R_e$ and $X_e$ on the magnitude of $\mathbf{V}_g$ for a particular $\mathbf{V}_L$ is obvious from Eq. (25.20) or Fig. 25.19. For increased values of $R_e$ or $X_e$, an increase in $\mathbf{V}_g$ is required for the same load voltage. For $R_e$ and $X_e = 0$, $\mathbf{V}_L$ and $\mathbf{V}_g$ are related by the turns ratio.

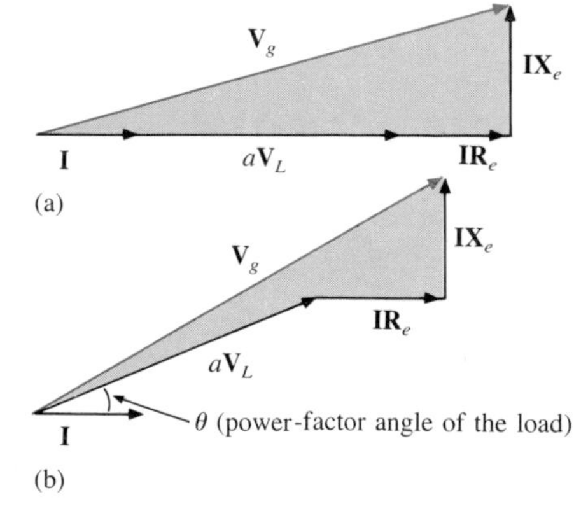


**FIG. 25.19**
*Phasor diagram for the iron-core transformer with (a) unity power-factor load (resistive) and (b) lagging power-factor load (inductive).*

---

**EXAMPLE 25.8.** For a transformer having the equivalent circuit of Fig. 25.20:

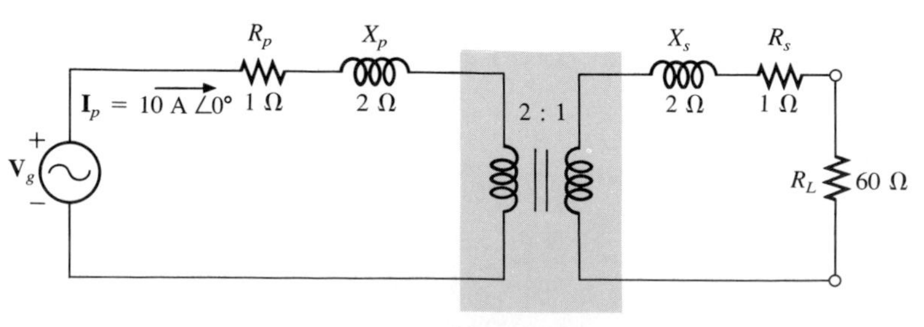

**FIG. 25.20**

a. Determine $R_e$ and $X_e$.
b. Determine $V_g$.

***Solutions:***

a. $R_e = R_p + a^2R_s = 1\ \Omega + (2)^2(1\ \Omega) = \mathbf{5\ \Omega}$
$X_e = X_p + a^2X_s = 2\ \Omega + (2)^2(2\ \Omega) = \mathbf{10\ \Omega}$

b. The transformed equivalent circuit appears in Fig. 25.21.

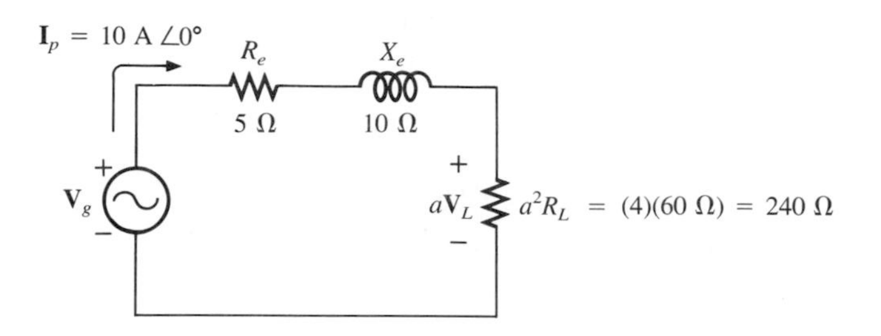

**FIG. 25.21**

$$aV_L = (I_p)(a^2R_L) = 2400\text{ V}$$

Thus,

$$V_L = \frac{2400\text{ V}}{a} = \frac{2400\text{ V}}{2} = 1200\text{ V}$$

and

$$\mathbf{V}_g = \mathbf{I}_p(R_e + a^2R_L + jX_e)$$
$$= 10\text{ A}(5\ \Omega + 240\ \Omega + j10\ \Omega) = 10\text{ A}(245\ \Omega + j10\ \Omega)$$
$$\mathbf{V}_g = 2450\text{ V} + j100\text{ V} = 2452.04\text{ V}\ \angle 2.34^\circ$$

and

$$V_g = \mathbf{2452.04\ V}$$

For $R_e$ and $X_e = 0$, $V_g = aV_L = 2400$ V. Therefore, it is necessary to increase the generator voltage by 52.04 V (due to $R_e$ and $X_e$) to obtain the same load voltage.

## 25.7 FREQUENCY CONSIDERATIONS

For certain frequency ranges, the effect of some parameters in the equivalent circuit of the iron-core transformer of Fig. 25.15 can be neglected. Since it is convenient to consider a low-, mid-, and high-frequency region, the equivalent circuits for each will now be introduced and briefly examined.

For the low-frequency region, the reactance ($2\pi fL$) of the primary and secondary leakage reactances can be ignored, and the reflected equivalent circuit will appear as shown in Fig. 25.22(a). The magnetizing inductance must be included, since it appears in parallel with the secondary reflected circuit. As the frequency approaches zero, the

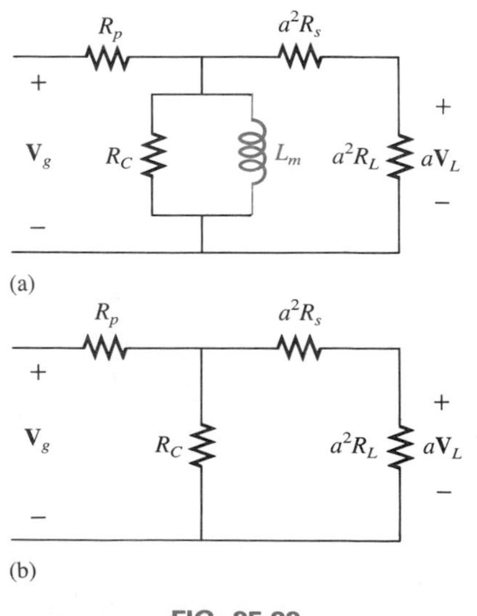


**FIG. 25.22**

*(a) Low-frequency reflected equivalent circuit; (b) mid-frequency reflected circuit.*

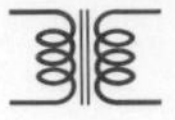

reactance of the magnetizing inductance will reduce in magnitude, causing a reduction in the voltage across the secondary circuit. For $f = 0$ Hz, $L_m$ is ideally a short circuit, and $V_L = 0$. As the frequency increases, the reactance of $L_m$ will eventually be sufficiently large compared with the reflected secondary impedance to be neglected. The mid-frequency reflected equivalent circuit will then appear as shown in Fig. 25.22(b). Note the absence of reactive elements, resulting in an *in-phase* relationship between load and generator voltages.

For higher frequencies, the capacitive elements and primary and secondary leakage reactances must be considered, as shown in Fig. 25.23.

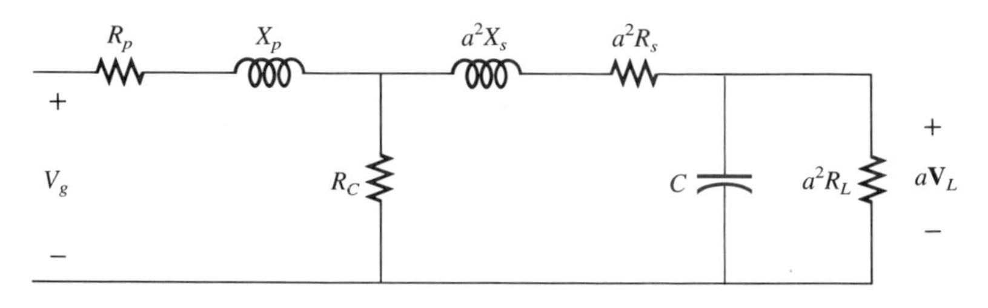


**FIG. 25.23**
*High-frequency reflected equivalent circuit.*

For discussion purposes, the effects of $C_w$ and $C_s$ appear as a lumped capacitor $C$ in the reflected network of Fig. 25.23; $C_p$ does not appear since the effect of $C$ will predominate. As the frequency of interest increases, the capacitive reactance ($X_C = 1/2\pi fC$) will decrease to the point that it will have a shorting effect across the secondary circuit of the transformer, causing $V_L$ to decrease in magnitude.

A typical transformer-frequency response curve appears in Fig. 25.24. For the low- and high-frequency regions, the primary element

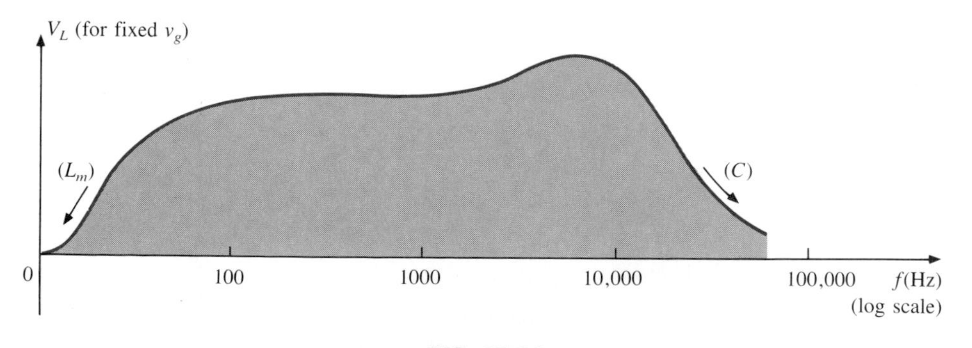


**FIG. 25.24**
*Transformer-frequency response curve.*

responsible for the drop-off is indicated. The peaking that occurs in the high-frequency region is due to the series resonant circuit established by the inductive and capacitive elements of the equivalent circuit. In the peaking region, the series resonant circuit is in, or near, its resonant or tuned state.

The network discussed in some detail earlier in this chapter was for the high mid-frequency region.

## 25.8 AIR-CORE TRANSFORMER

As the name implies, the air-core transformer does not have a ferromagnetic core to link the primary and secondary coils. Rather, the coils are placed sufficiently close to have a mutual inductance that will establish the desired transformer action. In Fig. 25.25, current direction and polarities have been defined for the air-core transformer. Note the presence of a mutual inductance term $M$, which will be positive in this case, as determined by the dot convention.

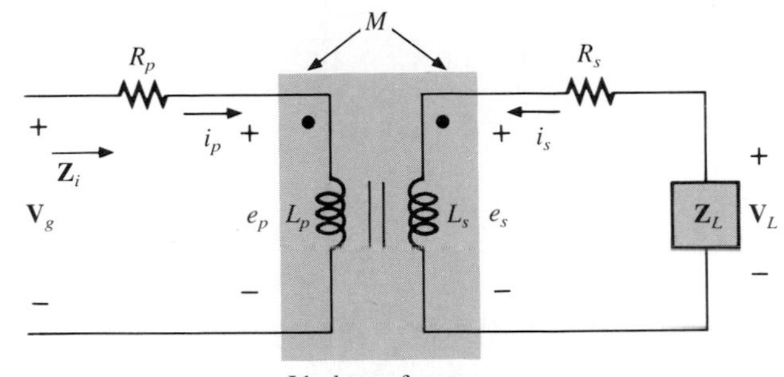


**FIG. 25.25**
*Air-core transformer equivalent circuit.*

From past analysis in this chapter, we now know that

$$e_p = L_p \frac{di_p}{dt} + M \frac{di_s}{dt} \qquad (25.21)$$

for the primary circuit.

We found in Chapter 12 that for the pure inductor, with no mutual inductance present, the mathematical relationship

$$v_1 = L_1 \frac{di_1}{dt}$$

resulted in the following useful form of the voltage across an inductor:

$$\mathbf{V}_1 = \mathbf{I}_1 \mathbf{X}_{L_1} \quad \text{where } \mathbf{X}_{L_1} = \omega L_1 \angle 90^\circ = j\omega L_1$$

Similarly, it can be shown, for a mutual inductance, that

$$v_1 = M \frac{di_2}{dt}$$

will result in

$$\mathbf{V}_1 = \mathbf{I}_2 \mathbf{X}_m \quad \text{where } \mathbf{X}_m = \omega M \angle 90^\circ = j\omega M \qquad (25.22)$$

Equation (25.21) can then be written (using phasor notation)

$$\mathbf{E}_p = \mathbf{I}_p\mathbf{X}_{L_p} + \mathbf{I}_s\mathbf{X}_m \tag{25.23}$$

and

$$\mathbf{V}_g = \mathbf{I}_p\mathbf{R}_p + \mathbf{I}_p\mathbf{X}_{L_p} + \mathbf{I}_s\mathbf{X}_m$$

or

$$\mathbf{V}_g = \mathbf{I}_p(\mathbf{R}_p + \mathbf{X}_{L_p}) + \mathbf{I}_s\mathbf{X}_m \tag{25.24}$$

For the secondary circuit,

$$\mathbf{E}_s = \mathbf{I}_s\mathbf{X}_{L_s} + \mathbf{I}_p\mathbf{X}_m \tag{25.25}$$

and

$$\mathbf{V}_L = \mathbf{I}_s\mathbf{R}_s + \mathbf{I}_s\mathbf{X}_{L_s} + \mathbf{I}_p\mathbf{X}_m$$

or

$$\mathbf{V}_L = \mathbf{I}_s(\mathbf{R}_s + \mathbf{X}_{L_s}) + \mathbf{I}_p\mathbf{X}_m \tag{25.26}$$

Substituting

$$\mathbf{V}_L = -\mathbf{I}_s\mathbf{Z}_L$$

into Eq. (25.26) results in

$$0 = \mathbf{I}_s(\mathbf{R}_s + \mathbf{X}_{L_s} + \mathbf{Z}_L) + \mathbf{I}_p\mathbf{X}_m$$

Solving for $\mathbf{I}_s$, we have

$$\mathbf{I}_s = \frac{-\mathbf{I}_p\mathbf{X}_m}{\mathbf{R}_s + \mathbf{X}_{L_s} + \mathbf{Z}_L}$$

and, substituting into Eq. (25.24), we obtain

$$\mathbf{V}_g = \mathbf{I}_p(\mathbf{R}_p + \mathbf{X}_{L_p}) + \left(\frac{-\mathbf{I}_p\mathbf{X}_m}{\mathbf{R}_s + \mathbf{X}_{L_s} + \mathbf{Z}_L}\right)\mathbf{X}_m$$

Thus, the input impedance is

$$\mathbf{Z}_i = \frac{\mathbf{V}_g}{\mathbf{I}_p} = \mathbf{R}_p + \mathbf{X}_{L_p} - \frac{\mathbf{X}_m^2}{\mathbf{R}_s + \mathbf{X}_{L_s} + \mathbf{Z}_L}$$

or, defining

$$\mathbf{Z}_p = \mathbf{R}_p + \mathbf{X}_{L_p} \qquad \mathbf{Z}_s = \mathbf{R}_s + \mathbf{X}_{L_s} \qquad \mathbf{X}_m = j\omega M$$

we have

$$\mathbf{Z}_i = \mathbf{Z}_p - \frac{(+j\omega M)^2}{\mathbf{Z}_s + \mathbf{Z}_L}$$

and

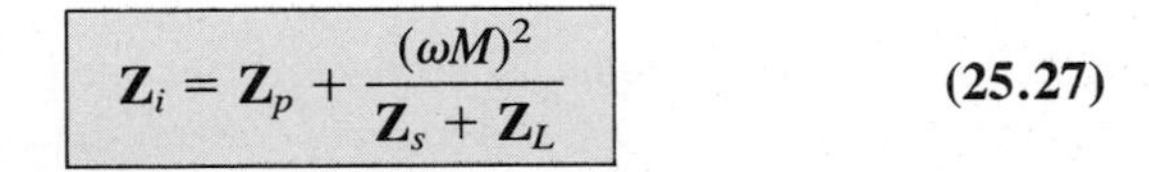

$$\mathbf{Z}_i = \mathbf{Z}_p + \frac{(\omega M)^2}{\mathbf{Z}_s + \mathbf{Z}_L} \qquad \textbf{(25.27)}$$

The term $(\omega M)^2/(\mathbf{Z}_s + \mathbf{Z}_L)$ is called the *coupled impedance*. Note that it is independent of the sign of $M$. Consider also that since $(\omega M)^2$ is a constant with 0° phase angle, if $\mathbf{Z}_L$ is resistive, the resulting coupled impedance term will appear capacitive, due to division of $(\mathbf{R}_L + \mathbf{Z}_s)$ into $(\omega M)^2$. This resulting capacitive reactance will oppose the series primary inductance $L_p$, causing a reduction in $\mathbf{Z}_i$. Including the effect of the mutual term, the input impedance to the network will appear as shown in Fig. 25.26.

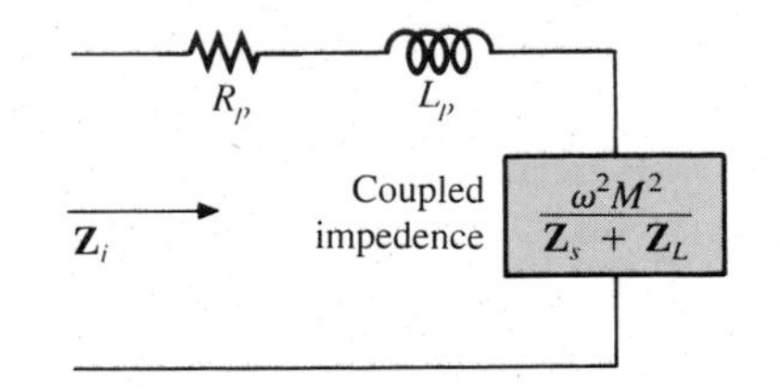


**FIG. 25.26**
*Input characteristics for the air-core transformer.*

**EXAMPLE 25.9.** Determine the input impedance to the air-core transformer in Fig. 25.27.

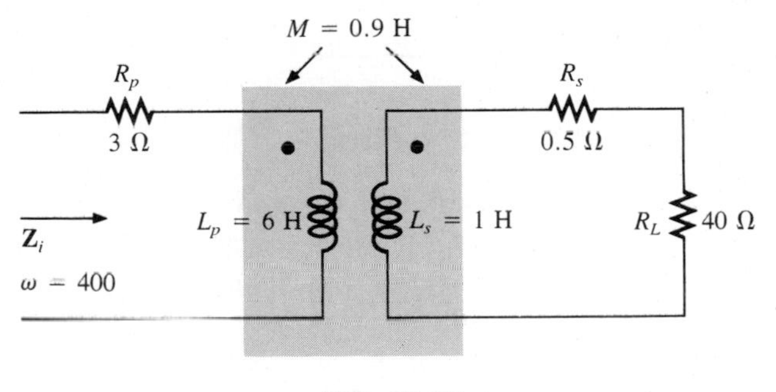


**FIG. 25.27**

***Solution:***

$$\mathbf{Z}_i = \mathbf{Z}_p + \frac{(\omega M)^2}{\mathbf{Z}_s + \mathbf{Z}_L}$$

$$= R_p + jX_{L_p} + \frac{(\omega M)^2}{R_s + jX_{L_s} + R_L}$$

$$= 3\ \Omega + j2.4\ \text{k}\Omega + \frac{((400\ \text{rad/s})(0.9\ \text{H}))^2}{0.5\ \Omega + j400\ \Omega + 40\ \Omega}$$

$$\cong j2.4\ \text{k}\Omega + \frac{129.6 \times 10^3}{40.5 + j400}$$

$$= j2.4\ \text{k}\Omega + \frac{129.6 \times 10^3}{402.05\ \angle 84.22^\circ}$$

$$= j2.4\ \text{k}\Omega + 322.4\ \Omega\ \angle -84.22^\circ$$

$$= j2.4\ \text{k}\Omega + (0.0325\ \text{k}\Omega - \underbrace{j0.3208\ \text{k}\Omega)}_{\text{capacitive}}$$

$$= 0.0325\ \text{k}\Omega + j(2.40 - 0.3208)\ \text{k}\Omega$$

$$\mathbf{Z}_i = \mathbf{32.5\ \Omega + \mathit{j}2079\ \Omega} = R_i + jX_{L_i} = \mathbf{2079.25\ \Omega\ \angle 89.10^\circ}$$

## 25.9 THE TRANSFORMER AS AN ISOLATION DEVICE

The transformer is frequently used to isolate one portion of an electrical system from another. By *isolation,* we mean the absence of any direct physical connection. As a first example of its use as an isolation device, consider the measurement of line voltages on the order of 40,000 V (Fig. 25.28).

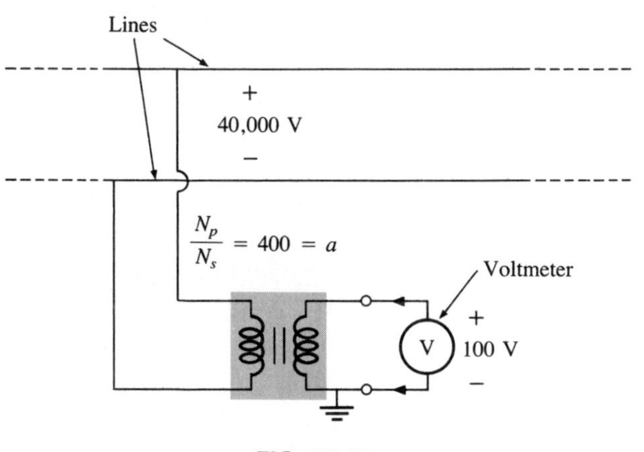


FIG. 25.28

To apply a voltmeter across 40,000 V would obviously be a dangerous task due to the possibility of physical contact with the lines when making the necessary connections. By including a transformer in the transmission system as original equipment, one can bring the potential down to a safe level for measurement purposes and can determine the line voltage using the turns ratio. Therefore, the transformer will serve both to isolate and to step down the voltage.

As a second example, consider the application of the voltage $v_x$ to the vertical input of the oscilloscope (a measuring instrument) in Fig. 25.29.

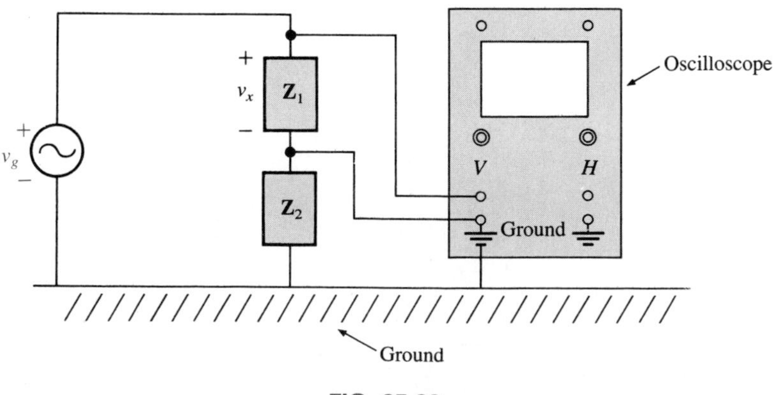


FIG. 25.29

If the connections are made as shown and the generator and oscilloscope have a common ground, the impedance $\mathbf{Z}_2$ has been effectively

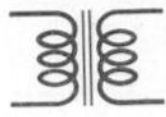



shorted out of the circuit by the ground connection of the oscilloscope. The input voltage to the oscilloscope will therefore be meaningless as far as the voltage $v_x$ is concerned. In addition, if $\mathbf{Z}_2$ is the current-limiting impedance in the circuit, the current in the circuit may rise to a level that will cause severe damage to the circuit. If a transformer is used as shown in Fig. 25.30, this problem will be eliminated, and the input voltage to the oscilloscope will be $v_x$.

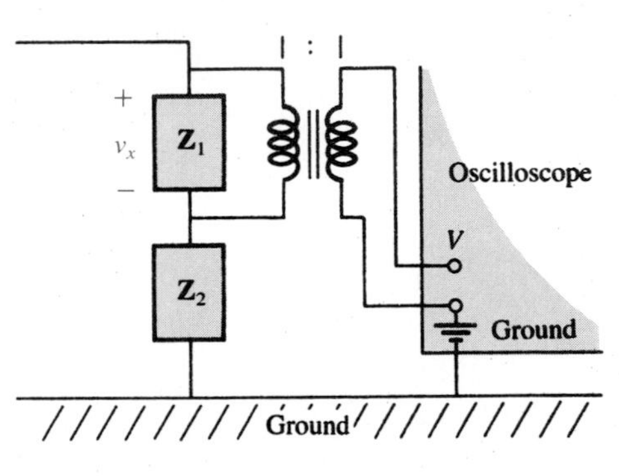


FIG. 25.30

## 25.10 NAMEPLATE DATA

A typical power transformer rating might be the following:

5 kVA, 2000/100 V, 60 Hz

The 2000 V or the 100 V can be either the primary or the secondary voltage; that is, if 2000 V is the primary voltage, then 100 V is the secondary voltage, and vice versa. The 5 kVA is the apparent power ($S = VI$) rating of the transformer. If the secondary voltage is 100 V, then the maximum load current is

$$I_L = \frac{S}{V_L} = \frac{5000 \text{ VA}}{100 \text{ V}} = 50 \text{ A}$$

and if the secondary voltage is 2000 V, then the maximum load current is

$$I_L = \frac{S}{V_L} = \frac{5000 \text{ VA}}{2000 \text{ V}} = 2.5 \text{ A}$$

The transformer is rated in terms of the apparent power rather than the average power for the reason demonstrated by the circuit of Fig. 25.31.

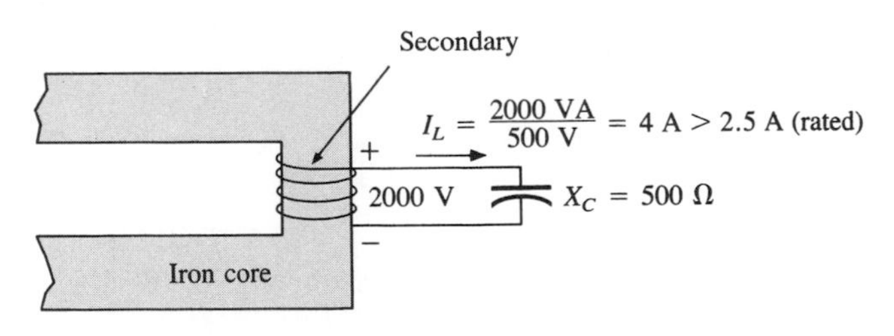


FIG. 25.31

Since the current through the load is greater than that determined by the apparent power rating, the transformer may be permanently damaged. Note, however, that since the load is purely capacitive, the average power to the load is zero. The wattage rating would therefore be meaningless regarding the ability of this load to damage the transformer.

The transformation ratio of the transformer under discussion can be either of two values. If the secondary voltage is 2000 V, the transformation ratio is $a = N_p/N_s = V_g/V_L = 100 \text{ V}/2000 \text{ V} = 1/20$, and the transformer is a step-up transformer. If the secondary voltage is 100 V, the transformation ratio is $a = N_p/N_s = V_g/V_L = 2000\text{ V}/100\text{ V} = 20$, and the transformer is a step-down transformer.

The rated primary current can be determined simply by applying Eq. (25.15):

$$I_1 = \frac{I_2}{a}$$

which is equal to [2.5 A/(1/20)] = 50 A if the secondary voltage is 2000 V, and (50 A/20) = 2.5 A if the secondary voltage is 100 V.

To explain the necessity for including the frequency in the nameplate data, consider Eq. (25.10a):

$$E_p = 4.44 f_p N_p \Phi_m$$

and the *B-H* curve for the iron core of the transformer (Fig. 25.32).

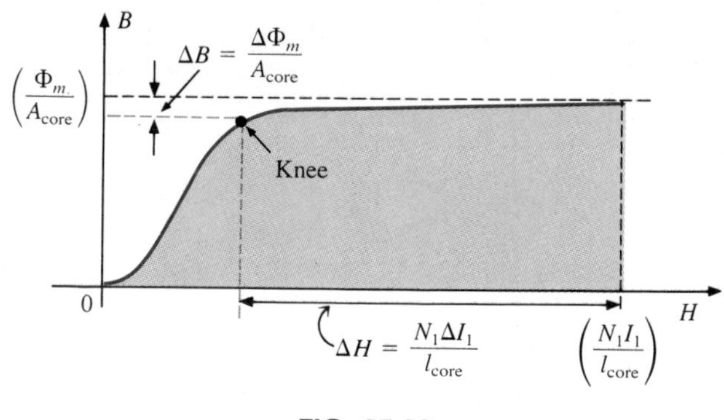

FIG. 25.32

The point of operation on the *B-H* curve for most transformers is at the knee of the curve. If the frequency of the applied signal should drop, and $N_p$ and $E_p$ remain the same, then $\Phi_m$ must increase in magnitude, as determined by Eq. (25.10a):

$$\Phi_m\uparrow = \frac{E_p}{4.44 f_p\downarrow N_p}$$

Note on the *B-H* curve that this increase in $\Phi_m$ will cause a very high current in the primary, resulting in possible damage of the transformer.

## 25.11 TYPES OF TRANSFORMERS

Transformers are available in many different shapes and sizes. Some of the more common types include the power transformer, audio transformer, I-F (intermediate-frequency) transformer, and R-F (radio-frequency) transformer. Each is designed to fulfill a particular requirement in a specific area of application. The symbols for some of the basic types of transformers are shown in Fig. 25.33.

The method of construction varies from one transformer to another. Two of the many different ways in which the primary and secondary

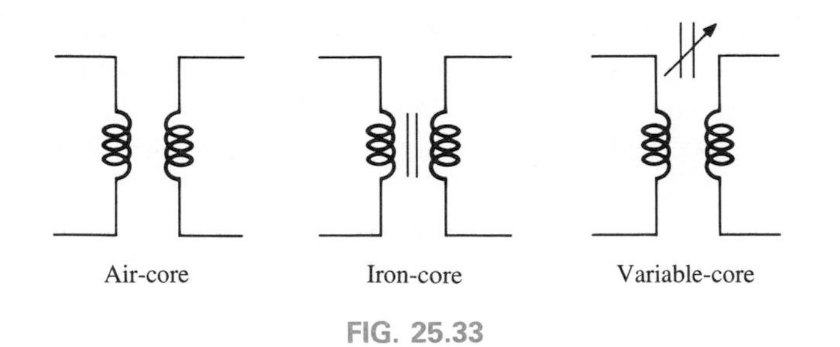


FIG. 25.33

coils can be wound around an iron core are shown in Fig. 25.34. In either case, the core is made of laminated sheets of ferromagnetic material separated by an insulator to reduce the eddy current losses. The sheets themselves will also contain a small percentage of silicon to

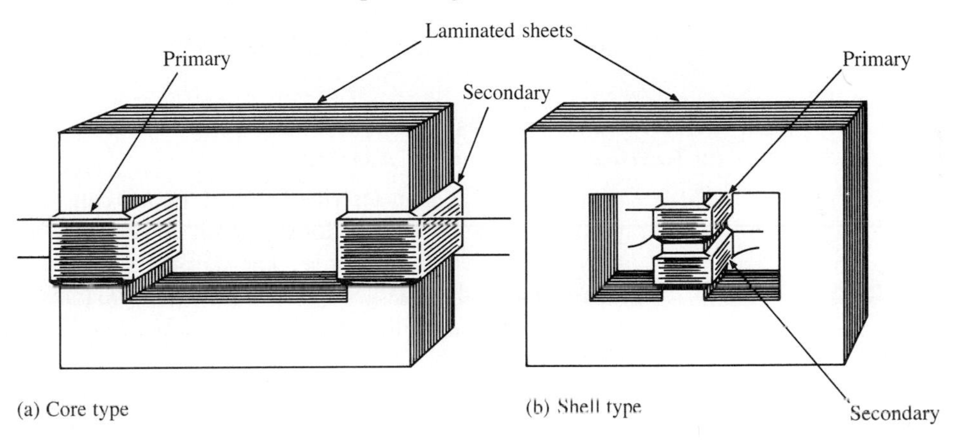

FIG. 25.34

increase the electrical resistivity of the material and further reduce the eddy current losses.

A shell-type power transformer with its schematic representation is shown in Fig. 25.35.

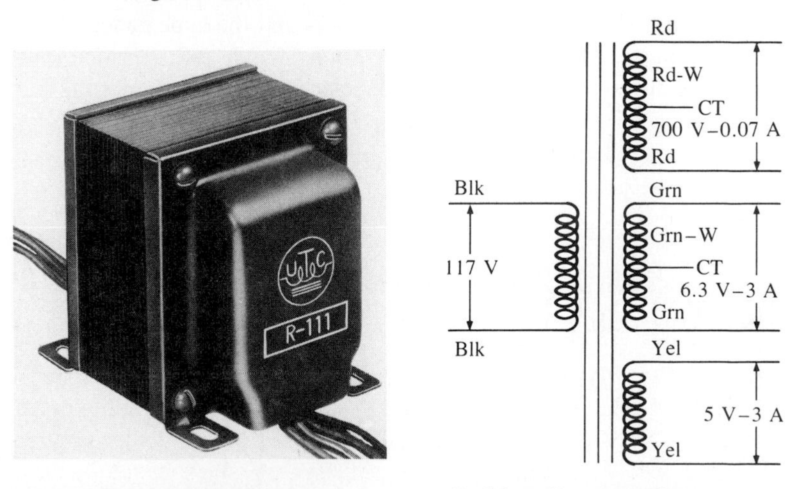


(a) Shell-type power transformer.

(b) Schematic representation.

FIG. 25.35

*(Courtesy of United Transformer Co.)*

The *autotransformer* [Fig. 25.36(b)] is a type of power transformer which, instead of employing the two-circuit principle (complete isolation between coils), has one winding common to both the input and output circuits. The induced voltages are related to the turns ratio in the

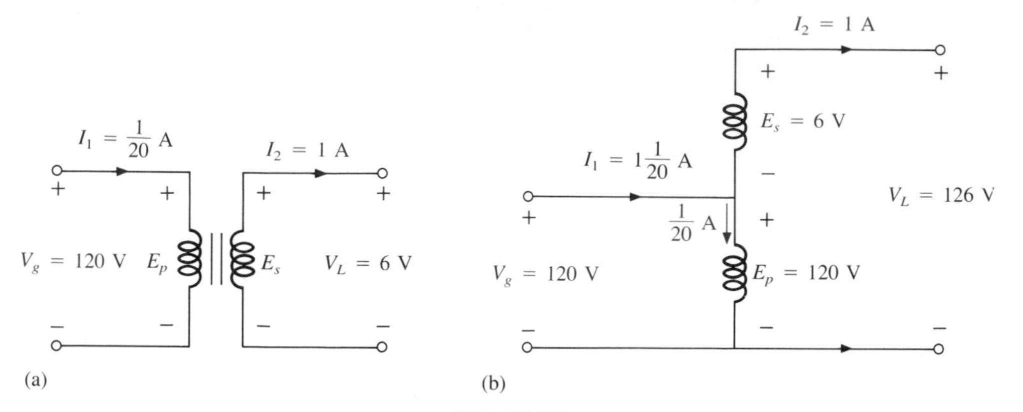


**FIG. 25.36**
*(a) Two-circuit transformer; (b) autotransformer.*

same manner as that described for the two-circuit transformer. If the proper connection is used, a two-circuit power transformer can be employed as an autotransformer. The advantage of using it as an autotransformer is that a larger apparent power can be transformed. This can be demonstrated by the two-circuit transformer of Fig. 25.36(a). It is shown in Fig. 25.36(b) as an autotransformer.

For the two-circuit transformer, note that $S = (\frac{1}{20}\text{ A})(120\text{ V}) = 6\text{ VA}$, while for the autotransformer, $S = (1\frac{1}{20}\text{ A})(120\text{ V}) = 126\text{ VA}$, which is many times that of the two-circuit transformer. Note also that the current and voltage of each coil are the same as those for the two-circuit configuration. The disadvantage of the autotransformer is obvious: loss of the isolation between the primary and secondary circuits.

A dual-in-line pulse transformer package appears in Fig. 25.37 for use with integrated circuits and printed circuit board applications. Note the availability of four isolated transformers and the appearance of the dot convention. The data for one such unit include a 2:1 primary-to-secondary turns ratio; a leakage inductance of 0.50 μH; a coupling capacitance of 7 pF; a primary dc resistance of 0.19 Ω; and a secondary resistance of 0.13 Ω.

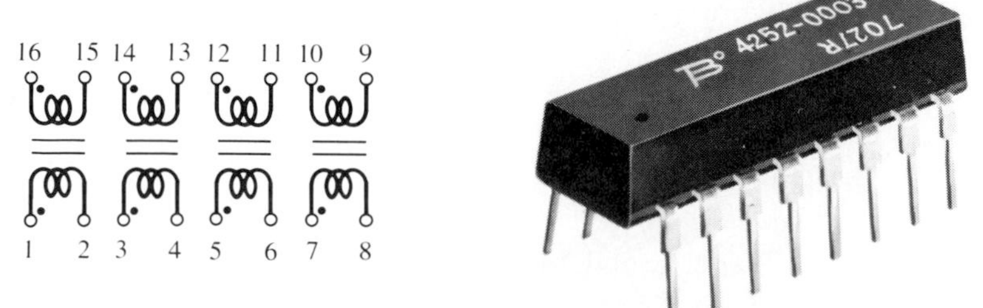


**FIG. 25.37**
*Dual-in-line pulse transformer package. (Courtesy of Bourns®, Inc.)*

A miniature radio transformer designed for direct mounting on a printed circuit board appears in Fig. 25.38.

FIG. 25.38
(*Courtesy of Microtran Company, Inc.*)

## 25.12 TAPPED AND MULTIPLE-LOAD TRANSFORMERS

For the center-tapped (primary) transformer of Fig. 25.39, where the voltage from the center tap to either outside lead is defined as $E_p/2$, the relationship between $E_p$ and $E_s$ is

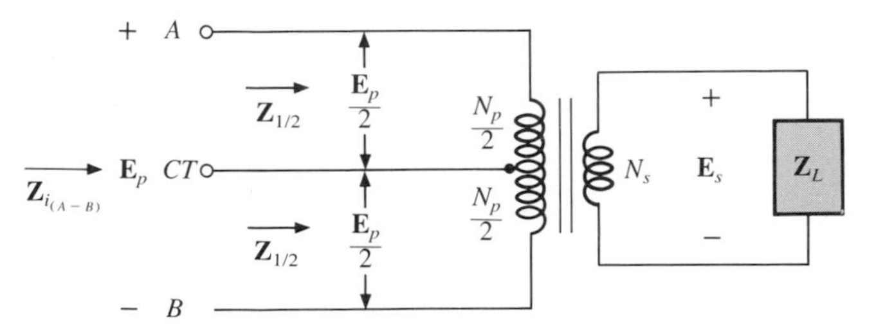


FIG. 25.39
*Ideal transformer with a center-tapped primary.*

$$\frac{\mathbf{E}_p}{\mathbf{E}_s} = \frac{N_p}{N_s} \tag{25.28}$$

For each half-section of the primary,

$$\mathbf{Z}_{1/2} = \left(\frac{N_p/2}{N_s}\right)^2 \mathbf{Z}_L$$

$$\mathbf{Z}_{1/2} = \frac{1}{4}\left(\frac{N_p}{N_s}\right)^2 \mathbf{Z}_L$$

and

$$\mathbf{Z}_{1/2} = \frac{1}{4}\mathbf{Z}_i \tag{25.29}$$

as indicated in Fig. 25.39.

For the multiple-load transformer of Fig. 25.40, the following equations apply:

$$\frac{\mathbf{E}_1}{\mathbf{E}_2} = \frac{N_1}{N_2} \qquad \frac{\mathbf{E}_1}{\mathbf{E}_3} = \frac{N_1}{N_3} \qquad \frac{\mathbf{E}_2}{\mathbf{E}_3} = \frac{N_2}{N_3} \tag{25.30}$$

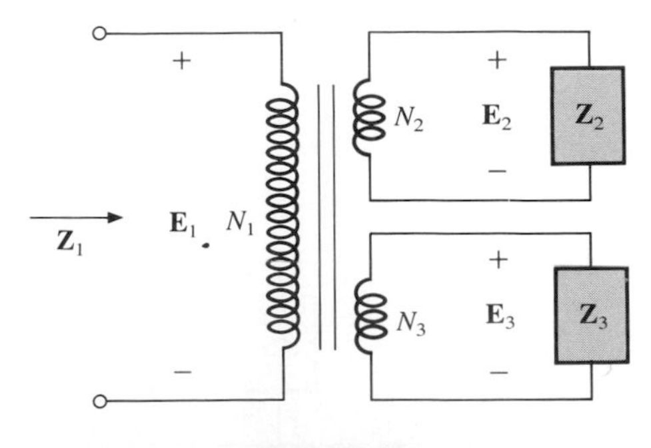


FIG. 25.40
*Ideal transformer with multiple loads.*

The total input impedance can be determined by first noting that for the ideal transformer, the power delivered to the primary is equal to the power dissipated by the load; that is,

$$P_1 = P_{L_2} + P_{L_3}$$

and, for resistive loads ($\mathbf{Z}_1 = R_1$, $\mathbf{Z}_2 = R_2$, and $\mathbf{Z}_3 = R_3$),

$$\frac{E_1^2}{R_1} = \frac{E_2^2}{R_2} + \frac{E_3^2}{R_3}$$

or, since

$$E_2 = \frac{N_2}{N_1}E_1 \quad \text{and} \quad E_3 = \frac{N_3}{N_1}E_1$$

then

$$\frac{E_1^2}{R_1} = \frac{[(N_2/N_1)E_1]^2}{R_2} + \frac{[(N_3/N_1)E_1]^2}{R_3}$$

and

$$\frac{E_1^2}{R_1} = \frac{E_1^2}{(N_1/N_2)^2R_2} + \frac{E_1^2}{(N_1/N_3)^2R_3}$$

Thus,

$$\frac{1}{R_1} = \frac{1}{(N_1/N_2)^2R_2} + \frac{1}{(N_1/N_3)^2R_3} \qquad \textbf{(25.31)}$$

indicating that the load resistances are reflected in parallel.

For the configuration of Fig. 25.41 with $E_2$ and $E_3$ defined as shown, Eqs. (25.30) and (25.31) are applicable.

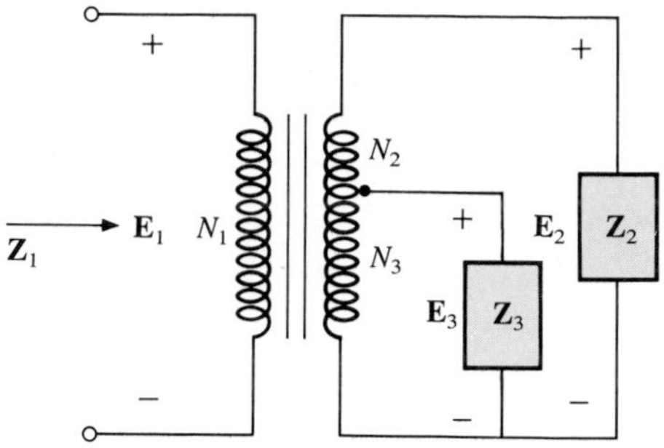

FIG. 25.41
*Ideal transformer with a tapped secondary and multiple loads.*

## 25.13 NETWORKS WITH MAGNETICALLY COUPLED COILS

For multiloop networks with magnetically coupled coils, the mesh-analysis approach is most frequently applied. A firm understanding of the dot convention discussed earlier should make the writing of the equations quite direct and free of errors. Before writing the equations for any particular loop, first determine whether the mutual term is positive or negative, keeping in mind that it will have the same sign as that for the other magnetically coupled coil. For the two-loop network of Fig. 25.42, for example, the mutual term has a positive sign since the current through each coil leaves the dot. For the primary loop,

$$\mathbf{E}_1 - \mathbf{I}_1\mathbf{Z}_1 - \mathbf{I}_1\mathbf{Z}_{L_1} - \mathbf{I}_2\mathbf{Z}_m - \mathbf{Z}_2(\mathbf{I}_1 - \mathbf{I}_2) = 0$$

where $M$ of $\mathbf{Z}_m = \omega M \angle 90°$ is positive and

$$\mathbf{I}_1(\mathbf{Z}_1 + \mathbf{Z}_{L_1} + \mathbf{Z}_2) - \mathbf{I}_2(\mathbf{Z}_2 - \mathbf{Z}_m) = \mathbf{E}_1$$

Note in the above that the mutual impedance was treated as if it were an additional inductance in series with the inductance $L_1$ having a sign determined by the dot convention and the voltage across which is determined by the current in the magnetically coupled loop.

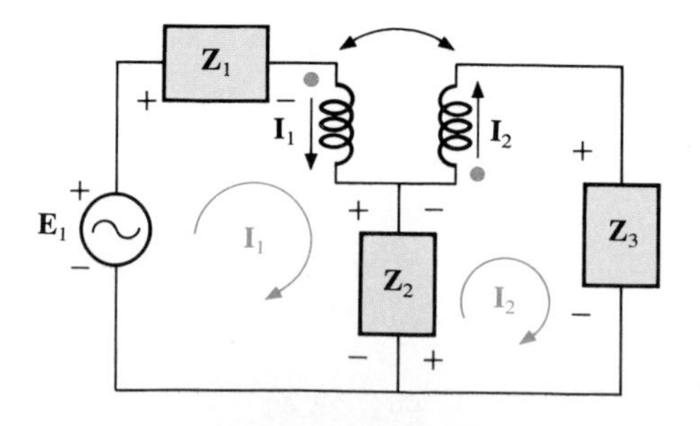


FIG. 25.42

For the secondary loop,

$$-\mathbf{Z}_2(\mathbf{I}_2 - \mathbf{I}_1) - \mathbf{I}_2\mathbf{Z}_{L_2} - \mathbf{I}_1\mathbf{Z}_m - \mathbf{I}_2\mathbf{Z}_3 = 0$$

or

$$\mathbf{I}_2(\mathbf{Z}_2 + \mathbf{Z}_{L_2} + \mathbf{Z}_3) - \mathbf{I}_1(\mathbf{Z}_2 - \mathbf{Z}_m) = 0$$

For the network of Fig. 25.43, we find a mutual term between $L_1$ and $L_2$ and $L_1$ and $L_3$ labeled $M_{12}$ and $M_{13}$, respectively.

For the coils with the dots ($L_1$ and $L_3$), since each current through the coils leaves the dot, $M_{13}$ is positive for the chosen direction of $I_1$ and $I_3$. However, since the current $I_1$ leaves the dot through $L_1$, and $I_2$ enters the dot through coil $L_2$, $M_{12}$ is negative. Consequently, for the input circuit,

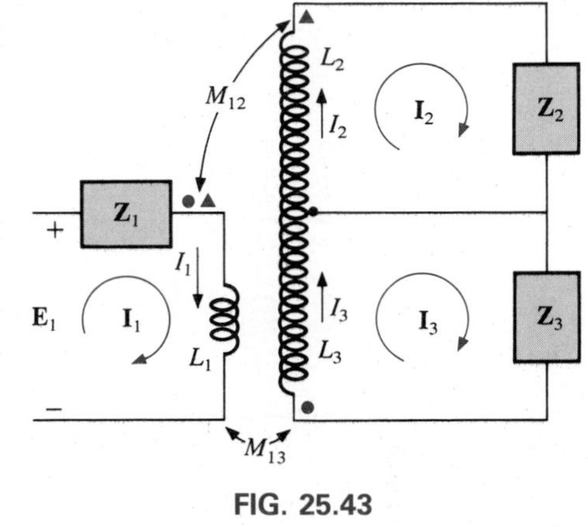

FIG. 25.43

$$\mathbf{E}_1 - \mathbf{I}_1\mathbf{Z}_1 - \mathbf{I}_1\mathbf{Z}_{L_1} - \mathbf{I}_2(-\mathbf{Z}_{m_{12}}) - \mathbf{I}_3\mathbf{Z}_{m_{13}} = 0$$

or

$$\mathbf{E}_1 - \mathbf{I}_1(\mathbf{Z}_1 + \mathbf{Z}_{L_1}) + \mathbf{I}_2\mathbf{Z}_{m_{12}} - \mathbf{I}_3\mathbf{Z}_{m_{13}} = 0$$

For loop 2,

$$-\mathbf{I}_2\mathbf{Z}_2 - \mathbf{I}_2\mathbf{Z}_{L_2} - \mathbf{I}_1(-\mathbf{Z}_{m_{12}}) = 0$$
$$-\mathbf{I}_1\mathbf{Z}_{m_{12}} + \mathbf{I}_2(\mathbf{Z}_2 + \mathbf{Z}_{L_2}) = 0$$

and for loop 3,

$$-\mathbf{I}_3\mathbf{Z}_3 - \mathbf{I}_3\mathbf{Z}_{L_3} - \mathbf{I}_1\mathbf{Z}_{m_{13}} = 0$$

or

$$\mathbf{I}_1\mathbf{Z}_{m_{13}} + \mathbf{I}_3(\mathbf{Z}_3 + \mathbf{Z}_{L_3}) = 0$$

In determinant form,

$$\begin{aligned}
\mathbf{I}_1(\mathbf{Z}_1 + \mathbf{Z}_{L_1}) &- \mathbf{I}_2\mathbf{Z}_{m_{12}} && + \mathbf{I}_3\mathbf{Z}_{m_{13}} &&= \mathbf{E}_1 \\
-\mathbf{I}_1\mathbf{Z}_{m_{12}} &+ \mathbf{I}_2(\mathbf{Z}_2 + \mathbf{Z}_{L_2}) && + 0 &&= 0 \\
\mathbf{I}_1\mathbf{Z}_{m_{13}} &+ 0 && + \mathbf{I}_3(\mathbf{Z}_3 + \mathbf{Z}_{13}) &&= 0
\end{aligned}$$

## 25.14 COMPUTER ANALYSIS

### BASIC

One of the obvious advantages of BASIC is that all the equations that would be employed in a longhand analysis of a network can also be applied in a BASIC program. The analysis would also follow the same logical order. An additional advantage is that new models for devices or systems do not have to be developed; the analysis can proceed in the same manner as applied to earlier systems.

### PSPICE

In PSPICE, however, the correct modeling of a device is a very important element of the input file if the results are to have any meaning whatsoever. In fact, to insure the model is correct, it must first be tested

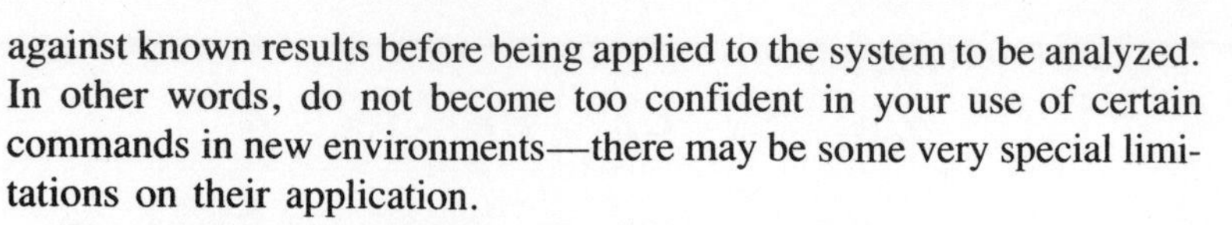

against known results before being applied to the system to be analyzed. In other words, do not become too confident in your use of certain commands in new environments—there may be some very special limitations on their application.

For transformers, a model for the ideal device requires the use of two controlled sources and an independent sensing source as shown in Fig. 25.44.

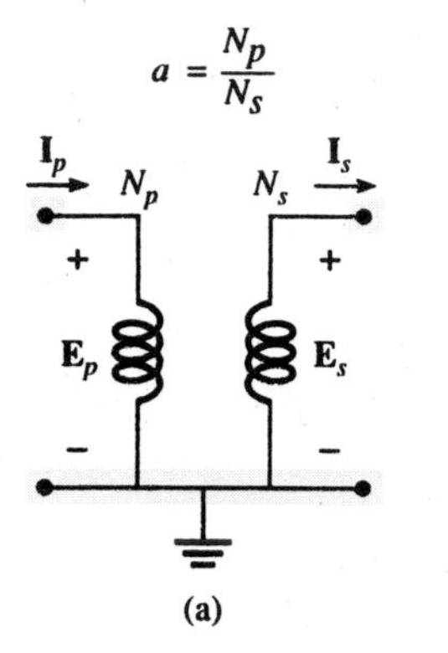


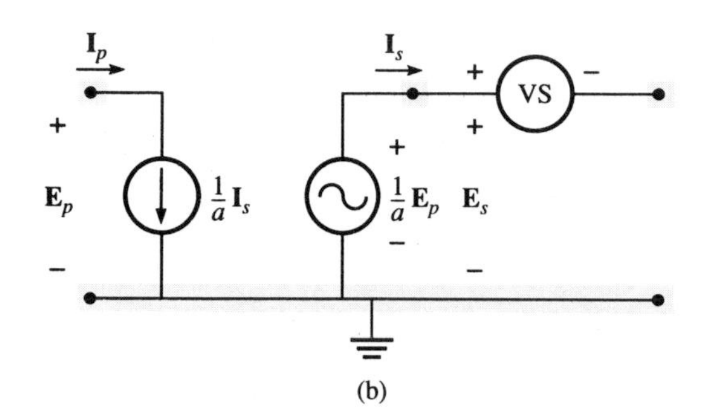


FIG. 25.44
*(a) Ideal transformer; (b) PSPICE model.*

The multiplying factor for both controlling sources is one over the turns ratio ($1/a$). Figure 25.44 reveals that the secondary voltage is $1/a$ times the primary voltage and the primary current is $1/a$ times the secondary current. The use of both a controlled current and voltage source includes the important voltage and current relationships for a transformer ($E_p = aE_s$, $I_s = aI_p$). The independent source VS is required to define the direction of $I_s$ for the CCCS. The controlling voltage for the VCVS is simply the primary voltage of the transformer. There is no need to include the impedance relationships ($Z_i = a^2 Z_L$, etc.) since they are already incorporated in the voltage and current relationships of the transformer.

As a first example let us analyze the basic transformer configuration of Fig. 25.45. The transformation ratio is

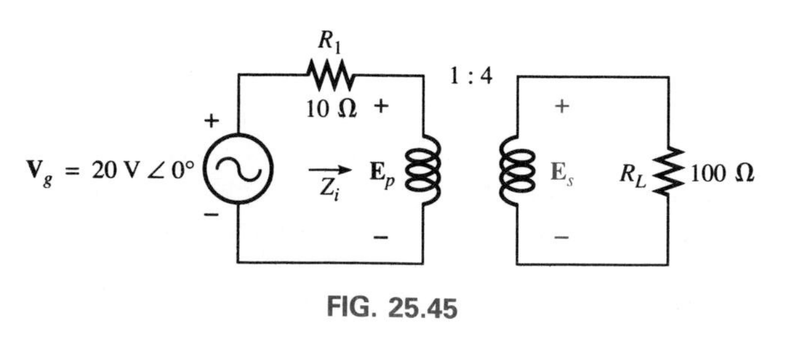


FIG. 25.45

$$a = \frac{N_p}{N_s} = \frac{1}{4} < 1$$

revealing that it is a step-up transformer.

Before assigning nodes and writing the input file, let us first analyze the network in the longhand fashion to provide a result for comparison with the output file.

$$\begin{aligned} Z_i &= a^2 Z_L \\ &= \left(\frac{1}{4}\right)^2 100\ \Omega \\ &= 6.25\ \Omega \end{aligned}$$

and

$$E_p = \frac{(6.25\ \Omega)(20\ \text{V})}{6.25\ \Omega + 10\ \Omega} = 7.692\ \text{V}$$

with

$$\begin{aligned} E_s &= \frac{1}{a}E_p = \frac{1}{(\frac{1}{4})}(7.692\ \text{V}) = 4(7.692\ \text{V}) \\ &= 30.77\ \text{V} \end{aligned}$$

and

$$V_L = E_s = \mathbf{30.77\ V}$$

The nodes are defined in Fig. 25.46. The reference node is carried through to the output circuit to establish a common ground. If preferred,

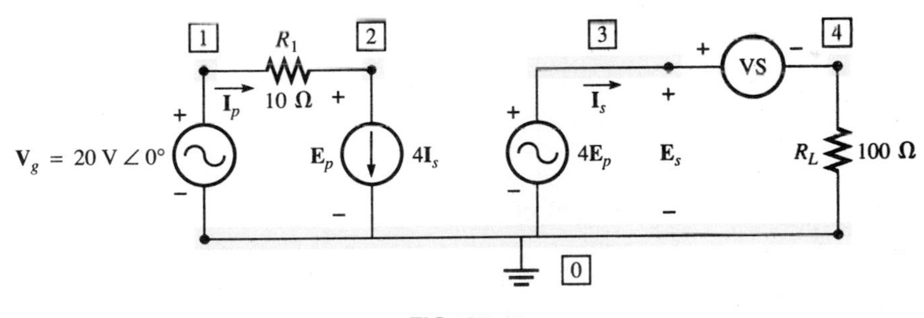


**FIG. 25.46**

isolation between the input and output circuits can be maintained by assigning a different node label for the base of the output circuit and connecting the two bases by a large impedance of perhaps 1E30 Ω to simulate an open circuit. In any event, a common ground is required for the entire system.

The input file of Fig. 25.47 employed the ''names'' PRI and SEC to distinguish between sources and requested the magnitude of the voltage across the load resistor, as calculated before. Note for both controlled sources that the multiplying factor is $1/a = 1/(\frac{1}{4}) = 4$ and that a frequency of 1 kHz was chosen to establish an .AC analysis. The output file reveals an exact match between the longhand and computer solutions.

```
************************ Evaluation PSpice (January 1989) ******* 09:15:15 *******

Chapter 25 - Transformer

****     CIRCUIT DESCRIPTION

****************************************************************************


VG 1 0 AC 20V 0
R1 1 2 10
FPRI 2 0 VS 4
VS 3 4 0
ESEC 3 0 2 0 4
RL 4 0 100
.AC LIN 1 1KH 1KH
.PRINT AC VM(RL)
.OPTIONS NOPAGE
.END


****     AC ANALYSIS                    TEMPERATURE =   27.000 DEG C


 FREQ        VM(RL)

  1.000E+03   3.077E+01
```

FIG. 25.47

The second transformer network to be analyzed appears in Fig. 25.20 of the text. In Example 25.8 the primary current established a load voltage of 1200 V with the requirement that $\mathbf{V}_g =$ 2452.04 V ∠2.34°. In this exercise we will apply a source voltage $\mathbf{V}_g$ = 2452.04 V ∠2.34° and note whether a load voltage of 1200 V results as in the example.

The nodes are chosen as in Fig. 25.48 with the inductor values determined at a frequency of 60 Hz. In this case, $a = N_p/N_s = 2/1 = 2$ re-

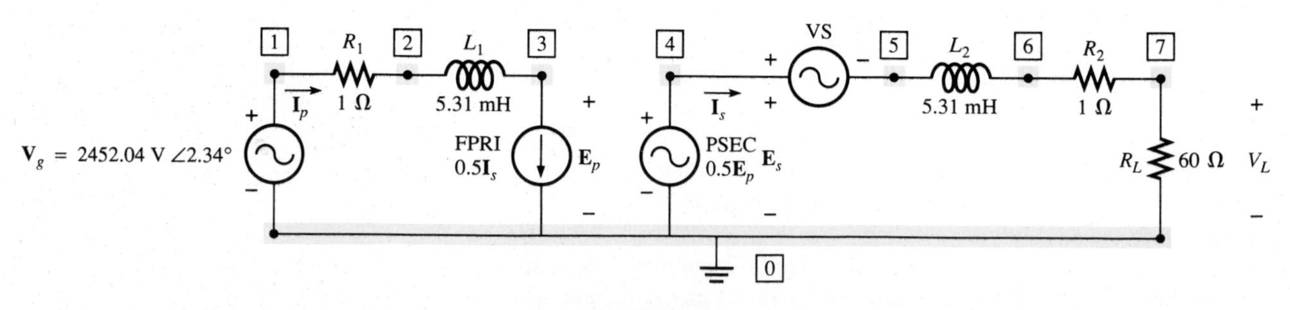


FIG. 25.48

sulting in a multiplying factor of $1/a = 1/2 = 0.5$ as appearing in the input file of Fig. 25.49. The output file reveals an exact match of 1200 V.

```
************************ Evaluation PSpice (January 1989) ******* 09:19:22 *******

Chapter 25 - Transformer (Example 25.8)

****      CIRCUIT DESCRIPTION

******************************************************************************

VG 1 0 AC 2452.04V 2.34
R1 1 2 1
L1 2 3 5.31MH
FPRI 3 0 VS 0.5
VS 4 5 0
ESEC 4 0 3 0 0.5
L2 5 6 5.31MH
R2 6 7 1
RL 7 0 60
.AC LIN 1 60 60
.PRINT AC VM(RL)
.OPTIONS NOPAGE
.END

****      AC ANALYSIS                    TEMPERATURE =   27.000 DEG C

  FREQ          VM(RL)

  6.000E+01    1.200E+03
```

FIG. 25.49

## PROBLEMS

### SECTION 25.2 Mutual Inductance

1. For the air-core transformer of Fig. 25.50:
   a. Find the value of $L_s$ if the mutual inductance $M$ is equal to 80 mH.
   b. Find the induced voltages $e_p$ and $e_s$ if the flux linking the primary coil changes at the rate of 0.08 Wb/s.
   c. Find the induced voltages $e_p$ and $e_s$ if the current $i_p$ changes at the rate of 0.3 A/ms.
2. a. Repeat Problem 1 if $k$ is changed to 1.
   b. Repeat Problem 1 if $k$ is changed to 0.2.
   c. Compare the results of parts (a) and (b).
3. Repeat Problem 1 for $k = 0.9$, $N_p = 300$ turns, and $N_s = 25$ turns.

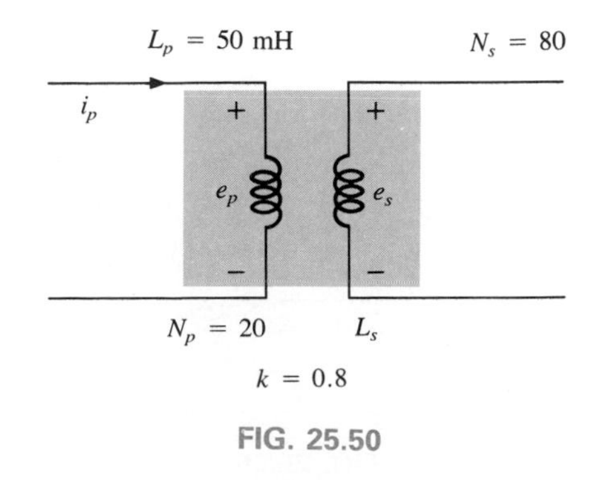


FIG. 25.50

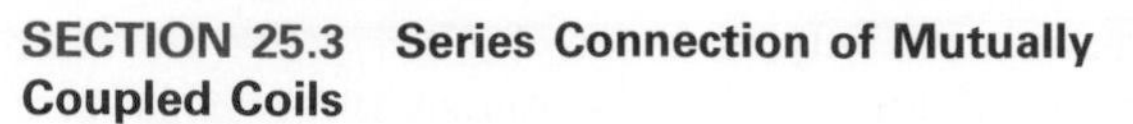

## SECTION 25.3 Series Connection of Mutually Coupled Coils

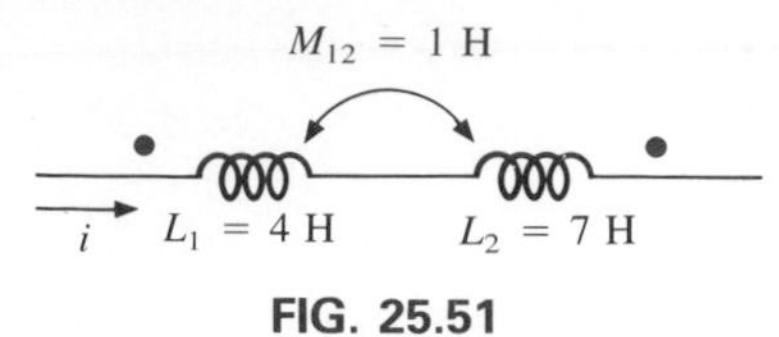


FIG. 25.51

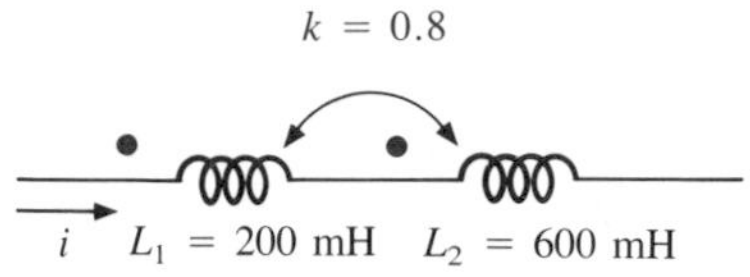

4. Determine the total inductance of the series coils of Fig. 25.51.

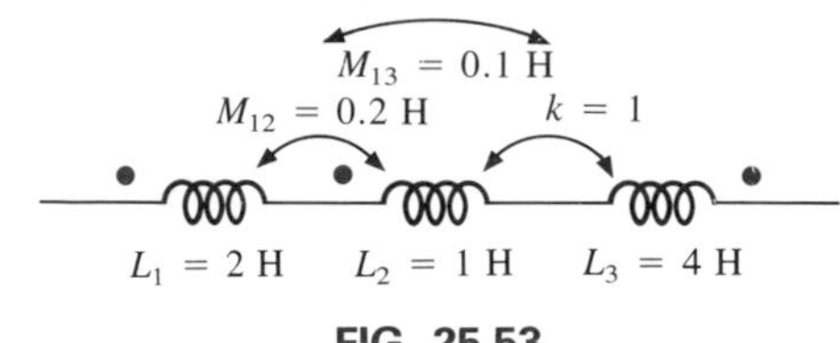


FIG. 25.52

5. Determine the total inductance of the series coils of Fig. 25.52.

FIG. 25.53

6. Determine the total inductance of the series coils of Fig. 25.53.

$M_{13} = 0.1$ H, $M_{12} = 0.2$ H, $k = 1$, $L_1 = 2$ H, $L_2 = 1$ H, $L_3 = 4$ H

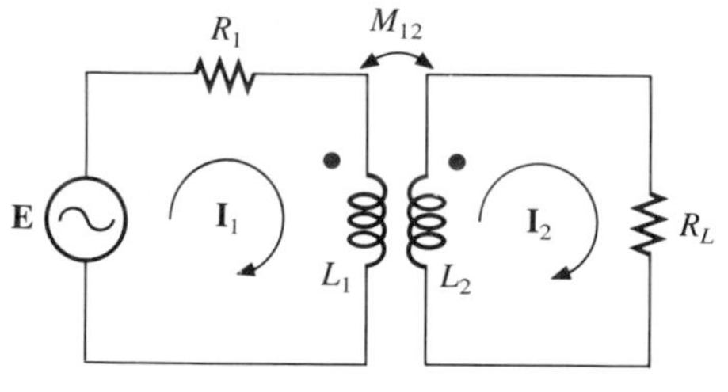


FIG. 25.54

7. Write the mesh equations for the network of Fig. 25.54.

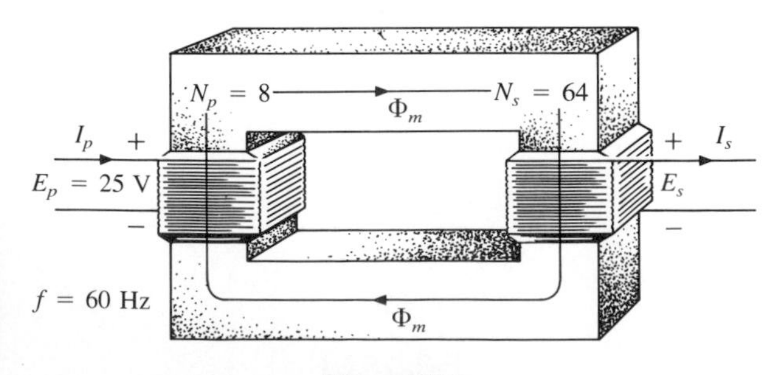


FIG. 25.55

## SECTION 25.4 The Iron-Core Transformer

8. For the iron-core transformer ($k = 1$) of Fig. 25.55:
   **a.** Find the magnitude of the induced voltage $E_s$.
   **b.** Find the maximum flux $\Phi_m$.

9. Repeat Problem 8 for $N_p = 240$ and $N_s = 30$.

10. Find the applied voltage of an iron-core transformer with a secondary voltage of 240 and $N_p = 60$ with $N_s = 720$.

11. If the maximum flux passing through the core of Problem 8 is 12.5 mWb, find the frequency of the input voltage.

## SECTION 25.5 Reflected Impedance and Power

12. For the iron-core transformer of Fig. 25.56:
    a. Find the magnitude of the current $I_L$ and the voltage $V_L$ if $a = 1/5$, $I_p = 2$ A, and $Z_L = 2\text{-}\Omega$ resistor.
    b. Find the input resistance for the data specified in part (a).

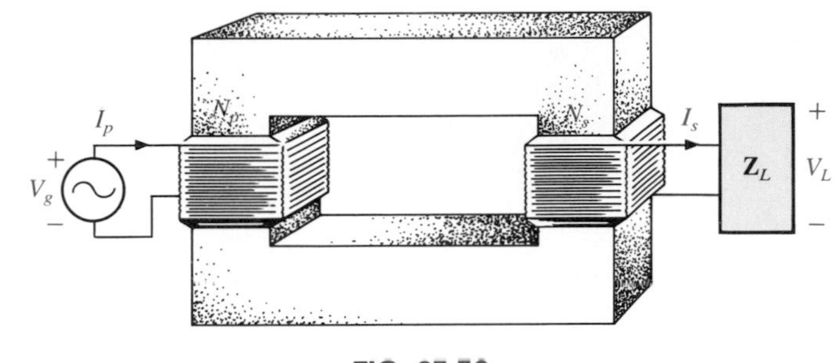


FIG. 25.56

13. Find the input impedance for the iron-core transformer of Fig. 25.56 if $a = 2$, $I_p = 4$ A, and $V_g = 1600$ V.

14. Find the voltage $V_g$ and the current $I_p$ if the input impedance of the iron-core transformer of Fig. 25.56 is 4 Ω and $V_L = 1200$ V with $a = 1/4$.

15. If $V_L = 240$ V, $Z_L = 20\text{-}\Omega$ resistor, $I_p = 0.05$ A, and $N_s = 50$, find the number of turns in the primary circuit of the iron-core transformer of Fig. 25.56.

16. a. If $N_p = 400$, $N_s = 1200$, and $V_g = 100$ V, find the magnitude of $I_p$ for the iron-core transformer of Fig. 25.56 if $Z_L = 9\ \Omega + j12\ \Omega$.
    b. Find the magnitude of the voltage $V_L$ and the current $I_L$ for the conditions of part (a).

17. a. For the circuit of Fig. 25.57, find the transformation ratio required to deliver maximum power to the speaker.
    b. Find the maximum power delivered to the speaker.

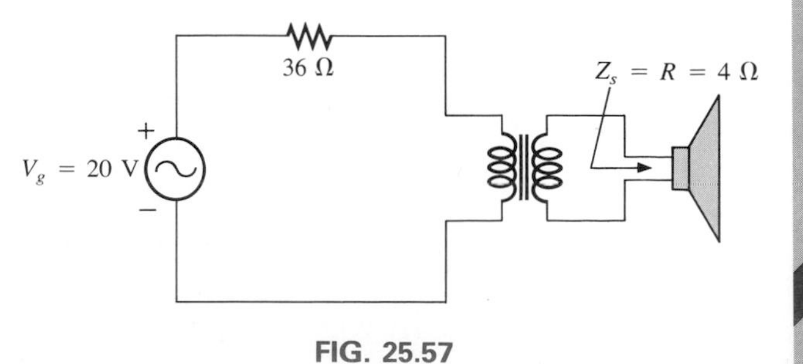


FIG. 25.57

## SECTION 25.6 Equivalent Circuit (Iron-Core Transformer)

**18.** For the transformer of Fig. 25.58, determine
- **a.** the equivalent resistance $R_e$.
- **b.** the equivalent reactance $X_e$.
- **c.** the equivalent circuit reflected to the primary.
- **d.** the primary current for $\mathbf{V}_g = 50\text{ V}\angle 0°$.
- **e.** the load voltage $V_L$.
- **f.** the phasor diagram of the reflected primary circuit.
- **g.** the new load voltage if we assume the transformer to be ideal with a 4:1 turns ratio. Compare the result with that of part (e).

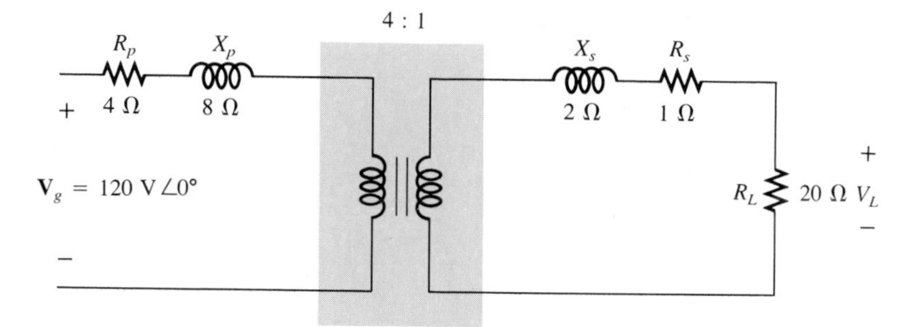

FIG. 25.58

**19.** For the transformer of Fig. 25.58, if the resistive load is replaced by an inductive reactance of 20 Ω:
- **a.** Determine the total reflected primary impedance.
- **b.** Calculate the primary current.
- **c.** Determine the voltage across $R_e$, $X_e$, and the reflected load.
- **d.** Draw the phasor diagram.

**20.** Repeat Problem 19 for a capacitive load having a reactance of 20 Ω.

## SECTION 25.7 Frequency Considerations

**21.** Discuss in your own words the frequency characteristics of the transformer. Employ the applicable equivalent circuit and frequency characteristics appearing in this chapter.

## SECTION 25.8 Air-Core Transformer

**22.** Determine the input impedance to the air-core transformer of Fig. 25.59. Sketch the reflected primary network.

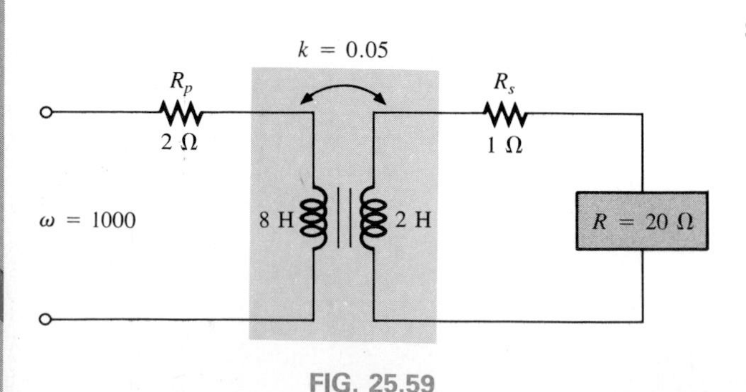


FIG. 25.59

## SECTION 25.10 Nameplate Data

**23.** An ideal transformer is rated 10 kVA, 2400/120 V, 60 Hz.
  **a.** Find the transformation ratio if the 120 V is the secondary voltage.
  **b.** Find the current rating of the secondary if the 120 V is the secondary voltage.
  **c.** Find the current rating of the primary if the 120 V is the secondary voltage.
  **d.** Repeat parts (a) through (c) if the 2400 V is the secondary voltage.

## SECTION 25.11 Types of Transformers

**24.** Determine the primary and secondary voltages and currents for the autotransformer of Fig. 25.60.

## SECTION 25.12 Tapped and Multiple-Load Transformers

**25.** For the center-tapped transformer of Fig. 25.39 where $N_p = 100$, $N_s = 25$, $\mathbf{Z}_L = \mathbf{R} = 5\ \Omega \angle 0°$, and $\mathbf{E}_p = 100\ \text{V} \angle 0°$:
  **a.** Determine the load voltage and current.
  **b.** Find the impedance $\mathbf{Z}_i$.
  **c.** Calculate the impedance $\mathbf{Z}_{1/2}$.

**26.** For the multiple-load transformer of Fig. 25.40 where $N_1 = 90$, $N_2 = 15$, $N_3 = 45$, $\mathbf{Z}_2 = \mathbf{R} = 8\ \Omega \angle 0°$, $\mathbf{Z}_3 = \mathbf{R}_L = 5\ \Omega \angle 0°$, and $\mathbf{E}_1 = 60\ \text{V} \angle 0°$:
  **a.** Determine the load voltages and currents.
  **b.** Calculate $\mathbf{Z}_1$.

**27.** For the multiple-load transformer of Fig. 25.41 where $N_1 = 120$, $N_2 = 40$, $N_3 = 30$, $\mathbf{Z}_2 = \mathbf{R} = 12\ \Omega \angle 0°$, $\mathbf{Z}_3 = \mathbf{R}_C = 10\ \Omega \angle 0°$, and $\mathbf{E}_1 = 120\ \text{V} \angle 60°$:
  **a.** Determine the load voltages and currents.
  **b.** Calculate $\mathbf{Z}_1$.

## SECTION 25.13 Networks with Magnetically Coupled Coils

**28.** Write the mesh equations for the network of Fig. 25.61.

**29.** Write the mesh equations for the network of Fig. 25.62.

## SECTION 25.14 Computer Analysis

### PSPICE

***30.** Write the input file to determine the magnitude and phase of the voltage $\mathbf{V}_L$ for the network of Fig. 25.58.

***31.** Write the input file for the network of Fig. 25.9 using the equations developed in the example to establish a network model. Print out the voltage $\mathbf{V}_L$ for $\mathbf{E}_1 = 100\ \text{V} \angle 0°$, $R_L = 100\ \Omega$, $R_1 = 10\ \Omega$, $L_1 = 2\ \text{H}$, $L_2 = 5\ \text{H}$, $M = 0.9\ \text{H}$, and $f = 60\ \text{Hz}$.

***32.** Write the input file to determine the input impedance $\mathbf{Z}_i$ for the air-core transformer of Fig. 25.27. Use the technique of earlier chapters to find the impedance level and use Eqs. (25.24) and (25.26) to establish a network model for the system.

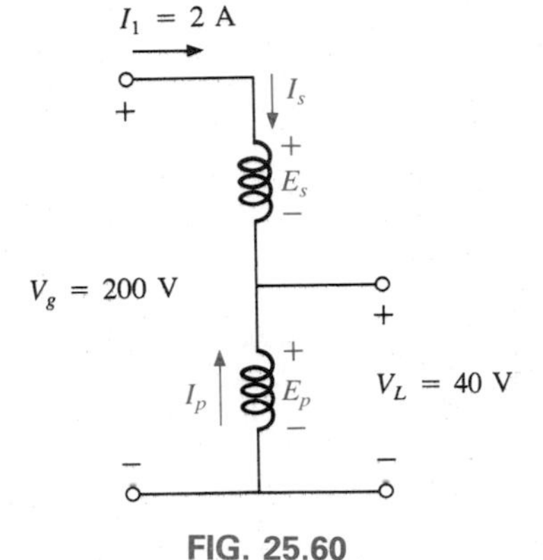

FIG. 25.60

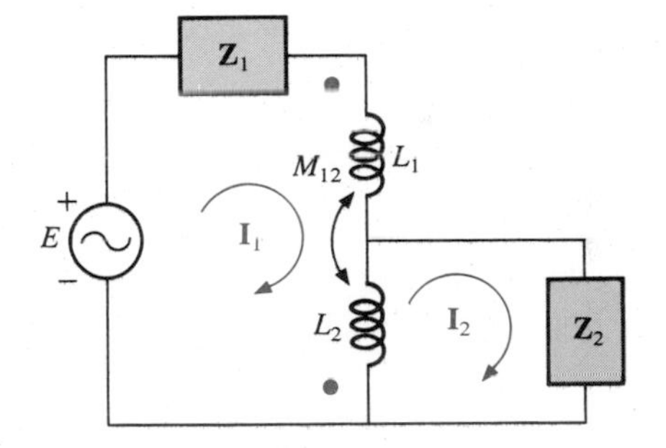

FIG. 25.61

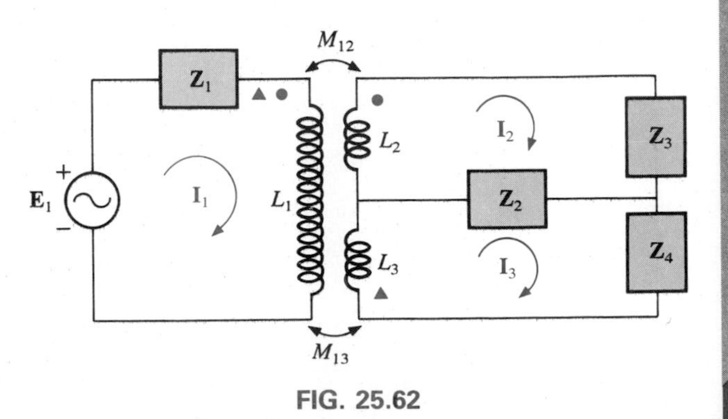


FIG. 25.62

### BASIC

33. Write a program to provide a general solution to the problem of impedance matching as defined by Example 25.6. That is, given the speaker impedance and the internal resistance of the source, determine the required turns ratio and the power delivered to the speaker. In addition, calculate the load and source current and the primary and secondary voltages. The source voltage will have to be provided with the other parameters of the network.
34. Given the equivalent model of an iron-core transformer appearing in Fig. 25.20, write a program to calculate the magnitude of the voltage $\mathbf{V}_g$.
35. Given the parameters of Example 25.9, write a program to calculate the input impedance in polar form.

## GLOSSARY

**Autotransformer** A transformer with one winding common to both the primary and the secondary circuits. A loss in isolation is balanced by the increase in its kilovolt-ampere rating.

**Coefficient of coupling** ($k$) A measure of the magnetic coupling of two coils which ranges from a minimum of zero to a maximum of 1.

**Dot convention** A technique for labeling the effect of the mutual inductance on a net inductance of a network or system.

**Leakage flux** The flux linking the coil that does not pass through the ferromagnetic path of the magnetic circuit.

**Loosely coupled** A term applied to two coils that have a low coefficient of coupling.

**Multiple-load transformers** Transformers having more than a single load connected to the secondary winding or windings.

**Mutual inductance** The inductance that exists between magnetically coupled coils of the same or different dimensions.

**Nameplate data** Information such as the kilovolt-ampere rating, voltage transformation ratio, and frequency of application that is of primary importance in choosing the proper transformer for a particular application.

**Primary** The coil or winding to which the source of electrical energy is normally applied.

**Reflected impedance** The impedance appearing at the primary of a transformer due to a load connected to the secondary. Its magnitude is controlled directly by the transformation ratio.

**Secondary** The coil or winding to which the load is normally applied.

**Step-down transformer** A transformer whose secondary voltage is less than its primary voltage. The transformation ratio $a$ is greater than 1.

**Step-up transformer** A transformer whose secondary voltage is greater than its primary voltage. The magnitude of the transformation ratio $a$ is less than 1.

**Tapped transformer** A transformer having an additional connection between the terminals of the primary or secondary windings.

**Transformation ratio** ($a$) The ratio of primary to secondary turns of a transformer.

# 26

# Two-Port Parameters (z, y, h)

## 26.1 INTRODUCTION

In the broad spectrum of courses to follow, there will be an increasing need for the ability to *model* both devices and systems and employ these models in the analysis and synthesis of combined and enlarged systems.

This chapter will be limited to the configuration most frequently subject to the modeling technique: the two-port network shown in Fig. 26.1.

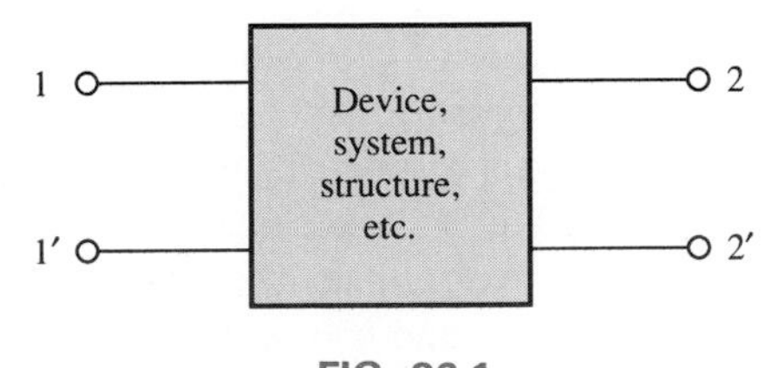


FIG. 26.1
*Two-port configuration.*

Note that in Fig. 26.1 there are two ports of entry or interest, each having a pair of terminals. For some devices, the two-port configuration of Fig. 26.1 may appear as shown in Fig. 26.2(a). The block diagram of Fig. 26.2(a) simply indicates that terminals 1′ and 2′ are in common—a particular case of the general two-port network. A single-port and a multiport network appear in Fig. 26.2(b). The former has been analyzed throughout the text, while the characteristics of the latter will be left for a more advanced course.

The primary purpose of this chapter is to develop a set of equations (and, subsequently, networks) that will allow us to model the device or system appearing within the enclosed structure of Fig. 26.1. That is, we will be able to establish a network that will display the same terminal

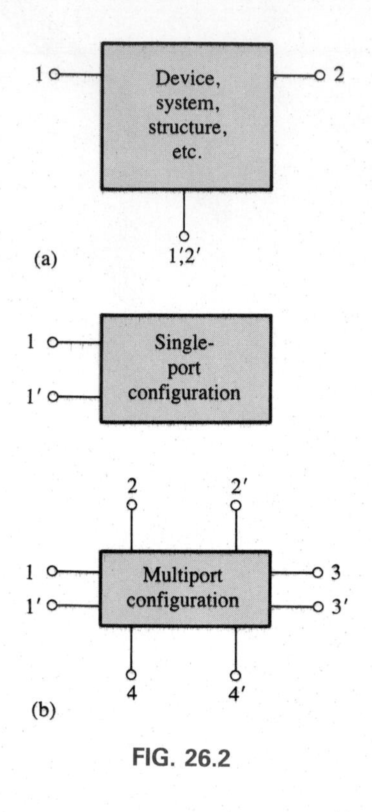


FIG. 26.2

characteristics as those of the original system, device, and so on. In Fig. 26.3, for example, a transistor appears between the four external terminals. Through the analysis to follow, we will find a combination of network elements that will allow us to replace the transistor by a network that will behave very much like the original device for a specific set of operating conditions. Methods such as mesh and nodal analysis can then be applied to determine any unknown quantities. The models, when reduced to their simplest forms as determined by the operating conditions, can also provide very quick estimates of network behavior without a lengthy mathematical derivation. In other words, someone well-versed in the use of models can analyze the operation of large, complex systems in short order. The results may be only approximate in most cases, but this quick return for a minimum of effort is often worthwhile.

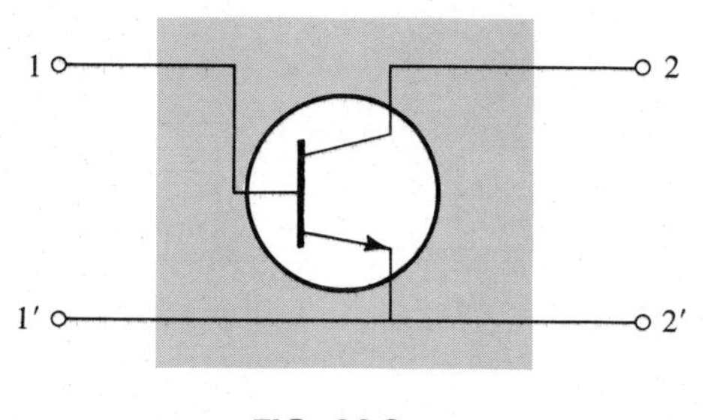

FIG. 26.3

The following analysis may initially appear very mathematical and devoid of any practical meaning. However, you will note that the parameters are determined by the ratio of electrical quantities under very specific network conditions. The resulting quantities also have a terminology that will be applied very frequently in the electronics course to follow. The analysis of this chapter will be limited to linear (fixed-value) systems with bilateral elements. Three sets of parameters will be developed for the two-port configuration, referred to as the *impedance* (**z**), *admittance* (**y**), and *hybrid* (**h**) parameters. A table will be provided at the end of the chapter relating the three sets of parameters.

## 26.2 IMPEDANCE (z) PARAMETERS

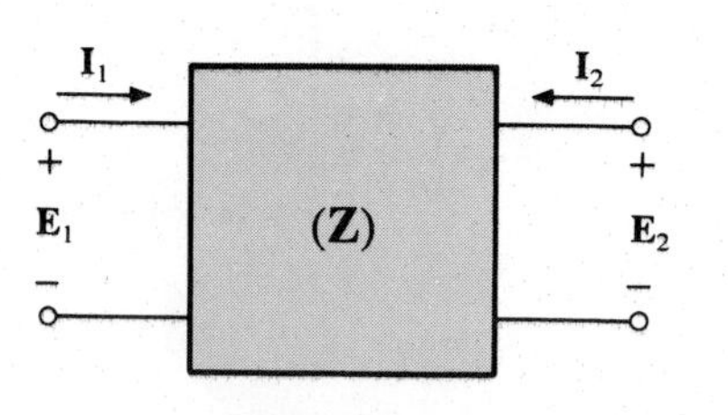


FIG. 26.4
*Two-port impedance parameter configuration.*

For the two-port configuration of Fig. 26.4, four variables are specified. For most situations, if any two are specified, the remaining two variables can be determined. The four variables can be related by the following equations:

$$\mathbf{E}_1 = \mathbf{z}_{11}\mathbf{I}_1 + \mathbf{z}_{12}\mathbf{I}_2 \qquad \textbf{(26.1a)}$$

$$\mathbf{E}_2 = \mathbf{z}_{21}\mathbf{I}_1 + \mathbf{z}_{22}\mathbf{I}_2 \qquad \textbf{(26.1b)}$$

The *impedance parameters* $\mathbf{z}_{11}$, $\mathbf{z}_{12}$, and $\mathbf{z}_{22}$ are measured in ohms.

To model the system, each impedance parameter must be determined. They are determined by setting a particular variable to zero.

### $\mathbf{z}_{11}$

For $\mathbf{z}_{11}$, if $\mathbf{I}_2$ is set to zero, as shown in Fig. 26.5, Eq. (26.1a) becomes

$$\mathbf{E}_1 = \mathbf{z}_{11}\mathbf{I}_1 + \mathbf{z}_{12}(0)$$

and

$$\mathbf{z}_{11} = \left.\frac{\mathbf{E}_1}{\mathbf{I}_1}\right|_{\mathbf{I}_2 = 0} \quad \text{(ohms, } \Omega) \tag{26.2}$$

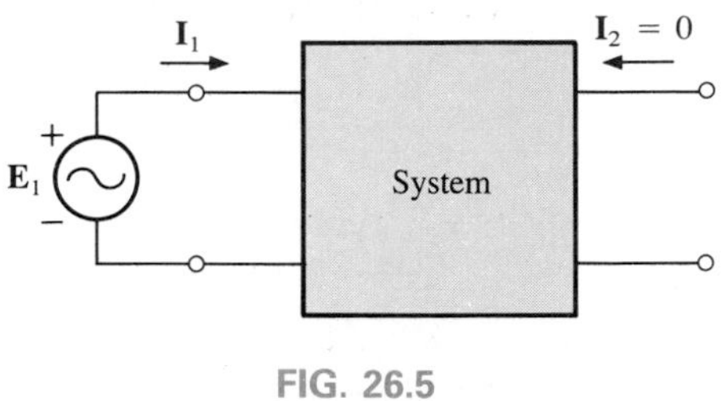


FIG. 26.5
$\mathbf{z}_{11}$ *determination.*

Equation (26.2) reveals that with $\mathbf{I}_2$ set to zero, the impedance parameter is determined by the resulting ratio of $\mathbf{E}_1$ to $\mathbf{I}_1$. Since $\mathbf{E}_1$ and $\mathbf{I}_1$ are both input quantities, with $\mathbf{I}_2$ set to zero, the parameter $\mathbf{z}_{11}$ is formally referred to in the following manner:

$\mathbf{z}_{11}$ = *open-circuit, input-impedance parameter*

### $\mathbf{z}_{12}$

For $\mathbf{z}_{12}$, $\mathbf{I}_1$ is set to zero, and Eq. (26.1a) results in

$$\mathbf{z}_{12} = \left.\frac{\mathbf{E}_1}{\mathbf{I}_2}\right|_{\mathbf{I}_1 = 0} \quad (\Omega) \tag{26.3}$$

For most systems where input and output quantities are to be compared, the ratio of interest is usually that of the output quantity divided by the input quantity. In this case, the *reverse* is true, resulting in the following:

$\mathbf{z}_{12}$ = *open-circuit, reverse-transfer impedance parameter*

The term *transfer* is included to indicate that $\mathbf{z}_{12}$ will relate an input and output quantity (for the condition $\mathbf{I}_1 = 0$). The network configuration for determining $\mathbf{z}_{12}$ is shown in Fig. 26.6.

For an applied source $\mathbf{E}_2$, the ratio $\mathbf{E}_1/\mathbf{I}_2$ will determine $\mathbf{z}_{12}$ with $\mathbf{I}_1$ set to zero.

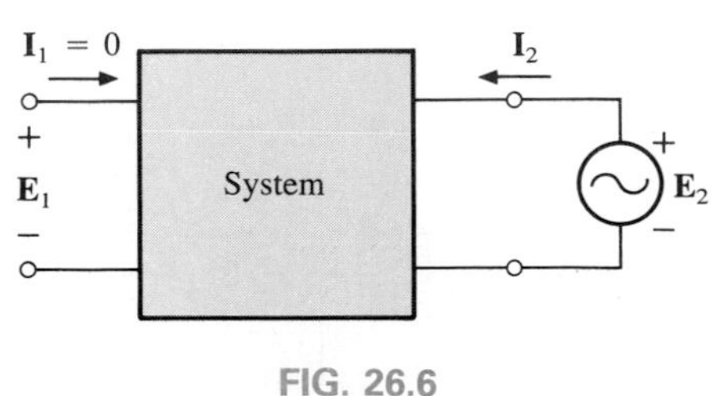


FIG. 26.6
$\mathbf{z}_{12}$ *determination.*

### $\mathbf{z}_{21}$

To determine $\mathbf{z}_{21}$, set $\mathbf{I}_2$ to zero and find the ratio $\mathbf{E}_2/\mathbf{I}_1$ as determined by Eq. (26.1b). That is,

$$\mathbf{z}_{21} = \left.\frac{\mathbf{E}_2}{\mathbf{I}_1}\right|_{\mathbf{I}_2 = 0} \quad (\Omega) \tag{26.4}$$

In this case, input and output quantities are again the determining variables, requiring the term *transfer* in the nomenclature. However,

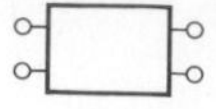

the ratio is that of an output to an input quantity, so the descriptive term *forward* is applied, and

$\mathbf{z}_{21}$ = *open-circuit, forward-transfer impedance parameter*

The determining network is shown in Fig. 26.7. For an applied voltage $\mathbf{E}_1$, it is determined by the ratio $\mathbf{E}_2/\mathbf{I}_1$ with $\mathbf{I}_2$ set to zero.

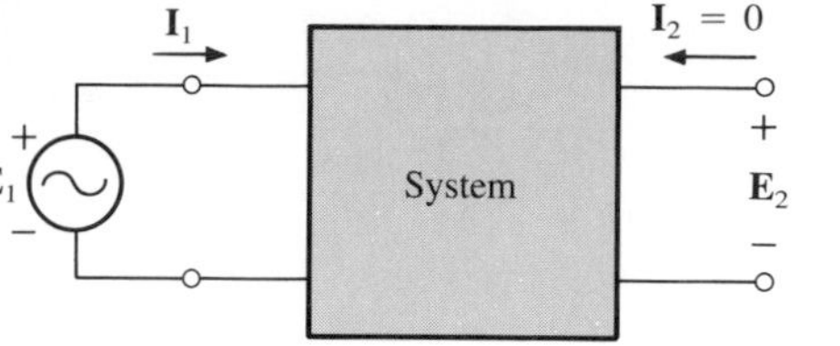


FIG. 26.7
$\mathbf{z}_{21}$ *determination.*

## $\mathbf{z}_{22}$

The remaining parameter, $\mathbf{z}_{22}$, is determined by

$$\mathbf{z}_{22} = \left.\frac{\mathbf{E}_2}{\mathbf{I}_2}\right|_{\mathbf{I}_1 = 0} \quad (\Omega) \tag{26.5}$$

as derived from Eq. (26.1b) with $\mathbf{I}_1$ set to zero. Since it is the ratio of the output voltage to the output current with $\mathbf{I}_1$ set to zero, it has the terminology

$\mathbf{z}_{22}$ = *open-circuit, output-impedance parameter*

The required network is shown in Fig. 26.8. For an applied voltage $\mathbf{E}_2$, it is determined by the resulting ratio $\mathbf{E}_2/\mathbf{I}_2$ with $\mathbf{I}_1 = 0$.

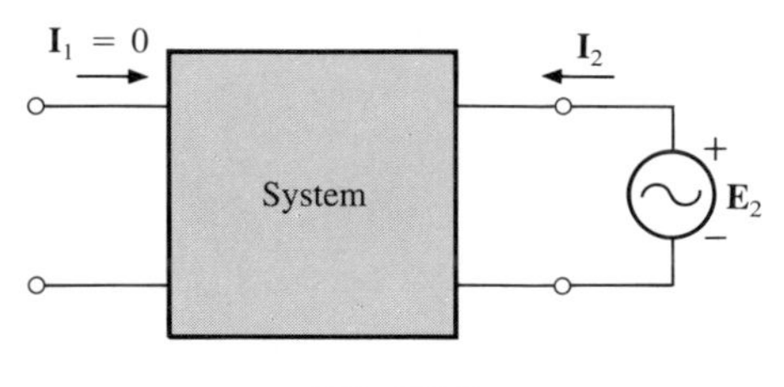


FIG. 26.8
$\mathbf{z}_{22}$ *determination.*

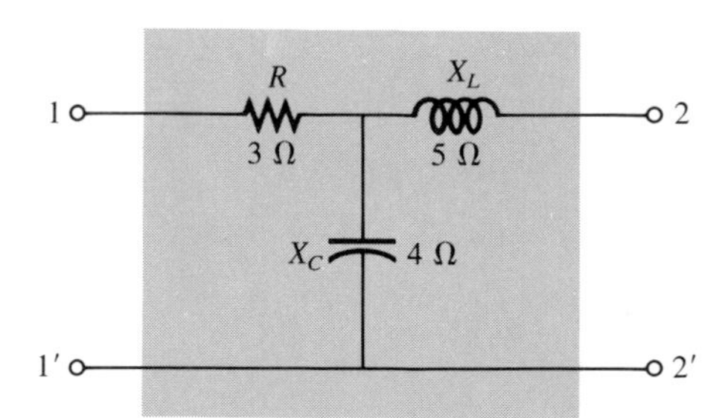


FIG. 26.9
*T configuration.*

**EXAMPLE 26.1.** Determine the impedance (**z**) parameters for the T network of Fig. 26.9.

***Solution:*** For $\mathbf{z}_{11}$, the network will appear as shown in Fig. 26.10, with $\mathbf{Z}_1 = 3\ \Omega\ \angle 0°$, $\mathbf{Z}_2 = 5\ \Omega\ \angle 90°$, and $\mathbf{Z}_3 = 4\ \Omega\ \angle -90°$:

$$\mathbf{I}_1 = \frac{\mathbf{E}_1}{\mathbf{Z}_1 + \mathbf{Z}_3}$$

Thus,

$$\mathbf{z}_{11} = \left.\frac{\mathbf{E}_1}{\mathbf{I}_1}\right|_{\mathbf{I}_2 = 0}$$

and

$$\mathbf{z}_{11} = \mathbf{Z}_1 + \mathbf{Z}_3 \tag{26.6}$$

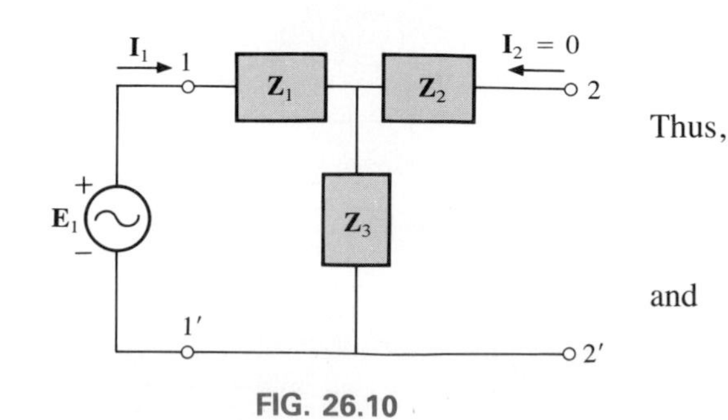


FIG. 26.10

For $\mathbf{z}_{12}$, the network will appear as shown in Fig. 26.11:

$$\mathbf{E}_1 = \mathbf{I}_2\mathbf{Z}_3$$

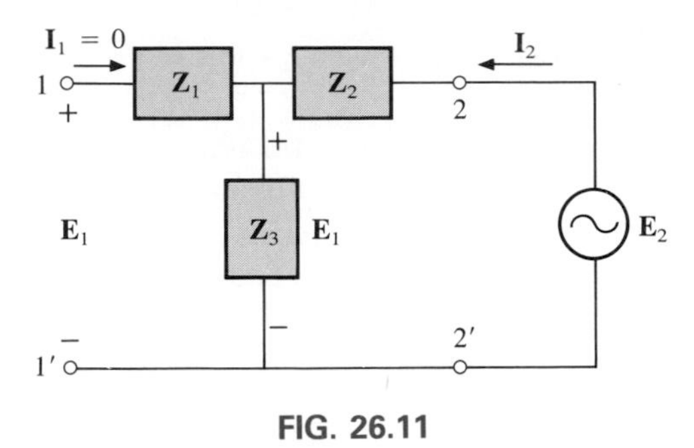


FIG. 26.11

Thus,

$$\mathbf{z}_{12} = \left.\frac{\mathbf{E}_1}{\mathbf{I}_2}\right|_{\mathbf{I}_1 = 0} = \frac{\mathbf{I}_2\mathbf{Z}_3}{\mathbf{I}_2}$$

and

$$\boxed{\mathbf{z}_{12} = \mathbf{Z}_3} \qquad \textbf{(26.7)}$$

For $\mathbf{z}_{21}$, the required network appears in Fig. 26.12:

$$\mathbf{E}_2 = \mathbf{I}_1\mathbf{Z}_3$$

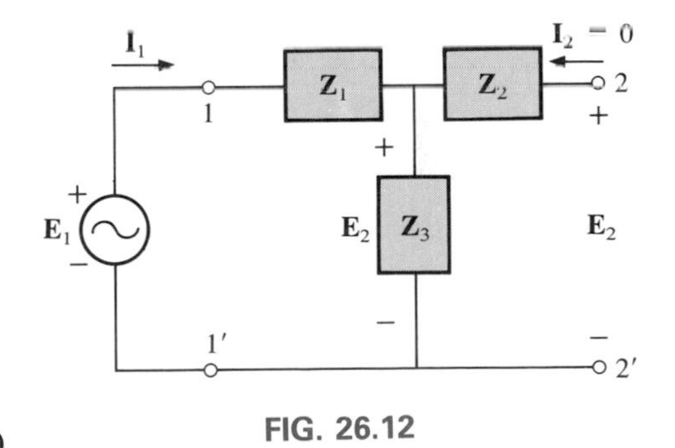


FIG. 26.12

Thus,

$$\mathbf{z}_{21} = \left.\frac{\mathbf{E}_2}{\mathbf{I}_1}\right|_{\mathbf{I}_2 = 0} = \frac{\mathbf{I}_1\mathbf{Z}_3}{\mathbf{I}_1}$$

and

$$\boxed{\mathbf{z}_{21} = \mathbf{Z}_3} \qquad \textbf{(26.8)}$$

For $\mathbf{z}_{22}$, the determining configuration is shown in Fig. 26.13:

$$\mathbf{I}_2 = \frac{\mathbf{E}_2}{\mathbf{Z}_2 + \mathbf{Z}_3}$$

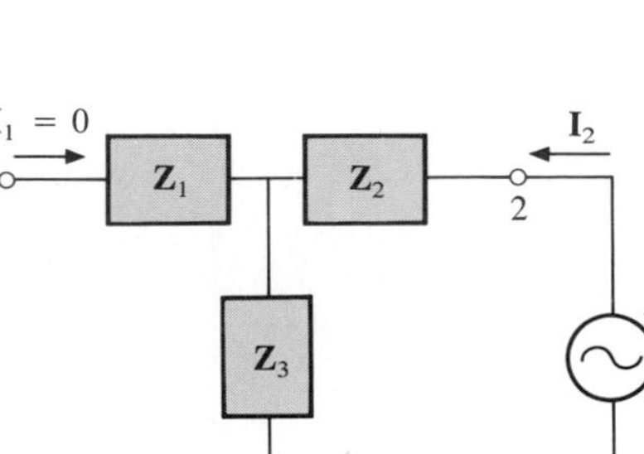

FIG. 26.13

Thus,

$$\mathbf{z}_{22} = \left.\frac{\mathbf{E}_2}{\mathbf{I}_2}\right|_{\mathbf{I}_1 = 0} = \frac{\mathbf{I}_2(\mathbf{Z}_2 + \mathbf{Z}_3)}{\mathbf{I}_2}$$

and

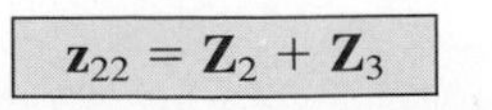

$$\boxed{\mathbf{z}_{22} = \mathbf{Z}_2 + \mathbf{Z}_3}$$

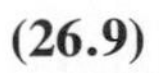
**(26.9)**

Note that for the T configuration, $\mathbf{z}_{12} = \mathbf{z}_{21}$. For $\mathbf{Z}_1 = 3\ \Omega\ \angle 0°$, $\mathbf{Z}_2 = 5\ \Omega\ \angle 90°$, and $\mathbf{Z}_3 = 4\ \Omega\ \angle -90°$, we have

$$\mathbf{z}_{11} = \mathbf{Z}_1 + \mathbf{Z}_3 = \mathbf{3\ \Omega - \mathit{j}4\ \Omega}$$
$$\mathbf{z}_{12} = \mathbf{z}_{21} = \mathbf{Z}_3 = 4\ \Omega\ \angle -90° = \mathbf{-\mathit{j}4\ \Omega}$$
$$\mathbf{z}_{22} = \mathbf{Z}_2 + \mathbf{Z}_3 = 5\ \Omega\ \angle 90° + 4\ \Omega\ \angle -90° = 1\ \Omega\ \angle 90° = \mathbf{\mathit{j}1\ \Omega}$$

For a set of impedance parameters, the terminal (external) behavior of the device or network within the configuration of Fig. 26.1 is determined. An *equivalent circuit* for the system can be developed using the impedance parameters and Eqs. (26.1a) and (26.1b). Two possibilities for the impedance parameters appear in Fig. 26.14.

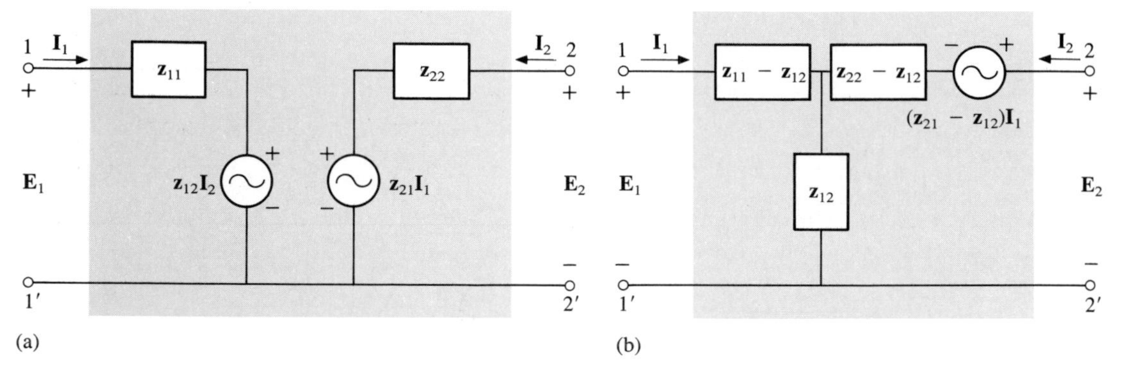


FIG. 26.14
*Two-port, **z**-parameter equivalent networks.*

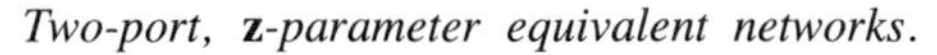

Applying Kirchhoff's voltage law to the input and output loops of the network of Fig. 26.14(a) results in

$$\mathbf{E}_1 - \mathbf{z}_{11}\mathbf{I}_1 - \mathbf{z}_{12}\mathbf{I}_2 = 0$$

and

$$\mathbf{E}_2 - \mathbf{z}_{22}\mathbf{I}_2 - \mathbf{z}_{21}\mathbf{I}_1 = 0$$

which, when rearranged, become

$$\mathbf{E}_1 = \mathbf{z}_{11}\mathbf{I}_1 + \mathbf{z}_{12}\mathbf{I}_2$$
$$\mathbf{E}_2 = \mathbf{z}_{21}\mathbf{I}_1 + \mathbf{z}_{22}\mathbf{I}_2$$

matching Eqs. (26.1a) and (26.1b).

For the network of Fig. 26.14(b),

$$\mathbf{E}_1 - \mathbf{I}_1(\mathbf{z}_{11} - \mathbf{z}_{12}) - \mathbf{z}_{12}(\mathbf{I}_1 + \mathbf{I}_2) = 0$$

and

$$\mathbf{E}_2 - \mathbf{I}_1(\mathbf{z}_{21} - \mathbf{z}_{12}) - \mathbf{I}_2(\mathbf{z}_{22} - \mathbf{z}_{12}) - \mathbf{z}_{12}(\mathbf{I}_1 + \mathbf{I}_2) = 0$$

which, when rearranged, are

$$\mathbf{E}_1 = \mathbf{I}_1(\mathbf{z}_{11} - \mathbf{z}_{12} + \mathbf{z}_{12}) + \mathbf{I}_2\mathbf{z}_{12}$$
$$\mathbf{E}_2 = \mathbf{I}_1(\mathbf{z}_{21} - \mathbf{z}_{12} + \mathbf{z}_{12}) + \mathbf{I}_2(\mathbf{z}_{22} - \mathbf{z}_{12} + \mathbf{z}_{12})$$

and

$$\mathbf{E}_1 = \mathbf{z}_{11}\mathbf{I}_1 + \mathbf{z}_{12}\mathbf{I}_2$$
$$\mathbf{E}_2 = \mathbf{z}_{21}\mathbf{I}_1 + \mathbf{z}_{22}\mathbf{I}_2$$

Note in each network the necessity for a current-controlled voltage source, that is, a voltage source the magnitude of which is determined by a particular current of the network.

The usefulness of the impedance parameters and the resulting equivalent networks can best be described by considering the system of Fig. 26.15(a), which contains a device (or system) for which the impedance parameters have been determined. As shown in Fig. 26.15(b), the

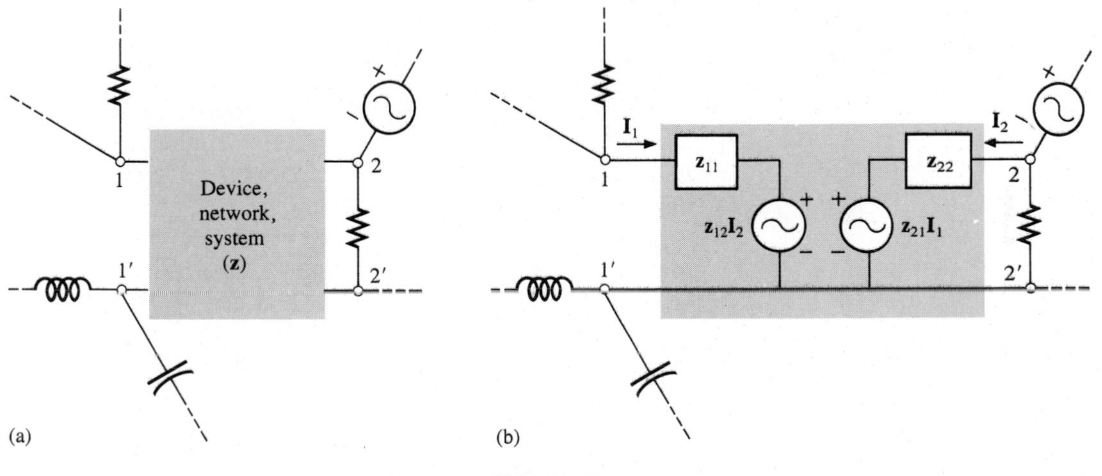

FIG. 26.15

equivalent network for the device (or system) can then be substituted, and methods such as mesh analysis, nodal analysis, and so on, can be employed to determine required unknown quantities. The device itself can then be replaced by an equivalent circuit and the desired solutions obtained more directly and with less effort than is required using only the characteristics of the device.

**EXAMPLE 26.2.** Draw the equivalent circuit in the form shown in Fig. 26.14(a) using the impedance parameters determined in Example 26.1.

***Solution:*** The circuit appears in Fig. 26.16.

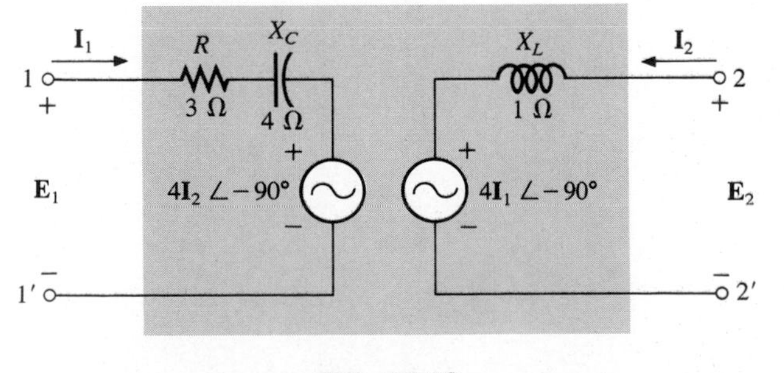


FIG. 26.16

## 26.3 ADMITTANCE (y) PARAMETERS

The equations relating the four terminal variables of Fig. 26.1 can also be written in the following form:

$$\mathbf{I}_1 = \mathbf{y}_{11}\mathbf{E}_1 + \mathbf{y}_{12}\mathbf{E}_2 \qquad \textbf{(26.10a)}$$

$$\mathbf{I}_2 = \mathbf{y}_{21}\mathbf{E}_1 + \mathbf{y}_{22}\mathbf{E}_2 \qquad \textbf{(26.10b)}$$

Note that in this case each term of each equation has the units of current, as compared to voltage for each term of Eqs. (26.1a) and (26.1b). In addition, the unit of each coefficient is siemens, compared with the ohm for the impedance parameters.

The impedance parameters were determined by setting a particular current to zero through an open-circuit condition. For the *admittance parameters* of Eqs. (26.10a) and (26.10b), a voltage is set to zero through a short-circuit condition.

The terminology applied to each of the admittance parameters follows directly from the descriptive terms applied to each of the impedance parameters. The equations for each are determined directly from Eqs. (26.10a) and (26.10b) by setting a particular voltage to zero.

### $\mathbf{y}_{11}$

$$\mathbf{y}_{11} = \frac{\mathbf{I}_1}{\mathbf{E}_1}\Bigg|_{\mathbf{E}_2 = 0} \quad \text{(siemens, S)} \qquad \textbf{(26.11)}$$

$\mathbf{y}_{11}$ = *short-circuit, input-admittance parameter*

The determining network appears in Fig. 26.17.

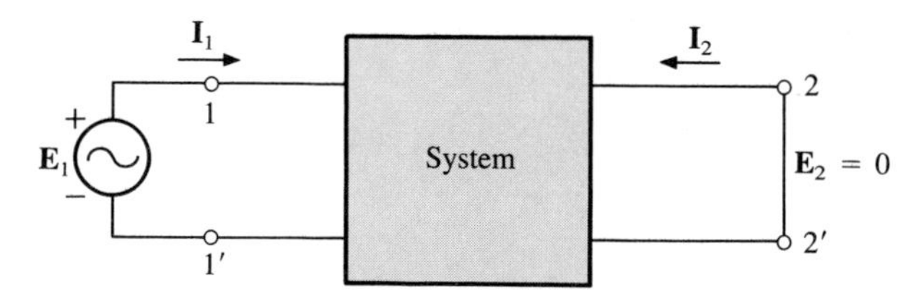


FIG. 26.17
$\mathbf{y}_{11}$ *determination.*

### $\mathbf{y}_{12}$

$$\mathbf{y}_{12} = \frac{\mathbf{I}_1}{\mathbf{E}_2}\Bigg|_{\mathbf{E}_1 = 0} \quad \text{(S)} \qquad \textbf{(26.12)}$$

$\mathbf{y}_{12}$ = *short-circuit, reverse-transfer admittance parameter*

The network for determining $\mathbf{y}_{12}$ appears in Fig. 26.18.

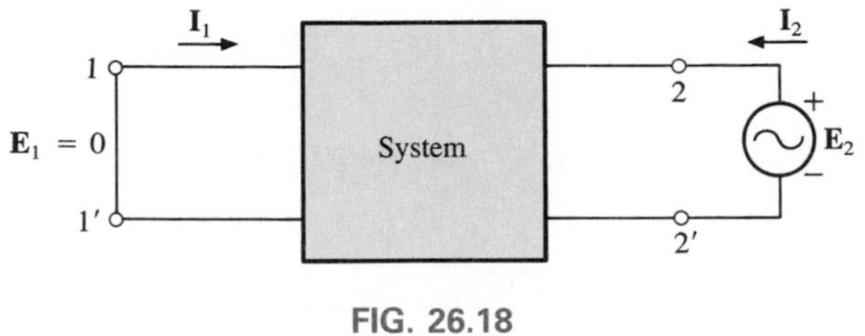


FIG. 26.18
$\mathbf{y}_{12}$ *determination.*

## $\mathbf{y}_{21}$

$$\mathbf{y}_{21} = \frac{\mathbf{I}_2}{\mathbf{E}_1} \Big|_{\mathbf{E}_2 = 0} \quad \text{(S)} \qquad (26.13)$$

$\mathbf{y}_{21}$ = *short-circuit, forward-transfer admittance parameter*

The network for determining $\mathbf{y}_{21}$ appears in Fig. 26.19.

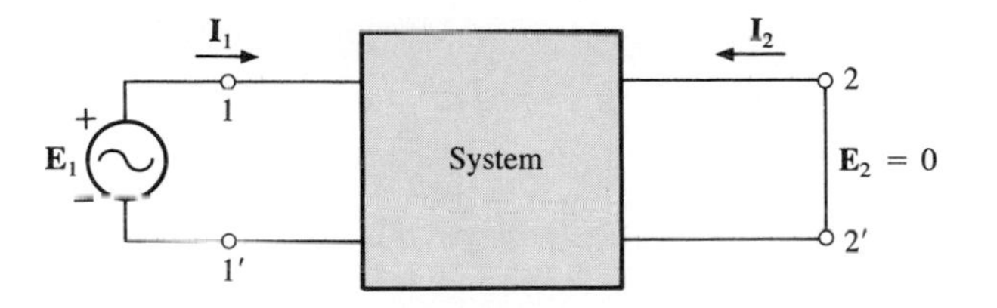


FIG. 26.19
$\mathbf{y}_{21}$ *determination.*

## $\mathbf{y}_{22}$

$$\mathbf{y}_{22} = \frac{\mathbf{I}_2}{\mathbf{E}_2} \Big|_{\mathbf{E}_1 = 0} \quad \text{(S)} \qquad (26.14)$$

$\mathbf{y}_{22}$ = *short-circuit, output-admittance parameter*

The required network appears in Fig. 26.20.

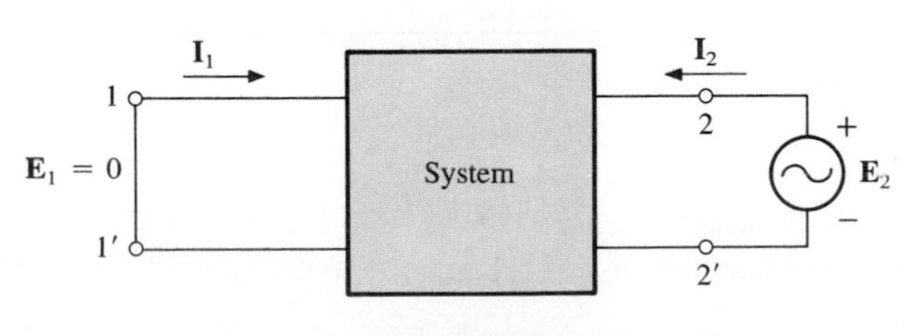


FIG. 26.20
$\mathbf{y}_{22}$ *determination.*

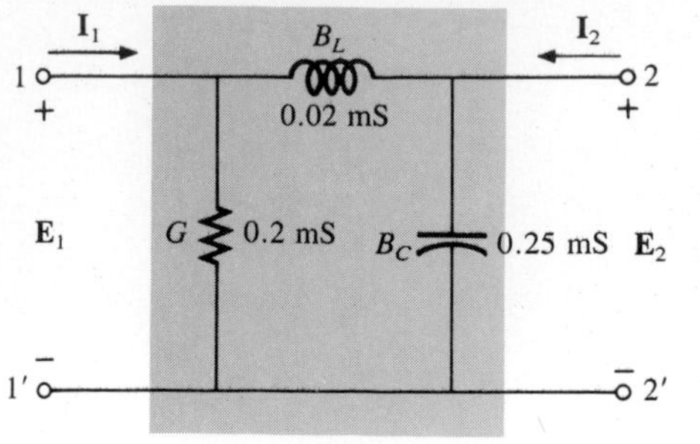


FIG. 26.21
*π network.*

**EXAMPLE 26.3.** Determine the admittance parameters for the $\pi$ network of Fig. 26.21.

***Solution:*** The network for $\mathbf{y}_{11}$ will appear as shown in Fig. 26.22, with

$$\mathbf{Y}_1 = 0.2 \text{ mS} \angle 0^\circ \quad \mathbf{Y}_2 = 0.02 \text{ mS} \angle -90^\circ \quad \mathbf{Y}_3 = 0.25 \text{ mS} \angle 90^\circ$$

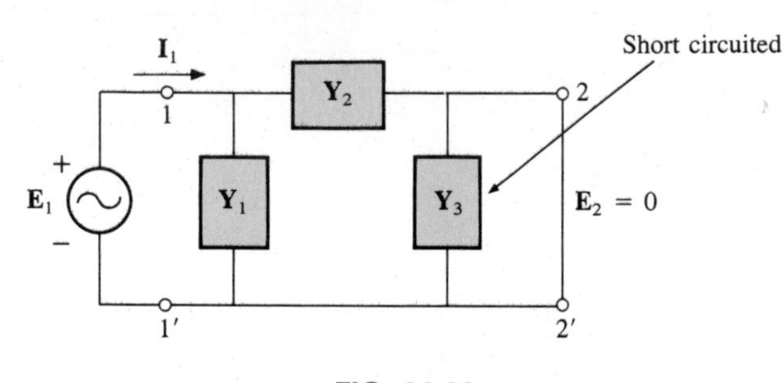

FIG. 26.22

We use

$$\mathbf{I}_1 = \mathbf{E}_1\mathbf{Y}_T = \mathbf{E}_1(\mathbf{Y}_1 + \mathbf{Y}_2)$$

Thus,

$$\mathbf{y}_{11} = \left.\frac{\mathbf{I}_1}{\mathbf{E}_1}\right|_{\mathbf{E}_2 = 0}$$

and

$$\boxed{\mathbf{y}_{11} = \mathbf{Y}_1 + \mathbf{Y}_2} \tag{26.15}$$

The determining network for $\mathbf{y}_{12}$ appears in Fig. 26.23. $\mathbf{Y}_1$ is short circuited; so $\mathbf{I}_{\mathbf{Y}_2} = \mathbf{I}_1$, and

$$\mathbf{I}_{\mathbf{Y}_2} = \mathbf{I}_1 = -\mathbf{E}_2\mathbf{Y}_2$$

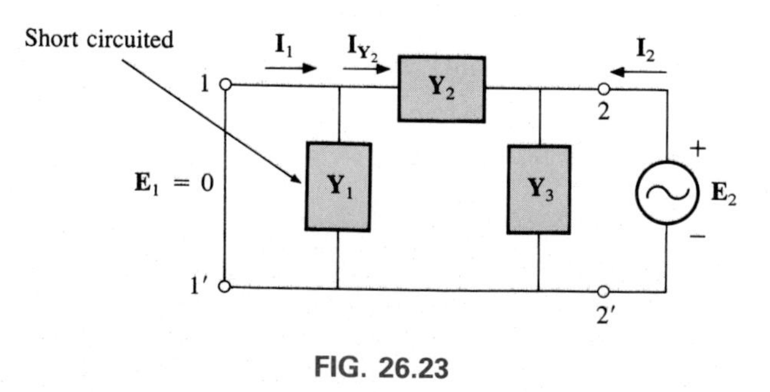


FIG. 26.23

The minus sign results because the defined direction of $\mathbf{I}_1$ in Fig. 26.23 is opposite to the actual flow direction due to the applied source $\mathbf{E}_2$; and

$$\mathbf{y}_{12} = \left.\frac{\mathbf{I}_1}{\mathbf{E}_2}\right|_{\mathbf{E}_1 = 0}$$

and

$$\boxed{\mathbf{y}_{12} = -\mathbf{Y}_2} \tag{26.16}$$

The network employed for $\mathbf{y}_{21}$ appears in Fig. 26.24. In this case, $\mathbf{Y}_3$ is short circuited, resulting in

$$\mathbf{I}_{\mathbf{Y}_2} = \mathbf{I}_2$$

and

$$\mathbf{I}_{\mathbf{Y}_2} = \mathbf{I}_2 = -\mathbf{E}_1\mathbf{Y}_2$$

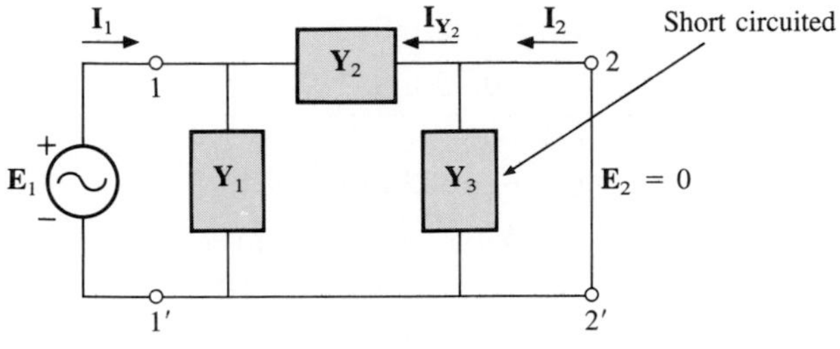

FIG. 26.24

Thus,

$$\mathbf{y}_{21} = \left.\frac{\mathbf{I}_2}{\mathbf{E}_1}\right|_{\mathbf{E}_2 = 0}$$

and

$$\boxed{\mathbf{y}_{21} = -\mathbf{Y}_2} \tag{26.17}$$

Note that for the $\pi$ configuration, $\mathbf{y}_{12} = \mathbf{y}_{21}$, which was expected, since the impedance parameters for the T network were such that $\mathbf{z}_{12} = \mathbf{z}_{21}$. A T network can be converted directly to a $\pi$ network using the Y-$\Delta$ transformation.

The determining network for $\mathbf{y}_{22}$ appears in Fig. 26.25:

$$\mathbf{Y}_T = \mathbf{Y}_2 + \mathbf{Y}_3$$

and

$$\mathbf{I}_2 = \mathbf{E}_2(\mathbf{Y}_2 + \mathbf{Y}_3)$$

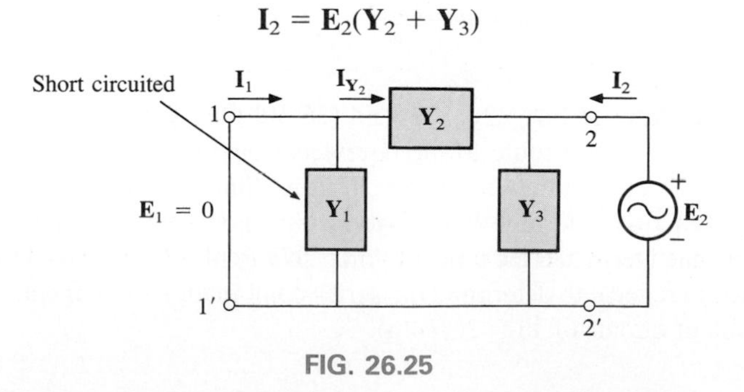


FIG. 26.25

Thus,

$$\mathbf{y}_{22} = \left.\frac{\mathbf{I}_2}{\mathbf{E}_2}\right|_{\mathbf{E}_1 = 0}$$

and

$$\boxed{\mathbf{y}_{22} = \mathbf{Y}_2 + \mathbf{Y}_3} \qquad \textbf{(26.18)}$$

Substituting values, we have

$$\mathbf{Y}_1 = 0.2 \text{ mS } \angle 0^\circ$$
$$\mathbf{Y}_2 = 0.02 \text{ mS } \angle -90^\circ$$
$$\mathbf{Y}_3 = 0.25 \text{ mS } \angle 90^\circ$$

$$\mathbf{y}_{11} = \mathbf{Y}_1 + \mathbf{Y}_2$$
$$= \mathbf{0.2\ mS} - \boldsymbol{j0.02}\ \mathbf{mS\ (ind.)}$$

$$\mathbf{y}_{12} = \mathbf{y}_{21} = -\mathbf{Y}_2 = -(-j0.02 \text{ mS})$$
$$= \boldsymbol{j0.02}\ \mathbf{mS\ (cap.)}$$

$$\mathbf{y}_{22} = \mathbf{Y}_2 + \mathbf{Y}_3 = -j0.02 \text{ mS} + j0.25 \text{ mS}$$
$$= \boldsymbol{j0.23}\ \mathbf{mS\ (cap.)}$$

---

Note the similarities between the results for $\mathbf{y}_{11}$ and $\mathbf{y}_{22}$ for the $\pi$ network compared with $\mathbf{z}_{11}$ and $\mathbf{z}_{22}$ for the T network.

Two networks satisfying the terminal relationships of Eqs. (26.10a) and (26.10b) are shown in Fig. 26.26. Note the use of parallel branches,

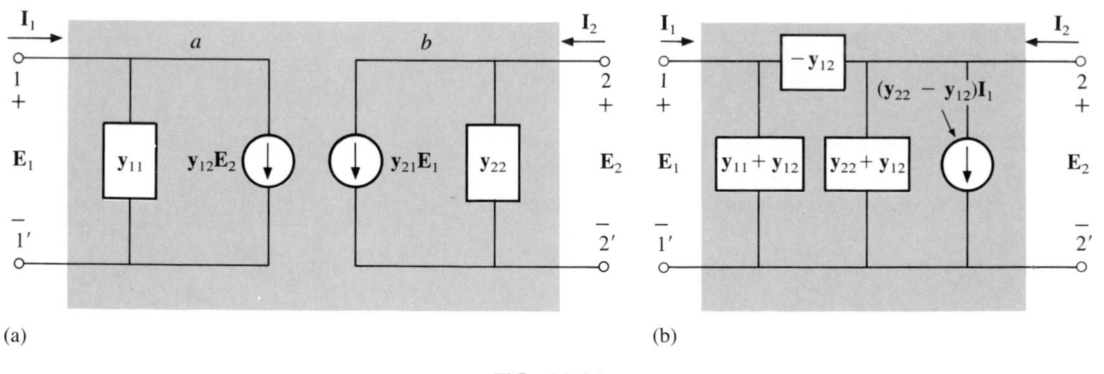


**FIG. 26.26**
*Two-port, **y**-parameter equivalent networks.*

since each term of Eqs. (26.10a) and (26.10b) has the units of current, and the most direct route to the equivalent circuit is an application of Kirchhoff's current law in reverse. That is, find the network that satisfies Kirchhoff's current law relationship. For the impedance parameters, each term had the units of volts, so Kirchhoff's voltage law was applied in reverse to determine the series combination of elements in the equivalent circuit of Fig. 26.14(a).

Applying Kirchhoff's current law to the network of Fig. 26.26(a), we have

$$\text{Node } a\text{: } \overbrace{\mathbf{I}_1}^{\text{entering}} = \overbrace{\mathbf{y}_{11}\mathbf{E}_1 + \mathbf{y}_{12}\mathbf{E}_2}^{\text{leaving}}$$
$$\text{Node } b\text{: } \mathbf{I}_2 = \mathbf{y}_{22}\mathbf{E}_2 + \mathbf{y}_{21}\mathbf{E}_1$$

which, when rearranged, are Eqs. (26.10a) and (26.10b).

For the results of Example 26.3, the network of Fig. 26.27 will result if the equivalent network of Fig. 26.26 is employed.

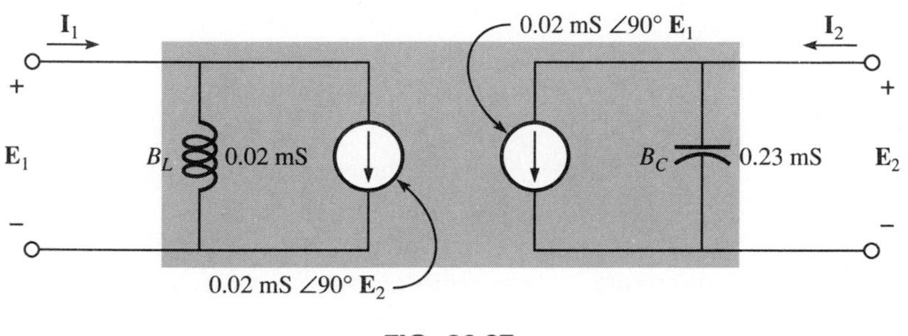


FIG. 26.27

## 26.4 HYBRID (h) PARAMETERS

The *hybrid* (**h**) *parameters* are employed extensively in the analysis of transistor networks. The term *hybrid* is derived from the fact that the parameters have a mixture of units (a hybrid set) rather than a single unit of measurement such as ohms or siemens, used for the **z** and **y** parameters, respectively. The defining hybrid equations have a mixture of current *and* voltage variables on one side, as follows:

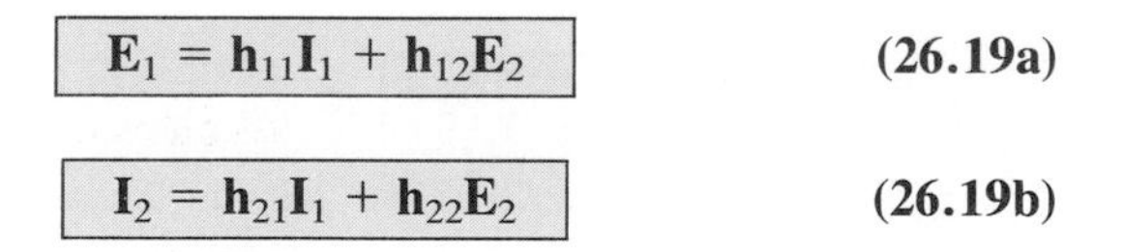

$$\mathbf{E}_1 = \mathbf{h}_{11}\mathbf{I}_1 + \mathbf{h}_{12}\mathbf{E}_2 \qquad (26.19a)$$

$$\mathbf{I}_2 = \mathbf{h}_{21}\mathbf{I}_1 + \mathbf{h}_{22}\mathbf{E}_2 \qquad (26.19b)$$

To determine the hybrid parameters, it will be necessary to establish both the short-circuit and the open-circuit condition, depending on the parameter desired.

### $\mathbf{h}_{11}$

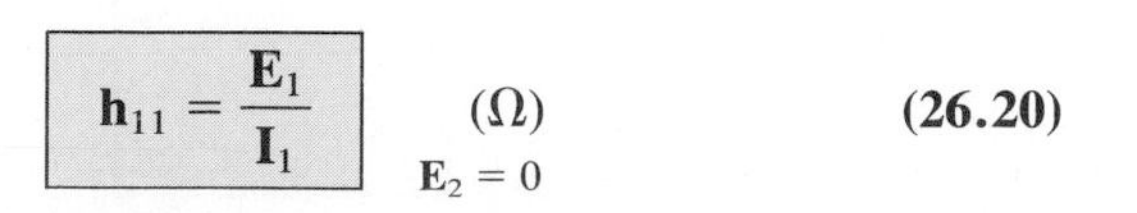

$$\mathbf{h}_{11} = \frac{\mathbf{E}_1}{\mathbf{I}_1} \bigg|_{\mathbf{E}_2 = 0} \quad (\Omega) \qquad (26.20)$$

$\mathbf{h}_{11}$ = *short-circuit, input-impedance parameter*

The determining network is shown in Fig. 26.28.

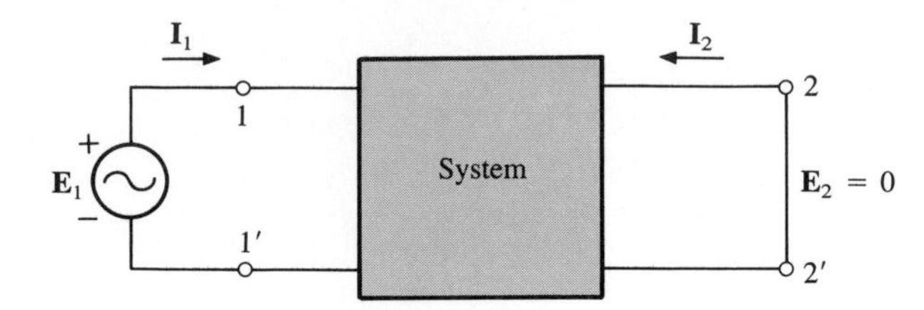


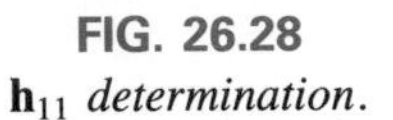
FIG. 26.28
$\mathbf{h}_{11}$ *determination.*

## $\mathbf{h}_{12}$

$$\mathbf{h}_{12} = \frac{\mathbf{E}_1}{\mathbf{E}_2} \Big|_{\mathbf{I}_1 = 0} \quad \text{(dimensionless)} \qquad \textbf{(26.21)}$$

$\mathbf{h}_{12}$ = *open-circuit, **r**everse-transfer voltage ratio parameter*

The network employed in determining $\mathbf{h}_{12}$ is shown in Fig. 26.29.

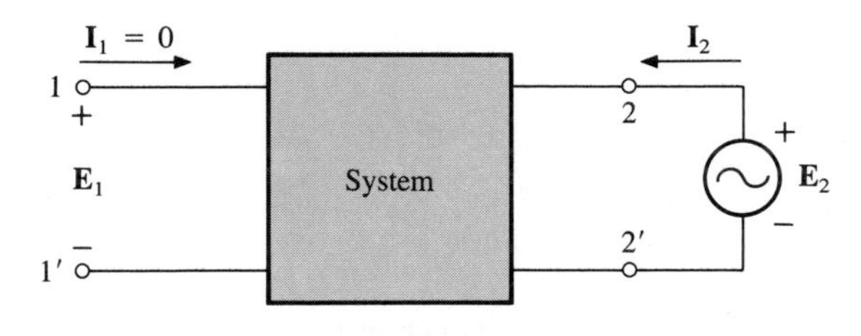


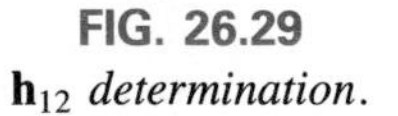
FIG. 26.29
$\mathbf{h}_{12}$ *determination.*

## $\mathbf{h}_{21}$

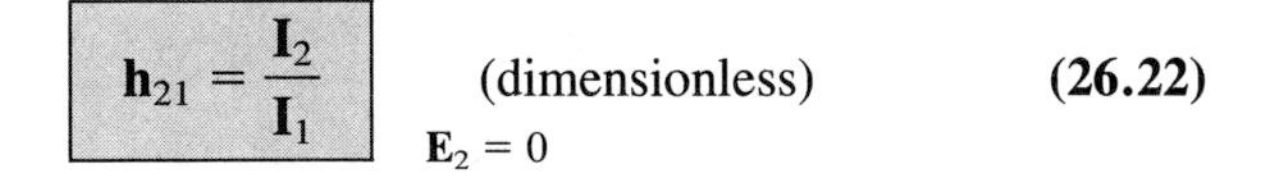

$$\mathbf{h}_{21} = \frac{\mathbf{I}_2}{\mathbf{I}_1} \Big|_{\mathbf{E}_2 = 0} \quad \text{(dimensionless)} \qquad \textbf{(26.22)}$$

$\mathbf{h}_{21}$ = *short-circuit, **f**orward-transfer current ratio parameter*

The determining network appears in Fig. 26.30.

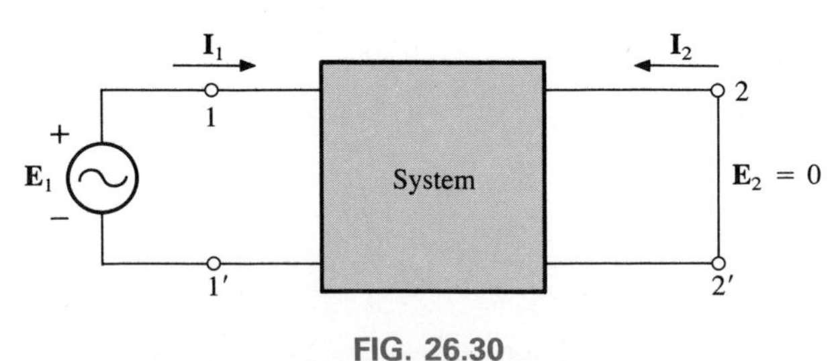


FIG. 26.30
$\mathbf{h}_{21}$ *determination.*

## $\mathbf{h}_{22}$

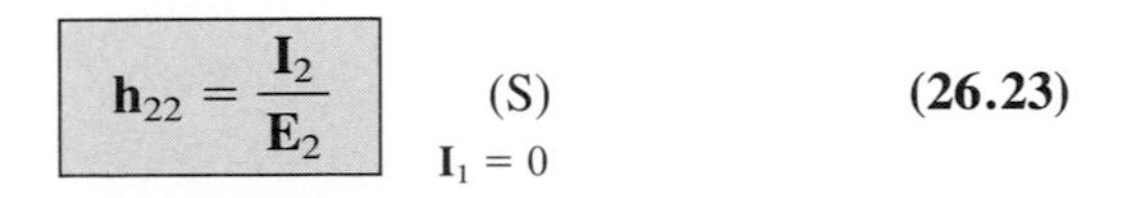

$$\mathbf{h}_{22} = \left.\frac{\mathbf{I}_2}{\mathbf{E}_2}\right|_{\mathbf{I}_1 = 0} \quad (S) \qquad \textbf{(26.23)}$$

$\mathbf{h}_{22}$ = *open-circuit, **output** admittance parameter*

The network employed to determine $\mathbf{h}_{22}$ is shown in Fig. 26.31.

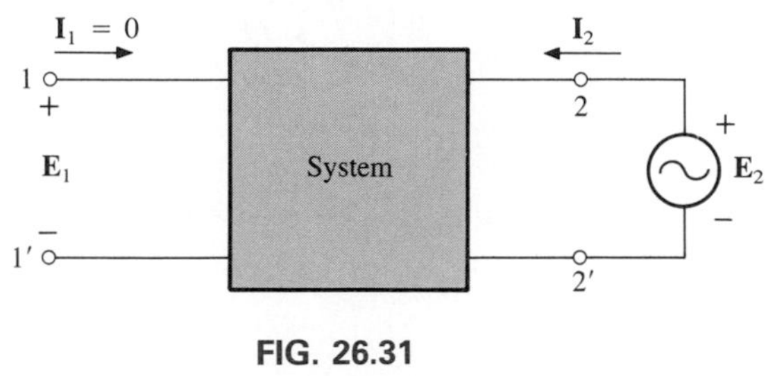


**FIG. 26.31**
$\mathbf{h}_{22}$ *determination.*

The subscript notation for the hybrid parameters is reduced to the following for most applications. The letter chosen is that letter appearing in boldface in the preceding description of each parameter:

$$\mathbf{h}_{11} = \mathbf{h}_i \qquad \mathbf{h}_{12} = \mathbf{h}_r \qquad \mathbf{h}_{21} = \mathbf{h}_f \qquad \mathbf{h}_{22} = \mathbf{h}_o$$

The hybrid equivalent circuit appears in Fig. 26.32. Since the unit of measurement for each term of Eq. (26.19a) is the volt, Kirchhoff's

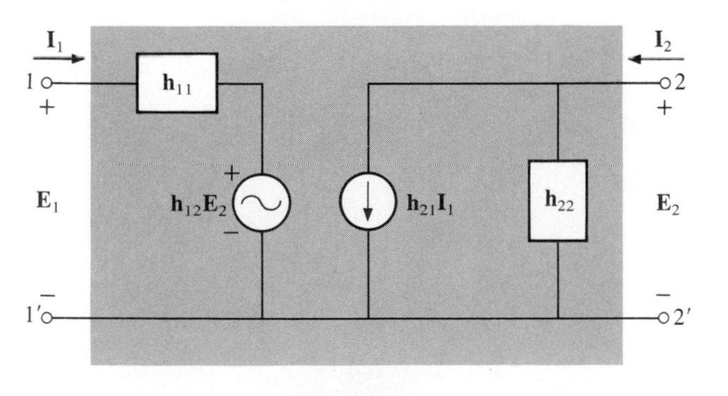


**FIG. 26.32**
*Two-port, hybrid-parameter equivalent network.*

voltage law was applied in reverse to obtain the series input circuit indicated. The unit of measurement of each term of Eq. (26.19b) has the units of current, resulting in the parallel elements of the output circuit as obtained by applying Kirchhoff's current law in reverse.

Note that the input circuit has a voltage-controlled voltage source whose controlling voltage is the output terminal voltage, while the output circuit has a current-controlled current source whose controlling current is the current of the input circuit.

**EXAMPLE 26.4.** For the hybrid equivalent circuit of Fig. 26.33:

a. Determine the current ratio (gain) $\mathbf{I}_2/\mathbf{I}_1$.
b. Determine the voltage ratio (gain) $\mathbf{E}_2/\mathbf{E}_1$.

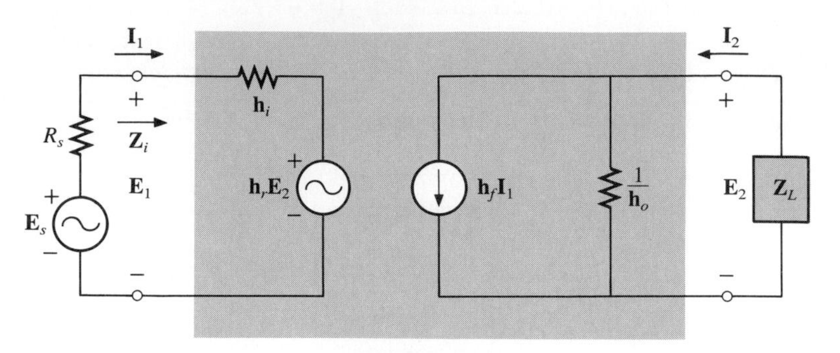


FIG. 26.33

***Solutions:***

a. Using the current divider rule, we have

$$\mathbf{I}_2 = \frac{(1/\mathbf{h}_o)\mathbf{h}_f\mathbf{I}_1}{(1/\mathbf{h}_o) + \mathbf{Z}_L} = \frac{\mathbf{h}_f\mathbf{I}_1}{1 + \mathbf{h}_o\mathbf{Z}_L}$$

and

$$A_i = \frac{\mathbf{I}_2}{\mathbf{I}_1} = \frac{\mathbf{h}_f}{1 + \mathbf{h}_o\mathbf{Z}_L} \tag{26.24}$$

b. Applying Kirchhoff's voltage law to the input circuit gives us

$$\mathbf{E}_1 - \mathbf{h}_i\mathbf{I}_1 - \mathbf{h}_r\mathbf{E}_2 = 0$$

and

$$\mathbf{I}_1 = \frac{\mathbf{E}_1 - \mathbf{h}_r\mathbf{E}_2}{\mathbf{h}_i}$$

Apply Kirchhoff's current law to the output circuit:

$$\mathbf{I}_2 = \mathbf{h}_f\mathbf{I}_1 + \mathbf{h}_o\mathbf{E}_2$$

However,

$$\mathbf{I}_2 = -\frac{\mathbf{E}_2}{\mathbf{Z}_L}$$

so

$$-\frac{\mathbf{E}_2}{\mathbf{Z}_L} = \mathbf{h}_f\mathbf{I}_1 + \mathbf{h}_o\mathbf{E}_2$$

Substituting for $\mathbf{I}_1$ gives us

$$-\frac{\mathbf{E}_2}{\mathbf{Z}_L} = \mathbf{h}_f\left(\frac{\mathbf{E}_1 - \mathbf{h}_r\mathbf{E}_2}{\mathbf{h}_i}\right) + \mathbf{h}_o\mathbf{E}_2$$

or

$$\mathbf{h}_i\mathbf{E}_2 = -\mathbf{h}_f\mathbf{Z}_L\mathbf{E}_1 + \mathbf{h}_r\mathbf{h}_f\mathbf{Z}_L\mathbf{E}_2 - \mathbf{h}_i\mathbf{h}_o\mathbf{Z}_L\mathbf{E}_2$$

and

$$\mathbf{E}_2(\mathbf{h}_i - \mathbf{h}_r\mathbf{h}_f\mathbf{Z}_L + \mathbf{h}_i\mathbf{h}_o\mathbf{Z}_L) = -\mathbf{h}_f\mathbf{Z}_L\mathbf{E}_1$$

with the result that

$$A_v = \frac{\mathbf{E}_2}{\mathbf{E}_1} = \frac{-\mathbf{h}_f\mathbf{Z}_L}{\mathbf{h}_i(1 + \mathbf{h}_o\mathbf{Z}_L) - \mathbf{h}_r\mathbf{h}_f\mathbf{Z}_L} \qquad \textbf{(26.25)}$$

**EXAMPLE 26.5.** For a particular transistor, $\mathbf{h}_i = 1\ \text{k}\Omega$, $\mathbf{h}_r = 4 \times 10^{-4}$, $\mathbf{h}_f = 50$, and $\mathbf{h}_o = 25\ \mu\text{s}$. Determine the current and the voltage gain if $\mathbf{Z}_L$ is a 2-kΩ resistive load.

***Solution:***

$$A_i = \frac{\mathbf{h}_f}{1 + \mathbf{h}_o\mathbf{Z}_L} = \frac{50}{1 + (25\ \mu\text{S})(2\ \text{k}\Omega)}$$

$$A_i = \frac{50}{1 + (50 \times 10^{-3})} = \frac{50}{1.050} = \mathbf{47.62}$$

$$A_v = \frac{-\mathbf{h}_f\mathbf{Z}_L}{\mathbf{h}_i(1 + \mathbf{h}_o\mathbf{Z}_L) - \mathbf{h}_r\mathbf{h}_f\mathbf{Z}_L}$$

$$= \frac{-(50)(2\ \text{k}\Omega)}{(1\ \text{k}\Omega)(1.050) - (4 \times 10^{-4})(50)(2\ \text{k}\Omega)}$$

$$A_v = \frac{-100 \times 10^3}{(1.050 \times 10^3) - (0.04 \times 10^3)} = -\frac{100}{1.01} = \mathbf{-99}$$

The minus sign simply indicates a phase shift of 180° between $\mathbf{E}_2$ and $\mathbf{E}_1$ for the defined polarities in Fig. 26.33.

## 26.5 INPUT AND OUTPUT IMPEDANCES

The input and output impedances will now be determined for the hybrid equivalent circuit and a **z** parameter equivalent circuit. The input impedance can always be determined by the ratio of the input voltage to the input current with or without a load applied. The output impedance is always determined with the source voltage or current set to zero. We found in the previous section that for the hybrid equivalent circuit of Fig. 26.33,

$$\mathbf{E}_1 = \mathbf{h}_i\mathbf{I}_1 + \mathbf{h}_r\mathbf{E}_2$$
$$\mathbf{E}_2 = -\mathbf{I}_2\mathbf{Z}_L$$

and

$$\frac{\mathbf{I}_2}{\mathbf{I}_1} = \frac{\mathbf{h}_f}{1 + \mathbf{h}_o\mathbf{Z}_L}$$

By substituting for $\mathbf{I}_2$ in the second equation (using the relationship of the last equation), we have

$$\mathbf{E}_2 = -\left(\frac{\mathbf{h}_f\mathbf{I}_1}{1 + \mathbf{h}_o\mathbf{Z}_L}\right)\mathbf{Z}_L$$

so the first equation becomes

$$\mathbf{E}_1 = \mathbf{h}_i\mathbf{I}_1 + \mathbf{h}_r\left(-\frac{\mathbf{h}_f\mathbf{I}_1\mathbf{Z}_L}{1 + \mathbf{h}_o\mathbf{Z}_L}\right)$$

and

$$\mathbf{E}_1 = \mathbf{I}_1\left(\mathbf{h}_i - \frac{\mathbf{h}_r\mathbf{h}_f\mathbf{Z}_L}{1 + \mathbf{h}_o\mathbf{Z}_L}\right)$$

Thus,

$$\mathbf{Z}_i = \frac{\mathbf{E}_1}{\mathbf{I}_1} = \mathbf{h}_i - \frac{\mathbf{h}_r\mathbf{h}_f\mathbf{Z}_L}{1 + \mathbf{h}_o\mathbf{Z}_L} \qquad \textbf{(26.26)}$$

For the output impedance, we will set the source voltage to zero but preserve its internal resistance $R_s$ as shown in Fig. 26.34.

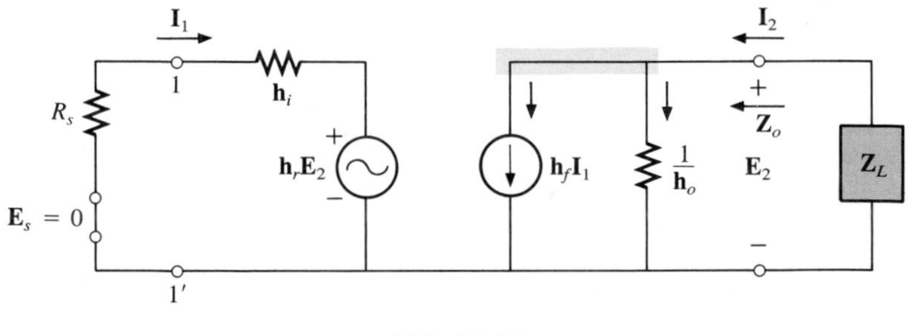

FIG. 26.34

Since

$$\mathbf{E}_s = 0$$

then

$$\mathbf{I}_1 = -\frac{\mathbf{h}_r\mathbf{E}_2}{\mathbf{h}_i + R_s}$$

From the output circuit,

$$\mathbf{I}_2 = \mathbf{h}_f\mathbf{I}_1 + \mathbf{h}_o\mathbf{E}_2$$

or

$$\mathbf{I}_2 = \mathbf{h}_f\left(-\frac{\mathbf{h}_r\mathbf{E}_2}{\mathbf{h}_i + R_s}\right) + \mathbf{h}_o\mathbf{E}_2$$

and

$$\mathbf{I}_2 = \left(-\frac{\mathbf{h}_r\mathbf{h}_f}{\mathbf{h}_i + R_s} + \mathbf{h}_o\right)\mathbf{E}_2$$

Thus,

$$\mathbf{Z}_o = \frac{\mathbf{E}_2}{\mathbf{I}_2} = \frac{1}{\mathbf{h}_o - \dfrac{\mathbf{h}_r\mathbf{h}_f}{\mathbf{h}_i + R_s}} \qquad \textbf{(26.27)}$$

**EXAMPLE 26.6.** Determine $\mathbf{Z}_i$ and $\mathbf{Z}_o$ for the transistor having the parameters of Example 26.5 if $R_s = 1\ \text{k}\Omega$.

***Solution:***

$$\mathbf{Z}_i = \mathbf{h}_i - \frac{\mathbf{h}_r\mathbf{h}_f\mathbf{Z}_L}{1 + \mathbf{h}_o\mathbf{Z}_L} = 1\ \text{k}\Omega - \frac{0.04\ \text{k}\Omega}{1.050}$$
$$= 1 \times 10^3 - 0.0381 \times 10^3 = \mathbf{961.9\ \Omega}$$

$$\mathbf{Z}_o = \frac{1}{\mathbf{h}_o - \dfrac{\mathbf{h}_r\mathbf{h}_f}{\mathbf{h}_i + R_s}} = \frac{1}{25\ \mu\text{S} - \dfrac{(4 \times 10^{-4})(50)}{1\ \text{k}\Omega + 1\ \text{k}\Omega}}$$
$$= \frac{1}{25 \times 10^{-6} - \dfrac{200 \times 10^{-4}}{2 \times 10^3}}$$
$$= \frac{1}{25 \times 10^{-6} - 10 \times 10^{-6}}$$
$$\mathbf{Z}_o = \frac{1}{15 \times 10^{-6}} = \mathbf{66.67\ k\Omega}$$

For the **z** parameter equivalent circuit of Fig. 26.35,

$$\mathbf{I}_2 = \frac{-\mathbf{z}_{21}\mathbf{I}_1}{\mathbf{z}_{22} + \mathbf{Z}_L}$$

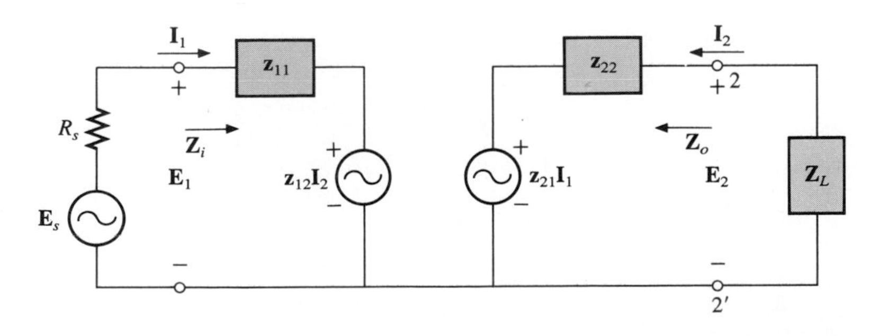


FIG. 26.35

and

$$\mathbf{I}_1 = \frac{\mathbf{E}_1 - \mathbf{z}_{12}\mathbf{I}_2}{\mathbf{z}_{11}}$$

or

$$\mathbf{E}_1 = \mathbf{z}_{11}\mathbf{I}_1 + \mathbf{z}_{12}\mathbf{I}_2 = \mathbf{z}_{11}\mathbf{I}_1 + \mathbf{z}_{12}\left(\frac{-\mathbf{z}_{21}\mathbf{I}_1}{\mathbf{z}_{22} + \mathbf{Z}_L}\right)$$

and

$$\mathbf{Z}_i = \frac{\mathbf{E}_1}{\mathbf{I}_1} = \mathbf{z}_{11} - \frac{\mathbf{z}_{12}\mathbf{z}_{21}}{\mathbf{z}_{22} + \mathbf{Z}_L} \tag{26.28}$$

For the output impedance, $\mathbf{E}_s = 0$, and

$$\mathbf{I}_1 = -\frac{\mathbf{z}_{12}\mathbf{I}_2}{R_s + \mathbf{z}_{11}} \quad \text{and} \quad \mathbf{I}_2 = \frac{\mathbf{E}_2 - \mathbf{z}_{21}\mathbf{I}_1}{\mathbf{z}_{22}}$$

or

$$\mathbf{E}_2 = \mathbf{z}_{22}\mathbf{I}_2 + \mathbf{z}_{21}\mathbf{I}_1 = \mathbf{z}_{22}\mathbf{I}_2 + \mathbf{z}_{21}\left(-\frac{\mathbf{z}_{12}\mathbf{I}_2}{R_s + \mathbf{z}_{11}}\right)$$

and

$$\mathbf{E}_2 = \mathbf{z}_{22}\mathbf{I}_2 - \frac{\mathbf{z}_{12}\mathbf{z}_{21}\mathbf{I}_2}{R_s + \mathbf{z}_{11}}$$

Thus,

$$\mathbf{Z}_o = \frac{\mathbf{E}_2}{\mathbf{I}_2} = \mathbf{z}_{22} - \frac{\mathbf{z}_{12}\mathbf{z}_{21}}{R_s + \mathbf{z}_{11}} \tag{26.29}$$

## 26.6 CONVERSION BETWEEN PARAMETERS

The equations relating the **z** and **y** parameters can be determined directly from Eqs. (26.1) and (26.10). For Eqs. (26.10a) and (26.10b),

$$\mathbf{I}_1 = \mathbf{y}_{11}\mathbf{E}_1 + \mathbf{y}_{12}\mathbf{E}_2$$
$$\mathbf{I}_2 = \mathbf{y}_{21}\mathbf{E}_1 + \mathbf{y}_{22}\mathbf{E}_2$$

The use of determinants will result in

$$\mathbf{E}_1 = \frac{\begin{vmatrix} \mathbf{I}_1 & \mathbf{y}_{12} \\ \mathbf{I}_2 & \mathbf{y}_{22} \end{vmatrix}}{\begin{vmatrix} \mathbf{y}_{11} & \mathbf{y}_{12} \\ \mathbf{y}_{21} & \mathbf{y}_{22} \end{vmatrix}} = \frac{\mathbf{y}_{22}\mathbf{I}_1 - \mathbf{y}_{12}\mathbf{I}_2}{\mathbf{y}_{11}\mathbf{y}_{22} - \mathbf{y}_{12}\mathbf{y}_{21}}$$

Substituting the notation

$$\Delta_y = y_{11}y_{22} - y_{12}y_{21}$$

we have

$$E_1 = \frac{y_{22}}{\Delta_y}I_1 - \frac{y_{12}}{\Delta_y}I_2$$

which, when related to Eq. (26.1a),

$$E_1 = z_{11}I_1 + z_{12}I_2$$

indicates that

$$z_{11} = \frac{y_{22}}{\Delta_y} \quad \text{and} \quad z_{12} = \frac{y_{12}}{\Delta_y}$$

and, similarly,

$$z_{21} = \frac{y_{21}}{\Delta_y} \quad \text{and} \quad z_{22} = \frac{y_{11}}{\Delta_y}$$

For the conversion of **z** parameters to the admittance domain, determinants are applied to Eqs. (26.1a) and (26.1b). The impedance parameters can be found in terms of the hybrid parameters by first forming the determinant for $I_1$ from the hybrid equations,

$$E_1 = h_{11}I_1 + h_{12}E_2$$
$$I_2 = h_{21}I_1 + h_{22}E_2$$

That is,

$$I_1 = \frac{\begin{vmatrix} E_1 & h_{12} \\ I_2 & h_{22} \end{vmatrix}}{\begin{vmatrix} h_{11} & h_{12} \\ h_{21} & h_{22} \end{vmatrix}} = \frac{h_{22}}{\Delta_h}E_1 - \frac{h_{12}}{\Delta_h}I_2$$

and

$$\frac{h_{22}}{\Delta_h}E_1 = I_1 - \frac{h_{12}}{\Delta_h}I_2$$

or

$$E_1 = \frac{\Delta_h I_1}{h_{22}} - \frac{h_{12}}{h_{22}}I_2$$

which, when related to the impedance parameter equation,

$$E_1 = z_{11}I_1 + z_{12}I_2$$

indicates that

$$z_{11} = \frac{\Delta_h}{h_{22}} \quad \text{and} \quad z_{12} = -\frac{h_{12}}{h_{22}}$$

The remaining conversions are left as an exercise. A complete table of conversions appears in Table 26.1.

**TABLE 26.1**
*Conversions between* **z**, **y**, *and* **h** *parameters.*

| FROM → / TO ↓ | z | y | h |
|---|---|---|---|
| **z** | $\begin{matrix} z_{11} & z_{12} \\ z_{21} & z_{22} \end{matrix}$ | $\begin{matrix} \frac{y_{22}}{\Delta_y} & \frac{-y_{12}}{\Delta_y} \\ \frac{-y_{21}}{\Delta_y} & \frac{y_{11}}{\Delta_y} \end{matrix}$ | $\begin{matrix} \frac{\Delta_h}{h_{22}} & \frac{h_{12}}{h_{22}} \\ \frac{-h_{21}}{h_{22}} & \frac{1}{h_{22}} \end{matrix}$ |
| **y** | $\begin{matrix} \frac{z_{22}}{\Delta_z} & \frac{-z_{12}}{\Delta_z} \\ \frac{-z_{21}}{\Delta_z} & \frac{z_{11}}{\Delta_z} \end{matrix}$ | $\begin{matrix} y_{11} & y_{12} \\ y_{21} & y_{22} \end{matrix}$ | $\begin{matrix} \frac{1}{h_{11}} & \frac{-h_{12}}{h_{11}} \\ \frac{h_{21}}{h_{11}} & \frac{\Delta_h}{h_{11}} \end{matrix}$ |
| **h** | $\begin{matrix} \frac{\Delta_z}{z_{22}} & \frac{z_{12}}{z_{22}} \\ \frac{-z_{21}}{z_{22}} & \frac{1}{z_{22}} \end{matrix}$ | $\begin{matrix} \frac{1}{y_{11}} & \frac{-y_{12}}{y_{11}} \\ \frac{y_{21}}{y_{11}} & \frac{\Delta_y}{y_{11}} \end{matrix}$ | $\begin{matrix} h_{11} & h_{12} \\ h_{21} & h_{22} \end{matrix}$ |

## 26.7 IMPACT OF VARIOUS PARAMETERS

The important parameters of any two-port electrical/electronics system are $A_v$, $A_i$, $Z_i$, and $Z_o$. A knowledge of all four often provides sufficient information to work effectively with the system in the design of a single- or multistage area of application.

The input impedance as defined by Fig. 26.36(a) is the impedance presented by the system to an applied voltage or current source. As shown in Fig. 26.36(b), the system can effectively be replaced by an

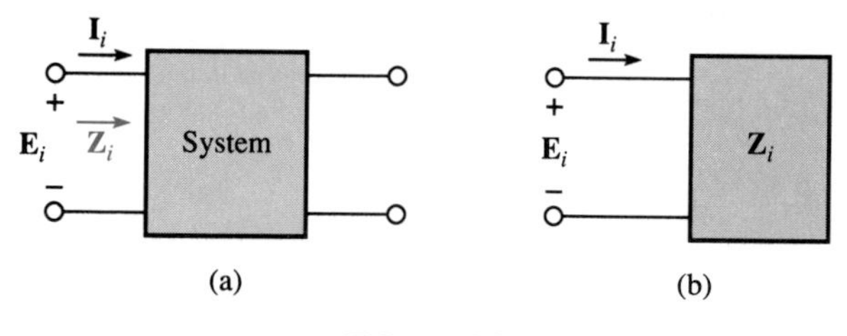


FIG. 26.36

impedance $\mathbf{Z}_i$ if the relationship between $\mathbf{E}_i$ and $\mathbf{I}_i$ is the only area of interest. Through Ohm's law we find that

$$\mathbf{E}_i = \mathbf{I}_i\mathbf{Z}_i \tag{26.30}$$

or

$$\mathbf{I}_i = \frac{\mathbf{E}_i}{\mathbf{Z}_i} \tag{26.31}$$

The input impedance can be determined experimentally by inserting a sensing resistor, as shown in Fig. 26.37, and applying a voltage $\mathbf{V}_g$. The input current can then be found using Ohm's law $\mathbf{I}_i = \mathbf{V}_R/\mathbf{R} = (\mathbf{V}_g - \mathbf{E}_i)/\mathbf{R}$ and the input impedance from $\mathbf{Z}_i = \mathbf{E}_i/\mathbf{I}_i$.

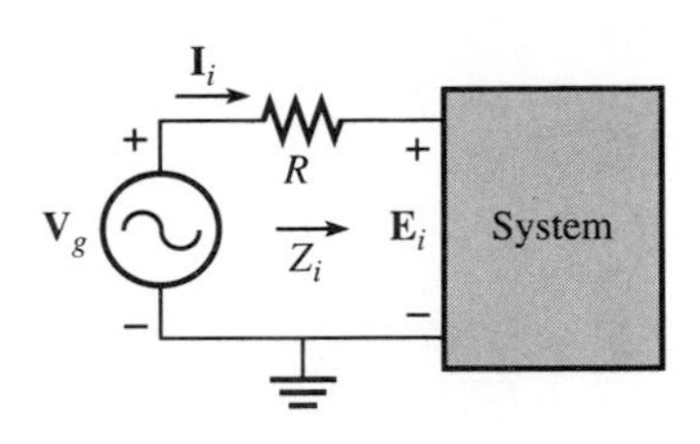

FIG. 26.37

As noted in earlier sections, the output impedance is determined with the applied source set to zero, as shown in Fig. 26.38. The output

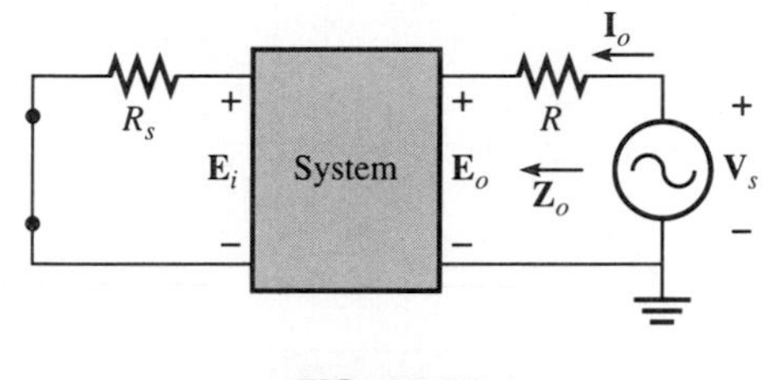

FIG. 26.38

impedance can then be determined experimentally by applying a source $\mathbf{V}_g$ and measuring the voltage $\mathbf{E}_o$. The result is that $\mathbf{I}_o = \mathbf{V}_R/\mathbf{R} = (\mathbf{V}_g - \mathbf{E}_o)/\mathbf{R}$ and $\mathbf{Z}_o = \mathbf{E}_o/\mathbf{I}_o$.

The voltage gain of a two-port system can be found under loaded or unloaded conditions. The current gain, however, requires a load impedance to establish an output current. The no-load voltage gain is defined by Fig. 26.39(a), whereas the loaded voltage and current gains are defined by Fig. 26.39(b). The voltage gain can be determined from $A_v = \mathbf{E}_o/\mathbf{E}_i$ and the current gain from $A_i = \mathbf{I}_o/\mathbf{I}_i$ with $\mathbf{I}_o = \mathbf{I}_L = \mathbf{V}_L/\mathbf{R}_L$ and $\mathbf{I}_i = \mathbf{V}_R/\mathbf{R} = (\mathbf{V}_g - \mathbf{E}_i)/\mathbf{R}$.

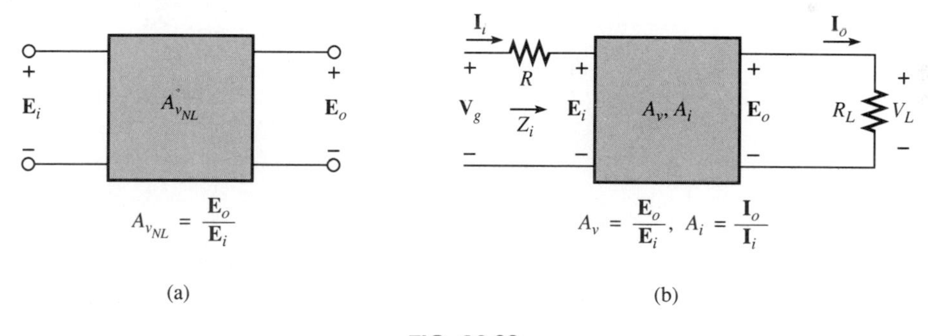


FIG. 26.39

For all two-port systems, keep in mind through the analysis to follow that the voltage gain and current gain are a function of the applied load as demonstrated by Eqs. (26.24) and (26.25) for the hybrid equivalent model. However, the output impedance $\mathbf{Z}_o$ and the no-load voltage gain $A_{v_{\text{NL}}}$ are independent of the applied load. This independence can be put to good use developing a Thevenin equivalent circuit for the output circuit as shown in Fig. 26.40.

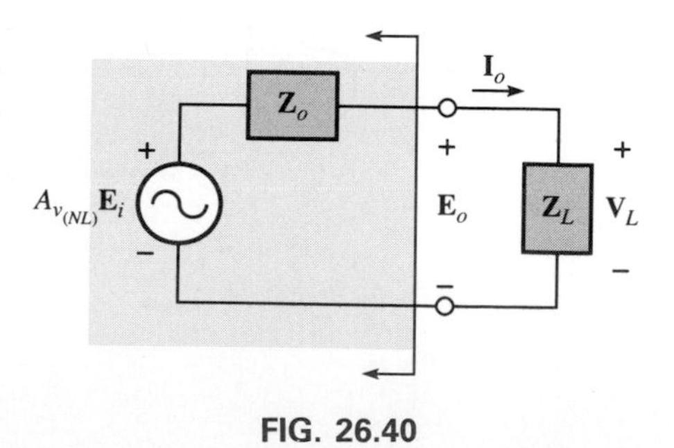


FIG. 26.40

Using Fig. 26.40 the load voltage can be determined using the voltage divider rule,

$$\mathbf{V}_L = \frac{\mathbf{Z}_L(A_{v_{\text{NL}}}\mathbf{E}_i)}{\mathbf{Z}_L + \mathbf{Z}_o}$$

and

$$\mathbf{V}_L = \frac{A_{v_{NL}}\mathbf{Z}_L\mathbf{E}_i}{\mathbf{Z}_L + \mathbf{Z}_o} \quad \textbf{(26.32)}$$

and the network voltage gain can be found from

$$A_v = \frac{\mathbf{E}_o}{\mathbf{E}_i} = \frac{\mathbf{V}_L}{\mathbf{E}_i} = \frac{\left(\dfrac{A_{v_{NL}}\mathbf{Z}_L\mathbf{E}_i}{\mathbf{Z}_L + \mathbf{Z}_o}\right)}{\mathbf{E}_i}$$

so that

$$A_v = \frac{A_{v_{NL}}\mathbf{Z}_L}{\mathbf{Z}_L + \mathbf{Z}_o} \quad \textbf{(26.33)}$$

The output current is

$$\mathbf{I}_o = \frac{A_{v_{NL}}\mathbf{E}_i}{\mathbf{Z}_o + \mathbf{Z}_L}$$

resulting in a current gain:

$$A_i = \frac{\mathbf{I}_o}{\mathbf{I}_i} = \frac{\dfrac{A_{v_{NL}}\mathbf{E}_i}{\mathbf{Z}_o + \mathbf{Z}_L}}{\dfrac{\mathbf{E}_i}{\mathbf{Z}_i}}$$

and

$$A_i = \frac{A_{v_{NL}}\mathbf{Z}_i}{\mathbf{Z}_o + \mathbf{Z}_L} \quad \textbf{(26.34)}$$

For two-port systems in cascade, as shown in Fig. 26.41, the important terminal parameters ($A_v$, $A_i$, $Z_i$, and $Z_o$) are affected by the loading of one stage on another. For example, the input impedance to the third

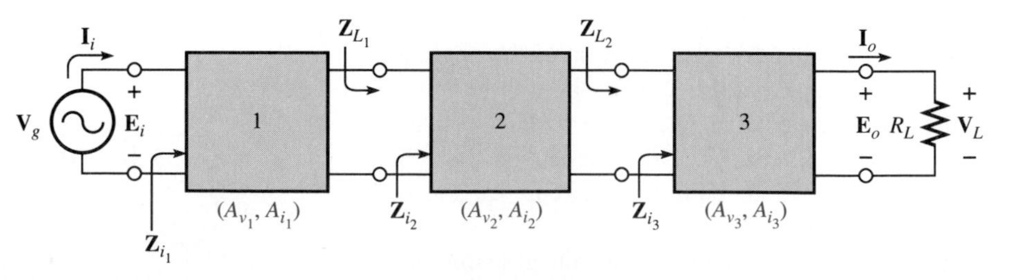


FIG. 26.41

stage is the load impedance for the second stage, and so on. There are, however, a number of equations that can be developed for the system of Fig. 26.41 if we define $A_v$, $A_i$, $Z_i$, and $Z_o$ to the levels of gain and impedance under loaded conditions.

For instance, the total voltage gain is determined by

$$A_{v_T} = A_{v_1} \cdot A_{v_2} \cdot A_{v_3} \tag{26.35}$$

For the total current gain:

$$A_{i_T} = A_{i_1} \cdot A_{i_2} \cdot A_{i_3} \tag{26.36}$$

The magnitude of the load voltage is determined by

$$V_L = I_o R_L$$

and the magnitude of the total voltage gain:

$$A_{v_T} = \frac{V_L}{V_g} = \frac{I_o R_L}{I_i Z_{i_1}}$$

so that

$$A_{v_T} = A_{i_T} \frac{R_L}{Z_{i_1}} \tag{26.37}$$

Eq. (26.37) permits determining the total voltage or current gain from the other quantity using the load and input impedance.

---

**EXAMPLE 26.7.** If the two-port system of Fig. 26.42(a) is employed with the source and load impedance of Fig. 26.42(b), determine:

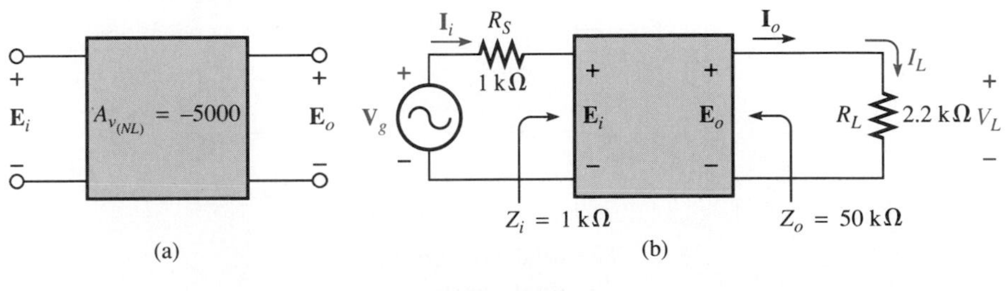


**FIG. 26.42**

a. The voltage gain $A_v = E_o/E_i$ for Fig. 26.42b.
b. The voltage gain $A_{v_T} = V_L/V_g$ for Fig. 26.42b.
c. The total current gain $A_{i_T} = I_o/I_i$.

***Solutions:***

a. Eq. (26.33):

$$A_v = \frac{E_o}{E_i} = \frac{A_{v_{NL}} R_L}{R_L + R_o} = \frac{(-5000)(2.2 \text{ k}\Omega)}{2.2 \text{ k}\Omega + 50 \text{ k}\Omega} = \mathbf{-210.73}$$

b. $V_L = E_o$
but

$$E_i = \frac{Z_i V_g}{Z_i + R_s} = \frac{(1 \text{ k}\Omega)(V_g)}{1 \text{ k}\Omega + 1 \text{ k}\Omega} = \frac{V_g}{2}$$

and

$$A_{v_T} = \frac{V_L}{V_g} = \left(\frac{V_L}{E_i}\right)\left(\frac{E_i}{V_g}\right) = \left(\frac{E_o}{E_i}\right)\left(\frac{E_i}{V_g}\right) = A_v\left(\frac{E_i}{V_g}\right)$$

$$= (-210.73)\left(\frac{1}{2}\right)$$

$$= \mathbf{-105.36}$$

c. Eq. (26.34):

$$A_i = \frac{I_o}{I_i} = \frac{A_{v_{NL}} Z_i}{Z_o + R_L} = \frac{-5000(1 \text{ k}\Omega)}{50 \text{ k}\Omega + 2.2 \text{ k}\Omega} = \mathbf{-95.79}$$

The negative sign in the solution for $A_{v_T}$ simply reveals that $\mathbf{V}_L$ and $\mathbf{V}_g$ are out of phase by 180°, as are $\mathbf{E}_o$ and $\mathbf{E}_i$, as specified for Fig. 26.42(a). The negative sign for $A_i$ reveals that $\mathbf{I}_o$ and $\mathbf{I}_L$ have a direction opposite to that indicated in Fig. 26.42(b)—in other words, a 180° phase shift between $\mathbf{I}_o$ and $\mathbf{I}_i$ for the directions indicated in Fig. 26.42(b).

**EXAMPLE 26.8.** For the two-stage system of Fig. 26.43, determine:

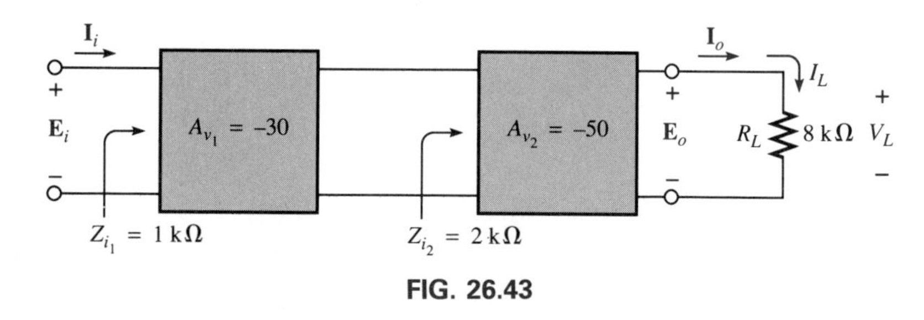


FIG. 26.43

a. The total voltage gain.
b. The total current gain.
c. The current gain of each stage.
d. The total current gain using the results of part (c). Compare to the results of part (b).

***Solutions:***

a. Eq. (26.35):

$$A_{v_T} = A_{v_1} \cdot A_{v_2} = (-30)(-50) = \mathbf{1500}$$

b. Eq. (26.37):

$$A_{i_T} = A_{v_T}\frac{Z_{i_1}}{R_L} = (1500)\frac{1\text{ k}\Omega}{8\text{ k}\Omega} = \mathbf{187.5}$$

c. $A_{i_1} = A_{v_1}\dfrac{Z_{i_1}}{Z_{i_2}} = (-30)\dfrac{1\text{ k}\Omega}{2\text{ k}\Omega} = \mathbf{-15}$

$A_{i_2} = A_{v_2}\dfrac{Z_{i_2}}{R_L} = (-50)\dfrac{2\text{ k}\Omega}{8\text{ k}\Omega} = \mathbf{-12.5}$

As a check:

$$A_{i_T} = A_{i_1} \cdot A_{i_2} = (-15)(-12.5) = 187.5 \qquad \text{(as above)}$$

## 26.8 COMPUTER ANALYSIS

The computer analysis of this chapter will be limited to a practice session in entering controlled sources into the PSPICE software package. The system to be analyzed is the loaded two-port hybrid equivalent model of Fig. 26.44. Note the insertion of a sensing voltage source to define the direction of $\mathbf{I}_1$ and the common reference point for both the input and output circuits. The desired output voltage $\mathbf{V}_L = \mathbf{E}_2$ will also be the controlling voltage for the VCVS of the input circuit.

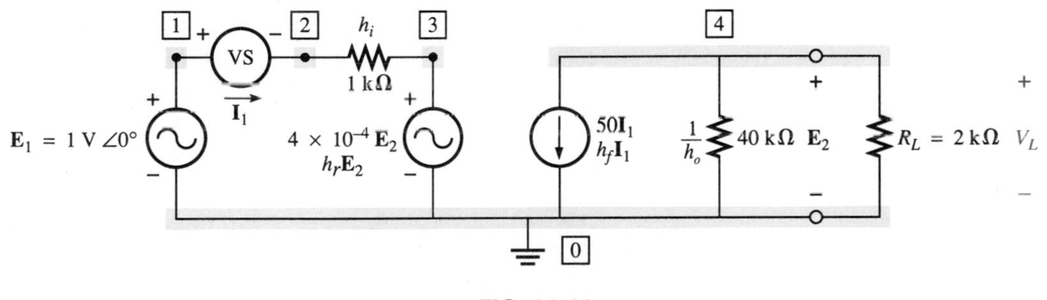


FIG. 26.44

Before examining the input and output files let us first determine the voltage gain using Eq. (26.25).

$$\begin{aligned}
A_v &= \frac{-h_fR_L}{h_i(1 + h_oR_L) - h_rh_fR_L} \\
&= \frac{-(50)(2\text{ k}\Omega)}{(1\text{ k}\Omega)(1 + (25 \times 10^{-6}\text{ S})(2\text{ k}\Omega)) - (4 \times 10^{-4})(50)(2\text{ k}\Omega)} \\
&= \frac{-100 \times 10^3}{(1\text{ k}\Omega)(1 + 50 \times 10^{-3}) - 40} = \frac{-100 \times 10^3}{1050 - 40} \\
&= \frac{-100 \times 10^3}{1010} = -99.01
\end{aligned}$$

and

$$A_v = \frac{\mathbf{E}_2}{\mathbf{E}_1} = \frac{\mathbf{V}_L}{\mathbf{E}_1}$$

so that

$$\mathbf{V}_L = A_v\mathbf{E}_1 = (-99.01)(1\text{ V }\angle 0°)$$
$$= \mathbf{99.01\text{ V }\angle 180°}$$

The input and output files appear in Fig. 26.45 with the required entries and the request for the load voltage. The output file provides an exact match with the solution just obtained.

```
************************ Evaluation PSpice (January 1989) ******* 17:27:55 *******

Chapter 26 - Two Port Networks - Hybrid Model

****     CIRCUIT DESCRIPTION

******************************************************************************


VI 1 0 AC 1V
VSENSE 1 2 0
R1 2 3 1K
ER 3 0 4 0 4E-4
FF 4 0 VSENSE 50
R2 4 0 40K
RL 4 0 2K
.AC LIN 1 1KH 1KH
.PRINT AC VM(4,0) VP(4,0)
.OPTIONS NOPAGE
.END

****     AC ANALYSIS                          TEMPERATURE =   27.000 DEG C


  FREQ          VM(4,0)       VP(4,0)


  1.000E+03    9.901E+01    1.800E+02
```

FIG. 26.45

The last PSPICE run of the section (and the text) will determine the output impedance for the network of Fig. 26.44 by setting the voltage source to zero volts, as shown in Fig. 26.46. A current source of 1 A is

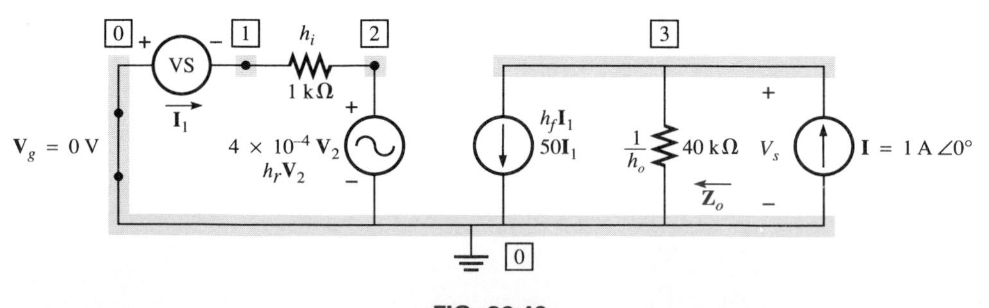


FIG. 26.46

applied, and the voltage $\mathbf{V}_s$ is determined; it has the same magnitude and angle as $\mathbf{Z}_o$, since

$$\mathbf{Z}_o = \frac{\mathbf{V}_s}{\mathbf{I}} = \frac{\mathbf{V}_s}{1\text{ A}\angle 0^\circ} = \frac{V_s}{1\text{ A}}\angle\theta_s - 0^\circ = V_s\,\Omega\angle\theta_s$$

Before reviewing the input and output files, let us again determine the solution using the longhand approach to check the computer result.

Eq. (26.27):

$$Z_o = \cfrac{1}{h_o - \cfrac{h_r h_f}{h_i + R_s}}$$

$$= \cfrac{1}{25 \times 10^{-6}\text{ S} - \cfrac{(4 \times 10^{-4})(50)}{1\text{ k}\Omega + 0}} = \cfrac{1}{25 \times 10^{-6}\text{ S} - \cfrac{200 \times 10^{-4}}{1\text{ k}\Omega}}$$

$$= \frac{1}{25 \times 10^{-6}\text{ S} - 20 \times 10^{-6}\text{ S}} = \frac{1}{5 \times 10^{-6}\text{ S}}$$

$$= \mathbf{200\text{ k}\Omega}$$

The input file of Fig. 26.47 include the applied current source of 1 A at zero degrees and uses a frequency of 1 kHz to establish the ac analy-

```
************************* Evaluation PSpice (January 1989) ******* 19:49:30 *******

Chapter 26 - Two Port Networks - Output Impedance

****     CIRCUIT DESCRIPTION

*********************************************************************************

VS 0 1 0
R1 1 2 1K
ER 2 0 3 0 4E-4
FF 3 0 VS 50
R2 3 0 40K
IS 0 3 AC 1 0
.AC LIN 1 1K 1K
.PRINT AC VM(3,0)
.OPTIONS NOPAGE
.END

****     AC ANALYSIS                          TEMPERATURE =   27.000 DEG C

  FREQ        VM(3,0)

   1.000E+03   2.000E+05
```

FIG. 26.47

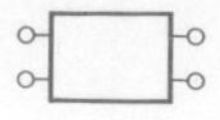

sis. In particular, note the ability to carry the reference node through to the sensing voltage source and the need to redefine the nodes of Fig. 26.44. The magnitude of the resulting voltage $\mathbf{V}_s$ is again an exact match with the above solution.

## PROBLEMS

### SECTION 26.2 Impedance (z) Parameters

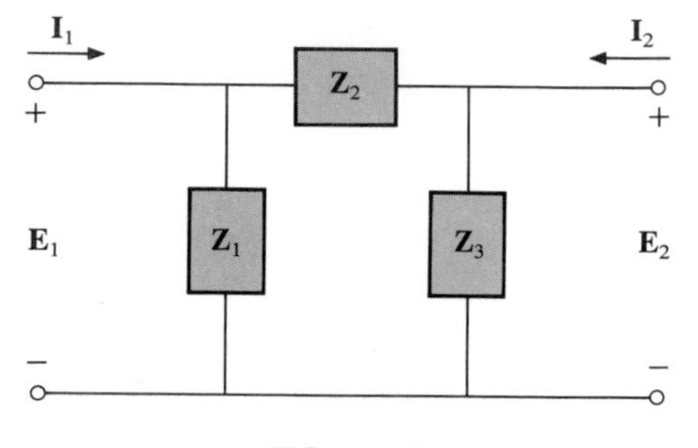


FIG. 26.48

1. **a.** Determine the impedance (**z**) parameters for the $\pi$ network of Fig. 26.48.
   **b.** Sketch the **z** parameter equivalent circuit (using either form of Fig. 26.14).

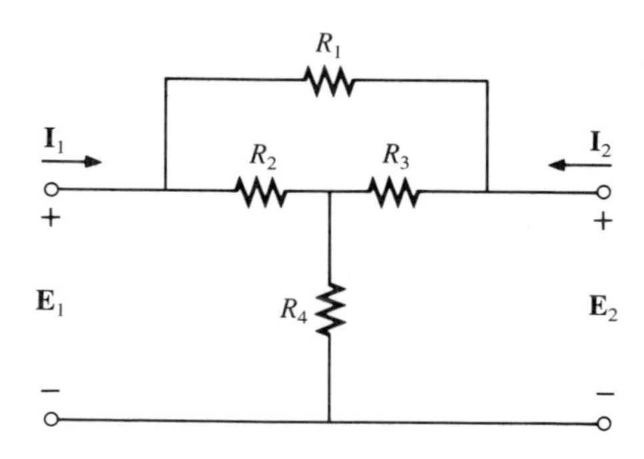


FIG. 26.49

2. **a.** Determine the impedance (**z**) parameters for the network of Fig. 26.49.
   **b.** Sketch the **z** parameter equivalent circuit (using either form of Fig. 26.14).

### SECTION 26.3 Admittance (y) Parameters

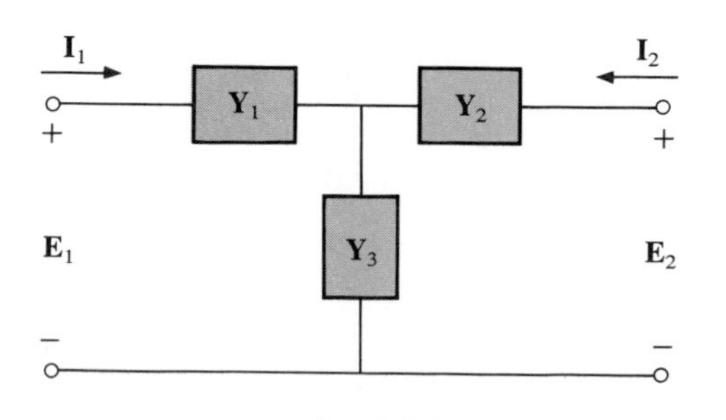


FIG. 26.50

3. **a.** Determine the admittance (**y**) parameters for the T network of Fig. 26.50.
   **b.** Sketch the **y** parameter equivalent circuit (using either form of Fig. 26.26).

4. **a.** Determine the admittance (**y**) parameters for the network of Fig. 26.51.
   **b.** Sketch the **y** parameter equivalent circuit (using either form of Fig. 26.26).

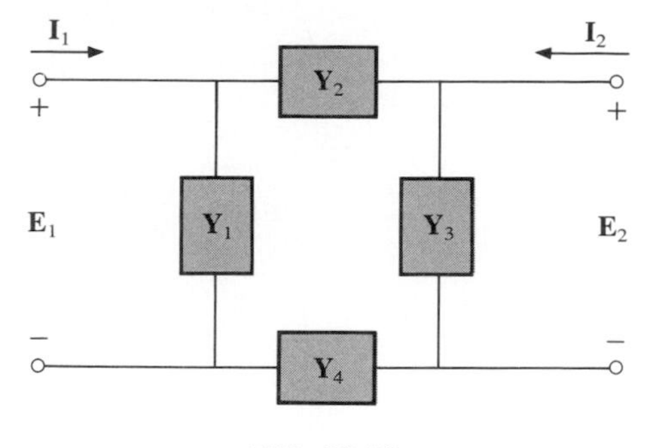


FIG. 26.51

## SECTION 26.4 Hybrid (h) Parameters

5. **a.** Determine the **h** parameters for the network of Fig. 26.48.
   **b.** Sketch the hybrid equivalent circuit.
6. **a.** Determine the **h** parameters for the network of Fig. 26.49.
   **b.** Sketch the hybrid equivalent circuit.
7. **a.** Determine the **h** parameters for the network of Fig. 26.50.
   **b.** Sketch the hybrid equivalent circuit.
8. **a.** Determine the **h** parameters for the network of Fig. 26.51.
   **b.** Sketch the hybrid equivalent circuit.
9. For the hybrid equivalent circuit of Fig. 26.52:
   **a.** Determine the current gain $A_i = \mathbf{I}_2/\mathbf{I}_1$.
   **b.** Determine the voltage gain $A_v = \mathbf{E}_2/\mathbf{E}_1$.

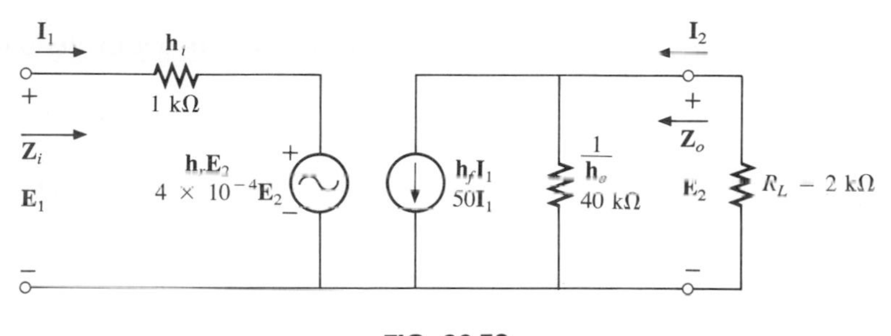

FIG. 26.52

## SECTION 26.5 Input and Output Impedances

10. For the hybrid equivalent circuit of Fig. 26.52:
    **a.** Determine the input impedance.
    **b.** Determine the output impedance.
11. Determine the input and output impedances for the **z** parameter equivalent circuit of Fig. 26.53.

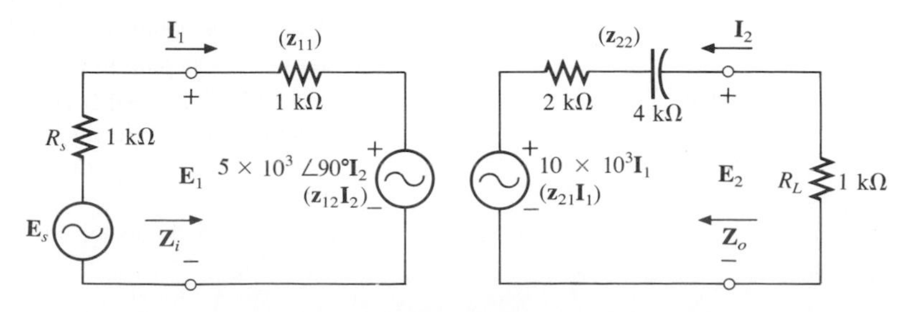


FIG. 26.53

12. Determine the expression for the input and output impedance of the **y** parameter equivalent circuit.

## SECTION 26.6 Conversion Between Parameters

13. Determine the **h** parameters for the following **z** parameters:

$$\mathbf{z}_{11} = 4\ \text{k}\Omega$$
$$\mathbf{z}_{12} = 2\ \text{k}\Omega$$
$$\mathbf{z}_{21} = 3\ \text{k}\Omega$$
$$\mathbf{z}_{22} = 4\ \text{k}\Omega$$

14. **a.** Determine the **z** parameters for the following **h** parameters:

$$\mathbf{h}_{11} = 1\ \text{k}\Omega$$
$$\mathbf{h}_{12} = 2 \times 10^{-4}$$
$$\mathbf{h}_{21} = 100$$
$$\mathbf{h}_{22} = 20 \times 10^{-6}\ \text{S}$$

**b.** Determine the **y** parameters for the hybrid parameters indicated in part (a).

## SECTION 26.7 Impact of Various Parameters

15. For the system of Fig. 26.36, if $\mathbf{E}_i = 120\ \text{V} \angle 0°$ and $\mathbf{I}_i = 6.2\ \text{A} \angle -10.8°$ determine the input impedance in rectangular coordinates. At a frequency of 60 Hz, determine the nameplate values of the parameters.

16. For the system of Fig. 26.37, determine $\mathbf{Z}_i$ if $v_g = 220 \sin 377t$ and $v_R = 120 \times 10^{-3} \sin(377t - 60°)$, with $R = 2.2\ \Omega$.

17. For the configuration of Fig. 26.38, determine $\mathbf{Z}_o$ if $v_g = 10 \sin 377t$ and $v_R = 40 \times 10^{-3} \sin 377t$, with $R = 100\ \Omega$ and $R_s = 1\ \text{k}\Omega$.

18. For the system of Fig. 26.40,
   **a.** Determine $Z_o$ if $Z_L = 1.2\ \text{k}\Omega$, $I_L = 5\ \text{mA}$, $A_{v_{NL}} = 1200$, and the applied voltage $E_i$ is 40 mV.
   **b.** Determine the loaded voltage gain $A_v = E_o/E_i$ for the parameters of part (a).
   **c.** Find the current gain $A_i = I_o/I_i$ for the parameters of part (a) and $Z_i = 1\ \text{k}\Omega$.
   **d.** Find the power gain using $A_p = A_v \cdot A_i$ and compare to the result obtained using $A_p = I_o^2 R_L / I_i^2 R_i$.

*19. For the system of Fig. 26.41, if $Z_{i_1} = 1\ \text{k}\Omega$, $Z_{i_3} = 2\ \text{k}\Omega$, $R_L = 2.2\ \text{k}\Omega$, $A_{v_1} = -12$, and $A_{v_3} = -32$,
   **a.** Determine $A_{v_2}$ if the total voltage gain is $-6912$.
   **b.** Find $Z_{i_2}$ using the results of part (a) and the fact that $A_{i_2} = -26$.
   **c.** Find $A_{i_1}$ and $A_{i_3}$ using the fact that $A_{i_2} = -26$ and $A_{i_3} = 2A_{i_1}$.

*20. For the system of Fig. 26.54,

a. Determine $Z_o$.

b. Find $A_{v_{NL}} = V_L/V_g$.

c. Sketch the Thevenin equivalent circuit of Fig. 26.40.

d. Calculate $V_L$ using the results of part (c) and the voltage divider rule.

e. Determine $I_o$ using the results of part (c).

f. Determine $V_L$ and $I_o$ for the network of Fig. 26.54, analyzing the network as it appears and compare with the results of parts (d) and (e).

g. Determine $A_{v_T} = V_L/V_g$.

h. Find $A_{i_T} = I_o/I_i$.

i. How does the no-load voltage gain of part (b) compare to the loaded gain of part (g)?

j. How does the current gain of part (h) compare with the multiplying factor of the controlled current source (100)? Why are they different?

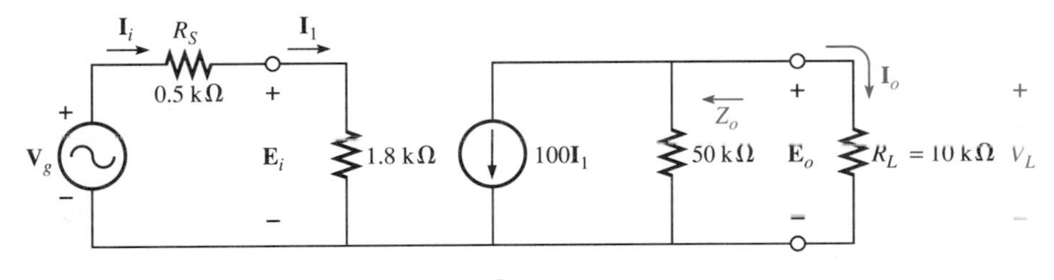

FIG. 26.54

## SECTION 26.8 Computer Analysis

### PSPICE

21. Write the input file to determine the input impedance for the system of Fig. 26.52.

*22. Write the input file to determine the output voltage and current for the system of Fig. 26.54 if $\mathbf{V}_g = 10\text{ V} \angle 0°$.

23. Write the input file to determine the output impedance for the system of Fig. 26.54.

24. Repeat Problem 23 for the input impedance.

### BASIC

25. Given a T configuration such as that shown in Fig. 26.9, write a program to generate the **z** and **y** parameters. Each branch of the T configuration can be a resistor, capacitor, or inductor.

26. Given the hybrid parameters of a system, write an equation to determine the current gain, voltage gain, input impedance, and output impedance.

27. Write a program to perform the conversions indicated in Table 26.1. That is, given the **z, y,** or **h** parameters, determine the **z, y,** or **h** parameters.

## GLOSSARY

**Admittance (y) parameters** A set of parameters, having the units of siemens, that can be used to establish a two-port equivalent network for a system.

**Hybrid (h) parameters** A set of mixed parameters (ohms, siemens, some unitless) that can be used to establish a two-port equivalent network for a system.

**Impedance (z) parameters** A set of parameters, having the units of ohms, that can be used to establish a two-port equivalent network for a system.

**Input impedance** The impedance appearing at the input terminals of a system.

**Output impedance** The impedance appearing at the output terminals of a system with the energizing source set to zero.

**Single-port network** A network having a single set of access terminals.

**Two-port network** A network having two pairs of access terminals.

# Appendixes

# Appendix A

## PSPICE

The PSPICE* software package employed throughout this text is derived from programs developed at the University of California at Berkeley during the early 1970s. Various versions of this program are provided by MicroSim Corp. for use on the IBM line of personal computers, the Macintosh II, SUN terminals, and on VAX terminals. SPICE is an acronym for Simulation Program with Integrated Circuit Emphasis. SPICE has undergone many changes since the early versions of circuit-analysis programs such as ECAP (Electrical Circuit Analysis Program) were developed. Although a number of companies have customized SPICE for their particular use, MicroSim has developed both a commercial package and a student version. The commercial package includes data on a large selection of popular integrated circuit (IC) and semiconductor units, whereas the student version is limited primarily to an introduction to SPICE and the range of areas of analysis. The commercial or professional versions employed by engineering companies can be costly, so MicroSim encourages duplication of the student version in the educational community. The programs of this text were all run on a student version to ensure that they will work in a classroom environment. PSPICE is one of a number of circuit-analysis and simulation programs that can be used to carry out either dc or ac analysis on essentially any type of circuit with practical elements. The circuit is first entered as a number of lines of text; then PSPICE is run to analyze the given circuit, providing an output file of the results obtained. The steps involved in performing an analysis include the following:

1. Using an editor or word processor program (one is supplied with the latest version of PSPICE) to create the circuit description file and list of output data desired.
2. Running the PSPICE analysis on the circuit entered in step 1.
3. Viewing the output file generated by PSPICE (it can be viewed using the browse feature on the latest PSPICE version).
4. Running PROBE if an output waveform is desired.

PROBE is a program provided with PSPICE designed to obtain a graphical display of the output data as appear on a graph or oscilloscope. It is part of the PSPICE package, but it requires additional calculations to develop a desired output display and therefore is best run on a system having a math coprocessor IC. Although any IBM computer can provide the lengthy calculations required by PROBE, a math coprocessor IC may be installed in any IBM computer to provide for faster mathematical calculations. The PROBE analysis will work with or without the math coprocessor, but it is much faster on a system with the IC installed. Commercial operations would invariably include one in their systems, but most student units do not. In the student environment



*PSPICE is a registered trademark of MicroSim Corp., Laguna Hills, California.

problems are generally short and the use of PROBE, while slow without a coprocessor, will nevertheless provide the desired output displays.

MicroSim Corp. has continued to update the PSPICE analysis program and their library of practical devices. The student version used in this text is dated January 1989, so newer versions will likely be forthcoming. It is expected that any of the available student versions will be able to analyze the types of circuits studied in an educational setting.

## BASIC SYSTEM REQUIREMENTS

PSPICE can be run on a suitably configured IBM PC or AT system, Macintosh II, SUN-3 or SUN-4 workstation, or on a VAX minicomputer or mainframe station. As example, PSPICE can be run under DOS (Version 3.3 or 4.0) on an IBM PC having at least 512K of RAM. MicroSim Corp. supplies the program on either 720-kilobyte (kb), 3.5-in. diskettes or 1.2 megabyte (Mb), 5.25-in. diskettes. These disks or copies of them can be used to run PSPICE. Although a hard disk is not necessary, it makes the overall operation much smoother. The PSPICE program, including a small version of the library for student use, and the PROBE program use over 1.25 Mb. If one uses a hard disk, the entire PSPICE package can be readily accessible from the hard disk. If desired, a floppy disk can then be used to store the various small programs and output files obtained using PSPICE. High-density floppy disks (720 kb, 1.2 Mb, or 1.4 Mb) can be used if a hard disk is available. Operation using only a 360-kb disk, while still possible, is terribly annoying, as one must constantly switch disks in and out to provide the computer with the more than 1.25 Mb of program needed.

PSPICE can be run using either a monochrome or color display, the latter being either CGA, EGA, or VGA. No difference occurs using the various displays for the resulting output file. A slight difference can be seen, however, using PROBE.

The latest version of PSPICE for the IBM includes an integrated input editor and SPICE analysis program, along with the ability to browse through the output file and to run PROBE to view any output desired. The editor permits typing in the SPICE input circuit information, correcting any typing errors that may have occurred or changing some of the information previously entered. Once the PSPICE analysis is started, it carries out the dc, ac, or other desired analysis of the given circuit, resulting in an output file (or PROBE data file). The output file will also show any errors the PSPICE analysis finds when trying to analyze the circuit. The user can then go back to the original input circuit file to make changes or corrections.

## OVERALL OPERATION WITH PSPICE

To use the PSPICE analysis program one must do the following:

1. First, identify the complete circuit, marking all node points and components (component name and value).

2. Second, use a text editor—the one provided with PSPICE or any other desired text editor (Professional Write, WORD, Wordperfect 5, etc.)—to enter a circuit file describing the given circuit.
3. Third, run the PSPICE analysis, which produces either an output file listing the results obtained or a PROBE data file that can be viewed using the PROBE program to display the results. Using the integrated PSPICE package, one can browse through the resulting output file to view the results obtained or to see what, if any, errors occurred when the circuit was analyzed.

## FILE NAMES

Files used with PSPICE have a number of standard three-letter extensions.

| | | |
|---|---|---|
| Input file describing circuit | .CIR | |
| Output file holding results | .OUT | |
| Probe data file to be run | PROBE.DAT | (only one such file may be present) |

The .CIR file is created by the person using the editor. In using PSPICE to edit a file, one merely enters a file name such as HW5.CIR or LAB3P2.CIR or just the file name HW5 or LAB3P2, with PSPICE then adding the default extension .CIR. In using a text editor, one must give the full name with .CIR extension to name the file.

When PSPICE runs a circuit analysis, it generates a second file with the extension .OUT. For example, running a PSPICE analysis on the input file HW5.CIR would result in a second file, HW5.OUT.

If the input file includes a line calling for PROBE, the output includes a file PROBE.DAT, which can then be run using PROBE to view the various output signals obtained from the circuit. When one stores the input and output data on a separate floppy disk, there can be many .CIR files (using different names) and many .OUT files (obtained from the PSPICE analysis), but there will be only one PROBE.DAT of the last file analyzed using PROBE analysis.

## GETTING A PRINTOUT

The easiest way to get a hard copy of the resulting output file is to use a DOS print command. To print the file HW5.OUT, for example, type the DOS command

```
PRINT HW5.OUT
```

and the full output file will be printed out on a number of pages. This output file includes a reference to the PSPICE version run to provide the analysis, the date and time when the analysis was run, a listing of the original input circuit file, and the resulting outputs called for in the input file.

If the PROBE feature was selected in the circuit file, a PROBE.DAT output file is generated. The analysis output file (.OUT) does not display a PROBE graphical output. PROBE must be selected and run to obtain the display. When a PROBE display is developed on the screen, a hard copy is easily obtained by choosing the hard-copy option (pressing function key F8).

## USING PROBE

The use of PROBE to provide graphical display of the output results starts by including a single line

```
.PROBE
```

in the input (.CIR) file. When PSPICE analysis is performed, an output file, PROBE.DAT, is generated (only one such file may be present, this being the last file generated). To display graphs, one merely runs the PROBE program (by typing in just the word PROBE or selecting the PROBE feature from the menu). Since only one PROBE.DAT file exists, the operation takes place without further reference to the name of the file to be viewed.

PROBE has its own menu of choices that allows the user to select various mathematical operations, variables to be displayed, the type of scaling to be used on either the *X*- or *Y*-axis, the number of variables to be displayed at one time, choice or linear of logarithmic axes, and when a hardcopy is to be printed out.

## USING AN INTEGRATED PSPICE PROGRAM

The present version of PSPICE is an integrated package including the editor, circuit analyzer, probe, and so on in a command shell format. The user carries out the various steps from one main command shell.

At the DOS prompt, enter the command PS to run the PSPICE program. The display now shows the boxed outline of the PSPICE Control Shell. To enter and analyze a new circuit proceed as follows:

1. Select *Files* from the top menu line. Pressing the ENTER key then highlights *Current*. This is a request for the name you wish to give the current circuit file. For example, enter PROB1. PSPICE will then assign the file name PROB1.CIR to the information you now type in to describe the circuit to be analyzed.
2. Next, select *Files* and then *Edit* to enter lines describing the circuit and analysis steps desired. Type in the lines of the circuit. When finished, press the Esc (escape) key to go back to the menu line choices. The program will ask whether a file should be created (to store the present circuit lines entered). Answer Y (for yes).

3. Using the top menu line, select *Analysis* to choose performing the desired analysis of the circuit. Press the Enter key for the choice *Start*.
4. If any error occurs, a message will alert you to the fact. If an error should occur, select the menu item *Files;* then choose *Browse* to look at the output file on the screen. Within the output file will be one or more messages indicating the line in error and the type of error. To correct the error, press Esc to get to the menu line; then continue with step 2 to edit in the required correction.
5. If errors do not surface, choose the menu items *Files* and then *Browse,* which will display the output file containing the desired analysis results.
6. To get a hard copy of the output file, choose the menu item *Quit,* then *DOS Command*. In the DOS command space, enter the DOS command PRINT *filename* (e.g., PRINT PROB3.OUT).
7. To run PROBE, select the menu item *Probe*. Probe will operate by placing a graph outline on the screen and then prompt you for which variables to display, the axes to use, and the like to obtain an output plot.

The preceding was a very brief description of PSPICE and the basic maneuvers associated with using the software package. Additional information and detail is available from the PSPICE manual and a growing list of publications.

# Appendix B

## CONVERSION FACTORS

| To Convert from | To | Multiply by |
|---|---|---|
| Btus | Calorie-grams | 251.996 |
| | Ergs | $1.054 \times 10^{10}$ |
| | Foot-pounds | 777.649 |
| | Hp-hours | 0.000393 |
| | Joules | 1054.35 |
| | Kilowatthours | 0.000293 |
| | Wattseconds | 1054.35 |
| Centimeters | Angstrom units | $1 \times 10^{8}$ |
| | Feet | 0.0328 |
| | Inches | 0.3937 |
| | Meters | 0.01 |
| | Miles (statute) | $6.214 \times 10^{-6}$ |
| | Millimeters | 10 |
| Circular mils | Square centimeters | $5.067 \times 10^{-6}$ |
| | Square inches | $7.854 \times 10^{-7}$ |
| Cubic inches | Cubic centimeters | 16.387 |
| | Gallons (U.S. liquid) | 0.00433 |
| Cubic meters | Cubic feet | 35.315 |
| Days | Hours | 24 |
| | Minutes | 1440 |
| | Seconds | 86,400 |
| Dynes | Gallons (U.S. liquid) | 264.172 |
| | Newtons | 0.00001 |
| | Pounds | $2.248 \times 10^{-6}$ |
| Electronvolts | Ergs | $1.60209 \times 10^{-12}$ |
| Ergs | Dyne-centimeters | 1.0 |
| | Electronvolts | $6.242 \times 10^{11}$ |
| | Foot-pounds | $7.376 \times 10^{-8}$ |
| | Joules | $1 \times 10^{-7}$ |
| | Kilowatthours | $2.777 \times 10^{-14}$ |
| Feet | Centimeters | 30.48 |
| | Meters | 0.3048 |
| Foot-candles | Lumens/square foot | 1.0 |
| | Lumens/square meter | 10.764 |
| Foot-pounds | Dyne-centimeters | $1.3558 \times 10^{7}$ |
| | Ergs | $1.3558 \times 10^{7}$ |
| | Horsepower-hours | $5.050 \times 10^{-7}$ |
| | Joules | 1.3558 |
| | Newton-meters | 1.3558 |

| To Convert from | To | Multiply by |
|---|---|---|
| Gallons (U.S. liquid) | Cubic inches | 231 |
| | Liters | 3.785 |
| | Ounces | 128 |
| | Pints | 8 |
| Gauss | Maxwells/square centimeter | 1.0 |
| | Lines/square centimeter | 1.0 |
| | Lines/square inch | 6.4516 |
| Gilberts | Ampere-turns | 0.7958 |
| Grams | Dynes | 980.665 |
| | Ounces | 0.0353 |
| | Pounds | 0.0022 |
| Horsepower | Btus/hour | 2547.16 |
| | Ergs/second | $7.46 \times 10^9$ |
| | Foot-pounds/second | 550.221 |
| | Joules/second | 746 |
| | Watts | 746 |
| Hours | Seconds | 3600 |
| Inches | Angstrom units | $2.54 \times 10^8$ |
| | Centimeters | 2.54 |
| | Feet | 0.0833 |
| | Meters | 0.0254 |
| Joules | Btus | 0.000948 |
| | Ergs | $1 \times 10^7$ |
| | Foot-pounds | 0.7376 |
| | Horsepower-hours | $3.725 \times 10^{-7}$ |
| | Kilowatthours | $2.777 \times 10^{-7}$ |
| | Wattseconds | 1.0 |
| Kilograms | Dynes | 980,665 |
| | Ounces | 35.2 |
| | Pounds | 2.2 |
| Lines | Maxwells | 1.0 |
| Lines/square centimeter | Gauss | 1.0 |
| Lines/square inch | Gauss | 0.1550 |
| | Webers/square inch | $1 \times 10^{-8}$ |
| Liters | Cubic centimeters | 1000.028 |
| | Cubic inches | 61.025 |
| | Gallons (U.S. liquid) | 0.2642 |
| | Ounces (U.S. liquid) | 33.815 |
| | Quarts (U.S. liquid) | 1.0567 |
| Lumens | Candle power (spher.) | 0.0796 |
| Lumens/square centimeter | Lamberts | 1.0 |
| Lumens/square foot | Foot-candles | 1.0 |

| To Convert from | To | Multiply by |
|---|---|---|
| Maxwells | Lines | 1.0 |
| | Webers | $1 \times 10^{-8}$ |
| Meters | Angstrom units | $1 \times 10^{10}$ |
| | Centimeters | 100 |
| | Feet | 3.2808 |
| | Inches | 39.370 |
| | Miles (statute) | 0.000621 |
| Miles (statute) | Feet | 5280 |
| | Kilometers | 1.609 |
| | Meters | 1609.344 |
| Miles/hour | Kilometers/hour | 1.609344 |
| Newton-meters | Dyne-centimeters | $1 \times 10^{7}$ |
| | Kilogram-meters | 0.10197 |
| Oersteds | Ampere-turns/inch | 2.0212 |
| | Ampere-turns/meter | 79.577 |
| | Gilberts/centimeter | 1.0 |
| Quarts (U.S. liquid) | Cubic centimeters | 946.353 |
| | Cubic inches | 57.75 |
| | Gallons (U.S. liquid) | 0.25 |
| | Liters | 0.9463 |
| | Pints (U.S. liquid) | 2 |
| | Ounces (U.S. liquid) | 32 |
| Radians | Degrees | 57.2958 |
| Slugs | Kilograms | 14.5939 |
| | Pounds | 32.1740 |
| Watts | Btus/hour | 3.4144 |
| | Ergs/second | $1 \times 10^{7}$ |
| | Horsepower | 0.00134 |
| | Joules/second | 1.0 |
| Webers | Lines | $1 \times 10^{8}$ |
| | Maxwells | $1 \times 10^{8}$ |
| Years | Days | 365 |
| | Hours | 8760 |
| | Minutes | 525,600 |
| | Seconds | $3.1536 \times 10^{7}$ |

# Appendix C

## DETERMINANTS

Determinants are employed to find the mathematical solutions for the variables in two or more simultaneous equations. Once the procedure is properly understood, solutions can be obtained with a minimum of time and effort and usually with fewer errors than when using other methods.

Consider the following equations, where $x$ and $y$ are the unknown variables and $a_1$, $a_2$, $b_1$, $b_2$, $c_1$, and $c_2$ are constants:

| Col. 1 | | Col. 2 | | Col. 3 |
|---|---|---|---|---|
| $a_1x$ | + | $b_1y$ | = | $c_1$ |
| $a_2x$ | + | $b_2y$ | = | $c_2$ |

**(C.1a)**
**(C.1b)**

It is certainly possible to solve for one variable in Eq. (C.1a) and substitute into Eq. (C.1b). That is, solving for $x$ in Eq. (C.1a),

$$x = \frac{c_1 - b_1y}{a_1}$$

and substituting the result in Eq. (C.1b),

$$a_2\left(\frac{c_1 - b_1y}{a_1}\right) + b_2y = c_2$$

It is now possible to solve for $y$, since it is the only variable remaining, and then substitute into either equation for $x$. This is acceptable for two equations, but it becomes a very tedious and lengthy process for three or more simultaneous equations.

Using determinants to solve for $x$ and $y$ requires that the following formats be established for each variable:

Col. 1 Col. 2 Col. 1 Col. 2

$$x = \frac{\begin{vmatrix} c_1 & b_1 \\ c_2 & b_2 \end{vmatrix}}{\begin{vmatrix} a_1 & b_1 \\ a_2 & b_2 \end{vmatrix}} \qquad y = \frac{\begin{vmatrix} a_1 & c_1 \\ a_2 & c_2 \end{vmatrix}}{\begin{vmatrix} a_1 & b_1 \\ a_2 & b_2 \end{vmatrix}} \tag{C.2}$$

First note that only constants appear within the vertical brackets and that the denominator of each is the same. In fact, the denominator is simply the coefficients of $x$ and $y$ in the same arrangement as in Eqs. (C.1a) and (C.1b). When solving for $x$, the coefficients of $x$ in the numerator are replaced by the constants to the right of the equal sign in Eqs. (C.1a) and (C.1b), whereas the coefficients of the $y$ variable are simply repeated. When solving for $y$, the $y$ coefficients in the numerator are replaced by the constants to the right of the equal sign and the coefficients of $x$ are repeated.

Each configuration in the numerator and denominator of Eqs. (C.2) is referred to as a *determinant* ($D$), which can be evaluated numerically in the following manner:

$$\text{Determinant} = D = \overset{\text{Col. 1 \quad Col. 2}}{\begin{vmatrix} a_1 & b_1 \\ a_2 & b_2 \end{vmatrix}} = a_1b_2 - a_2b_1 \tag{C.3}$$

The expanded value is obtained by first multiplying the top left element by the bottom right and then subtracting the product of the lower left and upper right elements. This particular determinant is referred to as a *second-order* determinant, since it contains two rows and two columns.

It is important to remember when using determinants that the columns of the equations, as indicated in Eqs. (C.1a) and (C.1b), be placed in the same order within the determinant configuration. That is, since $a_1$ and $a_2$ are in column 1 of Eqs. (C.1a) and (C.1b), they must be in column 1 of the determinant. (The same is true for $b_1$ and $b_2$.)

Expanding the entire expression for $x$ and $y$, we have the following:

$$x = \frac{\begin{vmatrix} c_1 & b_1 \\ c_2 & b_2 \end{vmatrix}}{\begin{vmatrix} a_1 & b_1 \\ a_2 & b_2 \end{vmatrix}} = \frac{c_1b_2 - c_2b_1}{a_1b_2 - a_2b_1} \tag{C.4a}$$

$$y = \frac{\begin{vmatrix} a_1 & c_1 \\ a_2 & c_2 \end{vmatrix}}{\begin{vmatrix} a_1 & b_1 \\ a_2 & b_2 \end{vmatrix}} = \frac{a_1c_2 - a_2c_1}{a_1b_2 - a_2b_1} \tag{C.4b}$$

---

**EXAMPLE C.1.** Evaluate the following determinants:

a. $\begin{vmatrix} 2 & 2 \\ 3 & 4 \end{vmatrix} = (2)(4) - (3)(2) = 8 - 6 = \mathbf{2}$

b. $\begin{vmatrix} 4 & -1 \\ 6 & 2 \end{vmatrix} = (4)(2) - (6)(-1) = 8 + 6 = \mathbf{14}$

c. $\begin{vmatrix} 0 & -2 \\ -2 & 4 \end{vmatrix} = (0)(4) - (-2)(-2) = 0 - 4 = \mathbf{-4}$

d. $\begin{vmatrix} 0 & 0 \\ 3 & 10 \end{vmatrix} = (0)(10) - (3)(0) = \mathbf{0}$

---

**EXAMPLE C.2.** Solve for $x$ and $y$:

$$\begin{aligned} 2x + \phantom{4}y &= 3 \\ 3x + 4y &= 2 \end{aligned}$$

***Solution:***

$$x = \frac{\begin{vmatrix} 3 & 1 \\ 2 & 4 \end{vmatrix}}{\begin{vmatrix} 2 & 1 \\ 3 & 4 \end{vmatrix}} = \frac{(3)(4) - (2)(1)}{(2)(4) - (3)(1)} = \frac{12 - 2}{8 - 3} = \frac{10}{5} = \mathbf{2}$$

$$y = \frac{\begin{vmatrix} 2 & 3 \\ 3 & 2 \end{vmatrix}}{5} = \frac{(2)(2) - (3)(3)}{5} = \frac{4 - 9}{5} = \frac{-5}{5} = \mathbf{-1}$$

***Check:***

$$\begin{aligned} 2x + y &= (2)(2) + (-1) \\ &= 4 - 1 = 3 \qquad \text{(checks)} \\ 3x + 4y &= (3)(2) + (4)(-1) \\ &= 6 - 4 = 2 \qquad \text{(checks)} \end{aligned}$$

**EXAMPLE C.3.** Solve for $x$ and $y$:

$$\begin{aligned} -x + 2y &= 3 \\ 3x - 2y &= -2 \end{aligned}$$

***Solution:*** In this example, note the effect of the minus sign and the use of parentheses to insure the proper sign is obtained for each product:

$$x = \frac{\begin{vmatrix} 3 & 2 \\ -2 & -2 \end{vmatrix}}{\begin{vmatrix} -1 & 2 \\ 3 & -2 \end{vmatrix}} = \frac{(3)(-2) - (-2)(2)}{(-1)(-2) - (3)(2)}$$

$$= \frac{-6 + 4}{2 - 6} = \frac{-2}{-4} = \mathbf{\frac{1}{2}}$$

$$y = \frac{\begin{vmatrix} -1 & 3 \\ 3 & -2 \end{vmatrix}}{-4} = \frac{(-1)(-2) - (3)(3)}{-4}$$

$$= \frac{2 - 9}{-4} = \frac{-7}{-4} = \mathbf{\frac{7}{4}}$$

**EXAMPLE C.4.** Solve for $x$ and $y$:

$$\begin{aligned} x &= 3 - 4y \\ 20y &= -1 + 3x \end{aligned}$$

***Solution:*** In this case, the equations must first be placed in the format of Eqs. (C.1a) and (C.1b):

$$\begin{aligned} x + 4y &= 3 \\ -3x + 20y &= -1 \end{aligned}$$

$$x = \frac{\begin{vmatrix} 3 & 4 \\ -1 & 20 \end{vmatrix}}{\begin{vmatrix} 1 & 4 \\ -3 & 20 \end{vmatrix}} = \frac{(3)(20) - (-1)(4)}{(1)(20) - (-3)(4)}$$

$$= \frac{60 + 4}{20 + 12} = \frac{64}{32} = \mathbf{2}$$

$$y = \frac{\begin{vmatrix} 1 & 3 \\ -3 & -1 \end{vmatrix}}{32} = \frac{(1)(-1) - (-3)(3)}{32}$$

$$= \frac{-1 + 9}{32} = \frac{8}{32} = \mathbf{\frac{1}{4}}$$

---

The use of determinants is not limited to the solution of two simultaneous equations; determinants can be applied to any number of simultaneous linear equations. First we will examine a shorthand method that is applicable to third order determinants only, since most of the problems in the text are limited to this level of difficulty. We will then investigate the general procedure for solving any number of simultaneous equations.

Consider the three following simultaneous equations:

| Col. 1 | | Col. 2 | | Col. 3 | | Col. 4 |
|---|---|---|---|---|---|---|
| $a_1x$ | + | $b_1y$ | + | $c_1z$ | = | $d_1$ |
| $a_2x$ | + | $b_2y$ | + | $c_2z$ | = | $d_2$ |
| $a_3x$ | + | $b_3y$ | + | $c_3z$ | = | $d_3$ |

in which $x$, $y$, and $z$ are the variables, and $a_{1,2,3}$, $b_{1,2,3}$, $c_{1,2,3}$, and $d_{1,2,3}$ are constants.

The determinant configuration for $x$, $y$, and $z$ can be found in a manner similar to that for two simultaneous equations. That is, to solve for $x$, find the determinant in the numerator by replacing column 1 with the elements to the right of the equal sign. The denominator is the determinant of the coefficients of the variables (the same applies to $y$ and $z$). Again, the denominator is the same for each variable.

$$x = \frac{\begin{vmatrix} d_1 & b_1 & c_1 \\ d_2 & b_2 & c_2 \\ d_3 & b_3 & c_3 \end{vmatrix}}{\begin{vmatrix} a_1 & b_1 & c_1 \\ a_2 & b_2 & c_2 \\ a_3 & b_3 & c_3 \end{vmatrix}}, \quad y = \frac{\begin{vmatrix} a_1 & d_1 & c_1 \\ a_2 & d_2 & c_2 \\ a_3 & d_3 & c_3 \end{vmatrix}}{D}, \quad z = \frac{\begin{vmatrix} a_1 & b_1 & d_1 \\ a_2 & b_2 & d_2 \\ a_3 & b_3 & d_3 \end{vmatrix}}{D}$$

$$D = \begin{vmatrix} a_1 & b_1 & c_1 \\ a_2 & b_2 & c_2 \\ a_3 & b_3 & c_3 \end{vmatrix}$$

A shorthand method for evaluating the third-order determinant consists simply of repeating the first two columns of the determinant to the right of the determinant and then summing the products along specific diagonals as shown below:

$$D = \begin{vmatrix} a_1 & b_1 & c_1 \\ a_2 & b_2 & c_2 \\ a_3 & b_3 & c_3 \end{vmatrix} \begin{matrix} a_1 & b_1 \\ a_2 & b_2 \\ a_3 & b_3 \end{matrix}$$

4(−) 5(−) 6(−)

1(+) 2(+) 3(+)

The products of the diagonals 1, 2, and 3 are positive and have the following magnitudes:

$$+a_1b_2c_3 + b_1c_2a_3 + c_1a_2b_3$$

The products of the diagonals 4, 5, and 6 are negative and have the following magnitudes:

$$-a_3b_2c_1 - b_3c_2a_1 - c_3a_2b_1$$

The total solution is the sum of the diagonals 1, 2, and 3 minus the sum of the diagonals 4, 5, and 6:

$$+(a_1b_2c_3 + b_1c_2a_3 + c_1a_2b_3) - (a_3b_2c_1 + b_3c_2a_1 + c_3a_2b_1) \quad \textbf{(C.5)}$$

*Warning:* **This method of expansion is good only for third-order determinants!** It cannot be applied to fourth- and higher-order systems.

---

**EXAMPLE C.5.** Evaluate the following determinant:

$$\begin{vmatrix} 1 & 2 & 3 \\ -2 & 1 & 0 \\ 0 & 4 & 2 \end{vmatrix} \Rightarrow \begin{vmatrix} 1 & 2 & 3 \\ -2 & 1 & 0 \\ 0 & 4 & 2 \end{vmatrix} \begin{matrix} 1 & 2 \\ -2 & 1 \\ 0 & 4 \end{matrix}$$

(−) (−) (−)

(+) (+) (+)

***Solution:***

$$[(1)(1)(2) + (2)(0)(0) + (3)(-2)(4)] - [(0)(1)(3) + (4)(0)(1) + (2)(-2)(2)]$$
$$= (2 + 0 - 24) - (0 + 0 - 8) = (-22) - (-8)$$
$$= -22 + 8 = \mathbf{-14}$$

---

**EXAMPLE C.6.** Solve for $x$, $y$, and $z$:

$$1x + 0y - 2z = -1$$
$$0x + 3y + 1z = +2$$
$$1x + 2y + 3z = 0$$

***Solution:***

$$x = \frac{\begin{vmatrix} -1 & 0 & -2 \\ 2 & 3 & 1 \\ 0 & 2 & 3 \end{vmatrix} \begin{matrix} -1 & 0 \\ 2 & 3 \\ 0 & 2 \end{matrix}}{\begin{vmatrix} 1 & 0 & -2 \\ 0 & 3 & 1 \\ 1 & 2 & 3 \end{vmatrix} \begin{matrix} 1 & 0 \\ 0 & 3 \\ 1 & 2 \end{matrix}}$$

$$= \frac{[(-1)(3)(3) + (0)(1)(0) + (-2)(2)(2)] - [(0)(3)(-2) + (2)(1)(-1) + (3)(2)(0)]}{[(1)(3)(3) + (0)(1)(1) + (-2)(0)(2)] - [(1)(3)(-2) + (2)(1)(1) + (3)(0)(0)]}$$

$$= \frac{(-9 + 0 - 8) - (0 - 2 + 0)}{(9 + 0 + 0) - (-6 + 2 + 0)}$$

$$= \frac{-17 + 2}{9 + 4} = \mathbf{-\frac{15}{13}}$$

$$y = \frac{\begin{vmatrix} 1 & -1 & -2 \\ 0 & 2 & 1 \\ 1 & 0 & 3 \end{vmatrix} \begin{matrix} 1 & -1 \\ 0 & 2 \\ 1 & 0 \end{matrix}}{13}$$

$$= \frac{[(1)(2)(3) + (-1)(1)(1) + (-2)(0)(0)] - [(1)(2)(-2) + (0)(1)(1) + (3)(0)(-1)]}{13}$$

$$= \frac{(6 - 1 + 0) - (-4 + 0 + 0)}{13}$$

$$= \frac{5 + 4}{13} = \mathbf{\frac{9}{13}}$$

$$z = \frac{\begin{vmatrix} 1 & 0 & -1 \\ 0 & 3 & 2 \\ 1 & 2 & 0 \end{vmatrix} \begin{matrix} 1 & 0 \\ 0 & 3 \\ 1 & 2 \end{matrix}}{13}$$

$$= \frac{[(1)(3)(0) + (0)(2)(1) + (-1)(0)(2)] - [(1)(3)(-1) + (2)(2)(1) + (0)(0)(0)]}{13}$$

$$= \frac{(0 + 0 + 0) - (-3 + 4 + 0)}{13}$$

$$= \frac{0 - 1}{13} = \mathbf{-\frac{1}{13}}$$

or from $0x + 3y + 1z = +2$,

$$z = 2 - 3y = 2 - 3\left(\frac{9}{13}\right) = \frac{26}{13} - \frac{27}{13} = \mathbf{-\frac{1}{13}}$$

***Check:***

$$
\begin{array}{r|l|l}
1x + 0y - 2z = -1 & -\dfrac{15}{13} + 0 + \dfrac{2}{13} = -1 & -\dfrac{13}{13} = -1 \ \checkmark \\[2ex]
0x + 3y + 1z = +2 & 0 + \dfrac{27}{13} + \dfrac{-1}{13} = +2 & \dfrac{26}{13} = +2 \ \checkmark \\[2ex]
1x + 2y + 3z = 0 & -\dfrac{15}{13} + \dfrac{18}{13} + \dfrac{-3}{13} = 0 & -\dfrac{18}{13} + \dfrac{18}{13} = 0 \ \checkmark
\end{array}
$$

The general approach to third-order or higher determinants requires that the determinant be expanded in the following form. There is more than one expansion that will generate the correct result, but this form is typically employed when the material is first introduced.

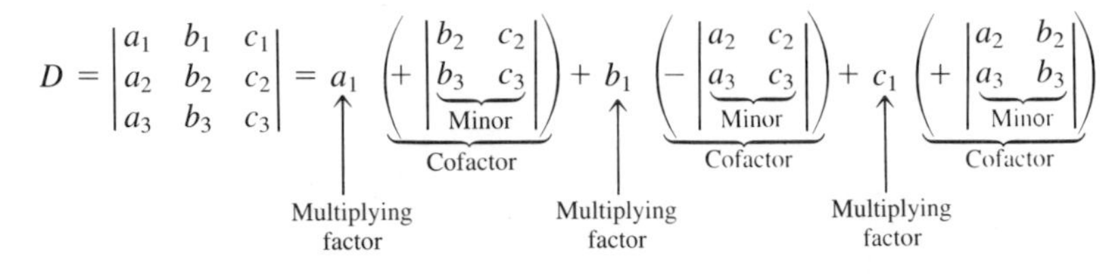


This expansion was obtained by multiplying the elements of the first row of $D$ by their corresponding cofactors. It is not a requirement that the first row be used as the multiplying factors. In fact, any *row* or *column* (not diagonals) may be used to expand a third-order determinant.

The sign of each cofactor is dictated by the position of the multiplying factors ($a_1$, $b_1$, and $c_1$ in this case) as in the following standard format:

$$
\begin{vmatrix}
+ \rightarrow & - & + \\
\downarrow & & \\
- & + & - \\
& & \\
+ & - & +
\end{vmatrix}
$$

Note that the proper sign for each element can be obtained by simply assigning the upper left element a positive sign and then changing sign as you move horizontally or vertically to the neighboring position.

For the determinant $D$, the elements would have the following signs:

$$
\begin{vmatrix}
a_1^{(+)} & b_1^{(-)} & c_1^{(+)} \\
a_2^{(-)} & b_2^{(+)} & c_2^{(-)} \\
a_3^{(+)} & b_3^{(-)} & c_3^{(+)}
\end{vmatrix}
$$

The minors associated with each multiplying factor are obtained by covering up the row and column in which the multiplying factor is located and writing a second-order determinant to include the remaining elements in the same relative positions that they have in the third-order determinant.

Consider the cofactors associated with $a_1$ and $b_1$ in the expansion of $D$. The sign is positive for $a_1$ and negative for $b_1$ as determined by the standard format. Following the procedure outlined above, we can find the minors of $a_1$ and $b_1$ as follows:

$$a_{1(\text{minor})} = \begin{vmatrix} \not{a}_1 & \not{b}_1 & \not{c}_1 \\ \not{a}_2 & b_2 & c_2 \\ \not{a}_3 & b_3 & c_3 \end{vmatrix} = \begin{vmatrix} b_2 & c_2 \\ b_3 & c_3 \end{vmatrix}$$

$$b_{1(\text{minor})} = \begin{vmatrix} \not{a}_1 & \not{b}_1 & \not{c}_1 \\ a_2 & \not{b}_2 & c_2 \\ a_3 & \not{b}_3 & c_3 \end{vmatrix} = \begin{vmatrix} a_2 & c_2 \\ a_3 & c_3 \end{vmatrix}$$

It was pointed out that any row or column may be used to expand the third-order determinant, and the same result will still be obtained. Using the first column of $D$, we obtain the expansion

$$D = \begin{vmatrix} a_1 & b_1 & c_1 \\ a_2 & b_2 & c_2 \\ a_3 & b_3 & c_3 \end{vmatrix} = a_1\left(+\begin{vmatrix} b_2 & c_2 \\ b_3 & c_3 \end{vmatrix}\right) + a_2\left(-\begin{vmatrix} b_1 & c_1 \\ b_3 & c_3 \end{vmatrix}\right) + a_3\left(+\begin{vmatrix} b_1 & c_1 \\ b_2 & c_2 \end{vmatrix}\right)$$

The proper choice of row or column can often effectively reduce the amount of work required to expand the third-order determinant. For example, in the following determinants, the first column and third row, respectively, would reduce the number of cofactors in the expansion:

$$D = \begin{vmatrix} 2 & 3 & -2 \\ 0 & 4 & 5 \\ 0 & 6 & 7 \end{vmatrix} = 2\left(+\begin{vmatrix} 4 & 5 \\ 6 & 7 \end{vmatrix}\right) + 0 + 0 = 2(28 - 30)$$

$$= \mathbf{-4}$$

$$D = \begin{vmatrix} 1 & 4 & 7 \\ 2 & 6 & 8 \\ 2 & 0 & 3 \end{vmatrix} = 2\left(+\begin{vmatrix} 4 & 7 \\ 6 & 8 \end{vmatrix}\right) + 0 + 3\left(+\begin{vmatrix} 1 & 4 \\ 2 & 6 \end{vmatrix}\right)$$

$$= 2(32 - 42) + 3(6 - 8) = 2(-10) + 3(-2)$$

$$= \mathbf{-26}$$

---

**EXAMPLE C.7.** Expand the following third-order determinants:

a. $$D = \begin{vmatrix} 1 & 2 & 3 \\ 3 & 2 & 1 \\ 2 & 1 & 3 \end{vmatrix} = 1\left(+\begin{vmatrix} 2 & 1 \\ 1 & 3 \end{vmatrix}\right) + 3\left(-\begin{vmatrix} 2 & 3 \\ 1 & 3 \end{vmatrix}\right) + 2\left(+\begin{vmatrix} 2 & 3 \\ 2 & 1 \end{vmatrix}\right)$$

$$= 1[6 - 1] + 3[-(6 - 3)] + 2[2 - 6]$$
$$= 5 + 3(-3) + 2(-4)$$
$$= 5 - 9 - 8$$
$$= \mathbf{-12}$$

b. $$D = \begin{vmatrix} 0 & 4 & 6 \\ 2 & 0 & 5 \\ 8 & 4 & 0 \end{vmatrix} = 0 + 2\left(-\begin{vmatrix} 4 & 6 \\ 4 & 0 \end{vmatrix}\right) + 8\left(+\begin{vmatrix} 4 & 6 \\ 0 & 5 \end{vmatrix}\right)$$

$$= 0 + 2[-(0 - 24)] + 8[(20 - 0)]$$
$$= 0 + 2(24) + 8(20)$$
$$= 48 + 160$$
$$= \mathbf{208}$$

---

# Appendix D

## COLOR CODING OF MOLDED MICA CAPACITORS (PICOFARADS)

RETMA *and standard* MIL *specifications*

| Color | Significant Figure | Decimal Multiplier | Tolerance ± % | Class | Temp. Coeff. PPM/°C Not More than | Cap. Drift Not More than |
|---|---|---|---|---|---|---|
| Black | 0 | 1 | 20 | A | ±1000 | ±(5% + 1 pF) |
| Brown | 1 | 10 | — | B | ±500 | ±(3% + 1 pF) |
| Red | 2 | 100 | 2 | C | ±200 | ±(0.5% + 0.5 pF) |
| Orange | 3 | 1000 | 3 | D | ±100 | ±(0.3% + 0.1 pF) |
| Yellow | 4 | 10,000 | — | E | +100 − 20 | ±(0.1% + 0.1 pF) |
| Green | 5 | — | 5 | — | — | — |
| Blue | 6 | — | — | — | — | — |
| Violet | 7 | — | — | — | — | — |
| Gray | 8 | — | — | I | +150 − 50 | ±(0.03% + 0.2 pF) |
| White | 9 | — | — | J | +100 − 50 | ±(0.2% + 0.2 pF) |
| Gold | — | 0.1 | — | — | — | — |
| Silver | — | 0.01 | 10 | — | — | — |

*Courtesy of Sprague Electric Co.*
*Note:* If both rows of dots are not on one face, rotate capacitor about axis of its leads to read second row on side or rear.

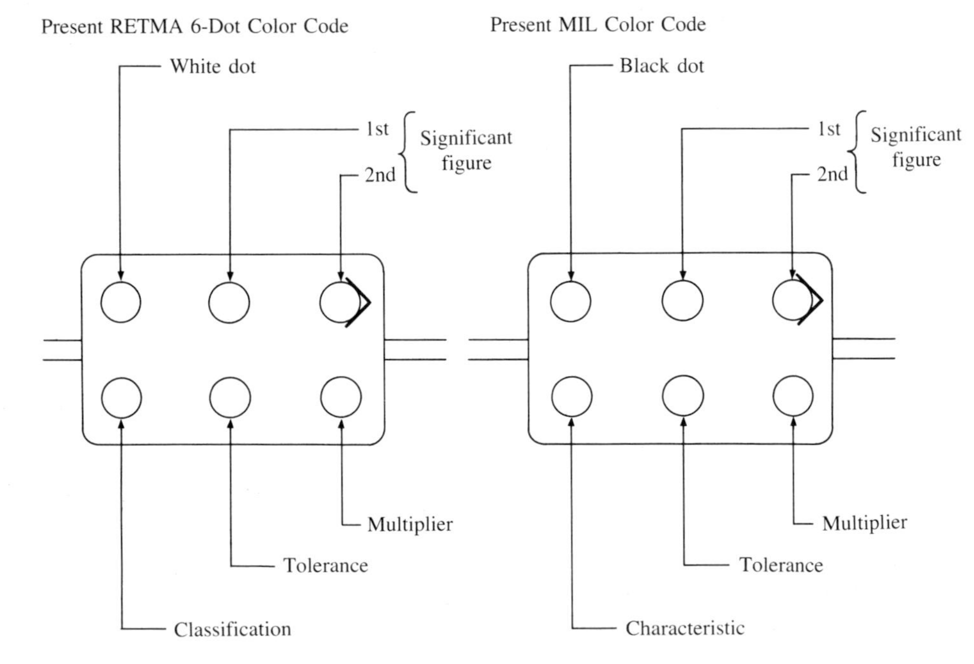


FIG. D.1

# Appendix E

## COLOR CODING OF MOLDED TUBULAR CAPACITORS (PICOFARADS)

| Color | Significant Figure | Decimal Multiplier | Tolerance ± % |
|---|---|---|---|
| Black | 0 | 1 | 20 |
| Brown | 1 | 10 | — |
| Red | 2 | 100 | — |
| Orange | 3 | 1000 | 30 |
| Yellow | 4 | 10,000 | 40 |
| Green | 5 | $10^5$ | 5 |
| Blue | 6 | $10^6$ | — |
| Violet | 7 | — | — |
| Gray | 8 | — | — |
| White | 9 | — | 10 |

*Courtesy of Sprague Electric Co.*

*Note:* Voltage rating is identified by a single-digit number for ratings up to 900 V and a two-digit number above 900 V. Two zeros follow the voltage figure.

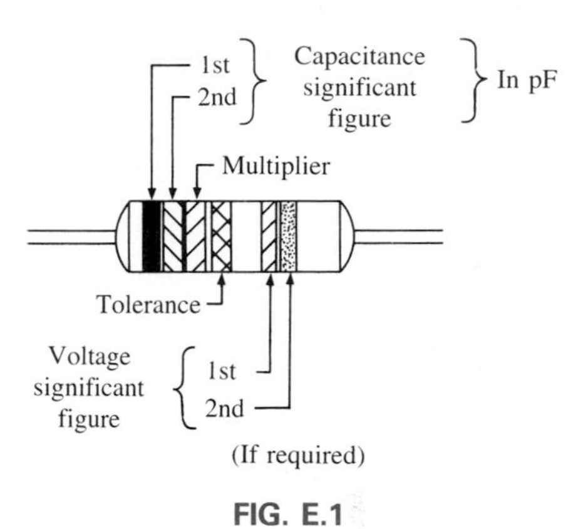


FIG. E.1

# Appendix F

## THE GREEK ALPHABET

| Letter | Capital | Lowercase | Used to Designate |
|---|---|---|---|
| Alpha | A | $\alpha$ | Area, angles, coefficients |
| Beta | B | $\beta$ | Angles, coefficients, flux density |
| Gamma | Γ | $\gamma$ | Specific gravity, conductivity |
| Delta | Δ | $\delta$ | Density, variation |
| Epsilon | E | $\epsilon$ | Base of natural logarithms |
| Zeta | Z | $\zeta$ | Coefficients, coordinates, impedance |
| Eta | H | $\eta$ | Efficiency, hysteresis coefficient |
| Theta | Θ | $\theta$ | Phase angle, temperature |
| Iota | I | $\iota$ | |
| Kappa | K | $\kappa$ | Dielectric constant, susceptibility |
| Lambda | Λ | $\lambda$ | Wavelength |
| Mu | M | $\mu$ | Amplification factor, micro, permeability |
| Nu | N | $\nu$ | Reluctivity |
| Xi | Ξ | $\xi$ | |
| Omicron | O | $o$ | |
| Pi | Π | $\pi$ | 3.1416 |
| Rho | P | $\rho$ | Resistivity |
| Sigma | Σ | $\sigma$ | Summation |
| Tau | T | $\tau$ | Time constant |
| Upsilon | Υ | $\upsilon$ | |
| Phi | Φ | $\phi$ | Angles, magnetic flux |
| Chi | X | $\chi$ | |
| Psi | Ψ | $\psi$ | Dielectric flux, phase difference |
| Omega | Ω | $\omega$ | Ohms, angular velocity |

# Appendix G

## MAGNETIC PARAMETER CONVERSIONS

| | SI (MKS) | | CGS | | English |
|---|---|---|---|---|---|
| $\Phi$ | webers (Wb)<br>1 Wb | = | maxwells<br>$10^8$ maxwells | = | lines<br>$10^8$ lines |
| $B$ | $Wb/m^2$<br><br>1 $Wb/m^2$ | = | gauss<br>($maxwells/cm^2$)<br>$10^4$ gauss | = | lines/in.$^2$<br><br><br>$6.452 \times 10^4$ lines/in.$^2$ |
| $A$ | 1 $m^2$ | = | $10^4$ $cm^2$ | = | 1550 in.$^2$ |
| $\mu_o$ | $4\pi \times 10^{-7}$ Wb/Am | = | 1 gauss/oersted | = | 3.20 lines/Am |
| $\mathscr{F}$ | $NI$ (ampere-turns,<br>At) 1 At | = | $0.4\pi NI$ (gilberts)<br>1.257 gilberts | | $NI$ (At)<br>1 gilbert = 0.7958 At |
| $H$ | $NI/l$ (At/m)<br>1 At/m | = | $0.4\pi NI/l$ (oersteds)<br>$1.26 \times 10^{-2}$ oersted | = | $NI/l$ (At/in.)<br>$2.54 \times 10^{-2}$ At/in. |
| $H_g$ | $7.97 \times 10^5 B_g$<br>(At/m) | | $B_g$ (oersteds) | | $0.313B_g$ (At/in.) |

# Appendix H

## MAXIMUM POWER TRANSFER CONDITIONS

Derivation of maximum power transfer conditions for the situation where the resistive component of the load is adjustable but the load reactance is set in magnitude*

For the circuit of Fig. H.1, the power delivered to the load is determined by

$$P = \frac{V_{R_L}{}^2}{R_L}$$

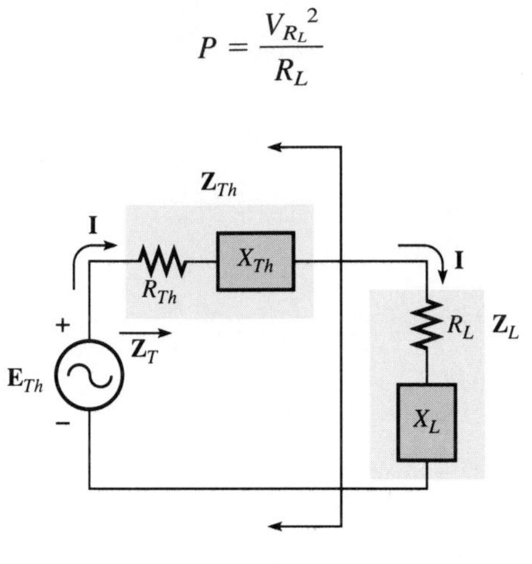

FIG. H.1

Applying the voltage divider rule:

$$\mathbf{V}_{R_L} = \frac{\mathbf{R}_L\mathbf{E}_{Th}}{\mathbf{R}_L + \mathbf{R}_{Th} + \mathbf{X}_{Th} + \mathbf{X}_L}$$

The magnitude of $\mathbf{V}_{R_L}$ is determined by

$$V_{R_L} = \frac{R_L E_{Th}}{\sqrt{(R_L + R_{Th})^2 + (X_{Th} + X_L)^2}}$$

and

$$V_{R_L}^2 = \frac{R_L^2 E_{Th}^2}{(R_L + R_{Th})^2 + (X_{Th} + X_L)^2}$$

with

$$P = \frac{V_{R_L}^2}{R_L} = \frac{R_L E_{Th}^2}{(R_L + R_{Th})^2 + (X_{Th} + X_L)^2}$$

---

*With sincerest thanks for the input of Professor Harry J. Franz of the Beaver Campus of Pennsylvania State University.

Using differentiation (calculus), maximum power will be transferred when $dP/dR_L = 0$. The result of the preceding operation is that

$$R_L = \sqrt{R_{Th}^2 + (X_{Th} + X_L)^2} \qquad \text{[Eq. (18.21)]}$$

The magnitude of the total impedance of the circuit is

$$Z_T = \sqrt{(R_{Th} + R_L)^2 + (X_{Th} + X_L)^2}$$

Substituting this equation for $R_L$ and applying a few algebraic maneuvers will result in

$$Z_T = 2R_L(R_L + R_{Th})$$

and the power to the load $R_L$ will be

$$P = I^2R_L = \frac{E_{Th}^2}{Z_T^2}R_L = \frac{E_{Th}^2R_L}{2R_L(R_L + R_{Th})}$$

$$= \frac{E_{Th}^2}{4\left[\dfrac{R_L + R_{Th}}{2}\right]}$$

$$= \frac{E_{Th}^2}{4R_{av}} \qquad \text{[Eq. (8.22)]}$$

with

$$R_{av} = \frac{R_L + R_{Th}}{2} \qquad \text{[Eq. (8.23)]}$$

# Appendix I

## ANSWERS TO SELECTED ODD-NUMBERED PROBLEMS

### Chapter 1

**5.** 3 h
**7.** CGS
**9.** MKS = CGS = 20°C,
K = SI = 273.15
**11.** 45.72 cm
**13.** **(a)** $15 \times 10^3$ **(b)** $30 \times 10^{-3}$
**(c)** $7.4 \times 10^6$ **(d)** $6.8 \times 10^{-6}$
**(e)** $402 \times 10^{-6}$ **(f)** $200 \times 10^{-12}$
**15.** **(a)** $10^{-1}$ **(b)** $10^{-4}$
**(c)** $10^9$ **(d)** $10^{-9}$
**(e)** $10^{42}$ **(f)** $10^3$
**17.** **(a)** $10^{-6}$ **(b)** $10^{-3}$
**(c)** $10^{-6}$ **(d)** $10^9$
**(e)** $10^{-16}$ **(f)** $10^{-1}$
**19.** 2 ms
**21.** 60 nF
**23.** 6000 mm
**25.** **(a)** 90 s **(b)** 144 s
**(c)** $50 \times 10^3$ $\mu$s
**(d)** 160 mm **(e)** 120 ns
**(f)** 41.898 days **(g)** 1.02 m
**27.** **(a)** 2.54 m **(b)** 1.219 m
**(c)** 26.7 N **(d)** 0.1348 lb
**(e)** 4921.26 ft
**(f)** 3.2187 m **(g)** 850.17 yds
**29.** $670.62 \times 10^6$ mph
**31.** 2.045 s
**33.** 67.20 days
**35.** 3600 quarters = $900
**37.** 345.6 m
**39.** 1.391 min.

### Chapter 2

**3.** **(a)** 18 mN **(b)** 2 mN **(c)** 0.18 mN
**5.** 0.01 m
**9.** 3.1 A
**11.** 90 C
**13.** 0.5 A
**15.** 1.194 A > 1 A (yes)
**17.** **(a)** 1.2484 million
**(b)** 0.9363 million, sol. = (a)
**19.** 252 J
**21.** 4 C
**23.** 3.533 V
**25.** 5 A
**27.** 25 h
**29.** 0.773 h
**31.** 60 Ah:40 Ah = 1.5:1, 50% more with 60 Ah
**33.** 545.45 mA, 129,598.92 kJ
**43.** 600 C

### Chapter 3

**1.** **(a)** 500 mils **(b)** 10 mils
**(c)** 4 mils **(d)** 1000 mils
**(e)** 240 mils **(f)** 3.937 mils
**3.** **(a)** 0.04 in. **(b)** 0.03 in.
**(c)** 0.2 in. **(d)** 0.025 in.
**(e)** 0.00278 in. **(f)** 0.009 in.
**5.** 73.33 Ω
**7.** 3.581 ft
**9.** **(a)** $R_{silver} > R_{copper} > R_{aluminum}$
**(b)** silver 9.9 Ω, copper 1.037 Ω, aluminum 0.17 Ω
**11.** **(a)** 21.71 $\mu\Omega$ **(b)** 35.59 $\mu\Omega$
**(c)** increases **(d)** decreases
**13.** 942.28 mΩ
**15.** **(a)** #8:1.1308 Ω, #18:11.493 Ω
**(b)** #18:#8 = 10.164:1 ≅ 10:1,
#18:#8 = 1:10.164 ≅ 1:10
**17.** **(a)** 1.087 mA/CM
**(b)** 1.384 kA/in.$^2$ **(c)** 3.6127 in.$^2$
**19.** **(a)** 21.71 $\mu$A **(b)** 33.87 $\mu$A
**21.** 0.15 in.
**23.** 2.409 Ω
**25.** 3.334 Ω
**27.** 0.046 Ω
**29.** **(a)** 40.29°C **(b)** −195.61°C
**31.** $\alpha_1 = 0.00393$ (Cu),
$\alpha_1 = 0.00391$ (Al)
**33.** **(a)** $\alpha_1 = 0.00365$ **(b)** 33.83 Ω
**35.** 142.86
**37.** **(a)** July, 1992 **(b)** 43.858 K/yr
**(c)** October, 1992
**41.** −30°C: 10.2 kΩ,
100°C: 10.15 kΩ
**43.** 6.5 kΩ
**47.** yes
**49.** yes
**51.** **(a)** 0.1566 S **(b)** 0.0955 S
**(c)** 0.0219 S
**57.** **(a)** 10 fc:3 kΩ, 100 fc:0.4 kΩ
**(b)** neg. **(c)** no—log scales
**(d)** −321.43 Ω/fc

### Chapter 4

**1.** 15 V
**3.** 4 kΩ
**5.** 72 mV
**7.** 54.55 Ω
**9.** 28.571 Ω
**11.** 1.2 kΩ
**13.** **(a)** 12.632 Ω **(b)** 4.1 MJ
**17.** 800 V
**19.** 1 W
**21.** **(a)** 57,600 J **(b)** $16 \times 10^{-3}$ kWh
**23.** 2 s
**25.** 196 $\mu$W
**27.** 4 A
**29.** 9.61 V
**31.** 144.058 Ω, 0.833 A
**33.** 70.71 mA, 1.414 kV
**35.** **(a)** 4.095 W **(b)** 19.78 Ω
**(c)** 88.45 kJ
**37.** 82.89%
**39.** 84.77%
**41.** **(a)** 1657.78 W **(b)** 15.07 A
**(c)** 19.38 A
**43.** 65.25%
**45.** 80%
**47.** **(a)** 17.9%
**(b)** 76.73%, 328.66% increase
**49.** **(a)** 1350 J
**(b)** *W* doubles, *P* the same
**51.** 6.67 h
**53.** **(a)** 50 kW **(b)** 240.38 A
**(c)** 90 kWh
**55.** 11.94¢

### Chapter 5

**1.** **(a)** 20 Ω, 3 A
**(b)** 1.63 MΩ, 6.135 $\mu$A
**(c)** 110 Ω, 318.2 mA
**(d)** 10 kΩ, 12 mA
**3.** **(a)** 16 V **(b)** 1.56 V
**5.** **(a)** 4.18 mA (CW)
**(b)** 0.388 A (CCW)
**7.** **(a)** 5 V **(b)** 70 V

**9.** 3.28 mA, 7.22 V
**11.** **(a)** 70.6 Ω, 85 mA (CCW),
$V_1 = 2.8045$ V,
$V_2 = 0.4760$ V,
$V_3 = 0.850$ V,
$V_4 = 1.870$ V
**(b)–(c)** $P_1 = 0.2384$ W,
$P_2 = 0.0405$ W,
$P_3 = 0.0723$ W,
$P_4 = 0.1590$ W
**(d)** all $\frac{1}{2}$ W
**13.** **(a)** 225 Ω, 0.533 A **(b)** 8 W
**(c)** 15 V
**15.** **(a)** 2 kΩ **(b)** 1 Ω
**17.** $V_{R_3} = 6$ V, $V_{R_2} = 42$ V,
$V_{R_1} = 12$ V
**19.** **a.** **(a)** $V_a = 4$ V, $V_b = -8$ V
**(b)** $V_a = 14$ V, $V_b = 4$ V
**(c)** $V_a = 13$ V, $V_b = -8$ V
**b.** **(a)** $V_{ab} = 12$ V **(b)** $V_{ab} = 10$ V
**(c)** $V_{ab} = 21$ V
**21.** **(a)** $V_a = 20$ V, $V_b = 26$ V,
$V_c = 35$ V, $V_d = -12$ V,
$V_e = 0$ V
**(b)** $V_{ab} = -6$ V, $V_{dc} = -47$ V,
$V_{cb} = 9$ V
**(c)** $V_{ac} = -15$ V, $V_{db} = -38$ V
**23.** 2 Ω
**25.** 100 Ω
**27.** 1.52%

## Chapter 6

**1.** **(a)** 2, 3, 4 **(b)** 2, 3 **(c)** 1, 4
**3.** **(a)** 6 Ω, 0.1667 S
**(b)** 1 kΩ, 1 mS
**(c)** 2.076 kΩ, 0.4817 mS
**(d)** 1.333 Ω, 0.75 S
**(e)** 9.948 Ω, 100.525 mS
**(f)** 0.6889 Ω, 1.4516 S
**5.** **(a)** 18 Ω **(b)** $R_1 = R_2 = 24$ Ω
**7.** 120 Ω
**9.** **(a)** 0.8571 Ω, 1.1667 S
**(b)** $I_S = 1.05$ A, $I_1 = 0.3$ A,
$I_2 = 0.15$ A, $I_3 = 0.6$ A
**(c)–(d)** $P_1 = 0.27$ W,
$P_2 = 0.135$ W, $P_3 = 0.54$ W,
$P_{\text{del.}} = 0.945$ W
**(e)** $R_1$, $R_2 = \frac{1}{2}$ W, $R_3 = 1$ W
**11.** **(a)** 66.67 mA **(b)** 225 Ω **(c)** 8 W
**13.** **(a)** $I_S = 7.5$ A, $I_1 = 1.5$ A
**(b)** $I_S = 9.6$ mA, $I_1 = 0.8$ mA
**15.** **(a)** 14.67 A **(b)** 256 W
**(c)** 14.67 A
**17.** **(a)** $I_1 = 3$ mA, $I_2 = 1$ mA
$I_3 = 1.5$ mA
**(b)** $I_2 = 4$ μA, $I_3 = 1.5$ μA,
$I_4 = 5.5$ μA, $I_1 = 6$ μA
**19.** **(a)** $I_1 = 4$ A, $I_2 = 8$ A
**(b)** $I_1 = 2$ A, $I_2 = 4$ A,
$I_3 = I_4 = 1$ A, $I_5 = \frac{4}{3}$ A
**(c)** $I_1 = 272.73$ mA,
$I_2 = 227.27$ mA,
$I_5 = 500$ mA,
$I_3 = 136.36$ mA,
$I_4 = 90.91$ mA
**(d)** $I_1 \cong 990$ mA, $I_2 \cong 9.90$ mA,
$I_3 \cong 900$ mA, $I_4 \cong 90$ mA,
$I_5 \cong 9$ mA, $I_6 \cong 0.9$ mA
**21.** 6.6 kΩ
**23.** $I_1 = 1.714$ A, $I_2 = 0.857$ A
**25.** **(a)** 6.13 V **(b)** 1.2776 mA
**(c)** 9 V **(d)** 1.6364 mA
**27.** **(a)** 4 V **(b)** 3.997 V **(c)** 3.871 V
**(d)** 3 V

## Chapter 7

**1.** **(a)** yes (KCL) **(b)** 3 A
**(c)** yes (KCL) **(d)** 4 V
**(e)** 2 Ω **(f)** 5 A
**(g)** $P_1 = 12$ W, $P_2 = 18$ W,
$P_{\text{del.}} = 50$ W
**3.** **(a)** 4 Ω
**(b)** $I_S = 9$ A, $I_1 = 6$ A, $I_2 = 3$ A
**(c)** 6 V
**5.** $I_1 = 6$ A, $I_2 = 16$ A, $I_3 = 0.8$ A,
$I = 22$ A
**7.** **(a)** 4 A
**(b)** $I_2 = 1.333$ A, $I_3 = 0.6665$ A
**(c)** $V_a = 8$ V, $V_b = 4$ V
**9.** **(a)** 5 Ω, 16 A
**(b)** $I_{R_2} = 8$ A, $I_3 = I_9 = 4$ A
**(c)** $I_8 = 1$ A **(d)** 14 V
**11.** **(a)** $V_G = 1.9$ V, $V_S = 3.65$ V
**(b)** $I_1 = I_2 = 7.05$ μA,
$I_D = 2.433$ mA **(c)** 6.268 V
**(d)** 8.02 V
**13.** **(a)** 2 A
**(b)** $I_1 = I_3 = \frac{2}{3}$ A, $I_8 = \frac{2}{9}$ A
**(c)** 1.037 W
**15.** **(a)** $I_2 = 1\frac{2}{3}$ A, $I_6 = 1\frac{1}{9}$ A,
$I_8 = 0$ A
**(b)** $V_4 = 10$ V, $V_8 = 0$ V
**17.** 12 V, 3 A
**19.** **(a)** 14 V **(b)** 9 A
**21.** **(a)** 24 A **(b)** 8 A
**(c)** $V_3 = 48$ V, $V_5 = 24$ V,
$V_7 = 16$ V
**(d)** $P_7 = 128$ W, $P_{\text{del.}} = 5760$ W
**23.** **(a)** 12 A **(b)** 0.5 A **(c)** 0.5 A
**(d)** 6 A
**25.** **(a)** $V_{ab} = 32$ V, $V_{bc} = 8$ V
**(b)** $V_{ab} = 31.51$ V, $V_{bc} = 8.49$ V
**(c)** 16.015 W **(d)** 16 W
**27.** 25 mA: $R_{\text{shunt}} \cong 2$ Ω,
50 mA: $R_{\text{shunt}} \cong 1$ Ω,
100 mA: $R_{\text{shunt}} \cong 0.5$ Ω
**29.** 5 V: $R_S = 4.9$ kΩ, 50 V:
$R_S = 499.9$ kΩ, 500 V:
$R_S = 499.9$ kΩ
**31.** **(a)** $R_{\text{series}} = 28$ kΩ
**(b)** FS: 0 Ω, $\frac{3}{4}$ FS: 10 kΩ,
$\frac{1}{2}$ FS: 30 kΩ, $\frac{1}{4}$ FS: 90 kΩ

## Chapter 8

**1.** 28 V
**3.** **(a)** $I_1 = 12$ A, $I_S = 11$ A
**5.** **(a)** 3 A, 6 Ω **(b)** 4.091 mA,
2.2 kΩ
**7.** **(a)** 8 A **(b)** 8 A
**9.** 2.4 A, 9.6 V
**11.** **(a)** 5.4545 mA, 2.2 kΩ
**(b)** 17.375 V **(c)** 5.375 V
**(d)** 2.443 mA
**13.** **(a)** CW: $I_{R_1} = 1.445$ mA;
down: $I_{R_3} = 9.958$ mA;
CCW: $I_{R_2} = 8.513$ mA
**(b)** CW: $I_{R_1} = 2.0316$ mA;
left: $I_{R_2} = 0.8$ mA;
CW: $I_{R_3} = 1.2316$ mA
**15.** **(b)** $E_1 - I_2(R_1 + R_2) - I_3R_1 = 0$,
$I_2R_2 - I_3(R_3 + R_4) + I_5R_4 =$
$0$, $I_3R_4 - I_5(R_4 + R_5) -$
$E_2 = 0$
**(d)** left: 63.694 mA
**17.** **(a)** CW: $I_{R_1} = -\frac{1}{7}$ A;
CW: $I_{R_2} = -\frac{5}{7}$ A
**(b)** CW: $I_{R_1} = -3.0625$ A;
CW: $I_{R_3} = 0.1875$ A
**19.** **(a)** CW: $I_{R_1} = 1.8701$ A;
CW: $I_{R_2} = -8.5484$ A;
$V_{ab} = -22.74$ V
**(b)** CW: $I_{R_2} = 1.274$ A;
CW: $I_{R_3} = 0.26$ A;
$V_{ab} = -0.904$ V
**21.** **(a)** $I_B = 63.02$ μA,
$I_C = 4.416$ mA,
$I_E = 4.479$ mA
**(b)** $V_B = 2.985$ V, $V_E = 2.285$ V,
$V_C = 10.285$ V
**(c)** 70.07

**23.** **(a)** CW: $I_{R_3} = 1.2059$ mA, $I_{R_4} = -0.4806$ mA, $I_{R_2} = -0.6206$ mA **(b)** CW: $I_{R_5} = -0.2385$ A, $I_{R_3} = -0.5169$ A, $I_{R_6} = -1.278$ A
**25.** **(a)** CW: $I_{R_1} = -\frac{1}{7}$ A, $I_{R_2} = -\frac{5}{7}$ A **(b)** CW: $I_{R_1} = -3.0625$ A, $I_{R_3} = 0.1875$ A
**27.** **(a)** CW: $I_{R_1} = 1.871$ A, $I_{R_2} = -8.548$ A **(b)** CW: $I_{R_2} = 1.274$ A, $I_{R_3} = 0.26$ A
**29.** check sol. 21
**31.** **(a)** CW: $I_{R_3} = 1.2059$ mA, $I_{R_4} = -0.4806$ mA, $I_{R_2} = -0.6206$ mA **(b)** CW: $I_{R_5} = -0.2385$ A, $I_{R_3} = -0.5169$ A, $I_{R_6} = -1.278$ A
**33.** **(a)** left: $V_1 = 8.077$ V, $V_2 = 9.385$ V **(b)** left: $V_1 = 4.8$ V, $V_2 = 6.4$ V
**35.** **(a)** left: $V_1 = -2.653$ V, $V_2 = 0.952$ V **(b)** left: $V_1 = 8.877$ V, right: $V_2 = 9.831$ V, center: $V_3 = -3.005$ V
**37.** **(a)** left: $V_1 = -5.311$ V, center: $V_2 = -0.6219$ V, right: $V_3 = 3.751$ V **(b)** top left: $V_1 = -6.917$ V, right: $V_2 = 12$ V, bottom left: $V_3 = 2.3$ V
**39.** **(b)** $V_{R_5} = 0.1967$ V **(c)** no
**41.** **(b)** $V_{R_5} = 0$ V, yes
**43.** **(a)** 3.33 mA **(b)** 1.177 A
**45.** **(a)** 133.33 mA **(b)** 7 A
**47.** **(b)** 0.256 mA

## Chapter 9

**1.** **(a)** CW: $I_{R_1} = \frac{5}{6}$ A, $I_{R_2} = 0$ A, CW: $I_{R_3} = \frac{5}{6}$ A **(b)** $E_1$: 5.33 W, $E_2$: 0.333 W **(c)** 8.333 W **(d)** no
**3.** **(a)** down: 4.4545 mA **(b)** down: 3.11 A
**5.** **(a)** 6 Ω, 6 V **(b)** 2 Ω: 0.75 A, 30 Ω: 0.1667 A, 100 Ω: 0.0566 A
**7.** (I) 2 Ω, 84 V (II) 1.579 kΩ, 1.149 V
**9.** (I) 45 Ω, 5 V (II) 2.055 kΩ, 16.772 V
**11.** 4.041 kΩ, 9.726 V
**13.** 14 Ω, 2.571 A
**15.** (I) 9.756 Ω, 0.95 A (II) 2 Ω, 30 A
**17.** (I) 10 Ω, 0.2 A (II) 4.033 kΩ, 2.9758 mA
**19.** **(a)** (I) 14 Ω, (II) 7.5 Ω **(b)** (I) 23.14 W, (II) 3.33 W
**21.** **(a)** (I) 9.756 Ω, (II) 2 Ω **(b)** (I) 2.20 W, (II) 450 W
**23.** 0 Ω
**25.** 500 Ω
**27.** 39.3 μA, 220 mV
**29.** 2.25 A, 6.075 V
**35.** **(a)** 0.357 mA **(b)** 0.357 mA **(c)** yes

## Chapter 10

**1.** $9 \times 10^3$ N/C
**3.** 70 μF
**5.** 50 V/m
**7.** $8 \times 10^3$ V/m
**9.** 937.5 μF
**11.** mica
**13.** **(a)** $10^6$ V/m **(b)** 4.96 μC **(c)** 0.0248 μF
**15.** 29,035 V
**17.** **(a)** 0.5 s **(b)** $20(1 - e^{-t/0.5})$ **(c)** $1\tau$: 12.64 V, $3\tau$: 19 V, $5\tau$: 19.87 V **(d)** $i_C = 0.2 \times 10^{-3}e^{-t/0.5}$, $v_R = 20e^{-t/0.5}$
**19.** **(a)** 5.5 ms **(b)** $100(1 - e^{-t/(5.5\times10^{-3})})$ **(c)** $1\tau$: 63.21 V, $3\tau$: 95.02 V, $5\tau$: 99.33 V **(d)** $i_C = 18.18 \times 10^{-3}e^{-t/(5.5\times10^{-3})}$, $v_R = 60e^{-t/(5.5\times10^{-3})}$
**21.** **(a)** 10 ms **(b)** $50(1 - e^{-t/(10\times10^{-3})})$ **(c)** $10 \times 10^{-3}e^{-t/(10\times10^{-3})}$ **(d)** $v_C \cong 50$ V, $i_C \cong 0$ A **(e)** $v_C = 50e^{-t/(4\times10^{-3})}$, $i_C = 25 \times 10^{-3}e^{-t/(4\times10^{-3})}$
**23.** **(a)** $80\ (1 - e^{-t/(1\times10^{-6})})$ **(b)** $0.8 \times 10^{-3}e^{-t/(1\times10^{-6})}$ **(c)** $v_C = 80e^{-t/(4.9\times10^{-6})}$, $i_C = 0.163 \times 10^{-3}e^{-t/(4.9\times10^{-6})}$
**25.** **(a)** 10 μs **(b)** 3 kA **(c)** yes
**27.** 1.386 μs
**29.** $R = 54.567$ kΩ
**31.** **(a)** $v_C = 15(1 - e^{-t/0.15})$, $i_C = 1.5 \times 10^{-3}e^{-t/0.15}$
**33.** 0–4 ms: 0.3 mA, 4–6 ms: 0.9 mA, 6–7 ms: 3 mA, 7–10 ms: 0 mA, 10–13 ms: −3.2 mA, 13–15 ms: 1.8 mA
**35.** 0 ms: $v_C = 0$ V, 4 ms: $v_C = 0$ V, 6 ms: $v_C = -8$ V, 16 ms: $v_C = +12$ V, 18 ms: $v_C = +12$ V, 20 ms: $v_C = 0$ V, 25 ms: $v_C = 0$ V
**37.** **(a)** $V_1 = 10$ V, $Q_1 = 60$ μC, $V_2 = 6.67$ V, $Q_2 = 40$ μC, $V_3 = 3.33$ V, $Q_3 = 40$ μC **(b)** $V_1 = 8$ V, $Q_1 = 9600$ pC, $V_2 = 16$ V, $Q_2 = 3200$ pC, $V_3 = 16$ V, $Q_3 = 6400$ pC, $V_4 = 16$ V, $Q_4 = 9600$ pC
**39.** **(a)** $100(1 - e^{-t/120\text{ ms}})$ **(b)** $42.857(1 - e^{-t/120\text{ ms}})$ **(c)** $57.143(1 - e^{-t/120\text{ ms}})$ **(d)** $100e^{-t/120\text{ ms}}$ **(e)** 433.44 ms
**41.** 8640 pJ
**43.** **(a)** 5 J **(b)** 0.1 C **(c)** 200 A **(d)** 10,000 W **(e)** 10 s

## Chapter 11

**1.** Φ: $5 \times 10^4$ maxwells, $5 \times 10^4$ lines, B: 8 gauss, 51.616 lines
**3.** **(a)** 0.04 T
**5.** $952.4 \times 10^3$ At/Wb
**7.** 2624.67 At/m
**9.** 2.133 A
**11.** **(a)** $N_1 = 60$ t **(b)** $13.34 \times 10^{-4}$ Wb/Am
**13.** 2.7432 A
**15.** 1.35 N
**17.** **(a)** 2.028 A **(b)** ≅ 2 N
**19.** $59.4 \times 10^{-4}$ Wb
**21.** **(a)** $T = 1.5(1 - e^{-H/700})$ **(c)** $H = -700 \log_e(1 - T/1.5)$ **(e)** Eq: 40.1 mA

## Chapter 12

**1.** 4.25 V
**3.** 14 turns
**5.** 15.65 μH
**7.** **(a)** 2.5 V **(b)** 0.3 V **(c)** 200 V

**9.** 0–3 ms: 0 V, 3–8 ms: 1.6 V, 8–13 ms: −1.6 V, 13–14 ms: 0 V, 14–15 ms: 8 V, 15–16 ms: −8 V, 16–17 ms: 0 V
**11.** 5 $\mu$s: 4 mA, 10 $\mu$s: −8 mA, 12 $\mu$s: 4 mA, 16 $\mu$s: 4 mA, 24 ms: 0 mA
**13.** **(a)** 2.27 $\mu$s
**(b)** $5.45 \times 10^{-3}(1 - e^{-t/2.27\ \mu s})$
**(c)** $v_L = 12e^{-t/2.27\ \mu s}$, $v_R = 12(1 - e^{-t/2.27\ \mu s})$
**(d)** $i_L$: $1\tau = 3.45$ mA, $3\tau = 5.179$ mA, $5\tau = 5.413$ mA, $v_L$: $1\tau = 4.415$ V, $3\tau = 0.598$ V, $5\tau = 0.081$ V
**15.** **(a)** $i_L = 0.882 \times 10^{-3}(1 - e^{-t/0.735\ \mu s})$, $v_L = 6e^{-t/0.735\ \mu s}$
**(b)** $i_L = 0.882 \times 10^{-3}e^{-t/0.333\ \mu s}$, $v_L = -13.23e^{-t/0.333\ \mu s}$
**17.** **(a)** $i_L = 1.333 \times 10^{-3}\ (1 - e^{-t/56\ ns})$, $v_L = 48e^{\ t/56\ ns}$
**(b)** 1.107 mA, 8.046 V
**19.** **(a)** $i_L = 3 \times 10^{-3}(1 - e^{-t/6.67\ \mu s})$, $v_L = 2.25e^{-t/6.67\ \mu s}$
**(b)** 1.896 mA, 0.304 V
**(c)** $i_L = 1.896 \times 10^{-3}e^{-t/3.33\ \mu s}$, $v_L = -0.304e^{-t/3.33\ \mu s}$
**21.** **(a)** 10 mH, 18 $\mu$F
**(b)** 25 mH, 18 $\mu$F
**23.** $I_1 = 4$ mA, $V_1 = 16$ V, $V_2 = 0$ V
**25.** $V_1 = 10$ V, $I_1 = 2$ A, $I_2 = 1.33$ A
**27.** $W_C = 360\ \mu$J, $W_L = 12$ J

## Chapter 13

**1.** **(a)** 10 ms **(b)** 2 **(c)** 100 Hz
**(d)** amplitude = 5 V, $V_{p\text{-}p} = 6.67$ V
**3.** 10 ms, 100 Hz
**5.** **(a)** 60 Hz **(b)** 100 Hz
**(c)** 29.41 Hz **(d)** 40 kHz
**7.** 0.25 s
**11.** **(a)** $\pi/4$ **(b)** $\pi/3$ **(c)** $\frac{2}{3}\pi$ **(d)** $\frac{3}{2}\pi$
**(e)** $0.989\pi$ **(f)** $1.228\pi$
**13.** **(a)** 3.14 rad/s
**(b)** $20.93 \times 10^3$ rad/s
**(c)** $1.57 \times 10^6$ rad/s
**(d)** 157 rad/s
**15.** **(a)** 120.06 Hz, 8.33 ms
**(b)** 1.34 Hz, 747.62 ms
**(c)** 955.41 Hz, 1.05 ms
**(d)** $9.95 \times 10^{-3}$ Hz, 100.5 s
**17.** 104.7 rad/s
**23.** 0.4755 A
**25.** 11.537°, 168.463°
**29.** **(a)** $v$ leads $i$ by 10°
**(b)** $i$ leads $v$ by 70°
**(c)** $i$ leads $v$ by 80°
**(d)** $i$ leads $v$ by 150°
**31.** **(a)** $v = 25 \sin(\omega t + 30°)$
**(b)** $i = 3 \times 10^{-3} \sin(6.28 \times 10^3 t - 60°)$
**33.** **(a)** 2 V **(b)** $\frac{1}{2}$ mV
**35.** **(b)** 0.5 V
**37.** **(a)** $2 \sin 377t$ **(b)** $100 \sin 377t$
**(c)** $84.84 \times 10^{-3} \sin 377t$
**(d)** $33.94 \times 10^{-6} \sin 377t$
**39.** 2.16 V
**41.** $G = 0$ V, $V_{\text{eff}} = 8.165$ V
**43.** **(a)** 17.76 V **(b)** 11.31 V

## Chapter 14

**3.** **(a)** $3770 \cos 377t$ **(b)** $-452.4 \sin 754t$
**(c)** $-7.85 \sin(157t - 10°)$
**(d)** $-500 \sin(20t - 150°)$
**5.** **(a)** $210 \sin 754t$
**(b)** $14.8 \sin(400t - 120°)$
**(c)** $42 \times 10^{-3} \sin(\omega t + 88°)$
**(d)** $28 \sin(\omega t + 180°)$
**7.** **(a)** 1.592 H **(b)** 2.654 H
**(c)** 0.8414 H
**9.** **(a)** $100 \sin(\omega t + 90°)$
**(b)** $8 \sin(\omega t + 150°)$
**(c)** $120 \sin(\omega t - 120°)$
**(d)** $60 \sin(\omega t + 190°)$
**11.** **(a)** $1 \sin(\omega t - 90°)$
**(b)** $0.6 \sin(\omega t - 70°)$
**(c)** $0.8 \sin(\omega t + 10°)$
**(d)** $1.6 \sin(377t + 130°)$
**13.** **(a)** $\infty\ \Omega$ **(b)** 530.79 Ω
**(c)** 265.39 Ω **(d)** 17.693 Ω
**(e)** 1.327 Ω
**15.** **(a)** 9.31 Hz **(b)** 4.66 Hz
**(c)** 18.62 Hz **(d)** 1.59 Hz
**17.** **(a)** $6 \times 10^{-3} \sin(200t + 90°)$
**(b)** $33.96 \times 10^{-3} \sin(377t + 90°)$
**(c)** $44.94 \times 10^{-3} \sin(374t + 300°)$
**(d)** $56 \times 10^{-3} \sin(\omega t + 160°)$
**19.** **(a)** $1334 \sin(300t - 90°)$
**(b)** $37.17 \sin(377t - 90°)$
**(c)** $127.2 \sin 754t$
**(d)** $100 \sin(1600t - 170°)$
**21.** **(a)** $C$ **(b)** $L = 254.78$ mH
**(c)** $R = 5\ \Omega$
**25.** 318.47 mH
**27.** 5.067 nF
**29.** **(a)** 0 W **(b)** 0 W **(c)** 122.5 W
**31.** 192 W
**33.** $40 \sin(\omega t - 50°)$
**35.** **(a)** $2 \sin(157t - 60°)$
**(b)** 318.47 mH **(c)** 0 W
**37.** **(a)** $i_1 = 2.828 \sin(10^4 t + 150°)$, $i_2 = 11.312 \sin(10^4 t + 150°)$
**(b)** $i_S = 14.14 \sin(10^4 t + 150°)$
**39.** **(a)** 5 ∠36.87° **(b)** 2.83 ∠45°
**(c)** 16.38 ∠77.66°
**(d)** 806.23 ∠82.87°
**(e)** 1077.03 ∠21.80°
**(f)** 0.00658 ∠81.25°
**(g)** 11.78 ∠−49.82°
**(h)** 8.94 ∠153.43°
**(i)** 61.85 ∠−104.04°
**(j)** 101.53 ∠−39.81°
**(k)** 4326.66 ∠123.69°
**(l)** $25.495 \times 10^{-3}$ ∠−78.69°
**41.** **(a)** 15.033 ∠86.19°
**(b)** 60.208 ∠4.76°
**(c)** 0.30 ∠88.09°
**(d)** 2002.5 ∠−87.14°
**(e)** 86.182 ∠93.73°
**(f)** 38.694 ∠−94°
**43.** **(a)** $11.8 + j7$ **(b)** $151.9 + j49.9$
**(c)** $4.72 \times 10^{-6} + j71$
**(d)** $5.2 + j1.6$ **(e)** $209.3 + j311$
**(f)** $-21.2 + j12$ **(g)** $7.03 + j9.93$
**(h)** $95.698 + j22.768$
**45.** **q.** 6 ∠−50° **r.** $0.2 \times 10^{-3}$ ∠140°
**s.** 109 ∠−230° **t.** 76.471 ∠−80°
**u.** 4 ∠0° **v.** 0.71 ∠−16.49°
**w.** $4.21 \times 10^{-3}$ ∠161.1°
**x.** 18.191 ∠−50.91°
**47.** **(a)** 100 ∠30° **(b)** 0.25 ∠−40°
**(c)** 70.71 ∠−90° **(d)** 29.69 ∠0°
**(e)** $4.242 \times 10^{-6}$ ∠90°
**(f)** $2.546 \times 10^{-6}$ ∠70°
**49.** $41.769 \sin(377t + 29.43°)$
**51.** $76.297 \sin(\omega t + 40.59°)$

## Chapter 15

**1.** **(a)** 6.8 Ω ∠0° **(b)** 754 Ω ∠90°
**(c)** 15.7 Ω ∠90°
**(d)** 265.25 Ω ∠−90°
**(e)** 318.47 Ω ∠−90°
**(f)** 200 Ω ∠0°
**3.** **(a)** $88 \times 10^{-3} \sin \omega t$
**(b)** $9045 \sin(377t + 150°)$
**(c)** $2547.02 \sin(157t - 50°)$

**5.** **(a)** 4.24 Ω ∠−45°
**(b)** 3.04 kΩ ∠80.54°
**(c)** 1617.56 Ω ∠88.33°

**7.** **(a)** 10 Ω ∠36.87°
**(c)** $\mathbf{I}$ = 10 A ∠−36.87°,
$\mathbf{V}_R$ = 80 V ∠−36.87°,
$\mathbf{V}_L$ = 60 V ∠53.13°
**(f)** 800 W **(g)** 0.8 lagging

**9.** **(a)** 1660.27 Ω ∠−73.56°
**(b)** 8.517 mA ∠73.56°
**(c)** $\mathbf{V}_R$ = 4.003 V ∠73.56°,
$\mathbf{V}_L$ = 13.562 V ∠−16.44°
**(d)** 34.09 mW, 0.283 leading

**11.** **(a)** 3.16 kΩ ∠18.43°
**(c)** 3.18 μF, 6.37 H
**(d)** $\mathbf{I}$ = 1.3424 mA ∠41.57°,
$\mathbf{V}_R$ = 4.027 V ∠41.57°,
$\mathbf{V}_L$ = 2.6848 V ∠131.57°,
$\mathbf{V}_C$ = 1.342 V ∠−48.43°
**(g)** 5.406 mW **(h)** 0.9487 lagging

**13.** **(a)** $\mathbf{V}_1$ = 8.94 V ∠223.43°,
$\mathbf{V}_2$ = 26.83 V ∠43.43°
**(b)** $\mathbf{V}_1$ = 112.92 V ∠12.432°,
$\mathbf{V}_2$ = 58.66 V ∠−139.936°

**15.** **(a)** $\mathbf{I}$ = 0.132 A ∠−38.552°,
$\mathbf{V}_R$ = 3.96 V ∠−38.552°,
$\mathbf{V}_C$ = 0.35 V ∠−128.552°
**(b)** 0.198 lagging **(c)** 0.523 W
**(g)** $R + jX_L$ = 30 Ω + $j$148.147 Ω

**17.** 31.34 Ω + $j$22.10 Ω

**21.** **(a)** $\mathbf{Y}_T$ = 0.0215 ∠0°,
$\mathbf{Z}_T$ = 47 Ω ∠0°
**(b)** $\mathbf{Y}_T$ = 5 mS ∠−90°,
$\mathbf{Z}_T$ = 200 Ω ∠90°
**(c)** $\mathbf{Y}_T$ = 1.667 S ∠90°,
$\mathbf{Z}_T$ = 0.6 Ω ∠−90°
**(d)** $\mathbf{Y}_T$ = 0.1014 S ∠−9.46°,
$\mathbf{Z}_T$ = 9.86 Ω ∠9.46°
**(e)** $\mathbf{Y}_T$ = 0.190 S ∠61.39°,
$\mathbf{Z}_T$ = 5.267 Ω ∠−61.39°
**(f)** $\mathbf{Y}_T$ = 0.338 mS ∠−9.546°,
$\mathbf{Z}_T$ = 2.959 kΩ ∠9.546°

**23.** **(a)** $\mathbf{Y}_T = G + jB_C$ = 0.171 S + $j$0.470 S
**(b)** $\mathbf{Y}_T = G + jB_C$ = 0.043 mS + $j$0.246 mS
**(c)** $\mathbf{Y}_T = G - jB_L$ = 0.013 mS − $j$0.022 mS

**25.** **(a)** 0.112 mS ∠26.57°
**(c)** $\mathbf{E}$ = 17.86 V ∠−6.57°,
$\mathbf{I}_R$ = 1.786 mA ∠−6.57°,
$\mathbf{I}_C$ = 0.893 mA ∠83.43°
**(f)** 31.9 mW **(g)** 0.893 leading

**27.** **(a)** $\mathbf{Y}_T$ = 0.885 S ∠−19.81°
**(c)** 531 μF, 5.31 mH
**(d)** $\mathbf{E}$ = 2.397 V ∠79.81°,
$\mathbf{I}_R$ = 1.998 A ∠79.81°,
$\mathbf{I}_L$ = 1.199 A ∠−10.19°,
$\mathbf{I}_C$ = 0.479 A ∠169.81°
**(g)** 4.79 W **(h)** 0.941 lagging

**29.** **(a)** $\mathbf{Y}_T$ = 0.110 S ∠65.77°
**(c)** 636.9 μF, 31.8 mH
**(d)** $\mathbf{E}$ = 25.03 V ∠60°,
$\mathbf{I}_S$ = 2.75 A ∠125.77°,
$\mathbf{I}_C$ = 5 A ∠150°,
$\mathbf{I}_R$ = 1.14 A ∠60°,
$\mathbf{I}_L$ = 2.503 A ∠−30°
**(g)** 28.59 W **(h)** 0.4091 leading

**37.** **(a)** 4.454 kΩ − $j$1.047 kΩ
**(b)** 17.481 Ω + $j$29.717 Ω

**39.** **(a)** $\mathbf{E}$ = 75.6 V ∠−70.11°,
$\mathbf{I}_R$ = 0.3436 A ∠−70.11°,
$\mathbf{I}_L$ = 12.04 mA ∠−160.11°
**(b)** 0.3401 leading **(c)** 25.9734 W
**(f)** 0.4748 A ∠19.63°
**(g)** $\mathbf{Z}_T$ = 25.72 Ω − $j$71.08 Ω

## Chapter 16

**1.** **(a)** 1.2 Ω ∠90° **(b)** 10 A ∠−90°
**(c)** 10 A ∠−90°
**(d)** $\mathbf{I}_2$ = 6 A ∠−90°,
$\mathbf{I}_3$ = 4 A ∠−90°
**(e)** 60 V ∠0°

**3.** **(a)** $\mathbf{Z}_T$ = 3.87 Ω ∠−11.817°,
$\mathbf{Y}_T$ = 0.258 S ∠11.817°
**(b)** 15.504 A ∠41.817°
**(c)** 3.985 A ∠82.826°
**(d)** 47.809 V ∠−7.174°
**(e)** 910.71 W

**5.** **(a)** 0.375 A ∠25.346°
**(b)** 70.711 V ∠−45° **(c)** 33.9 W

**7.** **(a)** 1.423 A ∠18.259°
**(b)** 26.574 V ∠4.763°
**(c)** 54.074 W

**9.** **(a)** $\mathbf{Y}_T$ = 0.099 S ∠−9.709°
**(b)** $\mathbf{V}_1$ = 20.4 V ∠30°,
$\mathbf{V}_2$ = 10.887 V ∠58.124°
**(c)** 1.933 A ∠11.109°

**11.** 33.201 A ∠38.889°

**13.** 139.707 mW

## Chapter 17

**3.** **(a)** $\mathbf{Z}$ = 21.93 Ω ∠−46.85°,
$\mathbf{E}$ = 10.97 V ∠13.15°
**(b)** $\mathbf{Z}$ = 5.15 Ω ∠59.04°,
$\mathbf{E}$ = 10.3 V ∠179.04°

**5.** **(a)** 5.15 A ∠−24.5°
**(b)** 0.442 A ∠143.48°

**7.** **(a)** 13.07 A ∠−33.71°
**(b)** 48.33 A ∠−77.57°

**9.** $-3.165 \times 10^{-3}$ **V** ∠137.29°

**11.** **(a)** $\mathbf{V}_1$ = 19.86 V ∠43.8°,
$\mathbf{V}_2$ = 8.94 V ∠106.9°
**(b)** $\mathbf{V}_1$ = 19.78 V ∠132.48°,
$\mathbf{V}_2$ = 13.37 V ∠98.78°

**13.** same as 11.(a)

**15.** $\mathbf{V}_1$ = 14.68 V ∠68.89°,
$\mathbf{V}_2$ = 12.97 V ∠155.88°

**17.** **(a)** $\mathbf{V}_1$ = 5.74 V ∠122.76°,
$\mathbf{V}_2$ = 4.04 V ∠145.03°,
$\mathbf{V}_3$ = 25.94 V ∠78.07°
**(b)** $\mathbf{V}_1$ = 15.13 V ∠1.29°,
$\mathbf{V}_2$ = 17.24 V ∠3.73°,
$\mathbf{V}_3$ = 10.59 V ∠−0.11°

**19.** $\mathbf{V}_L = 171.63 \times 10^3$ **I** ∠59.3°

**21.** **(a)** yes **(b)** 0 A **(c)** 0 V

**23.** $L_x$ = 5 mH, $R_x$ = 5 Ω

**27.** **(a)** 11.98 A ∠38.5°
**(b)** 7.02 A ∠20.56°

## Chapter 18

**1.** **(a)** 6.095 A ∠−32.115°
**(b)** 3.77 A ∠−93.8°

**3.** $i$ = 0.5 A + 1.581 sin($\omega t$ − 26.565°)

**5.** 6.261 mA ∠−63.43°

**7.** −22.09 V ∠6.34°

**9.** 19.62 V ∠53°

**11.** **(a)** $\mathbf{Z}_{Th}$ = 2.4 Ω ∠36.87°,
$\mathbf{E}_{Th}$ = 80 V ∠36.87°
**(b)** $\mathbf{Z}_{Th}$ = 5.263 kΩ ∠74.741°,
$\mathbf{E}_{Th}$ = 16.639 V ∠−33.69°

**13.** **(a)** $\mathbf{Z}_{Th}$ = 4.997 Ω ∠−38.663°,
$\mathbf{E}_{Th}$ = 77.139 V ∠50.412°
**(b)** $\mathbf{Z}_{Th}$ = 14.142 Ω ∠−45°,
$\mathbf{E}_{Th}$ = 11.571 V ∠36.336°

**15.** **(a)** $\mathbf{Z}_{Th}$ = 9 Ω ∠0°,
$\mathbf{E}_{Th}$ = 12 V + 24 V ∠0°
**(b)** 2.65 V ∠−83.66°

**17.** $\mathbf{Z}_{Th}$ = 5.1 kΩ ∠−11.31°,
$\mathbf{E}_{Th}$ = 10 **V**

**19.** $\mathbf{Z}_{Th}$ = 20 kΩ ∠0°,
$\mathbf{E}_{Th}$ = −3990 V ∠0°

**21.** $\mathbf{Z}_{Th}$ = 25 kΩ ∠0°,
$\mathbf{E}_{Th}$ = −1800 V ∠0°

**23.** $\mathbf{Z}_{Th}$ = 105 kΩ ∠0°,
$\mathbf{E}_{Th}$ = 315 V ∠0°

**25.** **(a)** $\mathbf{Z}_N$ = 21.312 Ω ∠32.196°,
$\mathbf{I}_N$ = 0.1 A ∠0°

(b) $\mathbf{Z}_N = 6.813\ \Omega\ \angle -54.228°$, $\mathbf{I}_N = 8.506\ \text{A}\ \angle 65.324°$

**27.** (a) $\mathbf{Z}_N = 9.66\ \Omega\ \angle 14.93°$, $\mathbf{I}_N = 2.15\ \text{A}\ \angle -42.87°$
(b) $\mathbf{Z}_N = 4.37\ \Omega\ \angle 55.67°$, $\mathbf{I}_N = 22.83\ \text{A}\ \angle -34.65°$

**29.** (a) $\mathbf{Z}_N = 9\ \Omega\ \angle 0°$, $\mathbf{I}_N = 1.333\ \text{A} + 2.667\ \text{A}\ \angle 0°$
(b) $2.65\ \text{V}\ \angle -83.66°$

**31.** $\mathbf{Z}_N = 5.1\ \text{k}\Omega\ \angle -11.31°$, $\mathbf{I}_N = -1.961 \times 10^{-3}\ \mathbf{V}\ \angle 11.31°$

**33.** $\mathbf{Z}_N = 5.1\ \text{k}\Omega\ \angle -11.31°$, $\mathbf{I}_N = 9.81\ \text{mA}\ \angle 11.31°$

**35.** $\mathbf{Z}_N = 6.63\ \text{k}\Omega\ \angle 0°$, $\mathbf{I}_N = 0.792\ \text{mA}\ \angle 0°$

**37.** (a) $\mathbf{Z}_L = 8.32\ \Omega\ \angle 3.18°$, 1198.2 W
(b) $\mathbf{Z}_L = 1.562\ \Omega\ \angle -14.47°$, 1.614 W

**39.** 40 kΩ, 25 W

**41.** (a) 9 Ω (b) 6.78 W

**43.** (a) 1.414 kΩ (b) 0.518 W

**47.** $25.77\ \text{mA}\ \angle 104.4°$

## Chapter 19

**1.** (a) $R$: 300 W, $L$: 0 W, $C$: 0 W
(b) $R$: 0 VAR, $L$: 900 VAR, $C$: 500 VAR
(c) $R$: 300 VA, $L$: 900 VA, $C$: 300 VA
(d) 300 W, 400 VAR($L$), 500 VA, 0.6 lagging
(f) 5 J
(g) $W_C = 1.327$ J, $W_L = 2.389$ J

**3.** (a) 1200 W, 1200 VAR($L$), 1697 VA, 0.7071 lagging
(c) $8.485\ \text{A}\ \angle -45°$

**5.** (a) 180 W, 0 VAR, 180 VA
(b) 0 W, 360 VAR($L$), 360 VA
(c) 580 W, 960 VAR($L$), 1121.61 VA, 0.52 lagging

**7.** (a) $R$: 300 W, $L$: 0 W, $C$: 0 W
(b) $R$: 0 VAR, $L$: 400 VAR, $C$: 250 VAR
(c) $R$: 300 VA, $L$: 400 VA, $C$: 250 VA
(d) 300 W, 150 VAR($L$), 335.41 VA, 0.89 lagging
(f) 8.33 J
(g) $W_C = 0.66$ J, $W_L = 1.06$ J

**9.** (a) $\mathbf{Z}_T = 2\ \Omega - j3.464\ \Omega$
(b) 5 kW

**11.** (a) 300 W, 400 VAR($C$), 500 VA, 0.6 leading
(b) $16.67\ \text{A}\ \angle 53.13°$
(d) 0 W: $X_C = 2.159\ \Omega$, 300 W: $R = 1.0796\ \Omega$, $X_L = 0.7197\ \Omega$

**13.** (a) 1100 W, 2366.26 VAR($C$), 2609.44 VA, 0.4215 leading
(b) $521.89\ \text{V}\ \angle -65.07°$
(c) 100 W: $R = 1743.38\ \Omega$, $X_C = 1307.53\ \Omega$, 1000 W: $R = 43.59\ \Omega$, $X_C = 99.88\ \Omega$

**15.** (a) 438 μF (b) 230 μF

**17.** (a) 37.441 kVA (b) 0.935 lagging
(c) 13.299 kVAR($L$) (d) 37.441 A
(e) 35 μF (f) 35 A

**19.** (a) $V$: 120 V, $I$: 5.5 A, $W$: 396 W
(b) $13.09\ \Omega + j17.46\ \Omega$

**21.** (a) 2.358 A, 22.24 W
(b) 2.07 A, 89.6 mH
(c) 28.9 W, 107.77 mH

## Chapter 20

**1.** (a) $\omega_s = 250$ rad/s, $f_s = 39.81$ Hz
(b) $\omega_s = 3535.53$ rad/s, $f_s = 562.98$ Hz
(c) $\omega_s = 21{,}880$ rad/s, $f_s = 3484.08$ Hz

**3.** (a) 40 Ω
(b) $\mathbf{I} = 10\ \text{A}\ \angle 30°$, $\mathbf{V}_R = 20\ \text{V}\ \angle 30°$, $\mathbf{V}_L = 400\ \text{V}\ \angle 120°$, $\mathbf{V}_C = 400\ \text{V}\ \angle -60°$
(d) 200 W, 4 kVAR($L$), 4 kVAR($C$)
(e) 20

**5.** (a) 400 Hz
(b) $f_1 = 5800$ Hz, $f_2 = 6200$ Hz
(c) 45 Ω (d) 375 mW

**7.** (a) 10 (b) 20 Ω
(c) 1.59 mH, 3.98 μF
(d) $f_1 = 1900$ Hz, $f_2 = 2100$ Hz

**9.** $R = 10\ \Omega$, 13,27 mH, 27.08 nF

**11.** (a) 1 MHz (b) 160 kHz
(c) 720 Ω, 0.7162 mH, 35.37 pF
(d) 56.25 Ω

**13.** (a) 159.155 kHz (b) 10 V
(c) $I_L = I_C = 0.1$ A (d) 12

**15.** (a) 104 Ω (b) 342.11 Ω
(c) $\mathbf{I}_C = 16.45\ \text{mA}\ \angle 90°$, $\mathbf{I}_L = 16.78\ \text{mA}\ \angle -78.69°$
(d) 0.796 mH, 76.56 nF
(e) 3.29, 6079.03 Hz

**17.** (a) 102.734 kHz
(b) $X_L = X_C = 51.64\ \Omega$
(c) 34.43 (d) 1509.7 Ω
(e) $\mathbf{E} = 15.1\ \text{V}\ \angle 0°$, $\mathbf{I}_C = 292.41\ \text{mA}\ \angle 90°$, $\mathbf{I}_L = 292.3\ \text{mA}\ \angle -88.34°$
(f) $Q_p = 29.24$, BW = 3.51 kHz

**19.** (a) 1581 Ω (b) 1581 Ω
(c) 15,735 Hz (d) 6.4 nF

**21.** $R_s = 100.44\ \text{k}\Omega$, $C = 1.268$ nF

**23.** (a) 3 kΩ (b) 100 (c) 11.764 kHz
(d) 211.764 V

**25.** (a) 251.65 kHz (b) 4.444 kΩ
(c) 14.05 (d) 17.91 kHz
(e) 20 μH: $f_s = 251.65$ kHz, $Q_l = 1.581$, $Z_{T_p} = 49.94\ \Omega$, $Q_p = 1.579$, BW = 159.373 kHz
(f) 2 μH: $f_s = 251.65$ kHz, $Q_l = 0.1581$, $Z_{T_p} = 0.5\ \Omega$, $Q_p = 0.158$, BW = 1.59 MHz
(g) $10^5$, $10^3$, 10 (h) yes

## Chapter 21

**1.** (a) 5 (b) −4 (c) 8 (d) −6
(e) 1.301 (f) 3.937 (g) 4.748
(h) −0.498

**3.** (a) 11.513 (b) −9.21 (c) 2.996
(d) 9.065

**5.** −0.699 (yes)

**7.** 1.431 (yes)

**9.** 25.12 W

**11.** 20.792 dBm

**13.** 251.785 mV

**15.** $10^3$

**17.** $f_c = 3617.16$ Hz,
$0.1f_c$: $0.995\ \angle -5.71°$,
$0.5f_c$: $0.894\ \angle -26.57°$,
$f_c$: $0.707\ \angle -45°$,
$2f_c$: $0.447\ \angle -63.43°$,
$10f_c$: $0.0995\ \angle -84.29°$

**19.** (a) $f_c = 15.915$ kHz

**21.** (a) low pass: $f_c = 795.77$ Hz, high pass: $f_c = 1989.44$ Hz
(b) BW = 1193.67 Hz, center = 1392.61 Hz

**23.** (a) $f_s = 100.658$ kHz, $V_{o_{max}} = 0.93$ V
(b) $Q_s = 18.37$, BW = 5,479.5 Hz

**25.** (a) 12.195
(b) BW = 410 Hz, $f_1 = 4795$ Hz, $f_2 = 5205$ Hz
(c) at $f_s$, $V_o = 0.024V_i$

**27.** band-stop-726.44 kHz, pass-band-2.013 MHz

**29.** **(a)** $f_1 = 6772.55$ Hz
**(c)** $\frac{1}{2}f_1$: $-6.7$ dB, $2f_1$: $-0.969$ dB
**31.** **(a)** $f_1 = 19.894$ kHz,
$f_2 = 1989.44$ Hz
**(b)** $f_1$: $-39.29°$, $f_2$: $-39.29°$
**33.** **(a)** $f_2 = 1591.55$ Hz,
$A_{v_{dB}} = -3.915$ dB;
$f_1 = 1768.4$ Hz,
$A_{v_{dB}} = -3.49$ dB
**(b)** $f_2$: 45°, $f_1$: 42°
**35.** **(a)** $f_2 = 79.577$ kHz,
$f < f_2$: $-1.938$ dB,
$f > f_2$: $-6$ dB/octave
**37.** $f_c = 2$ kHz, $f < f_c$: 0 dB,
$f > f_c$: $-6$ dB/octave
**39.** $f_1 = 159.15$ Hz, $f_2 = 318.31$ Hz,
$f_3 = 477.465$ Hz, $f < f_1$: 0 dB,
$f > f_1$: $+6$ dB, $f > f_2$: 12 dB,
$f > f_3$: 0 dB/octave

## Chapter 22

**1.** **(a)** positive-going **(b)** 2 V
**(c)** 0.2 ms **(d)** 6 V **(e)** 6.5%
**3.** **(a)** positive-going **(b)** 10 mV
**(c)** 3.2 ms **(d)** 20 mV **(e)** 3.4%
**7.** **(a)** 120 $\mu$s **(b)** 8.333 kHz
**(c)** maximum = 440 mV,
minimum = 80 mV
**9.** prf = 125 kHz,
duty cycle = 62.5%
**11.** 0 V
**13.** 18.88 mV
**15.** 117 mV
**17.** $v_C = 4(1 + e^{-t/20\text{ms}})$
**19.** $i_C = -8 \text{ mA}e^{-t}$
**21.** $i_C = 4 \text{ mA}e^{-t/0.2\text{ms}}$ **(a)** $5\tau = T/2$
**(b)** $5\tau = \frac{1}{5}(T/2)$ **(c)** $5\tau = 10(T/2)$
**23.** $0 - T/2$: $v_C = 20$ V, $T/2 - T$:
$v_C = 20e^{-t/\tau}$, $T - \frac{3}{2}T$:
$v_C = 20(1 - e^{-t/\tau})$
**25.** $\mathbf{Z}_p = 5.964 \text{ M}\Omega \angle -48.502°$,
$\mathbf{Z}_s = 0.663 \text{ M}\Omega \angle -48.502°$

## Chapter 23

**1.** **(a)** 120.1 V **(b)** 120.1 V
**(c)** 12.01 A **(d)** 12.01 A
**3.** **(a)** 120.1 V **(b)** 120.1 V
**(c)** 16.98 A **(d)** 16.98 A
**5.** **(a)** $\theta_2 = -120°$, $\theta_3 = 120°$
**(b)** $\mathbf{V}_{an} = 120 \text{ V} \angle 0°$,
$\mathbf{V}_{bn} = 120 \text{ V} \angle -120°$,
$\mathbf{V}_{cn} = 120 \text{ V} \angle 120°$
**(c)** $\mathbf{I}_{an} = 8 \text{ A} \angle -53.13°$,
$\mathbf{I}_{bn} = 8 \text{ A} \angle -173.13°$,
$\mathbf{I}_{cn} = 8 \text{ A} \angle 66.87°$
**(e)** 8 A **(f)** 207.85 V
**7.** $V_\phi = 127$ V, $I_\phi = 8.98$ A,
$I_L = 8.98$ A
**9.** **(a)** $\mathbf{E}_{AN} = 12.7 \text{ kV} \angle -30°$,
$\mathbf{E}_{BN} = 12.7 \text{ kV} \angle -150°$,
$\mathbf{E}_{CN} = 12.7 \text{ kV} \angle 90°$
**(b)** $\mathbf{I}_{an} = 11.107 \text{ A} \angle -97.54°$,
$\mathbf{I}_{bn} = 11.107 \text{ A} \angle -217.54°$,
$\mathbf{I}_{cn} = 11.107 \text{ A} \angle 22.46°$
**(c)** $\mathbf{I}_L = \mathbf{I}_\phi$
**(d)** $\mathbf{V}_{an} =$
$11{,}962.57 \text{ V} \angle -29.34°$,
$\mathbf{V}_{bn} =$
$11{,}962.57 \text{ V} \angle -149.34°$,
$\mathbf{V}_{cn} = 11{,}962.57 \text{ V} \angle 90.66°$
**11.** **(a)** 120.1 V **(b)** 208 V
**(c)** 13.364 A **(d)** 23.15 A
**13.** **(a)** $\theta_2 = -120°$, $\theta_3 = +120°$
**(b)** $\mathbf{V}_{ab} = 208 \text{ V} \angle 0°$,
$\mathbf{V}_{bc} = 208 \text{ V} \angle -120°$,
$\mathbf{V}_{ca} = 208 \text{ V} \angle 120°$
**(d)** $\mathbf{I}_{ab} = 9.455 \text{ A} \angle 0°$,
$\mathbf{I}_{bc} = 9.455 \text{ A} \angle -120°$,
$\mathbf{I}_{ca} = 9.455 \text{ A} \angle 120°$
**(e)** 16.376 A **(f)** 120.1 V
**15.** **(a)** $\theta_2 = -120°$, $\theta_3 = 120°$
**(b)** $\mathbf{V}_{ab} = 208 \text{ V} \angle 0°$,
$\mathbf{V}_{bc} = 208 \text{ V} \angle -120°$,
$\mathbf{V}_{ca} = 208 \text{ V} \angle 120°$
**(d)** $\mathbf{I}_{ab} = 86.67 \text{ A} \angle -36.87°$,
$\mathbf{I}_{bc} = 86.67 \text{ A} \angle -156.87°$,
$\mathbf{I}_{ca} = 86.67 \text{ A} \angle 83.13°$
**(e)** 150.11 A **(f)** 120.1 V
**17.** **(a)** $\mathbf{I}_{ab} = 15.325 \text{ A} \angle -73.30°$,
$\mathbf{I}_{bc} = 15.325 \text{ A} \angle -193.30°$,
$\mathbf{I}_{ca} = 15.325 \text{ A} \angle 46.7°$
**(b)** $\mathbf{I}_{Aa} = 26.54 \text{ A} \angle -103.31°$,
$\mathbf{I}_{Bb} = 26.54 \text{ A} \angle 136.68°$,
$\mathbf{I}_{Cc} = 26.54 \text{ A} \angle 16.69°$
**(c)** $\mathbf{E}_{AB} = 17{,}013.6 \text{ V} \angle -0.59°$,
$\mathbf{E}_{BC} =$
$17{,}013.77 \text{ V} \angle -120.59°$,
$\mathbf{E}_{CA} = 17{,}013.87 \text{ V} \angle 119.41°$
**19.** **(a)** 208 V **(b)** 120.09 V
**(c)** 7.076 A **(d)** 7.076 A
**21.** $V_\phi = 69.28$ V, $I_\phi = 2.89$ A,
$I_L = 2.89$ A
**23.** $V_\phi = 69.28$ V, $I_\phi = 5.77$ A,
$I_L = 5.77$ A
**25.** **(a)** 440 V **(b)** 440 V **(c)** 29.33 A
**(d)** 50.8 A
**27.** **(a)** $\theta_2 = -120°$, $\theta_3 = +120°$
**(b)** $\mathbf{V}_{ab} = 100 \text{ V} \angle 0°$,
$\mathbf{V}_{bc} = 100 \text{ V} \angle -120°$,
$\mathbf{V}_{ca} = 100 \text{ V} \angle 120°$
**(d)** $\mathbf{I}_{ab} = 5 \text{ A} \angle 0°$,
$\mathbf{I}_{bc} = 5 \text{ A} \angle -120°$,
$\mathbf{I}_{ca} = 5 \text{ A} \angle 120°$
**(e)** 8.66 A
**29.** **(a)** $\theta_2 = -120°$, $\theta_3 = 120°$
**(b)** $\mathbf{V}_{ab} = 100 \text{ V} \angle 0°$,
$\mathbf{V}_{bc} = 100 \text{ V} \angle -120°$,
$\mathbf{V}_{ca} = 100 \text{ V} \angle 120°$
**(d)** $\mathbf{I}_{ab} = 7.072 \text{ A} \angle 45°$,
$\mathbf{I}_{bc} = 7.072 \text{ A} \angle -75°$,
$\mathbf{I}_{ca} = 7.072 \text{ A} \angle 165°$
**(e)** 12.25 A
**31.** 2160 W, 0 VAR, 2160 VA,
$F_p = 1$
**33.** 7210.67 W, 7210.67 VAR($C$),
10,197.42 VA, 0.707 leading
**35.** 7.263 kW, 7.263 kVAR,
10.272 kVA, 0.707 lagging
**37.** 287.93 W, 575.86 VAR($L$),
643.83 VA, 0.4472 lagging
**39.** 900 W, 1200 VAR($L$), 1500 VA,
0.6 lagging
**41.** $\mathbf{Z}_\phi = 12.98\ \Omega - j17.31\ \Omega$
**43.** **(a)** 9237.6 V **(b)** 80 A
**(c)** 1276.8 kW **(d)** 0.576 lagging
**(e)** $\mathbf{I}_{Aa} = 80 \text{ A} \angle -54.83°$
**(f)** $\mathbf{V}_{an} = 7773.45 \angle -4.87°$
**(g)** $\mathbf{Z}_\phi = 62.52\ \Omega + j74.38\ \Omega$
**(h)** $F_p$ (entire system) = 0.576,
$F_p$ (load) = 0.643 (both
lagging)
**(i)** 93.98%
**45.** **(b)** $P_T = 5899.64$ W,
$P_{\text{meter}} = 1966.55$ W
**49.** **(a)** 120.09 V
**(b)** $I_{an} = 8.492$ A, $I_{bn} = 7.076$ A,
$I_{cn} = 42.465$ A
**(c)** 4928.5 W, 4928.53 VAR($L$),
6969.99 VA, 0.7071 lagging

## Chapter 24

**1.** (I) **a.** no **b.** no **c.** yes **d.** no
**e.** yes (II) **a.** yes **b.** yes **c.** no
**d.** no **e.** yes (II) **a.** yes **b.** yes
**c.** no **d.** yes **e.** yes (III) **a.** yes
**b.** yes **c.** no **d.** yes **e.** yes
(IV) **a.** no **b.** no **c.** yes **d.** yes
**e.** yes
**7.** **(a)** 19.04 V **(b)** 4.53 A

**9.** 71.872 W
**11.** **(a)** $i = 2 + 2.08 \sin(400t - 33.69°) + 0.5 \sin(800t - 53.13°)$
**(b)** 2.508 A
**(c)** $v_R = 24 + 24.96 \sin(400t - 33.69°) + 6 \sin(800t - 53.13°)$
**(d)** 30.092 V
**(e)** $v_L = 16.64 \sin(400t + 56.31°) + 8 \sin(800t + 36.87°)$
**(f)** 13.055 V **(g)** 75.481 W
**13.** **(a)** $i = 1.2 \sin(400t + 53.13°)$
**(b)** 0.848 A
**(c)** $v_R = 18 \sin(400t + 53.13°)$
**(d)** 12.73 V
**(e)** $v_C = 18 + 23.98 \sin(400t - 36.87°)$
**(f)** 24.73 V **(g)** 10.79 W
**15.** $v_o = 2.257 \times 10^{-3} \sin(377t + 93.66°) + 1.923 \times 10^{-3} \sin(754t + 1.64°)$
**17.** $i_s = 30 + 30.27 \sin(20t + 7.59°) + 0.5 \sin(40t - 30°)$

## Chapter 25

**1.** **(a)** 0.2 H
**(b)** $e_p = 1.6$ V, $e_s = 5.12$ V
**(c)** $e_p = 15$ V, $e_s = 24$ V
**3.** **(a)** 158.02 mH
**(b)** $e_p = 24$ V, $e_s = 1.8$ V
**(c)** $e_p = 15$ V, $e_s = 24$ V
**5.** 1.354 H
**7.** $\mathbf{I}_1(\mathbf{R}_1 + \mathbf{X}_{L_1}) + \mathbf{I}_2(\mathbf{X}_m) = \mathbf{E}_1$, $\mathbf{I}_1(\mathbf{X}_m) + \mathbf{I}_2(\mathbf{X}_{L_2} + \mathbf{R}_L) = 0$
**9.** **(a)** 3.125 V **(b)** 391.02 $\mu$Wb
**11.** 56.31 Hz
**13.** 400 Ω
**15.** $12{,}000t$
**17.** **(a)** $a = 3$ **(b)** 2.78 W
**19.** **(a)** 360.56 Ω ∠86.82°
**(b)** 332.82 mA ∠−86.82°
**(c)** $\mathbf{V}_{R_e} = 6.656$ V ∠−86.82°, $\mathbf{V}_{X_e} = 13.313$ V ∠3.18°, $\mathbf{V}_{X_L} = 106.5$ V ∠3.18°
**23.** **(a)** 20 **(b)** 83.33 A **(c)** 4.167 A
**(d)** $a = \frac{1}{20}$, $I_s = 4.167$ A, $I_p = 83.33$ A
**25.** **(a)** 25 V ∠0°, 5 A ∠0°
**(b)** 80 Ω ∠0° **(c)** 20 Ω ∠0°
**27.** **(a)** $\mathbf{E}_2 = 40$ V ∠60°, $\mathbf{I}_2 = 3.33$ A ∠60°, $\mathbf{E}_3 = 30$ V ∠60°, $\mathbf{I}_3 = 3$ A ∠60°
**(b)** $R_1 = 64.52$ Ω
**29.** $[\mathbf{Z}_1 + \mathbf{X}_{L_1}]\mathbf{I}_1 - \mathbf{Z}_{M_{12}}\mathbf{I}_2 + \mathbf{Z}_{M_{13}}\mathbf{I}_3 = \mathbf{E}_1$, $\mathbf{Z}_{M_{12}}\mathbf{I}_1 - [\mathbf{Z}_2 + \mathbf{Z}_3 + \mathbf{X}_{L_2}]\mathbf{I}_2 + \mathbf{Z}_2\mathbf{I}_3 = 0$, $\mathbf{Z}_{M_{13}}\mathbf{I}_1 - \mathbf{Z}_2\mathbf{I}_2 + [\mathbf{Z}_2 + \mathbf{Z}_4 + \mathbf{X}_{L_3}]\mathbf{I}_3 = 0$

## Chapter 26

**1.** **(a)** $\mathbf{z}_{11} = (\mathbf{Z}_1\mathbf{Z}_2 + \mathbf{Z}_1\mathbf{Z}_3)/(\mathbf{Z}_1 + \mathbf{Z}_2 + \mathbf{Z}_3)$, $\mathbf{z}_{12} = \mathbf{Z}_1\mathbf{Z}_3/(\mathbf{Z}_1 + \mathbf{Z}_2 + \mathbf{Z}_3)$, $\mathbf{z}_{21} = \mathbf{z}_{12}$, $\mathbf{z}_{22} = (\mathbf{Z}_1\mathbf{Z}_3 + \mathbf{Z}_2\mathbf{Z}_3)/(\mathbf{Z}_1 + \mathbf{Z}_2 + \mathbf{Z}_3)$
**3.** **(a)** $\mathbf{y}_{11} = (\mathbf{Y}_1\mathbf{Y}_2 + \mathbf{Y}_1\mathbf{Y}_3)/(\mathbf{Y}_1 + \mathbf{Y}_2 + \mathbf{Y}_3)$, $\mathbf{y}_{12} = -\mathbf{Y}_1\mathbf{Y}_2/(\mathbf{Y}_1 + \mathbf{Y}_2 + \mathbf{Y}_3)$, $\mathbf{y}_{21} = \mathbf{y}_{12}$, $\mathbf{y}_{22} = (\mathbf{Y}_1\mathbf{Y}_2 + \mathbf{Y}_2\mathbf{Y}_3)/(\mathbf{Y}_1 + \mathbf{Y}_2 + \mathbf{Y}_3)$
**5.** $\mathbf{h}_{11} = \mathbf{Z}_1\mathbf{Z}_2/(\mathbf{Z}_1 + \mathbf{Z}_2)$, $\mathbf{h}_{21} = -\mathbf{Z}_1/(\mathbf{Z}_1 + \mathbf{Z}_2)$, $\mathbf{h}_{12} = \mathbf{Z}_1/(\mathbf{Z}_1 + \mathbf{Z}_2)$, $\mathbf{h}_{22} = (\mathbf{Z}_1 + \mathbf{Z}_2 + \mathbf{Z}_3)/(\mathbf{Z}_1\mathbf{Z}_3 + \mathbf{Z}_2\mathbf{Z}_3)$
**7.** $\mathbf{h}_{11} = (\mathbf{Y}_1 + \mathbf{Y}_2 + \mathbf{Y}_3)/(\mathbf{Y}_1\mathbf{Y}_2 + \mathbf{Y}_1\mathbf{Y}_3)$, $\mathbf{h}_{21} = -\mathbf{Y}_2/(\mathbf{Y}_2 + \mathbf{Y}_3)$, $\mathbf{h}_{12} = \mathbf{Y}_2/(\mathbf{Y}_2 + \mathbf{Y}_3)$, $\mathbf{h}_{22} = \mathbf{Y}_2\mathbf{Y}_3/(\mathbf{Y}_2 + \mathbf{Y}_3)$
**9.** **(a)** 47.62 **(b)** −99
**11.** $\mathbf{Z}_i = 9{,}219.5$ Ω ∠−139.4°, $\mathbf{Z}_o = 29.07$ kΩ ∠−86.05°
**13.** $\mathbf{h}_{11} = 2.5$ kΩ, $\mathbf{h}_{12} = 0.5$, $\mathbf{h}_{21} = -0.75$, $\mathbf{h}_{22} = 0.25$ mS
**15.** $\mathbf{Z}_i = 19$ Ω $+ j3.63$ Ω, $R = 19$ Ω, $L = 9.629$ mH
**17.** 24.9 kΩ
**19.** **(a)** −18 **(b)** 2.89 kΩ

# Index